ADVANCED INORGANIC CHEMISTRY

ADVANCED INORGANIC CHEMISTRY

Volume I

For B.Sc. (Pass & Honours) Ist & IInd Year Students of Indian Universities

(Multiple Choice Questions with Hints and Explanations Added as per New U.G.C. Model Curriculum)

SATYA PRAKASH
D.Sc., F.A.Sc.
Formerly Professor of Chemistry
Allahabad University, Allahabad

G.D. TULI
M.Sc., Ph.D.
Formerly Principal, Shyam Lal College
University of Delhi, Delhi

S.K. BASU
M.Sc., Ph.D.
Formerly Professor of Chemistry
Banaras Hindu University, Varanasi

R.D. MADAN
M.Sc., Ph.D.
Reader, Department of Chemistry
N.R.E.C. College, Khurja

S. CHAND
PUBLISHING

S Chand And Company Limited
(ISO 9001 Certified Company)

S Chand And Company Limited

(ISO 9001 Certified Company)

Head Office: D-92, Sector–2, Noida – 201301, U.P. (India), Ph. 91-120-4682700

S. CHAND PUBLISHING

Registered Office: A-27, 2nd Floor, Mohan Co-operative Industrial Estate, New Delhi – 110 044, Phone: 011-49731800

www.**schandpublishing.com;** e-mail: **info@schandpublishing.com**

Marketing Offices:

Chennai : Ph: 23632120; chennai@schandpublishing.com

Guwahati : Ph: 2738811, 2735640; guwahati@schandpublishing.com

Hyderabad : Ph: 40186018; hyderabad@schandpublishing.com

Jalandhar : Ph: 4645630; jalandhar@schandpublishing.com

Kolkata : Ph: 23357458, 23353914; kolkata@schandpublishing.com

Lucknow : Ph: 4003633; lucknow@schandpublishing.com

Mumbai : Ph: 25000297; mumbai@schandpublishing.com

Patna : Ph: 2260011; patna@schandpublishing.com

First Edition 1944

Subsequent Editions and Reprints 1945, 50, 52, 54, 56, 58, 60, 61, 63, 64, 65, 66, 67, 68, 69, 70, 71, 74, 76, 79, 80, 81, 82, 83, 84, 85 (Twice), 87 (Twice), 88, 89, 91, 92, 93, 94, 96, 98, 99, 2000, 2001, 2003, 2005, 2006, 2008, 2009, 2010, 2011, 2012, 2013, 2014 (Twice), 2015, 2016 (Twice), 2017, 2018, 2019, 2020

Reprint 2021

Low Priced Students' Paperback Edition (LPSPE), 2022

Reprint 2025

ISBN: 978-93-550-1099-5 **Product Code:** H6AIC68CHEM10ENAA21L

PRINTED IN INDIA

By Vikas Publishing House Private Limited, Plot 20/4, Site-IV, Industrial Area Sahibabad, Ghaziabad – 201 010 and Published by S Chand And Company Limited, A-27, 2nd Floor, Mohan Co-operative Industrial Estate, New Delhi – 110 044.

Preface to the Ninteenth Edition

In this edition the style of the arrangement and representation of different chapters of the previous edition has been kept unchanged.

Main Features of this Edition

(*i*) The printing mistakes and errors reported to us, from time to time, by the students and fellow-teachers have been corrected.

(*ii*) At the end of each chapter, latest questions asked in the B.Sc. examination of Indian Universities have been given. The solutions of some typical questions are also given.

(*iii*) **Multiple choice questions for each chapter have been incorporated at the end of the book. The hints and explanations of the solution of these questions are also added.**

Acknowledgement

We are extremely grateful to Shri Ravindra Kumar Gupta (Managing Director), Shri Navin Joshi, (General Manager : Sales and Marketing), Shri R.S. Saxena (Advisor) and Shri Shishir Bhatnagar (Editor) for their whole-hearted cooperation in bringing out this new edition.

To Our Fellow-teachers and Students

We humbly request our students and teachers to send us their constructive criticism and suggestions to me in order to enable us to incorporate them in the next edition to make this book error-free and more useful from users point of view.

Dr. R.D. Madan

37, Bhagirathi Apartments

Sector – 9, Rohini

Delhi – 110085

Phone : 7860313

AUTHORS

PREFACE TO THE SEVENTEENTH EDITION

We feel happy to present this thoroughly revised and updated edition to our readers in a new format. In this edition, we have made major changes in the light of modern concepts, with the result that this edition has become almost a new book in all respects. Time to time many new topics have been introduced at the undergraduate level, and the subject-matter is too exhaustive, for us it was a difficult task to cover the text in one volume. Therefore, the subject-matter has been divided into two volumes. The present volume, i.e. Volume I is further divided into two parts. Part-I deals with *Basic Principles of Chemistry* and Part-II discusses the *Chemistry of Main Group Elements and Their Compounds: s- and p- Block Elements*. The book has been written in a simple language and subject-matter has been arranged systematically. Solved examples and solved numerical problems have been given in each chapter. To solve the difficult problems, hints have also been given, so that the students may not face any difficulty to solve them. Previous years' questions and objective-type questions have also been given, to acquaint the students with trend of questions they expect in the examinations as well other competitive examinations. We are sure, this edition will serve as a very useful textbook at the B.Sc. (Ist and IInd year) level of all the Indian Universities.

Its Volume II is also divided into four parts. Part I deals with *Coordination Chemistry*, Part II discusses *Transition Elements — d-Block Elements*, Part III is devoted on *Inner-Transition Elements— f-Block Elements* and Part IV explains some special topics —*Non-aqueous Solvents, Metallic Carbonyls* and *Nitrosyls, Pollution, Organometallic compounds, Chromatography*.

We invite constructive suggestions from the fellow teachers and students for further improvement of the book.

We express our gratitude to Shri Ravindra Kumar Gupta, Director, S. Chand & Company Ltd., who extended his full cooperation to us bring out this voluminous work in a short time.

AUTHORS

September, 1997

CONTENTS

PART I: Basic Principles of Chemistry

Part II : Chemistry of Main Group Elements and Their Compounds (s-and p-block elements)

Part I : Basic Principles of Chemistry

1

Quantum Mechanical Concepts of Atomic Structure : Rutherford's and Bohr's Atomic Models

Rutherford's α-particle scattering experiment : Discovery of nucleus

The discovery of electrons, protons and neutrons made it clear that an atom is made up of these three fundamental particles. In order to know how and where these particles are located in an atom, Rutherford (1911) performed α-particle scattering experiment, in which he produced α-particles ($_2^4He$) from a radioactive element, like radium, placed in a lead block and allowed them to strike a very thin gold foil. A movable circular screen, coated with ZnS was placed on the other side of the foil. This screen helped in detecting the direction in which α-particles moved, when they struck the foil.

Observations and conclusions

When α-particles were allowed to strike the thin gold foil, it was observed that (See Fig. 1.1).

(*i*) Some α-particles were deflected back through small angles (*small deflection*) and some through large angles (*large deflection*). This was explained by assuming that the atom contained a positively-charged part at its centre, and called it **nucleus.** It is this positively charged nucleus which repels the positively-charged α-particles to deflect back.

(*ii*) Most of the α-particles passed through the gold foil, without undergoing any deflection

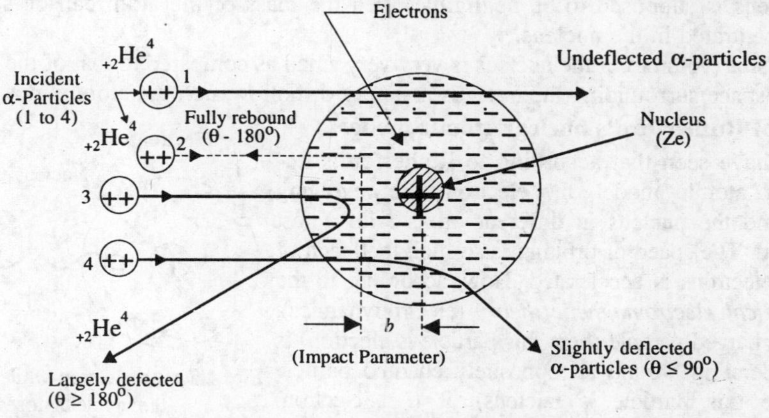

Fig. 1.1 : Three types of deflections of α-particles by a thin metal foil shown in a simple way.

from their original path (*No deflection state*). This was explained by saying that most of the space round the nucleus is empty, which allows the α-particles to pass undeflected through the foil. Thus an atom is not all solid, but *extraordinarily hollow.*

(*iii*) The number of of α-particles, which completely rebounded on hitting the gold foil and turn back to their original path, was very small. This was explained by assuming that the nucleus is very *dense*, *hard* and *rigid*.

Practically the whole mass of an atom is concentrated at its nucleus. Since the number of α-particles which come back to their original path is very small, the volume occupied by the nucleus is only a very minute fraction of the total volume of the atom.

What is nucleus?

The conclusions drawn from the observations of Rutherford's α-particle scattering experiment clearly established the fact that the nucleus is the central part of the atom which, due to the presence of protons in it, is positively-charged. It is very dense and rigid. Besides, the whole mass of the atom is concentrated in this nucleus.

Rutherford's Nuclear Model of Atom (Solar or planetary model of atom)

On the basis of the conclusions drawn from the α-particle scattering experiment, Rutherford gave a nuclear model of atom. According to this model :

(*i*) An atom consists of a positively-charged, very dense and rigid, nucleus, which is situated at the centre of the atom. This nucleus is composed of protons (positively-charged particles) and neutrons (neutral particles). It is due to the presence of protons that the nucleus is positively-charged.

(*ii*) Electrons (negatively-charged particles) are revolving round the nucleus, with high speed, in different orbits. Since the atom is electrically neutral, the number of electrons is equal to that of protons. The electrons keep on revolving in orbits round the nucleus, just as the planets revolve round the Sun. Thus, Rutherford's nuclear model of the atom bears a close resemblance with the Solar system, in which the Sun plays the role of the *nucleus* and the planets play the role of the revolving electrons. It is due to this close resemblance of Rutherford's nuclear model of the atom with the Solar system, that the electrons revolving round the nucleus are generally referred to as *planetary electrons* and Rutherford's nuclear model is also called *Solar model* or *Planetary model of the atom.*

(*iii*) Since the nucleus is made up of protons and neutrons, the mass of the nucleus is the sum of the masses of all protons and neutrons in it. If the mass of electrons revolving round the nucleus is supposed to be negligible, then the mass of the atom can be supposed to be concentrated in the nucleus.

(*iv*) The size (volume) of the nucleus is very very small as compared to that of the whole atom.

(*v*) The space surrounding the nucleus is empty or hollow, *i.e.*, the atom is not all solid.

Drawbacks of Rutherford's nuclear atomic model

(*i*) We have seen that according to Rutherford's nuclear atomic model, the electrons are revolving around the nucleus in different orbits with a fast speed. The speed of orbiting is so high that each of the electrons is accelerated. Now, according to the *classical electromagnetic theory* (electrodynamics), if a charged particle (here this particle is electron) is accelerating around an oppositely charged particle (here this particle is nucleus), it (*i.e.*, electron) would continuously radiate light radiations and hence its energy and speed both go on decreasing. Due to this, the electron will gradually come closer to the nucleus by a spiral path. The ultimate result of this

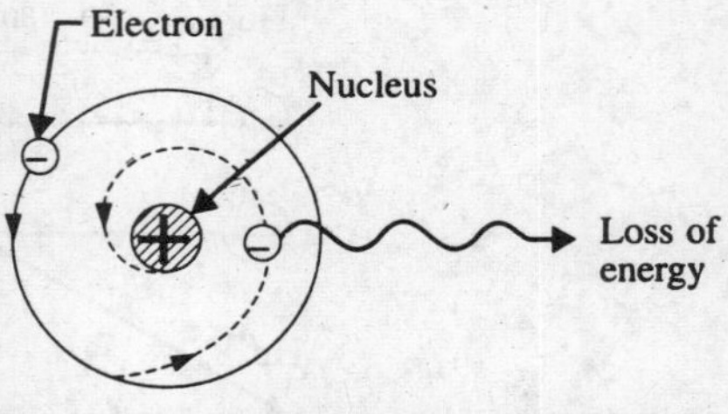

Fig. 1.2 : An electron while revolving round the nucleus loses energy and its speed is decreased, and ultimately the electron falls into the nucleus by a spiral path.

spiral path will be that the electron will fall into the nucleus, as shown in Fig. 1.2, thereby making the atom unstable, *i.e.* according to Rutherford's model, the atom should not exist. Thus, we find that the Rutherford's model does not explain the stability of the atom.

(*ii*) If the electron continuously radiates light radiations, its atomic spectrum should be continuous *i.e.* it should not contain lines of specific frequency / wave length. (*line spectrum*). In actual practice, however, the atomic spectrum is found to contain lines. Thus we see that Rutherford's model of the atom is also not able to explain the atomic 'line spectrum'.

What is atomic number (Z)

Atomic number (Z) of an element can be defined in various ways, *e.g.*,

Atomic number of an element can be defined as the number of positive charges on the nucleus of the atom of the element

OR

Atomic number of an element is also defined as the number of protons (p) present in the nucleus of the atom

OR

Atomic number of an element is also defined as the number of electrons (e) revolving round the nucleus in different orbits of the atom.

OR

Atomic number of an element is also defined as the serial number of its position in the periodic table, starting from hydrogen as the first element.

The above definitions show that $Z = p = e$.

Atomic number of an element is the fundamental property of the element (Mosley's law)

When cathode rays fall on an element of high m.pt. (called *anti-cathode* or *metal target)*, X-rays are emitted by it. When these rays are diffracted by the crystals of $K_4[Fe(CN)_6]$ (called *diffraction grating*), these rays are spread out into a spectrum, which can be photographed on a photographic plate. The spectrum, so obtained on the photographic plate, consists of various series of lines. When these series are arranged in the decreasing order of their frequency, we get K, L, M etc., series (See Fig. 1.3).

English scientist, Mosley (1913) obtained the X-ray spectrum of different elements and found that the frequency (v) of the lines of a given series is related with the atomic number (Z) of the element by the following relation.

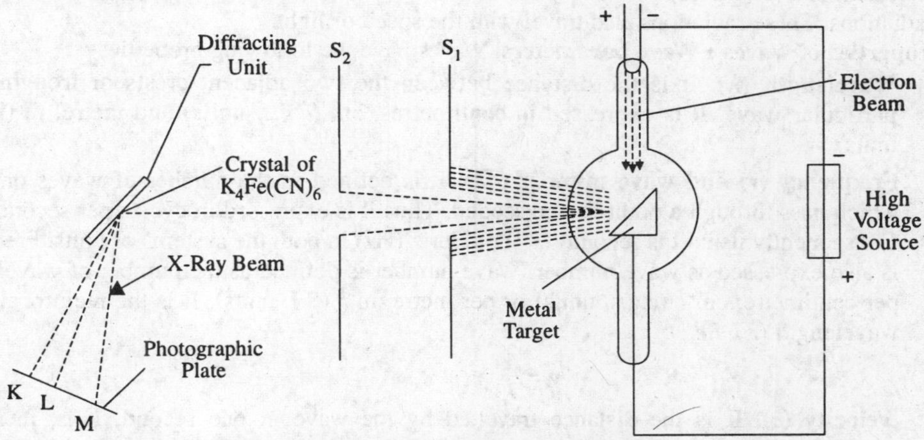

Fig. 1.3 : Splitting of *X*-rays into K, L, M etc. series.

$$\sqrt{v} = a\,(Z - b) \qquad ..(i)$$

Here a and b are constants. This equation is an equation of a straight line. When Mosley ploted a graph between $\sqrt{\text{frequency}} \times 10^8$ (on y-axis) of X-rays and the atomic number, Z (on x-axis) of the elements from which the X-rays have been obtained, he actually got a straight line (See Fig. 1.4).

He also plotted a graph between $\sqrt{\nu} \times 10^8$ and atomic weight of the elements, he *did not* get a straight line. Since the frequency of X-rays depends on the internal structure of the atom, the atomic number of an element must be the fundamental property of the atom, *i.e.*, the properties of the elements should be governed by the atomic number Z of the elements (*Mosley's law*) and not by atomic weight. This further means that the elements should be arranged in the Periodic Table, in the increasing order of their atomic number and not atomic weight, as suggested by Mendeleef in the beginning.

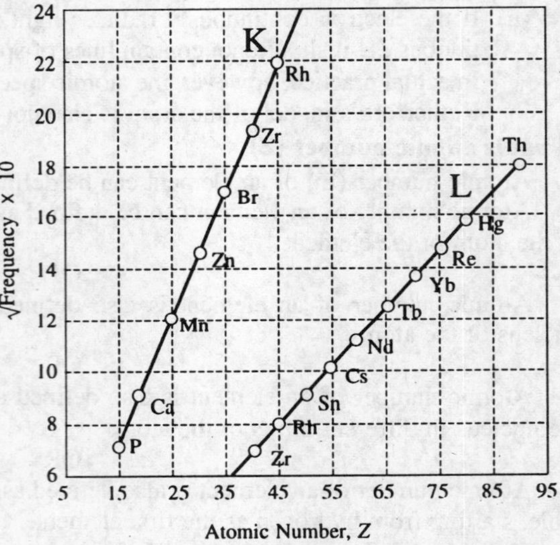

Fig. 1.4 : Relation between the frequency (ν) of the lines of a given series and the atomic number Z of the element.

What is mass number (A) ?

Mass number of an element is the sum of the protons (p) and neutrons (n) present in the nucleus, *i.e.*, $A = p + n$.

Mass number is a whole number.

Electromagnetic radiations or electromagnetic waves

In 1864 James Clark Maxwell showed that when an electric current is passed through a circuit (*e.g.*, through a tungsten filament), it radiates energy in the form of waves. The energy radiated is called *radiated energy* and the waves produced are called *electromagnetic waves* or *electromagnetic radiations.* These waves are so called because similar sort of waves can also be obtained by moving a charged body, as well as a magnet, *to* and *fro* in a magnetic field. These waves do not require any medium for propagation or transmission; *e.g.*, light and radio waves (which are electromagnetic waves) from the sun and stars reach the earth by travelling through non-material medium *i.e.*, through empty space. All these waves travel with the speed of light.

The radiations (or waves) emitted by the vibratory motion of an electron are also electromagnetic radiations. These radiations also travel with the speed of light.

Properties of waves : Wave parameters. Waves have the following properties :

1. **Wavelength (λ).** It is the distance between the two adjacent crests or troughs in a particular wave. It is expressed in centimetres, cm (*c.g.s.* units) and metre, m (in S.I. units)

2. **Frequency (ν) and wave number ($\bar{\nu}$).** It is defined as the number of waves or cycles which pass through a point in one second. Thus it is expressed in cycles per second (*cps*). Consequently its unit is second^{-1}(s^{-1}) or Hertz (Hz) in both the systems of units. Frequency is also expressed as wave number. Wave number is defined as the number of wavelengths per centimetre, cm^{-1} (c.g.s. units) or per metre, m^{-1} (S.I. units). It is the reciprocal of the wavelength (λ), *i.e.*,

$$\nu = \frac{1}{\lambda} \quad \ldots (i)$$

3. **Velocity (c).** It is the distance travelled by the wave in one second. It is, therefore, expressed in cm. s^{-1} (c.g.s. units) or ms^{-1} (*S.I.*units). All types of electromagnetic radiations have velocity equal to the velocity of light which is equal to 3×10^{10} cm s^{-1} or 3×10^8 ms^{-1}. It has been shown that c, ν and λ are related as :

$$c = \nu \times \lambda \quad \text{or} \quad \frac{1}{\lambda} = \frac{\nu}{c} \quad \ldots (ii)$$

On combining the relations (*i*) and (*ii*), we get : $\bar{\nu} = \frac{1}{\lambda} = \frac{\nu}{c}$ ... (*iii*)

Equation (*iii*) shows that wavelength (λ) and frequency (ν or $\bar{\nu}$) are inversely proportional to each other, *i.e.* for a given radiation, with the increase of its wavelength, its frequency decreases and vice-versa.

We know that, according to Planck's quantum theory of radiation, the energy (*E*) associated with the radiation is given by :

$$E = h\nu = \frac{hc}{\lambda} \quad \text{... } (iv)$$

Equation (*iv*) shows that energy of radiation increases with the increase of its frequency but decreases with the increase of its wavelength and vice versa.

Electromagnetic spectrum

A radiation consisting of one wavelength only is called *monochromatic radiation,* (*e.g.*, sun light) while that consisting of different wavelengths is known as *polychromatic radiation.* When a polychromatic radiation is passed through a prism and then allowed to fall on a screen, it gets splitted into different radiations, which have different wavelengths. The splitting of a polychromatic radiation into different radiations having different wavelengths is called *disperson,* which is due to the bending of different radiations through different angles. The pattern of radiations thus obtained is called *spectrum.* When sunlight is allowed to pass through a prism, the spectrum obtained is called *solar spectrum,* which consists of seven colours called VIBGYOR (V= violet, I = inidigo, B = blue, G = green, Y = yellow, O = orange, R = red). All the seven colours have different wavelengths (4000–7500 Å). Since these different wavelengths can be detected by human eye, the spectrum consisting of seven colours is also called *visible spectrum* or *visible region.* Since different colours are diffused, *i.e.,* the second colour starts before the first ends, there is no separation between different colours and hence the spectrum is called a *continuous spectrum.*

Bunsen and Kirchhoff (1859) showed that there also existed some other radiations, along with the radiations of VIBGYOR. The radiations like infrared radiations, microwaves and radiowaves are beyond red end of the spectrum and hence have longer wavelengths. The radiations like *ultra violet,* X-rays, γ-rays and cosmic rays lie on the other side of the violet end of the spectrum and have shorter wavelengths. The various radiations can be arranged as follows in the increasing order of their wavelengths:

Cosmic rays < γ- rays < X-rays < ultraviolet (U.V.) visible < infrared (I.R.) < microwaves < radiowaves .

The order shown here is called *electromagnetic spectrum.* Different radiations forming the electromagnetic spectrum have been shown in Fig. 1.5. The radiations lying in the visible region of the spectrum have their wavelengths in 4000-7500 Å region and these radiations are visible to the human eye. The wavelengths of different radiations

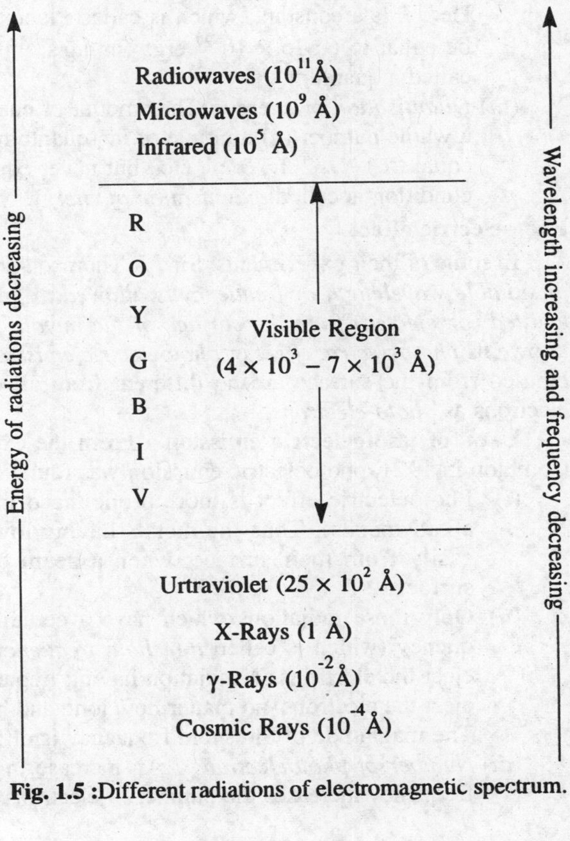

Fig. 1.5 :Different radiations of electromagnetic spectrum.

of the spectrum are : cosmic rays = 10^{-4} Å, γ-rays = 10^{-2} Å, X-rays = 1 Å, ultra violet = 2500 Å, visible = 6000 Å, near infrared = 10^4 Å, far infra-red = 10^5 Å microwaves = 10^9 Å, radiowaves = 10^{11} Å, electric waves = 3000 miles.

Maxwell's electromagnetic wave theory

Clark Maxwell (1856) postulated that : (*i*) light travels as waves of very small wavelength (*ii*) colour of light depends on its wavelength. (*iii*) energy is emitted or absorbed by a body continuously (*iv*) a black body emits the radiation of only one wavelength, irrespective of its temperature. (*v*) the energy of radiation is proportional to the intensity of radiation. This theory clearly explains the mode of propagation of electromagnetic radiation and optical phenomena like interference, diffraction, refraction, polarisation etc.; but is not able to explain black body radiations and photo-electric effect.

Planck's quantum theory of radiation

In order to explain the spectral distribution of energy emitted by a black body, Max Planck put forth (1901) his theory, which is called Planck's quantum theory of radiation. This theory is based on the following postulates :

(*i*) *Concept of quantum / photon :* The emission or absorption of energy (light as well as all other forms of electromagnetic radiations) by a body does not take place continuously, but discontinuously, *i.e.,* the emission or absorption of energy takes place in the form of small packets (bundles) of energy. Each packet of energy is called a *quantum* (plural is *quanta*). In case of light, each quantum is called *photon.* A photon can be defined as the smallest packet (unit) of energy wave, which can exist independently.

(*ii*) *Energy of a quantum or photon* : Each quantum is associated with a definite amount of energy (*E*) which is proportional to the frequency ν of the radiation, *i.e.,*

$$E \propto \nu \quad \text{or} \quad E = h\nu$$

Here *h* is a constant, which is called Planck's constant. The value of *h* has been found to be equal to 6.626×10^{-27} ergs (in c.g.s. units) and 6.626×10^{-34} Js (in S.I. units). $h\nu$ is called a quantum.

(*iii*) *Quantisation of energy :* The amount of energy emitted or absorbed (*i.e., E*) by a body is a whole number (*n*) multiple of $h\nu$ (quantum), *i.e.,* $E = nh\nu$. This equation means that *E* is equal to 1 $h\nu$, 2 $h\nu$, 3 $h\nu$, etc., but never equal to 0.8 $h\nu$, 1.3 $h\nu$, 3.5 $h\nu$, 4.89 $h\nu$ etc. This condition is called *quantisation of energy.*

Photoelectric effect

In some of their experiments, Sir J.J. Thomson and P. Lenard showed that *when a beam of light of suitable wavelength or frequency is allowed to fall on the surface of a metal, the electrons are emitted (or ejected) from the surface of the metal.* This phenomenon of emission of electrons is known as *photoelectric effect* or *photoelectric emission* (*photo* means *light*). Although the electrons emitted from the surface are no different from all other electrons, it is customary to call these electrons as *photo-electrons*.

Laws of photoelectric emission : From the experimental data collected by Richardson and Compton in 1921, photoelectric emission was found to be governed by the following laws :

(*i*) Photoelectric effect is more pronounced in metals having low ionisation energy (*e.g.,* alkali metals). Thus the metals having low ionisation energy emit the electrons quite easily from their surface, when a beam of light of suitable frequency falls on their surface.

(*ii*) Only those radiations which have a certain *critical* or *limiting minimum* value of frequency (which is called *threshold frequency* or *cut-off frequency*, represented as ν_0) can eject the electrons. A radiation having frequency smaller than threshold frequency cannot eject the electrons, no matter how long and how high in its intensity it falls on the surface. The magnitude of threshold frequency is different for different metals.

(*iii*) *Number of photoelectrons :* An increase in the intensity of the incident light of a given frequency increases the number of electrons emitted per second from the surface.

(*iv*) *Velocity and kinetic energy of photoelectrons.* The increase in the intensity of the incident light does not increase the velocity or the kinetic energy of the photo-electrons, no matter how long the light falls on the surface. Thus the velocity and kinetic energy of the emitted electrons, both, are indepedent of the intensity of the incident light. The magnitude of kinetic energy of the photo-electrons increases with an increase in the magnitude of the frequency of the incident radiation above the threshold frequency and hence a plot between the kinetic energy of the photo-electrons and the frequency of the incident radiation will be a straight line.

(*v*) Electrons are ejected even in high vacuum.

Explanation of photoelectric effect by Planck's quantum theory of radiation : Einstein (1905) explained photoelectric effect with the help of Planck's quantum theory of radiation. According to quantum theory of radiation, a photon of light of frequency, ν is associated with energy equal to $h\nu$. When a photon of light of frequency ν (ν is higher than threshold frequency) falls on a metal, some of the energy associated with the photon ($= h\nu$) is consumed to separate the electron from the surface and the remaining energy is imparted to the ejected electron, to give it certain velocity, say, equal to u. Due to this velocity, the emitted electron gains kinetic energy equal to $\frac{1}{2}mu^2$. The portion of energy which is consumed to separate the electron is equal to the binding energy of the electron. This energy is called *threshold eneryg* or *work function* and is equal to the product of h and threshold frequency (ν_0) of a photon of radiation falling on the metal surface. Thus, this energy is equal to $h\nu_0$. This is the minimum energy associated with the photon (of frequency ν_0) of the inicident radiation which is consumed to separate one electron from the metal surface. The above discussion shows that $h\nu$ is equal to the sum of $\frac{1}{2}mu^2$ (kinetic energy of one electron) and $h\nu_0$, *i.e.*, $h\nu = \frac{1}{2}mu^2 + h\nu_0$ ....(*i*)

$$\text{or} \begin{bmatrix}\text{Energy associated with one}\\ \text{photon of incident radiation}\\ \text{of frequency, } \nu\end{bmatrix} = \begin{bmatrix}\text{Minimum energy associated with the photon (of}\\ \text{threshold frequency, } \nu_0\text{), which is consumed}\\ \text{to separate one electron from the metal surface}\end{bmatrix} + \begin{bmatrix}\text{Kinetic energy of one}\\ \text{emitted electron}\end{bmatrix}$$

$$\text{Kinetic energy (K.E.) is given by K.E.} = \frac{\text{K.E. of electrons in J mol}^{-1}}{\text{Avogadro's number}}$$

Equation (*i*) was tested by Millikan. He determined kinetic energy, corresponding to a given incident light of frequency ν and also of frequency ν_o. With the help of these kinetic energy values, he calculated the value of h with the help of equation (*i*). This value was found to be equal to 6.570 $\times 10^{-34}$ Js which is in excellent agreement with the experimental value ($= 6.626 \times 10^{-34}$ Js).

Problem : *When sodium metal surface is exposed to radiation of 300 nm, the electrons with kinetic energy 1.68 × 10⁵ J mol⁻¹ are emitted. What is the minimum energy needed to remove an electron from the metal surface? What is the maximum wavelength that will cause a photo-electron to be emitted ?*

Solution. We know that :

$$h\nu = h\nu_0 + \text{kinetic energy of one emitted electron} = h \times \frac{c}{\lambda}$$

$$= \frac{(6.626 \times 10^{-34}\text{ Js}) \times (3 \times 10^8\text{ ms}^{-1})}{300 \times 10^{-9}\text{ m}} = 6.626 \times 10^{-19}\text{ J}$$

$$\text{Kinetic energy (K.E.) of one electron} = \frac{\text{K.E. of electrons in J mol}^{-1}}{\text{Avogadro's number}}$$

$$= \frac{1.68 \times 10^5\text{ J mol}^{-1}\text{ (given)}}{6.022 \times 10^{23}\text{ mol}^{-1}} = \mathbf{2.789 \times 10^{-19}\ J}$$

$h\nu_0 = E_{min} = h\nu - K.E. = (6.626 \times 10^{-19} - 2.789 \times 10^{-19})\ J = 3.837 \times 10^{-19}\ J$

Again, we know that $E_{min} = h\nu_0$ or $E_{min} = hc / \lambda$

or $(3.837 \times 10^{-19}\ J) = \dfrac{(6.626 \times 10^{-34}\ Js) \times (3 \times 10^8\ ms^{-1})}{\lambda}$

or $\lambda = \dfrac{(6.626 \times 10^{-34}\ Js) \times (3 \times 10^8\ ms^{-1})}{3.837 \times 10^{-19}\ J)} = \mathbf{5.181 \times 10^{-7}\ m}$

Bohr's Atomic Model

In order to remove the drawbacks of Rutherford's nuclear atomic model and to explain the origin of the production of spectrum of H and hydrogen-like species (*e.g.*, He^+, Li^{2+} etc.), Neil Bohr (1913) put forward an atomic model which is based on Planck's quantum theory of radiation.

Postulates of Bohr's atomic model

Main postulates of Bohr's atomic model are :

1. The electron, in an atom, behaves as a material particle and revolves round the nucleus in fixed concentric circular orbits or shells. These orbits are also called *energy shells* or *energy levels*, since the electron in these orbits is associated with a certain amount of energy. The energy associated with a certain energy level increases with the increase of its distance from the nucleus. Thus $E_1 < E_2 < E_3$ .... These energy levels are represented by capital letters: K, L, M, N etc., starting from the nucleus. The values of n (principal quantum number) for these levels are as : for K-shell (1st shell) n = 1, for L-shell (2nd shell) n = 2 etc.
2. As long as an electron revolves in a particular orbit, it neither absorbs nor emits energy. That is why energy levels are also called stationary states. The following two cases arise :
   (*a*) When an electron jumps from a lower energy level n_1 with energy E_1 to a higher energy level n_2 with energy E_2, the electron *absorbs* energy (radiation) in the form of photons. This absorbed energy is equal to $E_2 - E_1 = \Delta E$. According to quantum theory of radiation, $E_2 - E_1 = \Delta E = h\nu$ where ν = frequency of the radiation absorbed by the electron and h = Planck's constant.

   (*b*) When an electron falls down from a higher energy level n_2 with energy E_2 to a lower energy level n_1 with energy E_1, the electron *emits* (releases) energy in the form of photons. This emitted energy is equal to $E_2 - E_1 = \Delta E = h\nu$. The energy thus released appears as a spectral line in the emission spectrum.
3. The electron can move only in that orbit in which the angular momentum (= mvr) of the revolving electron is an integral multiple of $h / 2\pi$, *i.e.*,

$$mvr = n\frac{h}{2\pi}$$

where m = mass of the electron, v = velocity of the electron, r = radius of the orbit in which the electron is moving, n = an integral (called principal quantum number) which denotes the number of the orbit in which the electron is revolving and h = Planck's constant. The equation $mvr = nh / 2\pi$ means that the angular momentum of the revolving electron is quantised, *i.e.*, the magnitude of angular momentum is always a whole number and not in fraction. This is called th *principle of quantisation of angular momentum.*

Bohr's theory as applied to hydrogen atom and hydrogen-like species (one-electron species)

By applying his postulates and the classical laws of Physics, Bohr worked out a number of expressions for one-electron species (*e.g.*, H, He^{+1}, Li^{+2} etc.) These expressions give radius of the *nth* orbit, and velocity and energy of an electron revolving in *nth* orbit. He also derived an expression for the frequency (ν or v) and wavelength (λ) of the spectral line emitted (emission spectrum) when an electron fails from n_2 orbit (higher orbit) to n_1 orbit (lower orbit).

Expression for the radius of *nth* orbit of hydrogen atom and hydrogen-like species (one electron system)

Let us consider a hydrogen-like species (one electron system) with atomic number equal to Z. Suppose an electron of charge e is revolving round the nucleus. Obviously the charge on the nucleus is equal to Ze. Suppose r = distance between the revolving electron and nucleus (*i.e.*, r represents the radius of the orbit in which the electron is revolving), m = mass of the electron and v = tangential velocity of the revolving electron. Now the revolving electron is being acted upon simultaneously by the following two types of forces.

(*i*) **Electrostatic force of attraction (Centripetal force).** This force exeists between the nucleus and the revolving electron, this force tends to attract the electron towards the necleus. In *c.g.s* system of units. This force is equal to $Ze \times e / r^2 = Ze^2 / r^2$ while in SI system this force is equal to $\frac{1}{4\pi\epsilon_0} \times \frac{Ze^2}{r^2}$ where ϵ_0 is a constant which is called **permittivity** of the medium and whose value is 8.854×10^{-12} J^{-1} C^2 m^{-1}.

(*ii*) **Force of repulsion (Centrifugal force).** This force acts outwards from the nucleus and hence tries to keep the revolving electron away from its orbit. In both systems of units this force is equal to mv^2 / r.

Both the above forces act in opposite directions. In order that the electron may keep revolving in its orbit round the nucleus, Bohr assumed that these forces must be equal.

(*a*) **Calculation of r_n in c.g.s. units.** Here r_n is the radius of nth orbit of one-electron system. It has been shown above that both the forces acting on a revolving electron are equal, *i.e.*,

$$\frac{Ze^2}{r^2} = \frac{mv^2}{r} \quad \text{or} \quad v^2 = \frac{Ze^2}{mr} \qquad \text{... }(i)$$

Now since, according to one of the postulates of Bohr's model,

$$mvr = n \cdot \frac{h}{2\pi} \quad \text{or} \quad v^2 = \frac{n^2 h^2}{4\pi^2 m^2 h^2} \qquad \text{...}(ii)$$

From equations (*i*) and (*ii*), we get : $\frac{Ze^2}{mr} = \frac{n^2 h^2}{4\pi^2 m^2 h^2}$ or $r = \frac{n^2 h^2}{4\pi^2 me^2 Z}$ ... (*iii*)

If the radius of the nth orbit is represented by r_n, then equation (*iii*) can be written as :

$$r_n = \frac{n^2 h^2}{4\pi^2 m e^2 Z} \qquad \text{... }(iv)$$

On putting the values of h, π, m and e as $h = 6.626 \times 10^{-27}$ ergs, π (pi) = 3.14, $m = 9.108 \times 10^{-28}$ g and $e = 4.8 \times 10^{-10}$ e.s.u. in equation (*iv*) and using the relations erg = g cm^{2}s^{-2} and esu = g$^{1/2}$ cm $^{3/2}$ s^{-1}, we get the value of the radius of nth orbit of one-electron system with atomic number Z as given by

$$r_n = \frac{n^2 \times (6.624 \times 10^{-27} \text{ erg s})^2}{4 \times (3.14)^2 \times (9.108 \times 10^{-28} \text{ g}) \times (4.8 \times 10^{-10} \text{ esu})^2 \times Z}$$

$$= \frac{n^2 \times (6.624 \times 10^{-27})^2 \text{ erg}^2.\text{s}^2}{4 \times (3.14)^2 \times (9.108 \times 10^{-28}) \times (4.8 \times 10^{-10})^2 \times \text{g.esu}^2 \times Z}$$

$$= \frac{0.529 \times 10^{-8} \times n^2}{Z} \times \frac{(\text{g cm}^2 \text{ s}^{-2})^2 . \text{s}^2}{\text{g} . (\text{g}^{1/2} \text{ cm}^{3/2} \text{ s}^{-1})^2}$$

$$= \frac{0.529 \times 10^{-8} \times n^2}{Z} \times \frac{\text{g}^2 \text{ cm}^4 \text{ s}^{-4} \text{ s}^2}{\text{g} . \text{g} . \text{cm}^3 . \text{s}^{-2}} = \frac{0.529 \times 10^{-8} \times n^2}{Z} \text{ cm}$$

$$= \frac{0.529 \times n^2}{Z} \text{ Å} \qquad \text{... }(v)$$

(*b*) **Calculation of r_n in S.I. units.** On utilising the fact that centripetal and centrifugal forces acting on the revolving electron should be equal, we can write

$$\frac{1}{4\pi\varepsilon_0} \times \frac{Ze^2}{r^2} = \frac{mv^2}{r} \quad \text{or} \quad v^2 = \frac{Ze^2}{mr} \times \frac{1}{4\pi\varepsilon_0} \qquad \text{... (vi)}$$

From $mvr = n.\dfrac{h}{2\pi}$, we have $v^2 = \dfrac{n^2 h^2}{4\pi^2 m^2 r^2}$ ... (*viii*)

With the help of relations (*vi*) and (*vii*), we can write, $\dfrac{Ze^2}{mr} \times \dfrac{1}{4\pi\varepsilon_o} = \dfrac{n^2 h^2}{4\pi^2 m^2 r^2}$

or $$r = \frac{n^2 h^2 \times 4\pi\varepsilon_o}{4\pi^2 me^2 Z} = \frac{n^2 h^2 \varepsilon_o}{\pi me^2 Z}$$

Thus $$r_n = \frac{n^2 h^2 \varepsilon_o}{\pi me^2 Z} \qquad \text{...(viii)}$$

On putting the values of h, π, m, e and ε_0 as $h = 6.626 \times 10^{-34}$ Js, $\varepsilon_o = 8.85 \times 10^{-12}\ J^{-1}\ C^2\ m^{-1}$, $\pi = 3.14$, $m = 9.108 \times 10^{-31}$ kg, $e = 1.602 \times 10^{-19}$ C in equation (*viii*) and using the relation J = kg m^2s^{-2}, we get the value of the radius of *nth* orbit of one-electron system with atomic number Z as given by :

$$r_n = \frac{n^2 \times (6.626 \times 10^{-34}\ Js)^2 \times (8.85 \times 10^{-12}\ J^{-1}\ C^2\ m^{-1})}{3.14 \times (9.108 \times 10^{-31}\ kg) \times (1.602 \times 10^{-19}\ C)^2 \times Z}$$

$$= \frac{n^2 \times 5.29 \times 10^{-11} \times J^2 s^2\ J^{-1} C^2\ m^{-1}}{kg\ C^2 \times Z} = \frac{n^2 \times 5.29 \times 10^{-11} \times J\ s^2\ m^{-1}}{kg \times Z}$$

$$= \frac{5.29 \times 10^{-11} \times n^2 \times kg\ m^2\ s^{-2}\ s^2\ m^{-1}}{kg \times Z} = \frac{0.529 \times 10^{-10} \times n^2}{Z}\ m \qquad \text{... (ix)}$$

Bohr's radius [$(r_1)_H$ or r_H or a_0]

The radius of the first orbit (n = 1) of H-atom (Z = 1) is called Bohr's radius which is represented as $(r_1)_H$ or r_H or a_0. Its value can be obtained in both the systems of units as shown below :

(*a*) **In c.g.s. units,** we know that in c.g.s. system, r_n is given by :

$$r_n = \frac{h^2}{4\pi^2 me^2} \times \frac{n^2}{Z}$$

The value of $(r_1)_H$ is obtained by putting n = 1 and Z = 1 in the above equation. Thus :

$$(r_1)_H = \frac{h^2}{4\pi^2 me^2} \times \frac{(1)^2}{1}$$

On putting $h = 6.626 \times 10^{-27}$ erg s, $\pi = 3.14$, $m = 9.108 \times 10^{-28}$ g, $e = 4.8 \times 10^{-10}$ esu, and using the relations erg = g cm^2s^{-2} and esu = g $^{1/2}$ $cm^{3/2}$ s^{-1}, we get :

$$(r_1)_H = 0.529 \times 10^{-8}\ cm \quad \text{or} \quad (r_1)_H = 0.529\ Å$$

(*b*) **In S.I. units :** The value of r_n is given by $r_n = \dfrac{\varepsilon_0 h^2}{\pi me^2} \times \dfrac{n^2}{Z}$. Thus the value of $(r_1)_H$ is given by putting n = 1 and Z = 1 in the above equation. Hence $(r_1)_H = \dfrac{\varepsilon_0 h^2}{\pi me^2} \times \dfrac{(1)^2}{1} = \dfrac{\varepsilon_0 h^2}{\pi me^2}$

On putting $\varepsilon_0 = 8.85 \times 10^{-12}\ J^{-1}\ C^2\ m^{-1}$, $h = 6.626 \times 10^{-34}$ Js, $\pi = 3.14$, $m = 9.108 \times 10^{-31}$ kg, $e = 1.602 \times 10^{-19}$ C in the above equation and using the relation J = kg $m^2\ s^{-2}$, we get

$$(r_1)_H = 0.529 \times 10^{-10}\ m.$$

Relation between r_n and $(r_1)_H$

Here r_n is the redius of *nth* orbit of H-like species of atomic number equal to Z while $(r_1)_H$ is the radius of 1st orbit (n =1) of H–atom (Z = 1).

(*a*) **Relation in c.g.s. units.** The values of r_n and $(r_1)_H$ in c.g.s. units are given by :

$$r_n = \frac{h^2}{4\pi^2 me^2} \times \frac{n^2}{Z}, \quad (r_1)_H = \frac{h^2}{4\pi^2 me^2} \times \frac{(1)^2}{1}$$

On dividing, we get :

$$\frac{r_n}{(r_1)_H} = \frac{h^2}{4\pi^2 m e^2} \times \frac{n^2}{Z} \times \frac{4\pi^2 m e^2}{h^2} = \frac{n^2}{Z}$$

or $\quad r_n = (r_1)_H \times \frac{n^2}{Z}$ or $r_n = 0.529 \times 10^{-8} \times \frac{n^2}{Z}$ cm o $r_n = 0.529 \times \frac{n^2}{Z}$ Å ... (*i*)

All the above equations give us a relation between r_n and $(r_1)_H$ in c.g.s. units.

(*b*) **Relation in S.I. units.** The value of r_n and $(r_1)_H$ in S.I. units are given by the expressions :

$$r_n = \frac{\varepsilon_o h^2}{\pi m e^2} \times \frac{n^2}{Z}, \qquad (r_1)_H = \frac{\varepsilon_o h^2}{\pi m e^2} \times \frac{(1)^2}{1}$$

On dividing the two equations, we have : $\frac{r_n}{(r_1)_H} = \frac{\varepsilon_o h^2}{\pi m e^2} \times \frac{n^2}{Z} \times \frac{\pi m e^2}{\varepsilon_o h^2} = \frac{n^2}{Z}$

or $\quad r_n = (r_1)_H \times \frac{n^2}{Z}$ $\qquad$ or $\qquad r_n = 0.529 \times 10^{-10} \times \frac{n^2}{Z}$ m ... (*ii*)

The above two equations give us a relation between r_n and $(r_1)_H$ in S.I. system of units.

Problem. *Calculate radius of the 3rd orbit of H-atom.* **(Raj. B.Sc. 1983)**

Solution. We know that radius of *nth* orbit of one-electron species is given by :

$$r_n = \frac{0.529 \times 10^{-10} \times n^2}{Z} \text{ m} \qquad \text{(in S.I. units)}$$

On putting $n = 3$ and $Z = 1$, we get the radius of the 3rd orbit of H-atom given by :

$$(r_3)_H = \frac{0.529 \times 10^{-10} \times 3^2}{1} \text{ m} = \mathbf{4.761 \times 10^{-10} \text{ m}}$$

Expression for the velocity of an electron revolving in the *nth* orbit of H-atom and hydrogen-like species

(*a*) **In c.g.s. units.** We know that in case of a revolving electron the force of attraction $\left(= \frac{Ze^2}{r^2}\right)$ is exactly balanced by the centrifugal force $\left(= \frac{mv^2}{r}\right)$, *i.e.*, $\frac{mv^2}{r} = \frac{Ze^2}{r^2}$ or $v^2 = \frac{Ze^2}{mr}$ ...(*i*)

According to Bohr's quantum condition $mvr = n.\frac{h}{2\pi}$ or $\quad v = \frac{nh}{2\pi mr}$ ....(*ii*)

On dividing equation (*i*) by equation (*ii*), we get the velocity (v) given by :

$$\frac{v^2}{v} = \frac{Ze^2}{mr} \times \frac{2\pi mr}{nh} = \frac{2\pi Ze^2}{nh} \quad \text{or} \quad v_n = \frac{2\pi Ze^2}{nh}$$

On putting $\pi = 3.14$, $e = 4.8 \times 10^{-10}$ esu, $h = 6.626 \times 10^{-27}$ ergs and using the relations erg = g cm^2 s^{-2} and esu = g$^{1/2}$ cm $^{3/2}$ s^{-1}, we get :

$$v_n = \frac{2 \times 3.14 \times Z \times (4.8 \times 10^{-10})^2}{n \times 6.626 \times 10^{-27}} \text{ cm s}^{-1} = \frac{21.8368 \times 10^7 \times Z}{n} \text{ cm s}^{-1}$$

$$= \frac{2.1836 \times 10^8 \times Z}{n} \text{ cm s}^{-1} = \frac{0.2183 \times 10^9 \times Z}{n} \text{ cm s}^{-1}$$

(*b*) **In S.I. units.** In S.I. units, force of attraction is equal to $\frac{Ze^2}{4\pi\varepsilon_o r^2}$ but the centrifugal force is the same as it is in c.g.s. system $\left(= \frac{mv^2}{r}\right)$. Since these forces are equal, we have

$$\frac{mv^2}{r} = \frac{Ze^2}{4\pi\varepsilon_0 r^2} \quad \text{or} \quad v^2 = \frac{Ze^2}{4\pi\varepsilon_0 mr} \qquad \text{... (iii)}$$

According to Bohr's postulate, $mvr = n.\frac{h}{2\pi}$ or $v = \frac{nh}{2\pi mr}$ ... (*iv*)

On dividing equation (*iii*) by equation (*iv*), we have

$$\frac{v^2}{v} = \frac{Ze^2}{4\pi\varepsilon_0 mr} \times \frac{2\pi mr}{nh} = \frac{Ze^2}{2\varepsilon_o nh} \quad \text{or} \quad v_n = \frac{Ze^2}{2\varepsilon_0 nh}$$

On putting $e = 1.602 \times 10^{-19}$C, $\varepsilon_0 = 8.85 \times 10^{-12}$ $J^{-1}C^2 m^{-1}$ and $h = 6.626 \times 10^{-34}$ Js and using the relation J = kg $m^2 s^{-2}$, we have :

$$v_n = \frac{Z \times (1.602 \times 10^{-19} \text{ C})^2}{2 \times (8.85 \times 10^{-12} \text{ J}^{-1} \text{ C}^2 \text{ m}^{-1}) \times n \times (6.626 \times 10^{-34} \text{ J s})}$$

$$= \frac{0.2185 \times 10^7 \times Z}{n} \text{ ms}^{-1}$$

Number of revolutions made by an electron moving in *n*th orbit

Number of revolutions made by an electron moving with a velocity of v_n in *nth* orbit is givn by:

(*a*) **In c.g.s. units.** No. of revolutions $= \frac{v_n}{2\pi r_n}$

On putting the values of v_n and r_n as $v_n = \frac{0.2183 \times 10^9 \times Z}{n}$ cms^{-1} and

$r_n = \frac{0.529 \times 10^{-8} \times x^2}{Z}$ cm, we get :

$$\text{No. of revolutions} = \left[\frac{0.2183 \times 10^9 \times Z}{n} \text{ cm s}^{-1}\right] \times \left[\frac{Z}{2\pi \times 0.529 \times 10^{-8} \times n^2} \text{ cm}\right]$$

$$= \frac{0.2183 \times 10^9 \times Z^2}{2 \times 3.14 \times 0.529 \times 10^{-8} \times n^3} \text{ s}^{-1} = \frac{65.711 \times 10^{14} \times Z^2}{n^3} \text{ s}^{-1}$$

(*b*) **In S.I. units.** No. of revolutions $= \frac{v_n}{2\pi r_n}$

On putting the values of v_n and r_n as $v_n = \frac{0.2185 \times 10^7 \times Z}{n}$ ms^{-1} and

$r_n = \frac{0.529 \times 10^{-10} \times n^2}{Z}$ m, we get :

$$\text{No. of revolutions} = \left[\frac{0.2185 \times 10^7 \times Z}{n} \text{ ms}^{-1}\right] \times \left[\frac{Z}{2\pi \times 0.529 \times 10^{-10} \times n^2} \text{ m}\right]$$

$$= \frac{0.2185 \times 10^7 \times Z^2}{2 \times 3.14 \times 0.529 \times 10^{-10} \times n^3} \text{ s}^{-1} = \frac{65.711 \times 10^{14} \times Z^2}{n^3} \text{ s}^{-1}$$

Expression for the energy of an electron revolving in *n*th orbit of H-atom and hydrogen-like species (one-electron system)

Total energy (E_n) associated with an electron revolving in *nth* orbit, is the sum of its kinetic energy (K.E.) and potential energy (P.E.). The magnitude of K.E. in c.g.s. and S.I. units is equal to $\frac{1}{2}mv^2$. The magnitude of P.E. in c.g.s. units is equal to $-\frac{Ze^2}{r}$ while its magnitude in S.I. units is equal to $-\frac{Ze^2}{4\pi\varepsilon_0 r^2}$.

(*a*) **Calculation of E_n in c.g.s. units** : The sum of K.E. and P.E. is given by

$$E_n = \frac{1}{2}mv^2 - \frac{Ze^2}{r} \quad \text{... (i)}$$

We know that in case of a revolving electron the centripetal (Ze^2/r^2) and centrifugal forces ($= mv^2/r$) are equal to each other, *i.e.*,

$$\frac{mv^2}{r} = \frac{Ze^2}{r^2} \quad \text{or} - mv^2 = \frac{Ze^2}{r} \quad \text{or} \quad \frac{1}{2}mv^2 = \frac{Ze^2}{2r} \quad \text{... (ii)}$$

Putting the value of $\frac{1}{2}\, m\, v^2$ from equation (*ii*) in equation (*i*), we get :

$$E_n = \frac{Ze^2}{2r} - \frac{Ze^2}{r} = -\frac{Ze^2}{2r}$$

$$= -\frac{Ze^2}{2} \times \frac{4\pi^2\, m\, e^2\, Z}{n^2 h^2} = -\frac{2\pi^2\, m\, e^4\, Z^2}{n^2 h^2} \qquad \left(\because r = \frac{n^2 h^2}{4\pi^2 m e^2 Z}\right) \qquad \text{... (iii)}$$

On putting $\pi = 3.14$, $m = 9.108 \times 10^{-28}$ g, $e = 4.8 \times 10^{-10}$ esu and $h = 6.624 \times 10^{-27}$ ergs in equation (*iii*) and utilising the relations erg = g cm^2 s^{-2} and esu = g$^{1/2}$ cm$^{3/2}$ s^{-1}, we get :

$$E_n = \frac{2 \times (3.14)^2 \times (9.108 \times 10^{-28}\, g) \times (4.8 \times 10^{-10}\ \text{esu})^4 \times Z^2}{n^2 \times (6.624 \times 10^{-27}\ \text{erg s})^2}$$

$$= -\frac{2.18 \times 10^{-11} \times Z^2}{n^2} \times \frac{\text{g} \times (\text{esu})^4}{(\text{erg s})^2} = -\frac{2.18 \times 10^{-11} \times Z^2}{n^2} \times \frac{\text{g} \times (\text{g}^{1/2}\ \text{cm}^{3/2}\ \text{s}^{-1})^4}{(\text{g cm}^2\ \text{s}^{-2}\ \text{s})^2}$$

$$= -\frac{2.18 \times 10^{-11} \times Z^2}{n^2} \times \frac{\text{g} \times \text{g}^2\ \text{cm}^6\ \text{s}^{-4}}{\text{g}^2\ \text{cm}^4\ \text{s}^{-2}}$$

$$= -2.18 \times 10^{-11} \times \frac{Z^2}{n^2}\ \text{g cm}^2\ \text{s}^{-2} = -2.18 \times 10^{-11} \times \frac{Z^2}{n^2}\ \text{erg atom}^{-1}\ [\because \text{erg} = \text{g cm}^2\ \text{s}^{-2}]$$

Value of E_n in eV atom^{-1}. We have shown above that the value of E_n in erg atom^{-1} is given by the expression :

$$E_n = -2.18 \times 10^{-11} \times \frac{Z^2}{n^2}\ \text{erg atom}^{-1}$$

Now, since 1 erg = 6.2419×10^{11} eV, the above equation becomes

$$E_n = (-2.18 \times 10^{-11}) \times (6.2419 \times 10^{11}) \times \frac{Z^2}{n^2}\ \text{eV atom}^{-1}$$

or $$E_n = -13.6 \times \frac{Z^2}{n^2}\ \text{eV atom}^{-1}$$

Value of E_n in erg mol^{-1}. The value of E_n in erg mol^{-1} can be calculated as follows. We know that :

$$E_n = \frac{-2.18 \times 10^{-11} \times Z^2}{n^2}\ \text{erg atom}^{-1}$$

$$= \frac{(-2.18 \times 10^{-11}\ \text{erg atom}^{-1}) \times (\text{Avogadro's number}) \times Z^2}{n^2}$$

$$= \frac{(-2.18 \times 10^{-11}\ \text{erg atom}^{-1}) \times (6.022 \times 10^{23}\ \text{atoms mol}^{-1}) \times Z^2}{n^2}$$

$$= -\frac{13.1279 \times 10^{12} \times Z^2}{n^2}\ \text{erg mol}^{-1}$$

(*b*) **Calculation of E_n in S.I. units.** Sum of K.E. and P.E. is given by :

$$E_n = \frac{1}{2}\, mv^2 - \frac{Ze^2}{4\pi\, \varepsilon_0\, r} \qquad \text{... (vii)}$$

We have already said that for a revolving electron, the centripetal force (= $Ze^2 / 4\pi\, \varepsilon_0 r^2$) is equal to the centrifugal force (= mv^2/r), *i.e.*,

$$\frac{Ze^2}{4\pi\, \varepsilon_0\, r^2} = \frac{m\, v^2}{r} \quad \text{or} \quad \frac{Ze^2}{4\pi\, \varepsilon_o\, r} = m\, v^2 \quad \text{or} \quad \frac{1}{2}\, m\, v^2 = \frac{Ze^2}{8\pi\, \varepsilon_o\, r} \qquad \text{... (viii)}$$

Incorporating equation (*viii*) in equation (*vii*), we get :

$$E_n = \frac{Ze^2}{8\pi\, \varepsilon_o\, r} - \frac{Ze^2}{4\pi\, \varepsilon_o\, r} = -\frac{Ze^2}{8\pi\, \varepsilon_o\, r}$$

$$= -\frac{Ze^2}{8\pi\, \varepsilon_o} \times \frac{\pi\, me^2\, Z}{n^2 h^2 \varepsilon_o} \qquad \left[\because r = \frac{n^2 h^2 \varepsilon_0}{\pi\, m\, e^2\, Z}\right]$$

$$\text{or} \quad E_n = -\frac{m e^4 Z^2}{8 \varepsilon_0^2 n^2 h^2} \qquad \text{... (ix)}$$

On putting $m = 9.109 \times 10^{-31}$ kg, $e = 1.602 \times 10^{-19}$ C, $\varepsilon_0 = 8.85 \times 10^{-12}$ J^{-1}C^{2}m^{-1} and $h = 6.626 \times 10^{-34}$ Js and using the relation J = kg m^2 s^{-2}, we get the value of E_n of an electron revolving in the *nth* orbit of one-electron system with atomic number, Z given as :

$$E_n = -\frac{(9.108 \times 10^{-31}\text{ kg}) \times (1.602 \times 10^{-19}\text{ C})^4}{8 \times (8.85 \times 10^{-12}\text{ J}^{-1}\text{ C}^2\text{ m}^{-1}) \times (6.626 \times 10^{-34}\text{ Js})} \times \frac{Z^2}{n^2}$$

$$= -2.18 \times 10^{-18} \times \frac{Z^2}{n^2} \times \frac{\text{kg C}^4}{(\text{J}^{-1}\text{ C}^2\text{ m}^{-1})^2\ (\text{J s})^2}$$

$$= -2.18 \times 10^{-18} \times \frac{Z^2}{n^2} \times \frac{\text{kg} \times \text{C}^4}{\text{J}^{-1}\text{ C}^4\text{ m}^{-2}\text{ J}^2\text{ s}^2}$$

$$= -2.18 \times 10^{-18} \times \frac{Z^2}{n^2}\text{ kg m}^2\text{ s}^{-2} = -2.18 \times 10^{-18} \times \frac{Z^2}{n^2}\text{ J atom}^{-1} \quad (\because \text{J} = \text{kg m}^2\text{ s}^{-2})$$

Value of E_n in J mol^{-1} and kJ mol^{-1}. We have already seen that

$$E_n = -2.18 \times 10^{-18} \times \frac{Z^2}{n^2}\text{ J atom}^{-1} = (-2.18 \times 10^{-18}\text{ J atom}^{-1}) \times (\text{Avogadro's number}) \times \frac{Z^2}{n^2}$$

$$= (-2.18 \times 10^{-18}\text{ J atom}^{-1}) \times (6.022 \times 10^{23}\text{ atoms mol}^{-1}) \times \frac{Z^2}{n^2}$$

$$= -13.1279 \times 10^5 \times \frac{Z^2}{n^2}\text{ J mol}^{-1} = \frac{-13.1279 \times 10^5}{10^3} \times \frac{Z^2}{n^2}\text{ kJ mol}^{-1}$$

$$= -13.1279 \times 10^2 \times \frac{Z^2}{n^2}\text{ kJ mol}^{-1}$$

Energy of an electron in the first orbit of H-atom [$(E_1)_H$]

(*a*) **In c.g.s. units :** We know that the energy of an electron present in *n*th orbit of H-like species is given by:

$$E_n = -\frac{2\pi^2 m e^4}{h^2} \times \frac{Z^2}{n^2}$$

The value of $(E_1)_H$ is obtained by putting $n = 1$ and $Z = 1$ in the above equation. Thus,

$$(E_1)_H = -\frac{2\pi^2 m e^4}{h^2} \times \frac{(1)^2}{(1)^2}$$

On putting $\pi = 3.14$, $m = 9.108 \times 10^{-28}$ g, $e = 4.8 \times 10^{-10}$ esu, $h = 6.626 \times 10^{-27}$ erg s and using the relations erg = g cm^2 s^{-2} and esu = g $^{1/2}$ cm$^{3/2}$ s^{-1}, we get,

$(E_1)_H = -2.18 \times 10^{11}$ erg atom^{-1} or $(E_1)_H = -13.6$ eV atom^{-1}

or $(E_1)_H = -13.1279 \times 10^{10}$ erg mol^{-1}

(*b*) **In S.I. units:** The energy of an electron present in the *n*th orbit of H-like species is given by :

$$E_n = -\frac{m e^4}{8 \varepsilon_o^2 h^2} \times \frac{Z^2}{n^2}$$

The value of $(E_1)_H$ is obtained by putting $n = 1$ and $Z = 1$ in the above equation. Thus :

$$(E_1)_H = -\frac{m e^4}{8 \varepsilon_o^2 h^2} \times \frac{(1)^2}{(1)^2}$$

On putting $m = 9.108 \times 10^{-31}$ kg, $e = 1.602 \times 10^{-19}$C, $\varepsilon_0 = 8.85 \times 10^{-12}$ J^{-1}C^2 m^{-1}, $h = 6.626 \times 10^{-34}$ Js and using the relation J = kg m^{2}s^{-2}, we get the value of $(E_1)_H$ given by the expression :

$(E_1)_H = -2.18 \times 10^{-18}$ J atom^{-1} or $(E_1)_H = -13.1279 \times 10^5$ J mol^{-1}

or $(E_1)_H = -13.1279 \times 10^2$ kJ mol^{-1}

Relation between E_n and $(E_1)_H$

Here E_n is the energy of an electron revolving in *nth* orbit of H-like species with atomic number of Z while $(E_1)_H$ is the energy of the electron in 1*st* orbit ($n = 1$) in H-atom ($Z = 1$).

(*a*) **Relation in c.g.s. units.** We have already shown that the values of E_n and $(E_1)_H$ are given by :

$$E_n = -\frac{2\pi^2 m e^4}{h^2} \times \frac{Z^2}{n^2}, \quad (E_1)_H = -\frac{2\pi^2 m e^4}{h^2} \times \frac{(1)^2}{(1)^2}$$

On dividing, we get,

$$\frac{E_n}{(E_1)_H} = \frac{2\pi^2 m e^4 Z^2}{h^2 n^2} \times \frac{h^2}{2\pi^2 m e^4} = \frac{Z^2}{n^2}$$

$$\left.\begin{array}{llll} \text{or} & E_n = (E_1)_H \times \dfrac{Z^2}{n^2} & \text{or} & E_n = -2.18 \times 10^{-11} \times \dfrac{Z^2}{n^2} \text{ erg atom}^{-1} \\ \text{or} & E_n = -13.6 \times \dfrac{Z^2}{n^2} \text{ eV atom}^{-1} & \text{or} & E_n = -13.1279 \times 10^{10} \times \dfrac{Z^2}{n^2} \text{ erg mol}^{-1} \end{array}\right\} \quad ...(i)$$

All the above equations give us a relation between E_n and $(E_1)_H$ in c.g.s. units.

(*b*) **Relation in S.I. units :** The values of E_n and $(E_1)_H$ are given by the expressions :

$$E_n = -\frac{m e^4}{8\varepsilon_o^2 h^2} \times \frac{Z^2}{n^2}, \quad (E_1)_H = -\frac{m e^4}{8\varepsilon_o^2 h^2} \times \frac{(1)^2}{(1)^2}$$

On dividing we have : $$\frac{E_n}{(E_1)_H} = \frac{m e^4 Z^2}{8\varepsilon_0^2 h^2 n^2} \times \frac{8\varepsilon_0^2 h^2}{m e^4} = \frac{Z^2}{n^2}$$

$$\left.\begin{array}{llll} \text{or} & E_n = (E_1)_H \times \dfrac{Z^2}{n^2} & \text{or} & E_n = -2.18 \times 10^{-18} \times \dfrac{Z^2}{n^2} \text{ J atom}^{-1} \\ \text{or} & E_n = -13.1279 \times 10^5 \times \dfrac{Z^2}{n^2} \text{ J mol}^{-1} & \text{or} & E_n = -13.1279 \times 10^2 \times \dfrac{Z^2}{n^2} \text{ kJ mol}^{-1} \end{array}\right\} \quad ...(ii)$$

All the above equation give us a relation between E_n and $(E_1)_H$ in S. I. system of units.

To find out the order of $(E_1)_H$, $(E_2)_H$, $(E_3)_H$ etc.,

$(E_1)_H$, $(E_2)_H$, $(E_3)_H$ etc., are the energies associated with an electron moving in 1st ($n = 1$), 2nd ($n = 2$), 3*rd* ($n = 3$) etc. orbits of H-atom (Z = 1). The values of $(E_1)_H$, $(E_2)_H$, $(E_3)_H$ etc., are given by putting $n = 1, 2, 3$ etc., and $Z = 1$ in $E_n = -2.18 \times 10^{-11} \times \frac{Z^2}{n^2}$ erg. Thus :

$$(E_1)_H = -2.18 \times 10^{-11} \times \frac{(1)^2}{(1)^2} \text{ erg} = -2.18 \times 10^{-11} \text{ erg}$$

$$(E_2)_H = -2.18 \times 10^{-11} \times \frac{(1)^2}{(2)^2} \text{ erg} = -0.5448 \times 10^{-11} \text{ erg}$$

$$(E_3)_H = -2.18 \times 10^{-11} \times \frac{(1)^2}{(3)^2} \text{ erg} = -0.2421 \times 10^{-11} \text{ erg.}$$

The values of $(E_1)_H$, $(E_2)_H$, $(E_3)_H$ etc., as calculated above show that these go on increasing (*i.e.*, go on becoming less and less negative) as the electron moves away from the nucleus. Thus the magnitude of $(E_1)_H$, $(E_2)_H$ $(E_3)_H$etc., is in the order : $(E_1)_H < (E_2)_H < (E_3)_H$........

Why is the energy of an electron negative?

We have seen that the energy of an electron in *nth* orbit of one-electron system of atomic number equal to *Z* is given by :

$$E_n = -\frac{2\pi^2 m e^4}{h^2} \times \frac{Z^2}{n^2} \text{ (in c.g.s. units)}; \quad E_n = -\frac{m e^4}{8\varepsilon_0^2 h^2} \times \frac{Z^2}{n^2} \text{ (in S.I. units)}$$

These equations show that E_n in both the systems of units is negative. To explain it, consider an electron situated at an infinite distance from the nucleus so that there is no interaction between the electron and the nucleus. Now, as the electron approaches towards the nucleus, it does some work and some energy is released due to the attraction between the two oppositely-charged particles. Thus the energy of the electron decreases, *i.e.*, the energy of the elctron becomes negative as the electron moves away from the nucleus.

Expression for the ionisation energy of one-electron species (*e.g.*, H, He^+, Li^{2+} etc.) with atomic number of *Z*

Ionisation energy of a given species is the amount of energy involved in exciting an electron from an energy level $n = 1$ (lowest energy level) to infinity (*i.e.*, to an energy level having $n = \infty$), Thus if E_1 and E_∞ are the energies of an electron in energy level $n = 1$ and $n = \infty$, then :

∴ Ionisation energy of one-electron species with atomic number, $Z = E_\infty - E_1$

(*a*) **Calculation of ionisation energy in c.g.s. units** : E_1 and E_∞ in c.g.s. units are given by :

$$E_1 = -\frac{2\pi^2 m e^4}{h^2} \times \frac{Z^2}{(1)^2} = -2.18 \times 10^{-11} \times Z^2 \text{ erg}$$

$$E_\infty = -\frac{2\pi^2 m e^4}{h^2} \times \frac{Z^2}{(\infty)^2} = \text{Zero}$$

Thus ionisation energy of one-electron species with atomic number, $Z = E_1 - E_\infty$

$= -2.18 \times 10^{-11} \times Z^2 - \text{zero} \quad = -2.18 \times 10^{-11} \times Z^2 \text{ erg atom}^{-1}$

$= -13.6 \times Z^2 \text{ eV atom}^{-1} \quad = -13.1279 \times Z^2 \text{ erg mol}^{-1}$

Thus ionisation energy of H-atom ($Z = 1$) $= -13.6 \times 1^2 = -13.6$ eV

ionisation energy of He^+ ion ($Z = 2$) $= -13.6 \times 2^2 = -54.40$ eV

ionisation energy of Li^{2+} ion ($Z = 3$) $= -13.6 \times 3^2 = -122.40$ eV

(*b*) **Calculation of ionisation energy in S.I. units** : E_1 and E_∞ in S.I. units are given by :

$$E_1 = -\frac{m e^4}{8 \varepsilon_o^2 h^2} \times \frac{Z^2}{(1)^2} = -2.18 \times 10^{-18} \times Z^2 \text{ J atom}^{-1}$$

$$E_\infty = -\frac{m e^4}{8 \varepsilon_o^2 h^2} \times \frac{Z^2}{(\infty)^2} = \text{Zero}$$

∴ Ionisation energy of one-electron system with atomic number, $Z = E_1 - E_\infty$

$= -2.18 \times 10^{-18} \times Z^2 - \text{zero} \quad = \quad -2.18 \times 10^{-18} \times Z^2 \text{ J atom}^{-1}$

$= -13.1279 \times 10^5 \times Z^2 \text{ J mol}^{-1} = \quad -13.1279 \times 10^2 \times Z^2 \text{ kJ mol}^{-1}$

Thus :

Ionisation energy of H-atom ($Z = 1$) $= -2.18 \times 10^{-18} \times (1)^2 \text{ J atom}^{-1}$

Ionisation energy of He^+ ion ($Z = 2$) $= -2.18 \times 10^{-18} \times (2)^2 \text{ J atom}^{-1}$

Ionisation energy of Li^{2+} ion ($Z = 3$) $= -2.18 \times 10^{-18} \times (3)^2 \text{ J atom}^{-1}$

The above discussion showns that :

Ionisation energy of a H-like species with atomic number, Z

= Ionisation energy of H-atom × Atomic number of H-like species.

Achievements of Bohr's atomic model

Following are the achievements of Bohr's atomic model.

1. *Explanation of the stability of the atom : Removal of limitation of Rutherford's atomic model.* According to Bohr, the electron does not lose energy, as long as it revolves in a particular orbit. Also, it cannot jump from first orbit to lower orbit, as there is no orbit less than one. Thus, gradual loss of energy by the electron is not possible. Hence atom is stable.

2. *How is the emission spectrum produced ?* When an element is heated, its electrons absorb energy and jump to higher energy levels. In the higher energy levels, the electrons are unstable and hence jump back to the lower energy levels within 10^{-5} seconds in one or more strides and in this process, they release energy. The energy, thus released, gives the emission spectrum. Thus, we find that Bohr's atomic model explains very well the origion of the production of emission spectrum.

3. *How is the absorption spectrum produced ?* When an electromagnetic radiation is passed through the solution of a substance, some of its wavelengths are absorbed by the substance. The wavelengths absorbed are characteristic of that substance and cause the excitation of its electrons. The missing photons (from electromagnetic radiation) which are absorbed by the substance, appear as dark lines in the spectrum. Thus the Bohr's atomic model explains the origin of absorption spectrum also.

4. *How is emission spectrum of hydrogen produced ?* When an electric discharge is passed through hydrogen gas filled in a discharge tube at a very low pressure, bluish glow is obtained. When this glow is analysed by a prism, a group of spectral lines is obtained in the infrared, visible and ultraviolet region. This group of spectral lines is called the *emission spectrum of hydrogen.* This spectrum consists of several lines which can be grouped into five series, called *spectral series.* These series were named after the names of their discoverers, as shown below :–

(*i*) *Lyman series* (discovered by Lyman in 1915) which appears in the ultraviolet region of the spectrum. The wavelength of different spectral lines of this series (in Å) are 912, 931, 938, 950, 973, 1026, 1216

(*ii*) *Balmer series* (discovered by Balmer in 1835) which appears in the visible region. The wavelengths (in Å) of different spectral lines of this series are 3668, 3970, 4101, 4340, 4861, 6563.

(*iii*) *Paschen series* (discovered by Pashen in 1896) which appears in the near infrared region and the wavelength, (in Å) of different spectral lines of this series are 8207, 10050, 10938, 12818, 18751.

(*iv*) *Brackett series* (discovered by Brackett in 1922) which appears in the far infrared region and the wavelengths (in Å) of different spectral lines of this series are : 14590, 21660, 26280, 40570.

(*v*) *Pfund series* (discovered by Pfund in 1925) wich appears in the far infra-red region and the wavelength of different spectral lines of this series are (in Å) : 22810, 56360, 74800.

Bohr's atomic model explains the production of various spectral lines in the emission spectrum of hydrogen as follows :

When an electric discharge is passed through H-gas, filled in a discharge tube, at a very low pressure, the molecules of the gas break into atoms. These atoms absorb energy from the electric discharge and the single electron present in 1*st* energy level gets shifted to higher vacant energy levels 2, 3, 4 etc., depending on the amount of energy absorbed by the atoms of the gas. Now when the electron reaches in the higher energy levels, it becomes excited and hence unstable. Being unstable, it comes back to one of the lower energy levels or even to the lowest energy level ($n = 1$ level). In this process of shifting, the energy, which the electron had absorbed earlier is liberated by it, in the form of photon of light of specific frequency and wavelength, which appears as spectral lines in the emission spectrum. Thus we find that the emission spectrum of hydrogen is obtained due to the jump (*i.e.,* fall) of electron from the higher energy levels to lower energy levels. Since large number of jumps of the electron is possible, large number of spectral lines is obtained in the emission spectrum of H–atom. The various spectral lines of Lyman series are obtained when the electron jumps from *2nd, 3rd, 4th,* etc. energy levels (higher energy levels) to the *1st* energy level (ground state energy level). Thus the electronic jumps producing the spectral lines of Lyman series can be shown as $2 \rightarrow 1$, $3 \rightarrow 1$, $4 \rightarrow 1$ etc. Thus the lowest line in Lyman series corresponds to a drop of electron from *2nd* orbit to 1*st* orbit. The spectral lines of Balmer series originate when an electron jumps from 3*rd,* 4*th* etc., energy levels (higher energy levels) to *2nd* energy level (lower energy level). Thus the electronic jumps in case of Balmer series can be represented as $3 \rightarrow 2$, $4 \rightarrow 2$, $5 \rightarrow 2$ etc. In a similar manner the electronic jumps producing the spectral lines of other series can be shown as $4 \rightarrow 3$, $5 \rightarrow 3$, $6 \rightarrow 3$ etc. (Paschen series); $5 \rightarrow 4$, $6 \rightarrow 4$, $7 \rightarrow 4$ etc. (Brackett series); $6 \rightarrow 5$, $7 \rightarrow 5$ etc., (Pfund series)

To calculate the frequency (ν or $\bar{\nu}$) and wavelength (λ) of a spectral line emitted (emission spectrum) when an electron of H-atom falls from n_2 orbit (higher orbit) to n_1 orbit (lower orbit)

When an electron in H-atom falls from a higher orbit (n_2) to a lower orbit (n_1), H-emission spectrum is obtained. This spectrum consists of a large number of lines, which can be grouped into five series, namely Lyman series, Balmer series, Paschen series, Brackett series and Pfund series. When an electron falls from n_2 orbit to n_1 orbit, energy is released. This energy is equal to the difference of the energies of the electron in n_2 orbit and in n_1 orbit and is denoted by ΔE. Thus :

ΔE = Energy of the electron in n_2 orbit – Energy of the electron in n_1 orbit

or $\Delta E = E_2 - E_1$ ...(1)

Frequency of the spectral line can be expressed in S^{-1} or Hz and cm^{-1}(c.g.s. units) or m^{-1} (S.I. units).

(*a*) **Calculation of frequency in c.g.s. units.** (*i*) *Frequency ν in s^{-1} or Hz.* We know that in c.g.s. units, E_1 and E_2 are given by :

$$E_1 = -\frac{2\pi^2 m e^4}{h^2 n_1^2}, \quad E_2 = -\frac{2\pi^2 m e^4}{h^2 n_2^2}$$

$$\therefore \Delta E = E_2 - E_1 = -\frac{2\pi^2 m e^4}{h^2 n_2^2} - \left(-\frac{2\pi^2 m e^4}{h^2 n_1^2}\right) \text{ or } \Delta E = \frac{2\pi^2 m e^4}{h^2}\left[\frac{1}{n_1^2} - \frac{1}{n_2^2}\right] \quad ...(2)$$

Again we know that : $\Delta E = h\nu$ ...(3)

With the help of (2) and (3), we can write :

$$h\nu = \frac{2\pi^2 m e^4}{h^2}\left[\frac{1}{n_1^2} - \frac{1}{n_2^2}\right] \quad \text{or} \quad \nu = \frac{2\pi^2 m e^4}{h^3}\left[\frac{1}{n_1^2} - \frac{1}{n_2^2}\right] \quad ...(4)$$

On putting $\pi = 3.14$, $m = 9.108 \times 10^{-28}$g, $e = 4.8 \times 10^{-10}$ esu and $h = 6.626 \times 10^{-27}$ erg s and using the relations erg = g cm^2 s^{-2} and esu = $g^{1/2}$ $cm^{3/2}$ s^{-1}, we get the value of $2\pi^2 me^4 / h^3$ equal to 3.2906×10^{15} s^{-1} or Hz. Thus equation (4) becomes

$$\nu = 3.2906 \times 10^{15}\left[\frac{1}{n_1^2} - \frac{1}{n_2^2}\right] \text{ s}^{-1} \text{ or Hz} \quad ...(5)$$

(*ii*) *Frequency in cm^{-1} (wave number, $\bar{\nu}$):* The frequency of the spectral line in cm^{-1} is called wave number which is denoted as $\bar{\nu}$. $\bar{\nu}$ is given by :

$$\bar{\nu} = \frac{\nu}{c} \quad ...(6)$$

When the value of ν from equation (4) is put in equation (6), the latter [*i.e.*, equation (6)] reduces to

$$\bar{\nu} = \frac{2\pi^2 m e^4}{h^3 c}\left[\frac{1}{n_1^2} - \frac{1}{n_2^2}\right] \quad ...(7)$$

The term, $\frac{2\pi^2 m e^4}{h^3 c}$ in equation (7) is called *Rydberg constant of hydrogen atom* (in c.g.s. units) and is denoted by R_H. On putting $\pi = 3.14$, m = 9.108×10^{-28}g, $e = 4.8 \times 10^{-10}$ esu, $h = 6.626 \times 10^{-27}$ erg s, $c = 3 \times 10^{10}$ cm s^{-1} and using the relations erg = g cm^2 s^{-2} and esu = $g^{1/2}$ $cm^{3/2}$ s^{-1}, we get the value of $\frac{2\pi^2 m e^4}{h^3 c}$ equal to 1.0976×10^5 cm^{-1}. Thus equation (7) becomes :

$$\bar{\nu} = 1.0976 \times 10^5\left[\frac{1}{n_1^2} - \frac{1}{n_2^2}\right] \text{cm}^{-1} \quad \text{or} \quad \bar{\nu} = R_H\left[\frac{1}{n_1^2} - \frac{1}{n_2^2}\right] \text{cm}^{-1} \quad ...(8)$$

$(R_H = 1.0976 \times 10^5 \text{ cm}^{-1})$

Equation (4) is :

$$\nu = \frac{2\pi^2 m e^4}{h^3}\left[\frac{1}{n_1^2} - \frac{1}{n_2^2}\right] \quad \text{or } \nu = \frac{2\pi^2 m e^4 \times c}{h^3 c}\left[\frac{1}{n_1^2} - \frac{1}{n_2^2}\right] \quad ...(9)$$

We know that $\frac{2\pi^2 m e^4}{h^3 c}$ is Rydberg constant for H-atom (R_H) (equation 7). Thus equation (9) can be written as :

$$\nu = R_H \times c\left[\frac{1}{n_1^2} - \frac{1}{n_2^2}\right] \text{ s or Hz} \quad ...(10)$$

(b) Calculation of frequency in S.I. units. *(i) Frequency (ν) in s^{-1} or Hz.* In S.I. units E_1 and E_2 are given by :

$$E_1 = -\frac{m e^4}{8 \varepsilon_0^2 h^2 n_1^2} \quad , \quad E_2 = -\frac{m e^4}{8 \varepsilon_0^2 h^2 n_2^2}$$

$$\therefore \quad \Delta E = E_2 - E_1$$

$$= -\frac{m e^4}{8 \varepsilon_0^2 h^2 n_2^2} - \left(-\frac{m e^4}{8 \varepsilon_0^2 h^2 n_1^2}\right) = \frac{m e^4}{8 \varepsilon_0^2 h^2}\left[\frac{1}{n_1^2} - \frac{1}{n_2^2}\right] \quad \text{...(11)}$$

On putting the value of ΔE as given by equation (11) in $\Delta E = h\nu$, we get :

$$h\nu = \frac{m e^4}{8 \varepsilon_0^2 h^2}\left[\frac{1}{n_1^2} - \frac{1}{n_2^2}\right] \quad \text{or} \quad \nu = \frac{m e^4}{8 \varepsilon_0^2 h^3}\left[\frac{1}{n_1^2} - \frac{1}{n_2^2}\right] \quad \text{... (12)}$$

On putting $m = 9.108 \times 10^{-31}$ kg, $e = 1.602 \times 10^{-19}$ C, $\varepsilon_0 = 8.85 \times 10^{-12}$ J^{-1} C^2 m^{-1} and $h =$ 6.626×10^{-34} Js and utilising the relation $J = \text{kg m}^2 \text{ s}^{-2}$, we get the value of $\frac{m e^4}{8 \varepsilon_0^2 h^3}$ equal to 3.2906 $\times 10^{15}$ s^{-1}. Thus equation (12) becomes

$$\nu = 3.2906 \times 10^{15}\left[\frac{1}{n_1^2} - \frac{1}{n_2^2}\right] \text{ s}^{-1} \text{ or Hz} \quad \text{...(13)}$$

(ii) Frequency in m^{-1} (wave number, $\bar{\nu}$). Wave number, $\bar{\nu}$ is given by :

$$\bar{\nu} = \frac{\nu}{c} \quad \text{...(14)}$$

When the value of ν from equation (12) is put in equation (14), we get

$$\nu = \frac{m e^4}{8 \varepsilon_0^2 h^3 c}\left[\frac{1}{n_1^2} - \frac{1}{n_2^2}\right] \quad \text{...(15)}$$

The term, $\frac{m e^4}{8 \varepsilon_0^2 h^3 c}$ in equation (15) is called Rydberg constant of H-atom (in S.I. units). On putting $m = 9.108 \times 10^{-31}$ kg, $e = 1.602 \times 10^{-19}$ C, $\varepsilon_0 = 8.85 \times 10^{-12}$ J^{-1} $C^2 m^{-1}$, $h = 6.626 \times 10^{-34}$ Js, $c = 3 \times 10^8$ ms^{-1} and using relation $J = \text{kg ms}^{-2}$, we get the vlaue of $\frac{m e^4}{8 \varepsilon_0^2 h^3 c}$ equal to 1.0976×10^7 m^{-1}.

Thus equation (15) can be written as :

$$\bar{\nu} = 1.0966 \times 10^7\left[\frac{1}{n_1^2} - \frac{1}{n_2^2}\right] \text{m}^{-1} = R_H\left[\frac{1}{n_1^2} - \frac{1}{n_2^2}\right] \text{m}^{-1}; \; (R_H = 1.0976 \times 10^7 \text{ m}^{-1}) \text{...(16)}$$

We have shown above

$$\nu = \frac{m e^4}{8 \varepsilon_0^2 h^3}\left[\frac{1}{n_1^2} - \frac{1}{n_2^2}\right] \quad \text{[see equation (12)]}$$

$$\text{or} \quad \nu = \frac{m e^4 \times c}{8 \varepsilon_0^2 h^3 c}\left[\frac{1}{n_1^2} - \frac{1}{n_2^2}\right] \quad \text{... (17)}$$

We know that $\frac{m e^4}{8 \varepsilon_0^2 h^3}$ is Rydberg constant of H-atom (R_H). Thus equation (17) can also be written as :

$$\nu = R_H \times c\left[\frac{1}{n_1^2} - \frac{1}{n_2^2}\right] \text{ s}^{-1} \text{ or Hz} \quad \text{...(18)}$$

In the above equations n_1 and n_2 have different values for different series as shown below.

(*i*) For Lyman series $n_1 = 1$ and $n_2 = 2, 3, 4$ ....(*ii*) For Balmer series $n_1 = 2$ and $n_2 = 3, 4, 5$ ..... (*iii*) For Paschen series $n_1 = 3$ and $n_2 = 4, 5, 6$....... (*iv*) For Brackett series $n_1 = 4$ and $n_2 = 5, 6, 7$ .... (*v*) For Pfund series $n_1 = 5$ and $n_2 = 6, 7, 8$.......

Calculation of wave length (λ). $\lambda, \bar{v}, v$ and c are related as :

$$\frac{1}{\lambda} = \bar{\nu} = \frac{\nu}{c} \quad ...(19)$$

If the value of v as calculated from equations (4), (5), (10) (in c.g.s units), (12), (13), (18) (in S.I. units) is put in $\frac{1}{\lambda} = \frac{\nu}{c}$ ($c = 3 \times 10^{10}$ cm s^{-1} or 3×10^8 m s^{-1}), the value of λ is obtained in cm or m [$\frac{1}{\lambda} = \frac{\nu}{c} = \frac{s^{-1}}{cm\ s^{-1}}$ (in c.g.s units), $\frac{1}{\lambda} = \frac{\nu}{c} = \frac{s^{-1}}{ms^{-1}} = m^{-1}$ (in S.I. units)]. Similarly, since $\frac{1}{\lambda} = v$, the reciprocal of v as given by equations (7), (8) (in c.g.s. units), (15), (16) (in S.I. units) will give the value of λ.

Solved Examples

Example 1.1. *According to Bohr's theory, the electronic energy of hydrogen atom in the nth orbit is given by* $E_n = \frac{-21.76 \times 10^{-19}}{n^2}$ *J. Calculate the longest wavelength of light that would be needed to remove an electron from the third orbit of* He^+ *ion.* **(I.I.T. 1990)**

Solution: We know that the wavelength (λ) is given by :

$$\lambda = \frac{1}{\bar{\nu}} = \frac{1}{\nu/c} \quad \left(\because \bar{\nu} = \frac{\nu}{c}\right)$$

$$= \frac{c}{\nu} = \frac{c}{\Delta E/h} \quad (\because \Delta E = h\nu) \quad \text{or} \quad \lambda = \frac{ch}{\Delta E} \quad ... (i)$$

Since the energy of an electron in *nth* orbit of H-atom $= -\frac{21.76 \times 10^{-19}}{n^2}$ J, the energy of an electron in *n*th orbit of H-like species with atomic number Z $= -\frac{21.76 \times 10^{-19} \times Z^2}{n^2}$ J.

Thus the energy of an electron in *nth* orbit of He^+ ion having atomic number equal to 2(Z = 2).

$$= \frac{-21.76 \times 10^{-19} \times (2)^2}{n^2} \text{ J} = \frac{-87.04 \times 10^{-19}}{n^2} \text{ J}$$

We know that in order to have the longest wavelength, the electron will be promoted from *3rd* orbit to *4th* orbit. We, therefore, have

$$\Delta E = E_4 - E_3 = \left[-\frac{87.04 \times 10^{-19}}{4^2} - \left(-\frac{87.04 \times 10^{-19}}{3^2}\right)\right] \text{ J}$$

$$= 87.04 \times 10^{-19}\left(\frac{1}{9} - \frac{1}{16}\right) = 87.04 \times 10^{-19} \times 0.0486 \text{ J} = 4.230 \times 10^{-19} \text{ J}$$

$$\therefore \quad \lambda = \frac{ch}{\Delta E} = \frac{(3 \times 10^8 \text{ m.s}^{-1}) \times (6.624 \times 10^{-34} \text{ Js})}{4.230 \times 10^{-19} \text{ J}} = \mathbf{4.69 \times 10^{-7} m}$$

Example 1.2. *Calculate the wavelength in Å of the photon that is emitted when an electron in Bohr orbit with n = 2 returns to the orbit with n = 1 in H-atom. The ionisation potential of the ground state of H-atom is* 2.17×10^{-17} *erg.* **(I.I.T. 1982)**

Solution : We know that $\frac{1}{\lambda} = \bar{\nu} = \frac{\nu}{c}$ and $\Delta E = h\nu$. Utilising these two relations we get $\Delta E = h.\frac{c}{\lambda}$ or $\lambda = \frac{ch}{\Delta E}$.

Now $\Delta E = E_2 - E_1 = \frac{2\pi^2 m e^4}{h^2}\left(\frac{1}{1^2} - \frac{1}{2^2}\right)$ erg $= \frac{2\pi^2 m e^4}{h^2} \times \frac{3}{4}$

Again we know that ionisation potential of H-atom $= \frac{2\pi^2 m e^4}{h^2} = 2.17 \times 10^{-11}$

$\therefore \quad \Delta E = 2.17 \times 10^{-11} \times \frac{3}{4} = 1.62 \times 10^{-11}$ erg

$$\therefore \quad \lambda = \frac{ch}{\Delta E} = \frac{(3 \times 10^{10}\ \text{cm.s}^{-1}) \times (6.62 \times 10^{-27}\ \text{erg.s})}{1.627 \times 10^{-11}\ \text{erg}} = 12.20 \times 10^{-6}\ \text{cm} = \mathbf{1220\ Å}$$

Example 1.3. *According to Bohr's theory, the energy of an electron in nth orbit of hydrogen and hidrogen-like species with atomic number Z is given by the expression:*

$$E_n = -\frac{2\pi^2 m e^4 Z^2}{n^2 h^2}$$

(a) Find the expressions for the energy of an electron in nth orbit of H–atom, He^+, Li^{2+} and Be^{3+} ions.

(b) Compare the frequencies and wavelengths of the radiation emitted by H-atom with those emitted by He^+, Li^{2+} and Be^{3+} ions.

Solution : *(a) Determination of the expressions.* Since He^+ (Z = 2, $1s^1$), Li^{2+} (Z = 3, $1s^1$) and Be^{3+} (Z = 4, $1s^1$) ions are hydrogen-like ions, the expressions for the energy of an electron in the *nth* orbit of these ions will be obtained by putting Z = 2,3 and 4 respectively in the given expression. Thus :

$$E_{He^+} = \frac{-2\pi^2 m e^4 \times (2)^2}{n^2 h^2} = -\frac{8\pi^2 m e^4}{n^2 h^2}$$

$$E_{Li^{2+}} = \frac{-2\pi^2 m e^4 \times (3)^2}{n^2 h^2} = -\frac{18\pi^2 m e^4}{n^2 h^2}$$

$$E_{Be^{3+}} = \frac{-2\pi^2 m e^4 \times (4)^2}{n^2 h^2} = -\frac{-32\pi^2 m e^4}{n^2 h^2}$$

(b) Comparison of frequencies. We know that : $E = h\nu$ or $\nu = \frac{E}{h}$.

Utilising this relation, we can write the expressions for the frequencies as follows

$$\nu_H = \frac{E_H}{h} = \frac{-2\pi^2 m e^4}{n^2 h^3}; \qquad \nu_{He^+} = \frac{E_{He^+}}{h} = \frac{-8\pi^2 m e^4}{n^2 h^3}$$

$$\nu_{Li^{2+}} = \frac{E_{Li^{+2}}}{h} = \frac{-18\pi^2 m e^4}{n^2 h^3}; \qquad \nu_{Be^{3+}} = \frac{E_{Be^{3+}}}{h} = \frac{-32\pi^2 m e^4}{n^2 h^3}$$

$$\text{Thus } \frac{\nu_H}{\nu_{He^+}} = \frac{-2\pi^2 m e^4}{n^2 h^3} \times \frac{n^2 h^3}{-8\pi^2 m e^4} = \frac{1}{4}$$

$$\frac{\nu_H}{\nu_{Li^{2+}}} = \frac{-2\pi^2 m e^4}{n^2 h^3} \times \frac{n^2 h^3}{-18\pi^2 m e^4} = \frac{1}{9}$$

$$\frac{\nu_H}{\nu_{Be^{3+}}} = \frac{-2\pi^2 m e^4}{n^2 h^3} \times \frac{n^2 h^3}{-32\pi^2 m e^4} = \frac{1}{16}$$

(c) Comparison of wavelengths. We know that $\lambda = \frac{1}{\nu}$ Thus, utilising this relationship, we can write:

$$\lambda_H = \frac{1}{\nu_H} = \frac{h}{E_H} = \frac{n^2 h^3}{-2\pi^2 m e^4}; \qquad \lambda_{He^+} = \frac{n^2 h^3}{-8\pi^2 m e^4}$$

$$\lambda_{Li^{2+}} = \frac{n^2 h^3}{-18\pi^2 m e^4} \qquad \lambda_{Be^{3+}} = \frac{n^2 h^3}{-32\pi^2 m e^4}$$

$$\text{Thus : } \frac{\lambda_H}{\lambda_{H^+}} = \frac{4}{1}, \quad \frac{\lambda_H}{\lambda_{Li^{2+}}} = \frac{9}{1} \text{ and } \frac{\lambda_H}{\lambda_{Be^{3+}}} = \frac{16}{1}$$

Example 1.4. *Calculate the frequency of the spectral line emitted when the electron in n = 3 in hydrogen atom de-excites to ground state (Rydberg constant = 109737 cm^{-1}).* **(R.E.E. 1985)**

Solution : We know that when the value of Rydberg constant (*R*) is given in cm^{-1}, wave number of the spectral line is calculated first. $\bar{v}$ is given by :

$$v = R\left[\frac{1}{n_1^2} - \frac{1}{n_2^2}\right] = 109737 \times \left[\frac{1}{1^2} - \frac{1}{3^2}\right] \text{cm}^{-1} = 109737 \times \frac{8}{9} = 97544 \text{ cm}^{-1}$$

Now frequency v of the spectral line is given by:

$$v = c \times \bar{v} = (3 \times 10^{10} \text{ cm.s}^{-1}) \times (97544 \text{ cm}^{-1}) = \mathbf{292532 \times 10^{10}\ s^{-1}}$$

Example 1.5. *The electron energy (E) in H-atom is given by:* $E = \dfrac{-21.7 \times 10^{-12}}{n^2}$ *erg. Calculate the energy required to remove an electron completely from the n = 2 orbit. What is the longest wavelength (in cm) of light that can be used to cause this transition (R = 109686 cm^{-1}).* **(I.I.T. 1984)**

Solution: (*i*) We know that energy required is given by :

Energy required = Energy of the electron at infinite distance – Energy of the electron in $n = 2$ orbit.

$$= E_\infty - E_2$$

$$= \left[\frac{-21.7 \times 10^{-12}}{(\infty)^2} - \left(\frac{-21.7 \times 10^{-12}}{(2)^2}\right)\right] \text{erg}$$

$$= \frac{21.7 \times 10^{-12}}{(2)^2} \text{erg} = \frac{5.425 \times 10^{-12}}{(\text{v})} \text{erg}$$

(*ii*) The wave number ($\bar{v}$) of the transmitted radiation is given by :

$$\bar{v} = R\left[\frac{1}{2^2} - \frac{1}{\infty^2}\right] \text{cm}^{-1} = 109686 \times \frac{1}{4} \text{cm}^{-1} = 27421.5 \text{ cm}^{-1}$$

$$\text{Now } \lambda = \frac{1}{\bar{\text{v}}} = \frac{1}{27421.5 \text{ cm}^{-1}} = 0.0000364 \text{ cm}$$

$$= 0.0000364 \times 10^5 \times 10^{-5} \text{ cm} = \mathbf{3.64 \times 10^{-5}\ cm}$$

Example 1.6. *Calculate the frequency, energy and wavelength of the radiation corresponding to the spectral line of the lowest frequency in Lyman series in the spectrum of H-atom. Also calculate the energy for the corresponding line in the spectra of Li^{2+} ion. Given that $R_H = 1.09678 \times 10^7 m^{-1}$, $c = 3 \times 10^8$ ms^{-1} and $h = 6.625 \times 10^{-34}$ Js.* **(Roorkee 1991)**

Solution : We know that $\dfrac{1}{\lambda} = \bar{\text{v}} = R_H\left(\dfrac{1}{n_1^2} - \dfrac{1}{n_2^2}\right)$

Since the lowest line in Lyman series corresponds to a drop of the electron from *2nd* orbit to *1st* orbit, $n_1 = 1$ and $n_2 = 2$. Thus the above equation becomes :

$$\frac{1}{\lambda} = (1.09678 \times 10^7 \text{ m}^{-1})\left[\frac{1}{1^2} - \frac{1}{2^2}\right] = (1.09678 \times 10^7 \text{ m}^{-1} \times \frac{3}{4}$$

$$\therefore \quad \lambda = \frac{4}{3 \times 1.09678 \times 10^7 \text{ m}^{-1}} = \mathbf{1.215 \times 10^{-7}\ m}$$

Frequency (*v*) is given by :

$$v = \frac{c}{\lambda} = \frac{3 \times 10^8 \text{ ms}^{-1}}{1.215 \times 10^{-7} \text{ m}} = \mathbf{2.46 \times 10^{15}\ s^{-1}}$$

The energy, $(\Delta E)_H$ for the radiation corresponding to 2 → 1 transition in H-atom is given by :

$$(\Delta E)_H = \frac{2\pi^2 m e^4}{h^2}\left(\frac{1}{n_1^2} - \frac{1}{n_2^2}\right) \quad \text{... } (i)$$

Now since $R_H = \frac{2\,\pi_3\; m\,e^4}{h^3 c} = \frac{2\,\pi^2\;me^4}{h^2.\;hc}, \quad \frac{2\,\pi^2\;m\,e^4}{h^2} = R_H \times hc$

On putting $\frac{2\,\pi^2\;m\,e^4}{h^2} = R_H \times hc$ in equation (1), we get :

$$(\Delta E)_H = R_H \times hc\left(\frac{1}{n_1^2} - \frac{1}{n_2^2}\right)$$

$$= (1.09678 \times 10^7 \text{m}^{-1}) \times (6.625 \times 10^{-34} \text{ Js}) \times (3 \times 10^8 \text{ ms}^{-1}) \times \frac{3}{4} = \mathbf{16.34 \times 10^{-19}\ J}$$

Energy corresponding to the radiation in the spectra of Li^{2+} ion is given by :

$$(\Delta E)_{Li^{+2}} = (\Delta E)_H \times (\text{Atomic number of Li-atom})^2$$

$$= (16.34 \times 10^{-19} \text{ J}) \times 3^2 = \mathbf{147.06 \times 10^{-19}\ J}$$

Example 1.7. *Electromagnetic radiation of wavelength 242 nm is just sufficient to ionise sodium atom. Calculate the ionisation energy of Na-atom.* **(Roorkee 1992)**

Solution : Ionisation energy of Na-atom is the energy associated with that electromagnetic radiation whose wavelength is 242 nm. Thus we should determine the energy associated with this radiation.

$$E = h\nu = h\,.\,\frac{c}{\lambda}$$

$$= \frac{(6.626 \times 10^{-34} \text{ Js}) \times (3 \times 10^8 \text{ ms}^{-1})}{242 \times 10^{-9} \text{ m}} = \mathbf{0.082 \times 10^{-17}\ J}$$

Example 1.8. *Calculate the velocity (cm/sec) of an electron placed in the third orbit of H-atom. Also calculate the number of revolutions per second that this electron makes around the nucleus.* **(Roorkee 1987)**

Solution : (*i*) We know that the velocity of an electron moving in *n*th orbit is given by :

$$v_n = \frac{0.2183 \times 10^9 \times Z}{n} \text{ cms}^{-1}$$

On putting $Z = 1$ and $n = 3$, we get,

$$v_3 = \frac{0.2183 \times 10^9 \times 1}{3} = \mathbf{0.072766 \times 10^9\ cm\ s^{-1}}$$

(*ii*) Number of revolutions made by the moving electron in *nth* orbit is equal to $\frac{65.711 \times 10^{14} \times Z^2}{n^3}$ s^{-1}.

Thus the number of revolutions made by the electron revolving in *3rd* orbit.

$$= \frac{65.711 \times 10^{14} \times 1^2}{3^3} \text{ s}^{-1} = \mathbf{2.43 \times 10^{14}\ s^{-1}}$$

Lmitations of Bohr's theory

(*i*) Bohr's theory fails to explain the origin of the spectra given by multi-electron species. It only explains the origin of the spectrum given by one-electron species like H, He^+, Li^{2+}, Be^{3+} etc.

(*ii*) According to Bohr, circular orbits, in which the electrons revolve, are planar. But modern research has shown that an electron moves round the nucleus in the three dimentional space (*x, y, z* axes).

(*iii*) *No explanation for Zeeman and Stark effects*. It has been observed that when an excited atom, which gives a line emission spectrum, is put in magnetic field, its spectral lines are further split up into still thinner (*i.e.*, closely-spaced) lines. This type of splitting of spectral lines into a number of closely-spaced lines is called *Zeeman effect.* Another similar splitting of the spectral lines into a number of thinner lines is also observed when the excited atom is placed in an electric field. This type of splitting is called *Stark effect*. Bohr's theory has no explanation for any of these effects.

(*iv*) *Dual nature of electron and Heisenberg's uncertainty principle.* We have seen that Bohr had assumed that an electron behaves as a *material particle* of small mass moving round the nucleus in fixed circular orbits. But de Broglie suggested that *an electron has a dual nature, i.e., an electron, like light, behaves both as a material particle and as a wave.*

Further, Bohr's assumption that the electron revolves round the nucleus in circular orbits situated at fixed distance from the nucleus and with a fixed velocity is against Heisenberg's uncertainty principle.

(*v*) *No explanation for fine structure.* When a certain single line in the spectrum is seen with the help of a powerful microscope, it is found to split into a number of closely-spaced lines. Bohr's theory has no explanation for the splitting of a single line, since, according to Bohr, one and only one spectral line is produced when an electron jumps from one orbit to the other. The spectrum consisting of closely-spaced lines is called *fine structure* or *multiple structure.*

Sommerfeld's Extension of Bohr's Atomic Model : Bohr-Sommerfeld's Atomic Model

To account for the *fine structure* in the spectrum of hydrogen atom, Sommerfeld extended Bohr's theory in 1915 and suggested *that the moving electron might describe, in addition to the circular orbits, elliptical orbits as well, with the nucleus situated at one of the focii of the ellipse.* The two velocities *viz radial velocity* (which is along the radius vector) and *transverse* or *angular* velocity (which is perpendicular to radius vector) of an electron moving in an elliptical orbit, give rise to *radial momentum* and *angular* (or *azimuthal*) *momentum,* which, according to Summerfeld, are equal to $n_r \cdot h / 2\pi$ and $n\phi . h / 2\pi$ respectively. Here the integrals n_r and n_ϕ are two new quantum numbers which are called *radial quantum number* and *azimuthal quantum number* respectively n_r and n_ϕ are related to Bohr's quantum number, n as $n = n_r + n_\phi$. n and n_ϕ are related to the geometry of the ellipse as :

$$\frac{n}{n_\varphi} = \frac{n_r + n_\varphi}{n_\varphi} = \frac{\textit{Length of the major axis } (a)}{\textit{Length of the minor axis } (b)} \quad \text{... } (i)$$

Now let us consider the following possibllities :

(1) *When* $n_\phi = 0$. In this case the length of the minor axis (*b*) of the ellipse will also become equal to zero. In other words, the electronic path will be a straight line, passing through the nucleus, which is impossible. Thus n_ϕ *cannot be equal to zero.*

(2) *When* $n_\phi > n$. In this case the length of the minor axis (*b*) will be greater than the length of the major axis (*a*). Thus n_ϕ *cannot be greater than n* .

(3). *When* $n_\phi = n$. In this case *a* becomes equal to *b* and the orbit becomes circular. Thus circular orbit is a special case of elliptical orbit.

(4) *When* $n_\phi < n$. In this case the orbit will remain elliptical.

Thus, when $n = 4$, $n_\phi = 1, 2, 3, 4$. Corresponding to these four values of n_ϕ for $n = 4$, we have four possible orbits which are designated as 4_1, 4_2, 4_3 and 4_4 respectively. The subscripts represent the values of n_ϕ corresponding to $n = 4$. Quite obviously, one orbit (4_4 orbit) is circular, while the other three orbits *viz.* 4_3, 4_2 and 4_1 are elliptical.

Applying the principle of quantisation of momentun to the electron moving in an elliptical orbit, Sommerfeld deduced the energy of the electron in the hydrogen atom as :

$$E_n = -\frac{2\pi^2 e^4 m}{(n_\varphi + n_r)^2 h^2}$$

Thus, we find that the energy depends not only on the principal quantum number, *n* but also to some extent on the azimuthal quantum number, n_ϕ. The frequency of the radiation emitted as a result of transition of an electron from a level of energy E_2 to a level of energy E_1 ($E_2 > E_1$) would, therefore, vary, depending on the various values of n_ϕ and as a result a group of closely-spaced lines would appear instead of a single line in the spectrum of hydrogen atom or hydrogen like atoms.

The concepts advanced by Sommerfeld evidently postulate existence of sub-energy level for electrons in each of the principal energy levels of the atom.

Weakness of Bohr–Sommerfeld's atomic model

Experimental observations and theoretical treatment, based on wave mechanics show that the different values of azimuthal quantum number are : 0, 1, 2, 3, ..., $(n - 2)$, $(n - 1)$ (total *n* values) and not 1, 2, 3, ...$(n-1)$, *n* (total *n* values). The new quanntum number beginning with zero and ending at $(n - 1)$ has been represented by *l* in order to avoid the confusion with n_ϕ. Thus *l* is related with n_ϕ and n by the relation: $n_\phi = l + 1$ where $l = 0, 1, 2, ...(n-2), (n-1)$. With the help of this relation, relation (*i*) becomes as :

$$\frac{n}{l+1} = \frac{a}{b}$$

Sommerfeld's theory cannot give the correct number of lines observed in fine structure. Moreover, it gives no information about the relative intensities of the *fine lines*. Further, the exact definition of position and momentum is contrary to the uncertainty principle.

University Questions

Rutherford's atomic model

1. Explain Rutherford's model of the atom. What were its defects? Discuss Bohr's and Sommerfeld's modifications to remove three defects. *(Vikram 87, Bundelkhand 87)*

Bohr's atomic model

2. (*a*) What is Ritz combination principle? (*b*) Explain fully the term "Quantisation of Energy". *(Kanpur 92)*
3. Discuss the main postulates of Bohr's atomic model. *(Guwahati 89, Agra 93, E.C.C. Allahabad 94)*
4. (*a*) Derive an expression in C.G.S. unit for the radii of various Bohr's orbit in H-atom. (*b*) Calculate the frequency of the line in the emission spectrum of H-atom when the electrons in the atoms are in 3*rd* energy level. *(Tripura 91, Guwahati 91)*
5. (*a*) How will you explain the existance of various lines in H-spectrum with the help of Bohr's model of atomic structure. *(Banglore 92)*
   (*b*) What were the limitations of the Borh's model. *(G.N.D. 92)*
6. Derive an expression for the radius of any orbit in H-atom. *(Guwahati 89)*
7. What are the limitations of Bohr's Sommerfeld's model? *(Kanpur 93)*
8. Discuss photoelectric effect. What conclusion do you draw from it ? *(Guwahati 91)*

Numerical problems

1. Calculate the energy of transition involving $n_1 = 6$ to $n_2 = 3$ in H-atom ($R = 109, 737.32\ cm^{-1}$, $h = 6.63 \times 10^{-34}$ J). *(Guwahati 89)*
2. Calculate wavelength associated with an electron moving with a velocity of $1 \times 10^8\ cm\ s^{-1}$. Mass of electron = 9.1×10^{-28} g. **(Ans:** 7.273×10^{-8} cm) *(Bangalore 92)*
3. In a H-atom, an electron jumps from a third orbit to the first orbit. Find out the frequency and wavelength of the spectral line. *(Roorkee 89)* **(Ans:** $2.926 \times 10^{15}\ s^{-1}$)
4. Calculate the longest wavelength of light that will be needed to remove an electron from the 3rd orbit of He^+ ion. **[Ans:** 2.053×10^{-7} m = 2053 Å] *(IIT 1990)*
5. Calculate the frequency, energy and wavelength of the radiation corresponding to the spectral line of lowest frequency in Lyman series in the spectrum of H-atom. Also calcualte the energy for the corresponding line in the spectrum of Li^{2+} ion. Given $R_H = 1.09678 \times 10^7 m^{-1}$, $c = 3.0 \times 10^8\ ms^{-1}$ and $h = 6.625 \times 10^{-34}$ Js. *(Roorkee 1991)* **[Ans:** 1.636×10^{-18} J]
6. Estimate the difference in energy between 1*st* and 2*nd* Bohr_orbit for a H-atom. At what minimum atomic number, a transition from $n = 2$ to $n = 1$ energy level would result in the emission of X-rays with $\lambda = 3.0 \times 10^{-8}$ m? Which H-atom like species does this atomic number correspond to ? *(IIT 1993)* **[Ans:** Z = 2, He^+]
7. Electromagnetic radiations of wavelength 242 nm in just suffecient to ionise the sodium atom. Calculate the ionisation energy of sodium in kJ mol^{-1}. Given that $h = 6.62 \times 10^{-34}$ Js. *(Roorkee 1989, 92)* **(Ans:** 4.936×10^2 kJ mol^{-1})
8. Write the demerits of Bohr's atomic model. *(Nagpur 2003)*

2

Wave Mechanical Concept of Atomic Structure

de Broglie's concept of dual character of matter : de Broglie's wave equation

Einstein in 1905 suggested that light shows dual character, *i.e.*, light behaves both as a material particle as well as a wave. de Broglie in 1923 extended Einstein's view and said that all forms of matter like electrons, protons, neutorns, atoms, molecules etc. also show dual character. He further said that wavelength (λ) of a moving particle of mass m and moving with a velocity v is given by : $\lambda = h / mv$ (h = Planck's constant). This equation is called *de Broglie's wave equation* and λ is called *de Borglie's wave length.*

Derivation of de Broglie's equation

We know that the energy (E) of a photon is given by : $E = h\nu = hc / \lambda$ (Planck's equation) or by $E = mc^2$ (Einstein's equation). Here ν = frequency, c = velocity of light. λ = wavelength, m = mass of photon. On combining the above two equations we get $hc / \lambda = mc^2$ *or* $h/\lambda = mc$ or $\lambda = h/mc$. Thus for another particle moving with a velocity v, $\lambda = h/mv = h/p$ where $p = mv$ = momentum of the particle.

Experimental verification of de Broglie's equation

If an electron of charge e is accelerated by an applied potential, V (in *e.s.* units), its kinetic energy (K.E.) is given by : K.E. = Ve. Also the magnitude of K.E. of an electron moving with a velocity v is also given by : K.E $= \frac{1}{2} mv^2$. On combining these two relations we have K.E. $= Ve = \frac{1}{2} mv^2$ or $v = \sqrt{2\, Ve/m}$.

Now on putting the value of v obtained as above in de Broglie's equation ($\lambda = h / mv$), we get $\lambda = h / \sqrt{2\, Ve/m}$. On putting $h = 6.624 \times 10^{-34}$ Js; $e = 1.602 \times 10^{-19}$ C; $m = 9.108 \times 10^{-31}$ kg, the above equation becomes: $\lambda = \frac{12.26}{\sqrt{\text{V volts}}}$ Å (*de Broglie's equation*)

If a potential difference of 10 volts is applied, the wavelength, λ of the electron-wave emerging out would be equal to 3.877 Å. Similarly, if the potential difference is varied between 10 and 10,000 volts, λ should vary between 3.877 Å and 0.1226 Å. It is well known that X-rays have the wavelengths in this range.

Experimental verification of wave nature of electrons by Davison and Germer experiment

Wave nature of electrons was confirmed experimently by Davison and Germer (1927-28). They produced a beam of electrons from a heated tungsten filament and accelerated this beam by passing it through an electric field. The accelerated beam was allowed to fall on a nickel crystal which acted as a grating. The electrons got scattered by the crystal and were received on a photographic plate on which the electrons produced diffraction rings (Fig. 2.1). These rings were found to be similar to those produced by X-rays, when they are diffracted by the same nickel crystal. Since X-rays have wave character, electrons also should have a wave character.

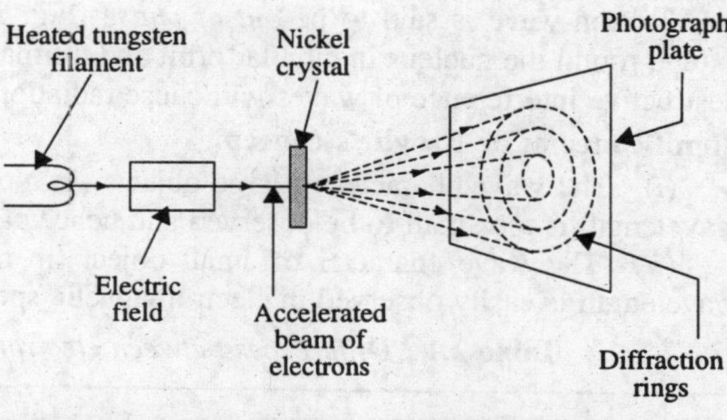

Fig. 2.1 : Davisson and Germer's experiment to verify the wave nature of electrons.

Confirmation of quantisation of angular momentum (Bohr's theory) by de Broglie's concept

Bohr's principle of quantisation of angular momentum of a moving electron can be derived from de Broglie's concept of wave nature of electron. This can be done as follows :

According to Bohr's principle of quantisation of angular momentum of a moving electron (Bohr had assumed an electron as a material particle), an electron can move only in that orbit in which the angular momentum ($= mvr$) is an integral whole number multiple of $h/2\pi$, *i.e.* $mvr = n \,.\, h/2\pi$ (*Bohr's equation*), where m = mass of the electron, v = velocity of the electron, r = radius of the orbit in which the electron is revolving, n = an integral whole number which can have values 1, 2 etc., and h = Planck's constant.

Now according to de Broglie an electron behaves as a standing or stationary wave which extends round the nucleus in a circular orbit (Fig. 2.2). If the two ends of the electron-wave meet to give a regular series of crests and troughs, the electron-wave is said to be *in phase i.e.* there is constructive interference of electron waves and the electron motion has a character of a standing wave or *non-energy radiating motion.* [Fig. 2.2 (*a*)]. Whatever be the path of the electron-wave round the nucleus, it is a necessary condition to get an electron-wave in phase so that the circumference of the Bohr's orbit ($= 2\pi r$) is equal to the whole number multiple of the wavelength (λ) of the electron wave, *i.e.*

$$2\pi r = n.\lambda \quad \text{or} \quad \lambda = \frac{2\pi r}{n} \qquad \text{... }(i)$$

Here n is a whole number which denotes the number of wave lengths associated with an electron-*wave extending round the nucleus.*

Now according to de Broglie, the wave length λ of an electron which behaves as a wave is given by

$$\lambda = \frac{h}{mv} \qquad (\textit{de Broglie's equation}) \qquad \text{... }(ii)$$

On comparing equations (*i*) and (*ii*) we get :

$$\frac{2\pi r}{n} = \frac{h}{mv} \quad \text{or} \quad mvr = n\,.\,\frac{h}{2\pi} \qquad (\textit{Bohr's equation})$$

This equation which is called Bohr's eqation clearly shows that *Bohr's assumption, namely the angular momentum* ($= mvr$) *of an electron is a whole number multiple of $h/2\pi$ is also confirmed only if an electron is supposed to behave as a wave as suggested by de Broglie.* In other words de Broglie's and bohr's concepts are in agreement with each other.

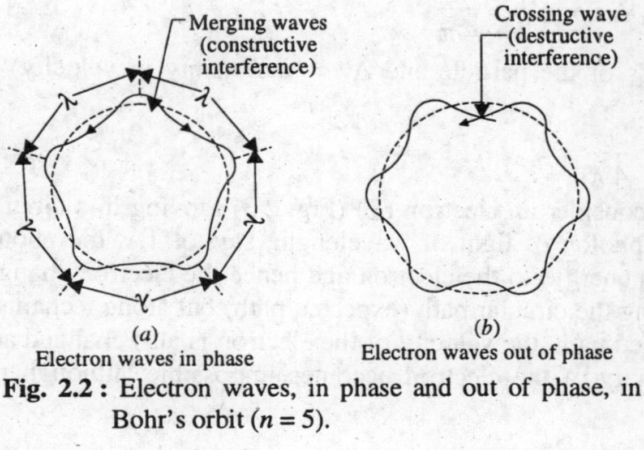

Fig. 2.2 : Electron waves, in phase and out of phase, in Bohr's orbit ($n = 5$).

If mvr is not a whole number multiple of $h/2\pi$ or if the circumference of the Bohr's orbit ($= 2\pi r$) is bigger or smaller than $n\lambda$, the electron-wave is said to be *out of phase* [Fig. 2.2 (*b*)] and hence the electron-wave will not extend round the nucleus in circular orbit and ultimately will come out of the Bohr's orbit, *i.e.*, the destructive interference of waves will cause radiation of energy.

Significance of de Broglie's concept

(*i*) The wave character of large objects (in motion) has no practical significance, since their wavelength is too small to be observed and hence cannot be measured.

(*ii*) The wave character of small object (in motion) has practical significance, since their wavelength is easily observed in electromagnetic spectrum.

Table 2.1 : *Differences between electromagnetic waves and matter waves*

Electromagnetic waves	Matter waves
1. According to clark Maxwell, these waves have electriç as well as magnetic field associated with them and hence are called electromagnetic waves. The electric and magnetic fields and the direction of propagation of waves are perpendicular to each other.	1. According to de Broglie the moving microscopic particles, because of their wave nature, are associated with waves which are called matter waves. These waves are not associated with electric and magnetic fields.
2. These waves can pass through vacuum.	2. These waves cannot pass through vacuum.
3. All electromagnetic waves travel with the velocity of light (= 3×10^{10} $cm.s^{-1}$)	3. These waves travel with different velocity as these are associated with different particles. The velocity of these waves is less than that of light.
4. These have physical significance, since their wavelengths can easily be measured.	4. These waves for large objects have no physical significance as their wavelengths are too small and hence cannot be measured.

Heisenberg's uncertainty principle

According to this principle, it is not possible to determine simulteneously and precisely both the position and momentum(or velocity) of a microscopic moving particle like electron, proton etc.

Mathematically this principle can be expressed as:

$$\Delta x \times \Delta p \geq \frac{h}{4\pi} \quad ...(i)$$

Here Δx = uncertainty involved in the position Δp = uncertainty in momentum and h = Planck's constant. The sign $\geq$ means that the product of Δx and Δp is equal or grater than $h/4\pi$ but never less than $h/4\pi$. The above relation shows Δx and Δp are inversely proportional to each other which means that if Δx is very small (*i.e.*, if the position of the particle is determined more or less exactly), Δp would be large (*i.e.*, uncertainty involved in the momentum will be large).

Similarly, if an attempt is made to measure the momentum (velocity) of the particle exactly, the uncertainty in the measurement of its position becomes large.

Generally the relation (*i*) is written as: $\Delta x \times \Delta p = \frac{h}{4\pi}$ ... (*ii*)

On putting $\Delta p = m \times \Delta v$ (m = mass of the particle and Δv = uncertainty in velocity) in equation (*ii*), we get,

$$\Delta x \times (m \times \Delta v) = \frac{h}{4\pi} \quad ...(iii)$$

Illustration of the principle : Let us consider an electron (*E*) (Fig. 2.3) moving in a circular path. When this electron colloides with a photon of light of wavelength, say, of 1Å, the photon transfers some of its energy (due to its high energy) to the electron and hence the electron changes its path, *i.e.* the electron does not move along the circular path (expected path) but along a changed path as shown in Fig. 2.3. Due to the changed path, the velocity of the electron is also changed and hence the determination of the exact velocity of the electron becomes impossible, although the measurement of the position or location of the electron is certain, since it can be observed by means of a microscope. On the other hand, if the electron is collided with a photon of light of wavelength greater than Å, the photon, due to its low energy, is not able to change the circular path of the electron and hence the velocity of the electron is not changed. Thus, the determination of the exact velocity of the electron is possible, but the position or location of the electron cannot be determined accurately. Thus, we see that the exact determination of the position and velocity (or momentum) of the electron cannot be made simultaneously.

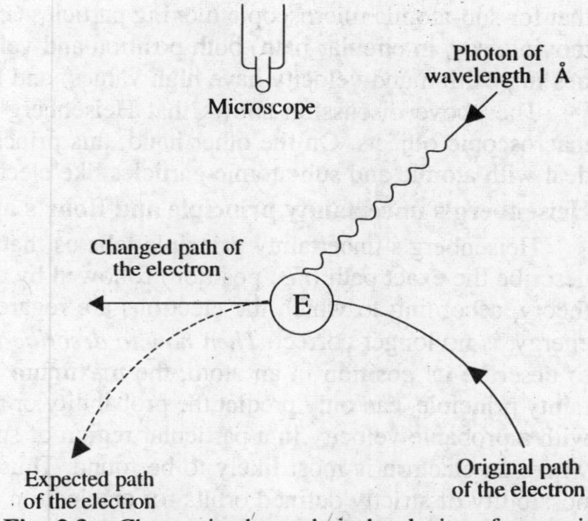

Fig. 2.3 : Change in the path and velocity of a moving electron by the impact of a photon of light of wavelength equal to 1Å.

Significance of the principle for macroscopic objects and atomic / sub-atomic particles

In order to discuss this topic, let consider the following two examples.

Problem 1. A cricket ball weighs 200 g. If the uncertainty in its position is 5 pm, what is the uncertainty in the velocity of the ball.

Solution : According to Heisenberg's uncertainty principle,

$$\Delta x \times \Delta p = \frac{h}{4\pi} \quad \text{or} \quad \Delta x \times (m \times \Delta v) = \frac{h}{4\pi} \quad (\because \Delta p = m \times \Delta v)$$

$$\text{or} \quad (5 \text{ pm}) \times (200 \text{ g}) \times \Delta v = \frac{6.626 \times 10^{-34} \text{ Js}}{4 \times 3.14}$$

$$\text{or} \quad \Delta v = \frac{0.05275 \times 10^{-34} \text{ Js}}{(5 \times 10^{-12}\, m) \times (200 \times 10^{-3} \text{ kg})} = 0.5275 \times 10^{-22} \times \frac{\text{Js}}{\text{m} \times \text{kg}}$$

$$= 5.275 \times 10^{-23} \times \frac{\text{kg m}^2 \text{ s}^{-2} \text{ s}}{\text{m kg}} \quad (\because \text{J} = \text{kg m}^2 \text{ s}^{-2}) = 5.275 \times 10^{-23} \text{ m.s}^{-1}$$

Problem 2. According to Bohr's theory of H-atom, the velocity of an electron in the 1st orbit is 2.183×10^6 ms^{-1}. If the uncertainty in the position of the electron is 5 pm, what will be the uncertainty in its velocity?

Solution : We know that : $\Delta x \times \Delta p = \frac{h}{4\pi}$ or $\Delta x \times (m \times \Delta v) = \frac{h}{4\pi}$

$$\text{or} \quad (5 \text{ pm}) \times (9.108 \times 10^{-31} \text{ kg}) \times \Delta v = \frac{6.626 \times 10^{-34} \text{ Js}}{4 \times 3.14}$$

$$\text{or} \quad (5 \times 10^{-12} \text{ m}) \times (9.108 \times 10^{-31} \text{ kg}) \times \Delta v = 0.5275 \times 10^{-34} \text{ Js}$$

$$\text{or} \quad \Delta v = \frac{0.5275 \times 10^{-34} \text{ Js}}{(5 \times 10^{-12} \text{ m}) \times (9.108 \times 10^{-31} \text{ kg})} \quad \text{or} \quad \Delta v = 1.158 \times 10^7 \text{ ms}^{-1}$$

Discussion of the results

The value of the uncertainty in velocity, as calculated in example 1 shows that this value is small (*i.e.*, negligible) in comparison to the velocity, which an object of this size can have. This means that for *macroscopic objects* (bigger objects) both position and velocity have precise values (*i.e.*, uncertainties in position and velocity are very small) and hence can be simultaneously determined to a reasonable degree of accuracy.

The value of the uncertainty in velocity, as calculated in example 2, indicates that this value is greater than the velocity of the electron in the *1st* orbit of H-atom (= 2.183×10^6 ms^{-1}). This means that for sub-atomic microscopic moving particles (*i.e.*, for extremely small particles like an electron moving, say, in circular path) both position and velocity cannot have precise value (*i.e.*, uncertainties in position and velocity have high values) and hence cannot be determined with accuracy.

The above discussion shows that Heisenberg's uncertainty principle has no significance for macroscopic objects. On the other hand, this principle has very important consequences when we deal with atomic and sub-atomic particles like electrons, protons etc.

Heisenberg's uncertainty principle and Bohr's atomic model

Heisenberg's uncertainty principle tells us that, due to the wave nature of electron, we cannot describe the exact path (*i.e.*, position) followed by an electron in an atom at any time. Thus, Bohr's theory, according to which the electrons are regarded as having fixed position (orbits) of definite energy, is no longer correct. *Then how to describe the position of an electron in an atom?* In order to describe its position in an atom, the maximum that we can do, is that we, according to uncertainty principle, can only predict the probability or relative chance of finding or locating an electron with a probable velocity in a particular region of space round the nucleus, *i.e.*, we can only predict where an electron is most likely to be found. Thus, we see that uncertainty principle rules out the possibility of strictly defined orbits for an electron, as suggest by Bohr.

Scattering of X-rays—Compton effect

In 1923 during the course of his experiments on the scattering of X-rays by solid matter (called *scattering substance* or *scatterer),* A.H. Compton found that when X-rays of high energy are allowed to fall on solid matter like carbon block or some other light element, an electron is ejected from the carbon block and X-rays are scattered from their original path. In this process a photon from the incident X-rays collides with the loosely bound electron in the carbon block (scatterer)and, after receiving an impact from the photon, the electron which was initially at rest gains some velocity and hence moves below the direction of the incident X-rays, while the photon is deflected above. Thus, we see that X-rays are scattered from their original path as a result of the collision between the photon and the electron. The collision between the photon and the electron can be regarded as a collision between two balls—one moving and the other at rest.It has been found that the scattered X-rays have longer wavelength (*i.e.* lower frequency or lower energy) than the incident X-rays. If the wavelengths of the incident X-rays and the scattered X-rays are λ, λ' respectively, then hc/λ and hc/λ' are the energies associated with the photon of the incident and scattered X-rays respectively. Thus the decrease in energy is equal to $hc/\lambda - hc/\lambda'$, since $hc/\lambda > hc/\lambda'$. This energy ($= hc/\lambda - hc/\lambda'$), is transferred to the electron at rest and therefore, the electron gains some velocity and moves from its path in this process. *This phenomenon in which there is a change in wavelength of the scattered X-rays is called Compton effect.* The angle through which the incident X-rays are deflected from their original path is known as *angle of scattering* and is denoted by θ. Thus θ is the angle between the incident and the scattered X-rays (See Fig. 2.4).

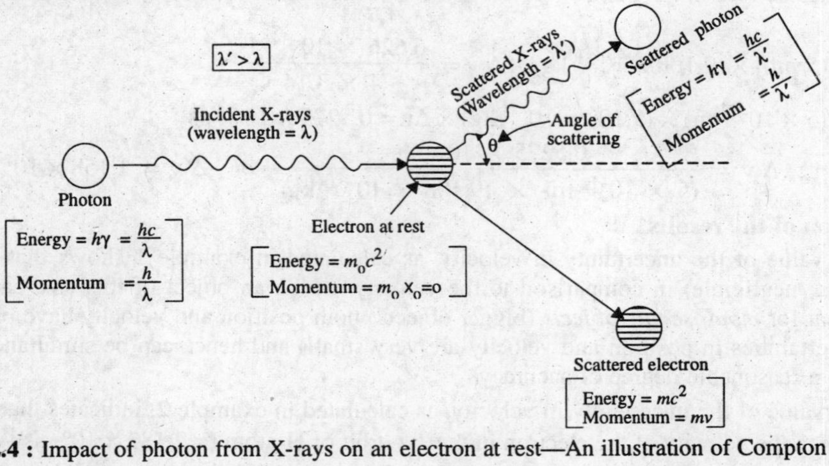

Fig. 2.4 : Impact of photon from X-rays on an electron at rest—An illustration of Compton effect

By applying the laws of conservation of energy and of momentum, and by assuming that X-rays are composed of photons each of which has energy equal to $h\nu$, Compton showed that the wavelength, λ' of the scattered X-rays and the wavelength, λ of the incident X-rays are related to each other by the relation:

$$\lambda' - \lambda = \frac{h}{mc}(1 - \cos\theta) \qquad ...(i)$$

where h is the Planck's constant, m is the rest mass of the electron and c is the velocity of light. The difference in λ' and λ [*i.e.*, $(\lambda' - \lambda)$] is called *Compton wavelength shift* and is denoted by $\Delta\lambda$. Thus :

$$\textit{Increase in wavelength} = \Delta\lambda = \lambda' - \lambda = \frac{h}{mc}(1 - \cos\theta). \qquad ...(ii)$$

This equation clearly shows that $\Delta\lambda$ depends on the magnitude of the angle of scattering, θ and is produced as a result of the collision between the photon of incident X-rays and electron at rest (photon-electron collison). This equation also shows that $\Delta\lambda$ is indepedent of the nature of the scattering substance and the wavelength of the incident X-rays. The term h/mc is called *Compton wavelength* whose value, in S.I. units, is equal to 2.424×10^{-12} metre as shown below :

$$\frac{h}{mc} = \frac{(6.626 \times 10^{-34} \text{ Js})}{(9.108 \times 10^{-31} \text{ kg}) \times (3 \times 10^{8} \text{ ms}^{-1})} = \frac{6.626 \times 10^{-11} \text{ Js}}{(9.108 \times 3) \times \text{kg ms}^{-1}}$$

$$= 0.2424 \times 10^{-11} \times \frac{\text{kg m}^2 \text{ s}^{-2} \text{ s}}{\text{kg m s}^{-1}} \qquad [\because \text{J} = \text{kg m}^2 \text{ s}^{-2}]$$

or $$\frac{h}{mc} = 0.2424 \times 10^{-11} \text{ m}$$

Thus eqation (*ii*) can be written as: $\Delta\lambda = \lambda' - \lambda = 0.2424 \times 10^{-11} (1 - \cos\theta) \text{ m}$ ...(*iii*)

Equation (*iii*) can be used to calculate the value of λ' (wavelength of the scattered beam of X-rays), if λ (wavelength of the incident beam of X-rays) and angle (through which the incident beam gets scattered) are known.

Some other calculations (*a*) *Energy of the incident photon* (*E*). It is given by:

$$E = h\nu = \frac{h}{\lambda/c} = \frac{hc}{\lambda} = \frac{(6.626 \times 10^{-34} \text{ J s}) \times (3 \times 10^{8} \text{ ms}^{-1})}{\lambda}$$

or $$E = \frac{19.878 \times 10^{-26} \text{ Jm}}{\lambda} \qquad ...(iv)$$

Thus if λ (in metres) is known , the value of E (in joules) can be calculated with the help of the above equation.

(*b*) *Energy of the scattered photon* (E'). It is givn by:

$$E' = h\nu' = \frac{h}{\lambda'/c} = \frac{hc}{\lambda'} = \frac{(6.626 \times 10^{-34} \text{ Js}) \times (3 \times 10^{8} \text{ ms}^{-1})}{\lambda'}$$

$$= \frac{19.878 \times 10^{-26} \text{ Jm}}{\lambda'} \qquad ...(v)$$

Thus, if λ' (in metres) is known, the value of E' (in joules) can be calculated by using the above equation.

Schrodinger's Wave Equation

We have seen that according to de Broglie, electron has a dual character, *i.e.* the electron behaves as a material particle and as a wave. In 1926, Ervin Schrodinger (an Austrian physicist) said that if an electron behaves as a wave, there must be a wave equation which should be able to describe the wave motion of the electron.

Schrodinger proposed that, since an electron behaves as a wave, it should obey the same equation of motion which all other known types of waves obey. On the basis of this simple idea, he derived an equation which describes the wave motion of an electron-wave propagating in three dimensions (*x*, *y* and *z* axes) in space. This wave equation is called Schrodinger's wave eqation. This equation can be written in many forms one of which is given below:

$$\frac{\partial^2\psi}{\partial x^2} + \frac{\partial^2\psi}{\partial y^2} + \frac{\partial^2\psi}{\partial z^2} + \frac{8\pi^2 m}{h^2}(E-V)\psi = 0 \quad \text{... }(i)$$

or $$\left(\frac{\partial^2}{\partial x^2} + \frac{\partial^2}{\partial y^2} + \frac{\partial^2}{\partial z^2}\right)\psi + \frac{8\pi^2 m}{h^2}(E-V)\psi = 0$$

or $$\nabla^2\psi + \frac{8\pi^2 m}{h^2}(E-V)\psi = 0 \quad \text{...}(ii)$$

Here $\nabla^2 = \left(\frac{\partial^2}{\partial x^2} + \frac{\partial^2}{\partial y^2} + \frac{\partial^2}{\partial x^2}\right)$. ∇^2 is called *laplacian mathematical operator.*

A mathematical operator such as $\int$, $\frac{d}{dx}$, $\sqrt{\ }$ etc., is always accompanied by a mathamatical quantity or a mathematical function upon which it operates. For example in $\sqrt{2}$, $\sqrt{\ }$ is a mathematical operator and 2 is the mathematical quantity on which it (*i.e.,* $\sqrt{\ }$) operates. Similarly in $\Delta^2\psi$, Δ^2 is also a mathematical operator and ψ is a mathematical function on which Δ^2 operates. Thus $\Delta^2\psi$ should not mean that Δ^2 is multiplied by ψ. *x, y,* and *z* are the three space coordinates of the electron with respect to nucleus (0, 0, 0).

ψ (greek letter *psi*) is a mathematical function which is called wave function. ψ represents the amplitude of the electron-wave at various points surrounding the nucleus. Since ψ is an amplitude function of certain of space coordinates *x, y* and *z,* it should be written as $\psi(x, y, z)$. For the sake of convenience $\psi(x,y,z)$ is generally written as ψ only.

$\frac{\partial^2\psi}{\partial x^2}$ is double differentiation of $\psi(x, y, z)$ with respect to *x* only keeping *y* and *z* constant;

$\frac{\partial^2\psi}{\partial y^2}$ is double differentiation of $\psi(x, y, z)$ with respect to *y* only keeping *x* and *z* constant and

$\frac{\partial^2\psi}{dz^2}$ is double differentiation of $\psi(x, y, z)$ with respect to *z* only keeping *x* and *y* constant .

m is the mass of the electron, *h* is the Planck's constant and *E* is the total energy (*i.e.,* kinetic energy + potential energy, *V*) of the electron.

Derivation of Schrodinger's wave equation

In driving the wave equation, Schrodinger assumed that the electronic waves are similar to stationary waves (stationary waves are those which are produced by the vibrations of a stretched string). He considered an electron-wave of amplitude ψ. If *x* is the displacement of the wave in *x* direction and λ is its wavelength, then ψ is given by (*A* is a constant):

$$\psi = A\sin\frac{2\pi x}{\lambda} \quad \text{...}(i)$$

On differentiating equation (*i*), we get :

$$\frac{\partial\psi}{\partial x} = \left(A\cos\frac{2\pi x}{\lambda}\right)\left(\frac{2\pi}{\lambda}\right) \quad \text{or} \quad \frac{\partial\psi}{\partial x} = \frac{2\pi A}{\lambda}\cos\frac{2\pi x}{\lambda} \quad \text{...}(ii)$$

On differentiating equation (*ii*), we have:

$$\frac{\partial^2\psi}{\partial x^2} = \frac{2\pi A}{\lambda}\left(-\sin\frac{2\pi x}{\lambda}\right)\left(\frac{2\pi}{\lambda}\right) = -\frac{4\pi^2}{\lambda^2}\left(A\sin\frac{2\pi x}{\lambda}\right)$$

$$= -\frac{4\pi^2}{\lambda^2}\cdot\psi \quad \left[\because \psi = A\sin\frac{2\pi x}{\lambda} \text{ from eqn. } (i)\right] \quad \text{...}(iii)$$

This is the wave equation for unidimensional stationary wave vibrating along *x*-axis only. When equation (*iii*) is extended to three dimensions, we replace $\frac{\partial^2\psi}{\partial x^2}$ by $\left(\frac{\partial^2\psi}{\partial x^2} + \frac{\partial^2\psi}{\partial y^2} + \frac{\partial^2\psi}{\partial z^2}\right)$ and hence equation (*iii*) becomes :

$$\frac{\partial^2\psi}{\partial x^2}+\frac{\partial^2\psi}{\partial y^2}+\frac{\partial^2\psi}{\partial z^2}=-4\pi^2\psi\times\frac{1}{\lambda^2}$$

or $$\frac{\partial^2\psi}{\partial x^2}+\frac{\partial^2\psi}{\partial y^2}+\frac{\partial^2\psi}{\partial z^2}=-4\pi^2\psi\times\frac{m^2v^2}{h^2}\quad\left[\because \lambda=\frac{h}{mv}\ \textit{(from de Broglie's relation)}\right] \quad ...(iv)$$

The above equation is a wave equation which describes the motion of an electron in three dimensions.

Now the total energy (*E*) of the electron is the sum of its kinetic energy ($\frac{1}{2}mv^2$) and potential energy (*V*). Thus :

$$E=\frac{1}{2}mv^2+V \quad \text{or} \quad \frac{1}{2}mv^2=(E-V) \quad \text{or} \quad v^2=\frac{2(E-V)}{m}$$

Substituting the above value of v^2 in equation (*iv*), we get

$$\frac{\partial^2\psi}{\partial x^2}+\frac{\partial^2\psi}{\partial y^2}+\frac{\partial^2\psi}{\partial z^2}=-\frac{4\pi^2\psi m^2}{h^2}\times\frac{2(E-V)}{m}=-\frac{8\pi^2 m(E-V)\psi}{h^2}$$

or $$\frac{\partial^2\psi}{\partial x^2}+\frac{\partial^2\psi}{\partial y^2}+\frac{\partial^2\psi}{\partial z^2}+\frac{8\pi^2 m}{h^2}(E-V)\psi=0$$

or $$\left(\frac{\partial^2}{\partial x^2}+\frac{\partial^2}{\partial y^2}+\frac{\partial^2}{\partial z^2}\right)\psi+\frac{8\pi^2 m}{h^2}(E-V)\psi=0$$

or $$\nabla^2\psi+\frac{8\pi^2 m}{h^2}(E-V)\psi=0 \quad ...(v)$$

Other forms of Schrodinger's wave equation

We have seen that Schrodinger wave equation is written as :

$$\left(\frac{\partial^2}{\partial x^2}+\frac{\partial^2}{\partial y^2}+\frac{\partial^2}{\partial z^2}\right)\psi+\frac{8\pi^2 m}{h^2}(E-V)\psi=0$$

On multiplying both the sides of the above equation by $-\frac{h^2}{8\pi^2 m}$, we get :

$$-\frac{h^2}{8\pi^2 m}\left(\frac{\partial^2}{\partial x^2}+\frac{\partial^2}{\partial y^2}+\frac{\partial^2}{\partial z^2}\right)\psi-E\psi+V\psi=0$$

or $$-\frac{h^2}{8\pi^2 m}\left(\frac{\partial^2}{\partial x^2}+\frac{\partial^2}{\partial y^2}+\frac{\partial^2}{\partial z^2}\right)\psi+V\psi=E\psi$$

or $$\left[-\frac{h^2}{8\pi^2 m}\left(\frac{\partial^2}{\partial x^2}+\frac{\partial^2}{\partial y^2}+\frac{\partial^2}{\partial z^2}\right)+V\right]\psi=E\psi$$

or $$\left[-\frac{h^2}{8\pi^2 m}\nabla^2+V\right]\psi=E\psi \quad ...(i)$$

The expression *viz* $\left[-\frac{h^2}{8\pi^2 m}\nabla^2+V\right]$ in eqn. (*i*) is called *Hamiltonian operator* and is written as $\hat{H}$. Thus equation (*i*) can also be written as

$$\hat{H}\psi=E\psi \quad ...(ii)$$

The Hamiltonian operator ($\hat{H}$) consists of two parts *viz.*, the kinetic energy part $\left(-\frac{h^2}{8\pi^2 m}\nabla^2\right)$ and the potential energy part (*V*). Equation (*i*) and (*ii*) are other forms of Schrodinger's wave equation.

To convert Cartesian coordinates (x, y, z) into polar coordinates (r, θ, ϕ)

Let us consider an electron at point A within an atom, with its nucleus at the origin (0, 0, 0). Let the coordinates of this electron be x, y, z in space (see Fig. 2.5). Let AK be the perpendicular on xy plane and KL, KN, AM be the perpendiculars on x, y, and z axes respectively and θ and ϕ be the angles, as shown in Fig. 2.5. Let AO = r, OM = z, ON = y and OL = x. It is obvious from the figure that AM = OK and ON = KL = y.

Now in Δ AMO, we have:

$$\cos\theta = \frac{OM}{AO} = \frac{z}{r} \quad \text{or} \quad z = r\cos\theta \qquad ...(i)$$

$$\sin\theta = \frac{AM}{AO} = \frac{OK}{r} \quad \text{or} \quad OK = r\sin\theta \qquad ...(ii)$$

In Δ KOL, we have : $\cos\phi = \frac{OL}{OK} = \frac{x}{r\sin\theta}$ [From eqn. (*ii*)]

or

$$x = r\sin\theta\cos\phi \qquad ...(iii)$$

$$\sin\phi = \frac{KL}{OK} = \frac{y}{r\sin\theta} \quad \text{or } y = r\sin\theta\sin\phi \qquad ...(iv)$$

Equations (*i*), (*ii*) and (*iv*) can be used to convert the cartesian coordinates namely z, x and y respectively into polar coordinates r, θ and ϕ.

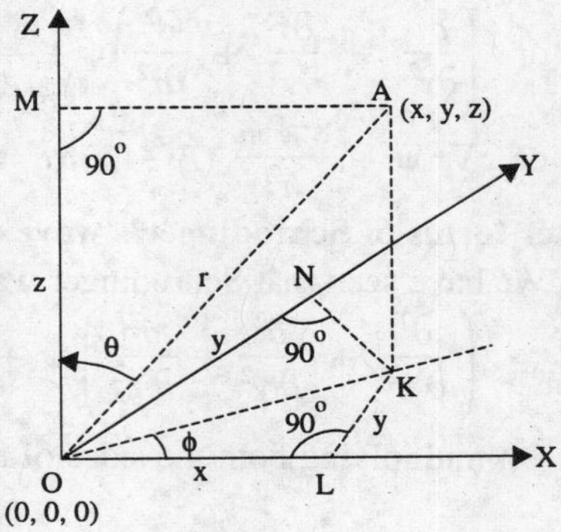

Fig. 2.5 : Correlation between Cartesian coordinates (x, y, z) and polar coordinates (r, θ, ϕ). The following relations may be noted AO = r, OL = x, ON = y = KL, OM = z.

To write Schrodinger's wave equation for H-atom in Cartesian and polar coordinates

We know that in case of H-atom, a single electron having charge equal to $-e$ moves round the nucleus with charge equal to $+e$. Thus potential energy (V) of the electron in H-atom is equal to $\frac{(-e)\times(+e)}{r} = \frac{-e^2}{r}$.

(*a*) *Wave equation in Cartesian coordinates* (x, y, z).

On putting $V = -\frac{e^2}{r}$ in Schrodinger wave equation, we get:

$$\left(\frac{\partial}{\partial x^2} + \frac{\partial}{\partial y^2} + \frac{\partial}{\partial z^2}\right)\psi + \frac{8\pi^2 m}{h^2}\left(E + \frac{e^2}{r}\right)\psi = 0 \quad ...(i)$$

This is Schrodinger wave equation for H-atom in Cartesian coordinates (x, y, z). In this equation the wave function (ψ) is a function of x, y and z.

(*b*) *Wave equation in polar coordinates* (r, θ, ϕ). When Cartesian coordinates (x, y, z) appearing in equation (*i*) are converted into polar coordinates (r, θ, ϕ) with the help of the relations $z = r\cos\theta$, $y = r\sin\theta\sin\phi$ and $x = r\sin\theta\cos\phi$, equation (*i*) gets converted into equation (*ii*) which is Schrodinger wave equation for H-atom in polar coordinates.

$$\left[\frac{\partial}{\partial r}\left(r^2\frac{\partial}{\partial r}\right) + \frac{1}{\sin\theta}\frac{\partial}{\partial\theta}\left(\sin\theta\frac{\partial}{\partial\theta}\right) + \frac{1}{\sin^2\theta}\frac{\partial^2}{\partial\phi^2}\right]\psi + \frac{8\pi^2 m r^2}{h^2}\left(E + \frac{e^2}{r}\right)\psi = 0 \quad ...(ii)$$

In equation (*ii*) ψ (wave function) is a function of polar coordinates (r, θ, ϕ) instead of Cartesian coordinates (x, y, z).

To find out the solution of Schrodinger wave equation for H-atom

We have shown above that Schrodinger wave equation for H-atom in polar coordinates is written as :

$$\left[\frac{\partial}{\partial r}\left(r^2\frac{\partial}{\partial r}\right)+\frac{1}{\sin\theta}\frac{\partial}{\partial\theta}\left(\sin\theta\frac{\partial}{\partial\theta}\right)+\frac{1}{\sin^2\theta}\frac{\partial^2}{\partial\phi^2}\right]\psi+\frac{8\pi^2 m r^2}{h^2}\left(E+\frac{e^2}{r}\right)\psi=0 \quad \text{... } (i)$$

A detailed analysis has shown that the determination of the value of ψ appearing in equation (*i*) above can be made only when ψ is written in the form

$$\psi(r,\theta,\phi) = R(r)\,\Theta(\theta)\,\Phi(\phi) \quad \text{... } (ii)$$

In equation (*ii*), R(*r*) is a function which depends on *r* only (*r* is the distance of the electron from the nucleus) but is indepedent of the other two polar coordinates *viz*., θ and ϕ. $\Theta(\theta)$ is a function depedent on θ only and is indepedent of *r* and ϕ. Similarly $\Phi(\phi)$ is a function depedent on ϕ only but is independent of *r* and θ.

The function R(*r*) is called *radial wave function* while the other two wave functions *viz.* $\Theta(\theta)$ and $\Phi\ \phi$ are called *angular wave function.* $\psi(r, \theta, \phi)$ is called *complete* or *total wave function.* Equation (*ii*) shows that the complete wave function of an orbital in equal to the product of radial wave function, R(*r*) and angular wave functions *viz*., $\Theta(\theta)$ and $\Phi(\phi)$. The product, $[\Theta(\theta)\times\Phi(\phi)]$ gives us the shape of the orbital.

By writting $\psi(r, \theta, \phi)$ equal to $R(r)\,\Theta(\theta)\,\Phi(\phi)$ in equation (*i*), this equation is converted into a complicated equation (not given here). This complicated equation can be split into the following three equations namely (*a*), (*b*), and (*c*).

$$(a)\qquad \frac{1}{R}\cdot\frac{\partial}{\partial r}\left(r^2\frac{\partial R}{\partial r}\right)+\frac{8\pi^2 m r^2}{h^2}\left(E+\frac{e^2}{r}\right)=l\,(l+1)$$

This equation is a function of *r* only and on solution gives the value of *R*(*r*). When this equation is solved for *R*(*r*), we get many values of *R*(*r*). Each value of *R*(*r*), besides depending on *r*, also depends on two constants *n* and *l*. *R*(*r*) is, therefore, generally written as $R_{n,l}(r)$ or simply as $R_{n,l}$ for the sake of simplicity. *n* and *l* are called quantum numbers, since they represent the quantisation of definite physical properties about the electron in the atom. For each permitted combination of *n* and *l*, equation (*a*) provides a distinct solution of *R*, *i.e.*, for different values of *n* and *l*, different mathematical equations involving *r* are found. These are represented by the symbols $R_{1,0}$ ($n = 1, l = 0$), $R_{2,0}$ ($n = 2, l = 0$), $R_{2,1}$ ($n = 2, l = 1$) and so on.

The quantum number, *n* is analogous to the principal quantum number suggested by Bohr in his atomic model (Bohr's atomic model) and can have integral values of 1, 2, 3 ...*n*. *n* determines the energy of an electron present in main shells. The quantum number, *l* is similar to the azimuthal quantum number of Bohr-Sommerfeld's theory and can have all possible whole number values from 0 to $(n-1)$, *i.e.*, $l = 0, 1, 2, 3, ...(n-1)$. $l = 0, 1, 2, 3$ and is designated by *s, p, d, f* etc., letters respectively.

The values of radial wave function, $R_{n,l}$ for *1s*, *2s*, and *2p* states for H-atom are given below:

For 1s state ($n = 1, l = 0$), $R_{1,0}$ or $R_{1s} = 2\left[\frac{Z}{a_0}\right]^{3/2} e^{-\sigma/2}$

For 2s state ($n = 2, l = 0$), $R_{2,0}$ or $R_{2s} = 2^{-3/2}\left[\frac{Z}{a_o}\right]^{3/2}(2-\sigma)\,e^{-\sigma/2}$

For 2p state ($n = 2, l = 1$), $R_{2,1}$ or $R_{2p} = 2^{-1}\times 6^{-1/2}\left[\frac{Z}{a_o}\right]^{3/2}\sigma\times e^{-\sigma/2}$

In the above expressions a_0 = Bohr's radius, Z = atomic number and $\sigma = \frac{Zr}{n\,a_o}$.

$$(b)\qquad \frac{\sin\theta}{\Theta}\frac{\partial}{\partial\theta}\left(\sin\theta\frac{\partial\Theta}{\partial\theta}\right)+l\,(l+1)\sin^2\theta = m^2.$$

This equation is a function of Θ only and on solution gives the value of $\Theta(\theta)$. When this equation is solved for Θ (θ), we get many values of Θ (θ). Each value of Θ (θ), besides depending on the angle θ, also depends on two constants l and m. Θ (θ) is, therefore, generally represented as $\Theta_{l,m}$ (θ) or simply as $\Theta_{l,m}$. The vaule of m depends on the value of l. For a given value of l, $m = 0, \pm 1, \pm 2 ... \pm l$. Thus m has $(2l + 1)$ values for a given value of l. m is called magnetic quantum number. If the values of l and m are known for a given orbital, the value of the angular wave function, $\Theta_{l,m}$ for that orbital can be written as shown below:

(*i*) $\Theta_{0,0}\ (l = 0, m = 0) = \dfrac{\sqrt{2}}{2} = \dfrac{1}{\sqrt{2}}$ (*s*-orbital)

(*ii*) $\Theta_{1,0}\ (l = 1, m = 0) = \sqrt{\dfrac{3}{2}} \cos\theta$; $\Theta_{1,\pm1}\ (l = 1, m = \pm 1) = \sqrt{\dfrac{3}{2}} \sin\theta$ (*p*-orbitals)

(*iii*) (*d*-orbitals):

$$\Theta_{2,0}\ (l = 2,\ m = 0) = \sqrt{\frac{5}{8}}\ (3\cos^2\theta - 1)$$

$$\Theta_{2,\pm1}\ (l = 2,\ m = \pm 1) = \sqrt{\frac{15}{4}}\ (\sin\theta\ \cos\theta)$$

$$\Theta_{2,\pm2}\ (l = 2,\ m = \pm 2) = \sqrt{\frac{15}{16}}\ \sin^2\theta$$

(*c*) $\dfrac{1}{\Phi} \cdot \dfrac{\partial^2 \Phi}{\partial \phi^2} = m^2$

This equation is a function of m only and on solution gives the value of Φ (ϕ). When this equation is solved for Φ (ϕ), we get many values of Φ (ϕ). Each value of Φ (θ), besides depending on the angle ϕ, also depends on the constant m. Hence the angular wave function Φ (ϕ) is also represented as $\Phi_m(\phi)$ *or* Φ_m. m fixes the direction of the angular momentum and hence fixes the direction of the orbital in space.

If the value of magnetic quantum number (m) for a given orbital is known, the value of Φ_m for the orbital can be writtten as shown below :

$$\Phi_0\ (m = 0) = \frac{1}{\sqrt{2\pi}}\ e^{o} = \frac{1}{\sqrt{2\pi}}$$

$$\Phi_{+1}\ (m = +1) = \frac{1}{\sqrt{2\pi}}\ e^{+i\phi} = \frac{1}{\sqrt{2\pi}}\ (\cos\phi + i\sin\phi)$$

$$\Phi_{-1}\ (m = -1) = \frac{1}{\sqrt{2\pi}}\ e^{-i\phi} = \frac{1}{\sqrt{2\pi}}\ (\cos\phi - i\sin\phi)$$

$$\Phi_{+2}\ (m = +2) = \frac{1}{\sqrt{2\pi}}\ e^{+2i\phi} = \frac{1}{\sqrt{2\pi}}\ (\cos 2\phi + i\sin 2\phi)$$

$$\Phi_{-2}\ (m = -2) = \frac{1}{\sqrt{2\pi}}\ e^{-2i\phi} = \frac{1}{\sqrt{2\pi}}\ (\cos 2\phi - i\sin 2\phi)$$

Complete wave function

We have said above that the radial wave function, R(r) is also written as $R_{n,l}$. Simlarly the angular wave functions *viz.*, Θ (θ) and Φ (ϕ) are also written as $\Theta_{l,m}$ and Φ_m respectively. Thus the equation $\psi(r, \theta, \phi) = R(r)\ \Theta(\theta)\ \Phi(\phi)$ can also be written as :

$$\psi_{n,l,m} = R_{n,l}\ \Theta_{l,m}\ \Phi_m$$

In this equation $\psi_{n,l,m}$ is the complete wave function. This equation shows that the complete wave function, besides depending on r, θ and ϕ, also depends on the three quantum numbers namely n,l and m. Each permitted combination of n,l and m gives a distinct wave function and hence a distinct orbital.

The above discussion shows that if the values of *n, l* and *m* of an orbital are known, the value of complete wave function can be written as shown below:

$1s$ orbital ($n = 1, l = 0, m = 0$) $= R_{1,0}\,\Theta_{0,0}\,\phi_0$

$2p_z$ orbital ($n = 2, l = 1, m = 0$) $= R_{2,1}\,\Theta_{1,0}\,\phi_0$

$2p_x$ orbital ($n = 2, l = 1, m = \pm 1$) $= \frac{1}{\sqrt{2}} R_{2,1}\,\Theta_{1,\pm 1}\,(\phi_{+1} + \phi_{-1})$

$2p_y$ orbital ($n = 2, l = 1, m = \pm 1$) $= \frac{1}{\sqrt{2}} R_{2,1}\,\Theta_{1,\pm 1}\,(\phi_{+1} - \phi_{-1})$

Eigen wave function and eigen value

Being a differential equation of second order, Schrodinger wave equation has many solutions for **ψ**, *i.e.*, when we solve Schrodinger wave equation for ψ, we get many values of ψ. Some of these values are imaginary, while others are real values. The imaginary values of ψ cannot be accepted. Only those values of ψ which give definite and acceptable value of the total energy (E) of the electron are acceptable. These acceptable values of the wave function ψ are called *eigen (acceptable) wave functions.* The value of *E* given by a particular eigen (acceptable) wave function is called *eigen value.* Thus every value of wave function is not an eigen (acceptable) wave function.

Characteristics of an eigen (acceptable) wave function

We have said above that all the values of ψ obtained by solving Schrodinger wave equation are not acceptable. The value of ψ which is acceptable is called eigen (acceptable) wave function. Only that value of ψ is an eigen (acceptable) wave function which has the following characteristics: (*i*) The function must be finite (*ii*) It should be single valued, *i.e.*, it should not have more than one value at a point in space (*iii*) It should be continuous, *i.e.*, if the function is plotted in space, it should not show any break in the plot (*iv*) $\frac{\partial^2\psi}{\partial x^2}$, $\frac{\partial^2\psi}{\partial y^2}$ and $\frac{\partial^2\psi}{\partial z^2}$ must be continuous functions of *x.y* and *z* respectively (*v*) The function must be normalised, *i.e.*, it should obey the relation $\int_{-\infty}^{+\infty} \psi^2\, dr = 1$ where *dr* is a small volume element and $\psi^2 dr$ represents the probability of finding the electron in the small volume element *dr*.

Concept of electron charge cloud and atomic orbital

Let us consider an electron of H-atom. This electron revolves round the nucleus and the nucleus remains fixed at its position. Let the position of the electron round the nucleus be represented by a dot. Because of the wave nature of the electron, it can be found anywhere round the nucleus, *i.e.* the probability of its being found is nowhere zero. The maximum concentration of the dots would be closest to the nucleus and would go on decreasing as the distance of the electron from the nucleus increases in all directions. In other words we can say that the probability of finding the electron will be maximum closest to the nucleus and will go on decreasing on the distance of the electron from the nucleus increases. The region in three dimensional space in which there is a probability of finding the electron is called *electron charge cloud* (See Fig. 2.6). The region in three dimensional space around the nucleus of an atom where the probability of finding the electron(s) is maximum is called an atomic orbital.

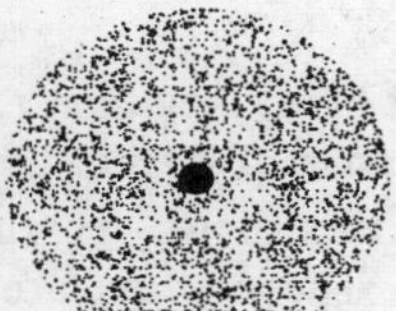

Fig. 2.6 : Electron charge cloud of an electron in H-atom.

Differences between an orbit and an orbital

The differences between an orbit and an orbital are given in Table 2.2.

Table 2.2 : ***Differences between an orbit and an orbital***

Orbit	Orbital
(*i*) As postulated by Bohr, an orbit is a definite circular path at a definite distance from the nucleus in which the electron revolves round the nucleus. Orbits are numbered as 1, 2, 3, 4......from the nucleus and are designated by the capital letters K, L, M, N....etc.	(*i*) As postulated by wave nature of an electron, an orbital is defined as a three-dimensional region or space around the nucleus within which the probability of finding an electron with a certain energy is maximum.
(*ii*) An orbit indicates an exact position or location of an electron in an atom. For example, according to Bohr's theory, an electron of hydrogen is found at a distance of 0.529 Å from the nucleus.	(*ii*) An orbital does not specify the definite or exact position of an electron in an atom, since the electron, due to its wave-nature, can not be found at a fixed distance from the nucleus.
(*iii*) There is a certainty about the movement of an electron in an orbit.	(*iii*) Due to uncertainty principle, there is no certainty about the movement of an electron in an orbital.
(*iv*) It represents the planar motion of the electron.	(*iv*) It represents the three-dimentional motion of the electron round the nucleus.
(*v*) The maximum number of electrons in an orbit is equal to $2n^2$ where *n* is the number of the orbit.	(*v*) An orbital cannot accommodate more than two electrons. These two electrons must have opposite spins.
(*vi*) Orbits are circular in shape.	(*vi*) Orbitals have different shapes, *e.g.*, *s*-orbital is spherically symmetrical, *p* orbitals are dumb-bell shaped etc.

Significance of ψ and ψ^2

The wave function (ψ), by itself, has no significance, since it cannot represent the probability of finding an electron in a small volume round the nucleus. This is because of the fact that the wave funcion ψ can have positive as well as negative value, whereas the probability of finding the electron can be zero or positive but can never be negative. The square of ψ (*i.e.*, ψ^2) always has positive value and hence measures the probability of finding the electron.

Electron probability function, D

The probability or chance of finding an electron in space round the nucleus is called *electron probability function,* D. The value of D for an extremely small spherical shell of radius *r* and thickness *dr* round the nucleus is given by :

$$D = \psi^2 \times \text{volume of the spherical shall} = \psi^2 \times 4\pi r^2\, dr$$

Thus, the electron probability between $r = 0$ and $r = r$ would be equal to :

$$\int_{r=0}^{r=r} \psi^2\, 4\,\pi r^2\, dr$$

Plot of $R_{n,l}$ against *r* and its relation with the electron probability density round a point at a distance of r from the nucleus

Plot of $R_{n,l}$ versus *r* for the electron present in 1*s*, 2*s*, and 3*s* orbitals are given in Fig. 2.7. These plots are called *probability curves.* The following points may be noted from these curves:

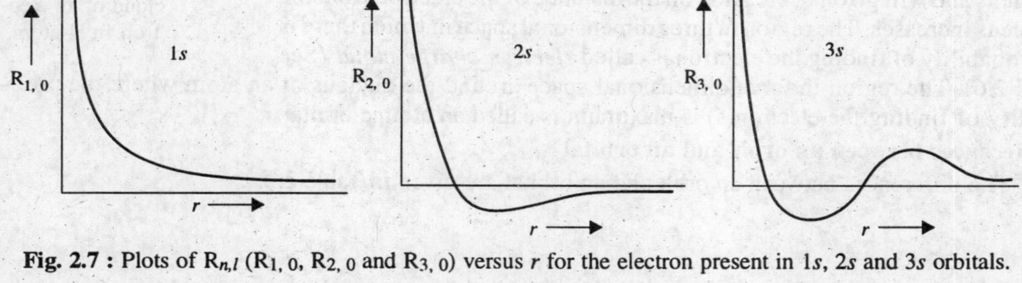

Fig. 2.7 : Plots of $R_{n,l}$ ($R_{1,0}$, $R_{2,0}$ and $R_{3,0}$) versus *r* for the electron present in 1*s*, 2*s* and 3*s* orbitals.

(*i*) The value of $R_{n,l}$ is maximum at $r = 0$, *i.e.* the probability of finding the electron at the nucleus (at nucleus $r = 0$) is maximum. This conclusion cannot be true, since the actual probability of finding the electron at the nucleus is zero. Thus we see that $R_{n,l}$ cannot be related with the probability of electron-density of an electron.

(*ii*) $R_{n,l}$ has both positive as well as negative values as is evident from the probability curves for 2*s* and 3*s* orbitals. The negative values of probability density cannot be possible.

Plots of $R^2_{n,l}$ against *r* and its relation with the electron probability density around a point at a distance of *r* from the nucleus.

Plots of $R^2_{n,l}$ versus *r* for the electron present in 1*s*, 2*s*, and 3*s* orbitals are given in Fig. 2.8

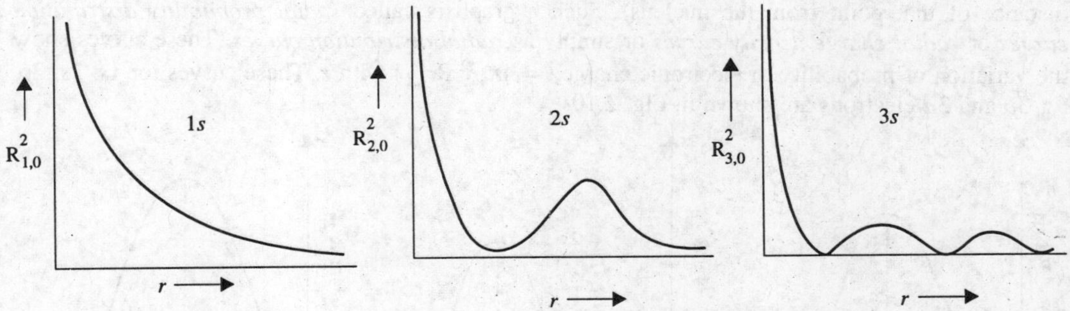

Fig. 2.8 : Plots of $R^2_{n,l}$ ($R^2_{1,0}$, $R^2_{2,0}$ and $R^2_{3,0}$) versus *r* for the electron present in 1*s*, 2*s*, and 3*s* orbitals.

The following points may be noted from these plots:

(*i*) $R^2_{n,1}$ has positive values only.

(*ii*) The magnitude of $R^2_{n,l}$ decreases as *r* increases. The value of $R^2_{n,l}$ is maximum at $r = 0$ which is not possible. Thus, although the values of $R^2_{n,l}$ are positive, $R^2_{n,l}$, like $R_{n,l}$, cannot be related with the probibility density of the electron.

The above discussion shows that neither $R_{n,l}$ nor $R^2_{n,l}$ can be directly related with the probability of finding an electron at a point distant *r* from the nucleus.

Radial probability of finding the electron in a small shell of thickness dr at a distance of *r* from the nucleus.

Let us consider two thin concentric shells. (spheres) of radius *r* and (r + *dr*) from the nucleus (N)which is supposed to be situated at the centre, as shown in Fig. 2.9. Let the volume of the space enclosed between the two shells be dv. Obviously dv is given by :

$$dv = \frac{4}{3}\pi (r + dr)^3 - \frac{4}{3}\pi r^3$$

$$= \frac{4}{3}\pi [r^3 + dr^3 + 3r^2 dr + 3 r dr^2 - r^3]$$

$$= \frac{4}{3}\pi [dr^3 + 3r^2dr + 3r dr^2]$$

since dr^3 and dr^2 are very small terms, these can be supposed to be equal to zero and hence the above equation becomes.

$$dv = \frac{4}{3}\pi \times 3r^2dr = 4\pi r^2dr$$

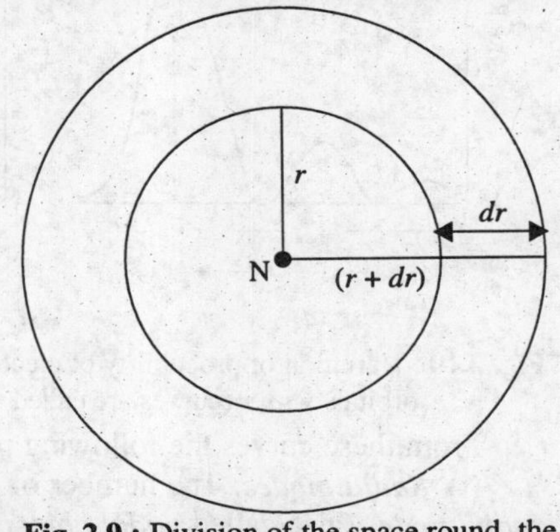

Fig. 2.9 : Division of the space round the nucleus (represented as N,) into a large number of thin concentric shells of very small thickness equal to *dr*.

The probability of finding the electron within this small radial shell of thickness *dr* around the nucleus is called *radial probability of finding the electron.* It is given by the product of $R^2_{n,l}$ and *dv.* Thus :

Radial probability of finding the electron within the small radial shell of thickness of *dr*

$$= R^2_{n,l} \times dv = R^2_{n,l} \times 4\pi r^2 dr$$

Radial probability distribution between $r = 0$ and $r = r$ and radial distribution curves for 1*s*, 2*s*, 2*p*, 3*s*, 3*p* and 3*d* electrons

We have shown above that the radial probability of finding an electron within the small radial shell of thickness dr is equal to $R^2_{n,l} \times 4\pi r^2 dr$ where r = inner radius of one of the thin concentric shells. Thus radial probability distribution between $r = 0$ and $r = r$ would be equal to

$$\int_{r=r}^{r=0} R^2_{n,l}\ 4\,\pi\, r^2\, dr$$

The above discussion shows that the radial probability distribution or radial probability of charge density of electron at a point may be obtained by plotting $4\pi r^2 R^2_{n,l}$ against r. (Here r is the distance of that point from the nucleus). Such a graph is called *radial probability distribution curves* or *radial charge density curves* or simply as *radial distribution curves*. These curves show the variation of probability of electronic charge ($= 4\pi r^2 R^2_{n,l}$) with r. These curves for 1*s*, 2*s*, 2*p*, 3*s*, 3*p* and 3*d* electrons are shown in Fig. 2.10.

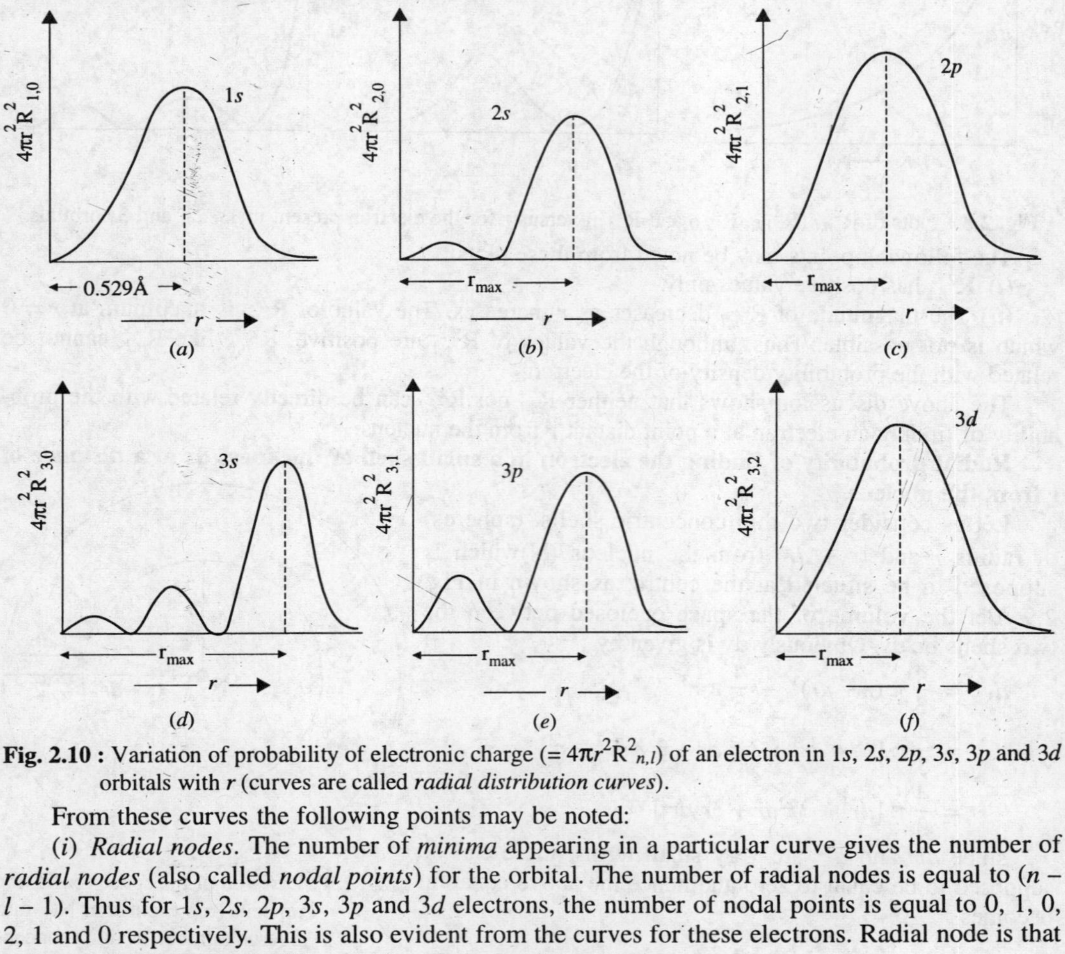

Fig. 2.10 : Variation of probability of electronic charge ($= 4\pi r^2 R^2_{n,l}$) of an electron in 1*s*, 2*s*, 2*p*, 3*s*, 3*p* and 3*d* orbitals with r (curves are called *radial distribution curves*).

From these curves the following points may be noted:

(*i*) *Radial nodes*. The number of *minima* appearing in a particular curve gives the number of *radial nodes* (also called *nodal points*) for the orbital. The number of radial nodes is equal to $(n - l - 1)$. Thus for 1*s*, 2*s*, 2*p*, 3*s*, 3*p* and 3*d* electrons, the number of nodal points is equal to 0, 1, 0, 2, 1 and 0 respectively. This is also evident from the curves for these electrons. Radial node is that point at which the probability density of electronic charge is zero.

That the number of radial nodes for *3s* orbital is two implies that the spherical *3s*-orbital has two concentric spherical shells within it at certain radial distances where the probability of finding the electron is zero. (See Fig. 2.11.)

Due to greater number of radial nodes for *3s* orbital (=2) than that for *1s* (number of nodes =

0) and 2*s* (number of nodes = 1) orbitals, this orbital (*i.e.* 3*s* orbital) is bigger in size and more diffused than both these orbitals.

(*ii*) Curves start from the origin and the areas covered by the envelopes of a particular curve go on increasing from left to right so that the area of the last envelop is maximum.

(*iii*) *Radius of maximum probability,* r_{max}. The peak of the last envelop of each curve gives the distance from the nucleus where the probability of finding the electron is maximum. This distance is called radius of maximum probability and has been denoted by r_{max} in each curve. In case of H-atom which contains only 1*s* orbital, the value of r_{max} is 0.529Å.

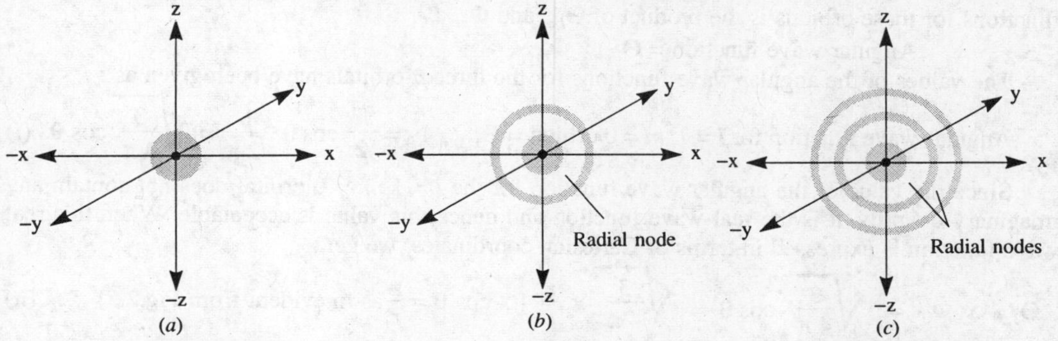

(*a*) 1*s* orbital (No radial node) (*b*) 2*s* orbital (Number of radial nodes = 1) (c) 3*s* orbital (Number of radial nodes = 3)

Fig. 2.11: Electron-population pictures for the electrons in 1*s*, 2*s* and 3*s* orbitals. The thick dot shown at the centre represents nucleus.

This value is in full agreement with the Bohr's radius calculated by Bohr (See *Bohr's atomic model*). As in evident, the values of r_{max} for a 3*s*, 3*p*, and 3*d* electron are in the order $(r_{max})_{3s} > (r_{max})_{3p} > (r_{max})_{3d}$

(*iv*) *Penetrating power of* 3*s*, 3*p*, 3*d electrons* . The smallest peak appearing in the curve for 3*s* electron is more closer to the nucleus than a similar peak for the 3*p* electron. This indicates that the probability of the occurence of a 3*s*-electron nearest to the nucleus is greater than that for the 3*p*-electrons. This means that a 3*s*-electron has greater penetrating power than a 3*p*-electron. Similarly it can be shown that a 3*p*-electron is more penetrating than a 3*d*-electron.. Thus the order of penetrating power of 3*s*, 3*p* and 3*d* electron is as 3*s* > 3*p* > 3*d*.

To write the value of angular wave function, $\Theta_{l,m} \times \phi_m$ for s, p and d orbitals and to deduce their shape from these values.

We have said earlier that the angular wave function of an orbital is the product of $\Theta_{l,m}$ and Φ_m where *l* and *m* are azimuthal and magnetic quantum numbers for the orbital.We have already given the values of $\Theta_{0,0}$, $\Theta_{1,0}$, $\Theta_{1,\pm1}$, $\Theta_{2,0}$, $\Theta_{2,\pm1}$, $\Theta_{2,\pm2}$ (page 36), Φ_0, Φ_{+1} , Φ_{-1}, Φ_{+2} and Φ_{-2} (page 36). With the help of these values, the values of $\Theta_{l,m} \times \Phi_m$ for *s* ($l = 0$, $m = 0$), *p* ($l = 1$, $m = 0, \pm 1$) and *d* ($l = 2$, $m = 0, \pm1, \pm2$) orbitals can be written as shown below. With the help of these values the shape of the orbitals can be predicted.

s-orbitals . We know that for an *s*-orbital $l = 0$. and hence $m = 0$. Thus the angular wave function for *s*-orbital can be written as $\Theta_{0,0} \times \Phi_0$ whose value is given by :

$$\Theta_{0,0} \times \Phi_0 = \frac{1}{\sqrt{2}} \times \frac{1}{\sqrt{2\pi}} = \frac{1}{2\sqrt{\pi}}$$

This euqation shows that $\Theta_{0,0}\Phi_0$ is indepednet of the angles θ and ϕ and hence *s*-orbital is symmetrical over all the directions, *i.e.* *s*-orbital has symmetrically spherical shape and is non-directional (Fig. 2.12).

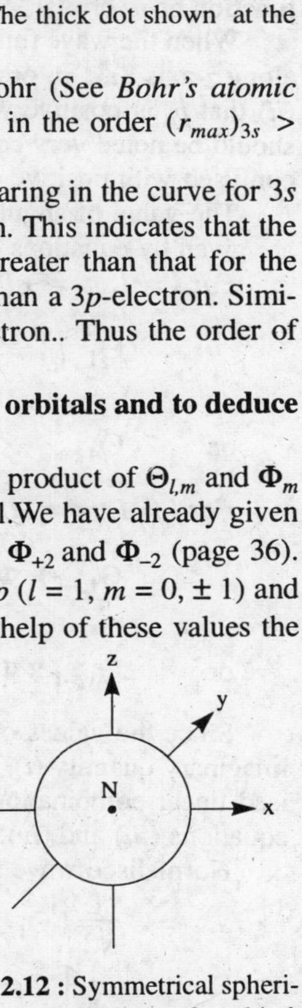

Fig. 2.12 : Symmetrical spherical shape of *s*-orbital. *N* represents nucleus.

Thus an *s*-orbital is usually represented by a circle. The higher

s-orbitals (*e.g.* $2s$, $3s$ etc.) are also non-directional and have symmetrically spherical shape. Greater the value of n for a given s-orbital, greater is the size of s-orbital. Thus 2s orbital (n = 2) is larger in size than $1s$ orbital (n = 1). The size of an s-orbital also depends on the number of the radial nodes for the orbital. Greater is the number of radial nodes, larger is the size of the orbital. Also see "*Radial nodes*" discussed earlier.

p-orbitals. For p-orbital $l = 1$ and hence $m = 0, +1, -1$ (three values). Three values of m for $l = 1$ indicate that a p-orbital has three sub-orbitals. The permitted combinations of l and m values for these three sub-orbitals are $l = 1, m = 0$; $l = 1, m = +1$ and $l = 1, m = -1$. The angular wave functions for these orbitals is the product of $\Theta_{l,m}$ and Φ_m, *i.e.*

Angular wave function = $\Theta_{l,m} \times \Phi_m$

The values of the angular wave functions for the three p-orbitals have been given as :

$$\text{Angular wave function for } l = 1, m = 0 \text{ orbital} = \Theta_{1,0}\ \Phi_0 = \frac{\sqrt{3}}{2}\cos\theta\,\frac{1}{\sqrt{2\pi}} = \sqrt{\frac{3}{4\pi}}\ \cos\theta \quad ...(i)$$

Since the value of the angular wave function for the $l = 1$, $m = 0$ orbital does not contain any imaginary quantity, it is the real wave function and hence this value is acceptable. When this real wave function is expressed in terms of Cartesian coordinates, we get :

$$\Theta_{1,0} \times \Phi_0 = \sqrt{\frac{3}{4\pi}}\ \cos\theta = \sqrt{\frac{3}{4\pi}} \times \frac{z}{r} \quad [\because \cos\theta = \frac{z}{r} \text{ as in evident from Fig.2.5}] \quad ...(ii)$$

It is obvious from equation (*ii*) that $\Theta_{1,0} \times \Phi_0$ is a function of z and hence is known as p_z wave function or p_z orbital.

When the wave function p_z is plotted in three dimensional space, we get dum-bell shaped curve along z-axis , *i.e.* p_z orbital has dum-bell shape as shown in Fig. 2.13 (*a*). It can be seen from eqn. (*ii*) that p_z is positive when z is positive and negative when z is negative (See Fig. 2.13). Here it should be noted very carefully that + and – signs are geometric signs only and hence should not be confused with positive or negative charges.

The value of angular wave functions for the sub-orbitals *viz.* $l = 1$, $m = +1$ and $l = 1$, $m = -1$ are given by equations (*iii*) and (*iv*) respectively.

For l = 1, m = +1 orbital

$$\Theta_{1,+1} \times \Phi_{+1} = \sqrt{\frac{3}{4}}\ \sin\theta \times \frac{1}{\sqrt{2\pi}}\ (\cos\phi + i\sin\phi)$$

$$\text{or} \quad \Theta_{1,+1} \times \Phi_{+1} = \sqrt{\frac{3}{8}\pi}\ \sin\theta\ (\cos\phi + i\sin\phi) \quad ...(iii)$$

For l = 1, m = –1 orbital

$$\Theta_{1,-1} \times \Phi_{-1} = \sqrt{\frac{3}{4}}\ \sin\theta \times \frac{1}{\sqrt{2\pi}}\ (\cos\theta - i\sin\phi)$$

$$\text{or} \quad \Theta_{1,-1} \times \Phi_{-1} = \sqrt{\frac{3}{8\pi}}\ \sin\theta\ (\cos\phi - i\sin\phi) \quad ...(iv)$$

Since the values of the angular wave functions as represented by equations (*iii*) and (*iv*) have imaginary quantity (i), these values are not acceptable. The real values are obtained by the normalised linear combinations (addition and subtraction) of the angular wave functions represented by equations (*iii*) and (*iv*). Thus :

Normalised wave function obtaned by the *addition* of (*iii*) and (*iv*)

$$= \frac{1}{\sqrt{2}}[\Theta_{1,+1}\ \Phi_{+1} + \Theta_{1,-1}\ \Phi_{-1}]$$

$$= \frac{1}{\sqrt{2}}\left[\sqrt{\frac{3}{8\pi}}\ \sin\theta\ (\cos\theta + i\sin\phi) + \sqrt{\frac{3}{8\pi}}\ \sin\theta\ (\cos\phi - i\sin\phi)\right]$$

$$= \sqrt{\frac{3}{4\pi}} \sin\theta \,.\, \cos\phi = \sqrt{\frac{3}{4\pi}} \cdot \frac{x}{r} \;[\because \sin\theta \,.\, \cos\phi = \frac{x}{r} \text{ as is evident from Fig. 2.5]} \quad ...(v)$$

Obviously the wave function represented by equation (*v*) varies with x and hence is called p_x wave function or p_x orbital.

When p_x wave function is plotted in three dimensional space, we get dumb-bell shaped curve along x-axis, *i.e.*, p_x orbital has dumb-bell shape as shown in Fig. 2.13 (*b*). It is obvious from equation (*v*) that p_x would be positive when x is positive and would be negative when x is negative.

Normalised wave function obtained by the ***subtraction*** of (*iii*) and (*iv*).

$$= \frac{1}{\sqrt{2}}[\Theta_{1,+1}\,\Phi_{+1} - \Theta_{1,-1} \times \Phi_{-1}]$$

$$= \frac{1}{\sqrt{2}}\left[\sqrt{\frac{3}{8\pi}} \sin\theta\;(\cos\phi + i\sin\phi) - \sqrt{\frac{3}{8\pi}} \sin\theta\;(\cos\phi - i\sin\phi)\right]$$

$$= \sqrt{\frac{3}{4\pi}} \sin\theta.\, \sin\phi = \sqrt{\frac{3}{4\pi}} \cdot \frac{y}{r} \;[\because \sin\theta \,.\, \sin\phi = \frac{y}{r} \text{ as is evident from Fig. 2.5]} \quad ...(vi)$$

Eq. (*vi*) shows that the wave function represented by this equation varies with y only and hence is called p_y wave function or p_y orbital.

A plot of p_y wave function in three dimensional space gives dum-bell shaped curve along y-axis. Thus p_y- orbital has dumbell shape as shown in Fig. 2.13 (*c*).

The above discussion shows that the values of angular wave functions of $2p_z$, $2p_x$ and $2p_y$ orbitals can be written as :

$$2p_z\;(n = 2, l = 1, m = 0) \;= \Theta_{1,0}\,\Phi_0$$

$$2p_x\;(n = 2, l = 1, m = \pm 1) = \frac{1}{\sqrt{2}}[\Theta_{1,+1}\,\Phi_{+1} + \Theta_{1,-1}\,\Phi_{-1}]$$

$$= \frac{1}{\sqrt{2}}\,\Theta_{1,+1}\,(\Phi_{+1} + \Phi_{-1}) \qquad (\because \Theta_{1,+1} = \Theta_{1,-1})$$

$$2p_y(n = 2, l = 1, m = \pm 1) \;= \frac{1}{\sqrt{2}}[\Theta_{1,+1}\,\Phi_{+1} - \Theta_{1,-1}\,\Phi_{-1}] = \frac{1}{\sqrt{2}}\,\Theta_{1,\pm 1}\,(\Phi_{+1} - \Phi_{-1})$$

In connection with *p*-orbitals, the following points must be noted carefully :

(*i*). Since the three axes *viz.* *x,y* and *z* axex are perpendicular to each other, the three *p*-orbitals (*i.e.* p_z, p_x and p_y) are also perpendicular to each other.

(*ii*). Each of the three *p*-orbitals has two egg-shaped lobes one on each side of the nucleus. The probability of finding the electron in both the lobes is equal. Both the lobes are separated by a nodal plane which passes through the nucleus. The electron density on this nodal plane and on the nucleus is zero. The angular part of the wave function, $\psi(\theta, \phi)$ has a + sign on one lobe and a – sign on the opposite lobe, although $\psi^2(\theta, \phi)$ will be positive on both the lobes.

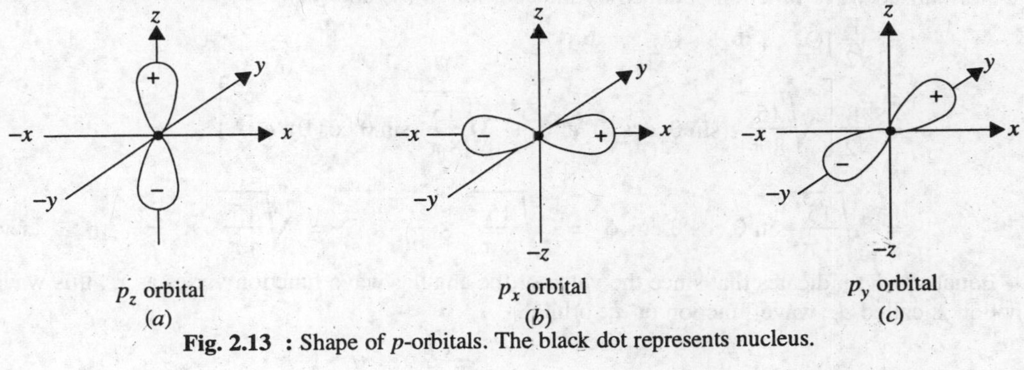

Fig. 2.13 : Shape of *p*-orbitals. The black dot represents nucleus.

(iii) Size of orbitals. Greater is the value of n (n = principal quantum number), larger is the size of the p-orbital. Thus $3p$ orbital ($n = 3$) is larger in size than $2p$ orbital ($n = 2$), although the shapes of both the orbitals are the same. Larger orbital occupies more space than the smaller orbital.

(iv) Energy of the three p-orbitals is the same *i.e.* all the three p-orbitals are ***degenerate.***

d-orbitals. For d-orbitals, $l = 2$ and hence $m = 0, \pm 1, \pm 2$ (five values). Five values of m for $l = 2$ (d-orbital) mean that a d-orbital has five sub-orbitals. The permitted combinations of l and m values for these five d sub-orbitals are $l = 2, m = 0$; $l = 2, m = +1$; $l = 2, m = -1$; $l = 2, m = +2$ and $l = 2, m = -2$.

The values of the angular wave functions for these d sub-orbitals are given below :

Angular wave function for l = 2, m = 0 orbital. It is given by :

$$\Theta_{2,0} \times \Phi_0 = \sqrt{\frac{5}{8}}\,(3\cos^2\theta - 1) \times \frac{1}{\sqrt{2\pi}}$$

$$= \sqrt{\frac{5}{16\pi}}\,(3\cos^2\theta - 1) = \sqrt{\frac{5}{16\pi}}\left(3 \times \frac{z^2}{r^2} - 1\right)$$

or
$$= \sqrt{\frac{5}{16\pi}} \times \frac{3z^2 - r^2}{r^2} \qquad \text{...}(i)$$

Since the value of $\Theta_{2,0} \times \Phi_0$ as given by equation (*i*) does not contain imaginary quantity, it is the real angular wave function. Thus, this value is acceptable. It is obvious from equation (*i*) that since $\Theta_{2,0} \times \Phi_0$ varies as z^2, this wave function is called $d_z{}^2$ wave function or $d_z{}^2$ orbital.

Angular wave functions for l = 2, m = +1 and l = 2, m = –1 orbitals. Angular wave functions for these orbitals are given by equations (*ii*) and (*iii*) respectively.

$$\Theta_{2,+1}\,\Phi_{+1} = \sqrt{\frac{15}{4}}\,\sin\theta\,\cos\theta \times \frac{1}{\sqrt{2\pi}}\,e^{+i\phi}$$

$$= \sqrt{\frac{15}{8\pi}}\,\sin\theta\cos\theta \times e^{-i\phi} \qquad \text{...}(ii)$$

$$\Theta_{2,-1}\,\Phi_{-1} = \sqrt{\frac{15}{4}}\,\sin\theta\cos\theta \times \frac{1}{\sqrt{2\pi}}\,e^{-i\phi}$$

$$= \sqrt{\frac{15}{8\pi}}\,\sin\theta\cos\theta\; e^{-i\phi} \qquad \text{...}(iii)$$

Since the values of both the above angular wave functions contain imaginary quantity i, these values cannot be accepted. The real values of these angular wave functions are obtained by the normalised linear combinations (addition and subtraction) of the angular wave functions represented by equations (*ii*) and (*iii*). Thus :

Normalised wave function obtained by the addition of (*ii*) and (*iii*)

$$= \frac{1}{\sqrt{2}}\,[\Theta_{2,+1}\,\Phi_{+1} + \Theta_{2,-1}\,\Phi_{-1}]$$

$$= \frac{1}{\sqrt{2}}\left[\sqrt{\frac{15}{8\pi}}\,(\sin\theta\cos\theta)\,e^{+i\phi} + \sqrt{\frac{15}{8\pi}}\,(\sin\theta\cos\theta)\,e^{-i\phi}\right]$$

$$= \sqrt{\frac{15}{4\pi}}\,\sin\theta\cos\theta\cos\phi = \sqrt{\frac{15}{4\pi}} \times \frac{x}{r} \times \frac{z}{r} = \sqrt{\frac{15}{4\pi}} \times \frac{xz}{r^2} \qquad \text{...}(iv)$$

Equation (*iv*) indicates that since the value of the angular wave function varies as xz, this wave function is called d_{xz} wave function or d_{xz} orbital.

Normalised wave function obtained by the *subtraction* of (*ii*) and (*iii*)

$$= \frac{1}{\sqrt{2}}[\Theta_{2,+1}\ \Phi_{+1} - \Theta_{2,-1}\ \Phi_{-1}]$$

$$= \frac{1}{\sqrt{2}}\left[\sqrt{\frac{15}{4\pi}}\ (\sin\theta\ \cos\theta)\ e^{+i\phi} - \sqrt{\frac{15}{4\pi}}\ (\sin\theta\cos\theta)\ e^{-i\phi}\right]$$

$$= \sqrt{\frac{15}{4\pi}}\ \sin\theta\cos\theta\sin\phi = \sqrt{\frac{15}{4\pi}} \times \frac{y}{r} \times \frac{z}{r} = \sqrt{\frac{15}{4\pi}} \times \frac{yz}{r^2} \qquad ...(v)$$

Equation (v) indicates that since the value of angular wave function varies with yz, this wave function is called d_{yz} wave function or d_{yz} orbital.

Angular wave functions for $l = 2$, $m = +2$ and $l = 2$, $m = -2$ orbitals. Angular wave functions for these orbitals are given by equations (*vi*) and (*vii*) respectively.

$$\Theta_{2,+2}\quad \Phi_{+2} = \sqrt{\frac{15}{16}}\ \sin^2\theta \times \frac{1}{\sqrt{2\pi}}\ e^{+2i\phi} = \sqrt{\frac{15}{32\pi}}\sin^2\theta\ \ e^{+2i\phi} \qquad ...(vi)$$

$$\Theta_{2,-2}\quad \Phi_{-2} = \sqrt{\frac{15}{16}}\ \sin^2\theta \times \frac{1}{\sqrt{2\pi}}\ e^{-2i\phi} = \sqrt{\frac{15}{32\pi}}\sin^2\theta\ \ e^{-2i\phi} \qquad ...(vii)$$

Since the values of both the above angular wave functions contain the imaginary quantity i, these values are not acceptable. The real values are obtained by the normalised linear combination of (*vi*) and (*vii*). Thus ;

Normalised wave function obtained by the addition of (*vi*) and (*vii*)

$$= \frac{1}{\sqrt{2}}[\Theta_{2,+2}\ \Phi_{+2} + \Theta_{2,-1}\ \Phi_{-2}]$$

$$= \frac{1}{\sqrt{2}}\left[\sqrt{\frac{15}{32\pi}}\ \sin^2\theta\ e^{-2i\phi} + \sqrt{\frac{15}{32\pi}}\sin^2\theta\ e^{-2i\phi}\right]$$

$$= \sqrt{\frac{15}{16\pi}}\ \sin^2\theta\ \cos 2\phi = \sqrt{\frac{15}{16\pi}}\ \sin^2\theta\ (\cos^2\phi - \sin^2\phi)$$

$$= \sqrt{\frac{15}{16\pi}}\ (\sin^2\theta\cos^2\phi - \sin^2\theta\sin^2\phi) = \sqrt{\frac{15}{16\pi}}\left(\frac{x^2}{r^2} - \frac{y^2}{r^2}\right) = \sqrt{\frac{15}{16\pi}} \times \frac{x^2 - y^2}{r^2} \qquad ...(viii)$$

Since the d wave function represented by equation (*viii*) varies as $x^2 - y^2$, this wave function is called $d_{x^2-y^2}$ function or $d_{x^2-y^2}$ orbital.

Normalised wave function obtained by the subtraction of (*vi*) and (*vii*)

$$= \frac{1}{\sqrt{2}}[\Theta_{2,+2}\ \Phi_{+2} + \Theta_{2,-2}\ \Phi_{-2}]$$

$$= \frac{1}{\sqrt{2}}\left[\sqrt{\frac{15}{32\pi}}\ \sin^2\theta\ \ e^{-2i\phi} - \sqrt{\frac{15}{32\pi}}\sin^2\theta\ \ e^{-2i\phi}\right]$$

$$= \sqrt{\frac{15}{16\pi}}\ \sin^2\theta\ \sin 2\phi = \sqrt{\frac{15}{16\pi}}\ \sin^2\theta\ .\ 2\sin\phi\ \cos\phi$$

$$= \sqrt{\frac{15}{4\pi}}\ \sin\theta\sin\phi\ \sin\theta\cos\phi = \sqrt{\frac{15}{4\pi}} \times \frac{y}{r} \times \frac{x}{r} = \sqrt{\frac{15}{4\pi}} \times \frac{xy}{r^2} \qquad ...(ix)$$

Since d-wave function represented by eqn. (*ix*) varies with xy, this wave function is called d_{xy} wave function or d_{xy} orbital.

The above discussion shows that the real angular wave functions of d-orbitals are d_{z^2}, d_{xz}, d_{yz}, $d_{x^2-y^2}$ and d_{xy}.

When the five d-wave functions are plotted in three dimensional space, we get the solid curves (three dimensional curves shown in Fig. 2.14. These are three dimensional curves because of the fact that they depend on r ($r = \sqrt{x^2 + y^2 + z^2}$).

All the five d-orbitals have been divided into the following three catagories :

(*i*) d_{xy}, d_{yz} *and* d_{zx} *orbitals.* Each of these orbitals has a double dum-bell shape, *i.e.* each of these orbitals contains four lobes. The four lobes of these orbitals lie in xy, yz and zx planes respectively and are concentrated *in between their respective coordinate axes*, making an angle of 45° with the axes. For example d_{xy} orbital has its four lobes between x and y axes. The quantity xy will be positive when both x and y are positive or both are negative. On the other hand, xy will be negative when either x or y is negative. Thus + and – signs occur in the alternate lobes of d_{xy} orbital. The four lobes of d_{yz} and d_{zx} orbitals also have +, –, +, – signs. (See Fig. 2.14).

(*ii*) $d_{x^2-y^2}$ *orbital.* This orbital, like d_{xy}, d_{yz} and d_{zx} orbitals, also has a double dum-bell shape with four lobes. These four lobes lie in xy plane and are concentrated *along x and y axes.* since, whatever be the sign of x, x^2 will always be positive, the signs in the lobes on x-axis will always be positive. On the other hand. the signs in the lobes on y-axis will always be negative, since whatever be the sign of y, $-y^2$ will always be negative. (See Fig. 2.14).

(*iii*) d_{z^2} *orbital.* This orbital has a dum-bell symmetrical shape with two lobes along the z-axis and a concentric collar (also called *ring, torus, lobe* or *tyre*) round the nucleus in the xy plane. The dum-bell shaped part has a positive geometric sign in both the lobes because, whatever be the sign of z (*i.e.* positive or negative), z^2 will always be positive. On the other hand, the collar in xy plane has a negative sign. The probability of finding the electron in this orbital is along z-axis (See Fig. 2.14).

Size of d-orbitals. d-orbitals belonging to all the main energy levels have similar shape. But their size will be different. The size depends on the number of radial nodes (No. of radial nodes = $n - l - 1$). Greater is the number of radial nodes, larger is the size of the orbital. For example, since $5d$ orbital (number of radial nodes = 5 – 2 – 1 = 2) has greater number of radial nodes than 4d orbital (No. of radial nodes = 4 – 2 – 1 = 1), this orbital has larger size. An orbital belonging to a higher main energy level has larger size than the orbital belonging to lower main energy level.

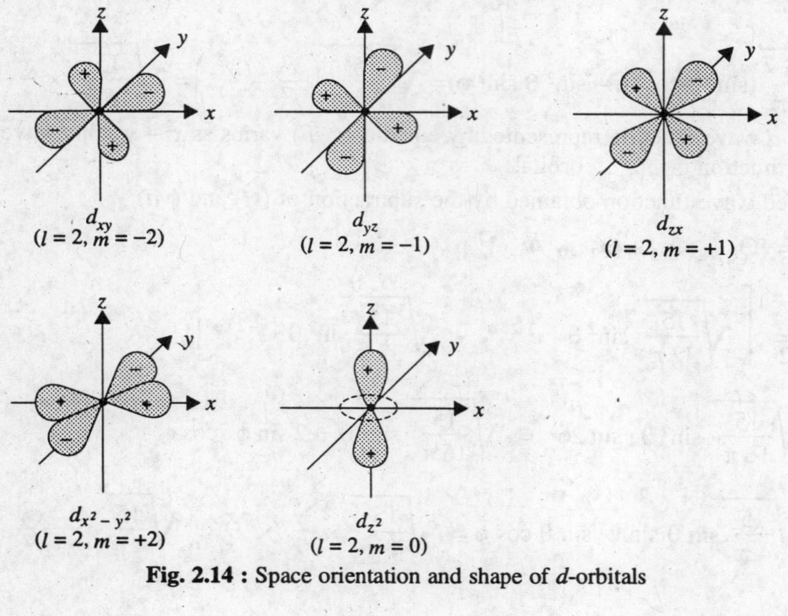

Fig. 2.14 : Space orientation and shape of *d*-orbitals

Quantum Numbers

These are the integral numbers which describe the energy of an electron in an orbit, the position (distance) of the electron from the nucleus (radius of the orbit), shape and number of orientations of an electron cloud (orbital) round its own axis and the direction of the spinning of the electron round its own axis. These are four numbers which are *principal quantum number (n), azimuthal quantum number (l), magnetic quantum number (m) and spin quantum number (s).* Thus, we see that the four quantum numbers give the complete description of an electron in an atom.

The three quantum numbers *viz n, l* and *m* are obtained when Schrodinger wave equation is solved for the wave function, ψ. Spin quantum number (*s*) arises due to the spinning of the electron about its own axis. These numbers are discussed below :

1. **Principal quantum number (*n*)**

This quantum number gives the following informations :

(*i*) *n* represents the number of the shell (orbit) or main energy level in which the electron revolves round the nucleus. *n* = 1, 2, 3, 4 etc. represent 1*st* (K), 2*nd* (L), 3*rd* (M), 4*th* (N) etc. shells.

(*ii*) *n* gives the distance (*r*) of an electron from the radius (radius of the orbit) given by the equation $\frac{h^2}{4\pi^2 me^2} \times \frac{n^2}{Z}$ (*c.g.s.* units) or $\frac{\varepsilon_o h^2}{\pi me^2} \times \frac{n^2}{Z}$ (*S.I.* units). This equation shows that as the value of *n* increases, the electron gets farther away from the nucleus.

(*iii*) *n* gives the energy (*E*) of an electron in an orbit, given by the equation $-\frac{2\pi^2 me^4}{h^2} \times \frac{Z^2}{n^2}$ (*c.g.s.* units) or $-\frac{me^4}{8\varepsilon_0^2 h^2} \times \frac{Z^2}{n^2}$ (*S.I.* units). This equation indicates that, due to negative sign, the magnitude of the energy associated with an orbit increases with the increase of the value of *n*.

(*iv*) *n* also gives the maximum number of electrons that may be accommodated in a given shell. According to Bohr-Bury scheme, the maximum number of electrons in *n*th shell = $2n^2$. Thus 1*st* (*n* = 1), 2*nd* (*n* = 2), 3*rd* (*n* = 3), 4*th* (*n* = 4) shells contain $2 \times 1^2 = 2$, $2 \times 2^2 = 8$, $2 \times 3^2 = 18$, $2 \times 4^2 = 32$ electrons respectively.

2. **Azimuthal quantum number (*l*).**

A portion of the energy of an electron is associated with the orbital motion of the electron round the nucleus. This orbital motion is described by the orbital angular momentum of the electron.

Azimuthal quantum number determines the magnitude of the orbital angular momentum and hence is also called *orbital angular momentum quantum number.* This quantum number explains the appearance of a group of closely-spaced lines seen in hydrogen spectrum and is designated by *l*. The possible values which *l* can have depend on the value of *n*. *l* may have all the possible whole number values from 0 to (*n* – 1), *i.e.* *l* = 0, 1, 2, 3........(*n* – 1). Thus for a given value of *n*, total number of *l* values is equal to *n*. Each value of *l* represents a particular sub-shell within the principal shell. *l* = 0, 1, 2, 3 represent *s,p,d* and *f* sub-shells respectively in a principal shell. The symbols *s,p,d* and *f* have been derived from the words *sharp, principal, diffuse* and *fundamental* respectively, which have been used to indentify the spectral lines in the atomic spectra of different atoms. Total number of *l* values give us the total number of sub-shells within a given main shell, *e.g.* if *n* = 2 (2*nd* main shell), *l* = 0, 1(two values). Two values of *l* imply that there are two sub-shells namely 2*s* and 2*p* in 2*nd* shell. In 2*s* sub-shell 2 indicates the number of the main shell (*n* = 2) and *s* indicates that *l* = 0. In 2*p* sub-shell *n* = 2 and *l* = 1. For *n* = 3 (third main shell), *l* = 0,1,2 (three values). Three values of *l* for 3*rd* main shell mean that this shell has three sub-shells, which are designated as 3*s* (*n* = 3, *l* = 0), 3*p*(*n* = 3, *l* = 1) and 3*d* (*n* = 3, *l* = 2). Similarly for *n* = 1 shell (1*st* shell), *l* = 0. This means that 1*st* shell has only one sub-shell *viz.* 1*s* (*n* = 1, *l* = 0).

The above discussion shows that :

(*i*) 1*st* ($n = 1$), 2*nd* ($n = 2$) and 3*rd* ($n = 3$) shells have *one* (1*s*), two (2s, 2*p*) and *three* (3*s*, 3*p*, 3*d*) sub-shells respectively.

(*ii*) Total number of sub-shells in a given shell is equal to the number of the main shell (*n*).

(*iii*) Maximum number of electrons, that may be accommodated in a given sub-shell, is equal to $2(2l + 1)$. Thus *s, p, d, f* sub-shells can have $2(2 \times 0 + 1) = 2$, $2(2 \times 1 + 1) = 6$, $2(2 \times 2 + 1) = 10$ and $2(2 \times 3 + 1) = 14$ electrons respectively.

3. **Zeeman effect and magnetic quantum number (*m*) or orientation quantum number**

The angular momentum of an electron is a vector quantity and hence has magnitude as well as direction. The magnitude of angular momentum, as already said, is determined by *l*. The direction of the anglural momentum is determined by magnetic quantum number (*m*).

Zeeman, in 1896, observed that when the source of a spectrum is placed in a strong magnetic field, each spectral line is further split into a number of separate lines. This phenomena is called *Zeeman effect*. If the spectrum is viewed in the same direction in which the magnetic field has been applied, each spectral line gets split up into two distinct lines. Both these lines are found to be circularly polarised in opposite directions. On the other hand, when the spectrum is viewed in a direction at right angles to the magnetic field, each spectral line is found to split up into three distinct lines. The middle line is found to be polarised at right angles to it.

The spectral lines obtained by the splitting of a single spectral line differ only slightly in energy from one another and also from the original spectral line which is seen in the spectrum in the absence of the magnetic field.

Magnetic quantum number has been introduced to account for the presence of additional spectral lines which are seen when a magnetic field is applied on the source of the spectrum.

If the total angular momentum of an electron is $h / 2\pi$, then the component of the total angular momentum in the direction of magnetic field in equal to $m \times h/2\pi$. If this component is quantised, the total angular momentum, $h/2\pi$ can have only a definite number of possible inclinations with respect to the direction of the magnetic field. The number of possible inclinations of the total angular momentum is given by the total number of *m* values. Possible values which *m* can have, depend on the value of *l*. *m* may have values which range from $-l$ through 0 to + l, *i.e.* $m = 0, \pm 1, \pm 2, ... \pm l$. Thus for a given value of *l*, total number of *m* values is equal to $(2l + 1)$. Each value of *m* represents a particular orbital within a sub-shell. Total number of *m* values give us total number of orbitals within a given sub-shell. For example :

(*a*) If $l = 0$ (*s* sub-shell), $m = 0$ (one value). One value of *m* indicates that *s* sub-shell has only one orbital. This, in other words, means that there is only one possible orientation of electrons in *s* sub-shell.

(*b*) When $l = 1$(*p* sub-shell), $m = 0, \pm 1$ (three values). Three values of *m* imply that *p* sub-shell has three orbitals which are perpendicular to each other and point towards *x*, *y* and *z*-axes. It is for this reason that these orbitals are designated as p_x, p_y and p_z. These three orbitals are *triply degenerate*. Three values of *m* ($= 0, \pm 1$) also mean that *p* sub-shell has three orientations in space.

(*c*) When $l = 2$ (*d* sub-shell), $m = 0, \pm 1, \pm 2$ (five values). This means that *d* sub-shell has five orbitals which are d_{xy}, d_{yz}, d_{zx}, $d_{x^2-y^2}$ and d_{z^2}. Thes five orbitals show *five fold degeneracy.*

(*d*) On the same lines of argument, it can be shown that *f* sub-shell ($l = 3$) has seven orientations in space corresponding to seven values of $m(= 0, \pm 1, \pm 2, \pm 3)$, *i.e.*, *f* sub-shell contains seven orbitals which have complicated shape. These seven orbitals show *seven-fold degeneracy.*

The above discussion indicates that there is only one *s*-orbital, three *p*-orbitals, five *d*-orbitals and seven *f*-orbitals, whatever be the principal energy level. Each orbital can have a maximum number of two electrons in it.

4. Spin quantum number(s)

The three quantum numbers *viz. n, l* and *m* are obtained when Schrodinger wave equation for H-atom is solved for ψ. Spin quantum number is not obtained in the solution of the wave equation. This quantum number arises due to the fact that the electron, while moving round the nucleus, also rotates or spins about its own axis (called *self-rotattion* or *spinning of the electron*) either in a clock-wise or in an anticlock-wise direction. Now since the electron is a charged particle (negatively-charged particle) and is spinning about its own axis, it would behave as a small magnet. Consequently, the spinning of the electron will create a magnetic moment, which is called *spin magnetic moment.* Since the electron can spin either in a clock-wise or in an anticlock-wise direction, the spin magnetic moment can be oriented either up or down. Thus corresponding to the clock-wise and anticlock-wise spinning of the electron, maximum two values of *s* are possible. since the quantum numbers differ from one another only by integers, *s* can have a value equal to +1/2 or – 1/2. Value equal to + 1/2 ($s = +1/2$) is used to indicate the clock-wise direction of the spin of the electron. This type of spin is known as α-spin and is generally represented by an upward arrow (↑). On the other hand, value equal to –1/2 ($s = -1/2$) indicates the anticlock direction of the spin. This type of spin is called β-spin and is represented by a downward arrow (↓).

Relation between different quantum numbers is given in Table 2.3.

Table 2.3 : ***Relations between different quantum numbers.***

Principal quantum number or main shell (*n*)	Values of *l* and the number of sub-shells in the main shell = Number of *l* values	Total number of electrons in a given sub-shell = 2 (2*l* + 1)	Values of *m* and the number of orbitals in a sub-shell = Number of *m* values	Maximum number of electrons in orbitals = 2 × No. of orbitals	Maximum number of electrons in main shell = $2n^2$
1 (1*st* shell or K-shell)	0 (1*s* sub-shell)	2 (2 × 0 + 1) = 2	0 (One value-one orbital *viz.* 1*s* orbital)	2 × 1 = 2	$2 \times 1^2 = 2$
2 (2nd shell or L shell)	0 (2*s* sub-shell)	2 (2 × 0 + 1) = 2	0 (One value—one orbital *viz* 2*s* orbital)	2 × 1 = 2	$2 \times 2^2 = 8$ or 2 + 6 = 8
	1 (2*p* sub-shell)	2(2 × 1 + 1) = 6	0, ± 1 (Three vlaues : three orbitals *viz.* $2p_x$, $2p_y$ and $2p_z$)	2 × 3 = 6	
3 (3rd shell or M shell)	0 (3*s* sub-shell)	2 (2 × 0 + 1) = 2	0 (One value-one orbital viz 3s orbital)	2 × 1 = 2	$2 \times 3^2 = 18$ or 2 + 6 + 10 = 18
	1 (3*p* sub-shell)	2 (2 × 1 + 1) = 6	0, ± 1 (Three values : three orbitals *viz* $3p_x$, $3p_y$ and $3p_z$)	2 × 3 = 6	
	2 (3*d* sub-shell)	2 (2 × 2 + 1) = 10	0, ± 1, ± 2 (Five values : five orbitals viz $3d_{xy}$, $3d_{z^2}$ $3d_{yz}$, $3d_{zz}$, $3d_{x^2-y^2}$)	2 × 5 = 10	

Pauli's Exclusion Principle

This principle states, that *the values of all the four quantum numbers for the two electrons residing in the same orbital like s, p_x, p_y, p_z, d_{xy}, d_{yz} etc. cannot be the same, i.e. for the two electrons residing in the same orbital, the values of three quantum numbers at the most may be the same but* *the value of the fourth quantum number namely spin quantum (s) for the two electrons must necessarily be different.*

This principle implies that *for one electron s will be +1/2 while for the other electron it will be –1/2. In other words, this means that two electrons residing in the same orbital of an atom have opposite spins,* since this state with opposite spins (↑↓) gives lower energy to the system, while the state with parallel spins (↑↑) gives higher energy to the system. According to thermodynamics, a system with lower energy has greater stability. Thus an orbital having two electrons with opposite spins (↑↓) has lower energy and hence is more stable than that containing the two electrons with paired spins (↑↑). $s = +1/2$ indicates the clock-wise direction of the spinning of one electron while $s = -1/2$ stands to indicate the anticlock-wise direction of the spinning of other electron.

Illustration of the principle

In order to illustrate the principle let us consider the following two different arrangements of the two electrons in 1*s* orbital. The values of the four quantum numbers for the two electrons in both the arrangements are given below :

Arrangement I		*Arrangement II*	
↑	↓	↑	↑
$n = 1$	$n = 1$	$n = 1$	$n = 1$
$l = 0$	$l = 0$	$l = 0$	$l = 0$
$m = 0$	$m = 0$	$m = 0$	$m = 0$
$s = +\frac{1}{2}$	$s = -\frac{1}{2}$	$s = +\frac{1}{2}$	$s = +\frac{1}{2}$

Here + 1/2 is the value of the spin quantum number (*s*) for the electron having *upward spin* (↑) and –1/2 is the value of the spin quantum number, for the electron having *downward spin* (↓). The values of different quantum numbers, as given above, for the two arrangments, make it clear that in arrangement (*I*) the values of the four quantum numbers for both the electrons are not the same, since the value c different for the two electrons. On the other hand, in arrangement (*II*) the values of the four quantum numbers for the two electrons are the *same*. Thus arrangement (*II*) is against Pauli's exclusion principle and hence is not allowable. In other words, it is arrangement (*II*) which is not a stable arrangement.

The above discussion shows that if two electrons are to be placed in an orbital like p_x, p_y, p_z, d_{xy}, d_{yz} etc., they should be placed in such a way that they should have opposite spins (↑↓) and not the same spins (↑↑ or ↓↓)

Application of Pauli's exclusion principle

With the help of this principle, the maximum number of electrons that can be accommodated in an orbital, sub-shell and main shell can be calculated as shown below :

(*i*) **Ist shell** (*K-shell* with *n = 1*). For this shell $n = 1$. The values of l and m for $n = 1$ are $l = 0$ and $m = 0$, s can have a value either equal to +1/2 or –1/2. Thus these different values of *n, l m,* and *s* give the **two** sets of the values of four quantum numbers, keeping in view Pauli's exclusion principle. (See Table 2.4).

Since the number of permitted sets of four quantum numbers is two for 1*st* shell ($n = 1$), this shell contains at the maximum two electrons which are contained in 1*s* orbital.

(*ii*) **2nd shell** (*L-shell* with $n = 2$). For this shell $n = 2$. For $n = 2$, the different values of *l, m* and *s* give the **eight** sets of quantum numbers which indicate that *2nd* shell contains eight electrons of which two are contained by 2*s* sub-shell while the remaining six are contained in 2*p* sub-shell. Six electrons in 2*p* sub-shell are distributed in $2p_x$, $2p_y$, and $2p_z$ orbitals as $2p_x^2$, $2p_y^2$, $2p_z^2$ in which the superscript indicates the number of electrons contained in the orbital.

(*iii*) **3rd and 4th shells** (*M* and *N shells*). On the same lines of arguement it can be shown that 3*rd* shell ($n = 3$) can contain at the maximum 18 electrons (2 in 3*s* sub-shell, 6 in 3*p* sub-shell and 10 in 3*d* sub-shell) and 4*th* shell ($n = 4$) can have 32 electrons (2 in 4*s* sub-shell, 6 in 4*p* sub-shell, 10 in 4*d* sub-shell and 14 in 4*f* sub-shell).

Whatever has been said above, has been shown in Table 2.4.

Table 2.4 : *To determine the number of elctrons present in orbitals and main shells with the help of Pauli's exclusian principle.*

Main shell	Combination number	Different values of *n*, *l*, *m* and *s*	Electron with upward / downward spin	Total number of electrons in orbitals and sub-shells	Total number of electrons in main shell
1*st* shell or K-shell (n = 1)	1	$n = 1, l = 0, m = 0, s = +\frac{1}{2}$	↑ (upward spin)	2 ($1s^2$) (↑↓)	2 or $2 \times 1^2 = 2$
	2	$n = 1, l = 0, m = 0, s = -\frac{1}{2}$	↓ (downward spin)		
2nd shell or L-shell (n=2)	1	$n = 2, l = 1, m = 0, s = +\frac{1}{2}$	↑ (upward spin)	2 ($2s^2$) (↑↓)	2 + 2 +2 + 2 = 8 or $2 \times 2^2 = 8$
	2	$n = 2, l = 1, m = 0, s = -\frac{1}{2}$	↓ (downward spin)		
	3	$n = 2, l = 1, m = 0, s = +\frac{1}{2}$	↑ (upward spin)	2 ($2p_z^2$) (↑↓)	
	4	$n = 2, l = 1, m = 0, s = -\frac{1}{2}$	↓ (downward spin)		
	5	$n = 2, l = 1, m = +1, s = +\frac{1}{2}$	↑ (upward spin)	2 ($2p_y^2$) (↑↓)	
	6	$n = 2, l = 1, m = +1, s = -\frac{1}{2}$	↓ (downward spin)		
	7	$n = 2, l = 1, m = -1, s = +\frac{1}{2}$	↑ (upward spin)	2 ($2p_x^2$) (↑↓)	
	8	$n = 2, l = 1, m = -1, s = -\frac{1}{2}$	↓ (down ward spin)		

Table 2.4 makes the following three rules evident: (*i*) *An orbital* like *s*, p_x, p_y, p_z, d_{xy}, d_{yz} *etc., can have at the most two electrons and these two electrons must have opposite spins, i.e.*, one electron should have the value of spin quantum number(*s*) equal to +1/2 and the other equal to –1/2, so that both the electrons residing in the same oribital, may have different sets of four quantum numbers, in accordance with Pauli's exclusion principle. (*ii*) *Maximum number of electrons, that can be accommodated* by *a shell with principal quantum number, n is equal to* $2n^2$. (*iii*) since the maximum capacity to contain the electrons is two, total number of electrons in *s,p,d* and *f* sub-shells is $2 \times (2l+1) = 2 \times (2 \times 0 + 1) = 2$, $2 \times (2 \times 1 + 1) = 6$, $2 \times (2 \times 2 + 1) = 10$ and $2 \times (2 \times 3 + 1) = 14$ respectively. These rules have been used in determining the electronic configuration of atoms of the elements.

Shielding (or screening) Effect and Effective Nuclear Charge

The electrons residing in the shells between the nucleus and the valence-shell are called *intervening electrons.* These electrons reduce the force of attraction between the nucleus and the outer-most shell electrons. The reduction produced in the force of attraction due to the presence of intervenining electrons is called *shielding* (or *screening) effect.* With the decrease in the force of attraction caused by the shielding effect of intervening electrons, the actual nuclear charge (which is equal to the atomic number, Z of the element) is decreased by the quantity, σ (sigma) which is called *screening constant.* The decreased nuclear charge which is obviously equal to (Z – σ) is called *effective nuclear charge* and is denoted by Z_{eff}. Thus :

$$Z_{eff} = Z - \sigma \quad \text{or} \quad Z_{eff} = Z_{actual} - \sigma \qquad ...(i)$$

σ is a measure of the extent to which the intervening electrons screen the outer-most shell electron from the nuclear pull on it. Equation (*i*) suggests that *effective nuclear charge* (Z_{eff}) *is defined as the actual nuclear charge (Z or* Z_{actual} *) minus the screening constant* (σ) *produced by the electrons residing between the nucleus and the outer-most shell electron* (intervenning electrons).

Factors affecting the magnitude of σ and Z_{eff} and their variation in the periodic table :

Following are the important factors which affect the magnitude of σ and Z_{eff} and predict their variation in the periodic table.

(*i*) *Number of interveening electrons.* Greater is the number of electrons interveening between the nucleus and the outer most shell (*i.e.*, intervening electrons), more will be the magnitude of σ and hence the magnitude of Z_{eff} will decrease ($Z_{eff} = Z_{actual} - \sigma$) to a greater extent. When we move down a group, the number of interveening electrons increases and hence the magnitude of σ also increases. The increase in the value of σ decreases the value of Z_{eff}. Thus on going down a group, the magnitude of Z_{eff} goes on decreasing.

(*ii*) *Size of the atom.* With the increase in the size of the atom, Z_{eff} decreases. Thus:

(*a*) Since atomic size increases in going down a group, Z_{eff} *decreases* in the same direction.

(*b*) Since the size of atoms decreases as we move along a period from left to right, Z_{eff} *increases* in the same direction.

Slater's rules for calculating σ and Z_{eff}

These rules are given at **A** and **B** discussed below :

A. The value of σ for an electron residing in an *ns* or in one of the *np* orbitals of *n*th shell ot an atom or an ion is the sum of the following contributions:

(*a*) Each of the remaining electrons in the *n*th shell makes a contribution of 0.35 to the value of σ.

(*b*) Each of the electrons in (*n* – 1)th shell makes a contribution of 0.85.

(*c*) Each of the electrons in the next inner shells makes a contribution of 1.0.

(*d*) There will be no contribution to the value of σ by the electrons residing in orbitals having higher value of their principal quantum number than the orbital containing the electron for which σ is being calculated. [See example below].

(*e*) If σ is being calculated for an electron of 1*s* orbital, there will be a contribution of 0.30 from other single electron in 1*s* orbital [See example below].

Thus σ for an electron residing in *ns* or *np* orbital

= 0.35 × [Number of the remaining electrons in *nth* shell] + 0.85 × [No. of electrons in (*n* – 1)*th* shell] + 1.0 × [Number of electrons in the inner shells].

The application of the above formula can be understood by the following examples.

Example 2.1 : *Calculate σ and Z_{eff} for 4s electron in (i) Mn(Z = 25) (ii) Cu (Z = 29) and (iii) K(Z = 19).*

Solution. (*i*) $Mn_{25} = \underbrace{1s^2, 2s^2p^6}_{(10)} \quad \underbrace{3s^2p^6d^5}_{(13)} \quad \underbrace{4s^2}_{(2)}$

σ for $4s^1$ electron in Mn atom = 0.35 × [No. of the remaining electrons in 4*s* orbital] + 0.85 × [No. of electrons in 3*s*, 3*p* and 3*d* orbitals] + 1.0 × [No. of electrons in the inner shells] = 0.35 × 1 + 0.85 × 13 + 1.0 × 10= 21.40.

∴ Z_{eff} experienced by $4s^1$ electron = 25 – 21.40 = 3.60

(*ii*) $Cu_{29} = \underbrace{1s^2, 2s^2p^6}_{(10)} \quad \underbrace{3s^2p^6d^{10}}_{(18)} \quad \underbrace{4s^1}_{(1)}$

σ for $4s^1$ electron in Cu atom = 0.35 × 0 + 0.85 × 18 + 1.0 × 10 = 25.30

∴ Z_{eff} experienced by $4s^1$ electron = 29 – 25.30 = 3.70

(*iii*) $K_{19} = \underbrace{1s^2, 2s^2p^6}_{(10)} \quad \underbrace{3s^2p^6}_{(8)} \quad \underbrace{4s^1}_{(1)}$

σ for $4s^1$ electron of K atom = 0.35 × 0 + 0.85 × 8 + 1.0 × 10 = 16.80

∴ Z_{eff} experienced by $4s^1$ electron = 19 – 16.80 = 2.20

Example 2.2. *Calculate Z_{eff} for the last electron in (i) Cl-atom (Z = 17) and (ii) Cl^- ion.*

Solution (*i*) Cl_{17} atom = $\underbrace{1s^2}_{(2)}$ $\underbrace{2s^2p^6}_{(8)}$ $\underbrace{3s^2p^5}_{(7)}$

σ for $3p^1$ electron (last electron) of Cl-atom

$= 0.35 \times 6 + 0.85 \times 8 + 1.0 \times 2 = 10.90$

$\therefore$ Z_{eff} exprienced by $3p^1$ electron of Cl-atom = 17 – 10.90 = 6.10

(*ii*) Cl^-_{17} ion = $\underbrace{1s^2}_{(2)}$ $\underbrace{2s^2p^6}_{(8)}$ $\underbrace{3s^2p^6}_{(8)}$

σ for $3p^1$ electron (last electron) of Cl^- ion

$= 0.35 \times 7 + 0.85 \times 8 + 1.0 \times 2 = 11.25$

$\therefore$ Z_{eff} experienced by $3p^1$ electron of Cl^- ion = 17 – 11.25 = 5.75

Example 2.3. *Calculate the Z_{eff} for the last electron in (i) Na atom (Z = 11) and (ii) Na^+ ion.*

Solution (*i*) Na_{11} atom = $\underbrace{1s^2}_{(2)}$ $\underbrace{2s^2p^6}_{(8)}$ $\underbrace{3s^1}_{(1)}$

σ for $3s^1$ electron (last electron)of Na-atom = $0.35 \times 0 + 0.85 \times 8 + 1.0 \times 2 = 8.80$

$\therefore$ Z_{eff} experienced by $3s^1$ electron of Na atom = 11 – 8.80 = 2.20

(*ii*) Na^+_{11} ion = $\underbrace{1s^2}_{(2)}$ $\underbrace{2s^2p^6}_{(8)}$

σ for $2p^1$ electron (last electron) of Na^+ ion = $0.35\times 7 + 0.85 \times 2 = 4.15$

$\therefore$ Z_{eff} experienced by $2p^1$ electron of Na^+ ion = 11 – 4.15 = 6.85

Example 2.4. *Calculate the effective nuclear charge for 1s electron in (i) nitrogen atom (Z = 7) and (ii) helium atom (Z = 2).*

Solution. (*i*) N_7 = $\underbrace{1s^2}_{(2)}$ $\underbrace{2s^2p^3}_{(5)}$

There will be no contribution from the five electrons present in 2*s* and 2*p* orbitals to the value of σ which is to be calculated for $1s^1$ electron. Thus :

σ for $1s^1$ electron of N-atom = $5 \times 0 + 1 \times 0.30 = 0.30$

$\therefore$ Z_{eff} experienced by $1s^1$ electron of N-atom = 7 – 0.30 = 6.70

(*ii*) He_2 = $\underbrace{1s^2}_{(2)}$

σ for $1s^1$ electron of He-atom = $1 \times 0.30 = 0.30$

$\therefore$ Z_{eff} experienced by $1s^1$ electron in He-atom = 2 – 0.30 = 1.70

Example 2.5. *Calculate Z_{eff} eperienced by an electron residing in K and L -shell in* Li_3, Be_4, B_5, C_6, N_7, O_8, F_9 *and* Ne_{10} *atoms.*

Solution. (*i*) Li_3 = $\underset{2}{\underbrace{\overset{K}{1s^2}}}$ $\underset{2}{\underbrace{\overset{L}{2s^1}}}$

σ for $2s^1$ electron (L-shell electron) = $0.35 \times 0 + 0.85 \times 2 = 1.70$

$\therefore$ Z_{eff} experienced by $2s^1$ electron (L-shell electron) = 3 – 1.70 = 1.30

σ for $1s^1$ electron (K-shell electron) = $1 \times 0.30 = 0.30$

$\therefore$ Z_{eff} felt by $1s^1$ electron (K-shell electron) = 3 – 0.30 = 2.70

(*ii*) Be_4 = $\underset{2}{\underbrace{\overset{K}{1s^2}}}$ $\underset{2}{\underbrace{\overset{L}{2s^2}}}$

σ for $2s^1$ electron (L-shell electron) = 0.35 × 1 + 0.85 × 2 = 2.05

∴ Z_{eff} experienced by $2s^1$ electron (L-shell electron)= 4 – 2.05 = 1.95

σ for $1s^1$ electron (K-shell electron) = 1 × 0.30 = 0.30

∴ Z_{eff} felt by $1s^1$ electron (K-shell electron) = 4 – 0.30 = 3.70

$$(iii)\ B_5 = \overset{K}{\underbrace{1s^2}_{2}}\quad \overset{L}{\underbrace{2s^2p^1}_{3}}$$

σ for $2p^1$ electron (L-shell electron) = 0.35 × 2 + 0.85 × 2 = 2.40

∴ Z_{eff} experienced by $2p^1$ electron (L-shell electron) = 5 – 2.40 = 2.60

σ for $1s^1$ electron (K-shell electron) = 1 × 0.30 = 0.30

∴ Z_{eff} experienced by $1s^1$ electron (K-shell electron) = 5 – 0.30 = 4.70

$$(iv)\ C_6 = \overset{K}{\underbrace{1s^2}_{2}}\quad \overset{L}{\underbrace{2s^2p^2}_{4}}$$

σ for $2p^1$ electron (L-shell electron) = 0.35 × 3 + 0.85 × 2 = 2.75

∴ Z_{eff} experienced by $2p^1$ electron (L-shell electron) = 6 – 2.75 = 3.25

σ for $1s^1$ electron (K-shell electron) = 1 × 0.30 = 0.30

∴ Z_{eff} experienced by $1s^1$ electron = 6 – 0.30 = 5.70

$$(v)\ N_7 = \overset{K}{\underbrace{1s^2}_{2}}\quad \overset{L}{\underbrace{2s^2p^3}_{5}}$$

σ for $2p^1$ electron (L-shell electron) = 0.35 × 4 + 0.85 × 2 = 3.10

∴ Z_{eff} experienced by $2p^1$ electron (L-shell electron) = 7 – 3.10 = 3.90

σ for $1s^1$ electron (K-shell electron) = 1× 0.30 = 0.30

∴ Z_{eff} experienced by $1s^1$ electron = 7 – 0.30 = 6.70

$$(vi)\ O_8 = \overset{K}{\underbrace{1s^2}_{2}}\quad \overset{L}{\underbrace{2s^2p^4}_{6}}$$

σ for $2p^1$ electron (L-shell electron) = 0.35 × 5 + 0.85 × 2 = 3.45

∴ Z_{eff} experienced by $2p^1$ electron = 8 – 3.45 = 4.55

σ for $1s^1$ electron (K-shell electron) = 1 × 0.30 = 0.30

∴ Z_{eff} experienced by $1s^1$ electron = 8 – 0.30 = 7.70

$$(vii)\ F_9 = \overset{K}{\underbrace{1s^2}_{2}}\quad \overset{L}{\underbrace{2s^2p^5}_{7}}$$

σ for $2p^1$ electron (L-shell electron) = 0.35 × 6 + 0.85 × 2 = 3.80

Z_{eff} experienced by $2p^1$ electron = 9 – 3.80 = 5.20

∴ σ for $1s^1$ electron (K-shell electron) = 1 × 0.30 = 0.30

∴ Z_{eff} experienced by $1s^1$ electron = 9 – 0.30 = 8.70

$$(viii)\quad Ne_{10} = \overset{K}{\underbrace{1s^2}_{2}}\quad \overset{L}{\underbrace{2s^2p^6}_{8}}$$

σ for $2p^1$ electron (L-shell electron) = 0.35 × 7+ 0.85 × 2 = 4.15

∴ Z_{eff} experienced by $2p^1$ electron = 10 – 4.15 = 5.85

σ for $1s^1$ electron (K-shell electron) = $1 \times 0.30 = 0.30$

$\therefore$ Z_{eff} experienced by $1s^1$ electron = $10 - 0.30 = 9.70$

B. The value of σ for an electron residing in $(n-1)d$ orbitals of $(n-1)th$ shell of an atom or an ion is the sum of the following contributions :

(*a*) There will be no contribution to the value of σ by the electrons residing in *ns* orbital.

(*b*) Each of the remaining electrons present in $(n-1)d$ orbitals, makes a contribution of 0.35.

(*c*) Each of the electrons present in the $(n-1)s$, $(n-1)p$ orbitals and inner shells [*i.e.* 1*st*, 2*nd*, 3*rd*, ... $(n-2)th$ shells] makes a contribution of 1.0. Thus :

σ for a $(n-1)d$ electron = $0.35 \times$ [No. of the electrons in $(n-1)d$ orbitals] + $1.0 \times$ [No. of electrons in $(n-1)s$, $(n-1)p$ orbitals and inner shells]

The application of the above formula can be understood by the following example.

Example 2.6. *Calculate σ and Z_{eff} for 3d electron in (i) Mn(Z= 25) and (ii) Cu (Z = 29).*

Solution : (*i*) Mn = $1s^2, 2s^2p^6, 3s^2p^6d^5, 4s^2$. On arranging the various orbitals as per Slater's rules, we get:

$$\underbrace{(1s^2), (2s^2p^6), (3s^2p^6)}_{(18)} \quad \underbrace{(3d^5)}_{(5)} \quad \underbrace{4s^2}_{(2)}$$

$4s^2$ electrons will not contribute to the value of σ for a $3d$ electron [See rule 2 (*a*)] Thus:

σ for a $3d$ electron in Mn atom = $0.35 \times$ [No. of the remaining electrons in $3d$ orbitals] + $1.0 \times$ [No. of electrons in $1s$, $2s$, $2p$, $3s$ and $3p$ orbitals] = $0.35 \times 4 + 1.0 \times 18 = 19.40$

$\therefore$ Z_{eff} experienced by a $3d$ electron of Mn atom = $25 - 19.40 = 5.60$

(*ii*) Cu = $1s^2, 2s^2p^6, 3s^2p^6d^{10}, 4s^1$

On arranging the various orbitals of Cu as per Slater's rules, we have :

$$\underbrace{(1s^2)\ (2s^2p^6), (3s^2p^6)}_{(18)} \quad \underbrace{3d^{10}}_{(10)} \quad \underbrace{4s^1}_{(1)}$$

Here $4s^1$ electron will not contribute to the value of σ for a $3d$ electron in Cu-atom. Thus:

σ for a $3d$ electron of Cu = $0.35 \times 9 + 1.0 \times 18 = 21.15$

$\therefore$ Z_{eff} experienced by a $3d$ electron = $29 - 21.15 = 7.85$

To calculate the effective nuclear charge felt at the periphery of an atom or an ion.

The effective nuclear charge (Z_{eff}) felt at the periphery of an atom or an ion is equal to the atomic number of the element minus the screening constant (σ) for all the electrons present in the electronic configuration of the atom or ion under consideration. This is illustrated by the following example.

Example 2.7. *Calculate Z_{eff} at the periphery of (i) N-atom (Z = 7) (ii) Cr-atom (Z = 24) and (iii) O^{2-} ion (Z = 8)*

Solution. (*i*) Electronic configuration of N-atom is $\underbrace{1s^2}_{(2)} \quad \underbrace{2s^2p^3}_{(5)}$

$\therefore$ σ for all the electrons present in the electronic configuration of N-atom $= 0.35 \times 5 + 0.85 \times 2 = 3.45$

$\therefore$ Z_{eff} at the periphery of N-atom = $7 - 3.45 = 3.55$

(*ii*) Cr = $\underbrace{1s^2, 2s^2p^6}_{(10)} \quad \underbrace{3s^2p^6d^5}_{(13)} \quad \underbrace{4s^1}_{(1)}$

$\therefore$ σ for all the electrons in Cr-atom = $0.35 \times 1 + 0.85 \times 13 + 1.0 \times 10 = 21.40$

$\therefore$ Z_{eff} at the periphery of Cr-atom = $24 - 21.40 = 2.60$

(*iii*) O = $1s^2, 2s^2p^4$; $\quad O^{2-} = \underbrace{1s^2}_{(2)} \quad \underbrace{2s^2p^6}_{(8)}$

$\therefore$ σ for all the electrons in O^{2-} ion = $0.35 \times 8 + 0.85 \times 2 = 4.5$

$\therefore$ Z_{eff} at the periphery of O^{2-} ion = $8 - 4.5 = 3.5$

Applications of Slater's rules and concept of effective nuclear charge

Slater's rules and the concept of effective nuclear charge have been used to explain the following :

1. *Why is 4s orbital filled earlier than 3d orbitals in potassium atom* (Z = 19)? We know that the configuration of Ar (Z = 18) which is the last element of 3*rd* period of the periodic table is $1s^2,2s^2p^6, 3s^2p^6$. Thus 3*rd* shell is not completely filled in Ar atom, since 3*d* orbitals remain vacant in it. After 3*p* orbitals have been filled completely in Ar, the 19th electron in K (Z = 19) does not enter 3*d* orbitals; rather it goes to 4*s* orbital. Why 4*s* orbital is filled in preference to 3*d* orbitals can be explained as follows:

The two configurations that are theoretically possible for K-atom are :

$$(a)\ K\ (Z = 19) = \underbrace{1s^2,2s^2p^6, 3s^2p^6}_{(18)}\ \underbrace{4s^1}_{(1)}$$

$$(b)\ K(Z = 19) = \underbrace{1s^2,2s^2p^6, 3s^2p^6}_{(18)}\ \underbrace{3d^1}_{(1)}$$

In example 1 we have calculated the value of Z_{eff} experienced by $4s^1$ electron of K-atom [configuration (*a*)] equal to 2.20. The value of Z_{eff} experienced by 3*d* electron of K-atom [configuration (*b*)] can be calculated as follows:

σ for 3*d* electron in structure (*b*) = $0.35 \times 0 + 1.0 \times 18 = 18$

$\therefore$ Z_{eff} experienced by 3*d* electron = 19 – 18 = 1.0

Since Z_{eff} for $4s^1$ electron is greater than that for $3d^1$ electron, the attraction between $4s^1$ electron and the nucleus is greater than that between the $3d^1$ electron and nucleus of K-atom. Thus $1s^2,2s^2p^6, 3s^2p^6, 4s^1$ configuration would be more stable than $1s^2, 2s^2p^6, 3s^2p^6d^1$ configuration. In other words 4*s* orbital is filled earlier than 3*d* orbital.

2. *Formation of cations from the isolated gaseous atoms of transitien metals.* When transition metals are converted into cations, it is *ns* electrons, and not $(n-1)d$ electrons, which are removed first from the isolated gaseous atoms of transition metals. In order to explain this fact let us consider the formation of Mn^+ ion from Mn atom (Z = 25). The configuration of Mn is $1s^2,2s^2p^6, 3s^2p^6d^5, 4s^2$.

In example 1, we have seen that Z_{eff} experienced by $4s^1$ electron in this atom is equal to 3.60 while according to example 6, Z_{eff} for $3d^1$ electron is equal to 5.60. Since Z_{eff} for $3d^1$ electron is greater than that for $4s^1$ electron, the attraction between 3*d* electron and the nucleus of Mn atom is greater than the attraction between 4*s* electron and the nucleus, Consequently the removal of a 4*d* electron from an isolated gaseous atom of manganese is more difficult than the removal of an electron from 4*s* orbital, *i.e.* the removal of a 4*s* electron is easier than the removal of a 3*d* electron. Thus, in general, we can say that it is *ns* electron, that is lost first by a transition metal to convert itself into a cation.

3. *A cation is smaller in size than its parent atom.* See "Atomic and ionic radii" discussed under the chapter on "*Periodic Properties*".

4. *An anion is larger in size than its parent atom..* See "*Atomic and ionic* radii" under the chapter on "*Periodic Properties*".

5. *Variation of atomic and ionic radii of the atoms of representative elements in a period.* See "Atomic and ionic radii" discussed in the chapter on "*Periodic Properties*".

6. *Variation of electronegativity values in a period and a group of representative elements.* See "*Electronegativity*" discussed in the chapter on "*Periodic Properties*".

7. *Variation in the values of successive ionisation energies of a given element.* The successive ionisation energies (IE_1, IE_2, IE_3 etc.) of a given element (*M*) increases in the order $IE_1 < IE_2 < IE_3 < \ldots$. This order has been explained on the basis of the concept of effective nuclear charge experienced by the last electron in M, M^+, M^{2+} etc. For the explanation see "*Ionisation energy*" discussed under the chapter on "*Periodic Properties*"

University Questions

1. What values are assigned to quantum numbers *n, l* and *m* for (*i*) 2*s* orbital (*ii*) 2*p* orbital (*iii*) 4*d* orbital ? *(VBS Purvanchal 2002)*
2. State and explain Pauli's exclusion principle. *(Nagpur 2002)*
3. Explain radial probability distribution curve for 1*s* electron. *(Nagpur 2002)*
4. Find effective nuclear charge on an electron in valence shell of calcium. (Z of Ca = 20). *(Nagpur 2002)*
5. What do you understand by the term quantum number ? What is the significance of principal quantum number and azimuthal quantum number ? Which quantum number determines the orientation of an atomic orbital ? *(Nagpur 2002)*
6. (*a*) Write a note on physical significance of ψ and Ψ^2.
   (*b*) Calculate the wave length of an electron (mass = 9.1×10^{-31} kg) travelling with one-third speed of light ($h = 6.624 \times 10^{-34}$ Js, $c = 2.998 \times 10^8$ ms^{-1}).
   (*c*) What is significance of four quantum numbers ? *(Delhi 2002)*

Answer (*b*) $$\lambda = \frac{h}{mc} = \frac{6.624\times10^{-34}\ \text{Js}}{(9.1\times10^{-31}\ \text{kg})\times\frac{1}{3}\times(2.998\times10^{8}\ \text{ms}^{-1})}$$

$$= \frac{6.624\times10^{-34}\ \text{kg}\,.\,\text{m}^2.\,\text{s}^{-2}.\,\text{s}}{(9.1\times10^{-31})\times\left(\frac{1}{3}\times2.998\times10^{8}\right)\text{kg ms}^{-1}} \qquad (\because J = \text{kg m}^2\text{s}^{-2})$$

$$= \frac{6.624\times10^{-34}}{9.1\times10^{-31}\times0.999\times10^{8}}\ \text{m}$$

$$\mathbf{= 0.7284 \times 10^{-11}\ m.}$$

7. Describe main features of *s, p, d* and *f* orbitals. How the Pauli's exclusion principle is utilised in determining the maximum number of electrons in the above orbitals ? *(VBS Purvanchal 2002)*
8. Which of the following arrangements of three electrons in three degenerate orbitals *viz* p_x, p_y and p_z is correct ? *(VBS Purvanchal 2002)*

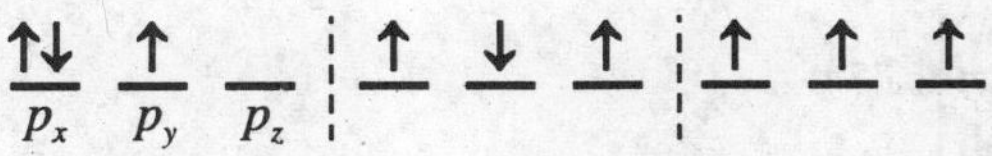

(*a*) (*b*) (*c*)

9. What is the difference between an orbit and an orbital ? *(VBS Purvanchal 2003)*
10. (*a*) What is meant by the polarisation of an ion ? State Fajan's rules of polarisation of ions.
    (*b*) Calculate the screening constant of silicon (atomic no. 14) with Slater's rule.
    (*c*) The value of ionization potential of hydrogen atom is 13.6 e.v. What would be the value of IIIrd I.P. of lithium atom ? *(Rohilkhand 2003)*
11. (*a*) Fourth shell may contain upto 32 electrons. Explain this fact with the help of quantum numbers.
    (*b*) State and discuss Pauli's exclusion principle. *(Rohilkhand 2003)*
12. Write the Schrodinger equation for H atom in polar form and separate the ϕ -equation and solve to obtain acceptable $\Phi(\phi)$ wave equation. *(Gujrat 2003)*
13. (*a*) Define orbital.
    (*b*) What are *n, l* and m values for $2p_x$ and $3p_y$ electrons ?
    (*c*) Explain briefly time independent Schrödinger wave equation. *(Delhi 2003)*
14. The first, second, third and fourth shells of an atom may contain maximum upto 2, 8, 18 and 32 electrons, respectively. Explain this arrangement in terms of quantum number. *(Rohilkhand 2003)*
15. (*a*) Calculate screening constant for elements with atomic number 11 and 17.
    (*b*) State Pauli's exclusion principle. *(Nagpur 2003)*

16. Which quantum number does not follow from the Schrodinger's equation and why ? *(VBS Purvanchal 2003)*

17. What is effective nuclear charge ? What relation does it have with shielding phenomenon ? *(Delhi 2003)*

18. (*a*) Define Screening Effect.

(*b*) Write Schrodinger wave equation for hydrogen atom and explain each term involved in it.

(*c*) What is meant by radial probability distribution curve ? Discuss distribution curves for 1*s* and 2*s* orbitals. *(Nagpur 2003)*

Shielding effect and Slater's rules

19. What do you understand by shielding effect? Explain the factors affecting it. Give its applications. *(Meerut 91)*

20. What is screening constant? How does it affect the value of ionisation potential of an element? *(Meerut 94, Avadh 93)*

21. Explain " Effective nuclear charge" and "effective atomic number" *(E.C.C. Allahabad 94)*

22. Write a note on "Slater's Rules." *(Gwahati 91)*

Numerical Problems

23. Calculate the uncertainty in the position (Δ_x) of an electron, if $\Delta v = 0.1\%$, $v = 2.2 \times 106\ ms^{-1}$, $m = 9.1 \times 10^{-31}$ kg and $h = 6.6 \times 10^{31}$ Js) *(Delhi 89)*

24. Determine the value of all the four quantum numbers of the 7*th* electron in N-atom (Z = 7), 15*th* electron in Cl–atom (Z = 17) and last electron in Sn-atom (Z = 50) *(Kanpur 92)*

25. Determine the values of all the four quantum numbers of 19*th* electron in Cr-atom, 8th electron in F-atom and last electron in Ar-atom (Z for Cr = 24, F = 9 and Ar = 18). *(Kanpur 93)*

26 Calculate the wave-length of an electron, moving with a velocity of $2.5 \times 10^{-7}\ ms^{-1}$ ($h = 6.63 \times 10^{-34}$ Js, $m = 9.11 \times 10^{-31}$ kg) *(Agra 92)*

27. Calculate the wave length for the motion of an electron of mass 9.1×10^{-31} kg, moving with a velocity of $1.2 \times 10^{5}\ ms^{-1}$ ($h = 6.625 \times 10^{-34}$ Js) *(Gwahati 91)*

28. (*a*) What is Schrödinger wave equation?

(*b*) Explain why 1*s* orbital is spherically symmetrical?

(*c*) Differentiate between an orbit and an orbital.

(*d*) Derive Schrödinger wave equation and define all the terms involved in it. *(GND Amritsar, 2004)*

3

Periodic Table

Mendeleef's Periodic Law and Mendeleef's Periodic Table

Attempts were made to classify the elements in a number of ways. Most noteworthy of these is that made by Russian chemist, Mendeleef (*men-de-h-lay-eff*), who gave a law, known as *Mendeleef's Periodic Law,* which states as : *The properties of elements are a periodic function of their atomic weights, i.e., if the elements are arranged in the increasing order of their atomic weights, the properties of the elements (i.e., similar elements) are repeated after definite regular intervals or periods.* Working on this law, Mendeleef arranged the elements in the increasing order of their atomic weights, in the form of a table, which is known as *Mendeleef's Periodic Table,* after his name. In this table, the elements are arranged in *groups* (or *columns*) and *periods* (or *rows*).

Defects of Mendeleef's Periodic Table

Main defects of Mendeleef's periodic table are given below :

(*i*) *Anomalous position of hydrogen* . Mendeleef has placed hydrogen in group IA,along with alkali metals, but at the same time, hydrogen also resembles halogens. Thus,the position of hydrogen is still a matter of dispute.

(*ii*) *Similar pairs of elements,placed in different groups,while dissimilar elements placed in the same group*. Many pairs of elements, which have similar properties with each other, have been placed in different groups. Examples of such pairs of elements are : Cu (IB)-Hg (IIB), Ag (IB)-Tl (IIIA), Ba (IIA)-Pb (IV A) etc. Many elements, which have dissimilar properties have been placed in the same group, *e.g.* coinage metals (Cu, Ag and Au) have been placed along with alkali metals (IA). Similarly Mn (VII B) has been placed with halogens (VII A).

(*iii*) *Position of lanthanides and actinides.* Lanthanides (elements with atomic number 58 to 103) and actinides (elements with atomic number 90 to 103) have not been placed in the main body of the Periodic Table. Rather these have been given a separate place at the bottom of the Periodic Table.

(*iv*) *Position of four anomalous pairs of elements.* Mendeleef's periodic table contains 4 pairs of elements, in which the element with lower atomic weight has been placed after the element with higher atomic weight. This is against Mendeleef's periodic law. These pairs are : Ar (Z = 18, at. wt = 40) – K(Z = 19, at. wt. 39.0), Co (Z = 27, at.wt. = 59.9) – Ni (Z = 28, at.wt. = 58.6), Te (Z = 52, at.wt. = 127.6) – I (Z = 53, at.wt = 126.9) and Th (Z = 90, at.wt. = 323.12) – Pa (Z = 91, at.wt. = 231).

(*v*) *Position of isotopes.* If the elements are arranged in the order of their increasing atomic weights, it is not possible to accomodate the large number of *isotopes* in the periodic table.

(*vi*) *Group does not represent valency.* Excepting osmium, elements placed in group eight do not show a valency of 8. Also the elements lying in the middle of long periods show two or more valencies *e.g.* Cr, Mn etc.

Mosely's Periodic Law: Modern Periodic Law

Mosley showed in 1911 that, since the physical and chemical properties of an element depend on the number of electrons and their arrangement in different orbitals of the atom, the classification of the elements should be based on the number of these electrons (*i.e.* atomic number) and their arrangement in different orbitals. This idea led Mosley to predict that most of the defects of Mendeleef's periodic table disappear, if the basis of classification of elements is changed to atomic number, in place of atomic weight. Accordingly, Mosley put forward Modern Periodic Law, which states as follows:

The properties of elements are a periodic function of their atomic number, i.e., if the elements are arranged in the increasing order of their atomic numbers, the properties of the elements (i.e. similar elements) are repeated after definite regular intervals or periods.

Long form of Periodic Table or Modern Periodic Table

This is a periodic table (given by Bohr), in which the arrangement of the elements has been made on the basis of atomic number of the elements, which is a more fundamental property of the elements. In this table, the arrangement of elements has been made in accordance with Mosley's periodic law, *i.e.* the elements have been grouped in the increasing order of their atomic number. This table has been shown on page 61. This table is also called Bohr's periodic table, since it is based on Bohr's scheme of classification of elements into four types, depending on the number of incomplete shells in the atom.

The vertical columns of the periodic table are called *groups* or *families* or simply *columns,* while the horizontal rows, shown in the table, are called *periods* or simply *rows.*

Groups. As already stated, the vertical columns are called groups. There are 18 vertical columns in the long form of the periodic table. In terms of electronic configurations of the atoms of the elements, a group consists of elements whose atoms have the same outer-most electronic configuration. There are two systems of numbering the groups. According to the recommendations of International Union of Pure and Applied Chemistry (IUPAC), the groups are numbered as 1, 2, 3, ...17, 18 while according to another system, the groups are named as IA (for 1), IIA (2), III B(3), IV B(4), VB (5), VIB (6), VII B(7), VIII B(8, 9, 10), IB (11), IIB (12), IIIA (13), IVA (14), VA (15), VIA (16), VIIA (17), VIII A or Zero (18). Both the systems of numbering the groups have been given in the periodic table. The elements of group IA (1) and IIA (2) are called *s-block elements* and those of groups IIIA (13), IVA (14), VA (15), VIA (16), VIIA (17) and VIII A(18) are called *p-block elements. s*-and *p-block elements* taken together are called *main group elements.* The elements of groups IA (1), IIA (2), IIIA (13), IVA (14), VA (15), VIA (16) and VIIA (17) are called *representative elements* and those of group VIII A (18) are called *noble gases.* The elements belonging to groups IIIB (3), IVB (4), VB (5), VIB (6), VII B(7), VIII B (8, 9, 10), IB (11) and IIB (12) (ten vertical columns) are called *d-block* or *transition elements.* There are two rows of elements which have been placed separately at the bottom of the periodic table. Each row consists of 14 elements. Fourteen elements of *Ist* row *viz* Ce_{58} to Lu_{71} are called *4f-block elements* or *lanthanides* while 14 elements of the *2nd* row *viz* Th_{90} to Lw_{103} are called *5f-block elements* or *actinides. 4f* and *5f* block elements are collectively called *f-block elements* or *inner-transition elements* or *rare earth metals.*

Periods. We have said above that the horizontal rows are called periods. There are seven periods in all. 7*th* period (last period) is an incomplete period, since it contains only 20 elements although it is expected to contain 2 + 10 + 14 + 6 = 32 elements. In each period the elements have been placed in the increasing order of their atomic number. First element of each period is on alkali metal and the last element is a noble gas. 1*st*, 2*nd*, and 3*rd* periods which contain 2, 8 and 8 elements respectively are called *short periods* while 4*th*, 5*th* and 6*th* periods which contain 18, 18 and 32 elements are called *long periods* . The elements of 2*nd* and 3*rd* periods were called typical elements by Mendeleef, since he thought that these are the true representatives of their respective groups. But, now since it has been found that the elements of 2*nd* period resemble more closely with the elements placed diagonally opposite to them than with the elements of their own groups, the elements of 2*nd* period are now not considered as trrue representative elements. Now a days the elements of 3*rd* period *viz* Na, Mg, Al, Si, P, S and Cl are called *typical elements.*

Table 3.1 : *Long form of Periodic Table (also called Extended or Bohr's or Modern Periodic Table)*

s-block elements (Representative elements): groups IA–II A

d-block elements (Transition elemnts): groups IIIB–IIB

p-block elements: groups III A–VIII A; Representative elements: III A–VII A; Noble gases: VIII A

Periods and No. of elements ↓	IA 1	II A 2	IIIB 3	IVB 4	VB 5	VIB 6	VIIB 7	VIII B 8	VIII B 9	VIII B 10	IB 11	IIB 12	III A 13	IV A 14	V A 15	VI A 16	VII A 17	VIII A 0 18
1 (2) (H_1 - He_2)	1 H																	2 He
2 (8) (Li_3 - Ne_{10})	3 Li	4 Be											5 B	6 C	7 N	8 O	9 F	10 NE
3 (8) (Na_{11} - Ar_{18})	11 Na	12 Mg											13 Al	14 Si	15 P	16 S	17 Cl	18 Ar
4 (18) (K_{19} - Kr_{36})	19 K	20 Ca	21 Sc	22 Ti	23 V	24 Cr	25 Mn	26 Fe	27 Co	28 Ni	29 Cu	30 Zn	31 Ga	32 Ge	33 As	34 Se	35 Br	36 Kr
5 (18) (Rb_{37} - Xe_{54})	37 Rb	38 Sr	39 Y	40 Zr	41 Nb	42 Mo	43 Tc	44 Ru	45 Rh	46 Pd	47 Ag	48 Cd	49 In	50 Sn	51 Sb	52 Te	53 I	54 Xe
6 (32) (Cs_{55} - Rn_{86})	55 Cs	56 Ba	57 La	72 Hf	73 Ta	74 W	75 Re	76 Os	77 Ir	78 Pt	79 Au	80 Hg	81 Tl	82 Pb	83 Bi	84 Po	85 At	86 Rn
7 (19) (Fr_{87} - Lw_{103})	87 Fr	88 Ra	89 Ac	104 Ku	105 Ha	106 Unh												

f-block elements (Inner-transition Elements)

Lanthanide series →	58 Ce	59 Pr	60 Nd	61 Pm	62 Sm	63 Eu	64 Gd	65 Tb	66 Dy	67 Ho	68 Er	69 Tm	70 Yb	71 Lu
Actinide series →	90 Th	91 Pa	92 U	93 Np	94 Pu	95 Am	96 Cm	97 Bk	98 Cf	99 Es	100 Fm	101 Md	102 No	103 Lw

(*a*) 1*st period (short period)* consists of two elements *viz.* H_1 and He_2.

(*b*) 2*nd period (short period)* contains 8 elements namely Li_3 to Ne_{10}.

(c) 3*rd period (short period)* also contains 8 elements *viz* Na_{11} to Ar_{18} .

(*d*) 4*th period (long period)* has 18 elements which are K_{19} to Kr_{36}. The first two elements (K_{19} and Ca_{20}) are *s-block* elements, further ten elements *viz* Sc_{21} to Zn_{30} are *d-block* elements and the last six elements(Ga_{31} to Kr_{36}) are *p*-block element.

(*e*) 5*th period (long period)* also has 18 elements which are Rb_{37}, Sr_{38} (*s*-block elements), Y_{39} to Cd_{48} (*d*-block elements) and In_{49} to Xe_{54} (*p*-block elements).

(*f*) 6*th period* (*long period*) contains 32 elements which are Cs_{55}, Ba_{56} (s-block elements), La_{57} (*d*-block element), Ce_{58} to Lu_{71} (*f*-block elements: 14 elements), Hf_{72} to Hg_{80} (*d*-block elements : nine elements), Tl_{81} to Rn_{86} (*p*-block elements : six elements)

(*g*) At present 7*th period (incomplete period)* contains only 20 elements which are Fr_{87}, Ra_{88} (*s*-block elements), Ac_{89} (*d*-block element), Th_{90} to Lw_{103} (*f*-block elements : 14 elements) and Ku_{104}, Ha_{105}, Unh_{106} (*d*-block elements).

Superiority of Long Form of Periodic Table Over Mendeleef's Table

Long form of periodic table is superior to Mandeleef's periodic table in the following respects:

(*i*) The classification of the elements in the long form of periodic table has been made on the basis of atomic number of the elements, which is a more fundamental property of the elements.

(*ii*) The classification of the elements in this table has been made on the basis of the valence-shell configurations of the elements, *i.e.*, the elements having the same valence-shell configurations have been placed in the same group, *e.g.*, the elements having ns^1 and ns^2 as their valence-shell configurations have been placed in the extreme left portion of the periodic table [group IA (alkali metals) and group IIA (alkaline earth metals)], the elements having $ns^2.np^{1\text{-}6}$ as their valence-shell configurations have been placed in the extreme right portion of the periodic table (group IIIA to zero group), the elements having $(n-1)d^{1-10}.\ ns^{0,1,2}$ as their valence-shell configuration (transitional elements) have been placed just before III A group, the elements having the differentiating electron in their $(n\text{–}2)\,f$ orbitals (*i.e.* lanthanides and actinides) have been placed between alkaline earth metals and transition elements (note the position of the elements *viz* Sc_{21},Y_{39}, La_{57} and Ac_{89} (group IIIB).

(*iii*) The elements of the two sub-group have been placed separately and thus dissimilar elements do not fall together.

(*iv*) This periodic table is easier to remember, understand and reproduce.

Defects of Long Form of Periodic Table

(*i*) The problem of the position of hydrogen still remains unsolved.

(*ii*) Like Mendeleefs' table, it fails to accommodate the lanthanides and actinides in the main body of the table.

Periodicity of Property and Magic Numbers

The term, periodicity of properties, indicates that the elements with similar properties reappear at certain regular intervals of atomic number in the periodic table. In order to understand the concept of periodicity of properties, we may consider the properties of the elements of groups I A (*Alkali metals*), zero (*Inert gases*) and VII A (*Halogens*). The examination of the properties of these elements will show that the elements belonging to the same group have similar properties. In other words, we can say that the atomic number intervals, at which the elements with similar properties reappear, are 2, 8, 8, 18, 18, 32 and 32 *i.e.*, we have to pass 2, 8, 8, 18, 18, 32, and 32 elements before we come across an element with similar properties. *The repetition of the elements with similar properties, in the order of increasing atomic number, as in the periodic table, is called periodicity of properties* and the numbers 2, 8, 18 and 32 are called *magic numbers*.

Cause of Periodicity of Properties.

What is the cause of periodicity of properties, *i.e.* why do the elements placed in the same group have similar properties? In order to answer this question, we may consider the valence-shell electronic configurations, for example, of the elements of groups IA, zero and VII A given as :

Group IA	*Group zero*	*Group VIIA*
H_1 .... $1s^1$	He_2 .... $1s^2$	
Li_3 .... $2s^1$	Ne_{10} ..... $2s^2p^6$	F_9 .... $2s^2p^5$
Na_{11} ..... $3s^1$	Ar_{18} ..... $3s^2p^6$	Cl_{17} ...... $3s^2p^5$
K_{19} .... $4s^1$	Kr_{36} ...... $4s^2p^6$	r_{35} .... $4s^2p^5$
Rb_{37} ..... $5s^1$	Xe_{54} ..... $5s^2p^6$	I_{53} .... $5s^2p^5$
Cs_{55}....... $6s^1$	Rn_{86} ..... $6s^2p^6$	At_{85}....... $6s^2p^5$

An examination of the valence-shell electronic configurations of these elements shows, that the elements showing periodicity of properties *i.e.* the elements belonging to the same group, have the same valence-shell electronic configuration. Thus, we see that in any group, similar valence-shell electronic configurations of the elements reoccur after certain regular intervals of atomic number and it is this reocurrence of similar valance-shell electronic configurations of the elements at certain regular intervals of atomic number, which becomes the cause of periodicity of properties.

According to modern periodic law, the properties of the elements (*i.e.* atomic properties)like atomic volume, atomic and ionic radii, ionisation potential, electron affinity electronegativity etc. are periodic functions of their atomic number. Here, the word *periodic* means that *with the rise of atomic number of the elements in the same period or group, there is a gradual variation (i.e. increase or decrease) in a particular property of the element.* These properties depend on the electronic configuration (particularly valence-shell configuration) of the atoms of the elements.

University Questions

1. Discuss the basis and salient features of the long form (or Modern) of periodic-table. *(Meerut 95; ECC Allahabad 94)*
2. Discuss the defects in Mendeleef's periodic table and explain how the modern periodic table removes them. *(Delhi Hons. Sub. 1990)*
3. Explain, why there are two elements in 1*st* period, eight elements in 2*nd* and 3*rd* periods, and eighteen elements in 4th period, of the periodic table? *(Raj. PMT 80)*
4. (*a*) What do you understand by the term "*periodicity of properties*"? What is the cause of periodicity of properties?
   (*b*) Write a brief note on "Typical Elements". *(Meerut, 96)*
5. Write a note on defects of the extended form of periodic table. *(Meerut 93)*
6. Give an account of the modern periodic table. State how it is more useful than Mendeleef's periodic table ? *(Lucknow 2002)*

4

Electronic Configuration of Elements

What are complete and valence-shell electronic configurations of an atom?

The distribution of electrons in various shells, sub-shells and orbitals, in an atom of an element, is called its *complete electronic configuration* or simply *electronic configuration.*

The distribution of electrons in various sub-shells and orbitals of the valence-shell (or outermost shell) of an atom is called its *valance-shell configuration.*

Representation of electronic configuration

The electronic configuration of an atom is written in terms of nl^x notation, where n = 0 (s-sub-shell), 1 (p sub-shell), 2 (d sub-shell) and 3 (f sub-shell). In this notation, the super-script x, written at the top of l, indicates the number of electrons present in the sub-shell, given by l, while n, written to the left of l^x indicates the number of the shell, to which the sub-shell denoted by l belongs. For example $3p^2$, in which $n = 3$, $x = 2$ and $l = 1$ (p sub-shell) indicates, that two electrons ($x = 2$) are present in p sub-shell ($l = 1$), which belongs to 3rd main shell ($n = 3$).

Difference between electronic configuration and electronic arrangement of a given atom or ion.

In order to understand the difference between these two terms, let us consider the gaseous Ti^{2+} ion. The valence-shell electronic configuration of this ion is $3d^2$. This configuration does not tell us about the manner in which the two electrons of Ti^{2+} ion are distributed, in a set of five degenerate $3d$ orbitals. It has been shown through permutations and combinations, that there are as many as 45 different ways in which the two electrons in the five $3d$ orbitals are distributed. In other words, we say that the electronic configuration, namely $3d^2$, can have 45 different electronic arrangements. Sixteen of these arrangements are given below :

(1)	↑	↑	___	___	___	(2)	↑	↓	___	___	___
(3)	↓	↓	___	___	___	(4)	↓	↑	___	___	___
(5)	↓	___	↑	___	___	(6)	↑↓	___	___	___	___
(7)	___	↑	↑	___	___	(8)	___	___	↑	↑	___
(9)	↑	___	___	↑	___	(10)	___	↑↓	___	___	___
(11)	___	↑	___	↑	___	(12)	___	___	↑	___	↑
(13)	↑	___	___	___	↑	(14)	___	___	↑↓	___	___
(15)	___	↑	___	___	↑	(16)	___	___	___	↑	↑

Symmetrical and unsymmetrical electronic charge distribution in orbitals

When the electronic charge, in orbitals, is distributed uniformly or evenly in all directions, this electronic charge is called symmetrical or uniform electronic charge distribution. On the other hand, when the electronic charge is not distributed uniformly or evenly, this electronic charge is called unsymmetrical or non-uniform electronic charge distribution. Whether the electronic charge distribution in orbitals is symmetrical or unsymmetrical, depends on the shape of the orbitals and the way in which the electrons are distributed in various orbitals. Let us consider the electronic charge distribution in *s*, *p* and *d* orbitals as discussed below.

s-orbital. Since *s*-orbital is spherical in shape, the electronic charge distribution in this orbital would be uniform in all directions, whether this orbital is singly filled (s^1 configuration) or doubly-filled (s^2 configuration).

p-orbitals. The distribution of electronic charge distribution in the electronic configuration having the three *p*-orbitals, either half-filled ($p_x^1\ p_y^1\ p_z^1$ configuration) or completely filled ($p_x^2\ p_y^2\ p_z^2$ configuration) is symmetrical. The configurations *viz* p^1, p^2, p^4 and p^5 have unsymmetrical distribution of electronic charge.

d-orbitals. The distribution of electronic charge in the electronic configuration having the five *d*-orbitals, either half-filled ($d_{xy}^1\ d_{yz}^1\ d_{zx}^1\ d_{z^2}^1\ d_{x^2-y^2}^1$ configuration) or completely-filled ($d_{xy}^2\ d_{yz}^2\ d_{zx}^2\ d_{x^2-y^2}^2\ d_{z^2}^2$ configuration) is symmetrical. The configurations *viz.* d^2, d^3, d^4, d^6, d^7, d^8 and d^9 have unsymmetrical distribution of electronic charge.

Relation between the electronic charge distribution of an electronic arrangement and its stability

It has been observed that the electronic arrangement having symmetrical electronic charge distribution, has lower energy and hence higher stability, than the electronic arrangement having unsymmetrical electronic charge distribution. This shows that since half-filled (*e.g.*, p^3, d^5 and f^7 orbitals) and completely-filled (*e.g.* p^6, d^{10} and f^{14} orbitals) orbitals have symmetrical electronic charge distribution, these orbitals are more stable than other orbitals.

What is an exchange pair of electrons ?

With the help of advanced mathematics, it can be shown that when two electrons, belonging to a system and having parallel spins [upward (↑) or downward (↓)], exchange their positions in space, without causing any change in the electronic arrangement of the system, the energy of the system decreases. Such a pair of electrons is called "exchange pair of electrons". An exchange pair of electrons can be defined as that pair of electrons, which contains electrons with parallel spins. Number of exchange pairs is equal to the number of ways of selecting two electrons, out of the electrons having parallel spins, in a given electronic arrangement. For a system containing electrons with parallel spins, the number of possible sets, each containing two electrons with parallel spins, represents the number of exchange pairs of electrons. Greater is the number of unpaired electrons with parallel spins, greater would be the number of exchange pairs of electrons.

To calculate the number of exchange pairs of electrons in a given electronic arrangement

The method used for calculating the number of exchange pairs of electrons consists of the following steps :

(*i*) Find out the number of electrons having upward spins (↑). Let this number be denoted by *m*. The number of exchange pairs for these *m* electrons is equal to :

$$^mC_2 = \frac{m\,!}{2\,(m-2)\,!} = \frac{m \times (m-1) \times (m-2) \ldots\ldots 2 \times 1}{2 \times (m-2) \times (m-3) \times \ldots\ldots 2 \times 1}$$

(*ii*) Similarly, find out the number of electrons having downward spins (↓). Let this number be denoted by *n*. The number of exchange pairs of electrons for these *n* electrons is equal to :

$$^nC_2 = \frac{n\,!}{2\,(n-2)\,!} = \frac{n \times (n-1) \times (n-2) \ldots\ldots 2 \times 1}{2 \times (n-2) \times (n-3) \times \ldots\ldots 2 \times 1}$$

(*iii*) Total number of exchange pairs of electrons in the given electronic arrangement, which contains *m* electrons with upward spins and *n* electrons with down-ward spins, is equal to the sum of $^mC_2$ and $^nC_2$. Thus :

Total number of exchange pairs of electrons

$$= \ ^mC_2 + {}^nC_2 \ = \ \frac{m\,!}{2\,(m-2)\,!} + \frac{n\,!}{2\,(n-2)\,!}$$

$$= \frac{m \times (m-1) \times (m-2) \times .....2 \times 1}{2 \times (m-2) \times (m-3) \times ..... 2 \times 1} + \frac{n \times (n-1) \times (n-2) \times .....2 \times 1}{2 \times (n-2) \times (n-3) \times ..... 2 \times 1}$$

Relation between the number of exchange pairs of electrons and stability of an electronic arrangement

Greater is the number of exchange pairs of electrons, present in a given electronic arrangement, greater would be the stability associated with the arrangement.

Exchange energy (E) (*stablising energy*)

We have already said, under the discussion of exchange pairs of electrons, that the presence of exchange pairs of electrons decreases the energy of the arrangement. The energy decrease per exchange pair of electrons, is called "exchange energy" E. It has a negative sign, since it tends to decrease the energy of the system. Thus E stabilises the system. E is, therefore, also called *stabilising energy.*

To calculate exchange energy for n exchange pairs of electrons

If an electronic arrangement contains n exchange pairs of electrons, exchange energy of this arrangement is equal to the product of $-E$ and n, where E is the exchange energy for one exchange pair of electrons. Thus :

Exchange energy for n exchange pairs of electron $= -nE$.

Thus the energy of the electronic arrangement, having n exchange pairs of electrons, is decreased by nE. The decrease is shown by putting a negative sign before nE.

Relation between the number of exchange pairs of electrons, exchange energy and stablity of the electronic arrangement

If we write many electronic arrangements for a given electronic configuration, then out of these electronic arrangements, the arrangement having the maximum number of electrons, with parallel spins, would have maximum number of exchange pairs of electrons and hence would have maximum exchange, energy. Consequently, such an arrangement would be the most stable. Whatever has been said above can be summarised by saying that, greater is the number of unpaired electrons, with the same spins, in an electronic arrangement, greater would be the number of exchange pairs of electrons and hence greater would be the amount of exchange energy of the arrangement and also greater would be the stability, associated with the arrangement.

Pairing energy (*P*) (*Destabilising energy*)

When two electrons are made to pair up in the same orbital, with opposite spins (*i.e.,* as ↑↓), energy is required. This energy is called "pairing energy". The presence of electron pairs, in a given electronic arrangement, increases the energy of the arrangement and hence the arrangement becomes destablished (unstable). Hence, pairing energy is shown with positive sign and is also called *destabilising energy.* Pairing energy, corresppnding to one electron pair of electrons is denoted by P.

To calculate pairing energy for *n* electron pairs in a given electronic arrangement

If an electronic arrangement has n electron pairs, the pairing energy of this arrangement is equal to the product of $+P$ and n where P is the pairing energy for one electron pair. Thus :

Pairing energy for n electron pairs $= +nP$.

Thus, the electronic arrangement, having no electron pair, has no pairing energy.

Relation between stability and the sum of exchange energy and pairing energy

We have already said that exchange energy of a given system stabilises it, *i.e.,* greater is the value of exchange energy, greater would be the stability of the system. On the other hand, pairing energy destabilises the system, *i.e.,* greater the value of pairing energy, greater would be the unstability of the system. This shows that the overall stability of a given system depends on the sum of exchange energy and pairing energy.

Promotional energy (Δ E)

The energy required to promote an electron from lower energy orbital to a higher energy orbital is called "promotional energy". It is denoted by ΔE and has positive sign.

Relation between stability and promotional and pairing energies

Let us consider the following two electronic arrangements for the six electrons in 4*s* (lower energy orbital) and 3*d* (higher energy orbital) orbitals in the valence-shell of Cr-atom ($Z = 24$).

In configuration (*a*) (Fig. 4.1) the two electrons are present in 4*s* orbital (paired electrons). Since this arrangement has only one pair of electrons, its pairing energy = P and promotional energy (ΔE) = 0. Thus, as $P > \Delta E$ ($\Delta E = 0$), arrangement (*a*) is unstable.

The arrangement (*b*) does not contain any pair of electrons, its pairing energy, $P = 0$. Now since arrangement (*b*) has been obtained from arrangement (*a*), by promoting one electron from a lower energy orbital (4*s* orbital) to higher energy orbitals (3*d* orbital), this arrangement has promotional energy, which is equal to ΔE. Thus, since $\Delta E > P$ (P is equal to zero), this arrangement is more stable than arrangement (*a*) [See Fig. 4.1].

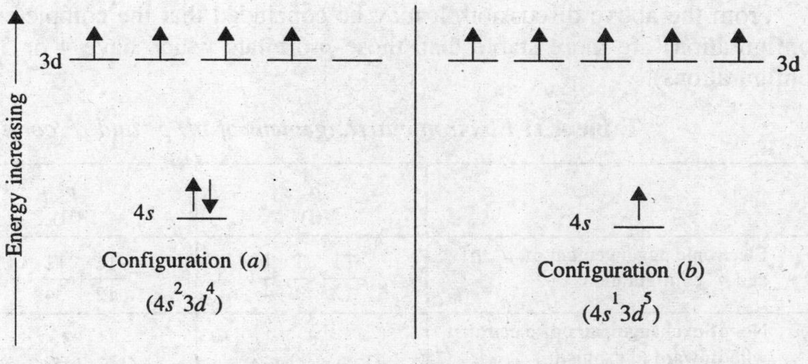

Fig. 4.1 : Relation between stability and promotional and pairing energies.

Comparison between the stability of p^1, p^2 and p^3 configurations

On the basis of Hund's rule, the electronic arrangements of p^1, p^2 and p^3 configurations can be represented by the electronic arrangements (*a*), (*b*) and (*c*) given below:

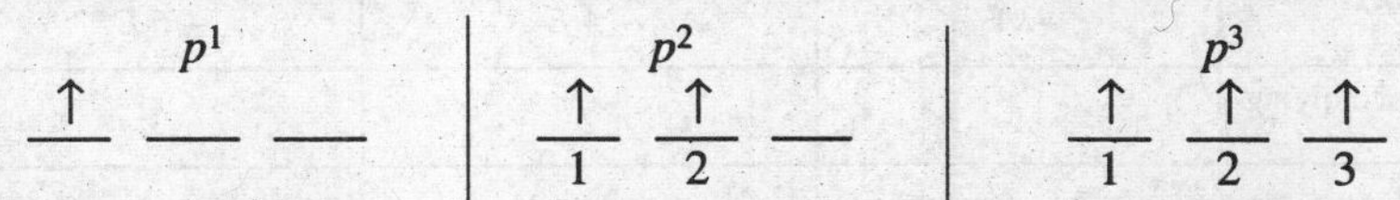

Arrangement (a) *Arrangement (b)* *Arrangement (c)*

With the help of the following concepts, it can be shown that arrangement (*c*) is the most stable.

(1) *Concept of exchange energy.* Since arrangement (*a*) has only one electron, it has no exchange pair of electrons and hence no exchange energy ($E = 0$).

Arrangement (*b*) has one exchange pair of electrons *viz* 1–2. Consequently the exchange energy of this arrangement is equal to $-E$.

Arrangement (*c*) has three exchange pairs of electrons, which are 1–2, 1–3 and 2–3. Thus the exchange energy of this arrangement is equal to $-3E$.

If we do not consider the negative sign associated with E terms, we find that the exchange energy of these arrangements is in the order (*c*) > (*b*) > (*a*). This order shows that arrangement (*c*) is the most stable arrangement. Thus we can conclude, that half filled *p*-orbitals (p^3 configuration) are more stable than other *p*-orbitals which contain 1 electron (p^1 configuration) or 2 electrons (p^2 configuration).

(2) *Concept of symmetrical and unsymmetrical distribution of electrons in orbitals.* According to this concept, the degenerate orbitals, having symmetrical distribution of electrons in them, have lower energy and hence are more stable than those orbitals, which have unsymmetrical distribution of electrons in them. Now since the distribution of three electrons in *p*-orbitals in arrangement (*c*) is symmetrical and that of one electron and two electrons in arrangements (*a*) and (*b*) respectively, is unsymmetrical, p^3 configuration is more stable than each of the two configurations *viz.*, p^1 and p^2.

Comparison between the stability of p^4, p^5 and p^6 configurations

On the basis of Hund's rule, the electronic arrangement of p^4, p^5 and p^6 configurations can be shown as depicted in Table.4.1. Let these be denoted as (I), (II) and (III) respectively. Number of exchange pairs of electrons, with upward and downward spins and decrease in exchange energy etc. are also given in the 1*st* column of this table. From this table, it may be seen that the net energy change in the given electronic arrangements (I), (II) and (III) are ($-3E$ + P), ($-4E + 2P$) and ($-6E + 3P$) respectively. Since the energy change is maximum in arrangement (III), this arrangement is the most stable. The symmetrical (uniform) arrangement of the six electrons in the three *p*-orbitals further adds to the stability of p^6 configuration.

From the above discussion, it may be concluded that the completely-filled *p*-orbitals (*i.e.*, p^6 configuration) are more stable than those *p*-orbitals which have 4 or 5 electrons (*i.e.*, p^4 and p^5 configurations).

Table 4.1: *Electronic arrangement of p^4, p^5 and p^6 configurations.*

	p^4 (I)	p^5 (II)	p^6 (III)
i) Electronic aggangement of p^4, p^5 and p^6 configurations	↑↓ ↑ ↑ 12 3 4	↑↓ ↑↓ ↑ 12 34 5	↑↓ ↑↓ ↑↓ 12 34 56
ii) No. of exchange pairs of electrons with upward (↑) spins (*a*)	3 (1–3, 1–4, 3–4)	3 (1–3, 1–5, 3–5)	3 (1–3, 1–5, 3–5)
iii) No. of exchange pairs of electrons with downward (↓) spins (*b*)	0	1 (2–4)	3 (2–4, 2–6, 4–6)
iv) Total number of exchange pairs of electrons, $n = (a + b)$	3 + 0 = 3 (1–3, 1–4, 3–4)	3 + 1 = 4 (1–3, 1–5, 3–5, 2–4)	3 + 3 = 6 (1–3, 1–5, 3–5, 2–4, 2–6, 4–6)
v) Decrease in exchange energy $= -E \times n$	–3 E	$-4E$	$-6E$
vi) Number of electron pairs occupying the same orbital (*m*)	1	2	3
vii) Increase in pairing energy $= +P \times m$	$+P$	$+2P$	$+3P$
viii) Net energy change = $(-nE + mP)$	$-3E + P$	$-4E + 2P$	$-6E + 3P$

Solved Examples

Example 4.1. *The three electronic arrangements, which can be written for p^3 configuration, are given below:*

p^3	p^3	p^3
↑↓ ↑ __ x y z	↑ ↓ ↑ x y z	↑ ↑ ↑ x y z
Arrangement (a)	*Arrangement (b)*	*Arrangement (c)*

(i) Calculate the number of exchange pairs of electrons in each of the arrangements.

(ii) Calculate the exchange energy for each arrangement and predict which of the three arrangements is the most stable.

(iii) Calculate the pairing energy in each case.

(iv) Taking into consideration (exchange energy + pairing energy), work out the most stable electronic arrangement.

Solution. (*i*) *Calculation of the number of exchange pairs. Arrangement* (*a*). Since the number of electrons having upward spins is equal to 2 ($m = 2$) and that of electrons having downward spins is equal to 1 ($n = 1$), total number of exchange pairs of electrons in this arrangement.

$$= {}^2C_2 + {}^1C_2 = \frac{2!}{2\,(2-2)\,!} + 0 = \frac{2 \times 1}{(2 \times 0)\,!} + 0 = \frac{2 \times 1}{2 \times 1} + 0 = 1$$

Arrangement (*b*). Here $m = 2$ and $n = 1$, and hence the number of exchange pairs of electrons $= {}^2C_2 + {}^1C_2 = 1$.

Arrangement (*c*). Here $m = 3$ and $n = 0$. Hence the number of exchange pairs of electrons $= {}^3C_2 + {}^0C_2 = \dfrac{3\,!}{2\,(3-2)\,!} + 0 = \dfrac{3 \times 2 \times 1}{2 \times 1} + 0 = 3.$

(*ii*) *Calculation of exchange energy.* Exchange energy is equal to $-E \times$ number of exchange pairs of electrons. Thus the exchange energy for arrangements (*a*), (*b*) and (*c*) would be equal to $-E \times 1 = -E$, $-E \times 1 = -E$ and $-E \times 3 = -3E$ respectively.

It may be seen from these values that of the three arrangements, arrangement (*c*) is associated with maximum value of exchange energy (do not consider negative sign) and hence is the most stable.

(*iii*) *Calculation of pairing energy.* It is only arrangement (*a*) which has *one electron pair,* whose two electrons are present in the same orbital and have opposite spins. The pairing energy of this arrangement $= +P \times 1 = +P$. Each of the remaining two arrangements has only *unpaired electrons* and hence the pairing energy of each arrangement is equal to zero.

(*iv*) *To find out the most stable arrangement.* The sum of exchange energy and pairing energy for these arrangements is equal to $-E + P$, $-E + 0$ and $-3E + 0$ respectively. Due to the presence of one electron pair, arrangement (*a*) has pairing energy equal to $+P$ which destabilises this arrangement. Thus arrangement (*a*) would be unstable. Among arrangements (*b*) and (*c*), since (*c*) has greater value of exchange energy (=3*E*) than (*b*) (=*E*), arrangement (*c*) would be more stable than arrangement (*b*) [(*c*) > (*b*)]. Thus the order of stability of the given arrangements is (*c*) > (*b*) > (*a*). Consequently arrangement (*c*) is the most stable.

Example 4.2. *Some of the possible electronic arrangements for the valence-shell electronic configuration of S-atom are given below :*

	3*s*	3*p*			4*s*
(*a*) $3s^2 3p^4 4s^0$	↑↓	↑↓	↑	↑	—
(*b*) $3s^2 3p^4 4s^0$	↑↓	↑↓	↑↓	—	—
(*c*) $3s^1 3p^5 4s^0$	↑	↑↓	↑↓	↑	—
(*d*) $3s^2 3p^3 4s^1$	↑↓	↑	↑	↑	↑
(*e*) $3s^1 3p^4 4s^1$	↑	↑↓	↑	↑	↑

(*i*) *Find out the number of exchange pairs of electrons for each of these arrangements.*

(*ii*) *Find out the total amount of energy (exchange energy + pairing energy + promotional energy) associated with each of these arrangements.*

(*iii*) *Which of the given electronic arrangements represents the valence-shell electronic configuration of S-atom in its ground state?*

Solution. (*i*) *To find out the number of exchange pairs of electrons. Arrangement* (*a*). Since this arrangement contains 4 electrons with upward spins and 2 electrons with downward spins, total number of exchange pairs of electrons present in this arrangement :

$$= {}^4C_2 + {}^2C_2 = \frac{4!}{2\,(4-2)\,!} + 1 = \frac{4 \times 3 \times 2 \times 1}{2 \times 2 \times 1} + 1 = 6 + 1 = 7$$

Arrangement (*b*). Here number of electrons with upward spins and downward spins = 3 each. Thus number of exchange pairs of electrons

$$= {}^3C_2 + {}^3C_2 = \frac{3\,!}{2\,(3-2)\,!} + \frac{3\,!}{2\,(3-2)\,!} = 3 + 3 = 6$$

Arrangement (*c*). Here number of electrons with upward spins (n_u) = 4 and the number of electrons having downward spins (n_d) = 2. Thus number of exchange pairs of electrons $= {}^4C_2 + {}^2C_2 = 6 + 1 = 7$

Arrangement (*d*). Here $n_u = 5$ and $n_d = 1$. Hence, the number of exchange pairs of electrons =

$${}^5C_2 + {}^1C_2 = \frac{5\,!}{2\,(5-2)\,!} + 0 = \frac{5 \times 4 \times 3 \times 2 \times 1}{2 \times 3 \times 2 \times 1} + 0 = 10$$

Arrangement (*e*). Here $n_u = 5$ and $n_d = 1$. Hence the number of exchange pairs of electrons $= {}^5C_2 + {}^1C_2 = 10 + 0 = 10$.

(ii) To find out the total amount of energy : Sum of the exchange energy, pairing energy and promotional energy is the total energy associated with a given arrangement. Thus

Total energy of arrangement (*a*) *i.e.*, (E_a) $= -7E + 2P$
Total energy of arrangement (*b*) *i.e.*, (E_b) $= -6E + 3P$
Total energy of arrangement (*c*) *i.e.*, (E_c) $= -7E + 2P + \Delta E\ (3s \rightarrow 3p)$
Total energy of arrangement (*d*) *i.e.*, (E_d) $= -10E + P + \Delta E\ (3p^1 \rightarrow 4s)$
Total energy of arrangement (*e*) *i.e.*, (E_e) $= -10E + P + \Delta E\ (3s^1 \rightarrow 4s)$

ΔE (promotional energy) shown in E_c, E_d and E_e terms is due to the excitation of one electron from 3*s* to 3*p*, from 3*p* to 4*s* and from 3*s* to 4*s* orbitals.

(iii) To find out the electronic configuration of S-atom. Obviously, the electronic arrangement which is the most stable, will be the electronic configuration of S-atom. Taking into consideration, the exchange energy, pairing energy and promotional energy of each of the given electronic arrangements, we can predict which is the most stable arrangement.

In electronic arrangements (*c*), (*d*) and (*e*), the magnitude of ΔE (promotional energy) corresponding to the electronic excitation from 3*s* to 3*p*, from 3*p* to 4*s* and from 3*s* to 4*s* orbitals respectively, has been found to greater than the sum of exchange energy and pairing energy of these arrangements and hence the electronic excitations *viz* $3s \rightarrow 3p$, $3p \rightarrow 4s$ and $3s \rightarrow 4s$ are energetically not possible and consequently these arrangements are ruled out.

The exchange energy (stabilising energy) of arrangement (*a*) is greater than that of arrangement (*b*). Also at the same time, the pairing energy (destabilising energy) of arrangement (*a*) is less than that of arrangement (*b*). Both these factors suggest that arrangement (*a*) is more stable than arrangement (*b*). Thus, out of the given five arrangements, it is only arrangement (*a*) which is the most stable. This arrangement is the electronic configuration of S-atom.

Example 4.3. *Assuming the energy difference between 4s and 3d orbitals to be small, the following two electronic arrangements have been proposed for Cu-atom (Z = 29).*

	4*s*	3*d*				
1. $4s^2\ 3d^9$	↑↓	↑↓	↑↓	↑↓	↑↓	↑↓
2. $4s^1\ 3d^{10}$	↑	↑↓	↑↓	↑↓	↑↓	↑↓

Which of the above two arrangements is more stable?

Solution. The answer to this question can be given by using any of the following two concepts.

(*a*) *Concept of symmetrical and unsymmetrical electronic charge distribution in orbitals.* Let us consider the above two configurations as follows :

Configuration 1. In this configuration, although the electronic charge distribution in 4*s* orbital is symmetrical, the charge distribution in 3*d* orbitals is unsymmetrical. Thus the electronic charge distribution in various orbitals of configuration 1 is unsymmetrical.

Configuration 2. In this configuration, the electronic charge distribution in 4*s*, as well as 3*d* orbitals, is symmetrical and hence the charge distribution in configuration 2 is symmetrical .

Due to the fact that electronic charge distribution in configuration 2 is symmetrical, this configuration has less energy than configuration 1. Consequently, configuration 2 is more stable than configuration 1, *i.e.*, the actual configuration of Cu is represented by configuration 2 and not by configuration 1.

(*b*) *Concept of exchange energy, pairing energy and promotional energy.* Let us calculate the sum of these energy terms for both the configurations.

Configuration 1. Since the number of electrons having upward spins (↑) and of those with downward spins (↓) is equal to 6 and 5 respectively, the number of exchange pairs of electrons present in this configuration is equal to ${}^6C_2 + {}^5C_2 = \dfrac{6\,!}{2\,(6-2)\,!} + \dfrac{5\,!}{2\,(5-2)\,!} = 15 + 10 = 25$. Hence exchange energy associated with this configuration $= -25E$. Since the number of pairs of electrons =

5, the pairing energy of this configuration = 5*P*. Thus the energy associated with configuration 1 (E_1) is equal to the sum of –25*E* and 5*P*, *i.e.*, $E_1 = -25E + 5P$

Configuration 2. Since the number of electrons having upward spins (↑) and that of the electrons having downward spins (↓) is equal to 6 and 5 respectively, the number of exchange pairs of electrons present in this configuration is equal to $^6C_2 + {}^5C_2 = 15 + 10 = 25$. Hence the exchange energy of this configuration is equal to –25*E*. Since the number of pairs of electrons = 5, pairing energy = 5*P*. Now since configuration 2 has been obtained from configuration 1 by promoting one electron from 4*s* orbital (lower energy orbital) to 3*d* orbitals (higher energy orbital), this configuration has also promotional energy equal to Δ*E*. Thus the total energy associated with configuration 2 (E_2) is equal to the sum, $(-25E + 5P + \Delta E)$, *i.e.*, $E_2 = -25E + 5P + \Delta E$.

As the difference in energy of 4*s* and 3*d* orbitals is very small, $\Delta E \approx P$ and hence $E_2 = -25E + 5P + P = -25E + 6P$. Thus $E_2 > E_1$ and hence on the basis of the energy consideration configuration 2 should be less stable than configuration 1. But actually configuration 2 is more stable than configuration 1.

The above discussion shows that the concept of exchange energy, pairing energy and promotional energy cannot be used to explain greater stability associated with configuration 2. It is only the concept of symmetrical and unsymmetrical electronic charge distribution which can explain the higher stability of configuration 2.

Rules for writting electronic configuration of an atom

The rules used for writing the electronic configuration of the atoms of the elements in the ground state are given below :

1. Maximum number of electrons in a shell. According to *Bohr-Bury scheme,* the maximum number of electrons, that a shell can hold, is equal to $2n^2$ where *n* is the number of the shell (*i.e.*, principal quantum number). Thus, according to *Bohr-Bury scheme,* the maximum number of electrons in 1*st* (*n* = 1), 2*nd* (*n* = 2), 3*rd* (*n* = 3) and 4*th* (*n* = 4) shell is equal to $2 \times 1^2 = 2$, $2 \times 2^2 = 8$, $2 \times 3^2 = 18$ and $2 \times 4^2 = 32$ respectively.

2. Maximum number of electrons in a sub-shell. We have seen during the study of azimuthal quantum number, that the maximum number of electrons in a sub-shell is equal to 2(2*l* + 1) where *l* = 0, 1, 2, or 3 for *s, p, d* or *f* sub-shells respectively. Thus, the maximum number of electrons in *s* (*l* = 0), *p* (*l* = 1), *d* (*l* = 2) and *f* (*l* = 3) sub-shells is equal to 2(2 × 0 + 1) = 2, 2(2 × 1 + 1) = 6, 2(2 × 2 + 1) = 10 and 2(2 × 3+1) = 14 respectively.

3. Aufbau principle. Aufbau is a German word which means *build up* or *construction.* It is for this reason that this principle is also often called *building up principle* or *construction principle.*It is pronoun-ced as *of bow*. This principle gives us a sequence in which various orbitals are filled with electrons. This principle states that *The orbitals are filled up with electrons*

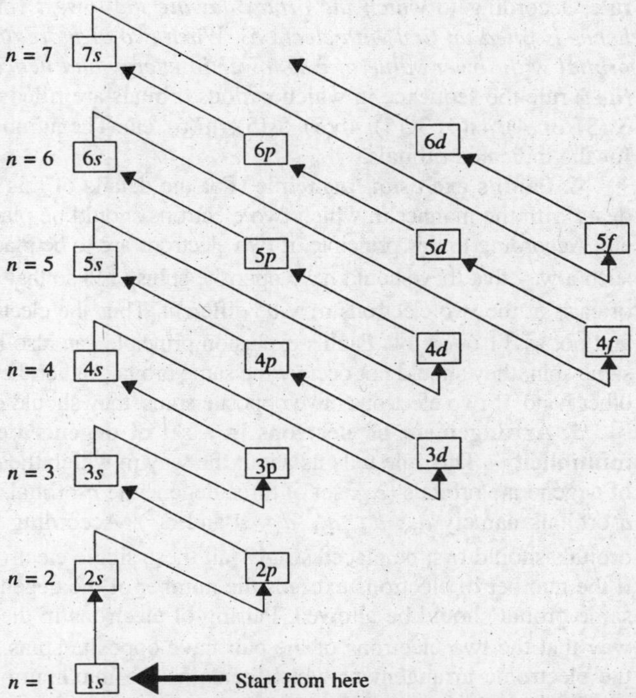

Fig. 4.2. The sequence in which various orbitals are filled up with electrons.

in the increasing order of their energy. The relative order of energy of various orbitals of an atom is as :

$1s < 2s < 2p < 3s < 3p < 4s < 3d < 4p < 5s < 4d < 5p < 6s < 4f < 5d < 6p < 7s < 5f < 6d < 7p$ $< 8s$.....,showing that the various orbitals are filled with electrons in the same order.

The sequence given above can be represented as shown in Fig. 4.2.

Limitations (Deviations) of Aufbau principle. There are some elements, in which the extra electron, while entering the orbital, does not obey Aufbau principle, as shown by the following examples:

(*i*) *Configuration of La atom (Z = 57).* We know that, according to Aufbau principle, the configuration of Ba(Z = 56) atom is $[Xe]6s^2$ where [Xe] indicates the configuration of Xe(Z = 54) atom which is 2, 8, 18, 18, 8. Now, after 6*s* orbital has been completely filled at Ba, the next electron, according to Aufbau principle, must enter 4*f* orbital in La atom to give $[Xe]4f^1 5d^0 6s^2$ configuration to La atom; but it has been observed that the extra electron enters 5*d* orbital (instead of 4*f* orbital) to give $[Xe]4f^0 5d^1 6s^2$ configuration to La atom. Why the extra electron goes to 5*d* orbital, instead of 4*f* orbital, in La atom, has been explained on the basis that at La, 5*d* and 4*f* orbitals are supposed to have almost the same energy and hence the extra electron is free to occupy any of these two orbitals.

(*ii*) *Configuration of Ac atom (Z = 89).* We know that, according to Aufbau principle, the configuration of Ra (Z = 88) atom is $[Rn]\ 7s^2$ where [*Rn*] is the configuration of Rn (Z = 86) atom which is 2, 8, 18, 32, 18, 8. Now after 7*s* orbital has been completely filled at Ra, the next electron, according to Aufbau principle, must enter 5*f* orbital in Ac atom; but it has been observed that the extra electron enters 6*d* orbital (instead of 5*f* orbital) to give $[Rn]\ 5f^0\ 6d^1.7s^2$ configuration to Ac atom. Why the extra electron goes to 6*d* orbital, instead of 5*f* orbital in Ac atom has been explained on the basis that in Ac, 6*d* and 5*f* orbitals are supposed to be almost of the same energy and hence the extra electron is free to go to any of these two orbitals.

(n+1) Rule. The relative order of energy of different orbitals (or the sequence in which various orbitals are filled up with electrons) can also be determined with the help of (*n*+*l*) value for a given orbital (*n* = principal quantum number, *l* = azimuthal quantum number). This rule is called (*n*+*l*) rule, according to *which the orbital having the lowest value of* (*n*+*l*) *has the lowest energy and hence is filled up first with electrons. When two or more orbitals have the same value of (n+l), the orbital with lower value of n is lower in energy and hence is filled up first with electrons. Using* (*n*+*l*) rule the sequence in which various orbitals are filled with electrons is as : 1*s*(1), 2*s*(2), 2*p*(3), 3*s*(3), 3*p*(4), 4*s*(4) 3*d*(5), 4*p*(5), 5*s*(5), 4*d*(6), etc. The numbers given in parentheses are (*n*+*l*) values for the indicated orbitals.

4. Pauli's exclusion principle (For the details of this principle, see chapter 2). This principle deals with the manner in which two electrons should be placed in an orbital like $s, p_x, p_y, p_z, d_{xy}, d_{yz}$ etc. According to this principle, if two electrons are to be placed in an orbital, they should be placed in such a way that they should have opposite spins (↑↓) so that the value of at least one quantum number of each of the two electrons may be different. Thus the electrons in an orbital should be placed as ↑↓ and not as ↑↑ or as ↓↓. Pauli's exclusion principle can also be stated as that if two electrons have the same spin, they should not occupy the same orbital; rather they should occupy different orbitals. On the other hand, if two electrons have opposite spins, they should occupy the same orbital.

5. Arrangement of electrons in a set of degenerate orbitals : Hund's rule of maximum multiplicity : This rule tells us about the way in which the electrons should be placed in a given set of degenerate orbitals (*e.g.* set of three degenerate *p*-orbitals *viz.*, p_x, p_y and p_z; set of five degenerate *d*-orbitals namely d_{xy}, d_{yz}, d_{zx}, $d_{x^2-y^2}$ and $d_z{}^2$). According to this rule, the electrons in degenerate orbitals should first be placed **singly** (all these single electrons should have the same spin) and then, if the number of electrons exceeds the number of the degenerate orbitals, pairing of electrons in the same orbital should be allowed. Pairing of electrons in the same orbital should be made in such a way that the two electrons of the pair have opposite spins (↑↓). This rule is based on the fact that the electronic arrangement which contains the maximum number of unpaired electrons with the same spin is the most stable .

Thus, according to this rule, the arrangement of three electrons in three degenerate orbitals *viz.*, p_x, p_y and p_z should be shown as ↑ ↑ ↑ and not as ↑↓ ↑ __ or as ↑ ↓ ↑.

Illustration and explanation of the rule. Let us again consider the distribution of three electrons in three degenerate orbitals of p sub-shell, **namely** p_x, p_y and p_z. There are 20 different ways of arranging the three electrons in p_x, p_y and p_z orbitals. Only three of these arrangements are shown below :

p^3			p^3			p^3		
↑↓	↑		↑	↓	↑	↑	↑	↑
x	*y*	*z*	*x*	*y*	*z*	*x*	*y*	*z*
Arrangement (a)			*Arrangement (b)*			*Arrangement (c)*		

According to Hund's rule of maximum multiplicity, arrangement (*c*) is the most stable of all the three arrangements, since it contains maximum number of unpaired electrons with the same spin. The stability of arrangement (*c*) can be explained with the help of the following concepts :

(*i*) *Concept of exchange pairs of electrons and exchange energy.* Since arrangement (*c*) has the maximum number of unpaired electrons having parallel spins, this arrangement has the maximum number of exchange pairs of electrons [this number for arrangements (*a*), (*b*) and (*c*) electrons has been found to be equal to 1, 1 and 3 respectively]

Now, since arrangement (*c*) has the maximum number of exchange pairs of electrons, this arrangement would be associated with maximum amount of exchange energy [This energy for the given arrangements is equal to $-E$, $-E$ and $-3E$ respectively] and hence this arrangement would be the most stable arrangement. The above discussion supports Hund's rule, which says that the electronic arrangement, having the maximum number of unpaired electrons with parallel spin, is the most stable.

(*ii*) *Concept of exchange energy plus pairing energy.* Before using this concept, it should be understood that the exchange energy of a system tends to stabilise it while the pairing energy tends to destablise the system. Now, on calculation we find that the sum of exchange energy (*E*) and pairing energy (*P*) for the given arrangements is equal to $-E + P$, $-E + 0$ and $-3E + 0$ respectively. Due to the presence of one electron pair, arrangement (*a*) has pairing energy equal to $+P$ which destabilises this arrangement. Thus arrangement (*a*) would be unstable. Among the arrangements (*b*) and (*c*), since (*c*) has greater value of exchange energy (= 3*E*) than (*b*) (= *E*), arrangement (*c*) is more stable than (*b*) [(*c*) > (*b*)]. Thus again we see that arrangement (*c*) is the most stable. The above discussion again confirms the rule that the electronic arrangement having the maximum number of unpaired electrons with the same spin is the most stable.

6. Rule of stability of half filled and completely-filled orbitals: This rule says that half-filled degenerate orbitals (*e.g.* p^3, d^5 and f^7 orbitals)and completely filled degenerate orbitals (*e.g.* p^6, d^{10} and f^{14} orbitals) have less energy and hence are more stable than other orbitals (*e.g.*, p^1, p^2, p^4, p^5; d^1, d^2, d^3, d^4, d^7, d^8, d^9 etc.) and hence give more stable electronic configuration to the atom.

Illustration and explanation of the rule. Let us consider the valence-shell electronic configuration of Cr-atom (Z = 24) which contains six electrons. These six electrons are distributed in 4*s* and 3*d* orbitals which are present in the valence-shell. The distribution of these six electrons in 4*s* and 3*d* orbitals can be shown by a number of different electronic arrangments. Out of these, only the following two electronic arrangements will be considered here.

$4s^2$	$3d^4$	$4s^1$	$3d^5$
↑↓	↑ ↑ ↑ ↑ __	↑	↑ ↑ ↑ ↑ ↑
Arrangement (a)		*Arrangement (b)*	

According to the rule of stability of half filled and completely-filled orbitals, configuration (*b*) is more stable than configuration (*a*) and hence the correct configuration of Cr-atom is $4s^1 3d^5$ and not $4s^2 3d^4$. This fact has been explained by the following concepts.

(*i*) *Concept of symmetrical and unsymmetrical electronic charge distribution in orbitals.* In configuration(*a*), since 3*d*-orbitals are neither half-filled nor completely-filled, the electronic charge distribution in these orbitals is unsymmetrical. On the other hand, the electronic charge distribution in 4*s* orbital is symmetrical. Thus the electronic charge distribution in various orbitals in configuration (*a*) is unsymmetrical.

In configuration (*b*), since 3*d* and 4*s* orbitals are half-filled, the electronic charge distribution in these orbitals in symmetrical.

Due to the symmetrical electronic charge distribution in configuration (*b*), this configuration has less energy than configuration (*a*). Consequently configuration (*b*) is more stable than configuration (*a*). In other words we say that Cr-atom prefers to have configuration (*b*) and not configuration (*a*).

(2) *Concept of exchange energy, pairing energy and promotional energy.* Let us calculate the sum of these three energy terms in each configuration.

Configuration (*a*). Since the number of electrons having upward spins (↑) and of those having downward spins (↓) is equal to 5 and 1 respectively, the number of exchange pairs of electrons present in this configuration is equal to $^5C_2 + {}^1C_2 = \frac{5\,!}{2!\,(5-2)!} + 0 = \frac{5 \times 4 \times 3 \times 2 \times 1}{2 \times 3 \times 2 \times 1}$ $= 10 + 0 = 10$. Hence exchange energy associated with this configuration $= -10E$. Since the number of pairs of electrons = 1, the pairing energy of this configuration $= P$. Thus the energy associated with configuration (*a*) (E_a) is equal to the sum of exchange energy ($= -10E$) and pairing energy ($= P$), *i.e.*, $E_a = -10E + P$.

Configuration (*b*). Since the number of electrons with upward spins (↑) and with downward spins (↓) is equal to 6 and zero, the number of exchange pairs of electrons present in this configuration is equal to $^6C_2 + {}^0C_2 = \frac{6\,!}{2!\,(6-2)\,!} + 0 = \frac{6 \times 5 \times 4 \times 3 \times 2 \times 1}{2 \times 4 \times 3 \times 2 \times 1} + 0 = 15$. Hence exchange energy of this configuration $= -15E$. Since this configuration does not have any pair of electrons, pairing energy is equal to zero. As configuration (*b*) has been obtained from configuration (*a*) by promoting one electron from 4*s* orbital (lower energy orbital) to 3*d*-orbitals(higher energy orbitals), this configuration has also promotional energy equal to ΔE. Thus the energy associated with configuration (*b*) (E_b) is equal to the sum of exchange energy, pairing energy and promotional energy. Thus

$$E_b = -15E + 0 + \Delta E = -15E + \Delta E$$

Now since the difference in energy of 4*s* and 3*d* orbitals in very small, $\Delta E \approx P$ and hence $E_b = -15E + P$. If negative sign associated with *E* is not considered, then we can say that configuration (*b*) is associated with greater amount of exchange energy (=15*E*) than configuration (*a*) which has exchange energy equal to 10*E*. Now since pairing energy is the same in both configurations, $E_b > E_a$ and hence configuration (*b*) is more stable than configuration (*a*).

Applications. The concept that half-filled and completely-filled orbitals are comparatively more stable than other orbitals has been used to explain the observed electronic configuration of many elements like Cr_{24} (explained above), Cu_{29}, Mo_{42}, Pd_{46}, Ag_{47}, Au_{79} Gd_{64}, Cm_{96} etc. The expected (configurations obtained on the basis of aufbou principle) and observed configurations of these elements are given below. The observed configurations have been obtained by promoting one electron from a lower energy orbital to a higher energy orbital (in case of Pd_{46}, both the electrons of 5*s* orbital (lower energy orbital) are shifted to 4*d* orbitals (higher energy orbitals).

Element	*Expected configuration*	*Observed configuration*	*Element*	*Expected configuration*	*Observed configuration*
Cu_{29}	[Ar] $3d^9 4s^2$	[Ar] $3d^{10} 4s^1$	Au_{79}	[Xe] $4f^{14} 5d^9 6s^2$	[Xe] $4f^{14} 5d^{10} 6s^1$
Mo_{42}	[Kr] $4d^4 5s^2$	[Kr] $4d^5 5s^1$	Gd_{64}	[Xe] $4f^8 5d^0 6s^2$	[Xe] $4f^7 5d^1 6s^2$
Pd_{42}	[Kr] $4d^8 5s^2$	[Kr] $4d^{10} 5s^0$	Cm_{97}	[Rn] $5f^8 6d^0 7s^2$	[Rn] $5f^7 6d^1 7s^2$
Ag_{47}	[Kr] $4d^9 5s^2$	[Kr] $4d^{10} 5s^1$			

Electronic configuration of the elements, according to their position in the periods of the periodic table

When we proceed from left to right in a period of the long form of periodic table, the extra electron occupies the orbital having higher energy, *i.e.*, the orbitals are filled up with electrons in the increasing order of their energy (*Aufbau principle*). According to Aufbau principle, the orbitals which are filled in different periods are shown below :

1st period	H_1, He_2 .....$1s^{1,2}$	5th period	Rb_{37}, Sr_{38} ..... $5s^{1,2}$ Y_{39} ..... Cd_{48} ..... $4d^{1-10}$ In_{49} ..... Xe_{54} ..... $5p^{1-16}$
2nd period	Li_3, Be_4 ..... $2s^{1,2}$ B_5 ..... Ne_{10} ..... $2p^{1-6}$	6th period	Cs_{55}, Ba_{56}, ... $6s^{1,2}$ La_{57} .....$5d^1$ Ce_{58} ..... Lu_{71} ..... $4f^{2-14}$ Hf_{72} ..... Hg_{80} ..... $5d^{2-10}$ Tl_{81} ..... Rn_{86} ..... $6p^{1-6}$
3rd period	Na_{11}, Mg_{12}, ... $3s^{1,2}$ Al_{13} .... Ar_{18} ..... $3p^{1-6}$	7th period (incomplete period)	Fr_{87}, Ra_{88} ..... $7s^{1,2}$ Ac_{89} ..... $6d^1$ Th_{90} ..... Lw_{103} ..... $5f^{0-14}$ Ku_{104}, Ha_{105} ..... $6d^{2,3}$
4th period	K_{19}, Ca_{20} ..... $4s^{1,2}$ Sc_{21} ..... Zn_{30} ..... $3d^{1-10}$ Ga_{31} ..... Kr_{36} ..... $4p^{1-6}$		

Elements of 1st period ($1s^{1\text{-}2}$). This period contains two elements *viz* H_1 and He and hence two electrons are added upto the end of this period. It is $1s$ orbital into which the two electrons enter. Thus in H_1, the single electron goes to $1s$ orbital to give $1s^1$ or 1 configuration to H_1atom. Second electron also goes to $1s$ orbital and thus the configuration of He_2 becomes $1s^2$ or 2. This configuration is also represented as $[He]_2$ or simply as [He]. Here it may be noted that $1s$ orbital becomes completely filled at the end of this period. The configuration of H_1 and He_2 are given below :

Hydrogen, H_1 $= 1s^1$ or 1 ; Helium, He_2 $= 1s^2$ or 2 = $[He]_2$ or [He]

Since the differentiating electron in both these elements enters *ns* orbital ($1s$ orbital), both these elements are *s*-block elements.

Elements of 2nd period ($2s^{1\text{-}2}$, $2p^{1\text{-}6}$). This period consists of 8 elements which are Li_3, Be_4 (two *s*-block elements) and B_5 to Ne_{10} (six p-block elements). Noble gas present in the 1*st* period is He_2 whose configuration is 2 which is also represented as $[He]_2$ or [He] . In this configuration $1s$ orbital is completely-filled and hence, according to aufbou principle, further eight electrons in eight elements (Li_3 to Ne_{10}) should enter $2s$ and $2p$ orbitals.

Electronic configuration of Li_3 and Be_4 (s-block elements). After the filling of $1s$ orbital at He_2, 3*rd* electron of Li_3 goes to $2s$ orbital (aufbou principle) and this orbital becomes completely filled at Be_4. Thus the configuration of these two elements are as shown below :

Lithium, Li_3 = [He] $2s^1$ = 2, 1 ; Beryllium, Be_4 = [He]$2s^2$ = 2, 2

Since in both these elements, the extra electron enters *ns* orbital (which is $2s$ orbital in this case), these elements are called *s*-block elements. Obviously the valence shell configuration of these elements is $2s^1$ (for Li_3) and $2s^2$ (for Be_4).

Configuration of the six elements viz., B_5 to Ne_{10} (p-block elements). After the filling of $2s$ orbital at Be_4, the next six electrons in these six elements occupy $2p$ sub-shell. Thus this sub-shell has one electron ($2p^1$) at B_5 and has six electrons ($2p^6$ -completely-filled) at Ne_{10}.

Thus the configuration of B_5 is [He] $2s^1$ and that of Ne_{10} is [He]$2s^2$ $2p^6$. The configuration of Ne_{10} is also represented as $[Ne]_{10}$ or [Ne]. Thus $[Ne]_{10}$ or [Ne] =[He] $2s^22p^6$ = 2, 8. The configurations these six elements are given below:

Boron, B_5 = [He] $2s^22p^1$ = 2, 3 ; Carbon, C_6 = [He] $2s^22p^2$ = 2, 4
Nitrogen, N_7 = [He] $2s^22p^3$ = 2, 5; Oxygen, O_8 = [He] $2s^22p^4$ = 2, 6
Fluorine, F_9 = [He] $2s^22p^5$ = 2, 7; Neon, Ne_{10} = [He] $2s^22p^6$ = 2, 8 = $[Ne]_{10}$ or [Ne]

Since in the given six elements, the differentiating electron enters *np* orbitals (Here these are $2p$ orbitals), these elements are called *p*-block elements. The two *s*-block elements (Be_4 and B_5) and five *p*-block elements (B_5 to F_9) are called *representative* (or *normal) elementss,* since they contain 1, 2, 3, 4, 5, 6 and 7 electrons respectively in their last shell (2*nd* shell), which is incompletely filled.

Elements of 3rd period ($3s^{1-2}$, $3p^{1-6}$)This period, like 2*nd* period, contains 8 electrons *viz* Na_{11}, Mg_{12} (two *s*-block elements) and Al_{13} to Ar_{18} (six *p*-block elements). Noble gas present in 2*nd* period is Ne_{10} whose configuration is 2, 8 which can also be represented as $[Ne]_{10}$ or [Ne]. In this configuration 2*p* orbitals are completely-filled and hence, according to aufbou principle, the orbitals which should be filled in the eight elements of this period are 3*s* and 3*p*.

Electronic configuration of Na_{11} and Mg_{12} (s-block elements). After the filling of 2*p* orbitals in Ne_{10}, 11*th* electron of Na_{11} goes to the *s*-orbital of the higher shell (*i.e.* 3*s* orbital) and this orbital becomes completely-filled at Mg_{12}. Thus the configuration of these two elements are as given below :

Sodium, Na_{11} = $[Ne]3s^1$ = 2, 8, 1 ; Magnesium, Mg_{12} = $[Ne]\ 3s^2$ = 2, 8, 2

Since in both of these elements, the differentiating electron goes to *ns* orbital (which is 3*s* orbital in this case) these elements are called *s*-block elements. Obviously, the valence-shell configuration of these *s*-block elements is $3s^1$ (for Na_{11}) and $3s^2$ (for Mg_{12}).

Configuration of the six elements viz., Al_{13} to Ar_{18} (p-block elements). After the filling of 3*s* orbital at Mg_{12}, the next six electrons in the next six elements enter 3*p* orbitals. Thus at Al_{13}, 3*p* sub-shell has one electron while this sub-shell contains six electrons in Ar_{18}. Thus the configuration of Al_{13} is $[Ne]\ 3s^2 3p^1$ and that of Ar_{18} is $[Ne]\ 3s^2\ 3p^6$. The configuration of Ar_{18} *viz.*, $[Ne]3s^2 3p^6$ can also be written as $[Ar]_{18}$ or [Ar]= 2, 8, 8. The configurations of these six elements are as shown below :

Aluminium, Al_{13} = $[Ne]3s^2 3p^1$ = 2, 8, 3; Silicon, Si_{14} = $[Ne]3s^2 3p^2$ = 2,8,4
Phosphorus, P_{15} = $[Ne]3s^2 3p^3$ = 2,8, 5; Sulpher, S_{16} = $[Ne]3s^2 3p^4$ = 2,8,6
Chlorine, Cl_{17} = $[Ne]3s^2 3p^5$ = 2,8, 7; Argon, Ar_{18} = $[Ne]3s^2 3p^6$ = 2,8,8 = $[Ar]_{18}$ or [Ar]

Since in the given six elements, the extra electron goes to *np* orbitals. (Here these are 3*p* orbitals), these elements are called *p*-block elements.

The electronic configurations of the two *s*-block eleeements(Na_{11} and Mg_{12}) and five *p*-block elements (Al_{13} to Cl_{17}) show that the last shell in these seven elements is incompletely-filled, because this shell (*i.e.*, 3*rd* shell) contains 1(in Na_{11}), 2(in Mg_{12}), 3(in Al_{13}), 4(in Si_{14}), 5(in P_{15}), 6(in S_{16}) and 7(in Cl_{17}) electrons. Thus, these seven elements are *representative* or *normal elements.*

Elements of 4th period ($4s^{1-2}$, $3d^{1-10}$, $4p^{1-6}$).This period contains 18 elements *viz* K_{19}, Ca_{20} (two *s*-block elements), Sc_{21} to Zn_{30} (ten *d*-block elements) and Ga_{31} to Kr_{36} (six *p*-block elements) We know that the noble gas present at the end of 3*rd* period is Ar_{18} whose electronic configuration is 2, 8, 8 or 2, 8, $3s^2p^6$. This configuration can also be written as $[Ar]_{18}$ or [Ar]. This configuration indicates that the last sub-shell which is completely-filled at Ar_{18} is 3*p*. Thus, according to aufbau principle, the sub-shells which should be filled with electrons in this period are 4*s*, 3*d, and* 4p, since this period contains 18 elements.

Electronic configuration of K_{19} and Ca_{20} (s-block elements). After the completion of 3*p* orbitals at Ar_{18}, 19*th* electron of K_{19} enters 4*s* orbital and this orbital becomes completely-filled at Ca_{20}. Thus the configurations of these two elements are as shown below :

Potassium, K_{19} = $[Ar]\ 4s^1$ = 2, 8, 8, 1; Calcium, Ca_{20} = $[Ar]\ 4s^2$ = 2, 8, 8, 2

Since in both the elements the extra electron enters *ns* orbital (which is 4*s* orbital in this case), these elements are also called *s-block elements.* Obviously the valence-shell configuration of these *s*-block elements can be written as $4s^1$(for K_{19}) and $4s^2$ (for Ca_{20}).

Note. After 3*p* orbitals have been filled completely at Ar_{18} (which is the noble gas of 3*rd* period), the 19th electron in K_{19} goes to 4*s* orbital instead of 3*d* orbitals. Why 4*s* orbital is filled earlier to 3*d* orbitals in K_{19} has been explained **under** "*Slater's Rules*".

Configuration of the ten elements viz Sc_{21} to Zn_{30} (*d-block elements*). We have seen above that the configuration of Ca_{20} is {Ar} $4s^2$ or 2, 8, 8, 2 or 2, 8, $3s^2p^6d^0$,$4s^2$ in which 3*d* orbitals are vacant ($3d^0$ configuration). Now, as we proceed from Sc_{21} to Zn_{30}, the differentiating electron in all the ten elements enters the vacant 3*d* sub-shell(aufbau principle), but in Cr_{24} and Cu_{29} one electron from

$4s$-orbital gets shifted to $3d$ orbitals (shifting of an electron from a higher shell to a lower shell). Thus, we find that in Cr_{24} and Cu_{29}, $4s$ orbital contains one electron (instead of two) and $3d$ orbitals contain 5($3d^5$) and 10($3d^{10}$) electrons, and not 4 ($3d^4$) and 9($3d^9$) electrons as expected. The shifting of one electron from $4s$ orbital to $3d$ orbitals in Cr_{24} makes $3d$ orbitals half-filled [Cr_{24} = [Ar] $4s^1 3d^5$] and the shifting of one $4s$ electron to $3d$ orbitals in Cu_{29} makes these orbitals completely filled [Cu_{29} = [Ar] $4s^1 3d^{10}$].

Again, we know that half-filled and completely-filled $3d$ orbitals are more stable than other $3d$ orbitals. Thus, due to the presence of half-filled $3d$ orbitals ($3d^5$ configuration), the observed configuration of Cr_{24} [*viz* [Ar] $4s^1 3d^5$] is more stable than its expected configuration [*viz.* [Ar] $4s^2\ 3d^4$]. Similarly, due to the presence of completely-filled $3d$ orbitals ($3d^{10}$ configuration), the observed configuration of Cu_{29} [*viz* [Ar] $4s^1 3d^{10}$] is more stable than its expected configuration [*viz.*, [Ar] $4s^2 3d^9$].

Table 4.2 : *Expected and observed (actual) configurations of the atoms of 3d transition series elements (1st transition series elements). Electronic configuration of* Ca_{20} *is also given.*

3d-transition series elements	Expected configurations	Observed configurations
Calcium, Ca_{20}	$[Ar]4s^2$ = 2, 8, 8, 2	$[Ar]4s^2$ = 2, 8, 8, 2
Scandium, Sc_{21}	$[Ar]4s^2\ 3d^1 = [Ar]3d^1\ 4s^2$ = 2, 8, (8+1), 2 = 2, 8, 9, 2	$[Ar]4s^2\ 3d^1 = [Ar]\ 3d^1 4s^2$ = 2, 8, (8 +1), 2 = 2, 8, 9, 2
Titanium, Ti_{22}	$[Ar]\ 4s^2 3d^2 = [Ar]\ 3d^2 4s^2$ = 2, 8, (8+2), 2 = 2, 8, 10, 2	$[Ar]\ 4s^2 3d^2 = [Ar]\ 3d^2 4s^2$ = 2, 8, (8+2), 2= 2, 8, 10, 2
Vanadium, V_{23}	$[Ar]4s^2 3d^3 = [Ar]3d^3 4s^2$ = 2, 8, (8+3), 2 = 2, 8, 11, 2	$[Ar]4s^2 3d^3 = [Ar]3d^3 4s^2$ = 2, 8, (8+3), 2 = 2, 8, 11, 2
Chromium, Cr_{24}	$[Ar]4s^2 3d^4 = [Ar]3d^4 4s^2$ = 2, 8, (8+ 4), 2 = 2, 8, 12, 2	$[Ar]\ 4s^1 3d^5 = [Ar]3d^5\ 4s^1$ = 2, 8, (8+5), 1 = 2, 8, 13, 1
Manganese, Mn_{25}	$[Ar]\ 4s^2 3d^5 = [Ar]3d^5 4s^2$ = 2, 8, (8+5), 2 = 2, 8, 13, 2	$[Ar]\ 4s^2 3d^5 = [Ar]3d^5 4s^2$ = 2, 8, (8+5), 2 = 2, 8, 13, 2
Iron, Fe_{26} (Ferrum)	$[Ar]4s^2 3d^6 = [Ar]3d^6 4s^2$ = 2, 8, (8+6), 2 = 2, 8, 14, 2	$[Ar]4s^2 3d^6 = [Ar]3d^6 4s^2$ = 2, 8, (8+6), 2 = 2, 8, 14, 2
Cobalt, Co_{27}	$[Ar]4s^2 3d^7 = [Ar]\ 3d^7 4s^2$ = 2, 8, (8+7), 2 = 2, 8, 15, 2	$[Ar]4s^2 3d^7 = [Ar]\ 3d^7 4s^2$ = 2, 8, (8+7), 2 = 2, 8, 15, 2
Nickel, Ni_{28}	$[Ar]4s^2 3d^8 = [Ar]^3 3d^8 4s^2$ = 2, 8, (8+8), 2 = 2, 8, 16, 2	$[Ar]4s^2 3d^8 = [Ar]3d^8 4s^2$ = 2, 8, (8+8), 2 = 2, 8, 16, 2
Copper, Cu_{29} (Cuprum)	$[Ar]4s^2 3d^9 = [Ar]\ 3d^9 4s^2$ = 2, 8, (8+9), 2 = 2, 8, 17, 2	$[Ar]4s^1 3d^{10} = [Ar]3d^{10} 4s^1$ = 2, 8, (8+10), 1 = 2, 8, 18, 1
Zinc, Zn_{30}	$[Ar]4s^2 3d^{10} = [Ar]\ 3d^{10} 4s^2$ = 2, 8, (8+10), 2 = 2, 8, 18, 2	$[Ar]4s^2 3d^{10} = [Ar]\ 3d^{10} 4s^2$ = 2, 8, (8+10), 2 = 2, 8, 18, 2

The observed configurations, as given in Table. 4.2., show that these elements have *four* shells, since they lie in 4*th* period of the periodic table. The distribution of electrons in these four shells can be shown as :

Shells :	*1st*	*2nd*	*3rd*	*4th*
Electrons in various shells	2 $1s^2$	8 $2s^2p^6$	9 to 18 $3s^2p^6d^{1-10}$	1 or 2 $4s^{1-2}$
	Completely-filled shells (two)		Partially-filled shells (two)	

The distribution of electrons shown above makes the following points evident.

(*i*) The configuration of the atoms of these ten elements can be written as 2, 8, 9 to 18, 1 or 2 or 2, 8, $3s^2p^6d^{1-10}$, $4s^{1-2}$ or [Ar] $3d^{1-10}\ 4s^{1-2}$. It is obvious, that the last two shells (*i.e.* 4*th* and 3*rd* shells) are *partially-filled,* since they contain 1 to 2 and 9 to 18 electrons respectively. The first two shells are *completely-filled.* Thus, these ten elements are called *transition elements.*

(*ii*) Last shell (*i.e.* 4*th* shell) in the elements (excepting Cr_{24} and Cu_{29}) contains 2 electrons ($4s^2$ configuration). Cr_{24} and Cu_{29} have one electron ($4s^1$ configuration).

(*iii*) The penultimate shell (*i.e.*, 3*rd* shell) has 9 to 18 electrons ($3s^2\,p^6d^{1-10}$ configuration) *i.e.*, it is 3*d* sub-shell which is sucessively filled up with electrons as we move from Sc_{21} to Zn_{30}. It is due to the successive filling of 3*d* sub-shell that these elements are called 3*d series elements* or 1*st transtian series elements.* Since 3*d* sub-shell is (*n* – 1) *d* sub-shell, these ten elements, according to definition of *d*-block elements, are called *d*-block elements.

(*iv*) All the elements have 2 and 8 electrons in 1*st* and 2*nd* shells respectively. Hence these shells are completely-filled.

(*v*) Valence-shell configuration of the atoms of these ten elements can be written as $3d^{1-10}$ $4s^{1-2}$.

Configuration of the six elements viz., Ga_{31} to Kr_{36} (p-block elements). After the filling of 3*d* orbitals at Zn_{30}, the next six electrons in the next six elements *viz.*, Ga_{31} to Kr_{36} enter 5*p* orbitals. In Ga_{31}, 5*p* sub-shell has one electron while the same sub-shell contains six electrons at Kr_{36}. Thus the configuration of Ga_{31} and Kr_{36} become as [Ar] $4s^23d^{10}4p^1$ and [Ar] $4s^23d^{10}4p^6$ respectively. The configuration of Kr_{36} is also represented as $[Kr]_{36}$ or [Kr]. Thus [Ar] $4s^23d^{10}4p^6$ = [Kr] = 2, 8, 18, 8. The configurations of these six elements are as given below :

Gallium, Ga_{31} = [Ar] $4s^23d^{10}4p^1$ = [Ar] $3d^{10}4s^24p^1$ = 2, 8, (8+10), 3 = 2, 8, 18, 3
Germanium, Ge_{32} = [Ar] $4s^23d^{10}4p^2$ = [Ar] $3d^{10}4s^24p^2$ = 2, 8, (8+10), 4 = 2, 8, 18, 4
Arsenic, As_{33} = [Ar] $4s^23d^{10}4p^3$ = [Ar]$3d^{10}4s^24p^3$ = 2, 8, (8+10), 5 = 2, 8, 18, 5
Selenium, Se_{34} = [Ar] $4s^23d^{10}4p^4$ = [Ar]$3d^{10}4s^24p^4$ = 2, 8, (8+10), 6 = 2, 8, 18, 6
Bromine, Br_{35} = [Ar] $4s^23d^{10}4p^5$ = [Ar]$3d^{10}4s^24p^5$ = 2, 8, (8+10), 7 = 2, 8, 18, 7
Krypton, Kr_{36} = [Ar] $4s^23d^{10}4p^6$ = [Ar]$3d^{10}4s^24p^6$ = 2, 8, (8+10), 8 = 2, 8, 18, 8 = $[Kr]_{36}$ or [Kr]

Since in the above six elements the extra electron goes to *np* orbitals (here these are *4p* orbitals), these elements are called *p-block elements.*

The electronic configuration of the *two s*-block elements (K_{19} and Ca_{20}) and *five p*-block elements (Ga_{31} to Br_{35}) show that the last shell in these seven elements is incompletely-filled because this shell (*i.e.*, 4*th* shell) contains 1 (in K_{19}), 2(in Ca_{20}), 3(in Ga_{31}), 4 (in Ge_{32}), 5(in As_{33}), 6(in Se_{34}) and 7 (in Br_{35}) electrons. Thus these seven elements are *representative (normal) elements.*

Elements of 5th period ($5s^{1-2}$, $4d^{1-10}$, $5p^{1-6}$). This period contains 18 elements *viz* Rb_{37}, Sr_{38} (two *s-block elements*), Y_{39} to Cd_{48} (ten *d*-block elements) and In_{49} to Xe_{54} (six *p*-block elements). We know that the noble gas present at the end of 4*th* period *is* Kr_{36} whose electronic configuration is 2, 8, 18, 8 or 2, 8, 18, $4s^2p^6d^0$ or [Kr]. This configuration shows that the last sub-shell which is completely-filled at Kr_{36} is 4*p*. Thus, according to aufbau principle, the sub-shells which should be filled with electrons in this period are 5*s*, 4*d* and 5*p*, since this period contains 18 elements.

Electronic configuration of Rb_{37} and Sr_{38} (s-block elements). After the completion of 4*p* orbitals at Kr_{36}, 37th electron of Rb_{37} goes to 5*s* orbital and this orbital gets completely-filled at Sr_{38}. Thus configurations of these two elements are as given below:

Rubedium, Rb_{37} = [Kr] $5s^1$ = 2, 8, 18, 8, 1 ; Strontium, Sr_{38} = [Kr] $5s^2$ = 2, 8, 18, 8, 2

Since in both the elements, the extra electron enters *ns* orbital (here *ns* orbital is 5*s* orbital), these elements are also called *s*-block elements. Obviously the valence-shell configuration of these *s*-block elements can be written as $5s^1$ (for Rb_{37}) and $5s^2$ (for Sr_{38}).

Note: After 4*p* orbitals have been filled completely at Kr_{36} (which is the noble gas of 4*th* period), the 37*th* electron in Rb_{37} enters the 5*s* orbital and not 4*d* orbitals. Why 5*s* is filled in preference to 4*d* orbitals in Rb_{37} has been explained under the study of "*Slater's Rules*"

Configuration of the ten elements viz Y_{39} to Cd_{48} (d-block elements). We have shown above that the configuration of Sr_{38} is [Kr] $5s^2$ *or* 2, 8, 18, $4s^2p^6d^0$, $5s^2$ in which 4*d* orbitals are vacant. Now, as we move from Y_{39} to Cd_{48}, the extra electron in all the ten elements enters the vacant 4*d* sub-shell (aufbau principle), but in the five elements *viz* Nb_{41}, Mo_{42}, Ru_{44}, Rh_{45} and Ag_{47} one electron from $5s$ orbital gets shifted to $4d$ orbitals (shifting of an electron from a higher shell to a lower shell) and in Pd_{46}, both $5s$ electrons are shifted to $4d$ orbitals. Thus we find that in the above five elements $5s$ orbital contains only one electron (instead of two) ($5s^1$) and in Pd_{46}, $5s$ orbital is vacant ($5s^0$).

Why the shifting of one or both electrons from $5s$ orbital to $4d$ orbitals takes place in case of Mo_{42}, Pd_{46} and Ag_{47} has been explained by saying that this type of shifting makes $4d$ orbitals either half-filled (as in case Mo_{42}) or completely-filled (as in case of Pd_{46} and Ag_{47}), and again we know that half-filled $4d$ orbitals ($4d^5$) are more stable than $4d^4$ orbitals (in Mo_{42}) and similarly completely-filled $4d$ orbitals ($4d^{10}$ configuration) are more stable than $4d^8$ (in Pd_{46}) and $4d^9$ (in Ag_{47}) orbitals. Thus due to the presence of half-filled $4d$ orbitals, the observed configuration of Mo_{42} (*viz.* [Kr] $5s^1 4d^5$) is more stable than its expected configuration (viz. [Kr] $5s^2\ 4d^4$). Similarly, because of the presence of completely-filled 4d orbitals, the observed configuration of Pd_{46} and Ag_{47} are more stable than their respective expected configurations.

Why the shifting of one of the two electrons of $5s$ orbital to $4d$ orbitals takes place in case of Nb_{41}, Ru_{44} and Rh_{45} has been explained on the basis of the nuclear-electron and electron–electron forces, the discussion of which is beyond the scope of this book.

Table 4.3: *Expected and observed (actual) configuration of the atoms of 4d transition series elements (2nd transition series elements). Electronic configuration of Sr_{38} in also given.*

$4d$ series element	Expected configuration	Observed configuration
Strontium, Sr_{38}	[Kr] $5s^2$ = 2, 8, 18, 8, 2	[Kr]$5s^2$ = 2, 8, 18, 8, 2
Yttrium, Y_{39}	[Kr]$5s^2\ 4d^1$ = [Kr] $4d^1 5s^2$ = 2, 8 18, (8 + 1) , 2 = 2, 8, 18, 9, 2	[Kr]$5s^2\ 4d^1$ = [Kr]$4d^1\ 5s^2$ = 2, 8, 18 (8 + 1), 2 = 2, 8, 18, 9, 2
Zirconium, Zr_{40}	[Kr] $5s^2 4d^2$ = [Kr] $4d^2\ 5s^2$ = 2, 8, 18, (8 + 2), 2 = 2, 8, 18, 10, 2	[Kr] $5s^2 4d^2$ = [Kr]$4d^2\ 5s^2$ = 2, 8, 18, (8 + 2), 2 = 2, 8, 18, 10, 2
Niobium, Nb_{41}	[Kr] $5s^2\ 4d^3$ = [Kr] $4d^3\ 5s^2$ = 2, 8, 18, (8 + 3), 2 = 2, 8, 18, 11, 2	[Kr]$5s^1\ 4d^4$ = [Kr] $4d^4\ 5s^1$ = 2, 8, 18 (8 + 4), 1 = 2, 8, 18, 12, 1
Molybdenum, Mo_{42}	[Kr] $5s^2\ 4d^4$ = [Kr] $4d^4\ 5s^2$ = 2, 8, 18, (8 + 4), 2 = 2, 8, 18, 12, 2	[Kr]$5s^1\ 4d^5$ = [Kr] $4d^5\ 5s^1$ = 2, 8, 18, (8 + 5), 1 = 2, 8, 18, 13, 1
Technetium, Tc_{43}	[Kr] $5s^2\ 4d^5$ = [Kr] $4d^5\ 5s^2$ = 2, 8, 18, (8 + 5), 2 = 2, 8, 18, 13, 2	[Kr] $5s^2\ 4d^5$ = [Kr] $4d^5\ 5s^2$ = 2, 8, 18, (8 + 5), 2 = 2, 8, 18, 13, 2
Ruthenium, Ru_{44}	Kr] $5s^2\ 4d^6$ = [Kr] $4d^6\ 5s^2$ = 2, 8, 18, (8 + 6), 2 = 2, 8, 18, 14, 2	[Kr] $5s^1\ 4d^7$ = [Kr] $4d^7\ 5s^1$ = 2, 8, 18, (8 + 7), 1 = 2, 8, 18, 15, 1
Rhodium, Rh_{45}	[Kr] $5s^2\ 4d^7$ = [Kr] $4d^7\ 5s^2$ = 2, 8, 18, (8+7), 2 = 2, 8, 18, 15, 2	[Kr] $5s^1\ 4d^8$ = [Kr] $4d^8\ 5s^1$ = 2, 8, 18, (18 + 8), 1 = 2, 8, 18, 16, 1
Palladium, Pd_{46}	[Kr] $5s^2\ 4d^8$ = [Kr] 4d$^8\ 5s^2$ = 2, 8, 18, (8 + 8), 2 = 2, 8, 18, 16, 2	[Kr] $5s^0\ 4d^{10}$ = [Kr] $4d^{10}\ 5s^0$ = 2, 8, 18, (8 + 10), 1 = 2, 8, 18, 18, 1
Silver, Ag_{47}	[Kr] $5s^2\ 4d^9$ = [Kr] $4d^9\ 5s^2$ = 2, 8, 18, (8 + 9), 2 = 2, 8, 18, 17, 2	[Kr] $5s^1\ 4d^{10}$ = [Kr] $4d^{10}\ 5s^1$ = 2, 8, 18, (8 + 8), 1 = 2, 8, 18, 18, 1
Cadmium, Cd_{48}	[Kr] $5s^2\ 4d^{10}$ = [Kr] $4d^{10}\ 5s^2$ = 2, 8, 18, (8 + 10), 2 = 2, 8, 18, 18, 2	[Kr] $5s^2\ 4d^{10}$ = [Kr] $4d^{10}\ 5s^2$ = 2, 8, 18, (8 + 10), 2 = 2, 8, 18, 18, 2

· The observed configurations, as given in Table 4.3, show that these elements have *five* shells, since they are located in *5th* period of the periodic table. The distribution of electrons in these five shell can be shown as :

Shells :	1*st*	2*nd*	3*rd*	4*th*	5*th*
Electrons in various shells	2 $1s^2$	8 $2s^2p^6$	18 $3s^2p^6d^{1-0}$	9 to 18 $4s^2p^6d$	0 to 2 $5s^{0-2}$
	Completely-filled shells (three)			Partially-filled shells (two)	

The distribution of electrons shown above makes following points evident :

(*i*) The configuration of the atoms of these ten elements can be represented as 2, 8, 18, 9 to 18, 0 to 2 or 2, 8, 18, $4s^2p^6d^{1-10}$, $5s^{0-2}$ or [Kr] $4d^{1-10}$ $5s^{0-2}$. It is obvious that the last two shells (*i.e.*, *5th* and *4th* shells) are partially-filled, since they contain 0 to 2 and 9 to 18 electrons respectively. The first three shells are *completely filled.* Thus, these ten elements are called *transition elements.*

(*ii*) Last shell (*i.e.*, 5th shell) contains *no* electron (in Pd_{46}), *one* electron (in Nb_{41}, Ru_{44}, Rh_{45} and Ag_{47}) and *two* electrons in the remaining five elements.

(*iii*) The penultimate shell (*i.e.* *4th* shell) has 9 to 18 electrons ($4s^2p^6d^{1-10}$ configuration), *i.e.* it is *4d* sub-shell which is progressively filled up with electrons as we move from Y_{39} to Cd_{48}. It is due to the successive filling of *4d* sub-shell that these elements are called *4d series elements* or *2nd transition series elements.* Since 4*d* sub-shell is (*n*–1)*d* sub-shell, these ten elments, according to definition of *d*-block elements, are called *d*-block elements.

(*iv*) All the ten elements have 2, 8 and 18 electrons in their 1*st*, 2*nd* and 3*rd* shells. Hence these shells are completely-filled.

(*v*) The valence-shell configuration of the atoms of these elments can be written as $4d^{1-10}5s^{0-2}$.

Configuration of the six elements viz In_{49} to Xe_{54} (p-block elements). We have shown above that the configuration of Cd_{48} is [Kr] $5s^2$ $4d^{10}$ or 2, 8, 18, 18, 2 which shows that 4*d* orbitals are completely-filled in this element. After the completion of 4*d* orbitals at Cd_{48}, the next six electrons in the next six elements *viz* In_{49} to Xe_{54} occupy 5*p* orbitals. In In_{49} 5*p* orbitals contain only one electron only while at Xe_{54} these orbitals contain six electrons ($5p^6$ configuration). Thus the configuration of In_{49} is [Kr] $5s^24d^{10}5p^1$ or 2, 8, 18, 18, 3 and that of Xe_{54} is [Kr] $5s^2$ $4d^{10}5p^6$ or 2, 8, 18, 18, 8. The configuration of Xe_{54} is also represented as $[Xe]_{54}$ or simply as [Xe]. The configurations of the above six elements are given below :

p-block elements

Indium, In_{49} = [Kr]$5s^24d^{10}5p^1$ = [Kr] $4d^{10}5s^25p^1$ = 2, 8, 18, (8 + 10), 3 = 2, 8, 18, 18, 3

Tin, Sn_{50} (Stannum) = [Kr] $5s^2$ $4d^{10}5p^2$ = [Kr] $4d^{10}5s^25p^2$ = 2, 8, 18, (8 + 10), 4
= 2, 8, 18, 18, 4

Antimony, Sb_{51} (Stibnium) = [Kr] $5s^2$ $4d^{10}5p^3$ = [Kr] $4d^{10}5s^25p^3$ = 2, 8, 18, (8 + 10), 5
= 2, 8, 18, 18, 5

Tellurium, Te_{52} = [Kr] $5s^24d^{10}5p^4$ = [Kr]$4d^{10}5s^25p^4$ = 2, 8, 18, (8 + 10), 6
= 2, 8, 18, 18, 6

Iodine, I_{53} = [Kr]$5s^24d^{10}5p^5$ = [Kr] $4d^{10}5s^25p^5$ = 2, 8, 18 (8 + 10), 7 = 2, 8, 18, 18, 7

Xenon, Xe_{54} = [Kr] $5s^24d^{10}5p^6$ = [Kr] $4d^{10}5s^25p^6$ = 2, 8, 18, (8 +10), 8 = 2, 8, 18, 18, 8
= $[Xe]_{54}$ or [Xe]

Since in these six elements the additional electron goes to *np* orbitals (Here these are 5*p* orbitals), these elements are called *p*-block elements. Obvisouly the valence-shell configuration of these elements can be represented as $5s^2p^{1-6}$. Among these six elements Xe_{54} is a noble gas and its last shell is completely-filled, since it contains 8 electrons (saturated shell).

The electronic configuration of the two *s*-block elements (Rb_{37} and Sr_{38}) and five *p*-block elements (In_{49} to I_{53}) indicate that the last shell of these seven elements is incompletely filled, becasue this shell (*i.e.* *5th* shell) contains 1 (in Rb_{37}), 2 (in Sr_{38}), 3 (in In_{49}), 4 (in Sn_{50}), 5 (in Sb_{51}), 6 (in Te_{52}) and 7 (in I_{53}) electrons. Thus these seven elements are *representative* (or *normal*) *elements.*

Elements of 6th period ($6s^{1-2}$, $5d^1$, $4f^{2-14}$, $5d^{2-10}$, $6p^{1-6}$). This period contains 32 elements viz

Cs_{55}, Ba_{56} (two *s*-block elements) La_{57} (*d*-block element), Ce_{58} to Lu_{71} (fourteen *f*-block elements), Hf_{72} to Hg_{80} (nine *d*-block elements) and Tl_{81} to Rn_{86} (six *p*-block elements). Thus we see that this period consists of two *s*-block, ten *d*-block, fourteen *f*-block and six *p*-block elements.

We have shown above that the electronic configuration of Xe_{54} (which is a noble gas of 5*th* period) is [Kr] $4d^{10}5s^25p^6$ or 2, 8, 18, 18, 18, 8 or 2, 8, 18, $4s^2p^6d^{10}f^0$, $5s^2p^6d^0$. This configuration is also represented as $[Xe]_{54}$ or simply as [Xe]. This configuration shows that the last orbitals, which are completely-filled in Xe_{54}, are 5*p*. Thus, according to aufbau principle, the orbitals which are filled with electrons in 6*th* period are 6*s*, 4*f*, 5*d* and 6*p*, since this period contains 32 elements.

Electronic Infiguration of Cs_{55} and Ba_{56} (s-blcok elements). After the completion of 5*p* orbitals at Xe_{54}, 55*th* electron of Cs_{55} enters the next higher shell *viz* 6*th* shell. This electron occupies 6*s* orbital and this orbital becomes completely-filled at Ba_{56}. Thus the configurations of these two elements are as shown below :

Cesium, Cs_{55}= $[Xe]6s^1$ = 2, 8, 18, 18, 8, 1 ; Barium, Ba_{56}= $[Xe]6s^2$ = 2, 8, 18, 18, 8, 2

Since in both the elements, the extra electron enters *s*-orbital of the last shell (here it is 6*s* orbital), these elements are *s*-block elements. Obviously, the valence shell configuration of these two elements can be written as $6s^1$ and $6s^2$ respectively.

Electronic configuration of La_{57} (d-block element) : After the completion of 6*s* orbital at Ba_{56} ($6s^2$ configuration), the next electron (*i.e.* 57*th* electron) of La_{57}, according to aufbau principle, must enter 4*f* orbitals (the squance of filling of the orbitals is 1*s*, 2*s*, 2*p*, 3*s*, 3*p*, 4*s*, 3*d*, 4*p*, 5*s*, 4*d*, 5*p*, 6*s*, 4*f*, 5*d*, 6*p*, 7*s*, 5*f*, 6*d*) to give [Xe] $6s^2\,4f^1\,4d^0$ configuration to La_{57} ; but actually what happens is that the extra electron of La_{57} enters 5*d* orbitals,(instead of 4*f* orbitals) to give [Xe] $6s^2\,4f^0\,5d^1$ configuration to La_{57} atom, in which 4*f* orbitals are vacant and 5*d* orbitals have one electron. Why the extra electron goes to 5*d* orbitals, instead of 4*f* orbitals in La_{57} atom, has been explained on the basis that at La_{57} atom, 5*d* and 4*f* orbitals are supposed to have almost the some energy ($4f \approx 5d$) and hence the additional electron is free to occpupy any of these orbitals. Since in La_{57} the differentiating electron occupies $(n-1)d$ orbitals (5*d* orbitals), it is a *d*-block element or an element of 5*d* transition series.

Electronic configuration of the fourteen elements viz Ce_{58} to Lu_{71} (called 4f-block elements or lanthanides). All these elements follow La_{57} in the periodic table and hence are called lanthanides. We have shown above that the configuration of La_{57} is [Xe] $4f^0\,5d^16s^2$, in which 5*d* sub-shell is singly-filled ($5d^1$ configuration) and 4*f* sub-shell is vacant ($4f^0$ configuration). This configuration shows that as we proceed from Ce_{50} to Lu_{71} (fourteen lanthanides), the additional electron should enter the vacant 4*f* sub-shell and 5*d* sub-shell should remain singly-filled ($5d^1$ configuration) in all the 14 lanthanides, as it is in La_{57}. Thus the expected configuration of the atoms of lathanides should be those in which 5*d* sub-shell is singly-filled and 4*f* sub-shell is progressively filled up with electrons as we move from Ce_{58} to Lu_{71} in the series. In other words we can say that the expected configuration of the atoms of lanthanides should be $[Xe]4f^{1-14}5d^16s^2$.

But, since the energies of 5*d* and 4*f* orbitals are closely similar in all the lanthanides, excepting Gd_{64} and Lu_{71}, $5d^1$ electron gets shifted to 4*f* orbitals and hence 5*d* orbitals remain vacant. The presence of one electron in 5*d* orbitals in Gd_{64} and Lu_{71} gives half-filled (in Gd_{64}) and completely-filled (in Lu_{71}) 4*f* orbitals, which are more stable. Expected as well as observed (actual) electronic configugration of lanthanides are given in Table. 4.4.

Table 4.4: *Expected and observed (actual) electronic configuration of tanthanide atoms. Configuration of La_{57} is also given.*

Lanthanide elements	*Expected configuration*	*Observed configuration*
Lanthanum, La_{57}	$[Xe]6s^2, 5d^0\,4f^1$ = [Xe] $4f^1\,5d^0\,6s^2$ = 2, 8, 18, (18 + 1), 8, 2 = 2, 8, 18, 19, 8, 2	[Xe] $6s^2\,5d^1\,4f^0$ = [Xe] $4f^05d^16s^2$ = 2, 8, 18, (18 + 0), (18 +1), 2 = 2, 8, 18, 18, 9, 2

Cerium, Ce_{58}	[Xe] $6s^2 5d^1 4f^1$ = [Xe] $4f^1 5d^1 6s^2$ = 2, 8, 18, (18+1), (8+1), 2 = 2, 8, 18, 19, 9, 2	[Xe] $6s^2 5d^0 4f^2$ = [Xe] $4f^2 5d^0 6s^2$ = 2, 8, 18 (18 + 2), 8, 2 = 2, 8, 18, 20, 8, 2
Prseodymium, Pr_{59}	[Xe] $6s^2 5d^1 4f^2$ = [Xe] $4f^2 5d^1 6s^2$ = 2, 8, 18, (18 + 2), 9, 2 = 2, 8, 18, 20, 9, 2	[Xe] $6s^2 5d^0 4f^3$ = [Xe] $4f^3 5d^0 6s^2$ = 2, 8, 18, (18+3), 8, 2 = 2, 8, 18, 21, 8, 2
Neodymium, Nd_{60}	[Xe] $6s^2 5d^1 4f^3$ = [Xe] $4f^3 5d^1 6s^2$ = 2, 8, 18, (18+3), 9, 2 = 2, 8, 18, 21, 9, 2	[Xe] $6s^2 5d^0 4f^4$ = [Xe] $4f^4 5d^0 6s^2$ = 2, 8, 18, (18+4), 8, 2 = 2, 8 ,18, 22, 8, 2
Promethium, Pm_{61}	[Xe] $6s^2 5d^1 4f^4$ = [Xe] $4f^4 5d^1 6s^2$ = 2, 8, 18, (18+4), 9, 2 = 2, 8, 18, 22, 9, 2	[Xe] $6s^2 5d^0 4f^5$ = [Xe] $4f^5 5d^0 6s^2$ = 2, 8, 18, (18+5), 8, 2 = 2, 8, 18, 23, 8, 2
Samarium, Sm_{62}	[Xe]$6s^2 5d^1 4f^5$ = [Xe] $4f^5 5d^1 6s^2$ = 2, 8, 18, (18 + 5), 9, 2 = 2, 8, 18, 23, 9, 2	[Xe] $6s^2 5d^0 4f^6$ = [Xe] $4f^6 5d^0 6s^2$ = 2, 8, 18, (18 + 6), 8, 2 = 2, 8, 18, 24, 8, 2
Europium, Eu_{63}	[Xe] $6s^2 5d^1 4f^6$ = [Xe]$4f^6 5d^1 6s^2$ = 2, 8, 18, (18 + 6), 9, 2* = 2, 8, 18, 24, 9, 2	[Xe] $6s^2 5d^0 4f^7$ = [Xe] $4f^7 5d^0 6s^2$ = 2, 8, 18, (18 + 7), 8, 2 = 2, 8, 18, 25, 8, 2
Gadolinium Gd_{64}	[Xe] $6s^2 5d^1 4f^7$ = [Xe] $4f^7 5d^1 6s^2$ = 2, 8, 18, (18 + 7), 9, 2 = 2, 8, 18, 25, 9, 2	[Xe] $6s^2 5d^1 4f^7$ = [Xe] $4f^7 5d^1 6s^2$ = 2, 8, 18 (18 + 7), 9, 2 = 2, 8, 18, 25, 9, 2
Terbium, Tb_{65}	[Xe] $6s^2 5d^1 4f^8$ = [Xe] $4f^8 5d^1 6s^2$ = 2, 8, 18, (18 + 8), 9, 2 = 2, 8, 18, 26, 9, 2	[Xe]$6s^2 5d^0 4f^9$ = [Xe]] $f^9 5d^0 6s^2$ = 2, 8, 18, (18 + 9), 8, 2 = 2, 8, 18, 27, 8, 2
Dysprosium, Dy_{66}	[Xe] $6s^2 5d^1 4f^9$ = [Xe] $4f^9 5d^1 6s^2$ = 2, 8, 18, (18 + 9), 9, 2 = 2, 8, 18,. 27, 9, 2	[Xe] $6s^2 5d^0 4f^{10}$ = [Xe] $4f^{10} 5d^0 6s^2$ = 2, 8, 18, (18 + 10), 8, 2 = 2, 8, 18, 28, 8, 2
Halmium, Ho_{67}	[Xe] $6s^2 5d^1 54f^{10}$ = [Xe] $4f^{10} 5d^1 6s^2$ = 2, 8, 18, (18 + 10), 9, 2 = 2, 8, 18, 28, 9, 2	[Xe] $6s^2 5d^0 4f^{11}$ = [Xe] $4f^{11} 5d^0 6s^2$ = 2, 8, 18, (18 +11), 8, 2 = 2, 8, 18, 29, 8, 2
Erbium, Er_{68}	[Xe] $6s^2 5d^1 4f^{11}$ = [Xe] $4f^{11} 5d^1 6s^2$ = 2, 8, 18, (18 + 11), 9, 2 = 2, 8, 18, 29, 9, 2	[Xe] $6s^2 5d^0 4f^{12}$ = [Xe] $4f^{12} 5d^0 6s^2$ = 2, 8, 18, (18 + 12), 8, 2 = 2, 8, 18, 30, 8, 2
Thulium, Tm_{69}	[Xe] $6s^2 5d^1 4f^{12}$ = [Xe] $4f^{12} 5d^1 6s^2$ = 2, 8, 18, (18 + 12), 9, 2 = 2, 8, 18, 30, 9, 2	[Xe] $6s^2 5d^0 4f^{13}$ = [Xe] $4f^{13} 5d^0 6s^2$ = 2, 8, 18, (18 + 13), 8, 2 = 2, 8, 18, 31, 8, 2
Ytterabium Yb_{70}	[Xe] $6s^2 5d^1 4f^{13}$ = [Xe] $4f^{13} 5d^1 6s^2$ = 2, 8, 18, (18 + 13), 9, 2 = 2, 8, 18, 31, 9, 2	[Xe] $6s^2 5d^0 4f^{14}$ = [Xe] $4f^{14} 5d^0 6s^2$ = 2, 8, 18, (18 + 14), 8, 2 = 2, 8, 18, 32, 8, 2
Lutetium Lu_{71}	[Xe] $6s^2 5d^1 4f^{14}$ = [Xe] $4f^{14} 5d^1 6s^2$ = 2, 8, 18, (18 +14), 9, 2 = 2, 8, 18, 32, 9, 2	[Xe] $6s^2 5d^1 4f^{14}$ = [Xe] $4f^{14} 5d^1 6s^2$ = 2, 8, 18 (18 + 14), 9, 2 = 2, 8, 18, 32, 9, 2

These configurations show that these elements have six shells, since they lie in 6th period of the priodic table. The distribution of electrons in these six shells can be shown as :

Shells :	1*st*	2*nd*	3*rd*	4*th*	5*th*	6*th*
Electrons in various shells	2 $1s^2$	8 $2s^2p^6$	18 $3s^2p^6d^{10}$	20 to 32 $4s^2p^6d^{10}f^{2-14}$	8 or 9 $5s^2p^6d^{0-1}$	2 $6s^2$
	←— completely-filled —→			←— Partially - filled —→		

The distribution of electrons shown above makes the following points evident:

(*i*) The configuration of the atoms of lanthanides can be written as 2, 8, 18, 20 to 32, 8 or 9, 2 or 2, 8, 18, $4s^2p^6d^{10}f^{2-14}$, $5s^2p^6d^{0-1}$, $6s^2$ or $[Xe]4f^{2-14}5d^{0-1}6s^2$. It is obvious that the last three shells (*i.e.*, 6*th*, 5*th* and 4*th* shells) are *partially-filled*, since they contain 2, 8 to 9 and 20 to 32 electrons respectively. First three shells are completely filled (1st shell = $2 \times 1^2 = 2$, 2*nd* shell = $2 \times 2^2 = 8$, 3*rd* shell = $2 \times 3^2 = 18$). Thus lanthanides can be defined as those elements whose last three shells are partially-filled while the first three shells are completely-filled.

(*ii*) Last shell (*i.e.* 6*th* shell) in each element has 2 electrons ($6s^2$ configuration).

(*iii*) Penultimate shell (*i.e.* 5*th* shell) has 8 or 9 electrons ($5s^2p^6d^{0-1}$ configuration). On the basis of the number of electrons present in 5*d* sub-shell, these elements have been grouped into the following two categories :

(*a*) Cd_{64} (7th element of the series) and Lu_{71} (14*th* element) have one electron in 5*d* sub-shell ($5d^1$ configuration and hence there are electrons in 5*th* shell of these elements ($5s^2p^6d^1$ configuration).

(*b*) The remaining elements have no electron in 5*d* sub-shell ($5d^0$ configuration). It is because of the presence of empty 5*d* sub-shell that 5*th* shell of the remaining elements has 8 electrons.

Important notes (*i*) In Eu_{63} (6*th* element of the series) and Yb_{70} (13*th* element), $5d^1$ electron is shifted to 4*f* sub-shell, since this type of shifting makes 4*f* sub-shell half-filled in Eu_{63} and completely filled in Yb_{70} and thus both these elements attain the configurations (observed configurations) which are more stable than their expected configurations.

(*ii*) In Gd_{64} (7*th* element), $5d^1$ electron is not shifted to $4f^7$ sub-shell, since the shifting gives [Xe] $6s^2\ 5d^0\ 4f^8$ configuration to Gd_{64}. This configuration is not as stable as [Xe] $6s^2\ 5d^1\ 4f^7$.

(*iii*) In Lu_{71} (14*th* element) the shifting of $5d^1$ electron to 4*f* sub-shell is not possible, since 4*f* sub-shell is already filled to its maximum capacity in Lu_{71}.

The above discussion makes it evident that Gd_{64} and Lu_{71} have 9 electrons in their 5*th* shell ($5s^2p^6d^1$ configuration) while other elements have 8 electrons ($5s^2\ p^6d^0$ configuration).

(*iv*) The ante-penultimate shell (*i.e.* 4*th* shell) contains 20 to 32 electrons ($4s^2p^6d^{10}f^{2-14}$ configuration). Thus 4*s*, 4*p* and 4*d* sub-shells are completely-filled but 4*f* sub-shell is being progressively filled up with electrons as we move in the series from Ce_{58} to Lu_{71}. It is due to the successive (progressive) filling of 4*f* sub-shell that these elements are called 4*f*-block elements. *Thus 4f block elements can be defined as those elements in which the last electron goes to 4f sub-shell.*

(*v*) Valence-shell configuration of the atoms of 14 lanthanides can be written as $4f^{2-14}\ 5d^{0-1}\ 6s^2$.

Electronic configuration of the nine elements viz Hf_{72} to Hg_{80} (d-block elements). We have seen that the configuration of Lu_{71} is $[Xe]6s^25d^14f^{14}$ in which 6*s* and 4*f* sub-shell are completely-filled while 5*d* sub-shell contains only one electron. Experiments have shown that as we proceed from Hf_{72} to Hg_{80}, the extra electron in all the nine elements goes to the 5*d* sub-shell (which is already having one electron at Lu_{71}) but in the *two* elements viz Pt_{78} and Au_{79}, one electron from 6*s* orbital gets shifted to 5*d* orbitals (shifting of an electron from a higher shell to lower shell). Thus, we find that in these two elements 6*s* orbital contains one electron only (instead of two), since one electron is shifted to 5*d* sub-shell.

Why the shifting of one electron from 6*s* orbital to 5*d* orbitals takes place in case of Au_{79} has been explained by saying that such a shifting produces ten electrons, instead of nine, in 5*d* orbitals ($5d^{10}$ configuration) and again we know that 5*d* orbitals hoving 10 electrons (completely-filled 5*d* orbitals) are more stable than those having 9 electrons (partially-filled 5*d* orbitals). Thus [Xe] $6s^15d^{10}4f^{14}$ configuration is more stable than [Xe] $6s^25d^94f^{14}$ configuration. The experimental (*i.e.* observed) configuration of Pt_{78} *viz* [Xe] $6s^15d^94f^{14}$ has been explained on the basis of nuclear-electron and electron-electron forces.

Table 4.5 : *Expected and observed (actual) configurations of the atoms of 5d transition series elements (3rd transition series elements). Configuration of lanthanides (Ce_{58} to Lu_{71}) are also given.*

5*d* series elements	Expected configuration	Observed configuration
Lanthanum, La_{57}	$[Xe]6s^2 5d^0 4f^1$ = $[Xe]\ 4f^1 5d^0 6s^2$ = 2, 8, 18, (18 + 1) 8, 2 = 2, 8, 18, 19, 8, 2	$[Xe]\ 6s^2 5d^1 4f^0$ = $[Xe]\ 4f^0 5d^1 6s^2$ = 2, 8, (18 + 0), (8 + 1), 2 = 2, 8, 18, 18, 9, 2
Cerium Ce_{58}	$[Xe]6s^2 5d^1 4f^1$ = $[Xe]\ 4f^1 5d^1 6s^2$ = 2, 8, 18 (18 + 1), (8 + 1), 2 = 2, 8, 18, 19, 9, 2	$[Xe]\ 6s^2 5d^0 4f^2$ = $[Xe]\ 4f^2 5d^0 6s^2$ = 2, 8, 18, (18 + 2), 8, 2 = 2, 8, 18, 20, 8, 2
Lutetium, Lu_{71}	$[Xe]6s^2 5d^1 4f^{14}$ = $[Xe]\ 4f^{14} 5d^1 6s^2$ = 2, 8, 18, (18 + 14), (8 + 1), 2 = 2, 8, 18, 32, 9, 2	$[Xe]\ 6s^2 5d^1 4f^{14}$ = $[Xe]\ 4f^{14} 5d^1 6s^2$ = 2, 8, 18, (18 + 14), (8 + 1), 2 = 2, 8, 18, 32, 9, 2
Hafnium, Hf_{72}	$[Xe]6s^2 5d^2 4f^{14}$ = $[Xe]\ 4f^{14} 5d^2 6s^2$ = 2, 8, 18, 32, (8 + 2), 2 = 2, 8, 18, 32, 10, 2	$[Xe]\ 6s^2 5d^2 4f^{14}$ = $[Xe]\ 4f^{14} 5d^2 6s^2$ = 2, 8, 18, 32, (8 + 2), 2 = 2, 8, 18, 32, 10, 2
Tantalum, Ta_{73}	$[Xe]6s^2 5d^3 4f^{14}$ = $[Xe]\ 4f^{14} 5d^3 6s^2$ = 2, 8, 18, 32, (8 + 3), 2 = 2, 8, 18, 32, 11, 2	$[Xe]\ 6s^2 5d^3 4f^{14}$ = $[Xe]\ 4f^{14} 5d^3 6s^2$ = 2, 8, 18, 32, (8 + 3), 2 = 2, 8, 18, 32, 11, 2
Tungsten, W_{74} (Wolfram)	$[Xe]6s^2 5d^4 4f^{14}$ = $[Xe]\ 4f^{14} 5d^4 6s^2$ = 2, 8, 18, 32, (8 + 4), 2 = 2, 8, 18, 32, 12, 2	$[Xe]\ 6s^2 5d^4 4f^{14}$ = $[Xe]\ 4f^{14} 5d^4 6s^2$ = 2, 8, 18, 32, (8 + 4), 2 = 2, 8, 18, 32, 12, 2
Rhenium, Re_{75}	$[Xe]6s^2 5d^5 4f^{14}$ = $[Xe]\ 4f^{14} 5d^5 6s^2$ = 2, 8, 18, 32, (8 +5), 2 = 2, 8, 18, 32, 13, 2	$[Xe]\ 6s^2 5d^5 4f^{14}$ = $[Xe]\ 4f^{14} 5d^5 6s^2$ = 2, 8, 18, 32 (8+5), 2 = 2, 8, 18, 32, 13, 2
Osmium Os_{76}	$[Xe]6s^2 5d^6 4f^{14}$ = $[Xe]\ 4f^{14} 5d^6 6s^2$ = 2, 8, 18, 32, (8 + 6), 2 = 2, 8, 18, 32, 14, 2	$[Xe]\ 6s^2 5d^6 4f^{14}$ = $[Xe]\ 4f^{14} 5d^6 6s^2$ = 2, 8, 18, 32, (8 + 6), 2 = 2, 8, 18, 32, 14, 2
Iridium, Ir_{77}	$[Xe]6s^2 5d^7 4f^{14}$ = $[Xe]\ 4f^{14} 5d^7 6s^2$ = 2, 8, 18, 32, (8 + 7), 2 = 2, 8, 18, 32, 15, 2	$[Xe]\ 6s^2 5d^7 4f^{14}$ = $[Xe]\ 4f^{14} 5d^7 6s^2$ = 2, 8, 18, 32, (8 + 7), 2 = 2, 8, 18, 32, 15, 2
Platinum, Pt_{78}	$[Xe]6s^2 5d^8 4f^{14}$ = $[Xe]\ 4f^{14} 5d^8 6s^2$ = 2, 8, 18, 32, (8 + 8), 2 = 2, 8, 18, 32, 16, 2	$[Xe]\ 6s^1 5d^9 4f^{14}$ = $[Xe]\ 4f^{14} 5d^9\ 6s^1$ = 2, 8, 18, 32, (8 + 9), 1 = 2, 8, 18, 32, 17, 1
Gold, Au_{79}	$[Xe]6s^2 5d^9 4f^{14}$ = $[Xe]\ 4f^{14} 5d^9\ 6s^2$ = 2, 8, 18, 32, (8 + 9), 2 = 2, 8, 18, 32, 17, 2	$[Xe]\ 6s^1 5d^{10} 4f^{14}$ = $[Xe]\ 4f^{14} 5d^{10} 6s^1$ = 2, 8, 18, 32, (8 + 10), 1 = 2, 8, 18, 32, 18, 1
Mercury, Hg_{80}	$[Xe]6s^2 5d^{10} 4f^{14}$ = $[Xe]\ 4f^{14} 5d^{10} 6s^2$ = 2, 8, 18, 32, (8 + 10), 2 = 2, 8, 18, 32, 18, 2	$[Xe]\ 6s^2 5d^{10} 4f^{14}$ = $[Xe]\ 4f^{14} 5d^{10} 6s^2$ = 2, 8, 18, 32, (8 + 10), 2 = 2, 8, 18, 32, 18, 2

The observed configurations, as given in Table 4.5, show that these elements have six shells, since these elements are located in 6*th* period of the periodic table. The distribution of electrons in these six shells can be shown as :

Shells :	*1st*	*2nd*	*3rd*	*4th*	*5th*	*6th*
Electrons in various shells	2 $1s^2$	8 $2s^2p^6$	18 $3s^2p^6d^{10}$	32 $4s^2p^5d^{10}f^{14}$	9 to 18 $5s^2p^6d^{1-10}$	1 or 2 $6s^{1-2}$
	← Completely-filled (four shells) →				Partially-filled ← (two shells) →	

The distribution of electrons, shown above, makes the following points evident :

(*i*) The configuration of the atoms of these nine elements can be represented as 2, 8, 18, 32, 9 to 18, 1 to 2 or 2, 8, 18, 32, $5s^2p^6\ d^{1-10}$, $6s^{1-2}$ or [Xe]$4f^{14}5d^{1-10}\ 6s^{1-2}$. It is obvious that the last two shells (*i.e.*, *6th* and *5th* shells) are partially-filled, since they contain 1 to 2 and 9 to 18 electrons respectively. The first four shells are completely-filled. On this basis, these nine elements are *transition elements.*

(*ii*) Last shell (*i.e.* *6th* shell) contains one electron in Pt_{78} and Au_{79}, and *two* electrons in the remaining seven elements.

(*iii*) The penutimate shell (*i.e.* *5th* shell) has 9 to 18 electrons ($5s^2p^6d^{1-10}$ configuration), *i.e.* it is 5*d* sub-shell which is progressively-filled up with elelctrons as we move from Hf_{72} to Hg_{80}. It is due to the successive filling of 5*d* sub-shell that these elements are called 5*d series elements* or 3*rd transition series elements.* Since 5*d* sub-shell is (*n*–1) *d* sub-shell, these nine elements, according to the definition of *d*-block elements, are called *d*-block elements.

(*iv*) All the nine elements have 2, 8, 18, and 32 electrons in their *1st*, *2nd*, *3rd* and *4th* shells respectively. Hence these shells are completely filled. Here it may be noted that *4th* shell of La_{57} contains only 18 electron ($4s^2p^6d^{10}f^0$) while the nine elements *viz* Hf_{72} to Hg_{80} have 32 electrons in their *4th* shell ($4s^2p^6d^{10}f^{14}$). The presence of 14 extra electrons in due to the lanclusion of 14 ianthanides (Ce_{58} to Lu_{71}) between La_{57} and Hf_{72}.

(*v*) The valence-shell configuration of the atoms of these elements can be written as $4f^{14}5d^{1-10}6s^{1-2}$.

Configurtion of the six elements viz., Tl_{81} to Rn_{86} (p-block elements). In the next six elements namely Tl_{81} to Rn_{86}, 6*p* orbitals are progressively filled with one electron at each step in these elements, as we move from Tl_{81} to Rn_{86}. Thus 6*p* orbitals have one electron at Tl_{81} ($6p^1$ configuration), while these orbitals are completely-filled at Rn_{86} ($6p^6$ configuration). Consequently the configuration of Tl_{81} and Rn_{86} are [Xe] $6s^24f^{14}5d^{10}6p^1$ or 2, 8, 18, 32, 18, 3 and [Xe] $6s^24f^{14}5d^{10}6p^6$ or 2, 8, 18, 32, 18, 8 respectively. The configuration of Rn_{86} is also represented as $[Rn]_{86}$ or [Rn]. The configurations of the above six elements can thus be written as shwon below :

p-block elements (↑↓)

Thallium, Tl_{81} = [Xe] $6s^2\ 4f^{14}5d^{10}6p^1$ = [Xe] $4f^{14}5d^{10}6s^26p^1$
= 2, 8, 18, (18 + 14), (8 + 10), 3 = 2, 8, 18, 32, 18, 3

Lead, Pb_{82} = [Xe] $6s^2\ 4f^{14}5d^{10}6p^2$ = [Xe] $4f^{14}5d^{10}6s^26p^2$
= 2, 8, 18, (18 + 14), (8 + 10), 4 = 2, 8, 18, 32, 18, 4

Bismuth, Bi_{83} = [Xe] $6s^2\ 4f^{14}5d^{10}6p^3$ = [Xe] $4f^{14}5d^{10}6s^26p^3$
= 2, 8, 18, (18 + 14), (8 + 10), 5 = 2, 8, 18, 32, 18, 5

Polomium, Po_{84} = [Xe] $6s^2\ 4f^{14}5d^{10}6p^4$ = [Xe] $4f^{14}5d^{10}6s^26p^4$
= 2, 8, 18, (18 + 14), (8 + 10), 6 = 2, 8, 18, 32, 18, 6

Astatine, At_{85} = [Xe] $6s^2\ 4f^{14}5d^{10}6p^5$ = [Xe] $4f^{14}5d^{10}6s^26p^5$
= 2, 8, 18, (18 + 14), (8 + 10), 7 = 2, 8, 18, 32, 18, 7

Radon Rn_{86} = [Xe] $6s^2\ 4f^{14}5d^{10}6p^6$ = [Xe] $4f^{14}5d^{10}6s^26p^6$
= 2, 8, 18 (18 +14), (8 +10), 8 = 2, 8, 18, 32, 18, 8 = $[Rn]_{86}$ or [Rn]

Since, in the above six elements, the additional electron enters *np* orbitals. (Here these are 6*p* orbitals), these elements are called *p*-block elements. Obviously the valence-shell configuration of these elements can be written as $6s^24f^{14}5d^{10}6p^{1-6}$. Among these six elements Rn_{86} is a noble gas and its last shell is completely-filled, since it contians 8 electrons (saturated shell).

The electronic configurations of the atoms of two *s*-block elements (*viz,* Cs_{55} and Ba_{56}) and *five* *p*-block elements (*viz,* Tl_{81}, Pb_{82}, Bi_{83}, Po_{84} and At_{85}) indicate that the last shell of these seven elements is incompletely-filled with electrons, since this shell (6th shell) contains 1 (in Cs_{55}), 2 (Ba_{56}), 3 (Tl_{81}), 4(Pb_{82}), 5(Bi_{83}), 6(Po_{84}) and 7(At_{85}) electrons, Thus, these seven elements are *representative* (or *normal*) *elements.*

Elements of 7th period ($7s^{1-2}$, $6d^1$, $5f^{0-14}$, $6d^{2-4}$). This period is an incomplete period and at present, contains 20 elements viz, Fr_{87}, Ra_{88} (two *s*-block elements), Ac_{89} (*d*-block element), Th_{90} to Lu_{103} (fourteen *f*-block elements) and Ku_{104}, Ha_{105} and Unh_{106} (three *d*-block elements). Thus this peirod contains two *s*-block, four *d*-block and fourteen *f*-block elements.

We have seen above that the electronic configuration of Rn_{86} (which is a noble gas of 6*th* period) is $[Xe]4f^{14}5d^{10}6s^26p^6$ or 2, 8, 18, 32, 18, 8 or 2, 8, 18, 32, $5s^2p^6d^{10}f^0$, $6s^2p^6d^0$. This configuration is also represented as $[Rn]_{86}$ or simply as [Rn]. This configurtion shows that Rn_{86} contains six shells and its last shell has 8 electrons ($6s^2p^6d^0$—saturated shell).

Thus, we find that the last orbitals which are completely-filled in Rn_{86} are 6*p*. Consequently, according to aufbou principle, the orbitals which should be filled with electrons in 7*th* period are 7*s*, 5*f* and 6*d*.

Electronic configuration of Fr_{87} and Ra_{88} (s-block elements). After the completion of 6*p* orbitals at Rn_{86}, 87*th* electron of Fr_{87} enters the next higher shell viz 7*th* shell. This electron goes to 7*s* orbital and this orbital becomes complete at Ra_{88}. The configuration of these two elements are as depicted below :

Francium Fr_{87} = $[Rn]\ 7s^1$ = 2, 8, 18, 32, 18, 8, 1
Radium, Ra_{88} = $[Rn]7s^2$ = 2, 8, 18, 32, 18, 8, 2
} *s-block elements*

Since, in both the elements, the differentiating electron goes to *s*-orbital of the last shell (here it is 7*s* orbital), these elements are *s*-block elements. It may be seen that the valence-shell configuration of these elements can be written as $7s^1$ and $7s^2$ respectively.

Electronic configuration of Ac_{89} (d-block elements). After the completion of 7*s* orbital at Ra_{88}, the next electron (*i.e.,* 89*th* electron) of Ac_{89}, according to aufbau principle, must enter 5*f* orbitals to give $[Rn]_{86}7s^25f^16d^0$ configuration to Ac_{89}; but actually the extra electron of Ac_{89} occupies 6*d* orbitals (instead of 5*f* orbitals) to give $[Rn]\ 7s^25f^06d^1$ configuration to Ac_{89} atom in which 5*f* orbitals are vacant and 6*d* orbitals have one electron. The reason of why the extra electron goes to 5*d* orbitals instead of 5*f* orbitals has already been given under the discussion of the configuraton of La_{57}. Since in Ac_{89} the differentiating electron goes to $(n-1)$ *d*-orbitals (here these are 6*d* orbitals), it is a *d*-block element or an element of 6*d* transition series.

Electronic configuration of the fourteen elements viz Th_{90} to Lw_{103} (called *actinides or 5f block elements*). All these elements follow Ac_{89} in the periodic table and hence are called actinides. We have already said earlier that the configuration of Ac_{89} is $[Rn]5f^06d^17s^2$ in which 6*d* sub-shell is singly-filled ($6d^1$ configuration) and 5*f* sub-shell is *empty* ($5f^0$ configuration). This configuration shows that as we move from one element to the next one in 14 actinides, the extra electron should enter the vacant 5*f* sub-shell and 6*d* sub-shell should remain singly-filled ($6d^1$ configuration) in all the 14 actinides. Thus, the expected configurations of the atoms of actinides should contain one electron in 6*d* sub-shell and 5*f* sub-shell should be progressively-filled up with electrons as we move from Th_{90} to Lw_{103} in the series. Experiments have, however, shown that the actual (observed) configurations of many actinides are different from their expected configurations. These configurations, along with their expected configurations, are given in Table 4.6.

Table 4.6 : *Expected and observed (actual) electronic configuration of actinide atoms. The configuration of Ac_{89} in also given.*

Acitinide elements	Expected configuration	Observed configuration
Actinium, Ac_{89}	$[Rn]\ 7s^26d^05f^1 = [Rn]5f^16d^07s^2$ = 2, 8, 18, 32, (18 + 1), (8 + 0) 2 = 2, 8, 18, 32, 19, 8, 2	$[Rn]\ 7s^26d^15f^0 = [Rn]\ 5f^06d^17s^2$ = 2, 8, 18, 32, (18 + 0), (8 + 1), 2 = 2, 8, 18, 32, 18, 9, 2
Thorium, Th_{90}	$[Rn]\ 7s^26d^15f^1 = [Rn]5f^16d^17s^2$ = 2, 8, 18, 32, (18+1), (8+1), 2 = 2,8, 18, 32, 19, 9, 2	$[Rn]\ 7s^26d^25f^0 = [Rn]\ 5f^06d^27s^2$ = 2, 8, 18, 32, (18+0), (8+2), 2 = 2, 8, 18, 32, 18, 10, 2

Protactinium, Pa_{91}	[Rn] $7s^2 6d^1 5f^2$ = [Rn]$5f^2 6d^1 7s^2$ = 2, 8, 18, 32, (18+2), (8+1), 2 = 2, 8, 18, 32, 20, 9, 2	[Rn] $7s^2 6d^1 5f^2$ = [Rn] $5f^2 6d^1 7s^2$ = 2, 8, 18, 32, (18+2), (8+1), 2 = 2, 8, 18, 32, 20, 9, 2
Uranium, U_{92}	[Rn] $7s^2 6d^1 5f^3$ = [Rn]$5f^3 6d^1 7s^2$ = 2, 8, 18, 32, (18+3), (8+1), 2 = 2, 8, 18, 32, 21, 9, 2	[Rn] $7s^2 6d^1 5f^3$ = [Rn] $5f^3 6d^1 7s^2$ = 2, 8, 18, 32, (18+3), (8+1),2 = 2, 8, 18, 32, 21, 9, 2
Neptumium, Np_{93}	[Rn] $7s^2 6d^1 5f^4$ = [Rn]$5f^4 6d^1 7s^2$ = 2, 8, 18, 32, (18+4), (8+1), 2 = 2, 8, 18, 32, 22, 9, 2	[Rn] $7s^2 6d^1 5f^4$ = [Rn] $5f^4 6d^1 7s^2$ = 2, 8, 18, 32, (18+4), (8+1), 2 = 2, 8, 18, 32, 22, 9, 2
Plutonium, Pu_{94}	[Rn] $7s^2 6d^1 5f^5$ = [Rn]$5f^5 6d^1 7s^2$ = 2, 8, 18 32 (18+5), (8+1), 2 = 2, 8, 18, 32, 23, 9, 2	[Rn] $7s^2 6d^0 5f^6$ = [Rn] $5f^6 6d^0 7s^2$ = 2, 8, 18, 32, (18+6), (8+0), 2 = 2, 8, 18, 32, 24, 8, 2
Americium, Am_{95}	[Rn] $7s^2$ $6d^1$ $5f^6$ = [Rn] $5f^6$ $6d^1$ $7s^2$ = 2, 8, 18, 32, (18 + 6), (8 + 1), 2 = 2, 8, 18, 32, 24, 9, 2	[Rn] $7s^2$ $6d^0$ $5f^7$ = [Rn] $5f^7$ $6d^0$ $7s^2$ = 2, 8, 18, 32, (18 + 7), 8, 2 = 2, 8, 18, 32, 25, 8, 2
Curium, Cm_{96}	[Rn] $7s^2 6d^1 5f^7$ = [Rn]$5f^7 6d^1 7s^2$ = 2, 8, 18, 32, (18 + 7), (8+1), 2 = 2, 8, 18, 32, 25, 9, 2	[Rn] $7s^2 6d^1 5f^7$ = [Rn] $5f^7 6d^1 7s^2$ = 2, 8, 18, 32, (18 + 7), (8 +1), 2 = 2, 8, 18, 32, 25, 9, 2
Berkelium, Bk_{97}	[Rn] $7s^2 6d^1 5f^8$ = [Rn]$5f^8 6d^1 7s^2$ = 2, 8, 18, 32, (18+8), (8 +1), 2 = 2, 8, 18, 32, 26, 9, 2	[Rn] $7s^2 6d^0 5f^9$ = [Rn] $5f^9 6d^0 7s^2$ = 2, 8, 18, 32 (18 + 9), 8, 2 = 2, 8, 18, 32, 27, 8, 2
Califonium, Cf_{98}	[Rn] $7s^2 6d^1 5f^9$ = [Rn]$5f^9 6d^1 7s^2$ = 2, 8, 18, 32, (18+9), (8+1), 2 = 2, 8, 18, 32, 27, 9, 2	[Rn] $7s^2 6d^0 5f^{10}$ = [Rn] $5f^{10} 6d^0 7s^2$ = 2, 8, 18, 32, (18+10), 8, 2 = 2, 8, 18, 32, 28, 8, 2
Einstenium Es_{99}	[Rn] $7s^2 6d^1 5f^{10}$ = [Rn]$5f^{10} 6d^1 7s^2$ = 2, 8, 18, 32, (18+10), (8+1), 2 = 2, 8, 18, 32, 28, 9, 2	[Rn] $7s^2 6d^0 5f^{11}$ = [Rn] $5f^{11} 6d^0 7s^2$ = 2, 8, 18, 32, (18+11), 8, 2 = 2, 8, 18, 32, 29, 8, 2
Fermium, Fm_{100}	[Rn] $7s^2 6d^1 5f^{11}$ = [Rn]$5f^{11} 6d^1 7s^2$ = 2, 8, 18, 32, (18+11), (8+1), 2 = 2, 8, 18, 32, 29, 9, 2	[Rn] $7s^2 6d^0 5f^{12}$ = [Rn] $5f^{12} 6d^0 7s^2$ = 2, 8, 18, 32, (18+12), 8, 2 = 2, 8, 18, 32, 30, 8, 2
Mendelevium, Md_{101}	[Rn] $7s^2 6d^1 5f^{12}$ = [Rn]$5f^{12} 6d^1 7s^2$ = 2, 8, 18, 32, (18 +12), 9, 2 = 2, 8, 18, 32, 30, 9, 2	[Rn] $7s^2 6d^0 5f^{13}$ = [Rn] $5f^{13} 6d^0 7s^2$ = 2, 8, 18, 32, (18+13), 8, 2 = 2, 8, 18, 32, 31, 8, 2
Nobelium, No_{102}	[Rn] $7s^2 6d^1 5f^{13}$ = [Rn]$5f^{13} 6d^1 7s^2$ = 2, 8, 18, 32, (18+13), 9, 2 = 2, 8, 18, 32, 31, 9, 2	[Rn] $7s^2 6d^0 5f^{14}$ = [Rn] $5f^{14} 6d^0 7s^2$ = 2, 8, 18, 32, (18+14), 8, 2 = 2, 8, 18, 32, 32, 8, 2
Lawrencium Lw_{103} or Lr_{103}	[Rn] $7s^2 6d^1 5f^{14}$ = [Rn]$5f^{14} 6d^1 7s^2$ = 2, 8, 18, 32, (18+14), 9, 2 = 2, 8, 18, 32, 32, 9, 2	[Rn] $7s^2 6d^1 5f^{14}$ = [Rn] $5f^{14} 6d^1 7s^2$ = 2, 8, 18, 32, (18+14), 9, 2 = 2, 8, 18, 32, 32, 9, 2

These configurations show that these elements have seven shells, since they lie in 7*th* period of the periodic table. The distribution of electrons in these seven shells can be shown as :

Shells :	1*st*	2*nd*	3*rd*	4*th*	5*th*	6*th*	7th
Electrons in various shells	2 $1s^2$	8 $2s^2p^6$	18 $3s^2p^6d^{10}$	32 $4s^2p^6d^{10}f^{14}$	18 to 32 $5s^2p^6d^{10}f^{0-14}$	8 or 10 $6s^2p^6d^{0-2}$	2 $7s^2$
	⟵ Completely-filled (four shells) ⟶				⟵ Partially-filled ⟶ (three shells)		

The distribution of electrons, shown above, makes the following points evident.

(*i*) The configuration of the atoms of actinides can be written as 2, 8, 18, 32, 18 to 32, 8 to 10, 2 or 2, 8, 18, 32, $5s^2p^6d^{10}f^{0-14}$, $6s^2p^6d^{0-2}$, $7s^2$ or [Rn] $5f^{0-14}6d^{0-2}7s^2$. The last three shells (*i.e.*, 7*th*, 6*th* and 5*th* shells) are *partially-filled,* since they contain 2, 8 to 10 and 18 to 32 electrons respectively. First four shells are *completely-filled* (1*st* shell = $2 \times 1^2 = 2$, 2*nd* shell = $2 \times 2^2 = 8$, 3*rd* shell = $2 \times 3^2 = 18$, 4*th* shell = $2 \times 4^2 = 32$). Thus *actinides can be defined as those elements whose last three shells are partially-filled and the first four shells are completely-filled.*

(*ii*) Last shell (*i.e.* 7*th* shell) in each element contains two electrons ($7s^2$ configuration).

(*iii*) Penultimate shell (*i.e.* 6*th* shell) has 8 to 10 electrons ($6s^2p^6d^{0-2}$ configuration). On the basis of the number of electrons present in 6*d* sub-shell, these elements have been grouped into the following three catagories :

(*a*) Th_{90} (1*st* element of the series) has two electrons in 6*d* sub-shell ($6d^2$ configuration) and hence there are 10 electrons in 6*th* shell of this element. The presence of two electron in 6*d* sub-shell in due to the shifting of 5*f* electron to 5*d* sub-shell (shifting of an electron from a lower sub-shell to a higher sub-shell).

(*b*) Pa_{91} (2nd element), U_{92} (3*rd* element) Np_{93} (4*th* element), Cm_{96} (7*th* element) and Lw_{103} (14*th* element) have one electron in 6*d* sub-sehll ($6d^1$ configuration) and hence these elements contain 9 electrons in their 6*th* shell.

(*c*) The remaining elements have *no* electron 6*d* sub-shell ($6d^0$ configuration). It is due to the presence of *empty* 6*d* sub-shell that 6*th* shell of these elements contain 8 electrons. 6*d* sub-shell is empty because $6d^1$ electron is shifted to 5*f* sub-shell (shifting of an electron from a higher sub-shell to a lower sub-shell).

(*iv*) In Am_{95} (6*th* element of the series) and No_{102} (13*th* element), $6d^1$ electron is shifted to 5*f* sub-shell, since this type of shifting makes 5*f* sub-shell half-filled in Am_{95} and completely-filled in No_{102} and thus both these elements attain the configuratons (observed configurations) which are more stable than their expected configurations.

(*v*) In Cm_{96} (7*th* element), $6d^1$ electron is not shifted to $5f^7$ sub-shell because the shifting gives $[Rn]7s^2\ 6d^0\ 5f^8$ configuration to Cm_{96}. This configuration is not as stable as $[Rn]7s^26d^15f^7$.

(*vi*) In Lw_{103} (14*th* element) the shifting of $6d^1$ electron to 5*f* sub-shell is not possible, since 5*f* sub-shell is already filled to its maximum capacity in Lw_{103}.

(*vii*) The ante-penultimate shell (*i.e.* 5*th* shell) contains 18 to 32 electrons ($5s^2p^6d^{10}f^{0-14}$ configuration). Thus 5*s*, 5*p* and 5*d* sub-shells are completely-filled but 5*f* sub-shell is being progressively filled up with electrons as we proceed in the series from Th_{90} to Lw_{103}. It is due to the successive (progressive) filling of 5*f* sub-shell that these elements are called 5*f*-block elements. Thus *5f-block elements can be defined as those elements in which the last electron goes to 5f sub-shell.*

(*viii*) Valence-shell configuration of the atoms of 14 actinides can be written as $5f^{0-14}\ 6d^{0-2}\ 7s^2$.

Electronic configuration of Ku_{104}, Ha_{105} and Unh_{106} (d-block elements). We have said above that the configuration of Lw_{103} is $[Rn]5f^{14}6d^17s^2$ in which 7*s* and 5*f* sub-shells are completly-filled while 6*d* sub-shell contains only one electron. Next three electrons in the next three elements *viz.* Ku_{104}, Ha_{105} and Unh_{106} enter the singly-filled 6*d* sub-shell. Thus the configurations of these three elements can be written as :

Kurchatovium, Ku_{104} = $[Rn]\ 7s^26d^25f^{14}$ = $[Rn]5f^{14}6d^27s^2$
= 2, 8, 18, 32, (18 + 14), (8 + 2), 2 = 2, 8, 18, 32, 32, 10, 2

Halnium, Ha_{105} = $[Rn]\ 7s^26d^35f^{14}$ = $[Rn]5f^{14}6d^37s^2$
= 2, 8, 18, 32, (18 + 14), (8 + 3), 2 = 2, 8, 18, 32, 32, 11, 2

Un-nilhexium, Unh_{106} = $[Rn]\ 7s^26d^45f^{14}$ = $[Rn]5f^{14}6d^47s^2$
= 2, 8, 18, 32, (18+14), (8 + 4), 2 = 2, 8, 18, 32, 32, 12, 2

Note : Ku is also called by other names like *Rutherfordium* (Rf) and *un-nilquandiun* (Unq). Ha is also called *Neielsborium* (Ns) and *un-nilpentium* (Unp)

The configurations of Ac_{89}, Ku_{104}, Ha_{105} and Unh_{106}, as given above, show that the configurations of all these elements can be represented by a general expression 2, 8, 18, 32, $5s^2p^6d^{10}f^{0-14}$, $6s^2p^6d^{1-4}$, $7s^2$ or 2, 8, 18, 32 18 to 32, 9 to 12, 2. This configuration shows that :

(*a*) Last two shells (*i.e.* 7*th* and 6*th* shells) are partially-filled, since they contain 2 (7*th* shell) and 9 to 12 (6*th* shell) electrons. The distribution of electrons in 6*th* shell ($6s^2p^6d^{1-4}$) shows that 6*s* and 6*p* sub-shells are completely-filled but 6*d* sub-shell is filled up with electrons successively as we proceed from Ac_{89} to Unh_{106}. Thus since the extra electron goes to (*n*–1)*d* orbitals (here these are 6*d* orbitals), all the four elements are *d*-block elements or 6*d* transition series elements (4*th* transition series).

(*b*) Ac_{89} contains 18 electrons in its 5*th* shell ($5s^2p^6d^{10}f^0$) while Ku_{104}, Ha_{105} and Unh_{106} have 18 + 14 = 32 electrons in this shell ($5s^2p^6d^{10}f^{14}$). The presence of 14 extra electrons in 5*th* shells is due to the inclusion of 14 actinides (Th_{90} to Lw_{103}) between Ac_{89} and Ku_{104}.

(*c*) The valence-shell configuration for Ac_{89} is $5f^06d^17s^2$ while for Ku_{104}, Ha_{105} and Unh_{106} the valence shell configuration is $5f^{14}6d^27s^2$, $5f^{14}6d^37s^2$ and $5f^{14}6d^47s^2$ respectively.

The electronic configurations of all 106 elements discussed above have been given in Table 4.7.

To write the electronic configuration of the atom of a given element

Before learning the technique of writting the electronic configuration of the atom of an element, we should memorise the following points.

(*i*) When we proceed from left to right in a period of the peiodic table, the atomic number of the elements (*i.e.* the number of electrons in various shells of the atom) increases by one unit at each next element. At each next element one more electron is added to the appropriate orbital. The order in which various orbitals are successively filled with electrons in various periods is given below. This order is based on aufbau principle, which states that orbitals are filled up with electrons in the increasing order of their energy.

(*ii*) The electronic configurations of the atoms of the elements of zero group (noble gases) are :

He_2 (1*st* peiod) = 2 or $[He]_2$ (one shell)
Ne_{10} (2*nd* peirod) = 2, 8 or $[Ne]_{10}$ (Two shells)
Ar_{18} (3*rd* period) = 2, 8, 8 or $[Ar]_{18}$ (Three shells)
Kr_{36} (4*th* period) = 2, 8, 18, 8 or $[Kr]_{36}$ (Four shells)
Xe_{54} (5*th* period) = 2, 8, 18, 18, 8 or $[Xe]_{54}$ (Five shells)
Rn_{86} (6*th* period) = 2, 8, 18, 32, 18, 8 or $[Rn]_{86}$ (Six shells)

There are two methods which are used to write the electronic configuration of the atom of a given element.

Method 1. According to this method, the electronic configuration of the atom of a given element is obtained by adding the extra electrons to the configuration of noble gas belonging to the previous period. The addition of extra electrons to various orbitals is made according to aufbau principle. The number of extra electrons, which have to added to various orbitals, is equal to the atomic number of the element minus the atomic number of the noble gas lying in the previous period, *e.g.* in case of At_{85} (6*th* period), the number of extra electrons, which have to be added to the configuration of Xe_{54} (This noble gas belongs to 5*th* period) is 85 – 54 = 31. These 31 electrons have to be added to 6*s* 5*d*, 4*f* and 6*p* orbitals (aufbau principle). Thus, the electronic configuration of At_{85} (No. of electrons = 54 + 31 = 85) is :

$$[Xe]_{54} \text{ (54 electrons)} + 6s^2\ 5d^{10}\ 4f^{14}\ 6p^5 \text{ (31 electrons)}$$

Now, since the electronic configuration of Xe_{54} = 2, 8, 18, 18, 8 (*five shells*), the electronic configuration of At_{85} = 2, 8, 18, 18, 8 + $6s^25d^{10}4f^{14}6p^5$ = 2, 8, 18, (18 + $4f^{14}$), (8 + $5d^{10}$), ($6s^2$ + $6p^5$) = 2, 8, 18, 32, 18, 7 (six shells).

Method 2. This method is based on the fact that the electrons present in a given element are introduced into various orbitals in the order 1*s*, 2*s*, 2*p*, 3*s*, 3*p*, 4*s*, 3*d*, 4*p*, 5*s*, 4*d*, 5*p*, 6*s*, $5d^1$ (in La_{57}), $4f^{2-14}$ (in Ce_{58} to Lu_{71}), $5d^{2-10}$ (in Hf_{72} to Hg_{80}), 6*p* (*in* Tl_{81} to Rn_{86}), 7*s* (Fr_{87} Ra_{88}), $6d^1$ (Ac_{89}), $5f^{0-14}$ (Th_{90} to Lw_{103}), $6d^{2-3}$ (Kw_{104}, Ha_{105}). This order is called aufbau principle. This method can be illustrated by determining the electronic configuration of Pd_{46}, Ta_{73} and Pt_{78}.

Table 4.7 : *Electronic configurations of the atoms of elements.*

← *s*-Block Elements → (Representative Elements) | ← *d*- Block Elements (Transition Elements) →

Period	IA (1)	II A (2)	(3) III B	(4) IV B	(5) V B	(6) VI B	(7) VII B	(8)	(9) VIII
1	H$_1$ $1s^1$								
2	Li$_3$ [He] $2s^1$ 2, 1	Be$_4$ [He] $2s^2$ 2, 2							
3	Na$_{11}$ [Ne] $3s^1$ 2, 8, 1	Mg$_{12}$ [Ne] $3s^2$ 2, 8, 2							
4	K$_{19}$ [Ar] $4s^1$ 2, 8, 8, 1	Ca$_{20}$ [Ar] $4s^2$ 2, 8, 8, 2	Sc$_{21}$ [Ar] $3d^1\ 4s^2$ 2, 8, 9, 2	Ti$_{22}$ [Ar] $3d^2\ 4s^2$ 2, 8, 10, 2	V$_{23}$ [Ar] $3d^3\ 4s$ 2, 8, 11, 2	Cr$_{24}$ [Ar] $3d^5\ 5s^1$ 2, 8, 13, 1	Mn$_{25}$ [Ar] $3d^5\ 4s^2$ 2, 8, 13, 2	Fe$_{26}$ [Ar] $3d^6\ 4s^2$ 2, 8, 14, 2	Co$_{27}$ [Ar] $3d^7\ 4s^2$ 2, 8, 15, 2
5	Rb$_{37}$ [Kr] $5s^1$ 2, 8, 18, 8, 1	Sr$_{38}$ [Kr] $5s^2$ 2, 8, 18, 8, 2	Y$_{39}$ [Kr] $4d^1\ 5s^2$ 2, 8, 18, 9, 2	Zr$_{40}$ [Kr] $4d^2\ 5s^2$ 2, 8, 18, 10, 2	Nb$_{41}$ [Kr] $4d^4\ 5s^1$ 2, 8, 19, 12, 1	Mo$_{42}$ [Kr] $4d^5\ 5s^1$ 2, 8, 18, 13, 1	Tc$_{43}$ [Kr] $4d^5\ 5s^2$ 2, 8, 18, 13, 2	Ru$_{44}$ [Kr] $4d^7\ 5s^1$ 2, 8, 18, 15, 1	Rh$_{45}$ [Kr] $4d^8\ 5s^1$ 2, 8, 18, 16, 1
6	Cs$_{55}$ [Xe]$6s^1$ 2, 8, 18, 18, 8, 1	Ba$_{56}$ [Xe] $6s^2$ 2, 8, 18, 18, 8, 2	La$_{57}$ [Xe] $4f^0$ $5d^1\ 6s^2$ 2, 8, 18, 18, 9, 2	Hf$_{72}$ [Xe] $4f^{14}$ $5d^2\ 6s^2$ 2, 8, 18, 32, 10, 2	Ta$_{73}$ [Xe] $4f^{14}$ $5d^3\ 6s^2$ 2, 8, 18, 32, 11, 2	W$_{74}$ [Xe] $4f^{14}$ $5d^4\ 6s^2$ 2, 8, 18, 32, 12, 2	Re$_{75}$ [Xe] $4f^{14}$ $5d^5\ 6s^2$ 2, 8,18, 32, 13, 2	Os$_{76}$ [Xe] $4f^{14}$ $5d^6\ 6s^2$ 2, 8, 18, 32, 14, 2	Ir$_{77}$ [Xe] $4f^{14}$ $5d^7\ 6s^2$ 2, 8, 18, 32, 15, 2
7	Fr$_{87}$ [Rn] $7s^1$ 2, 8, 18, 32, 18, 8, 1	Ra$_{88}$ [Rn] $7s^2$ 2, 8, 18, 32, 18, 8, 2	Ac$_{89}$ [Rn] $5f^0$ $6d^1\ 7s^2$ 2, 8, 18, 32, 18, 9, 2	Ku$_{104}$ [Rn] $5f^{14}$ $6d^2\ 7s^2$ 2, 8, 18, 32, 32, 10, 2	Ha$_{105}$ [Rn] $5f^{14}$ $6d^3\ 7s^2$ 2, 8, 18, 32, 32, 11, 2	Unh$_{106}$ [Rn] $5f^{14}$ $6d^4\ 7s^2$ 2, 8, 18, 32, 32, 12, 2			

← *f*- Block Elements (Inner-transition Elements) →

	Ce$_{58}$ [Xe] $4f^2$ $5d^0\ 6s^2$ 2, 8, 18, 20, 8, 2	Pr$_{59}$ [Xe] $4f^3$ $5d^0\ 6s^2$ 2,8, 18, 21, 8, 2	Nd$_{60}$ [Xe] $4f^4$ $5d^0\ 6s^2$ 2, 8, 18, 22, 8,2	Pm$_{61}$ [Xe] $4f^5$ $5d^0\ 6s^2$ 2, 8, 18, 23, 8,2	Sm$_{62}$ [Xe] $4f^6$ $5d^0\ 6s^2$ 2, 8, 18, 24, 8,2	Eu$_{63}$ [Xe] $4f^7$ $5d^0\ 6s^2$ 2, 8, 18, 25, 8, 2
Actinides	Th$_{90}$ [Rn] $5f^0$ $6d^2\ 7s^2$ 2, 8, 18, 32, 18, 10, 2	Pa$_{91}$ [Rn] $5f^2$ $6d^1\ 7s^2$ 2, 8, 18, 32, 20, 9, 2	U$_{92}$ [Rn] $5f^3$ $6d^1\ 7s^2$ 2, 8, 18, 32, 21, 9, 2	Np$_{93}$ [Rn] $5f^4$ $6d^1\ 7s^2$ 2, 8, 18, 32, 22, 9, 2	Po$_{94}$ [Rn] $5f^6$ $6d^0\ 7s^2$ 2, 8, 18, 32, 24, 8, 2	Am$_{95}$ [Rn] $5f^7$ $6d^0\ 7s^2$ 2, 8, 18, 32, 25, 8, 2

s-and p-block elements are collectively called main group elements

← *p*– Block Elements (Representative Elements) →

(10)	(11)	(12)	(13)	(14)	(15)	(16)	(17)	Noble gases VIII A or zero (18)
			(13) III A	(14) IV A	(15) V A	(16) VI A	(17) VII A	He_2 $1s^2$ 2 or [He]
→			B_5 [He] $2s^2\ 2p^1$ 2, 3	C_6 [He] $2s^2\ 2p^2$ 2, 4	N_7 [He] $2s^2\ 2p^3$ 2, 5	O_8 [He] $2s^2\ 2p^4$ 2, 6	F_9 [He] $2s^2\ 2p^5$ 2, 7	Ne_{10} [He] $2s^2\ 2p^6$ or [Ne] 2, 8
(10) →	(11) IB	(12) II B	Al_{13} [Ne] $3s^2\ 3p^1$ 2, 8, 3	Si_{14} [Ne] $3s^2\ 3p^2$ 2, 8, 4	P_{15} [Ne] $3s^2\ 3p^3$ 2, 8, 5	S_{16} [Ne] $3s^2\ 3p^4$ 2, 8, 6	Cl_{17} [Ne] $3s^2\ 3p^5$ 2, 8, 7	Ar_{18} [Ne] $3s^2$ $3p^6$ or [Ar] 2, 8, 8
Ni_{28} [Ar] $3d^8\ 4s^2$ 2, 8, 16, 2	Cu_{29} [Ar] $3d^{10}\ 4s^1$ 2, 8, 18, 1	Zn_{30} [Ar] $3d^{10}\ 4s^2$ 2, 8, 18, 2	Ga_{31} [Ar] $3d^{10}$ $4s^2\ 4p^1$ 2, 8, 18, 3	Ge_{32} [Ar] $3d^{10}$ $4s^2\ 4p^2$ 2, 8, 18, 4	As_{33} [Ar] $3d^{10}$ $4s^2\ 4p^3$ 2, 8, 18, 5	Se_{34} [Ar] $3d^{10}$ $4s^2\ 4p^4$ 2, 8, 18, 6	Br_{35} [Ar] $3d^{10}$ $4s^2\ 4p^5$ 2, 8, 18, 7	Kr_{36} [Ar] $3d^{10}\ 4s^2$ $4p^6$ or [Kr] 2, 8, 18, 8
Pd_{46} [Kr] $4d^{10}\ 5s^0$ 2, 8, 18, 18, 0	Ag_{47} [Kr] $4d^{10}\ 5s^1$ 2, 8, 18, 18, 1	Cd_{48} [Kr] $4d^{10}\ 5s^2$ 2, 8, 18, 18, 2	In_{49} [Kr] $4d^{10}$ $5s^2\ 5p^1$ 2, 8, 18, 18, 3	Sn_{50} [Kr] $4d^{10}$ $5s^2\ 5p^2$ 2, 8, 18, 18, 4	Sb_{51} [Kr] $4d^{10}$ $5s^2\ 5p^3$ 2, 8, 18, 18, 5	Te_{52} [Kr] $4d^{10}$ $5s^2\ 5p^4$ 2, 8, 18, 18, 6	I_{53} [Kr] $4d^{10}$ $5s^2\ 5p^5$ 2, 8, 18, 18, 7	Xe_{54} [Kr] $4d^{10}\ 5s^2$ $5p^6$ or [Xe] 2, 8, 18, 18, 8
Pt_{78} [Xe] $4f^{14}$ $5d^9\ 6s^1$ 2, 8, 18, 32, 17, 1	Au_{79} [Xe] $4f^{14}$ $5d^{10}\ 6s^1$ 2, 8, 18, 32, 18, 1	Hg_{80} [Xe] $4f^{14}$ $5d^{10}\ 6s^2$ 2, 8, 18, 32, 18, 2	Tl_{81} [Xe] $4f^{14}$ $5d^{10}\ 6s^2\ 6p^1$ 2, 8, 18, 32, 18, 3	Pb_{82} [Xe] $4f^{14}$ $5d^{10}\ 6s^2\ 6p^2$ 2, 8, 18, 32, 18, 4	Bi_{83} [Xe] $4f^{14}$ $5d^{10}\ 6s^2\ 6p^3$ 2, 8, 18, 32, 18, 5	Po_{84} [Xe] $4f^{14}$ $5d^{10}\ 6s^2\ 6p^4$ 2, 8,18, 32, 18, 6	At_{85} [Xe] $4f^{14}$ $5d^{10}\ 6s^2\ 6p^5$ 2, 8, 18, 32, 18, 7	Rn_{86} [Xe] $4f^{14}\ 5d^{10}\ 6s^2$ $6p^6$ or [Rn] 2, 8, 18, 32, 18, 8

→

Gd_{64} [Xe] $4f^7\ 5d^1$ $6s^2$ 2, 8, 18, 25, 9, 2	Tb_{65} [Xe] $4f^9\ 5d^0$ $6s^2$ 2, 8, 18, 27, 8, 2	Dy_{66} [Xe] $4f^{10}$ $5d^0\ 6s^2$ 2, 8, 18, 28, 8, 2	Ho_{67} [Xe] $4f^{11}$ $5d^0\ 6s^2$ 2, 8, 18, 29, 8, 2	Er_{68} [Xe] $4f^{12}$ $5d^0\ 6s^2$ 2, 8, 18, 30, 8, 2	Tm_{69} [Xe] $4f^{13}$ $5d^0\ 6s^2$ 2, 8, 18, 31, 8,2	Yb_{70} [Xe] $4f^{14}$ $5d^0\ 6s^2$ 2, 8, 18, 32, 8, 2	Lu_{71} [Xe] $4f^{14}$ $5d^1\ 6s^2$ 2, 8, 18, 32, 9, 2
Cm_{96} [Rn] $5f^7\ 6d^1$ $7s^2$ 2, 8, 18, 32, 25, 9, 2	Bk_{97} [Rn] $5f^9\ 6d^0$ $7s^2$ 2, 8, 18, 32, 27, 8, 2	Cf_{98} [Rn] $5f^{10}$ $6d^0\ 7s^2$ 2, 8, 18, 32, 28, 8, 2	Es_{99} [Rn] $5f^{11}$ $6d^0\ 7s^2$ 2, 8, 18, 32, 29, 8, 2	Fm_{100} [Rn] $5f^{12}$ $6d^0\ 7s^2$ 2, 8, 18, 32, 30, 8, 2	Md_{101} [Rn] $5f^{13}$ $6d^0\ 7s^2$ 2, 8, 18, 32, 31, 8, 2	No_{102} [Rn] $5f^{14}$ $6d^0\ 7s^2$ 2, 8, 18, 32, 32, 8, 2	Lw_{103} [Rn] $5f^{14}$ $6d^1\ 7s^2$ 2, 8, 18, 32, 32, 9, 2

Pd_{46}. This element has 46 electrons. Thus, the electronic configuration of this element should be : $1s^2$, $2s^2$, $2p^6$, $3s^2$, $3p^6$, $4s^2$, $3d^{10}$, $4p^6$, $5s^2$, $4d^8$. By spectroscopic studies, it has been shown that both the electrons from 5*s* orbital get shifted to 4*d* orbitals so that 5*s* orbital becomes vacant ($5s^0$ configuration) and 4*d* orbitals become completely-filled ($4d^{10}$ configuration). Thus the actual configuration of Pd_{46} is : $1s^2$, $2s^2$, $2p^6$, $3s^2$, $3p^6$, $4s^2$, $3d^{10}$, $4p^6$, $5s^0$, $4d^{10}$ = $1s^2$, $2s^2p^6$, $3s^2p^6d^{10}$, $4s^2p^6d^{10}$,$5s^0$ = 2, 8, 18, 18, 0.

Ta_{73}. This element has 73 electrons. Thus, the electronic configuration of this elements, is : $1s^2$, $2s^2$, $2p^6$, $3s^2$, $3p^6$, $4s^2$, $3d^{10}$, $4p^6$, $5s^2$, $4d^{10}$, $5p^6$, $6s^2$, $5d^1$, $4f^{14}$, $5d^2$ = $1s^2$, $2s^2p^6$, $3s^2p^6d^{10}$, $4s^2p^6d^{10}$,f^{14} $5s^2p^6d^3$, $6s^2$ = 2, 8, 18, 32, 11, 2.

Pt_{78}. This element has 78 electrons and hence its electronic configuration should be : $1s^2$, $2s^2$, $2p^6$, $3s^2$, $3p^6$, $4s^2$, $3d^{10}$, $4p^6$, $5s^2$, $4d^{10}$, $5p^6$, $6s^2$, $5d^1$, $4f^{14}$, $5d^7$ = $1s^2$, $2s^2p^6$, $3s^2p^6d^{10}$, $4s^2p^6d^{10}$, f^{14}, $5s^2$ p^6d^8, $6s^2$ = 2, 8, 18, 32, 16, 2.

It has ben found experimentally that one electron of 6*s* orbital gets shifted to 5*d* orbitals and hence 6*s* orbital becomes singly-filled and 5*d* orbitals contain 9 electrons ($5d^9$ configuration) instead of 8 ($5d^8$ configuration). Thus the configuration of Pt_{78} is : 2, 8, 18, 32, $5s^2p^6d^9$, $6s^1$ = 2, 8, 18, 32, 17, 1.

To write the valence-shell configuration of a given element

The technique used for writing the valence-shell configuration consists of the following setps :

(*a*) Find out the difference between the atomic number of the given element and that of the noble gas present in the previous period. For example the difference between Pb_{82} (6*th* period) and Xe_{54} (5*th* period) = 82 – 54 = 28. Let this difference be denoted by *x*. This difference denotes the number of electrons present in the valence-shell of the given element.

(*b*) Now find out the sub-shells which are to be filled up with *x* electrons.

(*c*) Then fill up these sub-shells with *x* electrons. This gives the valence-shell configuration of the given element.

Example 4.4. *Find out the valence-shell configuration of the atom of the following elements : Te_{52}, La_{57}, Pr_{59}, Pt_{78}, Pb_{82}, Ac_{89}, U_{92} and Ha_{105}.*

Solution : (*i*) **Te_{52}** : This element is present in 5*th* period and $x = Te_{52} - Kr_{36} = 16$.

Hence valence-shell configuration of $Te_{52} = 5s^2\ 4d^{10}\ 5p^4 = 4d^{10}\ 5s^2\ 5p^4$

(*ii*) **La_{57}** : Here $x = La_{57} - Xe_{54} = 3$

Hence valence-shell configuration of Li_{57} (6*th* period) = $6s^25d^1 = 5d^16s^2$

(*iii*) **Pr_{59}** : Here $x = Pr_{59} - Xe_{54} = 5$

∴ Expected valence-shell configuration of Pr_{59} (6*th* period) = $6s^25d^14f^2$

The actual valence-shell configuration is $6s^25d^0\ 4f^3 = 4f^35d^06s^2$

(*iv*) **Pt_{78}** : Here $x = Pt_{78} - Xe_{54} = 24$

Expected valence-shell configuration of Pt_{78} (6*th* period) = $6s^25d^14f^{14}5d^7 = 6s^25d^84f^{14}$

But the actual valence-shell configuration = $6s^15d^94f^{14} = 4f^{14}5d^96s^1$

(*v*) **Pb_{82}** : Here $x = Pb_{82} - Xe_{54} = 28$

∴ Valence-shell configuration = $6s^25d^14f^{14}5d^96p^2 = 6s^25d^{10}4f^{14}6p^2 = 4f^{14}5d^{10}6s^26p^2$

(*vi*) **Ac_{89}** : Here $x = Ac_{89} - Rn_{86} = 3$

∴ Valence-shell configuration of Ac_{89} (7*th* period) = $7s^26d^1 = 6d^17s^2$

(*vii*) **U_{92}** : Here $x = U_{92} - Rn_{86} = 6$

∴ Valence-shell configuration of $U_{92} = 7s^26d^15f^3 = 5f^36d^17s^2$

(*viii*) **Ha_{105}** : Here $x = U_{105} - R_{86} = 11$

∴ Valence-shell configuration of $Ha_{105} = 7s^26d^15f^{14}6d^2 = 7s^26d^35f^{14} = 5f^{14}6d^37s^2$

To write the electronic configuration of an ion

In order to write the electronic configuration of a given ion proceed as follows :

1. Write the complete or valence-shell configuration of the element from which the ion has been derived.

2. (*a*) In case of a cation, remove the electron (*s*), equal to the number of positive charges on the cation from the outer-most shell of the parental atom. This gives the electronic configuration of the cation.

(*b*) In case of an anion, add electron (*s*),equal to the number of negative charges on the anion to the outer-must shell of the parental atom. This gives the configuration of the anion.

Example 4.5. *Write the complete and valence-shell configuration of Pb^{2+} ion (Z = 82)*

Solution : Configuration of Pb-atom (Z = 82) = $[Xe]_{54}\ 6s^2 5d^{10} 4f^{14} 6p^2 = [Xe]_{54}\ 4f^{14} 5d^{10} 6s^2 6p^2$

$\therefore$ Configuration of Pb^{2+} ion = $[Xe]_{54}\ 4f^{14} 5d^{10} 6s^2$

Valence-shell configuration of Pb-atom = $4f^{14} 5d^{10} 6s^2 6p^2$

$\therefore$ Valence-shell configuration of pb^{2+}ion = $4f^{14} 5d^{10} 6s^2$

Example 4.6. *Write down complete and valence-shell configuration of O^{2-} ion (Z = 8)*

Solution : Configuration of 0-atom (*Z* = 8) = $[He]_2\ 2s^2 2p^4$

$\therefore$ Configuration of O^{2-} ion = $[He]_2 2s^2\ 2p^6$

Valence-shell configuration of O-atom (Z = 8) = $2s^2 2p^4$

$\therefore$ Valence-shell configuration of O^{2-} ion = $2s^2 2p^6$

Classification of Elements

Elements present in the periodic table, have been classified in a number of ways. Some classfications have been shown below :

Classification of elements in the periodic table

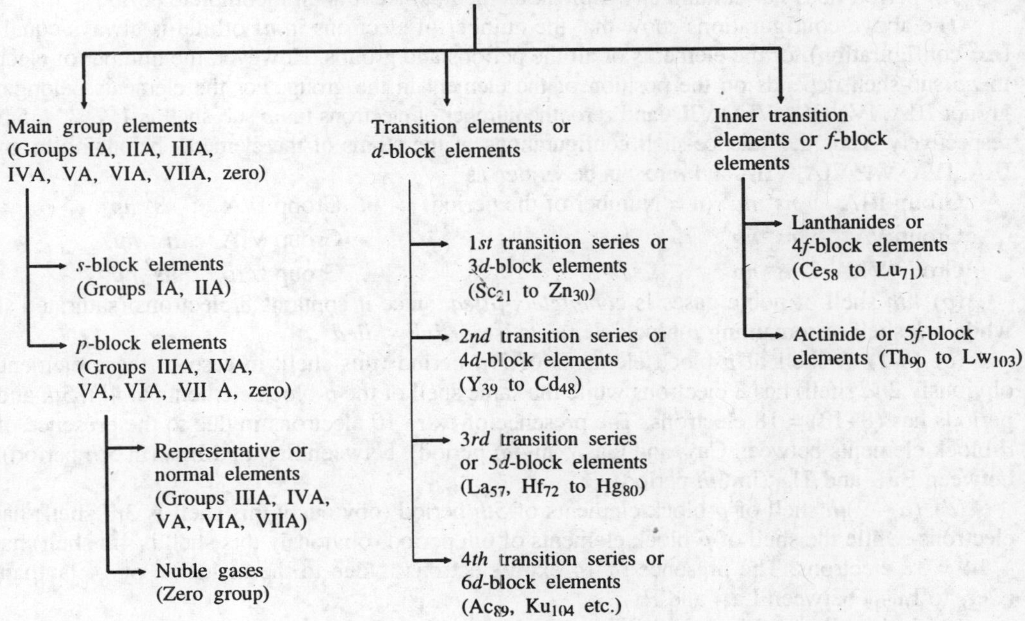

Classification of the elements into four blocks.

The elements have been classified, on the basis of the nature of the sub-shell into which the differentiating electron (last electron) enters, as we proceed from one element to the next in a given period. On this basis, the elements have been grouped into the following four blocks.

I. *s*-block elements. An inspection of the electronic configurations of the atoms of the elements of groups IA and IIA shows that the last electron in these elements enters *ns* orbital (*s*-orbital) of the last shell. Here *n* also denotes the number of the period in which the element is present and hence these elements are called *s*-block elements. Thus *s-block elements are defined as those* *elements in which the last electron goes to ns orbital* or *in which ns orbital is being filled up successively with electrons. s*-blcok elements are located at the extreme left of the periodic table (groups IA and IIA) and consist of metals (highly electro-positive elements). The following points may be noted.

(*a*) The valence-shell configuration of *s*-block elements can be written as ns^1 (for the elements of group IA) or ns^2 (for the elements of group IIA).

(*b*) In *s*-block elements, the outer-most shell (*i.e. nth* shell or last shell) is partially filled.

II. p-block elements. If the electronic configurations of the atoms of the elements belonging to groups IIIA, IVA, VA, VI A, VII A and zero are examined, it would be found that the last electron in these elements enters *np* sub-shell and hence these elements are called *p*-block elements. Thus *p*-block elements are defined as those elements in which the last electron enters *np* sub-shell (*p*-sub-shell of the last shell) or in which *np* sub-shell is progressively filled with electrons. These elements are located at the extreme right of the periodic table and consist of some metals, all non-metals, all metalloids and noble gases. The following points may be noted :

(*a*) The valence-shell configuration of the atoms of the elements of different periods can be written as :

2nd period : $2s^2\,2p^{1-6}$; *3rd* period : $3s^2\,3p^{1-6}$

4th period : $4s^2\,3d^{10}\,4p^{1-6}$ or $3d^{10}\,4s^2\,4p^{1-6}$

5th period : $5s^2\,4d^{10}\,5p^{1-6}$ or $4d^{10}\,5s^2\,5p^{1-6}$

6th period : $6s^2\,4f^{14}\,5d^{10}\,6p^{1-6}$ or $4f^{14}\,5d^{10}\,6s^2\,6p^{1-6}$

7th period does not contain any *p*-block element, since it is an incomplete period.

The above configurations show that the number of electrons in *ns* orbital is always equal to 2 (ns^2 configuration) for the elements of all the periods and groups. However, the number of electrons in *np* sub-shell depends on the position of the element in the group. For the elements belonging to groups IIIA, IVA, VA, VIA, VIIA and zero, the number of electrons in *np* sub-shell is 1, 2, 3, 4, 5 and 6 respectively. Thus the valence-shell configurations of the atoms of the elements belonging to groups IIIA, IVA, VA, VIA, VIIA and zero can be written as :

Group IIIA : $ns^2\,np^1$ (*n* = Number of the period) Group IVA : $ns^2\,np^2$

Group VA : $ns^2\,np^3$ Group VIA : $ns^2\,np^4$

Group VIIA : $ns^2\,np^5$ Group zero : $ns^2\,np^6$

(*b*) *nth* shell of noble gases is *completely-filled*, since it contains 8 electrons (saturated shell) while *nth* shell of remaining *p*-block elements is *partially-filled.*

(*c*) (*n*-1) *th* shell of *p*-block elements of *3rd* period (this shell, in case of these elements, is obviously *2nd* shell) has 8 electrons while the same shell of the *p*-block elements of *4th*, *5th* and *6th* periods has (8+10) = 18 electrons. The presence of extra 10 electrons in due to the presence of ten *d*-block elements between Ca_{20} and Ga_{31} (in *4th* period), between Sr_{38} and In_{49} (in *5th* period) and between Ba_{56} and Tl_{81} (in *6th* period).

(*d*) (*n* – 2)*th* shell of *p*-block elements of *5th* period (obviously this shell is *3rd* shell) has 18 electrons, while the shell of *p*-block elements of *6th* period (obviously this shell is *4th* shell) has 18 + 14 = 32 electrons. The presence of 14 extra electrons is due to the inclusion of 14 lanthanides (Ce_{58} to Lu_{71}) between La_{57} and Hf_{72}.

(*e*) *2nd* shell of *p*-block elements of *2nd* period has 3 electrons, while the same shell of the elements of other periods has 8 electrons.

(*f*) 1*st* shell of all the *p*-block elements has 2 electrons.

III. *d*-block elements or transition elements. The elements, namely Sc_{21} to Zn_{30} (10 elements of *4th* period), Y_{39} to Cd_{48} (10 elements of *5th* period), La_{57}, Hf_{72} to Hg_{80} (10 elements of *6th* period), Ac_{89}, Ku_{104}, Ha_{105}, Unh_{106} (4 elements of 7*th* period which is an incomplete period) are called *d*-block elements or transition elements.

From the electronic configurations of the atoms of these elements, the following points may be noted :

(*i*) In the atoms of these elements, the last (differentiating) electron enters (*n* – 1) *d* orbitals (*i.e.*

d-orbitals of the penultimate shell). Thus *d-block elements can be defined as those elements in which the last electron enters (n – 1)d orbitals or in which (n – 1)d orbitals are progressively filled up with electrons.*

(*ii*) *d*-block elements are located between *s*-and *p*-block elements, *i.e.* these elements are situated in the middle of the periodic table. All these elements are metals.

(*iii*) *d*-block elements are present in *4th* (Sc_{21} to Zn_{30}), *5th* (Y_{39} to Cd_{48}), *6th* (La_{57}, Hf_{72} to Hg_{80}) and *7th* (Ac_{89}, Ku_{104} to Unh_{106}) periods of the periodic table. *d*-block elements of periods *4th, 5th, 6th* and *7th* are called *3d*-series, *4d*-series, *5d*-series and *6d*-series elements respectively.

(*iv*) The last two shells of these elements are incomplete (*i.e.* partially-filled). *Thus d-block elements can also be defined as those elements whose two outer-most shells are partially-filled.* The last shell (*nth* shell) of Pd_{48} has *no* electron (ns^0 configuration), of Cr_{24}, Cu_{29}, Nb_{41}, Mo_{42}, Ru_{44}, Rh_{45}, Ag_{47}, Pt_{78} and Au_{79} has *one* electron (ns^1 configuration) and of the remaining elements has *two* electrons (ns^2 configuration).

The penultimate shell [*i.e.* (*n* – 1)*th* shell] of 3*d*, 4*d* and 5*d* series elements has 9 to 18 electrons [$(n-1)s^2p^6d^{1\text{-}10}$ configuration] and this shell of 6*d* series elements has 9 to 12 electrons [$(n-1)s^2p^6d^{1\text{-}4}$ configuration]

(*v*) The valence-shell configuration of the elements of *3d, 4d, 5d* and *6d* series can be written as:

3d series elements : $3d^{1\text{-}10}\,4s^{1,2}$ *4d series elements :* $4d^{1\text{-}10}5s^{0,1,2}$

5d series elements : $4f^{0,14}5d^{1\text{-}10}6s^{1,2}$ *6d series elements :* $5f^{0,14}\,6d^{1\text{-}4}\,7s^2$

(*vi*) Although Cu_{29}, Ag_{47} and Au_{79} have completely-filled (*n* – 1) *d* orbitals, yet these elements are considered as transition elements, because these elements show many properties which are similar to those of *d*-block elements, *e.g.* (*a*) Cu^{2+} and Au^{3+} have partially-filled (*n* – 1) *d* orbitals (Cu^{2+}= $3d^9$, Au^{3+}= $5d^8$) (*b*) Cu^+, Cu^{2+}, Ag^+, Au^+ and Au^3 form many complex compounds (*c*) Cu and Au show variable oxidation states etc.

(*vii*) According to the defination of *d*-block elements, Zn_{30}, Cd_{48} and Hg_{80} should not be placed along with *d*-block elements because (*n* – 1) *d* orbitals in the atoms of these elements and in M^{2+} cations (Zn Cd and Hg show only +2 oxidation state in their compounds *i.e.* these elements form only Zn^{2+}, Cd^{2+} and Hg^{2+} ions) are completely-filled. Thus some chemists regard these elements as *non-transition elements.* The non-transition nature of these elements is also confirmed by their many properties which are different from those of transition elements, *e.g.* these elements are diamagnetic, donot show variable oxidation states, most of their compounds are white, M^{2+} ions derived from these elements have completely-filled (*n* – 1) *d* orbitals etc. Like transition elements these elements in +2 oxidation state form complex compounds and it is due to this property that these elements are regarded as transition elements. In forming complex compounds, empty *s* and *p* orbitals of the outer-most shell of M^{2+} ions accept the electron pairs donated by the ligands and thus form complex compounds.

Classification of transition elements into 3d, 4d, 5d and 6d series elements and their electronic configuration. Depending on whether the differentiating electron goes to *3d, 4d, 5d* or *6d* sub-shell, transition elements (*i.e. d*-block elements) have been grouped into the following *four* series.

(*a*) *3d series elements* (*1st transition series : 4th period*). This series contains ten elements *viz.* $Sc_{21}$to Zn_{30}, These elements are present in *4th* period of the periodic table and hence have *four* shells. The configuration of the atoms of these elements have already been given in Table 4.2. The following points may be noted from these configurations:

(*i*) The configurations of these elements can be represented by the following Table :

Shells :	1*st*	2*nd*	3*rd*	4*th*
No. of electrons	2 2	8 8	9 to 18 $3s^2p^6d^{1\text{-}10}$	1 or 2 $4s^{1,2}$
	← First two shells → (completely-filled)		Outer-most two shells ← (partially-filled)⟶	

The above configuration shows that the last two shells (*i.e.*, 4*th* and 3*rd*) are partially-filled, since they contain 1 or 2 and 9 to 18 electrons respectively. First two shells are completely-filled. It is also evident from the above configuration that the valence-shell configuration of all the ten elements can be written as $3d^{1\text{-}10}\ 4s^{1,2}$

(*ii*) The expected configurations of Cr_{24} and Cu_{29} are different from their actual configurations as shown below :

Element	*Expected configuration*	*Actual configuration*
Cr_{24}	$[Ar]3d^4 4s^2$	$[Ar]3d^5 4s^1$
Cu_{29}	$[Ar]3d^9 4s^2$	$[Ar]3d^{10} 4s^1$

It may be seen that in both the elements *viz* Cr_{24} and Cu_{29} one electron from 4*s* orbital gets shifted to 3*d* sub-shell (shifting of an electron from an outer-orbital to an inner-orbital). This type of shifting of an electron makes 3*d* sub-shell half filled ($3d^5$) in Cr_{24} and completely-filled ($3d^{10}$) in Cu_{29}. Half-filled and completely-filled 3*d* orbitals are comparatively more stable than $3d^4$ and $3d^9$ orbitals. Thus $[Ar]3d^5\ 4s^1$ configuration (actual configuration of Cr_{24}) is more stable than $[Ar]\ 3d^4 4s^2$ configuration (expected configuration of Cr_{24}). Similarly $[Ar]3d^{10}4s^1$configuration (actual configuration of Cu_{29}) is more stable than $[Ar]3d^9\ 4s^2$ configuration which is the expected configuration of Cu_{29}.

(*b*) *4d series elements (2nd transition series : 5th period)*. This series contains ten elements *viz* Y_{39} to Cd_{48}. These elements are present in *5th* period of the periodic table and hence have *five* shells. The configurations of the atoms of these elements have already been given in Table 4.3. The following points may be noted from these configurations:

(*i*) The configurations of these elements can be represented by the following Table :

Shells :	*1st*	*2nd*	*3rd*	*4th*	*5th*
No. of electrons	2	8	18	9 to 18	0 or 2
	2	8	18	$4s^2p^6d^{1\text{-}10}$	$5s^{0\text{-}2}$
	← First two shells → (completely-filled)			← Outer-most two shells→ (partially-filled)	

The above configuration shows that the last two shells (*i.e.* 5*th* and 4*th*) are partially-filled, since they contain 0 to 2 and 9 to 18 electrons respectively. First three shells are completely-filled. It is also evident from the above configuration that the valence-shell configuration of all the ten elements can be written as $4d^{1\text{-}10}\ 5s^{0\text{-}2}$

(*ii*) The expected configurations of the elements *viz* Nb_{41}, Mo_{42}, Ru_{44}, Rh_{45}, Pd_{46} and Ag_{47} (six elements) are different from their respective actual configuration as shown below:

Element	*Expected configuration*	*Actual configuration*
Nb_{41}	$[Kr]4d^3 5s^2$	$[Kr]4d^4 5s^1$
Mo_{42}	$[Kr]4d^4 5s^2$	$[Kr]4d^5 5s^1$
Ru_{44}	$[Kr]4d^6 5s^2$	$[Kr]4d^7 5s^1$
Ru_{45}	$[Kr]4d^7 5s^2$	$[Kr]4d^8\ 5s^1$
Pd_{46}	$[Kr]4d^8 5s^2$	$[Kr]4d^{10} 5s^1$
Ag_{47}	$[Kr]4d^9 5s^2$	$[Kr]4d^{10} 5s^1$

It may be seen from the observed (actual) configurations that 5*s* orbital in Nb_{41}, Mo_{42}, Ru_{44}, Rh_{45}, and Ag_{47} contains *one* electron ($5s^1$ configuration). This is because of the reason that one electron from 5*s* orbital is shifted to 4*d* sub-shell (shifting of an electron from a higher shell to a lower shell). The shifting of one 5*s* electorn to 4*d* sub-shell in case of Mo_{42} produces half-filled 4*d* sub-shell ($4d^5$ configuration) which is comparatively more stable than $4d^4$ configuration. In case of Ag_{47}, the shifting of one 5*s* electron to *4d* sub-shell makes this sub-shell completely-filled ($5d^{10}$ configuration). Completely filled *5d* orbitals ($5d^{10}$) are more stable than $5d^9$ orbitals.

In case of Pd_{46} atom, 5*s* orbital is vacant, since both of its electrons get shifted to *4d* sub-shell so that the number of electrons present in this sub-shell becomes 10 instead of 8 ($4d^8 \rightarrow 4d^{10}$). $4d^{10}$ configuration is more stable than $4d^8$ configuration, since this sub-shell is completely-filled. The shifting of *5s* electron to *4d* sub-shell in case of other elements (*e.g.* Nb_{41}, Ru_{44} and Rh_{45}) has been explained on the basis of nuclear-electron and electron-electron forces.

(*c*) *5d series elements (3rd transition series: 6th period).* La_{57} (one element), Hf_{72} to Hf_{80} (nine elements) are the members of this series. These elements are lying in *6th* period of the period table and hence have six shells. In connection with the electronic configurations of these elements as given in Table 4.5, the following points may be noted:

(*i*) The configurations of the atoms of these ten elements can be represented by the following general expression:

Shells :	*1st*	*2nd*	*3rd*	*4th*	*5th*	*6th*
No. of electrons	2 2	8 8	18 18	18 or 32 $4s^2p^6d^{10}, f^{0\text{-}14}$	10 to 18 $5s^2p^6d^{2\text{–}10}$	1 or 2 $6s^{1\text{–}2}$
	←—— First four shells ——→ (completely-filled)				← Outer-most two shells → (partially-filled)	

The above configuration shows that the last two shells (*i.e. 6th* and *5th* shells) are partially-filled, since they contain 1 or 2 and 10 to 18 electrons respectively. First four shells are completely-filled. It is also evident from the above configuration that the valence-shell configuration of all the ten elements can be written as $4f^{0,14}\ 5d^{1\text{-}10}\ 6s^{1\text{-}2}$

(*ii*) It may also be noted from the above configuration that La_{57} has 18 electrons in its *4th* shell ($4s^2p^6d^{10}f^0$ configuration) while the same shell of the remaining nine elements has 14 extra electrons (*i.e.* 18 + 14 = 32 electrons) ($4s^2p^6d^{10}f^{14}$ configuration). The presence of 14 extra electrons is due to the presence of 14 lanthanides (Ce_{58} to Lu_{71}) between La_{57} and Hf_{72}.

Unique position of La_{57}. Whether La_{57} should be treated as a member of d-block elements or inner-transition elements can not be decided with certainty as shown below:

(*a*) *La_{57} is a member of d-block elements.* Since the last electron (*i.e.* 57*th* electron) of La_{57} enters (*n* – 1)*d* (*i.e. 5d*) sub-shell, it is treated as one of the members of *d*-block elements.

(*b*) *La_{57} is treated as a member of inner-transition elements.* The configuration of La_{57} is 2, 8, 18, 18, 9, 2 which shows that its three outer-most shells are partially-filled. Thus according to the definition of inner-transition elements, La_{57} should be put with inner-transition elements.

(*iii*) The configuration of 5*th* shell ($5s^2p^6d^{1\text{-}10}$) shows that, since the differentiating electron enters *5d* sub-shell, these elements are called *5d series elements.*

(*iv*) The expected configuration of the two elements *viz* Pt_{78} and Au_{79} are different from their respective actual configurations as shown below;

Element	*Expected configuration*	*Actual configuration*
Pt_{78}	$[Xe]4f^{14}5d^8 6s^2$	$[Xe]4f^{14}5d^9 6s^1$
Au_{79}	$[Xe]4f^{14}5d^9 6s^2$	$[Xe]4f^{14}5d^{10} 6s^1$

In these elements, *6s* orbital has one electron only. This is because of the fact that one electron from *6s* orbital gets shifted to *5d* sub-shell (shifting of an electron from a higher shell to a lower shell).

In case of Au_{79}, the shifting of one *6s* electron to *5d* sub-shell makes *5d* sub-shell completely-filled ($5d^{10}$ configuration) and completely-filled *5d* sub-shell is more stable than other *5d* sub-shells. Thus $[Xe]4f^{14}5d^{10}6s^1$ configuration of Au_{79} is more stable than its expected configuration *viz* $[Xe]4f^{14}5d^9 6s^2$. The configuration of Pt_{78} has been explained on the basis of nuclear-electron and electron-electron forces.

(*d*) *6d series elements [4th transition series : 7th* period (*incomplete period*)]. At present Ac_{89}, Ku_{104} to Unh_{106} (*four elements*) are the members of this series. These elements are present in *7th* period (incomplete period) and hence contain *seven* shells. The electronic configurations of the atoms of these elements make the following points evident:

(*i*) The configurations of the atoms of the above four elements can be represented by the following general expression :

Shells :	1st	2nd	3rd	4th	5th	6th	7th
No. of electrons	2 2	8 8	18 18	32 32	18 or 32 $5s^2p^6d^{10}f^{0-14}$	9 to 12 $6s^2p^6d^{1-4}$	2 $7s^2$
	← First five shells (completely-filled) →					← Outer-most two shells (partially-filled) →	

The above configuration shows that the last two shells (*i.e. 7th* and *6th shells*) are partially-filled, since they contain 2 and 9 to 12 electrons respectively. First five shells are completely-filled. It is also evident from the configuration that the valence-shell configuration of these four elements can be represented as $5f^{0,14}\,6d^{1-4}\,7s^2$:

(*ii*) It may also be noted that Ac_{89} has 18 electrons in its *5th* shell ($5s^2p^6d^{10}f^0$) while the same shell in the remaining three elements has 14 extra electrons (*i.e.* 18+14 =32 electrons) ($5s^2p^6d^{10}f^{14}$). The presence of 14 extra electrons is because of the presence of 14 actinides (Th_{90} to Lw_{103}) between Ac_{89} and Ku_{104}.

(*iii*) The configuration of *6th* shall ($6s^2p^6d^{1-4}$)shows that, since the extra electron in all these elements enters *6d* sub-shell, these elements are called *6d series elements.*

Formation of cations from the isolated gaseous atoms of transition metals. When transition metals are converted into cations, it is *ns* electrons, and not $(n-1)\,d$ electrons which are lost first by the isolated gaseous atoms of transition metals. This fact can be explained on the basis of the concept of effective nuclear charge. For this see "*Applications of Slater's Rules*" discussed in chapter 3.

IV. *f*-block elements or inner-transition elements. Fourteen elements *viz* Ce_{58} to Lu_{71} are called lanthanides, since they follow La_{57} in the periodic table. Similarly 14 elements namely Th_{90} to Lw_{103} are called actinides, since they follow Ac_{89} in the periodic table. The electronic configurations of all these 28 elements are given in Tables 4.4 and 4.6. These configurations show that the extra electron enters into *4f* sub-shell in lanthanides and into *5f* sub-shell in actinides. Now since lanthanides have six shells and actinides have seven shells, *4f* and *5f* sub-shells are called $(n-2)f$ sub-shells. Thus, since the extra electron goes to $(n-2)f$ sub-shell, these 28 elements are called *f*-block elements. *f-block elements can be defined as those elements in which the differentiating electron enters (n – 2)f sub-shell or in which (n – 2)f sub-shell is being filled up successively with electrons.* Since these elements lie within the transition series elements, *f*-block elements are also called inner-transition elements.

Last three shells namely *nth*, *(n-1)th* and *(n-2)th* contain 2 (ns^2 configuration), 8 to 10 [$(n-1)s^2p^6d^{0-2}$ configuration] and 18 to 32 [$(n-2)s^2p^6d^{10}f^{0-14}$ configuration] electrons respectively and hence are partially-filled while the remaining inner-shells (three in lanthanides and four in actinides) are completely filled. Thus inner-transition elements can also be defined as those elements whose outer-most three shells are partially-filled and the remaining shells are completely-filled Obviously the valence-shell configuration of these elements is $(n-2)f^{0-14}\,(n-1)d^{0-2}ns^2$.

Depending on whether the differentiating electron enters 4*f* sub-shell or 5*f* sub-shell, inner transition elements can be grouped into the following two categories.

(*i*) *4f-block elements. 4f-block elements are those in which the differentiating electron goes to 4f sub-shell or in which 4f sub-shell is successively filled up with electrons.* Obviously lanthanides (Ce_{58} to Lu_{71}) are 4*f*-block elements. These elements follow La_{57}. These are present in *6th* period and hence have six shells. The distribution of electrons in six shells is 2, 8, 18, 20 to 32, 8 or 9, 2 or 2, 8, 18, $4s^2p^6d^{10}f^{2-14}$, $5s^2p^6d^{0-1}$, $6s^2$, or [Xe]$4f^{2-14}5d^{0-1}\,6s^2$ which shows that the outer-most three shells are partially-filled and the three inner-shells are completely-filled. Valance-shell configuration can obviously be written as $4f^{2-14}5d^{0-1}6s^2$. For details see the configuration of the elements of *6th* period.

(*ii*) *5f-block elements. 5f-block elements are those elements in which the extra electron occupies 5f sub-shell or in which 5f sub-shell is progressively filled up with electrons.* Obviously actinides (Th_{90} to Lw_{103}) are *5f* block elements. These elements follow Ac_{89}. These are located in *7th* period and hence have seven shells. The distribution of electron in the seven shells is as 2, 8,

18, 32, 18 to 32, 8 to 10, 2 or 2, 8, 18, 32, $5s^2p^6d^{10}f^{0-14}$, $6s^2p^6d^{0-2}$, $7s^2$ or $[Rn]5f^{0-14}6d^{0-2}7s^2$. This configuration shows that the outer-most three shells are partially-filled while the inner four shells are completely-filled. Obviously valence-shell configuration can be written as $5f^{0-14}6d^{0-2}7s^2$. For details see the configurations of the elements of 7*th* period.

Classification of the elements on the basis of the number of completely-filled and partially-filled shells

The elements have also been classified on the basis of the number of completely-filled and partially-filled shells present in the configuration of the atoms of the elements. This basis of classification gives the following four types of elements.

I. Noble gases (Elements of zero group). The elements namely He_2, Ne_{10}, Ar_{18}, Kr_{36}, Xe_{54} and Rn_{86} (six elements) are called noble gases. The configurations of these elements show that all the shells are *completely-filled. The last shell has 8 electrons (ns^2p^6 configuration) and hence is called saturated shell.* It should be noted that although He_2 is a member of noble gases, it does not contain 8 electrons in its last shell. Rather, it has 2 electrons in *1s* orbital ($1s^2$ configuration).

II. Representative or normal elements. The elements of groups IA, IIA, IIIA, IVA, VA, VIA and VIIA are called *representative* or *normal elements*. The configurations of the atoms of these elements show that the *last shell is partially filled*, since this shell contains 1 (in elements of group IA), 2 (elements of group IIA), 3 (group IIIA), 4 (group IVA), 5 (group VA), 6 (group VIA) and 7 (group VIIA) electrons. The inner-shells are completely-filled. The valence-shell configuration of these elements belonging to different groups can be written as :

Group IA	:	ns^1	*Group IIA*	:	ns^2
Group IIIA	:	ns^2p^1	*Group IVA*	:	ns^2p^2
Croup VA	:	ns^2p^3	*Group VIA*	:	ns^2p^4
Group VIIA	:	ns^2p^5			

Here *n* indicates the number of the period to which the element belongs. The elements belonging to groups IA, IIA, IIIA,IVA,VA, VIA, VIIA and zero (noble gases) are collectively called **main group elements.**

III. Transition elements. The elements namely Sc_{21} to Zn_{30} (ten elements of *4th* period) Y_{39} to Cd_{48} (10 elements of *5th* period), La_{57}, Hf_{72} to Hg_{80} (10 elements of *6th* period), Ac_{89}, Ku_{104}, Ha_{105} and Unh_{106} (4 elements of *7th* period which is an incomplete period) are called transition elements. The electronic configurations of the atoms of these elements show that *two outer-most shells are partially filled while the inner shells the completely-filled.*

IV. Inner-transition elements. 14 elements *viz* Ce_{58} to Lu_{71} are called lanthanides and 14 elements namely Th_{90} to Lw_{103} are known as actinides. All these 28 elements are collectively called

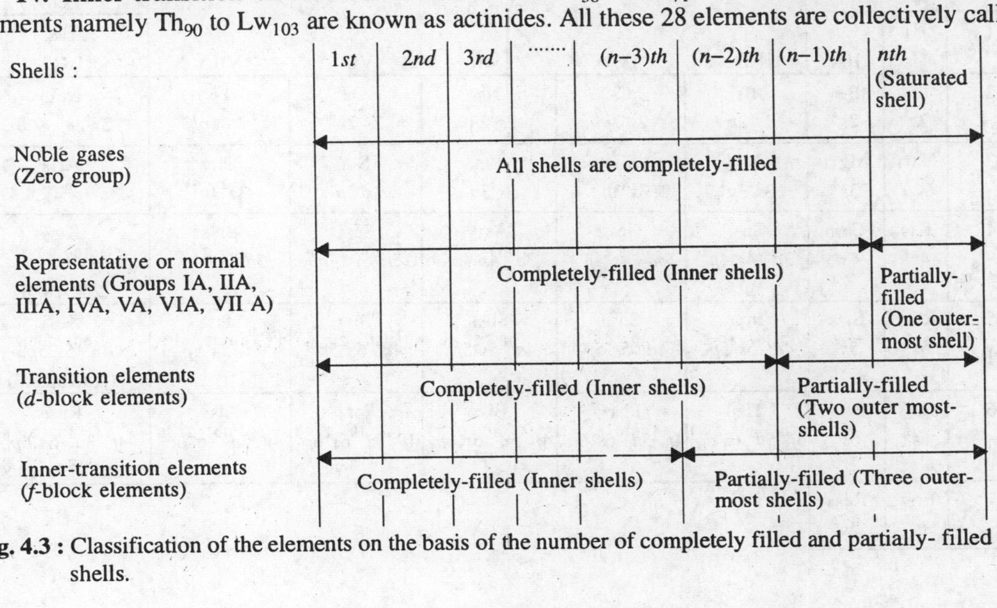

Fig. 4.3 : Classification of the elements on the basis of the number of completely filled and partially- filled shells.

inner-transition elements. Lanthanides have six shells, since they are present in *6th* period of the periodic table. On the other hand actinides have seven shells, since they are located in *7th* period of the periodic table. The electronic configurations of the atoms of lanthanides and actinides are given in Tables 4.4 and 4.6 respectiely. An inspection of their configurations shows that *the outer-most three shells of these elements are partially-filled while the inner-shells are completely-filled.* Thus in general we can say that in these elements *nth*, (*n*-1)*th* and (*n*-2)*th* shells (three outer-most shells) are partially-filled while 1*st*, 2*nd*, 3*rd*....(*n*-3)*th* shells (inner shells) are completely-filled.

Whatever has been said above has been summarised in Fig. 4.3.

Main group elements

The elements belonging to groups IA, IIA (*s*-block elements), IIIA, IVA, VA, VIA, VIIA and zero (*p*-block elements) are called main group elements. Thus the representative or normal elements (*i.e.* elements of groups IA, IIA, IIIA, IVA, VA, VIA and VIIIA) and zero group elements (noble gases) are called main group elements. s-block elements are present in all the seven periods of the periodic table while *p*-block elements are present only in *2nd, 3rd, 4th, 5th* and *6th* periods. Thus *1st* and *7th* periods donot have any *p*-block elements. As stated above *s*-block elements are present in groups IA and IIA while *p*-block elements are present in groups IIIA, IVA, VA, VIA, VIIA and Zero.

Elecronic configuration of main group elements according to their position in different periods of the periodic table

Electronic configuration of the atoms of main group elements are given in Table 4.8. These configurations make the following points evident.

(*i*) 1*st* period contains only *s*-block elements which are H_1 and He_2. The electronic configuration of these elements is $1s^1$ and $1s^2$. 1st period does not contain any *p*-block elements.

(*ii*) *2nd* and remaining periods contain *s*-block as well as *p*-block elements. The valence-shell configuration of *s*-block elements of *2nd, 3rd, 4th, 5th, 6th* and *6th* periods can be represented as $ns^{1\text{-}2}$ where *n* is the number of the period.

Table. 4.8: *Electronic configuration of the atoms of main group elements. Here* $[He]_2 = 2$; $[Ne]_{10} = 2, 8$; $[Ar]_{18} = 2, 8, 8$; $[Kr]_{36} = 2, 8, 18, 8$; $[Xe]_{54} = 2, 8, 18, 18, 8$; $[Rn]_{86} = 2, 8, 18, 32, 18, 8$

Period number	s-block elements		Representative Elements — p-block elements					Noble gases
	IA	IIA	IIIA	IVA	VA	VIA	VIIA	Zero
1	H1 $1s^1$							He2 $1s^2$(2)
2 $[He]_2$ +	Li3 $2s^1$	Be4 $2s^2$	B5 $2s^22p^1$	C6 $2s^22p^2$	N7 $2s^22p^3$	O8 $2s^22p^4$	F9 $2s^22p^5$	Ne10 $2s^22p^6$ (2,8)
3 $[Ne]_{10}$ +	Na11 $3s^1$	Mg12 $3s^2$	Al13 $3s^23p^1$	Si14 $3s^23p^2$	P15 $3s^23p^3$	S16 $3s^23p^4$	Cl17 $3s^23p^5$	Ar18 $3s^23p^6$ (2,8,8)
4 $[Ar]_{18}$ +	K19 $4s^1$	Ca20 $4s^2$	Ga31 $3d^{10}4s^24p^1$	Ge32 $3d^{10}4s^24p^2$	As33 $3d^{10}4s^24p^3$	Se34 $3d^{10}4s^24p^4$	Br35 $3d^{10}4s^24p^5$	Kr36 $3d^{10}4s^24p^6$ (2,8, 18, 8)
5 $[Kr]_{36}$ +	Rb37 $5s^1$	Sr38 $5s^2$	In49 $4d^{10}5s^25p^1$	Sn50 $4d^{10}5s^25p^2$	Sb51 $4d^{10}5s^25p^3$	Te52 $4d^{10}5s^25p^4$	I53 $4d^{10}5s^25p^5$	Xe54 $4d^{10}5s^25p^6$ (2,8,18,18,8)
6 $[Xe]_{54}$ +	Cs55 $6s^1$	Ba56 $6s^2$	Tl81 $4f^{14}5d^{10}6s^26p^1$	Pb82 $4f^{14}5d^{10}6s^26p^2$	Bi83 $4f^{14}5d^{10}6s^26p^3$	Po84 $4f^{14}5d^{10}6s^26p^4$	At85 $4f^{14}5d^{10}6s^26p^5$	Rn86 $4f^{14}3d^{10}6s^26p^6$ (2,8,18,32,18,8)
7 $[Rn]_{86}$ +	Fr87 $7s^1$	Ra88 $7s^2$						

(*iii*) The valence-shell configuration of the atoms of *p*-block elements belonging to *2nd* and *3rd* periods can be expressed as $2s^22p^{1\text{-}6}$ and $3s^23p^{1\text{-}6}$ respectively. The valence shell configurations of the atoms of *p*-block elements belonging to *4th* and *5th* periods can be represented as $3d^{10}4s^24p^{1-6}$ and $4d^{10}5s^25p^{1\text{-}6}$ respectively. The presence of extra ten electrons in *3d* or *4d* sub-shell is due to the presence of ten *3d*-block elements viz Sc_{21} – Zn_{30} between Ca_{20} and Ga_{31} (*4th* period) and ten *4d*-block elements *viz* Y_{39} – Cd_{48} between Sr_{38} and In_{49} (*5th* period). The valence-shell configuration of the atoms of *p*-block elements belonging to *6th* period can be expressed as $4f^{14}5d^{10}6s^26p^{1\text{-}6}$.

The presence of additional fourteen *4f* electrons is due to the inclusion of 14 lanthanides (*4f*-block elements *viz* Ce_{58} – Lu_{71}) between La_{57} and Hf_{72}. Similarly the presence of ten *5d* electrons is due to the inclusion of ten *5d*-block elements *viz* La_{57}, Hf_{72} to Hg_{80} between Ba_{56} and Tl_{81}. *7th* period does not contain any *p*-block elements.

Electronic configuration of main group elements according to their position in different groups of the periodic table

We have stated above that the elements present in groups IA, IIA, IIIA, IVA, VA, VIA, VIIA and zero [*noble gases*] are main group elements. Here we shall discuss the electronic configuration of the atoms of the elements belonging to the above groups.

Group IA (*s*-block elements). The configuration of these elements can be represented as [Noble gas]ns^1 where *n* is the number of the valence-shell or period number. The valence-shell is preceded by noble gas configuration (s^2p^6), *i.e.* their (*n* – 1)th shell contains 8 electrons. It is evident from the above configuration that the last electron in these elements enters *ns* orbital. Consequently the elements of group IA are *s*-block elements.

Group IIA (*s*-block elements). The configuration of these elements can be expressed as [Noble gas]ns^2 where *n* represents valence-shell or period number. Their valence-shell is also preceded by noble gas configuration. The last electron in these elements goes to *ns* orbital. The elements of group IIA are, therefore, *s*-block elements.

Group IIIA (*p*-block elements). The outer-most shell (*i.e. nth* shell) of the atoms of the elements of this group contain three electrons, two in *ns* and one in *np* orbitals. Ga_{31} and In_{49} contain extra $3d^{10}$ and $4d^{10}$ electrons respectively while Tl_{81} contains additional $4f^{14}$ and $5d^{10}$ electrons in their respective valence-shell. Thus the valence-shell configuration of B_6 and Al_{13} are $2s^22p^1$ and $3s^23p^1$ respectively but that for Ga_{31}, In_{49} and Tl_{81} is $3d^{10}4s^24p^1$, $4d^{10}5s^25p^1$ and $4f^{14}5d^{10}6s^26p^1$ respectively. These configurations show that the last electron in these elements occupies *np* orbitals and hence the elements of this group are *p*-block elements.

On the basis of the electronic configuration of the penultimate shell, these elements can be divided into two groups, namely (*i*) 1st group of elements consists of B and Al elements whose penultimate shell have *inert gas type configuration* (s^2 or s^2p^6 configuration) (B has 2 electrons and *Al* has 8 electrons in their penultimate shell) (*ii*) *2nd* group consists of *Ga*, *In* and *Tl* whose penultimate shells have *pseudo inert gas type configuration* ($s^2p^1d^{10}$ configuration). This difference in electronic configuration is responsible for the difference in properties of the elements of both the groups and also of boron and aluminium.

Group IVA (*p*-block elements). The electronic configuration of the atoms of the elements of this group show that their outer-most shell has four electrons, two in *ns* and two in *np* orbitals. Thus the general valences-shell configuration of these elements can be written as ns^2np^2. Since Ge_{32} and Sn_{50} are present in *4th* and *5th* periods respectively, their valence shell also contains ten electrons in *3d* and *4d* orbitals respectively. Similarly, Pb_{82} which is present in *6th* period also contains in its valence shell 14 electrons in *4f* orbitals and 10 electrons in *5d* orbitals. Penultimate shell of carbon contains s^2 electrons (saturated), of Si contains s^2p^6 electrons (saturated), of Ge contains $s^2p^6d^{10}$ electrons (saturated) while of Sn and Pb contain $s^2p^6d^{10}$ electrons (unsaturated). This shows why C differs from Si and these two (*i.e.* C and Si) from the other members of this sub-group.

Group VA (*p*-block elements). The electronic configurations of these elements indicate that the outer-most shell contains 5 electrons, two in *ns* and three in *np* orbitals. Thus the general configuration of these elements can be expressed as ns^2np^3. Three electrons present in *np* orbitals are equally distributed among the three *np* orbitals as np_x^1, np_y^1 and np_z^1 in accordance with Hund's rule, *i.e.* the three *np* orbitals are singly-filled. Since *np* orbitals are half-filled, these elements are fairly stable and not so reactive, *e.g.* nitrogen behaves as an almost inert element. It may be seen from the electronic configuration of these elements that the penultimate shells [*i.e.* (*n*-1)*th* shell] of N and P contain 2 and 8 electrons respectively. Thus the penultimate shells of both these elements are saturated shells. On the other hand, the penultimate shells of As, Sb and Si have 18 electrons. This shell in As is said to be saturated shell while in Sb and Bi this shell is unsaturated.

Group VIA (*p*-block elements). The outer-most shell of these elements has six electrons which means that the general outer-electronic configuration can be represented as ns^2np^4 or as ns^2 $np_x^2\ np_y^1\ np_z^1$. This configuration shows that these elements have two half-filled *np* orbitals which are responsible for chemical bonding with other elements. The valence-shell configuration also indicates that the elements of group VIA are *p*-block elements.

The penultimate shell in oxygen atom contains s^2 group (*saturated*), in sulphur s^2p^6 (*saturated*), in selenium $s^2p^6d^{10}$ (*saturated*), whereas in tellurium and polonium $s^2p^6d^{10}$ (*unsaturated*). This shows that oxygen differs from sulphur and these two elements from the remaining elements of this group. We thus conclude that the elements of this group are *p*-block elements.

Group VII A (*p*-block elements). The electronic configuration of the atoms of these elements show that these elements have the general electronic configuration as ns^2np^5 or $ns^2\ np_x^2\ np_y^2\ np_z^1$ which shows that there is only one half-filled *np* obital which is reponsible for chemical bonding between these and other elements. Since the halogens have only one electron less than eight electrons present in the adjascent noble gas, they have a very strong tendency to gain one electron to attain the stable noble gas configuration (ns^2np^6). Thus halogens are amongest the most reactive elements in inorganic chemistry.

The shell previous to the outermost shell contains in F the groups s^2 (*saturated*), in Cl s^2p^6 (*saturated*), in Br $s^2p^6d^{10}$ (*saturated*) while in I and At $s^2p^6d^{10}$ (*unsaturated*). This shows why F differs from Cl and these two from the remaining elements. Since the additional electron in these elements enters the p-orbital of the outermost shell, these are called p-block elements of the long form of the periodic table.

Zero group: Noble gases (*p*-block elements). The electronic configurations of the atoms of these elements show that all of them, except He_2, have stable $ns^2\ np^6$ configuration in their outer (valence) shell. He_2 has $2s^2$ configuration which, like $ns^2\ np^6$ configuration, is also a stable configuration. In these elements all the shells are completely-filled.

Conclusion. In all *s*-block elements (group IA and IIA) and in those *p*-block elements which belong to groups IIIA, IVA, VA, VIA and VIIA, the last shell (*i.e., nth* shell) is *partially-filled*, while all the inner shells are completely filled. In case of zero group elements, all the shells are *completely-filled*. Thus, in all main group elements the inner-shells are *completely-filled*. The general valence-shell configuration for *s*- and *p*-block elements can be written as $ns^{1\text{-}2}$ and $ns^2np^{1\text{-}6}$ respectively. Here *n* is the number of the last shell or the number of the period in which the element is present.

University Questions

Aufbau principle

1. How does Aufbau principle help us to understand the change in electronic configuration from element to element in the series K → Kr. *(Kanpur, 92; Kalyani, 89; Calicut 87; Rahelkhand, 92; Purvanchal, 93;' Gwahati, 91)*
2. What is the difference between energy level diagram of H-atom and that of multi-electron atom? How would you accout for the difference? *(M.S. Baroda 90)*

Hund's rule of maximum multiplicity

3. Explain the concept of "multiplicity". What is the justification of Hund's rule of maximum multiplicity? *(Agra 93; Kanpur 92; Gwahati 91)*

4. Write a note on Aufbau principle. *(Lucknow 2002)*

5. Give electronic configuration of group IA and group IIA elements. *(Nagpur 2002)*

6. Write the electronic configuration of the elements with atomic number 24, 29, 58, 64 and 96. *(VBS Purvanchal 2002)*

7. Write electronic configuration of Cr (24) and Cu (29). *(Nagpur 2002)*

8. (*a*) Write the electronic configuration of La(z = 57) and Ce(z = 58).

(*b*) What are the advantages of classifying the elements into *s*, *p*, *d* and *f* blocks in the periodic table? *(Delhi 2002)*

9. Correct the electronic configurations given below and give reasons for your answer :

(*a*) $1s^2, 2s^2\ 2p_x^2\ 2p_y^0\ 2p_z^0$ (*b*) $1s^2, 2s^2\ 2p^6\ 3s^2\ 3p^4, 4p^1$ *(VBS Purvanchal 2003)*

10. On the basis of electronic configuration, which of the following ions has maximum number of unpaired electrons ?

(*a*) Mg^{2+} (*b*) Ti^{3+} (*c*) V^{3+} (*d*) Fe^{2+} *(VBS Purvanchal 2003)*

11. (*a*) What is A U F B A U principle ? Use this principle to explain the filling of electrons in 4s-orbital before 3d-orbitals.

(*b*) Write the most stable electronic configuration using *s*, *p*, *d* notation of atom with atomic number 21, 31 and decide the position of each in periodic table indicate (*i*) The block (*ii*) The number of unpaired electrons for each. *(Rohilkhand 2003)*

12. (*a*) Elements A, B and C have atomic numbers 11, 29 and 35 respectively. Write their electronic configuration and classify them as *s*, *p* and *d* blocks.

(*b*) How many unpaired electrons are there in Cr^{2+} and Fe^{2+} ions ? *(Rohilkhand 2003)*

13. (*a*) What is Aufbua principle ? Give the sequence in which energy levels are filled. Write the electronic configuration of elements having atomic numbers 24 and 29.

(*b*) State Hund's rule of maximum multiplicity. Give suitable example.

(*c*) Write the general valence shell electronic configuration of IA and IIA group elements. *(Nagpur 2003)*

14. (*a*) Write the electronic configuration of F^-, Al^{3+}, Mn, Cu

(Given : Z of F = 9, Al = 13, Mn = 25, Cu = 29).

(*b*) What are s-block elements ? Write electronic configuration of IA group elements. *(Nagpur 2003)*

5

Atomic Properties : Periodic Properties

Introduction

Before proceeding to understand the topics, like chemical bonding and geometry (shape) of molecules and ions, it is necessary to have an adequate knowledge of the properties of the atoms of the elements. These properties are called **atomic properties.** These properties show a variation, when we proceed from left to right, in a period, and from top to bottom, in a group of the periodic table. These properties are, therefore, also called **periodic properties.**

The atomic properties, which we shall discuss in this chapter, are atomic and ionic radii, ionisation energies, electron affinity, eletronegativity, metallic (eletropositive) and non-metallic (electronegative) character, density, atomic volume, melting and boilong points, oxidising and reducing properties and oxidation states. These properties will be discussed in connection with main group elements.

Size of Atoms and Ions (Atomic and Ionic Radii)

Radius of an atom and of an ion

The size of an atom or an ion is represented by the magnitude of the radius of the atom (atomic radius) or ion (ionic radius). In ordinary sense, the radius of an atom or an ion is defined as the distance between the nucleus of the atom or ion and the electron cloud of the outer-most shell. If an atom or an ion is assumed to be a sphere, then atomic or ionic radius is the radius of the sphere.

The absolute size of an atom or an ion cannot be defined in an exact manner, due to the following reasons :

(*i*) The exact position occupied by an electron in an atom cannot be defined with certanity, since, according to wave mechanical picture of an electron, the electron may be very close to the nucleus at one time and may be far away from the nucleus at the other time. This means that the exact size of an atom cannot be defined.

(*ii*) The probability of finding the electron round the nucleus in an atom is influenced by the presence of other atoms in its environment. This results in that the size of the atom changes with the change of the atoms surrounding the atom.

The above discussion shows that it is not possible to measure the radius of an isolated atom or ion. We, therefore, measure the inter-nuclear distance between the nuclei of the atoms bounded together by covalent bond(s) or between the nuclei of the ions in an ionic crystal. It is these inter-nuclear distances which are used to calculate the atomic and ionic radii.

In spite of the above limitations, we have three operational concepts to define the atomic and ionic radius. These are *covalent radius*, *van der Waals radius* and *ionic raidus*.

Definition of covalent radius of atom A in A_2 molecule (Homo-nuclear diatomic molecule)

Covalent radius of atom *A* (abbreviated as r_A) in A_2 type covalent molecule, in which two

A-atoms are linked together by a single covalent bond (*A*—*A* bond), is defined as half of the distance between the nuclei of the two *A*-atoms. Thus if $d_{A\text{–}A}$ is the distance between the nuclei of the two *A*-atoms (this distance is called *bond distance* or *inter-nuclear distance* or *bond length* of A_2 molecule), then the covalent radius of atom *A* (*i.e.* r_A) is given by :

$$r_A \quad = \quad \frac{d_{A-A}}{2} \qquad ...(i)$$

Examples (*i*) Since the distance between the nuclei of the two H-atoms in H_2 molecule has been found to be equal to 0.74Å, covalent radius of H-atom = 0.74/2 = 0.37Å.

(*ii*) Since the inter-nuclear distances in Cl_2 and Br_2 molecules are 1.98 Å and 2.28 Å respectively, the covalent radii of Cl and Br atoms would be equal to 1.98/2 = 0.99 Å and 2.28/2 = 1.14 Å respectively.

Since the two *A*-atoms in A_2 molecule are linked together by a single non-polar covalent bond (*A*—*A* bond), the covalent radius discussed above is called *single bond covalent radius* or simply *normal covalent radius.* Single bond covalent radii of main group elements are given in Table. 5.1.

Table 5.1: *Single bond covalent radii and ionic radii (Å) of the atoms of main group elements. The figures given in bracket are the van der Waal's radii (1Å = 100 pm).*

Period number	I A	II A	III A	IV A	V A	VI A	VII A	Zero
1	H 0.37 2.08 (–1) (1.2)							He (0.93)
2	Li 1.23 0.68 (+1)	Be 0.90 0.31 (+2)	B 0.85 0.20 (+3)	C 0.77 0.15 (+4) 2.60 (–4)	N 0.75 0.11 (+5) 1.71 (–3) (1.5)	O 0.73 0.09 (+6) 1.42 (–2) (1.40)	F 0.72 1.36 (–1) (1.35)	Ne (1.31)
3	Na 1.54 0.95 (+1)	Mg 1.36 0.65 (+2)	Al 1.43 0.50 (+3)	Si 1.11 0.41 (+4) 2.71 (–1)	P 1.06 0.34 (+5) 2.12 (–3) (1.85)	S 1.02 0.29 (+6) 1.84 (–2) (1.80)	Cl 0.99 0.26 (+7) 1.81 (–1) (1.80)	Ar (1.75)
4	K 2.03 1.33 (+1)	Ca 1.74 0.99 (+2)	Ga 1.35 0.62 (+3) 1.13 (+1)	Ge 1.22 0.53 (+4) 0.93 (+2)	As 1.20 0.47 (+5) 2.22 (– 3) (2.00)	Se 1.16 0.42 (+6) 1.98 (–2) (1.90)	Br 1.14 0.39 (+7) 1.95 (–1) (1.95)	Kr (1.89)
5	Rb 2.16 1.48 (+1)	Sr 1.91 1.13 (+2)	In 1.67 0.81 (+3) 1.32 (+1)	Sn 1.41 0.71 (+4) 1.12 (+2)	Sb 1.40 0.62 (+5) 2.45 (–3) (2.20)	Te 1.36 0.56 (+6) 2.21 (–2) (2.10)	I 1.33 0.50 (+7) 2.16 (–1) (2.15)	Xe (2.09)
6	Cs 2.35 1.69 (+1)	Ba 1.98 1.35 (+2)	Tl 1.70 0.95 (+3) 1.40 (+1)	Pb 1.47 0.84 (+4) 1.20 (+2)	Bi 1.46 0.74 (+5) 1.20 (+3)	Po 1.46 — —	At —	Rn (2.14)

Calculation of covalent radius of atom A or B in AB molecule (Hetero-nuclear diatomic molecule)

Depending on whether the electronegativity difference (x_A – x_B) is very small or appreciably large, the following two cases arise :

(*i*) If the electronegativity difference ($x_A - x_B$) is *very small*, then the inter-nuclear distance between the centres of atom *A* and *B*, (d_{A-B}) in *AB* molecule is equal to the sum of the covalent radii of atoms *A* and *B*, *i.e.*

$$d_{A-B} = r_A + r_B \quad ...(ii)$$

If d_{A-B} and r_A are known, the covalent radius of atom *B* (*i.e.* r_B) can be calculated with the help of equation (*ii*).

Example. *Covalent radius of C-atom is 0.77 Å and the inter-nuclear distance between carbon and hydrogen atoms is CH_4 molecule is 1.14 Å. Find the covalent radius of H-atom.*

Solution. On using equation (*ii*), we get

$$1.14 = 0.77 + r_H$$

$$\therefore \quad r_H = 1.14 - 0.77 = \mathbf{0.37\ Å}$$

This value is the same as that obtained by spectroscopic method.

(*ii*) If the difference between the electronegativities of atoms *A* and *B* is *appreciably large*, then equation (*ii*) doed not hold good, *i.e.*, d_{A-B} is not equal to the sum of r_A and r_B; rather d_{A-B} has been found to be less than the sum of r_A and r_B. For example in HF, HCl, HBr, HI and $(CH_3)_3N$ molecules, the experimental values of d_{A-B}, in each molecule, is less than ($r_A + r_B$), as shown below :

In HF molecule, $d_{H-F} = r_H + r_F = 0.37 + 0.72 = 1.09$ Å

Experimental value of $d_{H-F} = 0.92$ Å

In HCl molecule, $d_{H-Cl} = r_H + r_{Cl} = 0.37 + 0.99 = 1.36$ Å

Experimental value of $d_{H-Cl} = 1.28$ Å

In HBr molecule, $d_{H-Br} = r_H + r_{Br} = 0.37 + 1.14 = 1.51$ Å

Experimental value of $d_{H-Br} = 1.42$ Å

In HI molecule, $d_{H-I} = r_H + r_I = 0.37 + 1.33 = 1.70$ Å

Experimental value of $d_{H-I} = 1.61$ Å

In $(CH_3)_3N$ molecule, $d_{N-C} = r_N + r_C = 0.75 + 0.77 = 1.52$ Å

Experimental value of $d_{N-C} = 1.47$ Å

It is obvious from the above calculations that the values of d_{A-B} obtained from equation (*ii*) are less than the experimental values of d_{A-B}.

The shortening of d_{A-B} (bond length) is due to the presence of ionic character in *A* – *B* bond. This ionic character is produced because of the difference in electronegativities of *A* and *B* atoms. If *A* is more electronegative than *B*, the electron pair forming the covalent bond between *A* and *B* would be shifted nearer towards more electronegative *A*-atom. This type of shifting of electron would accumulate negative charge on atom *A* and an equal (in magnitude) positive charge on atom *B*. This type of accumulation of opposite charges, on the two atoms viz. *A* and *B* would bring these atoms closer to each other and hence the bond distance (d_{A-B}) between them would get shortened. Greater is the electronegativity difference ($x_A - x_B$), greater would be the shortening of the bond distance and smaller would be the values of r_A and r_B.

Schomaker and Stevenson (1941) proposed that d_{A-B} in molecules of *AB* type in which the electronegativity difference ($x_A - x_B$) is large is given by the following equation.

$$d_{A-B} = r_A + r_B - 0.09\ (x_A - x_B) \quad ...(iii)$$

Here r_A and r_B are in Å and are equal to $d_{A-A/2}$ and $d_{B-B/2}$ respectively [see equation (*i*)]. If r_A and r_B are expressed in *pm*, then equation (*iii*) becomes equation (*iv*), since 1Å = 100 pm.

$$d_{A-B} = r_A + r_B - 9\ (x_A - x_B) \quad ...(iv)$$

Equation (*iii*) or (*iv*) can be used to calculate d_{A-B} in HF, HCl, HBr, HI and $(CH_3)_3N$ molecules and then these values can be compared with their experimental values.

In HF molecule, $d_{H-F} = r_H + r_F - 0.09\ (x_F - x_H)$

$= 0.37 + 0.72 - 0.09\ (4 - 2.1) = 0.919$ Å

Experimental value of $d_{H-F} = 0.92$ Å

In HCl molecule, d_{H-Cl} = $r_H + r_{Cl} - 0.09\ (x_{Cl} - x_H)$
= $0.37 + 0.99 - 0.09\ (3.0 - 2.1) = 1.36 - 0.09 \times 0.9 = 1.279$ Å

Experimental value of d_{H-Cl} = 1.28 Å

In HBr molecule, d_{H-Br} = $r_H + r_{Br} - 0.09\ (x_{Br} - x_H)$
= $0.37 + 1.14 - 0.09 \times (2.8 - 2.1)$
= $1.51 - 0.09 \times 0.7 = 1.447$ Å

Experimental value of d_{H-Br} = 1.42 Å

In HI molecule, d_{H-I} = $r_H + r_I - 0.09\ (x_I - x_H)$
= $0.37 + 1.33 - 0.09 \times (2.5 - 2.1)$
= $1.70 - 0.036 = 1.664$ Å

Experimental value of d_{H-I} = 1.61 Å

In $(CH_3)_3N$ molecule, d_{N-C} = $r_N + r_C - 0.09\ (x_N - x_C)$
= $0.75 + 0.77 - 0.09\ (3.0 - 2.5) = 1.475$ Å

Experimental value of d_{N-C} = 1.47 Å

The above claculations show that the values of d_{A-B} obtained from equation (*iii*) are almost the same as the experimental values of d_{A-B}. Thus equation (*iii*) or (*iv*) gives satisfactory results.

It is obvious from the above discussion that the radius of atom A (*i.e.*, r_A) in *AB* molecule can be calculated by using equation (*iii*) or (*iv*), if r_B and $(x_A - x_B)$ are known.

Double-bond and triple-bond covalent radii

If the two *A*-atom in A_2 molecule are linked together by a double bond, *A* = *A* (*e.g.*, two O-atoms in O_2 molecule are linked as O = O), the covalent radius of *A* atom is called its *double bond covalent radius*. Similarly if the two *A* atoms in A_2 molecule are attached together by a tripple bond, *A* ≡ *A* (*e.g.*, two N-atoms in N_2 molecule are attached together as N ≡ N), the covalent radius of *A* atom is called its *tripple bond covalent radius*. Double and triple bone covalent radii of some elements are given below :

Double bond covalent radii (Å) :	B = B,	C = C,	N = N	O = O
	0.76	0.67	0.60	0.55
Triple bond covalent radii (Å) :	B ≡ B,	C ≡ C,	Si ≡ Si,	N ≡ N
	0.68	0.60	1.07	0.55

Relation between single bond, double bond and triple bond covalent radii of an atom

It may be seen from the above values, that double and triple bond covalent radii of the atom of a given element are shorter than its single bond covalent radius. This is because of the fact that in the formation of a single covalent bond, only sigma orbitals overlap, while in the formation of a double or a triple bond, sigma as well as pi orbitals overlap. The additional overlapping of pi orbitals bring the two atoms closer to each other and hence bond length in reduced. The reduction in bond length reduces the double bond and triple bond covalent radius. We can conclude from this discussion that the magnitude of covalent radius of a given element decreases with the increase of the multiplicity or bond order of the bonds between the atoms, *i.e.*, these radii of a given element decrease in the order : single bond covalent radius > double bond covalent radius > triple bond covalent radius.

Example. Since C – C, C = C and C ≡ C bond distances are equal to 1.54 Å, 1.34 A and 1.20 Å respectively, single bond covalent radius, double bond covalent radius and triple bond covalent radius of carbon atom is equal to 1.54/2 = 0.77 Å, 1.34/2 = 0.67 Å and 1.20/2 = 0.60 Å respectively.

Covalent radii of metallic atoms : Metallic radii

It is not possible to determine the covalent radii of metals, since most of the metals *do not* form covalent compounds, excepting metallic hydrides and organo-metallic compounds. However, the radii of metallic atoms can be determined from the atomic values of metallic phases (Atomic volumes of metal atoms can be determined from the metallic phases, by dividing their atomic masses by their respective densities). The radii thus obtained are called *metallic radii.* It has been found that the covalent radii of metallic atoms as determined from metallic hydrides and organometallic compounds and the metallic radii, as determined from atomic volumes of metallic phases have almost similar values. It is, therefore, immaterial whether we call the radii of the metallic atoms as metallic radii or metallic covalent radii.

Atomic radius

The size of an atom, in its free state, is called its atomic radius. In many cases, it has been found that the atomic radius of an atom and its size determined from a non-polar single covalent bond (*i.e.* covalent radius) are almost the same. Therefore, the terms covalent radii and atomic radii have been used synonymously by many authors.

Factors affecting the magnitude of covalent radii of elements

The following factors are important :

1. Effective nuclear charge (Z_{eff}). With the increase in the magnitude of Z_{eff} felt by the electron of the outer-most shell, the force of attraction between the nucleus and the outer-most shell electron increases and hence the electron cloud of the element moves closer to the nucleus. This ultimately results in that the covalent radius of the atom of the element decreases. This factor has been used to explain why the covalent radii generally decrease, as we move from left to right, in a given period of main group elements (See the variation of covalent radii of main group elements in a period).

2. Number of shells or principle quantum number. Principal quantum number of an element represents the total number of shells present in that element. With the increase in the number of shells or principal quantum number, the outer-most shell electron(s) in the atom gets farther and farther away from the nucleus and hence the covalent radius increases. This factor has been used to explain why the covalent radii of the elements increase, as we move from top to bottom, in a given group of main group elements (See the variation of covalent radii of main group elements in a group).

3. Multiplicity of the bonds between the atoms. We have already seen that the magnitude of covalent radius decreases with the increase of the multiplicity of the bonds between the atoms. The reason for this has already been given.

Variation of covalent radii in main group elements

(*a*) **Variation in a period.** Generally speaking, the covalent radii decrease, in moving from left to right in a given period of the main group elements. This variation can be explained by considering, as an example, the covalent radii of the elements of 2*nd* period, as given in Table 5.2. In this table, the value of effective nuclear harge (Z_{eff}) felt by $2s^1$ electron (L-shell electron) in Li_3 and Be_4 and $2p^1$ electron (L-shell electron) in B_5, C_6, N_7, O_8, F_9 and Ne_{10} has also been given. These values have been calculated with the help of Slater's rules. This table shows that, as we proceed from Li_3 to F_9, the magnitude of Z_{eff} felt by L-shell electron goes on increasing and hence L-shell electrons (also K-shell electrons) are pulled closer and closer to the nucleus. Consequently the covalent radii of these elements go on decreasing, in moving from Li_3 to F_9. When we reach Ne_{10}, its Z_{eff} increases and hence its size should be smaller than that of F_9. But actually the size of Ne_{10} is greater than that of F_9. The larger size of Ne_{10}, as compared to that of F_9, has been explained as follows. Since the forces holding the atoms together in noble gases (*i.e.* inter-atomic forces) are only van der Waals forces, which are very weak forces, the radii of the atoms of noble gases are in fact van der Waals radii, which are larger than the covalent radii of the atoms of halogens. Thus Ne_{10} has bigger size than F_9.

Table 5.2 : *Covalent radii of the elements of 2nd period.* The radius of Ne_{10} is its van der Waals radius.

Elements of 2*nd* period	Li_3	Be_4	B_5	C_6	N_7	O_8	F_9	Ne_{10}
Covalent radii (Å)	1.23	0.85	0.52	0.77	0.75	0.73	0.72	1.31
Electronic configuration	$1s^2, 2s^1$	$1s^2, 2s^2$	$1s^2, 2s^2\,p^1$	$1s^2, 2s^2\,p^2$	$1s^2, 2s^2\,p^3$	$1s^2, 2s^2\,p^4$	$1s^2, 2s^2\,p^5$	$1s^2, 2s^2\,p^6$
Z_{eff} felt by the outer-most shell electron shown in bracket.	1.30 ($2s^1$)	1.95 ($2s^1$)	2.60 ($2p^1$)	3.25 ($2p^1$)	3.90 ($2p^1$)	4.55 ($2p^1$)	5.20 ($2p^1$)	5.85 ($2p^1$)

(*b*) **Variation in a group.** The covalent radii increase, in moving from top to bottom, in a given group of the main group elements. This variation can be explained by considering, as an example, the covalent radii of the elements of group IA as given below :

Elements of group IA :	Li	Na	K	Rb	Cs
Covalent radii (Å) :	1.23	1.54	2.03	2.16	2.35
Number of shells :	2	3	4	5	6
Z_{eff} felt by ns^1 electron :	increasing →				

As we proceed from Li to Cs, the number of shells or principal quantum number and Z_{eff} felt by ns^1 electron, both increase. The increase in the number of shells tends to increase the covalent radii while the increase in Z_{eff} tends to decrease the covalent radii. The effect caused by the increase of Z_{eff} is cancelled by the effect caused by the increae in the number of shells and hence the effect in the increase in the number of shells on covalent radii dominates. This results in that the covalent radii go on increasing, as we move down the group.

Definition of van der Waals radius

In order to get the definition of van der Waals radius, let us consider a crystal of iodine (*i.e.*, iodine in the solid state). The crystal of iodine contains a number of I_2 molecules. These I_2 molecules are not bonded to each other by chemical bonds (*strong bonds*) but are bonded together by *weak forces*, which are van der Waals forces. As a result of these weak forces, I_2 molecules in the crystal of iodine, remain at an equilibirium distance from each other, but donot form any chemical bonds between them.

Now, let us consider two adjacent I_2 molecules of iodine crystal (See Fig. 5.1). These two I_2 molecules are not bounded together by any chemical bonds, but by weak van der Waals forces. The distance between the nuclei of the two adjacent I-atoms (these atoms belong to two different I_2 molecules and hence are not bonded to each other) belonging to two adjascent I_2 molecules is called *van der Waals inter nuclear distance* and half of this distance is called van der Waals radius of iodine atom. Thus, van der Waaks radius of an element can be defined as half of the inter nuclear distance (called van der Waals inter nuclear distance) between the nuclei of the adjascent atoms, belonging to two neighbouring molecules of that element, in the solid state.

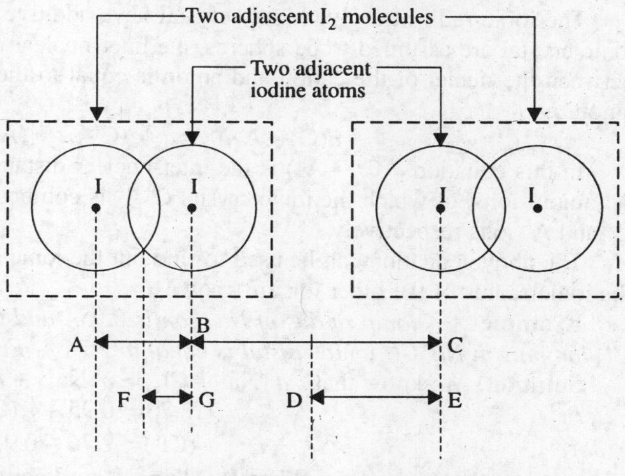

Fig. 5.1: van der Waals radius of iodine atom. AB = Covalent inter-nuclear distance = 2.66 Å. FG = AB/2 = 2.66/2 = 1.33 Å = covalent radius. BC = van der Waals inter-nuclear distance = 4.30 Å. DE = BC/2 = 4.30/2 = 2.15 Å = van der Waals radius of iodine atom.

Covalent and van der Waals radii, both are used for the atoms of covalent compounds. The van der Waals radii donot apply to metals. The magnitude of van der Waals radius depends on the effective packing size of the atom and is not a fixed quantity. The van der Waals radii of some elements have already been given in Table 5.1.

Comparison between covalent radius and van der Waals radius of an element

It may be seen from Fig 5.1 that since van der Waals inter nuclear distance represents the distance between the nuclei of two adjascent I-atoms of two neighbouring I_2 molecules, which are non-bonded together and covalent inter nuclear distance represents the distance between the two nuclei of two I-atoms, which are bonded together in a single I_2 molecule by a covalent bond (strong bond), van der Waals inter nuclear distance is larger than the covalent inter nuclear distance. In other words, van der Waals radius of an element is larger than its covalent radius. For example the van der Waals radii of N, P As , Sb, O, S Se, Te and halogens are larger than their covalent radii (See Table 5.1). This fact also explains why van der Waals forces are weaker than covalent bonds.

Comparison of the radii of noble gas atoms with those of halogen atoms

Although the effective nuclear charge felt by np^1 electron of noble gas atoms is greater than that felt by the same electron of halogen atoms, the size of noble gas atoms is larger (not smaller as expected) than that of halogen atoms. The larger size of the atoms of noble gases has been explained in the following manner.

We know that the atoms of noble gas in the solid state are linked together by *weak forces* (Van der Waals forces) while the atoms of halogens are bonded together by *strong forces* (chemical bonds). Consequently, the radii of noble gas atoms are larger than those of halogen atoms.

Comparison between the anionic radius and van der Waals radius of an element

We have said above that van der Waals radius of an element is larger than its covalent radius, but it is generally close to the radius of the anion (anionic radius) of that element. For example the van der Waals radii of N, P, As, Sb, O, S, Se, Te and halogens (X) are close to the radii of their anions *viz.* N^{3-}, P^{3-} , As^{3-}, Sb^{3-}, O^{2-}, S^{-2}, Se^{2-} and X^- respectively.

Definition of ionic radius

Ionic radius is the radius of an ion (cation or anion) in an ionic crystal. It may be defined as the distance between the nucleus of the ion and the point upto which the nucleus exerts its attractive force on the electron cloud of the ion. This definition shows that beyond the distance equal to the ionic radius the nucleus will not exert any attractive force on the electron cloud. Ionic radii of some ions (cations and anions) are given in Table 5.1.

Additive property of ionic radii

The ionic radii of a given ionic crystal have additive property which means that if ions of an ionic crystal are assumed to be spheres, the inter-nuclear distance (also called inter-ionic distance) between the nuclei of the cation and anion is equal to the sum of the radius of cation and that of anion, *i.e.*

$$d(C^+ - A^-) = r(C^+) + r(A^-) \qquad ...(i)$$

In this equation $d(C^+ - A^-)$ is the inter-nuclear distance between the nuclei of C^+ (cation) and A^-(anion) ions, of which the ionic crystal C^+A^- is composed, and $r(C^+)$ and $r(A^-)$ are the radii of C^+ and A^- ions respectively.

The above equation can be used to find out the ionic radius of cation/anion, if $d(C^+ - A^-)$ and the ionic radius of the other ion are known.

Example *Ionic radius of Na^+ ion is 0.95Å and the inter-nuclear distance between Na^+ – Cl^- ion pair in Na^+Cl^- ionic crystal is equal to 2.76 Å. Find the ionic radius of Cl^- ion.*

Solution. We know that $d(Na^+ - Cl^-) = r(Na^+) + r(Cl^-)$

or $2.76 = 0.95 + r(Cl^-)$

$\therefore$ $r(Cl^-) = 2.76 - 0.95 = \mathbf{1.81\ Å}$

Comparison between the radii of I^+, I^0 and I^- and conclusions drawn from it

The given species have the same nuclear charge which is equal to the atomic number of iodine (= 53), but have different number of electrons present in them (I^- = 53 + 1 = 54, I° = 53 and I^+ = 53 – 1 = 52). We know that greater is the total number of electrons (*e*) present in a given species, lesser is the force of attraction (*f*) between the nucleus and the total number of electrons in the species. With the decrease in the magnitude of force of attraction, the electron cloud of the atom or ion goes farther away from the nucleus and hence the radius (*r*) of the atom or ion increases. Thus *e*, *f* and *r* are related as :

$$e \propto \frac{1}{f} \propto r$$

Thus, among the given species, since the number of electrons decreases as $I^- > I^\circ > I^+$, the radii of these species would also decrease in the same order, *i.e.*, the radii would be in the order $I^- > I^\circ > I^+$. From this order we can draw the following conclusions :

(*i*) *The radii of various species having the same nuclear charge but different oxidation states decrease with the increase in the oxidation number of the species.* Thus since the oxidation number of the given species increases as $-1 \rightarrow 0 \rightarrow +1$, their sizes are in the order $I^- > I^\circ > I^+$. This order is based on the fact that with the increase in the oxidation number of the species, the number of electrons decreases and hence their size also decreases.

The above discussion shows that if we are given two or more cations of the same element, then the size of these cations will decrease with the increase in their positive oxidation number. Thus Pb^{4+} ion is smaller in size than Pb^{2+} (Pb^{2+} = 1.20 Å and Pb^{4+} = 0.84 Å).

Similarly, if we are given two or more anions of the same element (e.g. M^-, M^{2-}, M^{3-} etc.), then the size of these anions will increase with the increase in their negative oxidation number. Hence the sizes of M^-, M^{2-} and M^{3-} anions are in the order $M^- < M^{2-} < M^{3-}$.

(*ii*) *An anion is larger in size than its parent atom.* This conclusion is also explained by saying that, since the anion has greater number of electrons than its parent atom, it is larger in size than its parent atom.

(*iii*) *A cation is smaller in size than its parent atom.* This is because of the reason that the cation has smaller number of electrons and hence is smaller in size than its parent atom.

Variation of ionic radii of iso-electronic ions

Iso-electronic ions are the cations or **anions** of different atoms. These ions have the same number of electrons and same electronic configuration. Examples of iso-electronic ions are C^{-4} N^{3-}, O^{-2}, F^-, Na^+, Mg^{2+}. Al^{3+}, Si^{4+}, P^{5+}, S^{6+}, Cl^{7+} etc. Each of these ions has 10 electrons and the electronic configuration of each ion is $1s^2$, $2s^2p^6$.

The size of iso-electronic ions decreases with the increase in their nuclear charge (nuclear charge is equal to the atomic number with plus sign). This is because of the fact that with the increase in the nuclear charge, the force of attraction between the nucleus and the same number of electrons increases and hence the size of iso-electronic ions decreases. This fact is supported by the values of the size of the above iso-electronic ions as given below.

Iso electronic ions	*Number of electrons and electronic configuration*	*Nuclear charge (= + Z)*	*Size of the ions (Å)*
C^{-4}	10 ($1s^2$, $2s^2p^6$)	+ 6	2.60
N^{-3}	10 ($1s^2$, $2s^2p^6$)	+ 7	1.71
O^{-2}	10 ($1s^2$, $2s^2p^6$)	+ 8	1.42
F^-	10 ($1s^2$, $2s^2p^6$)	+ 9	1.36
Na^+	10 ($1s^2$, $2s^2p^6$)	+ 11	0.95
Mg^{2+}	10 ($1s^2$, $2s^2p^6$)	+ 12	0.65
Al^{3+}	10 ($1s^2$, $2s^2p^6$)	+ 13	0.50
Si^{4+}	10 ($1s^2$, $2s^2p^6$)	+ 14	0.41
P^{5+}	10 ($1s^2$, $2s^2p^6$)	+ 15	0.34
S^{6+}	10 ($1s^2$, $2s^2p^6$)	+ 16	0.29
Cl^{7+}	10 ($1s^2$, $2s^2p^6$)	+ 17	0.26

Variation of ionic radii in main group elements

(*a*) **Variation in a period.** Since the size of the atoms of the elements decreaes as we move from left to right in a given period, the cations and anions of the elements also decrease in size in the same direction (See Table 5.1).

(*b*) **Variation in a group.** Since the size of the atom of the elements increases as we move down a group of main group elements, the cations and anions derived from these elements also increase in size in the same direction (See Table 5.1).

It may be noted from Table 5.1 that the increase in ionic radii amongst the first two pairs of ions is larger than that in the pairs of the last two elements. For example the increase in the radii of the pairs of alkai metal cation is given as $Na^+ - Li^+$ pair = 0.95 – 0.60 = 0.35, $K^+ - Na^+$ pair = 1.33 – 0.95 = 0.38, $Rb^+ - K^+$ pair = 1.48 – 1.33 = 0.15, and $Cs^+ - Rb^+$ pair = 1.69 – 1.48 = 0.21. These values show that the increase in cationic radius is quite large, as we move from Li^+ to Na^+ and from Na^+ to K^+, but the increase is not so large, as we move from K^+ to Rb^+ and from Rb^+ to Cs^+. This behaviour is because of the presence of ten elements of 1*st* transition series (Sc_{21} to Zn_{30}) between K^+ and Rb^+ ions. As these transistion series elements enter between K^+ and Rb^+ ion, their increasing nuclear charge tends to cause the contraction in the size of the atoms and ions, which follow them. Thus, the ions which follow the transition series elements have smaller size.

Experimental methods for the determination of ionic radii

The following two methods are important.

1. **Lande's method (1920).** This was the first attempt to determine the radii of ions. Lande assumed that since Li^+ ion is very small and I^- ions are very large in size, I^- ions would be touching each other in Li^+I^- ionic crystal and hence half of $I^- - I^-$ distance, $d\,(I^- - I^-)$ would be equal to the radius of I^- ion, $r\,(I^-)$ [See Fig. 5.2 (*a*)] Thus :

$$r\,(I^-) \;=\; \frac{d\,(I^- - I^-)}{2} = \frac{426}{2} = 213\ pm$$

In the halides, like NaI, KI etc., which contain larger cations as compared to Li^+ cation (in LiI), halide ions would *not* be touching each other. Rather cation and anion will be touching each other and hence the sum of the radii of the cation and the anion would be equal to the inter-nuclear distance between the cation and anion. Thus, in case of K^+I^- ionic crystal, $d\,(K^+ - I^-) = r(K^+) + r(I^-)$. On putting $r\,(I^-)$ = 213 *pm* (obtained as above) and $d\,(K^+ - I^-)$ = 353 *pm*, we get $r\,(K^+)$ = 353 – 213 = 140 *pm*. Similarly the values of the ionic radii of other alkali metal cations and halide ions can be determined with the help of the following relationship.

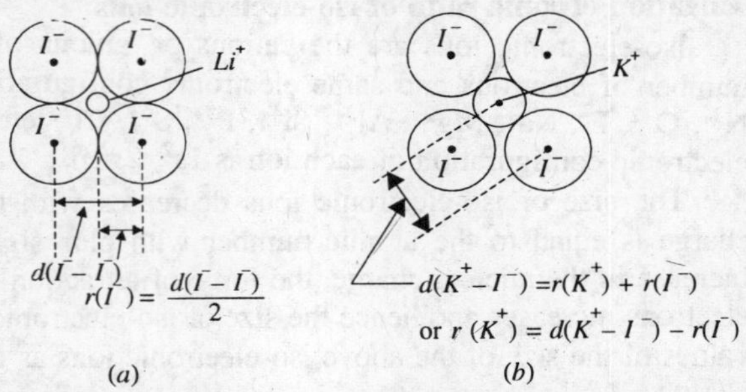

Fig. 5.2. Determination of (*a*) radius of I^- ion (*b*) radius of K^+ ion by Lande's method.

$$d\,(M^+ - X^-) = r\,(M^+) + r\,(X^-)$$

On substituting the value of $d\,(M^+ - X^-)$ and the radius of one ion in the above equation, the radius of the other ion can be calculated.

Other workers like Bragg and Goldschmidt have used this method to estimate the radius of O^{-2} ion (= 135 *pm*) from O — O distance in silicates and the ionic radii of over 80 other ions.

2. **Pauling's method : Determination of univalent and crystal radii.** Pauling has calculated the radii of ions in four alkali metal halides viz. NaF, KCl, RbBr and CsI. The ion pairs of each of these halides are iso-electronic, *i.e.*, they contain equal number of electrons as shown here : Na^+F^- = 2, 8 (Ne-type configuration) ; K^+Cl^- = 2, 8, 8 (Ar-type configuration); Rb^+Br^- = 2, 8, 18, 8 (Kr type configuration); Cs^+F^- = 2, 8, 18, 18, 8 (Xe type configuration). His method is based on the following assumptions :

(*i*) The cations and anions of the ionic crystal are in contact with each other and hence the sum of the radii of these ions is equal to inter-ionic distance between them, *i.e.*, for M^+X^- ionic crystal the following relation holds good.

$$d\,(M^+ - X^-) = r\,(M^+) + r\,(X^-) \qquad ...(i)$$

(*ii*) The radius of an ion [*i.e.* $r\,(M^+)$ or $r\,(X^-)$] having noble gas electronic configuration (ns^2 or ns^2p^6 configuration) is inversely proportional to the effective nuclear charge felt at the perifery of the ion. Thus for M^+ and X^- ions, we have :

$$r\,(M^+) = \frac{C}{(Z_{eff})_{M^+}} = \frac{C}{(Z)_M - (\sigma)_{M^+}} \quad ...(a)$$

$$r\,(X^-) = \frac{C}{(Z_{eff})_{X^-}} = \frac{C}{(Z)_X - (\sigma)_{X^-}} \quad ...(b)$$

In equations (*a*) and (*b*), C is a constant of proportionality whose value depends on the electronic configuration of the ion *viz.*, M^+ or X^-. σ (sigma) is screening constant for M^+ and X^- ions. $(Z_{eff})_{M^+}$ and $(Z_{eff})_{X^-}$ are the effective nuclear charges of M^+ and X^- ions. These are given by:

$$(Z_{eff})_{M^+} = (Z)_M - (\sigma)_{M^+}$$

$$(Z_{eff})_{X^-} = (Z)_X - (\sigma)_{X^-}$$

Now put the values of $r\,(M^+)$ and $r\,(X^-)$ from equations (*a*) and (*b*) in equation (*i*) to get:

$$d\,(M^+ - X^-) = \frac{C}{(Z)_M - (\sigma)_{M^+}} + \frac{C}{(Z)_X - (\sigma)_{X^-}} \quad ...(ii)$$

This equation can be used to calculate the value of C, if the value of other terms are given. The value of C as calculated from equation (*ii*) can be put in equations (*a*) and (*b*) to get the value of $r\,(M^+)$ and $r\,(X^-)$ respectively.

Illustrative Example. *Calculate the value of uni-valent radii of Na^+ and F^- ions, if the internuclear distance between Na^+ and F^- ions in Na^+F^- ionic crystal is equal to 231 pm.*

Solution. We have shown above that:

$$d\,(M^+ - X^-) = \frac{C}{(Z)_M - (\sigma)_{M^+}} + \frac{C}{(Z)_X - (\sigma)_{X^-}}$$

On applying this equation to Na^+F^- ionic crystal, we get:

$$d\,(Na^+ - F^-) = \frac{C}{(Z)_{Na} - (\sigma)_{Na^+}} + \frac{C}{(Z)_F - (\sigma)_{F^-}}$$

or

$$231 = \frac{C}{11 - (\sigma)_{Na^+}} + \frac{C}{9 - (\sigma)_{F^-}} \quad ...(iii)$$

Now,

$$(\sigma)_{Na^+}\,(e = 10 = 1s^2, 2s^2 2p^6 = 2, 8) = 0.35 \times 8 + 0.85 \times 2 = 4.5$$

$$(\sigma)_{F^-}\,(e = 10 = 1s^2, 2s^2 2p^6 = 2, 8) = 0.35 \times 8 + 0.85 \times 2 = 4.5$$

On putting the values of $(\sigma)_{Na^+}$ and $(\sigma)_{F^-}$ as calculated above in equation (*iii*), we get:

$$231 = \frac{C}{11 - 4.5} + \frac{C}{9 - 4.5} = \frac{C}{6.5} + \frac{C}{4.5}$$

This equation gives $C = 614.25$

Now put $C = 614.25$ in equations (*a*) and (*b*) to get:

$$r\,(Na^+) = \frac{614.25}{11 - 4.5} = \frac{614.25}{6.5} = \mathbf{94.5\ pm\ (Answer)}$$

$$r\,(F^-) = \frac{614.25}{9 - 4.5} = \frac{614.25}{4.5} = \mathbf{136.5\ pm\ (Answer)}$$

The experimental value of the inter-nuclear distance between the ions of an ionic crystal, containing univalent cations—univalent anions has been found to be equal or nearly equal to the sum of the uni-valent radii of univalent cations and uni-valent anions. On the other hand, the experimental value of the inter-nuclear distance between the ions of an ionic crystal containing poly-valent cations and poly-valent anions has been found to greater than the sum of the uni-valent radii of the poly-valent cations and poly-valent anions. For example the *experimental* value of $d(Mg^{2+} - O^{2-})$ in $Mg^{2+}\,O^{2-}$ ionic crystal is equal to 205 *pm*, which is less than the sum of the universal univalent radii of Mg^{2+} and O^{2-} ions [r_u of Mg^{2+} = 81.9 *pm* and r_u of O^{2-} = 175.5 and hence $(r_u)Mg^{2+} + (r_u)O^{2-} = 81.9 + 175.5 = 257.4$ *pm*].

Concept of crystal radius (r_c)

The above discussion shows that the uni-valent radii are very large. In order to remove the discrepency, Pauling has introduced the concept of crystal radius (r_c). The magnitude of r_c for a c-valent ion is related to the uni-valent radius (r_u) as :

$$r_c = r_u \times C^{-\frac{2}{n-1}} \qquad ...(i)$$

Here, n is the Born exponent, whose value depends on the electronic configuration of the ion. The value of n for the ions having helium, neon, argon, krypton and radon type configuration is 2, 7, 9, 10 and 12 respectively. Thus the values of r_c for Mg^{+2} and O^{-2} ions are given by :

r_c for Mg^{+2} ion $= 81.9 \times (2)^{-2/(7-1)} = 62$ *pm*

r_c for O^{-2} ion $= 175.5 \times (2)^{-2/(7-1)} = 140$ *pm*.

The sum of r_c of Mg^{+2} and O^{-2} ions = 62 + 140 = 202 *pm*, which is very near to the observed value of d ($Mg^{2+} - O^{2-}$) = 205 *pm*.

Variation of atomic and ionic radii of the elements in different groups

Here we shall consider the variation of atomic and ionic radii of the elements of the following groups :

Group IA (Alkali metals). As the alkali metals are the first elements in their respective periods, they have minimum effective nuclear charge and hence have the largest atomic radii in their respective periods. Down the group, the atomic radii increase due to the increase in the number of shells. Alkali metals have a strong tendency to lose ns^1 electron and hence form M^+ cations. The size of these cations also increases down the group. The size of these cations is smaller than that of the parental element, M, *i.e.* $M^+ < M$.

Group II A (Alkaline earth metals). Both atomic and ionic radii of these elements increase down the group. This increase is due to the increase in the number of shells. Atomic and ionic radii of these elements are lower as compared to those of alkali metals. The lower values of alkaline earth metals are due to the greater nuclear charge in these elements. Due to greater nuclear charge the electrons are attracted more towards the nucleus. It is on account of smaller radii of alkaline earth metals that these metals are harder, have higher densities and higher m.pts. than the elements of groups IA (alkali metals).

Alkaline earth metals can lose ns^2 electron to form M^{2+} cations. The size of these cations also increase down the group. As expected, the size of these cations are smaller than that of their parental element, M, *i.e.* $M^{2+} < M$.

Group III A. Down the group, atomic radii are expected to increase, due to the increase in the number of shells. However, as we move from Al to Ga, the atomic radius, instead of increasing, slightly decreases (Al = 1.43 Å, Ga = 1.35 Å).

Explanation. We know that the valence-shell configurations of Al and Ga are $4s^2\ 4p^1$ and $3d^{10}\ 4s^2\ 4p^1$ respectively. Since $3d$ orbitals present in the configuration of Ga are large in size, these orbitals screen the nucleus poorly and hence the effective nuclear charge in Ga becomes some what more than that in Al. Therefore, the atomic radius of Ga becomes slightly less than that of Al. In and Tl show the expected increasing trend in atomic radii. As expected, the radii of M^{3+} and M^+ cations also increase down the group.

Group IV A. The radii of the atoms of the elements of this group increase with the increase in atomic number as we move down the group from carbon to lead. The radii of M^{4+} and M^{2+} cations also show an increasing trend down the group.

Group VA. Due to the increase in the number of shells down the group, the size of the atoms (M) of these elements and of the ions *viz.* M^{3-} and M^{5+} increase. As expected the order of the size of M, M^{3-} and M^{5+} is as $M^{3-} > M > M^{5+}$.

Group VI A. The radii of the atoms (M) of the elements of oxygen family increase down the group. The radii of the ions also show an increasing trend. The radii of M, M^{2-} and M^{6+} ions are in the order $M^{2-} > M > M^{6+}$.

Group VII A (Halogens). As the halogens are the last elements in their respective periods, they have the maximum effective nuclear charge and hence have the smallest atomic radii in their respective periods. Down the group, the atomic radii increase due to the increase in the number of shells. Halogens have a strong tendency to accept an electron to form halide ions, X^-. The size of these halide ions also increases down the group. The size of the halide ions is longer than that of their parental halogen atoms (X), *i.e.* $X^- > X$.

Size of noble gas atoms. In the crystals of noble gases, the atoms of noble gases are not linked together by covalent bonds (strong bonds) but only by van der Waals forces (attractive forces) which are weak forces. Therefore, the atomic radii of noble gas atoms are actually van der Waals radii and not covalent radii. van der Waals radii represent the distance between the two adjascent noble gas atoms in the solid state. As we go down the group, the van der Waals radius increases due to the additon of new electronic shells.

Conclusion. The study of the variation of covalent and ionic radii in periods and groups of the main group elements, we have explained above, shows that these values generally decrease in moving from alkali metals to halogens and increase in moving down the group. Some irregular trends in these values have also been observed. Explanation of the regular and irregular trends of these values has already been given earlier.

The atomic radii of noble gas atoms are actually van der Waals radii and not covalent radii. These are larger than even the covalent radii of halogen atoms.

Ionisation Energy

What is ionisation energy?

Ionisation energy (IE), often mistakenly called "ionisation potential" of an element, is defined as the amount of energy required to remove an electron from the valence-shell of the isolated gaseous atom of the element, to convert the atom into the gaseous cation. Thus :

$$M(g) - e^- \dashrightarrow M^+ (g), \Delta H = \text{Energy required} = \text{IE of M atom} \qquad ... (i)$$

(*Endothermic reaction*)

Since IE is the energy required, it is represented with a positive sign, according to our usual convension of thermodynamics. The removal of an electron from a gaseous atom (called ionisation of the atom) is an *endothermic process*, since it requires energy.

Units used for expressing IE values.

IE values are expressed in electron volts per atom (eV/atom), kilo calories per mole(Kcal/mole) or kilojoules per mole (kJ/mole). These energy units are related as : 1 eV/atom = 96.4705 kJ/mole = 23.06 Kcal/mole. The IE values of *s*-and *p*-block elements are given in Table 5.3.

Table 5.3. *Ionisation energy values (in kJ/mole) of s-and p-block elements. These values have been taken from pages 40-42, Inorganic chemistry : Principles of Structure and Reactivity : 2nd edition, James E. Huheey.*

I A	II A	III A	IV A	V A	VI A	VII A	Zero
H 1312.0							He 2372.3
Li 520.3	Be 899.5	B 800.6	C 1086.4	N 1402.3	O 1314.0	F 1681.0	Ne 2080.3
Na 495.8	Mg 737.7	Al 577.6	Si 786.5	P 1011.8	S 999.62	Cl 1251.1	Ar 1520.5
K 418.9	Ca 589.8	Ga 578.8	Ge 762.2	As 944.0	Se 940.9	Br 1139.9	Kr 1350.7
Rb 403.0	Sr 549.5	In 558.3	Sn 708.6	Sb 831.6	Te 869.3	I 1008.4	Xe 1170.4
Cs 375.7	Ba 502.9	Tl 589.3	Pb 715.5	Bi 703.3	Po 812.0	At —	Rn 1037.0

Successive ionisation energies : Second, third etc. ionisation energies

IE defined by equation (*i*) is called first ionisation energy (IE_1), since it corresponds to the removal of one electron only. IE_2 is the amount of energy required to remove an electron from $M^+(g)$ ion, IE_3 is the energy required to remove an electron from M^{2+} (g) ion. In this way IE_4. IE_5 etc., can also be defined. Thus :

$$M\ (g) - e^- \longrightarrow M^+\ (g),\ \Delta H = +\ IE_1$$

$$M^{+}(g) - e^{-} \longrightarrow M^{2+}(g), \Delta H = + IE_2$$
$$M^{2+}(g) - e^{-} \longrightarrow M^{3+}(g), \Delta H = + IE_3$$

All the above processes can also be written as :

$$M(g) \xrightarrow[-e^{-}]{+IE_1} M^{+}(g) \xrightarrow[-e^{-}]{+IE_2} M^{2+}(g) \xrightarrow[-e^{-}]{+IE_3} M^{3+}(g)$$

IE_2, IE_3, IE_4 etc. are called *successive ionisation energies* of the neutral gaseous atom. It has been observed that for a given element M(g), the magnitude of successive ionisation energies are in the increasing order : $IE_1 < IE_2 < IE_3 <$.....

Explanation. The above order of successive ionisation enrgies can be explained as follows :

(*i*) *Explanation based on the size of the atom or cation.* The magnitude of ionisation energy increases with the decrease in the size of the atom or cation from which the electron is being removed. Thus, since the size of $M(g)$, $M^{+}(g)$, $M^{2+}(g)$, $M^{3+}(g)$....decreases in the order $M(g) > M^{+}(g) > M^{2+}(g) > M^{3+}(g)$...., the successive ionisation energies increase in the order $IE_1 < IE_2 < IE_3 < IE_4 <$ .....

(*ii*) *Explanation based on the effective nuclear charge experienced by the electron to be removed.* Effective nuclear charge (Z_{eff}) increases as an electron is removed from a gaseous atom or a gaseous cation. For example the values of Z_{eff} experienced by $2p^1$ electron in N ($N = 2s^2 2p^3$), N^+ ($N^+ = 2s^2 2p^2$) and N^{2+} ($N^{2+} = 2s^2\ 2p^1$) are in the increasing order (N = 3.9, N^+ = 4.25, N^{2+} = 4.60) (the students can themselves calculate these values with the help of Slater's rules). With the increase in the value of Z_{eff} from $N \rightarrow N^+ \rightarrow N^{2+}$, the values of successive iomisation energies also increase as shown in table 5.4.

Table 5.4: *Successive ionisation energies (kJ/mole) of some elements*

Element [M (*g*)] with electronic configuration	IE_1 [$M(g) \xrightarrow{-e^-} M^+(g)$]	IE_2 [$M^+(g) \xrightarrow{-e^-} M^{2+}(g)$]	IE_3 [$M^{2+}(g) \xrightarrow{-e^-} M^{3+}(g)$]
Li_3 ($1s^2$, $2s^1$)	520.3	7298.1	11814.9
Be_4 ($1s^2$, $2s^2$)	899.5	1757.1	14848.7
B_5 ($1s^2$, $2s^2p^1$)	800.6	2427.0	3659.8
C_6 ($1s^2$, $2s^2p^2$)	1086.4	2352.6	4620.5
N_7 ($1s^2$, $2s^2p^3$)	1402.3	2856.1	4578.1
O_8 ($1s^2$, $2s^2p^4$)	1314.0	3388.2	5300.4
F_9 ($1s^2$, $2s^2\ p^5$)	1681.0	3374.2	6050.4
Ne_{10} ($1s^2$, $2s^2p^6$)	2080.7	3952.3	6122.0

Important note

When an element, in its gaseous state, M(g) loses two electrons, to form a divalent cation [M^{2+}(g)], the energy required in this conversion, is equal to the *2nd* ionisation energy of the element. For example in the process,

$$Ca(g) \longrightarrow Ca^{2+}(g) + 2e^{-}$$
$$\text{Energy required} = +\ 2nd \text{ ionisation energy of Ca} = +\ 1146 \text{ kJ mol}^{-1}$$

Here (+) sign indicates the absorption of energy in the above process. It may be noted that *2nd* ionisation energy of an element is never equal to double the *1st* ionisation energy of that element.

Factors affecting the magnitude of ionisation energy

Following are the important factors which affect the magnitude of ionisation energy of the elements.

1. Nuclear charge. For the species having different charges, the force of attraction between the nucleus and the outer-most shell electron increases with the increase of nucleus charge and ence more energy is required to remove the outer-most shell electron from the atom. Thus with the ncrease of nuclear charge, the magnitude of ionisation energy also increases. This factor has been used to explain the following :

(*a*) Ionisation potential of Li^+ ion ($1s^2$) is greater than that of He-atom ($1s^2$) [Li^+ = 7298.1

KJ/mole, He = 2372.3 KJ/mole], although the electron to be removed belongs to the same orbital (*viz.* 1*s* orbital) in both the species. Greater value of ionisation potential for Li^+ ion is explained as follows. Both the species have different nuclear charges (Li^+ = 3, He = 2), but the number of electrons in both the species is the same (Li^+ = 3 – 1 = 2, He = 2). Due to greater nuclelar charge on Li+ ion, the force of attraction between the nucleus and the outer-most shell electron cloud increases and hence the value of ionisation potential also increases.

(*b*) With a few exceptions, the ionisation potentials of representative elements increase on proceeding from left to right in a period. For explanation see "Variation of IE values in main group elements" discussed on page 116.

2. Size of the atom or ion. Greater is the size of an atom or an ion, more far is the outer-most shell electron from the nucleus and hence lesser is the force of attraction between the nucleus and the outer-most shell electron and lesser amount of energy will be required to remove the outer-most shell electron. Consequently lesser will be the ionisation energy of the atom or its ion. Thus the magnitude of ionisation energy of an atom or an ion is inversely proportional to its size. This factor has been used to explain the following :

(*a*) The magnitude of 1*st* (IE_1), 2*nd* (IE_2), 3rd (IE_3) etc. ionisation potentials of an element, M(*g*) are in the order : $IE_1 < IE_2 < IE_3 <$ .... Since the size of M(*g*), $M^+(g)$ $M^{2+}(g)$ etc. is in the order : M(g) > $M^+(g)$ >$M^{2+}(g)$ > ...., the ionisation potentials of these species should be in the reverse order *i.e.* $IE_1 < IE_2 < IE_3 <$..... .

(*b*) On proceeding from top to bottom in a group of the representative elements of the periodic table, ionisation potential values decrease. For explanation, see page 117.

3. Principal quantum number (*n*). Greater is the value of *n* for the valence-shell electron of an atom, farther away this electron will be from the nucleus and hence lesser will be the force of attraction between the nucleus and the valence-shell electron. This means that lesser amount of energy will be required to remove the valence-shell electron. Thus with the increase of the principal quantum number (*n*) of the orbital from which the electron is to be removed, the magnitude of the ionisation energy decreases. This factor has been used to explain the following :

(*a*) For Mg-atom ($1s^2$, $2s^2p^6$, $3s^2$) the value of IE_3 is much greater than those of IE_1 and IE_2 both as shown in bracket (IE_1 = 737.7, IE_2 = 1450.7, IE_3 = 7732.8 KJ/mole). This can be explained as follows. IE_1 and IE_2 are ionisation energies required to remove an electron from 3s-orbital which has higher value of *n* (= 3) while IE_3 is the ionisation energy required to remove an electron from 2*p*-orbital which has a lower value of *n* (= 2). With the decrease of the value of principal quantum number of the orbital from which the electron is to be removed, the value of ionisation potential increases. Consequently the value of IE_3 is much greater than those of IE_1 and IE_2 both.

(*b*) IE_3 value for Be-atom ($1s^2$, $2s^2$) is much greater than IE_3 value of B-atom ($1s^2$, $2s^2p^1$) (IE for Be = 14848.7 KJ/mole, IE_3 for B-atom = 3659.8 KJ/mole). IE_3 for Be-atom is the energy required to remove an electron from 1*s*-orbital which has a lower value of *n* (=1) while IE_3 for B-atom is the energy required to remove an electron from 2*s* orbital which has a higher value of *n* (=2). With the increase of the value of *n* of the orbital from which the electron is to be removed, the value of IE_3 decreases. Thus IE_3 for B-atom is much lower than IE_3 for Be-atom.

(*c*) When we pass from a noble gas (zero group) to an alkali metal (group IA) belonging to the next period, a large decrease in the value of ionisation potential is observed, as is evident from the values given in Table 5.3. This large decrease has been explained by saying that as we come from a noble gas to an alkali metal, the value of *n* for the orbital from which electron is to be removed increases and hence the value of IE decreases.

4. Shielding effect. We have already said that, due to shielding effect produced by intervenning electrons, the force of attraction between the nucleus and the outer-most shell electron decreases and hence the ionisation energy also decreases. Thus with the increase of shielding effect, the magnitude of ionisation energy decreases. This factor has also been used to explain why a large decrease in the values of ionisation potential is observed, when we pass from a noble gas to an alkali metal. In alkali metals the electron to be removed from *ns* orbital is effectively screened or shielded by the electrons present in (*n* – 1) *th* shell, while in noble gases (ns^2p^6) the electron to be removed from *np* orbitals is *not effectively* shielded by the remaining five electrons residing in *np* orbitals. Due to greater shielding effect in alkali metals, ionisation potentials of alkali metals become smaller than those of noble gases.

5. Energy of the orbital from which the electron is to be removed. The relative order of energy of *s*, *p*, *d* and *f* orbitals of a given *nth* shell is as $ns < np < nd < nf$. This order implies that the relative ease with which the electron residing in these orbitals can be removed will be as $ns < np < nd < nf$. This order clearly indicates that to remove an electron from *f*-orbital of a given shell will be the easiest while to remove the same from *s-orbital* belonging to the same shell will be the most difficult, *i.e.* the amount of energy required to remove an electron from *f*-orbital is minimum while that required to remove an electron from *s*-orbital is maximum. In other words, other factors remaining the same, the ionisation energies of an electron in *s, p, d* and *f*-orbitals are in the order : $ns > np > nd > nf$.

6. Half-filled and completely filld orbitals. According to Hund's rule, half-filled (ns^1, np^3, nd^5) or completely filled (ns^2, np^6, nd^{10}) orbitals are comparatively more stable and hence more energy is needed to remove an electron form such orbitals. This means that the ionisation energy of an atom having half-filled or completely filled orbitals in its ground state electronic configuration is relatively higher than that expected normally from its position in the periodic table.

The relative order of the stability of p^3, p^6, d^5 and d^{10} configurations has been found as $d^5 < p^3 < d^{10} << p^6$. The concept of the high stabilty associated with s^2 configuration has been used to explain why the IE values of the first three elements of group II A (*i.e.* Be, Mg and Ca) are greater than those of the first three elements of group III A lying in the same period respectively (*i.e.* B, Al and Ga). Note that Sr and Ba (elements of group II A) have less IE values than In and Tl (elements of group III A) respectively. The extra stability associated with p^3 configuration has been employed to explain why IE values of the first three elements of group VA (*i.e.* N, P and As) are greater than those of the first three elements of group VI A lying in the same period (*i.e.* O, S and Se). Here also Sb and Bi have less IE than Te and Po (elements of group VI A) respectively. Noble gases have the highest IE values in their respective periods, since the atoms of noble gas elements have very stable octet of electrons (ns^2p^6 configuration) in their valence-shell.

Variation of IE values in main group elements

(*a*) **Variation in a period.** On proceeding from left to right in a period of main group elements of the periodic table, the nuclear charge increases by +1 at each next element. The increase in the nuclear charge increases the electrostatic force of attraction between the nucleus and the valence-shell electron and hence the magnitude of IE increases. Thus, on proceeding from left to right, the IE values, in general, increase with the successive increase in the magnitude of nuclear charge. Consequently, if we consider the IE values of the elements of 2*nd* period, these should increase from Li to Ne, *i.e.* the order should be Li < Be < B < C < N < O < F < Ne. The experimental values, as given in Table 5.3, however, show that the actual order is Li < B < Be < C < O < N < F < Ne. This order can be explained as follows : since the completely-filled 2*s* orbital in Be atom (Be = $2s^2$) is more stable than $2s^2p^1$ configuration of the next element *viz*., B -atom (B = $2s^2p^1$) and half-filled 2*p* orbitals of N-atom (N = $2s^2p^3$) are more stable than $2s^2p^4$ configuration of the next element *viz*., O-atom (O = $2s^2p^4$), both these elements (*i.e.* Be and N) have greater IE values than B and O-atoms respectively. Thus the actual order of IE values of the elements of 2*nd* period becomes Li < Be > B < C < N > O < F < Ne or Li < B < Be < C < O < N < F < Ne.

Similarly, it can be shown that the actual order of IE values of the elements of 3*rd* period is : Na <Al < Mg < Si < S < P < Cl < Ar. Greater values of Mg and P as compared to those of Al and S respectively have been explained in the same manner in which these values of Be and N as compared to B and O respectively have been explained.

It may be seen that in each period, alkali metals have the lowest IE values while the noble gases have the highest IE values. The highest values for noble gases can be explained on the basis of the fact that greater amount of energy is required to remove an electron from the highly stable 8-electron configuration (ns^2p^6 configuration).

(*b*) **Variation in a group.** On descending a group, IE values of the elements go on decreasing. The following two explanations have been suggested :

(*i*) On proceeding down the group, the size of the atoms and nuclear charge, both increase simultaneously. The increase in size of the atoms tends to decrease the magnitude of ionisation energies, while the increase in nuclear charge tends to increase the magnitude of ionisation energies. It appears that the increase in the size of atoms dominates the increase in the nuclear charge and hence ultimately IE values of the elements gradually go on decreasing down the group.

(*ii*) As we move down a group, the number of inner-shell electrons (*i.e.* intervening electrons) increases and hence the magnitude of shielding effect caused by these electrons on the valence-shell electron increases. With the increase of the magnitude of shielding effect, the ionisation energy decreases.

Variation of ionisation energy of the elements of different groups

Here, we shall discuss the variation of ionisation energy of the elements of main group elements, belonging to groups IA, IIA, IIIA, IVA, VA, VIA, VIIA and noble gases.

Group IA elements (Alkali metals). Since the atoms of these elements have the largest radii in their respective periods, these elements have the lowest values of their ionisation energy. In fact, in each period of the periodic table, the alkali metals have the lowest values of their ionisation energy. On going down the goup from Li to Cs, the ionisation energy decreases, as expected.

Formation of M^+ and M^{2+} ions. Since the ionisation energy values of alkali metals are the lowest, these metals can easily lose their ns^1 electron and can thus form M^+ ions quite easily. Since M^+ ion has stable noble gas congiguration *viz.* ns^2 (for Li^+) or ns^2np^6 (for Na^+, K^+, Rb^+, Cs^+) in the valence-shell, the energy required to pull out an electron from M^+ ion to convert this ion into M^{2+} ion is very high and hence under ordinary conditions, M^+ ion is not converted into M^{2+} ion. Consequently the chemistry of alkali metals is the chemistry of M^+ ions and these metals form ionic compounds which contain alkali metals as M^+ ions.

Alkali metals also form covalent diatomic molecules of A_2 type (A = Li, Na etc.) in which the two A atoms are linked together by a covalent bond. Li also forms organo metallic compounds which are covalent in nature.

Group II A elements (Alkaline earth metals). As expected, the first and second ionisation energy values decreases on proceeding from Be to Ba down the group. However, the ionisation energy of Ra is slightly greater than that of Ba. This abnormal behaviour of Ra cannot be explained satisfactorily.

Elements :	Be	Mg	Ca	Sr	Ba	Ra
1st ionisation energy (KJ mol^{-1})	*899.5*	*737.7*	*589.8*	*549.5*	*502.9*	*509.0*
2nd ionisation energy (KJ mol^{-1})	*1757.1*	*1450.7*	*1146.0*	*1064.0*	*965.0*	*979.0*

Formation of M^+, M^{2+} and M^{3+} ions. Since the ionisation energies of alkaline earth metals are low, they should lose one electron from *ns* orbital easily and hence form M^+ ions. However, in actual practice, these metals prefer to form M^{2+} ions and not M^+ ions. This abnormal behaviour of the metals can be explained by studying the energy change involved in the conversion of CaCl (*s*) to $CaCl_2(s)$. This conversion is represented as:

$$2CaCl\ (s) \longrightarrow CaCl_2\ (s) + Ca(s)$$

In this conversion heat equal to (–160 + S) KJ mol^{-1} is released. Here S = heat of sublimation of calcium. Since the energy is a negative quantity, the conversion of CaCl(*s*) to $CaCl_2(s)$ is spontaneous, *i.e.* Ca prefers to form Ca^{2+} ion and not Ca^+ ion. The greater stability of $CaCl_2$ (which consists of Ca^{2+} and Cl^- ions) over CaCl (which consists of Ca^+ and Cl^- ions) is also confirmed by the fact that the lattice energy of $CaCl_2$ (= 2200 KJ mol^{-1}) is greater than that of CaCl (= 720 KJ mol^{-1}). The metals of this group never form M^{3+} cations, since the value of 3*rd* ionisation potential is very high as third electron has to be removed from a highly stable ns^2 or ns^2p^6 configuration of M^{2+} cation.

Tendency of Be-atom to form ionic compounds. Since the 2*nd* Ionisation energy of Be-atom is the highest (= 1757.1 KJ mol^{-1}), it does not lose the two electrons from *ns* orbital readily to form

Be^{2+} ion. Accordingly, Be does not form ionic compounds readily. Rather, its compounds with O_2, Cl_2, N_2, S, etc. are more covalent in nature than the corresponding compounds of Mg, Ca, Sr and Ba.

Elements of group III A. IE values (in KJ/mol) of the elements of group III A are as B = 800.6, Al = 577.6, Ga = 578.8, In = 558.3 and Tl = 589.3. These values show that IE decreases sharply from B to Al, remains practically the same in Ga, again decreases from Ga to In and again shows a marginal increase from In to Tl, which is the last element of this group. Thus we see that these values change in an *oscillating* manner.

Explanation. The IE of B is the highest in the group, since its atomic radius is the smallest (B = 0.85 Å, Al = 1.43 Å, Ga = 1.35 Å, In = 1.67 Å, Tl = 1.70 Å) and the shielding of the nuclear charge by the intervenning $1s^2$ electrons is the lowest (B = $1s^2$, $2s^2p^1$). The sharp decrease from B to Al is due to an appreciable increase in atomic radius and effective shielding of the nuclear charge by the intervenning $2s^2$ and $2p^6$ electrons (Al = 2, $2s^2p^6$, $3s^2p^1$). In case of Ga, there are ten $3d$ electrons in its inner shell *viz.* $3rd$ shell (Ga = 2, 8, $3s^2p^6d^{10}$, $4s^2p^1$). Since d-electrons shield the nuclear charge less effectively than the s-and p- electrons, the outer electron *i.e.* ($4p^1$ electron) in Ga is held fairly strongly by the nucleus. Consequently the IE value of Ga remains almost the same as that of Al. In case of In, since the number of d-electrons *viz.*, $4d$ electrons (In = 2, 8, 18, $4s^2p^6d^{10}$, $5s^2p^1$) is the same as in $3d$ orbitals of Ga, their screening effect also remains the same. However, since the size increases as we move from Ga to In, the IE value decreases fom Ga to In. The last element *viz.*, Tl has fourteen $4f$ electrons (Tl = 2, 8, 18, $4s^2p^6d^{10}f^{14}$, $5s^2p^6d^{10}$, $6s^2p^1$). Since the shielding effect of f-electrons is much smaller than that of d-electrons, and also since there is only a slight increase in atomic size from In to Tl, the outer electrons in Tl are held quite firmly and hence the value of IE shows a marginal increase, when we proceed from In to Tl.

Although the nuclear charge of these elemens is *larger* and their size is *smaller* than that of the elements of Group II A, the first ionisation energies of these elements are lower than those of the elements of Group II A. This is due to the fact that the p-electrons (Group III A) are less *penetrating* and more shielded than s-electrons (Group II A), *i.e.*, p-electrons, being at a greater distance from the nucleus than the s-electrons, are held *less tightly* with the nucleus and hence can be removed more easily than the s-electrons.

The second and third ionisation energies are considerably higher because the successive electrons are to be removed from s-orbital.

In this group, since the sum of IE_1, IE_2 and IE_3 is very high for boron, boron is not able to exist as B^{3+} ion in its compounds. In fact this sum, for boron is the highest, because of its extremely small size and because of the fact that its valence-electrons are not effectively shielded from the nuclear charge by the intervening electrons. However, as we move down the group, the total ionisation energies decrease, but not in a regular order.

Tendency of boron atom to form ionic compounds. Values of IE_2 and IE_3 for the elements of this group are given as : (in KJ mol^{-1})

Elements :	B	Al	Ga	In	Tl
IE_2 :	2427	1803	1813	1623	1954
IE_3 :	3659.8	2723	2942	2682	2855

IE_3 value of boron is so high that this element has little tendency to form B^{3+} ion. Accodingly, B does not form ionic compounds readily. Most of the compounds of boron are covalent in nature. Due to comparatively smaller value of IE_3 for other elements, these elements have relatively higher tendency to give M^{3+} ions (M = Al, Ga, In) under suitable conditions. Thus Al, Ga and In prefer to form ionic compounds in which these elements are present as Al^{3+}, Ga^{3+} and In^{3+} respectively.

Elements of group IVA. As expected, the values of IE of the first four elements of group IV A (*i.e.* C, Si, Ge and Sn) show regular decrease from C to Sn, but the value of IE for Pb (last element of this group) shows a marginal increase as we move form Sn to Pb (C = 1086.4 KJ mol^{-1}, Si = 786.5, Ge = 762.2, Sn = 708.6, Pb = 715.5).This marginal increase is because of the following two factors :

(*i*) Fourteen 4*f* electrons present in the inner configuration of Pb (Pb = 2, 8, 18, $4s^2p^6d^{10}f^{14}$; $5s^2p^6d^{10}$, $6s^2p^2$) shield the nuclear charge less effectively than the 5*d* electrons.

(*ii*) There is only a marginal increase in the atomic radius as we move from Sn to Pb (Sn = 1.41 Å, Pb = 1.47 Å).

It is because of the above two-factors that the ionisation energy of Pb, instead of decreasing, shows a marginal increase.

Elements of group VA . As expected, the ionisation energy values of these elements decrease symmetrically as we move down the group from N to Bi (N = 1402.3 KJ mol^{-1}, P = 1011.8, As = 944.0, Sb = 831.6, Bi = 703.3).

Elements of group VI A. As we move down the group from O to Po, the ionisation energy values decrease systematically. This successive decrease takes place as expected (O = 1314.0 KJ mol^{-1}, S = 999.62, Se = 940.9, Te = 869.3, Po = 812.0).

Elements of group VII A (Halogens). Since the atoms of these elements have the smallest radii in their respective periods, these elements have the largest values of their ionisation energy.

In fact, in each period of the periodic table, the halogens have the highest values (except noble gases). The high values indicate that these elements have very little tendency to lose electrons. On going down the group from F to I, the ionisation energy decrease as expected. The last element namely iodine has comparatively very low value and hence has a tendency to lose one or more electrons to form the cations like I^+ and I^{3+}. Since F has the highest value of ionisation energy, it is not able to lose an electron to form F^+ cation.

Noble gases. The ionisation energies of noble gases are very high. These values are higher than even those of halogens. Their high values are due to the presence of stable ns^2np^6 configuration in their valence shell. These values decrease with the increase in van der Waals radii of the atoms of these elements when we move down the group from He to Rn.

Conclusion. From the values of ionisation energy, as given in Table 5.3, it may be seen that these values decrease regularly, as we move down the groups (groups IA and IIA) and increase across a period of *s*-block elements, but some *p*-block elements show irregular trends in these values. For example, the ionisation energy values of the elements of group IIIA and IVA donot decrease regularly, on moving down these groups. Similarly, these values for the elements of 2*nd* and 3*rd* periods donot increase regularly in moving from left to right. The ionisation energy values for the elements of 2*nd* period are in the order : Li < Be > B < C < N > O < F. The explanation of this unexpected variation has already been given at the appropriate place.

Ionisation energies of iso-electronic species

We know that (*i*) the size (*r*) of iso-electronic species decrease with the increase of their atomic number (Z), *i.e.* $r \propto \frac{1}{Z}$ or $Z \propto \frac{1}{r}$. (*ii*) the magnitude of ionisation energy (IE) of an atom decreases with the increase of its size (*r*), *i.e.* $\text{IE} \propto \frac{1}{r}$. On combining these two relations, we can write $IE \propto Z$, *i.e.* the ionisation energies of iso-electronic species increase with the increase of the atomic number. For example the ionisation energies of iso-electronic species viz. Ca^{2+} (Z = 20), Cl^- (Z = 17), K^+ (Z = 19) and S^{2-} (Z = 16) are in the order : $S^{2-} < Cl^- < K^+ < Ca^{2+}$. All these species contain 18 electrons.

To find out the order of the second IE values of the elements of 2nd period

Due to the successive increase in the magnitude of the nuclear charge from Li to Ne, the second IE values of the elements of 2*nd* period or the first IE values of $M^+(g)$ cations should increase in the same direction, *i.e.*, these values should be in the order: $Li^+ < Be^+ < B^+ < C^+ < N^+ < O^+ < F^+ < Ne^+$. The experimental values (corresponding to $M(g)^+ \rightarrow M(g)^{2+}$ process) given in KJ/mole (Li^+ = 7298.1, Be^+ = 1757.1, B^+ = 2427.0, C^+ = 2352.6, N^+ = 2856.1, O^+ = 3388.2, F^+ = 3374.2, Ne^+ 3952.3), however, suggest that the actual order is $Be^+ < C^+ < B^+ < N^+ < F^+ < O^+ < Ne^+ < Li^+$. This order can be explained by considering the valence-shell configurations of M^+

cations given as : $Li^+ = 1s^2$, $Be^+ = 2s^1$, $B^+ = 2s^2$, $C^+ = 2s^2p^1$, $N^+ = 2s^2p^2$, $O^+ = 2s^2p^3$, $F^+ = 2s^2p^4$ and $Ne^+ = 2s^2\ p^5$. Now, since the value of principal quantum number for the 1*s* orbital in Li^+ ion is minimum (= 1), maximum amount of energy is required to remove an electron from this ion, *i.e.* IE value for Li^+ ion is the highest. Due to the extra stability associated with the completely-filled 2*s*-orbital in B^+ ion ($2s^2$) and half-filled 2*p* orbitals in O^+ ion ($2s^2p^3$), these two ions have higher IE values than their next ions, namely C^+ ($2s^2p^1$) and F^+ ($2s^2p^4$) ions respectively. Thus the actual order of IE values of M^+ cations or the IE_2 values of M-atoms, as given above, is well explained.

Difference between ionisation potential and electrode potential of a metal

Main difference between these two properties of a metal is that the ionisation potential is concerned with the formation of a gaseous metal ion from the metal atom in the solid state.

$$M(s) \quad - e^- \xrightarrow{\text{Ionisation potential}} M^+ (g)$$

On the other hand, the electrode potential (oxidation potential) is concerned with the formation of a hydrated metal ion from the metal atom in the solid state.

$$M(s) \quad - e^- \xrightarrow{\text{oxidation electrode potential}} M^+ (aq)$$

Since the electrode potential of a metal depends on its sublimation, hydration and ionisation energies, the order in which electrode potential of metals are shown in the electrochemical series may not necessarily be the same as their ionisation energies.

Electron Affinity

What is electron affinity (EA) ?

Electron affinity (EA) of an element is defined as the amount of energy released when an electron is added to the valence-shell of an isolated gaseous atom of that element, to convert the atom into gaseous uni-negative ion (anion). Thus :

(*i*) $X^\circ(g) + e^- \rightarrow X^-(g)$, ΔH = Energy released = EA of X-atom (*Exothermic process*) ...(*i*)

The addition of an electron to an atom is thus an exothernic process. Since EA is the energy involved in adding an electron to an atom, it is also called *enthalpy of electron attachment.* Again, since EA is the energy released, it should be represented with a negative sign, according to our usual thermodynamic convention, but it is unfortunate that EA values are normally represented with a positive sign. EA, defined as above, is called *first electron affinity* (EA_1), since it is related with the addition of only one electron to a neutral gaseous atom. Since $(EA)_1$ represents the energy released, $(EA)_1$ is also called *exothermic electron affinity.*

In various calculations, the value for electron affinity is used with a negative sign, since electron affinity represents the energy released. For example, in Born-Haber cycle which is used for calculating the lattice energy of NaCl ionic crystal, the electron affinity of chlorine is used with a negative sign. (See Born-Haber cycle).

Just as ionisation energy is the measure of the tendency of an atom to change into a cation, electron affinity of an atom is the measure of the tendency of the atom to change into an anion. Greater is the amount of energy released in reaction (*i*) shown above, greater is the electron affinity of the atom, X. In other words, electron affinity of an atom gives a measure of the tightness with which the aditional electron is held by that atom.

Example. Let us consider the electron affinity of chlorine atom. Electron affinity of chlorine atom (= 348.8 KJ mol^{-1}) in the gaseous state is defined as —

$$Cl\ (g) + e^- \rightarrow Cl^-\ (g),\ \Delta H \;=\; \text{Electron affinity of Cl } (g)$$
$$=\; \text{Energy released} = +\ 348.8\ \text{KJ mol}^{-1}$$

Electron affinity of Cl (*g*) atom is the energy released when 1 mole or 6.02×10^{23} atoms or 35.5 g (atomic weight of chlorine = 35.5) of chlorine in the gaseous state are converted into $Cl^-(g)$ ions.

EA values of some elements are given below in Table 5.5.

Table 5.5 : *Electron affinity (KJ/g-atom) of s-and p-block elements*

I A	II A	III A	IV A	V A	VI A	VII A	Zero
H 72.77							He ≤ 0
Li 59.8	Be ≤ 0 (= – 0.6 eV)	B 26.8	C 122.3	N ≤ 0	O 141.1	F 328.0	Ne ≤ 0
Na 52.7	Mg ≤ 0	Al 42.6	Si 133.6	P 71.7	S 200.43	Cl 348.8	Ar ≤ 0
K 48.36	Ca ≤ 0	Ga 29.0	Ge 116.0	As 77.0	Sc 194.97	Br 324.6	Kr ≤ 0
Rb 46.89	Sr ≤ 0	In 29.0	Sn 121.0	Sb 101.0	Te 190.16	I 295.4	Xe ≤ 0
Cs 45.50	Ba ≤ 0	Tl 29.0	Pb 35.2	Bi 91.4	Po 183	At 270.9	Rn ≤ 0

Relation between EA of X(*g*) atom and IE of X^-(*g*) ion

We have seen that EA of X(*g*) atom is defined by the exothermic process, $X^o(g) + e^- \rightarrow X^-(g)$, ΔH (energy released) = EA of X(*g*) atom, while IE of $X^-(g)$ ion is defined by the endothermic process, $X^-(g) \rightarrow X^o(g) + e^-$, ΔH (energy required) = IE of X^- (*g*) ion. These two equations show that :

$$-\text{EA of } X(g) \text{ atom} = +\text{IE of } X^-(g) \text{ ion}$$

Since these reactions are opposite to each other, the magnitude of EA of X(*g*) atom and IE of X^- (*g*) ion are the same, but have opposite signs, because one energy is that which is released and the other is that which is required.

Second electron affinity (EA_2)

Second electron affinity (EA_2) of a gaseous element, M(*g*) is the amount of energy required to add an electron to the uni-negative gaseous ion of that element [*i.e.* $M^-(g)$ ion]. Thus :

$$M(g) + e^- \rightarrow M^-(g), \quad \Delta H = \text{Energy released} = +EA_1$$
$$M^-(g) + e^- \rightarrow M^{2-}(g); \quad \Delta H = \text{Energy required} = -EA_2$$

Since EA_1 (energy released) is shown with a positive sign (+), EA_2 (energy required) is shown with a negative sign (–).

The above equations show that the process defining EA_1 is an *exothermic process,* while that defining EA_2 is an *endothermic process.*

EA_1 and EA_2 values for O, S and Se are given below :

Oxygen : EA_1 = 141.13 KJ mol^{-1}, EA_2 = – 844.0 KJ mol^{-1}

Sulphur : EA_1 = 200.43 KJ mol^{-1}, EA_2 = – 333.0 KJ mol^{-1}

Selenium : EA_1 = 194.97 KJ mol^{-1}, EA_2 = – 420 KJ mol^{-1}

EA_2 is always greater than EA_1.

EA_2 represents energy required. Why ?

When an electron is added to a $M^-(g)$, energy is required (and not liberated), since the electron is being added against the electrostatic repulsion, existing between the negative charge on the electron being added and the negative charge on $M^-(g)$ ion. Thus, in the process of adding one electron to $M^-(g)$ ion, energy is required and not liberated, as in case of adding one electron to M(*g*) atom. Consequently the process defining EA_2 is as endothermic process.

Example 5.4 : *Calculate the magnitude of energy involved in the conversion of (a) O(g) atom to $O^{2-}(g)$ ion (b) S(g) atom to $S^{2-}(g)$ ion. Predict whether the given conversion is an endothermic or exothermic process.*

Solution. (*a*) The conversion of O-atom to $O^{2-}(g)$ ion can be shown as :

$$O(g) + 2e^- \rightarrow O^{2-}(g)$$

The above reaction actually takes place through the following two steps.

$$O(g) + e^- \rightarrow O^-(g); \quad EA_1 = +141.1 \text{ KJ mol}^{-1}$$
$$O^-(g) + e^- \rightarrow O^{2-}(g); \quad EA_2 = -844.0 \text{ KJ mol}^{-1}$$

$$O(g) + 2e^- \rightarrow O^{2-}(g), \Delta H = +EA_1 + (EA_2) = +141.1 - 844.0 = -702.9 \text{ KJ mol}^{-1}$$

Since ΔH has negative sign, the conversion of O(*g*) to O^{2-} (*g*) requires energy, equal to 702.9 KJ mol^{-1}, *i.e.*, the conversion of O(*g*) to $O^{2-}(g)$ is an endothermic process.

(*b*)
$$S(g) + e^- \rightarrow S^-(g), EA_1 = +200.43 \text{ KJ mol}^{-1}$$
$$S^-(g) + e^- \rightarrow S^{2-}(g), EA_2 = -333.00 \text{ KJ mol}^{-1}$$

$$S(g) + 2e^- \rightarrow S^{2-}(g), \Delta H = +EA_1 + (-EA_2) = +200.43 - 338.00 = -\mathbf{132.57 \text{ KJ mol}^{-1}}$$

Thus the conversion of S(*g*) to S^{2-} (*g*) is also an endothermic process.

Example. *The amount of energy released when one million of atoms of iodine in the vapour state are converted to I^- ions is 4.9×10^{-13} J according to the reaction.*

$$I(g) + e^- \rightarrow I^-(g).$$

Express the electron affinity of iodine in terms of KJ mol^{-1} and ev per atom.

Solution. Electron affinity of iodine is the amount of energy released when one mole or 6.02 × 10^{23} atoms of iodine in the gaseous state are converted into $I^-(g)$ ions. Thus :

∵ Amount of energy released, when 1 million or 10^6 atoms of iodine are converted into $I^-(g)$ ions = 4.9×10^{-13} J

∴ Amount of energy released when 6.02×10^{23} atoms of iodine are converted into $I^-(g)$

$$\text{ions} = \frac{4.9 \times 10^{-13}}{10^6} \times 6.02 \times 10^{23} \text{ J} = 29.5 \times 10^4 \text{ J} = \mathbf{295 \text{ KJ mol}^{-1}}$$

Now, since 96.3 KJ mol^{-1} = 1 eV $atom^{-1}$

$$\therefore \quad 295 \text{ KJ mol}^{-1} = \frac{1 \times 295}{96.3} \text{ eV atom}^{-1} = \mathbf{3.06 \text{ eV atom}^{-1}}$$

Example 5.6. *The electron affinity of chlorine is 349 KJ mol^{-1}. How much energy in KJ is released when 1 g. of chlorine is converted completely into $Cl^-(g)$ ions ?*

Solution. Electron affinity of chlorine is the amount of energy released when 1 mole or 35.5 g of chlorine in the gaseous state are converted into $Cl^-(g)$ ions. Thus :

Amount of energy released when 35.5 g of chlorine are converted into $Cl^-(g)$ ions = 349 KJ mol^{-1}

∴ Amount of energy released when 1 g of chlorine is converted into $Cl^-(g)$ ions

$$= \frac{349}{35.5} \text{ KJ mol}^{-1} = \mathbf{9.83 \text{ KJ mol}^{-1}}$$

Important note.

When two atoms of an element, in the gaseous state [2X(*g*)], gain two electrons, to form two monovalent anions [$2X^-(g)$], the energy released in this process is equal to the double of the electron affinity energy of that element. For example in the process,

$$2Cl(g) + 2e^- \rightarrow 2Cl^-(g)$$

Energy released =– 2 × Electron affinity energy of Cl-atom

$$= -2 \times 348.8 \text{ KJ mol}^{-1} = -697.6 \text{ KJ mol}^{-1}$$

Here (–) sign indicates the *release* of energy in the above process.

Factors affecting the magnitude of electron affinity.

Important factors, which affect the magnitudes of electron affinity values of elements, are discussed below :

1. Size of the atom. In case of smaller atoms, the attraction of the nucleus, for the electron to be added, is stronger. Thus, smaller is the size of the atom, greater is its electron affinity. (See electron affinity values of halogens).

2. Nuclear charge. Greater is the magnitude of nuclear charge of elements (along a period), stronger is the attraction of its nucleus for the electron to be added. Thus, with the increase in the magnitude of nuclear charge, electron affinity also increases.

3. Electronic configuration of the atom. How are the values of EA of elements affected by the electronic configuration of the atoms of the elements, can be explained by considering the following examples :

(*a*) *EA values of the elements of groups II A and II B.* The atoms of the elements of group II A have completely-filled *ns* orbital (ns^2 configuration) and those of the elements of group IIB have completely-filled $(n-1)$ *d* and *ns* orbitals [$(n-1)\ d^{10}.\ ns^2$ configuration] and hence the atoms of the elements of both these groups, either do not accept the electron being added to them or have a very little tendency to accept the electron. When they donot accept the electron, they show zero EA values.Even if they accept the electron, they do not liberate energy; rather they absorb some energy in the process of adding the electron to them, *i.e.* they have negative (*i.e.* endothermic) values of electron affinity. The elements of group IIA have completely-filed ns^2 orbitals and hence are reluctant to accept an electron. Therefore, they have negative (*i.e.* less than zero) electron affinity values. Similarly, since Zn, Cd and Hg have completely-filled *ns* and $(n-1)$ *d* orbitals [$(n-1)\ d^{10}$ ns^2 configuration], they show reluctance to accept an electron and hence have negative EA values (Zn = – 47 KJ/mole, Cd = – 32, Hg = – 61).

(*b*) *EA values of nitrogen and phosphorus.* EA values of N and P atoms are very low. This is because of the presence of half-filled *np* orbitals in their valence-shell (N = $2s^2p^3$, P = $3s^2p^3$). These half-filled orbitals, being very stable, have very little tendency to accept any extra electron to be added to them. Thus, EA values of N and P are very small.

(*c*) *EA values of noble gases.* Noble gases have stable ns^2p^6 configuration (helium has $1s^2$ configuration) and hence the atoms of these gases, either donot accept any extra electron, or have a very little tendency to accept the electron. Consequently these elements have either close to zero or slightly negative (corresponding to the endothermic process) EA values.

Variation of electron affinity in main group elements of the periodic table

(*a*) Variation in a period. On moving from alkali metals to halogens in a period, the size of the atoms decreases and the nuclear charge increases. Both these factors increase the force of attraction between the nucleus and the electron being added and hence the atom has a greater tendency to attract the electron towards itself. Thus *the electron affinity values go on increasing (generally) when we move from alkali metals to halogens.* From this it follows that in each period, alkali metals which lie to the extreme left of the periodic table, have the lowest values of electron affinity and the halogens which lie at the extreme right of the periodic table, have the highest values of electron affinity. Their highest values are because of the fact that, on gaining an extra electron, the halogen atoms attain the stable noble gas configuration (ns^2p^6 configuration).

(*b*) Variation in a group. On moving down a group, both the size of the atom and the nuclear charge increase. The increase in atomic size tends to decrease the electron affinity values, while the increase in nuclear charge tends to increase the electron affinity values. The net result is that the effect produced by the progressive increase of the atomic size outweighs the effect produced by the progressive increase in nuclear charge and *consequently the electron affinity goes on decreasing, as we move from top to bottom in a group.* The decrease in electron affinity values in a group is evident from the values (*i*) for the elements of group IA (H > Li > Na > K> Rb > Cs) (*ii*) for the elements namely S, Se, Te and Po (S > Se > Te > Po — group VI A) and (*iii*) for the elements *viz* Cl, Br, I, At (Cl > Br > I > At - Group VII A). It may be noted that the electron affinity value of O is less than that of S (O < S) in group VI A and similarly the electron affinity value of F is less than that of Cl (F < Cl) in group VIIA. Thus the actual order of electron affinity values for the elements of groups VIA and VIIA is O < S > Se > Te > Po or S > O > Se > Te > Po (group VII A) and F < Cl > Br > I > At or Cl > F > Br > I > At (group VII A). Why the electron affinity values of O and F are less than those of S and Cl respectively, has been explained as follows :

Due to small size of O and F atoms, as compared to the size of S and Cl atoms respectively, the addition of an extra electron to O and F atoms produces high electron density round them. The high electron density produced in this way, screens the nucleus and hence nuclear charge is decreased, *i.e.*, the repulsion between the electrons already present in the relatively compact $2p$ orbitals of the valence-shell of these atoms [4 in O-atom (O = $2s^2p^4$) and 5 in F-atom (F = $2s^2p^5$)] and the extra electron being added to these atoms to get O^- and F^- ions increase. Due to the increased (*i.e.* strong) electron-electron repulsion, O and F atoms show lesser tendency to attract an electron towards them to form O^- and F^- ions respectively and hence the electron affinity values of these elements become less than those of S and Cl atoms respectively.

Variation of electron affinity values of elements of different groups

Here we shall consider the variation of electron affinity of the elements of groups IA, IIA, IIB, III A, IVA, VA, VIA, VIIA and zero (noble gases).

Elements of group IA (Alkali metals). Since the atomic radii of the atoms of alkali metals increase as we move down the group, the EA values of these elements decrease in the same direction.

Elements of group IIA and II B. See page 123.

Elements of groups IIIA and IVA. EA values of the elements of these groups do not show the expected decreasing trend, as we move down the group. No satisfactory explanation for this unsystematic trend is available.

Elements of group VA. EA values of these elements do not show a regular trend. EA values of N and P are very low. The reason for their low values has already been given on page 123.

Elements of group VIA. EA values decrease gradually, as we move down the group from S to Po. EA value of O-atom is lower than that of S-atom although it is expected to be higher. The reason for this abnormal behaviour of O-atom has already been given on page 123.

Second electron affinity (EA_2) for O, S and Se atoms is also known. For detailed discussion, see page 121.

Elements of group VIIA (Halogens). Since the atomic radii of the atoms of halogens are the smallest in each period, these elements have the maximum tendency to gain an electron to form $X^-(g)$ ion. Thus, halogens have the highest EA values. Moreover, a halogen atom, $X(g)$, after gaining an electron, acquires stable noble gas configuration (ns^2np^6). EA values decrease gradually as we move down the group from Cl to I (Cl > Br > I). EA value of F is lower than that of Cl (F < Cl). Thus the actual order of EA values of halogens is : F < Cl > Br > I. The reason for the unexpected lower value of F, as compared to that of Cl, has already been given above.

Noble gases. See page 123.

Condition for the formation of an ionic compound by the combination of a metal with a non-metal : A combined application of ionisation potential and electron affinity

We know that the elements lying at the extreme left of the periodic table (*i.e.* metals of groups IA and IIA) have the lowest values of ionisation energy and hence can form cations easily by losing one or more electrons. The elements lying at the extreme right of the periodic table (*i.e.* non-metals of groups VIA and VIIA) have the highest values of electron affinity and hence can form anions readily by accepting one or more electrons. Now a large number of cations and anions thus formed, due to the electrostatic force of attraction between them, combine together and form an ionic crystal. This clearly shows that the essential condition for the formation of an ionic crystal by the combination of two elements, is that the element forming the cationic part of the ionic crystal should have low value of ionisation energy and the element forming the anionic part of the ionic crystal, should have high value of electron affinity. Consequently, the combination of a metal and a non-metal gives an ionic compound.

Determination of electron affinity of an element.

Electron affinity of an element, say Cl, can be determined by Born-Haber cycle. With the help of this cycle the following equation has been deduced for the determination of lattice energy of NaCl ionic crystal.

$$-(\Delta H_{for})_{NaCl} = (\Delta H)_{Na} + (I.P.)_{Na} + \frac{1}{2}(\Delta H_{disso})_{Cl_2} - (E.A)_{Cl} - (\Delta H_{lat})_{NaCl}$$

Obviously, electron affinity of Cl, $(EA)_{Cl}$ can be determined, if other energy terms namely heat of formation of NaCl, heat of sublimition of Na, ionisation potential of Na, heat of dissociation of Cl_2 and lattice energy of NaCl are known.

Electronegativity

Definition and methods (scales) of calculating electronegativity

The term electronegativity has been defined in different ways by different chemists and hence each of them has proposed his own definition and method to calculate (or measure) it. Different definitions and methods for its calculation are given below :

1. Pauling's approach (1932). According to Pauling, electronegativity of an atom of an element in a molecule is its relative tendency (or power or ability) to attract the shared electron pair towards itself.

Pauling calculated electronegativity values of elements by a method in which he considered the formation of AB covalent molecule by the combination of $\frac{1}{2}A_2$ and $\frac{1}{2}B_2$ molecules. Thus :

$$\frac{1}{2}A_2 + \frac{1}{2}B_2 \longrightarrow AB \qquad ...(i)$$

Now, since A_2, B_2, and AB are covalent molecules, they can also be represented as A–A, B–B and A–B respectively. Thus, equation (*i*) can also be written as :

$$\frac{1}{2}(A-A) + \frac{1}{2}(B-B) \longrightarrow A-B \qquad ...(ii)$$

He further said that bond dissociation energy of A–B covalent bond (E_{A-B}) is always greater than the geometric mean of the bond dissociation energies of A–A (E_{A-A}) and B–B (E_{B-B}) covalent bonds, *i.e.*

$$E_{A-B} > (E_{A-A} \times E_{B-B})^{1/2}$$

The difference between E_{A-B} and $(E_{A-A} \times E_{B-B})^{1/2}$ is called *ionic resonance energy* of A – B bond and is denoted by Δ_{A-B}. Thus :

$$\Delta_{A-B} = E_{A-B} - (E_{A-A} \times E_{B-B})^{1/2}$$

or

$$(\Delta_{A-B})^{1/2} = [E_{A-B} - (E_{A-A} \times E_{B-B})^{1/2}]^{1/2} \qquad ...(iii)$$

Pauling further suggested that the square root of Δ_{A-B} [*i.e.* $(\Delta_{A-B})^{1/2}$] is a measure of the ionic character in A – B covalent bond. With the increase in the magnitude of $(\Delta_{A-B})^{1/2}$, the amount of ionic character in A – B covalent bond also increases, *i.e.*

$$(\Delta_{A-B})^{1/2} \propto \text{Amount of ionic character in A – B bond} \qquad ...(a)$$

Now, the difference in electronegativity values of A and B atoms (*i.e.* $x_A - x_B$, if $x_A > x_B$) also increases with the increase in the amount of ionic character in A – B bond, *i.e.*

$$(x_A - x_B) \propto \text{Amount of ionic character in A – B bond} \qquad ...(b)$$

On combining (*a*) and (*b*) relations, we get :

$$(x_A - x_B) \propto (\Delta_{A-B})^{1/2} \quad \text{or} \quad (x_A - x_B) = K(\Delta_{A-B})^{1/2} \qquad ...(iv)$$

Here, K is a constant. On putting the value of Δ_{A-B} from equation (*iii*), equation (*iv*) becomes.

$$(x_A - x_B) = K[E_{A-B} - (E_{A-A} \times E_{B-B})^{1/2}]^{1/2} \qquad ...(v)$$

If bond dissociation energies (*i.e.* E_{A-B}, E_{A-A} and E_{B-B}) are expressed in Kcal (and not in KJ), the value of K has been found to be equal to 0.182. Thus equation (*v*) becomes

$$(x_A - x_B) = 0.182\,[E_{A-B} - (E_{A-A} \times E_{B-B})^{1/2}]^{1/2} \qquad ...(vi)$$

Equation (*vi*) is called Pauling's equation. Pauling assumed an orbitrary value of electronegativity for H-atom to be equal to 2.2 and determined the values for other elements with the help of equation (*vi*). These values for *s*-and *p*-block elements, as calculated by Pauling, are given in Table 5.6. The disadvantage of this method is that bond dissociation energies are not known with any degree of accuracy for many solid elements.

Table 5.6 : ***Pauling's electronegativity values of main group elements***

I A	II A	III A	IV A	V A	VI A	VII A	Zero
H 2.1							He 0
Li 1.0	Be 1.5	B 2.0	C 2.5	N 3.0	O 3.5	F 4.0	Ne 0
Na 0.9	Mg 1.0	Al 1.5	Si 1.8	P 2.1	S 2.5	Cl 3.0	Ar 0
K 0.8	Ca 1.2	Ga 1.6	Ge 2.0	As 2.0	Se 2.4	Br 2.8	Kr 0
Rb 0.8	Sr 1.0	In 1.7	Sn 1.9	Sb 1.8	Te 2.1	I 2.4	Xe 0
Cs 0.7	Ba 0.9	Tl 1.8	Pb 2.3	Bi 1.7	Po 2.0	At 2.2	Rn 0

Example. *Calculate the electronegativity of carbon from the data given as E_{H-H} = 104.2 Kcal/mole, E_{C-C} = 83.1 Kcal/mol, E_{C-H} = 98.0 Kcal/mole and x_H = 2.1.*

Solution. From Pauling's equation [equation (*vi*)] we have:

$$x_A - x_B = 0.182\,[\,E_{A-B} - (E_{A-A} \times E_{B-B})^{1/2}]^{1/2}$$

Thus, $$x_C - x_H = 0.182\,[\,E_{C-H} - (E_{C-C} \times E_{H-H})^{1/2}]^{1/2}$$

or $$x_C - 2.1 = 0.182\,[98.8 - (83.1 \times 104.2)^{1/2}]^{1/2}$$

$$= 0.182\,[98.8 - 93.05]^{1/2} = 0.182 \times 2.39 = 0.43$$

$$\therefore \quad x_C = 0.43 + 2.10 = \mathbf{2.53}$$

2. Allred and Rochow's scale (1958). Allred and Rochow suggested that electronegativity of an element, A can be calculated with the help of the following equation :

$$(x_A)_{AR} = \frac{0.359 \times Z_{eff}}{r^2} + 0.744 \quad ...(i)$$

Here, $(x_A)_{AR}$ = electronegativity of element A on Allred and Rochow's scale, Z_{eff} = effective nuclear charge at the periphery of element A and r = radius of the element (A) in Å.

Now, we know that :

$$Z_{eff} = Z_A - \sigma \quad ...(ii)$$

With the help of equation (*ii*), equation (*i*) becomes,

$$(x_A)_{AR} = \frac{0.359 \times (Z_A - \sigma)}{r^2} + 0.744 \quad ...(iii)$$

The value of electronegativity of *s*- and *p*-block elements, as claculated by equation (*iii*), are given in Table 5.7.

Table 5.7: *Electronegativity values of s- and p-block elements, as calculated by Allred and Rochow.*

IA	II A	III A	IV A	V A	VI A	VII A	Zero
H 2.20							He 3.2
Li 0.97	Be 1.47	B 2.01	C 2.50	N 3.07	O 3.50	F 4.10	Ne 5.1
Na 1.01	Mg 1.23	Al 1.47	Si 1.74	P 2.06	S 2.44	Cl 2.83	Ar 3.3
K 0.91	Ca 1.04	Ga (III) 1.82	Ge (IV) 2.02	As (III) 2.20	Se 2.48	Br 2.74	Kr 3.1
Rb 0.89	Sr 0.99	In 1.78	Sn (IV) 1.72	Sb 1.82	Te 2.01	I 2.21	Xe 2.4
Cs 0.86	Ba 0.97	Tl (III) 1.44	Pb (IV) 1.55	Bi 1.67	Po 1.76	At 1.90	Rn —

Example. *Calculate the electronegativity of lead following Allred-Rochow procedure. Use the following data : Covalent radius of lead = 1.53 Å and atomic number of lead = 82.*

Solution. We know that the electronic configuration of Pb (Z = 82) is

$$\underbrace{1s^2,\ 2s^2p^6,\ 3s^2p^6d^{10},\ 4s^2p^6d^{10}f^{14}}_{(60)},\ \underbrace{5s^2p^6d^{10}}_{(18)},\ \underbrace{6s^2p^2}_{(4)}.$$

According to Slater's rules, we have :

σ for all the electrons present in the electronic configuration of Pb-atom = 4 × 0.35 + 18 × 0.85 + 60 × 1.0 = 76.70.

$\therefore$ Z_{eff} at the periphery of Pb-atom = $Z_{Pb} - \sigma$ = 82 – 76.70 = 5.30

$$\therefore\quad x_{Pb} = \frac{0.359 \times 5.3}{(1.53)^2} + 0.744 = \mathbf{1.55}$$

3. **Mulliken's approach.** According to Mulliken, the average of ionisation energy (IE) and electron affinity (EA) of an atom is the measure of its electronegativity. He proposed that the electronegativity of an atom can be calculated with the help of the following relation.

$$x_A = 0.374\left[\frac{(IE)_A + (EA)_A}{2}\right] + 0.17$$

or

$$x_A = 0.187\,[(IE)_A + (EA)_A] + 0.17 \qquad ...(i)$$

In this equation x_A = electronegativity of atom A, $(IE)_A$ = ionisation energy of atom A and $(EA)_A$ = electron affinity of atom A. The above equation holds good when IE and EA both are measured in electron volts.

Example. *Calculate electronegativity of fluorine atom, given that ionisation energy of F-atom = 17.41 eV and electron affinity of F-atom = 3.45 eV*

Solution. According to equation (*i*), we have

$$x_F = 0.187\,[17.41 + 3.45] + 0.17 = \mathbf{4.07}$$

Factors affecting the magnitude of electronegativity

Important factors which influence the magnitude of electronegativity of elements are discussed below :

1. Size of the atom. Smaller atom has greater tendency to attract the shared electron pair towards itself and hence has greater electronegativity than a smaller atom. This factor has been used to explain the variation of electronegativity values of elements in a group and in a period of the periodic table (see below).

2. Oxidation state of the element. We know that if an atom shows many positive oxidation states, the size of the atom decreases with the increase in its oxidation state. Thus the size of Fe, Fe^{+2}, and Fe^{+3} is in the order : Fe > Fe^{+2} > Fe^{+2}.

Again, it is also known that smaller is the size of an atom or an ion, greater is its electronegativity. Thus the electronegativity of an element which shows many oxidation states is directly proportional to the magnitude of its positive oxidation state. Consequently the electronegativity of the above species is in the increasing order : Fe (= 1.80) < Fe^{+2} (= 1.83) < Fe^{+3} (= 1.96). Another example to show the variation of electronegativity of an atom with the change in its oxidation state is $\overset{+1}{HClO}$ _ $\overset{+5}{HClO_3}$ pair. Since Cl-atom is in higher oxidation state in $HClO_3$ (Cl = +5) than in HClO (Cl = + 1), Cl-atom in $HClO_3$ shows more electronegativity (*i.e.* electron pair-attracting power) than in HClO.

Due to the fact that an anion is larger in size than its parental atom, the anion has less tendency to attract the shared electron pair towards itself and hence has lower value of electronegativity, *e.g.*, F^- ion has lower electronegativity than F atom (F^- = 0.8, F = 4.0).

3. Effective nuclear charge of the element. We have already seen that according to Allred-Rochow's formula, the electronegativity of an element is proportional to the magnitude of effective nuclear charge at the periphery of the atom of that element (Z_{eff}). Thus with the increase in the magnitude of Z_{eff} of an element, the electronegativity of that element also increases. This factor has been used to explain the variation of electronegativity of elements in a group and in a period (see below).

4. Electropositive character of elements. Electropositive character of an element is opposite of its electronegative character, *i.e.*, the elements which are highly electropositive are weakly electronegative. In other words, the elements which are highly electropositive have low values of electronegativity (*e.g.* alkali metals). Similarly, the elements which have very weak electropositive character (*i.e.* non-metallic character) have very high values of electronegativity (*e.g.* halogens). In general we can say that the electronegativity of an element is inversely proportional to its eletropositive character. This factor has been used to explain the variation of electronegativity of elements in a group and in a period of the periodic table (see below).

5. Ionisation energy and electron affinity. We have already seen that, according to Mulliken, the average of ionisation energy and electron affinity of an atom is the measure of its electronegativity. Thus higher is the value of ionisation energy and electron affinity of an atom, higher is the electronegativity of that atom. Higher ionisation energy of an atom means that it is difficult to remove the most loosely bonded electron from the atom which leads us to expect that the electron affinity of that atom will also be greater. *Thus the atoms of the elements which have higher values of ionisation energy and electron affinity also have higher values of electronegativity.* For example, the elements of group VII A (halogens) which have the highest ionisation energies and electron affinities also have the highest values of electronegativity. Similarly, the elements of group IA (alkali metals) which have the lowest ionisation energies and electron affinities have the lowest values of electronegativity.

This factor has been used to explain the variation of electronegativity values of elements in a group and a period of the periodic table (see below).

6. Type of hybridisation undergone by the central atom. The electronic charge in hybrid orbitals of an atom in a molecule which has greater magnitude of *s*-character remains closer to the nucleus of that atom. This means that the electronegativity of an atom having hybrid orbital with greater magnitude of *s*-character would by high. As an example, let us see how the electronegativity of *C*-atom in methane $\left[CH_4 \text{ or } {}^{H}_{H}{>}C{<}^{H}_{H}\right]$, ethylene $\left[C_2H_4 \text{ or } {}^{H}_{H}{>}C{=}C{<}^{H}_{H}\right]$ and acetylene [C_2H_2 or H–C ≡ C–H] varies with the change of the type of hybridisation undergone by C-atom in these molecules. We know that in these molecules, C-atom is sp^3, sp^2 and sp hybridised respectively. Again it is also known that *s*-character in *sp*, sp^2 and sp^3 hybrid orbitals is 25%, 33% and 50% respectively. With the increase in the percentage of *s*-character in hybrid orbitals, electronegativity of C-atom also increases. Thus electronegativity of C-atom in CH_4, C_2H_4 and C_2H_2 is in the following increasing order :

$$CH_4 < C_2H_4 < C_2H_2$$

With the help of this factor, we can explain the following :

(i) Comparison of the acidic character of methane (CH_4), ethylene (C_2H_4) and acetylene (C_2H_2). With the increase in the electronegativity of C-atom or with the increase in the magnitude of s-character in hybrid orbitals from CH_4 to C_2H_4 to C_2H_2, the electron pair-accepting power of C-atom (*i.e.* acidic character) in these molecules also increases in the same direction, *i.e.* the acidic character of these molecules is in the order :

$$CH_4 < C_2H_4 < C_2H_2$$

The above order shows that CH_4 is neutral, C_2H_4 is slightly acidic while C_2H_2 is considerably acidic. The acidic character of C_2H_4 and C_2H_2 molecules can also be explained as follows : Due to greater electronegativity of C-atom in C_2H_4 and C_2H_2 molecules, the electron pair involved in the formation of C–H bond is shifted more towards C–atom ($C^{\delta-}$—$H^{\delta+}$) and hence when C–H bond gets broken, H-atom is obtained as H^+ ion. The formation of H^+ ion explains the acidic nature of C_2H_4 and C_2H_2 molecules. Since C-atom shows the least electronegativity in CH_4 molecule, H-atom remains neutral and hence this molecule shows neutral character.

(ii) Basic character of amines. Let us consider aniline, [$C_6H_5NH_2$ or 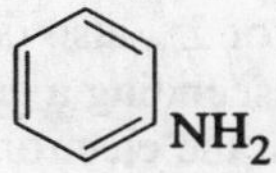], pyridine [C_5H_5N or 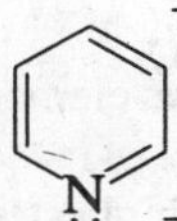] and $R-C\equiv N$: molecules. Obviously N–atom in these molecules is sp^3, sp^2 and sp hybridised. *s*-character in these hybrid orbitals is 25%, 33% and 50% respectively. We have seen above that with the increase in the magnitude of *s*-character in hybrid orbitals, the acidic character of the concerned molecules also increases. In other words we can say that with the increase in the magnitude of *s*-character in hybrid orbitals, the basic character of the molecules (*i.e.* electron pair-donating power of the central atom) decreases. Thus the basic character of the above molecules would be in the following order :

$$C_6H_5NH_2 > C_5H_5N > R-C\equiv N.$$

7. Nature of other atoms with which the concerned atom is attached in the molecule. The electronegativity of an atom in two or more molecules also depends on the nature of the atoms with which the atom is attached. As an example let us explain why the electronegativity of C-atom is different in CH_3I and CF_3I molecules. In CH_3I molecule, C-atom is less electronegative than I-atom. This creates a partial positive charge on C–atom. Thus CH_3I should be formulated as $[CH_3]^+I^-$ or $[C^{-2}H_3^+]^+ I^-$. In CF_3I molecule, because of the presence of F-atom (which is the most electronegative of all the atoms) C–atom becomes more electronegative than I-atom. This creates a partial negative charge on C–atom. Thus CF_3I should be formulated as $I^+[C^{+2}F_3^-]^-$. Thus it may be seen from this discussion that C-atom in CH_3I acquires smaller positive charge (= – 2) than in CF_3I (C = +2). In other words we can say that C–atom in CF_3I is more electronegative than in CH_3I. This factor has been used to explain why different products are obtained when CH_3I and CF_3I react with $Na^+[Mn(CO)_5]^-$ or are hydrolysed by H_2O.

(a) Reaction of $Na^+[Mn(CO)_5]^-$ with CH_3I and CF_3I. We have seen above that , due to the fact that C-atom has different electronegativity in CH_3I and CF_3I molecules, these molecules can be formulated as $[CH_3]^+I^-$ and $I^+[CF_3]^-$ respectively. When CH_3I reacts with $Na^+[Mn(CO)_5]^-$, the nucleophile, $[Mn(CO)_5]^-$ attacks on the positive site *viz.* $[CH_3]^+$ of $[CH_3]^+I^-$ and $[CH_3]^+[Mn(CO)_5]^-$ is obtained.

$$[CH_3]^+I^- + Na^+[Mn(CO)_5]^- \rightarrow [CH_3]^+[Mn(CO)_5]^- + Na^+I^-$$

On the other hand when CF_3I reacts with $Na^+ [Mn(CO)_5]^-$, the nucleophile attacks the positive site *viz.* I^+ of $I^+ [CF_3]^-$ and $I^+ [Mn (CO)_5]^-$ is produced.

$$2I^+ [CF_3]^- + Na^+ [Mn (CO)_5]^- \rightarrow NaI + C_2F_6 + I^+ [Mn(CO)_5]^-$$

(*b*) *Hydrolysis of CH_3 I and CF_3 I.* The hydrolysis of CH_3I and CF_3I by H_2O gives (CH_3OH + HI) and (CF_3H + IOH) respectively. The formation of these products can also be explained by the arguments used for (*a*) above.

$$[CH_3]^+I^- + H^+[OH]^- \rightarrow [CH_3]^+[OH]^- + H^+I^-; \; I^+[CF_3]^- + H^+[OH]^- \rightarrow H^+[CF_3]^- + I^+[OH]^-$$

Variation of electronegativity in a group of s- and p- block elements

In going down a group of *s*-and *p*-block elements, electronegativity values of the elements decrease. The decrease can be explained on the bais of any of the following facts :

(*i*) Smaller atom has greater tendency to attract the shared electron pair towards itself, *i.e.* smaller atom has greater electronegativity. Thus, with the increase in the size of the atoms down the group, the electronegativity of the elements decreases.

(*ii*) The number of electrons residing in the inner shells also increases. These elements screen the last electron. With the increase in the number of inner electrons, the magnitude of screening constant (σ) increases and hence the magnitude of Z_{eff} decreases ($Z_{eff} = Z_{actual} - \sigma$). The decrease in the magnitude of Z_{eff} also decreases the electronegativity values.

(*iii*) On descending a group, the electropositive character of the elements increases. This results in a decrease in the electronegativity values.

(*iv*) Ionisation energy and electron affinity both decrease as the group is descended. With the decrease of these quantities, the electronegativity values also decrease.

Variation of electronegativity of the elements of different groups

Here we shall discuss the variation of the electronegativity among the elements of groups belonging to main group elements.

Group IA. (Alkali metals). Since the atoms of alkali metals have the highest atomic size in their respective periods, alkali metals have the lowest tendency to attract electrons towards them and have the lowest electronegativities in their respective periods. Cesium has the lowest electronegativity of all the elements in the periodic table. It is due to the lowest electronegativity values for alkali metals that when these elements combine with the elements having high electronegativities (*e.g.* halogens), ionic compounds are obtained.

Group IIA. Like the elements of group IA, the elements of group IIA also have low electronegativity. As expected electronegativity values decrease on moving down the group. The value of Ca(= 1.2) is an exception.

Group IIIA. The members of this group are more electronegative as compared to those of alkaline earth metals group (group II A) due to their smaller atomic size. On moving down the group the electronegativity values donot show the exprected decrease.

Group IVA. Electronegativity values of the elements of this group show irregulalr trends. These values change in an oscillating manner, since electronegativity decreases from C to Si (C = 2.5, Si = 1.8) increases from Si to Ge (Ge = 2.0), again decreases from Ge to Sn (Sn = 1.9) and finally increases from Sn to Pb (Pb = 2.3). This type of variation appears to be due to the filling of *d*-orbitals in Ge and Sn and *d* and *f* orbitals in Pb (Ge = 2, 8, $3s^2p^6d^{10}$, $4s^2p^2$; Sn = 2, 8, 18, $4s^2p^6d^{10}$, $5s^2p^2$; Pb 2, 8, 18, 32, $5s^2p^6d^{10}f^{14}$, $5s^2p^6d^{10}$, $6s^2p^2$).

Groups VA and VIA. As expected, the electronegativity values of the elements of both the groups decrease gradually, as we move down the group. This decrease is because of the increase in atomic size of the elements which leads to the lesser attraction of the nucleus for the electrons. The gradual decrease in electronegativity, values down the group, indicates a gradual change from non-metallic to metallic character of the elements. Thus first element of both the groups viz. N and O respectively is exclusively a non-metal while the last element namely Bi or Po is definitely a metal. (See metallic and non-metallic character of the elements of groups VA and VIIA). Oxygen happens to be the second most electronegative element (O = 3.5) in the periodic table after fluorine.

Group VIIA (Halogens). Since the atoms of halogens have the least atomic size in their respective periods, halogens have the maximum tendency to attract the electrons towawrds them and hence have the maximum electronegativities in their respective periods. Fluorine has the maximum electronegativity (= 4.0) of all the elements in the periodic table. It is due to the highest electronegativity values for halogens that when these elements combine with the elements having low electronegativities (*e.g.* alkali metals and alkalinie earth metals), ionic compounds are obtained. These values decrrease gradually, on moving down the group, due to the increase in the atomic size. The gradual decrease down the group indicates a gradual decrease in the *non-metallic character of halogens* from F to I. Thus, the first element *viz.* fluorine, shows only non-metallic character, while the last element, iodine, shows non-metallic, as well as, metallic character (see non-metallic character of halogens). Large values of electronegativity of halogens indicate that *X* atoms have a strong tendency to form X^- ions.

Noble gases. The valence-shell configuration of the atoms of noble gases is ns^2np^6, which is very stable and hence these atoms have no tendency to accommodate any electron. Consequently noble gases have been assigned zero electronegativity values.

Problem. *Explain why the electronegativity of Si is lower than that of Ge.*

Solution. It has been observed that Ge – X bond (*X* = a halogen atom) can easily be reduced to Ge—H by nascent hydrogen (Zn/HCl) but Si – X bond cannot be reduced. (Ge – X + 2H → Ge – H + HX). This observation shows that Si – X bond is stronger than Ge – X bond. This, in turn, means that Si – X bond is more polar than Ge – X bond. This is possible only when electronegativity difference between X and Si is more than that between X and Ge $[(x_X - x_{Si}) > (x_X - x_{Ge})]$, *i.e.* the electronegativity of Si is lower than that of Ge.

Variation of electronegativity in a period of s- and p-block elements

In going from left to right in a period of *s*-and *p*-block elements (from alkali metals to halogens), electronegativity values increase, with the increase in atomic number. This increase can be explained on the basis of any of the following facts.

(*i*) Effective nuclear charge (Z_{eff}) increases, as we move from left to right in a period. With the increase in the magnitude of Z_{eff}, the magnitude of electronegativity also increases along a period from left to right. Thus, electronegativity values increase with the increase in atomic number, as we move from left to right along a period.

(*ii*) We know, that on moving from left to right in a period, the electropositive character of the elements goes on decreasing, *i.e.* the electronegative character increases in the same direction. This means that as we move from left to right in a period, the electronegativity values go on increasing.

(*iii*) On moving from left to right in a period there is a decrease in the size of the atoms. Smaller atoms have greater tendency to attract the electrons towards themselves *i.e. smaller atoms have higher electronegativity values.*

(*iv*) On moving from left to right in a period, there is an increase of ionisation energy and electron affinity of the elements. The atoms of the elements, which have higher value of ionisation energies and electron affinities, also have higher electronegativities.

The variation of electronegativity from alkali metals to halogens, as discussed above, makes it evident that in each period, alkali metals have the minimum values of electronegativity and halogens have the maximum values. Consequently, when alkali metals (which have the lowest electronegativity values) combine with halogens (which have the largest electronegativity values), ionic compounds are obtained. Noble gases have been assigned zero electronegativity values.

Conclusion. The elements of both the groups *viz,* IA and IIA, which are *s*-block elements, show the expected decreasing order in their electronegativity values, when we move down each of these groups. On the other hand, the elements of groups IIIA and IVA which are *p*-block elements show irregular trends in their electronegativity values on moving down the groups. The elements of the remaining groups *viz.* VA, VIA and VIIA (all are *p*-block elements) follow the expected decreasing trend.

The variation of electronegativity of s- and p- block elements in a period and in a group as discussed above clearly shows that Cs, which is the bottom element of the least electronegative elements (elements of group IA), has the minimum electronegativity, in the whole periodic table, while F, which is the top element of the most electronegative elements (elements of group VIIA), has the maximum electronegativity, in the whole periodic table. Other elements have intermediate values between those of Cs and F.

Applications of electronegativity

Some important facts, that can be explained on the basis of the concept of electronegativity, are given below.

1. To predict the nature of bond : Non-polar covalent bond, polar covalent bond and ionic bond. With the help of the electronegativity difference, $(x_A - x_B)$ between two atoms *viz.*, A and B, we can predict whether A–B bond would be non-polar covalent bond, polar covalent bond or ionic bond. See "*Polar and non-polar covalent bonds*" discussed in chapter 10.

2. To calculate the percentage of ionic character in A–B polar covalent bond in AB molecule. For this see "*Polar and non-polar covalent bonds*" discussed in chapter 10.

3. Explanation of the variation of bond angles. With the help of the concept of electronegativity, the variation in B–A–B bond angles of different molecules of AB_x type can be explained. For details see "*Factors affecting the magnitude of bond angles*", discussed in chapter 10.

4. Calculation of bond length, d_{A-B} in a hetero-diatomic molecule of AB type. If AB molecule has some ionic character and the radii and electronegativities of atoms A and B are represented as r_A, r_B (radii) and x_A, x_B (electronegativities), then d_{A-B} (in *pm* units) is given by Schomaker and Stevenson's equation (1941) *viz.*

$$d_{A-B} = r_A + r_B - 9\,(x_A - x_B) \qquad ...(i)$$

Here, the values of r_A and r_B are in *pm* units and hence d_{A-B} is also in *pm* units. If r_A and r_B are in Å, then d_{A-B} will be obtained in Å units and the above equation will take the form :

$$d_{A-B} = r_A + r_B - 0.09\,(x_A - x_B) \qquad ...(ii)$$

For details, see "*Bond Length*" (Chapter 10)

5. Explanation of diagonal relationship. The elements of 2*nd* period of the periodic table show similarities in properties with the elements lying to their right in 3*rd* period (see below). This is called diagonal relationship.

Elements in 2nd period : Li Be B C N O F

Elements in 3rd period : Ne Mg Al Si P S Cl

(Li → Mg, Be → Al, B → Si, C → P, N → S, O → Cl)

Why the two elements placed diagonally opposite to each other in 2*nd* and 3*rd* periods of the periodic table show similarities in properties (diagonal relationship), can also be explained on the basis of the concept of electronegativity. For details see "*Diagonal relationship*", discussed in chapter 10.

6. Calculation of enthalpy (heat) of formation of a molecule. With the help of electronegativity values of the elements, we can calculate the heat of formation of a molecule. Pauling has suggested the following equation for calculating the heat of formation of a given molecule.

$$\Delta H_f = 23\,\Sigma\,(x_A - x_B)^2 - 55.4\,x_N - 26.0\,n_O$$

Here Σ = number of bonds in the molecule [*e.g.*, $\Sigma = 1$ for HCl (H–Cl) and $\Sigma = 2$ for $BeCl_2$ (Cl – Be – Cl)], x_A and x_B = electronegativities of atoms A and B of which the molecule is composed, n_N and n_O = number of nitrogen and oxygen atoms present in the molecule. Values of ΔH_f obtained from Pauling's equation are only approximate and are in Kcal/mole.

Example 5.10. *Using Pauling's equation, calculate the heat of formation of $BeCl_2$ molecule. Given that $x_{Be} = 1.6$ and $x_{Cl} = 3.2$.*

Solution. Since $BeCl_2$ has two Be – Cl bonds, $\Sigma = 2$. Also, since there are no nitrogen and oxygen atoms present in $BeCl_2$ molecule, $n_N = 0$ and $n_O = 0$. Thus :

$$\Delta H_f \text{ of } BeCl_2 = 23 \times 2\,(3.2 - 1.6)^2 - 0 - 0. = 117.75 \text{ Kcal/mole.}$$

The experimental value of ΔH_f is 122 Kcal/mole.

7. (*a*) To explain the variation in the acidic character of the hydrides of the elements of the

same period. As an example, let us consider the acidity of CH_4, NH_3 H_2O and HF moleculeus which are the hydrides of the elements of *2nd* period. It has been observed that the acidity of the given molecules increases, as we move from CH_4 to HF ($CH_4 < NH_3 < H_2O < HF$), *i.e.* the acidity increases, as we move from left to right, in the period. The above order has been explained on the basis of the electronegativity values of the central atom of the molecules. For details see the chapter on "*Modern Concept of Acids and Bases*".

(*b*) To explain the variation in the acidic character of the oxy-acids of the elements of the same group and in the same oxidation state. As an example, let us consider perhalic acids of Cl, Br and I (group VII A) *viz.* $HClO_4$, $HBrO_4$ and HIO_4. The acidic strength of these acids decreases as the electronegativity of halogen atom (central atom) decreases as we move from Cl to I ($HClO_4 > HBrO_4 > HIO_4$). For explanation, see chapter on "*Modern Concepts of Acids and Bases*".

(*c*) To predict the acidic / basic character of hydroxy compounds. As an example, let us explain whether CsOH and IOH molecules will show acidic or basic character in aqueous solution. The values of electronegativity of different elements present in these molecules and those of $x_O - x_{Cs}$, $x_O - x_H$ and $x_O - x_I$ are given below :

$$\underset{0.8}{Cs} \overset{2.6}{—} \underset{3.4}{O} \overset{1.2}{—} \underset{2.2}{H}\ ; \qquad \underset{2.7}{I} \overset{0.7}{—} \underset{3.4}{O} \overset{1.2}{—} \underset{2.2}{H}$$

In Cs–O–H molecule, since ($x_O–x_{Cs}$) (= 2.6) > ($x_O - x_H$) (= 1.2), Cs–O bond would be more polar than O–H bond and hence the ionisation of CsOH in aqueous solution will take place at Cs – O bond in Cs–O–H and $[OH(aq)]^-$ ions will be produced.

$$Cs \ \vdots\!— O - H(aq) \rightarrow [Cs\ (aq)]^+ + [OH\ (aq)]^-$$

Due to the production of $[OH(aq)]^-$ ions, CsOH molecule will show basis character in aquecus solution. On similar lines of argument it can be shown that $Be(OH)_2$ will behave as a base.

In I–O–H molecule, since ($x_O - x_H$) (=1.2) > ($x_O - x_I$) (= 0.7), O–H bond would be more polar than O–I and hence the ionisation of IOH in aqueous solution will take place at O–H bond in I–O–H and $[H(aq)]^+$ ions will be obtained.

$$I — O \!—\!\vdots\ H(aq) \rightarrow [IO(aq)]^- + [H(aq)]^+$$

Due to the production of $[H(aq)]^+$ ions, IOH molecule will behave as an acid in aqueons solution.

Electropositive (Metallic) and Electronegative (Non-metallic) Character of Elements

Electropositive character of an element, in its gaseous state, is defined as its tendency to lose one or more electrons to undergo oxidation and thus to form gaseous cation.

$$M(g) \rightarrow M^+(g) + e^-$$

Ionisation energy (IE) of an element is the measure of its tendency to lose electrons and hence of its electropositive character. The elements having low values of ionisation energy are highly electropositive while those having high values of ionisation energy are highly electronegative.

In other words we can say, that with the decrease in the values of ionisation energy, the electropositive (metallic) character of the elements increases and the electronegative (non-metallic) character increases.

Variation of metallic and non-metallic character of the elements in the periodic table

(*a*) Variation in a group. Since the ionisation energy values decrease on decending a given group, the metallic character of elements increases and non-metallic character decreases in the same direction. This variation in best seen in the elements of groups IVA and VA.

(*b*) Variation in a period. Since the ionisation potential values of the elements increase on proceeding from left to right in a period, the metallic character of the elements decreases in the same direction. For example, when we proceed from left to right in *3rd* period, we find that the first two elements *viz.*, Na and Mg are strongly electropositive, Al also has metallic character while Si is a semi-metal (*i.e.*, Si shows metallic as well as non-metallic character). The last three elements

P, S, and Cl are non metals. Thus, when we proceed from alkali metals to halogens in a given period, the non-metallic character of the elements goes on increasing *i.e.,* alkali metals, which lie at the extreme left of the periodic table, show least non-metallic character (*i.e.,* these are metals) while the halogens, which lie at the extreme right of the periodic table, show maximum non-metallic or electronegative character, *i.e.*, halogens are non-metals.

The gradual decrease in the metallic character or the gradual increase in the non-metallic character of the elements, on proceeding from left to right in a period, is confirmed by the fact that the oxides of the elements of a given period become *more* and *more* acidic, as we proceed from left to right in the period, *e.g.* the oxides of the elements of 3*rd* period show this trend, as given here : Na_2O (*strongly basic*), MgO (*less basic*), Al_2O_3 (*amphoteric*), SiO_2 (*feebly acidic*), P_2O_5 (*acidic*), SO_3 (*more acidic*), Cl_2O_7 (*most acidic*).

Classification of the elements into metals, non-metals and semi-metals (metalliods)

We have said above that with the increase in the ionisation energy of the elements of the same period, the metallic character of the elements decreases or non-metallic character increases when we move from left to right in the period. Consequently, the elements located mostly in the lower left portion of the periodic table have strongly metallic character while those located in the top right portion of the periodic table have strongly non-metallic character. The elements like B, Si, Ge, As, Sb etc. which lie in-between the strongly metallic (metals) and strongly non-metallic (non-metals) show characteristics of both metals and non-metals, *i.e.* these elements are semi-metals or metalloids (See Fig. 5.3).

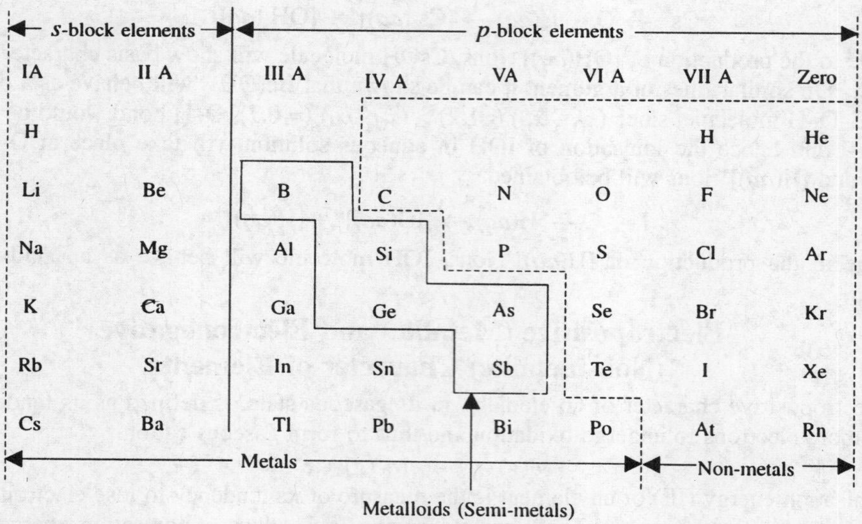

Fig. 5.3 : Division of main group elements (*s*-and *p*-block elements) into metals, non-metals and metalloids.

Defects of the above classification. Here it may be pointed out that, because of the existence of different allotropic forms of some elements, the above classification of the elements as metals, non-metals and metalloids is somewhat enigmatic. For example :

(*a*) White tin (or β–tin) is soft, ductile and good conductor of electricity. It has the usual metallic lustre. Consequently, white tin has metallic character. On the other hand, grey tin (or α–tin) is a poor conductor of electricity and hence lacks metallic properties. This allotropic form of tin resembles Si and Ge (which are semi-metals) and, therefore, has semi-metallic character.

(*b*) White phosphorus, which is composed of P_4 units and red phosphorus donot conduct elecricity and hence have non-metallic character. On the other hand, black phosphorus, which has a sheet-like structure, conducts electricity to a limited extent and hence behaves as a semi-metal.

(*c*) Se exists in two allotropic forms. The *red form* consists of eight-membered rings (Se_8) and is non-metallic, while the *grey form* has a chain structure. It has electrical properties and hence is metallic in nature.

Evidences supporting the electropositive character shown by metals

We have already said that metals have a strong tendency to lose one or more electrons and hence show strongly electropositive character. The following properties of metals support the electropositive character of metals.

(*i*) *Metallic oxides are basic in nature, i.e.*, when metallic oxides are disolved in H_2O, alkalies are produced. The reaction between oxide ion (O^{2-}) with H_2O produces hydroxyl ions (OH^- ions) which make the solution alkaline, *e.g.*

$$2Na - 2e^- \rightarrow 2\,Na^+; \quad O + 2e^- \rightarrow O^{2-}; \quad O^{2-} + H_2O \rightarrow 2OH^-$$

(*ii*) *Metals liberate* H_2 *gas from acids*. The production of H_2 gas from H_2SO_4 by the action of Mg on it, can be considered as an example.

$$H_2SO_4 \rightarrow 2H^+ + SO_4^{2-}, \quad Mg - 2e^- \rightarrow Mg^{2+}$$

$$2H^+ + 2e^- \rightarrow H_2, \quad Mg^{2+} + SO_4^{2-} \rightarrow MgSO_4$$

(*iii*) *Metals act as reducing agents*. The reducing property of Na-metal to reduce Cl_2 to Cl^- ion, for example, can be explained as follows :

$$Na - e^- \rightarrow Na^+, \quad \frac{1}{2}Cl_2 + e^- \rightarrow Cl^-$$

On adding these two equations we get :

$$Na \text{ (reducing agent)} + \frac{1}{2}Cl_2^0 \text{ (oxidising agent)} \rightarrow Na^+ \text{ (oxidising agent)} + Cl^- \text{ (reducing agent)}$$

Variation of electropositive (metallic) and electronegative (non-metallic) character of elements of different groups

Electropositive (metallic) character of group IA elements. Since the ionisation energies of these elements are low, these elements have strong electropositive or metallic character. Since ionisation energy decreases down the group, the electropositive character increases on moving down the group from Li to Cs.

Alkali metals are so highly electropositive that they emit their valence-shell electron (ns^1 electron) even when exposed to light (photoelectric effect). This property makes alkali metals useful in photoelectric cells. Cs and K are particularly used for this purpose.

Electropositive character of group IIA elements. Ionisation energies of these elements are low and hence these metals are fairly strongly electropositive, but since the ionisation energies of these elements are higher than those of alkali metals (group IA), alkaline earth metals (group II A) are not as strongly electropositive as the alkali metals. For example, unlike alkali metals, these metals are not able to emit electrons on exposure to light.

Since ionisation energy decreases down the group, the electropositive character increases on moving down the group. It is due to the gradual increase in the electropositive character of the elements, that the hydroxides of these elements become more and more basic, as we descend the group from Be to Ba. $Be[OH]_2$ is not at all basic; in fact it is amphoteric, since it reacts with acids to form salts and with alkalies to give beryllates (BeO_2^{2-})

$$Be(OH)_2 + 2HCl \rightarrow BeCl_2 + 2H_2O$$

$$Be(OH)_2 + 2NaOH \rightarrow Na_2BeO_2 + 2H_2O$$

The hydroxides of other metals are basic. Their basic character increases on moving down the group. Thus $Mg(OH)_2$ is *weakly basic* and $Ba(OH)_2$ is the *strongest base*.

Electropositive (metallic) character of group III A elements. Since the ionisation energy of B is more than that of Al, the former has less metallic character than the latter. As a matter of fact, B is regarded as a semi-metal (metalloid). It has more properties of a non-metal. It has high ionisation energy and is also a poor conductor of electricity. Al and the remaining elements of this group have metallic character. Their metallic character is more or less the same, since the ionisation energy values of these elements are slightly different from each other.

Metallic and non-metallic character of group IVA elements. The increase in metallic character of the elements as we move down a group, is best seen in the elements of this group. Carbon is a non-metal, Si and Ge are semi-metals (Electrical properties of Si are like those of a semi-metal) and Sn and Pb have distinctly metallic character in their physical and chemical properties. Thus, we see that as we move from carbon to lead, there is a change from non-metallic to metallic character of the elements. The first element *viz.,* carbon is a non-metal, while the last two elements, namely Sn and Pb, are distinctly metals.

Metallic and non-metallic character of group VA elements. The change from non-metallic character to the metallic character of the elements in the same group, which is seen in group IVA elements is also exhibited equally well in the elements of group VA. Thus, N and P (first-two elements) are exclusively non-metals, As and Sb are partly non-metals and partly metals (semi-metals) and the last element *viz.,* Bi is definitely a metal. The variation (*i.e.,* increase decrease) of the properties like (*i*) physical state of the elements (*ii*) electrical and thermal conductivity of the elements (*iii*) hardness and density (*iv*) stability of hydrides of these elements (*v*) acid character and solubility of oxides (*vi*) strength and solubility of oxy-acids (*vii*) hydrolysis of halides, confirms the fact that there is a transition from non-metallic to metallic character from N to Bi in group VA.

Metallic and non-metallic character of group VIA elements. The first four elements (O, S, Se and Te) are non-metals. The non-metallic character of O and S is stronger than that of Se and Te, *i.e.*, O and S are distinctly non-metallic and some metallic character begins to develop in Se and Te. The last element namely Po is a distincly metal. However, this element is radioactive and short-lived. The change from non-metallic character to metallic character in this group is also confirmed by the fact that the electronegativity values decrease gradually as we move down the group.

Non-metallic (or basic) character of halogens (group VII A). We have already said in this chapter that, in general, the metallic (*i.e.* basic) character of the elements increases on descending a group (See page 133) while the same decreases on proceeding from left to right in a period (see page 133).

Thus, among the elements of a given period, the halogens should be expected to show the minimum basic character and among halogens themselves astatine (the last member of the halogen family) should be the most basic in character. Unfortunately no evidence is available which may prove the basic character of astatine. Iodine which is the next heaviest halogen shows maximum basic properties in halogen family elements. Note that *iodine is not a metal.* A progressive increase in the basic character from F to I is due to the factors *viz* (*i*) Electronegativity decreases from F to I (*ii*) Electron affinity decreases from Cl to I (*iii*) Ionisation potential *decreases* from F to I.

Flourine is the most electronegative element and as such it does not show any basic character at all, *i.e.* it has no tendency to from positive ions in any of its known compounds. The non-existence of the oxy-acids of F further confirms our belief that F is unable to bear a positive charge. Since iodine has the lowest ionisation energy, it exhibits maximum basic character, *i.e.* iodine can lose electrons most readily and can thus form *unipositive iodine* (I^+) or *tripositive iodine* (I^{3+}).

Compounds containing cationic iodine. Following are the important compounds containing unipositive iodine (I^+) or tripositive iodine (I^{3+}).

(*i*) *Iodine chloride (ICl) and iodine bromide (IBr).* See chapter 29.

The electrolysis studies made on IBr suggest that this compound ionises as :

$$IBr \rightleftharpoons I^+ + Br^-$$

This mode of ionisation shows that IBr contains iodine as I^+.

(*ii*) *Iodine nitrate (INO_3) and iodine sulphate (I_2SO_4).* Solutions of these compounds are obtained by passing a solution of H_2SO_4 and of HNO_3 respectively in absolute alcohol through exchanger containing sorbed I^+ ion.

$$2I^+R^- + H_2SO_4 \longrightarrow I_2SO_4 + 2HR$$

$$I^+R^- + HNO_3 \longrightarrow INO_3 + HR$$

When the aqueous solutions of these compounds are electrolysed, iodine is liberated at cathode, proving that both of these salts contain uni-positive iodine, I^+.

(iii) Complex compounds of $[Ipy_2]^+X^-$ type ($X^- = NO_3^-, ClO_4^-$). In these compounds I^+ ion gets stablished by coordination. These compounds are obtained by treating Ag(I) salt with I_2 in chloroform, in presence of excess of pyridine (py). For example :

$$AgNO_3 + I_2 \text{ (in chloroform)} + 2py \text{ (excess)} \longrightarrow \underset{\text{ppt.}}{AgI} + [Ipy_2] NO_3$$

Complexes of this type have the following properties. We shall consider $[Ipy_2]NO_3$ as an example.

(a) Reaction with acidified KI solution. When $[Ipy_2]NO_3$ reacts with acidified KI solution, I_2 is liberated.

$$I^+ \text{ (From complex compound)} + I^- \text{ (From KI)} \longrightarrow I_2$$

I_2 liberated in the above reaction can be quantitatively estimated with sod. thiosulphate.

$$2Na_2S_2O_3 + I_2 \longrightarrow Na_2S_4O_6 + 2NaI$$

(b) Electrolysis. Electrolysis of $[Ipy_2]NO_3$ in chloroform liberates iodine at the cathode.

$$2I^+ \text{ (From the complex compound)} + 2e^- \longrightarrow I_2 \text{ (At cathode)}$$

These reactions show the presence of I^+ ion in the complex compound.

(c) Hydrolysis. The complex compound is hydrolysed by water as follows :

$$5[Ipy_2] NO_3 + 3H_2O \longrightarrow 5HNO_3 + 10py + HIO_3 + 2I_2$$

(d) Reaction with NaOH. Treatment of the complex compound with NaOH gives the base, (Ipy_2) OH which is converted into its anhydride.

$$[Ipy_2] NO_3 + NaOH \longrightarrow \underset{\text{Base}}{[Ipy_2] OH} + NaNO_3$$

$$\underset{\text{Base}}{2[Ipy_2]OH} \longrightarrow \underset{\text{Anhydride of the base}}{Ipy - O - Ipy} + H_2O + 2py$$

(e) Reaction with phenol (C_6H_5OH) gives tri-iodophenol.

$$C_6H_5OH + 3[Ipy_2] NO_3 \longrightarrow \underset{\text{Tri-iodophenol}}{C_6H_2I_3OH} + 3HNO_3 + 6py$$

(f) Noble metals like Au are dissolved in chloroform solution of the complex compound.

$$3[Ipy_2] NO_3 + Au \longrightarrow Au^{3+} + 3NO_3^- + \frac{1}{2} I_2 + 6py$$

In $[Ipy_2] NO_3$ and other similar compounds, I^+ appears to be two-coordinate and hence the complex, $[Ipy_2]^+ NO_3^-$ should be formulated as $[pyIpy]^+ NO_3^-$.

(iv) Iodine perchlorate, $I(ClO_4)_3.3H_2O$. It is formed as yellowish-green needles by the action of O_3 on a cooled solution of I_2 in anhydrous $HClO_4$.

$$I_2 + 6HClO_4 + O_3 \longrightarrow 2I(ClO_4)_3 3H_2O$$

When the aqeous solution of this compound is electrolysed, iodine is liberated at cathode.

$$I(ClO_4)_3 \longrightarrow I^{3+} + 3ClO_4^-$$

$$2I^{3+} + 6e^- \longrightarrow I_2 \text{ (At cathode)}$$

(v) Iodine triacetate, $I(CH_3COO)_3$. It is obtained by oxidising I_2 in acetic anhydride, $(CH_3.CO)_2O$ with fuming HNO_3.

It is a crystalline colourless substance, stable in cold but decomposes at 100°C. The saturated solution of the compound in acetic anhydride conducts electricity and during electrolysis I_2 is liberated at the cathode (of silver), forming AgI.

$$(CH_3COO)_3 + I \rightleftharpoons I^{3+} + 3CH_3COO^-$$

$$I^{3+} + 3e^- + Ag \longrightarrow AgI \text{ (At silver cathode)}$$

(vi) Iodine trichloride, ICl_3. For the preparation, properties and uses of this compound see chapter 29.

Conclusion. The discussion, made above, shows that the elements of group IA and IIA (*s*-block elements), which lie at the extreme left of the periodic table, are highly electro-positive (*i.e.* these elements have metallic character), while the elements of group VIIA (halogens — *p* block elements), which lie at the extreme right of the periodic table are highly electronegative. Thus the elements of groups IA and IIA have a strong tendency to lose one or two electrons to undergo oxidation ($M \rightarrow M^+ + e^-$, $M \rightarrow M^{2+} + 2e^-$) while the elements of group VIIA have a strong tendency to accept an electron to undergo reduction ($X + e^- \rightarrow X^-$). Some of the *p*-block elements like B, Si, Ge, As, Sb etc. which lie in-between strongly electropositive (elements of groups IA and IIA) and strongly electronegative (elements of groups VII A) elements are metalloids.

Density

Density of elements in the solid state

Densities (g cm^{-3}) of the main group elements are given in Table 5.8. Density of an element in the solid state (metallic element) depends on the type of packing of metal atoms in the lattice of the metal and on the size of the atoms. The elements having close packing of atoms in their lattices and large size of the atoms have low densities. It may be seen from table 5.8 that the densities of main group elements donot show any regular trend as we move from left to right in a period.

Densities generally increase as we move down a group of the main group elements. The increase down the group shows that the atoms of the elements are packed less and less closely as we move down the group.

Table 5.8: *Densities (in g. cm^{-3}) of the main group elements in the solid state.*
**Densities in liquid state, **Graphite*

I A	II A	III A	IVA	VA	VIA	VIIA	Zero
Li 0.53	Be 1.84	B 2.34	C** 2.26	N* 0.81	O* 1.14	F* 1.11	Ne* 1.20
Na 0.97	Mg 1.74	Al 2.70	Si 2.49	P 1.82	S 2.7	Cl* 1.56	Ar* 1.40
K 0.86	Ca 1.55	Ga 5.91	Ge 5.32	As 5.72	Se 2.79	Br * 3.12	Kr * 2.60
Rb 1.53	Sr 2.54	In 7.31	Sn 7.30	Sb 6.62	Te 6.24	I 4.94	Xe* 3.06
Cs 1.90	Ba 3.75	Tl 11.85	Pb 11.40	Bi 9.80	Po 9.2	At —	Rn —

Since alkali metals (group IA) have close packing of metal atoms in their lattices and have the largest size of their atoms in each period, they have the lowest densities. These values gradually increase down the group. Potassium has lower density than sodium, *i.e.* potassium in lighter than sodium. This is due to an unusual increase in the atomic size of potassium.

Densities of alkaline earth metals (group IIA) decrease slightly from Be to Ca and then increase considerably upto Ba. The density of these metals is considerably higher than that of the corresponding alkali metals, *i.e.* these metals are considerably denser than the corresponding alkali metals. Higher values of density of alkaline earth metals are due to the fact that the elements of group IIA can be packed more tightly due to their greater *nuclear charge* and *smaller size*.

Densities of the elements of other groups increase on moving down the group.

Atomic Volumes of Elements

The ratio of gram atomic mass and density is known as the "atomic volume" of the element. The atomic volumes (in cm^3) of main group elements are given in Table 5.9. Since the atomic radii increase in moving down the group, atomic volumes also increase in the same direction. Atomic volumes show no regular trends in a period. Atomic volumes of the elements of groups IA, IIA, IIIA, IVA, VIA and VIIA increase down the group. These values for the elements of group VA (excepting P and As) also increase down the group. The value for As is exceptionally low.

Table 5.9 : *Atomic volume (in cm^3) of main group elements*

I A	II A	III A	IV A	V A	VI A	VII A
Li 12.97	Be 4.90	B 4.62	C 3.4	N 17.3	O 14.0	F 17.1
Na 23.68	Mg 13.97	Al 10.0	Si 11.4	P 17.0	S 15.5	Cl 18.7
K 45.36	Ca 25.91	Ga 11.8	Ge 13.6	As 13.1	Se 16.5	Br 23.5
Rb 55.80	Sr 34.54	In 15.1	Sn 16.3	Sb 18.4	Te 20.5	I 25.7
Cs 69.95	Ba 36.72	Tl 17.2	Pb 18.27	Bi 21.3	Po 22.7	At —
	Ra 38.00					

Melting and Boiling Points of Elements

The melting and boiling points of main group elements are influenced by the electropositive (*i.e.,* metallic) character of the elements. As the metallic character increases, the melting and boiling points generally decrease. Since the metallic character of the elements increases on moving down the group, the melting and boiling points generally decrease in the same direction.

The melting and boiling points of the elements also depend on whether the forces between the atoms (inter-atomic forces) of the elements are weak or strong. In a crystal lattice having bigger atoms, the atoms are packed loosely and hence the inter-atomic forces are weak. The weak forces decrease the melting and boiling points of the elements. On the other hand in a crystal having smaller atoms, the atoms are packed together closely and hence the inter-atomic forces are strong. The strong forces increase the melting and boiling points. Thus, when we proceed down a group, the size of the atoms increases and hence the inter-atomic forces become weaker and weaker and the melting and boiling points of the elements, therefore, decrease. Thus, in a group, the melting and boiling points generally decrease, as we go down from top to bottom. The downward decreasing order of the melting and boiling points of the elements of the same group can be illustrated by considering the elements of group IA (See Table 5.10).

The melting and boiling points generally increase across a period and attain a maximum value in the middle and, thereafter, decrease until noble gas element is reached. This variation can be illustrated by considering the melting and boiling points of the elements of 2*nd* and 3*rd* periods. In 2*nd* period, the melting and boiling points increase from Li to C and then decrease from N to Ne. The boiling point of boron is, however, comparatively low. The last four elements of 2*nd* period (*viz.* N, O, F and Ne) are gases at ordinary temperature. Similarly in 3*rd* period, the melting and boiling points increase from Na to Si and thereafter decrease from P to Ar. However, the melting and boiling points of sulphur are comparatively high. (See table 5.10).

Table 5.10 : *Melting and boiling points of the main group elements*

I A	II A	III A	IV A	V A	VI A	VII A	Noble gases
H — —							He –272.1 –268.9
Li 181 1330	Be 1277 2770	B 2030 2550	C 3727 4830	N –210 –195.8	O –218.8 –183.0	F –218.6 –188.1	Ne –246.6 –246.0
Na 97.8 888	Mg 650 1107	Al 659 2450	Si 1410 2680	P* (white) 44.1 280	S 119 444.6	Cl –101.0 –34.6	Ar –189.3 –185.8
K 64.0 756	Ca 838 1440	Ga 28.8 2237	Ge 937.5 2830	As 817 (36 atm) 613 (sublimes)	Se 217 685	Br 59.5 2.94	Kr –157.2 –153.3
Rb 39.0 688	Sr 768 1380	In 156.4 2000	Sn 231.8 2270	Sb 630.5 1440	Te 449.5 489.8	I 185.2 3.76	Xe –118.8 –108.1
Cs 28.5 669	Ba 714 1640	Tl 303 1457	Pb 327.4 1725	Bi 271.3 20	Po 254 962	At — —	Rn –71 –62

Variations of melting and boiling points of the elemetns of different groups

Here, we shall consider the variation of melting and boiling points of the elements of groups IA, IIA, IIIA, IVA, VA, VIA, VIIA and noble gases.

Group IA (Alkali metals). Since the atoms of alkali metals have large size, the inter-atomic forces in the crystal lattice of these elements are quite low and hence the melting and boiling points of these elements are also low. These decrease down the group due to the weaking of the inter-atomic forces. The weaking of inter-atomic forces is due to the increase in atomic size.

Group IIA (Alkaline earth metals). The atoms of the elements of this group have smaller size than those of alkali metals (group IA). Consequently the atoms of these elements are more closely-packed in their crystal lattice and have, therefore, higher melting and boiling points than alkali metals. The trend in the values of melting and boiling points of these elements is not regular as we move down the group.

Group IIIA. The melting points decrease from B to Ga and then show an increasing trend. The deviation in the melting points may be attributed to the structural changes. The boiling points decrease regularly from B to Tl.The low value of melting point of Ga is explained on the basis that its structure consists of Ga_2 molecules.

Group IVA. Since the metallic character of the elements of this group increases on moving from C to Pb, the melting and boiling points decrease in the same direction. However, the melting point of Sn is exceptionally low and the boiling point of Ge is exceptionally high. The decrease in melting points as well as in boiling points along the group indicates that inter-atomic forces also decrease in the same direction. The melting and boiling points of C and Si are notably *high* because of the tendencies of these elements to form *giant molecules.*

Group VA. The melting points of the elements, excepting Sb and Bi, increase down the group. The boiling points of these elements (excepting Bi) show the increasing trend. The boiling point of Bi is unusually low. Low value of the melting point of Bi is due to the presence of weak forces of attraction between the atoms of this element in the solid state.

Group VIA. Both the melting and boiling points increase down the group. This is due to the increase in the atomic size as a result of which the van der Waals forces of attraction increase. However, the values for Po are less than those of Te.

Explanation. In Po, the inert pair effect is maximum due to the maximum nuclear charge. As a result, the valence *s*-electron pair (*i.e.* $6s^2$ pair) in Po, is less available, as compared to that in Te (*i.e.,* $5s^2$ pair) and the magnitude of the van der Waals forces of attraction in Po will also be less. The large difference in melting and boiling points between O and S is attributed, in part, to the fact that whereas oxygen exists in the form of O_2 molecules, sulphur exists as S_8 molecules.

Group VIIA (Halogens). Melting and boiling points both show an increasing trend, as we move down the group from F to I. This indicates that the attractive forces (van der Waals forces) between molecules become progressively more prominent, as the molecules increase in size and mass. This effect is also reflected in the physical state of halogens. F and Cl are gases at ordinary temperatures, Br is a heavy liquid, while I is a solid.

Noble gases. The melting as well as boiling points, of noble gases are very low. Their low values are due to the fact that the forces holding together the atoms of the gas, in the liquid and solid state, are van der Waals forces, which are weak forces. As we move down the group, these forces increase due to the addition of new electronic shells and hence the melting and boiling points, both increaase from He to Rn. The melting and boiling points of He are the lowest of any known element.

Electrode Potentials

The values of standard reduction electrode potential, E^o_{red} (in volts) of some important main group elements are given in Table 5.11.

Table 5.11 : E^o_{red} *values (in volts) of some important main group elements*

I A	II A	III A	IV A	V A	VI A	VII A
H^+ / H 0.0						
Li^+/Li –3.05	Be^{2+}/Be –1.85	B^{3+}/B –0.87			O_2/OH^- +0.44	$F_2/2F^-$ +2.85
Na^+/Na –2.71	Mg^{2+}/Mg –2.37	Al^{3+}/Al –1.68			S/S^{2-} +0.49	$Cl_2/2Cl^-$ + 1.36
K^+/K –2.92	Ca^{2+}/Ca –2.87	Ga^{3+}/Ga –0.52		As^{3+}/As +0.25		$Br_2/2Br^-$ +1.06
Rb^+/Rb –2.92	Sr^{2+}/Sr –2.89	In^{3+}/In – 0.34	Sn^{2+}/Sn – 0.14	Sb^{3+}/Sb +0.21		$I_2/2I^-$ +0.53
Cs^+/Cs –2.92	Ba^{2+}/Ba –2.90	Tl^+/Tl – 1.34	Pb^{2+}/Pb – 0.13	Bi^{3+}/Bi +0.23		

From these values the following points may be noted :

(*i*) The elements of groups IA, IIA, IIIA and IVA have negative values of E^o_{red}, while the elements of groups VA, VIA and VIIA have positive values.

(*ii*) As we proceed in a period from left two right, E^o_{red} values increase. For example, when we proceed in a period from the elements of group IA to those of group IVA, E^o_{red} values become less and less negative (*i.e.* these values increase) and when we move in the period from the elements of group VA to those of group VIIA, E^o_{red} values become more and more positive (*i.e.,* these values again increase).

(*iii*) In group IA, Li has the highest negative value. E^o_{red} values of the elements of group IIA become more and more negative as we move down the group.

(*iv*) Halogens have the largest positive values of E^o_{red}.

(*v*) In group IIIA, the value of standard reduction potential for boron is quite small. This is because of the fact that due to the large quantity of hydration energy associated with B^{3+} ion, boron normally does not form B^{3+} ion in aqueous solution. Since the values of standard reduction potential for all the elements are negative, these elements show reducing property. This property increases from B to Al, but decreases from Ga to In.

Oxidising and Reducing Properties of Elements in Aqueous Solution

The elements, which have a strong tendency to accept electrons, to undergo reduction in aqueous solution, are called "oxdising agents", while those which have a tendency to lose electrons, to undergo oxidation in aqueous solution, are called "reducing agents".

The elements of groups IA, IIA, IIIA and IVA are reducing agents, while those of groups VA, VIA and VIIA are oxidisng agents. As we proceed in a period from the elements of groups IA to those of group IVA, the reducing power of elements decreases. On proceeding in a period from the elements of group VA to those of group VIIA, the oxidising power of the elements increases (See chapter 18).

Recucing property of alkali metals, alkaline earth metals and transition metals, and oxidising property of halogens in aqueous solution, have been discussed in detail in chapter 18.

It may be seen that oxidising and reducing characters of the elements generally increases, as we move down a group of main group elements.

Oxidation States of Elements

Oxidation states shown by main group elements are given in table 5.12. The following points may be noted :

(*a*) Oxidation states of s-block elements. All the elements belonging to *s*-block show only one positive oxidation state, which is equal to the total number of electrons present in their valence-shell or equal to their group number. Thus, the elements of groups IA (alkali metals) and IIA (alkaline earth metals) exhibit oxidation state of +1 and +2 respectively.

(*b*) Positive oxidation states shown by p-block elements of groups IIIA, IVA and VA. Like *s*-block elements, the lighter *p*-block elements of groups IIIA (*e.g.*, B, Al), IVA (*e.g.*, C, Si) and VA (*e.g.*, N, P), show a positive state, which is equal to the sum of electrons present in the valence-shell of these elements or equal to their group number, G, *e.g.*, B and Al = +3 (group III A, $ns^2\ np^1$); C and Si = + 4 (group IVA, ns^2np^2) and N and P = +5(group VA, $ns^2\ np^3$). The *heavier p*-block elements of these groups show two types of oxidation states. The higher positive oxidation state is equal to the number of electrons in their valence-shell or group number (G) and the lower positive oxidation state is equal to (G –2).

The lower oxidation state becomes more and more stable, as we move down the group, while the higher oxidation state becomes less and less stable, on proceeding from top to bottom in a given group of *p*-block elements. This can be explained on the basis of inert pair effect, as discussed in chapter 10. Thus, we see that in case of the heavier elements of group IIIA *viz.*, Ga, In and Tl, +1 oxidation state (lower oxidation state) becomes more and more stable as we move in the series $Ga^+ \rightarrow In^+ \rightarrow Tl^+$ while + 3 oxidation state (higher oxidation state) becomes less and less stable in the series $Ga^{3+} \rightarrow In^{3+} \rightarrow Tl^{3+}$. In other words, we can say that the compounds of Tl(+1) (last element of group IIIA) in which the inert pair effect is maximum are more stable than those of Tl(+3). From the above study it can be concluded that the last element of each of the groups viz IIIA, IVA and VA shows reluctance to exist in higher oxidation state and prefers to exist in lower oxidation state.

Table 5.12: *Various oxidation states shown by main group elements*

I A (ns^1)	II A (ns^2)	III A ($ns^2\ np^1$)	IV A ($ns^2\ np^2$)	V A ($ns^2\ np^3$)	VI A ($ns^2\ np^4$)	VII A ($ns^2\ np^5$)
H +1						
Li +1	Be +2	B +3	C +4	N –3, –2, –1, +1, +2, +3, +4	O –1, –2, +2 (F_2O)	F –1
Na +1	Mg +2	Al +3	Si +4	P –3, +3, +4, +5	S –2, +2, +4, +6	Cl –1, +1, +3, +4, +5, +6, +7
K +1	Ca +2	Ga +1, +3	Ge +2, +4	As –3, +3, +5	Se –2, +2, +4, +6	Br –1, +1, +3, +4, +5, +6
Rb +1	Sr +2	In +1, +3	Sn +2, +4	Sb –3, +3, +5	Te –2, +2, +4, +6	I –1, +1, +3, +4, +5, +6, +7
Cs +1	Ba +2	Tl +1, +3	Pb +2, +4	Bi +3, +5	Po +2, +4	

(*c*) Variation of oxidation states in a period. Positive oxidation state increases with increasing atomic number on moving across the period, from left to right. On the other hand, negative oxidation state decreases in the same direction.

Oxidation states of elements of different groups.

Oxidation states shown by group IA elements. All the elements of this group show an oxidation state of +1, in all their ionic compounds. This oxidation state is equal to their group number or equal to the electrons in *s*-orbital of their valence-shell. These elements cannot show an oxidation state higher than +1, because, due to high second ionisation energy, the second electron from the valence-shell cannot be removed, *i.e.*,

$$M^o \xrightarrow{-e^-} M^+ + e^- \text{ (Possible)}; \quad M^+ \xrightarrow{-e^-} M^{2+} + e^- \text{ (Not possible)}$$

It may be seen from the electronic configuration of M^+ cation, that since these cations have no unpaired electrons, the compounds (of course, ionic compounds) having M^+ cations are diamagnetic and colourless. All the compounds of alkali metals are colourless, with the exception of those whose anions are coloured, *e.g.,* permanganates and dichromates.

Oxidation states of group IIA elements. All the elements of this group show an oxidation state of +2, in all their ionic compounds.

$$\underset{2,\ 2}{\overset{o}{Be}} \xrightarrow{-2e^-} \underset{2\ (s^2)}{Be^{2+}} + 2e^-$$

$$\underset{\ldots\ldots,\ 8,\ 2}{\overset{o}{M}} \xrightarrow{-2e^-} \underset{\ldots.,\ 8\ (s^2p^6)}{\overset{2+}{M}} + 2e^-$$

(M = Mg, Ca
Sr, Br, Ra)

Since the electronic configuration of Be^{2+} ion is s^2 and the valence-shell configuration of other M^{2+} ions is s^2p^6, these configurations are noble gas configurations and hence are very stable. Consequently, the removal of another electron from M^{2+} ions is not very easy. These elements, therefore, cannot show an oxidation state greater than +2. Thus, we see that alkaline earth metals show an oxidation state equal to their group number or equal to the number of electrons present in *s* orbital of their valence-shell.

Oxidation states of group IIIA elements. Oxidation states shown by the elements of this group are B = +3, Al = +3, Ga = +3, +1, In = +3, +1 and Tl = +3, +1. These oxidation states show that B and Al show consistantly +3 oxidation state (called higher oxidation state or group oxidation state) in their ionic compounds, while the remaining elements show +1 (called lower oxidation state) as well as +3 oxidation states. On the basis of inert pair effect (see chapte 10), it can be explained that in heavier elements *viz*., Ga, In and Tl, +1 oxidation state becomes more and more stable, as we move down the group ($Ga^+ < In^+ < Tl^+$) while +3 oxidation state becomes less and less stable in the same direction ($Ge^{3+} > In^+ > Tl^{3+}$). This shows that the last element, namely Tl of this group, shows reluctance to exist in higher oxidation state (= +3); rather it prefers to exist in lower oxidation state (= +1).

$GaCl_2$ is believed to be a dimeric molecule, Ga_2Cl_4, which has the structure as $Ga^+ [Ga^{3+} Cl_4]^-$, contains Ga in +1 and +3 oxidation state.

Oxidation states of the elements of group IVA. (*a*) *Positive oxidation states.* Oxidation states exhibited by these elements are C= + 4; Si = + 4, Ge = + 2, + 4; Sn = +2, + 4; Pb = + 2, + 4. These oxidation states show that C and Si show consistantly + 4 oxidation state (called higher oxidation state or group oxidation state) while the remaining three elements namely Ge, Sn and Pb (heavier elements) show + 2 (lower oxidation state) as well as + 4 oxidation state.

With the help of inert pair effect, it can be shown that in heavier elements *viz* Ge, Sn and Pb the stability of + 2 oxidation state (lower oxidation state) increases and that of + 4 oxidation state (higher oxidation state) decreases, as we move down the group, *i.e.*, $Ge^{2+} < Sn^{2+} < Pb^{2+}$; $Ge^{4+} > Sn^{4+} > Pb^{4+}$. From this discussion, it can be concluded that the last element *viz* Pb prefers to show lower oxidation state (= +2) and shows reluctance to exhibit higher oxidation state (= +4).

(*b*) *Negative oxidation states.* Since these elements have low values of electronegativity, they donot have much tendency to form negative ions like M^{4-}. However, carbon forms carbide (C^{-4}), acetylide (C_2^{-2}) and allylide (C_3^{-4}) ions in methanides, acetylides and allylides, which are ionic carbides. The structure of C_2^{-2} and C_3^{-4} ions are $\left[\ \ddot{C} \equiv \ddot{C}\ \right]^{2-}$ and $\left[\ :\ddot{C} = C = \ddot{C}:\ \right]^{4-}$ respectively. Examples of carbides containing C^{-4}, C_2^{-2}, and C_2^{-4} ions are $(Al^{+3})_4\ (C^{-4})_3$ or Al_4C_3, $(Be^{+2})_4\ (C^{-4})_2$ or Be_2C ; $(Ca^{+2})_2\ (C_2^{-2})^2$ or CaC_2 ; $(Mg^{+2})_4\ (C_3^{-4})_2$ or Mg_2C_3.

Oxidation states shown by the elements of group VA. Oxidation states shown by these elements are: N = + 1 (N_2O), + 2(NO), + 3 (N_2O_3), + 4 (N_2O_4), + 5 (N_2O_5), –1 (NH_2OH), –2(N_2H_4), –3 (NH_3), $-\frac{1}{3}$ (HN_3); P = –3, +3, +4, +5, As = –3, +3, +5; Sb = –3, +3, +5; Bi = +3 +5.

(*a*) *+3 and +5 oxidation states.* For the discussion of +3 and +5 oxidation states, see "*Inert pair Effect*" (Chapter 10).

(*b*) *Negative oxidation states.* It may be seen that the last element *viz.*, Bi, hardly exhibits negative oxidation state.

Oxidation states shown by the elements of group VIA. Different oxidation states shown by the elements of this group are O = –1, –2, +2 (F_2O); S, Se and Te = –2, +2, + 4; + 6, Po = + 2, + 4. The valence-shell electronic configuration of the atoms of these elements ($ns^2\ np^4$) suggests that these elements tend to gain or share two electrons, to acquire noble gas configuration ($ns^2\ np^6$) and thus show an oxidation state of –2 or +2 depending on whether the element with which the elements of this group combine is less electronegative or more, than these elements. Since O-atom has high electronegativity, it shows only negative oxidation states. Oxygen shows –1 oxidation state in H_2O_2 (H = +1, O = –1) and in the derivatives of H_2O_2 (*e.g.*, Na_2O_2). With the exception of F_2O molecule in which O-atom shows +2 state and F shows –1 oxidation state, in all other compounds O-atom exhibits –2 oxidation state. Since the electronegativity decreases as we move from S to Po (heavier elements), the tendency of these elements to show –2 oxidation state decreases and that to show +2 oxidation state increases down the group. Thus Po which is the last element in this group does not show –2 oxidation state. Rather this element shows +2 oxidation state. For Po +4 oxidation state is more stable than +6 oxidation state due to inert pair effect.

Oxidation states of halogens (group VIIA elements). Different oxidation states shown by halogens are F = –1; Cl = –1, + 1, + 3, + 4, + 5, + 6, + 7; Br = –1, + 1, + 3, + 4, + 5, + 6; I = –1, +1, + 3, + 4, + 5, + 6, + 7. The valence-shell electronic configuration of the atoms of these elements ($ns^2\ np^5$) suggests that these elements tend to gain or share one electron to attain noble gas configuration ($ns^2\ np^6$) and thus show an oxidation state of –1 or +1 depending on whether the element with which the halogen atoms combine is less electronegative or more so than the halogen atom. For example since each of the halogens is more electronegative than hydrogen, the oxidation state of halogen in HF, HCl, HBr and HI molecules is equal to –1. On the other hand, since the electronegativity of Cl, Br and I in less than that of O-atom, the oxidation state of these elements in HClO, HBrO and HIO molecules is +1 (Oxidation state of hydrogen in each molecule is equal to +1).

Since F is the most electronegative element, it exhibits only negative oxidation state equal to –1 and does not show any positive oxidation state. The remaining halogens show –1 oxidation state and a number of positive oxidation states like +1, +3, + 4, + 5, + 6 and +7.

University Questions

1. Arrange the following atoms/ions in the increasing order of their size. Justify your answer with reasons.
   (*i*) F^-, O^{2-}, C^{-4}, Ne and N^{3-} ***(Kanpur 93)***

(*ii*) Na^+, Mg^{2+}, Al^{3+}, F^-, O^{2-}, N^{3-}, C^{-4} *(Lucknow 94)*

(*iii*) C^{-4}, N^{-3}, O^{-2} and F^- *(Delhi 89)*

2. Which atom or ion, in each of the following pairs, has smaller size and why ?

(*i*) Fe^{3+}, Fe^{2+} *(Allahabad 91)*

(*ii*) Cl, Cl^- *(Allahabad 91)*

(*iii*) (*a*) Na, Na^+ (*b*) Be, Mg *(E.C.C. Allahabad 94)*

3. (*a*) Explain ionic radii with example. *(Purvanchal; 93)*

(*b*) How do atomic and ionic radii change within a group and a period of the periodic table? *(Agra 92; Rohilkhand 91)*

4. (*a*) Write a note on "Ionic radii and Crystal radii". *(Meerut 90)*

(*b*) Why the size of cation is smaller, as compared to that of the atom. *(Meerut, 96)*

5. Explain the following, giving appropriate reasons for your answer :

(*i*) The atomic radius decreases with the increasing atomic number in a period. *(Gwahati 91, Meerut 93)*

(*ii*) Na-atom is larger than both Li and Mg. *(Ravishankar 93)*

(*iii*) The ionic radii of Na^+ and Cu^+ ions are almost the same. *(Meerut 94)*

6. By using Pauling's method, calculate the ionic radii of K^+ and Cl ions in KCl ionic crystal. Given that $d_{K^+-Cl^-} = 3.14$ Å. *(Delhi 89)*

Ionisation energy

7. Define ionisation energy. What are the factors affecting it? Explain its periodicity in the periodic table. *(Ravishankar 93)*

8. (*a*) Distinguish between ionisation potential and ionic potential. *(Lucknow 94)*

(*b*) Why 2*nd* ionisation potential is higher than the first? *(Meerut 96, Jodhpur, 89)*

9. (*a*) Establish a relation between size and ionisation energy of an atom. *(Gwahati 91)*

(*b*) Arrange them in the decreasing order of their ionisation energies: Be, B, N, O and F.*(Meerut 1996, 95)*

10. Three elements have the electronic configuration : $1s^2, 2s^2\,2p^6, 3s^2\,3p^6$; $1s^2, 2s^2\,2p^6, 3s^2$ and $1s^2, 2s^2\,2p^6, 3s^2\,3p^6, 4s^1$. The first ionisation energies of these elements (not in the same order) are 418.9, 737.7 and 1520.7 KJ mol^{-1} and their atomic radii are 1.91, 1.36 and 2.03 Å. Match the appropriate values of ionisation energy and atomic radius to each configuration. *(Lucknow 94)*

11. Ionisation energies of five elements A, B, C, D and E are listed below :

Atom	Ionisation energies (Kcal/mol)		
	I_1	I_2	I_3
A	300	549	920
B	99	734	1100
C	118	1091	1652
D	176	347	1848
E	497	947	1500

Two elements are non-metals and three are metals. One of the elements is a noble gas. Identify by letter (*a*) the noble gas, (*b*) the reactive non-metal, (*c*) the most reactive metal, (*d*) the atom which exhibits a common oxidation state of +2, (*e*) the atom(s) which exhibit a common oxidation state of +1. *(Lucknow 93)*

12. Explain the following, giving appropriate reasons.

(*i*) Ionisation potential of N is greater than that of O. *(Meerut 93, Gauhati 91: Jodhpur 89; Bundelkhand 88)*

(*ii*) Second ionisation potential is always greater than the first. *(Meerut 94)*

(*iii*) First ionisation energy of Be is higher than that of B. *(Ravishankar 93, Bahuguna 93)*

(*iv*) Ionisation energy of P is greater than that of S. *(Delhi 89)*

(*v*) Be and N in the 2*nd* period and Mg and P in the 3*rd* periof of the periodic table have slightly higher iomisation energies than expected. *(Delhi 1989)*

(*vi*) First ionisation potential of C-atom is greater than that of B-atom, whereas the reverse is true for the second ionisation potential. *(I.I.T. 1989)*

(*vii*) The ionisation of Mg to Mg^{2+} requires three times as much energy as Mg to Mg^+. *(Delhi 90)*

(*viii*) Magnetism oxide is formulated as $Mg^{2+} O^{2-}$ and not as $Mg^{+} O^{-}$. *(Delhi 90)*
(*ix*) Ionisation energy of transition metals donot show any appreriable variation. *(Bahuguna 93)*
(*x*) The valence-shell electronic configuration of both Ca and Zn is $4s^2$. Explain why first ionisation energy of Ca is lower than that of Zn. *(Lucknow 94)*
(*xi*) First ionisation energies (in electron volts) of Cl (= 13.01), Br (= 11.84) and I (= 10.44) are lower than that of H (= 13.60), yet halogen cations (X^+), similar to H^+, are unknown in solution. *(Lucknow 93)*

13. In the following pairs, which element has lower value of ionisation energy and why?
(*i*) Na or K (*ii*) C or O (*iii*) Br or I (*iv*) Ca or Be
(*v*) O or N (*vi*) Cl or I (*vii*) Be or B *(Allahabad 91, Meerut 88)*

14. Which of the following elements has the lowest value of ionisation potential ?
(*i*) Na, F, I, Cs (*ii*) N, O, F

15. Which of the following pairs of elements has higher value of ionisation energy ? Give appropriate reason for your answer.
(*i*) K or Cu (*ii*) Be or B (*iii*) I or Br (*iv*) Mg or Al
(*v*) Cl or F (*vi*) K or Ca (*vii*) Cr or Mn (*viii*) S or P
(Jodhpur 89, Meerut 88, Avadh 83, Kanpur 82)

16. Among the alkali metals and alkaline earth metals, which has the highest value of ionisation potential? *(Meerut 83)*

17. (*i*) Arrange the following elements in the increasing order of their ionisation potential.
(*a*) LI, Be, B, (*b*) N, O, F (*c*) Li, B, C, O, N, Be, F (*d*) Na, K, Li, Cs, Rb *(REE 85)*
(*ii*) Arrange the following electronic configurations in the increasing order of ionisation potentials
(*a*) $1s^2$ (*b*) $1s^2, 2s^2p^1$ (*c*) $1s^2, 2s^1$ (*d*) $1s^2, 2s^2 2p^5$ (*e*) $1s^2, 2s^2 2p^3$
(*iii*) Arrange C, N, O and F in the decreasing order of their 2nd ionisation potential values. *(I.I.T. 1981)*

18. Select the correct answer :
(*i*) The first ionisation potentials (in electron volts) of N and O « atoms are respectively given by :
(*a*) 14.6, 13.6 (*b*) 13.6, 14.6 (*c*) 13.6, 13.6 (*d*) 14.6, 14.6 *(Roorkee 87)*
(*ii*) Na cannot exhibit + 2 oxidation state, because its (*a*) $IE_2 < IE_1$ (*b*) $IE_2 >> IE_1$ (*c*) $IE_2 = IE_1$, (*d*) $IE_2 = 0$ *(Andhra 90)* **[Ans : (*b*)]**

Electron affinity

19. (*a*) Justify the following order of electron affinity of the halogens : Cl > Br > I *(Kanpur 93)*
(*b*) Arrange the Be, B, N, O and F in the increasing order of electron affinities. *(Meerut, 95)*

20. (*a*) Explain the meaning of "Electron affinity" *(Agra 92)*
(*b*) Explain how electron affinity of elements vary in a period and a group of the periodic table. *(Agra 91)*

21. Explain the following, giving appropriate reasons :
(*i*) The electron affinity values of Be, Mg and noble gases are zero and those of N (= 0.02 eV) and P (= 0.80 eV) are very low. *(Delhi Hons. 87)*
(*ii*) EA value of Cl is higher than that of F although the electronegativity value of these elements is in the reverse order. *(Bahuguna 93)*
(*iii*) EA value of noble gases are zero and those of halogens are the highest in each period. *(Raj. 86)*
(*iv*) Although second electron affinity values of elements are negative, yet stable oxide ion (O^{2-}) and sulphide ion (S^{2-}) are formed. *(Jodhpur 89)*
(*v*) Electron affinity of an atom increases in the 3*rd* period as we move from Na to Cl. Mg (Z = 12) and P (Z = 15) are, however, exceptions. *(Meerut 92; Agra 91)*
(*vi*) Electron affinity of N is lower than that of carbon. *(Agra 88)*
(*vii*) The formation of F^- (*g*) from F(*g*) is exotheromic while that of O^{2-}(*g*) from O(*g*) is endothemic. *(Meerut 92, Delhi 90)*
(*viii*) Electron affinites of N and Be are very low. *(Gwahati 91)*
(*ix*) Electron affinity value increases from N to F in the periodic table. *(Meerut 89)*

22. What is electronegativity of an element ? Explain Pauling's scale of electronegativity. Give the periodicity of electronegativity. *(Nagpur 2002)*

23. Define ionisation potential. Which factors influence the values of ionisation potential ? How does it vary in a group of *s* and *p* block elements ? *(VBS Purvanchal 2002)*

24. What is ionisation potential of an element. Explain why IE_1 of N is higher than that of O? *(Nagpur 2002)*

25. Write a note on Electron affinity. *(Lucknow 2002)*

26. Define ionic and covalent radii. *(Nagpur 2002)*

27. Explain why the size of the cation is smaller and that of the anion is bigger than its corresponding neutral atom ? *(Delhi 2002)*

28. Explain why IE_1 of carbon is greater than that of boron whereas reverse is true for IE_2 ? *(VBS Purvanchal 2003)*

29. Suggest the reason for the decrease in first ionization potential from N to O and P to S. *(Rohilkhand 2003)*

30. (*a*) Which of the following elements would have maximum electron affinity value ? Give reason. Na, I, Cl, Ar.

(*b*) Which of the following iso-electronic anions has the largest radius ? Give reason. N^{-3}, O^{-2}, F^-, C^{-4}

(*c*) What is electronegativity ? Briefly explain basis of Pauling scale of electronegativity. State the trends in the variation of electronegativity in group and in periods. *(Rohilkhand 2003)*

31. (*a*) Arrange the following in the increasing order of their size. Justify your answer with reasons : F^-, O^{2-}, C^{-4}, Ne and N^{3-}

(*b*) Define ionization potential. Why the second ionization potential of an element is higher than the first ?

(*c*) Explain the meaning of electron affinity. Why is the electron affinity of fluorine less than that of chlorine, although fluorine has higher electronegativity than chlorine ?

(*d*) The first ionization energy of Be is greater than that of Li but the position is reversed in the case of second ionization energy. Explain. *(Rohilkhand 2003)*

32. (*a*) Define ionisation energy. State the factors on which it depends.

(*b*) Explain the following :

(*i*) O^{2-} ion is larger in size than that ion.

(*ii*) $[CuCl_4]^{2-}$ exists whereas $[CuI_4]^{2-}$ does not.

(*iii*) First IE of B is less than H_4 of Be.

(*iv*) IE_2 of Na is much higher than its IE_1.

(*v*) Alkali metal cations are generally smaller in size than their corresponding atoms.

(Delhi 2003)

Answer (*b*) (*ii*). Since Cl^- ion is smaller in size than I^- ion, the electrostatic attraction between Cu^{2+} and smaller Cl^- ions is greater than that between Cu^{2+} and larger I^- ions.

33. Given that IE_1 of Al is less than that of Mg. What about IE_2 of Al compared to that of Mg ? Why ? *(Delhi 2003)*

34. (*a*) Define ionisation potential. What are the factors affecting ionisation potential ? Give its trend across the period and down the group of periodic table.

(*b*) Define ionic and covalent radii.

(*c*) Draw p-p axial and p-p lateral overlaps.

(*d*) Draw the resonating structures of HCl molecule.

(*e*) Discuss Pauling's scale of electronegativity. (Nagpur 2003)

35. (*a*) Define electronegativity.

(*b*) Give reasons for the following :

(*i*) The second ionisation energy of sodium is very high as compared to its first ionisation energy.

(*ii*) First ionisation energy of nitrogen is higher than that of oxygen.

(*c*) Define the following and explain their trends across the period and down the group in periodic table : (*i*) Atomic radius and (*ii*) Electron affinity. *(Nagpur 2003)*

36. (*a*) Differentiate between covalent reactions and Van der Waal's radius.

(*b*) What are iso-electronic ions? Arrange the given ions in order of their decreasing size : F^-, O^{2-}, Mg^{2+} and Na^+.

(*c*) What is screening effect? How does it govern the ionisation energy of an atom? *(GND Amritsar 2004)*

6

Radioactivity

Radioactivity or natural radioactivity

The nuclei of naturally occuring heavy elements like U, Th, Ra and Po are unstable and keep on emitting spontaneously invisible rays or radiations *viz* α, β or γ-rays and give more stable elements. These heavy elements, which emit α, β or γ-rays are called *radioactive elements* and the property of emitting these rays is called *radioactivity of the elements.* The word "radioactivity" means "*Ray-emitting-activity*".

It is the nucleus of the atom of an element which spontaneously disintegrates, (*i.e.* disintegrates by itself) to emit α, β or γ-rays. The rays emitted by a radioactive element are called *radioactive rays.*

Thus radioactivity can also be defined as the phenomenon in which the nucleus of the atom of an element undergoes spontaneous and uncontrollable disintegration (or decay) and emits α, β, or γ-rays. Actually the definition given above is the definition of natural radioactivity.

Examples of natural radioactivity are given below:

$${}^{238}_{92}U \longrightarrow {}^{234}_{90}Th + {}^{4}_{2}He \qquad {}^{234}_{90}Th \longrightarrow {}^{234}_{91}Pa + {}^{0}_{-1}e$$

$${}^{226}_{88}Ra \longrightarrow {}^{222}_{86}Th + {}^{4}_{2}He$$

The spontaneous emission of radioactive rays from the nucleus of a radioactive element is called *radioactive disintegration or radioactive decay of the element.*

Types of radioactivity

Radioactivity is of the following two types.

(*i*) *Natural radioactivity.* Radioactivity discussed above is, in fact, natural radioactivity. In this type of radioactivity, the heavy elements like U, Th, Ra etc. emit spontaneously α, β or γ-rays and new elements are produced (also see above).

(*ii*) *Artificial or induced radioactivity.* Artificial radioactivity is a process in which a stable (non-radioactive) nucleus is changed into an unstable (radioactive) nucleus by bombarding it with appropriate atomic projectiles like α-particles, neutrons, protons, deutrons etc.

As an example, let us mention the experiment performed by Iren Curie and her husband, Frederick Jiolet (1934). They showed that when ${}^{27}_{13}Al$ isotope which is non-radioactive is bombarded with α-particles, it is converted into ${}^{30}_{15}P$ isotope (a radioactive isotope) with the emission of a neutorn.

$$\underset{\text{Stable (Non-radioactive)}}{{}^{27}_{13}Al} + {}^{4}_{2}He \longrightarrow \underset{\text{Unstable (Radioactive)}}{{}^{30}_{15}P} + {}^{1}_{0}n$$

${}^{30}_{15}P$ obtained as above has been found to be radioactive, *i.e.*, radioactivity is produced in ${}^{30}_{15}P$ isotope. This type of radioactivity is called artificial or induced radioactivity, since it has been produced by artificial means. Being radioactive, $^{30}_{15}P$ isotope gets disintegrated into a stable $^{30}_{14}Si$ isotope with the emission of a positron $\left(^{0}_{+1}e\right)$

$$\underset{\text{Unstable (Radioactive)}}{^{30}_{15}P} \longrightarrow \underset{\text{Stable (non-radioactive)}}{^{30}_{14}Si} + {}^{0}_{+1}e \ (positron)$$

More examples showing artificial radioactivity are given below:

(*i*) $\underset{\text{Non-radioactive}}{^{10}_{5}B + {}^{4}_{2}He} \longrightarrow \underset{\text{Radioactive}}{^{13}_{7}N} + {}^{1}_{0}n \ (neutron)$

$\underset{\text{Radiactive}}{^{13}_{7}N} \longrightarrow {}^{13}_{6}C + {}^{0}_{+1}e \ (positron)$

(*ii*) $^{11}_{5}B + {}^{4}_{2}He \longrightarrow {}^{14}_{6}C + {}^{1}_{1}H \ (proton)$

$^{14}_{6}C \longrightarrow {}^{14}_{7}N + {}^{0}_{-1}e \ (\beta\text{-}particle)$

(*iii*) $^{12}_{6}C + {}^{1}_{1}H \longrightarrow {}^{13}_{7}N$

$^{13}_{7}N \longrightarrow {}^{13}_{7}C + {}^{0}_{+1}e \ (positron)$

(*iv*) $^{14}_{7}N + {}^{4}_{2}He \longrightarrow {}^{17}_{9}F + {}^{1}_{0}n \ (neutron)$

$^{17}_{9}F \longrightarrow {}^{17}_{8}O + {}^{0}_{+1}e \ (positron)$

(*v*) $^{23}_{11}Na + {}^{2}_{1}H \longrightarrow {}^{24}_{11}Na + {}^{1}_{1}H \ (proton)$

$^{24}_{11}Na \longrightarrow {}^{24}_{12}Mg + {}^{0}_{+1}e \ (positron)$

(*vi*) $^{23}_{11}Na + {}^{4}_{2}He \longrightarrow {}^{26}_{13}Al + {}^{1}_{0}n \ (neutron)$

$^{26}_{13}Al \longrightarrow {}^{26}_{12}Mg + {}^{0}_{+1}e$

(*vii*) $^{27}_{13}Al + {}^{1}_{0}n \longrightarrow {}^{24}_{11}Na + {}^{4}_{2}He$

$^{24}_{11}Na \longrightarrow {}^{24}_{12}Mg + {}^{0}_{+1}e$

(*viii*) $^{24}_{12}Mg + {}^{4}_{2}He \longrightarrow {}^{27}_{14}Si + {}^{1}_{0}n$

$^{27}_{14}Si \longrightarrow {}^{27}_{13}Al + {}^{0}_{+1}e$

Differences between natural and artificial radioactivity

The differences between natural and artificial radioactivity are shown in Table 6.1.

Table 6.1: *Differences between natural and artificial radioactivity.*

Natural radioactivity	Artificial radioactivity
(*i*) Natural radioactivity is spontaneous, since in natural radioactivity, the nuclei of the heavy atoms disintegrate on their own accord, forming slightly lighter and more stable nuclei and emitting α, β or γ-radiations.	(*i*) Artificial radioactivity is not spontaneous, since in it the nuclei of the atoms have to be bombarded by fast moving particles like α-particles, neutrons, protons, deutrons etc.
(*ii*) Natural radioactivity is uncontrollable and hence it cannot be slowed down or accelerated by any means.	(*ii*) Artificial radioactivity can be controlled by controlling the speed of the bombarding particles used for bringing about the artificial radioactivity.
(*iii*) Natural radioacitivity is usually shown by heavy elements.	(*iii*) Artificial radioactivity can be induced even in light elements.

Units of radioactivity

The standard unit of radioactivity is *curie* (c) which is defined as the mass of a radioactive element which produces 3.7×10^{10} disintegrations per second. Thus *millicurie* (*mc*) and *microcurie* (μc) are the masses of a radioactive element that produce respectively $(3.7 \times 10^{10})/ 10^3 = 3.7 \times 10^7$ and $(3.7 \times 10^{10}) / 10^6 = 3.7 \times 10^4$ disintegrations per second. On this basis the statement that P-32, a beta-emitter, has an activity of 50 millicuries per gram means that each gram of P-32 present in any radioactive material has $50 \times 3.7 \times 10^7$ disintegration per second. Another unit of radioactivity is *rutherford* (*r*) which is defined as the mass of a radioactive substance which produces 10^6 disintegrations per second. Thus *millirutherford* (*mr*) and *microrutherford* (μ*r*) are the masses of a radioactive element that produces respectively $10^6 / 10^3 = 10^3$ and $10^6 / 10^6 = 1$ disintegrations per second. Thus we see that *c* and *r* are related as $1c / 1r = 3.7 \times 10^{10} / 10^6$ *or* $1c = 3.7 \times 10^4 r$

Detection and measurement of radioactivity

The radioactivity of a radioactive substance is detected and measured by instruments like Geiger-Muller (G-M) counter, (Fig.6.1), Wilson Cloud Chamber, scintillation counters and dosi-meters. The detection of radioactivity by G-M counter depends on the ionisation of a gas in a tube. Radioacitive substance is allowed to enter a tube where it produces the gas which undergoes ionisation in the tube.

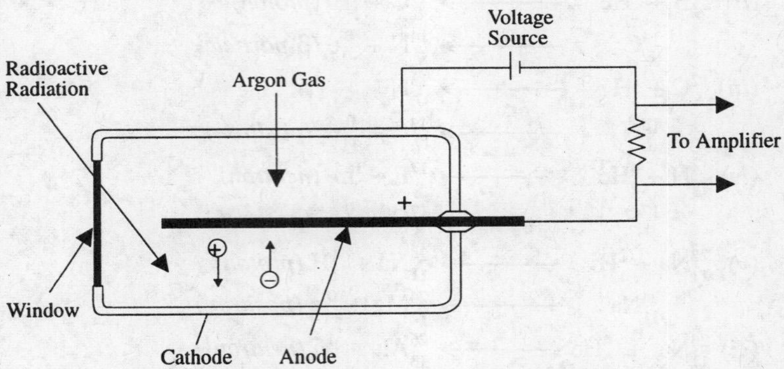

Fig. 6.1 : Geiger-Muller counter used for the detection and measurement of radioactivity of a substance.

Types of radioactive rays

In 1904, Rutherford and his co-workers placed a small piece of radium (a radioactive substance) in thick lead box having a small opening and got a thin beam of radioactive rays by using a slit. Now they passed the beam of invisible rays comming from radium between the parallel positive and negative plates of a strong electric field and found that the beam splitted into three different rays which are : (*i*) *Alpha* (α) *rays* which are deflected towards the negative plate (*ii*) *Beta* (β) *rays* which are deflected towards the positive plate (*iii*) *Gamma* (γ) *rays* which are not deflected at all. It is the nucleus of the radioactive element that is unstable and hence emits, α, β or γ-rays.

Nature, mass, charge (atomic number) and representation (symbol) of α, β and γ-rays

(*a*) α-particle is a helium nucleus which contains two protons and two neutrons and hence its mass (or mass number) is equal to 2 + 2 = 4 and its atomic number is 2 (due to the presence of two protons). Thus an α-particle is represented as ${}^{4}_{+2}He$ or ${}^{4}_{2}He$ which shows that an α-particle has a positive charge equal to +2, *i.e.*, α-particle is a *positively-charged particle.* Positive charge equal to +2 means that the atomic number of an α-particle is equal to +2 or simply 2.

The mass of an α-particle is equal to 4 *amu* = $4 \times 0.166 \times 10^{-26}$ kg.

(*b*) A β-particle has a negative charge equal to –1. Thus a β-particle is a *negatively charged particle.* –1 charge means that the atomic number of β-particle is equal to –1.

The mass of a β-particle is equal to $\frac{1}{1840}$ *amu* = 0.0005482 *amu* = $0.0005482 \times 0.166 \times 10^{-26}$ kg.

Since the mass in *amu* is so small (= 0.0005482 *amu*) that the mass of a β-particle is taken to be equal to zero in nuclear reactions. Thus a β-particle has a negative charge equal to –1 and no mass, *i.e.*, β-particle can be represented as ${}_{-1}^{0}\beta$. Now we know that since an electron, like a β-particle, also has –1 charge and no mass, a β-particle can also be regarded as an electron and hence can also be represented as ${}_{-1}^{0}e$.

The fact that a β-particle is a massless particle with atomic number equal to –1 can be explained as follows :

When the nucleus of the atom of a radioactive element loses a β-particle, one of the neutrons (${}_{0}^{1}n$) present in the nucleus of the atom is changed into a proton (${}_{+1}^{1}p$ or ${}_{1}^{1}p$) and a β-particle. Let the mass number of β-particle be x and atomic number equal to y. Thus :

$${}_{0}^{1}n \longrightarrow {}_{1}^{1}p + {}_{y}^{x}\beta$$

Obviously in order to balance the mass numbers we have $1 = 1+x$ *or* $x = 0$ and on balancing the atomic numbers we have $0 = 1+y$ *or* $y = -1$. Thus a β-particle is a massless particle ($x = 0$) with atomic number equal to $-1 (y = -1)$

(*c*) A γ-ray has no charge and no mass. This means that a γ-ray is a chargeless (neutral) and massless ray. Thus a γ-ray can be represented as ${}_{0}^{0}\gamma$. γ-rays are high frequency electromagnetic waves like X-rays and have very short wavelength.

Other properties of α, β and γ-rays

Nature, mass and charge of these rays have already been discussed. Here we are giving the properties like speed, penetration power, ionising power etc. of these rays in a tabular form (See Table 6.2).

Table 6.2 : *Comparison between some properties of α, β and γ-rays*

Property	α-rays (${}_{+2}^{4}He$ or ${}_{2}^{4}He$)	β-rays (${}_{-1}^{0}\beta$ or ${}_{-1}^{0}e$)	γ-rays (${}_{0}^{0}\gamma$)
(*i*) Penetration power	Least (1)	More than α-particles (100)	Greatest (10000)
(*ii*) Effect of electric and magnetic fields	Deflected towards the negative plate/pole	deflected towards the positive plate/pole	Not deflected
(*iii*) Ionising power	Greatest (10000)	Less (100)	Least (1)
(*iv*) Kinetic energy	High	Small	Nil
(*v*) Effect on photographic plate	The plate becomes black	Plate becomes black	Plate becomes black
(*vi*) Effect on ZnS and barium platinocyanide	Luminosity is produced	Less luminosity is produced	Least luminosity is produced.

Comparison between a β-particle and an electron

(*a*) Both the particles are negatively charged. Both have a negative charge equal to –1. Thus the atomic number of a β-particle is equal to –1.

(*b*) Both are massless particles ($m = 0$)

(*c*) When an atom of an element loses an electron, a cation of the same element is formed and hence the composition of the nucleus of the parent element remains the same. On the other hand, when the nucleus of an element loses a β-particle a new neutral element (called daughter element) is botained. Due to the formation of a new element, the composition of the nucleus of the parent element gets changed. The daughter element has one more proton but one less neutron than the parent element. Thus the sum of the protons and neutrons in the daughter element remains the same as in the parent element. Due to the formation of one more proton, the atomic number of the daughter element is increased by one unit.

Soddy-Fajans and Russel group displacement law (1913)

This law tells us about the nature and the new periodic position of the element produced when a radioactive element loses an α-particle or a β-particle, or a positron or gains an electron. The element losing or gaining the particle is called *parent element* and the element which is produced after the loss or gain of the particle is called *daughter element* or *daughter product.* This law can be studied under the following heads.

***(i)* Emission of an alpha particle, α or $^{4}_{2}He$.** We know that an α-particle is a helium nucleus whose mass number is 4 and nuclear charge (*i.e.*, atomic number) is +2. Thus this particle is represented as $^{4}_{+2}He$ or simply as $^{4}_{2}He$. When a radioactive element emits an α-particle ($^{4}_{2}He$), the mass number of the daughter element is decreased by 4 units (since the mass number of helium nucleus is 4) and atomic number gets decreased by 2 units (since the charge or atomic number of helium nucleus is +2). Due to the decrease in atomic number by two units, the daughter element gets shifted two places to the left in the periodic table. For example :

$$\underset{\text{Group VIA}}{^{215}_{84}Po_{131}} \xrightarrow{-^{4}_{2}He} \underset{\text{Group IVA}}{^{211}_{82}Pb_{129}} + {^{4}_{2}He}$$

Explanation. We know that a helium nucleus ($^{4}_{2}He$) has mass number equal to 4 and atomic number equal to 2. Thus this nucleus contains 2 protons and 4 – 2 = 2 neutrons ($^{4}_{2}He_2$). Thus the above nuclear reaction can be written as

$$^{215}_{84}Po_{131} \xrightarrow{-^{4}_{2}He_2} {^{211}_{82}Pb_{129}} + {^{4}_{2}He_2}$$

The presence of two protons and two neutrons in helium nucleus implies that when the nucleus of $^{215}_{84}Po_{131}$ loses an α-particle, 2 protons and 2 neutrons are lost by it, *i.e.*, the number of protons (or atomic number) in the daughter element becomes 84 – 2 = 82 and the number of neutrons becomes equal to 131 – 2 = 129 and consequently the mass number becomes equal to 82 + 129 = 211. Thus the loss of 2 protons implies that the atomic number of the daughter element is two units less than that of the parent element while the loss of (2 protons + 2 neutrons) suggests that the mass number of the daughter element becomes four units less than that of the perent element.

***(ii)* Emission of a beta particle, β or $^{0}_{-1}e$.** We know that a β-particle is an electron which has mass number equal to zero and its charge or atomic number is –1. Thus it is represented as $^{0}_{-1}e$ or β. When a radioactive element emits one $^{0}_{-1}e$ (or β-particle), the mass number of the daughter element remains the same (since the mass number of $^{0}_{-1}e$ is zero) but its atomic number is increased by one unit (since the charge or atomic number of $^{0}_{-1}e$ is –1). Due to the increase in atomic number by one unit, the newly-formed element gets shifted one place to the right in the periodic table. For example :

$$\underset{\substack{\text{(Radioactive)}\\\text{(Group IA)}}}{^{24}_{11}Na_{13}} \xrightarrow{-^{0}_{-1}e} \underset{\substack{\text{(Stable)}\\\text{(Group IIA)}}}{^{24}_{12}Mg_{12}} + {^{0}_{-1}e} \text{ or } \overline{\beta}$$

$$\underset{\text{(Group Zero)}}{^{23}_{10}Ne_{13}} \xrightarrow{-^{0}_{-1}e} \underset{\text{(Group IA)}}{^{24}_{12}Na_{12}} + {^{0}_{-1}e} \text{ or } \overline{\beta}$$

Explanation. We know that nucleus does not contain ${}_{-1}^{0}e$ (β-particle) as such. Then how is $\overline{\beta}$-particle produced in the nuclear reactions like those as given above? In order to explain its production, it is assumed that one of the neutrons (${}_{0}^{1}n$) present in the parent element is converted into a proton (${}_{1}^{1}H$ or ${}_{1}^{1}p$) and in this conversion process, one ${}_{-1}^{0}e$ (β-particle) and one anti-neutrino (${}_{0}^{0}\nu$) are also produced, *i.e.*,

$$ {}_{0}^{1}n \rightarrow {}_{1}^{1}H \text{ or } {}_{1}^{1}p + {}_{-1}^{0}e \text{ or } \beta + {}_{0}^{0}\nu \qquad \ldots (i) $$

Thus in the nuclear reaction,

$$ {}_{11}^{24}Na_{13} \xrightarrow{-{}_{-1}^{0}e} {}_{12}^{24}Mg_{12} + {}_{-1}^{0}e $$

one of the 13 neutrons is converted into a proton and hence the newly-formed element has 13 – 1 = 12 neutrons and 11 + 1 = 12 protons and in this process one ${}_{-1}^{0}e$ (β-particle) and one anti-neutrino (${}_{0}^{0}\nu$) are also produced. Thus the newly-formed element is ${}_{12}^{24}Mg_{12}$ and hence the above reaction should be written as :

$$ {}_{11}^{24}Na_{13} \longrightarrow {}_{12}^{24}Mg_{12} + {}_{-1}^{0}e \text{ or } \beta + {}_{0}^{0}\nu $$

The discussion made above shows that since mass number of the parent element and of the daughter element are the same, these two elements are isobars to each other. Thus we can say that the emission of one or more β-particles (or ${}_{-1}^{0}e$) by a radioactive element produces its isobar(s).

β decay is common over the entire range of nuclides and amongst the naturally occuring heavy radioactive nuclides and in the fission products (see *Nuclear Fission*).

Equation (*i*) is balanced in so far as angular momentum values of different particles is concerned as shown below:

$$ \begin{array}{lccccccccc} \text{Equation :} & {}_{0}^{1}n & \longrightarrow & {}_{1}^{1}H & - & {}_{+1}^{0}e & + & {}_{0}^{0}\nu \\ \text{Angular momentum values :} & +\dfrac{h}{4\pi} & & +\dfrac{h}{4\pi} & - & \dfrac{h}{4\pi} & + & \dfrac{h}{4\pi} \end{array} $$

If ${}_{0}^{0}\nu$ is not supposed to be emitted in the above equation, then angular momentum values are not balanced as shown below:

$$ \begin{array}{lccccc} \text{Equation :} & {}_{0}^{1}n & \longrightarrow & {}_{1}^{1}H & + & {}_{-1}^{0}e \\ \text{Angular momentum values :} & +\dfrac{h}{4\pi} & & +\dfrac{h}{4\pi} & - & \dfrac{h}{4\pi} \end{array} $$

Formation of isotopes and isobars. When a radioactive element emits one α-particle (${}_{2}^{4}He$) and two β-particles (${}_{-1}^{0}e$), the daughter element has the same atomic number as the parent element and hence these two elements are isotopes to each other. Thus the combined emission of one α-particle and two β-particles produces an isotope of the parent element, *e.g.*

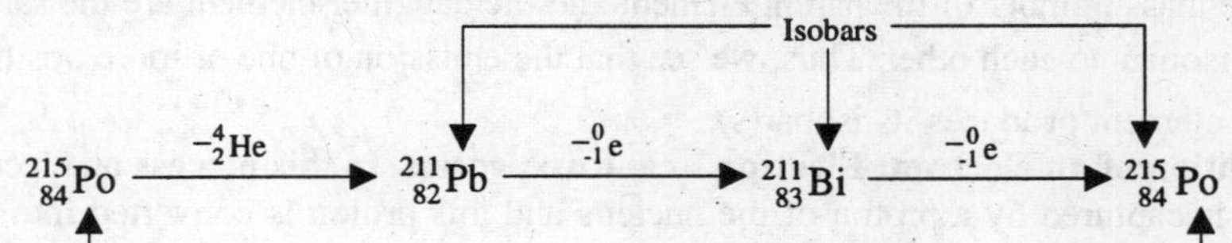

It may be seen from the above nuclear reactions that the emission of one or more β-particles (${}_{-1}^{0}e$) by a radioactive element produces its isobar(s).

(*iii*) Emission of a positron, ${}_{+1}^{0}e$. We know that a positron is an anti-particle of an electron (${}_{-1}^{0}e$), *i.e.*, a positron has mass number equal to zero (its actual mass = 0.0005486 *amu*) and its charge or atomic number is +1. Thus positron is represented as ${}^{0}_{+1}e$. This particle was discovered by Iren Curie and her husband (name-Frederick Jiolet) in 1934. They bombarded ${}^{27}_{13}Al$ nuclide (non-radioactive isotope) with α-particles so that it was converted into radioactive ${}^{30}_{15}P$ isotope with the emission of a neutron (${}^{1}_{0}n$).

$$ {}^{27}_{13}Al + {}^{4}_{2}He \longrightarrow {}^{30}_{15}P \text{ (Radioactive)} + {}^{1}_{0}n $$

Being radioactive ${}^{30}_{15}P$ disintegrates to give stable ${}^{30}_{14}Si$ isotope and a positron (${}^{0}_{+1}e$)

$$ {}^{30}_{15}P \text{ (Radioactive)} \longrightarrow {}^{30}_{14}Si \text{ (Stable istope)} + {}^{0}_{+1}e \textit{ (positron)} $$

It is evident from the above nuclear reaction that when a radioactive element emits a positron, the mass number of the daughter element remains the same (since the mass number of ${}^{0}_{+1}e$ is zero) but its atomic number is decreased by one unit (since the charge or atomic number of ${}^{0}_{+1}e$ is +1). Due to the decrease in the atomic number by one unit, the newly-formed element gets shifted one place to the left in the periodic table. This fact is also evident form the following nuclear reactions:

$$ \underset{\text{(Group IIA)}}{{}^{23}_{12}Mg_{11} \text{ (Radioactive)}} \xrightarrow{-{}^{0}_{+1}e} \underset{\text{(Group IA)}}{{}^{23}_{11}Na_{12} \text{ (Stable)}} + {}^{0}_{+1}e \textit{ (positron)} $$

$$ \underset{\text{(Group IA)}}{{}^{22}_{11}Na_{11} \text{ (Radioactive)}} \xrightarrow{-{}^{0}_{+1}e} \underset{\text{(Group zero)}}{{}^{22}_{10}Ne_{12} \text{ (Stable)}} + {}^{0}_{+1}e \textit{ (positron)} $$

Explanation: How is the positron produced in a nuclear reaction has been explained by assuming that one of the protons (${}^{1}_{1}p$ or ${}^{1}_{1}H$) present in the nucleus of the parent element is converted into a neutron (${}^{1}_{0}n$) and in this process one positron (${}^{0}_{+1}e$) and one neutrino (${}^{0}_{0}\nu$) are also produced, *i.e.*,

$$ {}^{1}_{1}H \rightarrow {}^{1}_{0}n + {}^{0}_{+1}e + {}^{0}_{0}\nu $$

Thus in the nuclear reaction

$$ {}^{23}_{12}Mg_{11} \xrightarrow{-{}^{0}_{+1}e} {}^{23}_{11}Na_{12} + {}^{0}_{+1}e, $$

one of the 12 protons is converted into a neutron and hence the daughter element has 12 – 1= 11 protons and 11+ 1 = 12 neutrons and in this process one positron (${}^{0}_{+1}e$) and one neutrino (${}^{0}_{0}\nu$) are also produced. Thus the daughter element is ${}^{23}_{11}Na_{12}$ and hence the above equation should be written as :

$$ {}^{23}_{12}Mg_{11} \longrightarrow {}^{23}_{11}Na_{12} + {}^{0}_{+1}e + {}^{0}_{0}\nu $$

Since the mass number of the parent element and the daughter element are the same, these two elements are isobars to each other. Thus, we see that the emission of one or more positrons (${}^{0}_{+1}e$) by a radioactive element produces its isobar(s).

***(iv)* Addition of an electron: Electron - capture process.** In this process an electron from the K or L shell is captured by a proton of the nucleus and this proton is converted into a neutron. In this process a neutrino (${}^{0}_{0}\nu$) is also produced.

$$ {}^{1}_{1}H + {}^{0}_{-1}e \longrightarrow {}^{1}_{0}n + {}^{0}_{0}\nu $$

The following nuclear reactions are electron-capture reactions.

$$ {}^{55}_{26}Fe_{29} + {}^{0}_{-1}e \longrightarrow {}^{55}_{25}Mn_{30} + {}^{0}_{0}\nu \quad \text{...}(i) $$

$$ {}^{131}_{56}Ba_{75} + {}^{0}_{-1}e \longrightarrow {}^{131}_{55}Cs_{76} + {}^{0}_{0}\nu $$

In reaction (*i*), one of the 26 protons captures an electron and is converted into a neutron and in this process one neutrino is also produced. This the number of protons (*i.e.* atomic number) of the daughter element is decreased by one unit and the number of neutrons is increased by one unit. Consequently the mass number remains the same. Thus we see that the process of electron-capture is similar to that reaction in which a radioactive element emits a positron ($_{+1}^{0}e$). Other examples of electron capture are.

$$_{37}^{82}Rb_{45} \xrightarrow{+\ _{-1}^{0}e} {}_{36}^{82}Kr_{46}\ ; \qquad _{79}^{194}Au_{115} \xrightarrow{+\ _{-1}^{0}e} {}_{78}^{194}Pt_{116}$$

Solved Examples

Example 6.1. *$_{90}^{234}Th$ disintegrates to give $_{82}^{206}Pb$ as the final product. How many alpha and beta particles are emitted during this process?* **(I.I.T 1986)**

Solution. If the number of α-and β particles which are emitted from $_{90}^{234}Th$ is *x* and *y* respectively, then the formation of $_{82}^{206}Pb$ can be represented as :

$$_{90}^{234}Th \xrightarrow{-x\,_{2}^{4}He} {}_{90-2x}^{234-4x}A \xrightarrow{-y\,_{-1}^{0}e} {}_{90-2x+y}^{234-4x}Pb = {}_{82}^{206}Pb$$

Thus $234 - 4x = 206$ or $x = 7$ and $90 - 2x+y = 82$

or $90 - 2 \times 7 + y = 82$ or $y = 6$

Thus 7 α and 6β particles are emitted.

Example 6.2. *In the following natural radioactive series where only the first and last elements are given, calculate the number of α-and β-particles emitted in each case.*

(a) $_{92}^{238}U \longrightarrow {}_{82}^{206}Pb$ *(b)* $_{90}^{232}Th \longrightarrow {}_{82}^{208}Pb$ *(c)* $_{92}^{235}U \longrightarrow {}_{82}^{207}Pb$

Solution. Apply the same procedure to find the number of α-and β-particles emitted in each reaction [**Ans:** *(a)* α = 8, β = 6 *(b)* α = 6, β = 4 *(c)* α = 7, β = 4]

Definition of half-life (or half-life period, $t_{1/2}$ or $t_{0.5}$) of a radioactive substance

The time in which half of the total amount (or total number of atoms) of a radioactive substance gets disintegrated is called half-life of that substance.

OR

Half-life of a radioactive substance is that time in which half of the total amount (or total number of atoms) of that substance disintegrates.

OR

The time in which the radioactivity of a given substance falls to half of its orignial value is called half-life of that substance.

The above discussion shows that in half-life of a radioactive substance, half of the total amount of that substance undergoes disintegration and the remaining half remains undisintegrated (*i.e.*, radioactive). Half-life is denoted by the symbol $t_{1/2}$ or $t_{0.5}$.

To calculate the amount of a radioactive substance left undisintegrated or disintegrated in n half-lives

Suppose that a radioactive substance undergoes disintegration. The initial amount of the substance taken is N_0 and after *t* time the amount of the substance that remains undisintegrated is N. Thus the amount that has disintegrated in *t* time is equal to (N_0 – N). Let the half-life of the substance be $t_{0.5}$. If *t* time is supposed to be composed of *n* half-lives, then:

$$t = n \times t_{0.5} \quad \text{or} \quad n = t / t_{0.5} \qquad ...(i)$$

Obviously:

(*i*) After one half-life (*i.e.* after 1 × $t_{0.5}$), the amount of the substance that would remain undisintegrated will be one half of its initial (original) amount (*i.e.*, 1/2 N_0). Thus the weight of the substance that undergoes disintegration $= N_0 - \frac{1}{2}N_0 = \frac{1}{2}N_0$

(*ii*) After two half-lives (*i.e.*, after $2 \times t_{0.5}$), the amount of the substance that would remain undisintegrated will be one-fourth of its initial amount [*i.e.*, $\frac{1}{4}N_0 = \left(\frac{1}{2}\right)^2 N_0$]. Thus the weight of the substance that undergoes disintegration $= N_0 - \frac{1}{4}N_0 = \frac{3}{4}N_0$.

(*iii*) After three half-lives (*i.e.*, after $3 \times t_{0.5}$), the amount of the substance that remains as such will be one-eighth of its initial amount [*i.e.*, $\frac{1}{8} N_0 = \left(\frac{1}{2}\right)^3 N_0$]. Thus the amount of the substance that is decomposed $= N_0 - \frac{1}{8}N_0 = \frac{7}{8}N_0$

(*iv*) After four half-lives (*i.e.*, after $4 \times t_{0.5}$), the amount of the substance that remains undecomposed will be one-sixteenth of its initial amount [*i.e.*, $\frac{1}{16}N_0 = \left(\frac{1}{2}\right)^4 N_0$]. Thus the amount of the substance that gets disintegrated $= N_0 - \frac{1}{16} N_0 = \frac{15}{16} N_0$.

The above discussion shows that after *n* half-lives (*i.e.*, after $n \times t_{0.5}$ or *t*), the amount of the substance that remains undisintegrated will be equal to $\left(\frac{1}{2}\right)^n N_0$ and hence the amount of the substance that gets disintegrated $= N_0 - \left(\frac{1}{2}\right)^n N_0 = \left[1 - \left(\frac{1}{2}\right)^n\right] N_0$. Thus :

Amounts of the radioactive substance that remains undisintegrated after *n* half-lives or *t* time ($t = n \times t_{0.5}$)

$$= N = N_0\left(\frac{1}{2}\right)^n = N_0\left(\frac{1}{2}\right)^{t/t_{0.5}} \qquad ...(ii)$$

Amount of the radioactive substance that gets disintegrated in *n* half lives or *t* time ($t = n \times t_{0.5}$) $= N_0 - N$.

$$= N_0 - N_0\left(\frac{1}{2}\right)^n = N_0\left[1 - \left(\frac{1}{2}\right)^n\right] = N_0\left[1 - \left(\frac{1}{2}\right)^{t/t_{0.5}}\right] \qquad ...(iii)$$

Equation (*iii*) gives the amount of the product formed by the disintegration of the radioactive substance.

Examples (*i*) Suppose we start with 24 atoms ($= N_0$) of a radioactive substance, then in one half-life ($1 \times t_{0.5}$)the number of atoms that remain undisintegrated would be equal to $\frac{1}{2}N_0 = \frac{1}{2} \times 24 = 12$ and the number of atoms that undergo disintegration would be equal to $N_0 - \frac{1}{2} N_0 = 24 - \frac{24}{2} = 24 - 12 = 12$. Similarly in the two half-lives ($2 \times t_{0.5}$), the number of atoms undisintegrated $= \left(\frac{1}{2}\right)^2 N_0 = \left(\frac{1}{2}\right)^2 \times 24 = 6$ and the number of atoms disintegrated $= N_0 - \left(\frac{1}{2}\right)^2 N_0 = 24 - 6 = 18$. In this way the number of atoms left undisintegrated and the number of atoms undergone disintegration in the subsequent half-lives can also be found out (See Fig. 6.2).

Time = 0
No of atoms = 24 = N_0
(black circles)

After one $t_{0.5}$ $(1 \times t_{0.5})$

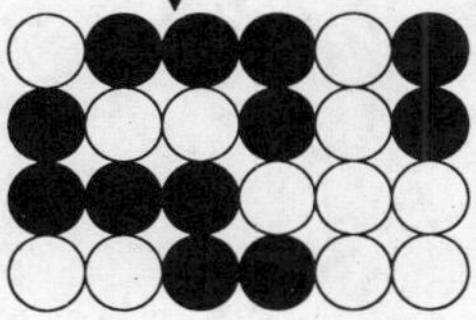

No. of atoms undisintegrated (black circles)

$$= \frac{1}{2} \times N_0 = \frac{1}{2} \times 24 = 12$$

No. of atoms disintegrated (white circles)

$$= N_0 - \frac{1}{2} N_0 = 24 - \frac{1}{2} \times 24 = 12$$

After two $t_{0.5}$ $(2 \times t_{0.5})$

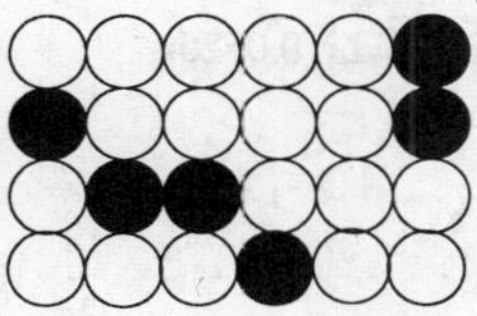

No. of atoms undisintegrated $= \left(\frac{1}{2}\right)^2 N_0.$

$$= \left(\frac{1}{2}\right)^2 \times 24 = 6 \text{ (black circles)}$$

No. of atoms disintegrated $= N_0 - \left(\frac{1}{2}\right)^2 N_0 = 24 - 6 = 18$ (white circles)

After three $t_{0.5}$ $(3 \times t_{0.5})$

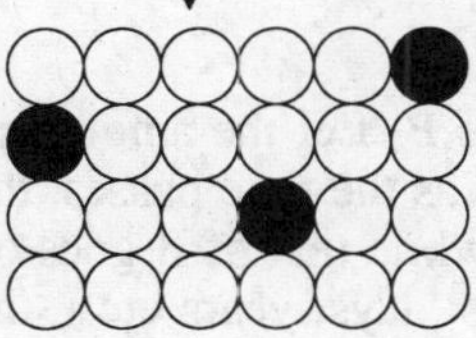

No. of atoms undisintegrated $= \left(\frac{1}{2}\right)^3 N_0$

$$= \left(\frac{1}{2}\right)^3 \times 24 = 3 \text{ (black circles)}$$

No. of atoms disintegrated $= N_0 - \left(\frac{1}{2}\right)^3 N_0 = 24 - 3 = 21$ (white circles)

So on

Fig. 6.2: To calculate the number of atoms of a readioactive substance left undisintegrated or disintegrated in *n* half-lives

(*ii*) ${}^{128}_{53}I$ isotope has half-life period = 25 minutes. Suppose we start with $1g(= N_0)$ of this isotope and allow it to undergo disintegration for a period of 100 minutes which are equal to 4 half-lives. Obviously :

Amount of ${}^{128}_{53}I$ left undisintegrated after 25 minutes (one half-life = 1 × 25 = 25)

$$= \frac{1}{2} N_0 = \frac{1}{2} \times 1 = 0.50 \text{ g}$$

Amount of ${}^{128}_{53}I$ left undisintegrted after 50 minutes (two half-lives = 2 × 25 = 50)

$$= \left(\frac{1}{2}\right)^2 N_0 = \left(\frac{1}{2}\right)^2 \times 1 = 0.25 \text{ g}$$

Amount of ${}^{128}_{53}I$ left undisintegrated after 75 minutes (three half-lives = 3 × 25 = 75)

$$= \left(\frac{1}{2}\right)^3 N_0 = \left(\frac{1}{2}\right)^3 \times 1 = 0.125\text{g}$$

Amount of $^{128}_{53}I$ left undisintegrated after 100 minutes (4 half-lives = 4 × 25 = 100)

$$= \left(\frac{1}{2}\right)^4 \times 1 = 0.0625g$$

All the above points have been shown in Fig. 6.3.

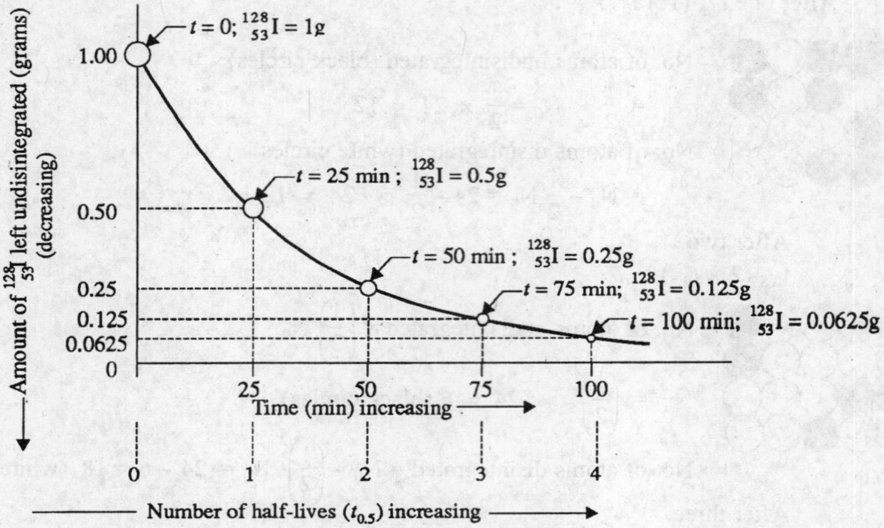

Fig. 6.3 : Radioactive decay of $^{128}_{53}I$.

Disintegration constant or decay constant (K)

Suppose a radioactive element A disintegrates into another substance B. Let the amount or the number of nuclei of A in the beginning (*i.e*, at $t = 0$ time) be N_0. Now as the time passes, the element A disintegrates and hence the amount of A goes on decreasing while that of B goes on increasing. Suppose that after t time (which may be in seconds, minutes, days, years etc.), the amount of A left undisintegrated is N *i.e.*, $(N_0 - N)$ is the amount of A that gets disintegrated into B after time t. Thus :

$$A \xrightarrow{\text{Disintegration}} B$$

		A	B
At $t = 0$ time, amounts of A and B	=	N_0	0
After t time, amounts of A and B	=	N	$(N_0 - N)$
		(Undisintegrated)	(Disintegrated)

Now if a small amount, dN of A gets disintegrated into B in a small time dt, then the rate of disintegration (*i.e.*, rate of decrease) of A into B is equal to – dN / dt which is proportional to the amount of A left undisintegrated (N), *i.e.*,

$$-\frac{dN}{dt} \propto N \quad \text{or} \quad -\frac{dN}{dt} = KN$$

(K is a constant of proportionality which is called *disintegration or decay constant).*

$$\text{or} \quad -\frac{dN}{N} = K \cdot dt \qquad (i)$$

In equation (*i*) negative sign shows that the amount of A is decreasing with time. Now if dt = 1 second, minute etc, then equation (*i*) becomes

$$-\frac{dN}{N} = K \qquad (ii)$$

Equation (*ii*) shows that *disintegration constant (K) can be defined as the fraction of the total amount of the radioactive substance* $\left(= \frac{dN}{N}\right)$ *which disintegrates in unit time.* K is constant for a particular radioactive element, *i.e.,* different elements have different values of K. The magnitude of K is not changed with the change of temperature, concentration, pressure and other physical factors. K is expressed in time^{-1} units (*i.e.,*in s^{-}, min^{-1}, hrs^{-1}; days^{-1}, yrs^{-1}).

Integrating equation (*i*) over limits N_0 (lower limit) and N (higher limit) (for left hand side) and 0 and *t* (for right hand side), we get :

$$\int_{N_0}^{N} \frac{dN}{N} = -\int_0^t K\,dt$$

$$\ln \frac{N}{N_0} = -Kt \qquad (iii)$$

$$\text{or} \quad \frac{N}{N_0} = e^{-Kt} \quad \text{or} \quad N = N_0\, e^{-Kt} \qquad (iv)$$

Equation (*iii*) can also be written as :

$$2.303 \log_{10} \frac{N}{N_0} = -Kt \quad \text{or} \quad -2.303 \log_{10} \frac{N}{N_0} = Kt \quad \text{or} \quad 2.303 \log_{10} \frac{N_0}{N} = Kt$$

$$\text{or} \quad t = \frac{2.303}{K} \log_{10} \frac{N_0}{N}$$

$$\text{or} \quad t = \frac{2.303}{K} \log_{10} \frac{\text{Amount of the radioactive subsaence before its decay}}{\text{Amount of the radioactive substance after its decay for time t}}$$

$$\text{or} \quad t = \frac{2.303}{K} \log_{10} \frac{\text{Amount of the substance in the beginning}}{\text{Amount of the substance left undisintegrated after time t}} \qquad (v)$$

Equation (*v*) is the same as is obtained for a first order reaction. Hence, rate of a radioactive disintegration is a first order reaction. This equation shows that the complete disintegration of a radioactive substance takes place in infinite time.

Another definition of disintegration constant

We have seen that , $N = N_0 . e^{-Kt}$

Now put $K = \frac{1}{t}$ in the above equation to get :

$$N = N_0 . e^{-\frac{1}{t} . t} = N_0 . e^{-1}$$

$$\text{or} \quad N = \frac{N_0}{e} = \frac{N_0}{2.718} \qquad (\because e = 2.718)$$

$$\text{or} \quad \boxed{N = 0.37\, N_0 = 37\% \text{ of } N_0.} \qquad ...(vi)$$

Equation (*vi*) shows that the *disintegration constant (K) can also be defined as the reciprocal of that time in which the amount of a radioactive substance reduces to 37% of its initial amount.*

To find out an expression for half-life period: Relation between half-life period ($t_{0.5}$ or $t_{1/2}$) and disintegration constant (K)

It is evident from the definition of half-life period that an expression for half-life period can be obtained by putting $t = t_{0.5}$ and $N = N_0 / 2$ either in equation (A) or in equation (B).

$$N = N_0\, e^{-Kt} \qquad ....(A)$$

On putting $N = N_0/2$ and $t = t_{0.5}$ in this equation, we get:

$$\frac{N_0}{2} = N_0\, e^{-K . t_{0.5}}$$

$$\text{or} \quad \frac{1}{2} = e^{-K . t_{0.5}} \quad \text{or} \quad \ln\left(\frac{1}{2}\right) = \ln\,[e^{-K . t_{0.5}}] = -K . t_{0.5}$$

$$\text{or} \quad 2.303 \log_{10}\left(\frac{1}{2}\right) = -K . t_{0.5} \quad \text{or} \quad -2.303 \log_{10} 2 = K . t_{0.5}$$

or $\frac{2.303 \log_{10} 2}{K} = t_{0.5}$ or $t_{0.5} = \frac{2.303 \times 0.301}{K}$

or $$t_{0.5} = \frac{0.6932}{K} \quad \text{... (vii)}$$

$$t = \frac{2.303}{K} \log_{10} \frac{N_0}{N} \quad \text{....(B)}$$

On putting $t = t_{0.5}$ and $N = N_0/2$, we get :

$$t_{0.5} = \frac{2.303}{K} \log_{10} \frac{2N_0}{N_0} = \frac{2.303}{K} \log 2$$

or $t_{0.5} = \frac{2.303 \times 0.301}{K}$ or $$t_{0.5} = \frac{0.6932}{K} \quad (viii)$$

Equations (*vii*) and (*viii*) are the same and give us a relation between half-life period ($t_{0.5}$) and disintegration constant (K) of a radioactive substance. Each of these equations shows that $t_{0.5}$ depends only on the magnitude of K and not on the weight or the number of the atoms of the radioactive substance or any other factor. $t_{0.5}$ is inversely proportional to K. $t_{0.5}$ for different substances is different, *i.e.*, half-life is fixed for a particular substance.

To find out an expression showing the relation between t and $t_{0.5}$

We have already shown that :

$$t = \frac{2.303}{K} \log_{10} \frac{N_0}{N} \quad \text{and} \quad t_{0.5} = \frac{0.6932}{K}$$

When K is eliminated from the above two equation, we get:

$$\frac{2.303}{t} \log_{10} \frac{N_0}{N} = \frac{0.6932}{t_{0.5}} \quad \text{or} \quad \frac{2.303 \times t_{0.5}}{0.6932} \log_{10} \frac{N_0}{N} = t$$

or $$t = \frac{2.303 \times t_{0.5}}{0.6932} \log_{10} \frac{N_0}{N} \quad (ix)$$

This equation gives us a relation between t and $t_{0.5}$.

Average life period (t_{av})

The disintegration of a radioactive substance goes on indefinitely and the toal period for complete disintegration of any radioactive sample is infinity which is meaningless. Therefore, a new term *viz.* average life period has been introduced. Average life period (t_{av}) of a radioactive substance is the reciprocal of its disintegration constant (K), *i.e.*

$$t_{av} = \frac{1}{K}$$

Relation between t_{av} and $t_{0.5}$

We know that $$t_{av} = \frac{1}{K} \quad (i)$$

We also know that $t_{0.5} = \frac{0.6932}{K}$ or $\frac{1}{K} = \frac{t_{0.5}}{0.6932} = t_{0.5} \times 1.44$

or $$t_{av} = \frac{1}{K} = t_{0.5} \times 1.44 \quad (ii)$$

Equation (*ii*) gives us a relation between t_av and $t_{0.5}$ of a given radioactive material. This equation shows that the value of $t_{0.5}$ of a radioactive element is smaller than t_{av} and t_{av} is directly proportional to $t_{0.5}$.

Example 6.3. *The half-life of radon is 3.8 days. After how many days will one twentieth of radon sample be left over?* **(I.I.T 1981, MLNR 1976)**

Solution. Here the amount of radon left undisintegrated $= \frac{1}{20}$*th* of its initial amount

or $$N = \frac{1}{20} \times N_0 \quad \text{or} \quad \frac{N_0}{n} = 20$$

Using the equation *viz.* $t = \frac{2.303 \times t_{0.5}}{0.693} \times \log \frac{N_0}{N}$, we get

$$t = \frac{2.303 \times 3.8}{0.693} \times \log 20 = \frac{2.303 \times 3.8}{0.693} \times \log (10 \times 2)$$

$$= \frac{2.303 \times 3.8}{0.693} [\log 10 + \log 2]$$

$$= \frac{2.303 \times 3.8}{0.693} [1 + 0.301] \qquad (\because \log 2 = 0.301)$$

$$= \textbf{16.42 days}$$

Example 6.4. *Half-life period of* $^{125}_{53}I$ *is 60 days. What percentage of the original radioactivity would be present after 180 days?* *(Roorkee Entrance 1980)*

Solution . Here the original amount of $^{125}_{53}$I isotope (N_0) = 100 and N is to be found out. We know that :

$$N = N_0 \left(\frac{1}{2}\right)^n = N_0 \left(\frac{1}{2}\right)^{t/t_{0.5}} \qquad (\because t = n \times t_{0.5})$$

or $$N = 100 \times \left(\frac{1}{2}\right)^{180/60} \quad \text{or} \quad N = 100 \times \left(\frac{1}{2}\right)^3 = 12.5\%$$

∴ % of the original radioactivity present after 180 days = 12.5%

Example 6.5. *Half life of* ^{226}Ra *is 1580 years. How many grams will be left undisintegrated from 1.0 g. of the isotope after 4740 years.*

Solution. We know that the amount of the isotope left undisintegrated $= N = N_0 \left(\frac{1}{2}\right)^n = N_0 \left(\frac{1}{2}\right)^{t/t_{0.5}}$ $(\because t = n \times t_{0.5})$

Here N_0 = 1g, t = 4740 yrs, $t_{0.5}$ = 1580 yrs

$$\therefore N = 1 \times \left(\frac{1}{2}\right)^{4740/1580} = \left(\frac{1}{2}\right)^3 = \frac{1}{8} = 0.125$$

Thus the amount of Ra left undisintegrated after 4740 yrs = **0.125 g.**

Example 6.6. *Half-life of* ^{210}Po *is 140 days. Calculate the number of days after which* $\frac{1}{4}$*g of* ^{210}Po *will be left undisintegrated from 1g of the isotope.*

Solution. Here $N = N_0 \left(\frac{1}{2}\right)^{t/t_{0.5}}$

or $$\frac{1}{4} \times 1 = 1 \times \left(\frac{1}{2}\right)^{t/140} \quad \text{or} \quad \frac{t}{140} = 2 \quad \text{or} \quad t = \textbf{280 days}$$

Example 6.7. *One gram of* $^{198}_{79}Au$ *($t_{0.5}$ = 65 hrs) decays by β-emission to produce stable mercury. Answer the following (i) Write the nuclear reaction for the process. (iii) How much mercury will be present after 260 hrs?* *(Roorkee Entrance 1984)*

Solution. (*i*) $^{198}_{79}Au \xrightarrow{-\beta \text{ or } -\,^{0}_{-1}e} \,^{198}_{80}Hg + \,^{0}_{-1}e$

$N_0 = 1g \qquad (N_0 - N) = (1 - N)$

(*ii*) Amount Au left undisintegrated after 260 hrs

$$= N = N_0 \times \left(\frac{1}{2}\right)^{t/t_{0.5}} = 1 \times \left(\frac{1}{2}\right)^{260/65} = 1 \times \left(\frac{1}{2}\right)^4 = \frac{1}{16}$$

$\therefore$ Amount of Hg obtained = $(N_0 - N)g = 1 - \frac{1}{16} = \mathbf{\frac{15}{16}\ g}$

Example 6.8. *Radioactivity is a first order process. Radioactive carbon in a wood sample decays with a half life of 5770 years. What is the rate constant (in yrs^{-1}) for the decay? What fraction would remain after 11540 years?* **(I.I.T. 1984)**

Solution. $t_{0.5}$ = 5770 yrs, t = 11540 yrs.

(*i*) Rate constant, K is given by :

$$K = \frac{0.6932}{t_{0.5}} = \frac{0.6932}{5770 \text{ yrs.}} = 1.2 \times 10^{-4} \text{ yrs}^{-1}$$

$$(ii)\ N = N_0\left(\frac{1}{2}\right)^n = N_0\left(\frac{1}{2}\right)^{t/t_{0.5}} \quad \text{or} \quad N = 1 \times \left(\frac{1}{2}\right)^{11540/5770} = \left(\frac{1}{2}\right)^2 = \frac{1}{4}$$

Thus the fraction of the radioactive carbon that is left undisintegrated after 11540 yrs = $\frac{1}{4}$ = 25% of the initial amount of carbon.

Example 6.9. *A certain radioisotope $^{A}_{Z}X$ (half life = 10 days) decays to give $^{A-4}_{Z-2}Y$. If 1.0g atom of $^{A}_{Z}X$ is kept in a sealed vessel, how much helium will accumulate in 20 days? Express the result in cm^3 at STP.* **(Roorkee Entrance 1987)**

Solution. The decay can be shown as :

$$^{A}_{Z}X \xrightarrow{-\,^{4}_{2}He} \,^{A-4}_{Z-2}Y + \,^{4}_{2}He$$

$N_0 \qquad (N_0 - N) = (1 - N)$

$$\text{Now } N = N_0\left(\frac{1}{2}\right)^{t/t_{0.5}} = 1 \times \left(\frac{1}{2}\right)^{20/10} = \left(\frac{1}{2}\right)^2 = \frac{1}{4}$$

$\therefore$ Amount of $^{4}_{2}He$ produced = $N_0 - N = 1 - \frac{1}{4} = \frac{3}{4}$

$\therefore$ Volume of $\frac{3}{4}$g atom of $^{4}_{2}He$ at STP = $\frac{3}{4} \times 22400$ cm^3 = **16800 cm^3**

Example 6.10. *A radioactive nuclide decays to form a stable nuclide. Its half-life time is 3 min. What fraction of 1 gm sample of the nuclide will remain radioactive after 9 min.* **(Roorkee 1982)**

Solution We know that :

$$N = N_0\left(\frac{1}{2}\right)^n \quad \text{or} \quad N = N_0\left(\frac{1}{2}\right)^{t/t_{0.5}} \qquad (\because t = n \times t_{0.5})$$

$$= N_0\left(\frac{1}{2}\right)^{9/3} = N_0\left(\frac{1}{2}\right)^3 = 1 \times \left(\frac{1}{2}\right)^3 \quad \text{or} \quad N = \frac{1}{8}$$

Thus $\frac{1}{8}$ gram of the sample will remain undisintegrated or radioactive.

Example 6.11. *In a sample of radioactive substance, there are 10^6 radioactive nuclei. Its* *half-life period is 20 seconds. How many nuclei will be left after 10 seconds?*

Solution. We know that

$$N = N_0\left(\frac{1}{2}\right)^n = N_0\left(\frac{1}{2}\right)^{t/t_{0.5}} \qquad (\because t = n \times t_{0.5})$$

$$= 10^6 \times \left(\frac{1}{2}\right)^{10/20} = 10^6 \times \left(\frac{1}{2}\right)^{1/2} \quad \text{or} \quad N = \mathbf{10^6 \times 0.707}$$

Example 6.12. *The half-life period of radium is 1590 yrs. After how many years will one gram of the pure radium (i) reduce to one centigram (ii) lose one centigram.*

Solution. (*i*) Initial amount = 1gm.

Amount remained undisintegrated = 1 centigram = $\frac{1}{100}$ gram = .01 gm

$$\therefore \quad t = \frac{2.303 \times t_{0.5}}{0.693}\log\frac{N_0}{N} = \frac{2.303 \times 1590}{0.693} \times \log\frac{1}{0.01}$$

$$= \frac{2.303 \times 1590}{0.693}\log 100 = \frac{2.303 \times 1590 \times 2}{0.693} = \mathbf{10567.8\ yrs}$$

(*ii*) Initial amount = 1 gram

Amount that remained undisintegrated = 1 – 0.01 = 0.99 gram

$$\therefore \quad t = \frac{2.303 \times 1590}{0.693}\log\frac{1}{0.99} = \mathbf{23.03\ yrs}$$

Radioactive equilibrium : Law of successive disintegration

Let a radioactive substance A disintegrate to form a product B which further disintegrates to form the product C. Thus

Successive disintegration :	A $\longrightarrow$	B $\longrightarrow$	C
No. of atoms :	N_A	N_B	N_C
Disintegration constant :	K_A	K_B	K_C
Half-life period :	$(t_{0.5})_A$	$(t_{0.5})_B$	$(t_{0.5})_C$

The stage at which the rate of decay of A is equal to the rate of decay of B, the rate of decay of B is equal to that of C and so on is called radioactive equilibrium. At this stage the amount of B in the sample remains constant.

Rate of decay of A = Rate of decay of B = Rate of decay of C

$$\text{or} \quad -\frac{dN_A}{dt} = -\frac{dN_B}{dt} = -\frac{dN_C}{dt} \quad \text{or} \quad K_AN_A = K_BN_B = K_CN_C$$

On considering the equilibrium A → B, we get:

$$K_AN_A = K_BN_B \quad \text{or} \quad \frac{N_A}{N_B} = \frac{K_B}{K_A} \qquad (i)$$

Now since $\frac{K_B}{K_A} = \frac{(t_{0.5})_A}{(t_{0.5})_B}$, we can also write

$$\boxed{\frac{N_A}{N_B} = \frac{K_B}{K_A} = \frac{(t_{0.5})_A}{(t_{0.5})_B}} \qquad (ii)$$

Suppose a radioactive sample contains two isotopes namely A and B. If the percentage of these isotopes present in the sample is *x* % and y % respectively, then:

$$\frac{\text{No. of moles of A isotope}}{\text{No. of moles of B isotope}} = \frac{(t_{0.5})_A}{(t_{0.5})_B} = \frac{K_B}{K_A}$$

or $$\frac{x/\text{mass number of A istope}}{y/\text{mass number of B isotope}} = \frac{(t_{0.5})_A}{(t_{0.5})_B} = \frac{K_B}{K_A}$$

or $$\frac{\text{x}}{\text{Mass number of A isotope}} \times \frac{\text{Mass number of B isotope}}{\text{y}} = \frac{(t_{0.5})_A}{(t_{0.5})_B} = \frac{K_B}{K_A}$$

Comparison between radioactive equilibrium and chemical equilibrium

A chemical equilibrium is reversible, since its direction *can be changed* by making a change in temperature, pressure or concentration of the species involved in the reaction. On the other hand, a radioactive equilibrium is irreversible and hence its direction *cannot be changed* by making a change in temperature, pressure etc.

Example 6.13. *A radioactive sample contains 15.8% ^{238}U and 5.3 × 10^{-6} % ^{226}Ra. If half-life of ^{226}Ra is 1590 years, what is the decay constant of ^{238}U?*

Solution. We know that

$$\frac{\text{No. of moles of U–238 }(N_{238})}{\text{No. of moles of Ra} - 226\ (N_{226})} = \frac{15.80/238}{5.3 \times 10^{-6}/226}$$

or $$\frac{N_{238}}{N_{226}} = \frac{15.80}{238} \times \frac{226}{5.3 \times 10^{-6}} = 2.83 \times 10^6$$

Again we know that:

$$\frac{N_{238}}{N_{226}} = \frac{(t_{0.5})_{238}}{(t_{0.5})_{226}} \quad \text{or} \quad 2.83 \times 10^6 = \frac{(t_{0.5})_{238}}{1590}$$

$$\therefore \quad (t_{0.5})_{238} = 2.83 \times 10^6 \times 1590 = 4.4997 \times 10^9 \text{ years}$$

$$\text{Now } K_{238} = \frac{0.693}{(t_{0.5})_{238}} = \frac{0.693}{4.4997 \times 10^9 \text{ years}}$$

$$\text{or } K_{238} = 1.54 \times 10^{-10} \text{ years}^{-1}$$

Acitivity of a radioactive substance ($-\frac{dN}{dt}$ or A)

Activity of a radioactive substance is defined as the number of disintegrations which the substance undergoes per second. Thus the unit of activity is disintegrations per second (*dps*). The activity is denoted by $-\frac{dN}{dt}$ or A which, as we have already seen, is equal to KN, *i.e.*,

Activity of a radioactive substance (A) or the number of disintegrations undergone by the substance in one second $= KN = \frac{0.693}{t_{0.5}} \times N$ ...(*i*)

In the above equation K = disintegration constant of the radioactive substance, N = number of atoms or nuclei of the substance left undisintegrated and $t_{0.5}$ = half-life of the substance.

Calculation of N

Suppose we have *x* grams of a radioactive element whose mass number is B. The number of atoms (N) present in *x* grams of this element is given by:

Thus activity (A) of *x* grams of an element is given by :

$$N = \frac{\text{Avogadro's number} \times \text{Mass of the element}}{\text{Mass number of the element}}$$

$$\text{Activity (A)} = K \times \frac{\text{Avogadro's number} \times \text{Mass of the element}}{\text{Mass number of the element}} \text{ s}^{-1}$$

or $$A = K \times \frac{6.022 \times 10^{23} \times x}{\text{B}} \text{ s}^{-1}$$

or $\quad A = \frac{0.693}{t_{0.5}} \times \frac{6.022 \times 10^{23} \times x}{B} \; s^{-1}$

The above equation gives us the number of disintegrations that x g of the substance undergoes in *one second.* If the number of disintegrations undergone by x g of the substance in t time is to be found out, the following relation is used.

No. of disintegrations undergone by x g of the substance in t time

$$= KNt = \frac{0.693}{t_{0.5}} \times \frac{(6.022 \times 10^{23}) \times \text{mass}}{\text{mass number}} \times t = \frac{0.693}{t_{0.5}} \times \frac{(6.022 \times 10^{23}) \times x}{B} \times t$$

We have already shown that:

$$t = \frac{2.303}{K} \log_{10} \frac{\text{Amount of the radioactive substance in the beginning}}{\text{Amount of the radioactive substance left undisintegrated after time t}}$$

Now since the activity shown by a radioactive substance is directly proportional to its amount, the above equation can also be written as :

$$t = \frac{2.303}{K} \log_{10} \frac{\text{Activity shown by the radioactive substance in the beginning}}{\text{Activity shown by the radioactive substance after time t}}$$

or $t = \frac{2.303}{K} \log_{10} \frac{Ao}{A}$ ...(*ii*)

Equation (*ii*) can also be written as equation (*iii*) or equation (*iv*)

$A = A_0 \, e^{-kt}$ ...(*iii*)

$A = A_0 \left(\frac{1}{2}\right)^n$ ...(*iv*)

In equation (*iv*) n in the number of half-lives ($t_{0.5}$) of which the total time (t) is composed, *i.e.*

$$t = n \times t_{0.5} \quad \text{or} \quad n = (t \,/\, t_{0.5})$$

Example 6.14. *Half-life of radium (atomic mass 226) is 1580 yrs. Show that 1 g of radium gives 3.70×10^{10} disintegrations per second.*

Solution. We know that the activity (A) of a radioactive substance is given by:

$$A = KN = \frac{0.693}{(t_{0.5})} \times N$$

$$= \frac{0.693}{1580 \times 365 \times 24 \times 60 \times 60 \text{ s}} \times \frac{\text{Avogadro's number} \times 1 \text{ gram}}{\text{Mass number}}$$

$$= \frac{0.693 \times 6.022 \times 10^{23} \times 1}{1580 \times 365 \times 24 \times 60 \times 60 \times 226} \text{ dis. s}^{-1}$$

or $\quad A = \mathbf{3.70 \times 10^{10}}$ **dis.** $\mathbf{s^{-1}}$

Example 6.15. *The half-life of the nuclide Rn^{220} is 54.5 s. What mass of this nucleus is equivalent to 1 millicurie (mci)?*

Solution. Activity of Rn^{220} = 1 mci = 3.7×10^7 dis s^{-1}

Let the mass of Rn^{220} = x gram

$$\therefore 3.7 \times 10^7 = \frac{0.693}{54.5} \times \frac{6.022 \times 10^{23} \times x}{220}$$

$$\therefore \quad x = \frac{3.7 \times 10^7 \times 54.5 \times 220}{0.693 \times 6.022 \times 10^{23}} \text{ g} = 1.06 \times 10^{-12} \text{ g} = \mathbf{1.06 \times 10^{-15}\ kg}$$

Example 6.16. *Calculate the alpha activity in terms of disintegrations per minute (dpm) of 10^{-3}g. sample of Pu^{239}. Given that $t_{0.5}$ = 24, 300 yrs and 1 yr = 3.15×10^7 s*

Solution. Activity of Pu^{239} is given by:

$$A = KN = \frac{0.693}{t_{0.5}} \times \frac{6.022 \times 10^{23} \times \text{mass of } Pu^{239}}{\text{Mass number}}$$

$$= \frac{0.693 \times 60}{24300 \times 3.15 \times 10^7 \text{ m}} \times \frac{6.022 \times 10^{23} \times 10^{-3}}{239} = \mathbf{1.43 \times 10^8 \text{ dis. m}^{-1}}$$

Example 6.17. *U^{235} yields 4770 disintegrations min^{-1} mg^{-1}. Calculate its decay constant and half-life (1 year = 3.15×10^7 s).*

Solution. We know that

$$A = K \times N \quad \text{or} \quad A = K \times \frac{(6.022 \times 10^{23}) \times \text{mass}}{\text{Mass number}}$$

or $$4770 \text{ d.m}^{-1}\text{mg}^{-1} = K \times \frac{(6.022 \times 10^{23} \text{ g}) \times 10^{-3} \, g}{235}$$

or $$\frac{4770}{60} \text{ d.s}^{-1} \text{ mg}^{-1} = K \times \frac{(6.022 \times 10^{23}) \times 10^{-3}}{235} \quad \therefore \quad K = \frac{4770 \times 235 \text{ s}^{-1}}{60 \times 6.022 \times 10^{23} \times 10^{-3}}$$

$$= \frac{4770 \times 235 \times (3.15 \times 10^7)}{60 \times 6.022 \times 10^{23} \times 10^{-3}} \text{ yr}^{-1} = 9.78 \times 10^{-10} \text{ yr}^{-1}$$

$$\therefore \quad t_{0.5} = \frac{0.693}{K} = \frac{0.693}{9.78 \times 10^{-10} \text{ yr}^{-1}}$$

$$= \frac{0.693}{9.78} \times 10^{10} \text{ yr} = \frac{69.3}{9.78} \times 10^8 \text{ yr} \quad \text{or} \quad t_{0.5} = \mathbf{7.08 \times 10^8 \text{ y}}$$

Example 6.18. *The activity of a sample of titanium is decreased by 5.5% over a period of one year. What is the half-life of titanium?*

Solution. We know that

$$t = \frac{2.303}{K} \log \frac{Ao}{A} \qquad 1\text{y} = \frac{2.303}{K} \log \frac{100}{100 - 5.5}$$

or $$1\text{y} = \frac{2.303}{K} \log \frac{100}{94.5} \quad \text{or} \quad K = \frac{2.303}{1\,y} \log \frac{100}{94.5} = 0.0566 \text{ y}^{-1}$$

Again we know that

$$t_{0.5} = \frac{0.693}{K} = \frac{0.693}{0.0566 \, y^{-1}} \quad \text{or} \quad t_{0.5} = \mathbf{12.24 \text{ y}}$$

Example 6.19. *Calculate the weight of ^{14}C (half-life = 5720 yrs) atoms which give 3.7×10^{10} disintegrations per second.* **(Roorkee 1986)**

Solution. If the weight of ^{14}C be x gram, then :

$$3.70 \times 10^7 \text{ s}^{-1} = \frac{0.693}{5720 \times 365 \times 24 \times 60 \times 60 \text{ s}} \times \frac{(6.022 \times 10^{23}) \times x}{14}$$

or $$3.70 \times 10^7 \text{ s}^{-1} = 3.83 \times 10^{-12} \times 6.022 \times 10^{23} \times \frac{x}{14} \text{ s}^{-1}$$

$$\therefore \quad x = \frac{14 \times 3.70 \times 10^7}{3.83 \times 10^{-12} \times 6.022 \times 10^{23}} \text{ g} = \mathbf{0.224 \times 10^{-3} g}$$

Example 6.20. *Half-life of Co^{60} is 5.27 yrs. Find out the alpha activity (dpm) of a 2.0 g sample of the isotope.*

Solution. $$\text{Activity(A)} = KN = \frac{0.693}{t_{0.5}} \times \frac{\text{Avogadro's number} \times \text{mass}}{\text{mass number}}$$

$$= \frac{0.693}{5.27 \times 365 \times 24 \times 60 \, m} \times \frac{(6.022 \times 10^{23}) \times 2}{60}$$

$$= \mathbf{5.028 \times 10^{15} \text{ dpm}}$$

Example 6.21. *A sample of a radioactive isotope showed an activity of 8640 counts per minute at one time and 7620 counts per minute after 1 hr. Calculate the disintegration constant?*

Solution. We know that

$$t = \frac{2.303}{K}\log_{10}\frac{A_o}{A} \quad \text{or} \quad K = \frac{2.303}{t}\log_{10}\frac{A_o}{A}$$

$$= \frac{2.303}{1 \times 60 \text{ min}} \times \log_{10}\frac{8640}{7620} = 0.0383 \times \log 1.19 \text{ min}^{-1} = \mathbf{2.09 \times 10^{-3} min^{-1}}$$

Example 6.22. *The rate of decay of Co^{60} changes from an initial value of 3112 dpm to 980 dpm in 42 days. Compute the value of half-life of Co^{60}.*

Solution. We know that

$$t = \frac{2.303}{K}\log_{10}\frac{A_o}{A} \quad \text{or} \quad t = \frac{2.303 \times (t_{0.5})}{0.693}\log_{10}\frac{A_o}{A} \left[\because K = \frac{0.693}{t_{0.5}}\right]$$

$$42 \text{ d} = \frac{2.303 \times (t_{0.5})}{0.693}\log_{10}\frac{3112}{980} = \frac{2.303 \times t_{0.5}}{0.693} \times 0.5018$$

$$\therefore \quad t_{0.5} = \frac{42 \times 0.693}{2.303 \times 0.5018} \text{ d} = \frac{29.106}{1.1556} = \mathbf{25.18 \text{ d}}$$

Example 6.23. *A count rate-meter is used to measure the activity of a given sample. At one time the meter shows 4750 counts min^{-1}. Five minutes later it shows 2700 counts min^{-1}. Find the decay constant and half-life of the sample.* *(Roorkee 86, MLNR 85)*

Solution. (*i*) We know that

$$t = \frac{2.303}{K}\log_{10}\frac{A_o}{A} \quad \text{or} \quad 5 = \frac{2.303}{K}\log_{10}\frac{4750}{2700}$$

$$\text{or} \quad K = \frac{2.303}{5}\log\frac{4750}{2700} = 0.4606 \times \log 1.759 \quad \text{or} \quad K = 0.4606 \times 0.2455 = \mathbf{0.1130 \text{ min}^{-1}}$$

$$(ii) \quad t_{0.5} = \frac{0.6932}{K} = \frac{0.6932}{0.1130 \text{ min}^{-1}} = \mathbf{6.1345 \text{ min.}}$$

Example 6.24. *The activity of a radioactive isotope falls to 12.5% in 90 days. Compute the half-life and decay constant of the isotope.* *(MLNR 1986)*

Solution. We know that

$$t = \frac{2.303}{K}\log_{10}\frac{A_o}{A}$$

$$\text{or} \quad K = \frac{2.303}{t}\log_{10}\frac{A_o}{A} = \frac{2.303}{90 \text{ d}}\log_{10}\frac{100}{12.5}$$

$$= 0.0255 \times \log_{10}(2)^3 = 0.0255 \times 3 \times \log 2$$

$$= 0.0255 \times 3 \times 0.301 = \mathbf{0.02312 \text{ d}^{-1}}$$

$$\text{Now } t_{0.5} = \frac{0.693}{k} = \frac{0.693}{0.0231} \text{ d} = \mathbf{30.10 \text{ d}}$$

Example 6.25. *The activity of a radioactive substance falls to 87.5% of its initial value after 5 yrs. Compute the half-life period of the substance. After what time the activity will fall by 87.5% [log 87.5 = 1.9420 and log 12.5 = 1.0969]*

Solution. We know that

$$K = \frac{2.303}{t}\log\frac{A_o}{A} = \frac{2.303}{5 \text{ yrs}} \times \log\frac{100}{87.5}$$

$$= 0.4606 \times [\log 100 - \log 87.5] \text{yrs}^{-1} = 0.4606 \times [2 - 1.9420] \text{yrs}^{-1}$$

$$= 0.4606 \times 0.0580 \text{ yrs}^{-1} = 0.02671 \text{ yrs}^{-1}$$

$$\therefore \quad t_{0.5} = \frac{0.693}{0.026} = \mathbf{26.65 \text{ yrs}}$$

Now let the time after which the activity falls by 87.5% be *t*, then

$$t = \frac{2.303}{K}\log\frac{A_o}{A} = \frac{2.303}{0.0267}\log\frac{100}{(100 - 87.5)} = \frac{2.303}{0.0267} \times \log\frac{100}{12.5} \text{yrs}$$

$$= 86.254 \times [\log 100 - \log 12.5] \text{yrs} = 86.254 \times [2 - 1.0969] \text{ yrs}$$

$= 86.254 \times 0.9031$ yrs. = **77.89 yrs**

Example 6.26. *The radioactive decay rate of a radioactive element is found to be 10^3 disintegrations per second at a certain time. If the half-life of the element is one second, find the decay rate after 1 second and after 3 seconds.* *(I.I.T. 1983)*

Solution. We know that

$$(i)\quad N = N_0\left(\frac{1}{2}\right)^n = N_0 \times \left(\frac{1}{2}\right)^{t/t_{0.5}} = 10^3 \times \left(\frac{1}{2}\right)^{1/1} = 10^3 \times \frac{1}{2} = \mathbf{500}$$

$$(ii)\quad N = N_0\left(\frac{1}{2}\right)^{t/t_{0.5}} = 10^3 \times \left(\frac{1}{2}\right)^{3/1} = 10^3 \times \frac{1}{8} = \mathbf{125}$$

Transmutation or disintegration of elements

Transmutation of an element is a process in which the element is converted into another element. This process may be of the following two types.

(i) **Natural or spontaneous transmutation or disintegration.** In this transmutation, a *radioactive element* is spontaneously converted into another element by emitting α, β or γ-rays. Examples of natural transmutation are given below. In these examples ^{226}Ra, ^{238}U, ^{234}Th and ^{213}Bi are radioactive elements and hence undergo spontaneous disintegration.

$$^{226}_{88}Ra \longrightarrow {}^{222}_{86}Rn + {}^{4}_{2}He \quad ; \quad ^{238}_{92}U \longrightarrow {}^{234}_{90}Th + {}^{4}_{2}He$$

$$^{234}_{90}Th \longrightarrow {}^{234}_{91}Pa + {}^{0}_{-1}e \quad ; \quad ^{213}_{83}Bi \longrightarrow {}^{213}_{84}Po + {}^{0}_{-1}e$$

(ii) **Artificial transmutation or artificial disintegration.** In this transmutation, *a stable* (non-radioactive) element is converted into another *stable element* by bombarding it with bombarding particles like α-particles, protons, deuterons etc. The bombarding particles are called by many other names like *incidents, atomic projectiles, bullets* etc. Examples are

$$^{10}_{5}B + {}^{4}_{2}He \longrightarrow {}^{13}_{6}C + {}^{1}_{1}H \quad ; \quad ^{23}_{11}Na + {}^{4}_{2}He \longrightarrow {}^{26}_{12}Mg + {}^{1}_{1}H$$

$$^{27}_{13}Al + {}^{4}_{2}He \longrightarrow {}^{30}_{14}Si + {}^{1}_{1}H \quad ; \quad ^{32}_{16}S + {}^{4}_{2}He \longrightarrow {}^{35}_{17}Cl + {}^{1}_{1}H$$

$$^{39}_{19}K + {}^{4}_{2}He \longrightarrow {}^{40}_{20}Ca + {}^{1}_{1}H$$

Discovery of artificial transmutation

It was for the first time in 1919 when Rutherford showed that when nitrogen atom $^{14}_{7}N$ (which is a stable element) is bombarded by energised α-particles, it is converted into $^{17}_{8}O$ atom and proton ($^{1}_{1}H$) is emitted. $^{17}_{8}O$ is also a stable element. $^{14}_{7}N$ nucleus is called *target nucleus* while $^{17}_{8}O$ is called *recoil nucleus*.

$$\underset{\text{Stable (non-radioactive)}}{^{14}_{7}N + {}^{4}_{2}He} \longrightarrow \underset{\text{Stable (non-radioactive)}}{^{17}_{8}O + {}^{1}_{1}H} \qquad ...(i)$$

Explanation of reaction (i). Rutherford proved by experiments that in reaction (*i*) α-particle is captured by $^{14}_{7}N$ nucleus and unstable intermediate nucleus namely $^{18}_{9}F$ is formed. $^{18}_{9}F$, being unstable, disintegrates spontaneously to give $^{17}_{8}O$ and $^{1}_{1}H$. Thus reaction (*i*) can be shown as:

$$\underset{\text{Stable nucleus}}{^{14}_{7}N + {}^{4}_{2}He} \longrightarrow \underset{\text{Unstable intermediate nucleus}}{^{18}_{9}F} \longrightarrow \underset{\text{Stable nucleus}}{^{17}_{8}O + {}^{1}_{1}H}.$$

Some important particles

Before we study nuclear reactions, it is essential for us to know the symbol, mass number, charge (atomic number) etc. of some particles which we shall be using again and again during our discussion. The following are important particles.

(i) **Alpha particle** $\left[{}^{4}_{+2}He \text{ or } {}^{4}_{2}He\right]$. It is a helium nucleus which contains 2 protons and 2

neutrons. Thus its symbol is ${}^{4}_{+2}He$ or ${}^{4}_{2}He$, showing that its mass number (A) is 4 and charges (atomic number) on it are +2, *i.e.*, A = 4 = 2*p* + 2*n* and Z = 2.

(*ii*) **Beta particle** $\left[{}^{0}_{-1}\beta \text{ or } {}^{0}_{-1}e\right]$. It is an electron and hence is represented by the symbol ${}^{0}_{-1}\beta$ or ${}^{0}_{-1}e$. Its mass number (A) = 0 and charges or atomic number = –1. The nucleus of an atom does not contain a β-particle or an electron as such. It is produced when a neutron (${}^{1}_{0}n$) is converted into a proton $\left({}^{1}_{1}H \text{ or } {}^{1}_{1}p\right)$.

$${}^{1}_{0}n \longrightarrow {}^{1}_{1}H \text{ (or } {}^{1}_{1}p) + {}^{0}_{-1}e \text{ or } {}^{0}_{-1}\beta \text{ (}\beta\text{-particle)}$$

(*iii*) **γ-ray** $\left[{}^{0}_{0}\gamma\right]$. It is an electromagnetic radiation of very short wave length. Its mass number and charges (atomic number) on it both are *zero*. Thus γ-ray is neutral and massless ray and hence is represented as ${}^{0}_{0}\gamma$.

(*iv*) **Deuteron** $\left[{}^{2}_{+1}H \text{ or } {}^{2}_{1}H \text{ or } {}^{2}_{+1}D \text{ or } {}^{2}_{1}D\right]$. It is the nucleus of heavy hydrogen atom called deuterium. Its mass number (A) = 2 and atomic number (Z) = 1.

(*v*) **Triton** $\left[{}^{3}_{+1}H \text{ or } {}^{3}_{1}H \text{ or } {}^{3}_{+1}T \text{ or } {}^{3}_{1}T\right]$. It is the nucleus of heavy hydrogen atom called tritium. It mass number (A) = 3 and atomic number(Z) = 1.

(*vi*) **Neutrino** $\left({}^{0}_{0}\nu\right)$. It is a nuclear particle having no mass and charge. Thus its mass number and atomic number both are equal to zero. The mass of this particle is equal to zero. Actual mass of this particle is 0.00002 with respect to O – 16 scale.

It was assumed by Pauli that this particle is emitted when a neutron (${}^{1}_{0}n$) is converted into a proton (${}^{1}_{1}H$) and a β-particle (${}^{0}_{-1}e$).

$${}^{1}_{0}n \longrightarrow {}^{1}_{1}H + {}^{0}_{-1}e + {}^{0}_{0}\nu$$

The above equation is balanced in so far as angular momentum values of different particles is concerned, as shown below:

$$\text{Equation : } {}^{1}_{0}n \longrightarrow {}^{1}_{1}H + {}^{0}_{-1}e + {}^{0}_{0}\nu$$

$$\text{Angular momentum values : } +\frac{1}{2}\cdot\frac{h}{2\pi} \quad + \frac{1}{2}\cdot\frac{h}{2\pi} - \frac{1}{2}\cdot\frac{h}{2\pi} + \frac{1}{2}\cdot\frac{h}{2\pi}$$

If ${}^{0}_{0}\nu$ is supposed to be emitted in the above reaction, angular momentum values are not balanced as shown below:

$$\text{Equation : } {}^{1}_{0}n \longrightarrow {}^{1}_{1}H + {}^{0}_{-1}e$$

$$\text{Angular momentum values : } +\frac{1}{2}\cdot\frac{h}{2\pi} \quad +\frac{1}{2}\cdot\frac{h}{2\pi} \quad -\frac{1}{2}\cdot\frac{h}{2\pi}$$

(*vii*) **Neutron** (${}^{1}_{0}n$). In 1931 Chadwick showed that when beryllium nucleus, ${}^{9}_{4}Be$ is bombarded by α-particles (${}^{4}_{2}He$), penetrating rays are obtained. He studied the characteristics of these rays and suggested that these rays are composed of *neutral particles* whose mass is equal to that of H-atom (*i.e.*, equal to 1) These particles were called *neutrons* by him and denoted as ${}^{1}_{0}n$.

$${}^{9}_{4}Be + {}^{4}_{2}He \longrightarrow {}^{12}_{6}C + {}^{1}_{0}n$$

Neutrons can also be obtained by many other nuclear reactions.

(*viii*) **Proton** $\left({}^{1}_{+1}H \text{ or } {}^{1}_{1}H \text{ or } {}^{1}_{1}p\right)$. It is the nucleus of ordinary hydrogen atom. Its mass number is 1 and charge equal to +1. This particle is present in the nucleus of the atom of all elements. Its presence in the nucleus was realised by its production in the nuclear reaction in which lighter elements like ${}^{6}_{3}Li$, ${}^{23}_{11}Na$, ${}^{31}_{15}P$ etc. are bombarded by deuteron (${}^{2}_{1}D$).

$${}^{6}_{3}Li + {}^{2}_{1}D \longrightarrow {}^{7}_{3}Li + {}^{1}_{1}H;\ {}^{23}_{11}Na + {}^{2}_{1}D \longrightarrow {}^{24}_{11}Na + {}^{1}_{1}H;\ {}^{31}_{15}P + {}^{2}_{1}D \longrightarrow {}^{32}_{15}P + {}^{1}_{1}H$$

and in Rutherford's artificial transmutation reaction (1919) in which he bombarded $^{14}_{7}N$ nucleus by α-particles

$$^{14}_{7}N + ^{4}_{2}He \longrightarrow ^{17}_{8}O + ^{1}_{1}H$$

(*ix*) **Positron.** ($^{0}_{+1}e$). It is a counterpart of an electron which has +1 charge and negligible mass. (Actual mass = 0.0005486 amu). Thus it is a positive electron. This particle was discovered by Iren Curie and her husband (name-Frederick Jiolet) in 1934. They bombarded Al-27 isotope (non-radioactive isotope) with α-particles so that it was converted into radioactive P-30 isotope with the emission of a neutron.

$$^{27}_{13}Al + ^{4}_{2}He \longrightarrow ^{30}_{15}P + ^{1}_{0}n$$

Being radioactive, P-30 disintegrates to give stable ^{30}Si isotope and positron ($^{0}_{+1}e$).

$$^{30}_{15}P \longrightarrow \underset{\text{Stable isotope}}{^{30}_{14}Si} + ^{0}_{+1}e \text{ (positron)}$$

Positron has also been obtained in many nuclear reactions. (See "*artificial radioactivity*")

Some important projectiles and projectile accelerators

α-particle ($^{4}_{+2}He$ or $^{4}_{2}He$), proton ($^{1}_{+1}H$ or $^{1}_{1}H$), deuteron ($^{2}_{+1}H$ or $^{2}_{1}H$), neutron ($^{1}_{0}n$) etc. are important fundamental projectiles which are generally used in the transmutation of different elements. Since α-particles have positive charge equal to +2, they are repelled by the nucleus when they hit the positively-charged nucleus and hence donot prove to be good projectiles. It has been seen that hardly one α-particle in a million may hit the nucleus. Protons and deuterons which have less positive charge (+1) than an α-particle (positive charge = +2) are better projectiles than α-particle, Neutrons ($^{1}_{0}n$) are the most effective projectiles, since they donot contain any charge.

The positively-charged α-particles, protons and deuterons can be made to act as more effective projectiles, if their velocity is accelerated. For this purpose, several instruments have been developed. These instruments are called *projectile accelerators*. Some such instruments are *cyclotron* (made by E.O. Lowrence at the university of California, U.S.A), *Van de Greef electrostatic generator, betatron, electron and proton synchotron etc.* In cyclotron the projectile particles are made to follow a spiral path under the combined high frequency a.c. voltage of the order of 10^4 to 2×10^4 volts. The particles finally leave the instrument with a velocity of about 40,000 Km. per second.

Artificial transmutation reactions induced by different bombarding projectiles

Artificial transmutation reactions induced by different bombarding particles may be of the following types:

1. Artificial transmutation reactions induced by α-particles ($^{4}_{2}He$). These reactions may be of the following types:

(*a*) (α, *p*) *reactions*. Example are:

$$^{14}_{7}N + ^{4}_{2}He \longrightarrow ^{17}_{8}O + ^{1}_{1}H \quad ; \quad ^{32}_{16}S + ^{4}_{2}He \longrightarrow ^{35}_{17}Cl + ^{1}_{1}H$$

$$^{10}_{5}B + ^{4}_{2}He \longrightarrow ^{13}_{6}C + ^{1}_{1}H \quad ; \quad ^{23}_{11}Na + ^{4}_{2}He \longrightarrow ^{26}_{12}Mg + ^{1}_{1}H$$

$$^{27}_{13}Al + ^{4}_{2}He \longrightarrow ^{30}_{14}Si + ^{1}_{1}H \quad ; \quad ^{39}_{19}K + ^{4}_{2}He \longrightarrow ^{40}_{20}Ca + ^{1}_{1}H$$

(*b*) (α, *n*) *reactions*. Examples are:

$$^{9}_{4}Be + ^{4}_{2}He \longrightarrow ^{12}_{6}C + ^{1}_{0}n \quad ; \quad ^{40}_{18}Ar + ^{4}_{2}He \longrightarrow ^{43}_{20}Ca + ^{1}_{0}n$$

$$^{7}_{3}Li + ^{4}_{2}He \longrightarrow ^{10}_{5}B + ^{1}_{0}n$$

(*c*) (α, γ) *reactions*. One example of such reactions is given below:

$$^{7}_{3}Li + ^{4}_{2}He \longrightarrow ^{11}_{5}B + ^{0}_{0}\gamma$$

2. Artificial transmutation reactions induced by protons ($^{1}_{1}H$). These reactions may be of the following types:

(*a*) (*p*, α) *reactions*. Examples are:

$$^{7}_{3}Li + ^{1}_{1}H \longrightarrow ^{4}_{2}He + ^{4}_{2}He \quad ; \quad ^{6}_{3}Li + ^{1}_{1}H \longrightarrow ^{4}_{2}He + ^{4}_{3}He$$

$$^{19}_{9}F + ^{1}_{1}H \longrightarrow ^{16}_{8}O + ^{4}_{2}He \quad ; \quad ^{14}_{7}N + ^{1}_{1}H \longrightarrow ^{11}_{6}C + ^{4}_{2}He$$

$$^{9}_{4}Be + ^{1}_{1}H \longrightarrow ^{6}_{3}Li + ^{4}_{2}He$$

(*b*) (*p*, *n*) *reactions.* Examples are given below:

$$^{18}_{8}O + ^{1}_{1}H \longrightarrow ^{18}_{7}F + ^{1}_{0}n \quad ; \quad ^{58}_{28}Ni + ^{1}_{1}H \longrightarrow ^{58}_{29}Cu + ^{1}_{0}n$$

$$^{65}_{29}Cu + ^{1}_{1}H \longrightarrow ^{65}_{30}Zn + ^{1}_{0}n \quad ; \quad ^{23}_{11}Na + ^{1}_{1}H \longrightarrow ^{23}_{12}Mg + ^{1}_{0}n$$

(*c*) (*p*, γ) *reactions.* Examples are given below:

$$^{7}_{3}Li + ^{1}_{1}H \longrightarrow ^{8}_{4}Be + ^{0}_{0}\gamma \quad ; \quad ^{27}_{13}Al + ^{1}_{1}H \longrightarrow ^{28}_{14}Si + ^{0}_{0}\gamma$$

$$^{12}_{6}C + ^{1}_{1}H \longrightarrow ^{13}_{7}N + ^{0}_{0}\gamma$$

(*d*) (*p*, *D*) *reactions.* Examples are:

$$^{4}_{2}Be + ^{1}_{1}H \longrightarrow ^{8}_{4}Be + ^{2}_{1}H \text{ or } ^{2}_{1}D \quad ; \quad ^{7}_{3}Li + ^{1}_{1}H \longrightarrow ^{6}_{3}Li + ^{2}_{1}H$$

3. Artificial transmutation reactions induced by deuterons ($^{2}_{1}H$ or $^{2}_{1}D$). These reactions may be of the following types:

(*a*) (*D*, α) *reactions*: $^{6}_{3}Li + ^{2}_{1}H \longrightarrow ^{4}_{2}He + ^{4}_{2}He$ (D, 2 α type)

$$^{16}_{8}O + ^{2}_{1}H \longrightarrow ^{14}_{7}N + ^{4}_{2}He \quad ; \quad ^{40}_{20}Ca + ^{2}_{1}H \longrightarrow ^{38}_{19}K + ^{4}_{2}He$$

$$^{34}_{16}Si + ^{2}_{1}H \longrightarrow ^{32}_{15}P + ^{4}_{2}He$$

(*b*) (*D*, *p*) *reactions* : $^{31}_{15}P + ^{2}_{1}H \longrightarrow ^{32}_{15}P + ^{1}_{1}H$

$$^{23}_{11}Na + ^{2}_{1}H \longrightarrow ^{24}_{11}Na + ^{1}_{1}H \quad ; \quad ^{56}_{27}Co + ^{2}_{1}H \longrightarrow ^{57}_{27}Co + ^{1}_{1}H$$

$$^{6}_{3}Li + ^{2}_{1}H \longrightarrow ^{7}_{3}Li + ^{1}_{1}H$$

(*c*) (*D*, *n*) *reactions:* $^{12}_{6}C + ^{2}_{1}H \longrightarrow ^{13}_{7}N + ^{1}_{0}n$

$$^{209}_{83}Bi + ^{2}_{1}H \longrightarrow ^{210}_{84}Po + ^{1}_{0}n \quad ; \quad ^{2}_{1}H + ^{2}_{1}H \longrightarrow ^{4}_{2}He + ^{1}_{0}n$$

(*d*) (*D*, *T*) *reactions*: $^{7}_{3}Li + ^{2}_{1}H \longrightarrow ^{6}_{3}Li + ^{3}_{1}H$ or $^{3}_{1}T$

$$^{31}_{15}P + ^{2}_{1}H \longrightarrow ^{30}_{15}P + ^{3}_{1}T \quad ; \quad ^{9}_{4}Be + ^{2}_{1}D \longrightarrow ^{8}_{4}Be + ^{3}_{1}T$$

4. Artificial transmutation reactions induced by neutrons ($^{1}_{0}n$). Examples are:

(*a*) (*n*, γ) *reactions* : $^{27}_{13}Al + ^{1}_{0}n \longrightarrow ^{28}_{13}Al + ^{0}_{0}\gamma$

$$^{238}_{92}U + ^{1}_{0}n \longrightarrow ^{239}_{92}U + ^{0}_{0}\gamma \quad ; \quad ^{85}_{37}Rb + ^{1}_{0}n \longrightarrow ^{86}_{37}Rb + ^{0}_{0}\gamma$$

(*b*) (*n*,α) *reactions:* $^{6}_{3}Li + ^{1}_{0}n \longrightarrow ^{3}_{1}H + ^{4}_{2}He$

$$^{10}_{5}B + ^{1}_{0}n \longrightarrow ^{7}_{3}Li + ^{4}_{2}He \quad ; \quad ^{27}_{13}Al + ^{1}_{0}n \longrightarrow ^{24}_{11}Na + ^{4}_{2}He$$

(*c*) (*n*,*p*) *reactions:* $^{11}_{5}B + ^{1}_{0}n \longrightarrow ^{11}_{4}Be + ^{1}_{1}H$

$$^{35}_{17}Cl + ^{1}_{0}n \longrightarrow ^{35}_{16}S + ^{1}_{1}H$$

(*d*) (*n*, *xn*) *reactions* : $^{235}_{92}U + ^{1}_{0}n \longrightarrow ^{139}_{56}Ba + ^{94}_{36}Kr + 3\,^{1}_{0}n$

$$^{39}_{16}K + ^{1}_{0}n \longrightarrow ^{38}_{19}K + 2\,^{1}_{0}n$$

(*e*) (*n, T*) *reactions* : $^{11}_{5}B + ^{1}_{0}n \longrightarrow ^{4}_{4}Be + ^{3}_{1}T$

$^{6}_{3}Li + ^{1}_{0}n \longrightarrow ^{4}_{2}He + ^{3}_{1}T$; $^{14}_{7}N + ^{1}_{0}n \longrightarrow ^{12}_{6}C + ^{3}_{1}T$

5. Artificial transmutation reactions induced by tritons ($^{3}_{1}H$ or $^{3}_{1}T$). Examples are:

(*a*) (*T,p*) *reactions* $^{59}_{27}Co + ^{3}_{1}H \longrightarrow ^{61}_{27}Co + ^{2}_{1}H$

(*b*) (*T,d*) *reactions* $^{6}_{3}Li + ^{3}_{1}H \longrightarrow ^{7}_{3}Li + ^{2}_{1}H$; $^{63}_{29}Cu + ^{3}_{1}H \longrightarrow ^{64}_{29}Cu + ^{2}_{1}H$

(*c*) (*T,n*) *reactions* $^{32}_{16}S + ^{3}_{1}H \longrightarrow ^{34}_{17}Cl + ^{1}_{0}n$

(*d*) (*T,α*) *reactions* $^{7}_{3}Li + ^{3}_{1}H \longrightarrow ^{6}_{2}He + ^{4}_{2}He$

6. Artificial transmutation reactions induced by γ-rays ($^{0}_{0}\gamma$). Nuclear reactions induced by γ-rays or photons are called *photo disintegration reactions.* Examples are:

(*a*) (γ, *n*) *reactions* $^{2}_{1}H + ^{0}_{0}\gamma \longrightarrow ^{1}_{1}H + ^{1}_{0}n$; $^{9}_{4}Be + ^{0}_{0}\gamma \longrightarrow ^{8}_{4}Be + ^{1}_{0}n$

$^{31}_{15}P + ^{0}_{0}\gamma \longrightarrow ^{30}_{15}P + ^{1}_{0}n$

(*b*) (γ, *p*) *reactions* $^{25}_{12}Mg + ^{0}_{0}\gamma \longrightarrow ^{24}_{11}Na + ^{1}_{1}H$

(*c*) (γ, 2*p*, *n*) *reactions* $^{27}_{13}Al + ^{0}_{0}\gamma \longrightarrow ^{24}_{11}Na + 2^{1}_{1}H + ^{1}_{0}n$

Example 6.27. *State the reactions including the compound nuclei for the following processes:* (*a*) $^{106}Pd\ (\alpha,\ p)\ ^{109}Ag$ (*b*) $^{6}Li\ (^{3}He,\ n)\ ^{8}B$ (*c*) $^{31}P\ (d,\ p)^{32}P$ (*d*) $^{9}Be\ (d,\ 2p)^{9}Li$ (*e*) $^{12}C\ (p,\ \gamma)^{13}N$ (*f*) $^{27}Al\ (\alpha,\ p)^{30}$ P.

Solution. We have:

(*a*) $^{106}_{46}Pd + ^{4}_{2}He \longrightarrow \left[^{110}_{48}Cd\right] \longrightarrow ^{109}_{47}Ag + ^{1}_{1}H$

(*b*) $^{6}_{3}Li + ^{3}_{2}He \longrightarrow \left[^{9}_{5}B\right] \longrightarrow ^{8}_{5}B + ^{1}_{0}n$

(*c*) $^{31}_{15}P + ^{2}_{1}H \longrightarrow \left[^{33}_{16}S\right] \longrightarrow ^{32}_{15}P + ^{1}_{1}H$

(*d*) $^{9}_{4}Be + ^{2}_{1}H \longrightarrow \left[^{11}_{5}B\right] \longrightarrow ^{9}_{3}Li + 2^{1}_{1}H$

(*e*) $^{12}_{6}C + ^{1}_{1}H \longrightarrow \left[^{13}_{7}N\right] \longrightarrow ^{13}_{7}N + ^{0}_{0}\gamma$

(*f*) $^{27}_{13}Al + ^{4}_{2}He \longrightarrow \left[^{31}_{15}P\right] \longrightarrow ^{30}_{15}P + ^{1}_{0}n$

Example 6.28. *Construct equations for the following conversions :* (*a*) $^{59}_{27}Co\ (n,\ \gamma)\ ^{60}Co$ (*b*) $^{14}_{7}N\ (n,\ p)\ ^{14}C$ (*c*) $^{9}_{4}Be\ (\alpha,\ n)\ ^{12}C$ (*d*) $^{7}_{3}Li\ (\alpha,\ \gamma)\ ^{11}B$ (*e*) $^{96}_{42}Mo\ (d,\ n)\ ^{97}Tc$ (*f*) $^{144}_{62}Sn\ (n,\gamma)\ ^{145}Sm$ (*g*) $^{209}_{83}Bi\ (\alpha,\ 2n)\ ^{211}At$ (*h*) $^{232}_{90}Th\ (^{12}C,\ 4n)\ ^{240}Cm$.

Solution. (*a*) $^{59}_{27}Co + ^{1}_{0}n \longrightarrow ^{60}_{27}Co + ^{0}_{0}\gamma$

(*b*) $^{14}_{7}N + ^{1}_{0}n \longrightarrow ^{14}_{6}C + ^{1}_{1}H$ (*c*) $^{9}_{4}Be + ^{4}_{2}He \longrightarrow ^{12}_{6}C + ^{1}_{0}n$

(*d*) $^{7}_{3}Li + ^{4}_{2}He \longrightarrow ^{11}_{5}B + ^{0}_{0}\gamma$ (*e*) $^{96}_{42}Mo + ^{2}_{1}H \longrightarrow ^{97}_{43}Tc + ^{1}_{0}n$

(*f*) $^{144}_{62}Sm + ^{1}_{0}n \longrightarrow ^{145}_{62}Sm + ^{0}_{0}\gamma$ (*g*) $^{209}_{83}Bi + ^{4}_{2}He \longrightarrow ^{211}_{85}At + 2^{1}_{0}n$

(*h*) $^{234}_{90}Th + ^{12}_{6}C \longrightarrow ^{240}_{96}Cm + 4^{1}_{0}n$

Example 6.29. *The nuclide* $^{247}_{99}Es$ *can be obtained by bombardment of* $^{238}_{92}U$ *in a reaction that emits five neutrons. Identify the bombarding particle.*

Solution. Let the bombarding particle be $^{a}_{b}X$. Therefore :

$^{238}_{92}U + ^{a}_{b}X = ^{247}_{99}Es + 5^{1}_{0}n$ ∴ $238 + a = 247 + 5$

or $a = (247 + 5) - 238 = 14$; $92 + b = 99$ ∴ $b = 99 - 92 = 7$

Thus the bombarding particle is $^{14}_{7}N$.

Example 6.30. *Fill in the blanks :*

(*a*) $^{14}_{7}N + ^{4}_{2}He \longrightarrow ^{17}_{8}O +$ ....

(*b*) $^{7}_{3}Li + ^{1}_{1}H \longrightarrow 2$ ....

(*c*) $^{10}_{5}B + ^{2}_{1}H \longrightarrow ^{11}_{6}C +$ ....

(*d*) $^{35}_{17}Cl + ^{1}_{0}n \longrightarrow ^{35}_{16}S +$ ....

Solution. (*a*) $^{14}_{7}N + ^{4}_{2}He \longrightarrow ^{17}_{8}O + \boxed{^{1}_{1}H}$

(*b*) $^{7}_{3}Li + ^{1}_{1}H \longrightarrow 2\,\boxed{^{4}_{2}He}$

(*c*) $^{10}_{5}B + ^{2}_{1}H \longrightarrow ^{11}_{6}C + \boxed{^{1}_{0}n}$

(*d*) $^{35}_{17}Cl + ^{1}_{0}n \longrightarrow ^{35}_{16}S + \boxed{^{1}_{1}H}$

Example 6.31. Express the following reactions in shorthand notation.

(*a*) $^{6}_{3}Li + ^{1}_{0}n \longrightarrow ^{4}_{2}He + ^{3}_{1}H$; (*b*) $^{31}_{15}P + ^{2}_{1}H \longrightarrow ^{32}_{15}P + ^{1}_{1}H$

(*c*) $^{238}_{92}U + ^{1}_{0}n \longrightarrow ^{239}_{93}Np + ^{0}_{-1}e$; (*d*) $^{253}_{99}Es + ^{4}_{2}He \longrightarrow ^{256}_{101}Md + ^{1}_{0}n$

(*e*) $^{27}_{13}Al + ^{1}_{0}n \longrightarrow ^{26}_{13}Al + 2^{1}_{0}n$

Solution. (*a*) $^{6}_{3}Li\ (n, \alpha)\ ^{3}_{1}H$; (*b*) $^{31}_{15}P\ (d, p)\ ^{32}_{15}P$

(*c*) $^{238}_{92}U\ (n, e)\ ^{239}_{93}Np$; (*d*) $^{253}_{99}Es\ (\alpha, n)\ ^{256}_{101}Md$

(*e*) $^{27}_{13}Al\ (n, 2n)\ ^{26}_{13}Al$

Example 6.32. *Complete the following reactions, giving the mass number and atomic number wherever necessary.*

(*a*) $^{235}U + n \longrightarrow$ .... $+ \gamma$; (*b*) $^{238}U + ^{4}He \longrightarrow$ .... $+ n$

(*c*) $^{243}Pu \longrightarrow ^{243}Am + _{-1}e$; (*d*) $^{239}_{94}Pu + n \longrightarrow ^{243}Pu + \gamma$

Solution. : (*a*) $^{235}_{92}U + ^{1}_{0}n \longrightarrow ^{236}_{92}U + ^{0}_{0}\gamma$; (*b*) $^{238}_{92}U + ^{4}_{2}He \longrightarrow ^{241}_{94}Pu + ^{1}_{0}n$

(*c*) $^{243}_{94}Pu \longrightarrow ^{243}_{95}Am + ^{0}_{-1}e$; (*d*) $^{239}_{94}Pu + 4^{1}_{0}n \longrightarrow ^{243}_{94}U + 4^{0}_{0}\gamma$

Applications of artificial transmutation reactions (artificial radioactivity)

Some of the applications of artificial transmutation reactions are discussed below:

1. Discovery of new fundamental particles and their uses. The study of transmutation reactions has enabled us to discover new fundamental particles like α-particles, protons, deuterons, neutrons, tritons, γ-rays, positrons etc. The discovery of protons ($^{1}_{1}H$), neutrons ($^{1}_{0}n$) and positron ($^{0}_{+1}e$) has been shown by the following nuclear reactions:

$^{14}_{7}N + ^{4}_{2}He \longrightarrow ^{17}_{8}O + ^{1}_{1}H$ *(Rutherford's transmutation reaction-1919)*

$^{6}_{3}Li + ^{2}_{1}H \longrightarrow ^{7}_{3}Li + ^{1}_{1}H$

$^{9}_{4}Be + ^{4}_{2}He \longrightarrow ^{12}_{6}C + ^{1}_{0}n$ *(Chadwick's reaction-1931)*

$^{27}_{13}Al + ^{4}_{2}He \longrightarrow \underset{\text{Radioactive (unstable)}}{^{30}_{15}P} + ^{1}_{0}n$

$^{30}_{15}P \longrightarrow \underset{\text{Stable}}{^{30}_{14}Si} + ^{0}_{+1}e$

} *Experiment performed by Curie and Frederick Jiolet (1934)*

The discovery of these particles has added to our knowledge of the composition of the nucleus. For example:

(*a*) The scattering of α-particles seen by Rutherford in his α-particle scattering experiment has proved the existence of positively charged nucleus at the centre of the atom of all elements. The extent of defection of α-particles has established that the atomic number of an element is numerically equal to the nuclear charge carried by the nucleus.

(*b*) When lighter elements like ^{6}Li, ^{23}Na, ^{31}P etc. are bombarded by means of deuterons ($^{2}_{1}H$), protons ($^{1}_{1}H$) are produced.

$$^{6}_{3}Li + ^{2}_{1}H \longrightarrow ^{7}_{3}Li + ^{1}_{1}H \quad ; \quad ^{23}_{11}Na + ^{2}_{1}H \longrightarrow ^{24}_{11}Na + ^{1}_{1}H$$

$$^{31}_{15}P + ^{2}_{1}H \longrightarrow ^{32}_{15}P + ^{1}_{1}H$$

The production of protons confirms the presence of these particles in the nucleus.

(*c*) The bombardment of $^{27}_{13}Al$ isotope by α-particles produces $^{30}_{15}P$ which is radioactive and hence disintegrates to give a stable isotope of $^{30}_{14}Si$ with the emission of positrons ($^{0}_{+1}e$).

$$^{27}_{13}Al + ^{4}_{2}He \longrightarrow ^{30}_{15}P + ^{1}_{0}n \quad ; \quad ^{30}_{15}P \longrightarrow ^{30}_{14}Si + ^{0}_{+1}e$$

The production of positrons in this way reveals the presence of these particles in the nucleus.

2. Preparation of isotopes of various elements. Examples showing the use of artifical transmutation reactions for the preparation of isotopes of elements are given below:

(*a*) *Preparation of missing elements.* When Mendeleef arranged the elements in the form of a periodic table (Mendeleef periodic table), he left four gaps in it. These gaps were left for the elements with atomic numbers 43, 61, 85 and 87, since these elements were not known at that time. He called these elements as *missing elements.* After a few years, radioactive isotopes of these elements were synthesised by using artificial transmutation reactions and were characterised. These elements were named as *technetium, Tc(Z = 43), promethium,* Pm(Z = 61), *astatine,* At (Z = 85) *and francium,* Fr (Z = 87) respectively. The synthesis of these elements *is* given below:

Technetium, $^{99}_{43}Tc$

$$^{98}_{42}Mo + ^{1}_{0}n \longrightarrow \underset{\text{Radioactive}}{^{99}_{42}Mo} + ^{0}_{0}\gamma$$

$$^{99}_{42}Mo \longrightarrow ^{99}_{43}Tc + ^{0}_{-1}e$$

Promethium, $^{148}_{61}Pm$

$$^{148}_{60}Nd + ^{1}_{1}H \longrightarrow ^{148}_{61}Pm + ^{1}_{0}n$$

Astatine, $^{211}_{85}At$

$$^{209}_{83}Bi + ^{4}_{2}He \longrightarrow ^{211}_{85}At + 2^{1}_{0}n$$

Francium, $^{223}_{87}Fr$

$$^{226}_{88}Ra + ^{1}_{0}n \longrightarrow \underset{\text{Radioactive}}{^{227}_{88}Ra} + ^{0}_{0}\gamma$$

$$^{227}_{88}Ra \longrightarrow \underset{\text{Radioactive}}{^{227}_{89}Ac} + ^{0}_{-1}e \quad (\beta\text{-particle})$$

$$^{227}_{89}Ac \xrightarrow{1\%} ^{223}_{87}Fr + ^{4}_{2}He$$

$$^{227}_{89}Ac \xrightarrow{99\%} ^{227}_{90}Th + ^{0}_{-1}e$$

The missing elements whose preparation is given above have been identified and have been given proper place in the periodic table. $_{87}Fr$ has been placed along with alkali metals in group IA, $_{43}Tc$ has been placed in group VII B under $_{25}Mn$ (transition elements), $_{85}At$ has been put in group VIIA along with halogens and $_{61}Pm$ has been placed along with lanthanides.

(*b*) *Preparation of trans-uranic elements.* Thirteen elements with atomic numbers 93 to 105 are called *trans-uranic elements,* since these elements come after $_{92}U$ in actinide series. By changing the target element and the bombarding projectiles, various isotopes of these elements have been obtained in sub-microgram quantities (10^{-5}g to 10^{-6}g) by artificial transmutation reactions and have been characterised. The synthesis of these elements is shown below:

Target element		*Bombarding projectile*		*Product*
$^{238}_{92}U$	+	$^{2}_{1}H$	$\longrightarrow$	$^{238}_{93}Np$ (*Neptunium*) + $2^{1}_{0}n$
$^{239}_{92}U$	+	$^{4}_{2}He$	$\longrightarrow$	$^{240}_{94}Pu$ (*Plutonium*) + $2^{1}_{0}n$
$^{239}_{94}Pu$	+	$^{4}_{2}He$	$\longrightarrow$	$^{241}_{95}Am$ (*Americium*) + $^{1}_{1}H$ + $^{1}_{0}n$
$^{239}_{94}Pu$	+	$^{4}_{2}He$	$\longrightarrow$	$^{240}_{96}Cm$ (*Curium*) + $3^{1}_{0}n$
$^{244}_{96}Cm$	+	$^{4}_{2}He$	$\longrightarrow$	$^{245}_{97}Bk$ (*Berkelium*) + $^{1}_{1}H$ + $2^{1}_{1}n$
$^{238}_{92}U$	+	$^{12}_{6}C$	$\longrightarrow$	$^{245}_{98}Cf$ (*Californium*) + $5^{1}_{1}n$
$^{238}_{92}U$	+	$^{14}_{7}N$	$\longrightarrow$	$^{247}_{99}Es$ (*Einsteinium*) + $5^{1}_{0}n$
$^{238}_{92}U$	+	$^{16}_{8}O$	$\longrightarrow$	$^{250}_{100}Fm$ (*Fermium*) + $4^{1}_{0}n$
$^{253}_{99}Es$	+	$^{4}_{2}He$	$\longrightarrow$	$^{256}_{101}Md$ (*Mendeleevium*) + $^{1}_{0}n$
$^{246}_{96}Cm$	+	$^{13}_{6}C$	$\longrightarrow$	$^{251}_{102}No$ (*Nobelium*) + $8^{1}_{0}n$
$^{252}_{98}Cf$	+	$^{10}_{5}B$	$\longrightarrow$	$^{257}_{103}Lw$ (*Lawrencium*) + $5^{1}_{0}n$
$^{242}_{94}Pu$	+	$^{22}_{10}Ne$	$\longrightarrow$	$^{260}_{104}Ku$ (*Kurchatovium*) + $4^{1}_{0}n$
$^{249}_{98}Cf$	+	$^{15}_{7}N$	$\longrightarrow$	$^{260}_{105}Ha$ (*Hahnium*) + $4^{1}_{0}n$

(*c*) *Preparation of radioactive and non-radioactive isotopes: Discovery of isotopes.* Artificial transmutation reactions have been used to prepare many radioactive isotopes (*e.g.* Na-24, Co-60, P-30, P-32 etc.) which have been used as radioactive tracers in various processes met with in medical field, agriculture, industry, analytical chemistry etc. The preparation of some radioactive isotopes are given below. These are the examples of induced radioactivity.

Non-radioactive (stable) isotope		*Bombarding projectile*		*Radioactive (unstable) isotope*
$^{23}_{11}Na$	+	$^{2}_{1}H$	$\longrightarrow$	$^{24}_{11}Na$ + $^{1}_{1}H$
$^{27}_{11}Al$	+	$^{1}_{0}n$	$\longrightarrow$	$^{24}_{11}Na$ + $^{4}_{2}He$
$^{59}_{27}Co$	+	$^{1}_{0}n$	$\longrightarrow$	$^{60}_{27}Co$
$^{27}_{13}Al$	+	$^{4}_{2}He$	$\longrightarrow$	$^{30}_{15}P$ + $^{1}_{0}n$
$^{32}_{16}S$	+	$^{1}_{0}n$	$\longrightarrow$	$^{32}_{15}P$ + $^{1}_{1}H$
$^{14}_{7}N$	+	$^{1}_{0}n$	$\longrightarrow$	$^{14}_{6}C$ + $^{1}_{1}H$
$^{10}_{5}B$	+	$^{4}_{2}He$	$\longrightarrow$	$^{13}_{7}N$ + $^{1}_{0}n$

Transmutation reactions have also been used to convert stable isotope of an element into stable isotope of another element. For example :

Stable isotope (Non-radio-active isotope)		*Bombarding projectile*		*Stable isotope (Non-radioactive isotope)*
$^{14}_{7}N$	+	$^{4}_{2}He$	$\longrightarrow$	$^{17}_{8}O$ + $^{1}_{1}H$
$^{10}_{5}B$	+	$^{4}_{2}He$	$\longrightarrow$	$^{13}_{6}C$ + $^{1}_{1}H$
$^{23}_{11}Na$	+	$^{4}_{2}He$	$\longrightarrow$	$^{26}_{12}Mg$ + $^{1}_{1}H$
$^{27}_{13}Al$	+	$^{4}_{2}He$	$\longrightarrow$	$^{30}_{14}Si$ + $^{1}_{1}H$
$^{32}_{16}S$	+	$^{4}_{2}He$	$\longrightarrow$	$^{35}_{17}Cl$ + $^{1}_{1}H$
$^{39}_{19}K$	+	$^{4}_{2}He$	$\longrightarrow$	$^{40}_{20}Ca$ + $^{1}_{1}H$

3. Release of atomic energy in nuclear fission and nuclear fusion reactions. A large amount of energy (called atomic energy or nuclear energy) is realeased in nuclear fission and nuclear fusion processes. This energy has been put to peaceful uses like the production of electricity or it can be used for the destructive purposes of making an atom bomb.

Natural and artificial radioactive series

The isotopes of naturally occuring heavy elements like $^{232}_{90}Th$, $^{235}_{92}U$, $^{238}_{92}U$ etc. are unstable and hence are constantly emitting α and β particles and giving rise to the formation of new radioactive elements. Being radioactive, these new elements spontaneously disintegrate and give a series of elements until a stable (*i.e.*, non-radioactive) element (usually lead) is obtained. *The series of elements thus obtained by the successive disintegration of the new radioactive elements is known as radioactive series.* The series obtained by the emission of α and β particles from $^{232}_{90}Th$, $^{238}_{92}U$ and $^{235}_{92}U$ are called *natural radioactive series.* Obviously there are three natural radioactive series. In addition to the natural radioactive series, there is another series which has been obtained by emission of α and β particles from the artificially prepared $^{237}_{93}Np$ isotope. This series is called *artificial radioactive series* or *neptunium-237* or (4*n*+1) *series.*

Natural radioactive series

As already said, the series obtained by the emission of various α and β particles from $^{232}_{90}Th$, $^{238}_{92}U$ and $^{235}_{92}U$ are called natural radioactive series. Obviously these series are of three types as shown below:

(*i*) *Thorium-232 or 4n series* ($^{232}_{90}Th$ – $^{208}_{82}Pb$). This series starts from $^{232}_{90}Th$ (parent element) and ends at $^{208}_{82}Pb$ (end element) which is a stable element. The parent element *viz.* $^{232}_{90}Th$ decays into $^{228}_{88}Ra$ by emitting a α-particle. $^{228}_{88}Ra$ emits a β-particle to form $^{228}_{89}Ac$ which again emits a β-particle to become $^{228}_{90}Th$. This disintegration process goes on proceeding continuously until stable end product *viz.* $^{208}_{82}Pb$ is obtained (See Table 6.3). The mass number of different members of this series is equal to 4*n* where *n* is a whole number whose value is different for different members of the series. The value of *n* for the parent element ($^{232}_{90}Th$) and end element ($^{208}_{82}Pb$) is 232/4 = 58 and 208/4 = 52 respectively. If the mass number of different members is divided by 4, the remainder obtained is equal to zero. Number of α and β particles emitted in the conversion of Th to Pb can be calculated with the help of the equations $232 - 4x = 208$ and $90 - 2x + y = 82$. These equations give *x* (number of α-particles) = 6 and *y* (number of β-particles) = 4. Half-life period of each member is given in bracket (y = years, d = days, h = hours, m = minutes and s = seconds).

Table 6.3: *Thorium-232 or 4n series (natural radioactive series [$^{232}_{90}Th$ – $^{208}_{82}Pb$ series]*

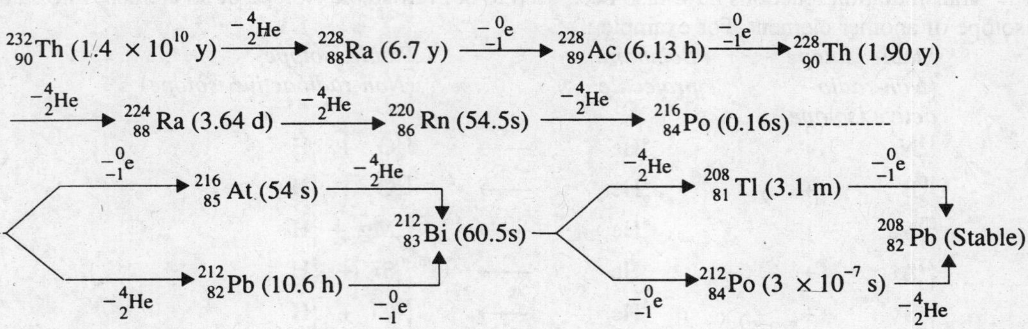

(*ii*) *Uranium-238 or (4n+2) series* [$^{238}_{92}U$ – $^{206}_{82}Pb$]. This series has $^{238}_{92}U$ as its parent element and $^{206}_{82}Pb$ as its final (stable) product (See Table 6.4). The mass number of different members is (4*n*+2) which, on being divided by 4, gives 2 as the remainder. This means that the value of *n* for ^{238}U and ^{206}Pb

is equal to $\frac{238 - 2}{4} = 59$ and $\frac{206 - 2}{4} = 51$ respectively. The number of α and β particles emitted by $^{238}_{92}U$ to form $^{206}_{82}Pb$ can be calculated by the equations $238 - 4x = 206$ and $92 - 2x + y = 82$. These equations give x (*i.e.*, number of α-particles) = 8 and y (*i.e.*, number of β-particles) = 6.

Table 6.4: *Uranium–238 or (4n+2) series (Natural radioactive series) [$^{238}_{92}U - ^{206}_{82}Pb$ series]*

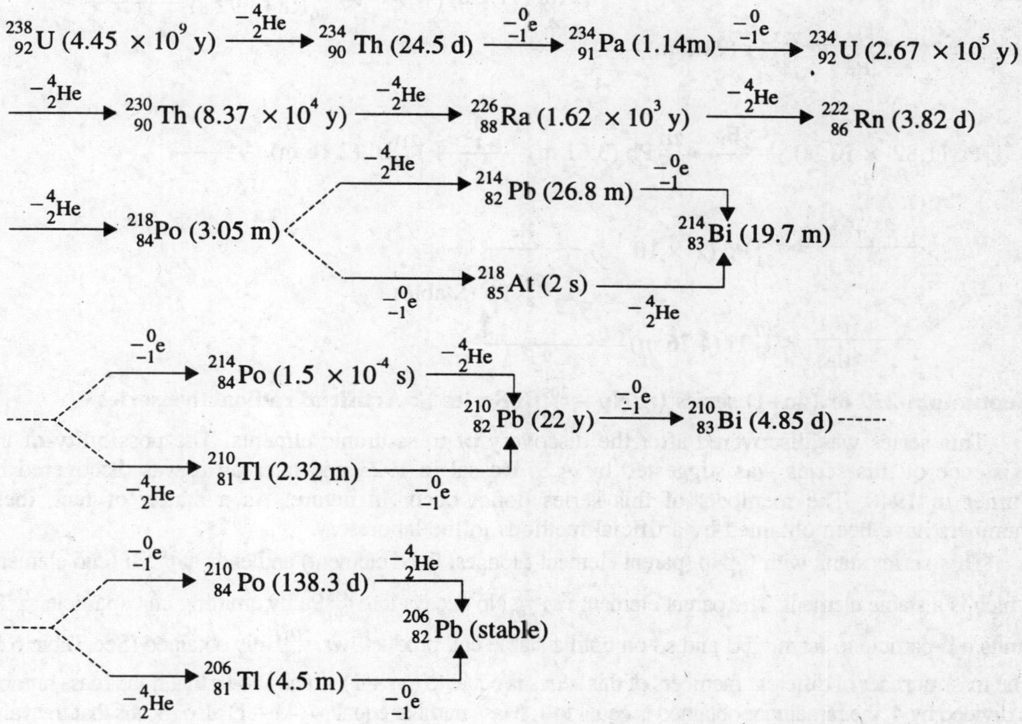

(*iii*) *Uranium-235 or (4n + 3) series [$^{235}_{92}U - ^{207}_{82}Pb$ series].* This series starts with $^{235}_{92}U$ and ends at $^{207}_{82}Pb$ which is a stable element. If the mass number of each member of this series is divided by 4, then 3 is obtained as a remainder. This means that the value of n for ^{235}U and ^{207}Pb is equal to $\frac{235 - 3}{4} = 58$ and $\frac{207 - 3}{4} = 51$ respectively. The number of α and β particles emitted in the series can be calçulated by using the equations $235 - 4x = 207$ and $92 - 2x + y = 82$. On solving, we get $x = 7$ and $y = 4$, where x and y are the number of α and β particles. (See table 6.5).

Similarities between the natural radioactive series

(*i*) In all the three natural radioactive series, sometimes a product is formed which disintegrates in a branching way by emitting one α and one β particle and thus produces two elements. These two elements are then converted into a common element when one of them emits a α-particle and the other emits a β-particle.

(*ii*) In all the series an element with atomic number 86 ($_{86}Rn$) is produced. This element has the properties of an inert gas and is called *emanation.*

(*iii*) The end product obtained in each series is a stable isotope of lead which may be $^{206}_{82}Pb$, $^{207}_{82}Pb$ or $^{208}_{82}Pb$.

Table 6.5: *Uranium -235 or (4n+3) series (Natural radioactive series) [$^{235}_{92}U$ – $^{209}_{82}Pb$ series]*

$$^{235}_{92}U\ (7.07 \times 10^{8}\ y) \xrightarrow{-^{4}_{2}He} {}^{231}_{90}Th\ (24.6h) \xrightarrow{-^{0}_{-1}e} {}^{231}_{91}Pa\ (3.2 \times 10^{4}\ y) \xrightarrow{-^{4}_{2}He} {}^{227}_{89}Ac\ (21.7y) \text{----}$$

$$\text{----} \xrightarrow{-^{0}_{-1}e} {}^{227}_{90}Th\ (18.9\ d) \xrightarrow{-^{4}_{2}He} {}^{223}_{88}Ra\ (11.2\ d)$$

$$\text{----} \xrightarrow{-^{4}_{2}He} {}^{223}_{87}Fr\ (21.0\ m) \xrightarrow{-^{0}_{-1}e} {}^{223}_{88}Ra\ (11.2\ d)$$

$$^{223}_{88}Ra\ (11.2\ d) \xrightarrow{-^{4}_{2}He} {}^{219}_{86}Rn\ (3.92\ s) \xrightarrow{-^{4}_{2}He}$$

$$^{215}_{84}Po\ (1.82 \times 10^{-3} s) \xrightarrow{-^{4}_{2}He} {}^{211}_{82}Pb\ (3.61\ m) \xrightarrow{-^{0}_{-1}e} {}^{211}_{83}Bi\ (2.16\ m)\ \text{--------}$$

$$\text{------} \xrightarrow{-^{0}_{-1}e} {}^{211}_{84}Po\ (2 \times 10^{-1}\ s) \xrightarrow{-^{4}_{2}He} {}^{207}_{82}Pb\ (\text{Stable})$$

$$\text{------} \xrightarrow{-^{4}_{2}He} {}^{207}_{81}Tl\ (4.76\ m) \xrightarrow{-^{0}_{-1}e} {}^{207}_{82}Pb\ (\text{Stable})$$

Neptunium-237 or (4n+1) series ($^{237}_{93}Np$ – $^{209}_{83}Bi$ Series) : Artificial radioactive series

This series was discovered after the discovery of trans-uranic elments. The possibility of the existence of this series was suggested by A.S. Russal in 1923 and later on it was discovered by Turner in 1940. The members of this series donot occur in nature. As a matter of fact, these members have been obtained by artificial methods in the laboratory.

This series starts with $^{237}_{93}Np$ (parent element : longest lived element) and ends at $^{209}_{83}Bi$ (end element) which is a stable element. The parent element *viz.* $^{237}_{93}Np$ decays into $^{233}_{91}Pa$ by emitting an α-particle. $^{233}_{91}Pa$ emits a β-particle to form $^{233}_{92}U$ and so on until a stable end product *viz.* $^{209}_{83}Bi$ is obtained (See Table 6.6). The mass number of different members of this series is equal to $(4n + 1)$ which shows that if the mass number is divided by 4, the ramainder obtained is equal to 1. Mass number equal to $(4n + 1)$ also means that the value of n for ^{237}Np and ^{209}Bi is equal to $\frac{237\ -\ 1}{4} = 59$ and $\frac{209\ -\ 1}{4} = 52$ respectively. The members of this series are not found in nature, since the half-life period of its longest-lived member *viz* ^{237}Np ($= 2.2 \times 10^{9}$ years) is much less than the age of the earth ($= 5 \times 10^{9}$ years). The number of α and β particles emitted in the transmutation of $^{237}_{93}Np$ to $^{209}_{83}Bi$ can be calcualted with the help of the equations $237 - 4x = 209$ and $93 - 2x + y = 83$ which give $x = 7$ and $y = 4$ where x and y denote the number of α and β–particles.

Table 6.6: *Neptunium or (4n+1) series (Artificial radioactive series) [$^{237}_{93}Np$ – $^{209}_{83}Bi$ series]*

$$^{237}_{93}Np\ (2.2 \times 10^{6}\ y) \xrightarrow{-^{4}_{2}He} {}^{233}_{91}Pa\ (27.0d) \xrightarrow{-^{0}_{-1}e} {}^{233}_{92}U\ (1.62 \times 10^{5}\ y) \xrightarrow{-^{4}_{2}He} {}^{229}_{90}Th\ (7000y)$$

$$\xrightarrow{-^{4}_{2}He} {}^{225}_{88}Ra\ (14.8d) \xrightarrow{-^{0}_{-1}e} {}^{225}_{89}Ac\ (10.0d) \xrightarrow{-^{4}_{2}He} {}^{221}_{87}Fr\ (4.8\ m) \xrightarrow{-^{4}_{2}He} {}^{217}_{85}At\ (0.018\ s)$$

$$\xrightarrow{-^{4}_{2}He} {}^{213}_{83}Bi\ (47\ m) \xrightarrow{-^{0}_{-1}e} {}^{213}_{84}Po\ (4.2 \times 10^{-6}\ s) \xrightarrow{-^{4}_{2}He} {}^{209}_{82}Pb\ (3.3h) \xrightarrow{-^{0}_{-1}e} {}^{209}_{83}Bi\ (\text{stable})$$

$$^{213}_{83}Bi\ (47\ m) \xrightarrow{-^{4}_{2}He} {}^{209}_{81}Tl\ (2.2m) \xrightarrow{-^{0}_{-1}e} {}^{209}_{82}Pb\ (3.3h)$$

Summary

Summary of all the four radioactive series is given below:

Series	Parent element of the series	End element of the series	No. of particles emitted α	No. of particles emitted β	Value of *n* for the parent element of the series	Value of *n* for the end element of the series
(*A*) Natural radioactive series						
(*i*) Thorium-232 or 4*n* series $\left[{}^{232}_{90}Th - {}^{208}_{82}Pb \text{ series}\right]$	${}^{232}_{90}Th$	${}^{208}_{82}Pb$	6	4	$\frac{232}{4} = 58 \left[{}^{232}_{90}Th\right]$	$\frac{208}{4} = 52 \left[{}^{208}_{82}Pb\right]$
(*ii*) Uranium-238 or (4*n* + 2) series $\left[{}^{238}_{92}U - {}^{206}_{82}Pb \text{ series}\right]$	${}^{238}_{92}U$	${}^{206}_{82}Pb$	8	6	$\frac{238-2}{4} = 59 \left[{}^{238}_{92}U\right]$	$\frac{206-2}{4} = 51 \left[{}^{206}_{82}Pb\right]$
(*iii*) Uranium-235 or (4*n* + 3) series $\left[{}^{235}_{92}U - {}^{207}_{82}Pb \text{ series}\right]$	${}^{235}_{92}U$	${}^{207}_{82}Pb$	7	4	$\frac{235-3}{4} = 58 \left[{}^{235}_{92}U\right]$	$\frac{207-3}{4} = 51 \left[{}^{207}_{22}Pb\right]$
(*B*) Artificial radioactive series : Neptunium-237 or (4*n* + 1) series $\left[{}^{237}_{93}Np - {}^{209}_{83}Bi \text{ series}\right]$	${}^{237}_{93}Np$	${}^{209}_{83}Bi$	7	4	$\frac{237-1}{4} = 59 \left[{}^{237}_{93}Np\right]$	$\frac{209-1}{4} = 52 \left[{}^{209}_{83}Bi\right]$

Use of radioactive series: Discovery of isotopes and isobars

For the first time, it was during the study of radioactive series that the existence of isotopes and isobars was realised. Thus radioactive series became a source of the discovery of isotopes and isobars. In 1919, Thomson showed that it is not only radioactive elements that give isotopes; non-radioactive elements (stable isotopes) also exist as isotopes.

University Questions

1. (*a*) Define radioactivity. *(Devi Ahilya, 1992; Meerut 95; Omania, 93)*
(*b*) Compare α, β and γ rays in terms of mass, charge and penetrating power. *(Rohelkhand, 90)*

2. What is group displacement law? How does it throw light on the idea of radioactive isotopes? *(M.S. Baroda 91, Osmania 93, Meerut 95)*

3. Explain what is meant by "Radioactive Equilibirium"? How does it differ from chemical equilibirium? *(Bombay 90, Osmania 93, Meerut 97).*

4. Write a note on:
(*i*) (*a*) Artificial transmutation (*b*) Health hazards of radiations *(Bahuguna, 92)*
(*ii*) (*a*) Artificial radioactivity (give examples) *(Agra, 92)*
(*b*) Transmutation of Elements *(Meerut, M.Sc., 94)*
(*iii*) (*a*) Atomic disintegration *(Agra, 92)*
(*b*) Decay constant (give examples) *(Meerut, 94)*
(*iv*) Carbon-dating technique *(Bahuguna, 92)*

5. What is meant by average-life period of a radioactive element and how is it related with half-life period of that element? *(Meerut M.Sc., 94)*

6. State and explain the law of radioactive disintegration. What do you understand by the half-life period of a radioactive element? Show that half-life is independent of the amount of the substance. *(Meerut, 97)*

7. Prove that half-life period of a radioactive element is gien by 0.693/k where K is the decay constant. *(Meerut, 94)*

8. Complete the following equations and identify X, Y, Z, Q and R.

(*i*) ${}^{24}_{11}Na \longrightarrow X + {}^{0}_{-1}e$ (*ii*) ${}^{24}_{11}Na + {}^{1}_{0}n \longrightarrow {}^{14}Y + {}^{1}_{1}Z$

(*iii*) ${}^{27}_{14}Si \longrightarrow {}^{27}_{13}Q + {}^{0}_{+1}e$ (*iv*) $R + {}^{4}_{2}He \longrightarrow {}^{13}_{7}N + {}^{1}_{0}n$ *(Agra ,92)*

9. The amount of C-14 in a peice of wood is found to be one-sixth of its amount in a fresh piece of wood. Calculate the age of the wood. *(Bahuguna, 92)*

10. At radioactive equilibirium the ratio between atoms of the two radioactive elements A and B was found to be 3.1×10^9. If the half-life period of element A is 2×10^9 years, what is the half-life period of the element B? *(Meerut M.Sc., 94)*

11. Complete the following nuclear reactions:

(*i*) ${}_7N^{14}$ + ... $\longrightarrow {}_8O^{17} + {}_1H^1$ (*ii*) ${}_{15}P^{30} \longrightarrow {}_{14}Si^{30}$ + ...

(*iii*) ${}_{13}Al^{27} + {}_0n^1 \longrightarrow$ .... $+ {}_2He^4$ (*iv*) ${}_4Be^9 + {}_2He^4 \longrightarrow$ .... $+ {}_0n^1$

(*v*) .... $+ {}_0n^1 \longrightarrow {}_5B^{11} + {}_2He^4$ *(Meerut, 93)*

12. The radioactivity of a radioactive isotope falls to 12.5% in 90 days. Calculate the half-life period and disintegration constant of the radio isotope. *(Meerut, 93)*

13. The half-life period of a radioactive element is 1600 years. Calculate the disintegration constant. *(Meerut, 96)*

14. Complete the follwing :

(*i*) ${}_7N^{14}$ + ... $\longrightarrow {}_6C^{11} + {}_2He^4$ (*ii*) ${}_{13}Al^{27}$ + ... $\longrightarrow {}_{12}Mg^{27} + {}_1H^1$

(*iii*) ${}_{13}Al^{27}$ + ... $\longrightarrow {}_{15}P^{30} + {}_0n^1$ (*iv*) ${}_{11}Na^{24} \longrightarrow {}_{12}Mg^{24}$ + ...

(*v*) ${}_4Be^9 + {}_2He^4 \longrightarrow {}_6C^{12}$ + ... *(Meerut, 96)*

15. How does the number of protons in a nucleus change when an α and then a β particle is emitted from the nucleus? *(Meerut, 95)*

16. What is artificial transmutation? Give the nuclear reactions induced by neutron, proton and α-particles. *(Osmania, 93)*

17. Ac^{227} has a half-life of 220 yrs with respect to radioactive decay. The decay follows two parallel paths, one leading to ${}^{227}Th$ and the other to ${}^{223}Fr$. The percentage yields of these daughter nuclides are 2.0 and 98.0 respectively. What are the decay constant (λ) for each of the separate paths. *(I.I.T., 96)*

7

Isotopes, Isotones, Isobars and Nuclear Isomers

Isotopes

What are isotopes?

Isotopes are the atoms of the same elements which have the same atomic number (Z) or the same number of protons (p) but different mass numbers (A). For example ^1_1H ($A = 1, Z = 1$), ^2_1H or ^2_1D ($A = 2, Z = 1$) and ^3_1H or ^3_1T ($A = 3, Z = 1$) are the isotopes of hydrogen.

Representation of an isotope

The atomic number of an isotope is written on the lower side of the symbol of the element whereas the mass number is written on the upper side of the symbol. For example an isotope of carbon whose atomic number is 6 and mass number is 12 is written as $^{12}_6C$.

Characteristics of isotopes

(*i*) Since the atomic numbers (Z) of the isotopes of an element are the same, they contain the same number of protons (p) in the nucleus and same number of electron (e) revolving round the nucleus in different orbits ($Z = p = e$).

(*ii*) Since the mass numbers (A) of isotopes are different, the sum of protons and neutrons in the nucleus is also different ($A = n + p$).

Now, since, as said above, the number of protons is the same, the number of neutrons is different so that A may be different.

On the basis of the above two properties the composition of the isotopes of hydrogen, carbon and oxygen is given in Table 7.1.

Table 7.1: *Composition of some isotopes*

Element	Isotopes	Atomic number (Z) = No. of protons (p) = No. of electrons (e)	Mass number, $A = n + p$	No. of neutrons, $n = A - p$
Hydrogen	(*a*) ordinary hydrogen or protium, ^1_1H	1	1	$1 - 1 = 0$
	(*b*) Heavy hydrogen or deuterium, ^2_1H or ^2_1D.	1	2	$2 - 1 = 1$
	(*c*) Tritium, ^3_1H or ^3_1T	1	3	$3 - 1 = 2$
Carbon	(*a*) $^{12}_6C$	6	12	$12 - 6 = 6$
	(*b*) $^{14}_6C$	6	14	$14 - 6 = 8$
Oxygen	(*a*) $^{16}_8O$	8	16	$16 - 8 = 8$
	(*b*) $^{17}_8O$	8	17	$17 - 8 = 9$
	(*c*) $^{18}_8O$	8	18	$18 - 8 = 10$

(*ii*) We have already said that the isotopes have different mass numbers (A). This means that the isotopes have different masses. The difference in their masses is due to the different number of protons present in their nuclei.

(*iii*) Since the isotopes of a given element have different number of neutrons (*i.e.*, different mass numbers or atomic masses), they show different physical properties. (*e.g.* densities, melting points, boiling points etc.). On the other hand, since the isotopes have the same atomic number or the same number of electrons, they have the same electronic configurations and hence have the same chemical properties.

(*iv*) Isotopes have different radioactive properties, since the composition of their nuclei is different.

(*v*) All the isotopes are placed in the same group of the periodic table, since they have the same atomic number.

Discovery of isotopes

In 1919, Soddy for the first time, showed that many radioactive elements exist as isotopes, but later on, Thomson showed that isotopy was found not only in radioactive elements but also existed in the non-radioactive elements. At present isotopes of both types of elements are known. Also see *"Natural and artificial radioactive series."*

Types of isotopes

Radioactive (unstable) as well as non-radioactive (stable) elements give isotopes. Thus isotopes are of the following two types:

(*i*) *Radioactive or unstable isotopes.* The nuclei of such isotopes are unstable and hence undergo spontaneous disintegration to produce α, β, and γ rays. The isotopes of heavy elements are generally radioactive. Examples of such isotopes are ^{12}C, ^{24}Na, ^{30}P, ^{32}P, ^{60}Co, ^{74}As, ^{131}I etc. Hydrogen gives three isotopes which are ^{1}H, ^{2}H and ^{3}H. It is only ^{3}H (tritium) which is radioactive.

(*ii*) *Non-radioactive or stable isotopes.* The nuclei of these isotopes are stable and hence donot undergo disintegration. The isotopes of light elements are generally non-radioactive. Examples are ^{206}Pb, ^{207}Pb, ^{208}Pb etc. Carbon gives three isotopes *viz.* C–12, C–13 and C–14. C–12 and C–13 are non-radioactive isotopes.

Separation of isotopes

Since the isotopes of a given element have the same chemical properties, they cannot be separated from each other by chemical methods. However, since the isotopes have different physical properties, they can be separated by physical methods. Some of the physical methods which have been used for the separation purpose are discussed below.

(*i*) **Gaseous diffusion method.** We know that, according to Grahm's law of diffusion, the rate of diffusion of a gas is inversely proportional to the square root of its density or molecular wieght. Thus, if a gaseous mixture containing two gaseous isotopes is allowed to pass through the pores of a porous membrane, then, according to Grahm's law of diffusion, the isotope with lower mass number (*i.e.*, lighter isotope) will diffuse through the membrane more quickly than the isotope having higher mass number (*i.e.*, heavier isotope). The lighter isotope gets more and more concentrated in the faster diffusing portions of the gas, while the heavier isotope gets more and more concentrated in the slower diffusing portions of the gas left behind.

This method has been used for the separation of gaseous ^{20}Ne and ^{22}Ne isotopes of neon and ^{16}O and ^{18}O isotopes of oxygen.

Efficiency of this method. The efficiency of this method depends on the ratio of the mass numbers of the two isotopes which have to be separated from each other. The larger is the ratio between the mass numbers of the two isotopes, higher is the efficiency of the process, *i.e.*, as the ratio between the mass numbers of the two isotopes or the ratio between the rates of diffusion of the two gaseous isotopes increases, the number of stages of diffusion process required to bring about the separation decreases and hence the separation becomes relatively easier. We know that the ratio between the rates of diffusion of two isotopes of hydrogen (*viz.* $^{2}_{1}H$ and $^{1}_{1}H$) is $\sqrt{2/1}$:

$\sqrt{1/1} = 1.414 : 1$ and that of the two isotopes of neon ($^{21}_{10}Ne$ and $^{20}_{10}Ne$) is $\sqrt{21/20} : \sqrt{20/20} = 1.02 : 1$. Since the ratio for the isotopes of hydrogen is greater than that for the isotopes of neon, the separation of the isotopes of hydrogen is relatively easier than the separation of the isotopes of neon.

(*ii*) **Thermal diffusion method.** This method was used by Clausius to separate the isotopes of chlorine *viz.* ^{35}Cl and ^{37}Cl and those of uranium, ^{235}U (less abundent isotope) and ^{238}U (move abundent isotope).

The apparatus used in this method consists of a vertical cyclindrical tube which is 3 m long and 1 cm in diameter. The tube is heated electrically to a temperature of 600° C or above from its inside by means of a thin Pt-wire enclosed in a porcelain tube. On the outside, the tube is kept cool by circulating cold water around it. (Fig. 7.1). Now when the gaseous mixture containing the two isotopes is passed through the space around the hot porcelain tube, the molecules of the heavier isotope get accumulated preferentially in the cooler part (*i.e.*, in the bottom) of the tube while the molecules of the lighter isotope get collected in the hotter part (*i.e.*, at the top) of the tube. Thus we see that the heavier isotope is carried downwards by the convection currents while the lighter isotope is carried upwards by the same currents. This process is repeated many times to achieve a complete separation of the isotopes.

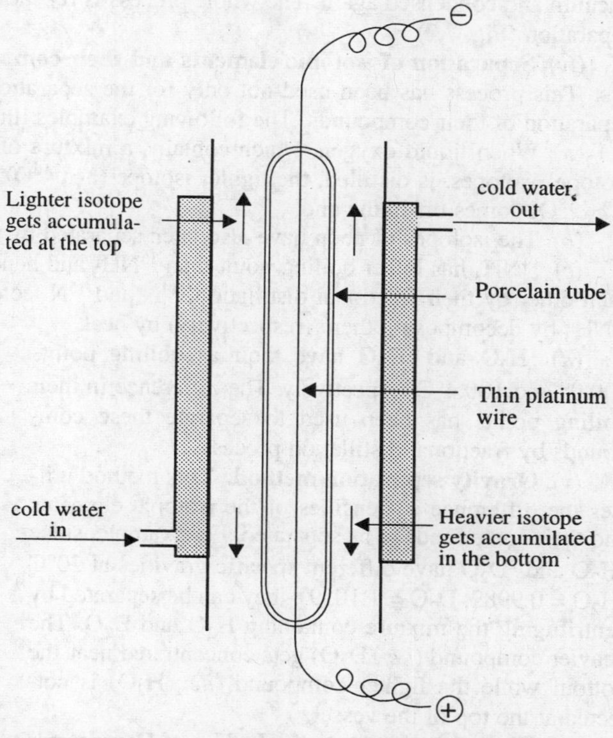

Fig. 7.1: Separation of isotopes by thermal diffusion method.

(*iii*) **Fractional evaporation process**. The process has been used for the seperation of the two isotopes of Hg. In this process Hg is allowed to undergo fractional evaporation at a temperature lying between 40°C and 60°C in vacuum. The vapours of the lighter isotope of Hg are condensed on a surface cooled in liquid air and placed a little above the evaporating Hg. After sometime, the liquid Hg left behind consists mostly of heavier isotope. This mercury is taken out by opening the stopcock at the bottom. The frozen Hg collected above is melted, evaporated under

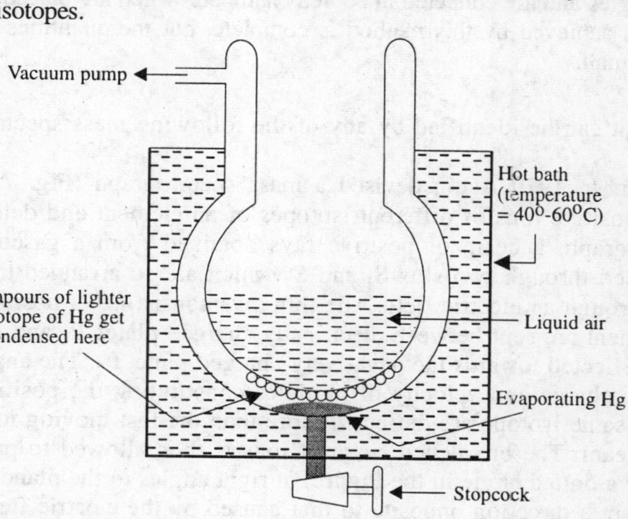

Fig. 7.2: The separation of the two isotopes of Hg by fractional evaporation process.

vacuum and condensed again. The whole process is repeated several times to achieve the complete separation (Fig. 7.2).

(*iv*) **Separation of isotopic elements and their compounds by fractional distillation process.** This process has been used not only for the separation of isotopic elements but also for the separation of their compounds. The following examples illustrate the use of this process.

(*a*) When liquid oxygen, which contains a mixture of ^{16}O (lighter isotope) and ^{18}O (heavier isotope) isotopes, is distilled, the lighter isotope (*i.e.*, ^{16}O) comes off first and the heavier isotope (*i.e.*, ^{18}O) comes off in the end.

(*b*) The isotopes of neon have also been separated by this process.

(*c*) $^{14}NH_3$ has lower boiling point than $^{15}NH_3$ and hence these compounds are separated from each other by their fractional distillation. ^{14}N and ^{15}N isotopes can be recovered from $^{14}NH_3$ and $^{15}NH_3$ by decomposing them (respectively) by heat.

(*d*) ^1H_2O and ^2D_2O have their as boiling points 100.0° and 100.4°C respectively. The difference in their boiling points has been used to separate these compounds by fractional distillation process.

(*v*) **Gravity separation method.** This method utilises the difference in densities of the isotopic elements and their compounds to be separated. For example, since ^1H_2O and ^2D_2O have different specific gravities at 20°C (H_2O = 0.9982, D_2O = 1.1059), they can be separated by centrifuging the mixture containing H_2O and D_2O. The heavier compound (*i.e.*, D_2O) gets concentrated near the bottom while the lighter compound (*i.e.*, H_2O) is collected at the top of the vessel.

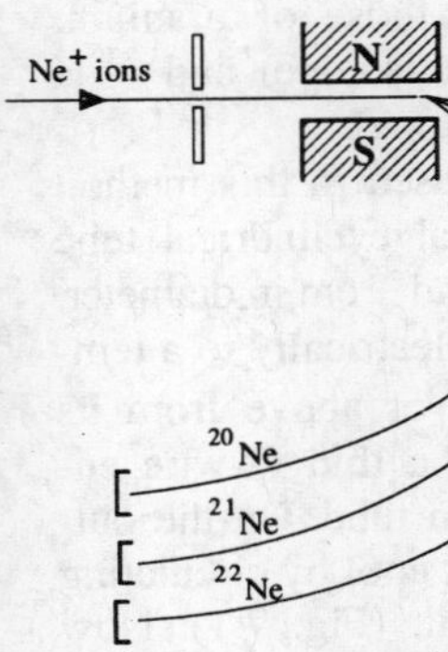

Fig. 7.3 : Separation of the three isotopes of neon (^{20}Ne, ^{21}Ne and ^{22}Ne) by electromagnetic method making the use of Dempster's mass spectrograph.

(*vi*) **Electromagnetic method : Use of Dempster's mass spectrograph.** This method is based on the principle of Dempster's mass spectrograph. This method has been used for the separation of the three isotopes of neon. The beam of ions of the three isotopes of neon namely ^{20}Ne ^{21}Ne and ^{22}Ne as obtained in Dempster's mass spectrograph is passed between the poles of a magnet (Fig. 7.3). The different isotopes get defflected at different angles and are collected in cooled chambers which are placed in appropriate positions. The separation achieved by this method is complete, but the quantities of different isotopes obtained are very small.

Indentification of isotopes

Different isotopes of an element can be identified by any of the following mass spectrographs.

1. **Aston's mass spectrograph**. In 1919 Aston devised a mass spectrograph (Fig. 7.4) which can accurately sort out the positive ions of different isotopes of an element and determine their masses. In this spectrograph, a beam of positive rays obtained from a gaseous element in a discharge tube is passed through two slits S_1 and S_2 which are so arranged that only a thin ribbon of rays passes through an electric field. The thin rays consisting of positive ions of various isotopes of the element are sent between electrically charged plates P_1 and P_2. Here the beam of positive ions is deflected towards the negatively-charged plate, P_2. The angle of deflection depends on the mass, charge and velocity of particles constituting the positive rays. The slow moving ions of the same isotope are deflected more than the fast moving ions which cause a broadening of the beam. The broadened beam of ions is then allowed to pass through a magnetic field (shown by a dotted circle in the Figure) at right angles to the plane of the charged plates and is thus sent in a direction opposite to that caused by the electric field. Here also the slow moving particles (heavier particles) undergo maximum deflection. The strengths of electric and magnetic fields are so adjusted that all the ions of the same mass are

focussed at the same point on the photographic plate where a sharp line or spot is obtained. The developed plate is known as **mass spectrogram.** Each line or spot seen on the photographic

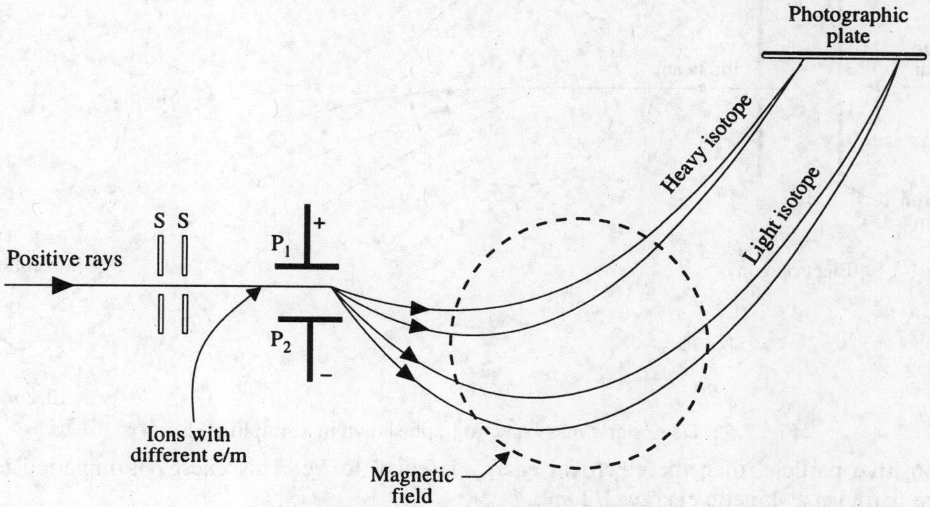

Fig. 7.4 : Aston's mass spectrograph.

plate shows the existence of separate isotopes. Further, the intensity of the line or spot in comparison with the lines of other isotopes gives the relative abundence of this particular isotope. This point is evident from the fact that an examiniation of the Aston's mass spectrogram of a sample of ordinary neon shows that the sample consists of two lines (see Fig. 7.5) which indicate that neon in a mixture of ^{20}Ne and ^{22}Ne isotopes. The relative intensities of these lines show that the relative abundance of these iostopes is as $^{20}Ne : {}^{22}Ne$ = 9 : 1. Similarly the mass spectrogram of a sample of chlorine has two lines, indicating the presence of ^{35}Cl and ^{37}Cl isotopes (Fig. 7.5). The relative intensities of these lines show that the relative abundance of these isotopes is as : $^{35}Cl : {}^{37}Cl = 3 : 1$.

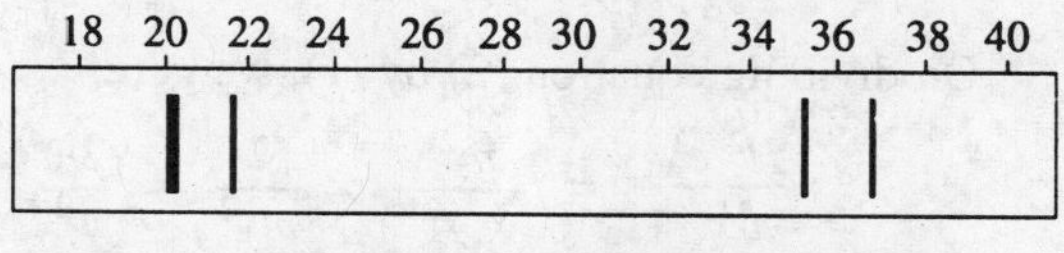

Fig. 7.5: Mass spectrogram of neon and chlorine.

Aston's mass spectrograph not only helps in identifying the isotopes present in an element but also helps in determining the actual atomic weight of a given element as is clear from the following example.

Problem. *A sample of ordinary neon is found to consist of* ^{20}Ne, ^{21}Ne *and* ^{22}Ne *isotopes in the precentage :* $^{20}Ne = 90.22\%$, $^{21}Ne = 0.26\%$ *and* $^{22}Ne = 8.82\%$. *Calculate the actual atomic weight of neon.*

Solution. Actual atomic weight of neon.

$$= \frac{90.92 \times 20 + 0.26 \times 21 + 8.82 \times 22}{90.92 + 0.26 + 8.82} = \frac{1818.40 + 5.46 + 194.04}{100} = \frac{2017.90}{100}$$

$$= 20.1790 \approx \mathbf{20.18}$$

2. **Dempster's mass spectrograph.** This spectrograph is shown in Fig. 7.6. In this apparatus a slow stream of gas or vapour of the sample is allowed to pass in between two perforated plates. Between these plates the stream is bombarded with the help of high-energy electrons which are produced by an electric gun. The gas atoms (A) are thus ionised and are converted into monopositive ions (A^+). Now when a potential of 500–2000 volts is applied between the perforated accelerating plates, the positively-charged ions are strongly attracted towards the negatively-charged plate. the positively-charged ions moving with accelerated speed then enter the magnetic field at right angles to their path and are thus made to move in a circular path.

Principle. If V is the potential applied between the accelerating plates and *e* is the charge on

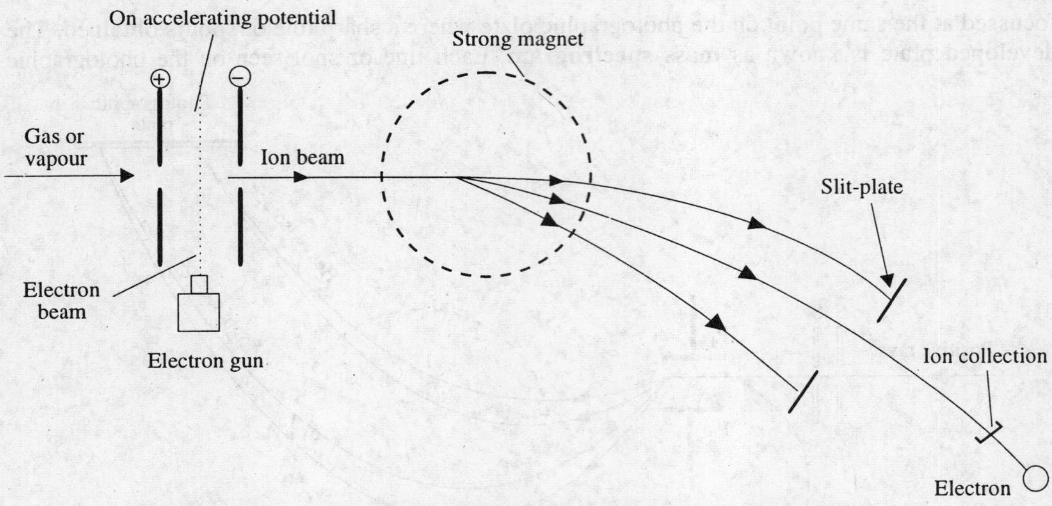

Fig. 7.6 : Dempster's mass spectrograph shown in a simplified form.

each positive particle, then the electrical energy is equal to Ve. This energy is imparted to the positive particles as kinetic energy, 1/2 mv^2, *i.e.,*

$$Ve = \frac{1}{2} mv^2 \qquad ...(1)$$

Here *m* is the mass and *v* is the velocity of the particle.

If the strength of the magnetic fields is H, then the magnetic force acting on the particle is equal to HeV which is again equal to centrifugal force, mv^2/r (r = radius of the circular path). Thus:

$$HeV = \frac{mv^2}{r} \quad \text{or} \quad H^2e^2 = \frac{m^2v^2}{r^2} \qquad ...(2)$$

On dividing equation (2) by (1), we get :

$$\frac{H^2e^2}{Ve} = \frac{m^2v^2}{r^2} \times \frac{2}{mv^2} = \frac{2m}{r^2} \quad \text{or} \quad e/m = \frac{2V}{H^2r^2} \qquad ...(3)$$

In the above equation [eqn. (3)], *e*, being the unit electrical charge, is constant. The radius (r) of the circular path is constant for a given spectrograph. Now, if the strength of the magnetic field (H) is kept constant, then *m* is proportional to 1/V, *i.e.,* $m \propto \frac{1}{V}$. Thus, by adjusting the accelerating potential (V), positively-charged particles of mass *m* are made to fall on the collector plate. Each particle sets up a minute electric current which passes to the electrometer. The strength of the current thus gives the relative abundance of the individual particles (*i.e.,* isotopes). By comparing the current strength with an experiment performed with C – 12 ion, the mass numbers of the various isotopes can be determined.

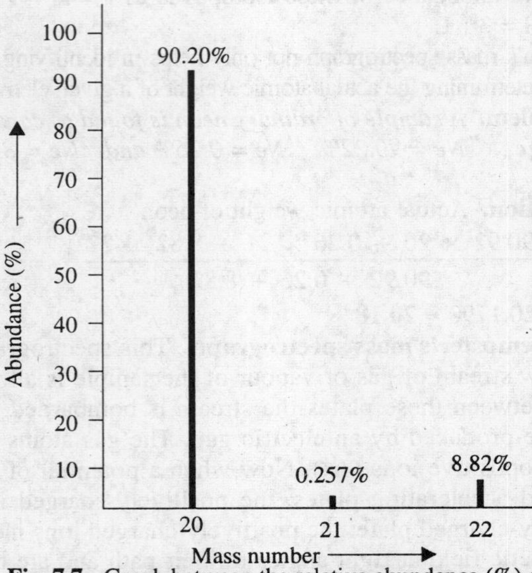

Fig. 7.7: Graph between the relative abundance (%) of the three isotopes of neon against their mass numbers.

In the modern mass spectrograph

each ion strikes a detector, the ion current is amplified and fed to a recorder. The recorder makes a graph between the abundance and mass number of each isotope. The heights of the peaks gives the relative abundance of the isotopes of the element. Such a graph of neon isotopes is given in Fig. 7.7 which shows that the three isotopes namely ^{20}Ne, ^{21}Ne and ^{22}Ne are in the relative proportion 90.20%, 0.257%, and 8.82% respectively.

Production of an isotope by the emission of one α–particle and two β-particles

When a radioactive element loses an α-particle ($^{4}_{2}He$) and then two β-particles ($^{0}_{-1}e$), the element formed has the same atomic number as the parent element but its mass number is less by 4 units. This fact is evident from the following nuclear reactions:

(*a*) $^{232}_{90}Th \xrightarrow{-^{4}_{2}He} {}^{228}_{88}Ra \xrightarrow{-^{0}_{-1}e} {}^{228}_{89}Ac \xrightarrow{-^{0}_{-1}e} {}^{228}_{90}Th$

(*b*) $^{238}_{92}U \xrightarrow{-^{4}_{2}He} {}^{234}_{90}Th \xrightarrow{-^{0}_{-1}e} {}^{234}_{91}Pa \xrightarrow{-^{0}_{-1}e} {}^{234}_{92}U$

(*c*) $^{218}_{84}Po \xrightarrow{-^{4}_{2}He} {}^{214}_{82}Pb \xrightarrow{-^{0}_{-1}e} {}^{214}_{83}Bi \xrightarrow{-^{0}_{-1}e} {}^{214}_{84}Po$

Since the atomic numbers of the parent element and of the daughter element are the same, they are isotopes to each other. Thus we see that the emission of one α-particle and two β-particles by a radioactive element produces its isotope.

Isotopes of hydrogen

It has been established by means of mass spectrograph that hydrogen has three isotopes with mass number 1, 2 and 3. These isotopes are discussed below:

(i) Hydrogen or protium ($^{1}_{1}H$ or H). It is represented as $^{1}_{1}H$ in which the numbers 1 and 1 indicate the atomic number and mass number (*i.e.,* atomic mass or atomic weight) of this isotope. Its exact mass number is, however, 1.0081. Its nucleus consists of one proton and no neutron. It has one electron in 1s energy level.

(ii) Heavy hydrogen or deuterium ($^{2}_{1}H$ or $^{2}_{1}D$). Its actual mass number is 2.0147. Its nucleus consists of one proton and one neutron. It has also one electron in 1*s* energy level.

(iii) Tritium ($^{3}_{1}H$ or $^{3}_{1}T$). Its mass number is 3.0170. Its nucleus consists of one proton and two neutrons. It has one electron in 1s energy level.

It has been estimated that a sample of ordinary hydrogen contains 99.994% of H_2 and 0.0156% of D_2. 10^7 molecules of ordinary hydrogen contain only one molecule of tritium. The cause of the presence of tritium in ordinary hydrogen in traces only is that the nucleus of tritium is radioactive and hence unstable.

Since the quantities of D_2 and T_2 present in natural hydrogen are exceedingly small, the properties of natural hydrogen are substantially those of protium.

Structure of isotopes

The structure of the three isotopes of hydrogen in terms of the number of electrons (*e*), protons (*p*) and neutrons (*n*) is shown pictorially in Fig. 7.8.

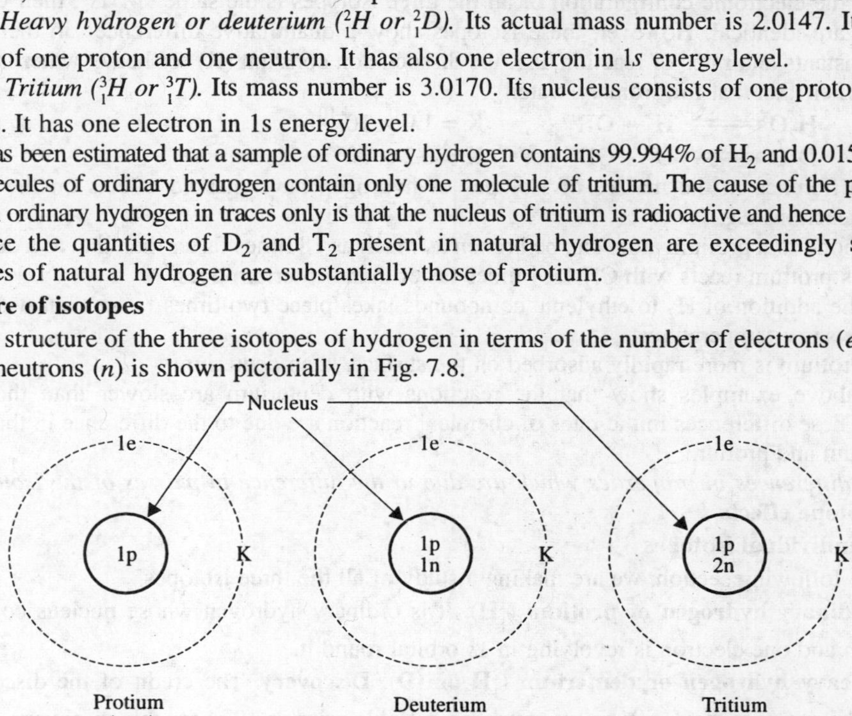

Fig. 7.8: Structure of the three isotopes of hydrogen in terms of the number of electrons, protons and neutrons.

Properties of isotopes

Some of the properties of three isotopes of hydrogen are given in table 7.2.

Table 7.2: *Some properties of the three isotopes of hydrogen*

Property	Ordinary hydrogen or protium	Heavy hydrogen or deuterium	Tritium
(i) Symbol	^1_1H or H	^2_1H or ^2_1D	^3_1H or ^3_1T
(ii) Atomic number, Z	1	1	1
(iii) Mass number, *A* (atomic mass or atomic weight)	1 (Exact value = 1.0081)	2 (Exact value = 2.0147)	3 (Exact value = 3.0170)
(iv) No. of electrons, $e = Z$	1	1	1
(v) No. of protons, $p = Z$	1	1	1
(vi) No. of neutrons, $n = (A - Z)$	$1 - 1 = 0$	$2 - 1 = 1$	$3 - 1 = 2$
(vii) Electronic configuration	$1s^1$	$1s^1$	$1s^1$
(viii) Atomicity	2	2	2
(ix) Molecular formula	H_2	D_2	T_2
(x) Relative abundance of their diatomic molecules in ordinary hydrogen	99.984%	0.156%	10–15%
(xi) Stability	Stable	Stable	Radiactive and hence unstable

Isotope effects

Since the electronic configuration of all the three isotopes is the same *viz.* $1s^1$, their chemical properties are identical. However, these isotopes show a quantitative difference in their equilibirium constants and rates of reaction, *e.g.* the dissociation constant (K) for heavy water is significantly smaller than that for ordinary water.

$$H_2O \rightleftharpoons H^+ + OH^- \quad ; \quad K = 1.0 \times 10^{-10}$$
$$D_2O \rightleftharpoons D^+ + OD^- \quad ; \quad K = 0.2 \times 10^{-4}$$

The difference in the masses of isotopes of hydrogen have marked effects on the rates of chemical reactions *e.g.*

(i) A bond to a protium atom can be broken as much as 18 times faster than that to a deuterium atom. Thus protium reacts with Cl_2 13.4 times faster than deuterium does.

(ii) The addition of H_2 to ethylenic compounds takes place two times faster than with deuterium at the same temperature.

(iii) Protium is more rapidly adsorbed on the surfaces than deuterium.

The above examples show that the reactions with deuterium are slower than those with protium. These differences in the rates of chemical reaction are due to the difference in the masses of deuterium and protium.

Such differences in properties which are due to the difference in masses of the isotopes are termed **isotope effects.**

Study of individual isotopes

In the following section, we are making a study of all the three isotopes.

I. Ordinary hydrogen or protium ($^1_1$H). It is ordinary hydrogen whose nucleus consists of one proton and one electron is revolving in 1*s* orbital round it.

II. Heavy hydrogen or deuterium ($^2_1$H or $^2_1$D). Discovery. The credit of the discovery of heavy hydrogen goes to *Urey* who was awarded a Noble prize in 1934 for this discovery.

In 1931 *Birge* and *Manzel* reported that the atomic weight of hydrogen as determined by mass spectrograph was 1.0076 while that determined by chemical methods such as displacement of hydrogen from H_2SO_4 by Zn metal, was 1.0078. In order to explain this discrepency in atomic weights, they suggested that ordinary hydrogen contains a very small proportion of an isotope of mass 2.

About the same time, *Allison* and *his coworkers* also reported that ordinary hydrogen is a mixture of two isotopes of hydrogen. The heavier isotope has mass 2 and is present in the ratio of 1 part in about 5000 parts.

Encouraged by the clues and the calculations shown above *Urey, Brickwedde* and *Murphy* evaporated 4 litres of liquid hydrogen under reduced pressure and at a temperature of freezing hydrogen. On concentrating the residue thus obtained, they examined the spectrum and found that it contained very faint lines. They assumed that the faint lines were due the heavy isotopes of hydrogen with atomic wieght 2. Aftewards the masses of light and heavy isotopes of hydrogen were more accurately determined as 1.0081 and 2.0147 respectively. They named this heavier isotope as **deuterium** and gave the symbol **D** to it.

Preparation. Deuterium is prepared by the following methods:

(1) By separating deuterium from the mixture of protium and deuterium. Water is decomposed by metals to give a gas which consists of varying amounts of H_2, HD and D_2. Since hydrogen, being lighter than deuterium, is liberated faster than deuterium from water, the first few fractions of the gas are richer in hydrogen while the last fractions contain greater proportion of deuterium. The final fraction of the gas is collected and as electric discharge is passed through it, whereby the atoms of hydrogen and deuterium combine to give a mixture containing H_2 and D_2 molecules. On electrolysis HD also gives H_2 and D_2 molecules.

Now from the mixture of H_2 and D_2, we can separate D_2. The methods used for the separation of D_2 from the mixture are given below :

(a) Diffusion process. The separation of deuterium from the mixture of deuterium and protium (*i.e.,* hydrogen) can be effected by the diffusion process. This process *makes the use of difference in the rates of diffusion of the two isotopes, viz.* protium ($_1H^1$) and deuterium ($_1D^2$). Since protium is lighter, it diffuses more readily than deuterium. Although this method is slow and tedious, yet it gives spectroscopically pure deuterium, *i.e.,* the spectrum of deuterium obtained by this method shows no trace of ordinary hydrogen.

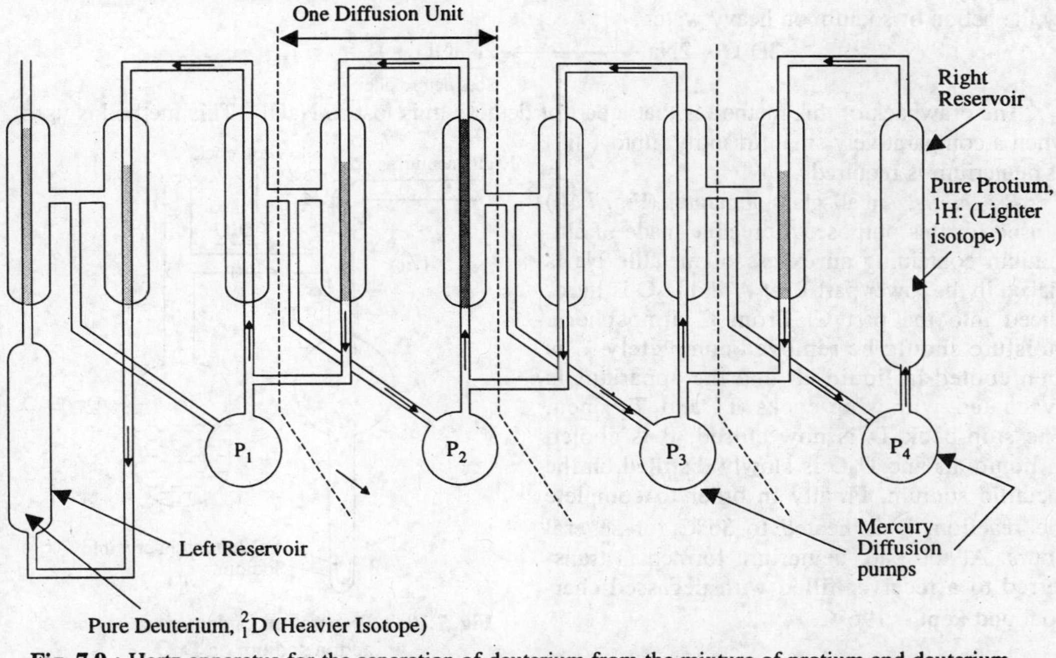

Fig. 7.9 : Hertz apparatus for the separation of deuterium from the mixture of protium and deuterium.

The process is carried out under reduced pressure since lower the pressure, the greater is the efficiency of the process.

The apparatus used in this process consists of a number of porous diffusion units which are called *Hertz diffusion units* (This method is due to Hertz) and are placed in series as shown in Fig. 7.9. The apparatus is called *Hertz apparatus.* Each unit contains a porous membrance shown by dotted portion in the figure and is connected with mercury diffusion pump (P_1, P_2, P_3.....). With the help of these diffusion pumps the mixture of the two isotopes is allowed to circulate into the diffusion units at the reduced pressure.

As the mixture of the two isotopes is allowed to pass into the diffusion units under reduced pressure, say, from left to right, with the help of mercury diffusion pumps, the lighter isotope *viz.* protium diffuses faster and its proportion increases on the right side, *i.e.*, it is collected in the right reservoir (placed on the right). On the other hand, the heavier isotope *viz.* deuterium is left behind and is collected into the left reservoir (placed on the left). This process is repeated a number of times to get pure deuterium.

(b) Fractional disillation of liquid hydrogen isotopes process. The mixture of the two isotopes is first of all liquefied by cooling and compression. The liquid so obtained is then slowly distilled under reduced pressure and at the temperature of freezing hydrogen. On concentrating the residue, enriched deuterium is obtained from which pure deuterium can be obtained by the fractional diffusion process as discussed in (*a*).

(c) Adsorption on charcoal process. This method is based on the fact that protium gets adsorbed on the surface of coconut charcoal more rapidly than deuterium. Thus, when a mixture of protium and deuterium is kept in contact with coconut charcoal at the temperature of liquid air for a few hours, most of protium gets adsorbed on the surface of the charcoal while deuterlum is collected.

(2) From heavy water. Ordinary water contains about 1 part in 6000 parts of what is known as *heavy water*, the molecule of which consists of two heavy hydrogen atoms and one oxygen atom (D_2O). Heavy water is obtained by the electrolysis of an aqueous alkali solution, between nickel electrodes, 20 litres of the solution yielding about 0.5 c.c. of pure heavy water. Deuterium can be obtained from this heavy water, D_2O by using any of the following methods:

(*a*) *By the action of sodium on D_2O.* Fairly pure sample of heavy hydrogen may be obtained by the action of sodium on heavy water

$$2D_2O + 2Na \longrightarrow \underset{\text{Sod. deutroxide}}{2NaOD} + D_2$$

The drawback of this method is that a part of deuterium is lost as NaOD. This method is used when a comparatively small quantity (upto 1 lit.) of deuterium is required.

Procedure. An all-glass apparatus (Fig. 7.10) is used for this purpose. A crucible made of aluminium containing an excess of metallic Na is placed in the lower part B of A and D_2O is introduced into the part C. From C atmospheric moisture should be removed completely. C is then cooled in liquid air and the apparatus is evacuated with stop-cocks T_1 and T_2 open. The stop-cock T_2 is now closed, B is cooled in liquid air and D_2O is slowly distilled on the metallic sodium. Finally in order to complete the reaction, B is heated to 35°C for several hours. At the end, deuterium formed is transferred to a receiver filled with degassed charcoal and kept – 196°C.

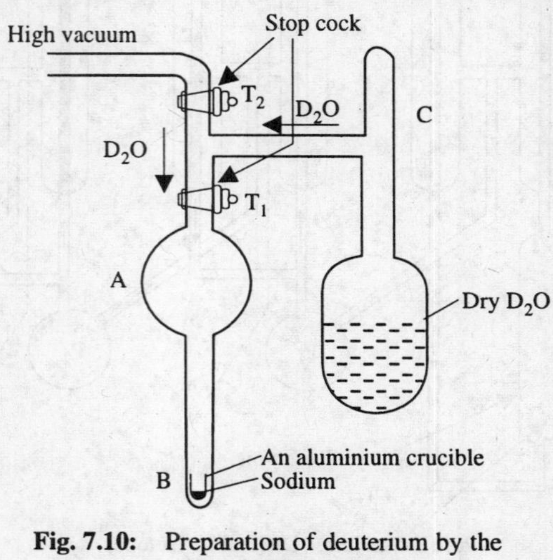

Fig. 7.10: Preparation of deuterium by the action sodium on D2O.

(b) By the action of magnesium on D_2O. When large quantities of deuterium are required, reaction between Mg and D_2O has been found to be more suitable.

$$D_2O + Mg \longrightarrow D_2 + MgO$$

Procedure. In this method, D_2O is slowly evaporated from a Pyrex flask that has been previously evacuated to 10^{-4} mm. The vapour passes up a reaction column, set up vertically on the top of the flask. The column is charged with magnesium shavings and is heated to about 500° by the means of a furnace. The gas of deuterium produced is very pure and the yield is quantitative.

(c) By the action of uranium on D_2O. This method not only gives deuterium but also gives as a method for its storage as *uranium deuteride*, UD_3. UD_3 when heated gives D_2 of high purity. The reactions are:

(*i*) *Preparation of D_2* : $2D_2O + U \longrightarrow UO_2 + 2D_2$.

(*ii*) *Storage of UD_3* : $3D_2 + 2U \longrightarrow 2UD_3$.

(*iii*) *Decomposition of UD_3* : $2UD_3 \longrightarrow 3D_2 + 2U$.

With the help of reaction (*iii*) we can get D_2 in any desired quantity.

(d) By the decomposition of D_2O by heated tungsten filament. This method has been given by *Forkas* and *Farkas* and is useful when small volumes of D_2 are required. The apparatus consists of a decomposition chamber provided with a tungsten filament (See Fig. 7.11).

One side of the chamber is connected with a small receiver containing heavy water and the other to a pump. Before beginning the experiment the apparatus is heated to about 1000°C and the tungsten filament to 2000°C as to remove the whole of the absorbed water. The whole apparatus is connected to a high vacuum pump during this preliminary treatment. The receiver containing heavy water is heated and vapours of D_2O so obtained are led to the decomposition chamber where they are decomposed to D_2 by the electrically heated tungsten filament.

$$D_2O\ (\textit{Vapours}) + 2W\ (\textit{Heated}) \rightarrow W_2O + D_2$$

W_2O is volatile and is subsequently condensed in the U-tube following the decomposition chamber and D_2 is carried further for collection. It is stated that a wire of tungsten 0.2 mm in thickness and 20 cm long ordinarily gives 80 ml of D_2O.

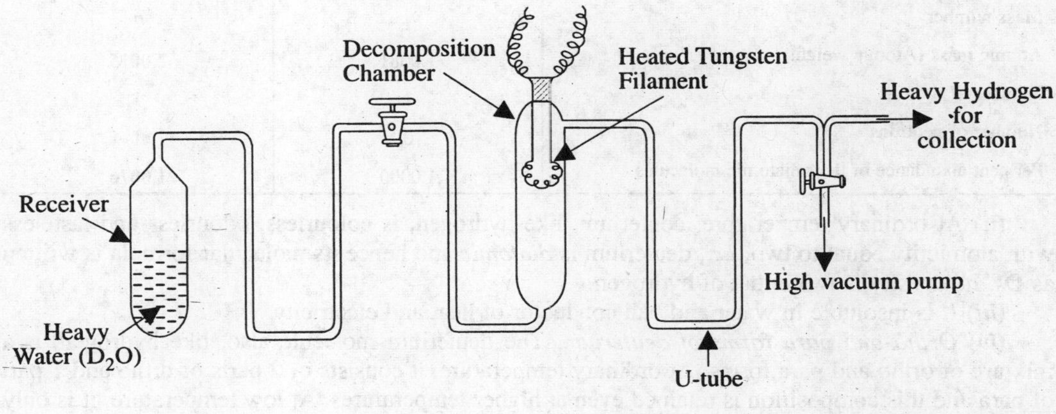

Fig. 7.11 : Preparation of heavy hydrogen by the decomposition of heavy water (D_2O) by heated tungsten filament.

(e) By the decomposition of D_2O by heated Zn turnings. When D_2 is required in large quantities, D_2O is decomposed by heated Zn turning to get D_2.

$$Zn\ (\textit{Heated}) + D_2O\ (\textit{Vapours}) \rightarrow ZnO + D_2$$

(f) By the electrolysis of D_2O. D_2 can also be prepared by the electrolysing pure D_2O acidified with 35% deuteriumsulphuric acid (D_2SO_4) in a specially designed all-glass apparatus. The apparatus is evacuated before electrolysis and a low current of 5 amperes is passed between platinum electrodes when D_2 is set free at the cathode and O_2 at the anode.

$$2D_2O \rightarrow \underset{\text{(At cathode)}}{2D_2} + \underset{\text{(At anode)}}{O_2}$$

Deuterium obtained by this method may contain small quantities of O_2 and D_2O vapour. It is purified by heating it over platinised asbestos, followed by drying with liquid air. The yield at 5 amp. is reported to be 2 litres per hour.

Physical properties. (*i*) Physical properties of deuterium are some what different from those of ordinary hydrogen. Some of the physical constants of the two isotopes are given in Table 7.3.

Table 7.3 : *Physical constant of ordinary (protium) and heavy (deuterium) hydrogen*

Property	Ordinary hydrogen (Protium, H_2)	Heavy hydrogen (Deuterium, D_2)
Symbol	H	D
Boiling point (°C)	– 252.6	– 249.5
Melting point (°C) (Freezing point)	– 259.1	– 254.5
Triple point (°C)	– 259.1	– 254.3
Latent heat of fusion (cal/mole)	28.0	52.3
Heat of dissociation (cal/mole)	104,500	14,900
Heat of vaporisation at triple point (cal/g.mole)	219.7	308
Molecular volume of solid (c.c.)	26.15	23.17
Density of liquid at 25°C	1.0000	1.0076
Vapour pressure at – 259.1°C (mm of Hg)	5.4	5.8
Atomicity	2	2
Molecular formula	H_2	D_2
Mass number	1	2
Atomic mass (Atomic weight)	1.0081	2.0015
Number of protons	1	1
Number of neutrons	0	1
Per cent abundance of their diatomic molecules	1.0000	1.0076

(*ii*) At ordinary temperature, deuterium, like hydrogen, is colourless, odourless and tasteless with atomictity equal to two, *i.e.*, deuterium is *diatomic* and hence its molecular formula is written as D_2. Its density is twice that of hydrogen.

(*iii*) It is insoluble in water and bad conductor of heat and electricity.

(*iv*) *Ortho and para forms of deuterium.* The deuterium molecule also, like hydrogen, is a mixture of ortho and para forms. At ordinary temperature, it consists of 2 parts of ortho and 1 part of para and this composition is retained even at higher temperatures. At low temperature, it is only ortho form which predominates while in normal hydrogen the para form predominates at ordinary termperature.

Chemical properties. The chemical properties of deuterium are similar to those of hydrogen, the only difference being that deuterium reacts more slowly and less completely as compared to hydrogen. Some of the chemical properties of deuterium are given below:

(*i*) *Occlusion.* Deterium is occluded by metallic surfaces more slowly than hydrogen.

(*ii*) *Action of oxygen.* It is combustible and burns in O_2 at 560° C to form D_2O.

$$2D_2 + O_2 \rightarrow \underset{\text{Heavy water}}{2D_2O}$$

(*iii*) *Action of water.* It reacts with water vapour at 200–400° C in presence of zinc oxide or alumina to form a mixture of deuterium oxide and hydrogen.

$$H_2O + D_2 \rightleftharpoons D_2O + H_2$$

(*iv*) *Action with halogens.* It combines with halogens to form deuterium halides.

$$D_2 + Cl_2 \xrightarrow{\text{In light}} 2DCl \quad \text{(Deuterium chloride)}$$

$$D_2 + Br_2 \xrightarrow{\text{On heating}} 2DBr \quad \text{(Deuterium bromide)}$$

$$D_2 + F_2 \xrightarrow{\text{In cold and dark}} 2DF \quad \text{(Deuterium fluoride)}$$

(*v*) *Combination with nitrogen.* Deuterium combines with nitrogen in the presence of a catalyst to form deutero-ammonia or heavy ammonia.

$$3D_2 + N_2 \longrightarrow 2ND_3$$

(*vi*) *Combination with metals.* It combines with alkali metals at 300°C to form deutrides which closely resemble metallic hydrides in their chemical properties.

$$2Na + D_2 \longrightarrow 2NaD \ \textit{(Sodium deuteride)}$$

$$NaD + H_2O \longrightarrow NaOH + HD$$

(*vii*) *Reaction with unsaturated hydrocarbons.* Deuterium gives addition reactions with unsaturated hydrocarbons. The reaction, however, is slower as compared with addition reaction with hydrogen.

$$\underset{\text{Ethylene}}{CH_2{=}CH_2} + D_2 \xrightarrow[260^{\circ}]{Ni} \underset{\text{Dideutero ethane}}{CH_2D{-}CH_2D}$$

(*viii*) *Exchange reactions.* under suitable conditions, it is possible for deuterium to take the place of hydrogen in some compounds, *e.g.*, at high temperature deuterium reacts with H_2, NH_3, H_2O and CH_4 and slowly exchanges hydrogen atoms of these molecules. The exchange of hydrogen by deuterium may take place partially or completely.

$$2HN_3 + 3D_2 \rightleftharpoons \underset{\text{Deuterium ammonia}}{2ND_3} + 3H_2 \ ; \quad H_2O + D_2 \rightleftharpoons \underset{\text{Deuterium oxide}}{D_2O} + H_2$$

$$CH_4 + 2D_2 \longrightarrow \underset{\text{Deuterium methane}}{CD_4} + 2H_2$$

Some exchange reactions take place in presence of hydrogenation catalyst such as finely divided nickel, palladium or platinum black.

$$C_6H_6 + 3D_2 \xrightleftharpoons[\text{black}]{\text{Platinum}} \underset{\text{Deutero benzene}}{C_6D_6} + 3H_2$$

Exchange reactions occur with great readiness when they involve ions in solutions, *e.g.*, in alkaline solution both acetone and acetyl acetone exchange all of their hydrogen atoms for deuterium because of their ionisation.

$$CH_3{-}\overset{O}{\overset{\|}{C}}{-}CH_3 \xrightarrow{\text{Alkali}} CH_3{-}\overset{OH}{\overset{|}{C}}{=}CH_2 \xrightleftharpoons[-3H_2]{+3D_2} CD_3{-}\overset{OD}{\overset{|}{C}}{=}CD_2$$

$$\rightleftharpoons \underset{\text{Deutro acetone}}{CD_3{-}\overset{O}{\overset{\|}{C}}{-}CD_3}$$

Similarly, acetylene undergoes deuterium exchange in alkaline solution because of the weakly acidic character of C—H bond in acetylene.

$$C_2H_2 \xrightleftharpoons{\text{Alkali}} \underset{\substack{\text{Deutero}\\ \text{acetylene}}}{C_2D_2} + H_2$$

(ix) Reducing property. At 700°C, AgCl is reduced to metallic silver, when deuterium is passed through its solution.

$$D_2 + 2AgCl \longrightarrow 2Ag + 2DCl$$

Uses. (*i*) It has been found as a source of *deuterons* which are capable of initiating several nuclear transformations through the formation of compound nucleus at relatively low energies. Some of nuclear transformations produced by deuterons are given below:

(a) (D, α) reactions:

$$_3Li^6 + {}_1D^2 \longrightarrow [_4Be^8] \longrightarrow {}_2He^4 + {}_2He^4$$
$$_6O^{16} + {}_1D^2 \longrightarrow [_9F^{18}] \longrightarrow {}_7N^{14} + {}_2He^4$$

(b) (D, p) reactions :

$$_{15}P^{31} + {}_1D^2 \longrightarrow [_{16}S^{33}] \longrightarrow {}_{15}P^{32} + {}_1H^1$$
$$_{11}Na^{22} + {}_1D^2 \longrightarrow [_{12}Mg^{32}] \longrightarrow {}_{11}Na^{24} + {}_1H^1$$

(c) (D, n) reactions:

$$_6Cl^{12} + {}_1D^2 \longrightarrow [_7N^{14}] \longrightarrow {}_7N^{13} + {}_0n^1$$
$$_1T^3 \text{ (or } {}_1H^3) + {}_1D^2 \longrightarrow [_2He^5] \longrightarrow {}_2He^4 + {}_0n^1 + \text{energy}$$

The energy liberated is used in hydrogen bomb.

(d) (D, T) reactions:

$$_3Li^7 + {}_1D^2 \longrightarrow [_4Be^9] \longrightarrow {}_3Li^6 + {}_1T^3$$
$$_{15}P^{31} + {}_1D^2 \longrightarrow [_{16}S^{33}] \longrightarrow {}_{15}P^{30} + {}_1T^3$$

(*ii*) It is used in the preparation of hydrogen bomb.

(*iii*) It has been used in the preparation of deuterium compounds which are very useful, *e.g.* D_2O is used as a *moderator* in nuclear reactors to slow down the speed of fast moving neutrons.

(*iv*) Deuterium and its compounds are used as a *tracer* in the study of mechanism of various chemical and metabolic reactions taking place in the body, *e.g.*, with the help of deuterium it has been shown that butyric acid (a component of butter) is never stored in the body and is an immediate source of energy which at once gets consumed while stearic acid (a component of lard) is not used up at once and hence is useful in building up reverse stock in the body.

III. Tritium (^3_1H or ^3_1T). Occurrence. It is an extremely rare isotope of hydrogen. It has been established that 10^{17} molecules of ordinary hydrogen contain only traces of tritium, *i.e.*, only 10^{-15}% of ordinary hydrogen is tritium. The cause of the presence of tritium in ordinary hydrogen only in traces is that the nucleus of tritium is radioactive and hence unstable. Evidently it is very difficult to separate tritium from ordinary hydrogen. 10^{17} parts of ordinary water contain about 7 parts of tritium.

In nature, the traces of tritium are present. These are formed continuously in the upper atmosphere by the action of neutrons (produced by the action of cosmic radiations on the atmospheric gases) on atmospheric nitrogen.

$$_7N^{14} + {}_0n^1 \longrightarrow {}_6C^{12} + {}_1T^3$$

Preparation. (*i*) *Oliphant, Harteek* and *Rutherford* (1934) obtained tritium by bombarding compounds such as D_3PO_4, ND_4Cl, $(ND)_2SO_4$ etc., with high energy deuterons.

$$\underset{\text{Deuterium}}{_1D^2} + \underset{\text{Deuteron}}{_1D^2} \longrightarrow \underset{\text{Tritium}}{_1T^3} + {}_1H^1$$

This is the first method of its preparation which led to its discovery.

(*ii*) *By the nuclear reactions produced by deuterons.*

$$_4Be^9 + {}_1D^2 \longrightarrow {}_4Be^8 + {}_1T^3 \quad ; \quad {}_8Li^7 + {}_1D^2 \longrightarrow [_4Be^9] \longrightarrow {}_3Li^6 + {}_1T^3$$

$$_{47}Ag^{109} + {}_1D^2 \longrightarrow [_{48}Cd^{111}] \longrightarrow {}_{47}Ag^{106} + {}_1T^3$$

$$_{15}P^{31} + {}_1D^2 \longrightarrow [_{16}S^{33}] \longrightarrow {}_{15}P^{30} + {}_1T^3$$

(*iii*) *By neutron bombardment of boron and nitrogen* (*i.e.*, neutron bombarding against saturated aqueous solution of boric acid and ammonium nitrate):

$$_5B^{11} + {}_0n^1 \longrightarrow [_5B^{12}] \longrightarrow {}_4Be^9 + {}_1T^3$$

$$_7N^{14} + {}_0n^1 \longrightarrow [_7N^{15}] \longrightarrow {}_5C^{12} + {}_1T^3$$

(*iv*) *By slow neutron bombardment of lithium:*

$$_3Li^6 + {}_0n^1 \longrightarrow {}_1T^3 + {}_2He^4$$

This is probably the best method of obtaining tritium.

Properties. (*i*) Because of the unfavourable neutron-to-proton ratio, tritium is a radioactive isotope with its half-life period of 12.4 years. It emits low energy β-particles and is changed into a lighter isotope of helium, $_2He^3$.

$$_1T^3 \longrightarrow {}_2He^3 + {}_{-1}e^0 \text{ (β–particle)}$$

The radioactive decay of tritium is the source of the traces of $_2He^3$ in the atmosphere.

(*ii*) Some of its physical constants are given below: Boiling point = – 248.1°C; Melting point (Triple point) = –252.4°C; Heat of sublimation at triple point = 303.0 cals/mole; Heat of vaporisation at normal boiling point = 333 cals/mole; Vapour pressure at the triple point = 162.00 mm.

(*iii*) Its atomicity is 2 and hence its molecular formula is T_2.

(*iv*) Its chemistry is very nearly identical with that of hydrogen. However, reactions with tritium are still slower than those with deuterium.

(*v*) *Action of H_2.* It combines with H_2 on heating (300 – 400°C) to form tritium hydride, HT.

$$H_2 + T_2 \longrightarrow 2HT$$

HT reacts with HCl to give tritium chloride, TCl.

$$HT + HCl \longrightarrow TCl + H_2$$

(*vi*) *Action of O_2.* When burnt in O_2, it forms tritium oxide, T_2O.

$$2T_2 + O_2 \longrightarrow 2T_2O$$

Uses. (*i*) It has the possibility of use as an artificial radioactive tracer in chemistry, medicine and biology.

(*ii*) It is used in nuclear fusion reactions such as :

$$_1D^2 + {}_1H^3 \longrightarrow {}_2He^4 + {}_0n^1 + \text{Energy}$$

In this reaction huge amount of energy is liberated which is used in hydrogen bomb.

(*iii*) It is a source of tritons which are used in several nuclear transformations. For example:

(*a*) (T, p) *reactions :*

$$_{27}Co^{59} + {}_1T^3 \longrightarrow {}_{27}Cu^{64} + {}_1H^1 \text{ (or } {}_1p^1)$$

(*b*) (T, D) *reactions :*

$$_3Li^6 + {}_1T^3 \longrightarrow {}_3Li^7 + {}_1D^2$$

$$_{29}Cu^{63} + {}_1T^3 \longrightarrow {}_{29}Cu^{64} + {}_1D^2$$

(*c*) (T, n) *reactions :*

$$_{16}S^{32} + {}_1T^3 \longrightarrow {}_{17}Cl^{34} + {}_0n^1$$

(*d*) (T, α) *reactions :*

$$_3Li^7 + {}_1T^3 \longrightarrow {}_2He^6 + {}_2He^4$$

(*iv*) It is also of potential use in thermo-nuclear process.

(*v*) Tritium hydride, TH, is a very important compound of tritium. It is used in the preparation of a number of organic compounds containing this radioactive isotope.

Uses of radioactive isotopes as radioactive tracers

Many radioactive elements or their compounds are used in tracing various processes met with in medical field, agriculture, industry, analytical chemistry etc. In this technique which is known as *tracer technique,* a radioactive isotope or its suitable compound is introduced at one point of a system and its movement is traced by measuring its radioactivity at different parts of the system.

Such isotopes or their compounds are called *radioactive tracers.* Let us consider an example in medical field. In order to find out if blood is circulating to a wound or not, a radiactive isotope is introduced into the body and after suitable time, some quantity of the blood is taken from the wound and its radioactivity is measured by means of a Geiger-Muller Counter.

In order to know the mechanism of a chemical reaction, a tracer element or its compound is added in small quantity to the reactants of the chemical reaction.

Radioactive isotopes are used as tracer in the following fields.

I. In medical field : Diagonosis and treatment of deseases. Many radioactive isotopes have been used in the diagnosis (detection) and treatment of deseases. The treatment of deseases by the use of radioactive isotopes is called *radiotherapy.* Examples showing the use of radioactive isotopes in the detection and treatment of deseases are given below:

(*a*) Radioactive Na-24 tracer which emits γ-rays has been used to detect the exact position of the tumor and blood clot in the human body. Radioactive Na-24 is injected into the body of the patient or given orally in the form of radioactive Na^{24} Cl which contains Na–24 tracer. In $Na^{24}Cl$, Na^{24} is called *labelled* or *tagged* element. When $Na^{24}Cl$ moves in the body, it accumulates in the area of the tumor or blood clot. The exact position of $Na^{24}Cl$ is found with the help of γ-ray detector which gives the exact position of the tumor or blood clot.

(*b*) Radioactive I-131 tracer emits β-rays and has been used to detect the proper functioning of the thyroid gland. This isotope has been used for the treatment of the cancer of thyroid. Radioactive I-131 destroys the abnormal tissues in the thyroid gland and in this way checks their further growth.

(*c*) Radioactive Co-50 which emits high energy γ-rays has been used for testing deeply-seated cancer growth.

(*d*) Radioactive Co-60 and radioactive Au-198 emit γ-rays and have been used for the treatment of cancer. When γ-rays emitted by Co-60 are directed at the cancerous tumor in the human body, the cancerous cells get burnt.

(*e*) Radioactive P-30 which is β-ray emitter is used for skin deseases.

(*f*) Radioactive P-32 is used to get relief from leukaemia.

II. In industry. Some uses of radioactive isotopes in industry are discussed below:

(*a*) γ-rays produced by radioactive Co-60 have been used as catalyst in the manufacture of C_2H_5Br from ethylene and HBr.

(*b*) C-15 has been used as a tracer in studying mechanisms involved in many industrial processes like alkylation, polymerisation, catalytic synthesis etc.

(*c*) Radioactive isotopes are used to determine the surface area of solids, the thickness of thin films and the sheets of paper, steel, rubber etc.

(*d*) Radioactive isotopes are used to test the wear and tear inside the car engine.

(*e*) Radioactive isotopes are used in detecting and preventing the corrosion of metals.

(*f*) Radioactive isotopes are used to detect the leakage or crack in the underground oil pipes, gas pipes and water pipes. To do this a solution of the radioactive substance is introduced in the pipeline. At the place of crack or leakage in the pipeline, the radioactive solution leaks out which is indicated by the radioactive detector.

III. In agriculture. Some uses of radioactive isotopes in agriculture are given below:

(*a*) Radioactive P-32 has been used to know that exact ratio between the amount of phosphorus taken up by the plant from the soil and that taken by it from the added phosphate fertiliser. This is done by adding some quantity of radioactive P-32 (which is a β-emitter) to phosphate fertiliser.

(*b*) When the crops are exposed to γ-rays emitted by Co-60, the yield of crops is both quantitatively and qualitatively increased.

IV. In biological field. Radioactive isotopes have been used to study the mechanism of various biological processes. Examples are given below:

(*a*) The mechanism of the movement of ions across the cell membrance can be studied with the help of radioactive isotopes.

(*b*) *Photosynthesis process.* This process takes place in plants. In this process plants take CO_2 and prepare sugar in presence of H_2O and O_2 gas is evolved.

$$6CO_2 + 6H_2O \rightarrow C_6H_{12}O_6 \text{ (sugar)} + 6O_2 \quad ...(i)$$

Whether O_2 gas evolved in the above reaction has come from CO_2 or from H_2O or from both has been settled by the use of O-18 isotope (O-18 is not radioactive). For this H_2O is made to react with CO_2^{18} containing O-18 and oxygen evolved is analysed with mass spectrometer. The analysis shows that oxygen produced in the above reaction does not contain O-18. This means that oxygen evolved comes from H_2O and not from CO_2^{18}. O-18 present in CO_2^{18} is consumed to form sugar ($C_6H_{12}O_6^{18}$) and H_2O^{18}. Consequently the process of photosynthesis represented by equation (*i*) should be correctly written as:

$$6CO_2^{18} + 12\,H_2O \rightarrow C_6H_{12}O_6^{18} \text{ (sugar)} + 6O_2 + 6H_2O^{18}.$$

V. In analytical chemistry. The use of radioactive isotopes in analytical chemistry can be understood by considering the following examples:

(*a*) *Precipitation of Co (II) from its salt solution.* Co (II) can be precipitated from its salt solution by using both β-nitroso-α-naphthol (I) and α-nitroso-β-naphthol (II) as precipitating agents. Which of these two compounds is a better precipiating agent for Co (II) can be settled with the help of Co (II) salt labelled with Co-60 isotope. For this Co(II) salt solution containing Co-60 which is β-emitter is treated with each of the two precipitating agents and the activity of the β-particles emitted by Co-60 is measured in the filltrate. The measurement shows that 0.17 mg of cobalt per litre is left in the filtrate when (I) is used as a precipitant while 1.5 mg cobalt per litre is left in the filtrate when (II) is used as precipitant. Thus β-nitroso-α-naphthol is a better precipitant for Co (II).

(*b*) *Determination of solubility of $PbSO_4$ in water.* Radioactive lead (Pb*) has been used to determine the solubility of $PbSO_4$ (this salt contains ordinary lead) in water. For this, a known quantity of Pb* is mixed with ordinary lead (Pb) in a known proportion. The entire amount consisting of Pb* and Pb is dissolved in HNO_3 so that we get a solution containing mixture of $Pb^*(NO_3)_2$ and $Pb(NO_3)_2$. To this solution is added H_2SO_4 whereby Pb^*SO_4 and $PbSO_4$ are precipitated.

$$Pb^* + Pb \xrightarrow{HNO_3} \underbrace{Pb^*(NO_3) + Pb(NO_3)_2}_{\text{Solution}} \xrightarrow{H_2SO_4} \underbrace{Pb^*SO_4 + PbSO_4}_{\text{Precipitate}}$$

After filtration, the radioactivity of the solution is measured and thus the amount of Pb^*SO_4 present in solution is calculated. This, of course, gives the amount of Pb* present in solution. If the ratio of the two varieties of lead originally taken is known, the amount of Pb present in solution can be calculated and hence the solubility of $PbSO_4$ in water can also be calculated.

(*c*) Radioactive isotopes have also been used in *activation analysis* and *isotopic dilution analysis.*

VI. In controlling the production of insects. For this purpose, large number of male insects are exposed to radiations obtained from radioactive substances. These radiations make them sterilised and hence they are not able to produce other insects when they mate with female insects. Malaria is brought under control by reducing the number of mosquitoes by this technique. The production of other types of insects or pests can also be controlled by this technique.

VII. In the determination of the age of a dead matter like wooden piece, peat, fossil, dead animal etc. by radio-carbon dating method. This method was developed by Willard F. Libby, who was awarded the Noble prize for this method. This method is called *radio-carbon dating method.*

Principle of the method. Carbon dioxide found in the atmosphere is a mixture of $^{12}CO_2$ (which is non-radioactive and found in great abundance = 98.89%) and $^{14}CO_2$ which is radioactive and found in traces. $^{14}CO_2$ is formed by the following reactions taking place in the upper portion of the atmosphere.

$^{14}_{7}N$ (present in the atmosphere) + $^{1}_{0}n$ (produced from the cosmic rays from the sun) → $^{14}_{6}C$ (radioactive) + $^{1}_{1}H$

$$^{14}_{6}C + O_2 \text{ (in the atmosphere)} \rightarrow {}^{14}_{6}CO_2 \text{ (radioactive)}$$

$^{12}CO_2$ and $^{14}CO_2$ are present in a fixed ratio in the atmosphere. ($^{12}CO_2$ + $^{14}CO_2$) mixture is assimilated by plants and through them by animals. As long as the living organism is alive, the ratio of the amounts of $^{14}CO_2$ and $^{12}CO_2$ (or $^{14}C/^{12}C$ ratio) in it remains almost constant and is the same as it is in the atmosphere. Now when the living organism dies, ^{14}C present in the dead organism begins to decay (since ^{14}C is radioactive) by emitting β-particles ($^{0}_{-1}e$) [$^{14}_{6}C \rightarrow {}^{14}_{7}N + {}^{0}_{-1}e$, $(t_{0.5})_{C-14}$ = 5568 yrs.] while ^{12}C which is non-radioactive remains intact in the organism (*i.e.*, ^{12}C does not decay). Due to the decay of ^{14}C, the amount of ^{14}C in the dead organism decreases. The decrease in the amount of ^{14}C also decreases $^{14}C/^{12}C$ ratio in the dead organism, *i.e.*, $^{14}C/^{12}C$ ratio in the dead organism becomes less than that in the living species. If $^{14}C/^{12}C$ ratios in the living organism and in the dead organism are determined by experiments and half-life period of ^{14}C is known, then the time (*t*) elapsed since the living body had died can be calculated by using equation (*ii*) which has been derived form equation (*i*).

We know that:

$$N = N_0 e^{-Kt} \qquad ...(i)$$

Here N = $^{14}C/^{12}C$ ratio in the dead organism, N_0 = $^{14}C/^{12}C$ ratio in the living or fresh organism, *t* = age of the organism and K = disintegration constant for ^{14}C. Equation (*i*) can also be written as:

$$\frac{N}{N_0} = e^{-Kt} \quad \text{or} \quad \ln\frac{N}{N_0} = -Kt$$

$$\text{or} \quad 2.303 \log_{10}\frac{N}{N_0} = -t \times \frac{0.693}{(t_{0.5})_{C-14}}$$

$$\text{or} \quad t = -\frac{2.303 \times (t_{0.5})_{C-14}}{0.693} \times \log_{10}\frac{N}{N_0}$$

$$\text{or} \quad t = \frac{2.303 \times (t_{0.5})_{C-14}}{0.693} \times \log_{10}\frac{N_0}{N}$$

$$\text{or} \quad t = \frac{2.303 \times (t_{0.5})_{C-14}}{0.693} \times \log_{10}\frac{^{14}C/^{12}C \text{ ratio in the living organism}}{^{14}C/^{12}C \text{ ratio in the dead organism}} \qquad ...(ii)$$

We have said above that when a plant or an animal dies, ^{12}C present in the species, being non-radioactive, does not decay and hence the amount of ^{12}C in the fresh and dead species remains the same. Consequently, in place of equation (*ii*), equation (*iii*) can also be used for calculating the age of the organism.

$$t = \frac{2.303 \times (t_{0.5})_{C-14}}{0.693} \times \log_{10}\frac{\text{Amount of } ^{14}C \text{ in the living organism}}{\text{Amount of } ^{14}C \text{ in the dead organism}} \qquad ...(iii)$$

Now we know that the amount of ^{14}C found in the organism is directly proportional to its activity, *i.e.* amount of ^{14}C is proportional to the number of disintegrations per minute (*dpm*) or disintegrations per second (*dps*) given by ^{14}C. Thus equation (*iii*) can also be written as equation (*iv*).

$$t = \frac{2.303 \times (t_{0.5})_{C-14}}{0.693} \times \log_{10}\frac{\text{Activity of } ^{14}C \text{ in the living organism}}{\text{Activity of } ^{14}C \text{ in the dead organism}} \qquad ...(iv)$$

Solved Examples

Example 7.1. *A piece of wood was found to have $^{14}C/^{12}C$ ratio 0.7 times that in the living plant. Calculate the period when the plant died. Given half-life of ^{14}C = 5760 years.*

Solution. We know that the age of the wood is given by:

$$\text{Age of the wood (t)} = \frac{2.303 \times \text{half-life of } ^{14}C}{0.693} \times \log_{10}\frac{^{14}C/^{12}C \text{ ratio in the living plant}}{^{14}C/^{12}C \text{ ratio in the dead plant}}$$

Now as given in the question, $^{14}C/^{12}C$ ratio in the wood (dead plant) = 0.7 × $^{14}C/^{12}C$ ratio in the living plant.

or $$\frac{^{14}C/^{12}C \text{ ratio in the living plant}}{^{14}C/^{12}C \text{ ratio in the dead plant (wood)}} = \frac{1}{0.7}$$

Thus $$t = \frac{2.303 \times 5760}{0.693} \times \log \frac{1}{0.7} = \frac{2.303 \times 5760}{0.693} \times \log \frac{10}{7}$$

$$= \frac{2.303 \times 5760 \times 0.155}{0.693} = \textbf{2970 years}$$

Example 7.2. *The ^{14}C to ^{12}C ratio in a certain piece of wood is 13% that of the atmosphere. Calculate the age of wood, given that half-life of ^{14}C = 5580 years.*

Solution. It is given that: $$\frac{^{14}C/^{12}C \text{ ratio in the living plant}}{^{14}C/^{12}C \text{ ratio in the dead plant (wood)}} = \frac{100}{13}$$

$$\therefore \quad \text{Age of the wood (t)} = \frac{2.303 \times (t_{0.5}) \text{ of } ^{14}C}{0.693} \times \log \frac{100}{13} = \frac{2.303 \times 5580}{0.693} \times \log \frac{100}{13}$$

$$= \frac{2.303 \times 5580 \times 0.8161}{0.693} = \textbf{16430 years}$$

Example 7.3. *A piece of charcoal from the ruins of a settlement in Japan was found to have $^{14}C/^{12}C$ ratio that was 0.617 times that found in living organism. How old is this piece of charcoal? Given the half-life of $^{14}_{6}C$ = 5770 years and log 1.62 = 0.210.* **(B.I.T. Ranchi 1989)**

Solution. $$\text{Age of the charcoal (t)} = \frac{2.303 \times t_{0.5}}{0.693} \times \log \frac{^{14}C/^{12}C \text{ ratio in the living wood}}{^{14}C/^{12}C \text{ ratio in the dead plant (wood)}}$$

$$= \frac{2.303 \times 5770}{0.693} \times \log \frac{1.000}{0.617} = \frac{2.303 \times 5770}{0.693} \times \log 1.62$$

$$= \frac{2.303 \times 5770}{0.693} \times 0.210 = \textbf{4026.76 years}$$

Example 7.4. *The amount of ^{14}C isotope in a piece of wood is found to be one-sixth of its amount present in a fresh piece of wood. Calculate the age of the wood, given that half-life of ^{14}C = 5577 years.* **(Raj. B.Sc. 1985)**

Solution. The age (t) of the wood is given by:

$$t = \frac{2.303 \times \text{half-life of } ^{14}C}{0.693} \times \log_{10} \frac{\text{Amount of } ^{14}C \text{ in the fresh wood}}{\text{Amount of } ^{14}C \text{ in the dead wood}}$$

$$= \frac{2.303 \times 5577}{0.693} \times \log_{10} \frac{N_0}{N_0/6} = \frac{2.303 \times 5577}{0.693} \log_{10} 6$$

$$= \frac{2.303 \times 5577}{0.693} \times 0.7782 = \textbf{14422.9 years}$$

Example 7.5. *The amount of ^{14}C in a sample of wood is found to be one-fourth of its amount present in a fresh piece of wood. Calculate the age of the wood ($t_{0.5}$ = 5577 years).* **(Pune B.Sc. 1986)**

Solution. We know that the age of the wood is given by:

$$\text{Age of wood (t)} = \frac{2.303 \times \text{half-life of } ^{14}C \ (t_{0.5})}{0.693} \times \log_{10} \frac{\text{Amount of } ^{14}C \text{ in the fresh wood } (N_0)}{\text{Amount of } ^{14}C \text{ in the dead wood } (N_t)}$$

or $$t = \frac{2.303 \times 5577}{0.693} \log_{10} \frac{N_0}{N_0/4} \text{ years}$$

$$= \frac{2.303 \times 5577}{0.693} \log_{10} 4 = \mathbf{11154 \text{ years}}$$

Example 7.6. *A wooden statue was found to contain 10.1% as much ^{14}C as a piece of fresh wood. Calculate the age of wooden statue. (Half-life of ^{14}C = 5570 yrs.).*

Solution. The age of the wooden statue (t)

$$= \frac{2.303 \times \text{half-life of } ^{14}C}{0.693} \times \log \frac{\text{Amount of } ^{14}C \text{ in the fresh wood}}{\text{Amount of } ^{14}C \text{ in the wooden statue}}$$

$$= \frac{2.303 \times 5570}{0.693} \times \log \frac{100}{10.1} = 18510.40 \times \log 9.90 = \mathbf{18500 \text{ years}}$$

Example 7.7. *A piece of wood recovered in an excavation has 33.32% as much ^{14}C as ordinary wood has today. When did this piece get burried ? ($t_{0.5}$ = 5770 years).*

Solution. The time for which the wooden piece was burried (t)

$$= \frac{2.303 \times t_{0.5}}{0.693} \log \frac{\text{Amount of } ^{14}C \text{ in the fresh wood}}{\text{Amount of } ^{14}C \text{ in the dead wood}}$$

$$= \frac{2.303 \times 5770}{0.693} \log \frac{100}{33.32} = \frac{2.303 \times 5770}{0.693} \log 3.00 = \mathbf{9148.4 \text{ years}}$$

Example 7.8. *The activity of the hair of an Egyptian mummy is 7.0 disintegrations min^{-1}, g^{-1} of carbon. Find the age of the mummy. Given that $t_{0.5}$ of ^{14}C = 5770 years and disintegration rate of fresh sample of ^{14}C = 14 min^{-1} g^{-1}.*

Solution. We know that the age of the mummy is given by:

$$\text{Age of the mummy (t)} = \frac{2.303 \times \text{half-life of } ^{14}C}{0.693} \times \log_{10} \frac{\text{Activity of } ^{14}C \text{ in fresh sample of mummy}}{\text{Activity of } ^{14}C \text{ in hair}}$$

$$= \frac{2.303 \times 5770}{0.693} \times \log \frac{14}{7} = \frac{2.303 \times 5770}{0.693} \times \log 2$$

$$= \frac{2.303 \times 5770}{0.693} \times 0.3010 = \frac{0.693 \times 5770}{0.693} = \mathbf{5770 \text{ years.}}$$

This example shows that if the activity of a piece of old wood is half of the activity of the piece of the fresh wood, then the age of the wood is equal to the half-life period of ^{14}C.

Example 7.9. *A sample of preserved wood was found to have specific activity of 4.11 disintegrations min^{-1} g^{-1}, while a fresh sample was found to contain a specific activity of 15.3 disintegrations min^{-1} g^{-1}. How old is the preserved wood? (Half life of ^{14}C = 5570 years).*

Solution. Age of the preserved wood (t)

$$= \frac{2.303 \times t_{0.5}}{0.693} \log \frac{\text{Activity of } ^{14}C \text{ in fresh sample of wood}}{\text{Activity of } ^{14}C \text{ in the sample of preserved wood}}$$

$$= \frac{2.303 \times 5570}{0.693} \times \log \frac{15.3}{4.11} = 18510.40 \times \log 3.72 = \mathbf{1.06 \times 10^4 \text{ years}}$$

Example 7.10. *A sample of wooden artifact is found to undergo nine $^{14}_{6}C$ disintegrations per minute per gram of carbon. What is the approximate age of artifact? The half-life of $^{14}_{6}C$ is 5730 years and radioactivity of wood recently cut down is 15 disintegrations per minute per gram of carbon.* **(Dhanbad 1989)**

Solution. Age of artifact (t)

$$= \frac{2.303 \times t_{0.5}}{0.693} \log \frac{\text{Activity of wood recently cut down}}{\text{Activity of wooden artifact}}$$

$$= \frac{2.303 \times 5730}{0.693} \times \log \frac{15}{9} = \mathbf{4230.2\ years}$$

VIII. In the determination of the age of an ore, earth or rock by rock-dating (also called radioactive dating) method. There are three methods:

(*a*) *$^{238}U/^{206}Pb$ ratio method.* As an example, let us determine the age of a rock containing U–238. The ore is analysed and analysis shows that the ore contains U–238 and Pb-206 isotopes. The presence of U-238 and Pb-206 isotopes in the ore indicates that the ore contained *only U-238 before its decay* and after its decay for time *t* (which is the age of the ore or rock) the ore contains Pb-206 and U-238. Pb-206 present in the ore is supposed to have formed by the decay (disintegration) of some quantity of U-238. The remaining quantity of U-238 remains undisintegrated (*i.e.*, remains as such). Now if it is supposed that the weight of U-238 which was present in the ore before its decay into Pb-206 is N_0 and the weight of U-238 which has remained undisintegrated even after the formation of Pb-206 is N, then the age of the ore (*t*) is given by equation (*i*) or (*ii*) in which K_{238} and $(t_{0.5})_{238}$ are *disintegration constant* and *half-life period* of U-238 isotope respectively and *t* is the age of the ore

$$N = N_0 \cdot e^{-K_{238} \cdot t} \quad ...(i)$$

or $$\frac{N}{N_0} = e^{-K_{238} \cdot t} \quad \text{or} \quad \ln \frac{N}{N_0} = -K_{238}.t$$

or $$2.303 \log_{10} \frac{N}{N_0} = -K_{238}.t \quad \text{or} \quad -2.303 \log_{10} \frac{N}{N_0} = K_{238}.t$$

or $$t = -\frac{2.303}{K_{238}} \cdot \log_{10} \frac{N}{N_0}$$

$$t = \frac{2.303}{K_{238}} \cdot \log_{10} \frac{N_0}{N}$$

or $$t = \frac{2.303 \times (t_{0.5})_{238}}{0.693} \cdot \log_{10} \frac{N_0}{N} \quad ...(ii)$$

Calculation of N_0 and N. Suppose the analysis of U-238 ore shows that the ore contains Pb^{206} grams of Pb-206 and U^{238} grams of U-238 that has been left undisintegrated after time *t*. Since all Pb^{206} grams of Pb-206 have been obtained by the disintegration of U-238, weight of U-238 which was present in the ore (or rock) before its decay into Pb-206 will be equal to the weight of U-238 which, on its disintegration, gave Pb^{206} grams of Pb-206 plus the weight of U-238 which has been found in the ore by its analysis *i.e.*, which has remained undisintegrated (= U^{238}).

Now, since 206 grams of Pb-206 are given by the decay of 238 grams of U-238,

Pb^{206} grams of Pb-206 must have been given by the decay of $\left(\frac{238}{206} \times Pb^{206}\right)$ grams of U–238 *i.e.*, the weight of U-238 which decays to give Pb^{206} grams of Pb-206 isotope = $\left(\frac{238 \times Pb^{206}}{206}\right)$ grams.

Hence the weight of U-238 which was present in the ore before its decay into Pb-206 = $\left(\frac{238 \times Pb^{206}}{206} + U^{238}\right) = N_0$.

Since, U^{238} is the weight of U-238 that has remained undisintegrated, $N = U^{238}$.

On putting the values of N_0 and N in equation (*ii*), we get:

$$\left.\begin{aligned} t &= \frac{2.303 \times (t_{0.5})_{238}}{0.693} \log_{10} \frac{\frac{238 \times Pb^{206}}{206} + U^{238}}{U^{238}} \\ \text{or} \quad t &= \frac{2.303 \times (t_{0.5})_{238}}{0.693} \log_{10} \left(1 + \frac{238}{206} \cdot \frac{Pb^{206}}{U^{238}}\right) \end{aligned}\right\} \quad ...(iii)$$

(*b*) *$^{235}U/^{207}Pb$ ratio method.* If we consider an ore or a rock containing U-235, then the isotope of lead which is obtained by its decay is Pb-207 and hence the age (*t*) of such an ore is given by equation (*iv*).

$$t = \frac{2.303 \times (t_{0.5})_{235}}{0.693} \log_{10} \frac{\frac{235 \times Pb^{207}}{207} + U^{235}}{U^{235}}$$

$$\text{or} \quad t = \frac{2.303 \times (t_{0.5})_{235}}{0.693} \log_{10} \left[1 + \frac{235}{207} \cdot \frac{Pb^{207}}{U^{235}}\right] \qquad ...(iv)$$

(*c*) *$^{238}U/^{235}U$ ratio method.* If an ore contains both the isotopes of uranium *viz.* U-238 and U-235, then the age (*t*) of the ore or earth can be calculated as follows:

On applying equations, $N = N_0.e^{-Kt}$ for U–238 and U-235 isotopes, we get:

For U-238 isotope $N_{238} = (N_0)_{238} e^{-K_{238} \cdot t}$

For U-235 isotope $N_{235} = (N_0)_{235} e^{-K_{238} \cdot t}$

On dividing the above two equations, we get : $\frac{N_{238}}{N_{235}} = \frac{(N_0)_{238} \cdot e^{-K_{238} \cdot t}}{(N_0)_{235} \cdot e^{-K_{235} \cdot t}}$

Now, since initially the two isotopes were in equal abundance and that no isotopic separation has occured through physical or chemical changes, $(N_0)_{238} = (N_0)_{235}$ and hence the above equation becomes :

$$\frac{N_{238}}{N_{235}} = \frac{e^{-K_{238} \cdot t}}{e^{-K_{235} \cdot t}} = e^{-K_{238} \cdot t + K_{235} t}$$

$$\text{or} \quad \frac{N_{238}}{N_{235}} = e^{t(K_{235} - K_{238})} \qquad \text{or} \qquad \ln \frac{N_{238}}{N_{235}} = t(K_{235} - K_{238})$$

$$\text{or} \quad 2.303 \log_{10} \frac{N_{238}}{N_{235}} = t\left[\frac{0.693}{(t_{0.5})_{235}} - \frac{0.693}{(t_{0.5})_{238}}\right]$$

From this equation *t* can be calculated, if N_{238}/N_{235} ratio and half-life periods of U-238 and U-235 isotopes are known.

Example 7.11. *A sample of uranium ore was found to contain 0.25 g of Pb-206 and 1.605 g of U-238. Assuming that all Pb-206 had accumulated due to the decay of U-238, find out the age of the ore (half-life of U-238 = 4.5 × 10^9 years).*

Solution. We know that:

∵ 206 g. of Pb-206 are given by the decay of 238 g of U-238

∴ 0.25 g of Pb-206 must have been given by the decay of $\frac{238}{206} \times 0.25$ g of U-238.

$$= \frac{238 \times 0.25}{206} \text{ g.}$$

Thus total weight of U-238 which was present in the ore before its decay into Pb-206 = weight of U-238 which gave 0.25 g of Pb-206 + weight of U-238 which has been found in the ore by its analysis

$$= \left[\frac{238 \times 0.25}{206} + 1.605\right] g = (0.2888 + 1.605) \text{ g} = 1.8938 \text{ g} = N_0$$

$$N = 1.6050 \text{ g.}$$

$$\therefore \quad \text{Age of the ore } (t) = \frac{2.303 \times (t_{0.5})_{U^{238}}}{0.693} \log_{10} \frac{N_0}{N}$$

$$= \frac{2.303 \times (t_{0.5})_{U^{238}}}{0.693} \log_{10} \frac{1.8938}{1.6050}$$

$$= \frac{2.303 \times 4.5 \times 10^9}{0.693} \log_{10} 1.1799$$

$$\therefore \quad t = \mathbf{1.075 \times 10^9 \ years}$$

Example 7.12. *A uranite sample (ore of uranium) was found on analysis to contain 0.214 g. of lead for every gram of uranium-238. Estimate the data when the mineral was formed in the earth's crust (Half-life U-238 = 4.5 × 10^9 years). Assume that Pb-206 results only from radioactive disintegration of U-238.*

Solution. We know that:

$\because$ 206 g. of ^{206}Pb are obtained from the decay of 238 g. of ^{238}U

$\therefore$ 0.214 g. of ^{206}Pb must have been obtained by the decay of $\frac{238 \times 0.214}{206}$ g of ^{238}U

$= (1.1553 \times 0.214)$ g. $= 0.2472$ g

$\therefore$ Total weight of ^{238}U which was present in the ore before its decay = 0.2471 + 1 = 1.2471g.

$\therefore$ Age of the mineral (t)

$$= \frac{2.303 \times (t_{0.5})_{U^{238}}}{0.693} \times \log \frac{1.2471}{1.0000} = \frac{2.303 \times 4.5 \times 10^9}{0.693} \log 1.2471$$

$$= \mathbf{1.4 \times 10^9 \ years}$$

Example 7.13. *$_{92}U^{238}$ by successive radioactive decays change to $_{82}Pb^{206}$. A sample of uranium ore was analysed and found to contain 1.0 g. of U^{238} and 0.1 g of Pb^{206}. Assuming that all the pb^{206} had accumulated due to the decay of U^{238}, find out the age of the ore (half-life of U^{238} = 4.5 × 10^9 years).* **(Roorkee Entrance 1990)**

Solution. We know that:

$\because$ 206 g. of Pb^{206} are given by the decay of 238 g of U^{238}

$\therefore$ 0.1 g. of Pb^{206} will have been given by the decay of $\frac{238 \times 0.1}{206}$ g of U^{238}

$= 1.1553 \times 0.1 = 0.11553$g.

Total weight of U^{238} which was present in the ore before its decay = 0.11553 + 1 = 1.11553g

$$\therefore \text{ Age of the ore (t)} = \frac{2.303 \times (t_{0.5})_{238}}{0.693} \log \frac{1.11553}{1.0000}$$

$$= \frac{2.303 \times 4.5 \times 10^9}{0.693} \log \frac{1.11553}{1.0000} = \mathbf{0.703 \times 10^9 \ years}$$

Example 7.14. *In a sample of uranium, the ratio by weight of Pb^{206} to U^{238} is 0.2. Find the age of the mineral. The decay constant of U^{238} is $1.54 \times 10^{-10} y^{-1}$.*

Solution. We know that:

$\because$ 206g. of ^{206}Pb are given by the decay of 238g of U^{238}

$\therefore$ 0.2g. of ^{206}Pb must have been given by the decay of $\left(\frac{238}{206} \times 0.2\right)$ g of ^{238}U

$= 1.1553 \times 0.2 = 0.23106$ g

$\therefore$ Total weight of U^{238} which was present in the ore before its decay

$= 0.23106 + 1.0 = 1.23106$

Let the half-life of the ore be $(t_{0.5})_{U^{238}}$ years,then

$$K = \frac{2.303}{(t_{0.5})_{U^{238}}} \times \log \frac{1.23106}{1.00000}$$

$$\therefore \quad (t_{0.5})_{U^{238}} = \frac{2.303}{1.54 \times 10^{-10}} \log 1.231 = \mathbf{1.58 \times 10^9 \ years}.$$

Example 7.15. *Analysis of a sample of uranium ore showed that it contained 0.28 gms. of*

Pb^{206} and 1.70 gms. of U^{238}. If all Pb^{206} had accumulated due to the decay of U^{238}, calculate the age of the ore (half-life period of U^{238} = 4.5 × 10^9 years). **(Delhi B.Sc. Hons 1979)**

Solution. Do it yourself.

Example 7.16. *The ratio of mass of Pb^{206} to the mass of U^{238} in a certain rock specimen is found to be 0.5. Assuming that the rock originally contained no lead, estimate its age. Half-life of U^{238} = 4.5 × 10^9 years.*

Solution. weight of Pb^{206} : weight of U^{238} = 0.5 : 1

0.5g. of Pb^{206} had been given by $\frac{238 \times 0.5}{206}$ g of U^{238}

$$= 1.1553 \times 0.5 = 0.57765 \text{ g}$$

∴ Total weight of U^{238} which was present in the ore before its decay $= 0.57765 + 1.0 = 1.57765\text{g}$

$$\therefore \quad \text{Age of the rock (t)} = \frac{2.303 \times (t_{0.5})_{U^{238}}}{0.693} \log \frac{1.57765}{1.0}$$

$$= \frac{2.303 \times 4.5 \times 10^9}{0.693} \log 1.57765$$

$$= \frac{2.303 \times 4.5 \times 10^9}{0.693} \times 0.1761 = \mathbf{2.63 \times 10^9 \text{ years}}$$

Example 7.17. *The half of ^{238}U decomposes to ^{206}Pb in 4.5 × 10^8 years. What will be the age of a rock that contains equal weights of both.* **(B.I.T. 1986)**

Solution. Let the weights of ^{238}U and ^{206}Pb isotopes present in the rock be *a* gram (both have the same amount).

Now we know that:

∵ 206g. of ^{206}Pb are given by the decay of 238g of ^{238}U

∴ *a* g of ^{206}Pb must have been given by the decay of $\left(\frac{238}{206} \times a\right)$ g of ^{238}U

Thus total weight of ^{238}U which was present in the ore before its decay $= \left(\frac{238}{206} \times a + a\right) g = \frac{444\,a}{206}$ g

$$\therefore \quad \text{Age of the rock (t)} = \frac{2.303 \times (t_{0.5})_{U^{238}}}{0.693} \log_{10} \frac{\frac{444\,a}{206}}{a}$$

$$= \frac{2.303 \times 4.5 \times 10^8}{0.693} \times \log_{10} \frac{444\,a}{206\,a} = \mathbf{5.004 \times 10^8 \text{ years}}$$

Example 7.18. *The half-life periods of U^{238} and U^{235} isotopes ore 4.5 × 10^9 years and 7.1 × 10^8 years respectively. The relative abundance of these isotopes at present is found to be 99.2% and 0.8% respectively. Calculate the age of the earth.*

Solution. *For U^{238} isotope* $N_{238} = (N_0)_{238}\, e^{-K_{238}\, t}$

For U^{235} isotope $N_{235} = (N_0)_{235}\, e^{-K_{235}\, t}$

On dividing the above two equations, we get :

$$\frac{N_{238}}{N_{235}} = \frac{(N_0)_{238}\, e^{-K_{238}\, t}}{(N_0)_{235}\, e^{-K_{235}\, t}}$$

Now since the two isotopes are assumed to be present in the same ratio when the earth was formed, $(N_0)_{238} = (N_o)_{235}$ and hence the above equation reduces to:

$$\frac{N_{238}}{N_{235}} = \frac{e^{-K_{238}\, t}}{e^{-K_{235}\, t}} = e^{-K_{238}\,.\,t + K_{235}\,.\,t} \quad \text{or} \quad \frac{N_{238}}{N_{235}} = e^{t\,(K_{235} - K_{238})}$$

or $\ln\left(\frac{N_{238}}{N_{235}}\right) = t\,(K_{235} - K_{238})$ or $2.303 \times \log_{10}\frac{N_{238}}{N_{235}} = t\left[\frac{0.693}{(t_{0.5})_{235}} - \frac{0.693}{(t_{0.5})_{238}}\right]$

or $2.303 \times \log\frac{99.2}{0.8} = t\left[\frac{0.693}{7.1 \times 10^8} - \frac{0.693}{4.5 \times 10^9}\right]$

or $2.303 \times \log 124 = t\,[0.0976 \times 10^{-8} - 0.0154 \times 10^{-8}]$

or $2.303 \times 2.0934 = t \times 0.0822 \times 10^{-8}$

or $t = \frac{2.303 \times 2.0934}{0.0822 \times 10^{-8}} = \mathbf{58.6508 \times 10^8}$ **years**

Example 7.19. *The isotopes U^{238} and U^{235} occur in nature in the ratio 140:1. Assuming that at the time of earth's formation they were present in equal ratio. Make an estimate of the age of earth. The half-lives of U^{238} and U^{235} are 4.5×10^9 years and 7.13×10^8 years respectively.*

(Roorkee Entrance 1983)

Solution:

For U-238 isotope : $N_{238} = (N_0)_{238}\, e^{-K_{238}\cdot t}$

For U-235 isotope : $N_{235} = (N_0)_{235}\, e^{-K_{235}\cdot t}$

On dividing these equations, we get :

$$\frac{N_{238}}{N_{235}} = \frac{(N_0)_{238}\, e^{-K_{238}\cdot t}}{(N_0)_{235}\, e^{-K_{235}\cdot t}}$$

Now since the two isotopes were present in the same ratio when the earth was formed, $(N_0)_{238} = (N_0)_{235}$ and hence we have:

$$\frac{N_{238}}{N_{235}} = \frac{e^{-K_{238}\cdot t}}{e^{-K_{235}\cdot t}} = e^{-K_{238}\cdot t + K_{235}\, t} \quad \text{or} \quad \frac{N_{238}}{N_{235}} = e^{t\,(K_{235} - K_{238})}$$

$$\ln\left(\frac{N_{238}}{N_{235}}\right) = t\,(K_{235} - K_{238})$$

$$2.303 \log_{10}\frac{N_{238}}{N_{235}} = t\left[\frac{0.693}{(t_{0.5})_{235}} - \frac{0.693}{(t_{0.5})_{238}}\right]$$

$$= t\left[\frac{0.693}{7.13 \times 10^8} - \frac{0.693}{4.5 \times 10^9}\right] = t\,[0.0971 \times 10^{-8} - 0.1540 \times 10^{-9}]$$

or $2.303 \log_{10}\frac{N_{238}}{N_{235}} = t \times 0.0817 \times 10^{-8}$

or $2.303 \times \log_{10} 140 = t \times 0.0817 \times 10^{-8}$

$\therefore \quad t = \frac{2.303 \times \log_{10} 140}{0.0817 \times 10^{-8}} = \frac{2.303 \times 2.1037}{0.0817 \times 10^{-8}} = \mathbf{59.30 \times 10^8}$ **years**

Isotones

Definition and composition

Isotones are the atoms of different elements which have the same number of neutrons (*n*), *i.e.*, isotones have the same value of (A – Z) where A = mass number and Z = atomic number. For example $^{77}_{33}As_{44}$ and $^{78}_{34}Se_{44}$ are isotones to each other, since they have the same number of neutrons as shown below:

For $^{77}_{33}As$, $Z = p = e = 33$, $A = 77$ and $n = A - Z = 77 - 33 = 44$

For $^{78}_{34}Se$, $Z = p = e = 34$, $A = 78$ and $n = A - Z = 78 - 34 = 44$

More examples of isotones are given below:

(*i*) ${}^{30}_{14}Si_{16}$ ($n = 30 - 14 = 16$), ${}^{31}_{15}P_{16}$ ($n = 31 - 15 = 16$) and ${}^{32}_{16}S_{16}$ ($n = 32 - 16 = 16$)

(*ii*) ${}^{3}_{1}H_2$ ($n = 3 - 1 = 2$) and ${}^{4}_{2}He_2$ ($n = 4 - 2 = 2$)

(*iii*) ${}^{226}_{88}Ra_{138}$ ($n = 226 - 88 = 138$) and ${}^{227}_{89}Ac_{138}$ ($n = 227 - 89 = 138$)

Isotones have different properties.

Isobars

What are isobars?

Isobars are the atoms of different elements which have the same mass number, A (A = $n + p$) but different atomic number (Z) *i.e.* the isobars have the same value of the sum of protons and neutrons. Examples of some stable isobaric *pairs* and isobars *triods* are given below:

Isobaric pairs : ${}^{36}_{16}S$, ${}^{36}_{20}Ar$; ${}^{40}_{18}Ar$, ${}^{40}_{20}Ca$; ${}^{46}_{20}Ca$, ${}^{46}_{22}Ti$; ${}^{113}_{48}Cd$, ${}^{113}_{49}In$; ${}^{123}_{51}Sb$, ${}^{123}_{52}Te$ etc.

Isobaric triads : ${}^{50}_{22}Ti$, ${}^{50}_{23}V$, ${}^{50}_{24}Cr$; ${}^{124}_{50}Sn$, ${}^{124}_{52}Te$, ${}^{124}_{54}Xe$; ${}^{40}_{18}Ar$, ${}^{40}_{19}K$, ${}^{40}_{20}Ca$ etc.

Characteristics of isobars

(*i*) Since the mass number (A) of the isobars are the same, the sum of neutrons (n) and protons (p) which is equal to A (A = $n + p$) is also the same in all the isobaric elements.

(*ii*) since the atomic numbers (Z) of the isobars are different, the number of protons and the number of neutrons in the nucleus are also different. The number of electrons is also different.

On the basis of the above two properties, the composition of the isobars *viz.* ${}^{40}_{18}Ar$, ${}^{40}_{19}K$ and ${}^{40}_{20}Ca$ can be shown as ${}^{40}_{18}Ar_{22}$, ${}^{40}_{19}K_{21}$ and ${}^{40}_{20}Ca_{20}$ respectively. In these, the numbers given at the top are the mass numbers (A), those given at the left are the number of protons (p), atomic number (Z) or number of electrons while those shown at the right indicate the number of neutrons (n) (n = A – Z).

(*iii*) No stable isobars have been found among the light nuclei.

(*iv*) With the exception of isobaric pairs having A = 113 and 123, all other stable isobaric pairs have even mass numbers, even atomic number (or protons) and even number of neutrons.

The atomic numbers of such pairs differ by two units. Note that the atomic numbers for the isobaric pair having A = 113 and 123 (odd number) differ by one unit only.

(*v*) Since isobars are different elements, they have different physical as well as chemical properties.

(*vi*) since isobars have different atomic numbers, they are placed in different groups of the periodic table.

Production of isobar(s) by the emission of one or more β-particles

It has been be seen under the study of Soddy-Fajans' group displacement law that when a radioactive element loses β-particle (${}^{0}_{-1}e$), the daughter element formed has the same mass number as the parent element. Thus the emission of one or more β-particles by a radioactive element produces its isobar(s). In the following example ${}^{228}_{88}Ra$, ${}^{228}_{89}Ac$ and ${}^{228}_{90}Th$ are isobars to each other, since they have the same mass number (= 228).

$$ {}^{228}_{88}Ra \xrightarrow{-{}^{0}_{-1}e} {}^{228}_{89}Ac \xrightarrow{-{}^{0}_{-1}e} {}^{228}_{90}Th $$

Mirror nuclei.

Mirror nuclei are a pair of isobars in which the number of protons and neutrons (or the number of neutrons and protons) differs by one and are interchanged, *i.e.* in mirror - nuclei $|p - n| = 1$. Examples of some mirror nuclei are ${}^{13}_{6}C_7$ — ${}^{13}_{7}N_6$, ${}^{15}_{7}N_8$ — ${}^{15}_{8}O_7$, ${}^{23}_{11}Na_{12}$ — ${}^{23}_{12}Mg_{11}$, ${}^{39}_{19}K_{20}$ — ${}^{39}_{20}Ca_{19}$ etc.

Isobaric isotopes or nuclear isomers

Isobaric isotopes or nuclear isomers are the atoms of the same element which have the same atomic number and same mass number but have different radioactive properties. The phenomenon of the existence of nuclear isomers is called nuclear isomerism.

Nuclear isomerism was first discovered in UX_2 and UZ nuclei. Both these nuclei are the atoms of the same element namely protoactinium and each has a mass number of 234 and atomic number 91 ($^{234}_{91}Pa$). These nuclei are formed when $UX_1(^{234}_{90}Th)$ emits two β-particles. When 99.65% of $^{234}_{90}Th$ emits a β-particle, ($^{0}_{-1}e$), UX_2 ($^{234}_{91}Pa$) is obtained and when the remaining amount (*i.e.*, 0.35%) of $^{234}_{90}Th$ emits a β-particle, UZ ($^{234}_{91}Pa$) is obtained (See Fig. 7.12).Both these nuclei have different values of their $t_{0.5}$ (UX_2 = 1.18m, UZ = 6.7h).UX_2 and UZ nuclei are called nuclear isomers. Nuclear isomers have the following characteristics:

(*i*) Nuclear isomers are in the meta-stable state

(*ii*) They have different half-lives and emit different radiations.

(*iii*) They come to the ground state by the emission of γ-rays.

(*iv*) These have the same mass number and same atomic number, *i.e.*, the nuclei of nuclear isomers are made up of the same number of protons and neutrons which differ in energy states due to variable arrangement of these particles in shell structure of the nucleus.

Nuclear isomerism has also been detected in artificial radioactive substances. Many isomeric pairs have been produced by bombarding radio-nuclies with neutrons.

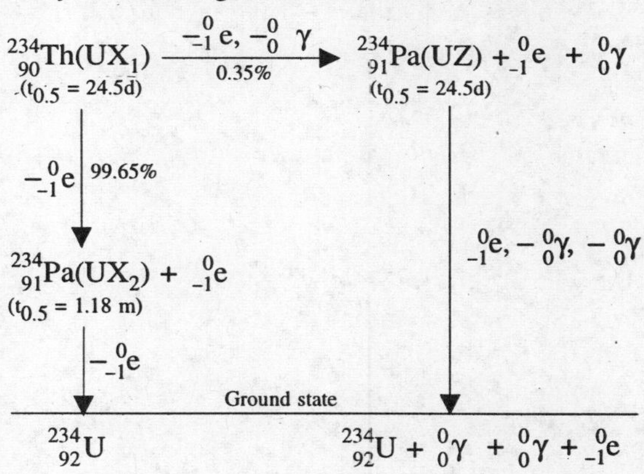

Fig. 7.12 : Formation of UZ and UX_2 isomeric pairs from UX_1

University Questions

1. What are rapid-active isotopes? Mention important applications of isotopes in the field of Medicine and Industry. *(Meerut, M.Sc. 89, 93; Behrampur, 91, Madurai 87, Agra, 92)*
2. (*a*) How isotopes are prepared? *(Madras, 87)*
   (*b*) State the principle of thermal diffusion as a technique for isotopic separation. *(Gauhati, 91; Kerela 91)*
3. (*a*) What is sotopic effect? How has it been utilised in the separation of isotopes? *(Meerut, 92)*
   (*b*) Explain, group displacement law is radio-activity. *(Gauhati, 91)*
4. Illustrate with an example the application of isotopes in (*a*) elucidation of a reaction mechanism (*b*) photo-synthesis in plants. *(Lucknow, 93; Meerut, 92)*
5. (*a*) What is the difference between isotope and isotone? Explain with examples. *(Meerut M.Sc., 89)*
   (*b*) What is nuclear isomerism? Give examples. *(Poona, 92)*
6. Why chemical methods are not used for the separation of isotopes? Give Aston's mass spectrograph method for the separation of isotopes. *(Lucknow, 93)*

7. Discuss the advantages and disadvantages of radioactive isotopes as compared with stable isotopes for the purpose of tracer chemistry and explain the tracer technique involved in medicine. *(Meerut M.Sc., 91)*

8. Write a note on:

(*i*) Isotopes *(Meerut. 89 S)*
(*ii*) Tracer technique *(Himachal 87)*
(*iii*) Separation and detection of isotopes. *(Meerut M.Sc., 93)*
(*iv*) Isobars *(Garhwal. 88)*
(*v*) Applications of radioactive isotopes. *(Meerut 90)*
(*vi*) Nuclear disintegration *(Meerut M.Sc.. 91)*
(*vii*) Isotopes and isobars. *(Gauhati 91)*
(*viii*) Isotones *(Gauhati 91)*

9. What is meant by radio-carbon dating? How it has been used to determine the age of a wooden piece? The amount of ${}^{14}_{6}C$ in a piece of antique wood piece is found to be one-sixth of a fresh piece of wood. Calculate the age of antique wooden piece. Given $t_{1/2}$ of ${}^{14}_{6}C$ = 5577 yrs. It is assurred that the radioactivity is entirely due to C-14. *(Goa, 90; Lucknow 93)*

10. A sample of U^{238} (half-life = 4.5×10^9 yrs) ore is found to contain 23.8 g of U^{238} and 20.6g of Pb^{206}. Calculate the age of the ore (log 2 = 0.30). *(Roorkee Entrance, 96)*

11. Discuss the uses of isotopes as tracers. *(Gujrat 2003)*

12. Name the isotopes of hydrogen. *(Nagpur 2003)*

8

Nuclear Chemistry : Chemistry of the Nucleus

Discovery of nucleus

Rutherford's α-particle scattering experiment showed that the atom of an element contains a positively-charged hard core at its centre. This positively-charged central part of the atom is called *nucleus.*

What is nucleus?

The conclusions drawn from the observations of Rutherford's α-particle scattering experiment show that nucleus is the central part of the atom which is positively-charged, dense and rigid. The whole mass of the atom is concentrated in it. The positive charge on the nucleus is due to the presence of protons in it.

Properties of nucleus

(i) Size/volume of the nucleus. α-particle scattering experiment performed by Rutherford has shown that the size (or volume) of the nucleus of an atom is very small as compared to that of the whole atom. Experiments on scattering of neutrons have shown that the radius (r) of the nucleus (nuclear radius) can be given by the expression:

$$r = R_0 A^{1/3}$$

Here R_0 is a constant whose value is 1.5×10^{-15} m and A = mass number of the element.

The radius of the nucleus is of the order of 10^{-14} m while that of an atom is of the order of 10^{-10} m. The volume of the nucleus is about $1/10^{12}$th part of that of the atom.

(ii) Density of the nucleus. It has been found that the density of the nuclear matter is the same in all nuclei. Its value is very high and is of the order of 2.0×10^{17} kg/m^3. Density of the nuclear matter of an atom is given by:

$$\text{Density} = \frac{\text{Nuclear mass}}{\frac{4\pi}{3} \times (\text{nuclear radius})^3} \text{ kg / m}^3$$

Obviously the nuclear mass should be in kg and nuclear radius should be in m. The density of the nucleus has been found to be 10^{12} times that of the atom.

(iii) Composition, charge and mass of the nucleus. Nucleus is composed of protons and neutrons. Protons are positively-charged while neutrons are neutral. It is due to the presence of positively-charged protons that the nucleus is positively-charged. The number of positive charges on a nucleus is equal to the number of protons in it. Since the nucleus is made up of protons and neutrons, the mass of a nucleus is the sum of the masses of all protons and neutrons in it.

In stable nuclei, the distance between protons and neutrons is about $2f - 3f$ (f indicates *fermi,* $1f = 10^{-13}$cm). The electrons present in an atom are revolving round the nucleus in different orbits. Since the number of electrons (electrons are negatively-charged) is equal to that of protons, the atom is electrically neutral.

We have said above that the nucleus of the atom of an element contains protons and neutrons. In cases where it is not necessary to differentiate between protons and neutrons, these particles are collectively c 'led *nucleons*.

(*iv*) *Emission of α, β, γ etc. particles from the nucleus.* Here we shall discuss the nature of the newly-formed element (called daughter element) when the nucleus of the atom of a radioactive element loses α, β, γ etc. particles.

(*a*) *Emission of α or β particle.* For this see "*Soddy-Fajan and Russel group displacement law* (1913)".

(*b*) *Emission of γ-rays. γ-rays* are secondary radiations. After the emission of α or β particles, the resulting nucleus becomes excited. After a definite time interval, this excited nucleus changes to a more stable state with the emission of extra energy in the form of γ-rays.

Since γ-rays are electromagnetic waves having no mass and no atomic number ($^0_0\gamma$), there is no change in the mass number and no change in the atomic number of the daughter element when the nucleus of a radioactive element emits a γ-ray. This shows that during γ-ray emission the newly-formed element has the same mass number and the same atomic number, as the original element.

$$^A_ZM \xrightarrow{-^0_0\gamma} {}^A_ZM + {}^0_0\gamma$$

(*c*) *Emission of a positron* ($_{+1}^{\ 0}e$). When the nucleus of a radioactive element emits a positron, which has zero mass (= 0.000548 *amu*) and +1 charge, the mass number of the daughter element remains the same, but its atomic number is decreased by one unit. For example.

$$^{120}_{51}Sb \xrightarrow{-^{\ 0}_{+1}e} {}^{120}_{50}Sn + {}^{\ 0}_{+1}e$$

Explanation. Anderson (1932) assumed that the nucleus has no positron and hence, during the emission of this particle, a proton (^1_1H) present in the nucleus is converted into a neutron (^1_0n) and a positron ($^{\ 0}_{+1}e$)

$$^1_1H \longrightarrow {}^1_0n + {}^{\ 0}_{+1}e$$

Due to the conversion of a proton into a neutron and positron, the daughter element has one less protons but one more neutrons than the parent element.

What is atomic number (Z)?

We know that the nucleus of the atom of an element is positively-charged, since it contains positivly-charged protons. The number of positive charges on the nucleus is equal to the number of protons in it. Chadwick showed by experiments that the number of positive charges on the nucleus of an atom is equal to the atomic number of the element. Thus atomic number of an element can be defined as the number of positive charges on the nucleus of the atom of that element.

Now since the number of positive chrges on the nucleus is equal to the number of protons in it, atomic number of an element is also equal to the number of protons residing in the nucleus.

Since the atom as whole is electrically neutral, the number of protons (positively-charged particles) in the nucleus is equal to the number of electrons (negatively-charged particles) revolving in different orbits round the nucleus. Consequently, atomic number of an element is also equal to the number of electrons revolving in orbits round the nucleus. Thuss if p = number of protons present in the nucleus, e = number of electrons revolving in orbits round the nucleus and Z = atomic number of the element, then

$$Z = p = e$$

Here it should be noted that the atomic number of an element is equal to the number of electrons only in a *neutral atom* and *not in an ion*, because only a neutral atom contains equal number of protons and electrons. On the other hand, since an ion is formed either by the removal or by the addition of one or more electrons from (to) a neutral atom, the ion has less or more electrons than protons.

When a chemical reaction takes place, the number of electrons change, but that of protons remains the same. Thus, during a chemical reaction, the atomic number of an element does not change, *i.e.*, it remains the same.

Atomic number of an element is also defined as the serial number of its position in the periodic table, starting from hydrogen as the first element.

What is mass number(A)?

We know that the nucleus of the atom of an element is composed of protons and neutrons. The sum of the number of protons (p) and neutrons (n) present in the nucleus of the atom of an element is called mass number (A) of that element, *i.e.*,

$$A = p + n \quad ...(i)$$

Since protons and neutrons are collectively called *nucleons*, the mass number (A) of an element is equal to the total number of nucleons present in the nucleus of the atom of that element. Obviously *mass number of an element is a whole number.*

Since the number of protons(p) in an atom is equal to the atomic number (Z) of the element, equation(i) can also be written as

$$A = Z + n \quad ...(ii)$$

Atomic mass (atomic weight) and its relation with mass number

Atomic mass of an element is equal to the sum of the masses of all protons and all neutrons present in the nucleus, *i.e.*,

Atomic mass or atomic weight= Mass of all protons + Mass of all neutrons

Let the number of protons and neutrons in the nucleus be p and n respectively. Now since mass of a proton = 1 *amu* and mass of a neutron = 1 *amu*, then :

$$\text{Atomic mass or atomic weight} = \text{Mass of } p \text{ protons} + \text{Mass of } n \text{ neutrons}$$
$$= p \times 1 + n \times 1 = (p + n)\ amu = \text{Mass number}$$

Thus the atomic mass (atomic weight) of an element, which has no isotopes, is equal to its mass number. Now since mass number is a whole number, the atomic mass of the elements having no isotopes is also a whole number and not a fraction.

Atomic mass (atomic weight) of an element is a figure which indicates the number of times by which one atom of that element is heavier than 1/12th (one-twelfth) part of a ^{12}C-atom which has been assigned an atomic mass of 12 units. For example the atomic mass of Mg is 24 which indicates that one atom of Mg is 24 times heavier than 1/12th part of a ^{12}C atom. Atomic mass of an element expressed in grams is called *gram atomic mass*. Thus if atomic mass of carbon is 12, then gram atomic mass of carbon will be 12 grams or 12g.

What is atomic mass unit (*amu*)?

The above discussion shows that one atomic mass unit (1 *amu*) is euqal to 1/12th of the mass of a carbon-12 atom, *i.e.*,

$$1\ a.m.u = 1/12\text{th of the mass of } ^{12}C \text{ atom.}$$

Reason for the fractional atomic mass (atomic weight) of an element having isotopes

We have said above that the atomic mass (atomic weight) of an element having no isotope is generally a whole number. However, in case of an element which has isotopes, its atomic mass (atomic weight) is generally a fractional quantity, since atomic mass of such an element is the average of the atomic masses of all the isotopes given by that element. Thus we find that the fractional atomic mass (atomic weight) of an element is due to the existence of its isotopes having different mass numbers or atomic masses.

To calculate the atomic mass (atomic weight) of an element having isotopes of different mass numbers (*isotopic mass*)

Let us consider the isotopes of chlorine as an example. This element has two isotopes *viz.* ^{35}Cl and ^{37}Cl. If the abundance of these isotopes in nature is a% and b% respectively, then

Atomic mass (or atomic weight) of Cl–atom

$$= \frac{\% \text{ of abundance of } ^{35}\text{Cl isotope} \times \text{mass number of } ^{35}\text{Cl} + \% \text{ of abundance of } ^{37}\text{Cl isotope} \times \text{mass number of } ^{37}\text{Cl}}{\text{Sum of the \% of abundance of the two isotopes}}$$

$$= \frac{a \times 35 + b \times 37}{a + b} \quad ...(i)$$

Now if $a + b = 100$, the above equation becomes :

Atomic mass (or atomic weight) of Cl-atom $= \dfrac{a \times 35 + b \times 37}{100}$

$$= \frac{a \times 35 + (100 - a) \times 37}{100} \quad ...(ii)$$

Solved Examples

Example 8.1. *Natural chlorine has two isotopes viz. ^{35}Cl and ^{37}Cl. The percentage of the abundance of these isotopes is 75% and 25% respectively. Calculate the atomic mass (atomic weight) of Cl-atom.*

Solution. On using relation (*ii*) given above we have

$$\text{Atomic mass of Cl-atom} = \frac{\% \text{ of the abundance of } ^{35}\text{Cl isotope} \times \text{mass of } ^{35}\text{Cl isotope} + \% \text{ of the abundance of } ^{37}\text{Cl isotope} \times \text{mass of } ^{37}\text{Cl isotope}}{100}$$

$$= \frac{75 \times 35 \times 25 \times 37}{100} = \mathbf{35.50}$$

Example 8.2. *The relative atomic mass of copper is 63.5. Calculate the percentage of the isotopes $^{63}_{29}Cu$ and $^{65}_{29}Cu$ present in it.*

Solution. If the percentage of the first isotope is supposed to be equal to x, then the percentage of the second isotope will be $(100 - x)$. The relative atomic mass of 63.5 will be given by

$$63.5 = \frac{x \times 63 + (100 - x) \times 65}{x + (100 - x)}$$

Solve this equation and find out the value of x [$x = 75\%$ and $(100 - x) = 25\%$]

Thus the % of the first isotope is 75 and that of the second isotope is 25.

Example 8.3. *The two isotopes of hydrogen having mass numbers 1 and 2 have atomic weights equal to 1.007825 and 2.01473 as determined by mass spectrograph. These isotopes are present in ordinary hydrogen in the ratio of 6400 to 1. Determine the mean-atomic weight of H-atom.*

Solution. The mean atomic weight of H-atom

$$= \frac{6400 \times 1.007825 + 1 \times 2.01473}{6400 + 1} = \mathbf{1.0080}$$

Example 8.4. *Naturally occuring boron consists of two isotopes whose atomic weights are 10.01 and 11.01. The atomic weight of boron is 10.181 Calculated the % of each isotope in natural boron.* **(I.I.T. 1982)**

Solution. Atomic weight of boron

$$= \frac{x \times 10.01 + (100 - x) \times 11.01}{x + (100 - x)}$$

$$\text{or} \quad 10.81 = \frac{x \times 10.01 + (100 - x) \times 11.01}{100} \qquad \therefore \; x = 20$$

Thus the % of the first isotope = 20 and that of the other isotope = (100 – 20) = 80.

Example 8.5. *Naturally occuring carbon consists of two isotopes namely ^{12}C and ^{13}C. What are the percentage abundance of these isotopes in a sample of carbon whose atomic weight is 12.01112. Assume that the nuclidic masses of ^{12}C and ^{13}C are 12.0 and 13.00334 amu respectively.* **(B.I.T. 1989).**

Solution. Suppose that the % abundance of ^{12}C isotope = x

∴ % abundance of ^{13}C isotope = $(100 - x)$

Now we know that

Atomic weight of carbon

$$= \frac{\%\text{ abundance of } ^{12}C \times \text{mass of } ^{12}C + \%\text{ abundance of } ^{13}C \times \text{mass of } ^{13}C}{100}$$

$$12.01112 = \frac{x \times 12.0 + (100 - x) \times 13.00334}{100}$$

$$\therefore \quad x = 98.89\%$$

Thus % of ^{12}C = **98.89%** and % ^{13}C = (100 – 98.89)% = **1.11%**

Example 8.6. *A sample of ordinary neon is found to consist of ^{20}Ne, ^{21}Ne and ^{22}Ne isotopes in the percentage 90.92%, 0.26% and 8.82% respectively. Calculate the actual atomic weight of neon.*

Solution. Actual atomic weight of neon

$$= \frac{\%\text{ of } ^{20}Ne \times \text{mass number of } ^{20}Ne + \%\text{ of } ^{21}Ne \times \text{mass number of } ^{21}Ne + \%\text{ of } ^{22}Ne \times \text{mass number of } ^{22}Ne}{100}$$

$$= \frac{90.22 \times 20 + 0.26 \times 21 + 8.82 \times 22}{100} = \mathbf{20.18}$$

Example 8.7. *^{14}N and ^{15}N present in ordinary nitrogen are in the ratio $^{14}N/^{15}N$ = 272. Calculate the atomic weight of nitrogen.*

Solution. Atomic weight of nitrogen

$$= \frac{14 \times 272 + 15 \times 1}{272 + 1} = \mathbf{14}$$

It should be noted that although the atomic masses of a large number of elements are fractional, yet they are generally written as the nearest whole number for the sake of convenience. For example, though the exact atomic mass of hydrogen is 1.008, it is written as 1. Similarly though the exact atomic mass of oxygen is 15.9994, it is written as 16 to simplify the calculations.

Forces which operate inside the nucleus of an atom

We know that the nucleus of an atom contains protons (positively-charged particles) and neutrons (neutral particles) which are collectively called *nucleons* or *nuclear particles.* There are two types of forces which operate inside the nucleus of an atom. These are *electrostatic force* and *nuclear force* (or *strong force*).

(*i*) *Electrostatic force (Repulsive force).* The force acting between two charged particles is called electrostatic force depending on whether the two charged particles have the same or different charges, this force may be repulsive electrostatic force or attractive electrostatic force. Since the nucleus of an atom contains similarly-charged (positively-charged) protons, the electrostatic force acting between them is *repulsive electrostatic force.* This force causes repulsion between the various protons present in the nucleus and hence makes the nucleus unstable. The magnitudes of this force is inversely proportional to the distance between the protons. Thus, this force becomes large in magnitude (*i.e.*, more effective or strong), when the distance between the positively-charged protons is small.

(*ii*) *Nuclear force (Attractive force).* Nuclear force is that which, without considering the nature of the charges on the protons and neutrons, causes attraction between protons and protons, between protons and neutrons and between neutrons and neutrons. Thus the nuclear force is always an attractive force and hence is also called *nuclear force of attraction* or *attractive nuclear force.* Due to its attractive nature, it is a strong force and holds together the nucleons in the nucleus, and thus makes the nucleus stable. This force becomes large in magnitude (*i.e.*, effective or strong) when the various protons and neutrons (nuclear particles) in a nucleus are very close to each other. Nuclear forces fall rapidly to zero at distances greater than 1.4×10^{-13}cm. and are referred to as *short range forces.* These forces donot obey the inverse square law.

A problem. Let us explain why small nuclei (light nuclei) are stable while big nuclei (heavy nuclei) are unstable.

We have said above that there are two types of forces which operate in the nucleus of an atom. These are electrostatic force of repulsion and nuclear force of attraction. The former makes the nucleus unstable while the latter makes the nucleus stable.

If an atom of an element is small (*i.e.*, the element has small mass number), its nucleus will also be small. Due to small size of the nucleus, the various protons and neutrons in it are very very close to each other and hence the nuclear force of attraction between the nuclear particles is large. Also, at the same time, due to the small distance between the various protons, the electrostatic force of repulsion existing between positively-charged protons is also large. But, since the nuclear force of attraction is much greater than the electrostatic force of repulsion, small nucleus attains stability. Thus, we see that, inspite of the existence of electrostatic force of repulsion between the positively-charged protons, nuclei of small atoms are stable, because the nuclear force of attraction existing between various nuclear particles is greater in magnitude than the electrostatic force of repulsion.

If an atom is big. (*i.e.*, the element has a large mass number like $^{235}_{92}U$), its neucleus would also be big and hence the distance between various nuclear particles in it would be comparatively large. Due to the large distance between the various nuclear particles, the nuclear force of attraction between these particles would decrease (*i.e.*, would become weak). At the same time, due to the large distance between the protons the electrosatic force of repulsion existing between protons would also decrease. But it has been found that the magnitude of decrease in nuclear force of attraction is more than the magnitude of decrease in electrostatic force of repulsion and hence, in big nuclei, the electrostatic force of repulsion becomes slightly greater than the nuclear force of attraction, *i.e.*, the electrostatic force of repulsion dominates. This results in that the nucleus becomes unstable and hence disintegrates by emitting certain particles. ^{235}U is an example of big nucleus in which the electrostatic force of repulsion is greater than the nuclear force of attraction and hence ^{235}U nucleus is very unstable due to which it can easily be fissioned (broken) into smaller nuclei. The fission of ^{235}U nucleus can be done by means of slow moving neutrons (neutrons having low energy).

$$\underset{\substack{\text{Big nucleus}\\\text{(Heavy nucleus)}}}{^{235}_{92}U} + \underset{\substack{\text{Slow moving}\\\text{neutrons}}}{^{1}_{0}n} \xrightarrow{\text{Fission}} \underbrace{^{139}_{56}Ba + {}^{94}_{36}Kr}_{\substack{\text{Smaller nuclei}\\\text{(Medium weight nuclei)}}} + 3^{1}_{0}n + \text{Energy}$$

How are nuclear forces produced in the nucleus of an atom : Meson exchange theory of the origin of nuclear forces

In 1939 Hideki Yukawa, a Japenese physicist put forward meson exchange theory to explain the origin of nuclear forces which operate in the nucleus of an atom and keep the nucleons bound together. According to this theory the protons ($_{+1}p$) and neutorns ($_{o}n$) present in the nucleus are surrounded by positively-charged pi mesons ($_{+1}\pi$ mesons) and negatively-charged pi mesons ($_{-1}\pi$ mesons).

Positive pi mesons interact with the neutrons and convert them into protons

$$\text{Neutron } (_{0}n) \quad + \quad \text{Positive } \pi \text{ meson } (_{+1}\pi) \quad \rightarrow \quad \text{Proton } (_{+1}p) \qquad ...(i)$$

Similarly negative pi mesons interact with the protons and convert them into neutrons.

$$\text{Proton } (_{+1}p) \quad + \quad \text{Negative } \pi \text{ meson } (_{-1}\pi) \quad \rightarrow \quad \text{Neutron } (_{0}n) \qquad ...(ii)$$

This type of interaction takes place, for a part of time. The interactions shown by equations (*i*) and (*ii*) can be combined together and shown by a single equation *viz.* equation (*iii*).

$$_{0}n \underset{+\,_{-1}\pi}{\overset{+\,_{+1}\pi}{\rightleftharpoons}} {}_{+1}p \qquad ...(iii)$$

The interaction shown by equation (*iii*) indicates that $_{+1}\pi$ and $_{-1}\pi$ mesons oscillate between the neighbouring protons and neutrons (*i.e.*, nucleons) present in the nucleus, *i.e.*, there is a continuous exchange of $_{+1}\pi$ and $_{-1}\pi$ mesons (this exchange is of back and forth type exchange) between the neighbouring protons and neutrons. This type of exchange of mesons between the two neighbouring protons and neutrons produces attraction between these particles and thus keeps them bound together in the nucleus.

How the exchange of a $_{+1}\pi$ meson or a $_{-1}\pi$ meson between a neutron and a proton produces attraction between them can be understood by a simple example. Consider two dogs who are trying to snatch a piece of bone from each other. In this attempt, the bone is rapidly exchanged between the two dogs. Since neither of the dogs is able to get the piece, the latter goes on oscillating continuously between them and thus keeps the two dogs *bound together*.

Modern researches have shown that in addition to the exchange of $_{-1}\pi$ and $_{+1}\pi$ mesons between a neutron and a proton as shwon by equation (*iii*), there is also an exchange of neutral pi mesons ($_0\pi$) between two protons and between two neutrons.

$$_{+1}p \underset{+\,_0\pi}{\overset{+\,_0\pi}{\rightleftharpoons}} {}_{+1}p\,, \qquad {}_0n \underset{+\,_0\pi}{\overset{+\,_0\pi}{\rightleftharpoons}} {}_0n$$

Nuclear stability

It has been observed that some isotopes (*i.e.* nuclides or nuclei) are stable while others are unstable (*i.e.*, radioactive). Stable nuclides are not able to undergo spontaneous disintegration. This is due to the stability of their nuclei. The stability of the nucleus of an atom is called nuclear stability. Radioactive nuclei undergo spontaneous disintegration.

Factor affecting the nuclear stability

Following are some important factors on which the nuclear stability depends, *i.e.*, with the help of these factors we can predict whether a given nuclide is stable or radioactive.

(*i*) Even and odd number of protons (*p*) and neutrons (*n*)

(*ii*) Neutron - to - proton ratio (*n/p* ratio) ; (*iii*) Packing fraction (*f*)

(*iv*) Binding energy per nucleon ($\overline{B}$) ; (*v*) Nuclear shell model : Magic numbers.

All these factors are discussed below:

Even and odd number of protons (*p*) and neutrons (*n*)

At present, more than 900 nuclides are known. Out of these only 272 nuclides are stable while others are radioactive. On the basis of even and odd number of protons and neutrons, these 272 nuclides have been grouped into four classes shown in Table 8.1. This table shows that the number of even-*p*-odd-*n* and odd-*p*-even-*n* type nuclei are nearly the same in number. The nuclei of odd-*p*-odd-*n* type are only four and hence are not found in nature. These nuclei are limited only to light nuclei. The maximum number of stable nuclei are of even-*p*-even-*n* type. 85% of the stable isotopes found in earth's crust are of even-*p*-even-*n* type, while the remaining (15%) isotopes are of odd-*p*-even - *n* type. The presence of even –*p* even –*n* type isotopes in the maximum percentage in the earth's crust shows that the stable nuclei have a tendency to from *p-p* and *n-n* pairs.

Table 8.1 : *Classification of stable nuclides (isotopes) on the basis of even and odd number of protons (p) and neutrons (n)*

Nuclide type	No of protons (*p*)	No. of neutrons (*n*)	Mass number (A) = (*p* + *n*)	No. of stable nuclides or isotopes	Examples
i) Even-*p*-even-*n* nuclides	Even	Even	Even	160	$^{4}_{2}He$, $^{24}_{12}Mg$, $^{208}_{82}Pb$, $^{16}_{8}O$, $^{28}_{14}Si$, $^{56}_{26}Fe$, $^{40}_{20}$ Ca etc.
ii) Even-*p*-odd-*n* nuclides	Even	Odd	Odd	56	$^{17}_{8}O$, $^{25}_{12}Mg$, $^{57}_{26}Fe$ etc.
iii) Odd-*p*-even-*n* nuclides	Odd	Even	Odd	52	$^{7}_{3}Li$, $^{19}_{9}F$, $^{63}_{29}Cu$, $^{27}_{13}Al$, $^{23}_{11}Na$, $^{39}_{19}K$, $^{31}_{15}P$ etc.
iv) Odd-*p*-odd-*n*- nuclides	Odd	Odd	Even	4	$^{2}_{1}H$, $^{6}_{3}Li$, $^{10}_{5}B$, $^{14}_{7}N$ only.
Total number of nuclides = 272					

Neutron-to-proton ratio (*n/p* ratio)

Excepting light elements and ordinary hydrogen ($^{1}_{1}H$), all other elements contain both neutrons and protons. Table 8.2 contains *n/p* ratio values for some stable nuclides. This table shows that *n/p* value for these stable nuclides is equal to 1 or greater than 1 ($n/p > 1$). The value of *n/p* ratio for light stable nuclides up to $^{40}_{20}Ca_{20}$ is equal to 1 and for other nuclides (*i.e.*, heavy nuclides) *n/p* ratio value is greater than 1 ($n/p > 1$).

A graph has been drawn between the number of neutrons (*n*) and number of protons (*p*) present in the stable nuclides (or nuclei) of various elements. This graph is shown in Fig. 8.1 and is called *Serge chart.*

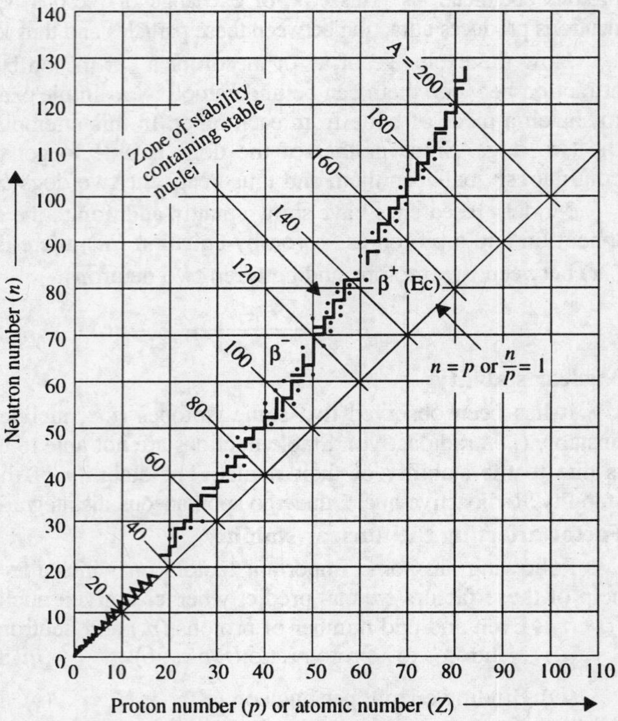

Fig. 8.1: *Graph (called Serge chart) between the number of protons (p) and the number of neutrons (n) in stable nuclei.*

The line shown in the chart is an imaginary line at which $n = p$, *i.e.*, $n/p = 1$. The strip or the zone shown in the figure is called *belt* (or *zone*) of stability, since most of the stable nuclei (*i.e.*, non-radioactive nuclei) lie within this zone. This strip goes on widening as the number of protons (or atomic number of the elements) increases. The nuclei whose *n/p* value lies above or below this belt are radioactive and hence spontaneously disintegrate to give stable nuclei. The following three cases may be studied.

Table 8.2 : *n/p ratio values for some stable nuclides.*

Nuclides :	$^{2}_{1}H_1$	$^{20}_{10}Ne_{10}$	$^{40}_{20}Ca_{20}$	$^{64}_{30}Zn_{34}$	$^{90}_{40}Zr_{50}$	$^{120}_{50}Sn_{70}$	$^{150}_{60}Nd_{90}$	$^{202}_{80}Hg_{122}$
Z or p =	1	10	20	30	40	50	60	80
n =	1	10	20	34	50	70	90	122
n/p ratio =	1.00	1.00	1.00	1.13	1.25	1.40	1.50	1.53
	← Light nuclides → (n/p = 1.00)			← Heavy nuclides → (n/p > 1)				

I. The nuclei lying *above* the belt of stability are richer in neutrons and hence they disintegrate in such manner that one of their neutrons is converted into a proton [$^{1}_{0}n \rightarrow {}^{1}_{1}H$ or ${}^{1}_{1}p + {}^{0}_{-1}e$ (β-particle)], *i.e.*, such nuclei emit a β-particle ($^{0}_{-1}e$), since it is this decay process in which a neutron is converted into a proton. After emitting a β-particle the newly-formed nuclide has its *n/p* value lower than its parent nuclide and hence approaches closer to the zone of stability and becomes stable. The following examples illustrate this point.

(*a*) $^{24}_{11}Na_{13} \xrightarrow{-\,^{0}_{-1}e} \,^{24}_{12}Mg_{12} + \,^{0}_{-1}e$

Radioactive Stable

or $11p + 13n \longrightarrow 11p + 12n + (1p + \,^{0}_{-1}e)$

or $11p + 13n \longrightarrow 12p + 12n + \,^{0}_{-1}e$

$n/p = 13/11 = 1.18 \longrightarrow n/p = 12/12 = 1.00$

(*b*) $^{14}_{6}C_8 \xrightarrow{-\,^{0}_{-1}e} \,^{14}_{7}N_7 + \,^{0}_{-1}e$ ($\overline{\beta}$-particle)

Radioactive Stable

or $6p + 8n \longrightarrow 6p + 7n + (1p + \,^{0}_{-1}e)$

or $6p + 8n \longrightarrow 7p + 7n + \,^{0}_{-1}e$

$n/p = 8/6 = 1.33 \longrightarrow n/p = 1/1 = 1.00$

(*c*) $^{23}_{10}Ne_{13} \xrightarrow{-\,^{0}_{-1}e} \,^{23}_{11}Na_{12} + \,^{0}_{-1}e$

Radioactive Stable

or $10p + 13n \longrightarrow 10p + 12n + (1p + \,^{0}_{-1}e)$

or $10p + 13n \longrightarrow 11p + 12n + \,^{0}_{-1}e$

$n/p = 13/10 = 1.30 \longrightarrow n/p = 12/11 = 1.09.$

II. The nuclei lying *below* the zone of stability are deficient in neutrons and hence disintegrate in such a way that one of their protons is converted into a neutron. The conversion of a proton into a neutron can be done by any of the following two ways:

(*a*) *Emission of a positron* : $^{1}_{1}H \rightarrow \,^{1}_{0}n + \,^{0}_{+1}e$ (*positron*)

(*b*) *Electron-capture process* : $^{1}_{1}H + \,^{0}_{-1}e$ (*electron*) $\rightarrow \,^{1}_{0}n$

Thus the nuclei lying below the belt of stability disintegrate in such a way that they either emit a positron or undergo electron-capture process.

In examples (*i*) and (*ii*) given below a proton is converted into a neutron and a positron is emitted while in examples (*iii*), (*iv*) and (*v*) a proton is converted into a neutron by capturing an electron.

(*i*) $^{23}_{12}Mg_{11} \xrightarrow{-\,^{0}_{+1}e} \,^{23}_{11}Na_{12} + \,^{0}_{-1}e$ (*positron*)

Radioactive Stable

$12p + 11n \longrightarrow (11p + 1n + \,^{0}_{+1}e) + 11n$

or $12p + 11n \longrightarrow 11p + 12n + \,^{0}_{+1}e$

$n/p = 11/12 = 0.91 \longrightarrow n/p = 12/11 = 1.09$

(*ii*) $^{22}_{11}Na_{11} \xrightarrow{-\,^{0}_{+1}e} \,^{22}_{10}Ne_{12} + \,^{0}_{+1}e$ (*positron*)

Radioactive Stable

$11p + 11n \longrightarrow (10p + 1n + \,^{0}_{+1}e) + 11n$

or $11p + 11n \longrightarrow 10p + 12n + \,^{0}_{+1}e$

$n/p = 11/11 = 1.00 \longrightarrow n/p = 12/10 = 1.20$

(*iii*) $^{55}_{26}Fe_{29} \xrightarrow{-\,^{0}_{+1}e} \,^{55}_{25}Mn_{30} + \,^{0}_{0}\nu$

$26p + 29n \longrightarrow (25p + n) + 29n$

or $26p + 29n \longrightarrow 25p + 30n$

$n/p = 29/26 = 1.11 \longrightarrow n/p = 30/25 = 1.20$

$$(iv)\ {}^{82}_{37}Rb_{45} \xrightarrow{+\ {}^{0}_{-1}e} {}^{82}_{36}Kr_{46}$$

$n/p = 45/37 = 1.21 \longrightarrow n/p = 46/36 = 1.27$

$$(v)\ {}^{194}_{79}Au_{115} \xrightarrow{+\ {}^{0}_{-1}e} {}^{194}_{78}Pt_{116}$$

$n/p = 115/79 = 1.45 \longrightarrow n/p = 116/78 = 1.48$

After emitting a positron or capturing an electron, the newly-formed element has higher value of *n/p* ratio and comes closer to the zone of stability. Thus the newly-formed element is more stable.

III. ${}^{208}_{82}Pb_{126}$ and ${}^{209}_{83}Bi_{126}$ are the heaviest stable nuclei. Other nuclei having higher number of protons (*i.e.,* atomic number) or neutrons disintegrate by α (${}^{4}_{2}He$), ${}^{0}_{+1}e$ or ${}^{0}_{-1}e$ decay or by fission process.

The above discussion shows that an optimum value of *n/p* ratio also determines the stability of a given nuclide.

Packing Fraction (*f*)

Mass spectrograph has enabled us to determine the extact value of the mass of a given isotope (called *isotopic mass* or *actual atomic mass* of the isotope). It has been found that the isotopic mass as determined by the spectrograph is not the same as the *mass number* of the isotope which is a whole number, because it is the sum of the protons and neutrons. When the difference namely (isotopic mass – mass number) × 10^4 is divided by mass number, we get what is known as packing fraction (*f*) of the isotope. Thus

$$\text{Packing fraction } (f) = \frac{\text{Isotopic mass} - \text{Mass number}}{\text{Mass number}} \times 10^4 = \frac{M - A}{A} \times 10^4$$

Since the mass number is the sum of the protons (*p*) and neutrons (*n*) present in the nucleus of the isotope (called *nucleons*), the above equation can also be written as

$$\text{Packing fraction } (f) = \frac{\text{Isotopic mass} - \text{Mass number}}{\text{Nucleons}} \times 10^4$$

$$\text{or} \quad f = \frac{M - A}{n + p} \times 10^4$$

Relation between the stability of a nucleus and its packing fraction value

The stability of a nuclide depends on the value of its packing fraction as shown below:

(*i*) If the value of packing fraction of a nuclide is a *negative quantity* , then the nuclide would be stable. For example, since the values of packing fraction of ${}^{16}O$ and ${}^{40}Ar$ nuclides are negative (${}^{16}O$ = – 3.6875, ${}^{40}Ar$ = – 9.404), both these nuclei are *stable.* Negative value of packing fraction is obtained when M < A.

(*ii*) If a nuclide has a *low positive value* for its packing fraction, then the nuclide would be *stable.* For example, since ${}^{12}C$ and ${}^{4}He$ both have low positive values for their packing fraction (${}^{12}C$ = +3.1666, ${}^{4}He$ = + 6.5), both these nuclides are stable. The nuclei having *very high positive values* of packing fraction are *unstable* and hence such nuclei undergo spontaneous disintegration. Positive value of *f* is obtained when M > A.

Variation of packing fraction with mass number and relation between the packing fraction and nuclear stability

The plot shown in Fig. 8.2 is a graph of packing fraction against the mass number of various isotopes. This graph shows how the magnitude of packing fraction changes with the change of mass number.

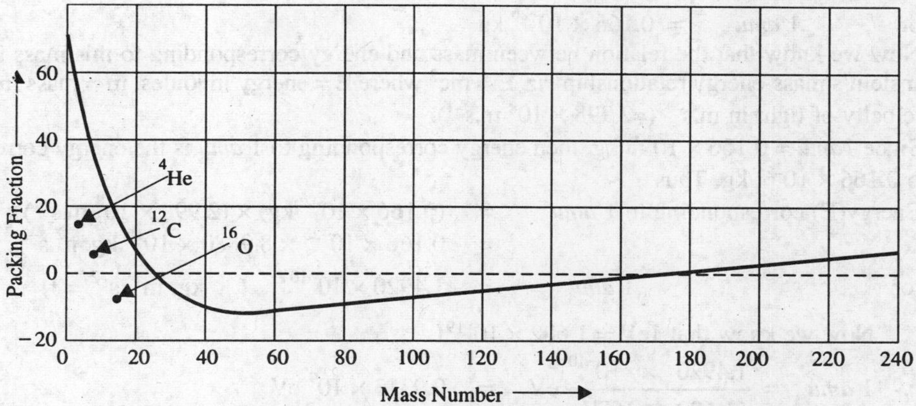

Fig. 8.2 : *Variation of packing fraction of elements with their mass number.*

From this graph the following points may be noted:

(*i*) $^{12}_{6}C$ and $^{4}_{2}He$ have low *positive* values for their packing fraction ($^{12}_{6}C$ = +3.1666, $^{4}_{2}He$ = + 6.5) and hence these are stable nuclei.

(*ii*) ^{16}O has negative value of its packing fraction (= –3.6875) and hence its nucleus is stable.

(*iii*) The transition elements like Mo, Tc, Ru, Rh, Pd, Ag which have mass numbers in the neighbourhood of 45 have the lowest negative values for their packing fraction and hence the nuclei of these isotopes are highly stable.

(*iv*) For the elements having mass numbers more than 190, the value of packing fraction is positive (low value) and becomes more positive as the mass number increases. This indicates that the stability of these isotopes goes on decreasing with the increase in their mass numbers.

(*vi*) Elements having mass numbers more than 230 have very high positive values for their packing fraction and hence are highly unstable, *i.e.*, such elements are radioactive and hence undergo disintegration spontaneously.

Problem : *Calculate the packing fraction of argon isotope, $^{40}_{18}Ar$. Mass of the isotope of argon = 39.962384 amu. What does the value of packing fraction imply?*

Solution. We know that the packing fraction of $^{48}_{18}Ar$ isotope (*f*)

$$= \frac{\text{Mass of } ^{48}_{18}Ar - \text{Mass number of } ^{40}_{18}Ar}{\text{Mass number of } ^{40}_{18}Ar} \times 10^4$$

$$= \frac{39.962384 - 40}{40} \times 10^4 = -9.404$$

Negative value of *f* indicates that $^{40}_{18}Ar$ is a stable isotope.

Calculation of energy (in MeV) which is equivalent to 1 *amu*

We know that

1 atomic mass unit (1 *amu*) = 1/12*th* of the mass of a $^{12}_{6}C$ atom

Now since mass of a $^{12}_{6}C$ atom $= \frac{12}{\text{Avogadro's number}}$ g $= \frac{12}{6.022 \times 10^{23}}$ g = 12 *amu*

Thus since 12 *amu* $= \frac{12}{6.022 \times 10^{23}}$ g,

or 1 *amu* $= \frac{1}{12} \times \frac{12}{6.022 \times 10^{23}} g = \frac{1}{6.022 \times 10^{23}}$ g

$$= 0.166 \times 10^{-23} \text{ g} = \frac{0.166 \times 10^{-23}}{10^3} \text{ kg}$$

or $\quad 1\ amu = 0.166 \times 10^{-26}$ kg.

Now we know that the relation between mass and energy corresponding to this mass is given by Einstein's mass-energy relationship *viz.* $E = mc^2$ where E = energy in joules, m = mass in kg and c = velocity of light in $m.s^{-1}$ ($=2.998 \times 10^8\ m.s^{-1}$).

Since 1*amu* = 0.166×10^{-26} kg, then energy corresponding to 1 *amu* is the energy corresponding to 0.166×10^{-26} kg. Thus

$$\text{Energy(E) corresponding to 1 } amu = (0.166 \times 10^{-26}\text{kg}) \times (2.998 \times 10^8\ \text{m.s}^{-1})^2$$
$$= 0.166 \times 10^{-26} \times 8.9880 \times 10^{16}\ \text{kg.m}^2.\text{s}^{-2}$$

or
$$1\ amu = 1.4920 \times 10^{-10}\text{J} \quad (\because \text{kg. m}^2.\ \text{s}^{-2} = \text{J})$$

Now we know that 1eV = 1.602×10^{-19}J

$$\therefore\ 1\ amu = \frac{1.4920 \times 10^{-10}}{1.602 \times 10^{-19}}\ \text{eV} = 0.9313 \times 10^9\ \text{eV}$$
$$= \frac{0.9313 \times 10^9}{10^6}\ \text{MeV} \quad (\because 1\ \text{MeV} = 10^6\ \text{eV})$$
$$= 0.9313 \times 10^3\ \text{MeV} = 931.3\ \text{MeV}$$

i.e., 1 *amu* = 931.3 MeV

For the sake of convenience, in our calculations we shall be using the relation:

$$1 amu = 931.50\ \text{MeV}$$

Mass Defect (ΔM)

Since the nucleus of an atom contains protons and neutrons and the mass of electrons (which are present in different shells outside the nucleus) is negligible, almost the entire mass of the atom is concentrated in the nucleus. If the mass of the electrons is also considered, then it has been found that the actual mass (M) of an atom as determined by mass spectrograph (*i.e.,* atomic mass of the isotope or atom) is always less than the sum of the masses (M′) of all protons, all neutrons and all electrons present in the atom *i.e.,* M < M′. The difference in masses *viz* (M′ – M) is called *mass defect, mass loss, packing loss, mass deficit* or *mass decrement* (ΔM). Thus:

Mass defect = (Masses of all protons + masses of all neutrons + masses of all electrons) – Actual atomic mass of the atom

or $\quad \Delta M = M' - M.$

When nucleons (*i.e.,* neutrons and protons) combine to form the nucleus of an atom, mass equal to (M′ – M) is lost.

Let us calculate the mass defect for an isotope namely ^A_ZX in which A is its mass number and Z is its atomic number (nuclear charge). We know that since Z = *p* = *e*, A = *n* + *p* and *n* = A – Z, the atom of the given isotope contains Z protons, Z electrons and (A – Z) neutrons. Now if the mass of one proton, one electron and one neutron is m_p, m_e and m_n respectively, then

Sum of the masses of Z protons, Z electrons and (A – Z) neutrons (M′) = $Z.m_p + Z.m_e + (A - Z)\ m_n$

or
$$M' = Z(m_p + m_e) + (A - Z)m_n \quad ...(i)$$

We know that $(m_p + m_e)$is the sum of the mass of one proton and mass of one electron. Again we know that since, ^1_1H has one proton and one electron but has no neutron, the mass of H-atom is the sum of mass of one proton and mass of one electron, *i.e.,*

$$\text{Mass of H-atom } (m_H) = (m_p + m_e) \quad ...(ii)$$

With the help of equation (*ii*), equation (*i*) reduces to:

$$M' = Z.m_H + (A - Z)m_n \quad ...(iii)$$

Now if the actual atomic mass of ^A_ZX isotope as determined by mass spectrograph in M, then mass defect (Δ M) of this isotope will be given by:

Mass defect (ΔM) = $M' - M$

or $\Delta M = [Z.m_p + Z.m_e + (A - Z)m_n] - M$ or $\Delta M = [Z(m_p + m_e) - (A - Z)m_n] - M$

or $\Delta M = [Z.m_H - (A - Z)m_n] - M$ ...(*iii*)

Some important values and relations

(*i*) Mass of one proton (m_p) = 1.007277 *amu*

(*ii*) Mass of one neutron (m_n) = 1.008665 *amu*

(*iii*) Mass of one electron (m_e) = 0.0005486 *amu*

(*iv*) Mass of a helium nucleus, ^4_2He = 4.0026 *amu*

(*v*) Mass of a deuterium nucleus, ^2_1H = 2.01471 *amu*

(*vi*) Mass of one H-atom ^1_1H (m_H) = Mass of one proton + Mass of one electron

= 1.0072770 + 0.0005486 = 1.0078256 *amu*

(*vii*) Avogadro's number = 6.022×10^{23}

(*viii*) 1 *amu* = 0.166×10^{-26} kg

(*ix*) velocity of light (c) = 3.0×10^{10} cm.s^{-1} = $\frac{3.0 \times 10^{10}}{100}$ m.s^{-1} = 3.0×10^8 m.s^{-1}

(*x*) J = kg.m^2.s^{-2}

(*xi*) 1 *amu* = 1.4920×10^{-10} J = 931.5 MeV

(xii) 1 eV = 10^{-6} MeV = 23.06 Kcal = 1.602×10^{-19} J = 1.6021×10^{-12} erg

(*xiii*) 1 MeV = 1.602×10^{-16} kJ

Example 8.8. *Calculate the mass defect of oxygen atom, $^{16}_8O$ which has a mass of 15.994910 amu. Given that mass of a neutron = 1.008665 amu, mass of a proton = 1.007277 amu and mass of an electron = 0.0005486 amu.*

Solution. $^{16}_8O$ contains 8 protons, 8 neutrons and 8 electrons. Thus

mass of 8 protons = 8×1.007277 = 8.058216 *amu*

mass of 8 neutrons = 8×1.008665 = 8.069320 *amu*

mass of 8 electrons = 8×0.0005486 = 0.004388 *amu*

∴ M′ = 8.058216 + 8.069320 + 0.004388 = 16.131924 *amu*

M = 15. 994910 *amu* (given)

∴ Mass defect (ΔM) = 16.131924 – 15.994910 = ***0.137014 amu***

Example 8.9. *Calculate the mass defect in helium atom, ^4_2He which has a mass of 4.00260 amu. Given that mass of a neutron 1.008665 amu and mass of one H-atom = 1.007825 amu.*

Solution. ^4_2He contains 2 protons, 2 electrons and 2 neutrons.

M′ = mass of 2 protons + mass of 2 electrons + mass of 2 neutrons

= 2 × (mass of one proton + mass of one electron) + mass of 2 neutrons

= 2 × mass of a H-atom + mass of 2 neutrons

= (2 × 1.007825 + 2 × 1.008665) *amu* = (2.01560 + 2.017330) *amu* = 4.032980 *amu*

M = 4.002600 *amu* (given)

∴ ΔM = (4.032980 – 4.002600) *amu* = **0.030300 amu**

Example 8.10. *An element with mass number 15 and isotopic mass 15.00486 amu has mass defect of 0.124043 amu. If mass of a proton and a neutorn is 1.008145 amu and 1.008986 amu respectively, find out the atomic number of the element. Mass of the electron can be neglected.*

Solution. Let the element be represented as $^{15}_aX$. Obviously the nucleus of the atom of this element has *a* protons and (15 – *a*) neutrons

Now, M′ = Mass of *a* protons + Mass of (15 – *a*) neutrons

= [*a* × 1.008145 + (15 – *a*) × 1.008986] *amu*

M = 15.00486 *amu* (given)

$\therefore$ Mass of defect (ΔM) = M′ – M

or $0.124043 = 1.008145\,a + 1.008986 \times (15 - a) - 15.00486$

On solving this equation we get $a = 7$

Binding energy of a nucleus or neclear binding energy (B) and its calculation

We have said above (see *Mass Defect*) that when necleons (*i.e.* neutrons and protons) combine together to form the necleus of an atom, some mass is lost. The mass lost is called mass defect (ΔM). The mass is lost in the process of combination of nucleons to form the nucleus, since this mass is converted into energy, *i.e.*, when protons and neutrons combine together to form the nucleus, mass lost is transformed into energy which is released in the process. The energy released in this way is called **binding energy** of the nucleus or **nuclear binding energy (B).** Thus the energy released in the process of combination of protons and neutrons to form the nucleus is called binding energy of the nucleus.

It should be evident from the above discussion that when a nucleus is broken up into its constituents neutrons and protons, energy is needed and this energy is equal to the boinding energy of the nucleus.

If the mass lost in the formation of a nucleus by the combination of neutrons and protons is ΔM *amu (i.e., if the mass defect is* ΔM *amu), then the energy into which this lost mass is converted is equal to* (ΔM × 931.5)MeV, since, as we have already shown that 1 *amu* = 931.5 MeV.

Thus, we find that when the mass defect (*i.e.*, the mass lost) is equal to ΔM *amu*, then the energy released in the formation of a nucleus by the combination of neutrons and protons is equal to ΔM × 931.5 MeV. This is the binding energy (B) of the nucleus, *i.e.*,

Binding energy of the nucleus (B) = *Mass defect,* ΔM (in *amu*) × *931.5 MeV*

The above relation shows that if the mass defect is 1 *amu.* then binding energy is equal to 1 × 931.5 MeV = 931.5 MeV.

Variation of the nuclear binding energy with mass number and its relation with the nuclear stability

The study of the variation of the nuclear binding energy with mass numbers of isotopes can be made from the plot shown in Fig. 8.3.

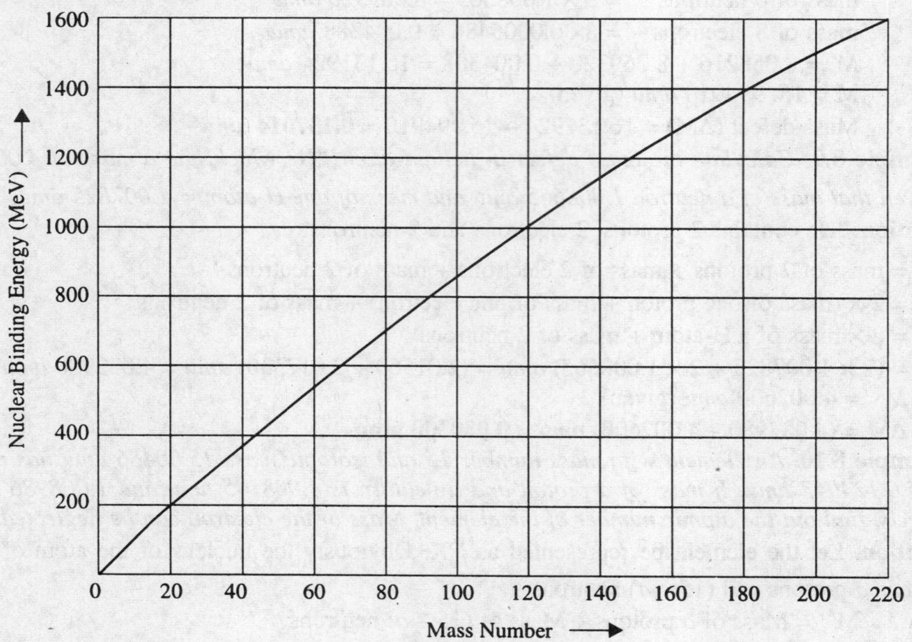

Fig. 8.3 : Variation of nuclear binding energy (in MeV) with mass number.

This plot is a graph between the nuclear binding energy (in MeV) of a number of isotopes and their corresponding mass number. This graph shows that as the mass number of the isotopes increases, the magnitude of the nuclear binding energy also increases. However, for the isotopes having high mass numbers, the increase in the magnitude of nuclear binding energy is small.

In case of two or more isotopes having the *same* mass numbers, the nucleus of the isotope having higher value of B is more stable than the nucleus of the isotope having lower value of B.

The stability of the nuclei of the isotopes having *different* mass numbers is generally compared with the help of the value of their binding energies per nucleon ($\overline{B}$).

Binding energy per nucleon ($\overline{B}$)

As the name suggests, the binding energy of a nucleus (B) divided by the sum of the protons (*p*) and neutrons (*n*) (*i.e.*, nucleons) present in the nucleus is called binding energy per nucleon ($\overline{B}$). Thus $\overline{B}$ is given by

$$\text{Binding energy per nucleon } (\overline{B}) = \frac{\text{Binding energy of the nucleons (B)}}{\text{No. of nucleons in the nucleus}}$$

or

$$\overline{B} = \frac{\text{Mass defect } (\Delta M) \times 931.5}{\text{No. of nucleons in the nucleus}} \text{ MeV}$$

or

$$\overline{B} = \frac{\Delta M \times 931.5}{p + n} \text{ MeV} = \frac{\Delta M \times 931.5}{A} \text{ MeV}$$

Here A is the mass number which is equal to $(p + n)$.

Variation of binding energy per nucleon ($\overline{B}$) with mass number (A) and its relation with nuclear stability

How the magnitude of $\overline{B}$ varies with the mass numbers (A) of the isotopes can be studied with the help of the plot shown in Fig. 8.4.

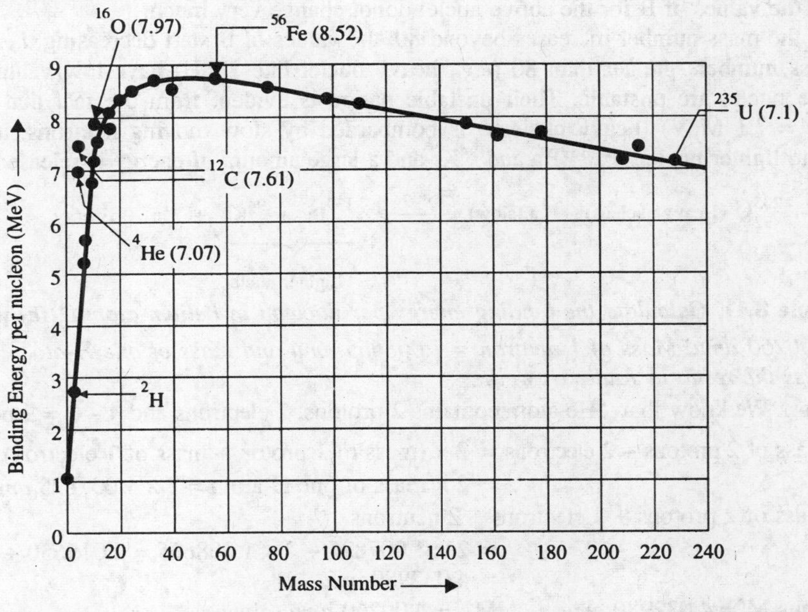

Fig. 8.4 : Variation of binding energy per nucleon (in MeV) with mass number of different nuclei.

This plot is a graph between the binding energy per nucleon (in MeV) of a number of isotopes and their corresponding mass numbers. This plot is called *binding energy curve*. Binding energy per nucleon is a measure of the stability of the nucleus. Greater is the magnitude of binding energy per nucleon, greater is the stability attained by the nucleus or greater is the force holding the nucleons together in the nucleus. The largest values of $\overline{B}$ are the characteristics of the most stable nuclei. From the binding energy curve the following points may be noted.

(*i*) $\overline{B}$ values of the nuclei-having *very low mass numbers* (*i.e.,* lighter nuclei like ${}_1^1H$, ${}_1^2H$, ${}_1^3H$) are very small and hence these nuclei are unstable. Being unstable they combine together (fusion) to give heavy nuclei and a huge amount of energy is also liberated. Examples of nuclear fusion reactions are given below:

$${}_1^1H + {}_1^2H \xrightarrow{\text{Fusion}} {}_2^3He + 5.50 \text{ MeV}$$

$${}_1^2H + {}_1^3H \longrightarrow {}_2^4He + {}_0^1n + 17.6 \text{ MeV}$$

$$4{}_1^1H \longrightarrow {}_2^4He + 2{}_{+1}^{0}e \text{ (positron)} + 24.64 \text{ MeV}$$

$${}_1^2H + {}_1^2H \longrightarrow {}_2^3He + {}_0^1n + 3.3 \text{ MeV}$$

(*ii*) The nuclei whose mass numbers are multiples of 4 or multiples of helium nucleus and which have equal number of protons and neutrons (*i.e.,* light nuclei like ^4He, ^{12}O, ^{16}O, ^{20}Ne, ^{24}Mg, ^{28}Si) show a rapid increase in their $\overline{B}$ values. ${}_4^8Be$ is an exception. It splits into two alpha particles (${}_4^8Be \longrightarrow 2{}_2^4He$). $\overline{B}$ values for the above said light nuclei are *high* (${}_2^4He$ = 7.0747 MeV, ${}_6^{12}C$ = 7.61833 MeV, ${}_8^{16}O$ = 7.976 MeV) and hence these nuclei are stable.It may also be concluded that helium nucleus has quite stable structure and this is the reason why α-particles are emitted by many radioactive elements.

(*iii*) At mass number 56, the value of $\overline{B}$ becomes maximum (= 8.52 MeV). This maximum value is for ${}_{26}^{56}Fe$ isotope. This value shows that the nucleus of iron is exceptionally stable and hence iron is found in large abundance in nature.

(*iv*) The plot for the nuclei having mass numbers in the range 60 – 80 is almost flat which means that the values of $\overline{B}$ for the above nuclei donot change very much.

(*v*) As the mass number increases beyond 80, the values of $\overline{B}$ start decreasing, *i.e.,* the nuclei having mass numbers greater than 80 (*e.g.,* heavy nuclei like 235U) have low values of $\overline{B}$ and hence these nuclei are unstable. Their unstable nature is evident from the fact that when ^{235}U nucleus ($\overline{B}$ = 7.1 MeV) (heavy nucleus) is bombarded by slow moving neutrons, it is splitted (fission) into lighter nuclei *viz.* ^{139}Ba and ^{94}Kr and a large amount of energy is released.

$${}_{92}^{235}U \text{ (Heavy nucleus)} + {}_0^1n \text{ (slow)} \xrightarrow{\text{Fission}} \underbrace{{}_{56}^{139}Ba + {}_{36}^{94}Kr}_{\text{Lighter nuclei}} + 3{}_0^1n + \text{Energy}$$

Example 8.11. *Calculate the binding energy per nucleon in helium atom (${}_2^4He$) which has a mass of 4.00260 amu. Mass of 1 neutron = 1.008665 amu and mass of one H-atom = 1.007825 amu. Express the result in Joules as well.*

Solution. We know that ${}_2^4He$ atom contains 2 protons, 2 electrons and 4 – 2 = 2 neutrons.

Mass of 2 protons + 2 electrons = 2 × (mass of 1 proton + mass of 1 electron)
= 2 × mass of one H-atom = 2 × 1.007825 *amu.*

∴ Mass of 2 protons + 2 electrons + 2 neutrons
= 2 × 1.007825 + 2 × 1.008665 = 2.015650 + 2.017330
= 4.032980 *amu*

Thus M′ = 4.032980 *amu* ; M = 4.00260 *amu* (given)

∴ Mass defect (ΔM) = M′ – M = 4.032980 – 4.002600 = 0.030380 *amu*

Since 1 *amu* = 931.5 MeV, 0.030380 *amu* = 0.030380 × 931.5 MeV = 28.298 MeV

Thus binding energy of helium nucleus (B) = 28.298 MeV

$\therefore$ Binding energy per nucleon $(\overline{B})$ $= \dfrac{28.298}{4} = \mathbf{7.0745\ MeV}$

$= 7.0745 \times 10^6\ eV = 7.0745 \times 10^6 \times 1.602 \times 10^{-19} J$

$= \mathbf{11.33 \times 10^{-13}\ J}$

Example 8.12. *The observed mass of $^{56}_{26}Fe$ is 55.9375 amu. Using the masses of one proton and one neutron as 1.00732 amu and 1.00866 amu respectively , calculate the binding energy per necleon in MeV and in Joules. Also calculate the energy evolved when one mole of $^{56}_{26}Fe$ atoms are formed from protons and neutrons. Neglect the mass of electrons.*

Solution. The nucleus of $^{56}_{26}Fe$ contains 26 protons and 56 – 26 = 30 neutrons.

Mass of 26 protons + mass of 30 neutrons (M′)

$= 26 \times 1.00732 + 30 \times 1.00866$

$= 26.19032 + 30.2698 = 56.45012$ *amu*

Now M = 55.9375 *amu* (given)

Mass defect (ΔM) $= M' - M = 56.45012 - 55.9375 = 0.51262$ *amu*

Binding energy (B) $= \Delta M \times 931.5\ MeV = 0.51262 \times 931.5\ MeV = 477.50553\ MeV$

$\therefore$ Binding energy per nucleon $(\overline{B}) = \dfrac{477.50553}{56}\ MeV = 8.5268844\ MeV$

$= 8.5268844 \times 1.602 \times 10^{-13} J = 1.3660068 \times 10^{-12}\ J$

Energy released when one mole of $^{56}_{26}Fe$ atoms are formed from protons and neutrons

$= [56 \times 8.5268844 \times \text{Avogadro's number}]\ MeV$

$= 56 \times 8.5268844 \times 6.022 \times 10^{23}\ MeV$

$= 2875.5382 \times 10^{23}\ MeV.$

Nuclear shell model: Magic numbers

This model was proposed by Maria Mayer, Engene Wigner and Hans Jensen in 1950 which earned them noble prize in 1963.

According to this model:

1. Each nucleon moves in its orbit within the nucleus, independently of all other nucleons. It is for this reason that this model is also known as *Independent particle model* or *single particle model.*

2. Each nucleon is acted upon by an average field produced by the action of other nucleons.

3. The neutrons as well as protons pair amongst themselves (*i.e.,* a neutron forms a pair with another neutron and similarly a proton forms a pair with another proton) in the nucleus and these paired nucleons are packed into separate shells (called *nuclear shells)* within the nucleus. A nuclear shell which contains a specific number of neutrons and / or protons in it is called a filled shell and this filled shell is more stable than other shells. These specific numbers are 2, 8, 20, 28, 50, 82 and 126 which are called **magic numbers**. These magic numbers indicate that the nucleons exist in pairs in the nucleus. The nuclei having either protons or neutrons or both equal to the magic numbers given above are called **magic nuclei** and these nuclei are very stable. Being stable, these nuclei are also called **noble nuclei.** The greater stability associated with these nuclei is confirmed by the following facts:

(*i*) These nuclei are found in greater abundance in nature as compared to other isotopic forms of the same element.

(*ii*) These nuclei cannot capture a neutron because the nuclear shells in their respective nuclei are already filled with nucleons and hence cannot contain an extra neutron. The nuclei having the magic number (*i.e.,* magic nuclei) have been classified into the following two groups:

(*a*) Nuclei in which only protons or neutrons are magic numbers. Such nuclei are called *single magic number nuclei*. Examples of such nuclei are $^{17}_{8}O$ ($p = 8$, $n = 17 - 8 = 9$), $_{50}Sn$ ($p = 50$), $^{207}_{82}Pb$ ($p = 82$, $n = 207 - 82 = 125$) and $^{209}_{83}Bi$ ($p = 83$, $n = 209 - 83 = 126$).

(*b*) Nuclei in which protons and neutrons both are magic numbers. Such nuclei are called *double magic number nuclei.* Examples are $^{4}_{2}He$ ($p = 2, n = 2$), $^{16}_{8}O$ ($p = 8, n = 8$), $^{40}_{20}Ca$ ($p = 20, n = 20$), $^{208}_{82}Pb$ ($p = 82, n = 126$)

When a nuclide contains protons or neutrons or both equal to any of the magic numbers, we say that this nuclide has closed or filled shell. Such a nuclide gives a number of stable isotopes or isotones, *e.g.* $_{50}Sn$ ($p = 50$ which is a magic number) has ten stable isotopes having their mass numbers equal to 112, 114, 115, 116, 117, 118, 120, 122 and 124. $_{20}Ca$ ($p = 20$ which is a magic number) has six stable isotopes. The maximum number of stable isotones (isotones are those nuclei which have the same number of neutrons) occur for those nuclei which have $n = 50$ or 82. These numbers are also magic numbers. The nuclei having 6, 14, 28 or 40 protons or neutrons are less stable than those nuclei which have protons or neutrons or both equal to magic numbers. The numbers namely 6, 14, 28 and 40 are called *semi-magic numbers*. The nuclei having protons or neutrons other than magic numbers or semi-magic numbers are comparatively less stable. This model has successfully explained the nuclear stability, spin, magnetic moment etc. The nuclei *viz.* $^{2}_{1}H$, $^{6}_{3}Li$, $^{10}_{6}B$ and $^{14}_{7}N$ are not covered under the scheme of this model.

UNIVERSITY QUESTIONS

Binding energy, binding energy curve and stability of the nucleus

1. What is meant by binding energy per nucleon? How does it vary with atomic number of the element? Discuss the importance of binding energy curve in the release of nuclear energy. *(Rohilkhand 90; Meerut M.Sc. 91, 92, 93)*
2. Discuss the various theories given for the stability of the nucleus. *(M.Sc. Baroda 90; Lucknow 93)*
3. Establish relation between the binding energy, mass nucleus and mass defect of the nucleus. *(M.Sc. Baroda 91; Allahabad 91; Berhampur 91; Kurukshetra 92)*
4. How is nuclear stability related to packing fraction and binding energy. Explain. *(Osmania 93; E.C.D. Allahabad 94)*
5. Discuss the stability of nucleus under the following heads (*a*) Binding energy per nucleon (*b*) Neutron-proton ratio and (*c*) Odd-even rule. *(Lucknow 95)*

Numerical prblems

6. Calculate the mass defect and binding energy of $_{2}He^{4}$. Given that mass of $_{2}He^{4}$ = 4.00390 *amu*, mass of proton = 1.00758 *amu*, mass of neutron = 1.00893 and mass of electron = 0.0005486 *amu*. *(Allahabad 91)*
7. Calculate the mass defect and binding energy per nucleon of $_{6}C^{12}$ which has isotopic mass equal to 12.0 *amu*. Given that mass of neutron = 1.008671 *amu*, mass of hydrogen = 1.00787 *amu*, 1 *amu* = 931.5 MeV. *(Osmania 93; Delhi 85)*
8. Calculate the binding energy per nucleon in $^{35}_{17}Cl$ nucleus. Given that mass of $_{0}n^{1}$ = 1.008930 *amu*, mass of $_{1}H^{1}$ = 1.007825 *amu*, mass of $_{17}Cl^{35}$ = 34.9800 *amu*, 1 *amu* = 931 MeV. *(Delhi 88)*
9. The atomic mass of a nucleus is 6.017 *amu*. Calculate its total binding energy (Mass of $^{1}_{1}p$ = 1.00727 *amu*, mass of $^{1}_{0}n$ = 1.008665 *amu*). *(Kurukshetra 92)* **[Ans.: 2876 MeV]**
10. Calculate the binding energy of a deuteron nucleus, given that mass of neutron = 1.008665 *amu*, mass of proton = 1.007825 *amu* and mass of deuteron nucleus = 2.014103 *amu*. *(Meerut M.Sc., 94)*
11. Calculate the relative atomic mass of an element which consists of the following isotopes with the indicated relative abundance.

Isotope	*Isotopic mass*	*Natural abundance*
1	28.0	92.0
2	29.0	5.0
3	30.0	3.0

(Gauhati 89) **[Ans.: 28.11]**

12. Calculate the mass defect and binding energy in MeV for the isotope $_{17}Cl^{35}$. Given that 1 amu = 931.5 MeV, mass of electron = 0.0005486 amu, mass of proton = 0.007277 amu and mass of neutron = 1.008665 amu. *(Delhi 2003)*

9

Nuclear Reactions

What are nuclear reactions?

These are the reactions in which one element is converted into the other either by emitting α or β particles (*spontaneous disintegration*) or by bombarding it with suitable bombarding particle (*artificial transmutation of elements*). In these reactions, the nucleus of the original element undergoes a change in its composition, *i.e.*, during a nuclear reaction the number of nucleons present in the nucleus of the original element changes. Thus in a nuclear reaction, new atoms are formed. In other words we can say that in nuclear reactions transformation or transmutation of atoms takes place.

Differences between chemical reactions and nuclear reactions

We have said above that in a nuclear reaction, the composition of the nucleus of the original element changes and hence new elements are formed. On the other hand, in a chemical reaction, the composition of the nucleus remains unchanged; only the rearrangement of the outer-most electrons of the atoms takes place, *i.e.*, in a chemical reaction only a rearrangement of atoms is involved and hence no new atoms or elements are formed.

In a nuclear reaction the amount of energy absorbed or released is a million times greater than that absorbed or released in a chemical reaction, *i.e.*, in a chemical reaction the amount of energy absorbed or released is *small* while in a nuclear reaction the amount of energy absorbed or released is *large* which is because of the reason that in a nuclear reaction the gain (absorption of energy) or loss (release of energy) in mass is ***small but measurable.*** The mass gained or lost is converted into energy in accordance with Einstein's mass-energy relationship *viz.* $E = mc^2$ where E = energy absorbed or released, m = mass gained or lost, c = velocity of light.

The differences between chemical reactions and nuclear reactions are summarised in Table 9.1.

Table 9.1 : ***Differences between the chemical reactions and nuclear reactions***

Chemical reactions	Nuclear reactions
(*i*) In these reactions only the outer-most electrons of the participating atoms take part and the nuclei of the atoms remain unaffected.	(*i*) In these reactions the nuclei of the atoms undergo a change in their composition.
(*ii*) The chemical identity of an element remains the same even after the chemical reaction, *i.e.*, no new elements are formed.	(*ii*) In these reactions one element gets changed into the other element or elements, *i.e.*, new elements are formed.
(*iii*) Usually a small amount of energy is abosorbed or released in these reactions.	(*iii*) A large amount of energy is abosorbed or released in these reactions.
(*iv*) These reactions are greatly affected by a change in temperature, pressure etc.	(*iv*) These reactions are not affected by a change in temperature, pressure etc.
(*v*) These reactions can be reversed under suitable conditions.	(*v*) These reactions can not be reversed.
(*vi*) All the isotopes of an element give the same type of chemical reactions.	(*vi*) All the isotopes of an element give different types of nuclear reactions.

Classification of nuclear reactions

Nuclear reactions have been classified into the following categories.

Classification I. This classification is based on the fact whether the nuclear reaction takes place spontaneously (*i.e.*, by itself) or is brought about by artificial means. This classification gives the following two types of nuclear reactions:

(*i*) Nuclear reactions in which a radioactive element (unstable element) spontaneously disintegrates into another element by emitting α, β or γ-rays. Thus in these reactions, a single nucleus (*i.e.*, a single radioactive element) is involved. Examples of such reactions are:

$$^{238}_{92}\text{U (Radioactive)} \longrightarrow \,^{234}_{90}\text{Th} + \,^{4}_{2}\text{He } (\alpha\text{–particle})$$

$$^{234}_{90}\text{Th (Radioactive)} \longrightarrow \,^{234}_{91}\text{Pa} + \,^{0}_{-1}\text{e } (\beta\text{-particle})$$

These reactions are studied under the title "*Natural Radioactivity*".

(*ii*) In these reactions, a stable element (non-readioactive element) is bombarded with a suitable bombarding particle (*e.g.* α-particle, proton, deuteron, neutron etc.) and a new element is obtained. In addition to new element, one of the particles like neutron, proton, deuteron, α-particle, γ-ray etc. is also emitted. The newly-formed element may or may not be stable. Thus two cases arise:

(*a*) If the new element is *unstable* (radioactive), it disintegrates by itself and gives a stable element, *e.g.*

$$\underset{\text{Stable (non-radioactive)}}{^{27}_{13}\text{Al}} + \underset{\text{Bombarding particle}}{^{4}_{2}\text{He}} \longrightarrow \underset{\text{Unstable (radioactive)}}{^{30}_{15}\text{P}} + \,^{1}_{0}\text{n}$$

$$^{30}_{15}\text{P} \longrightarrow \underset{\text{Stable (non-radioactive)}}{^{30}_{14}\text{Si}} + \,^{0}_{+1}\text{e } (\textit{positron})$$

Such types of reactions are studied under "*Artificial or induced radioactivity*"

(*b*) If the new element is *stable* (non-radioactive), it does not disintegrate, *i.e.*, no more radioactivity is observed. Such types of reactions are studied under "*Artificial transmutation* or *artificial disintegration of elements*" Example is:

$$\underset{\text{Stable}}{^{14}_{7}\text{N}} + \underset{\text{Bombarding particle}}{^{4}_{2}\text{He}} \longrightarrow \underset{\text{Stable}}{^{17}_{8}\text{O}} + \,^{1}_{1}\text{H}$$

$$^{17}_{8}\text{O (Stable)} \longrightarrow \text{No disintegration.}$$

Classification II. This classification is based on the nature of the bombarding particle (or projectile) used in the nuclear reaction. The particles which are generally used as bombarding projectiles are α-particle ($^{4}_{2}$He), proton ($^{1}_{1}$H), deuteron ($^{2}_{1}$H or $^{2}_{1}$D), neutron ($^{1}_{0}$n), triton ($^{3}_{1}$H or $^{3}_{1}$T), γ-ray ($^{0}_{0}\gamma$). Thus these reactions may be those which are induced by α-particles, protons, deuterons, neutrons, tritons and γ-rays. Examples of such reactions have already been given under "*Artificial transmutation reactions induced by different bombarding projectiles*"

Classification III. Nuclear reactions are classified on the basis of overall energy transformation that occurs during the nuclear change. On this basis the reactions are of the following types.

(*i*) *Capture reactions*. These are the nuclear reactions in which bombarding particle is captured by the target nucleus with the emission of γ-ray photons. For example :

$$^{7}_{3}\text{Li} + \,^{4}_{2}\text{He} \longrightarrow \,^{11}_{5}\text{B} + \,^{0}_{0}\gamma \quad ; \quad ^{7}_{3}\text{Li} + \,^{1}_{1}\text{H} \longrightarrow \,^{8}_{4}\text{Be} + \,^{0}_{0}\gamma$$

$$^{12}_{6}\text{C} + \,^{1}_{1}\text{H} \longrightarrow \,^{13}_{7}\text{N} + \,^{0}_{0}\gamma \quad ; \quad ^{26}_{12}\text{Mg} + \,^{1}_{0}\text{n} \longrightarrow \,^{27}_{13}\text{Al} + \,^{0}_{0}\gamma$$

$$^{27}_{13}\text{Al} + \,^{1}_{0}\text{n} \longrightarrow \,^{28}_{13}\text{Al} + \,^{0}_{0}\gamma \quad ; \quad ^{238}_{92}\text{U} + \,^{1}_{0}\text{n} \longrightarrow \,^{239}_{92}\text{U} + \,^{0}_{0}\gamma$$

The process in which the nucleus of a radioactive element captures an electron (*i.e.*, addition of an electron to a nucleus) is called *electron-capture process*. For this see "*Soddy-Fajan and Russel Group Displacement Law*".

(ii) Particle-particle reactions. In these reactions, the bombarding particle is absorbed by the target nucleus to form the unstable compound nucleus whose half-life period is 10^{-14} to 10^{-12} seconds. This compound nucleus, being unstable, breaks down (*i.e.*, disintegrates) to form the stable nucleous. Examples are : (compound nucleus has been shown in bracket)

$$^{24}_{11}\text{Na} + ^{4}_{2}\text{He} \longrightarrow \left[^{28}_{13}\text{Al}\right] \longrightarrow ^{26}_{13}\text{Al} + 2\,^{1}_{0}\text{n}$$
$$^{27}_{13}\text{Al} + ^{1}_{0}\text{n} \longrightarrow \left[^{28}_{13}\text{Al}\right] \longrightarrow ^{28}_{13}\text{Al} + ^{0}_{0}\gamma$$
$$^{27}_{12}\text{Mg} + ^{1}_{1}\text{H} \longrightarrow \left[^{28}_{13}\text{Al}\right] \longrightarrow ^{24}_{11}\text{Na} + ^{4}_{2}\text{He}$$
$$^{26}_{12}\text{Mg} + ^{2}_{1}\text{H} \longrightarrow \left[^{28}_{13}\text{Al}\right] \longrightarrow ^{27}_{12}\text{Mg} + ^{1}_{1}\text{H}$$

(iii) Spallation reactions. In these reactions, a high speed projectile having high energy (energy = 400 MeV or more than this value) chips a fragment off from a heavy nucleus and a large number of light particles are emitted. The daughter nucleus has atomic number 10-20 units less and mass number 30 units less than the parent nucleus.

Examples. High energy α-particle, $^{4}_{2}$He (energy = 400 MeV) converts Cu-63 nucleus into Cl-37 nucleus and releases 14 protons ($14\,^{1}_{1}$H) and 16 neutrons ($16\,^{1}_{1}$n).

$$^{63}_{29}\text{Cu} + ^{4}_{2}\text{He} \longrightarrow ^{37}_{17}\text{Cl} + 14\,^{1}_{1}\text{H} + 16\,^{1}_{0}\text{n}$$

Other examples are :

$$^{75}_{33}\text{As} + ^{2}_{1}\text{H} \longrightarrow ^{56}_{55}\text{Mn} + 9\,^{1}_{1}\text{H} + 12\,^{1}_{0}\text{n}$$
$$^{118}_{50}\text{Sn} + ^{1}_{1}\text{H} \longrightarrow ^{72}_{31}\text{Ga} + 10\,^{4}_{2}\text{He} + 7\,^{1}_{0}\text{n}$$

(v) Fission reactions. See pages (230-239).

(vi) Fusion reactions. See pages (239-242).

Classification IV. This classification is based on the fact whether energy is absorbed or released in the nuclear reaction. Thus this classification gives the following two types.

(i) Endo-ergic nuclear reactions. In these reactions energy is absorbed by the reactants. In these reactions, the sum of the masses of the reactants is *less* than the sum of the masses of the products, *i.e.,* in these reactions some mass is gained or increased. The mass increased is converted into energy which is absorbed in the reaction. Thus the absorption of energy is due to the increase in mass of the products. Energy absorbed in a given nuclear reaction is given by Einstein mass-energy equation *viz.* $E = mc^2$ where E = energy absorbed m = mass gained or gain (or increase) in mass and c = velocity of light.

If the increase in mass is in *amu*, then energy absorbed

= Increase in mass × 931.5 MeV

= [sum of the masses of the products – sum of the masses of the reactants] × 931.5 MeV.

Let us calculate the amount of energy *absorbed* in Rutherford's artificial transmutation reaction *viz.*

$$^{14}_{7}\text{N} + ^{4}_{2}\text{He} + \text{Q}\ (\textit{energy absorbed}) \longrightarrow ^{17}_{8}\text{O} + ^{1}_{1}\text{H}$$

Masses of the reactants = mass of $^{14}_{7}$N nucleus + mass of a helium nucleus, $^{4}_{2}$He

= (14.0031 + 4.0026) *amu* = 18.0057 *amu*

Masses of the products = mass of $^{17}_{8}$O nucleus + mass of a H-atom

= (16.9991 + 1.0078) *amu* = 18.0069 *amu*

∴ Increase in mass = 18.0069 – 18.0057 = 0.0012 *amu*

∴ Energy absorbed = 0.0012 × 931.5 MeV= **1.117 MeV.**

Since energy is absorbed in the given reaction, the given reaction is an endo-ergic nucleur reaction

(ii) Exo-ergic nuclear reactions. In these reactions energy is released. The energy released in these reactions is called *nuclear energy* or *atomic energy*. In these reactions, the sum of the masses of the products is *less* than the sum of the masses of the reactants, *i.e.,* in these reactions *some mass*

is lost. The mass lost is converted into energy which is released in the reaction. Thus the release of energy is due to the loss in mass of the reactants. Energy released in a given nuclear reaction is given by Einstein mass-energy equation *viz.* $E = mc^2$ where E = energy released m = mass lost or loss in mass and c = velocity of light. If mass lost is in *amu*, then energy released

= Mass lost or loss in mass × 931.5 MeV = [(sum of the masses of the reactants) – (sum of the masses of the products)] × 931.5 MeV.

Let us calculate the amount of energy released in the following reactions which are exo-ergic nuclear reactions.

(a) Nuclear fission of ^{235}U nucleus by slow neutrons. The calculations show that energy equal to 200.27 MeV is released by the fission of one atom of ^{235}U nucleus. For details see *"Nuclear Fission"*.

(b) Fusion of two deuterium nuclei (^2_1H) to form a helium nucleus (^4_2He). This reaction can be shown as:

$$^2_1H + ^2_1H \longrightarrow ^4_2He + Q \text{ (energy released)}$$

Masses of the reactants	=	2 × mass of one deuterium nucleus
	=	2 × 2.01471 *amu* = 4.02942 *amu*
Mass of the product	=	mass of helium nucleus = 4.0026 *amu.*
Mass lost or loss in mass	=	(4.02942 – 4.00260) *amu* = 0.02682 *amu*
Energy released	=	0.02682 × 931.5 MeV= **24.9828 MeV**

Bohr's theory of nuclear reaction

According to Bohr's theory, the bombarding part. :le combines with the target nucleus to form an intermediate compound nucleus which is in the excited state and hence is unstable. This unstable intermediate compound nucleus has small life period (10^{-14} to 10^{-12} seconds) and hence breaks up to give the final product, called recoil nucleus. The nature of the final product depends on the amount of energy associated with the unstable intermediate compound nucleus. For example when $^{27}_{13}Al$ (target nucleus) is bombarded with protons ^1_1H (bombarding particle), an unstable intermediate compound nucleus, $[^{28}_{14}Si]$ is formed which breaks up in any of the following ways depending on the amount of energy associated with the intermediate coumpound nucleus.

$$^{27}_{13}Al + ^1_1H \rightarrow [^{28}_{14}Si] \longrightarrow \begin{cases} ^{27}_{14}Si + ^1_0n \\ ^{24}_{12}Mg + ^4_2He \\ ^{24}_{11}Na + ^1_0n + 3\,^1_1H \end{cases}$$

Nuclear Fission

What is nuclear fission?

Nuclear fission is a nuclear reaction in which an unstable heavier nucleus, on being bombarded by slow moving neutrons, splits up (*i.e.*, gets fissioned) into two lighter stable nuclei of almost equal size with the liberation of a large amount of heat energy.

Discovery of nuclear fission.

In 1939 German scientists, Otto Hahn and F. Strassman found that when $^{235}_{92}U$ nucleus is bombarded with slow neutrons, it gets splitted into two lighter nuclei (called *fission products* or *fragments*) namely $^{141}_{56}Ba$ and $^{92}_{36}Kr$ with the liberation of three neutrons and a large amount of heat energy which is called *fission energy* or *atomic energy.*

$$^{235}_{92}U + ^1_0n\ (slow) \xrightarrow{\text{Fission}} \underbrace{^{141}_{56}Ba + ^{92}_{36}Kr}_{\text{Fission products}} + 3\,^1_0n + \text{Huge amount of energy} \quad ...(i)$$

Other nuclei that can undergo nuclear fission reaction

It is not only ^{235}U nucleus that undergoes nuclear fission. ^{233}U, ^{238}U, ^{232}Th and ^{239}Pu nuclei are also liable to fission. Uranium found in nature contains only 0.7% of ^{235}U isotope and 99.3% of ^{238}U isotope. ^{235}U is fissionable while ^{238}U cannot undergo fission. Therefore, when natural uranium without removing ^{238}U from it, is put to fission process, ^{235}U nuclei alone undergo fission but the chain reaction is not set up because the secondary neutrons produced in the fission of ^{235}U nuclei are absorbed by ^{238}U nuclei without causing any further fission reaction.

$$^{238}_{92}U + {}^{1}_{0}n \longrightarrow {}^{239}_{92}U \longrightarrow {}^{239}_{93}Np + {}^{0}_{-1}e \longrightarrow {}^{239}_{94}Pu + 2\,{}^{0}_{-1}e$$

Thus ^{238}U nucleus is converted into ^{239}Pu nucleus which can easily be separated chemically from the uranium residues of the atomic pile. ^{239}Pu nucleus, like ^{235}U nucleus, can undergo nuclear fission very easily even by slow moving neutrons. Thus we find that ^{239}Pu nucleus can also be used for the production of fission energy.

^{232}Th nucleus is changed to ^{233}U nucleus *via* neutron capture and subsequent β-emission. U-233 is fissionable.

$$^{232}_{90}Th + {}^{1}_{0}n \longrightarrow {}^{233}_{90}Th \longrightarrow {}^{233}_{91}Pa + {}^{0}_{-1}e \longrightarrow {}^{233}_{92}U + 2\,{}^{0}_{-1}e$$

Lighter nuclei like Bi, Cu, Au, Ta and rare earths also undergo fission by the impact of some projectiles with higher energies (50 – 450 MeV).

$$^{63}_{29}Cu + {}^{1}_{1}H\ (\text{proton}) \longrightarrow {}^{24}_{11}Na + {}^{39}_{19}K + {}^{1}_{0}n$$

Nature of fission products

In nuclear fission reaction (*i*), we have shown that the fission products consist of a pair of stable isotopes *viz.* $^{92}_{36}$Kr and $^{141}_{56}$Ba. But the analysis of the reaction products has shown that the nuclear reaction can also take place in many other modes and hence pairs of other isotopes are also obtained as shown below:

	Lighter isotopes		*Heavier isotopes*		*Neutrons*
$^{235}_{92}U + {}^{1}_{0}n \longrightarrow$	$^{90}_{37}Rb$	+	$^{144}_{55}Cs$	+	$2\,{}^{1}_{0}n$
	$^{98}_{40}Zn$	+	$^{136}_{52}Te$	+	$2\,{}^{1}_{0}n$
	$^{92}_{36}Kr$	+	$^{141}_{56}Ba$	+	$3\,{}^{1}_{0}n$
	$^{96}_{36}Kr$	+	$^{137}_{56}Ba$	+	$3\,{}^{1}_{0}n$
	$^{90}_{36}Kr$	+	$^{143}_{56}Ba$	+	$3\,{}^{1}_{0}n$
	$^{87}_{35}Br$	+	$^{146}_{57}La$	+	$3\,{}^{1}_{0}n$
	$^{97}_{39}Y$	+	$^{135}_{53}I$	+	$4\,{}^{1}_{0}n$
	$^{94}_{38}Sr$	+	$^{140}_{54}Xe$	+	$2\,{}^{1}_{0}n$
	$^{95}_{38}Sr$	+	$^{139}_{54}Xe$	+	$2\,{}^{1}_{0}n$

Which particular pair of stable isotopes is formed actually depend on chance. The pairs of fission products that have been obtained in different fission reactions have been grouped into two categories : one category consists of those *lighter nuclei* whose mass numbers are between 80 and 100 and their atomic numbers are in the range *viz.* 35–42; the other category contains *heavier nuclei* having mass numbers and atomic numbers in the range *viz.* 130–150 and 51–58 respectively.

Why different pairs of isotopes are obtained in different fission reactions is because of the fact that the immediate isotopes originally formed are sometimes unstable and hence disintegrate, emitting β-particles ($^{0}_{-1}$e), to form stable isotopes of other elements with higher atomic number. For example:

(*i*) $$^{140}_{54}Xe \xrightarrow{-{}^{0}_{-1}e} {}^{140}_{55}Cs \xrightarrow{-{}^{0}_{-1}e} {}^{140}_{56}Ba \xrightarrow{-{}^{0}_{-1}e} {}^{140}_{57}La \xrightarrow{-{}^{0}_{-1}e} {}^{140}_{58}Ce\ (\textit{stable isotopes})$$

$$(ii)\ {}^{90}_{40}Zr \xrightarrow{-{}^{0}_{-1}e} {}^{98}_{41}Nb \xrightarrow{-{}^{0}_{-1}e} {}^{98}_{42}Mo$$

$$(iii)\ {}^{136}_{52}Te \xrightarrow{-{}^{0}_{-1}e} {}^{136}_{53}I \xrightarrow{-{}^{0}_{-1}e} {}^{136}_{54}Xe$$

Such reactions are called β-*active decay chains.* Total number of identified nuclides obtained from the fission of ^{235}U nucleus is about 300. Out of these, 200 are β-emitters.

Yield of fission products: Bohr yield curve

The yield of fission products depends an their mass number. It has been found with the help of Bohr yield curve that the isotopes with mass numbers 95 and 140 have maximum yield, *i.e.*, the isotopes with mass numbers 95 and 140 dominate the fission products. Bohr yield curve is obtained by plotting the logrithm of the percentage yield of different isotopes (*y*-axis) against the mass numbers of these isotopes (*x*-axis).The yields of the various nuclides are known as *Bohr yields.*

Production of neutrons

The number of neutrons obtained along with fission products depends on the mode in which the fission reaction takes place and on the energy of the neutron that has been used to induce the fission of ^{235}U nucleus.

Amount of energy liberated in the fission of one atom of ^{235}U nucleus

We have already said that in the fission of ^{235}U nucleus by slow neutrons, energy is released, *i.e.*, the fission of ^{235}U nucleus is an *exo-ergic reaction.* The energy released is called *nuclear energy / atomic energy* or *fission energy.*

$$\underbrace{{}^{235}_{92}U + {}^{1}_{0}n}_{\text{Reactants}} \longrightarrow \underbrace{{}^{141}_{56}Ba + {}^{92}_{36}Kr + 3\,{}^{1}_{0}n}_{\text{Products}} + \textit{Energy}$$

The release of energy takes place because of the fact that in this reaction some mass is lost, since the sum of the masses of the reactants is more than the sum of the masses of the products. The mass lost is converted into energy which is released in the reaction. The energy released is calculated with the help of Einstein's mass-energy relationship *viz.* $E = mc^2$ where E = energy released, m = mass lost or loss in mass and c = velocity of light. If mass lost is in *amu*, then energy released

= Mass lost or loss in mass (in *a.m.u*) × 931.5 MeV

= [(sum of the masses of the reactants) – (sum of the masses of the products)] × 931.5 meV.

In our case:

Masses of the reactants = mass of ^{235}U + mass of one ${}^{1}_{0}n$

= 235.044 + 1.008 = 236.052 *amu.*

Masses of the products = mass of ^{141}Ba + mass of ^{92}Kr + 3 × mass of one ${}^{1}_{0}n$

= 91.905 + 140.908 + 3 × 1.008 = 235.837 *amu.*

∴ Mass lost or loss in mass = 236.052 – 235.837 = 0.215 *amu*

∴ Energy released = 0.215 × 931.5 MeV = **200.27 MeV**

Thus, we see that 200. 27 MeV is the energy that is obtained by the fission of one atom of ^{235}U nucleus.

Amount of energy released by the fission of 1 gram of ^{235}U

First of all let us find out the number of atoms present in 1 gram of ^{235}U. We know that 235.0 g (atomic weight of uranium) of ^{235}U contains atoms equal to Avogadro's number = 6.022×10^{23}

$$\therefore \quad 1 \text{ g. of } {}^{235}U \text{ contains} = \frac{6.022 \times 10^{23}}{235} \text{ atoms}$$

We have already seen that the energy released by the fission of 1 atom of ^{235}U.

= 200.27 MeV

$$\therefore \quad \text{Energy released by the fission of } \frac{6.022 \times 10^{23}}{235} \text{ atoms of } {}^{235}U$$

$$= \frac{6.022 \times 10^{23}}{235} \times 200.27 \text{ Mev}$$

$$= \frac{6.022 \times 10^{23} \times 200.27 \times 10^{6} \times 1.602 \times 10^{-19}}{235} \text{ Joules}$$

$$= \frac{6.022 \times 10^{23} \times 200.27 \times 10^{6} \times 1.602 \times 10^{-19}}{235 \times 10^{3}} \text{ kJ} = 8.22 \times 10^{7} \text{ kJ}$$

Thus, when 1 gram of ^{235}U is subjected to fission, 8.22×10^7 kJ of energy is produced. This amount of energy can be produced only after burning about 2.5 metric tons of good quality coal. Thus 1 gram of ^{235}U is equivalent to 2.5 metric tones of coal in respect of energy.

It has been found that the heat energy obtained by the fission of one atom of ^{235}U is about 200 million times more than the heat energy obtained by burning one molecule of natural gas (methane)

Peaceful uses of atomic (nuclear) energy

Nuclear energy is used (*i*) for generating electricity, (*ii*) for the treatment of deseases like cancer, (*iii*) for the improvement of agriculture and industry.

Explanation of nuclear fission of U-235 nucleus by liquid drop model

Bohr and Wheeler have suggested that the nuclear fission of U-235 can be explained with the help of *liquid drop model*, since an atomic nucleus has many similarities with a liquid drop. A liquid drop has a spherical shape due to its forces of surface tension. When a sufficiently large amount of energy is applied on the drop, the forces of surface tension are destroyed and hence the spherical shape of the drop becomes elliptical and then dumbell. This dumbell shape ultimately breaks into two parts of spherical shape.

Now let us see what happens when a slow moving neutron attack a U-235 nucleus. Before the attack of the neutron, the shape of U-235 remains spherical, since the protons and neutrons present in the nucleus keep held together very stongly. Now when a sufficiently large amount of energy is applied on U – 235 nucleus by bombarding it with slow moving neutron, the shape of the nucleus gets elongated and hence its spherical shape becomes elliptical. Due to large amount of energy which is being applied on U-235 nucleus, the elliptical shape ultimately becomes dumbell. Dumbell shape has a waist in it and both the parts of this shape have positive charge. This dumbell shape is called *critical shape*. Now since the two parts of the dumbell shape have similar (positive) charges, they repell each other and ultimately U-235 nucleus is broken (nuclear fission) into two nuclei (*viz.* Ba-139 and Kr-94) each of which has a spherical shape (Fig 9.1). In this reaction three neutrons are also emitted and a large amount of energy is also released.

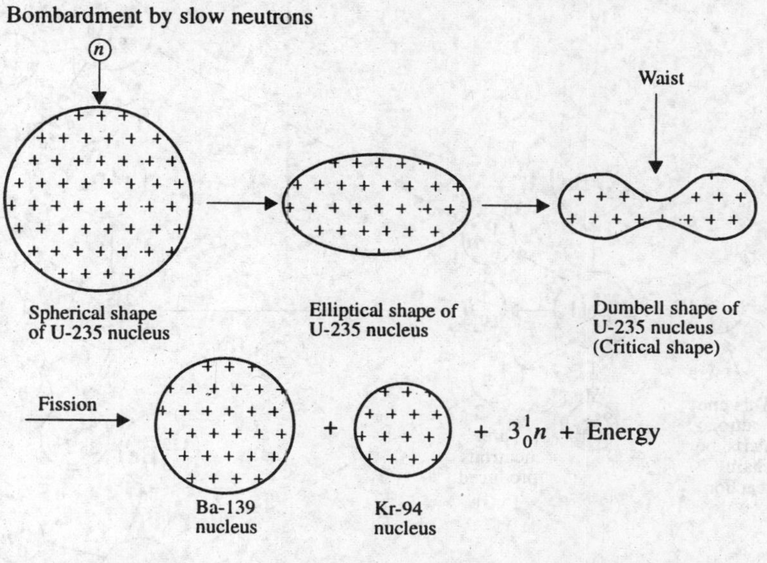

Fig. 9.1: Explanation of nuclear fission of U-235 nucleus by liquid drop model.

Types of fission reaction

We have seen that when a U-235 nucleus is bombarded by means of a slow moving neutron, it gets fissioned into two smaller nuclei namely $^{139}_{56}Ba$ and $^{94}_{36}Kr$ and three neutrons (secondary neutrons) and a large amount of energy are also produced. Depending on whether all the three secondary neutrons or only one of them is allowed to cause the fission of other U-235 nuclei (if a considerable number of U-235 nuclei is taken for the fission in the beginning), we have two types of nuclear fission reaction. These are *(i) uncontrolled or explosive fission reaction* and *(ii) controlled or critical fission reaction.*

(i) Uncontrolled or explosive fission reaction. If none of the three neutrons produced in the fission (secondary neutrons) escapes out of the air but causes the fission of other U-235 nuclei, then each of these three neutrons causes the fission of another U-235 nucleus and produces three neutrons and heat energy. Thus the three neutrons cause the fission of three more U-235 nuclei and produce 3 × 3 = 9 neutrons along with a lot of more heat energy. These nine neutrons cause the fission of 9 more U-235 nuclei, producing 9 × 3 = 27 neutrons and yet greater amount of energy (see Fig. 9.2). This process of fission of U-235 nucleus goes on like an unending chain of fission reaction and ultimately an uncontrollable amount of heat energy is produced in a very short time. This energy cannot be used for a peaceful purpose (like the production of electricity from this energy); rather it causes explosion in a few moments (atom bomb). Since this reaction takes place at a very fast rate, it cannot be controlled and hence is called uncontrolled fission reaction. Since

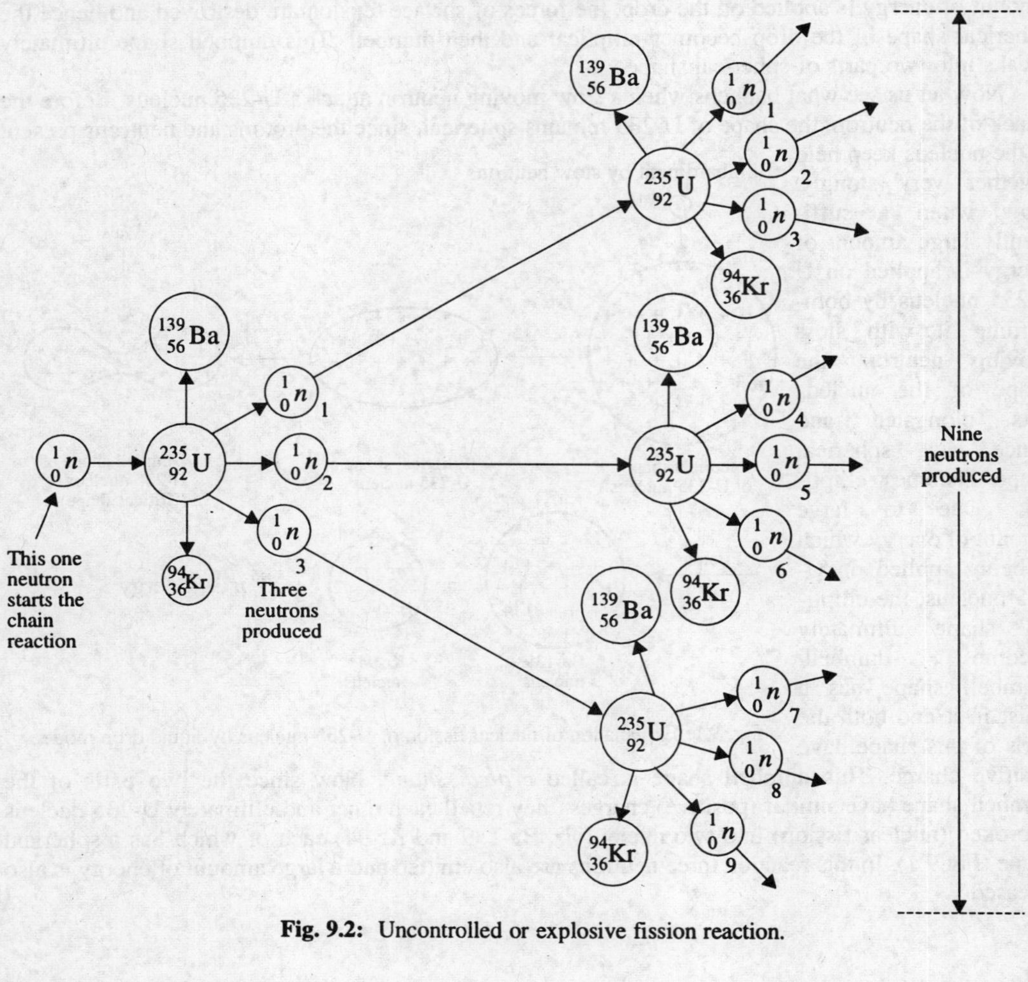

Fig. 9.2: Uncontrolled or explosive fission reaction.

this reaction produces a tremendous (uncontrolled) amount of heat energy which causes explosion, this reaction is also called explosive fission reaction and takes place in atom bomb.

The above discussion shows that an explosive nuclear fission reaction can be defined as that fission reaction in which all the neutrons produced during the fission of one U-235 nucleus are deliberately allowed to cause the fission of other U-235 nuclei leading to the production of untrollable amount of heat energy which produces explosion.

If it is assumed that one neighbouring U-235 nucleus gets fissioned by one secondary neutron in one second, then calculations have shown that 10^{20} U-235 nuclei would be fissioned in one minute only, producing an unmanageable amount of heat energy.

(*ii*) *Controlled or critical fission reaction.* If two of the three secondary neutrons are absorbed by cadmium or boron rods (controlling rods) before they can cause the fission of other U-235 nuclei, then only one neutron is left which causes the fission of one other U-235 nucleus. This fission reaction proceeds at a steady and slow rate and hence a controlled amount (manageable amount) of energy is produced. Since this fission reaction takes place at a *steady* and *slow rate*, it can be controlled and hence is called controlled fission reaction. This type of reaction takes place in nuclear reactor. Since it produces controlled quantity of heat energy, this energy can be converted into electricity with the help of atomic power plant. This reaction is also called critical fission reaction, since this reaction takes place when U-235 has critical mass.The first controlled fission reaction was carried out by Fermi on 2nd December, 1941.

If it is assumed that one neighbouring U-235 nucleus gets fissioned by one secondary neutron in one second, then it has been found that only sixty U-235 nuclei would be fissioned in one minute, producing a manageable (controlled) amount of heat energy.

Chain reaction

A chain reaction is that in which the particle which initiates the reaction is also produced in the reaction and this particle makes the reaction proceed like an unending chain. Thus a chain reaction is a self-sustaining or self-propagating process. The fission of U-235 nucleus by slow moving neutrons is also a chain reaction since this reaction is started by neutrons and the same particles are also produced in it. The neutrons thus produced cause the fission of other U-235 nuclei and hence the fission of U-235 nucleus goes on like an unending chain, *i.e.*, the fission of U-235 nucleus becomes self-sustaining or self-propagating process.

Three different ways in which the fission chain reaction can take place

The value of neutron-multiplication factor (K) is given by the expression:

$$K = \frac{\text{No. of neutrons produced per fission}}{\text{No. of neutrons producing fission}}$$

Depending on whether the value of K is equal to, smaller or greater than 1, a fission chain reaction can take place in the following three different ways:

(*i*) When K = 1, the number of neutrons remains constant and a controlled fission chain reaction takes place at a steady rate as in atomic reactors and a controlled amount of heat energy is produced. This energy can be put to peaceful purposes. This reaction occurs when U-235 (or any other fissionable material) has mass equal to its **critical mass.** Critical mass of a fissionable material is its that minimum mass which can undergo chain reaction (fission reaction). If the mass is less than the critical mass, chain reaction will not occur.

(*ii*) When K < 1, some of the secondary neutrons escape into the air and decrease in number with time and hence the fission chain reaction cannot be sustained, *i.e.*, this reaction is not a chain reaction. This reaction occurs when U-235 has sub-critical mass, *i.e.*, U-235 a mass less than its critical mass. Sub-critical mass is quite safe, since in this case no chain reaction occurs.

(*iii*) When K > 1 , the number of neutrons emitted increases and hence the rate of fission chain reaction also increases rapidly, and an uncontrolled reaction occurs. This reaction produces an uncontrollable amount of heat energy in a fraction of a second which leads to explosion as in atomic bomb. This reaction occurs when U-235 has **super-critical** mass, *i.e.*, U-235 has mass greater than its critical mass.

Atomic bomb or fission bomb

First atomic bomb was made in 1945. An atomic bomb consists of several thousands of pieces of U-235 isotope. The mass of each of the pieces should be less than the critical mass of U-235 isotope but the sum of the masses of all the pieces should be greater than the critical mass of U-235 isotope. Between each of the pieces explosive material like trinitrotoulene (TNT) is also placed. Now when TNT is fired, the different pieces of U-235 rapidly come closer together and form a large piece of U-235 whose mass, as said above, is greater than the critical mass of U-235 isotope. Now neutrons obtained from Ra—Be source are allowed to strike the large piece of U-235. This results in that a rapid uncontrolled chain reaction starts. This reaction takes place at an uncontrollable speed, leading to an explosion and hence to the production of an unmanageable quantity of heat energy.

The amount of energy liberated in an atom bomb is equal to that produced by the detonation of about 20,000 tons of TNT. This energy raises the temperature to the order of 10,000,000°C (10^7 °C) and more and produces a pressure of several millions atmosphere. The explosion also produces a violent and intense blast of visible, ultraviolet, X-ray and γ-ray radiations which cause blinding flash.

When fission occurs, the fragments fly apart with tremendous speeds. They collide with each other and then kinetic energy is changed into heat energy. The radioactive fragments formed scatter over wide areas and get deposited on the surrounding objects. These deposited radioactive fragments and different types of radiations have very bad effect on living and non-living objects.

Nuclear Reactor or Atomic Reactor

What is a nuclear reactor?

A nuclear reactor is a kind of furnace in which controlled fission of a radioactive material like U-235 takes place and a manageable amount of nuclear energy (atomic energy) is produced at a steady and slow rate. Nuclear energy thus produced is in the form of heat energy. This heat energy can be converted into electricity in a nuclear power plant. Atomic reactor is also called **atomic pile.**

Components of a nuclear reactor and their functions

A nuclear reactor consists of a moderator, controlling rods, fuel element and coolant. All these parts are enclosed in a chamber which is called *protective screen*. These parts and their functions are discussed below:

How to slow down the speed of secondary neutrons: use of moderators

We know that the fission of U-235 nucleus takes place only when this nucleus is bombarded by slow moving neutrons. We also know that the fission of U-235 nucleus by slow neutrons produces 3 neutrons (called secondary neutrons) which have fairly high speed (fast neutrons). Since the secondary neutrons are fast neutrons, they cannot cause fission of other U-235 nuclei effectively. So, in order to enable these neutrons to cause fission of other U-235 nuclei more effectively, the speed of these neutrons must be slowed down. The speed of neutrons is slowed down by using the substances like pure carbon (graphite), heavy water etc. inside the reactor. These substances are called moderators. Thus a moderator can be defined as a substance which slows down the speed of secondary neutrons produced in the reactor to a level at which they (*i.e.*, neutrons) can cause the fission of other U-235 nuclei effectively. When high speed neutrons collide with the moderator, they lose some of their kinetic energy and thus their speed is slowed down. The slow neutrons thus obtained can cause the fission of new U-235 nuclei. The slowing of the speed of the secondary neutrons makes the fission reaction proceed under control (controlled fission reaction). The reactor in which a graphite core is used as a moderator is called *atomic pole.* The graphite core contains a number of holes in which uranium rods (fuel element) and cadmium or boron rods (which act as controlling rods) are inserted. The reactor in which D_2O is used as a moderator in a tank is called *swimming pool reactor.* Uranium rods (fuel element) are suspended into the tank.

How to decrease the number of excess neutrons (secondary neutrons) produced in the fission of ^{235}U nucleus: use of controlling rods

The essential condition for making the fission of U-235 nucleus proceed smoothly at a controlled rate (controlled fission reaction) is that the excess number of neutrons produced in the reaction must be removed, since it is these excess neutrons which make the nuclear fission of U-235 go out of control (uncontrolled or explosive fission reaction) and proceed at a very fast rate, producing a tremendous amount of energy which can not be used for peaceful purposes.

In order to remove or regulate the excess number of neutrons, cadmium or boron rods are used. These rods are called *controlling rods*, since they control the number of excess neutrons. These rods absorb the excess neutrons and thus remove them. With the removal of excess neutrons, the fission reaction proceeds at steady speed. If the controlling rods are fully inserted into the fuel element, they will absorb all the generated neutrons and hence the reaction will stop. Under this condition the nuclear fire will be extinguished. This is called *shut down position of the reactor.* On the other hand, if the controlling rods are completely withdrawn from the fuel element (*i.e.,* taken out of the reactor), all the generated neutrons will be available for the further fission of other U-235 nuclei and hence the fission reaction will take place at a faster speed, *i.e.*, the fission process will go out of controll and hence a tremendous amount of heat energy will be produced at a fast rate in a short time. This energy will obviously cause explosion.

The above discussion shows that the length of controlling rods in the fuel element should be adjusted in such a way that the fission reaction proceeds under control at a steady speed. Under this condition a controlled amount of heat will be produced at a steady rate. The length of the controlling rods inside the fuel element is adjusted from out side the reactor. If the number of neutrons has to be decreased, the controlling rods should be pushed further into the reactor so that they can absorb the excess neutrons. On the other hand, if the number of neutrons is to be increased, the controlling rods should be pulled out of the reactor.

Whatever has been said above can be summarised by saying that the controlling rods regulate the number of neutrons that can cause the further nuclear fission of U-235 nucleus and thus controls the rate of fission and ultimately controls the amount of heat energy obtained per unit time. This controlled amount of heat energy can be put to useful purposes, *i.e.,* it can be converted into electricity.

Fuel element

The fuel element which is used in a nuclear reactor as a fissionable material is ^{235}U isotope. Natural uranium consists chiefly of two isotopes *viz.* ^{238}U (99.3%) and ^{235}U (0.7%) of which only ^{235}U (lighter isotope) is fissionable. ^{238}U (heavier isotope) can undergo nuclear fission reaction by *very fast neutrons* .

Hence ^{235}U should first be separated from ^{238}U and then employed in the reactor. But since it is very costly, U-238 enriched with U-235 is used in the reactor. Enriched uranium contains both the isotopes of uranium *viz.* U-235 and U-238, but it is U-235 isotope alone which undergoes nuclear fission and hence acts as the real fuel element. Enriched uranium is used in the form of long rods or plates. These uranium rods are inserted into the graphite core (which acts as a moderator) in such a way that cadmium or boron rods (control rods) lie in between the uranium rods.

Coolant or heat transfer agent

Since the energy produced in the reactor is in the form of heat (heat energy), it must be quickly transferred to heat exchanger. The transfer of heat energy to the heat exchanger is done by circulating a coolant in a pipe which is connected to both the reactor and the heat exchanger, and forms a coil inside the heat exchanger. A coolant must have high boiling point and high specific heat. The coolant is pumped into the reactor by means of a pump. The above discussion shows that a coolant can be defined as that substance which transfers the heat energy produced in the reactor to the heat exchanger.

In most successful reactors, *liquid sodium metal* is used as the coolant which transfers the heat energy to the heat exchanger in a short time. In some reactors, *CO_2 gas* and even *water* are used as coolants. *Heavy water, organic compounds (e.g. polyphenyls), air* etc. have also been used as coolants.

Shield or protective screen or protective chamber

All the above components *viz.* moderator, controlling rods and fuel element are enclosed in a chamber which has thick walls made of cement and concrete. These walls stop the nuclear radiations (produced in the reactor) to go out of the reactor and thus save the human beings and animals from their dangerous effects on them. Thus we see that the concrete chamber acts as a shield or protective screen and is called *protective chamber.*

Uses of a nuclear reactor

(*i*) Radioactive isotopes of various elements are produced. These isotopes have been found to be very useful in pure science, biology and medicine.

(*ii*) Heat energy produced in the fission of U-235 nucleus by slow neutrons taking place in the atomic reactor has been used to generate electricity.

(*iii*) It produces fissionable material like plutonium which is used in atom bomb.

(*iv*) It is also used for the production of fast neutrons that are needed for nuclear bombardment.

Nuclear Power Plant

To get electricity from atomic (nuclear) energy with the help of a nuclear power plant

We know that in nuclear reactor, controlled nuclear fission of U-235 nucleus takes place by means of slow neutrons and a manageable amount of nuclear energy (atomic energy) is produced at a steady and slow rate. Nuclear energy thus obtained is in the form of heat energy. This heat energy is converted into electricity (electrical energy) by means of a nuclear power plant. The constitution and working of a nuclear power plant are discussed below:

Constitution of a nuclear power plant

The nuclear power plant consists of a nuclear reactor which is enclosed in protective chamber which is called *shield / protective screen* or *protective chamber.* The nuclear reactor is connected with a heat exchanger by a pipe which forms a coil inside it. The heat exchanger has water in its lower portion, which is connected to a steam turbine. The shaft of the steam turbine is connected to an electric generator or a dynamo. The steam turbine is also connected with a condenser through which cold water is circulated.

We know that a nuclear reactor consists of a graphite core which acts as moderator. The graphite core has a number of holes into which cadmium or boron rods are inserted. These rods act as controlling rods. The length of these rods which are to be inserted into the graphite core (*i.e.,* reactor) can be adjusted from outside the reactor. The adjustment can be done by pushing the controlling rods further into the reactor or by pulling them out of the reactor.

The rods of uranium enriched with U-235 are inserted into the graphite core in such a way that cadmium or boron rods lie in between the uranium rods. These uranium rods act as fuel element.

Liquid sodium is used as a coolant (heat transfer agent) which circulates in a pipe. This pipe is connected to both the nuclear reactor and the heat exchanger and forms a coil in the heat exchanger.

All the above parts *viz.* graphite core (moderator), cadmium or boron rods (controlling rods), uranium rods (fuel element) are enclosed in a chamber which has thick walls made of cement and

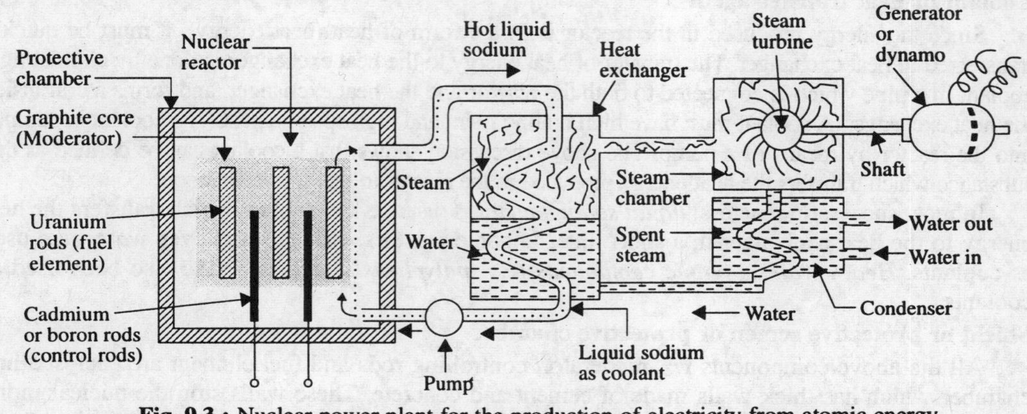

Fig. 9.3 : Nuclear power plant for the production of electricity from atomic energy.

concrete and acts as a shield or protective screen. This chamber is, therefore, called protective chamber (see Fig. 9.3).

Working of a nuclear power plant

First of all graphite core (moderator) is placed in the nuclear reactor and then cadmium or boron rods (controlling rods) are completely inserted into the holes of the graphite core so that these rods may absorb all the neutrons being emitted by the fission of U-235 nucleus and the fission chain reaction may not start at once. After that, rods of uranium enriched with U-235 (fuel element) are inserted into the graphite core in such a way that the cadmium or boron rods may lie in between the uranium rods. The reactor is then sealed.

To start the nuclear reactor, the cadmium or boron rods are pulled so much out of the reactor that a correct number of neutrons are available in the reactor (other neutrons are absorbed by cadmium or boron rods) and hence a controlled fission reaction takes place in the reactor. This controlled fission reaction produces a manageable amount of heat energy at a steady and slow rate. Now, since the energy produced in the reactor is in the form of heat (heat energy), it is quickly transfered to the heat exchanger by circulating liquid sodium (which acts as a coolant) in a pipe which is connected to both reactor and heat exchanger, and forms a coil in the heat exchanger. The coolant is sent into the reactor with the help of a pump. When the coolant reaches inside the reactor, it absorbs heat energy produced in the reactor and becomes extremely hot. This extremely hot coolant is passed into the coil inside the heat exchanger. By gaining heat, water starts boiling and super-heated and highly pressurised steam is produced in the heat exchanger and transfers its heat to water contained in the lower part of the heat exchanger. This steam is passed into steam turbine. Since the steam is super-heated and has high pressure, it drives the steam turbine. When steam turbine rotates, the coil of an electric generator or dynamo (fixed to the shaft of turbine) also rotates and produces electricity. The used steam (*i.e.*, spent steam) coming out of the steam chamber is passed into a condenser where this steam gets cooled and is converted into water. This water is again sent to the heat exchanger for the formation of fresh steam.

Major research laboratories of atomic energy in India

Bhabha Atomic Research Centre (BARC), Bombay is the major research laboratory of atomic energy in India. Other research laboratories are located in Gulmarg in Kashmir and in Calcutta in West Bengal.

Atomic power stations in India

At present there are four atomic power stations (atomic power plants) in India. These are : (*i*) Tarapur atomic power station at Tarapur (Maharashtra), (*ii*) Rajasthan atomic power station at Rana Pratap Sagar near Kota (Rajsthan), (*iii*) Madras atomic power station at Kalpakkam (Tamil Nadu), (*iv*) Narora atomic power station at Narora (Uttar Pradesh).

Nuclear Fusion

(Also called *Atomic Fusion*)

What is nuclear fusion?

Nuclear fusion is a nuclear reaction in which lighter nuclei combine together (*i.e.*, are fused together) to form a single heavy and more stable nucleus and a large amount of energy is also released (called *fusion energy*). For example the following nuclear reaction which shows the formation of one helium nucleus, ${}_2^4He$ (heavy nucleus) by the combination (fusion) of two deuterons, ${}_1^2H$ (light nuclei) is a fusion reaction. In this reaction energy equal to 23.75 MeV is also released.

$$ {}_1^2H + {}_1^2H \xrightarrow{\text{Fusion}} {}_2^4He + 23.7 \text{ MeV.} $$

More examples of nuclear fusion reactions

The following nuclear reactions are the examples of fusion reactions:

$$ 4\,{}_1^1H \longrightarrow {}_2^4He + 2\,{}_{+1}^{0}e + 24.64 \text{ MeV.} $$

$$ {}_1^1H + {}_1^1H \longrightarrow {}_1^2H + {}_{+1}^{0}e + 0.42 \text{ MeV.} $$

$$^{2}_{1}H + ^{1}_{1}H \longrightarrow ^{3}_{2}He + 5.50 \text{ MeV.}$$
$$^{13}_{6}C + ^{1}_{1}H \longrightarrow ^{14}_{7}N + 7.55 \text{ MeV.}$$
$$^{2}_{1}H + ^{2}_{1}H \longrightarrow ^{3}_{2}He + ^{1}_{0}n + 3.3 \text{ MeV.}$$
$$^{2}_{1}H + ^{3}_{1}H \longrightarrow ^{4}_{2}He + ^{1}_{0}n + 17.6 \text{ MeV.}$$

Conditions necessary for fusion reaction

(*i*) When the two lighter nuclei are brought closer to each other to combine together to form the heavy nucleus, the protons present in their respective nuclei repel each other. Hence in order to overcome the repulsive forces, both the lighter nuclei should be given high kinetic energy, *i.e.*, the nuclei should be fused at high temperature. Thus nuclear fusion takes place only when the temperature is very high (4×10^6 degrees celsius). Since nuclear fusion reactions require very high temperature, these reactions are also called *thermo-nuclear reactions* or *temperature- dependent nuclear reactions*

(*ii*) High pressure is also essential so that collisions are frequent.

High temperature and high pressure both can be obtained only by the explosion of an atomic bomb. Therefore, on the earth, fusion reaction can take place during the explosion of an atom bomb. However, in the sun and stars high temperature is already there.

Energy released in the nuclear fusion reactions

We have already said that in a nuclear fusion reaction, energy is released, *i.e.* nuclear fusion reaction is an *exo-ergic raction.* The energy released is called *fusion energy.*

The release of energy takes place because of the fact that in this reaction some mass is lost, since the sum of the masses of the reactants is greater than the sum of the masses of the products. The mass lost is converted into energy which is released in the reaction. The energy is calculated with the help of Einstein's mass energy relationship *viz.* $E = mc^2$ (E = energy, m = mass and c = velocity of light). If mass lost is in *amu,* then energy released

= Mass lost or loss in mass (in *amu*) × 931.5 MeV

= [(sum of the masses of the reactants) – (sum of the masses of the products)] × 931.5 MeV.

The magnitude of energy released in a nuclear fusion reaction is so high that scientists have not as yet been able to control its magnitude, *i.e.*, in a nuclear fusion reaction an uncontrolled amount of energy is obtained and hence this energy cannot be used for a peaceful purpose (*e.g.* for the production of electricity). Note that the magnitude of energy obtained in the fission of ^{235}U nucleus by slow neutrons is also high but it can be converted into a controlled amount in a nuclear reactor and then can be converted into electricity.

Amount of energy released in the formation of one $^{4}_{2}He$ nucleus by the fusion of four $^{1}_{1}H$ nuclei

We know that the formation of one $^{4}_{2}He$ nucleus by the fusion of four $^{1}_{1}H$ nuclei can be shown as

$$4\,^{1}_{1}H \longrightarrow ^{4}_{2}He + 2\,^{0}_{+1}e \text{ (positrons)}$$

Mass of the reactants = 4 × mass of one H-atom

= 4 × 1.0078256 *amu* = 4.0313024 *amu*

Mass of the products = mass of He nucleus + 2 × mass of one positron

= (4.0026 + 2 × 0.0005486) *amu* = 4.0036972 *amu*

∴ Mass lost = (4.0313024 – 4.0036972) *amu* = 0.0276052 *amu*

∴ Energy released in the formation of one helium nucleus ($^{4}_{2}He$) = 0.0276052 × 931.5 MeV

= **25.7140 Mev**

Energy released in the formation of one mole of helium nucleus from four moles of hydrogen

$= 25.7140 \times 6.022 \times 10^{23}$ Mev $= 25.7140 \times 6.022 \times 10^{23} \times 1.602 \times 10^{-16}$KJ

$= 248.0692 \times 10^{7}$ KJ $=$ $\mathbf{24.806 \times 10^{8}}$ **KJ**

Energy released in the fusion of one gram of hydrogen

$$= \frac{24.806 \times 10^8}{4 \times 1.0078} \text{KJ} = \mathbf{6.153 \times 10^8\ KJ}$$

Here it may be noted that the burning of 1 g. of coal in air produces only 0.350 KJ of heat energy. Obviously the difference between 6.153×10^8 and 0.350 is very high.

Origin of the energy of the sun and stars (stellar energy or solar energy)

The temperature in the centre of the sun is extremely high (about 2×10^7 Kelvin). Sun has been providing us continuously heat and light for the last billion of years and will continue to do so for a similar another period, yet its temperature has never decreased nor it is likely to decrease even in the future. Now some questions arise : How is the energy produced in the sun? Why does the temperature of the sun not decrease?

It was in 1932 when Hans Bethe answered the above questions by proposing that the production of heat in the sun is due to the fusion reaction which is going on inside the sun (or stars) all the time.

Analysis of the spectrum of the light emitted by the sun has shown that the sun contains a tremendous quantity of $^1_1$H-atoms (*i.e.*, protons, $^1_1$H). According to Bethe, four hydrogen nuclei (4^1_1H) combine together (*i.e.*, are fused together) at high temperature prevailaing in the sun and form one $^4_2$He nucleus (heavier nucleus) and two positrons ($^0_{+1}$e), and a large amount of heat and light energy is also produced.

$$4\,^1_1\text{H} \xrightarrow[\text{in the sun}]{\text{Fusion at high temperature}} {}^4_2\text{He} + 2\,^0_{+1}\text{e} + 24.64 \text{ MeV.}$$

It is this liberated energy which produces extremely high temperature in the sun and is also supplied in the form of heat and light by the sun to the earth. The above discussion shows that the sun may be regarded as a big thermonuclear furnace. Here it should be noted that the conversion of four $^1_1$H nuclei into one $^4_2$He nucleus which is the source of the solar energy is so slow even at such a high temperature prevailing in the sun that it takes several million years for the conversion of one gram of hydrogen into helium. Why the sun produces energy at high rate is because of the enormous size of the sun.

The above hydrogen-helium fusion reaction does not take place through a single step; rather it occurs through the following two types of cycles :

(i) Proton-proton cycle or (p – p) cycle. This cycle proceeds through the following three steps. In each step energy is released.

(*a*) $\left[{}^1_1\text{H} + {}^1_1\text{H} \rightarrow {}^2_1\text{H} + {}^0_{+1}\text{e} + 0.42 \text{ MeV}\right] \times 2$

(*b*) $\left[{}^2_1\text{H} + {}^1_1\text{H} \rightarrow {}^3_2\text{He} + 5.50 \text{ MeV}\right] \times 2$

(*c*) ${}^3_2\text{He} + {}^3_2\text{He} \rightarrow {}^4_2\text{He} + 2\,^1_1\text{H} + 12.80 \text{ MeV}$

On adding (*a*) × 2, (*b*) × 2 and (*c*), we get :

$$4\,^1_1\text{H} \longrightarrow {}^4_2\text{He} + 2\,^0_{+1}\text{e} + (0.84 + 11.00 + 12.80)$$

or $$4\,^1_1\text{H} \longrightarrow {}^4_2\text{He} + 2\,^0_{+1}\text{e} + 24.64 \text{ MeV.}$$

(ii) Carbon-nitrogen cycle or (C–N) cycle. This cycle operates through the followong six steps. In each step energy is also released

$${}^{12}_6\text{C} + {}^1_1\text{H} \longrightarrow {}^{13}_7\text{N (unstable)} + 1.94 \text{ MeV.}$$

$${}^{13}_7\text{N} \longrightarrow {}^{13}_6\text{C} + {}^0_{+1}\text{e} + 1.20 \text{ MeV.}$$

$${}^{13}_6\text{C} + {}^1_1\text{H} \longrightarrow {}^{14}_7\text{N} + 7.55 \text{ MeV.}$$

$${}^{14}_7\text{N} + {}^1_1\text{H} \longrightarrow {}^{15}_8\text{O (unstable)} + 7.29 \text{ MeV.}$$

$${}^{15}_8\text{O} \longrightarrow {}^{15}_7\text{N} + {}^0_{+1}\text{e} + 1.74 \text{ MeV.}$$

$$^{15}_{7}N + ^{1}_{1}H \longrightarrow ^{12}_{6}C + ^{4}_{2}He + 4.96 \text{ MeV}.$$

On adding all the above six equation we get:

$$4^{1}_{1}H \longrightarrow ^{4}_{2}He + 2\,^{0}_{+1}e + (1.94 + 1.20 + 7.55 + 7.29 + 1.74 + 4.96)$$

$$\text{or} \quad 4^{1}_{1}H \longrightarrow ^{4}_{2}He + 2\,^{0}_{+1}e + 24.68 \text{ MeV}.$$

In (C – N) cycle chain reactions $^{12}_{6}C$ acts as a sort of catalyst, since this istotope is regenerated at the end of these reactions. This cycle does not contribute much to the production of solar energy, since the temperature at the sun's centre (of the order of 2×10^7 Kalvin) is too low for this reaction to proceed.

In certain more massive stars the temperature prevailing is many times higher than the solar temperature. At these high temperatures (C – N) cycle chain reactions proceed very well and is the principal source of the energy of the massive stars.

An interesting problem

We have said above that the heat energy of the sun is generated by the fusion of four $^{1}_{1}H$ nuclei into one $^{4}_{2}He$ nucleus. Thus in the production of heat energy $^{1}_{1}H$ nuclei present in the sun are being consumed continuously due to their conversion into $^{4}_{2}He$ nuclei. Now here a question arises. If the consumption of $^{1}_{1}H$ nuclei goes on continuously, then a day may come when all the stock of hydrogen in the sun will be consumed and no more fusion reaction will take place and hence no energy will be obtained.

The answer to the above question has been given as follows:

When all the $^{1}_{1}H$ nuclei are exhausted by their conversion into $^{4}_{2}He$ nuclei, the temperature and pressure in the sun will decrease and the sun will shrink in size under its own gravitational attraction. When the sun has considerably shrunk in its size, the temperature in the sun will again become high—so high that $^{4}_{2}He$ nuclei will fuse together to form $^{12}_{6}C$ isotope and a tremendous amount of energy will be liberated ($3^{4}_{2}He \rightarrow ^{12}_{6}C$ + Energy). In this way sun will regain its lost sun-shine and will become as bright as ever.

Hydrogen bomb

Hydrogen bomb consists of (deuterium + tritium) core, which is surrounded by an atomic bomb assembly. When atomic bomb is detonated, explosion takes place and extremely high temperature and pressure are attained. High temperature and high pressure thus obtained make $^{2}_{1}H$ and $^{3}_{1}H$ nuclei to fuse together to produce helium nucleus ($^{4}_{2}He$) and an uncontrolled amount of energy is also generated.

$$^{2}_{1}H(\text{Deuterium}) + ^{3}_{1}H\,(\text{Tritium}) \xrightarrow{\text{Fusion}} ^{4}_{2}He + {}_{0}n^{1} + 17.6 \text{ MeV}.$$

This reaction takes place at a very high speed which cannot be controlled, *i.e.*, an uncontrolled fusion reaction takes place in hydrogen bomb. Since atomic bomb is used to produce high temperature and high pressure which are essential for the fusion of $^{2}_{1}H$ and $^{3}_{1}H$ nuclei to produce $^{4}_{2}He$ nucleus, this bomb is named as *trigger.*

Comparison between hydrogen bomb and atomic bomb

(*i*) We have already said that in an atomic bomb, uncontrolled fission reaction takes place. For the occurrence of such a reaction the mass of the fissionable material (like ^{235}U nucleus) should be greater than its critical mass. On the other hand, in a hydrogen bomb any amount of the fusible materials can be taken.

(*ii*) The amount of energy released in a hydrogen bomb is several times greater than that produced due to the explosion of atomic bomb.

Comparison between nuclear fission reactions and nuclear fusion reactions

(*A*) **Similarity.** The only similarity between these two types of reactions is that the cause of the production of energy is the same and that is loss in mass (mass defect), *i.e.*, energy is released because of the fact that the sum of the masses of the products is less than the sum of the masses of the reactants. The mass which is lost is converted into energy and this energy is generated in the reaction.

(*B*) **Dissimilarities.** (*i*) Nuclear fission is a nuclear reaction in which a heavier nucleus splits up into two lighter nuclei. On the other hand nuclear fusion reaction is that nuclear reaction in which two lighter nuclei combine together (fuse together) to form a heavier nucleus.

(*ii*) Nuclear fission is a chain reaction while nuclear fusion is not so, since the product formed in this reaction is a single nucleus which is stable.

(*iii*) Nuclear fission is carried out by using slow moving neutrons while nuclear fusion reaction takes place at high tempurature and high pressure, since these are essential for overcoming the repulsive forces, existing betwen the protons present in the lighter nuclei which are fusing together to form the heavy nucleus.

(*iv*) The amount of energy produced in nuclear fusion reaction is much more than that generated in a fission reaction, since the loss in mass in a fusion reaction is greater than that in a fission reaction.

(*v*) The products (*e.g.* $^{139}_{56}Ba$ and $^{94}_{36}Kr$) obtained in a nuclear fission reaction are radioactive and emit harmful radioactive radiations. Thus, these products are harmful. So the problem of the disposal of these harmful products arises. The products (*e.g.* $^{4}_{2}He$) obtained in the fusion reaction are not radioactive and hence are harmless.

University Questions

1. (*a*) What is a nuclear reaction? How are they classified?
(*b*) Describe the different types of nuclear reactions, giving examples. *(Gauhati 89, Agra 92, Meerut M.Sc. 93)*

2. (*a*) Explain the differences between chemical and nuclear reactions. *(Meerut M.Sc. 92, 94)*
(*b*) Discuss with examples important nuclear reactions caused by neutrons. *(Meerut M.Sc. 93)*

3. (*a*) Energy is released in nuclear fusion as well as in nuclear fission. Explain why? *(Meerut 90, 93, 95; Utkal 92)*
(*b*) Show that nuclear fission is a chain reaction. *(Rohilkhand 90)*

4. (*a*) Explain the liquid drop model for nuclear structure. *(Devi Ahilya 92)*
(*b*) Very high temperatures exist around the sun. Why? *(Lucknow, 95)*

5. Select the correct answer :
(*i*) Nuclear reaction is possible from the following particle: (*a*) Alpha particle (*b*) Electron (*c*) Positron (*d*) All the above.
(*ii*) Isotope used in atom bomb is (*a*) U^{238} (*b*) U^{237} (*c*) U^{235} (*d*) all the above. *(Ravishankar 93)*
(*iii*) $^{27}_{13}Al$ is a stable isotope. $^{24}_{13}Al$ is expected to disintegrate by (*a*) α-emission (*b*) β-emission (*c*) positron emission (*d*) proton emission. *(I.I.T. 96)*

6. What is the Q value (in ergs) of the reaction ^{14}N (α, p) ^{17}O? The sum of ^{14}N and ^{4}He masses is 18.005678 amu and sum of ^{17}O and ^{1}H masses is 18.006958 *amu*. *(Meerut M.Sc. 93)*

10

Different Types of Chemical Bonds

What is a chemical bond ?

A chemical bond is defined as the attractive force that holds various constituent particles (atoms, ions or molecules) in different chemical species.

Why do atoms combine together? — Cause of chemical bonding.

The atoms combine together for the following reasons:

1. **Concept of lowering of energy of combining atoms.** When two atoms combine together to form a bond, there is an overall decrease in the potential energy of the combining atoms, *i.e.*, a system having bonded atoms has lower energy than that having the unbonded atoms. This implies that the system of bonded atoms having lower energy is more stable than that of unbonded atoms having higher energy. *From this it, therefore, follows that the process of chemical bonding between the atoms decreases the energy of the combining atoms and gives rise to the formation of a system which has lower energy and hence has greater stability.*

2. **Octet rule or rule of eight : Electronic theory (or octet theory) of valency.** Lewis, Kossel and Longmuir (1916) tried, for the first time, to explain *why atoms combine together* on the basis of the electronic configuration of noble gasses as given here : He (Z = 2) → 2; Ne (Z = 10) → 2, 8; Ar (Z = 18) → 2, 8, 8; Kr (Z = 36) → 2, 8, 18, 8; Xe (Z = 54) → 2, 8, 18, 18, 8; Rn (Z = 86) → 2, 8, 18, 32, 18, 8. They assumed that since the atoms of noble gases do not normally react with other atoms to form compounds, it is reasonable to assume that the outermost shell configuration of the atoms of noble gases is a stable configuration of 8 electrons which they called an *octet.* They also concluded that the two electrons in case of helium (called *duplet*) is also as stable as an octet present in other noble gases. *Since the octet of electrons is so stable in the gases, one can reasonably assume that when atoms of other elements combine to form a molecule, the electrons in their outermost orbits are arranged between themselves in such a way that they achieve an octet of electrons which is stable and thus a chemical bond is established between the atoms.*

The tendency of the atoms to have eight electrons in their outermost shell is known as **octet rule** or **rule of eight** Since helium atom has only two electrons, this rule is called **doublet rule** or **rule of two** in case of helium. Octet rule was given in the form of a theory which is known as *octet theory of valency* or *electronic theory of valency* which states that:

In the formation of a chemical bond atoms interact with each other by losing, gaining or sharing of electrons so as to acquire a stable outer-shell oj eight electrons.

The main points of this theory can be summarised as follows:

(*i*) Atom with 8 electrons in the outer most shell (2 in case of helium) are chemically stable and hence are incapable of chemical combination.

(*ii*) An atom having less than 8 electrons in its outer-most shell is chemically active and hence has a tendency to combine with other atoms. The atoms possessing less than 4 electrons in their ultimate shell usually tend to lose them, while those having more the 4 electrons in the outermost shell tend to gain the electrons during the chemical combination or bond formation to attain stable configuration of the nearest inert gas.

(*iii*) Atoms combine chemically as a result of transferring of electrons from the outer-most shell of one atom to that of the other or by sharing one, two or three electron pairs between the valence-shell of both the combining atoms. The transfer of electrons or sharing of electron pairs gives a stable configuration of 8 electrons to the valence-shell of both the atoms.

(*iv*) The tendency of an atom for transference or sharing its electron pairs is a measure of its chemical activity.

Types of bonds

Main types of bonds are *ionic bond* (also called *electrovalent bond*), *covalent bond* and *coordinate covalent bond.* In addition to these bonds, *hydrogen bond, odd electron bond, metallic bond* and *van der Waals forces (inter-molecular forces)* are also known. In this chapter we shall discuss all these bonds.

Ionic Bond

What is ionic bond?

An ionic bond (also called electrovalent bond) is that bond which is formed when one or more electrons from the valence-shell of an atom (usually a highly electropositive element) are completely transferred to the valence-shell of another atom (usually a highly electronegative element). Thus an ionic bond is established between the two atoms whose electronegativity difference is very high.

The atom that loses electrons becomes positively-charged (cation) while the atom that accepts electrons becomes negatively-charged (anion). The cations and anions thus obtained are then attracted towards each other by the electrostatic force of attraction and are thus linked together by an ionic bond. Thus an ionic bond can also be defined as the electrostatic force of attraction existing between the cations and anions of an ionic compound. In the formation of an ionic bond or ionic compound energy is released. The cations and anions obtained by the loss or gain of electrons attain the nearest noble gas configuration.

Formation of ionic bonds in some ionic compounds

How an ionic bond is formed in ionic compounds can be explained by considering the formation of NaCl, MgO, CaF_2 and Al_2O_3 molecules. The formation of each of these molecules consists of three steps shown below:

NaCl (*a*) $Na\ (2, 8, 1) \xrightarrow{-e^-} Na^+\ (2, 8) + e^-$

(*b*) $Cl\ (2, 8, 7) \xrightarrow{+e^-} Cl^-\ (2, 8, 8)$

(*c*) $Na^+\ (2, 8) + Cl^-\ (2, 8, 8) \longrightarrow Na^+Cl^-$

MgO (*a*) $Mg\ (2, 8, 2) \xrightarrow{-2e} Mg^{2+} + 2e^-$

(*b*) $O\ (2, 6) \xrightarrow{+2e} O^{2-}\ (2, 8)$

(*c*) $Mg^{2+}\ (2, 8) + O^{2-}\ (2, 8) \longrightarrow Mg^{2+}\ O^{2-}$

CaF_2 (*a*) $Ca\ (2, 8, 8, 2) \xrightarrow{-2e^-} Ca^{2+}\ (2,8,8) + 2e^-$

(*b*) $F\ (2, 7) \xrightarrow{+e^-} F^-\ (2,8)$

(*c*) $Ca^{2+}(2, 8, 8) + 2F^-\ (2, 8) \longrightarrow Ca^{2+}.2F^-$ or $Ca^{2+}(F^-)_2$

Al_2O_3 (*a*) $Al\ (2, 8, 3) \xrightarrow{-3e^-} Al^{3+}\ (2, 8) + 3e^-$

(*b*) $O\ (2, 6) \xrightarrow{+2e} O^{2-}\ (2, 8)$

(*c*) $2Al^{3+}\ (2, 8) + 3O^{2-}\ (2, 8) \longrightarrow (Al^{3+})_2\ (O^{2-})_3$

Favourable conditions for the formation of an ionic bond in M^+X^- ionic crystal

We know that the formation of the ionic crystal, M^+X^- takes place through the following steps:

(*i*) When energy equal to ionisation energy of atom M is supplied to it, it loses an electron and is converted into M^+ cation.

$$M^+ + \text{Ionisation energy of M} \rightarrow M^+ + e^-$$

(*ii*) Now the atom X takes up the electron given up by M and is converted into X^- anion. In the formation of X^- ion, energy equal to the electron affinity of X is released.

$$X^+ + e^- \rightarrow X^- + \text{Electron affinity of X.}$$

(*iii*) Now M^+ and X^- ions combine together due to the electrostatic force of attraction existing between them and form M^+X^- ionic crystal. In this process, energy equal to lattice energy of M^+X^- ionic crystal is released.

$$M^+ + X^- \rightarrow M^+X^- \text{ (ionic crystal)} + \text{Lattice energy of } M^+X^-$$

Different steps shown above indicate that the favourable conditions for the formation of an ionic bond or an ionic compound can be described as follows:

(*i*) *Low ionisation energy of atom M*. The ionisation energy of atom M should be low so that this atom may lose the electron readily to form the cation, M^+.

(*ii*) *High electron affinity of atom X*. The electron affinity of atom X should be high so that this atom may accept the electron (lost by M) readily to form the anion, X^-.

Since the elements lying on the extreme left of the periodic table (*i.e.*, metals of groups IA and IIA) have the lowest ionisation energies and the elements lying on the extreme right of the periodic table (*i.e.*, non-metals of groups VIA and VIIA) have the highest values of electron affinity, the formation of ionic bond is favoured between metals and non-metals, *i.e.*, the combination of metals and non-metals gives ionic compounds.

(*iii*) *High value of lattice energy of M^+X^- ionic crystal.* Greater is the force of attraction between the constituent ions in an ionic solid, greater is the stability of the ionic solid. Greater is the magnitude of lattice energy of an ionic solid, greater is the electrostatic force of attraction between the ions (lattice energy is the direct measure of the electrostatic force of attraction between the ions) and hence greater is the stability of the ionic solid. Thus in order to get strong ionic bond in an ionic solid, its lattice energy should be very high.

(*iv*) *Electronegativity difference between M and X atoms very high.* The electronegativity difference between M and X atoms should be very high so that complete transfer of electron(s) form M (less electronegative atom) to X (more electronegative atom) may be possible.

General properties of ionic compounds

See "*Properties of ionic crystals*" discussed in the chapter on "*Structure of Solids.*"

Energy changes involved in the formation of ionic compounds

The energy changes that take place in the formation of ionic solids can be studied by considering the formation of NaCl(*s*) ionic crystal. For this see "*Born-Haber Cycle*" discussed under "*Lattice Energy*" in the chapter on "*Structure of Solids*".

Electrovalency or electrovalence

In the formation of electrovalent compounds, the number of electrons lost by an atom of an element to form the cation or gained by the atom to form an anion, both having the nearest inert gas configuration (ns^2p^6) is known as the electrovalency or electrovalence of that element. In other words, electrovalency of an element is equal to the number of positive or negative charges acquired by the atom of that element in forming cation or anion respectively by losing or gaining electron(s). Thus electrovalency is either positive electrovalency or negative electrovalency.

Variable electrovalency (oxidation states) shown by p-block and d-block elements

There are many *p*-block (*e.g.* Ga, In, Tl, Ge, Sn, Pb etc.) and *d*-block (*e.g.* Fe, Cu etc.) elements which form cations with variable positive electrovalency. The phenomenon in which the elements show more than one electrovalency is called variable electrovalency.

(*a*) *p-block elements.* The heavier elements of groups IIIA (*e.g.* Ga, In, Tl) and IVA (Ge, Sn, Pb) show variable electrovalency as shown below:

Group IIIA : Ga = +3, +1; In = +3, +1; Tl = +3, +1
Group IVA : Ge = +4, +2; Sn = +4, +2; Pb = +4, +2

These variable oxidation states can be explained on the basis of "*Inert Pair Effect*" discussed on page 247.

(*b*) *d-block elements.* There are many transition elements like Fe, Cu etc. which form cations with variable positive electrovalancy. For example Fe shows two positive oxidation states *viz.* +2 (*ferrous*, Fe^{2+}) and +3 (*ferric*, Fe^{3+}). The reason for variable oxidation states is that $(n-1)d$ and *ns* orbitals in the atoms of these elements have almost the same energy and hence the electrons can be lost from $(n-1)d$ orbitals as easily as they can be lost from *ns* orbitals. Thus oxidation states of +2, + 3 and + 4 shown by Ti (2, 8, $3s^2p^6d^2$, $4s^2$) can be explained as follows. This element shows +2 oxidation state when both of its *4s* electrons are lost [Ti (2, 8, $3s^2p^6d^2$, $4s^2$) – $2e^-$ $\rightarrow$ Ti^{2+} (2, 8, $3s^2p^6d^2$)]. It shows +3 oxidation state when both 4*s* and one *3d* electrons are lost [Ti(2, 8, $3s^2p^6d^2$,$4s^2$] – $3e^-$ $\rightarrow$ Ti^{3+} (2, 8, $3s^2p^6d^1$)] and shows +4 oxidation state when both *4s* and both *3d* electrons are lost [Ti(2, 8, $3s^2p^6d^2$, $4s^2$] - $4e^-$ $\rightarrow$ Ti^{4+} (2, 8, $3s^2p^6$).

Inert Pair Effect

What is inert pair effect?

Some oxidation states shown by the elements of groups IIIA, IVA and VA (*p*-block elments) are given below.

Group IIIA (ns^2p^1) $\rightarrow$ B = +3; Al = +3; Ga = +3, +1; In = +3, +1, Tl = +3, +1
Group IVA (ns^2p^2) $\rightarrow$ C = +4; Si = +4; Ge = +4, +2; Sn = +4, +2; Pb = +4, +2
Group VA (ns^2p^3) $\rightarrow$ N = +3, P = +3, +5; As = +3, +5; Sb = +3, +5; Bi = +3, +5

From these oxidation states it may be seen that the elements with higher atomic numbers [*e.g.* Ga, In, Tl (group IIIA); Ge, Sn, Pb (group IVA) and As, Sb, Bi (group VA)] show two types of oxidation states (variable oxidation states) which are given below:

(*i*) *Higher oxidation states (Group oxidation states, G):* These are equal to the group number of the elements or the number of electrons present in the valence-shell or in *ns* and *np* orbitals. These oxidation states are called *group oxidation states* and are represented as G. These oxidation states are obtained when all the electrons in the valence-shell (*i.e.*, all the electrons in the *ns* and *np* orbitals) are involved in bond formation.

(*ii*) *Lower oxidation states.* These are equal to group oxidation state minus two = (G – 2). These result when only the electrons in the *np* orbitals are involved in bond formation and ns^2 electron pair shows reluctance to participate in the bond formation. The reluctance of ns^2 electron pair to participate in bond formulation is called inert pair effect.

It may be seen that the group oxidation state (G) of an element is higher than its oxidation state of (G – 2).

Cause of inert pair effect

In order to explain the cause of inert pair effect let us consider, as an example, the valence-shell electronic configuration of the atoms of the elements of group IVA which is ns^2np^2. Since the two *np* electrons are farther away from the nucleus than the two *ns* electrons,the former (*i.e.*, 2 np^2 electrons) can easily be removed or be excited to the higher orbitals than the latter (*i.e.*, ns^2 electrons) and hence the two *ns* electrons are thought to be inert, *i.e.* ns^2 electrons remain paired and donot take part in bonding especially in the heaviest element namely lead. The above discussion shows that lead forms Pb^{2+} more easily than Pb^{4+}, *i.e.*, lead prefers to exist in +2 oxidation state (lower oxidation state) and shows reluctance to exist in +4 oxidation state (higher oxidation state).

Variation of inert pair effect in a group

Inert pair effect increases (*i.e.*, becomes more and more effective) as we move down the group. Thus this effect is absent in the elements situated in the beginning of a group and becomes prominent in the heavier elements of the group. This means that as we move down the group, ns^2 electron pair present in the cations having lower oxidation states becomes more and more inert and hence tends to remain paired. Thus as we move down the group, ns^2 electron pair, due to its stability, is not easily lost to form the cation having higher oxidation state.

Consequences or applications of inert pair effect: Explanation of the relative stability of various oxidation states of the heavier elements of groups IIIA, IVA and VA

The relative stability of the two types of oxidation states shown by the heavier elements of groups IIIA, IVA and VA can be explained with the help of this effect. This is shown below:

1. Explanation of relative stability of +1 and +3 oxidation states of the elements of group IIIA. Since inert pair effect is absent in B and Al which are situated at the beginning of this group, both these elements lose all their three ns^2p^1 electrons to form B^{+3} and Al^{+3} cations. (As a matter of fact, boron atom, due to its small size and high electron density, forms covalent compounds and hence does not give B^{+3} ion). Thus B and Al show consistantly +3 oxidation state (group oxidation state) while the heavier elements namely Ga, In and Tl which show inert pair effect exhibit +1 (lower oxidation state) and +3 (higher oxidation state) oxidation states. Since inert pair effect in Ga^+ and In^+ ions both is not effective to any great extent, $4s^2$ electron pair in Ga^+ ion ($Ga = 4s^2p^1$, $Ga^+ = 4s^2p^0$) and $5s^2$ electron pair in In^+ ion ($In = 5s^2p^1$, $In^+ = 5s^2p^0$) is not stable or inert and is, therefore, lost easily to convert these monovalent cations into more stable trivalent cations namely Ga^{+3} and In^{+3} respectively. Thus Ga^{+3} and In^{+3} , both are more stable than Ga^+ and In^+ respectively ($Ga^{+3} > Ga^{+1}$, $In^{+3} > In^{+1}$). In other words we can say that Ga and In are more stable in their group oxidation states (+3). Being less stable, Ga (+1) and In (+1) compounds are readily oxidised to more stable Ga (+3) and In (+3) compounds and in this conversion they reduce other substances. Consequently Ga and In in their lower oxidation states (= +1) act as reducing agents.

Now since the inert pair effect in Tl^+ ion, which is the last element (heaviest element) of this group, takes place to the maximum extent, $6s^2$ electron pair of Tl^+ ion ($Tl = 6s^26p^1$, $Tl^+ = 6s^26p^0$) is very stable or inert and hence is not lost easily to convert Tl^+ to Tl^{3+}. Thus Tl^+ ion is more stable than Tl^{3+} ion ($Tl^+ > Tl^{3+}$). Being less stable, Tl(+3) compounds are readily reduced to Tl(+1) compounds and in this conversion they oxidise some other substance. Consequently Tl(+3) compounds act as oxidising agents while Tl(+1) compounds act as reducing agents. In other words Tl in its higher oxidation state (= +3) acts as an oxidising agent and in its lower oxidation state (= +1) behaves as a reducing agent.

Now since the inert pair effect increases from Ga to Tl, the stability of Ga^+, In^+ and Tl^+ also increases as : $Ga^+ < In^+ < Tl^+$ while that of Ga^{+3}, In^{+3}, and Tl^{+3} decreases in the same direction ($Ga^{+3} > In^{3+} > Tl^{+3}$). Thus we find that the stability of lower oxidation states (= +1) increases and that of higher oxidation states (= +3) decreases as we move down in the series Ga $\longrightarrow$ In $\longrightarrow$ Tl. This discussion clearly shows that the most common oxidation state (*i.e.*, most stable oxidation state) shown by B, Al, Ga and In is +3 (group oxidation state) while that for Tl is +1 (lower oxidation state). In other words we can say that the last element *viz.* Tl of this group shows reluctance to exist in higher oxidation state (= +3); rather it prefers to exist in lower oxidation (= +1).

2. Explanation of the relative stability of +2 and +4 oxidation states of the elements of group IVA. Carbon and silicon both show consistantly +4 oxidation state (group oxidation state) while the heavier elements *viz.* Ge, Sn and Pb show +2 (lower oxidation state) as well as +4 (higher oxidation state) oxidation states. On the same lines of arguments as used for the elements of group IIIA, the following conclusions can be drawn in connection with the relative stability of various oxidation states shown by the elements of this group.

(*i*) C and Si both show uniformly +4 oxidation state.

(*ii*) Ge^{4+} and Sn^{4+} both are more stable than Ge^{2+} and Sn^{2+} respectively ($Ge^{2+} < Ge^{4+}$ and $Sn^{2+} < Sn^{4+}$). In other words Ge and Sn both are more stable in their group oxidation state (= +4 state) and hence Ge (+2) and Sn (+2) compounds are good reducing agents as is evident from the following reducing properties shown by Ge (+2) and Sn (+2) compounds.

$$M^{2+} \text{ (M = Ga, Sn)} + 2Fe^{3+} \longrightarrow 2Fe^{2+} + M^{4+}$$

$$2M^{2+} + [2Cr^{+6} O_4]^{2-} + 16H^+ \longrightarrow 2Cr^{3+} + 8H_2O + 3M^{4+}$$

$$Sn^{+2} S + (NH_4)_2 S_2 \longrightarrow (NH_4)_2 Sn^{+4} S_3$$

$$2Sn^{2+}Cl_2 + 2I_2 \longrightarrow Sn^{+4} Cl_4 + SnI_4$$

$$3 Sn^{+2} + \left[Cr_2^{+6} O_7\right]^{2-} + 14H^+ \longrightarrow 3Sn^{+4} + 2Cr^{+3} + 7H_2O$$

(*iii*) Pb(+2) compounds are more stable than Pb(+4) compounds ($Pb^{2+} > Pb^{4+}$) and hence

Pb(+4) compounds are oxidising agents as is evident from the following oxidising properties shown by Pb(+4) compounds:

$$2Mn^{+2}SO_4 + 5Pb^{+4}O_2 + 6HNO_3 \rightarrow 2HMn^{+7}O_4 + 2Pb^{+2}SO_4 + 3Pb^{+2}(NO_3)_2$$

Or $$2Mn^{+2} + 5Pb^{+4}O_2 + 4H^+ \rightarrow 2[Mn^{+7}O_4]^- + 5Pb^{+2} + 2H_2O$$

$$2Cr^{+3}(OH)_3 + 10KOH + 3Pb^{+4}O_2 \rightarrow 2K_2Cr^{+6}O_4 + 2K_2Pb^{+2}O_2 + 8H_2O$$

$$Pb^{+4}O_2 + 4H^+Cl^- \rightarrow Pb^{+2}Cl_2 + 2H_2O + Cl_2^0$$

(*iv*) The stability of lower oxidation state (+2) increases ($Ge^{+2} < Sn^{+2} < Pb^{+2}$) and that of higher oxidation state (+4) decreases ($Ge^{+4} > Sn^{+4} > Pb^{+4}$) as we move down the group from Ge to Pb. The increasing order of +2 state is supported by the fact that only a few compounds of Ge^{+2} are known while Pb^{+2} forms many compounds. Similarly the decreasing order of +4 state is evident from the fact that very few compounds of Pb^{+4} have been obtained and they are unstable, $GeBr_4$, GeI_4, GeS_2, Ge_3N_4, $SnBr_4$, SnI_4, SnS_2 Sn_3N_4 etc. in which Ge and Sn both are in +4 oxidation state are stable compounds.

(*v*) The most common (*i.e.*, the most stable) oxidation states of C, Si, Ge and Sn is +4 while that of Pb is +2.

3. Explanation of the relative stability of +3 and +5 oxidation states of the elements of group VA. We know that the valence-shell configuration of the atoms of these elements is ns^2p^3 which suggests that all these elements show +3 and +5 oxidation states. +3 state is caused when only three *np* electrons are involved in chemical bonding and +5 state is obtained when all the three *np* and both *ns* electrons are involved. On the basis of inert pair effect it can be shown that the stability of +3 oxidation state (lower oxidation state) increases and that of +5 (higher oxidation state) decreases as we move down the group from N to Bi. The last element *viz.* Bi prefers to show +3 oxidation state and shows reluctance to exhibit +5 oxidation state. Thus Bi in +3 oxidation state is more stable than when it is in +5 oxidation state. Consequently Bi forms only trihalides and not pentahalides.

Summary.

The above discussion shows that, due to inert pair effect, the lighter elements of groups IIIA, IVA and VA prefer to show higher oxidation state (which is equal to group oxidation state, G) and have reluctance to exhibit the lower oxidation state [which is equal to (G – 2)]. On the other hand, the heaviest elements *viz.* Tl, Pb and Bi of these groups respectively prefer to show lower oxidation state and have reluctance to exhibit the higher oxidation state.

Covalent Bond

Introduction

According to Lewis-Langmuir (1919), a covalent bond is that bond which is formed between the two combining atoms by the mutual sharing of one, two or three electron pairs between them. By the mutual sharing of electron pair(s) each of the two combining atoms attains stable noble gas configuration. One electron of each electron pair is contributed by one atom while the other electron is contributed by the other atom. Thus both the combining atoms have equal claim on the shared electron pair.

The covalent bond formed by the sharing of *one, two* or *three* electron pairs between the participating atoms is called *single* (represented as —, *e.g.* H – H, Cl – Cl), *double* (designated as =, *e.g.* O = O) and *triple* (designated as ≡, e.g. N ≡ N) covalent bond respectively. Double and triple covalent bonds are collectively called *multiple covalent bonds.* The formation of single covalent bond in H_2, double bond in O_2 and triple bond in N_2 molecules can be illustrated as follows:

H× · H or H—H, $\overset{\times\times}{\underset{\times\times}{O}}\,{}^{\times}_{\times}\!:\overset{\cdot\cdot}{\underset{\cdot\cdot}{O}}$ or $\underset{\times\times}{\overset{\times\times}{O}}=\underset{\cdot\cdot}{\overset{\cdot\cdot}{O}}$, $\overset{\times\times}{N}\,\vdots\!:\overset{\times\times}{N}$ or $\overset{\times\times}{N}\equiv\overset{\times\times}{N}$

The compounds containing covalent bonds are called *covalent compounds*.

Factors favouring the formation of covalent bond between A and B atoms

The factors which favour the formation of a covalent bond between the two atoms A and B are given below:

(i) A and B atoms should have higher values of ionisation potential. A and B atoms should have higher values of ionisation potential so that none of them should be able to lose electron(s) and form cations. Thus, when the elements having high values of ionisation potential combine together, they form covalent bond between them (*e.g.* Cl — Cl, F — F etc.).

(ii) Electronegativity of A and B atoms should be equal or slightly different from each other. When two identical atoms (*i.e.*, the atom having the same electronegativity) combine together, a covalent bond is formed between them, since they have the same electronegativity. Due to the same electronengativity of the atoms, the electrons pair being shared by both the atoms is attracted equally by those atoms and hence lies exactly at the mid point of the distance between the two identical linked atoms. Thus a symmetrical distribution of the electron pair between the two atoms is obtained. This symmetrical distribution gives rise to the formation of a purely covalent bond between the two atoms. When the electronegativities of A and B are slightly different from each other, a polar covalent bond is formed between them. If B is more electronegative than A, polar covalent bond is represented as $A^{\delta+} - B^{\delta-}$ and if the electronegativity of A is more than that of B, the polar covalent bond is represented as $A^{\delta-} — B^{\delta+}$.

(*iii*) Colvalent bond is formed between A and B atoms when they equally share one (single bond), two (double bond) or three (triple bond) pairs of electrons with each other. Thus for the formation of covalent bonds between the two atoms, electron pairs must be shared equally between them.

(*iv*) In terms of overlap of atomic orbitals, a covalent bond is obtained by the overlap of only those atomic orbitals of the valence-shell which contain unpaired electrons only and these unpaired electrons must have opposite spins. If the electrons have the same (*i.e.*, parallel) spins, due to the repulsion between them, the overlap of orbitals will increase the potential energy of the system and the system will become unstable, *i.e.*, no bond will be established.

Important properties of covalent compounds

Some of the properties of covalent compounds are summarised below:

(*i*) *Physical state*: The covalent compounds are usually made up of discrete molecules which are held together by van der Waals forces which are very weak. Due to the weak nature of these forces, the covalent compounds usually exist in all the three states, namelly solids (*e.g.* I_2), liquids (*e.g.* Br_2) and gases (e.g. Cl_2) at ordinary temperature.

(*ii*) *Crystal structure*. The crystals of covalent compounds are of three types.

(*a*) Those in which the molecules are small and held together by van der Waals forces. Organic compounds and sulpher (S_8) are the examples of such type of crystals. Since the van der Waals forces are very weak, very small work is required to separate these molecules and hence the crystals of this type are soft, easily fusable and volatile.

(*b*) Those in which every atom is united with the other by covalent links forming infinite three dimensional covalent structures called giant molecule. In these molecules strong forces operate in all the directions and hence these compounds are very hard and have high melting points. Diomond and silica are the examples of such type of covalent compounds.

(*c*) Those which consits of separate layers. Graphite is an example of such type of covalent crystals.

(*iii*) *Melting and boiling points.* Due to the existance of very weak van der Waals forces, energy required to break these weak forces is low and hence the melting points and boiling points of covalent compounds, excepting those which form giant molecules, are comparatively low.

(*iv*) *Electrical conductivity.* When an electric current is passed through a covalent compound in its liquid or fused state, due to the absence of free electrons and ions in them, the conduction of electric current does not take place and hence these compounds act as bad conductors of electricity. Graphite which is a net work covalent compound is an exception, since, due to the presence of free electrons in its structure, it is a good conductor of electricity.

(*v*) *Solubility in polar and non-polar solvents.* Most of the covalent compounds are insoluble in polar solvents, since covalent malecules do not contain ions and hence do not interact with the dipolar solvent molecules 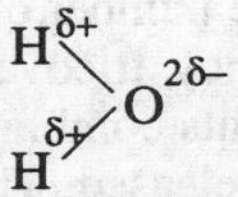 to form a hydrated ions. Due to the non-formation of hydrated ions, no hydration energy is released. Covalent compounds are soluble in non polar solvents (*e.g.* C_6H_6, CCl_4 etc.), because they have got merely the same molecular structures, *i.e.*, because both of them are of covalent nature. High polymeric covalent compounds and those forming gaint molecules are virtually insoluble in any solvent.

(*vi*) *Moleculer reactions.* The aqueous solution of covalent compounds donot contain any ions and hence donot undergo ionic reactions as ionic compounds do. These compounds, however, undergo moleculer reactions which are very slow and need a control of temperature and pressure.

(*vii*) *Soft and waxy.* Covalent compounds are soft and waxy, since they consist of separate molecules.

(*viii*) *Directional nature of covalent bonds, geometry and isomerism in covalent compounds.* Covalent bonds are directed in particular directions in space, *i.e.*, covalent bonds are directional and rigid. It is due to the directional nature of covalent bonds that covalent compounds have definite geometry (*e.g.* linear, trigonal planar, tetrahedral etc.) and hence show space isomerism. Mainly organic compounds which are covalent show isomerism.

Covalency and Maximum Covalency of an Element

What is covalency or covalence of an element?

Covalency of an element in its covalent compounds or ions is the total number of covalent and coordinate bonds which one atom of that element can form with the neighbouring atoms. In other words, covalency of an element is the number of electrons contributed by one atom of the element in forming covalent bonds with the neighbouring atoms. For example:

(*a*) Covalency of H-atom in each of the molecules *viz.* H_2(H—H), H_2O (H—O—H) and HCl (H—Cl) is equal to 1.

(*b*) Covalency of N-atom in NH_3 molecule 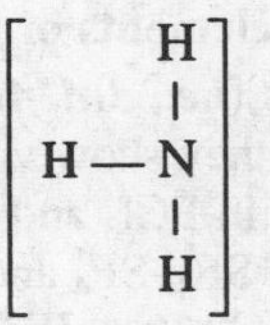 is 3 but its covalency in NH_4^+ ion, 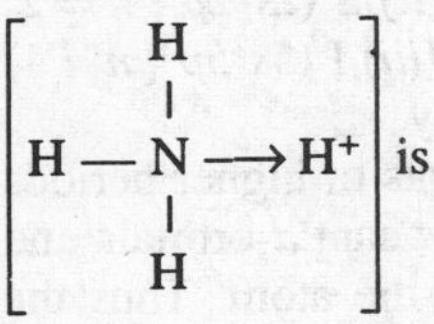 is 4.

(*c*) Covalency of O-atom in O_2 (O = O) and H_2O (H – O – H) molecules is equal to 2.

Covalency of hydrogen and of the elements of 2nd period

The covalency of H ($1s^1$, n = 1), N ($2s^22p^3$, n = 3), O($2s^22p^4$, n = 2) and F ($2s^22p^5$, n = 1) is equal to the number of unpaired electrons (n) present in their respective valence-shell. Thus the covalency of H (in H X, H_2, NH_3 etc.), N(in N_2, NH_3, NF_3 etc.), O (in O_2, H_2O etc.) and I (F_2, HF etc.) is equal to 1, 3, 2, and 1 respectively as shown below:

	1s				
H-atom ($1s^1$)	↑				(n = 1, covalency = 1)
	2s		2p		
N-atom ($2s^22p^3$)	↑↓	↑	↑	↑	(n = 3, covalency = 3)
O-atom ($2s^22p^4$)	↑↓	↑↓	↑	↑	(n = 2, covalency = 2)
F-atom ($2s^2$ $2p^5$)	↑↓	↑↓	↑↓	↑	(n = 1, covalency = 1)

Covalency of Be ($2s^2$, n = 0), B ($2s^22p^1$, n = 1) and C ($2s^22p^2$, n = 2) atoms is equal to 2 (*e.g.* BeF_2, $BeCl_2$, BeH_2 etc.), 3(e.g. BF_3, BCl_3 etc.) and 4 (*e.g.* CH_4, CCl_4 etc.) respectively, although the number of unpaired electrons in their valence-shell is equal to 0, 1 and

2. Thus we see that the covalency of Be, B and C-atoms is not equal to the number of unpaired electrons as in case of N, O and F atoms. In order to explain the dicovalency, tricovalency and tetracovalency of Be, B and C-atoms, it is assumed that when these elements combine with the elements like F, Cl, H etc. to form BeF_2, BF_3, CH_4 etc; some energy is released. This energy promotes one electron from 2*s* orbital to the vacant $2p_x$, $2p_y$ or $2p_z$ orbital, making two, three or four unpaired electrons available for the formation of two, three and four covalent bonds.

	2s	2p (x y z)	
(*i*) (*a*) Be-atom in the ground state ($2s^2\ 2p^0$)	↑↓	— — —	
(*b*) Be-atom in the excited state ($2s^1\ 2p^1$)	↑	↑ — —	(n = 2, covalency = 2)
(*ii*) (*a*) B-atom in the ground state ($2s^2\ 2p^1$)	↑↓	↑ — —	
(*b*) B-atom in the excited state ($2s^1\ 2p^2$)	↑	↑ ↑ —	(n = 3, covalency = 3)
(*iii*) (*a*) C-atom in ground state ($2s^2\ 2p^2$)	↑↓	↑ ↑ —	
(*b*) C-atom in the excited state ($2s^1\ 2p^3$)	↑	↑ ↑ ↑	(n = 4, covalency = 4)

3 and C-atoms having the configurations shown at (*a*) are said to be in the *ground (or normal) state* and the configurations in this state are called *ground state* (or *normal state*) *configurations* while these atoms having the configurations shown at (*b*) are said to be in the *excited state* and the configurations in this state are called *excited state configurations.*

Variable covalency shown by the elements of higher periods

The elements of higher periods (*i.e.*, *3rd*, *4th* etc) donot show a fixed covalency. Rather they show more than one covalency, *i.e.*, they show variable covalency. For example (*i*) P ($3s^2 3p^3$, n = 3) shows covalencies equal to 3 and 5 in PCl_3 and PCl_5 molecules respectively. (*ii*) S ($2s^2 2p^4$, n = 2) exhibits covalencies of 2,4, and 6 in SF_2, SF_4 and SF_6 molecules respectively. (*iii*) I ($5s^2 5p^5$, n = 1) has covalencies of 1, 3, 5 and 7 in ICl, ICl_3, IF_5 and IF_7 molecules respectively.

Explanation of variable covalency. The variable covalency of the elements of higher periods arises due to the unpairing of *s*-and *p* - electrons by promoting them into the vacant *d*-orbitals and thus increasing the number of unpaired electrons (n) in the valence-shell of the atom. Thus the covalencies of P(= 3, 5), S (=2, 4, 6) and *I* (= 1, 3, 5, 7) can be explained as follows:

	ns	*np*	*nd*
(*i*) (*a*) P-atom in the ground state ($3s^2\ 3p^3\ 3d^0$)	↑↓	↑ ↑ ↑ (n = 3, covalency = 3)	— — —
(*b*) P-atom in the excited state ($3s^1\ 3p^3\ 3d^1$)	↑	↑ ↑ ↑ (n = 5, covalency = 5)	↑ — —
(*ii*) (*a*) S-atom in the ground state ($3s^2\ 3p^4\ 3d^0$)	↑↓	↑↓ ↑ ↑ (n = 2, covalency = 2)	— — —
(*b*) S-atom in the 1st excited state ($3s^2\ 3p^3\ 3d^1$)	↑↓	↑ ↑ ↑ (n = 4, covalency = 4)	↑ — —
(*c*) S-atom in the 2nd excited state ($3s^1\ 3p^3\ 3d^2$)	↑	↑ ↑ ↑ (n = 6, covalency = 6)	↑ ↑ —

	5s	5p	5d
(*iii*) (*a*) I-atom in the ground state ($5s^2\ 5p^5\ 5d^0$)	↑↓	↑↓ ↑↓ ↑ (n = 1, covalency = 1)	— — —
(*b*) I-atom in the 1st excited state ($5s^2\ 5p^4\ 5d^1$)	↑↓	↑↓ ↑ ↑ (n = 3, covalency = 3)	↑ — —
(*c*) I-atom in the 2nd excited state ($5s^2\ 5p^3\ 5d^2$)	↑↓	↑ ↑ ↑ (n = 5, covalency = 5)	↑ ↑ —
(*d*) I-atom in the 3rd excited state ($5s^1\ 5p^3\ 5d^3$)	↑	↑ ↑ ↑ (n = 7, covalency = 7)	↑ ↑ ↑

It may be seen from the above discussion that variable covalency is shown only by those elements whose atoms contain vacant *d*- orbitals in their valence-shell so that the unpairing of *s*-and *p*-electrons by promoting them to the vacant *d*-orbitals may be possible. The elements having no *d*-orbitals (*i.e,* elements of 2*nd* period) donot exhibit variable covalency.

Maximum covalency of an element

Maximum covalency of an element is defined as the maximum number of covalent bonds which an atom of that element can form with the neighbouring atoms. For example the maximum covalency shown by Al, Si, P, S and I atoms is equal to 6, 6, 5, 6 and 7 which is exhibited by these elements in AlF_6^{3-}, SiF_6^{2-}, PF_5, SF_6 and IF_7 respectively.

Possible maximum covalency that an element can have in its covalent molecules or ions: Rule of maximum covalency

Sidgwick has given a rule which is called rule of maximum covalency or covalency maximum rule. According to rule, the possible maximum covalency that an element can be expected to exhibit in its covalent molecules/ions depends on the number of the period in which the element is present. According to this rule the maximum covalency of the elements of different periods is as follows:

For H-atom (1*st* period) = 2 ; For the elements of 2*nd* period = 4

For the elements of 3*rd* and 4*th* periods = 6; For the elements of higher periods = 8

It may be noted that it is not essential for an element to show its possible maximum covalency in accordance with the rule, *e.g.* although I-atom belongs to 5*th* period, it does not form any covalent molecule or ion in which it may exhibit the expected maximum covalency of 8. The actual maximum covalency that I-atom can show is equal to 7. This covalency of 7 is shown by I-atom in IF_7 molecule.

Explanation of the rule of maximum covalency

(*a*) **Maximum covalency of hydrogen atom.** Since H-atom has only one *s*-orbital (1*s* orbital) and one unpaired electron, it can share only *two* electrons in making covalent bond. Thus H-atom has a maximum convalency equal to 2 / 2 = 1.

(*b*) **Maximum covalency of the elements of 2nd period.** Since the elements of 2*nd* period have only four orbitals namely 2*s*, $2p_x$, $2p_y$ and $2p_z$ in their valence-shell, they can share a maximum of 8 electrons in making covalent bonds, *i.e,. their maximum covalency is equal to* 8/2 = 4 only. In showing this covalency, the paired electrons in 2*s* orbital are promoted to the vacant $2p_x$, $2p_y$ or $2p_z$ orbitals and thus all the electrons become unpaired. In showing this maximum covalency the element undergoes sp^3 hybridisation. For example Be, B, C and N (central atom) in BeF_4^-, BH_4^-, CH_4 and NH_4^+ molecules / ions show a covalency of 4 which is the maximum covalency of these elements and undergo sp^3 hybridisation as shown below:

(*i*) Be-atom in the ground state ($2s^2\ 2p^0$)

Be-atom in the excited state ($2s^1\ 2p^1$)

Be-atom in sp^3 hybridised state in BeF_4^{2-} ion

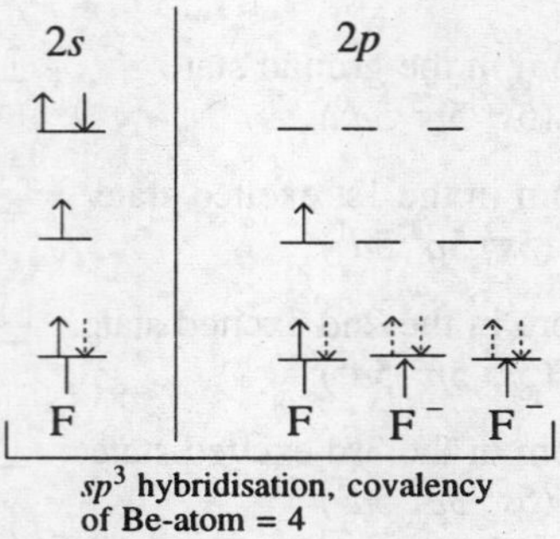

(*ii*) B-atom in the ground state ($2s^2\ 2p^1$)

B-atom in the excited state ($2s^1\ 2p^2$)

B-atom in sp^3 hybridised state in BH_4^- ion

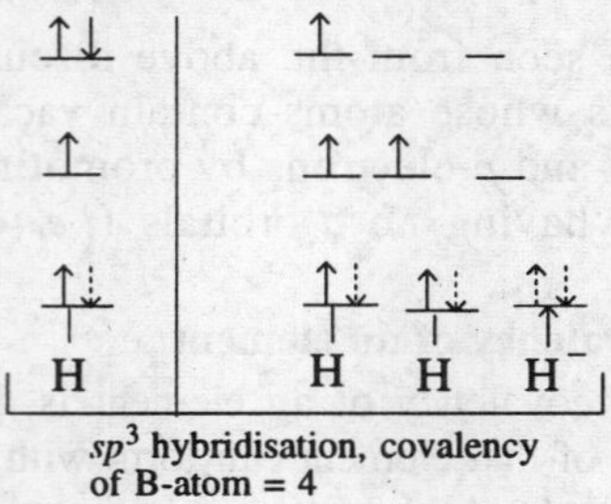

(*iii*) C-atom in the ground state ($2s^2\ 2p^2$)

C-atom in the excited state ($2s^1\ 2p^3$)

C-atom in sp^3 hybridised state in CH_4 molecule

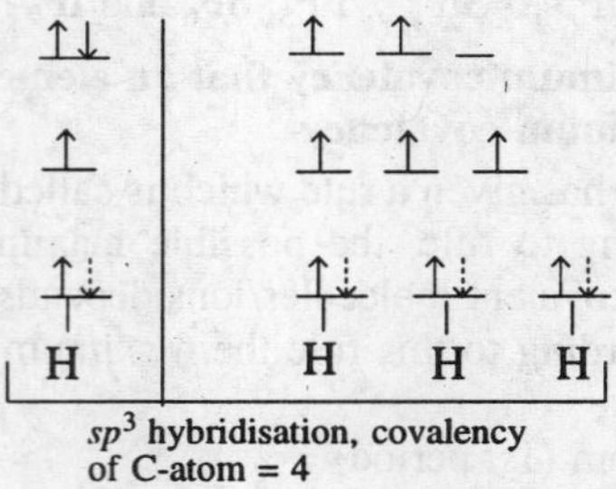

(*iv*) N-atom in the ground state ($2s^2\ 2p^3$)

N-atom in sp^3 hybridised state in NH_4^+ ion

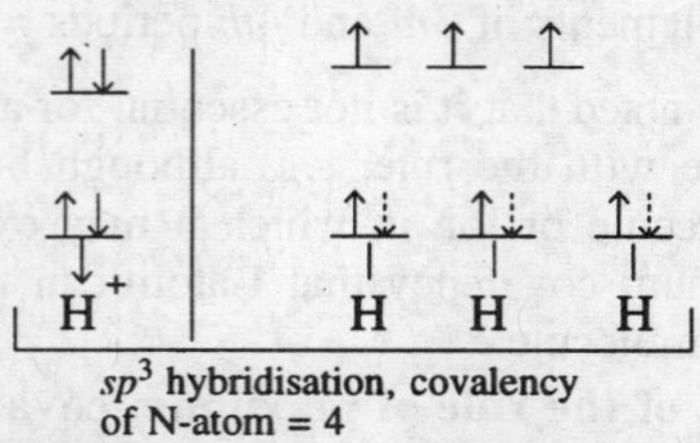

(*c*) **Maximum covalency of the elements of 3rd and higher periods.** We have seen that the elements of *2nd* period show a maximum covalency of *4*. In contrast, the maximum covalency shown by the elements of 3*rd* and higher periods is *more than 4*. Their maximum covalency which is greater than 4 is due to the fact that one, two or three *nd* orbitals (total number of *nd* orbital is five) of the valence-shell of these elements also get involved in sp^3d, sp^3d^2 or sp^3d^3 hybridisation respectively. For example:

Al-atom in the ground state ($3s^2\ 3p^1\ 3d^0$)

Al-atom in the excited state ($3s^1\ 3p^1\ 3d^1$)

Al-atom in sp^3d^2 hybridised state in AlF_6^{3-} ion

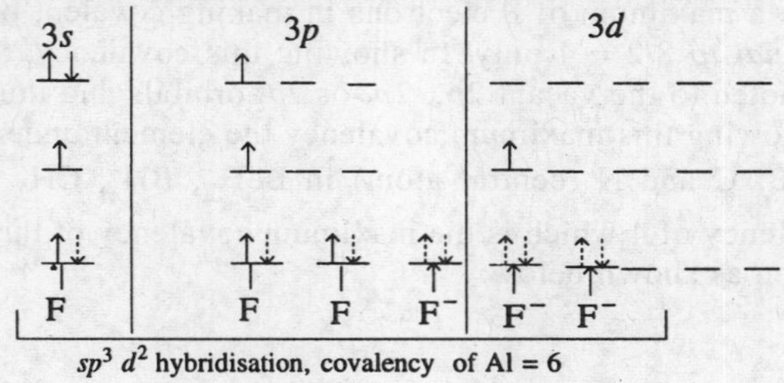

(*i*) The maximum covalency of B-atom (2*nd* period) is 4 which is seen in BH_4^- ion but that of Al-atom (3*rd* period) is 6 which is seen in AlF_6^{3-} ion. This covalency arises due to sp^3d^2 hybridisation.

(*ii*) The maximum covalency of C-atom is equal to 4 which is found in CH_4 molecule but the maximum covalency of Si-atom is 6 which is found in SiF_6^{2-} ion. This covalency arises because of sp^3d^2 hybridisation of Si-atom.

Si-atom in the ground state ($3s^2\ 3p^2\ 3d^0$)

Si-atom in the excited state ($3s^1\ 3p^3\ 3d^0$)

Si-atom in sp^3d^2 hybridised state in SiF_6^{2-} ion

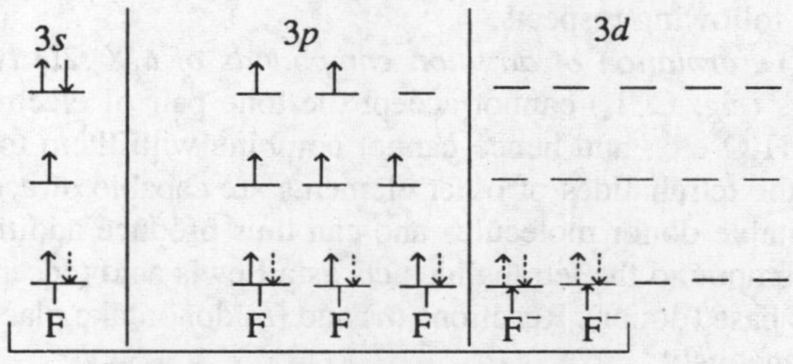

sp^3d^2 hybridisation, covalency of Si-atom = 6

(*iii*) The maximum covalency of N (2*nd* period) is 4 which is exhibited in NH_4^+ ion, but that of P-atom (3*rd* period) is 6 which is shown by this element in PF_6^- ion which arises due to sp^3d^2 hybridisation.

P-atom in the ground state ($3s^2\ 3p^3\ 3d^0$)

P-atom in the excited state ($3s^1\ 3p^3\ 3d^1$)

P-atom in sp^3d^2 hybridised state in PF_6^- ion

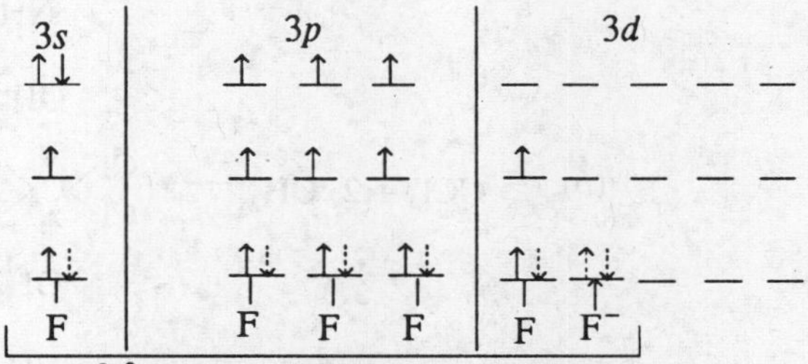

sp^3d^2 hybridisation, covalency of P-atom = 6

Applications of the rule of maximum covalency

A number of facts can be explained with the help of the rule of maximum covalency. Some applications are discussed below:

1. Aluminium can form hexafluo-aluminate ion, AlF_6^{3-} but boron cannot form hexa fluo-borate ion, BF_6^{3-}. Al-atom, being a member of 3*rd* period, contains vacant 3*d* orbitals in its valence-shell (Al = $3s^23p^13d^0$). One of the two electrons of 3*s*-orbital is shifted to the vacant 3*p* orbital.

Al-atom in the ground state ($3s^2\ 3p^1\ 3d^0$)

Al-atom in the excited state ($3s^1\ 3p^2\ 3d^0$)

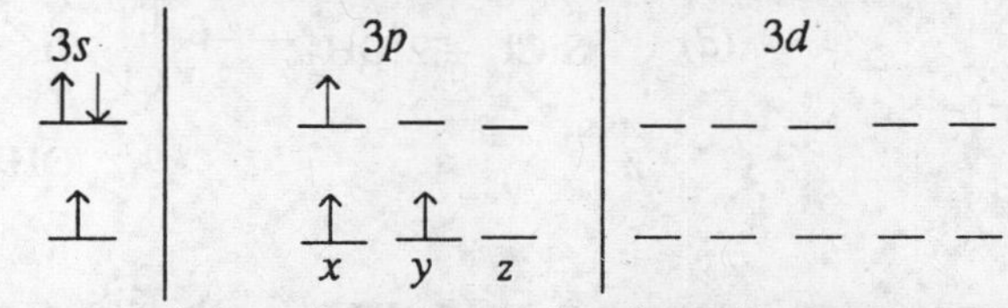

The singly-filled three orbitals (3*s*, $3p_x$ and $3p_y$) form three Al—F bonds while the three vacant orbitals ($3p_z$ and two 3*d* orbitals) accept the lone pair of electrons donated by three F^- ions and form three $F^- \rightarrow Al$ coordinate bonds. In the formation of AlF_6^{3-}, Al-atom undergoes sp^3d^2 hybridisation.

Al-atom in sp^3d^2 hybridisation in AlF_6^{3-} ion

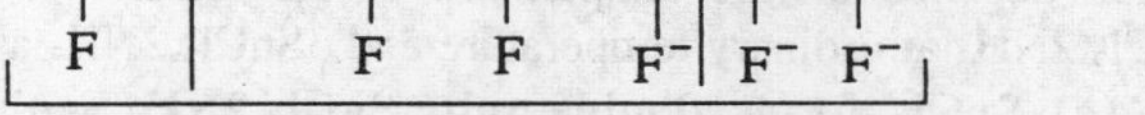

sp^3d^2 hybridisation, covalency of Al-atom = 6

Thus, we see that it is due to the presence of vacant *d*-orbitals (3*d* orbitals) in the valence-shell of Al-atom that this atom accepts the lone pair of electrons and AlF_6^{3-} ion is formed. Since

B-atom has only-four orbitals and no *d*-orbitals (B = $2s^2$, $2p^1_x\ 2p_y^0\ 2p_z^0$), this atom cannot accommodate the electron pairs donated by three F^- ions and hence cannot form BF_6^{3-} ion. However, this atom can form BF_4^- in which its maximum covalency of 4 is completely satisfied.

2. Differences between the tetrahalides of carbon and tetrahalides of other elements of group IVA. The tetrahalides of carbon differ from the tetrahalides of other elements of this group in the following respects.

(*i*) *Formation of addition compounds of $MX_4.2D$ type* (D = donor molecule). Carbon tetrahalides (*e.g.* CCl_4) cannot accept the lone pair of electrons donated by the donor molecules like NH_3, H_2O etc., and hence cannot combine with them to form addition compounds. On the other hand the tetrahalides of other elements are capable of accepting the lone pair of electrons donated by suitable donor molecules and can thus produce addition compounds. In the formation of addition compound the tetrahalide acts as a Lewis acid (acceptor) and the donor molecule behaves as a Lewis base (donor). Reactions (*a*) and (*b*) donot take place while the remaining reactions take place spontaneously.

(*a*) $CCl_4 + 2{:}\ NH_3 \longrightarrow$ 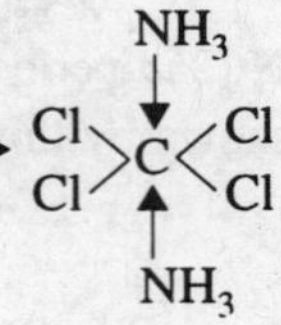 or $CCl_4\ .\ 2NH_3$

(*b*) $CCl_4 + 2\ {:}\ddot{O}H_2 \longrightarrow$ 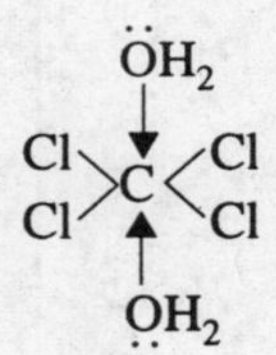 or $CCl_4\ .\ 2H_2O$

(*c*) $SiF_4 + 2\ {:}\ NH_3 \longrightarrow$ 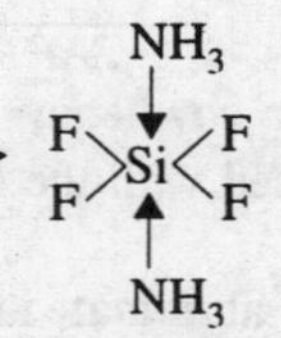 or $SiF_4\ .\ 2NH_3$

(*d*) $SiCl_4 + 2\ \ddot{O}H_2 \longrightarrow$ 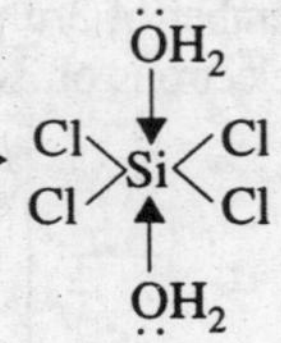 or $SiCl_4\ .\ 2H_2O$

(*e*) $SnF_4 + {:}NH_3 \longrightarrow$ 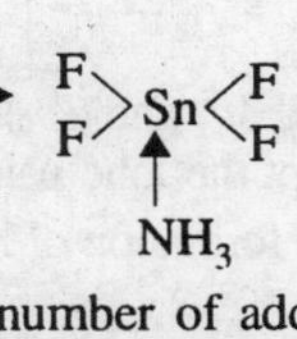 or $SnF_4\ .\ NH_3$

(*f*) $SnCl_2$ combines with NH_3 to form a number of addition compounds the composition of which depends on the temperature of the reaction. For example $SnCl_2$ gives $SnCl_2.NH_3$ and $SnCl_2.2NH_3$ at ordinary temperature and $3SnCl_2.2NH_3$ above 375K.

(*g*) $SnCl_4$ forms double salt, $SnCl_4.2NH_3$ with NH_3 [$SnCl_4 + 2NH_3 \rightarrow SnCl_4.2NH_3$] which can be sublimed without decomposition. It also gives crystalline double salts with substances like N_2O, $NOCl$, PCl_5 etc.

(ii) Formation of complex hexahalide ions of $[MX_6]^{2-}$ type. Carbon tetrahalides cannot accept the electron pairs donated by other halide ions and hence cannot combine with them to produce complex halide ions. Thus the following reaction does not take place.

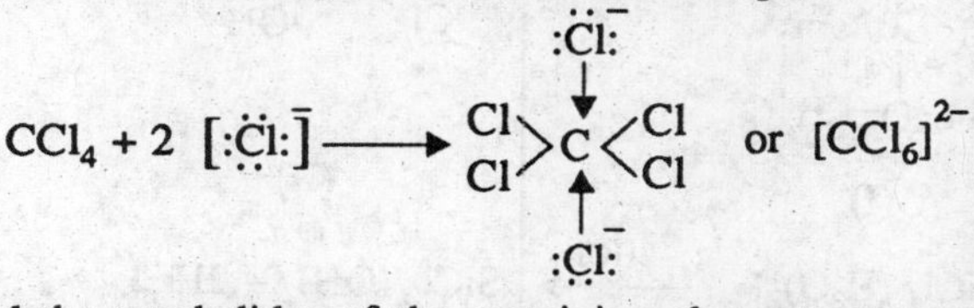

On the other hand the tetrahalides of the remaining elements can accept the electron pairs donated by the halide ions and hence can combine with them to give complex halide ions. Thus each of the following reactions takes place spontaneously.

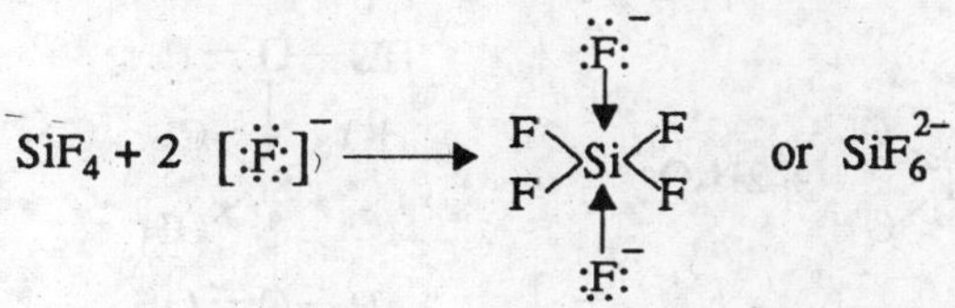

$$GeX_4 + 2\,[:\ddot{X}:]^- \longrightarrow GeX_6^{2-} \quad [X{=}F, Cl]$$

$$SnX_4 + 2\,[:\ddot{X}:]^- \longrightarrow SnX_6^{2-} \quad [X{=}F, Cl, Br, I]$$

$$PbCl_4 + 2\,[:\ddot{Cl}:]^- \longrightarrow PbCl_6^{2-}$$

(iii) Hydrolysis. While tetrahalides of carbon are not hydrolysed by water under normal conditions, the tetrahalides of other elements are readily hydrolysed. For example the hydrolysis of $SiCl_4$ gives silicic acid, $Si(OH)_4$ or H_4SiO_4 and HCl while CCl_4 cannot be hydrolysed to form H_4CO_4 and HCl *i.e.*, reaction (*B*) cannot take place

(*A*) $SiCl_4 + 4H.OH \longrightarrow Si(OH)_4$ or $H_4SiO_4 + 4HCl$ (This reaction takes place)

(*B*) $CCl_4 + 4\,H.OH \longrightarrow C(OH)_4$ or $H_4CO_4 + 4HCl$ (This reaction does not take place)

The hydrolysis of $SiCl_4$ as represented by equation (*A*) actually takes place through the following steps:

(*a*) $SiCl_4$ combines with two H_2O molecules (donor molecules) and gives a 6-coordinated coordination compound, [$SiCl_4$ $(H_2O)_2$](I)

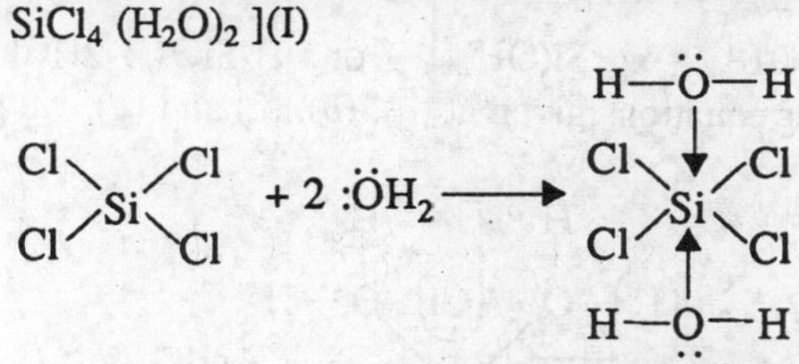

6-coordinated coordination compound (I)

or $\qquad SiCl_4 + 2H_2O \rightarrow [SiCl_4\,(H_2O)_2](I)$

(*b*) 6-coordination compound (I) formed in reaction (*a*) above loses two HCl molecules and gives $SiCl_2(OH)_2$ (II).

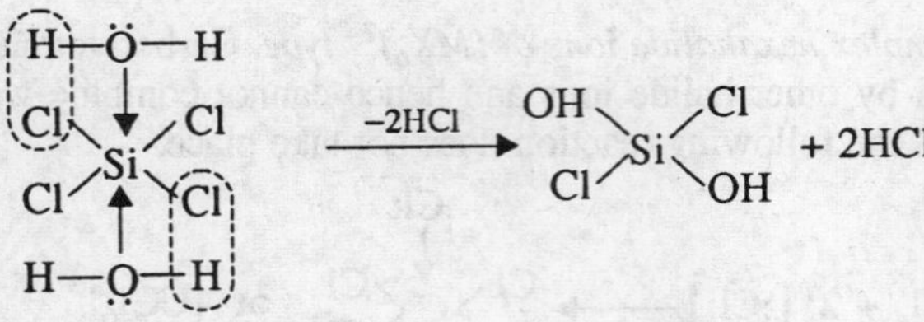

(I) (II)

or $[SiCl_4(H_2O)_2] \xrightarrow{-2HCl} SiCl_2(OH)_2 + 2HCl$
(I) (II)

(*c*) $SiCl_2(OH)_2$ (II) formed in reaction (*b*) above adds two H_2O molecules and gives 6-coordinated coordination compound, $[SiCl_2(OH)_2(H_2O)_2]$ (III).

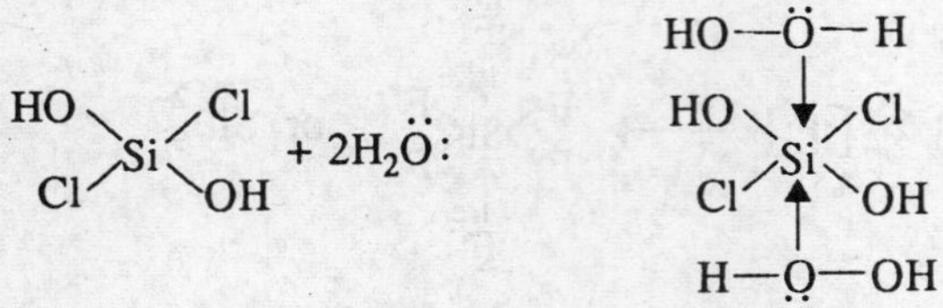

(II)

6-coordinated coordination compound, $[SiCl_2(OH)_2(H_2O)_2]$
(III)

or $SiCl_2(OH)_2 + 2H_2O \rightarrow [SiCl_2(OH)_2(H_2O)_2]$
(II) (III)

(*d*) 6-coordinated coordination compound (III) formed in reaction (*c*) loses two HCl molecules and produces $Si(OH)_4$ or H_4SiO_4.

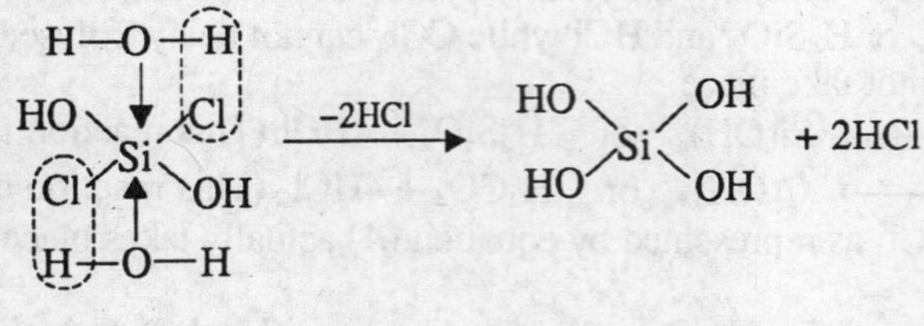

(III)

or $[SiCl_2(OH)_2(H_2O)_2] \rightarrow Si(OH)_4$ or $H_4SiO_4 + 2HCl$

On combining all the four equations given at (*a*), (*b*), (*c*) and (*d*), we get the final equation as:

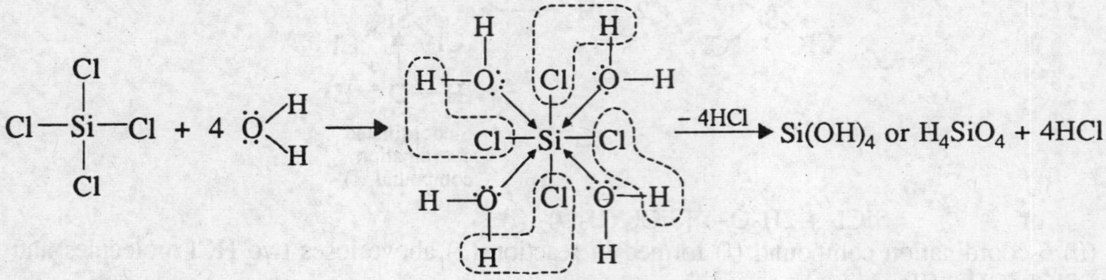

or $SiCl_4 + 4H_2O \rightarrow SiCl_4(H_2O)_4 \xrightarrow{-4HCl} Si(OH)_4$ or $H_4SiO_4 + 4HCl$

Explanation of the differences. The above differences between carbon tetrahalides and the tetrahalides of other elements can be explained as follows:

(a) Explanation based on the rule of maximum covalency. We know that, according to the rule of maximum covalency, the maximum covalency of carbon is 4 (tetracovalency) while maximum covalency of silicon is six (hexacovalency). Thus C-atom cannot have a covalency more than 4. In $CCl_4.2NH_3$, $CCl_4.2H_2O$ (addition compounds), CCl_6^{2-} (complex halide ion) the covalency of carbon is 6 which, according to the rule of maximum covalency, is not possible. Consequently CCl_4 molecule can form neither addition compounds nor complex halide ions. It is due to the non formation of 6-coordinated coordination compound, $[CCl_4(H_2O)_2]$ during the hydrolysis of CCl_4 that CCl_4 resists hydrolysis.

In $SiCl_4$, Si-atom shows a covalency of 4 which is not its maximum covalency. Thus in order to achieve its maximum covalency of 6, $SiCl_4$ can combine with two NH_3, two H_2O molecules and two Cl^- ions to give $SiCl_4.2NH_3$, $SiCl_4.2H_2O$ (addition compounds) and $SiCl_6^{2-}$ ion (complex halide ion) respectively. It is due to the formation of 6-coordinated coordination compound, $[SiCl_4\ (H_2O)_2]$ that $SiCl_4$ can easily be hydrolysed to give $Si(OH)_4$ or H_4SiO_4 and HCl as the final products of hydrolysis.

(b) Explanation based on the absence or presence of suitable vacant orbitals. In CCl_4 molecule, the central C-atom in its excited state has four singly-filled orbitals in its valence-shell ($C = 2s^1\ 2p_x^{\ 1}\ 2p_y^{\ 1}\ 2p_z^{\ 1}$) which are completely filled with electrons after forming four C—Cl bonds in CCl_4 molecule. Thus there is no vacant orbital in the valence-shell of C-atom which may accommodate the lone pairs of electrons donated by donor molecules (*e.g.* NH_3, H_2O etc.) or donor halide ions to form $CCl_4.2NH_3$, $CCl_4.2H_2O$, CCl_6^{2-} respectively. Also at the same time, since C-atom belongs to 2*nd* period of the periodic table, it has no *d*-orbitals which may accommodate the lone pairs of electrons. Thus we find that it is due to the absence of *d*-orbitals in the valence-shell of C-atom that the formation of addition compounds or complex halide ions is not possible.

In case of $SiCl_4$ molecule, the central Si-atom, being a member of 3*rd* period, has two vacant 3*d* orbitals in its valence-shell ($Si = 3s^2\ 3p_x^1\ 3p_y^1\ 3p_z^0\ 3d^0$). These two vacant 3*d* orbitals ($3d_{x^2-y^2}$, $3d_{z^2}$) can accommodate the lone pairs of electrons donated by the donor molecules or halide ions. In the formation of these species Si-atom is sp^3d^2 hybridised.

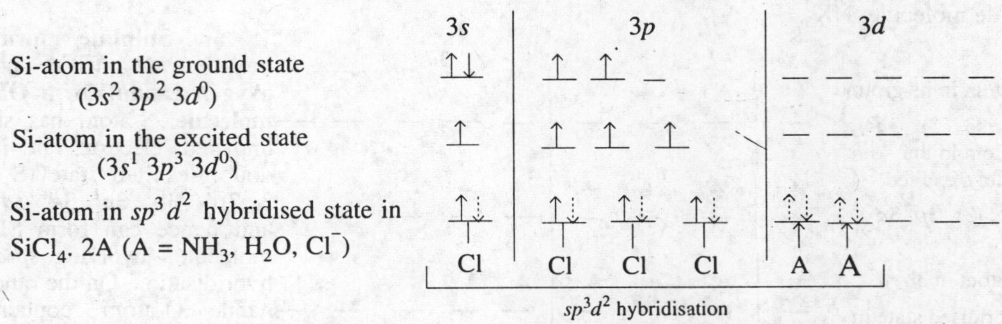

3. Phosphorus can form PX_6^- ions (X = F, Cl etc.) while nitrogen cannot form NX_6^- ions. See the formation of AlF_6^{3-} and non-formation of BF_6^{3-} ions discussed on page 255.

It can also be explained that nitrogen trihalides cannot form the anions like NX_5^{2-} while the trihalides of other elements of group VA can form the anions, MX_5^{2-}.

4. Phosphorus forms both trihalides (PX_3) and pentahalides (PX_5) but nitrogen can form only trihalides (NX_3) and not pentahalides (NX_5). Phosphorus, being an element of 3*rd* period, contains vacant 3*d* orbitals in its valence-shell ($P = 3s^2\ 3p^3\ 3d^0$). When P-atom uses only three unpaired electrons present in 3*p* orbitals, trihalide of this element (PX_3) is formed. In the formation of this molecule P-atom in its ground state is sp^3 hybridised.

When one of the two electrons of 3*s* orbital is transferred to 3*d* orbitals, P-atom gets five unpaired electrons which explain the formation of PX_5 molecule. In the formation of this molecule, P-atom in its excited state is sp^3d hybridised.

The valence-shell electronic configuration of N-atom ($2s^22p^3$) indicates that it has only three unpaired electrons and hence N-atom can form NX_3 molecule. In the formation of this molecule,

N-atom is sp^3 hybridised.

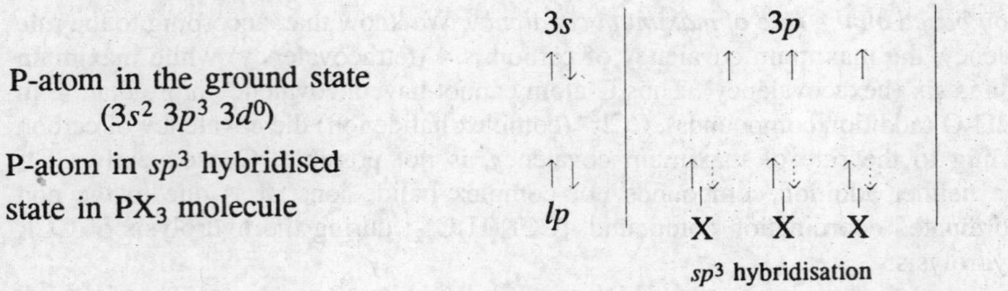

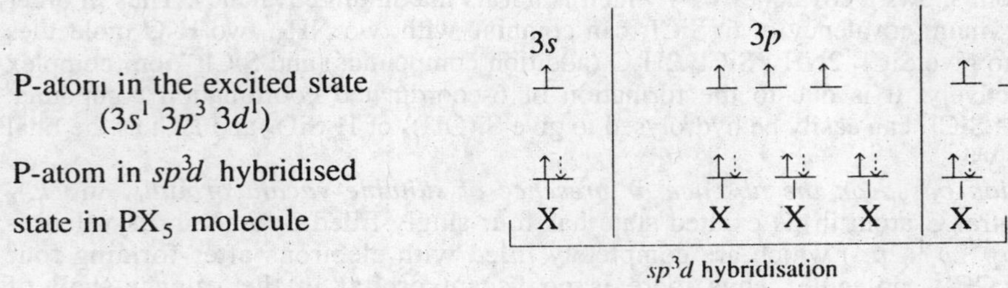

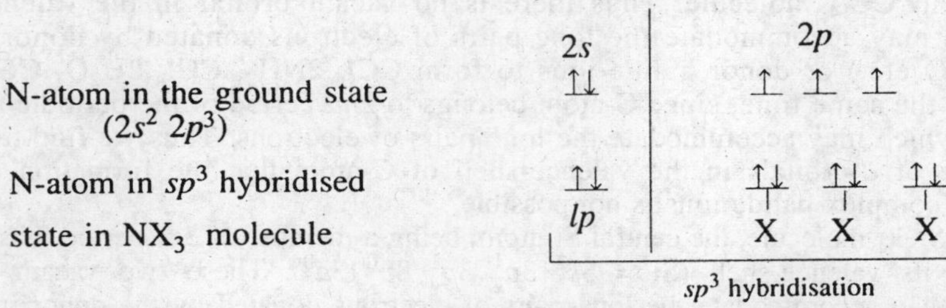

Since N-atom has no *d*-orbitals, one electron from 2*s* orbital cannot be promoted to *d*-orbitals and hence five unpaired electrons cannot be made available. Thus N-atom cannot form pentahalide molecule, NX_5.

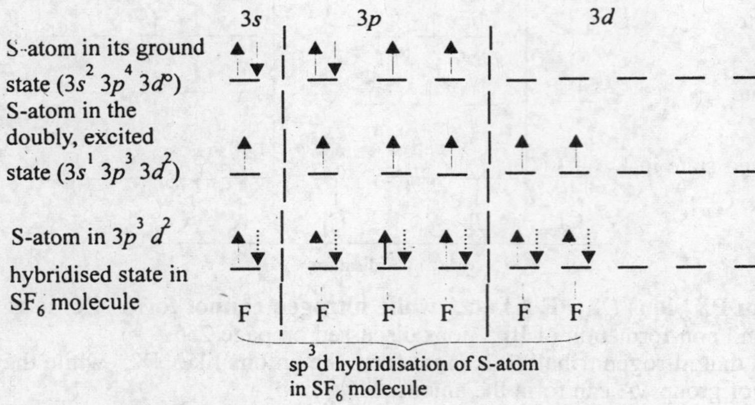

5. Sulphur forms hexafluoride (SF_6) but oxygen cannot form OF_6 molecule. S-atom has six unpaired electrons in its doubly-excited state (S = $3s^1 3p_x^1\ 3p_y^1\ 3p_z^1\ 3d^1\ 3d^1$) and hence can form SF_6 molecule, involving sp^3d^2 hybridisation. On the other hand, O-atom contains only two unpaired electrons and no *d*-orbitals (O = $2s^2 2p_{x2}\ {}^{py}1\ {}^{2p}{}_{z1}$). Thus O-atom can form only OF2 molecule but not OF_6. In the formation of OF_2 molecule O-atom (central atom) is sp^3 hybridised.

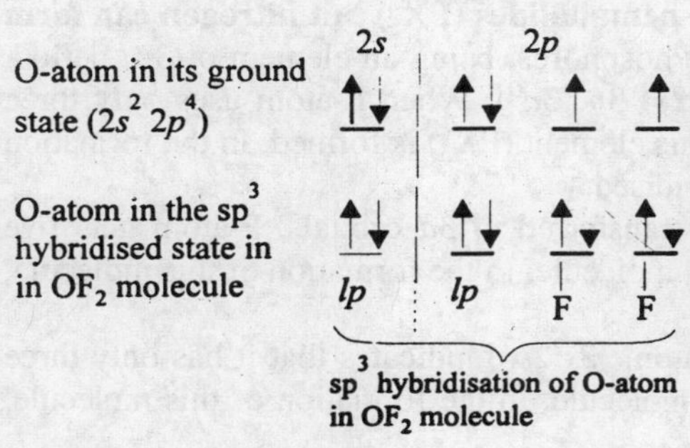

6. SF_6 and SeF_6 cannot be hydrolysed while TeF_6 can be hydrolysed. Although the central atom of each of the molecules has *d*-orbitals in its valence-shell, it is only TeF_6 which is reactive and hence can be hydrolysed by water; SF_6 and SeF_6 are stable and hence cannot be hydrolysed. Their different behaviour towards hydrolysis can be explained as follows:

S and Se atoms belong to *3rd* and *4th* periods respectively and hence both these elements have maximum covalency equal to six which is satisfied in their hexafluorides (SF_6 and SeF_6). Thus these elements donot use any of their *d*-orbitals for accepting lone pairs of electrons from O-atom of H_2O molecules. Te-atom, being a member of *5th* period (higher period), is expected to have maximum covalency equal to eight. Note that in TeF_6, Te-atom has covalency of six. In order to attain maximum covalency of eight, TeF_6 adds two H_2O molecules to form an unstable 8-covalent intermediate, $TeF_6 . 2H_2O$. This intermediate loses two HF molecules and gives $TeF_4(OH)_2$. The process of the addition of two H_2O molecules followed by the process of elimination of two HF molecules stops when we finally get orthotelluric acid, $Te(OH)_6$ or H_6TeO_6.

$$TeF_6 \xrightarrow{+2H_2O} TeF_6.2H_2O \xrightarrow{-2HF} TeF_4(OH)_2$$

$$\xrightarrow{+2H_2O} TeF_4(OH)_2.2H_2O \xrightarrow{-2HF} TeF_2(OH)_4$$

$$\xrightarrow{+2H_2O} TeF_2(OH)_4.2H_2O \xrightarrow{-2HF} Te(OH)_6 \text{ or } H_6TeO_6$$

All the above steps can be combined together to get the following final equation.

$$TeF_6 + 6H_2O \longrightarrow Te(OH)_6 \quad \text{or} \quad H_6TeO_6 + 6HF$$

Covalency of the elements of group IVA

Although the valence-shell configuration of atoms of these elements ($ns^2\ np_x^1\ np_y^1\ np_z^0$) indicates that, since these elements have two unpaired elements ($n = 2$) they should show bicovalency in their covalent compounds, but experiments have proved that these elements form covalent compounds in which they exhibit tetracovalency. In order to explain the tetracovalency, it is assumed that one of the paired electrons from ns^2 orbital gets shifted to the vacant $2p_z$ orbital (excited state) so that the number of unpaired electrons becomes four which explains the tetracovalency of these elements. Four orbitals, each containing one electron, hybridise together and give a set of four equivalent sp^3 hybrid orbitals, which can be used to form four σ bonds with the adjacent atoms. Let us take the case of carbon atom.

Carbon atom in its ground state : $2s^2\ 2p_x^1\ 2p_y^1\ 2p_z^0$ $(n = 2)$

Carbon atom in its excited state : $2s^1\ 2p_x^1\ 2p_y^1\ 2p_z^1$ $(n = 4)$

Carbon atom in sp^3 hybridised state : $\underbrace{(sp^3)\ (sp^3)^1\ (sp^3)^1 (sp^3)^1}_{sp^3 \text{ hybridisation}}$ $(n = 4)$

Si, Ge, Sn and Pb which belong to higher periods exhibit 4 as well as 6 covalencies. The exhibition of higher covalencies (= 6) by these elements is due to the presence of vacant *d*-orbitals in the valence-shell of the atoms of these elements. The exhibition of higher covalency can also be explained on the basis of the "Rule of maximum covalency". For example Si-atom shows a covalency of 6 in $SiCl_4.2NH_3$, $SiCl_4.2H_2O$ (addition compounds), SiF_6^{2-} (complex ion) etc. *d*-orbitals, being vacant, accept the lone pairs of electrons donated by the donor molecules or ions which act as Lewis bases. The formation of $SiCl_4.2H_2O$ has already been shown on page 259. $SiCl_4.2NH_3$, $SiCl_4.2H_2O$, SiF_6^{2-} are octahedral species and hence are formed by sp^3d^2 hybridisation.

Covalency of group VA elements

Since N-atom has only three unpaired electrons and no *d*-orbitals in its valence-shell, this element shows a covalency of 3 (*e.g.* NH_3). This covalency corresponds to sp^3 hybridisation. When this element uses its lone pair of electrons and all the three unpaired electrons in bonding,

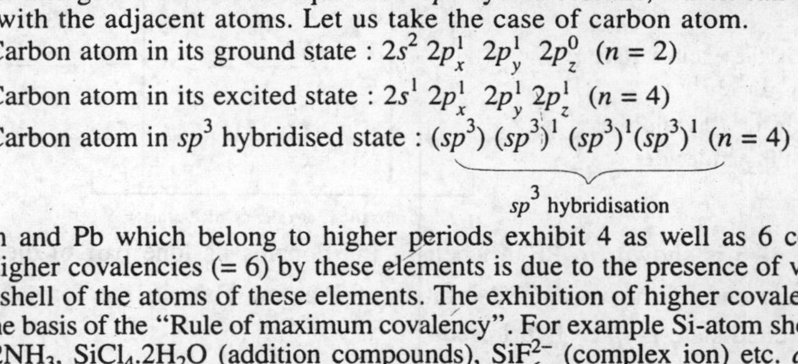

it shows a covalency of 4 (*e.g.* NH_4^+ ion) which is the maximum covalency of this element (See "*Rule of maximum covalency*"). This covalency also corresponds to sp^3 hybridisation.

The remaining elements of this group, due to the presence of *d*-orbitals in their valence shell, can exhibit many covalencies like 3(*e.g.* PF_3), 4 (*e.g.* PCl_4^+), 5(*e.g.*PF_5) and 6(*e.g.* PF_6^-). As an example let us see how P-atom shows covalency of 3, 4, 5 and 6 in PF_3, PCl_4^+, PF_5 and PF_6^- respectively.

The covalency of 3 (in PF_3) is shown by P-atom when this element is in the ground state and this covalency corresponds to sp^3 hybridisation.

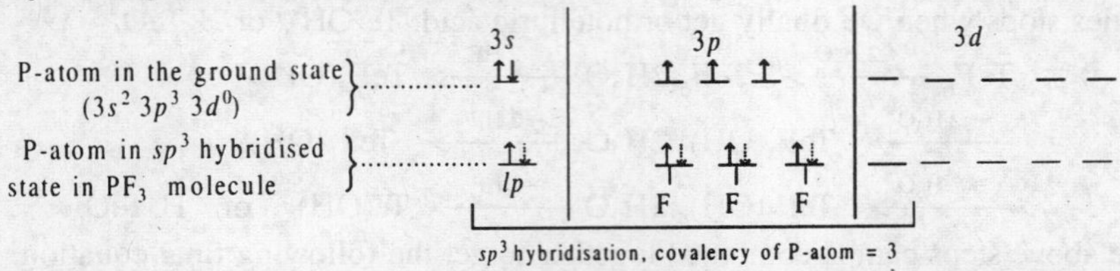

sp^3 hybridisation, covalency of P-atom = 3

Covalency of 4 (in PCl_4^+) is shown when P atom uses all its three unpaired electrons and one pair of electrons residing in 3*s* orbital. This covalency also corresponds to sp^3 hybridisation.

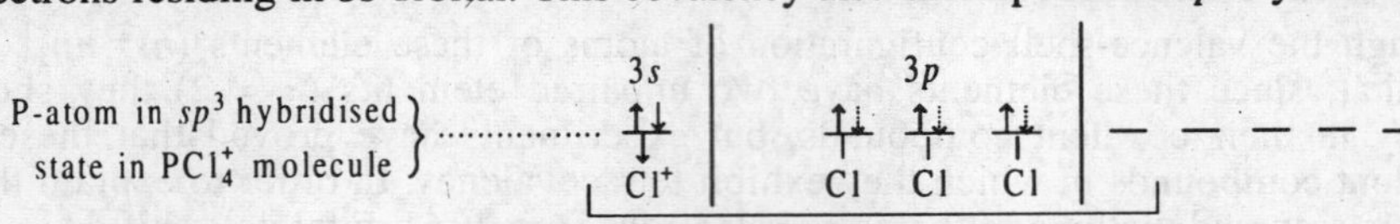

sp^3 hybridisation, covalency of P-atom = 4

Covalency of 5 is caused (in PF_5) when one of the paired electrons in 3*s* orbital is shifted to 3*d* orbitals so that five unpaired electrons are obtained. In showing this covalency, P-atom is sp^3d hybridised.

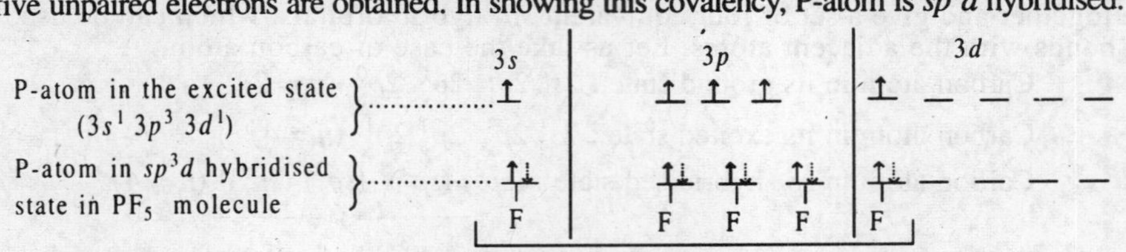

sp^3d hybridisation, covalency of P-atom = 5

Covalency of 6 is shown (in PF_6^-) when F^- ion donates its lone pair of electrons which is accommodated in one of the vacant 3*d* orbitals of P-atom. Thus in the formation of PF_6^- ion, P-atom in its excited state is sp^3d^2 hybridised.

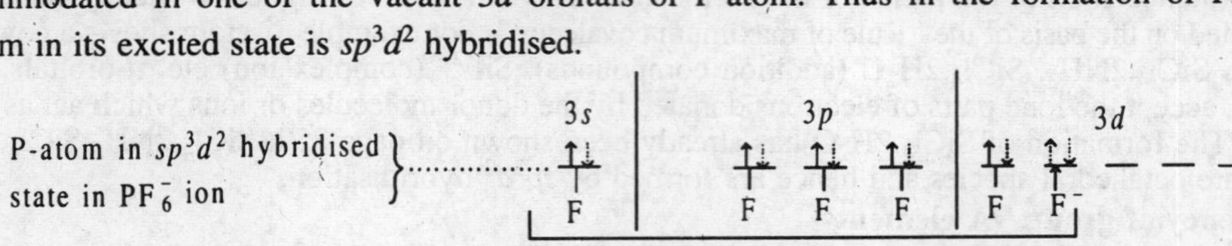

sp^3d^2 hybridisation, covalency of P-atom = 6

It may be seen from the above discussion that N and P both show a covalency of 4 when the lone pair of electrons residing in 2*s* (in N-atom) or 3*s* (in P-atom) forms a dative bond with Lewis acid (*e.g.* H^+, Cl^+ etc.). Thus :

Lewis base		*Lewis Acid*		
H_3N :	+	H^+	→	$H_3N \rightarrow H^+$ *or* NH_4^+
Cl_3P:	+	Cl^+	→	$Cl_3P \rightarrow Cl^+$ *or* PCl_4^+

It may also be seen from the above discussion that N-atom can show a covalency of 3 and not of 5, while other elements can show both types of covalencies. Thus N can form the compounds like NF_3 but not the compounds like NF_5, NCl_5 etc. while other elements can form both types of compounds (PF_3, PF_5, AsF_3, AsF_5 etc.)

Reason. The reason for the non-existence of the molecules like NF_5, NCl_5 etc. can be explained on the basis of valence-shell configuration of N and other elements.

The valence-shell configuration of N(N = $2s^2p^3$) and other elements (P = $3s^2p^3d^0$; As = $4s^2p^3d^0$; Sb = $5s^2p^3d^0$; Bi = $6s^2p^3d^0$) show that the valence-shell of N-atom contains only *s* and *p* orbitals and no *d*-orbitals, while other elements have *s*, *p* and vacant *d*-orbitals in their valence-shells. Thus, due to the absence of *d*-orbitals, N-atom is not able to undergo sp^3d hybridisation and hence cannot form two additional N—F bonds in NF_5 molecule by the overlap of singly-filled sp^3d hybrid orbital on N-atom and singly-filled $2p$ orbital on F-atom. On the other hand, other elements, due to the presence of vacant *d*-orbitals in their valence-shell, can undergo sp^3d hybridisation and can thus form five $sp^3d(A) - 2p(X)$ σ-bonds (A = P, As, Sb; X = F, Cl) in AX_5 type molecules.

Covalency of group VIA elements

Since O-atom has only two-unpaired electrons and no *d*-orbitals in its valence-shell, this element shows a covalency of 2 only (*e.g.* OF_2), *i.e.*, this element cannot show higher covalencies. This covalency corresponds to sp^3 hybridisation of O-atom.

O-atom in the ground state ($2s^2\ 2p^4$)

O-atom in sp^3 hybridised state in OF_2 molecule

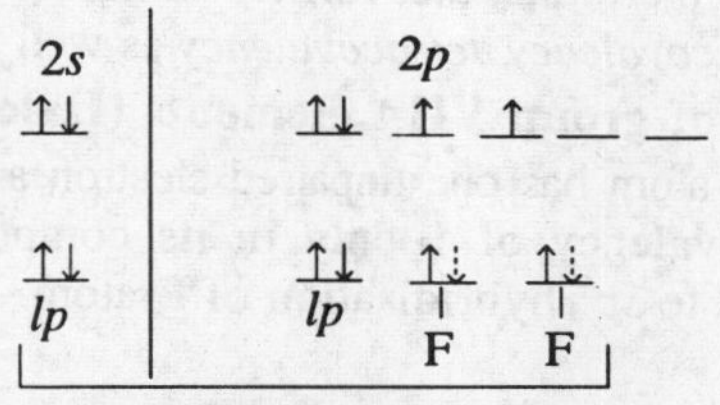

sp^3 hybridisation, covalency of O-atom = 2

The remaining elements of this group, due to the presence of *d*-orbitals, can exhibit many covalencies like 2 (*e.g.* SF_2), 4 (*e.g.* SF_4) and 6 (*e.g.* SF_6). As an example let us see how S-atom shows covalency of 2, 4 and 6 in SF_2, SF_4 and SF_6 molecules respectively. The covalency of 2 is shown by S-atom when this element is in the ground state and this covalency corresponds to sp^3 hybridisation. The covalency of 4 is caused when one of the paired electrons in $3p$ orbitals is shifted to $3d$ orbitals and thus we get four unpaired electrons. In showing this covalency S-atom is sp^3d hybridised. The covalency of 6 is shown when $3s$ electrons also become unpaired, thus making six unpaired electrons available for bonding. In showing this covalency S-atom is sp^3d^2 hybridised.

S-atom in the ground state ($3s^2\ 3p^4\ 3d^0$)

S-atom in sp^3 hybridised state in SF_2 molecule

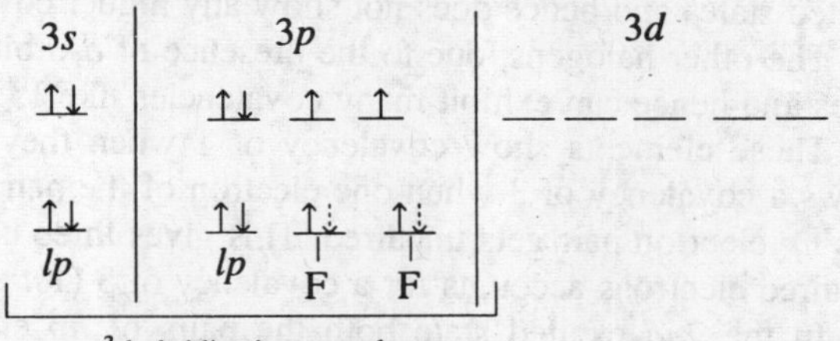

sp^3 hybridisation, covalency of S-atom = 2

S-atom in the 1*st* excited state ($3s^2\ 3p^3\ 3d^1$)

S-atom in sp^3d hybridised state in SF_4 molecule

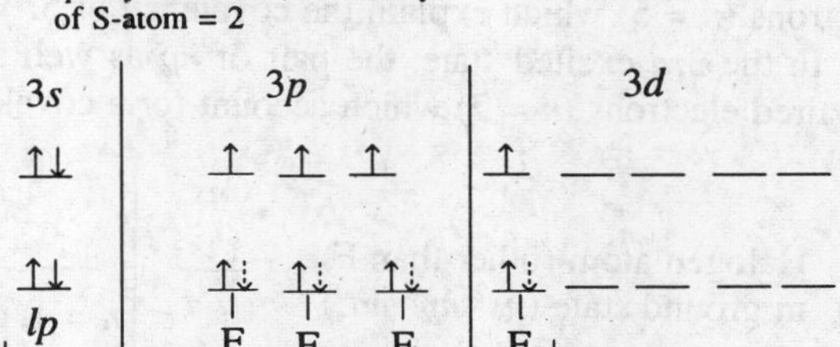

sp^3d hybridisation, covalency of S-atom = 4

S-atom in the 2*nd* excited state ($3s^1\ 3p^3\ 3d^2$)

S-atom in sp^3d^2 hybridised state in SF_6 molecule

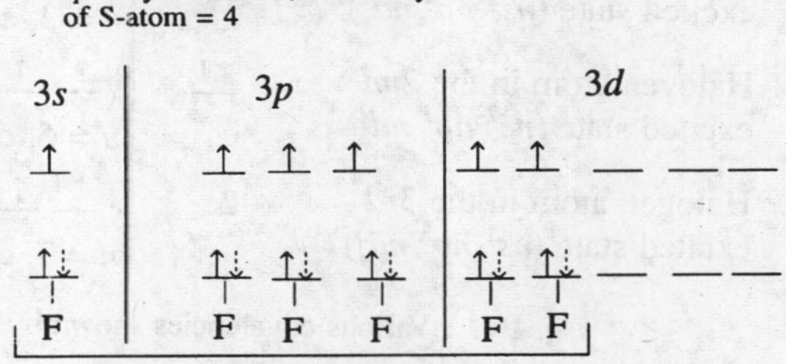

$sp^3\ d^2$ hybridisation, covalency of S-atom = 6

We have stated above that O-atom cannot show covalencies greater than 2 while other elements can show higher covalencies (e.g. 4, 6) as well. The abnormal behaviour of O-atom can be explained as follows:

O-atom belongs to *2nd* period and hence has only 2*s* and 2*p* orbitals ($O = 2s^2 2p^4$) and no *d*-orbitals. So, in case of O-atom, neither 2*p* nor 2*s* electrons can be promoted to *d*-orbitals to get more number of unpaired electrons. However, in order to get more number of unpaired electrons, 2*p* or 2*s* electrons may be excited to go to 3*s* orbital, but, since too much energy is needed for this type of excitation of the electron, the electrons in 2*p* or 2*s* orbitals donot get unpaired.O-atom, therefore, does not show any covalency which may be higher than 2. The above discussion shows that oxygen never shows more than *dicovalency* in its covalent compounds while other members of this group may show *dicovalency, tetracovalency* as well as *hexacovalency.*

Covalency of group VIIA elements (Halogens)

Since F-atom has one unpaired electron and no *d*-orbitals in its valence-shell, this element shows a covalency of 1 only in its compounds like HF, ClF, BrF, IF etc. This covalency corresponds to sp^3 hybridisation of F-atom.

	2*s*	2*p*		
F-atom in the ground state ($2s^2\ 2p^5$)	↑↓	↑↓	↑↓	↑
F-atom in sp^3 hybridised state in AF molecule (A = H, Cl, Br, I)	↑↓ *lp*	↑↓ *lp*	↑↓ *lp*	↑↓ A
	sp^3 hybridisation, covalency of F-atom = 1			

Due to the absence of *d*-orbitals in the valence-shell of F-atom, this element can not have excited states and hence does not show any higher covalencies.

The other halogens, due to the presence of *d*-orbitals in their valence-shell, can have excited states and hence can exhibit many covalencies like 1, 3, 5 and 7.

These elements show covalency of 1 when they are in their ground state. A halogen atom shows a covalency of 3 when one electron of the pair of *np* electrons is shifted to *nd* orbitals and thus *np* electron pair gets unpaired. This gives three unpaired electrons ($n = 3$). The presence of 3 unpaired electrons accounts for a covalency of 3 (*1st* excited state).

In the *2nd* excited state both the pairs of *np* electrons get unpaired, providing 5 unpaired electrons ($n = 5$) which explain the covalency of 5.

In the *3rd* excited state, the pair of *np* as well as *ns* electrons get unpaired, providing seven unpaired electrons ($n = 7$) which account for a covalency of 7 (See Fig. 10.1).

	ns	np	*nd*
Halogen atom (other than F) in ground state ($ns^2\ np^5\ nd^0$)	↑↓	↑↓ ↑↓ ↑ n = 1, covalency = 1	_ _ _ _ _
Halogen atom in the first excited state ($ns^2\ np^4\ nd^1$)	↑↓	↑↓ ↑ ↑ n = 3, covalency = 3	↑ _ _ _ _
Halogen atom in the *2nd* excited state ($ns^2\ np^3\ nd^2$)	↑↓	↑ ↑ ↑ n = 5, covalency = 5	↑ ↑ _ _ _
Halogen atom in the *3rd* excited state ($ns^1\ np^3\ nd^3$)	↑	↑ ↑ ↑ n = 7, covalency = 7	↑ ↑ ↑ _ _

Fig. 10.1 : Various covalencies shown by a halogen atoms other than F-atom.

Failure (Limitations) of octet rule

Although, according to octet theory, the central atom in most of the covalent molecules attains 8 electrons, yet there are several covalent molecules in which the central atom does not obey octet rule, *i.e.*, the central atom has either less than 8 electrons (*incomplete octet*) or more than eight electrons (*expansion of octet*) in its outer-most shell. This fact is evident from the structure of some covalent molecules given below. The number of electrons present in the valence-shell of each atom is shown in the bracket.

(*i*) 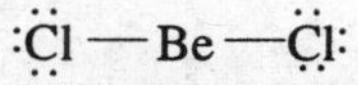

$BeCl_2$ molecule (Be=4, Cl=8)

(*ii*) 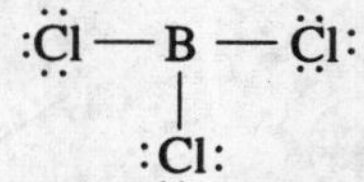

BCl_3 molecule (B = 6, Cl = 8)

(*iii*) 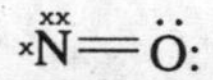

NO molecule (N=7, O=8)

(*iv*) 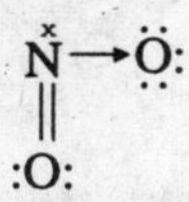

NO_2 molecule (N=7, O=8)

(*v*) 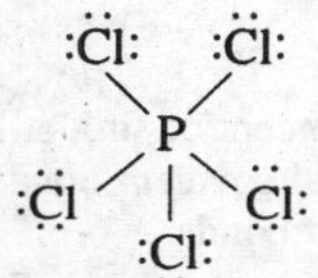

PCl_5 molecule (P=10, Cl=8)

(*vi*) 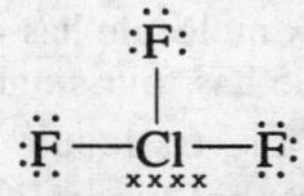

ClF_3 molecule (Cl=10, F=8)

(*vii*) 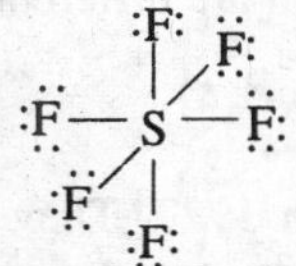

SF_6 molecule (S=12, F=8)

(*viii*) 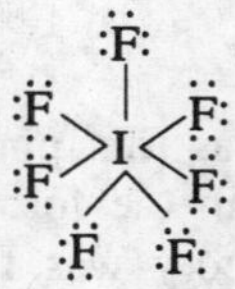

IF_7 molecule (I=14, F=8)

(*ix*) 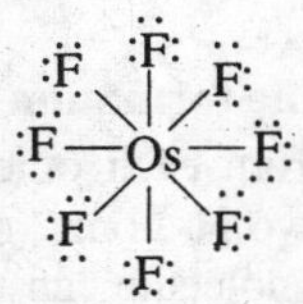

OsF_8 molecule (Os=16, F=8)

(*x*) 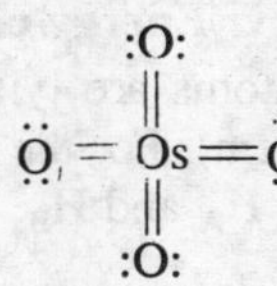

OsO_4 molecule (Os=16, O=8)

Explanation of the failure of octet rule in the molecules having central atom with more than 8 electrons

PCl_5, ClF_3, SF_6, IF_7, OsF_8, OsO_4 etc. are the covalent molecules whose central atom has more than 8 electrons in its outer-most shell. The structure of such molecules has been explained with the help of the following two concepts:

(*i*) *Sidgwick's rule of maximum covalency* (Aslo called *Maximum covalency rule*). This rule has been already discussed on page 253.

(*ii*) *Sugden's concept of singlet bonds.* Sugden did not agree with the rule of maximum covalency proposed by Sidgwick. He, however, suggested that the octet rule is never disobeyed. According to him, the outer-most shell of the central atom in molecules like PCl_5, SF_6 etc. has eight electrons, if it is assumed that in PCl_5 molecule three Cl-atoms are linked to the central P-atom by three normal covalent bonds (two-electron bonds) while each of the remaining two Cl-atoms is attached to P-atom by a special type of bound which is produced by the one-sided sharing of only one electron between P-and Cl-atoms (Fig. 10.2). In this way octet of electrons is maintained by P-atom. This special type of bond which is produced by the one-sided sharing of only one electron between the atoms is called *singlet bond, singlet linkage, single electron linkage, half bond* or simply *singlet.* This bond is a special type of co-ordinate bond because the single electron shared is from one atom (central atom) to the other atom. This bond is represented by a half arrow (↿) whose head points from the donor (*i.e.*, P-atom) to the accepter (*i.e.*, Cl-atom). On the basis of the concept of singlet bonds, the structure of PCl_5 can be shown as given at (*a*) in Fig. 10.2. Obviously, PCl_5 molecule has two singlet bonds and three normal covalent bonds. The structure of SF_6 molecule as shown at (*b*) of the same figure indicates that it has four singlet bonds and two normal covalent bonds. It may be noted that, after the formation of singlet linkages, both atoms in both the molecules attain an octet of electrons.

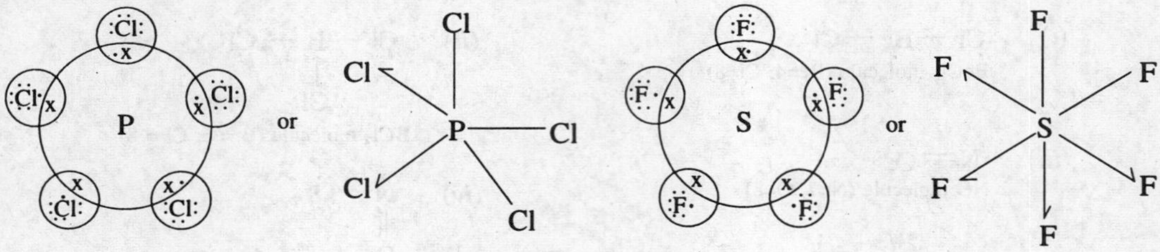

(*a*) (*b*)

Fig. 10.2 : Structure of PCl_5 and SF_6 molecules on the basis of the concept of singlet linkages (*a*) PCl_5 molecule has two singlet linkages and three normal covalent bonds (*b*) SF_6 molecule has four singlet linkages and two normal covalent bonds.

The fact that PCl_5 molecule has two types of bonds (as suggested by Sugden) has been supported by the fact that when PCl_5 is heated, it decomposes to give PCl_3 and Cl_2. The formation of these compounds shows that PCl_5 has two types of bonds which are different from each other.

Valence Bond Theory

Introduction

This theory was first put forward by Heitler and London in 1927. They gave a theoretical treatment to explain the formation of a covalent bond and energy changes taking place in H_2 molecule. This theory was further extended by Pauling and Slater (1931) to account for the directional characteristics of covalent bond.

Heitler-London theory: Energy changes taking place during the formation of H_2 molecule

Suppose two H-atoms are lying at a very large distance from each other (*i.e.*, there is no interaction between them). For the sake of convenience let the two H-atoms (nuclei) of H_2 molecule be represented as H_A and H_B. The covalent structure of H_2 molecule can thus be represented as $\overset{1}{H_A}\ \overset{2}{H_B}$. In this structure, the numbers 1 and 2 indicate two electrons. Electron 1 is associated with H_A nucleus and electron 2 is associated with H_B nucleus. If the wave functions of the orbitals containing these electrons are $\psi_A(1)$ and $\psi_B(2)$ respectively, then the combined wave function of H_2 molecule (ψ) having the above covalent structure is given by:

$$\psi = \psi_A(1).\psi_B(2) \qquad ...(i)$$

Now if a graph is drawn between the energy and the inter-nuclear distance, a curve like that shown in Fig. 10.3 (*a*) is obtained. The solution of equation (*i*) shows that when the two H-atoms are at a very large distance from each other (*i.e.*, there is no interaction between them), the energy of the system is zero, and when the atoms are brought closer and closer to each other, the energy of the system starts decreasing (*i.e.*, the energy becomes negative) till a certain minimum value is reached. The curve (*a*) shows that this minimum value is obtained at – 24KJ/mole (in the subsequent discussion we shall not use negative sign with energy terms) which is obtained at inter-nuclear distance equal to 0.90Å. At this point where E = 24 KJ/mole and r = 0.90Å, the two H-atoms are said to be bonded together. Consequently this minimum value of energy (= 24 KJ/mole) is called the bond (or bonding) energy and the inter-nuclear distance (= 0.90Å) at which this energy is obtained is called bond length (r_0).

This value of energy (= 24 KJ/mole) is very very low when it is compared with the experimental value which, as is evident from the dotted curve, is equal to 458 KJ/mole. This value is obtained at an inter-nuclear distance equal to 0.74 Å.

Since the difference between the experimental value of energy and bonding energy obtained by using equation (*i*) is very large (= 458 – 24 = 434 KJ/mole), the total (combined) wave function (ψ) as given by equation (*i*) is not the correct wave function for H_2 molecule. Thus in order to get the correct value of the combined wave function, some other points should also be considered. The following points are worth consideration. These points will improve the value of the total wave function so that we may obtain a correct value of energy which may be close to the experimental value of energy.

(i) Exchange of electrons: In the above discussion, the covalent structure of H_2 molecule has been represented as $\overset{1}{H}_A \overset{2}{H}_B$ in which electron 1 has been supposed to be solely associated with nucleus H_A and electron 2 is associated with nucleus H_B. This assumption is wrong, because when two H-atoms come closer to each other and a bond is formed between them, electrons 1 and 2 have equal freedom to get them associated with any of the two hydrogen nuclei. Thus in order to get the correct wave function of H_2 molecule, the following both covalent structures of this molecule should be considered.

$$\underset{\text{(I)}}{\overset{1}{H}_A \overset{2}{H}_B} \qquad\qquad \underset{\text{(II)}}{\overset{2}{H}_A \overset{1}{H}_B}$$

In structure (I), electron 1 is associated with nucleus H_A and electron 2 is associated with nucleus H_B. In structure (II) the position of electrons has been interchanged. The wave functions of these structures are $\psi_A(1).\ \psi_B(2)$ and $\psi_A(2).\ \psi_B(1)$ respectively.

According to the law of linear combination, the true wave function for H_2 molecule can be obtained by combining linearly the wave functions of structures (I) and (II), *i.e.*, the true wave function of H_2 molecule is obtained by combining linearly the wave functions $\psi_A(1).\ \psi_B(2)$ and $\psi_A(2).\ \psi_B(1)$. The linear combination of these wave functions can be made in the following two ways:

(*a*) When the combination of $[\psi_A(1).\psi_B(2)]$ and $[\psi_A(2).\ \psi_B(1)]$ takes place in a symmetric way, we get symmetric wave function (ψ_b) which is equal to the sum of the above two wave functions, *i.e.*,

$$\psi_b = [\psi_A(1).\psi_B(2)] + [\psi_A(2).\ \psi_B(1)] \qquad \text{....}(ii)$$

Symmetric wave function (ψ_b) is that whose value does not change, if the numbers 1 and 2 are interchanged. Since symmetric wave function is perfectly covalent, ψ_b is also called *covalent wave function* and hence is represented as ψ_{cov}. Thus:

$$\psi_b = \psi_{cov} = [\psi_A(1).\psi_B(2)] + [\psi_A(2).\ \psi_B(1)] \qquad \text{....}(iii)$$

As is evident from curve (*b*), the energy of H_2 molecule as calculated from equation (*iii*) comes out to be equal to 303 KJ/mole (fall in energy) which is obtained at an inter-nuclear distance of 0.80Å. Since ψ_b leads to a fall of energy (= 303 KJ/mole), the bonding between H-atoms is possible, *i.e.*, ψ_b represents a bonding (or attractive) state. The difference (303 – 24 = 279 KJ/mole) is called *exchange energy* of H_2 molecule, since it arises from the concept of exchange of electrons between the two hydrogen nuclei. Exchange energy of a diatomic molecule is obtained due to the exchange of electrons between the two atomic nuclei.

(*b*) When the combination of $[\psi_A(1)\psi_B(2)]$ and $[\psi_A(2)\ \psi_B(1)]$ takes place in an asymmetric way, we get asymmetric wave function (ψ_*) which is equal to the difference of the above two wave functions, *i.e.*,

$$\psi_* = [\psi_A(1).\psi_B(2)] - [\psi_A(2).\ \psi_B(1)] \qquad \text{....}(iv)$$

Asymmetric wave function (ψ_*) is that whose value changes to – ψ_*, if the number 1 and 2 are interchanged. The energy of H_2 molecule as calculated from equation (*iv*) comes out to be a positive quantity [see curve(*c*)]; *i.e.*, ψ_* leads to an increase in energy and hence the system becomes unstable. This enegy is obtained when the electron spin of the two H-atoms are parallel (↑↑) and they are brought nearer to each other.

(ii) Screening effect of electrons. We have seen above that the bonding energy as calculated from the equation (*ii*) or (*iii*) which has been obtianed after considering the exchange of electrons is equal to 303 KJ/mole. This value is still smaller than the experimental value (= 458 KJ/mole). This shows that the wave function ψ_b or ψ_{cov}) represented by equation (*ii*) or (*iii*) is not the correct wave function for H_2 molecule. Thus in order to get the correct wave function for H_2 molecule, the

screening effect of electrons should also be considered. When two H-atoms come closer to each other, the electron of one H-atom shilelds the electron of the other H-atom from the nucleus and hence the electrons donot feel the full positive charge of the nucleus. Consequently the actual positive charge felt by an electron is less than the nuclear charge and is called effective nuclear charge. Now when we use the value of effective nuclear charge to improve the value of ψ_b or ψ_{cov} wave function given by equation (*ii*) or (*iii*), bonding energy equal to 365 KJ/mole is obtained as shown in curve (*d*). This value is obtained at an inter-nuclear distance equal to 0.76Å.

(iii) Ionic character of H—H bond. We have said above that after considering the screening effect of electrons, the bonding energy is equal to 365 KJ/mole. This value is still smaller than the experimental value (= 458 KJ/mole). This shows that the wave function which has been obtained after considering the screening effect of electrons is not the correct wave function of H_2 molecule. In order to get the correct wave function, it has been suggested that the following ionic structures of H_2 molecule should also be considered.

$$\overset{1,2}{H_A}\ H_B \quad \text{or} \quad H_A^-\ H_B^+ \qquad \text{(III)} \qquad\qquad H_A\ \overset{1,2}{H_B} \quad \text{or} \quad H_A^+\ H_B^- \qquad \text{(IV)}$$

In structure (III), both the electrons (shown as 1 and 2) are attached with H_A nucleus while in structure (IV) both electrons reside on H_B nucleus. The wave functions representing these structures are $\psi_A(1)\ \psi_A(2)$ and $\psi_B(1)\ \psi_B(2)$ respectively. Thus the total ionic wave function, ψ_{ionic} may be given by:

$$\psi_{ionic} = \psi_A(1)\psi_A(2) + \psi_B(1)\psi_B(2) \qquad ...(v)$$

The ionic structures shown above increase the force of attraction between the two H-atoms. This attracion leads to an increase (in the negative side of the graph) in the bond energy. Thus we see that both the structures (*i.e.*, covalent as well as ionic) of H_2 molecule contribute to the actual wave function of H_2 molecule. Thus the combined wave function of this molecule (ψ_{H_2}) is given by combining ψ_{cov} and ψ_{ionic} *i.e.*,

$$\psi_{H_2} = \psi_{cov} + \psi_{ionic}$$

On putting the value of ψ_{cov} and ψ_{ionic} from equations (*iii*) and (*v*), we get:

$$\psi_{H_2} = [\psi_A(1).\psi_B(2)+\psi_A(2).\ \psi_B(1)] + \lambda[\psi_A(1)\ \psi_A(2) + \psi_B(1)\ \psi_B(2)] \qquad ...(vi)$$

Here λ is called the *mixing coeficient or weighting factor* and represents the magnitude of the contribution made by the ionic structures to the total wave function of H_2 molecule. The value of λ has been found to be equal to 0.17. As is evident from curve(*e*), the energy of H_2 molecule calculated by using equation (*vi*) comes out to be equal to 388 KJ/mole which is obtained at an inter-nuclear distance of 0.75Å. This value (= 388 KJ/mole) is closer to the experimental value (= 458 KJ/mole) and the value of inter-nuclear distance (= 0.75Å) is also very close to the experimental value (= 0.74Å. This discussion shows that when both the structures namely covalent as well as ionic contribute to the true structure of H_2 molecule, the bonding energy is equal to 388 KJ/mole.

We have seen that when H_2 molecule is supposed to have only covalent structures *viz.* (*I*) and (II), the bonding energy is equal to 303 KJ/mole. When covalent as well as ionic structures contribute to the true structure of H_2 molecule, the bonding energy is equal to 388 KJ/mole. The difference (388 – 303 = 85 KJ/mole) is due to the contribution made by the ionic structure. This difference is called *ionic-covalent resonance energy.*

The above discussion shows that H_2 molecule can neither be represented by covalent structures $\left[\ \overset{1}{H_A}\text{—}\overset{2}{H_B} \text{ and } \overset{2}{H_A}\text{—}\overset{1}{H_B}\ \right]$ nor by ionic structures $\left[\ \overset{1,2}{H_A}\ H_B \text{ or } \overset{-}{H_A}\ \overset{+}{H_B} \text{ and } H_A\ \overset{1,2}{H_B} \text{ or } \overset{+}{H_A}\ \overset{-}{H_B}\ \right]$. The true structure of this molecule is said to be something which is in between the above four structures, *i.e.*, H—H bond in H_2 molecule is partly covalent and partly ionic.

Bond energies and bond lengths for various wave functions of H_2 molecule are summed up in Table 10.1 and have been shown in Fig. 10.3.

Table 10.1 : *Bond energies and bond lengths for various wave functions of H_2 molecule.*

Structure of H_2 molecule	Combined wave function	Curve shown in Fig. 10.3	Bond energy (KJ/mole)	Bond length (inter-nuclear distance) (Å)
$\overset{1}{H_A}\overset{2}{H_B}$	$\psi = \psi_A(1)\ \psi_B(2)$	(*a*)	24	0.90
$\overset{1}{H_A}\overset{2}{H_B}$, (I) $\overset{2}{H_A}\overset{1}{H_B}$ (II)	$\psi_b = \psi_{cov} = [\psi_A(1)\ \psi_B(2)] + [\psi_A(2)\ \psi_B(1)]$ $\psi_* = [\psi_A(1)\ \psi_B(2)] - [\psi_A(2)\ \psi_B(1)]$	(*b*) (*c*)	303 positive value	0.80 ——
	ψ_{cov} with screening effect of electrons	(*d*)	365	0.76
$\overset{1}{H_A} - \overset{2}{H_B}$, (I) $\overset{2}{H_A} - \overset{1}{H_B}$ (II) Covalent structures $\overset{1,2}{H_A}\ H_B$ or $\overset{-}{H_A}\ \overset{+}{H_B}$ (III) $H_A\ \overset{1,2}{H_B}$ or $\overset{-}{H_A}\ \overset{+}{H_A}$ (IV) Ionic structure	$\psi_{H_2} = \psi_{cov} + \lambda_{ionic}$ $= [\psi_A(1)\ \psi_B(2) + \psi_A(2)\ \psi_B(1)] + \lambda[\psi_A(1)\ \psi_A(2) + \psi_B(1)\ \psi_B(2)]$	(*e*)	388	0.75
Experimental value		dotted curve	458	0.74

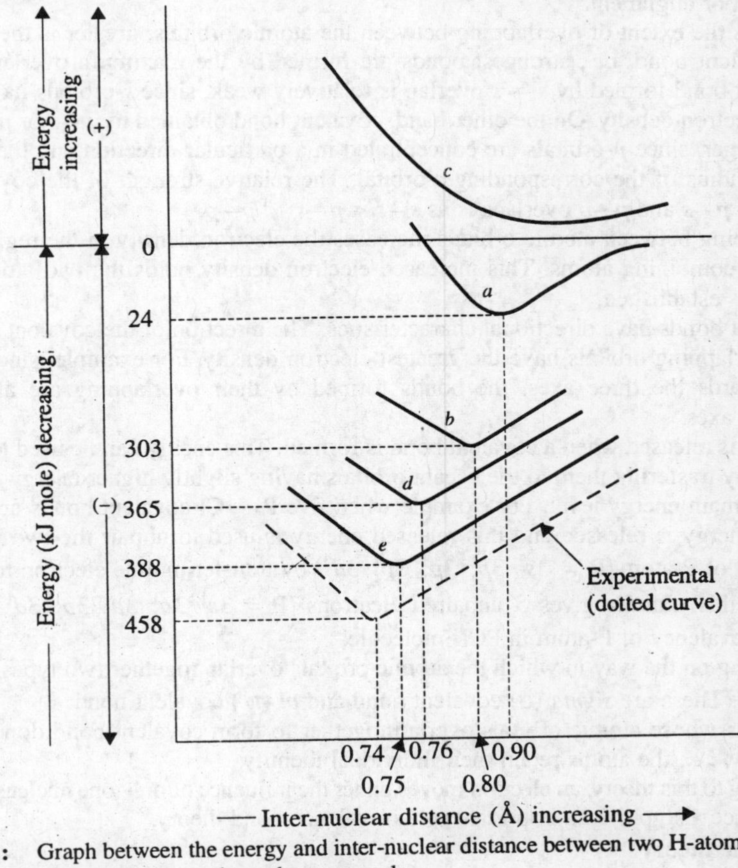

Fig. 10.3 : Graph between the energy and inter-nuclear distance between two H-atoms of H_2 molecule. This graph has not been drawn to scale.

Pauling-Slater's Theory: Orbital Overlap Theory of Covalent Bond

Introduction

As already said, valence bond theory proposed by Heitler and London, was modified by Pauling and Slater in 1931 to account for the formation and directional characteristics of covalent bond. Thus we can say that Pauling-Slater's theory is the modified form of Heitler-London theory. Since, according to Pauling - Slater's theory, covalent bond is obtained when the atomic orbitals of the valence-shell (containing unpaired electrons only) of the atoms overlap together, this theory is called *orbital overlap theory.*

Postulates of Pauling - Slater's theory

(*i*) This theory explains the formation and directional characteristics of covalent bond.

(*ii*) A covalent bond is formed by the overlap of pure atomic orbitals of the atoms between which the bond is being formed. The overlapping takes place only between those atomic orbitals which belong to the valance-shell of each atom and contain unpaired electrons with opposite spins. If the spins of the electrons are the same, orbitals, due to the repulsion between electrons, will increase the energy of the system and hence the system will become unstable, *i.e.*, no bond will be formed.

(*iii*) The atomic orbitals of the valence-shell which have paired electrons (completely-filled orbitals) donot participate in the overlapping process and hence donot contribute to the bond formation. The electron pairs in these orbitals are, therefore, called *non-bonding electron pairs* or *lone pairs* of electrons.

(*iv*) The orbitals overlapping together should have the same symmetry, *i.e.*, they should have proper orientation or alignment.

(*v*) Greater is the extent of overlapping between the atomic orbitals, greater is the strength of the resulted covalent bond, *i.e.*, strongest bonds are formed by the maximum overlap of atomic orbitals. Covalent bond formed by *s* — *s* overlap is relatively weak, since *s*-orbitals have spherical distribution of electron density. On the other hand, covalent bond obtained by *p—s* or *p—p* overlap is relatively stronger, since *p*-orbitals are concentrated in a particular direction and their lobes are longer than the radius of the corresponding *s*-orbital. The relative strength of the covalent bonds obtained by *s—s*, *p—s* and *p—p* overlaps is as $s—s < p—s < p—p$.

(*vi*) Overlapping between atomic orbitals increases the electron density in the region between the nuclei of the combining atoms. This increased electron density holds the two atoms together and thus a bond is established.

(*vii*) Covalent bonds have directional characteristics. The direction of the covalent bond is that in which the overlapping orbitals have the greatest electron density. For example, since *p*-orbitals are directed towards the three axes, the bonds formed by their overlapping are also directed towards the three axes.

(*viii*) Energy is released when a covalent bond is formed. This energy can be used to unpair the paired electrons by trasfering them to the vacant orbitals having slightly higher energy and belonging to the same main energy level. For example when five P — Cl covalent bonds are formed in PCl_5 molecule, energy is released and this released energy is used to unpair the two paired electrons in 3*s* orbital of P-atom $\left(P = 3s^2\ 3p_x^1\ 3p_y^1\ 3p_z^1\ 3d^0\right)$ by transfering one electron to the vacant 3*d* orbital. Thus this transfer gives 5 unpaired electrons $\left(P = 3s^1\ 3p_x^1\ 3p_y^1\ 3p_z^1\ 3d^1\right)$ which explains the pentacovalency of P-atom in PCl_5 molecule.

(*ix*) Depending on the way in which the atomic orbitals overlap together two types of covalent bonds are formed. These are *sigma* (σ) covalent bond and *pi* (π) covalent bond.

(*x*) The atoms whose atomic orbitals overlap together to form covalent bond donot lose their individual identity, *i.e.*, the atoms retain their individual identity.

(*xi*) According to this theory, an electron moves under the influence of only one nucleus of an atom.

(*xii*) Resonance also plays an important role in valence bond theory.

Limitations of the theory

(*i*) We have said that the atomic orbitals which are completely-filled donot participate in the process of overlapping and hence donot contribute to bond formation. However, this seems to be exaggeration, since the nucleus of an approaching atom is bound to affect the electron-waves of nearly all orbitals.

(*ii*) According to valence bond theory the structures of B_2 and O_2 molecules are : B — B: and $:\ddot{O}=\ddot{O}:$ which indicate that these molecules have all the electrons in the paired state and hence should be diamagnetic. But experiments have shown that both these molecules are paramagnetic to the extent of two unpaired electrons. Valence bond theory is unable to explain the parmagnetic character of these molecules.

(*iii*) According to valence bond theory, the covalent bonds are obtained by the overlap of pure atomic orbital of the atoms constituting the molecule. The formation of bonds in this way does not explain the observed properties (*e.g.* bond lengths, bond angles etc.) of many molecules. For example, according to valence-bond theory, all the four C—H bonds in CH_4 molecule should be of unequal length, but actually these have been found to be of equal length. In case of H_2O molecule, HOH bond angle should be equal to 90°, but actually the bond angle is 104.5°.

Sigma (σ) Bond

What is sigma(σ) bond?

A covalent bond which is formed between two atoms by the overlap of their two half-filled orbitals belonging to their valence-shell along a line joining the nuclei of both the atoms (*i.e.*, along the inter-nuclear axis, bond axis or molecular axis, as it is called) is called a σ-bond. In other words, σ-bond is produced by the *head-to-head overlap* (head-to-head overlap is also called *end-to-end overlap, head-on overlap, end-on overlap, axial overlap* or *linear overlap*) of the two half-filled orbitals belonging to the valence-shell of the two combining atoms. Thus if *x*-axis is assumed to be the molecular axis, then σ-bond is produced by $s-s$ overlap (as in H_2 molecule), $s-p_x$ overlap [as in NH_3, H_2O and HX (X = F, Cl, Br or I) molecules] and p_x-p_x overlap (as in X_2 molecules). A sigma bond is also produced when two hybrid orbitals overlap together along the inter-nuclear axis.

Examples to show the formation of σ-bond

Here we shall see how σ-bond is formed in H_2, NH_3, H_2O, HF and F_2 molecules.

(*i*) *H_2 molecule (s—s overlap)*. The electronic configuration of H-atom is $1s^1$. 1*s*-orbitals of both H-atoms which have one unpaired electron each overlap with each other along their axes and form one H—H σ-bond (Fig. 10.4).

(*ii*) *NH_3 molecule (s—p overlap)*. The valence-shell electronic configuration of N-atom is $2s^2$, $2p_x^1$, $2p_y^1$, $2p_z^1$. All the three *p*-orbitals on N-atom, each having one electron, overlap with three 1*s* orbitals of three H-atoms along their axes and form three N—H σ-bonds (Fig. 10.5). 2*s*-orbital of N-atom which has paired electrons does not participate in the overlapping process and hence the electron pair residing in 2*s*-orbital does not contribute to bond formation.

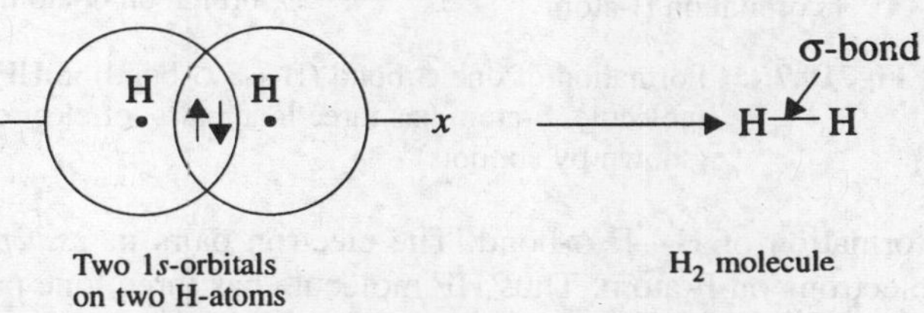

Fig. 10.4: Formation of one σ-bond (H—H σ-bond) in H_2 molecule by the overlap of two 1*s*-orbitals (s—s overlap)

This pair of electrons is called *non-bonding electron pair* or *lone pair of electrons.* Thus NH_3 molecule has one lone pair of electrons on N-atom.

Now since p_x, p_y and p_z orbitals are at right angle to each other, each of the three H—N—H bond angle in NH_3 molecule must be equal to 90°. The actual angle determined experimentally is equal to 107° instead of 90°. The increase in bond angle is due to the repulsion between different electron pairs.

(*iii*) *H_2O molecule (s—p overlap)*. The valence-shell configuration of O-atom is $2s^2$, $2p_x^1$,

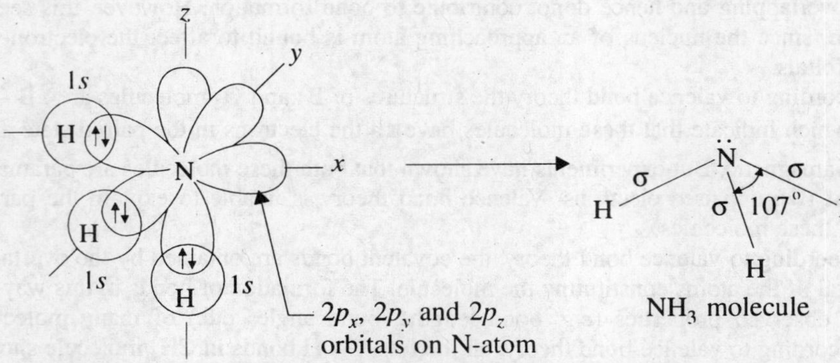

Fig 10.5: Formation of three σ-bonds (N—Hσ-bonds) in NH_3 molecule by three *s-p* overlaps. Two dots shown at N-atom represent a lone pair of electrons which is present in 2*s* orbital of N-atom.

$2p_y^2$, $2p_z^1$. Two 1*s*-orbitals on two H-atoms overlap with $2p_x$ and $2p_z$ orbitals (both having unpaired electron) along their axes and form two O—H σ-bonds (Fig. 10.6). The electron pairs in 2*s* and $2p_y$ orbitals reside as lone pairs of electrons on O-atom. Thus H_2O molecule has two lone pairs of electrons on O-atom.

Now since p_x and p_y orbitals are perpendicular to each other, each of the two H—O—H bond angles in H_2O molecule must be equal to 90°. However, the actual bond angle determined experimentally is 104.5° instead of 90°. The increase in bond angle is due to the repulsion between different electron pairs.

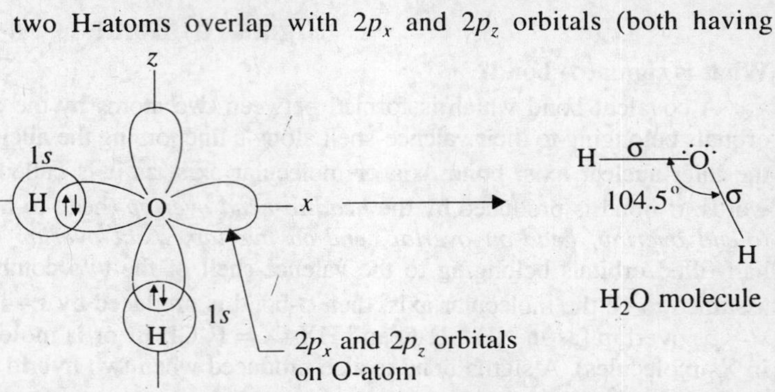

Fig. 10.6 : Formation of two σ–bonds (O—Hσ–bonds) in H_2O molecule by two *s—p* overlaps. O-atom has two lone pairs of electrons (shown by four dots)

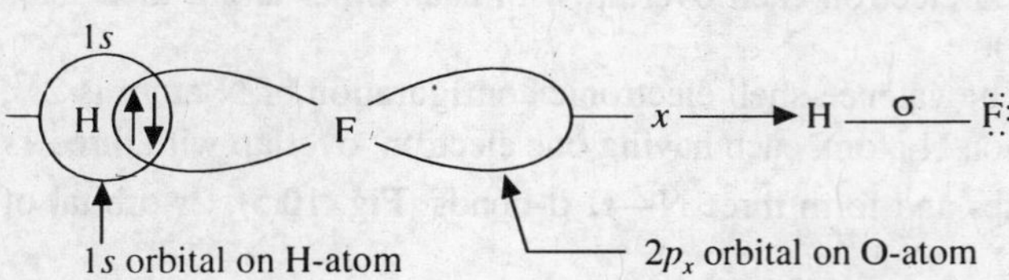

Fig. 10.7: Formation of one σ-bond (H—F σ-bond) in HF molecule. F-atom has three lone pairs of electrons (shown by six dots)

(*iv*) *HF molecule (s — p overlap).* The valence-shell configuration of F-atom is $2s^2$, $2p_x^1$, $2p_y^2$, $2p_z^2$. 1*s*-orbital of H-atom overlaps with the singly- filled $2p_x$ orbital on F-atom along the *x*-axis and gives rise to the formation of H—F σ-bond. The electron pairs in 2*s*, $2p_y$ and $2p_z$ orbitals reside as lone pairs of electrons on F-atom. Thus HF molecule has three lone pairs of electrons on F-atom. (Fig.10.7).

(*v*) *F_2 molecule (p—p overlap).* The valence-shell configuration of F-atom is $2s^2$, $2p_x{}^1$ $2p_y{}^2$ $2p_z{}^2$. $2p_x$ orbitals (which have one unpaired electron) on two F-atoms overlap together along the *x*-axis and form one F—F σ-bond. The electron pairs in 2*s*, $2p_y$ and $2p_z$ orbitals of both F-at-

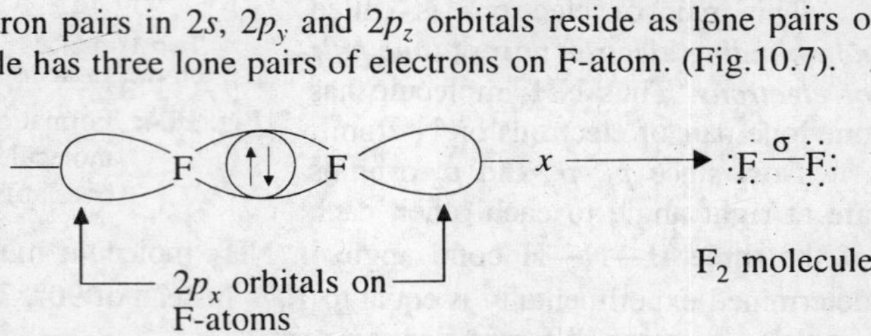

Fig. 10.8 : Formation of one σ-bond (F—F σ-bond) in F_2 molecule by *p-p* overlap. Each F-atom in F_2 molecule has three lone pairs of electrons.

oms act as lone pairs of electrons. Thus each F-atom in F_2 molecule has three lone pairs of electrons on it (Fig. 10.8).

Pi (π) Bond

What is pi (π) bond?

A covalent bond which is formed between two atoms by the overlap of their two-half-filled orbitals belonging to their valence-shell along a line perpendicular to the inter-nuclear axis is called a π-bond. In other words, π-bond is produced by the *side-to-side overlap* (side -to-side overlap is also called *side-wise overlap,* or *lateral overlap*) of the two half-filled orbitals belonging to the valence-sheil of the two combining atoms. Thus if *x*-axis is assumed to be the molecular axis, then π-bond is produced by $p_y - p_y$ and $p_z - p_z$ overlaps.

Examples to show the formation of π-bond

Here we shall see how π-bond is formed in O_2 and N_2 molecules.

O_2 molecule. The valence-shell configuration of O-atom is $2s^2$, $2p_x^1$ $2p_y^1$ $2p_z^2$, showing th: the atom has two halt-filled *p*-orbitals namely $2p_x$ and $2p_y$. If *x*-axis is assumed to be the molecular axis, then the overlap between two $2p_x$ orbitals (each orbital has one unpaired electron) on two O-atoms along the *x*-axis gives rise to the formation of one O—O σ-bond while the overlap of two $2p_y$ orbitals (each is having one unpaired electron) along a line perpendicular to the *x*-axis (molecular axis) gives one O—O π bond. Thus O_2 molecule has two O-atoms linked together by one σ-bond and one π-bond, *i.e.*, O_2 molecule has a double bond (Fig. 10.9). The electron pairs present in 2*s* and $2p_z$ orbitals reside as lone pairs on both O-atoms. This discussion shows that the structure of O_2 molecule can be written as $:\ddot{O} = \ddot{O}:$. In this structure each O-atom is surrounded by an octet of electrons, *i.e.*, this structure conforms to the octet rule.

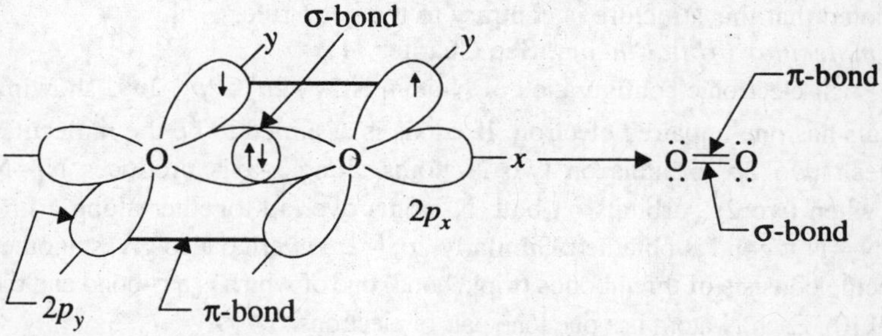

Fig. 10.9 : Formation of one σ-bond (O — O σ-bond) and one π-bond (O — O π-bond) in O_2 molecule by two *p*—*p* overlaps.

Now let us see how the above structure of O_2 molecule can account for the following properties of O_2 molecule.

(*i*) O_2 has paramagnetic character corresponding to the presence of two unpaired electrons.

(*ii*) O_2 molecule has high bond dissociation energy which is equal to 494.0 KJ/mol.

Since the above structure of O_2 molecule does not contain any unpaired electrons, it cannot account for the paramagnetic behaviour of O_2 molecule, although, due to the presence of O = O double bond, it accounts for the high bond dissociation energy of O_2 molecule.

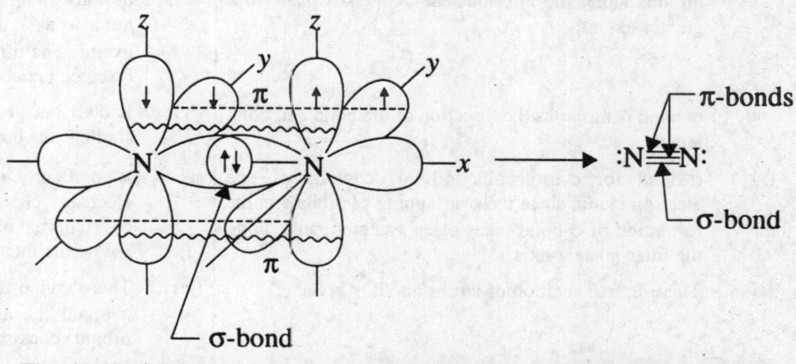

Fig. 10.10 : Formation of one σ-bond (N—N σ-bond) and two π-bonds (N—N π-bonds) in N_2 molecule by $2p_x - 2p_x$, $2p_y - 2p_y$ and $2p_z - 2p_z$ overlaps respectively.

In addition to the above structure $:\ddot{O} = \ddot{O}:$, the following structures have also been proposed for O_2 molecule.

(*a*) $:\ddot{\underset{\cdot}{O}} — \ddot{\underset{\cdot}{O}}:$ This structure contains one 2-electron bond and each O-atom has one unpaired electron. Thus, due to the presence of two unpaired electrons, this structure can account for the paramagnetic character of O_2 molecule. Since the two O-atoms are linked by only one single covalent bond, this structure can not explain the high bond dissociation energy of O_2 molecule.

(*b*) $:O \overset{\cdots}{\underset{\cdots}{—}} O:$ This structure was suggested by Pauling and has one 2-electron bond (normal covalent bond) and two 3-electron bonds. Each of the 3-electron bond contains one unpaired electron. Thus, due to the presence of two unpaired electrons this structure can account for the paramagnetic nature of O_2 molecule. This structure also accounts for the high bond dissociation energy, since a 3-electron bond imparts additional strength equal to one-half the strength of a 2-electron bond. It may be noted that this structure is contrary to the octet rule.

(*c*) *Structure based on molecular orbital theory.* See Chapter 11.

***N_2* mlecule.** The valence-shell electronic configuration of N-atom is $2s^2$, $2p_x^1$, $2p_y^1$, $2p_z^1$ showing that each of the three *p*-orbitals has one unpaired electron. If *x*-axis is assumed to be the molecular axis, then the overlap between two $2p_x$ orbitals on two N-atoms along *x*-axis produces N—N σ-bond. On the other hand, when two $2p_y$ orbitals on both N-atoms overlap together along a line perpendicular to *x*-axis, then N—N π bond is obtained. Similarly $2p_z - 2p_z$ overlap also gives another N—N π-bond. Thus N_2 molecule consists of three bonds (triple bond) one of which is a σ-bond and the other two are π-bonds. (Fig. 10.10). Each N-atom has one lone pair of electrons.

Comparison between sigma (σ) and pi (π) bonds

The comparison between sigma (σ) and pi (π) bonds has been shown in Table 10.2.

Table 10.2: *Comparison between sigma (σ) and pi (π) bonds.*

	Sigma (σ) bond		Pi (π) bond
(*i*)	It is obtained when two pure atomic orbitals or hybrid orbitals overlap with each other along the inter-nuclear axis.	(*i*)	It is obtained when two pure atomic orbitals overlap with each other along a line perpendicular to the inter-nuclear axis.
(*ii*)	If *x*-axis is assumed to be the inter-nuclear axis (molecular axis), then *s–s*, *s–*p_x and $p_x - p_x$ overlaps produce σ-bond.	(*ii*)	If *x*-axis is assumed to be the inter-nuclear axis, then $p_y - p_y$ and p_z–p_z overlaps produce π-bond.
(*iii*)	σ-bond is a stronger bond, since the overlap of orbitals along the inter-nuclear axis takes place to a greater extent.	(*iii*)	π-bond is a weaker bond because the overlap of orbitals along a line perpendicular to the inter-nuclear axis does not take place to a greater extent. The already present σ-bond restricts the distance between the atoms.
(*iv*)	σ-bond determines the direction of the bond and bond length.	(*iv*)	It does not decide the direction of the bond but shortens the bond length due to multiple bonding.
(*v*)	σ-bond (or σ-molecular orbital) consists of one electron cloud, since the overlapping of orbitals in the formation of σ-bond takes place symmetrically along the inter-nuclear axis.	(*v*)	π-bond (or π-molecular orbital) consists of two electron clouds (electron lobes), since the overlapping of orbitals takes place above and below the inter-nuclear axis.
(*vi*)	There is free rotation of atoms about σ-bond.	(*vi*)	There cannot be a free rotation of atoms about π-bond, because a π-bond (or π-molecular orbital) consists of two electron clouds.
(*vii*)	*s*-orbitals take part in the formation of a σ-bond.	(*vii*)	*s*-orbitals never take part in the formation of a π-bond.
(*viii*)	One lobe of *p*-orbital taking part in the formation of a σ-bond stretches while the other lobe reduces in size.	(*viii*)	Both lobes of *p*-orbital taking part in π-bond formation remain of the same size.

To predict the nature of bond: Non-polar covalent bond, polar covalent bond and ionic bond

Let us consider a AB molecule which contains A—B bond. Let the electronegativity of atom A (x_A) be greater than than of atom B (x_B), *i.e.*, $x_A > x_B$. With the help of the known value of electronegativity difference, ($x_A - x_B$), we can predict whether A—B bond will be non-polar covalent bond, polar covalent bond or ionic bond. The following three cases arise:

(*a*) *When* ($x_A - x_B$) *is equal to zero, i.e.*, when the electronegativity values of the two atoms are the same or when two identical or similar atoms (homo-nuclear atoms) are linked together (e.g. H—H, F—F bonds), the bond formed between the two identical atoms is purely covalent. This bond does not contain any polarity or ionic character and hence is called *non-polar covalent bond.* Why this bond does not contain any ionic character has been explained as follows:

Due to the same electronegativity values of the two atoms, the electron pair being shared by both the atoms is attracted equally by these atoms and hence lies exactly at the mid point of the distance between the two identical linked atoms. In terms of wave mechanics we say that the shared electron pair spends equal time round both the atoms, since the atoms have the same electronegativity values. The symmetrical distribution of the electron pair round the two identical atoms does not lead to the development of any charge on any of the two atoms and hence the bond between the two identical atoms remains non-polar. From the above discussion it follows that a *non-polar covalent bond may be defined as a bond existing between two atoms having the same electronegativity values (i.e., identical or homo-nuclear atoms) having no charge seperation between them.*

(*b*) *When* ($x_A—x_B$) *is relatively small* (*e.g.* in O—H, Cl—H bonds), the bond formed between A and B is covalent which has some ionic character or polarity, *i.e.*, the bond is polar covalent. The formation of polar covalent bond which is represented as $A^{\delta-} - B^{\delta+}$ can be explained as follows. Due to the small difference in electronegativity values between the two atoms, the electron pair being shared by the two atoms is attracted slightly towards the more electronegative A-atom and hence does not lie exactly at the mid point of the distance between the two atoms. In terms of wave mechanics we say that the shared electron pair spends more time near A-atom. The unsymmetrical distribution of electron pair between A and B leads to the development of a partial negative charge (represented as δ^-) on more electronegative A-atom and equal positive partial charge (= δ^+) on less electronegative B-atom. Thus the polar covalent bond formed between A (more electronegative atom) and B (less electronegative atom) is represented as $A^{\delta-} - B^{\delta+}$. The molecule containing a polar covalent bond is called a *dipole,* since the more electronegative atom at one end of the molecule carries a negative (partial) charge and the less electronegative atom at the other end of the molecule carries a positive (partial) charge which is equal in magnitude to that of the negative charge. Thus the formation of a polar covalent bond between H (less electronegative) and F (more electronegative) in HF molecule can be shown as $H^{\delta+}—F^{\delta-}$, indicating that HF molecule is acting as a dipole. From the above discussion it follows that *a polar covalent bond may be defined as a covalent bond existing between two atoms having slightly different electronegativities (i.e., different atoms) and having charge separation between them.*

(*c*) *When* ($x_A - x_B$) *is very large,* complete transfer of an electron from atom B to atom A takes place and hence the bond between A and B would be purely ionic which can be represented as B^+A^- (*e.g.* Li^+F^-).

The above discussion shows that polar covalent bond is intermediate between a pure covalent bond and a pure ionic bond.

Factors on which the amount of polarity of a polar covalent bond depends

The factors on which the polarity of a given polar covalent bond depends are given below:

1. *Electronegativity difference between the linked atoms.* Greater is the difference between the electronegativity of A and B atoms in A—B bond, greater is the amount of ionic character (polarity) in A—B bond. For example, since the electronegativity difference between X (X = F,Cl, Br, I) and H atoms in HF, HCl, HBr and HI molecules decreases in the order : HF > HCl > HBr > HI, the

amount of ionic character in H—X bond also decreases in the same order, *i.e.*, HF is the most polar ionic molecule and HI is the least so.

2. *Dipole moment of the polar molecule.* Greater is the value of experimental dipole moment of a polar molecule, greater is the amount of ionic character in the polar covalent bond in the polar molecule. For example, since the experimental dipole moment values of HF (= 0.640×10^{-29} C.m), HCl (= 0.34×10^{-29}C.m), HBr (= 0.263×10^{-29} C.m.) and HI (= 0.127×10^{-29} C.m.) molecules decrease in the order, HF > HCl > HBr > HI, the amount of ionic character in H—X (X = F, Cl, Br, I) bonds in these molecules also decreases in the same order, *i.e.*, HF is the most polar molecule and HI is the least so.

Table10.3 *Differences between polar and non-polar covalent bonds.*

	Polar covalent bond		Non-polar covalent bond
(*i*)	A polar covalent bond is that covalent bond which has some amount of polar or ionic character (*i.e.*, polarity	(*i*)	A non-polar covalent bond is nothing but a normal covalent bond which does not have any amount of polar character.
(*ii*)	This bond is established between the two atoms which have small difference in their electronegativity values.	(*ii*)	This bond is established between two atoms which have no difference in their electronegativity values, *i.e.*, this bond is established between two atoms which have the same electronegativity or is established between two similar atoms.
(*iii*)	Due to the difference in electronegativity values between the two atoms, the electron pair constituting the polar covalent bond is not equally attracted or shared by the two atoms.	(*iii*)	Due to no difference in electronegativity values of the two atoms, the electron pair constituting the non-polar covalent bond is equally attracted or shared by the two atoms.
(*iv*)	Due to unequal sharing of the electron pair, the more electronegative atom acquires a partial negative charge on it while the less electronegative atom acquires a partial positive (equal) charge.	(*iv*)	Due to the equal sharing of the electron pair, neither of the two atoms acquires any charge on it.
(*v*)	A polar covalent bond between A and B atoms in AB molecule is represented as $\overset{\delta+}{A}—\overset{\delta-}{B}$ (when $x_A < x_B$) or $\overset{\delta-}{A}—\overset{\delta+}{B}$ (when $x_A > x_B$)	(*v*)	A non-polar covalent bond which is a normal covalent bond between two A atoms in A_2 molecule is represented as A — A.
(*vi*)	$\overset{\delta+}{A} — \overset{\delta-}{B}$ or $\overset{\delta-}{A} — \overset{\delta+}{B}$ acts as a dipole.	(*vi*)	A — A does not act as a dipole.
(*vii*)	When a polar molecule is placed between two charged parallel plates, the molecule orients in such a way that the positive end of its dipole is directed towards the negatively-charged plate and the negative end of its dipole is directed towards the positively-charged plate. As a result of this, the intensity of the charge on each plate is diminished.	(*vii*)	When a non-polar molecule is placed between two charged parallel plates, the molecule does not undergo any orientation in the electric field.
(*viii*)	Due to charge separation in polar molecules these have higher melting and higher boiling points.	(*viii*)	Since there is no charge separation in non polar molecules, these molecules have low melting and low boiling points.

How to distinguish between a polar and a non-polar molecule

A molecule which has purely covalent bond is called *non-polar molecule* while that which contains polar covalent bond is called *polar molecule.* There are two methods to distinguish between a polar and a non-polar molecule.

(*i*) *Dipole moment method.* For this see "*Applications of dipole moment values*".

(*ii*) *A simple experimental method.* A polar molecule can be distinguished from a non-polar molecule by a simple experiment in which we place the substance under the examination between two charged parallel plates. If the substance is polar (*e.g.* HCl), the molecules of this substance will orient in such a manner that the positive ends of the molecules are directed towards the negatively-charged plate and the negative ends are directed towards the positively-charged plate. The orientation of the molecules in this manner results in that the intensity of charge on each plate will diminish.

If the substance is non-polar, the molecules of this substance will not orient when the substance is placed between the two charged parallel plates.

Calculation of the percentage of ionic character in A—B polar covalent bond in AB molecule

The following two methods have been proposed for this purpose.

1. Electronegativity difference method. The amount of ionic character in A—B polar covalent bond indicates the extent to which the electron pair has shifted towards the more electronegative atom (extent of charge separation) in the bond. Greater is the electronegativity difference between the bonded atoms, A and B [*i.e.*, $(x_A - x_B)$, if $x_A > x_B$], greater will be the charge separation between the bonded atoms and hence greater will be the magnitude of ionic character in the bond. Several empirical equations have been proposed to calculate the amount of ionic character in A—B polar covalent bond. All these equations make use of electronegativity difference, $(x_A - x_B)$. Two of these equations are given below:

(*a*) *Pauling's equation.* According to Pauling, amount of ionic character is given by:

Amount of ionic character in A—B bond = $[1 - e^{0.25\,(x_A - x_B)}]$ % ...(*i*)

With the help of this equation, Pauling established the following relation between $(x_A - x_B)$ and percentage of ionic character.

$(x_A - x_B)$	=	0.6	1.0	1.4	1.7	2.0	2.4	3.0	3.2
% of ionic character	=	9	22	39	51	63	76	91	92

The above data show that when (*a*) $(x_A - x_B) = 1.7$, A—B bond has 50% ionic character and 50% covalent character. (*b*) $(x_A - x_B) < 1.7$, A—B bond has ionic character less than 50% and covalent character more than 50%, *i.e.*, *A — B* bond is more covalent than ionic in character. (*c*) $(x_A - x_B) > 1.7$, A – B bond has ionic character more than 50% and covalent character less than 50%, *i.e.*, A – B bond is more ionic than covalent in character.

Pauling suggested the following alternative empirical equation also for calculating the ionic character in A—B bond.

Ionic character in A—B bond = $\left[18\,(x_A - x_B)^{1.4}\right]$ % ...(*ii*)

Problem : *On the basis of Pauling's equation [equation (ii)], calculate the percentage of ionic character in HF molecule, given that* $x_F = 4.0$, $x_H = 2.2$.

Solution. $x_F - x_H = 4.0 - 2.2 = 1.8$

Ionic character in HF molecule = $18 \times (1.8)^{1.4}\% = 44.12\%$

This value (= 44.12%) is quite close to that (= 45%) obtained by Pauling from dipole moment of HF molecule.

(*b*) *Hannay and Smith equation.* These two chemists proposed the following equation to calculate the ionic character

Ionic character in A—B bond = $[16(x_A - x_B) + 3.5\,(x_A - x_B)^2]\%$ ....(*iii*)

Problem : *On the basis of Hanny and Smith equation, calculate the percentage of ionic character in gaseous HF, HCl, HBr and HI molecules, given that* $x_F = 4.0$, $x_{Cl} = 3.2$, $x_{Br} = 3.0$, $x_I = 2.7$ *and* $x_H = 2.1$.

Solution. $x_F - x_H = 4.0 - 2.2 = 1.8$; $x_{Cl} - x_H = 3.2 - 2.2 = 1.0$; $x_{Br} - x_H = 3.0 - 2.2 = 0.8$; $x_I - x_H = 2.7 - 2.2 = 0.5$

On putting the above values in Hannay - Smith's equation, we get :

Ionic character in HF molecule = $[(16 \times 1.8 + 3.5 \times (1.8)^2]\%$ = 40.14%

Ionic character in HCl molecule = $[16 \times 1.0 + 3.5 \times (1)^2]\%$ = 19.5%

Ionic character in HBr molecule = $[16 \times 0.8 + 3.5\,(0.8)^2]\%$ = 15.04%

Ionic character in HI molecule = $[16 \times 0.5 + 3.5 \times (0.5)^2]\%$ = 8.875%

The percentage of ionic character in gaseous HF, HCl, HBr and HI molecules as calculated above shows that their ionic character is in the order : HF(40.14%) > HCl(19.5%) > HBr (15.04%) > HI(8.875%). Since the percentage of ionic character in each of these molecules is less than 50%, these molecules in their gaseous state are *mostly covalent* and only *partly ionic*. The predominant covalent nature of these molecules is confirmed by the facts that these compounds have low melting and boiling points and in their pure state are bad conductors of electricity.

2. Dipole moment method. According to this method the percentage of ionic character in A—B bond in AB molecule is equal to

$$\frac{100 \times \mu_{exp}}{(\mu_{cal})_{ionic}}\% = \frac{100 \times \mu_{exp}}{d_{A-B} \times \text{Electronic charge}}\%$$

Here μ_{exp} = Experimental (observed or actual) value of dipole moment of AB molecule

$(\mu_{cal})_{ionic}$ = calculated (theoretical value) of dipole moment of AB molecule when this molecule is assumed to be completely ionic. The value of $(\mu_{cal})_{ionic}$ is calculated as follows:

$(\mu_{cal})_{ionic}$ = Distance between the centres of the positive and negative ends of $A^{\delta-}$—$B^{\delta+}$ dipole × Electronic charge.

= d_{A-B} × Electronic charge.

Thus if d_{A-B} and μ_{exp} are known, the percentage of ionic character in A – B bond can be calculated by using the above relation. For details see "*Applications of dipole moment*".

Dipole Moment

What is dipole moment?

We have seen under the study of polar covalent bond that if A—B bond in a AB molecule is polar (*i.e.*, AB is polar molecule), then AB molecule behaves as a dipole, since, it has two poles. The dipole can be represented as $A^{\delta-} - B^{\delta+}$ (when $x_A > x_B$) or $A^{\delta+} - B^{\delta-}$ (when $x_B < x_A$). Since the molecule as a whole is a neutral, one of the two ends of the dipole has a positive charge and the other end has a negative charge whose magnitude is the same as that of positive charge. These two charges are separated by a definite distance which is called bond length. If q is the charge (positive or negative) at each end of the dipole and l is the distance between the positive and negative charges *i.e.* l is the bond length), then the product of q and l is called dipole moment (μ) of AB molecule. Thus:

$$\mu = q \times l$$

Representation of dipole moment

Sidgwick has suggested that dipole moment of a molecule is represented by an arrow with a crossed tail. The arrow is directed from the positive end to negative end of the dipole. For example the dipole moment of HCl molecule can be represented as depicted in the margin (Fig. 10.11).

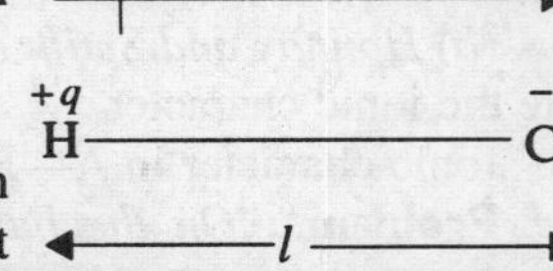

Fig. 10.11: Representation of dipole moment in HCl molecule.

Origin of dipole moment of a polar molecule

Dipole moment of a polar molecule arises due to the difference in electronegativities of the two atoms bonded together by a polar covalent bond. Dipole moment of a polar molecule measures the degree of polarity of that molecule. Thus greater is the value of dipole moment of a given polar molecule, greater is the magnitude of polarity associated with the molecule.

Units of dipole moment

If in a molecule an electron is assumed to be placed at a distance of 1Å from an equally positive charge, then the dipole moment (μ) of the molecule is given by:

μ = charge on the electron × Distance between the electron and the positive charge.

Now the charge on the electron in c.g.s. units is 4.8×10^{-10} electrostatic unit (*e.s.u*) while the charge on the electron in S.I. units is equal to 1.602×10^{-19} coulomb (C). Thus :

(*a*) μ *in c.g.s. units* $= (4.8 \times 10^{-10}\ esu) \times 1$ Å

$= (4.8 \times 10^{-10}\ esu) \times (10^{-8}\ \text{cm}) = 4.8 \times 10^{-18}\ esu.\text{cm}$.

Now the quantity 10^{-18} *esu.cm* is called one debye (D), *i.e.*, 10^{-18} *esu*.cm = 1D

$\therefore \quad \mu = 4.8$ D

Thus we find that in c.g.s units μ is expressed in debye (D).

(*b*) μ in *S.I. units* $= (1.602 \times 10^{-19}\ C) \times (10^{-8} \times 10^{-2}\ \text{m}) = 1.602 \times 10^{-29}\ C.\text{m}$.

Thus we find that in S.I. units μ is expressed in Coulomb meter (*C*.m) unit.

Relation between Debye (D) and Coulomb-meter (C.m).

We have seen above that the dipole moment of a molecule in which electron is assumed to be placed at a distance of 1Å from an equally positive charge is equal to 4.8 D (in *c.g.s. units*) or 1.602×10^{-29} C.m (in *S.I.*units). Thus :

$$4.8\text{D} = 1.602 \times 10^{-29}\ C.\text{m.} \quad \text{or} \quad 1\text{D} = \frac{1.602 \times 10^{-29}}{4.8}\ \text{C.m.}$$

$$\text{or} \quad 1\text{D} = 0.3337 \times 10^{-29}\ C.\text{m.}$$

$$\text{Also} \quad 1.602 \times 10^{-29}\ C.\text{m.} = 4.8\text{D} \quad \text{or} \quad 1\ \text{C.m.} = \frac{4.8}{1.602 \times 10^{-29}}\ \text{D}$$

$$\text{or} \quad 1\ \text{C.m} = 2.9962 \times 10^{29}\ \text{D}$$

Problem. *Calculate the dipole moment values (in S.I. units) of HF, HCl, HBr and HI molecules whose H–X bond lengths are 0.92 Å, 1.27 Å, 1.41 Å and 1.61 Å respectively.*

Solution. By definition :

Dipole moment of HX molecule, μ_{HX} = Charge on an electron (in coulomb) × H—X bond distance (in meter)

Thus :

$$\mu_{HF} = (1.602 \times 10^{-19}\ \text{C}) \times (0.92 \times 10^{-10}\ \text{m}) = 1.47 \times 10^{-29}\ \text{Cm.}$$

$$\mu_{HCl} = (1.602 \times 10^{-19}\ \text{C}) \times (1.27 \times 10^{-10}\ \text{m}) = 2.03 \times 10^{-29}\ \text{Cm.}$$

$$\mu_{HBr} = (1.602 \times 10^{-19}\ \text{C}) \times (1.41 \times 10^{-10}\ \text{m}) = 2.25 \times 10^{-29}\ \text{Cm.}$$

$$\mu_{HI} = (1.602 \times 10^{-19}\ \text{C}) \times (1.61 \times 10^{-10}\ \text{m}) = 2.57 \times 10^{-29}\ \text{Cm.}$$

Comparison between calculated and experimental value of dipole moment of a given molecule.

The value of dipole moment calculated with the help of the equation:

$$\mu = q \times l$$

is called calculated value of dipole moment. The value of dipole moment obtained experimentally is called experimental or observed value of dipole moment. It has been seen that the the calculated value of dipole moment of a given molecule is different from the experimental value of its dipole moment. Experimental values of dipole moment (in *C*.m.) are given as HF = 0.640×10^{-29}, HCl = 0.343×10^{-29}, HBr = 0.263×10^{-29}, HI = 0.127×10^{-29}, CO = 0.033×10^{-29}, NO = 0.023×10^{-29} $H_2O = 0.610 \times 10^{-29}$, $H_2S = 0.340 \times 10^{-29}$, $SO_2 = 0.53 \times 10^{-29}$, $NO_2 = 0.097 \times 10^{-29}$, $NH_3 = 0.494 \times 10^{-29}$, $PH_3 = 0.183 \times 10^{-29}$, $NF_3 = 0.07 \times 10^{-29}$.

Bond moment

A poly-atomic molecule (*e.g.* NH_3, H_2O, CH_4, CCl_4, CH_3Cl etc.) contains a number of bonds. For example NH_3 molecule has three N—H bonds and CH_3Cl molecule has three C—H bonds and one C—Cl bond. Each bond in a given molecule has a definite value of its bond moment which is called *bond moment.* The moment of a bond does not depend on the nature of the molecule. Values of moments of some bonds are given below. Note that A—A bond moment is equal to zero.

H — F	H — Cl	H — Br	H — I	H — O,
1.9D	1.1D	0.8D	0.4D	1.5D
H—N	H—C	C — C	C — F	C — Cl
1.3D	0.4D	0.0D	1.4D	1.5D
C — Br	C — I	C = O		
1.4D	1.2D	2.3D		

Dipole moment of A_2 (non-polar molecule) and AB type molecules

(*a*) A molecule of A_2 type (*e.g.* H_2, O_2, etc.) is non-polar and contains one A—A bond. This molecule has zero dipole moment, since both A atoms have the same electronegativity values (Note that dipole moment of a molecule arises due to the difference in electronegativity values of the two atoms bonded together).

(*b*) A molecule of AB type (*e.g.* HF, HCl, HBr, HI etc.) which contains A—B bond has a certain value of dipole moment, since A and B atoms have different values of their electronegativity. The dipole moment of this type of molecule is equal to the dipole moment of A—B bond. Thus the dipole moment values of HF, HCl, HBr and HI molecules are the same as the dipole moment values of H—F, H—Cl, H—Br and H—I bonds, *i.e.*, $\mu_{HF} = 1.9D$, $\mu_{HCl} = 1.1D$, $\mu_{HBr} = 0.8$ D and $\mu_{HI} = 0.4D$.

Additive property of dipole moments and dipole moment of a molecule containing two or more bonds (poly-atomic molecule)

Dipole moments are vectorial quantities, *i.e.*, they can be added. This property has been utilised to calculate the net dipole moment of a molecule which contains two or more bonds (poly-atomic molecule). The net dipole moment of a molecule is equal to the vectorial sum of the dipole moments of various bonds present in the molecule. For example, since H_2O molecule has two identical O—H bonds, the net dipole moment of H_2O molecule which has V-shape (unsymmetrical shape) is the vector sum of 1.5D and 1.5D (1.5 D is the value of the bond moment of O—H bond) which comes out to be equal to 1.48D. Thus the dipole moment of H_2O molecule is 1.48 D.

If the poly-atomic molecule has symmetrical shape (*e.g.* CH_4, CCl_4, BCl_3 etc.), then the dipole moments of various bonds (which are identical bonds) present in the symmetrical molecule are oriented in opposite directions and hence are cancelled by each other. Thus the net dipole moment of symmetrical molecules comes out to be equal to zero.

Applications of dipole moment values

Following are the important applications of dipole moment values:

1. To distinguish between polar and non-polar molecules. A molecule which has purely covalent bond is called non-polar molecule while that which contains polar covalent bond is called polar molecule. On the basis of the dipole moment value of a given molecule, it is possible to predict whether the given molecule is polar or non-polar in nature. The molecules having some value of dipole moment (*e.g.* HCN, OCS, NH_3, PF_3, AsF_3, H_2O, H_2S, SO_2, NO_2, O_3, N_2O etc.) are polar while those having zero value of dipole moment (*e.g.* CO_2, BeF_2, H_2, N_2, CH_4, SF_6, CS_2, BF_3, SO_3, CF_4, CCl_4 etc.) are non-polar. Greater is the value of dipole moment of a polar molecule, greater is the amount of ionic character in the polar covalent bond in the polar molecule. For example the calculated values of dipole moment (calculated from the equation, $\mu = q \times l$) of HF, HCl, HBr and HI molecules are (in *C*.m.) 0.64×10^{-29}, 0.34×10^{-29}, 0.263×10^{-29} and 0.127×10^{-29} respectively. Since these values decrease in the order HF > HCl > HBr > HI, the amount of ionic character in H—X bonds (X = F, Cl, Br, I) in the given molecules also decreases in the same order, *i.e.*, HF is the most polar molecule and HI is the least so.

2. To calculate the percentage of ionic character in the polar covalent bond in a given molecule. We have already said that the dipole moment of a polar molecule arises due to the difference in electronegativity values of the two atoms bonded together. Thus greater is the electronegativity difference between the bonded atoms in a polar molecule, greater is the value of dipole moment. Again we know that greater is electronegative difference, greater is % of ionic character in the bond linking the two atoms. This discussion shows that dipole moment values of a given polar molecule can aslo be used to calculate the percentage of ionic character in the covalent bond in that molecule. The percentage of ionic character is calculated with the help of the following equation.

$$\text{Percentage of ionic character in A — B bond in AB molecule} = \frac{100 \times \mu_{exp}}{\mu_{ionic}} \% \quad ...(i)$$

Here, μ_{exp} = experimental (observed or actual) value of dipole moment of AB molecule, μ_{ionic}

= calculated (theoretical) value of dipole moment of AB molecule when this molecule is assumed to be *completely ionic, i.e.*, an electron is completely transfered from less electronegative atom to more electronegative atom. Thus μ_{ionic} represents the dipole moment value for the ionic structure of AB molecule which can be represented as A^+B^- (when $x_B > x_A$) or as $A^- B^+$(when $x_A > x_B$). It is obvious from the above discussion that μ_{ionic} can also be represented by many other notations like $\mu_{A^+B^-}$ (when $x_B > x_A$), $\mu_{B^+A^-}$ (when $x_A > x_B$), $(\mu_{cal})_{A^+B^-}$, $(\mu_{cal})_{B^+A^-}$ etc.

The value of $(\mu_{cal})_{A^+B^-}$ is calculated as follows :

In c.g.s. units.

$(\mu_{cal})A^+B^-$ = Distance (in cm) between the centres of the positive and negative ends of $A^{\delta+} - B^{\delta-}$ dipole × Electronic charge (in *e.s.u.*)

$= d_{A-B}$ (in cm) × (4.8×10^{-10}*e.s.u*)

In S.I. units

$(\mu_{cal})_{A^+B^-}$ = Distance (in meter) between the centres of the positive and negative ends of $A^{\delta+} - B^{\delta-}$ dipole × Electronic charge (in Coulomb)

$= d_{A-B}$ (in m) × (1.602×10^{-19}C).

The calculation of the percentage of ionic character in a given bond by using equation (*i*) can be illustrated by the following problems :

Problems : *H—X bond length in HF, HCl, HBr and HI molecules are 0.92Å, 1.27Å, 1.41Å and 1.61Å, and the experimental value of dipole moment of these molecules is 0.64×10^{-29}C.m., 0.343×10^{-29} C.m., 0.263×10^{-29} C.m. and 0.127×10^{-29} C.m. respectively. Calculate the percentage of ionic character in H-X bond in each molecule.*

Solution. We know that:

$$\% \text{ of ionic character in H—X bond in HX molecule} = \frac{100 \times (\mu_{exp})_{HX}}{(\mu_{cal})_{H^+X^-}} \%$$

The above formula can be used, if $(\mu_{cal})_{H^+X^-}$ is known. The calculation of $(\mu_{cal})_{H^+X^-}$ in S.I. units is done by the following formula

$(\mu_{cal})_{H^+X^-}$ = Distance (in m) between the centres of the positive and negative ends of the dipole, $H^{\delta+} - X^{\delta-}$ × Electronic charge (in Coulomb)

$= d_{H-X}$ (in m) × (1.602×10^{-19}C)

Thus the value of $(\mu_{cal})_{H^+X^-}$ for each of the given molecules is given by (1Å = 10^{-8}cm, 1 cm = 10^{-2}m)

$$(\mu_{cal})_{H^+F^-} = (0.92 \times 10^{-8} \times 10^{-2}\text{m}) \times (1.602 \times 10^{-19}\text{C}) = 1.47 \times 10^{-29} \text{ C.m.}$$

$$(\mu_{cal})_{H^+Cl^-} = (1.27 \times 10^{-10}\text{m}) \times (1.602 \times 10^{-19}\text{C}) = 2.03 \times 10^{-29} \text{ C.m}$$

$$(\mu_{cal})_{H^+Br^-} = (1.41 \times 10^{-10}\text{m}) \times (1.602 \times 10^{-19}\text{C}) = 2.25 \times 10^{-19} \text{ C.m}$$

$$(\mu_{cal})_{H^+I^-} = (1.61 \times 10^{-10}\text{m}) \times (1.602 \times 10^{-19}\text{C}) = 2.57 \times 10^{-29} \text{ C.m}$$

Therefore, the % of ionic character in HF, HCl, HBr and HI molecules is given by the following expressions:

$$\% \text{ of ionic character in HF molecule} = \frac{0.64 \times 10^{-29}}{1.47 \times 10^{-29}} \times 100 = \mathbf{43.5\ \%}$$

$$\% \text{ of ionic character in HCl molecule} = \frac{0.343 \times 10^{-29} \times 100}{2.03 \times 10^{-29}} = \mathbf{16.8\ \%}$$

$$\% \text{ of ionic character in HBr molecule} = \frac{0.263 \times 10^{-29} \times 100}{2.25 \times 10^{-29}} = \mathbf{11.6\ \%}$$

$$\% \text{ of ionic character in HI molecule} = \frac{0.127 \times 10^{-29} \times 100}{2.57 \times 10^{-29}} = \mathbf{4.9\ \%}$$

Thus, we see that the % of ionic and covalent character in HF, HCl, HBr and HI molecules in their gaseous state is respectively : HF 43.5%, 56.5%; HCl = 16.8%, 83.2%; HBr = 11.6%,

88.4%; HI = 4.9%, 95.1%. Since in these molecules the percentage of covalent character is greater than that of ionic character, these molecules in their gaseous state are *mostly covalent* and *only partly ionic.* The predominant covalent nature of these molecules is confirmed by the facts that these compounds have low melting and boiling points and in their pure state these compounds are bad conductors of electricity. The percentage of ionic character in these molecules is in the order HF (= 43.5%) > HCl (=16.8%) > HBr (= 11.6%) > HI (= 4.9%)

Problem : *In NaCl ionic crystal, bond length is 2.36Å and the experimental dipole moment of this molecule is 8.5D. Calculate the percentage of ionic charecter in Na—Cl bond in the given molecule.*

Solution. Expected (calculated) dipole moment,

$(\mu_{cal})Na^+Cl^-$ = Bond length in NaCl crystal (in cm) × Electronic charge (in *e.s.u.*)

= $(2.36 \times 10^{-8}\text{cm}) \times (4.8 \times 10^{-10}\ esu) = 11.328 \times 10^{-18}$ cm. *esu*

= 11.328D ($\because 10^{-18}$cm.*esu* = 1D)

Observed dipole moment, $(\mu_{exp})_{NaCl} = 8.5D$.

Thus the % of ionic character in Na—Cl bond in NaCl molecule $= \dfrac{100 \times 8.5}{11.328}\% = 75.035\%$

Hence bond in NaCl has 75 per cent ionic character.

Problem : *The calculated value of dipole moment of Li I (= 11.5D) in the gaseous state is more than its experimental value (= 3.25D). Explain why?*

Solution. We know that the calculated value of dipole moment is based on complete transfer of one electron from Li to I atom. The difference in the two values of dipole moment can be explained on the basis of the fact that in Li I, the electron charge cloud of I^- ion is polarised by Li^+ ion. This results in that a partial overlap between the filled valence-orbital of I^- ion and empty valence - orbital of Li^+ ion takes place and hence the flow of electric charge occurs from I^- to Li^+ ion. This flow of electric charge decreases the dipole moment of Li—I bond. Thus the observed dipole moment value of Li I is less than its calculated value.

3. To predict the shape of a given molecule. The dipole moment value of a given molecule can be used to predict whether the given molecule has symmetrical or unsymmetricl shape.

(i) Structure of homo-nuclear diatomic molecules of A_2 type. Molecules of this type (*e.g.* H_2, C_2, N_2 O_2, Cl_2 etc) have zero dipole moment and hence have symmetrical dumbell stape (A—A).

(ii) Structure of hetero-nuclear diatomic molecules of AB type. Molecules of this type (*e.g.* HF, HCl, HBr, HI, CO, NO, ClF etc.) have some value of dipole moment and hence have symmetrical dumbbell shape.

(iii) Structure of tri-atomic molecules (*a*)Triatomic molecules like CO_2, $BeCl_2(g)$, $HgCl_2$ etc. have zero dipole moment and hence are linear in shape (symmetrical shape). Let us consider CO_2 molecule as an example. In this molecule (O = C = O) both carbon—oxygen bonds are polar but the overall molecule is non-polar because of its symmetrical linear geometry. It is due to its symmetrical geometry that the dipole moment of one C = O bond which is equal to 2.3D is cancelled by the dipole moment of the other C = O bond lying on the opposite side. Thus the resultant dipole moment of CO_2 molecule becomes zero. On the same lines of argument it can be explained that the dipole moment of $BeCl_2(g)$ and $HgCl_2$ molecules are also zero and hence these molecules, like CO_2 molecule, have linear geometry (see Fig. 10.12).

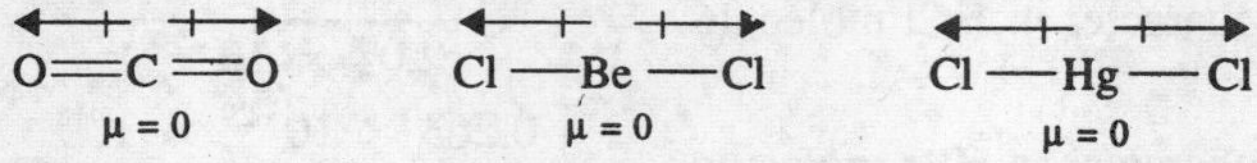

Fig. 10.12: Linear geometry of CO_2, $BeCl_2$ (g) and $HgCl_2$ molecules.

(*b*) Tri-atomic molecules like H_2O, H_2S, SO_2 etc. have some value of their dipole moment and hence have angular shape (unsymmetrical shape). As an example let us consider H_2O molecule as an example. This molecule has two indentical O—H bonds. Due to the unsymmetrical structure of the molecule, the dipole moments of two O—H bonds donot cancel each other and hence the net dipole moment of H_2O molecule is not zero but it has some value which has been found to be equal to 1.84D. This value is the vector sum of dipole moments of two O—H bonds (each O—H bond has its dipole moment value equal to 1.5 D). Also see Fig. 10.13.

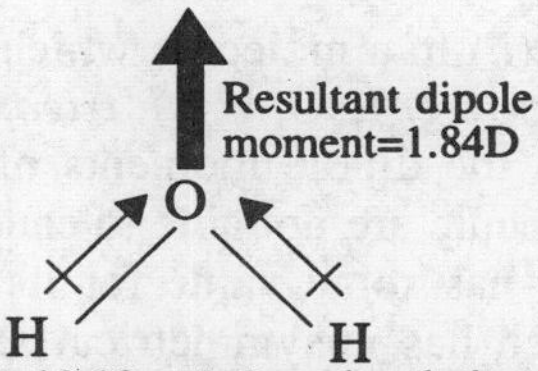

Fig. 10.13 : H_2O molecule has dipole moment equal to 1.84D which is the vector sum of the dipole moments of the two O—H.

(*iv*) *Structure of tetra-atomic molecules* (*a*) Tetra-atomic molecules like BF_3, BCl_3, SO_3 etc. have zero dipole moment and hence have symmetrical trigonal geometry. Each of these molecules has three identical bonds. Let us consider BF_3 molecule as an example. This molecule has three B —F polar bonds but the overall molecule is non-polar because of its symmetrical geometry. Due to the symmetrical geometry, the dipole moments of the three B—F bonds are concelled by each other and hence the resultant dipole moment of BF_3 molecule becomes equal to zero (see Fig.10.14). On similar lines, zero dipole moment of BCl_3 and SO_3 molecules can also be accounted for.

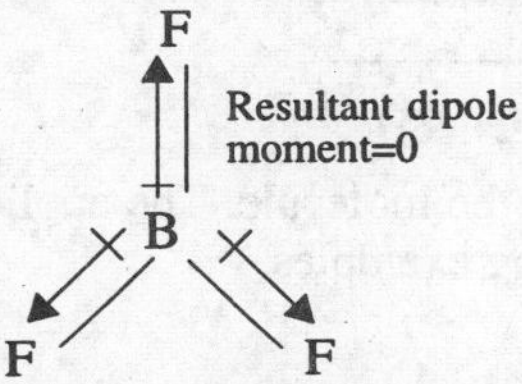

Fig. 10.14 : BF_3 molecule has zero dipole moment, since the dipole moments of the three B—F bonds cancel, each other.

(*b*) The tetra-atomic molecules like NH_3, NF_3, PH_3, PCl_3 etc. have same value of their dipole moment and hence these molecules have trigonal pyramidal shape (unsymmetrical shape). Each of these molecules has three identical bonds. Although these bonds are identical, due to the unsymmetrical shape of the molecule, the dipole moments of the three bonds are not able to cancell each other and hence the net dipole moment of these molecules is not equal to zero, *i.e.*, they have some value of dipole moment. This value is the vector sum of the dipole moments of the three bonds (See Fig. 10.15).

(*v*) *Structure of penta-atomic molecules (a).* Penta-atomic molecules like CCl_4, CF_4, SiF_4, CH_4 etc. have zero dipole moment and hence have symmetrical tetrahedral geometry. Each of these molecules has four identical bonds. Each of these bonds is polar but the overall molecule is non-polar due to its symmetrical shape. Due to symmetrical shape the dipole moments of the four bonds are concelled by each other and hence the net dipole moment of each of the molecules becomes equal to zero. (See Fig. 10.16).

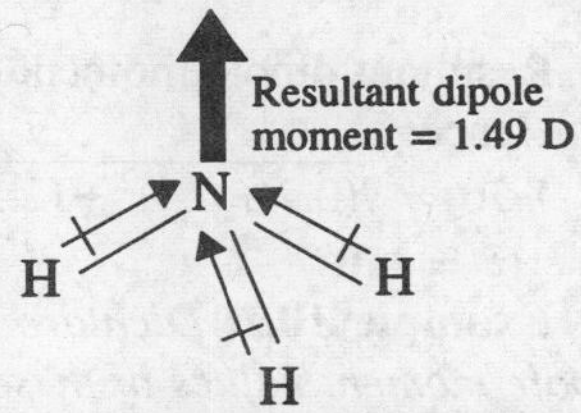

Fig. 10.15 : NH_3 molecule has dipole moment equal to 1.49 D which is the vector sum of the dipole moments of the three N—H bonds.

(*b*) Penta-atomic molecules like CH_3Cl has some value of dipole moment and hence has unsymmetrical tetrahedral structure. This has been explained as follows. This molecule has three C—H bonds and one C—Cl bond. Since all the four bonds are not identical, the dipole moments of these individual bonds are not able to cancel each other and hence the molecule has some value of its dipole moment and it has unsymmetrical tetrahedral structure. (See Fig. 10.17).

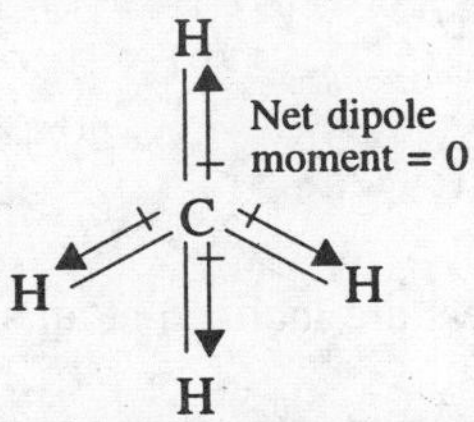

Fig. 10.16 : CH_4 molecule has zero dipole moment, since the dipole moments of the four C—H bonds are cancelled by each other.

Summary. The above discussion can be summerised as:

(*i*) In a molecule which has symmetrical shape like *dumbbell*, (*e.g.* C_2, H_2, N_2, O_2, Cl_2, Br_2, HF, HCl, HBr, HI, CO, NO, ClF etc.), *linear* (*e.g.* CO_2, $BeCl_2(g)$, $HgCl_2$), *trigonal planar* (*e.g.* BF_3, BCl_3, SO_3 etc.), *regular tetrahedral* (*e.g.* CH_4, CCl_4, CF_4 SiF_4etc.), *trigonal bipyramidal* (*e.g.* PF_5, PCl_5 (*g*)), *square planar* (*e.g.* XeF_4), *regular octahedral* (*e.g.* SF_6, WF_6), *pentagonal bipyramidal* (IF_5) the dipole moments of the various bonds present in the molecule cancel each other and hence the net dipole moment of the molecule becomes equal to zero. Thus a molecule which has a symmetrical shape has no (zero) dipole moment.

(*ii*) In a molecule which has unsymmetrical shape like *angular* (*e.g.* H_2O, H_2S), *trigonal pyramidal* (*e.g.* NH_3, NF_3, PH_3 etc.) the dipole moments of the various bonds present in the molecule are not able to cancel each other and hence the molecule has some value for its dipole moment. Thus a molecule which has unsymmetrical shape has a certain value of dipole moment.

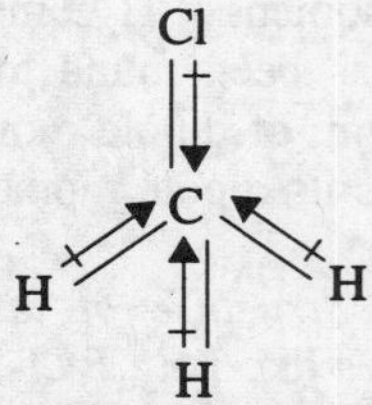

Fig. 10.17 : CH_3Cl molecule has some value of dipole moment, since due to the unsymmetrical shape of CH_3Cl molecule, the dipole moments of the four bonds are not able to cancel each other.

4. To calculate the bond angle between two bonds in a molecule. The angle subtended by two axes of overlaps of orbitals or bonds is called bond angle. The bond dipole moment values have been used to calculate the bond angle between two bonds. Bond angle (α) between two bonds in a given molecule is given by the expression:

$$\text{Net dipole moment of the molecule} = \sqrt{m_1^2 + m_2^2 + 2m_1\, m_2 \cos\alpha}$$

in which m_1 and m_2 are the bond moments of various bonds present in the molecule. The application of the above formula can be illustrated by considering the following examples.

Solved Examples

Example 10.1. *Calculate the HOC bond angle in CH_3OH molecule in which the bond moments are given as m_{H-C} = 0.4 D, m_{C-O} = 0.7 D and m_{H-O} = 1.5 D. Resultant dipole moment of CH_3OH molecule is 1.56D.*

Solution. The structure of CH_3OH molecule is given in the margin. In this structure α is HOC bond angle. It is evident from the structure that m_1 = 0.4 + 0.7 = 1.1D, m_2 = 1.5D.

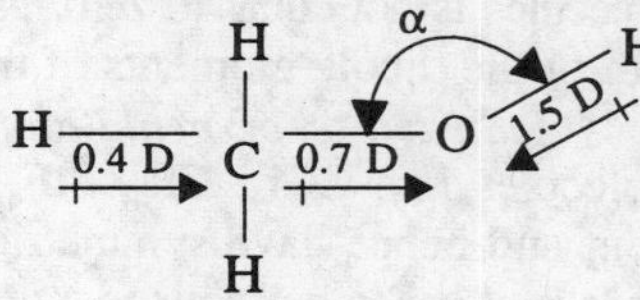

Resultant dipole moment of CH_3OH = 1.56D

Hence:

$$1.56 = \sqrt{(0.7 + 0.4)^2 + (1.5)^2 + 2 \times (0.7 + 0.4) \times 1.5 \times \cos\alpha}$$

$$\alpha = \mathbf{110°}$$

Example 10.2. *Dichloro benzene ($C_6H_4Cl_2$) has three isomers viz. ortho, meta and para. Dipole moment values of these isomers are 2.6D, 1.5D and 0.0D respectively. Calculate the angle between two C — Cl bonds in each isomer.*

Solution. In each isomer m_{C-Cl} = 1.5D and hence

$$\mu = \sqrt{(m_{C-Cl})^2 + (m_{C-Cl})^2 + 2 \times (m_{C-Cl})\,(m_{C-Cl}) \cos\alpha}$$

In ortho isomer : $2.6 = \sqrt{(1.5)^2 + (1.5)^2 + 2 \times 1.5 \times 1.5 \times \cos\alpha}$

$\therefore \quad \alpha = \mathbf{60°}$

In meta isomer : $1.5 = \sqrt{(1.5)^2 + (1.5)^2 + 2 \times 1.5 \times 1.5 \times \cos\alpha}$

$\therefore \quad \alpha = \mathbf{120°}$

In para isomer: $0.0 = \sqrt{(1.5)^2 + (1.5)^2 + 2 \times 1.5 \times 1.5 \times \cos\alpha}$

$\therefore \quad \alpha = \mathbf{180°}$

Thus, we find that both C—Cl bonds in ortho, meta and para isomers are inclined at an angle of 60°, 120° and 180° respectively.

Example 10.3. *Calculate the HOH bond angle in H_2O molecule whose dipole moment is equal to 1.84D. Bond moment value for O—H bond is 1.5D.*

Solution. H_2O molecule has two O—H bonds. Let them be inclined at an angle of α. Then:

$$1.84 = \sqrt{(1.5)^2 + (1.5)^2 + 2 \times 1.5 \times 1.5 \cos\alpha}$$

$\therefore \quad \alpha = \mathbf{104.5°}$

Thus in H_2O molecule two O—H bonds are inclined at an angle of 104.5° with each other.

5. To explain the behaviour of solutions and gases. The dipole moment values also help us in explaining the behaviour of solutions and gases which is generally different from their behaviour in ideal states.

6. To explain the solubility of substances. The study of dipole moment values suggests that polar substances can dissolve better in polar solvents and most ionic compounds can dissolve in water.

7. To distinguish between cis-and trans-isomers. It has been found by experiments that *cis*-isomer has higher dipole moment value than the *trans*-isomer.

8. To ascertain the orientation in organic molecules. This point can be understood by considering the isomers of dichlorobenzene molecule ($C_6H_4Cl_2$). This molecule has three isomers *viz. ortho, meta* and *para.* In *para*-isomer, the two C—Cl bond moments are opposite to each other and hence are cancelled. Thus the dipole moment of *para*-isomer is zero and hence this molecule is a *symmetrical molecule. Para*-isomer is purely non-polar. In case of *ortho*-and *meta*-isomers, the two C—Cl bond moments are not able to cancel each other and hence both these isomers have definite values of dipole moment and are, therefore, *assymetric molecules.* Since *ortho* isomer has greater value of dipole moment (= 2.6 D) than the *meta* isomer (= 1.5 D), the former has greater assymnetry than the latter.

Factors affecting the magnitude of dipole moment

Following are important factors:

1. Magnitude of polarity of the molecule. Greater is the magnitude of polarity associated with a molecule, greater is the value of dipole moment of the molecule.

2. Electronegativity difference between the two atoms of the molecule. If the electronegativity difference between the two atoms of a molecule is zero (*i.e.*, if the molecule is of A_2 type), the dipole moment of the molecule would also be equal to zero. On the other hand, if the electronegativity difference has some value, the molecule would also have some value of its dipole moment.

3. Symmetrical / asymmetrical shape of the molecule. A molecule having symmetrical shape has no dipole moment while that having asymmetrical shape has some value of its dipole moment. For details see "*Applications of dipole moment values*".

A problem. *The dipole moment of NH_3 molecule is considerably higher than that of NF_3 even though the electronegativity difference between the atoms in the respective bonds are almost identical. Explain.* (*Delhi B.Sc. Hons. 1982*)

Solution. Due to greater electronegativity of F-atom as compared to N-atom, the directions of the three dipoles of the three N—F bonds in NF_3 molecule are towards F-atom ($\overset{+\!\!\longrightarrow}{\text{N—F}}$) and hence oppose the effect of the lone pair of electrons on N-atom. Thus the resultant dipole moment of NF_3 molecule is towards F-atom. On the other hand, the directions of the three dipoles of the three N—H bonds in NH_3 molecule are towards N-atom ($\overset{+\!\!\longrightarrow}{\text{H—N}}$) and hence are added to the effect of the lone pair of electrons on N-atom. Thus the resultant dipole moment of NH_3 molecule is towards the lone pair of electrons. Consequently, dipole moment of NH_3 molecule becomes greater than that NF_3 molecule (see Fig. 10.18).

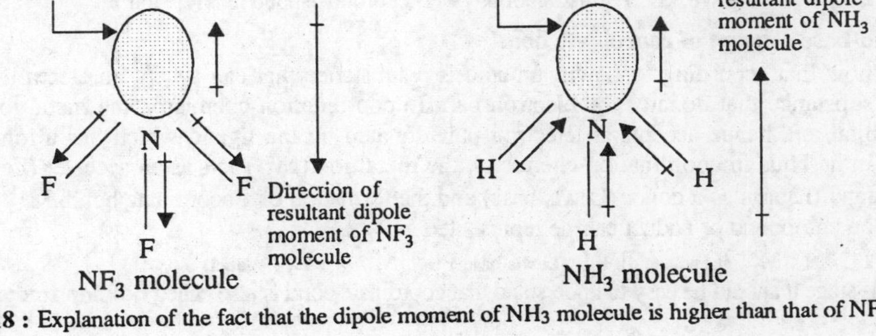

Fig. 10.18 : Explanation of the fact that the dipole moment of NH_3 molecule is higher than that of NF_3 molecule.

Coordinate Bond

Introduction.

According to G.A. Parkins (1921), a coordinate bond is that bond which is formed between the two atoms by the sharing of two electrons (one electron pair) between them, both the electrons being donated entirely by only one atom. Thus in the formation of this bond, one atom donates an electron pair while the other atom accepts it. The atom donating the electron pair is called *donor* while the atom accepting it is called *acceptor.* The atom acting as a donor must have one unused pair of electrons (called lone pair of electrons) which may be donated by it to the acceptor atom The acceptor atom must have an empty orbital to accept the lone pair of electrons. Coordinate bond is represented as A → B where A acts as a donor and B behaves as an acceptor.

In the formation of a coordinate bond, a partial positive charge is developed on donor atom and an equal partial negative charge is developed on the acceptor atom. Thus the coordinate bond, A → B should actually be represented as $A^{\delta+}$—$B^{\delta-}$. Due to the development of partial charges, the coordinate bond becomes slightly polar and hence is also called ***semi-polar bond*** or ***dative bond.*** Coordinate bond has the properties of both covalent bond and ionic bond and, therefore, has also been called *co-ionic* bond by Polymer (1944). Coordinate bond is a union of one electrovalent and one covalent bond. It is important to note that a coordinate bond once formed, becomes identical with a covalent bond and hence it is indistinguishable from a covalent bond.

Formation of a coordinate bond by orbital overlap theory

According to orbital overlap theory, a coordinate bond is formed when a completely filled orbital (*i.e.,* an orbital having a lone pair of electrons) of a atom overlaps with an empty orbital of the other atom.

Example. Ammonium in (NH_4^+) is formed by the coordination of NH_3 molecule and H^+ ion ($NH_3 + H^+ \rightarrow NH_4^+$). We know that on NH_3 molecule N-atom is sp^3 hybridised. Three sp^3 hybrid orbitals are singly-filled and hence are used in forming three (N—H) σ-bonds. The fourth sp^3 hybrid orbital has a lone pair of electrons. H^+ ion has 1*s* orbital which is vacant. In the formation of NH_4^+ ion vacant 1*s* orbital of H^+ ion overlaps with the completely-filled (lone pair of electrons) sp^3 hybrid orbital of N-atom of NH_3 molecule and forms N→ H^+ coordinate bond. (See Fig.10.19). The coordinate bond formed is correctly represented as sp^3(N) → 1*s* (H^+) coordinate bond.

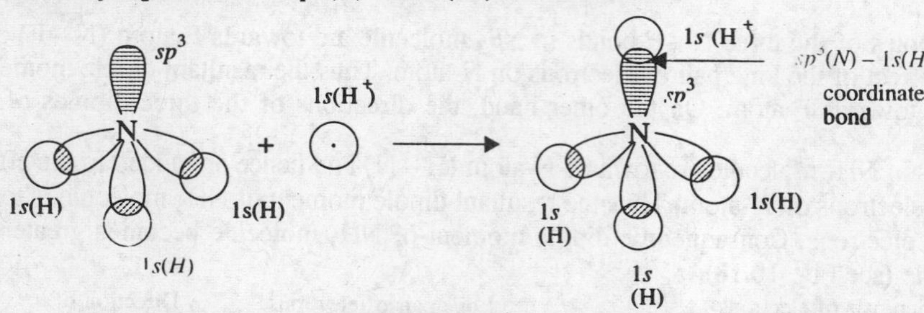

Fig. 10.19 : Formation of N → H^+ coordinate bond in NH_4^+ ion.

Lewis acid-base concept of coordinate bond

We know that according to Lewis, an acid is a substance that can accept an electron pair and base is a substance that donates the electron pair. In coordination chemistry, the metal ion has an empty orbital and hence accepts an electron pair donated by the ligand which has a lone pair of electrons in it. Thus in coordination chemistry, the metal ion (M^{n+}) acts as an accepter (Lewis acid) and the ligand (L) acts as a donor (Lewis base) and the formation of a coordinate bond, $M^{n+} \leftarrow L$ in a coordination compound or adduct can be represented as:

$$M^{n+} \text{ (Lewis acid)} + L \text{ (Lewis base)} \rightarrow [M^{n+} \leftarrow L] \text{ (Adduct)}$$

At this stage it should be easy to understand that coordinate bond is also called **donnor-acceptor bond.**

Structures of some molecules and ions contaning coordinate bond

The structures of some molecules/ions having coordinate bond are given below:

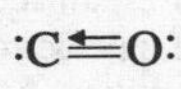

CO molecule

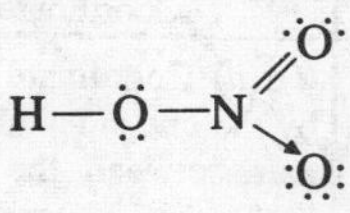

HNO_3 molecule

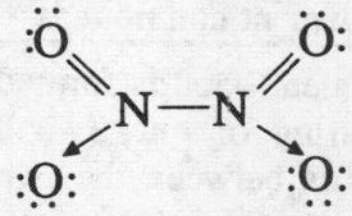

Dinitrogen tetroxide molecule (N_2O_4)

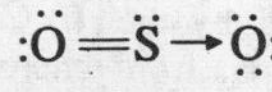

SO_2 molecule

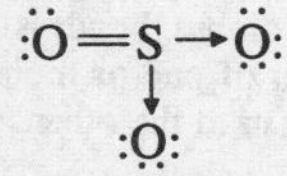

SO_3 molecule

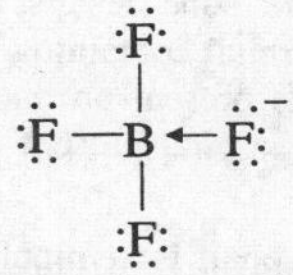

Fluoroborate ion (BF_4^-)

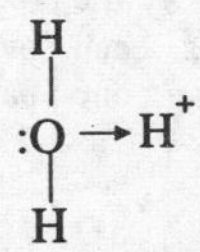

Hydronium ion (H_3O^+)

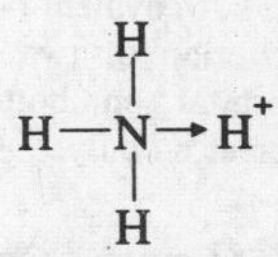

Ammonium ion (NH_4^+)

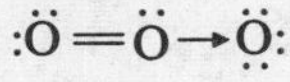

O_3 molecule

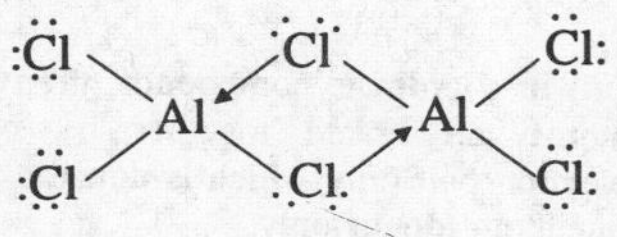

Anhydrous Al_2Cl_6

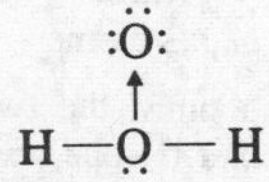

Hydrogen peroxide H_2O_2

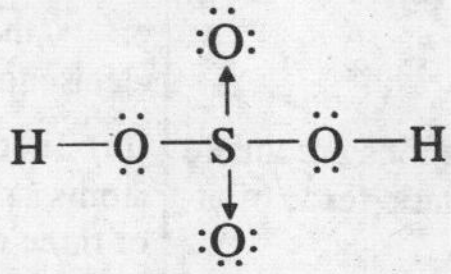

H_2SO_4 molecule

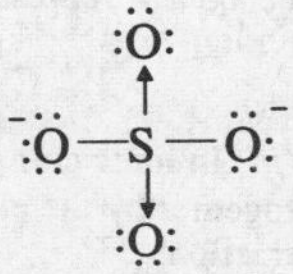

SO_4^{2-} ion

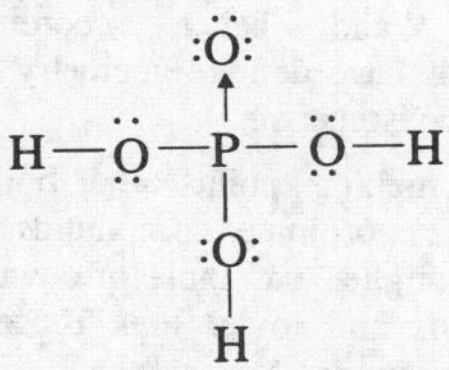

H_3PO_4 molecule

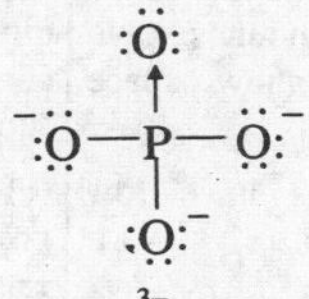

PO_4^{3-} ion

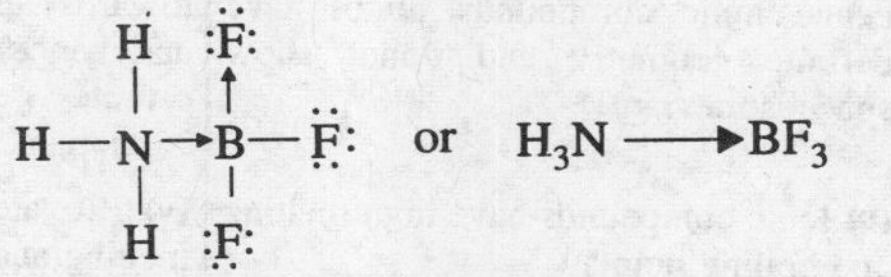

$NH_3 \cdot BF_3$ addition compound

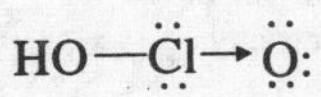

Chlorous acid, $HClO_2$

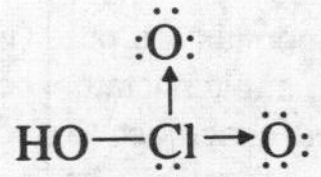

Chloric acid, $HClO_3$

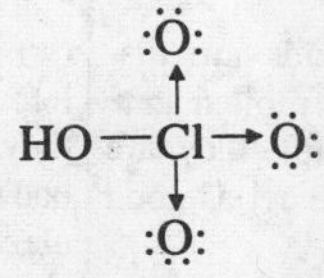

Perchloric acid, $HClO_4$

Characteristics of coordinate compounds

Some characteristics of coordinate compounds are summerised below:

(*i*) Once a cordinate is formed, it becomes identical to a normal covalent bond and hence it is difficult to differentiate it form a covalent bond.

(*ii*) *Space isomerism.* Coordinate bond is rigid and has directional properties. It is for this reason that the coordinate bond compounds have definite geometry and exhibit space isomerism.

(*iii*) *Solubility.* Coordinate compounds are soluble in organic solvents (non-polar solvents) and insoluble in H_2O (polar solvent).

(*iv*) *Electrical conductivity.* Coordinate compounds donot dissociate into ions when dissolved in H_2O or melted and hence donot conduct electricity, *i.e.*, these compounds behave as non-conductors of electricity.

(*v*) *Melting and boiling points:* Since a coordinate bond is a union of one electrovalent and one covalent bond, the volatility of these compounds lies between that of covalent and ionic compounds. Thus their melting and boiling points are relatively higher than those of the covalent compounds and lower than those of ionic compounds.

(*vi*) *Stability.* Coordinate compounds are stable like covalent compounds with the exception of those compounds in which the donor and acceptor atoms are capable of indepednet existence. Such compounds are called *molecular compounds.*

Table 10.4 : *Differences between ionic, covalent and coordinate compounds (or bonds)*

Ionic compounds	Covalent compounds	Coordinate compounds
(*i*) (*a*) Ionic bond is formed by the transfer of one or more electrons from one atom to the other.	(*i*) (*a*) Covalent bond is formed by the mutual sharing of one, two or three electron pairs between the two linked atoms. One electron of the electron pair is donated by one atom and the other electron is donated by the other atom.	(*i*) (*a*) Coordinate bond is established by the one-sided sharing of an electron pair between the linked atoms, *i.e.*, both the electrons of the shared electron pair are donated by one atom (donor) only.
(*b*) In the formation of an ionic bond, no overlap of atomic orbitals takes place.	(*b*) A covalent bond is obtained by the overlap of two atomic orbitals, each orbital on both atoms having one electron only.	(*b*) A coordinate bond results by the overlap of an orbital containing a lone pair of electrons present on one atom with the empty orbital present on the other atom.
(*ii*) Ionic bond is highly polar and hence is represented as A^+B^- (*e.g.* Na^+Cl^-)	(*ii*) Covalent bond may be polar ($A^{\delta-}$ - $B^{\delta+}$ or non-polar (A—A), depending on the relative values of electronegativity of A and B atoms	(*ii*) Coordinate bond is semipolar and hence is represented as $A^{\delta+} \rightarrow B^{\delta-}$.
(*iii*) In ionic compounds ions are linked together by the electrostatic force of attraction.	(*iii*) In covalent compounds, the two atoms are linked together by one, two or three electron pairs which are shared by both the linked atoms.	(*iii*) In coordinate compounds, the two atoms are linked together by one electron pair only which is donated by one atom (donor)only.
(*iv*) Ionic bonds are non-directional and hence ionic compounds donot have definite geometry and donot show space isomerism.	(*iv*) Covalent bonds are directed towards specific directions in space and hence covalent compounds have difinite geometry and show space isomerism.	(*iv*) Like covalent bonds, co-ordinate bonds are also directed in specific directions and hence coordinate compounds have definite geometry and show space isomerism.
(*v*) Ionic compounds have high melting and boiling points	(*v*) Covalent compounds have low melting and boiling points.	(*v*) The melting points and boiling points of coordinate compounds are relatively higher than those of covalent compounds and lower than those of ionic compounds.
(*vi*) When ionic compounds are dissolved in H_2O or are in the molten state (*i.e.*, fused or melted state), they are broken into ions and hence act as good conductors of electricity.	(*vi*) When covalent compounds are put into water or are in the molten state, they donot give ions and hence act as bad conductor (non-conductor) of electricity (graphite is an exception, since it is a good conductor).	(*vi*) Like covalent compounds coordinate compounds also behave as non-conductors of electricity.
(*vii*) Ionic compounds consist of ions.	(*vii*) Covalent compounds consist of individual molecules.	(*vii*) Like covalent compounds, co-ordinate compounds also consist of individual molecules.
(*viii*) Generally these compounds are crystalline solids.	(*viii*) These compounds exist as gases, liquids and solids.	(*viii*) These compounds exist as gases, liquids and solids.
(*ix*) These are soluble in polar solvents (*e.g.* H_2O) but insoluble in non-polar (organic) solvents (*e.g.*, C_6H_6, CCl_4 etc.).	(*ix*) These are insoluble in polar solvents but soluble in non-polar solvents.	(*ix*) Like covalent compounds these are insoluble in polar solvent but soluble in non-polar solvents.
(*x*) These compounds undergo ionic reactions in solution which are very fast.	(*x*) These compounds undergo moleculer reactions which are very slow.	(*x*) These compounds, like covalent compounds, undergo molecular reactions which are very slow.

Concept of Resonance

What is resonance?

It has been observed that certain molecules and ions can be represented by more than one Lewis structures but none of these structures is able to explain the properties like bond length, bond energy etc. of the molecules / ions. For example O_3 molecule can be represented by the following two Lewis structures:

$$:\overset{-}{\underset{\cdot\cdot}{\ddot{O}}} - \overset{+}{\ddot{O}} = \ddot{O}: \qquad\qquad :\ddot{O} = \overset{+}{\ddot{O}} - \overset{-}{\underset{\cdot\cdot}{\ddot{O}}}:$$

Structure (I) Structure (II)

The above structures show that since O_3 molecule has one $\overset{+}{O} = O$ double bond and one $\overset{+}{O}$—O single bond, both oxygen-oxygen bond lengths in O_3 molecule should be different, *i.e.*, the length of $\overset{+}{O} = O$ double bond should be equal to 121 pm and that of $\overset{+}{O}$—O single bond should be equal to 148 pm. Experiments have, however, shown that both oxygen—oxygen bonds are of equal length (= 126 pm). Thus, we find that the experimental value of oxygen—oxygen bond length (which is equal to 126 pm) is intermediate between O—O single bond length (= 148 pm) and O = O double bond length (= 121 pm).

In order to explain the discrepancy between the two oxygen—oxygen bond length in O_3 molecule, it has been suggested that none of the above structures is the correct structure of O_3 molecule; rather the correct structure is that which resonates between the above two structures, *i.e.*, the true structure of O_3 molecule is a *resonance hybrid* of the above two structures. These structures are called *contributing* or *resonating structures* of O_3 molecule. The resonating structures are also called *resonance forms* or *canonical forms*.The resonance between the two or more resonating structures of a given molecule or ion is represented by putting a double-headed arrow (↔) between the various resonating structures. It may be noted that the resonating structures have no physical reality and hence donot exist. It is only resonance hybrid which exists in reality. The above description can be summed up as follows:

When a molecule or ion can be represented by more than one Lewis structures, none of these structures is able to explain the properties of the molecule or ion and the true structure of the molecule/ion is a resonance hybrid of these various structures which are called resonating structures and this phenomenon is called *resonance.*

To draw the structure of resonance hybrid of O_3 molecule

Although the resonance hybrid of O_3 mole (or any other molecule or ion) is imaginary, its structure can be represented as shown in Fig.10.20. In this figure the dotted lines show that π-electron cloud of the π-bond in O = O double bond is not localised between any two particular O-atoms. Rather it is spread over both O-atoms, *i.e.*, π-electron cloud is delocalised.

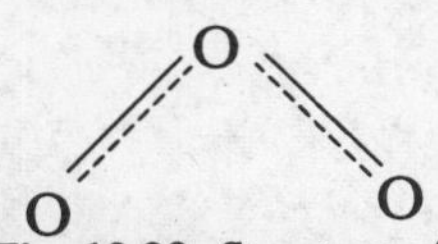

Fig. 10.20: Structure of the resonance hybrid of O_3 molecule.

Characteristics of resonating structures

(*i*) None of the resonating structures can completely describe all the properties of the given species and hence none of these structures represents the actual structure of the species. The actual structure is said to resonate between the resonating structures and hence is supposed to be a resonance hybrid of these various resonance structures, *i.e.*, the actual structure has the characteristics of all the resonating structures. The actual structure is not a mixture of the resonance forms and is different from all the contributing structures. The actual structure cannot be depicted by a pictorial representation on a paper. The resonating forms are imaginary and none of them has its indepedent or real existence. Also these forms donot mean that the actual structure of the given species exists for a certain fraction of time as one form and for another fraction of time as the other form.

(*ii*) All the resonating structures contribute to the actual structure of the given species, although the contribution made by each form is not the same. The magnitude of contribution actually depends on the structure of that form. For example:

(*a*) Of the following three resonating structures of HCl molecule, the contribution of structure (III) to the actual structure of this molecule is poor, since Cl-atom is much more electronegative than H-atom and hence cannot carry positive charge.

$$H—\ddot{\underset{\cdot\cdot}{Cl}}: \qquad H^{+}\,[:\ddot{\underset{\cdot\cdot}{Cl}}:]^{-} \qquad [\ddot{H}\ ]^{-}\quad [\ddot{\underset{\cdot\cdot}{Cl}}:\ ]^{+}$$

(I) (II) (III)

(*b*) Among the following three resonating structures of H_2 molecule, the ionic structures (II and III structures) are less stable than the covalent structure (structure I), since the two electrons present on the same H-atom in ionic structures would repel each other and hence the ionic structures would become unstable.

$$\underset{\text{(I)}}{H_A—H_B} \qquad \underset{\text{(II)}}{[\ddot{H}_A]^- [H_B]^+} \qquad \underset{\text{(III)}}{[H_A]^+ [\ddot{H}_B]^-}$$

Thus, we see that the covalent structure contributes more to the actual structure of H_2 molecule. In order to distinguish between two H-atoms, they have been represented as H_A and H_B.

(*iii*) Resonance forms have the same number of unpaired electrons (*n*) so that a continuous change from one bond type to the other may occur. For example:

(*a*) Structure (VI) for benzene has two unpaired electrons while structures (I) to (V) have none. Structure (VI) is, therefore, not to be accepted as a resonance form (Fig. 10.21).

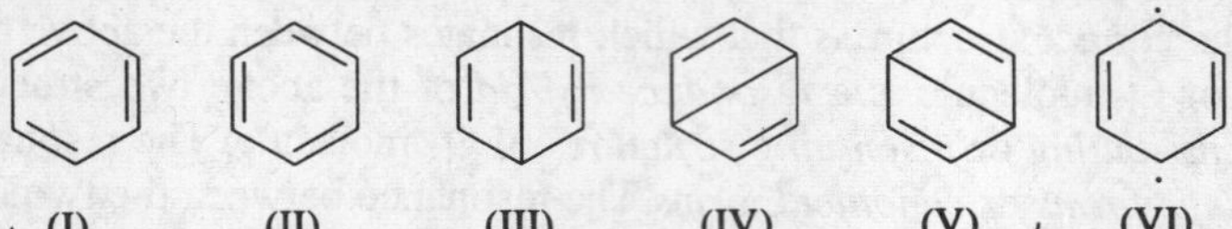

(I) (II) (III) (IV) (V) — $n = 0$; (VI) $n = 2$

Fig. 10.21 : Six structures for benzene molecule, C_6H_6.

(*b*) Structure (IV) for O_2 molecule has two unpaired electrons while structures (I) to (III) donot have any unpaired electrons. Thus structure (IV) cannot be taken as resonating structure. Note that in structure (IV) each O-atom has only 7 electrons instead of 8 and hence is less stable than (I), (II) and (III) structures.

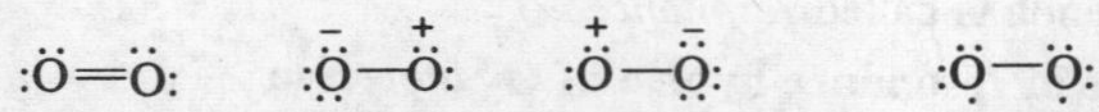

(I) (II) (III) — $(n = 0)$; (IV) $(n = 2)$

(*iv*) Resonance forms have *almost* the same energy, *i.e.*, there is not much difference in the energy of resonance forms of a given species. Thus a resonance form having much higher energy than each of the remaining resonance forms cannot be taken as a contributing structure. For example for benzene out of the structures given above in Fig. 10.21 structure (VI) has higher energy than each of the remaining five structures and hence is ruled out. This structure is also ruled out on the basis of the fact that it has two unpaired electrons while the other structures have none. Due to the presence of long bonds in the ring, structures III to V have more energy than structures I and II and hence are not taken as the contributing structures. Thus benzene is supposed to be a resonance hybrid of structures (I) and (II) as shown in Fig. 10.22.

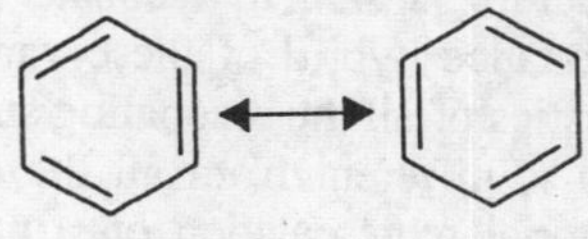

(I) (II)

Fig. 10.22: Benzene (C_6H_6) is a resonance hybrid of the given resonating structures.

(*v*) In the contributing structures negative charge should be shown on electronegative atom and positive charge should be shown on electro-positive element. For example theoretically three resonance forms for HF molecule are possible. These are H—F (covalent form), H^+F^- and H^-F^+(both ionic forms). Out of the ionic forms, H^-F^+ form is not feasible, since in this structure negative charge resides on electropositive element (*viz.* H-atom) and positive charge resides on electronegative element (F-atom) H^-F^+ will have very high energy and hence is unimportant. Thus the actual structure of HF molecule is supposed to be resonance hybrid of the first two structures (H — H $\leftrightarrow$ H^+F^-).

(*vi*) In contributing structures, similar charges should not reside on adjacent atoms while dissimilar charges should reside on the adjacent atoms. For example.

(*a*) For nitrous oxide molecule (N_2O) the following resonating structures can be written. Since in structure (III), + + charges (similar charges) reside on N and O atoms which are adjacent to each other, this structure is not to be accepted.

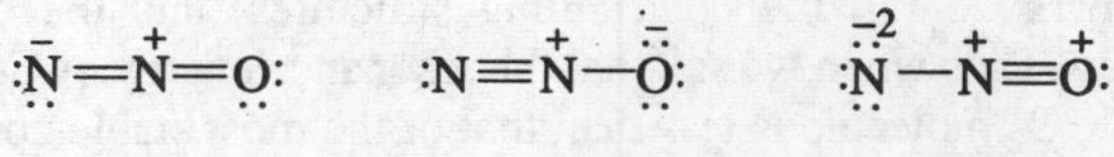

(I) (II) (III)

(*b*) For hydrazoic acid molecule (HN_3), structure (III) is unstable because in this structure positive charges reside on the adjacent N-atoms.

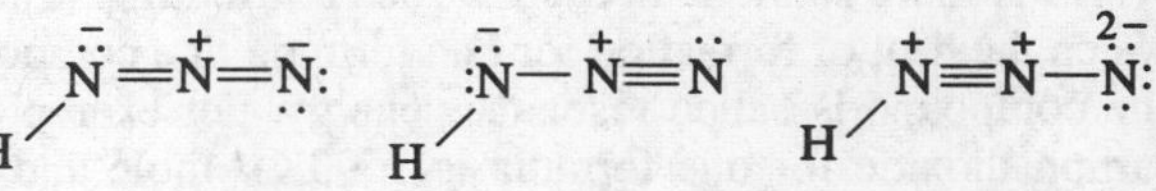

(I) (II) (III)

(*vii*) The relative position of all the atoms in each of the resonating structures should be the same; the position of electrons can, however, change among the bonding atoms. If we change the position of atoms, we are not writing the structure of the same compound but doing so far an isomeric species. For example:

(*a*) The relative position of N, N and O atoms in the various resonating structures of N_2O molecule should remain as N — N — O. If any of the contributing forms has the relative position as N—O—N, that form is not to be accepted as a contributing form.

(*b*) The relative position of S and two O-atoms in the contributing forms of SO_2 molecule should be as O — S — O and not as S — O — O. Thus, :S → Ö = Ö : cannot be a resonance form of SO_2 molecule.

(*viii*) Greater is the number of covalent bonds in a contributing structure, greater is the stability of that structure. For example the ionic contributing structures [*i.e.*, structures (*II*), (III) and (IV)] for BF_3 molecule are more stable due to the presence of additional (extra) B — F covalent bond in them. The presence of extra B — F bond reduces the energy of the species and hence makes it more stable. The actual bond order of B — F bond in BF_3 is more than 1. It has 54% π character.

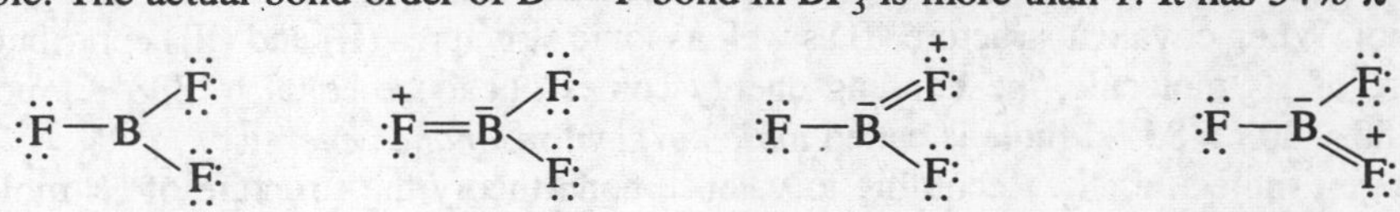

(I) (II) (III) (IV)

(*ix*) The larger is the number of contributing structures, the greater is the stability of the molecule.

Resonance Energy

Suppose a given molecule or an ion can be represented by three resonating structures *viz.* I, II and III. Let the energies of these structures be E_1, E_2 and E_3 respectively. Obviously the difference in their energies would be very small. Now suppose that their energies are in the order $E_3 < E_1 < E_2$. Obviously, since structure III has the lowest energy, it is the most stable contributing structure. Let the energy determined experimentally for the spe-

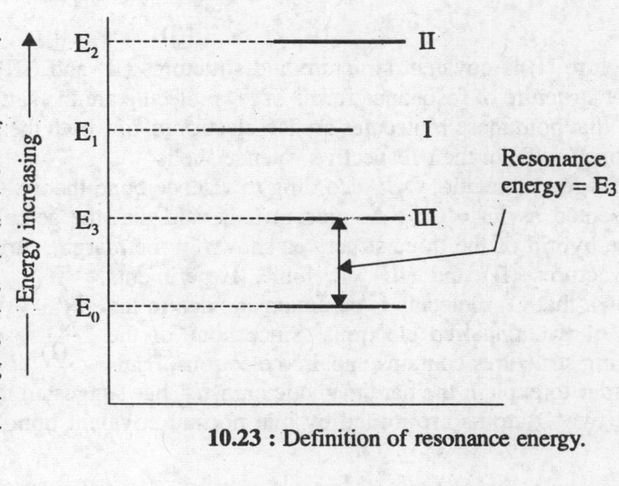

10.23 : Definition of resonance energy.

cies be E_0. The difference between E_3 and E_0 (*i.e.*, $E_3 - E_0$) is called resonance energy of the given species (See Fig 10.23). Thus resonance energy of a given species is the difference between the energy of its most stable contributing structure (*i.e.*, obviously such a contributing structure will be having the lowest energy of all the contributing structures) and the energy of the species determined experimentally. Thus when we say that the resonance energy of CO_2 is 126 KJ/mol, this means that the energy of CO_2 molecule is less than that of the most stable contributing structure of this molecule by an amount equal to 126 KJ. In other words we can say that CO_2 molecule is 126 KJ more stable than the most stable contributing structure of CO_2 molecule. The above discussion shows that since the energy of the resonance hybrid is less than that of each of the contributing structures, resonance hybrid is more stable than each of the contributing structures.

The difference between the heat of formation for one formula of a compound and the observed heat of formation for the compound is called resonance energy. For example since the calculated heat of formation of carbon dioxide for one formula is 1590KJ/ mole and the observed heat of formation is 1464 KJ/mol, the resonance energy of CO_2 is equal to 1590 – 1464 = 126 KJ / mol.

Resonance structures of some common molecules and ions

1. H_2 molecule. During the discussion of Heither-London theory of H_2 molecule, we have said that H_2 molecule can neither be represented by the pure covalent structure (structure I) nor by the ionic structures (structures II and III). The true structure of H_2 molecule is considered to be a resonance hybrid of the above three structures

$$\underset{(I)}{H_A - H_B} \leftrightarrow \underset{(II)}{[\ddot{H}_A]^- [H_B]^+} \leftrightarrow \underset{(III)}{[H_A]^+ [\ddot{H}_B]^-}$$

In order to differentiate the two H-atoms, they have been represented as H_A and H_B. Among the above three contributing structures, ionic structures are less stable than the covalent structure, since the two electrons present on the same H-atom in ionic structures would repel each other and hence the ionic structures would become unstable. The resonance hybrid is more stable than each of the three resonating structures and explains all the properties of H_2 molecule. The actual structure cannot be drawn on paper.

When H_2 molecule is supposed to have only covalent structure (I), the bonding energy is equal to 303 KJ/mol. When covalent structure (I) as well as ionic structures (II) and (III) contribute to the true structure of H_2 molecule, the bonding energy comes out to be equal to 388 KJ/mole. The difference 388 – 303 = 85 KJ/mole is called *ionic covalent resonance energy.*

2. Nitrogen molecule, N_2. According to valence bond theory the structure of N_2 molecule is represented as : N≡N:. This structure is not able to explain the properties of N_2 molecule and hence its actual structure is supposed to be a resonance hybrid of the following resonating structures:

$$\underset{(I)}{:N{\equiv}N:} \longleftrightarrow \underset{(II)}{:\overset{-}{\ddot{N}}{=}\overset{+}{\ddot{N}}} \longleftrightarrow \underset{(III)}{\overset{+}{\ddot{N}}{=}\overset{-}{\ddot{N}}:}$$

Structure (I) is covalent structure and structures (II) and (III) are ionic structures. It may be noted that structure of resonance forms of N_2 molecule are the same as those of CO molecule. The reason is that both these molecules are iso-electronic, *i.e.*, both these species have the same number of electrons (= 10) in their respective valence shells.

3. Oxygen molecule, O_2. According to valence bond theory, the structure of O_2 molecule can be represented as $:\dot{O} = \dot{O}:$. At present it is said that the actual structure of O_2 molecule is a resonance hybrid of the three structures shown in the margin. Structure (I) is a covalent structure while structures (II) and (III) are ionic. Experiments have shown that O_2 molecule is paramagnetic due to the presence of two unpaired electrons, Since none of the contributing structures contains unpaired electrons, Pauling, in order to explain the paramagnetic chracter, has suggested that this molecule has a structure in which two O-atoms are joined by one normal covalent bond and two three-electron bonds.

$$\underset{(I)}{:\ddot{O}{=}\ddot{O}:} \longleftrightarrow \underset{(II)}{:\overset{-}{\ddot{O}}{-}\overset{+}{O}:} \longleftrightarrow \underset{(III)}{:\overset{+}{O}{-}\overset{-}{\ddot{O}}:}$$

Thus according to Pauling, the structure of O_2 molecule is $:O \overset{\cdot\cdot\cdot}{-} \overset{\cdot\cdot\cdot}{-} O:$ in which each of the two three-electron bonds has one unpaired electron. Thus this structure explains the presence of two unpaired electrons in O_2 molecule. It may be noted that Pauling's structure does not obey octet rule.

4. Oxides of carbon. CO and CO_2 the most important oxides given by carbon. Both these molecules show resonance which has been discussed in chapter 15.

5. Oxides of nitrogen. Oxides of nitrogen like nitrous oxide (N_2O), nitric oxide in the gaseous state (NO) and nitrogen dioxide (NO_2) show resonance. It has been discussed in chapter 15.

6. Oxides of sulphur SO_2 and SO_3 are the most important oxides formed by sulphur. Both these molecules show resonance which has been discussed in chapter 15.

7. Oxy anions of carbon, nitrogen and sulphur. Oxy-anions like CO_3^{2-}, NO_2^-, NO_3^-, SO_4^{2-}etc., have various resonance forms and thus exhibit resonance. The resonance of each of these ions has been discussed in chapter 15.

Polarisation of Ions and Fajans' Rules

What is polarisation of ions?

In the formation of an ionic molecule, when two oppositely-charged ions (*i.e.*, cation and anion) of the ionic molecule come closer to each other, the cation attracts the electron charge cloud of the outer-most shell of the anion towards itself and hence the symmetrical shape of the anion gets distorted (*i.e.*, deformed or polarised). The phenomenon in which the symmetrical shape of the anion gets deformed by the approach of the cation closer to it is called polarisation (or distortion or deformation) of the anion. As an example the distortion of I^- ion (anion) by Li^+ ion (cation) in LiI ionic crystal has been shown in Fig. 10.24.

The ability of a cation to polarise an anion is called its *polarising power* or *polarising ability.* It may be noted that it is not only the anion that gets polarised; the cation is also polarised by the anion, but due to its smaller size, its electron cloud remains strongly held to the nucleus and hence is not attracted much towards the anion, *i.e.*, the cation is not polarised to an appreciable extent by the anion. We, therefore, generally donot consider the polarisation of a cation by an anion. The tendency of an anion to get polarised by a cation is called its *polarisability.*

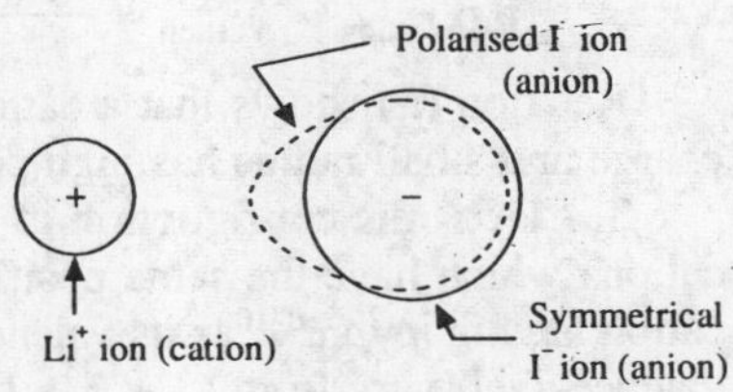

Fig. 10.24: Polarisation of I^- ion by Li^+ ion in LiI ionic molecule.

Factors affecting the magnitude of polarising power of a cation

Followinng are important factors which influence the magnitude of the polarising power of cations

1. Magnitude of positive charge on the cation. Higher is the positive charge on the cation, more strongly it attracts the outermost shell electron cloud of the anion towards itself and hence polarises a given anion more strongly, *i.e.*, with the increase in the magnitude of the positive charge on the cation, the polarising power of the cation also increases. In other words we can say that the polarising power of a cation is directly proportional to the magnitude of positive charge on it.

Examples : (*a*) Let us see how the polarising power of cations varies when we proceed from NaCl to $AlCl_3$ via $MgCl_2$. Since the magnitude of positive charge on the cation in these molecules increases as +1 (in NaCl) $\rightarrow$ +2 (in $MgCl_2$) $\rightarrow$ +3 (in $AlCl_3$), the polarising power of the cation to polarise the common Cl^- ion also increases from Na^+ to Al^{3+}, *i.e.*, the polarising power of the cations increases as $Na^+ < Mg^{2+} < Al^{3+}$.

(*b*) Now let us consider $SnCl_2$ and $SnCl_4$ molecules in which the same element (*viz.* Sn) has different positive charges on it. In $SnCl_2$ molecule, Sn has +2 positive charges while in $SnCl_4$ this element has + 4 positive charges. Because of higher positive charge, Sn^{4+} ion polarises Cl^- ion in

$SnCl_4$ more effectively than Sn^{2+} ion polarises the same Cl^- ion in $SnCl_2$ molecule, *i.e.*, Sn^{4+} ion has higher polarising power than Sn^{2+} ion ($Sn^{4+} > Sn^{2+}$).

2. Size of the cation. Smaller is the size of a cation, more strongly it attracts the outer-most shell electron cloud of the anion towards itself and hence greater is its polarising power to polarise a given anion, *i.e.*, with the decrease in the size of a cation, the polarising power of the cation increases. In other words we can say that the polarising power of a cation is inversely proportional to its size.

Examples: (*a*) Let us see how the polarising power of the alkali metal cations *viz.* Li^+, Na^+, K^+, Rb^+ and Cs^+ in their chlorides (LiCl, NaCl, KCl, RbCl and CsCl) varies on proceeding down the group. Since the size of the cations increases from Li^+ to Cs^+($Li^+ < Na^+ < K^+ < Rb^+ < Cs^+$), their polarising power to polarise Cl^- ion which is present in all the chlorides decreases in the same order, *i.e.*, the polarising power of the cations is in the order : $Li^+ > Na^+ > K^+ > Rb^+ > Cs^+$

(*b*) Let us again consider $SnCl_2$ (Sn = +2) and $SnCl_4$ (Sn = + 4) molecules. Because of the lower positive charge on Sn in $SnCl_2$ molecule Sn^{2+} ion has larger size than Sn^{4+} ion. Due to smaller size, Sn^{4+} ion polarises Cl^- ion in $SnCl_4$ molecule more strongly than Sn^{2+} ion polarises the same anion in $SnCl_2$ molecule, *i.e.*, Sn^{4+} ion has higher polarising power ($Sn^{4+} > Sn^{2+}$).

On combining factors (*i*) and (*ii*) discussed above, we can say that the polarising power of a cation is directly proportional to the magnitude of the positive charge on it, but inversely proportional to its size. Thus greater is the value of [charge on the cation / radius of the cation], greater is the polarising power of the cation and *vice-versa, i.e.*,

$$(\text{p.p.})_{\text{cation}} \propto \frac{\text{Positive charge on the cation}}{\text{Radius of the cation}} \qquad ...(i)$$

The ratio between the positive charge on the cation and its radius is called **ionic potential** of the cation (ϕ_{cation}). Thus relation (*i*) can also be written as:

$$(\text{p.p.})_{\text{cation}} \propto \phi_{\text{cation}} \propto \frac{\text{Positive charge on the cation}}{\text{Radius of the cation}} \qquad ...(ii)$$

Relation (*ii*) shows that a cation having high value of ionic potential or having high positive charge and small radius has high polarising power.

3. Electronic configurtion of the cation. If we have to compare the polarising power of two cations which have the same positive charge on them and the same or almost same size, then the cation having $ns^2p^6d^{1-10}$ as its valence-shell configuration [obviously the number of electrons in the valence-shell vary from (2 + 6 + 1) to (2 + 6 + 10) or 9 to 18] will polarise the given anion to a greater extent than the cation having ns^2p^6 as its valence-shell configuration (Number of electrons in the valence-shell = 2 + 6 = 8). $ns^2p^6d^{10}$ configuration is called *pseudo-inert gas configuration* while ns^2p^6 configuration is known as *inert gas* or *noble gas configuration.* Thus we see that a cation containing 18 electrons in its valenc-shell has greater polarising power than the cation containing 8 electrons.

Reason. We know that *d*-electrons shield the nucleus poorly as compared to *s*- and *p*- electrons. Thus *d*-electrons present in an 18-electron cation has poorer shielding of the nucleus than the *s*- and *p*- electrons present in 8-electron cation. Consequently the 18-electron cation would have more greater polarising power than the 8-electron cation.

Examples (*a*) Although Cu^+ and Na^+ ions have almost the same size (Cu^+ = 96 pm, Na^+ = 95 pm) and the same positive chrge (= +1), Cu^+ ion polarises Cl^- ion in CuCl more strongly than Na^+ ion can polarise Cl^- ion in NaCl. Greater polarising power of Cu^+ ion compared to that of Na^+ ion is because of the fact that Cu^+ ion has $3s^2p^6d^{10}$ configuration (18 electrons) while Na^+ ion has $2s^2p^6$ configuration (8 electrons). $3d^{10}$ electrons of Cu^+ ion shield the nuclear charge less effectively than the $2s^2$ and $2p^6$ electrons of Na^+ ion.

(*b*) Although Hg^{2+} (2, 8, 18, 32, $5s^2p^6d^{10}$) and Ca^{2+}(2, 8, $3s^2p^6d^0$) have the same charge and nearly the same size, yet Hg^{2+} ion has more polarising power than Ca^{2+} ion.

Factors affecting the polarisability of an anion

Following are the important factors:

1. Magnitude of negative charge on the anion. Higher is the negative charge on the anion, more easily or more strongly its outer-most shell electron cloud is attracted by the cation towards itself and hence the anion is polarised more strongly by a given cation, *i.e.*, with the increase in the magnitude of the negative charge on the anion, the polarisability of the anion increases. In other words we can say that the polarisability of an anion is directly proportional to the magnitude of negative charge on it.

Examples (*a*) Let us cosider the hydrides namely CH_4, NH_3, H_2O and HF. Since the magnitude of negative charge on the anions in these molecules decreases from – 4 (in CH_4) to –1 (in HF), the ease with which the anion is polarised by H^+ cation also decreases from C^{-4} to F^-, *i.e.*, the polarisability of the anions is in the decreasing order *viz.* $C^{-4} > N^{-3} > O^{-2} > F^-$. This order shows that H^+ ion cannot polarise F^- ion to any significant extent while it polarises C^{-4}, N^{-3} and O^{-2} ions to a great extent.

(*b*) The anions such as As^{3-} and P^{3-} are easily polarisable because of the high negative charge on them.

2. Size of the anion. Larger is the size of the anion, more easily or more strongly its outer-most shell electron cloud is attracted by the cation towards itself and hence greater is the polarisability of the anion, *i.e.,* with increase in the size of the anion, the polarisability of the anion also increases. In other words we can say that the polarisability of an anion is directly proportional to its size.

Examples (*a*) Let us consider the hydrides of halogens namely HF, HCl, HBr and HI. Since the size of the halide ions increases from F^- to I^- ($F^- < Cl^- < Br^- < I^-$), the polarisability of these ions also increases in the same order, *i.e.*, in HF molecule F^- ion is not polarised by H^+ ion to any significant extent while in HI molecule I^- ion gets polarised by H^+ ion to the maximum extent.

(*b*) The anions such as I^- and Se^{2-} are easily polarisable because of their large size.

The study of the above two factors shows that the anion having high negative charge and large size is more easily polarised by a given cation, since in an anion having high negative charge and large size the attraction between the nucleus of the anion and its electron cloud is decreased and hence the anion is easily distorted by a cation.

Fajans' rules of covalency

We have said above that in an ionic molecule the cation polarises the anion. If the cation has high polarising power, it attracts the electron charge cloud of the anion towards itself more strongly and this results in that the electron charge clouds of both the ions overlap together. This overlap decreases the ionic character between the cation and the anion in the molecule. Consequently, depending on the polarising power of the cation, the ionic bond between the cation and anion becomes either partially covalent or purely covalent. Greater is the polarising power of a cation, greater will be the amount of covalent character produced in the ionic molecule.The above discussion shows that all those factors which increase the polarising power of a cation or polarisability of an anion would make the ionic bond in the ionic molecule either partially covalent or purely covalent.

In 1924, Fajans told that there are certain conditions which, if fulfilled by the cation and anion of a given ionic molecule, make the ionic bond partially covalent (polar covalent) or purely covalent. These condition are called Fajans' rules which are given below:

1. The cation should have high positive charge on it. Higher is the positive charge on the cation in a given ionic molecule, higher will be its polarising power and hence greater will be the magnitude of covalent character produced in the ionic bond.

2. The cation should be small in size. Smaller is the cation in its size, greater is its polarising power and hence greater is the amount of covalent character produced in the ionic molecule.

3. The cation should have $ns^2p^6d^{10}$ configuration. We have already said that a cation having $ns^2p^6d^{10}$ configuration has greater polarising power than the cation having ns^2p^6 configuration. Greater polarising power produces covalent character in the ionic bond. Thus CuCl is covalent while NaCl is ionic. The covalent character of CuCl is because of the fact that Cu^+ ion which has $3s^2p^6d^{10}$ configuration polarises Cl^- ion to a greater extent than Na^+ ($2s^2p^6$) polarises Cl^- ion in NaCl.

4. The anion should have high negative charge on it. Greater is the magnitude of negative charge on the anion, more strongly it gets polarised by a given cation and hence the ionic bond gains covalent character. Thus higher is negative charge on the anion, greater is the magnitude of the covalent character produced in the ionic bond.

5. The anion should be large in size. Larger is the size of the anion, more strongly it is polarised by a given cation and consequently the covalent character increases.

Summary of Fajans' rules

Fajans' rules can be summarised in the following words: The formation of covalent bond in an ionic molecule is favoured when the cation has high positive charge, small size and $ns^2p^6d^{10}$ configuration while the anion has high negative charge and large size.

Applications of Fajans' rules: Consequences of polarisation

With the help of Fajans' rules a number of properties (*e.g.* covalent / ionic character, melting points, solubility etc.) of the compounds can be explained. Here we shall consider the following types of compounds.

1. Halides having cations of different elments with different positive charge on them. Let us consider NaCl, $MgCl_2$ and $AlCl_3$. In these chlorides, since the magnitude of positive charge on the cation increases from +1 (in NaCl) to +3 (in $AlCl_3$), the polarising power of these cations to polarise Cl^- ion increases as $Na^+ < Mg^{2+} < Al^{3+}$. With the increase in the polarising power of the cations, the covalent character in the given chlorides also increases in the same order (NaCl < $MgCl_2$ < $AlCl_3$). $AlCl_3$ is essentially covalent in character. It is due to the increasing order of the covalent character that the melting points of these chlorides decreases as : NaCl > $MgCl_2$ > $AlCl_3$. (NaCl = 800°C, $MgCl_2$ = 712°C, $AlCl_3$ = sublimes at 180°C).

2. Halides having cations of the same element with different positive charge on them. It has been found that if a given main group element forms two chlorides, the chloride of the element with higher positive charge is covalent while that of the element with lower positive charge is ionic. Being colvalent, the higher chloride is insoluble in water and has lower melting point. All these informations regarding the chloride pairs *viz.* TlCl — $TlCl_3$, $SnCl_2$ — $SnCl_4$ and $PbCl_2$ — $PbCl_4$ are given below:

Chloride pairs	Charges on the cation and their radii	Covalent / ionic character, m.pt. and solubility.
(*i*) TlCl—$TlCl_3$	Tl^+ = +1, 164 pm, Tl^{+3} = +3, 95 pm	TlCl = ionic, m.pt = 430° $TlCl_3$ = covalent, m.pt = 45°C
(*ii*) $SnCl_2$ —$SnCl_4$	Sn^{2+} = +2, 112 pm Sn^{4+} = +4, 71 pm	$SnCl_2$ = ionic, solid, soluble in water, m.pt = 246°C $SnCl_4$ = covalent, liquid, insoluble in water but soluble in C_6H_6, non-conductor of electricity, m.pt = – 33°C
(*iii*) $PbCl_2$ —$PbCl_4$	Pb^{2+} = +2, 133 pm Pb^{4+} = + 4, 84 pm	$PbCl_2$ = ionic, m.pt = 501°C $PbCl_4$ = covalent, m.pt = –15°C

In order to explain the difference in bonding , solubility and melting point, let us consider $SnCl_4$ — $SnCl_2$ pair as an example. In this pair, since Sn^{4+} ion has higher positive charge and is smaller in size than Sn^{2+} ion, Sn^{4+}ion polarises Cl^- ion in $SnCl_4$ molecule more strongly than Sn^{2+} ion polarises Cl^- ion in $SnCl_2$ molecule and hence $SnCl_4$ becomes covalent while $SnCl_2$ is ionic. The covalent character of $SnCl_4$ is supported by the fact that it is a covalent liquid which is soluble in C_6H_6 and is a non-conductor of electricity. The ionic character of $SnCl_2$ is shown by its being a solid which is soluble in H_2O. Being covalent, $SnCl_4$ has lower melting point than $SnCl_2$ ($SnCl_4$ = –33°C, $SnCl_2$ = 246°C).

On the same lines of argument it can be explained that $PbCl_4$ and $TlCl_3$ are covalent while $PbCl_2$ and TlCl are ionic. Being covalent $PbCl_4$ and $TlCl_3$ have lower melting points than $PbCl_2$ and TlCl respectively.

3. Halides having cations of different elements with the same positive charge on them but with different size. (*a*) *LiCl, NaCl, KCl, RbCl, CsCl.* Lithium metal gives largely covalent compounds, especially halides, while the other alkali metals form largely ionic halides. As an example, let us take the case of chlorides of alkali metals. With the successive increase in the size of M^+ cation (all the cations have the same positive charge on them) from Li^+ to Cs^+, the polarising power of these cations to polarise Cl^- ion decreases in the same direction and hence the covalent character of the chlorides of alkali metals also decreases from LiCl to CsCl (LiCl > NaCl > KCl > RbCl > CsCl). Thus LiCl has the maximum covalent character while other chlorides are less covalent (*i.e.*, more ionic). Being the most covalent, LiCl has the lowest melting point of all the chlorides (LiCl = 613°C, NaCl = 800°C, KCl = 776°C, RbCl = 715°C, CsCl = 625°C). Now the melting points of other chlorides should increase as NaCl < KCl < RbCl < CsCl. The melting points, however, decrease from NaCl to CsCl. This decrease is due to the successive decrease in the lattice energy values from NaCl to CsCl (NaCl = 770.3 KJ/mole, KCl = 701.2, RbCl = 682.4, CsCl = 629.7). Thus we see that the concept of polarisation (Fajans' rules) is not able to explain the variation in the melting points of the chlorides of alkali metals. Being the most covalent, LiCl is insoluble in H_2O but soluble in organic solvents (*e.g.* alcohol and pyridine). On the othe hand, other chlorides, being less covalent (*i.e.*, more ionic) are insoluble in organic solvents. NaCl, being ionic, is soluble in water.

(*b*) *$BeCl_2$, $MgCl_2$, $CaCl_2$, $SrCl_2$* and *$BaCl_2$*. Beryllium gives *largely covalent compounds,* especially halides, while other alkaline earth metals form largely ionic halides. As an example let us consider the case of the chlorides of alkaline earth metals. Due the successive increase in the size of M^{+2} cations from Be^{+2} to Ba^{+2} on moving down the group, the covalent character of MCl_2 molecules decreases in the same direction ($BeCl_2$ > $MgCl_2$ > $CaCl_2$ > $SrCl_2$ > $BaCl_2$). Thus $BeCl_2$ has the maximum covalent character and other chlorides are less covalent (*i.e.*, more ionic) This order is also supported by the fact that the melting points of these chlorides increase in the same direction as shown in the bracket ($BeCl_2$ = 405°C, $MgCl_2$ = 712°C, $CaCl_2$ = 772°C, $SrCl_2$ = 872°C, $BaCl_2$ = 960°C).

4. Halides of coinage metals and the corresponding halides of alkali metals. It has been observed that the halides of coinage metals are covalent, have lower melting points and are insoluble in water. On the other hand the corresponding halides of alkali metals (whose positive charge and the size is the same as those of coinage metals), have higher melting points and are soluble in water. All these informations are given below:

Halide pairs	Valence shell configuration and radius of the cation	Covalent / ionic character, m.pt. and solubility
(*i*) CuCl, NaCl	$Cu^+ = 3s^2p^6d^{10}$, 96 pm $Na^+ = 2s^2p^6$, 95 pm	CuCl — covalent and insoluble, m.pt. = 430°C NaCl — ionic and soluble, m.pt = 800°C
(*ii*)AgCl, KCl	$Ag^+ = 4s^2p^6d^{10}$, 126 pm $K^+ = 3s^2p^6$, 133 pm	AgCl — covalent and insoluble, m.pt. = 455°C KCl — ionic and soluble, m.pt. = 776° C
(*iii*) AuCl, RbCl	$Au^+ = 5s^2p^6d^{10}$, 137 pm $Rb^+ = 4s^2p^6$, 148 pm	AuCl — Covalent and insoluble, m.pt = 170°C RbCl — ionic and soluble, m.pt. = 715°C.

Why the halides of coinage metals are covalent and those of alkali metals are ionic can be explained by considering CuCl – NaCl pair as an example. Since Cu^+ ion has $3s^2p^6d^{10}$ configuration (pseudo-inert gas configuration) and Na^+ ion has $2s^2p^6$ configuration (inert gas configuration), Cu^+ ion polarises Cl^- ion in CuCl more strongly than Na^+ ion polarises the same anion in NaCl. Due to greater extent to which Cl^- ion gets polarised by Cu^+ ion, CuCl becomes covalent and NaCl is ionic. Being covalent, CuCl has lower melting point than NaCl and is insoluble in water while NaCl (which is ionic) is soluble. It is also due to the covalent character of the halides of coinage metals that for these halides the observed inter-nuclear distances are more close to the sum of the atomic (or covalent) radii of the atoms, while in case of alkali metal halides the observed inter-nuclear distances are close to the sum of the ionic radii of the atoms.

5. Compounds having anions of different elements with different negative charge on them. Here we shall consider the variation of covalent character in the ionic bond in the molecules namely $CH_4(C = -4)$, NH_3 $(N = -3)$, H_2O $(O = -2)$ and $HF(F = -1)$. In all these molecules H^+ acts as a cation. With the decrease in the magnitude of negative charge on the anion from – 4 (in CH_4) to –1 (in HF), the ease with which the anion gets polarised by the cation (H^+) decreases from C^{-4} to F^-. with the decrease of polarisation, the covalent character in the ionic bond in the given molecules also decreases, *i.e.,* the covalent character is in the order : $CH_4 > NH_3 > H_2O > HF$.

6. Halides of a given element. Here we shall consider the following halides.

(*i*) *HF, HCl, HBr and HI.* These are the halides of hydrogen. As we go down the group from F^- to I^- ion, the size of these anions (these anions have the same negative charge on them) increases ($F^- < Cl^- < Br^- < I^-$) and hence the extent to which these anions get polarised by H^+ ion also increases in the same direction. With the increase in the polarisability of the halide ions from F^- to I^-, the covalent character of H—X bond or HX molecules also increases in the same direction. Thus HF is the least covalent (maximum ionic character) and HI is the most covalent (minimum ionic character). This order of the covalent character of HX molecules can also be proved by applying Hannay and Smith equation to these molecules (see "*Electronegativity*" in chapter 5).

(*ii*) *LiF, LiCl, LiBr, LiI.* As we go down the group from F^- ion to I^- ion, the size of these anions (these anions have the same negative charge on them) increases ($F^- < Cl^- < Br^- < I^-$) and hence the extent to which these anions get polarised by Li^+ ion also increases in the same direction ($F^- < Cl^- < Br^- < I^-$). With the increase in the polarisability of the halide ions from F^- ion to I^- ion, the covalent character of the halides also increases in the same direction, *i.e.,* the covalent character of the halides increases as LiF < LiCl < LiBr < LiI, showing that LiF has the least covalent character (*i.e.,* maximum ionic character) and LiI has the maximum covalent character (*i.e.,* least ionic character). Due to the successive increase in the covalent character on proceeding from LiF to LiI, the melting points of these halides show the expected decrease from LiF to LiI as shown in bracket (LiF = 870°C, LiCl = 613°C, LiBr = 547°C, LiI = 446°C). Being the least covalent (most ionic) in the series, LiF should be soluble in H_2O and LiI which has the maximum covalent character should be insoluble in H_2O. But the fact is that LiF, due to its highest value of lattice energy, is insoluble and LiI which has the minimum value of lattice energy is soluble. Values of lattice energy (in KJ / mole) are LiF = 1034, LiCl = 840.1, LiBr = 781.2, LiI = 718.2.

7. Sulphides and oxides of a given metal. Let us consider M_2O and M_2S molecules which are oxide and sulphide of a given monovalent metal cation M^+. Since S^{-2} ion is larger in size than O^{-2} ion, S^{-2} ion gets polarised by M^+ cation to a greater extent as compared to O^{-2} ion by the same cation. Thus M_2S is more covalent than M_2O ($M_2S > M_2O$). Being more covalent, M_2S is less soluble in H_2O than M_2O. In general the sulphides are less soluble in water than the oxides of the same metal. Sulphides being covalent, are soluble in orgainc solvents like alcohol.

8. Hydrides of the elements of groups VA and VIA. Here we shall consider the hydrides namely NH_3, PH_3, AsH_3, SbH_3, BiH_3 (group VA) and H_2O, H_2S, H_2Se and H_2Te (group VIA).

(*i*) *NH_3 , PH_3, AsH_3, SbH_3 and BiH_3.* Due to the successive increase in the size of negatively-charged A^{3-} ion ($A^{3-} = N^{3-}$, P^{3-}, As^{3-}, Sb^{3-} and Bi^{3-}) from N^{3-} to Bi^{3-}, the covalent character of the given hydrides increases as we pass from NH_3 to BiH_3 ($NH_3 < PH_3 < AsH_3 < SbH_3 < BiH_3$). This order shows that NH_3 is the least covalent and BiH_3 is the most covalent.

(*ii*) *H_2O, H_2S, H_2Se, H_2Te.* On the same lines of argument as we have used in case of NH_3,

PH_3, AsH_3 etc. molecules, we can explain that the covalent character of the given hydrides increases as we move down the group. ($H_2O < H_2S < H_2Se < H_2Te$). Thus H_2O is the least covalent and H_2Te is the most so. Being least covalent, H_2O is the most stable and hence dissociates only at a high temperature (=2000°C) while H_2Te which is the most covalent is extremely unstable and hence dissociates even at ordinary temperature.

9. Monoxides of alkali metals, M_2O (M = Li, Na, K, Rb, Cs). All the alkali metals form monoxides of M_2O type. All these oxides are ionic oxides and hence are represented as $M_2^+O^{2-}$. Because of the smallest size of Li^+ ion, this ion polarises the electron cloud of O^{2-} ion to the maximum extent. This allows partial overlap between the orbital of Li^+ and O^{2-} ions and hence some covalent character in Li—O bond; apart from this ionic character is also introduced in Li_2O molecule. Thus Li_2O has covalent character while other monoxides are almost entirely ionic in character. Due to covalent character in Li_2O, this molecule has extra stability.

10. Cause of diagonal relationship. The similarity in properties existing between the diagonal pairs of elements like Li—Mg, Be—Al and B—Al is called diagonal relationship. The cause of this relationship has been given on the basis of the polarising power (ionic potential) of the cations of the elements of these pairs. For the details of this application see "Diagonal Relationship" discussed in this chapter.

11. Decomposition temperature of metallic carbonates. When a metallic carbonate undergoes decomposition, its CO_3^{2-} ion gets split into CO_2 and O^{2-} ion. Thus the decomposition of a metallic carbonate gives CO_2 and metallic oxide, *e.g.*

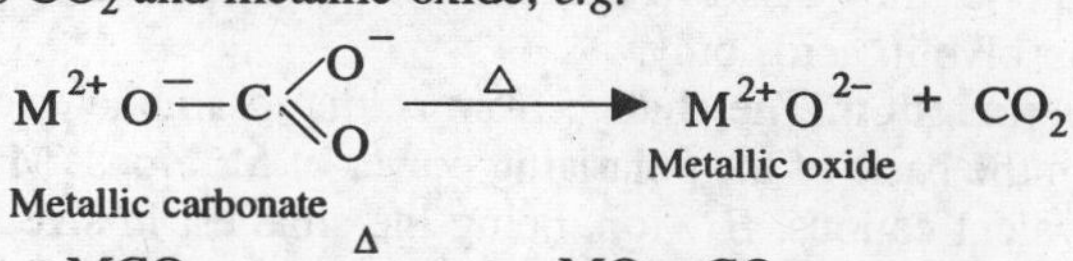

or $\quad MCO_3 \xrightarrow{\Delta} MO + CO_2$

The temperature at which a given carbonate gets decomposed depends on the polarising power of the cation. The cation with greater polarising power will distort the electron cloud of the neighbouring O-atom of CO_3^{2-} ion very effectively and hence CO_3^{2-} ion would get split into CO_2 and O^{2-} ion more easily, *i.e.*, the decomposition temperature of a metallic carbonate having the cation with greater polarising power will take place at a lower temperature. This point is illustrated by the following examples:

(*a*) *Decomposition of alkaline earth metal carbonates*. We know that the size of M^{2+} ions (M^{2+} = Be^{2+}, Mg^{2+}, Ca^{2+}, Sr^{2+} and Ba^{2+}) increases as we move down the group. With the increase in the size of M^{2+} ions, the polarising power of M^{2+} ions decreases down the group and hence the temperature at which MCO_3 molecule gets decomposed into MO and CO_2 increases. Thus $BeCO_3$ decomposes at the lowest temperature while $BaCO_3$ decomposes at the highest temperature [Decomposition temperatures (in °C) are given as $BeCO_3$ = 100, $MgCO_3$ = 400, $CaCO_3$ = 900, $SrCO_3$ = 1290, $BaCO_3$ = 1360). Due to the successive increase in the decomposition temperatures, the thermal stability of these molecules increases as $BeCO_3 < MgCO_3 < CaCO_3 < SrCO_3 < BaCO_3$.

(*b*) *Decomposition of $CdCO_3$, $HgCO_3$ and $CaCO_3$*. Decomposition temperatures of these molecules are $CdCO_3$ ~ 350°C, $HgCO_3$ ~350°C and $CaCO_3$ = 900°C. These values show that $CdCO_3$ and $HgCO_3$ molecules decompose at much lower temperature than $CaCO_3$. This is explained as follows. The valence-shell electronic configurations of Cd^{2+}, Hg^{2+} and Ca^{2+} ions are as Cd^{2+} = $4s^2p^6d^{10}$, Hg^{2+} = $5s^2p^6d^{10}$ and Ca^{2+} = $3s^2p^6$. Due to the presence of *d*-electrons in Cd^{2+} and Hg^{2+} ions, these ions have greater polarising power then Ca^{2+} ion which does not have any *d*-electron in its valence-shell. Thus Cd^{2+} and Hg^{2+} ions distort the electron charge cloud of O-atom of CO_3^{2-} ion more easily than Ca^{2+} ion does. Thus CO_3^{2-} ion of $CdCO_3$ and $HgCO_3$ molecules split into CO_2 and MO (M = Cd, Hg) at lower temperature while $CaCO_3$ will give CO_2 and CaO at higher temperature.

12. Several salts of lithium are hydrated whereas the salts of other alkali metals are not hydrated. This abnormal behaviour of lithium salts can be explained as follows: Since Li^+ ion is of the smallest size, it has the highest charge / size ratio and hence has the highest polarising power of all the alkali metal cations (M^+ ions). When Li^+ ion comes in contact with H_2O molecules, it, due to its highest polarising power, polarises the electron charge cloud of H_2O molecules more effectively than each of the remaining alkali metal ions. Consequently, Li^+ ion is hydrated by water molecules to the maximum extent. It is due to the extensive hydration of Li^+ ion that several salts of this ions are hydrated (*e.g.* $LiClO_4.3H_2O$), whereas the salts of other alkali metal ions are not hydrated.

13. Variation of the acidic, amphoteric and basic character of the oxides and hydroxides of the elements of group IIIA. All the elements of group IIIA form sesquioxides of the general formula, M_2O_3. Examples are B_2O_3, Al_2O_3, Ga_2O_3, In_2O_3 and Tl_2O_3. These elements also form trihydroxides of $M(OH)_3$ type. The oxides may be considered to have been obtained by the dehydration of $M(OH)_3$ molecules

$$2M(OH)_3 \rightarrow M_2O_3 + 3H_2O$$

It has been found that, on moving down the group, there is a gradual change from acidic to amphoteric and then to basic character of the oxides and hydroxides. Thus:

$$\underbrace{B_2O_3 \text{ and } B(OH)_3,}_{\text{Weakly acidic}} \quad \underbrace{Al_2O_3, Al(OH)_3, Ga_2O_3, Ga(OH)_3,}_{\text{Amphoteric}} \quad \underbrace{In_2O_3, In(OH)_3, Tl_2O_3, Tl(OH)_3}_{\text{Basic}}$$

Being acidic, B_2O_3 and $B(OH)_3$ both dissolve in alkalies only. Since Al_2O_3 , $Al(OH)_3$, Ga_2O_3 and $Ga(OH)_3$ are amphoteric, they dissolve both in acids and alkalies. In_2O_3, $In(OH)_3$, Tl_2O_3 and $Tl(OH)_3$, being basic, dissolve in acids only.

Explanation of the variation. The change from acidic to amphoteric and then to basic character can be explained on the basis of the polarising power of M^{3+} ions ($M^{3+} = B^{3+}, Al^{3+}, Ga^{3+}, In^{3+}$ and Tl^{3+}). Of all the trivalent cations, B^{3+} ion, being the smallest in size, has the highest positive charge density (charge/size) and hence highest polarising power. When $B(OH)_3$ is placed in water, O-atom of H_2O molecule gets coordinated with B-atom of $B(OH)_3$ molecule and thus a hydrated species, $\substack{H \\ H}\!>O \rightarrow B(OH)_3$ is formed. In this hydrated species, B^{3+}ion, due to its high polarising power, attracts the electrons of the coordinated O-atom towards itself. Coordinated O-atom, in turn, pulls the σ-electron pair of O—H bond in $\substack{H \\ H}\!>O$ molecule towards itself suffeciently strongly to rupture the O—H bond. The rupture of O—H bond makes H^+ ions available. H^+ ions thus obtained make the aqueous solution of $B(OH)_3$ feebly acidic; *i.e.*, the solution of $B(OH)_3$ in water acts as a weak acid.

$$H_2O + B(OH)_3 \rightarrow \substack{H \\ H}\!>O \rightarrow B(OH)_3 \rightarrow [B(OH)_4]^- + H^+$$

It is due to the acidic nature of $B(OH)_3$ that it reacts with NaOH solution to produce sodium meta borate, $Na^+BO_2^-.2H_2O$

$$B(OH)_3 + H_2O \rightarrow [B(OH)_4]^- + H^+$$
$$NaOH \rightarrow Na^+ + OH^-$$
$$\overline{B(OH)_3 + H_2O + NaOH \rightarrow Na^+[B(OH)_4]^- + H_2O}$$

or $$B(OH)_3 + NaOH \rightarrow Na^+ [B(OH)_4]^- \text{ or } Na^+BO_2^-. 2H_2O$$

Since the size of M^{3+} ions increases as we move down the group from B^{3+} to Tl^{3+}, the polarising power of the ions decreases in the same direction. Thus , since polarising power of Al^{3+} and Ga^{3+} ions is less than that of B^{3+} ion, the capacity of these ions to rupture O— H bond present in $\substack{H \\ H}\!>O \rightarrow Al(OH)_3$ and $\substack{H \\ H}\!>O \rightarrow Ga(OH)_3$ to produce H^+ ions is relatively small and hence $Al(OH)_3$ and $Ga(OH)_3$ both, instead of being acidic, are amphoteric. Since In^{3+} and Tl^{3+} ions are even larger than B^{3+}, Al^{3+} and Ga^{3+} ions, their polarising power is so small they cannot rupture

O—H bond in $\overset{H}{\underset{H}{}}>O \rightarrow In(OH)_3$ and $\overset{H}{\underset{H}{}}>O \rightarrow Tl(OH)_3$ to produce H^+ ions. As a matter of fact, $In(OH)_3$ and $Tl(OH)_3$ undergo disproportionation in water to give OH^- ions which make the solution basic.

$$\overset{H}{\underset{H}{}}>O + In\,(OH)_3 \rightarrow \left[\overset{H}{\underset{H}{}}>O \rightarrow In\,(OH)_3\right] \rightarrow \left[\overset{H}{\underset{H}{}}>O \rightarrow In(OH)_2\right]^+ + OH^-$$

and so on.

$$\overset{H}{\underset{H}{}}>O + Tl(OH)_3 \rightarrow \left[\overset{H}{\underset{H}{}}>O \rightarrow Tl(OH)_3\right] \rightarrow \left[\overset{H}{\underset{H}{}}>O \rightarrow Tl(OH)_2\right]^+ + OH^-$$

and so on.

The acidic / basic character of the oxides can also be explained in a similar manner.

Bond parameters

Important characteristics of a bond are *bond length, bond energy* and *bond angle.* These are called *bond parameters.*

Bond Length

Bond length or bond distance in covalent molecules (also called inter-nuclear distance)

The atoms in a molecule are always vibrating with respect to each other and hence it is not possible to determine the exact distance between the nuclei of the atoms. We, therefore, talk about an average distance between the atoms. This distance is called *bond length* (also called *bond distance* or *inter-nuclear distance*). Thus bond length can be defined as the average distance between the nuclei of the two atoms bonded together by covalent bond (s).

According to molecular orbital theory, a covalent bond between two atoms is formed by the overlap of their atomic orbitals. The overlapping of their atomic orbitals brings the two atoms closer to each other (force of attraction). During the process of overlapping a stage is reached when the force of attraction between the two atoms is balanced by the force of repulsion between the nuclei of the two atoms. At this stage the equilibrium distance between the two atomic nuclei is called bond length or bond distance.

Determination of bond length in covalent molecules

Bond length of a molecule can be determined by various methods like electronegativity method, electron diffraction method, X-ray diffraction method, neutron diffraction method and molecular spectra method. Determination of bond length can be studied under the following headings.

A. Determination of bond length (d_{A-A}) in A_2 type molecule. We have seen under the study of covalent radius that the inter-nuclear distance between the nuclei (d_{A-A}) of two A-toms linked together by a single covalent bond in A_2 molecule, is given by

$$d_{A-A} = 2r_A \quad ...(i)$$

where r_A is the covalent radius of atom A. This equation shows that in A_2 type molecule, d_{A-A} is equal to double the radius of atom A (r_A). Thus if r_A is known, d_{A-A} can be calculated with the help of the above equation.

B. Determination of bond length (d_{A-B}) in AB type molecule. The following two cases can be studied.

(*i*) If the electronegativity difference ($x_A - x_B$) is *very small,* then d_{A-B} is equal to the sum of the covalent radii of atoms A and B, *i.e.,*

$$d_{A-B} = r_A + r_B \quad ...(ii)$$

If r_A and r_B are known, d_{A-B} can be calculated with the help of equation (*ii*)

(*ii*) If the electronegativity difference, ($x_A - x_B$) is appreciably large, then the experimental value of d_{A-B} is not equal to the sum of r_A and r_B; rather the experimental value of d_{A-B} is less than the sum, ($r_A + r_B$). Schomaker and Stevenson (1941) have shown that when the electronegativity difference, ($x_A - x_B$) is very large, then d_{A-B} is given by equation (*iii*) or (*iv*) which is called Schomaker and Stevenson's equation.

$$d_{A-B} = r_A + r_B - 0.09\ (x_A - x_B) \quad ...(iii)$$

In this equation r_A and r_B are in Å and hence d_{A-B} is also obtained in Å. If r_A and r_B are in pm, then equation (*iii*) should be written as (1Å = 100 pm)

$$d_{A-B} = r_A + r_B - 9\ (x_A - x_B) \quad ...(iv)$$

Equation (*iii*) and (*iv*) show that greater is the value of $(x_A - x_B)$, greater will be the decrease in the magnitude of d_{A-B}. The determination of d_{A-B} by using equation (*iii*) or (*iv*) is called *electronegativity method.*

Factors affecting the magnitude of bond length in covalent molecules

Following are the factors which affect the magnitude of bond length in covalent molecules.

(*i*) *Electronegativity difference between A and B atoms in AB molecule or polar character of A—B bond.* If the electronegativity difference $(x_A - x_B)$ between A and B atoms in AB molecule is *very large,* then A—B bond becomes polor ($A^{\delta-} - B^{\delta+}$, if A atom has more electronegativity than B atom) because more electronegative atom acquires partial negative charge and less electronegative atom acquires partial positive charge (equal in magnitude) on it. This type of accumulation of opposite charges on the two atoms brings these atoms nearer to each other and hence d_{A-B} is shortened. Greater is the electronegativity difference $(x_A - x_B)$, greater is the shortening of d_{A-B}.

Schomaker and Stevenson (1941) have proposed that when $(x_A - x_B)$ is very large, then d_{A-B} is given by: $d_{A-B} = r_A + r_B - 0.09\ (x_A - x_B)$. Here r_A and r_B are the radii (in Å) of A and B.

(*ii*) *Bond order or bond multiplicity.* Bond order or bond multiplicity is defined as the number of covalent bonds formed between two atoms in a given molecule. It is represented as B.O. It has been observed that the bond length between two atoms (which may be the same or different) bonded together by a triple bond(*i.e.,* A ≡ A or A ≡ B) is smaller than that between the same atoms bonded by a double bond (*i.e.,* A = A or A = B). Similarly the bond length betwen two atoms bonded together by a double bond is smaller than that betwen the same atoms bonded by a single bond (*i.e.,* A — A or A — B). Thus we find that the bond length of A ≡ A, A = A and A — A bonds is in the order A ≡ A < A = A < A — A. This order indicates that with the decrease of bond order (or multiplicity) between the two atoms, the bond length between the atoms increases. This conclusion is supported by the following example.

Carbon — carbon bond:	C ≡ C	C = C	C — C
Bond order:	3	2	1
Bond length (Å)	1.20	1.34	1.54

Explanation. We know that in the formation of a single bond the overlapping takes place between one pair of orbitals only, while in the formation of a double bond the overlapping takes place between two pairs of orbitals and in the formation of a triple bond the overlapping occurs between three pairs of orbitals. Thus with the increase in the number of pairs of orbitals overlapping together, the bond order also increases. With the increase in bond order, the degree of overlapping of orbitals also increases. The increase in the degree of overlapping of orbitals brings the atoms nearer to each other and hence bond length decreases. Thus bond order and bond length are related as :

$$\text{Bond order} \propto \frac{1}{\text{Bond length}}$$

(*iii*) *Type of hybridisation.* With the increase in the percentage of *s*-character in a hybrid orbital, bond length decreases. For example with the increase in the percentage of *s*-character in sp^3, sp^2 and *sp* hybrid orbitals formed in ethane, ethylene and acetylene, C—H and C—C bond lengths decrease as shown below:

Molecule	Type of hybridisation	% of *s*-character in hybrid orbitals	C—H bond length (pm)	C—C bond length (pm)
Ethane (H_3C-CH_3)	sp^3	25	109.3	154
Ethylene ($H_2C = CH_2$)	sp^2	33.3	108.7	134
Acetylene ($HC \equiv CH$)	*sp*	50	105.7	120

Inter-ionic distance in ionic solids

In an ionic crystal, the distance between the nuclei of the two ions (*i.e.*, cation and anion) is called *inter-ionic distance.*

Determination of inter-ionic distance in ionic solids

In an ionic crystal of M^+X^- type in which M^+ and X^- ions are in close contact with each other, the inter-nuclear distance between the nuclei of the two ions is equal to the sum of the radius of M^+ and X^- ions, *i.e.*

$$d\,(M^+ - X^-) = r\,(M^+) + r\,(X^-)$$

where $d\,(M^+ - X^-)$ is the distance between the nuclei of M^+ and X^- ions and $r\,(M^+)$ and $r\,(X^-)$ are their ionic radii.

Bond Energy

Bond formation energy of A — A bond in A_2 molecule and A — B bond in AB molecule

A—A and A—B bond is formed in A_2 and AB molecule respectively when two A-atoms or A and B atoms combine together. In the formation of A — A or A — B bond some amount of energy is released. This energy is called *bond formation energy* or *simply bond energy* or *bond enthalpy.*

$$\underbrace{A + A}_{\text{Two atoms}} \rightarrow \underbrace{A - A}_{A_2\text{ molecule}} + \text{Energy released (called } \textit{bond formation energy}\text{)}$$

$$\underbrace{A + B}_{\text{Two atoms}} \rightarrow \underbrace{A - B}_{\text{AB molecule}} + \text{Energy released (called } \textit{bond formation energy}\text{)}$$

Bond formation energy is shown with negative sign (*release of energy*)

Thus bond formation energy can be defined as the amount of energy released when one mole (one Avogadro's number) of bonds are formed between the constituent atoms (in the gaseous state) of a molecule of A_2 or AB type. It is obvious that the bond formation process is an exothermic process.

Bond dissociation energy of A—A bond in A_2 molecule and A—B bond in AB molecule

Bond dissociation energy of A—A bond in A_2 molecule and A—B bond in AB molecule is the energy required to break A—A or A—B bond so that A_2 and AB molecules are broken into their constituent atoms, *i.e.*,

$$\underbrace{A—A}_{A_2\text{ molecule}} + \text{Energy required (} \textit{bond dissociation energy}\text{)} \rightarrow \underbrace{A + A}_{\text{Two atoms}}$$

$$\underbrace{A—B}_{\text{AB molecule}} + \text{Energy required (} \textit{bond dissociation energy}\text{)} \rightarrow \underbrace{A + B}_{\text{Two atoms}}$$

Bond dissociation energy is shown with a *positive sign (absorption of energy).*

The definition of bond formation and bond dissociation energy as given above shows that both these energies for A—A bond (in A_2 molecule) and A—B bond (in AB molecule) are equal but are represented with opposite signs.

Bond dissociation energies (D_1, D_2, D_3 etc.) of a poly-atomic molecule of AB_n type

The molecule of AB_n type has *n* A—B bonds. For example $H_2O\left(\begin{matrix}H\\H\end{matrix} > O\right)$, NH_3 $\left(\begin{matrix}H\\H\end{matrix}\!\!>N—H\right)$ and $CH_4\left(\begin{matrix} & H & \\ & | & \\ H— & C & —H \\ & | & \\ & H & \end{matrix}\right)$ molecules have two O—H, three N—H and four C—H bonds respectively. Since AB_n molecule has *n* A—B bonds, *n* bond dissociation energies are required to break *n* A—B bonds. Thus the number of bond dissociation energies for a given poly-atomic molecule of AB_n type is always equal to the number of A—B bonds present in the molecule. Thus since H_2O, NH_3 and CH_4 molecules have two O—H, three N—H and four C—H bonds respectively, these molecules have 2, 3 and 4 bond dissociation energies. Two O—H bond dissociation energies corresponding to the breaking of two O—H bonds in H_2O molecule are given below:

(*a*) H — O — H(*g*) → H(*g*) + O — H(*g*), D_1 = + 500.8 KJ/mole

(*b*) O—H(*g*) → O(*g*) + H(*g*), D_2 = + 424.7KJ/mole

Thus two bond dissociation energies of H_2O molecule are D_1 = 500.8 KJ/mole and D_2 = 424.7 KJ/mole. It may be seen that these values are not equal.

Heat of atomisation of a gaseous substance

Heat required to break one mole of a gaseous substance into its constituent gaseous atoms is called heat of atomisation of the substance. For example heat of atomisation of $H_2O(g)$, $CH_4(g)$ and $C_2H_6(g)$ (ethane) are 925.5 KJ/mole, 397.2 Kcal/mole and 674.7 Kcal/mole. These values are shown with a positive sign, since they represent the heats absorbed.

$H_2O(g) \rightarrow 2H(g) + O(g)$, Heat of atomisation of $H_2O(g)$ = +925.5 KJ/mole

$CH_4(g) \rightarrow C(g) + 4H(g)$, Heat of atomisation of $CH_4(g)$ = +397.2 Kcal/mole

$C_2H_6(g) \rightarrow 2C(g) + 6H(g)$, Heat of atomisation of $C_2H_6(g)$ = +674.7 Kcal/mole.

Relation between bond dissociation energies and heat of atomisation of a molecule

Let us consider H_2O(g) molecule. Two bond dissociation energies (D_1 and D_2) of H_2O (g) molecule can be represented by the following equations:

H—O—H(*g*) → H(g) + O—H(*g*), D_1 = +500.8 KJ/mole ...(*i*)

O—H(*g*) → O(g) + H(*g*), D_2 = +424.7KJ/mole ...(*ii*)

Add: H—O—H(*g*) → 2H(*g*) + O(*g*), Δ H = $+D_1 + D_2$

= + 500.8 + 424.7 = + 925.5 KJ/mole ...(*iii*)

Equation (*iii*) represents heat of atomisation of $H_2O(g)$ molecule. Thus we can say that heat of automisation of a molecule is equal to the sum of the bond dissociation energies corresponding to the breaking of all the bonds present in the molecule. Thus heat of atomisation of H_2O is equal to (D_1+D_2) = +500.8 + 524.7 = + 925.5 KJ/mole.

Average bond dissociation energy of a bond in a gaseous molecule

Let us consider $H_2O(g)$ molecule. This molecule has two equivalent O—H bonds ($\begin{matrix}H\\H\end{matrix}$ > O).

Heat of atomisation of $H_2O(g)$ molecule has been found to be equal to 925.5KJ/mole, *i.e.*.

$H_2O(g) \rightarrow 2H(g) + O(g)$, Heat required in the reaction = Heat of atomisation of $H_2O(g)$ molecule = + 925.5 KJ/mole.

If this value is divided by total number of O—H bonds present in $H_2O(g)$ molecule, we get average bond dissociation energy for one O—H bond in $H_2O(g)$ molecule, *i.e.*,

Average bond dissociation energy of O—H bond in $H_2O(g)$ molecule

$$= \frac{\text{Heat of atomisation of } H_2O\text{ (g) molecule}}{\text{Total number of O–H bonds in } H_2O\text{(g) molecule}} = \frac{925.5}{2} = 462.5 \text{ kJ/mole} \quad ...(i)$$

462.5 KJ/mole represents the average energy that is needed to break one mole of O—H bond in $H_2O(g)$ molecule.

Thus average bond dissociation energy of a bond in a gaseous molecule can be defined as the atomisation energy of the molecule, divided by the total number of bonds present in the molecule.

We have already said that heat of atomisation of a gaseous molecule is equal to the sum of successive bond dissociation energies of the molecule. Thus:

Average bond dissociation energy of A—B bond in $AB_n(g)$ molecule

$$= \frac{\text{Sum of successive A–B bond dissociation energies}}{\text{Total number of A–B bonds in } AB_n\text{ (g) molecule}}$$

Relation (*i*) shows that if the total number of bonds (*n*) in a given gaseous molecule and the average bond dissociation energy of the bond (ΔH_{av}) are known, then heat of atomisation of the molecule will be equal to the product of *n* and ΔH_{av}.

Example 10.4. *Calculate the heat of atomisation of CH_4 molecule, if bond enthalpy of C—H bond = 99.3 KCal.*

Solution. Heat of atomisation of $CH_4(g)$ molecule

= Number of C—H bonds × Enthalpy of C — H bond. = 4 × 99.3 = **397.2 KCal**

Methods for the determination of heat of atomisation of a gaseous molecule and average bond dissociation energy of a bond in the molecule

The methods which are generally used for this purpose are *thermochemical method, electron absorption analysis, mass spectroscopic analysis, magnetic susceptibility measurements, magnetic resonance method, molar refraction method* and *kinetic method.* Here, we shall discuss thermochemical (calorimetric) method. This method can be illustrated by the following examples:

Example 10.5. *Find out the heat of atomisation of $CH_4(g)$ molecule and average bond dissociation energy for C—H bond in this molecule by using the following thermochemical data:*

(*i*) $C(s) + 2H_2(g) \rightarrow CH_4(g)$, $\Delta H = -17.9$ *Kcal/mole*

(*ii*) $C(s) \rightarrow C(g)$, $\Delta H = +170.9$ *Kcal/mole*

(*iii*) $2H_2(g) \rightarrow 4H(g)$, $\Delta H = +208.4$ *Kcal/mole*

Solution. Equation (*i*) can also be written as:

$CH_4(g) \rightarrow C(s) + 2H_2(g)$, $\Delta H = +17.9$ Kcal/mole

Now add this equation to equations(*ii*) and (*iii*) to get:

$CH_4(g) \rightarrow C(g) + 4H(g)$, $\Delta H = +17.9 + 170.9 + 208.4 = +397.2$ Kcal/mole

∴ Heat of atomisation of $CH_4(g)$ molecule = +397.2 Kcal/mole

Since CH_4 has four C—H bonds, average bond dissociation energy for one C—H bond = +397.2/4 = +99.3 Kcal/mole.

Example 10.6. *Calculate the bond energy of C—C bond in ethane molecule, $C_2H_6(g)$ from the following data (bond enthalpy of C—H bond = 99 Kcal/mole)*

(*i*) $2C(s) + 3H_2(g) \rightarrow C_2H_6(g)$, $\Delta H = -20.30$ *Kcal*

(*ii*) $2C(s) \rightarrow 2C(g)$, $\Delta H = +341.8$ *Kcal*

(*iii*) $3H_2(g) \rightarrow 6H(g)$, $\Delta H = +312.6$ *Kcal*

Solution. In order to find out the bond energy of C—C bond in $C_2H_6(g)$ molecule we must find out the heat of atomisation of this molecule. Equation(*i*) can also be written as:

$C_2H_6(g) \rightarrow 2C(s) + 3H_2(g)$, $\Delta H = +20.3$ Kcal

Now add this equation to equations (*ii*) and (*iii*) to get:

$C_2H_6(g) \rightarrow 2C(g) + 6H(g)$, $\Delta H = 20.3 + 341.8 + 312.6 = 674.7$ Kcal

∴ Heat of atomisation of $C_2H_6(g)$ = 674.7 Kcal.

Now, we know that C_2H_6 molecule has one C—C bond and six C—H bonds

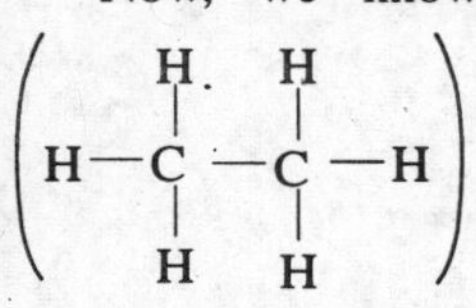

Thus

Heat required to break one C—C bond and six C—H bonds in C_2H_6 molecule = 674.7 Kcal

Now since bond energy of C—H bond = 99Kcal(given), bond energy of one C—C bond

= 674.7 – 6 × 99 = **+ 80.7 Kcal**

Applications of bond energies

The most important application of bond energies is that these can be used to evaluate the heat of reaction of a given process. This application is evident from the following example.

Example 10.7. *Calculate the heat of reaction of the following process:*

$$CH_4(g) + 4F_2(g) \rightarrow CF_4(g) + 4HF(g),$$

given that bond enthalpies of C—H, F—F, C—F and H—F bonds are 99.3, 38.0, 116.0 and 136.0 Kcal/mole respectively.

Solution. In the given reaction four C—H bonds and four F—F bonds are broken down while four C—F bonds and four H—F bonds are formed. Hence heat of reaction of the above process will be equal to the sum of bond dissociation energies of all the broken bonds (these energies have to be shown with a positive sign) and bond formation energies for all the bonds that have formed (these energies have to be shown with a negative sign). Thus:

Heat of reaction = [Bond dissociation energies for four C—H bonds + bond dissociation energies for four F—F bonds] + [Bond formation energies for four C—F bonds + bond formation energies for four H—F bonds]

= $[4 \times (+99.3) + 4 \times (+38)] + [4 \times (-116) + 4 \times (-136)]$

= **– 458.8 Kcal/mole.**

Factors on which the magnitude of bond energy depends

Following are the factors on which the magnitude of bond energy depends.

(*i*) The magnitude of bond energy changes with the change of the nature of the atoms present in the neighbouring bonds. Hence bond energy of a given bond changes with the change of the molecule in which it is present.

(*ii*) Bond energy of a bond depends on its exact situation in the molecule, *i.e.*, bond energy changes with the change of the geometry of the molecule.

(*iii*) Bond energy also changes with the polarity of the bond. Hence ionic and covalent bonds have different bond energies.

Comparison between the bond dissociation energies of a single bond, double bond and triple bond and relation between bond order (B.O), bond length and bond energy.

It has been found that a triple bond (*e.g.* C ≡ C, C≡O etc.) has higher bond dissociation energy than a double bond (*e.g.* C = C, C = O etc.) which, in turn, has higher bond dissociation energy than a single bond (*e.g.* C—C, C—O etc.). Thus the bond dissociation energy of a triple bond, double bond and single bond is in the order:

Triple bond dissociation energy > Double bond dissociation energy > Single bond dissociation energy.

or Bond dissociation energy of A ≡ A bond (B.O. = 3) > Bond dissociation energy of A = A bond (B.O. = 2) > bond dissociation energy of A — A bond (B.O = 1).

The above order clearly shows that with the decrease of bond order (B.O.), the bond dissociation energy also decreases.

Explanation. We have already said during the discussion of bond length that in the formation of triple or double bond, the two atoms come closer to each other and hence bond length decreases. The decrease in bond length makes the bond stronger and hence bond dissociation energy increases. Thus

$$\text{Bond order} \propto \text{Bond energy} \quad ...(i)$$

We know that:

$$\text{Bond order} \propto \frac{1}{\text{Bond length}} \quad ...(ii)$$

On combining the above two relations, we get:

$$\text{Bond order} \propto \frac{1}{\text{Bond length}} \propto \text{Bond energy}$$

Bond disociation energies of double and triple bonds are called *multiple bond energies.*

Calculation of bond dissociation energy (E_{A-B}) of A—B bond in AB molecule by Pauling's formula

Pauling has suggested that the bond energy (E_{A-B}) of A—B bond (in electron volts) in AB molecule can be calculated by using his formula which is given below:

$$x_A - x_B = 0.208\left[E_{A-B} - \sqrt{E_{A-A} \times E_{B-B}}\right]^{1/2}$$

Here x_A and x_B are electronegativities of A and B atoms and E_{A-B}, E_{A-A} and E_{B-B} are bond energies of A—B, A—A and B—B bonds in AB, A_2 and B_2 molecules respectively. Value of E_{A-B} can be calculated, if the values of other terms appearing in the quation are known.

Bond Angles

What is bond angle?

The angle between the bonding electron pairs in a given molecule or an ion is called bond angle. For example HCH bond angle in CH_4 molecule is equal to 109.5°.

Factors affecting the magnitude of bond angles

Following are the important factors which affect the magnitude of bond angles.

1. Number of lone pairs of electrons on the central atom of a given molecule or an ion. How the number of lone pairs of electrons on the central atom affects the magnitude of bond angle can be understood by considering the following examples.

(a) Comparison of the bond angles of the nitrogen group hydrides with those of oxygen group hydrides. The bond angles of the hydrides of the elements of nitrogen and oxygen group elements are given in the margin. Bond angle of CH_4 is also given for comparison. It may be seen that the bond angles in CH_4, NH_3 and H_2O molecules are in the order : $CH_4 > NH_3 > H_2O$. This order can be explained as follows:

CH_4 >	NH_3 >	H_2O
109.5°	107.5°	104.5°
	PH_3 >	H_2S
	93.5°	92.2°
	AsH_3 >	H_2Se
	91.8°	91.0°
	SbH_3 >	H_2Te
	91.3°	89.4°

Due to the absence of lone pairs of electrons on C-atom in CH_4 molecule, HCH bond angle in this molecule is equal to the expected tetrahedral angle (= 109.5°). In NH_3 molecule, N-atom has one lone pair of electrons on it. Here *(lp – bp)* repulsion is greater than the *(bp – bp)* repulsion. The greater magnitude of *(lp – bp)* repulsion makes three bonding electron pairs come closer to each other and hence HNH bond angle reduces from 109.5° (in CH_4) to 107.5° in NH_3 molecule. O-atom in H_2O molecule has two lone pairs of electrons and two bonding pairs of electron. Here the repulsion between different types of electron pair decreases in the order : *(lp – lp)*> *(lp – bp)* > *(bp – bp)*. It is due to this order that the lone pairs in H_2O molecule tend to be as far apart from each other as possible and hence HOH bond angle in H_2O molecule gets reduced compared to that in NH_3 molecule (NH_3 = 107.5°, H_2O = 104.5°). Thus we see that with the increase in the number of lone pairs of electrons on the central atom, the bond angle decreases.

On similar lines of argument it can be shown that the bond angles of the hydrides of nitrogen group elements are greater than those of the hydrides of oxygen group elements lying in the same period. Thus $PH_3 > H_2S$, $AsH_3 > H_2Se$ and $SbH_3 > H_2Te$.

(b) Comparison between ONO bond angles in NO_2^+ and NO_2^- ions. The structure of these ions is given below:

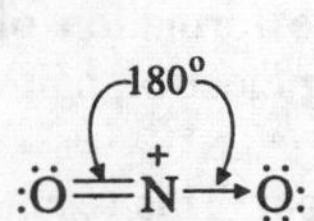

NO_2^+ ion (called nitronium or nitryl ion)

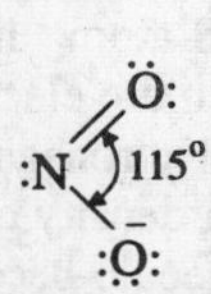

NO_2^- ion (called nitrite ion)

In NO_2^+ ion, N-atom is *sp* hybridised and hence this ion has linear shape. ONO bond angle is equal to 180° which is the expected bond angle for a linear geometry.

In NO_2^- ion, N-atom is sp^2 hybridised. Due to the presence of one lone pair of electrons which resides in one of the three sp^2 hybrid orbitals, the shape of NO_2^- ion becomes *angular* or *V-shaped* (and not trigonal planar). Corresponding to sp^2 hybridisation, ONO bond angle in NO_2^- ion should be equal to 120°. Actually, ONO bond angle in this ion has been found to be equal to 115° which is less than 120°. The decrease in bond angle is due to the presence of one lone pair of electrons on N-atom. This lone pair occupies more space than the bonding electron pairs and hence the bonding electron pairs are pushed closer to each other. This results in that the bond angle between the bonding electron pairs decreases.

Since N-atom in NO_2^+ ion has no lone pair of electrons, its bond angle is the same as is expected for a linear geometry.

(*c*) *Cl — N — O bond angle in NOCl molecule.* Electron dot structure of NOCl molecule is $^{Cl}_{O}\!\!\gg\!N\!:$ which shows that N-atom is sp^2 hybridised. One of the three sp^2 hybrid orbitals has one lone pair of electrons. Due to sp^2 hybridisation, the expected value of Cl—N—O bond angle should be 120°. In fact the bond angle has been found to be equal to 116° which is less than 120°. The decrease in bond angle is because of the fact that the lone pair of electrons on N-atom makes the bonding electron pairs involved in N—Cl and N—O σ-bonds come closer together and hence Cl—N—O bond angle derceases from 120° to 116°. It is also due to the decrease in bond angle that the given molecule does not have the expected trigonal planar geometry; rather it has angular geometry.

2. Electronegativity or size of atom A (central atom) in molecules of AB_x type. The effect of this factor can be seen when we compare the B—A—B bond angles of different molecules of AB_x type in which atom A changes but atom B remains the same as we move from one molecule to the other in a given set of molecules. The molecules given in the following three sets are the examples of AB_x type molecules. Their B—A—B bond angles are also given. Electronegativity value of the central atom A is given in bracket.

Set 1 (A= N, P, As, Sb; B = H)	Set 2 (A = N, P As, Sb; B = F)	Set 3 (A = O, S, Se, Te; B = H)
NH_3 = 107.5 (N = 3.0)	NF_3 = 102° (N = 3.0)	H_2O = 104.5° (O = 3.5)
PH_3 = 93.5° (P = 2.1)	PF_3 = 97° (P = 2.1)	H_2S = 92.2° (S = 2.5)
As H_3 = 91.8° (As = 2.0)	AsF_3 = 96° (As = 2.0)	H_2Se = 91.0° (Se = 2.4)
SbH_3 = 91.3° (Sb = 1.8)	SbF_3 = 88° (Sb = 1.9)	H_2Te = 89.5° (Te = 2.1)

It may be seen from the above table that as we move from one molecule to the other in a given set, the B—A—B bond angle decreases with the decrease of the electronegativity or with the increase of the size of the central atom A.

Explanation. In order to explain the variation of bond angles, let us explain the change in bond angle when we proceed from H_2O to H_2S as an example. Since electronegativity of O-atom is greater than that of S-atom, the bonding electron pair used in the formation of O—H bond is shifted more towards more electronegative O-atom than the bonding electron pair of S—H bond is shifted towards less electronegative S-atom.

Thus the electron density on O-atom is more than that on S-atom. This results in that (*bp-bp*) repulsion in H_2O molecule becomes greater than the (*bp-bp*) repulsion in H_2S molecule. Thus we see that, on proceeding from OH_2 molecule to SH_2 molecule, (*bp-bp*) repulsion decreases. Due to the decrease in (*bp-bp*) repulsion bond angle in SH_2 molecule also decreases, *i.e.,* H—S—H bond angle becomes less than H—O—H bond angle.

On the basis of the above explanation which we have used for one pair of molecules, we can show why the bond angles decrease when we move from H_2O to H_2Te in the given set.

3. Electronegativity or size of atom B in the molecules of AB_x type. The effect of this factor can be seen when we compare the B—A—B bond angles of different molecules of AB_x type in which atom A (central atom) remains the same but atom B changes as we move from one molecule to the other in a given set of molecules. The molecules given in the following four sets are the examples of AB_x type molecules. Their B—A—B bond angles are also given. Electronegativity of atom B is given in bracket.

Set 1 (A = N; B = F, H)	Set 2 (A = P; B = Cl Br, I)	Set 3 (A = As; B = Cl, Br, I)	Set 4 (A = O; B = F, Cl)
NF_3 = 102° (F = 4.0)	PCl_3 = 100° (Cl = 3.0)	$AsCl_3$ = 98.4° (Cl = 3.0)	OF_2 = 103.2° (F = 4.0)
NH_3 = 107.5° (H = 2.1)	PBr_3 = 101.5° (Br = 2.8) PI_3 = 102° (I = 2.5)	As Br_3 = 100.5° (Br = 2.8) As I_3 = 101.5° (I = 2.5)	OCl_2 = 111° (Cl = 3.0)

It may be seen from the above table that as we move from one molecule to the other in a given set of molecules, the B—A—B bond angle increases with the decrease of the electronegativity or with the increase of the size of atom B.

Explanation. As an example, we shall consider the variation in bond angle in OF_2 — OCl_2 pair. Since electronegativity of F-atom is greater than that of Cl-atom, the bonding electron pair of O—F bond would shift more towards F-atom than the bonding electron pair of O—Cl bond would shift towards Cl-atom. Thus the electron density on F-atom in OF_2 molecule would be greater than the electron density on Cl-atom in OCl_2 molecule . This results in the *(bp-bp)* repulsion in OF_2 molecule becomes less than the *(bp - bp)* repulsion in OCl_2 molecule. Thus, we seen that as we proceed from OF_2 molecule to OCl_2, *(bp-bp)* repulsion increases. Due to the increase in *(bp-bp)* repulsion, bond angle in OCl_2 molecule also increases, *i..e.,* Cl—O—Cl bond angle becomes greater than F—O—F bond angle.

Conclusion. The study of the effect of electronegativity of atom A or atom B in molecules of AB_x type on B—A—B bond angle as made above shows that *the bond angle increases with the increase in the magnitude of (bp – bp) repulsion and vice-versa.*

4. Presence of multiple bonds in a molecule. The bond angle of a molecule having multiple bond or bonds has been found to be different from that which is expected on the basis of the geometry of the molecule. In order to illustrate this point, we shall consider bond angles in some trigonal and tetrahedral molecules (given below) which have multiple bonds.

Trigonal planar molecules (Expected bond angle = 120°)

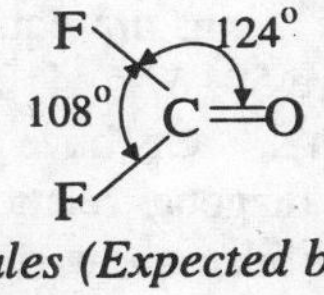

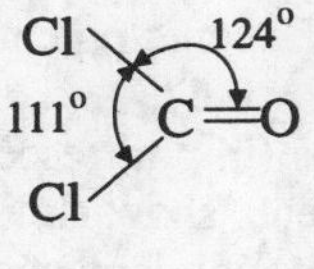

Tetrahedral molecules (Expected bond angle 109.5°)

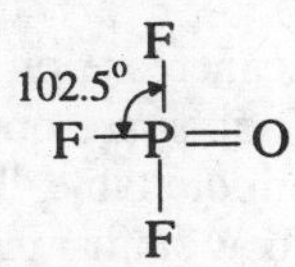

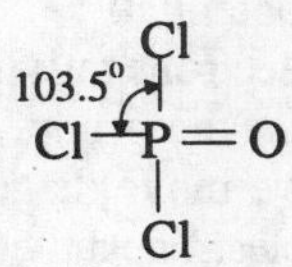

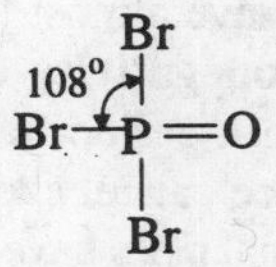

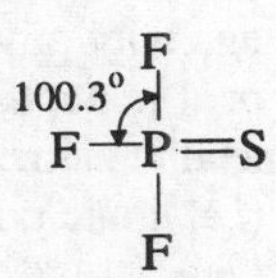

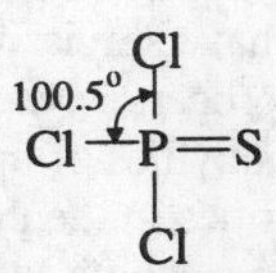

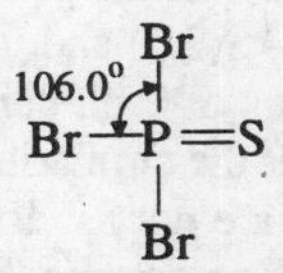

We know that the expected bond angle in tetrahedral and trigonal planar molecules is 109.5° and 120° respectively. Bond angle shown in each molecule indicates that its value is less than that which is expected for the geometry of the molecule. It may also be seen that the angle between the multiple bond and the other single bonds in the molecule is greater than the angle between two single bonds of the molecule.

Explanation. We know that the presence of a π-bond in a molecules decreases the bond length. This means that a double bond or a triple bond is considerably fatter than a single bond and hence repels the other single bonds. This results in that the bonding electron pairs involved in the formation of single bonds come closer to each other and hence X A X (A = central atom, X = F, Cl, Br) bond angle falls below that expected from the geometry of the molecule.

Diagonal Relationship or Diagonal Similarity

What is diagonal relationship?

It has been observed that the elements of 2*nd* period of the periodic table show similarity in properties with the elements which are lying to their right in 3*rd* period or are lying diagonally opposite to them. Thus, Li—Mg, Be—Al, B—Si, C—P, N—S and O—Cl pairs of elements (These pairs are called *diagonal pairs*) show many similar properties. The similarity in properties existing between Li—Mg, Be—Al, B—Si etc. pairs is called **diagonal relationship** or **diagonal similarity.**

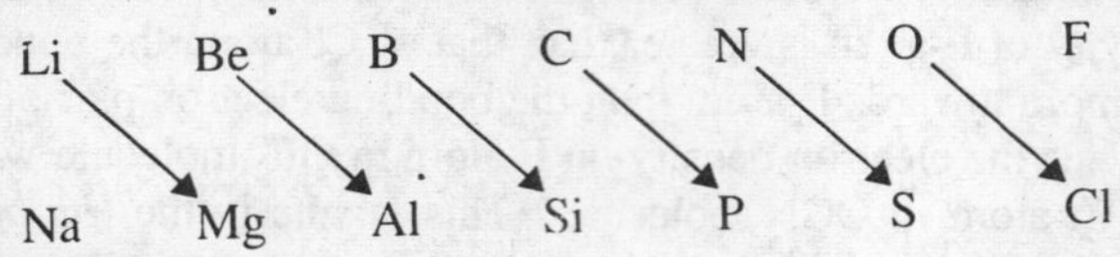

Cause of diagonal relationship

Diagonal relationship existing between a pair of elements can be explained on the basis of polarising power concept or on the basis of electronegativity concept.

1. Polarising power concept (Ionic potential concept). The ratio, cationic charge / cationic size, is a measure of the polarising power of a given cation. This ratio is also called *ionic potential.* Since on moving from left to right accross a period, the cationic charge increases and cationic size decreases, the magnitude of polarising power (*i.e.,* ionic potential) of the cations (cationic charge/cationic size ratio) increases (Fajans' rules). On the other hand, on moving down a group, since the cationic charge remains the same and cationic size increases, the polarising power of the cations decreases (Fajans' rules). Since these two variations of the polarising power of cations (one along the period and the other down the group) are opposite to each other, they partly cancel (balance) each other when we move diagonally in the periodic table. This results in that the elements placed diagonally opposite to each other have similar polarising power and hence show similar properties. The ionic potentials of Be^{2+}, Al^{3+}, B^{3+} and Si^{4+} ions are equal to 2/59 = 0.033, 3/67.5 = 0.044, 3/41 = 0.073 and 4/54 = 0.074 respectively (Here 59, 67.5, 41 and 54 are the ionic radii of the given ions (in pm) respectively). Now since Be^{2+} and Al^{3+} ions have almost similar ionic potential, these elements show similar properties, *i.e.,* they show diagonal relationship because they have almost the same ionic potential. Similarly B^{3+} and Si^{4+} ions also show diagonal relationship because they have almost the same ionic potential.

2. Electronegativity concept. The electronegativities of elements increase on moving from left to right in a period and decrease down the group. These two variations are opposite to each other and hence cancel each other when we move diagonally in the periodic table. Thus results in that the diagonal pairs have almost similar electronegativities, and hence show similar properties. Thus, since the elements of the pairs namely Li(1.0) – Mg(1.2), Be (1.5) – Al (1.5), B(2.0) – Si(1.8), C(2.5) – P(2.1), N(3.0) – S(2.5) and O(3.5) – Cl(3.1) have similar values of electronegativity (given in bracket), the two elements of each pair show similar properties, *i.e.,* they show diagonal relationship. It may be emphasised that the similarities in electronegativities of the diagonal non-metallic elements (*e.g.* C—P, N—S, O—Cl) are not as close as the similarities in the polarising powers (*i.e.,* ionic potentials) of the diagonal metallic elements (*e.g.,* Li—Mg, Be—Al etc.)

Diagonal relationship between Li and Mg.

The diagonal relationship between Li and Mg is due to two reasons : (*i*) Both the elements have almost the same atomic and ionic radii [Li = 134 pm, Mg = 145pm; Li^+ = 90 pm(for C.N. = 6), Mg^{2+} = 86 pm (for C.N. = 6)] (*ii*) Both the elements have almost similar values for their electronegativity (Li= 1.00, Mg = 1.20).

Following are the important points which justify the similarities between Li and Mg.

1. *Formation of hydrated salts.* Both the elements give many hydrated salts like $LiCl.2H_2O$, $MgCl_2, 6H_2O$; $LiNO_3.H_2O$, $Mg(NO_3)_2.6H_2O$ etc.

2. (*a*) *Action of heat on nitrates, carbonates and hydroxides.* These compounds of both the metals decompose on heating to given the corresponding oxides of the metals.

Nitrates
$$4LiNO_3 \xrightarrow{\Delta} 2Li_2O + 4NO_2 + O_2$$
$$2Mg(NO_3)_2 \xrightarrow{\Delta} 2MgO + 4NO_2 + O_2$$

Carbonates
$$LiCO_3 \xrightarrow{\Delta} Li_2O + CO_2$$
$$MgCO_3 \xrightarrow{\Delta} MgO + CO_2$$

Hydroxides
$$2LiOH \xrightarrow{\Delta} Li_2O + H_2O$$
$$Mg(OH)_2 \xrightarrow{\Delta} MgO + H_2O$$

(*b*) *Action of heat on formates.* Formates of both the metals, on heating, give metallic carbonates and formaldehyde (HCHO)

$$2(HCOO)Li \xrightarrow{\Delta} Li_2CO_3 + HCHO$$

$$(HCOO)_2Mg \xrightarrow{\Delta} MgCO_3 + HCHO$$

3. *Action of N_2, Cl_2, carbon, O_2 and H_2O on the metals.* (*a*) Both the metals combine with N_2, on heating, to form the nitrides *viz.* Li_3N and Mg_3N_2. Both these nitrides are readily hydrolysed by H_2O and form the metallic hydroxides and liberate NH_3.

$$Li_3N + 3H_2O \longrightarrow 3LiOH + NH_3$$
$$Mg_3N_2 + 6H_2O \longrightarrow 3Mg(OH)_2 + 2NH_3$$

(*b*) Both the metals combine with dry Cl_2, on heating to form LiCl and $MgCl_2$ respectively. Both these chlorides have many similar properties, *e.g.* (*i*) Both the chlorides are deliquescent solids (*ii*) Both are soluble in alcohol and pyridine, showing that both these chlorides are covalent in character. (*iii*) Both have water of crystallisation (*e.g.* $LiCl.2H_2O$, $MgCl_2.6H_2O$). (*iv*) Both undergo hydrolysis in hot water.

(*c*) On heating, both metals combine with carbon and give the carbides. Li gives acetylide type carbide ($HC \equiv CH$) while Mg gives allylide type carbide ($H_2C = C = CH_2$). Thus these carbides contain C_2^{-2} and C_3^{-4} ions and hence are formulated as $Li_2^+(C_2)^{-2}$ or Li_2C_2 and $Mg_4^{+2}(C_3)_2^{-4}$ or Mg_2C_3.

(*d*) Li and Mg ribon, when burnt in O_2 or air, form the normal oxides namely Li_2O and MgO. Both these oxides are sparingly soluble in water .

(*e*) Both the metals react with H_2O slowly and form the metallic hydroxides and H_2 gas is liberated.

$$2Li + 2H_2O \longrightarrow 2LiOH + H_2 \quad ; \quad Mg + 2H_2O \longrightarrow Mg(OH)_2 + H_2$$

Both these hydroxides are weak bases and decompose on heating to give the matallic oxides and H_2O as already shown.

4. *Solubility of fluorides, carbonates, oxalates, phosphates and bicarbonates.* Fluorides, carbonates, oxalates and phosphates of both the metals are *sparingly soluble* in water while the bicarbonates are *freely soluble.*

5. Both the metals are hard metals

6. Both metals can be melted in dry air without losing their lusture (brightness).

Diagonal relationship between Be and Al.

The diagonal relationship between Be and Al is due to the reasons: (*i*) The ionic potentials (= ionic charge/ionic radius) of Be^{+2} and Al^{+3} ions are almost the same (Be^{+2} = 2 / 59 = 0.033, Al^{+3} =

3/67.5 = 0.044. Here 59 and 67.5 are the radii (in pm) of Be^{+2} and Al^{+3} ions respectively for C.N. = 6 (*ii*) The values of standard reduction potentials are almost the same ($E^0_{Be^{+2}/Be} = -1.70$ V, $E^0_{Al^{+3}/Al} = -1.67$ V) (*iii*) The electronegativity values of both the elements are the same (Be = 1.5, Al = 1.5) (*iv*) Both the metals have weakly electropositive character (*v*) The covalent radii of both the elements are almost the same (Be = 125 pm, Al = 130 pm).

Following are the important points which justify the simlarities between Be and Al.

1. *Occurrence of the metals in nature.* In nature both the metals are found as beryl, $3BeO.Al_2O_3.6SiO_2$.

2. *Extraction of the metals.* Both the metals can be extracted by the electrolytic reduction of their oxides

3. *Action of HNO_3 and alkalies.* (*a*) Both the metals react with HNO_3 and form a layer of protective oxide on their surface. This layer of oxide makes the metals passive for the further action of the acid on them.

(*b*) Both the metals react with boiling alkalies to form soluble *beryllate* (BeO_2^{-2}) and *aluminate* (AlO_2^-) and H_2 is evolved.

$$Be + 2OH^- \longrightarrow BeO_2^{-2} + H_2; \quad 2Al + 2OH^- + 2H_2O \longrightarrow 2AlO_2^- + 3H_2$$

4. *Formation of halides, nitrides, carbides and oxides.* (*a*) Both the metals combine with halogens and form the halides *viz.* BeX_2 and AlX_3 respectively. These halides have many similar properties, *e.g.* (*i*) they have low melting points; (*ii*) they are hydrolysed by H_2O; (*iii*) the anhydrous halides (excepting fluorides) are covalent in nature and hence are soluble in organic solvents; (*iv*) they behave as Lewis acids; (*v*) $BeCl_2$ and $AlCl_3$ both have dimeric (bridged) structure as shown below.

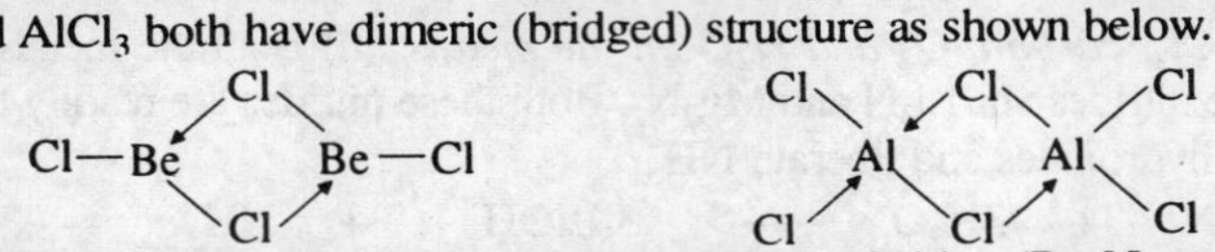

(*b*) Both the metals combine with N_2 and form their nitrides (Be_3N_2 and AlN). Both these nitrides are decomposed by H_2O to form metallic hydroxides and NH_3 gas is evolved.

$$Be_3N_2 + 6H_2O \longrightarrow 3Be(OH)_2 + 2NH_3 \quad , \quad AlN + 3H_2O \longrightarrow Al(OH)_3 + NH_3$$

(*c*) Both the metals form methanide type carbides ($HC \equiv CH$) and hence contain C^{-4} ion. Thus these carbides can be formulated as $Be_4^{+2}\ C_2^{-4}$ or Be_2C and $Al_4^{+3}\ C_3^{-4}$ or Al_4C_3 and hence evolve CH_4 on hydrolysis.

$$Be_2C + 4H_2O \longrightarrow CH_4 + 2Be(OH)_2 \quad ; \quad Al_4C_3 + 12H_2O \longrightarrow 3CH_4 + 4Al(OH)_3$$

(*d*) Both the metals burn in O_2 and form oxides (BeO and Al_2O_3). Both these oxides are amphoteric in nature, *i.e.*, they react with both acids and alkalies to form the salts

$$BeO + 2HCl \longrightarrow BeCl_2 + H_2O \quad ; \quad BeO + 2NaOH \longrightarrow Na_2\,BeO_2 + 2H_2O$$

$$Al_2O_3 + 6HCl \longrightarrow 2AlCl_3 + 3H_2O \quad ; \quad Al_2O_3 + 2NaOH \longrightarrow 2NaAlO_2 + H_2O$$

5. *Formation of complexes by Be^{+2} and Al^{+3} ions.* Both these ions have a strong tendency to combine with ligands and form chelated complexes like $[Be(H_2O)_4]^{+2}$, $[BeF_4]^{-2}$, $[Be(C_2O_4)_2]^{-2}$; $[Al(H_2O)_6]^{+3}$, $[AlF_6]^{-3}$, $[Al(C_2O_4)_3]^{-3}$ etc.

Diagonal relationship between B and Si.

Due to the following reasons, these elements show diagonal relationship with each other (*i*) The electronegativity values of both the elements are almost the same (B = 2.0, Si = 1.8). (*ii*) Both the elements have almost identical values of their ionic potentials (B^{+3} = 3/41 = 0.073, Si^{+4} = 4/54 = 0.074).

The following properties justify the diagonal relationship between B and Si.

1. *Preparation of the elements.* Both the elements can be obtained by reducing their oxides with Mg-metal.

$$B_2O_3 + 3Mg \longrightarrow 2B + 3MgO \quad ; \quad SiO_2 + 2Mg \longrightarrow Si + 2MgO$$

2. *Occurrence of the elements.* None of these elements occurs in the free state in nature, rather they are found in the form of their oxy compounds namely borates and silicates respectively.

3. *Non-metallic nature.* Both the elements are non-metals and hence are bad conductors of heat and electricity.

4. *Existence of allotropic forms.* Both the elements exist in two allotropic forms *viz.* amorphous and crystalline. Crystalline forms of both the elements are hard.

5. *Density, electronegativity, ionisation potential and boiling points.* Both the elements have almost the same density (B = 2.34 g/cc, Si = 2.35 g/cc), electronegativity (B = 2.0, Si = 1.8) and ionisation potential values (B = 800.6 KJ / mole, Si = 786.4 KJ / mole). Their boiling points are very high and close to each other (B = 2550°C, Si = 2680°C).

6. *Tendency to form covalent compounds.* Most of the compounds of these elements are covalent in nature.

7. *Formation of B^{+3} and Si^{+4} cations.* B and Si normally donot form stable cations namely B^{+3} and Si^{+4}.

8. *Formation of oxides.* Both the elements form stable oxides *viz.* B_2O_3 and SiO_2 which can be prepared either by the action of O_2 on the element at high temperature or by exposing their hydrides *viz.* B_2H_6 and SiH_4 to air

$$B_2H_6 + 3O_2 \longrightarrow B_2O_3 + 3H_2O \quad ; \quad SiH_4 + 2O_2 \longrightarrow SiO_2 + 2H_2O$$

These oxides have many similar properties *e.g.* (*a*) Both the oxides are sparingly soluble solid compounds having high melting points.

(*b*) Both the oxides are weakly acidic in character, since in water they give boric acid (H_3BO_3) and meta silicic acid (H_2SiO_3) which are weak acids.

$$B_2O_3 + 3H_2O \longrightarrow H_3BO_3 \quad ; \quad SiO_2 + H_2O \longrightarrow H_2SiO_3$$

These oxides also seem to have *amphoteric character,* since they also react with NaOH to form stable borate (BO_3^{-3}) and meta silicate (SiO_3^{-2}) and with acids (*e.g.* HF) they give their salts

$$B_2O_3 + 6NaOH \longrightarrow 2Na_3BO_3 + 3H_2O \quad ; \quad SiO_2 + 2NaOH \longrightarrow Na_2SiO_3 + H_2O$$

$$B_2O_3 + 6HF \longrightarrow 2BF_3 + 3H_2O \quad ; \quad SiO_2 + 4HF \longrightarrow SiF_4 + 2H_2O$$

(*c*) Both the oxides combine with metallic oxides, on heating, to form metaborate and metasilicate, *e.g.*

$$B_2O_3 + CoO \xrightarrow{\Delta} Co(BO_2)_2 \quad \textit{(Cobalt meta borate)}$$

$$SiO_2 + CaO \xrightarrow{\Delta} CaSiO_3 \quad \textit{(Calcium meta silicate)}$$

(*d*) Both the oxides can be reduced to their element with the help of Mg-metal (Chemical equations have already been given).

(*e*) Both the oxides, when heated with carbon, in an electric furnace, gives carbides.

$$2B_2O_3 + 7C \xrightarrow{\Delta} B_4C + 6CO \ ; \quad SiO_2 + 3C \xrightarrow{\Delta} SiC + 2CO$$

9. *Formation of sulphides* (B_2S_3 and SiS_2). Both the elements combine with sulphur at elevated temperature to form sulphides. These sulphides are attacked by H_2O with the liberation of H_2S gas.

$$B_2S_3 + 6H_2O \longrightarrow 2H_3BO_3 + 3H_2S \quad ; \quad SiS_2 + 4H_2O \longrightarrow H_4SiO_3 + 2H_2S$$

10. *Formation of halides.* Halides of both the elements can be obtained by similar methods, *e.g.*

(*i*) Heated mixture of $(B_2O_3 + 3C) + 3Cl_2 \longrightarrow 2BCl_3 + 3CO$

Heated mixture of $(SiO_2 + 2C) + 2Cl_2 \longrightarrow SiCl_4 + 2CO$

(*ii*) $B_2O_3 + 6HF \longrightarrow 2BF_3 + 3H_2O$

$SiO_2 + 4HF \longrightarrow SiF_4 + 2H_2O$

(*iii*) B or Si + vapours of halogen $(X_2) \xrightarrow{\text{Burn}} BX_3$ or SiX_4.

The halides have many similar properties, *e.g.* (*a*) Halides of both the elements are covalent in nature. These act as Lewis acids and hence form adducts with donor molecules like pyridine (*py*) (*e.g.* BX_3 .*py*, SiX_4.2*py*)

(*b*) BCl_3 and $SiCl_4$.both are liquids and get hydrolysed by H_2O and form H_3BO_3 and H_2SiO_3 respectively.

$$BCl_3 + 3H_2O \longrightarrow H_3BO_3 + 3HCl \quad ; \quad SiCl_4 + 3H_2O \longrightarrow H_2SiO_3 + 4HCl$$

(*c*) BCl_3 and $SiCl_4$ both get reduced to B_2H_6 and SiH_4 respectively by $LiAlH_4$.

$$4BCl_3 + 3LiAlH_4 \longrightarrow 3LiCl + 3AlCl_3 + 2B_2H_6$$
$$SiCl_4 + LiAlH_4 \longrightarrow LiCl + AlCl_3 + SiH_4$$

(*d*) BF_3 and SiF_4 both are colourless fuming gases and get hydrolysed by H_2O to form the complex acids namely $H[BF_4]$ (called hydrofluoboric acid) and $H_2[SiF_6]$ (called fluosilicic acid)

$$4BF_3\ (excess) + 3H_2O \longrightarrow H_3BO_3 + 3H[BF_4]$$
$$3SiF_4\ (excess) + 3H_2O \longrightarrow H_2SiO_3 + 2H_2[SiF_6]$$

11. *Formation of oxy acids.* Both the elements form oxy acids (ortho acids) namely ortho boric acid (H_3BO_3) and orthosilicic acid (H_4SiO_4). These can be prepared by the action of H_2O on the sulphides of the elements.

$$B_2S_3 + 6H_2O \longrightarrow 2H_3BO_3 + 3H_2S \quad ; \quad SiS_2 + 4H_2O \longrightarrow H_4SiO_4 + 2H_2S$$

Both these acids are weak acids (K_A for $H_3BO_3 = 6.0 \times 10^{-10}$ and K_A for $H_4SiO_4 \approx 10^{-10}$) and form volatile esters namely tetraethyl ortho borate, $(C_2H_5)_3BO_3$ and tetraethyl orthosilicate, $(C_2H_5)_4SiO_4$ respectively when heated with alcohol in presence of conc.H_2SO_4.

$$H_3BO_3 + 3C_2H_5OH \xrightarrow{\text{Conc. } H_2SO_4} (C_2H_5)_3\ BO_3 + 3H_2O$$
$$H_4SiO_4 + 4C_2H_5OH \xrightarrow{\text{Conc. } H_2SO_4} (C_2H_5)_4SiO_4 + 4H_2O$$

12. *Action of alkalies on the elements:* Both the elements are attacked by fused alkalies to form borate (BO_3^{3-}) and meta silicate (SiO_3^{2-}).

$$2B + 6NaOH \longrightarrow 2Na_3BO_3 + 3H_2 \quad ; \quad Si + 2NaOH + H_2O \longrightarrow Na_2SiO_3 + 2H_2.$$

13. *Formation of borides and silicides:* Both the elements combine with metals and form borides (B^{3-}) and silicides (Si^{4-}) respectively, *e.g.*

$$2B + 3Mg \longrightarrow Mg_3B_2 \quad ; \quad Si + 2Mg \longrightarrow Mg_2Si$$

Borides and silicides both are decomposed by dil. acids like HCl to give volatile hydrides namely B_2H_6 and SiH_4 respectively.

$$Mg_3B_2 + 6HCl \longrightarrow 3MgCl_2 + B_2H_6 \quad ; \quad Mg_2Si + 4HCl \longrightarrow 2MgCl_2 + SiH_4$$

14. *Formation of boranes and silanes :* Both the elements can form their hydrides which are called boranes (*e.g.* B_2H_6, B_4H_{10} etc.) and silanes (*e.g.* SiH_4, Si_2H_6). B_2H_6 and SiH_4 can be prepared either by the action of dil HCl on Mg_3B_2 or Mg_2Si or by reducing BCl_3 or $SiCl_4$ with $LiAlH_4$. Chemical equations have already been given above. B_2H_6 and SiH_4 both are covalent in nature, and catch fire on exposure to air forming B_2O_3 and SiO_2 respectively (For chemical equation see above). These hydrides are also hydrolysed by H_2O to give H_3BO_3 and H_2SiO_3 respectively.

$$B_2H_6 + 6H_2O \longrightarrow 2H_3BO_3 + 6H_2 \quad ; \quad SiH_4 + 3H_2O \longrightarrow H_2SiO_3 + 4H_2.$$

B_2H_6 and SiH_4 both are strong reducing agents as is evident from the following reactions.

$$B_2H_6 + Cl_2 \longrightarrow C_2H_5Cl + HCl \quad ; \quad SiH_4 + Cl_2 \longrightarrow SiH_3Cl + HCl$$

15. *Formation of nitrides(N^{3-}):* Both the elements form nitrides, when N_2 gas is passed over heated elements or over heated mixture of the oxides (B_2O_3 or SiO_2) and carbon.

$$B_2O_3 + 3C + N_2 \longrightarrow 2BN + 3CO\uparrow \quad ; \quad 3SiO_2 + 6C + 2N_2 \longrightarrow Si_3N_4 + 6CO$$

Both the nitrides react with steam to liberate NH_3

$$BN + 3H_2O\ (steam) \longrightarrow H_3BO_3 + NH_3 \quad ; \quad Si_3N_4 + 9H_2O\ (steam) \longrightarrow 3H_2SiO_3 + 4NH_3.$$

16. *Formation of carbides*: Both the elements form carbides namely B_4C and SiC which can be obtained by heating the element or its oxide with carbon. For chemical equations see above. These carbides are covalent in nature. These are hard substances and hence are used as abrasives.

17. *Formation of complex acids:* Both the elements form the complex acids *viz.* $H[BF_4]$and $H_2[SiF_4]$ which can be obtained by the action of H_2O on BF_3 and SiF_4 respectively

$$4BF_3\ (excess) + 3H_2O \rightarrow H_3BO_3 + 3H[BF_4]$$
$$3SiF_4\ (excess) + 3H_2O \rightarrow H_2SiO_3 + 2H_2[SiF_4]$$

Odd Electron Bonds and Odd Electron Molecules / Ions

Introduction.

There are certain molecules and ions (*e.g.* H_2^+, He_2^+, N_2^+, O_2, O_2^-, NO, NO_2, ClO_2 etc.) which are paramagnetic. In order to explain their paramagnetic character, it has been suggested by moleculer orbital theory that these species also contain, in addition to normal covalent bonds, one or more odd-electron bonds. Odd-electron bonds are those which are formed by the sharing of odd number of electrons (*i.e.*, 1, 3, 5 or 7 electrons) between the two atoms.

Thus odd-electron bonds may be one-electron bond, three-electron bond etc. One-electron bond results by the sharing of one electron between the two combining atoms. Similarly three-electron bond is established by the sharing of three electrons between the atoms. The species having odd-electron bonds are called odd-electron species.

Examples of odd-electron molecules/ions

(*i*) *Homo-nuclear and heteronuclear diatomic molecule / ions* H_2^-, H_2^+, He_2^+, N_2^+, O_2, O_2^- and O_2^+ (homo-nuclear diatomic species) and NO and NO^- (hetero-nuclear diatomic species) are the important examples. Their structure, stability (dissociation energy), bond length, magnetic properties (number of unpaired electrons) etc. have been discussed in chapter 11. (Molecular orbital theory) and summarised in Table 10.5.

Table 10.5 : *Examples of some odd-electron molecules/ions*
(n = No. of unpaired electrons, B.O. = bond order)

Molecule / ion	B.O.	Structure	Nature and number of bonds	*n*
Diatomic molecules/ions				
H_2^- ion (3)	0.5	$[H \dots H]^-$	One 3-electron bond	1
H_2^+ ion (1)	0.5	$[H \cdot H]^+$	One 1-electron bond	1
He_2^+ ion (3)	0.5	$[He \dots He]^+$	One 3-electron bond	1
N_2^+ ion (9)	2.5	$[N \stackrel{\cdots}{=} N]^+$	Two 2-electron bonds One 3-electron bond	1
O_2 molecule (12)	2.0	$:O \stackrel{\cdots}{\underset{\cdots}{-}} O:$	One 2-electron bond Two 3-electron bonds	2
Superoxide ion, O_2^- (13)	1.5	$[:O \stackrel{\cdots}{-} O:]^-$	One 2- electron bond One 3- electron bond	1
Oxygenyl ion, O_2^+ (11)	2.5	$[:O \stackrel{\cdots}{=} O:]^+$	Two 2-electron bonds One 3-electron bond	1
NO molecule (11)	2.5	$:N \stackrel{\cdots}{=} O:$	Two 2-electron bonds One 3-electron bond	1
NO^- ion (12)	2.0	$[:N \stackrel{\cdots}{\underset{\cdots}{-}} O:]^-$	One 2-electron bond Two 3-electron bonds	2
Triatomic molecule				
NO_2 molecule (17)	3.5	$:O \stackrel{\cdots}{-} N = O:$	Three 2-electron bonds One 3-electron bond	1
ClO_2 and BrO_2 molecules (19)	3.5	$:O \stackrel{\cdots}{-} X = O:$	Three 2-electron bonds One 3-electron bonds	1
Tetratomic molecule				
ClO_3 molecule (25)	—	$\ddot{O}$: ‖ $:\ddot{O}=Cl\cdot$ ‖ $:O:$	Six 2-electron bonds	1

(*ii*) *Triatomic molecules of AO_2 type* (A = N, Cl, Br). Here we shall consider the structure of dioxides of N, Cl and Br (NO_2, ClO_2 and BrO_2). NO_2 molecule has $5 + 2 \times 6 = 17$ electrons (odd number of electrons) in its valence-shell while ClO_2 and BrO_2, both have $7 + 2 \times 6 = 19$ electrons (odd number of electrons) in their respective shell. Thus all the three molecules are odd-electron molecules. Each of these molecules has angular (V-shaped) structure. O—N—O bond angle in NO_2 molecule = 132° and O—Cl—O angle in ClO_2 molecule = 118°.

According to molecular orbital theory, each of the molecules is an odd-electron molecule, since it contains three 2-electron bonds and one 3-electron bond (See Fig. 10.25). Thus the bond order of each of the molecules is equal to 3.5. Bond order of 0.5 is due to the presence of one 3-electron bond. It is due to the presence of one unpaired electron in 3-electron bond that these molecules show paramagnetic character. In case of NO_2 molecule the unpaired electron present in 3-electron bond is *more localised* on the central N-atom and hence when two NO_2 molecules are brought in contact with each other, they combine together and form nitrogen tetroxide, $O_2N — NO_2$ or N_2O_4. The linking of two NO_2 molecules in N_2O_4 molecule takes place through N — N σ-bond. Thus NO_2 molecule has a tendency to dimerise to form N_2O_4 molecule.

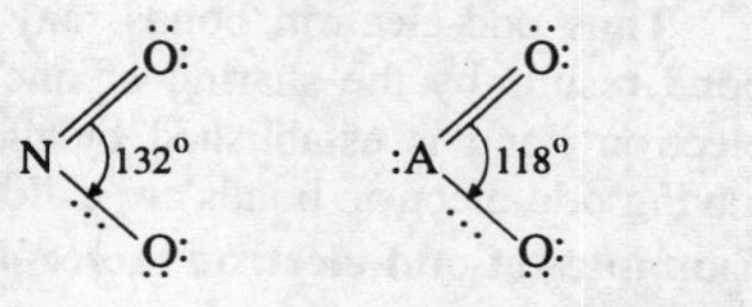

Fig. 10.25: Angular structures of NO_2 molecule [shown at *(a)*], ClO_2 and BrO_2 molecules [shown at *(b)*].

Like NO_2 molecule, ClO_2 molecule is also paramagnetic and reactive, and has marked colour. But, unlike NO_2, ClO_2 has no tendency to dimerise to form Cl_2O_4. This is perhaps due to the fact that the unpaired electron present in 3-electron bond in ClO_2 molecule is *less localised* on the central Cl-atom. The presence of 3-electron bond accounts for the shorter Cl — O bond lengths (Cl — O bond length = 1.49Å).

(iii) Tetratomic molecules: Structure of ClO_3 and Cl_2O_6 molecules. Like NO_2, ClO_2 and BrO_2, ClO_3 molecule is also an odd-electron molecule, since it contains odd number of electrons (=7 + 6 × 3 = 25) in its valence-shell. Thus ClO_3 molecule has one unpaired electron which is localised on the central Cl-atom. Due to the presence of one unpaired electron on Cl-atom, ClO_3 shows paramagnetic character and when two ClO_3 molecules come in contact with each other they combine together to form dimeric molecule, $(ClO_3)_2$ or Cl_2O_6 which can be represented as $O_3Cl—ClO_3$. Thus Cl—Cl bond in Cl_2O_6 molecule is formed by the sharing of two unpaired electrons on two Cl-atoms of two ClO_3 molecules. Thus we find that in Cl_2O_6 molecule all the electrons become paired and hence this molecule is diamagnetic. The above discussion shows that the structures of ClO_3 and Cl_2O_6 molecules can be shown as depicted in Fig. 10.26.

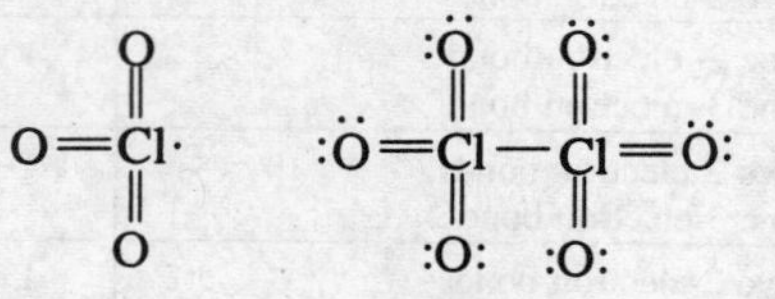

Fig. 10.26: Structure of ClO_3 [shown at *(a)*] and Cl_2O_6 [shown at *(b)*] molecules.

Properties of odd electron bonds/molecules or ions

(i) One-and three-electron bonds are weaker than a normal covalent bond (two-electron bond) and hence the bond length of one-electron bond is greater than that of two-electron bond, *e.g.* H—H bond length in H_2 molecule [H · · H or H—H] is 0.74Å while H—H bond length in H_2^+ ion, $[H \cdot H]^+$ which contains one-electron bond is 1.06Å. The bond length of a three-electron bond is intermediate between that of a double bond and a triple bond.

(ii) Odd electron bond is generally formed between the atoms which have similar or almost similar electronegativity. Odd electron bonds are not stable, if the electronegativity difference between the atoms involved is large.

(iii) Molecules having odd electron bonds are paramagnetic, coloured and very reactive, *e.g.* NO_2 is paramagnetic, brown in colour and so reactive that it dimerises to form N_2O_4. The paramagnetic character of such molecules is because of the presence of one (*e.g.* NO, N_2^+, O_2^-, O_2^+, He_2^+, NO_2, ClO_2) or two (*e.g.* NO^-, O_2)unpaired electron in three-electron bond(s).

Hydrogen Bond

What is hydrogen bond?

Hydrogen bond or hydrogen bonding is defined as the electrostatic force of attraction between

H-atom (which is already covalently bonded to a highly electronegative atom like N, O or F in a molecule) and a highly electronegative atom present in a molecule of the same substance or present in a molecule of different substance or present within the same molecule. Thus H-bond is formed either between the molecules of the same substance (see Fig. 10.27) or between the molecules of different substances (see Fig. 10.28) or between two atoms present in the same molecule (see Fig. 10.29). H-bond is represented by a dotted line.

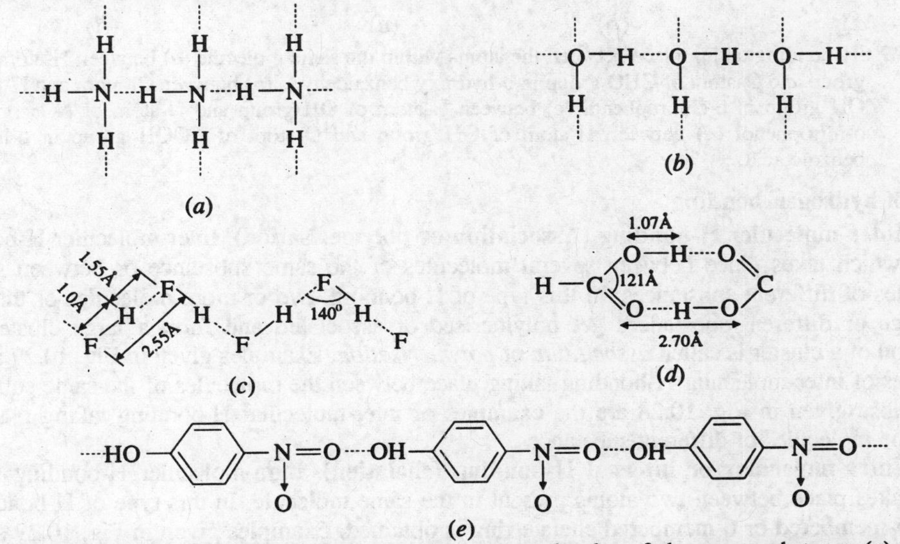

Fig. 10.27: H-bonding taking place between two or more molecules of the same substance (*a*) between several molecules of NH_3 : Formation of $(NH_3)_x$ cluster (*b*) between several molecules of H_2O: Formation of $(H_2O)_x$ cluster (*c*) between several molecules of HF: Formation of $(HF)_x$ cluster (*d*) between two molecules of formic acid (HCOOH), leading to the formation of a dimer, $(HCOOH)_2$. (*e*) between several molecules of *p*-nitrophenol.

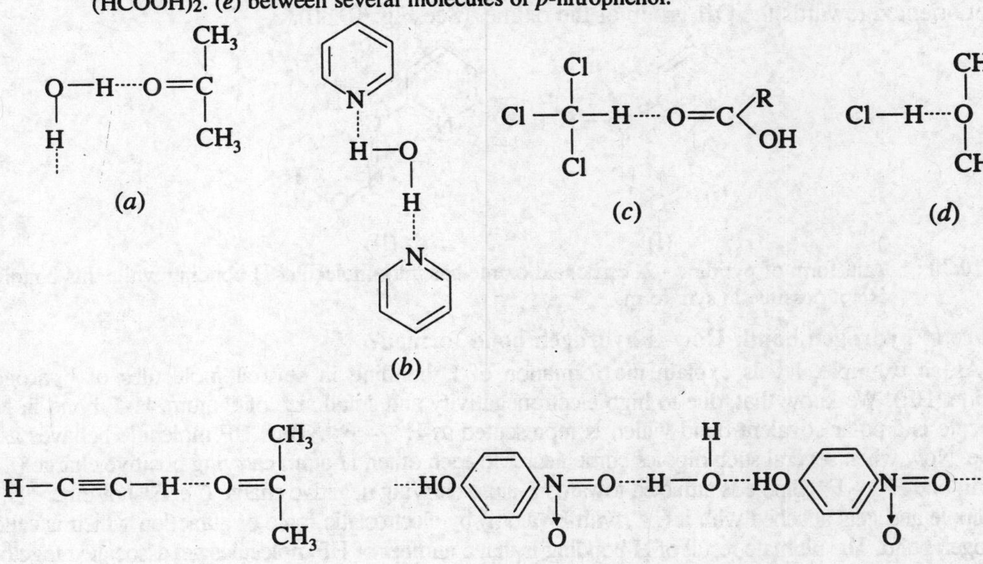

Fig. 10.28: H-bonding taking place between the molecules of different substances (*a*) between water (H_2O) and acetone ($CH_3CO.CH_3$) (*b*) between pyridine (C_5H_5N) and water (H_2O) (*c*) between chloroform ($CHCl_3$] and fatty acid (RCOOH) (*d*) between HCl and dimethyl ether ($H_3C.O.CH_3$). (*e*) between acetylene (HC ≡ C—H) and acetone ($CH_3.CO.CH_3$) (*f*) between *p*-nitrophenol (OH – C_6H_4. NO_2) and water (H_2O).

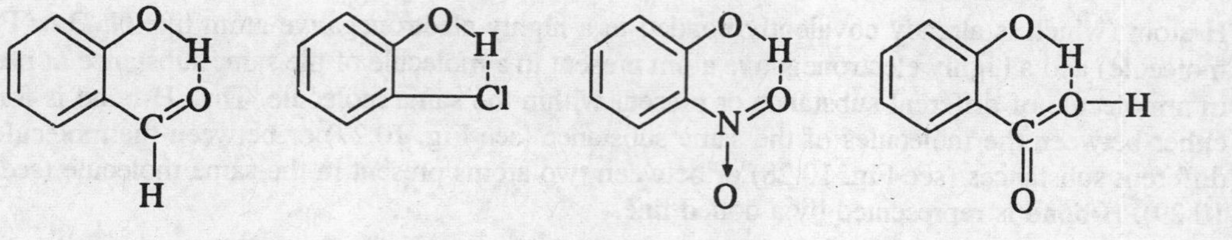

(a) (b) (c) (d)

Fig. 10.29 : H-bonding taking place between the atoms within the same molecule (*a*) between H-atom of OH group and O-atom of CHO group in o-hydroxy benzaldehyde (*b*) between Cl-atom and H-atom of OH group in o-chlorophenol (*c*) between H-atom of OH group and O-atom of NO_2 group in o-nitrophenol (*d*) between H-atom of OH group and O-atom of COOH group in o-hydroxy benzoic acid.

Types of hydrogen bonding

1. Inter-moleculer H-bonding (Association or polymerisation). Inter moleculer H-bonding is that which takes place between several molecules of the same substance or between several molecules of different substances. In this type of H-bonding, two or more molecules of the same substance or different substances get polymerised or associated and form a large cluster. The formation of a cluster is called *association* or *polymerisation*. Examples given in Fig. 10.27 are the examples of inter-molecular H-bonding taking place between the molecules of the same substance while those given in Fig. 10.28 are the examples of inter-molecular H-bonding taking place between the molecules of different substances.

2. Intra-molecular or internal H-bonding (chelation). Intra-moleculer H-bonding is that which takes place between two atoms present in the same molecule. In this type of H-bonding, a planar 5-membered or 6-membered chelate ring is obtained. Examples given in Fig. 10.29 are the examples of intra-molecular H-bonding. In this type of H-bonding, the interacting atoms should be placed in such a way that there is minimum strain during the closure of the ring. Thus the *anti* form of pyridine-2-carboxaldoxime (I) contains intra-molecular H-bonding while the *syn form* (II) can not have intra-molecular H-bonding, since the lone pair of electrons on N-atom in the pyridine ring is not oriented towards the OH group of the oxime. (See Fig. 10.30).

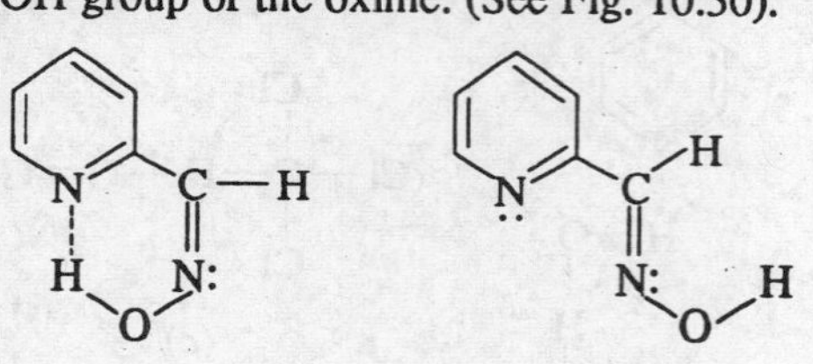

(I) (II)

Fig. 10.30 : Anti form of pyridine - 2- carboxald oxime has intra-molecular H-bonding while this bonding is not possible in syn. form.

Nature of hydrogen bond: How is hydrogen bond formed?

As an example, let us explain the formation of H-bonding in several molecules of hydrogen fluoride (HF). We know that, due to high electronegativity and small size of F-atom, H–F bond in HF molecule is a polar covalent bond which is represented as $H^{\delta+} - F^{\delta-}$. Thus HF molecule behaves as a dipole. Now when several such dipoles come nearer to each other, H-atom carrying positive charge (*i.e.*, $H^{\delta+}$) in one $H^{\delta+} - F^{\delta-}$ dipole is attracted towards F-atom carrying negative charge (*i.e.* $F^{\delta-}$) of other $H^{\delta+} - F^{\delta-}$ dipole and gets attached with it (*i.e.*, with $F^{\delta-}$ atom) by electrostatic force of attraction which is called hydrogen bond. The ultimate result of H-bonding is that a number of HF molecules get associated together and form a large cluster of molecules which can be represented as $(HF)_x$. Thus the formation of H-bonds between several HF molecules [*i.e.*, the formation $(HF)_x$ cluster] can be shown as:

$$\cdots\overset{\delta+}{H}-\overset{\delta-}{\ddot{F}}:\cdots\overset{\delta+}{H}-\overset{\delta-}{\ddot{F}}:\cdots\overset{\delta+}{H}-\overset{\delta-}{\ddot{F}}:\cdots\overset{\delta+}{H}-\overset{\delta-}{\ddot{F}}:$$

On the basis of the above discussion we can say that hydrogen bonding is simply dipole-dipole attraction. This dipole-dipole attraction is stronger than other dipole-dipole attractions.

Conditions for the formation of hydrogen bonds

Following conditions must be satisfied for the formation of H-bonds in compounds.

(*i*) *Presence of highly electronegative atom.* The molecule having H-bonds should have a highly electronegative atom like N, O or F directly linked to H-atom by a covalent bond.

(*ii*) *Presence of the atom with small size.* The highly electronegative atom should be of small size so that B—H bond (B = N, O or F) may be highly polar ($B^{\delta+}$—$H^{\delta-}$) and a strong interaction between several such dipoles may occur. This explains that, although O and Cl both have almost the same electronegativity (O = 3.5, Cl = 3.0), O-atom, due to its small size (O= 0.73Å, Cl = 0.99Å) forms H-bonds while Cl-atom does not do so. On the same lines of argument, it can be explained that O and N can form H-bonds but S and P cannot do so. Their electronegativity values and covalent radii are as N = 3.0, 0.76Å; P = 2.1, 1.06Å; O = 3.5, 0.73Å and S = 2.5, 1.02Å.

Strength of H-bond

H-bond is a weak bond, since it is merely electrostatic force of attraction and not a chemical bond. The strength of a H-bond increases with the increase of the electronegativity of the atom attached with H-atom by a covalent bond in N—H · · · N, O—H· · ·O and F—H· · ·F. Now since the electronegativity of N, O and F is in the order N < O < F, the strength of H-bond in N—H· · ·N (*e.g.* in NH_3 molecules), O–H · · · O (*e.g.* in H_2O molecules) and F—H· · ·F (*e.g.* in HF molecules) is also in the same order, *i.e.* N—H· · ·N (= 8.37KJ/mole) < O—H· · ·O(= 29.29 kJ/mole) < F—H· · ·F (= 41.85 KJ/mole). H-bond is much stronger than van der Waals forces but weaker than a covalent bond and an ionic bond. The order of strength is, therefore, as : van der Waals forces < H-bond < covalent bond < ionic bond.

Effect of hydrogen bonding on the physical properties of the compounds containing hydrogen bonds

Although H-bond is a weak bond, it affects a large number of physical properties of hydrogen bonded componds. Some of these properties are given below:

1. Variation in the boiling points of the binary hydrogen compounds of the elements of groups VA, VIA and VIIA. Molecular weights and boiling points (°C) of binary hydrogen compounds of the elements of groups VA, VIA and VIIA are given in Table 10.6.

Table 10.6 : *Variation of the boiling points (°C) of binary hydrogen compounds of the elements of groups VA, VIA and VII A.*

Group V A			Group VI A			Group VII A		
Hydride	Mol. Wt.	B.P. (°C)	Hydride	Mol. Wt.	B.P. (°C)	Hydride	Mol. Wt.	B.P. (°C)
NH_3	17	– 34.5	H_2O	18	+100.0	HF	20	+19.4
PH_3	34	– 87.5	H_2S	33	– 60.7	HCl	36.5	– 85.0
AsH_3	78	– 62.4	H_2Se	81	– 41.5	HBr	81	– 67.1
SbH_3	125	– 18.4	H_2Te	130	– 1.8	HI	128	– 35.0

We know that with the increase in the molecular weight of the hydrides in each group, the magnitude of van der Waals forces of attraction also increases in the same direction (*i.e.*, downwards). Due to the increase in the magnitude of van der Waals forces the boiling points of the hydrides should also increase downwards, *i.e.*, the boiling points should be in the order 1*st* member < 2*nd* member < 3*rd* member < 4*th* member. But we see from the table that as we move from the 1*st* member to the 2*nd* member, the boiling point, instead of increasing, decreases and then the boiling points show the expected increasing trend on moving from 2*nd* to 4*th* member.

Explanation. Why the boiling point of the 1*st* member is higher than that of 2*nd* member in each group can be explained on the basis of the fact that since N, O and F have much more electronegativity than H atom and the electronegativity values of other elements are close to that of

H-atom (H = 2.1, N = 3.0, P= 2.1; As = 2.0, Sb = 1.9, O=3.5, S = 2.5, Se = 2.5, Te = 2.1, F = 4.0, Cl = 3.0, Br = 2.8 and I = 2.5), NH_3, H_2O and HF molecules undergo inter-molecular hydrogen bonding while the occurence of this type of hydrogen bonding is absent in other molecules. It is because of H-bonding that $(NH_3)_x$, $(H_2O)_x$ and $(HF)_x$ clusters are formed. In order to break H-bonds holding NH_3, H_2O and HF molecules in their clusters more heat energy is required and hence boiling points of these molecules are higher than those of PH_3, H_2S and HCl respectively.

2. Melting points, boiling points, heat of fusion and heat of vaporisation of hydrogen-bonded compounds and their methyl derivatives. When H-atom of hydrogen bonded compounds are replaced, one by one, by CH_3 groups, we get methyl derivatives. The extent of H-bonding decreases in methyl derivatives. With the decrease in the extent of H-bonding, the melting points, boiling points, heat of fusion and heat of vaporisation for methyl derivatives (non hydrogen-bonded compounds) also decrease. This fact is evident from the melting points, boiling points, heat of fusion and heat of vaporisation of H_2O or H–OH, CH_3OH and $(CH_3)_2O$ given as : *Melting points* (°C): H_2O = 0.0, CH_3OH = –97.8, $(CH_3)_2O$ = –138.5; *boiling points* (°C): H_2O = 100.0, CH_3OH = 64.7, $(CH_3)_2O$ = –23.7; *heat of fusiom* (K J mol^{-1}): H_2O = 6.0, CH_3OH = 2.2, $(CH_3)_2O$ = 4.9; *heat of vaporisation* (KJ mol^{-1}); H_2O = 40.6, CH_3OH = 35.6, $(CH_3)O$ = 21.5.

Hydrogen-bonded compounds namely H_2SO_4, H_3PO_4 and H_3BO_3 donot boil but decompose at 340°, 213°C and 185°C respectively with the loss of water. On the other hand the methyl derivatives of these compounds *viz.* $(CH_3)_2SO_4$, $(CH_3)_3PO_4$ and $(CH_3)_3BO_3$ which are non hydrogen-bonded compounds boil at lower temperatures which are 188°C, 193°C and 65°C respectively. The pairs namely HF–CH_3F and HN_3–CH_3N_3 also show the similar trend in their boiling points (HF = 19.5°C, CH_3F = –78°C; HN_3 = 37°C, CH_3N = 20°C).

3. Comparison between the boiling points of CH_4, NH_3, H_2O and HF. The boiling points of these hydrides are in the order of : CH_4 (= –161°C) < NH_3 (= –34.5°C) < HF = (= +19.4°C) < H_2O (= +100°C).

Explanation. Due to very low electronegativity of C-atom (=2.5), there is no possibility of the formation of $(CH_4)_x$ cluster. In fact, in CH_4 several molecules are linked together by van der Waals forces which are very weak, weaker even than H-bonds. It is due to the presence of weak van der Waals forces in methane molecules that methane has the lowest boiling point. Now, due to the high electronegativity and small size N, O and F atoms, all the three molecules *viz.* NH_3, H_2O and HF have H-bonding. The way in which H-bonds are formed in these molecules as shown in Fig. 10.27, suggests that, due to the presence of one lone pair of electrons on N-atom and three H-atoms attached covalently with N-atom, each NH_3 molecule can form four H-bonds with other four NH_3 molecules. Similarly, due to the presence of the lone pairs of electrons on O-atom and two H-atoms attached covalently with O-atom, each H_2O molecule, like NH_3 molecule, can form four H-bonds with other four H_2O molecules. But, since F-atom, in HF molecule, has three lone pairs of electrons and one H-atom attached with it, each HF molecule can form only two H-bonds with two other HF molecules. Thus the strength of H-bonding, which is directly proportional to the number of H-bonds formed, in NH_3, H_2O and HF molecules should be in the order of NH_3 > H_2O > HF. Now since N-atom has the lowest value of electronegativity of all the three atoms (N = 3.0, O = 3.5, F = 4.0), the strength of H-bonding in NH_3 molecule becomes minimum of all the three molecules. Now since the strength of two H-bonds in HF molecule is less than that of four H-bonds in H_2O, the order of the strength of H-bonding in NH_3, H_2O and HF molecules is NH_3 < HF < H_2O. The boiling points of these hydrides are also in this order. Now since CH_4 has the lowest boiling point, as said above, the order of the boiling points of all the four hydrides is in the order : CH_4 < NH_3 < HF < H_2O.

4. Solubility of organic compounds. The solubility of organic compounds in H_2O is due to the formation of H-bonds between H_2O molecules and organic compounds. Alcohols are soluble in H_2O but alkanes are not: This is because of the reason that alkane molecules are not capable of forming H-bonds with H_2O molecules and hence are not miscible with H_2O. On the other hand alcohol molecules can form H-bonds with H_2O molecules and hence are miscible with H_2O.

Alcohols having alkyl groups with three carbon atoms are freely soluble in H_2O. With the increase in the number of C-atoms, the solubility decreases, *i.e.*, it is only lower alcohols (*e.g.* CH_3OH, C_2H_5OH, C_3H_7OH etc.)which are freely soluble in H_2O. $(CH_3)_2O$ forms H-bonds with H_2O (H–O–H $\cdots$ $O<\begin{smallmatrix}CH_3\\CH_3\end{smallmatrix}$) and hence is completely miscible with H_2O whereas $(CH_3)_2S$ which is not capable of forming H-bonds with H_2O is partially miscible. Benzene (C_6H_6) is partially miscible in H_2O while pyridine (C_5H_5N) is miscible in all proportions due to strong H-bond formation through the lone pair of electrons residing on N-atom in C_5H_5N molecule.

The solubility of some compounds in non-aqueous solvents like $CHCl_3$, HCl, $(CH_3)_2CO$ etc. can also be explained on the basis of H-bonding. For example (*a*) organic fatty acids (RCOOH) dissolve in chloroform ($CHCl_3$) due to the formation of H-bonds between polar H-atom of C — H bond in chloroform and O-atom of COOH group. 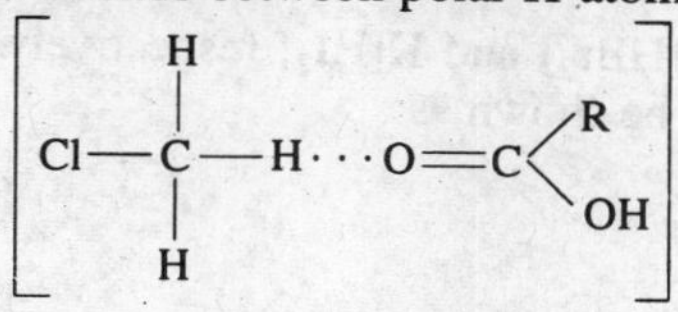 (*b*) HCl dissolves in $(CB_3)_2$ due to 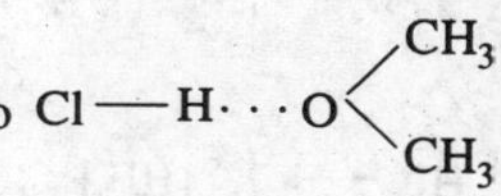 bond formation. (c) acetylene (H–C ≡ C–H) dissolves in acetone ($CH_3.CO.CH_3$) due to H—C ≡ C—H$\cdots$ $O<\begin{smallmatrix}CH_3\\CH_3\end{smallmatrix}$ bond formation. Here it should be noted that acetylene is not soluble in H_2O, since H-bonds present in H_2O molecules are stronger than those formed between H-atoms of acetylene molecule and O-atoms of H_2O molecule. Thus acetylene is not able to break the H-bonds present in H_2O molecule and hence is '*squeezed out*, *i.e.*, acetylene does not dissolve in H_2O.

5. Viscosity of liquids. Due to inter-molecular H-bonding, the attraction between molecules of hydrogen-bonded liquids is increased. This increase in the attraction decreases the tendency of the liquids to flow smoothly. For example since glycerol (CH_2OH–CHOH–CH_2OH) has three OH groups and C_2H_5OH has only one OH group, glycerol forms more number of H-bonds, *i.e.*, more glycerol molecules are associated together by inter-molecular H-bonding and hence its viscosity is greater than that of C_2H_5OH(C_2H_5OH = 5.97 millipoise, CH_2OH–CHOH–CH_2OH = 10^4 millipoise).

On similar grounds it can be explained why the viscosity of H_2O is greater than that of C_2H_5OH (H_2O = 10.05 millipoise). Similarly the viscosity of CH_2OH—CHOH—CH_2OH is greater than that of $(CH_3)_2O$ (= 2.33 millipoise). It is due to intermoleculer H-bondig between several H_2SO_4 molecules that the acid has high b.pt and is highly viscous. Its highly viscous nature makes the acid a syrupy liquid.

6. Physical state of the hydrides of the elements of groups VA, VIA and VIIA. First hydride of each of these groups is a liquid at ordinary temperature while the remaining hydrides are gases. We have already said that, since the central atom of each of the first hydride has high electronegativity and small size, these molecules unite together through inter-molecular H-bonding and form a cluster or an associated molecule. It is due to the formation of the cluster that these hydrides exist as liquids at ordinary temperature. On the other hand, since the central atom of the remaining hydrides have low electronegativity and bigger size, they are not capable of uniting together to form the cluster *i.e.*, these hydrides exist as gases under normal conditions.

7. Formation of hydrogen difluoride ion, HF_2^-. This ion is formed by the combination of F^- ion and HF molecule.

$$F^- + HF \rightarrow HF_2^-$$

In the formation of HF_2^- ion, F^- ion, because of its high electronegativity, gets attached with H-atom of HF (H—F) molecule by hydrogen bond. Thus the formation of HF_2^- ion can be shown as:

$$:\ddot{\underset{..}{F}}:^- + H—\ddot{\underset{..}{F}}: \longrightarrow \left[:\ddot{\underset{..}{F}}:\cdots H—\ddot{\underset{..}{F}}:\right]^- \text{ or } HF_2^-$$

Due to smaller electronegativity of Cl^-, Br^- and I^- ions, these ions cannot combine with H-atom of HCl, HBr and HI molecules by H-bond and hence HCl_2^-, HBr_2^- and HI_2^- ions are not formed. It is due to the existence of HF_2^- ion that HF can combine with KF (This salt is F^- ion donor) and form $K^+[HF_2]^-$ salt (acid salt). On the other hand, due to the non-formation of HCl_2^-, HBr_2^- and HI_2^- ions, HCl, HBr and HI molecules cannot combine with KCl, KBr and KI to give $K[HCl_2]$, $K[HBr_2]$ and $K[HI_2]$ respectively. The formation of $K[HF_2]$ by the combination of HF and KF can be shown as :

$$KF \longrightarrow K^+ + :\ddot{\underset{..}{F}}:^-$$

$$:\ddot{\underset{..}{F}}:^- + H—\ddot{\underset{..}{F}}: \longrightarrow \left[:\ddot{\underset{..}{F}}: \cdots H—\ddot{\underset{..}{F}}:\right]^- \text{ or } HF_2^-$$

On adding we get :

$$KF + HF \longrightarrow K^+[HF_2]^-$$

8. Structure of molecular crystals. There are many molecular crystals in which the individual molecules are connected together through H-bonding. The formation of H-bonding gives chain structure (*e.g.* linear chains in HCN, zig-zag chains in HF, CH_3OH, formic acid etc.), sheet structure (*e.g.* α-oxalic acid) or three-dimensional net work (*e.g.* KH_2PO_4) to the crystals. In some cases the crystals contain units consisting of dimeric or trimeric H-bonded species. Zig-zag structure of solid HF has already been shown at (*c*) of Fig. 10.27 and the sheet structure of α-oxalic acid is shown in Fig. 10.31.

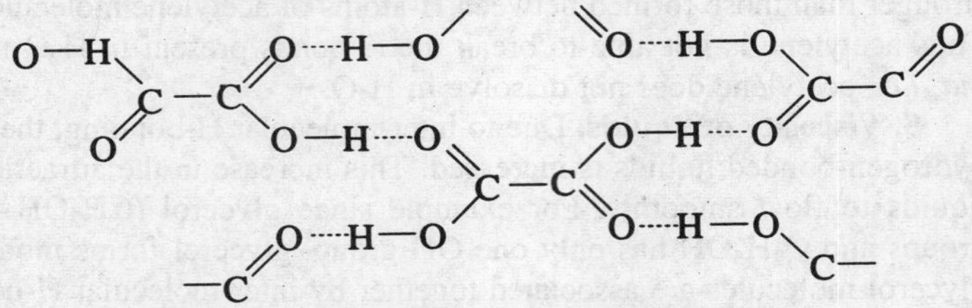

Fig. 10.31 : Sheet structure of α-oxalic acid

9. Comparison between the structures of the crystals of NH_4F, NH_4Cl, NH_4Br and NH_4I. The crystals of NH_4Cl, NH_4Br and NH_4I have CsCl structure below 184.3°C, 137.8°C and –17.6°C respectively, but above these temperatures the crystals of these halides have NaCl structure. The radius ratio value of NH_4F (= 0.92) suggests that the crystals of this molecule should have CsCl structure. Instead, NH_4F crystals have wurtzite structure (zinc sulphide structure) in which each NH_4^+ ion is tetrahedrally surrounded by F^- ions and H-bonds are formed between the F^- ions and H-atom of NH_4^+ ions. H-bonds thus formed lock NH_4^+ ions in position. Thus we see that the difference in structure of NH_4F as compared to that of other halides is due to the formation of H-bonds in NH_4F. H-bonds are formed only in NH_4F and not in other halides, since F^- has the highest electronegativity and smallest size.

10. Comparison between the properties of o, m and p-isomers of an aromatic compound. Since the two groups in m– and *p*–isomers of a given aromatic compound are far away from each other, these isomers show inter-molecular H-bonding, *i.e.*, several molecules of these isomers get polymerised or associated to form a cluster. The inter-molecular H-bonding taking place between severals molecules of *p*-nitrophenol has already been shown in Fig. 10.27 (*e*). On the other hand, in *o*-isomer the two groups are nearer to each other and hence this isomer shows intra-molecular (*i.e.*, internal) H-bonding which does not lead to the formation of any cluster as in case of *m*-and *p*-isomers. The intra-molecular H-bonding taking place in *o*-nitrophenol has already been shown in Fig. 10.29 (*c*). Due to the formation of a cluster in *m*-and *p*-isomers, these isomers have higher melting points and boiling points than the *o*-isomer. Due to higher boiling point of *m*-and *p*-isomers, these have lower volatility than the *o*-isomer. Due to greater volatility of *o*-isomer than *p*-isomer, the former can easily be separated from the latter by steam distillation.

m- and *p*-isomers are more soluble in water, since the molecules of these isomers can associate with H_2O molecules by inter-molecular H-bonding. The formation of inter-molecular H-bonding between *p*-nitrophenol and water molecules has already been shown in Fig. 10.28 (*f*). On the other hand, the formation of intra-moleculer H-bonding in *o*-isomer prevents the formation of inter-moleculer H-bonding between *o*-isomer and H_2O molecules and this results in the decrease in the solubility of *o*-isomer. All the above facts have been shown for *o*–, *m*– and *p*-nitrophenols in Table 10.7.

Table 10.7: *Physical properties of o-, m- and p-nitrophenols.*

Compound	m.pt. (°C)	b.pt (°C)	Volatility	Solubility in H_2O
o-nitrophenol	45	217	highly volatile	less soluble
m-nitrophenol	96	290	less volatile	more soluble
p-nitrophenol	114	245	non-volatile	more soluble

Other compounds in which intra-molecular H-bonding affects their melting points, boiling points. volatility and solubility as in *o*-nitro-phenol are *o*-chlorophenol and *o*-hydroxy benzoic acid.

11. Monomeric or polymeric nature of HF, HCl, HBr and HI molecules. In case of HCl, HBr and HI, H-bonding does not take place and hence these molecules are not able to form associated molecules, $(HX)_n$ where X = Cl, Br or I, *i.e.*, these molecules exist as *monomeric molecules*.

HF vapour is *monomeric* above 353 K (1 atom). However, at higher pressures and lower temperatures liquid as well as gaseous HF molecule, due to H-bonding, exists as a *polymerric* (associated) molecule, $(HF)_n$ (n = 2 to 6). X-ray diffraction study has shown that, due to H-bonding, solid HF possesses a *zig-zag* chain structure in which H–F = 1.00Å, H- - -F = 1.55Å and HFH angle = 140°. [See Fig. 10.27 (*c*)]. H-bonds have been shown by dotted lines. HCl, HBr and HI are not able to form zig-zag chains.

The anhydrous acid is probably a mixture of H_2F_2, H_3F_3, H_4F_4 and H_5F_5.

van der Waals Forces (Inter-molecular Forces)

Introduction

It is now known with certainty that there exist forces of attraction between the molecules of polar as well as non-polar substances. The magnitude of these forces is maximum in case of solids and decreases as we move form solid → liquid → gas. These forces are called *inter-molecular forces*. The existence of these forces in gases was first recognised by van der Waals and hence inter-molecular forces are also termed as van der Waals forces. The crystals whose constituent particles are held together in position by van der Waals forces are called *molecular crystals.*

Origin of van der Waals forces.

van der Waals forces existing between the molecules of polar and non-polar substances are due to the presence of different types of interactions which are discussed below:

1. Dipole-dipole interactions. These intractions are present in polar molecules. The gases like NH_3, SO_2, HF, HCl etc. are the

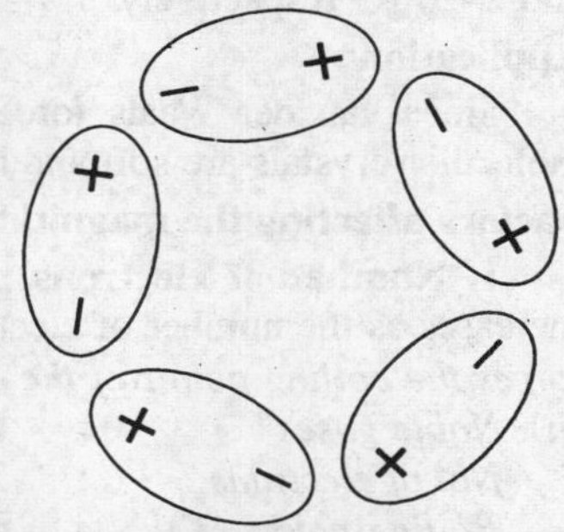

Fig. 10.32 : Dipole-dipole interaction between polar molecules.

examples of polar molecules. Although polar molecules are neutral, they have permanent dipole moments, *i.e.*, a polar molecule behaves as a dipole. van der Waals forces existing between the polar molecules are due to the electrostatic interactions between the positive end of one dipole and the negative end of the other dipole. Thus van der Waals forces between the polar molecules are due to dipole-dipole interactions (see Fig. 10.32). The magnitude of dipole-dipole interaction depends on the dipole moment of the molecule concerned. Greater is the dipole moment, greater is the magnitude of dipole-dipole interaction. It is due to the presence of dipole-dipole interactions that the gases can be liquified.

2. Instantaneous dipole-induced dipole interactions: London forces. These interactions are present in non-polar molecules. Cl_2, I_2, O_2, N_2 and mono-atomic molecules like He, Ne, Ar etc. are the examples of non-polar molecules. Non-polar molecules have no permanent dipole moments. In 1930, Fritz London explained the origin of van der Waals forces existing between non-polar molecules in the solid and liquid state with the help of quantum (wave) mehanics. We know that the electrons surrounding the nucleus in the atom of each non-polar molecule is symmetrically distributed round the nucleus. According to London, these electrons are in continuous and rapid motion with respect to the nucleus and hence it is possible that at any instant the electron density (negative charge) round the nucleus may be concentrated in one region and the nucleus (positive charge) may be concentrated in the other region of the same molecule. This results in an unsymmetrical distribution of the electron density round the nucleus and hence the non-polar molecule becomes an *instantaneous* or *temperary dipole*. When this instantaneous dipole comes closer to another non-polar molecule, the other non-polar molecule developes an *induced dipole* in it. The two dipoles *viz.*, instaneous dipole and induced dipole are then attracted towards each other by electrostatic forces which are called **London forces** (see Fig. 10.33). These forces are called **dispersive** or **dispersion forces** because the phenomenon of dispersion of light is also connected with these dipoles.

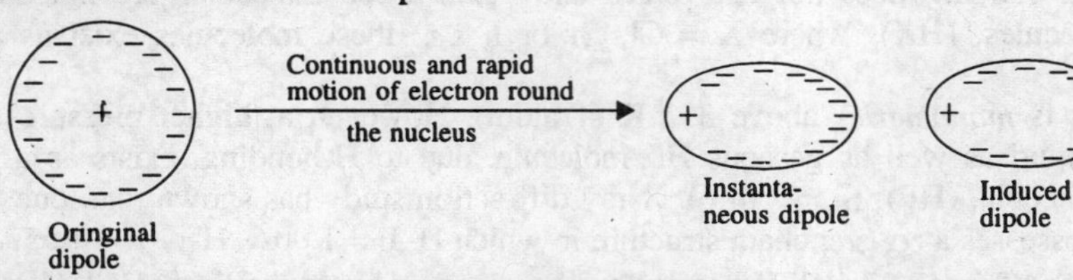

Fig. 10.33 : Instantaneous dipole–induced dipole interactions between non-polar molecules : London forces.

For a pair of adjacent molecules, London forces vary inversely as r^7 (= distance between the pair of molecules). Thus the magnitude of these forces is very small. The above discription shows that the van der Waals forces between non-polar molecules is exclusively due to London forces.

Characteristics of van der Waals forces

(*i*) These forces are only short range forces. In the simplest molecules they are important only upto 10^{-7} cm

(*ii*) These forces are much weaker than covalent and hydrogen bonds. The energies per molecule involved in covalent bond, hydrogen bond and van der Waals attraction are 200–8400, 10–40 and 2–20 KJ respectively.

Applications

Since van der Waals forces are easily overcome, the condensed gas readily vaporises and molecular crystals are soft and have low melting points.

Factors affecting the magnitide of van der Waals forces

1. Number of electrons present in a molecule. The magnitude of van der Waals forces increases as the number of electrons in the *molecules increase. With the increase in van der Waals forces the boiling points of the molecules also increase. For example:*

(*a*) *Noble gases*	:	He	Ne	Ar	Kr	Xe	Rn
No. of electrons	:	2	10	18	36	54	86
Boiling points (K)	:	4.1	27	67.2	121	165	211.2

(b) Molecules	:	CH_4	SiH_4	GeH_4
No. of electrons	:	10	18	36
Boiling points(K)	:	112	161	183

2. Molecular size or molar mass. Larger is the molecular size or roughly larger is the molar mass, the stronger is the van der Waals force and hence boiling point also increases. Thus boiling point of ethane, C_2H_6 is higher than that of methane, CH_4 (C_2H_6 = 185K, CH_4 = 112K). Similarly the boiling point of C_2F_6 is higher than that of CF_4 (C_2F_6 = 195K, CF_4 = 145K).

Metallic Bond

See "*Metallic Crystals*" discussed in the chapter on "*Structure of Solids*".

Universtiy Questions

Ionic bond and electrovalency

1. What is an ionic bond? What are the factors favouring an ionic bond formation? *(Meerut 94)*
2. Write a note on "Elecctovalency and variable electrovalency". Exaplain your answer with examples. *(Agra 92, Kanpur, 92)*
3. Under what conditons does ionic valency pass into co-valency? How are the properties of ionic compounds affected by lattice energy? *(Allahabad 1991)*
4. Explain the following
   (*i*) Ionic compounds are normally solids and soluble in polar solvents, having high dielectric constant. *(Manipur 81)*
   (*ii*) NaCl is electrolyte but CCl_4 is not? *(Utkal 81, Punjabi 82)*
   (*iii*) Ionic compounds, like solid NaCl, does not conduct electricity, whereas molten (fused) NaCl does. *(Venkateswara, 81; Agra 88)*
   (*iv*) Electrovalent compounds are mostly solids and have higher m.pts. than covalent compounds. *(Manipur, 81; Kanpur 85)*
   (*v*) Ionic compounds melt at a higher temperature as compared to covalent compounds.

Inert pair effect

5. What is inert pair effect?. Discuss this effect with special reference to VA group elements *(ECC Allahabad 94, Meerut 90, 93, Agra 92)*
6. (*a*) What is inert pair effect and how is the variable valency of Pb and Tl explained on this basis? *(Lucknow 95, Avadh 93)*
   (*b*) How does inert pair effect affect the chemical nature of some of the elements? Give examples. *(Lucknow 95, Meerut 96)*
7. Explain the follwing:
   (*i*) Thallous compounds are more stable than thallic compounds
   (*ii*) Boron forms no compound in the unipositive state while Tl^+ compounds are quite stable. *(M.D. Rohtak 1984)*
   (*iii*) The lower oxidation states become more stable progressively as we descend down in group IIIA and IVA of the periodic table. *(Delhi 91)*
   (*iv*) Boron forms BCl_3 and not BCl. *(Raj. 86)*
   (*v*) Tl(+1) compounds are more stable as compared to Tl(+3) compounds. *(Meerut 94)*
   (*vi*) Tin forms divalent compounds more easily than silicon.
   (*vii*) On descending the group IVA the stability of +IV oxidation state decreases and that of +II oxidation state increases. *(Delhi B.Sc. Hons, 1986, Punjab 86, Pune 86)*
   (*viii*) PbO_2 is an oxiding agent while SnO_2 is not. *(Agra 85, Raj. 85)*
   (*ix*) Sn^{4+} is not more stable than Sn^{2+} whereas Pb^{2+} is more stable than Pb^{4+}
   (*x*) Ge(II) is a strong reducing agent and Ge (IV) is not. Pb(IV) is an oxidising agent and Pb(II) is not. *(Delhi Hons., 1980)*
   (*xi*) $PbCl_2$ is more stable than $PbCl_4$ *(Delhi 1976)*
   (*xii*) Compounds of Bi(+III) are more stable than Bi(+V). *(Meerut 96)*
   (*xiii*) Which of the following halides is least stable and has doutful existence?
   (*a*) CI_4 (*b*) GeI_4 (*c*) SnI_4 (*d*) PbI_4 *(IIT 96)*

Covalent bond, covalency and rule of maximum covalency

8. Explain the meaning of covalent bond. Illustrate with examples. *(Agra 92; Meerut 92)*

9. What are the factors favouring the formation of a covalent bond between two atoms A and B as against ionic bond. *(Bahuguna 93)*

10. Differentiate between (*a*) ionic bond and covalent bond (*b*) polar and non-polar covalent bond, giving examples. *(Meerut 94, 96)*

11. Explain why covalent compounds are often gases, liquids or solids with low melting points. *(E.C.C. Allahabad 94)*

12. Give the reaction, if any, of H_2O on $SiCl_4$ and CCl_4. Why do they differ in their behaviour? *(Himachal 1986)*

13. Which has greater reactivity and why SF_6 or TeF_6? *(Punjabi 83; Punjab 86)*

14. (*a*) Explain the meaning of variable covalency.

(*b*) How would you account for it? What is the maximum covalency that an element can exhibit. *(Kanpur 83)*

15. Explain the following, giving appropriate reasons:

(*i*) $SiCl_4$ or SiF_4 is easily hydrolysed by H_2O but CCl_4 or CF_4 is not, although Si—Cl bond is stronger than C—Cl. *(Raj., 87, Allahabad 87, 94, Meerut 88; Delhi Hons, 86, 87)*

OR

Saturated carbon compounds are not reactive while the corresponding silicon compounds are reactive. *(Agra 85; Raj. 85)*

(*ii*) CCl_4 does not act as Lewis acid while $SiCl_4$ and $SnCl_4$ do so. *(Lucknow 94)*

(*iii*) PCl_5 is stable while NCl_5 cannot be prepared. *(Gauhati 91)*

OR

N-atom does not form pentahalides whereas P-and Sb both form penta halides *(Gorakhpur, 84, 88; Delhi, 80, Meerut 91)*

OR

P, As and Sb exhibit 5-covalent state whereas N-atom exhibits only 3-covalent state. *(Himachal 84)*

(*iv*) NF_3 is known but NF_5 is not known. *(Delhi 84)*

(*v*) OF_6 does not exist while SF_6 exists. *(Punjab 85; Raj. 86)*

(*vi*) $[CF_6]^{2-}$ ion does not exist while $[SiF_6]^{2-}$ ion exists.

(*vii*) $[NF_6]^-$ ion does not exist but $[PF_6]^-$ ion exists.

(*viii*) Boron does not show a covalency beyond four. *(Delhi Hons. Sub., 86)*

(*ix*) The formula of telluric acid is H_6TeO_6 or $Te(OH)_6$. The analogous sulphuric acid, $S(OH)_6$ does not exist. *(Delhi Hons. 87)*

(*x*) SF_6 is inert whereas SF_4 hydrolyses even though the average bond energy in SF_4 is greater than that in SF_6. *(Pune 86; Delhi Hons., 87)*

(*xi*) Although the central atom in the hexafluorides SF_6, SeF_6 and TeF_6 belongs to the same group, it is only TeF_6 which can be hydrolysed by H_2O; SF_6 and SeF_6 both remain unreactive. *(E.C.C. Allahabad 94)*

(*xii*) Chlorine atom has only one unpaired electron, even then it shows the covalency of 1, 3, 5 and 7. *(Agra 88)*

Singlet linkages

16. Explain the formation of singlet linkages in chemical compounds giving suitable examples. *(Meerut 90)*

Valence bond theory

17. Give an account of valence bond theory of bonding in H_2 molecule. Point out some main limitations of valence bond theory. *(E.C.C. Allahabad 94)*

18. Explain the structure of the following pairs of molecules in terms of valence bond theory:

(*a*) CO_2 and SO_2 (*b*) CCl_4 and $SiCl_4$ (*c*) BF_3 and NH_3 (*d*) H_2O and H_2S. *(Meerut M.Sc., 84)*

Heitler-London (electron pair) theory

19. Explain fully Heither-London theory and give important characteristics of covalent compounds. *(Agra 92)*

20. Explain fully Heitler-London theory and limitations of valence bond theory. *(Meerut 92)*

Pauling Slater's Theory

21. Discuss in brief the Pauling Slater's theory of valency. *(Meerut 90)*

Lewis structures.

22. Write down the electronic structures of N_2H_4, PCl_5, NH_2OH, H_3PO_4 and N_3H. *(Jodhpur 90)*

23. Draw the structures of the following : SF_4, H_6TeO_6, $(NPF_2)_3$, $2SbF_5$, ICl_4^-, P_4S_7. *(Delhi Hons. 89)*

Sigma and pi-bonds

24. What is pi bond? Explain the formation of pi bond in ethylene and acetylene molecules. *(Pondichery 92)*

25. Differentiate between sigma and pi-bonds. Why is a σ bond stronger than a π bond. *(Gauhati 91; Meerut 93)*

Polar and non-polar covalent bonds

26. Explain electronegativity. How is the percentage character of a bond related with the electronegativity difference of the bonded atoms? *(Avadh 93)*

27. Arrange the following bonds in the increasing order of their covalent character : C—Cl, B—Cl, N—Cl, Cl—Cl and B—F. *(Lucknow 94)*

28. Write a note on :

(*i*) Polar and non-polar covalent bond. *(Agra, 91, Meerut 91, 93)*

(*ii*) Percentage ionic character of covalent bonds. *(Bahuguna 93)*

29. Explain partial ionic character of covalent bond with the help of suitable examples. *(Meerut 89, 91)*

30. How has the concept of electronegativity been used to predict the nature of a bond. *(Meerut 92)*

31. The electronegativity values of Sn and Cl are 1.8 and 3.0 respectively. Predict the nature of bond between the two atoms *(Delhi 89)*

Hint : Since the electronegativity difference between the two atoms is large, these atoms will form ionic bond between them.

32. Calculate the % ionic character in Cs—F bond in CsF molecule. The electronegativity values for Cs and F are 0.7 and 4.0 respectively. Predict the nature of CsF molecule. *(Delhi 89)*

33. Explain the following:

(*i*) CsF is the most ionic compound. *(Raj PMT 1979)*

(*ii*) HCl in the gaseous state is covalent while in the aqueous solution it is ionic.

Answer. Due to the electronegativity difference between H and Cl atoms, these atoms in the gaseous HCl acquire partial positive and negative charges and become $H^{\delta+}$ and $Cl^{\delta-}$ and consequently H — Cl covalent bond becomes polar bond ($H^{\delta+} - Cl^{\delta-}$). Now when gaseous HCl molecule is dissolved in H_2O, the polar molecule of H_2O

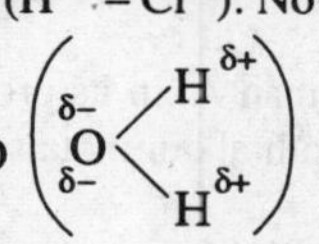

surround the $H^{\delta+}$ and $Cl^{\delta-}$ ions and split $H^{\delta+}$— $Cl^{\delta-}$ bond into $H^{\delta+}$ and $Cl^{\delta-}$ ions. Thus aqueous solution of HCl is ionic.

(*iii*) Cs—F bond in CsF is more ionic than Li—F bond in LiF, yet the melting point of CsF is less than that of LiF. *(Meerut M.Sc. 93)*

(*iv*) HCl in the gaseous state is covalent but in the aqueous solution it is ionic.

Dipole moment

34. What is dipole moment? What is debye? Describe one method to measure the dipole moment. How is it useful in the elucidation of molecular structure? *(M.D. Rohtak 90)*

35. (*a*) What is dipole moment? Give any two methods for its determination.

(*b*) "Dipole moment values are sometimes used to distinguish between *cis* and *trans* isomers". Substantiate this statement. *(Pondichery 92)*

36. Explain the following :

(*i*) Dipole moment of BF_3 is zero. *(Saurashtra 92)*

(*ii*) The dipole moment of NH_3 is greater than that of NF_3. *(Saurashtra 92)*

(*iii*) SO_2 has dipole moment while CO_2 does not. *(Delhi 87)*

(*iv*) Dipole moments of CO_2 and SO_2 molecules are zero and 1.60D respectively. *(Karela 87)*

(*v*) Dipole moment of CH_4 is zero. *(Madras 85)*
(*vi*) The dipole moment of CH_4 is zero but CH_2Cl_2, CH_3Cl and $CHCl_3$ have some dipole moment.
(*vii*) H_2O molecule is a polar molecule.
(*viii*) CH_3Cl has dipole moment, although it has a planar structure (symmetrical structure).
(*ix*) PF_3 has no dipole moment but IF_5 has.
(*x*) BF_3 has no dipole moment but NF_3 has. *(Delhi Hons., 77)*
(*xi*) $BeCl_2$ has zero dipole moment while H_2S has some value of dipole moment.

Answer : $BeCl_2$ has zero dipole moment, since this molecule has symmetrical shape (linear shape). On the other hand, since H_2S has angular shape, it has some value of dipole moment.

(*xii*) Carbon — oxygen bond is polar but CO_2 molecule is non-polar. *(Delhi 89)*
(*xiii*) Water has a dipols moment of 1.85 debye while the dipole moment of CO_2 molecule is zero. *(Kalyani 1991; Madurai Kamraj 1987)*
(*xiv*) The dipole moments of HF, HCl, HBr and HI decrease in the order :
HF (= 1.91D) > HBr (= 1.80 D) > HCl (= 1.08 D) > HI (= 0.42 D) *(Kerela 87)*

37. Arrange the following molecules in the increasing order of their dipole moment values (Give reasons): CH_4, CH_3Cl, CH_2Cl_2, $CHCl_3$ and CCl_4. *(Kalyani 91)*

38. The dipole moment of KCl is 3.336×10^{-29} coulomb metre which indicates that it is highly polar molecule. The inter-atomic distance between K^+ and Cl^- in this molecule is 2.6×10^{-10}m. Calculate the dipole moment of KCl molecule, if there were opposite charges of one fundamental unit located at each nucleus. Calculate the percentage of ionic character of KCl. *(IIT 93)*
[Ans : *4.165×10^{-29} cm, 80%***]**

39. Calculate the percentage of ionic character in HCl molecule when the bond distance in thus molecule is 1.275 and the observed dipole moment value is 1.03 D. *(Bharathidasan 92)*

Diagonal relationship

40. Expalin what you understand by "Diagonal Relationship". *(Allahabad 91; Lucknow 93; Agra, 92; Meerut 94)*

41. Explain, with examples and reasons, the diagonal relationship between the following pairs:
(*i*) Li and Mg *(Meerut 90; Allahabad 91)*
(*ii*) Be and Al *(Agra 92)*
(*iii*) B and Si *(Meerut 94)*

42. Explain the following:
(*i*) Li shows greater resemblance with Mg than with other alkali metals. *(Meerut 1981, 82, 83, 84, Lucknow 81. 82)*
(*ii*) Boron shows similarities with Si *(Jodhpur 85)*

OR

Boron is more similar to silicon than Al in its properties. *(Agra 89)*
(*iii*) Li does not show resemblance with alkali metals and shows similarities with Mg. *(Raj, 87, 87 supp.)*
(*iv*) Li shows diagonal relationship with Mg. *(Agra 84; REE 85)*
(*v*) Some elements show diagonal relationship in the periodic table *(Jodhpur 85)*
(*vi*) Li resembles Mg in its chemical properties even then it is placed with alkali metals. *(Delhi B.Sc. Hons., 86; Agra 88)*
(*vii*) Be resembles more with Al than with other alkaline earth metals. *(Punjab 85)*

43. Explain the following:
(*i*) Bond length of covalent bonds in SO_2 and SeO_2 are much shorter than expected. *(Delhi 87)*
(*ii*) C — C bond length decreases in the order C_2H_6, C_2H_4 and C_2H_2. *(Raj, 79)*
(*iii*) Bond length between Si and Cl is shorter than that between Si and Br. *(Gauhati 91)*

44. State whether the following statements are true or false.
(*a*) C—C bond is longer than C = C bond. (True)
(*b*) Bond lengths decrease in the order : C—C > C = C > C ≡ C. (True) *(MLNR 1983)*

45. The C—C and Si—C inter-atomic distances are 1.54 and 1.88 Å respectively. Calculate the atomic radius of Si-atom. **[Ans.** *1.11***]**

Bond energy

46. Which of the following pairs of species would you expect to have higher bond energy? (*i*) F_2 , F_2^+

(ii) NO, NO^- (iii) BN, BO (iv) NF, NO (v) Be_2, Be_2^+. **[Ans : (i)]**

47. Write a note on "*Bond energy and average bond energy*". *(Raj. 87)*

48. Discuss the factor which affect bond energy.

49. Explain the meaning of (*a*) Bond energy (*b*) Bond dissociation energy (*c*) Average bond energy.

50. Give Pauling's formula used for calculating energy of A—B bond in AB molecule.

51. State whether the following statement is true or false. The dissociation energy of C = C bond is less than of C—C bond. *(MLNR 1983)*
[Ans: False]

Bond angle

52. Explain, the following, giving proper reasons for your answer;

(i) Bond angles in CH_4, NH_3 and H_2O molecules decrease from 109.5° to 107.3° to 104.5°. *(Raj. 79)*

(ii) OSO bond angle in SO_2 molecule is less than 120°.

(iii) Bond angle in PH_3 is smaller as compared to that in NH_3.

(iv) Bond angle in OF_2 is less than that in OCl_2. *(Delhi 81, 82; Raj, 82; Avadh 83)*

(v) Bond angles decrease from H_2O, H_2S, H_2Se and TeH_2. *(GND 1985)*

(vi) Bond angle is CH_4 is 109.5° and in H_2O it is 104.5°. *(Raj. 87)*

(vii) Although O and S are the member of the same group of the periodic table, HOH in H_2O is 104° 27′ while HSH in H_2S is only 92°. *(Gorkhpur 1984; Rohelkhand 91)*

(viii) Though in the molecules of NH_3 and H_2O there is sp^3 hybridisation, even then the bond angles are 107° and 104.5° respectively. *(Agra 88)*

(ix) HXH (X = O, S, Se) bond angles in the hydrides H_2O, H_2S and H_2Se are 104.27°, 92.16° and 91° respectively. *(Delhi B.Sc. Hons., 1986)*

(x) HNH and HOH bond angles in NH_3 and H_2O molecules are 106.45° and 104.37° respectively. *(Raj.B.Sc. 84)*

(xi) Bond angle in the hydrides NH_3, PH_3 & AsH_3 decreases as : $NH_3 > PH_3 > AsH_3$. *(Lucknow. 94)*

53. Give the bond angles of the following BCl_3, H_2O, NH_3, BeF_2, C_2H_2, C_2H_4 and CH_4 *(Kanpur 80)*

[**Ans**: BCl_3 = 120°, H_2O = 105°, NH_3 = 107°, BeF_2 = 180°, C_2H_2 = 180°, C_2H_4 = 180°, C_2H_4 = 120°, CH_4 = 109°]

54. Arrange the following in the increasing order of bond angle. Give reasons for your answer. PCl_3, PBr_3 and PI_3. *(Delhi 89)*

55. Which of the following molecules has the highest bond angle : H_2O, CO_2, SO_2 & NO_2 *(Ravishankar 93)*

Coordinate bond

56. Write a note on "Lewis acid-base concept of coordinate bond". *(Agra 92)*

Resonance

57. Write a note on:

(i) Resonance in inorganic compounds *(Agra 92; Gauhati 91)*

(ii) Resonance energy *(Bahuguna 93; Gauhati 91)*

Fajans' rules

58. Write short note on.

(i) Fajans' rules of covalency, *(Meerut 91, Gauhati 91, Agra 92; E.C.C Allahabad 94; Bahuguna 93)*

(ii) Polarisation of covalent molecules *(Agra 92)*

(iii) Polarisability *(Kanpur 93)*

(iv) Polarisation *(Meerut 93)*

(v) Polarising power of ions. *(Meerut 91)*

59. (*a*) How does the polarising power of cations vary in a period and a group of the long form of the periodic table? Illustrate your answer with suitable examples and appropriate reasons.

(*b*) Place the halide ions in the increasing order of their polarisability. *(Rohilkhand 91)*

60. Describe the factors favouring the formation of covalent bond in reference to Fajans' rules. *(Ravishankar 93)*

61. Discuss the factors affecting the polarising power of ions. *(Kanpur 93)*

62. On the basis of Fajans' rules explain which compound of each of the following pairs is more covalent CuO or CuS, AgCl or NaCl, LiCl or KCl, $SnCl_2$ or $SnCl_4$ and $AlCl_3$ or $GaCl_3$. *(Meerut 91)*

63. What is meant by polarisation of ions? Explain the variation of polarisability of an anion by a given cation in a period and in a group. *(Gauhati 91)*

64. Explain the following, giving appropriate reason:

(*i*) LiCl is not soluble in H_2O but soluble in organic solvents like alcohol. *(Raj. 85)*

(*ii*) Lithium salts are slightly covalent while sodium salts are ionic. *(Meerut 84)*

(*iii*) (*a*) LiCl is covalent
(*b*) LiCl is more covalent than NaCl. *(Kurukhetra, 82; Bundelkhand, 88)*

(*iv*) In spite of the fact that Li is an alkali, lithium halides are covalent in nature. *(Raj. 85)*

(*v*) Li forms many covalent compounds. *(Raj. 84)*

(*vi*) LiCl is soluble in organic solvents while the chlorides of other alkali metals are not. *(Meerut 83)*

(*vii*) LiI is soluble in water while LiF is not. *(Jodhpur 87)*

(*viii*) Beryllium forms many covalent compounds. *(Raj. 85)*

(*ix*) Beryllium halides are covalent while magnesium halides are ionic. *(Raj. 86)*

(*x*) The m.pt. of NaCl is 1073 K while that of $AlCl_3$ is 463K. *(Bharthiar 86)*

(*xi*) NaCl is ionic but NaI is predominantly covalent.

(*xii*) NaCl melts at a much higher temperature than $AlCl_3$.

(*xiii*) $AlCl_3$ is essentially covalent whereas AlF_3 is predominantly ionic. *(Delhi 87)*

(*xiv*) AlF_3 is a high melting solid while $AlCl_3$ is a low melting volatile solid. *(M.D.Rohtak 84)*

(*xv*) Boron has a tendency to form covalent compounds. *(Delhi 87, 89)*

(*xvi*) The boiling points of the halides of Al are : AlF_3 = 1270°C, $AlCl_3$ = 180°C, $AlBr_3$ = 265°C, AlI_3 = 382°C. *(Kurukshetra 81, 82)*

(*xvii*) $PbCl_4$ is essentially covalent while $PbCl_2$ is largely ionic.

(*xviii*) The solubility of $CaCO_3$ is less than that of K_2CO_3.

(*xix*) The m.pt. of $HgCl_2$ is less than that of $CaCl_2$ ($HgCl_2$ = 276°C, $CaCl_2$ = 722°C)

(*xx*) Silver halides (*e.g.* AgF, AgCl, AgBr) are more covalent than the potassium halides (KF, KCl, KBr)

(*xxi*) The solubility of silver halides decreases in the order : AgF AgCl AgBr AgI *(Raj. 86 S)*

(*xxii*) The ionic compounds having highly charged cations and anions are insoluble in water. *(Raj. 86 S)*

(*xxiii*) Boron does not form ionic compounds, although its first ionisation energy may suggest the formation of B^+ ion. *(Raj. 86)*

(*xxiv*) (*a*)Oxides are more covalent than fluoride.
(*b*) CuCl is more covalent than NaCl.
(*c*) Covalent character decreases from $BeCl_2$ to $RaCl_2$.

(*xxv*) AgCl is white in colour whereas AgI is yellow. *(Meerut 91)*

(*xxvi*) Polarising power of Cu^+ ion is greater than that of Na^+ ion. *(Gauhati 91)*

Odd-electron bond

65. Write a note on "Odd electron compounds and odd electron bonds". *(Rohilkhand 91, Agra 92)*

Hydrogen bond

66. Write a note on

(*i*) Hydrogen bond and its applications
(Meerut, 90; Kanpur, 93; Bahuguna, 93; Gauhati, 91;Avadh, 93; Agra, 92)

(*ii*) Different classes of hydrogen bond. *(E.C.C. Allhabad 94)*

(*iii*) Inter-molecular and intra-molecular hydrogen bond. *(Nagpur 93)*

67. Give a brief account of hydrogen bonding in inorganic compounds. *(Gauhati91)*

68. How does hydrogen bonding affect the viscosity of a liquid? *(Nagpur 93)*

69. Explain the following:

(1) NH_3 has higher b.pt than PH_3 *(I.I.T. 1989)*

(2) NH_3 is highly soluble in water.

Answer: NH_3 is highly soluble in H_2O, since with H_2O molecules, NH_3 molecules combine through H-bonding and give rise to the formation of (I) and (II).

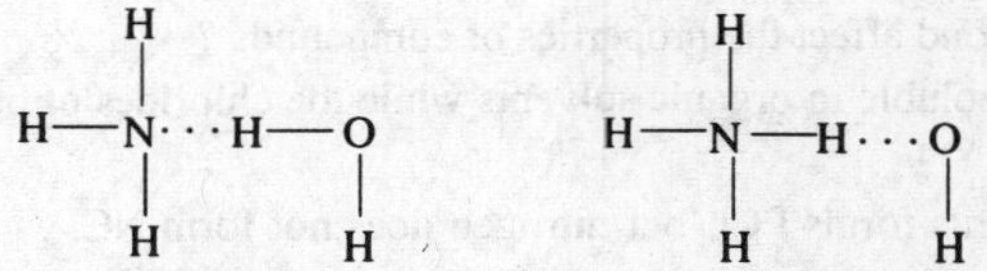

(I) (II)

Aqueous solution of NH_3 contains 46.2% of uncombined ammonia, 52.4% of hydrated amonia (H_3N ......H—O—H) and 1.4% of NH_4^+ and OH^- ions.

(3) NH_4OH is a weak base in aqueous solutions. *(Meerut, 84S)*
(4) NH_3 has exceptionally high m.pt. and b.pt. as compared to those of the hydrides of the remaing elements of group VA. *(Delhi, 85)*
(5) H_2O is a liquid while H_2S is a gas. *(Gauhati, 91; Ravishankar, 93)*
(6) H_2O boils at much higher temperature than H_2S. *(Meerut, 91)*
(7) Boiling point of H_2O is higher than that of H_2Se.
(8) Boiling point of wter is higher than that of other hydrides of the elements of group VIA. *(Andhra, 86)*
(9) H_2O and HF have abnormally high boiling points. *(Bahuguna, 93)*
(10) Boiling point of HF is higher than that of HCl. *(Ravishankar, 93)*
(11) Hydrofluoric acid is polymeric while other halogen acids have monomeric formula, HX (X = Cl, Br, I).
(12) A chlorine derivative corresponding to KHF_2 is not possible. *(Calcutta B.Sc., 86)*
(13) Anhydrous fluoride is a liquid whereas as other halogen acids are gases at room temperature. *(Meerut, 84)*
(14) F^- ion can combine with HF to form HF_2^- ion but other halide ions (Cl^-, Br^- and I^-) can not combine with HCl, HBr, and HI molecules to form HCl_2^-, HBr^- and HI_2^- ions.
(15) Acid salts only of hydrofluoric acid are known.
(16) HF has the highest b.pt. of all the halogen acids.
(17) Why does HF have abnormally high b.pt. ?

70. What are limitations of valence bond theory ? *(Nagpur 2002)*

71. Define the terms : (*i*) Bond length (*ii*) Bond energy. *(Nagpur 2002)*

72. Explain why H_2O is a liquid and H_2S is a gas. *(VBS Purvanchal 2002)*

73. Explain why CsF is the most ionic. *(VBS Purvanchal 2002)*

74. Write a short note on Fajan's rules. *(Nagpur 2002)*

75. Write a note on Dipole Moment. *(Lucknow 2002)*

76. (*a*) What is intra-molecular Hydrogen Bonding ?
(*b*) Discuss the effect of hydrogen bonding on m. pt. and b. pt. of the substance. *(Nagpur 2002)*

77. Write a detailed note on : (*i*) odd-electron molecules (*ii*) Van der Waal's forces. *(Lucknow 2002)*

78. Draw the figures for p-p axial and p-p heteral overlaps. *(Nagpur 2002)*

79. What is an inert pair effect ? *(Nagpur 2002)*

80. What is an odd electron molecule ? Draw the structure of NO_2 molecule. *(Nagpur 2002)*

81. (*a*) Explain with proper reasoning, why NF_3 has zero dipole moment but NH_3 has high dipole moment.
(*b*) Methane (CH_4) is very slightly soluble in water, but methanol is highly soluble. Explain.
(*c*) Why is triple bond shorter than a double bond ?
(*d*) Why is boiling point of NH_3 higher than that of PH_3 ?
(*e*) Write a note an Metallic Bond. *(Delhi 2002)*

82. According to Fajan's theory, the tendency towards covalency is dependent on ionic size and charge. Discuss the usefulness of this theory by giving suitable examples. *(Delhi 2002)*

83. (*a*) Explain why beryllium compounds are covalent.
(*b*) Explain diagonal relationship of Be with Al.
(*c*) What is inert pair effect ? Explain with reference to Sn and Pb compounds. *(Delhi 2002)*

84. What are the essential conditions for the formation of ionic bond ? *(VSB Purvanchal 2003)*

85. How does hydrogen bond affect the properties of compounds ? *(VBS Purvanchal 2003)*

86. Explain why Li Cl is soluble in organic solvents while the chlorides of other alkali metals are not. *(VBS Purvanchal 2003)*

87. Explain why phosphorus forms PCl_5 but nitrogen does not form NCl_5. *(VBS Purvanchal 2003)*

88. (*a*) What is the percentage ionic character in HCl bond if the electronegativities of H and Cl are 2.1 and 3.5 e.v. respectively ?

(*b*) Write short note on Hydrogen Bond.

(*c*) Explain : (*i*) AlF_3 is ionic while $AlCl_3$ is covalent (*ii*) At ordinary temperature H_2O is liquid, but H_2S is a gas. *(Rohilkhand 2003)*

89. (*a*) What is meant by polarising power of a cation and polarizability of an anion ?

(*b*) TI (III) compounds are oxidising agents while Ga (I) compounds are reducing agents. Explain.

(*c*) What do you mean by odd electron bond ? Explain giving two examples.

(*d*) What is an ionic bond ? Give conditions for the formation of ionic bonds. *(Rohilkhand 2003)*

90. Explain the bonding in H_2, molecule on the basis of valence bond theory. Obtain the following relation :

$\psi_{V.B} = \psi_{covalent} + \lambda\psi_{ionic}$. *(Gujrat 2003)*

91. (*a*) How are Be and Al are related with each other ? Explain giving suitable example S.

(*b*) Write note on (*i*) Fajan's rules (*ii*) Metallic bond.

(*c*) What kind of forces must be overcome to : (*i*) boil water (*ii*) melt NaCl (*iii*) melt iron (*iv*) sublime dry ice. *(Delhi 2003)*

Answer (*c*) (*i*) Vender Waals' forces of attraction (*ii*) Ionic electrostatic forces (*iii*) Molecular bonding forces (*iv*) Binding energy.

92. Explain why $AlCl_3$ is covalent but AlF_3 is not. *(Delhi 2003)*

93. (*a*) What is polarisation ? Write a note on Fajan's rules. Which of the following cations will have greater polarising power ? (*i*) Na^+ or Mg^{2+} (*ii*) Cu^{2+} or Ca^{+2}.

(*b*) Define the terms : bond order and bond energy.

(*c*) What are the limitations of valence bond theory ?

(*d*) Draw potential energy diagram for formation of hydrogen molecule. *(Nagpur 2003)*

94. (*a*) What is hydrogen bonding ? How does it affect the melting point and boiling point of the compounds ?

(*b*) What is metallic bond ? Explain free electron theory of metals.

(*c*) Define : (*i*) Bond length, and (*ii*) Bond energy. *(Nagpur 2003)*

95. (*a*) Discuss the effect of intermolecular and intramolecular hydrogen bonding on boiling point of a liquid.

(*b*) What are odd electron molecules ? Give the structure of NO_2 molecule.

(*c*) Apply VSEPR theory to explain why the H–N–H bond angle in NH_3 is 107° and the F–Cl–F bond angle in ClF_3 is 87°. *(Nagpur 2003)*

96. Define resonance. What are important rules for writing resonance structure ? *(Nagpur 2002)*

97. How is the percentage ionic character of a bond related with the electronegativity of the bonded atoms ? Calculate the percentage ionic character of HCl, if the electronegativities for H and Cl are 2.1 and 3.0 respectively. *(VBS Purvanchal, 2002)*

98. An electronegative element X has outer most electronic configuration $3s^2 3p^5$ and an electropositive element Y has outer most configuration $4s^1$. (*i*) Name both elements (*ii*) Write position in the periodic table (*iii*) What would be the nature of the two elements : metallic or non-metallic ? (*iv*) What type of bond will be formed, if X and Y combine ? *(VBS Purvanchal 2002)*

99. (*a*) What is electronegativity and how does it differ from electron affinity ?

(*b*) How is electronegativity related with the type of bond formed between atoms. *(VBS Purvanchal 2003)*

100. Explain why $PbCl_2$ is more stable than $PbCl_4$. *(Bhopal 2004)*

101. (*a*) What is inert pair effect?

(*b*) Arrange the following compounds in order of their increasing covalent character : NaF, AlN, MgO and SiC. *(GND Amritsar 2004)*

11

Molecular Orbital Theory of Simple Diatomic Molecules and Ions

Introduction

Molecular orbital theory (MOT) is another approach to explain the formation of covalent bonds in molecules or ions. This theory was mainly developed by Hund and Mulliken in 1932 and later by Lennerd Jones and Coulson. This theory is also sometimes called **Hund-Mulliken theory.** This theory is based on the *Linear Combination of Atomic Orbitals* (LCAO) of the atoms constituting the molecule or ion and hence is also called **LCAO–MO theory.**

According to this theory, all the electrons in a given molecule or ion are considered to be present in molecular orbitals. Thus these electrons move under the influence of all the nuclei of all the constituent atoms making the molecule or ion.

Salient Features of Molecular Orbital Theory

(*i*) When two isolated atomic orbitals (or their wave functions) residing on two atoms and having similar energy and same symmetry combine linearly with each other by linear combination of atomic orbitals (LCAO) method (see below), two molecular orbitals are obtained. The two atomic orbitals combining linearly may be of the same two atoms of a homo-nuclear diatomic molecule or ion (*e.g.*, H_2, H_2^+, H_2^-, N_2, O_2, O_2^+, O_2^-, O_2^{2-}, etc.) or may be of two different atoms of a hetero-nuclear diatomic molecule or ion (*e.g.*, CO, CN^-, NO, NO^+, NO^- etc.).

When three atomic orbitals of similar energy and same symmetry combine with one another, three molecular orbitals are obtained. Thus we can conclude that the number of molecular orbitals obtained is always equal to the number of atomic orbitals combining together.

(*ii*) *Two different ways of linear combination of atomic orbitals (or their wave functions) and formation of bonding and antibonding molecular orbitals.* Suppose that the wave functions of two atomic orbitals on two atoms A and B of AB type molecule (hetero-diatomic molecule) are represented as ψ_A and ψ_B respectively. There are two different ways in which ψ_A and ψ_B can combine linearly with each other. These two different ways are discussed below:

(*a*) *Additive overlap (also called positive overlap or ++ overlap) and formation of bonding molecular orbital.* In this type of linear combination, the positive lobe (*i.e.*, the lobe having + sign) of ψ_A overlaps with the positive lobe of ψ_B and a molecular orbital is formed. This molecular orbital has lower energy than each of the two isolated atomic orbital wave functions *viz.* ψ_A and ψ_B (decrease in energy) and hence the formation of this molecular orbital produces attraction between the two nuclei of A and B atoms. This attraction results in the establishment of a stable chemical bond between A and B. Since this molecular orbital leads to the formation of a stable bond, it is called *bonding molecular orbital* which is represented as ψ^b.

(*b*) *Subtractive overlap (also called negative or + – overlap) and formation of anti bonding* *molecular orbital.* In this type of linear combination the positive lobe of ψ_A overlaps with the negative lobe (*i.e.,* the lobe having – sign) of ψ_B and a molecular orbital is formed. This molecular orbital has higher energy than each of the two atomic orbital wave functions (ψ_A and ψ_B) (increase in energy) and hence the formation of this molecular orbital produces repulsion between the two nuclei of A and B. This repulsion opposes the formation of any bond between the nuclei. Since this molecular orbital opposes the formation of any bond between the nuclei, it is called *antibonding molecular orbital* which is represented as ψ^*.

Here it should be understood clearly that + and – signs of the lobes are only geometric signs of the atomic orbital wave function and should not be confused with positive (+) and negative (–) charges.

Above discussion shows that the formation of bonding molecular orbital wave function (ψ^b) and anti-bonding molecular orbital wave function (ψ^*) by the linear combination of two atomic orbital wave functions *viz.*, ψ_A and ψ_B can be represented by the following equations:

$$\psi^b = +\psi_A + \psi_B \quad (++\text{ overlap}) \qquad ...(i)$$

$$\psi^* = +\psi_A - \psi_B \quad (+-\text{ overlap}) \qquad ...(ii)$$

When we show the contributions made by ψ_A and ψ_B in ψ^b and ψ^* molecular orbitals, equations (*i*) and (*ii*) are written as:

$$\psi^b = \frac{1}{\sqrt{2}}(\psi_A + \psi_B) \qquad ...(iii)$$

$$\psi^* = \frac{1}{\sqrt{2}}(\psi_A - \psi_B) \qquad ...(iv)$$

Obviously the sum of the squares of the coefficients of atomic orbital wave functions ψ_A and ψ_B appearing in the bonding molecular orbital wave function is equal to unity $\left[\left(\frac{1}{\sqrt{2}}\right)^2 + \left(\frac{1}{\sqrt{2}}\right)^2 = 1\right]$. Similarly this sum for antibonding molecular orbital is also equal to unity.

The relative order of the energy of ψ_A, ψ_B, ψ^b and ψ^* can be shown pictorially (See Fig. 11.1).

The above discussion can be summarised by stating that when two atomic orbital wave functions having similar energy and same symmetry combine together linearly, two molecular orbital wave functions are obtained. One of these molecular orbital wave functions is bonding and the other is antibonding. Bonding molecular orbital wave function has lower energy than antibonding orbital wave function.

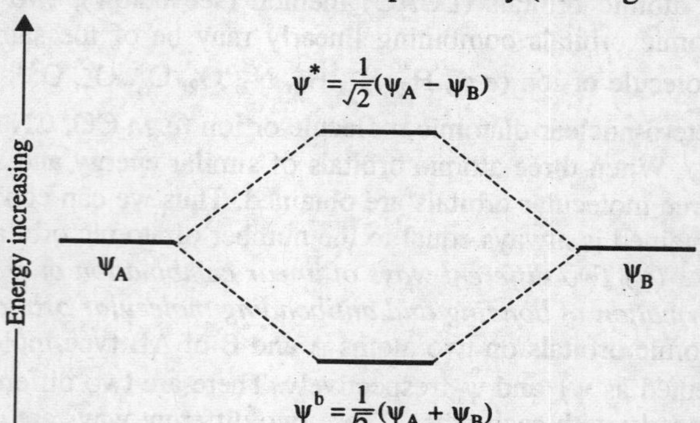

Fig. 11.1 : Relative order of the energy of ψ_A, ψ_B, ψ^b and ψ^*.

Variation of electron charge density with inter-nuclear distance in H_2 molecule

For the sake of simplicity let the two H-atoms of H_2 molecule be designated as H_A and H_B. Let the wave functions of 1*s* orbital of these atoms be represented as ψ_{H_A} and ψ_{H_B}. Obviously the positive overlap of ψ_{H_A} and ψ_{H_B} will give bonding molecular orbital, $\psi^b_{H_2}$ and the negative overlap of ψ_{H_A} and ψ_{H_B} will produce antibonding molecular orbital, $\psi^*_{H_2}$. Thus

$$\psi^b_{H_2} = \psi_{H_A} + \psi_{H_B} \qquad ...(i)$$

$$\psi^*_{H_2} = \psi_{H_A} - \psi_{H_B} \qquad ...(ii)$$

On squaring equation (*i*), we get:

$$\left(\overset{b}{\psi}_{H_2}\right)^2 = \psi_{H_A}^2 + \psi_{H_B}^2 + 2\,\psi_{H_A}\,\psi_{H_B} \qquad \text{... }(iii)$$

The probability of finding the electron in a particular region or at a particular point is called *electron probability density* and is directly proportional to the *electron charge density.* In equation (*iii*) $\left(\overset{b}{\psi}_{H_2}\right)^2$ denotes electron probability density or electron charge density in the bonding molecular orbital $\left(\overset{b}{\psi}_{H_2}\right)^2$ while $\psi_{H_A}^2$ and $\psi_{H_B}^2$ indicate the electron charge density in the isolated (*i.e.,* uncombined) atomic orbitals *viz.* ψ_{H_A} and ψ_{H_B}. It may be seen from equation (*iii*) that $(\overset{b}{\psi}_{H_2})^2$ is greater than $(\psi_{H_A}^2 + \psi_{H_B}^2)$ by an amount equal to $2\psi_{H_A}\psi_{H_B}$. This means that the electron charge density in the bonding molecular orbital $(\overset{b}{\psi}_{H_2})$ is greater by $2\psi_{H_A}\psi_{H_B}$ than the sum of the electron charge densities of the isolated atomic orbitals, ψ_{H_A} and ψ_{H_B}. The excess or increase in electron charge density $(=2\psi_{H_A}\psi_{H_B})$ occurs in the region lying in between the two positively charged nuclei. The increase in electron charge density between the two positive nuclei shields (*i.e.*, screens) the two nuclei from mutual repulsion, *i.e.,* due to the increase in electron charge density, the two nuclei are attracted towards each other. Thus the formation of bonding molecular orbital produces attraction between the two nuclei and hence leads to the establishment of a stable chemical bond. The attraction between the nuclei (*i.e.*, the formation of a stable bond) leads to a decrease in the energy of the bonding molecular orbital in the event of its being occupied by electrons. In other words, the bonding molecular orbital has lower energy than each of the isolated atomic orbitals from which this molecular orbital is derived (See Fig. 11.2). At this stage it should be easy to understand that since $\overset{b}{\psi}_{H_2}$ leads to the formation of a state bond, it is called bonding molecular orbital.

On squaring equation (*ii*), we get:

$$\left(\overset{*}{\psi}_{H_2}\right)^2 = \psi_{H_A}^2 + \psi_{H_B}^2 - 2\,\psi_{H_A}\,\psi_{H_B} \qquad \text{...}(iv)$$

In this equation $(\overset{*}{\psi}_{H_2})^2$ represents the electron charge density in the antibonding molecular orbital $(\overset{*}{\psi}_{H_2})$. As is evident from equation (*iv*), $\left(\overset{*}{\psi}_{H_2}\right)^2$ is smaller than $(\psi_{H_A}^2 + \psi_{H_B}^2)$ by an amount equal to $2\psi_{H_A}\psi_{H_B}$. This means that electron charge density $\left(\overset{*}{\psi}_{H_2}\right)^2$ the antibonding molecular orbital $(\overset{*}{\psi}_{H_2})$ is smaller by $2\psi_{H_A}\,\psi_{H_B}$ than the sum of the electron charge densities of the uncombined atomic orbitals, ψ_{H_A} and ψ_{H_B}. The decrease in the electron charge density $(= 2\psi_{H_A}\,\psi_{H_B})$ occurs in the region lying in between the two positively charged nuclei. The decrease in electron charge density between the nuclei means that there is no shielding of the nuclei, *i.e.*, due to the decrease in electron charge density between the nuclei, these are repelled from each other. Thus the formation of antibonding molecular orbital produces repulsion between the two nuclei and hence opposes the formation of any bond between the nuclei. The repulsion between the nuclei leads to an increase in the energy of the antibonding molecular orbital in the event of its being occupied by electrons. In other words, the antibonding molecular orbital has higher energy than each of the isolated atomic orbitals from which this molecular orbital is derived (See Fig. 11.2). Since $\overset{*}{\psi}_{H_2}$ molecular orbital opposes the formation of any bond, it is called antibonding molecular orbital.

A filled bonding molecular orbital of H_2 molecule has high electron density in the region lying between the nuclei H_A and H_B while a filled antibonding molecular orbital has zero electron density at the middle point of the internuclear distance between H_A and H_B. This fact can be illustrated by drawing a graph between the internuclear distance between H_A and H_B and the electron charge densities in the isolated atomic orbitals (*i.e.*, $\psi_{H_A}^2$ and $\psi_{H_B}^2$), bonding molecular orbital $\left[\left(\psi_{H_2}^{b}\right)^2\right]$ and antibonding molecular orbital $\left[\left(\psi_{H_2}^{*}\right)^2\right]$. This graph is called *electron density diagram.* The variation of electron charge density in bonding molecular orbital $\left[\left(\psi_{H_2}^{b}\right)^2\right]$ and in isolated atomic orbitals $[(\psi_{H_A})^2$ and $(\psi_{H_B})^2]$ with internuclear distance between the nuclei of H_A and H_B has been shown by solid and broken lines respectively in Fig. 11.2 (*a*). It is clear from this figure that in bonding molecular orbital the electron charge density or probability of finding an electron is maximum in the region lying between the nuclei of H_A and H_B atoms. It is due to this maximum electron density that the two nuclei are attracted towards each other and a stable bond is established between them.

The variation of electron charge density in antibonding molecular orbital $\left[\left(\psi_{H_2}^{*}\right)^2\right]$ and in isolated atomic orbitals $[(\psi_{H_A})^2$ and $(\psi_{H_B})^2]$ with internuclear distance has been shown by solid

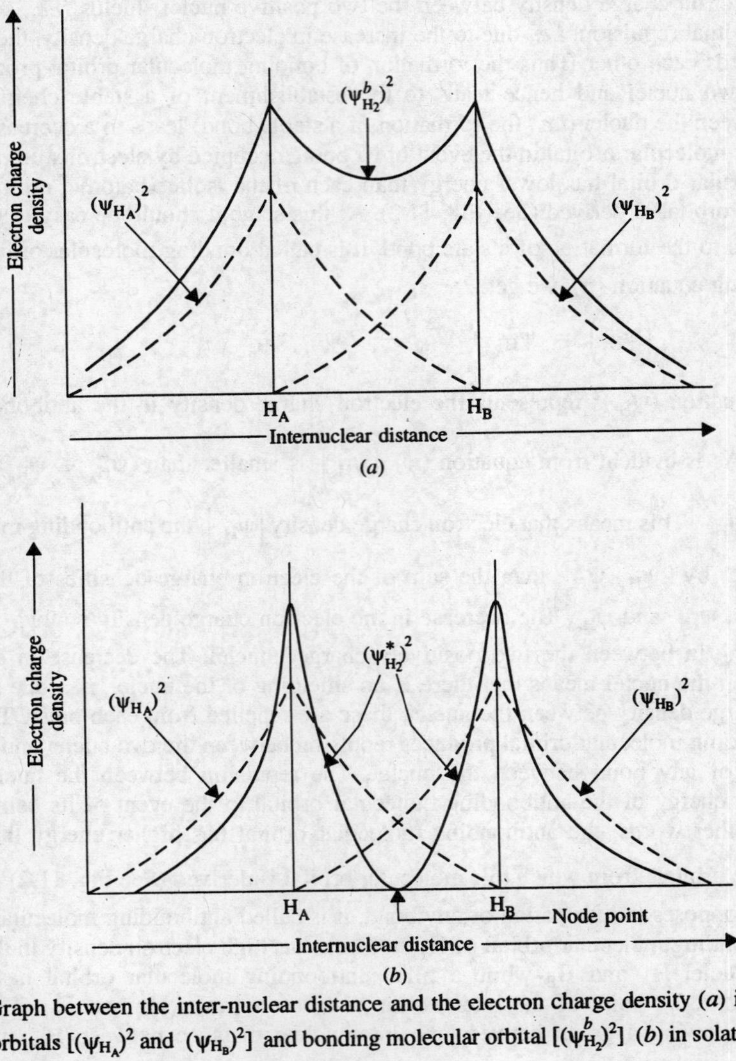

Fig. 11.2 : Graph between the inter-nuclear distance and the electron charge density (*a*) in isolated atomic orbitals $[(\psi_{H_A})^2$ and $(\psi_{H_B})^2]$ and bonding molecular orbital $[(\psi_{H_2}^{b})^2]$ (*b*) in solated atomic orbitals and antibonding molecular orbital $[(\psi_{H_2}^{*})^2]$.

and broken lines respectively in Fig. 11.2 (*b*). It can be seen from this figure that the probability of finding the electron decreases in the region between the nuclei and becomes zero at the mid-point of the internuclear distance. Now the two nuclei repel each other and hence no bond is formed. The middle point at which the electron density is zero is called **node point.**

Molecular orbitals formed by the combination of two *s*-orbitals

When two *s*-orbitals overlap with each other, two sigma molecular orbitals are obtained. One of these is sigma bonding molecular orbital (designated as σ_s^b) and the other is sigma antibonding molecular orbital (represented as σ_s^*). σ_s^b molecular orbital is obtained when the lobe of one *s*-orbital with + sign overlaps with that of the other *s*-orbital with + sign. Such type of overlap of two atomic orbitals is called + + overlap or addition of orbitals. σ_s^* molecular orbital results when the lobe of one *s*-orbital with + sign overlaps with that of the other *s*-orbital with – sign. Such type of overlap of two atomic orbitals is called + – overlap or subtraction of orbitals. When + + overlap takes place, we say that the wave functions of the two *s*-orbitals are being added and hence + + overlap is also called addition overlap. + – overlap occurs when on *s*-orbital with + sign is added to another *s*-orbital with a – sign or when one *s*-orbital with + sign is subtracted from another *s*-orbital with + sign. The formation of σ_s^b and σ_s^* molecular orbitals has been shown in Fig. 11.3.

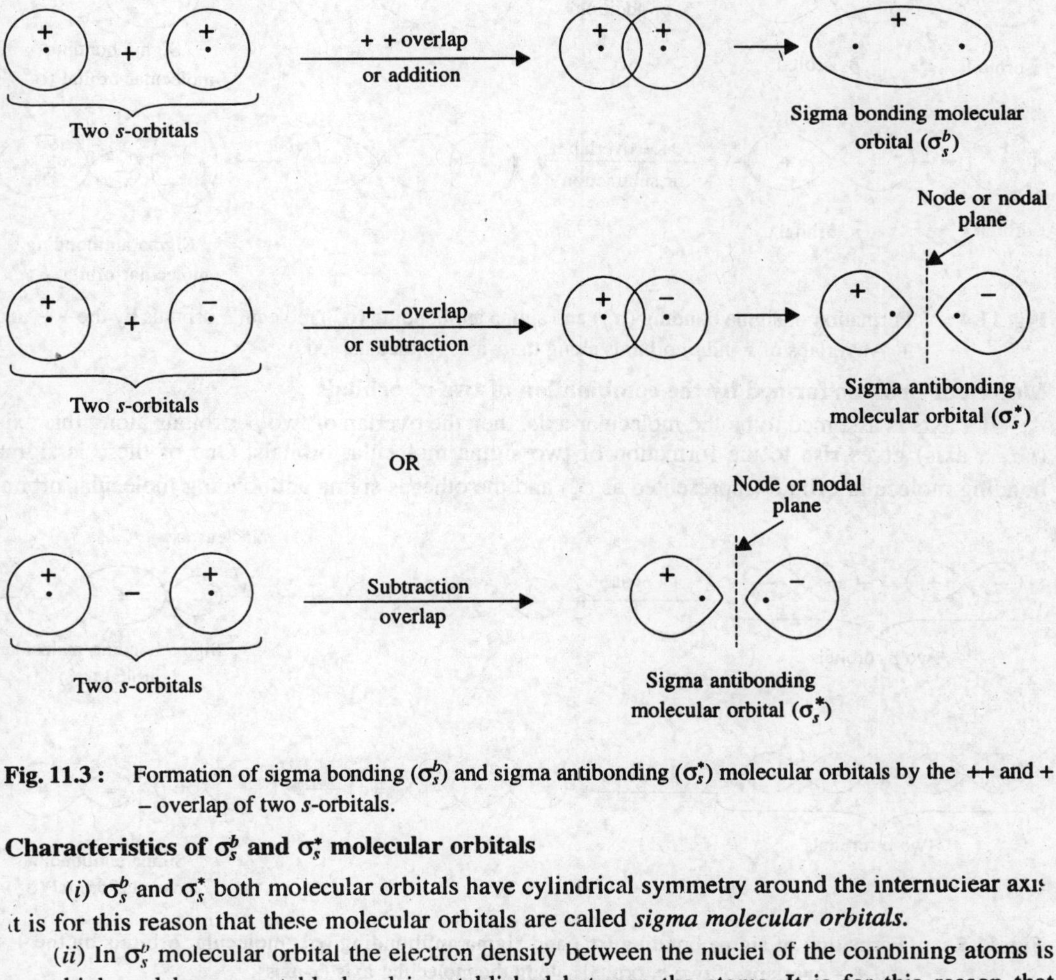

Fig. 11.3 : Formation of sigma bonding (σ_s^b) and sigma antibonding (σ_s^*) molecular orbitals by the ++ and + – overlap of two *s*-orbitals.

Characteristics of σ_s^b and σ_s^* molecular orbitals

(*i*) σ_s^b and σ_s^* both molecular orbitals have cylindrical symmetry around the internuclear axis it is for this reason that these molecular orbitals are called *sigma molecular orbitals.*

(*ii*) In σ_s^b molecular orbital the electron density between the nuclei of the combining atoms is very high and hence a stable bond is established between the atoms. It is for this reason that

σ_s^b molecular orbital is called *bonding* sigma molecular orbital. On the other hand, in σ_s^* molecular orbital the electron density between the nuclei is negligible and hence the nuclei fly apart. Thus the formation of σ_s^* molecular orbital does not lead to the formation of a stable bond between the combining atoms. It is why σ_s^* molecular orbital is called *antibonding* sigma molecular orbital.

(*iii*) The energy of σ_s^b molecular orbital is less than that of both the combining *s*-orbitals while the energy of σ_s^* molecular orbital is greater.

(*iv*) σ_s^b molecular orbital has no node while σ_s^* molecular orbital has one node.

Molecular orbitals formed by the combination of *s* and p_x orbitals

If *x*-axis is assumed to be the molecular axis (molecular axis is also called internuclear axis or bond axis), then the overlap of one *s* and one p_x orbital along *x*-axis gives rise to the formation of two molecular orbitals, one of which is sigma bonding molecular orbital (represented as σ_{sp}^b) and the other is sigma antibonding molecular orbital (represented as σ_{sp}^*). σ_{sp}^b is obtained when + + overlap takes place and σ_s^* is obtained when + – overlap occurs (See Fig. 11.4). These molecular orbitals are formed in HF, HCl etc., molecules.

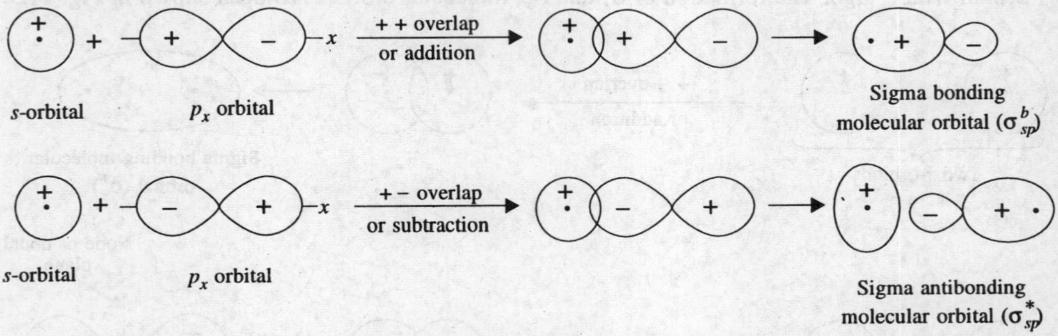

Fig. 11.4 : Formation of sigma bonding (σ_{sp}^b) and sigma antibonding (σ_{sp}^*) molecular orbitals by the + + and + – overlaps of *s* and p_x orbitals along the *x*-axis (molecular axis)

Molecular orbitals formed by the combination of two p_x orbitals

If *x* axis is assumed to be the molecular axis, then the overlap of two p_x orbitals along this axis (*i.e.*, *x*-axis) gives rise to the formation of two sigma molecular orbitals. One of these is sigma bonding molecular orbital (represented as σ_x^*) and the other is sigma antibonding molecular orbital

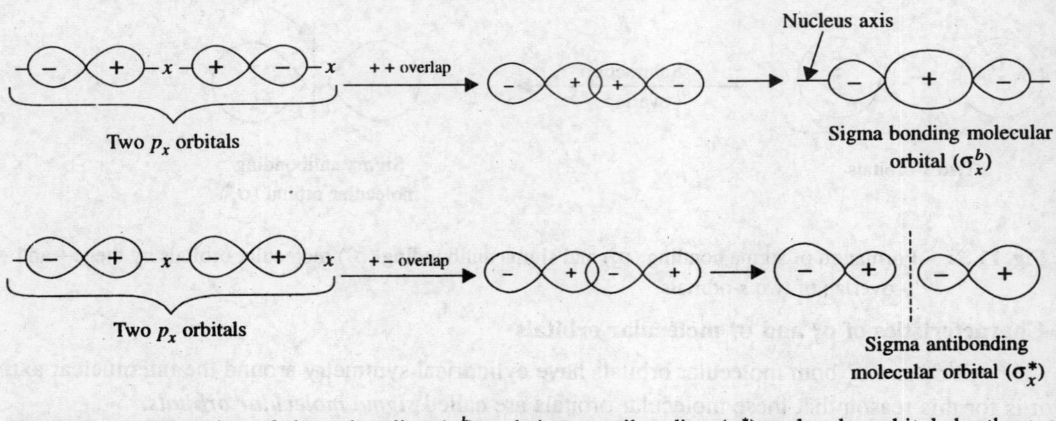

Fig. 11.5 : Formation of sigma bonding (σ_x^b) and sigma antibonding (σ_x^*) molecular orbitals by the + + and + – overlaps of two p_x orbitals along the molecular axis (*x*-axis).

(represented as σ_x^*). σ_x^b molecular orbital is obtained when + + overlap between two p_x orbitals takes place along x-axis while σ_x^* molecular orbital results when + – overlap takes place along x-axis (see Fig. 11.5).

The overlap between two s-orbitals, between s-and one p_x orbital and between two p_x orbitals along the molecular axis is called *head-to-head, head-on, end-on* or *linear overlap.*

Characteristics of σ_x^b and σ_x^* molecular orbitals

Since σ_x^b and σ_x^* molecular orbitals, like σ_s^b and σ_s^* molecular orbitals are also sigma type molecular orbitals, σ_x^b and σ_x^* molecular orbitals have the same characteristics as σ_s^b and σ_s^* molecular orbitals have.

Molecular orbitals formed by the combination of two p_y or two p_z orbitals

When two p_y or two p_z orbitals overlap together along a line perpendicular to the molecular axis, two pi molecular orbitals are formed in each case. One of these molecular orbitals is pi bonding molecular orbital and the other is pi antibonding molecular orbital. For example when two p_z orbitals overlap together, two pi molecular orbitals are obtained. One of these is pi bonding moleculer orbital (π_z^b) and the other is a pi antibonding molecular orbital (π_z^*). π_z^b moleculer orbital is obtained when the lobes of the two p_z orbitals with + + sign overlap (+ + overlap) along the line perpendicular to the molecular axis (See Fig. 11.6). σ_z^* molecular orbital is obtained when the lobes of two p_z orbitals with + – sign overlap (+ – overlap) along the line perpendicular to the molecular axis (Fig. 11.6).

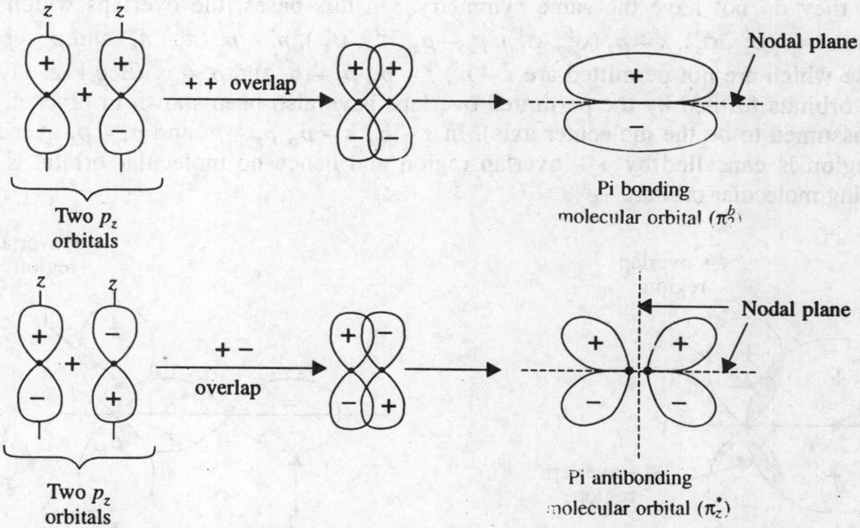

Fig. 11.6 : Formation of pi bonding (π_z^b) and pi antibonding (π_z^*) molecular orbitals by the + + and + – overlap of two p_z orbitals along a line perpendicular to the nuclear axis (x-axis).

Similarly when two p_y orbitals overlap together in two different ways (+ + overlap and + – overlap), two pi molecular orbitals (π_y^b and π_y^*) are obtained. π_y^b is called bonding pi molecular orbital while π_y^* is called antibonding pi molecular orbital.

The overlap between two p_y or between two p_z atomic orbitals along a line perpendicular to the molecular axis is called *side-to-side, sidewise, sideway* or *lateral* overlap.

Characteristics of π-molecular orbitals

(*i*) Pi molecular orbitals are not symmetrical about the molecular axis and hence are not sigma molecular orbitals.

(*ii*) The electron density in π_z^b or π_y^b molecular orbital is zero on a plane that contains the nuclear axis (nodal plane). The electron density is concentrated in the region above and below this nodal plane. π_z^* or π_y^* molecular orbital has the least energy in the region between the nuclei of the two atoms.

(*iii*) π_y^b and π_z^b molecular orbitals have the same energy and hence are called degenerate orbitals. Similarly π_y^* and π_z^* are also degenerate orbitals.

Necessary conditions to be satisfied by atomic orbitals to combine together to form molecular orbitals

The necessary conditions which are essential to enable atomic orbitals to overlap effectively together to form molecular orbitals are discussed below:

(*i*) The combining atomic orbitals must have the same or almost the same energy. On this basis, in the formation of a homonuclear diatomic molecule $1s$–$1s$ overlap is possible but $1s - 2s$ and $2s - 2p$ overlaps are not permitted, since the energy of $1s$ orbital is less than that of $2s$ orbital. Similarly the energy of $2s$ orbital is less than that of $2p$ orbital.

(*ii*) The combining atomic orbitals must be able to overlap to the maximum extent to form molecular orbitals, since greater is the extent of overlap between the combining atomic orbitals, greater will be the electron density in the region between the nuclei of the combining atoms and hence stronger will be the bond between them.

(*iii*) The combining atomic orbitals must have the same symmetry about the molecular axis. This means that the atomic orbitals having the same or roughly the same energy will not overlap if they do not have the same symmetry. On this bases, the overlaps which can be allowed are $s - s$ (σ_s^b , σ_s^*), $s - p_x$ (σ_{sp}^b, σ_{sp}^*), $p_x - p_x$ (σ_x^b , σ_x^*), $p_y - p_y$ (π_y^b , π_y^*) and $p_z - p_z$ (π_z^b, π_z^*), while those which are not permitted are $s - p_y$, $s - p_z$, $p_x - p_y$ and $p_x - p_z$ (See Fig. 11.7). The molecular orbitals formed by the permitted overlaps have also been shown in bracket. (*x*-axis has been assumed to be the moleculer axis). In $s - p_y$, $s - p_z$, $p_x - p_y$ and $p_x - p_z$ overlaps + + overlap region is cancelled by + – overlap region and hence no molecular orbital is formed (non-bonding molecular orbital).

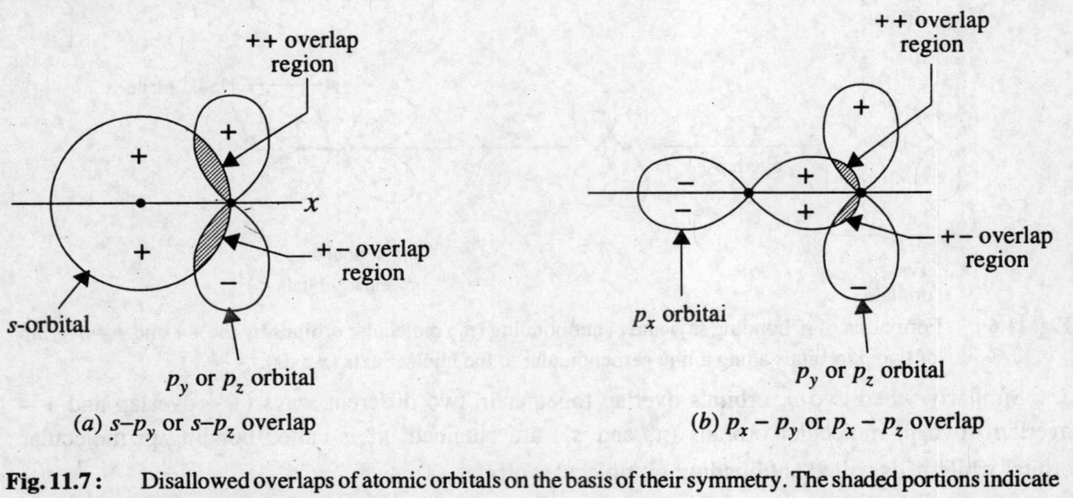

(*a*) s–p_y or s–p_z overlap (*b*) $p_x - p_y$ or $p_x - p_z$ overlap

Fig. 11.7 : Disallowed overlaps of atomic orbitals on the basis of their symmetry. The shaded portions indicate the overlap regions.

Comparison between (1) bonding and antibonding molecular orbitals and (2) sigma and pi molecular orbitals

These comparisons have been shown in Tables 11.1 and 11.2 respectively.

Table 11.1 *Comparison between bonding and anti-bonding molecular orbitals.*

Bonding molecular orbitals	Antibonding molecular orbitals
(*i*) Bonding molecular orbitals are obtained when two atomic orbitals combine together in such a way that their lobes with + + sign overlap (+ + overlap), *i.e.*, bonding molecular orbitals are obtained when the wave functions of the two atomic orbitals are added.	(*i*) Antibonding molecular orbitals are obtained when two atomic orbitals combine together in such a way that their lobes with + – sign overlap (+ – overlap), *i.e.*, antibonding molecular orbitals are obtained when the wave functions of the two atomic orbitals are subtracted.
(*ii*) Since the repulsion between the nuclei in bonding molecular orbital is minimum, the energy of bonding molecular orbital is less than that of both the combining atomic orbitals and hence the formation of bonding molecular orbital leads to the formation of a stable bond (attraction) between the combining atoms.	(*ii*) Since the repulsion between the nuclei in antibonding molecular orbital is maximum, the energy of antibonding molecular orbital is more than that of both the combining atomic orbitals and hence the formation of an antibonding molecular orbital produces repulsion between the combining atoms
(*iii*) The probability of finding the electron in the region between the nuclei of the combining atoms is very high and hence a stable molecule is formed.	(*iii*) The probability of finding the electron in the region between the nuclei of the combining atoms is negligible. The electron density is concentrated only on one nucleus at any particular moment.
(*iv*) Each electron in a bonding molecular orbital contributes to attraction between the two combining atoms.	(*iv*) Each electron in an antibonding molecular orbital contributes to repulsion between the combining atoms.
(*v*) The difference in energy between the molecular orbital and the combining atomic orbitals is called *stabilisation energy.*	(*v*) The difference in energy between the molecular orbital and the combining atomic orbitals is called *de-stablisation energy.*

Table 11.2 : *Comparison between sigma and pi molecular orbitals.*

Sigma molecular orbitals	Pi molecular orbitals
(*i*) σ-molecular orbital is obtained when + + or + – portions of two atomic orbitals overlap with each other along the internuclear axis. + + overlap gives σ-bonding molecular while + – overlap gives σ-antibonding molecular orbital.	(*i*) π-molicular orbital is obtained when ++ or + – portions of the two atomic orbitals overlap with each other along a line perpendicular to the internuclear axis. ++ overlap gives π-bonding molecular orbital while + – overlap gives π-antibonding molecular orbital.
(*ii*) Sigma bonding as well as sigma antibonding molecular orbitals have cylindrical symmetry around the internuclear axis. It is for this reason that these molecular orbitals are called sigma molecular orbitals.	(*ii*) Since π-molecular orbitals are concentrated perpendicular to the molecular axis, these molecular orbitals do not have cylindrical symmetry around the molecular axis, *i.e.*, these molecular orbitals are antisymmetric about the internuclear axis.
(*iii*) σ_x^b and σ_x^* molecular orbitals donot have the same energy, *i.e.*, these molecular orbitals are not degenerate.	(*iii*) π_y^b and π_z^b molecular orbitals have the same energy and hence are degenerate orbitals. Similarly π_y^* and π_z^* are also degenerate.
(*iv*) σ-molecular orbital can be rotated freely about the bond axis.	(*iv*) π-molecular orbital cannot be rotated freely about the bond axis.
(*v*) σ-molecular orbital is a stronger molecular orbital, since it results by the greater overlap of orbitals along the nuclear axis.	(*v*) π–molecular orbital is a weaker molecular orbital since it results by the lesser overlap of orbitals along a line perpendicualr to the molecular axis.
(*vi*) The formation of a σ-molecular orbital decides the direction and length of the bond (bond length).	(*vi*) The formation of a π-molecular orbital does not decide the direction of the bond but shortens the bonds distance. The formation of two π-molecular orbitals shortens (as in a triple bond) the bond distance to a greater extent than the formation of one π-molecular orbital (as in a double bond).

Comparison between atomic orbitals and molecular orbitals

Similarities. (*i*) The molecular orbitals in a molecule or an ion have different energy and different shape and size as atomic orbitals in an atom have.

(*ii*) The filling of atomic orbitals with electrons takes place in accordance with Aufbou principle,

Pauli's exclusion principle and Hund's rule of maximum multiplicity. Molecular orbitals are also filled with electrons according to these principles.

(*iii*) Like atomic orbitals, a molecular orbital can be represented by a wave function ψ. ψ of an atomic orbital is called atomic orbital wave function and that for a molecular orbital is called molecular orbital wave function.

(*iv*) Each atomic orbital wave function in an atom is associated with a set of quantum numbers which determine the energy and shape of the atomic orbital. Similarly each molecular orbital in a molecule is also associated with a set of quantum numbers which determine the energy, shape and size of molecular orbital.

Dissimilarities

Atomic orbital	Molecular orbital
(*i*) An atomic orbital is defined as the region is space round one nucleus in an atom in which there is maximum probability of finding the electron.	(*i*) A molecular orbital is defined as the region of space round two or more nuclei in a molecule.
(*ii*) An atomic orbital belongs to one nucleus only and hence is *monocentric.*	(*ii*) A molecular orbital belongs to two or more atoms (polycentric) *i.e.*, a molecular orbital is considered to be a charged cloud having all the nuclei of the whole molecule.

Bond order or bond multiplicity

Bond order of a molecule or an ion is the measure of the strength or stability of the bond. Numerically bond order (B.O.) is equal to half of the difference between the total number of electrons present in bonding molecular orbitals (n_b) and the total number of electrons present in the antibonding molecular orbitals (n_a). Thus

$$\text{B.O.} = (n_b - n_a)/2$$

In common practice, the electrons residing in the molecular orbitals formed only from the orbitals of the valence-shell are considered for determining the bond order. The following informations are given by the magnitude of bond order of a given molecule or an ion.

(*i*) *Number of covalent bonds between the atoms and structure of the species.* The number of covalent bonds between the two atoms of a given diatomic species is equal to the B.O. of that species *i.e.*, the species having B.O. = 1 implies that the two atoms in the species are linked by a single covalent bond. B. O. = 2 means that the two atoms are joined by a double bond (= bond) while B.O. = 3 indicates the presence of a triple bond (≡ bond) between the atoms. Thus since B.O. of B_2, C_2 and N_2 molecules is equal to 1, 2, and 3 respectively, the number of covalent bonds by which the two atoms of these molecules are linked together is 1, 2 and 3 respectively. Consequently the structures of these molecules can be written as : B – B:, :C = C: and :N≡ N: respectively.

Bond order equal to zero implies that the species does not exist. For example, since bond order for He_2, Be_2 and Ne_2 molecules is equal to zero, these molecules are incapable of existence, *i.e.*, these species exist as monoatomic species (*i.e.*, in the atomic state).

Bond order equal to 0.5 suggests the presence of one odd-electron bond between the atoms. For example, since bond order for H_2^+ ion is 0.5, the two H-atoms in this ion are linked together by one *one-electron bond* which is an odd electron bond. Thus the structure of H_2^+ ion can be written as [H · H]$^+$. Bond order of H_2^- and He_2^+ ions is also equal to 0.5. This bond order implies that the two H-atoms in these ions are linked together by one *three-electron bond* (odd electron bond) and hence the structure of these ions are [H ··· H]$^-$ and [He ··· He]$^+$ respectively.

Bond orders of N_2^+ and O_2^+ ions are 2.5 which suggests that the two atoms in these ions are linked together by two normal covalent bonds and one three-electron bond (odd electron bond). Thus the structure of these ions are [N $\overset{\cdot\cdot\cdot}{=}$ N:]$^+$ and [:O $\overset{\cdot\cdot\cdot}{=}$ O:]$^+$ respectively. Bond order equal to 1.5 for O_2^- ion suggests the presence of one normal covalent bond and one *three-electron bond* between the two O-atoms. Thus the structure of O_2^- ion is [:$\underset{\cdot\cdot}{\text{O}}$ $\overset{\cdot\cdot\cdot}{—}$ $\underset{\cdot\cdot}{\text{O}}$:]$^-$. The presence of odd-electron bond explains the paramagnetic character of the species.

(*ii*) *Bond dissociation energy , stability and bond length.* Greater is the value of bond order, greater is the value of bond dissociation energy and hence greater is the stability of the bond or the species, *i.e.*, the species with higher value of bond order will have higher value of bond dissociation energy and hence would be more stable than the species having lower value of bond order. For example, since the values of bond order of O_2^+, O_2, O_2^- and O_2^{2-} are in the decreasing order : O_2^+ (B.O. = 2.5) > O_2 (B.O. = 2) > O_2^- (B.O. = 1.5) > O_2^{2-} (B.O. = 1), the bond dissociation energies and the stability of both of these species will also be in the decreasing order as shown below (bond dissociation energy values are given in brackets): O_2^+ (= 642 kJ/mole) > O_2 (= 494 kJ/mole) > O_2^- (= 394 kJ/mole) > O_2^{2-} (= 210 kJ/mole). Obviously the order of stability is $O_2^+ > O_2 > O_2^- > O_2^{2-}$.

Higher is the value of bond order, shorter is the bond length. Since the bond orders of O_2^+, O_2, O_2^- and O_2^{2-} are in the decreasing order as shown above, O – O bond lengths are in the increasing order as shown below (bond lengths are given in bracket) : O_2^+ (=112 pm) < O_2 (=121 pm) < O_2^- (=125 pm) < O_2^{2-} (=149 pm).

Molecular orbital configurations of H_2, H_2^+, H_2^-, He_2 and He_2^+ species (Homo-nuclear diatomic species)

We know that each of the two atoms of which the given species are composed consists of only $1s$ orbital. $1s$ orbital on one atom combines with $1s$ orbital on the other similar atom and gives rise to the formation of two sigma molecular orbitals. One of these molecular orbitals is a sigma bonding molecular orbital (σ_{1s}^b) and the other is a sigma antibonding molecular orbital (σ_{1s}^*). In a simplified way, the formation of these molecular orbitals can be shown as :

$1s$ atomic orbital on one atom $+1s$ atomic orbital on the other identical atom

$$\xrightarrow[\text{(addition of orbitals)}]{++\text{ overlap}} \sigma_{1s}^b \text{ molecular orbital}$$

$1s$ atomic orbital on one atom + $1s$ atomic orbital on the other identical atom

$$\xrightarrow[\text{(Subtraction of orbitals)}]{+-\text{ overlap}} \sigma_{1s}^*$$

The energy of these molecular orbitals is in the order : $\sigma_{1s}^b < \sigma_{1s}^*$. σ_{1s}^b molecular orbital has lower energy than either of the two $1s$ orbitals of two atoms from which it is formed while σ_{1s}^* molecular orbital has higher energy than each of the $1s$ atomic orbitals. Since σ_{1s}^b molecular orbital has lower energy than σ_{1s}^* molecular orbital, it will be filled up first with electrons. Keeping in mind that the electronic configuration of H, H^+, H^-, He and He^+ are H $=1s^1$, $H^+ = 1s^0$, $H^- = 1s^2$, He $= 1s^2$ and $He^+ = 1s^1$, the molecular orbital configurations of the given species can be written as:

H_2 molecule ($1s^1 + 1s^1$ = 2 electrons) = $(\sigma_{1s}^b)^2\ (\sigma_{1s}^*)^0$

H_2^+ cation ($1s^1 + 1s^0$ = 1 electrons) = $(\sigma_{1s}^b)^1\ (\sigma_{1s}^*)^0$

H_2^- anion ($1s^1 + 1s^2$ = 3 electrons) = $(\sigma_{1s}^b)^2\ (\sigma_{1s}^*)^1$

He_2 molecule ($1s^2 + 1s^2$ = 4 electrons) = $(\sigma_{1s}^b)^2\ (\sigma_{1s}^*)^2$

He_2^+ cation ($1s^2 + 1s^1$ = 3 electrons) = $(\sigma_{1s}^b)^2\ (\sigma_{1s}^*)^1$

Molecular orbital energy level diagram of H_2 and H_2^- are given below in Figs. 11.8 and 11.9 respectively. In a similar way these diagrams for other species can also be drawn.

In the molecular orbital diagram, the molecular orbitals have been shown in the centre and atomic orbitals which combine to form molecular orbitals have been shown at the two extremes. Atomic orbitals have been shown at the same energy level while σ_{1s}^{b} molecular orbital has been shown lower and σ_{1s}^{*} molecular orbital has been shown higher than the corresponding atomic orbitals.

(*i*) *Bond order and related properties.* Obviously bond order of the given species are as : H_2 = (2 – 0)/2 = 1, H_2^+ = 1 – 0/2 = 0.5, H_2^- = (2 – 1)/2 = 0.5, He_2 = (2 – 2)/2 = 0 (zero) and He_2^+ = (2 – 1)/2 = 0.5. Since the bond order of H_2 molecule (= 1) is greater than that of H_2^- ion (= 0.5), H_2 molecule has higher value of its bond dissociation energy (= 458 KJ / mole) than H_2^+ ion (= 269 KJ/mole).

Thus the bond in H_2 molecule is stronger than that in H_2^+ ion and hence the bond length in H_2 molecule is smaller (= 0.74 Å) than that in H_2^+ ion (1.04 Å). Bond dissociation energy and bond length in He_2^+ ion are 230 KJ / mole and 1.08 Å respectively.

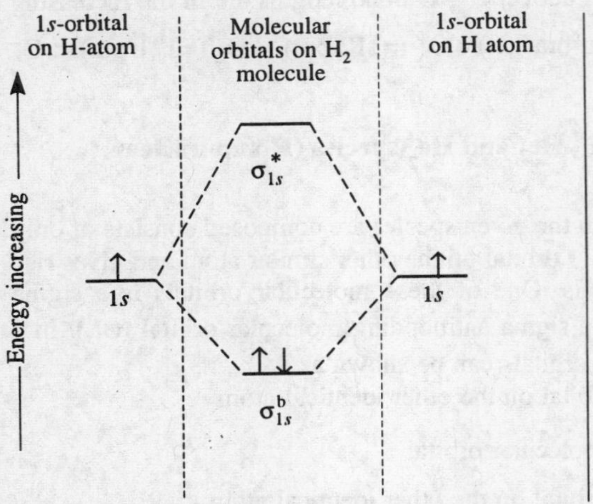

Fig. 11.8 : Molecular orbital energy level diagram for H_2 molecule.

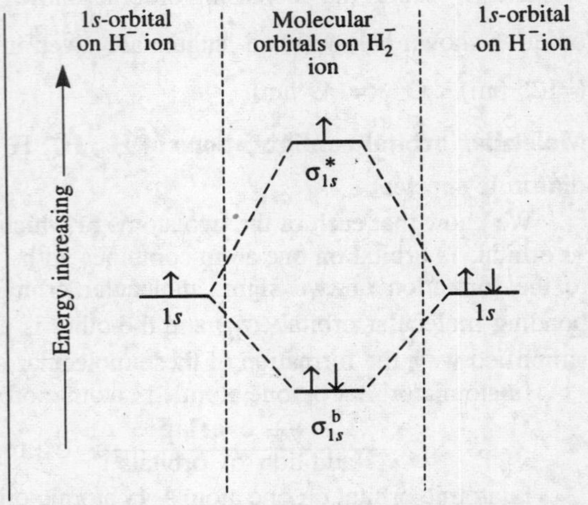

Fig. 11.9 : Molecular orbital energy level diagram for H_2 ion.

Bond order equal to zero for He_2 molecule implies that this molecule does not exist, *i.e.*, helium element exists as a monoatomic species.

(*ii*) *Magnetic properties and structure.* Since each of the species *viz* H_2^+, H_2^- and He_2^+ has unpaired electron (the number of unpaired electrons is one in each case), these are paramagnetic. On the other hand, H_2 and He_2 molecules have only paired electrons and hence are diamagnetic. Bond order equal to one for H_2 molecule indicates that the two H-atoms in H_2 molecule are linked together by σ-bond (H – H). In order to explain the paramagnetic nature and bond order equal to 0.5 for H_2^+, H_2^- and He_2^+ ions, the concept of the formation of odd-electron bond between the atoms has been suggested. An odd-electron bond is that which is formed by the sharing of an odd number (*e.g.* 1, 3, 5 or 7) of electrons between the atoms. Thus, on the basis of this concept, it has been assumed that H_2^+, H_2^- and He_2^+ ions consists of *one one-electron bond, one three-electron bond* and *one three-electron bond* respectively. Consequently the structure of these ions can be written as $[H{\cdot}H]^+$, $[H{\cdots}H]^-$ and $[He{\cdots}He]^+$ respectively. One-electron bond in H_2^+ ion arises due to the presence of one electron in σ_{1s}^{b} molecular orbital while three-electron bond in H_2^- and He_2^+ ions is formed by an electron pair residing in σ_{1s}^{b} molecular orbital and one unpaired electron in σ_{1s}^{*} molecular orbital. The presence of an odd electron bond in a molecule or an ion gives a bond order equal to 0.5 to the species.

Formation of molecular orbitals in a homo-nuclear diatomic species of A_2 type

Here we shall see how molecular orbitals are formed in a homonuclear diatomic species of A_2 type (*e.g.*, Li_2, Be_2, B_2, C_2, N_2, N_2^+, O_2, O_2^+, O_2^-, O_2^{2-}, F_2, Ne_2) in which both A-atoms belong to second period of the periodic table. Since each atom of A_2 type species belongs to 2*nd* period, each atom has 1*s*, 2*s*, $2p_x$ $2p_y$ and $2p_z$ atomic orbitals (five atomic orbitals). Now five atomic orbitals on atom A combine with the corresponding atomic orbitals of the other similar atom A and ten molecular orbitals are obtained. These molecular orbitals are σ_{1s}^b, σ_{1s}^*, σ_{2s}^b, σ_{2s}^* σ_x^b, σ_x^*, π_y^b, π_y^*, π_z^b and π_z^*. In the simplified way the formation of these ten molecular orbitals can be shown as : (AO denotes atomic orbital)

1*s* AO of atom A+ 1*s* AO of another atom A —
- ++ overlap (Addition) → σ_{1s}^b
- +− overlap (Subtraction) → σ_{1s}^*

2*s* AO of atom A+ 2*s* AO of another atom A —
- ++ overlap (Addition) → σ_{2s}^b
- +− overlap (Subtraction) → σ_{2s}^*

$2p_x$ AO of atom A+ $2p_x$ AO of another atom A —
- ++ overlap (Addition) → σ_x^b
- +− overlap (Subtraction) → σ_x^*

$2p_y$ AO of atom A+ $2p_y$ AO of another atom A —
- ++ overlap (Addition) → π_y^b
- +− overlap (Subtraction) → π_y^*

$2p_z$ AO of atom A+ $2p_z$ AO of another atom A —
- ++ overlap (Addition) → π_z^b
- +− overlap (Subtraction) → π_z^*

In the formation of these molecular orbitals *x*-axis has been assumed to be the molecular axis. It is due to this assumption that the overlap of two $2p_x$ orbitals on two atoms gives sigma molecular orbitals (*viz.*, σ_x^b and σ_x^*) and not pi molecular orbitals.

Relative order of the energy of ten molecular orbitals

Spectroscopic studies have shown that the relative order of the energy of the ten molecular orbitals in Li_2, Be_2, B_2, C_2 and N_2 molecules and in the ions, if any, derived from these molecules is as :

$$\sigma_{1s}^b < \sigma_{1s}^* < \sigma_{2s}^b > \sigma_{2s}^* < \pi_y^b = \pi_z^b < \pi_x^b < \pi_y^* = \pi_z^* < \sigma_x^*$$

while the relative order of energy of these molecular orbitals in O_2,F_2 and Ne_2 molecules and in the ions formed by these molecules is as:

$$\sigma^b_{1s} < \sigma^*_{1s} < \sigma^b_{2s} > \sigma^*_{2s} < \sigma^b_x < \pi^b_y = \pi^b_z < \pi^*_y = \pi^*_z < \sigma^*_x$$

Both the orders given above have been depicted in Fig. 11.10.

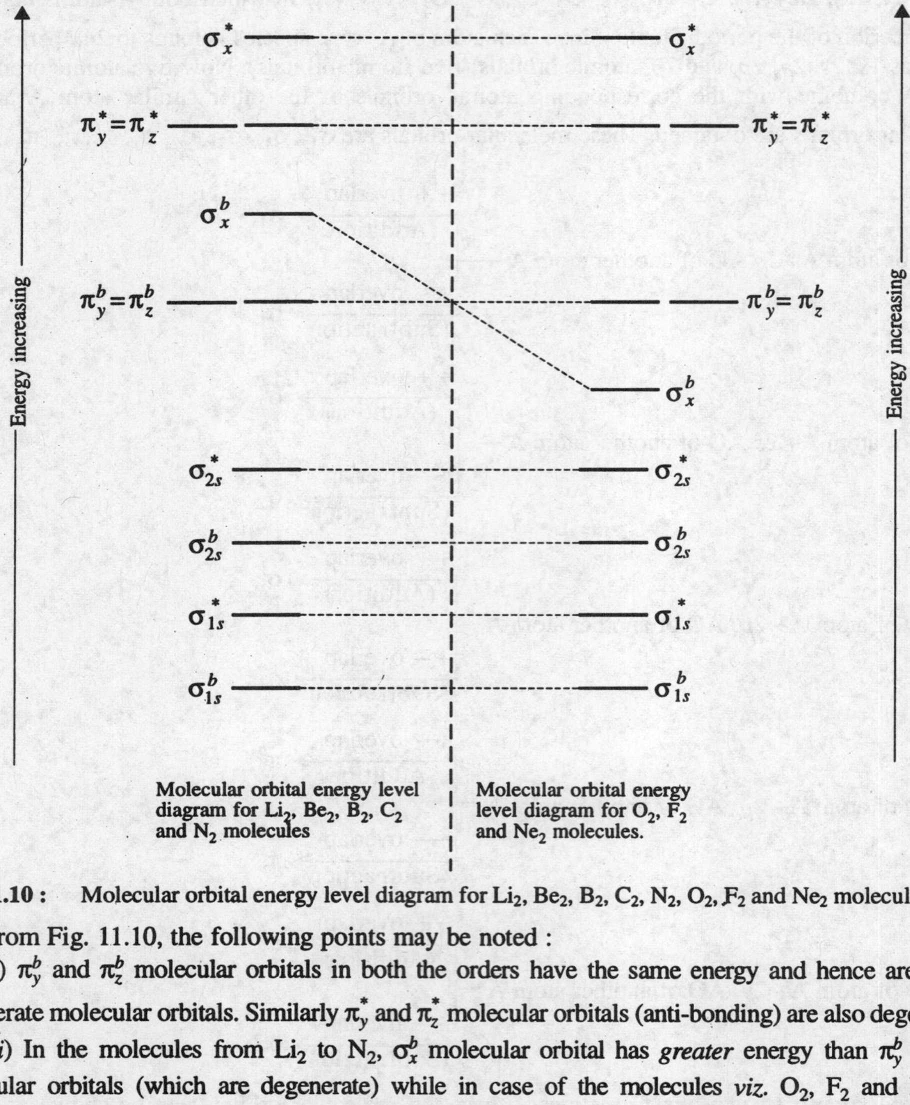

Fig. 11.10 : Molecular orbital energy level diagram for Li_2, Be_2, B_2, C_2, N_2, O_2, F_2 and Ne_2 molecules.

From Fig. 11.10, the following points may be noted :

(*i*) π^b_y and π^b_z molecular orbitals in both the orders have the same energy and hence are called degenerate molecular orbitals. Similarly π^*_y and π^*_z molecular orbitals (anti-bonding) are also degenerate.

(*ii*) In the molecules from Li_2 to N_2, σ^b_x molecular orbital has *greater* energy than π^b_y and π^b_z molecular orbitals (which are degenerate) while in case of the molecules *viz.* O_2, F_2 and Ne_2 σ^b_x molecular has lower energy than each of the orbitals namely π^b_y and π^b_z. The difference in the relative order of the energy of these molecular orbitals in due to the fact that in case of B_2, C_2, N_2 etc., species, 2*s* and $2p_x$ orbitals have very small difference in their energies and hence they mix together. On the other hand, the energy difference between 2*s* and $2p_x$ orbitals in O_2, F_2, Ne_2 etc., species is very large and hence their mixing together is not possible. It may be noted that out of the three 2*p* orbitals, only $2p_x$ orbital can mix with 2*s* orbital, since these orbitals have the same symmetry.

Rules for filling the molecular orbitals with electrons

(*i*) Various molecular orbitals are filled up with electrons in the increasing order of their energy, *i.e.*, the molecular orbital having the lowest energy is filled up first while that having the highest energy is filled up in the end. (*Aufbau principle*)

(*ii*) The maximum number of electrons that a molecular orbital can accomodate in it is two.

These two electrons must have opposite spins, *i.e.,* if one electron has spin equal to +1/2, the other should have spin equal to – 1/2.

(*iii*) In molecular orbitals having the same energy (degenerate orbitals), the pairing of electrons takes place when each of the molecular orbitals has at least one electron. It means that whenever two molecular orbitals having the same energy are available, electrons prefer to remain unpaired (*Hund's rule*).

Determination of the molecular orbital configuration of A_2 type species

Here we shall learn to write the molecular orbital configuration of A_2 type species in which both A-atoms belong to 2nd period of the periodic table. We have already seen that ten molecular orbitals are formed in A_2 type species by the overlap of ten atomic orbitals. Since 1*s* atomic orbitals are deeply seated below the valence-shell, molecular orbitals *viz.* $\overset{b}{\sigma}_{1s}$ and $\overset{*}{\sigma}_{1s}$ which are obtained by the combination of 1*s* orbitals do not participate in the bond formation in A_2 species, *i.e.,* $\overset{b}{\sigma}_{1s}$ and $\overset{*}{\sigma}_{1s}$ are considered to be non-bonding molecular orbitals. Consequently, while considering the molecular orbital configuration of A_2 species, we omit $\overset{b}{\sigma}_{1s}$ and $\overset{*}{\sigma}_{1s}$ molecular orbitals and hence write only those molecular orbitals which are obtained by the combination of atomic orbitals belonging to the valence-shell only (valence-shell orbitals). Valence-shell orbitals of an element of 2nd period are 2*s*, $2p_x$, $2p_y$ and $2p_z$ (four orbitals).

Now let us find out one by one, the molecular orbital configurations of A_2 type molecules / ions where A is an element of 2nd period of the periodic table.

1. Molecular orbital configuration of Li_2 molecule (1 + 1 = 2 electron). The valence-shell configuration of Li-atom is $2s^1$. 2*s* orbital of one Li-atom overlaps with 2*s* orbital of the other Li-atom and gives two molecular orbitals, which are $\overset{b}{\sigma}_{2s}$ and $\overset{*}{\sigma}_{2s}$. Since the energy of $\overset{b}{\sigma}_{2s}$ is less than that of $\overset{*}{\sigma}_{2s}$, the two electrons present in the valence-shell of Li_2 molecule ($2s^1 + 2s^1 = 2$) will occupy $\overset{b}{\sigma}_{2s}$ molecular orbital and $\overset{*}{\sigma}_{2s}$ molecular orbital will remain vacant. Thus the molecular orbital configuration of Li_2 molecule can be writtten as:

Li_2 molecule (2 electrons) = $\left(\overset{b}{\sigma}_{2s}\right)^2 \left(\overset{*}{\sigma}_{2s}\right)^0$

Molecular orbital energy level diagram of Li_2 molecule is shown in Fig. 11.11.

It may be seen from the molecular orbital configuration of Li_2 molecule that bond order of this molecule is equal to (2 – 0)/2 = 1 which implies that there is one Li—Li-bond in Li_2 molecule. Due to the fact that both the electrons present in $\overset{b}{\sigma}_{2s}$ molecular orbital are paired, Li_2 molecule is diamagnetic. The bond dissociation energy of Li_2 molecule is about 105 KJ/mole and Li — Li bond length is equal to 2.67 Å.

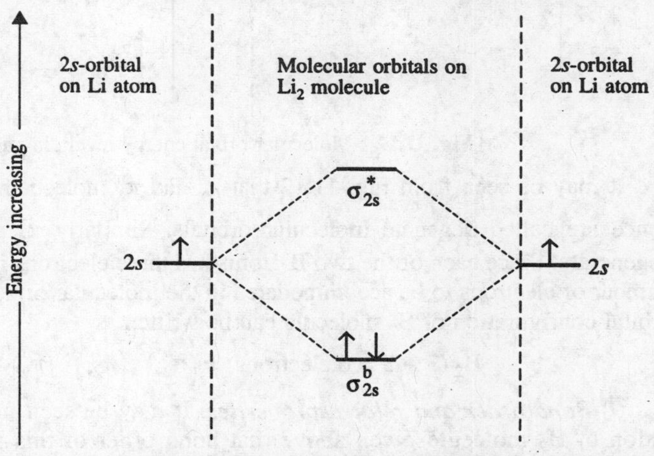

Fig. 11.11 : Molecular orbital energy level diagram Li_2 molecule.

2. Molecular orbital configuration of Be_2 molecule (2 + 2 = 4 electrons). The valence-shell configuration of Be-atom is $2s^2$. 2*s* orbital of one Be-atom overlaps with 2*s* orbital of the other Be-atom and gives two molecular orbitals which are $\overset{b}{\sigma}_{2s}$ and $\overset{*}{\sigma}_{2s}$. Both of these molecular orbitals will be filled with electrons. Thus molecular orbital configuration of Be_2 molecule is $\left(\overset{b}{\sigma}_{2s}\right)^2 \left(\overset{*}{\sigma}_{2s}\right)^2$.

Bond order is equal to (2 – 2)/2 = 0 (zero). Zero value of bond order implies that Be_2 molecule does not exist.

3. Molecular orbital configuration of B_2 molecule (3 + 3 = 6 electrons). The valence-shell configuration of B-atom is $2s^2\ 2p_x^1,\ 2p_y^0,\ 2p_z^0$ (four orbitals). The overlap of four atomic orbitals of one B-atom with the identical four atomic orbitals of the other B-atom gives eight molecular orbitals which are σ_{2s}^b, σ_{2s}^*, π_y^b, π_z^b, π_x^b, π_y^*, π_z^* and σ_x^*. The relative order of the energy of these molecular orbitals has been shown in the molecular orbital energy level diagram depicted in Fig. 11.12.

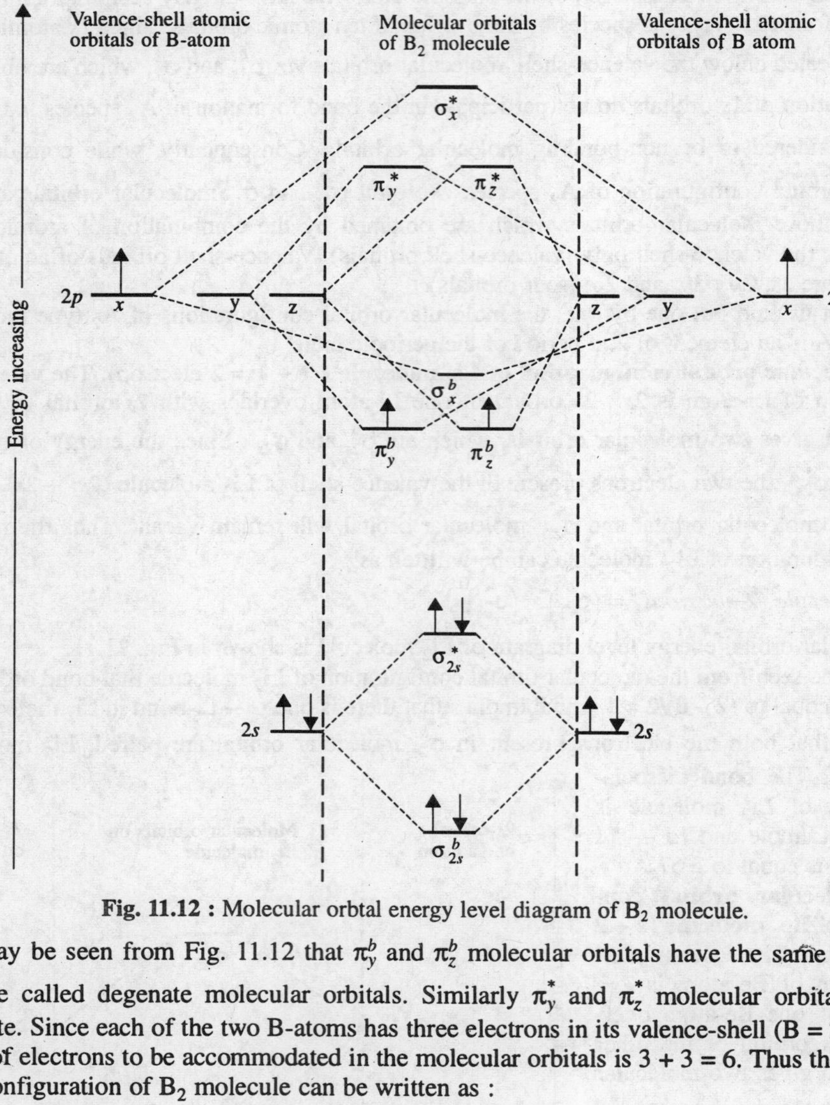

Fig. 11.12 : Molecular orbtal energy level diagram of B_2 molecule.

It may be seen from Fig. 11.12 that π_y^b and π_z^b molecular orbitals have the same energy and hence are called degenate molecular orbitals. Similarly π_y^* and π_z^* molecular orbitals are also degenerate. Since each of the two B-atoms has three electrons in its valence-shell (B = $2s^2p^1$), total number of electrons to be accommodated in the molecular orbitals is 3 + 3 = 6. Thus the molecular orbital configuration of B_2 molecule can be written as :

$$B_2\ (3 + 3 = 6\ \text{electrons}) = \left(\sigma_{2s}^b\right)^2 \left(\pi_{2s}^*\right)^2 \left(\pi_y^b\right)^1 \left(\pi_z^b\right)^1.$$

(*i*) *Bond order and related properties.* It may be seen from the molecular orbital configuration of B_2 molecule given above that bond order of this molecule is (4 – 2) / 2 = 1 which indicates that the two B-atoms in B_2 molecule are linked together by a single covalent bond

(:B—B: band). B—B bond dissociation energy of B_2 molecule is 289 KJ/mole and B—B bond length is 1.59 Å.

(*ii*) *Magnetic properties.* Since B_2 molecule has two unpaired electrons, it is paramagnetic to the extent of two unpaired electrons. It may be recalled that according to valence bond theory (VBT), the structure of B_2 molecule is : B — B: which shows that each B-atom has one electron pair and hence, according to VBT, B_2 molecule is diamagnetic. Thus we find that VBT is unstable to explain the paramagnetic character of B_2 molecule. Molecular orbital can well explain its paramagnetic character as shown above.

4. Molecular orbtial configuration of C_2 molecule (4 + 4 = 8 electrons). The valence-shell configuration of C-atom is $2s^2$, $2p_x^1$, $2p_y^1$, $2p_z^0$ (four orbitals). The overlap of four atomic orbitals of one C-atom with the identical four atomic orbitals of the other C-atom gives eight molecular orbitals which are $\overset{b}{\sigma}_{2s}$, $\overset{*}{\sigma}_{2s}$, $\overset{b}{\pi}_y$, $\overset{b}{\pi}_z$, $\overset{b}{\sigma}_x$, $\overset{*}{\pi}_y$, $\overset{*}{\pi}_z$, $\overset{*}{\sigma}_x$. The relative order of the energy of these molecular orbitals is as : $\overset{b}{\sigma}_{2s} < \overset{*}{\sigma}_{2s} < \overset{b}{\pi}_y = \overset{b}{\pi}_z < \overset{b}{\sigma}_x < \overset{*}{\pi}_y = \overset{*}{\pi}_z < \overset{*}{\sigma}_x$. Total number of electrons which occupy the molecular orbitals is eight (4 + 4 = 8). Thus the molecular orbital configuration of C_2 molecule is as :

$$C_2\ (4 + 4 = 8 \text{ electrons}) = \left(\overset{b}{\sigma}_{2s}\right)^2 \left(\overset{*}{\sigma}_{2s}\right)^2 \left(\overset{b}{\pi}_y\right)^2 \left(\overset{b}{\pi}_z\right)^2$$

The molecular orbital configuration shows that bond order of C_2 molecule is (6 – 2)/2 = 2 which indicates that the two C-atoms in C_2 molecule are linked together by a double bond (:C = C:). Since C_2 molecule does not have any unpaired electrons, it is diamagnetic, as observed experimentally. Bond dissociation energy of C_2 molecule is 627.9 KJ/mole and bond length is 1.31 Å.

5. Molecular orbital configurations of N_2 and N_2^+ species (N_2 = 5 + 5 = 10 electrons, N_2^+ = 5 + 4 = 9 electrons). The valence-shell configuration of N-atom is $2s^2$ $2p_x^1$, $2p_y^1$, $2p_z^1$ which shows that N-atom has 5 electrons in its valence-shell. Thus the total number of electrons present in the valence shell of N_2 and N_2^+ is 5 + 5 = 10 and 5 + 5 – 1 = 9 respectively. The molecular orbital configurations of these species are given below:

$$N_2 \text{ molecule (10 electrons)} = \left(\overset{b}{\sigma}_{2s}\right)^2 \left(\overset{*}{\sigma}_{2s}\right)^2 \left(\overset{b}{\pi}_y\right)^2 \left(\overset{b}{\pi}_z\right)^2 \left(\overset{b}{\sigma}_x\right)^2$$

$$N_2^+ \text{ ion (9 electrons)} = \left(\overset{b}{\sigma}_{2s}\right)^2 \left(\overset{*}{\sigma}_{2s}\right)^2 \left(\overset{b}{\pi}_y\right)^2 \left(\overset{b}{\pi}_z\right)^2 \left(\overset{b}{\sigma}_x\right)^1$$

Molecular orbital energy level diagram of N_2 molecule is shown in Fig.11.13. In a similar manner this diagram for N_2^+ ion can also be drawn.

(*a*) *Bond order and related properties.* It may be seen from the molecular orbital configurations given above that the bond orders (B.O.) of these species are as: B.O. of N_2 molecule = (8 – 2)1/2 = 3 and B.O. of N_2^+ ion = (7 – 2)/2 = 2.5. Thus since the bond order of N_2 is greater than that of N_2^+, (*i*) length at N—N bond in N_2 molecule is shorter than that in N_2^+ ion (N—N bond length in N_2 molecule = 1.10 Å, N—N bond length in N_2^- ion = 1.12 Å) (*ii*) dissociation energy of N_2 molecule is greater than that at N_2^+ ion (N_2 = 940 KJ/mole, N_2^+ = 828 KJ/mole). Thus N_2 molecule is more stable than N_2^+ ion.

(*b*) *Magnetic properties and structure.* It is evident from the molecular orbital configurations that N_2 molecule has all the electrons in the paired state and hence is diamagnetic. On the other hand N_2^+ ion has one unpaired electron and hence is paramagnetic to the extent of one unpaired electron. B.O. equal to 3 for N_2 molecule suggests that this molecule has three normal covalent bonds and hence its structure can be represented as :N≡N:. In order to explain the paramagnetic character (due to the presence of one unpaired electron) and bond order of 2.5 of N_2^+ ion, it has been

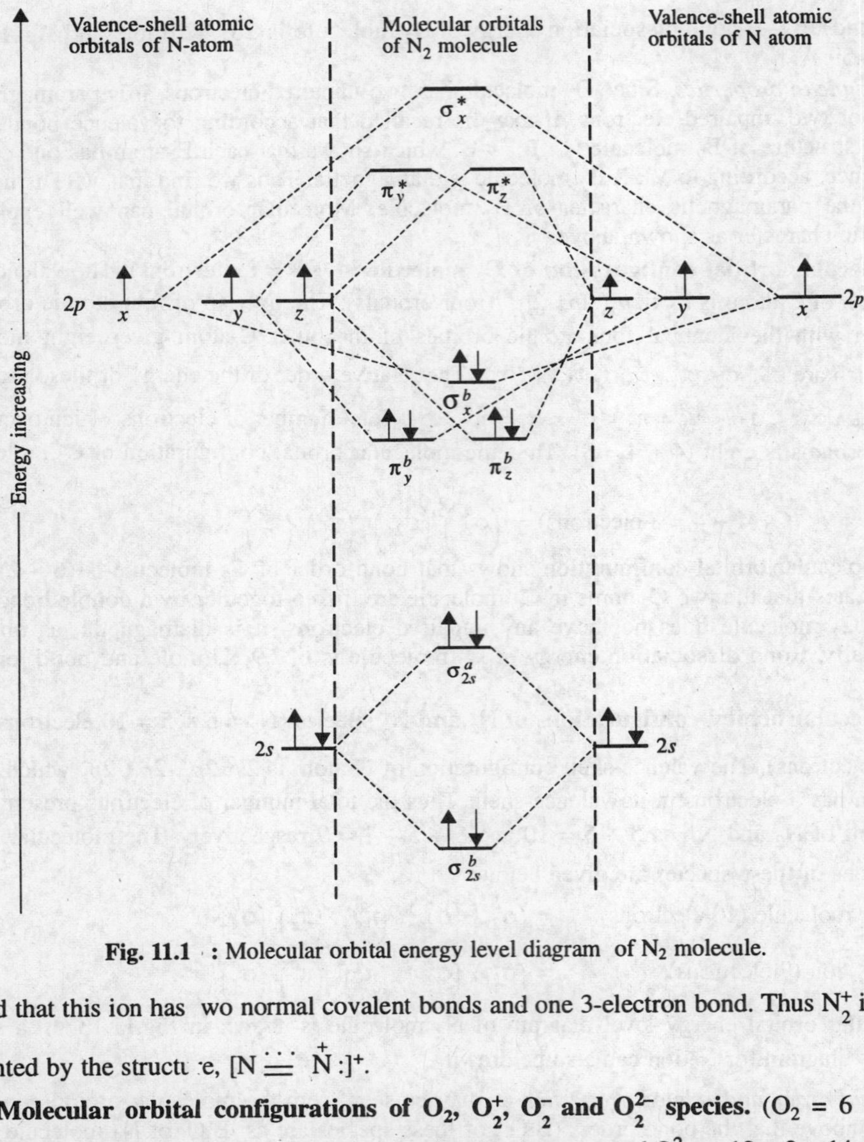

Fig. 11.1 : Molecular orbital energy level diagram of N_2 molecule.

assumed that this ion has wo normal covalent bonds and one 3-electron bond. Thus N_2^+ ion can be represented by the structu :e, $[N \stackrel{\cdots}{=} \overset{+}{N}:]^+$.

6. Molecular orbital configurations of O_2, O_2^+, O_2^- and O_2^{2-} species. (O_2 = 6 + 6 = 12 electrons, O_2^+ = 12 – 1 = 11 electrons, O_2^- = 12 + 1 = 13 electrons and O_2^{2-} = 12 + 2 = 14 electrons). The molecular orbital configurations of these species are given below :

O_2 molecule (12 electrons) $= (\sigma_{2s}^b)^2 (\sigma_{2s}^*)^2 (\sigma_x^b)^2 (\pi_y^b)^2 (\pi_z^b)^2 (\pi_y^*)^1 (\pi_z^*)^1$

Oxygenyl ion, O_2^+ (11 electrons) $= (\sigma_{2s}^b)^2 (\sigma_{2s}^*)^2 (\sigma_x^b)^2 (\pi_y^b)^2 (\pi_z^b)^2 (\pi_y^*)^1$

Superoxide ion, O_2^- (13 electrons) $= (\sigma_{2s}^b)^2 (\sigma_{2s}^*)^2 (\sigma_x^b)^2 (\pi_y^b)^2 (\pi_z^b)^2 (\pi_y^*)^2 (\pi_z^*)^1$

Peroxide ion, O_2^{2-} (14 electrons) $= (\sigma_{2s}^b)^2 (\sigma_{2s}^*)^2 (\sigma_x^b)^2 (\pi_y^b)^2 (\pi_z^b)^2 (\pi_y^*)^2 (\pi_z^*)^2$

Molecular orbital energy level diagram of O_2 molecule is shown in Fig. 11.14. In a similar manner these diagrams for other species can also be drawn.

(*a*) *O_2 molecule.* Bond order of this molecule is equal to (8 – 4)/2 = 2 which implies that the

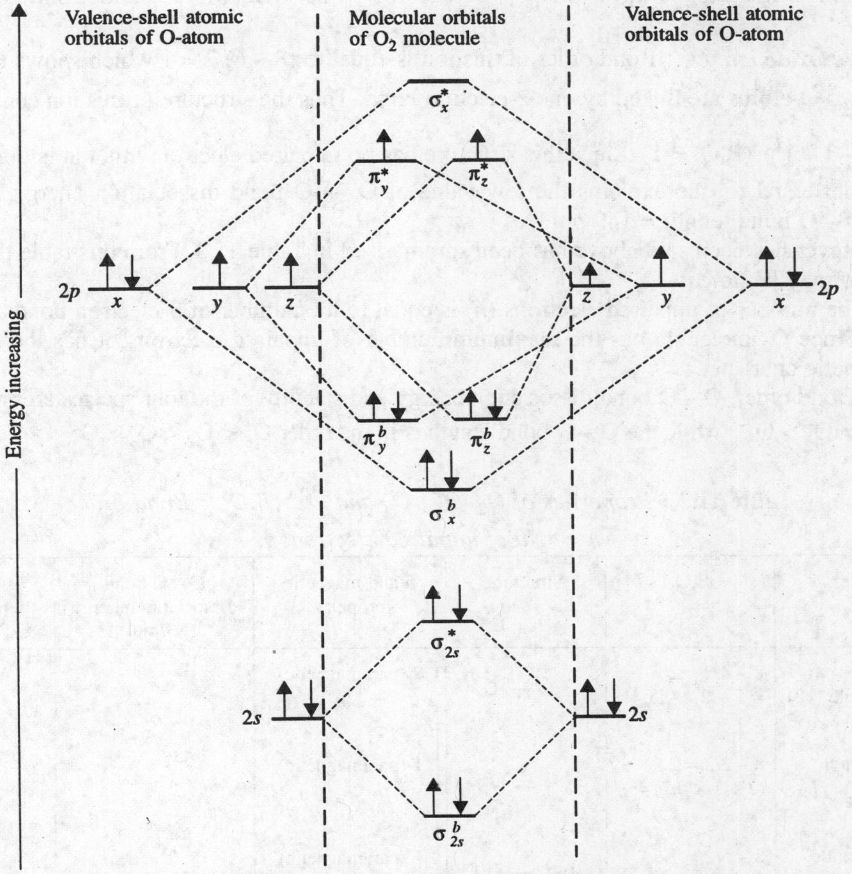

Fig. 11.14 : Molecular orbital energy level diagram O_2 molecule.

two O-atoms in O_2 molecule are joined together by one 2-electron bond and two 3-electron bonds, since a 3-electron bond gives a bond order equal to 0.5. Thus the structure of O_2 molecule can be written as :O $\overset{\cdots}{\underset{\cdots}{—}}$ O: (B.O. = 1 + 0.5 × 2 = 2). Since each 3-electron bond has one unpaired electron, this structure accounts for the paramagnetic character of O_2 molecule. These two unpaired electrons reside in antibonding molecular orbitals *viz*., π_y^* and π_z^*. Bond order equal to 2 also explains the high O – O bond dissociation energy of O_2 molecule (= 494 KJ mol). O – O bond length = 121 pm.

(*b*) *Oxygenyl ion, O_2^+*. Bond order of this ion is equal to (8 – 3)/2 = 2.5 which means that the two O-atoms in this ion are linked together by two 2-electron bonds and one 3-electron bond. Thus the structure of ion can be written as [:O $\overset{\cdots}{=}$ O:]$^+$ (B.O. = 2 + 1 × 0.5 = 2.5). 3-electron bond contains one unpaired electron which explains the paramagnetic character of this ion. Bond order equal to 2.5 also explains the high value of O — O bond dissociation energy of O_2^+ ion (= 642 KJ mol^{-1}). O — O bond length in this ion is equal to 112 pm.

(*c*) *Superoxide ion, O_2^-*. Bond order of this ion is equal to (8 – 5)/2 = 1.5 which suggests that the two O-atoms in this ion are bonded together by one 2-electron bond and one 3-electron bond. Thus the structure of this ion can be written as [:Ọ $\overset{\cdots}{—}$ Ọ:]$^-$ (B.O. = 1 + 1 × 0.5 = 1.5). The presence of one unpaired electron in 3-electron bond accounts for the paramagnetic nature of the ion. Bond dissociation energy of the ion = 394 KJ mole^{-1} and bond distance = 125 pm.

(*d*) *Peroxide ion, O_2^{2-}*. Bond order of this ion is equal to (8 – 6)/2 = 1 which shows that in this ion the two O-atoms are linked by one 2-electron bond. Thus the structure of this ion can be shown as [:Ö — Ö :]$^{2-}$ (B.O. = 1). Since this structure has no unpaired electron, this ion is diamagnetic. Bond order equal to one explains the low value of O — O bond dissociation energy (210.0 KJ mol^{-1}). O—O bond length = 149 pm.

Whatever has been said above has been summarised in Table 11.3. From this table the following points may be noted:

(*i*) The number of unpaired electrons (*n*) is equal to the number of 3-electron bonds.

(*ii*) Since O_2 molecule has the maximum number of unpaired electrons, it has the maximum paramagnetic character.

(*iii*) Bond order, O—O bond dissociation energy and stability of the four species are in the order $O_2^+ > O_2 > O_2^- > O_2^{2-}$ while the O—O bond length is in the order $O_2^+ < O_2 < O_2^- < O_2^{2-}$.

Table 11.3 : *Properties of O_2, O_2^+, O_2^- and O_2^{2-} (B.O. = Bond order, n = No of unpaired electrons).*

Species	B.O.	Structure	*n* and magnetic property	O – O bond dissociation energy (KJ/mol)	O – O bond length (pm)
(*i*) O_2 molecule (12)	8 – 4/2 = 2 = 1 + 2 × 0.5 = 2	:O $\overset{\cdots}{\underset{\cdots}{—}}$ O:	2 (paramagnetic)	494	121
(*ii*) Oxygenyl ion, O_2^+ (11)	8 – 3/2 = 2.5 (2 + 1 × 0.5) = 2.5	[:O $\overset{\cdots}{=}$ O:]$^+$	1 (paramagnetic)	642	112
(*iii*) Superoxide ion, O_2^- (13)	8 – 5/2 = 1.5 (1 + 1 × 0.5) = 1.5	[:Ö $\overset{\cdots}{—}$ Ö:]$^-$	1 (paramagnetic)	394	125
(*iv*) Peroxide ion, O_2^{2-} (14)	8 – 6/2 = 1	[:Ö — Ö:]$^{2-}$	0 (diamagnetic)	210	149

7. Molecular orbital configuration of F_2 molecule (7 + 7 = 14 electrons). The valence-shell configuration of F-atom is $2s^2\ 2p_x^2\ 2p_y^2\ 2p_z^1$ (four atomic orbitals). The overlap of four atomic orbitals of one F-atom with the identical four atomic orbitals of the other F-atom gives eight molecular orbitals which are $\overset{b}{\sigma}_{2s}$, $\overset{*}{\sigma}_{2s}$, $\overset{b}{\pi}_y$, $\overset{b}{\pi}_z$, $\overset{b}{\sigma}_x$, $\overset{*}{\pi}_y$, $\overset{*}{\pi}_z$ and $\overset{*}{\sigma}_x$. The relative order of the energy of these molecular orbitals is $\overset{b}{\sigma}_{2s} < \overset{*}{\sigma}_{2s}, \overset{b}{\sigma}_x < \overset{b}{\pi}_y = \overset{b}{\pi}_z < \overset{*}{\pi}_y = \overset{*}{\pi}_z < \overset{*}{\sigma}_x$. Total number of electrons which have to be accommodated in these eight molecular orbitals is 7 + 7 = 14. Thus the molecular orbital configuration of F_2 molecule is

$$F_2\ (7 + 7 = 14 \text{ electrons}) = \left(\overset{b}{\sigma}_{2s}\right)^2 \left(\overset{*}{\sigma}_{2s}\right)^2 \left(\overset{b}{\sigma}_x\right)^2 \left(\overset{b}{\pi}_y\right)^2 \left(\overset{b}{\pi}_z\right)^2 \left(\overset{*}{\pi}_y\right)^2 \left(\overset{*}{\pi}_z\right)^2$$

This configuration shows that the bond order of F_2 molecule is (8 – 6)/2 = 1, which indicates that the two F-atoms in F_2 molecule are linked together by a single covalent bond (:F̤̈ – F̤̈ :). Since all the electrons in the molecular orbital configuration of F_2 are paired, F_2 molecule is diamagnetic. F—F bond dissociation energy is equal to 155 KJ/mole and F—F bond length is 1.42 Å.

8. Molecular orbital configuration of Ne_2 molecule (8 + 8 = 16 electrons). The valence-shell configuration of Ne-atoms is $2s^2\ 2p_x^2\ 2p_y^2\ 2p_z^2$ (eight electrons). Ne_2 molecule has 8 + 8 = 16 electrons which are accommodated in eight molecular orbitals. Thus the molecular orbital configuration of Ne_2 molecule is

$$Ne_2 \text{ molecule } (8 + 8 = 16 \text{ electrons}) = \left(\sigma_{2s}^b\right)^2 \left(\sigma_{2s}^*\right)^2 \left(\sigma_x^b\right)^2 \left(\pi_y^b\right)^2 \left(\pi_z^b\right)^2 \left(\pi_y^*\right)^2 \left(\pi_z^*\right)^2 \left(\sigma_x^*\right)^2$$

Thus bond order is equal to (8 – 8)/2 = 0 (zero) which implies that Ne_2 molecule is incapable of existence, *i.e.*, Ne_2 molecule is not formed.

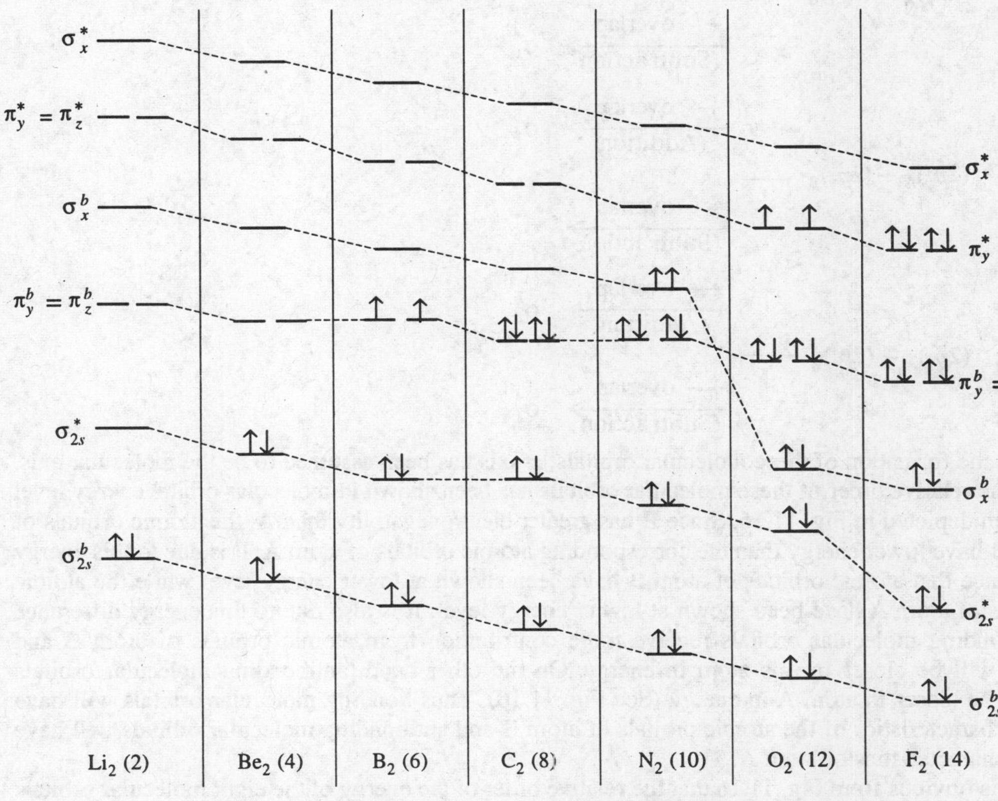

Fig. 11.15 : Molecular orbital configuration of diatomic elements.

Summary

Molecular orbital energy level diagrams of Li_2, Be_2, B_2, C_2, N_2, O_2 and F_2 molecules have been shown in a single diagram (Fig. 11.15).

Formation of molecular orbitals in a heteronuclear diatomic species of AB type

Here we shall see how molecular orbitals are formed in a hetero-nuclear diatomic species of AB type (*e.g.*, CO, CN, CN^-, NO, NO^- etc.) in which A and B atoms belong to 2nd period of the periodic table and electronegativity of A is less than that of B. Since both the atoms belong to 2nd period, each of them has $2s$, $2p_x$, $2p_y$ and $2p_z$ atomic orbitals (4 atomic orbitals) in its valence-shell. If atomic orbitals on atom A are represented $(2s)_A$, $(2p_x)_A$, $(p_y)_A$ and $(2p_z)_A$ and those on atom B are represented as $(2s)_B$, $(2p_x)_B$, $(2p_y)_B$ and $(2p_z)_B$, the overlap of the four atomic orbitals on atom A with the corresponding four atomic orbitals of atom B gives eight molecular orbitals which are σ_{2s}^b, σ_{2s}^*, π_y^b, π_z^b, σ_x^b, σ_x^*, π_y^*, π_z^* and σ_z^*. In a simplified way, the formation of these eight molecular orbitals can be shown as :

$(2s)_A + (2s)_B$ — ++ overlap (Addition) → σ_{2s}^{b}
 — +− overlap (Subtraction) → σ_{2s}^{*}

$(2p_x)_A + (2p_x)_B$ — ++ overlap (Addition) → σ_{x}^{b}
 — +− overlap (Subtraction) → σ_{x}^{*}

$(2p_y)_A + (2p_y)_B$ — ++ overlap (Addition) → σ_{y}^{b}
 — +− overlap (Subtraction) → σ_{y}^{*}

$(2p_z)_A + (2p_z)_B$ — ++ overlap (Addition) → σ_{z}^{b}
 — +− overlap (Subtraction) → σ_{z}^{*}

In the formation of these molecular orbitals, *x*-axis has been assumed to be the molecular axis.

The relative order of these molecular orbitals has been shown in molecular orbital energy level diagram depicted in Fig. 11.16. Since B has greater electronegativity than A, the atomic orbitals of atom B have lower energy than the corresponding atomic orbitals of atom A. It is due to this energy difference that atomic orbitals of atom B have been shown at lower energy level while the atomic orbitals of atom A have been shown at higher energy level. It is also due to this energy difference that bonding molecular orbitals receive more contribution from atomic orbitals of atom B and hence will be closer to this atom in energy. On the other hand, antibonding molecular orbitals would be closer to atom A in energy (See Fig. 11.16). Thus bonding molecular orbitals will have more characteristics of the atomic orbitals of atom B and antibonding molecular orbitals will have more characteristics of atom A.

It is obvious from Fig. 11.16 that the relative order of the energy of the eight molecular orbitals is $\sigma_{2s}^{b} < \sigma_{2s}^{*} < \pi_y^{b} = \pi_z^{b} < \sigma_x^{b} < \pi_y^{*} = \pi_z^{*} < \pi_x^{*}$.

Molecular orbital configuration of CO molecule

CO molecule is a heteronuclear diatomic molecule of AB type and is composed of one C-atom and one O-atom. Since these atoms have 4 ($C = 2s^2p^2$) and 6 ($O = 2s^2p^4$) electrons in their valence-shell respectively, CO molecule has 4 + 6 = 10 electrons in its valence-shell. These ten electrons are distributed in various molecular orbitals as $\left(\sigma_{2s}^{b}\right)^2, \left(\sigma_{2s}^{*}\right)^2, \left(\pi_y^{b}\right)^2 = \left(\pi_z^{b}\right)^2, \left(\sigma_x^{b}\right)^2,$ $\left(\pi_y^{*}\right)^0 = \left(\pi_z^{*}\right)^0, \left(\sigma_x^{*}\right)^0$. This is the molecular orbital configuration of CO molecule. Obviously bond order of this molecule is equal to (8 – 2)/2 = 3 which implies that C-and O-atoms in CO molecule are linked by a triple bond [:C≡ O:]. One of the three bonds is a σ-bond and the other two are π-bonds. Bond dissociation energy has been found to be equal to 1067 KJ/mole and C—O bond length is equal to 1.14 Å. One of the two electron pairs residing in σ_{2s}^{b} and σ_{2s}^{*} molecular orbitals is located on C-atom while the other electron pair resides on O-atom.

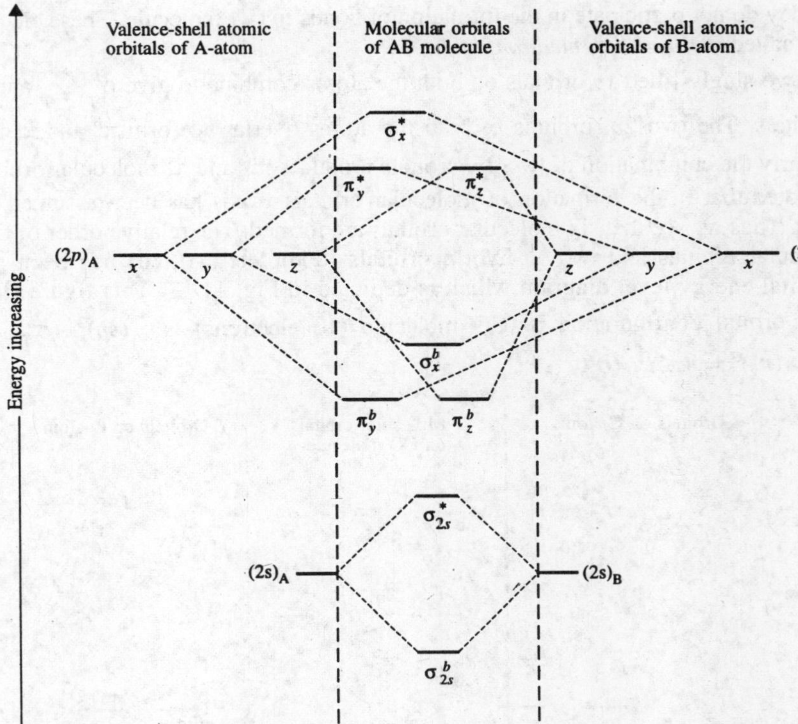

Fig. 11.16 : Molecular orbital energy level diagram of a heteronuclear diatomic species of AB type.

These two electron pairs act as non-bonding electron pairs (*i.e.*, as *lps*). The electron pair present in σ_x^b molecular orbital makes a σ-bond while the two electron pairs present in π_y^b and π_z^b molecular orbitals make two π-bonds between C- and O-atoms. CO acts as a ligand in various complexes known as *carbonyl complexes.* The presence of a non-bonding electron pair on C-atom accounts for the σ-donating property of CO molecule while the presence of a vacant π_y^* or π_z^* molecular orbital accounts for its π-accepting property. As is evident from the molecular orbital configuration, CO molecule has no unpaired electron and hence is diamagnetic in character.

Molecular orbital configuration of CO molecule suggested by Coulson

According to Coulson, before the formation of molecular orbitals, carbon and oxygen atoms undergo *sp* hybridisation as shown below :

C-atom in the ground state $(2s^2p^2)$: $2s^2\ 2p_x^1\ 2p_y^1\ 2p_z^0$

C-atom in *sp* hybridised state : $\underbrace{(sp)_C^2\ \ (sp)_C^1}_{\text{Two } sp \text{ hybrid orbitals}}\ 2p_y^0\ 2p_z^1$

Similarly *sp* hybridisation of O-atom can be shown as:

O-atom in the ground state $(2s^2p^4)$: $2s^2\ 2p_x^1\ 2p_y^2\ 2p_z^1$

O-atom in *sp* hybridised state : $(sp)_O^2\ \ (sp)_O^1\ 2p_y^2\ 2p_z^1$

In case of carbon, *sp* hybrid orbital containing the electron pair has higher energy than the hybrid orbital containing only one electron. On the other hand, in case of oxygen atom the situation is reversed. $2p_y$ and $2p_z$ orbitals of each of the atoms have the same energy.

The electron pairs residing in *sp* hybrid orbitals of both the atoms remain as lone pairs on these atoms, since they do not participate in the formation of bonds in CO molecule. These electron pairs are, therefore, called *non-bonding electron pairs.*

Now the two singly-filled *sp* orbitals on both the atoms combine to give σ^b_{sp-sp} and σ^*_{sp-sp} molecular orbitals. The two $2p_y$ orbitals of both the atoms overlap to form π^b_y and π^*_y molecular orbitals. Similarly the combination of two $2p_z$ orbitals produces π^b_z and π^*_z molecular orbitals. Here it should be noted that in the formation of molecular orbitals *x*-axis has been assumed to be the molecular axis. Thus we see that six molecular orbitals are formed. The relative order of the energy of these molecular orbitals and two *sp* hybrid orbitals (completely filled) has been shown in molecular orbital energy level diagram which is depicted in Fig. 11.17. This figure shows that the molecular orbital configuration of CO molecule (10 electrons) is : $(sp)^2_O$ $(\sigma^b_{sp-sp})^2$, $(\pi^b_y) = (\pi^b_z)^2$, $(sp)^2_C$, $(\pi^*_y)^0 = (\pi^*_z)^0$, $(\sigma^*_{sp-sp})^0$.

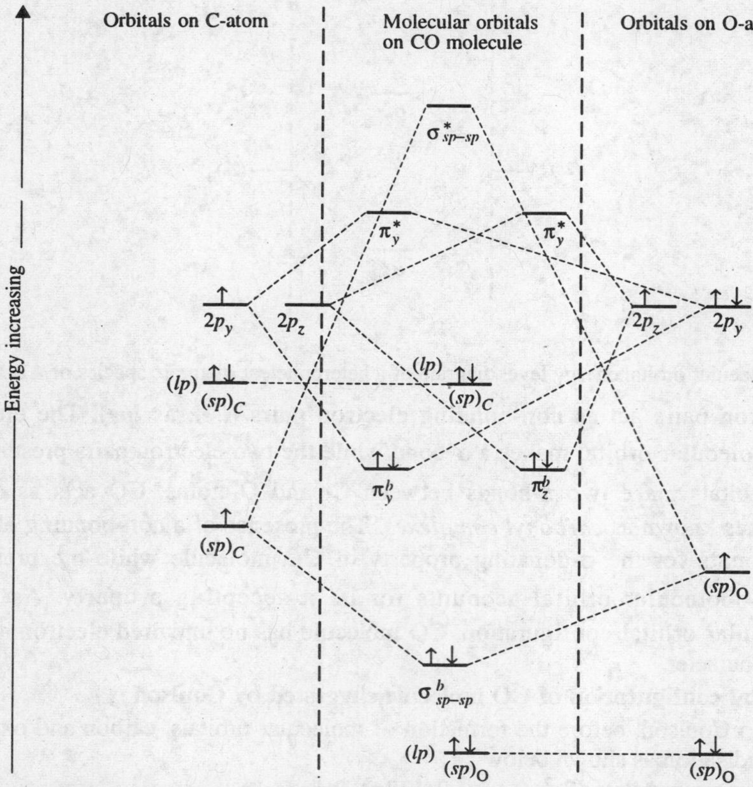

Fig. 11.17 : Molecular orbital energy level diagram of CO molecule as suggested by Coulson.

Molecular orbital configuration given above indicates that the bond order of CO = (6 – 0) / 2 = 3. Thus carbon and oxygen atoms are linked together by a triple bond which is composed of one *sigma* and two *pi* bonds.

sp hybrid orbital shown at the bottom of the energy level diagram belongs to O-atom and has a *lp* of electrons in it. Similarly *sp* hybrid orbital shown at the 4th place in the energy level diagram belongs to C-atom and, like *sp* hybrid orbital of O-atom, also has a *lp* of electrons. Thus we see that carbon and oxygen both have one *lp* of electrons each which is contained in their respective *sp* hybrid orbital. π^b_y and π^b_z molecular orbitals are degenerate and are at right angles to each other. Similarly, π^*_y and π^*_z molecular orbitals are also degenerate and at right angles to each other. Note that π^b_y and π^b_z molecular orbitals are completely filled while π^*_y and π^*_z molecular orbitals are empty. The pictorial representation of completely-filled *sp* hybrid orbitals and of vacant π_y^*/π_z^* molecular orbital has been given in Fig. 11.18.

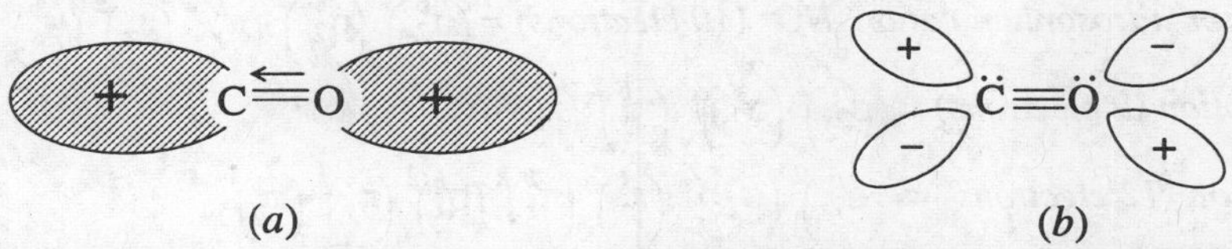

Fig. 11.18 : (*a*) Pictorial representation of completely-filled *sp* hybrid orbitals on carbon and oxygen atoms in CO molecule. One orbital is on carbon atom and the other is on oxygen atom. Smaller lobes of these orbitals have not been shown for the sake of convenience (*b*) Pictorial representation of vacant π_y^* molecular orbital. Similar is another empty π_z^* molecular orbital. These two molecular orbitals are at right angles to each other. Two dots shown on each atom indicates a lone pair of electrons contained in *sp* hybrid orbital of each atom.

Tendency of CO molecule to act as a terminal ligand in metallic carbonyl

We have said above that, according to Coulson, in CO molecule, carbon and oxygen both have one *lp* of electrons each in their respective *sp* hybrid orbitals. When CO molecule acts as a terminal ligand (called *terminal carbonyl group*) in metallic carbonyls, the attachment of this group takes place with one metallic atom (M) through the *lp* located in *sp* hybrid orbital of C-atom (and not of O-atom, since C-atom is less electronegative than O-atom) of CO molecule. The electron pair residing in *sp* hybrid orbital of C-atom is donated to the appropriate vacant hybrid orbital of the metallic atom [*e.g.*, sp^3 hybrid orbital of Ni-atom in $Ni(CO)_4$, dsp^3 hybrid orbital of Fe-atom in $Fe(CO)_5$ and of Co-atom in $Co_2(CO)_8$ (in solution), d^2sp^3 hybrid orbital in $Cr(CO)_6$, $Mn_2(CO)_{10}$, $Co_2(CO)_8$ (in solid state) etc.] and OC → M bond is formed.

We have also said above that CO molecule also has vacant π_y^* and π_z^* molecular orbitals. Being vacant, these molecular orbitals can accept electron pairs donated by the appropriate *filled* orbitals of metallic atoms to produce M → CO π-bonds. M → CO π-bond is also called *back donation, back-bonding* or *back coordination.* Due to the acceptor property of π_y^* and π_z^* molecular orbitals, these are also called *acceptor molecular orbitals*.

Molecular orbital configuration of CN and CN⁻ species

Valence-shell of CN and CN^- species contain 4 + 5 = 9 and 4 + 5 + 1 = 10 electrons respectively. Molecular orbital configurations of these species are given below:

$$\text{CN (9 electrons)} = (\sigma_{2s}^b)^2 (\sigma_{2s}^*)^2 (\pi_y^b)^2 (\pi_z^b)^2 (\sigma_x^b)^1$$

$$\text{CN}^- \text{(10 electrons)} = (\sigma_{2s}^b)^2 (\sigma_{2s}^*)^2 (\pi_y^b)^2 (\pi_z^b)^2 (\sigma_x^b)^2 (\pi_y^*)^0$$

(*i*) *Bond order and related properties.* It may be seen from the molecular orbital configuration of CN and CN^- that the bond orders of these species are (7–2)/2 = 2.5 (for CN) and (8 – 2)/2 = 3 (for CN^-). Since CN^- has higher bond order than CN, it has shorter bond length and larger bond dissociation energy than CN.

(*ii*) *Magnetic properties and structure.* As is evident from the molecular orbital configurations that CN has one unpaired electron and hence is paramagnetic. On the other hand, CN^- has all the electrons in the paired state and hence is diamagnetic. Due to the presence of one unpaired electron and bond order equal to 2.5, CN is assumed to have two normal covalent bonds and one 3-electron bond. Thus CN has the structure [C$\stackrel{\cdot\cdot\cdot}{=}$N:]. CN^- has the structure $[:C \equiv N:]^-$.

Molecular orbital configuration of NO, NO⁺, NO²⁺ and NO⁻ species

Total number of electrons present in the valence-shell of these species is 5 + 6 = 11 (in NO), 5 + 6 – 1 = 10 (in NO^+), 5 + 6 – 2 = 9 (in NO^{2+}) and 5 + 6 + 1 = 12 (in NO^-). The molecular orbital configurations of these species are:

Nitric oxide, NO (11 electrons) = $(\sigma_{2s}^b)^2 (\sigma_{2s}^*)^2 (\pi_y^b)^2 (\pi_z^b)^2 (\sigma_x^b)^2 (\pi_y^*)^1$

Nitrosyl or nitrosonium cation, NO^+ *(10 electrons)* = $(\sigma_{2s}^b)^2 (\sigma_{2s}^*)^2 (\pi_y^b)^2 (\pi_z^b)^2 (\sigma_x^b)^2$

NO^{2+} *cation (9 electrons)* = $(\sigma_{2s}^b)^2 (\sigma_{2s}^*)^2 (\pi_y^b)^2 (\pi_z^b)^2 (\sigma_x^b)^1$

NO^- *anion (12 electrons)* = $(\sigma_{2s}^b)^2 (\sigma_{2s}^*)^2 (\pi_y^b)^2 (\pi_z^b)^2 (\sigma_x^b)^2 (\pi_y^*)^1 (\pi_z^*)^1$

(*a*) *Bond order and bond lengths.* It is obvious from the molecular orbital configurations given above that bond order (B.O.) of NO = (8 – 3)/2 = 2.5, B.O. of NO^+ = (8 – 2)/2 = 3, B.O. of NO^{2+} = (7 – 2)/2 = 2.5 and B.O. of NO^- = (8 – 4)/2 = 2. Since the bond order of NO^+ ion is the highest, it has the shortest bond length. Similarly NO^- ion which has the lowest bond order has the largest bond length.

(*b*) *Magnetic properties and structure.* It is obvious from the molecular orbital configurations that NO, NO^- and NO^{2+} contain unpaired electrons (NO = 1, NO^{2+} = 1, NO^- = 2) and hence are paramagnetic. On the other hand, all the electrons in NO^+ ion are paired and hence this ion is diamagnetic. In order to explain the bond order of 2.5 of NO molecule and its paramagnetic character, it has been assumed that this molecule has two normal covalent bonds (one σ-bond and one π-bond) and one 3-electron bond, since a 3-electron bond gives a bond order equal to 0.5. Thus the structure of NO molecule can be written as [$\overset{\times}{\underset{\times}{\cdot}}$N $\overset{\times\,\cdot\,\cdot}{=}$ O:]. This structure agrees well with small value of its dipole moment (= 0.02 × 10^{-29} C.m) and the observed nitrogen—oxygen bond length (= 1.14 Å). Since the unpaired electron resides in an antibonding molecular orbital (*viz.*, π_y^*), this molecule has no tendency to dimerise with another NO molecule to form N_2O_2, *i.e.*, NO molecule does not show the properties of an odd-electron molecule. NO molecule acts as a ligand and forms complexes which are known as nitrosyl complexes. NO molecule acts as a σ-and π-donor and as a π-acceptor. Bond dissociation energy of NO molecule is equal to 667.8 KJ / mole.

In order to explain the bond order of 3 and diamagnetic character of NO^+ ion, it has been suggested that the ion has a triple bond and both the atoms have one lone pair of electrons each. N–O bond length has been found to be equal to 106 pm. Thus the structure of this ion is [:N ≡ O$\overset{\times}{\underset{\times}{}}$]$^+$.

In order to explain the paramagnetic character (which is due to the presence of two unpaired electrons) and bond order of 2 in NO^- ion, it has been suggested that this ion has one normal covalent bond and two 3-electron bonds. The presence of two 3-electron bonds gives a bond order equal to 2 × 0.5 = 1 to the ion. Thus the structure of this ion can be written as [:N $\overset{\cdot\times\times}{\underset{\cdot\times\times}{-}}$ O$\overset{\times}{\underset{\times}{}}$]$^+$. N – O bond length is equal to 124 pm.

Molecular orbital configuration of NO^+ ion suggested by Coulson

Since NO^+ ion and CO molecule both have the same number of electrons (=10 electrons), they are iso-electronic and hence have the same Coulson's molecular orbital configuration (see Coulson's molecular orbital configuration of CO molecule). Thus the molecular orbital configuration of NO^+ ion can be written as :

$$(sp)_O^2,\ (\sigma_{sp-sp}^b)^2,\ (\pi_y^b)^2 = (\pi_z^b)^2,\ (sp)_N^2,\ (\pi^*)^0 = (\pi_z^*)^0,\ (\sigma_{sp-sp}^*)^0$$

This configuration shows that bond order of the ion = (6 – 0)/2 = 3. Thus NO^+ ion has a triple bond and both the atoms have one *lp* of electrons each, *i.e.*, the structure is : N ≡ $\overset{+}{O}$:

Molecular orbital configuration of HF molecule

This molecule is composed of one H and one F-atom. The configuration of H-atom is $1s^1$ (one electron) while the valence shell configuration of F-atom is $2s^2\ 2p_x^1\ 2p_y^2\ 2p_z^2$ (seven electrons). In the formation of molecular orbitals in HF molecule, 1*s* (H) – $2p_y$ (F) and 1*s* (H) – $2p_z$ (F) overlaps do not take place, since the + + overlap region is cancelled by the + – overlap region and hence no molecular orbitals are formed (non-bonding molecular orbitals). Thus only singly-filled $2p_x$ orbital on F-atom overlaps with the singly-filled 1*s* orbital on H-atom and forms two sigma molecular orbitals which are designated as σ^b_{sp} and σ^*_{sp} . The molecular orbital energy level diagram of HF molecule can thus be shown as given in Fig. 11.19.

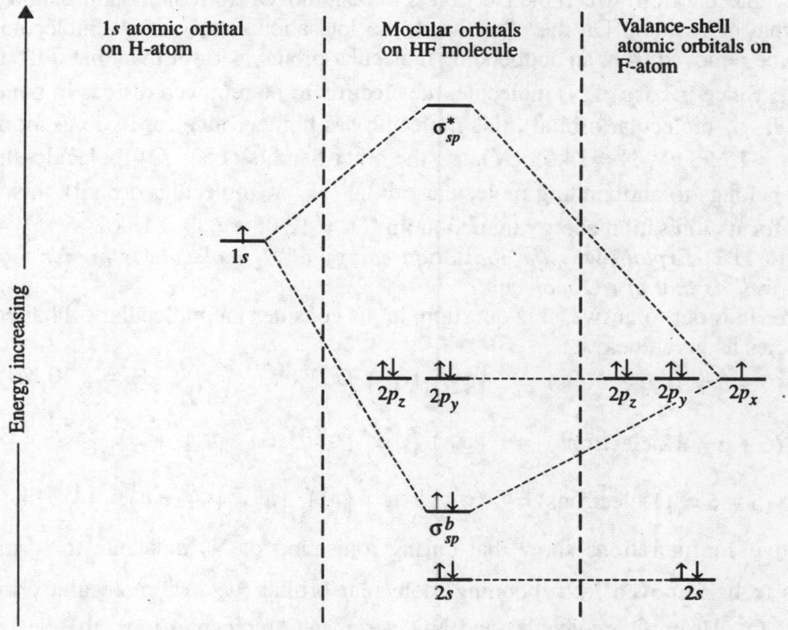

Fig. 11.19 : Molecular orbital energy level diagram of HF molecule.

The two electrons (one from $1s$ orbital of H-atom and one from $2p_x$ orbital of F-atom) occupy σ^b_{sp} molecular orbital since the energy of this molecular orbital is less than that of σ^*_{sp} molecular orbital. The remaining six electrons of F-atom (two in $2s$, two in $2p_y$ and two in $2p_z$ atomic orbitals) remain in their respective atomic orbitals. The shape of σ^b_{sp} molecular orbital is not symmetrical as shown in Fig. 11.20 since the energies of $1s$ (H) and $2p_x$ (F) atomic orbitals are not the same. Molecular orbital energy level diagram shows that the molecular orbital configuration of HF molecule can be written as :

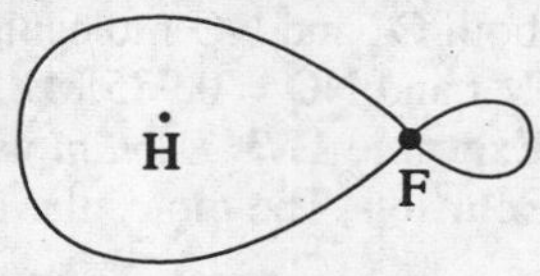

Fig. 11.20 : Unsymmetrical shape of σ^b_{sp} molecular orbital obtained by the overlap of $1s$ orbital on H-atom with the $2p_x$ orbital on F-atom in HF molecule.

$$\text{HF} = (2s)^2\,(\sigma^b_{sp})^2\,(2p_y)^2\,(2p_z)^2\,(\sigma^*_{sp})^0$$

Thus B.O.= (2–0)/2 = 1 (H – F)

Solved Examples

Example 11.1. *Explain why the ionisation energy of N_2 molecule is higher than that of N-atom, but the ionisation energy of O_2 molecule is lower than that of O-atom?*

Solution. The molecular orbital configurations of N_2 and O_2 molecules are given below:

$$N_2 \text{ (10 electrons)} = \left(\sigma^b_{2s}\right)^2 \left(\sigma^*_{2s}\right)^2 \left(\pi^b_y\right)^2 = \left(\pi^b_z\right)^2 \left(\sigma^b_x\right)^2$$

$$O_2 \text{ (12 electrons)} = \left(\sigma^b_{2s}\right)^2 \left(\sigma^*_{2s}\right)^2 \left(\sigma^b_x\right)^2 \left(\pi^b_y\right)^2 = \left(\pi^b_z\right)^2 \left(\pi^*_y\right)^1 = \left(\pi^*_z\right)^1$$

We know that the electrons residing in bonding molecular orbitals of a diatomic molecule to A_2 type are closer to the nucleus than the electrons residing in atomic orbitals of A atom while the electrons in antibonding molecular orbitals of A_2 molecule are for away from the nucleus than the electrons present in atomic orbitals of A atom. Thus the energy required to remove an electron from the outermost bonding molecular is greater in magnitude than that which is required to remove an electron from an atomic orbital of the atom. Similarly the energy required to remove an electron from the outermost antibonding molecular orbital is less in magnitude than that which is needed to remove an electron from an atomic orbital of an atom. In other words, we can say that the ionisation energy of A_2 molecule from which the electron is to be removed from the bonding molecular orbital is higher than that of A atom. On the other hand, the ionisation energy of A_2 molecule in which the electron is to be removed from an antibonding molecular orbital is lower than that of A atom.

Example. Since in case of N_2 molecule, the electron to be removed resides in bonding molecular orbital (*viz.* σ_x^b molecular orbital), this molecule has higher value for its ionisation energy than N-atom. (N_2 = 15.58 eV, N = 14.53 eV). On the other hand, since in O_2 molecule, the electron to be removed belongs to antibonding molecular orbital (*viz.* π_z^* molecular orbital), this molecule has lower value for its ionisation energy than O-atom (O_2 = 12.08 eV, O = 13.62 eV).

Example 11.2. *Explain why the ionisation energy of N_2 molecule is higher than that of O_2 molecule as well as that of NO molecule?*

Solution. In order to answer this question, let us consider the molecular orbital configurations of these species as given below :

$$N_2\ (5 + 5 = 10 \text{ electrons}) = \left(\sigma_{2s}^b\right)^2 \left(\sigma_{2s}^*\right)^2 \left(\pi_y^b\right)^2 \left(\pi_z^b\right)^2 \left(\sigma_x^b\right)^2, \text{ I.E.} = 1.50 \text{ MJ}$$

$$O_2\ (6 + 6 = 12 \text{ electrons}) = \left(\sigma_{2s}^b\right)^2 \left(\sigma_{2s}^*\right)^2 \left(\sigma_x^b\right)^2 \left(\pi_y^b\right)^2 \left(\pi_z^b\right)^2 \left(\pi_y^*\right)^1 \left(\pi_z^*\right)^1, \text{ I.E.} = 1.18 \text{ MJ}$$

$$NO\ (5 + 6 = 11 \text{ electrons}) = \left(\sigma_{2s}^b\right)^2 \left(\sigma_{2s}^*\right)^2 \left(\pi_y^b\right)^2 \left(\pi_z^b\right)^2 \left(\sigma_x^b\right)^2 \left(\pi_y^*\right)^1, \text{ I.E.} = 0.985 \text{ MJ}$$

The above configurations show that during ionisation of N_2 molecule to form N_2^+ ion, the electron has to be removed from bonding molecular orbital (*viz.*, σ_x^b molecular orbital) while in order to get O_2^+ (from O_2 molecule) and NO^+ (from NO molecule) ions, the electron has to be removed in both the cases from the antibonding molecular orbital. Again it is known that to remove an electron from a bonding molecular orbital requires more energy while to remove an electron from an antibonding molecular orbital less energy is needed. Thus N_2 has greater ionisation energy than both O_2 and NO molecules. Ionisation energies of these molecules are as N_2 = 1.50 MJ, O_2 = 1.18 MJ and NO = 0.985 MJ.

Example 11.3. *Explain why N_2 molecule is more stable than NO molecule?*

Solution. The molecular orbital configurations of N_2 and NO molecules are given below:

$$N_2 \text{ molecule } (10 \text{ electrons}) = \left(\sigma_{2s}^b\right)^2 \left(\sigma_{2s}^*\right)^2 \left(\pi_y^b\right)^2 \left(\pi_z^b\right)^2 \left(\sigma_x^b\right)^2$$

$$NO \text{ molecule } (11 \text{ electrons}) = \left(\sigma_{2s}^b\right)^2 \left(\sigma_{2s}^*\right)^2 \left(\pi_y^b\right)^2 \left(\pi_z^b\right)^2 \left(\sigma_x^b\right)^2 \left(\pi_y^*\right)^1$$

The molecular orbital configurations given above show that bond orders of N_2 and NO molecules are (8 – 2)/2 = 3 and (8 – 3)/2 = 2.5 respectively. Since bond order of N_2 molecule is greater than that of NO molecule, N_2 molecule is more stable than NO molecule. This fact is confirmed by the dissociation energy and bond distance values of these molecules which are given as N_2 = 945.6 KJ/mole, N–N bond distance = 1.10 Å, NO = 667.8 KJ/mole, N – O bond distance = 1.14 Å

Comparison between valence bond theory (VBT) and molecular orbital theory (MOT)

Similarities. (*i*) Both the theories explain the formation of covalent bond.

(*ii*) According to both the theories, a covalent bond (VBT) or a molecular orbital (MOT) is formed when two atomic orbitals having the same symmetry and same energy overlap with each other.

(*iii*) According to both the theories, a covalent bond (VBT) or a molecular orbital (MOT) has directional characteristics.

(*iv*) Both the theories predict that, when covalent bond is formed, an increase in the electron density between the nuclei takes place. This increased electron density holds the two atoms together and thus a bond is established.

Dissimilarities

VBT	MOT
(*i*) According to VBT, an electron moves under the influence of only one nucleus of an atom.	(*i*) According to MOT, an electron moves under the influence of two or more nuclei (atoms) of a molecule.
(*ii*) The atoms retain their individual identity.	(*ii*) The atoms lose their identity.
(*iii*) VBT predicts the existence of antibonding orbitals in terms of repulsive forces.	(*iii*) MOT predicts the formation of bonding as well as antibonding molecular orbitals.
(*iv*) VBT fails to explain the paramagnetic character of B_2 and O_2 molecules. Each of these molecules is paramagnetic due to the presence of two unpaired electrons.	(*iv*) MOT can well explain the paramagnetic character of B_2 and O_2 molecules.
(*v*) Resonance plays an important role in VBT.	(*v*) Resonance does not play any role in MOT.
(*vi*) VBT starts with individual atoms and considers the interaction between them.	(*vi*) MOT starts with nuclei of the constituent atoms.
(*vii*) In VBT, only half-filled orbitals of the valence shell take part in bonding and completely-filled orbitals remain non-bonding orbitals.	(*vii*) In MOT, all the atomic orbitals, whether completely filled, partially-filled or vacant, overlap to form molecular orbitals, provided they have proper symmetry and comparable energy.
(*viii*) In the original VBT, the wave function does not give any weightage to ionic structures.	(*viii*) MOT inherently gives equal weightage to covalent and ionic structures.
(*ix*) VBT is less exact than MOT and the calculations involved in VBT are less tedious than those of MOT.	(*ix*) MOT is more exact than VBT. Calculations involved in MOT are much more tedious than those involved in VBT.

Summary : Postulates (or features or basic principles) of molecular orbital theory

Whatever has been studied so far in connection with molecular orbital theory can be summarised as follows. This summary gives us the basic principles of molecular orbital theory.

(*i*) When two atoms of a homonuclear diatomic species (A_2 type species) or a heteronuclear diatomic species (AB type species) come closer to each other to form a bond between them, the valence-shell atomic orbitals of one atom overlap with the corresponding valence-shell atomic orbitals of the other atom and form a set of new orbitals which are called molecular orbitals. Thus A_2 or AB type species is supposed to consist of many molecular orbitals. After the molecular orbitals have been formed, the atomic orbitals lose their identity.

The number of molecular orbitals formed is equal to the number of atomic orbitals overlapping together. Thus, if two atomic orbitals combine, two molecular orbitals are formed. One of these two molecular orbitals is called bonding molecular orbital, since it has lower energy than the atomic orbital of each of the two isolated atoms. On the other hand, the other molecular orbital is called antibonding molecular orbital, because it has higher energy than the atomic orbital of each of the two isolated atoms. The bonding molecular orbital has high electron density in the region between the nuclei. The accumulation of electron density in the region between the nuclei explains the formation of a stable bond between the atoms. In an antibonding molecular orbital, the probability of finding the electron in between the nuclei is almost equal to zero, since the electron density is concentrated at the far ends of the molecule or ion. Because of no electron density between the nuclei, the nuclei are repelled from each other and hence no bond is formed.

(*ii*) When the overlapping lobes of the two atomic orbitals have + + signs, we get bonding molecular orbital. On the other hand, when the overlapping lobes of the combining atomic orbitals have + – signs, antibonding molecular orbitals are obtained.

(*iii*) The molecular orbitals formed by the combination of atomic orbitals may be sigma (σ) molecular orbitals or pi (π) molecular orbitals. If atomic orbitals overlap along the nuclear axis, sigma molecular orbitals are produced, and if the overlap of atomic orbitals takes place along a line perpendicular to the nuclear axis, pi molecular orbitals are obtained.

(*iv*) The inner orbitals of the individual atoms do not participate in the formation of molecular orbitals.

(*v*) In order that effective overlap of atomic orbitals may take place to give fruitful molecular orbitals, they should have the same or almost the same energy and same symmetry.

(*vi*) Molecular orbitals of a molecule or an ion have different energy, shape and size.

(*vii*) Each molecular orbital is described by a wave function, ψ which is called molecular orbital wave function. The molecular orbital wave function is associated with a set of quantum numbers which determine the energy, shape and size of the molecular orbital.

(*viii*) The square of the molecular orbital wave function (*i.e.*, ψ^2) denotes the electron probability density or electron charge density in the molecular orbital wave function, ψ.

(*ix*) All the electrons present in the valence-shell of all the atoms making the molecule or ion occupy the molecular orbitals in accordance with Aufbau principle, Pauli's exclusion principle and Hund's rule of maximum multiplicity.

(*x*) Each electron in molecular orbital belongs to all the nuclei present in the molecule or ion *i.e.*, each electron in a molecular orbital is considered to be moving along the entire molecule or ion under the influence of all the nuclei of all the atoms constituting the molecule or ion. In other words we can say that a molecular orbital is considered to be a charged cloud having all the nuclei of the whole molecule, *i.e.*, molecular orbital is *polycentric.*

University Questions

1. Draw moelcular orbital, energy bond diagrams of N_2, F_2, O_2 and Cl_2 molecules. Write their molecular configuration and discuss their stability and bond order. *(Himachal, 92, Agra, 92; Bangalore, 93)*
2. Draw the molecular orbital energy level diagram of F_2, HF and CO molecules. *(Gauhati 91, Saurashtra, 91)*
3. How does molecular orbital theory (MOT) explain the difference in the reactivity of N_2, O_2 and F_2? *(Bahuguna, 93)*
4. Give molecular orbital structures of H_2, H_2^+, He_2^+ and He^+. *(Lucknow, 94)*
5. What are sigma and pi molecular orbitals. Give their characteristics. *(Saurashtra, 91)*
6. (*a*) Write an account of molecular orbital theory applied to homo-diatomic molecules (*b*) How does bond order explain the stability of a molecule (*c*) How can you deduce bond order from molecular orbital theory (*d*) Calculate the bond or of He_2^+, O_2^-, O_2^{2-}, NO, F_2 and CO. *(Meerut, 91, 92, Lucknow, 1995)*
7. (*a*) Show with a neat diagram that the charge density between two bonded nuclei is greater than that between two non-bonded atoms. (*b*) Differentiate between atomic orbital and molecular orbital. *(Meerut, 93, 94)*
8. Describe Coulson's molecular orbital diagram of CO molecule. *(Nagpur, 93)*
9. Fill in blanks : When N_2 goes to N_2^+, the N—N bond distance............, ar.d when O_2 goes to O_2^+, the O—O bond distance.................. . *(I.I.T., 96)*
10. What is molecular orbital theory? Explain paramagnetism of O_2 molecule by this theory. *(Ravishankar, 93)*

11. Explain the following on the basis of molecular orbital theory :

(*i*) (*a*) Hydrogen forms diatomic molecule and helium remains monoatomic.
(*b*) The bond order of O_2 is two and that of H_2 is one.
(*c*) O_2 molecule is paramagnetic while O_2^{2-} ion is diamagnetic *(Meerut, 96)*
(*d*) He^+ ion is less stable than H_2^+ ion
(*e*) CO is diamagnetic while NO is paramagnetic. *(Gauhati 91; Lucknow, 93)*

(*ii*) Liq. O_2 sticks to the poles of a magnet while liq. N_2 dos not do so. *(E.C.C. Allahabad 94)*

(*iii*) He_2 and Be_2 molecules do not exist. *(Meerut, M.Sc., 94)*

(*iv*) O_2 molecule is paramagnetic while N_2 molecule is diamagnetic. *(Meerut, 93)*

(*v*) B_2 molecule is stable whereas Be_2 molecule is unstable. *(Meerut, 91)*

(*vi*) The bond order in O_2 molecule is two and that of N_2 molecule is three. *(Meerut, 91)*

(*vii*) The reactivities of N_2, O_2 and F_2 molecules are different. *(Himachal, 92)*

(*viii*) The bond strength of NO^+, NO and NO^- decreases in the order : $NO^+ > NO > NO^-$ *(Lucknow, 94)*

(*ix*) F_2 molecule is diamagnetic while O_2 molecule is paramagnetic. *(Gorakhpur, 87)*

(*x*) B_2 molecule is paramagnetic in nature. *(Punjab, 86)*

(*xi*) Bond length in O_2^+ ion is greater than that in O_2 molecule but that in N_2^+ ion is less than that in N_2 molecule. *(GND, 84)*

(*xii*) A helium atom does not form any bond with another helium atom but its bond with H or He^+ has been detected. *(Raj., 86)*

(*xiii*) H_2 molecule is more stable than H_2^+ ion. *Raj. 86)*

(*xiv*) The ionisation energy of O_2 molecule is less than that of O_2^+ ion.

12. Write short note on:

(*i*) Bonding and anti bonding molecular orbitals *(Agra, 92; Bahuguna, 93; E.C.C. Allahabad, 94, Lucknow, 94)*

(*ii*) Bond order *(Agra, 92; E.C.C. Allahabad, 94)*

(*iii*) Molecular orbitals. *(Agra, 92)*

(*iv*) Bonding and non-bonding molecular orbitals. *(Meerut, 91)*

13. Explain why He_2 molecule does not exist. *(Lucknow 2002)*

14. (*i*) Draw molecular orbital configurations of the species O_2, O_2^+, O_2^- and O_2^{2-} and calculate their bond order.

(*ii*) Explain why O_2 is paramagnetic whereas O_2^{2-} is diamagnetic.

(*iii*) Arrange O_2, O_2^+, O_2^- and O_2^{2-} in the increasing order of their stability, bond length and bond dissociation energy y *(Lucknow 2002)*

15. (*a*) What do you mean by the term linear combination of atomic orbitals ?
(*b*) What are the conditions necessary for the formation of molecular orbitals ?
(*c*) Differentiate between bonding and antibonding molecular orbitals.
(*d*) Explain the formation of oxygen molecule on the basis of molecular orbital theory. Explain magnetic character of oxygen molecule and calculate its bond order. *(Nagpur 2002)*

16. (*a*) Draw molecular orbital energy level diagram for NO and find its bond order.
(*b*) Explain why O_2 is paramagnetic. *(Delhi 2002)*

17. Obtain the delocalized molecular orbitals for BeH_2. Give neat sketches. *(Gujrat 2003)*

18. (*a*) Construct MO diagram for O_2 molecule and answer the following : (*i*) Calculate the bond order in O_2 and O_2^+ ion. (*ii*) What would be the magnetic character of O_2 molecule ?
(*b*) What are necessary conditions for an effective combination of two atomic orbitals to form molecular orbitals ?
(*c*) Distinguish between bonding and anti bonding molecular orbitals.

19. (*a*) What is LCAO approximation ? Draw the shapes of bonding and antibonding M.Os. formed by 1*s*-orbitals.
(*b*) Write M.O. configuration for Li_2 and B_2 molecules.
(*c*) Explain in detail, the Coulson's model of CO molecule. Calculate its bond order.
(*d*) Explain electron-charge density curves for bonding and antibonding M.Os. *(Nagpur 2003)*

12

Valence-shell Electron Pair Repulsion (VSEPR) Theory : Shape of Molecules and Ions

This theory is very useful to predict the geometry or shape of a number of simple covalent molecules and ions of non-transition elements. This theory was proposed for the first time by Sidgwick and Powell in 1940 and was developed by Gillespie and Nyholm in 1957. In the following discussion the electron pairs used in the formation of σ- and π-bonds have been represented as *σ-bps* and *π-bps* respectively. The electron pairs which are not used up in the formation of any bond are called *lone pairs of electrons* or *unshared electron pairs* and are represented as *lps*.

Assumptions or postulates of VSEPR theory

This theory makes the following assumptions:

1. Spatial arrangement of the central atom's valence-shell electron pairs. We know that in the Lewis structure (electron dot structure) of a given molecule or ion, only the valence-shell electrons of the central atom participate in the formation of various bonds. On the basis of the nature of the central atom's valence-shell surrounding the central atom, the species can be grouped into the following categories :

(*a*) Species whose central atom is surrounded by σ-bonding electron pairs (*σ-bps*) only. Examples of such species are BeF_2, BF_3, CH_4, PF_5, SF_6, IF_7 etc.

(*b*) Species whose central atom is surrounded by *σ-bps* as well as by lone pairs of electrons (*lps*). Examples of such species are $SnCl_2$, NH_3, H_2O, SF_4, ClF_3, XeF_2, IF_5, XeF_4, XeF_6 etc.

(*c*) Species whose central atom is surrounded by *σ-bps* as well as by *π-bps*. *lps* are absent. Examples of such species are CO_2 (O = C = O), HCN (H—C≡ N), H_2C_2 (H—C≡ C—H), etc.

(*d*) species whose central atom is surrounded by *σ-bps*, *lps* and *π-bps*. Examples are SO_2 (O = $\ddot{S}$ → O), ClO_3^- $\left[O \leftarrow \ddot{Cl} \rightarrow O \text{ with } O^- \text{ bonded below Cl} \right]$ etc.

The type of spatial arrangement (*e.g.*, linear, trigonal planar, tetrahedral etc.) the electron pairs surrounding the central atom depends on the sum of *σ-bps* and *lps* surrounding the central atom. The presence of *π-bps* does not influence the spatial arrangement of the electron pairs. The spatial arrangement takes place in such a way that the electron pairs occupy their positions in space as far away from each other as possible because in this case the electrostatic repulsion between the electron pairs is reduced to minimum. The system with minimum repulsive forces is the most stable.

It has been observed that when (*σ-bps* + *lps*) is equal to 2, 3, 4, 5, 6 and 7, the spatial orientation of these electron pairs round the central atom is *linear* (angle = 180°), *trigonal planar* (angle = 120°), *tetrahedral* (angle = 109.5°), *trigonal bipyramidal* (angles = 120°, 90°), *octahedral (angles* = 90°, 90°) and *pentagonal bipyramidal* (angles = 72°, 90°). The orientation of σ-*bps* for the molecules whose central atom is surrounded only by σ-*bps* and no *lp* is shown in Fig. 12.1 in which A is the central atom and B are the surrounding atoms with which A (central atom) is linked by σ-*bps* only.

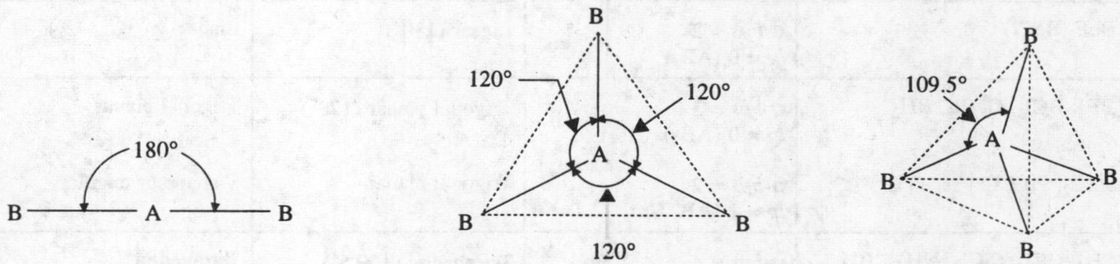

(*a*) Linear arrangement of two σ-*bps* (σ-*bps* = 2, *lp*=0)

(*b*) Trigonal planar arrangement of three σ-*bps* (σ-*bps* = 3, *lp* = 0)

(*c*) Tetrahedral arrangement of four σ-*bps* (σ-*bps* = 4, *lp* = 0)

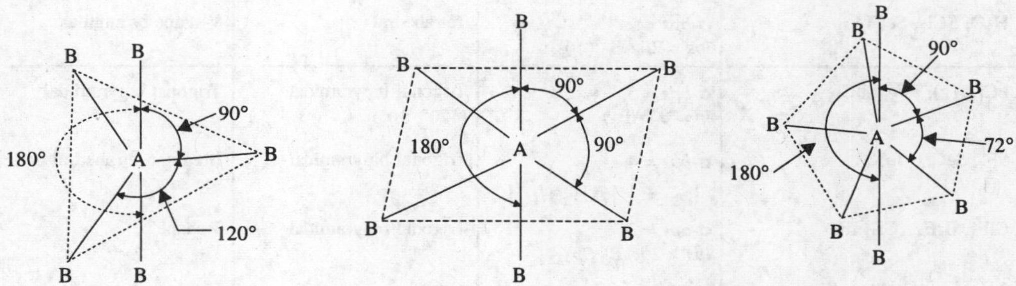

(*d*) Trigonal bipyramidal arrangement of five σ-*bps* (σ-*bps* = 5, *lp* = 0)

(*e*) Octahedral arrangement of six σ-*bps* (σ-*bps* = 6, *lp* = 0)

(*f*) Pentagonal bipyramidal arrangement of seven σ-*bps* (σ-*bps* = 7, *lp* = 0)

Fig. 12.1 : Spatial arrangement of 2, 3, 4, 5, 6 and 7 σ-bonding electron pairs (σ-*bps*) round the central atom (A). The central atom is not surrounded by any lone pair of electrons (*lps*). B are the atoms with which central atom (A) is linked by σ-*bps*.

2. Geometry of the molecule/ion and bond angle. (*a*) While determining geometry of a given molecule or ion on the basis of VSEPR theory, the presence of π-*bps* surrounding the central atom of the species should not be considered, since these electron pairs donot influence the geometry of the molecule or ion. It is only σ-*bps* and *lps* surrounding the central atom which decide the shape of the species and hence only these electron pairs (*i.e.*, σ-*bps* and *lps*) should be considered.

(*b*) *Effect of lps on the geometry and bond angle of a given species.* If the central atom is surrounded purely by σ-*bps*, then the shape (geometry) of the molecule or ion is the same as the spatial arrangement of the σ-*bps* round the central atom. However, if the central atom is surrounded by σ-*bps* as well as by *lps*, then the geometry of the molecule or ion gets distorted from its expected geometry and becomes different from the spatial arrangement of (σ-*bps* + *lps*) (See Table 12.1). Due to the presence of *lps*, bond angle also becomes different from the expected bond angle between the σ-*bps*. For example, since C-atom (central atom) in CH_4 molecule is surrounded only by four σ-*bps*, it has the expected tetrahedral shape with the expected H—C—H bond angle of 109.5° (expected angle). On the other hand, since N–atom (central atom) in NH_3 molecule is surrounded by three σ-*bps* and one *lp*, it does not have the tetrahedral shape; rather it has pyramidal shape with H—N—H bond angle = 107.5°.

Table 12.1 : *Effect of lone pairs of electrons (lps) on the geometry of a given species.*

No. of (σ-*bps* + *lps*) surrounding the central atom and examples of molecules/ions having these electron pairs	No. of σ-*bps* and *lps*. Type of molecule/ion is given in bracket	Spatial arrangement of (σ-*bps* + *lps*). Bond angles are given in parentheses	Actual shape (geometry) of the molecule/ion
2. BeF_2,$BeCl_2$	σ-*bps* = 2 *lp* = 0 (AB_2)	Linear (180°)	Linear
3. BF_3, BCl_3, $GaCl_3$, BH_3	σ-*bps* = 3 *lp* = 0 (AB_3)	Trigonal planar (120°)	Trigonal planar
$SnCl_2$, Pb Cl_2	σ-*bps* = 2 *lp* = 1 [AB_2 (*lp*)]	Trigonal planar	V-shape or angular
4. CH_4, CCl_4, $SiCl_4$, NH_4^+, BH_4^-	σ-*bps* = 4 *lp* = 0 (AB_4)	Tetrahedral (109.5°)	Tetrahedral
NH_3, PH_3, PX_3	σ-*bps* = 3 *lp* = 1 [AB_3 (*lp*)]	Tetrahedral	Trigonal pyramidal
H_2O, SCl_2, $SeCl_2$	σ-*bps* = 2 *lps* = 2 [AB_2 $(lp)_2$]	Tetrahedral	V-shape or angular
5. PCl_5, (g), PF_5, $SbCl_5$	σ-*bps* = 5 *lp* = 0 (AB_5)	Trigonal bipyramidal (120°, 90°)	Trigonal bipyramidal
$5F_4$, $SeCl_4$, TeX_4	σ-*bps* = 4 lps = 1 [AB_4 (*lp*)]	Trigonal bipyramidal	Irregular tetrahedral or see-saw
ClF_3, BrF_3, ICl_3, IF_3	σ-*bps* = 3 *lps* = 2 [AB_3 $(lp)_2$]	Trigonal bipyramidal	T-shape
XeF_2, ICl_2^-	σ-*bps* = 2 *lps* = 3 [AB_2 $(lp)_3$]	Trigonal bipyraminal	Linear
6. SF_6, TeF_6, $(SbF_6)^{2-}$, $(SbF_5)^{2-}$ etc.	σ-*bps* = 6 *lp* = 0 (AB_6)	Octahedral (90°, 90°)	Octahedral
IF_5, BrF_5, ClF_5, $(SbF_5)^-$	σ-*bps* = 5 *lp* = 1 [AB_5 (*lp*)]	Octahedral	Square pyramidal
XeF_4, ICl_4^{-1}	σ-*bps* = 4 *lps* = 2 [AB_4 $(lp)_2$	Octahedral	Square planar
7. IF_7	σ-*bps* = 7 *lp* = 0 (AB_7)	Pentagonal bipyramidal (72°, 90°)	Pentagonal bipyramidal
XeF_6	σ-*bps* = 6 *lp* = 1 [AB_6 (*lp*)]	Pentagonal bipyramidal	Distorted octahedral

3. Order of the repulsion between electron pairs. σ–*bps* are considered to be localised between the two atoms linked together by covalent bond while a *lp* is held by only one atom, *i.e.*, bonding electron pair is under the influence of two atoms and a *lp* is under the influence of one atom only *viz.*, central atom. Thus we expect that the orbital containing a *lp* is more spread out in space in comparison to that (*i.e.*, orbital) which contains a bonding electron pair. Consequently a *lp* will exert more repulsion on the bonding electron pair as compared to the repulsion exerted by a bonding electron pair on the other bonding electron pair, *i.e.*,

$$(lp - bp) > (bp - bp) \quad ...(i)$$

On similar lines of argument it can be shown that :

$$(lp - lp) > (lp - bp) \quad ...(ii)$$

On combining the above two relations we get :

$$(lp - lp) > (lp - bp) > (bp - bp) \qquad ...(iii)$$

Relation (*iii*) gives us the order of the magnitude of repulsion between electron pairs.

The repulsions between the electron pairs at 90° are stronger than that between the electron pairs at 120°. These repulsions are the weakest (lowest) when the electron pairs are at 180°.

To determine the shape of molecules and ions whose central atom is surrounded by σ-*bps* only and no *lps*

In the following discussion we shall explain the shape of molecules and ions with the help of VSEPR theory. We shall consider the following types of molecules/ions.

1. Species of AB_2 type : Linear species. The following examples are considered.

(*a*) *$BeCl_2$ molecule in the vapour phase.* The Lewis structure of this molecule is Cl—Be—Cl in which the central atom (Be-atom) is surrounded by two σ-*bps* and no *lp*. These two σ-*bps* will occupy their positions as far apart as possible so that the electrostatic repulsion between them is reduced to minimum. This is possible only when these bonding electron pairs are at 180° about the central atom, *i.e.*, the bonding electron pairs occupy positions on the opposite sides of Be-atom forming Cl—Be—Cl angle equal to 180°.

Thus $BeCl_2$ molecule in the vapour phase has linear geometry

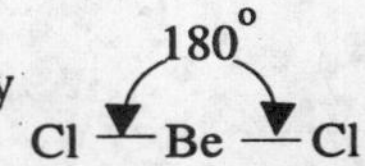

(*b*) *CO_2 molecule.* Lewis structure of this molecule is O = C = O. Since the central atom (C-atom) in this molecule is surrounded by two σ-*bps* and no *lp* (π-*bps* have not been considered), this molecule, like $BeCl_2$ molecule, has linear geometry.

Other examples of linear molecules are hydrogen cyanide (H — C≡ N) and acetylene (H — C ≡ C — H).

2. Species of AB_3 type: Trigonal planar species. The following examples may be considered.

(*a*) *BCl_3 molecule.* Lewis structure of this molecule is Cl—B—Cl (with a third Cl bonded below B) which shows that B-atom (central atom) is surrounded by three σ-*bps* and no *lp*. This molecule can have any of the two shapes shown in Fig. 12.2. T-shaped geometry has two *bp-bp* repulsions at 90° and one *bp-bp* repulsion at 180°, while trigonal geometry has three *bp-bp* repulsions at 120°. Consequently in trigonal structure the repulsive forces between the electron pairs are reduced to minimum and hence this structure is the most probable structure of BCl_3 molecule.

BF_3, $GaCl_3$ and BH_3 molecules also have trigonal planar geometry.

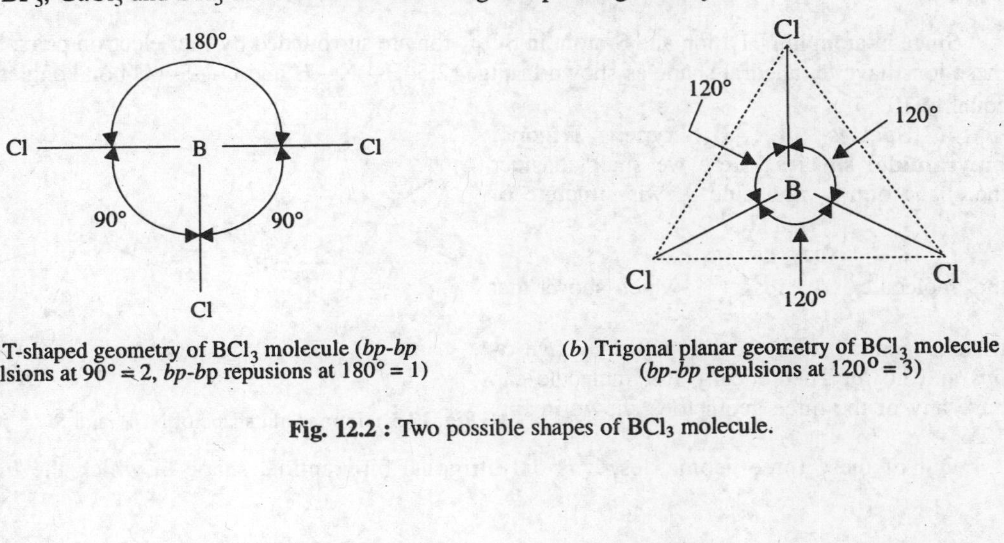

(*a*) T-shaped geometry of BCl_3 molecule (*bp-bp* repulsions at 90° = 2, *bp-bp* repusions at 180° = 1)

(*b*) Trigonal planar geometry of BCl_3 molecule (*bp-bp* repulsions at $120^o = 3$)

Fig. 12.2 : Two possible shapes of BCl_3 molecule.

(*b*) *SO_3 molecule*. Lewis structure of this molecule as given alongside (Fig. 12.3) suggests that S-atom (central atom) is linked with one O-atom by a double bond and with two O-atoms by coordinate (covalent) bonds. Now since S-atom is surrounded by three **σ-*bps*** and no *lp* (π-bonds have not been considered), SO_3 molecule, like BCl_3 molecule, has trigonal geometry as shown in Fig. 12.3.

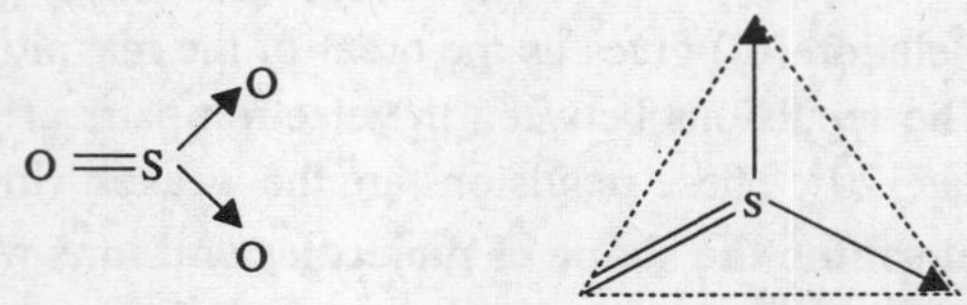

Lewis structure of SO_3 molecule — Trigonal planar geometry of SO_3 molecule

Fig. 12.3 : Lewis structure and trigonal planar shape of SO_3 molecule.

3. **Species of AB_4 type: Tetrahedral species.** The following examples may be considered:

(*a*) *CH_4 molecule*. Lewis structure of this molecule is 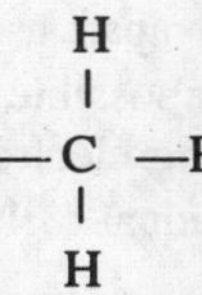 which shows that C-atom (central atom) is surrounded by four **σ-*bps*** and no *lp*. This molecule can have any of the two structures shown in Fig. 12.4.

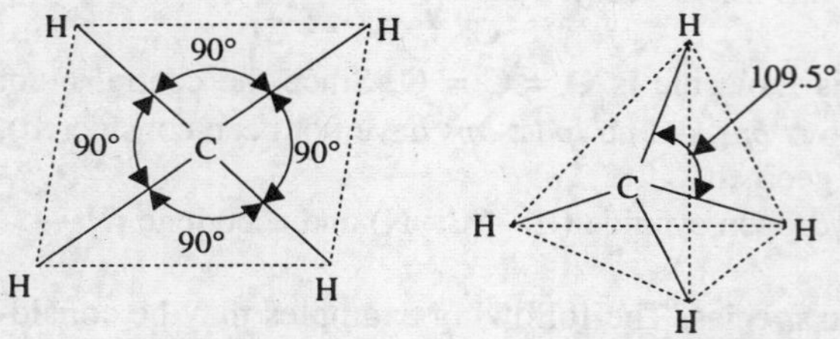

(*a*) Square planar geometry. *bp-bp* repulsions at 90°=4

(*b*) Tetrahedral shape *bp-bp* repulsions at 90° = Zero

Fig. 12.4 : Two possible structures of CH_4 molecule.

Clearly, since the repulsive interactions in tetrahedral shape are minimum, this shape is the most probable shape of CH_4 molecule. A tetrahedron is a solid figure which contains four faces each of which is an equilateral triangle. Each of the four H—C—H bond angles is equal to 109.5°.

CCl_4 and $SiCl_4$ molecules aso have tetrahedral geometry.

(*b*) *NH_4^+ and SO_4^{2-} ions*. Lewis structure of these ions can be written as :

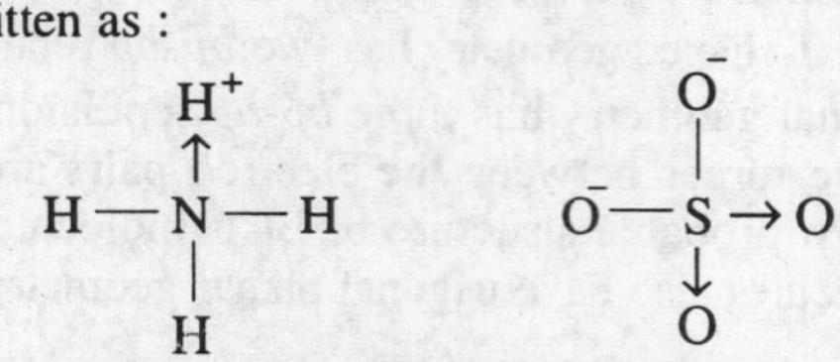

Since N-atom in NH_4^+ ion and S-atom in SO_4^{2-} ion are surrounded by four electron pairs, both these ions have tetrahedral shape as shown in Fig. 12.5. H—N—H and O—S—O bond angles are equal to 109.5°.

4. **Species of AB_5 type: Trigonal bipyramidal species.** Here we shall consider the shape of PF_5 molecule. Lewis structure of this molecule is 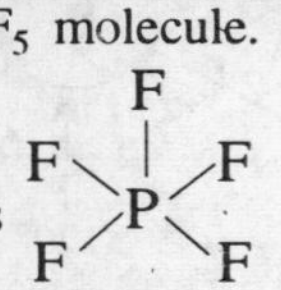 which shows that P-atom (central atom) is surrounded by five σ-*bps* and no *lp*. Theoretically PF_5 molecule can have any of the three geometries shown in Fig. 12.6.

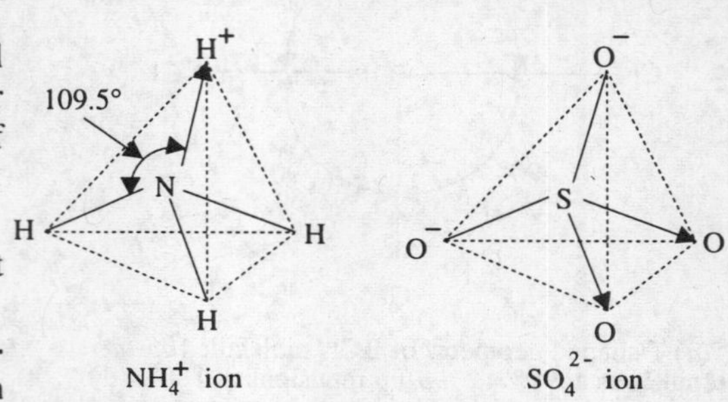

NH_4^+ ion — SO_4^{2-} ion

Fig. 12.5 : Tetrahedral shape of NH_4^+ and SO_4^{2-} ions.

Out of these three geometries, it is only trigonal bipyramidal shape in which the *bp-bp* repulsions are minimum and hence this geometry is the most probable geometry of PF_5 molecule.

Other molecules which have trigonal bipyramidal shape are PCl_5 (*g*), $SbCl_5$ etc.

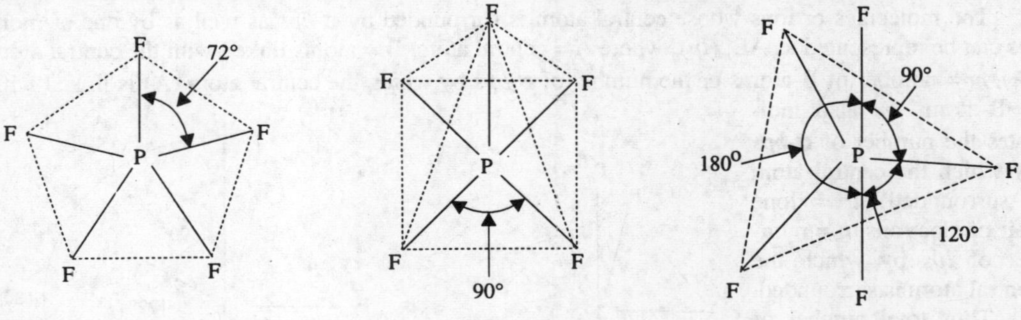

(*a*) Planar pentagonal structure (bond angle = 72°, *bp-bp* repulsions at 72° = 5)

(*b*) Square pyramidal geometry (bond angle = 90°, repulsions at 90° = 8)

(*c*) Trigonal bipyramidal geometry. (*bp-bp*) repulsions at 90° = 6

Fig. 12.6 : Three possible geometries of PF_5 molecule.

5. Species of AB_6 type: Octahedral species. SF_6 and TeF_6 are the examples of molecules which have octahedral shape. Here we shall consider the geometry of SF_6 molecule. Lewis structure of this molecule is  in which S-atom (central atom) is surrounded by six σ-*bps* and no *lp*. This molecule can have any of the three structures shown in Fig. 12.7.

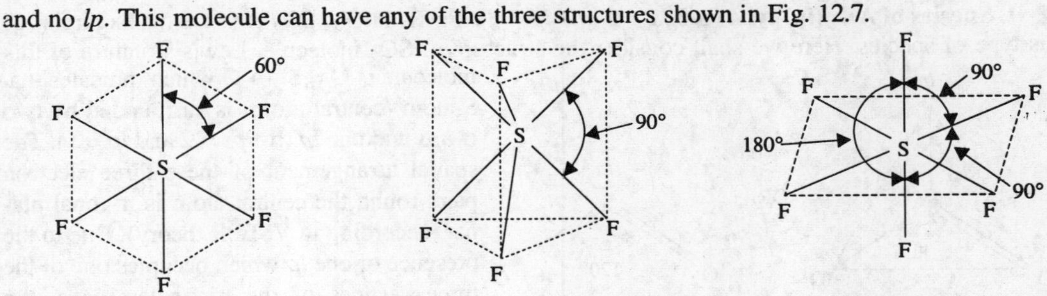

(*a*) Planar hexagonal geometry. *bp-bp* repulsions at 60° = 6. These repulsions are strong.

(*b*) Trigonal prismatic geometry. *bp-bp* repulsions at 90° = 6. These repulsions are very strong.

(*c*) Octahedral geometry. *bp-bp* repulsions at 90° = 12.

Fig. 12.7: Three possible geometries of SF_6 molecule.

Out of these structures octahedral structure is the most stable. Each of F—S—F-bond angle is equal to 90°.

6. Species of AB_7 type : Pentagonal bipyramidal species. IF_7 is an example of AB_7 type species. So here we shall consider the structure of this molecule. Lewis structure of this molecule is  which shows that I-atom (central atom) is surrounded by seven σ-*bps* and no *lp*.

Thus according to VSEPR theory this molecule has pentagonal bipyramidal geometry shown in Fig. 12.8. In this figure, F_a and F_b are the fluorine atoms lying in the pentagonal plane and at the axes of the pentagon respectively.

To determine the shape of molecules and ions whose central atom is surrounded by σ-*bps* as well as by one or more *lps*.

The molecules or ions whose central atom is surrounded by σ-*bps* as well as by one or more *lps* can be represented as $AB_m(lp)_n$ where A = central atom, B = atoms linked with the central atom (A), *m* = number of B-atoms or the number of σ-*bps* by which the central atom (A) is linked with *m* B-atoms (*m* also indicates the number of σ-*bps* by which the central atom is surrounded), *lp* = lone pair of electrons, *n* = number of *lps* by which the central atom is surrounded.

Thus total number of electron pairs which surround the central atom is equal to (*m* + *n*). While determining the geometry of the molecules or ions having one or more π-bonds, it should be remembered that the presence of π-*bps* surrounding the central atom does not influence the geometry of the molecule/ion, We should, therefore, not consider the presence of π-bonds. Let us consider the structure of the species of the following types:

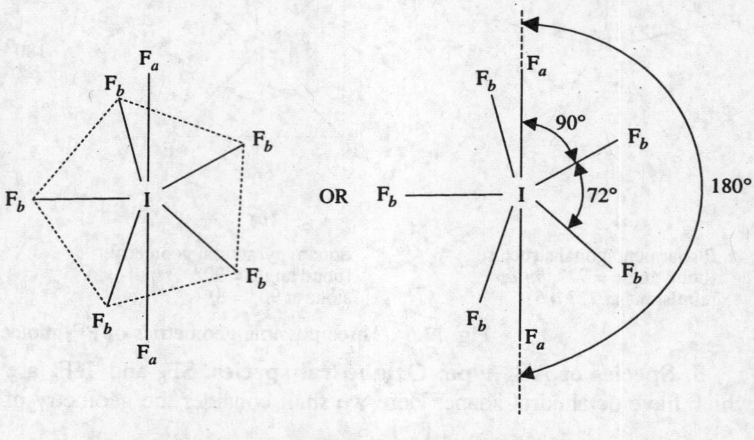

Fig. 12.8 : Pentagonal bipyramidal structure of IF_7 molecule.

1. Species of AB_2 (lp) type (σ-*bps* = *2 and lp* = *1*). $SnCl_2$ $PbCl_2$, SO_2 etc., are the examples of this type of species. Here we shall consider the structure of SO_2 molecule. Lewis structure of this molecule is O = S → O which indicates that S-atom (central atom) is surrounded by two σ-*bps* and one *lp* (σ-*bps* = 2 and *lp* = 1). The spatial arrangement of these three electron pairs round the central atom is trigonal planar (according to VSEPR theory). Due to the presence of one *lp* which occupies one of the three vertices of the triangular plane, the shape of SO_2 molecule is not trigonal planar but it is angular with O—S—O bond angle equal to 120°. (See Fig. 12.9).

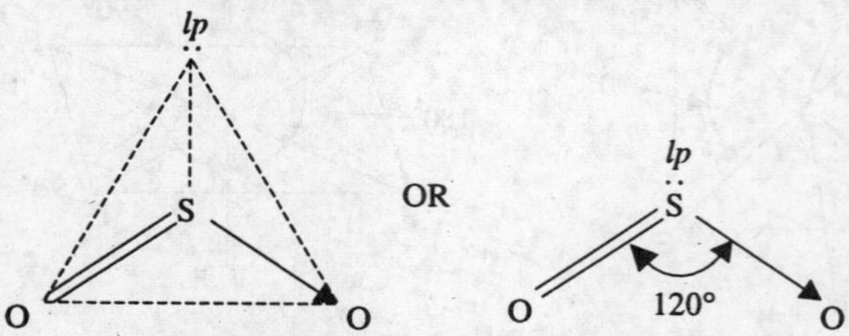

Fig. 12.9 : Angular shape of SO_2 molecule.

2. Species of AB_3 (*lp*) type. (*m* = 3, *n* = 1, *i.e.*, σ-*bps* = 3, *lp* = 1). NH_3, PH_3, PCl_3, RNH_2, R_2N, R_3N, ClO_3^- etc., are the examples of this type of species. Here we shall consider the structure of NH_3 and PCl_3 molecules.

(*a*) *Structure of NH_3 molecule.* Lewis structure of this molecule is H—$\ddot{N}$—H (with H bonded below N) which shows that since N-atom (central atom) is surrounded by 4 electron pairs (σ-*bps* = 3 and *lp* = 1), the spatial arrangement of these four electron pairs round N-atom (central atom) is tetrahedral (according to one of the postulates of VSEPR theory). Due to the presence of one *lp* at one of the four vertices of the tetrahedron, the shape of NH_3 molecule is not tetrahedral but it gets distorted and becomes trigonal pyramidal as shown in Fig. 12.10 (*a*).

H — N — H bond angle in NH_3 molecule. Although the orientation of 4 electron pairs (σ-*bps* = 3, *lp* = 1) round N-atom is tetrahedral, H—N—H bond angle is not equal to the expected

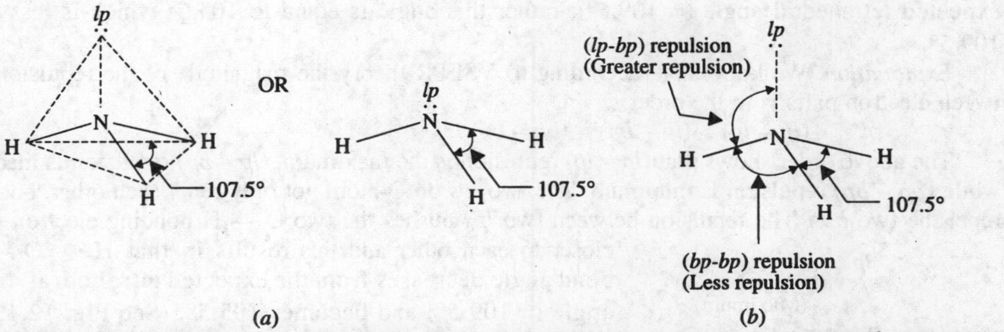

Fig. 12.10 *(a)* **: Trigonal pyramidal shape of NH_3 molecule.** *lp* **indicates a lone pair of electrons.** *(b)* **To explain why H—N—H bond angle in NH_3 is less than the tetrahedral bond angle (=109.5°).** *lp* **indicates a lone pair of electrons on N-atom.**

tetrahedral angle (109.5°); rather this angle is 107.5° which is less than 109.5°. Why H—N—H bond angle is less than 109.5° has been explained as follows:

We know that, according to VSEPR theory, repulsion between a ***lp*** **and a** ***bp*** **is greater than that between a** ***bp*** **and a** ***bp, i.e.,***

$$(lp - bp) > (bp - bp)$$

It is due to the greater magnitude of (*lp* – *bp*) repulsion in comparison to that of (*bp* – *bp*) repulsion that the two N — H bonding electron pairs are pushed closer to each other and the result is that H — N — H bond angle decreases from the expected tetrahedral bond angle (= 109.5°) and becomes 107.5° [See Fig. 12.10 (*b*)].

(b) Structure of PCl_3 molecule. **The structural formula of this molecule is $Cl-\ddot{P}-Cl$ (with Cl below P) which shows that the central atom (P-atom) is surrounded by three σ-*bps* and one *lp*. Thus, like NH_3 molecule, PCl_3 molecule also has trigonal pyramidal structure and the observed Cl—P—Cl bond angle is 100°. (See Fig. 12.11).**

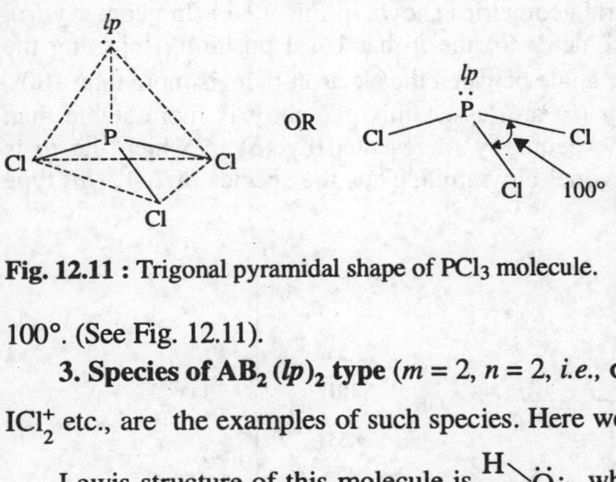

Fig. 12.11 : Trigonal pyramidal shape of PCl_3 molecule.

3. Species of AB_2 $(lp)_2$ type (m = 2, n = 2, *i.e.*, σ-*bps* = 2 and *lps* = 2). H_2O, SCl_2, $SeCl_2$, NH_2^-, ICl_2^+ etc., are the examples of such species. Here we shall consider the structure of H_2O molecule.

Lewis structure of this molecule is $H_2\ddot{O}:$ (H—Ö:—H, bent) which shows that since O-atom (central atom) is surrounded by 4 electron pairs (σ-*bps* = 2 and *lps* = 2), the spatial orientation of 4 electron pairs round O-atom is tetrahedral (according to VSEPR theory). Due to the presence of two *lps* located at the two tetrahedral positions, the shape of H_2O molecule gets distorted from tetrahedral shape and becomes angular (V-shape or bent) as shown in Fig. 12.12.

H — O — H bond angle in H_2O molecule. Although the special orientation of 4 electron pairs (σ-*bps* = 2, *lps* = 2) round O-atom is tetrahedral, H — O — H bond angle is not equal to the

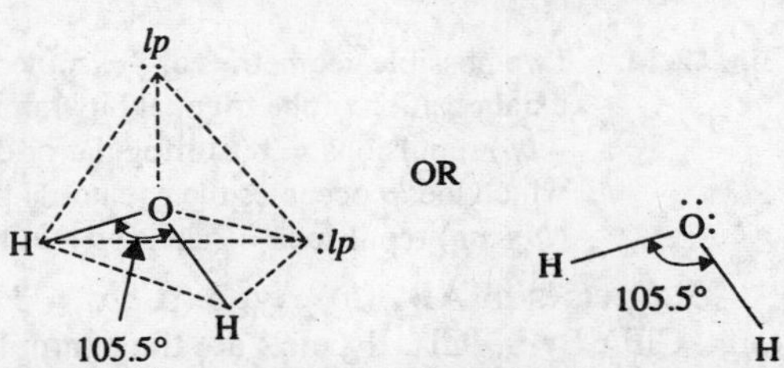

Fig. 12.12 : Angular (V-shaped or bent) structure of H_2O

expected tetrahedral angle (= 109.5°); rather this angle is equal to 105.5° which is less than 109.5°.

Explanation. We know that, according to VSEPR theory the magnitude of the repulsion between electron pairs is in the order:

$$(lp - lp) > (lp - bp) > (bp - bp)$$

The above order shows that $(lp - lp)$ repulsion is the maximum, $(lp - bp)$ repulsion is medium while $(bp - bp)$ repulsion is minimum. The two *lps* on O-atom not only repel each other, but also repel the two *bps*. The repulsion between two *lps* pushes the two O — H bonding electron pairs closer to each other and this results in that H — O — H bond angle decreases from the expected tetrahedral bond angle (= 109.5°) and becomes 105.5°. (See Fig. 12.13).

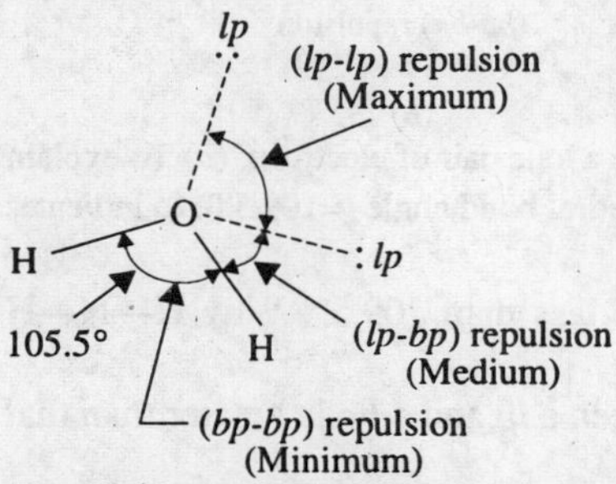

Fig. 12.13 : To explain why H — O — H bond angle in H_2O molecule is less than the tetrahedral bond angle.

4. Species of AB_4 (*lp*) type (σ-*bps* = 4 and *lp* = 1). SF_4, $TeCl_4$, $SeCl_4$ etc., are the examples of this type of species. Here we shall discuss the structure of SF_4 molecule.

Lewis structure of this molecule is 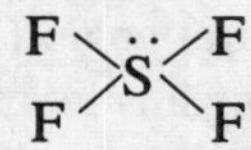 which shows that since S-atom (central atom) is surrounded by five electron pairs (σ-*bps* = 4 and *lp* = 1), the spatial arrangement of these electron pairs round S-atom is trigonal bipyramidal. Due to the presence of one *lp* of electrons, SF_4 molecule has *irregular* or *distorted tetrahedral* (also called *see-saw*) geometry in which the *lp* can occupy either one of the two axial positions or one of the three equatorial (also called basal) positions. Thus theoretically SF_4 molecule can have any of the two distorted tetrahedral geometries shown in Fig. 12.14. In geometry (*a*), the *lp* occupies the axial position while in geometry (*b*) the *lp* has basal position. Rejecting the electron pair–electron pair repulsions where the angle between the electron-pairs is more than 100°, the $(lp - bp)$ repulsions involved in geometry (*b*) are lesser, this geometry is more stable than geometry (*a*). Thus SF_4 molecule has see-saw geometry represented by (*b*) in which the *lp* is situated at one of the basal positions of the trigonal bipyramid. Thus the species of AB_4 (*lp*) type have geometry (*b*).

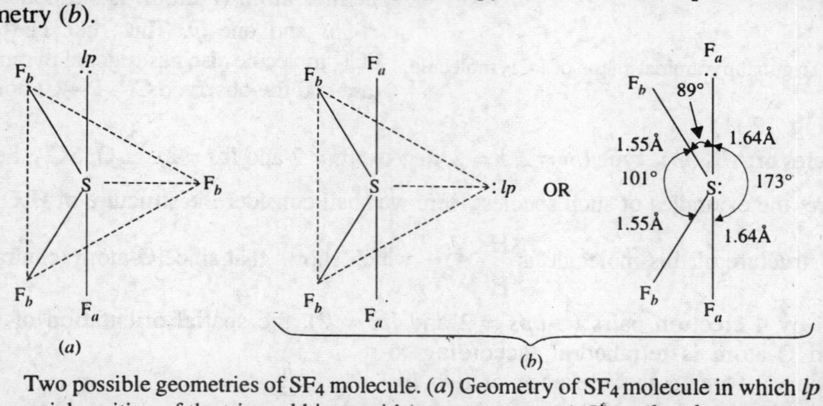

Fig. 12.14. Two possible geometries of SF_4 molecule. (*a*) Geometry of SF_4 molecule in which *lp* occupies the axial position of the trigonal bipyramid (wrong geometry). Here $(lp - bp)$ repulsions = 3 and (bp $- bp)$ repulsions = 3 (*b*) Irregular or distorted tetrahedral (see-saw) geometry of SF_4 molecule in which the *lp* occupies the equatorial position of the trigonal bipyramid (correct geometry). Here $(lp - bp)$ repulsions = 2 and $(bp - bp)$ repulsions = 4.

5. Species of AB_3 $(lp)_2$ type (σ-*bps* = 3 and *lps* = 2). Interhalogen compounds of XY_3 type (*e.g.*, ClF_3, BrF_3, ICl_3, IF_3 etc.) are the examples of AB_3 $(lp)_2$ type species. In XY_3 interhalogens X is the bigger halogen atom and Y is the smaller halogen atom. X is the central atom in these molecules. As an example let us discuss the structure of ClF_3 molecule. Lewis structure of this molecule is 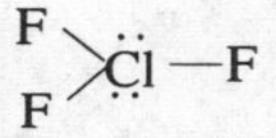 which indicates that since Cl-atom (central atom) is surrounded by five electron pairs (σ-*bps* = 3 and *lps* = 2), the spatial orientation of these 5 electron pairs round Cl-atom is trigonal bipyramidal. Due to the presence of two *lps*, ClF_3 molecule has *slightly bent T-shaped* structure.

Theoretically, depending on the positions occupied by the two *lps*, ClF_3 molecule can have any of the three geometries namely (*a*), (*b*) and (*c*) shown in Fig. 12.15.

If we do not consider the electron pair-electron pair repulsions where the angle between them is greater than 100°, we find that geometry (*c*) in which both the *lps* occupy the equatorial positions of the trigonal bipyramid involves minimum repulsion. Thus geometry (*c*) which is called *bent T-shaped geometry* is the most stable geometry for ClF_3 molecule, *i.e.*, ClF_3 molecule or any other molecule or ion of AB_3 $(lp)_2$ type has *bent T-shaped geometry*.

Each of the F—Cl—F (F_a — Cl — F_b) bond angles is equal to 87.5°. The basal Cl—F (Cl — F_b) bond length is equal to 1.60Å while each of the two axial Cl—F (Cl — F_a) bond lengths is equal to 1.70°Å (see Fig. 12.15).

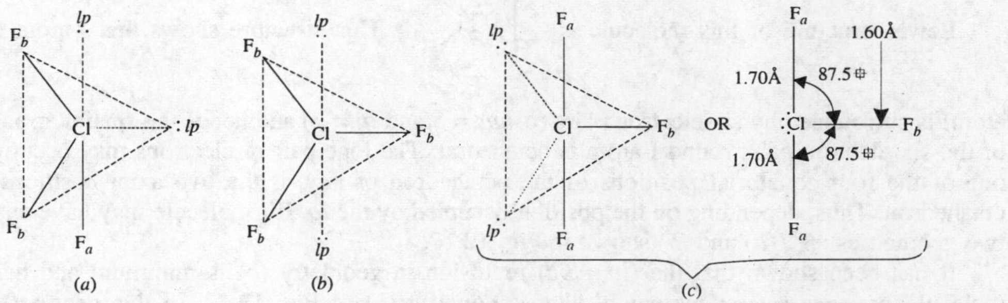

Fig. 12.15 : Three possible geometries of ClF_3 molecule. (*a*) In this geometry one *lp* is at the equatorial position and the other *lp* is at the axial position of the trigonal bipyramid (wrong geometry). Here (*lp* – *lp*) repulsions = 1, (*lp* – *bp*) repulsions = 3 and (*bp* – *bp*)repulsions = 2. (*b*) In this geometry both the *lps* occupy the axial positions of the trigional bipyramid (wrong geometry). Here (*lp* – *lp*) repulsions = 0, (*lp* – *bp*) repulsions = 6 and (*bp* – *bp*) repulsions = 0. (*c*) Bent T-shaped geometry in which both *lps* occupy the equatorial positions of the trigonal bipyramid (correct geometry). Here (*lp* – *lp*) repulsions = 0, (*lp* – *bp*) repulsions = 4 and (*bp* – *bp*) repulsions = 2.

6. Species of AB_2 $(lp)_3$ type (σ–*bps* = 2 and *lps* = 3). XeF_2, ICl_2^- etc. are the examples of AB_2 $(lp)_3$ type species. As an example let us consider the geometry of XeF_2 molecule.

Lewis structure of this molecule is F—Xe—F which shows that since Xe-atom (central atom) is surrounded by five electron pairs (σ-*bps* = 2 and *lps* = 3), the spatial arrangement of 5 electron pairs round the central atom (Xe-atom) is trigonal bipyramidal. Due to the presence of three *lps*, XeF_2 molecule assumes linear shape with F—Xe—F bond angle equal to 180°.

Depending on the positions occupied by the three *lps*, XeF_2 molecule can have any of the three structures shown in Fig. 12.16. The number of electron pair-electron pair repulsions shown in each geometry indicates that this repulsion in geometry (*c*) is minimum and hence this geometry is the most stable. In other words we can say that all the three *lps* occupy the equatorial positions of the trigonal bipyramid and hence XeF_2 or any other species of AB_2 $(lp)_3$ type has linear shape in which the length of each of Xe — F bonds is equal to 2.0 Å.

7. Species of AB_5 (*lp*) type (σ–*bps* = 5 and *lp* = 1). IF_5, BrF_5, ClF_5, $[SbF_5]^{2-}$ etc., are the examples of AB_5 (*lp*) type species. As an example let us discuss the geometry of IF_5 molecule.

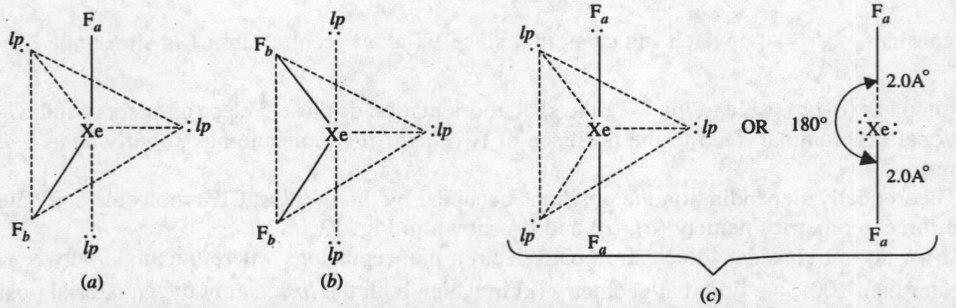

Fig. 12.16 : Three possible geometries of XeF_2 molecule. (*a*) In this geometry one *lp* is at the axial position and the two *lps* are at the equatorial positions of the trigonal bipyramid (wrong geometry). Here (*lp* – *lp*) repulsions = 2, (*lp* – *bp*) repulsions = 3 and (*bp* – *bp*) repulsions = 1 (*b*) In this geometry one *lp* is at the equatorial position and the two *lps* are at the axial positions (wong geometry). Here (*lp* – *lp*) repulsions = 2, (*lp* – *bp*) repulsions = 4 and (*bp* – *bp*) repulsions = 0. (*c*) Linear shape of XeF_2 molecule in which all the three *lps* occupy the equatorial positions (correct geometry). Here (*lp* – *lp*) repulsions = 0 and (*lp* – *bp*) repulsions = 6.

Lewis structure of this molecule is 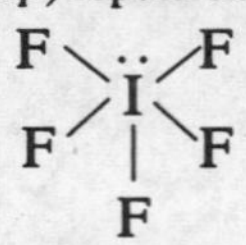 This structure shows that I-atom (central atom) is surrounded by six electron pairs (σ-*bps* = 5 and *lp* = 1) and hence the spatial arrangement of the six electron pairs round I-atom is octahedral. The lone pair of electrons may occupy either one of the four equatorial positions of the octahedron or any of the two axial positions of the octahedron. Thus, depending on the position occupied by the *lp*, IF_5 molecule may have any of the two geometries *viz.*, (*a*) and (*b*) shown in Fig. 12.17.

It has been shown that the (*lp* – *bp*) repulsion in geometry (*b*) is minimum and hence IF_5 molecule *assumes square pyramidal shape* shown at (*b*) of Fig. 12.17. In this geometry the *lp* occupies the axial position.

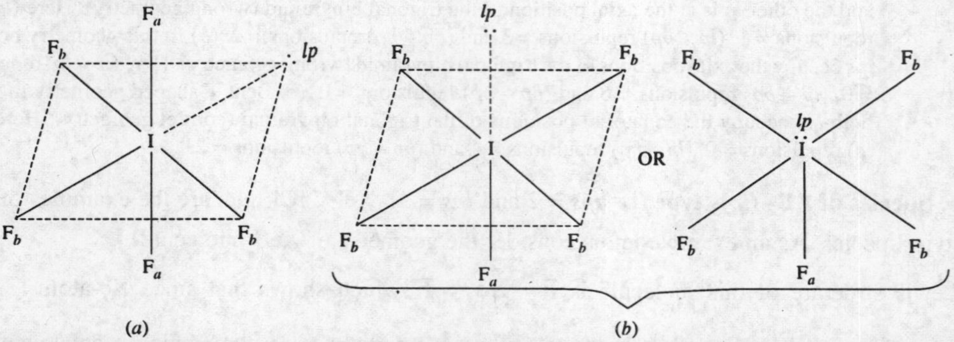

Fig. 12.17 : Two possible geometries of IF_5 molecule. (*a*) In this geometry *lp* occupies the equatorial position of the octahedtron (wrong geometry). (*b*) Square pyramidal geometry of IF_5 molecule in which the *lp* occupies the axial position (correct geometry).

8. Species of AB_4 (*lp*)$_2$ type (σ-*bps* = 4 and *lps* = 2) XeF_4, ICl_4^- etc. are the examples of this type of species. As an example let us consider the geometry of XeF_4 molecule.

Lewis structure of this molecule is 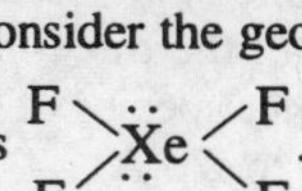 . This structure clearly shows that since Xe-atom (central atom) is surrounded by six electron pairs (σ-*bps* = 4 and *lps* = 2), spatial arrangement of these six electron pairs round Xe-atom is octahedral.

Depending on the positions occupied by the two *lps*, XeF_4 molecule can have any of the three geometries (*a*), (*b*) and (*c*) depicted in Fig. 12.18. It has been observed that electron pair-electron pair repulsions in geometry (*c*) in which the two *lps* occupy the axial positions of the octahedron are minimum and XeF_4 has square planar geometry in which each of the two F_b—Xe—F_b bond angles is equal to 90° and each of the four Xe—F_b bond lengths is equal to 1.95Å.

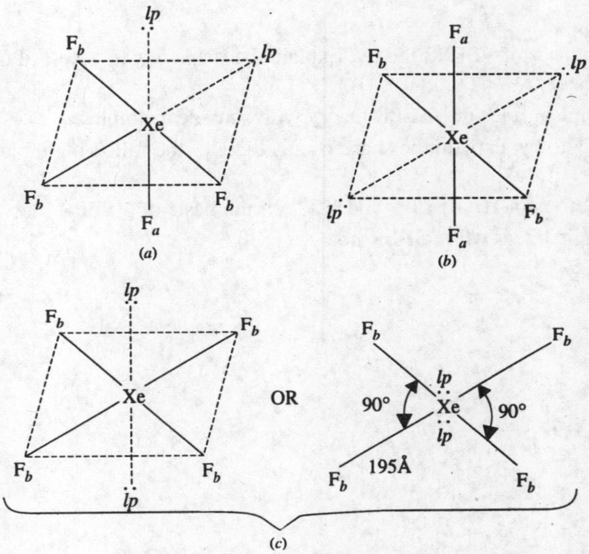

Fig 12.18 : Three different geometries of XeF_4 molecule. (*a*) Geometry of XeF_4 molecule in which one *lp* is at the equatorial position and the other *lp* is at the axial position of the octahodron. (wrong geometry). Here (*lp* — *lp*) repulsions = 1, (*lp* — *bp*) repulsions = 6 and (*bp* — *bp*) repulsions = 5. (*b*) Geometry of XeF_4 molecule in which both *lps* occupy the axial positions (wrong geometry) (*c*) Square planar geometry of XeF_4 molecule in which both *lps* occupy the axial positions (correct geometry). Here (*lp* – *lp*) repulsions = 0, (*lp* – *bp*) repulsions = 8 and (*bp* – *bp*) repulsions = 4.

University Questions

1. Describe the basic ideas of VSEPR theory. Explain the application of this theory for predicting the shapes of the molecules namely BCl_3, NH_3, H_2O and SF_6. *(Kerela, 1991)*
2. In CH_4, NH_3 and H_2O molecules the bond angle is decreasing. Explain giving reasons. *(Gujarat, 1992)*
3. (*a*) What will be the shape of the following molecules on the basis of VSEPR model : ClF_3, XeF_2, PCl_5 and SF_6. *(Nagpur, 1993)*

   (*b*) Explain the shape of BF_3 molecule on the bais of VSEPR theory. *(Meerut, 96)*
4. Using VSEPR theory, predict and draw the shapes of the following molecules : BF_3, H_2O, PCl_5, SF_6, NH_3, XeF_4, IF_3, $SiCl_4$, SO_3 and XeO_3. *(Delhi, 1980, 81; Bharathiar, 85)*
5. What are the limitations of VSEPR theory? *(Delhi, B.Sc. Hons., 1982)*
6. Write a note on Valence-shell Electron Pair Repulsion (VSEPR) theory. On the basis of this theory discuss the shapes of BrF_3 and PCl_5. *(Allahabad, 1988; Lucknow, 93)*
7. How may bonding and non-bonding electron pairs are there in the valence-shell of the central atoms in SF_4, BrF_5 and XeF_5 molecules? *(Allahabad, 1988)*
8. Using the VSEPR theory, predict the structure of each of the following and indicate whether the bond angles are likely to be distorted from the theoretical value : $BeCl_2$, H_2O and $SnCl_2$. *(Delhi, 89)*
9. Write Lewis dot formula of HOCl, BF_3 and NH_4^+. *(Nagpur, 1987)*
10. Explain basic features of VSEPR theory. Draw Lewis structures of BCl_3, NH_3 and ICl_3, and deduce their geometry. *(Allahabad, 89)*

11. Describe the structures of H_2O, NH_3 and CH_4 molecules on the basis of VSEPR theory. Explain why the bond angles are different in the three molecules. *(Andhra, 87)*

12. Discuss the shape of NH_3, CH_4 and PCl_3 molecules on the basis of VSEPR theory. *(Kashmir, 1988)*

13. (*a*) Discuss the basic postulates of VSEPR theory. State some of its limitations.

(*b*) Give reasons for the following on the basis of VSEPR theory. The H—C—H, H—N—H and H—O—H bond angles in CH_4, NH_3 and H_2O are 109.5°, 107.3° and 104.5° respectively. *(Utkal, 91; Lucknow, 95)*

14. Describe the basic ideas of VSEPR theory. Explain the shape of ClF_3, XeF_2, PCl_5 and SF_4 molecules on the basis of this theory. *(VBS Purvanchal 2002)*

15. Explain stereochemical rules of VSEPR theory with suitable examples. *(Nagpur 2002)*

16. Using V S E P R Theory, explain the shape of any three of the following molecules (ion) : (*i*) ClF_3 (*ii*) SF_4 (*iii*) NH_3 (*iv*) H_3O^+ *(Rohilkhand 2003)*

17. Discuss the geometries of H_3O^+, ClF_3 and ICl_2^- on the basis of VSEPR theory. *(Rohilkhand 2003)*

18. Explain why BeF_2 is linear while SF_2 is not. *(Delhi 2003)*

13

Hybridisation and its Different Types

Hybridisation

How to explain the bicovalency, tricovalency and tetracovalency of Be, B and C-atoms with the help of the concept of promotion of electron(s)

The valence-shell electronic configuration of Be, B and C-atoms in their ground state are given below (n = No. of unpaired electrons).

$$Be\ (2s^2\ 2p^0) \rightarrow 2s^2\ 2p_x^0\ 2p_y^0\ 2p_z^0 \quad (n = 0)$$
$$B\ (2s^2\ 2p^1) \rightarrow 2s^2\ 2p_x^1\ 2p_y^0\ 2p_z^0 \quad (n = 1)$$
$$C\ (2s^2\ 2p^2) \rightarrow 2s^2\ 2p_x^1\ 2p_y^1\ 2p_z^0 \quad (n = 2)$$

A look at these configurations shows that Be-atom has no unpaired electron (n = 0) and, therefore, should have *zero covalency, i.e.,* Be-atom should not be able to form any compound. Similarly since B-atom has one unpaired electron (n = 1), it should have *monocovalency, i.e.,* it should form the compounds like BH, BF, BCl etc. Similarly C-atom has two unpaired electrons (n = 2) and hence should show *bicovalency, i.e.,* this element should form the compounds like CH_2, CF_2, CCl_2 etc. But contrary to the above expectations, we find that beryllium is bicovalent in its compounds (*e.g.*, BeH_2, BeF_2 etc.), boron shows tricovalency in the compounds like BH_3, BF_3 etc., and carbon is always tetracovalent in its compounds.

In order to explain the observed bicovalency, tricovalency and tetracovalency of Be, B and C atom, the concept of promotion of electron from the lower energy orbital to the vacant higher energy orbital has been introduced. According to this concept, it is assumed that Be, B and C-atoms acquire excited states before overlapping with the appropriate orbitals of the surrounding atoms to make σ-bonds. It is assumed that one of the two paired electrons present in 2s orbital is promoted to the vacant 2p orbital (promotion of an electron from a lower energy orbital to a higher energy orbital). In this state the elements are said to be in their excited states. The excited states of Be, B and C-atoms are shown below :

$$Be\ (2s^1 2p^1) \rightarrow 2s^1\ 2p_x^1\ 2p_y^0\ 2p_z^0 \quad (n = 2)$$
$$B\ (2s^1 2p^2) \rightarrow 2s^1\ 2p_x^1\ 2p_y^1\ 2p_z^0 \quad (n = 3)$$
$$C\ (2s^1 2p^3) \rightarrow 2s^1\ 2p_x^1\ 2p_y^1\ 2p_z^1 \quad (n = 4)$$

In the excited state these elements have 2, 3 and 4 unpaired electrons. When an electron is promoted from a lower energy orbital to a higher energy orbital, energy is *required (absorbed)* and the energy required for the promotion of the electron is compensated by the energy released during bond formation.

To explain the formation of two (Be—F) σ-bonds in BeF_2 molecule by the overlap of pure atomic orbitals.

We have seen above that Be-atom in its excited state has two half-filled orbitals namely 2s and 2p_x. These orbitals make an axial overlap with two half-filled 2p_x orbitals on two F-atoms (F = 2s^2,

$2p_x^1, 2p_x^2, 2p_z^2$) as shown in Fig. 13.1 and give rise to the formation of Be (2*s*)— F($2p_x$) and Be ($2p_x$) – F ($2p_x$) σ-bonds in BeF_2 molecule.

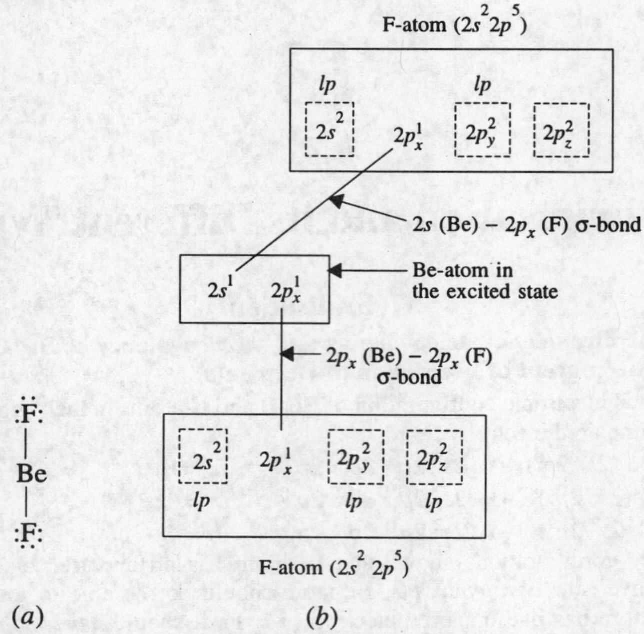

Fig. 13.1 : (*a*) Electron dot structure of BeF_2 molecule (*b*) Formation of two Be — F σ–bonds by the overlap of pure atomic orbitals on Be and F-atoms. *lp* represents lone pair of electrons in F-atom.

It has been observed that since the energy required (increase in energy) for promoting an electron from the completely-filled 2*s* orbital to the vacant $2p_x$ orbital is less than the energy released (decrease in energy) in the formation of two Be—F σ-bonds in BeF_2 molecule, BeF_2 molecule is *energetically stable*. But when we consider the stability/strength or bond lengths of Be (2*s*) — F ($2p_x$) and Be ($2p_x$)—F ($2p_x$) bonds, a difficulty arises. Since 2*s* and $2p_x$ orbitals have different energies, Be (2*s*) — F ($2p_x$) and Be ($2p_x$) – F ($2p_x$) bonds in BeF_2 should have different energies and stability / strength. Due to the difference in energies of the two bonds, these bonds should have different bond lengths. This is against the experimental facts which have shown that both the bonds in BeF_2 molecule have equal energies and have equal bond lengths. Then how to explain the identical length of both the bonds. In order to explain the identical length of both the bonds in BeF_2 molecule, the concept of hybridisation has been introduced.

To explain the formation of two identical (Be—F) σ-bonds in BeF_2 molecule by the overlap of hybrid orbitals with atomic orbitals : Definition of hybridisation

According to the concept of hybridisation, the formation of two identical (Be —F) σ-bonds in BeF_2 molecule is assumed to take place through the following steps:

1. An electron from the completely-filled 2*s*-orbital is promoted to the vacant $2p_x$ orbital so that Be-atom is obtained in its excited state ($2s^1$ $2p_x^1$).

2. The singly-filled 2*s* and $2p_x$ orbitals in the excited Be-atom mix together and give rise to the formation of *two* new orbitals which have identical energy and shape. These orbitals are called *hybrid orbitals* and the process of mixing the atomic orbitals to form hybrid orbitals is called *hybridisation.* Thus hybridisation can be defined as a process in which two or more atomic orbitals of the valence-shell of an atom (central atom of the molecule or ion) either in its ground state or in its excited state mix (merge or combine or overlap) together and give rise to the formation of new orbitals which are called *hybrid orbitals* or *hybridised orbitals.* The atom which undergoes hybrid-

isation is said to be *hybridised.* The number of hybrid orbitals obtained is always equal to the total number of atomic orbitals mixing together and the energy of all the hybrid orbitals is the same.

Since in the formation of two hybrid orbitals in Be-atom one *s* and one *p* orbital (p_x orbital) mix together, this hybridisation is called *sp hybridisation* and the two hybrid orbitals are called *sp hybrid orbitals.* Here Be-atom in its excited state is said to be *sp hybridised.*

Since the lobes of a *p*-orbital can combine with an *s*-orbital in two different ways, two *sp* hybrid orbitals are obtained when one *s* and one *p* orbital are combined together (see Fig. 13.2).

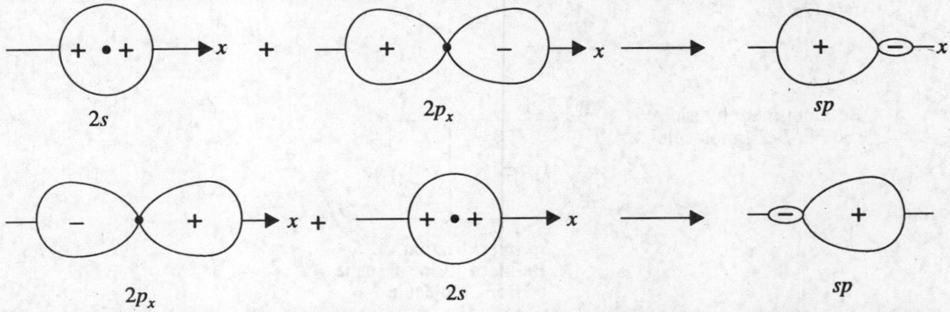

Fig. 13.2 : Formation of two *sp* hybrid orbitals in case of Be-atom by the combination of 2*s* and 2p_x orbitals.

The two *sp* hybrid orbitals have equivalent energy and are directed in opposite directrons, *i..e.,* these hybrid orbitals are inclined to each other at an angle of 180°. Each of these contains one electron. Two *sp* hybrid orbitals which have been shown separately in Fig. 13.2 have been shown in the combined state in Fig. 13.3 given in the margin.

Fig. 13.3 : Two *sp* hybrid orbitals shown in the combined way. These hybrid orbitals are oriented in opposite directions and make an angle of 180° between them.

3. Now the two singly-filled *sp* hybrid orbitals of Be-atom overlap with the two singly-filled 2p_x orbital on two F-atoms (F = $2s^2$, $2p_x^1$ $2p_y^2$ $2p_z^2$) and two (Be — F) σ-bonds are obtained (see Fig. 13.4).

The formation of (Be—F) σ-bonds that has been shown in Fig. 13.4 can also be shown in simplified way as depicted in Fig. 13.5.

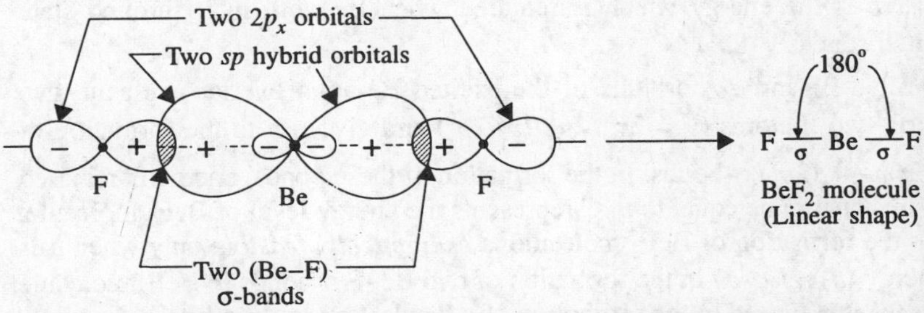

Fig. 13.4 : Formation of two Be–F σ bonds by the overlap between singly-filled two *sp* hybrid orbitals on Be-atom and two singly-filled 2p_x orbitals on two F-atoms.

Since the two *sp* hybrid orbitals are of equivalent energy, both (Be – F) σ-bonds are also of equal energy / strength or stability and hence are of equal bond length. Obviously (Be – F) σ-bonds are Be (*sp*) – F (2p_x) σ-bonds. Since the two *sp* hybrid orbitals are directed in opposite directions (*i.e.,* they are inclined to each other at an angle of 180°), both (Be – F) σ bonds also point in opposite directions and BeF_2 molecule has *linear shape* (F – Be – F) with FBeF bond angle equal to 180°.

Energy changes taking place during the formation of two (Be – F) σ bonds in BeF_2 molecule

We have seen that the formation of two Be – F σ-bonds in BeF_2 molecule can take place in the following two ways:

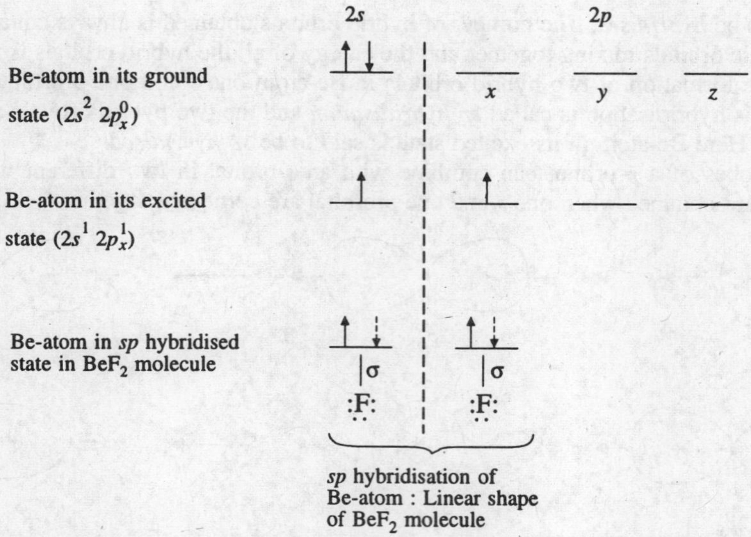

Fig. 13.5 : Formation of two Be—F σ bonds in BeF_2 molecule by the overlap of two half-filled *sp* hybrid orbitals with two half-filled $2p_x$ orbitals on two F-atom. Solid arrow (↑) indicates the electron in each *sp* hybrid orbital while the dotted arrow (↓) indicates the electron present in $2p_x$ orbital on F-atom. Dots shown round F-atoms indicate the electron pairs present in $2s$, $2p_y$ and $2p_z$ orbitals of F-atoms (F = $2s^2$, $2p_x^1\ 2p_y^2\ 2p_z^2$).

I. According to this way, the formation of two (Be—F) σ bonds takes place through the following steps :

(*i*) We know that the valence-shell electronic configuration of Be-atom in its ground state is $2s^22p_x^0$. In Fig. 13.6. A represents the energy level of Be-atom in its ground state. First of all, one electron from $2s$ orbital is promoted to the vacant $2p_x$ orbital so that the valence-shell configuration of Be-atom becomes $2s^12p_x^1$. When Be-atom has this configuration, Be-atom is said to be in the *excited state*. In Fig. 13.6 B represents the energy level of Be-atom in the excited state. In the process of promotion of a $2s$ electron to $2p_x$ orbital, energy is required (increase in energy). Let this energy be equal to *a*, *i.e.*, *a* is the energy which is required when Be-atom in its ground state ($2s^22p_x^0$) goes to the excited state ($2s^1\ 2p_x^1$).

(*ii*) Now the singly-filled $2s$ and $2p_x$ orbitals of the excited Be-atom overlap with the two singly-filled $2p_x$ orbitals on two F-atoms (F = $2s^2\ \ 2p_x^1\ 2p_y^2\ 2p_z^2$) and give rise to the formation of Be($2s$) – F ($2p_x$) and Be($2p_x$) – F ($2p_x$) σ-bonds. In the formation of these bonds, energy is released (decrease in energy). Let this energy be equal to *b*. C represents the energy level of BeF_2 molecule. Here it may be noted that the formation of BeF_2 molecule is *energetically feasible* only when *b* is greater than *a*, *i.e.*, the energy (*b*) *released* in the formation of two Be–F σ-bonds in BeF_2 molecule is greater than that which (*a*) is *required* in the promotion of a $2s$ electron to $2p$ orbital.

II. In this way, the formation of two (Be—F) σ-bonds takes place through the following steps :

(*i*) Be-atom in its ground state ($2s^2\ 2p_x^0$) goes to the excited state ($2s^12p_x^1$). Obviously energy required for this process is equal to *a*.

(*ii*) Now the singly-filled $2s$ and $2p_x$ orbitals present in the excited Be-atom get hybridised to form two equivalent *sp* hybrid orbitals each of which contains one electron. The process of the formation of hybrid orbitals also requires some energy. Let this energy be equal to *c*. D represents the energy level of *sp* hybrid orbitals.

(*iii*) Now both the singly-filled *sp* hybrid orbitals of Be-atom overlap with the singly-filled $2p_x$ orbitals of two F atoms (F = $2s^2$, $2p_x^1\ 2p_y^2\ 2p_z^2$) and give rise to the formation of two Be (*sp*) – F

($2p_x$) σ-bonds. In the formation of these bonds, energy is released (decrease in energy). Let this energy be equal to *d*.

It may be seen from Fig. 13.6 that *d* is greater than *b*, *i.e.*, the energy released or decreased in the formation of Be – F σ-bonds by the use of *sp* hybrid orbitals is more than that which is released or decreased in the formation of Be – Fσ-bonds by the use of 2*s* and $2p_x$ orbitals. In a general way we can say that σ-bonds formed by the use of hybrid orbitals of the atom are stronger and hence more stable than those which are formed by the use of pure atomic orbitals only. The energy which makes ($sp - 2p_x$) σ bonds stronger than ($2s - 2p_x$) and ($2p_x - 2p_x$) σ bonds is equal to ($d - b$).

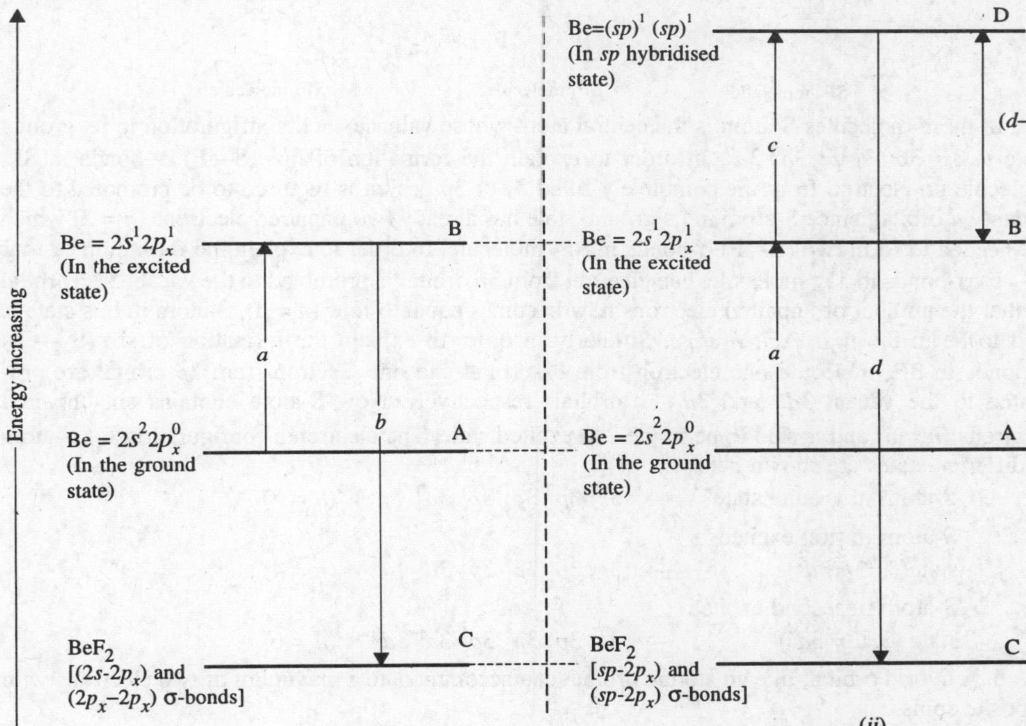

Fig. 13.6 : (*i*) Energy changes taking place during the formation of Be–F σ-bonds in BeF_2 molecule by the use of *s*- and *p* orbitals only (*ii*) Energy changes taking place in the formation of Be–F σ-bonds in BeF_2 molecule by the use of *sp* hybrid orbitals.

Salient features of the phenomenon of hybridisation: Rules of hybridisation

1. It is the valence-shell atomic orbitals of the central atom or ion of a given molecule / ion which mix together to give the hybrid orbitals, *i.e.*, it is the valence-shell atomic orbitals of the central atom which get hybridised. After the central atom has undergone hybridisation and has formed a certain number of hybrid orbitals, we say that the central atom is sp, sp^2, sp^3, sp^3d, sp^3d^2 etc., hybridised.

2. The atomic orbitals of the central atom of a given molecule or ion which participate in the formation of π-bonds remain excluded from the hybridisation process. For example in CO_2 molecule (O = C = O) the singly-filled 2*s* and $2p_x$ orbitals of C-atom in its excited state (C-atom in its excited state has the valence-shell electronic configuration as $2s^1\ 2p_x^1,\ 2p_y^1\ 2p_z^1$) mix together to form two *sp* hybrid orbitals while the singly-filled $2p_y$ and $2p_z$ orbitals remain excluded from the hybridisation process, since these orbitals make two (C — O) π-bonds.

3. In a few cases, empty atomic orbitals are also involved in hybridisation.

4. Atomic orbitals having lone pairs of electrons (*completely-filled orbitals*) are also involved

in the hybridisation, if the central atom of the given molecule or ion also has one or more lone pairs of electrons.

5. If required, one or more electrons from an orbital in the ground state valence-shell configuration of the central atom of a given molecule or ion are promoted to the ***next empty higher energy orbital*** belonging to the same main energy level. In order to illustrate this point let us consider the formation of two, four and six (S—F) σ-bonds in SF_2, SF_4 and SF_6 molecules. These molecules have the following electron dot structures (Lewis structures).

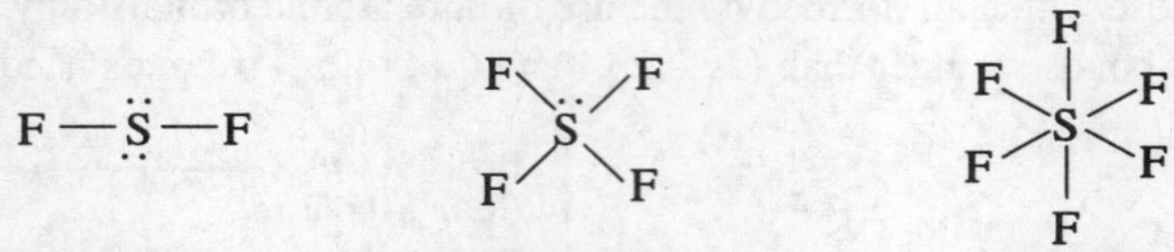

SF_2 molecule SF_4 molecule SF_6 molecule

In these molecules S-atom is the central atom whose valence-shell configuration in its ground state is $3s^2\ 3p^2\ 3p^1p^1\ 3d^0\ 3d^0$. In order to explain the formation of *two* (S—F) σ–bonds in SF_2 molecule no electron from the completely-filled 3*s* or 3*p* orbital is required to be promoted to the vacant 3*d* orbital, since S-atom in its ground state has already two unpaired electrons ($n = 2$) which are needed to form two (S—F) σ-bonds in SF_2 molecule. In order to explain the formation of four (S—F) σ-bonds in SF_4 molecule, one electron from 3*p* orbital is promoted to the vacant $3d_z{}^2$ orbital so that the number of unpaired electrons now becomes equal to four ($n = 4$). S-atom in this state is said to be in the ***first excited state***. Similarly in order to explain the formation of six (S — F) σ-bonds in SF_6 molecule one electron from 3*s* orbital and one electron from 3*p* orbital are promoted to the vacant $3d_{z^2}$ and $3d_{x^2-y^2}$ orbitals respectively. Now S-atom contains six unpaired electrons ($n = 6$) and is said to be in the 2*nd* excited state. The electronic configurations of S-atom in different states are shown below :

S-atom in ground state	→	$3s^2\ 3p_x^2\ 3p_y^1\ 3p_z^1\ 3d^0$	($n = 2$)
S-atom in first excited state($3s^23p^33d^1$)	→	$3s^2\ 3p_x^1\ 3p_y^1\ 3p_z^1\ 3d^1$	($n = 4$)
S-atom in second excited state ($3s^13p^33d^2$)	→	$3s^1\ 3p_x^1\ 3p_y^1\ 3p_z^1\ 3d^1\ 3d^1$	($n = 6$)

6. A hybrid orbital, like an atomic orbital, can accommodate a maximum of two electrons with opposite spins.

7. Hybrid orbitals generally from either σ-bonds or contain lone pairs of electrons. They do not form *pi* bonds.

8. The number of hybrid orbitals formed is equal to the total number of pure atomic orbitals which combine together to form hybrid orbitals.

9. The electronic charge in hybrid orbitals is concentrated more in one direction. This property makes the hybrid orbitals to overlap with the atomic orbitals to a greater extent and hence give stronger bonds.

10. Each hybrid orbital has one larger lobe and one smaller lobe (Fig. 13.7). The larger lobe gives a better overlapping with the orbital of another atom. The resulting bond is, therefore, stronger. Generally, smaller lobes are not shown because they do not extend sufficiently far from the nucleus to participate in bond formation. A hybrid orbital has electron density concentrated in the bigger lobe.

11. The hybridised orbitals have equivalent energies and identical size and shape, *i.e.,* hybrid orbitals are degenerate. The sum of the energies of all the hybrid orbitals is equal to the sum of the energies of various orbitals which are being mixed to form the hybrid orbitals, *i.e.,* the process of hybridisation does not involve any loss or gain of energy.

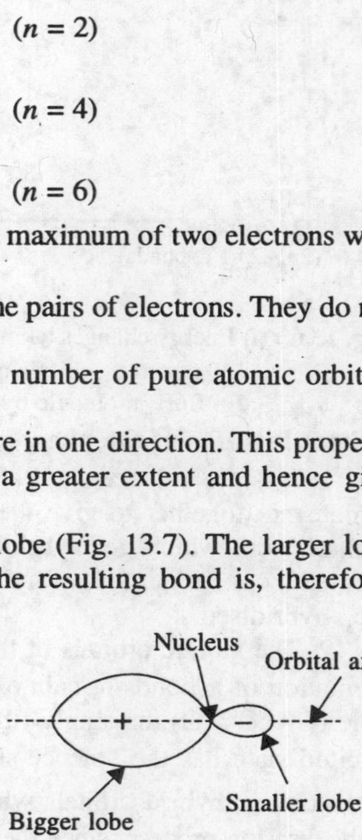

Fig. 13.7 : Representation of a hybrid orbital

12. Atomic oritals can mix together effectively to give hybrid orbitals only when they are of almost same energy and same symmetry.

13. Hybrid orbitals are directed in space in specific directions round the central atom. The direction of orientation in space depends on the number of hybrid orbitals formed or on the sum of σ-*bps* and *lps* surrounding the central atom. It has been found that if the number of hybrid orbitals or the sum of σ-*bps* and *lps* is equal to 2, 3, 4, 5, 6 and 7, then the spatial orientation (arrangement) of these hybrid orbitals or (σ-*bps* + *lps*) round the central atom is *linear, trigonal planar, tetrahedral* or *square planar, trigonal bipyramidal, octahedral* and *pentagonal bipyramidal respectively.*

14. The shapes of hybrid orbitals are different from those of the atomic orbitals from which the hybrid orbitals are obtained by mixing them.

15. The energy released (decrease in energy) in the formation of σ-bonds by the use of hybrid orbitals is more than that which is released in the formation of σ-bonds by the use of pure atomic orbitals. σ-bonds formed by the use of hybrid orbitals of the atom are, therefore, stronger and hence more stable than those which are formed by the use of pure atomic orbitals only.

16. The type of hybridisation undergone by the central atom / ion in a given molecule or ion depends on the sum of σ-*bps* and *lps* surrounding the central atom / ion, *e.g.*, if the sum of σ-*bps* and *lps* is 2, then the central atom / ion undergoes *sp* hybridisation. Similarly for the sum equal to 3, 4, 5, 6 and 7, the central atom undergoes sp^2, sp^3, sp^3d, sp^3d^2 and sp^3d^3 hybridisation respectively.

17. It is only the sum of σ-*bps* and *lps* surrounding the central atom which decides the shape of a given molecule or ion. The shape of the molecule/ion does not depend on the π-bonds present in the molecule/ion.

18. In the formation of a coordinate bond between the central atom(donor atom) and the surrounding atom (acceptor atom), the electron pair residing in the hybrid orbital of the central atom is donated to the appropriate vacant orbital of the surrounding atom.

Types of Hybridisation : Types of Hybrid Orbitals

sp, sp^2, sp^3, sp^3d, sp^3d^2, sp^3d^3 etc., are the important types of hybridisation. The central atom or ion in a given molecule / ion can undergo any of these hybridisations.

I. *sp* hybridisation : Linear hybridisation

In this hybridisation one *s* and one *p* orbital (If axis is supposed to be the molecular *x*-axis, *p*-orbital will be p_x-orbital) of the valence-shell of the central atom of a given molecule or ion combine together and give rise to the formation of a set of two equivalent hybrid orbitals, which are called *sp* hybrid orbitals. Since the lobes of a *p*-orbital can combine with an *s*-ortital in two different ways as shown in Fig. 13.8, two *sp*-hybrid orbitals are obtained. Two *sp* hybrid orbitals which have been shown separately in Fig. 13.8 have already been shown in Fig. 13.3. in the combined way.

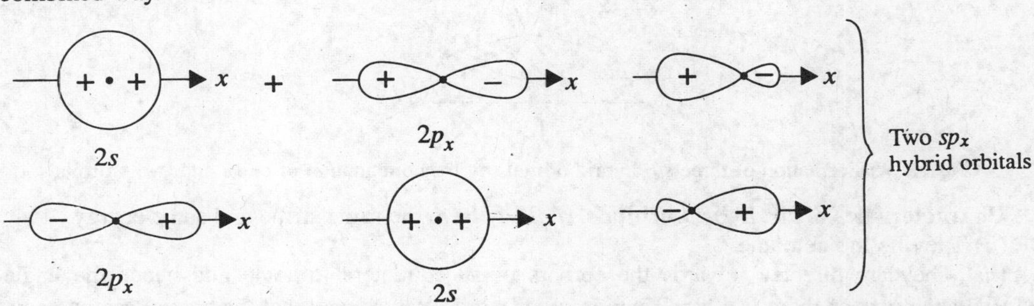

Fig. 13.8 : Formation of two *sp* hybrid orbitals by the combination of one *s* and one p_x orbital.

Since the two lobes of each of the three *p*-orbitals are oriented along the axes in opposite directions, the mixing of one *s* with any of the three *p*-orbitals would generate two *sp* hybrid orbitals.

Characteristics of sp-hybrid orbital. (*i*) *sp* hybrid orbitals are equivalent and symmetrical hybrid orbitals

(*ii*) These have the same shape and same energy.

(*iii*) These are collinear, *i.e.*, they lie in one axis in a straight line and hence the angle between them is 180° and their spatial arrangement is linear *i.e.*, they are directed in a straight line in opposite directions.

(*iv*) *s*-orbital is *spherical* and *p*-orbital is *pear-shaped* but *sp* hybrid orbital which results from the mixing of one *s* and one *p* orbital is *oval-shaped.*

II. *sp*² hybridisation: Trigonal planar hybridisation

In this hybridisation one *s*- and any two *p*-orbitals of the valence-shell of the central atom of a given molecule or ion combine together and give a set of three equivalent hybrid orbitals which are called sp^2 or *trigonal hybrid* orbitals (see Fig. 13.9).

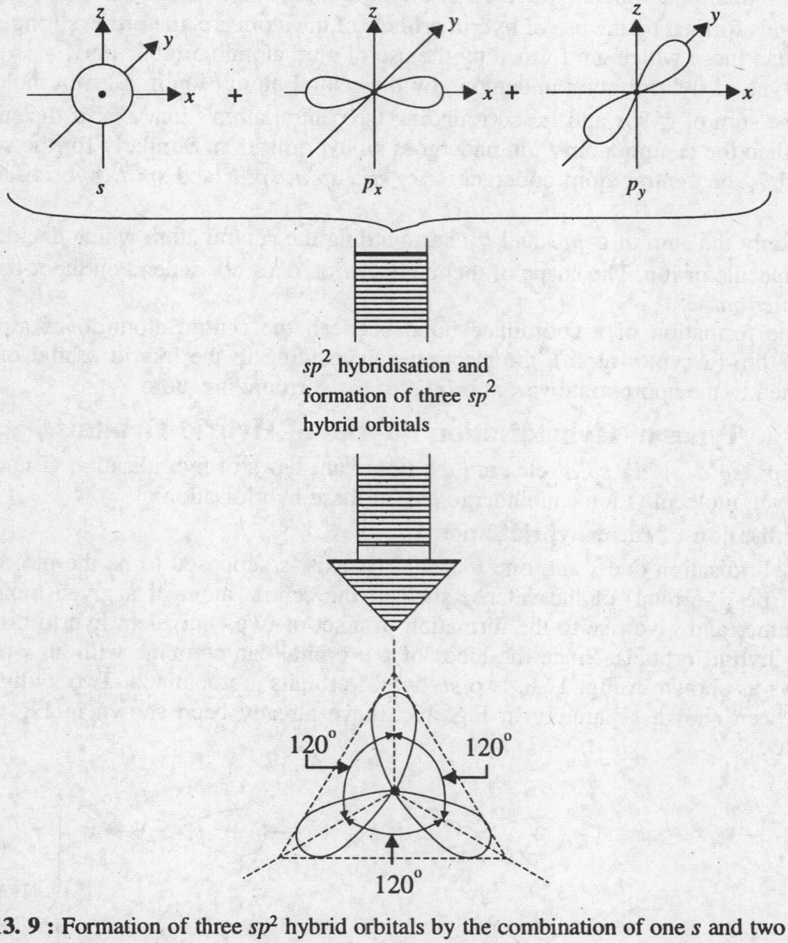

Fig 13. 9 : Formation of three sp^2 hybrid orbitals by the combination of one *s* and two *p* orbitals

Characteristics of sp^2 hybrid orbitals. (*i*) They have the same shape and same energy.

(*ii*) They lie in one plane.

(*iii*) They are directed towards the corners of an equilateral triangle and hence the angle between each pair of them is 120°. Thus, in other words, we can say that the spatial arrangement of the three sp^2 hybrid orbitals is *trigonal.*

(*iv*) The wave functions of the three sp^2 hybrid orbitals are given below:

$$\psi_{sp^2}\ (1) \quad = \frac{1}{\sqrt{3}}\,\psi_{2s} + \sqrt{\frac{2}{3}}\ \psi_{2p_x}$$

$$\psi_{sp^2}\ (2) \quad = \frac{1}{\sqrt{3}}\,\psi_{2s} - \frac{1}{\sqrt{6}}\,\psi_{2p_x} + \frac{1}{\sqrt{2}}\,\psi_{2p_y}$$

$$\psi_{sp^2}\ (3) \quad = \frac{1}{\sqrt{3}}\,\psi_{2s} - \frac{1}{\sqrt{6}}\,\psi_{2p_x} - \frac{1}{\sqrt{2}}\,\psi_{2p_y}$$

(*v*) The sp^2 hybrid orbitals which result from the combination of one *s* and two *p*-orbitals are *less oval* than the *sp* hybrid orbitals because the contribution of the two *pear-shaped p*-orbitals to the formation of sp^2 hybrid orbitals is greater. The shape of sp^2 hybrid orbitals, therefore, tends more towards a pear.

Types of species whose central atom undergoes sp^2 hybridisation

The central atom of the species (molecules or ions) whose structure can be explained on the basis of sp^2 hybridisation is surrounded by *three* electron pairs, *i.e.*, the sum of *σ-bps* and *lps* surrounding the central atom is equal to three. Maximum number of lone pairs that may be present in sp^2 hybrid orbitals is only *one* and hence these species may be of two types *viz.*, AB_3 and AB_2 (*lp*).

Since (*σ-bps* + *lps*) = 3, the central atom (A) undergoes sp^2 hybridisation to form a set of three equivalent sp^2 hybrid orbitals. Each of these hybrid orbitals overlaps in head-to-head manner with the appropriate atomic orbitals on B-atoms and forms three (in AB_3 molecules) and two [in AB_2(*lp*) molecules] A—B σ-bonds. The central atom is placed at the centre of the equilateral triangle while the three corners of the triangle are occupied by three B-atoms or two B-atoms and one lone pair of electrons as the case may be.

1. AB_3 type species : Trigonal planar species. Species of this type can be represented as B—A—B which shows that the
                                                                                     |
                                                                                     B
central atom (A) is surrounded by three *σ-bps* (*σ-bps* = 3) and no *lp* (*lp* = 0), *i.e.*, *σ-bps* + *lps* = 3 + 0 = 3. These species have trigonal planar geometry (see Fig. 13.10).

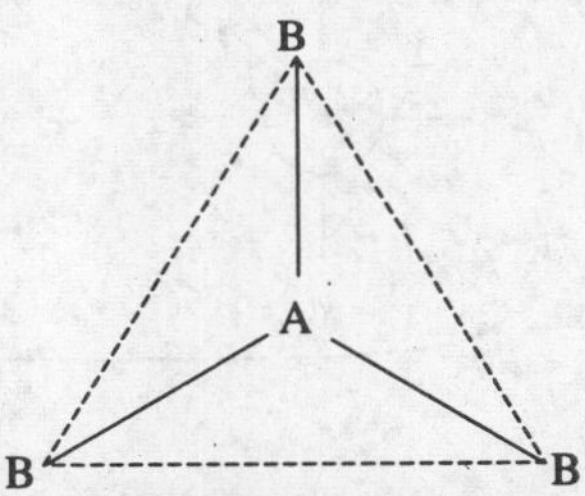

Fig 13.10 : Trigonal planar geometry of AB3 type molecule or ion.

Examples of molecules/ions which are of AB_3 type and have trigonal planar geometry are BX_3 (X = F, Cl, Br), BO_3^{3-}, BH_3, BeF_3^-, SO_3 etc.

2. AB_2 (*lp*) type species : Angular or V-shaped species. These species can be represented as B—$\ddot{A}$—B which shows that the central atom (A) is surrounded by two *σ-bps* and one *lp*. According to VSEPR theory, lone pair of electrons can occupy any of the three sp^2 hybrid orbitals, since all the three hybrid orbitals are equivalent. Thus, although the spatial orientation of the three sp^2 hybrid orbitals round the central atom is trigonal planar, due to the presence of one *lp* in one sp^2 hybrid orbital, the shape of AB_2 (*lp*) molecule/ion gets distorted and becomes angular (V-shaped structure) (see Fig. 13.11). Examples of molecules/ions having angular geometry are $SnCl_2(g)$, $PbCl_2$ etc.

III. $d_{x^2-y^2}\,sp^2$ hybridisation : Square planar molecules/ions

If the central atom or ion of a molecule/ion is surrounded by four σ-bonding electron pairs (*σ-bps* = 4), the shape of the molecule/ion is either square planar or tetrahedral. Here we shall consider square planar shape only. This shape arises because of dsp^2 hybridisation of the central metal atom or ion of the molecule / ion. Examples of molecules / ions which have square planar geometry are $[Ni^{2+}(CN)_4]^{2-}$, $[Ni^{2+}(dmg)_2]^0$, $[Cu^{2+}(NH_3)_4]^{2+}$, $[Cu^{2+}(big\ H)_2]^{2+}$, $[Cu^{2+}(big)_2]^0$, $[Cu^{2+}(en)_2]^{2+}$, $[Pt^{2+}Cl_4]^{2-}$, $[Pt^{2+}(NH_3)_4]^{2+}$, $[Pt^{2+}(PPh)_2Cl_2]^0$, $[Pd^{2+}Cl_4]^{2-}$, $[Pt^{2+}(gly)_2]^0$, $[Ni^{2+}(NH_3)_4]^{2+}$, $[Cu^{2+}(py)_4]^{2+}$, $[Ni^{2+}(big\ H)_2]^{2+}$.

The term "*dsp^2 hybridisation*" means that when $d_{x^2-y^2}$, *s* and two *p*-orbitals (say p_x and p_y) are mixed together, four hybrid orbitals are generated. These hybrid orbitals are called *dsp^2 square* *planar hybrid orbitals* and the mixing of the orbitals is called dsp^2 *hybridisation.* These hybrids are co-planar and lie in xy plane. Each pair of these four hybrids is inclined at an angle of 90°.

Four dsp^2 hybrid orbitals can also be regarded as having been formed by adding an orbital to three sp^2 hybrid orbitals which, we know, lie in the same plane and are inclined to each other at 120°. If sp^2 hybrid orbitals are supposed to be formed by the combination of s, p_x and p_y orbitals, then the three sp^2 hybrids lie in xy plane. In order to get four square planar dsp^2 hybrids, we have to mix three sp^2 hybrids to one such atomic orbital which is located in the same plane in which sp^2 hybrids are located (*i.e.*, xy plane) and whose lobes are directed towards the four corners of the square plane (*i.e.*, the lobes of the orbital are directed along x and y axes).

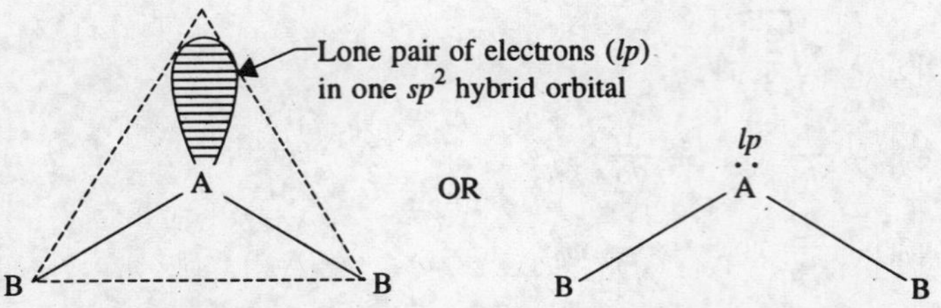

Fig. 13.11 : Angular geometry of AB_2 (lp) type species.

By inspection we find that no p-orbital has such characteristics. However, $d_{x^2-y^2}$ orbital has these characteristics, since this orbital is located in xy plane and its four lobes are directed along x and y axes. Thus when $d_{x^2-y^2}$ orbital is mixed with the three trigonal planar sp^2 hybrid orbitals, four square planar dsp^2 hybrids are obtained.

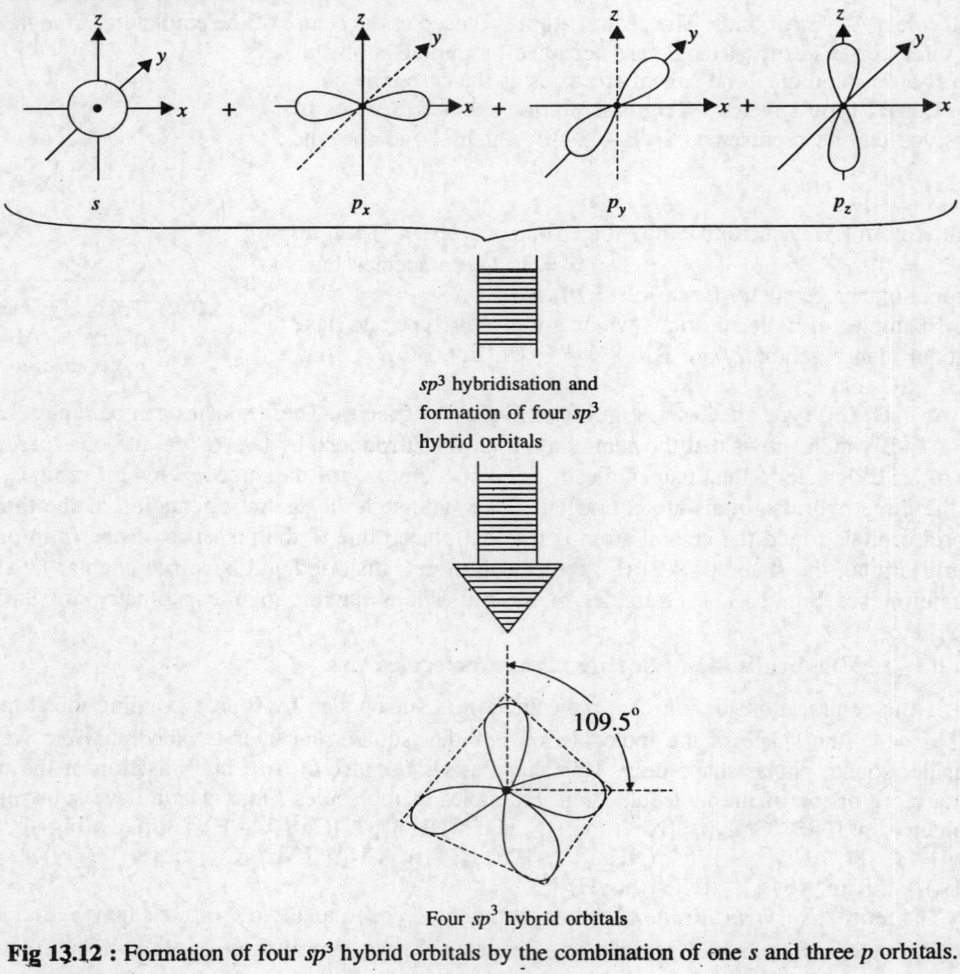

Fig 13.12 : Formation of four sp^3 hybrid orbitals by the combination of one s and three p orbitals.

How dsp^2 hybridisation of the central atom or ion has been used to explain the formation of square planar complex ions like $[Ni(CN)_4]^{2-}$, $[Cu^{2+}(NH_3)_4]^{2+}$ etc. has been discussed under "*Valence Bond Theory of coordination compounds*".

IV. sp^3 hybridisation: Tetrahedral hybridisation

The term "sp^3" hybridisation means that one *s* and all the three *p*-orbitals (p_x, p_y and p_z orbitals) of the valence-shell of the central atom of a given molecule / ion combine together and form a set of four hybrid orbitals which are called sp^3 or *tetrahedral hybrid orbitals* (see Fig. 13.12).

Four sp^3 hybrid orbitals can also be regarded as having been formed by adding one atomic orbital to three sp^2 hybrid orbitals which, we know, lie in the same plane (co-planar) and are inclined to each other at 120°. If sp^2 hybrid orbitals are supposed to be formed by the combination of *s*, p_x and p_y orbitals, then the three sp^2 hybrids lie in *xy* plane. In order to get four sp^3 tetrahedral hybrid orbitals, we have to mix three sp^2 hybrids to one such atomic orbital which is located above and below the *xy* plane. Obviously, since sp^2 hybrids lie in *xy* plane, such atomic orbital is p_z orbital. As soon as we mix p_z atomic orbital with the three sp^2 hybrid orbitals, these hybrids (*i.e.*, sp^2 hybrids) are pressed below the *xy* plane and four tetrahedral sp^3 hybrid orbitals are generated (See Fig. 13.13). Each pair of these four sp^3 hybrids is inclined at an angle of 109.5°. The mixing of p_z and sp^2 orbitals is called sp^3 hybridisation.

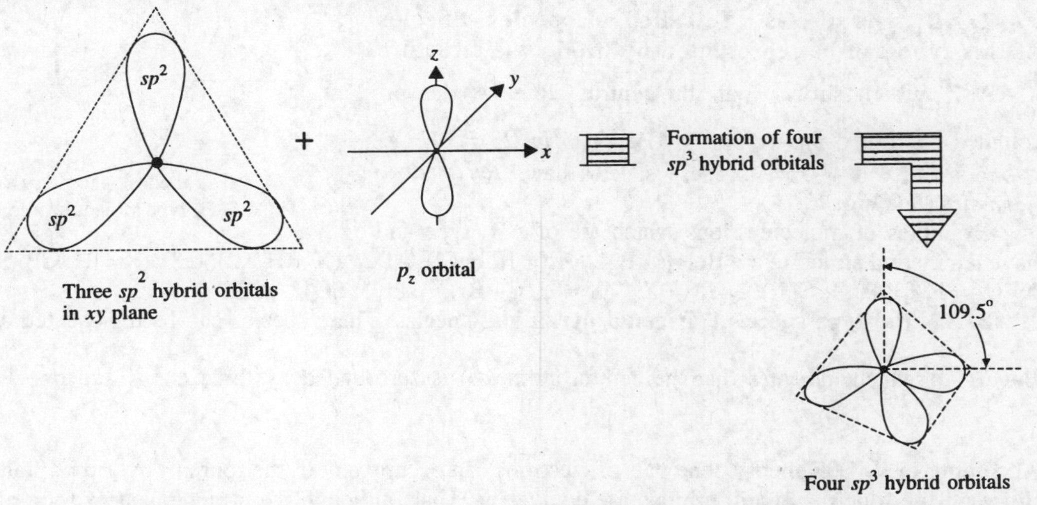

Fig. 13.13 : Formation of four sp^3 hybrid orbitals by the combination of three sp^2 hybrid orbitals and one p_z orbital.

Characteristics of sp^3 hybrid orbitals. (*i*) sp^3 hybrid orbitals are equivalent and symmetrical hybrid orbitals.

(*ii*) These have the same shape and same energy.

(*iii*) They are directed towards the four corners of a regular tetrahedron and the angle between each pair of them is 109.5°. This angle is called *tetrahedral angle.*

(*iv*) Each of the four sp^3 hybrid orbitals has one-fourth *s*-character and three-fourth *p*-character, *i.e.*, *s*-character = 25% and *p*-character = 75%.

(*v*) The wave function of each *sp* hybrid orbital is written as:

$$\Psi_{sp}(1) = \frac{1}{\sqrt{2}}\Psi_{2s} + \frac{1}{\sqrt{2}}\Psi_{2p} = \frac{1}{\sqrt{2}}(\Psi_{2s} + \Psi_{2p})$$

$$\Psi_{sp}(2) = \frac{1}{\sqrt{2}}\Psi_{2s} - \frac{1}{\sqrt{2}}\Psi_{2p} = \frac{1}{\sqrt{2}}(\Psi_{2s} - \Psi_{2p})$$

(*vi*) In case of sp^3 hybrid orbitals which result from the mixing of one *s* and three *p*-orbitals, since the contribution of the three *p*-orbitals predominates over the contribution of one *s* orbital, the shape of sp^3 hybrid orbital is practically the same as that of the parent *p*-orbitals except for the fact that the lobes in sp^3 hybrid orbitals are somewhat more spread and are somewhat shorter in length than the pure *p*-orbitals.

Types of species whose central atom undergoes sp^3 hybridisation

The central atom of the species (molecules or ions) whose structure can be explained on the basis of sp^3 hybridisation is surrounded by four electron pairs, *i.e.*, the sum of σ-*bps* and *lps* surrounding the central atom is equal to 4. Maximum number of *lps* that may be present in sp^3 hybrid orbitals is *three* and hence these species may be of four types *viz.*, AB_4, AB_3 (*lp*), AB_2 $(lp)_2$ and AB $(lp)_3$.

Since (σ-*bps* + *lps*) = 4, the central atom (A) undergoes sp^3 hybridisation to form a set of four equivalent sp^3 hybrid orbitals. Each of these hybrid orbitals overlaps in head-to-head manner with the appropriate atomic orbitals on B-atoms and forms four (in AB_4 molecules), three [in AB_3 (*lp*) molecules], two [in AB_2 $(lp)_2$ molecules] and one [in AB $(lp)_3$ molecules] A—B σ-bonds. The central atom (A) is placed at the centre of the tetrahedron while the four corners of the tetrahedron are occupied by B-atoms or B-atoms and one, two or three lone pairs of electrons, as the case may be.

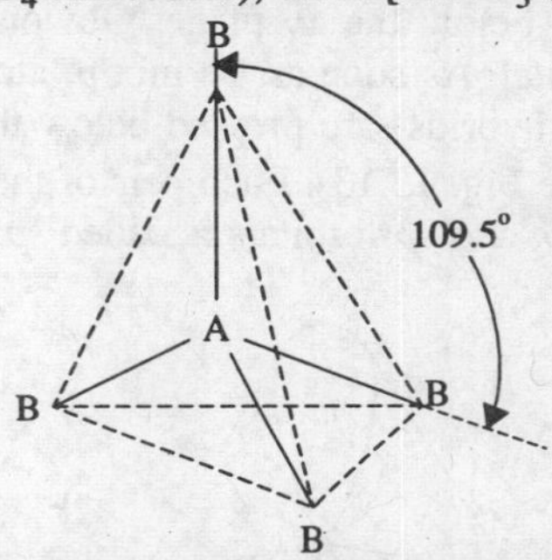

Fig. 13.14 : Tetrahedral geometry of AB4 type molecule or ion.

1. AB_4 type species : Tetrahedral species. Species of this type can be represented by the Lewis structure $\genfrac{}{}{0pt}{}{B}{B}\!>A<\!\genfrac{}{}{0pt}{}{B}{B}$ which shows that the central atom (A) is surrounded by four σ-*bps* (σ-*bps* = 4) and *no lp* (*lp* = 0), *i.e.*, σ-*bps* + *lps* = 4 + 0 = 4. These species have *tetrahedral geometry* (See Fig. 13.14).

Examples of molecules/ions which are of AB_4 type and have tetrahedral structure are BeH_4^{2-}, BX_4^- (X = H, F, Cl), AlX_4^- (X = H, Cl), CX_4 (X = H, Cl), Si X_4 (X = F, Cl, Br, I, CH_3), NH_4^+, PX_4^+ (X = H, Cl, Br), $AsPh_4^+$, SO_4^{2-} etc.

2. AB_3 (*lp*) type species : Trigonal pyramidal species. These species can be represented as

$$\begin{array}{c} \ddot{\text{A}} \\ \text{B—}\;\;\;\text{—B} \\ | \\ \text{B} \end{array}$$

which indicates that the central atom (A) is surrounded by three σ-*bps* and one *lp*.

According to VSEPR theory, lone pair of electrons can occupy any of the four sp^3 hybrid orbitals. Since all the four sp^3 hybrid orbitals are equivalent. Thus, although the arrangement of four sp^3 hybrid orbitals in space round the central atom is tetrahedral, due to the presence of one *lp* in one sp^3 hybrid orbital, the shape of AB_3 (*lp*) species gets distorted and becomes distorted tetrahedral which is called *trigonal pyramidal* or *simply pyramidal shape* (See Fig. 13.15). The central A atom is placed at the centre of the pyramid and B atoms form its base. The lone pair of electrons (*lp*) occupies the apex of the pyramid.

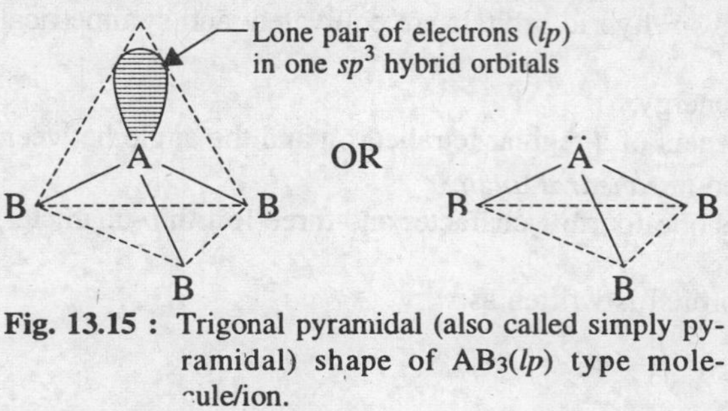

Fig. 13.15 : Trigonal pyramidal (also called simply pyramidal) shape of AB3(*lp*) type molecule/ion.

Example of molecules/ions having trigonal pyramidal structural are NX_3 (X = H,F), PX_3 (X = H, F, Cl, Br, I), AsX_3 (X = H, F, Cl, Br, I), $SbCl_3$, $BiBr_3$, H_3O^+, SbF_3^+ SeF_3^+, Te Me_3^+ etc.

3. AB_2 $(lp)_2$ type species: Angular (V-shaped) species. These species can be represented by the structure, B — $\ddot{\underset{\cdot\cdot}{A}}$ — B. This structure shows that the central atom (A) is surrounded by two σ-*bps* and two *lps*. According to VSEPR theory, lone pairs of electrons can occupy any two sp^3 hybrid orbitals. Thus, although the spatial orientation of the four sp^3 hybrid orbitals round the central atom is tetrahedral, due to the presence of two *lps* in two sp^3 hybrid orbitals, the shape of AB_2 $(lp)_2$ molecule/ion is *angular* (V-shaped structure) as shown in Fig. 13.16.

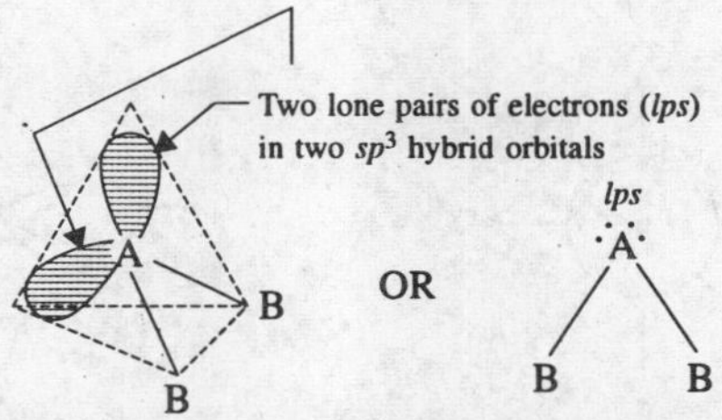

Fig 13.16 : Angular geometry (V-shaped structure) of AB_2 $(lp)_2$ molecule/ion.

Examples of molecules/ions having angular geometry are NH_2^- (amide ion), H_2O, OX_2, H_2S, SCl_2, SO_2, H_2Se, $SeCl_2$, triatomic interhalogen cations (*e.g.* ICl_2^+, BrF_2^+, ClF_2^+, IBr_2^+, IF_2^+ etc.) .

4. AB $(lp)_3$ type species: Linear species. These species can be represented as $:\ddot{\underset{..}{A}}$ —B which shows that the central atom (A) is surrounded by one σ-*bp* and three *lps*. According to VSEPR theory, lone pairs occupy any three sp^3 hybrid orbitals. Thus although the spatial arrangement of four sp^3 hybrid orbitals round the central atom (A) is tetrahedral, due to the presence of three *lps* in three sp^3 hybrid orbitals, the shape of AB $(lp)_3$ molecule/ion is linear as shown in Fig. 13.17.

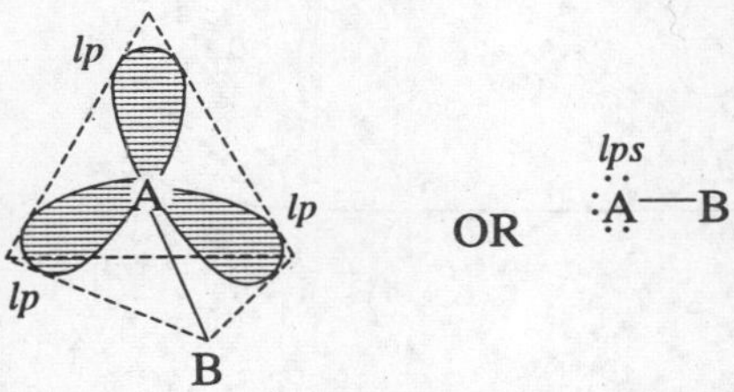

Fig. 13.17 : Linear shape of AB $(lp)_3$ type species.

Inter halogen molecules of XY type (*e.g.*, ClF, BF, IF, ICl, IBr etc.) have linear shape which can be explained on the basis of sp^3 hybridisation of their central atom.

V. $sp^3d_{z^2}$ or sp^3d hybridisation: Trigonal bipyramidal hybridisation

In this type of hybridisation, one *s*, three *p* and one *d* (d_{z^2}) orbitals of the valence-shell of the central atom of a given molecule/ ion combine together and give rise to the formation of a set of five hybrid orbitals which are called sp^3d or *trigonal bipyramidal hybrid orbitals* (See Fig. 13.18).

Characteristics of sp^3d hybrid orbitals. (*i*) These have the same shape and same energy.

(*ii*) They are directed towards the five corners of a regular pentagonal bipyramid, *i.e.*, the arrangement of the five hybrid orbitals in space is *pentagonal bipyramidal*. Due to their pentagonal bipyramidal arrangement they donot lie in one plane. Three hybrid orbitals which have been represented by *b* (called *basal* or *equatorial* hybrid orbitals) are lying in the equilateral (equatorial) triangle while the remaining two which have been represented by *a* (called *axial* or *apical* or *polar* hybrid orbitals) are lying above and below on the axis passing through the centre of the triangular base of the trigonal bipyramid. Thus the angle between each pair of the basal hybrid orbitals is of 120°, that between the two axial hybrid orbitals is of 180° and that between the axial and basal hybrid orbitals is equal to 90°. The three basal hybrid orbital are equivalent and similarly the two axial hybrid orbitals are also equivalent.

Types of species whose central atom undergoes sp^3d hybridisation

The species (molecules or ions) whose structure can be explained on the basis of sp^3d hybridisation is surrounded by *five* electron pairs, *i.e.*, the sum of σ-*bps* and *lps* surrounding the central atom is equal to 5. Maximum number of *lps* that may be present in sp^3d hybrid orbitals is *three* and hence these species may be of four types namely AB_5, AB_4 (*lp*), AB_3 $(lp)_2$ and AB_2 $(lp)_3$.

Since (σ-*bps* + *lps*) = 5, the central atom (A) undergoes sp^3d hybridisation to form a set of five sp^3d hybrid orbitals. Each of these hybrid orbitals overlaps in head-to-head manner with the appropriate atomic orbitals on B-atoms and forms five [in AB_5 molecules], four [in AB_4 (*lp*) molecules], three [in AB_3 $(lp)_2$ molecules] and two [in AB_2 $(lp)_3$ molecules] A – B σ-bonds. The central atom (A) is placed at the centre of the equilateral triangular base of the trigonal bipyramid while the five corners of the bipyramid are occupied by B-atoms or B-atoms and one, two or three lone pairs of electrons, as the case may be.

1. AB_5 type species : Trigonal bipyramidal species. Species of this type can be represented

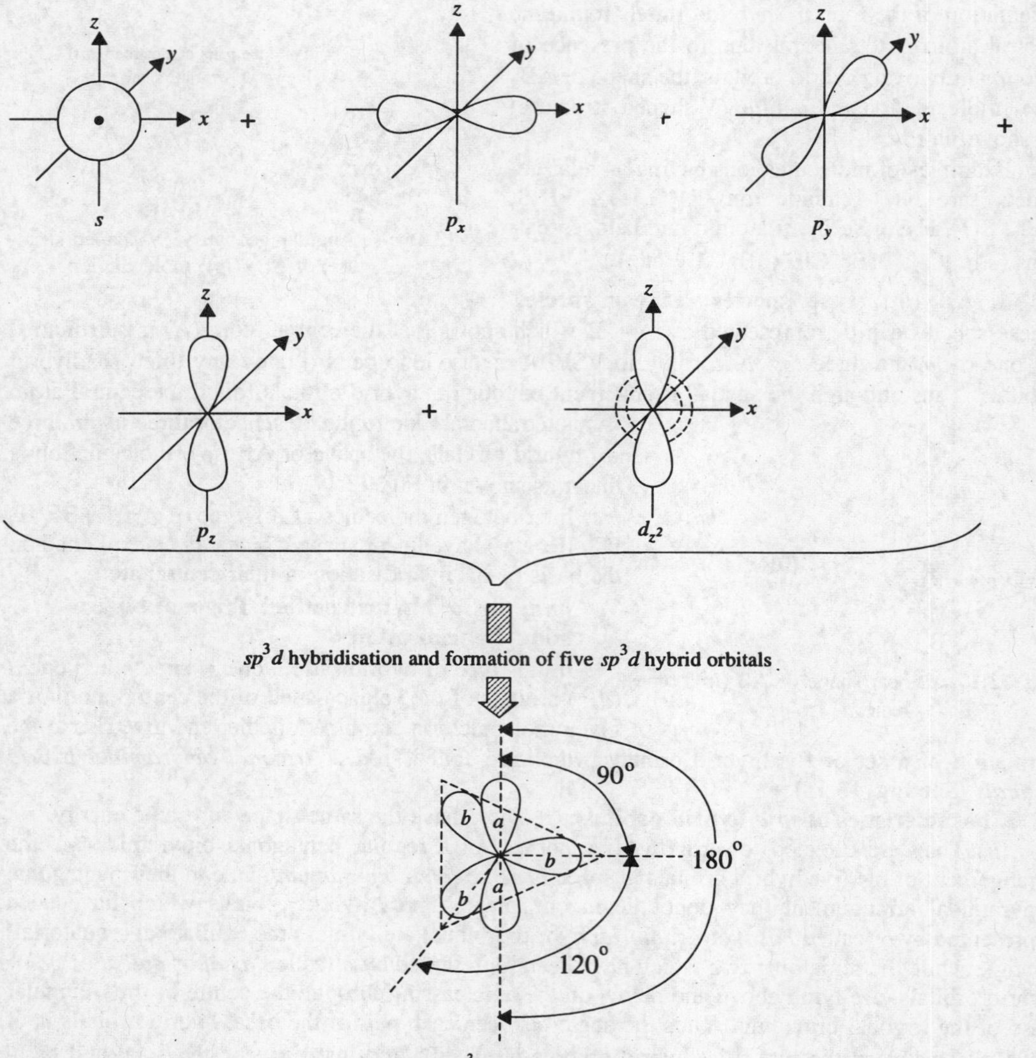

Fig. 13.18 : Formation of sp^3d hybrid orbitals by the combination of one s and three p and one d (d_{z^2}) orbitals.

as $\begin{matrix} B \\ B \end{matrix}\!> \underset{\overset{|}{B}}{A} <\!\begin{matrix} B \\ B \end{matrix}$ which shows that the central atom A is surrounded by five σ-*bps* (σ-*bps* = 5) and no *lp* (*lp* = 0), *i.e.*, σ-*bps* + *lps* = 5 + 0 = 5. These species have *trigonal bipyramidal structure* (see Fig. 13.19).

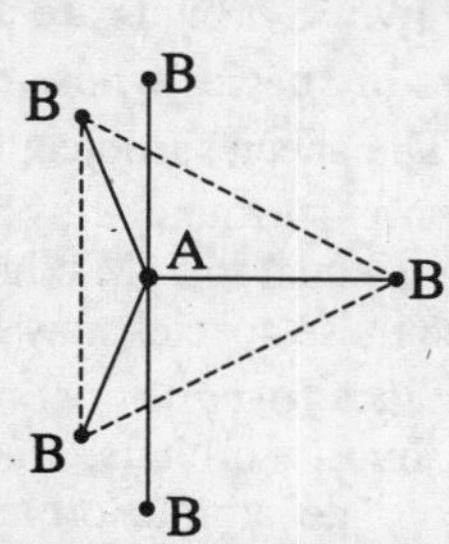

Fig. 13.19: Trigonal bipyramidal structure of AB_5 type molecule or ion.

Examples of species which are of AB_5 type and have trigonal bipyramidal structure are PF_5, PCl_5 (*g*), AsFs, SbF_5.

2. AB_4 (*lp*) type species : Distorted tetrahedral (or see-saw) species. These species can be represented by the electron dot structure, $\begin{matrix} B \\ B \end{matrix}\!> \ddot{A} <\!\begin{matrix} B \\ B \end{matrix}$ which clearly shows that the central atom (A) is surrounded by four σ-*bps* and one *lp*. According to VSEPR theory, lone pair of electrons occupies one of the three basal sp^3d hybrid orbitals, since in this case (*lp* – *bp*) repulsions at 90° are lesser and the *lp* can occupy larger volume in the basal hybrid orbital. Thus the structure of AB_4 (*lp*) molecule/ion having its *lp* in the basal hybrid orbital is the most stable. Thus although the arrangement of five sp^3d hybrid orbitals in space is trigonal bipyramidal, due to the presence of one lone pair of electrons in the basal hybrid orbital, the shape of AB_4 (*lp*) species gets distorted and becomes *distorted tetrahedral* or *see-saw* (See Fig. 13.20).

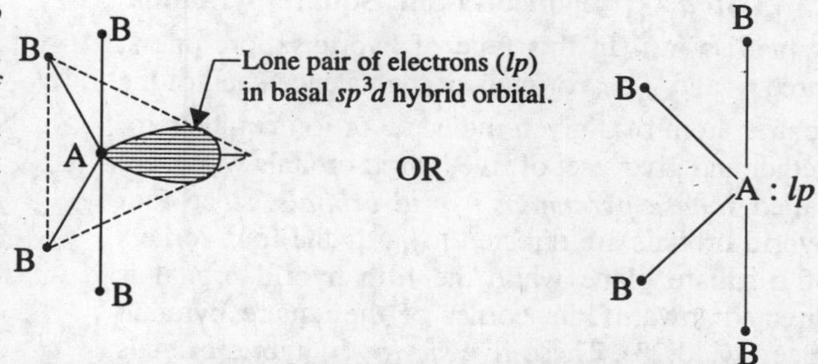

Fig. 13.20 : Distorted tetrahedral or see-saw structure of AB_4 (*lp*) molecule/ion.

Examples of molecules/ions having see-saw structure are SF_4, SeF_4, $TeCl_4$, TeF_4, $TeBr_4$, $SnCl_4^{2-}$, penta-atomic interhalogen cations (*e.g.* IF_4^+, ClF_4^+, BrF_4^+) etc.

3. AB_3 (*lp*)$_2$ type species : Bent T-shaped species. These species can be represented by the electron dot structure, $B-\ddot{\underset{..}{A}}<^{B}_{B}$ which shows that the central atom (A) is surrounded by three σ-*bps* and two *lps*. According to VSEPR theory, the two *lps* prefer to occupy the two basal (equilateral) sp^3d hybrid orbitals, since in this case the repulsions between the electron pairs where the angle between them is less than 100° are minimum and the *lps* can occupy larger volume in the basal hybrid orbitals. Thus the structure of AB_3 (*lp*)$_2$ type molecule/ion having both the *lps* in the basal hybrid orbitals is the most stable, *i.e.*, although the arrangement of five sp^3d hybrid orbitals in space is trigonal bipyramidal, due to the presence of two *lps* residing in the basal hybrid orbitals, the shape of AB_3 (*lp*)$_2$ species gets distorted and becomes *bent T-shape* (see Fig. 13.21).

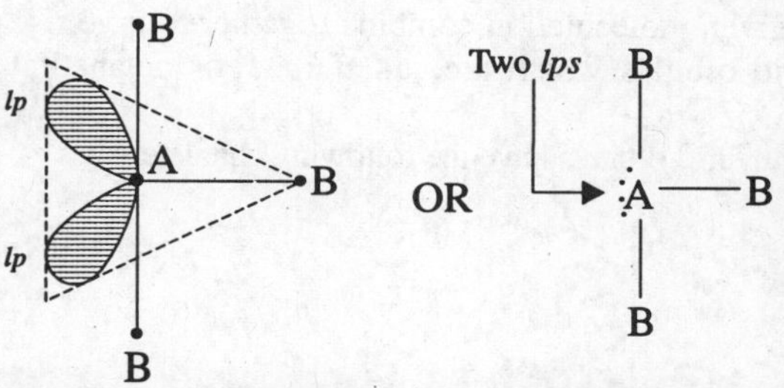

Fig. 13.21: Bent T-shaped structure of AB_3 (*lp*)$_2$ type molecule/ion

Examples of molecules/ions having bent T-shaped structure are interhalogen compounds of XY_3 type (*e.g.*, ClF_3, IF_3, BrF_3 etc.), XeF_3^+ etc.

4. AB_2(*lp*)$_3$ type species: Linear species. These species can be represented by the structure $B-\ddot{\underset{..}{A}}-B$ which shows that the central atom (A) is surrounded by two σ-*bps* and three *lps*. According to VSEPR theory, the three *lps* occupy the three basal (equatorial) sp^3d hybrid orbitals, since in this case the repulsion between the two electron pairs inclined at an angle less than 100° is minimum. Thus the structure of AB_2 (*lp*)$_3$ type species having the three *lps* in the basal hybrid orbitals is the most stable, *i.e.*, although the orientation of the five sp^3d hybrid orbitals in space is trigonal bipyramidal, due to the presence of three *lps* in the basal hybrid orbitals, the shape of AB_2(*lp*)$_3$ species becomes linear (See Fig. 13.22).

Example of molecules/ions having linear structure are XeF_2, KrF_2, triatomic interhalogen anions (*e.g.*, ICl_2^-, I_3^-, IBr^-, ClF_2^-, $IBrCl^-$) etc.

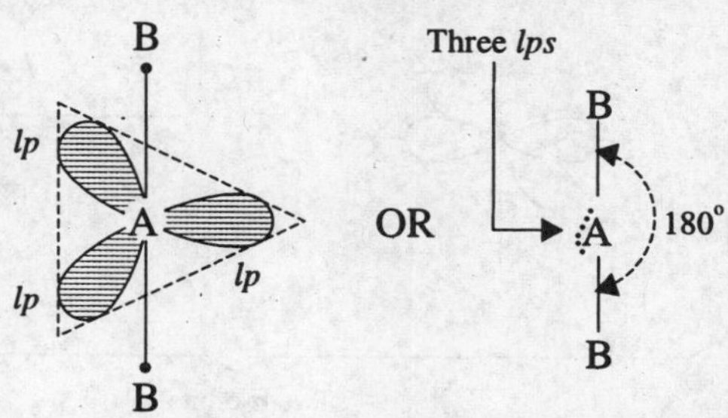

Fig. 13.22: Linear shape of AB_2 (*lp*)$_3$ type species.

VI. $sp^3d_{x^2-y^2}$ hybridisation: Square pyramidal hybridisation. In this type of hybridisation, one *s*, three *p*- and $d_{x^2-y^2}$ orbitals of the valence-shell of the central atom of a given molecule or ion combine together and give a set of five hybrid orbitals which are called *square pyramidal hybrid orbitals*, since four hybrid orbitals are directed towards the four corners of a square plane while the fifth hybrid orbital is directed towards the corner of the square pyramid (See Fig.13.23). These five $sp^3d_{x^2-y^2}$ hybrid orbitals do not lie in the same plane.

The molecules of AB_5 type whose central atom(A) undergoes $sp^3d_{x^2-y^2}$ hybridisation have square pyramidal geometry (See Fig. 13.23).

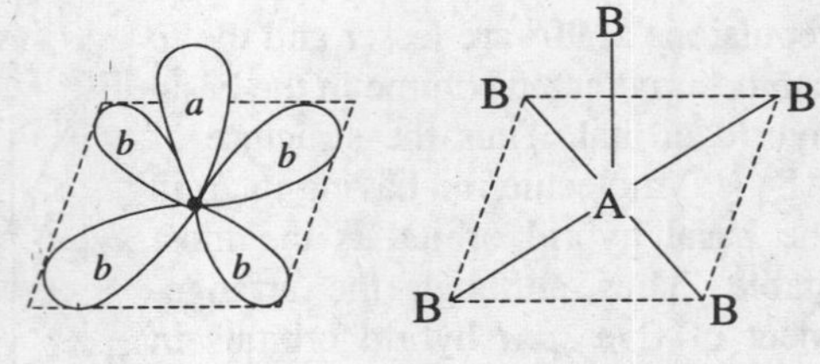

(*a*) (*b*)

Fig. 13.23: (*a*) Five $sp^3d_{x^2-y^2}$ hybrid orbitals formed by the combination of ones three *p* and one $d_{x^2-y^2}$ orbitals *a* indicates the axial hybrid orbital while *b* indicates the hybrid orbitals lying in the square plane. (*b*) square pyramidal geometry of AB_5 molecule whose central atom (A) undergoes $sp^3d_{x^2-y^2}$ hybridisation.

VII. $sp^3d_{x^2-y^2}\ d_{z^2}$ or sp^3d^2 hybridisation : Octahedral hybridisation. In this type of hybridisation, one *s*, three *p* and two *d* orbitals (d_{z^2} and $d_{x^2-y^2}$ orbitals) of the valence-shell of the central atom of a given molecule/ion combine together to give rise to the formation of a set of six equivalent hybrid orbitals which are called sp^3d^2 or octahedral hybrid orbitals. (See Fig. 13.24).

Characteristics of sp^3d^2 hybrid orbitals. These hybrid orbitals have the following characteristics :

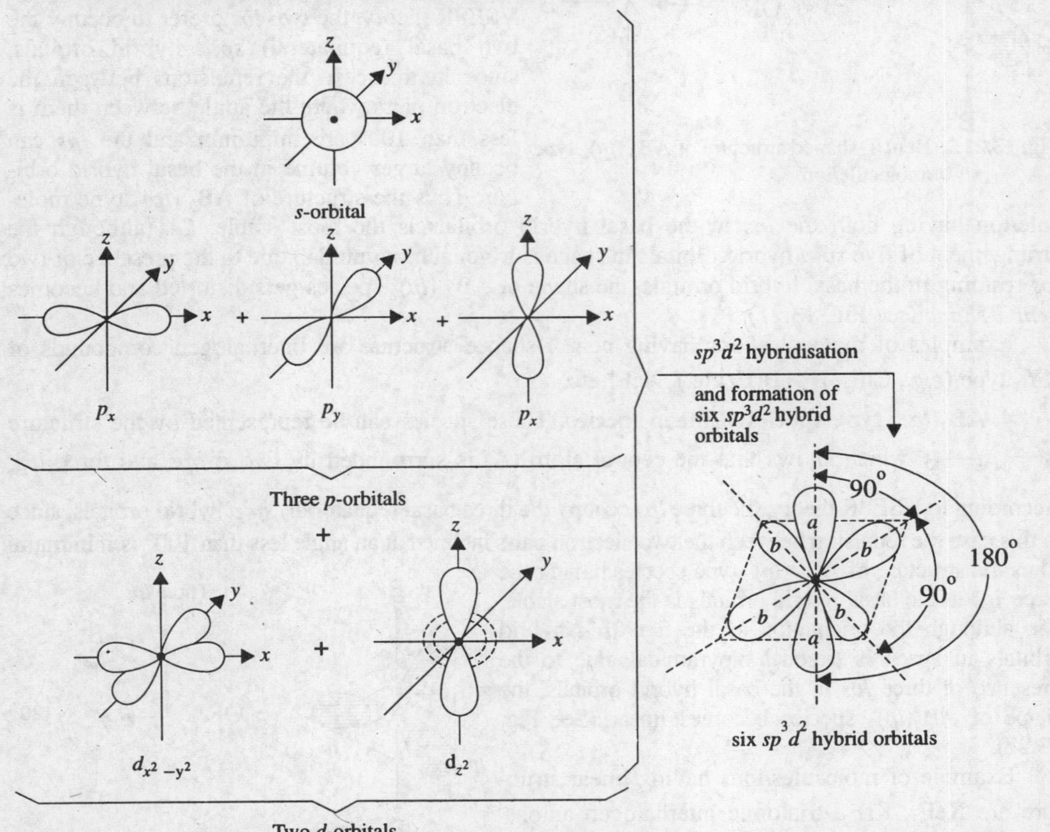

Fig. 13.24 : Formation of six sp^3d^2 hybrid orbitals by the combination of one *s*, three *p* and two *d* ($d_{x^2-y^2}$ and d_{z^2}) orbitals.

(*i*) These have the same shape and same energy.

(*ii*) They are directed towards the six corners of a regular (symmetrical) octahedron, *i.e.*, the arrangement of the six hybrid orbitals in space is octahedral. Due to their octahedral arrangement, they do not lie in one plane. Four hybrid orbitals which been represented by *b* (called *basal* or *equatorial hybrid orbitals*) are lying in square plane while the remaining two which have been represented by *a* (called *axial hybrid orbitals*) are lying in a plane which is perpendicular to that containing the basal hybrid orbitals, *i.e.*, the axial hybrid orbitals lie above and below on the axis passing through the centre of the square base of the octahedron. Thus the angle between each pair of basal hybrid orbitals is of 90°, that between the two axial hybrid orbitals is of 180° and that between the axial and basal hybrid orbitals is equal to 90°. The four basal hybrid orbitals are equivalent and similarly the two axial hybrid orbitals are also equivalent.

Types of species whose central atom undergoes sp^3d^2 hybridisation

The central atom of the species (molecules or ions) whose structure can be explained on the basis of sp^3d^2 hybridisation is surrounded by *six* electron pairs *i.e.*, the sum of σ-*bps* and *lps* surrounding the central atom is equal to six. Maximum number of *lps* that may be present in sp^3d^2 hybrid orbitals is *two* and hence these species can be grouped into *three* types namely AB_6, AB_5 (*lp*) and AB_4 $(lp)_2$. Since (σ-*bps* + *lps*) = 6, the central atom(A) undergoes sp^3d^2 hybridisation and forms a set of six sp^3d^2 hybrid orbitals. Each of these hybrid orbitals makes a head-to-head overlap with the appropriate atomic orbitals of B-atoms and forms six [in AB_6 molecules], five [in AB_5 (*lp*)] and four [in AB_4 $(lp)_2$] A – B σ-bonds. The central atom A is situated at the centre of the square base of the octahedron while the six corners of the octahedron are being occupied by B-atoms or by B-atoms and one or two lone pairs of electrons, as the case may be.

1. AB_6 type species: Octahedral species. Species of this type can be represented by the electron dot structure,

$$\begin{matrix} & B & \\ & | & \\ {}^{B}_{B}> & A & <{}^{B}_{B} \\ & | & \\ & B & \end{matrix}$$

which shows that the central atom, A is surrounded by six σ-*bps* (σ-*bps* = 6) and no *lp* (*lp* = 0), *i.e.*, σ-*bps* + *lp* = 6 + 0 = 6. These species have octahedral geometry (See Fig. 13.25).

Examples of the species which are of AB_6 type and have octahedral structure are $[AlF_6]^{3-}$, $[GaCl_6]^{3-}$, $[InCl_6]^{3-}$, $[TlCl_6]^{3-}$, $[SnCl_6]^{2-}$, $[PF_6]^-$ $[PCl_6]^-$, $[AsF_6]^-$, $[SbCl_6]^-$, $[BiCl_6]^-$ SF_6, TeF_6, SeF_6, $[IF_6]^+$ etc.

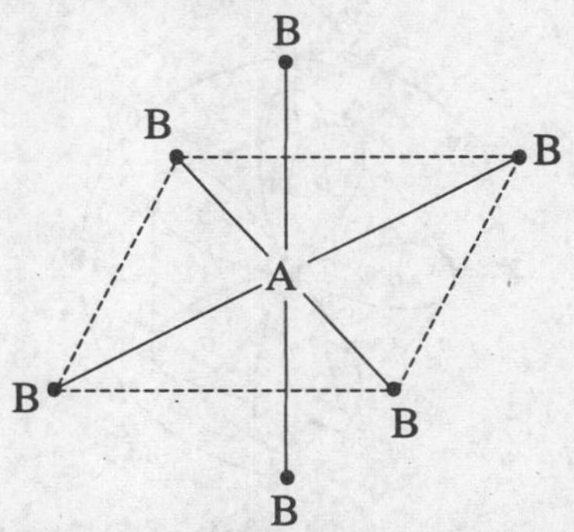

Fig. 13.25: Octahedral structure of AB_6 type species

2. AB_5 (*lp*) type species : Square pyramidal species.

These species can be represented as $\begin{matrix} {}^{B}_{B}> & \ddot{A} & <{}^{B}_{B} \\ & | & \\ & B & \end{matrix}$ which shows that the cantral atom (A) is surrounded by five σ-*bps* (σ-*bps* = 5) and one *lp* (*lp* = 1). According to VSEPR theory, lone pair of electrons occupies one of the two axial sp^3d^2 hybrid orbitals, since in this case (*lp*—*bp*) repulsions are minimum and hence AB_5 (*lp*) species attains maximum stability. Thus, although the arrangement of six sp^3d^2 hybrid orbitals in space is octahedral, due to the presence of one *lp* in the axial hybrid orbital, the shape of AB_5 (*lp*) species gets distorted and becomes *square pyramidal* (See Fig. 13.26).

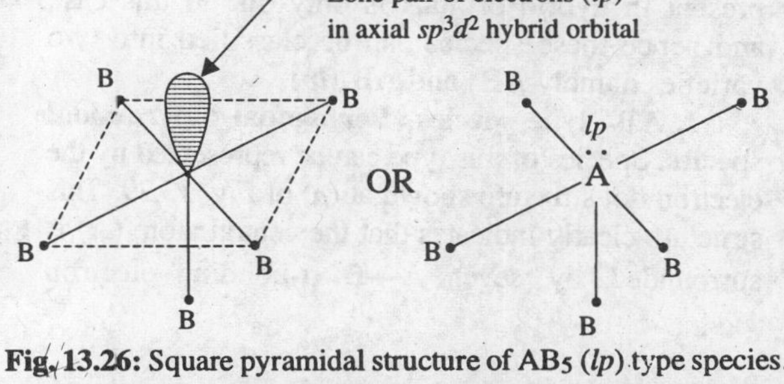

Fig. 13.26: Square pyramidal structure of AB_5 (*lp*) type species.

Examples of molecules/ions having square pyramidal structure are IF_5, BrF_5, ClF_5, $[SbF_5]^{2-}$, $[SbCl_5]^{2-}$, $[TeCl_5]^-$, $[XeF_5]^+$ etc.

3. AB_4 *(lp)*$_2$ type species : Square planar species. These species have their electron dot structure as $\begin{matrix} B \\ B \end{matrix} > \ddot{\underset{..}{A}} < \begin{matrix} B \\ B \end{matrix}$. It is obvious from this structure that the central atom (A) is surrounded by four σ-*bps* and two *lps*. According to VSEPR theory, the two *lps* occupy the two axial sp^3d^2 hybrid orbitals so that (*lp—bp*) repulsion becomes minimum and hence AB_4 $(lp)_2$ molecule attains maximum stability. Thus we see that, although the spatial orientation of the six sp^3d^2 hybrid orbitals is octahedral, due to the presence of two *lps* in the axial hybrid orbitals, the shape of $AB_4(lp)_2$ molecule/ion becomes *square planar* (See Fig. 13.27).

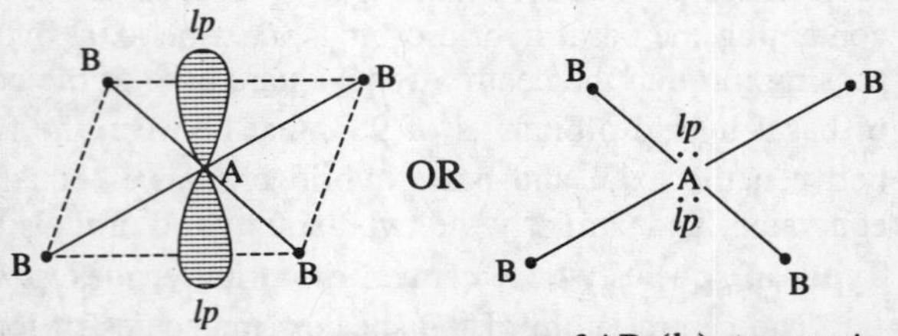

Fig. 13.27 : Square planar geometry of $AB_4(lp)_2$ type molecule/ion.

Examples of molecules / ion having square planar geometry are XeF_4, penta-atomic interhalogen anions (*e.g.*, BrF_4^-, IF_4^-, I_5^-, ICl_4^-) etc.

VIII. sp^3d^3 hybridisation : Pentagonal bipyramidal hybridisation

In this hybridisation one *s*, three *p* and three *d* (d_{xy}, d_{yz} and d_{zx}) orbitals combine together and give a set of seven hybrid orbitals which are called sp^3d^3 or pentagonal bipyramidal hybrid orbitals (See Fig. 13.28).

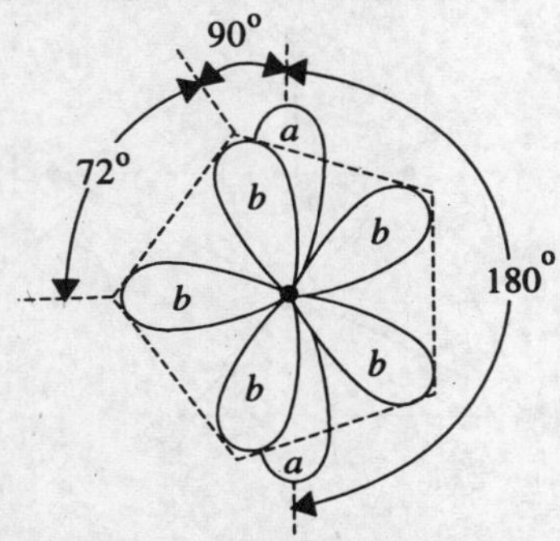

Fig. 13.28: Seven sp^3d^3 hybrid orbitals

sp^3d^3 hybrid orbitals have the following characteristics :

(*i*) These have the same shape and same energy.

(*ii*) They are directed towards the seven corners of a regular (symmetrical) pentagonal bipyramid, *i.e.*, the arrangement of the seven hybrid orbitals in space is pentagonal bipyramidal. Due to their pentagonal bipyramidal arrangement, they do not lie in one plane. Five hybrid orbitals (which have been represented by *b* and hence are called *basal hybrid orbitals*) are lying in the pentagonal plane while the remaining two which have been represented by *a* (called *axial hybrid orbitals*) are lying above and below on the axis passing through the centre of the pentagonal base of the pentagonal bipyramid. Thus the angle between each pair of basal hybrid orbitals is of 360/5 = 72°, that between two axial hybrid orbitals is of 180° and that between the axial and basal hybrid orbitals is equal to 90°. The five basal hybrid orbitals are equivalent and similarly the two axial hybrid orbitals are also equivalent.

In the species whose structure can be explained on the basis of sp^3d^3 hybridisation, the sum of σ-*bps* and *lps* surrounding the central item is always equal to 7. Maximum number of *lps* that may be present in hybrid orbitals is only one in this case and hence these species can be classified into two varieties namely AB_7 and $AB_6(lp)$.

1. AB_7 type species : Pentagonal bipyramidal species. Species of this type can be represented by the electron dot structure shown at (*a*) of Fig. 13.29. This structure clearly indicates that the central atom (A) is surrounded by seven A—B σ-bonding electron

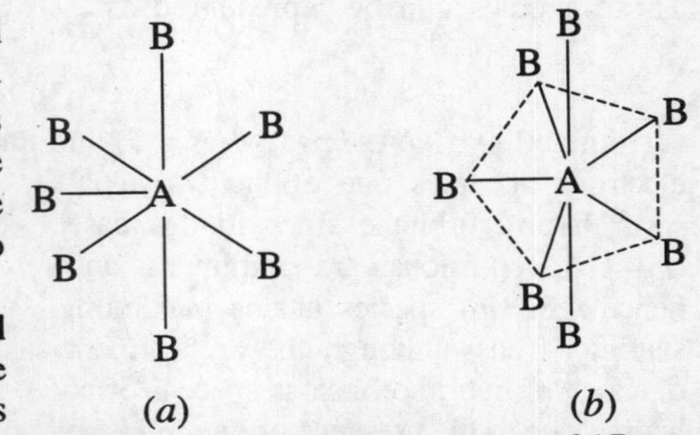

Fig. 13.29: (*a*) Electron dot structure of AB_7 type molecule (*b*) Pentagonal bipyramidal geometry of AB_7 type molecule.

pairs (σ-*bps* = 7) and no lone pair of electrons (*lp* = 0). Thus, since σ-*bps* + *lps* = 7 + 0 = 7, A-atom (central atom) in AB_7 molecule is sp^3d^3 hybridised and this molecule has pentagonal bipyramidal geometry shown at (*b*) of Fig. 13.29. IF_7 and ReF_7 are the examples of molecules of AB_7 type.

2. AB_6(*lp*)type species : Distorted octahedral geometry. Species of this type can be represented by the electron dot structure shown in Fig. 13.30. This structure shows that the central atom (A) is surrounded by six A—B σ-bonding electron pairs (σ-*bps* = 6) and one lone pair of electrons (*lp* = 1). Thus the central A-atom is sp^3d^3 hybridised. Theoretically the lone pair of electron should occupy one of the five basal positions of the pentagonal bipyramid, *i.e.,* the lone pair of electrons should be directed towards one of the five corners of the pentagonal base and hence AB_6(*lp*) molecule/ion should have *distorted pentagonal bipyramidal* structure shown at (*a*) of Fig. 13.31. However, experimentally it has been proved that structure (*a*) is wrong. The correct structure which AB_6(*lp*) molecule/ion has is *distorted octahedral.* In this structure the six corners of the octahedron are occupied by six B-atoms and the lone pair of electrons (*lp*) is directed either towards the mid-point of an edge of the octahedron or directed towards the centre of one of the trigonal eight faces of the octahedron [See Fig. 13.31. (*b*)].

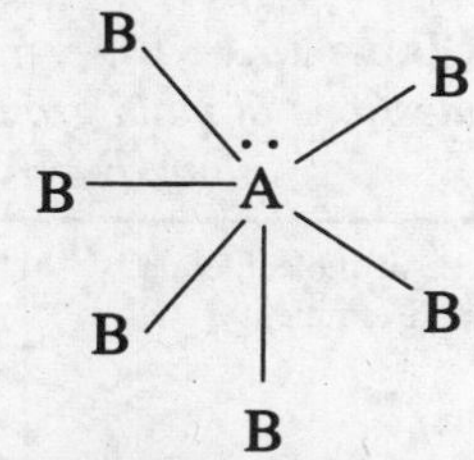

Fig. 13.30: Lewis structure (dot structure) of AB_6 (*lp*) molecule.

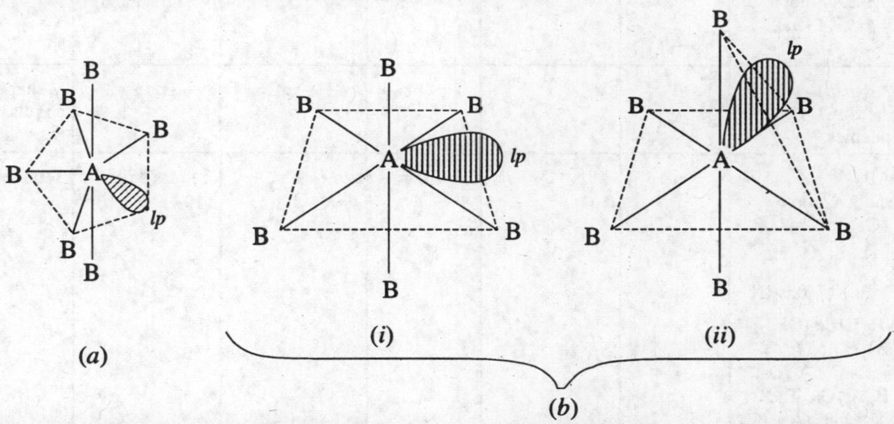

Fig. 13.31 : (*a*) Distorted pentagonal bipyramidal structure of AB6 (*lp*) molecule/ion. (Wrong structure) (*b*) Distorted octahedral structure of AB6 (*lp*) molecule/ion having a *lp* directed towards the mid point of an edge of the octahedron [see (*i*)] or directed towards the centre of one of the trigonal eight faces of the octahedron [see (*ii*)].

XeF_6, IF_6^-, $TeCl_6^{2-}$, TeF_6^{2-}, SbX_6^{3-} etc. are the examples of AB_6 (*lp*) species which have distorted octahedral structure.

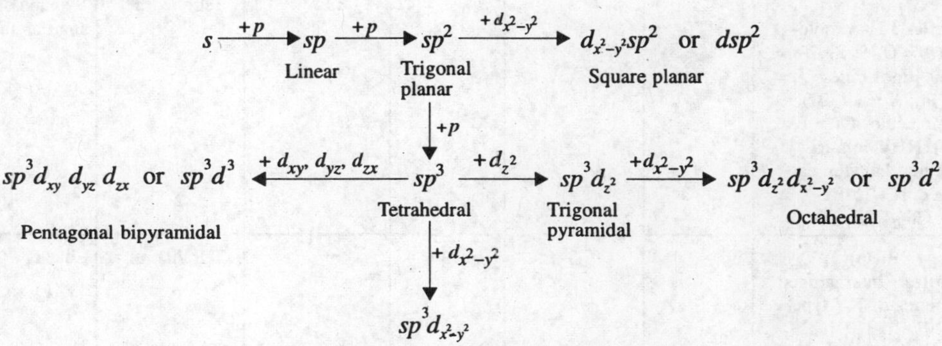

Fig. 13.32: Formation of different hybrid orbitals and their orientation.

Summary

The formation of different hybrid orbitals and their orientation round the central atom are given in Fig. 13.32 and Table 13.1.

Table 13.1 : *Different types of hybridisation undergone by the central atom of simple molecules/ions of main group elements. These molecules/ions contain only sigma bonds. A = Central atom, B = Atoms surrounding the central atom, lp = lone pair of electron.*

Type of molecule/ion with examples	No. of σ-*bps* surrounding the central atom	No. of *lps* surrounding the central atom	(σ-*bps* + *lps*)	Type of hybridisation undergone by the central atom	Spatial arrangement of (σ-*bps* + *lps*) round the central atom and bond angle(s)	Actual shape (geometry) of the molecule/ion
AB_2 **Examples :** BeX_2 molecules (X = H, F, Cl)	2	0	2 + 0 = 2	*sp* ($s + p_x$)	Linear (180°)	Linear
AB_3 **Examples :** BX_3 molecules (X = H, F, I, Br), Bo_3^{3-}, BeF_3^-	3	0	3 + 0 = 3	sp^2($s + p_x + p_y$)	Trigonal planar (120° each)	Trigonal planar
AB_2 (*lp*) **Examples :** $SnCl_2$ (g), $PbCl_2$ molecules	2	1	2 + 1 = 3	sp^2	— do —	V-shape or angular or bent shape
AB_4 **Examples :** AX_4 molecules (A = C, Si; X = (H F, Cl, Br, I), BeX_4^{2+} ions, (X = H, F), BeX_4^{2-} ions (X = H_2O, NH_3), $[Be(OH)_2]^{2-}$ ion, $[Be(ox)_2]^{2-}$ ion (*ox* = oxalate ion, $C_2O_4^{2-}$), AX_4^- ions (A = B, Al, Ga, Tl; X = H, F), AX_4^+ ions (A = N, P, As; X = H, Cl, Br, I, Ph)	4	0	4 + 0 = 4	sp^3 ($s + p_x + p_y + p_z$)	Tetrahedral (109.5° each)	Tetrahedral
AB_3 (*lp*) **Examples :** AX_3 molecules (A = N, P, As, Sb, Bi; X = H, F, Cl, Br, I), H_3O^+ ion	3	1	3 + 1 = 4	sp^3	Tetrahedral	Trigonal pyramidal
AB_2 $(lp)_2$ **Examples :** H_2A molecules (A = O, S, Se, Te, Po), AX_2 molecules (A = S, Te, Po; X = Cl, Br), X_2O molecules (X = F, Cl, Br), NH_2^- ion, triatomic interhalogen cations (*e.g.* ICl_2^+, IBr_2^+, BrF_2^+, ClF_2^+, IF_2^+ etc.)	2	2	2 + 2 = 4	sp^3	Tetrahedral	V-shape or angular shape
AB $(lp)_3$ **Examples :** Interhalogen compounds of XY type (e.g. ClF, BrF, IF, ICl, IBr)	1	3	1 + 3 = 4	sp^3	Tetrahedral	Linear

contd.

AB_5 **Examples :** AX_5 molecules in the gaseous and liquid state (A = P, As, Sb; X = F, Cl, Br), $SnCl_5^-$	5	0	5 + 0 = 5	sp^3d $(s + p_x + p_y + p_z + d_{z^2})$	Trigonal bipyramidal (120°, 180°, 90°)	Trigonal bipyramidal
AB_4 (lp) **Examples :** AX_4 molecules (A = S, Se, Te, Po; X = F, Cl, Br, I), R_2 Se X_2, $SnCl_4^{2-}$, penta-atomic interhalogen cations (e.g., IF_4^+, ClF_4^+, BrF_4^+ etc.)	4	1 (occupies one of the equatorial sp^3d hybrid orbitals)	4 + 1 = 5	sp^3d	Trigonal bipyramidal	Distorted tetrahedral or see-saw.
AB_3 $(lp)_2$ **Examples :** Interhalogen compounds of XY_3 type (*e.g.*, ClF_3, BrF_3, IF_3), XeF_3^+.	3	2 (occupy the two equatorial sp^3d hybrid orbitals)	3 + 2 = 5	sp^3d	Trigonal bipyramidal	Bent T-shaped structure
AB_2 $(lp)_3$ **Examples :** XeF_2 triatomic interhalogen anions (*e.g.* ICl_2^-, IBr_2^-, $BrCl_2^-$, ClF_2^-, I_3^-, $ClBrI^-$ etc.)	2	3 (occupy all the three equatorial sp^3d hybrid orbitals)	2 + 3 = 5	sp^3d	Trigonal bipyramidal	Linear shape
AB_6 **Examples :** AF_6 molecules (A = S, Se, Te), AX_6^{3-} ions (A = Al, Ga, In, Tl; X = F, Cl), AX_6^{2-} ions (A = Si, Ge, Sn; X = F, Cl), AX_6^- ions (A = P, As, Sb, Bi, X = F, Cl) hepta-atomic interhalogen cations (*e.g.* IF_6^+, BrF_6^+ etc.)	6	0	6 + 0 = 6	sp^3d^2 $(s + p_x + p_y + p_z + d_{z^2} + d_{x^2-y^2})$	Octahedral (90°, 180°, 90°)	Octahedral
AB_5 (lp) **Examples :** Interhalogen compounds of XY_5 type (*e.g.* IF_5, BrF_5, ClF_5), TeX_5^- ions (X = F,Cl), Sb X_5^{2-} (X = F, Cl), XeF_5^+.	5	1 (occupies one of the two axial sp^3d^2 hybrid orbitals)	5 = 1 = 6	sp^3d^2	Octahedral	Square pyramidal
AB_4 $(lp)_2$ **Examples :** XeF_4 molecule, penta-atomic interhalogen anions (e.g. ICl_4^-, BrF_4^-, ClF_4^- etc.)	4	2 (occupy both the axial sp^3d^2 hybrid orbitals)	4 + 2 = 6	sp^3d^2	Octahedral	Square planar
AB_7 **Examples :** Interhalogen compounds of XY_7 type (*e.g.* IF_7)	7	0	7 + 0 = 7	sp^3d^3 $(s + p_x + p_y + p_z + d_{xy} + d_{yx} + d_{zx})$	Pentagonal bipyramidal (72°, 180°, 90°)	Pentagonal bipyramidal
AB_6 (lp) **Examples :** XeF_6, SbX_6^{3-} (X = F, Cl), TeX_6^{2-} (X = F, Cl), hepta-atomic interhalogen anions (*e.g.*, IF_6^-, BrF_6^- etc.)	6	1	6 + 1 = 7	sp^3d^3	Pentagonal bipyramidal	Distorted octahedral

University Questions

1. (*a*) What do you understand by "orbital hybridistion"?
   (*b*) Explain different types of hybridisation with suitable examples. *(Meerut, 91)*
2. Draw schematic representation of the following hybrid orbitals : *sp*, sp^2 and sp^3. *(Nagpur 85; Bharat Hidasan, 92)*
3. What is hybridisation? Show how the concept of hybrid orbitals gives a better description of the geometry of molecules. *(Kerala, 87; Jodhpur, 89; Behrampur, 91; Rohelkhand, 91; North-East Hill., 92; Agra, 89; Garhwal, 88)*
4. Explain sp^3 hybridisation. *(Behrampur, 91)*
5. Write down the rules of hybridisation of orbitals. *(Raj., 86; Jodhpur, 89)*
6. Discribe different types of hybrid orbitals that can be formed by an atom with *s*, *p* and *d* orbitals in its valence-shell. What are the shapes which each of the hybrid orbitals has. *(Delhi, 81; Agra, 81, MD Rohtak, 82; Calcutta, 82)*
7. Write a note on hybridisation. *(Agra, 92; Gauhati, 91)*
8. Discuss *sp*, sp^2, sp^3, sp^3d, dsp^2, dsp^3 and sp^3d^2 types of hybridisation, giving one examples of each. *(Rajasthan, 87; Garhwal, 88)*
9. Giving examples, explain the concept of hybridisation with reference to sp^3 and sp^3d hybridisation. *(Delhi, 88)*
10. Name the types of hybrid orbitals which may be formed by an atom with only *s*- and *p*-orbitals in its valence-shell. Draw the shapes of hybrid orbitals so formed. *(Delhi, 89)*
11. State the geometric arrangement of bonds produced by a set of sp^3 hybrid orbitals. *(Poona, 86)*
12. Discuss the type of hybridisation that exists in the following geometrical shapes of molecules (*i*) Tetrahedral (*ii*) Planar (*iii*) Octahedral *(Bundelkhand, 81; Mysore, 82)*
13. Using one example of each, discuss how the geometry of molecules is related with the orbital hybridisation in respect of sp^2, sp^3 and sp^3d^2 hybridisation. *(Rajasthan, 87)*
14. What is the expected geometry for the following types of hybridisation : *sp*, sp^2, sp^3, sp^3d and sp^3d^2 ? *(Rohelkhand, 91)*
15. Explain the diagrammatic representation of *sp*, sp^2 and sp^3 hybridisations. *(Ravishankar, 93)*
16. Explain the following, giving appropriate reasons:
    (*i*) Hybrid orbitals form comparatively stronger bonds. *(Ravishankar, 93)*
    (*ii*) PF_5 is known but NF_5 is unknown. *(E.C.C. Allahabad, 94)*
    (*iii*) PCl_5 is known but PH_5 is unknown. *(E.C.C Allahabad, 94)*
17. What is hybridization ? Discuss the following types of hybridization giving one example of each : (*i*) sp^3 and (*ii*) sp^3d^2. *(Nagpur 2002)*
18. Explain the structure of the following pairs of molecules in terms of hybridization theory (*i*) CO_2, H_2O (*iii*) BF_3, NH_3 *(VBS Purvanchal 2002)*
19. Give the geometry of the following types of hyridised orbitals : *sp*, sp^2, sp^3, sp^3d and sp^3d^2. Give one example of inorganic molecules for each type of hybridisation. *(VBS Purvanchal 2002)*
20. In CH_4, NH_3 and H_2O bond angles are 109.5°, 107.3° and 104.5° respectively. Give reasons for this decrease. *(VBS Purvanchal 2002, 2003)*
21. Explain the structure of $BeCl_2$, BCl_3 and CCl_4 molecules on the basis of hybridisation theory. *(VBS Purvanchal 2003)*
22. What do you mean by hybridization ? Explain with examples taking the cases of sp^3d and sp^3d^2 hybridization. *(Rohilkhand 2003)*
23. What is hybridisation ? Discuss sp^2, sp^3d and sp^3d^2 hybridisation with suitable examples. *(Nagpur 2003)*

14

Structure and Shape of Molecules/Ions Containing only Sigma Bonds

Structure and Shape of Molecules

Introduction

Here we shall consider the structure and geometry of those molecules in which the central atom is an element of non transition element (main group elements) and the surrounding elements with which the central atom is linked by two or more σ-bonds is hydrogen or halogens. The structure of interhalogen compounds (*e.g.*, ClF, BrF, ICl, ClF_3, IF_3, IF_5, BrF_5, IF_7 etc.) has also been discussed in this section. Depending on the nature of the molecule, the central atom (in the ground state or in the excited state) is *sp*, sp^2, sp^3 or sp^3d etc., hybridised. The molecules whose central atom does not contain any lone pair of electrons have the expected geometry while the molecules whose central atom is surrounded by one or more lone pairs of electrons have distorted shape.

1. Geometry and structure of BeX_2 molecules (X = H, F, Cl)

All these molecules have *linear geometry* (X — Be — X) which arises from *sp* hybridisation of the central Be-atom in its excited state (Be-atom in its excited state has its valence-shell configuration as $2s^1\ 2p_x^1$). Both Be—X σ-bonds have the same energy/stability and hence are of the same length. They are inclined to each other at an angle equal to 180°, *i.e.*, X Be X bond angle = 180°. How Be—F σ-bonds are formed in BeF_2 molecule by the use of *sp* hybrid orbitals has already been explained in chapter 13.

2. Geometry of BX_3 molecules (X = H, F, Cl, Br)

All these molecules have *trigonal planar geometry* which results from sp^2 hybridisation of the central B-atom. As an example let us discuss the geometry of BF_3 molecule. Lewis structure of this molecule is given at (*a*) of Fig. 14.1. This structure shows that B-atom (central atom) is surrounded by three σ-*bps* and *no lone pair of electrons*. Thus, since σ-*bps* + *lps* = 3 + 0 = 3, B-atom is sp^2 hybridised in BF_3 molecule.

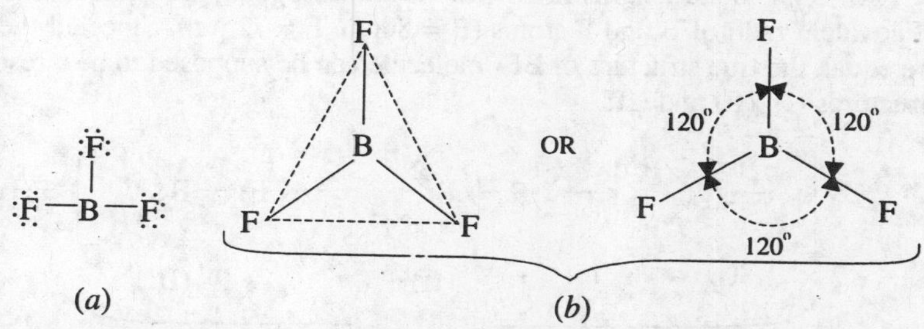

Fig. 14.1 : (*a*) Lewis structure of BF_3 molecule (*b*) Trigonal planar geometry of BF_3 molecule.

B-atom in ground state ($2s^2\ 2p^1$) : $2s^2\ 2p_x^{\ 1}\ 2p_y^{\ 0}\ 2p_z^{\ 0}$

B-atom in the excited state ($2s^1\ 2p^2$) : $2s^1\ 2p_x^{\ 1}\ 2p_y^{\ 1}\ 2p_z^{\ 0}$

B-atom in sp^2 hybridised state in BF_3 molecule : $\underbrace{(sp^2)^1\ (sp^2)^1\ (sp^2)^1}_{\text{Three } sp^2 \text{ hybrid orbitals}}\ 2p_z^{\ 0}$

BF_3 molecule has three B—F σ-bonds each of which is formed by the overlap between the singly-filled sp^2 hybrid orbital on B-atom and singly-filled $2p_x$ orbital on F-atom (F = $2s^2\ 2p_x^{\ 1}\ 2p_y^{\ 2}\ 2p_z^{\ 2}$). Due to sp^2 hybridisation and no lone pair of electrons on B-atom, BF_3 molecule has the expected trigonal planar shape shown at (*b*) of Fig. 14.1. Each of the three F B F bond angles is equal to 120°. sp^2 hybridisation given above indicates that the vacant $2p_z$ orbital on B-atom does not participate in sp^2 hybridisation process and hence remains unused. Completely-filled $2p_z$ orbital of any of the three F-atoms (F = $2s^2$, $2p_x^{\ 1}\ 2p_y^{\ 2}\ 2p_z^{\ 2}$) donates its electron pair to the vacant and unused $2p_z$ orbital of B-atom and gives rise to the formation of F($2p_z^{\ 2}$) → B($2p_z^{\ 0}$) coordinate π-bond [see hybridisation and also Fig. 14.2]. Thus we find that an additional F → B π-bond is established between B and F atoms when the vacant $2p_z$ orbital of B-atom accepts the electron pair from the filled $2p_z$ orbital on any F-atom. Since F → B π-bond is formed by the overlap of two *p*-orbitals, it is $(p - p)\pi$ or $(p\pi - p\pi)$ bond. The formation of F → B π-bond is called *back bonding* or *dative bonding*. This back bonding creates eight electrons (octet of electrons or inert gas configuration) round B-atom in BF_3 molecule. In this way B-atom which is surrounded by only six electrons (sextant) in BF_3 molecule gets surrounded by eight electrons (octet) and thus compensates its deficiency for two electrons. The same thing happens with other BX_3 molecules as well.

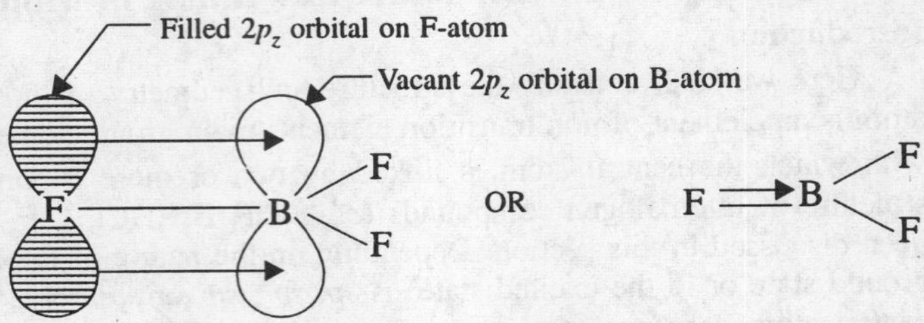

Fig. 14.2 : Formation of F ($2p_z^{\ 2}$) → B($2p_z^{\ 0}$) coordinate π-bond in BF_3 molecule.

The above discussion shows that since $2p_z$ orbital of B-atom is filled by the pair of electrons donated by F-atom, the electron deficiency of boron atom gets compensated and, therefore, the Lewis acid character of BF_3 molecule decreases.

All B—F bond lengths in BF_3 molecule are equal (=130 pm) which are shorter than the sum of covalent radii of B and F atoms (B = 80pm, F = 72 pm). Since all the three B—F bond lengths are equal, the true structure of BF_3 molecule can be supposed to be a resonance hybrid of the three structures (I), (II) and (III).

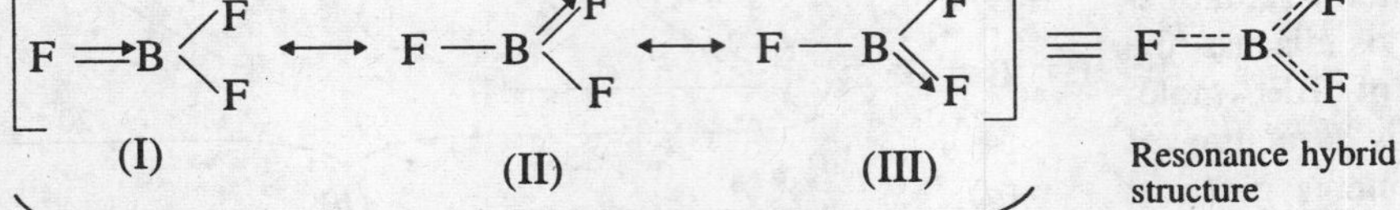

(I) (II) (III) Resonance hybrid structure

Three resonating structures

The dotted lines in the resonance hybrid structure indicate that *p*—*p* π-bond is not localised between any two particular B and F atoms; rather it is spread over the whole molecule. Thus π-bond is a delocalised π-bond.

Formation of (Cl—B) and (Br—B) π-bonds in BCl_3 and BBr_3 molecules. The overlap between 2*p* orbital (filled) on F-atom and 2*p* orbital (vacant) on B-atom which gives rise to the formation of F(2*p*) → B(2*p*) π bond is **easily** possible and hence very effective, since 2*p* orbitals of both the atoms belong to the same energy level. In case of BCl_3 molecule, the formation of Cl → B π-bond takes place due to the overlap between 3*p* orbital (filled) of Cl-atom and 2*p* orbital (vacant) on B-atom. Similarly in case of BBr_3 molecule, the formation of Br → B π bond takes place due to the overlap between 4*p* orbital (filled) on Br-atom and 2*p* orbital (vacant) on B-atom. But 3*p* (Cl) — 2*p* (B) overlap in BCl_3 molecule and 4*p* (Br) — 2*p* (B) overlap in BBr_3 molecule would not be very effective, since there is an appreciable difference in the energy levels of 3*p* and 2*p* orbitals (in BCl_3 molecule) and 4*p* and 2*p* orbitals (in BBr_3 molecule).

3. Structure and geometry of $SnCl_2$ (g) and $PbCl_2$ molecules

Both these molecules have *angular* (or *bent* or *T-shaped*) *structure* which arises due to sp^3 hybridisation of the central atom. As an example let us discuss the geometry of $SnCl_2(g)$ molecule whose Lewis structure is shown in the margin. From this structure it is evident that the central Sn-atom is surrounded by two *σ-bps* and *one lone pair of electrons*. Thus Sn-atom in $SnCl_2$ (*g*) molecule is sp^3 hybridised as shown below :

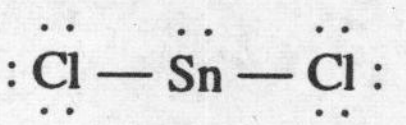

Lewis structure of $SnCl_2$ (g) molecule

Sn-atom in ground state ($5s^25p^2$) : $5s^2\ 5p_x^1\ 5p_y^1\ 5p_z^0$

Sn-atom in sp^2 hybridised state in $SnCl_2$ (g) molecule : $\underbrace{\boxed{(sp^2)^2}_{lp}\ (sp^2)^1\ (sp^2)^1}_{\text{Three } sp^2 \text{ hybrid orbitals}}\ 2p_z^0$

$SnCl_2$ molecule has two Sn — Cl σ-bonds. Each of these is formed by the overlap between the singly-filled sp^2 hybrid orbitals on Sn-atom and singly-filled $3p_x$ orbital on Cl-atom (Cl = $3s^2\ 3p_x^1\ 3p_y^2\ 3p_z^2$).

Although the spatial arrangement of the three electron pairs (two bonding electron pairs and one lone pair) round the central Sn-atom is *trigonal planar*, due to the presence of one lone pair of electrons in one sp^2 hybrid orbital, the expected *trigonal planar shape* of $SnCl_2$ molecule gets distorted and becomes *angular shape* which has been shown in Fig. 14.3.

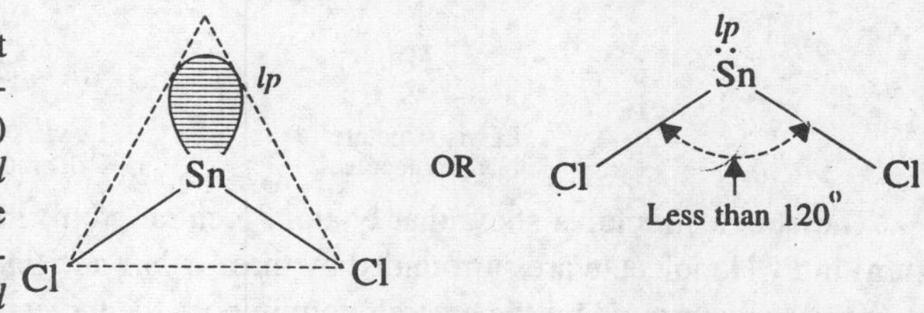

Fig. 14.3 : Angular geometry (V-shaped or bent structure) of $SnCl_2(g)$ molecule.

According to VSEPR theory, since (*lp –bp*) repulsion is greater than (*bp–bp*) repulsion in $SnCl_2$ molecule, there is some distortion in the molecule. It is due to this distortion that Cl Sn Cl bond angle becomes slightly less than 120° and the molecule assumes *V-shape* (*angular* or *bent shape*).

4. Structure and geometry of the molecules of AX_4 type (A = C, Si; X= H, F, Cl, Br, I)

All these molecules have *symmetrical tetrahedral shape* which is due to sp^3 hybridisation of the central atom in its excited state configuration. As an example let us discuss the geometry of CH_4 molecule. In this molecule, C-atom (central atom) is surrounded by four *σ-bps* as is evident from the electron dot structure of this molecule [Fig. 14.4 (*a*)] and hence is sp^3 hybridised.

C-atom in ground state ($2s^2\ 2p^2$) : $2s^2\ 2p_x^1\ 2p_y^1\ 2p_z^0$

C-atom in the excited state ($2s^12p^3$): $2s^1\ 2p_x^1\ 2p_y^1\ 2p_z^1$

C-atom in sp^3 hybridised state in CH_4 molecule : $\underbrace{(sp^3)^1\ (sp^3)^1\ (sp^3)^1\ (sp^3)^1}_{\text{Four } sp^3 \text{ hybrid orbitals}}$

CH_4 molecule has four C—H σ-bonds. Each of these bonds is formed by the overlap of singly-filled sp^3 hybrid orbital on C-atom with the singly-filled 1s orbital of H-atom (H = $1s^1$). According to VSEPR theory, since C-atom is sp^3 hybridised and has no lone pair of electrons on it, CH_4 molecule has symmetrical tetrahedral shape with HCH bond angle equal to 109.5°. [See Fig. 14.4 (*b*)].

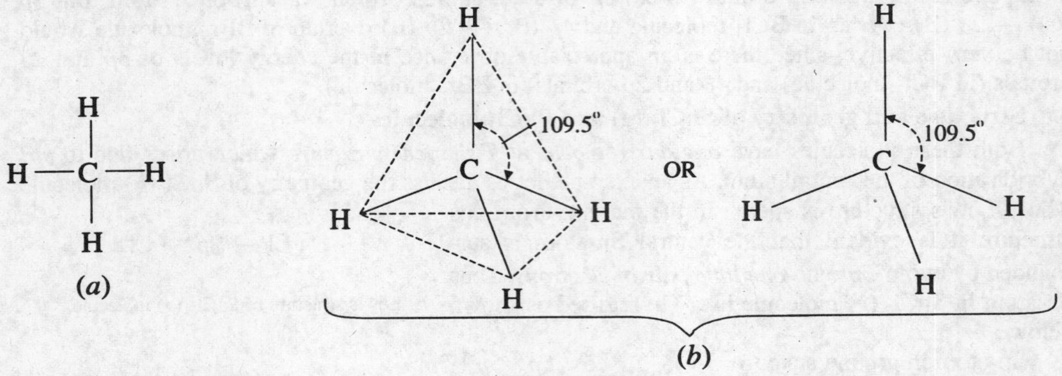

Fig 14.4 : (*a*) Electron dot structure of CH_4 molecule (*b*) Tetrahedral geometry of CH_4 molecule.

Structure and geometry of the molecules of AX_3 type (A = N, P, As, Sb, Bi; X = H, F, Cl Br, I)

All these molecules have trigonal pyramidal structure. This structure arises because of sp^3 hybridisation of the central atom. Let us consider the structures of NH_3 and PCl_3 molecules. Lewis structure of these molecules are given below.

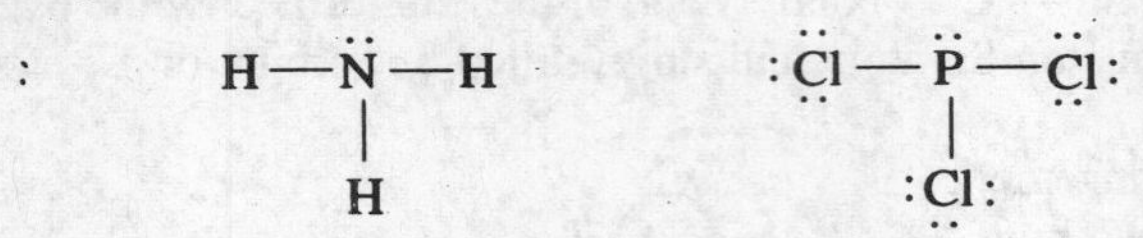

Lewis structure of NH_3 molecule   Lewis structure of PCl_3 molecule

The above structures show that N-atom (central atom) in NH_3 molecule and P-atom (central atom) in PCl_3 molecule are surrounded by three *σ-bps* and *one lone pair of electorons.* Thus since (*σ-bps* + *lps*) surrounding the central atom in both the molecules is equal to 3+1 = 4, central atom in these molecules is sp^3 hybridised.

N-atom in ground state ($2s^2\ 2p^3$) : $2s^2\ 2p_x^1\ 2p_y^1\ 2p_z^1$

N-atom in sp^3 hybridised state in NH_3 molecule : $\boxed{(sp^3)^2}\ (sp^3)^1\ (sp^3)^1\ (sp^3)^1$

P-atom in ground state ($3s^2 3p^3$) : $3s^2\ 3p_x^1\ 3p_y^1\ 3p_z^1$

P-atom in sp^3 hybridised state in PCl_3 molecule : $\boxed{(sp^3)^2}\ (sp^3)^1\ (sp^3)^1\ (sp^3)^1$

NH_3 molecule has three N — H σ-bonds. Each of these is formed by the overlap between the singly-filled sp^3 hybrid orbital on N-atom and singly-filled 1s orbital of H-atom. PCl_3 molecule has three P—Cl σ-bonds each of which is obtained through the overlap between the singly-filled sp^3 hybrid orbitals of P-atom and singly-filled $3p_x$ orbital of Cl-atom (Cl = $3s^2$, $3p_x^1\ 3p_y^2\ 3p_z^2$).

Although the spatial arrangement of four electron pairs (three bonding electron pairs and one lone pair of electrons) round the central N or P atom in NH_3 or PCl_3 molecule is tetrahedral, due to the presence of one lone pair of electrons in one sp^3 hybrid orbital, the expected tetrahedral shape of the molecule gets distorted and becomes trigonal pyramidal with N-atom (or P-atom) at its centre.

The three corners of the trigonal base are occupied by three H-atoms (in NH_3) or by three Cl-atoms (in PCl_3). The lone pair of electrons occupies the apex of the pyramid (see Fig. 14.5).

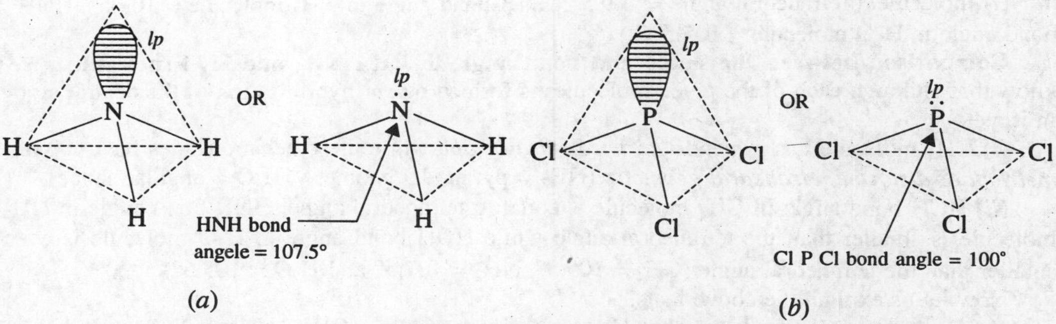

Fig. 14.5 : Trigonal pyramidal shape of NH_3 [shown at (*a*)] and PCl_3 [shown at (*b*)] molecules.

According to VSEPR theory, since (*lp-bp*) repulsion is greater than (*bp-bp*) repulsion, each of the three N—H/P—Cl bonds comes closer to each other and hence each HNH and ClPCl bond angle becomes less than the tetrahedral angle (= 109.5°). Thus in case of NH_3 molecule, HNH bond angle has been found to be equal to 107.5° which is less than the tetrahedral angle. Similarly ClPCl bond angle in PCl_3 molecule is equal to 100° which is less than the tetrahedral angle.

6. Structure and geometry of H_2A (A = O, S, Se, Te, Po) and AX_2 (A = S, Te, Po; X = Cl, Br) type molecules

All these molecules have *angular* (*V-shaped*) *structure* which arises due to sp^3 hybridisation of the central atom in the molecules and due to the presence of two lone pairs of electrons (*lps* = 2) in two sp^3 hybrid orbitals. As an example let us discuss the geometry of H_2O molecule. The electron dot structure of H_2O molecule as given at (*a*) of Fig. 14.6 shows that the central O-atom in H_2O molecule is surrounded by two σ-*bps* and two *lps* (*i.e.*, σ-*bps* + *lps* = 4). Thus, since σ-*bps* + *lps* = 4, O-atom (central atom) is sp^3 hybridised in H_2O molecule.

O-atom in ground state ($2s^2\ 2p^4$) : $2s^2\ 2p_x^2\ 2p_y^1\ 2p_z^1$

O-atom in sp^3 hybridised state in H_2O molecule : $\underbrace{\underset{lp}{\boxed{(sp^3)^2}}\ \underset{lp}{\boxed{(sp^3)^2}}\ (sp^3)^1\ (sp^3)^1}_{\text{Four } sp^3 \text{ hybrid orbitals}}$

H_2O molecule has two O—H σ-bonds. Each of these bonds results from the overlap of singly-filled sp^3 hybrid orbital with the singly-filled 1*s* orbital on H-atom.

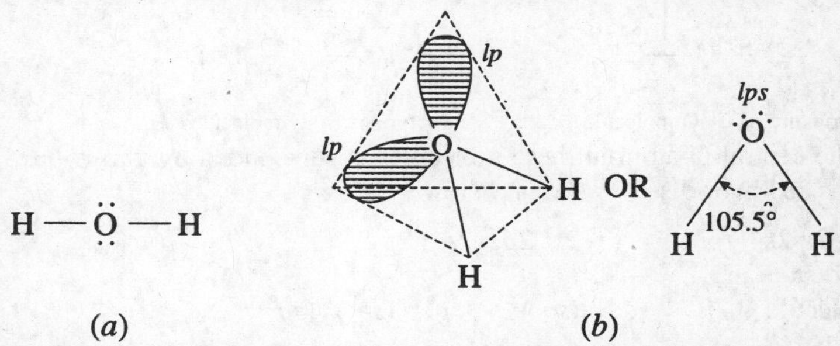

Fig. 14.6 : (*a*) Electron dot structure of H_2O molecule. (*b*) Angular geometry of H_2O molecule.

Although the spatial orientation of four sp^3 hybrid orbitals round the central O-atom in H_2O molecule is tetrahedral, due to the presence of two lone pairs of electrons in two sp^3 hybrid orbitals, the expected tetrahedral shape of H_2O gets distorted and assumes *angular shape* (See Fig. 14.6).

We know that the central O-atom in H_2O molecule is surrounded by two lone pairs of electrons (*lps*) and two bonding electron pairs (*bps*). Now since, according to VSEPR theory,

$$(lp - lp) \text{ repulsion} > (lp - bp) \text{ repulsion} > (bp - bp) \text{ repulsion},$$

there is a greater distortion in H_2O molecule than in NH_3 molecule. Consequently, HOH bond angle in H_2O molecule decreases by a greater amount from tetrahedral angle than HNH bond angle in NH_3 molecule. (Tetrahedral angle = 109.5°, HNH bond angle in NH_3 molecule = 107.5°, HOH bond angle in H_2O molecule = 105.5°C.)

Comparison between the shape and bond angle in CH_4, NH_3 and H_2O molecules. We know that although each of the given molecules is formed by sp^3 hybridisation of the central atom in it, yet:

(*i*) CH_4 molecule has *symmetrical tetrahedral shape* while the other molecules have *unsymmetrical* or *distorted tetrahedral geometry* (NH_3 = pyramidal geometry, H_2O = angular shape).

(*ii*) HCH bond angle in CH_4 molecule is equal to tetrahedral angle, HNH bond angle in NH_3 molecule is smaller than the tetrahedral angle while HOH bond angle in H_2O molecule is even smaller than the tetrahedral angle (CH_4 = 109.5°, NH_3 = 107.5° and H_2O = 105.5°).

Now let us explain the above facts.

Explanation of (i). We know that C-atom (central atom) in CH_4 molecule is surrounded by *σ-bps* only (four) and *no lp* (*lp* = 0), while N–atom in NH_3 molecule is surrounded by *three σ-bps* and *one lp* (*lp* = 1) and O-atom in H_2O molecule is surrounded by two *σ-bps* and two *lps* (*lps* = 2). Since C-atom has no *lp* on it, CH_4 molecule does not undergo any distortion and hence has the expected symmetrical tetrahedral geometry. On the other hand the central N-atom in NH_3 and O-atom in H_2O have one and two *lps* respectively, these molecules undergo distortion and hence their shapes get distorted, becoming pyramidal and angular respectively.

Explanation of (ii). With the increase in the number of *lps* on the central atom, the distortion in the molecule becomes greater and hence bond angle becomes smaller. Thus HCH bond angle in CH_4 molecule (*lp* = 0) is equal to tetrahedral angle (=109.5°), HOH bond angle in H_2O molecule (*lps* = 2) is smaller than HNH bond angle in NH_3 molecule (*lp* = 1) (H_2O=105.5°, NH_3 = 107.5°).

Bond angles in some other angular molecules involving sp^3 hybridisation. Bond angles in some molecules which have angular geometry because of sp^3 hybridisation of the central atom are as: HOH bond angle in H_2O molecule = 105.5°, HSH in H_2S = 92.5°, HSeH in H_2Se = 90.0°, HTeH in H_2Te = 89.0°, FOF in OF_2= 105°, ClOCl in Cl_2O = 110°, BrOBr in Br_2O > 110°, ClSCl in SCl_2 = 103° and BrTeBr in $TeBr_2$ = 98°. It may be noted that, as expected, in molecules namely H_2O, H_2S, H_2Se, H_2Te, OF_2, SCl_2 and $TeBr_2$, the bond angle is less than the tetrahedral angle (= 109.5°), but in the remaining molecules (*viz.*, Cl_2O and Br_2O) the bond angle is greater than the tetrahedral angle.

7. Structure and geometry of monoxides of halogens, X_2O (X = F, Cl, Br)

Fluorine, chlorine and iodine form monoxides which are formulated as OF_2 (difluorine oxide), Cl_2O (dichlorine oxide) and Br_2O (dibromine oxide). The electron dot structure of these oxides can be written as :

:F̤̈ — Ö̤ — F̤̈:	:C̤̈l — Ö̤ — C̤̈l:	:B̤̈r — Ö̤ — B̤̈r:
Structure of OF_2 molecule	Structure of Cl_2O molecule	Structure of Br_2O molecule

These structures show that the central O-atom in these molecules is surrounded by two *σ-bps* and two *lps* and hence this atom is sp^3 hybridised as shown below :

O-atom in ground state ($2s^2 2p^4$) : $2s^2\ 2p_x^2\ 2p_y^1\ 2p_z^1$

O-atom in sp^3 hybridised state in OX_2 (X = F, Cl, Br) molecule : $(sp^3)^2\ \ (sp^3)^2\ (sp^3)^1\ (sp^3)^1$

Four sp^3 hybrid orbitals

It may be seen from Fig. 14.7 that two of the four sp^3 hybrid orbitals contain lone pair of electrons and the remaining two are half-filled and hence are used in forming two (O—X) σ -bonds with two halogen atoms (X).

Although the central O-atom is sp^3 hybridised in each oxide, due to the presence of two lone pairs of electrons on O-atom, the expected tetrahedral shape of OX_2 molecule gets distorted and becomes *angular* (*V-shaped structure*) as given in Fig. 14.7.

XOX bond angle in F_2O, Cl_2O and Br_2O molecules. We have seen above that two of the four sp^3 hybrid orbitals on the central O-atom are occupied by two lone pairs of electrons. Due to the presence of lone pairs of electrons, XOX bond angle in all the three oxides is different from the tetrahedral angle (= 109.5°) as shown below:

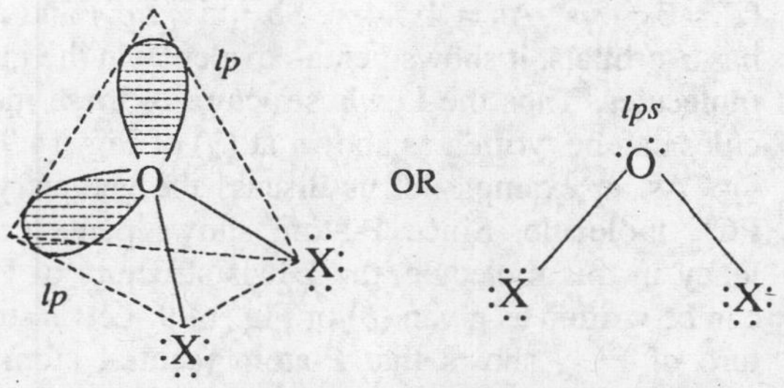

Fig. 14.7 : Angular shape (V-shaped structure) of OX_2 molecule (X = F, Cl and Br).

FOF bond angle in F_2O molecule = 105°
ClOCl bond angle in Cl_2O molecule = 110°
BrOBr bond angle in Br_2O molecule = > 110°.

F—O bond length in OF_2 molecule is equal to 1.409Å and Cl—O bond length in Cl_2O molecule is equal to 1.70 Å.

Explanation of bond angles. In F_2O molecule, since F-atom is more electronegative than O-atom (F = 4.0, O = 3.5), the bonding electron pairs lie nearer to F-atom. This results in that (*bp–bp*) repulsion becomes less than (*lp–lp*) repulsion and hence FOF bond angle is *less* than 109.5° (tetrahedral angle), being equal to 105°.

In Cl_2O molecule, since O-atom is more electronegative than Cl-atom (O = 3.5, Cl = 3.0), the bonding electron pairs lie closer to O-atom and hence (*bp – bp*) repulsion is *greater* than (*lp – lp*) repulsion and ClOCl bond angle is greater than 109.5°, being equal to 110°.

In Br_2O molecule, since Br-atom is less electronegative than Cl-atom (Br = 2.8, Cl = 3.0), the bonding electron pairs will be drawn more towards O-atom and hence (*bp – bp*) repulsion is much greater than (*lp – lp*) repulsion. Thus BrOBr bond angle is even more than 110°. (See Fig. 14.8).

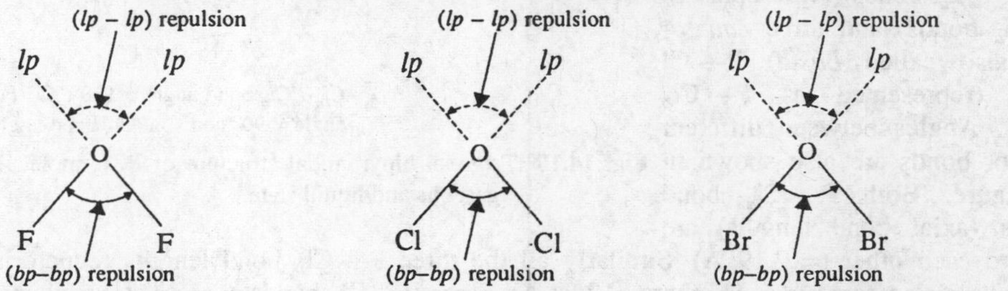

Here (*bp-bp*) repulsion < (*lp–lp*) repulsion and hence FOF bond angle <109.5° (=105°)

Here (*bp–bp*) repulsion > (*lp–lp*) repulsion and hence ClOCl bond angle >109.5° (=110°)

Here (*bp–bp*) repulsion >> (*lp–lp*) repulsion and hence Br O Br bond angle > 110°

Fig. 14.8: XOX bond angles in F_2O, Cl_2O and Br_2O molecules.

8. Structure and geometry of dihalides of S, Se and Te

As an example let us consider the structure and geometry of SCl_2 molecule. The electron dot structure of this molecule is $Cl—\ddot{\underset{\cdot\cdot}{S}}—Cl$. Since S-atom (central atom) is surrounded by two *σ-bps* and two *lps*, S-atom is sp^3 hybridised in this molecule. Since two of the four sp^3 hybrid orbitals are occupied by two *lps*, Cl S Cl bond angle decreases from the tetrahedral angle (= 109.5°) to 103° and the shape becomes angular. The bond angle in $TeBr_2$ is 98°.

9. Structure and geometry of the molecules of AX_5 type (A = P, As, Sb; X = F, Cl, Br) in the gaseous and liquid state

In the given molecules, since the central atom (A-atom) has five electrons in its valence-shell

($P = 3s^2\ 3p^3$, As = $4s^2\ 4p^3$, Sb = $5s^2\ 5p^3$) and also has *d*-orbitals, it shows penta-covalency in the given molecules. Thus the Lewis structure of these molecules can be written as shown at (*a*) of Fig. 14.9.

As an example let us discuss the geometry of PCl_5 molecule. Since P-atom shows penta covalency in this molecule, the Lewis structure of PCl_5 can be written as given (*b*) of Fig. 14.9. Lewis structure of PCl_5 shows that P-atom (central atom) is surrounded by five *σ-bps* and no lone pair of electrons and hence is sp^3d hybridised as shown below:

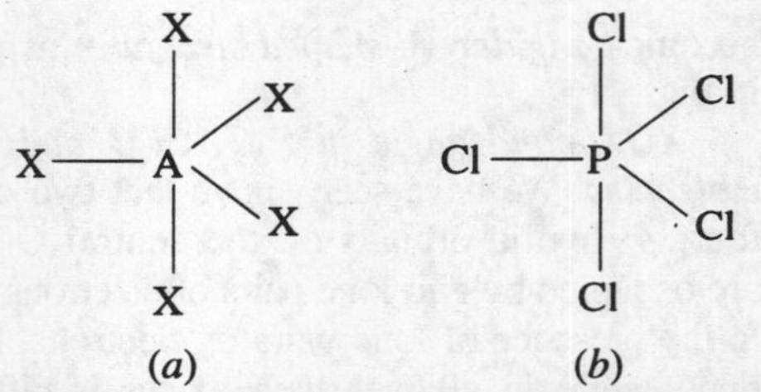

Fig. 14.9 : (*a*) Lewis structure of AX_5 molecule (A = P, As, Sb) (*b*) Lewis structure of PCl_5 molecule.

P-atom in ground state ($3s^23p^3d^0$) : $3s^2\ 3p_x^{\ 1}\ 3p_y^{\ 1}\ 3p_z^{\ 1}\ 3d^0$

P-atom in excited state ($3s^13p^33d^1$) : $3s^1\ 3p_x^{\ 1}\ 3p_y^{\ 1}\ 3p_z^{\ 1}\ 3d^1$

P-atom in sp^3d hybridised state in PCl_5 molecule: $\underbrace{(sp^3d)^1\ (sp^3d)^1\ (sp^3d)^1\ (sp^3d)^1\ (sp^3d)^1}_{\text{Five } sp^3d \text{ hybrid orbitals}}$

PCl_5 molecule has five P— Cl σ-bonds. Each of these bonds results by the overlap between the singly-filled sp^3d hybrid orbital on P-atom and singly-filled $3p_x$ orbital on Cl-atom (Cl = $3s^2\ 3p_x^1\ 3p_y^2\ 3p_z^2$). Since there is no lone pair of electrons in any of the five sp^3d hybrid orbitals, PCl_5 molecule has the expected *trigonal bipyramidal shape* which has been shown in Fig. 14.10. PCl_5 molecule has *two axial* P— Cl bonds (represented as P— Cl_a bonds) and three *equatorial* (also called *basal*) P— Cl bonds (represented as P— Cl_b bonds). Angles between different types of bonds are also shown in the figure. Both P— Cl_a bond lengths (axial bond lengths) are equal to each other (= 2.19 Å). Similarly all the three P— Cl_b bond lengths (equatorial bond lengths) are also equal to each other (= 2.04 Å). Since P— Cl_a bonds are less strong than P— Cl_b bonds, P— Cl_a bond lengths are longer than the P— Cl_b bond lengths and hence have considerable ionic character. The difference between the axial and equatorial bond lengths confirms the fact that the five sp^3d hybrid orbitals are *non-equivalent.*

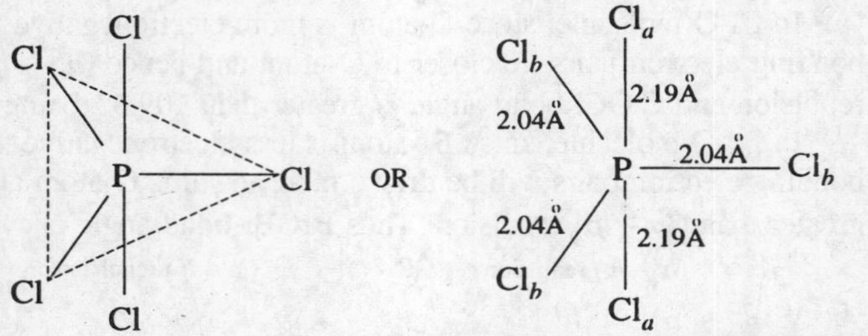

$Cl_a\ PCl_a$ bond angle = 180°, $Cl_a\ PCl_b$ bond angle = 90° and Cl_bPCl_b bond angle = 120°

Fig. 14.10 : Trigonal bipyramidal structure of PCl_5 molecule in the gaseous and liquid state.

10. Structure of PCl_5 and PBr_5 molecules in the solid state

X-ray studies of these molecules in the solid state have shown that these molecules have ionic structures which can be represented as $[PCl_4]^+\ [PCl_6]^-$ and $[PBr_4]^+\ Br^-$ respectively. The cations *viz.* $[PCl_4]^+$ and $[PBr_4]^+$ both have *tetrahedral geometry* which arises because of sp^3 hybridisation of the central P-atom. $[PCl_6]^-$ ion has *octahedral geometry* which results from sp^3d^2 hybridisation of the central P-atom. For the details of the structure of these ions see the subsequent pages.

11. Structure and geometry of the molecules of AX_4 type (A = S, Se, Te, Po; X = F, Cl, Br, I)

All these molecules have *distorted tetrahedral* or *see-saw geometry* which arises due to sp^3d hybridisation of the central atom in the molecule (S, Se, Te or Po atom as the case may be) and due to the presence of one lone pair of electrons in one of the equatorial hybrid orbitals.

As an example let us discuss the geometry of SF_4 molecule in which S-atom is the central atom. Since S-atom has six electrons in its valence-shell (S = $3s^2$, $3p_x^{\ 2}\ 3p_y^{\ 1}\ 3p_z^{\ 1}$), the electron-dot structure of SF_4 molecule can be written as given at (*a*) of Fig. 14.11.

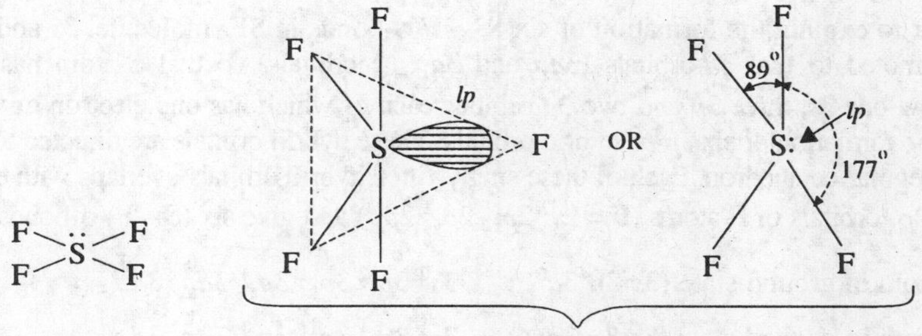

(*a*) (*b*)

Fig.14.11 : (*a*) Electron dot structure of SF_4 molecule (*b*) Distorted tetrahedral or see-saw geometry of SF_4 molecule.

The electron dot structure shows that the central S-atom in SF_4 molecule is surrounded by four σ-*bps* and one *lp*. Consequently S-atom in its excited state (S = $3s^2\ 3p_x^{\ 1}\ 3p_y^{\ 1}\ 3p_z^{\ 1}\ 3\overset{1}{d}_{z^2}$) undergoes sp^3d hybridisation so that four half-filled and one completely-filled hybrid orbital may be available.

S-atom in ground state ($3s^2\ 3p^4\ 3d^0$) : $3s^2\ 3p_x^{\ 2}\ 3p_y^{\ 1}\ 3p_z^{\ 1}\ 3\overset{0}{d}_{z^2}$

S-atom in excited state ($3s^2\ 3p^3\ 3d^1$) : $3s^2\ 3p_x^{\ 1}\ 3p_y^{\ 1}\ 3p_z^{\ 1}\ 3\overset{1}{d}_{z^2}$

S-atom in sp^3d hybridised state in SF_4 molecule. : $\boxed{(sp^3d)^2}$ $(sp^3d)^1\ (sp^3d)^1\ (sp^3d)^1\ (sp^3d)^1\ (sp^3d)^1$

lp

Five sp^3d hybrid orbitals

SF_4 molecule has four S–F σ-bonds. These bonds result by the overlap between the singly-filled four sp^3d hybrid orbitals and four singly-filled $2p_x$ orbitals on for F-atoms (F = $2s^2\ 2p_x^{\ 1}\ 2p_y^{\ 2}\ 2p_z^{\ 2}$). Thus each S–F σ-bond is a S (sp^3d) – F($2p_x$) σ-bond.

Position occupied by the lone pair and shape of SF_4 molecule. sp^3d hybridisation scheme given above indicates that one of the hybrid orbitals contains a lone pair of electrons. It should be noted that the hybrid orbital containing the lone pair of electrons is the equatorial hybrid orbital in the trigonal bipyramid and not the axial hybrid orbital, since the lone pair present in the equatorial hybrid orbital experiences comparatively *less repulsion* than when it is in the axial hybrid orbital and hence the molecule is more stable in this case. Thus we find that although the spatial arrangement of five hybrid orbitals round the central S-atom is trigonal bipyramidal, due to the presence of a lone pair of electrons in one of the three equatorial hybrid orbitals, the expected trigonal bipyramidal shape of SF_4 molecule gets distorted and becomes *distorted tetrahedral* or *see-saw* with the bond angles equal to 89° and 177° instead of the expected angles of 90° and 180° respectively.

12. Geometry of molecules of AF_6 type (A = S, Se and Te)

Molecules of AF_6 type have *regular octahedral geometry* which results from sp^3d^2 hybridisation of the central A-atom (which may be S, Se or Te) in doubly excited state. As an example let us consider the geometry of SF_6 molecule. Ground state electronic configuration of S-atom (central atom) is $3s^2\ 3p^2\ 3p^1\ 3p^1$. Electron dot structure of this molecule as given at (*a*) of Fig. 14.12 shows that S-atom is bonded to six F-atoms by six S—F σ-bonds and the central atom (*i.e.,* S-atom) has no lone pair of electrons on it.

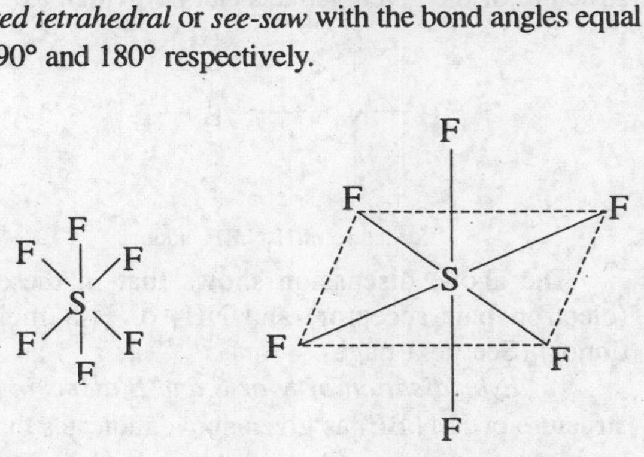

(*a*) (*b*)

Fig.14.12. (*a*) Electron dot structure of SF_6 molecule (*b*) Octahedral geometry of SF_6 molecule.

Thus in order to explain the formation of six S — F σ-bonds in SF_6 molecule, 3*s* and 3*p* electrons must be promoted to two 3*d*-orbitals ($3d_{z^2}$ and $3d_{x^2-y^2}$ orbitals) so that S-atom has six unpaired electrons. Now one 3*s*, three 3*p* and two 3*d* orbitals each of which has one electron mix together and give rise to the formation of six sp^3d^2 hybrid orbitals. These hybrid orbitals are directed towards the six corners of a regular octahedron. Each of these singly-filled hybrid orbitals overlaps with each of the six singly-filled $2p_x$ orbitals of F-atoms (F = $2s^2\ 2p_x^1\ 2p_y^2\ 2p_z^2$) and give six (S—F) σ-bonds

S-atom in ground state ($3s^23p^43d^0$) : $3s^2\ 3p_x^2\ 3p_y^1\ 3p_z^1\ 3d_{z^2}^0\ 3d_{x^2-y^2}^0$

S-atom in excited state ($3s^13p^33d^2$) : $3s^1\ 3p_x^1\ 3p_y^1\ 3p_z^1\ 3d_{z^2}^1\ 3d_{x^2-y^2}^1$

S-atom in sp^3d^2 hybridised state in SF_6 molecule : $\underbrace{(sp^3d^2)^1\ (sp^3d^2)^1\ (sp^3d^2)^1\ (sp^3d^2)^1\ (sp^3d^2)^1\ (sp^3d^2)^1}_{\text{Six } sp^3d^2 \text{ hybrid orbitals}}$

It is due to sp^3d^2 hybridisation and due to the absence of any lone pair of electrons on S-atom that SF_6 molecule has octahedral shape as shown at (*b*) of Fig. 14.12.

13. Formation of $H_3N \rightarrow BF_3$ and $H_2S \rightarrow BF_3$ adducts

$H_3N.BF_3$ consists of H_3N and BF_3 molecules. Similarly $H_2S.BF_3$ is composed of H_2S and BF_3 molecules. We know that in BF_3 moleucle (F—B(F)—F) B-atom (central atom) is sp^2 hybridised. In sp^2 hybridisation, $2p_z$ orbital of B-atom remains unhybridised and hence is vacant. Due to the presence of vacant $2p_z$ oribital on B-atom. BF_3 molecule is an *electron-deficient molecule*. In NH_3 molecule (H—$\ddot{N}$(H)—H) N-atom (central atom) is sp^3 hybridised. One of the four sp^3 hybrid orbitals on N-atom (central atom) contains a lone pair of electrons. Similarly in H_2S molecule (H, H > $\ddot{S}$:), S-atom (central atom) is also sp^3 hybridised. Two sp^3 hybrid orbitals contain a lone pair of electrons each. When BF_3 molecule combines with NH_3 or H_2S molecule, the vacant $2p_z$ orbital on B-atom of BF_3 accepts the lone pair of electrons donated by NH_3/H_2S molecule and forms N → B coordinate bond in $H_3N \rightarrow BF_3$ and S→ B coordinate bond in $H_2S \rightarrow BF_3$. Thus the structure of the given adducts can be written as :

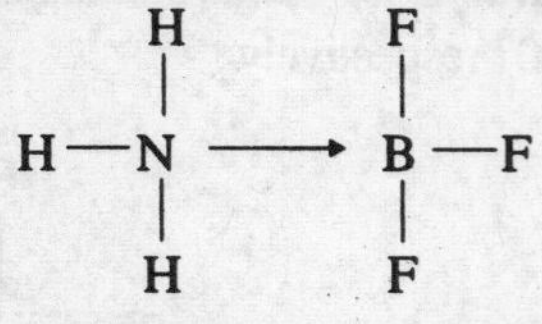

Structure of $H_3N.BF_3$ adduct

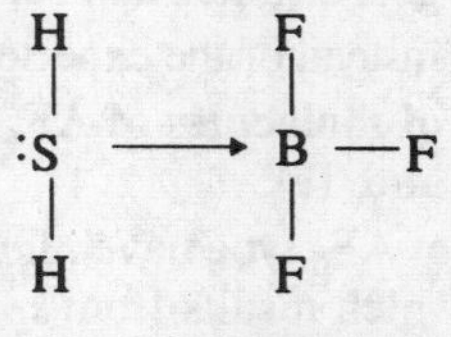

Structure of $H_2S.BF_3$ adduct

The above discussion shows that in these adducts, BF_3 molecule acts as a Lewis acid (electron pair acceptor) and NH_3 or H_2S molecule behaves as a Lewis base (electron pair donor) (See next page).

sp^3 hybridisation of N or S and B atoms in the adducts and formation of different bonds. The structure of $H_3N.BF_3$ as given above indicates that N and B both are surrounded by four σ-*bps* and hence each atom is sp^3 hybridised in this adduct. N → B coordinate bond is formed by the overlap of completely-filled sp^3 hybrid orbital on N-atom with the vacant sp^3 hybrid orbital on B-atom. Thus N→ B coordinate bond is N(sp^3) → B (sp^3) coordinate bond. Each of the three N—H

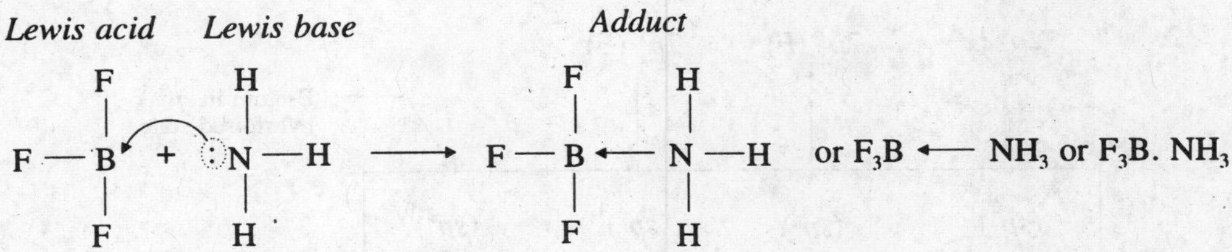

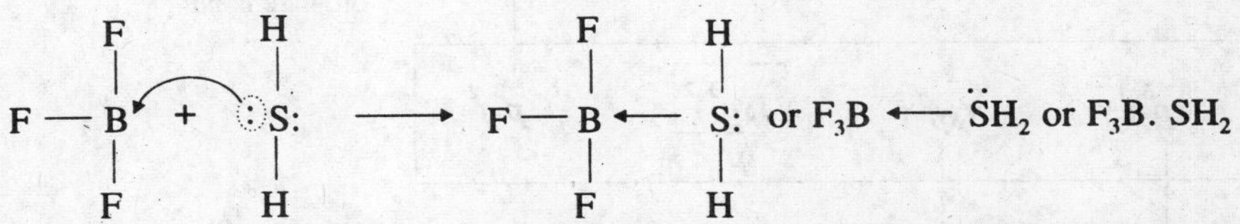

σ–bonds results by the overlap of singly-filled sp^3 hybrid orbital of N-atom with the singly-filled $1s$ orbital of H-atom. Each of the three B—F σ-bonds results when singly-filled sp^3 hybrid orbital of B-atom overlaps with the singly-filled $2p_x$ orbital on F-atom (see Fig. 14.13).

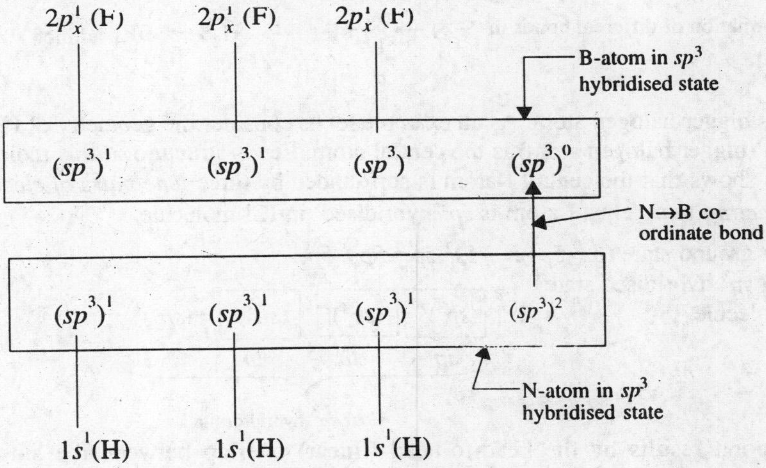

Fig. 14.13 : Formation of different bonds in 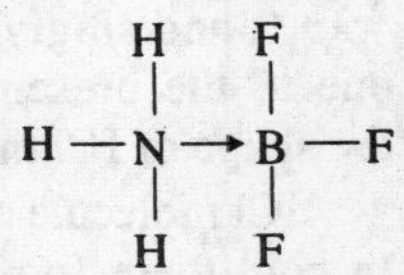 or $H_3N \rightarrow BF_3$ adduct

The above discussion shows that in $H_3N \rightarrow BF_3$ adduct, N-atom is tetrahedrally surrounded by four atoms (three H-atoms and one B-atom). Similarly B-atom is also tetrahedrally surrounded by four atoms (three F-atoms and one N-atom).

On similar lines, the formation of different bonds in $H_2S \rightarrow BF_3$ adduct can also be explained [See Fig. 14.14].

14. Structure and geometry of interhalogen compounds

While considering the geometry of interhalogen compounds the larger halogen atom is treated as the central atom and it is this central atom (bigger halogen atom) which is hybridised. The structure of various types of interhalogen compounds is discussed as follows:

(*i*) **Structure and geometry of interhalogen compounds of XY type.** Examples of such compounds are ClF, BrF, IF, BrCl, ICl and IBr. All these molecules have *linear geometry* which arises because of sp^3 hybridisation of the central halogen

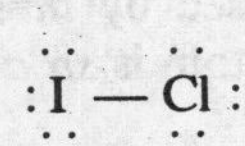

Lewis structure of ICl molecule

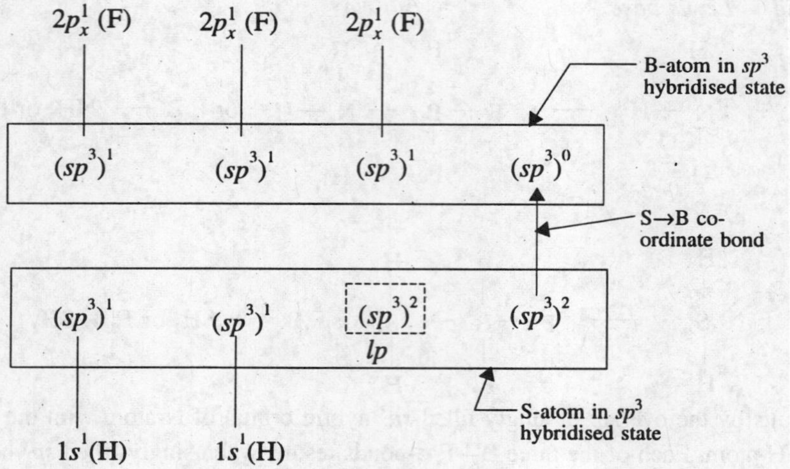

Fig. 14.14 : Formation of different bonds in 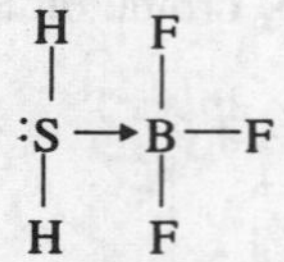 or $H_2\ddot{S} \rightarrow BF_3$ adduct

atom which is bigger halogen atom. As an example let us consider the geometry of ICl molecule in which I-atom (bigger halogen atom) is the central atom. Lewis structure of this molecule as given in the margin shows that the central I-atom is surrounded by three *lone pairs of electrons* and *one σ-bonding electron pair*. Thus I-atom is sp^3 hybridised in ICl molecule.

I-atom in ground state ($5s^2\ 5p^5$): $5s^2\ 5p_x^{\ 2}\ 5p_y^{\ 2}\ 5p_z^{\ 1}$

I-atom in sp^3 hybridised state in ICl molecule : $(sp^3)^2$ $(sp^3)^2$ $(sp^3)^2$ $(sp^3)^1$
lp lp lp
Four sp^3 hybrid orbitals

I–Cl σ-bond results by the head-to-head (linear) overlap between the singly-filled $3p_x$ orbital on Cl-atom (Cl = $3s^2\ 3p_x^{\ 1}\ 3p_y^{\ 2}\ 3p_z^{\ 2}$) and singly filled sp^3 hybrid orbital on I-atom. It is due to the presence of three lone pairs of electrons that the shape of ICl molecule is *linear.*

ICl molecule exists in two forms namely α and β forms. In both these forms, ICl molecule exists as zig-zag chains shown in the margin (Fig. 14.15).

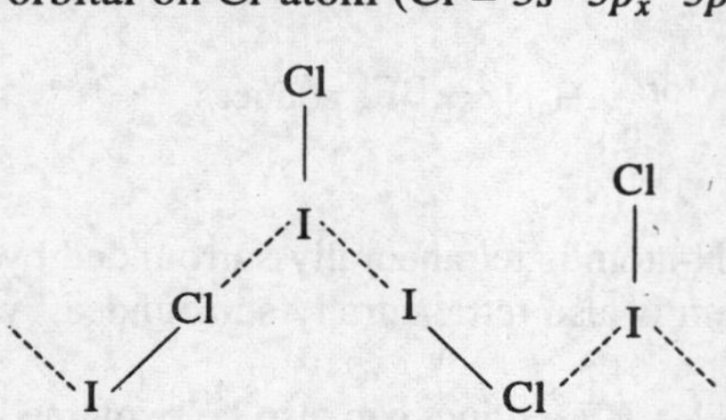

Fig. 14.15 : Zig-zag chains of ICl molecule in its α and β forms.

(*ii*) **Structure and geometry of interhalogen compounds of XY_3 type**. Examples of such compounds are ClF_3, BrF_3 and IF_3. All these molecules have bent *T-shaped structure* which is due to sp^3d hybridisation of the central bigger halogen atom. As an example let us discuss the geometry of ClF_3 molecule in which Cl-atom which is the bigger halogen atom is the central atom. Lewis structure of ClF_3 molecule which is given in the margin shows that the central Cl-atom is surrounded by *three σ-bps* and *two lone pairs of electrons* and hence Cl-atom is sp^3d hybridised in ClF_3 molecule.

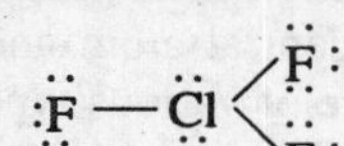

Lewis structure of ClF_3 molecule

Cl-atom in ground state ($3s^2\ 3p^5\ 3d^0$)	:	$3s^2\ 3p_x^{\ 2}\ 3p_y^{\ 2}\ 3p_z^{\ 1}\ 3d_{z^2}^{0}$
Cl-atom in excited state ($3s^2\ 3p^4\ 3d^1$)	:	$3s^2\ 3p_x^{\ 2}\ 3p_y^{\ 1}\ 3p_z^{\ 1}\ 3d_{z^2}^{1}$
Cl-atom in sp^3d hybridised state in ClF_3 molecule	:	$(sp^3d)^2$ $(sp^3d)^2$ $(sp^3d)^1$ $(sp^3d)^1$ $(sp^3d)^1$ *lp* *lp* Five sp^3d hybrid orbitals

Each of the three Cl—F σ-bonds results by the overlap of singly-filled sp^3d hybrid orbital on Cl-atom and singly-filled $2p_x$ orbital of F-atom ($F = 2s^2\ 2p_x^{\ 1}\ 2p_y^{\ 2}\ 2p_z^{\ 2}$).

Position of two lps and shape of ClF_3 molecule. It may be seen from the discussion of the structure of molecules/ions of $AB_3(lp)_2$ type that the two *lps* occupy the basal positions of the trigonal bipyramid. (See page 373) Due to the presence of *lps* in the basal position, the expected trigonal bipyramidal shape of ClF_3 molecule gets distorted and assumes *slightly bent T-shaped geometry*. Each of the two F Cl F bond angles is equal to 87.5°. The basal Cl—F (Cl—F_b) bond length is equal to 1.60Å while each of the two axial Cl—F (Cl—F_a) bond lengths is equal to 1.70 Å (See Fig. 14.16).

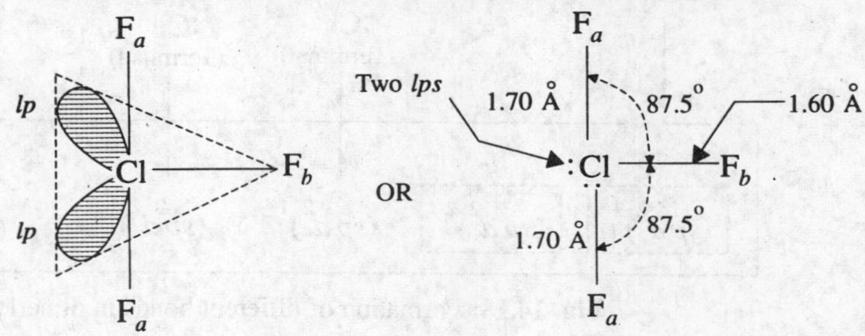

(*a*) (*b*)

Fig. 14.16 : Bent T-shaped structure of ClF_3 molecule.

Each of the two F Br F bond angles in BrF_3 molecules is equal to 86.12°. The basal Br—F bond length is 1.72 Å and each of the two axial Br—F bond lengths is equal to 1.81Å.

Structure of $(ICl_3)_2$ dimeric molecule. IF_3, ClF_3 and BrF_3 are monomeric T-shaped molecules but ICl_3 is a dimeric molecule, $(ICl_3)_2$ in the *solid state*. $(ICl_3)_2$ has a chlorine-bridged structure in which Cl-atom makes a bridge between two ICl_2 units (see Fig. 14.17). It may be seen from the structure that each I-atom is bonded to four Cl-atoms and hence is surrounded by four σ-*bps* and two *lps*. Since the sum of σ-*bps* and *lps* = 4 + 2 = 6, each I-atom in its excited state is sp^3d^2 hybridised.

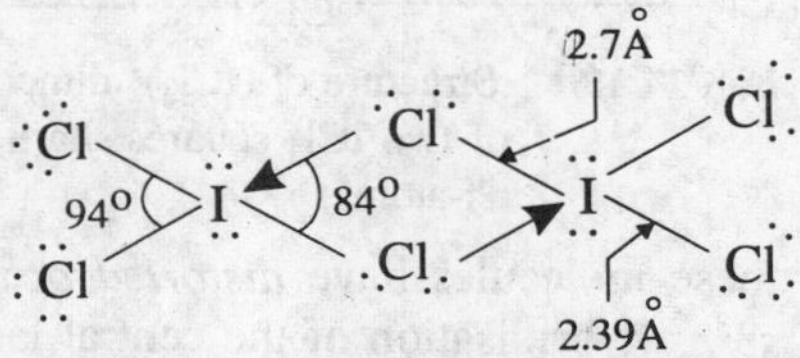

Fig. 14.17 : Chlorine-bridged structure of $(ICl_3)_2$ dimer.

I-atom in ground state ($5s^2\ 5p^5\ 5d^0$)	:	$5s^2\ 5p_x^{\ 2}\ 5p_y^{\ 2}\ 5p_z^{\ 1}\ 5d_{z^2}^{0}\ 5d_{x^2-y^2}^{0}$
I-atom in excited state ($5s^2\ 5p^4\ 5d_{z^2}^{1},\ 5d_{x^2-y^2}^{0}$)	:	$5s^2\ 5p_x^{\ 2}\ 5p_y^{\ 1}\ 5p_z^{\ 1}\ 5d_{z^2}^{1}\ 5d_{x^2-y^2}^{0}$
I-atom in sp^3d^2 hybridised state in ICl_4 unit :		$(sp^3d^2)^2$ $(sp^3d^2)^2$ $(sp^3d^2)^1$ $(sp^3d^2)^1$ $(sp^3d^2)^1$ $(sp^3d^2)^0$ *lp* *lp* Six sp^3d^2 hybrid orbitals

Two lone pairs (*lp*) of electrons are residing in the axial sp^3d^2 hybrid orbitals of the octahedron. Each of the three Cl—I σ-bonds results from the overlap between the singly-filled $3p_x$ orbital of Cl-atom and singly-filled sp^3d^2 hybrid orbital on I-atom while Cl → I coordinate bond arises by the donation of an electron pair on bridging Cl-atom to the vacant sp^3d^2 hybrid orbital on I-atom.

sp^3d^2 hybridisation of another I-atom and the formation of bonds between this atom and the four Cl-atoms in another ICl_4 unit can also be explained in a similar way. Formation of 8 bonds between two I-atoms and six Cl-atoms in dimeric molecule, $(ICl_3)_2$ has been shown in Fig. 14.18.

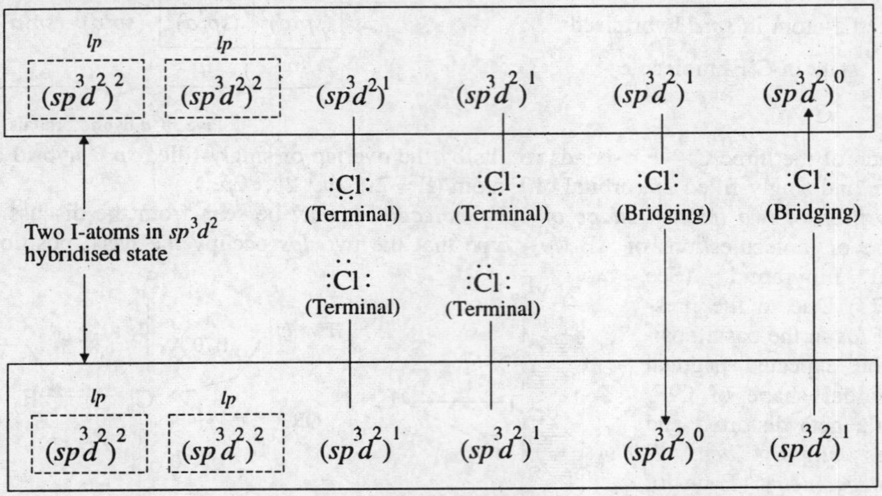

Fig. 14.18: Formation of different bonds in dimeric $(ICl_3)_2$ molecule.

Due to the presence of two *lps* in the axial positions of the octahedron, each ICl_4 unit has square planar structure. In the structure of $(ICl_3)_2$ molecule, two ICl_4 squares share two bridging Cl-atoms. The centre of each square is occupied by I-atom. Thus the structure of $(ICl_3)_2$ dimer can be shown as given in Fig. 14.19.

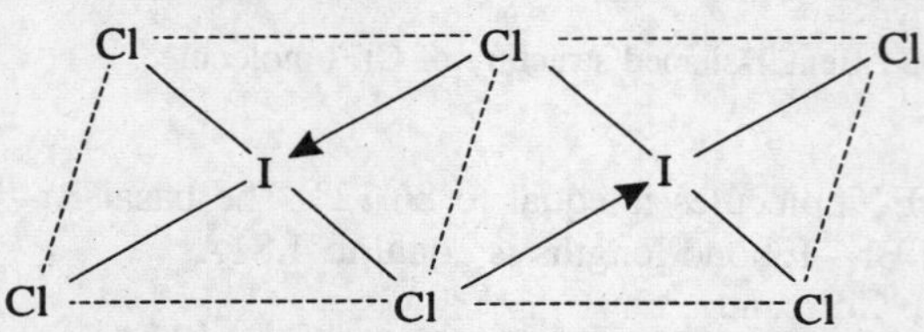

Fig. 14.19 : Structure of $(ICl_3)_2$ dimer which consists of two ICl_4 squares sharing two bridging Cl-atoms.

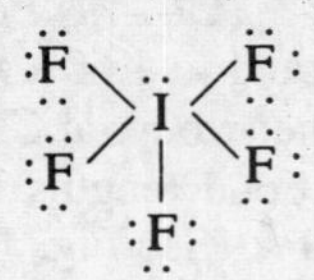

Lewis structure of IF_5 molecule

(*iii*) **Structure and geometry of interhalogen compounds of XY_5 type.** Examples of such compounds are ClF_5, BrF_5 and IF_5. All these molecules have *distorted octahedral (i.e., square pyramidal) structure* which arises from sp^3d^2 hybridisation of the central larger halogen atom. As an example let us discuss the shape of IF_5 molecule in which I-atom (bigger halogen atom) is the central atom. Lewis structure of this molecule is given in the margin. This structure clearly shows that the central I-atom is surrounded by five σ-*bps* and *one lone pair of electrons* and hence I-atom is sp^3d^2 hybridised in IF_5 molecule.

I-atom in ground state
$(5s^2\, 5p^3\, 5d^0)$: $5s^2\, 5p_x^{\,2}\, 5p_y^{\,2}\, 5p_z^{\,1}\, 5d_{z^2}^{0}\, 5d_{x^2-y^2}^{0}$

I-atom in the excited state
$(5s^2\, 5p^3\, 5d^1\, 5d^1)$: $5s^2\, 5p_x^{\,1}\, 5p_y^{\,1}\, 5p_z^{\,1}\, 5d_{z^2}^{1}\, 5d_{x^2-y^2}^{1}$

I-atom in sp^3d^2 hybridised state
in IF_5 molecule : $\boxed{(sp^3d^2)^2}\, (sp^3d^2)^1\, (sp^3d^2)^1\, (sp^3d^2)^1\, (sp^3d^2)^1\, (sp^3d^2)^1$

lp

Six sp^3d^2 hybrid orbitals

How are I–F σ-bonds formed ? Out of the six sp^3d^2 hybrid orbitals, one axial hybrid orbital contains a lone pair of electrons while the remaining five hybrid orbitals contain one electron each.

The lone pair of electrons occupies the axial hybrid orbital, since in this case ($lp - bp$) repulsion is minimum and hence ClF_5 molecule attains maximum stability. Each of the five F—I σ-bonds results by the overlap of singly-filled sp^3d^2 hybrid orbital on Cl-atom and singly-filled $2p_x$ orbital of F-atom ($F = 2s^2\ 2p_x^1\ 2p_y^2\ 2p_z^2$).

Shape of IF_5 molecule. Although the spatial arrangement of six electron pairs (*viz.*, *bps* = 5 and *lp* = 1) round the central I-atom in space is octahedral, due to the presence of one *lp* of electrons in the axial hybrid orbital the shape of IF_5 molecule is not octahedral, rather the shape gets distorted and hence IF_5 molecule assumes *square pyramidal shape.*

A recent crystal structure study of this molecule made by Burbank Jones shows that all the four basal F-atoms (F_b-atoms) are slightly displaced upwards from the base of the square pyramid and hence IF_5 molecule assumes the structure shown in Fig. 14.20. F_aIF_b bond angle has been found to be equal to 81.9°. I—F_a bond distance (= 1.73Å) is appreciably shorter than the I—F_b bond distance (= 1.87Å).

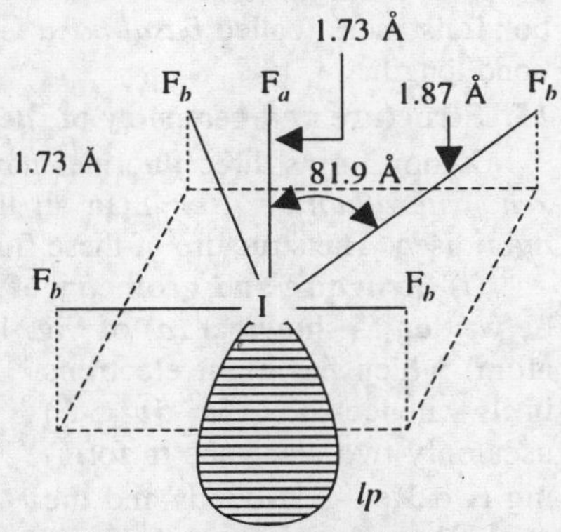

Fig. 14.20 : Square pyramidal structure of IF_5 molecule.

On the basis of similar arguments it can be shown that ClF_5 and BrF_5 molecules also have pyramidal shape.

(*iv*) **Structure and geometry of XY_7 type interhalogen compounds.** IF_7 is the only interhalogen compound of this type. The Lewis structure of this molecule is given at (*a*) of Fig. 14.21. This structure shows that the central I-atom (bigger halogen atom) is surrounded by seven σ-*bps* and hence I-atom in its excited state ($5s^1\ 5p^3\ 5d^3$) is sp^3d^3 hybridised in IF_7 molecule.

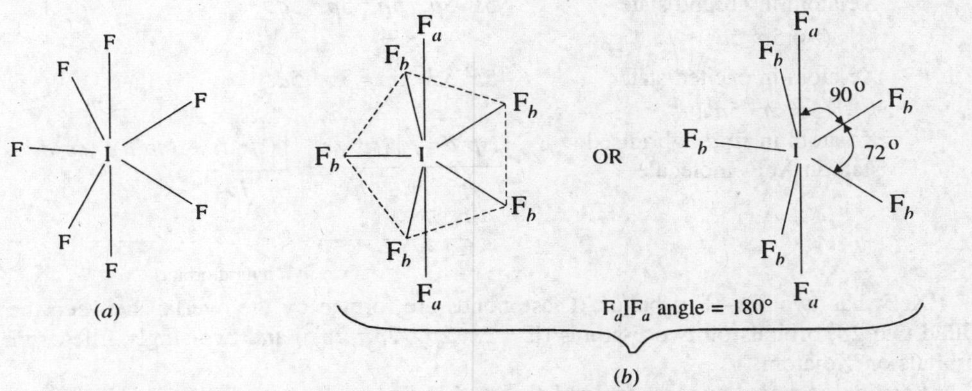

Fig. 14.21 : (*a*) Lewis structure of IF_7 molecule (*b*) Pentagonal bipyramidal geometry of IF_7 molecule.

I-atom in ground state ($5s^2\ 5p^5\ 5d^0$)	:	$5s^2\ 5p_x^2\ 5p_y^2\ 5p_z^1\ 5d^0\ 5d^0\ 5d^0$
I-atom in excited state ($5s^1\ 5p^3\ 5d^3$)	:	$5s^1\ 5p_x^1\ 5p_y^1\ 5p_z^1\ 5d^1\ 5d^1\ 5d^1$
I-atom in sp^3d^3 hybridised state in IF_7 molecule	:	$\underbrace{(sp^3d^3)^1\ (sp^3d^3)^1\ (sp^3d^3)^1\ (sp^3d^3)^1\ (sp^3d^3)^1\ (sp^3d^3)^1\ (sp^3d^3)^1}_{\text{Seven } sp^3d^3 \text{ hybrid orbitals}}$

Each of the seven I—F σ-bonds is formed by the overlap between the singly-filled sp^3d^3 hybrid orbital on I-atom and singly-filled $2p_x$ orbital on F-atom ($F = 2s^2\ 2p_x^1\ 2p_y^2\ 2p_z^2$). Since there is no lone pair of electrons in any of the hybrid orbitals, IF_7 molecule has the expected pentagonal bipyramidal shape which has been shown at (*b*) of Fig. 14.21. IF_7 molecule has two axial I—F bonds (represented as I — F_a bonds) and five equatorial (also called *basal*) I—F bonds (represented as I—F_b bonds). Angles between different types of bonds are also shown in the figure. Both the I—F_a bond distances (called *axial bond lengths*) are equal to each other. Similarly all the five I—F_b bond distances (called *basal bond lengths*) are also equal. I—F_a bond length is greater than I—F_b bond length.

15. Structure and geometry of the fluorides of xenon

Xenon forms three fluorides namely *xenon difluoride* (XeF_2), *xenon tetrafluoride* (XeF_4) and *xenon hexafluoride* (XeF_6). In all these fluorides Xe-atom (central atom) in its excited state is hybridised. The structure of these fluorides is discussed below on the basis of hybridisation.

(*i*) **Structure and geometry of XeF_2 molecule.** Electron dot structure of this molecule can be written as shown at (*a*) of Fig. 14.22. It is evident from this structure that Xe-atom (central atom) which has eight electrons in its valence-shell ($5s^2\ 5p^6\ 5d^0$) uses only two electrons in forming two Xe—F σ-bonds and the remaining six electrons remain as three lone pairs of electrons on Xe-atom. It is also evident from the electron dot structure that the central Xe-atom is surrounded by two *σ-bps* and three *lps* and hence Xe-atom is sp^3d hybridised in XeF_2 molecule. Three hybrid orbitals have lone pairs of electrons each while the remaining two have one electron each.

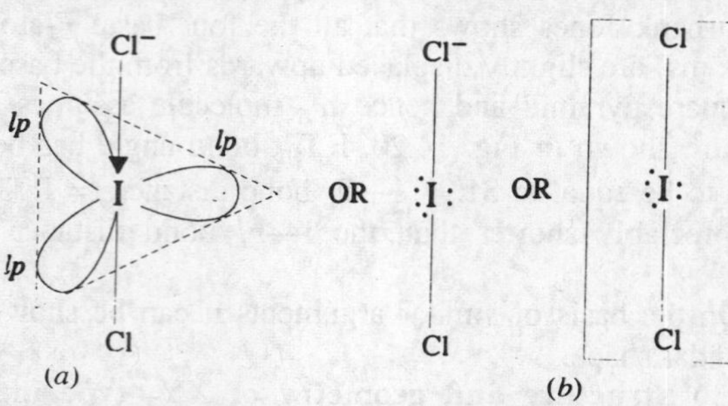

Fig.14.22 : (*a*) Electron dot structure of XeF_2 molecule (*b*) Linear shape of XeF_2 molecule.

Xe-atom in ground state ($5s^2\ 5p^6\ 5d^0$)	:	$5s^2\ 5p_x^2\ 5p_y^2\ 5p_z^2\ 5d_{z^2}^0$
Xe-atom in excited state ($5s^2\ 5p^5\ 5d^1$)	:	$5s^2\ 5p_x^2\ 5p_y^2\ 5p_z^1\ 5d_{z^2}^1$
Xe-atom in sp^3d hybridised state in XeF_2 molecule	:	$(sp^3d)^2$ $(sp^3d)^2$ $(sp^3d)^2$ $(sp^3d)^1$ $(sp^3d)^1$ lp lp lp Five sp^3d hybrid orbitas

XeF_2 has two Xe—F σ-bonds. These bonds are formed by the overlap between the singly-filled two $2p_x$ orbitals on two F-atoms (F = $2s^2\ 2p_x^1\ 2p_y^2\ 2p_z^2$) and two singly-filled sp^3d hybrid orbitals on Xe-atom.

Position occupied by three lps and shape of XeF_2 molecule. According to VSEPR theory, the three *lps* occupy the equatorial hybrid orbitals because in this case the electron pairs repulsion is minimum and XeF_2 molecule attains the maximum stability. Thus we see that although the spatial arrangement of five sp^3d hybrid orbitals round the central Xe-atom is trigonal bipyramidal, the shape of XeF_2 molecule, due to the presence of three *lps* in the equatorial positions, gets distorted and becomes linear (See Fig. 14.22).

(*ii*) **Structure and geometry of XeF_4 molecule.** Electron dot structure of this molecule as given at (*a*) of Fig. 14.23 discloses the fact that Xe-atom (central atom) which has 8 electrons in its valence shell ($5s^2\ 5p^6\ 5d^0$) uses only 4 electrons in forming four Xe — F σ-bonds and the remaining two electron pairs (*i.e.*, 4 electrons) remain lone pairs (*lps* = 2). It is also obvious from the dot structure that the central Xe-atom is surrounded by 4 *σ-bps* and 2 *lps* and hence Xe-atom undergoes sp^3d^2 hybridisation in XeF_4 molecule. Two hybrid orbitals have a lone pair of electrons each while the remaining four hybrid orbitals contains one electron each.

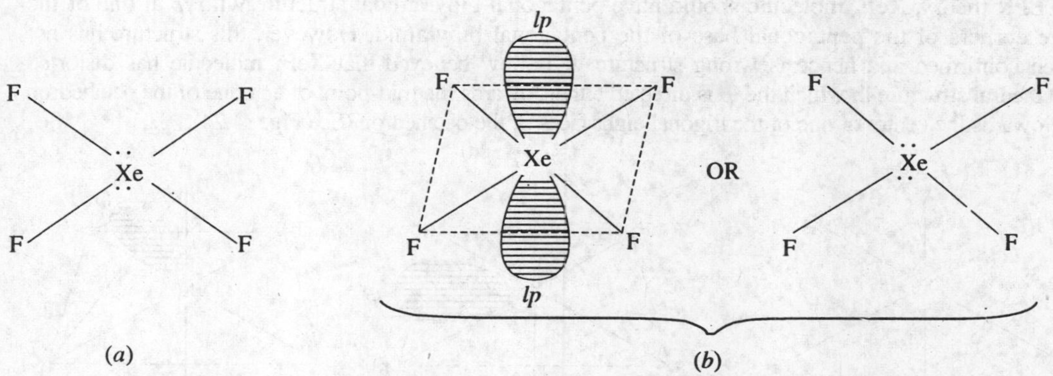

Fig. 14.23. (*a*) Electron dot structure of XeF_4 molecule (*b*) Square planar geometry of XeF_4 molecule.

Xe-atom in ground state ($5s^2\ 5p^6\ 5d^0$) : $5s^2\ 5p_x^2\ 5p_y^2\ 5p_z^2\ 5d_{z^2}^0\ 5d_{x^2-y^2}^0$

Xe-atom in excited state ($5s^2\ 5p^4\ 5d^2$) : $5s^2\ 5p_x^2\ 5p_y^1\ 5p_z^1\ 5d_{z^2}^1\ 5d_{x^2-y^2}^1$

Xe-atom in sp^3d^2 hybridised state in XeF_4 molecule : $\underbrace{\boxed{(sp^3d^2)^2}_{lp}\ \boxed{(sp^3d^2)^2}_{lp}\ (sp^3d^2)^1\ (sp^3d^2)^1\ (sp^3d^2)^1\ (sp^3d^2)^1}_{\text{Six } sp^3d^2 \text{ hybrid orbitals}}$

XeF_4 has four Xe—F σ-bonds. These bonds are obtained when the singly-filled four sp^3d^2 hybrid orbitals overlap with four singly-filled $2p_x$ orbitals of four F-atoms (F = $2s^2\ 2p_x^1\ 2p_y^2\ 2p_z^2$).

Position occupied by two lps and shpae of XeF_4 molecule. According to VSEPR theory, the two lps **occupy the axial hybrid orbitals because in this case the electron pairs repulsion is** minimum and XeF_4 molecule gains maximum stability. Thus, according to VSEPR theory, although the spatial orientation of six electron pairs (4σ-*bps* and 2*lps*) round the central Xe-atom is octahedral, the shape of XeF_4 molecule, due to the presence of two *lps* in the axial positions, gets distorted and becomes *square planar* as shown in Fig. 14.23 (*b*).

iii) **Structure and geometry of XeF_6 molecule.** In this molecule Xe is the central atom Electron dot structure of this molecule is given at (*a*) of Fig. 14.24. This structure shows that Xe-atom (central atom) uses only six electrons in forming six Xe—F σ- bonds and the remaining two electrons (*i.e.,* one pair of electrons) remain as a non bonding electron pair on Xe-atom. (lp = 1). Thus Xe-atom is surrounded by *six bps* and *one lp* and hence Xe-atom in its excited state (Xe = $5s^2\ 5p^3\ 5d_{xy}^1\ 5d_{yz}^1\ 5d_{zx}^1$) undergoes sp^3d^3 hybridisation in XeF_6 molecule. One hybrid orbital contains the lone pair of electrons and the remaining six hybrid orbitals contain one unpaired electron each.

Xe-atom in ground state ($5s^2\ 5p^6\ 5d^0$) : $5s^2\ 5p^6\ 5d_{xy}^0\ 5d_{yz}^0\ 5d_{zx}^0$

Xe-atom in excited state ($5s^2\ 5p^3\ 5d^3$) : $5s^2\ 5p_x^1\ 5p_y^1\ 5p_z^1\ 5d_{xy}^1\ 5d_{yz}^1\ 5d_{zx}^1$

Xe-atom in sp^3d^3 hybridised state in XeF_6 molecule : $\underbrace{\boxed{(sp^3d^3)^2}_{lp}\ (sp^3d^3)^1\ (sp^3d^3)^1\ (sp^3d^3)^1\ (sp^3d^3)^1\ (sp^3d^3)^1\ (sp^3d^3)^1}_{\text{Seven } sp^3d^3 \text{ hybrid orbitals}}$

XeF_6 molecule has six Xe—F σ-bonds. These bonds are obtained when the singly-filled six sp^3d^3 hybrid orbitals overlap with six singly-filled $2p_x$ orbitals of six F-atoms (F = $2s^2\ 2p_x^1\ 2p_y^2\ 2p_z^2$).

Position occupied by the lone pair of electrons (lp) and shape of XeF_6 molecule. According to

VSEPR theory, XeF_6 molecule should have pentagonal bipyramidal structure with *lp* at one of the five corners of the pentagonal base of the pentagonal bipyramid. However, this structure has not been confirmed and hence is wrong structure. It is now believed that XeF_6 molecule has distorted octahedral structure in which the *lp* is directed either towards the mid-point of an edge of the octahedron or towards the centre of one of the trigonal eight faces of the octahedron (See Fig. 14.24).

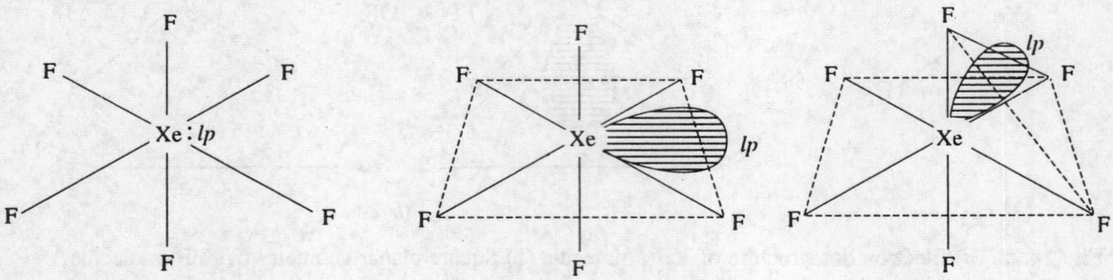

(*a*) (*b*) (*c*)

Fig. 14.24 : (*a*) Electron dot structure of XeF_6 molecule (*lp* indicates the lone pair of electrons on Xe-atom) (*b*) Distorted octahedral structure of XeF_6 molecule having a *lp* directed towards the mid point of an edge of the octahedron (*c*) Distorted octahedral structure of XeF_6 molecule having a *lp* pointing towards the centre of one of the eight trigonal faces of the octahedron.

Structure and Shape of Ions

1. Geometry of ortho borate ion, BO_3^{3-}

Three negative charges on BO_3^{3-} ion are distributed on three O-atoms, *i.e.*, each O-atom in BO_3^{3-} ion is negatively-charged. Thus the electron dot structure of this ion can be written as shown in the margin. This structure shows that the central B-atom is surrounded by three σ-*bps* and no lone pair of electrons and hence B-atom is sp^2 hybridised. Consequently BO_3^{3-} ion has trigonal planar structure with each OBO bond angle equal to 120°.

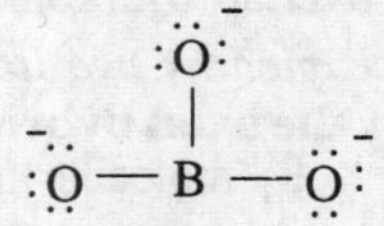

Electron dot structure of BO_3^{3-} ion

B-atom in ground state ($2s^2 2p^1$)	:	$2s^2\ 2p^1$
B-atom in excited state ($2s^1\ 2p^2$)	:	$2s^1\ 2p_x^{\ 1}\ 2p_y^{\ 1}\ 2p_z^{\ 0}$
B-atom in sp^2 hybridised state BO_3^{3-} ion	:	$(sp^2)^1\ (sp^2)^1\ (sp^2)^1$

Three singly-filled sp^2 hybrid orbitals

Each of the three ($B—O^-$) σ-bonds is formed by the overlap between the singly-filled sp^2 hybrid orbital of B-atom and singly-filled $2p$ orbital of O^- ($O^- = 2s^2\ 2p^2\ 2p^2\ 2p^1$).

2. Geometry of beryllium trifluoride ion, BeF_3^-

This ion has *trigonal planar geometry* corresponding to sp^2 hybridisation of Be-atom (central atom). This ion can be regarded as having been formed by the combination of one BeF_2 molecule with one F^- ion.

$$BeF_2 \quad + \quad F^- \longrightarrow \quad [BeF_3]^-$$

In BeF_2 molecule (F—Be—F), Be-atom (central atom) is *sp* hybridised. In *sp* hybridisation two *p*-orbitals of Be-atom are excluded from hybridisation and hence remain vacant. Thus BeF_2 molecule can accept an electron pair donated by F^- ion to form $F_2Be \leftarrow \ddot{\underset{..}{F}}:^-$ or BeF_3^- ion. Thus in the formation of BeF_3^- ion, BeF_2 molecule behaves as a Lewis acid and F^- ion acts as a Lewis base.

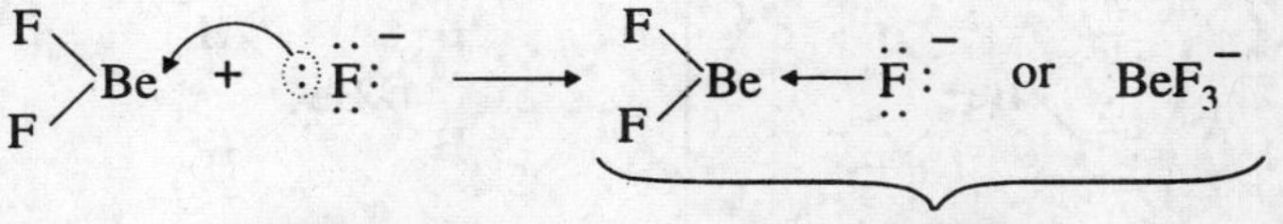

Lewis acid Lewis base Adduct

The structure of BeF_3^- ion as given above shows that the central Be-atom is surrounded by three σ-*bps* and hence is sp^2 hybridised as shown below :

Be-atom in ground state ($2s^2\ 2p^0$)	:	$2s^2 2p^0$
Be-atom in excited state ($2s^1\ 2p^1$)	:	$2s^1\ 2p_x^1\ 2p_y^0\ 2p_z^0$
Be-atom in sp^2 hybridised state in BeF_3^- ion	:	$(sp^2)^1\ (sp^2)^1\ (sp^2)^0$ Three sp^2 hybrid orbitals

Each of the two (Be—F) σ-bonds is obtained by the overlap between singly-filled sp^2 hybrid orbital of Be-atom and singly-filled 2*p* orbital of F-atom, and $F^- \rightarrow Be$ coordinate bond results by the donation of an electron pair from F^- ion to the vacant sp^2 hybrid orbital. Due to sp^2 hybridisation, BeF_3^- ion has *trigonal planar geometry.*

3. Geometry of ions of BeX_4^{2-} type (X = H, F)

Examples of the ions of this type are BeH_4^{2-} and BeF_4^{2-} ions. Both these ions have tetrahedral geometry due to sp^3 hybridisation of Be-atom (central atom). These ions can be regarded as having been formed by the combination of BeF_2 molecule with $2F^-$ ions and BeH_2 molecule with $2H^-$ ions respectively. Thus.

$$BeF_2 + 2F^- \rightarrow BeF_4^{2-}\ ; \qquad BeH_2 + 2H^- \rightarrow BeH_4^{2-}$$

In BeF_2 (F — Be — F) and BeH_2 (H — Be — H) molecules, Be-atom is *sp* hybridised. In *sp* hybridisation two *p*-orbitals of Be-atom remain unhybridised and hence vacant. Thus BeF_2 molecule can accept two electron pairs donated by two $2F^-$ ions to form 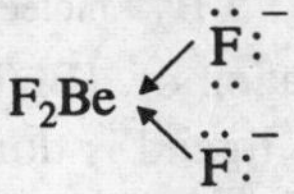 or BeF_4^{2-} ion.

Similarly BeH_2 molecule can accept two electron pairs donated by two H^- ions to form 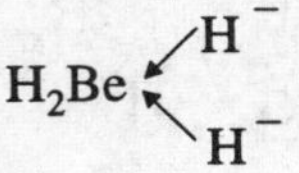 or BeH_4^{-2} ion.

Lewis acid Lewis base Adduct

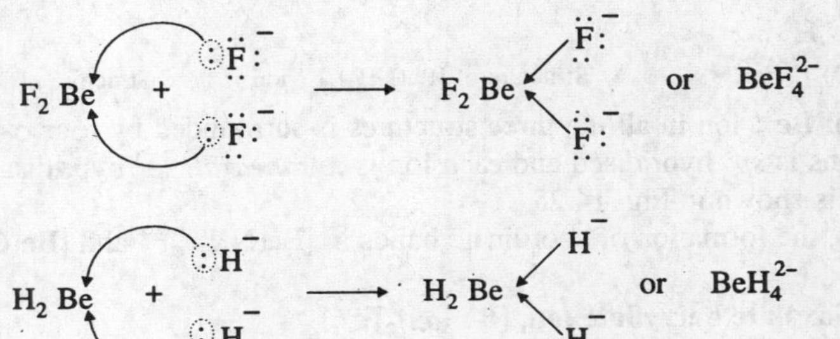

The structures of BeF_4^{2-} and BeH_4^{2-} as given below show that Be-atom is sp^3 hybridised.

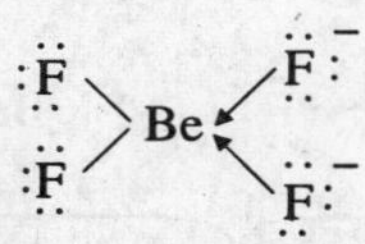 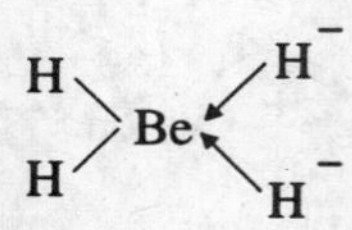

sp^3 hybridisation of Be-atom in these two ions can be shown as :

Be-atom in ground state ($2s^2\ 2p^0$) : $2s^2\ 2p^0$

Be-atom in excited state ($2s^1\ 2p^1$) : $2s^1\ 2p_x^1\ 2p_y^0\ 2p_z^0$

Be-atom in sp^3 hybridised state in Be X_4^{2-} ion (X = H, F) : $\underbrace{(sp^3)^1\ (sp^3)^1\ (sp^3)^0\ (sp^3)^0}_{\text{Four } sp^3 \text{ hybrid orbitals}}$

Each of the (Be—X) σ-bonds is formed in its usual way while each of the $X^- \rightarrow$ Be coordinate bonds is established by the donation of an electron pair from X^- ($X^- = H^-$ or F^-) to the vacant sp^3 hybrid orbital of Be-atom.

4. Geometry of $[Be(H_2O)_4]^{2+}$, $[Be(NH_3)_4]^{2+}$ and $[Be(OH)_4]^{2-}$ ions

In these ions Be atom is present as Be^{2+} ion whose valence-shell electronic configuration is $2s^0\ 2p_x^0\ 2p_y^0\ 2p_z^0$. All these ions have *tetrahedral shape* which arises due to sp^3 hybridisation of Be^{2+} ion. Each of these ions can be regarded as having been formed by the combination of Be^{2+} ion and four ligands.

$$Be^{2+} + 4H_2O \rightarrow [Be(H_2O)_4]^{2+} ; \quad Be^{2+} + 4NH_3 \rightarrow [Be(NH_3)_4]^{2+}$$
$$Be^{2+} + 4OH^- \rightarrow [Be(OH)_4]^{2-}$$

The structures of H_2O, NH_3 and OH^- (ligands) as given below show that O-atom in H_2O molecule N-atom in NH_3 molecule and O-atom in OH^- ion has two, one and three lone pairs of electrons respectively.

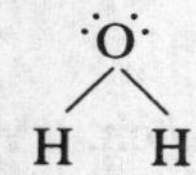

H_2O molecule

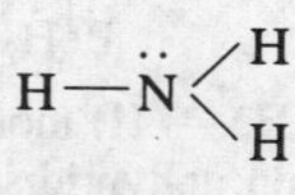

NH_3 molecule

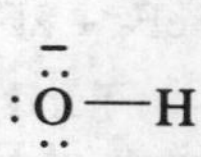

OH^- ion

In the formation of $[Be(H_2O)_4]^{2+}$, $[Be(NH_3)_4]^{2+}$ and $[Be(OH)_4]^{2+}$ ions, one lone pair of electrons on the ligand is donated to Be^{2+} ion to form four L $\rightarrow Be^{2+}$ co-ordinate bonds (L = H_2O, NH_3, OH^-). Thus the structure of these ions can be written as :

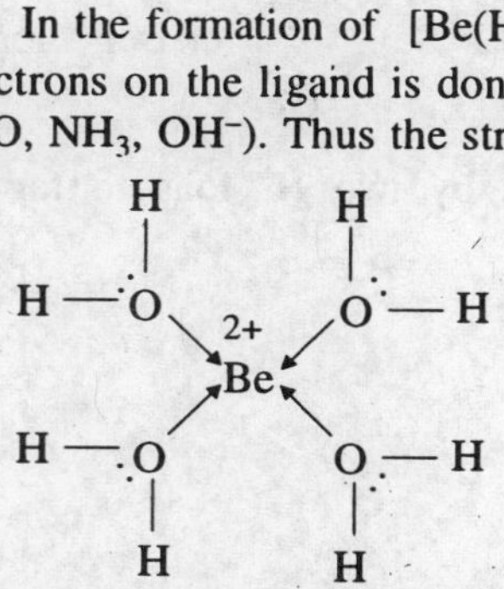

Structure of $[Be(H_2O)_4]^{2+}$ ion

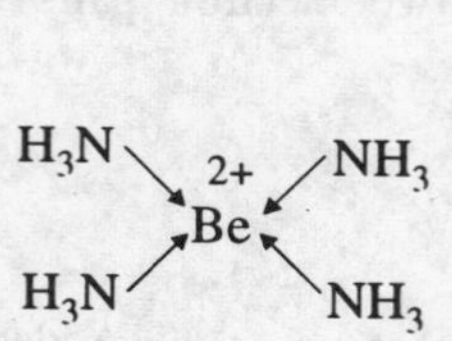

Structure of $[Be(NH_3)_4]^{2+}$ ion

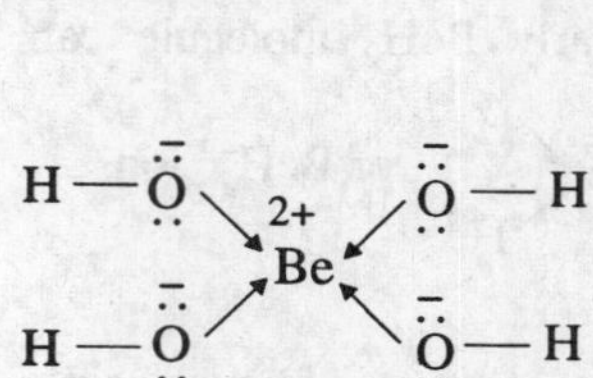

Structure of $[Be(OH)_4]^{2+}$ ion

Since the central Be^{2+} ion in all the three structures is surrounded by four σ-*bps*, Be^{2+} ion in each of the given ions is sp^3 hybridised and each ion is *tetrahedral*. sp^3 hybridisation of Be^{2+} ion in $[Be(H_2O)_4]^{2+}$ ion is shown in Fig. 14.25.

On similar lines, the formation of coordinate bonds in $[Be(NH_3)_4]^{2+}$ and $[Be(OH)_4]^{2-}$ ions can be shown.

5. Geometry of dioxalate berryllate ion, [Be $(ox)_2]^{2-}$

In this ion, ox^{2-} represents oxalate ion whose structure is given below.

Be^{2+} ion in its ground state ($2s^0\ 2p^0$)

Be^{2+} ion in sp^3 hybridised state in $[Be(H_2O)_4]^{2+}$ ion

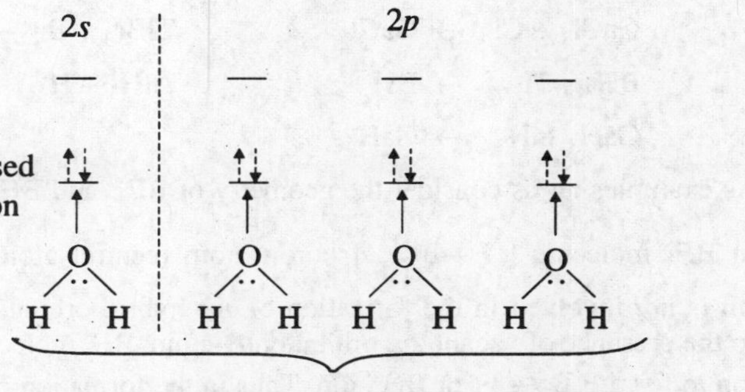

sp^3 hybridisation : Tetrahedral shape of $[Be(H_2O)_4]^{2+}$ ion

Fig. 14.25 : sp^3 **hybridisation of** Be^{2+} **ion in** $[Be(H_2O)_4]^{2+}$ **ion.**

$[Be\,(ox)_2]^{2-}$ ion has *tetrahedral shape* which is due to sp^3 hybridisation of Be^{2+} ion. In the given ion, Be-atom is present as Be^{2+} ion, since each of the two ligand ions attached with Be-atom has two negative charges on it. The given ion can be regarded as being formed by the combination of one Be^{2+} ion and two oxalate ions.

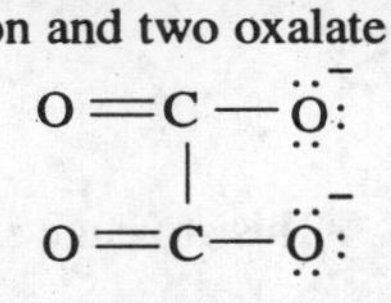

Structure of oxalate ion, $(ox)^{2-}$ (Ligand ion)

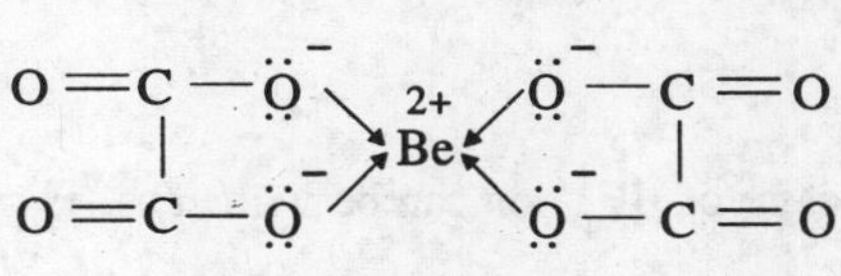

Structure of $[Be\,(ox)_2]^{2-}$ ion

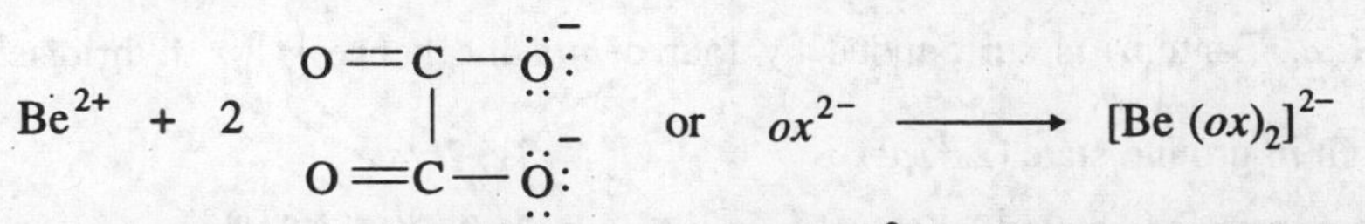

Each of the two negatively-charged O-atoms of each ox^{2-} ion has three lone pairs of electrons on it. One of these lone pairs of each O-atom is donated to Be^{2+} ion to form four $Be \leftarrow \ddot{O}:$ coordinate bonds in $[Be(ox)_2]^{2-}$ ion. Thus the structure of $[Be\,(ox)_2]^{2-}$ ion can be written as shown above. This structure shows that Be^{2+} ion is surrounded by four σ-*bps* and hence Be^{2+} ion is sp^3 hybridised as shown below:

Be-atom in ground state ($2s^2\ 2p^0$)	:	$2s^2\ 2p^0$
Be^{2+} ion in ground state ($2s^0\ 2p^0$)	:	$2s^0\ 2p_x^{\ 0}\ 2p_y^{\ 0}\ 2p_z^{\ 0}$
Be^{2+} ion in sp^3 hybridised state in $[Be\,(ox)_2]^{2-}$ ion	:	$(sp^3)^0\ (sp^3)^0\ (sp^3)^0\ (sp^3)^0$ (Four vacant sp^3 hybrid orbitals)

Each of the $ox \rightarrow Be^{2+}$ coordinate bond is formed by donation of an electron pair of negatively-charged O-atom of $C_2O_4^{2-}$ ion (ox^{2-}) to the vacant sp^3 hybrid orbitals of Be^{2+} ion.

6. Geometry of the ions of AX_4^- type (A = B, Al, Ga, Tl; X = H, F, Cl, Br)

Examples of such ions are BH_4^-, BF_4^-, AlH_4^-, AlF_4^-, GaH_4^-, $GaCl_4^-$, $TlBr_4^-$ etc. All these ions have *tetrahedral geometry* which is due to sp^3 hybridisation of the central atom. All these ions can be regarded as having been formed by the combination of trihalide and one halide ion or by the combination of trihydride with one hydride ion. Thus:

$$BF_3 + F^- \rightarrow BF_4^- \ ; \qquad AlF_3 + F^- \rightarrow AlF_4^-$$

$GaCl_3 + Cl^- \rightarrow GaCl_4^-$; $TlBr_3 + Br^- \rightarrow TlBr_4^-$

$BH_3 + H^- \rightarrow BH_4^-$; $AlH_3 + H^- \rightarrow AlH_4^-$

$GaH_3 + H^- \rightarrow GaH_4^-$

As examples let us consider the geometry of BF_4^- and BH_4^- ions.

In BF_3 molecule [F—B<^F_F] boron atom (central atom) is sp^2 hybridised. $2p_z$ orbital of B-atom is *not* involved in the formation of sp^2 hybrid orbitals (three) and hence remains *vacant.* Due to the presence of vacant $2p_z$ orbital on B-atom, BF_3 molecule can accept an electron pair from a F^- ion to form $F_3B \leftarrow F^-$ or BF_4^- ion. Thus in the formation of BF_4^- ion, BF_3 molecule acts as an electron pair acceptor (Lewis acid) and F^- ion behaves as an electron pair donor (Lewis base).

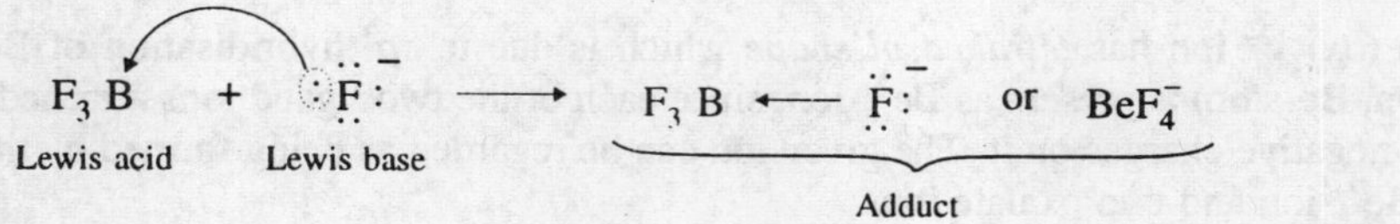

Thus the structure of BF_4^- ion can be written as 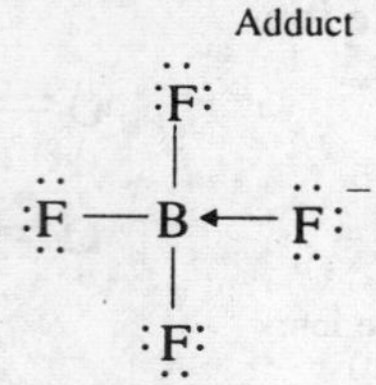 . This structure shows that the central atom (*i.e.*, B-atom) is surrounded by four σ-*bps* and hence is sp^3 hybridised as shown below:

B-atom in ground state ($2s^22p^1$)	:	$2s^2\ 2p^1$
B-atom in excited state ($2s^12p^2$)	:	$2s^12p_x^{\ 1}\ 2p_y^{\ 1}\ 2p_z^{\ 0}$
B-atom in sp^3 hybridised state in BF_4^- ion	:	$(sp^3)^1\ (sp^3)^1\ (sp^3)^1\ (sp^3)^0$ — Four sp^3 hybrid orbitals

Due to sp^3 hybridisation, BF_4^- ion has tetrahedral geometry.

The tetrahedral geometry of AlF_4^-, $GaCl_4^-$ and $TlBr_4^-$ ions can also be explained on the same lines.

The formation of BH_4^- 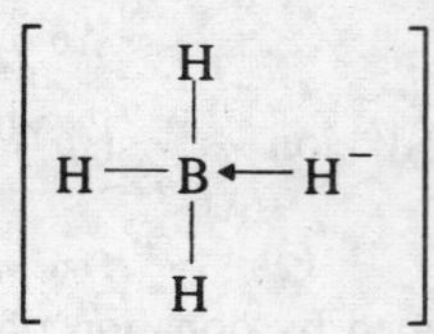 ion can be shown as:

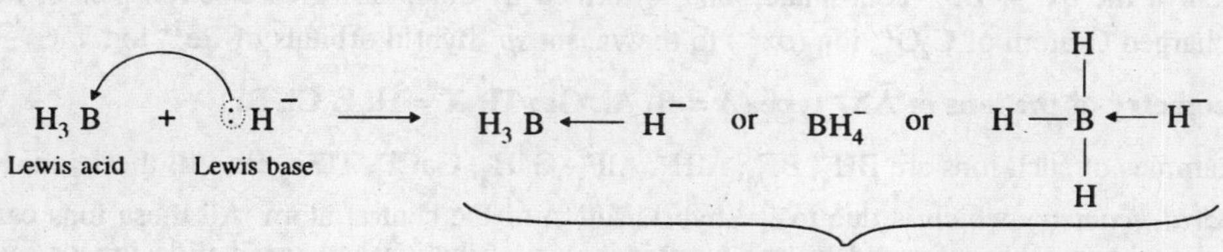

sp^3 hybridisation scheme shown above for BF_4^- ion can also be shown in a similar way for BH_4^- ion.

7. Geometry of AX_4^+ ions (A = N, P, As; X = H, Cl, Br, I, Ph)

NH_4^+ (ammonium ion), PH_4^+ (phosphonium), PCl_4^+, PBr_4^+ and PI_4^+ ions are the examples of this type of ions. All these ions can be regarded as having been formed by the donation of an electron pair residing on the central atom of AX_3 molecules to X^+ ions. Thus in the formation of AX_4^+ ion AX_3 molecule acts as a Lewis base and X^+ ion behaves as a Lewis acid.

Lewis base Lewis acid Electron dot structure of the ion formed

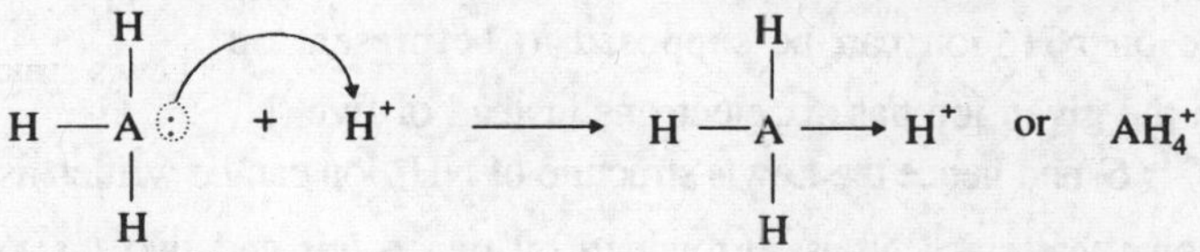

(A = N, P)

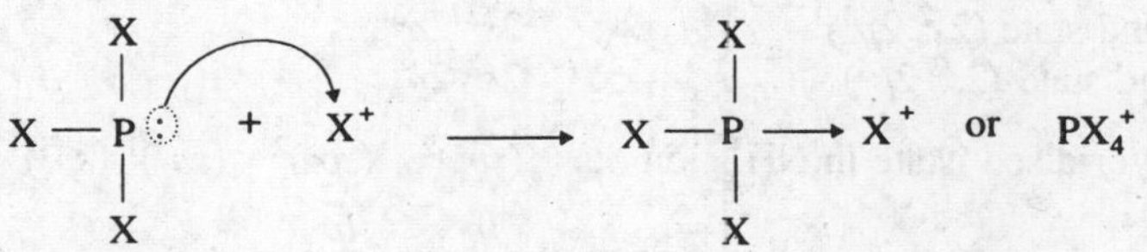

(X = Cl, Br, I)

The electron dot structures of AX_4^+ ions as given above show that the central atom (A-atom) in each ion is surrounded by four σ-*bps* and hence is sp^3 hybridised. sp^3 hybridisation of A-atom is shown below :

A-atom (A = N, P) in its ground state (ns^2np^3) : $ns^2\ np_x^1\ np_y^1\ np_z^1$

A-atom in sp^3 hybridised state : $\underbrace{(sp^3)^2 (sp^3)^1 (sp^3)^1 (sp^3)^1}_{\text{Four } sp^3 \text{ hybrid orbitals}}$

(A → X^+) coordinate bond is formed when the electron pair residing in one of the four sp^3 hybrid orbitals of A atom is denoted to the vacant 1*s* orbital of H^+ ion or vacant *np* orbital of Cl^+, Br^+ or I^+ ion. After the formation of (A → X^+) coordinate bond, this bond becomes identical to (A–X) normal covalent bond. Thus all the four bonds in AX_4^+ ion are identical. PCl_4^+ ion is found in PCl_5 molecule in the solid state, since in this state PCl_5 molecule is represented as $[PCl_4]^+$ $[PCl_6]^-$. It is due to sp^3 hybridisation of the central A-atom and no lone pair of electrons that all these ions have the expected tetrahedral shape.

8. Geometry of H_3O^+ (hydronium) ion

This ion can be regarded as having been formed by the donation of an electron pair residing on the central O-atom of H_2O molecule to the vacant 1*s* orbital of H^+ ion. Thus in the formation of H_3O^+ ion, H_2O acts as a Lewis base and H^+ ion acts as a Lewis acid.

Lewis base Lewis acid

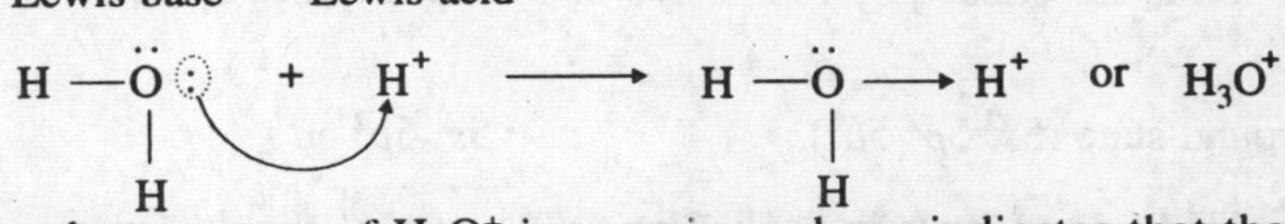

The electron dot structure of H_3O^+ ion as given above indicates that the central O-atom is surrounded by three σ-*bps* and one *lp*. Consequently O-atom in H_3O^+ ion is sp^3 hybridised as shown below:

O-atom in ground state ($2s^22p^4$) : $2s^2\ 2p_x^2\ 2p_y^1\ 2p_z^1$

O-atom in sp^3 hybridised state in H_3O^+ ion : $\underbrace{\underset{lp}{(sp^3)^2}\ \underset{lp}{(sp^3)^2}\ (sp^3)^1\ (sp^3)^1}_{\text{Four } sp^3 \text{ hybrid orbitals}}$

Each (O—H) σ-bond is formed by the overlap between singly-filled sp^3 hybrid orbital on O-atom and singly-filled $1s$ orbital of H-atom ($H = 1s^1$) while O → H^+ coordinate is established by the donation of electron pair residing in sp^3 hybrid orbital to the vacant $1s$ orbital of H^+ ($H^+ = 1s^0$). Due to sp^3 hybridisation and presence of one lone pair of electrons, H_3O^+ ion does not have tetrahedral geometry. Rather it has *trigonal pyramidal shape.*

9. Geometry of amide ion, NH_2^-

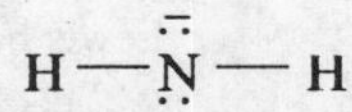

Lewis structure of NH_2^-

The negative charge on NH_2^- ion can be supposed to be present on N-atom. Thus N-atom in the given ion has six electrons instead of five ($N^- = 2s^2 2p^3 = 5$, $N^- = 2s^2\,2p^4 = 6$) and hence the Lewis structure of NH_2^- ion can be written as given in the margin. The structure shows that since N^- is surrounded by two σ-*bps* and two *lps*, N^- ion is sp^3 hybridised as shown below:

N-atom in ground state ($2s^2\,2p^3$) : $2s^2\,2p^3$

N^- ion in ground state ($2s^2\,2p^4$) : $2s^2\,2p^4$

N^- ion in sp^3 hybridised state in NH_2^- ion : $(sp^3)^2$ $(sp^3)^2$ $(sp^3)^1$ $(sp^3)^1$
lp *lp*

Four sp^3 hybrid orbitals

It is due to the presence of two lone pairs of electrons on N^- ion that NH_2^- ion has *angular geometry.* It is due to the presence of two lone pairs of electrons on N-atom that NH_2^- ion is capable of acting as a bridging ligand. When it functions as a bridging ligand, it donates both of its lone pairs of electrons residing in two sp^3 hybrid orbitals to two metal atoms.

10. Geometry of $SnCl_4^{2-}$ ion

This ion has sea-saw geometry which arises due to sp^3d hybridisation of Sn-atom (central atom) and due to the presence of one lone pair of electrons in one of the equatorial sp^3d hybrid orbital. This ion can be regarded as having been obtained by combining one $SnCl_2$ molecule and two Cl^- ions, *i.e.,*

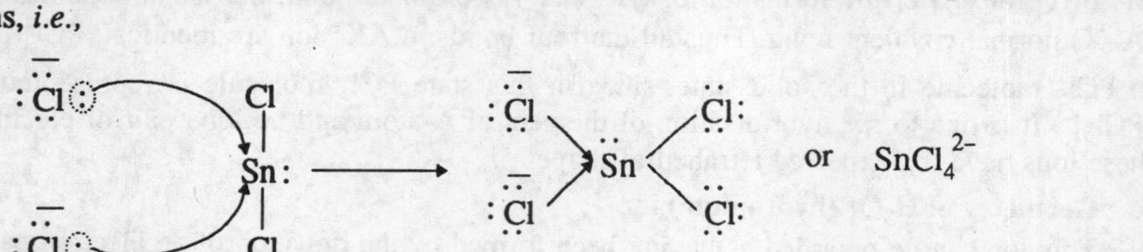

The electron dot structure of $SnCl_4^{2-}$ ion as given above makes it evident that the central Sn-atom in $SnCl_4^{2-}$ ion is surrounded by four σ-*bps* and one *lp*. Hence Sn atom is sp^3d hybridised as shown below:

Sn-atom in ground state ($5s^2\,5p^2\,5d^0$) : $5s^2\,5p^2\,5d^0_{z^2}$

Sn-atom in sp^3d hybridised state in $SnCl_4^{2-}$ ion. : $(sp^3d)^2$ $(sp^3d)^1$ $(sp^3d)^1$ $(sp^3d)^0$ $(sp^3d)^0$
lp

Five sp^3d hybrid orbitals

(Cl^- → Sn) coordinate bond is obtained by the donation of an electron pair from $3p$ orbital of Cl^- ion ($Cl^- = 3s^2\,3p^2\,3p^2\,3p^2$) to the vacant sp^3d hybrid orbitals on Sn-atom. Due to the presence of lone pair of electrons in sp^3d hybrid orbital, the shape of $SnCl_4^{2-}$ ion is *sea-saw.*

11. Geometry of AX_6^- ions (A = P, As, Sb, Bi; X = F, Cl)

All these ions have octahedral geometry which results from sp^3d^2 hybridisation of the central atom. These ions may be regarded as having been formed by the combination of pentahalides and one halide ion. Thus :

$$PCl_5 + Cl^- \rightarrow PCl_6^- \quad ; \quad AsF_5 + F^- \rightarrow AsF_6^-$$
$$SbF_5 + F^- \rightarrow SbF_6^- \quad ; \quad BiCl_5 + Cl^- \rightarrow BiCl_6^-$$

As an example let us consider the structure of PCl_6^- ion. We know that in PCl_5 molecule, P-atom (central atom) is sp^3d^2 hybridised. One *d*-orbital ($3d_{x^2-y^2}$ orbital) is not involved in sp^3d^2 hybridisation and hence remains vacant. Due to the presence of vacant $3d_{x^2-y^2}$ orbital, PCl_5 molecule can accept one electron pair donated by one Cl^- ion to form $Cl_5P \leftarrow Cl^-$ or PCl_6^- ion. Thus in the formation of PCl_6^- ion, PCl_5 molecule acts as a Lewis acid and Cl^- ion behaves as a Lewis base.

Lewis base		Lewis acid		Ion formed
Cl_5P	+	$:\ddot{Cl}:^-$	$\longrightarrow$	$Cl_5P \leftarrow Cl^-$ or PCl_6^-

Thus the structure of PCl_6^- **ion can be written as given in the margin. This structure shows that the central atom (P-atom) is surrounded by six σ-*bps* and hence is sp^3d^2 hybridised as shown below :**

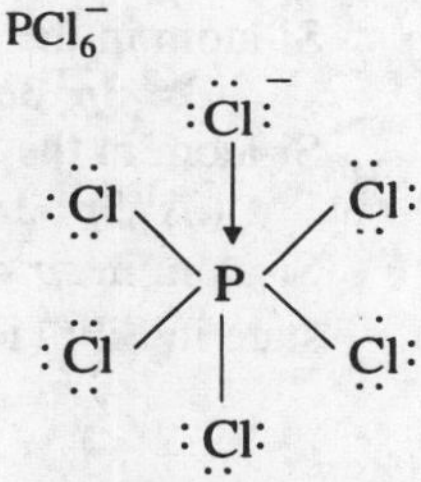

Lewis structure of PCl_6^- ion

P-atom in the ground state ($3s^2\ 3p^3\ 3d^0$) : $3s^2\ 3p^3\ 3d^0$

P-atom in the excited state ($3s^1\ 3p^3\ 3d^1\ 3d^1\ 3d^0$) : $3s^1\ 3p_x^1\ 3p_y^1\ 3p_z^1\ 3d_{z^2}^1\ 3d_{x^2-y^2}^0$

P-atom in sp^3d^2 hybridised state in PCl_6^- ion : $(sp^3d^2)^1\ (sp^3d^2)^1 (sp^3d^2)^1\ (sp^3d^2)^1\ (sp^3d^2)^1\ (sp^3d^2)^0$

($Cl^- \rightarrow P$) coordinate bond is formed by the donation of an electron pair on Cl^- ion ($Cl^- = 3s^2\ 3p^2\ 3p^2\ 3p^2$) to the vacant sp^3d^2 hybrid orbital of P-atom. PCl_6^- ion has the expected octahedral shape. This ion is found in the solid state of PCl_5, since this molecule in the solid state is represented as $[PCl_4]^+\ [PCl_6]^-$.

12. Geometry of AX_6^{2-} ions (A = Si, Ge, Sn; X= F, Cl). All these ions have octahedral geometry which results from sp^3d^2 hybridisation of the central A-atom. These ions can be regarded as having been formed by the combination of one AX_4 molecule and two X^- ions.

Thus:

$$AX_4\ (A = Si, Ge, Sn;\ X = F, Cl) + 2X^- \rightarrow AX_6^{2-}$$

As an example let us explain the geometry of SiF_6^{2-} ion. We know that in SiF_4 molecule $\left[\begin{matrix}F \\ F\end{matrix} > Si < \begin{matrix}F \\ F\end{matrix}\right]$, Si-atom (central atom) is sp^3 hybridised. Two $3d$ orbitals ($3d_{z^2}$ and $3d_{x^2-y^2}$) are not involved in the formation of sp^3 hybrid orbitals (four) and hence remain vacant. Due to the presence of vacant two $3d$ orbitals, SiF_4 molecule can accept two electron pairs donated by two F^- ions to form $F_4\,Si \begin{matrix}\swarrow F^- \\ \nwarrow F^-\end{matrix}$ or SiF_6^{2-} ion. Thus in the formation of SiF_6^{2-} ion, SiF_4 molecule acts as an electron pair acceptor (*Lewis acid*) and each of the two F^- ions behaves as an electron pair donor (*Lewis base*).

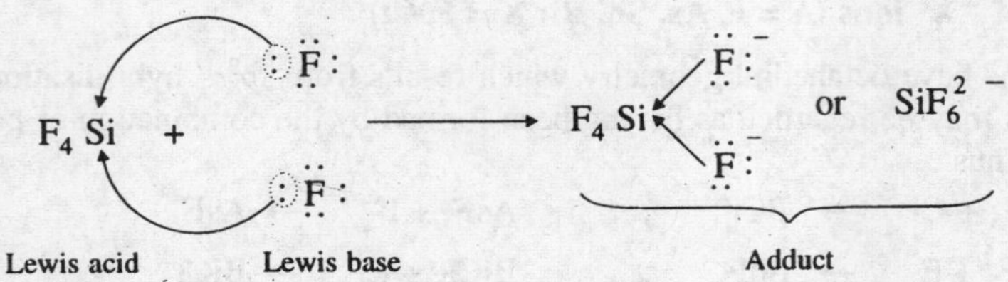

Thus the structure of SiF_6^{2-} ion can be written as 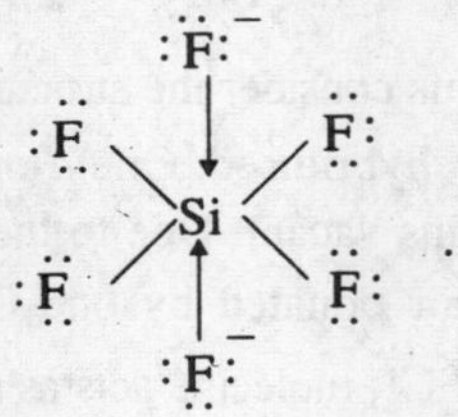 . This structure shows that the central atom (Si-atom) is surrounded by six σ-*bps* and hence is sp^3d^2 hybridised as shown below:

Si-atom in the ground state $(3s^2\,3p^2\,3d^0)$: $3s^2 3p^2\, 3d^0\, 3d^0$

Si-atom in the excited state $(3s^1\,3p^3\,3d^0)$: $3s^1\, 3p^3\, 3d^0\, 3d^0$

Si-atom in sp^3d^2 hybridised state in SiF_6^{2-} ion : $\underbrace{(sp^3d^2)^1\,(sp^3d^2)^1\,(sp^3d^2)^1\,(sp^3d^2)^1\,(sp^3d^2)^0\,(sp^3d^2)^0}_{\text{Six } sp^3d^2 \text{ hybrid orbitals}}$

Due to sp^3d^2 hybridisation SiF_6^{2-} ion has octahedral geometry. Both ($F^- \rightarrow$ Si) coordinate bonds are formed by the donation of electron pair of F^- ion to the vacant sp^3d hybrid orbital. On the same lines the octahedral geometry of $SnCl_6^{2-}$ and GeF_6^{2-} ions can also be explained.

13. Geometry of AX_6^{3-} ions (A = Al, Ga, In, Tl; X = F, Cl)

All these ions have octahedral shape which results from sp^3d^2 hybridisation of the central A-atom. All these ions can be regarded as having been formed by the combination of one AX_3 molecule and three X^- ions (halide ions). Thus :

$AlF_3 + 3F^- \rightarrow AlF_6^{3-}$, $\quad GaCl_3 + 3Cl^- \rightarrow GaCl_6^{3-}$

$InCl_3 + 3Cl^- \rightarrow InCl_6^{3-}$, $\quad TlCl_3 + 3Cl^- \rightarrow TlCl_6^{3-}$

As an example let us consider the geometry of AlF_6^{3-} ion. We know that in AlF_3 [$\frac{F}{F}{>}Al — F$], Al-atom (central atom) is sp^2 hybridised. In sp^2 hybridisation one $3p$ and two $3d$ orbitals of Al-atom remain unhybridised and hence can accept three electron pairs donated by F^- ions to form AlF_6^{3-} ion (The structure of AlF_6^{3-} ion is shown in the margin). Thus in the formation of AlF_6^{3-} ion, AlF_3 molecule acts as a Lewis acid and three F^- ions behave as Lewis base.

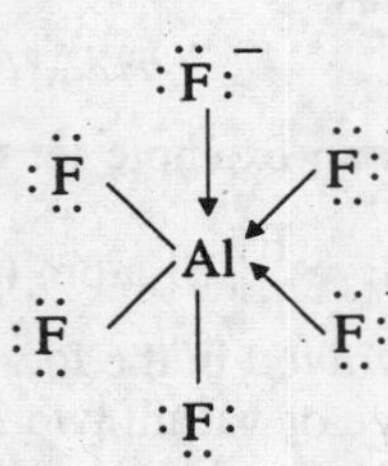

The structure of AlF_6^{3-} ion as given in the margin suggests that Al-atom (Central atom) in AlF_6^{3-} ion is surrounded by 6 σ-bps and hence is sp^3d^2 hybridised as shown below:

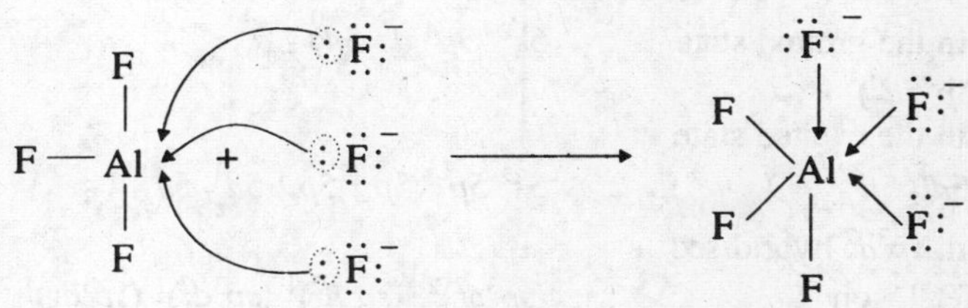

or $AlF_3 + 3\ :\ddot{F}:^- \longrightarrow AlF_6^{3-}$

Lewis base — Lewis acid — Adduct

Al-atom in ground state' $(3s^2\,3p^1 3d^0)$	:	$3s^2\ 3p^1\ 3d^0$
Al-atom in the excited state $(3s^1\,3p^2\,3d^0)$	:	$3s^1\ 3p^2\ 3d^0\ 3d^0$
Al-atom in sp^3d^2 hybridised state in AlF_6^{3-} ion	:	$\underbrace{(sp^3d^2)^1\ (sp^3d^2)^1\ (sp^3d^2)^1\ (sp^3d^2)^0\ (sp^3d^2)^0\ (sp^3d^2)^0}_{\text{Six } sp^3d^2 \text{ hybrid orbitals}}$

Vacant three sp^3d^2 hybrid orbitals form three ($F^- \rightarrow Al$) coordinate bonds.

14. Geometry of TeX_5^- ions (X = F, Cl).

Both these ions have *square pyramidal geometry.* This shape arises because of sp^3d^2 hybridisation of the central Te-atom and the presence of one lone pair of electrons in one of two axial sp^3d^2 hybrid orbitals. These ions can be supposed to have been formed by the combination of one TeX_4 molecule and one X^- ion. Thus :

$$Te\ X_4\ (X = F, Cl) + \quad X^- \longrightarrow TeX_5^-$$

As an example let us consider the geometry of $TeCl_5^-$ ion. We know that electron dot structures of $TeCl_4$ molecule and $TeCl_5^-$ ion can be written as :

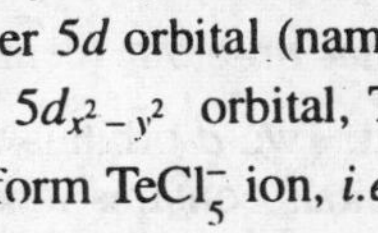

Structure of $TeCl_4$ molecule

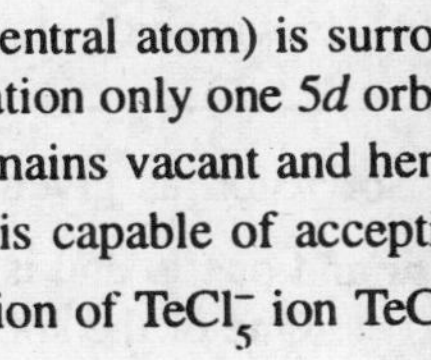

Structure of $TeCl_5^-$ ion

In the structure of $TeCl_4$ molecule, since Te-atom (central atom) is surrounded by four *σ-bps* and one *lp*, this atom is sp^3d hybridised. In sp^3d hybridisation only one 5*d* orbital (*viz.*, $5d_{z^2}$ orbital) is involved and the other 5*d* orbital (namely $5d_{x^2-y^2}$) remains vacant and hence is unused. Due to the presence of vacant $5d_{x^2-y^2}$ orbital, $TeCl_4$ molecule is capable of accepting one electron pair donated by Cl^- ion to form $TeCl_5^-$ ion, *i.e.*, in the formation of $TeCl_5^-$ ion $TeCl_4$ molecule acts as a Lewis acid and Cl^- ion behaves as a Lewis base.

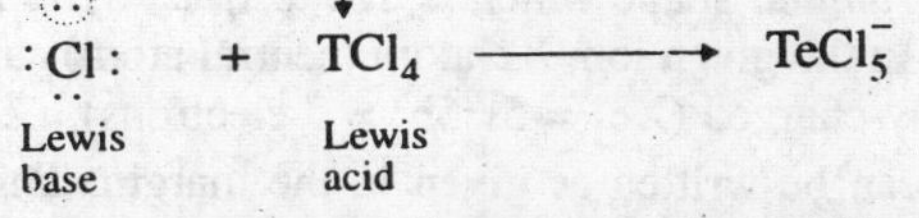

Thus the structure of $TeCl_5^-$ ion can be written as given above. This structure shows that since the central Te-atom is surrounded by five *σ-bps* and one *lp*, central atom is sp^3d^2 hybridised as shown below:

Te-atom in the ground state $(5s^2 5p^4 5d^0)$: $5s^2\ 5p^4\ d_{z^2}^0\ 5d_{x^2-y^2}^0$

Te-atom in the excited state $(5s^2\ 5p^3\ 5d_{z^2}^1\ 5d_{x^2-y^2}^0)$: $5s^2\ 5p_x^1\ 5p_y^1\ 5p_z^1\ 5d_{z^2}^1\ 5d_{x^2-y^2}^0$

Te-atom in sp^3d^2 hybridised state in $TeCl_5^-$ ion : $\underbrace{\underset{lp}{\boxed{(sp^3d^2)^2}}\ (sp^3d^2)^1\ (sp^3d^2)^1\ (sp^3d^2)^1\ (sp^3d^2)^1\ (sp^3d^2)^0}_{\text{Six } sp^3d^2 \text{ hybrid orbitals}}$

$(Cl^- \rightarrow Te)$ co-ordination bond is formed in its usual way.

15. Geometry of SbX_5^{2-} ions (X = F, Cl)

Both these ions have *square pyramidal* structure which results due to sp^3d^2 hybridisation of the central Sb-atom and presence of one lone pair of electrons in one of the two axial sp^3d^2 hybrid orbitals. These ions can be regarded as having been formed by the combination of one molecule of trihalide of Sb with two halide ions, *i.e.*,

$$SbF_3 + 2F^- \rightarrow SbF_5^{2-}, \qquad SbCl_3 + 2Cl^- \rightarrow SbCl_5^{2-}$$

As an example let us consider the structure of SbF_5^{2-} ion. the Lewis structure of SbF_3 molecule and SbF_5^{2-} ion can be written as shown below:

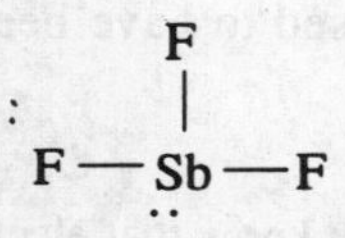

Structure of SbF_3 molecule

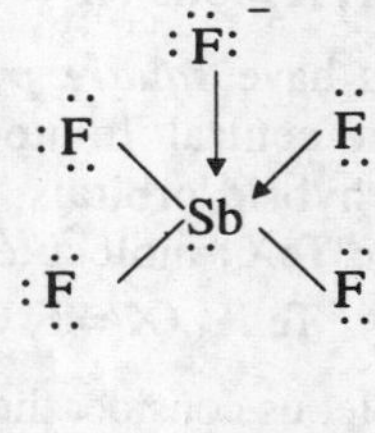

Structure of SbF_5^{2-} ion

In the structure of SbF_3 molecule, since Sb-atom (central atom) is surrounded by three ***σ-bps*** and *one lone pair of electrons*, this atom is sp^3 hybridised and both *d*-orbitals (*i.e.*, $5d_{z^2}$ and $5d_{x^2-y^2}$ orbitals) remain excluded from hybridisation and hence are vacant. Due to the presence of two vacant *d* orbitals, SbF_3 molecule can accept two electron pairs donated by two F^- ions to form SbF_5^{2-} ion.

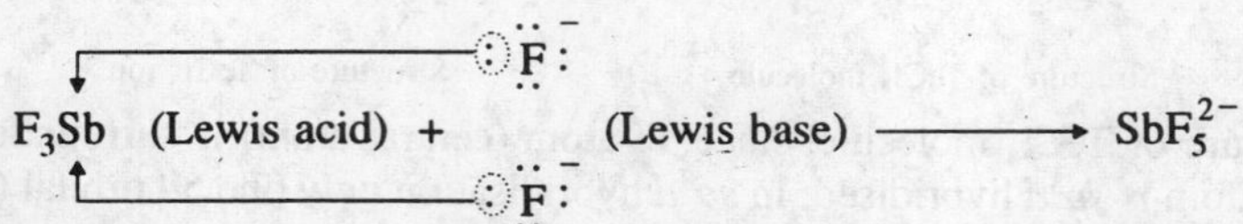

The structure of SbF_5^{2-} ion as given above shows that Sb-atom (central atom) in this ion is surrounded by 5 ***σ-bps*** and one ***lp*** and is sp^3d^2 hybridised. Two *d*-orbitals used in sp^3d^2 hybridisation are $5d_{z^2}$ and $5d_{x^2-y^2}$. One of the six sp^3d^2 hybrid orbitals contains lone pair of electrons.

16. Geometry of XeF_5^+ ion

This ion has *square pyramidal* shape which is due to sp^3d^2 hybridisation of the central atom. In the given ion, Xe-atom (central atom) can be supposed to be positively-charged ($Xe^+ = 5s^25p^5$ = 7 electrons), *i.e.*, the structure of XeF_5^+ ion can be written as given in the margin. This structure indicates that Xe^+ ion is surrounded by five ***σ-bps*** and one ***lp***.

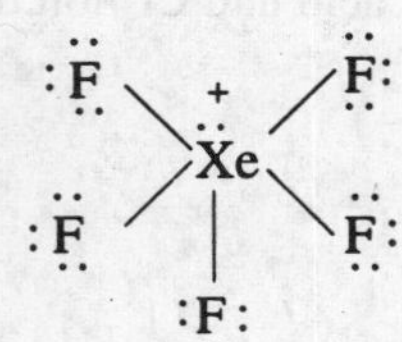

Structure of XeF_5^+ ion

Thus Xe^+ ion is sp^3d^2 hybridised. One of the axial sp^3d^2 hybrid orbital contains a lone pair of electrons. It is due to the presence of lone pair of electrons that XeF_5^+ ion has square pyramidal geometry.

17. Geometry of SbX_6^{3-} ions (X = F, Cl)

Both these ions have distorted octahedral geometry which arises because of sp^3d^3 hybridisation of the central Sb-atom and the presence of one lone pair of electrons in one sp^3d^3 hybrid orbitals. SbX_6^{3-} ion can be regarded as having been formed by the combination of one SbX_3 molecule and three X^- ions. Thus the formation of SbF_6^{3-} ion can be represented as:

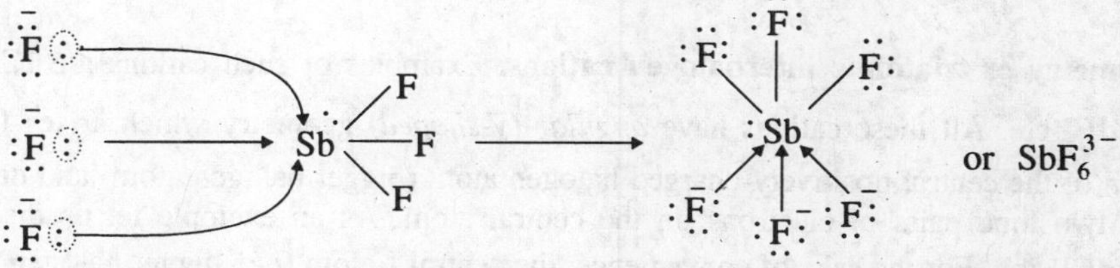

The Lewis structure (electron dot structure) of SbF_6^{3-} ion as given above shows that the central Sb-atom is surrounded by six σ-*bps* and one *lp*. Consequently Sb-atom (central atom) in this ion is sp^3d^3 hybridised and due to the presence of one *lp* in one of the seven sp^3d^3 hybrid orbitals, the shape of SbF_6^{3-} ion is *distorted octahedral.* sp^3d^3 hybridisation of Sb-atom has been shown below:

Sb-atom in ground state ($5s^2\ 5p^3\ 5d^0$) : $5s^2\ 5p^3\ 5d_{xy}^0\ 5d_{yz}^0\ 5d_{zx}^0$

Sb-atom in sp^3d^3 hybridised state in SbF_6^{3-} ion : $\underbrace{\boxed{(sp^3d^3)^2}}_{lp}\ (sp^3d^3)^1\ (sp^3d^3)^1\ (sp^3d^3)^1\ (sp^3d^3)^0\ (sp^3d^3)^0\ (sp^3d^3)^0$

Seven sp^3d^3 hybrid orbitals

Each of three ($F^- \rightarrow Sb$) coordinate bond is formed by the donation of an electron pair from $2p$ orbital of F^- ion ($2s^2\ 2p^2\ 2p^2\ 2p^2$) to the vacant sp^3d^3 hybrid orbital.

18. Geometry of TeX_6^{2-} ions (X = F, Cl)

Both these ions have distorted octahedral geometry which is due to sp^3d^3 hybridisation of Te-atom (central atom) and due to the presence of one *lp* of electrons in one sp^3d^3 hybrid orbitals. TeX_6^{2-} ion can be supposed to have been obtained by the combination of one TeX_4 molecule and two X^- ions. Thus the formation of TeF_6^{2-} ion can be represented as :

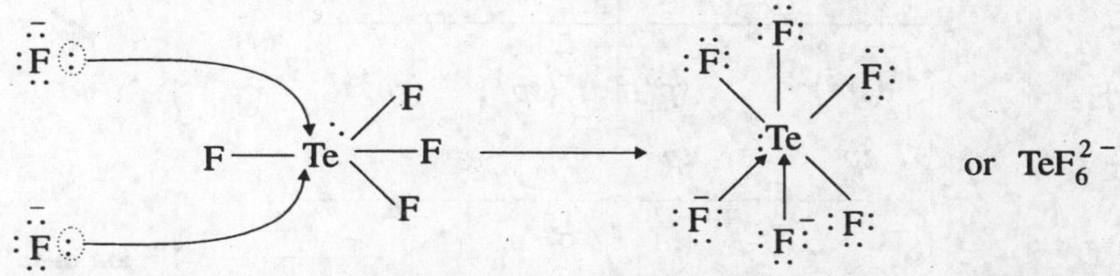

The electron dot structure of TeF_6^{2-} ion given above shows that the central Te-atom is surrounded by six σ-*bps* and one *lp*. Consequently Te-atom in this ion is sp^3d^3 hybridised and due to the presence of one *lp* in one of the seven sp^3d^3 hybrid orbitals the shape of TeF_6^{2-} ion is distorted octahedral.

19. Structure of interhalogen ions (Polyhalide ions)

While determining the geometry of a given interhalogen cation, the positive charge on the interhalogen cation is supposed to be present on the central halogen atom. In case of interhalogen anions, the negative charge on the anion is supposed to be present on one of the halogen atoms attached with the central halogen atom. Thus in ICl_2^+ ion (cation), I-atom (central atom is always bigger halogen atom) is supposed to be positively-charged ($I^+ = 5s^2\ 5p^4$) and hence the Lewis structure (electron dot structure) of this ion is $:\ddot{Cl} — \overset{+}{\ddot{\underset{..}{I}}} — \ddot{Cl}:$. In ICl_2^- ion (anion), one of the two Cl-atom is supposed to be negatively-charged and hence the electron dot structure of this ion is $:\ddot{\underset{..}{Cl}} — \ddot{\underset{..}{I}} \leftarrow \overset{..-}{\underset{..}{Cl}}:$.

(i) Geometry of triatomic interhalogen cations. Examples of such cations are ICl_2^+, IBr_2^+, BrF_2^+, ClF_2^+, IF_2^+ etc. All these cations have *angular (V-shaped)* geometry which arises from sp^3 hybridisation of the central positively-charged halogen atom (bigger halogen atom) and due to the presence of two lone pairs of electrons on the central atom. As an example let us discuss the geometry of ICl_2^+ ion. For the sake of convenience, the central I-atom (*i.e.*, bigger halogen atom) is supposed to be positively-charged. Thus the electron-dot structure of ICl_2^+ ion can be written as shown at (*a*) of Fig. 14.26. Since the central I-atom is surrounded by two *σ-bps* and two *lps*, I^+ ion is supposed to be sp^3 hybridised as shown below.

I-atom in the ground state ($5s^2\ 5p^5$) : $5s^2\ 5p_x^2\ 5p_y^2\ 5p_z^1$

I^+ ion in the ground state ($5s^2\ 5p^4$) : $5s^2\ 5p_x^2\ 5p_y^1\ 5p_z^1$

I^+ ion in sp^3 hybridised state in ICl_2^+ ion : $\underbrace{\underset{lp}{(sp^3)^2}\ \ \underset{lp}{(sp^3)^2}\ \ (sp^3)^1\ (sp^3)^1}_{\text{Four } sp^3 \text{ hybrid orbitals}}$

Formation of both I^+ – Cl bonds in ICl_2^+ ion has been shown at (*b*) of Fig. 14.26.

28

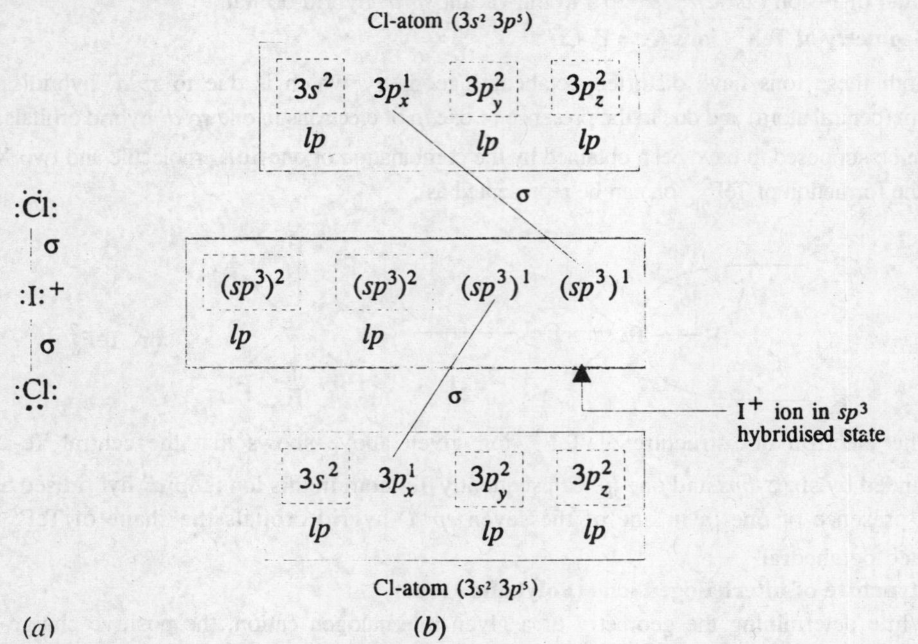

Fig. 14.26 : (*a*) Lewis structure of ICl_2^+ ion (*b*) sp^3 hybridisation of I^+ in ICl_2^+ ion and formation of different bonds.

Although the spatial arrangement of four sp^3 hybrid orbitals is tetrahedral, due to the presence of two lone pairs of electrons located in two sp^3 hybrid orbitals, ICl_2^+ ion assumes *angular shape* (*V-shaped structure*) as shown in Fig. 14.27.

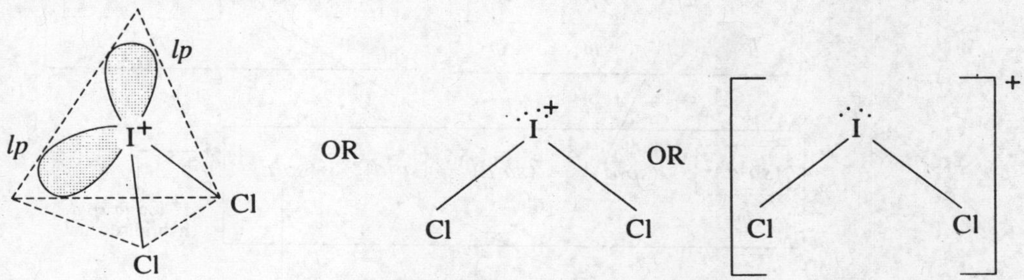

Fig. 14.27 : Angular (V-shaped) structure of $[ICl_2]^+$ ion.

Bond angles and bond lengths in various ionic species containing ICl_2^+, ClF_2^+ ions are given below:

Ionic species	Bond angles	Bond lengths
$[ICl_2]^+$ $[AlCl_4]^-$	ClICl = 96.7°	I—Cl = 228 pm
$[ICl_2]^+$ $[SbCl_4]^-$	ClICl = 92.5°	I—Cl = 231 pm
$[ClF_2]^+$ $[AsF_6]^-$	FClF = 103.2°	F—Cl = 154 pm
$[ClF_2]^+$ $[SbF_6]^-$	FClF = 95.9°	F—Cl = 158 pm
$[BrF_2]^+$ $[SbF_6]^-$	FBrF = 93.5°	F—Br = 169 pm

(*ii*) Geometry of penta-atomic interhalogen cations. Examples of such cations are IF_4^+, ClF_4^+, BrF_4^+ etc. All these ions have *distorted tetrahedral* or *see-saw structure.* This structure arises due to sp^3d hybridisation of the central positively-charged halogen atom (bigger halogen atom) and due to the presence of one lone pair of electrons on the central atom. As an example let us consider the geometry of IF_4^+ ion. For the sake of convenience, the central I-atom (*i.e.*, bigger halogen atom) is supposed to be positively-charged. Thus the Lewis structure of IF_4^+ ion can be written as shown at (*a*) of Fig. 14.28. Since the central I-atom is surrounded by four σ-*bps* and one *lp*, I^+ ion is said to be sp^3d hybridised as shown below:

I-atom in ground state ($5s^2\ 5p^5\ 5d^0$) : $5s^2\ 5p^5\ 5d_{z^2}^0$

I^+ ion in ground state ($5s^2\ 5p^4\ 5d^0$) : $5s^2\ 5p_x^2\ 5p_y^1\ 5p_z^1\ 5d_{z^2}^0$

I^+ ion in the excited state ($5s^2\ 5p^3\ 5d^1$) : $5s^2\ 5p_x^1\ 5p_y^1\ 5p_z^1\ 5d_{z^2}^1$

I^+ ion in sp^3d hybridised state in IF_4^+ ion. : $(sp^3d)^2$ $(sp^3d)^1$ $(sp^3d)^1$ $(sp^3d)^1$ $(sp^3d)^1$

lp (under $(sp^3d)^2$); Five sp^3d hybrid orbitals

sp^3d hybridisation scheme shown above indicates that IF_4^+ ion is a $AB_4(lp)$ type species. Formation of four (I^+ – F) σ- bonds has been shown at (*b*) of Fig. 14.28.

Position occupied by lone pair of electrons and shape of IF_4^+ *ion.* We have already said under the discussion of the structure of $AB_4(lp)$ type molecules/ions that the lone pair of electrons (*lp*) prefers to occupy the basal hybrid orbital. Due to the presence of one *lp* in the basal position, the shape of IF_4^+ ion is no longer trigonal bipyramidal but gets distorted and

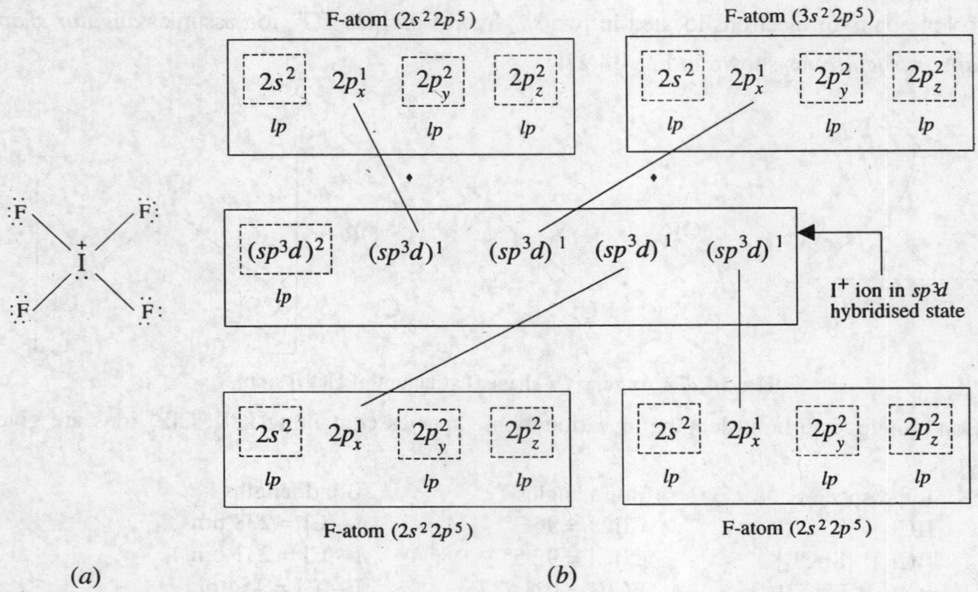

Fig. 14.28 : (*a*) Lewis structure of IF_4^+ ion. (*b*) sp^3d hybridisation of I^+ ion in IF_4^+ ion and formation of different bonds.

becomes *distorted tetrahedral* or *see-saw* (see Fig. 14.29) although the spatial arrangement of five sp^3d hybrid orbitals round the central I^+ ion is trigonal bipyramidal .

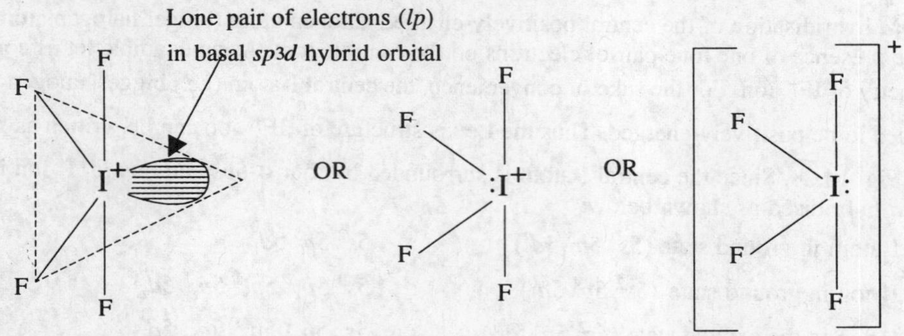

Fig. 14.29 : Distorted tetrahedral or see-saw structure of IF_4^+ ion.

***(iii)* Geometry of hepta-atomic interhalogen cations.** Examples of such cations are IF_6^+, BrF_6^+ etc. All these ions have *octahetral structure* corresponding to sp^3d^2 hybridisation of the central positively-charged halogen atom (bigger halogen atom) and due to the absence of any lone pair of electrons on the central atom. As an example let us consider the geometry of IF_6^+ ion. For the sake of simplicity, the central I-atom (*i.e.*, bigger halogen atom) is supposed to be positively-charged. Thus Lewis structure of IF_4^+ ion can be written as shown in the margin. Since the central I-atom is surrounded by six σ-*bps* and *no lone pair of electrons,* I^+ ion is said to be sp^3d^2 hybridised as shown below:

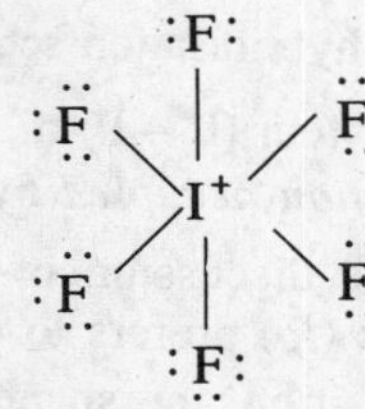

I-atom in ground state ($5s^2\ 5p^5\ 5d^0$)	:	$5s^2\ 5p^5\ 5d_{z^2}^0\ 5d_{x^2-y^2}^0$
I^+ ion in ground state ($5s^2\ 5p^4\ 5d^0$)	:	$5s^2\ 5p^4\ 5d_{z^2}^0\ 5d_{x^2-y^2}^0$
I^+ ion in doubly-excited state ($5s^1\ 5p^3\ 5d^2$)	:	$5s^1\ 5p^3\ 5d_{z^2}^1\ 5d_{x^2-y^2}^1$
I^+ ion in sp^3d^2 hybridised state in IF_6^+ ion	:	$\underbrace{(sp^3d^2)^1\ (sp^3d^2)^1\ (sp^3d^2)^1\ (sp^3d^2)^1\ (sp^3d^2)^1\ (sp^3d^2)^1}_{\text{Six singly-filled } sp^3d^2 \text{ hybrid orbitals}}$

sp^3d^2 hybridisation scheme given above shows that IF_6^+ ion has no lone pair of electrons and hence this ion has octahedral geometry which has been shown in Fig. 14.30.

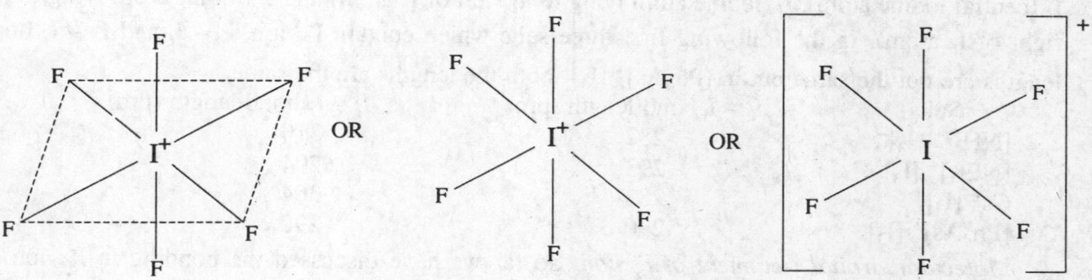

Fig. 14.30 : Octahedral structure of IF_6^+ ion.

***(iv)* Structure of tri-atomic interhalogen anions.** Examples of the anions of this type are ICl_2^-, IBr_2^-, $BrCl_2^-$, ClF_2^-, I_3^-, $ClBrI^-$ etc. All these ions have *linear structure* which is due to sp^3d hybridisation of the central halogen atom and the presence of three lone pairs of electrons on the central halogen atom. As examples let us discuss the structure and geometry of ICl_2^- and I_3^- ions.

(a) Structure and geometry of ICl_2^- ion. The negative charge on this ion may be supposed to reside on one of the two Cl^- atoms (smaller halogen atoms). Thus the Lewis structure of this ion can be written as given in the margin. This structure shows that the central I-atom is surrounded by two *σ-bps* and *three lone pairs of electrons* and hence I-atom is sp^3d hybridised. sp^3d hybridisation of I-atom is shown below:

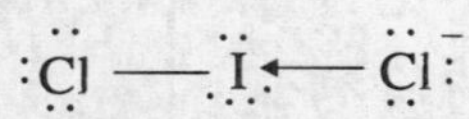

Lewis structure of ICl_2^-

I-atom in ground state ($5s^2 5p^5 5d^0$)	:	$5s^2\ 5p^5\ 5d_{z^2}^0$
I-atom in sp^3d hybridised state in ICl_2^- ion.	:	$\underbrace{\underset{lp}{(sp^3d)^2}\ \underset{lp}{(sp^3d)^2}\ \underset{lp}{(sp^3d)^2}\ (sp^3d)^1\ (sp^3d)^0}_{\text{Five } sp^3d \text{ hybrid orbitals}}$

The sp^3d hybrid orbital which is vacant forms $Cl^- \rightarrow I$ coordinate bond. sp^3d hybridisation scheme shows that ICl_2^- ion is a $AB_2\ (lp)_3$ type species.

Position of three lps and shape of ICl_2^- ion. We have already said under the discussion of the structure of $AB_2(lp)_3$ type species that the three *lps* prefer to sit at the basal position of the trigonal bipyramid, since in this case the repulsions between different electron pairs are minimum and *lps* have a larger volume to occupy in the basal positions. Due to the presence of three *lps* in the basal positions, the shape of ICl_2^- ion gets distorted and becomes linear (See Fig. 14.31).

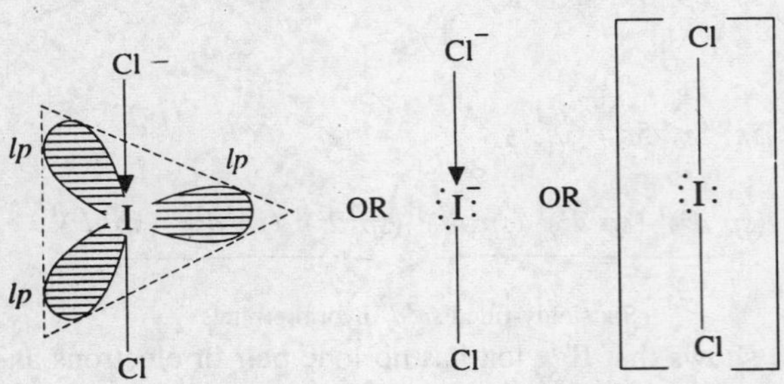

Fig. 14.31: Linear shape of ICl_2^- ion.

(*b*) *Structure of* I_3^-. This ion is called triiodide ion. Like ICl_2^- ion, I_3^- ion also has linear structure ($:\ddot{I}_l — :\ddot{I}_c \leftarrow \ddot{I}_r:$) in which I_c atom undergoes sp^3d hybridisation. In order to differentiate the three iodine atoms, they have been designated as I_c (central iodine atom), I_l (iodine atom lying to the left of I_c atom) and I_r (iodine atom lying to the right of I_c atom). In the following first three salts which contain I_3^- ion, $I_c – I_l$ and $I_c – I_r$ bond lengths are not the same but in $[Ph_4As]^+ [I_3]^-$ both the lengths are the same.

Salts	$I_c – I_l$ bond length (pm)	$I_c – I_r$ bond length (pm)
$[NH_4]^+ [I_3]^-$	282	310
$[NEt_4]^+ [I_3]^-$	293	294
$Cs^+ [I_3]^-$	283	304
$[Ph_4As]^+ [I_3]^-$	290	290

Molecular orbital treatment of I_3^- *ion.* So far we have discussed the bonding in I_3^- ion by hybridisation concept. The bonding in I_3^- ion can also be described with the help of molecular orbital theory, if we take linear combination of the two singly-filled 5*p* orbitals of two I-atoms ($I = 5s^2\ 5p^2\ 5p^2\ 5p^1$) and one completely-filled 5*p* orbital of the central I^- ion ($I^- = 5s^2\ 5p^2\ 5p^2$

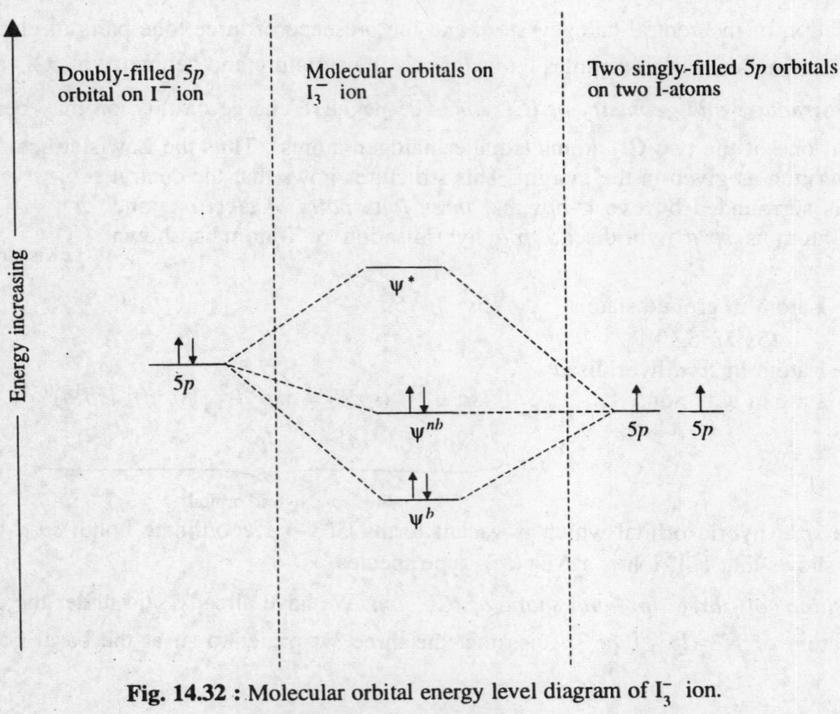

Fig. 14.32 : Molecular orbital energy level diagram of I_3^- ion.

$5p^2$). The linear combination of the three atomic orbitals gives three molecular orbitals which are bonding molecular orbital (ψ^b), non-bonding molecular orbital (ψ^{nb}) and antibonding molecular orbital (ψ^*). The molecular orbital energy level diagram of I_3^- ion is shown in Fig. 14.32.

It should be noted that for the sake of simplicity, all the mixing coefficients have been taken to be equal to unity. The two electrons involved in bond formation are contained in bonding molecular orbital (ψ^b) which spreads out on either side of the negatively-charged central iodine atom and thus two equivalent bonds are formed.

(v) Structure of penta-atomic interhalogen anions. Examples of the anions of this type are ICl_4^-, BrF_4^-, I_5^-. All these ions have *square planar geometry* which arises due to sp^3d^2 hybridisation of the central halogen atom and due to the presence of two lone pairs of electrons on this atom. As an example let us discuss the geometry of ICl_4^- ion. The negative charge on this ion is supposed to be present on one of the four Cl-atoms. Thus the Lewis structure of this ion can be written as given in the margin. This structure clearly shows that the central I-atom is surrounded by four σ-*bps* and two *lps*. Thus, I-atom is sp^3d^2 hybridised. One of the six sp^3d^2 hybrid orbitals is vacant and hence accepts an electron pair donated by Cl^- ion and thus $Cl^- \rightarrow I$ coordinate bond is established.

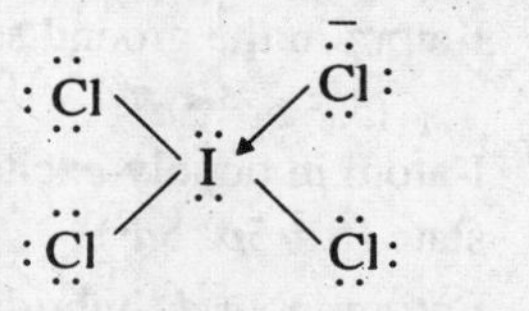

Lewis structure of ICl_4^- ion

I-atom in ground state ($5s^25p^55d^0$) : $5s^2\ 5p^5\ 5d_{z^2}^0\ 5d_{x^2-y^2}^0$

I-atom in excited state ($5s^25p^4\ 5d^1$) : $5s^2\ 5p^4\ 5d_{z^2}^1\ 5d_{x^2-y^2}^0$

I-atom in sp^3d^2 hybridised state in ICl_4^- ion : $\underbrace{\underset{lp}{(sp^3d^2)^2}\ \underset{lp}{(sp^3d^2)^2}\ (sp^3d^2)^1\ (sp^3d^2)^1\ (sp^3d^2)^1\ (sp^3d^2)^0}_{\text{Six } sp^3d^2 \text{ hybrid orbitals}}$

sp^3d^2 hybridisation scheme given above shows that ICl_4^- ion is a $AB_4\ (lp)_2$ type species.

Position of the two lone pairs of electrons and shape of ICl_4^- ion. Experiments have shown that the two *lps* prefer to sit in the axial positions of the octahedron, since in this case (*lp* – *bp*) repulsion is minimum and a stable structure is obtained by ICl_4^- ion. Thus, although the arrangement of six electron pairs (σ-*bps* = 4 and *lp* = 2) round the central I-atom is octahedral, due to the presence of two *lps* in the axial position, the shape of ICl_4^- ion becomes *square planar* as shown in Fig. 14.33.

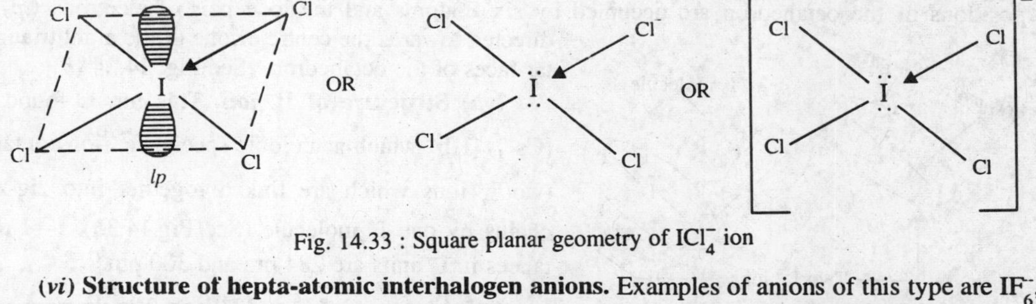

Fig. 14.33 : Square planar geometry of ICl_4^- ion

(vi) Structure of hepta-atomic interhalogen anions. Examples of anions of this type are IF_6^-, BrF_6^- etc. These ions have *distorted octahedral structure* which is because of sp^3d^3 hybridisation of the central halogen atom and presence of one lone pair of electrons on the central atom. As an example let us discuss the geometry of IF_6^- ion. The negative charge on IF_6^- ion may be supposed to be present on one of the six F-atoms (smaller halogen atom). Thus the Lewis structure of IF_6^- ion is that which is given in the margin. This structure shows that the central I-atom is surrounded by six σ *bps* and one lone pair of electrons and hence I-atom is supposed to be sp^3d^3 hybridised. One of the seven hybrid orbitals is vacant and hence accepts an electron pair donated by F^- ion ($F^- = 2s^2\ 2p^2\ 2p^2\ 2p^2$) to form $F^- \rightarrow I$ coordinate bond. sp^3d^3 hybridisation of I-atom can be shown as follows:

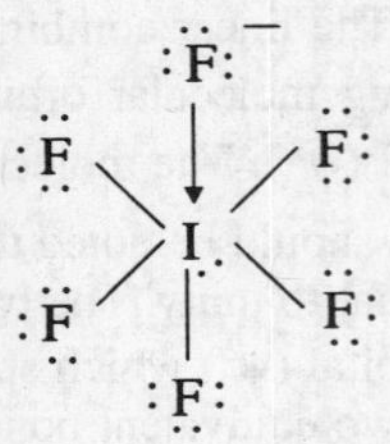

Structure of IF_6^- ion

I-atom in the ground state ($5s^2\ 5p^5\ 5d^0$) : $5s^2\ 5p^5\ 5d^0_{xy}\ 5d^0_{yz}\ 5d^0_{zx}$

I-atom in doubly-excited state ($5s^2\ 5p^3\ 5d^2$) : $5s^2\ 5p^3\ 5d^1_{xy}\ 5d^1_{yz}\ 5d^0_{zx}$

I-atom in sp^3d^3 hybridised state in IF_6^- ion : $\underbrace{\underset{lp}{(sp^3d^3)^2}\ (sp^3d^3)^1\ (sp^3d^3)^1\ (sp^3d^3)^1\ (sp^3d^3)^1\ (sp^3d^3)^1\ (sp^3d^3)^0}_{\text{seven } sp^3d^3 \text{ hybrid orbitals}}$

sp^3d^3 hybridisation scheme given above indicates that IF_6^- ion is a AB_6 (*lp*) type species.

Theoreticaly the lone pair of electrons should occupy one of the five basal positions of the pentagonal base of the pentagonal bipyramid, *i.e.*, the lone pair of electrons should be directed towards one of the five corners of the pentagonal base and hence IF_6^- ion should have *distorted pentagonal bipyramidal structure* shown at (*a*) of Fig. 14.34. However, experiments have shown that distorted pentagonal bipyramidal structure of IF_6^- ion is wrong. At present it is believed that IF_6^- ion has *distorted octahedral* structure in which the six positions of the octahedron are occupied by six F-atoms and the lone pair of electrons (*lp*) is directed towards the centre of one of the eight triangular faces of the octahedron. [See Fig. 14.34 (*b*)].

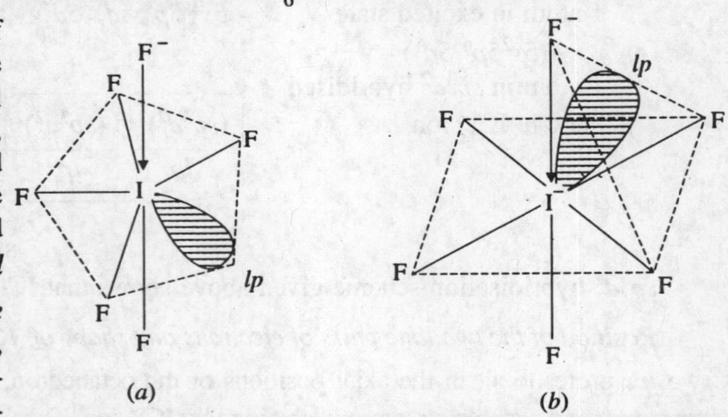

Fig. 14.34 : (*a*) Distorted pentagonal bipyramidal structure of IF_6^- Ion (wrong structure) (*b*) Distorted octahedral structure of IF_6^- ion (correct structure).

(*vii*) **Structure of I_8^{2-} ion.** This ion is found in $[Cs^+]_2\ [I_8]^{2-}$ which is an ionic species. I_8^{2-} ion contains two I_3^- ions which are linked together into zig-zag chains by one I_2 molecule (See Fig.14.35). I—I distances in I_3^- units are 284 pm and 300 pm.

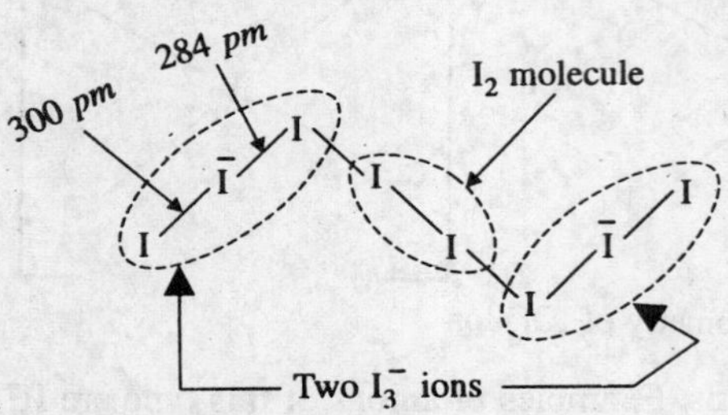

Fig. 14.35 : Zig-zig chain structure of I_8^{2-} ion in $[Cs^+]_2\ [I_8]^{2-}$

(*viii*) **Structure of I_3^+ and I_5^+ ions.** The structure of I_3^+ ion is bent (See Fig. 14.36) as expected (*cf* SCl_2) and that of I_5^+ ion is said to be a resonance hybrid of the two structures shown in Fig. 14.37.

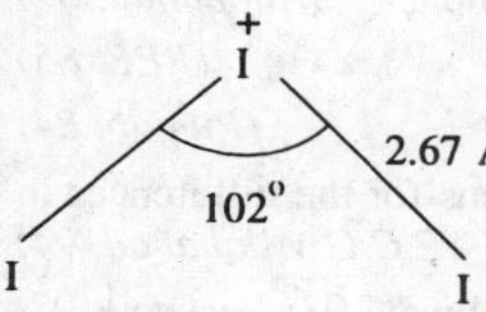

Fig. 14.36 : Bent Structure of I_3^+ ion

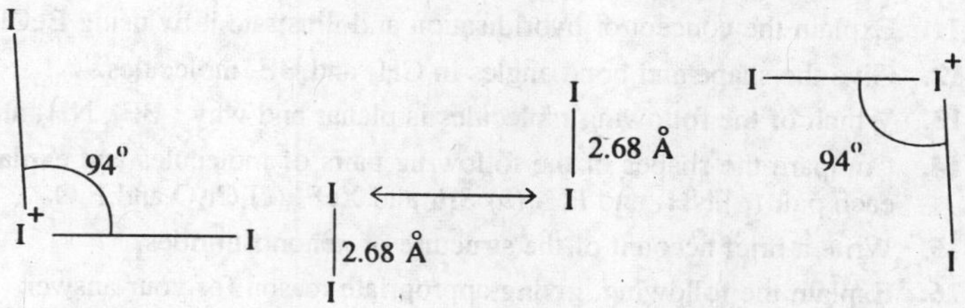

Fig. 14.37 : Resonance in I_5^+ ion

University Questions

1. Give the structure of all the four types of interhalogen compounds. *(Rohilakhand, 1991; Ravishankar, 1993)*
2. Discuss the structure of IF_7. *(Delhi, 89; Nagpur, 1993)*
3. Discuss the types of hybridisation and geometrical shapes of PCl_5 and SF_6 molecules. *(Gauwahati, 1991)*
4. Explain with examples the role of lone pairs of electrons on a central atom in determining the shape of an ion or a molecule. *(Gauwahati, 1991)*
5. Give the structure of a halide of xenon in its six oxidation state. *(Ravishakar, 93)*
6. What type of hybridisation is taking place in the following compounds? Discuss their shapes: $BeCl_2$, BCl_3, NH_3, PCl_5 and SF_6. *(Ravishankar, 1993)*
7. CH_4, NH_3 and H_2O represent sp^3 hybridisation but their bond angles are different. Explain. *(Ravishankar, 1993)*
8. Indicate the different states of hybridisation and deduce the geometry of the following isolated molecules:
   - *(i)* $BeCl_2$ *(Bundelkhand, 88; Kanpur, 93)*
   - *(ii)* BeH_2 *(Avadh, 93)*
   - *(iii)* BeF_2 *(Manglore, 86)*
   - *(iv)* BF_3, BH_3 *(Agra, 89; Kanpur, 93; Purvanchal, 93; Lucknow, 93)*
   - *(v)* BCl_3 *(Rohilkhand, 91)*
   - *(vi)* $SnCl_2$ *(Rajasthan, 84)*
   - *(vii)* CH_4 *(Kanpur, 93)*
   - *(viii)* NH_4Cl *(Utkal, 86)*
   - *(ix)* NH_3 *(Purvanchal, 93)*
   - *(x)* H_2O *(Tripura, 91; Purvanchal, 93, Lucknow, 93)*
   - *(xi)* SCl_2 *(Pune, 85)*
   - *(xii)* PCl_5 *(Agra, 89; Purvanchal, 93)*
   - *(xiii)* SF_4 *(Allahabad, 88; Purvanchal, 93)*
   - *(xiv)* ClF_3 *(Meerut M.Sc., 94; Purvanchal, 93; Lucknow, 93)*
   - *(xv)* XeF_2 *(Agra, 89)*
   - *(xvi)* SF_6 *(Meerut, 90; Gauwahati, 91; Lucknow, 93)*
   - *(xvii)* IF_5 *(Meerut M.Sc., 89)*
   - *(xix)* XeF_4 *(Allahabad, 88)*
   - *(xx)* Trimethyl aluminium, $Al(CH_3)_3$ *(Lucknow, 93)*
   - *(xxi)* OF_2 *(Meerut, 90)*
9. Discuss the shape of the following ions on the basis of hybridisation:
   - *(i)* ICl_4^- *(Delhi, 88; Meerut, 94)*
   - *(ii)* ICl_2^- *(Delhi, 88; Meerut, 94)*
   - *(iii)* BeF_4^{2-} *(Agra, 89; Lucknow, 94)*
   - *(iv)* AlH_4^- *(Lucknow, 93)*

10. What is the state of hybridisation of the central atom in the following molecules : $BeCl_2$, BCl_3, CCl_4 and PF_5. *(Bharathidasan, 92)*

11. Explain the concept of hybridisation and illustrate it by using $BeCl_2$ as an example. *(Allahabad, 89)*

12. Give the shape and bond angles in CH_4 and BF_3 molecules. *(REE, 85)*

13. Which of the following molecules is planar and why : BF_3, NH_3 and PCl_5? *(Punjab, 84)*

14. Compare the shapes of the following pairs of molecules and explain the reasons for the differences in each pair (*a*) NH_3 and PH_3 (*b*) SF_4 and XeF_4 (*c*) Cl_2O and F_2O. *(E.C.C. Allahabad, 94)*

15. Write a brief account of the structure of xenon fluorides. *(Avadh, 93; Lucknow, 93)*

16. Explain the following, giving appropriate reason for your answer.

(*i*) He and Be have ns^2 configuration in their outer-most orbit. However, Be forms compounds while He does not do so.

(*ii*) The central atoms in CH_4, NH_3 and H_2O molecules contain the same number of electrons (equal to eight) but their geometries are different.

(*iii*) PCl_5 exists while NCl_5 does not exist. *(Rohilkhand, 91)*

(*iv*) H_2O has bent structure. *(Lucknow, 85)*

(*v*) The molecules namely CH_4, NH_3 and H_2O all involve sp^3 hybridisation of the central atom but the bond angles in them are 109°28′, 107°18′ and 104°30′ respectively. *(Nagpur, 87)*

(*vi*) CH_4, NH_3 and H_2O molecules have tetrahedral geometry, yet their bond angles are different. *(Jodhpur, 89)*

(*vii*) XeF_2 is linear while SF_2 molecule is not. *(Punjab, 86)*

(*viii*) The shape of trisilyamine molecule, $(SiH_3)_3N$ is planar while that of trimethyl amine molecule, $(CH_3)_3N$ is pyramidal. *(Delhi Hons., 86; Allahabad, 89)*

(*ix*) BeF_2 molecule is linear but SF_2 molecule is angular. *(Punjab, 85)*

(*x*) $BeCl_2$ molecule is linear and the angle between Be—Cl bonds is 180°. *(Agra, 88)*

(*xi*) PCl_5 molecule has trigonal bipyramidal structure. *(Meerut, 94)*

(*xii*) BeF_2 molecule is linear while H_2O molecule is V-shaped. *(Rohilkhand, 90)*

(*xiii*) NH_3 molecule is pyramidal while NH_4^+ ion is tetrahedral. *(Gauwahati, 91)*

(*xiv*) CH_4, H_2O and NH_3 contain the same number of electrons but their shapes are different. *(Gauwahati, 91)*

(*xv*) Al forms AlF_6^{3-} ion but B does not form BF_6^{3-} ion *(Lucknow, 93)*

(*xvi*) According to electronic configuration H_2O molecule should be linear and NH_3 molecule should be trigonal planar but actually both the molecules have unsymmetrical tetrahedral structure. *(Bahuguna, 93)*

(*xvii*) NH_3 molecule has pyramidal structure. *(Meerut, 93)*

17. Among the following species identify the iso-structural pairs.

NF_3, NO_3^-, BF_3, H_3O^+, HN_3

(*a*) [NH_3, NO_3^-] and [BF_3, H_3O^+]

(*b*) [NF_3, HN_3] and [NO_3^-, BF_3]

(*c*) [NF_3, H_3O^+] and [NO_3^-, BF_3]

(*d*) [NF_3, H_3O^+] and [HN_3, BF_3] *(I.I.T., 96)*

18. (*a*) Write a note on structure of boric acid and its titration against alkalies.

(*b*) Give the structures of the following (any *six*) : (*i*) Permonosulphuric acid (*ii*) $HClO_4$ (*iii*) ClF_3 (*iv*) IF_7 (*v*) PCl_5 (*vi*) H_3PO_3 (*vii*) $H_2S_2O_8$ (*viii*) Hydrazine. *(Delhi 2002)*

19. Discuss the structure of XeF_2 with the help of valence bond theory. *(Delhi 2002)*

20. What is meant by hybridization of atomic orbitals ? Write down the shape of any three of the following molecules and indicate the hybrid orbitals used by central atom in them : (*i*) $BeCl_2$ (*ii*) BCl_3 (*iii*) SF_6 (*iv*) PCl_5 *(Rohilkhand 2003)*

21. Explain the geometries of BF_3 and PCl_5 molecules on the basis of hybridisation. *(Nagpur 2003)*

22. Find out the number of lone pairs and bond pairs present in the following : (*i*) NH_3 (*ii*) H_2O (*iii*) SF_4 (*iv*) ClF_3. *(Nagpur 2003)*

15

Structure and Shape of Molecules/Ions Containing Sigma as well as Pi Bonds

Introduction

The presence of one or more pi bonds (π-bonds) in the electron dot structure (Lewis structure) of a given molecule/ion does not affect the type of hybridisation undergone by the central atom/ion. In other words, the presence of one or more π-bonds does not affect the shape of the molecule or ion. In order to know the nature of hybridisation undergone by the central atom/ion, we have to find out only the sum of *sigma bonding electron pairs (σ-bps)* and *lone pairs of electrons (lps)* surrounding the central atom/ion. Here we shall consider the structure (bonding) and geometry of the following molecules/ions.

Oxides of carbon

Carbon monoxide (CO) and carbon dioxide (CO_2) are important oxides of carbon. Here we shall consider the structure and geometry of these oxides.

1. Structure and geometry of CO molecule

Following electron dot structures have been proposed for CO molecule.

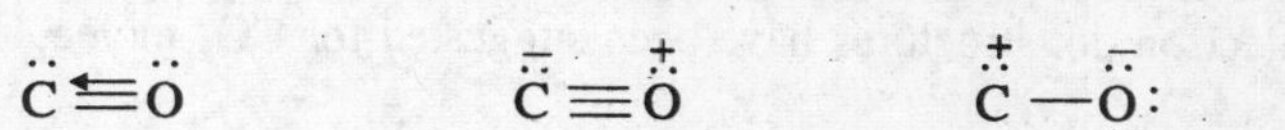

Structure (I) **(Covalent structure)** **Structure (II)** **Structure (III)**

(Ionic structures)

The following points may be noted from these structures :

(*i*) In structures (I) and (II), both the atoms obey octet rule while in structures (III) only O-atom obeys octet rule.

(*ii*) In all the three structures, since the central C-atom is surrounded by one sigma bonding electron pair and one lone pair of electrons, this atom is *sp* hybridised in each of these structures, *i.e.*, the valence-shell configuration of C-atom in *sp* hybridised state in different structures can be written as :

C-atom in structure (I) $\rightarrow (sp)^2\ (sp)^1\ 2p_y^1\ 2p_z^0$

$\bar{C}$ in structure (II) $\rightarrow (sp)^2\ (sp)^1\ 2p_y^1\ 2p_z^1$

$\overset{+}{C}$ in structure (III) $\rightarrow (sp)^2\ (sp)^1\ 2p_y^0\ 2p_z^0$

(*iii*) In structures (I) and (II), O-atom is *sp* hybridised but in structure (III) O-atom is sp^3 hybridised.

O-atom in structure (I) $\rightarrow$ $(sp)^2\ (sp)^1\ 2p_y^1\ 2p_z^2$

$\overset{+}{\text{O}}$ in structure (II) $\rightarrow$ $(sp)^2\ (sp)^1\ 2p_y^1\ 2p_z^1$

$\overset{-}{\text{O}}$ in structure (III) $\rightarrow$ $(sp^3)^2\ (sp^3)^1\ (sp^3)^1\ (sp^3)^2$

On the basis of the type of hybridisation undergone by C–and O–atoms in structure (I), the formation of different bonds in this structure can well be explained.

Resonance in CO molecule. X-ray diffraction study of CO molecule has shown that carbon–oxygen bond length in this molecule is equal to 1.13 Å. This value is neither equal to the calculated values of C – O single bond length (= 1.43 Å) nor C = O double bond length (= 1.22 Å). This value is, however, very close to the calculated value of C ≡ O triple bond length (= 1.10 Å). In order to account for the discrepancy between the experimental value of carbon-oxygen bond length and the calculated values of C – O single bond and C = O double bond lengths, it is assumed that none of the above three structures represents the actual structure of CO molecule. Rather, the real structure is supposed to be resonance hybrid of the following contributing structures.

$$:\text{C}\equiv\text{O}: \longleftrightarrow :\overset{-}{\text{C}}\equiv\overset{+}{\text{O}}: \longleftrightarrow :\overset{+}{\text{C}}-\overset{-}{\ddot{\text{O}}}:$$

(I) (II) (III)

Since the experimental value of carbon-oxygen bond length (= 1.13 Å) is very close to the calculated value of C ≡ O triple bond length (= 1.10 Å), it is supposed that the bond between carbon and oxygen atoms in CO molecule is approximately a triple bond, *i.e.*, out of the above resonating structures, (I) and (II) forms are more important. Since in structure (II), there is a positive charge on O-atom and a negative charge on C-atom, CO molecule should have appreciable dipole moment from oxygen to carbon $\left(\overrightarrow{\overset{+}{\text{O}}-\text{C}}\right)$. But resonance structure (III) has dipole moment from carbon to oxygen $\left(\overrightarrow{\overset{+}{\text{C}}-\overset{-}{\text{O}}}\right)$. This reduces the polarity of the molecule. Therefore, as expected, CO molecule has very low value of dipole moment (= 0.12 D).

2. Structure and geometry of CO_2 molecule

Following electron dot structures have been suggested for CO_2 molecule.

$$:\ddot{\text{O}}=\text{C}=\ddot{\text{O}}: \qquad :\overset{-}{\underset{\cdot\cdot}{\ddot{\text{O}}}}-\text{C}\equiv\overset{+}{\text{O}}$$

Structure (I) (*Covalent structure*) Structure (II) (*Ionic structure*)

The following points may be noted from these structures:

(*i*) All the three atoms in each structure are surrounded by an octet of electrons. Thus both the structures obey octet rule.

(*ii*) Structure (I) is covalent while structure (II) ionic.

(*iii*) Since central carbon atom in both the structures is surrounded by two σ-*bps* and no lone pair of electrons, it is *sp* hybridised in both the structures.

It may be noted that (C–O) σ-bonds are obtained by [sp(C) – $2p_x$(O)] overlap while (C – O) π bonds result from [2*p* (C) – 2*p*(O)] overlap, *i.e.*, π–bond is a (*p* – *p*)π-bond.

Due to *sp* hybridisation of the central carbon atom and absence of any lone pair of electrons on this atom, CO_2 molecule has *linear shape*. Linear shape is also confirmed by the fact that dipole moment of CO_2 molecule is zero.

Resonance in CO_2 molecule. We have said above that CO_2 molecule has linear shape. Experiments have shown that both carbon-oxygen bond lengths in this molecule are equal, each bond being equal to 1.15 Å. This value of bond length is shorter than the calculated C = O double bond length (= 1.22 Å) and larger than the calculated C ≡ O triple bond length (= 1.10 Å). In order to account for the discrepancy between the experimental and calculated values of bond lengths, it has been suggested that none of the above two structures is the true structure of CO_2 molecule. The actual structure is said to be a resonance hybrid of the following contributing structures:

$$:\ddot{O}_a{=}C{=}\ddot{O}_b: \longleftrightarrow :\overset{-}{\ddot{\underset{..}{O}}}_a - C \equiv \overset{+}{\ddot{O}}_b \longleftrightarrow \overset{+}{\ddot{O}}_a \equiv C - \overset{-}{\ddot{\underset{..}{O}}}_b:$$
$$\text{(A)} \qquad\qquad \text{(B)} \qquad\qquad \text{(C)}$$

Structure (A) is covalent while structures (B) and (C) are ionic structures. The dipoles of structures (B) and (C) are equal and act in opposite directions so that the resultant dipole moment of the actual structure becomes zero. Observed dipole moment of CO_2 molecule is also equal to zero. Thus it can be concluded that structures (B) and (C) make equal and opposite contributions to the overall structure of this molecule. The shortening of the carbon-oxygen bond lengths mentioned above is due to the stabilisation acquired through resonance.

Comparision between the structure of CO_2 and SiO_2 molecules

Although the central atom in both the oxides belongs to the same group *viz.*, IV A, these molecules have different structures as is evident from Fig. 15.1.

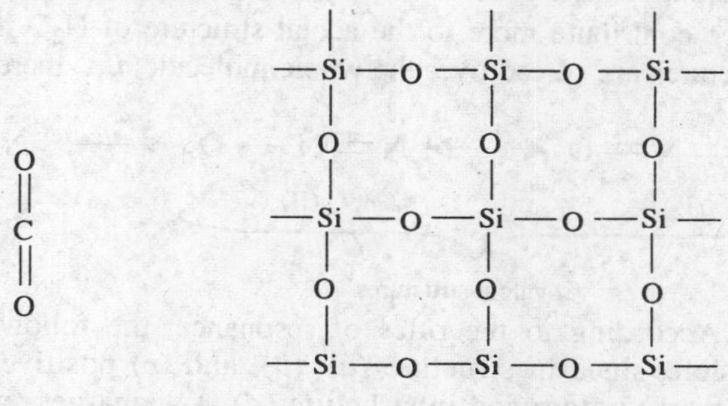

(*a*) (*b*)

Fig. 15.1 : (*a*) Linear monomeric structure of CO_2 molecule (*b*) Infinite three–dimensional polymeric structure of SiO_2 molecule.

Fig. 15.1 shows that while CO_2 molecule has a monomeric linear structure (O = C = O), SiO_2 molecule does not have O = Si = O structure; rather it has an infinite three dimensional polymeric structure in which a large number of SiO_4 tetrahedrons are linked together. The linking of various SiO_4 tetrahedrons gives a giant (polymeric) structure to SiO_2 molecule. In each SiO_4 tetrahedron, Si-atom is attached with four O-atoms by Si—O σ-bonds. In the structure of CO_2 molecule C–atom is attached with each of the two O-atoms by one C—O σ-bond and one C—O π-bond while in the structure of SiO_2 molecule each Si-atom is attached with each of the four O-atoms only by one Si—O σ-bond and Si — O π-bond is absent. Thus, we see that in the structure of SiO_2 molecule, Si–atom is not able to form Si — O π-bond. The cause of the nonformation of Si — O π-bond in SiO_2 molecule has been given as follows.

Explanation. We have already stated that in CO_2 molecule C—O π-bond results by the overlap between singly-filled 2*p* orbital of C-atom and singly-filled 2*p* orbital on O-atom, *i.e.*, C — O π-bond is a 2*p* (C) – 2*p* (O) or (*p* — *p*) π -bond. Since the size and energy of 2*p* orbitals of C and O-atoms are nearly the same, they overlap effectively and form (*p* — *p*) π-bond in CO_2 molecule. On the other hand in SiO_2 molecule, since the 3*p* orbitals of Si-atom and 2*p* orbitals of O-atom have appreciable difference in their size and energy, they cannot overlap effectively and hence cannot form 3*p*(Si) – 2*p*(O) π-bond. The tetracovalency of Si-atom in SiO_2 molecule is satisfied by forming four Si—O σ-bonds while the tetracovalency of C-atom in CO_2 molecule is satisfied by forming two C = O double bonds with two O-atoms. In the giant structure of silica, since each O-atom is attached with two Si-atoms, each Si-atom has only half share of each of the four O-atoms attached with it and consequently the ratio of Si to O-atom in silica molecule is as $1 : 1/2 \times 4 = 1:2$. Thus the formula of silica is SiO_2.

CO_2 is a gas while SiO_2 is a solid of great hardness and high melting point. The difference in these properties is because of the difference in the structure of CO_2 and SiO_2 molecules. Since the giant structure of SiO_2 is very stable, great amount of energy is needed to break Si—O bonds present in the structure of SiO_2 molecule. This explains the fact that SiO_2 is a solid with high hardness and high melting point. Because of the monomeric structure of CO_2 molecule, this oxide exists as a gas.

Oxides of nitrogen and ions derived from them

Here we shall consider the structure of N_2O, NO, NO_2, N_2O_4, N_2O_3 and N_2O_5 molecules and NO^- and NO_2^+ ions.

1. Structure and shape of nitrous oxide molecule, N_2O

Molecular spectroscopic study of this molecule has shown that the three atoms in this molecule are arranged in such a way that two N-atoms are directly linked together, *i.e.*, the three atoms are arranged in the order N—N—O and not as N—O—N. It has been suggested that the true structure of N_2O molecule is a resonance hybrid of the following four structures two of which [*viz.* structures (I) and (II)] are covalent while the remaining two [*i.e.*, structures (III) and (IV)] are ionic and highly polar. All these structures obey octet rule. N_2O molecule has a small value of its dipole moment equal to 0.16 D. This small value of dipole moment can be explained only on the basis of structures (III) and (IV), since the dipole moments in these structures act in opposite directions. Thus out of these four structures, (III) and (IV) structures (ionic structures) are more important and hence contribute more to the actual structure of N_2O molecule. + and – charges shown in ionic structures are spread over the whole molecule, *i.e.*, there is delocalisation of the charge.

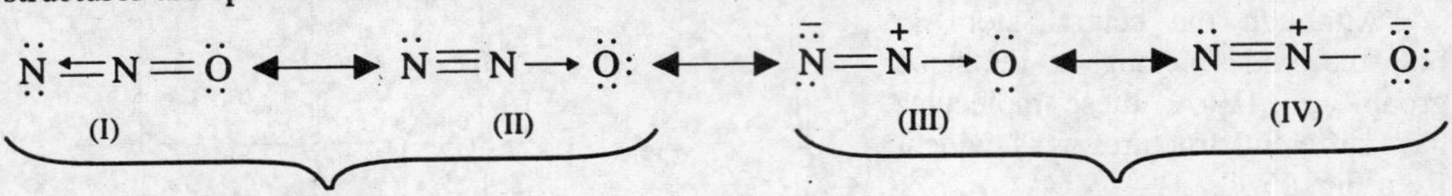

(I) (II) (III) (IV)

Covalent structures — Ionic structures

According to the rules of resonance, the following structures of N_2O molecule are not possible, since in structure (*a*), (*b*), and (*c*) positive charge is present on the more electronegative O-atom and in structure (*d*), + + charges (which are like charges) are present on the adjacent N and O-atoms.

(*a*) (*b*) (*c*) (*d*)

The following points may be noted from the structures namely (I), (II), (III) and (IV):

(*i*) All the structures obey octet rule, since the valence-shell of each atom in each structure is surrounded by 8 electrons (octet of electrons).

(*ii*) Out of structures (III) and (IV), structure (IV) is more favourable, since O-atom (more electronegative element) is having negative charge on it in this structure.

(*iii*) Since the central N-atom or N^+ ion in each structure is surrounded by two bonding electron pairs (σ-*bps* = 2) and no lone pair of electron (*lp* = 0), this atom is *sp* hybridised in each structure. (N — O) or (N—N) σ-bonds in all the structures are obtained by the overlap between the half-filled *sp* hybrid orbital on the central N-atom and half-filled 2*p* orbital on O-atom or N-atom as the case may be. π-bond is obtained by (*p*—*p*) overlap. N → N or N → O coordinate bond is obtained when *sp* hybrid orbital donates its electron pair to the vacant 2*p* orbital of N-atom (N → N coordinate bond) or of O-atom (N → O coordinate bond).

Shape of N_2O molecule. Since the central N-atom or N^+ ion in each structure is surrounded by two σ-*bps* and *no lp*, each of the structures has a *linear shape*. In ionic structures, nitrogen-nitrogen and nitrogen-oxygen bond lengths are 1.13 Å and 1.19 Å respectively.

2. Structure of nitric oxide molecule, NO

(*a*) Structure of NO molecule in the gaseous state. NO molecule has 11 electrons (N = 5, O = 6) (odd number of electrons) in its valence-shell and hence it is not possible to write a structure which may contain all the eleven electrons in the paired state. It is, therefore, suggested that NO molecule has one unpaired electron which may be present either on N-atom or on O-atom. Thus NO molecule is an odd-electron molecule which can be represented by any of the following two structures.

$$:\dot{N}=\ddot{O}: \qquad :\overset{-}{\ddot{N}}=\overset{+}{\dot{O}}:$$

Structure (I) (Covalent structure) — Structure (II) (Ionic structure)

Structure (I). In this structure, since N-atom is surrounded by one σ-bonding electron pair and one lone pair of electrons, it is *sp* hybridised. Similarly O-atom which is surrounded by one σ-bonding electron pair and two lone pairs of electron is sp^2 hybridised. In their hybridised state the valence-shell configuration of N and O atoms can be written as:

N-atom in *sp* hybridised state : $(sp)^2\ (sp)^1\ 2p_y^1\ 2p_z^1$

O-atom in sp^2 hybridised state : $(sp^2)^2\ (sp^2)^2\ (sp^2)^1\ 2p_z^1$

In structure (I) (N—O)σ-bond is obtained by [*sp* (N)–sp^2 (O)] overlap while (N—O) π-bond is formed by [2*p* (N) – 2*p* (O)] overlap, *i.e.*, π-bond is a (*p*—*p*) π-bond. Unpaired electron on N-atom is residing in 2*p* orbital of this element while lone pair of electrons is residing in *sp* orbital. Both lone pairs of electrons of O-atom are present in sp^2 hybrid orbitals.

sp hybridisation of N-atom in NO molecule shown as above indicates that the structure of NO molecule can be shown as depicted in Fig. 15.2.

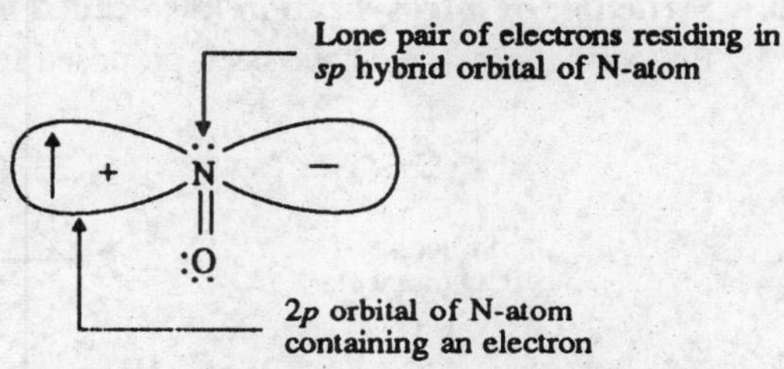

Fig. 15.2 : Structure of NO molecule. The lone pair present on N-atom resides in *sp* hybrid orbital while the unpaired electron resides in 2*p* orbital of N-atom.

Structure II. In this structure, $\overset{-}{N}$ in sp^2 hybridised and $\overset{+}{O}$ is *sp* hybridised. The valence-shell configurations of $\overset{-}{N}$ and $\overset{+}{O}$ in sp^2 and *sp* hybridised state respectively are given below:

$\overset{-}{N}$ in sp^2 hybridised state : $(sp^2)^2\ (sp^2)^2\ (sp^2)^1\ 2p_z^{\ 1}$

$\overset{+}{O}$ in *sp* hybridised state : $(sp)^2\ (sp)^1\ 2p_y^{\ 1}\ 2p_z^{\ 1}$

$(\overset{-}{N} - \overset{+}{O})$σ-bond is obtained by [$sp^2$ $(\overset{-}{N})$ – *sp* $(\overset{+}{O})$] overlap and $(N — \overset{+}{O})$π -bond is established by [$2p(\overset{-}{N})$ – $2p(\overset{+}{O})$] overlap, *i.e.*, π -bond is a (*p*—*p*)π -bond. Lone pairs of electrons on $\overset{-}{N}$ and $\overset{+}{O}$ are present in their respective sp^2 hybrid orbitals. Unpaired electron on $\overset{+}{O}$ is present in its 2*p* orbital.

Location of odd-electron and paramagnetic character of NO. We have seen above that the unpaired electron (called odd electron) on N-atom in structure (I) and on $\overset{+}{O}$ in structure (II) resides in 2*p* orbital. As a matter of fact the unpaired electron is not localised on N–atom or on $\overset{+}{O}$, but is spread over the whole molecule. It is due to the presence of unpaired electron that gaseous NO molecule shows paramagnetic character. Due to the presence of an odd electron in gaseous NO molecule, this molecule should behave as an odd electron molecule. In fact, it does not show the properties of an odd electron molecule.

Shape of NO molecule. It is due to *sp* hybridisation of N–atom that NO molecule ($:\dot{N}=\ddot{O}:$) has *linear shape.*

Resonance in NO molecule. Nitrogen-oxygen bond distance in the gaseous NO molecule has been found to be equal to 114 pm. This distance is intermediate between the N = O double bond distance (=115 pm) and N ≡ O triple bond distance (= 105 pm) In order to explain nitrogen—oxygen bond distance equal to 114 pm, it has been assumed that the true structure of NO mole in the gaseous state is a resonance hybrid of structures (I) and (II).

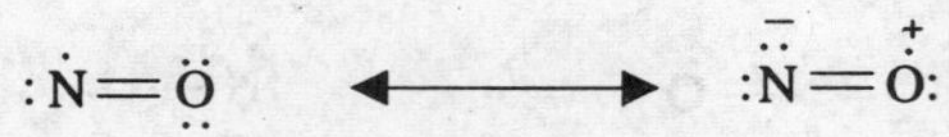

Structure (I) (Covalent structure) Structure (II) (Ionic structure)

(b) Structure of NO molecule in the solid and liquid state. In the solid and liquid state, two NO molecules associate together and form a rectangular loose dimer which has the structure shown in the margin. In the formation of this dimer the two unpaired electrons become paired and hence the dimer shows diamagnetic property. Thus nitric oxide in the solid and liquid state is diamagnetic while the same in the gaseous state is paramagnetic.

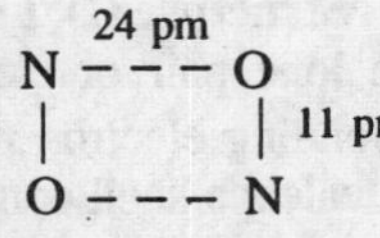

Nitric oxide dimer

3. Structure of nitrosyl cation (also called nitrosonium cation), $\overset{+}{NO}$

Following structures have been proposed for this ion.

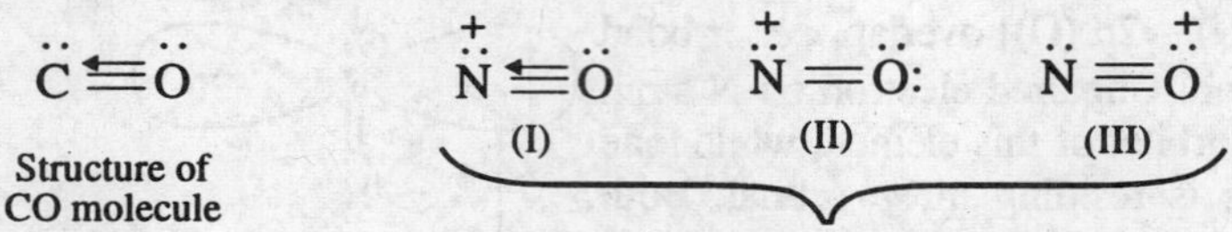

Structure of CO molecule — Structures of NO^+ ion

Structure I. When NO molecule loses an electron, it becomes $\overset{+}{NO}$ ion. This ion is iso-electronic with CO molecule, since both the species have the same number of electrons equal to 10. Thus $\overset{+}{NO}$ and CO have the same structure as shown above. In structure (I), $\overset{+}{N}$ [$\overset{+}{N} = 2s^2\ 2p^2$] is *sp* hybridised [$\overset{+}{N} = (sp)^2\ (sp)^1\ 2p_y^1\ 2p_z^0$]. ($\overset{+}{N}$ — O) σ-bond is formed by *sp* ($\overset{+}{N}$) – 2*p* (O) overlap. One ($\overset{+}{N}$ — O) π -bond is obtained by $\overset{+}{N}$ (*p*) — O (*p*) overlap while (O→ $\overset{+}{N}$) coordinate bond is formed by the donation of $2p^2$ electron pair of O-atom to the vacant orbital of $\overset{+}{N}$. Lone pair of electrons on $\overset{+}{N}$ resides in its *sp* hybrid orbital.

Structure II. This structure can be supposed to have been obtained when one unpaired electron residing on N-atom in NO molecule is removed.

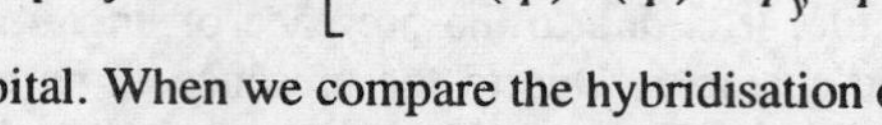

Structure of NO molecule — Structures of NO^+ ion (II)

In the structure of NO molecule, N–atom is *sp* hybridised. Similarly in the structure of $\overset{+}{NO}$ ion [structure (II)], $\overset{+}{N}$ is also *sp* hybridised. $\left[\overset{+}{N} = (sp)^2\ (sp)^1\ 2p_y^1\ 2p_z^0\right]$. Lone pair of electrons on $\overset{+}{N}$ resides in its *sp* hybrid orbital. When we compare the hybridisation of $\overset{+}{N}$ in structures (I) and (II), we find that $\overset{+}{N}$ in both the structures is *sp* hybridised and lone pair of electrons is present in one of the two *sp* hybrid orbitals.

Structure III. In this structure N-atom is *sp* hybridised [$N = (sp)^2(sp)^1\ 2p_y^1\ 2p_z^1$]. Lone pair of electrons on N-atom is present in one of the two *sp* hybrid orbitals. (N—O) σ-bond is obtained by [N(*sp*) – $\overset{+}{O}$ (2*p*)] overlap while both (N — $\overset{+}{O}$) π -bonds are (*p* — *p*) π -bonds.

4. Structure of $\overset{-}{NO}$ ion

This ion is found in bent metal nitrosyls which are considered to be complex compounds containing $\overline{\text{N}}\text{O}$ ion as a ligand. Negative charge on NO ion can be considered to be present on N-atom and hence electron dot structure of this ion can be written as $:\ddot{\overline{\text{N}}} = \ddot{\text{O}}:$ which shows that $\overline{\text{N}}$ whose configuration is $2s^2\ 2p^4$ is sp^2 hybridised [$\overline{\text{N}} = (sp^2)^2\ (sp^2)^2\ (sp^2)^1\ 2p_z^1$]. Two lone pairs of electrons on $\overline{\text{N}}$ are situated in two sp^2 hybrid orbitals. ($\overline{\text{N}}$ — O) σ-bond is obtained by $\overline{\text{N}}$ (sp^2) — O (p) overlap while ($\overline{\text{N}}$ — O) π-bond results by $\overline{\text{N}}$ (p) – O(p) overlap.

5. Structure of nitrogen dioxide molecule, NO_2

Like NO, NO_2 molecule also has an odd number of electrons (5 + 2 × 6 = 17) in its valence-shell and hence has one unpaired electron which resides on N-atom. It is due to the presence of one unpaired electron on N-atom that NO_2 molecule shows paramagnetic character. Thus NO_2 molecule is an odd-electron molecule and its structure can be represented by any of the following two *angular structures.*

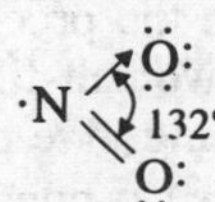

Structure (A)
(Covalent structure)

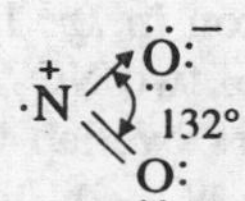

Structure (B)
(Ionic structure)

In each of these structures, N-atom (central atom) is sp^2 hybridised.

N-atom in sp^2 hybridised state in structure (A) : $(sp^2)^2\ (sp^2)^1\ (sp^2)^1\ 2p_z^1$

$\overset{+}{\text{N}}$-atom in sp^2 hybridised state in structure (B) : $(sp^2)^1\ (sp^2)^1\ (sp^2)^1\ 2p_z^1$

It is due to sp^2 hybridisation of the central N-atom that NO_2 molecule has *angular shape* (*V–shaped structure*) with ONO bond angle equal to 132°. It may also be noted that this angle is larger than the expected trigonal planar angle (= 120°) corresponding to sp^2 hybridisation. It may be seen from sp^2 hybridisation scheme that in both the structures unpaired electron resides in one of the three sp^2 hybrid orbitals. It is due to the presence of this unpaired electron that NO_2 molecule behaves as an odd electron molecule, *i.e.*, NO_2 molecule is colourless and paramagnetic. It may also be noted that N–O π -bond in both the structures is a (p—p) π-bond.

sp^2 hybridisation scheme indicates that the structure of NO_2 molecule can be represented by any of the two structures given in Fig. 15.3. In these structures sp^2 hybrid orbital of N or $\overset{+}{\text{N}}$ which contains one unpaired electron has also been shown.

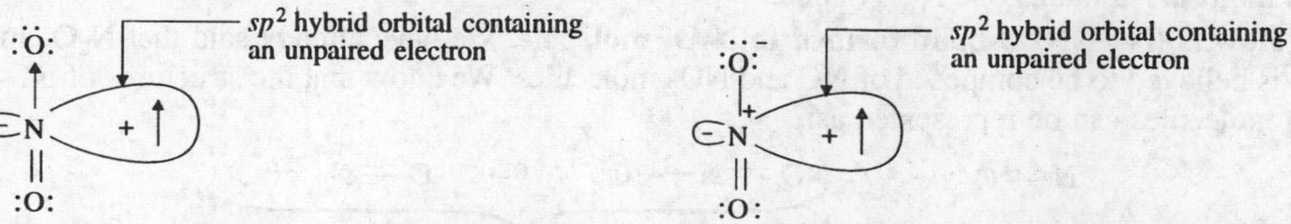

Fig 15.3 : Structure of NO_2 molecule. sp^2 hybrid orbital on N or N^+ contain one unpaired electron.

Resonance in NO_2 molecule. The experimental value of nitrogen-oxygen bond distnace in NO_2 molecule is 1.19 Å. This value is intermediate between the calculated N—O single bond distance (= 1.36 Å) and N = O double bond distance (= 1.15 Å). In order to explain the discrepancy between the experimental and calculated nitrogen-oxygen bond distances, it has been suggested that the true structure of NO_2 molecule is a resonance hybrid of the following four structures. (I) and (II) are covalent structures while (III) and (IV) are ionic structures.

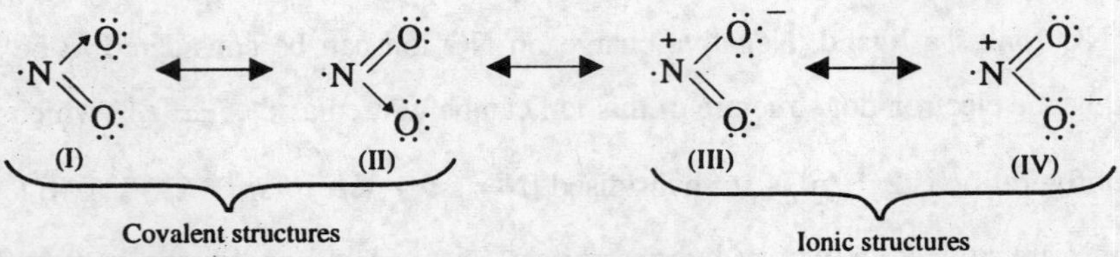

6. Structure of dinitrogen tetraoxide molecule, N_2O_4

When two NO_2 molecules dimerise together, we get N_2O_4 molecule. Thus the structure of N_2O_4 molecule can be represented by any of the following two structures:

How is N—N bond formed? We have seen above that in NO_2 molecule, N-atom (central atom) has one unpaired electron in sp^2 hybrid orbital. In the process of the formation of N_2O_4 molecule by the dimerisation of two NO_2 molecules, the two half-filled sp^2 hybrid orbitals on two N-atoms of two NO_2 molecules overlap with each other and give rise to the formation of N—N bond in N_2O_4 molecule. Thus N — N bond is a $(sp^2–sp^2)$ σ-bond (See Fig. 15.4). It is due to $(sp^2–sp^2)$ overlap that the two unpaired electrons in two sp^2 hybrid orbitals become paired. Thus N_2O_4 molecule has no unpaired electrons and hence shows diamagnetic character.

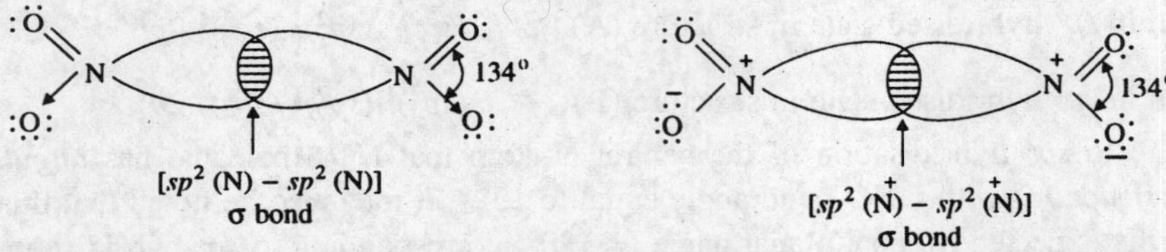

Fig. 15.4 : Formation of N — N σ-bond in N2O4 molecule by the overlap of two half-filled sp^2 hybrid orbitals on two N-atoms.

7. Structure of dinitrogen trioxide molecule, N_2O_3 (N_2O_3 is also called nitrogen sesquioxide)

Since in the liquid and gaseous state, N_2O_3 molecule dissociates to give nitric oxide (NO) and nitrogen dioxide (NO_2) ($N_2O_3 \rightleftharpoons NO + NO_2$), N_2O_3 molecule is believed to contain NO and NO_2 molecules which are linked together by a N—N bond existing between two N-atoms belonging to NO and NO_2 molecules respectively, *i.e.*, N_2O_3 molecule can be represented as ON – NO_2 or N_2O_3. This molecule contains one N—N bond.

How is (N — N) σ-bond formed in N_2O_3 molecule. We have already said that N_2O_3 molecule is believed to be composed of NO and NO_2 molecules. We know that the structures of NO and NO_2 molecules can be represented as:

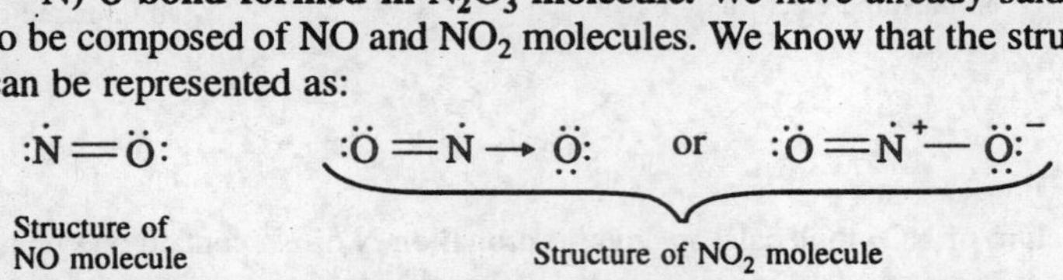

In NO molecule N-atom is *sp* hybridised and the unpaired electron present on N-atom resides in $2p_z$ orbital of this atom. The pair of electrons present on N-atom resides in one of the two *sp* hybrid orbitals.

In case of NO_2 molecule, N/$\overset{+}{N}$ atom is sp^2 hybridised and the unpaired electron present on this atom resides in one of the three sp^2 hybrid orbitals.

When half-filled $2p_z$ orbitals of N-atom belonging to NO molecule overlaps with the half-filled sp^2 hybrid orbital on N/$\overset{+}{N}$ atom of NO_2 molecule, $[2p_z\,(N) - sp^2\,(N)]$ σ-bond is obtained. Thus (N — N) σ-bond results by $(2p_z - sp^2)$ overlap (See Fig. 15.5).

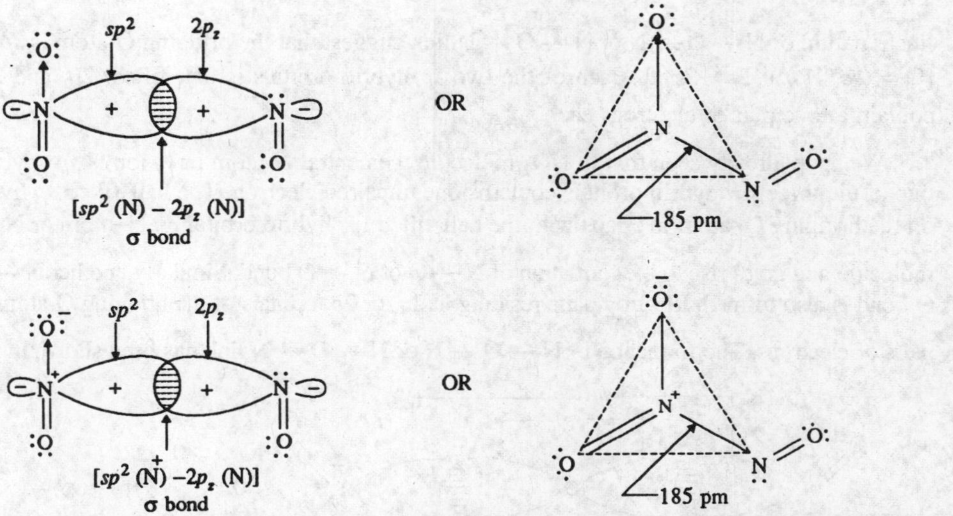

Fig 15.5 : Formation of (N—N) or (N—N^+) σ-bond in N_2O_3 molecule by the overlap between half-filled sp^2 hybrid orbital on N/N^+ atom of NO_2 molecule and half–filled $2p_z$ orbital on N-atom in NO molecule.

Since the two unpaired electrons on two N-atoms become paired in forming N—N bond in N_2O_3 molecule, N_2O_3 molecule has no unpaired electron and hence shows diamagnetic character. N—N bond length has been found to be equal to 185 pm.

The structure of N_2O_3 molecule shown in Fig. 15.5 indicates that it has a planar structure with one planar N-atom and one angular N-atom. Planar N-atom comes from NO_2 molecule and hence is sp^2 hybridised. On the other hand the angular N-atom comes from NO molecule and is, therefore, *sp* hybridised. Planar N-atom has been shown at the centre of the triangle while the angular N-atom occupies one of the vertices of the triangle. Lone pair of electrons shown at the angular N-atom resides in one of the two *sp* hybrid orbitals.

Tracer studies made by Liefer (1940) have shown that when normal NO (*i.e.* ^{14}NO) and NO_2 enriched with ^{15}N isotope (*i.e.*, $^{15}NO_2$) are mixed together, a rapid exchange between ^{14}N and ^{15}N atoms takes place.

$$^{14}NO + {}^{15}NO_2 \rightleftharpoons {}^{15}NO + {}^{14}NO_2$$

This rapid exchange of N-atoms indicates that the two N-atoms in N_2O_3 molecule have equivalent position, *i.e.*, the structure of N_2O_3 molecule can be shown as depicted in the margin. This structure has also been confirmed by the infra–red spectroscopic studies of the crystals of N_2O_3.

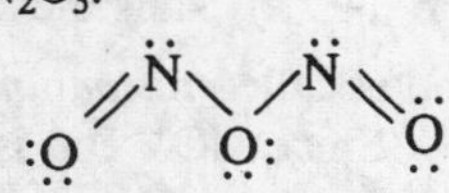

8. Structure of dinitrogen pentoxide molecule, N_2O_5

(a) Structure of N_2O_5 molecule in the gaseous state. In the *gaseous state*, N_2O_5 molecule can be regarded as being formed by combining two NO_2 molecules through one O-atom which acts as a bridge between two N-atoms of two NO_2 molecules. Thus the structure of N_2O_5 molecule can be written as shown below:

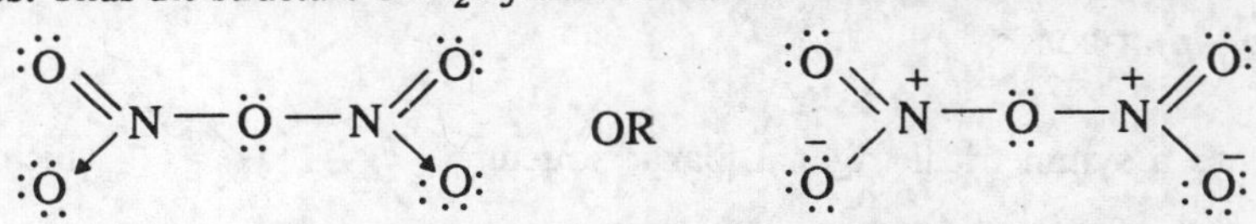

Symmetrical structure of N_2O_5 molecule in the gaseous state.

How is N—O—N or $\overset{+}{N}$—O—$\overset{+}{N}$ link formed in N_2O_5 molecule? The symmetrical structure of N_2O_5 molecule given above shows that N—O—N or $\overset{+}{N}$—O—$\overset{+}{N}$ link is almost *linear*. The linear nature of N—O—N or $\overset{+}{N}$—O—$\overset{+}{N}$ link suggest that the bridging O-atom is *sp* hybridised [O = $(sp)^1$ $(sp)^1$ $2p_y^2$ $2p_z^2$]. Each of the two *sp* hybrid orbitals is half-filled. $2p_y$ and $2p_z$ orbitals contain a lone pair of electrons each.

We have already seen that in NO_2 molecule the central N-atom or $\overset{+}{N}$ ion is sp^2 hybridised and one of the three sp^2 hybrid orbitals contains one unpaired electron. The half-filled *sp* hybrid orbital on the bridging O–atom overlaps with the half-filled sp^2 hybrid orbital on N–atom or N ion of NO_2 molecule and gives rise to the formation of N—O or $\overset{+}{N}$—O bond. Similarly another N—O or $\overset{+}{N}$—O bond is also formed. Electron pairs residing in $2p_y$ or $2p_z$ orbitals of the bridging O-atom act as lone pairs of electrons. The formation of N—O—N or $\overset{+}{N}$—O—$\overset{+}{N}$ link has been shown in Fig. 15.6.

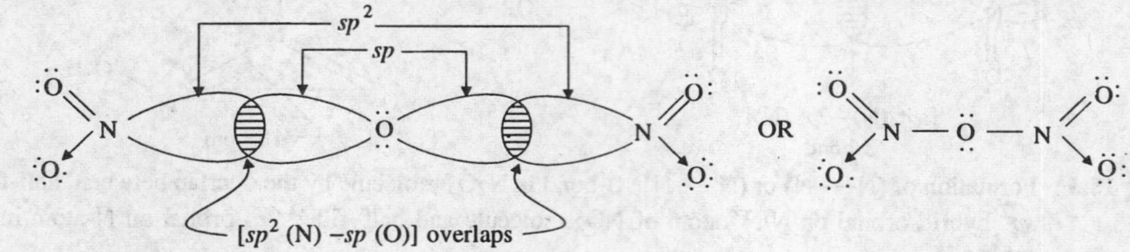

(*a*)

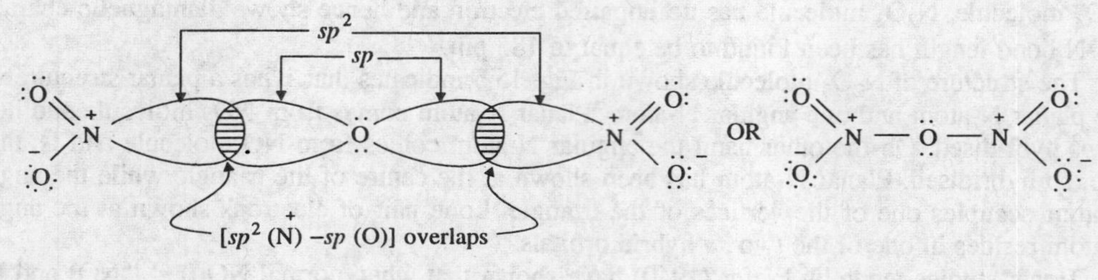

(*b*)

Fig 15.6 : Formation of N—O—N or $\overset{+}{N}$ — O—$\overset{+}{N}$ links in N_2O_5 molecule.

(*b*) Structure of N_2O_5 molecule in the solid state. The gaseous N_2O_5 changes to solid state at low temperature. X-ray studies have shown that in the solid state, N_2O_5 molecule behaves as an ionic compound which consists of *nitronium* (also called *nitryl* or *nitrosyl*) ion (NO_2^+) and *nitrate* ion (NO_3^-). Thus the structure of N_2O_5 molecule in the solid can be represented as $[NO_2]^+$ $[NO_3]^-$ and is called *nitronium nitrate.*

Since NO_2^+ ion is isoelectronic with CO_2 molecule [Both species have the same number of electrons (= 16)], this ion, like CO_2 molecule, has a *linear structure* (:O = $\overset{+}{N}$ = O :) in which $\overset{+}{N}$ is *sp*-hybridised. [$\overset{+}{N}$ = $(sp)^1$ $(sp)^1$ $2p_y^1$ $2p_z^1$]. N—O bond length is equal to 115 pm. In $\overset{+}{N}O_2$ ion both π-bonds are (*p* – *p*) π-bonds.

NO_3^- ion has a symmetrical trigonal planar structure, 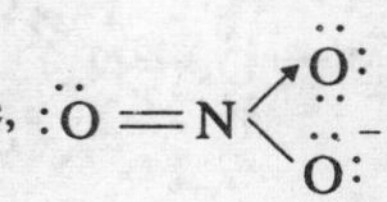 which shows that N-atom is sp^2 hybridised [N = $(sp^2)^2$ $(sp^2)^1$ $(sp^2)^1$ $2p_z^1$]. Thus N → O coordinate bond is obtained by the donation of the electron pair residing in sp^2 hybrid orbital of N-atom to the vacant 2*p* orbital of O-atom. N—O π bond is a (*p*—*p*) π -bond.

The above discussion shows that the structure of N_2O_5 molecule in the solid can be represented as:

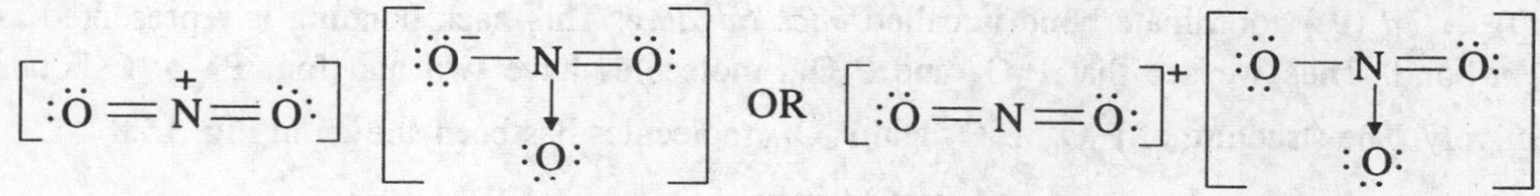

OR $\left[NO_2\right]^+\left[NO_3\right]^-$

Oxides of phosphorus

The well known oxides formed by phosphorus are phosphorus trioxides (P_2O_3), phosphorus tetroxide (P_2O_4) and phosphorus pentoxide (P_2O_5). The vapour density of phosphrous trioxide and the lowering of the freezing point of its solution in benzene suggest that this oxide exists as a dimer, $(P_2O_3)_2$ or P_4O_6. The vapour density of phosphorus tetroxide at 1800°C indicates that this oxide also exists as a dimer, $(P_2O_4)_2$ or P_4O_8. Similarly the vapour density of phosphorus pentoxide at 1400°C corresponds to the dimeric formula, $(P_2O_5)_2$ or P_4O_{10}. Each of these oxides has tetrahedral shape. The four corners of the tetrahedron are occupied by four P-atoms. Six O-atoms in all the three oxides make a bridge between two P-atoms of P_4 tetrahedron. These six O-atoms are called bridging O-atoms. Thus we find that each P-atom is attached covalently with three bridgning O-atoms and each bridging O-atom is covalently linked with two P-atoms. In P_4O_8 molecule, the two extra O-atoms (called *terminal* O-atoms) are attached with two P-atoms by P $\rightleftarrows$ O bonds while in P_4O_{10} molecule the four terminal O-atoms are attached with four P-atoms by P $\rightleftarrows$ O bonds. P_4O_6 molecule has no terminal O-atom.

In all the three oxides, each P-atom is sp^3 hybridised [P = $(sp^3)^2$ $(sp^3)^1$ $(sp^3)^1$ $(sp^3)^1$ $3d^0$]. It may be noted from sp^3 hybridisation scheme of P-atom given above that one sp^3 hybrid orbital contains a lone pair of electrons while the remaining three sp^3 hybrid orbitals have one electron each. One 3*d* orbital of P-atom which is vacant remains unhybridised. The three singly-filled sp^3 hybrid orbitals of one P-atom overlaps with the singly-filled 2*p* orbitals of three different bridging O-atoms to give three P—O σ bonds. The electron pair residing in one of the four sp^3 hybrid orbitals acts as a lone pair (*lp*) of electrons on P-atom in P_4O_6 and P_4O_8 molecules. Each of the four P-atoms in P_4O_6 molecule has a *lp* of electrons while only two P-atoms in P_4O_8 have a *lp* of electrons each.

How are P $\rightleftarrows$ O bonds formed in P_4O_8 and P_4O_{10} molecules? We have said above that in P_4O_6 molecule each P-atom has one lone pair (*lp*) of electrons in its sp^3 hybrid orbital. In P_4O_8 molecule, two P-atoms donate the *lps* residing in sp^3 hybrid orbital to the *vacant* 2*p* orbitals of two terminal O-atoms (O = $2s^2\ 2p^0\ 2p^2\ 2p^2$) and form two [sp^3 (P) → *p* (O)] coordinate bonds. In P_4O_{10} molecule, all the four P-atoms donate the *lps* siding in sp^3 hybrid orbitals to the vacant *p*-orbitals of four terminal O-atoms to give rise to the formation of four [sp^3 (P) → 2*p* (O)] coordinate bonds. The formation of [sp^3 (P) → 2*p* (O)] coordinate bond has been shown in Fig. 15.7.

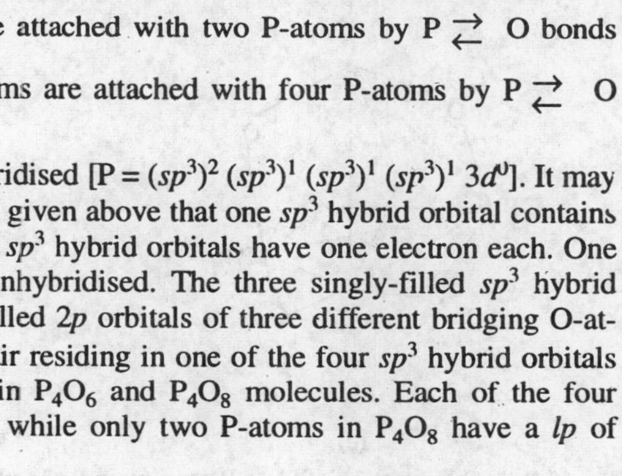

Fig. 15.7 : Formation of P → O bonding in P4O8 and P4O10 molecules.

In P_4O_8 and P_4O_{10} molecules P → O bond length has been found to be much shorter than the length of other P— O bonds in the same molecules. In order to explain the shorter value of P → O bond length it is assumed that in these molecules the filled 2*p* orbital of the terminal O-atom makes a lateral overlap with the vacant 3*d* orbital of P-atom and thus [2*p* (O) → 3*d* (P)] coordinate bond is formed. The donation of an electron pair residing in $2p$ orbital of the terminal O-atom to the vacant $3d$ orbital of P-atom is called back donation of electron and the formation of [$2p$ (O) $\rightarrow$ $3d$ (P)] coordinate bond is called *back bonding*. This back bonding is represented as (p–d) π-bond. Thus we see that P_4O_8 and P_4O_{10} molecules have two and four P $\rightleftarrows$ O-bonds respectively. The structure of P_4O_6, P_4O_8 and P_4O_{10} molecules has been shown in Fig. 15.8

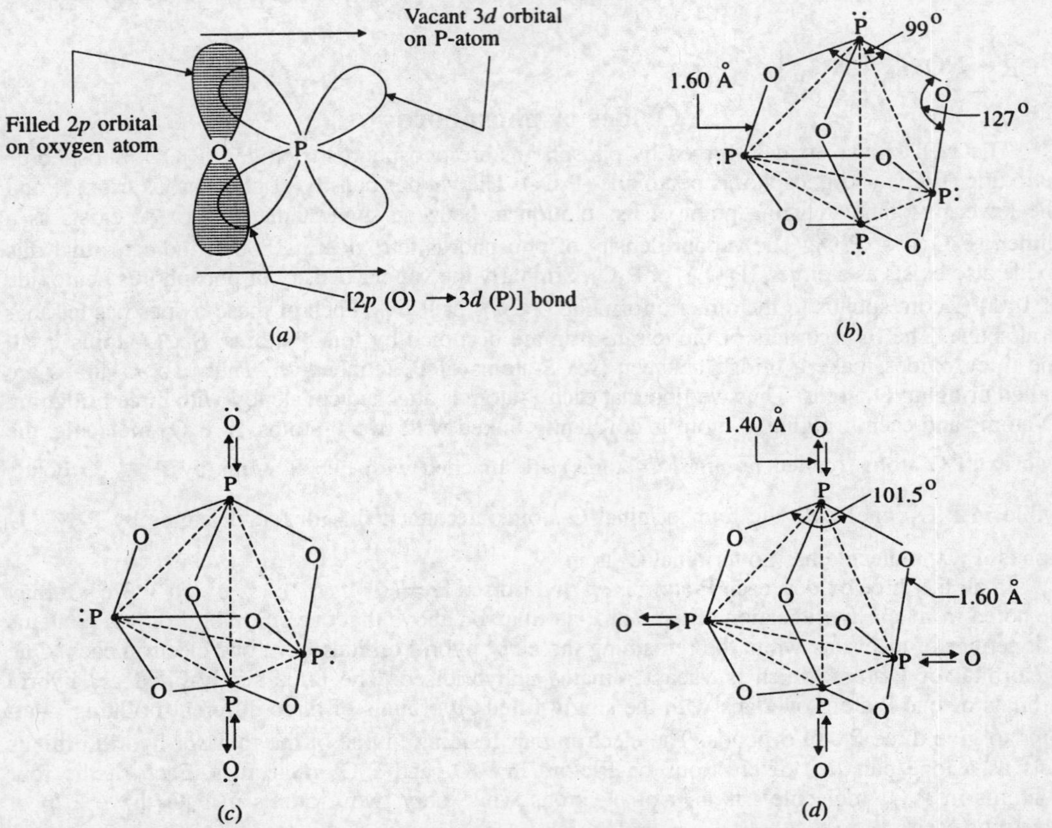

Fig. 15.8 : (*a*) Formation of O → P bonding in P_4O_8 and P_4O_{10} molecules (*b*) Structure of phosphorus trioxide molecule (P_4O_6) (*c*) Structure of phosphorus tetroxide molecule (P_4O_8) (*d*) structure of phosphorus pentoxide molecule (P_4O_{10}).

Oxides of MO_2 type (M = S, Se, Te, Po)/ in the gaseous and solid state

Here we shall discuss the structure of the following oxides:

1. Structure and geometry of SO_2 molecule.

In the gaseous as well as in the solid state, SO_2 exists as discrete molecule which can be represented by any of the following three structures. Structures (I) and (II) are *covalent structures* while structure (III) is an *ionic structure.*

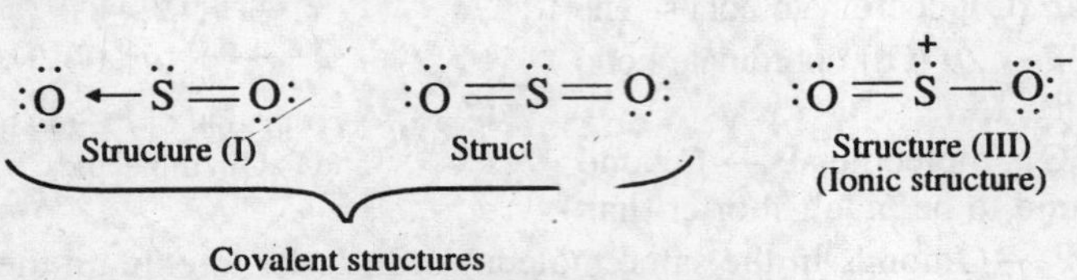

The following points may be noted from these structures:

(*i*) Structures (I) and (III) obey octet rule, since all the three atoms (one S-atom and two O-atoms) are surrounded by 8 electrons in these structures. Structure (II) *does not* obey octet rule because S-atom is surrounded by 10 electrons while each of the two O-atoms is surrounded by 8 electrons.

(*ii*) Since the central S-atom in all the three structures is surrounded by two *σ-bps* and one lone pair (*lp*) of electrons, this atom is sp^2 hybridised. sp^2 hybridisation of S-atom in these structures is shown below:

sp^2 hybridisation of S-atom in structure (I) : $(sp^2)^2\ (sp^2)^2\ (sp^2)^1\ 3p_z^1$

sp^2 hybridisation of S–atom in structure(II) : $(sp^2)^2\ (sp^2)^1\ (sp^2)^1\ 3p_z^1\ 3d^1$

sp^2 hybridisation of S^+ in structure (III) : $(sp^2)^2\ (sp^2)^1\ (sp^2)^1\ 3p_z^1$

In structure (I), (S—O) π-bond is a (*p–p*)π-bond. In structure (II) one (S—O) π-bond is a (*p—p*) π-bond while the other (S—O) π-bond is a (*d* – p) π-bond. In order to form (*d* – *p*) π-bond, 3*d* and 2*p* orbitals make as *side wise overlap* with each other. This type of sidewise overlap between *d* and *p* orbitals brings S and O-atoms closer to each other and thus accounts for the shorter length of S – O bond. The formation of (*p—p*) and (*d—p*) π-bonds in SO_2 molecule has been shown in Fig. 15.9. In structure (III), (S—O) π-bond is a (*p—p*) π-bond.

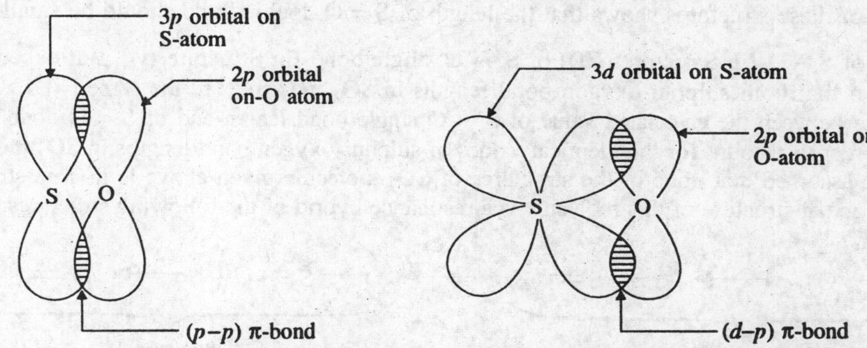

Fig. 15.9 : Formation of (*p–p*) and (*d–p*)π-bonds in SO_2 molecule (O = S = O).

Shape of SO_2 molecule and OSO bond angle. It is due to the presence of one lone pair of electrons in one of the three sp^2 hybrid orbitals and sp^2 hybridisation of S or S^+ atom that SO_2

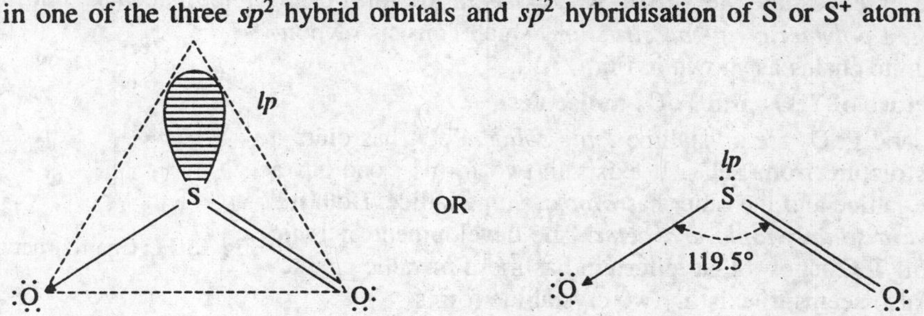

(I)

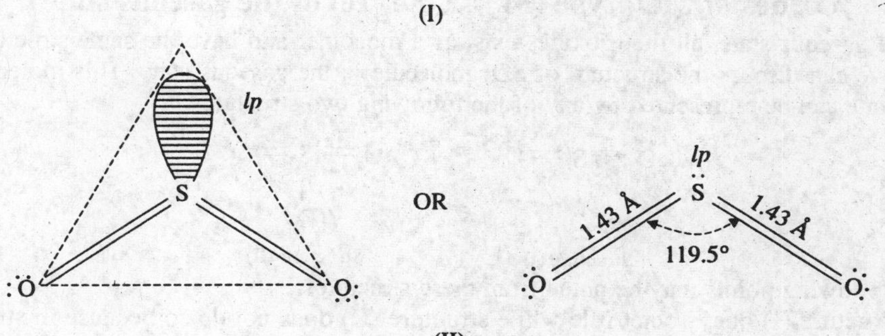

(II)

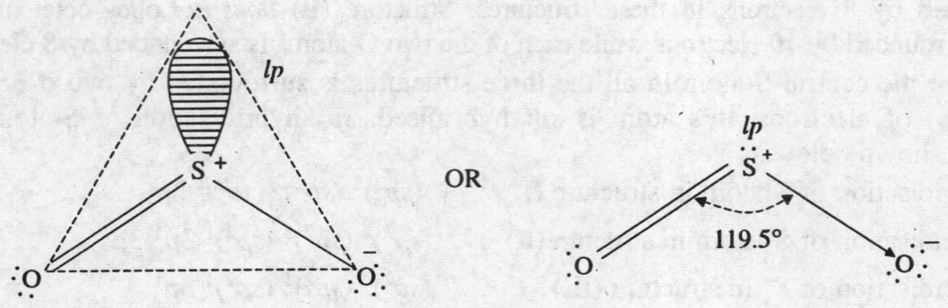

(III)

Fig. 15.10 : Angular shape of SO_2 molecule.

molecule has *angular* (*V-shaped*) structure. Since (*lp–bp*) repulsion is greater than (*bp–bp*) repulsion, OSO bond angle gets slightly reduced from the expected trigonal planar angle of 120° to 119.5° (See Fig. 15.10).

Resonance in SO_2 molecule. We have seen above that the structures (I) and (II) obey octet rule. Each of these structures shows that the length of S = O double bond should be smaller than the length of $\overset{+}{S} - \overset{-}{O}$ [in Structure (III)] or S → O single bond [in Structure (I)]. But experiments have shown that both sulphur-oxygen bond lengths in SO_2 molecule are the same (=1.43 Å) and intermediate between the calculated value of S — O single bond length and of S = O double bond length. In order to account for the identical values of sulphur-oxygen bond lengths in SO_2 molecule, it has been assumed that none of the structures of SO_2 molecule given above is its true structure. Rather the actual structure of this molecule is a resonance hybrid of the following structures:

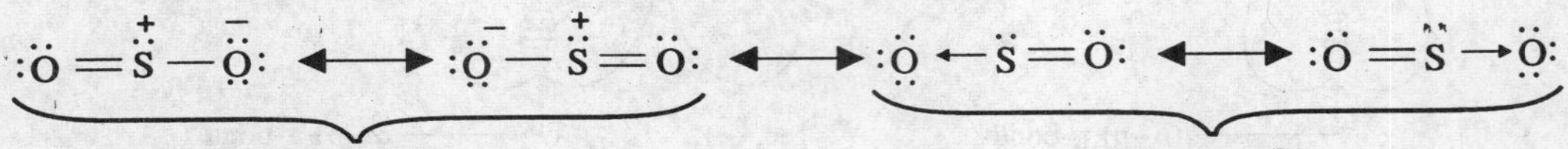

Structure (III) Structure (I)

2. Structure of SeO_2 molecule

In the gaseous state, SeO_2 like SO_2, exists in the form of discrete molecules, but in the solid state it has a *polymeric zig-zag structure* which consists of non-planar infinite chains as shown in Fig. 15.11.

3. Structure of TeO_2 and PoO_2 molecules.

TeO_2 and PoO_2 are crystalline *ionic solids*. TeO_2 has quite a different structure from SeO_2. It exists in two forms : one has a *rutile* type lattice and the other has *brookite* type lattice. Both the forms have *distinctly ionic character*. The development of ionic character in TeO_2 shows that tellurium has more metallic character. PoO_2 also seems to exist in two crystalline forms.

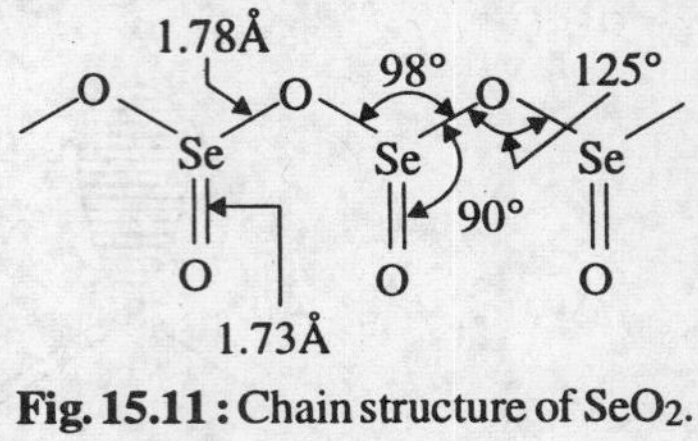

Fig. 15.11 : Chain structure of SeO_2.

Oxides of MO_3 type (M = S, Se, Te) in the gaseous state

In the gaseous state, all the trioxides exist as a monomer and have the same structure. As an example we can discuss the structure of SO_3 molecule in the gaseous state. This molecule in the gaseous state can be represented by any of the following two structures.

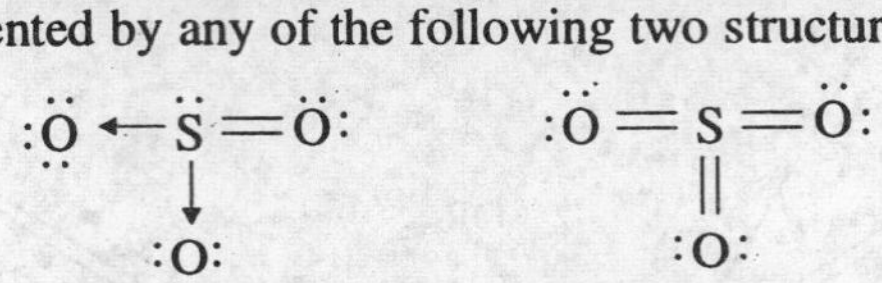

Structure (I) Structure (II)

The following points may be noted from these structures :

(*i*) Structure (I) obeys octet rule while structure (II) does not do so because in structure (I) each atom is surrounded by an octet of electrons. In structure (II) the central S-atom is surrounded by 12 electrons while each of the three O-atoms is surrounded by 8 electrons (octet of electrons).

(*ii*) Since the central S–atom in both the structures is surrounded by three *σ-bps* and no lone pair of electrons (*lps* = 0), this atom is sp^2 hybridised in both the structures.

S-atom in structure (I) : $(sp^2)^2\ (sp^2)^2\ (sp^2)^1\ 3p_z^1$

S-atom in structure (II) : $(sp^2)^1\ (sp^2)^1\ (sp^2)^1\ 3p_z^1\ 3d^1\ 3d^1$

In structure (I), (S — O) π-bond is a (*p—p*) π-bond. In structure (II), one (S — O) π-bond is a (*p—p*) π-bond while each of the remaining two (S—O) π -bonds is a (*p—d*) π-bond.

Shape of SO_3 molecule and OSO bond angle. It is due to sp^2 hybridisation of the central S-atom and absence of any lone pair of electrons on the central S-atom that SO_3 molecule has the expected *trigonal planar geometry* (See Fig. 15.12). Each of the three OSO bond angles is equal to 120° which is the expected angle for a *trigonal planar geometry.*

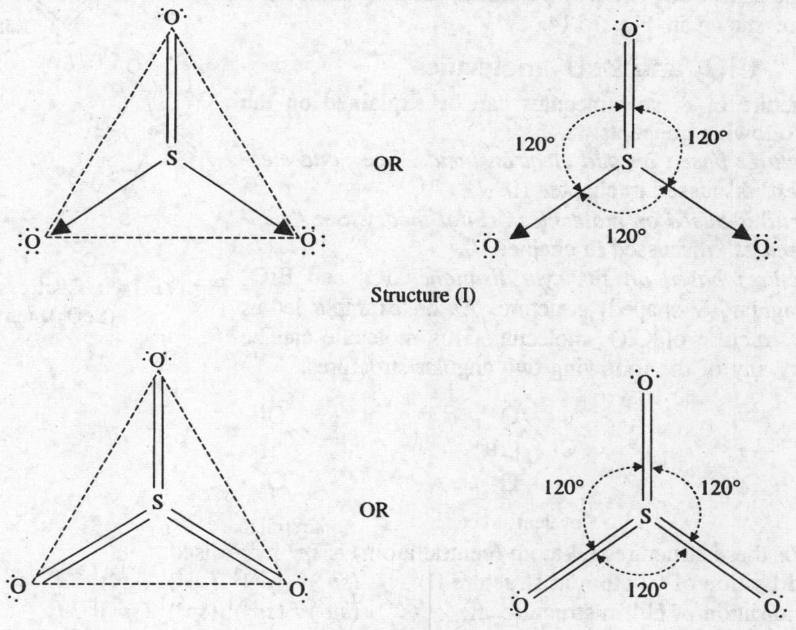

Structure (II)

Fig. 15.12 : Trigonal planar shape of SO_3 molecule.

Resonance in SO_3 molecule. We have seen above that it is only structure (I) which obeys octet rule. Experiments have shown that each of the three sulphur-oxygen bond length in this structure is equal to 1.42 Å. This value is intermediate between the calculated value of S — O single bond length and of S = O double bond length. In order to account for the observed value of sulphur-oxygen bond length in SO_3 molecule, it has been suggested that the true structure of SO_3 cannot be represented by structure (I), but it is a resonance hybrid of the following contributing structures :

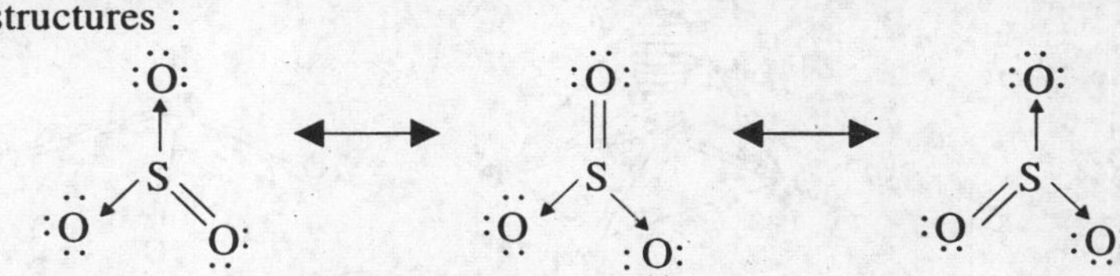

Oxides of MO_3 type (M = S, Se) in the solid state

In the solid state, α-form of SO_3 molecule gets polymerised and gives a trimer, $(SO_3)_3$ which

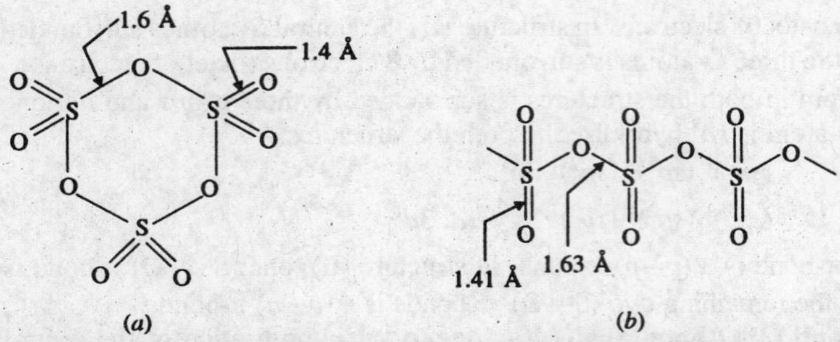

Fig. 15.13 : (*a*) Ring (cyclic) trimeric structure of α–form of SO_3 molecule in the solid state (*b*) chain (or layer) structure of β and γ-forms of SO_3 molecule in the solid state.

has a ring (cyclic) structure shown at (*a*) of Fig. 15.13. On the other hand, β- and γ-forms of SO_3 solid have linear layer (or chain) structure shown at (*b*) of Fig. 15.13. This structure consists of a number of SO_3 chains linked together. These chains are produced when many SO_4 tetrahedrons get linked together by the sharing of O-atom.

In the solid state, SeO_3 exists as a tetramer, $(SeO_3)_4$ which has a cyclic structure shown in Fig. 15.14.

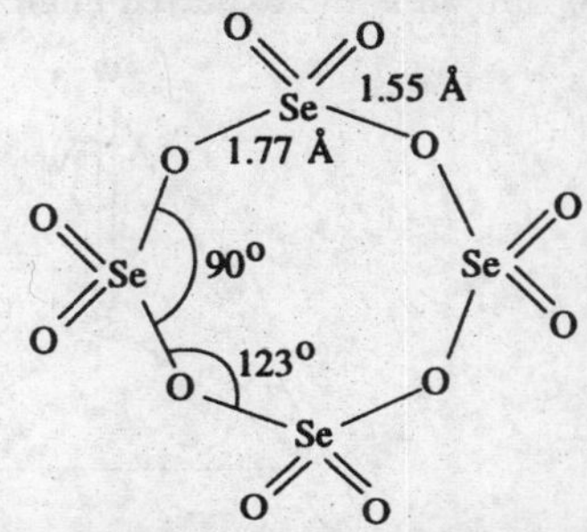

Fig. 15.14 : Cyclic structure of $(SeO_3)_4$ tetramer.

ClO_2 and BrO_2 molecules

The structure of these molecules can be explained on the basis of the following concepts:

(*a*) *Structures based on odd-electron bonds.* See *"odd-electron molecules"* discussed in chapter 10.

(*b*) *Structures based on molecules orbital theory.* See *"odd-electron molecules"* discussed in chapter 10.

(*c*) *Structures based on sp^3 hybridisation.* ClO_2 and BrO_2 both have angular (V-shaped) structure. As an example let us consider the structure of ClO_2 molecule. This molecule can be represented by any of the following two angular structures:

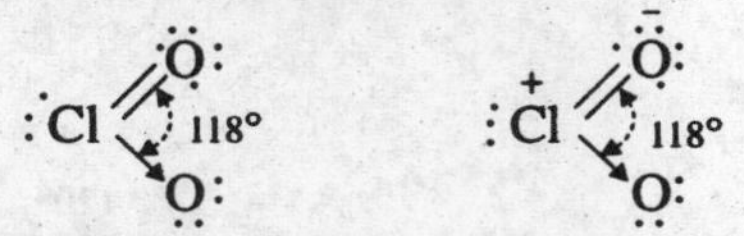

Covalent structure **Ionic structure**

In each of these structures Cl-atom (central atom) is sp^3 hybridised.

sp^3 hybridisation of Cl-atom in structure (I) : $(sp^3)^2\ (sp^3)^2\ (sp^3)^1\ (sp^3)^1\ 3d^1$

sp^3 hybridisation of Cl^+ in structure (II) : $(sp^3)^2\ (sp^3)^1\ (sp^3)^1\ (sp^3)^1\ 3d^1$

It is due to sp^3 hybridisation and the presence of one lone pair of electrons and one unpaired electron in two sp^3 hybrid orbitals that ClO_2 molecule has angular geometry with OClO bond angle equal to 118°. It may be noted that Cl – O π-bond in both the structures is (*p–d*) π -bond.

Oxides of xenon

The following oxides will be considered:

1. Structure of xenon trioxide molecule, XeO_3

This molecule can be presented by the Lewis structure shown at (*a*) of Fig. 15.15. This

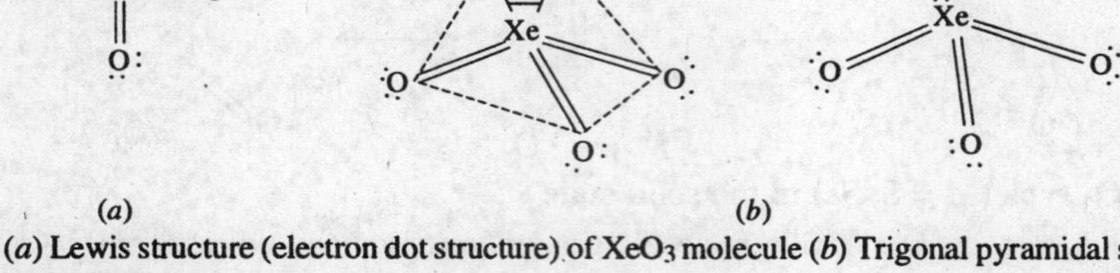

Fig. 15.15 : (*a*) Lewis structure (electron dot structure) of XeO_3 molecule (*b*) Trigonal pyramidal shape of XeO_3 molecule.

structure shows that Xe-atom (central atom) is surrounded by three σ-bonding electron pairs and has one lone pair of electrons on it. Hence Xe-atom in its excited state is sp^3 hybridised. [Xe = $(sp^3)^2$ $(sp^3)^1$ $(sp^3)^1$ $5d^1$ $5d^1$ $5d^1$]. Each of the three (Xe – O) π-bonds is a (*p–d*) π-bond. Due to the presence of one lone pair of electrons in one of the four sp^3 hybrid orbitals, XeO_3 molecule, instead of having tetrahedral shape, has trigonal pyramidal geometry as shown at (*b*) of Fig. 15.15.

2. Structure of xenon tetroxide molecule, XeO_4

This molecule can be represented by the Lewis structure shown at (*a*) of Fig. 15.16. This structure shows that Xe-atom (central atom) is surrounded by four σ-bonds and has no lone pair of electrons on it. Hence Xe-atom in its excited state is sp^3 hybridised in XeO_4 molecule. [Xe = $(sp^3)^1$ $(sp^3)^1$ $(sp^3)^1$ $(sp^3)^1$ $3d^1$ $3d^1$ $3d^1$ $3d^1$]. Each of the (Xe–O) π-bonds is a (*d – p*)π-bond. Due to sp^3 hybridisation and no lone pair of electrons, XeO_4, molecule has the expected tetrahedral geometry shown at (*b*) of Fig. 15.16.

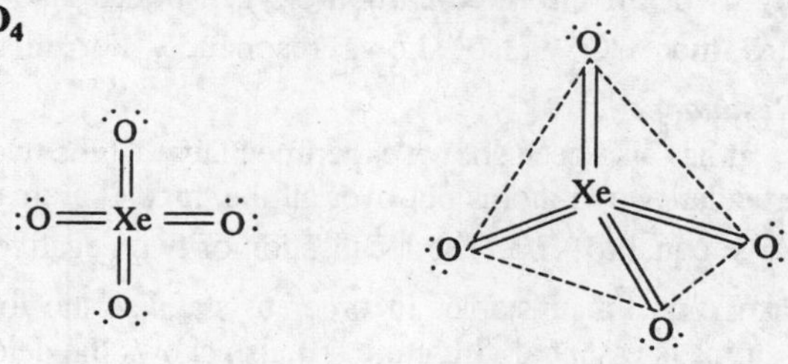

(*a*) (*b*)

Fig. 15.16 : (*a*) Electron dot structure (Lewis structure) of XeO_4 molecule (*b*) Tetrahedral shape of XeO_4 molecule.

Structure of carbonic acid molecule (H_2CO_3) and carbonate ion (CO_3^{2-})

Since H_2CO_3 is a dibasic acid, its structure has two OH groups attached directly with the central carbon atom. Thus the electron dot structure of H_2CO_3 molecule is as given at (I) below. The electron dot structure of CO_3^{2-} ion is derived from that of H_2CO_3 molecule by removing two H-atoms of two OH groups. Thus the electron dot structure of CO_3^{2-} ion can be written as given at (II).

Structure of H_2CO_3 molecule (I) Structure of CO_3^{2-} ion (II)

Structure of CO_3^{2-} given above shows that the central C-atom is surrounded by three σ-bonds and no lone pair of electron (*i.e.*, σ–*bps* = 3, *lp* = 0). Thus carbon atom in its excited state ($2s^12p^3$) is sp^2 hybridised [C = $(sp^2)^1$ $(sp^2)^1$ $(sp^2)^1$ $2p_z^1$. In CO_3^{2-} ion, (C — O) π -bond is a (*p*—*p*) π-bond. It is due to sp^2 hybridisation that CO_3^{2-} ion has trigonal planar geometry as shown in Fig. 15.17.

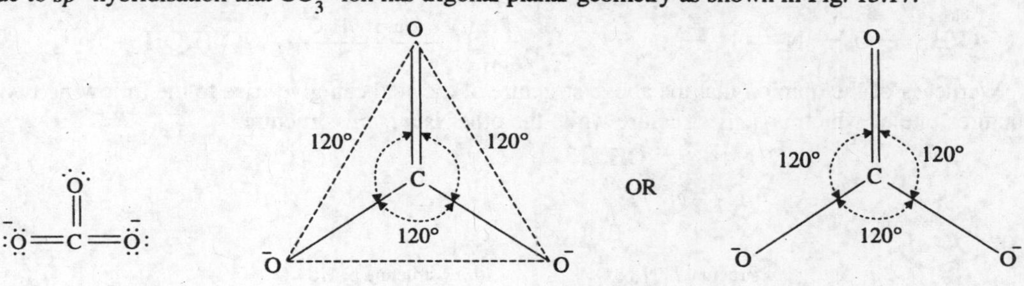

(*a*) (*b*)

Fig. 15.17 : (*a*) Electron dot structure of CO_3^{2-} ion (*b*) Trigonal planar structure of CO_3^{2-}

Delocalisation of π-bond and negative charges and resonance in CO_3^{2-} ion. X-ray examination of CO_3^{2-} ion has indicated that all the three carbon-oxygen bonds are of the same length (= 1.31 Å). This fact indicates that π-bond (π-electron cloud) is not localised between carbon and any particular O-atom, rather, it is spread over all the three O-atoms, *i.e.*, π-electron cloud is delocalised. In order to show the delocalisation of the π- electron cloud, the structure shown in Fig. 15.18 has been suggested for CO_3^{2-} ion. In this structure, the dotted lines show the presence of delocalised π-electron cloud in this ion. It is due to the presence of delocalised π-electron cloud that the bond order of C—O bonds in this ion is equal to 4/3 = 1.33. The similarity in the lengths of all the three carbon-oxygen bonds also indicates that CO_3^{2-} also shows resonance. For this see "*Resonance*".

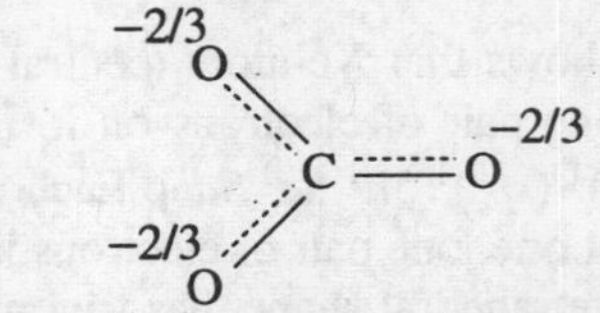

Fig. 15.18 : Structure of CO_3^{2-} ion representing the delocalisation of π-electron cloud.

It has also been shown experimentally that the two negative charges (= −2) are not localised on any particular two O-atoms but over all the three O-atoms. Thus each O-atom can be regarded as having a charge equal to −2/3. The localisation of two negative charges on two particular O-atoms in CO_3^{2-} ion makes this ion unstable. In order to stabilise the ion, the delocalisation of charge on all the three O-atoms is essential. Structure (II) also shows the delocalisation of negative charges on O-atoms.

Oxyacids and oxy anions of nitrogen

Here we shall discuss the structure and geometry of $H_2N_2O_2$, HNO_2, HNO_3 and HNO_4 molecules and oxy-anions derived from them.

1. Structure of hyponitrous acid molecule, $H_2N_2O_2$

The determination of the molecular weight of the acid suggest that the acid has the molecular formula as $H_2N_2O_2$. This acid has a structure in which two N-atoms are attached with each other by a double bond and each N-atom is attached with one OH group. Thus the acid has the structure, $H\ddot{O}—\ddot{N}=\ddot{N}—\ddot{O}H$. The presence of two OH groups in the acid confirms the dibasic nature of the acid. The above structure of the acid is supported by the following reactions shown by the acid.

(*i*) The aqueous solution of $H_2N_2O_2$ decomposes, on heating, to give N_2O and H_2O.

$$H_2N_2O_2 \xrightarrow{\Delta} N_2O + H_2O$$

(*ii*) The esters [*e.g.*, $(C_2H_5)_2N_2O_2$] and the salts (*e.g.*, $Na_2N_2O_2$) of the acid, on heating, give N_2.

$$(C_2H_5)_2\,N_2O_2 \xrightarrow{\Delta} C_2H_5OH + CH_3CHO + N_2$$

$$3Na_2N_2O_2 \xrightarrow{\Delta} 2NaNO_2 + 2Na_2O + 2N_2$$

(*iii*) The diethyl ester, $(C_2H_5)_2N_2O_2$, on being reduced with sodium amalgam and H_2O, gives C_2H_5OH and N_2.

$$C_2H_5—O\overset{H}{|}N=N\overset{H}{|}O—C_2H_5 + 2H \xrightarrow{Na\text{–}Hg/H_2O} 2C_2H_5OH + N_2$$

Werner is of the opinion that the above structure of the acid can give rise to the following two structures one of which is a *cis* structure while the other is a *trans* structure.

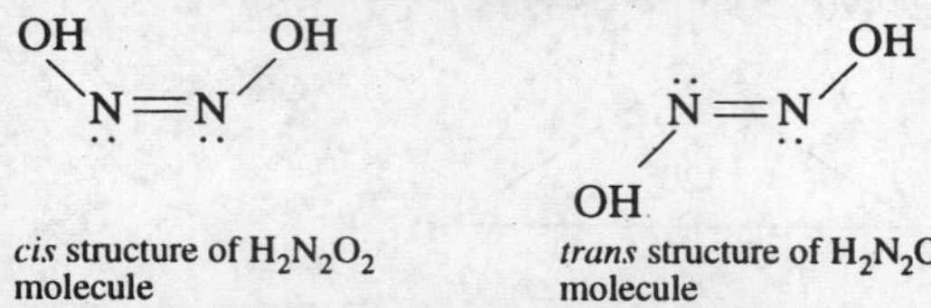

cis structure of $H_2N_2O_2$ molecule   *trans* structure of $H_2N_2O_2$ molecule

Since the dipole moment of dimethyl ester of the acid, $(C_2H_5)_2N_2O_2$ has been found to be equal to zero, the acid has *trans* structure.

2. Structure of hyponitrite ion, $N_2O_2^{2-}$

The infrared spectrum of Na, Hg and Ag salts of hyponi-

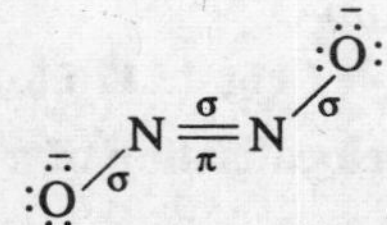

Fig. 15.19 : Trans-structure of hyponitrite ion, $N_2O_2^{2-}$.

trous acid ($H_2N_2O_2$) have shown that $N_2O_2^{2-}$ ion has *trans* structure in which two O-atoms are lying on opposite sides of N = N bond [Fig. 15.19]. Each N-atom is sp^2 hybridised, since round each N-atom (central atom) σ-*bps* + *lps* = 2 + 1 = 3.

N-atom in its ground state : $2s^2\ 2p_x^1\ 2p_y^1\ 2p_z^1$

N–atom in sp^2 hybridised state: $(sp^2)^2\ (sp^2)^1\ (sp^2)^1\ \ 2p_z^1$

One of the three sp^2 hybrid orbitals contains a lone pair of electrons. The other two sp^2 hybrid orbitals, each containing one electron, make N—N and N—O σ-bonds. Finally, the two unhybridised $2p_z$ orbitals on two N-atoms ovelap and make N—N π-bond.

3. Structure of nitrous acid molecule HNO_2

HNO_2 gives two types of organic derivatives which are *alkyl nitrites* (R — O — N = O) and *nitro alkanes* (R—N(=O) → O). Thus corresponding to the structure of these derivatives, HNO_2 molecule can be given the following two structures.

(I) Acid structure (II) Nitro structure

It may be seen from these structures that in acid structure, N-atom is attached with H-atom through O-atom while in nitro structure, N-atom is linked with H-atom directly. Microwave spectroscopic studies have shown that HNO_2 molecule in the gaseous state has acid structure (I).

HNO_2 shows reactions which support both the above structures of HNO_2 molecule. These reactions are given below :

Reactions supporting acid structure (I). (*i*) HNO_2 is a monobasic acid and hence gives only one type of metallic salts. The presence of one OH group attached directly with N-atom (central atom) confirms the monobasic nature of the acid.

(*ii*) When HNO_2 is treated with a primary alkyl alcohol (ROH), H-atom of HNO_2 is replaced by alkyl group (R) and alkyl nitrite (RONO) is obtained. The structure of alkyl nitrite has been found as R — $\ddot{O}$ — $\ddot{N}$ = $\ddot{O}$:. This structure is confirmed by the fact that when alkyl nitrite is reduced by Sn/HCl, ROH and NH_2OH are obtained. These reactions can be shown as follows and support structure (I) for HNO_2 molecule.

$$R — \boxed{OH + H} — O — N{=}O \xrightarrow{-H_2O} R — O — N{=}O + H_2O$$
Alkyl nitrite

$$\underset{\text{Alkyl nitrite}}{R — O — N{=}O} \xrightarrow[\text{(Sn/HCl)}]{+\ 4H} ROH\ +\ NH_2OH \qquad ...(i)$$

(*iii*) When HNO_2 is treated with aliphatic primary amine (RNH_2), primary alcohol (ROH) and N_2 are obtained.

$$R — N\boxed{H_2 + O} = N — O — H \xrightarrow{-H_2O} R — \boxed{N = N} — O — H \xrightarrow{-N_2} R — OH + N_2$$

or $R —NH_2 + O = N — OH \longrightarrow R — OH + H_2O + N_2$

Reactions supporting nitro structure (II). Nitro structure shows that N-atom is attached directly with H-atom. The following reaction confirms the direct attachment of N-atom with H-atom in HNO_2 molecule.

When ethyl bromide (C_2H_5Br) is treated with $AgNO_2$, nitroethane ($C_2H_5NO_2$) is formed. $C_2H_5NO_2$, on being reduced with tin and conc. HCl, gives ethyl amine ($C_2H_5NH_2$). Now, since

N-atom is attached directly with C_2H_5 group in $C_2H_5NO_2$ molecules, N-atom in $AgNO_2$ molecule should also be directly attached with Ag atom, *i.e.*, N-atom in HNO_2 molecule is also attached directly with H-atom.

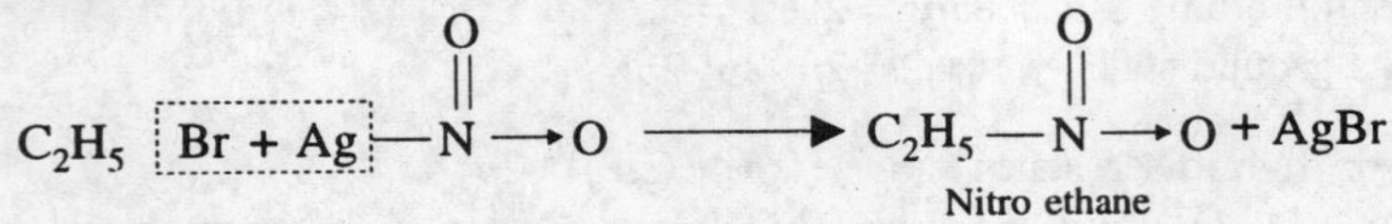

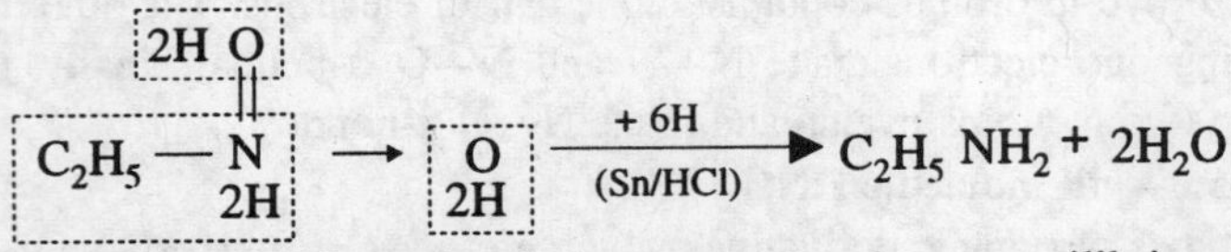

Now the structure of HNO_2 molecule is supposed to be an equilibrium mixture of the above two structures (tautomeric structures), *i.e.*,

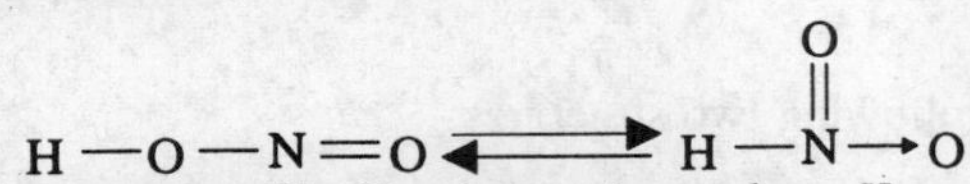

Parameters in acid structure (I). Bond lengths and bond angles that have been determined in structure (I) are shown in Fig. 15.20.

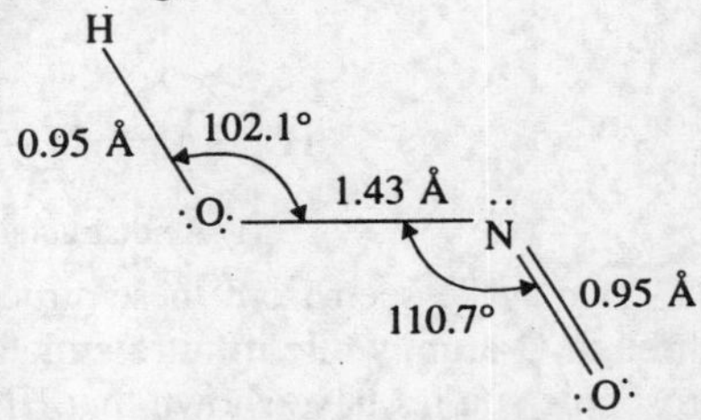

Fig. 15.20 : Parameters in acid structure of HNO_2 molecule.

4. Structure of nitrite ion, NO_2^-

We have already seen that the structure of HNO_2 molecule can be represented by any of the two structures namely *acid structure (I)* and *nitro structure (II)*. When H-atom is removed from these structures, we get the structure of NO_2^- ion. Thus NO_2^- ion can have any of the two structures namely *structure* (A) and *structure* (B) which have been derived from *structures* (I) and (II).

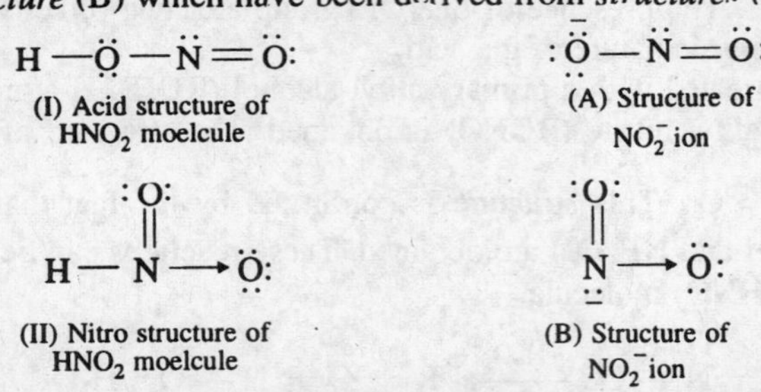

(I) Acid structure of HNO_2 moelcule

(A) Structure of NO_2^- ion

(II) Nitro structure of HNO_2 moelcule

(B) Structure of NO_2^- ion

In structure (A), since N-atom is surrounded by two *σ-bps* and one *lp*, it is sp^2 hybridised in this ion [N = $(sp^2)^2\,(sp^2)^1\,(sp^2)^1\,2p_z^1$]. One electron pair residing in one sp^2 hybrid orbitals act as a non-bonding electron pair on N-atom while the two singly-filled sp^2 hybrid orbitals make two N—O σ-bonds. $2p_z$ orbital of N-atom which is not involved in sp^2 hybridisation and contains one electron makes (N — O) π-bond. This π-bond is a $2p$ (N) – $2p$(O) or (*p*—*p*) π-bond. It is due to

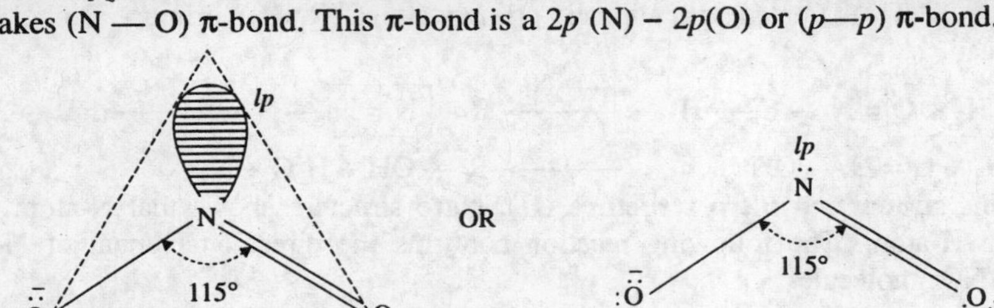

Fig. 15.21 : Angular shape of NO_2^- ion.

sp^2 hybridisation of N-atom and the presence of one *lp* on it that NO_2^- ion has angular shape as shown in Fig. 15.21.

In structure (B), N-atom is negatively-charged and hence contains six electrons in its valence-shell ($\overline{N} = 2s^22p^4$). Both O-atoms attached with $\overline{N}$ atom are neutral. Since $\overline{N}$ ion is surrounded by two σ-*bps* and one lone pair of electrons, it is sp^2 hybridised [$\overline{N} = (sp^2)^2\ (sp^2)^2\ (sp^2)^1\ 2p_z^1$]. One electron pair residing in one sp^2 hybrid orbital acts as a lone pair of electrons on $\overline{N}$ ion while the other electron pair is donated to the vacant 2*p*-orbital of O-atom and makes a $\overline{N} \rightarrow O$ coordinate bond. ($O = 2s^2\ 2p_x^2\ 2p_y^2\ 2p_z^0$). Half-filled $2p_z$ orbital on the other O-atom makes a (N — O) π-bond. This π-bond is a 2*p* (N) – 2*p* (O) π-bond [(*p* — *p*) π-bond]. It is due to sp^2 hybridisation of N^- ion and the presence of one lone pair of electrons on N^- that NO_2^- ion has angular geometry (See Fig. 15:22].

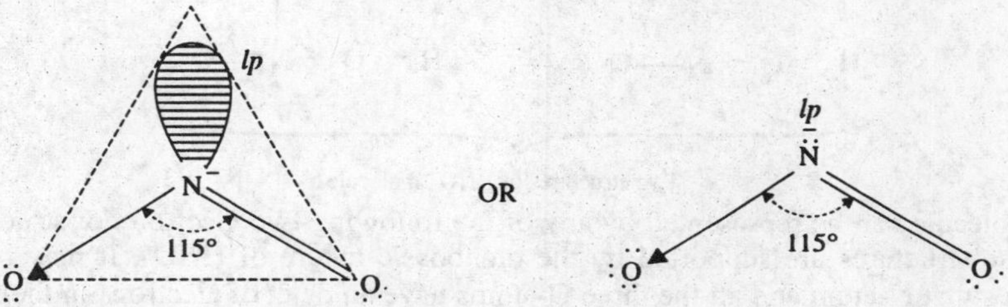

Fig. 15.22 : Angular shape of NO_2^- ion.

Position of π-electron pair cloud and angular shape of NO_2^- ion : Resonance in NO_2^- ion. π-electron pair cloud is spread (delocalised) over both O-atoms. The occurance of sp^2 hybridisation and the presence of one electron pair on N-atom both indicate that NO_2^- ion has angular shape shown in Fig.15.22. X-ray and Raman studies of crystalline $AgNO_2$ and $NaNO_2$ have also shown that NO_2^- ion has angular shape with ONO bond angle equal to 115°. Due to the delocalisation π-electron pair cloud over both O-atoms, both nitrogen-oxygen bond distances are equal (= 122 pm). This bond distance is intermediate between single N — O bond distance (= 136 pm) and double N = O bond distance (= 115 pm). The equivalence of the two nitrogen-oxygen bond distances suggests that the actual structure of NO_2^- ion is a resonances hybrid of structures (A) and (B).

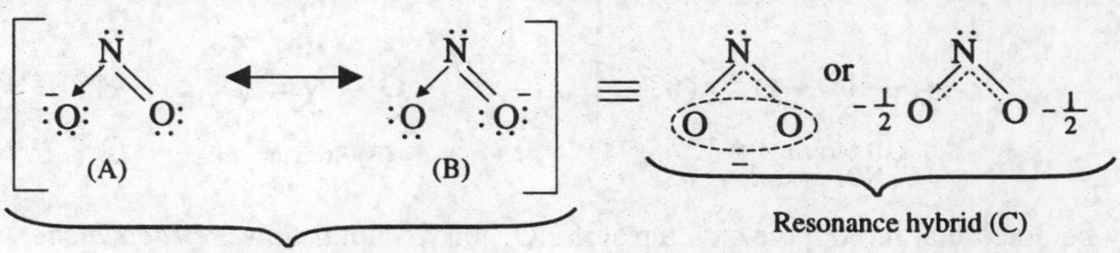

In order to show that delocalisation of π-electron cloud over both O-atoms, both canonical forms can be represented by a single structure namely (C) which is called resonance hybrid of structures (A) and (B). In structure (C), the dotted lines represent the delocalisation of π-electron cloud and each O-atom carries – 1/2 charge.

ONO bond angle in NO_2^- ion. Since NO_2^- ion is formed by sp^2 hybridisation of N-atom, ONO bond angle should be equal to 120° which is the expected bond angle for sp^2 hybridisation. But actually ONO bond angle in NO_2^- ion has been found to be equal to 115°. The decrease in the bond angle from the expected value is due to the presence of a lone pair of electrons on N-atom. This lone pair occupies more space than the bonding electron pair and hence the bonding electron pairs come closer to each other. This results in the decreases in the bond angle.

5. Structure of nitric acid molecule, HNO_3

HNO_3 is a monobasic acid and hence contains one OH group attached directly with N-atom (central atom). Thus the structure of HNO_3 molecule can be represented by the formula (HO) NO_2. Spectroscpic and electron diffraction studies have shown that HNO_3 molecule in the *vapour state* has *trigonal planar structure* with various bond angles and bond lengths as shown in Fig. 15.23.

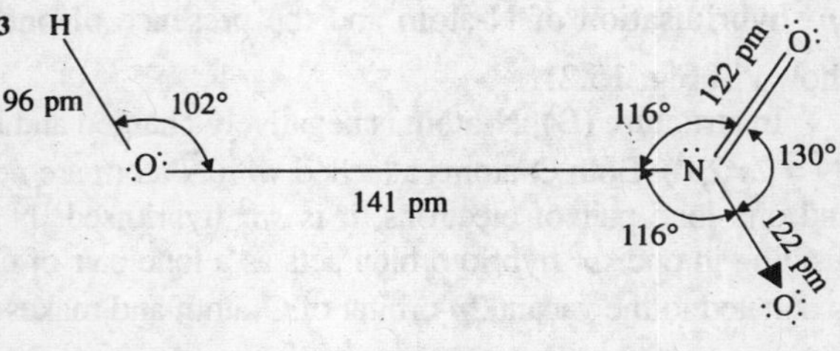

Fig. 15.23 : Trigonal planar structure of HNO_3 molecule.

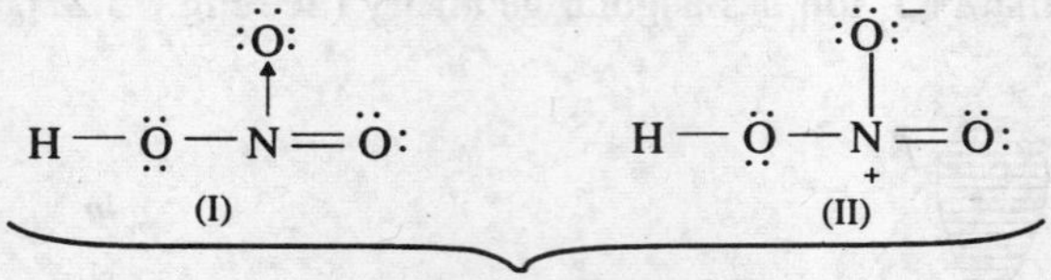

· Two structures of HNO_3 molecule

HNO_3 molecule can be represented by any of the following two electron dot structures:

Both these structures are supported by the monobasic nature of HNO_3. It may be seen from these structures that N-atom and all the three O-atoms have an octet of electrons in their outer shell. H-atom has two electrons in its shell.

6. Structure of nitrate ion, NO_3^-

Corresponding to structures (I) and (II) which are the structures of HNO_3 molecule, NO_3^- ion can be represented by the structures (A) and (B) respectively.

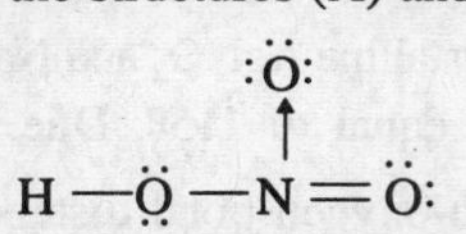

(I) Structure of HNO_3 moelcule

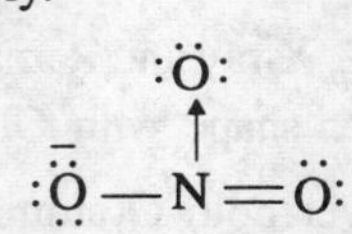

(A) Structure of NO_3^- ion

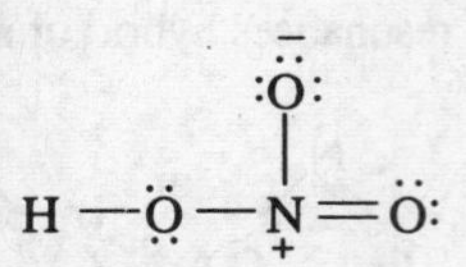

(II) Structure of HNO_3 moelcule

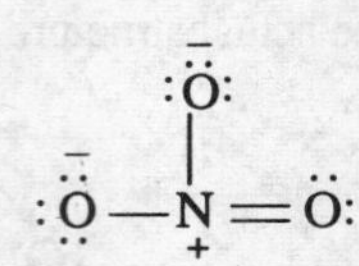

(B) Structure of NO_3^- ion

It may be noted that in both the structures of NO_3^- ion, N-atom shows *tetracovalency* and both the structures obey octet rule. Since in both the structures, the central N-atom is surrounded by *three σ-bps* and no *lp* (*i.e., σ-bps* + *lps* = 3 + 0 = 3), N-atom is sp^2 hybridised in both these structures. Due to sp^2 hybridisation, NO_3^- ion has *triangular planar geometry* (Fig. 15.24) in which all the three introgen-oxygen bonds are of equal length (= 1.22 Å) and each of the three O—N—O bond angles is equal to 120° which is the expected angle for sp^2 hybridisation.

In structure (A) sp^2 hybridisation of N-atom can be shown as : $(sp^2)^2\ (sp^2)^1\ (sp^2)^1\ 2p_z^1$. Completely-filled sp^2 hybrid orbital on N-atom donates its electron pair to the vacant 2*p* orbital of O atom and forms a N → O coordinate bond. The head-to-head overlap of two singly-filled sp^2 hybrid orbitals with the two singly-filled 2*p* orbitals of O^- and O gives N — O^- and N — O σ-bonds. N — O π-bond results by the sidewise overlap of singly-filled $2p_z$ orbital on N-atom with the singly-filled $2p_z$ orbital on O-atom.

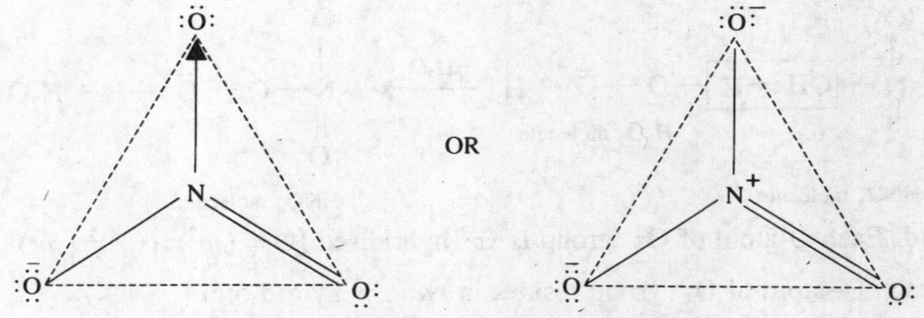

Fig. 15.24 : Trigonal planar geometry of NO_3^- ion.

In structure (B), sp^2 hybridisation of $\overset{+}{N}$ ($\overset{+}{N} = 2s^2\ 2p^2$) can be shown as $(sp^2)^1\ (sp^2)^1\ (sp^2)^1\ 2p_z^1$. Three $\overset{+}{N} - O$ bonds are obtained by the head-to-head overlap between three sp^2 hybrid orbitals and half-filled $2p$ orbitals on three O-atoms. ($\overset{+}{N} - O$) π-bond is obtained by $2p\ (\overset{+}{N}) - 2p$ (O) sidewise overlap. Thus π-bond is a ($p\pi$ — $p\pi$) π-bond.

Resonance in NO_3^- ion. Structures (A) and (B) show that nitrogen-oxygen bond distances should be of unequal length but experiments have shown that all the three nitrogen-oxygen bond lengths are equal (= 1.22 Å). This value is intermediate between the calculated N—O single bond distance (= 1.36 Å) and N = O double bond distance (= 1.15 Å). In order to explain the equivalence in the three bond lengths it has been suggested that none of the above two structures is the actual structure of NO_3^- ion. Rather the true structure is a resonance hybrid of the following contributing structures. Contributing structures from (*i*) to (*vi*) correspond to structure (A) while the contributing structures (*vii*) to (*ix*) correspond to structure (B).

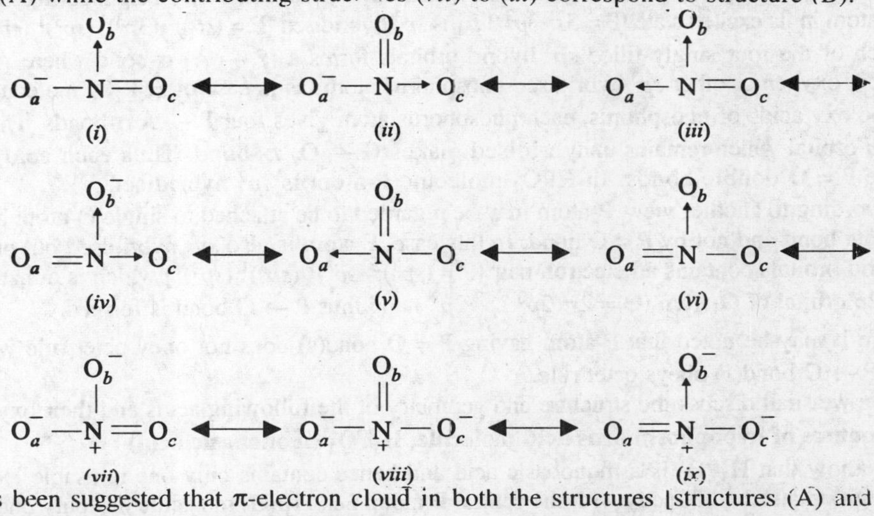

It has been suggested that π-electron cloud in both the structures [structures (A) and (B)] is spread (delocalised) over all the three O-atoms.

7. Structure of peroxynitric acid (HNO_4) and peroxynitrate ion (NO_4^-)

Peroxynitric acid may be regarded as having been obtained by removing one H_2O molecule from the combination of one molecule of nitric acid (HNO_3) and one molecule of hydrogen peroxide (H_2O_2).

The structure of HNO_4 molecule suggests that the structure of peroxynitrate ion (NO_4^-) can be written as given in the margin. In the structure of HNO_4 molecule and NO_4^- ion, N-atom (central atom) is sp^2 hybridised. [N = $(sp^2)^2\ (sp^2)^1\ (sp^2)^1\ 2p_z^1$]. (N—O) π-bond is a

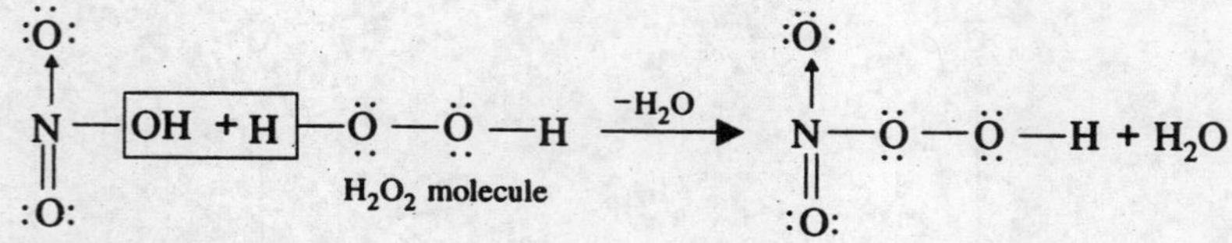

HNO_3 molecule H_2O_2 molecule HNO_4 molecule

$(p\text{–}p)$ π-bond. Each O-atom of O_2^{2-} group is sp^3 hybridised $[O = (sp^3)^2\,(sp^3)^2\,(sp^3)^1\,(sp^3)^1]$. The two *lps* on each O–atom of O_2^{2-} group resides in two sp^3 hybrid orbitals of O-atom. Due to sp^2 hybridisation of N-atom (central atom) the shape of HNO_4 and NO_4^- are *trigonal planar*. The structures of HNO_4 molecule and NO_4^- ion contain one peroxy group, O_2^{2-} (— O — O —).

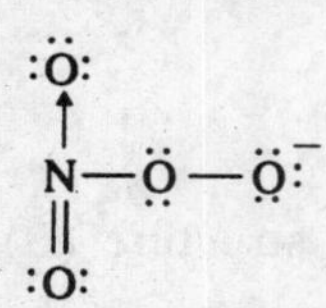

Structure of peroxy nitrate ion, NO_4^-

Oxy acids and oxy anions of phosphorus

While considering the structure of oxyacids and oxy anions, the following points must be kept in mind.

(*i*) Total number of OH groups attached directly with P-atom(s) is equal to the basicity of the oxy acid. H-atom of an OH groups is ionisable.

(*ii*) H-atoms other than those present in OH groups are attached directly with P-atoms and these H-atoms are responsible for the reducing property of the acids. Greater is the number of H-atoms attached directly with P-atom, greater is the reducing property of the acid. Thus the oxy acid which shows reducing property also contains one or more P—H bonds. H-atoms of these bonds are not ionisable. Examples of such acids are H_3PO_2, H_3PO_3 and HPO_2.

(*iii*) X-ray study and other studies of the acids and their salts have shown that each P-atom of the acid (excepting HPO_2) and in the oxy anions is tetrahedrally surrounded by neighbouring atoms, *i.e.*, each P-atom in its excited state $[P = 3s^1\,3p^3\,3d^1]$ is sp^3 hybridised $[P = (sp^3)^1\,(3p^3)^1\,(sp^3)^1\,(sp^3)^1\,3d^1]$.

Each of the four singly-filled sp^3 hybrid orbitals forms a (P — A) σ-bond where A may be hydrogen, oxygen or other sp^3 hybridised phosphorus atom (as in case of $H_4P_2O_6$ molecule). Thus in all the oxy acids of phosphorus, each phosphorus atom gives four P — A σ-bonds. The singly-filled $3d$ orbital which remains unhybridised makes (P— O) π–bond. Thus each acid has one or more P = O double bonds. In HPO_2 molecule P-atom is sp^2 hybridised.

According to another view, P-atom may be regarded to be attached to single O-atom by P→ O coordinate bond and not by P = O bond. In this case, P-atom is also sp^3 hybridised but one of the sp^3 hybrid orbitals contains an electron pair $[P = (sp^3)^2\,(sp^3)^1\,(sp^3)^1\,(sp^3)^1]$ which is donated to the vacant $2p$ orbital of O-atom $(O = 2s^2\,2p_x^0\,2p_y^2\,2p_z^2)$ and thus P → O bond is formed.

Here it may be noted that P-atom having P = O bond(s) does not obey octet rule while that having P → O bond(*s*) obeys octet rule.

Here we shall discuss the structure and geometry of the following acids and their oxy anions.

1. Structure of hypophosphorus acid molecule, H_3PO_2 (monobasic acid)

We know that H_3PO_2 is a monobasic acid and hence contains only one ionisable H-atom or one OH group attached directly with the central P-atom, *i.e.*, H_3PO_2 molecule has only one P—OH bond. This acid is a reducing agent and hence contains P—H bonds (two). On the basis of these properties of this acid, the electron dot structure of H_3PO_2 molecule can be written as shown below:

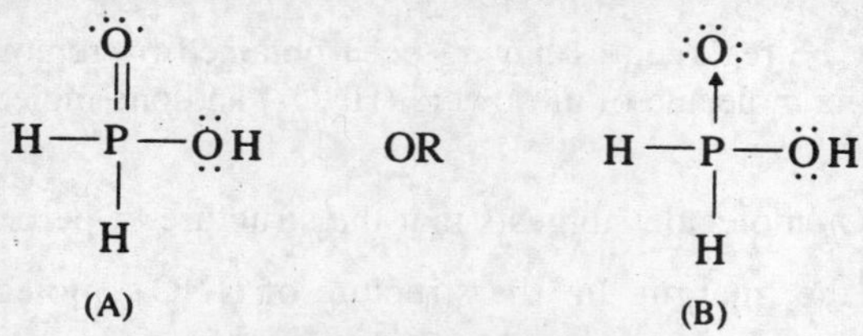

(A) OR (B)

Electron dot structure of H_3PO_2 molecule

It is obvious from the electron dot structure that P-atom (central atom) forms four σ-bonds with the neighbouring O-atoms and has no *lp* of electrons on it. Consequently P-atom is sp^3 hybridised in both the structures. In structure (A), P-atom undergoes sp^3 hybridisation in its excited state (P = $3s^1\ 3p_x^1\ 3p_y^1\ 3p_z^1\ 3d^1$) while in structure (B) P-atom undergoes sp^3 hybridisation in the ground state (P = $3s^2\ 3p_x^1\ 3p_y^1\ 3p_z^1$). Due to sp^3 hybridisation and no *lp* of electrons on P-atom H_3PO_2 molecule has tetrahedral shape which has been shown in Fig. 15.25.

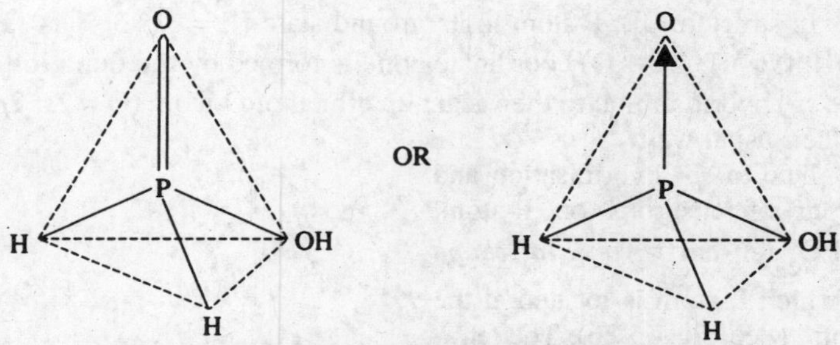

Fig. 15.25 : Tetrahedral structures of H_3PO_2 molecule.

The structure of H_3PO_2 molecule as given above is also confirmed by the fact that when two molecules of H_3PO_2 are heated, H_3PO_4 and PH_3 are obtained.

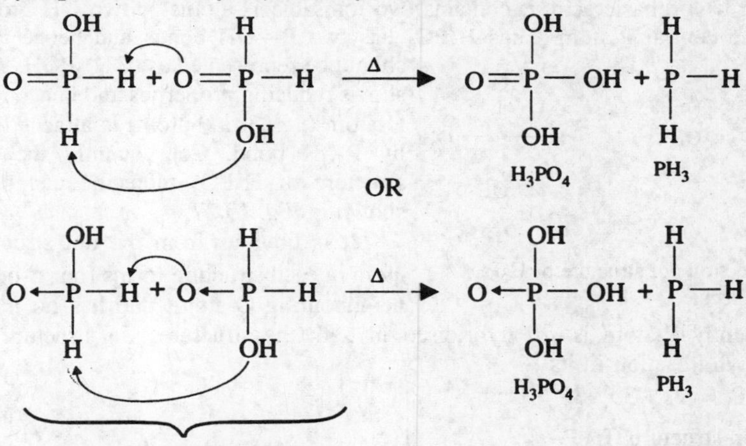

Two molecules of H_3PO_2

2. Structure of dihydrogen hypophosphite ion, $H_2PO_2^-$

Since H_3PO_2 is a monobasic acid, it ionises as $H_3PO_2 \rightleftharpoons H^+ + H_2PO_2^-$. $H_2PO_2^-$ ion is called *dihydrogen hypophosphite ion.* The electron dot structure of this ion is either (I) or (II).

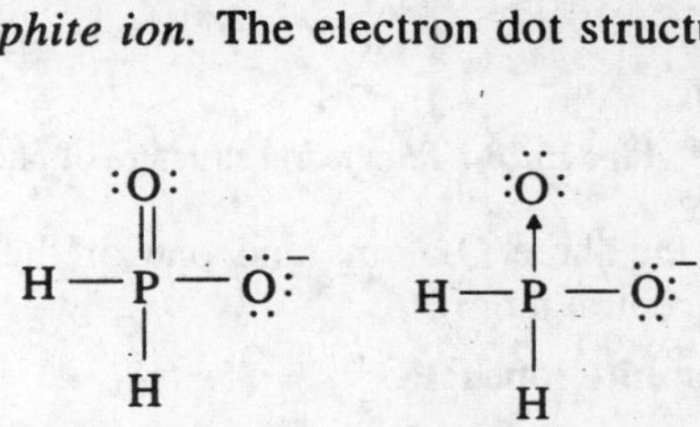

(I) (II)

In structure (I), P-atom is sp^3 hybridised in its excited state (P = $3s^1\ 3p^3\ 3d^1$). sp^3 hybridisation gives four singly-filled sp^3 hybrid orbitals and one singly-filled $3d$ orbital [P = $(sp^3)^1\ (sp^3)^1\ (sp^3)^1\ (sp^3)^1\ 3d^1$]. σ-bonds are formed in their usual way. (P — O) π-bond is formed by the sidewise overlap of half-filled $3d$ orbital on P-atom and half-filled $2p$ orbital on O-atom. Thus (P — O) π bond is a $3d$ (P) – $2p$ (O) or (d–p) π -bond.

In structure (II), P-atom in its ground state [P = $3s^2$ $3p^3$] is sp^3 hybridised [P = $(sp^3)^2$ $(sp^3)^1$ $(sp^3)^1$ $(sp^3)^1$]. (P → O) coordinate bond is formed by the donation of an electron pair present in one sp^3 hybrid orbital to the vacant $2p$ orbital of O-atom (O = $2s^2$ $2p^2$ $2p^2$ $2p^0$). σ-bonds are formed in their usual way.

Due to sp^3 hybridisation and no *lp* of electrons on P-atom, $H_2PO_2^-$ ion has *tetrahedral shape* in which P-atom is located at the centre (See Fig. 15.26). The tetrahedral shape has also been confirmed by the Raman spectral and X-ray study of the salts containing $H_2PO_2^-$.

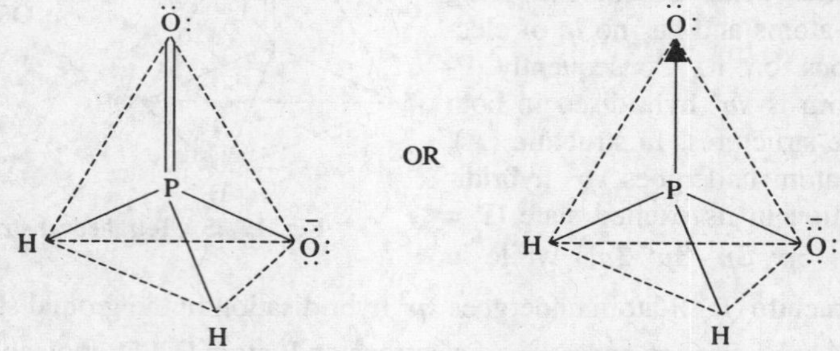

Fig. 15.26 : Tetrahedral shape of $H_2PO_2^-$ ion.

3. Structure of phosphorus acid, H_3PO_3 (Dibasic acid)

Since H_3PO_3 is dibasic acid, it contains two ionisable H-atoms or two OH groups attached directly with the central P-atom. Thus H_3PO_3 has two P—OH bonds and hence this molecule should be represented as $(OH)_2POH$. This acid also shows reducing properties and hence the remaining H-atom (*i.e.*, 3rd H-atom) is attached with P-atom by P—H bond. Consequently the electron dot structure of H_3PO_3 molecule can be written as shown in Fig. 15.27.

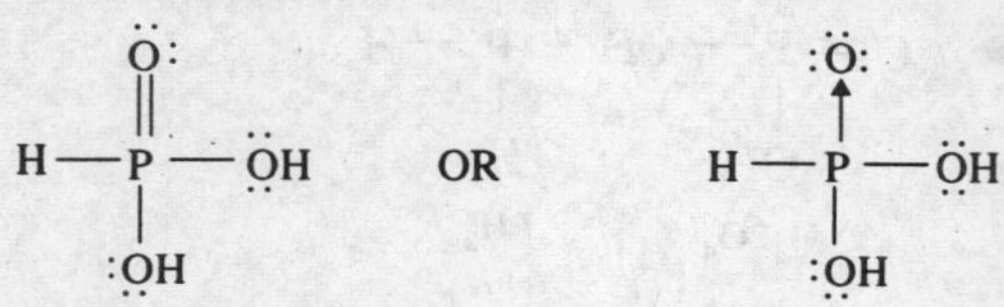

(A) (B)

Fig. 15.27 : Electron dot structure of H_3PO_3 molecule.

It is obvious from the dot structures that P-atom in each structure forms four σ-bonds with the neighbouring O-atoms and has no *lp* of electrons on it. Consequently P-atom is sp^3 hybridised in both the structures. In structure (A) P-atom undergoes sp^3 hybridisation in its excited state (P = $3s^1$ $3p_x^1$ $3p_y^1$ $3p_z^1$ $3d^1$) while in structure (B) P–atom undergoes sp^3 hybridisation in its ground state (P = $3s^2$ $3p_x^1$ $3p_y^1$ $3p_z^1$). Due to sp^3 hybridisation and no *lp* of electrons on P-atom, H_3PO_3 molecule has *tetrahedral shape* which has been shown in Fig. 15.28. X-ray study of the crystalline acid has shown that the three O-atoms and one of the three H-atoms are arranged tetrahedrally around the P-atom (central atom).

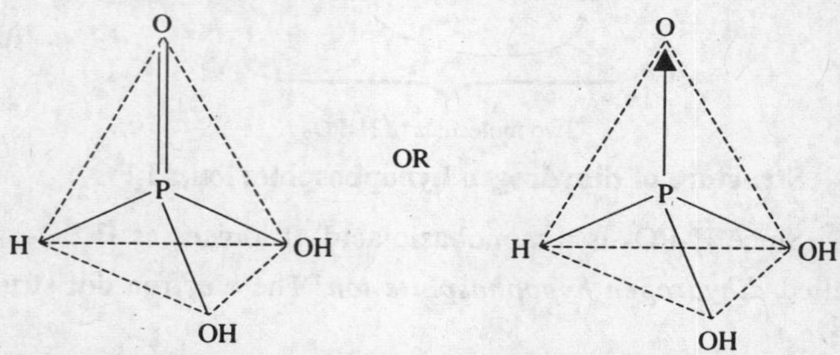

Fig. 15.28 : Tetrahedral structure of phosphorus acid molecule, H_3PO_3.

4. Structure of hydrogen phosphite ion, HPO_3^{2-}

The electron dot structures of H_3PO_3 molecule (which is the parental acid of HPO_3^{2-} ion) as given above indicates that the electron dot structure of HPO_3^{2-} ion can be written as given below in Fig. 15.29.

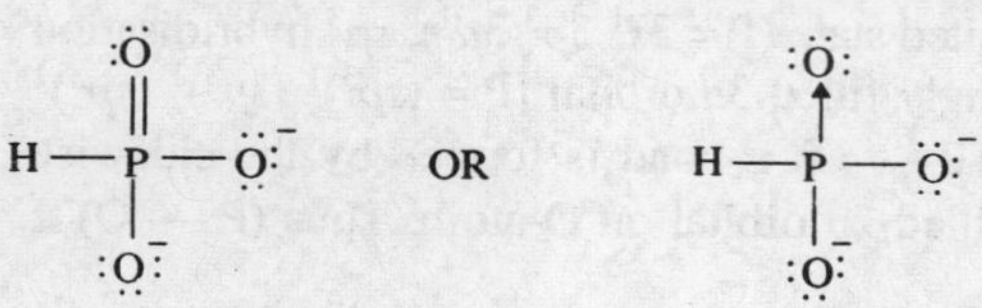

(I) (II)

Fig. 15.29 : Electron dot structures of HPO_3^{2-} ion.

In structure (I) P-atom in its excited state [P = $3s^1$ $3p^3$ $3d^1$] is sp^3 hybridised [P = $(sp^3)^1$ $(sp^3)^1$ $(sp^3)^1$ $(sp^3)^1$ $3d^1$]. (P – O) π-bond is a $3d$ (P) – $2p$

(O) or $(d-p)$ π-bond. In structure (II) P-atom in its ground state $[P = 3s^2\ 3p^3]$ is sp^3 hybridised $[P = (sp^3)^2\ (sp^3)^1\ (sp^3)^1\ (sp^3)^1]$. $(P \rightarrow O)$ coordinate bond is obtained when one sp^3 hybrid which is completely-filled donates its electron pair to the vacant $2p$ orbital of O-atom $(O = 2s^2\ 2p^2\ 2p^2\ 2p^0)$.

Due to sp^3 hybridisation and the absence of any lone pair of electrons on P-atom, HPO_3^{2-} ion has the expected *tetrahedral shape* in which P-atom (central atom) occupies the central position of the tetrahedron (See Fig. 15.30).

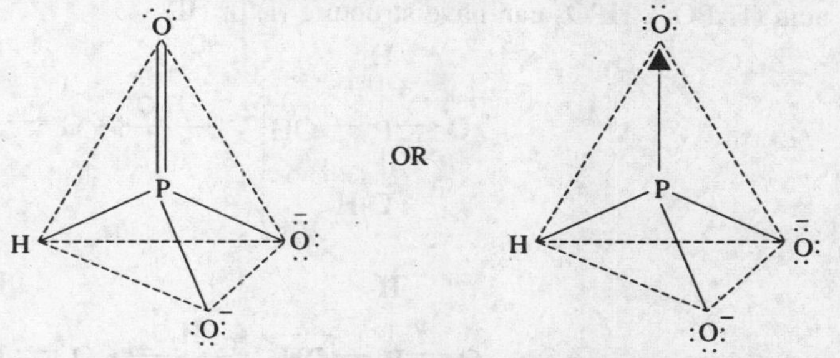

Fig. 15.30 : Tetrahedral shape of hydrogen phosphite ion, HPO_3^{2-}.

5. **Structure of di(pyro) phosphorus acid molecule ($H_4P_2O_5$) (Dibasic acid)**

Since $H_4P_2O_5$ is dibasic ($H_4P_2O_5 \rightleftharpoons 2\overset{+}{H} + H_2P_2O_5^{2-}$) and is obtained by eliminating one H_2O molecule from two molecules of phosphorus acid (H_3PO_3), its structure is (I) or (II).

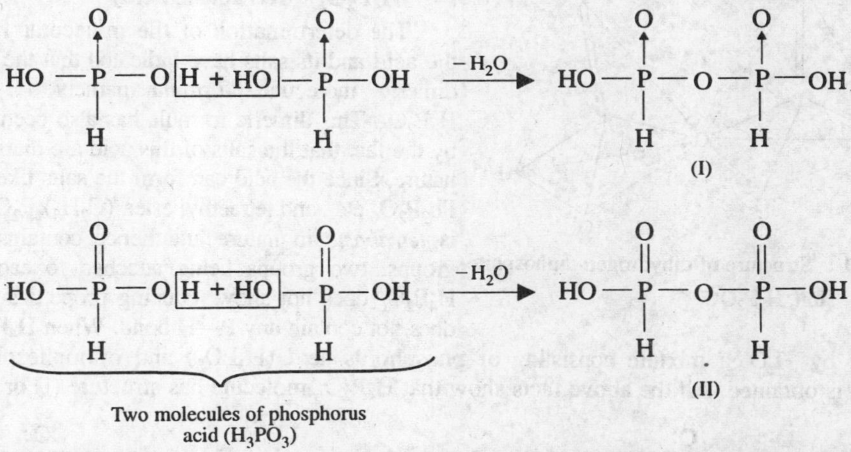

The presence of two OH groups attached with P-atom confirms the dibasic nature of $H_4P_2O_5$.

6. **Structure of dihydrogen di(pyro) phosphite ion, $H_2P_2O_5^{2-}$**

We have already said that $H_4P_2O_5$ is a dibasic acid and hence ionises as $H_4P_2O_5 \rightleftharpoons 2H^+ + H_2P_2O_5^{2-}$. $H_2P_2O_5^{2-}$ ion is called *dihydrogen pyrophosphite ion.* The electron dot structure of this ion is (I) or (II).

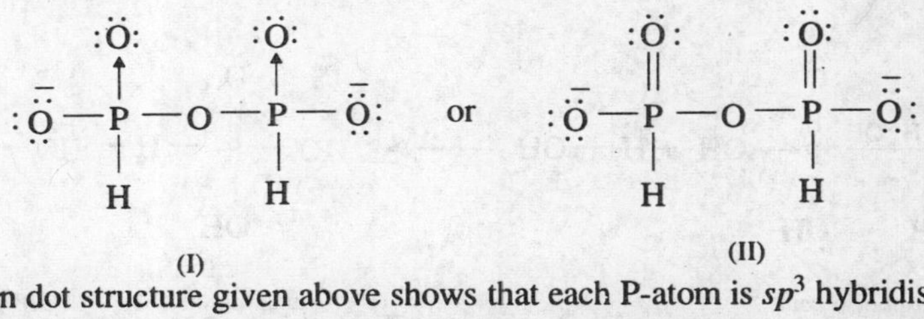

The electron dot structure given above shows that each P-atom is sp^3 hybridised. The structure of $H_2P_2O_5^{2-}$ ion can be obtained by joining two HPO_3 tetrahedrons in such a way that O-atom is common to both tetrahedrons. (see Fig. 15.31).

7. Structure of metaphosphorus acid molecule, HPO_2.

Since this acid is obtained by eliminating one H_2O molecule from one molecule of phosphorus acid (H_3PO_3), HPO_2 can have structure (I) or (II).

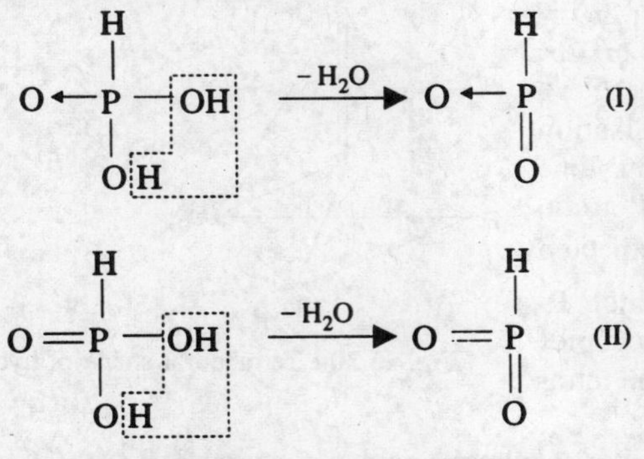

One molecule of phosphorus acid (H_3PO_3)

Metaphosphorus acid (HPO_2)

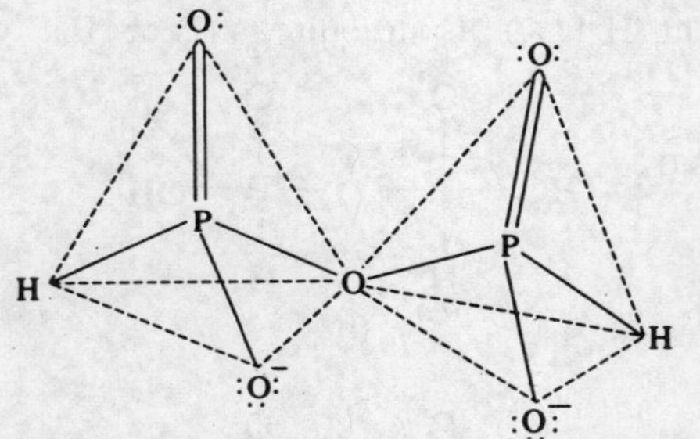

Fig. 15.31 : Structure of dihydrogen diphosphite ion, $H_2P_2O_5^{2-}$.

8. Structure of hypophosphoric acid molecule ($H_4P_2O_6$) (Tetrabasic acid)

The determination of the molecular formula of the acid and its salts have indicated that the acid has a dimeric molecular formula namely $(H_2PO_3)_2$ or $H_4P_2O_6$. This dimeric formula has also been supported by the fact that the salts of this acid are diamagnetic in nature. Since the acid can form the salts like $Na_4P_2O_6$, $Pb_2P_2O_6$ etc., and tetraethyl ester, $(C_2H_5)_4P_2O_6$, the acid is *tetrabasic* in nature and hence contains four OH groups, two groups being attached to each P-atom. $H_4P_2O_6$ does not show reducing properties and hence does not contain any P—H bond. When $H_4P_2O_6$ is hydrolysed by H_2O, a mixture consisting of phosphorus acid (H_3PO_3) and orthophosphoric acid (H_3PO_4) is obtained. All the above facts show that $H_4P_2O_6$ molecule has structure (I) or (II).

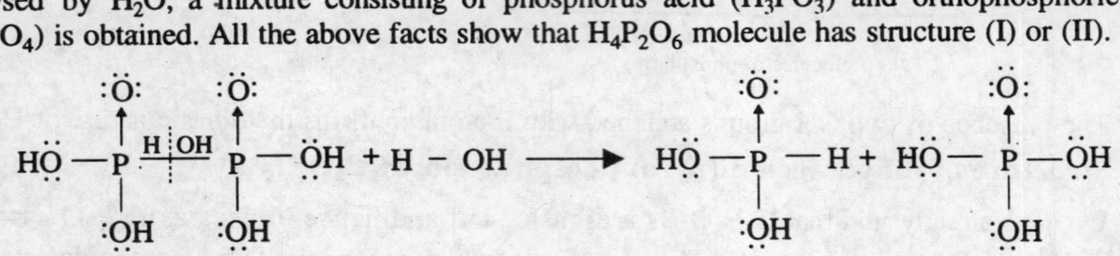

Hypophosphoric acid ($H_4P_2O_6$) (I)

Phosphorus acid (H_3PO_3)

Orthophosphoric acid (H_3PO_4)

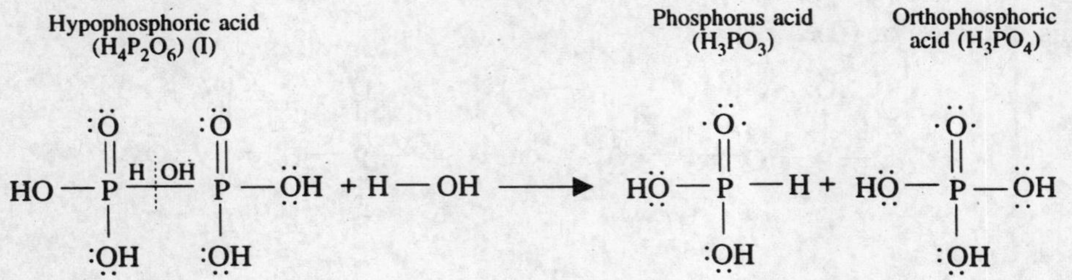

Hypophosphoric acid ($H_4P_2O_6$) (II)

Phosphorus acid (H_3PO_3)

Orthophosphoric acid (H_3PO_4)

The above structures of $H_4P_2O_6$ have been confirmed by the Raman spectrum and X-ray studies of the diammonium salt of the acid, $(NH_4)_2H_2P_2O_6$. Each P-atom is sp^3 hybridised. P – P bond results by the overlap of two singly-filled sp^3 hybrid orbitals, one on each P-atom.

9. Structure of hypophosphate ion, $P_2O_6^{4-}$

We have already said that $H_4P_2O_6$ is a tetrabasic acid and hence it can also be represented as $H_4^+ [P_2O_6]^{4-}$. $P_2O_6^{4-}$ ion is called *hypophosphate ion.* This ion can have structure (I) or (II).

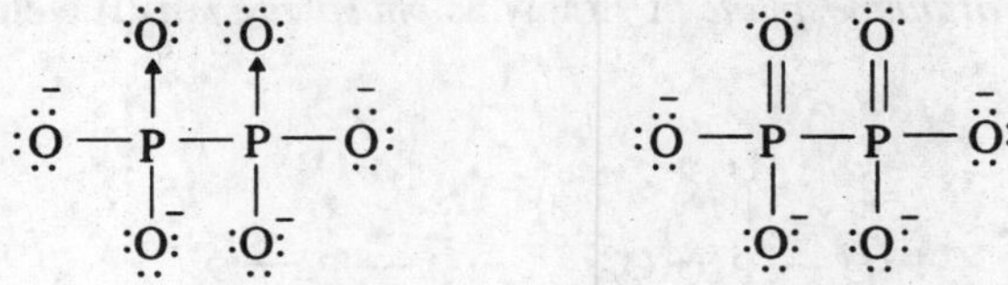

(I) (II)

Structure of $P_2O_6^{4-}$ ion

In this structure each P-atom in its excited state [P = $3s^1\ 3p^3\ 3d^1$] is sp^3 hybridised [P = $(sp^3)^1\ (sp^3)^1\ (sp^3)^1\ 3d^1$]. P – P bond is obtained when singly-filled sp^3 hybrid orbitals on both P-atoms overlap with each other.

10. Structure of orthophosphoric (also calledphosphoric) acid molecule, H_3PO_4 (Tribasic acid)

We know that H_3PO_4 is a tribasic acid. This indicates that H_3PO_4 molecule has three OH groups attached directly with P-atom, *i.e.*, H_3PO_4 molecule has three P – OH bonds and hence can be represented by the formula, $(OH)_3PO$.

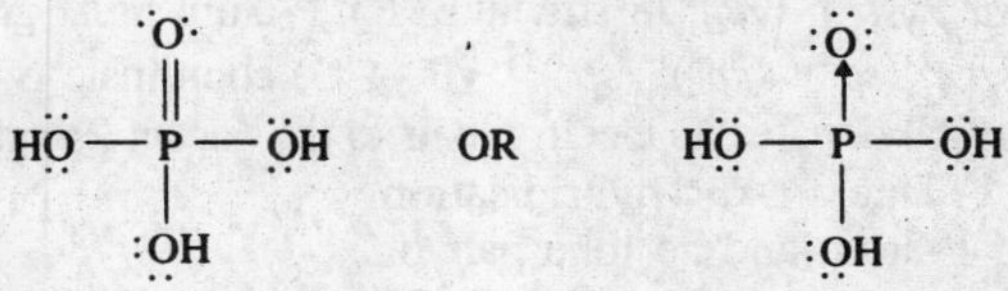

(A) (B)

Fig. 15.32 : Electron dot structure of H_3PO_4 molecule.

This acid does not show reducing property and hence does not contain any P – H bond. On the basis of the above facts, it can be said that H_3PO_4 molecule has the electron dot structure shown in Fig. 15.32. The presence of three OH groups is also confirmed by the fact the H_3PO_4 forms triethyl phosphate, $(C_2H_5O)_3OP$ which is an ester.

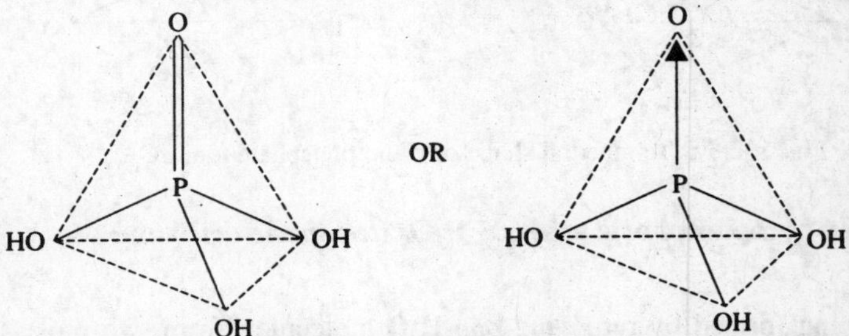

Fig. 15.33 : Tetrahedral shape of orthophosphoric acid molecule, H_3PO_4.

It may be seen from the above structure that P-atom in structure (A) does not obey octet rule, since it is surrounded by 10 electrons (instead of 8 electrons). On the other hand P-atom in structure (B) is surrounded by 8 electrons and hence obeys octet rule.

It is obvious from the above electron dot structure that P-atom in each structure forms four σ-bonds with the neighbouring O-atoms and has no *lp* of electrons on it. Consequently P-atom is sp^3 hybridised in both the structures. In structure (A), P-atom undergoes sp^3 hybridisation in its excited state (P = $3s^1\ 3p_x^1\ 3p_y^1\ 3p_z^1\ 3d^1$), but in structure (B) P-atom undergoes sp^3 hybridisation in the ground state (P = $3s^2\ 3p_x^1\ 3p_y^1\ 3p_z^1$). Due to sp^3 hybridisation and no *lp* of electrons on P-atom, H_3PO_4 molecule has tetrahedral shape which has been shown in Fig. 15.33.

The solid and concentrated solution of orthophosphoric acid has various H_3PO_4 tetrahedral molecules connected together by H–bonds (Fig. 15.34). The existence of H-bonds between different tetrahedral H_3PO_4 molecules increases the viscosity of the concentrated

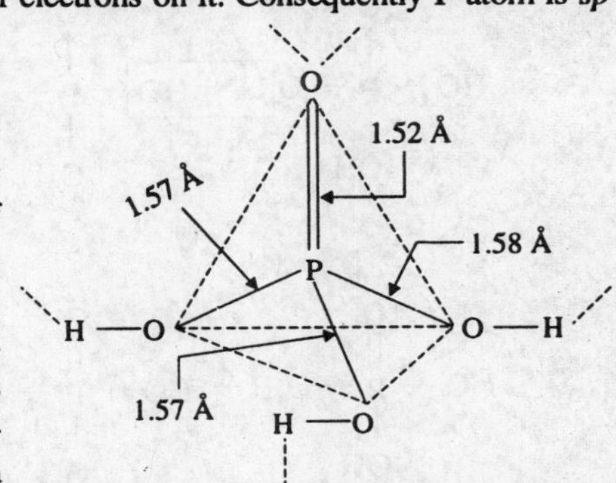

Fig. 15.34 : Structure of H_3PO_4 in solid and concentrated solution form.

solution of the acid and thus accounts for the syrupy nature of the acid. H-bonds have been shown by dotted lines.

11. Structure of orthophosphate ion, PO_4^{3-}

We have said above that H_3PO_4 is a tribasic acid and hence ionises as $H_3PO_4 \rightleftharpoons 3H^+ + PO_4^{3-}$, PO_4^{3-} ion is called *orthophosphate* or simply as *phosphate ion.* The electron dot structure of this ion is (A) or (B).

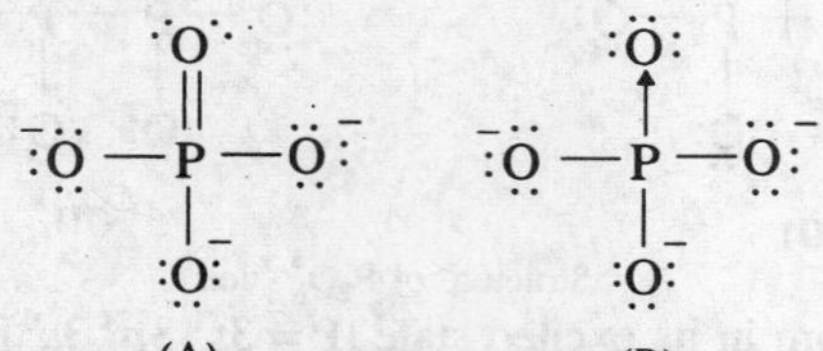

(A) (B)

In structure (A), P-atom in its excited state [P = $3s^1\ 3sp^3\ 3d^1$] is sp^3 hybridised [P = $(sp^3)^1\ (sp^3)^1\ (sp^3)^1 (sp^3)^1\ 3d^1$]. In structure (B), P-atom in its ground state [P = $3s^2\ 3p^3$] is also sp^3 hybridised [P = $(sp^3)^2\ (sp^3)^1\ (sp^3)^1\ (sp^3)^1$]. (P → O) coordinate σ-bond is obtained when completely-filled sp^3 hybrid orbital donates its electron pair to the vacant 2*p* orbital of O-atom (O = $2s^2\ 2p^2\ 2p^2\ 2p^0$).

Due to sp^3 hybridisation of P-atom and no lone pair of electrons on the central P-atom in both the structures, PO_4^{3-} ion has tetrahedral shape which has been shown in Fig. 15.35. Tetrahedral shape has also been confirmed by the Raman spectral and X-ray study of the salts containing PO_4^{3-} ion.

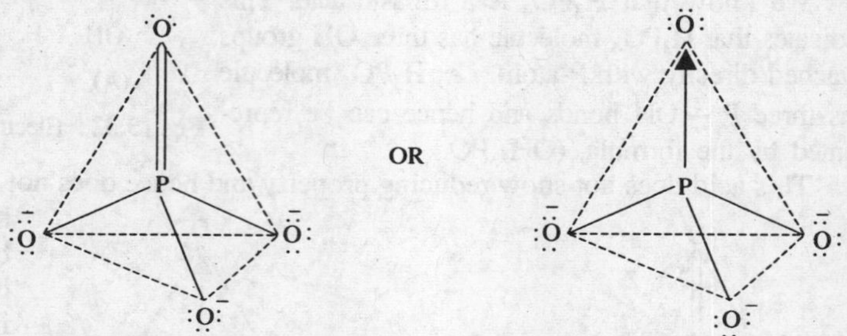

Fig. 15.35 : Tetrahedral shape of ortho phosphate ion, PO_4^{3-}.

12. Diphosphoric acid (also called pyrophosphoric acid), $H_4P_2O_7$ (*tetrabasic acid*) and diphosphate ion, $P_2O_7^{4-}$

$H_4P_2O_7$ can be regarded as being formed by removing *one* H_2O molecule "from two mole cules of H_3PO_4.

$$2H_3PO_4 \xrightarrow{-H_2O} H_4P_2O_7$$

Thus the structure of $H_4P_2O_7$ molecule and $P_2O_7^{2-}$ ion can be derived as follows:

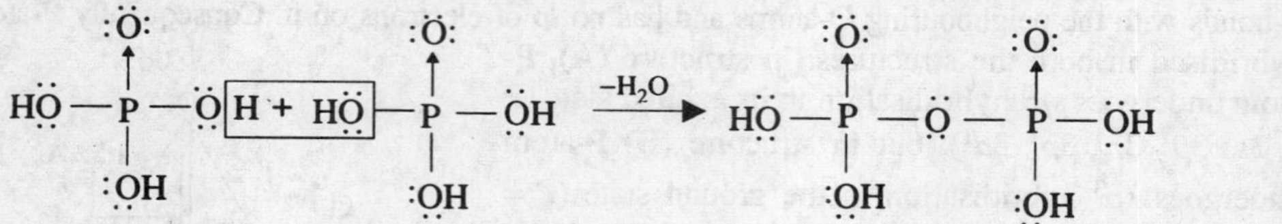

OR

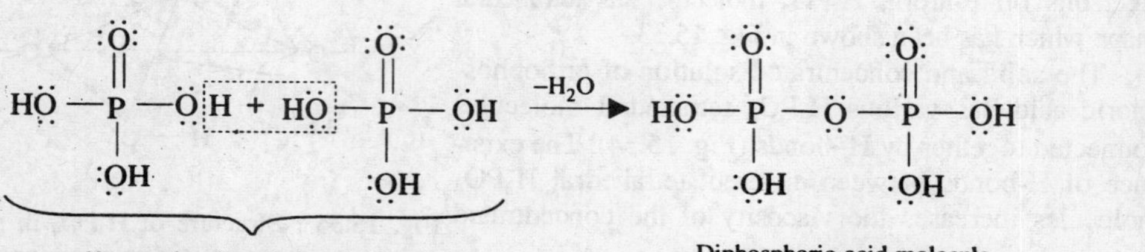

Two molecules of phosphoric acid (H_3PO_4)

Diphosphoric acid molecule ($H_4P_2O_7$)

The presence of four OH groups attached with P-atom confirms the tetrabasic nature of $H_4P_2O_7$.

Obviously the structure of pyrophate ion ($P_2O_7^{2-}$) can be written as:

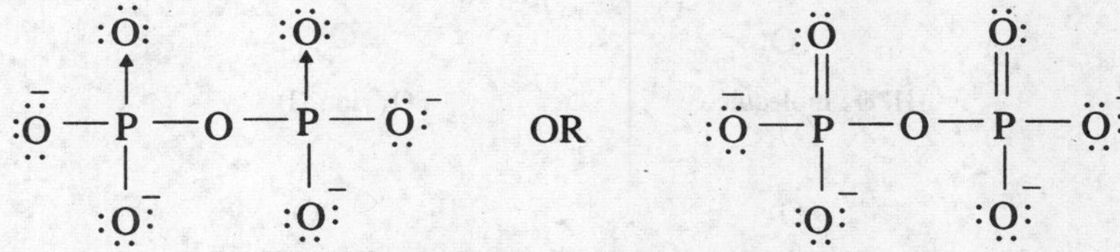

Structure of pyro (di) phosphate ion, $P_2O_7^{4-}$

The electron dot structures of $H_4P_2O_7$ molecule and $P_2O_7^{4-}$ ion as given above show that each P-atom is sp^3 hybridised. The structure of $P_2O_7^{4-}$ ion can be obtained by joining two PO_4 tetrahedrons in such a way that O-atom is common to both tetrahedrons (See Fig. 15.36).

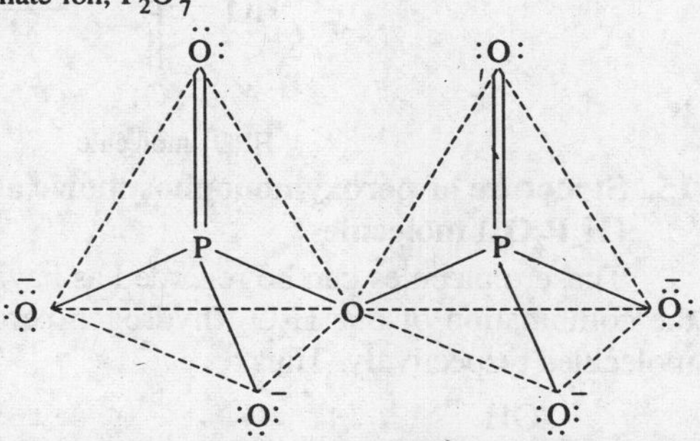

Fig. 15.36 : Structure of diphosphate (also called pyrophosphate) ion, $P_2O_7^{4-}$.

13. Structure of metaphosphoric acid molecule, HPO_3 (Monobasic)

Since metaphosphoric acid molecule (HPO_3) is regarded as having been obtained by removing one H_2O molecule from one molecule of H_3PO_4, HPO_3 molecule has structure (I) or (II).

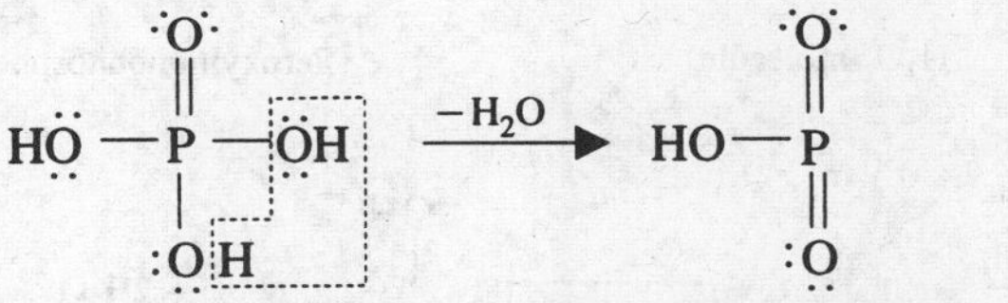

One molecule of H_3PO_4 — Structure of HPO_3 molecule (I)

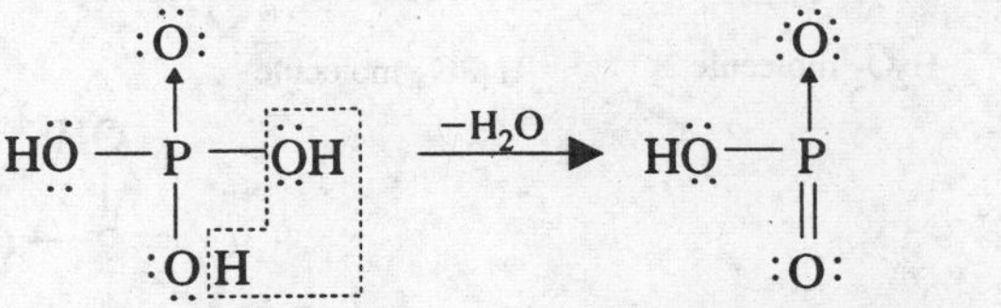

One molecule of H_3PO_4 — Structure of HPO_3 molecule (II)

The presence of one OH group attached directly with P-atom in HPO_3 molecule confirms the monobasic nature of the acid. In HPO_3 molecule P-atom is sp^2 hybridised and hence this molecule has *trigonal planar shape.*

14. Structure of metaphosphate ion PO_3^-

Since HPO_3 is a monobasic acid, it ionises as $HPO_3 \rightleftharpoons H^+ + PO_3^-$. PO_3^- ion is called metaphosphate ion. This ion has structure (I) or (II) shown below:

In this ion P-atom is sp^2 hybridised and hence PO_3^- ion has *triangular planar* shape. One (P – O) π-bond is a P (3*p*) – O(2*p*) or (*p* – *p*) π-bond while the other (P – O) π-bond is a P(3*d*) – O(2*p*) or (*d* – *p*)π-bond.

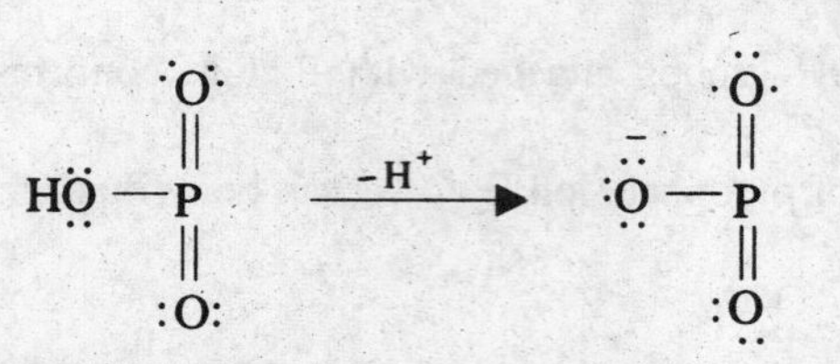

HPO_3 molecule $\qquad$ PO_3^- ion (I)

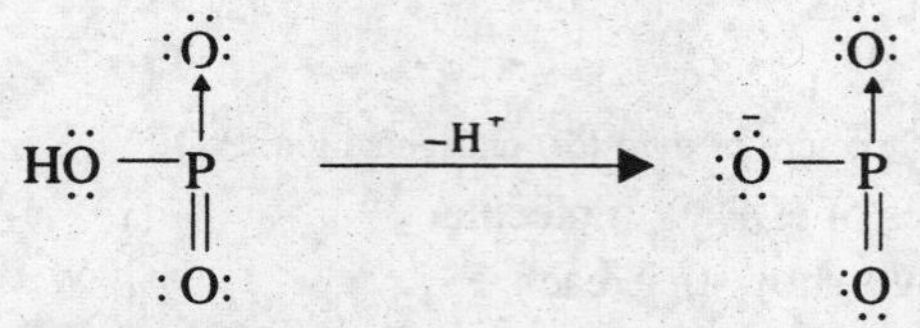

HPO_3 molecule $\qquad$ PO_3^- ion (II)

15. Structure of peroxymonophosphoric acid (H_3PO_5) and peroxydiphosphoric acid ($H_4P_2O_8$) molecule

These molecules can be regarded as having been obtained by removing one H_2O molecule by the combination of one H_2O_2 (hydrogen peroxide) and one or two H_3PO_4 (orthophosphoric acid) molecules respectively. Thus :

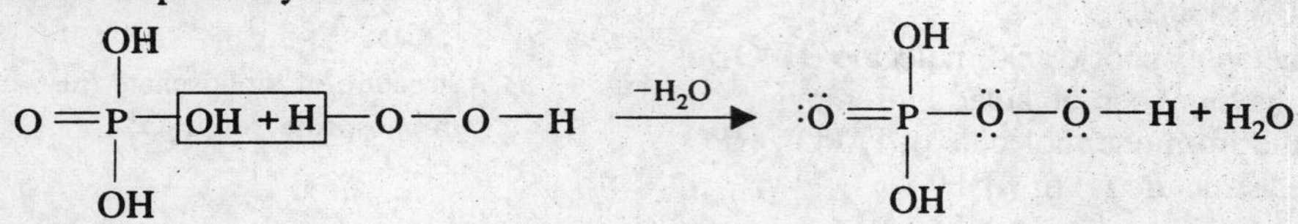

H_3PO_4 molecule $\qquad$ H_2O_2 molecule $\qquad$ Peroxymonophosphoric acid (H_3PO_5)

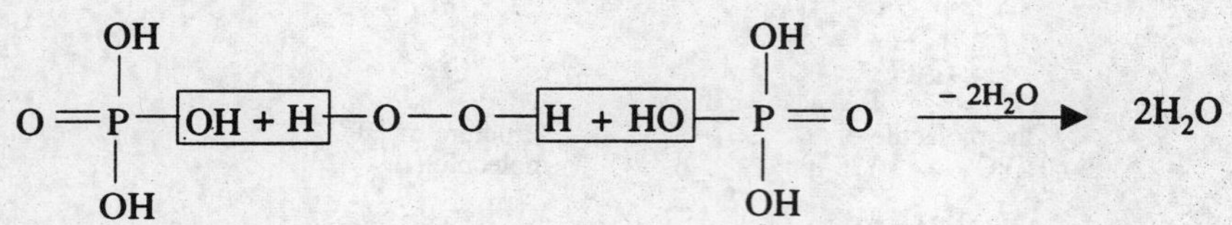

H_3PO_4 molecule $\qquad$ H_2O_2 molecule $\qquad$ H_3PO_4 molecule

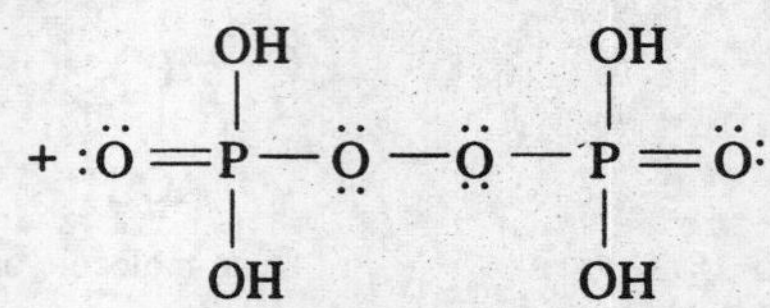

Peroxydiphosphoric acid ($H_4P_2O_8$)

The structures of H_3PO_5 and $H_4P_2O_8$ molecules as given above indicate that each P-atom in its excited state [$P = 3s^1\ 3p^3\ 3d^1$] and each O-atom in its ground state [$O = 3s^2\ 3p^4$] is sp^3 hybridised.

$$P = (sp^3)^1\ (sp^3)^1\ (sp^3)^1\ (sp^3)^1\ 3d^1$$

$$O = (sp^3)^2\ (sp^3)^2\ (sp^3)^1\ (sp^3)^1$$

Two lone pairs of electrons on O-atom reside in two sp^3 hybrid orbitals. Each (O – P) π-bond is obtained by $3d$ (P)–$2p$ (O) overlap, *i.e.*, (O – P) π-bond is a $(d-p)$ π -bond. (P – O) σ-bond is due to sp^3 (P) – sp^3 (O) overlap. Similarly (O – O) σ-bond is also due to sp^3 (O) – sp^3 (O) overlap. The structure of both peroxyphosphoric acids contains one peroxy group, O_2^{2-} (– O – O –).

Oxy acids and oxy anions of sulphur

Here we shall consider the structure and geometry of the following oxy acids and oxy anions derived from them.

1. Structure of sulphoxylic acid molecule, H_2SO_2

It is a dibasic acid and hence contains two OH groups attached with the central S-atom. Thus H_2SO_2 molcule can be represented by the structure, $H\ddot{\underset{\cdot\cdot}{O}} - \ddot{\underset{\cdot\cdot}{S}} - \ddot{\underset{\cdot\cdot}{O}}H$. This structure shows that S-atom is sp^3 hybridised and H_2SO_2 molecule has angular geometry.

2. Structure of sulphurous acid molecule, H_2SO_3

Following two structures have been suggested for H_2SO_3 molecule.

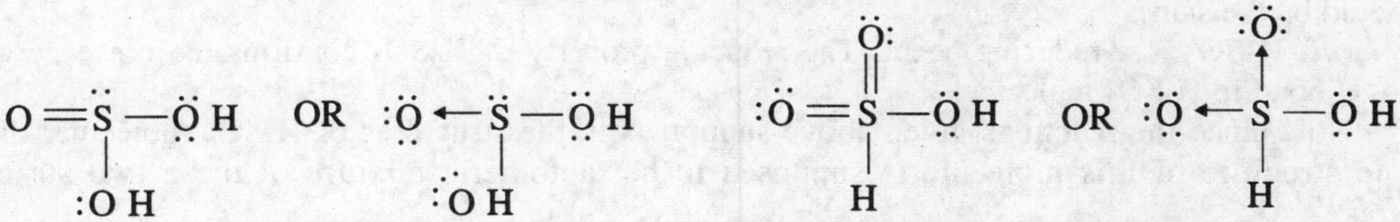

(I) Symmetrical structure (II) Unsymmetrical structure

Reactions supporting the symmetrical structure (I). (*i*) H_2SO_3 is obtained by the hydrolysis of thionyl chloride, $OSCl_2$ (which has symmetrical structure) by two molecules of water.

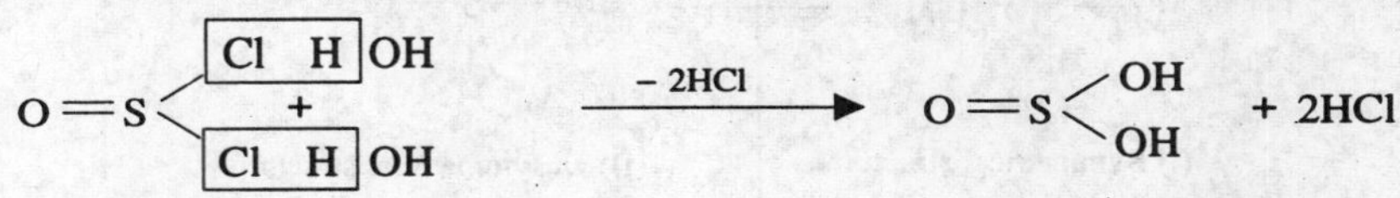

$(OSCl_2)$ (H_2SO_3)

(*ii*) When $OS.Cl_2$ is treated with C_2H_5OH and then with NaOH, Na_2SO_3 is obtained. If it is accepted that the incoming group takes the position of the outgoing group, Na_2SO_3 should have the symmetrical structure as shown below :

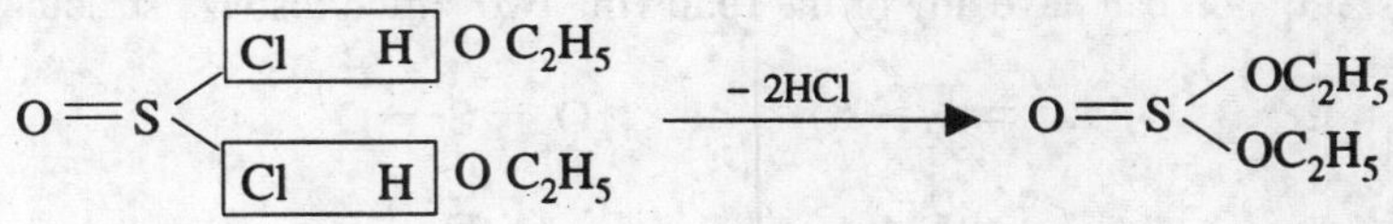

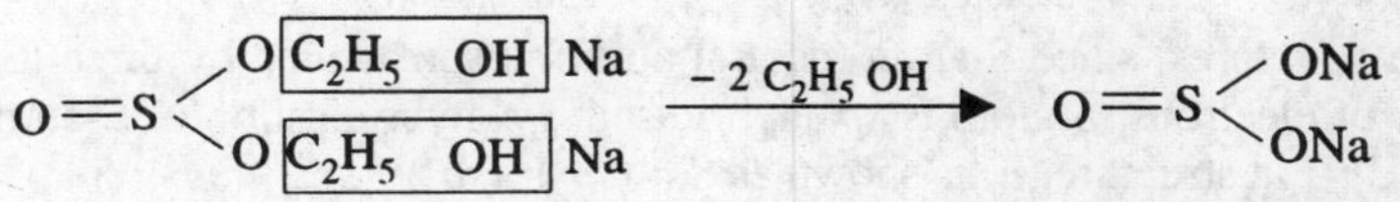

(Na_2SO_3)

$(OSCl_2)$

(*iii*) H_2SO_3 is a dibasic acid and gives two types of salts *e.g.*, $NaHSO_3$ and Na_2SO_3. Its dibasic nature indicates that H_2SO_3 molecule has symmetrical structure which contains two OH groups attached directly with S-atom.

Reactions supporting the unsymmetrical structure (II). (*i*) $NaKSO_3$ can be obtained either by the action of $NaHSO_3$ on KOH or by the action of $KHSO_3$ on NaOH.

$$NaHO_3 + KOH \longrightarrow NaKSO_3 + H_2O$$

$$KHSO_3 + NaOH \longrightarrow KNaSO_3 + H_2O$$

It has been observed that $NaKSO_3$ and $KNaSO_3$ formed in these reactions have different structures. This indicates that if H_2SO_3 molecule has symmetrical structure, $NaKSO_3$ and $KNaSO_3$ should have the same structure. Thus H_2SO_3 has an unsymmetrical structure and the above two salts have the following structure:

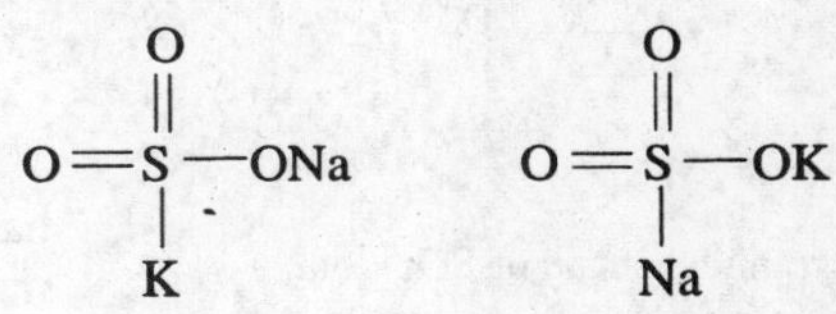

(*ii*) Diethyl sulphite, $(C_2H_5)_2SO_3$ can be obtained either by the action of $OS.Cl_2$ on C_2H_5OH or by the action of Na_2SO_3 on C_2H_5I.

$$OS.Cl_2 + 2C_2H_5OH \longrightarrow (C_2H_5)_2\ SO_3\ (\text{b.pt.} = 161°C) + 2HCl$$

$$NaSO_3 + 2C_2H_5I \longrightarrow (C_2H_5)_2\ SO_3\ (\text{b.pt.} = 213.4°C) + 2NaI$$

It has been observed that diethyl sulphites formed in the above reactions are different from each other, since the former has b.pt. = 161°C and the latter has b.pt. = 213.4°C. This indicates that if H_2SO_3 molecule has symmetrical structure, diethyl sulphites formed in both the above reactions should be the same.

(*iii*) H_2SO_3 is a reducing agent. The reducing property of H_2SO_3 confirms the presence of S—H bond in H_2SO_3 molecule.

Now since the reactions given above support both the structures of H_2SO_3 molecule, the true structure of this molecule is supposed to be tautomeric mixture of these two structures.

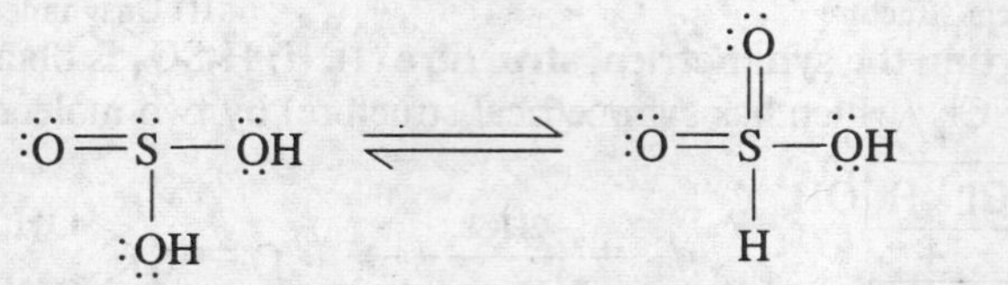

(I) Symmetrical structure (II) Unsymmetrical structure

3. Structure of sulphite ion, SO_3^{2-}

X-ray study of $Na_2S_2O_3$ ionic crystal has shown that SO_3^{2-} ion has a structure which can be derived by removing two H-atoms from two OH groups of the symmetrical structure of H_2SO_3 molecule. Thus SO_3^{2-} ion can have any of the following two structures *viz.*, structures (*a*) and (*b*).

Structure (*a*) Structure (*b*)

In both these structures, since S-atom (central atom) is surrounded by three (S – O) σ-bonds and one lone pair of electrons, S-atom is sp^3 hybridised. sp^3 hybridisation of S-atom in its excited state [S = $3s^2\ 3p^3\ 3d^1$] in structure (*a*) is shown as S = $(sp^3)^2\ (sp^3)^1\ (sp^3)^1\ (sp^3)^1\ 3d^1$. It may be noted that $3d$ orbital does not participate in sp^3 hybridisation and remains as such. This orbital which is singly-filled overlaps with the half-filled $2p$ orbital of O-atoms and gives (S—O) π -bond. Thus π-bond is a (*d–p*) π -bond. (S—O) σ-bonds are obtained in the usual manner.

Although the spatial arrangement of four electron pairs (three σ-*bps* and one *lp*) round the central S-atom is tetrahedral, due to the presence of one *lp* on S-atom and greater magnitude of (*lp–bp*) repulsion than (*bp–bp*) repulsion the expected tetrahedral shape of SO_3^{2-} ion gets distorted and becomes *trigonal pyramidal* (distorted *tetrahedral shape*) (See Fig. 15.37).

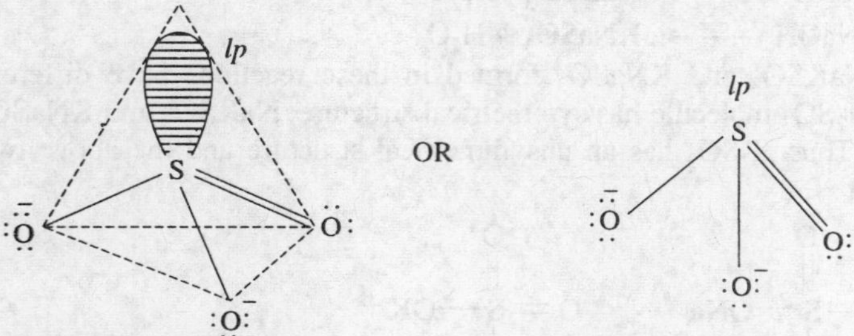

Fig. 15.37 : Trigonal pyramidal shape of SO_3^{2-} ion.

4. Structure of hyposulphurous acid (also called hydro sulphurous acid or dithionous acid) molecule, $H_2S_2O_4$

Following three structures have been suggested for $H_2S_2O_4$ molecule.

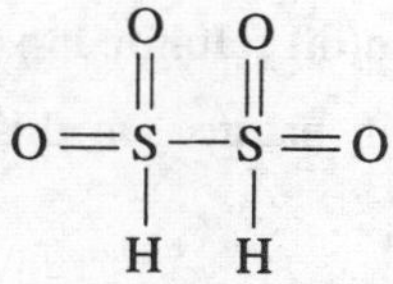

(I)

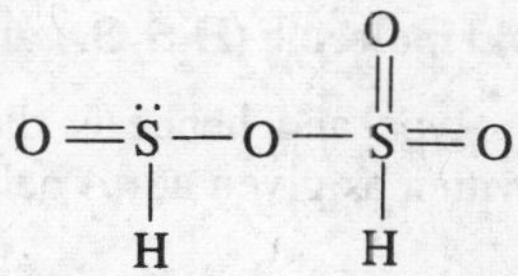

(II)

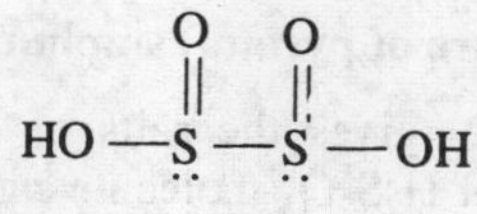

(III)

Structures (I) and (II) explain the reducing property of the acid while structure (III) explains the dibasic nature of the acid. The acid gives the salts of $M_2S_2O_4$ type (*e.g.*, $Na_2S_2O_4$, called sold.hyposulphite).

5. Structure and geometry of hyposulphite ion, $S_2O_4^{2-}$

Since hyposulphurous acid ($H_2S_2O_4$) is a dibasic acid and has the structure (*a*), $S_2O_4^{2-}$ ion has structure (*b*).

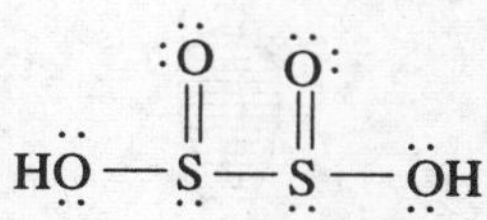

(*a*) Structure of $H_2S_2O_4$ molecule

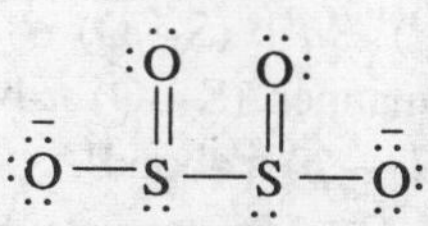

(*b*) Structure of $S_2O_4^{2-}$ ion

Since in the structure of $S_2O_4^{2-}$ ion, each S-atom is surrounded by three σ-*bps* and one *lp*, S-atom in its excited state [S = $3s^2\ 3p^3\ 3d^1$] is sp^3 hybridised [S = $(sp^3)^2\ (sp^3)^1\ (sp^3)^1\ (sp^3)^1\ 3d^1$]. One of the four sp^3 hybrid orbitals contains a lone pair of electrons while other three sp^3 hybrid orbitals are having one electron each. (S—S) σ-bond is obtained by the overlap of two singly-filled sp^3 hybrid orbitals on both S-atoms. ($\overline{O}$—S) and (O—S) σ-bonds are obtained by the overlap of singly-filled sp^3 hybrid orbitals of S-atom with the singly-filled 2*p* orbital of the surrounding two O-atoms. (S—O) π-bond results by *d*(S) – *p*(O) overlap.

6. Structure of thiosulphurous acid molecule ($H_2S_2O_2$) and thiosulphite ion ($S_2O_2^{2-}$)

$H_2S_2O_2$ is a dibasic acid and hence has structure (*a*). Obviously $S_2O_2^{2-}$ ion can be represented by structure (*b*).

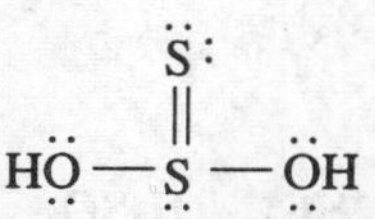

(*a*) Structure of thiosulphorus acid molecule ($H_2S_2O_2$)

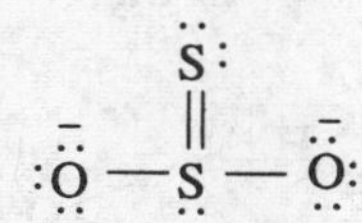

(*b*) Structure of $S_2O_2^{2-}$ ion

Since the central S-atom in $S_2O_2^{2-}$ ion is surrounded by three σ-*bps* and one one pair of electrons, S-atom in excited state [S = $3s^2\ 3p^3\ 3d^1$] is sp^3 hybridised state [S = $(sp^3)^2\ (sp^3)^1\ (sp^3)^1\ (sp^3)^1\ 3d^1$]. It may be noted that one sp^3 hybrid orbital contains one lone pair of electrons (*lp*). (S—O^-) σ-bonds are obtained in the usual manner. (S—S) π-bond results by the overlap of *p*-and *d*-orbitals of two S-atoms, *i.e.*, (S—S) π-bond is a (*p*—*d*) π-bond.

Although the spatial arrangement of four sp^3 hybrid orbitals in space is tetrahedral, due to the presence of one *lp* in one of the hybrid orbitals, $S_2O_2^{2-}$ ion has trigonal pyramidal shape shown in Fig. 15.38.

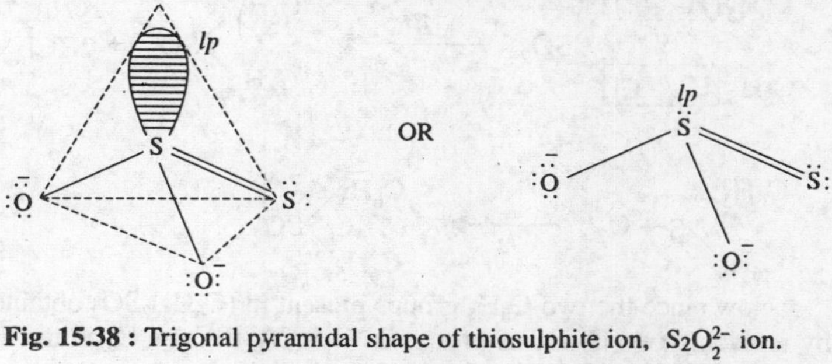

Fig. 15.38 : Trigonal pyramidal shape of thiosulphite ion, $S_2O_2^{2-}$ ion.

7. Structure of pyro(di) sulphurous acid molecule ($H_2S_2O_5$) and pyro(di) sulphite ion ($S_2O_5^{2-}$)

$H_2S_2O_5$ gives the salts of $K_2S_2O_5$ type and hence is dibasic in nature. On this basis the structure of $H_2S_2O_5$ molecule can be written as given at (*a*) below.

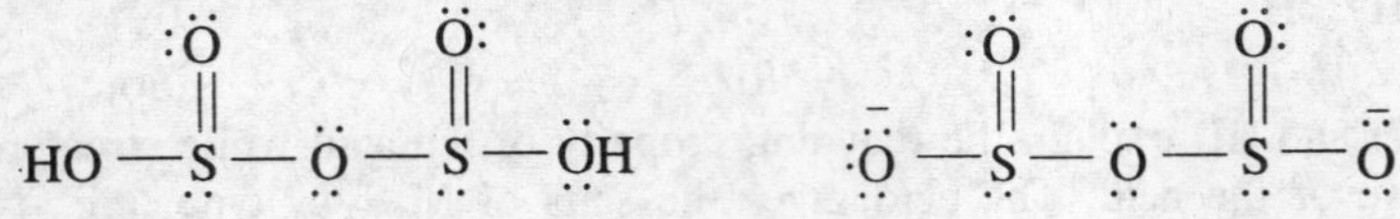

(*a*) Structure of ($H_2S_2O_5$) molecule. (*b*) Structure of $S_2O_5^{2-}$ ion.

Obviously the structure of $S_2O_5^{2-}$ ion is that which is shown at (*b*). Structure of $S_2O_5^{2-}$ as given above shows that each S-atom is surrounded by three σ-*bps* and one *lp*. Thus each S-atom in the excited state [S = $3s^2\ 3p^3\ 3d^1$] is sp^3 hybridised. [S = $(sp^3)^2\ (sp^3)^1\ (sp^3)^1 (sp^3)^1\ 3d^1$]. (S—O) σ- bonds are formed in their usual manner. (S—O) π-bond is a S (*d*) – O (*p*) π -bond [(*d* — *p*) π -bond].

The structure of $S_2O_5^{2-}$ ion consists of two SO_3 tetrahedrons which are joined together through O-atom (See Fig. 15.39).

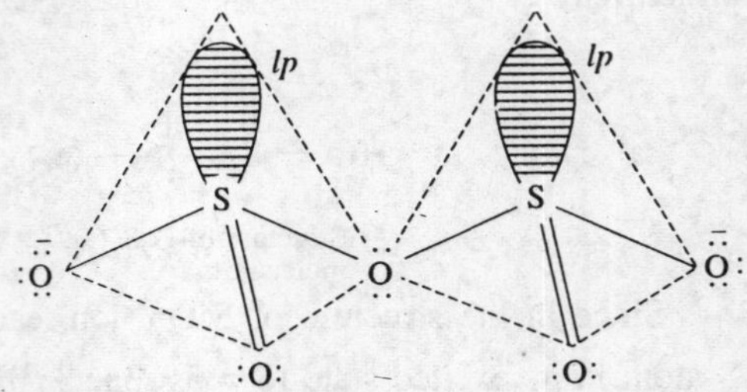

Fig. 15.39 : Structure of $S_2O_5^{2-}$ ion which consists of two SO_3 tetrahedrons joined together through a common O-atom.

8. Structure of sulphuric acid molecule, H_2SO_4

It is a dibasic acid and hence contains two OH groups which are attached directly with S-atom. Thus H_2SO_4 can also be represented by the formula, $(OH)_2SO_2$. The presence of two OH groups in the molecule is confirmed by the fact that when H_2SO_4 reacts with PCl_5, two OH groups present in H_2SO_4 or $(OH)_2SO_2$ molecule are replaced by two Cl-atoms and sulphuryl chloride (Cl_2SO_2) is obtained. When Cl_2SO_2 thus obtained is treated with H_2O, H_2SO_4 is obtained. These reactions can be explained only when it is assumed that H_2SO_4 molecule contains two OH groups.

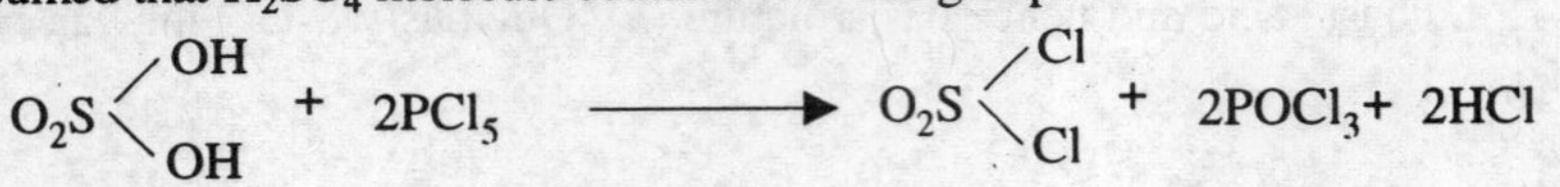

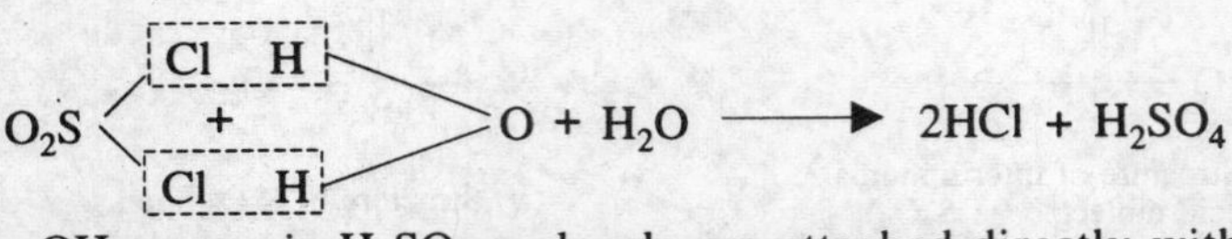

That two OH groups in H_2SO_4 molecule are attached directly with S-atom is also confirmed by the fact that when sulphuryl chloride ($O_2S.Cl_2$) reacts with C_6H_6, diphenyl sulphone [$O_2S\ (C_6H_5)_2$] is obtained. Diphenyl sulphone is also formed by the oxidation of diphenyl sulphide [$(C_6H_5)_2S$] in which two C_6H_5 groups are directly attached to S-atom.

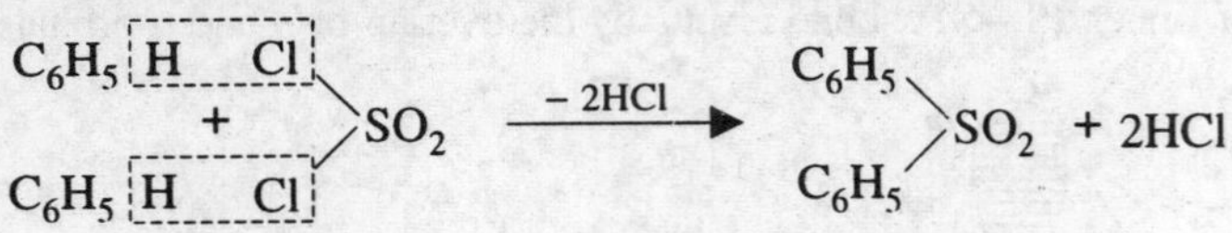

... (*i*)

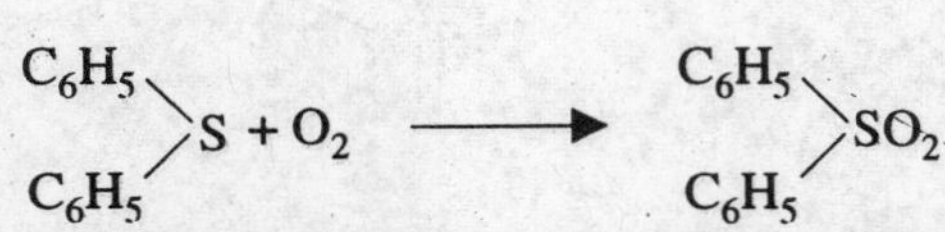

Now since the two C_6H_5 groups present in $(C_6H_5)_2SO_2$ obtained by reaction (*i*) are introduced by replacing two Cl-atoms present in $O_2S.Cl_2$, the two Cl-atoms in $O_2S.Cl_2$ and consequently two OH groups in H_2SO_4 must also be directly linked with S-atom.

On the basis of the above facts, any of the following two structures can be suggested for H_2SO_4 molecule.

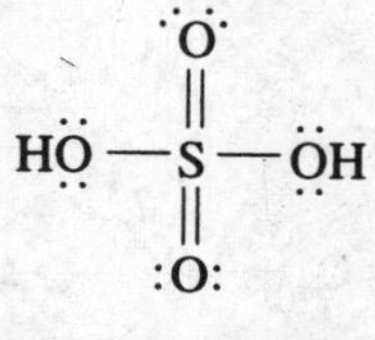

(I) (Based on hexacovalency of S-atom)

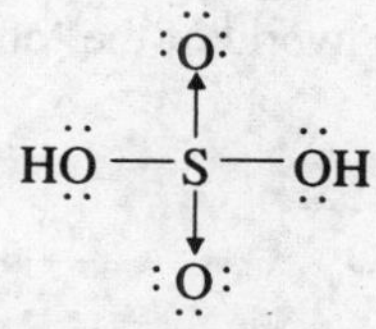

(II) (Based on tetracovalency of S-atom)

Now it is assumed that the structure of H_2SO_4 molecule is a resonance hybrid of the above two structures namely (I) and (II). Thus:

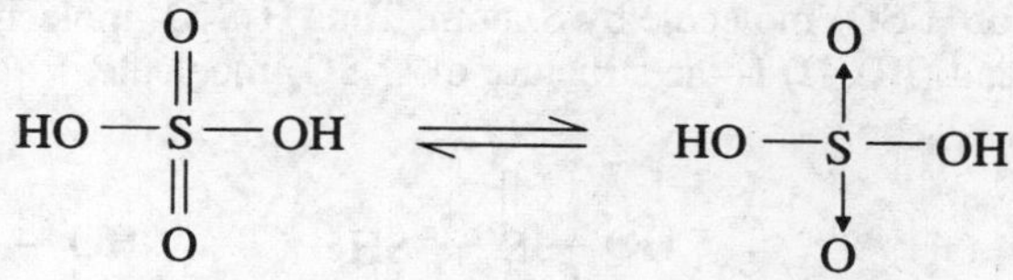

(I) (II)

9. Structure of sulphate ion, SO_4^{2-}

X-ray analysis of the ionic crystals containing SO_4^{2-} ion has shown that SO_4^{2-} ion has a structure that can be obtained by removing two H-atoms from two OH groups of structures (I) or (II) of H_2SO_4 molecule (shown above). Thus SO_4^{2-} ion can have any of the structures namely (A) and (B) shown below:

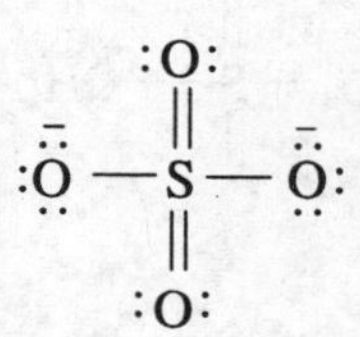

(A)

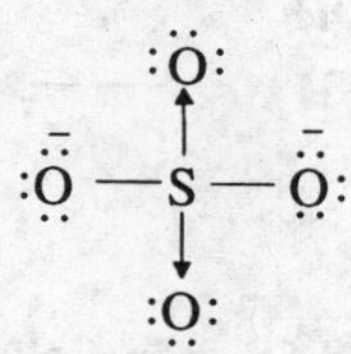

(B)

Both these structures are Lewis structures of SO_4^{2-} ion. It may be noted that S-atom (central atom) in structure (A) is surrounded by 12 eletrons (6 electron pairs) while S-atom in structure (B) surrounded by 8 electrons (4 electron pairs). Thus it is structure (B) which obeys octet rule. In both these structures since S-atom is surrounded by four σ-bps, S-atom (central atom) in its excited state [S= $3s^1\ 3p^3\ 3d^2$] is sp^3 hybridised. sp^3 hybridisation of S-atom in the excited state of S-atom in structure (A) is shown as S = $(sp^3)^1\ (sp^3)^1\ (sp^3)^1\ (sp^3)^1\ 3d^1\ 3d^1$. It may be noted that all the four (S—O) σ-bonds are obtained by the overlap between one sp^3 hybrid orbital on S-atom and $2p$ orbital on O-atom while the two (S—O) π -bonds are obtained by $3d$ (S) – $2p$ (O) overlap. Thus each (S—O) π-bond is a (d — p) π-bond.

It is due to sp^3 hybridisation of S-atom and no lone pair of electrons on S-atom, SO_4^{2-} ion has the expected *tetrahedral shape* (See Fig. 15.40).

Resonance in ion. In SO_4^{2-} ion all the four sulphur–oxygen bond distances are

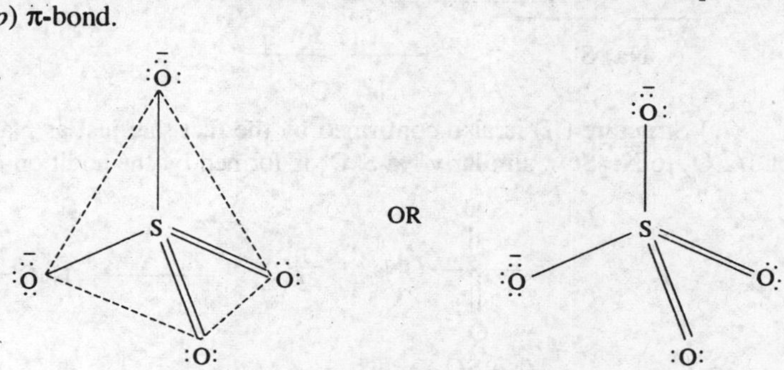

Fig. 15.40 : Tetrahedral shape of SO_4^{2-} ion.

equal (= 1.50 Å). This bond length is intermediate between the calculated bond length for single S — O bond (= 1.70 Å) and for double S = O bond (=1.49 Å). This indicates that the true structure of SO_4^{2-} ion is a resonance hybrid of the following structures:

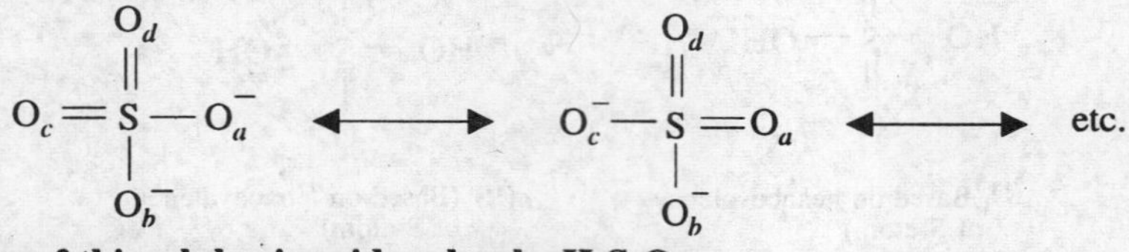

10. Structure of thiosulphuric acid molecule, $H_2S_2O_3$

The structure of $H_2S_2O_3$ molecule can be regarded as having been obtained by replacing one O-atom in the structure of H_2SO_4 molecule by S-atom. Thus $H_2S_2O_3$ molecule can have any of the two structures *viz.*, (II) and (III). (I) is the structure of H_2SO_4 moelcule.

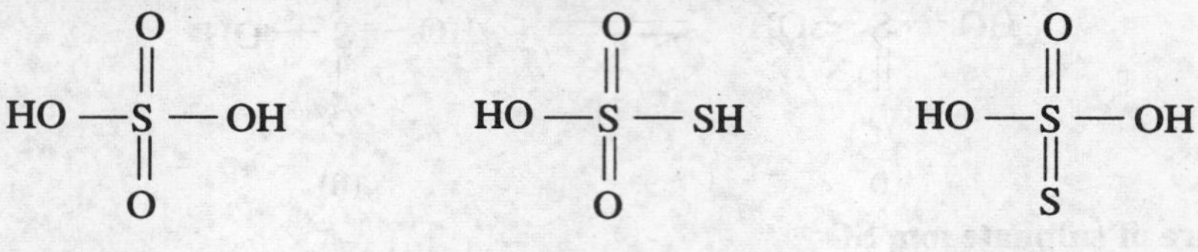

(I) (II) (III)

Structure (II). The following reactions support structure (II) for $H_2S_2O_3$ molecule.

(*i*) When $H_2S_2O_3$ is hydrolysed by H_2O, H_2S and H_2SO_4 are obtained. The formation of these products support structure (II) for $H_2S_2O_3$ molecule.

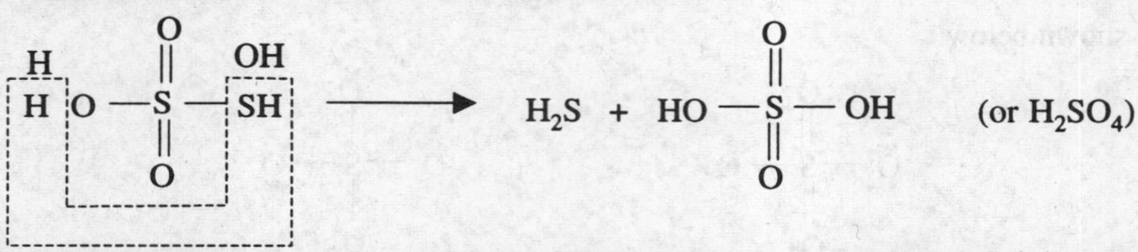

Similarly when $Ag_2S_2O_3$ is hydrolysed, Ag_2S and H_2SO_4 are obtained. This also supports structure (II).

(*ii*) $H_2S_2O_3$ can be obtained by the action of phosphorus pentasulphide (P_2S_5) on H_2SO_4. This implies the presence of SH group in $H_2S_2O_3$ molecule.

$$H_2SO_4 + P_2S_5 \longrightarrow H_2S_2O_3 + \ldots\ldots\ldots\ldots$$

(*iii*) The formation of $Na_2S_2O_3$ by the action of iodine on a mixture of Na_2S and Na_2SO_3 solutions (Spring's method) also supports structure (II) for $Na_2S_2O_3$ or $H_2S_2O_3$ molecule.

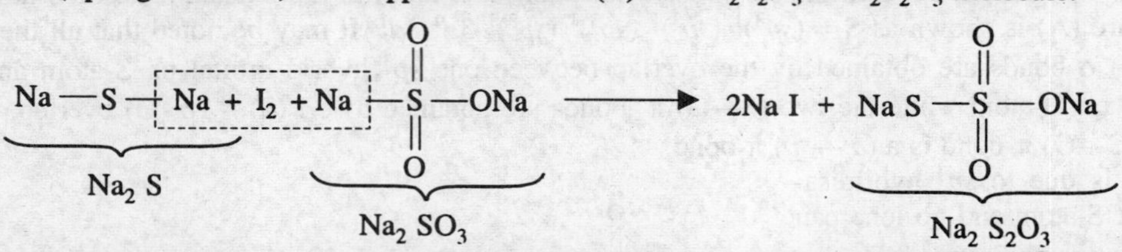

(*iv*) Structure (II) is also confirmed by the fact that just as Na_2SO_4 is formed by the addition of 1/2 O_2 to Na_2SO_3, similarly $Na_2S_2O_3$ is formed by the addition of S to Na_2SO_3.

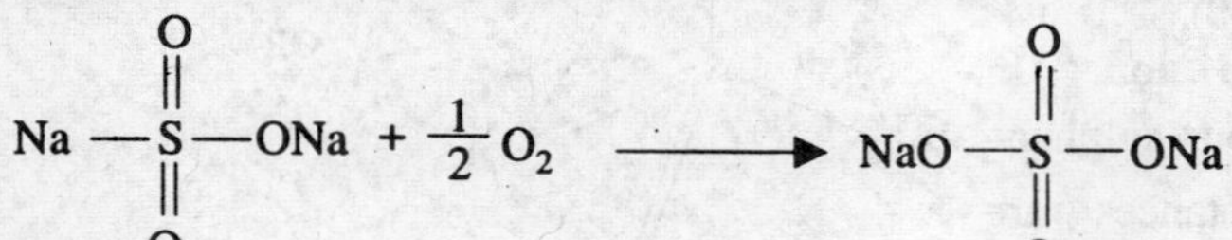

(Na_2SO_3) (Na_2SO_4)

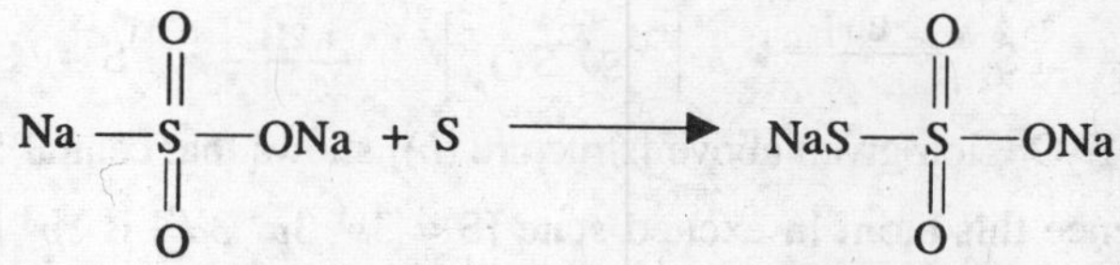

(Na_2SO_3) ($Na_2S_2O_3$)

(*v*) $Na_2S_2O_3$ reacts with C_2H_5Br to give ethyl derivative (A) which, on hydrolysis, gives ethylmercaptan (C_2H_5SH).

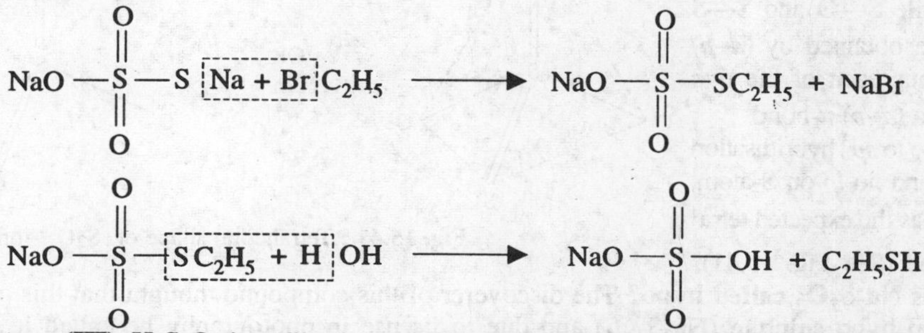

Structure (III). $H_2S_2O_3$ is a dibasic acid (but forms only one type of salts) and hence contains two OH groups attached directly with the central S-atom. The dibasic nature of the acid supports structure (III) for $H_2S_2O_3$ molecule.

Now it is assumed that structures (II) and (III) are in equilibrium with each other and hence the structure of $H_2S_2O_3$ molecule is a tautomeric form of these two structures:

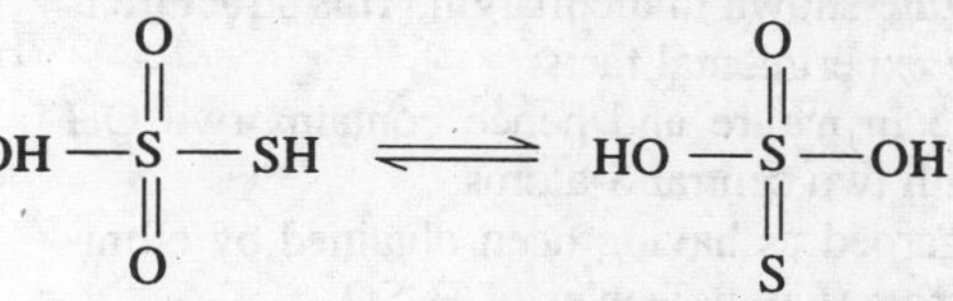

(II) (III)

11. Structure of thiosulphate ion $S_2O_3^{2-}$

The structure of $S_2O_3^{2-}$ ion is obtained by removing both H-atoms of OH groups from the structure of $H_2S_2O_3$ molecule. Thus $S_2O_3^{2-}$ ion has structure (*b*). It may be seen that $S_2O_3^{2-}$ ion has the same structure as SO_4^{2-} ion (see below):

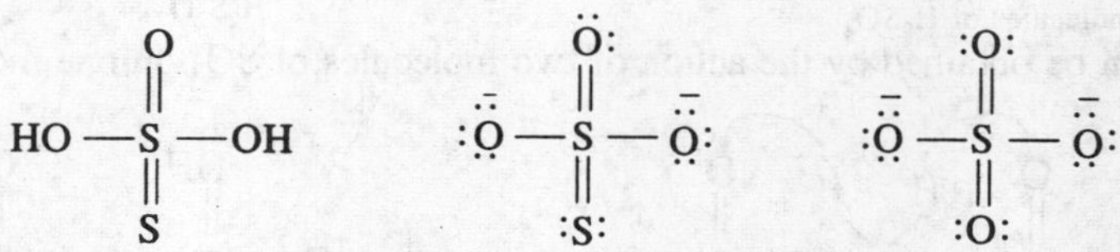

(*a*) Structure of $H_2S_2O_3$ molecule (*b*) Structure of $S_2O_3^-$ ion (Lewis structure) (*c*) Structure of SO_4^{2-} ion (Lewis structure)

In $S_2O_3^{2-}$ ion, S—S bond length = 2.01 Å and S—O bond length = 1.47 Å. These bond lengths suggest the presence of some amount of S—S and S—O π-bondings in ion. In $S_2O_3^{2-}$ ion, the two S-atoms are not equivalent. The non-equivalent nature of the two S-atoms has been confirmed by the tracer studies in which it has been observed that when the solution containing active thiosulphate ion $[^{35}S\ ^{32}S\ O_3]^{2-}$ obtained by boiling non-radioactive sulphite $[^{32}SO_3]^{2-}$ with radioactive sulphur (^{35}S) is treated with an acid (H^+), a precipitate of free sulphur is obtained and SO_2 gas is evolved. The whole radioactivity is found in free sulphur and not in the evolved SO_2. If both S-atoms in $S_2O_3^{2-}$ ion were equivalent, radioactivity should have been distributed between both the products namely S and SO_2 gas.

$$\left[{}^{32}SO_3\right]^{2-} + {}^{35}S \xrightarrow{\text{Boil}} \left[{}^{35}S\,{}^{32}SO_3\right]^{2-} \xrightarrow{+2H^+} {}^{35}S + H_2O + {}^{32}SO_2$$

Lewis structure of $S_2O_3^{2-}$ ion given above [structure (*b*)] shows that central S-atom is surrounded by four σ–*bps* and hence this atom in excited state [S = $3s^1\ 3p^3\ 3d^2$] is sp^3 hybridised [S =$(sp^3)^1\ (sp^3)^1\ (sp^3)^1\ (sp^3)^1\ 3d^1 3d^1$]. Three S—O and one S—S σ-bonds are obtained by (sp^3–*p*) overlap while S—O and S—S π-bonds are obtained by (*d–p*) overlap. Thus each of the two π-bonds is a (*d–p*) π-bond.

It is due to sp^3 hybridisation of S-atom and no *lp* on S-atom, $S_2O_3^{2-}$ ion has the expected tetrahedral shape (See Fig. 15.41).

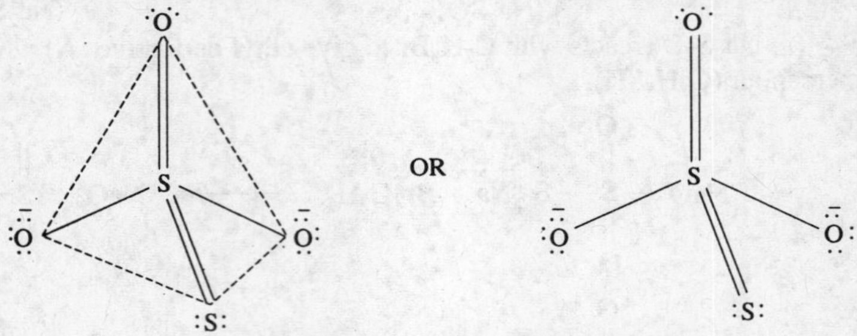

Fig. 15.41 : Tetrahedral shape of $S_2O_3^{2-}$ ion.

Why is $Na_2S_2O_3$ called hypo? The discoverer of this compound thought that this compound was sodium hypo sulphite ($Na_2S_2O_4$) and due to its use in photography he called it simply as *hypo*. But later on, the structure of this compound showed that it is sodium thiosulphate ($Na_2S_2O_3$). For this compound the word hypo is still in use by photographers.

12. Structure of pyrosulphuric acid (also called disulphuric acid or fuming sulphuric acid or Oleum) molecule, $H_2S_2O_7$

This acid has the structure shown in the margin. This structure is supported by the following experimental facts:

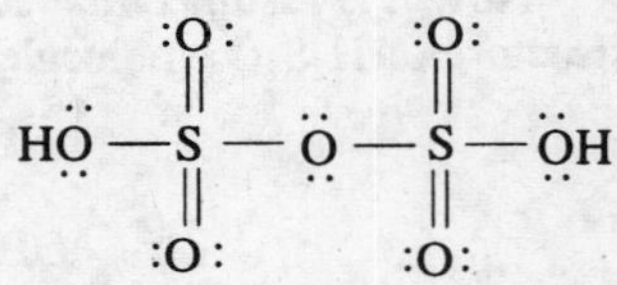

Structure of $H_2S_2O_7$ molecule

(*i*) The acid is dibasic in nature and hence contains two OH groups attached directly with two central S-atoms.

(*ii*) $H_2S_2O_7$ can be regarded as having been obtained by eliminating one H_2O molecule from two molecules of H_2SO_4.

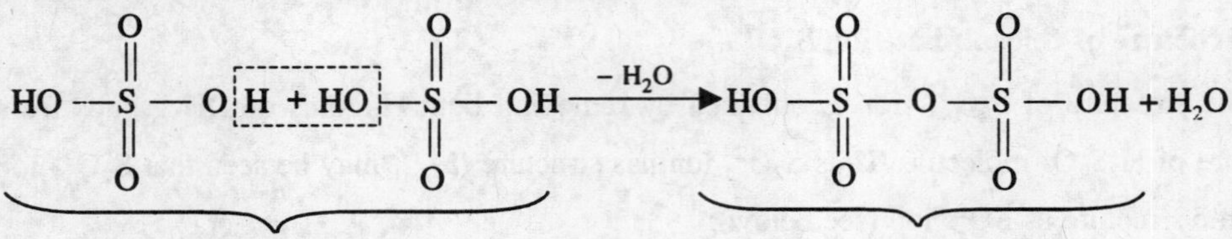

Two molecules of H_2SO_4 $\qquad$ $H_2S_2O_7$

(*iii*) $H_2S_2O_7$ can be obtained by the action of two molecules of SO_3 on one molecule of H_2O.

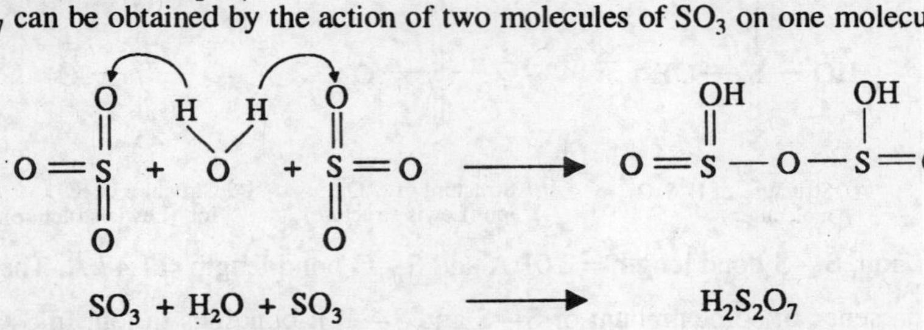

or $\qquad SO_3 + H_2O + SO_3 \longrightarrow H_2S_2O_7$

13. Structure of pyrosulphate ion, $S_2O_7^{2-}$

The structure of this ion is that which has been shown in the margin. It is obvious from this structure that each S-atom is surrounded by four σ-*bps* and hence each S-atom in its excited state [S = $3s^1\ 3p^3\ 3d^2$] is sp^3 hybridised [S = $(sp^3)^1\ (sp^3)^1\ (sp^3)^1\ (sp^3)^1\ 3d^1\ 3d^1$].

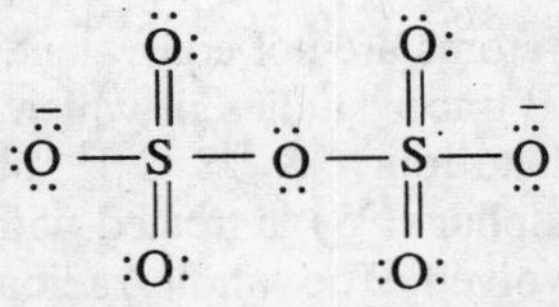

Structure of $S_2O_7^{2-}$ ion

It may be seen from sp^3 hybridisation scheme that two $3d$ orbitals donot participate in the hybridisation process. Each of the four sp^3 hybrid orbitals contains one electron. These hybrid orbitals make (S—O) σ-bonds by overlapping with the singly-filled $3p_x$ orbital on each O-atom. S—O π bonds are $3d$ (S) – $2p_x$ (O) π-bonds.

In the structure of $S_2O_7^{2-}$ ion, two SO_4 tetrahedral units are linked together by sharing one O-atom (See Fig. 15.42).

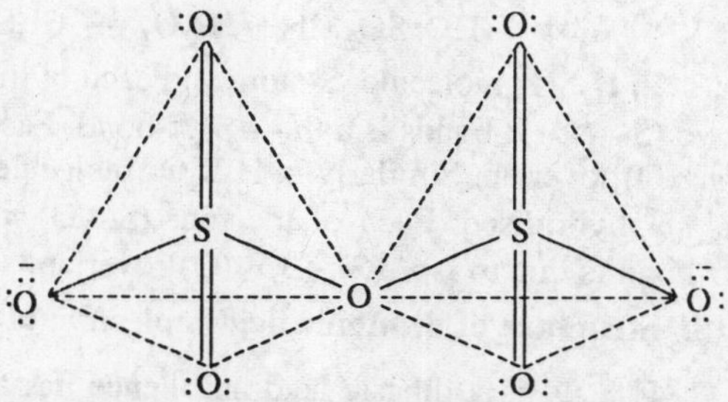

Fig. 15.42 : Structure of pyrosulphate ion, $S_2O_7^{2-}$.

14. Structure of peroxymonosulphuric acid (H_2SO_5) and peroxydisulphuric acid ($H_2S_2O_8$) molecules

These molecules can be regarded as having been obtained by removing one H_2O molecule by the combination of one H_2O_2 and one or two H_2SO_4 molecules respectively. Thus:

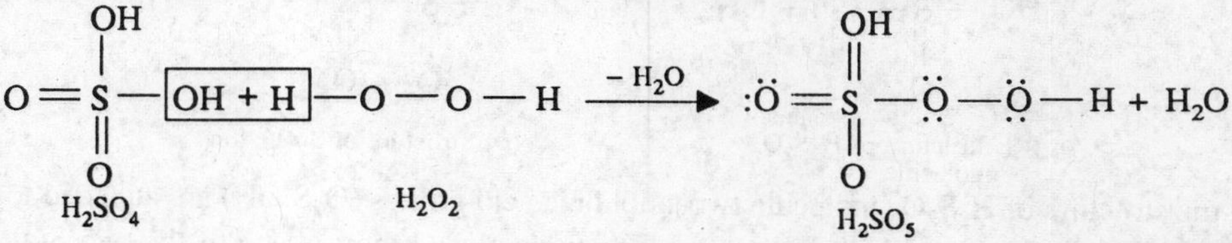

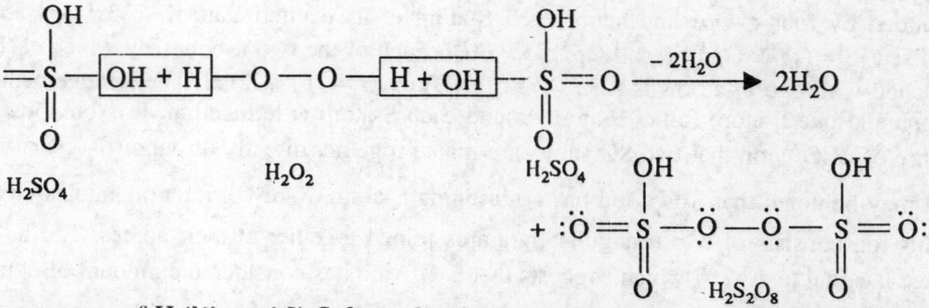

The structures of H_2SO_5 and $H_2S_2O_8$ molecules as given above are confirmed by the method of their preparation. H_2SO_5 can be obtained by the action of one mole of anhydrous hydrogen peroxide (H_2O_2) on one mole of chlorosulphonic acid [$ClSO_2(OH)$]. Similarly $H_2S_2O_8$ can be obtained by the action of one mole of anhydrous hydrogen peroxide (H_2O_2) on two moles of chlorosulphonic acid [$2Cl\ SO_2(OH)$]. This method of preparation of the acids shows that both these acids contain one peroxy linkage and hence the structures can be shown as:

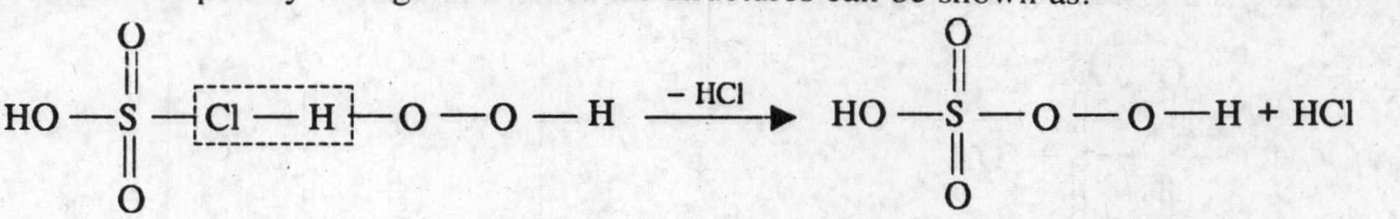

or $$HO.\ SO_2 . Cl + H_2O_2 \xrightarrow{-HCl} H_2SO_5 + HCl$$

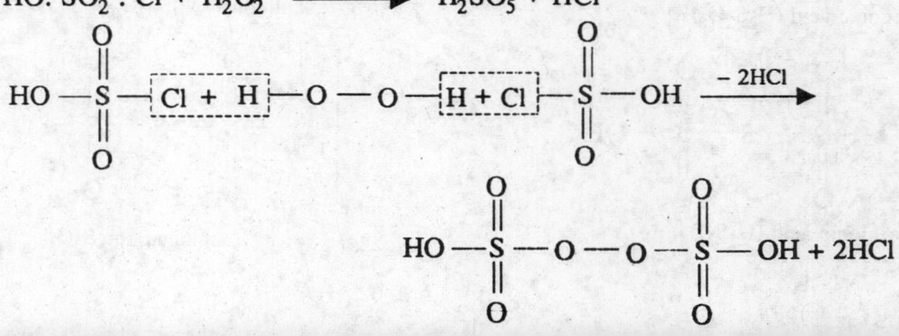

or $HO.SO_2.Cl + H_2O_2 + Cl.SO_2.OH \xrightarrow{-2HCl} H_2S_2O_8 + 2HCl$

In H_2SO_5 molecule, S-atom and each of the two central O-atoms is sp^3 hybridised. Each of the two (S—O) π-bonds is a (*d* — *p*) π-bond. Each [S — O (central)] σ-bond is obtained by [sp^3(S) – sp^3 (O)] overlap. Similarly in $H_2S_2O_8$ molecule, both S-atoms and each of the two central O-atoms is sp^3 hybridised. Each of the four (S—O) π-bonds is (*d* — *p*) π -bond. Each [S—O (central)] σ-bond is due to [sp^3 (S) – sp^3 (O)] overlap.

15. Structure of dithionic acid molecule ($H_2S_2O_6$) and dithionate ion ($S_2O_6^{2-}$)

$H_2S_2O_6$ is a dibasic acid and hence has two OH groups. One OH group is attached directly with each S-atom. Thus the structure of $H_2S_2O_6$ molecule can be written as given at (*a*) below. Obviously the structure of $S_2O_6^{2-}$ ion can be represented as given at (*b*).

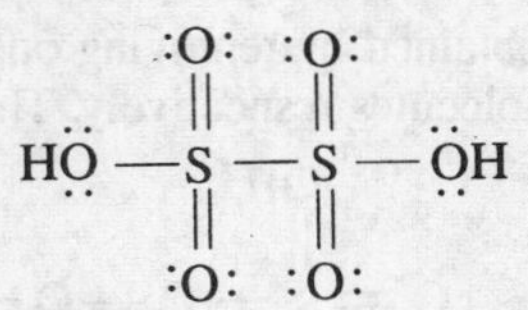

(*a*) Structure of $H_2S_2O_6$ molecule

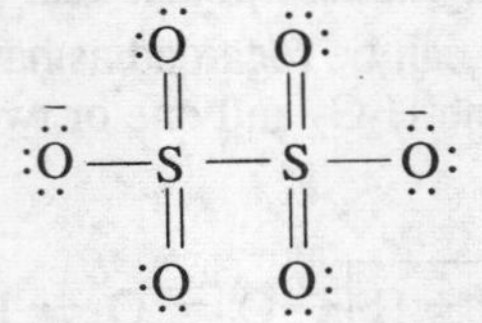

(*b*) Structure of $S_2O_6^-$ ion

In the structure of $H_2S_2O_6$ molecule two sulphonic acid groups ($O_2S.OH$) are attached together directly through a (S—S) σ-bond. In both the structures given above, each of the two S-atoms is surrounded by four ***σ–bps*** and hence each S-atom in its excited state [S = $3s^1$ $3p^3$ $3d^2$] is sp^3 hybridised [S = $(sp^3)^1$ $(sp^3)^1$ $(sp^3)^1$ $(sp^3)^1$ $3d^1$ $3d^1$]. Each of the two π-bonds by which each S-atom is attached with two O-atoms is a 3*d* (S) – 2*p* (O) or (*d* — *p*) π-bond. The arrangement of three O-atoms and one S-atom (other S-atom) round each S-atom is tetrahedral. It is obvious from the structure of $S_2O_6^{2-}$ ion that two SO_3 units are linked together directly through (S—S) σ-bond.

It may be noted that $S_2O_6^{2-}$ ion has no bridging S-atom. X-ray diffraction studies have shown that this ion consists of two triangular pyramids joined together at their apices. The S—O bond distance is equal to 1.43 Å which suggests that $S_2O_6^{2-}$ ion has considerable amount of double bond character.

16. Structure of polythionic acid molecules ($H_2S_nO_6$) and polythionate ions ($S_nO_6^{2-}$)

In diothionic acid molecule ($H_2S_2O_6$) two sulphonic acid groups ($O_2S.OH$) are attached together directly while in polythionic acid molecules (*e.g.*, $H_2S_3O_6$, $H_2S_4O_6$, $H_2S_5O_6$ and $H_2S_6O_6$) two $O_2S.OH$ groups are linked together through one or more S-atoms. Thus the structure of polythionic acid molecules and polythionate ions can be written as given below:

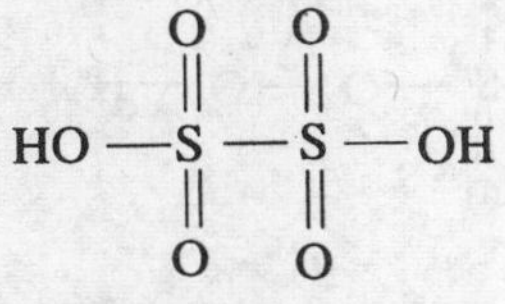

Dithionic acid ($H_2S_2O_6$)

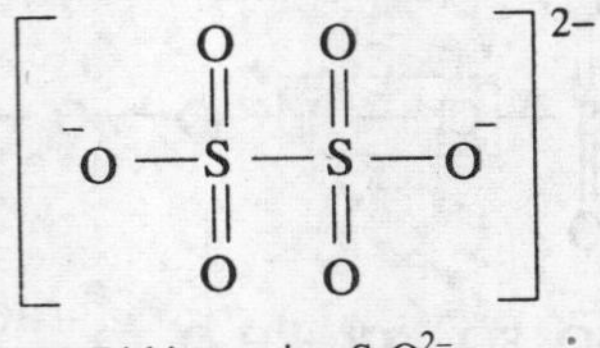

Dithionate ion $S_2O_6^{2-}$

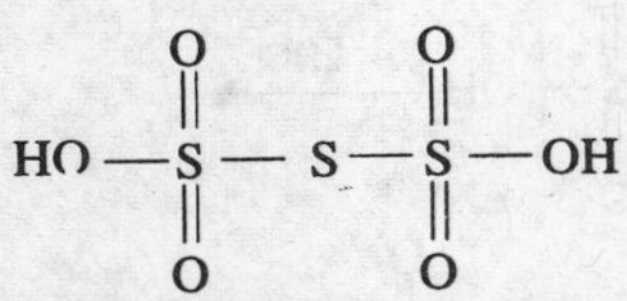

Trithionic acid ($H_2S_3O_6$)

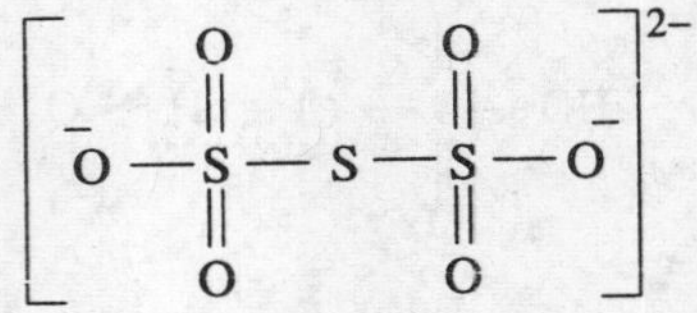

Trithionate ion $S_3O_6^{2-}$

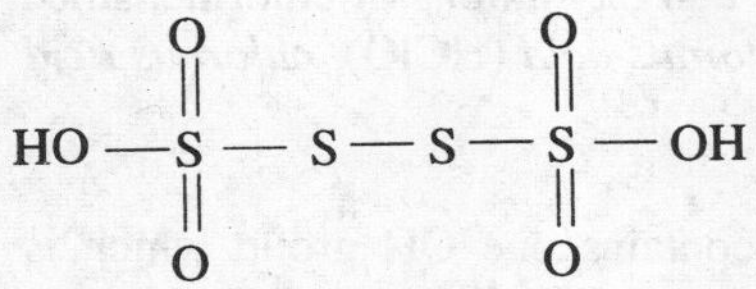

Tetrathionic acid ($H_2S_4O_6$)

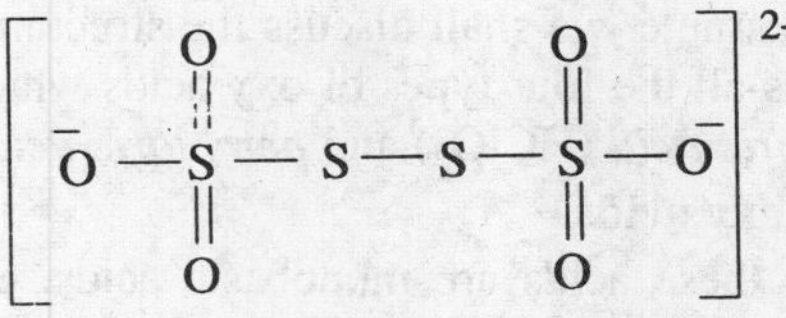

Tetrathionate ion ($S_4O_6^{2-}$)

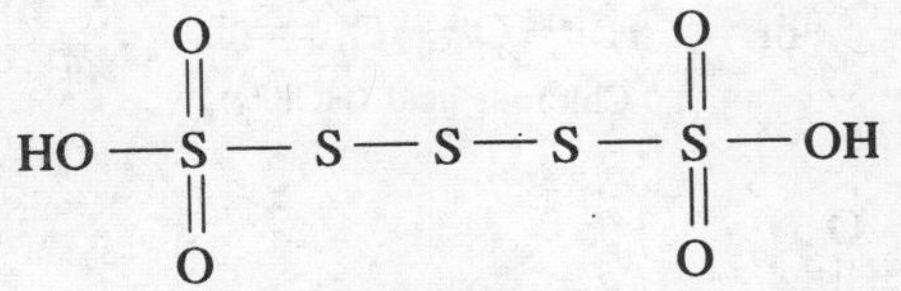

Pentathionic acid ($H_2S_5O_6$)

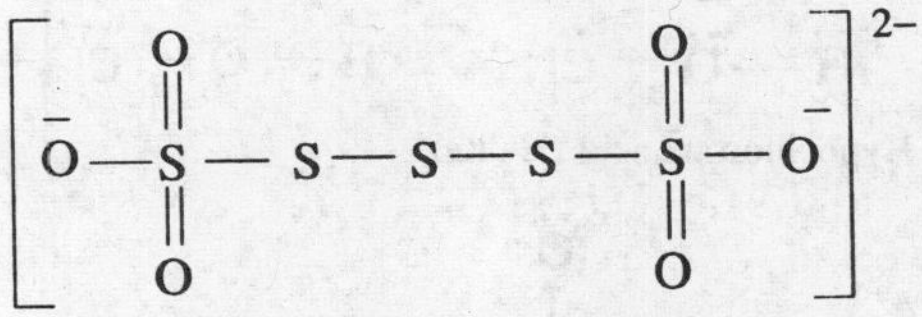

Pentathionate ion ($S_5O_6^{2-}$)

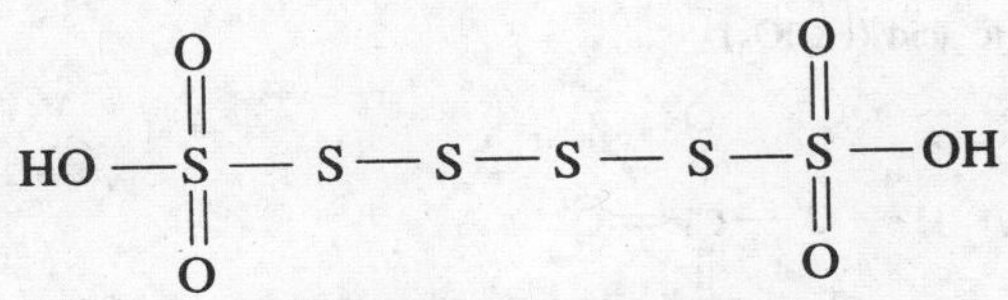

Hexathionic acid ($H_2S_6O_6$)

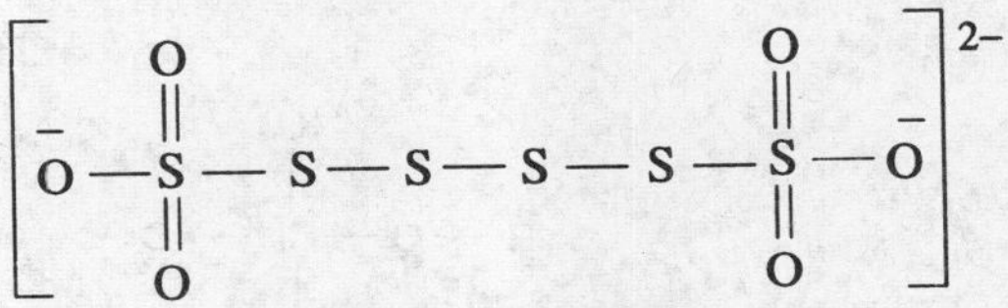

Hexathionate ion ($S_6O_6^{2-}$)

The structures given above make the following points evident:

(*i*) These structures contain two types of S-atoms *viz.*, (*a*) *two terminal S-atoms* which are situated at the end of S-atoms chain and are surrounded tetrahedrally by three O-atoms and one bridging S-atom. (*b*) 1, 2, 3, or 4 *bridging S-atoms* which make a bridge between the two terminal S-atoms. These S-atoms make a chain.

(*ii*) Each terminal S-atom is surrounded by three O-atoms and one of the bridging S-atoms. Thus each terminal S-atom is sp^3 hybridised.

The above discussion shows that the structure of polythionate ion, $S_nO_6^{2-}$ (n = 3 to 6) can be shown as given in Fig. 15.43.

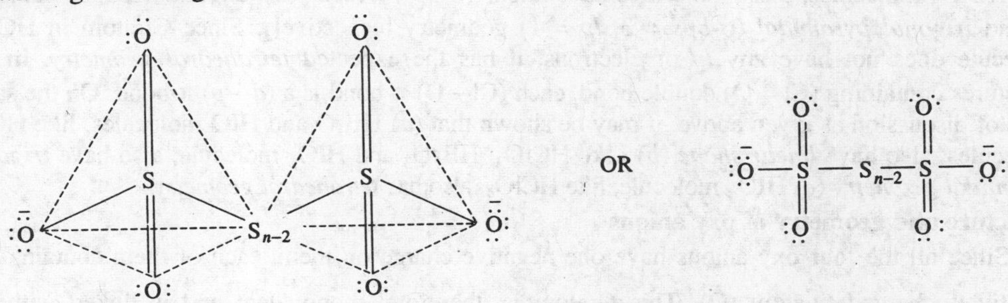

Fig. 15.43 : Structure of polythionate ion, $S_nO_6^{2-}$ or $[S_2\ S_{n-2}\ O_6]^{2-}$ (n = 3 to 6).

Oxy acids and oxy anions of Cl, Br and I

Excepting F, all the remaining three halogens *viz.*, Cl, Br and I form oxy acids. These acids formed by Cl, Br and I are given below : *Chlorine* → HClO, $HClO_2$, $HClO_3$, $HClO_4$; *Bromine* → HBrO, $HBrO_2$; *Iodine* → HIO, HIO_3, HIO_4. These acids can be grouped into four types *viz.*, (*i*) *HOX type acids* (X = Cl, Br, I). These are called *hypohalous acids* (*ii*) $HClO_2$ is called *chlorous acid*. Bromine and iodine do not form $HBrO_2$ and HIO_2 acids (*iii*) HXO_3 *type acids* (X = Cl, Br, I). These are called *halic acids* (*iv*) HXO_4 *type acids* (X = Cl, I). These are called *perhalic acids.*

The acids of one type have the same structure, *e.g.*, acids of HXO type have linear geometry.

Here, as an example, we shall discuss the structure of oxy acids and oxy anions of chlorine, since chlorine forms all the four types of oxy acids which are *hypochlorous acid* ($HClO$), *chlorous acid* ($HClO_2$), *chloric acid* ($HClO_3$) and *perchloric acid* ($HClO_4$).

Structure of oxy acids

Since all these acids are monobasic acids, each of them contains one OH group which is directly attached with the central Cl-atom. Thus electron dot structures (Lewis structures) of these acids can be written as:

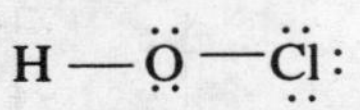

Hypochlorous acid ($HClO$)

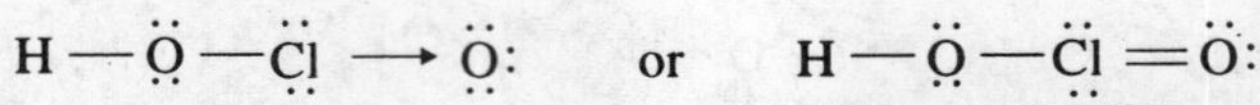

Chlorous acid ($HClO_2$)

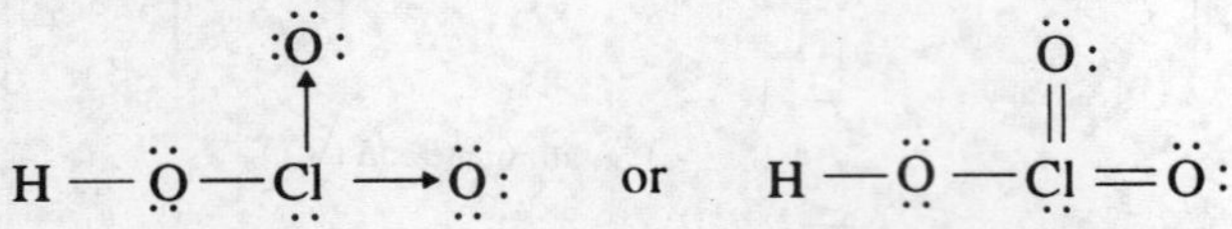

Chloric acid ($HClO_3$)

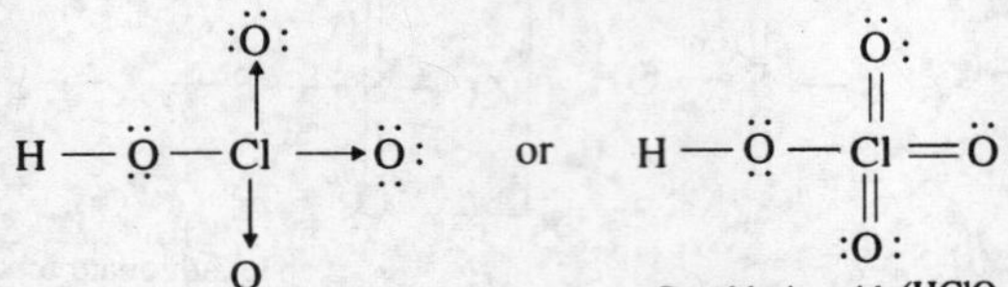

Penchloric acid ($HClO_4$)

It may be noted from the above structures that the structures containing Cl → O coordinate bonds obey octet rule while those having Cl = O double bonds do not obey this rule.

Due to monobasic nature of these acids, these acids can also be represented by the formulae (OH)Cl (*hypochlorous acid*), (OH)ClO (*chlorous acid*), (HO)ClO_2 (*chloric acid*) and (OH)ClO_3 (*perchloric acid*).

Since the sum of *σ-bps* and lone pair of electrons (*lps*) surrounding the central Cl-atom in all these acids is equal to 4, Cl-atom is sp^3 hybridised in each of them. Due to sp^3 hybridisation and presence of 3, 2 and 1 lone pairs of electrons on Cl-atom (central atom) of $HClO$, $HClO_2$ and $HClO_3$ molecules, these molecules have *linear* (*σ-bp* = 1, *lps* = 3), *angular* (*σ-bps* = 2, *lps* = 2) and *trigonal pyramidal* (*σ-bps* = 3, *lp* = 1) geometry respectively. Since Cl-atom in $HClO_4$ molecule does not have any *lp* of electrons, it has the expected *tetrahedral geometry.* In the structures containing (Cl = O) double bond, each (Cl—O) π-bond is a (*d*—*p*) π-bond. On the same lines of discussion as given above, it may be shown that (*a*) HBrO and HIO molecules, like HClO molecules, also have *linear shape* (*b*) Like $HClO_3$, $HBrO_3$ and HIO_3 molecule, also have *trigonal pyramidal geometry* (*c*) HIO_4 molecule, like $HClO_4$, also has *tetrahedral geometry*.

Structure and geometry of oxy anions

Since all the four oxy anions have one negative charge on them, each of them contains one negatively-charged O-atom ($\bar{O}$). This O-atom is, therefore, monovalent and is linked with the central Cl-atom by a single (Cl—$\bar{O}$) σ-bond. The remaining O-atoms are neutral and are attached with the Cl-atom either by a (Cl = O) double bond or by a (Cl → O) coordinate bond.

1. Hypochlorite ion, ClO^-. The electron dot structure of this ion can be represented as

$\cdot\ddot{Cl}$ — : $\ddot{O}$ which shows that this structure obeys octet rule and in it Cl-atom is linked with

O by one (Cl — O) σ-bond and has three *lps* on it (*σ-bps* = 1, *lps* = 3). Thus, since Cl-atom is surrounded by one *σ-bp* and three *lps*, this atom is sp^3 hybridised [$Cl = (sp^3)^2\ (sp^3)^2\ (sp^3)^2\ (sp^3)^1$]. It is obvious from sp^3 hybridisation scheme given above that three hybrid orbitals are completely-filled while fourth hybrid orbital contains one electron. This hybrid orbital overlaps with the with the half-filled $2p_x$ orbital of $\bar{O}$ ion ($\bar{O}$ = $2s^2\ 2p_x^1\ 2p_y^2\ 2p_z^2$) and forms (Cl – O) σ-bond. It is due to the presence of three *lps* on Cl-atom that Cl$\bar{O}$ ion does not have tetrahedral geometry but has *linear shape* (See Fig. 15.44). Cl—O bond distance in Cl$\bar{O}$ ion is equal to 1.70 Å and Cl—O bond energy is 209 KJ/mole.

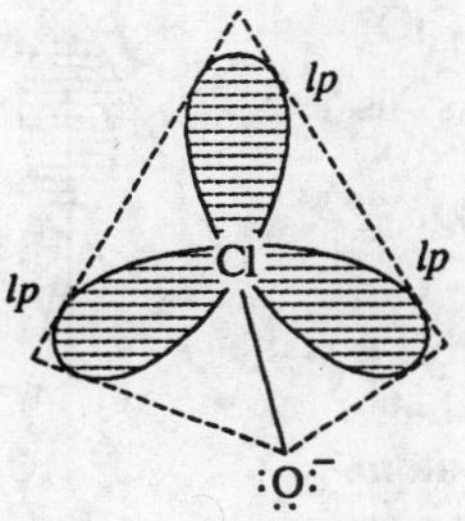

OR :C̈l — Ö:

Fig. 15.44 : Linear shape of ClO^- ion.

2. Chlorite ion, ClO_2^- . This ion can have any of the following two electron dot structures:

$:\bar{\ddot{O}} — \ddot{Cl} \rightarrow \ddot{O}:$ Structure (*a*)

$:\bar{\ddot{O}} — \ddot{Cl} = \ddot{O}:$ Structure (*b*)

The above structures show that one O-atom is negatively-charged and hence is attached with Cl-atom (central atom) by a single (Cl—$\bar{O}$) σ-covalent bond while the other O-atom which is neutral is bonded to Cl-atom either by a Cl → O coordinate bond [Structure (*a*)] or by a (Cl = O) double bond [structure (*b*)]. Cl = O double bond consists of one (Cl — O) σ-bond and one (Cl—O) π-bond. Since Cl-atom in both the structures is surrounded by two σ-bonds and two *lps*. Cl-atom in its excited state [Cl = $3s^2\ 3p^4\ 3d^1$] is sp^3 hybridised in both the structures. sp^3 hybridisation of Cl-atom in structure (*b*) is as Cl = $(sp^3)^2\ (sp^3)^2\ (sp^3)^1\ (sp^3)^1\ 3d^1$.

It is obvious from sp^3 hybridisation scheme that π-bond is formed by the overlap between half-filled $3d$ orbital (this orbital remains excluded from sp^3 hybridisation) and half-filled $2p$ orbital on O-atom [O = $2s^2\ 2p_x^1\ 2p_y^1\ 2p_z^2$]. Thus (Cl — O) π-bond is a (*d* — *p*) π-bond.

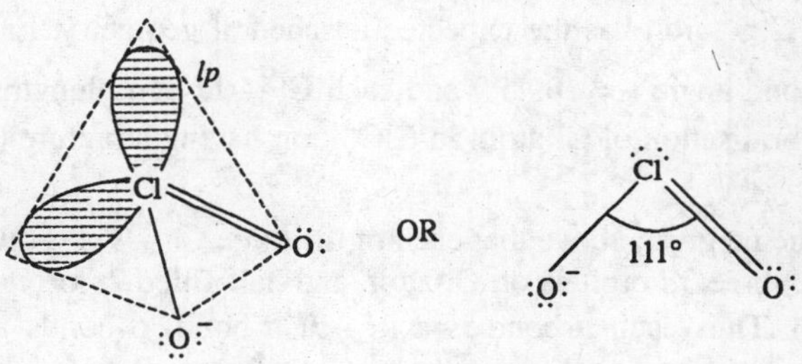

Fig. 15.45 : Angular shape of ClO_2^- ion.

Although the spatial arrangement of four electron pairs (2 σ-*bps* + 2 *lps*) round the central Cl-atom is tetrahedral (due to sp^3 hybridisation), due to the presence of two *lps*, the shape of ClO_2^- ion gets distorted and becomes angular. (See Fig. 15.45).

3. Chlorate ion, ClO_3^- . This ion can have any of the following two electron dot structures:

$:\bar{\ddot{O}} — \ddot{Cl} \rightarrow \ddot{O}:$ (with $\uparrow$:Ö: on Cl) Structure (*a*)

$:\bar{\ddot{O}} — \ddot{Cl} = \ddot{O}:$ (with $\|$:O: on Cl) Structure (*b*)

The above structures indicate that one O-atom is negatively-charged and hence is linked with Cl-atom (central atom) by a single (Cl — $\bar{O}$) σ-covalent bond while the remaining two O-atoms are neutral and are, therefore, bonded to Cl-atom either by Cl → O coordinate bond [Structure (*a*)] or by a Cl = O double bond [Structure (*b*)]. Obviously Cl = O bond is composed of one (Cl — O) σ-bond and one (Cl — O) π-bond. Since Cl-atom (central atom) in both the structures is surrounded by 3σ–*bps* + 1*lp* , this atom in its excited state [Cl = $3s^2\ 3p^3\ 3d^2$] is sp^3 hybridised. sp^3

hybridisation of Cl-atom in ClO_3^- ion having structure (*b*) is as $(sp^3)^2\ (sp^3)^1\ (sp^3)^1\ (sp^3)^1\ 3d^1\ 3d^1$. It is obvious from sp^3 hybridisation scheme shown above that both (Cl — O) π-bonds result from the overlap between half-filled two 3*d* orbitals on Cl-atom and half-filled 2*p* orbitals on two O-atoms [O = $2s^2\ 2p_x^1\ 2p_y^2\ 2p_z^1$].

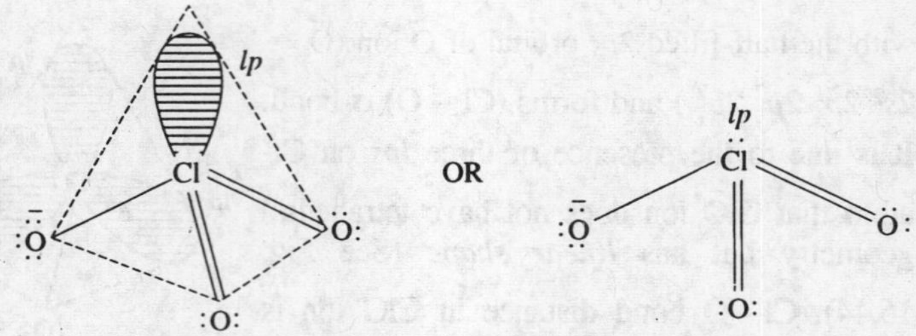

Fig. 15.46 : Trigonal pyramidal shape of ClO_3^- ion.

Thus each π-bond is a (*d*—*p*) π-bond. σ-bonds are obtained in the usual manner.

Although the spatial arrangement of four sp^3 hybrid orbitals is tetrahedral, due to (*lp*—*bp*) and (*bp*—*bp*) repulsions, the shape of ClO_3^- ion is not tetrahedral; rather this shape changes and becomes *trigonal pyramidal* (See Fig. 15.46).

4. Perchlorate ion, ClO_4^-. This ion can have any of the following electron dot structures. It may be seen that it is only structure (*a*) which obeys octet rule.

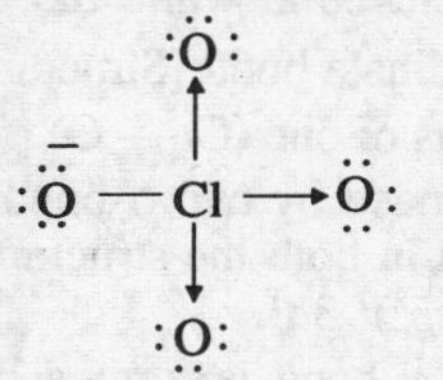

Structure (*a*)

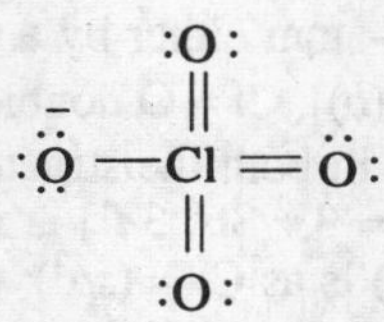

Structure (*b*)

Since in both the structures the number of σ-bonds by which Cl-atom (central atom) is linked with O-atoms is 4 and there is no lone pair of electrons on it, the central atom is sp^3 hybridised in these structures. Due to sp^3 hybridisation, ClO_4^- ion has the expected tetrahedral geometry. Each OClO bond angle is equal to tetrahedral bond angle (= 109.5°) and each Cl — O bond length is 1.45Å. Bond energy is 364 KJ/mol. sp^3 hybridisation of Cl-atom in ClO_4^- ion having structure (*b*) is as Cl = $(sp^3)^1\ (sp^3)^1\ (sp^3)^1\ (sp^3)^1\ 3d^1\ 3d^1\ 3d^1$.

It is obvious from sp^3 hybridisation scheme given above that each of the three Cl—O π -bonds results from the overlap between half-filled three 3*d* orbitals on Cl-atom and half-filled 2*p* orbitals on three O-atoms (O = $2s^2\ 2p_x^1\ 2p_y^2\ 2p_z^1$). Thus each π-bond is a (*d*—*p*) π-bond. σ-bonds are obtained in the usual manner.

It is due to sp^3 hybridisation and the absence of lone pair of electrons on Cl atom, ClO_4^- ion has the expected tetrahedral shape [See Fig. 15.47].

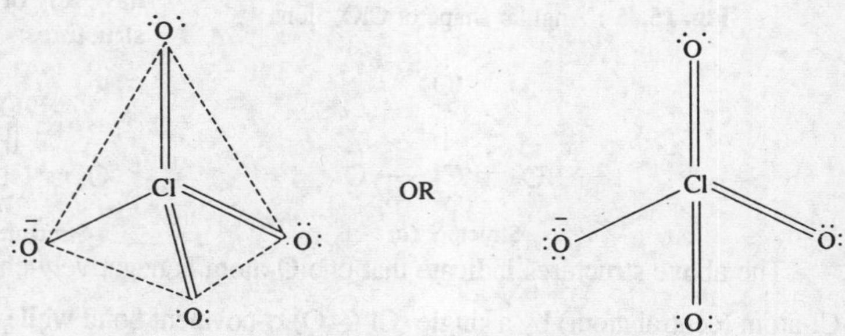

Fig. 15.47 : Tetrahderal shape of ClO_4^- ion.

On the same lines of discussion as given above it may be shown that:

(*a*) Like ClO^- ion, BrO^- and IO^- ions, also have *linear shape*.

(*b*) BrO_3^- and IO_3^- ions, like ClO_3^- ion also have *trigonal pyramidal geometry*.

(*c*) Like ClO_4^- ion, IO_4^- ion has *tetrahedral shape.*

It may be noted that BrO_2^- and IO_2^- ions are not formed.

Oxyhalides of nitrogen

Nitrogen forms two types of oxyhalides *viz.*, *Nitrosyl halides*, $[NO]^+ [X]^-$ (X = F, Cl, Br) and *Nitryl halides*, $[NO_2]^+ [X]^-$ (X = F, Cl). Here we shall discuss the structure of *nitrosyl chloride*, NO^+Cl^- and *nitryl chloride*, $NO_2^+ Cl^-$.

1. Structure of nitrosyl chloride, NOCl

Electron diffraction studies have shown that NOCl molecule can have any of the following two structures:

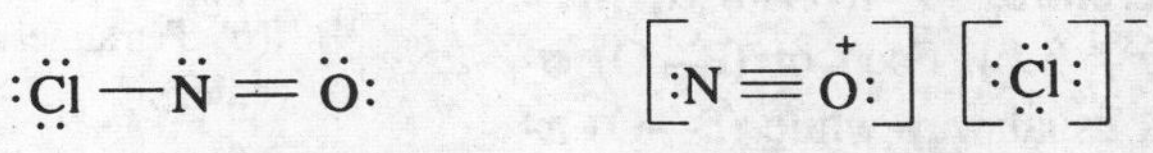

Covalent structure Ionic structure

In covalent structure, since N-atom [N = $2s^2\ 2p^3$] is surrounded by two σ-*bps* and one *lp*, this atom is sp^2 hybridised [N = $(sp^2)^1\ (sp^2)^1\ (sp^2)^1\ 2p_z^1$]. Since one of the three sp^2 hybrid orbitals is occupied by a lone pair of electrons (*lp*), NOCl molecule has distorted trigonal (*i.e., angular* or *V-shaped*) geometry in which ClNO bond angle is equal to 115° (instead of 120°). Cl—N and N = O bond distances are 1.95 Å and 1.14 Å respectively (See Fig. 15.48). The dipole moment of covalent structure has been measured in CCl_4 solution and has been found to be equal to 1.83D. This high value suggests the presence of some ionic character in NOCl molecule. N – O π-bond is a (*p*—*p*) π-bond. The true structure of NOCl molecule is supposed to be a resonance hybrid of its covalent and ionic structures.

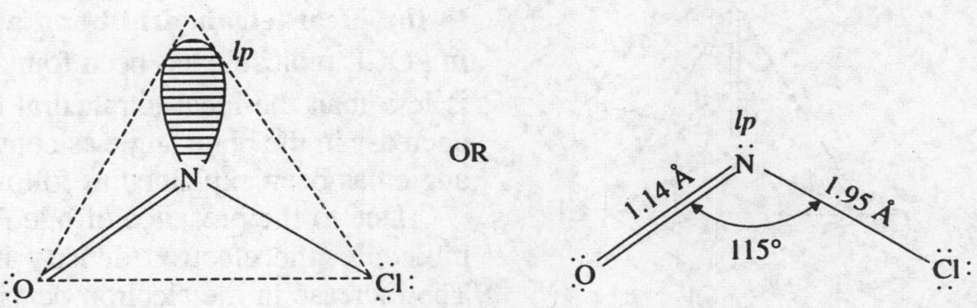

Fig. 15.48 : Angular geometry of nitrosyl chloride molecule, NOCl.

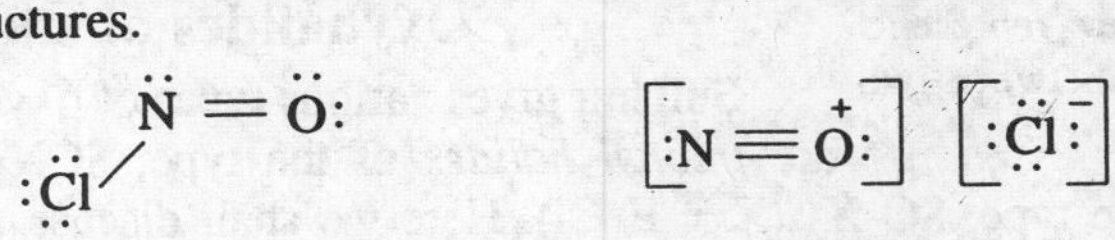

Covalent structure Ionic structure

2. Structure of nitryl chloride molecule, NO_2Cl

The electron dot structure of this molecule is 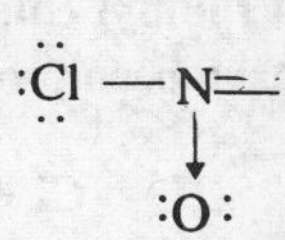 . This structure shows that since the central N-atom is surrounded by three s-bps. it is sp^2 hybridised [N = $(sp^2)^2\ (sp^2)^1\ (sp^2)^1\ 2p_z^1$] N → O co-ordinate bond results by the donation of the electron pair residing in one sp^2 hybrid orbitals to the vacant 2*p* orbital of O-atom (O = $2s^2\ 2p^2\ 2p^2\ 2p^0$). N – O π-bond is a (*p*—*p*) π-bond.

Due to sp^2 hybridisation of the central N-atom and absence of any *lp* on this atom, NO_2Cl molecule has the expected *trigonal planar geometry* with ClNO and ONO bond angles equal to 120°.

Oxytrihalides of phosphorus (also called phosphoyl trihalides), POX3

Phosphorus forms oxytrihalides which are of POX_3 type. These are *phosphorus oxytrifluoride*

(POF_3), *phosphorus oxytrichloride* ($POCl_3$) and *phosphorus oxytribromide* ($POBr_3$). All these molecules have tetrahedral shape which results from sp^3 hybridisation of P-atom in its excited state. As an example let us discuss the structure and geometry of $POCl_3$ molecule.

Electron dot structure of $POCl_3$ molecule as shown at (*a*) of Fig. 15.49 indicates the P-atom (central atom) is surrounded by four σ-*bps* and no lone pair of electrons. P-atom in its excited state [P = $3s^1\ 3p^3\ 3d^1$] is therefore, sp^3 hybridised [P = $(sp^3)^1\ (sp^3)^1\ (sp^3)^1\ (sp^3)^1\ 3d^1$]. Each of (P—O) σ-bonds is formed in its usual way while (P—O) π-bond results by (*d* — *p*) overlap, *i.e.*, (P—O) π-bond is a (*d* — *p*) π-bond. It is due to sp^3 hybridisation of the central P-atom and absence of lone pair of electrons on P-atom that $POCl_3$ molecule has tetrahedral shape shown at (*b*) of Fig. 15.49.

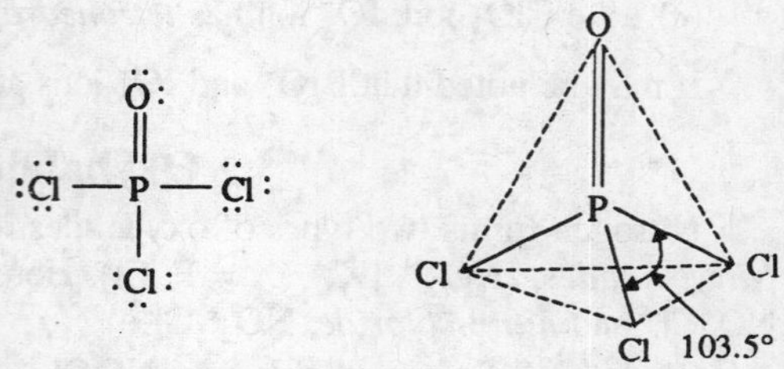
(*a*) (*b*)

Fig. 15.49 : (*a*) Electron dot structure of phosphorus oxytrichloride ($POCl_3$) molecule. (*b*) Tetrahderal shape of $POCl_3$ molecule.

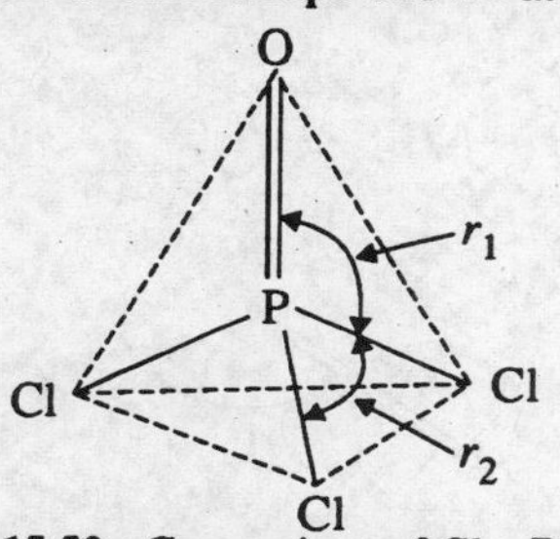

Fig. 15.50 : Comparison of Cl—P—Cl bond angle in $POCl_3$ molecule (= 103.5°) with the tetrahedral bond angle (= 109.5°). r_1 represents repulsion between O = P and P—Cl bonds while r_2 represents the repulsion between two P—Cl bonds.

Cl—P—Cl bond angle in $POCl_3$ molecule as compared to the ideal tetrahedral bond angle. Cl—P—Cl bond angle in $POCl_3$ molecule has been found to be equal to 103.5° which is less than the ideal tetrahedral bond angle (= 109.5°). The decrease in the bond angle as compared to the tetrahedral bond angle has been explained as follows:

Due to the presence of one P = O double bond in $POCl_3$ molecule, the electron density in P = O bond is increased. The increase in the electron density in P = O bond makes the repulsion between P = O and P—Cl bonds greater than that between two P—Cl bonds (*i.e.*, $r_1 > r_2$) and hence O = P—Cl bond angle is increased and Cl — P — Cl bond angle is decreased as compared to the ideal tetrahderal bond angle (= 109.5°) (Fig.15.50).

Oxyhalides of sulphur

Sulphur gives various types of oxyhalides. Important ones are *thionyl halides* of the type, SOX_2 (X = F, Cl, Br) and *sulphuryl halides* of the type, SO_2X_2 (X = F, Cl). Here we shall discuss the structure of *thionyl chloride* ($SOCl_2$) and *sulphuryl chloride* (SO_2Cl_2) molecules only.

1. Structure and geometry of thionyl chloride molecule, $SOCl_2$

Following three structures have been proposed for $SOCl_2$ molecule.

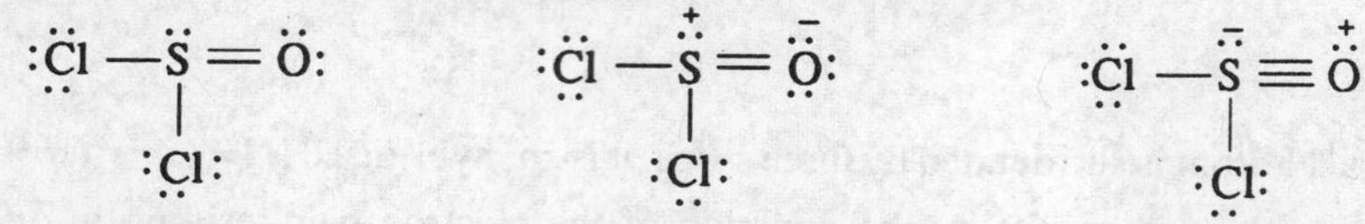

Structure (I) Structure (II) Structure (III)

Each of the above structures shows that the central S-atom is surrounded by three σ-*bps* and one lone pair of electrons (*lp* = 1) and hence this atom is sp^3 hybridised.

In structure (I), S-atom in its excited state [S = $3s^2\ 3p^3\ 3d^1$] is sp^3 hybridised [S = $(sp^3)^2\ (sp^3)^1\ (sp^3)^1\ (sp^3)^1\ 3d^1$]. Each of the

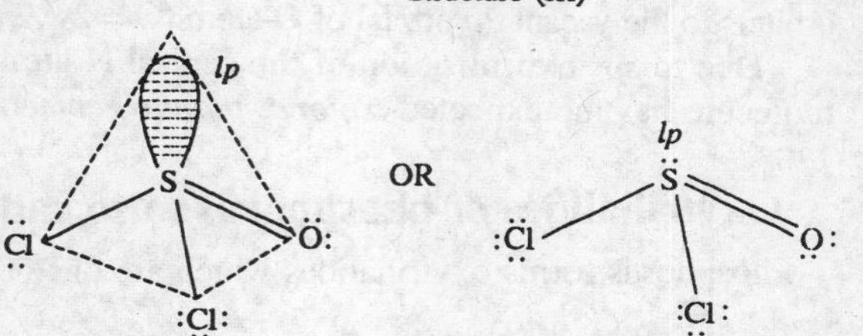

Fig. 15.51 : Trigonal pyramidal shape of $SOCl_2$ molecule.

(S — A) σ-bonds (A = Cl or O) is formed in its usual way while (S — O) π-bond is obtained by the overlap between the singly-filled 2*p* orbital of O-atom with half-filled 3*d*-orbital of S-atom. Thus (S — O) π-bond is a (*p* — *d*) π-bond.

Due to sp^3 hybridisation of the central S-atom and presence of one lone pair of electrons (*lp*) on this atom, $SOCl_2$ molecule has distorted tetrahedral or *trigonal pyramidal geometry* as shown in Fig. 15.51.

2. Structure and geometry of sulphuryl chloride molecule, SO_2Cl_2

This molecule can be represented by any of the following two electron dot structures:

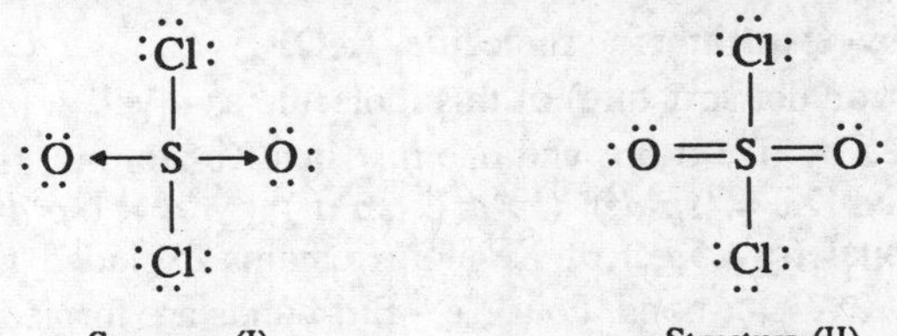

Structure (I) Structure (II)

These structures show that:

(*i*) In structure (I) all the atoms are surrounded by 8 electrons and hence obey octet rule while in structure (II) S-atom is surrounded by 12 electrons and other atoms are surrounded by 8 electrons.

(*ii*) S-atom (central atom) is surrounded by four σ-*bps* and no *lp* (*lp* = 0) and hence this atom in excited state [S = $3s^2\ 3p^3\ 3d^2$] is sp^3 hybridised. sp^3 hybridisation of S-atom in structure (II) gives four singly-filled sp^3 hybrid orbitals and two unhybridised 3*d* orbitals [S = $(sp^3)^1\ (sp^3)^1\ (sp^3)^1\ (sp^3)^1\ 3d^1\ 3d^1$]. In structure (II), each S — Cl and each S — O σ-bond is formed in usual way and each of the two (S — O) π-bonds is obtained by S (3*d*) – O (*sp*) overlap. Thus (S — O) π -bond is a (*d* — *p*) π - bond. It is due to sp^3 hybridisation and no lone pair of electrons on S-atom that SO_2Cl_2 molecule has the expected tetrahedral structure (Fig. 15.52).

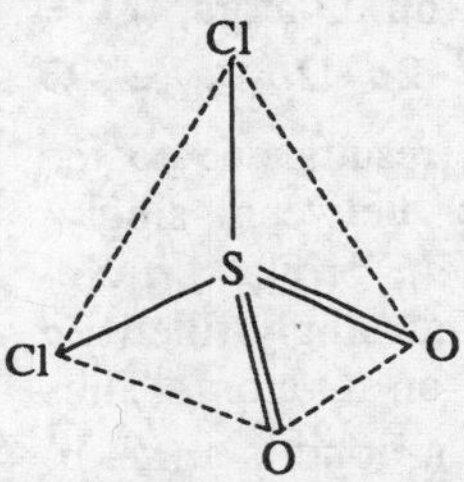

Fig. 15.52 : Tetrahedral geometry of $SO_2\ Cl_2$ molecule.

Oxyfluorides of xenon

Xenon gives a number of oxyfluorides like $XeOF_2$, $XeOF_4$, XeO_2F_2, XeO_3F_2 and XeO_2F_4. Their structure and geometry are discussed below:

1. Structure of xenon oxydifluoride molecule $XeOF_2$

The electron dot structure (Lewis structure) of this molecule is given at (*a*) of Fig. 15.53. It is obvious from this structure that Xe-atom (central atom) is surrounded by three σ-*bps* and two lone pairs of electrons (*lps* = 2) and hence Xe-atom in its excited state [Xe = $5s^2\ 5p^4\ 5d^1\ 5d^1$] is sp^3d hybridised [Xe = $(sp^3d)^2\ (sp^3d)^2\ (sp^3d)^1\ (sp^3d)^1\ (sp^3d)^1\ 5d^1$]. It must be noted that one 5*d*-orbital (say $5d_{x^2-y^2}$ orbital) of Xe-atom remains unhybridised. This orbital is used in forming Xe—O π-bond. Two Xe — F σ-bonds are formed by the overlap between the singly-filled two sp^3d hybrid orbitals and two singly-filled $2p_x$ orbitals on two F-atoms (F = $2s^2\ 2p_x^1\ 2p_y^2\ 2p_z^2$) while Xe — O σ-bond is obtained by the overlap between the singly-filled sp^3d

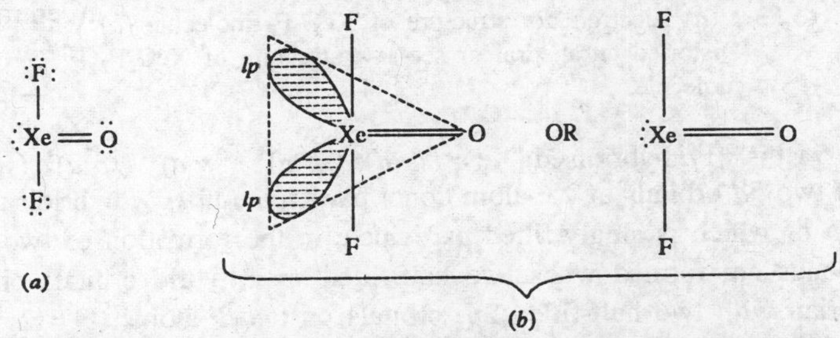

Fig. 15.53 : (*a*) Electron dot structure of $XeOF_2$ molecule, (*b*) Bent T-shaped structure of $XeOF_2$ molecule.

hybrid orbital and singly-filled $2p$ orbital of O-atom ($O = 2s^2\ 2p_x^1\ 2p_y^1\ 2p_z^2$). (Xe—O) π-bond results due to the overlap between singly-filled $5d_{x^2-y^2}$ orbital on Xe-atom and singly-filled $2p$-orbital on O-atom. Thus Xe—O π-bond is a (d—p) π-bond.

Although the spatial arrangement of five electron pairs round the central Xe-atom is *trigonal bipyramidal*, due to the presence of two lone pairs of electrons in the equatorial hybrid orbitals, the shape of $XeOF_2$ molecule gets distorted and becomes *bent T-shaped geometry* (See Fig. 15.53). It is because of the presence of lone pairs of electrons in the equatorial position that the repulsions between electron pairs become minimum.

2. Structure of xenon oxytetrafluoride molecule, $XeOF_4$

Lewis structure (electron dot structure) of this molecule as given at (a) of Fig. 15.54 suggests that Xe-atom is surrounded by five σ-*bps* and one *lp*. Thus Xe-atom in its excited state [$Xe = 5s^2\ 5p^3\ 5d^3$] is sp^3d^2 hybridised [$Xe = (sp^3d^2)^2\ (sp^3d^2)^1\ (sp^3d^2)^1\ (sp^3d^2)^1\ (sp^3d^2)^1\ (sp^3d^2)^1\ 5d^1$]. It must be noted that one $5d$-orbital (say $5d_{xy}$) of Xe-atom remains excluded from hybridisation. This orbital is used in forming Xe—O π-bond. Four Xe—F σ-bonds are formed by the overlap between singly-filled four sp^3d^2 hybrid orbitals and four singly-filled $2p_x$ orbital on four F-atoms ($F = 2s^2$, $2p_x^1\ 2p_y^1\ 2p_z^2$) while Xe—O σ-bond results by the overlap between the singly-filled sp^3d^2 hybrid orbital and singly-filled $2p$ orbital on O-atom ($O = 2s^2, 2p_x^1\ 2p_y^1\ 2p_z^2$). Xe—O π-bond results due to the overlap between singly-filled $5d_{xy}$ orbital on Xe-atom and singly-filled $2p$ orbital on O-atom. Thus Xe—O π-bond is a (d—p) π-bond. Although the spatial arrangement of six electron pairs round the central Xe-atom is *octahedral*, due to the presence of one lone pair of electrons in the axial position of the octahedron, the shape of $XeOF_4$ molecule is distorted and becomes *square pyramidal* (See Fig. 15.54) in which four F-atoms are co-planar, since they lie in the same plane (square plane). The lone pair of electrons lies opposite to O-atom which is bonded to Xe-atom by a double bond. This arrangement of lone pair of electrons minimises the repulsion between electron pairs and a stable molecule is obtained.

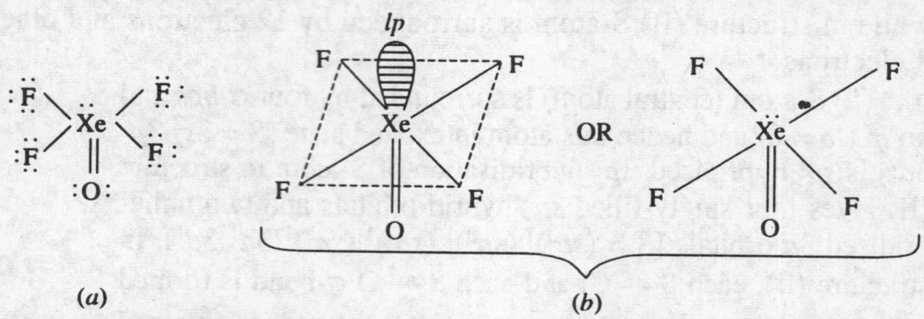

Fig. 15.54 : (a) Electron dot structure of $XeOF_4$ molecule, (b) Square pyramidal shape of $XeOF_4$ molecule.

3. Structure of xenon dioxydifluoride molecule, XeO_2F_2

Electron dot structure of this molecule is given at (a) of Fig. 15.55. This structure shows that Xe-atom (central atom) is surrounded by four σ-*bps* (two Xe—F σ-bonds and two Xe—O σ-bonds) and one *lp*. Thus Xe-atom in its excited state [$Xe = 5s^2\ 5p^3\ 5d^3$] is sp^3d hybridised [$Xe = (sp^3d)^2\ (sp^3d)^1\ (sp^3d)^1\ (sp^3d)^1\ (sp^3d)^1\ 5d^1\ 5d^1$]. It must be noted that two $5d$ orbitals of Xe-atom donot participate in sp^3d hybridisation process. These $5d$ orbitals each of which is singly-filled participate in the formation of two Xe—O π-bonds. Two Xe—F σ-bonds are formed when two half-filled sp^3d hybrid orbitals on Xe-atom make head-to-head overlap with two half-filled $2p_x$ orbitals on two F-atoms ($F = 2s^2\ 2p_x^1\ 2p_y^2\ 2p_z^2$). Two Xe—O

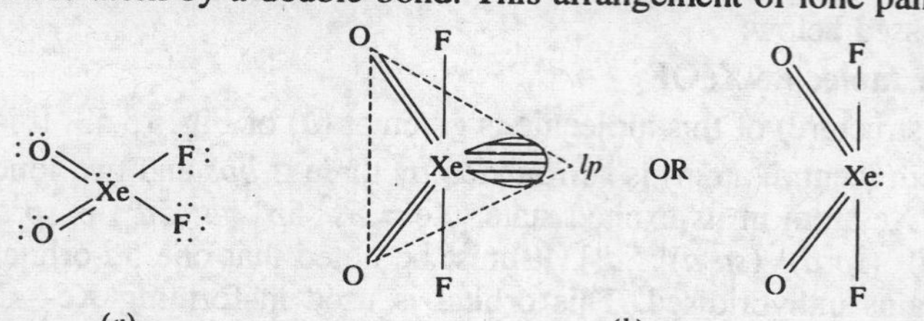

Fig. 15.55 : (a) Electron dot structure of XeO_2F_2 molecule (b) Distorted tetrahedral or see-saw structure of XeO_2F_2 molecule.

σ-bonds are formed when two half-filled sp^3d hybrid orbitals on Xe-atom make head-to-head overlap with two half-filled $2p_x$ orbitals on two F-atoms ($F = 2s^2\,2p_x^1\,2p_y^2\,2p_z^2$). Two Xe—O σ-bonds are obtained by the overlap between the singly-filled two sp^3d hybrid orbitals on Xe-atom with the singly-filled $2p_x$ orbitals on two O-atoms. Two Xe—O π-bonds are formed when two $5d$ orbitals on Xe-atom which did not participate in the hybridisation process overlap with the singly-filled $2p$ orbitals of two O-atoms. Thus each Xe—O π-bond is a $(d—p)$ π-bond.

Position occupied by lp in the trigonal bipyramid and shape of XeO_2F_2 molecule. We know that the five sp^3d hybrid orbitals are directed towards the five corners of a trigonal bipyramid. The *lp* occupies the hybrid orbital directed towards one of the corners of the trigonal base of the trigonal bipyramid *i.e.*, the *lp* occupies the equatorial hybrid orbital, since in this case electron pairs repulsion is minimum. Thus due to the presence of one *lp* in the equatorial hybrid orbital, the shape of XeO_2F_2 is no longer trigonal bipyramidal, but it is distorted *tetrahedral* or *see-saw* (See Fig. 15.55).

4. Structure of xenon trioxydifluoride molecule, XeO_3F_2

Electron dot structure of this molecule is given at (*a*) of Fig. 15.56. This structure suggests that Xe-atom (central atom) is surrounded by five σ-*bps* (three Xe—O σ-bonds and two Xe—F σ-bonds) and no *lp*. Thus Xe-atom in its excited state [$Xe = 5s^1\,3p^3\,5d^4$] is sp^3d hybridised [$Xe = (sp^3d)^1\,(sp^3d)^1\,(sp^3d)^1\,(sp^3d)^1\,(sp^3d)^1\,5d^1\,5d^1\,5d^1$]. Each of the five σ-bonds is formed by the overlap between half-filled sp^3d hybrid orbital and half-filled 2*p*-orbital F or O-atom. Each of (Xe—O) π-bonds results by $2p$ (O)—$5d$ (Xe) overlap, *i.e.*, each π-bond is a $(p—d)$ π-bond.

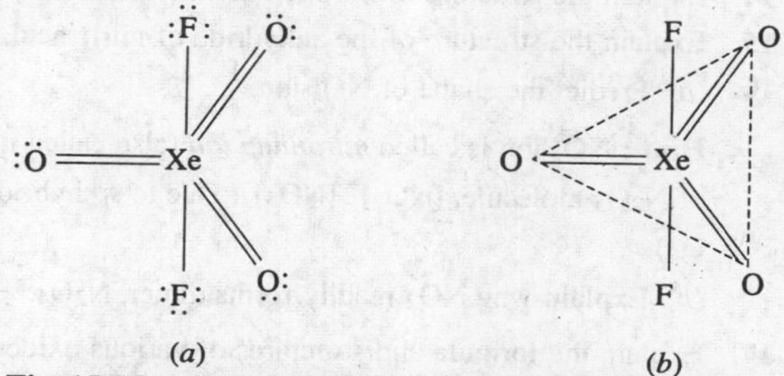

Fig. 15.56 : (*a*) Electron dot structure of XeO_3F_2 molecule (*b*) Trigonal bipyramidal geometry of XeO_3F_2 molecule.

It is due to sp^3d hybridisation of the central Xe-atom and due to the absence of any lone pair of electrons ($lp = 0$) in any of the five sp^3d hybrid orbitals that XeO_3F_2 molecule has the expected *trigonal bipyramidal shape* that has been shown at (*b*) of Fig. 15.56. In this shape the three doubly-bounded O-atoms are present at the equatorial positions of the trigonal bipyramid.

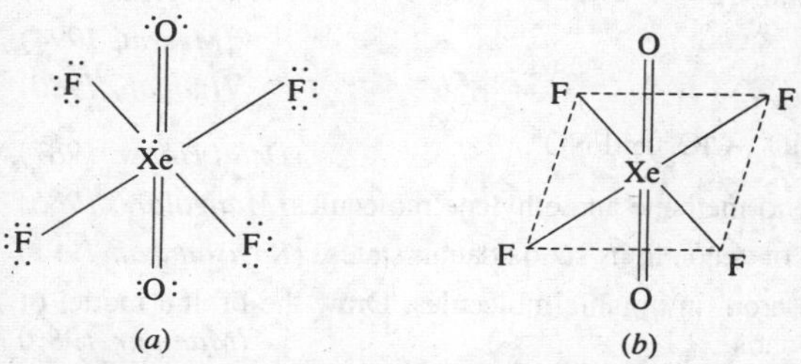

Fig. 15.57 : (*a*) Electron dot structure of XeO_2F_4 molecule (*b*) Octahedral geometry of XeO_2F_4 molecule.

5. Structure of xenon dioxytetrafluoride molecule, XeO_2F_4

Electron dot structure of this molecule is given at (*a*) of Fig. 15.57. This structure suggests that the central Xe-atom is surrounded by six σ-*bps* and no *lp*. Consequently Xe-atom in its excited state [$Xe = 5s^1\,5p^3\,5d^4$] is sp^3d^2 hybridised [$Xe = (sp^3d^2)^1\,(sp^3d^2)^1\,(sp^3d^2)^1\,(sp^3d^2)^1\,(sp^3d^2)^1\,(sp^3d^2)^1\,5d^1\,5d^1$]. Each σ-bond is formed by the overlap of half-filled sp^3d^2 hybrid orbitals with the half-filled $2p$ orbitals of F/O atom while (Xe—O) π-bond is obtained by [$2p$ (O) —$5d$ (Xe)] overlap, *i.e.*, each π bond is a $(p—d)$ π-bond.

Since Xe-atom (central atom) undergoes sp^3d^2 hybridisation and none of the six sp^3d^2 hybrid orbitals contains a lone pair of electrons, XeO_2F_4 has the expected *octahedral shape* in which the two doubly-bounded O-atoms are having axial position to each other [See Fig. 15.57 (*b*)].

University Questions

1. Discuss the structure of Caro's and Marshall's acids. *(Rohilkhand, 91; Nagpur, 93)*
2. Discuss the structure of $XeOF_4$. *(Nagpur, 1993)*
3. Discuss the structure of NO molecule. *(Nagpur, 1993)*
4. Discuss the structure of polythionic acids. *(Ravishankar, 1993)*

5. Explain why the shapes of ClO_4^- and ClO_3^- ions are different, although they have the same number of electron pairs round Cl-atom. *(Punjab, 1984)*
6. Discuss the structure of permonosulphuric acid. *(Bahuguna, 1992)*
7. Discuss the shape of CO_2 molecule. *(Purvanchal, 1993)*
8. Write down the electronic dot structure of (*a*) orthophosphoric acid (*b*) peroxy sulphuric acid. *(Avadh, 1993)*
9. Draw the electron-dot representations of oxy acids of chlorine. *(Rohilkhand, 1991)*
10. Give the electronic structures of any two oxy acids of sulphur *(Guahati, 91)*
11. Discuss the structure of various oxides of nitrogen. *(GND, 1982)*
12. Discuss the structures of N_2O_3 and N_2O_5 molecules. *(M.D.Rohtak, 1984)*
13. Draw the electron dot structures of the oxides of nitrogen *(Meerut, 1985)*
14. Explain the structure of nitrogen dioxide molecule. *(Delhi Hons., 1987; 1988)*
15. Explain the structure of the anhydride of nitric acid. *(GND, 1984 S)*
16. (*a*) Predict the shape of NO_2^+ ion. *(Delhi Hons., 1987)*

    Hint : NO_2^+ ion is called ***nitronium ion*** (also called nitryl or nitrosyl ion) and is present in the solid state of N_2O_5 molecule, $[NO_2]^+ [NO_3]^-$. Due to *sp* hybridisation of $\overset{+}{N}$, NO_2^+ ion has linear shape. ($:\ddot{O} = \overset{+}{N} = \ddot{O}:$).

    (*b*) Explain why NO_2 readily forms dimer, N_2O_4.
17. Explain the formula and structures of various oxides of phosphorus. *(Punjab, 1984; Delhi, 1987)*
18. Discuss the structure of phosphoric oxides. *(Lucknow, 1993)*
19. Write a note on the structure of oxides of sulphur. *(Delhi, 1986; Meerut, 1986; Calcutta, 1986)*
20. Discuss the shape of SO_2 molecule on the basis of hybridisation. *(Meerut, 1988)*
21. Discuss the structure of SO_3 molecule. *(Delhi, 1988)*
22. Give and explain the stereochemistry of solid SO_3. *(Delhi Hons., 1987)*
23. Write a note on "Structure of oxy-acids of sulphur". *(Meerut, 1994)*
24. Write the structure of sodium thiosulphate. *(Meerut, 1994)*
25. Discuss the structure of XeO_3 molecule. *(Punjab, 1980)*
26. Predict which of the following ions is planar: ClO_3^-, ClO_4^- and NO_3^-. *(Delhi Hons., 1987)*
27. Explain the term "hybridisation" with reference to methane and ethylene molecules.*(Mangalore, 1986)*
28. Give the structures of an oxide and an oxyhalide of xenon in its six oxidation states. *(Ravishankar, 1993)*
29. Explain sp^3, sp^2 and *sp* hybridisation of carbon atom in organic molecules. Draw the orbital model of acetylene and label the bonds as sigma and pi bonds. *(Manipur, 1989)*
30. Explain *sp* and sp^2 hybridisation with the help of simple organic molecules. Indicate the shape of the molecule in each case. *(North Eastern Hill 1992)*
31. Indicate the different states of hybridisation and deduce the shape of the following ions:

    (*i*) NO_3^- *(Pune, 1980; Kurukshetra, 1983)*

    (*ii*) ClO_4^- *(Madras, 1986; Delhi Hons., 1987; Delhi, 1988)*

    (*iii*) CO_3^{2-} *(Calcutta, 1985)*

    (*iv*) SO_4^{2-} *(Meerut, 1988)*

    (*v*) Oxy anions of HClO, $HClO_2$, $HClO_3$, and $HClO_4$. *(Lucknow, 1993)*

16

Structure of Solids

Classification of solids

Solids may be classified in the following two manners:

1st classification. This classification gives two categories of solids *viz. true solids* and *pseudo solids*. The comparison between the properties of these two types of solids is given as follows, in the tabular form:

True solids	Pseudo solids
(*i*) These have definite shape and volume. These donot lose their shape and volume even on long standing.	(*i*) These donot have definite shape and volume. On long standing, they lose their shape and volume under their own weights.
(*ii*) They are rigid and hence are not easily distorted by mild distorting forces.	(*ii*) These are less rigid and hence are easily distorted by even mild distorting forces.
(*iii*) They have sharp melting points.	(*iii*) They melt over a range of temperature. On heating, they usually first soften and then are converted into liquid state.
(*iv*) Examples : All crystalline solids like NaCl, KCl, Fe, Cu, S, sugar etc.	(*iv*) Examples : Glass and pitch.

2nd classification. This classification gives three categories of solids *viz., crystalline solids, non-crystalline* or *amorphous solids* and *polycrystalline solids.* Based on the nature of the bonding forces existing between the constituent particles of crystalline solids, these solids have been classified into five catagories, namely *ionic crystals, covalent crystals, molecular (van der Waals) crystals, metallic cyrstals* and *hydrogen bonded crystals.*

Comparison between crystalline and amorphous solids

The comparison between the properties of these two types of solids has been shown in Table 16.1.

Table 16.1 : *Comparison between the properties of crystalline and amorphous solids*

Crystalline solids	Amorphous solids
(*i*) The constituent particles (atoms, ions or molecules) of crystalline solids are arranged in three dimensional space in a definite and regular manner. These solids, therefore, have definite and regular geometry which extends throughout the crystal. Thus crystalline solids have *long range order.*	(i) The constituent particles of amorphous solids are *not* arranged in a regular manner and hence these solids have only *short range order.*
(*ii*) They have sharp (definite) melting points.	(*ii*) They donot have sharp melting points *i.e.,* they melt over a wide range of temperature. For example, when glass in heated, it softens and does not change abruptly into liquid state.

(*iii*) Crystalline solids can be broken into pieces, with *definite* surfaces, when cut with a knife.	(*iii*) Amorphous solids can be broken into pieces, with *irregular* surfaces, when cut with a knife.
(*iv*) Their properties like electrical conductivity, refractive index, thermal expansion etc. are *different* in different directions, *i.e.*, crystalline solids are *unisotropic* and this property is a strong evidence of the fact that the crystalline solids have an orderly arrangement of their different constituent particles.	(*iv*) Their properties like mechanical strength, refractive index, electrical conductivity etc., are the *same* in all the directions, *i.e.*, amorphous solids are *isotropic*.
(*v*) They *have* definite heat of fusion	(*v*) They *do not* have definite heat of fusion.
(*vi*) Crystalline solids are characterized by their symmetry, which is called crystal symmetry.	(*vi*) Amorphous solids donot possess symmetry.
(*vii*) Examples : NaCl, diamond, sugar, sulphur, graphite	(*vii*) Examples : Rubbers, certain plastics, glass, fused silica.

Polycrystalline solids

There are solids which occur as powder and resemble amorphous solids, but their individual particles have all the characteristics of crystalline solids. Such solids are called polycrystalline solids. These solids appear to be isotropic, even when each single crystal is an isotropic.

Types of crystal symmetry found in crystals : Elements of crystal symmetry

All the crystals of the same substance possess the same elements of symmetry. Three elements of symmetry are discussed below :

1. Centre of symmetry. It is such an imaginary point within the crystal that if a straight line is drawn through it, it will meet the surfaces of the crystal at equal distances on each side of the point. The centre of symmetry in cubical crystal like NaCl has been shown by A in Fig. 16.1 (*a*).

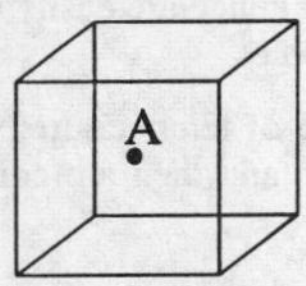

(*a*) Centre of symmetry in a cubical crystal

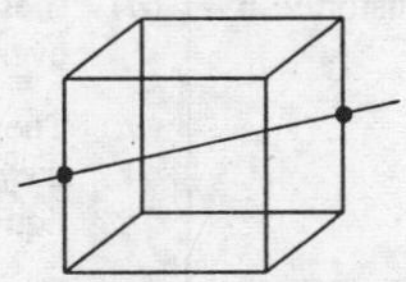

(*b*) Axis of two-fold symmetry (six)

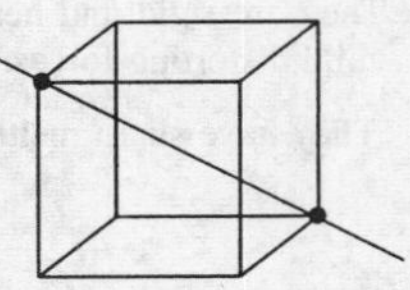

(*c*) Axis of three-fold symmetry (Four)

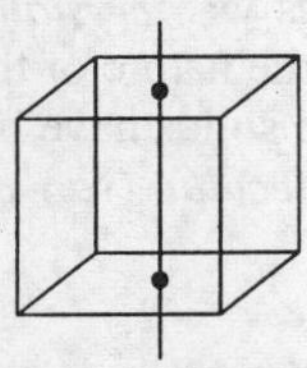

(*d*) Axis of four-fold symmetry (three)

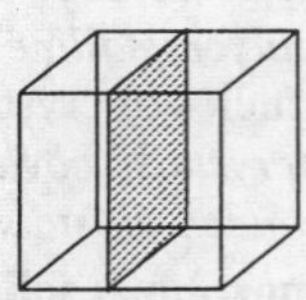

(*e*) Rectangular plane of symmetry (three)

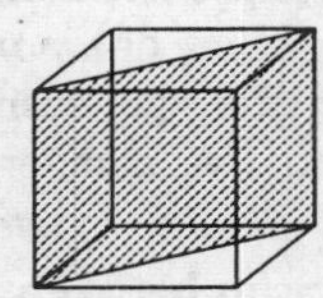

(*f*) Diagonal plane of symmetry (six)

Fig. 16.1 : Elements of crystal symmetry.

2. Axes of symmetry. It is such an imaginary line within the crystal that if the crystal is rotated about this line, the crystal will have the same appearance more than once during the course of a complete rotation. A cubical crystal like NaCl can have in all 13 axes of symmetry. These are:

(*a*) *Six axes of two-fold symmetry.* Each of these six axes is called *diad axis.* When the crystal is rotated about any of these axes, it presents the same appearance *two times* in a complete rotation. One of such axes is shown in Fig. 16.1 (*b*). Each of these three axes intersects at the middle points of the two opposite edges of the cube.

(*b*) *Four axes of three-fold symmetry.* Each of these axes is called *triad axis.* When the crystal is rotated about any of these axes, it presents the same appearance *three times* in a complete rotation. One of such axes is shown in Fig. 16.1 (*c*). Each of these four axes intersects at the two opposite solid angles (*i.e.*, corners) of the cube.

(*c*) *Three axes of four-fold symmetry.* Each of these three axes is called *tetrad axis*. When the crystal is rotated about any of these axes, it presents the same appearance *four times* in a complete rotation. One of such axes is shown in Fig. 16.1 (*d*). Each of these three axes intersects at the middle points of the two opposite faces of a cube. All the three axes are at right angles to one another.

3. Plane of symmetry. *It is an imaginary plane which divides a crystal into two such parts that one is the exact mirror image of the other.* In other words, a plane of symmetry divides the crystal into two identical and similarly placed halves. In all, there are *nine* planes of symmetry possessed by a cubical crystal like NaCl as shown below :

(*a*) *Three rectangular planes of symmetry.* One such rectangular plane of symmetry is shown in Fig. 16.1 (*e*). There are in all *three* such rectangular planes of symmetry which are at right angles to each other.

(*b*) *Six diagonal planes of symmetry*. One such diagonal plane of symmetry is shown in Fig. 16.1 (*f*). There are in all six such diagonal planes of symmetry which pass diagonally through the cube.

Definition of lattice points and crystal lattice

The constituent particles (atoms, ions or molecules) of a crystalline solid are arranged in a regular order. The position of these particles in the crystalline solid, relative to one another, is usually represented by points (•) which are called *lattice points* or *lattice sites.* The orderly or regular arrangement of infinite set of these lattice points is called *crystal lattice* or *space lattice* or simply *crystal.* Thus, a crystal lattice can be defined as an orderly arrangement of lattice points (*i.e.*, atoms, molecules or ions) which shows how the constituent particles present in the crystal lattice are arranged in three dimensional space. A three dimensional crystal lattice is shown in Fig 16.2.

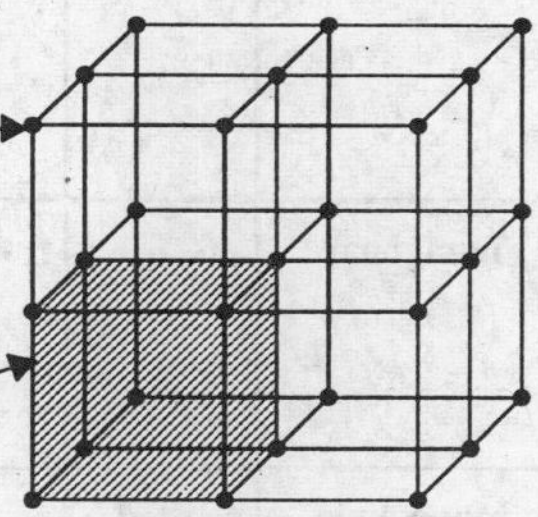

Fig. 16.2 : Definition of lattice points, crystal lattice and unit cell.

Definition of unit cell

If we carefully look at the crystal lattice, shown in Fig. 16.2, it will be found that the crystal lattice is composed of an infinite number of small units, which are adjacent to one another, in three dimensions. Here we have shown only eight small units. Each of these small units is called a unit cell. Thus, a unit cell can be defined as the smallest unit of the crystal lattice which, when repeated again and again, gives the entire crystal of the given substance. When a large number of unit cells of the crystal combine together and arrange themselves in a regular order, the entire crystal lattice is obtained. A unit cell of a crystal lattice has all those properties which the crystal lattice has. One of the eight unit cells of the crystal lattice, shown in Fig. 16.2, has been represented by shaded lines.

Seven crystal systems and 14 Bravais lattices : Different types of structures that a crystal can have

It can be shown geometrically that theoretically there can be 32 different combinations of elements of symmetry of a crystal. These are called **32 point groups** or **32 systems.** Some of these systems have been grouped into seven different categories which are known as seven **basic crystal systems.** These seven systems are *cubic, orthorhombic* (or *rhombic*), *tetragonal, monoclinic, triclinic, hexagonal and rhombohedral* (or *trigonal*).

The properties of these systems along with the types and number of space lattices present in each of the seven crystal systems are given in Table 16.2. A column wise study of the table is given below (Also see Fig 16.3).

Table 16.2 : *The seven crystal systems*

1st Name of the crystal system	*2nd* Parameters or dimensions of the unit cell		*3rd* Types of space lattices present in the crystal system and their total number	*4th* Minimum symmetry elements	*5th* Examples
	Intercepts	Crystal angles (Also called axial or interfacial angles)			
1. Cubic	$a = b = c$	$\alpha = \beta = \gamma = 90°$	(*i*) Simple (S) or primitive (P) cubic (SC) (*ii*) Body-centered (I) cubic (BCC) (*iii*) Face-centered (F) cubic (FCC) Total = 3	Four 3-fold	NaCl, KCl, CaF_2, $NaClO_2$, diamond, ZnS, Cu_2O, alums, Pb, Ag, Au, Hg.
2. Orthorhombic (or simply rhombic)	$a \neq b \neq c$	$\alpha = \beta = \gamma = 90°$	(*i*) Simple or primitive orthorhombic (*ii*) Body-centered orthorhombic (*iii*) End-centered orthorhombic (*iv*) Face-centered orthorhombic Total = 4	Three mutually perpendicular 2-fold	KNO_3, K_2SO_4, $BaSO_4$, $MgSO_4$, Mg_2SiO_4 α-sulphur, $PbCO_3$
3. Tetragonal	$a = b \neq c$	$\alpha = \beta = \gamma = 90°$	(*i*) Simple or primitive tetragonal (*ii*) Body-centered tetragonal Total = 2	One 4-fold	Sn, $NiSO_4$, SnO_2, TiO_2, Zr SiO_4, KH_2PO_4, $PbWO_4$
4. Monoclinic	$a \neq b \neq c$	$\alpha = \gamma = 90°$ $\beta \neq 90°$	(*i*) Simple monoclinic (*ii*) End-centered monoclinic Total = 2	One 2-fold	$Na_2SO_4 \cdot 10H_2O$, $Na_2B_4O_7 \cdot 10H_2O$, $CaSO_4 \cdot 2H_2O$, $FeSO_4$
5. Triclinic	$a \neq b \neq c$	$\alpha \neq \beta \neq \gamma \neq 90°$	Simple triclinic 1	One 1-fold	$CuSO_4 \cdot 5H_2O$, $K_2Cr_2O_7$, H_3BO_3
6. Hexagonal	$a = b \neq c$	$\alpha = \beta = 90°$ $\gamma = 120°$	Simple hexagonal 1	One 6-fold	AgI, graphite, Zn, quartz, ZnO, CdS, HgS, ice, PbI_2, beryl, Mg, Zn, Cd.
7. Rhombohedral (or trigonal)	$a = b = c$	$\alpha = \gamma = 90°$ $\beta = 90°$	Simple hexagonal 1 Total number of space lattices = 3 + 4 + 2 + 2 + 1 + 1 + 1 = 14	One 3-fold	$NaNO_3$, ICl, calcite, magnesite, Sb, Bi.

1st column. This column gives us the names of the seven crystal systems. The seven crystal systems correspond to the seven distinct types of unit cells (or sub-units). These unit cells are all parallelopipeds.

2nd column. The three sides of a unit cell are *crystallographic axes* or simply *crystal axes.* The angles between the three pairs of the three intersecting crystal axes are called *crystal angles interfacial* or *interaxial angles.* The interfacial angles of the unit cell are represented by α, β and γ. The lengths of the three intersecting axes are called *primitives* or *characteristic intercepts* on the axes. These are denoted by *a, b* and *c.* The intercepts (*i.e., a, b,* and *c*) and interfacial angles (*i.e.,* α, β and γ) together are called *parameters* of the unit cell. If we know the magnitude of intercepts and interfacial angles of a unit cell, we can determine the shape and actual size of the unit cell. But if we donot know the actual magnitude of the intercepts but only their ratio and the value of interfacial angles, then we can determine only the shape (form) of the unit cell and not its actual size. Following points may be noted from this column.

(*i*) *In cubic crystal system,* the lengths of the three intercepts are equal ($a = b = c$) and the three crystal axes are at right angles to each other (*i.e.,* $\alpha = \beta = \gamma = 90°$)

(*ii*) In *orthorhombic crystal system,* the lengths of the three intercepts are different ($a \neq b \neq c$) but each of the three crystal angles is equal to 90° ($\alpha = \beta = \gamma = 90°$)

(*iii*) In *tetragonal structure,* the length of the two intercepts are equal, but the length of the third intercept is different ($a = b \neq c$). Each of the three crystal angles is equal to 90° ($\alpha = \beta = \gamma = 90°$).

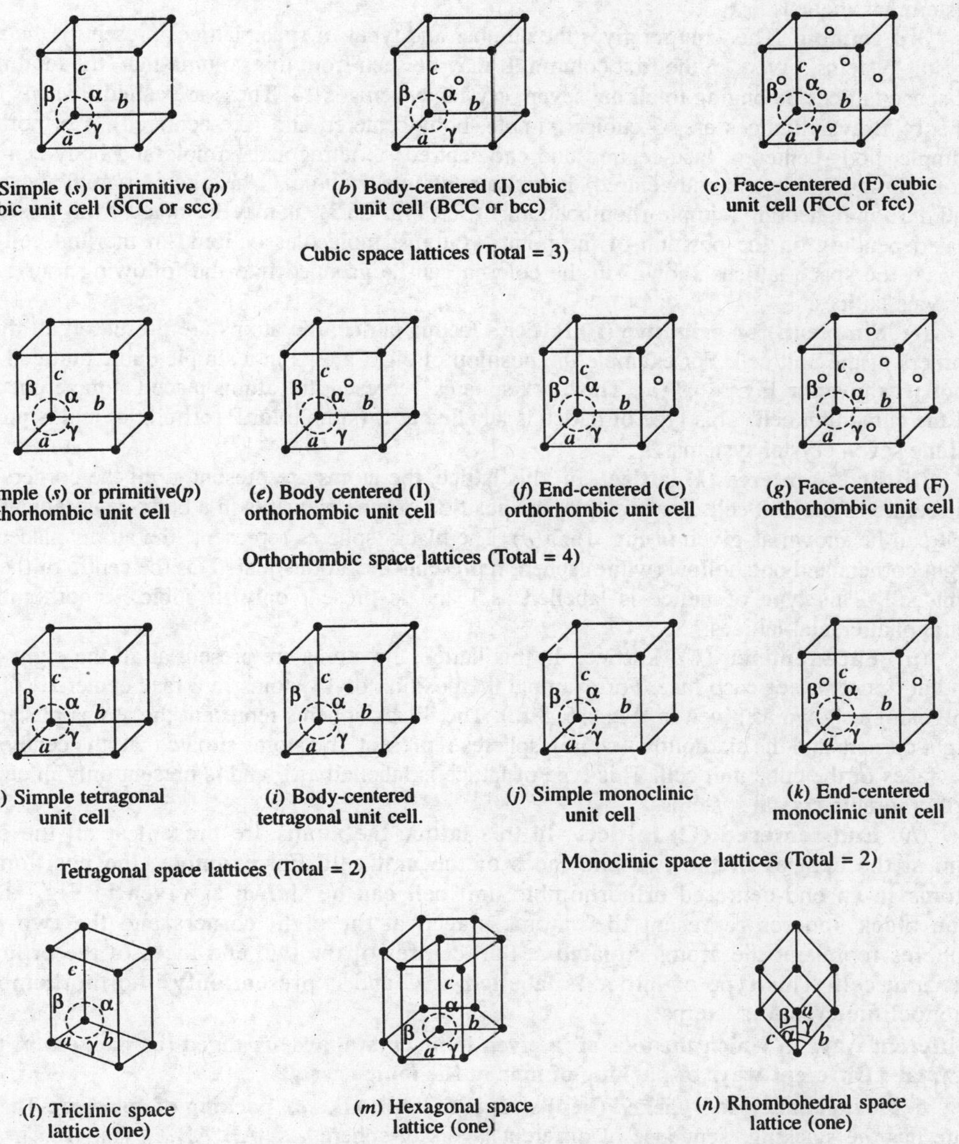

(*a*) Simple (*s*) or primitive (*p*) cubic unit cell (SCC or scc)
(*b*) Body-centered (I) cubic unit cell (BCC or bcc)
(*c*) Face-centered (F) cubic unit cell (FCC or fcc)

Cubic space lattices (Total = 3)

(*d*) Simple (*s*) or primitive(*p*) orthorhombic unit cell
(*e*) Body centered (I) orthorhombic unit cell
(*f*) End-centered (C) orthorhombic unit cell
(*g*) Face-centered (F) orthorhombic unit cell

Orthorhombic space lattices (Total = 4)

(*h*) Simple tetragonal unit cell
(*i*) Body-centered tetragonal unit cell.

Tetragonal space lattices (Total = 2)

(*j*) Simple monoclinic unit cell
(*k*) End-centered monoclinic unit cell

Monoclinic space lattices (Total = 2)

(*l*) Triclinic space lattice (one)
(*m*) Hexagonal space lattice (one)
(*n*) Rhombohedral space lattice (one)

Fig. 16.3 : Fourteen (3 + 4 + 2 + 2 + 1 + 1 + 1 = 14) Bravais lattices.

(*iv*) In *monoclinic arrangement,* all the three intercepts are of unequal length ($a \neq b \neq c$). Two crystal angles are equal to 90° ($\alpha = \beta = 90°$) but the third angle is not of 90° ($\beta \neq 90°$).

(*v*) The *triclinic structure* has complete lack of symmetry, since all the three intercepts are unequal in length ($a \neq b \neq c$) and similarly all the three crystal angles are of unequal magnitude and none of three is equal to 90° ($\alpha \neq \beta \neq \gamma \neq 90°$).

(*vi*) In *hexagonal structure,* there are three axes in one plane and a fourth axis perpendicular to the plane having the three axes. The three axes are inclined at an angle of 120° to each other. The intercepts along the three co-planer axes are equal, but the intercepts along the fourth axis are unequal.

(*vii*) In *rhomhohedral structure*, all the three intercepts are equal in length. Two axial angles are equal to 90°, while the third axial angle is not equal to 90°.

The above discussion shows that the cubic system has the greatest symmetry, whereas triclinic system has the least.

3rd column. This column gives the number and types of space lattices present in the various crystal systems, given in the first column. It may be seen from this column that the total number of space lattices belonging to all the seven crystal systems is 14. These are called *Bravais lattices* The 14 Bravais lattices are 3 cubic (simple, body-centered and face-centered), 4 orthorhombic (simple, body-centered, face-centred and end-centred), 2 tetragonal (simple and body-centred), 2 monoclinic (simple and end-centred) 1, triclinic (simple triclinic), 1 hexagonal (simple hexagonal) and 1 rhombohedral (simple rhombohedral) (See Fig. 16.3). It may be noted from this column that depending on the position of the points (atoms, molecules or ions) in the unit cell of the crystal the space lattices shown in the column can be grouped into the following four types of Bravais lattices.

(*i*) **Simple (S) or primitive (P) lattice.** In this lattice, the atoms are present at all the eight corners of the unit cell. For example the position of eight atoms in a simple cubic unit cell can be shown as given in Fig. 16.3 (*a*). The black spheres represent the atoms placed at the eight corners of the cubic unit cell. This type of lattice is labelled as S (simple) or P (primitive) and is present is all the seven crystal systems.

(*ii*) **Body-centered (I) lattice.** In this lattice, the atoms are present at all the corners and at the centre of the unit cell. For example, the position of nine elements in a body-centered cubic unit cell can be shown as given in Fig. 16.3 (*b*). The black spheres represent the atoms placed at the eight corners and one hollow (white) sphere represents the atom situated as the centre of the cubic unit cell. This type of lattice is labelled as I and is present only in cubic, orthorhombic and tetragonal crystal lattices.

(*iii*) **Face-centered (F) Lattice.** In this lattice, the atoms are present at all the eight corners and at the centre of each face. For example the position of 14 atoms in a face-centered cubic unit cell can be shown as given in Fig. 16. 3 (*c*). The black spheres represent the atoms placed at the eight corners and the six hollow (white) spheres represent the atoms situated at the centers of the six faces of the cube unit cell. This type of lattice is labelled as F and is present only in cubic and orthorhombic crystal systems.

(*iv*) **End-centered (C) lattice.** In this lattice the atoms are present at all the corners and at the centres of the two end faces of the unit cell. For example the position of 10 atoms in an end-centered orthorhombic unit cell can be shown as given in Fig. 16.3 (*k*). The black spheres represent the atoms placed at the eight corners and the two hollow spheres represent the atoms situated at the centres of the two end faces of the orthorhombic unit cell. This type of lattice is labelled as C and is present only in orthorhombic and monoclinic crystal systems.

Different ways in which the ions of a given ionic crystal are arranged (or packed) in the crystal : Different ways of packing of ions in the ionic crystals

1. Hexagonal closed-packed (hcp) structure (AB AB........ packing of spheres). This structure has the stacking sequence of different layers of spheres as AB AB... . (Fig 16.4). A close examination of this arrangement of different layers of spheres shows that it has *hexagonal close* *packing symmetry* and hence this arrangement is called *hexagonal closest packed (hcp) structure.* In this structure, each sphere is surrounded by 12 other nearest spheres and hence the coordination number of each sphere is 12.

2. Cubic (or cubical) close-packed (ccp) or face-centered cubic (fcc) structure (ABC ABC....... packing of spheres). This structure has the stacking sequence of different layers of spheres as ABC ABC .... This structure has cubic symmetry and hence is also called cubic close packed (ccp) structure. A closer study of ABC ABC .... arrangement shows that, in addition to the 8 spheres placed at the 8 corners of a cube, there is also one sphere at the centre of each of the six faces of the cube and hence this structure is also called face-centered cubic (fcc) structure (See Fig. 16.5). In this structure, each sphere is surrounded by 12 other nearest spheres and hence the coordination number of each sphere in *ccp* or *fcc* structure is 12.

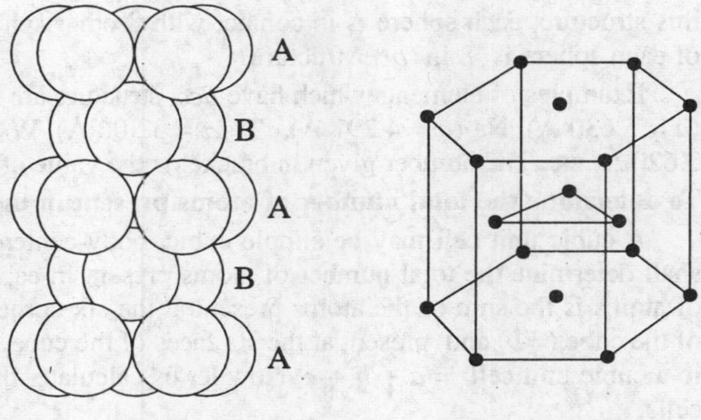

(*a*) (*b*)

Fig. 16.4 : AB AB.. system of cose packing of spheres or hexagonal close packing (*hcp*) of spheres.

Examples of elements which have cubic close-packed structure are Al (a = 4.049 Å), Cu (a = 3.615 Å), Au (a = 4.079 Å), γ-Fe (a = 3.591 Å), Pb (a = 4.951 Å), Ni (a = 3.524 Å), Pt (a = 4.086 Å), Ag (a = 4.086 Å) etc. The number given in bracket indicates the edge length (a) of the cubic unit cell.

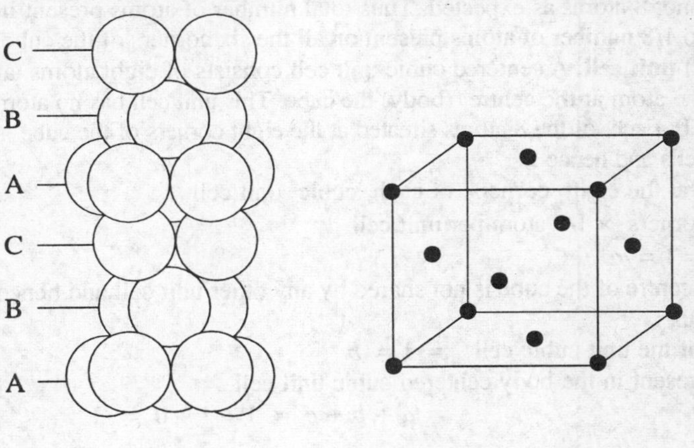

(*a*) (*b*)

Fig 16.5 : (*a*) ABC ABC ...... system of close packing of spheres (*b*) Face-centered cubic arrangement of spheres.

Common features of hcp and ccp (or fcc) structures. Following are common features of both these structures.

(*i*) Coordination number of each sphere in both the types of structure is 12, since each sphere is surrounded by six nearest spheres lying in the same layer, three spheres lying in the layer just above it and three spheres lying in the layer just below it.

(*ii*) Both types of structures have the same packing of efficiency which in equal to 74%, *i.e.,* 74% of the total volume of the unit cell is occupied by the spheres.

3. Body-centered cubic (bcc) structure. In this structure, there are eight spheres at 8 corners of a cube and one sphere is in the centre (or body) of the cube (See Fig. 16.6). In

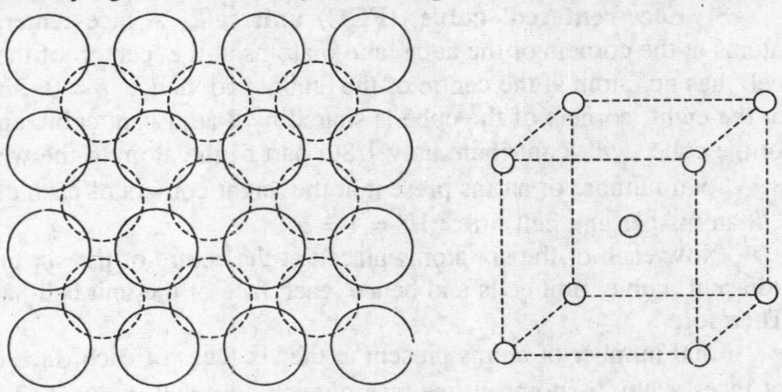

(*a*) (*b*)

Fig. 16.6 : Body-centred cubic arrangement of spheres.

this structure, each sphere is in contact with 8 other spheres and hence the coordination number of each sphere is 8 in *bcc* structure.

Examples of elements which have *bcc* structure are Ba (a = 5.025 Å), α-Fe (a = 2.867 Å), Rb (a = 5.630 Å), Na (a = 4.291 Å), Ti (a = 3.306 Å), W (a = 3. 115 Å), U (a = 3.474 Å), Zr (a = 3.620 Å) etc. The number given in bracket is the value of the edge length (a) of the cubic unit cell.

To determine the total number of atoms present in the cubic unit cell

A cubic unit cell may be simple cubic, body-centered cubic or face-centered cubic. Here we shall determine the total number of atoms present in each of these cubic unit cells. Total number of atoms is the sum of the atoms present at the six corners of the cube (= a), present at the centre of the cube (= b) and present at the six faces of the cube (= c). Thus total number of atoms present in a cubic unit cell $= a + b + c$. Now let us calculate this number is each of the above three unit cells.

1. Simple (primitive) cubic (SC) unit cell. Simple cubic unit cell consists of 8 atoms at the 8 corner of the cube. This unit cell has no atom at its centre and at any of its six faces (*i.e.*, $b = 0$ and $c = 0$). Since each of the 8 atoms situated at the 8 corners is shared by 8 adjacent cubic unit cells, each atom situated at the corner contributes only 1/8th part of an atom to the cubic unit cell. Thus the number of atoms contributed by 8 atoms is equal to $1/8 \times 8 = 1 = a$. Thus we seen that a simple cubic unit cell has only 1 atom and not 8 atoms as expected. Thus total number of atoms present in one simple cubic unit cell is equal to 1/8 number of atoms present on all the six corners of the cube.

2. Body-centered cubic (BCC) unit cell. A centered cubic unit cell consists of eight atoms at the eight corners of the cube and one atom at the centre (body) the cube. This unit cell has no atom at any of its six faces and hence $c = 0$. Each of the 8 atoms situated at the eight corners of the cube is shared by eight adjacent cubic unit cells and hence :

Total number of atoms present at the eight corners of each cubic unit cell

= No. of corners × 1/8 atom per unit cell

= $8 \times 1/8 = 1 = a$.

The single atom placed at the centre of the cube is not shared by any other unit cell and hence belongs entirely to the unit cell. Thus :

No. of atoms present at the centre of the unit cubic cell $= 1 = b$

Therefore, total number of atoms present in the body centered cubic unit cell

$$= a + b + c = 1 + 1 + 0 = 2$$

General formula is :

Total number of atoms present in one body-centered cubic unit cell

= 1/8 × No. of atoms placed at 8 corners of the cube + No. of atoms situated at the centre of the cube

$= 1/8 \times 8 + 1 = 1 + 1 = 2$

3. Face-centered cubic (FCC) unit cell. A face-centered cubic unit cell consists of 8 atoms at the corners of the cube and 6 atoms at the center of the six faces of the cube. This unit cell has no atom at the centre of the cube and hence $b = 0$. Since each of the 8 atoms situated at the eight corners of the cube is shared by 8 adjascent cubic unit cells, each atoms at the corner of the cube will contribute only 1/8th part of the atom to the whole unit cell. Therefore :

Total number of atoms present at the eight corners of each cubic unit cell = No. of corners × 1/8 atoms per unit cell $= 8 \times 1/8 = 1 = a$

Now each of the six atoms placed at the centre of the six faces of the cube is shared by two adjacent cubic unit cells and hence each face of the unit cell has only 1/2 atom (and not one). Therefore :

Total number of atoms present at the six faces of each face-centered cubic unit cell = No. of faces × No. of atoms at one face of each unit cell $= 6 \times 1/2 = 3 = c$

Thus total number of atoms present in face-centered cubic unit cell

$$= a + b + c = 1 + 0 + 3 = 4.$$

General formula is :

Total number of atoms present in one face-centered cubic unit cell = 1/8 × No. of atoms placed at 8 corners of the cube + 1/2 × No. of atoms placed at the six centers of the six faces of the cube

$$= \frac{1}{8} \times 8 + \frac{1}{2} \times 6 = 4$$

Example 16.1. *A unit cell consists of a cube in which there are anions (B) at each corner and one at the centre of the unit cell and cations (A) at the centre of each face. How many cations and anions make up the unit cell? What is the simplest formula for this compound ?*

Solution : No. of anions (B) present in the unit cell = $1/8 \times 8 + 1 \times 1$ = **1 + 1 = 2**

No. of cations (A) present in the unit cell = $6 \times 1/2 = 3$

∴ Formula of the compound = $\mathbf{A_3B_2}$

To find out a relation between the edge length (*a*) of a cubic unit cell and the radius (*r*) of the spherical atom present in it.

We know that a cubic unit cell may be a simple cubic unit cell, face-centered cubic unit cell or a body centered cubic unit cell.

(*i*) *Relation between a and r in a simple cubic unit cell.* We know that a simple cubic unit cell has 8 atoms at its eight corners. Suppose the length of each side of the cube is *a* and the radius of each atom (atomic radius) is *r*. Eight atoms lying at 8 corners have been represented by hollow spheres in Fig 16.7. Let us consider two nearest atoms, A and B. Since A and B atoms are nearest neighbours, they touch each other and hence :

$$a = r_A + r_B \quad ...(i)$$

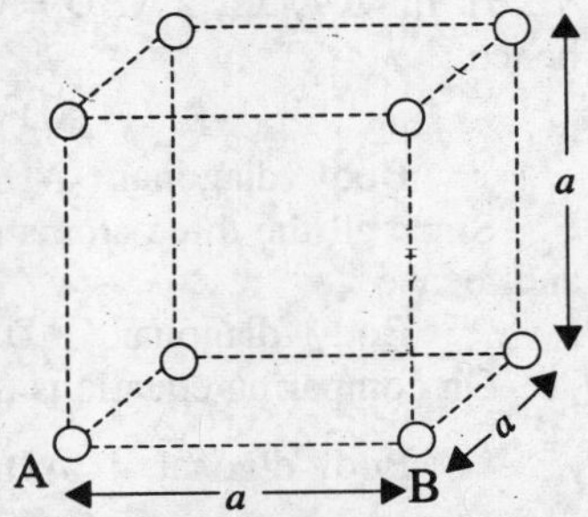

Fig. 16.7 : Calculation of radius of an atom (atomic radius) in a simple cubic unit cell.

Now since A and B atoms are identical atoms, $r_A = r_B = r$, and hence equation (*i*) becomes :

$$a = 2r \quad ...(ii)$$

Equation (*ii*) gives us a relation between *a* and *r* in a simple cubic unit cell.

(*ii*) *Relation between a and r in a face-centered cubic unit cell.* We know that a face-centered cubic unit cell has 8 atoms at its 8 corners and 6 atoms at the centre of its six faces [See Fig. 16.8 (*a*)]. Suppose the length of each side of the cube is *a* and the radius of each atom is *r.* It is easy to see from the figure that AB is one of the face diagonals of the cubic unit cell. In Δ ABC, ∠ ACB = 90° and hence

$$AB^2 = AC^2 + CB^2$$
$$= a^2 + a^2 = 2a^2$$

∴ Face diagonal, $AB = \sqrt{2}\, a$ ... (*i*)

Since all the three atoms, namely A, D and B are nearest neighbours. they touch each other and hence :

$$a = r_A + r_B$$

Now since A and B are identical atoms, $r_A = r_B = r$ and therefore $a = 2r$

Edge length (*a*) of the cubic unit cell = $2r$

Face diagonal, $AB = r_A + 2r_D + r_B$.

Since all the atoms are identical,

$$r_A = r_D = r_B = r$$

∴ Face diagonal, $AB = 4r$ ...(*ii*)

On comparing equations (*i*) and (*ii*), we get :

Face diagonal, $AB = \sqrt{2}\, a = 4r$ ...(*iii*)

Equation (*iii*) gives us a relation between *a* and *r* in a face-centered cubic unit cell.

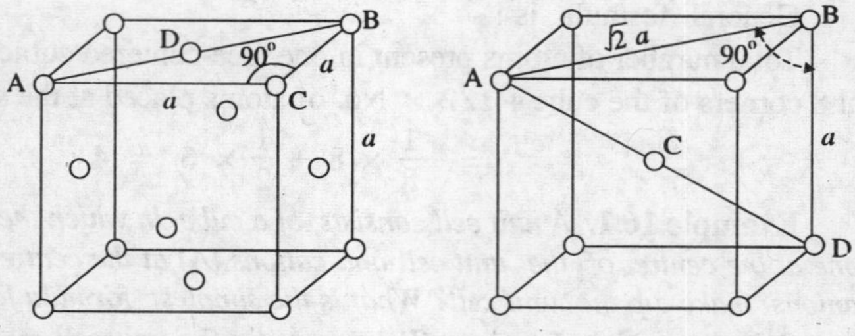

(a) Face-centred cubic unit cell (b) Body-centred cubic unit cell

Fig. 16.8 : (a) Calculation of the length of the face diagonal (AB) in a face-centered cubic unit cell (b) Calculation of the body diagonal (A D) in a body-centered cubic unit cell.

(iii) *Relation between a and r in a body-centered cubic unit cell.* We know that a body-centered cubic unit cell contains 8 atoms at its eight corners and one atom at its body (centre) [See Fig. 16.8 (b)]. Suppose the length of each side of the cube is a and the radius of each atom is r. It is easy to see from the figure that AD is a body diagonal and AB is a face diagonal. The length of face diagonal (AB), as we have shown above, is equal to $\sqrt{2}$ a (*i.e.*, AB = $\sqrt{2}$ *a*). In Δ ADB, $\angle$ABD = 0° and hence :

$$AD^2 = AB^2 + BD^2 = (\sqrt{2}a)^2 + a^2 = 3a^2$$

$\therefore$ Body diagonal, $AD = \sqrt{3}a$ ...(i)

Since all the three atoms namely A, C and D are nearest neighbours, they touch each other and hence :

Body diagonal, $AD = r_A + 2r_C + r_D = 4r$ ...(ii)

On comparing equations (i) and (ii), we get :

Body digonal, $AD = \sqrt{3}a = 4r$ ...(iii)

Body diagonal is also called *cross diagonal* or *long diagonal.* Equation (iii)gives a relation between a and r in a body-centered cubic unit cell.

Example 16.2. *In Li I crystal, I^- ions from cubical closest packing and Li^+ ions occupy octahedral holes. What is the relationship between the edge length of unit cell and the radius of I^- ions ? Calculate the ionic radius of Li^+ an I^- ions, if a = 600 pm.*

Solution. When we say that I^- ions form cubic closest packing, this means that I^- ions are present at all the eight corners and at the centre of each of the six faces of the cube. In this arrangement of I^- ions, the length of face diagonal (= $\sqrt{2}$ a) is equal to 4 times the radius of the anion (r_{I^-}), and the edge length (a) of the cube is equal to double the sum of the radius of Li^+ and I^- ions. Thus :

$$\sqrt{2}\,a = 4r_{I^-} \quad \text{... (i)}$$

$$a = 2\left(r_{Li^+} + r_{I^-}\right) \quad \text{... (ii)}$$

From equation (i), we have :

$$\sqrt{2} \times 600\ pm = 4r_{I^-}$$

or $$r_{I^-} = \frac{1.41 \times 600}{4} = \mathbf{211.5\ pm}$$

From equation (ii), we have :

$$600 = 2\left(r_{Li^+} + 211.5\right)$$

or $$r_{Li^+} = \frac{600}{2} - 211.5 = \mathbf{88.5\ pm}$$

To determine the relative density of packing of atoms in a cubic unit cell

Relative density of packing of a cubic unit cell is defined as the ratio of the volume occupied by the actual number of spherical atoms present in the cubic unit cell and the total volume of the cube. Suppose that the radius of the spherical atom is r and edge length of the cube is a . Also suppose that the unit cell has N atoms. Now since an atom is supposed to be a hard sphere, its volume is equal to $4/3\ \pi r^3$. Thus the volume occupied by n atoms = $n \times 4/3\ \pi r^3$. Now we know that volume of the cubic unit cell = a^3. Therefore :

Relative density of packing

$$= \frac{\text{Volume occupied by } n \text{ atoms present in the cubic unit cell}}{\text{Volume of the cubic unit cell of packing}}$$

$$= \frac{n \times 4/3\ \pi r^3}{a^3}$$

Again we know that a cubic unit cell may be simple cubic unit cell, body-centered cubic unit cell and face centred cubic unit cell. Here, we shall determine the relative density of packing in all these three unit cells.

1. To calculate the relative density of packing in a simple cubic (sc) unit cell. We know that although a simple cubic unit cell has eight atoms at its eight corners, yet the actual number of atoms present in the unit cell is not equal to 8 but is equal to $1/8 \times 8 = 1$. (*i.e.*, $n = 1$).

Volume of this one atom = $1 \times 4/3\ \pi r^3$.

We know that in sc unit cell r and a are related as :

$$r = \frac{a}{2}$$

$\therefore$ Volume of one atom = $4/3\ \pi\ (a/2)^3$

Volume of the the cube = a^3

$$\therefore \text{ Relative density of packing in sc unit cell } = \frac{\frac{4}{3}\pi\left(\frac{a}{2}\right)^3}{a^3} = \frac{\pi}{6} = = 0.504$$

This relation shows that the volume occupied by *one* atom present in a sc unit cell is $0.504 \times 100 = 50.4\ \%$ of the whole volume of the cubic unit cell and the volume that remains vacant (*i.e.*, unoccupied) is equal to $(100 - 50.41)\ \% = 49.6\ \%$

2. To calculate the relative density of packing in body-centered cubic (bcc) unit cell. We know that a body-centered cubic unit cell has 8 atoms at its eight corners and one atom in the body (*i.e.*, centre) of the unit cell. Thus the total number of atoms in the unit cell should be equal to $8 + 1 = 9$. But actually, as we know, the actual number of atoms present in the body-centered cubic unit cell is equal to $8 \times 1/8 = 1 + 1 = 2$ and not 9. Thus $n = 2$ and volume of 2 atoms = $2 \times 4/3\ \pi r^3$.

In a body-centered cubic unit cell r and a are related as :

$$r = \frac{\sqrt{3}\ a}{4}$$

$$\therefore \text{ Volume of 2 atoms } = \frac{2 \times 4 \times \pi}{3}\left(\frac{\sqrt{3}\ a}{4}\right)^3$$

Volume of the cube = a^3

$\therefore$ Relative density of packing in *bcc* cubic unit cell

$$= \frac{\frac{2 \times 4 \times \pi}{3}\left(\frac{\sqrt{3}\ a}{4}\right)^3}{a^3} = \frac{\sqrt{3}\ \pi}{8} = 0.680$$

The above relation shows that the volume occupied by *two* atoms present in the *bcc* cubic unit cell is $0.680 \times 100 = 68.\%$ of the whole volume of the cubic unit cell and the volume that remains vacant (*i.e.*, unoccupied) is equal to $(100 - 68)\% = 32\%$.

3. To calculate the relative density of packing in a face-centered cubic (fcc) unit cell [*Also called cubic close packing (ccp) arrangement of atoms*]. We know that a face-centered cubic unit cell has 8 atoms at its eight corners and 6 atoms at the six centres of its six faces. Thus the total number of atoms present in a face-centered cubic unit cell should be equal to 8 +

6 = 14. But, as we know, the actual number of atoms present in a face-centered cubic unit cell is equal to $8 \times 1/8 + 6 \times 1/2 = 1 + 3 = 4$ and not 14. Thus $n = 4$ and volume of 4 atoms $= \dfrac{4 \times 4 \times \pi r^3}{3}$.

In a face-centered cubic unit cell r and a are related as : $r = \dfrac{\sqrt{2}\,a}{4}$

$$\therefore \quad \text{Volume of 4 atoms} = \frac{4 \times 4 \times \pi}{3} \times \left(\frac{\sqrt{2}\,a}{4}\right)^3$$

$$\text{Volume of the cube} = a^3$$

∴ Relative density of packing in *fcc* cubic unit cell

$$= \frac{\dfrac{4 \times 4 \times \pi}{3} \times \left(\dfrac{\sqrt{2}\,a}{4}\right)^3}{a^3} = \frac{\sqrt{2}\,\pi}{6} = 0.740$$

The above relation shows that the volume occupied by four atoms present in the *fcc* cubic unit cell is $0.740 \times 100 = 74\ \%$ of the whole volume of the cubic unit cell and the volume that remains vacant (*i..e.,* unoccupied) is equal to $(100 - 74)\ \% = 26\ \%$.

To calculate the density of a crystal having cubic structure

The density of a crystal having cubic structure depends on whether the crystal has simple cubic structure, body-centered cubic structure or face-centered cubic structure. Density is given by :

$$\text{Density of the crystal } (d) = \frac{\text{Mass of the cubic unit cell}}{\text{Volume of the cubic unit cell}} \quad \text{...}(i)$$

$$\text{Mass of a cubic unit cell} = \text{No. of atoms present in a unit cell } (n) \times \text{Mass of a single atom } (m)$$

and,
$$\text{Mass of a single atom } (m) = \frac{\text{Atomic mass (M)}}{\text{Avogadro's number (N)}}$$

$$\therefore \quad \text{Mass of a cubic unit cell} = n \times \frac{\text{M}}{\text{N}} \quad \text{...}(ii)$$

If the edge length of the unit cell is a, then :

$$\text{Volume of the cubic unit cell} = a^3 \quad \text{...}(iii)$$

With the help of equations (*ii*) and (*iii*), equation (*i*) becomes :

$$\text{Density of the crystal } (d) = \frac{n \times \dfrac{\text{M}}{\text{N}}}{a^3}$$

or

$$d = \frac{n\text{M}}{a^3\text{N}} \quad \text{...}(iv)$$

Now we know that the number of atoms present in a simple cubic (SC) unit cell, body-centered cubic (*bcc*) unit cell and face-centered cubic (*fcc*) unit cell is 1, 2 and 4 respectively.

∴ Density of a crystal having SC structure ($n = 1$)

$$= \frac{1 \times \text{M}}{a^3\text{N}} = \frac{1 \times \text{M}}{a^3 \times 6.023 \times 10^{23}\ \text{mol}^{-1}}$$

Density of a crystal having *bcc* structure ($n = 2$)

$$= \frac{2 \times \text{M}}{a^3\text{N}} = \frac{2 \times \text{M}}{a^3 \times 6.023 \times 10^{23}\ \text{mol}^{-1}}$$

Density of a crystal having *fcc* structure ($n = 4$)

$$= \frac{4 \times \text{M}}{a^3\text{N}} = \frac{4 \times \text{M}}{a^3 \times 6.023 \times 10^{23}\ \text{mol}^{-1}}$$

The units of various quantities used in equation (*iv*) are :
Molar mass or atomic mass of the crystal (M) = $g.mol^{-1}$
Edge-length of the cube (a) = cm (Note the relation 1 pm = 10^{-10}cm)
Volume of the cubic unit cell (a^3) = cm^3
Avogadro's number (N) = 6.023×10^{23} mol^{-1}

$$\therefore \quad \text{Density} = \frac{g\,.\,mol^{-1}}{cm^3 \times mol^{-1}} = g/cm^3 = g.cm^{-3}$$

Example 16.3. *Calculate the value of the Avogadro's constant from the following data: density of NaCl = 2. 165 g. cm^{-3} and distance between Na^+ and Cl^- ions in NaCl = 281 pm.*

Solution. The unit cell of NaCl has face-centered cubic unit cell, *i.e.*, in the cubic unit cell of NaCl crystal every next lattice point is occupied by oppositely charged ions and every alternate lattice point is occupied by the same type of ions. Hence :

The edge length of the cubic unit cell of NaCl,

$$a = 2\,(r_{Na^+} + r_{Cl^-}) = 2 \times \text{(Distance between } Na^+ \text{ and } Cl^- \text{ ions)}$$
$$= 2 \times 281\ pm = 2 \times 281 \times 10^{-10}\ \text{cm.}$$
$$n = 4$$

Molar mass of NaCl (M) = $(23 + 35.5)\ g.mol^{-1} = 58.5\ g.mol^{-1}$
Density of NaCl (d) = $2.\ 165 g.cm^{-3}$
Hence, by using the expression

$$d = \frac{n\,M}{Na^3}$$

$$\text{or} \quad 2.165\ g.cm^{-3} = \frac{4 \times (58.5\ g\,.\,mol^{-1})}{N \times (2 \times 281 \times 10^{-10}\ cm)^3}$$

$$\therefore \quad N = \frac{4 \times (58.5\ g\,.\,mol^{-1})}{(2.165\ g\,.\,cm^{-3}) \times (2 \times 281 \times 10^{-10}\ cm)^3}$$
$$= \mathbf{6.09 \times 10^{23}\ mol^{-1}}$$

To calculate the total number of cubic unit cells and total number of atoms present in a given mass of a metal.

Let us consider w grams of a metal whose density is d cm^{-3}. Suppose w grams of the metal contain x cubic unit cells and the edge length of each cubic unit cell is a cm.

Volume at a unit cell = a^3 cm^3.
Volume of x grams of the metal.

$$= \frac{\text{Mass of the metal } (w)}{\text{Density of the metal } (d)} = \frac{w}{d}$$

$$\therefore \quad \text{No. of cubic unit cells } (x) \text{ present in } w \text{ grams of the metal } = \frac{w/d}{a^3}$$

$$\text{or} \quad x = \frac{w}{d\,a^3}$$

Now if n is the number of atoms of the metal present in the cubic unit cell, then :

$$\text{Total number of atoms present in } w \text{ grams of the metal } = n \times \frac{w}{da^3}$$

Example 16.4. *An element crystallisers in a structure having a fcc unit cell of an edge 200 pm. Calculate its density, if 200 g of this element contains 24 × 10^{23} atoms.*

Solution. We know that :

Total number of atoms present in w grams of the element

$$= n \times \frac{w}{d \times a^3}$$

or $$24 \times 10^{23} = 4 \times \frac{200\text{ g}}{d \times (200 \times 10^{-10}\text{ cm})^3}$$

or $$d = \frac{4 \times (200\text{ g})}{(200 \times 10^{-10}\text{ cm})^3 \times (24 \times 10^{23})}$$

$$= \frac{4 \times (200\text{ g})}{(8 \times 10^{-24}\text{ cm}^3) \times (24 \times 10^{23})}$$

or Density = **41.6 g cm^{-3}**

Interstitial sites (or interstitial holes or interstitial voids) in ionic crystals

We know that an ionic crystal is composed of cations and anion. Generally the cation is smaller in size than the anion. The bigger ions (*i.e.,* anions) have close-packed structure which may be either *hexagonal closest-packed (hcp) structure* or *cubical closest-packed (ccp) structure.* In this close-packed structure there is left some vacant space between the anions. This vacant space is is called **interstitial site** or **interstitial void** or **interstitial hole.** The vacant spaces (*i.e.,* interstitial sites) are occupied by the smaller ions (cations).

Types of interstitial sites

Depending on the number of the anions and the way in which they surround a cation in a given ionic crystal, we have the following four type of interstitial sites: *trigonal* or *triangular site* (coordination number of cation = 3), *tetrahedral site* (C.N. of cation = 4), *octahedral site* (C.N. of cation = 6) and *cubic site* (C.N. of cation = 8).

To determine the number of tetrahedral holes for each sphere present in a face-centered cubic (fcc) unit cell (ccp arrangement of spheres)

Let us consider a face-centered cubic unit cell shown in Fig 16.9. Obviously, in this unit cell, one sphere is placed at each of the 8 corners and one sphere is present at the centre of each of the six faces of the cube. We have already said that the total number of spheres present in the face-centered cubic unit cell

= No. of spheres contributed by 8 spheres placed at the eight corners
+ No. at sphere contributed by 6 spheres placed at the centre of the six faces.

$= 8 \times 1/8 + 6 \times 1/2 = 1 + 3 = 4$

The face-centered cubic unit cell, shown in Fig. 16.9, can be divided into eight small cubes. One of the eight small cubes has been shown as ABCDEFGH (cube shown by dark lines). Four spheres, namely A, C, G and F, which are situated at the alternate corners of the small cube, form a tetrahedron. Fifth sphere (J) placed at the centre of the cube or tetrahedron, occupies the tetrahedral site, created by the four spheres (A, C, G and F). Thus, we see that the small cube (ABCDEFGH) has one tetrahedral site. Now, since the entire face-centered cubic unit cell is divided into 8 small cubes, the total number of tetrahedral holes present in the entire unit cell is equal to $8 \times 1 = 8$. Now :

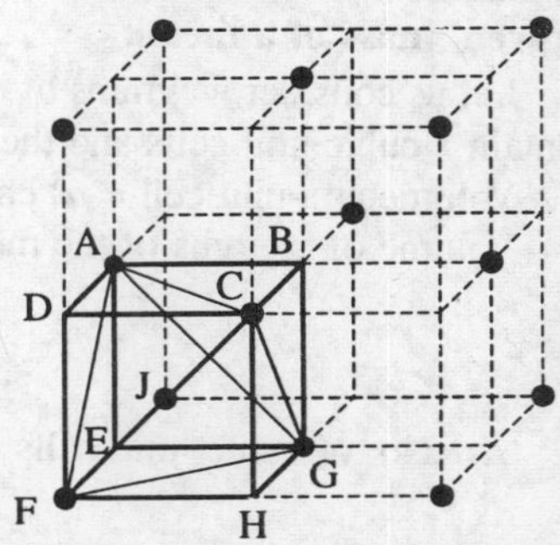

Fig.16.9 : Determination of the number of tetrahedral sites (holes) for each sphere present in a face-centered cubic unit cell.

$\because$ No. of tetrahedral holes for a spheres present in the unit cell = 8

$\therefore$ No. of tetrahedral holes for one sphere present in the unit cell = 8/4 = 2

or No. of tetrahedral holes for N spheres present in the unit cell = 2 N

Thus, we see that every sphere present in a face-centered cubic unit cell [also called cubic close-packed (ccp) structure] has two tetrahedral sites. For example, since in the structure of zinc blend (ZnS) crystal, S^{2-} ions are present at each of the 8 corners and at the centre of each of the 6 faces (*i.e.,* S^{2-} ions have *fcc* or *ccp* arrangement), there are two tetrahedral sites for every S^{2-} ion.

The above discussion also shows that the number of tetrahedral sites (holes) present in the face-centred cubic unit cell (= 8) in double the number of spheres present in the cubic unit cell (= 4).

To determine the number of octahedral holes for each sphere present in a face-centered cubic unit cell

Let us consider a face-centered cubic unit cell (cubic close packed) shown in Fig. 16.10. In this figure M denotes an intersection point at which the lines joining six spheres (A, B, C, D, E and F) situated at the centre of the six faces of the cubic unit cell intersect. This point is not occupied by any sphere and is surrounded octahedrally by six spheres, since the six spheres are placed at the six corners of an octahedron. This point, therefore, becomes the octahedral hole, when all the six spheres touch each other. Thus we see that the unit cell has one octahedral hole (intersection point, M) at its centre. In addition to the intersection point M (octahedral hole), the centre of each of the 12 edges of the unit cell also has an intersection point. Now since each edge is shared by 4 adjacent unit cell, the actual number of intersections (or octahedral holes) contributed by each edge = 1/4 and hence the total number of intersections (octahedral holes) contributed by 12 edges in equal to $12 \times 1/4 = 3$. Thus :

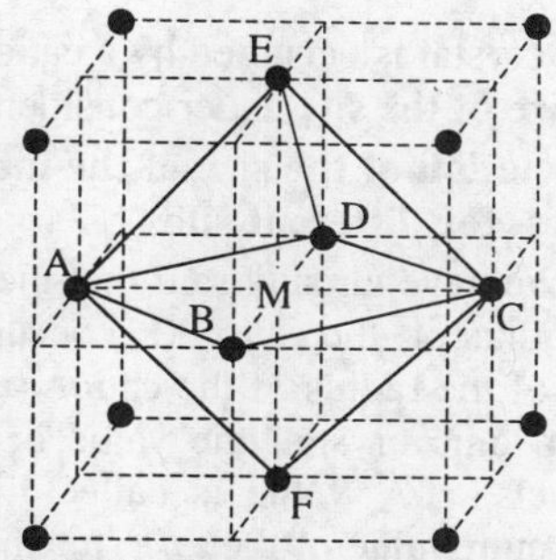

Fig. 16.10 : Determination of the number of octahedral holes for each sphere present in a face-centered cubic unit cell.

Total number of octahedral holes present in the entire face-centered cubic unit cell = No. of octahedral holes situated at the centre of the unit cell + No. of octahedral holes contribute by 12 edges of the unit cell = 1 + 3 = 4.

Now, we know that the total number of spheres present in the face-centered cubic unit cell = 4. Hence :

$\because$ No. of octahedral holes for 4 spheres present in the unit cell = 4

$\therefore$ No. of octahedral holes for 1 sphere present in the unit cell = 4/4 = 1

$\therefore$ No. of octahedral holes for N spheres present in the unit cell = 1 N

Thus, we see that every sphere present in a face-centered cubic unit cell [also called cubic close-packed (*ccp*) structure] has one octahedral site. For example, since, in the structure of NaCl crystal, Cl^- ions are situated at the 8 corners and at the centre of each of the six faces (*i.e.*, Cl^- ions have *fcc as ccp* arrangement), there is only one octahedral site for every Cl^- ion, *i.e.*, there is only one Na^+ ion for every Cl^- ion. The above discussion shows that the number of octahedral sites (or holes) present in the face-centered cubic unit cell (= 4) is equal to the number of spheres present in the cubic unit cell (= 4). The above study shows that the total number of octahedral holes created by the cubic close packed arrangement of anions is equal to the number of anions. For example, since the number of Cl^- ions in the unit cell of NaCl crystal is 4 and all these anions have close-packed arrangement, the number of octahedral holes in this closed-packed arrangement of Cl^- ions is also equal to 4.

Example 16.5. *In a cubic close packed structure of mixed oxides, the lattice is made up of oxide ions, one-eighth of tetrahedral voids are occupied by divalent ions (A^{2+}) while one-half of octahedral voids are occupied by trivalent ions (B^{3+}). What is the formula of the oxide ?.*

Solution. Since the oxide ions (O^{2-}) are in ccp arrangement, each O^{2-} ion has two tetrahedral sites and one octahedral site. It is given in the question that 1/8*th* of 2 tetrahedral sites are occupied by A^{2+} ions and 1/2 octahedral sites are occupied by B^{3+} ions. Thus :

No. of A^{2+} ions occupying $1/8 \times 2 = 1/4$ *tetrahedral sites* = $1/4\ A^{2+}$

No. of B^{3+} ions occupying $1/2 \times 1 = 1/2$ *octahedral site* = $1/2\ B^{3+}$

Thus one O^{2-} ion is combined with 1/4 A^{2+} ions 1/2 B^{3+} ions and hence the formula of the oxide is $1/4A^{2+}\ 1/2\ B^{3+}O^{2-}$ or $\mathbf{AB_2O_4}$.

Size of interstitial site

We have said above that the interstitial site left in between the closely-packed anions in an ionic crystal is occupied by a cation. The radius of the cation occupying the interstitial site is called the size of the site under consideration.

Calculation of the size of the interstitial site : Calculation of limiting radius ratio (r_+ / r_-) values for different sites

We have already said that the size of a given site created by the anions (in an ionic crystal) is the radius of the cation (r_+) occupying the given site. The value of r_+ is calculated in terms of the ratio of the radius of the cation (r_+) to that of the anions (r_-) which create the interstitial site. In a given ionic crystal, the value of r_+/r_- is calculated when there is anion-anion and anion-cation contacts. r_+/r_- value is called **limiting radius ratio value.** Limiting radius ratio value is the minimum value of the r_+/r_- for the cation that can fit into a given site.

Now, let us calculate limiting r_+/r_- value for each of the four sites mentioned above.

(*a*) Trigonal site and to calculate the limiting radius ratio (r_+/r_-) value for this site (C.N. of cation = 3). When three sphere which represent anions are placed at the corners of an equilateral triangle, a vacant space is left in between these spheres (anions). This vacant space is called **trigonal site.** This site is occupied by a cation in an ionic crystal having triangular structure. The radius of the cation is the size (radius) of the trigonal site.

In order to calculate the size of trigonal site, let us consider the unit cell of an ionic crystal having triangular (trigonal) structure. In this crystal the coordination number of the cation is three, *i.e.*, each cation is surrounded by three anions (trigonal arrangement of three anions round the central cation). Although the three anions are closely packed with each other (*i.e.* these ions touch each other), some empty space is left between them. This empty space, as we have already mentioned, is called trigonal or triangular site. This space (or site) is occupied by the central cation. The radius of this cation is called the size (radius) of the trigonal site. How this trigonal site is formed has been shown in Fig. 16.11 which shows that the anions (represented by bigger spheres having negative sign) lie at the vertices of the equilateral triangle EBC and the cation (represented by a smaller sphere with a positive sign) is situated at the centre of Δ EBC. Under this condition we say that the cation is occupying the trigonal site. In Fig. 16.11, for the sake of simplicity, the cations and anions have been shown *not* touching each other. Actually the cations and anions are touching one another, *i.e.*, cation-anion and anion-anion contacts are present in the ionic crystal.

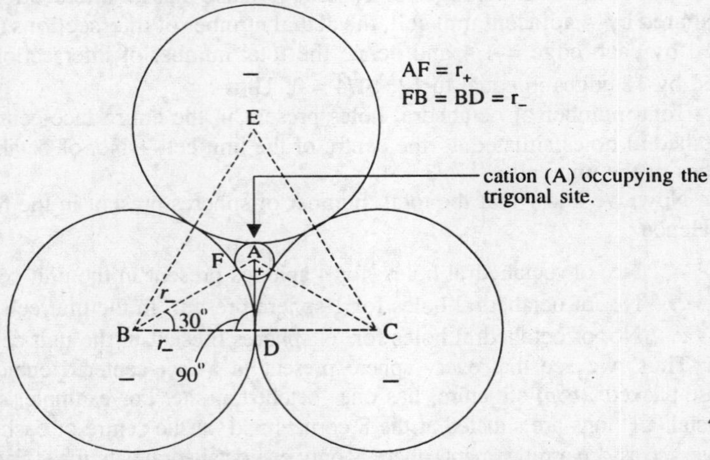

Fig. 16.11 : Trigonal arrangement of three anions (represented by bigger spheres having negative sign) round a cation, A (shown by smaller sphere with a positive sign) in an ionic crystal having trigonal structure. All the anions (B, E and C) have the same size (radius).

Now let r_+ and r_- be the radii of the cation and anion respectively. It is obvious from the figure that :

(*i*) In equilateral Δ EBC, BE = BC = CE = $r_- + r_- = 2r_-$. Also AF = r_+, BF = r_- and BD = r_-

(*ii*) Since A is the centre of Δ EBC (or A is the centre of the cation) and the cation is touching all the three anions, AB = AF + BF = $r_+ + r_-$.

Now in right angled Δ ABD, ∠ BDA = 90° and ∠ ABD = 30° and hence

$$\cos 30° = \frac{BD}{AB} = \frac{r_-}{r_+ + r_-} \quad \text{or} \quad \frac{\sqrt{3}}{2} = \frac{r_-}{r_+ + r_-}$$

$$\text{or} \quad \frac{2}{\sqrt{3}} = \frac{r_+ + r_-}{r_-} = \frac{r_+}{r_-} + 1 \qquad \text{or} \quad \frac{r_+}{r_-} = \frac{2}{\sqrt{3}} - 1 = \mathbf{0.155}$$

Thus the value of r_+/r_- in the limiting case of anion-anion and anion-cation contacts for a trigonal site is equal to 0.155. This value indicates that the radius ratio value for a cation to fit into a trigonal site should not be less than 0.155, *i.e.*, the lowest value of radius ratio for a cation having C.N. = 3 (trigonal site) is 0.155. That is why 0.155 is called the limiting radius ratio value.

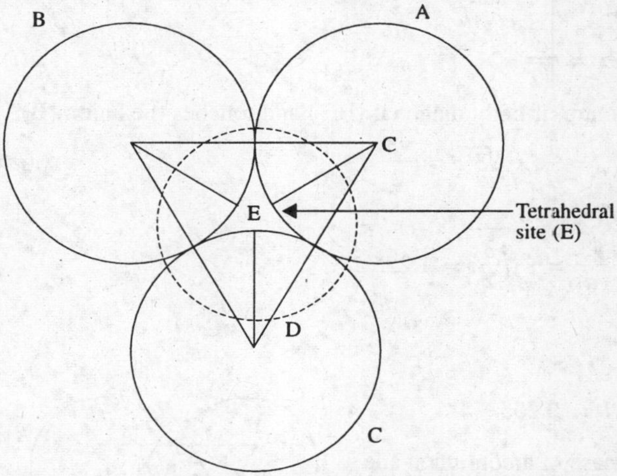

Fig. 16.12 : A tetrahedral site (E). All the four spheres (A, B, C and D) touch each othe and are of the same size.

(*b*) Tetrahedral site and to calculate the limiting radius ratio (r_+/r_-) value for this site (C.N. of cation = 4). Let us consider three spheres (A, B and C) touching each other and placed in the same plane (See Fig. 16.12). Now let us place another sphere, D (this sphere has been shown by dotted lines in Fig. 16.12) on the centre of the triangle obtained by joining the centres of the three spheres. Since the four spheres touch each other at one point only, a small empty space is left in between them. This small empty space has been denoted by E and is called *tetrahedral site*. Here it should be noted that all the four spheres are of equal size (radius).

In the unit cell of an ionic crystal having zinc blende (ZnS) type structure, a tetrahedral site can be created within the tetrahedron obtained by placing the four anions (*i.e.*, four spheres A, B, C and D) at the alternate corners of a cube in such away that they touch each other. (In zinc blende each Zn^{2+} ion is surrounded tetrahedrally by four S^{2-} ions). Tetrahderal site is created at the centre of the tetrahedron (ABCD) (Fig. 16.13). The cation (E) (shown by smaller sphere with a positive sign) is placed at the centre of the tetrahdron. Since the centre of the tetrahedron, centre of the cube and the centre of the body diagonal (BF) lies at the same point, the cation (E) is said to be placed at the centre of the body diagonal or at the centre of the tetrahedron. Under this condition we say that the cation is occupying the tetraheral site. Thus we see that the name "*tetrahderal site*" is used to indicate that the arrangement of the four anions (A, B, C and D) round the cation (E) is tetrahedral. Tetrahedral site, therefore, does not mean that the shape of the site is tetrahedral. The most stable arrangement of the ions present in the ionic crystal is that in which all the four anions placed at the corners of the cube touch each other as well as the cation put at the centre of the cube.

Now if a is the length of each side of the cube, then , as we know,

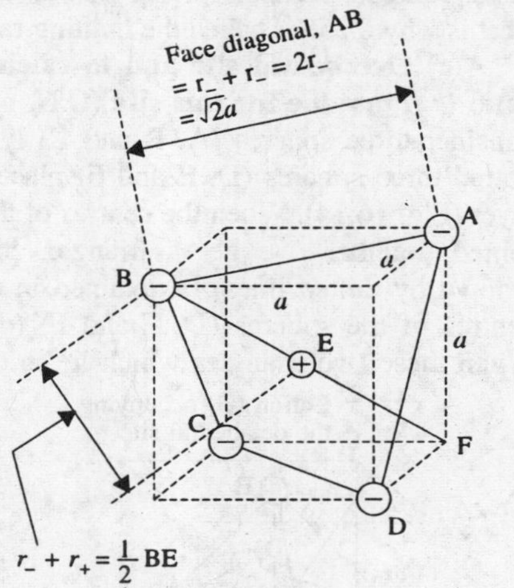

$$r_- + r_+ = \frac{1}{2}\,\text{BE}$$
$$= \frac{1}{2} \times \text{Body diagonal}$$
$$= \frac{1}{2} \times \sqrt{3}\, a = \frac{\sqrt{3}}{2} a$$

Fig. 16.13 : Tetrahedral arrangement (AB CDA is the *tetrahedron*) of four anions (represented by bigger spheres having negative sign) around a cation, E (shown by smaller sphere with a positive sign) in an ionic crystal having zinc blende (ZnS) type structure. All the anions namely A, B, C and D touch each other and have the same size (radius).

Face diagonal, AB $= \sqrt{2}\,a$ (see Fig. 16.13)

and Body diagonal, BF $= \sqrt{3}\,a$

Now, since the anions namely A and B touch each other,

$$r_- + r_- = \text{Face diagonal} = \sqrt{2}\,a$$

or

$$2r = \sqrt{2}\,a$$

or

$$r_- = \frac{\sqrt{2}}{2}a = \frac{a}{\sqrt{2}} \qquad ...(i)$$

Now, since the cation (E) lies at the centre of body diagonal (BF′) and touches the anion (B),

$$r_- + r_+ = \frac{1}{2} \times \text{BF} \quad \text{or} \quad r_- + r_+ = \frac{1}{2} \times \sqrt{3}\,a = \frac{\sqrt{3}}{2}a \qquad ...(ii)$$

On dividing equation (*ii*) by (*i*), we get :

$$\frac{r_- + r_+}{r_-} = \frac{\sqrt{3}}{2}a \times \frac{\sqrt{2}}{a} = \sqrt{\frac{3}{2}} = 1.255$$

or

$$1 + \frac{r_+}{r_-} = 1.255$$

or

$$\frac{r_+}{r_-} = 1.255 - 1 = \mathbf{0.255}$$

Thus the value of r_+/r_- in the limiting case of anion-anion and anion-cation contacts for a tetrahedral site is equal to 0.255. This value indicates that the radius ratio value for a cation to fit into a tetrahedral site should not be less than 0.255, *i.e.*, the lowest value for radius ratio for a cation having C.N. = 4 (tetrahedral site) is 0.255. That is why 0.255 is called the limiting radius ratio value.

(*c*) Octahedral site and to calculate the limiting radius ratio (r_+/r_-) value for this site (C.N. of the cation = 6). Let us consider three spheres (A, B and C) lying in one plane and the dotted three spheres (D, E and F) placed on them in the second layer (Fig. 16.14). When the centres of the spheres A, B and C are joined together, we get a triangle. Similarly another triangle (shown by dotted lines) is obtained in the other layer, when the centres of the spheres (D, E and F) (dotted sphere) are joined. When these two triangles which lie in different layers are superimposed one above the other, we get octahedron, which has been shown by a cross (×) in Fig. 16.14. In this position the apices of the two triangles point in opposite directions. Thus an octahedral site is created by two triangles whose apices are in opposite directions. In simple language, we can say that an octahedral site is created between the six spheres placed at the six vertices of a regular octahedron.

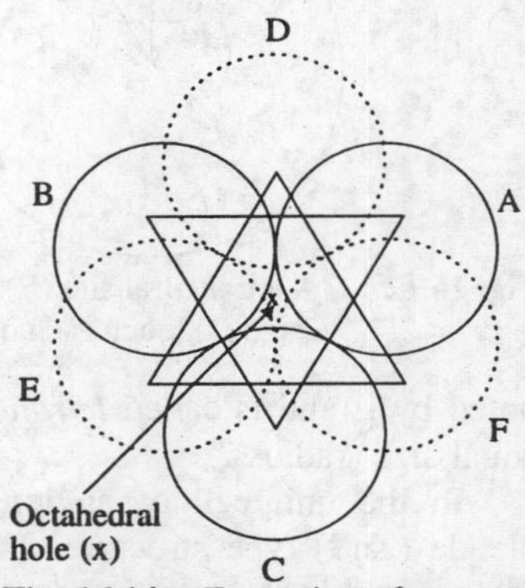

Fig. 16.14 : Formation of an octahedral hole (site) between six spheres. All the six spheres are of the same size (radius).

In the unit cell of an ionic crystal having rock salt (NaCl) type structure, an octahedral site is created by placing the six anions (shown by bigger spheres with negative sign) at the centre of the six faces of a cube (see Fig. 16.15) in such a way that they touch each other (closest packing of anions). The four anions namely A, B, C and D lie at the four corners of the square plane, one anion (*viz.*, E) lies above and one anion (*viz.*, F) lies below the square plane. Octahedral site is created at the centre of the square plane contain-

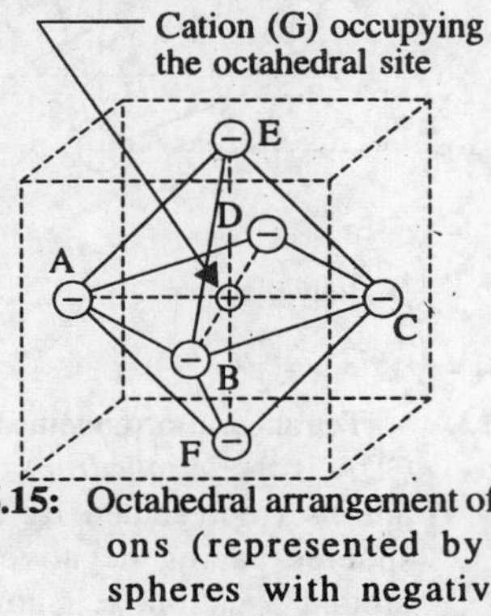

Fig. 16.15: Octahedral arrangement of six anions (represented by bigger spheres with negative sign) around a cation, G (shown by smaller sphere with a positive sign) in an ionic crystal having NaCl type structure. All the six anions have the same sze (radius).

ing the four anions. The cation (shown by a smaller sphere with a positive sign) is placed at the centre of the square. Under this condition, we say that the cation (G) is occupying the octahedral site created by the six anions placed at the six vertices of an octahedron. Here the cation is surrounded octahedrally by six anions. For the sake of simplicity, the cation and the six anions have been shown *not* touching each other. Actually they touch one another so that the most stable structure may be given to the ionic crystal.

In order to calculate the value of r_+/r_- for an octahedral site, let us consider only the square plane of the octahedron, which contains four anions (A, B, C and D) at its four corners and one cation (G) at its centre (octahedral site) (Fig.16.16). Let r_+ and r_- be the radii of the cation and each of the anions respectively. All the four anions touch each other and the cation simultaneously. The value of r_+/r_- (radius ratio) can be calculated by simple geometry as shown below. Let X and Z be the centres of the cation (G) and the anion (D) respectively. Let Y be the point at which the two anions *viz.*, C and D touch each other. Now join the points X, Y and Z to get the right angled Δ XYZ, in which ∠ XYZ = 90° and ∠ XZY = 45°.

Now since the central cation (G) is touching all the four anions, $XZ = XP + PZ = r_+ + r_-$.

$$\text{Now,} \quad \cos 45° = \frac{YZ}{XZ} = \frac{r_-}{r_+ + r_-}$$

$$\text{or} \quad \frac{1}{\sqrt{2}} = \frac{r_-}{r_+ + r_-}$$

$$\text{or} \quad \sqrt{2} = \frac{r_+ + r_-}{r_-} = \frac{r_+}{r_-} + 1$$

$$\text{or} \quad \frac{r_+}{r_-} = \sqrt{2} - 1 = 1.414 - 1 = \mathbf{0.414}$$

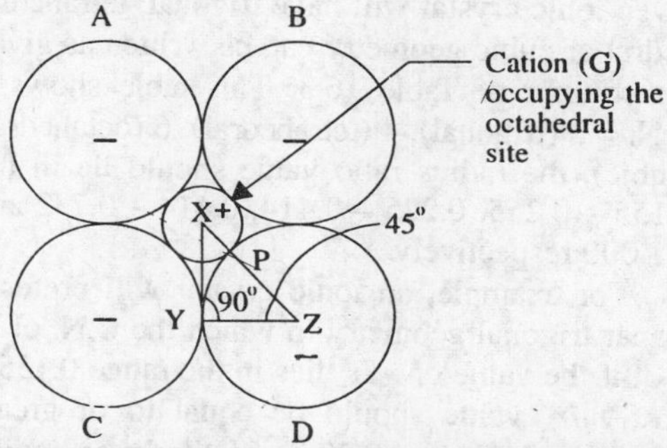

Fig. 16.16 : A cross section through an octahedral site. Only four anions (A, B, C, D) and the central cation (G) lying in the square of the octahedron have been shown. Here, $XP = r_+$, $PZ = YZ = r_-$, $XZ = r_+ + r_-$, ∠XZY = 45° and ∠XYZ = 90°.

Thus, we find that the value of r_+/r_- in the limiting case of anion-anion and anion-cation contacts for an octahedral site is equal to 0.414.

This value indicates that the radius ratio value for a cation to fit into an octahedral site should not be less than 0.414, *i.e.*, the lowest value of radius ratio for a cation having C.N. = 6 (octahedral site) is 0.414. That is why 0.414 is called the limiting radius ratio value.

(*d*) Cubic site and to calculate the limiting radius ratio (r_+/r_-) value for this site. (C.N. of cation = 8). Let us consider the unit cell of an ionic crystal having caesium chloride (CsCl) type crystal structure. (Examples are CsBr, CsI, TlCl, TlBr etc.). In this ionic crystal, the coordination number of the cation and of the anion is equal to 8, *i.e.*, each ion is surrounded by 8 ions (Cubic arrangement of ions) of opposite charge. Let us consider the cubic arrangement of 8 anions around a cation. The cubic arrangement of 8 anions is obtained by placing the 8 anions at the eight corners of a cube in such away that they touch each other and the cation is put at the body centre of the cube (See Fig. 16.17). In this condition, the cation is said to occupy the cubic interstitial site left in between the eight anions.

Suppose the length of each side of the cube is a and r_+ and r_- are the radii of the cation and anion respectively. Since the anions A and B touch each other,

$$AB = a = r_- + r_- = 2r_- \quad \text{or} \quad r_- = \frac{a}{2} \qquad ...(i)$$

We know that :

Body diagonal, $AC = \sqrt{3}\, a$

Now, since the cation lies at the centre of the body diagonal AC and touches both the anions *viz.*, A and C,

$$AC = r_- + 2r_+ + r_- = 2r_+ + 2r_- \quad \text{or} \quad \sqrt{3}\,a = 2r_+ + 2r_- \quad \text{or} \quad r_+ + r_- = \frac{\sqrt{3}\,a}{2} \qquad ...(ii)$$

On dividing equation (*ii*) by (*i*), we get :

$$\frac{r_+ + r_-}{r_-} = \frac{\sqrt{3}\,a}{2} \times \frac{2}{a} = \sqrt{3} \quad \text{or} \quad \frac{r_+}{r_-} + 1 = \sqrt{3} \quad \text{or} \quad \frac{r_+}{r_-} = \sqrt{3} - 1 = \mathbf{0.732}$$

This value indicates that the radius ratio value for a cation to fit into a cubic site should not be less 0.732, *i.e.*, the lowest value of radius ratio for a cation having C.N. = 8 (cubic site) is 0.732. That is why 0.732 is called the limiting radius ratio value.

To find out the range of radius ratio values for a given coordination number

We have seen above that the lowest value (limiting value) of radius ratio (r_+/r_-) for an ionic crystal having trigonal (C.N. = 3), tetrahedral (CN = 4), octahedral (C.N. = 6) and cubic (C.N. = 8) geometry is 0.155, 0.255, 0.414 an 0.732 respectively. Thus, the range of radius ratio values between which a given ionic crystal will have trigonal, tetrahedral, octahedral or cubic geometry can be written as given in the 1*st* column of Table 16.3. This table shows that for C.N. = 3 (trigonal), 4 (tetrahedral), 6 (octahedral) and 8 (cubic), the radius ratio value should lie in the range 0.155 – 0.255, 0.225 – 0.414, 0.414 – 0.732 and 0.732 – 1.000 respectively.

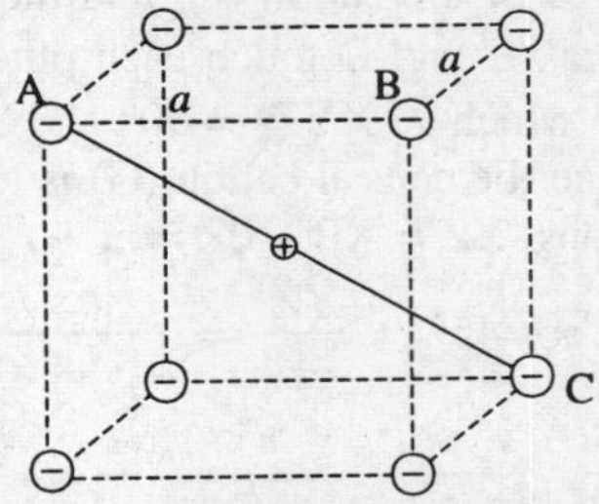

Fig. 16.17 : Cubic arrangement of eight anions (represented by bigger spheres having negative sign) round a cation (shown by smaller sphere with a positive sign) in an ionic crystal having caesium chloride (CsCl) type structure. All the eight anions have the same size (radius).

For example, an ionic crystal will prefer to have planar trigonal geometry in which the C.N. of each ion is 3, if the value of r_+/r_- lies in the range 0.155 – 0.225 (*i.e., r_+/r_-* value should be equal to or greater than 0.155 but less than 0.225). Similarly an ionic crystal would prefer to have NaCl type structure in which the C.N. of each ion is 6, if the value of r_+/r_- lies in the range 0.414 – 0.732 (*i.e., r_+/r_-* value should be equal to or greater than 0.414 but less than 0.732). If the value of r_+/r_- is less than 0.414, tetrahedral geometry would be preferred and if the value of r_+/r_- is more than 0.732, cubic geometry will be preferred. The above discussion shows that in a close-packed cubic arrangement the size of the cubic, octahedral and tetrahedral hole is in the order cubic hole > octahedral hole > tetrahedral hole.

Table 16.3 : *Ranges of radius ratio (r_+/r_-) values and geometry of the ionic crystals.*

Range of radius ratio (r_+/r_-) values for the given cation	C.N. of the cation in ionic crystal	Arrangement of anions round the cation (Geometry of ionic crystal)	Examples of ionic crystals
0.155 — 0.225	3	Planar trigonal	Boron oxide, BO_3^{3-}, BN
0.225 — 0.414	4	Tetrahedral	Ionic crystals having zinc blende (ZnS) type structure (e.g., ZnO, CuCl, CuBr, CuI) SiO_4^{4-}.
0.414 — 0.732	6	Octahedral	Ionic crystals having rock salt (NaCl) type structure (e.g., AgF, AgCl, AgBr).
0.732 — 1.000	8	Cubic	Ionic crystals having caesium chloride (CsCl) type structure (*e.g.*, CsBr, CsI, TlCl, TlBr).

Effect of the decrease/increase in the value of radius ratio (r_+/r_-) on the coordination number or structure of ionic crystal

Let us consider an ionic crystal in which a cation occupies the octahedral site created by six closely-packed anions. In Fig. 16.18, only four anions placed at the four corners of the square base of the octahedron have been shown. For the sake of simplicity, the anion lying above the square base and that lying below the square base have not been shown. The cation is situated at the centre of the square plane and hence occupies the octahedral site. Since the anions are closely-packed, the anions touch one another and the cation simultaneously and hence r_+/r_- value lies in the range 0.414 – 0.732. Now let us consider the following two cases :

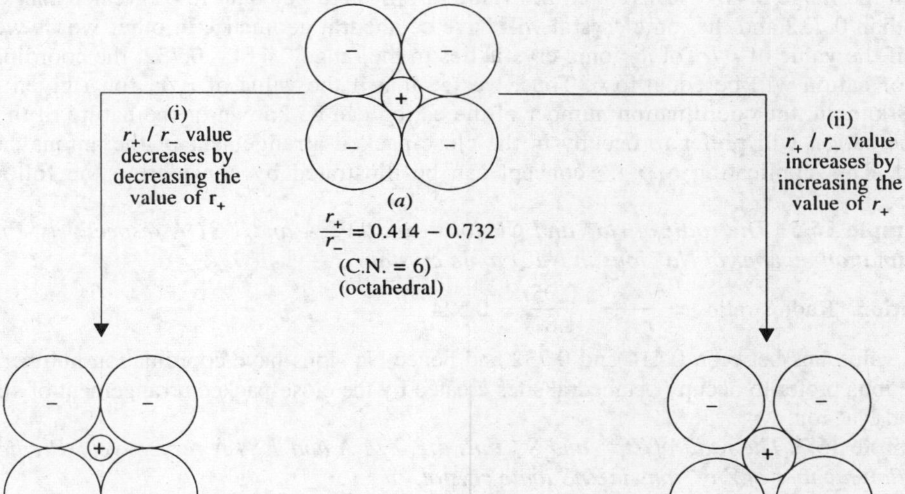

Fig. 16.18 : Effect of the change in the value of r_+/r_- on the coordination number or the structure of the ionic crystal.

(*i*) If we decrease the radius of the cation, the value of r_+/r_- ratio also decreases and the four anions touch one another but are not able to touch the cation as they do in structure (*a*). In order to make the anions to touch the cation, the anions will have to come closer and closer to each other and, therefore, the repulsion between them will increase. The increase in the repulsion between the anions will make structure (*a*) unstable. In order to get a stable structure, we will have to push away one or more anions so that the remaining anions touch the cation and a more stable structure is obtained. In other words, we can say that in order to get a stable structure, we will have to decrease the coordination number of the cation from 6 to 4. With the change of coordination number of the cation, the geometry of the ionic crystal will also change from octahedral to tetrahedral.

(*ii*) If we increase the radius of the cation, the value of r_+/r_- ratio also increases and the four anions touch the cation but are not able to touch one another as they do in structure (*a*), since they go away from each other. If the value of r_+/r_- ratio is further increased, the anions go farther and farther apart and structure (*a*) becomes unstable. In order to make the anions to touch one another, more anions have to be accommodated between the four anions. In other words, we can say that the coordination number of the central cation may increase from 6 to 8.

The above discussion shows that the decrease or increase in the value of r_+/r_- ratio changes the coordination number of the cation and hence changes the structure of the ionic crystal.

Calculation of coordination number of a cation in an ionic crystal : An application of radius ratio (r_+/r_-) concept.

We have already said that for a particular coordination number of a cation in an ionic crystal, there is a definite range of the radius ratio value for the ionic crystal (see Table 16.3). For example, if the coordination number of each ion in an ionic crystal is 6, the value of r_+/r_- will lie in the range 0.414 – 0.732, *i.e.*, the value of r_+/r_- will be equal to or greater than 0.414 but less than 0.732 and the ionic crystal will have octahedral geometry. In other words we can say that if the value of r_+/r_- of an ionic crystal lies in the range 0.414 – 0.732, the coordination number of cation will be equal to 6. Thus, we see that if the value of r_+/r_- for a given ionic crystal is known, the coordination number of the cation can be known or the nature of the site which the cation will prefer to occupy in the close-packed arrangement of the anions can be predicted. This application of r_+/r_- concept can be illustrated by considering the following examples.

Example 16.6. *The radii of Na^+ and Cl^- ions are 0.95 Å and 1.81 Å respectively. Predict the coordination number of Na^+ ions in NaCl ionic crystal.*

Solution. Radius ratio $= \frac{r_+}{r_-} = \frac{0.95}{1.81} = 0.524$

This value lies between 0.414 and 0.732 and hence Na^+ ions have coordination number of 6 and these ions prefer to occupy octahedral sites created by the close-packed arrangement of six Cl^- ions round Na^+ ion.

Example 16.7. *The radii of Zn^{2+} and S^{2-} ions are 0.74 Å and 1.84 Å respectively. Predict the coordination number of Zn^{2+} ions in ZnS ionic crystal.*

Solution. Radius ratio $= \frac{r_+}{r_-} = \frac{0.74}{1.84} = 0.40$

This value lies in the range 0.225 – 0.414 and hence Zn^{2+} ions have coordination number of 4 and these ions occupy tetrahedral sites in close-packed lattice of S^{2-} ions.

Example 16.8. *The radii of Cs^+ and Cl^- ions are 1.69 Å and 1.81 Å respectively. Predict the coordination number of Cs^+ ions in CsCl ionic crystal.*

Solution. Radius ratio $= \frac{1.69}{1.81} = 0.93$

Since r_+/r_- value lies between 0.732 and 1.000, coordination number of Cs^+ ions in CsCl ionic crystal is 8, *i.e.*, Cs^+ ions occupy cubic sites in the close-packed lattice of Cl^- ions.

Radius ratio values and the structure of some ionic crystals of AB type are given in Table 16.4.

The radius ratio (*i.e.*, r_+/r_-) values for the chlorides of alkali metals as given in the table show that the observed coordination numbers for Na, K, Rb and Cs are the same as expected from r_+/r_- values but the observed coordination number for Li is higher. The higher value for Li is due to the higher lattice energy of LiCl. The higher coordination number explains the stability of LiCl. It may be seen from the table that caesium halides crystallise with CsCl type crystal lattice having coordination number = 8, while the halides of other alkali metals crystallise with crystal lattices of NaCl type with coordination number = 6.

Table 16.4 : *Radius ratio values (r_+/r_- values) and structure of some ionic crystals of AB type (C.N. = coordination number)*

Ionic crystal	r_+ (Å)	r_- (Å)	r_+/r_-	C.N. of the cation expected from r_+/r_- value	Observed C.N. of the cation	Structure of the ionic crystal
LiCl	0.68	1.81	0.375	4	6	NaCl type structure (C. Ns. = 6 : 6)
NaCl	0.95	1.81	0.524	6	6	—do—
KCl	1.33	1.81	0.624	6	6	—do—
RbCl	1.48	1.81	0.817	8	6 or 8	NaCl or CsCl type structure (C. Ns. = 6 : 6 or 8 : 8)
CsCl	1.69	1.81	0.933	8	8	CsCl type structure (8 : 8)
NaI	0.95	2.16	0.439	6	6	NaCl type structure
RbI	1.48	2.16	0.546	6	6	NaCl type structure
CsI	1.69	2.16	0.782	8	8	CsCl type structure
CsBr	1.69	1.95	0.866	8	8	CsCl type structure
MgO	0.65	1.42	0.457	6	6	NaCl type structure
CuI	0.91	2.16	0.421	6	4	ZnS type structure (4 : 4)
ZnS	0.74	1.84	0.402	4	4	—do—
ZnSe	0.74	1.98	0.372	4	4	—do—

Properties of Solids

Here we shall discuss *electrical properties* (*electrical conductivity*), *magnetic properties* and *dielectric properties* of the solids. All these properties depend on the structure, nature of bonds and the composition of the solids.

Electrical conductivity of solids : Classification of solids on the basis of the value of their electrical conductivity

The electrical conductivity of a solid is defined as its ability to conduct electric current through it. Many solids conduct electricity while others donot. On the basis of the value of electrical conductivity of solids, these have been classified as (1) *good conductors* (or simply *conductors*) (2) *insulators* and (3) *semi conductors.*

1. Good conductors or conductors.

The electrical conductivity of good conductors is of the order of 10^8 ohm^{-1} cm^{-1} and they allow the maximum portion of the applied electric field to flow through them.

Electrical conductivity of metals. Metals are good conductors of electricity, since according to electron sea theory, the metallic crystals have mobile or delocalised (free) electrons. When an electric field is applied between the two ends of a metal, the mobile electrons begin to move towards the positive pole and the new electrons from negative pole take their position. Thus with the movement of electrons, the electric current starts to flow in the metal and the metal, therefore, acts as a good conductor of electricitiy.

It has been observed that the electrical conductivity of most of the metals decreases with the increase of temperature. This is because of the fact that with increase of temperature the positive metal ions also begin to vibrate and their motion hinders the free movement of the mobile electrons between the positive metal ions. Thus the vibrating positive metal cores create resistance to the flow of the electrons towards the positive pole and hence the movement of the electrons towards the positive pole falls and so does the electrical conductivity of the metals at higher temperature.

2. Poor conductors or insulators

The electrical conductivity of insulators is of the order 10^{-22} ohm^{-1} cm^{-1} and they donot practically allow electric current to flow through them. Many of the inorganic and organic solids are insulators.

Electrical conductivity of ionic solids (a). Ionic solids are poor or bad conductors of electricity, since they do not allow the electric current to pass through them. This is because of the fact that, due to the electrostatic force of attraction, the positive and negative ions present in the ionic crystals are fixed in their position and hence cannot move even if an external electric field is applied on them.

(*b*) Ionic solids in their fused (melted) state conduct electricity because the positive and negative ions now become free to move under the applied electric field.

(*c*) An ionic solid in water conducts electricity because it reacts with water and gives free ions in solution.

(*d*) The ionic solids having defects, conduct electricity because the ion moves from its lattice site to occupy a vacancy. This results is creating a new vacancy and another nearby ion moves into it and so on. This process continues and a vacancy, thereby, migrates from one end to the other end. Thus it conducts electricity across the whole of the ionic crystal.

Electrical conductivity of covalent solids : (*a*) At low temperature (0°K) covalent solids like silicon and germanium do not conduct electricity because the electrons present in the localised covalent bonds are fixed in their position and hence are not free to move. At high temperature, these elements act as semiconductors (see below).

(*b*) Because of its unique structure with free electrons, graphite conducts electric current.

(*c*) Hydrogen chloride dissolved in water (hydrochloric acid) conducts electricity, since it reacts with water to form H^+ ions.

3. Semiconductors

The electrical conductivity of a semiconductor at normal temperature is intermediate of a metal (good conductor) and an insulator. It is in the range of 10^{-9} to 10^2 ohm^{-1} cm^{-1}. The semiconductors allow only a portion of the applied electric field to flow through them.

Conversion of an insulator into a semiconductor : Classification of semiconductors

The crystals of silicon and germanium act as insulator at low temperature of 0°K. These elements can be made to behave as semiconductor either by heating them or by adding impurity atoms like phosphorus or boron to them. A semiconductor which is obtained by heating an insulator is called **instrinsic semiconductor,** while the semiconductor which is obtained by adding impurity atoms is called **extrinsic semiconductors.**

Electrical conductivity of silicon and germanium crystals

The crystals of silicon and germanium act as insulators at low temperature (0°K), but when their temperature is increased above 0°K, they act as semiconductors (*intrinsic semiconductors*). These elements also act as semiconductors (*extrinsic semiconductors*), when the impurity of phosphorus or boron is added to them.

Why silicon or germanium acts as an intrinsic semiconductor at high temperature, can be explained as follows, on the basis of the *concept of positive hole formation.*

Silicon and germanium have four electrons in their valence-shell (Si = $3s^2p^2$ and Ge = $4s^2p^2$). The crystal structure of these elements has shown that each atom of these elements is bounded by four covalent bonds, with the neighbouring four similar atoms of the same element. Thus, each atom is associated with 8 electrons and hence is relatively stable. At very low temperature of 0°K, the two electrons participating in the formation of a covalent bond, are tightly held in their position and hence cannot move freely (mobile electrons) as they can move in a metallic crystal. Consequently the crystals of silicon and germanium at low temperature act as insulators. Now when the temperature of the crystals is increased above 0°K, one of the covalent bonds gets broken and one of the two electrons in the covalent bond between the two silicon or germanium atoms moves through the crystal. The movement of the electron creates a positive hole at the site of the missing electron. The positive hole is vacant and hence an electron present in the nearby covalent bond moves into this vacant positive hole and thus a fresh positive hole is created in the nearby covalent bond.

This process continues and a positive hole, thereby, migrates from one end to the other end.

Thus, it conducts electricity across the whole of the crystal of silicon or germanium. The electrons migrate in one direction and the positive holes in the opposite direction, *i.e.*, the movement of positive holes in an electrical field is opposite to that of electrons. Thus we find that the presence of positive hole produces electrical conductivity in silicon/germanium element at high temperature, *i.e.*, it is due to the presence of positive hole in the crystal of silicon or germanium that this element acts as a conductor at high temperature. Since the conductivity value of these elements lies in between that of metals and insulators, these elements are said to act as semiconductors at high temperature.

Free electrons and positive holes are generally designated by the letters *e* and *h* respectively, while their concentrations are represented by *n* (concentration of free electrons) and *p* (concentration of positive holes). In pure crystals of silicon or germanium (pure crystals are those which donot contain any impurity) *n* and *p* are equal. Since the conductivity produced in the crystals of silicon or germanium, has been created, not by adding any external (foreign) substance (impurity), this conduction is called **intrinsic conduction** and silicon or germanium is called **intrinsic semiconductor.** We have already said that when an impurity of a foreign substance, like phosphorus or boron, is added to silicon (or germanium) atoms, these elements act as extrinsic semiconductors. This behaviour can be explained as follows :

(*a*) *Addition of the impurity of phosphorus atoms : n-type extrinsic semiconductors.* When a small quantity of phosphorus is added to extremely pure silicon (99.99% pure) by a process called **dopping,** some of the silicon atoms are substituted by phosphorus atoms. Each phosphorus atom forms four covalent bonds with the surrounding silicon atoms by using its four valence electrons and its fifth electron is left unused or unbonded. Thus there is an extra electron at each of the lattice points occupied by phosphorus atoms compared to the lattice points occupied by silicon atoms. There are as many extra electrons in the lattice points of phosphorus as there are phosphorus atoms (Fig. 16.19). Since the extra electrons on phosphorus atoms remain unused (unbonded), they move

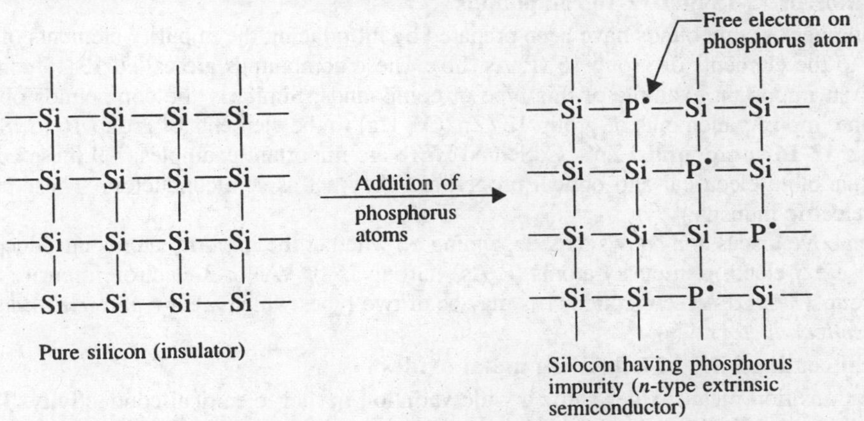

Fig. 16.19 : Representation of the formation of *n*-type extrinsic semiconductor.

freely through the crystal. These unused electrons can easily be excited to the empty conduction bonds of silicon (or germanium) crystal by the application of electric field. Thus we see that the unused electrons are the current carriers. Since phosphorus atom has 5 valence-electrons and silicon atom has 4 valence electrons, we can say that the impurity atoms have greater number of valence-electrons than the parent insulator atoms. When impurity atoms having more valence-electrons than the parent insulator atoms are introduced into them (*i.e.*, insulator atoms), we get ***n*-type extrinsic semiconductors** (*n* indicates negatively-charged free electrons). The mixture obtained by adding 5-electron impurity atom (*i.e.*, phosphorus atom) to 4-electron insulator atom (*i.e.*, silicon atom) is electrically neutral.

(*b*) *Addition of the impurity of boron atoms : p-type extrinsic semiconductors.* When a small quantity of boron is added to pure silicon, some silicon atoms are replaced by boron atoms. Each boron atom is linked with three Si atoms by normal two-electron covalent bonds, while it is linked with the other nearby silicon atom by one-electron bond. In one-electron bond there is an electron deficiency. This electron deficiency creates a positive hole or vacancy at the site where the electron is missing. There are as many positive holes as there are boron atoms. The unbonded electrons on silicon atoms move through the crystal by hobing one vacancy site to another (Fig. 16.20). The holes move opposite to the direction of flow of electrons. Since boron has three valence-electrons and silicon atom has four valence-electrons, we can say that the impurity atoms have less number of valence-electrons than the parent insulator atoms. When impurity atoms, having less number of valence-electrons than the parent insulator atoms, are introduced into them (*i.e.*, insulator atoms), we get ***p*-type extrinsic semiconductors** (*p*- indicates positive holes).

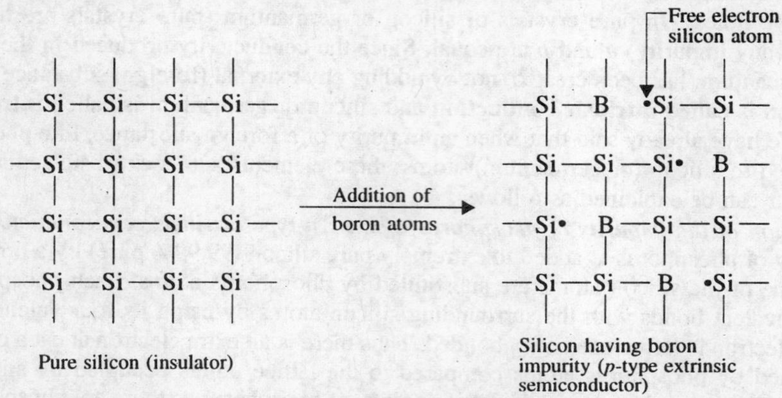

Fig. 16.20 : Representation of the formation of *p*-type extrinsic semiconductor.

Preparation of 13-15 and 12-16 compounds

A number of compounds have been prepared by introducing the impurity elements of group 13 (B, Ga) to the elements of group 15 (P, As, Sb). These compounds are called **13-15 compounds.** GaAs is an important example of this type of compounds. Similarly the compounds obtained by adding the impurity elements of group 12 (Zn, Cd, Hg) to the elements of group 16 (S, Se, Te) are known as **12-16 compounds.** ZnS, CdSe and HgTe are important examples. All these compounds have remarkable electrical and optical properties. They act as semiconductors. These are widely used in electric industry.

The above discussion shows that, depending on whether the impurity atoms introduced into an insulator are 5-electron impurity atoms (P, As) (Group 15 or VA) or 3-electron impurity atoms (B, Al) (Group 13 or III A) semiconductors may be of two types which are *n-type semiconductors* and *p-type semiconductors.*

Electrical conductivity of transition metal oxides

The transition metal oxides show a wide variation in their electrical conductivity. The difference in their electrical conductivity is due to the impurities and crystal imperfections in them. Some oxides are as good conductors as metals (*e.g.*, TiO, VO, CrO_3) while some are insulators (*e.g.*, TiO_2, V_2O_5, Cr_2O_3, MnO, Mn_2O_3, MnO_2, FeO, Fe_2O_3, CoO, NiO, CuO). There are also oxides which show a transition from metal to insulator at a certain temperature (*e.g.*, Ti_2O_3, V_2O_3, VO_2). It may be noted that monoxides namely TiO and VO are metallic while the monoxides *viz.*, MnO, FeO, CoO, NiO and CuO are insulators.

Super conductivity and super conducting material

It was in 1913 when Kammerligh Onnes observed for the first time that helium offered no resistance to the passage of electricity (*i.e.*, allowed free flow of electricity through it) at 4 K. This phenomenon is called *superconductivity* and the material (*e.g.*, helium) which offers no resistance to the passage of electricity through it at temperatures lying between 2 K and 5 K is called *super conducting material.* At this temperature the substance becomes diamagnetic and hence is repelled by magnets.

Soon after Onne's observation, it was found that most of the metals become super-conducting at temperatures lying between 2 K and 5 K. In 1986, Bednorz and Muller prepared a compound, $La_{1.5}$ $Ba_{1.5}$ Cu O_4 which behaved as a super conducting material at 30 K. Recent discoveries have shown that many mixed oxides become super conducting at fairly high temperatures. A few examples of such oxides are (temperature is given in bracket) $Ba_{0.7}$ $K_{0.3}$ BiO_3 (30 K), $La_{1.8}$ $Sr_{0.2}$ CuO_4 (40 K), YBa_2 Cu_3O_7 (90 K), Ba_2Ca_2 Sr_2 Cu_3O_{10} (105 K), Tl_2 Ca_2 Ba_2 Cu_3O_{10} (125 K). Much research is being done to find materials which may act as super conducting materials even at higher temperature.

Applications of super conducting materials.

Super conducting materials have applications in electronics, building super-magnets, levigation transportation and in power transmission.

Magnetic properties.

The electron, while revolving round the nucleus, has two types of motions namely *orbital motion* and *spin motion.* In orbital motion the electron revolves round the nucleus in various orbitals while in spin motion the electron spins round its own axis. Both these motions produce magnetic fields when a substance is placed between the poles of a horse shoe magnet. The magnetic field produced by the movement of the electrons present in the substance interacts with the external magnetic field and different substances show different behaviour when placed between the poles of a magnet.

On the basis of the behaviour shown by the substances when placed in a magnetic field, they have been classified as *paramagnetic substances, diamagnetic substances, ferromagnetic substances, anti-ferromagnetic substances* and *ferrimagnetic substances.*

Paramagnetic and diamagnetic substances.

The points showing differences between these two types of substances, are given below in Table 16.5.

Table 16.5 : *Differences between paramagnetic and diamagnetic substances.*

Paramagnetic substances	Diamagnetic substances
(*i*) The substances which, when placed in a magnetic field, increase the intensity of the magnetic field, are called paramagnetic substances.	(*i*) The substances which, when placed in a magnetic field, decrease the intensity of the magnetic field are called diamagnetic.
(*ii*) A paramagnetic substance permits more number of lines of force to pass through it (Fig. 16.21) and is, therefore, attracted by the external magnetic field. Consequently, a paramagnetic species tends to move to the region of the strongest field strength of the magnetic field when placed in a non-uniform magnetic field. Due to the attraction in the magnetic field, the paramagnetic substance sets itself with its length parallel to the lines of force (see Fig. 16.21 (*c*)).	(*ii*) A diamagnetic substance permits less number of lines of force to pass through it (Fig. 16.21) and is therefore, repelled by the external magnetic field. Thus a diamagnetic species tends to move to the weakest part when placed in a non-uniform magnetic field. Due to the repulsion in the magnetic field, the diamagnetic substance sets itself at right angles (*i.e.*, perpendicular) to the lines of force. [See Fig. 16.21 (*d*)].
(*iii*) The magnitude of paramagnetism varies with the temperature, *i.e.*, it is temperature-dependent.	(*iii*) The magnitude of diamagnetism does not vary with temperature, *i.e.*, it is temperature indepedent.
(*iv*) Paramagnetic character of a substance is due to the presence of unpaired electrons in it.	(*iv*) If an orbital contains two electrons, these electrons will have opposite spins and hence will have opposite orbital momentum. Due to the opposite orbital momentum, the substances containing only paired electrons are repelled by the magnetic field when placed in it, *i.e.*, the substances containing only paired electrons show diamagnetic character.
(*v*) When placed in a magnetic field, paramagnetic substance acquires a feeble magnetisation in the direction in which the magnetic field has been applied. When the magnetic field is removed, the paramagnetic material loses its magnetisation.	(*v*) When a diamagnetic material is placed in a magnetic field, it acquires feeble magnetisation in the direction opposite to that in which the magnetic field has been applied.

(*vi*) When a tube contaning paramagnetic substance is placed between the poles of a magnet and weighed, an increase in weight is observed. The increase in weight is due to the fact that the paramagnetic substance is attracted by the lines of force and the substance is pulled down into the gap between the poles. As a result of this attraction, the paramagnetic substance shows an increase in weight.	(*vi*) When a tube containing diamagnetic substance is placed between the poles of a magnet and weighed, a decrease in weight is observed. This is due to the reason that diamagnetic substance is repelled by the lines of force and hence shows a decrease in weight.
(*vii*) Examples of paramagnetic substances are Al, Cr, Mn, $CuSO_4$, *liq.* O_2, Fe, Co, Ni, Pt, $[Ni(H_2O)_6]Cl_2$, $[Mn(H_2O)_6]Cl_2$, NO, transition metal oxides (*e.g.*, TiO, Ti_2O_3, VO, VO_2, CuO etc.), transition metal ions etc.	(*vii*) Examples of diamagnetic substances are Zn, Bi, Sb, Cu, Au, H, alcohol, H_2O, quartz, NaCl, KCl, organic ligands, $[Zn(NH_3)_4]^{2+}$, $[ScF_6]^{3-}$, TiO_2, N_2, benzene etc.
(*viii*) A paramagnetic substance has some value of its magnetic moment.	(*viii*) A diamagnetic substance has zero value for its magnetic moment, since such a substance has no unpaired electrons.

Even paramagnetic substances show diamagnetism. Since diamagnetism is much weaker than paramagnetism (10 to 100 times weaker) and both act opposite to each other, it is difficult to experience diamagnetism in the substances which are paramagnetic or which have unpaired electrons.

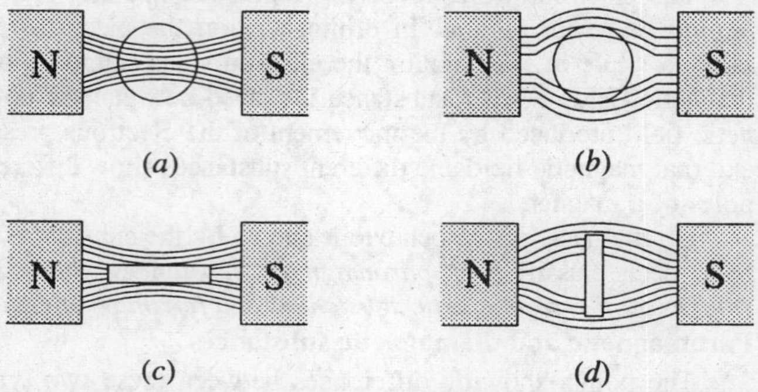

Fig. 16.21 : Behaviour of paramagnetic [(*a*) and (*c*)] and diamagnetic [(*b*) and (*d*)] substances in a magnetic field.

Why are the substances paramagnetic or diamagnetic ?

It is due to the presence of one or more unpaired electrons in the molecules (or atoms or ions) of a substance that the substance shows paramagnetic character. Similarly the presence of all the electrons in the paired state make the substance diamagnetic. This can be explained as follows :

We know that an electron is a charged particle and has orbital motion (*i.e.*, motion of an electron in orbitals) and spin motion (*i.e.*, spining of the electron about its own axis). Due to these two types of motions, an electron creates a magnetic field, *i.e.*, each unpaired electron in an atom, an ion or a moelcule of a substance may be regarded as a micromagnet which has a definite value of magnetic moment.

If a substance has many unpaired electrons, these electrons obviously will have the same spin and hence the magnetic moment of each electron will be equal and in the same direction. Thus the magnetic moment of each unpaired electron will *not* be cancelled by the magnetic moment of the other unpaired electrons present in the substance. Consequently if such a substance is placed in the magnetic field, it experiences attraction and hence shows paramagnetic character. The molecules (or atoms or ions) of a paramagnetic material has a definite value of magnetic moment.

Now, consider the case of a substance in which one or more orbitals contain paired electrons only. Since the two paired electrons residing in the same orbital have opposite spins, the spin magnetic moment of one electron will be equal and opposite to that of the other electron occupying the same orbital. This will result in that the magnetic moment of one electron will be cancelled by that of the other electron residing in the same orbital and consequently, if such a substance is placed in the magnetic field, it experiences repulsion and hence shows diamagnetic behaviour.

Ferromagnetic substances

When the paramagnetic substances like Fe, Co, Ni and CrO_2 at room temperature, are placed in an external magnetic field, their magnetic moments (electron spins) are aligned with respect to the external magnetic field. In case of these substances, the alignment of the magnetic moments does not undergo any change, even if the external magnetic field is removed, *i.e.*, unlike paramagnetic substances, these substances exhibit permanent magnetism even after the magnetic field is removed. Such substances are called ferromagnetic substances. In case of ferromagnetic substances the alignment of all magnetic moments is in the same direction (parallel direction) [Fig. 16.22 (*a*)] and hence such substances have some definite value of their magnetic moment. Thus *a ferromagnetic substance can be defined as that substance in which the alignment of all the magnetic moments is in the same direction* (parallel direction).

When a ferromagnetic material is placed in a magnetic field, it acquires a strong magnetism in the direction of the applied magnetic field. Thus a ferromagnetic substance shows more paramagnetic character than the paramagnetic substance. These substances are strongly attracted by the magnetic field. Once a ferromagnetic material is magnetised, it remains permanently so. Ferromagnetism, like paramagnetism, is also due to the presence of unpaired electrons in the ferromagnetic material. However, unlike paramagnetism, ferromagnetic materials show permanent magnetism even in the absence of magnetic field. Ferromagnetic substance are very important in technology. For example CrO_2 is used to make magnetic tapes for use in casette recorders. Other examples of ferromagnetic substances are steel, Mn,Cu_2MnSn alloy etc.

↑ ↑ ↑ ↑ ↑ ↑	↑↓ ↑↓ ↑↓	↑↑ ↓↑ ↑↓
(*a*) Alignment of magnetic moment vactors in a ferromagnetic substance	(*b*) Alignment of magnetic moment vactors in an anti-ferromagnetic substance	(*c*) Alignment of magnetic moment vactors in a ferrimagnetic substance

Fig. 16.22 : Alignment of magnetic moment vectors in ferromagnetic, anti-ferromagnetic and ferrimagnetic substances.

Anti-ferromagnetic substances

An anti-ferromagnetic substance is that in which the number of magnetic moment vectors aligned in one direction is equal to the number of magnetic moment vectors aligned in opposite direction [Fig. 16.22 (*b*)] and hence such a substance has zero value for its resultant (net) magnetic moment. In other words we can say that in an anti-ferromagnetic substance half of the magnetic moment vectors are aligned in one direction while the remaining half magnetic moment vectors are aligned in the opposite direction.

Transition metal oxides, namely V_2O_3, Cr_2O_3, MnO, Mn_2O_3, FeO, Fe_2O_3, CoO, Co_3O_4 and NiO, show anti-ferromagnetic character. Cupric acetate monohydrate is also an example of anti-feromagnetic substance.

Ferrimagnetic substances.

A ferimagnetic substance is that in which the number of magnetic moment vectors, aligned in parallel and antiparallel directions, is unequal [Fig. 16.22 (*c*)] and hence such a substance has some value of its resultant (net) magnetic moment. Examples of the substances, which show ferrimagnetic behaviour are Fe_3O_4 and ferrites having the general formula $M^{2+} Fe_2^{3+} O_4$ ($M^{2+} = Mg^{2+}, Cu^{2+}, Zn^{2+}$ etc.).

Effect of temperature on ferromagnetic, antiferromagnetic and ferrimagnetic substances

By the increase of temperature the ordered alignment of magnetic moment vectors in ferromagnetic, anti-ferromagnetic and ferrimagnetic substances get changed and hence the substances become paramagnetic. For example :

(*a*) Fe_3O_4 which is ferrimagnetic at ordinary temperature shows paramagnetic character at 850 K.

(*b*) V_2O_3 transforms from anti-ferromagnetic state to paramagnetic state at 150 K.

(*c*) NiO transforms from anti-ferromagnetic state to paramagnetic state at 395 K.

(*d*) CrO_2 transforms from ferromagnetic state to paramagnetic state at 395 K.

Dielectric Properties

We know that the electrons in an insulator are closely bound to individual atoms, or ions which constitute the insulator and, therefore, they donot migrate under the applied electric field. However, when an insulator is placed in an electrostatic field, the constituents of the insulator get polarised, *i.e.*, the electrons are displaced towards the positive charges and the nuclei are displaced towards the negative charges of the applied electrostatic field. Due to the shifting of charges, dipoles are created. The dipoles in a crystal may align in a number of ways in the field.

Depending on the way in which the dipoles in a crystal are aligned, the crystals have been classified into the following categories :

1. Piezoelectric crystals. Piezoelectric crystals are those in which the dipoles align themselves in such a way that the crystal may have some value for its dipole moment. When mechanical stress (pressure) is exerted on a piezoelectric crystal, electricity or electric polarity known as **piezoelectricity** is produced. This phenomenon is called **piezoelectric effect.** Electric current is produced due to the displacement of ions or dipoles. Conversely, when a peizoelectric crystal is placed in an electric field, the displacement of dipoles causes deformation in the crystal. Thus a piezoelectric crystal acts as mechanical-electrical transducers. Piezoelectric crystals are used as pick-ups in record players in which they produce electric signals by the application of pressure (*i.e.*, mechanical stress or strain). It is for this reason that piezoelectric crystals are also called **pick-up crystals.** Piezoelectic crystals may be classified into the following two categories.

(*a*) *Ferroelectric crystals.* In a ferroelectric crystal, the dipoles are permanently aligned even in the absence of electric field and the direction (orientation) of the dipoles can be changed (reversed) by applying an electric field. The phenomenon in which the orientation of dipoles in a peizoelectric crystal can be changed (reversed) by applying an electric field is called **ferroelectricity.** Common example of ferroelectric crystals are barium titanate ($BaTiO_3$), sodium potassium tartrate (Rochelle salt), potassium dihydrogen phosphate (KH_2PO_4) etc.

(*b*) *Anti-ferroelectric crystals.* In an anti-ferroelectric crystal, the dipoles are aligned in opposite directions, *i.e.*, the dipoles point up and down in the alternate polyhedra and hence the net dipole moment of the crystal is zero. Lead zirconate ($PbZrO_3$) is an anti-ferroelectric crystal.

2. Pyroelectric crystals. Pyroelectric crystals are those which produce small electric current when heated. This phenomenon is called **pyroelectricity** or **pyroelectric effect.**

Classification of crystals on the basis of the nature of particles present in their lattice

On the basis of the nature of the particles constituting the crystals, these have been clssified into the following four groups :

I. Ionic crystals

X-ray study of these crystals has shown that the constituent particles of which these crystals are made up are cations and anions. The cations and anions are held together by ionic bounds (*i.e.*, electrostatic attractions) and hence these crystals are called ionic crystals. Examples of ionic crystals are NaCl, LiF, $CaCl_2$, Li_2O, Na_2O etc.

Properties of ionic crystals

(*i*) **Crystal structure and physical state of ionic crystals.** The study of the structure of ionic crystals can be made under the following headings.

(*a*) *Constituent particles and nature of force holding them together.* The constituent particles of an ionic crystal are cations and anions which are held together by strong electrostatic force of attraction, known as *ionic bond.* Since the electrostatic force of attraction is strong, the ionic crystals exist as *solids.*

Ionic crystals may be of AB (*e.g.*, LiCl, NaCl, KBr, RbI, AgCl, AgBr, BeO, MgO, CaO, SrO, BaO, CsCl, CsBr, CsI, ZnS, HgS etc.), AB_2 (*e.g.*, CaF_2, SrF_2, ThO_2, TiO_2 etc.) and A_2B (*e.g.*, Li_2O, Na_2O etc.) type. There are many ionic crystals in which some of the atoms are held by covalent bonds to form the anions which have definite positions and orientations. Examples of such crystals are $CaCO_3$, $BaSO_4$ etc.

(*b*) *Nature of ionic bond, geometry of ionic crystals and possibility of space isomerism.* Ionic bonds existing in ionic crystals are *not rigid* and *non-directional. i.e.,* ionic bond extends equally in all directions. Due to the non-directional nature of ionic bonds, ionic crystals neither have any geometry nor show space (stereo) isomerism.

(*c*) *Arrangement (packing) of ions in an ionic crystal.* Since in most of the ionic crystals, anions are larger in size than the cations, anions have closest-packed arrangement which may be *cubic close-packed* (*ccp*-ABC ABC... sequence of layers) or *hexagonal close-packed* (*hcp* - AB AB A.... sequence of layers) arrangement. In this arrangement of anions, there are left some empty spaces in between the anions which are called *interstitial sites* or *interstitial holes.* These interstitial sites are of three types *viz.*, tetrahedral sites, octahedral sites and cubic sites. The relative order of the size of these holes is as :

cubic hole > octahedral hole > tetrahedral hole

Each of the anions having closest-packed arrangement has *two* tetrahderal sites or *one* octahedral site, *i.e.*, for the anions having closet-packed arrangement, the number of tetrahedral sites is twice the number of the anions but the number of octahedral sites is equal to the number of anions.

Some or all of the holes created by the closes-packed arrangement of anions are occupied by the cations present in the ionic crystal. The size of the interstitial hole or the radius of the cation which occupies the interstitial hole depends on the radius of the anions (see "*Radius Ratio*"). If we suppose the anions as hard spheres touching each other and if we know the radius of the anion (r_-), we can calculate the radius of the cation (r_+) that exactly fits into the interstitial site.

(*d*) *Coordination number of an ion (cation or anion) in an ionic crystal.* In an ionic crystal, the arrangement of ions is such that there is more electrostatic attraction between the oppositely-charged ions than there is electrostatic repulsion between the similarly-charged ions. It is for this reason that each ion in an ionic crystal is surrounded by the greatest possible number of oppositely-charged ions as its nearest neighbours. The number of nearest oppositely-charged ions by which a given ion is surrounded is called coordination number of the given ion. Thus the coordination number of a cation is the number of nearest anions by which the cation is surrounded. Similarly the coordination number at an anion is the number of nearest cations by which the anion is surrounded. The coordination numbers of positive and negative ions are the same, if there are equal number of cations and anions present in the ionic crystal (*e.g.,* NaCl, ZnS CsCl etc.). The coordination numbers of cations and anions are different, if the number of cations and anions present in the ionic crystal is different (*e.g.*, CaF_2, TiO_2, Na_2O etc.).

Coordination number of an ion depends on the nature of the interstitial site which the ion occupies. For example :

(*a*) A cation, in a tetrahedral site, is surrounded by 4 anions and hence its coordination number is 4 [See Fig. 16.23 (*a*) and the structure of zinc blende crystal, ZnS].

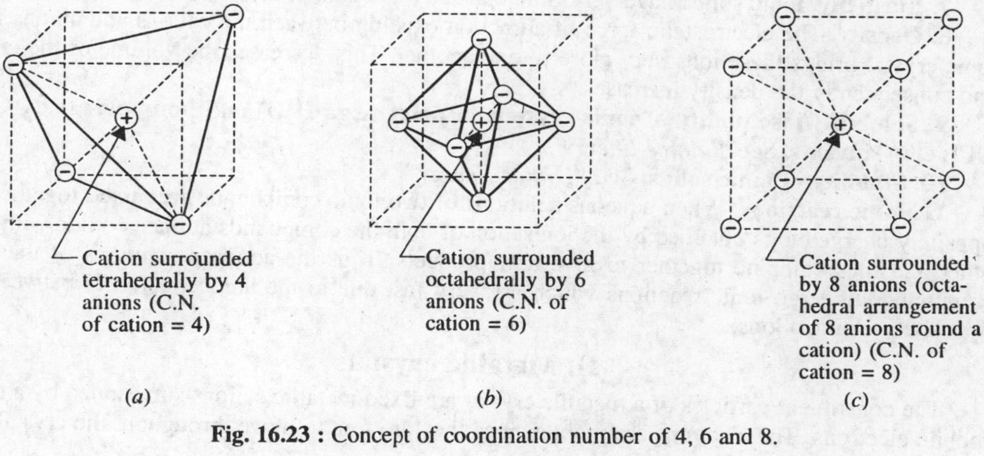

Fig. 16.23 : Concept of coordination number of 4, 6 and 8.

(*b*) A cation which occupies an octahedral site is surrounded by six anions and hence its coordination number is equal to six [See Fig. 16.23 (*b*) and the structure of NaCl crystal].

(*c*) A cation which occupies a cubic hole is surrounded by 8 anions and hence its cordination number is 8 [See Fig. 16.23 (*c*) and structure of CaF_2 crystal].

(*d*) The coordination number of each sphere in a hexagonal close-packed (*hcp*) arrangement of spheres is equal to 12.

Factors affecting the coordination number of ions. The above discussion shows that the coordination number of an ion depends on the following factors :

(*a*) *Arrangement of oppositely-charged ions round the given ion.* When the arrangement of oppositely-charged ions round a given ion is tetrahedral, octahedral or body-centered cubic, this means that the given ion is surrounded by 4, 6 or 8 oppositely charged ions and hence, the coordination number of the given ion is 4, 6 or 8 respectively.

(*b*) *Nature of interstitial site occupied by a cation.* When a cation occupies a tetrahedral, octahedral or cubic interstitial site. the coordination number of the cations is 4, 6 or 8 respectively.

(*c*) *Radius ratio* (r_+/r_-) *of the ionic crystal.* The coordination number of an ion in an ionic crystal also depends on the value of radius ratio (r_+/r_-). This has already been discussed.

(*d*) *Effect of temperature and pressure on the structure (or coordination number) of ionic crystals.* At ordinary temperatures and pressures, chlorides, bromides and iodides of Li, Na, K and Rb as well as the halides of Ag have NaCl type structure in which the coordination number of both cations and anions are 6 (6 : 6 ionic crystal). When pressure is applied on these ionic crystals, their crystal structure is changed from NaCl to CsCl in which the coordination number of both the ions is equal to 8 (8 : 8 ionic crystal). Thus we see that the application of pressure increases the coordination number.

On heating to about 760 K, the CsCl structure transforms to NaCl structure. Similarly NH_4Cl, NH_4Br and NH_4I which have CsCl type structure at ordinary temperature transform the CsCl structure to the NaCl structure at 457 K, 411 K and 253 K respectively.

2. Electrical nature of unit cell. The unit cell by which the crystal structure of an ionic crystal can be represented must be electrically neutral.

3. Melting and boiling points. See *"Lattice Energy"*.

4. Electrical properties (electrical conductivity) of ionic solids. See page 511.

5. Hardness and briteness. Ionic solids are highly hard and brittle, since the application of external force on the ionic crystals makes one layer of ions slip over the other layer along a plane so that the ions having similar charges in the two layers come closer to each other. This causes repulsion between the similarly-charged ions and ultimately this repulsion breaks the crystal, *i.e.*, the ionic crystals are brittle.

6. Lattice energy. Ionic crystals have high values of lattice energy.

7. Elasticity. Ionic solids have very little elasticity and cannot be easily bent.

8. Density. The electrostatic force of attraction existing between the cations and anions in an ionic crystal brings these ions very close to one another. This decreases the volume of the crystal and consequently the density increases.

9. Solubility/insolubility of ionic crystals in polar (e.g., H_2O) and non-polar (e.g., C_6H_6, CCl_4 etc.) solvents. See *"Lattice Energy"*.

10. Stability of ionic solids. See *"Lattice Energy"*.

11. Ionic reactions. When aqueons solutions of two ionic compounds are mixed together, the oppsitely-charged ions obtained by the ionisation of both the compounds at once collide with each other and thus combine together to give new products. Thus the aqueons solutions of the ionic compounds undergo ionic reactions which are very fast due to the attractive forces between the oppositely-charged ions.

II. Metallic crystals

The constituent particles of a metallic crystal are fixed metallic cations surrounded by a sea of mobile electrons. The electrons are mobile since they are free to move throughout the crystal. The metallic cations are in their positions and mobile electrons are in the interstices between them. The metallic cations and mobile electrons are held together by electrostatic force of attraction known as **mettalic bond** and hence these crystals are called metallic crystals.

Thus metallic bond can be defined as the force of attraction between the metallic cations and the mobile electrons and this force holds the metal atoms firmly together in the metallic crystal.

Examples of metallic crystals are metals (*e.g.*, Hg, W, alkali metals, Os, Cu, Bi, Fe, Au), alloys etc.

Factors favouring the formation of metallic bond

The factors favouring the formation of metallic bond are :

1. The ionisation energy of the metals should be low so that the valence-electrons of the metal are losely held by the nucleus and thus become mobile.

2. The metals should have low electronegativity so that they donot accept electrons and never form anions.

3. The number of vacant orbitals in a metal should be more than the number of valence-electrons in it so that the mobile electrons easily move into the vacant orbitals and the properties like thermal conductivity etc. of the metals are easily explained.

Nature of metallic bond

In order to explain the nature of metallic bonding the following theories have been put forward.

1. Electron-sea theory. This theory is also called *electron-gas* or *electron-cloud* theory and was proposed by Lorentz. According to this theory, since the ionisation energies of metals are very low, they readily lose their valence-electrons and form the positive metal ions, called *kernels*. The electrons given up by the metal atoms are not stationary but can move freely from one place to the other through the vacant orbitals. Thus these electrons are mobile or delocalised. According to this theory the force of attraction between the metal positive ions and the mobile electrons holds the metal atoms together in the metallic crystal.

Limitations of electron-sea theory. (*i*) This theory can not explain why tungsten has high melting point (= 3300° C) while Hg has a melting point equal to –39°C. (*ii*) It does not explain the high hardness of osmium metal. (*iii*) It fails to explain why some metals have greater electrical conductivity than others.

2. Valence bond theory. This theory is also called *resonance theory* and was suggested by Pauling. According to this theory, metallic bond is covalent in origin and the structure of a metal involves resonance of electron pair bonds (*i.e.*, covalent bonds) between each atom and its nearest neighbour. In case of Li-atom it has been observed that this atom (Li $\rightarrow 1s^2, 2s^1, 2p_x^0\ 2p_y^0\ 2p_z^0$) is surrounded by eight neighbouring Li-atoms. Since the single electron in 2*s* orbital cannot form electron pair bonds with eight neighbours, it is assumed that resonance takes place throughout the solid. Different resonance (contributing) forms can be written as :

Li — Li
Li — Li
	↔	Li Li
		| |
		Li Li
			↔	Li⁺ Li
				     |
				Li—Li⁻
					↔	Li Li⁺
						|
						Li⁻—Li
							↔	Li⁻—Li
								     |
								Li Li⁺
									↔	Li—Li⁻
										     |
										Li⁺ Li

(I)		(II)		(III)		(IV)		(V)		(VI)

The contributing structures *viz.*, (III), (IV), (V) and (VI) contain a negatively-charged Li-atom which is bounded to other two Li-atoms by two covalent bonds. These covalent bonds are called *resonating covalent bonds.* How these bonds are formed has been explained as follows.

The electronic configuration of Li-atom: $1s^2, 2s^1\ 2p_x^0, 2p_y^0, 2p_z^0$ shows that there are three vacant 2*p*-orbitals ($2p_x$, $2p_y$ and $2p_z$). Since the energy of these three 2*p*-orbitals is not very different from that of the valence orbital (*i.e.*, 2*s*-orbital), an electron from 2*s*-orbital of one Li-atom can easily be transferred to one of three 2*p* vacant orbitals of the other Li-atom to convert it into Li^- ion with the configuration $1s^2, 2s^1, 2p_x^1, 2p_y^0, 2p_z^0$. The Li-atom which is giving the electron from its 2*s*-orbital is converted into Li^+ ion with the configuration $1s^2$. This point can more clearly be understood by distinguishing the four Li-atoms as Li_a, Li_b, Li_c and Li_d. All Li-atoms have $1s^2, 2s^2$ configuration.

Now suppose Li_d atom gives its $2s^1$-electron to Li_b atom, so that Li_d becomes Li_d^+ (with configuration $1s^2$) and Li_b becomes Li_b^- (with configuration $1s^2, 2s^1, 2p_x^1, 2p_y^0, 2p_z^0$). Now $2s^1$ and $2p_x^1$ orbitals on Li_b^- ion form two *sp*-hybrid orbitals, which form two covalent bonds, one with Li_a and the other with Li_c atom, as shown below. In brackets we have shown the electronic configuration of the atom or ion.

$$\begin{array}{lcl} Li_a\,(1s^2, 2s^1) & \text{———————} & Li_b^-\,[1s^2, (sp)^1, (sp)^1, 2p_y^0, 2p_z^0] \\ & & \quad | \\ Li_d^+\,(1s^2) & & Li_c\,(1s^2, 2s^1) \end{array}$$

Thus we find that the requirement, in terms of valence bond theory of metallic bonding, for the formation of resonating covalent bonds is that *there should be orbitals, energetically not very different from the outermost occupied orbitals, in the configuration of the metallic atom.*

The two vacant orbitals namely $2p_y^0$ and $2p_z^0$ on Li_b^- can accept the conductivity electrons and thus contribute to metallic conduction. These two $2p$ vacant orbitals are termed *metallic orbitals.*

3. Molecular orbital theory (MOT) : Band model. The MOT of metals is commonly known as the band model. Since the smallest perceptible piece of a metal would contain about 10^{20} atoms, a large number of molecular orbitals (MOs) are formed from a large number of atomic orbitals (AOs).

Let us consider an idealised one-dimensional lattice. Imagine the step-by-step formation of a chain of Li-atoms, Li-Li-Li...Li-Li-Li. In the formation of this chain of Li-atoms first of all two Li-atoms will form a Li_2-moelcule. In the formation of this molecule the two 2*s*-AOs (each having one electron, Li → $2s^1$), one from each Li-atom, combine to give two MOs. The lower energy MO is completely filled with two electrons and the higher energy MO is vacant [Fig. 16.24 (*a*)]. Attaching another Li-atom to Li_2-molecule gives the linear Li_3 molecule (hypothetical molecule). Li_3 molecule is formed when three 2*s*-AOs combine to give three MOs [Fig. 16.24 (*b*)].

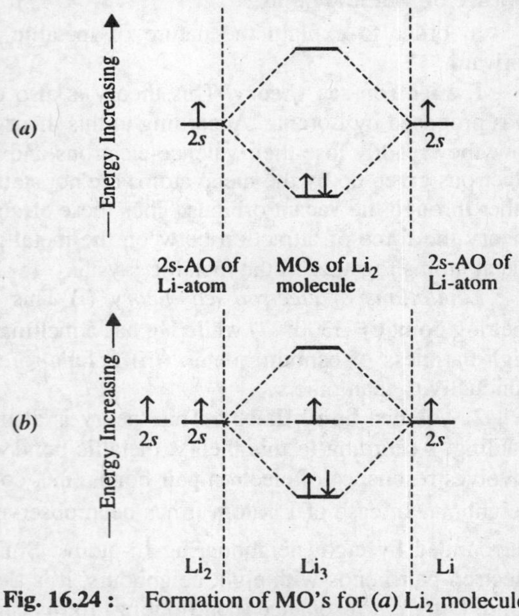

Fig. 16.24 : Formation of MO's for (*a*) Li_2 molecule (*b*) Li_3 molecule (hypothetical).

Now as the length of the chain increases by increasing the number of Li-atoms we get a large number of MOs closely spaced together. The number MOs (*i.e.,* energy levels) is the same as that of AOs responsible for their formation. As the number of Li-atoms increases, the energy levels get closer and closer and ultimately become continuous. Such a group of continuous energy levels is known as a *band* and consequently the molecular orbital theory is called *band model.*

Since 2*s*-AO of each Li-atom has only one electron, lower-half of the total MOs (or the lower-half of the band) formed is completely filled and the upper-half is vacant. The upper-half of the band has higher energy (*i.e.,* less stable) than the lower-half. Thus the band (which is called 2*s* band) in Li-Li-Li....Li-Li-Li chain is half-filled provided that we consider a three dimensional arrow of Li-atoms. Fig. 16.25 shows the schematic band structure of Li-metal. In this figure the empty 2*p*-AOs have also been shown.

Band model can readily explain the metallic properties. When a crystal of Li is placed in an electric field, a few electrons acquire energy and are excited to the higher vacant MOs. The higher energy electrons carry the current with them. The higher conductivity of metals is thus due to the presence of large number of higher energy vacant MOs. Thus, we see that the electrons in the lower energy MOs

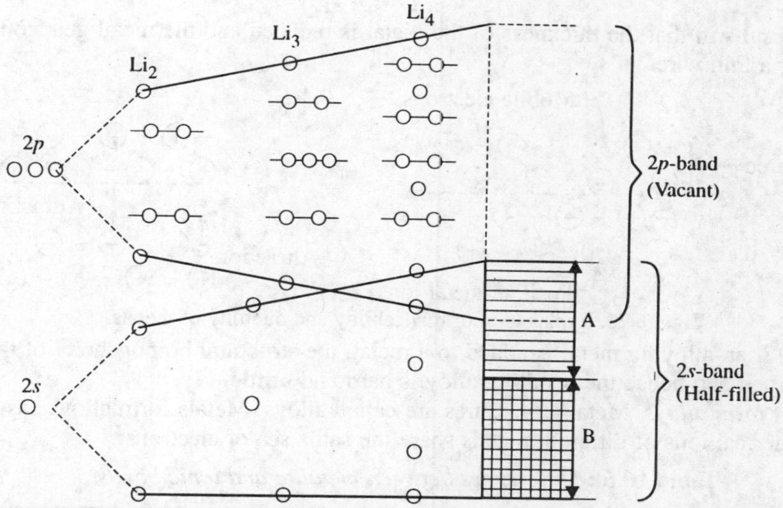

Fig. 16.25 : Schematic band structure of Li-metal. A is the upper half of the 2*s*-band (vacant) and B is the lower half (completely filled). Small circles indicate AOs and MOs.

are not localised. These delocalised electrons absorb and re-emit light and thus render the metal surface shiny and lustrous. Since the bonding electrons are not localised at any particular atom in the metal crystal, the lattice can be deformed easily which explains the malleability and ductility of metals.

Physical properties of metals and their explanation on the basis of electron-gas theory.

Some of the physical properties of metals and their explanation on the basis of electron-gas theory are given below :

1. *Metals are good conductors of electricity (Electrical conductivity of metals)*. For this see "*Electrical conductivity of solids*" discussed earlier.

2. *Metals are good conductors of heat (Thermal conductivity)*. It has been observed that the metals which are good conductors of electricity are also good conductors of heat. Thermal conductivity of metals can also be explained on the basis of mobile electrons present in the metallic lattice. When a piece of metal is heated at one end, the mobile electrons at this end absorb heat energy from this end and move very rapidly through the metallic lattice towards the cooler end. During this process they collide with adjacent electrons and thus transfer their heat energy to them. Thus the mobility of electrons allows heat transfer to the other end.

3. *Metals are opaque and have lusture or colour*. The light that falls on metals is absorbed completely by the electronic transition of the sea of electrons. Since no light is permitted to pass through the metal, it is opaque.

When freshly-cut or scratched, most of the metals possess metallic lusture, *i.e.,* they have a shining surface. The lustre of metals can also be explained on the basis of mobile electrons. When light falls on the surface of the metal, the incident light collides with the mobile electrons, which get excited. When these excited electrons revert to their original position, they give off energy in the form of light. This light appears to be reflected from the surface of the metal and hence the metal appears to possess its characteristic lustre.

4. *Metals are malleable or plastic and ductile*. Metals are malleable and ductile, *i.e.,* when they are beaten by a hammer, they are converted into sheets (*malleability*) and when they are drawn through a die, they are converted into thin wire (*ductility*).

When a metal is beaten, say by a hammer, one layer of positive metal ions moves along another layer through a distance and occupies new positions. Since the sea of electrons (*i.e.,* valence electrons) are mobile, they also move along with the positive metal ions and thus the structure of the metal does not alter but remains the same and the crystal structure is restorted (Fig.

16.26). This results in that the thickness of the metal is reduced and the metal gets converted into a thin sheet or a thin wire.

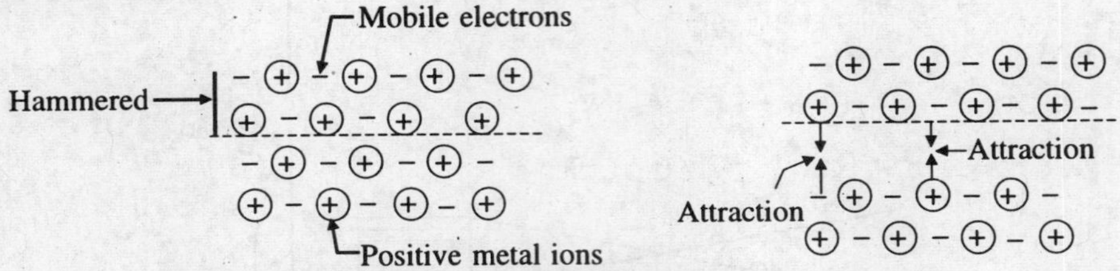

Fig. 16.26. Explanation of malleability and ductility of metals.

However, if an alloying metal is added to a metal, the structural homogeneity of the principal metal is disturbed and hence the resulting alloy is hard and brittle.

5. *Metals form alloys.* Metallic mixtures are called alloys. Metals form alloys easily. In these alloys, the spherical ions of different metals share the same sea of electrons.

Table 16.6 : *Differences between metallic and ionic bonds.*

Metallic bond	Ionic bond
1. This bond is formed by the simultaneous attractive interaction between the kernals (positively charged metal atoms) and mobile electrons in a metal crystal.	1. This bond is formed by the transfer of electrons between two dissimilar atoms.
2. Metals are malleable	2. Ionic compounds are brittle.
3. It is a weak bond.	3. It is a strong bond due to strong electrostatic force of attraction.
4. Metals are good conductors in the solid state.	4. Ionic compounds are bad conductors in the solid state.

Table 16.7 : *Differences between metallic bond and covalent bond.*

Metallic bond	Covalent bond
1. This bond is formed by the simultaneous attractive interaction between the kernals mobile electrons in a metal crystal.	1. This bond is formed by the sharing of electrons between two similar or dissimilar atoms.
2. It is a weak bond because mobile electrons are simultaneously attracted by a large number of nuclei.	2. It is a strong bond because the bonded electron pair (or pairs) is strongly attracted by nuclei.
3. Metals are usually solids (Hg is a liquid).	3. Covalent compounds are usually liquids and gases.
4. Metals are good conductors of electricity.	4. Genrally, covalent solids are bad conductors of electricity (graphite is an exception).
5. This bond is non-directional.	5. This bond is directional.

III. Atomic or covalant crystals

The constituent particles of a covalent crystal are neutral atoms of the same or different kind, which are held together by electron pair covalent bonds. It is due to the presence of covalent bonds that atomic crystals are also called **covalent crystals.** A covalent crystal consists of a large network of atoms, covalently linked to one another and hence covalent crystals are also called **covalent network crystals.**

Examples of covalent crystals are diamond, graphite, silicon carbide, boron nitride, silicon dioxide, rhombic sulphur, quartz etc.

Properties of covalent crystals.

In contrast to ionic crystals, the valencies in covalent crystals, are saturated and covalent bonds are oriented in definite directions. This leads to looser packing than in ionic and metallic crystals, *i.e.*, covalent crystals have *open structure*. However, the covalent bond is very strong and hence considerable amount of energy is required to break it. Thus, covalent crystals are usually very hard, possess high melting points and low volatility and are insoluble in most of the solvents. Generally, these solids are bad conductors of electricity (graphite is an exception). This is because they are not made up of ions and donot have any free electrons available for electric conduction, all the electrons being engaged in covalent bonding.

Diamond

Occurrence. Diamond is the purest crystalline allotropic form of carbon, occuring in nature in the free state. Diamonds are found in South Africa, Russia, Australia and Brazil. In nature, diamond occurs as octahedral crystals.

Preparation. The artificial diamonds were prepared by heating pure sugar charcoal and iron in a graphite crucible at about 3000°C in an electric furnace. At this high temperature, carbon was dissolved in iron. Now, when the molten mass was cooled rapidly by plunging it into a bath of molten lead and then treated with conc. HCl, crystals of artificial diamond were obtained.

Structure. X-ray analysis has shown that the crystal of diamond is made up of a number of tetrahedral units. Each tetrahedral unit contains C-atoms at its centre and this C-atom, is linked with other four C-atoms placed at the four corners of the tetrahedron. C-atoms of each tetrahedron are also linked by strong C—C covalent bonds with C-atoms of each of the neighbouring tetrahedrons (see Fig. 16.27). Each C-atom is sp^3 hybridised. C—C bond distance in diamond is equal to 1.54 Å, which is the normal distance of C—C covalent bond. Since diamond consists of a number of tetrahedral units, it is a giant molecule. Since each C-atom forms four covalent bonds, it produces a repetitive three-dimensional structure, using only covalent bonds.

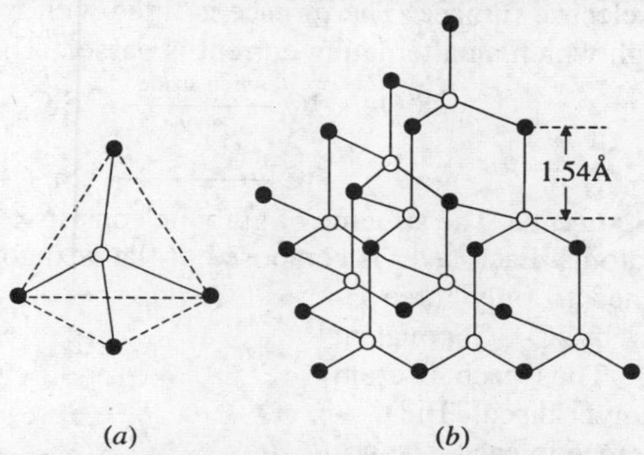

Fig. 16.27 : (*a*) Tetrahedral unit of C-atoms in diamond crystal. (*b*) Structure of diamond crystal in which a number of tetrahedral units are linked together. White carcles indicate the C-atom placed at the centre of each tetrahedron while black curles indicate carbon atoms placed at the four corners of each tetrahderon.

Silicon, germanium and grey tin also crystallise in the same way as diamond.

Properties. The structure of diamond, discussed above, explains the following properties of diamond.

(*i*) It is insoluble in all solvents.

(*ii*) Due to the giant molecular nature of the structure of diamond, it is a very hard substance. In fact diamond is the hardest natural substance known so far. Due to its great hardness, diamond is used for making glass cutters and rock borers.

(*iii*) Since all the four electrons of the valence-shell of each C-atom are used up in making four covalent bonds with other C-atoms, there is no free electron left in diamond crystal to carry electricity or heat in the crystal. Thus diamond is a bad conductor of electricity and heat.

(*iv*) Since, in order to break strong C—C covalent bonds present in diamond, a large amount of heat energy is needed, the melting point of diamond is very high.

(*v*) Its relative density is high (= 3.5).

(*vi*) It is transparent to X-rays.

(*vii*) It has high value of refractive index (= 2.45).

(*viii*) At red heat (900°C), diamond burns in air to form CO_2. It is slowly oxidised by $K_2Cr_2O_7$ and conc., H_2SO_4 at 200°C to form CO_2.

(*ix*) On heating with Na_2CO_3, it gives CO.

(*x*) It is stable in vacuum upto 1500°C but when heated to 1800 – 2000°C, it is converted into graphite.

(*xi*) It reacts with F_2 at 700°C to form CF_4 ($C + 2F_2 \xrightarrow{700°\,C} CF_4$).

(*xii*) The colourless diamonds are used as precious jewellery stones because of their ability to reflect and refract light.

(*xiii*) Diamonds are used for drawing of thin wires.

Graphite

Graphite is found in Ceylon.

Manufacture. Graphite is obtained by Acheson's process which consists of heating a mixture of sand (SiO_2) and coke anthracite in presence of iron oxide as catalyst at a temperature of 3000°C in an electric furnace. The furnace is fitted with two carbon electrodes joined by a thin carbon rod through which an alternating current is passed. The reaction between SiO_2 and coke proceeds as :

$$SiO_2 + 3C \xrightarrow[3000°\,C]{\text{Iron oxide}} SiC + 2CO$$

$$SiC \xrightarrow{\Delta} Si + C\ (\text{graphite})$$

Structure. The structue of graphite consists of a large number of flat parallel layers (or sheets) of C-atoms. Each layer is composed of flat hexagonal rings of C-atoms. In each layer, each C-atom is joined to only three C-atoms by C—C covalent bonds. Thus each C-atom is sp^2 hybridised. The C—C distance in each hexagonal ring is 1.42 Å and the distance between the two adjacent layers of C-atoms is 3.40 Å. Such a large distance between the two layers suggests that these layers are not joined together by covalent bonds. The layers are in fact joined together by weak van der Waals forces. The presence of hexagonal rings in each layer is supported by the fact that with alkaline $KMnO_4$ solution, graphite is oxidised to mellitic acid, $C_6(COOH)_6$ and oxalic acid. Mellitic acid on distillation with lime gives C_6H_6. Structure of graphite is given in Fig. 16.28.

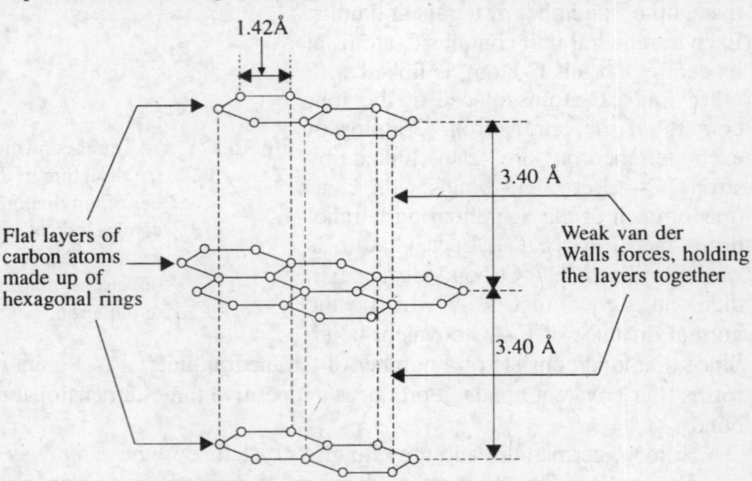

Fig. 16.28. Structure of graphite crystal.

Properties and uses. (*i*) Since the van der Walls forces, by which layers of C-atoms are held together, are very weak, these layers in graphite can easily slide over one another and hence graphite is a *soft substance* and has *lubricating properties*. Thus, graphite is used as a lubricant for the fast moving parts of machinery. Since graphite is non-volatile, it can also be used for lubricating the machinery parts which operate at a very high temperature. It is a greasy substance.

(*ii*) Since in each layer, each C-atom is linked to other three C-atoms only, one electron on each C-atom is free to carry heat and electricity in the structure of graphite. Thus, graphite is a good conductor of electricity and heat. Due to its good thermal and electrical conductivity, graphite is used for making carbon electrodes or grahite electrodes. The electrical conductivity increases with the increase of temperature.

(*iii*) Due to the fact that graphite is soft and black in colour, it is used for making lead pencils.

(*iv*) It is due to the strong C—C covalent bonding within the sheets that graphite has high melting point (= 3700°C) and hence is used for making refractory crucibles, which are used for melting metals in industry.

(*v*) At 700° – 800°C, graphite burns in air, to give CO_2.

(*vi*) It reacts with conc. HNO_3 or conc. H_2SO_4 below 100°C to form graphitic acid ($C_{11}H_4O_5$).

(*vii*) Graphite is used for lining and for making electrodes of electric furnaces.

(*viii*) Graphite is also used as a moderator, *i.e.*, for reducing the speed of fast moving neutrons in an atomic reactor.

Table 16.8 : *Differences between the properties of diamond and graphite.*

Diamond	Graphite
(*i*) It occurs in nature in the free state.	(*i*) It occurs naturally and is also manufactured artificially.
(*ii*) It is the hardest natural substance.	(*ii*) It is soft and greasy to touch.
(*iii*) It has high relative density (= 3.5).	(*iii*) Its relative density is 2.3.
(*iv*) It is transparent and has high refractive index (= 2.45). It is a non-conductor of heat and electricity.	(*iv*) It is black, opaque and a good conductor of heat and electricity.
(*v*) It burns in air at 900°C to give CO_2.	(*v*) It burns in air at 700°C-800°C to give CO_2.
(*vi*) It occurs as octahedral crystals.	(*vi*) It occurs as hexagonal crystals.
(*vii*) It is insoluble in all solvents.	(*vii*) It is insoluble in all ordinary solvents.

IV. Molecular crystals

The constituent particles of a molecular crystal are discrete or independent covalent molecules (or atoms in case of noble gases). These molecules may be polar molecules or non-polar molecules and are oriented in a manner to give the best packing. Thus molecular crystals may be polar molecular crystals or non-polar molecular crystals. The molecules in polar molecular crystals are held together by relatively weak polar forces and the molecules in non-polar molecular crystals are held together by weak van der Waals forces.

Examples of molecular crystals are iodine, ice (polar molecular crystal), sugar, noble gases (except helium), carbon dioxide (non-polar molecular crystal) etc.

Properties of molecular crystals. Since the forces holding the molecules in molecular crystals are weak van der Waals forces, these crystals are generally soft and have low melting points. The electrical conductivity is extremely low *i.e.*, these crystals are bad conductors of electricity. Polar molecular crystals are usually soluble in the polar solvents, while non-polar molecular crystals are usually soluble in non-polar solvents. The noble gases, which form molecular crystals (except helium) have cubic close packed structure. Solid H_2, N_2 and O_2 crystallise with hexagonal close-packed structure.

Polymorphism and polymorphic (or polymorphous) forms

The phenomenon by which a certain *crystalline compound* exists in two or more different crystalline forms is called *polymorphism* and the different crystalline forms are called *polymorphic (or polymorphous) forms* of the given compound.

Examples. (*i*) Calcium carbonate occurs naturally in two crystalline forms namely *calcite* (which has rhombohedral structure) and *aragonite* (orthorhombic structure). These two forms are the polymorphic forms of calcium carbonate.

(*ii*) As_2O_3 exists in two polymorphous forms which have rhombic and octahedral structure.

(*iii*) Sb_2O_3 exists in two polymorphous forms which have rhombic and octahedral structure.

Isomorphism and isomorphous forms

We have already said that As_2O_3 exists in two poly-morphous forms *viz.*, rhombic and octahedral. Similarly, Sb_2O_3 also has two poly-morphous forms which have rhombic and octahedral structure. Since the atoms in the rhombic forms of both the oxides are united in a similar way, these forms are said to be *isomorphous* with each other. Similarly, since the atoms in the octahedral forms of both the oxides are also united in a similar way, these forms are also said to be isomorphous with each other. The phenomenon by which the two substances have the same number of atoms, which are united in a similar way, is called *isomorphism.*

Table 16.9 : *Summary of the characteristics of various types of crystals.*

Characteristics	Ionic crystals	Metallic crystals	Atomic or covalent crystals	Molecular crystals
(*i*) Constituent particles	Cations and anions arranged in a regular pattern	Fixed metallic cations surrounded by a sea of mobile electrons	Neutral atoms covalently bounded with each other	Polar molecules or non polar molecules (atoms in case of noble gases)
(*ii*) Binding forces between the constituent particles	Strong electrostatic attraction between the cations and anions (Ionic bonds)	Electrostatic force of attraction between metallic cations and mobile electrons (metallic bond)	Covalent bonds between the neutral atoms	Weak polar forces in polar molecular crystals and van der waals forces in non-polar molecular crystals.
(*iii*) Melting points	High	Intermediate as well as high	Very high	Low
(*iv*) Hardness and brittleness	Hard and brittle	Low as well as high hardness, low brittleness	Usually very hard (Graphite is soft)	Polar molecular crystals are moderately hard, while non-polar molecular crystals are very soft.
(*v*) Electrical conductivity	High in the fused state and in solution	High	Usually low (Graphite is an exception).	Extremely low
(*vi*) Solubility (*a*) in polar solvents	Soluble	Insoluble	Insoluble	Polar molecular crystals are usually soluble in polar solvents.
(*b*) in non-polar solvents	Insoluble	Insoluble	Usually soluble	Non-polar molecular crystals are usually soluble in non-polar solvents.
(*vii*) Examples	NaCl, LiF, $CaCl_2$, ZnS	Metals (e.g. Cu, Fe, Au etc.) and alloys	Diamond, graphite, SiC, BN, SiO_2, rhombic sulphur etc.	Ice (polar molecular crystal), CO_2 (non-polar molecular crystal), iodine, sugar, noble gases.

Other examples of isomorphous substances

(*i*) $ZnSO_4.7H_2O$ and $FeSO_4.7H_2O$ are isomorphous with each other.

(*ii*) Chrome alum, $K_2SO_4.Cr_2(SO_4)_3$. $24H_2O$ is isomorphous with potash alum, K_2SO_4. $Al_2(SO_4)_3$. $24H_2O$.

(*iii*) The ionic compounds *viz.* NaF and MgO are also isomorphous substances, since they are made up of the ions which have identical electronic configuration.

Mitcherlich's Law of isomorphism

In 1820 Mitcherlich put forward a law, called *Mitcherlish's law of isomorphism* with regard to the isomorphous substances. It is stated as :

Isomorphous substances have an equal number of atoms which are united in a similar way, i.e. they have similar chemical formulae or constitution.

Properties of isomorphous substances. The chief characteristics of isomorphous substances are : (*i*) The crystals of these substances have the same shape. (*ii*) The crystals of one substance can be grown in a saturated solution of the other isomorphous substance. For example, if a crystal of $FeSO_4.7H_2O$ is suspended in a saturated solution of $ZnSO_4.7H_2O$, the former (*i.e.*, $FeSO_4$, $7H_2O$) begins to grow in size as the latter (*i.e.*, $ZnSO_4$, $7H_2O$) deposits on it. Similarly, if a crystal of $ZnSO_4.7H_2O$ is suspended in a saturated solution of $FeSO_4.7H_2O$, crystals of $ZnSO_4.7H_2O$ begin to grow in size as $FeSO_4.7H_2O$ deposits on the crystal of $ZnSO_4.7H_2O$. (*iii*) The elements forming the isomorphous substances have the same valency. For example the valency of Zn and Fe in $ZnSO_4.7H_2O$ and $FeSO_4.7H_2O$ which are isomorphous with each other is 2. (*iv*) They can form mixed crystals with each other, *i.e.*, when a solution containing two isomorphous substances is concentrated to the point of crystallisation, we get crystals, which are homogeneous and contain both the isomorphous substances in the ratio in which they were present in the solution.

Allotropy and allotropic forms or allotropes

The phenomenon by which an element exists in two or more different crystalline or amorphous forms is called *allotropy* and the different forms are called *allotropic forms* or *allotropes* of the given elements. Different allotropic forms of an element differ in their physical properties, but they have the same chemical properties.

Examples. Carbon, phosphorus, oxygen, sulphur etc. exist in many allotropic forms.

Cause of allotropy

The phenomenon of allotropy is due to the difference in (*i*) the arrangement of atoms or molecules in the crystal lattice, *e.g.*, diamond and graphite are the allotropic forms of carbon because they have different arrangement of carbon atoms in their crystal lattice, (*ii*) the number of the atoms in the molecule (*e.g.*, O_2 and O_3) (*iii*) the method of crystallisation, (*iv*) the amount of available energy.

Stable, unstable and metastable allotropes

The allotrope that can exist indefinitely in a given physical condition without changing even in part into the other allotrope is called a *stable allotrope* while the allotrope which can change into another allotrope is called an *unstable allotrope*. The allotrope which does not change into another allotrope under the conditions in which it is theoretically unstable is called a *metastable allotrope.*

Types of allotropy

Allotropy is of three types *viz.* enantiotropy, monotropy and dynamic allotropy. These have been discussed below :

1. Enantiotropy. Let us consider the two crystalline allotropic forms of sulphur namely rhombic sulphur (S_R) and monoclinic sulphur (S_M). S_R is stable at all the temperatures below 96°C while S_M is stable only at temperatures above 96°C. The temperature of 96°C is called transition temperature or transition point of the two forms of sulphur. At the transition temperature (= 96°C) both the allotropic forms of sulphur co-exist. These two allotropic forms are called *enantiotropes* or *enantiotropic forms* of sulphur and the phenomenon by which one form is stable at all temperatures below the transition temperature and the other form is stable only at the temperatures above the transition temperature is called *enantiotropy.* The conversion of S_R to S_M at the transition temperature of 96°C can occur in either direction and hence is reversible process. This can be shown as :

$$S_R \xrightleftharpoons{96^{\circ}C} S_M$$

Thus , we see that S_R and S_M show enantiotropy at 96°C. Similarly the enantiotropy shown by white and grey tin (transition temperature = 13°C) and shown by red and yellow forms of HgI_2 (transition temperature = 120°C) can be depicted by the following enantitropic changes :

$$\text{Grey tin} \xrightleftharpoons{13^{\circ}C} \text{White tin}$$

$$\text{Red HgI}_2 \xrightleftharpoons{120^{\circ}C} \text{Yellow HgI}_2$$

It should be noted that the transition temperature of the enantiotropes of an element is always less than the melting point of that element. At transition temperature, both enantiotropes have equal vapour pressures and same solubility in a given solvent.

2. Monotropy. Let us consider the two crystalline allotropic forms of phosphorus *viz.*, red phosphorus and white phosphorus. Red phosphorus is stable below the melting point of phosphorus while the white phosphorus is metastable. The phenomenon by which one of the two crystalline allotropic forms of a substance is stable below the melting point of the substance and the other is metastable or unstable is called *monotropy* and the two crystalline forms are called *monotropic* *forms.* Thus we see that red phosphorus and white phosphorus are the two monotropic forms of phosphorus. Similarly diamond and graphite are the monotropic forms of carbon. Graphite is stable below the melting point of carbon while diamond is metastable. The theoretical transition temperature of the substance showing monotropy is always higher than its melting point.

3. Dynamic allotropy. This is that form of allotropy in which dynamic equilibrium exists between the two allotropic forms, *i.e.*, both the forms change into one another exactly at the same rate. Both the forms are in the same physical state, *i.e.*, both the forms are either in the liquid state or in the gaseous state. The equilibrium composition changes with temperature.

Example. S_λ and S_μ are the dynamic allotropes of liquid sulphur. When sulphur is heated above its melting point, it melts into a pale yellow liquid known as S_λ. As the temperature is raised, S_λ becomes darker and darker and viscous. At this stage liquid sulphur obtained is called S_μ. These two forms are in equilibirium with each other and are liquid allotropes of sulphur.

Allotropes of carbon

Carbon exists in two allotropic forms *viz., crystalline and amorphous forms.* Diamond and graphite are the crystalline allotropic forms while animal charcoal, wood charcoal, sugar charcoal, lamp black, coal, coke and gas carbon are the amorphous forms.

Crystalline forms of carbon.

Diamond and graphite are the crystalline allotropic forms of carbon. These forms have already been discussed under "*Covalent crystals*".

Amorphous forms of carbon.

(1) Animal charcoal. It is obtained as a residue when destructive distillation of animal bones is done in cast iron retorts in a limited supply of air. It can absorb coloured substances from solutions. Due to this property animal charcoal is used for removing brown colour from the sugar-cane juice during the manufacture of sugar. It is also used for the manufacture of phosphorus and phosphoric acid.

(2) Wood charcoal. It is obtained as a residue when destructive distillation of wood is done in cast iron retorts in a limited supply of air. When burnt in O_2, CO_2 is formed. It is an extremely porous substance due to which it absorbs large volumes of gases. It is used in gas masks to remove poisonous gases. The gas masks are used by the workers in coal mines. It is used as a reducing agent in certain metallurgical processes. It is also used as a fuel and as a constituent of gun powder.

(3) Sugar charcoal. It is obtained as a black mass by dehydrating cane sugar with conc. H_2SO_4.

$$C_{12}H_{22}O_{11} \text{ (Sugar-white)} \xrightarrow[(-11\ H_2O)]{\text{Conc. } H_2SO_4} 12C \text{ (Sugar charcoal-black)}$$

When burnt in O_2, sugar charcoal produces CO_2.

(4) Lamp black. It is obtained by burning substances which are rich in carbon content such as kerosene oil, tar, turpentine oil or any other vegetable oil in a limited supply of air. The smoke so obtained is deposited on wet blankets hung in chamber. It is used for making black shoe polish, printer's ink, black paints, type writer ribbons and rubber tyers.

(5) Coal. It is formed in nature by slow decomposition of vegetable matter under the influence of heat, pressure and limited supply of air. This process of decomposition takes place in stages and in each stage a different form of coal is produced. Different forms of coal are *peat* containing 60% carbon, *lignite* containing 67% carbon, *bituminous* containing 88% carbon and *anthracite* containing about 95% carbon. Anthracite is the purest form of coal. Coal is used : (*i*) as a fuel in boilers, engines furnaces, (*ii*) in the manufacture of fuel gases such as coal gas, producer gas, water gas etc., (*iii*) in the manufacture of synthetic petrol, (*iv*) anthracite is used in the manufacture of graphite.

(6) Coke. It is obtained as a residue when destructive distillation of coal is done in cast iron retorts. It is the pure form of carbon containing about 80-95% carbon. Coke is used as a reducing agent in the metallurgical processes and also as a fuel.

(7) Gas carbon. It is a very pure form of carbon and is obtained as a deposit on the sides and roofs of the iron retorts in which the destructive distillation of coal is done. Gas carbon is very hard and is a good conductor of heat and electricity. It is used for making electrodes.

Allotropes of sulphur

Sulphur exists in a number of allotropic forms, which are (1) *crystalline sulphur* like rhombic or octahedral sulphur (α-sulphur) and monoclinic or prismatic sulphur (β-sulphur) (2) *liquid sulphur*, like λ and μ sulphur (3) *colloidal sulphur* (4) *plastic* or *γ-sulphur.* We will now discuss these forms one by one, as follows :

1. Rhombic and monoclinic sulphur. Both these forms exist in octa-atomic molecule of sulphur, S_8. S_8 molecule has a packered ring structure which contains 8-sulphur atom rings or zig-zag chains. In these rings or chains eight sulphur atoms are joined together by covalent bonds (See Fig. 16.29).

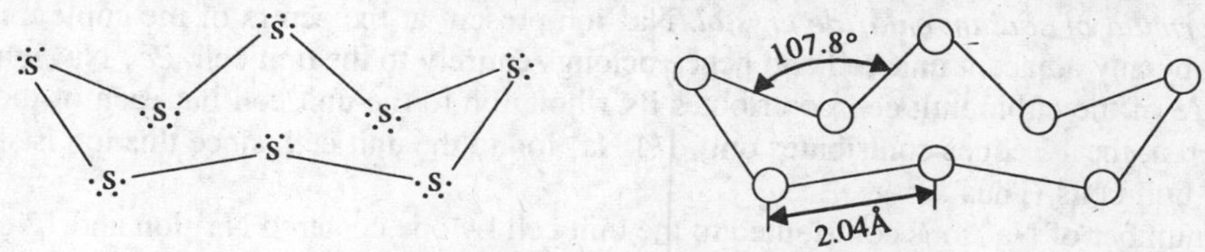

Fig. 16.29. Structure of sulphur S_8 molecule.

Such rings give molecular crystal in which the rings are held together by relatively weak polar forces. Consequently both the forms have low melting points. Rhombic sulphur is the stable form at room temperature. When it is heated at 96.0°, it is converted into monoclinic form. Rhombic sulphur has *octahedral shape* while monoclinic sulphur has long *needle like shape.*

2. S_λ and S_μ liquid allotropes of sulphur (Liquid sulphur). See "*Dynamic allotropy*".

3. Colloidal sulphur. Colloidal sulphur is frequently produced in chemical reactions involving sulphur compounds.

4. Plastic or γ-sulphur. When liquid sulphur is cooled rapidly, plastic sulphur or γ-sulphur is obtained. The nature of this form may vary since it depends on the proportion of λ-and μ-sulphur present in liquid sulphur.

Structure of AB type ionic crystals

Here we shall discusses the crystal structure of *sodium chloride* (rock salt, NaCl), *zinc blende* (ZnS), *wurtzite* (ZnS) and *Caesium chloride* (CsCl) crytals.

1. Structure of sodium chloride crystal (Also called *rock salt type structure*).

The cubic unit cell of NaCl crystal is shown in Fig. 16.30, in which the circles with negative

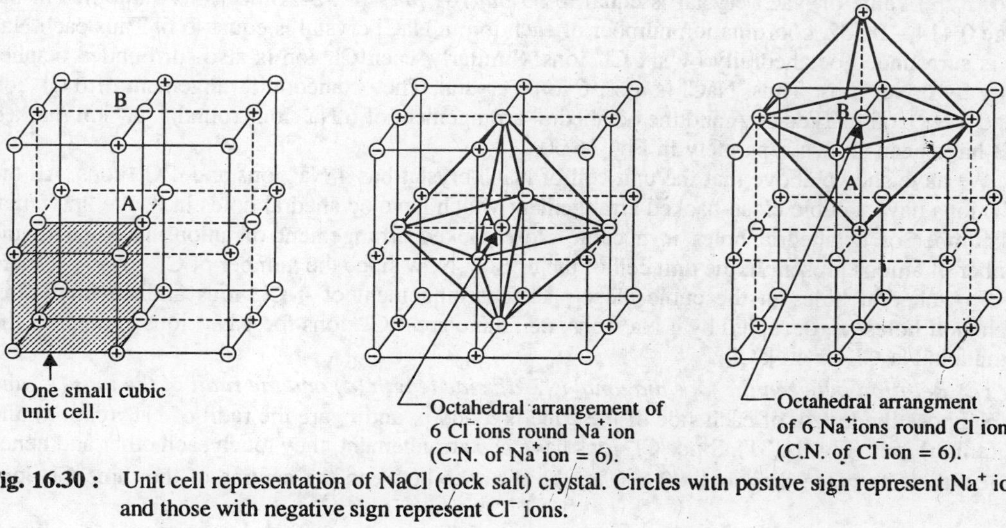

Fig. 16.30 : Unit cell representation of NaCl (rock salt) crystal. Circles with positve sign represent Na^+ ions and those with negative sign represent Cl^- ions.

sign represent Cl^- ions and the circles having positive sign represent Na^+ ions. The structure of NaCl crystal can be discussed under the following headings.

(*a*) *Arrangement of Na^+ and Cl^- ions.* In the unit cell of NaCl crystal, Cl^- ions have cubic close-packed (*ccp*) arrangement (ABC ABC.... sequence of layers), *i.e.*, in the cubic unit cell of NaCl crystal, Cl^- ions are present at all the eight corners of the cube and at the centre of each of the six faces. Thus there are $14Cl^-$ ions in the cubic unit cell. One Na^+ ion is present at the centre of the cubic unit cell and 12 Na^+ ions are present at the centre of 12 edges. Thus there are 13 Na^+ ions in the cubic unit cell. The above discussion shows that the cubic unit cell of NaCl crystal contains alternate Na^+ and Cl^- ions each of which has face-centered cubic arrangement. When this type of arrangement takes place, we say that Cl^- ions have *ccp* arrangement and Na^+ ions occupy all the available octahedral sites created by the *ccp* arrangement of Cl^- ions.

(*b*) *Formula of sodium chloride crystal.* Na^+ ion present at the centre of the cubic unit cell is not shared by any adjacent unit cell and hence belongs entirely to the unit cell, *i.e.,* Na^+ ion located at the centre of the cubic unit cell contributes its whole ion to the unit cell but each of the 12 Na^+ ions located at the 12 edges contributes only 1/4 Na^+ ion to the unit cell, since this ion is shared by 4 adjacent unit cells. Thus :

Total number of Na^+ ions contributed to the unit cell by one centered Na^+ ion and 12 edge Na^+ ions = $1 + 12 \times 1/4 = 1 + 3 = 4$.

The unit cell of sodium chloride crystal also contains 8 Cl^- ions at the eight corners (each of these 8 Cl^- ions is being shared by 8 surrounding unit cells) and 6 Cl^- ions at the centre of the six faces (each of these 6 Cl^- ions is being shared by 2 surrounding unit cells). Thus :

Total number of Cl^- ions contributed to the unit cell by 8 corner Cl^- ions and 6 Cl^- ions situated at the centre of the faces $= 8 \times 1/8 + 6 \times 1/2 = 1 + 3 = 4$.

Thus, we see that if the entire cubic unit cell of sodium chloride crystal is divided into eight small cubic unit cells, then each of these small cubic unit cells has 4 Na^+ ions and $4Cl^-$ ions which are present at the eight corners of the cube in an alternate manner, *i.e.,* each of the small cubic unit cell can be represented by the formula Na_4Cl_4 or NaCl. In other words we can say that each of the small cubic unit cell contains 4 molecules of NaCl (4 NaCl).

The above formula of sodium chloride crystal can also be deduced as follows: Since in the crystal of sodium chloride, Cl^- ions have *ccp* arrangement, each Cl^- ion is associated only with *one octahedral* hole and this hole is occupied by one Na^+ ion (cation). Thus there is only one Na^+ ion for one Cl^- ion in the formula of rock salt, *i.e.,* the formula of rock salt crystal is NaCl which shows that the stereo chemistry of the crystal is 1 : 1.

(*c*) *Coordination number of Na^+ and Cl^- ions and occupation of octahedral holes.* The radius ratio (r_+/r_-) value of NaCl crystal is equal to 95 *pm*/181 *pm* = 0.524. Since, this value lies in the range 0.414 – 0.732, coordination number of each ion in NaCl crystal is equal to 6. Thus each Na^+ ion is surrounded octahedrally by six Cl^- ions. Similarly, each Cl^- ion is also surrounded octahedrally by 6 Na^+ ions. Thus, NaCl is a 6 : 6 ionic crystal. The octahedral arrangement of 6 Cl^- ion round a Na^+ ion marked as A and the octahedral arrangement of 6 Na^+ ions round a Cl^- ion marked as B have been shown separately in Fig. 16.30.

We have shown above that the unit cell of NaCl crystal has 4 Na^+ ions and 4 Cl^- ions. All the 4 Cl^- ions have a cubic close-packed arrangement which have octahedral holes in it. We know that the number of octahedral holes in a cubic close-packed arrangement of anions is equal to the number of anions present in the unit cell of the crystal. Now since the number of Cl^- ions is 4, there are 4 octahedral holes in the cubic close-packed arrangement of 4 Cl^- ions and all there four octahedral holes are occupied by 4 Na^+ ions, *i.e.*, there are 4 Cl^- ions for 4 Na^+ ions and hence the formula is Na_4Cl_4 or 4 NaCl.

(*d*) *Relation between the face diagonal ($\sqrt{2}\,a$), side length (a) and the radii of the ions (r_+ and r_-).* Suppose the length of each side of the cube is *a* and r_+ and r_- are the radii of Na^+ (cation) and Cl^- (anion) ions (Fig. 16.31). Since Cl^- ions have *ccp* arrangement, they touch each other and hence the length of the face diagonal, $\sqrt{2}\,a$ (*i.e.*, the distance between the centres of the three Cl^- ions marked as 1, 2 and 3) is equal to the sum of $(r_-)_1$, $2(r_-)_2$ and $(r_-)_3$ where $(r_-)_1$, $(r_-)_2$ and $(r_-)_3$ are the radii of Cl^- ions designated as 1, 2 and 3 (See Fig. 16.31). Thus :

$$\text{Face diagonal} = \sqrt{2}\,a = (r_-)_1 + 2(r_-)_2 + (r_-)_3$$

Now, since, $(r_-)_1 = (r_-)_2 = (r_-)_3$, the above equation becomes

$$\text{Face diagonal} = \sqrt{2}\,a = 4r_-$$

The above relation shows that in NaCl ionic crystal (or in any ionic crystal having NaCl type structure), the length of face diagonal ($= \sqrt{2}\,a$) is equal to four times the radius of the anion.

It may be seen from Fig. 16.31 that since the ions namely 4 (anion), 5 (cation) and 6 (anion) touch each other, the length of the side of the cube (a) is given by :

$$\text{Length of the side} = a = (r_-)_4 + 2\,(r_+)_5 + (r_-)_6$$

or $$\text{Length of the side} = a = 2\,(r_+ + r_-)$$

This relation shows that the length of each side of the cubic unit cell (a) of NaCl ionic crystal is equal to double of the sum of the radii of the cation and anion.

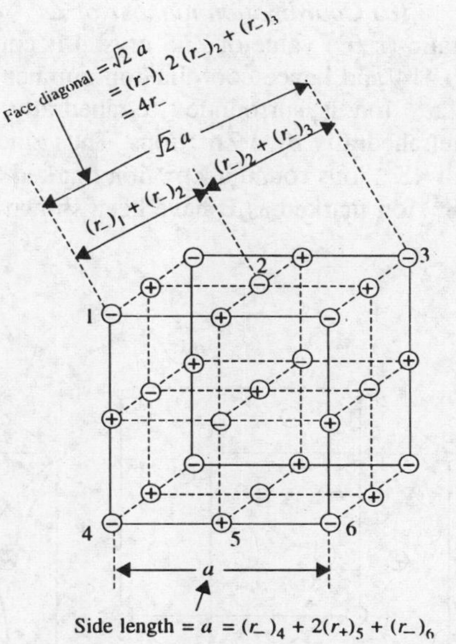

Fig. 16.31 : Relation between the face diagonal ($= \sqrt{2}\,a$), side length of the cube ($= a$) and radii of the ions (r_+ and r_-) in NaCl type ionic crystal.

Examples of ionic crystals, having NaCl type structure are halides of Li, Na, K and Rb (*e.g.*, LiCl, NaCl, KBr, RbI), AgCl, AgBr and oxides of alkaline earth metals (*e.g.*, BeO, MgO, CaO, SrO, BaO).

2. Structure of zinc blende (ZnS) crystal.

The cubic unit cell of zinc blende crystal is shown in Fig. 16.32 in which the circles with positive sign represent Zn^{2+} ions while those having negative sign represent S^{2-} ions. The structure of zinc blende crystal can be discussed as follows : (zinc blende is also called *sphalerite*).

(*a*) *Arrangement of Zn^{2+} and S^{2-} ions.* In the unit cell of zinc blende crystal, S^{2-} ions have cubic close-packed (*ccp*) arrangement, *i.e.*, S^{2-} ions are present at all the eight corners of the cube and at the centre of each of the six forces of the cubic unit cell. Thus there are 14 S^{2-} ions in the cubic unit cell.

The unit cell also contains 4 Zn^{2+} ions which are present within the unit cell. These ions (*i.e.*, Zn^{2+} ions) occupy half of the available tetrahedral sites created by the cubic close-packed arrangement of $4S^{2-}$ ions.

(*b*) *Formula of zinc blende crystal.* Since all the 4 Zn^{2+} ions are present within the unit cell, none of them is shared by any adjacent unit cell and hence each of these 4 Zn^{2+} ions belongs entirely to the unit cell, *i.e.*, all the 4 Zn^{2+} ions located within the unit cell contribute wholly to the unit cell. Thus total number of Zn^{2+} ions present in the unit cell is equal to 4.

The cubic unit cell of zinc blende also has 8 S^{2-} ions at its eight corners (each of these 8 S^{2-} ions is being shared by 8 adjacent unit cells) and 6 S^{2-} ions at the centre of the six faces (each of these 6 S^{2-} ions is being shared by 2 surrounding unit cells). Thus :

Total number of S^{2-} ions contributed to the unit cell by 8 corner S^{2-} ions and $6S^{2-}$ ions situated at the centre of the faces $= 8 \times 1/8 + 6 \times 1/2 = 1 + 3 = 4$.

Thus we see that each unit cell of zinc blende crystal contains 4 Zn^{2+} ions and 4 S^{2-} ions and hence the formula of this crystal is Zn_4S_4 or ZnS, *i.e.*, each unit cell of zinc blende crystal contains 4 molecules of ZnS.

The above formula of zinc blende crystal can also be deduced as follows. Since in the crystal of zinc blende, S^{2-} ions have cubic close-packed arrangement, each S^{2-} ion is associated with *two tetrahedral holes*. It is only one tetrahedral site which is occupied by *one* Zn^{2+} ion. Thus there is only one Zn^{2+} ion for one S^{2-} ion in the formula of zinc blende, *i.e.*, the formula of zinc blende crystal is ZnS which shows that the stereochemistry of the crystal is 1 : 1.

(*a*) *Coordination number of Zn^{2+} and S^{2-} ions and occupation of tetrahedral sites.* The radius ratio (r_+/r_-) value of ZnS crystal is equal to 74 *pm*/184 *pm* = 0.40 which lies in the range 0.225 – 0.414 and hence coordination number of each ion in zinc blende crystal is equal to 4. Thus each Zn^{2+} ion is surrounded tetrahedrally by $4S^{2-}$ ions. Similarly each S^{2-} ion is also surrounded tetrahedrally by 4 Zn^{2+} ions. Thus zinc blende is a 4 : 4 ionic crystal. The tetrahedral arrangement of $4S^{2-}$ ions round a Zn^{2+} ion marked as A and the tetrahedral arrangement of 4 Zn^{2+} ions round a S^{2-} ion marked as B have been shown separately in Fig. 16.32.

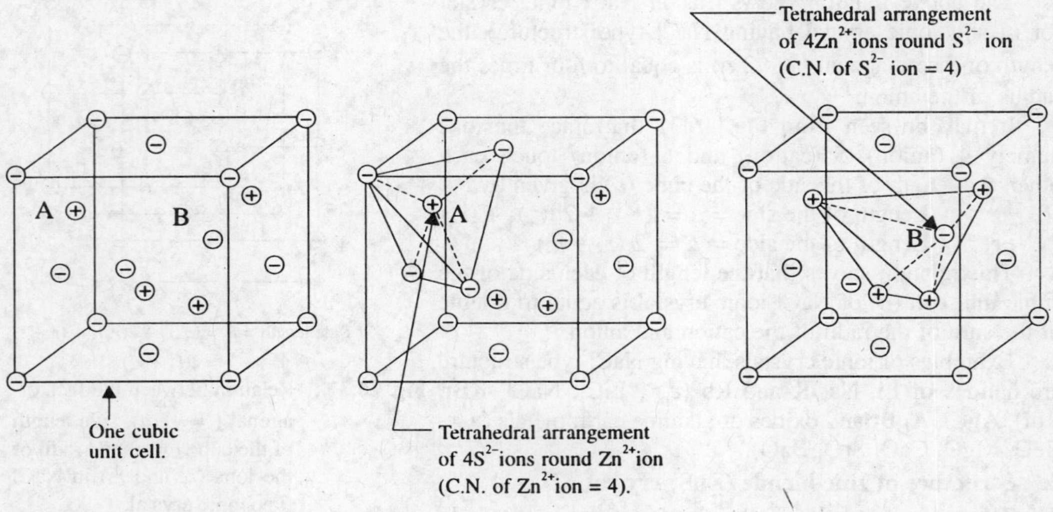

Fig. 16.32 : Unit cell representation of zinc blende (ZnS) crystal. Circles with positive sign represent Zn^{2+} ions while those with negative sign represent S^{2-} ions.

We have shown above that the unit cell of zinc blende has 4 Zn^{2+} ions and $4S^{2-}$ ions. All the $4S^{2-}$ ions have a cubic close-packed arrangement which has tetrahedral holes in it. We know that the number of tetrahedral holes in a cubic close-packed arrangement of anions is double the number of anions present in the unit cell of the crystal. Now since the number of S^{2-} ions is 4, there are 8 tetrahedral holes in the cubic close-packed arrangement of 4 S^{2-} ions. Since one Zn^{2+} ion can occupy only one tetrahedral hole, 4 Zn^{2+} ions occupy 4 tetrahedral holes only and the remaining 4 tetrahedral holes remain unoccupied, *i.e.*, out of 8 tetrahedral holes only 4 holes (alternate holes) are occupied by 4 Zn^{2+} ions. In other words we can say that Zn^{2+} ions occupy only half of the available tetrahedral holes which are created due to *ccp* arrangement of S^{2-} ions.

Examples of ionic crystals which have zinc blende type crystal structure are CuCl, CuBr, CuI, AlP, SiC, sulphides of Be, Cd and Hg (II), ZnO etc.

3. Structure of wurtzite (ZnS) crystal.

Zinc sulphide has two types of structures namely zinc blende and wurtzite. Zinc blende structure has already been explained. Here, we shall discuss the structure of wurtzite crystal. The unit cell of this crystal is shown in Fig. 16.33. From this figure the following points may be noted.

(*a*) *Packing of S^{2-} ions.* S^{2-} ions have hexagonal close-packed (*hcp*) arrangement, *i.e.* there is ABABA...sequence of layers. Hexagonal close-packed arrangement of S^{2-} ions produces tetrahedral sites in the arrangement. Each S^{2-} ion has two tetrahedral sites, one above and the other below it.

(*b*) *Occupation of tetrahedral sites.* We have said above that each S^{2-} ion has two tetrahedral sites in the lattice. It is only one of these two tetrahedral sites which is filled by one Zn^{2+} ion. Thus, only half of the total tetrahedral sites are filled with Zn^{2+} ions and hence the stereochemistry of wurtzite crystal is 1 : 1.

(*c*) *Coordination number of Zn^{2+} and S^{2-} ions.* Each Zn^{2+} ion is surrounded by $4S^{2-}$ ions which are disposed towards the four corners of a regular tetrahedron. Similarly each S^{2-} ion is also surrounded by $4Zn^{2+}$ ions. Thus the coordination number of each ion is 4 and hence wurtzite crystal is a 4 : 4 ionic crystal.

The radius ratio value is equal to 0.40 which suggests that coordination number is 4 and the wurtzite crystal has tetrahedral structure.

Examples of ionic crystals having wurtzite structure are AgI, ZnO, NH_4F, AlN etc.

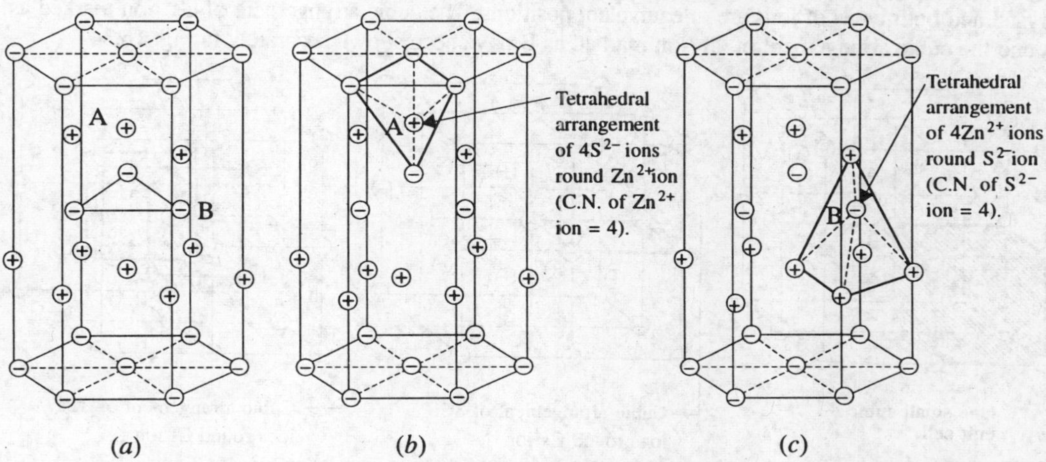

Fig. 16.33 : Crystal structure of wurtzite crystal (ZnS). Circles with positive sign represent Zn^{2+} ions while those with negative sign represent S^{2-} ions.

4. Structure of CsCl crystal

The cubic unit cell of the crystal of CsCl has been shown in Fig. 16.34. Circles with positive sign represent Cs^+ ions while the circles having negative sign represent Cl^- ions. The crystal structure of CsCl can be discussed under the following headings.

(a) Arrangement of Cs^+ and Cl^- ions. The unit cell of CsCl crystal contains 27 Cl^- ions and 8 Cs^+ ions. 8 Cl^- ions are present at the eight corners of the unit cell, 6 Cl^- ions are present at the centre of each of the six faces, 12 Cl^- ions are present at the centre of 12 edges and one Cl^- ion is located at the centre of the cubic unit cell. All the 8 Cs^+ ions are present within the unit cell.

(b) Formula of caesium chloride crystal and occupation of cubic interstitial sites. If the entire cubic unit cell of caesium chloride is divided into eight small cubic unit cells, then we may see that each of these small cubic unit cells has 8 Cl ions and one Cs^+ ion. 8 Cl^- ions are present at the eight corners of the cube (cubic arrangement of Cl^- ions) and Cs^+ ion is located at the centre of the cube. Since Cs^+ ion is placed in the body centre of the cube, it belongs entirely to the cube and hence contributes its whole ion to the cube. Since each of the 8 corner Cl^- ions is being shared by 8 adjacent cubes, 8 corner Cl^- ions contribute only $8 \times 1/8 = 1$ Cl^- ion to the cube. Thus each of the eight small cubes into which the whole unit cell is divided contains one Cs^+ ion and one Cl^- ion and hence the formula of caesium chloride is CsCl, *i.e.*, the stereochemistry of this compound is 1 : 1. Since Cl^- ions are present at the 8 corners of the cube, they are said to have cubic arrangement. Since Cs^+ ion is located at the centre of the cube, this ion is said to occupy the cubic hole created by the cubic arrangement of 8 Cl^- ions.

Some chemists say that since each small cube contains eight ions (8 Cl^- ions) at its eight corners and one ion (Cs^+ ion) at its body centre, this arrangement of ions (8 Cl^- ions and one Cs^+ ion) should be called *body-centered cubic arrangement*, but actual definition of body-centered cubic is that this arrangement should contain the *same ions* at the corners as well as at the body centre. Consequently, to call the arrangement of Cs^+ and Cl^- ions a body-centered cubic arrangement is misnomer.

(c) Coordination number of Cs^+ and Cl^- ions. The radius ratio (r_+/r_-) value for CsCl crystal is

equal to 160 *pm*/ 181 *pm* = 0.884. Since this value lies in the range 0.732 – 1,000, the coordination number of each ion in CsCl crystal is equal to 8. Thus each Cs^+ ion surrounded by 8 Cl^- ions which are disposed towards the eight corners of a cube. Similarly each Cl^- ion is also surrounded by 8 Cs^+ ions which are placed at the eight corners of a cube. Thus in the structure of CsCl crystal each ion is surrounded by eight oppositely-charged ions which have cubic arrangement. Thus CsCl is a 8 : 8 ionic crystal and both types of ions are in equivalent positions. The cubic arrangement of Cs^+ ion marked as A and the cubic arrangement of Cl^- ion marked as B have been shown separately in Fig. 16.34.

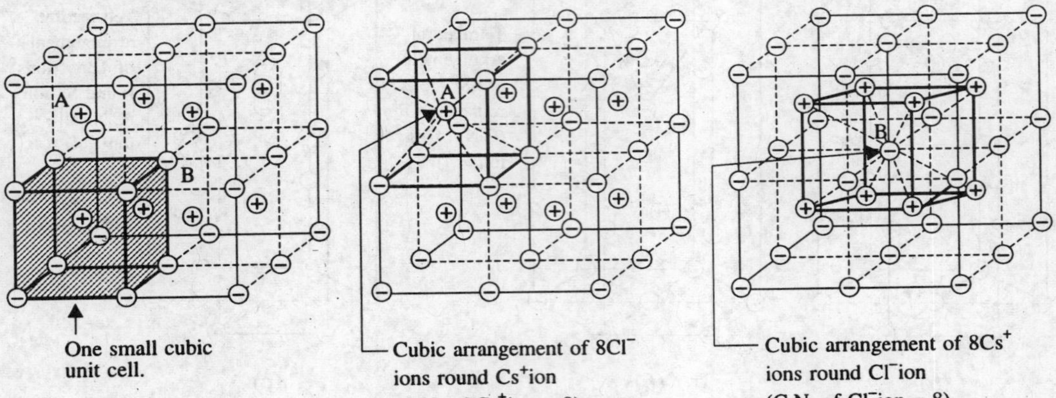

Fig. 16.34 : Unit cell representation of CsCl crystal. Circles with positive sign represent Cs^+ ions and those with negative sign represent Cl^- ions.

Examples of ionic crystals which have CsCl type crystal structure are CsBr, CsI, TlCl, TlBr etc.

Structure of AB_2 type ionic crystals

Here we shall discuss the crystal structures of *calcium fluoride* (CaF_2–called *fluorite structure*), *titanium dioxide* (TiO_2–called *rutile structure*) and *calcium carbide* (CaC_2).

1. Structure of calcium fluoride, CaF_2 crystal (*called fluorite type structure*)

The unit cell of CaF_2 crystal is shown in Fig. 16.35 in which Ca^{2+} and F^- ions have been represented by circles having positive sign and negative sign respectively. The salient features of the crystal structure of CaF_2 are as follows :

(*a*) *Arrangement of Ca^{2+} and F^- ions and occupation of tetrahedral sites.* Ca^{2+} ions have *ccp* arrangement, *i.e.*, Ca^{2+} ions are at all the eight corners and at all the six centres of the six faces of the cube. F^- ions occupy all the available tetrahedral sites.

(*b*) *Formula of calcium fluoride crystal.* Each of the Ca^{2+} ions has two tetrahedral sites and both these tetrahedral sites are occupied by two F^- ions. Thus there are two F^- ions for one Ca^{2+} ion, *i.e.*, the formula of calcium fluoride is CaF_2 which shows that the stereochemistry of the compound is 1 : 2.

(*c*) *Coordination number of Ca^{2+} and F^- ions.* The radius ratio (r_+/r_-) value of CaF_2 crystal is 99 *pm*/ 136 *pm* = 0.73. Since this value lies in the range 0.732–1.000, coordination number of Ca^{2+} ion in CaF_2 crystal is equal to 8.

Coordination number equal to 8 means that each Ca^{2+} ion (located at the body centre of the cube) is surrounded by 8 F^- ions which are placed at the 8 corners of the cube (cubic arrangement of 8F^- ions round Ca^{2+} ion). Each F^- ion (located at the centre of the cube) is surrounded by Ca^{2+} ions which are situated at the 4 corners of the tetrahedron (tetrahedral arrangement of 4 Ca^{2+} ions round F^- ion). Thus CaF_2 is a 8 : 4 ionic crystal. The cubic arrangement of 8F^- ions round Ca^{2+} ion marked as A and the tetrahedral arrangement of 4 Ca^{2+} ions round F^- ion marked as B are shown separately in Fig. 16.35.

BaF_2, SrF_2, PbF_2, HgF_2, CuF_2 also have CaF_2 type structure.

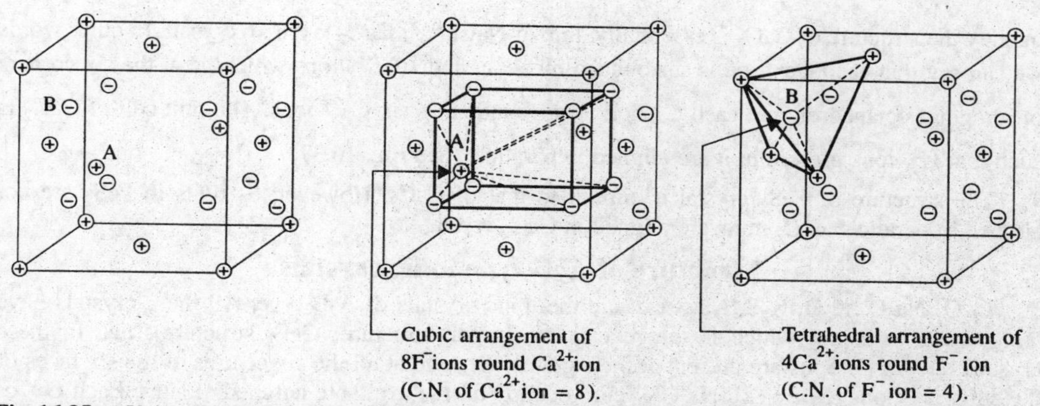

Fig. 16.35 : Unit cell representation of CaF_2 (fluorite structure) crystal circles with positive sign represent Ca^{2+} ions while those with negative sign represent F^- ions.

2. Structure of titanium dioxide (TiO_2) crystal (*Rutile structure*)

The unit cell of the crystal of TiO_2 has been shown in Fig.16.36. Circles with positive sign

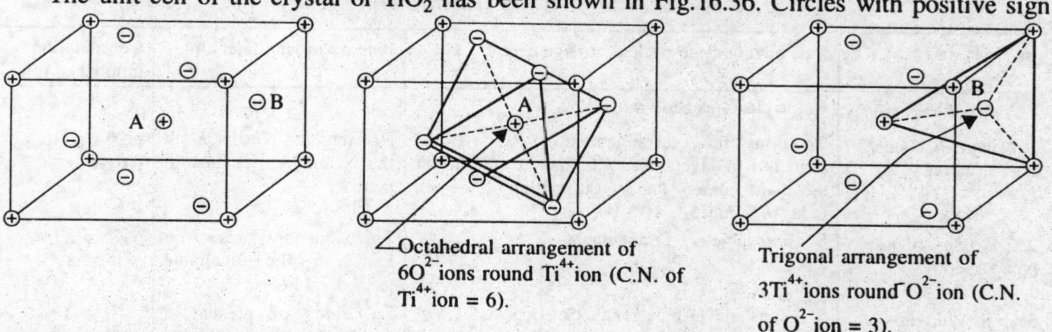

Fig. 16.36 : Crystal structure of titanium dioxide. TiO_2 (Rutile structure). Circles with positive sign represent Ti^{4+} ions and those with negative signs represent O^{2-} ions.

represent Ti^{4+} ions while the circles having negative sign represent oxide ions (O^{2-}). The crystal structure of TiO_2 molecule can be discussed under the following headings :

(*a*) *Arrangement of* Ti^{4+} *and* O^{2-} *ions in the unit cell.* Ti^{4+} ions are present at all the eight corners and at the body centre of a distorted cube. The cube is distorted because one of the axis of the cube is shorter than the other by 30%. Therefore, we can say that Ti^{4+} ions have a *body-centered distorted cubic arrangement.* Oxide ions (O^{2-} ions) occupy positions of *three fold coordination.*

(*b*) *Coordination numbers of* Ti^{4+} *and* O^{2-} *ions.* It may be seen from the figure that each Ti^{4+} ion is surrounded octahedrally by 6 O^{2-} ions and each O^{2-} ion is surrounded by three Ti^{4+} ions which are disposed towards the corners of an equilateral triangle. Thus the coordination numbers of Ti^{4+} and O^{2-} ions are 6 and 3 respectively, *i.e.*, TiO_2 crystal is a 6 : 3 ionic crystal.

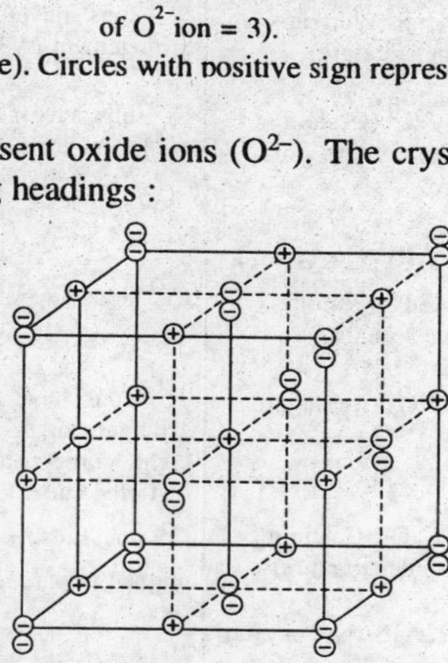

Fig. 16.37 : Structure of calcium carbide crystal (CaC_2). Pairs of circles having negative sign represent carbide ions (C_2^{2-} ions) and single circles with positive sign represent calcium ions (Ca^{2+} ions).

3. Structure of calcium carbide (CaC_2) and FeS_2 crystals

In CaC_2 crystal, the value of radius ratio, Ca^{2+}/C_2^{2-} has been found to be equal to 0.52 which is close to the value of radius ratio, Na^+/Cl^- in Na^+Cl^- crystal (= 0.524). Consequently the structure of CaC_2 is essentially that of NaCl, *i.e.*, CaC_2 is a 6 : 6 crystal. In other words we can say that each Ca^{2+} ion is surrounded by six carbide (C_2^{2-}) ions which lie at the six corners of a regular octahedron and each C_2^{2-} ion is surrounded by six Ca^{2+} ions. The unit cell of CaC_2 is such that C_2^{2-} ions present in it are aligned in parallels (See Fig. 16.37).

The structure of FeS_2 crystal is similar to that of CaC_2. However, S_2 units in FeS_2 are not aligned in parallels as C_2 units are aligned in CaC_2 crystal.

Structure of A_2B type ionic crystals

Li_2O, Na_2O, K_2O, K_2S, Na_2S etc. are the ionic crystals of A_2B type. All these crystals have *anti-fluorite structure* which is the reverse of fluorite structure (CaF_2 structure), *i.e.*, in these crystals, O^{2-} or S^{2-} ions are present at the eight corners and at all the six centres of the six faces of the cube. M^+ ions ($M^+ = Li^+, Na^+, K^+$ etc.) ocuppy all the available tetrahedral sites. Each O^{2-} or S^{2-} ion is surrounded by 8 M^+ ions while each M^+ ion is surrounded by $4O^{2-}$ or $4S^{2-}$ ions. Thus the coordination number of M^+ and O^{2-} ion (or S^{2-} ion) is 4 and 8 respectively and hence each of the given crystals is a 4 : 8 ionic crystal.

Table 16.10 : *Summary of the structure of ionic crystals.*

Ionic crystals	Ions having close-packed arrangement	Ions occupying interstitial sites	Coordination number of ions
I. AB type crystals	Cubic close-packed (*ccp*)		
1. NaCl (Rock salt type structure)	Cl^- ions have arrangement (ABC ABC.... sequence of layers) *i.e.*, Cl^- ions are present at the eight corners and at the centre of the six faces of a cube.	Na^+ ions are present in all the available octahedral holes.	$Na^+ = 6$ $Cl^- = 6$
2. ZnS (Zinc blende type structure)	S^{2-} ions have *ccp* arrangement	Zn^{2+} ions are present in half of the available tetrahderal holes.	$Zn^{2+} = 4$ $S^{2-} = 4$
3. ZnS (Wurtzite type structure)	S^{2-} ions have hexagonal close-packed (*hcp*) arrangement (ABABA.... sequence of layers).	Zn^{2+} ions are present in half of the available tetrahderal sites.	$Zn^{2+} = 4$ $S^{2-} = 4$
4. CsCl (Cesium chloride type structure)	Cl^- ions have cubic arrangement, *i.e.*, Cl^- ions are located at the eight corners of al simple cube.	Cs^+ ion is present in the cubic hole, *i.e.*, Cs^+ ion is located at the centre of the cube.	$Cs^+ = 8$ $Cl^- = 8$
II. AB_2 type crystal			
1. CaF_2 (Fluorite type structure)	Ca^{2+} ions have *ccp* arrangement.	F^- ions are present in all the available tetrahedral holes.	$Ca^{2+} = 8$ $F^- = 4$
2. TiO_2 (Rutile structure)	Ti^{4+} ions have a body-centred distorted cubic arrangement, *i.e.*, Ti^{4+} ions are present at the eight corners and at the body centre of a distorted cube	O^{2-} ions occupy positions of three-fold coordination.	$Ti^{4+} = 6$ $O^{2-} = 3$
3. CaC_2 (Calcium carbide structure)	C_2^{2-} ions have *ccp* arrangement. These ions are aligned in parallels.	Ca^{2+} ions occupy all the available octahedral holes.	$Ca^{2+} = 6$ $C_2^{2-} = 6$
III. A_2B type crystal			
Na_2O (anti-fluorite type structure)	O^{2-} ions have *ccp* arrangement.	Na^+ ions are present in all the available tetrahederal holes.	$Na^+ = 4$ $O^{2-} = 8$

Lattice Energy

What is lattice energy ?

We know that the cations and anions which constitute an ionic lattice (or solid) are considered to be hard spheres. When a large number of cations and anions are brought nearer to each other to form an ionic lattice, they arrange themselves in a regular pattern to attain a close-packed type structure. In the process of attaining this type of structure, energy is released and hence the system gets *stabilised* and ionic crystal is formed. The magnitude of energy released is called lattice energy of the ionic lattice or solid. Thus the *lattice energy of an ionic solid can be defined as the amount of energy released when one mole of the gaseous cations and one mole of the gaseous anions are brought closer to each other from an infinite distance to combine together to form one mole of the ionic crystal or solid.* Lattice energy is represented as H_{lat}. Since H_{lat} is the energy released, it is denoted with a negative sign. Thus, for Na^+Cl^- (*s*) ionic crystal $(H_{lat})_{NaCl(s)}$ is defined as :

$$Na^+ (g) + Cl^- (g) \longrightarrow Na^+ Cl^- (s), \text{ Energy released} = -(H_{lat})_{NaCl\,(s)}$$

Due to the release of energy, the formation of $Na^+ Cl^-$ (*s*) by the combination Na^+ (*g*) and Cl^- (*g*) is an *exothermic process.*

It follows from the above discussion that when one mole of $Na^+ Cl^-$ (*s*) ionic crystal is broken up into its constituent gaseous ions *viz.*, Na^+ (*g*) and Cl^- (*g*), energy will be required. The energy required in the above reaction has been found to be numerically equal to the lattice energy of $Na^+Cl^-(s)$. Thus lattice energy of an ionic crystal can also be defined *as the amount of energy needed to break one mole of the ionic solid into its constituent gaseous ions.* Thus :

$$Na^+ Cl^- (s) \longrightarrow Na^+ (g) + Cl^- (g)\ ; \quad \text{Energy required} = + (H_{lat})_{NaCl\,(s)}$$

Heat of formation of a compound (Also called formation energy).

Heat of formation of a compound is the amount of energy released or absorbed when one molecule of that compound is formed by the combination of its constituent atoms. Heat of formation is denoted as H_{for}. H_{for} released is shown with a negative sign while H_{for} absorbed is indicated with a positive sign.

Examples. (*a*) When one molecule of sodium chloride, NaCl (*s*) is formed by the combination of Na(*s*) and $\frac{1}{2}Cl_2$ (*g*), energy equal to 118.2 Kcal mol^{-1} is *released.* Thus the heat of formation of NaCl (*s*) is 118.2 Kcal mol^{-1}

$$Na\ (s) + 1/2\ Cl_2\ (g) \longrightarrow NaCl\ (s)\ ; \text{ Energy released} = -(H_{for})_{NaCl(s)} = -118.2 \text{ Kcal mol}^{-1}$$

(*b*) When one molecule of $NaCl_2$ (s) is formed by the combination of Na (*s*) and Cl_2 (g), energy equal to 501.7 Kcal mol^{-1} is *absorbed.* Thus the heat of formation of $NaCl_2$ is + 501.7 Kcal mol^{-1}.

$$Na(s) + Cl_2\ (g) \longrightarrow NaCl_2\ (s)\ ; \quad \text{Energy absorbed} = +(H_{for})_{NaCl_2\,(s)} = +501.7 \text{ Kcal mol}^{-1}$$

Relation between the stability and heat of formation of compounds

A compound having negative value for its heat of formation is stable while that having positive value is unstable. Thus, since NaCl (*s*) has negative value for its heat of formation (= – 118.2 Kcal mol^{-1}) and $NaCl_2(s)$ has positive value (= + 501.7 Kcal mol^{-1}), the former is stable while the latter is unstable.

Comparison between lattice energy and heat of formation

Lattice energy of an ionic compound is the energy released (negative sign) when one molecule of the ionic compound in the solid state is formed by the combination of its constituent ions in the gaseous state, while heat of formation of a compound is defined as the energy released (negative sign) or absorbed (positive sign) when one molecule of that compound (in any state) is formed by the combination of the constituent elements (in any state).

Different types of interactions taking place between the ions in M^{z+} X^{z-} ionic crystal

We know that M^{z+} X^{z-} ionic crystal contains numberless M^{z+} and X^{z-} ions which are arranged in an orderly manner. In this ionic crystal the following attractive and repulsive interactions between the ions take place :

(*i*) *Cation-anion attractive interactions.* These interactions take place between M^{z+} ion (cation) and all X^{z-} ions (anions) surrounding the cation. This type of interaction gives *attractive* (electrostatic) *potential energy* which is denoted as $(PE)_{att}$. The magnitude of this energy is given by :

$$(PE)_{att} = \frac{(z^+ e) \times x \times (-z^- e)}{r} = \frac{-x\, z^+ z^- e^2}{r}$$

Here z^+ and z^- = Number of charges on M^{z+} and X^{z-} ions respectively, e = charge on an electron (= 4.8×10^{-10} *esu*), r = inter-ionic distance between the ions, and x = number of X^{z-} ions (anions) surrounding a single M^{z+} ion (cation).

(*ii*) ***Cation-cation repulsive interactions.*** This type of interaction takes place between M^{z+} ion (cation) and all the remaining M^{z-} ions (cations) surrounding the single cation. This interaction gives *repulsive potential energy*, $(PE)_{rep}$ whose magnitude is given by :

$$(PE)_{rep} = \frac{(z^+ e) \times x \times (z^+ e)}{r} = \frac{x\,(z^+)^2 e^2}{r}$$

Here x is the number of M^{z+} ions surrounding a single M^{z+} ion.

(*iii*) ***Anion-cation attractive interactions.*** These interactions take place between X^{z-} ion (anion) and all M^{z+} ions (cations).

(*iv*) ***Anion-anion repulsive interactions.*** These interactions occur between X^{z-} ion (anion) and all the remaining X^{z-} ions (anions) surrounding the single anion.

(*v*) ***Repulsion between electron clouds of M^{z+} and X^{z-} ions.*** When M^{z+} and X^{z-} ions are brought nearer to each other from an infinite distance, they repel each other due to the overlap of their similarly-charged electron clouds. This repulsion gives repulsive potential energy, $(PE)_{rep}$. A number of expressions have been suggested for the magnitude of $(PE)_{rep}$. According to one expression given by Born and Lande, $(PE)_{rep}$ is inversely proportional to r^n. Thus, according to Born and Lande, $(PE)_{rep}$ is given by :

$$(PE)_{rep} = \frac{be^2}{r^n}$$

Here b is a proportionality constant which is called *Born Coefficient* or *repulsion coefficient.* This constant expresses the repulsion between the ions. n is another constant which is called *Born exponent.* The value of n may be obtained from compressibility K, since we are measuring the resistance between the ions when they are forced to come closer to each other.

$$K = \frac{18\, r^4}{Ae^2\,(n-1)}$$

The value of n for a given ion depends on its configuration or on the number of electrons present in it. The value of n for the ions having He-type (electrons = 2, *e.g.*, Li^+, Be^{2+}), Ne-type (electrons = 10, *e.g.*, Na^+, Mg^{2+}, O^{2-}, F^-), Ar-type (electrons = 18, *e.g.*, K^+, Ca^{2+}, S^{2-}, Cl^-, Cu^+), Kr-type (electrons = 36, *e.g.*, Rb^+, Br^-, Ag^+) and Xe-type (electrons = 54, *e.g.* Ca^{2+}, I^-, Au^+) configuration is 5, 7, 9, 10 and 12 respectively. The value of n for a given ion increases with the increase in electron density around it.

n-value for an ionic crystal is the average of n-values of the two constituent ions of which the ionic crystal is composed. For example, n-value for NaCl ionic crystal = 1/2 (n-value of Na^+ ion + n-value of Cl^- ion) = 1/2 (7 + 9) = 8. The value of n for an ionic crystal is always more than one.

Since repulsive potential energy which arises due to the overlap of electron clouds of ions varies inversely as r^n, it is enough to consider the overlap of the electron clouds of the nearest ions only in a given ionic crystal. For example since in the crystal of NaCl each Na^+ ion is surrounded by the six nearest

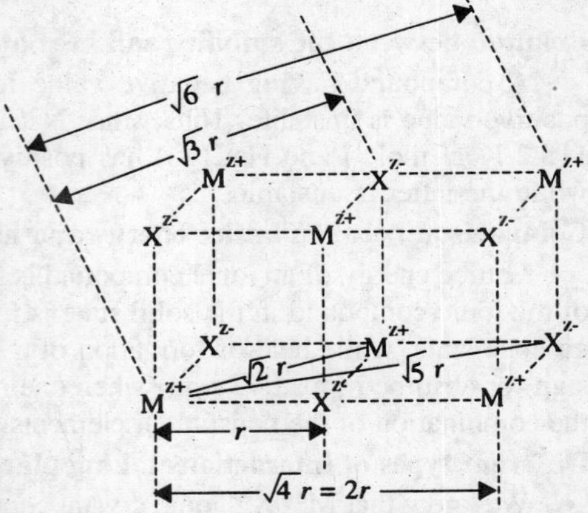

Fig. 16.38 : Inter ionic distance in $M^{z+} X^{z-}$ ionic crystal which has NaCl crystal structure. r is the length of the side of a cube.

Cl^- ions, the repulsive potential energy due to the overlap of the electron clouds of one Na^+ ion and $6Cl^-$ ions is given by :

$$(PE)_{rep} = \frac{6\,be^2}{r^n} = \frac{B}{r^n}$$

In the above equation $B = 6\,be^2$, *i.e.*, B is the product of the number of the nearest Cl^- neighbours of Na^+ ion and be^2.

To calculate the potential energy of a pair of ions separated by a distance of *r* within an ionic crystal ($M^{z+}X^{z-}$) having NaCl type crystal structure

By considering the interactions between M^{z+} and X^{z-} ions and the overlap between their electron clouds, we can calculate the potential energy of a pair of ions separated by a distance of *r* within the ionic crystal, $M^{z+}X^{z-}$ which has NaCl type crystal structure. This calculation involves the following steps :

(*i*) *To calculate potential energy when M^{z+} cation interacts with all the remaining M^{z+} cations and all X^{z-} anions.* Since $M^{z+}X^{z-}$ ionic crystal has Na^+Cl^- crystal structure, each M^{z+} cation is surrounded :

(*a*) by six X^{z-} anions at a distance of $\sqrt{1}\,r = r$, where *r* is the distance between M^{z+} cation and its nearest X^{z-} anions (See Fig. 16.38 and Fig. 16.39). The interaction of M^{z+} ion with 6 X^{z-} ions gives *attractive potential energy*, $(PE)_{att}$ whose magnitude is given by :

$$(PE)_{att} = \frac{-6z^+z^-e^2}{\sqrt{1}\,r} \quad ...(a)$$

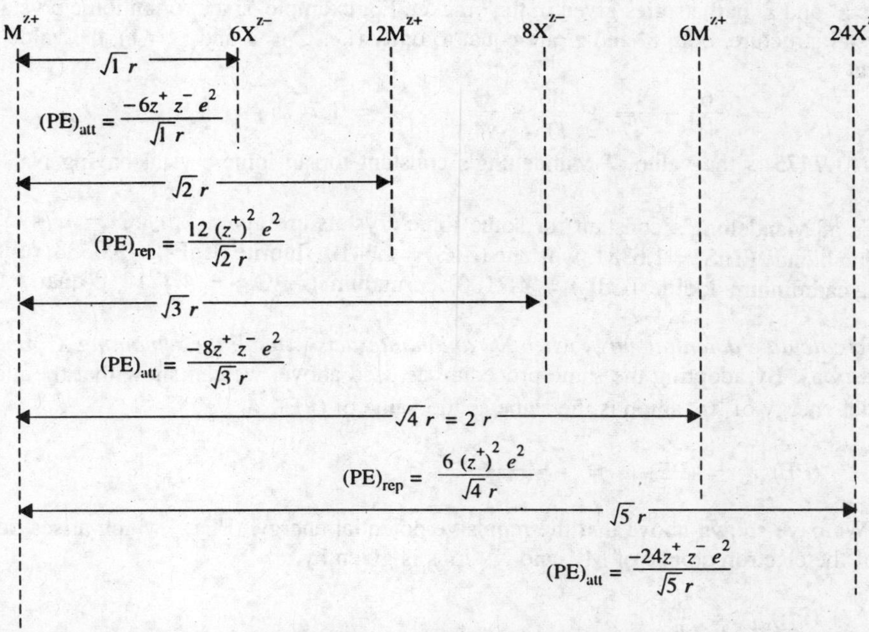

Fig. 16.39 : To calculate the potential energy when M^{z+} cation interacts with all the remaining M^{z+} cations and all X^{z-} anions.

(*b*) by 12 M^{z+} ions at a distance of $\sqrt{2}\,r$. The interaction between M^{z+} ion with 12 M^{z+} ions gives *repulsive potential energy*, $(PE)_{rep}$ whose magnitude is give by :

$$(PE)_{rep} = \frac{12\,(z^+)^2e^2}{\sqrt{2}\,r} \quad ...(b)$$

(*c*) by 8 X^{z-} ions at a distance of $\sqrt{3}\,r$. The interaction between M^{z+} ion and 8 X^{z-} ions gives *attractive potential energy*, $(PE)_{att}$ whose value is given by :

$$(PE)_{att} = \frac{-8z^+ z^- e^2}{\sqrt{3}\, r} \quad ...(c)$$

(*d*) by six M^{z+} ions which are placed at a distance of $\sqrt{4}\, r$ from M^{z+} ion. The interaction between M^{z+} ion with 6 M^{z+} ions gives *repulsive potential energy*, $(PE)_{rep}$ whose magnitude is given by :

$$(PE)_{rep} = \frac{6\,(z^+)^2\, e^2}{\sqrt{4}\, r} \quad ...(d)$$

When all the energy terms given by equations (*a*), (*b*), (*c*), (*d*) etc. are added, the sum thus obtained gives us the potential energy of M^{z+} cation, $(PE)_{M^{z+}}$ when the interaction of this cation (*i.e.*, M^{z+} cation) with 6 X^{z-}, 12 M^{z+}, 8 X^{z-}, 6 M^{z+} etc,. ions is considered. Thus :

$$(PE)_{M^{z+}} = \frac{-6\, z^+ z^- e^2}{\sqrt{1}\, r} + \frac{12\,(z^+)^2 e^2}{\sqrt{2}\, r} - \frac{8\, z^+ z^- e^2}{\sqrt{3}\, r} + \frac{6\,(z^+)^2 e^2}{\sqrt{4}\, r} - \ldots$$

$$= \frac{-z^+ z^- e^2}{r}\left(\frac{6}{\sqrt{1}} - \frac{12\, z^+}{\sqrt{2}\, z^-} + \frac{8}{\sqrt{3}} - \frac{6\, z^+}{\sqrt{4}\, z^-} + \ldots\right)$$

$$\text{or} \quad (PE)_{M^{z+}} = -\frac{z^+ z^- e^2 A}{r} \quad ...(i)$$

In equation (*i*) A has been used for the sum of all the terms given in the bracket above. A is called *Mandelung's constant*. The value of A for a given ionic crystal can be calculated by putting the values z^+ and z^- in the series given in the bracket. For example, since for an ionic crystal having NaCl crystal structure, both z^+ and z^- are equal to unity (*i.e.*, $z^+ = 1$ and $z^- = 1$), the value of A for this crystal

$$= \frac{6}{\sqrt{1}} - \frac{12}{\sqrt{2}} + \frac{8}{\sqrt{3}} - \frac{6}{\sqrt{4}} + \ldots = 1.7475$$

Thus, 1.7475 is the value of Mandelung's constant for an ionic crystal having NaCl crystal structure.

Values of Mandelung's constant for some ionic crystals are given as NaCl = 1.7475, CsCl = 1.7627, zinc blende (ZnS) = 1.6381, wurtzite (ZnS) = 1.6410, fluorite (CaF_2) = 5.0388, rulite (TiO_2) = 4.8160, cadminium iodide (CdI_2) = 4.7100, corundum (Al_2O_3) = 4.1718 β-quartz (SiO_2) = 4.4393.

(*ii*) *To calculate potential enrgy when X^{z-} anion interacts with all the remaining X^{z-} anions and all M^{z+} cations.* By adopting the same procedure as used above, we can show that the magnitude of potential energy of X^{z-} anion is the same as the value of $(PE)_{M^{z+}}$, *i.e.*,

$$(PE)_{X^{z-}} = (PE)_{M^{z+}} = -\frac{z^+ z^- e^2 A}{r} \quad ...(ii)$$

(*iii*) We have shown above that the repulsive potential energy, $(PE)_{rep}$ which arises due to the overlap of the electron clouds of M^{z+} and X^{z-} ions is given by :

$$(PE)_{rep} = \frac{B}{r^n} \quad ...(iii)$$

Thus total potential energy $[(PE)_{total}]$ of a pair of ions in $M^{z+} X^{z-}$ ionic crystal which has NaCl crystal structure is given by :

$$(PE)_{total} = 1/2\ [(PE)_{M^{z+}} + (PE)_{X^{z-}}] + (PE)_{rep}$$

$$= \frac{1}{2}\left[-\frac{z^+ z^- e^2 A}{r} - \frac{z^+ z^- e^2 A}{r}\right] + \frac{B}{r^n}$$

$$\text{or} \quad (PE)_{total} = \left[-\frac{z^+ z^- e^2 A}{r} + \frac{B}{r^n}\right] \text{ per ion pair} \quad ...(iv)$$

The factor 1/2 has been used since the addition of $(PE)_{M^{z+}}$ and $(PE)_{X^{z-}}$ takes the interaction in account twice. Equation (*iv*) gives in the potential energy of an ion-pair separated by a distance of *r*.

Estimation of the value of B in equation (iv). We known that at equilibrium distance between the ions in the crystal (r_0) the attractive and repulsive forces acting in the crystal are equal to each other and hence $(PE)_{total}$ becomes minimum at $r = r_0$, *i.e.*, when $r = r_0$, $d(PE)_{total}/dr = 0$. Thus :

$$\left[\frac{(PE)_{total}}{dr}\right]_{r=r_0} = \frac{d\left[-\frac{z^+ z^- e^2 A}{r} + \frac{B}{r^n}\right]}{dr} = 0$$

$$= \frac{z^+ z^- e^2 A}{r_0^2} - \frac{nB}{r^{n+1}} = 0$$

or $$\frac{z^+ z^- e^2 A}{r_0^2} = \frac{nB}{r^{n+1}}$$

$$B = \frac{z^+ z^- e^2 A\, r_0^{\,n-1}}{n} \quad ...(v)$$

Now put the value of B from equation (*v*) and $r = r_0$ in equation (*iv*) to get :

$$\left[(PE)_{total}\right]_{r=r_0} = -\frac{z^+ z^- e^2 A}{r_0} + \frac{z^+ z^- e^2 A\, r_0^{n-1}}{n \cdot r_0^n} = -\frac{z^+ z^- e^2 A}{r_0}\left(1 - \frac{1}{n}\right)$$

or $$\left[(PE)_{total}\right]_{r=r_0} = \frac{z^+ z^- e^2 A}{r_0}\left(\frac{1}{n} - 1\right) \text{ per ion pair (in cgs units)} \quad ...(vi)$$

To derive Born-Lande equation for the lattice energy of M^{z+} X^{z-} ionic crystal having NaCl crystal structure

Since the given ionic crystal has NaCl type crystal structure, the value of potential energy of a pair of ions placed at an equilibrium distance, r_0 from each other (at equlibrium distance the attractive and repulsive forces in the crystal are the same) is given by :

$$(PE)_{r=r_0} = \frac{z^+ z^- e^2 A}{r_0}\left(\frac{1}{n} - 1\right) \text{ per ion pair (in } cgs \text{ units)}$$

Again we know that lattice energy (U) is defined as the amount of energy released when one mole of an ionic crystal is formed by combining the constituent gaseous cations and gaseous anions separated from each other by an infinite distance. Thus if *N* is the Avogadro's number (N), then U is given by :

$$U = (PE)_{r=r_0} \times N$$

or $$U = \frac{z^+ z^- e^2 AN}{r_0}\left(\frac{1}{n} - 1\right) \text{ per ion pair (in } cgs \text{ units)}$$

Since *n* is always greater than 1, the value of U as given by the above equation is negative. The above equation is called *Born-Lande equation.*

Factors affecting the magnitude of lattice energy of ionic solids

We have shown above that lattice energy of an ionic solid is given by Born-Lande equation which is given above. This equation shows that the lattice energy of ionic solid depends on the following factors :

(*i*) *Charge on the cation (z^+) or on the anion (z^-) or the product of charges ($z^+ z^-$).* Greater is the charge on the cation (z^+) or on the anion (z^-) or greater is the product of electronic charges on the ions ($z^+ z^-$), greater is the magnitude of lattice energy (U). Thus the lattice energies of ionic solids containing monovalent-monovalent ions, divalent-monovalent ions and divalent-divalent ions are in the increasing order, *e.g.*, (values of lattice enrgies are in kJ/mole).

Na^+F^-	<	$Mg^{2+}F_2^-$	<	$Mg^{2+}O^{2-}$
914.2		2882.0		3895.0
K^+F^-	<	$Ca^{2+}F_2^-$	<	$Ca^{2+}O^{2-}$
812.1		2581.0		3520.0

(ii) Inter-ionic distance (r_0) between the ions or the size of the cation (r_+) or of the anion (r_-). Born-Lande equation shows that the lattice energy is inversely proportional to the inter-ionic distance (r_0) between the ions. Since r_0 is equal to the sum of r_+ and r_-, smaller is the size of the ions, smaller would be the value of r_0 and hence higher would be the magnitude of lattice energy. Thus smaller is the size of a cation for a given anion or of an anion for a given cation, smaller is the valur of inter-nuclear distance (*d*) and hence greater is the magnitude of force of attraction existing between the ions [See relations (*ii*) and (*iii*) given above] and consequently greater is the magnitude of lattice energy.

Examples. (*a*) Since the size of monovalent alkali metal cations (M^+) *increases* on moving down the group from Li^+ to Cs^+, the lattice energy of their fluorides (or of any other halides) *decreases* in the order LiF > NaF > KF > RbF > CsF.

Size of M^+ ions : $Li^+ < Na^+ < K^+ < Rb^+ < Cs^+$

——— Increasing ———→

Lattice energy of fluorides (kJ/mole)/of alkali metals : 1034 > 914.2> 812.1 > 780.3 > 743.9

——— Decreasing ———→

(*b*) Since the size of monovalent halide ions (X^-) *increases* on moving down the group from F^- to I^-, the lattice energy of the halides of a given alkali metal (say) *decreases* in the order LiF > LiCl > LiBr > LiI.

Size of X^+ ions : $F^- < Cl^- < Br^- < I^-$

——— Increasing ———→

Lattice energy of fluorides of Li (kJ/mole) : 1034 > 840.1> 781.2 > 718.2

——— Decreasing ———→

The study made at (*a*) and (*b*) above, shows that in the series of a particular halide of various alkali metals (*e.g.*, Li F, NaF, KF, RbF, CsF) or in the series of halides of a given alkali metals (*e.g.*, LiF, LiCl, LiBr, LiI), the lattice energy of the halides decreases with the increase in internuclear distance ($M^+ - X^-$ distance). The lattice energies of salts of alkali metals, having a given anion, decreases with the increase in atomic number of the alkali metals.

(*c*) The lattice energies of the oxides of alkaline earth metals also decrease on descending the group, since the size of Be^{2+}, Mg^{2+} etc., ions increase in the same direction. Thus :

Oxides :	BeO	MgO	CaO	SrO	BaO
Lattice energies of oxides (KJ/mole) :	4541	3895	3520	3325	3108
	——— Decreasing ———→				
Size of M^{2+} ions (Å) :	Be^{2+} 3.1	Mg^{2+} 6.5	Ca^{2+} 9.9	Sr^{2+} 1.13	Ba^{2+} 1.35
	——— Increasing ———→				

(*d*) Since the monovalent cations of alkali metals (M^+ ions) are larger in size than the divalent cations of alkaline earth metals (M^{2+} ions), the lattice energies of the salts (like fluorides) of M^+

ions are lower in magnitude than those of the corresponding salts of M^{2+} ions, *e.g.*, (values of lattice energies given are in kJ/mole).

NaF	<	MgF_2	;	KF	<	CaF_2
914.2		2882.0	;	812.1		2581.0

Conclusion. The study of the above two factors clearly show that in order to get high value of lattice energy, the size of the cation and (or) anion should be small and electric charges on the cation/anion should be high.

(*iii*) *Mandelung constant (A).* The magnitude of lattice energy is directly proportional to the value of A which depends on the coordination number of each ion and geometric arrangement of ions in the crystal lattice of the crystal.

Lattice energy of the sulphates, chlorides nitrates etc., of alkaline earth metals

Since the size of the divalent cations of alkaline earth metals increase as we move down the group from Be^{2+} to Ba^{2+} ions, the lattice energy of their sulphates, chlorides, nitrates etc., should decrease in the same direction. But it has been observed that their lattice energies are almost the same. The similarity in their lattice energy values is due to the fact that the anions *viz*., sulphates, chlorides, nitrates etc. present in these salts are very large in size.

To determine the magnitude of energy released in the reaction, Na(*s*) + 1/2 Cl_2 (*g*) → NaCl (*s*) by Born-Haber cycle : Determination of lattice energy of NaCl (*s*)

When sodium atom in the solid state, Na(*s*) combines with chlorine in the gaseous state, 1/2 Cl_2 (*g*), NaCl in the solid state, NaCl (*s*) is obtained and in this process, some energy is also released. This energy is called *heat of formation (or formation energy)* of NaCl and is denoted by $(H_{for})_{NaCl\,(s)}$. The release of energy is indicated by putting a negative sign before it (*Exothermic reaction*). Thus :

(A) $Na\,(s) + 1/2\,Cl_2\,(g) \rightarrow NaCl\,(s)$; Heat released = $-(H_{for})_{NaCl\,(s)}$

The value of $(H_{for})_{NaCl\,(s)}$ can be calculated, if it is realised that the formation of NaCl (*s*) as represented by equation (A) can also take place through the following five steps from (*a*) to (*e*).

(*a*) Solid sodium atom, Na(*s*) *absorbs* energy equal to its sublimation energy, $(H_s)_{Na\,(s)}$ and gets changed into gaseous sodium atom, Na(*g*). *(Endothermic reaction).*

$Na\,(s) \rightarrow Na\,(g)$; Energy absorbed = $+(H_s)_{Na\,(s)}$

(*b*) Gaseous sodium atom Na (*g*) obtained at (*a*) absorbs energy (*endothermic reaction*) equal to its ionisation energy, $(IE)_{Na\,(g)}$ and is converted into gaseous sodium cation, Na^+ (*g*). In this reaction one electron is lost by Na (*g*) atom.

$Na\,(g) \rightarrow Na^+\,(g) + e^-$; Energy absorbed = $+(IE)_{Na\,(g)}$

(*c*) 1/2 Cl_2 (*g*) molecule absorbs energy (endothermic reaction) equal to half of the dissociation energy of Cl_2 (*g*) molecule and is converted into gaseous chloride atom, Cl (*g*).

$1/2\,Cl_2\,(g) \rightarrow Cl\,(g)$; Heat absorbed = $+1/2\,(H_{diss})_{Cl_2\,(g)}$

(*d*) Cl (*g*) obtained as above accepts an electron lost by Na (*g*) in step (*b*) and is converted into gaseous chloride ion, Cl^-(*g*). In this process energy equal to electron affinity of Cl (*g*), $(EA)_{Cl\,(g)}$ is released (*Exothermic reaction*)

$Cl\,(g) + e^- \rightarrow Cl^-\,(g)$; Energy released = $-(EA)_{Cl\,(g)}$

(*e*) Now Na^+(*g*) [obtained in step (*b*)] and Cl^-(*g*) [obtained in step (*d*)] combine together and produce NaCl(*s*). In this combination process, energy equal to lattice energy of NaCl(*s*), $(H_{lat})_{NaCl\,(s)}$ is released (*Exothermic reaction*).

$Na^+\,(g) + Cl^-\,(g) \rightarrow NaCl\,(s)$; Heat released = $-(H_{lat})_{NaCl\,(s)}$

On adding equations given at (*a*), (*b*), (*c*), (*d*) and (*e*), we get :

$Na\,(s) + 1/2\,Cl_2\,(g) \rightarrow NaCl\,(s)$

This equation is the same as equation (A) which represents the formation of NaCl (s) by the combination of Na (s) and $\frac{1}{2}$ Cl_2 (g). Various steps given above can be represented in the form of a cycle known as Born-Haber cycle (See Fig. 16.40).

Since the sum of all the above reactions gives us reaction (A), the sum of all the energy terms corresponding to the reactions given at (a), (b), (c), (d) and (e) should be equal to $(H_{for})_{NaCl\,(s)}$, *i.e.*

$$-(H_{for})_{NaCl\,(s)} = +(H_s)_{Na\,(s)} + (IE)_{Na\,(g)} + 1/2\,(H_{diss})_{Cl_2\,(g)} - (EA)_{Cl\,(g)} - (H_{lat})_{NaCl\,(s)}$$

This equation can be used to calculate the value of $(H_{lat})_{NaCl(s)}$, if the values of other energy terms with appropriate sign are put in this equation. This equation has also been used to calculate the value of electron affinity of Cl-atom, $(EA)_{Cl}$.

Lattice energy (in kJ/mole) of alkali metal halides are given below :

	Li^+	Na^+	K^+	Rb^+	Cs^+
F^-	1034	914.2	812.1	780.3	743.9
Cl^-	840.1	770.3	701.2	682.4	629.7
Br^-	781.2	728.4	671.1	654.0	612.5
I^-	718.2	680.7	632.2	616.7	584.5

Lattice energy values (in kJ mol^{-1}) of some other ionic crystals are given below :

BeO = 4541, MgO = 3895, MgF_2 = 2882, MgS = 3254

CaO = 3520, CaF_2 = 2581, CaCl = 720, $CaCl_2$ = 2200

SrO = 3325, BaO = 3108, AgCl = 895, AgBr = 883, AgI = 870, ZnO = 4032.

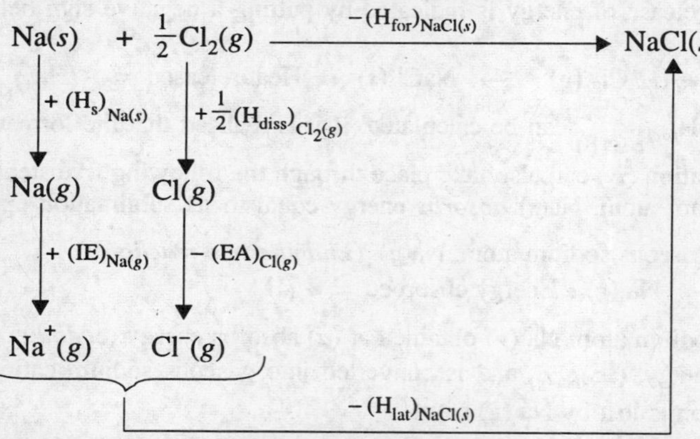

Fig. 16.40 : Born-Haber cycle to show the formation of NaCl (s). Positive sign (+) indicates absorption of heat energy and negative sign (–) indicates release of heat energy.

Example 16.9. *When a mole of crystalline NaCl is prepared from 1 gram atom of sodium and 0.5 mole of chlorine gas, 410 kJ of heat is produced. The heat of sublimation of Na metal is 108.8 kJ. The heat of dissociation of chlorine gas into atoms is 242.7 kJ, the ionisation energy of Na is 493.7 kJ and the electron affinity of chlorine is 368.2 kJ. Calculate the lattice energy of NaCl.* **(Delhi B.Sc. 1985)**

Solution. We know that lattice energy of NaCl (s) $(H_{lat})_{NaCl(s)}$ is given by :

$$-(H_{for})_{NaCl(s)} = +(H_s)_{Na} + (IE)_{Na} + 1/2\,(H_{diss})_{Cl_2} - (EA)_{Cl-} - (H_{lat})_{NaCl(s)}$$

or $$-410 = +108.8 + 493.7 + 1/2 \times 242.7 - 368.2 - (H_{lat})_{NaCl(s)}$$

or $$(H_{lat})_{NaCl(s)} = 1133.8 - 368.2 = \mathbf{765.65\ KJ\ mol^{-1}}$$

Example 16.10. *Use the following data for the given reactions and calculate the value* *of electron affinity of iodine (i)* $K(s) + \frac{1}{2} I_2 (s) \rightarrow KI (s)$, $\Delta H = -327.0$ *KJ/mol, (ii)* $K^+ (g) + I^- (g) \rightarrow KI (s)$, $\Delta H = -643.9$ *KJ/mol, (iii)* $K (s) \rightarrow K (g)$, $\Delta H = 90.0$ *KJ/mol, (iv)* $K (g) \rightarrow K^+ (g) + e^-$, $\Delta H = 415.0$ *KJ/mol (v)* $I_2 (s) \rightarrow 2I^- (g)$, $\Delta H = 213.0$ *KJ/mole.* **(Delhi B.Sc. Hons. 1987)**

Solution. We know that

$$-\left(\Delta H_{for}\right)_{KI} = +\left(\Delta H_{sub}\right)_K + (IP)_K + 1/2\left(\Delta H_{diss}\right)_{I_2} - (EA)_I - \left(\Delta H_{lat}\right)_{KI}$$

or $\qquad -327 = 90 + 415 + 1/2 \times 213 - (EA)_I - 643.9$

$\therefore \qquad (EA)_I = \mathbf{294.6\ kJ/mol.}$

Explanation of some properties of ionic crystals on the basis of their lattice energy values

Following are important properties of ionic solids which can be explained on the basis of lattice energy of the ionic solids.

1. Solubility/insolubility of ionic solids in polar (e.g., H_2O) and non-polar (e.g., C_6H_6, CCl_4 etc.) solvents

Whether a given ionic solid is soluble or insoluble in a polar (ionic) or a non-polar solvent depends on two factors *viz.*, (*a*) *lattice energy of the ionic solid* which keeps the constituent ions (*i.e.*, cations and anions) of the ionic solid held together tightly in the ionic crystal, and (*b*) *hydration energy of the ionic solid*, which is the energy released when the ions of the ionic solid obtained by dissolution of the ionic solid in ionic solvent (*e.g.*, H_2O) interact with water dipoles to form hydrated ions. Higher is the magnitude of lattice energy, more tightly the ions are held together in the ionic crystal and hence lesser is the tendency of the ionic solid to split into ions by dissolution and to interact with water dipoles to form hydrated ions, *i.e.*, lesser is the solubility of the ionic solid. Higher is the magnitude of hydration energy, greater will be the tendency of the ionic solid to break into ions in water and then to get hydrated by H_2O molecules and consequently the solubility of the solid *will increase*. Thus we see that lattice energy tends to oppose the dissolution of the ionic solid while the hydration energy tends to dissolve the ionic solid into water. On this basis we can say that if the hydration energy is greater than the lattice energy, the ionic solid will be soluble in polar solvent like water and *vice-versa*, since it is only greater magnitude of hydration energy as compared to lattice energy which can break the ionic solid into its constituent ions against the electrostatic force of attraction existing between the ions.

Ionic solids are not soluble in non-polar solvents, (*e.g.*, C_6H_6, CCl_4 etc.), since the ions of ionic solid donot interact with the solvent molecules (due to the non-polar nature of the solvent molecules) to form the hydrated ions and hence no hydration energies are obtained.

Examples (*a*). LiCl, Li Br, NaCl, NaBr etc., are the examples of ionic solids which are soluble in water, since their hydration energies are greater than their lattice energies as shown below : (The values are given in kJ/mole).

Hydration energies : LiCl = 883, Li Br = 854, NaCl = 775, NaBr = 741

Lattice energies : LiCl = 840.1, Li Br = 781.2, NaCl = 770.3, NaBr = 728.4

(*b*) The ionic solids like KCl, KBr and KI are insoluble in water, since their hydration energies are less than their lattice energies as shown below. (The values are given in kJ/mole).

Hydration energies : KCl = 686, KBr = 657, KI = 619

Lattice energies : KCl = 701.2, KBr = 671.1, KI = 632.2

Similarly, AgCl, $BaSO_4$, CaF_2 and AlF_3 are also insoluble in water, since their hydration energies are less than their lattice energies.

High value of dielectric constant of the solvent also favours the dissolution of the ionic solids, because it weakens the electrostatic force of attraction existing between the oppositely charged ions present in the ionic solids.

Comparison between the solubility of halides of a given metal. Let us compare the solubility of the halides given by Li-metal. We have said above that higher is the magnitude of lattice energy, lesser is the solubility of the ionic solid in water. Now since the lattice energies of the halides of Li decrease from LiF to LiI, the solubility of these halides is in the increasing order. Thus :

Halides of Li	:	Li F	LiCl	LiBr	LiI
Values of lattice energy (kJ/mole)	:	1034	840.1	781.1	718.2
		——— Decreasing ———→			
Solubility	:	——— Increasing ———→			

(LiF < LiCl < LiBr < LiI)

Thus, we see that Li F is insoluble in water while LiI is soluble in this solvent. In general we can say that the order of the solubility of the halides of a given metal can be written as fluorides < chlorides < bromides < iodides.

Comparison between the solubility of hydroxides and fluorides of alkaline earth metals. Since the size of the divalent alkaline earth metal cations (M^{2+} ions) increase on moving down the group from Be^{2+} to Ba^{2+}, the magnitude of the lattice energy of the hydroxides and fluorides of these metals decrease in the same direction. With the decrease in the lattice energy value, the solubility of these compounds increases. Thus $Be(OH)_2$ and $Mg(OH)_2$ are almost insoluble in water and hydroxides of other alkaline earth metals are slightly soluble. Similarly MgF_2 and CaF_2 are also insoluble in water.

Comparison between the solubility of M^+- salts with polyvalent ions (*e.g.* SO_4^{2-}, $C_2O_4^{2-}$, PO_4^{3-} etc.) and that of M^{2+}-salts with the same ions. Due to the smaller size of alkaline earth metal ions (M^{2+} ions) as compared to those of alkali metal cations (M^+ ions) of the same period, the lattice energies of M^{2+}-salts with polyvalent ions (*e.g.* SO_4^{2-}, $C_2O_4^{2-}$, PO_4^{3-} etc.) are higher than those of the corresponding to M^+-salts. Consequently M^{2+}- salts are less soluble than M^+-salts, *e.g.*, $CaCO_3$ is less soluble in water than K_2CO_3.

2. Melting points and boiling points of ionic solids

Since the electrostatic force of attraction existing between the cations and anions present in an ionic crystal are very strong, the lattice energies of ionic solids are very high. It is because of high values of their lattice energies that large amount of energy is required to separate the ions of the ionic solid and hence melting and boiling points of ionic solids are also very high. The melting points of ionic solids increase with the increase in the magnitude of their lattice energy and *vice-versa*, since the lattice energy is a measure of the strength of the attractive force existing between the oppositely-charged ions of the ionic solid.

Examples. (*a*) Since the lattice energies of the halides of Li decrease as LiF > LiCl > LiBr > LiI, the melting points of these halides also decrease in the same direction.

Halides of Li	:	LiF	LiCl	LiBr	LiI
Lattice energy (kJ/mole)	:	1034	840.1	781.2	718.2
		——— Decreasing ———→			
Melting points (°C)	:	870	613	597	446
		——— Decreasing ———→			

(*b*) Melting points of the chlorides of alkali metals (excepting LiCl) also decrease as NaCl > KCl > RbCl > CsCl, since the lattice energy values of these chlorides also decrease in the same direction.

Chlorides of alkali metals	:	LiCl	NaCl	KCl	RbCl	CsCl
Lattice energy (kJ/mole)	:	840.1	770.3	701.2	682.4	629.7
			——— Decreasing ———→			
Melting points (°C)	:	613	800	776	710	625
			——— Decreasing ———→			

(*c*) The melting point of AgCl (= 455°C) in higher than that of AgBr (= 434°C), since AgCl higher value of its lattice energy (Lattice energy of AgCl = 895 kJ/mole and that of AgBr = 883 kJ/mole).

3. Stability of ionic solids

Greater is the force of attraction between the oppositely-changed ions (constituent ions) of the ionic solid, greater is the stability of the ionic solid. Again, we know that greater is the magnitude of lattice energy of the ionic solid, greater is the force of attraction between the ions (lattice energy is the direct measure of the electrostatic force of attraction between the ions) and hence greater is the stability of the ionic solid. Thus an ionic solid having higher magnitude of lattice energy will be more stable than that having lower value of lattice energy.

Example. Since $CaCl_2$ has higher value for its lattice energy (= 2200 kJ mol^{-1}) than CaCl (= 720 kJ mol^{-1}), the former is more stable than the latter.

Solvation process and solvation energy

When a gaseous cation or a gaseous anion reacts with a solvent (abbreviated as *sol*) in the liquid state, solvated cation or solvated anion is obtained and in this process which is called *solvation process* energy is liberated. This liberated energy is called *solvation energy*. Thus :

$$M^+ (g) + \underset{\text{solvent}}{sol.} \longrightarrow \underset{\text{solvated cation}}{M^+ (sol)} + \text{solvation energy}$$

$$X^- (g) + sol \longrightarrow \underset{\text{solvated anion}}{X^- (sol)} + \text{solvation energy}$$

Due to the liberation of energy, solvation process is an *exothermic process*.

Hydration process and hydration energy

When a gaseous cation or a gaseous anion reacts with water, hydrated cation or hydrated anion is produced and in this process which is called *hydration process* energy is released. This released energy is called *hydration energy*. Thus :

$$M^+ (g) + H_2O\ (l) \text{ or } aq \longrightarrow \underset{\text{Hydrated cation}}{M^+ (aq)} + \underset{\text{energy}}{\text{Hydration}}$$

$$X^-(g) + H_2O\ (l) \text{ or } aq \longrightarrow \underset{\text{Hydrated anion}}{X^- (aq)} + \underset{\text{energy}}{\text{Hydration}}$$

Obviously, hydration process is *an exothermic process*.

Hydration energies of some ions and salts

Hydration energies (heat of hydration) of some ions and some salts are given below (in kJ/mole).

Ions

H^+ = 1091			
Li^+ = 531.4	Be^{2+} = 2494.0	Al^{3+} = 4694.5	F^- = 515.0
Na^+ = 422.6	Mg^{2+} = 1953.9		Cl^- = 381.0
K^+ = 338.9	Ca^{2+} = 1577.0		Br^- = 347.0
Rb^+ = 318.9	Sr^{2+} = 1443.0		I^- = 305.0
Cs^+ = 280.3	Ba^{2+} = 1352.0		

Salts

LiCl = 883,	LiBr = 854	
NaCl = 775,	NaBr = 741	
KCl = 686,	KBr = 657,	KI = 619

Factors affecting the magnitude of solvation (or hydration) energy.

Following factors are important :

1. Ionic potential (ϕ) of the ion. The charge-to-radius ratio of an ion is called its ionic potential (ϕ). Thus :

$$\text{Ionic potential } (\phi) = \frac{\text{Charge on the ion}}{\text{Radius of the ion}} \quad ...(i)$$

An ion with high value of ϕ will tend to form the solvated or hydrated ion more easily than the ion with a smaller value of ionic potential. In other words, we say that solvation energy or hydration energy increases with increasing ionic potential and *vice versa.* The relation (*i*) given above shows that :

(*a*) ϕ increases with the increase of the charge on the ion. Again we have said above that hydration energy also increases with the increase of the value of ϕ. Consequently hydration energy of an ion increases with the increase of the charge on it.

Example. The hydration energies of Na^+ (charge = +1), Mg^{2+} (charge = +2) and Al^{3+} (charge = +3) ions increases as Na^+ (= 422.6 kJ/mole) < Mg^{2+} (= 1953.9 kJ/mole) < Al^{3+} (= 4694.5 kJ/mole).

(*b*) ϕ decreases with the increase in the size of the ion. Thus, since hydration energy increases with the increase of ϕ, hydration energy also increases with the decrease of radius and *vice versa.*

Examples. (*i*) Since the size of alkali metal cations increases as $Li^+ < Na^+ < K^+ < Rb^+ < Cs^+$, the hydration energy of these cations are in the decreasing order as shown below :

Alkali metal cations	:	Li^+	Na^+	K^+	Rb^+	Cs^+
Radii (Å)	:	0.60	0.95	1.33	1.48	1.69
		Increasing ⟶				
Hydration energy	:	531.4	422.6	338.9	318.8	280.3
(kJ/mole)		Decreasing ⟶				

(*ii*) On similar lines of argument it can be shown that with the increase of the size of divalent cations of alkaline earth metals from Be^{2+} to Ba^{2+}, the hydration energies of these cations decrease down the group. The values of hydration energies (in kJ/mol) are :

Be^{2+} = 2494, Mg^{2+} = 1953.9, Ca^{2+} = 1577, Sr^{2+} = 1443 and Ba^{2+} = 1352.

(*iii*) On similar lines of argument it can be explained why hydration energy of halide ions decreases down the group as shown below :

Halide ions	:	F^-	Cl^-	Br^-	I^-
Radii (Å)	:	1.36	1.81	1.95	2.16
		Increasing ⟶			
Hydration energy	:	515	381	347	305
(kJ/mole)		Decreasing ⟶			

Hydration energies of K^+ and F^- ions. Although both these ions have the same size (K^+ = 1.33 Å, F^- = 1.36 Å), there is a considerable difference in the value of their hydration energy (K^+ = 338.9 kJ/mole, F^- = 515.0 kJ/mole). This difference is due to the fact that F^- ion interacts with the small $H^{\delta+}$ atom while K^+ ion has to interact with the much larger $O^{\delta-}$ atom of water dipoles. Thus more number of H-atoms can get around a F^- ion compared to O-atoms around the K^+ ion.

2. Dipole moment of the solvent. Greater is the dipole moment of the solvent, greater is the magnitude of solvation energy.

Variation of hydration energies of ions in the periodic table

We have already said that the tendency of ions to get hydrated decreases with the increase of their size, *i.e.,* the magnitude of hydration energy of ions decrease with the increase of their size (radius) and *vice versa.*

(*a*) **Variation of hydration energy in a group.** We know that since the radii of the ions derived from the atoms of the elements of a given group of *s*- and *p*-block elements increase on descending the group, the magnitude of the hydration energy of the ions decrease in the same direction.

Examples. (*i*) Since the radii of alkali metal cations increase down the group, the magnitude of their hydration energy decreases in the same direction, *i.e.*, the relative order of the hydration energies of alkali metal cations is as :

$$Li^+ > Na^+ > K^+ > Rb^+ > Cs^+ \quad \textit{(Decreasing order)}$$

(*ii*) Since the size of Mg^{2+} ion is less than that of Ba^{2+}, Mg^{2+} ion has higher value of its hydration energy than Ba^{2+} ion ($Mg^{2+} > Ba^{2+}$).

(*iii*) Since the radii of halide ions increase down the group, the magnitude of their hydration energy decreases in the same direction, *i.e.*, the relative order of the hydration energies of halide ions is as :

$$F^- > Cl^- > Br^- I^- \quad \textit{(Decreasing order)}$$

(*b*) Variation of hydration energy in a period. Since the radii of the ions derived from the atoms of the elements of a given period of *s*- and *p*-block elements decrease on moving from left to right, the magnitude of hydration energy of the ions increase in the same direction.

Example. Since the radii of Na^+, Mg^{2+} and Al^{3+} ions (period 3) decrease as $Na^+ > Mg^{2+} > Al^{3+}$, the hydration energies of these ions are in the increasing order, *i.e.*, the relative order of hydration energies of the these ions is as :

$$Na^+ < Mg^{2+} < Al^{3+} \quad \textit{(Increasing order)}$$

Explanation of some properties on the basis of hydration energy values

Following are the important properties which can be explained on the basis of hydration energy concept.

1. Size and electrical conductivity of hydrated alkali metal cations, [M (*aq*)]$^+$. We know that smaller is the size of an ion, greater is the degree of its hydration. Thus as we move down the group of alkali metals from Li^+ to Cs^+ ion, the size of the ion increases and hence the degree to which the ion gets hydrated decreases, *i.e.*, the degree of hydration of alkali metal ions is in the order :

$$Li^+ > Na^+ > K^+ > Rb^+ > Cs^+$$

The above order shows that Li^+ ion is hydrated to the maximum extent and Cs^+ ion gets hydrated to the minimum extent. Thus, Li^+ ion is surrounded by the maximum number of H_2O molecules and hence has the largest size when it is in the *aqueous solution*. This means that the hydrated Li^+ ion, $[Li\,(aq)]^+$ is the *largest in size*. On the other hand Cs^+ ion is surrounded by the minimum number of H_2O molecules and hence has the smallest size in *aqueous solution*. This means that hydrated Cs^+ ion, $[Cs\,(aq)]^+$ is the *smallest in size*. Thus we see that the size of alkali metal hydrated cations decreases down the group as shown below :

$$[Li\,(aq)]^+ > [Na\,(aq)]^+ > [K(aq)]^+ > [Rb(aq)]^+ > [Cs\,(aq)]^+ \quad \textit{(Decreasing order)}$$

The radii (Å) of the hydrated ions, M^+ (aq) have been found to be equal to $Li^+\,(aq) = 3.40$, $Na^+\,(aq) = 2.76$, $K^+\,(aq) = 2.32$, $Rb^+\,(aq) = 2.28$, $Cs^+\,(aq) = 2.28$.

Because of the fact that the tendency of alkali metal cations to get hydrated decreases down the group, the electrical conductivity (ionic conductance) of these cations in solution of infinite dilution is in the increasing order as shown below :

$$[Li\,(aq)]^+ < [Na\,(aq)]^+ < [K\,(aq)]^+ < [Rb\,(aq)]^+ < [Cs\,(aq)]^+ \quad \textit{(Increasing order)}$$

The above order shows that Li^+ ion in solution has the lowest and Cs^+ ion has the lighest mobality or ionic conductance in an electric field (electrical conductivity). The values of ionic conductance (ohm^{-1} cm^2 equiv.$^{-1}$) are as $Li^+\,(aq) = 33.5$, $Na^+\,(aq) = 43.5$, $K^+\,(aq) = 64.6$, $Rb^+\,(aq) = 67.5$, $Cs^+\,(aq) = 68.0$.

2. Formation of hydrated salts by alkali metals. Since the size of alkali metal monovalent cations (M^+ ions) increases down the group from Li^+ to Cs^+ with the increase in their atomic number, the extent to which these cations get hydrated decrease in the same direction. Thus all salts of Li^+ ion which get hydrated to the maximum extent are hydrated, many salts of Na^+ ion are also hydrated while those of Rb^+ and Cs^+ ions are very rarely hydrated.

3. Formation of hydrated salts by alkaline earth metals. Since the alkaline earth metal divalent cations (M^{2+} ions) are smaller in size than alkali metal monovalent cations (M^+ ions), the extent to which M^{2+} ions get hydrated is greater than that of M^+ ions. For example Mg^{2+}, Ca^{2+} and Ba^{2+} ions give a large number of hydrated salts which are more extensively hydrated than those given by M^+ ions. Example of such salts are $MgCl_2.6H_2O$, $Mg(NO_3)_2.6H_2O$, $MgSO_4.7H_2O$, $CaCl_2.6H_2O$, $BaCl_2.6H_2O$. On the other hand Na^+ ion gives a limited number of hydrated salts and Rb^+ gives very rarely hydrated salts.

4. Comparison between the reactivity of coinage metals and of alkali metals. The reactivity of a metal increases when its heat of sublimation and ionisation energy are low and heat of hydration is high Thus, since for coinage metals the heat of sublimation and ionisation energies both are very high and hydration energies are very low as compared with these values for alkali metals, coinage metals show less reactivity than alkali metals. Low reactivity of coinage metals is evident from the fact that these metals can be attacked only by strong oxidising agents like HNO_3, conc. H_2SO_4 etc. These metals attack neither water nor steam while alkali metals attack H_2O, liberating H_2 gas.

5. Solubility/insolubility of ionic solids in polar (*e.g.* H_2O) and non-polar (*e.g.* C_6H_6, CCl_4 etc.) solvents. For this see "*Lattice Energy*".

6. Solubility of the nitrates, sulphates, chlorides etc., of alkaline earth metals. We have already said that the solubility of an ionic salt in water depends on the difference in the magnitude of hydration energy and lattice energy of the ionic salt. Now since the lattice energies of nitrates, chlorides etc., of divalent alkaline earth metal cations (M^{2+}) are almost the same, the relative order of the solubility of these salts cannot be predicted on the basis of the difference in the magnitude of hydration energy and lattice enrgy. Consequently the relative order is determined by the magnitude of hydration energies of the gaseous M^{2+} ions only. Greater is the magnitude of hydration energy of M^{2+} cation, greater is the solubility of the salt of this ion. Thus, since the hydration energies of M^{2+} ions decrease down the group (Be^{+2} = 2494 kJ/mole, Mg^{+2} = 1953.9 kJ/mole, Ca^{2+} = 1577 kJ/mole, Sr^{+2} = 1443 kJ/mole, Ba^{+2} = 1352 kJ/mole), the solubility of the sulphates, nitrates, chlorides etc., also decreases in the same direction. $MgSO_4$ is soluble in H_2O while $BaSO_4$ is insoluble or has very low solubility.

7. Ca^{2+} ion is more stable than Ca^+ ion in aqueous solution. Heat of hydration of Ca^+ and Ca^{2+} ions are 320 kJ mol^{-1} and 1577 kJ mol^{-1} respectively.

$Ca^+ (g) + aq \rightarrow Ca+ (aq)$; Heat of hydration of Ca^+ (*g*) ion = 320 kJ mol^{-1}

$Ca^{2+}(g) + aq \rightarrow Ca^{2+}(aq)$; Heat of hydration of Ca^{2+} (*g*) ion = 1577 kJ mol^{-1}

Since hydration energy of Ca^{2+} ion is higher than that of Ca^+ ion, Ca^{2+} ion is more stable than Ca^+ ion in aqueous solution.

An ideal ionic crystal.

An ideal ionic crystal is that in which all the lattice points are occupied by the ions constituting the ionic crystal (See Fig. 16.41) and no lattice point is vacant. In an ideal ionic crystal the two-dimensional arrangement of the constituent ions of the crystal is perfect or ideal. Fig. 16.41 shows two-dimensional pattern of the ions (A^+ and B^+ ions) of a perfect ionic crystal, $A^+ B^-$.

A^+ B^- A^+ B^- A^+
B^- A^+ B^- A^+ B^-
A^+ B^- A^+ B^- A^+
B^- A^+ B^- A^+ B^-

Fig. 16.41 : Two dimentional arrangement of A^+ and B^- ions in a perfect ionic crystal, $A^+ B^-$.

The above description shows that *an ideal crystal is that which has the same unit cell and contains the same lattice points throughout the whole crystal.* It should be noted that such ideal crystals exist only at absolute zero (0 K) temperature. At any temperature greater than 0 K, the crystals have some departure from the ordered arrangement of the constituent particles.

Stoichiometric and non-stoichiometric crystals

Stoichiometric ionic crystals are those in which the ratio of positive and negative ions present in the compound are the same as required by the ideal formula of the compound. On the other hand, non-stoichiometric compounds are those in which the ratio of positive and negative ions present in the compound differ from that which is required by ideal chemical formula of the compound. For example, vandium (II) oxide is non-stoichiometric oxide, since it has the formula VO_x where $x = 0.6 - 1.3$. Similarly iron (II) oxide samples contain more oxygen atoms than iron atoms, while samples of zinc (II) oxide usually has excess of zinc atoms than oxygen atoms. In each case electrical neutrality is maintained.

Imperfections (defects) in crystals

We have already said that in an ideal ionic crystal the arrangement of constituent ions is perfect or ideal. The imperfections in crystals implies that the perfect arrangement of the constituent ions is destroyed due to the presence of cationic or anionic vacancies in the lattice points, dislocation of ions in the lattice, the presence of the same type of ions in the different oxidation states and the presence of impurities in the lattice points. In short we can say that the term disorder or imperfection (defect) in a crystal denotes the departure of the constituent particles of the crystal from its regular position in the lattice. The presence of defects in crystals affects many properties of the crystals. For example, the electrical conductivity and mechanical strength of the solids are changed to a large extent due to the presence of defects in crystals.

A brief description of different types of imperfections present in the crystalline solids is given in chart 1 :

Chart 1 : *Classification of imperfections or defects found in ionic crystals*

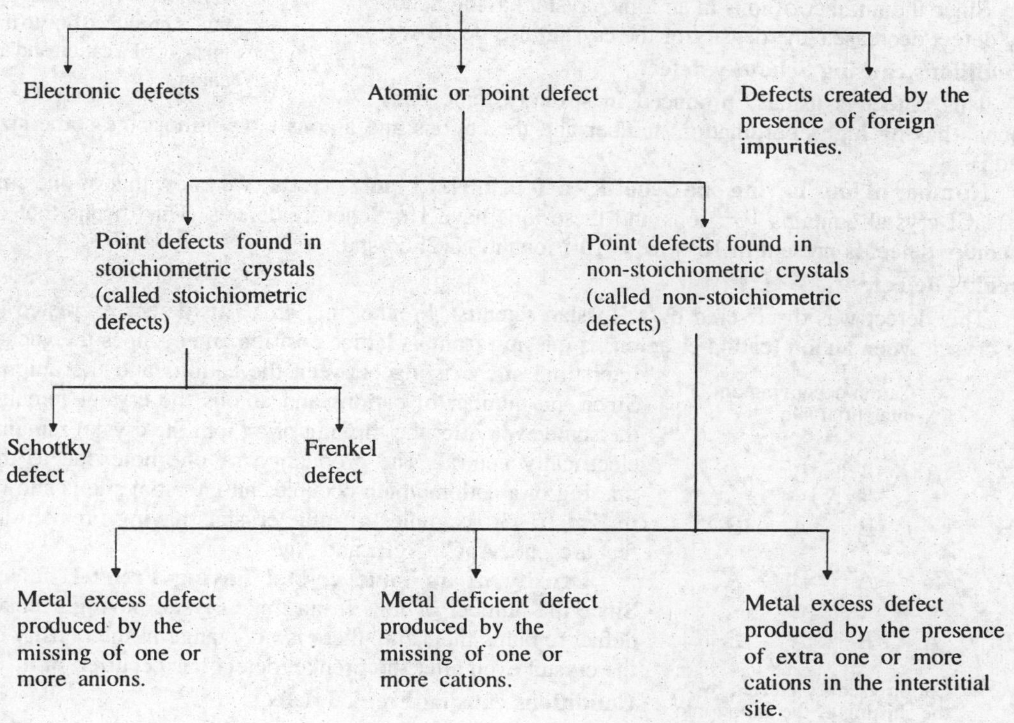

Atomic or point defects

Atomic or point defects are produced in an ionic crystal when an equal number of cations and anions are missing from their regular lattice positions and thus an equal number of cations and anion vacancies are produced in the structure of the crystal or when an ion (cation or anion) of an ionic crystal is missing and is shifted to a vacant interstitial site existing in the structure of the crystal.

Atomic or point defects may occur in stoichiometric as well as in non-stoichiometric crystals. The point defects found in stoichiometric crystals are called *stoichiometric defects* while the point defects found in non-stoichiometric crystals are known as *non-stoichiometric defects.*

Point defects found in stoichiometric crystals (Stoichiometric defects)

Point defects found in stoichiometric defects are of the following two types (*i*) *Schottky defect* (*ii*) *Frenkel defect.*

Schottky defect

This defect was discovered by a German scientist, Schottky in 1930. This defect is created in the crystal when the same number of cations and anions are missing from their positions and thus holes (called cation vacancy or anion vacency) are produced. Since the number of cations and anions that are left in the crystal even after the missing of cations and anions is the same, the crystal remains electrically neutral even after the missing of the cations and anions. The production of two holes, one due to the missing of a cation and the other due to the missing of an anion, is shown in Fig. 16.42. Examples of ionic crystals having Schottky defect are NaCl, KCl, KBr, CsCl etc.

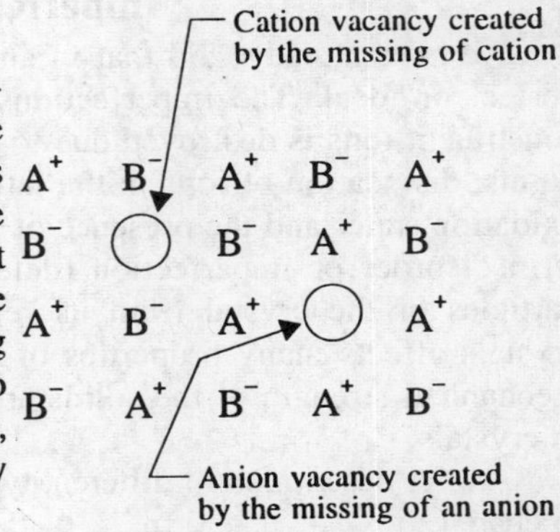

Fig. 16.42 : Schottky defect in an ionic crystal (A^+B^-). This defect is produced due to the missing of a cation and an anion.

Density of an ionic crystal having Schottky defect

Since the number of ions in an ionic crystal having Schottky defect decrease,s the density of the crystal also decreases.

Conditions causing Schottky defect

This defect is usually produced in strongly ionic compounds having high coordination number and the cations and anions have almost the some size (radii).

Number of ions having one Schottky defect in NaCl ionic crystal. We know that of one cm^3 of NaCl crystal contains 10^{22} ions and these ions have 10^6 Schottky defects. This means that of Schottky defect is present in $10^{22}/10^6 = 10^{16}$ ions in NaCl crystal.

Frenkel defect.

This defect was discovered by a Russian scientist, Frenkel in 1926. This defect is created in the crystal when an ion (cation or anion) is missing from its lattice position and occupes the vacant interstitial site existing between the cations and the anions. Since the number of cations and anions the crystal remains the same even after the missing of an ion, the crystal remains electrically neutral. The production of one hole due to the missing of a cation which occupies an interstial site is shown in Fig. 16.43. Examples of ionic crystals having Frenkel defect are ZnS, AgCl, AgBr and AgI.

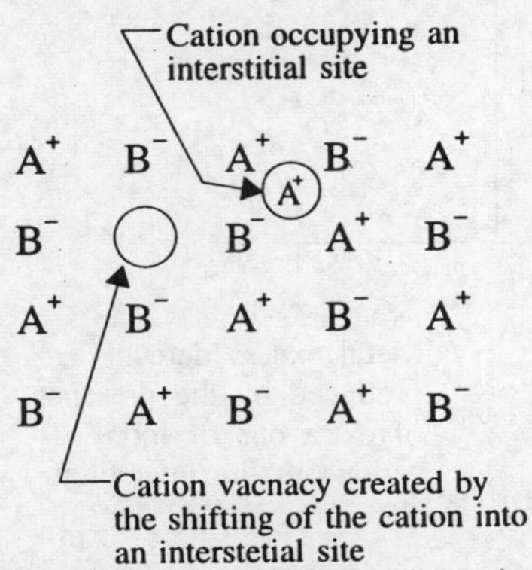

Fig. 16.43 : Frenkel defect in an ionic crystal (A^+B^-). This defect is produced due to the shifting of a cation into an interstial site.

Density of an ionic crystal having Frenkel defect. Since the number of ions in an ionic crystal having Frenkel defect remains the same, there is no change in the density of the crystal even after the Frenkel defect has occurred in it.

Conditions causing Frenkel defect

This defect is generally produced in compounds having low coordination number and the cation is smaller in size than the anion so that the cation may fit into the interstitial site. For example this defect occurs in AgCl, since, because of smaller size of Ag^+ ion, this ion can fit into the interstitial site.

Consequences of Schottky and Frenkel defects. (*i*) It is due to the presence of these defects that the crystal is able to conduct electricity through it when an electric field is applied on it. This tendency of the crystal can be explained as follows. When an electric field is applied on a crystal, having both types of defects, a nearby ion moves from its position to occupy a hole. This results in creating a new hole and another nearby ion moves into it and so on. This process continues and a hole, thereby, moves from one end to the other end. Thus it conducts electricity across the whole crystal.

(*ii*) Density of the crystals, having Schottky defects, decreases, as expected.

(*iii*) The presence of holes also decreases the lattice energy or the stability of the crystal. The presence of too many holes decreases the stability of the crystal to such an extent that a partial collapse of the lattice energy may take place.

(*iv*) We have seen that in a crystal having Schottky defect, the similar ions (charges) come closer to each other and hence the dielectric constant of the crystal increases.

Table 16.11 : *Differences between Schottky and Frenkel defects.*

Schottky defect	Frenkel defect
(*i*) The positive and negative ions are completely missing from their lattice positions. The number of cations and anions, which are missing, is the same.	(*i*) An ion (cation or anion) leaves its lattice position and occupies a vacant interstitial site existing between the cations and anions in the crystal.
(*ii*) The density of the crystal having this defect, decreases, *i.e.*, the presence of Schottky defect in a crystal decreases its density.	(*ii*) The density of the crystal having this defect remains the same; *i.e.*, the presence of Frenkel defect in a crystal does not change its density.
(*iii*) This defect occurs in highly ionic crystal, having high coordination numbers and the cations and the anions have almost the same size.	(*iii*) This defect is found in those ionic crystals which have low coordination number and the cation is smaller in size than the anion.

Point defects found in non-stoichiometric crystals (Non-stoichiometric defects)

Depending on whether the positive ions are present in excess (metal excess) or negative ions are in excess (metal deficiency) in a given non-stoichiometric crystal, there are two types of defects found in a non-stoichiometric crystal. These defects are *metal excess defects* and *metal deficient defects* respectively. Following are the important types of metal excess and metal deficient defect.

1. Metal excess defect, produced by the missing of one or more anions. This defect is produced when one or more anions are missing from their positions and thus anion vacancies are produced. In order to maintain the electrical neutrality of the crystal, these anion vacancies are occupied by the electrons, since the electrons like anions, are also negatively-charged (see Fig. 16.44). Thus, we see that although the crystal as a whole is electrically neutral, the number of metal ions (cations) is greater than that of the anions in the crystal, *i.e.*, there is an excess of metal ions in the crystal and it is for this reason that this defect is called **metal excess defect**. It is easy to understand that this defect is found in those crystals which have Schottky defects.

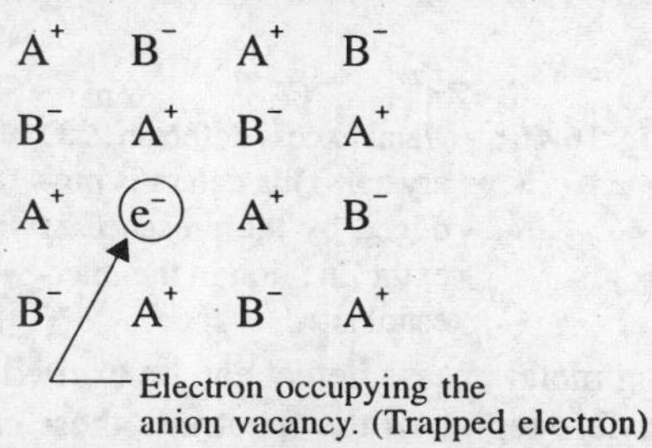

Fig. 16.44 : Metal excess defect produced by the missing of one anion (B^-) in an ionic crystal, A^+B^-.

Example. When alkali metal halide crystal is heated in the atmosphere of alkali metal vapours, the halide ions move into the surface of the metal (missing of anions) and combine with alkali metals atoms ($X^- + M \rightarrow MX + e^-$). The electrons thus produced get diffused into the crystal and occupy the holes (anion vacancies) produced by the missing of the halide ions. The electrons, which occupy the anion vacancy are called *F-centres* or *colour centres*. The presence of F-centres (or electrons occupying the interstitial site) in an ionic crystal modifies the properties of the crystal. For example :

(*a*) *Colour.* The presence of excess of potassium in KCl crystal makes the crystal *violet* and similarly the presence of excess of lithium in LiCl crystal gives *pink* colour to the crystal. Greater is the number of F-centres, greater is the intensity of colouration.

(*b*) Crystals having F-centres are *paramagnetic*, since the electrons occupying the interstitial sites are unpaired.

(*c*) When the crystals having F-centres are irradiated with light, they become photoconductor.

2. Metal excess defect produced by the presence of extra one or more cations in the interstitial site. This defect is produced when one or more cations occupy the interstitial site and in order to maintain the electrical neutrality of the crystal, equal number of electrons are trapped into other interstitial sites (See Fig. 16.45). Thus, we see that although the crystal as a whole is electrically neutral, the number of metal ions is greater than that of the anions, *i.e.*, there is an excess of metal ions in the crystal and hence this defect is called *metal excess defect.*

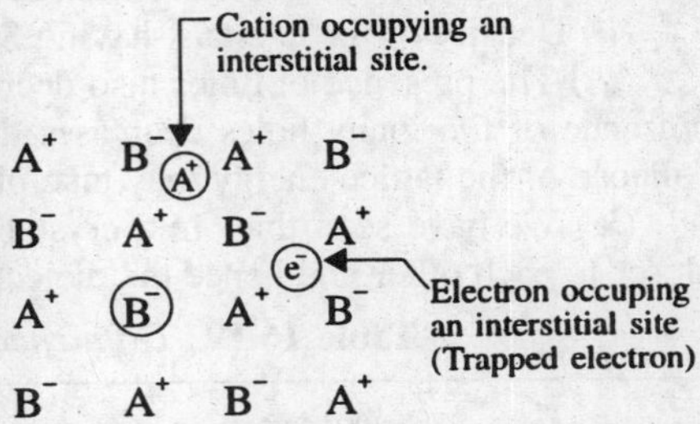

Fig. 16.45 : Metal excess defect produced by the presence of extra cation (A^+) in the interstial site in A^+ B^- ionic crystal.

It is easy to understand that this defect is found in those crystals which have Frenkel defects. The crystals having this type of defect have advanced electrical conductivities due to the presence of interstitial electrons.

Example. This defect is found in zinc oxide (ZnO). ZnO loses O_2 reversibly at high temperatures and becomes yellow in colour. The excess of Zn^{2+} ions are trapped into the interstitial sites and equal number of electrons are trapped into other interstitial sites to balance the electrical charge (See Fig. 16.46).

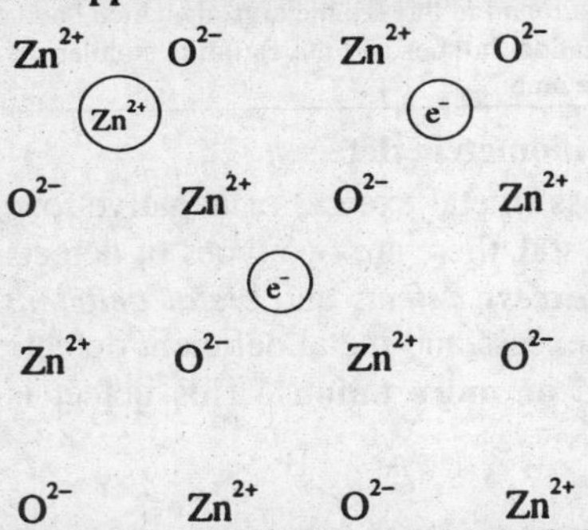

Fig. 16.46 : Metal excess defect in ZnO crystal. This defect is produced by the presence of extra Zn^{2+} ion in the interestial site.

Consequences of metal excess defects. (*a*) *Electrical conductivity of non-stoichiometric crystals.* The above discussion shows that in a non-stoichiometric crystals the metal excess defect is produced either by the missing of one or more anions or by the presence of extra one or more cations in the interstial sites. In both the ways the non-stoichiometric crystal having metal excess defect contains free electrons which are present either in the anionic vacancy or are trapped into the interstitial site. Since these electrons are mobile, the crystals having metal excess defect conduct electricity through them. Since the number of electrons is low, the crystals show low electrical conductivity and hence are called *n-type semiconductors*.

(*b*) *Colour of the non-stoichiometric crystals.* Since the free electrons present in the non-stoichiometric crystals having metal excess defect can be excited to higher energy levels by the absorption of radiations in the visible region of the spectrum, these crystals are generally coloured. For example non-stoichiometric NaCl is *yellow* and KCl is *lilac.*

3. Metal deficient defect produced by the missing of one or more cations. This defect is produced when one or more cations are missing from their positions and thus cation vacancies are produced. In order to maintain the electrical neutrality of the crystal, one or more nearby cations acquire extra positive charge and are thus converted into cations with higher oxidation state (See Fig. 16.47). Thus we see that although the crystal as a whole is electrically neutral, the number of metal ions (cations) is less than that of the anions, *i.e.*, there is deficiency of metal ions in the crystal and hence it is for this reason that this defect is called *metal deficient defect.* This defect is found in those ionic crystals whose positive ions show variable oxidation state.

Examples. Ferrous oxide, nickel oxide, ferrous sulphide (iron pyrites) etc., are the important examples of ionic crystals which have this type of defect. In the crystal of ferrous sulphide, one Fe^{2+} ion is missing and two Fe^{2+} ions acquire one extra positive charge so that the electrical neutrality of the sulphide crystal may be maintained. Thus in the crystal of iron pyrites one of the three Fe^{2+} ions is missing from its lattice site and two Fe^{2+} ions are converted into Fe^{3+} ions (See

Fig. 16.48). The conversion of Fe^{2+} ions into Fe^{3+} ions takes place due to the movement of one electron from Fe^{2+} ion to Fe^{3+} ion. It is due to the movement of electron(s) that the ionic crystals having this type of defect act as conductors. The movement of electrons from Fe^{2+} to Fe^{3+} ion produces a shiny colour in iron pyrites. Some samples of pyrites possess so much lusture that they are known as *fools gold.*

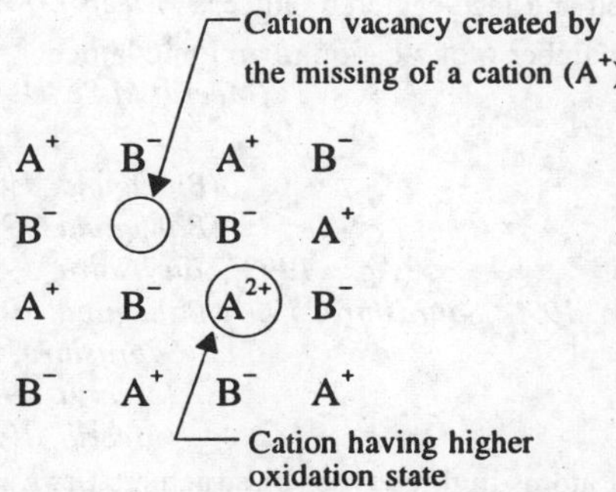

Fig. 16.47 : Metal deficient defect produced by the missing of one cation (A^+) in an ionic crystal, A^+B^-.

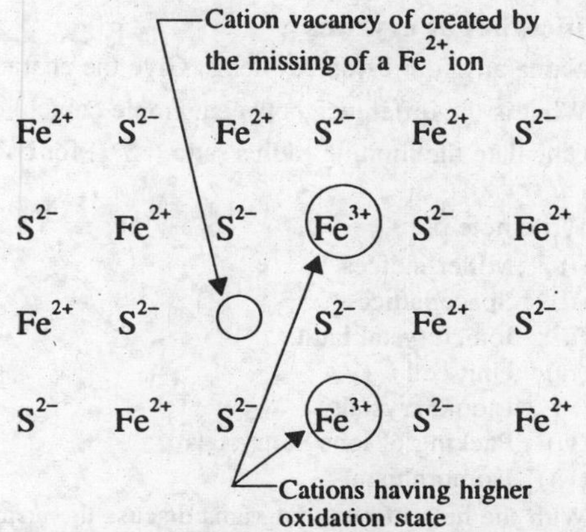

Fig. 16.48 : Metal deficient defect in the crystal of ferrous sulphide (iron pyrites).

Consequences of metal deficient defects. Crystals having metal deficient defects are semi-conductors. The conductivity is due to the movement of an electron from one cation to the other cation. When an electron moves from one cation, it is changed into a cation with higher positive oxidation state. The movement of a cation is also called the movement of a positive hole and the substances are called *p-semiconductors.*

Structure of ferrous oxide (FeO) crystal.

The crystal of FeO has rock salt (NaCl) structure, In this oxide, the ratio of Fe^{2+} and O^{2-} ions is not 1 : 1. Its composition has been found to be as $Fe_{0.95}$ O (Wustite). This composition is explained by saying that in the crystal of FeO, one Fe^{2+} ion is missing and two Fe^{2+} ions acquire positive charge to maintain the electrical neutrality of the crystal of FeO. Thus two Fe^{2+} ions are converted into Fe^{3+} ions. Also, see the structure of iron pyrites (FeS).

Normal spinel structure : Structure of Mg^{2+} Al_2^{3+} O_4 and Zn^{2+} Fe_2^{3+} O_4 (zinc ferrite) oxides.

Both these oxides adopt normal spinel structure in which the oxide ions (O^{2-} ions) have cubic close-packed (*ccp*) arrangement and the divalent cations (Mg^{2+} and Zn^{2+} ions) occupy one-eighth of the tetrahedral holes while trivalent cations (Al^{3+} and Fe^{3+} ions) occupy half of the octahedral holes.

Uses of ferrites.

Many of the ferrites ($Fe_2O_4^{2-}$) which have spinel structure are important magnetic materials and are used in telephones or memory loops in computers.

Inverse spinel structure : Structure of Fe_3O_4 or Fe^{2+} Fe_2^{3+} O_4 or FeO^{2+} Fe_2^{3+} O_3 (magnetite) and Mg^{2+} Fe_2^{3+} O_4 (magnesium ferrite)

Fe_3O_4 has the composition $Fe^{2+}OFe_2^{3+}O_3$ which shows that it has Fe^{2+} as well as Fe^{3+} ions. In the structure of Fe_3O_4 three Fe^{2+} ions have been replaced by two Fe^{3+} ions to balance the charge. In its structure, oxide ions (O^{2-} ions) have cubic close-packed (*ccp*) arrangement. Fe^{2+} ions (divalent cations) occupy octahedral holes while Fe^{3+} ions (trivalent cations) are equally distributed between octahedral and tetrahedral holes. Such a structure is called *inverse spinel structure.*

$MgFe_2O_4$ also has inverse spinel structure. The structure of $MgFe_2O_4$ can be obtained by replacing Fe^{2+} ion of Fe_3O_4 ($Fe_3O_4 \equiv Fe^{2+}O.\ Fe_2{}^{3+}O_3$) by Mg^{2+} ion. Thus, $MgFe_2O_4$ can be represented as $Mg^{2+}O.\ Fe_2{}^{3+}O_3$.

University Questions

Classification of crystals

1. Name any four crystal systems. Give the characteristics of any two of them. *(Meerut M.Sc., 1992)*

2. What is the difference between simple cubic, body-centered and face-centered lattices. *(Meerut, 1994)*

3. Calculate the limiting radius ratio (r_+/r_-) for coordination number 6 of a cation in an ionic lattice. *(Meerut M.Sc., 1992)*

4. Write note on :

(*i*) Miller indices *(Bahuguna, 1992)*
(*ii*) Space lattice *(Bahuguna, 1992)*
(*iii*) Ionic crystal lattice *(Agra 1992; Saurashtra, 1991)*
(*iv*) Unit cell *(MS Baroda, 1990; Saurashtra, 1991; Bahuguna, 1992)*
(*v*) Liquid crystals *(Saurashtra, 91)*
(*vi*) Packing of ions in crystals *(Meerut, 1990)*
(*vii*) Isomorphism *(Meerut, 1990)*

5. With the help of neat diagrams discuss the arrangement of atoms in three types of structures (*hcp, ccp, bcc*) in which metals generally crystallise. Why are some metals more malleable and ductile than others ? *(Meerut M.Sc., 1994)*

6. Give the rules to determine the number of points/atoms/ions in one unit cell and find out how many cations and anions are present in sodium chloride unit cell. *(M.S. Baroda, 1990)*

7. Draw unit cell of NaCl. *(Saurashtra, 1991)*

8. Explain the terms : axes of symmetry, plane of symmetry and centre of symmetry. *(Himachal, 1992)*

9. What are the typical properties of ionic crystal ? *(Pondicherry, 1992)*

10. Differentiate between the cubic close packing and hexagonal close packing of spheres. *(Punjab, 1992)*

11. How are the minimum values of radius ratio arrived at for various coordination numbers and what are their limits ? What type of crystal structures are associated with different coordination numbers ? *(Meerut M.Sc., 1991)*

12. Draw the diagrams to represent *f.c.c.* lattice and *b.c.c.* lattice. *(Guwahati, 1991; Bangalore, 92)*

13. A metallic element crystallises into a lattice containing a sequence of layers of ABABAB... . Any packing of spheres leaves out void in the lattice. What percentage by volume of this lattice is empty space ? *(I.I.T., 1996)*

14. What are the three types of cubic crystals ? How are they distinguished from each other ? Calculate the number of atoms per unit cell in each case. *(Utkal, 1992)*

15. Give a brief account of hexagonal close packed (*hcp*) structures of crystals. *(Usmania, 1993)*

16. Discuss "Limiting radius ratio" for coordination numbers 3, 4, 6 and 8. Discuss the types of crystal structures of ionic compounds of the type AX and AX_2. *(Lucknow, 1995)*

Conductors, non-conductors and semiconductors

17. Differentiate between conductors and non-conductors on the basis of molecular orbital theory of metallic bonding. *(Agra, 1992)*

18. What are semiconductors ? Why are they so called ? *(Guwahati, 1991)*

Crystal structure of some ionic crystals

19. Describe the crystal structure of caesium chloride. *(E.C.C. Allahabad, 1994)*

20. What will be the coordination number of each ion in NaCl, if the ionic radius of $Na^+ = 0.95$ Å and that of $Cl^- = 1.81$ Å. Predict the shape of NaCl crystal. *(Lucknow, 1993)*

21. How is an octahedral hole formed in a close-packed structure? *(Meerut M.Sc., 1994)*

22. Discuss the crystal structures of rutile and cesium chloride. *(Meerut M.Sc., 1992)*

23. Discuss the crystal structure of NaCl and ZnS. *(Meerut M.Sc. 1991, 94; Bahuguna, 1992)*

24. Silver iodide has a zinc blende structure. Draw a unit cell of silver iodide. What are the coordination number of Ag^+ and I^- an AgI ? How many formula units (AgI) are present in a unit cell ? *(Meerut M.Sc. 1993)*

25. What is coordination number ? What is the coordination number of each atom in a hexagonal closed packed structure ? *(Meerut M.Sc. 1994)*

Lattice energy, solvation energy, dissolution of salts and Born-Haber cycle

26. The energy involved in the dissolution of salts can be handled conveniently by Born-Haber cycle. Explain. *(E.C.C. Allahabad 1994)*
27. Explain with examples "Born-Haber cycle". *(Ruhelkhand 1990; Purvanchal 1993; Kurukshetra, 1992)*
28. What is lattice energy? Describe how lattice energy of ionic solids is evaluated by Born-Haber cycle? *(Utkal 1992)*
29. Explain lattice energy and solvation energy. *(Purvanchal 1993; Rohilkhand 1991; Meerut 1994)*
30. What is lattice energy ? What are the factors which control lattice energy ? *(Meerut M.Sc. 1993)*
31. In which case would you expect larger difference in the calculated and experimental lattice energy, LiBr or KF ? Why ? *(Meerut M.Sc. 1993; Delhi, 1986; Raj. 1986)*
32. What is Madelung constant ? Why do NaCl and MgO have the same value of this constant ? *(Meerut M.Sc. 1994)*
33. Calculate the limiting radius ratio (r_+/r_-) for tetrahedral coordination of a cation in an ionic lattice. *(Meerut M.Sc. 1993)*
34. Write a note on "Hydration Energy". *(Meerut 1990)*
35. Write a note on "Crystal Energy". *(Meerut 1990)*
36. Arrange the following in the increasing order of their lattice energy : CaO, RbBr, KCl, MgO and CsI. *(Meerut M.Sc. 1992)*
37. Describe an expression for the lattice energy of one mole of NaCl crystal. *(Meerut M.Sc. 1992)*
38. Why do the calculated and Born-Haber cycle lattice energies differ sometime ? Explain with two suitable examples. *(Meerut M.Sc. 1992)*
39. Which of the following halides has higher value of its lattice energy and why ? LiF or LiI. *(Bharathiar 1986)*
    (Ans : *Li F)*
40. Explain the following, giving reasons:
    (*i*) LiF has higher lattice energy than LiI. *(Bharathiar 1986)*
    (*ii*) Although the size of the bivalent cations of alkaline earth metals increase from Be^{2+} to Ba^{2+} down the group, the lattice energy of the sulphates, nitrates and chlorides of these metals are almost the same.
    (*iii*) Lattice energy of the fluorides of alkali metals decreases from LiF. (= 1032 kJ/mole) to CsF (= 726 kJ/mole) *(Delhi Hons. 1985)*
    (*iv*) Degree of hydration of alkali metal cations decreases from Li^+ ion to Cs^+ ion. *(Delhi Hons. 1985)*
    (*v*) Li^+ ion is most extensively hydrated among alkali metal cations. *(Delhi 1985)*
    (*vi*) Hydration energy of Na is greater than that of K. *(Guwahati 1991)*
    (*vii*) Solubility of the halides of a given metal increases in the order :
    Fluoride < Chloride < Bromide < Iodide
    (*viii*) The solubility of the hydroxides of alkaline earth metals increases in moving down the group. *(Delhi 1982)*
    (*ix*) $MgSO_4$, $7H_2O$ is soluble in water whereas $BaSO_4$ is insoluble. *(Raj. 1983)*
    (*x*) The solubility of the sulphates of alkaline earth metals decreases on descending the group from Be to Ba while the solubility of the fluorides of these metals increases on moving down the group. *(Delhi 1981, 1982, GND 1982, 1983)*
    (*xi*) $CaCO_3$ is less soluble in H_2O than K_2CO_3.
    (*xii*) $BaSO_4$ has a very low solubility in water.
    (*xiii*) LiI is soluble in H_2O while LIF is not. *(Jodhpur 1987)*
    Answer : LiF, due to its highest value of lattice energy among the halides of Li-metal, is insoluble in H_2O while LiI which has the minimum value of lattice energy is soluble in H_2O.
    (*xiv*) The melting points of calcium halides decreases in the order : $CaF_2 > CaCl_2 > CaBr_2 > CaI_2$. *(Jodhpur 1985; Punjab 1985)*

Numerical Problems

41. Calculate the lattice energy of NaI with the following data : The heat of formation of NaI = – 287.6 kJ mol^{-1}, sublimation and ionisation energies of Na = 108.7 and 493.8 kJ mol^{-1} respectively. The value of sublimation and dissociation energy of iodine is 106.6 kJ mol^{-1} and electron affinity of iodine atom = – 305.9 kJ mol^{-1}. *(Ravishankar, 93)*

42. Calculate the proton affinity of NH_3 [*i.e.*, enthalpy of the reaction $NH_3\ (g) + H^+(g) \rightarrow NH_4^+\ (g)$] and enthalpy of formation of $NH_4^+\ (g)$ from the following data : Lattice energy of $NH_4Cl\ (s)$ = – 162 Kcal mol^{-1}, enthalpy of formation of NH_4Cl = – 75 Kcal mol^{-1}, enthalpy of dissociation of $H_2\ (g)$ = 104 Kcal mol^{-1}, enthalpy of dissociation of $Cl_2\ (g)$ = 58 Kcal mol^{-1}, ionisation energy of H (*g*) = 314 Kcal mol^{-1}, electron affinity of Cl (*g*) = – 83 Kcal mol^{-1}, enthalpy of formation of $NH_3\ (g)$ = – 11 Kcal mol^{-1}. *(Meerut M.Sc., 1993)*

43. Assuming atoms as spheres of the same size, discuss the geometry of the solids formed by packing them in 3-dimensions and hence distinguish between hexagonal close packing (HCP) and cubic close packing (CCP). How are tetrahedral and octahedral holes generated in solids and do they differ from each other? *(Lucknow, 1994)*

44. When a mole of crystalline NaCl is prepared from 1 gram atom of Na and 0.5 mole of Cl_2 gas, 410 kJ of heat is produced. The heat of sublimation of Na metal is 108.8 kJ. The heat of dissociation of Cl_2 gas into atoms is 242.7 kJ, the ionisation energy of Na is 493.7 kJ and electron affinity of Cl is 368.2 kJ. Calculate the lattice energy of NaCl. *(Meerut M.Sc., 1994)*

45. Aluminium forms face-centred cubic crystals. The density of Al is 2.70 g/cm^3. Calculate the length of the side of the unit cell of aluminium (At. wt. of Al = 27) (**Ans.:** 4.054×10^8 cm) *(Pune, 1992)*

46. Gold has a face-centred cubic structure with a unit length 4.07 Å and density equal to 19.3 g/cm^3. Calcualte the Avogadro's number from these data (At. wt. of Au = 197) (**Ans.:** 6.05×10^{23}) *(Sauhashtra, 1992)*

47. Polunium crystallises in a simple cubic unit cell. It has atomic mass = 209 and density = 91.5 kg m^{-3}. What is the edge length of its unit cell ? (**Ans :** 15.59×10^{-8} cm) *(Delhi, 1992)*

48. Using the following data calculate the enthalpy of formation of CaF(*s*) and explain its stability. Estimated lattice energy of CaF (*s*) = – 795 kJ mol^{-1}, enthalpy of atomisation of Ca(*s*) = 201 kJ mol^{-1}, 1*st* ionisation energy of Ca (*g*) = 590 kJ mol^{-1}, enthalpy of atomisation of $F_2(g)$ = 79 kJ mol^{-1}, electron affinity of F (*g*) = – 335 kJ mol^{-1}, enthalpy of $CaF_2\ (s)$ = – 1243 kJ mol^{-1}. *(Meerut M.Sc., 1992)*

Mettalic Bond

49. What is meant by metallic bond ? Explain main theories of metallic bonding in brief. *(Agra, 1992; Utkal, 1992; Rohelkhand, 1991)*

50. Write a note on free electron theory of metallic bond. *(Usmania, 1993)*

51. Explain the following :

(*i*) Metals are good conductors of electricity. *(Raj., 1978)*

(*ii*) Hybrid orbitals are more suitable for metallic bonding. *(Raj., 1978, 1980)*

(*iii*) Copper is ductile and malleable but brass is tenacious.

Graphite and diamond

52. Compare the properties, structure and used of diamond and graphite. *(Meerut, 1993, 1994)*

53. "What is allotropy"? What is the difference in the spatial arrangement of carbon atoms in diamond and graphite ? *(Rohilkhand, 1991; Meerut, 1991)*

54. Based on their crystal structure give a comparative account of the properties of diamond and graphite. *(Meerut, 1990)*

55. Explain the following, giving appropriate reasons :

(*i*) Melting point of diamond is very high, although it is a covalent crystal. *(M.D. Rohtak, 1984)*

(*ii*) Graphite acts as an abbrasive and as a lubricant. *(I.I.T., 1981)*

(*iii*) Pure graphite shows intrinsic conductivity. *(Pune, 1986)*

(*iv*) Graphite is a good conductor of electricity while diamond is not. *(Raj. 1987; Meerut, 1988)*

(*v*) Diamond is a hard substance. *(Meerut, 1994)*

(*vi*) Diamond is a non-conductor, while graphite is a good conductor. *(Meerut, 1988 S)*

(*vii*) Unlike diamond, graphite is soft; but like diamond it has a high melting point. *(Allahabad, 1989)*

(*viii*) Diamond is a hard substance while graphite is soft and greasy solid. *(Meerut 1988)* (*ix*) Graphite forms Lamellar compounds but not diamond.

Defects in crystals

56. What are stoichiometric defects? How are they caused? Discuss the consequences of these defeats. *(Lucknow, 1993)*

57. What type of lattice defects are responsible for the existence of non-stoichiometric compounds of AB type ? Which defect causes the crystals to act as semiconductors ? *(Meerut M.Sc., 1993)*

58. What are crystal defects ? How are they caused ? Explain non-stoichiometric defects in crystals and its consequences. *(Agra, 1992; Guwahati, 1991)*

59. Wht is "Isomorphism" ? *(Utkal, 1992)*

60. What are "Non-stoichiometric defects" ? *(Lucknow, 1995)*

61. Define the term lattice energy of an ionic solid. Describe how is the lattice energy of sodium chloride calculated with the help of Born-Haber cycle. *(Nagpur 2002)*

62. Write a note on Born Haber Cycle. *(Lucknow 2002)*

63. Explain why graphite is soft and diamond is hard. *(Delhi 2002)*

64. What is allotropy ? Explain why graphite is a good conductor of electricity while diamond is not. *(VBS Purvanchal 2003)*

65. What is lattice energy ? How is it determined ? *(VBS Purvanchal 2003)*

66. Describe band theory of metals. How does it explain the behaviour of conductors, bad conductors and semiconductors ? *(VBS Purvanchal 2003)*

67. Graphite is a good conductor of electricity whereas diamond is a non-conductor of electricity. Explain. *(Rohilkhand 2003)*

68. (*a*) What is lattice energy ? How does the solubility of an ionic solid depend upon its lattice energy ?

(*b*) Discuss in details the Born-Haber cycle for calculating lattice energy.

(*c*) Discuss radius ratio rules for determining the shapes of ionic crystals. *(Rohilkhand 2003)*

69. Write a short note on metallic bond. *(Rohilkhand 2003)*

70. What is radius ratio rule ? How can the structure of ionic crystal be predicted with its help ? *(Nagpur 2003)*

71. (*a*) Define : (*i*) Lattice energy and (*ii*) Solvation energy.

(*b*) Write brief notes on : (*i*) Radius ratio rule (*ii*) Born-Haber cycle for calculating lattice energy.

(*c*) Discuss the structure of SiO_4^{4-} ion. *(Nagpur 2003)*

72. (*a*) Draw the crystal structure of NaCl.

(*b*) What are the limitations of radius ratio rule.

(*c*) Draw the crystal structure of CaF_2.

(GND Amritsar, 2004)

17

Redox Reactions

Oxidation number (O.N.) and oxidation state

In most of the cases, these two terms are used in the same sense, *i.e.*, these are synonymous (*i.e.* same) terms. Oxidation number of an atom is the apparent charge on the atom in its compounds or on its ion, while oxidation state of an atom is the oxidation number per atom of that element in a given compound or an ion, *i.e.*, the oxidatin state of an atom is equal to the total oxidation number of that atom in the molecule, divided by the number of the atoms present in the molecule. For example, since the total oxidation number of Cr-atom in $K_2Cr_2O_7$ molecule is + 12, oxidation state of Cr-atom, in this molecule, is equal to $+12/2 = +6$.

The O.N. of an atom in its ionic compound is equal to the number of positive or negative charges on its ion *e.g.*, the O.N. of Na and Cl atoms, in Na^+Cl^- ionic compound is equal to +1 and –1 respectively.

The O.N. of an atom, in its covalent compound, is equal to the charge which the atom appears to have acquired when the electron pairs being shared by the two linked atoms are counted according to the following arbitrary rules :

(*i*) When the sharing of electron pairs takes place between like (*i.e.*, same) atoms, the half of the shared electrons are counted along with the outermost shells of each of the two like atoms. On the basis of this rule, it can be shown that the O.N. of each Cl-atom in Cl_2 molecule is zero. Since in Cl_2 molecule the sharing of two electrons takes place between two like atoms *viz.* Cl-atoms, one electron will be counted with each Cl-atom, *i.e.*, each Cl-atom will have 7 electrons. Now since chlorine nucleus has +7 charges and 7 negative charges are also counted with the nucleus, the net charge which Cl-atom appears to have in Cl_2 molecule is zero. Thus, the O.N. of each Cl-atom in Cl_2 molecule is zero. Similarly, it can also be shown that the O.N. of each N-atom in N_2 molecule, of each P-atom in P_4, of each S-atom in S_2 etc., is also zero.

(*ii*) When the sharing of electrons takes place between two unlike (*i.e.*, different) atoms, all the shared electrons are counted along with the outermost shell electrons of the more electronegative atom. On the basis of this rule, it can be shown that O.N. of O-atom in H_2O, H_2O_2 and OF_2 molecules is –2, –1 and +2 respectively. In H_2O molecule (H—O—H) :

Charges on the nucleus of O-atom = + 8

Charges due to 2 electrons present in the inner-shell of O-atom = – 2

Charges due to 6 electrons in the outer-shell of O-atom = – 6

Charges due to 2 electrons of two H-atoms counted with O-atom = –2

∴ Apparent charges on O-atom = + 8 – 2 – 6 – 2 = – 2.

Thus the O.N. of O-atom in H_2O molecule = – 2.

H_2O_2 molecule (H—O—O—H) has two O—H bonds and one O—O bond. The electron pair being shared by O and H atoms in O—H bond will be counted with O-atom which is more electronegative than H-atom, but the electrons pair being shared by two O-atoms in O—O bond will be counted with each O-atom. Thus, the O.N. of each O-atom in H_2O_2 molecule is –1. In OF_2 molecule (F—O—F), both pairs of electrons responsible for the formation of two O—F bonds will be counted with F-atom, which is more electronegative than O-atom. Thus, O.N. of O-atom in OF_2 molecule is +2.

The following additional points may be noted in connection with oxidation number of an atom :

(*i*) O.N. of an atom may be zero or any whole number or fractional number, having plus (+) or minus (–) sign associated with it. For example, O.N. of each Cl-atom in Cl_2 molecule is zero, O.N. of C-atom in CO molecule is +2, O.N. of N-atom in N_2H is –1/3, O.N. of C-atom in CH_4 is – 4, O.N.of S in $Na_2S_4O_6$ is +5/2 etc.

(*ii*) The O.N. of an atom may be different in its different compounds. For example, O.N. of O-atom in H_2O, H_2O_2 and OF_2 molecules is –2, –1 and + 2 respectively.

(*iii*) In A → B coordinate bond, since the electron pair is donated by A, it appears that A has lost two electrons and hence its O.N. is +2 and consequently the O.N. of B will be –2.

General rules for calculating the O.N. of elements

Since it is laborious to draw the structure of all the molecules and then to find out the O.N. of the concerned element or elements, some general rules have been found out, which can be used to calculate the O.N. of the elements. These rules are given below :

1. O.N. of an element, in the free state or uncombined state (*i.e.*, elementry state), is zero, *e.g.*, O.N. of Cl in Cl_2, of P in P_4 and of S in S_8 is zero.

2. O.N. of an atom, in a monoatomic ion (*e.g.*, Al^{3+},Cl^- etc.), is equal to the number of positive or negative charges on the ion, *e.g.*, O.N. of Al-atom in Al^{3+}ion and of Cl-atom in Cl^- ion is +3 and –1 respectively.

3. The sum of the oxidation numbers of all the atoms, in a given molecule, is zero. This sum is equal to the sum of oxidation numbers of the individual atoms, each multiplied by the number of the atoms present in the molecule, *e.g.*, in Na_2SO_4 molecule, the sum of oxidation numbers = 2 × (+1) + 1 × (+ 6) + 4 × (–2) = 2 + 6 – 8 = 0.

4. In olyatomic ions [*e.g.*, $Fe(CN)_6]^{4-}$, $Cr_2O_7^{2-}$, SO_4^{2-}, NH_4^+etc.], the algebraic sum of O.N. of all the constituent atoms is equal to the number of positive or negative charges on the polyatomic ion, *e.g.*, in $Cr_2O_7^{2-}$ ion, the sum namely 2 × (+ 6) + 7 × (–2) is equal to –2 which is the charge on $Cr_2O_7^{2-}$ ion .

5. In compounds containing ions or radicals (*e.g.*, NO_3^-, NH_2^-, N^{3-},CN^-, OH^-, SCN^-, NH_4^+, $-CH_3^+$, NH_3^0, CO^0, NO^+ etc.), it is more convenient to use the O.N. of the ions or radicals, rather than to assign the oxidation numbers to the individual atoms, *e.g.*, O.N. of Cl-atom in both the compounds, namely NOCl (nitrosyl chloride) and CH_3Cl (methyl chloride) is –1, since O.N. of NO^+ (nitrosyl) and CH_2^+ (methyl) groups is +1. Similarly O.N. of Fe in sodium nitroprusside, Na_2 $[Fe(CN)_5 (NO)]$ is +2, since oxidation numbers of CN^- and NO^+ groups are –1 and +1 respectively.

6. O.N. of a non-metal is negative in its compound with metals or with hydrogen, *e.g.*, O.N. of N in both the molecules *viz.* Ca_3N_2 and NH_3 is –3.

7. O.N. of the metals of groups IA (alkali metals) and of group IIA (alkaline earth metals) is +1 and +2 respectively.

8. The O.N. of an atom is a variable quantity, since it may be different in different compounds. The fact is evident from the following examples :

(*i*) O.N. of H-atom is : (*a*) zero in free H-atom or H_2 molecule (*b*) +1 in compounds like H_2O_2, NH_2OH, HN_3, HX(X = F, Cl, Br, I), PH_3, NH_3, NH_4^+, PH_4^+, NH_4^-, NH^{2-}, H_2SO_4, SiH_4, CH_4, C_2H_6, C_3H_8, CH_3Cl, C_2H_4, C_2H_2, $CHCl_3$, HCN, $H_2C_2O_4$, H_2CO_3 etc. (*c*) –1 in metallic hydrides *e.g.*, CaH_2, NaH, $LiAlH_4$ etc.

(*ii*) O.N. of carbon atom is : (*a*) – 4 in CH_4, Al_4C_3, etc. (*b*) –3 in C_2H_6 (*c*) – 8/3 in C_3H_8 (*d*) – 2 in CH_3 Cl, C_2H_4 etc. (*e*) – 4/3 in Mg_2C_3 etc. (*f*) – 1 in CaC_2, C_2H_2 (*g*) zero in diamond, graphite, CH_2Cl_2, $C_6H_{12}O_6$ (glucose), $C_{12}H_{22}O_{11}$, CH_2O etc. (*h*) +2 in CO, $CHCl_3$, HSCN, HCN, CN^- etc. (*i*) +3 in $H_2C_2O_4$, $C_2O_4^{2-}$ etc. (*j*) + 4 in CO_2, H_2CO_3, CCl_4, Na_2CO_3, CS_2, CF_4 etc.

(*iii*) O.N. of N-atom is : (*a*) – 3 in NH_3, NH_4^+, AlN, SCN^-, CN^- etc. (*b*) – 2 in N_2H_4 etc. (*c*) –1 in NH_2OH etc. (*d*) –1/3 in HN_3, N_3^-, NaN_3 etc. (*e*) zero in free nitrogen atom or N_2 molecule (*f*) +1 in N_2O etc. (*g*) +2 in N_2O_2 or NO etc. (*h*) +3 in HNO_2, N_2O_3, NF_3 etc. (*i*) + 4 in NO_2, N_2O_4 etc. (*j*) +5 in HNO_3, N_2O_5 etc.

(*iv*) O.N. of oxygen atom is : (*a*) –2 in Cl_2O, H_2O, PbO_2, CO_3^{2-}, SO_2, $C_2O_4^{2-}$, HOX (X = Cl, Br, I), NH_2OH etc. (*b*) –1 in compounds containing peroxide ion, O_2^{2-} (*e.g.*, Na_2O_2, BaO_2, H_2O_2 etc.) (*c*) –1/2 in KO_2 (*d*) zero in free O-atom and O_2 and O_3 molecules (*e*) +1 in O_2F_2 (*f*) +2 in OF_2.

(*v*) O.N. of Si-atom is : (*a*) – 4 in SiH_4, Mg_2Si etc. (*b*) + 4 in SiO_2 and K_2SiO_3.

(*vi*) O.N. of P-atom is : (*a*) –3 in PH_3, PH_4^+, Ca_3P_2 etc. (*b*) zero in P_4 (*c*) +1 in H_3PO_2 (*d*) +3 in PX_3 (X = Cl, Br, I), P_2O_3, H_3PO_3 etc. (*e*) +5 in H_3PO_4, $H_4P_2O_7$, P_4O_{10}, PCl_5 etc.

(*vii*) O.N. of sulphur atom is : (*a*) –2 in compounds containing S^{-2} (*e.g.*, H_2S, ZnS etc.) and HS^- (*e.g.*, NaHS etc.) ions. (*b*) zero in S, S_4, S_8, SCN^-, $(CH_3)_2$ SO etc. (*c*) +2 in $Na_2S_2O_3$ etc. (*d*) + 2.5 in $Na_2S_4O_6$ etc. (*e*) +4 in SO_2, H_2SO_3, $SOCl_2$, $NaHSO_3$, SF_4 etc. (*f*) +1 in S_2Cl_2, S_2Br_2 etc. (*g*) + 6 in H_2SO_4, SO_2Cl_2, SO_3, SF_6, $H_2S_2O_7$ etc.

(*viii*) O.N. of Cl-atom is : (*a*) –1 in HCl, ICl, ICl_5, $SOCl_2$, SO_2Cl_2, CrO_2Cl_2, $K_2[PtCl_6]$, $H[AuCl_4]$, CCl_4 etc. (*b*) zero in Cl, Cl_2 (*c*) +1 in HClO, Cl_2O (*d*) +3 in $HClO_2$, ClF_3 etc. (*e*) +4 in ClO_2 etc. (*f*) +5 in $HClO_3$ (*g*) +7 in $HClO_4$, Cl_2O_7 etc.

(*ix*) In all the halides (*i.e.*, compounds of HX or MX type, where X is a halogen and M is a metal atom) the O.N. of the halogen is –1.

Solved Examples

Example 17.1. *Find out the oxidation number of (i) Cr in CrO_2Cl_2 (ii) Pt in $H_2[PlCl_6]$ (iii) U in $UO_2\ (NO_3)_2$ (iv) N in NH_4VO_3.*

Solution. Let the oxidation number of the required element be x. Then

(*i*) $x + 2 \times (-2) + 2 \times (-1) = 0$

or $x - 4 - 2 = 0$ or $x = +6$.

Thus O.N. of Cr atom is $CrO_2Cl_2 = +6$.

(*ii*) $2 \times (+1) + x + 6 \times (-1) = 0$.

or $2 + x - 6 = 0$ or $x = +4$.

Thus O.N. of Pt metal in $H_2[Pt\ Cl_6] = +4$.

(*iii*) In $UO_2\ (NO_3)_2$, O.N. of NO_3^- ion is –1 and hence $x + 2 \times (-2) + 2 \times (-1) = 0$.

or $x - 4 - 2 = 0$ or $x = +6$.

Thus, O.N. of U-metal in $UO_2\ (NO_3)_2 = +6$.

(*iv*) In $NH_4\ VO_3$, O.N. of VO_3^- ion is –1 and hence $x + 4 + (+1) + 1 \times (-1) = 0$.

or $x + 4 - 1 = 0$ or $x = -3$.

Thus O.N. of N-atom in $NH_4VO_3 = -3$.

Distinction between oxidation number and valency or covalency of an element

The main points of difference between oxidation number and valency have been given below in Table 17.1

Table 17.1

Valency or covalency	Oxidation number
(*i*) Valency is the combining capacity of an element which is equal to the number of H-atoms or double the number of O-atoms which combine with one atom of the element. Covalency of an element is the number of covalent bonds by which that element is linked with the surrounding elements.	(*i*) Oxidation number (O.N.) is the residual charge which an atom has or appears to have when the other atoms from the molecule are removed as ions by counting the shared electrons with more electronegative atoms.
(*ii*) It is a number only and does not carry any plus (+) or minus (–) sign. For example in CO_2 molecule (O = C = O), the covalency of C-atom is 4 and that of O-atom is 2.	(*ii*) It refers to the charge which can be positive, negative or zero. For example in CO_2 molecule, the oxidation number of C-atom is + 4 and that of O-atom is – 2 ($O^{-2} = C^{+4} = O^{-2}$)

Valency or covalency	Oxidation number
(*iii*) The valency of an element is always a whole number.	(*iii*) The O.N. of an element may also have fractional value, *e.g.*, O.N. of S-atom in $Na_2S_4O_6$ is +2.5.
(*iv*) Many elements like C, N and O exhibit constant valency.	(*iv*) Even the elements like C, N and O which exhibit constant valency can have variable oxidation numbers.

The differences given above can be seen from the examples given below :

(*i*) *Molecules :*

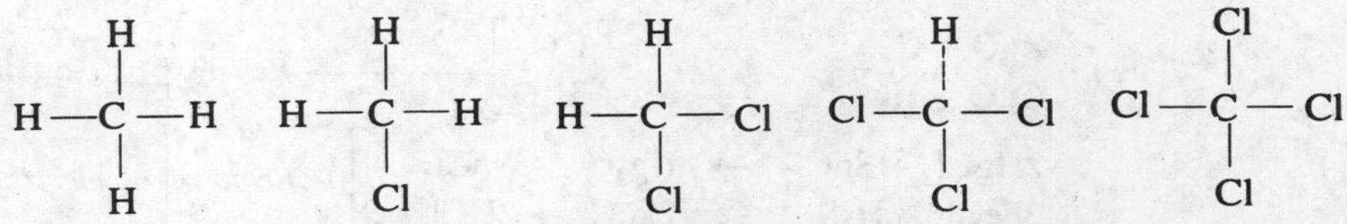

	(CH_4)	(CH_3Cl)	(CH_2Cl_2)	($CHCl_3$)	(CCl_4)
Covalency of C-atom =	4	4	4	4	4
O.N. of C-atom =	−4	−2	0	+ 2	+ 4

(*ii*) *Molecules :*

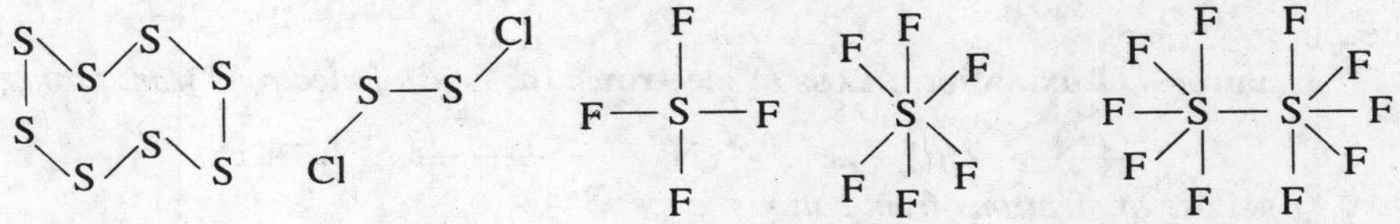

	(S_8)	(S_2Cl_2)	(SF_4)	(SF_6)	(S_2F_{10})
Covalency of S-atom =	0	2	4	6	6
O.N. of S-atom =	0	+1	+4	+ 6	+ 5

(*iii*) *Molecules :*

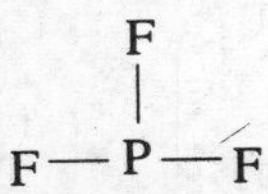

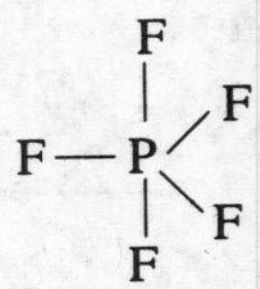

	(PF_3)	(PF_5)
Covalency of P-atom =	3	5
O.N. of P-atom =	+3	+5

Oxidation and Reduction

Various definitions have been given from time to time.

1. Old definition. Oxidation is defined as a process in which oxygen or any other electronegative element is added or hydrogen or any electropositive element is removed.

Examples :

$2Mg + O_2 \rightarrow 2MgO$
$C + O_2 \rightarrow CO_2$ .... *Addition of oxygen*

$Cu + Cl_2 \rightarrow CuCl_2$
$Hg_2Cl_2 + Cl_2 \rightarrow 2HgCl_2$ .... *Addition of electronegative element*

$3Fe + 4H_2O$ (*Steam*) $\rightleftharpoons Fe_3O_4 + 4H_2$
$Mg + H_2O$ (*Steam*) $\rightarrow MgO + H_2$ .... *Removal of hydrogen*

$H_2S + Cl_2 \rightarrow 2HCl + S$ ------------ *Removal of electropositive element*

Reduction is a process in which hydrogen or any electropositive element is added or oxygen or any electronegative element is removed.

Examples : $2C + H_2 \rightarrow C_2H_2$, $N_2 + 3H_2 \rightarrow 2NH_3$ ------------------ *Addition of hydrogen*

$HgCl_2 + Hg \rightarrow Hg_2Cl_2$, $CuCl_2 + Cu \rightarrow 2CuCl$ ------------------ *Addition of electropositive element*

$ZnO + C \rightarrow Zn + CO$, $CuO + H_2 \rightarrow Cu + H_2O$ -------------- *Removal of oxygen*

$2HgCl_2 + SnCl_2 \rightarrow Hg_2Cl_2 + SnCl_4$, $2FeCl_3 + H_2S \rightarrow 2FeCl_2 + 2HCl + S$ --- *Removal of electronegative element.*

2. Definition based on electron transfer : Electronic concept. According to the electronic concept, oxidation is a process in which an atom, an ion (cation or anion) or a molecule loses one or more electrons *i.e.*, electrons are released in an oxidation process. On the other hand, reduction is a process in which an atom, an ion or a molecule gains (accepts) one or more electrons, *i.e.*, electrons are consumed in a reduction process. Thus, in an oxidation process electrons are shown at the RHS of the reaction, while in a reduction process electrons are shown at the LHS of the reaction.

Examples of oxidation : Loss of electrons. (*a*) *Loss of electrons from atoms, e.g.,*

$Zn \xrightarrow{-2e^-} Zn^{2+} + 2e^-$; $Na \xrightarrow{-e^-} Na^+ + e^-$

(*b*) *Loss of electrons from cations, e.g.,*

$Fe^{2+} \xrightarrow{-e^-} Fe^{3+} + e^-$; $Sn^{2+} \xrightarrow{-2e^-} Sn^{4+} + 2e^-$

(*c*) *Loss of electrons from anions, e.g.,*

$2I^- \xrightarrow{-2e^-} I_2 + 2e^-$; $2S_2O_3^{2-} \xrightarrow{-2e^-} S_4O_6^{2-} + 2e^-$

$[Fe(CN)_6]^{4-} \xrightarrow{-e^-} [Fe(CN)_6]^{3-} + e^-$

(*d*) *Loss of electrons from molecules, e.g.* ,

$H_2 \xrightarrow{-2e^-} 2H^+$; $H_2O_2 \xrightarrow{-2e^-} O_2 + 2H^+ + 2e^-$

$2H_2O \xrightarrow{-4e^-} O_2 + 4H^+ + 4e^-$

Examples of reduction : Gain of electrons. (*a*) *Gain of electrons by atoms, e.g.,*

$F + e^- \rightarrow F^-$; $Cl + e^- \rightarrow Cl^-$

(*b*) *Gain of electrons by cations, e.g.,*

$Cu^{2+} + 2e^- \rightarrow Cu$; $Fe^{3+} + e^- \rightarrow Fe^{2+}$

$2H^+ + 2e^- \rightarrow H_2$; $2Hg^{2+} + 2e^- \rightarrow Hg_2^{2+}$

(*c*) *Gain of electrons by anions, e.g.,*

$MnO_4^- + 8H^+ + 5e^- \rightarrow Mn^{2+} + 4H_2O$

$Cr_2O_7^{2-} + 14H^+ + 6e^- \rightarrow 2Cr^{3+} + 7H_2O$

$2IO_3^- + 12H^+ + 10e^- \rightarrow I_2 + 6H_2O$

(*d*) *Gain of electrons by molecules, e.g.,*

$H_2O_2 + 2H^+ + 2e^- \rightarrow 2H_2O$

$O_2 + 4H^+ + 4e^- \rightarrow 2H_2O$

$2H_2O + 2e^- \rightarrow H_2 + 2OH^-$

$Cl_2 + 2e^- \rightarrow 2Cl^-$

3. Definition based on the change (increase or decrease) in oxidation number : Oxidation number concept. According to this concept, oxidation is defined as a process in which the oxidation number (O.N.) of the effective element present in the species, increases, while reduction is a process in which the O.N. of the effective element decreases.

Examples of oxidation. Increase in O.N. has been shown in bracket.

$Zn^0 \rightarrow Zn^{2+} + 2e^-$ $(0 \rightarrow +2)$

$Fe^{2+} \rightarrow Fe^{3+} + e^-$ $(+2 \rightarrow +3)$

$2I^- \rightarrow I_2^0 + 2e^-$ $(-1 \rightarrow 0)$

$2S_2O_3^{2-} \rightarrow S_4O_6^{2-} + 2e^-$ $(+2 \rightarrow +2.5)$
(S = +2) (S = +2.5)

$[Fe(CN)_6]^{4-} \rightarrow [Fe(CN)_6]^{3-} + e^-$ $(+2 \rightarrow +3)$
(Fe = +2) (Fe = +3)

$H_2O_2 \rightarrow O_2 + 2H^+ + 2e^-$ $(-1 \rightarrow 0)$
(O = –1) (O = 0)

Examples of reduction. Decrease in O.N. has been indicated in bracket.

$F^0 + e^- \rightarrow F^-$ $(0 \rightarrow -1)$

$Cu^{2+} + 2e^- \rightarrow Cu^0$ $(+2 \rightarrow 0)$

$MnO_4^- + 8H^+ + 5e^- \rightarrow Mn^{2+} + 4H_2O$ $(+7 \rightarrow +2)$
(Mn = +7) (Mn = + 2)

$H_2O_2 + 2H^+ + 2e^- \rightarrow 2H_2O$ $(-1 \rightarrow -2)$
(O = –1) (O = – 2)

$2H_2O + 2e^- \rightarrow H_2 + 2OH^-$ $(+1 \rightarrow 0)$
(H = +1) (H = 0)

$ClO_2 + 2H_2O + 5e^- \rightarrow Cl^- + 4OH^-$ $(+4 \rightarrow -1)$
(Cl = + 4) (Cl = –1)

Redox Reactions and Half Reactions

Redox reaction is that chemical reaction in which oxidation (*i.e.*, loss of electrons or increase in oxidation number) and reduction (*i.e.*, gain of electrons or decrease in oxidation number) take place simultaneously, *i.e.*, in a redox reaction one substance is oxidised and the other substance is reduced and hence oxidation and reduction proceed simultaneously.

A redox reaction is also called *oxidation-reduction reaction.* A redox reaction is composed of two reactions one of which represents only oxidation while the other represents only reduction. The reaction representing oxidation is called *oxidation half-reaction* while that representing reduction is called *reduction half-reaction.* Both these reactions are collectively called *half reactions.* Consider the following redox reaction.

Reduction (Mn = +7 → +2)

$$MnO_4^- + 16H^+ + 5C_2O_4^{2-} \rightarrow 2Mn^{2+} + 8H_2O + 10\,CO_2$$

(Mn = +7) (C = +3) (Mn = +2) (C = + 4)

Oxidation (C = +3 → + 4)

In the above reaction, since the oxidation number (O.N.) of C-atom is increased from +3 (in $C_2O_4^{2-}$) to + 4 (in CO_2), $C_2O_4^{2-}$ ions are said to have been oxidised to CO_2 molecules. Similarly, since the O.N. of Mn-atom is decreased from +7 (in MnO_4^-) to +2 (in Mn^{2+}), MnO_4^- ions are said to have been reduced to Mn^{2+} ions. Thus we find that MnO_4^- ions oxidise $C_2O_4^{2-}$ ions to CO_2 molecules and are themselves reduced to Mn^{2+} ions.

The above redox reaction is composed of the following two half-reactions, one of which is the oxidation half-reaction and the other is reduction half-reaction. The addition of these half-reactions gives the above redox reaction. The oxidation half-reaction represents the oxidation of $C_2O_4^{2-}$ ions to CO_2 molecules, while the reduction half-reaction represents the reduction of MnO_4^- ions to Mn^{2+} ions.

$$\left[\underset{(C=+3)}{C_2O_4^{2-}} \xrightarrow{\text{Oxidation}} \underset{(C=+4)}{2\,CO_2} + 2e^-\right] \times 5 \qquad \textit{(Oxidation half-reaction)}$$

$$\left[\underset{(Mn=+7)}{MnO_4^-} + 8H^+ + 5e^- \xrightarrow{\text{Reduction}} \underset{(Mn=+2)}{Mn^{2+}} + 4H_2O\right] \times 2 \qquad \textit{(Reduction half-reaction)}$$

$$\underset{(Mn=+7)}{2MnO_4^-} + 16H^+ + \underset{(C=+3)}{5C_2O_4^{2-}} \rightarrow \underset{(Mn=+2)}{2Mn^{2+}} + 8H_2O + \underset{(C=+4)}{10CO_2} \qquad \textit{(Redox reaction)}$$

It may be noted that in a redox reaction, the number of electrons lost by the substance oxidised is the same as that gained by the substance reduced. It may also be noted that in a redox reaction the total increase in oxidation number of the element oxidised is equal to the total decrease in oxidation number of the element reduced.

Oxidising Agent (Oxidant) and Reducing Agent (Reductant)

Oxidising agent is that species (*i*) which oxidises some other species and is itself reduced, (*ii*) which gives up oxygen or any other electronegative element, (*iii*) which accepts hydrogen or any other electropositive element, (*iv*) which gains one or more electrons (*Electronic concept*), *i.e.*, an oxidising agent acts as *electron-acceptor*, (*v*) in which the O.N. of an effective element decreases (*oxidation number concept*) and thus an oxidising agent gets reduced.

Reducing agent is that species (*i*) which reduces some other species and is itself oxidised, (*ii*) which accepts oxygen or any other electronegative element, (*iii*) which gives up (loses) hydrogen or any other electropositive element, (*iv*) which loses one or more electrons (electronic concept), *i.e.* a reducing agent acts as *electron-donor*, (*v*) in which the O.N. of an effective element increases and thus the reducing agent gets oxidised.

Illustration. Let us consider the following redox reaction. In this reaction, oxidising agent and reducing agent have been designated as O.A. and R.A. respectively.

$$\underset{(Cr=+6)}{K_2Cr_2O_7} + 7H_2SO_4 + \underset{(I=-1)}{6KI} \rightarrow \underset{(Cr=+3)}{Cr_2(SO_4)_3} + 4K_2SO_4 + 7H_2O + \underset{(I=0)}{3I_2} \qquad \textit{(Molecular equation)}$$

Reduction (Cr = +6 → +3)

$$\text{or} \quad \underset{\substack{(Cr=+6)\\(O.A.)}}{Cr_2O_7^{2-}} + 14H^+ + \underset{\substack{(I=-1)\\(R.A.)}}{6I^-} \longrightarrow \underset{\substack{(Cr=+3)\\\text{Reduction}\\\text{product}}}{2Cr^{3+}} + 7H_2O + \underset{\substack{(I=0)\\\text{Oxidation}\\\text{product}}}{3I_2} \qquad \textit{(Ionic equation)}$$

Oxidation (I = – 1 → 0)

In the above redox reaction, $Cr_2O_7^{2-}$ ions oxidise I^- ions to I_2 molecules and are reduced to Cr^{3+} ions, since the oxidation number (O.N.) of Cr-atom in $Cr_2O_7^{2-}$ ions is decreased from + 6 (in $Cr_2O_7^{2-}$ ions) to +3 (in Cr^{3+} ions) and that of I^- ions is increased from –1 (in I^- ions) to zero (in I_2

molecules). Thus, on the basis of oxidation number concept, $Cr_2O_7^{2-}$ ions act as oxidising agent and I^- ions behave as reducing agent. Cr^{3+} ions are the reduction product of $Cr_2O_7^{2-}$ ions, since these ions (*i.e.*, Cr^{3+} ions) are obtained by the reduction of $Cr_2O_7^{2-}$ ions. Similarly I_2 molecules are the oxidation product of I^- ions, because these molecules are obtained by the oxidation of I^- ions.

The ionic equation given above can be broken into the following two half-reactions.

$$\underset{\substack{(Cr = +6)\\(O.A.)}}{Cr_2O_7^{2-}} + 14H^+ + 6e^- \xrightarrow{\text{Reduction}} \underset{(Cr = +3)}{2Cr^{3+}} + 7H_2O \quad \text{--- (\textit{Reduction half-reaction})}$$

$$\underset{\substack{(I = -1)\\(R.A.)}}{6\,I^-} \xrightarrow{\text{Oxidation}} \underset{(I = 0)}{3I_2} + 6e^- \quad \text{--- (\textit{Oxidation half-reaction})}$$

$$Cr_2O_7^{2-} + 14H^+ + 6I^- \longrightarrow 2Cr^{3+} + 7H_2O + 3I_2 \quad \text{(\textit{Redox reaction})}$$

It may be seen from the above half-reactions that $Cr_2O_7^{2-}$ ions gain electrons (six electrons) while I^- ions lose the same number of electrons. Thus on the basis of electronic concept, $Cr_2O_7^{2-}$ ions act as oxidising agent and I^- ions behave as reducing agent. It may also be noted that the transfer of electrons takes place from the reducing agent to the oxidising agent.

Important oxidising agents and their reduction half-reactions

Hydrogen peroxide (H_2O_2), halogens (X_2), potassium dichromate ($K_2Cr_2O_7$), potassium chlorate ($KClO_3$), potassium permanganate ($KMnO_4$), nitric acid (HNO_3) etc., are important oxidising agents. Reduction half-reaction for each of these oxidising agents is given below. Reduction half-reaction gives us the products which are obtained when a given oxidising agent oxidises the reducing agent and is itself reduced. The products thus obtained are called *reduction products*. In this reaction, the oxidising agent accepts one or more electrons (given by the reducing agent) and is reduced, *i.e.*, in reduction half-reaction electron(s) are accepted by the oxidising agent or are added to the LHS of the reduction reaction. As an example, let us consider the reduction half-reaction of Br_2 gas, which is a good oxidising agent. When Br_2 gas is passed through KI solution, iodine (I_2) is liberated. In this reaction, Br_2 oxidises KI to I_2 and is itself reduced to KBr. Thus KBr (or Br^- ion) is the reduction product of Br_2.

$$\underset{(Br = 0)}{Br_2} + \underset{(I = -1)}{2KI} \rightarrow \underset{\substack{(Br = -1)\\\text{Reduction product}}}{2KBr} + \underset{(I = 0)}{I_2} \quad \text{(\textit{Redox reaction})}$$

$$\text{or} \quad Br_2 + 2I^- \rightarrow \underset{\text{Reduction product}}{2Br^-} + I_2 \quad \text{(\textit{Redox reaction})}$$

The above redox reaction is made up of the following two half-reaction :

$$\underset{(Br = 0)}{Br_2} + 2e^- \rightarrow \underset{(Br = -1)}{2Br^-} \quad \text{(\textit{Reduction half-reaction})}$$

$$\underset{(I = -1)}{2I^-} + \rightarrow \underset{(I = 0)}{I_2} + 2e^- \quad \text{(\textit{Oxidation half-reaction})}$$

$$Br_2 + 2I^- \rightarrow 2Br^- + I_2 \quad \text{(\textit{Redox reaction})}$$

Example of oxidising agents

(1) Ferric salts. Ferric salts are good oxidising agents, since when Fe^{3+} ion reacts with a reducing agent (*e.g.*, nascent hydrogen, H_2S, Sn^{2+} ion etc.), it gets reduced to Fe^{2+} ion, *e.g.*,

$$Fe^{3+} + \underset{(H = 0)}{H} \rightarrow Fe^{2+} + \underset{(H = +1)}{H^+}$$

$$2Fe^{3+} + \underset{(S = -2)}{H_2S} \rightarrow 2Fe^{2+} + 2H^+ + \underset{(S = 0)}{S}$$

$$2Fe^{3+} + Sn^{2+} \rightarrow 2Fe^{2+} + Sn^{4+}$$

The ionic equation representing the reduction of Fe^{3+} ion to Fe^{2+} ion can be written as :

$$Fe^{3+} + e^- \rightarrow Fe^{2+} \quad \textit{(Reduction half-reaction)}$$

(2) Hydrogen peroxide, H_2O_2. Hydrogen peroxide oxidises many substances. Depending on the medium in which the oxidation takes place, H_2O_2 gets reduced to different substances. For example.

(*a*) *In acidic medium*, H_2O_2 (O = –1) is reduced to H_2O (O = –2)

$$\underset{(O=-1)}{H_2O_2} \rightarrow \underset{(O=-2)}{H_2O} + O \quad \textit{(Non-ionic equation)}$$

$$\underset{(O=-1)}{H_2O_2} + 2H^+ + 2e^- \rightarrow \underset{(O=-2)}{2H_2O} \quad \textit{(Ionic equation)},\ E^0_A = +1.77\ V$$

(*b*) *In basic (alkaline) medium*, H_2O_2 (O = –1) gets reduced to OH^- ions (O = – 2)

$$\underset{(O=-1)}{H_2O_2} + 2e^- \rightarrow \underset{(O=-2)}{2OH^-},\quad E^0_B = +0.87\ V$$

(3) Halogens, X_2. All the halogens are oxidising agents. When they oxidise a certain substance, they themselves are reduced to the halides (X^-).

$$\underset{(X=0)}{X_2} + 2e^- \rightarrow \underset{(X=-1)}{2X^-}$$

(4) Acidified solution of patassium dichromate, $K_2Cr_2O_7$. In presence of H_2SO_4, $K_2Cr_2O_7$ acts as a strong oxidising agent. $K_2Cr_2O_7$ is reduced to $Cr_2(SO_4)_3$

$$\underset{(Cr=+6)}{K_2Cr_2O_7} + 4H_2SO_4 \rightarrow K_2SO_4 + \underset{(Cr=+3)}{Cr_2(SO_4)_3} + 4H_2O + 3O \quad \textit{(Non-ionic equation)}$$

or $Cr_2O_7^{2-} + 8H^+ \rightarrow 2Cr^{3+} + 4H_2O + 3O$ *(Ionic equation)*

or $Cr_2O_7^{2-} + 14H^+ + 6e^- \rightarrow 2Cr^{3+} + 7H_2O$ *(Ionic equation)*

(5) Potassium chlorate, $KClO_3$. Potassium chlorate is a powerful oxidising agent. It is reduced to KCl

$$KClO_3 \rightarrow KCl + 3O \quad \textit{(Non-ionic equation)}$$

or $\underset{(Cl=+5)}{ClO_3^-} + 6H^+ + 6e^- \rightarrow \underset{(Cl=-1)}{Cl^-} + 3H_2O$ *(Ionic equation)*

$KClO_3$ oxidises hot H_2SO_3 (S = + 4) to H_2SO_4 (S = + 6) and is itself reduced to KCl.

$$KClO_3 \rightarrow KCl + 3O$$

$$[H_2SO_3 + O \rightarrow H_2SO_4] \times 3$$

$$\underset{(Cl=+5)}{KClO_3} + \underset{(S=+4)}{H_2SO_3} \rightarrow \underset{(Cl=-1)}{KCl} + \underset{(S=+6)}{H_2SO_4} \quad \textit{(Molecular equation)}$$

$$\underset{(Cl=+5)}{ClO_3^-} + 6H^+ + 6e^- \rightarrow \underset{(Cl=-1)}{Cl^-} + 3H_2O \quad \textit{(Reduction half-reaction)}$$

$$3 \times [\underset{(S=+4)}{SO_3^{2-}} + H_2O \rightarrow \underset{(S=+6)}{SO_4^{2-}} + 2H^+ + 2e^-] \quad \textit{(Oxidation half-reaction)}$$

$$\underset{(Cl=+5)}{ClO_3^-} + \underset{(S=+4)}{3SO_3^{2-}} \rightarrow \underset{(Cl=-1)}{Cl^-} + \underset{(S=+6)}{3SO_4^{2-}} \quad \textit{(Ionic equation)}$$

(6) Copper sulphate, $CuSO_4$. Aqueous solution of $CuSO_4$ is a good oxidising agent, since Cu^{2+} ion oxidises a number of substances and is itself reduced to Cu^+ ion ($Cu^{2+} + e^- \rightarrow Cu^+$). For example when KI is added to a solution of $CuSO_4$, a precipitate of CuI mixed with iodine is obtained. The reaction involves two steps :

$$\underset{(Cu=+2)}{2CuSO_4} + \underset{(I=-1)}{2KI} \rightarrow K_2SO_4 + \underset{(Cu=+1)}{Cu_2SO_4} + \underset{(I=0)}{I_2}$$

$$Cu_2SO_4 + 2KI \rightarrow 2CuI\downarrow + K_2SO_4 \quad (\text{ppt})$$

$$\underset{(Cu=+2)}{2CuSO_4} + \underset{(I=-1)}{4KI} \rightarrow 2K_2SO_4 + \underset{(I=0)}{I_2} + \underset{(Cu=+1)}{2CuI\downarrow} \ (ppt)$$

The ionic equations corresponding to the above equationss can be written as :

$$2Cu^{2+} + 2I^- \rightarrow 2Cu^+ + I_2$$

$$2Cu^+ + 2I^- \rightarrow 2CuI \ (\text{ppt})$$

$$\underset{(Cu=+2)}{2Cu^{2+}} + \underset{(I=-1)}{4I^-} \rightarrow \underset{(Cu=+1)}{2CuI \ (\text{ppt})\downarrow} + \underset{(I=0)}{I_2}$$

(7) Potassium permanganate, $KMnO_4$. This substance acts as a strong oxidising agent in acidic (H_2SO_4), neutral and alkaline solution.

(a) In acidic medium. In acidic solution (H_2SO_4) $KMnO_4$ (MnO_4^-) is reduced to $MnSO_4$ (Mn^{2+}) *(Reduction product)*

$$\underset{(Mn=+7)}{2KMnO_4} + 3H_2SO_4 \rightarrow K_2SO_4 + \underset{(Mn=+2)}{2MnSO_4} + 3H_2O + 5O \quad \textit{(Non-ionic equation)}$$

or $2MnO_4^- + 6H^+ \rightarrow Mn^{2+} + 3H_2O + 5O$ *(Ionic equation)*

or $MnO_4^- + 8H^+ + 5e^- \rightarrow Mn^{2+} + 4H_2O$ *(Ionic equation)*

(b) In neutral or weakly alkaline solution. In neutral or weakly alkaline solution $KMnO_4$ (MnO_4^-) is reduced directly to MnO_2 (Reduction product)

$$\underset{(Mn=+7)}{2KMnO_4} + H_2O \rightarrow \underset{(Mn=+4)}{2MnO_2} + 2KOH + 3O \quad \textit{(Non-ionic equation)}$$

or $2MnO_4^- + H_2O \rightarrow 2MnO_2 + 2OH^- + 3O$ *(Ionic equation)*

or $MnO_4^- + 2H_2O + 3e^- \rightarrow MnO_2 + 4OH^-$ *(Ionic equation)*

(c) In strongly alkaline medium. In strongly alkaline solution $KMnO_4$ (MnO_4^-) is first reduced to potassium manganate, K_2MnO_4 (MnO_4^{2-}) and then to insoluble MnO_2.

$$\underset{(Mn=+7)}{2KMnO_4} + 2KOH \rightarrow \underset{(Mn=+6)}{2K_2MnO_4} + H_2O + O$$

$$2K_2MnO_4 + 2H_2O \rightarrow 2MnO_2 + 4KOH + 2O$$

$$2KMnO_4 + 2KOH + H_2O \rightarrow 2MnO_2 + 4KOH + 3O$$

or $\underset{(Mn=+7)}{2KMnO_4} + H_2O \xrightarrow{\text{Alkali}} \underset{(Mn=+4)}{2MnO_2} + 2KOH + 3O$ *(Non-ionic equation)*

OR

$$2MnO_4^- + 2OH^- \longrightarrow 2MnO_4^{2-} + H_2O + O$$

$$2MnO_4^{2-} + 2H_2O \longrightarrow 2MnO_2 + 4OH^- + 2O$$

$$\underset{(Mn=+7)}{2MnO_4^-} + H_2O \xrightarrow{\text{Alkali}} \underset{(Mn=+4)}{2MnO_2} + 2OH^- + 3O \quad \textit{(Ionic equation)}$$

OR

$$MnO_4^- + e^- \rightarrow MnO_4^{2-}$$

$$MnO_4^{2-} + 2H_2O + 2e^- \rightarrow MnO_2 + 4OH^-$$

$$\underset{(Mn=+7)}{MnO_4^-} + 2H_2O + 3e^- \rightarrow \underset{(Mn=+4)}{MnO_2} + 4OH^- \quad \textit{(Ionic equation)}$$

(8) Nitric acid, HNO_3. Oxidising properties of HNO_3 depend on whether the acid used is dilute or concentrated. For example :

(*i*) *Hot and conc. HNO_3* (N = +5) gets reduced to NO_2 (N = + 4). Thus NO_2 is the reduced product.

$$\underset{(N=+5)}{2HNO_3} \rightarrow \underset{(N=+4)}{2NO_2} + H_2O + O \quad (\textit{Non-ionic equation})$$

$$\underset{(N=+5)}{NO_3^-} + 2H^+ + e^- \rightarrow \underset{(N=+4)}{NO_2} + H_2O \quad (\textit{Ionic equation})$$

(*ii*) (*a*) *Dil. HNO_3* (50%) is reduced to NO (N = +2)

$$\underset{(N=+5)}{2HNO_3} \rightarrow \underset{(N=+2)}{2NO} + H_2O + 3O \quad (\textit{Molecular equation})$$

$$\underset{(N=+5)}{HNO_3} + 3H \rightarrow \underset{(N=+2)}{NO} + 2H_2O \quad (\textit{Molecular equation})$$

$$\underset{(N=+5)}{NO_3^-} + 4H^+ + 3e^- \rightarrow \underset{(N=+2)}{NO} + 2H_2O \quad (\textit{Ionic equation})$$

(*b*) *Dil. HNO_3* (20%) is reduced to N_2O (N = +1)

$$\underset{(N=+5)}{2HNO_3} \rightarrow \underset{(N=+1)}{N_2O} + H_2O + 4O \quad (\textit{Molecular equation})$$

$$\underset{(N=+5)}{2HNO_3} + 8H \rightarrow \underset{(N=+1)}{N_2O} + 5H_2O \quad (\textit{Molecular equation})$$

$$\underset{(N=+5)}{2NO_3^-} + 10H^+ + 8e^- \rightarrow \underset{(N=+1)}{N_2O} + 5H_2O \quad (\textit{Ionic equation})$$

(*iii*) (*a*) *Very dil. HNO_3* (6%) is reduced to NH_3 (N = –3) or NH_4NO_3 or NH_4^+ ions (N = –3)

$$\underset{(N=+5)}{HNO_3} + 8H \rightarrow \underset{(N=-3)}{NH_3} + 3H_2O$$

$$NH_3 + \underset{(N=+5)}{HNO_3} \rightarrow \underset{(N=-3)}{NH_4\,NO_3}$$

$$2HNO_3 + 8H \rightarrow 3H_2O + NH_4NO_3 \quad (\textit{Non-ionic equation})$$

$$\underset{(N=+5)}{NO_3^-} + 10H^+ + 8e^- \rightarrow \underset{(N=-3)}{NH_4^+} + 3H_2O \quad (\textit{Ionic equation})$$

(*b*) *Very dil.* HNO_3 is also reduced to H_2 (see the action of HNO_3 on Mg and Mn)

$$\underset{(H=+1)}{2HNO_3} + 2e^- \rightarrow \underset{(H=0)}{H_2} + 2NO_3^-$$

(9) Mercuric salts. Mercuric salts (Hg^{2+}) are first reduced to mercurous salts (Hg_2^{2+}) which in turn are reduced to free mercury (Hg^0)

$$2Hg^{2+} + 2e^- \rightarrow Hg_2^{2+}$$

$$Hg_2^{2+} + 2e^- \rightarrow 2Hg^0$$

(10) Hydrogen ions, H^+. These ions are reduced to free hydrogen, H_2 (H = 0)

$$\underset{(H=+1)}{2H^+} + 2e^- \rightarrow \underset{(H=0)}{H_2}$$

(11) Acidified MnO_2. MnO_2 (Mn = + 4) in presence of strong acids like conc. H_2SO_4 is reduced to Mn^{2+} ion

$$\underset{(Mn=+4)}{MnO_2} + 4H^+ + 2e^- \rightarrow \underset{(Mn=+2)}{Mn^{2+}} + 2H_2O$$

(12) Sodium hypochlorite, NaOCl. Since aqueous solution of NaOCl produces nascent oxygen, NaOCl acts as an oxidising agent.

$$\underset{(Cl = +1)}{NaOCl\,(aq)} \rightarrow \underset{(Cl = -1)}{NaCl\,(aq)} + O\,(aq) \quad \text{(Non-ionic equation)}$$

$$\underset{(Cl = +1)}{OCl^-} + 2H^+ + 2e^- \rightarrow \underset{(Cl = -1)}{Cl^-} + H_2O \quad \text{(Ionic equation)}$$

Important reducing agents and their oxidation half-reactions

Nitrous acid (HNO_2), sodium arsenite (Na_3AsO_3), oxalic acid ($H_2C_2O_4$), acidified solution of $FeSO_4$ etc., are oxidising agents. Oxidation half-reactions for each of these oxidising agents is given below. Oxidation half-reaction gives us the products which are obtained when a given reducing agent reduces an oxidising agent and is itself oxidised. The products thus obtained are called *oxidation products.* In this reaction, the reducing agent loses one or more electrons and is oxidised, *i.e.*, in oxidation half-reaction electron(s) are released by the reducing agent or are added to RHS of the oxidation reaction. As an example let us consider the oxidation half-reaction of HNO_2 which acts as a reducing agent towards strong oxidising agent like acidified solution of $KMnO_4$, acidified solution of K_2CrO_7, Br_2, H_2O_2 etc., when HNO_2 reacts with Br_2, HNO_2 is oxidised to HNO_3 and Br_2 is reduced to HBr. Thus HNO_2 acts as a reducing agent (R.A.) and Br_2 behaves as an oxidising agent (O.A.)

$$\underset{\substack{(Br = 0)\\ \text{O.A.}}}{Br_2} + H_2O + \underset{\substack{(N = +3)\\ \text{R. A.}}}{HNO_2} \rightarrow \underset{(N = +5)}{HNO_3} + \underset{(Br = -1)}{2HBr}$$

or $\quad Br_2 + H_2O + NO_2^- \rightarrow NO_3^- + 2H^+ + 2Br^-$

The above redox reaction is composed of the following two half-reactions.

$$\underset{(N = +3)}{NO_2^- + H_2O} \xrightarrow{\text{Oxidation}} \underset{(N = +5)}{NO_3^- + 2H^+ + 2e^-} \quad \text{(Oxidation half-reaction)}$$

$$\underset{(Br = 0)}{Br_2 + 2e^-} \xrightarrow{\text{Reduction}} \underset{(Br = -1)}{2Br^-} \quad \text{(Reduction half-reaction)}$$

$$Br_2 + H_2O + NO_2^- \rightarrow NO_3^- + 2H^+ + 2Br^- \quad \text{(Redox reaction)}$$

Examples of reducing agents

(1) Oxalic acid ($H_2C_2O_4$). $H_2C_2O_4$ or $C_2O_4^{2-}$ ion is oxidised to CO_2

$$H_2C_2O_4 + O \rightarrow CO_2 + H_2O$$

or $\quad \underset{(C = +3)}{C_2O_4^{2-}} \rightarrow \underset{(C = +4)}{2CO_2 + 2e^-}$

(2) Acidified solution of $FeSO_4$. It is oxidised to $Fe_2(SO_4)_3$.

$$2FeSO_4 + H_2SO_4 + O \rightarrow Fe_2(SO_4)_3 + H_2O$$

or $\quad Fe^{2+} \rightarrow Fe^{3+} + e^-$

(3) Sdium thiosulphate, $Na_2S_2O_3$. $S_2O_3^{2-}$ ion may be oxidised to various species as shown below :

(*a*) $S_2O_3^{2-}$ ion is oxidised to SO_4^{2-} ion

$$Na_2S_2O_3 + O \rightarrow Na_2SO_4 + S$$

or $\quad \underset{(S = +2)}{S_2O_3^{2-}} \rightarrow \underset{(S = +6)}{SO_4^{2-}} + S + 2H^+ + 2e^-$

(*b*) $S_2O_3^{2-}$ ion is oxidised to dithionate ion, $S_2O_6^{2-}$ (S = ÷5)

$$Na_2S_2O_3 + 3O \rightarrow Na_2S_2O_6$$

or $\quad \underset{(S = +2)}{S_2O_3^{2-}} + 3H_2O \rightarrow \underset{(S = +5)}{S_2O_6^{2-}} + 6H^+ + 6e^-$

(*c*) $S_2O_3^{2-}$ ion may be oxidised to tetrathionate ion, $S_4O_6^{2-}$ (S = +2.5)

$$\underset{(S = +2)}{2S_2O_3^{2-}} \rightarrow \underset{(S = 2.5)}{S_4O_6^{2-}} + 2e^-$$

(4) Stanuous chloride ($SnCl_2$). It is oxidised to stannic chloride, $SnCl_4$

$$Sn^{2+} \rightarrow Sn^{4+} + 2e^-$$

(5) Hydrogen sulphide, H_2S (S = –2). It is oxidised to sulphur (S = 0)

$$H_2S + O \rightarrow H_2O + S$$

or $$S^{2-} \rightarrow S + 2e^-$$

(6) Aqueous solution of SO_2 or sulphurous acid (H_2SO_3). It is oxidised to H_2SO_4.

(*a*) $SO_2 + H_2O + O \rightarrow H_2SO_4$

or $\underset{(S = +4)}{SO_2} + 2H_2O \rightarrow \underset{(S = +6)}{SO_4^{2-}} + 4H^+ + 2e^-$

(*b*) $H_2SO_3 + O \rightarrow H_2SO_4$

or $\underset{(S = +4)}{SO_3^{2-}} + H_2O \rightarrow \underset{(S = +6)}{SO_4^{2-}} + 2H^+ + 2e^-$

(7) Potassium iodide KI (*a*) Acidified solution of KI is oxidised to iodine, I_2

$$2KI + H_2SO_4 + O \rightarrow K_2SO_4 + H_2O + I_2$$

or $$\underset{(I = -1)}{I^-} \rightarrow \underset{(I = 0)}{I_2} + 2e^-$$

(*b*) An alkaline solution of KI is oxidised to potassium iodate, KIO_3

$$KI + 3O \rightarrow KIO_3$$

$$\underset{(I = -1)}{I^-} + 6OH^- \rightarrow \underset{(I = +5)}{IO_3^-} + 3H_2O + 6e^-$$

(8) Hydrogen peroxide, H_2O_2. In acidic as well as in basic medium, H_2O_2 (O = –1) gets oxidised to O_2 (O = 0)

$$H_2O_2 + O \rightarrow H_2O + O_2$$

$$\underset{(O = -1)}{H_2O_2} \rightarrow \underset{(O = 0)}{O_2} + 2H^+ + 2e^- \quad \text{(In acidic medium)}$$

$$\underset{(O = -1)}{H_2O_2} + 2OH^- \rightarrow \underset{(O = 0)}{O_2} + 2H_2O + 2e^- \quad \text{(In basic medium)}$$

Example 17.2. *Which of the following reactions is / are redox reactions? Give reason for your answer.*

(*i*) $CuSO_4 + 4NH_3 \rightarrow [Cu(NH_3)_4]SO_4$

(*ii*) $2CuSO_4 + 4KI \rightarrow 2CuI + 2K_2SO_4 + I_2$

(*iii*) $Na_2SO_4 + BaCl_2 \rightarrow BaSO_4 + 2NaCl$

(*iv*) $3Br_2 + 6NaOH \rightarrow NaBrO_3 + 5NaBr + 3H_2O$

(*v*) $NH_4NO_3 \xrightarrow{\Delta} N_2O + 2H_2O$

Solution. In equation (*ii*) the oxidation number of Cu decreases from +2 (in $CuSO_4$) to +1 (in CuI), while that of iodine increase from –1 (in KI) to zero (in I_2). Thus it is a redox reaction. Similarly in equation (*iv*), since the O.N. of bromine increases from zero (in Br_2) to +5 (in $NaBrO_3$) and decreases from zero (in Br_2) to –1 (in NaBr), it is also a redox reaction. In other reactions, the O.N. of other elements does not undergo any change and hence these reactions are not redox reactions.

Example 17.3. *Select oxidising and reducing agents in the following redox reaction and break this reaction into two half-reactions.*

$$3N_2H_4 + 2BrO_3^- \rightarrow 3N_2 + 2Br^- + 6H_2O$$

Solution. The given reaction can be written as : (R.A. = reducing agent, O.A. = oxidising agent)

Oxidation
(N = –2 → 0)

$$\underset{\substack{(N = -2) \\ \text{R.A.}}}{N_2H_4} + \underset{\substack{(Br = +5) \\ \text{O.A.}}}{2BrO_3^-} \longrightarrow \underset{(N = 0)}{3N_2} + \underset{(Br = -1)}{2Br^-} + 6H_2O$$

Reduction
(Br = + 5 → –1)

In the above reaction, since the O.N. of N increases from –2 (in N_2H_4) to zero (in N_2), N_2H_4 is being oxidised to N_2, *i.e.*, N_2H_4 is acting as reducing agent. Similarly, since the O.N. of bromine decreases from +5 (in BrO_3^-) to –1 (in Br^-), BrO_3^- ions are being reduced to Br^- ions, *i.e.*, BrO_3^- ions act as oxidising agent. N_2H_4 reduces BrO_3^- ions to Br^- ions. BrO_3^- ions act as oxidising agent, since these ions oxidise N_2H_4 to N_2.

The two half-reactions into which the given equation can be broken are :

$$3N_2H_4 \xrightarrow{\text{Oxidation}} 3N_2 + 12H^+ + 12e^- \quad \textit{(Oxidation half-reaction)}$$

$$2BrO_3^- + 12H^+ + 12e^- \xrightarrow{\text{Reduction}} 2Br^- + 6H_2O \quad \textit{(Reduction half-reaction)}$$

To calculate the equivalent weight of a given oxidising or reducing agent

The following two methods are used for calculating the equivalent weight a given oxidising or reducing agent.

1. Method based on the number of electron (n) gained by an oxidising agent or lost by a reducing agent. The equivalent weight of an oxidising agent is obtained by dividing the molecular weight of the oxidising agent by the number of electrons (*n*), which one molecule of the oxidising agent gains in its reduction half reaction. As an example, let us calculate the equivalent weight of $KMnO_4$ which acts as an oxidising agent in acid, neutral and alkaline medium. The ionic equations representing the oxidising property of $KMnO_4$ in these media can be written as :

$$MnO_4^- + 8H^+ + 5e^- \xrightarrow{\text{Reduction}} Mn^{2+} + 4H_2O \quad \textit{(In acid medium, } n = 5)$$

$$MnO_4^- + 2H_2O + 3e^- \xrightarrow{\text{Reduction}} MnO_2 + 4OH^- \quad \textit{(In neutral medium, } n = 3)$$

$$MnO_4^- + e^- \xrightarrow{\text{Reduction}} MnO_4^{2-} \quad \textit{(In strongly alkaline solution, } n = 1)$$

Obviously the equivalent weight of $KMnO_4$:

$$\textit{In acidic medium} = \frac{\text{Molecular weight of } KMnO_4}{5} = \frac{158}{5} = \mathbf{31.6}$$

$$\textit{In neutral solution} = \frac{158}{3} = \mathbf{52.66}$$

$$\textit{In alkaline medium} = \frac{158}{1} = \mathbf{158}$$

The above calculations show that the equivalent weight of $KMnO_4$, in acidic solution, is equal to 1/5*th* of its molecular weight; in neutral solution, it is equal to 1/3*rd* of its molecular weight, while in alkaline solution, it is equal to its molecular weight.

The equivalent weight of a reducing agent is obtained by dividing its molecular weight by the number of electrons (*n*), which one molecule of the reducing agent loses in its oxidation half reaction. As an example let us calculate the equivalent weight of $Na_2S_2O_3.5H_2O$, which acts as a reducing agent. The ionic equation representing the oxidising property of this substance can be written as :

$$2S_2O_3^{2-} \xrightarrow{\text{Oxidation}} S_4O_6^{2-} + 2e^-$$

or $\quad S_2O_3^{2-} \xrightarrow{\text{Oxidation}} \frac{1}{2} S_4O_6^{2-} + e^- \quad (n = 1)$

Here we see that since two molecules of $Na_2S_2O_3$ give two electrons, one molecule of this compound will give one electron. Thus $n = 1$ and hence :

Equivalent weight of $Na_2S_2O_3.5H_2O$ (reducing agent)

$$= \frac{\text{Molecular weight of } Na_2S_2O_3 \,.\, 5H_2O}{n}$$

$$= \frac{248}{1} = \mathbf{248}$$

2. Method based on the decrease or increase in the oxidation number of the effective element present in the oxidising or reducing agent. The equivalent weight of an oxidising agent is obtained by dividing the molecular weight of the oxidising agent by the *decrease* in the oxidation number that all the atoms of the effective element present in the oxidising agent undergo. Similarly the equivalent weight of a reducing agent is obtained by dividing the molecular weight of the reducing agent by the *increase* in the oxidation number that all the atoms of the effective element present in the reducing agent undergo. The method for calculating the decrease (in case of oxidising agent) or increase (in case of reducing agent) in the oxidation number of the effective element and then calculating the equivalent weight can be understood by considering the following examples :

(*i*) $K_2Cr_2O_7$ acts as an oxidising agent in acid medium.

$$\underset{\substack{(Cr = +6)\\ \textit{(Oxidising agent)}}}{Cr_2O_7^{2-}} + 14\,H^+ + 6e^- \xrightarrow{\text{Reduction}} \underset{\substack{(Cr = +3)\\ \textit{(Reduced product)}}}{2Cr^{3+}} + 7H_2O$$

It may be seen from the above reduction reaction that each Cr-atom in the oxidising agent ($Cr_2O_7^{2-}$ ion) carries +6 charges while Cr-atom in the reduced product (Cr^{3+} ion) has +3 charges. Now since the oxidising agent has two Cr-atoms, we must calculate the decrease in oxidation number undergone by two Cr-atoms in the above reaction. Thus :

Decrease in oxidation number undergone by two Cr-atoms $= 2 \times (+6) - 2 \times (+3) = 12 - 6 = 6$

$$\therefore \quad \text{Equivalent weight of } K_2Cr_2O_7 \textit{ (oxidising agent)} = \frac{\text{Molecular weight of } K_2Cr_2O_7}{6} = \frac{294.21}{6} = \mathbf{49.04}$$

(*ii*) I_2 acts as an oxidising agent as shown below :

$$\underset{\substack{(I = 0)\\ \textit{(Oxidising agent)}}}{I_2} + 2e^- \rightarrow \underset{\substack{(I = -1)\\ \textit{(Reduced product)}}}{2I^-}$$

In the above reaction, each I-atom in the oxidising agent (I_2) carries zero charge, while I-atom in the reduced product (I^- ion) has –1 charge. Now, since the oxidising agent has two I-atoms, we must calculate the decrease in oxidation number undergone by two I-atoms in the above reaction. Thus :

Decrease in oxidation number undergone by two I-atoms

$$= 2 \times 0 - 2 \times (-1) = 0 + 2 = 2$$

$$\therefore \quad \text{Equivalent weight of } I_2 \textit{ (oxidising agent)} = \frac{\text{Molecular weight of } I_2}{2} = \frac{253.80}{2} = \mathbf{126.9}$$

(*iii*) $Na_2S_2O_3.\ 5H_2O$ acts as a reducing agent

$$\underset{\substack{(S = +2)\\ \textit{(Reducing agent)}}}{2S_2O_3^{2-}} \xrightarrow{\text{Oxidation}} \underset{\substack{(S = +2.5)\\ \textit{(Oxidised product)}}}{S_4O_6^{2-}} + 2e^-$$

It is obvious from the above euqation that each S-atom in the reducing agent ($S_2O_3^{2-}$ ion) carries +2 charges while each S-atom in the oxidised product ($S_4O_6^{2-}$ ion) carries 2.5 charges. Now. since the reducing agent has two S-atoms, we must calculate the increase in the oxidation number undergone by two S-atoms in the above reaction. Thus :

Increase in the oxidation number undergone by two S-atoms

$$= 2 \times 2.5 - 2 \times 2 = 5 - 4 = 1$$

∴ Equivalent weight of $Na_2S_2O_3.5H_2O$ (*reducing agent*)

$$= \frac{\text{Molecular weight of } Na_2S_2O_3.5H_2O}{1}$$

$$= \frac{248.2}{1} = \mathbf{248.2}$$

(*iv*) Let us calculate the equivalent weight of $CuSO_4.5H_2O$ (oxidising agent) and KI (reducing agent) in the following redox reaction :

$2CuSO_4$	+	$4KI$	→	$2CuI$	+	I_2	+	$2K_2SO_4$
(Cu = +2)		(I = –1)		(Cu = +1)		(I = 0)		
(*Oxidising agent*)		(*Reducing agent*)		(*Reduced product*)		(*Oxidised product*)		

In the above reaction, $CuSO_4$ acts as an oxidising agent and KI behaves as a reducing agent, since the oxidation state of Cu decreases from +2 to +1 and that of I-atom increases from –1 to 0 (zero). Thus :

Decrease in oxidation number uncdrgone by one Cu-atom in oxidising agent ($CuSO_4.5H_2O$) = +2 – 1 = 1

∴ Equivalent weight of $CuSO_4.5H_2O$ (*oxidising agent*)

$$= \frac{\text{Molecular weight of } CuSO_4.5H_2O}{1}$$

$$= \frac{249.69}{1} = \mathbf{249.69}$$

Increase in oxidation number undergone by one I-atom in reducing agent (KI) = 0 – (–1) = 1

∴ Equivalent weight of KI (*reducing agent*)

$$= \frac{\text{Molecular weight of KI}}{1}$$

$$= \frac{166.0}{1} = \mathbf{166.0}$$

Thus we see that in the above redox reaction the equivalent weight of $CuSO_4.5H_2O$ (oxidising agent) and KI (reducing agent) is the same as their molecular weight.

Example 17.4. *In the reaction,* $I_2 + 2S_2O_3^{2-} \rightarrow 2I^- + S_4O_6^{2-}$, *the equivalent weight of iodine will be equal to : (a) its molecular weight (b) 1/2 of its molecular weight (c) 1/4 of its molecular weight (d) twice its molecular weight.* [**Ans** : (*a*)]. **[MLNR 1985]**

Example 17.5. *Calculate the equivalent weight of* $Na_2S_2O_3.5H_2O$ *and* $KBrO_3$ *in the following reactions:*

(*a*) $2S_2O_3^{2-} + I_2 \rightarrow S_4O_6^{2-} + 2I^-$

(*b*) $BrO_3^- + 6H^+ + 6e^- \rightarrow Br^- + 3H_2O$ **(I.I.T. 1979)**

[**Ans** : (*a*) $\frac{248.2}{1} = 248.2$ (*b*) $\frac{167}{6} = 27.8$]

Disproportionation Reactions (Also called self oxidation-reduction reactions or self redox reactions)

Disproportnation reaction is that redox reaction in which some part of a substance undergoes oxidation while the remaining part of that same substance undergoes reduction, *i.e.*, the same substance is oxidised as well as reduced and hence acts both as oxidising as well as reducing agent.

Examples of disproportionation reactions

Following are important examples of disproportionation reactions.

1. Disproportionation of Cl_2, Br_2 and I_2 by the action of NaOH. The given halogens undergo disproportionation in the following two ways. Here we shall consider the disproportionation of Cl_2 molecule, as an example. The behaviour of other halogens towards NaOH is the same as that of Cl_2.

(*a*) When Cl_2 (Cl = 0) reacts with a *cold and dilute solution* of NaOH (or any alkali), sodium chlorite, NaOCl (Cl = +1) and sodium chloride, NaCl (Cl = –1) are obtained.

$$\underset{(Cl\,=\,0)}{Cl_2} + \underset{\text{cold and dil. solution}}{2NaOH} \rightarrow \underset{(Cl\,=\,+1)}{NaOCl} + \underset{(Cl\,=\,-1)}{NaCl} + H_2O \quad \textit{(Non-ionic equation)}$$

$$\underset{(Cl\,=\,0)}{Cl_2} + 2OH^- \rightarrow \underset{(Cl\,=\,+1)}{OCl^-} + \underset{(Cl\,=\,-1)}{Cl^-} + H_2O \quad \textit{(Ionic equation)} \quad ...(i)$$

It is obvious from the above equation that the oxidation state (O.S.) of Cl-atom increases from zero (in Cl_2) to +1 (in NaOCl) and decreases to –1 (in NaCl). Thus Cl-aotm undergoes oxidation (increase in O.S.) as well as reduction (decrease in O.S.). Consequently, Cl-atom undergoes disproportionation reaction.

Equation (*i*) is composed of the following two half reactions :

$$\underset{(Cl\,=\,0)}{Cl_2} + 4OH^- \xrightarrow{\text{Oxidation}} \underset{(Cl\,=\,+1)}{2ClO^-} + 2H_2O + 2e^- \quad \textit{(Oxidation half-reaction)}$$

$$\underset{(Cl\,=\,0)}{Cl_2 + 2e^-} \xrightarrow{\text{Reduction}} \underset{(Cl\,=\,-1)}{2Cl^-} \quad \textit{(Reduction half-reaction)}$$

$$\underset{(Cl\,=\,0)}{2Cl_2} + \underset{(Cl\,=\,+1)}{4OH^-} \longrightarrow \underset{(Cl\,=\,+1)}{2ClO^-} + 2H_2O + \underset{(Cl\,=\,-\,1)}{2Cl^-}$$

Oxidation and reduction half-reactions given above, show that if we consider the disproportionation of two Cl_2 molecules, then one Cl_2 molecule (Cl = 0) is reduced to Cl^- ion (Cl = –1) and the other Cl_2 molecule is oxidised to ClO^- ion (Cl = + 1). In other words we can say that one Cl_2 molecule acts as an oxidising agent (O.A.) and the other Cl_2 molecule behaves as a reducing agent (R.A.). Thus equation (*i*) can also be written as :

Reduction

$$\underset{\substack{(Cl\,=\,0)\\ R.A.}}{Cl_2} + \underset{\substack{(Cl\,=\,0)\\ O.A.}}{Cl_2} + 4OH^- \longrightarrow \underset{(Cl\,=\,+1)}{2ClO^-} + 2H_2O + \underset{(Cl\,=\,-1)}{2Cl^-}$$

Oxidation

(*b*) When Cl_2 (Cl = 0) reacts with a *hot and concentrated solution* of NaOH (or any alkali), sodium chlorate, $NaClO_3$ (Cl = +5) and sodium chloride, NaCl (Cl = –1) are obtained.

$$\underset{(Cl\,=\,0)}{3Cl_2} + \underset{\text{Hot and concentrated solution}}{6NaOH} \rightarrow \underset{(Cl\,=\,+5)}{NaClO_3} + \underset{(Cl\,=\,-1)}{5NaCl} + 3H_2O$$

$$\text{or} \quad \underset{(Cl\,=\,0)}{3Cl_2} + 6OH^- \rightarrow \underset{(Cl\,=\,+5)}{ClO_3^-} + \underset{(Cl\,=\,-1)}{5Cl^-} + 3H_2O \quad ...(i)$$

Equation (*i*) is composed of the following half-reactions

$$Cl_2 + 12OH^- \xrightarrow{\text{Oxidation}} 2ClO_3^- + 6H_2O + 10e^- \quad (\textit{Oxidation half reaction})$$
$$(Cl = 0) \qquad\qquad (Cl = +5)$$

$$5 \times [Cl_2 + 2e^- \xrightarrow{\text{Reduction}} 2Cl^-] \quad (\textit{Reduction half-reaction})$$
$$(Cl = 0) \qquad\qquad (Cl = -1)$$

$$6\,Cl_2 + 12\,OH^- \rightarrow 2\,ClO_3^- + 6H_2O + 10\,Cl^- \quad ...(ii)$$
$$(Cl = 0) \qquad (Cl = +5) \qquad (Cl = -1)$$

or $3Cl_2 + 6\,OH^- \rightarrow ClO_3^- + 3H_2O + 5Cl^-$

Equation (*ii*) shows that if we consider the disproportionation reaction of six Cl_2 molecules, one Cl_2 molecule is oxidised to ClO_3^- ion and the remaining five Cl_2 molecules are reduced to Cl^- ion, *i.e.,*

Oxidation

$$\underset{\substack{(Cl = 0)\\ \text{R.A.}}}{Cl_2} + \underset{\substack{(Cl = 0)\\ \text{O.A.}}}{5Cl_2} + 12OH^- \longrightarrow \underset{(Cl = +5)}{2ClO_3^-} + 6H_2O + \underset{(Cl = -1)}{10Cl^-}$$

Reduction

2. Disproportionation of white phosphorus (P_4) in boiling caustic soda. When white phosphorus, $P_4(P = 0)$ is boiled with caustic soda, NaOH in an inert atmosphere, $PH_3(P = -3)$ is evolved and sodium hypophosphite, NaH_2PO_2 (P = +1) is also produced.

$$P_4 + 3NaOH + 3H_2O \rightarrow PH_3 + 3NaH_2PO_2 \quad (\textit{Non-ionic equation})$$
$$(P = 0) \qquad (P = -3) \qquad (P = +1)$$

or
$$P_4 + 3OH^- + 3H_2O \rightarrow PH_3 + 3H_2PO_2^- \quad (\textit{Ionic equation})$$
$$(P = 0) \qquad (P = -3) \qquad (P = +1)$$

The above ionic equation is composed of the following two half-reactions:

$$3 \times [P_4 + 8OH^- \xrightarrow{\text{Oxidation}} 4H_2PO_2^- + 4e^-] \quad (\textit{Oxidation half-reaction})$$
$$(P = 0) \qquad\qquad (P = +1)$$

$$P_4 + 12H^+ + 12e^- \xrightarrow{\text{Reduction}} 4PH_3 \quad (\textit{Reduction half-reaction})$$
$$(P = 0) \qquad\qquad (P = -3)$$

$$4P_4 + 24OH^- + 12H^+ \rightarrow 12H_2PO_2^- + 4PH_3$$

or $4P_4 + (12OH^- + 12H^+) + 12OH^- \rightarrow 12\,H_2PO_2^- + 4PH_3$

or $4P_4 + 12H_2O + 12OH^- \rightarrow 12H_2PO_2^- + 4PH_3$ ...(*i*)

or $P_4 + 3H_2O + 3OH^- \rightarrow 3H_2PO_2^- + PH_3$

Oxidation and reduction half-reactions, given above, clearly show that if we consider the disproportionation reaction undergone by four P_4 molecules, then one P_4 molecule is reduced to PH_3 and three P_4 molecules are oxidised to $H_2PO_2^-$ ion. In other words, we can say that one P_4 molecule acts as oxidising agent (O.A.) and three P_4 molecules behave as reducing agent (R.A.). Thus equation (*i*) can also be written as :

Reduction

$$\underset{\substack{\text{O.A.}\\(P=0)}}{P_4} + \underset{\substack{\text{R.A.}\\(P=0)}}{3P_4} + 12H_2O + 12OH^- \longrightarrow \underset{(P=+1)}{12H_2PO_2^-} + \underset{(P=-3)}{4PH_3}$$

Oxidation

3. Disproportionation of cuprous compounds in solution. The soluble cuprous compounds (*e.g.*, Cu_2SO_4, $CuCH_3COO$ etc.) disproportionate in solution with the formation of metallic copper and corresponding cupric salt.

$$\underset{(Cu=+1)}{Cu_2SO_4\,(aq)} \rightarrow \underset{(Cu=0)}{Cu\,(s)} + \underset{(Cu=+2)}{CuSO_4\,(aq)}$$

or

$$\underset{(Cu=+1)}{2Cu^+\,(aq)} \rightarrow \underset{(Cu=0)}{Cu\,(s)} + \underset{(Cu=+2)}{Cu^{2+}\,(aq)}$$

This reaction is composed of the following half-reactions :

$Cu^+\,(aq) \rightarrow Cu^{2+}(aq)$ (*Oxidation half-reaction*)

$Cu^+\,(aq) + e^- \rightarrow Cu^0\,(s)$ (*Reduction half-reaction*)

$Cu^+\,(aq) + Cu^+\,(aq) \rightarrow Cu^{2+}(aq) + Cu^0\,(s)$

This reaction shows that one Cu^+ (*aq*) ion is oxidised to Cu^{2+} (*aq*) ion while the other Cu^+ (*aq*) ion is reduced to metallic Cu(*s*), *i.e.*,

Reduction
($Cu = +1 \rightarrow 0$)

$$\underset{\substack{(Cu=+1)\\(\text{R.A.})}}{Cu^+\,(aq)} + \underset{\substack{(Cu=+1)\\(\text{O.A.})}}{Cu^+\,(aq)} \rightarrow \underset{(Cu=+2)}{Cu^{2+}\,(aq)} + \underset{(Cu=0)}{Cu\,(s)}$$

Oxidation
($Cu = +1 \rightarrow +2$)

When the insoluble Cu_2O (Cu = +1) is dissolved in dil. H_2SO_4, the product obtained is Cu (II) sulphate and not Cu (I) sulphate. Thus we see that in this reaction Cu^+ ion undergoes disproportionation to Cu^{2+} ion and Cu^0 metal.

$$\underset{(Cu=+1)}{Cu_2O} + H_2SO_4 \rightarrow \underset{(Cu=+2)}{CuSO_4} + \underset{(Cu=0)}{Cu} + H_2O$$

or

$$\underset{(Cu=+1)}{Cu_2O} + 2H^+ \rightarrow \underset{(Cu=+2)}{Cu^{2+}} + \underset{(Cu=0)}{Cu} + H_2O$$

4. Disproportionation of manganate ion (MnO_4^{2-}) in neutral and acidic solution. MnO_4^{2-} ion (green solution, Mn = +6) is stable in highly alkaline solution but on adding large quantity of water or on acidifying the solution by carbonic acid (by passing CO_2 into the solution), it readily undergoes disproportionation to produce $MnO_2(s)$ (Mn = + 4) and permanganate ion, MnO_4^- (pink colour, Mn = +7).

In neutral solution

$$\underset{(Mn=+6)}{3K_2MnO_4} + 2H_2O \rightarrow \underset{(Mn=+7)}{2KMnO_4} + \underset{(Mn=+4)}{MnO_2\,(s)} + 4KOH \quad (\textit{Non-ionic equation})$$

or

$$3MnO_4^{2-} + 2H_2O \rightarrow 2MnO_4^- + MnO_2(s) + 4OH^- \quad (\textit{Ionic equation})$$

In acidic solution

$3K_2MnO_4 + 2H_2O + 4CO_2 \rightarrow 2KMnO_4 + MnO_2(s) + 4KHCO_3$ (*Non-ionic equation*)

or $3MnO_4^- + 4H^+ \rightarrow 2MnO_4^- + MnO_2(s) + 2H_2O$ (*Ionic equation*)

The above ionic equation is composed of the following two half-reactions :

$2 \times [MnO_4^{2-} \rightarrow MnO_4^- + e^-]$ (*Oxidation half-reaction*)
(Mn = +6) (Mn = +7)

$MnO_4^{2-} + 4H^+ + 2e^- \rightarrow MnO_2(s) + 2H_2O$ (*Reduction half-reaction*)
(Mn = +6) (Mn = +4)

$3MnO_4^{2-} + 4H^+ \rightarrow 2MnO_4^- + MnO_2(s) + 2H_2O$ ...(*i*)

Oxidation and reduction half-reactions given above indicate that if we consider the disproportionation reaction of three MnO_4^{2-} ions, then two MnO_4^{2-} ions undergoes oxidation to MnO_4^- ion and one MnO_4^- ion undergoes reduction to MnO_2. Thus ionic equation (*i*) can be written as :

Oxidation
(Mn = +6 → +7)

$2MnO_4^{2-} + MnO_4^{2-} + 4H^+ \rightarrow 2MnO_4^- + MnO_2(s) + 2H_2O$
(Mn = +6) (Mn = +6) (Mn = +7) (Mn = +4)
R.A. O.A.

Reduction
(Mn = +6 → +4)

5. Disproportionation of H_2O_2 in presence of a catalyst. When a pinch of MnO_2 is added to a small quantity of H_2O_2 taken in a test tube, brisk evolution of O_2 takes place. In this reaction H_2O_2 (O = –1) gets decomposed into H_2O (O = –2) and O_2 (O = 0). MnO_2 acts as a catalyst.

$$2H_2O_2 \xrightarrow{MnO_2} 2H_2O + O_2$$
(O = –1) (O = –2) (O = 0)

This reaction is composed of the following half-reactions:

$H_2O_2 \xrightarrow{\text{Oxidation}} O_2 + 2H^+ + 2e^-$ (*Oxidation half-reaction*)
(O = –1) (O = 0)

$H_2O_2 + 2H^+ + 2e^- \xrightarrow{\text{Reduction}} 2H_2O$ (*Reduction half-reaction*)
(O = –1) (O = –2)

$2H_2O_2 \rightarrow 2H_2O + O_2$

6. Disproportionation of the aqueous solution of hypochlorous acid (HClO) and its salt.
(*a*) When the aqueous solution of HClO (Cl = +1) or its salt *is exposed to light,* it undergoes disproportionation to Cl_2 (Cl = 0) and chloric acid, $HClO_3$ (Cl = +5)

$5HClO \xrightarrow{\text{Light}} 2Cl_2 + HClO_3$ (*Non-ionic equation*)
(Cl = +1) (Cl = 0) (Cl = +5)

$5ClO^- + 4H^+ \rightarrow 2Cl_2 + ClO_3^- + 2H_2O$ (*Ionic equation*)
(Cl = +1) (Cl = 0) (Cl = +5)

The above ionic equation is composed of the following half-reactions :

$$\underset{(Cl = +1)}{ClO^-} + 2H_2O \xrightarrow{\text{Oxidation}} \underset{(Cl = +5)}{ClO_3^-} + 4H^+ + 4e^- \text{ (Oxidation half-reaction)}$$

$$2 \times [\underset{(Cl = +1)}{2ClO^-} + 4H^+ + 2e^- \xrightarrow{\text{Reduciton}} \underset{(Cl = 0)}{Cl_2} + 2H_2O] \text{ (Reduction half-reaction)}$$

$$5ClO^- + 4H^+ \rightarrow 2Cl_2 + ClO_3^- + 2H_2O$$

(*b*) When equeous solution of HClO (Cl = +1) or its salt *is heated,* it undergoes disproportionation to produce HCl (Cl = –1) and chloric acid, $HClO_3$ (Cl = +5)

$$\underset{(Cl = +1)}{3HClO} \xrightarrow{\text{Heat}} \underset{(Cl = -1)}{2HCl} + \underset{(Cl = +5)}{HClO_3} \quad \text{(Non-ionic equation)}$$

$$\underset{(Cl = +1)}{ClO^-} \xrightarrow{\text{Heat}} \underset{(Cl = -1)}{2Cl^-} + \underset{(Cl = +5)}{ClO_3^-} \quad \text{(Ionic equation)}$$

The above ionic equation is composed of the following two half reactions.

$$\underset{(Cl = +1)}{ClO^-} + 2H_2O \xrightarrow{\text{Oxidation}} \underset{(Cl = +5)}{ClO_3^-} + 4H^+ + 4e^- \quad \text{(Oxidation half-reaction)}$$

$$2 \times [\underset{(Cl = +1)}{ClO^-} + 2H^+ + 2e^- \xrightarrow{\text{Reduciton}} \underset{(Cl = -1)}{Cl^-} + H_2O] \quad \text{(Reduction half-reaction)}$$

$$\underset{(Cl = +1)}{3ClO^-} \longrightarrow \underset{(Cl = -1)}{2Cl^-} + \underset{(Cl = +5)}{ClO_3^-}$$

7. Disproportionation of chlorous acid ($HClO_2$) and its salts. When $HClO_2$ (Cl = +3) or its salts are heated, they undergo disproportionation into chloric acid, $HClO_3$ (Cl = +5) and HCl (Cl = –1)

$$\underset{(Cl = +3)}{3HClO_2} \xrightarrow{\Delta} \underset{(Cl = +5)}{2HClO_3} + \underset{(Cl = -1)}{HCl} \quad \text{(Non-ionic equation)}$$

$$\text{or } \underset{(Cl = +3)}{3ClO_2^-} \xrightarrow{\Delta} \underset{(Cl = +5)}{2ClO_3^-} + \underset{(Cl = -1)}{Cl^-} \quad \text{(Ionic equation)}$$

The above ionic equation is composed of the following two half-reactions:

$$2 \times [\underset{(Cl=+3)}{ClO_2^-} + H_2O \xrightarrow{\text{Oxidation}} \underset{(Cl = +5)}{ClO_2^-} + 2H^+ + 2e^-] \quad \text{(Oxidation half-reaction)}$$

$$\underset{(Cl = +3)}{ClO_2^-} + 4H^+ + 4e^- \xrightarrow{\text{Reduciton}} \underset{(Cl = -1)}{Cl^-} + 2H_2O \quad \text{(Reduction half-reaction)}$$

$$\underset{(Cl = +3)}{3ClO_2^-} \longrightarrow \underset{(Cl = +5)}{2ClO_3^-} + \underset{(Cl = -1)}{Cl^-}$$

8. Disproportionation of $KClO_3$. Potassium chlorate, $KClO_3$ undergoes disproportionation in the following two conditions :

(*a*) When $KClO_3$ (Cl = +5) is *heated at a high temperature,* it undergoes disproportionation into KCl (Cl= –1) and potassium perchlorate, $KClO_4$ (Cl = +7)

$$\underset{(Cl = +5)}{4KClO_3} \xrightarrow{\Delta} \underset{(Cl = -1)}{KCl} + \underset{(Cl = +7)}{3KClO_4} \quad \text{(Non-ionic equation)}$$

$$\text{or } \underset{(Cl = +5)}{4ClO_3^-} \longrightarrow \underset{(Cl = -1)}{Cl^-} + \underset{(Cl = +7)}{3ClO_4^-} \quad \text{(Ionic equation)}$$

The above ionic equation is composed of the following two half-reactions :

$$3 \times [ClO_3^- + H_2O + 2e^- \xrightarrow{\text{Oxidation}} ClO_4^- + 2H^+] \quad \textit{(Oxidation half-reaction)}$$
$$(Cl = +5) \qquad\qquad (Cl = +7)$$
$$ClO_3^- + 6H^+ \xrightarrow{\text{Reduciton}} Cl^- + 3H_2O + 6e^- \quad \textit{(Reduction half-reaction)}$$
$$(Cl = +5) \qquad\qquad (Cl = -1)$$

$$4ClO_3^- \longrightarrow 3ClO_4^- + Cl^-$$
$$(Cl = +5) \qquad (Cl = +7) \qquad (Cl = -1)$$

(*b*) When $KClO_3$ (Cl = +5) *is very gently heated* with conc. H_2SO_4, loud explosions are produced and chlorine dioxide, ClO_2(Cl = + 4) and potassium perchlorate, $KClO_4$ (Cl = +7) are obtained.

$$3KClO_3 + H_2SO_4 \rightarrow 2ClO_2 + KClO_4 + H_2O + K_2SO_4 \quad \textit{(Non-ionic equation)}$$
$$(Cl = +5) \qquad (Cl = + 4) \qquad (Cl = +7)$$

or $$3ClO_3^- + 2H^+ \rightarrow 2ClO_2 + ClO_4^- + H_2O \quad \textit{(Ionic equation)}$$
$$(Cl = +5) \qquad (Cl = +4) \qquad (Cl = +7)$$

The above ionic equation is composed of the following half-reactions:

$$ClO_3^- + H_2O \xrightarrow{\text{Oxidation}} ClO_4^- + 2H^+ + 2e^- \quad \textit{(Oxidation half-reaction)}$$
$$(Cl = +5) \qquad (Cl = +7)$$
$$2 \times [ClO_3^- + 2H^+ + e^- \xrightarrow{\text{Reduciton}} ClO_2 + H_2O] \quad \textit{(Reduction half-reaction)}$$
$$(Cl = +5) \qquad (Cl = +4)$$

$$3ClO_3^- + 2H^+ \longrightarrow ClO_4^- + ClO_2 + H_2O$$
$$(Cl = +5) \qquad (Cl = +7) \qquad (Cl = +4)$$

Comproportionation Reaction

This reaction is the reverse of disproportionation reaction, *i.e.*, in comproportionation reaction two species containing an element in two different oxidation states unite together to form a product in which the element is in an intermediate oxidation state, *e.g.*,

$$Ag^{2+}(aq) + Ag^0(s) \rightleftharpoons 2Ag^+(aq)$$
$$(Ag = +2) \qquad (Ag = 0) \qquad (Ag = +1)$$

E^0 value of the above reaction is positive (= +1.18 V), suggesting that the forward reaction is highly favoured thermodynamically.

Methods for balancing a given unbalanced redox reaction

There are two methods for balancing a given unbalanced redox reaction. These are *ion-electron method* or *method of half-reactions* and *oxidation number method.*

I. Ion-electron method or method of half-reactions

This method was developed in 1927 by Jatle and La Mer. Various steps involved in this method are given below :

(*i*) If the given unbalanced equation is in the molecular form, convert it into ionic form and then break this ionic equation into two half-reactions *viz.*, oxidation half-reaction and reduction half-reaction.

(*ii*) Balance both the half-reactions in the following way :

(*a*) Balance the number of atoms other than H and O by multiplying the RHS or LHS of the half-reaction by a suitable multiple.

(*b*) In order to balance H-atoms and to show the acidic medium of the reaction, add H^+ ions to the LHS of the half-reactions.

(*c*) In order to balance O-atoms, add the appropriate number of H_2O molecules to the side deficient in O-atoms.

(*d*) If the reaction takes place in the basic or alkaline medium, H^+ ions used in step (*b*) should be allowed to combine with the equal number of OH^- ions to produce the same number of H_2O molecules. Now add the resulted two equations, so that H^+ ions are cancelled out from both the equations.

(*e*) Balance the electrical charges on both sides by adding electrons (e^-) to the side deficient in negative charges.

(*f*) Multiply one or both the half-reactions by a suitable number, so that by adding the two half-reactions, the electrons are cancelled out.

(*iii*) The above six steps [steps from (*a*) to (*f*)] give both the half-reactions in the balanced state. Now add the two balanced half-reactions and cancel out the terms which are common to both sides.

Various steps given above can be illustrated by the following problems.

Example 17.6. *Balance the following equations by ion–electron method :*

(1) $FeCl_3 + SnCl_2 \rightarrow FeCl_2 + SnCl_4$

(2) $P_4 \rightarrow H_2PO_2^- + PH_3$ *(In basic medium)*

(3) $MnO_4^- + C_2O_4^{2-} \rightarrow Mn^{2+} + CO_2$ *(In acidic medium)*

(4) $I_2 \rightarrow IO_3^- + I^- + H_2O$ *(In alkaline medium)*

Solution (1) $FeCl_3 + SnCl_2 \rightarrow FeCl_2 + SnCl_4$

Prceed according to the following steps :

(*i*) Write the given molecular equation in the ionic form

$$Fe^{3+} + Sn^{2+} \rightarrow Fe^{2+} + Sn^{4+}$$

(*ii*) Break the above ionic equation into oxidation and reduction half reactions

$Sn^{2+} \rightarrow Sn^{4+}$ *(Oxidation half-reaction)* ...(*A*)

$Fe^{3+} \rightarrow Fe^{2+}$ *(Reduction half reaction)* ...(*B*)

(*iii*) In the oxidation half-reaction (*A*), the number of Sn-atoms is already balanced. In order to balance the electrical charges add $2e^-$ to the RHS of equation (*A*)

$$Sn^{2+} \rightarrow Sn^{4+} + 2e^- \qquad ...(C)$$

This is the balanced oxidation half–reaction

(*iv*) In the reduction half-reaction (*B*), the number of Fe atoms is already balanced. In order to balance electrical charges add $1e^-$ to the LHS of equation (*B*).

$$Fe^{3+} = + e^- \rightarrow Fe^{2+} \qquad ...(D)$$

This is the balanced reduction-half reaction.

(*v*) Now in order to get the same number of electrons in the balanced oxidation half reaction (*C*) and balanced reduction half reaction (*D*), multiply equation (*D*) by 2 and then add the two equations.

$$Sn^{2+} \rightarrow Sn^{4+} + 2e^-$$
$$[Fe^{3+} + e^- \rightarrow Fe^{2+}] \times 2$$
$$\overline{Sn^{2+} + 2Fe^{3+} \rightarrow Sn^{4+} + 2Fe^{2+}}$$

This is the balanced ionic equation.

(*vii*) Convert the above ionic equation into molecular form

$$SnCl_2 + 2FeCl_3 \rightarrow SnCl_4 + 2FeCl_2$$

This is the desired balanced molecular equation.

Solution (2) $P_4 \rightarrow H_2PO_2^- + PH_3$ *(In basic medium)*

Proceed through the following steps:

(*i*) It is evident from the given ionic equation that it is a disproportionation reaction, since P_4 (P = 0) is oxidised to $H_2PO_2^-$ (P = +1) as well as reduced to PH_3 (P = –3). Thus the given equation can be broken into the following oxidation and reduction half–reactions:

$P_4 \rightarrow H_2PO_2^-$ *(Oxidation half–reaction)* ...(*A*)

$$P_4 \rightarrow PH_3 \quad \textit{(Reduction half-reaction)} \quad ...(B)$$

(ii) To balance oxidation half-reaction. (*A*) In order to balance phosphorus atoms multiply the RHS of equation (*A*) by 4.

$$P_4 = 4H_2PO_2^- \quad ...(C)$$

(*b*) Now balance O–atoms by adding $8H_2O$ to the LHS of equation (*C*).

$$P_4 + 8H_2O \rightarrow 4H_2PO_2^- \quad ...(D)$$

(*c*) To balance H–atoms add $8H^+$ to the RHS of equation (*D*)

$$P_4 + 8H_2O \rightarrow 4H_2PO_2^- + 8H^+ \quad ...(E)$$

(*d*) Now since the reaction takes place in basic medium $8H^+$ ions should combine with $8OH^-$ to produce $8H_2O$ molecules

$$8H^+ + 8OH^- \rightarrow 8H_2O \quad ...(F)$$

(*e*) Now add equations (*E*) and (*F*) so that $8H^+$ ions are cancelled and we get :

$$P_4 + 8OH^- \rightarrow 4H_2PO_2^- \quad ...(G)$$

(*f*) To balance the electrical charges, add $4e^-$ to the RHS of equation (*G*)

$$P_4 + 8OH^- \rightarrow 4H_2PO_2^- + 4e^- \quad ...(H)$$

This is the balanced oxidation half–reaction.

(iii) To balance reduction half–reaction.(*a*) Multiply RHS of equation (*B*) by 4 so that the number of P– atoms becomes equal on both sides of the equation.

$$P_4 \rightarrow 4PH_3 \quad ...(I)$$

(*b*) Add $12H^+$ to the LHS of equation (*I*) to balance H–atoms

$$P_4 + 12H^+ \rightarrow 4PH_3 \quad ...(J)$$

(*c*) To balance the electrical charges add $12e^-$ to the LHS of equation (*J*)

$$P_4 + 12H^+ + 12e^- \rightarrow 4PH_3 \quad ...(K)$$

This equation is the balanced reduction half–reaction.

(iv) In order to make the number of electrons equal in the balanced oxidation and reduction half–reactions, multiply equation (*H*) by 3 and then add the resulted equation to equation (*K*).

$$[P_4 + 8OH^- \rightarrow 4H_2PO_3^- + 4e^-] \times 3$$

$$P_4 + 12H^+ + 12e^- \rightarrow 4PH_3$$

$$4P_4 + 24OH^- + 12H^+ \rightarrow 12H_2PO_2^- + 4PH_3$$

or $$4P_4 + (12\ OH^- + 12H^+) + 12OH^- \rightarrow 12H_2PO_2^- + 4PH_3$$

or $$4P_4 + 12H_2O + 12OH^- \rightarrow 12H_2PO_2^- + 4PH_3$$

or $$P_4 + 3H_2O + 3OH^- \rightarrow 3H_2PO_2^- + PH_3$$

This is the desired balanced equation.

Solution (3) $MnO_4^- + C_2O_4^{2-} \rightarrow Mn^{2+} + CO_2$ (*In acidic medium*)

Proceed according to the following steps :

(*i*) Break the given ionic equation into oxidation and reduction half reactions.

$$C_2O_4^{2-} \rightarrow CO_2 \quad \textit{(Oxidation half-reaction)} \quad ...(A)$$

(C = +3) (C = +4)

$$MnO_4^- \rightarrow Mn^{2+} \quad \textit{(Reduction half-reaction)} \quad ...(B)$$

(Mn = +7) (Mn = +2)

(ii) To balance the oxidation half–reaction (*A*). The oxidation half-reaction (*A*) can be balanced by using the following steps :

(*a*) To balance carbon atoms in the oxidation half-reaction (*A*), multiply RHS of this equation by two to get :

$$C_2O_4^{2-} \rightarrow 2CO_2 \qquad ...(C)$$

(*b*) In order to balance electrical charges in equation (*C*), add $2e^-$ to the RHS of this equation.

$$C_2O_4^{2-} \rightarrow 2CO_2 + 2e^- \qquad ...(D)$$

This is the balanced oxidation half-reaction.

(*iii*) *To balance the reduction half-reaction* (*B*). Reduction half-reaction (*B*) can be balanced by proceeding through the following steps :

(*a*) The number of Mn-atoms is already balanced.

In order to balance O-atoms add $4H_2O$ to the RHS of this equation.

$$MnO_4^- \rightarrow Mn^{2+} + 4H_2O \qquad ...(E)$$

(*b*) Since the given reaction takes place in acidic medium, in order to balance H-atoms in equation (E), add $8H^+$ to the LHS of this equation.

$$MnO_4^- + 8H^+ \rightarrow Mn^{2+} + 4H_2O \qquad ...(F)$$

(*c*) In order to balance the electrical charges, add $5e^-$ to the LHS of equation (E)

$$MnO_4^- + 8H^+ + 5e^- \rightarrow Mn^{2+} + 4H_2O \qquad ...(G)$$

This is the balanced reduction half-reaction.

(*iv*) In order to make the number of electrons equal in the balanced oxidation and reduction half-reactions, multiply equation (*D*) by 5 and equation (*G*) by 2 and then add the reoulted equations

$$\left[C_2O_4^{2-} \longrightarrow 2CO_2 + 2e^-\right] \times 5$$

$$\left[MnO_4^- + 8H^+ + 5e^- \longrightarrow Mn^{2+} + 2H_2O\right] \times 2$$

$$\overline{2MnO_4^- + 16H^+ + 5C_2O_4^{2-} \rightarrow 2Mn^{2+} + 10CO_2 + 8H_2O}$$

This is the balanced ionic equation.

Solution (4) $I_2 \rightarrow IO_3^- + I^- + H_2O$ (*In alkaline medium*)

Proceed through the following steps :

(*i*) It is evident from the given equation that the reaction shown in this equation is a disproportionation reaction, since I_2 is oxidised (to IO_3^-) as well as reduced (to I^-). Thus the given equation can be broken into the following two half-reactions.

$$I_2\ (I = 0) \rightarrow IO_3^-\ (I = +5) \quad (\textit{Oxidation half-reaction}) \qquad ..(A)$$

$$I_2\ (I = 0) \rightarrow I^-\ (I = -1) \quad (\textit{Reduction half-reaction}) . \qquad ..(B)$$

(*ii*) *To balance oxidation half-reaction.* (*a*) In order to balance the number of iodine atoms in both the sides of equation (*A*), multiply the RHS of this equation by 2.

$$I_2 \rightarrow 2IO_3^- \qquad ...(C)$$

(*b*) In order to balance O-atoms, add $6H_2O$ on the LHS of equation (*C*).

$$I_2 + 6H_2O \rightarrow 2IO_3^- \qquad ...(D)$$

(*c*) In order to balance H-atoms in equation (*D*), add $12H^+$ on the RHS of equation (*D*)

$$I_2 + 6H_2O \rightarrow 2IO_3^- + 12H^+ \qquad ...(E)$$

(*d*) Now since the oxidation of I_2 to IO_3^- takes place in alkaline medium, $12H^+$ ions combine with $12OH^-$ to produce $12H_2O$ molecules

$$12H^+ + 12OH^- \rightarrow 12H_2O \qquad ...(F)$$

(*e*) Add equation (*E*) and (*F*) so that $12H^+$ ions are cancelled from these equations

$$I_2 + 12OH^- \rightarrow 2IO_3^- + 6H_2O \qquad ....(G)$$

(*f*) Balance the electrical charges by adding $10e^-$ to the RHS of equation (*G*)

$$I_2 + 12OH^- \rightarrow 2IO_3^- + 6H_2O + 10e^- \quad ...(H)$$

This is the balanced oxidation half-reaction.

(*iii*) *To balance the reduction half reaction.* (*a*) Multiply the RHS of equation (*B*) by 2 to equalise the number of I-atoms on both sides of this equation.

$$I_2 \rightarrow 2I^- \quad ...(I)$$

(*b*) To balance the electrical charges add $2e^-$ to the LHS of equation (*I*) so that we get equation (*J*)

$$2e^- + I_2 \rightarrow 2I^- \quad ...(J)$$

This is the balanced reduction half-reaction.

(*iv*) Now multiply both the sides of equation (*J*) by 5 and then add the resulted equation to equation (*H*).

$$[2e^- + I_2 \rightarrow 2I^-] \times 5$$

$$I_2 + 12OH^- \rightarrow 2IO_3^- + 6H_2O + 10e^-$$

$$6I_2 + 12OH^- \rightarrow 10I^- + 2IO_3^- + 6H_2O$$

or
$$3I_2 + 6OH^- \rightarrow 5I^- + IO_3^- + 3H_2O$$

This is the desired balanced ion-electron equation.

II. Oxidation number method. The use of this method can be understood by considering the following examples:

Example 17.7. *Balance the following redox reaction by oxidation number method*

$$MnO_4^- + C_2O_4^{2-} \rightarrow Mn^{2+} + CO_2 \quad \textit{(in acidic medium)}$$

Solution. (*i*) In the given equation, MnO_4^- acts as an oxidising agent and $C_2O_4^{2-}$ behaves as a reducing agent. This is because of the fact that oxidation number (O.N.) of Mn is decreased from +7 (in MnO_4^-) to +2 (in Mn^{2+}) and that of carbon atom is increased from +3 (in $C_2O_4^{2-}$) to + 4 (in CO_2). The oxidation number of O-atom remains the same.

(*ii*) Increase in O.N. of one C-atom $= +4 - 3 = +1$

∴ Increase in O.N. of two C-atoms $= +1 \times 2 = +2$

Decrease in O.N. of one Mn-atom $= +7 - 2 = +5$

In order to make the total increase in O.N. equal to the total decrease in O.N., the oxidising agent (*i.e.*, MnO_4^-) should be multiplied by 2 and the reducing agent (*i.e.*, $C_2O_4^{2-}$) should be multiplied by 5. Thus the given equation should be written as:

$$2MnO_4^- + 5C_2O_4^- \rightarrow 2Mn^{2+} + 10CO_2$$

(*iii*) In order to balance O-atoms, add $8H_2O$ molecules to the RHS of the above equation to get :

$$2MnO_4^- + 5C_2O_4^{2-} \rightarrow 2Mn^{2+} + 10CO_2 + 8H_2O$$

(*iv*) In order to show the acidic medium of the reaction and to balance H-atoms, add $16H^+$ to the LHS of the above equation.

$$2MnO_4^- + 5C_2O_4^{2-} + 16H^+ \rightarrow 2Mn^{2+} + 10CO_2 + 8H_2O$$

This is the balanced equation, since the electrical charges on both sides are equal.

Example 17.8. *Balance the redox reaction,*

$$As_2S_3 + NO_3^- \rightarrow H_3AsO_4 + SO_4^- + NO \quad \textit{(in acidic medium)}$$

Solution.	As_2S_3	+ NO_3^-	→ H_3AsO_4	+ SO_4^{2-}	+ NO
	As = +3	N = +5	As = +5	S = +6	N = +2
	S = −2	O = −2	O = −2	O = −2	O = −2

(*i*) In the given equation, it is only O-atom whose O.N. remains the same while the O.Ns of other elements (As, S and N) change. It may be seen that O.Ns of As and S increase while the O.N. of N-atom decreases. Thus :

	Increase in O.N. of one As-atom	$= +5 - 3 = +2$
∴	Increase in O.N. of two As-atoms	$= +2 \times 2 = +4$
	Increase in O.N. of one S-atom	$= +6 - (-2) = +8$
∴	Increase in O.N. of three S-atoms	$= +8 \times 3 = +24$
∴	Total increase in O.N.	$= +4 + 24 = +28$
	Decrease in O.N. of N-atom	$= +5 - 2 = +3$

In order to make total increase in O.N. equal to the total decrease in O.N., multiply As_2S_3 by 3 and NO_3^- by 28. Thus the above equation becomes :

$$3As_2S_3 + 28NO_3^- \rightarrow 6H_3AsO_4 + 9SO_4^{2-} + 28NO$$

(*ii*) In order to balance O-atoms, add $4H_2O$ to the LHS of the above equation.

$$3As_2S_3 + 28NO_3^- + 4H_2O \rightarrow 6H_3AsO_4 + 9SO_4^{2-} + 28NO$$

(*iii*) In order to show acidic medium of the reaction and to balance H-atoms, add $10H^+$ to the LHS of the above equation.

$$3As_2S_3 + 28NO_3^- + 4H_2O + 10H^+ \rightarrow 6H_3AsO_4 + 9SO_4^{2-} + 28NO$$

This is the balanced equation.

Direct and indirect redox reactions

A redox reaction, in which oxidation and reduction take place in the *same vessel,* is called *direct redox reaction.* In this reaction the transfer of electrons from the reducing agent to the oxidising agent takes place *directly* and over a very short distance (generally within the molecular diameters). The decrease in energy during a direct redox reaction appears as *heat, i.e.,* in a direct redox reaction the chemical energy is converted into *heat energy.*

A redox reaction, in which oxidation and reduction take place in *different vessels,* is called *indirect redox reaction.* In this reaction, the transfer of electrons from the reducing agent to the oxidising agent takes place *indirectly* through the conducting wires. The decrease in energy in an indirect redox reaction appears as *electrical energy, i.e.*, in an indirect redox reaction the chemical energy is converted into *electrical energy.* The arrangement in which the indirect redox reaction is carried out, is called *electrochemical cell.* Electrical cell is also called *Galvanic cell* (after the name of *Luigi Galvani*) or *Voltaic cell* (after the name of *Alessandro Volta*).

Importance of redox reactions

In our daily life we come across many proceses, like rusting of iron articles, fading of colour of the clothes, burning of combustible substances (*e.g.*, cooking gas, wood, coal etc.). All these processes are redox processes.

Redox processes are also involved in a large number of industrial processes like electroplating, extraction of metals (*e.g.*, Na, Al etc.), bleaching of wood pulp, manufacture of NaOH etc.

Redox reactions also form the basis of electrochemical and electrolytic cells.

University Questions

1. Calculate O.N. of (*i*) C in HCN (*ii*) C in HNC (*iii*) Cr in CrO_5 (*iv*) S in H_2SO_5 (*v*) S in $H_2S_2O_8$.

Answer.

(*i*) **O.N. of C-atom in hydrocyanic acid or hydrogen cyanide molecule (HCN).** The structure of this molecule is $H-C\equiv\ddot{N}$. Since N-atom is linked with C-atom by three covalent bonds, N-atom has –3 charges on it. $(\overset{+}{H}-\overset{x}{C}\equiv\overset{-3}{N})$. Thus x is given by : $+1 + x - 3 = 0$, $\therefore\ x = +2.$

(*ii*) **O.N. of C-atom in hydroisocyanic acid molecule (HNC).** The structure of this molecule is $H-\ddot{N}=\ddot{C}$ or $H \rightleftharpoons \ddot{N}=C$. Since N-atom is linked with H and C atoms by three covalent bonds, this atom has – 3 charges on it. $(\overset{+1}{H}-\overset{-3}{N}=\overset{x}{C}:)$. Thus x is given by : $+1 - 3 + x = 0$ or $x = +2$

(*iii*) **O.N. of Cr atom in chromium diperoxide or chromic peroxide molecule, CrO_5 or $Cr(O_2)_2O$.** Usual method gives O.N. of Cr atom in CrO_5 molecule = + 10 as shown below :

$$x + 5 \times (-2) = 0 \text{ or } Cr = +10$$

This O.N. of Cr atom is wrong, since the maximum O.N. of Cr atom is only + 6 and cannot be more than this. This is because Cr atom has six unpaired electrons in its valence shell [Cr = $3d^5 4s^1$ (n = 6)]. Correct O.N. of Cr is calculated by considering the structure of CrO_5 molecule given in the margin. This structure shows that CrO_5 has two peroxide ions (O_2^{2-}) and one double-bonded O-atom. Each O-atom of O_2^{2-} ions has – 1 charge and double-bonded O-atom has – 2 charge. Thus O.N. of Cr atom (x) is given by :

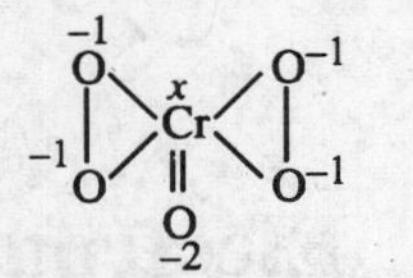

$$x + \underset{\text{(For double-bonded O-atom)}}{1 \times (-2)} + \underset{\text{(For O-atom in } O_2^{2-} \text{ ion)}}{4 \times (-1)} = 0 \quad \text{or} \quad x = +6$$

(*iv*) **O.N. of S-atom in Caro's acid or permonosulphuric acid, H_2SO_5.** Conventional method gives O.N. to S-atom in H_2SO_5 molecule = + 8 as shown below :

$2 \times (+1) + x + 5 \times (-2) = 0$ or $x = +8$. This O.N. is wrong, since S-atom has maximum O.N. = + 6.

Correct O.N. can be calculated by considering the structure of H_2SO_5 molecule as given in the margin. This structure shows that H_2SO_5 has one O_2^{2-} ion (peroxide ion) in which each O-atom has – 1 charge. Each of the remaining three O-atoms has – 2 charge. Each H-atom has + 1 charge. Thus

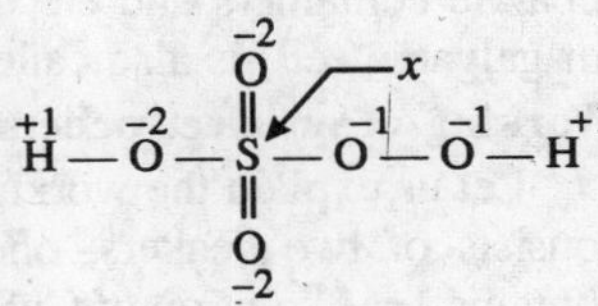

$$\underset{\text{(For H-atom)}}{2 \times (+1)} + \underset{\text{(For O-atom in } O_2^{2-} \text{ ion)}}{2 \times (-1)} + \underset{\text{(For other O-atoms)}}{3 \times (-2)} + x = 0$$

$\therefore \quad x = +6$

(*v*) **O.N. of S-atoms in per disulphuric acid or Marshall's acid ($H_2S_2O_8$).** Conventional method gives O.N. of S-atom in $H_2S_2O_8$ molecule equal to + 7 as shown below :

$$2 \times (+1) + 2 \times x + 8 \times (-2) = 0, \qquad \therefore x = +7$$

This O.N. is wrong, since S-atom has maximum O.N. = + 6. Correct O.N. can be calculated by considering the structure of $H_2S_2O_8$ molecule as given in the margin. This structure shows that $H_2S_2O_8$ has two O_2^{2-} (peroxide ion) ions in which each O-atom has – 1 charge. Each of the remaining six O-atoms has – 2 charge. Thus :

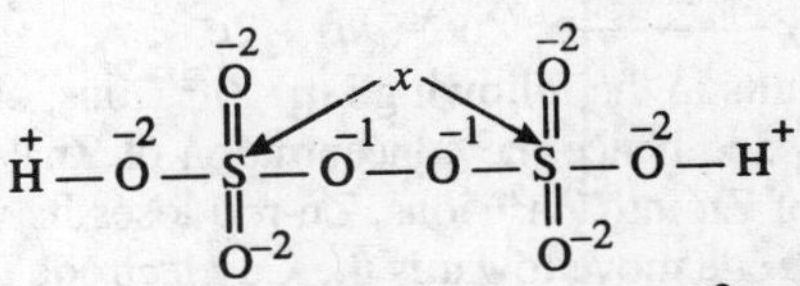

$$\underset{\text{(For two H-atoms)}}{2 \times (+1)} + \underset{\text{(For six O-atoms)}}{6 \times (-2)} + \underset{\text{(For two O-atoms in } O_2^{2-} \text{ ions)}}{2 \times (-1)} + 2 \times x = 0 \qquad x = +6$$

2. Evaluate equivalent weight of $KMnO_4$ for its reduction reaction in acidic, basic and neutral solutions. *(VBS Purvanchal 2002)*
3. What are redox reactions ? Discuss with examples the ion-electron method to balance the chemical equations. *(Lucknow 2002)*
4. How is the equivalent weight of an oxidizing and reducing agent calculated by using oxidation number concept ? Find the oxidation number of underlined elements in the following : (*i*) $\underline{Br}O_3^-$ (*ii*) $H_3\underline{P}O_4$ (*iii*) $H\underline{N}O_3$ (*iv*) $H_2\underline{S}O_3$ *(Nagpur 2002)*
5. Give only one example (equation only) for redox reaction. *(Rohilkhand 2003)*
6. (*a*) Find oxidation number of (*i*) Mn in $KMnO_4$ and (*ii*) C in $H_2C_2O_4$

   (*b*) Balance the following equation by oxidation number method in basic medium : $Br_2 \rightarrow BrO_3^- + Br^-$ *(Nagpur 2003)*

18

Electrode Potentials and Electrochemical Series

Electrochemical cell or galvanic cell

Electrochemical cell is a device in which oxidation and reduction reactions take place in two separate containers and electrical energy is produced during these reactions. Electrochemical cell (or galvanic cell) is also called **voltaic cell.**

Working of an electrochemical cell

Let us explain the working of a galvanic cell by taking the example of a Daniell cell. This cell consists of two beakers, one of which contains 1 M solution of $ZnSO_4$ and the other beaker contains 1 M solution of $CuSO_4$. A Zn-rod is dipped into $ZnSO_4$ solution, while a Cu-rod is dipped into $CuSO_4$ solution. These metallic rods are called **electrodes.**

The metallic rods are connected with an ammeter, by means of an insulated wire. A key is also inserted in the circuit. Ammeter is used to know the presence of electric current. The solutions in the two beakers are called **electrolyte solutions.** These solutions are connected together by an inverted U-tube, containing saturated solution of an *inert electrolyte* like KCl, KNO_3 or NH_4NO_3 (See Fig. 18.1). The U-tube is called **salt bridge.**

When the circuit is completed by inserting the key in the circuit, the needle of the ammeter shows a deflection. The deflection of the needle indicates that electric current is flowing in the external circuit.

Following observations are made :

(*i*) Zn (*s*) is oxidised to Zn^{2+} (*aq*) ions

$$Zn\ (s) \xrightarrow{\text{Oxidation}} Zn^{2+}\ (aq) + 2e^-$$

The above oxidation reaction results in the following : (*a*) Zn^{2+} ions, obtained by the oxidation of Zn, go into the solution of $ZnSO_4$ and hence the concentration of Zn^{2+} ions in $ZnSO_4$ solution increases (*b*) Due to the conversion of Zn into Zn^{2+} ions, Zn-rod loses its weight.

(*ii*) Electrons released at Zn-electrode move towards the Cu-electrode through the outer circuit. These electrons are accepted by Cu^{2+} (*aq*) ions of $CuSO_4$ solution, which are reduced to Cu (*s*).

$$Cu^{2+}\ (aq) + 2e^- \xrightarrow{\text{Reduction}} Cu\ (s)$$

The above reduction reaction results in the following : (*a*) Since Cu^{2+} (*aq*) ions of $CuSO_4$ solution are reduced to Cu (*s*), the concentration of Cu^{2+} ions in $CuSO_4$ solution decreases. (*b*) Cu obtained by reduction of Cu^{2+} (*aq*) ions gets deposited on Cu-rod and hence Cu-rod gains weight.

In the external circuit, electrons flow from Zn-rod to Cu-rod. Therefore, by convention, electric current flows in the opposite direction, *i.e.,* electric current flows from Cu-rod to Zn-rod.

Zn-electrode, where oxidation of Zn (*s*) to Zn^{2+} (*aq*) ions takes place or electrons are released, is called **anode** or **oxidation electrode,** while Cu-electrode, at which reduction of Cu^{2+} (*aq*) ions to Cu (*s*) takes place or electrons are accepted, is called **cathode** or **reduction electrode.**

Since electrons move from Zn-rod to Cu-rod, Zn-rod (*i.e.,* anode or oxidation electrode) is also called **negative terminal** and Cu-rod (*i.e.*, cathode or reduction electrode) is also called **positive terminal.** Sometimes, negative and positive signs are also put on the electrodes, to show the release and loss of electrons taking place on them.

The two containers, in which oxidation and reduction takes place, are called **half-cells.** The beaker containing Zn-rod dipped in $ZnSO_4$ solution is called **oxidation half-cell,** while the beaker having Cu-rod dipped in $CuSO_4$, is called **reduction half-cell.** Oxidation reaction taking place at Zn-electrode (anode) is called **oxidation half-cell reaction** and the reduction reaction taking place at Cu-electrode (cathode) is called **reduction half-cell reaction.** The sum of these half-cell reactions is called **cell reaction.** Thus :

Oxidation half-cell reaction	:	$Zn\ (s) \rightarrow Zn^{2+}\ (aq) + 2e^-$
Reduction half-cell reaction	:	$Cu^{2+}\ (aq) + 2e^- \rightarrow Cu\ (s)$
Cell reaction	:	$Zn\ (s) + Cu^{2+}\ (aq) \rightarrow Zn^{2+}\ (aq) + Cu\ (s)$

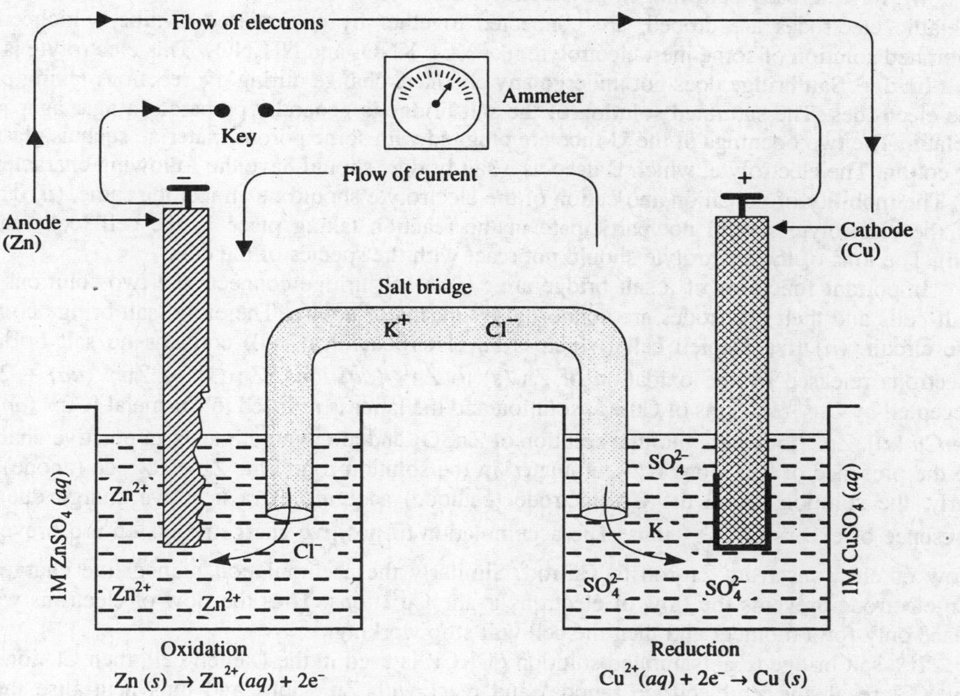

Fig. 18.1 : Working of a Daniell cell.

Cell reaction is a redox reaction which takes place in the cell.

In a simplified way, Daniell cell shown in Fig. 18.1, can be represented as depicted in Fig. 18.2.

Flow of electrons in a galvanic cell

We know that oxidation of metal to metal ions takes place at anode and in this process of oxidation electrons are generated. These electrons move towards the cathode through the external circuit. When these electrons reach cathode, they reduce the metal

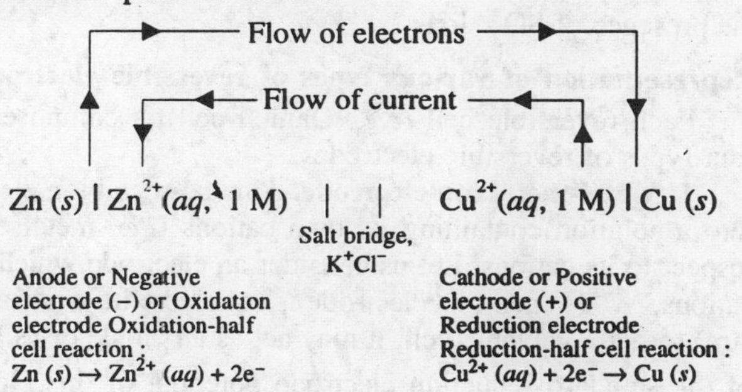

Fig. 18.2 : A simplified way of representing a Daniell cell.

ions into metal atoms. Thus we see, that in a galvanic cell, electrons flow through the external circuit from the negative electrode (anode) to positive electrode (cathode). Since the oxidation potential of anode is higher than that of cathode, we can say that the electrons, in the external circuit flow from the electrode of higher oxidation potential to the electrode of lower oxidation potential. For example, the flow of electrons in the Daniell cell, namely

$$\underset{\text{Anode}}{Zn\,(s)\,|\,Zn^{2+}\,(aq)}\;||\;\underset{\text{Cathode}}{Cu^{2+}\,(aq)\,|\,Cu\,(s)}$$

takes place from Zn (*s*) / Zn^{2+} (*aq*) (anode) to Cu^{2+} (*aq*) / Cu (*s*) (cathode) in the external circuit. This is due to the difference in the relative oxidation tendencies of Zn and Cu, to give their corresponding cations (Zn^{2+} and Cu^{2+} ions respectively) or due to the difference in the relative reduction tendencies of Zn^{2+} (*aq*) and Cu^{2+} (*aq*) ions to give Zn (*s*) and Cu (*s*) atoms respectively.

Salt bridge and its functions

We have already said that in an electrolytic cell the two electrolyte solutions, in which the metallic electrodes are dipped, are connected together by an inverted U-tube, which contains saturated solution of some inert electrolyte like KCl, KNO_3 and NH_4NO_3. This electrolyte is called **salt bridge.** Salt bridge does not undergo any chemical change during the reactions taking place at the electrodes. The saturated solution of the salt bridge is generally prepared in agar agar jelly or gelatin. The two openings of the U-tube are plugged with some porous material, such as glass wool or cotton. The electrolyte, which is used as a salt bridge, should have the following characteristics: (*i*) The mobility of the anion and cation of the electrolyte should be almost the same. (*ii*) The ions of the electrolyte should not participate in the reaction taking place in the cell (cell reaction). (*iii*) The ions of the electrolyte should not react with the species of the cell.

Important functions of a salt bridge are : (*i*) A salt bridge connects the two solutions of the half-cells and their electrodes are connected by means of a wire. Therefore, salt bridge completes the circuit. (*ii*) If a Daniell cell (which is an electrochemical cell) contains no salt bridge, the electrons released by the oxidation of Zn (*s*) to Zn^{2+} (*aq*) ions [Zn (*s*) $\rightarrow$ Zn^{2+} (*aq*) + $2e^-$] are accepted by Cu^{2+} (*aq*) ions of $CuSO_4$ solution and the latter is reduced to Cu-metal [Cu^{2+} (*aq*) + $2e^-$ $\rightarrow$ Cu (*s*)]. Zn^{2+} ions pass into the solution of $ZnSO_4$ and after sometime extra positive charge due to the presence of Zn^{2+} ions is accumulated in the solution round the Zn-electrode (anode). Similarly, the solution round the Cu-electrode (cathode) acquires extra negative charge due to the presence of excess of SO_4^{2-} ions. The accumulation of positive charge round Zn-rod prevents the flow of electrons from Zn-rod to Cu-rod. Similarly the accumulation of negative charge round Cu-electrode prevents the flow of electrons to the Cu^{2+} ions. Thus the flow of electrons will take place only for a moment and then the cell will stop working.

If a salt bridge (*e.g.*, saturated solution of KCl) is used in the Daniell cell, then Cl^- ions given by KCl reach the Zn-electrode (anode) and react with Zn^{2+} ions and thus neutralise the extra positive charge accumulated round Zn-rod due to the presence of Zn^{2+} ions. Similarly K^+ ions furnished by KCl reach Cu-electrode (cathode) and neutralise the negative charge created due to the presence of SO_4^{2-} ions.

Representation of various types of reversible electrodes

Each reversible cell (*e.g.*, Daniell cell) is composed of two reversible electrodes. There are four types of reversible electrodes.

1. Metal-metal ion electrode. A metal-metal ion electrode consists of a pure metal rod dipped into a solution containing its own cations (*i.e.*, metal cations). This electrode is reversible with respect to its cations. Let us consider an electrode which consists of a metal, M in contact with its cations, M^{n+}. When this electrode is coupled with a normal hydrogen electrode, $2H^+$ (*aq*) / H_2 (*g*, 1 atm) to get a galvanic cell, it may act as an anode or as a cathode depending on whether the value of the standard reduction electrode potential of this electrode (*i.e.*, $E^0_{M^{n+}/M}$ value) is negative or positive. Thus the following two cases arise :

(*a*) When the value of $E^0_{M^{n+}/M}$ is negative, the given electrode constitutes the anode of the resulted cell and the given electrode is represented as M(*s*) / M^{n+} (*aq*) in which the vertical line denotes the contact boundary between the solid metal and the solution phase.

M (*s*) / M^{n+} (*aq.*)	‖	$2H^+$ (*aq.*) ∣ H_2 (*g*, 1 atm)
Metal-metal ion electrode		Normal hydrogen electrode
(Anode or oxidation electrode)		(Cathode or reduction electrode)
$E^0_{M^{n+}/M}$ = Negative value		

At the given electrode oxidation of M (*s*) to Mn^+ (*aq*) takes place. Hence this electrode is also called *oxidation electrode.*

$$M(s) \rightarrow M^{n+}(aq) + ne^- \ (oxidation)$$

(*b*) When the value of $E^0_{M^{n+}/M}$ is positive, the given electrode constitutes the cathode of the resulted cell and the given electrode is represented as M^{n+} (*aq*)/M (*s*)

H_2 (*g* 1 atm) / $2H^+$ (*aq*)	‖	M^{n+}(*aq*) / M (*s*)
Normal hydrogen electrode		Metal-metal ion electrode
(Anode or oxidation electrode)		(Cathode or reduction electrode)

At the given electrode reduction of M^{n+} (*aq*) to M (*s*) takes place. Hence this electrode is also called *reduction electrode.*

$$M^{n+}(aq) + ne^- \rightarrow M(s) \ (Reduction)$$

2. Gas-inert metal electrodes. Examples are given below :

(*i*) *Hydrogen electrode* (H^+/ ½ H_2 electrode). In this electrode hydrogen gas at a given pressure is bubbled into the solution of an acid, say HCl, *i.e.*, in this electrode hydrogen gas at a given pressure is bubbled into a solution containing H^+ ions. Since hydrogen gas is non-conducting, Pt or some other metal which is not attacked by the acid and comes easily into equilibrium with hydrogen, is used for making electrical contact in the circuit. When the pressure of hydrogen gas is 1 atmosphere, hydrogen electrode is called normal (standard) hydrogen electrode, NHE.

(*a*) When a NHE is coupled with a metal-metal ion electrode, M (*s*) / M^{n+} (*aq*) which has a negative value for its standard reduction electrode potential (*i.e.*, $E^0_{M^{n+}/M}$ value is negative), this electrode acts as cathode and hence is represented as $2H^+$ (*aq*) / H_2 (*g*, 1 atm), Pt or simply as $2H^+$ (*aq*) / H_2 (*g*, 1 atm). Reduction of H^+ (*aq*) to H_2 (*g*) takes place at this electrode.

$$2H^+(aq) + 2e^- \rightarrow H_2(g) \ (Reduction)$$

When a NHE acts as cathode, its standard reduction electrode potential is assumed to be equal to zero with a positive sign, *i.e.*, $E^0_{2H^+/H_2} = +\,0.00$ V.

(*b*) When NHE is coupled with a metal-metal ion electrode M (*s*) / Mn^+ (*aq*) which has a positive value for its standard reduction electrode potential (*i.e.*, $E^0_{M^{n+}/M}$ value is positive), this electrode acts as anode and hence is represented as Pt, H_2 (*g*, 1 atm) / $2H^+$ (*aq*) or simply as H_2 (*g*, 1 atm) / $2H^+$ (*aq*). Oxidation of H^2 (*g*) to H^+ (*aq*) takes place at this electrode

$$H_2(g) \rightarrow 2H^+(aq) + 2e^- \ (Oxidation)$$

When a NHE acts as an anode, its standard reduction electrode potential is assumed to be equal to zero with a negative sign, *i.e.*, $E^0_{2H^+/H_2} = -\,0.00$ V.

The above description shows that hydrogen electrode is reversible with respect to H^+ ions.

(*ii*) *Chlorine electrode* [Pt, Cl_2 (*g*, 1 atm)/$2Cl^-$ (*aq*)]. In this electrode chlorine gas at a given pressure is bubbled into a solution of HCl, *i.e.*, in this electrode chlorine gas is bubbled through a solution containing Cl^- ions. Pt metal is used for making electrical contact in the circuit. This electrode is represented as Pt, Cl_2 (*g*, 1, atm)/$2Cl^-$ (*aq*). The reduction reaction taking place at this electrode is represented as :

$$Cl_2(g) + 2e^- \rightarrow 2Cl^-(aq) \ (Reduction\ reaction)$$

This reaction shows that this electrode is reversible with respect to Cl^- ions. The values of standard reduction electrode potentials of fluorine, chlorine, bromine and iodine electrodes are given below. These values are for the reduction reaction taking place at the electrode.

$$X_2\,(g) + 2e^- \rightarrow 2X^-\,(aq)$$
$$(X_2 = F_2, Cl_2, Br_2, I_2)$$

Electrode	*Reduction reaction*	E^0_{red} *(Volt)*
Pt, F_2 (*g*, 1 atm) / $2F^-$ (*aq*)	$F_2\,(g) + 2e^- \rightarrow 2F\,(aq)$	+ 2.87
Pt, Cl_2 (*g*, 1 atm) / $2Cl^-$ (*aq*)	$Cl_2\,(g) + 2e^- \rightarrow 2Cl^-\,(aq)$	+1.36
Pt, Br_2 (*g*, 1 atm) / $2Br^-$ (*aq*)	$Br_2\,(g) + 2e^- \rightarrow 2Br^-\,(aq)$	+ 1.06
Pt, I_2 (*g*, 1 atm) / $2I^-$ (*aq*)	$I_2\,(g) + 2e^- \rightarrow 2I^-\,(aq)$	+ 0.53

The values given above show as we proceed from F_2 to I_2 (*i.e.*, down group VIIA), the values of $E^0_{X_2(g)/2X^-(aq)}$ go on decreasing.

(*iii*) *Oxygen electrode (O_2/OH^- electrode)*. In this electrode, O_2 gas at a given pressure is bubbled into a solution containing OH^- ions. Pt metal is used for making electrical contact in the circuit. This electrode is represented as O_2 (*g*, 1 atm), Pt / OH^- (*aq*) or simply as O_2 (*g*) /OH^- (*aq*). The value of standard reduction electrode potential for this electrode is + 0.40 V, *i.e.*, + 0.40 V is the value of the standard reduction potential, corresponding to the following reduction reaction, in which O_2 gas (O = zero) is reduced to OH^- ions (O = –2).

$$\underset{(O\,=\,zero)}{O_2\,(g)} + 2H_2O + 4e^- \rightarrow \underset{(O\,=\,-2)}{4OH^-\,(aq)} \quad \textit{(Reduction reaction)}$$

Since the above equilibrium between O_2 gas and OH^- ions in solution cannot be established quickly, O_2 – OH^- electrode does not behave as a truely reversible electrode.

3. Metal-metal salt electrodes. These electrodes consist of a metal in contact with a sparingly soluble salt of the same metal dipped in the solution of a soluble salt having the same anion. Examples of such electrodes are Ag – AgCl and Hg – Hg_2Cl_2 (calomel) electrodes.

(*i*) *Ag-AgCl electrode.* This electrode consists of a Ag wire, coated with a layer of AgCl, which is inserted in the solution of a chloride (*e.g.*, KCl or HCl) of 1 M concentration. On the hydrogen scale, this electrode has the value for its standard reduction electrode potential equal to + 0.2225 V. Positive value indicates that when this electrode is coupled with a normal hydrogen electrode, this electrode constitutes the cathode of the resulted cell and hence reduction takes place at this electrode. At this electrode Ag^+ ions are reduced to Ag.

$$\underset{(Ag\,=\,+1)}{AgCl\,(s)} + e^- \rightarrow \underset{(Ag\,=\,0)}{Ag\,(s)} + Cl^-\,(aq)$$

Since this electrode acts as cathode, it is represented as : KCl (*aq*), AgCl (*s*) / Ag (*s*) or HCl (*aq*), AgCl (*s*) / Ag (*s*) or Cl^- (*aq*), AgCl (*s*) / Ag (*s*).

(*ii*) *Hg – Hg_2Cl_2 electrode* (*Also called calomel electrode*). This electrode consists of metallic mercury in contact with calomel (Hg_2Cl_2) which, in turn, is in contact with a solution of a chloride (*e.g.*, KCl). This electrode has positive value for its standard reduction electrode potential. This value depends on the concentration of KCl solution used as shown below at 25°C. The concentrations of KCl solution are given in parentheses.

E° values = + 0.3335 V (at 0.1 N), + 0.2810 V (at 1.0 N), + 0.2422 V (*Saturated solution*)

Positive values indicate that then calomel electrode is connected with a standard hydrogen electrode, this electrode constitutes the cathode of the resulted cell and hence reduction takes place at this electrode. At this electrode Hg_2^{2+} ions are reduced to Hg

$$\underset{(Hg_2\,=\,+2)}{Hg_2Cl_2\,(s)} + 2e^- \rightarrow \underset{(Hg\,=\,0)}{2Hg(l)} + 2Cl^-\,(aq)$$

Since calomel electrode acts as cathode, it is represented as : KCl (*aq*), Hg_2Cl_2 (*s*) / Hg (*l*) or Cl^- (*aq*), Hg_2Cl_2 (*s*)/Hg (*l*).

(*iii*) *Hg-Hg_2SO_4 electrode.* The value of standard reduction potential of this electrode is equal to + 0.6) V. Positive value implies that when this electrode is coupled with a NHE to get a galvanic cell, this electrode acts as a cathode and hence the reduction of Hg_2^{2+} ions to Hg takes place at this electrode

$$Hg_2SO_4\,(s) + 2e^- \rightarrow 2Hg\,(l) + 2SO_4^{2-}\,(aq)$$
$$(Hg_2 = +2) \qquad (Hg = 0)$$

Since the given electrode acts as cathode, it is represented as SO_4^{2-} (*aq*), Hg_2SO_4 (*s*) / Hg (*l*).

4. Oxidation-reduction (or redox) electrodes. The electrode consisting of an inert metal like Pt, Au etc., inserted into a solution containing the ions of the same element in different oxidation states is called redox electrode. The inert metal picks up the electrons and provides electrical contact of the electrodes (cf. gas electrodes). For example : Pt/Fe^{3+} (*aq*), Fe^{2+} (*aq*) which is a redox electrode is obtained by inserting Pt-wire into a solution containing Fe^{2+} and Fe^{3+} ions. The potential of a redox electrode arises due to the tendency of the ions in one of the oxidation states to change into the other more stable oxidation state. Thus the reaction taking place the above redox electrode can be written as :

$$Fe^{3+}\,(aq) + e^- \rightleftharpoons Fe^{2+}\,(aq)$$

Some redox electrodes, reduction reactions taking place on them and the values of standard reduction electrode potentials (E^o_{red} values) are given below in Table 18.1 :

Table 18.1 : *Some redox electrodes*

Redox electrode	Reduction reaction	E^o_{red} value (Volts)
Sn^{4+} (*aq*) / Sn^{2+} (*aq*), Pt	Sn^{4+} (*aq*) + $2e^- \rightarrow Sn^{2+}$ (*aq*)	+ 0.14
Fe^{3+} (*aq*) / Fe^{2+} (*aq*), Pt	Fe^{3+} (*aq*) + $e^- \rightarrow F^{2+}$ (*aq*)	+ 0.77
Co^{3+} (*aq*) / Co^{2+} (*aq*), Pt	Co^{3+} (*aq*) + $e^- \rightarrow Co^{2+}$ (*aq*)	+ 1.82
$8H^+$ (*aq*), MnO_4^- (*aq*) / Mn^{2+} (*aq*), Pt	MnO_4^- (*aq*) + $8H^+$ + $5e^-$ $\rightarrow Mn^{2+}$ (*aq*) + $4H_2O$	+ 1.51
Ce^{4+} / Ce^{3+} (*aq*), Pt	Ce^{4+} (*aq*) + $e^- \rightarrow Ce^{3+}$ (*aq*)	+ 1.61
$14H^+$ (*aq*), $Cr_2O_7^{2-}$ (*aq*)/ Cr^{3+} (*aq*) Pt	$Cr_2O_7^{2-}$ (*aq*) + $14H^+$ + $6e^- \rightarrow$ $2Cr^{3+}$ (*aq*) + 7 H_2O	+ 1.33

What is electrode potential (E_{elec})

Let us consider a metal-metal ion electrode which is obtained by immersing a metal rod (M) in its salt solution, *i.e.*, metal-metal ion electrode consists of a metal (M) immersed in the solution containing metal cations (M^{n+} ions). In this electrode there exists a separation of charges between the metal and its salt solution. This separation of charges results in that a net negative or positive charge is developed on the metal with respect to it solution. The negative or positive charge developed on the metal produces electrical potential difference between the metal and its solution. This electrical potential difference is called potential of the electrode or **electrode potential.** Thus electrode potential is a measure of the tendency of an electrode in a half-cell to gain or lose electrons. The unit of electrode potential is **Volt** (V). Electrode potential is generally represented as E_{elec}.

What is Volt ?

The volt, is the potential, required to flow one coulomb of electric current per second, through a uniform thread of mercury 106.3 cm long, containing 14.4521 g of mercury at 0°C.

Origin of electrode potential

How the electrode potential is generated, has been explained by Nernst by his theory, which is called *Nernst's theory*. Let us consider the following two examples :

1. Zn-metal plate placed in a solution containing Zn^{2+} ions. When a Zn-metal plate is placed in a solution containing Zn^{2+} ions (*e.g.*, Zn metal plate may be placed in 1 M solution of $ZnSO_4$), Zn atoms lose electrons and are oxidised to Zn^{2+} (*aq*) ions.

$$Zn\,(s) \rightarrow Zn^{2+}\,(aq) + 2e^-$$

Electrons thus generated get accumulated on the Zn-plate and hence make the plate negatively-charged. In order to neutralise the negative charge created on the surface of the metal plate, a layer of Zn^{2+} ions (Zn^{2+} ions are positively-charged) is formed round the surface of negatively-charged Zn-plate [Fig. 18.3 (*a*)]. In this way, an electrical layer is generated between the metal plate and the solution containing Zn^{2+} ions. This electrical layer develops a definite potential difference between the metal (Zn) and Zn^{2+} (*aq*) ions of the solution. This potential difference is called electrode potential of Zn (*s*)/$ZnSO_4$ (*aq*) or Zn (*s*) / Zn^{2+} (*aq*) electrode.

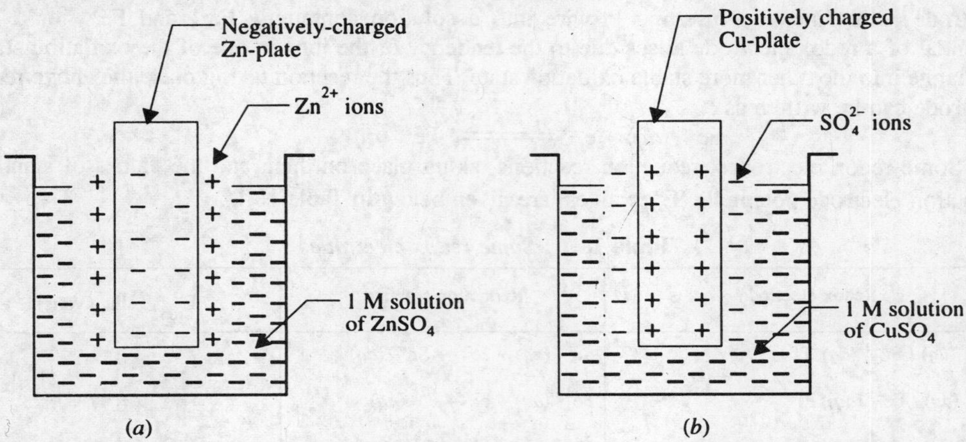

Fig. 18.3 : Origin of electrode potential in (*a*) Zn (*s*) / $ZnSO_4$ (*aq*) or Zn (*s*) / Zn^{2+} (*aq*) electrode (*b*) Cu (*s*)/ $CuSO_4$ (*aq*) or Cu (*s*) / Cu^{2+} (*aq*) electrode.

2. Cu-metal plate placed in a solution containing Cu^{2+} ions. When a Cu-metal plate is placed in a solution containing Cu^{2+} ions (*e.g.* Cu-metal plate may be placed in a solution of $CuSO_4$), Cu plate becomes positively charged and in order to neutralise the positive charge on the plate, a layer of SO_4^{2-} ions (these ions are negatively-charged) is formed round the surface of the positively-charged plate [Fig. 18.3 (*b*)]. In this case too, an electrical layer is generated between the metal plate and the solution containing Cu^{2+} ions. This electrical layer gives rise to a potential difference between the metal (Cu) and Cu^{2+} (*aq*) ions (or solution). This potential difference is called electrode potential of Cu (*s*) / $CuSO_4$ (*aq*) or Cu (*s*) / Cu^{2+} (*aq*) electrode.

What is standard electrode potential (E^o_{elec}) ?

Standard electrode potential can be defined in any of the following two ways :

(*i*) When the concentration of the ions and the pressure of the gaseous species appearing in the electrode reaction (half-cell reaction) are 1 mol dm^{-3} and 101.325 kPa = 1 atom respectively, the electrode potential is called standard electrode potential.

(*ii*) Standard electrode potential of an electrode can also be defined as its potential determined with respect to a normal hydrogen electrode. In simple words, standard electrode potential of a metal-metal ion electrode is defined as the potential difference between a normal hydrogen electrode and the metal-metal ion electrode obtained by immersing the metal in its salt solution of 1 M concentration at 25°C or 25 + 273 = 298 K. Standard electrode potential is designated as E^o_{elec}.

Types of electrode potential

Electrode potential may be of the following types :

1. Oxidation electrode potential $(E_{ox})_{elec}$. Oxidation potential of an electrode is for the oxidation reaction taking place at the electrode. Thus oxidation electrode potential is the potential of the electrode at which oxidation takes place and hence is a measure of the tendency of the electrode in a half-cell to get oxidised or to lose electrons. For example, standard oxidation potential of Zn (s)/ Zn^{2+} (aq) electrode, which is equal to + 0.76 V, is for the oxidation reaction (oxidation of Zn to Zn^{2+} ions) taking place at the electrode.

$$Zn\ (s) \rightarrow Zn^{2+}\ (aq) + 2e^-,\ E^o_{ox} \text{ or } E^o_{Zn/Zn^{2+}} = +\ 0.76\ V$$

2. Reduction electrode potential, $(E_{red})_{elec}$. Reduction potential of on electrode is for the reduction reaction taking place at the electrode. Thus reduction electrode potential is the potential of the electrode at which reduction takes place and hence is a measure of the tendency of the electrode in a half-cell to get reduced or to gain electrons. For example,standard reduction potential of Cu (s) / Cu^{2+} (aq) electrode, which is equal to + 0.34 V, is for the reduction reaction (reduction of Cu^{2+} ions to Cu) taking place at the electrode.

$$Cu^{2+}\ (aq) + 2e^- \rightarrow Cu\ (s),\ E^o_{red} \text{ or } E^o_{Cu^{2+}/Cu} = +\ 0.34\ V$$

Relation between the oxidation potential and reduction potential of a given electrode. Oxidation potential of an electrode, $(E_{ox})_{elec}$ is numerically equal to the reduction potential of the same electrode, $(E_{red})_{elec}$ with opposite sign, *i.e.,*

$$(E_{ox})_{elec} = -\ (E_{red})_{elec}$$

or

$$-\ (E_{ox})_{elec} = (E_{red})_{elec}$$

Similarly standard oxidation potential of an electrode, $(E^o_{ox})_{elec}$ is also numerically equal to the standard reduction potential of the same electrode, (E^o_{red}) with opposite sign, *i.e.,*

$$(E^o_{ox})_{elec} = -\ (E^o_{red})_{elec}$$

or

$$-\ (E^o_{ox})_{elec} = (E^o_{red})_{elec}$$

An important convention. According to the present convention, the value of electrode potential is expressed as **reduction electrode potential.** Thus, if we say that the value of standard electrode potential of Zn (s) / Zn^{2+} (aq) electrode is – 0.76 volt, this means that the given value is the value of the standard **reduction** electrode potential of Zn-electrode, *i.e.,* $E^o_{Zn^{2+}/Zn} = -\ 0.76$ Volt

3. Negative reduction electrode potential and positive reduction electrode potential. We have already said that, as a convention, the value of the electrode potential is expressed as reduction electrode potential. This reduction electrode potential may have a negative value or a positive value, depending on whether the given electrode acts as a negative terminal (anode or oxidation terminal) or as a positive terminal (cathode or reduction terminal) of the galvanic cell obtained by coupling the given electrode with a normal (or standard) hydrogen electrode (NHE).

Let us consider a galvanic cell which is obtained by connecting a normal hydrogen electrode (NHE), H_2 (g) 1 atm) / $2H^+$ (aq 1 M), Pt with a metal-metal ion electrode, M (s) / M^{n+} (aq, 1 M). Now the following two cases arise :

(*a*) If M (s) / M^{n+} (aq, 1 M) acts as anode and H_2 (g, 1 atm) / $2H^+$ (aq, 1 M), Pt acts as cathode, then the cell thus obtained can be represented as :

M (s) / M^{n+} (aq, 1 M)	‖	$2H^+$ (aq, 1 M) / H_2 (g, 1 atm), Pt
Metal-metal ion electrode (Anode or oxidation electrode)		NHE (Cathode or reduction electrode)

The two half-cell reactions can be written as :

Oxidation reaction at anode : $M\ (s) \rightarrow M^{n+}\ (aq) + ne^- \times 2$

Reduction reaction at cathode : $2H^+\ (aq) + 2e^- \rightarrow H_2\ (g) \times n$

Cell reaction : $2M\ (s) + 2nH^+\ (aq) \rightarrow 2M^{n+}\ (aq) + nH_2\ (g)$

Obviously the above cell will work only when E^o_{cell} comes out to be a positive quantity. Thus in order to get a positive value of E^o_{cell}, the value of $E^o_{M^{n+}/M}$ should be a negative quantity, since $E^o_{2H^+/H_2}$ = 0.0 volt and (–) × (–) = (+). Thus we can say that if in a galvanic cell of the above type, M (*s*) / M^{n+} (*aq*, 1 M) electrode acts as anode and H_2 (*g*, 1 atm)/$2H^+$ (*aq*, 1 M), Pt acts as cathode, the standard reduction electrode potential of M (*s*) / M^{n+} (*aq*, 1 M) electrode (anode) will have negative value, *i.e.*,

$$E^o_{M^{n+}/M} = \text{A negative quantity}$$

Since oxidation (release or loss of electrons) takes place at metal-metal ion electrode, this electrode has less electron accepting tendency than NHE and hence its standard reduction potential gets a negative sign.

(*b*) If M (*s*) / Mn^+ (*aq*, 1 M) acts as cathode and H_2 (*g*, 1 atm) / $2H^+$ (*aq*, 1 M), Pt acts as anode, then the cell thus obtained can be represented as :

Pt, H_2 (*g*, 1 atm)/ H^+ (*aq*,1 M) ‖	M^{n+} (*aq*, 1 M) / M (*s*)
NHE (anode or oxidation electrode)	Metal-metal ion electrode (cathode or reduction electrode)

The two half-cell reactions taking place at the two electrodes are :

Oxidation reaction at anode : $H_2 (g) \rightarrow 2H^+ (aq) + 2e^-$] × *n*

Reduction reaction at cathode : $M^{n+} (aq) + ne \rightarrow M (s)$] × 2

Cell reaction : $nH_2 (g) + 2M^{n+} (aq) \rightarrow 2nH^+ (aq) + 2M (s)$

$$\therefore \quad E^o_{cell} = (E^o_{red})_{cathode} - (E^o_{red})_{anode}$$

$$= E^o_{M^{n+}/M} - E^o_{2H^+/H_2}$$

Obviously, in order to get a positive value of E^o_{cell}, the value of $E^o_{M^{n+}/M}$ should be a positive quantity, since $E^o_{2H^+/H_2}$ = 0.0 volt. Thus we can say that if a galvanic cell of the above type, M (*s*) / M^{n+} (*aq*, 1 M) electrode acts as cathode and H_2 (*g*, 1 atm)/$2H^+$ (*aq*, 1 M), Pt acts as anode, the standard reduction electrode potential of M (*s*)/M^{n+} (*aq*, 1 M) electrode will have a positive sign, *i.e.*,

$$E^o_{M^{n+}/M} = \text{A positive quantity}$$

Since reduction (gain of electrons) takes place at metal-metal ion electrode, this electrode has greater electron accepting tendency than NHE and its standard reduction potential gets a positive sign.

The above discussion can be summarised by stating that the electrode having a negative value for its standard reduction electrode potential will constitute the negative terminal (anode) of the galvanic cell obtained by compling the given electrode with a NHE. Similarly an electrode having positive value for its standard reduction electrode potential will act as positive terminal (cathode) of the resulted galvanic cell. In other words we can say that an electrode at which oxidation takes place acts as anode and has negative value for its standard reduction electrode potential. Similarly an electrode at which reduction occurs acts as cathode and has positive value of its standard reduction electrode potential.

Determination of the standard reduction electrode potential of a given metal-metal ion electrode

Standard electrode potential of a given metal-metal ion electrode (half-cell) can be determined by a method which consists of the following steps :

(*i*) Constitute a galvanic cell by connecting the given electrode with a normal hydrogen electrode (NHE) through a salt bridge. Here NHE acts as a reference electrode.

(*ii*) Put a voltmeter in the circuit.

(*iii*) Connect the electrodes with each other and read the value given by voltmeter. This value is the value of standard e.m.f. of the cell which is obtained by coupling the given electrode with NHE.

This method can be understood by considering the following examples :

1. Determination of reduction electrode potential of Zn (*s*) / Zn^{2+} (*aq*, 1 M) electrode. Zn (*s*) / Zn^{2+} (*aq*, 1 M) electrode is obtained by immersing a zinc rod in 1 M solution of $ZnSO_4$ and a NHE contains 1 M HCl with Pt electrode surrounded by bubbles of hydrogen gas at 1 atm pressure. These two electrodes are connected together through a salt bridge of KNO_3 and thus a galvanic cell is obtained. A voltmeter is also put in the circuit. This voltmeter gives the value of standard e.m.f. of the resulted cell. When these electrodes are joined together to form the cell, a competition sets in between the electrodes to liberate electrons. Now since Zn has a greater tendency to liberate electrons than hydrogen, the electrons flow from Zn-electrode to hydrogen electrode and hence Zn-electrode acts as negative electrode (anode) and H-electrode acts as positive electrode (cathode). In other words we can say that since oxidation [oxidation of Zn (*s*) to Zn^{2+} (*aq*) ions] takes place at Zn-electrode and reduction [reduction of H^+ (*aq*) ions to H_2 (*g*)] occurs at H-electrode, Zn-electrode acts as oxidation electrode (anode) and NHE electrode acts as reduction electrode (cathode).

As soon as the two electrodes are joined together, voltmeter gives a reading of 0.76 V which is the value of standard e.m.f. of the resulted cell (*i.e.*, E^o_{cell} = 0.76 V). This value is obtained in the voltmeter, only when the +ive terminal of the voltmeter in connected to the +ive electrode (cathode) and –*ive* terminal of the voltmeter is connected to the –ive electrode (anode) of the cell. If the connection of the voltmeter to the electrodes is not made in the way mentioned above, no reading will be obtained in the voltmeter. This discussion shows that the cell obtained by coupling Zn (*s*) / Zn^{2+} (*aq*, 1 M) electrode to Pt, H_2 (*g*, 1 atm) / H^+ (*aq*, 1 M) electrode can be represented, as shown in Fig. 18.4.

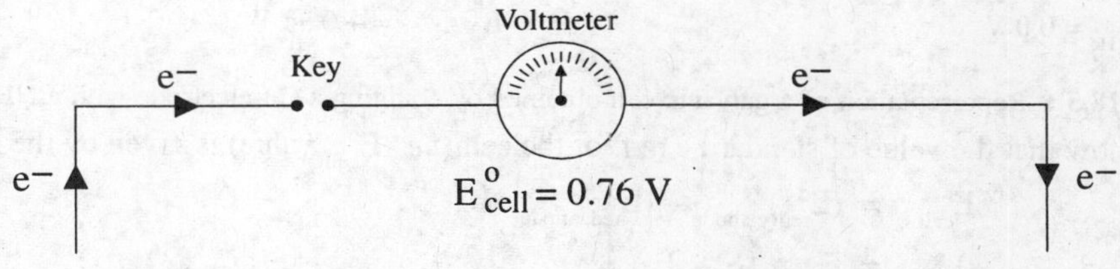

Zn (*s*)/Zn^{2+} (*aq*, 1 M) I K NO_3I	H^+ (*aq*, M) I H_2 (*g*, 1 atm), Pt NHE
Zn-electrode (Anode or oxidation or Negative electrode).	Cathode or reduction or (Positive electrode)
Oxidation reaction :	Reduction reaction
Zn (*s*) → Xn^{2+} (*aq*) + $2e^-$	$2H^+$ (*aq*) + $2e^-$ → H_2 (*g*)
$E^0_{Zn^{2+}/Zn}$ = – 0.76 V.	$E^0_{2H^+/H_2}$ = 0.0 V.

Fig. 18.4 : Representation of a galvanic cell obtained by coupling Zn-electrode with a NHE.

We know that the value of standard e.m.f. of the cell (*i.e.*, E^o_{cell} value) is given by the expression.

$$E^o_{cell} = (E^o_{red})_{cathode} - (E^o_{red})_{anode}$$

or $$0.76 = E^o_{2H^+/H_2} - E^o_{Zn^{2+}/Zn}$$

or $$0.76 = 0.0 - E^o_{Zn^{2+}/Zn}$$

or $$E^o_{Zn^{2+}/Zn} = (E^o_{red})_{anode} = -\,0.76 \text{ V (A negative quantity)} \quad ...(i)$$

2. Determination of reduction electrode potential of Cu (*s*)/Cu^{2+} (*aq*, 1 M) electrode. Cu (*s*) / Cu^{2+} (*aq*, 1 M) electrode is obtained by immersing a Cu-rod in 1 M solution of $CuSO_4$. This electrode is connected with a NHE through salt bridge of KCl and thus a galvanic cell is obtained. A voltmeter is also put in the circuit. This voltmeter gives the value of standard e.m.f. of the resulted cell. When Cu-electrode and NHE are connected together to form the cell, a competition sets in between the two electrode to release electrons, since hydrogen has greater tendency to liberate electrons than copper the electrons from hydrogen electrode to Cu-electrons from hydrogen electrode to Cu-electrode and hence NHE acts as negative electrode (anode) and Cu-electrode acts as positive electrode (cathode). In other words we can say that since oxidation [oxidation of H_2 (*g*) to H^+ (*aq*) ions] takes place at NHE and reduction [reduction of Cu^{2+} (*aq*) ions to Cu (*s*) metal] occurs at Cu-electrode, NHE acts as oxidation electrode (anode) and NHE acts as reduction electrode (cathode). As soon as the two electrodes are connected together voltmeter gives a reading of 0.34 V which is the value of standard e.m.f. of the resulted cell. (*i.e.*, E^o_{cell} = 0.34 V). The above discussion shows that the cell obtained by coupling Cu (*s*) / Cu^{2+} (*aq*, 1 M) electrode to Pt, H_2 (*g*, 1 atm) / H^+ (*aq*, 1 atm) can be represented as shown in Fig. 18.5.

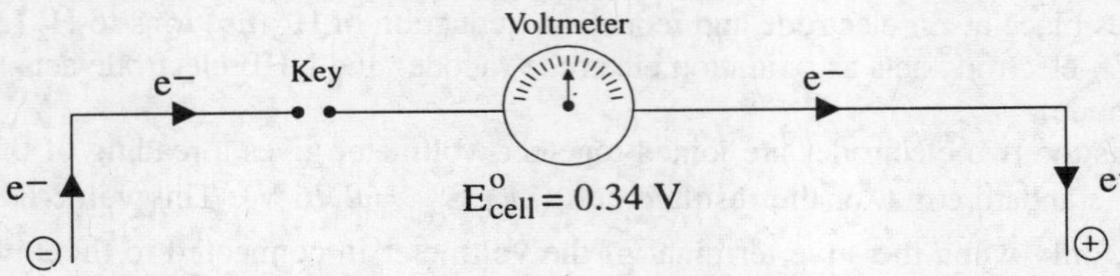

NHE Pt, H_2 (*g*, 1 atm) | H^+ *aq*, 1 M) | KCl |	Cu^{2+} (*aq*), | Cu (*s*)
Anode or oxidation electrode	Cu-electrode
(Negative electrode)	Cathode or reduction electrode (Positive electrode)
Oxidation Reaction :	Reduction reaction :
$H_2\ (g) \rightarrow 2H^+\ (aq) + 2e^-$	$Cu^{2+}\ (aq) + 2e^- \rightarrow Cu\ (s)$
$E^0_{2H^+/H_2} = 0.0$ V	$E^0_{Cu^{2+}/Cu} = +0.34$ V

Fig. 18.5 : Representation of a galvanic cell obtained by coupling a Cu-electrode with a NHE.

We know that the value of standard e.m.f. of the cell (*i.e.*, E^o_{cell} value) is given by the relation :

$$E^o_{cell} = (E^o_{red})_{cathode} - (E^o_{red})_{anode}$$

or $$0.34 = E^o_{Cu^{2+}/Cu} - E^o_{2H^+/H_2}$$

or $$0.34 = E^o_{Cu^{2+}/Cu} - 0.0$$

or $$E^o_{Cu^{2+}/Cu} = +0.34 \text{ V}$$

or $$E^o_{Cu^{2+}/Cu} = (E^o_{red})_{cathode} - = +0.34 \text{ (A positive quantity)} \qquad ...(ii)$$

Important conclusions

1. Relations (*i*) and (*ii*) given above, show that the value of standard reduction potential of a given electrode may be either negative or positive, depending on whether the given electrode acts as negative terminal (anode) or as a positive terminal (cathode) of the galvanic cell which is obtained by coupling the given electrode with a NHE. When the given electrode acts as a negative terminal (anode), the electrode has a negative value for its standard reduction potential. On the other hand when the given electrode acts as a positive terminal, the electrode has a positive value for its standard reduction potential. This conclusion can also be stated in the following words. If an electrode which has *negative value* for its standard reduction electrode potential is coupled with a NHE to get a galvanic cell, this electrode will act as a *negative terminal* (*anode*) of the resulted cell. On the other hand, if an electrode which has *positive value* for its standard reduction electrode potential is coupled with a NHE to get a galvanic cell, this electrode will act as a *positive terminal (cathode)* of the resulted cell.

2. The above discussion also shows that, if two electrodes are coupled together to get a galvanic cell, then the electrode having less value of its standard reduction electrode potential will constitute the negative terminal (*i.e.*, anode or left hand electrode) of the resulted cell and the

electrode having more value of its standard reduction electrode potential will constitute the positive terminal (*i.e.,* cathode or right hand electrode) of the cell. In other words we can say that the standard reduction potential of anode and cathode may have + +, – – or – + signs respectively. +(anode) – (cathode) combination is not allowed, since in this case $(E^o_{red})_{cathode}$, being negative, becomes less than $(E^o_{red})_{anode}$.

Example 18.1. *Represent the cell obtained by coupling the following pairs of metal-metal ion electrodes (i) Cu / Cu^{2+}, Ag / Ag^+ (ii) Zn / Zn^{2+}, Pb / Pb^{2+} and (iii) Zn / Zn^{2+}, Ag / Ag^+. The values of standard reduction potentials of these electrodes are as $E^o_{Cu^{2+}/Cu}$ = + 0.36 V, $E^o_{Ag^+/Ag}$ = + 0.80 V, $E^o_{Zn^{2+}/Zn}$ = – 0.76 V, $E^o_{Pb^{2+}/Pb}$ = – 0.13 V.*

Solution. (*i*) Since $E^o_{Ag^+/Ag}$ (= + 0.80 V) > $E^o_{Cu^{2+}/Cu}$ (= + 0.36 V), Ag^+/Ag electrode will constitute the cathode and Cu / Cu^{2+} will act as anode. Thus the cell obtained by coupling these two electrodes can be represented as :

$$\underset{\text{Anode}}{Cu / Cu^{2+}} \;||\; \underset{\text{Cathode}}{Ag^+ / Ag}$$

Similarly it can be shown that the cell obtained by coupling the other two pairs of electrodes can be represented as

(*ii*) $\underset{\text{Anode}}{Zn \mid Zn^{2+}} \;||\; \underset{\text{Cathode}}{Pb^{2+} \mid Pb}$

$E^o_{Zn^{2+}/Zn}$ (= – 0.76 V) < $E^o_{Pb^{2+}/Pb}$ (= – 0.13 V)

(*iii*) $\underset{\text{Anode}}{Zn \mid Zn^{2+}} \;||\; \underset{\text{Cathode}}{Ag^+ \;||\; Ag}$

$E^o_{Zn^{2+}/Zn}$ (= – 0.76 V) < $E^o_{Ag^+/Ag}$ (= + 0.80 V)

How to construct a galvanic cell for a given redox reaction or for a given pair of electrodes : Representation (or notation) of a galvanic cell

While constructing a galvanic cell, in which the given redox reaction takes place, we should keep the following points in mind :

(*i*) A galvanic cell consists of two electrodes, which are called half-cells. These two half-cells operate independently.

(*ii*) In a galvanic cell, the electrode at which oxidation takes place is called oxidation electrode or anode or negative electrode, while the electrode at which reduction takes place is called reduction electrode or cathode or positive electrode. Remember that in an electrolytic cell, anode is positive and cathode is negative.

(*iii*) In the representation of a galvanic cell, anode is written at the left hand side of the cell and cathode is written at the right hand side of the cell. It is for this reason that anode is also called left hand electrode (LHE) and cathode is called right hand electrode (RHE).

(*iv*) If the two electrolytic solutions being used in the two electrodes are in direct contact, a single full vertical line (|) is drawn between them.

(*v*) If the two electrolytic solutions are not in direct contact, but are connected through a salt bridge, two parallel vertical lines (||) are drawn between them.

(*vi*) The concentration of the two electrolytic solutions are indicated in the respective half-cells.

(*vii*) Whether the cell representation found out by us is correct or wrong, can be decided by calculating the value of e.m.f. of the cell, E^o_{cell}. If this value comes out to be a positive quantity, the cell representation is correct. The value of E^o_{cell} can be calculated by using the formula :

$$E^o_{cell} = (E^o_{red})_{cathode} - (E^o_{red})_{anode} \qquad ...(i)$$

Relation (*i*) given above shows that in order to get a positive value of E^o_{cell}, the value of $(E^o_{red})_{cathode}$ should be greater than the value of $(E^o_{red})_{anode}$. Thus, if we are given two electrodes and their standard reduction electrode potential values, the electrode having higher value (*i.e.*, more positive or less negative value) for its standard reduction electrode potential, will constitute the cathode and the electrode having lower value (*i.e.*, less positive or more negative value) for its standard electrode potential, will constitute the anode of the galvanic cell, obtained by coupling these two electrodes.

(*viii*) If we are given a galvanic cell and asked to tell whether the given cell will operate or not, we must calculate the value of E^o_{cell} of the given cell. If the value comes out to be positive, the given cell will operate and if the value of E^o_{cell} is a negative quantity, the given cell will not work. However, the cell obtained by interchanging the position of the electrodes will operate, since in this case the value of E^o_{cell} will be positive.

(*ix*) If we are given representation of a galvanic cell and asked to tell whether the cell representation is correct or wrong, we must calculate the value of E^o_{cell} of the given cell. If this value is a positive quantity, the given cell representation is correct. On the other hand, if the value of E_{cell} is negative, the given representation is wrong. The correct cell representation can be obtained by interchanging the position of electrodes, *i.e.*, anode and cathode of the wrong representation are made cathode and anode respectively.

(*x*) If we are given a redox reaction and asked to tell whether the given reaction will take place spontaneously or not (*i.e.*, whether the given reaction would be feasible or not), we must construct a galvanic cell in which the given redox reaction may take place and then find out the value of E^o_{cell} of the cell thus btained. If the value of E^o_{cell} is positive, the given redox reaction will proceed as written (*i.e.*, it will proceed in the forward direction). On the other hand, if the value of E^o_{cell} is negative, the given reaction will not proceed as shown (*i.e.*, it will not proceed in the forward direction but will proceed in the backward direction).

How to find out the number of electrons (*n*) involved in a cell reaction?

Suppose a cell reaction is represented as :

$$Ni\ (s) + 2Ag^+\ (aq) \rightarrow Ni^{2+}\ (aq) + 2Ag\ (s)$$

The method for finding out *n*, consists of the following steps :

(*i*) Break the given cell reaction into oxidation and reduction half-cell reactions.

Oxidation half-cell reaction (at anode)

$$Ni\ (s) \rightarrow Ni^{2+}\ (aq) + 2e^- \quad ...(a)$$

Reduction half-cell reaction (at cathode)

$$Ag^+\ (aq) + e^- \rightarrow Ag\ (s) \quad ...(b)$$

(*ii*) Equalise the number of electrons in both the half-cell reactions. For this, multiply equation (*b*) by 2 to get :

$$2Ag^+\ (aq) + 2e^- \rightarrow 2Ag\ (s) \quad ...(c)$$

(*iii*) Add equations (*a*) and (*c*) to get the cell reaction. Obviously, $n = 2$.

What is standard e.m.f. of a galvanic cell (E^o_{cell})?

When all the reactants and the products involved in the cell reaction of a galvanic cell are at unit concentration (one molar), the e.m.f. of the cell is called standard e.m.f. of the cell which is represented as E^o_{cell}. E^o_{cell} is also called **standard cell potential.**

Calculation of e.m.f. of a galvanic cell at standard conditions. Calculation of standard e.m.f. of a given galvanic cell (E^o_{cell})

The redox reaction taking place in a galvanic cell is called cell reaction. Cell reaction is obtained by adding the equations representing the oxidation reaction taking place at anode and reduction reaction taking place at cathode, and hence the value of E^o_{cell} is also obtained by adding the value of standard oxidation potential of anode $[(E^o_{oxi})_{anode}]$ and the value of standard reduction potential of cathode $[(E^o_{red})_{cathode}]$. Thus, for a Daniell cell,

$$\underset{\text{Anode}}{Zn\,(s)\,/\,Zn^{2+}\,(aq)} \quad \| \quad \underset{\text{Cathode}}{Cu^{2+}\,(aq)\,/\,Cu\,(s)}$$

Oxidation reaction at anode : $Zn\,(s) \rightarrow Zn^{2+}\,(aq) + 2e^-$

Reduction reaction at cathode : $Cu^{2+}\,(aq) + 2e^- \rightarrow Cu\,(s)$

Cell reaction : $Zn\,(s) + Cu^{2+}\,(aq) \rightarrow Zn^{2+}\,(aq) + Cu\,(s)$

$$E^o_{cell} = (E^o_{oxi})_{anode} + (E^o_{red})_{cathode} \qquad \text{...(i)}$$

$$= E^o_{Zn/Zn^{2+}} + E^o_{Cu^{2+}/Cu}$$

$$= 0.76 + 0.34 = 1.10\text{ V}$$

Now we know that:

$$(E^o_{oxi})_{anode} = -(E^o_{red})_{anode} \qquad \text{... (a)}$$

With the help of relation (*a*), equation (*i*) changes to :

$$E^o_{cell} = -(E^o_{red})_{anode} + (E^o_{red})_{cathode}$$

or $$E^o_{cell} = (E^o_{red})_{cathode} - (E^o_{red})_{anode} \qquad \text{...(ii)}$$

New since the value of $(E^o_{red})_{cathode}$ is higher than the value of $(E^o_{red})_{anode}$, equation (*ii*) can also be written as :

E^o_{cell} = Higher standard reduction electrode potential value – Lower standard reduction electrode potential value ...(*iii*)

Also we know that:

$$(E^o_{red})_{cathode} = -(E^o_{oxi})_{cathode} \qquad \text{...(b)}$$

With the help of relation (*b*), equation (*i*) becomes

$$E^o_{cell} = (E^o_{oxi})_{anode} - (E^o_{oxi})_{cathode} \qquad \text{...(iv)}$$

Now since the value of $(E^o_{oxi})_{anode}$ is higher than the value of $(E^o_{oxi})_{cathode}$, equation (*iv*) becomes

E^o_{cell} = Higher standard oxidation electrode potential value – Lower standard oxidation electrode potential value ...(*v*)

Conditions for getting positive value of E^o_{cell}

In order to make a galvanic cell work spontaneously, the value of its E^o_{cell} should be a positive quantity. We know that the value of E^o_{cell} can be calculated by any of the following two equations :

$$E^o_{cell} = (E^o_{red})_{cathode} - (E^o_{red})_{anode} \qquad \text{...(i)}$$

$$E^o_{cell} = (E^o_{oxi})_{anode} - (E^o_{oxi})_{cathode} \qquad \text{...(ii)}$$

Equations (*i*) and (*ii*) show that the value of E^o_{cell} would be positive only when the value of $(E^o_{red})_{cathode}$ is greater than the value of $(E^o_{red})_{anode}$ [see equation (*i*)] or the value of $(E^o_{oxi})_{anode}$ is greater than the value of $(E^o_{oxi})_{cathode}$ [See equation (*ii*).

Solved Examples

Example 18.2. *Calculate the e.m.f. of the following cell and predict whether the given cell representation is correct or wrong. If wrong, write the correct representation and correct cell reaction.*

$$Cu\,(s)\,/\,Cu^{2+}\,(aq)\,\|\,Zn^{2+}\,(aq)\,/\,Zn\,(s)$$

Given $E^o_{Cu^{2+}/Cu} = +0.34$ *V and* $E^o_{Zn^{2+}/Zn} = -0.76$ *V.*

Solution. Obviously the value of E^o_{cell} is given by :

$$E^o_{cell} = (E^o_{red})_{cathode} - (E^o_{red})_{anode}$$
$$= E^o_{Zn^{2+}/Zn} - E^o_{Cu^{2+}/Cu}$$
$$= -0.76 - (+0.34) = \mathbf{-1.10\ V}$$

Since the value of E^o_{cell} is a negative quantity, the given representation of the cell is wrong. Correct representation is obtained by interchanging the position of electrodes, *i.e.*, the correct representation is

$$Zn\ (s) \mid Zn^{2+}\ (aq) \parallel Cu^{2+}\ (aq) \mid Cu$$

Oxidation reaction at anode : $Zn\ (s) \rightarrow Zn^{2+}\ (aq) + 2e^-$

Reduction reaction at cathode : $Cu^{2+}\ (aq) + 2e^- \rightarrow Cu\ (s)$

Cell reaction : $Zn\ (s) + Cu^{2+}\ (aq) \rightarrow Zn^{2+}\ (aq) + Cu\ (s)$

This is the correct cell reaction.

Example 18.3. *Construct a galvanic cell in which the following redox reaction takes place :*

$$2Fe^{3+}\ (aq) + 2I^-\ (aq) \rightarrow Fe^{2+}\ (aq) + I_2\ (s)$$

Solution. The given redox reaction shows that I^- ions are being oxidised to I_2 molecules and Fe^{3+} ions are reduced to Fe^{2+} ions. Thus $I^-\ (aq)\ /\ I^2\ (s)$ electrode will constitute the oxidation terminal (anode) and $Fe^{3+}\ (aq)$, $Fe^{2+}\ (aq)$ / Pt will become the reduction terminal (cathode) of the galvanic cell. Thus the cell in which the given reaction takes place is represented as :

$$I^-\ (aq) \mid I^2\ (s) \quad \parallel \quad Fe^{3+}\ (aq),\ Fe^{2+}\ (aq) \mid Pt$$

Oxidation electrode (Anode) Reduction electrode (Cathode)

It is oxidation electrode at which oxidation half-reaction takes place and at reduction electrode, reduction half-reaction occurs.

Oxidation half-reaction at anode : $2I^-\ (aq) \rightarrow I_2\ (s) + 2e^-$

Reduction half-reaction at cathode : $[Fe^{3+}\ (aq) + e^-] \times 2$

Cell reaction : $2I^-\ (aq) + 2Fe^{3+}\ (aq) \rightarrow I_2\ (s) + 2Fe^{2+}\ (aq)$

Whether the cell representation is correct or wrong can be decided by calculating its e.m.f. value (*i.e.*, E^o_{cell} or E_{cell} value). This value is calculated with the help of the following expressions :

$$E^o_{cell} = (E^o_{red})_{cathode} - (E^o_{red})_{anode}$$

and $$E_{cell} = E^o_{cell} - \frac{0.0591}{n} \quad \textit{(Nernst equation)}$$

If the value of E^o_{cell} or E_{cell} comes out to be a positive quantity, the cell representation is correct. On the other hand, if the value of E^o_{cell} or E_{cell} is a negative quantity, then the cell representation is wrong. The correct representation is obtained by interchanging the position of the electrodes. In the present case :

$$E^o_{cell} = (E^o_{red})_{cathode} - (E^o_{red})_{anode} = E^o_{Fe^{3+}/Fe^{2+}} - E^o_{I_2/I^-} = 0.771 - 0.537 = \mathbf{+0.0234\ V}$$

is a positive quantity and hence the cell representation is correct.

Example 18.4. *Construct the cell in which the following redox reaction takes place :*

(i) $Zn\ (s) + Hg_2SO_4\ (s) \rightarrow ZnSO_4\ (aq) + 2Hg\ (l)$

(ii) $H_2\ (g) + 1/2O_2\ (g) \rightarrow H_2O\ (l)$

(iii) $K_2Cr_2O_7 + 6KI + 14HCl \rightarrow 8KCl + 2CrCl_3 + 3I_2 + 7H_2O$

(iv) $MnO_4^- + 5Fe^{2+} + 8H^+ \rightarrow Mn^{2+} + 5Fe^{3+} + 4H_2O.$

Solution. (*i*) From reaction (*i*) it is evident that Zn is oxidised to Zn^{2+} ions and Hg_2^{2+} ions (in Hg_2SO_4) are being reduced to Hg metal. Thus $Zn\ (s) \mid Zn^{2+}\ (aq)$ electrode constitutes the anode and $SO_4^{2-}\ (aq)$, $Hg_2SO_4\ (s) \mid Hg\ (l)$ acts as cathode and hence, the cell thus obtained, can be represented as :

$Zn\ (s)\ |\ Zn^{2+}\ (aq)\ ||\ SO_4^{2-}\ (aq),\ Hg_2SO_4\ (s)\ |\ Hg\ (l)$

Anode Cathode

The correctness of the answer can be verified by calculating the value of E^o_{cell} of the above cell. If this value is positive, the representation of the cell is correct. For the given cell,

$$E^o_{cell} = (E^o_{red})_{cathode} - (E^o_{red})_{anode}$$
$$= E^o_{SO_4^{2-},\ Hg_2SO_4/Hg} - E^o_{Zn^{2+}/Zn}$$
$$= +0.61 - (-0.6) = \mathbf{+1.37\ V}$$

(*ii*) Reaction (*ii*) can also be written as :

$$H_2\ (g) + \frac{1}{2}O_2\ (g) \rightarrow OH^-\ (aq) + H^+\ (aq)$$
$$(H = 0)\quad (O_2 = 0) \qquad (O = -2) \quad (H = +1)$$

This reaction shows that H_2 (H = 0) is being oxidised to H^+ (H = +1) and O_2 (O = 0) is being reduced to OH^- (O = –2). Thus the galvanic cell in which the given reaction will take place will be composed of Pt, H_2 (*g*)/$2H^+$ (*aq*) (anode) and Pt, O_2 (*g*)/OH^- (*aq*) (cathode) electrodes. Consequently, the galvanic cell can be represented as :

$Pt,\ H_2\ (g)\ /\ 2H^+\ (aq)\ ||\ O_2\ (g)\ /\ OH^-\ (aq),\ Pt$

Anode Cathode

$$E^o_{cell} = (E^o_{red})_{cathode} - (E^o_{red})_{anode}$$
$$= E^o_{Pt,\ O_2/OH^-} - E^o_{2H^+/H_2,\ Pt}$$
$$= 0.40 - 0.0 = \mathbf{+0.40\ V}$$

which is a positive quantity and hence the cell representation is correct.

(*iii*) The given reaction can also be written as :

$$Cr_2O_7^{2-} + 6I^- + 14H^+ \rightarrow 2Cr^{3+} + 7H_2O + 3I_2$$
$$(Cr = +6) \quad (I = -1) \qquad (Cr = +3) \quad (I = 0)$$

This reaction shows that since $Cr_2O_7^{2-}$ (Cr = + 6) is reduced to Cr^{3+} (Cr = +3), $14H^+$, $Cr_2O_7^{2-}$ (*aq*), Cr^{3+} (*aq*)/Pt electrode (It is a redox electrode) acts as *cathode.* On the other hand, since I^- (I = – 1) is being oxidised to I_2 (I = 0), Pt, I_2 (*g*) / $2I^-$ (*aq*) electrode acts as *anode.* Thus the cell in which the given redox reaction takes place, can be represented as :

$2I^-\ (aq)\ /\ I_2(g),\ Pt\ ||\ 14H^+,\ Cr_2O_7^{2-}\ (aq),\ Cr^{3+}\ (aq)/Pt$

Anode Cathode

$$E^o_{cell} = (E^o_{red})_{cathode} - (E^o_{red})_{anode}$$
$$= +1.33 - 0.53 = \mathbf{+0.80}$$

(*iv*) The given reaction shows that MnO_4^- (Mn = +7) is being reduced to Mn^{2+} (Mn = +2) while Fe^{2+} (Fe = +2) is being oxidised to Fe^{3+} (Fe = +3). Thus $8H^+$, MnO_4^- (*aq*), Mn^{2+}/Pt forms the *cathode* and Fe^{2+} (*aq*), Fe^{3+} (*aq*)/Pt (redox electrode) is acting as *anode.* Thus, the cell can be represented as :

$Pt/Fe^{2+}\ (aq),\ Fe^{3+}\ (aq)\ ||\ 8H^+,\ MnO_4^-\ (aq),\ Mn^{2+}/Pt$

Anode Cathode

$$E^o_{cell} = (E^o_{red})_{cathode} - (E^o_{red})_{anode}$$
$$= +1.51 - 0.77 = \mathbf{+0.74\ V}$$

Example 18.5. *Construct the galvanic cell by coupling the following pairs of electrodes. Justify your answer.*

(*i*) *Cu (s) / Cu^{2+} (aq), Ag^+ (aq) / Ag (s)*

(*ii*) *Zn (s) / Zn^{2+} (aq), Cu^{2+} (aq) / Cu (s)*

(iii) Mg (s) / Mg^{2+} (aq), Ni (s) / Ni^{2+} (aq)

Given that $E^o_{Cu^{2+}/Cu} = +0.34$ V, $E^o_{Ag^+/Ag} = +0.80$ V, $E^o_{Zn^{2+}/Zn} = -0.76$ V, $E^o_{Mg^{2+}/Mg} = -2.37$ V *and* $E^o_{Ni^{2+}/Ni} = -0.25$ V.

Solution. We know that the half-cell, having higher value (*i.e.,* more positive or less negative value) of E^o_{red}, constitutes the cathode and the half-cell, having lower value (*i.e.,* less positive or more negative value) of E^o_{red}, acts as the anode of the galvanic cell.

(*i*) Here since $E^o_{Ag^+/Ag}$ (= + 0.80 V) is more than $E^o_{Cu^{2+}/Cu}$ (= + 0.34 V), Ag/Ag^+ electrode will act as cathode and Cu/Cu^{2+} electrode will behave as anode. Thus, the cell may be represented as :

Cu (*s*) / Cu^{2+} (*aq*) || Ag^+ (*aq*) / Ag (*s*)

Anode Cathode

$$\therefore \quad E^o_{cell} = (E^o_{red})_{cathode} - (E^o_{red})_{anode}$$
$$= 0.80 + 0.34 = +\mathbf{1.10\ V}$$

Since E^o_{cell} is a positive quantity, the cell representation is correct.

(*ii*) Here, since $E^o_{Cu^{2+}/Cu}$ (= +0.31 V) > $E^o_{Zn^{2+}/Zn}$ (= – 0.76 V) = + Cu / Cu^{2+} will act as cathode and Zn / Zn^{2+} will act as anode and hence cell may be represented as :

Zn (*s*) / Zn^{2+} (*aq*) || Cu^{2+} (*aq*) / Cu (*s*)

Anode Cathode

$$E^o_{cell} = +0.31 - (-0.76) = \mathbf{+1.07\ V}$$

Since E^o_{cell} is positive, the cell notation is correct.

(*iii*) Since $E^o_{Ni^{2+}/Ni}$ (= – 0.25 V) > $E^o_{Mg^{2+}/Mg}$ (= – 2.37 V), Ni/Ni^{2+} will become the cathode and Mg/Mg^{2+} will act as anode. Thus the cell may be represented as :

Mg (*s*) / Mg^{2+} (*aq*) || Ni^{2+} (*aq*) / Ni (*s*)

Anode Cathode

$$E^o_{cell} = (E^o_{red})_{cathode} - (E^o_{red})_{anode}$$
$$= -0.25 - (-2.37) = +\mathbf{2.12\ V}$$

Since E^o_{cell} is positive, the cell representation is correct.

Example 18.6. *What will be spontaneous reaction, when the following half-cell reactions are combined together?*

(a) $Fe^{3+} + e^- \rightarrow Fe^{2+}$; $E^o = +0.77$ V

(b) $MnO_4^- + 8H^+ + 5e^- \rightarrow Mn^{2+} + 4H_2O$; $E^o = +1.49$ V

What is the value E^o_{cell} *?*

Solution. Since E^o_{red} value for half-cell reacion (*a*) is smaller than that for half cell reaction (*b*), oxidation will take place at half-cell reaction (*a*) and reduction will take place at half-cell reaction (*b*). We, therefore, write reaction (*a*) as oxidation reaction and reaction (*b*) as reduction reaction. Now, in order to make the number of electrons equal in both the reactions, we multiply equation (*a*) by five and then add it to equation (*b*).

(*a*) [$Fe^{2+} \rightarrow Fe^{3+} + e^-$ (*Oxidation reaction*)] × 5

(*b*) $MnO_4^- + 8H^+ + 5e^- \rightarrow Mn^{2+} + 4H_2O$ (*Reduction reaction*)

$$5Fe^{2+} + MnO_4^- + 8H^+ \rightarrow 5Fe^{3+} + Mn^{2+} + 4H_2O$$

This is the desired spontaneous reaction. Obviously,

$$E^o_{cell} = E^o_{MnO_4^-/Mn^{2+}} - E^o_{Fe^{3+}/Fe^{2+}}$$
$$= +1.49 - 0.77 = +\mathbf{0.72\ V}$$

Example 18.7. *The reaction, Zn (s) + Co^{2+} $\rightleftharpoons$ Co (s) + Zn^{2+} occurs in a cell. Write the electrode reactions and compute the standard e.m.f. of the cell. Given that* $E^o_{Zn/Zn^{2+}} = +0.76$ *V,* $E^o_{Co/Co^{2+}} = +0.28$ *V.* **(Roorkee 1988)**

Solution. The cell reaction, Zn (*s*) + Co^{2+} $\rightleftharpoons$ Co (*s*) + Zn^{2+} shows that since Zn is being oxidised to Zn^{2+} ions at anode and Co^{2+} ions are being reduced to Co at cathode, Zn (*s*)/Zn^{2+} electrode will constitute the anode of the cell and Co^{2+} / Co electrode will form the cathode of the cell. Thus, the cell can be represented as :

$$\underset{\text{Anode}}{Zn\,(s)\,/\,Zn^{2+}} \quad || \quad \underset{\text{Cathode}}{Co^{2+}\,/\,Co\,(s)}$$

$$E^o_{Zn^{2+}/Zn} = -0.76\text{ V} \quad E^o_{Co^{2+}/Co} = -0.28\text{ V}$$

$$\therefore\ E^o_{cell} = (E^o_{red})_{cathode} - (E^o_{red})_{anode}$$

$$= E^o_{Co^{2+}/Co} - E^o_{Zn^{2+}/Zn}$$

$$= -0.28 - (-0.76) = \mathbf{+0.48\ V}$$

Obviously the two half-cell reactions taking place at anode (oxidation) and cathode (reduction) can be written as :

Oxidation at anode : Zn (*s*) $\rightarrow$ Zn^{2+} +2e$^-$

Reduction at cathode : Co^{2+} + 2e$^-$ $\rightarrow$ Co (*s*)

Example 18.8. *In the voltanic cell, Cu, Cu^{2+} || Ag^+, Ag answer the following : (i) which electrode acts as anode and which electrode acts as cathode (ii) write the half cell reactions occurring at the two electrodes (iii) write the cell reaction (iv) calculate the standard e.m.f. of the cell. Given that* $E^o_{Cu/Cu^{2+}} = -0.33$ *V and* $E^o_{Ag/Ag^+} = -0.80$ *V.* **(Avadh B.Sc. 1981)**

Solution. (*i*) It is obvious from the cell representation that, Cu, Cu^{2+} electrode acts as anode and Ag^+, Ag electrode acts as cathode.

(*ii*) In this cell, Cu is being oxidised to Cu^{2+} ions at anode and Ag^+ ions are being reduced to Ag at cathode.

Hence the two half-cell reactions can be written as :

Oxidation at anode : Cu $\rightarrow$ Cu^{2+} + 2e$^-$ ...(*i*)

Reduction at cathode : 2Ag^+ + 2e$^-$ $\rightarrow$ 2Ag ...(*ii*)

(*iii*) Cell reaction is obtained by adding oxidation reaction taking place at anode and reduction reaction occurring at cathode. Thus on adding (*i*) and (*ii*) we get cell reaction as :

$$Cu + 2Ag^+ \rightarrow Cu^{2+} + 2Ag$$

(*iv*) $E^o_{cell} = (E^o_{red})_{cathode} - (E^o_{red})_{anode}$

$$= E^o_{Ag^+/Ag} - E^o_{Cu^{2+}/Cu}$$

$$= +0.80 - (+0.33) = \mathbf{+0.47\ V}$$

Electrochemical Series

The values of standard reduction potentials of many electrodes have been determined at 25°C or (25 + 273) = 298 K. These values have been arranged in their **increasing order,** in the form of a series, which is called **electrochemical series.** This series has been shown in Table 18.2. From electro-chemical series, the following points may be noted :

(*i*) The half-cell reactions (*i.e.,* reactions taking place at the given electrode) are written as reduction reactions and hence the species written on LHS of the reduction half-cell reactions are oxidised form (*i.e.,* oxidising agents) while the species written on RHS are reduced form (*i.e.,* reducing agents). Oxidising agents accept electrons and are reduced into reducing agents.

(*ii*) (*a*) Since the half-cell reactions are reduction reactions, the standard electrode potential value given in the series are standard reduction electrode potentials. The value of standard reduction electrode potential (E^o_{red}) and standard oxidation potential (E^o_{ox}) of an electrode are equal to each other but they have opposite sign, *i.e.,* if one has positive sign, the other has negative sign and *vice versa.* Thus :

$$+E^o_{red} = -E^o_{ox} \text{ and } +E^o_{ox} = -E^o_{red}$$

Example : $Li^+ (aq) + e^- \xrightarrow{\text{Reduction}} Li (s);$ $\quad E^o_{red} = E^o_{Li^+/Li} = -3.04$ V

$Li (s) \xrightarrow{\text{Oxidation}} Li^+ (aq) + e^-;$ $\quad E^o_{ox} = E^o_{Li/Li^+} = +3.04$ V

(*b*) Some standard reduction electrode potential values are negative, while others are positive. Negative value of standard reduction electrode potential of a given electrode, implies that this electrode will constitute negative terminal (anode) of the galvanic cell which is obtained by coupling the given electrode with a normal hydrogen electrode. On the other hand, the positive value of standard reduction potential of a given electrode implies, that this electrode will constitute positive terminal (cathode) of the galvanic cell, which is obtained by coupling the given electrode with a normal hydrogen electrode.

(*c*) As we proceed from top to bottom in the series (down the series), the values of standard reduction electrode potential go on increasing and hence become more and more positive ($- \rightarrow 0 \rightarrow +$). On proceeding from bottom to top, E^o_{red} values become more and more negative ($+ \rightarrow 0 \rightarrow -$).

(*iii*) The electrodes, having negative values for their standard reduction electrode potentials, have been placed *above* hydrogen electrode, while the electrodes, having positive values for their standard reduction electrode potentials, have been put *below* hydrogen electrode.

E^o_{red} values for the reduction of water, O_2 and halogen molecules (F_2, Cl_2, Br_2 and I_2 molecules) are given below. In the reduction reactions, hydrogen (H = 0) is reduced to H_2 molecule (H = 0) and O_2 molecule (O = 0) is reduced to H_2O molecule (O = –2) or OH^- ions (O = –2) In the reduction of halogen molecules, halide ions (F^-, Cl^-, Br^- and I^- ions) are formed

Oxidising agent		*Reducing agent*	*E^o_{red} values (in volts)*
$2H^+ + e^-$	→	H_2	$E^o_A = +0.00$ (1 M acid)
(H = +1)		(H = 0)	
$2H_2O + 2e^-$	→	$H_2 + 2OH^-$	$E^o_B = -0.83$ (1 M base)
(H = +1)		(H = 0)	
$2H_2O + 2e^-$	→	$H_2 + 2OH^-$	$E^o_N = -0.41$ (at pH = 7)
(H = +1)		(H = 0)	
$O_2 + 4H^+ + 4e^-$	→	$2H_2O$	$E^o_A = +1.23$ (1 M acid)
(O = 0)		(O = –2)	
$O_2 + 2H_2O + 4e^-$	→	$4OH^-$	$E^o_B = +0.44$ (1 M base)
(O = 0)		(O = –2)	
$O_2 + 4H^+ + 4e^-$	→	$2H_2O$	$E^o_N = +0.82$ (at pH = 7)
(O = 0)		(O = –2)	
$2O_2 + 4H^+ + 4e^-$	→	$2H_2O_2$	$E^o_A = +1.229$
$F_2 + 2e^-$	→	$2F^-$	$E^o_A = +2.87$
(F = 0)			
$Cl_2 + 2e^-$	→	$2Cl^-$	$E^o_A = +1.36$
(Cl = 0)			
$Br_2 + 2e^-$	→	$2Br^-$	$E^o_A = +1.06$
(Br = 0)			
$I_2 + 2e^-$	→	$2I^-$	$E^o_A = +0.54$
(I = 0)			

Table 18.2 : *Standard reduction electrode potentials of some electrodes at 25°C (=25 + 273 = 298 K) in acidic medium (electrochemical series)*

Electrode or couple	Electrode reduction reaction (Also called half-cell reduction reaction)				Standard reduction electrode potentials in volts (E^o_{red} values)
	Oxidising agent (Oxidised form)			Reducing agent (Reduced form)	
Li^+ / Li	Li^+ (*aq*)	$+ e^-$	$\longrightarrow$	$Li(s)$	– 3.05
M^+/M (M = K, Rb or Cs)	M^+	$+ e^-$	$\longrightarrow$	M	– 2.93
Ba^{2+} / Ba	Ba^{2+}	$+ 2 e^-$	$\longrightarrow$	Ba	– 2.90
Sr^{2+} / Sr	Sr^{2+}	$+ 2e^-$	$\longrightarrow$	Sr	– 2.89
Ca^{2+} / Ca	Ca^{2+}	$+ 2e^-$	$\longrightarrow$	Ca	– 2.76
Na^+ / Na	Na^+	$+ e^-$	$\longrightarrow$	Na	– 2.71
Mg^{2+} / Mg	Mg^{2+}	$+ 2e^-$	$\longrightarrow$	Mg	– 2.38
La^{3+} / La	La^{3+}	$+ 3e^-$	$\longrightarrow$	La	– 2.37
$\frac{1}{2} H_2 / H^-$	$\frac{1}{2} H_2$	$+ 2e^-$	$\longrightarrow$	H^-	– 2.23
Be^{2+} / Be	Be^{2+}	$+ 2e^-$	$\longrightarrow$	Be	– 1.70
Al^{3+} / Al	Al^{3+}	$+ 3e^-$	$\longrightarrow$	Al	– 1.67
B^{3+} / B	B^{3+}	$+ 3e^-$	$\longrightarrow$	B	– 0.87
Zn^{2+} / Zn	Zn^{2+}	$+ 2e^-$	$\longrightarrow$	Zn	– 0.76
Cr^{3+} / Cr	Cr^{3+}	$+ 3e^-$	$\longrightarrow$	Cr	– 0.74
Ga^{3+} / Ga	Ga^{3+}	$+ 3e^-$	$\longrightarrow$	Cr	– 0.52
H_3PO_3/H_3PO_2	H_3PO_3 (P = + 3)	$+ 2H^+ + 2e^-$	$\longrightarrow$	$H_3PO_2 + H_2O$ (P = + 1)	– 0.51
$CO_2 / H_2C_2O_4$	$2CO_2$ (C = + 4)	$+ 2H^+ + 2e^-$	$\longrightarrow$	$H_2C_2O_4$ (C = + 3)	– 0.49
Fe^{2+} / Fe	Fe^{2+}	$+ 2e^-$	$\longrightarrow$	Fe	– 0.41
Cr^{3+} / Cr	Cr^{3+}	$+ 3e^-$	$\longrightarrow$	Cr	– 0.41
Cd^{2+} / Cd	Cd^{2+}	$+ 2e^-$	$\longrightarrow$	Cd	– 0.40
In^{3+} / In	In^{3+}	$+ 3e^-$	$\longrightarrow$	In	– 0.34
Ni^{2+} / Ni	Ni^{2+}	$+ 2e^-$	$\longrightarrow$	Ni	– 0.25
H_3PO_4 / H_3PO_3	H_3PO_4 (P = +5)	$+ 2H^+ + 2e^-$	$\longrightarrow$	$H_3PO_3 + H_2O$ (P = +3)	– 0.20
Sn^{2+} / Sn	Sn	$+ 2e^-$	$\longrightarrow$	Sn	– 0.14
Pb^{2+} / Pb	Pb^{2+}	+ 2e	$\longrightarrow$	Pb	– 0.13
$H^+ / \frac{1}{2} H_2$	H^+	$+ e^-$	$\longrightarrow$	$\frac{1}{2} H_2$	± 0.000 (By definition)
$S_4O_6^{2-} / S_2O_3^{2-}$	$S_4O_6^{2-}$ (S = 2.5)	$+ 2e^-$	$\longrightarrow$	$2S_2O_3^{2-}$ (S = 2.0)	+ 0.09
Sn^{4+} / Sn^{2+}	Sn^{4+}	$+ 2e^-$	$\longrightarrow$	Sn^{2+}	+ 0.15
Cu^{2+}/Cu^+	Cu^{2+}	$+ e^-$	$\longrightarrow$	Cu^+	+ 0.15

SO_4^{2-} / H_2SO_3	$SO_4^{2-} + 4H^+$	$+ e^-$	$\longrightarrow$	$H_2SO_3 + H_2O$	+ 0.17
Cu^{2+} / Cu	Cu^{2+}	$+ 2e^-$	$\longrightarrow$	Cu	+ 0.34
S / H_2S	S	$+ 2H^+ + 2e^-$	$\longrightarrow$	H_2S	+ 0.44
Cu^+ / Cu	Cu^+	$+ e^-$	$\longrightarrow$	Cu	+ 0.52
I_2 / $2I^-$	I_2	$+ 2e^-$	$\longrightarrow$	$2I^-$	+ 0.54
H_3ASO_4 / H_3ASO_3	H_3AsO_4	$+ 2H^+ + 2e^-$	$\longrightarrow$	$H_3AsO_3 + H_2O$	+ 0.56
O_2 / H_2O_2	O_2 (O = 0)	$+ 2H^+ + 2e^-$	$\longrightarrow$	H_2O_2 (O = –1)	+ 0.68
Fe^{3+} / Fe^{2+}	Fe^{3+}	$+ e^-$	$\longrightarrow$	Fe^{2+}	+ 0.77
$\frac{1}{2} Hg_2^{2+}$ / Hg	$\frac{1}{2} Hg_2^{2+}$ (Hg = +1)	$+ e^-$	$\longrightarrow$	Hg (Hg = 0)	+ 0.80
Ag^+ / Ag	Ag^+	$+ e^-$	$\longrightarrow$	Ag	+ 0.80
Hg^{2+} / Hg	Hg^{2+}	$+ 2e^-$	$\longrightarrow$	Hg	+ 0.85
Br_2 / $2Br^-$	Br_2	$+ e^-$	$\longrightarrow$	$2Br^-$	+ 1.06
IO_3^- / I^-	IO_3^- (I = + 5)	+ e	$\longrightarrow$	$I^- + 3 H_2O$ (I = 0)	+ 1.09
IO_3^- / I^-	$IO_3 + 6H^+$ (1 = +5)	$+ 5e^-$	$\longrightarrow$	$\frac{1}{2} I + 3H_2O$ (I = 0)	+ 1.20
$\frac{1}{2}Cr_2O_7^{2-}$ / Cr^{3+}	$\frac{1}{2} Cr_2O_7^{2-}$	$+ 7H^+ + 3e^-$	$\longrightarrow$	$Cr^{3+} + \frac{7}{2} H_2O$ (Cr = +3)	+ 1.33
Cl_2 / $2Cl^-$	Cl_2	$+ 2e^-$	$\longrightarrow$	$2 Cl^-$	+ 1.36
HIO / $\frac{1}{2} I_2$	HIO (I = +1)	$+ H^+ + e^-$	$\longrightarrow$	$\frac{1}{2} I_2 + H_2O$ (I = 0)	+ 1.45
MnO_4^- / Mn^{2+}	MnO_4^- (Mn = +7)	$+ 8H^+ + 5e^-$	$\longrightarrow$	$Mn + 4H_2O$ (Mn = +2)	+ 1.51
BrO_3^- / $\frac{1}{2} Br_2$	BrO_3 (Br = +5)	$+ 6H^+ + 5e^-$	$\longrightarrow$	$\frac{1}{2} Br_2 + 3H_2O$ (Br = 0)	+1.52
HBrO / $\frac{1}{2} Br_2$	HBrO (Br = +5)	$+ H^+ +$ e–	$\longrightarrow$	$\frac{1}{2} Br_2 + H_2O$ (Br = 0)	+1.59
H_5IO_6 / IO_3^-	H_5IO_6 (I = +7)	$+ H^+ + 2e^-$	$\longrightarrow$	$IO_3^- + 3H_2O$ (I = 5)	+1.60
Ce^{4+} / Ce^{3+}	Ce^{4+}	$+ e^-$	$\longrightarrow$	Ce^{3+}	+1.61
HClO / $\frac{1}{2} Cl_2$	HClO (Cl = + 1)	$+ H^+ + e^-$	$\longrightarrow$	$\frac{1}{2} Cl_2 + H_2O$ (Cl = 0)	+1.64
$HClO_2$ / HClO	$HClO_2$ (Cl = +3)	$+ 2H^+ + 2e^-$	$\longrightarrow$	$HClO + H_2O$ (Cl = +1)	+ 1.64
H_2O_2 / H_2O	H_2O_2 (O = –1)	$+ 2H^+ + 2e^-$	$\longrightarrow$	$2H_2O$ (O = – 2)	+1.77
$\frac{1}{2} S_2O_8^{2-}$ / SO_4^{2-}	$\frac{1}{2} S_2O_8^{2-}$	$+ e^-$	$\longrightarrow$	SO_4^{2-}	+2.01
O_3 / O_2	O_3	$+ 2H^+ + 2e^-$	$\longrightarrow$	$O_2 + H_2O$	+2.07

$F_2 / 2F^-$	F_2	$+ 2e^-$	$\longrightarrow$	$2F^-$	+2.87
$\frac{1}{2}F_2$ / HF	$\frac{1}{2}F_2$ (F = 0)	$+ H^+ + e^-$	$\longrightarrow$	HF (F = –1)	+3.06

Applications of standard reduction electrode potential values : Applications of electrochemical series

Some of the applications of standard reduction electrode potential values, given in the electrochemical series, are discussed below :

I. To predict which half-cell will constitute the cathode and which half-cell will form the anode. If we are given two half-cells (electrodes) and the values of their standard reduction electrode potentials, with the help of these values, we can predict which half-cell will constitute the cathode and which half-cell will act as anode of the galvanic cell, which is obtained by coupling these two helf-cells. The half-cell having higher value (*i.e.*, more positive or less negative value) of its standard reduction electrode potential will form the cathode, while the half-cell, having lower value (*i.e.*, less positive or more negative value) for its standard reduction electrode potential, will act as anode of the galvanic cell.

II. To predict whether a given cell representation is correct or wrong. Whether a given cell representation is correct or wrong, can be decided with the help of E^o_{red} values of the two electrodes which constitute the galvanic cell. If the difference $[(E^o_{red})_{cathode} - (E^o_{red})_{anode}]$ [which is equal to E^o_{cell} of the cell, *i.e.*, $E^o_{cell} = (E^o_{red})_{cathode} - (E^o_{red})_{anode}]$ is a positive quantity, the cell representation is correct. If the value of $[(E^o_{red})_{cathode} - (E^o_{red})_{anode}]$ is a negative quantity, the cell representation is wrong.

III. To calculate the value of reduction electrode potential of a given electrode at non-standard conditions: To calculate the value of $(E_{red})_{elec}$ by using Nernst equation. With the help of the value of standard reduction electrode potential of a given electrode, we can also calculate the value of the reduction electrode potential of the same electrode when the species appearing in the reduction reaction taking place at the electrode are not in their standard conditions (Non-standard conditions). This value is calculated with the help of Nernst equation which is written as:

$$(E_{red})_{elec} = (E^o_{red})_{elec} - \frac{2.303\ RT}{nF} \log \frac{\text{Concentration of the products}}{\text{Concentration of the reactants}}$$

Here $(E^o_{red})_{elec}$ = Reduction electrode potential at non-standard conditions.

$(E^o_{red})_{elec}$ = Reduction electrode potential at standard conditions, *i.e.*, $(E^o_{red})_{elec}$ is the standard reduction electrode potential.

n = Number of electrons released in the electode reduction reaction

IV. To calculate the value of potential (e.m.f.) of a given cell at standard conditions : To calculate the value of E^o_{cell}. With the help of E^o_{red} values we can calculate the value of standard potential (standard e.m.f.) of a given galvanic cell (E^o_{cell} value). The value of E^o_{cell} is given by :

$$E^o_{cell} = (E^o_{red})_{cathode} - (E^o_{red})_{anode}$$

This equation is applied when the reactants and the products appearing in the cell reaction are in their standard conditions.

V. To calculate the value of potential (e.m.f.) of a given cell at non-standard conditions : To calculate the value of E_{cell} by using Nernst equation. If the reactants and the products appearing in the cell reaction are *not* in their standard states (non-standard conditions), the e.m.f. of the cell (E_{cell} value) is determined with the help of Nernst equation which is written as :

$$E_{cell} = E^o_{cell} - \frac{2.303\ RT}{nF} \log \frac{\text{Product of the concentration of the products}}{\text{Product of the concentration of the reactants}}$$

Here E_{cell} = e.m.f. of the cell when the reactants and products appearing in the cell reaction are not in their standard conditions (Non-standard conditions).

E^o_{cell} = e.m.f. of the cell when the reactants and the products are in their standard conditions (standard conditions). Thus E^o_{cell} is the standard e.m.f. of the cell.

n = Number of electrons involved in the cell reaction.

If the cell reaction (redox reaction) is represented as :

$$aA + bB \rightleftharpoons cC + dD, \text{ (}n = \text{No. of electrons involved),}$$

then Nernst equation for this reaction is written as :

$$E_{cell} = E^o_{cell} - \frac{2.303\ RT}{nF} \log \frac{[C]^c\ [D]^d}{[A]^a\ [B]^b}$$

VI. Oxidising and reducing property of substances in aqueous solution. A substance (element, molecule or ion) which has a tendency to accept (gain) electrons to undergo reduction in aqueous solution is said to act as an oxidising agent. On the other hand a substance which has a tendency to lose electrons to undergo oxidation in aqueous solution is said to act as a reducing agent.

The substance which has a positive value of E^o_{red} has a tendency to gain electrons to undergo reduction and hence is capable of acting as an oxidising agent in aqueous solution. On the other hand, a substance which has a negative value of E^o_{red} has a tendency to lose electrons to undergo oxidation and hence is capable of acting a reducing agent in aqueous solution.

Variation of oxidising power of oxidising agents and reducing power of reducing agents in electrochemical series. Higher is the vlaue of E^o_{red}, higher is the tendency of the substance to accept electrons to undergo reduction in aqueous solution and hence higher is the oxidising power of the substance. Since E^o_{red} values *increase* down the electrochemical series (– → 0 → +), the oxidising power of the oxidising agents also *increases* in the same direction. Thus Li^+ ion is the weakest oxidising agent and F_2 is the strongest oxidising agent. The increasing order of the oxidising power (or strength) of oxidising agents down the series shows that strong oxidising agents have large positive values of E^o_{red}.

Higher is the value of E^o_{red}, lower is the tendency of the substance to lose electrons to undergo oxidation in aqueous solution and hence lower is the reducing power of the substance. Since E^o_{red} values increase down the electrochemical series, the reducing power of the reducing agents decreases in the series. Thus Li is the strongest reducing agent and F^- ion is the weakest reducing agent. The decreasing order of the strength of reducing agents down the series shows that strong reducing agents have large negative values of E^o_{red}.

Periodic trends of oxidising and reducing power of main group elements. Values of E^o_{red} of main group elements are given in Table 18.3.

Table 18.3 : *E^o_{red} values (in volts) of some important main group elements*

IA	IIA	IIIA	IVA	VA	VIA	VIIA
H^+ / H 0.0						
Li^+ / Li – 3.05	Be^{2+} / Be – 1.85	B^{3+} / B – 0.87			O_2 / OH^- + 0.44	F_2 / $2F^-$ + 3.85
Na^+ / Na – 2.71	Mg^{2+} / Mg – 2.37	Al^{3+} / Al – 1.68			S/S^{2-} + 0.49	Cl_2 / $2Cl^-$ + 1.36
K^+ / K – 2.92	Ca^{2+} / Ca – 2.87	Ga^{3+} / Ga – 0.52		As^{3+}/ As +0.25		Br_2 / $2Br^-$ + 1.06
Rb^+ / Rb – 2.92	Sr^{2+} / Sr – 2.89	In^{3+} / In – 0.34	Sn^{2+}/Sn – 0.14	Sb^{3+} / Sb +0.21		I_2 / $2I^-$ + 0.53
Cs^+ / Cs – 2.92	Ba^{2+} / Ba – 2.90	Tl^+ / Tl – 0.34	Pb^{2+} / Pb – 0.13	Br^{3+} / Bi + 0.23		

These values make the following points evident :

1. Since the elements of groups IA, IIA, IIIA and IVA have negative values of E^o_{red}, these elements are reducing agents. On the other hand, the elements of group VA, VIA and VIIA have positive values and hence are oxidising agents.

2. Since as we proceed in a period from the elements of groups IA to those of group IVA, E^o_{red} values become less and less negative (*i.e.*, these values increase), the reducing power of the reducing agents decreases. Thus the first element of each period is the strongest reducing agent.

3. Since as we proceed in a period from the elements of group VA to those of group VIIA, E^o_{red} values become more and more positive (*i.e.*, these values increase), the oxidising power of the oxidising agents increases. Thus the last element of each period is the strongest oxidising agent.

4. *Reducing property of alkali metals in aqueous solution.* E^o_{red} values (in volts) for alkali metals (group IA) corresponding to the reduction reaction, $M^+ (aq) + e^- \rightleftharpoons M (s)$ are given in Table 18.3. Since these values for all the alkali metals are *large negative values,* all these metals are strong reducing agents in aqueous solution.

Reducing property of alkali metals is evident from the fact that these metals reduce H^+ ions to H_2 in aqueous solution, *i.e.*, these metals can react with compounds having acidic hydrogen atoms like H_2O, acetylene ($HC \equiv CH$), hydrogn halides etc.

$$2Li + 2H^+OH^- \rightarrow 2LiOH + H_2$$
$$2Li + HC \equiv CH \rightarrow LiC \equiv CLi + H_2$$
$$2Na + 2H^+Cl^- \rightarrow 2NaCl + H_2$$

Since Li^+ / Li couple has the lowest value of E^o_{red} (most negative value), Li is the strongest reducing agent.

Why is Li-metal the strongest reducing agent in aqueous solution? When an alkali metal, M (*s*) acts as a reducing agent in aqueous solution, it loses its ns^1 electron and is oxidised to the hydrated cation, $M^+ (aq)$

$$M (s) + aq \rightarrow M^+ (aq) + e^- \qquad ...(A)$$

Equation (*A*) represents the reducing property of alkali metals. The oxidation reaction given above actually takes place through the following three steps :

(*a*) In this step M (*s*) absorbs energy equal to the sublimation energy of M, $(H_s)_{M (s)}$ and is converted into gaseous state, M (*g*)

$$M (s) \rightarrow M (g) \text{ ; Heat absorbed } = + (H_s)_{M (s)}$$

(*b*) In this step, M (*g*) loses its ns^1 electron and is converted into $M^+ (g)$ cation. Energy equal to ionisation energy of M (*g*), $(IE)_{M(g)}$ is absorbed in this conversion.

$$M (g) \rightarrow M^+ (g) + e\text{; Heat absorbed} = + (IE)_{M (g)}$$

(*c*) $M^+ (g)$ ion formed in step (*b*) gets hydrated to form the hydrated ion $M^+ (aq)$. In this process of hydration, hydration energy of $M^+ (g)$, $(H_h)_{M^+ (g)}$ gets liberated.

$$M^+ (g) + aq \rightarrow M^+ (aq)\text{; Energy released} = - (H_h)_{M^+ (g)}$$

On adding equation given at (*a*), (*b*) and (*c*), we get :

$$M (s) + aq \rightarrow M^+ (aq) + e^-$$

This equation is the some as equation (A) which represents the reducing property of alkali metals in aqueous solution. Various steps, given above, can be represented in the form of a cycle, known as Born-Haber cycle (See Fig. 18.6).

Since the sum of all the above equations gives us reaction (A), the sum of all the energy terms corresponding to the re-

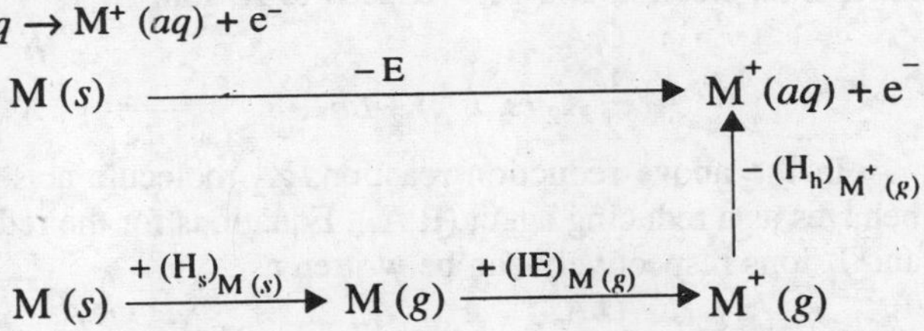

Fig. 18.6 : Born-Haber cycle to show the oxidation or reducing power of alkali metals in aqueous solution. Positive (+) sign indicates absorption of heat and negative (–) sign indicates release of heat energy.

actions given at (*a*), (*b*) and (*c*) should be equal to E, *i.e.*,

$$E = +(H_s)_{M\,(s)} + (IE)_{M\,(g)} - (H_h)_{M^+(g)}$$

Evidently the oxidation of M (*s*) to M^+ (*aq*) ion, (*i.e.*, reducing property of an alkali metal) will be high, if E has a large negative value. Now since sublimation energy of all the alkali metals is almost the same, the value of E depends on the difference between the values of ionisation energy and hydration energy of alkali metals. For Li-metal hydration energy (= – 531.4 kJ mol^{-1}) is greater than that of its ionisation energy (= + 520.3 kJ mol^{-1}) while for other metals hydration energy is less than their respective ionisation energy. Thus it is only for Li-metal that E is a negative value and for other alkali metals E is a positive quantity. Thus Li is the strongest reducing agent in aqueous solution. In other words, we can say that the maximum reducing power of Li-metal is because of maximum value of its hydration energy.

5. *Reducing property of alkaline earth metals in aqueous solution.* E^o_{red} values (in volts) for the alkaline earth metals (group IIA), corresponding to the reduction reaction, M^{2+} (*aq*) + $2e^-$ $\rightleftharpoons$ M (*s*), are given in Table 18.3. Since these values for alkaline earth metals are also large negative values, these metals, like alkali metals, are also reducing agents. Since E^o_{red} values decrease (*i.e.*, become more and more negative), as we move down the group from Be to Ba, the reducing power of the metals increases in the same direction.

Since E^o_{red} values of alkaline earth metals are greater (*i.e.*, less negative) than those of alkali metals, the former are weaker reducing agents than the latter.

6. *Reducing property of transition metals of 1st transition series.* E^o_{red} values (in volts) for these elements corresponding to reduction reaction, M^{2+} (*aq*) + $2e^-$ $\rightleftharpoons$ M (*s*) are given below :

Sc	Ti	V	Cr	Mn	Fe	Co	Ni	Cu	Zn
—	– 1.60	– 1.20	– 0.91	– 1.81	– 0.44	– 0.28	– 0.25	+ 0.34	– 0.76

Since E^o_{red} values for all the metals, except Cu, are negative, these elements can readily lose two electrons to form hydrated cation, M^{2+} (*aq*) in aqueous solution and hence act as good reducing agent in solution. Their reducing property is confirmed by the fact that these metals can liberate H_2 gas from a dil. acid like HCl, *i.e.*, these metals can reduce H^+ ions of HCl to H_2.

$$M + 2HCl \rightarrow MCl_2 + H_2 \quad \text{or} \quad M^o + 2H^+ \rightarrow M^{2+} + H_2$$

In actual practice, the rate at which most of these metals react with acids and liberate H_2 gas, is very slow. Some of the metals get protected from further attack of the acids due to the formation of a thin protective layer of their inert and sticky oxide on them. This layer prevents the acid to come in further contact of the metal, *e.g.*, when Cr-metal reacts with an acid, it gets coated with a thin protective layer of Cr_2O_3. This layer makes the metal unreactive and hence the metal can be used as a protective non-oxidising metal.

Since Cu has positive value of E^o_{red}, it is not able to liberate H_2 from an acid and hence does not act as a reducing agent.

7. *Relative order of oxidising power of halogen molecules (X_2) and reducing power of halide ions (X^-) in aqueous solution.* When X_2 molecule acts as an oxidising agent in aqueous solution, it accepts an electron and gets reduced to X^- ion.

O.A. *R.A.*

$$\frac{1}{2}X_2\,(s, l, g) + aq + e^- \rightleftharpoons X^-\,(aq)$$

In the above reduction reaction, X_2 molecule acts as an oxidising agent (O.A.) and X^- ion behaves as a reducing agent (R.A.). Equations for the reduction of F_2, Cl_2, Br_2 and I_2 to F^-, Cl^-, Br^- and I^- ions respectively can be written as :

O.A. *R.A.*

$$\frac{1}{2}F_2\,(g) + aq + e^- \rightleftharpoons F^-\,(aa) \qquad E^o_{red} = +2.85\text{ V}$$

$$\frac{1}{2}Cl_2\,(g) + aq + e^- \rightleftharpoons Cl^-\,(aq) \qquad E^o_{red} = +\,1.36\ V$$

$$\frac{1}{2}Br_2\,(l) + aq + e^- \rightleftharpoons Br^-\,(aq) \qquad E^o_{red} = +\,1.06\ V$$

$$\frac{1}{2}I_2\,(g) + aq + e^- \rightleftharpoons I^-\,(aq) \qquad E^o_{red} = +\,0.53\ V$$

Since E^o_{red} values are large positive values, X_2 molecules are capable of accepting an electron and hence act as oxidising agents. On the other hand, X^- ions are capable of losing an electron and hence act as reducing agents. Since E^o_{red} values decrease, as we move down the group from F_2 to I_2, the oxidising power of X_2 molecules decreases from F_2 to I_2 ($F_2 > Cl_2 > Br_2 > I_2$) and the reducing power of X^- ions increases from F^- to I^- ($F^- < Cl^- < Br^- < I^-$).

Thus F_2 is the strongest oxidising agent and I_2 is the weakest oxidising agent. F^- ion (or HF) is the weakest reducing agent and I^- ion (or HI) is the strongest reducing agent. As a matter of fact HF or F^- does not show reducing properties at all.

The increasing order of the reducing power of hydrogen halide molecules (HX) or halide ions (X^-) in aqueous solution can also be explained on the basis of the strength of H–X bond in HX molecules. We know that as we proceed from HF to HI, the size of the halogen atom (X) increases from F to I and hence H–X bond length also increases from H–F to H–I. (H–F = 0.92, H–Cl = 1.27, H–Br = 1.41 and H–I = 1.61). Due to the successive increase in H–X bond length, the strength of H–X bond decreases as H–F > H–Cl > H–Br > H–I (H–F = 574.0 kJ mol^{-1}, H–Cl = 428.1 kJ mol^{-1}, H–Br = 362.5 kJ mol^{-1}, H–I = 294.6 kJ mol^{-1}). Consequently the ease with which H–X bond gets broken to give H atom which is responsible for the reducing power of HX molecules increases as we move from HF to HI, *i.e.*, the reducing power of HX molecules is in the order HF < HCl < HBr < HI or $F^- < Cl^- < Br^- < I^-$.

Reactions showing that the oxidising power of halogens (X_2) is in the decreasing order ($F_2 > Cl_2 > Br_2 > I_2$) while the reducing power of hydrogen halides (HX) or halide ions (X^-) is in the increasing order ($F^- < Cl^- < Br^- < I^-$). Let us study the following redox reactions. In these reactions halogens oxidise hydrogen halides or halide ions to their corresponding halogen molecules and are themselves reduced to their corresponding halide ions. Thus halogens act as oxidising agents (O.A.) and haldie ions behave as reducing agents (R.A.).

	O.A.		R.A	
(i)	$F_2 + 2HCl$	→	$2HF + Cl_2$	
or	$F_2 + 2Cl^-$	→	$2F^- + Cl_2$	
(ii)	$F_2 + 2HBr$	→	$2HF + Br_2$	
or	$F_2 + Br^-$	→	$2F^- + Br_2$	
(iii)	$F_2 + 2HI$	→	$2HF + I_2$	
or	$F_2 + 2I^-$	→	$2F^- + I_2$	
(iv)	$Cl_2 + 2HF$	→	$2HCl + I_2$	*(Not possible)*
or	$Cl_2 + 2F^-$	→	$2Cl^- + F_2$	
(v)	$Cl_2 + 2HBr$	→	$2HCl + Br_2$	
or	$Cl_2 + 2Br^-$	→	$2Cl^- + Br_2$	
(vi)	$Cl_2 + 2HI$	→	$2HCl + I_2$	
or	$Cl_2 + 2I^-$	→	$2Cl^- + I_2$	
(vii)	$Br_2 + 2HF$	→	$2HBr + F_2$	*(Not possible)*
or	$Br_2 + 2F^-$	→	$2Br^- + F_2$	
(viii)	$Br_2 + 2HCl$	→	$2HBr + Cl_2$	*(Not possible)*
or	$Br_2 + 2Cl^-$	→	$2Br^- + Cl_2$	
(ix)	$Br_2 + 2HI$	→	$2HBr + I_2$	
or	$Br_2 + 2I^-$	→	$2Br^- + I_2$	

$$\left.\begin{array}{llll} (x) & I_2 + 2HF & \rightarrow & 2HI + F_2 \\ \text{or} & I_2 + 2F^- & \rightarrow & 2I^- + F_2 \end{array}\right\} \textit{(Not possible)}$$

$$\left.\begin{array}{llll} (xi) & I_2 + 2HCl & \rightarrow & 2HI + Cl_2 \\ \text{or} & I_2 + 2Cl^- & \rightarrow & 2I^- + Cl_2 \end{array}\right\} \textit{(Not possible)}$$

$$\left.\begin{array}{llll} (xii) & I_2 + 2HBr & \rightarrow & 2HI + Br_2 \\ \text{or} & I_2 + 2Br^- & \rightarrow & 2I^- + Br_2 \end{array}\right\} \textit{(Not possible)}$$

From the above reactions, the following point may be noted in connection with the oxidising power of halogens.

(*a*) Since F_2 can oxidise all the three halide ions *viz.*, Cl^-, Br^- and I^- ions to Cl_2, Br_2 and I_2 respectively, as is evident from reactions (*i*), (*ii*) and (*iii*) reactions, F_2 has the maximum oxidising power, *i.e.*, F_2 is the strongest oxidising agent among all the halogens.

(*b*) Since Cl_2 can not oxidise F^- ion to F_2 [see reaction (*iv*)] but can oxidise Br^- and I^- ions to Br_2 and I_2 respectively [see reactions (*v*) and (*vi*)], it has lower oxidising power than F_2 but higher than Br_2 and I_2 both.

(*c*) Reactions (*vii*) and (*viii*) show that since Br_2 can oxidise neither F^- nor Cl^- to F_2 and Cl_2 respectively, it has lower oxidising power than F_2 and Cl_2 both. On the other hand, since Br_2 can oxidise I^- ion to I_2, as is evident from reaction (*ix*), this halogen has greater oxidising power than I_2.

(*d*) Reactions (*x*), (*xi*) and (*xii*) show that I_2 cannot oxidise any of the halide ions and hence has the lowest oxidising power, *i.e.*, I_2 is the weakest oxidising agent, among all the halogens.

The above discussion can be summarised by saying that a halogen molecule can oxidise only those halide ions to their corresponding halogens, which lie below it, in the halogen family. Conversely, a halogen molecule cannot oxidise any of the halide ions, lying above it, in the family. Thus :

$$F_2 + 2X^- \rightarrow 2F^- + X_2 \quad (X^- = Cl^-, Br^-, I^-)$$

$$Cl_2 + 2F^- \rightarrow 2Cl^- + F_2 \quad \textit{(Not possible)}$$

$$Cl_2 + 2X^- \rightarrow 2Cl^- + X_2 \quad (X^- = Br^-, I^-)$$

$$Br_2 + 2X^- \rightarrow 2Br^- + X_2 \quad (X^- = F^-, Cl^-) \quad \textit{(Not possible)}$$

$$Br_2 + 2I^- \rightarrow 2Br^- + I_2$$

$$I_2 + 2X^- \rightarrow 2I^- + X_2 \quad (X^- = F^-, Cl^-, Br^-) \quad \textit{(Not possible)}$$

In connection with the reducing power of HX molecules or X^- ions, the following points may be noted :

(*a*) Since reactions (*iv*), (*vii*) and (*x*) do not occur as shown, HF molecule or F^- ion is not capable of reducing any of the halogens to their corresponding halide ions. Thus HF molecule or F^- ion does not show reducing properties at all.

(*b*) Since reaction (*i*) takes place spontaneously, HCl or Cl^- can reduce F_2 to F^- and hence has greater reducing power than HF or F^-. On the other hand, since Cl^- ion can reduce neither Br_2 to Br^- not I_2 to I^- as is evident from reactions (*viii*) and (*xi*), it is a weaker reducing agent than Br^- and I^- ions both.

(*c*) It is evident from reactions (*ii*) and (*v*) that Br^- ion can reduce F_2 to F^- and Cl_2 to Cl^-, this ion has greater reducing power than F^- and Cl^- ions. On the other hand, since Br^- ion is not able to reduce I_2 to I^- ion [see reaction (*xii*)], this ion has lower reducing power than I^- ion.

(*d*) Equations (*iii*), (*vi*) and (*ix*) show I^- ion can reduce all the three halogens to their corresponding halide ions and hence this ion has the maximum reducing power among all the halide ions. Thus HI or I^- is the strongest reducing agent.

The above discussion makes it evident that HF or F^- cannot reduce any of the halogens. HCl can reduce only F_2 to F^- but not Br_2 or I_2, HBr can reduce only F_2 and Cl_2 but not I_2 while HI can reduce all the halogens. Thus :

$$\left.\begin{array}{lll} & 2HF + X_2 & \rightarrow 2HX + F_2 \\ & (X_2 = Cl_2, Br_2, I_2) & \\ \text{or} & 2F^- + X_2 & \rightarrow 2X^- + F_2 \end{array}\right\} \textit{(Not possible)}$$

$$\begin{array}{lll} & 2HCl + F_2 & \rightarrow Cl_2 + 2HF \\ \text{or} & Cl^- + F^2 & \rightarrow Cl_2 + 2F^- \end{array}$$

$$\left.\begin{array}{lll} & 2HCl + X_2 & \rightarrow Cl_2 + 2HX \\ & (X_2 = Br_2, I_2) & \\ \text{or} & 2Cl^- + X_2 & \rightarrow Cl_2 + 2X^- \end{array}\right\} \textit{(Not possible)}$$

$$\begin{array}{lll} & 2HBr + X_2 & \rightarrow Br_2 + 2HX \\ & (X_2 = F_2, Cl_2) & \\ \text{or} & 2Br^- + X_2 & \rightarrow Br_2 + 2X^- \end{array}$$

$$\left.\begin{array}{lll} & 2HBr + I_2 & \rightarrow Br_2 + 2HI \\ \text{or} & 2Br^- + I_2 & \rightarrow Br_2 + 2I^- \end{array}\right\} \textit{(Not possible)}$$

$$\begin{array}{lll} & 2HI + X_2 & \rightarrow 2HX + I_2 \\ & (X_2 = F_2, Cl_2, Br_2) & \\ \text{or} & 2I^- + X_2 & \rightarrow 2X^- + I_2 \end{array}$$

An alternative explanation to account for the decreasing order of the oxidising power of halogens. We have stated above that when a halogen molecule oxidises a certain substance in aqueous solution, it accepts an electron and is reduced to hydrated halide ion, X^- (*aq*). Thus the oxidation reaction or oxidising property of a halogen molecule can be represented by the following reduction reaction.

$$\frac{1}{2} X_2\ (s,\ l,\ g) + e^- + aq \xrightarrow{\text{Reduction}} X^-\ (aq) \qquad \text{... (A)}$$

In this reduction reaction some amount of heat is *liberated.* This heat is denoted by E. The amount of E is a measure of the oxidising power of halogen molecule in aqueous solution. Larger is the negative value of E (energy liberated is given negative sign), greater of the tendency of the halogen molecule to accept an electron to undergo reduction to form X^- (*aq*) ion and hence greater is the oxidising power of the halogen molecule in aqueous solution. The values of E (in kJ mol^{-1}) for F_2 (*g*), Cl_2 (*g*), Br_2 (*l*) and I_2 (*s*) molecules have been found to be equal to – 763.5, – 608.3, – 575.1 and – 524.9 respectively. Since these value are in the order F_2 (*g*) > Cl_2 (*g*) > Br_2 (*l*) > I_2 (*s*) (negative sign has not been considered), the oxidising power of halogen molecules is also in the some order. Thus F_2 is the strongest oxidising agent and I_2 is the weakest oxidising agent.

Calculation of E for the reduction reaction, $\frac{1}{2} X_2\ (s,\ l,\ g) + e^- + aq \rightarrow X^-\ (aq)$, *where* X_2 = F_2 *(g),* Cl_2 *(g),* Br_2 *(l) and* I_2 *(s).* The value of E liberated in the reduction reaction (A) can be calculated by using the fact that this reaction takes place through the following steps :

(*a*) If X_2 molecule is a solid (*e.g.*, I_2 molecule), then the solid $\frac{1}{2}X_2$ (*s*) molecule *absorbs* energy equal to the half of *heat of fusion* of X_2 (*s*), $\frac{1}{2}\ (H_f)_{X_2\,(s)}$ and is changed into liquid state. Since the heat of fusion is the heat which is absorbed in the reaction, it is denoted by putting a positive sign before it, *i.e.*,

$$\frac{1}{2} X_2\ (s) \rightarrow \frac{1}{2} X_2\ (l); \qquad \text{Heat absorbed} = +\frac{1}{2}\ (H_f)_{X_2\,(s)}$$

$$(X_2 = I_2)$$

Since F_2 and Cl_2 are gases and Br_2 is a liquid at room temperature, heat of fusion would be equal to zero for these halogens.

(*b*) In this step, if the halogen is a liquid (*e.g.*, Br_2 molecule), then the liquid molecule, $\frac{1}{2}X_2\,(l)$ absorbs energy equal to half of the *heat of vaporisation* of $X_2\,(l)$ molecule, $\frac{1}{2}(H_v)_{X_2\,(l)}$ and is changed into gaseous state, *i.e.*,

$$\frac{1}{2}X_2\,(l) \rightarrow \frac{1}{2}X_2\,(g) \qquad \text{Heat absorbed} = +\frac{1}{2}(H_v)_{X_2\,(l)}$$

$(X_2 = Br_2)$

Since F_2 and Cl_2 are in the gaseous state, the heat of vaporisation for these halogens are equal to zero.

(*c*) Here, $\frac{1}{2}X_2\,(g)$ molecule obtained in step (*b*) absorbs energy equal to half of the X—X bond dissociation energy of $X_2\,(g)$ molecule, $\frac{1}{2}(H_d)_{X_2\,(g)}$ and gets changed to X (*g*) atom.

$$\frac{1}{2}X_2\,(g) \rightarrow X\,(g) \qquad \text{Heat absorbed} = +\frac{1}{2}(H_d)_{X_2\,(g)}$$

(*d*) In this step, X (*g*) atom obtained in step (*c*) accepts an electron (supplied by a reducing agent) and gets reduced to $X^-\,(g)$ ion. In this reduction reaction, energy equal to electron affinity of X (*g*) atom, $(EA)_{X\,(g)}$ is released. The release of energy is denoted by a negative sign.

$$X\,(g) + e^- \rightarrow X^-\,(g) \qquad \text{Heat released} = -(EA)_{X\,(g)}$$

(*e*) $X^-\,(g)$ ion obtained as above gets hydrated and is changed into hydrated ion, $X^-\,(aq)$. In this process of hydration, energy equal to hydration energy of $X^-\,(g)$ ion, $(H_h)_{X^-\,(g)}$ is released.

$X^-\,(g) + aq \rightarrow X^-\,(aq)$; Heat released $= -(H_h)_{X_2\,(g)}$

On adding the equations given at (*a*), (*b*), (*c*) (*d*) and (*e*), we get :

$$\frac{1}{2}X_2\,(s, l, g) + e^- + aq \rightarrow X^-\,(aq)$$

This equation is the same as equation (A) which represents the oxidising property of halogen in aqueous solution. Various steps given above can be represented in the form of a cycle known as Born-Haber cycle (See Fig. 18.7).

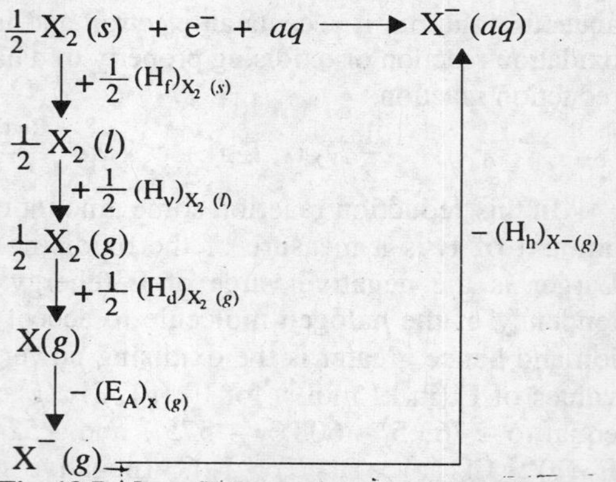

Fig. 18.7 : Born-Haber cycle to show the reducing or oxidising power of halogen molecule in aqueous solution. Positive (+) sign indicates absorption of heat and negative (–) sign indicates release of heat.

Since the sum of all the above reactions gives us reaction (A), the sum of all the energy terms corresponding to the reactions given at (*a*), (*b*), (*c*), (*d*) and (*e*) should be equal to E, *i.e.*,

$$E = +\frac{1}{2}(H_f)_{X_2(s)} + \frac{1}{2}(H_v)_{X_2(l)} + \frac{1}{2}(H_d)_{X_2(g)} - (EA)_{X(g)} - (H_h)_{X^-(g)}$$

If the values of the energy terms shown on the RHS of the above equation are put, the value of E for the reduction reaction (A) for each of the halogens can be calculated. The values obtained for each reaction have been shown in Table. 18.4

Table 18.4 : *To calculate the energy(E) released in the reduction reaction,* $\frac{1}{2}X_2\ (s, l, g) + e^- + aq \rightarrow X^-(aq)$ *where* $X_2 = F_2, Cl_2, Br_2$ *and* I_2*. Values of E are in kJ mol*$^{-1}$.

$\frac{1}{2}X_2\ (s, l, g) + e^- + aq \rightarrow X^-(aq)$ reaction	$+\frac{1}{2}(H_f)_{X_2(s)}$ *a*	$+\frac{1}{2}(H_v)_{X_2(l)}$ *b*	$+\frac{1}{2}(H_d)_{X_2(g)}$ *c*	$-(EA)_{X(g)}$ *d*	$-(Hh)_{X^-(g)}$ *e*	$E = a+b+c$
$\frac{1}{2}F_2(g) + e^- + aq \rightarrow F^-(aq)$	—	—	+159/2 = +79.5	– 328.0	– 515	– 763.5
$\frac{1}{2}Cl_2(g) + e^- + aq \rightarrow Cl^-(aq)$	—	—	+243 / 2 = +121.5	– 348.8	– 381	– 608.3
$\frac{1}{2}Br_2(l) + e^- + aq \rightarrow Br^-(aq)$	—	+30/2 = +15	+193/2 = 96.5	– 324.6	– 347	– 575.1
$\frac{1}{2}I_2(s) + e^- + aq \rightarrow I^-(aq)$	+16/2 = +8	+44/2 = +22	+151/2 = +75.5	– 295.5	– 305	– 524.9

From Table 18.4 the following points may be noted :

(*a*) The values of E have negative sign with them. Negative sign indicates that the energy is released when the halogen is reduced and thus acts as an oxidising agent. Negative sign also implies that halogens can easily accept the electron and hence can act as oxidising agent.

(*b*) Without considering negative sign, it can be said that E values decrease as we move from F_2 to I_2. It is due to the successive decrease in the values of E that oxidising power of halogens decreases from F_2 to I_2 ($F_2 > Cl_2 > Br_2 > I_2$).

(*c*) The value of E for F_2 is greater than that for Cl_2. This is because of the fact that the bond dissociation energy and electron affinity values for F_2 molecule are lower and heat of hydration for F^- ion is higher. Thus we can say that the stronger oxidising character of F_2 as compared to Cl_2 is due to its relatively lower bond dissociation energy, lower electron affinity and higher heat of hydration.

Oxidising properties shown by halogens. Here we shall discuss the following oxidising properties of halogens :

1. *Oxidation of metallic halides to halogens.* When a halogen reacts with the aqueous solution of an appropriate metallic halide (generally potassium or sodium halide), the halogen present in the halide is liberated. Thus in this reaction halide ion (X^-) is oxidised to halogen (X_2).

F_2 can liberate Cl_2, Br_2 and I_2 from the aqueous solution of KCl, KBr and KI respectively.

$$F_2 + 2KX \rightarrow 2KF + X_2 \quad (X = Cl, Br, I)$$

or $$F_2 + 2X^- \rightarrow 2F^- + X_2$$

Cl_2 can liberate Br_2 and I_2 from KBr and KI respectively but has no action on KF.

$$Cl_2 + 2KX \rightarrow 2KCl + X_2 \quad (X = Br, I)$$

or $$Cl_2 + 2X^- \rightarrow 2Cl^- + X_2$$

Br_2 can liberate I_2 from KI but has no action on KF and KCl

or $$Br_2 + 2KI \rightarrow 2KBr + I_2$$

$$Br_2 + 2I^- \rightarrow 2Br^- + I_2$$

I_2 is not able to liberate halogen from any of the halides *i.e.*, the following reaction is not possible

$$\left.\begin{array}{l} I_2 + 2KX \rightarrow 2KI + X_2 \\ (X = F^-, Cl^-, Br^-) \\ \text{or} \quad I_2 + 2X^- \rightarrow 2I^- + X_2 \end{array}\right\} \textit{(Not possible)}$$

The above discussion can be summarised by stating that a halide ion can be oxidised only by that halogen which lies above it in the halogen family. Thus :

(*a*) F^- ion cannot be oxidised to F_2 by any of the halogen namely Cl_2, Br_2 and I_2.

(*b*) Cl^- ion can be oxidised to Cl_2 only by F_2 and neither by Br_2 nor by I_2.

(*c*) Br^- ion can be oxidised to Br_2 only F_2 and Cl_2 but not by I_2.

(*d*) I^- ion can be oxidised to I_2 by each of the remaining three halogens *viz.*, F_2, Cl_2 and Br_2.

2. *Oxidation (decomposition) of water.* All the halogens can oxidise H_2O. Since the oxidising power of halogens decreases as $F_2 > Cl_2 > Br_2 > I_2$, the reactivity of halogens towards water also decreases from F_2 to I_2. Thus, since F_2 has the maximum oxidising power of all the halogens, it oxidises H_2O (O = –2) to O_2(O = 0) and O_3 (O = 0) *vigorously* and *even in the dark.*

(*a*) $2H_2O\ (l) + 2F_2\ (g) \rightarrow 4HF\ (aq) + O_2\ (g)$

or $2H_2O\ (l) + 2F_2\ (g) \rightarrow 4H^+\ (aq) + 4F^-\ (aq) + O_2\ (g)$; $\Delta H = -\ 795$ kJ mol^{-1}

(*b*) $3H_2O\ (l) + 3F_2\ (g) \rightarrow 6HF\ (aq) + O_3\ (g)$

or $3H_2O\ (l) + 3F_2\ (g) \rightarrow 6H^+\ (aq) + 6F^-\ (aq) + O_3\ (g)$.

Reaction (*a*) is an exothermic reaction, since energy (= 795 kJ mol^{-1}) is evolved in this reaction. This energy is quite large which means that F_2 is a very strong oxidising agent.

Cl_2 oxidises H_2O to O_2 only in the presence of sunlight while Br_2 oxidises H_2O only in presence of sunlight but slowly. HClO and HBrO formed are unstable and hence undergo decomposition to give ($HCl + O_2$) and ($HBr + O_2$) respectively.

$$2X_2\ (g, l) + 2H_2O(l) \rightarrow 2HX(aq) + 2HOX\ (aq)$$
$$(X_2 = Cl_2 \text{ or } Br_2)$$
$$2HOX(aq) \rightarrow 2HX\ (aq) + O_2\ (g)$$
$$2H_2(g) + 2H_2O(l) \rightarrow 4HX\ (aq) + O_2\ (g)$$
or
$$2X_2(g) + 2H_2O(l) \rightarrow 4H^+\ (aq) + 4X^-\ (aq) + O_2$$

The oxidation of H_2O to O_2 by iodine is a reversible and endothermic process.

$$2I_2\ (s) + 2H_2O\ (l) \rightleftharpoons 4HI\ (aq) + O_2(g)$$
or
$$2I_2\ (s) + 2H_2O\ (l) \rightleftharpoons 4H^+\ (aq) + 4I^-\ (aq) + O_2\ (g);\quad \Delta H = +\ 105 \text{ kJ mol}^{-1}$$

The endothermic nature of the process means that energy equal to 105 kJ mol^{-1} is needed for the forward reaction to occur. In fact, the reverse reaction takes place and O_2 oxidises I^- ion to I_2 molecule.

3. *Oxidation of other substances. Flourine oxidises :* (*a*) Chlorate, ClO_3^- (Cl = +5) and iodate, IO_3^- (I = +5) to perchlorate, ClO_4^- (Cl = +7) and periodate, IO_4^- (I = +7) respectively.

$$F_2 + KXO_3 + H_2O \rightarrow KXO_4^- + 2HF$$
$$(X = Cl, I) \qquad (X = +7)$$
or
$$F_2 + XO_3^- + H_2O \rightarrow XO_4 + 2F^- + 2H^+$$
$$(F = 0)\ (X = +5) \qquad (X = +7)\ (F = -\ 1)$$

(*b*) Bisulphite (*e.g.*, $KHSO_4$) to peroxy-disulphate, $K_2S_2O_8$

$$F_2 + 2KHSO_4 \rightarrow K_2S_2O_8 + 2HF$$
or
$$F_2 + 2KHSO_4 \rightarrow S_2O_8^{2-} + 2H^+ + 2F^-$$
$$(F = 0)\ (S = +\ 6) \qquad (S = +\ 7) \qquad (F = -1)$$

(*c*) Chromic salts (Cr = + 3) to dichromate, $Cr_2O_7^{2-}$ (Cr = +6)

$$2F_2 + 2Cr^{3+} + 7H_2O \rightarrow Cr_2O_7^{2-} + 14H^+ + 6F–$$
$$(F = 0)\ (Cr = +\ 3) \qquad (Cr = +\ 6) \qquad (F = -1)$$

Chlorine Oxidises: (*a*) Acidified solution of ferrous (Fe = +2) salts to ferric (Fe = + 3)

$$Cl_2 + 2Fe^{2+} \rightarrow 2Cl^- + 2Fe^{3+}$$

(*b*) H_2S (S = – 2) to sulphur (S = 0)

$$Cl_2 + S^{-2} \rightarrow 2Cl^- + S$$

(*c*) SO_3^{2-} (S = + 4) to SO_4^{2-} (S = + 6)

$$Cl_2 + SO_3^{2-} + H_2O \rightarrow 2Cl^- + SO_4^{2-} + 2H^+$$

(*d*) Thiosulphate, $S_2O_3^{2-}$ (S = + 2) to SO_4^{2-} (S = + 6)

$$Cl_2 + S_2O_3^{2-} + H_2O \rightarrow 2Cl^- + SO_4^{2-} + S + 2H^+$$

(*e*) Suspension of I_2 (I = 0) in water to iodic acid, HIO_3 (I = + 5)

or $5Cl_2 + 6H_2O + I_2 \rightarrow 10Cl^- + 12H^+$

(*f*) Aqueous solution of SO_2 (S = + 4) to H_2SO_4 (S = + 6)

$$Cl_2 + SO_2 + 2H_2O \rightarrow 2Cl^- + SO_4^{2-} + 4H^+$$

(*g*) Arsenite, AsO_3^{3-} (As = +3) to arsenate, AsO_4^{3-} (As = +5)

$$Cl_2 + AsO_3^{3-} + H_2O \rightarrow 2Cl^- + AsO_4^{3-} + 2H^+$$

(*h*) Potassium manganate, K_2MnO_4 (Mn = + 6) to permangnate, $KMnO_4$ (Mn = +7)

$$Cl_2 + 2MnO_4^{2-} \rightarrow 2Cl^- + 2MnO_4^-$$

(*i*) Vegetable coloured matter to colourless matter. When a vegetable coloured matter (*e.g.*, wood pulp, cotton fibres etc.) is treated with the aqueous solution of Cl_2, nascent oxygen produced by the action of Cl_2 on H_2O oxidises the coloured matter to colourless matter, called *oxidation product.*

$$Cl_2\,(g) + H_2O\,(l) \rightarrow 2HCl\,(aq) + O\,(aq)$$

$$\text{Coloured matter} + O \xrightarrow{\text{Oxidation}} \text{Colourless matter} \qquad (\textit{Oxidation product})$$

The formation of colourless product explains the bleaching action of Cl_2. The colourless matter is not reduced back in air to coloured matter. Thus the bleaching done by Cl_2 is *permanent.* Now let us compare this bleaching action of Cl_2 with that of SO_2. In case of SO_2, bleaching action takes place due to the reduction of coloured matter. The reduction is done by nascent hydrogen which is obtained by the action of H_2O on SO_2.

$$SO_2\,(g) + 2H_2O\,(l) \rightarrow H_2SO_4\,(aq) + 2H\,(aq)$$

$$\text{Coloured matter} + H \xrightarrow{\text{Reduction}} \text{Colourless matter } (\textit{Reduction matter})$$

In this case, the reduction matter (*i.e.*, colourless matter) is slowly oxidised in air to coloured matter. Thus the bleaching done by SO_2 is temporary.

Bromine oxidises : (*a*) H_2S (S = – 2) to S (S = 0)

$$Br_2 + S^{2-} \rightarrow 2Br^- + S$$

(*b*) SO_3^{2-} (S = + 4) to SO_4^{2-} (S = + 6)

$$Br_2 + SO_3^{2-} + H_2O \rightarrow 2Br^- + 2H^+ + SO_4^{2-}$$

(*c*) $S_2O_3^{2-}$ (S = + 2) to SO_4^{2-} (S = + 6)

$$Br_2 + S_2O_3^{2-} + H_2O \rightarrow 2Br^- + 2H^+ + SO_4^{2-} + S$$

(*d*) Aqueous solution of SO_2 (S = + 4) to H_2SO_4 (S = + 6)

$$Br_2 + SO_2 + 2H_2O \rightarrow 2Br^- + 4H^+ + SO_4^{2-}$$

(*e*) AsO_3^{3-} (As = +3) to AsO_4^{3-} (As = +5)

$$Br_2 + AsO_3^{3-} + H_2O \rightarrow 2Br^- + 2H^+ + AsO_4^{3-}$$

(*f*) Acidified solution of Fe (+2) salts to Fe (+3) salts.

$$Br_2 + 2Fe^{2+} \rightarrow 2Br^- + 2Fe^{3+}$$

(*g*) Nitrite, NO_2^- (N = +3) to nitrate, NO_3^- (N = +5)

$$Br_2 + NO_2^- + H_2O \rightarrow 2Br^- + 2H^+ + NO_3^-$$

(*h*) Vegetable coloured matter to colourless matter. For reactions see Cl_2.

Iodine oxidises : (*a*) AsO_3^{3-} (As = +3) to AsO_4^{3-} (As = +5)

$I_2 + AsO_3^{3-} + H_2O \rightarrow 2I^- + AsO_4^{3-} + 2H^+$

(*b*) As_2O_3 (As = +3) to As_2O_5 (As = +5)

$2I_2 + 2H_2O + As_2O_3 \rightarrow As_2O_5 + 4H^+ + 4I^-$

(*c*) $S_2O_3^{2-}$ (S = + 2) to tetrathionate ion, $S_4O_6^{2-}$ (S = + 2.5)

$I_2 + 2S_2O_3^{2-} \rightarrow 2I^- + S_4O_6^{2-}$

(*d*) Acidified solution (acidified with HCl) of $SnCl_2$ to $SnCl_4$.

$I_2 + Sn^{2+} \rightarrow 2I^- + Sn^{4+}$

(*e*) NO (N = +2) to HNO_3 (N = +5)

$3I_2 + 2NO + 4H_2O \rightarrow 6I^- + 2HNO_3 + 6H^+$

(*f*) H_2S (S = –2) to S (S = 0)

$I_2 + S^{2-} \rightarrow 2I^- + S$

(*g*) Aqueous solution of SO_2 (S = + 4) to H_2SO_4 (S = + 6)

$SO_2 + 2H_2O + I_2 \rightarrow 4H^+ + SO_4^{2-} + 2I^-$

(*h*) SO_3^{2-} to SO_4^{2-}

$I_2 + SO_3^{2-} + H_2O \rightarrow 2I^- + 2H^+ + SO_4^{2-}$

(*i*) NO_2^- to NO_3^-

$I_2 + NO_2^- + H_2O \rightarrow 2I^- + 2H^+ + NO_3^-$

Example 18.9. *The values of standard reduction electrode potentials (E^o_{red} values) of some half-cell reactions are given below : (O.A. = Oxidising agent, R.A. = Reducing agent).*

O.A.		*R.A.*	*E^o_{red} values (V)*
$I_2 + 2e^-$	$\longrightarrow$	$2I^-$	+0.54
$MnO_4^- + 8H^+ + 5e^-$	$\longrightarrow$	$Mn^{2+} + 4H_2O$	+1.52
$Fe^{3+} + 2e^-$	$\longrightarrow$	Fe^{2+}	+0.77
$Sn^{4+} + 2e^-$	$\longrightarrow$	Sn^{2+}	+0.1

Select the strongest reducing agent and strongest oxidising agent.

Solution. In the given reduction reactions, the species written on LHS accept electrons and are reduced to the species written on the RHS. Thus the species given on RHS act as reducing agents.

We know that the species which has the lowest E^o_{red} value acts as the strongest reducing agent $\left[E^o_{red} \propto \dfrac{1}{\text{Reducing power}}\right]$ while the species having the highest E^o_{red} value acts as the strongest oxidising agent [$E^o_{red} \propto$ Oxidising power]. Thus since the value of $E^o_{Sn^{4+}/Sn^{2+}}$ (= + 0.1 V) is the lowest, Sn^{2+} ion acts as the strongest reducing agent. Similarly, since the value of $E^o_{MnO_4^-/Mn^{2+}}$ (= +1.52 V) is the highest, MnO_4^- ion is the strongest oxidising agent.

Example 18.10. *On the basis of the following E° values, predict whether ferricyanide ion is stronger or weaker oxidising agent than ferric ion.*

$[Fe(CN)_6]^{4-} \rightarrow [Fe(CN)_6]^{3-} + e^-$; $E^o = -0.36$ V

$Fe^{2+} \rightarrow Fe^{3+} + e^-$; $E^o = -0.77$ V

Solution. The given reactions are oxidation reactions. On writing them as reduction reactions we get :

Oxidising agent		*Reducing agent*	
$(Fe(CN)_6)^{3-} + e^-$	$\xrightleftharpoons{\text{Reduction}}$	$[Fe(CN)_6]^{4-}$;	$E^o_{red} = +0.36$ V

Ferrocyanide ion
(Fe = +3)

$$Fe^{3+} + e^- \xrightleftharpoons{\text{Reduction}} Fe^{2+}\ ; \qquad E^o_{red} = +0.77\ V$$

Now since the value of standard reduction electrode potential for Fe^{3+} / Fe^{2+} couple is greater than the value of standard electrode reduction electrode potential for $[Fe(CN)_6]^{3-}$ / $[Fe(CN)_6]^{4-}$ couple, Fe^{3+} is stronger oxidising agent than $[Fe(CN)_6]^{3-}$. (Oxidising power $\propto E^o_{red}$).

Example 18.11. *The following reactions represent the reduction of IO_3^- ion into I^- ion in acidic and basic medium. Predict in which medium IO_3^- ion will act as a better oxidising agent.*

$$IO_3^- + 6H^+ + 6e^- \rightarrow I^- + 3H_2O;\ E^o = +0.907\ V \qquad \text{(Acidic medium)}$$

$$IO_3^- + 3H_2O + 6e^- \rightarrow I^- + 6OH^-;\ E^o = +0.260\ V \qquad \text{(Basic medium)}$$

Solution : *Oxidising agent* *Reducing agent*

$$IO_3^- + 6H^+ + 6e^- \xrightarrow{\text{Reduction}} I^- + 3H_2O; \quad E^o_{red} = +0.907\ V\ \textit{(Acidic medium)}$$

(I = +5) (I = –1)

$$IO_3^- + 3H_2O + 6e^- \xrightarrow{\text{Reduction}} I^- + 6OH^-; \quad E^o_{red} = +0.260\ V\ \textit{(Basic medium)}$$

(I = +5) (I = –1)

Since the value of for E^o_{red} for IO_3^- / I^- couple in acidic medium (= + 0.907 V) is greater than that (= + 0.26 V) for the same couple in basic medium, IO_3^- ion will act as stronger (better or more powerful) oxidising agent acidic medium.

Example 18.12. *The standard reduction potential values of some electrodes are given as Zn^{2+}/Zn = – 0.762 V, Cr^{2+}/Cr = – 0.740 V, $2H^+/H_2$ = +0.00 V, Fe^{3+}/Fe^{2+} = + 0.770 V. Which of these species is the most powerful reducing agent?* *(I.I.T. 1981)*

Solution. Let us write the reduction reactions so that oxidising agents are written on the LHS and reducing agents are shown on the RHS.

Oxidising agent			*Reducing agents*	E^o_{red} (V)
Zn^{2+}	$+ 2e^-$	$\rightleftharpoons$	Zn	– 0.762
Cr^{2+}	$+ 2e^-$	$\rightleftharpoons$	Cr	– 0.740
$2H^+$	$+ 2e^-$	$\rightleftharpoons$	H_2	0.00
Fe^{3+}	$+ e^-$	$\rightleftharpoons$	Fe^{2+}	+ 0.77

Since Zn^{2+}/Zn electrode has the lowest value of E^o_{red} (most negative value), Zn metal is the most powerful reducing agent.

Example 18.13. *Select the strongest reducing agent among the species namely K, Mg, Al, Ba and F^-.* *(MLNR 1984)*

Solution. Let us write reduction reactions, so that oxidising agents are shown on the LHS and reducing agents are written on the RHS. Values of E^o_{red} have been taken from the electrochemical series. (Table 18.2 ; page 609-10).

Oxidising agents			*Reducing agents*	E^o_{red} (V)
K^+	$+ e^-$	$\rightleftharpoons$	K	–2.93
Mg^{2+}	$+ 2e^-$	$\rightleftharpoons$	Mg	–2.38
Al^{3+}	$+3e^-$	$\rightleftharpoons$	Al	–1.67
Ba^{2+}	$+2e^-$	$\rightleftharpoons$	Ba	–2.90
F^2	$+2e^-$	$\rightleftharpoons$	$2F^-$	+2.87

Thus since E^o_{red} values increase as – 2.93 (for K^+/K couple) < – 2.90 (Ba^{2+} / Ba) < – 2.38 (Mg^{2+} / Mg) < – 1.67 (Al^{3+} / Al) < + 2.87(F_2 / $2F^-$), the reducing power of the elements decreases in the same direction *i.e.*, reducing power decreases as K > Ba > Mg > Al > F_2. Consequently K is the strongest reducing agent .

Example 18.14. *Arrange the following in the decreasing order of their tendency of being oxidised. Zn, Cu, Mg and Ag.*

Solution. We know that a metal which has a strong tendency of being oxidised acts as a strong reducing agent. Thus in order to find out the decreasing order of the tendency of the metals of being oxidised, we should find out the decreasing order of the reducing power of the metals.

For this, we should write reduction reactions in which oxidising agents should be shown on the LHS and reducing agents should be written on RHS. Values of E^o_{red} been taken from the electrochemical series.

Oxdising agent			*Reducing agent*	E^o_{red} *(V)*
Zn^{2+}	+	$2e^-$	$\rightleftharpoons$ Zn	– 0.76
Cu^{2+}	+	$2e^-$	$\rightleftharpoons$ Cu	+ 0.34
Mg^{2+}	+	$2e^-$	$\rightleftharpoons$ Mg	– 2.38
Ag^+	+	e^-	$\rightleftharpoons$ Ag	+ 0.80

Increasing order of E^o_{red} values : = – 2.38 < – 0.76 < + 0.34 < + 0.80.

Decreasing order of reducing power of reducing agents : Mg > Zn > Cu > Ag.

VII. Reactivity of metals. The metals which have large negative E^o_{red} values have a strong tendency to lose electrons to form hydrated cations (reducing agents) and hence are very reactive. On the other hand the metals which have low positive values of E^o_{red} are unreactive. For example since alkali metals and alkaline earth metals have large negative values of E^o_{red}, these are chemically very reactive. On the other hand Cu, Ag, and Pt-metals (Ru, Rh, Pd, Os, Ir, Pt) which have low positive values of E^o_{red} are quite unreactive or noble (Cu^+ / Cu = + 0.52 V, Ag^+ / Ag = + 0.80 V, Au^+ / Au = + 1.6 V, Ru^{2+} / Ru = 0.50 V, Rh^{2+} / Rh = + 0.60 V, Pd^{2+} / Pd = +1.20 V, Os^{2+} / Os = + 0.90 V, Ir^{2+} / Ir = + 1.0, Pt^{2+} / Pt = +1.20 V).

Since E^o_{red} values become more and more positive (– → 0 → +) as we move down the electrochemical series, the reactivity of the metals decreases in the same direction. Thus the metals lying on the top of the series are very reactive.

Example 18.15. *Arrange the following metals in the decreasing order of their reactivity : Al, Zn, Sn and Cu.*

Given that $E^o_{Al^{3+}/Al} = -1.60$ V, $E^o_{Zn^{2+}/Zn} = -0.74$ V, $E^o_{Sn^{4+}/Sn^{2+}} = +0.15$ V, $E^o_{Cu^{2+}/Cu} = +0.34$ V.

Solution. E^o_{red} values can be arranged as :

E^o_{red} values :	$E^o_{Al^{3+}/Al}$	$E^o_{Zn^{2+}/Zn}$	$E^o_{Sn^{4+}/Sn^{2+}}$	$E^o_{Cu^{2+}/Cu}$
	– 1.6 V	– 0.74 V	+ 0.15 V	+ 0.34 V

——————— Increasing ———————→

Reactivity of metals : ——————— Decreasing ———————→

Thus the reactivity of the given metals can arranged as : Al > Zn > Sn > Cu (*Decreasing order*).

VIII. Reactivity of non-metals. The non-metals which have a strong tendency to gain electrons are more reactive. Now we know that the tendency of a non-metal to gain electrons (*i.e.*, oxidising agent) increases with the increase in the value of standard reduction potential of non-metal. Hence reactivity of non-metals increases with the increase in the value of standard reduction electrode potential of the non-metal.

Example. Since the values of E^o_{red} of F_2, Cl_2, Br_2 and I_2 decreases from F_2 to I_2 ($F_2/2F^-$ = +2.87 V, Cl_2 / $2Cl^-$ = +1.36 V, Br_2 / $2Br^-$ = +1.06 V, I_2 / $2I^-$ = + 0.54 V), the reactivity of these halogens also decreases from F_2 to I_2, *i.e.*, F_2 is the most reactive while I_2 is least so.

IX. To find out the conditions of spontaneous occurrence of some redox reactions. A spontaneous reaction is that which proceeds by itself once it has been properly initiated, if necessary. All these reactions are spontaneous reactions which are feasible.

Here we shall find out the conditions for the spontaneous occurrence of redox reactions of the following types :

(A) *R.A.* + *O.A.* → *R.A.* + *O.A.*

$$\underset{\text{Metal}}{Zn\,(s)} + \underset{\text{Metallic salt}}{CuSO_4\,(aq)} \rightarrow Cu\,(s)\downarrow + ZnSO_4\,(aq)$$

or

$$\underset{\text{Displacing metal (R.A.)}}{Zn\,(s)} + \underset{\text{Metal being displaced (O.A.)}}{Cu^{2+}\,(aq)} \rightarrow \underset{\text{(R.A.)}}{Cu\,(s)\downarrow} + \underset{\text{(O.A.)}}{Zn^{2+}\,(aq)}$$

The above redox reaction gives us the following informations :

(*i*) When Zn-metal is added to the aqueous solution of $CuSO_4$, Cu present in $CuSO_4$ is displaced by Zn (*i.e.*, Cu is precipitated) and solution of $ZnSO_4$ is obtained. Thus Zn-metal acts as a *displacing metal.*

(*ii*) Zn-metal reduces Cu^{2+} ions (present in $CuSO_4$ solution) to free Cu-metal and Zn-metal itself is oxidised to Zn^{2+} ions. Thus Zn-metal acts as a reducting agent (R.A.) and Cu^{2+} ions behave as oxidising agent (O.A.).

In the given redox reaction, since Zn (*s*) is being oxidised to Zn^{2+} (*aq*) ions and Cu^{2+} (*aq*) ions are being reduced to Cu (*s*) metal, Zn (*s*) / Zn^{2+} (*aq*) electrode will constitute the anode and Cu^{2+} (*aq*) / Cu (*s*) electrode will form the cathode of galvanic cell which is obtained by coupling these two electrodes together. Thus the galvanic cell obtained can be represented as :

$$\underset{\text{Anode (oxidation electrode). Reducing agent (R.A.)}}{Zn\,(s)\,/\,Zn^{2+}\,(aq)} \;\|\; \underset{\text{Cathode (reduction electrode). Oxidising agent (O.A.)}}{Cu^{2+}\,(aq)\,/\,Cu\,(s)}$$

The reactions taking place at the two electrodes can be written as :

Oxidation reaction at anode : $Zn\,(s) \rightarrow Zn^{2+}(aq) + 2e^-$

Reduction reaction at cathode : $Cu^{2+}\,(aq) + 2e^- \rightarrow Cu\,(s)$

Cell reaction : $Zn\,(s) + Cu^{2+}\,(aq) \rightarrow Zn^{2+}\,(aq) + Cu\,(s)$

The cell reaction is the same as that given above.

The value of E^o_{cell} of the above cell is given by :

$$E^o_{cell} = (E^o_{red})_{cathode} - (E^o_{red})_{anode} \quad ...(i)$$

The above cell will function smoothly, if E^o_{cell} is a positive quantity. In order to make E^o_{cell} positive, the value of $(E^o_{red})_{cathode}$ should be higher than that of $(E^o_{red})_{anode}$, *i.e.*,

$$(E^o_{red})_{cathode} > (E^o_{red})_{anode} \quad ...(ii)$$

Now since reduction takes place at cathode and oxidation occurs at anode, the relation (*ii*) can also be written as :

$$(E^o_{red})_{O.A.} > (E^o_{red})_{R.A.} \quad ...(iii)$$

or

$$(E^o_{red})_{R.A.} < (E^o_{red})_{O.A.} \quad ...(iv)$$

or

$$(E^o_{red})_{\text{Displaying metal}} < (E^o_{red})_{\text{Metal being displaced}} \quad ...(v)$$

The above relations [relations (*i*), (*ii*), (*iii*), (*iv*) and (*v*)] give us the following rules :

(*i*) The given redox reaction would occur as written only when the value of standard e.m.f. of the cell (*i.e.*, E^o_{cell} value) in which the given redox reaction occurs is a positive quantity [See relation (*i*)]. If the value of E^o_{cell} is a negative quantity, the given reaction would proceed in the opposite direction *i.e.*, the reaction would occur from right to left.

(*ii*) The given redox reaction will occur spontaneously only when the value of standard reduction potential of cathode is higher than that of the standard reduction potential of anode [See relation (*ii*)]. Now since cathode acts as an oxidising agent and anode acts as a reducing agent, the value of standard reduction electrode potential of the oxidising agent (*i.e.*, metal being displaced) should be higher than that of the reducing agent (*i.e.*, displacing metal). In other words we can say that the value of E^o_{red} of the metal being displaced should be higher than that of the displacing metal.

(*iii*) The species which is being reduced (*i.e.*, which accepts electrons or acts as an oxidising agent) should have higher value of its standard reduction electrode potential while the species which is being oxidised (*i.e.*, which loses the electrons or acts as a reducing agent) should have lower value of its standard reduction potential.

(*iv*) We have already said that in a redox reaction, a reducing agent reacts with an oxidising agent. This redox reaction proceeds from left to right spontaneously only when the standard reduction electrode potential of the reducing agent is less than that of the oxidising agent. We know that in the reduction electrode reaction the species written on LHS acts as O.A. and the species written on the LHS acts as R.A. Now since the values of standard reduction electrode potential increase down the electrochemical series (*i.e.*, the values become more and more positive as we move down the series), R.A. of an electrode reduction reaction will react with the O.A. of only that electrode reaction which has greater value of its standard reduction electrode potential (*i.e.*, which lies below in the electrochemical series). The reaction between R.A. and O.A. produces another pair of R.A. and O.A. R.A. thus produced belongs to the electrode which has higher value of E^o_{red} (*i.e.*, lower electrode) and O.A., thus generated, belongs to the electrode which has lower value of E^o_{red} (*i.e.*, upper electrode). In simple words we can say that in the electrochemical series, a metal which can displace another metal from its salt solution lies above the metal to be displaced while a metal which cannot displace another metal from its salt solution lies below the metal to be displaced.

Example. The values of standard reduction electrode potential of some electrodes are given below. These electrodes have been shown in the same order in which they are placed in the electrochemical series, *i.e.*, E° values increase down the series.

O.A.		*R.A*	*E° values*
$Mg^{2+}(aq)$	→	$Mg(s)$	$E^o_{Mg^{2+}/Mg} = -2.37$ V
$Al^{3+}(aq) + 3e^-$	→	$Al(s)$	$E^o_{Al^{3+}/Al} = -1.66$ V
$Zn^{2+}(aq) + 2e^-$	→	$Zn(s)$	$E^o_{Zn^{2+}/Zn} = -0.76$ V
$Fe^{2+}(aq) + 2e^-$	→	$Fe(s)$	$E^o_{Fe^{2+}/Fe} = -0.44$ V
$Cu^{2+}(aq) + 2e^-$	→	$Cu(s)$	$E^o_{Cu^{2+}/Cu} = +0.34$ V

The increasing order of E^o_{red} values as given above show that :

(*a*) Mg (*s*) can reduce $Al^{3+}(aq)$, $Zn^{2+}(aq)$, $Fe^{2+}(aq)$ and $Cu^{2+}(aq)$ ions (oxidising agents).

Thus the following redox reactions in which Mg displaces Al^{3+}, Zn^{2+}, Fe^{2+} and Cu^{2+} ions from their salt solution occur spontaneously.

R.A.		*O.A.*		*O.A.*		*R.A.*
$3Mg(s)$	+	$2Al^{3+}(aq)$	→	$3Mg^{2+}(aq)$	+	$2Al(s)$
$Mg(s)$	+	$Zn^{2+}(aq)$	→	$Mg^{2+}(aq)$	+	$Zn(s)$
$Mg(s)$	+	$Fe^{2+}(aq)$	→	$Mg^{2+}(aq)$	+	$Fe(s)$
$Mg(s)$	+	$Cu^{2+}(aq)$	→	$Mg^{2+}(aq)$	+	$Cu(s)$

(*b*) Al (*s*) can reduce $Zn^{2+}(aq)$, $Fe^{2+}(aq)$ and $Cu^{2+}(aq)$ ions (oxidising agents), *i.e.*, Al (*s*) can displace Zn^{2+}, Fe^{2+} and Cu^{2+} ions from their salt solution.

R.A.		*O.A.*		*O.A.*		*R.A.*
$2Al(s)$	+	$3Zn^{2+}(aq)$	→	$2Al^{3+}(aq)$	+	$3Zn(s)$

$2Al\ (s) + 3Fe^{2+}\ (aq) \rightarrow 2Al^{3+}\ (aq) + 3Fe\ (s)$

$2Al\ (s) + 3Cu^{2+}\ (aq) \rightarrow 2Al^{3+}\ (aq) + 3Cu\ (s)$

(*c*) Zn (*s*) can reduce Fe^{2+} (*aq*) and Cu^{2+} (*aq*) ions (oxidising agents), *i.e.*, Zn (*s*) can displace Fe^{2+} and Cu^{2+} ions from their salt solution.

R.A.		*O.A.*		*O.A.*		*R.A.*
Zn (*s*)	+	Fe^{2+} (*aq*)	→	Zn^{2+} (*s*)	+	Fe (*s*)
Zn (*s*)	+	Cu^{2+} (*aq*)	→	Zn^{2+} (*s*)	+	Cu (*s*)

(*d*) Fe (*s*) can reduce only Cu^{2+} (*aq*) ion (oxidising agent), *i.e.*, Fe can displace only Cu^{2+} ions from the salt solution

	R.A.		*O.A.*		*O.A.*		*R.A*
	Fe (*s*)	+	Cu^{2+} (*aq*)	→	Fe^{2+} (*aq*)	+	Cu (*s*)
(B)	*O.A.*		*R.A.*		*R.A.*		*O.A*
	Cl_2	+	2KI (*aq*)	→	2KCl (*aq*)	+	I_2
	Halogen		Halide		Halide		Halogen
or	Cl_2	+	$2I^-$ (*aq*)	→	$2Cl^-$ (*aq*)	+	I_2
	Displacing halogen (O.A.)		Halogen being displaced (R.A.)		(R.A.)		(O.A.)

The above redox reaction gives the following informations :

(*i*) When Cl_2 reacts with aqueous solution of KI, iodine present in KI solution is displaced by Cl_2 and hence I_2 is liberated in the free state. Thus Cl_2 acts as a *displacing agent.*

(*ii*) Cl_2 oxidises I^- ions to I_2 and is itself reduced to Cl^- ions. Thus Cl_2 acts as an oxidising agent and I^- ions act as reducing agent.

In the given redox reaction, since I^- (*aq*) ions are being oxidised to I_2 and Cl_2 is being reduced to Cl^- (*aq*) ions, Pt, I^- (*aq*) / I_2 electrode will constitute the anode and Cl^- (*aq*) / Cl_2 (*g*), Pt electrode will form the cathode of the galvanic cell which is obtained by coupling these two electrodes together. Thus the galvanic cell obtained can be represented as :

Pt / I^- (*aq*)/ I_2	‖	Cl^- (*aq*) / Cl_2 (*g*) / Pt
Anode (oxidation electrode)		Cathode (Reduction electrode)
Reducing agent (R.A.)		Oxidising agent (O.A.)

The reactions taking place at the two electrodes can be written as :

Oxidation reaction at anode : $2I^- (aq) \rightarrow I_2 + 2e^-$

Reduction reaction at cathode : $Cl_2 (g) + 2e^- \rightarrow 2Cl^- (aq)$

Cell reaction : $2I^- (aq) + Cl_2 (g) \rightarrow I_2 + 2Cl^- (aq)$

This cell reaction is the same as the given redox reaction. The value of E^o_{cell} of the above cell is given by :

$$E^o_{cell} = (E^o_{red})_{cathode} - (E^o_{red})_{anode} \quad ...(i)$$

The above cell will work as shown, if E^o_{cell} is a positive quantity. In order to make E^o_{cell} positive, the value of $(E^o_{red})_{anode}$ should be greater than that of $(E^o_{red})_{anode}$, *i.e.*,

$$(E^o_{red})_{cathode} > (E^o_{red})_{anode} \quad ...(ii)$$

Now since reduction takes place at cathode and oxidation occurs at anode, the relation (*ii*) can also be written as :

$$(E^o_{red})_{O.A.} > (E^o_{red})_{R.A.} \quad ...(iii)$$

or $$(E^o_{red})_{R.A.} < (E^o_{red})_{O.A.} \quad ...(iv)$$

or $$(E^o_{red})_{\text{Halogen being displaced}} < (E^o_{red})_{\text{Displacing halogen}} \quad ...(v)$$

The above five relations give us the following rules :

(*i*) The given redox reaction will be possible only when the value of E^o_{cell} of the galvanic cell in which the given redox reaction takes place is positive. If the value of E^o_{cell} is negative, the given reaction will proceed from right to left.

(*ii*) The given redox reaction will occur spontaneously only when the value of $(E^o_{red})_{cathode}$ is greater than the value of $(E^o_{red})_{anode}$. Now since cathode acts as an oxidising agent and anode acts as a reducing agent, the value of $(E^o_{red})_{O.A.}$ (*i.e.,* displacing halogen) should be higher than $(E^o_{red})_{R.A.}$ (*i.e.,* halogen being displaced). In other words we can say that the value of E° of the displacing halogen should be higher than that of the halogen being displaced.

(*iii*) The halogen which is being reduced (*i.e.,* which accepts electrons or acts as an oxidising agent) should have higher value of its standard reduction electrode potential while the halogen which is being oxidised (*i.e.,* which loses the electrons or acts as a reducing agent) should have lower value of its standard reduction potential.

(*iv*) The values of standard reduction potential of X_2 / $2X^-$ electrodes decrease from F_2 to I_2 as shown in the margin. We have already said that the halogen which is displacing (*i.e.,* which is acting as an oxidising agent) another halogen from its salt solution should have higher value of its standard reduction electrode potential while the halogen which is being displaced (*i.e.,* which is acting as a reducing agent) should have lower value. Thus :

F_2 / $2F^-$	= + 2.87 V
Cl_2 / $2Cl^-$	= + 1.36 V
Br_2 / $2Br^-$	= + 1.06 V
I_2 / $2I^-$	= + 0.54 V

(*i*) Since fluorine has greater value of its standard reduction electrode potential than the remaining three halogens, it can displace all these halogens from the aqueous solution of metallic chlorides, bromides and iodides. *i.e.,* all the following reactions are possible.

	O.A.		R.A.		R.A.		O.A.	
(*a*)	F_2	+	2KCl	→	2KF	+	Cl_2	Possible
or	F_2	+	$2Cl^-$	→	$2F^-$	+	Cl_2	
(*b*)	F_2	+	2KBr	→	2KF	+	Br_2	
or	F_2	+	$2Br^-$	→	$2F^-$	+	Br_2	
(*c*)	F_2	+	2KI	→	2KF	+	I_2	
or	F_2	+	$2I^-$	→	2F	+	I_2	

(*ii*) Since chlorine has lower value of its standard reduction electrode potential than that of fluorine, it cannot displace fluorine from the aqueous solution of metallic fluorides, *i.e.,* reaction (*a*) is not possible. On the other hand, since chlorine has greater value of its standard reduction electrode potential than bromine and iodine both, it can displace both these halogens from the aqueous solution of metallic bromide and iodide, *i.e.,* reactions (*b*) and (*c*) are possible.

	O.A.		R.A.		R.A.		O.A.	
(*a*)	Cl_2	+	2KF	→	2KCl	+	F_2	Not Possible
or	Cl_2	+	$2F^-$	→	$2Cl^-$	+	F_2	
(*b*)	Cl_2	+	2KBr	→	2KCl	+	Br_2	Possible
or	Cl_2	+	$2Br^-$	→	$2Cl^-$	+	Br_2	
(*c*)	Cl_2	+	2KI	→	2KCl	+	I_2	
or	Cl_2	+	$2I^-$	→	$2Cl^-$	+	I_2	

(*iii*) Since bromine has lower value of its standard reduction electrode potential than fluorine and chlorine both, this halogen can displace neither fluorine nor chlorine from the aqueous solution of metallic fluorides and chlorides, *i.e.,* reactions (*a*) and (*b*) are not possible. On the other hand, since bromine has greater value of its standard reduction electrode potential than iodine, it can displace iodine from the aqueous solution of a metallic iodide, *i.e.,* reaction (*c*) possible.

	O.A.		R.A.		R.A.		O.A.	
(*a*)	Br_2	+	KF	→	2KBr	+	F_2	Not possible
or	Br_2	+	$2F^-$	→	$2Br^-$	+	F_2	
(*b*)	Br_2	+	2KCl	→	2KBr	+	Cl_2	
or	Br_2	+	$2Cl^-$	→	$2Br^-$	+	Cl_2	
(*c*)	Br_2	+	2KI	→	2KBr	+	I_2	Possible
or	Br_2	+	$2I^-$	→	$2Br^-$	+	I_2	

(*d*) Since iodine has lower value of its standard reduction electrode potential than each of the remaining three halogens, this halogen cannot displace F, Cl and Br from metallic fluorides, chlorides and bromides, *i.e.*, all the following reactions are impossible to occur.

(*a*)	I_2	+	2KF	→	$2KI + F_2$	Not possible
or	I_2	+	$2F^-$	→	$2I^- + F_2$	
(*b*)	I_2	+	2KCl	→	$2KI + Cl_2$	
or	I_2	+	$2Cl^-$	→	$2I^- + Cl_2$	
(*c*)	I_2	+	2KBr	→	$2KI + Br_2$	
or	I_2	+	$2Br^-$	→	$2I^- + Br_2$	

Example 18.16. *Iodine (I_2) and bromine (Br_2) are added to solution containing I^- and Br^- ions. What reaction would occur, if the concentration of each species is 1 M? The electrode potentials for the reactions are :*

$$I_2 + 2e^- \rightarrow 2I^-,\ E^\circ = +0.54\ V$$
$$Br_2 + 2e^- \rightarrow 2Br^-,\ E^\circ = +1.08\ V$$

Solution. The possible reaction of I_2 with Br^- ions and Br_2 with I^- ions may be represented as:

(*a*) $I_2 + 2Br^- \rightarrow Br_2 + 2I^-$ (*b*) $Br_2 + 2I^- \rightarrow I_2 + 2Br^-$

Now we have to predict which of the above two reactions occurs spontaneously.

Reaction. (*a*) In this redox reaction Br^- ions are being oxidised to Br_2 molecules and I_2 molecules are being reduced to I^- ions. Thus this reaction occurs as written, only when $E^\circ_{Br_2/2Br^-} <$ $E^\circ_{I_2/2I^-}$. Now since $E^\circ_{Br_2/2Br^-}$ (= +1.08 V) is not smaller than $E^\circ_{I_2/2I^-}$ (= + 0.54 V), reaction (*a*) will not proceed as shown.

(*b*) In this reaction, I^- ions are oxidised to I_2 molecules and Br_2 molecules are reduced to Br^- ions. Thus, this reaction proceeds from left to right only when $E^\circ_{I_2/2I^-} < E^\circ_{Br_2/2Br^-}$. Now since $E^\circ_{I_2/2I^-}$ (= + 0.54 V) is smaller than $E^\circ_{Br_2/2Br^-}$ (= + 1.08 V), reaction (*b*) will proceed as written.

(C)	2M Metal	+	2HX Acid	→	2MX Metallic salt	$+ H_2$
or	2M (*s*) Reducing agent (R.A.) Displacing metal	+	$2H^+$ (*aq*) Oxidising agent (O.A.)	→	$2M^+$ (*aq*)	$+ H_2$ (*g*)

The above reaction is a redox reaction. This reaction gives us the following informations :

(*i*) When a metal (M) is acted upon by an acid (HX), hydrogen of the acid is displaced by the metal and the metallic salt of the acid is formed, *i.e.*, H_2 gas is liberated. Thus meter (M) acts as a *displacing metal.*

(*ii*) When a metal (M) is treated with an acid (HX), H^+ ions present in the acid are reduced to H_2 gas by the metal and the metal is oxidised to M^+ ions. Thus M acts as a reducing agent and HX (or H^+ ions) acts as an oxidising agent.

In the given redox reaction, since the metal (M) is being oxidised to M^+ (*aq*) ions and H^+ (*aq*) ions are being reduced to H_2 (*aq*), M (*s*) / M^+ (*aq*) electrode will form the anode and H^+ (*aq*) / H_2

(*g*), Pt electrode will constitute cathode of the galvanic cell which is obtained by combining the above two electrodes together. Thus the galvanic cell obtained can be represented as :

$M\ (s) / M^+\ (aq)$	‖	$H^+\ (aq) / H_2\ (g)$, Pt
Anode (oxidation electrode)		Cathode (Reduction electrode)
Reducing agent (R.A.)		Oxidising agent (O.A.)

The reactions taking place at the two electrodes can be written as :

Oxidation reaction at anode : $[M\ (s) \rightarrow M^+\ (aq) = e^-] \times 2$

Reduction reaction at cathode : $2H^+\ (aq) + 2e^- \rightarrow H_2\ (g)$

Cell reaction : $2M\ (s) + 2M^+\ (aq) \rightarrow 2M^+\ (aq) + H_2\ (g)$

This cell reaction is the same as that redox reaction given above. The value of E^o_{cell} is given by:

$$E^o_{cell} = (E^o_{red})_{cathode} - (E^o_{red})_{anode} = E^o_{2H^+/H_2} - E^o_{M^+/M}$$

$$= (E^o_{red})_{O.A.} - (E^o_{red})_{R.A.} \qquad ...(i)$$

Now the galvanic cell shown above will work only when E^o_{cell} is positive. Thus in order to get a positive value of E^o_{cell}, the value of $(E^o_{red})_{O.A.}$ should be greater than that of $(E^o_{red})_{R.A.}$ Now since the value of $(E^o_{red})_{O.A.}$ or $E^o_{2H^+/H_2}$ = 0.0 V, the value of $(E^o_{red})_{R.A.}$ or $E^o_{M^+/M}$ should be less than zero, *i.e.*, $(E^o_{red})_{R.A.}$ should be negative quantity.

$$(E^o_{red})_{R.A.} < \text{Zero}$$

or $$(E^o_{red})_{R.A.} = \text{A negative quantity}$$

The above discussion shows that a metal can displace hydrogen from an acid, only when one of the following conditions is satisfied :

(*i*) The value of e.m.f. of the cell (*i.e.*, E^o_{cell} value) in which the given redox reaction occurs is a positive quantity. If the value of E^o_{cell} comes out to be a negative quantity, the given redox reaction would proceed in the opposite direction, *i.e.*, the reaction would take place from right to left.

(*ii*) The value of E^o_{red} of the reducing agent (*i.e.*, displacing metal) should be less than zero, *i.e.*, it should a negative quantity. In other words we can say that the displacing metal should lie above hydrogen in the electrochemical series. Thus all those metals which lie above hydrogen in the electrochemical series can displace hydrogen from an acid, since these metals have negative values for their E^o_{red}. The metals having positive value for their E^o_{red} are not able to displace hydrogen from an acid. Such metals lie below hydrogen in the electrochemical series.

(*iii*) The species which is being oxidised (*i.e.*, which loses uses electrons or acts as a reducing agent–displacing metal) should have lower value of its standard reduction potential while the species which is being reduced (*i.e.*, which accepts electrons or acts as an oxidising agent) should have higher value for its standard reduction electrode potential.

Example 18.17. *Will Fe (s) be oxidised to Fe^{2+} ions by reacting with 1.0 M HCl ? Given that $E^o_{Fe/Fe^{2+}} = +0.44$ V.* *(I.I.T. 1986)*

Solution. The oxidation of Fe (*s*) by 1.0 M HCl can be represented by the following redox reaction :

$$Fe\ (s) + 2HCl\ (aq) \rightarrow FeCl_2\ (aq) + H_2\ (g)$$

or $$Fe\ (s) + 2H^+\ (aq) \rightarrow Fe^{2+}\ (aq) + H_2\ (g)$$

This reaction will occur spontaneously when $E^o_{Fe^{2+}/Fe}$ is lower than $E^o_{2H^+/H_2}$. Now since $E^o_{2H^+/H_2}$ is equal to zero, $E^o_{Fe^{2+}/Fe}$ should be a negative quantity. Actually it is a negative quantity as shown below.

$$E^o_{Fe/Fe^{2+}} = +0.44\ \text{V} \quad \text{or} \quad E^o_{Fe^{2+}/Fe} = -0.44\ \text{V}$$

Thus Fe (*s*) will be oxidised to Fe^{2+} ions when this metal reacts with 1 M HCl.

A very important conclusion. The discussion made so far indicates that the essential condition to make a redox reaction to occur spontaneously is that the value of standard reduction potential of the species which is being oxidised (*i.e.*, the species which loses electrons or acts as a reducing agent) should be lower than that of the species which is being reduced (*i.e.*, the species which gains electrons or acts as an oxidising agent).

Example 18.18. *Predict whether the reaction,* $Sn^{4+} + 2Fe^{2+} \rightarrow Sn^{2+} + 2Fe^{3+}$ *will occur or not. Given that* $E^o_{Sn^{4+}/Sn^{2+}} = 0.1$ *V and* $E^o_{Fe^{3+}/Fe^{2+}} = 0.77$ *V.*

Solution. In the given redox reaction, Sn^{4+} ion is being reduced to Sn^{2+} by Fe^{2+} and Fe^{2+} ion is being oxidised to Fe^{3+} by Sn^{4+}. Thus Sn^{4+} acts as oxidising agent and Fe^{2+} acts as reducing agent. We know that the given redox reaction is possible only when the value of standard reduction potential of oxidising agent (*i.e.*, $E^o_{Sn^{4+}/Sn^{2+}}$ value) is higher than that of the reducing agent (*i.e.*, $E^o_{Fe^{3+}/Fe^{2+}}$ value). But here we find that the value of $E^o_{Sn^{4+}/Sn^{2+}}$ is not higher than that of $E^o_{Fe^{3+}/Fe^{2+}}$. Thus the given reaction will not proceed as written.

Example 18.19. *Will metallic tin reduce* Sn^{4+} *to* Sn^{2+} *? Given* $E^o_{Sn^{2+}/Sn} = -0.13$ *V,* $E^o_{Sn^{4+}/Sn^{2+}} = 0.15$ *V.*

Solution. If metallic tin (Sn) reduces Sn^{4+} to Sn^{2+}, the metallic tin itself is oxidised to Sn^{2+}. Thus :

$$\underset{\text{R.A.}}{Sn} + \underset{\text{O.A.}}{Sn^{4+}} \rightarrow Sn^{2+} + Sn^{2+}$$

Now we see that since

$$(E^o_{red})_{O.A.} > (E^o_{red})_{R.A.}$$

or $$E^o_{Sn^{4+}/Sn^{2+}} > E^o_{Sn^{2+}/Sn}$$

or $$+0.15 > -0.13,$$

The given redox reaction is possible, *i.e.*, metallic tin will reduce Sn^{4+} to Sn^{2+}.

Example 18.20. *Giving appropriate reasons explain whether Zn will be oxidised or not when placed in contact with a solution of* $NiSO_4$*. Given that* $E^o_{Ni^{2+}/Ni} = -0.25$ *V,* $E^o_{Zn^{2+}/Zn} = -0.76$ *V.*

(R.E.E. 1985)

Solution. The reaction between Zn and $NiSO_4$ solution can be written as :

$$Zn + NiSO_4 \rightarrow ZnSO_4 + Ni$$

or $$\underset{\text{R.A.}}{Zn\,(s)} + \underset{\text{O.A.}}{Ni^{2+}\,(aq)} \rightarrow Zn^{2+}\,(aq) + Ni\,(s)$$

The above reaction would be possible only when $(E^o_{red})_{O.A.} > (E^o_{red})_{R.A.}$ Now since $(E^o_{red})_{O.A.}$ or $E^o_{Ni^{2+}/Ni}$ (= – 0.25 V) is less negative than $(E^o_{red})_{R.A.}$ or $E^o_{Zn^{2+}/Zn}$ (= – 0.76), $E^o_{Ni^{2+}/Ni}$ has higher value than $E^o_{Zn^{2+}/Zn}$ and hence the given reaction is possible, *i.e.*, Zn will be oxidised when placed in contact with a solution of $NiSO_4$.

Example 18.21. *Each of the following redox reactions takes place spontaneously as shown. Arrange Mg, Zn, Cu and Ag in the decreasing order of their standard reduction electrode potentials*

$$Cu + 2Ag^+ \rightarrow Cu^{2+} + 2Ag$$
$$Mg + Zn^{2+} \rightarrow Mg^{2+} + Zn$$
$$Zn + Cu^{2+} \rightarrow Zn^{2+} + Cu$$

Solution. The given reactions can be written as :

(*i*) $$\underset{\text{Reducing agent (R.A.)}}{Cu} + \underset{\text{Oxidising agent (O.A.)}}{2Ag^+} \rightarrow Cu^{2+} + 2Ag$$

(*ii*) $Mg + Zn^{2+} \rightarrow Mg^{2+} + Zn$
R.A. O.A.

(*iii*) $Zn + Cu^{2+} \rightarrow Zn^{2+} + Cu$
R.A. O.A.

Reaction (*i*) is possible only when :

$$(E^o_{red})_{O.A.} > (E^o_{red})_{R.A.}$$

or $$E^o_{Ag^+/Ag} > E^o_{Cu^{2+}/Cu} \quad ...(a)$$

Reaction (*ii*) is possible only when :

$$(E^o_{red})_{O.A.} > (E^o_{red})_{R.A.}$$

or $$E^o_{Zn^{2+}/Zn} > E^o_{Mg^{2+}/Mg} \quad ...(b)$$

Reaction (*iii*) is possible only when :

$$(E^o_{red})_{O.A.} > (E^o_{red})_{R.A.}$$

or $$E^o_{Cu^{2+}/Cu} > E^o_{Zn^{2+}/Zn} \quad ...(c)$$

From relations (*a*), (*b*) and (*c*) we get,

$$E^o_{Ag^+/Ag} > E^o_{Cu^{2+}/Cu} > E^o_{Zn^{2+}/Zn} > E^o_{Mg^{2+}/Mg}$$

This is the required order.

Example 18.22. *Will Cu^+ ion disproprtionate in aqueous solution into Cu^{2+} and Cu^0 ? Given that $E^o_{Cu^+/Cu} = +0.52$ V and $E^o_{Cu^{2+}/Cu^+} = +0.153$ V.*

Solution. We know that when Cu^+ ion undergoes disproportionation in aqueous solution, some of its part is oxidised to Cu^{2+} (+1 → +2) and the remaining part is reduced to Cu° (+1 → 0). Thus the disproportionation of Cu^+ ion into Cu^{2+} and Cu° can be shown as

$$2Cu^+ \rightarrow Cu^{2+} + Cu^o \quad ...(i)$$

Now let us construct a galvanic cell whose cell reaction is (*i*). On breaking the cell reaction [reaction (*i*)] into two half-cell reactions we get :

Oxidation half-reaction anode : $Cu^+ \rightarrow Cu^{2+} + e^-$
Reduction half-reaction at cathode : $Cu^+ + e^- \rightarrow Cu^o$

Cell reaction : $2Cu^+ \rightarrow Cu^{2+} + Cu^o$

or

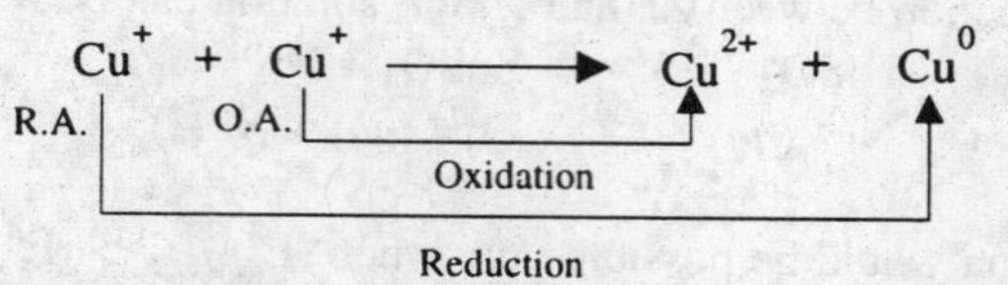

Thus the cell in which the above reaction takes place can be represented as:

Cu^+ / Cu^{2+} || Cu^+ / Cu^0
Anode Cathode

e.m.f. of he above cell is given by :

$$E^o_{cell} = (E^o_{red})_{cathode} - (E^o_{red})_{anode} = E^o_{Cu^+/Cu^0} - E^o_{Cu^{2+}/Cu^+}$$

$$= 0.52 - (+0.153) = \mathbf{+0.367\ V}$$

Since E^o_{cell} comes out to be a positive quantity, the reaction shown at (*i*) is feasible, *i.e.*, Cu^+ ion disproportionates in aqueous solution into Cu^{2+} and Cu^0.

An alternative method. The above problem can also be solved by an alternative method given below.

We have seen that the disproportionation Cu^+ ion into Cu^{2+} and Cu° can be represented as :

$$2Cu^+ \rightarrow Cu^{2+} + Cu^0$$

or $\underset{R.A.}{Cu^+} + \underset{O.A.}{Cu^+} \rightarrow \underset{O.A.}{Cu^{2+}} + \underset{R.A.}{Cu^0}$

This is a redox reaction in which Cu^+ ion acts both as R.A. and O.A. Thus this reaction is possible only when $(E^o_{red})_{R.A.} < (E^o_{red})_{O.A.}$ or $E^o_{Cu^{2+}/Cu^+} < E^o_{Cu^+/Cu^0}$. Now since $E^o_{Cu^{2+}/Cu^+} = +0.153$ V and $E^o_{Cu^+/Cu^0} = +0.52$ V, $E^o_{Cu^{2+}/Cu^+} < E^o_{Cu^+/Cu^0}$ and hence the above redox reaction occurs spontaneously.

Example 18.23. *Using the standard potential values, predict whether the reaction between the following can occur (a) Fe^{3+} (aq) and I^- (aq) (b) Ag^+ (aq) and Cu (s) (c) Fe^{3+} (aq) and Br^- (aq) (d) Ag (s) and Fe^{3+} (aq), (e) Br (s) and Fe^{2+} (aq) (f) Fe and Fe^{3+} (g) Sn^{2+} and Fe^{3+}.*

Solution. (*a*) The possible reaction that may take place between Fe^{3+} (*aq*) and I^- (*aq*) ions may be written as :

$$2Fe^{3+}(aq) + 2I^-(aq) \rightarrow 2Fe^{2+}(aq) + I_2(s) \qquad ...(i)$$

This reaction shows that since I^- ions are oxidised to I_2 molecules and Fe^{3+} ions are reduced to Fe^{2+} ions, the value of e.m.f. of the cell (E^o_{cell} value) in which the above redox reaction takes place is given by :

$$E^o_{cell} = E^o_{I^-/I_2} + E^o_{Fe^{3+}/Fe^{2+}} = -0.53 + 0.77 = \mathbf{+0.24}$$

Since the value of E^o_{cell} is positive, reaction (*i*) will proceed as shown, *i.e.*, the reaction between Fe^{3+} (*aq*) and I^- (*aq*) ions is feasible (spontaneous).

(*b*) The possible reaction between Ag^+ (*aq*) and Cu (*s*) can proceed as :

$$2Ag^+(aq) + Cu(s) \rightarrow 2Ag(s) + Cu^{2+}(aq)$$

which shows that Cu is being oxidised to Cu^{2+} ions and Ag^+ is being reduced to Ag. Thus:

$$E^o_{cell} = E^o_{Cu/Cu^{2+}} + E^o_{Ag^+/Ag} = -0.337 + 0.799 = \mathbf{+0.462\ V}$$

Since E^o_{cell} is positive, the reaction between Ag^+ (*aq*) and Cu (*s*) is possible as shown by the above equation.

(*c*) The possible reaction between Fe^{3+} (*aq*) and Br^- (*aq*) ions can be represented by the equation

$$2Fe^{3+}(aq) + 2Br^-(aq) \rightarrow 2Fe^{2+}(aq) + Br_2(l) \qquad ...(i)$$

This reaction clearly indicates that Br^- ions are oxidised to Br_2 molecules and Fe^{3+} ions are reduced to Fe^{2+} ions. Thus

$$E^o_{cell} = E^o_{Br^-/Br_2} + E^o_{Fe^{3+}/Fe^{2+}} = -1.076 + 0.771 = \mathbf{-\ 0.305\ V}$$

Since the value of E^o_{cell} is negative, the reaction between Fe^{3+} (*aq*) and Br^- (*aq*) ions will not proceed in the forward direction as shown by equation (*i*). However, this reaction will proceed in the backward direction as shown below by equation (*ii*), since in this case the value of E^o_{cell} comes out to be a positive quantity.

$$2Fe^{2+}(aq) + Br_2(l) \rightarrow 2Fe^{3+}(aq) + 2Br(aq) \qquad ...(ii)$$

$$E^o_{cell} = E^o_{Fe^{2+}/Fe^{3+}} + E^o_{Br_2/Br^-} = -0.771 + 1.1076 = 0.305\ V$$

(*d*) The possible reaction is :

$$Ag(s) + Fe^{3+}(aq) \rightarrow Ag^+(aq) + Fe^{2+}(aq)$$

This reaction shows that Ag is being oxidised to Ag^+ and Fe^{3+} is reduced to Fe^{2+}.

Thus :

$$E^o_{cell} = E^o_{Ag/Ag^+} + E^o_{Fe^{3+}/Fe^{2+}} = -0.799 + 0.771 = -0.028$$

Since the value of E^o_{cell} is negative, the reaction between Ag (*s*) and Fe^{3+} (*aq*) is not possible. However, the reverse reaction is feasible, since in this case E^o_{cell} is positive, as shown below.

$$Ag^+(aq) + Fe^{2+}(aq) \rightarrow Ag(s) + Fe^{3+}(aq)$$

$\therefore \quad E^o_{cell} = E^o_{Fe^{2+}/Fe^{3+}} + E^o_{Ag^+/Ag} = -0.771 + 0.799 = +0.028$

(*e*) The possible reaction is :

$$Br_2\ (aq) + Fe^{2+}\ (aq) \rightarrow Fe^{3+}\ (aq) + 2Br^-\ (aq)$$

Fe^{2+} ions are being oxidised to Fe^{3+} ions and Br_2 molecule is reduced to Br^- ion.

Thus :

$$E^o_{cell} = E^o_{Fe^{2+}/Fe^{3+}} + E^o_{Br_2/Br^-} = -0.771 + 1.076 = \mathbf{+0.305}$$

Since the value of E^o_{cell} is positive, the reaction shown above is feasible.

(*f*) The possible reaction that may take place between Fe and Fe^{3+} is that Fe reduces Fe^{3+} to Fe^{2+} which can be shown as :

$$Fe + 2Fe^{3+} \rightarrow 3Fe^{2+}$$

The above reaction shows that Fe is oxidised to Fe^{2+} (oxidation reaction) and Fe^{3+} is reduced to Fe^{2+} (reduction reaction). Thus the oxidation and reduction half-reactions the addition of which gives the above redox reaction can be written as :

Oxidation half-reaction : $Fe \rightarrow Fe^{2+} + 2e^-$

Reduction half-reaction : $[Fe^{3+} + e^- \rightarrow Fe^{2+}] \times 2,\ E^o_{Fe^{3+}/Fe^{2+}} = 0.771\ V$

Cell reaction : $Fe + 2Fe^{3+} \rightarrow 3Fe^{2+},\ E^o_{cell} = 0.440 + 0.771 = \mathbf{1.211\ V}$

Since the value of E^o_{cell} is positive, the above reaction will occur, *i.e.,* Fe will reduce Fe^{3+} to Fe^{2+}.

(*g*) The possible reaction that may take place between Sn^{2+} and Fe^{3+} is that Sn^{2+} reduces Fe^{3+} to Fe^{2+} and is itself oxidised to Sn^{4+} which can be shown as :

$$Sn^{2+} + 2Fe^{3+} \rightarrow 2Fe^{2+} + Sn^{4+}$$

The oxidation and reduction half-reactions into which the above redox reaction can be broken are :

Oxidation half-reaction : $Sn^{2+} \rightarrow Sn^{4+} + 2e^-$

Reduction half-reaction : $[Fe^{3+} + e^- \rightarrow Fe^{2+}] \times 2$

Cell reaction : $Sn^{2+} + 2Fe^{3+} \rightarrow 2Fe^{2+} + Sn^{4+},\ E^o_{cell} = -0.15 + 0.771 = \mathbf{+0.621\ V}$

Since the value of E^o_{cell} is positive, the redox reaction given above will proceed as shown, *i.e.,* Sn^{2+} will reduce Fe^{3+} to Fe^{2+}.

To fix the position of elements in the electrochemical series on the basis of the informations furnished by given redox reactions

The method used in fixing the position of elements in the electrochemical series is based on the following facts :

(*i*) The metals which can displace hydrogen from an acid lie above hydrogen in the electrochemical series (*i.e.,* such metals have negative values for their standard reduction electrode potential) while the metals which cannot displace hydrogen from an acid lie below hydrogen in the electrochemical series (*i.e.,* such metals have positive values for their standard reduction electrode potential).

(*ii*) In the electrochemical series, a metal which can displace another metal from its salt solution lies above the metal to be displaced while a metal which cannot displace another metal from its salt solution lies below the metal to be displaced.

Examples 18.24. *Study, the following redox reactions and arrange A, B, H, D and C in the form of electrochemical series:*

(*i*) $2HCl + 2A \rightarrow 2ACl + H_2$ (*ii*) $2HCl + B \rightarrow BCl_2 + H_2$

(*iii*) $BCl_2 + 2A \rightarrow 2ACl + B$ (*iv*) $CCl_2 + D \rightarrow DCl_2 + C$

(*v*) $D + HCl \rightarrow$ *No reaction takes place.*

Solution (*i*) Since A metal can displace H_2 gas from HCl, A should be placed above hydrogen in the electrochemical series, *i.e.*,

$$A \rightarrow H \qquad ...(a)$$

(*ii*) Since B metal can also displace hydrogen from HCl, B should be placed above hydrogen in the series, *i.e.*,

$$B \rightarrow H \qquad ...(b)$$

(*iii*) Reaction (*iii*) shows that since A metal can displace B metal from BCl_2, A metal should lie above B in the electrochemical series, *i.e.*,

$$A \rightarrow B \qquad ...(c)$$

(*iv*) Since D metal can displace C metal from CCl_2, D should be placed above C, *i.e.*,

$$D \rightarrow C \qquad ...(d)$$

(*v*) Since D cannot displace H_2 gas from HCl, D should be placed below hydrogen, *i.e.*,

$$H \rightarrow D \qquad ...(e)$$

The relations (*a*), (*b*) and (*c*) show that the position of A, B and H should be as :

$$A \rightarrow B \rightarrow H \qquad ...(f)$$

Relations (*d*), (*e*) and (*f*) indicate that :

$$A \rightarrow B \rightarrow H \rightarrow D \rightarrow C$$

Thus the position of A, B, H, D and C in the electrochemical series is A, B, H, D, C. In this the values of standard reduction electrode potentials increase from A → B . . . → C.

Example 18.25. *Study the following reactions and fix the position of H, A and B metals in the electrochemical series.*

(*i*) $A + H_2SO_4 \rightarrow ASO_4 + H_2$ (*ii*) $B + H_2SO_4 \rightarrow$ *No reaction.*

Solution. Reaction (*i*) shows that metal A can displace hydrogen from H_2SO_4 and hence A has negative value for its standard reduction electrode potential. In other words we can say that A metal lies above hydrogen in the electrochemical series. Reaction (*ii*) indicates that B metal is not able to displace hydrogen from H_2SO_4 and hence has positive value for its standard reduction electrode potential, *i.e.*, B metal lies below hydrogen in the electrochemical series. Thus the position of A, B and H in the electrochemical series can be shown as :

A ... E^o_{red} value – negative (–)

H ... E^o_{red} value – zero

B ... E^o_{red} value – positive (+)

Uses of standard e.m.f. value of a given galvanic cell : Uses of E^o_{cell} value

The value of standard e.m.f. of a given galvanic cell has been used to make a number of predictions some of which are discussed below :

1. To predict whether a given galvanic cell will work or not. If we are given a galvanic cell and its standard e.m.f. value (E^o_{cell} value), we can predict with the help of this value whether the given cell will operate or not. If the value of E^o_{cell} is positive, the given cell will operate. If the value of E^o_{cell} is negative, the cell will not work. However, the cell obtained by interchanging the position of the electrodes will work, since in that case the value of E^o_{cell} comes out to be a positive quantity.

Now we know that E^o_{cell} is given by equations (*i*) and (*ii*) given below :

$$E^o_{cell} = (E^o_{red})_{cathode} - (E^o_{red})_{anode} \qquad ...(i)$$

$$E^o_{cell} = (E^o_{oxi})_{anode} - (E^o_{oxi})_{cathode} \qquad ...(ii)$$

The above equations show that E^o_{cell} will be positive when

$(E^o_{red})_{cathode} > (E^o_{red})_{anode}$ [From equation (*i*)]

OR

$(E^o_{oxi})_{anode} > (E^o_{oxi})_{cathode}$ [From equation (*ii*)]

The above discussion shows that in order to make a galvanic cell to work properly, any of the following conditions must be fulfilled.

(*i*) E^o_{cell} should have a positive value

(*ii*) $(E^o_{red})_{cathode}$ > $(E^o_{red})_{anode}$

(*iii*) $(E^o_{red})_{anode}$ > $(E^o_{oxi})_{cathode}$

Example 18.26. *Predict whether the cell, Cu, Cu^{2+} || Zn^{2+}, Zn will operate or not ? Given that $E^o_{Cu/Cu^{2+}}$ = – 0.34 and $E^o_{Zn/Zn^{2+}}$ = + 0.76 V.*

Solution. E^o_{cell} is given by : $E^o_{cell} = (E^o_{red})_{cathode} - (E^o_{red})_{anode}$

$$= E^o_{Zn^{2+}/Zn} - E^o_{Cu^{2+}/Cu}$$

$$= -0.76 - (+0.34) = \mathbf{-1.10\ V}$$

Since E^o_{cell} is negative, the cell will not operate. If the position of the electrodes is interchanged, the new cell obtained can be represented as :

Zn, Zn^{2+}	||	Cu^{2+}, Cu
Anode		Cathode

$$E^o_{cell} = E^o_{Cu^{2+}/Cu} - E^o_{Zn^{2+}/Zn}$$

$$= +0.34 - (-0.76\ V) = \mathbf{+1.10\ V}$$

Since E^o_{cell} is positive, the above cell will work.

2. To predict whether the given cell representation is correct or wrong. If we are given the representation of a galvanic cell and its standard e.m.f. value (E^o_{cell} value), we can predict with the help of E^o_{cell} value whether the given cell representation is correct or wrong. The given cell representation would be correct, if any of the following conditions is fulfilled :

(*i*) Value of E^o_{cell} should be a positive quantity

(*ii*) $(E^o_{red})_{cathode} > (E^o_{red})_{anode}$

(*iii*) $(E^o_{oxi})_{anode} > (E^o_{oxi})_{cathode}$

Example 18.27. *Predict whether the following cell representation is correct or wrong. If wrong, write the correct representation. Given $E^o_{Cu^{2+}/Cu}$ = + 0.337 V and $E^o_{Zn^{2+}/Zn}$ = – 0.763 V.*

Cu (s) | Cu^{2+} (aq) || Zn^{2+} (aq) | Zn (s)

Solution. $E^o_{cell} = (E^o_{red})_{cathode} - (E^o_{red})_{anode} = -0.763 - 0.337 = \mathbf{-1.100\ V}$

Since the value of E^o_{cell} is negative, the cell representation is wrong, *i.e.*, this cell will not work. The correct cell representation is that in which the positions of the electrodes are interchanged, *i.e.*, anode of the wrong representation constitutes cathode and the cathode acts as anode. Thus the correct representation is :

Zn (*s*) | Zn^{2+} (*aq*)	||	Cu^{2+} (*aq*) | Cu (*s*)
Anode		Cathode

It may be seen that the value of E^o_{cell} of the above cell is a positive quantity as shown below :

$$E^o_{cell} = E^o_{Cu^{2+}/Cu} - E^o_{Zn^{2+}/Zn} = 0.337 - (-0.763) = \mathbf{+1.10\ V}$$

Positive value (= + 1.10 V) shows that the cell representation is correct.

3. To calculate the value of e.m.f. of a given galvanic cell (cell potential) at non-standard conditions by Nernst equation. e.m.f. of a given cell at standard conditions is represented as E^o_{cell} while e.m.f. of a given cell at non-standard conditions is represented as E_{cell}. With the help of the value of E^o_{cell} we can also calculate the value of E_{cell} by using Nernst equation. The value of E_{cell} of galvanic cell in which the redox reaction namely.

$$a\,A + b\,B \rightleftharpoons c\,C + d\,D$$

takes place is given by Nernst equation *viz.*,

$$E_{cell} = E^{o}_{cell} - \frac{2.303\,RT}{nF} \log \frac{\text{Product of the concentration of the products}}{\text{Product of the concentration of the reactants}}$$

or

$$E_{cell} = E^{o}_{cell} - \frac{2.303\,RT}{nF} \log \frac{[C]^c\ [D]^d}{[A]^a\ [B]^b}$$

4. To predict whether a given redox reaction will occur spontaneously or not. See "Applications of Electrochemical series".

To calculate the electrode potential of an electrode potential of an electrode at non-standard conditions : To write Nernst equation for an electrode at non-standard conditions.

Let us consider a metal-metal ion electrode which consists of metal (M) in contact with a solution containing M^{n+} cations. This electrode can be represented as M^{n+} (*aq*) / M (*s*) and the reduction reaction taking place at this electrode can be written as :

$$\underset{\text{Reactants (Oxidised form)}}{M^{n+}(aq) + ne^-} \rightarrow \underset{\text{Products (Reduced form)}}{M(s)} \ (\textit{Reduction reaction}) \qquad ...(A)$$

Here n is the number of electrons taken up by the electrode. If the concentration of M^{n+} (*aq*) ion is equal to 1 mol dm^{-3} (standard conditions) reduction electrode potential of the given electrode is called *standard reduction electrode potential* which is to represented as $E^{o}_{M^{n+}/M}$. If the concentration of M^{n+} (*aq*) ions is *not* equal to 1 mol dm^{-3} (*non-standard conditions*), the reduction electrode potential of the given electrode is represented as $E_{M^{n+}/M}$. Under the non-standard conditions, the value of $E_{M^{n+}/M}$ is calculated by Nernst equation which can be derived as follows :

If – ΔG is the decrease in free energy (negative sign denotes decrease in energy) accompanying the above reduction electrode reaction (A), ΔG° is the decrease in *standard* free energy of the reaction (Free energy is called standard free energy, when all the reactants and products appearing in the reaction are in their standard conditions of unit activity) and Q is the reaction quotient of the activities of the products and reactants at any stage of the above reaction, then

$$-\Delta G = -\Delta G^{\circ} - RT \ln Q \qquad ...(i)$$

Now we know that if $(E_{red})_{elec}$ and $(E^{o}_{red})_{elec}$ are the electrode reduction potential at non-standard conditions and electrode reduction potential at standard conditions (standard reduction electrode potential) of the given electrode M^{n+} (*aq*) / M (*s*), then :

$$-\Delta G = nF(E_{red})_{elec} \qquad ...(a)$$

and

$$-\Delta G^{\circ} = nF(E^{o}_{red})_{elec} \qquad ...(b)$$

Here F is the Faraday of electricity. Its value is constant and is equal to 96500 coulombs mol^{-1} (C mol^{-1}). With the help of relations (*a*) and (*b*) equation (*i*) becomes :

$$nF(E_{red})_{elec} = nF(E^{o}_{red})_{elec} - RT \ln Q$$

$$(E_{red})_{elec} = (E^{o}_{red})_{elec} - \frac{RT}{nF} \ln Q \qquad ...(ii)$$

Now we know that :

$$Q = \frac{\text{Activity of the products}}{\text{Activity of the reactants}} \qquad ...(c)$$

If the solution containing M^{n+} (*aq*) ions is dilute, then the activity of the products and reactants can be supposed roughly to be equal to their concentrations and hence relation (*c*) can be written as :

$$Q = \frac{\text{Concentration of the products}}{\text{Concentration of the reactants}} = \frac{[\text{Products}]}{[\text{Reactants}]} \qquad ...(d)$$

Now since the product in the electrode reaction is the reduced form and the reactant is the oxidised form, relation (*d*) can be written as :

$$Q = \frac{\text{Concentration of the reduced form}}{\text{Concentration of the oxidised form}} = \frac{[\text{Reduced form}]}{[\text{Oxidised form}]} \qquad ...(e)$$

With the help of relations (*c*), (*d*) and (*e*) equation (*ii*) can be written as :

$$(E_{red})_{elec} = (E^o_{red})_{elec} - \frac{RT}{nF}\ln\frac{\text{Activity of the products}}{\text{Activity of the reactants}}$$

$$= (E^o_{red})_{elec} - \frac{RT}{nF}\ln\frac{\text{Concentration of the products}}{\text{Concentrationof the reactants}}$$

$$= (E^o_{red})_{elec} - \frac{RT}{nF}\ln\frac{[\text{Products}]}{[\text{Reactants}]}$$

$$= (E^o_{red})_{elec} - \frac{RT}{nF}\ln\frac{[\text{Reduced Form}]}{[\text{Oxidised Form}]}$$

$$= (E^o_{red})_{elec} + \frac{RT}{nF}\ln\frac{[\text{Oxidised Form}]}{[\text{Reduced Form}]}$$

$$= (E^o_{red})_{elec} + \frac{2.303\ RT}{nF}\log\frac{[\text{Oxidised Form}]}{[\text{Reduced Form}]} \quad (\because \ln x = 2.303 \log x)$$

$$\therefore \quad E_{M^{n+}/M} = E^o_{M^{n+}/M} + \frac{2.303\ RT}{nF}\log\frac{[M^{n+}\ (aq)]}{[M\ (s)]}$$

All the above equations are different forms of Nernst equation. This equation gives us the value of the potential of the electrode, M^{n+} (*aq*) / M (*s*) at non-standard conditions at a temperature equal to *T* Kelvin or (*T* – 273)°C. While using the Nernst equation, it should be kept in mind that the electrode reaction should be written as a *reduction reaction*.

To write Nernst equation at 25°C or (25 + 273) = 298 K : To calculate the value of $(E_{red})_{elec}$ at 298 K

We have seen that Nernst equation for an electrode reduction reaction,

$$M^{n+}\ (aq) + ne^- \rightarrow M\ (s)$$

at a temperature equal to *T* Kelvin is given by :

$$(E_{red})_{elec} = (E^o_{red})_{elec} - \frac{2.303\ RT}{nF}\log\frac{[\text{Products}]}{[\text{Reactants}]} \quad \text{... } (i)$$

Obviously Nernst equation at 25°C or (25 + 273) = 298 K can be obtained by putting R = 8.31 JK^{-1} mol^{-1} = 8.31 C.V. K^{-1} (since J = C.V.), T = 298 K and F = 96,500 C mol^{-1} in equation (*i*). Thus Nernst equation at 298 K can be written as :

$$(E_{red})_{elec} = (E^o_{red})_{elec} - \frac{2.303 \times (8.31\ \text{C.V. K}^{-1}\ \text{mol}^{-1}) \times 298\ \text{K}}{n \times 96500\ \text{C mol}^{-1}}\log\frac{[\text{Products}]}{[\text{Reactant}]}$$

$$= (E^o_{red})_{elec} - \frac{0.05912}{n}\log\frac{[\text{Products}]}{[\text{Reactants}]}$$

$$\text{or} \quad E_{M^{n+}/M} = E^o_{M^{n+}/M} - \frac{0.05912}{n}\log\frac{[M\ (s)]}{[M^{n+}\ (aq)]}$$

The above equation gives the value of $E^o_{M^{n+}/M}$ at 298 K.

To write Nernst equation at 25°C for a hydrogen electrode at non-standard conditions

Reduction reaction taking place at a hydrogen electrode is written as :

$$\underset{\text{Reactants (Oxidised form)}}{2H^+\ (aq) + 2e^-} \rightarrow \underset{\text{Products (Reduced form)}}{H_2\ (g)}$$

The number of electrons involved in the reaction is equal to 2 (*i.e.*, *n* = 2). Obviously at 25°C, the value of reduction potential for the above reduction reaction is given by :

$$E_{2H^+/H_2} = E^o_{2H^+/H_2} - \frac{0.05912}{n}\log\frac{[\text{Products}]}{[\text{Reactants}]}$$

$$= 0.0 - \frac{0.05912}{2}\log\frac{p_{H_2}}{[H^+]^2}$$

$$= +\frac{0.0591 \times 2}{2}\log[H^+]$$

or $$E_{2H^+/H_2} = 0.0591 \log[H^+] \quad ...(i)$$

Now we know that :

$$pH = -\log[H^+]$$

or $$\log[H^+] = -pH \quad ...(i)$$

With the help of equationn (*ii*), equation (*i*) becomes

$$E_{2H^+/H_2} = 0.0591(-pH)$$

or $$E_{2H^+/H_2} = -0.0591 \text{ pH (at 25°C)} \quad ...(iii)$$

To calculate the e.m.f. of a galvanic cell (cell potential) at non-standard conditions: To write Nernst equation for a cell reaction at non-standard conditions.

As an example let us consider the following cell :

$$\underset{\text{Anode}}{Zn(s) / Zn^{2+}(aq)} \quad || \quad \underset{\text{Cathode}}{Cu^{2+}(aq) / Cu(s)}$$

Obviously the given cell is composed of two electrodes *viz.*, Zn (*s*) / Zn^{2+} (*aq*) (anode) and Cu^{2+} (*aq*) / Cu (*s*) (cathode). The cell reaction can be obtained by adding oxidation reaction at anode and reduction reaction at cathode.

Oxidation reaction at anode : $Zn(s) \rightarrow Zn^{2+}(aq) + 2e^-$

Reduction reaction at cathode : $Cu^{2+}(aq) + 2e^- \rightarrow Cu(s)$

Cell reaction : $\underbrace{Zn(s) + Cu^{2+}(aq)}_{\text{Reactants}} \rightarrow \underbrace{Zn^{2+}(aq) + Cu(s)}_{\text{Products}}$

The value of cell potential (E_{cell}) of the given cell is given by :

$$E_{cell} = (E_{red})_{cathode} - (E_{red})_{anode}$$

$$E_{cell} = E_{Cu^{2+}/Cu} - E_{Zn^{2+}/Zn} \quad ...(i)$$

The reduction reactions taking place at the electrodes are :

$$Zn^{2+}(aq) + 2e^- \rightarrow Zn(s)\ (n = 2) \quad \textit{(At anode)} \quad ...(a)$$

$$Cu^{2+}(aq) + 2e^- \rightarrow Cu(s)\ (n = 2) \quad \textit{(At cathode)} \quad ...(b)$$

Thus the values of $E_{Zn^{2+}/Zn}$ and $E_{Cu^{2+}/Cu}$ can be written as follows, by using Nernst equation for an electrode.

$$E_{Zn^{2+}/Zn} = E^o_{Zn^{2+}/Zn} - \frac{2.303\,RT}{2F}\log\frac{[Zn(s)]}{[Zn^{2+}(aq)]} \quad \text{[For reduction reaction (a)]}$$

$$E_{Cu^{2+}/Cu} = E^o_{Cu^{2+}/Cu} - \frac{2.303\,RT}{2F}\log\frac{[Cu(s)]}{[Cu^{2+}(aq)]} \quad \text{[For reduction reaction (b)]}$$

Put the values of $E_{Zn^{2+}/Zn}$ and $E_{Cu^{2+}/Cu}$ in equation (*i*) to get the value of E_{cell}, then, equation (*ii*) becomes

$$E_{cell} = \left[E^o_{Cu^{2+}/Cu} - \frac{2.303\,RT}{2F}\log\frac{[Cu(s)]}{[Cu^{2+}(aq)]}\right] - \left[E^o_{Zn^{2+}/Zn} - \frac{2.303\,RT}{2F}\log\frac{[Zn(s)]}{[Zn^{2+}(aq)]}\right]$$

$$= \left(E^o_{Cu^{2+}/Cu} - E^o_{Zn^{2+}/Zn}\right) - \frac{2.303\,RT}{2F}\log\frac{[Zn^{2+}(aq)]\,[Cu(s)]}{[Zn(s)]\,[Cu^{2+}(aq)]}$$

Now since $\left(E^o_{Cu^{2+}/Cu} - E^o_{Zn^{2+}/Zn}\right) = E^o_{cell}$, the above equation becomes :

$$E_{cell} = E^o_{cell} - \frac{2.303\,RT}{2F}\log\frac{[Zn^{2+}(aq)]\,[Cu(s)]}{[Zn(s)]\,[Cu^{2+}(aq)]}$$

$$E_{cell} = E^o_{cell} - \frac{2.303\,RT}{2F}\log\frac{\text{Product of the concentration of the products in the cell reaction}}{\text{Product of the concentration of the reactants in the cell reaction}} \quad ...(ii)$$

At 298 K, equation (*ii*) becomes :

$$E_{cell} = E^o_{cell} - \frac{2.303 \times (8.31 \text{ C.V.J. K}^{-1} \text{ mol}^{-1}) \times 298 \text{ K}}{2 \times 96500 \text{ C mol}^{-1}} \log \frac{\text{Product of the concentration of the products}}{\text{Product of the concentration of the reactants}}$$

$$E_{cell} = E^o_{cell} - \frac{0.05912}{2} \log \frac{\text{Product of the concentration of the products}}{\text{Product of the concentration of the reactants}} \quad ...(iii)$$

To write Nernst equation for a general cell reaction (*Redox reaction*)

Suppose the reaction taking place in a galvanic cell is represented by (*i*) in which the number of electrons involved in *n*.

$$a\text{ A} + b\text{ B} \rightleftharpoons c\text{ C} + d\text{ D} \quad ...(i)$$

Nernst equation for this cell reaction in written as :

$$E_{cell} = E^o_{cell} - \frac{2.303 \text{ RT}}{nF} \log \frac{\text{Product of the concentration of the products}}{\text{Product of the concentration of the reactants}}$$

$$E_{cell} = E^o_{cell} - \frac{2.303 \text{ RT}}{nF} \log \frac{[C]^c [D]^d}{[A]^a [B]^b} \quad ...(ii)$$

At 298 K, equation (*ii*) becomes :

$$E_{cell} = E^o_{cell} - \frac{0.05912}{n} \log \frac{[C]^c [D]^d}{[A]^a [B]^b} \quad ...(iii)$$

Example 18.28. *Choose the correct answer out of the following five answers. The reduction potential of a half-cell consisting of a platinum wire dipped into a solution of 0.01 M Sn^{2+} ions and 0.001 M Sn^{4+} ions at 25°C is : (A) $E^o_{oxid} + 0.059$ (B) $E^o_{oxid} - 0.059$ (C) E^o_{red} (D) $E^o_{red} - 0.059/2$ (E) $E^o_{red} + 0.059/2$.*

Solution. The reduction reaction taking place at the given electrode is represented as :

$$Sn^{4+} (0.001 \text{ M}) + 2e^- \rightarrow Sn^{2+} (0.01 \text{ M})$$

According to Nernst equation, the reduction electrode potential, E^o_{red} or $E^o_{Sn^{4+}/Sn^{2+}}$ at 25°C is given by :

$$E_{red} = E^o_{red} - \frac{0.059}{2} \log \frac{[Sn^{2+}]}{[Sn^{4+}]}$$

$$= E^o_{red} - \frac{0.059}{2} \log \frac{0.01}{0.001}$$

$$= E^o_{red} - \frac{0.059}{2} \log 10$$

$$\text{or } E_{red} = E^o_{red} - \frac{0.059}{2}$$

Thus answer (*D*) is the correct answer.

Example 18.29. *Calculate E^o_{cell} and E_{cell} for the cell which is represented as Sn / Sn^{2+} (1.0 M) || Pb^{2+} (0.001 M) / Pb. Given that $E^o_{Sn^{2+}/Sn} = -0.14$ V and $E^o_{Pb^{2+}/Pb} = -0.13$ V. What do you infer from the value of E_{cell} ?*

Solution. The given cell is :

$$\underset{\text{Anode}}{Sn / Sn^{2+}(1.0 \text{ M})} \quad || \quad \underset{\text{Cathode}}{Pb^{2+} (0.001 \text{ M}) / Pb}$$

$$E^o_{cell} = (E^o_{red})_{cathode} - (E^o_{red})_{anode}$$

$$= E^o_{Pb^{2+}/Pb} - E^o_{Sn^{2+}/Sn} = -0.13 - (-0.14) = 0.01 \text{ V.}$$

Oxidation half-reaction at anode : $Sn \rightarrow Sn^{2+} + 2e^-$
Reduction half-reaction at cathode : $Pb^{2+} + 2e^- \rightarrow Pb$

Cell reaction : $Sn + Pb^{2+} \rightarrow Sn^{2+} + Pb$

Since the number of electrons involved in the cell reaction is 2 (*i.e.*, $n = 2$), E_{cell} is given by Nernst equation *viz.*,

$$E_{cell} = E^o_{cell} - \frac{0.05912}{n} \log Q$$

$$= 0.01 - \frac{0.0592}{2} \log \frac{[Sn^{2+}]\ [Pb]}{[Sn]\ [Pb^{2+}]}$$

$$= 0.01 - 0.0295 \times \log \frac{(1.0) \times (1)}{(1) \times (0.001)}$$

$$= 0.01 - 0.0295 \times \log 10^3$$

$$= 0.01 - 0.0295 \times 3 = 0.01 - 0.0886$$

$$\therefore \quad E_{cell} = -0.078 \text{ V}$$

Since E_{cell} is a negative quantity, the given cell will not operate. However, it will operate, if the position of the electrodes is changed with each other *i.e.*, it will work, if it is represented as :

$$\underset{\text{Anode}}{Pb / Pb^{2+} (0.001 \text{ M})} \parallel \underset{\text{Cathode}}{Sn^{2+} (1.0 \text{ M}) / Sn}$$

In this case the cell reaction would be

$$Pb + Sn^{2+} \rightarrow Pb^{2+} + Sn$$

Example 18.30. *Write Nernst equation for the cell Ni (s) / Ni^{2+} (aq, 0.1 M) || Ag^+ (aq, 0.1 M) Ag (s) and also find its cell potential. Given $E^o_{Ag^+/Ag}$ = + 0.80 V and $E^o_{Ni^{2+}/Ni}$ = – 0.25 V.*

Solution. (*a*) The representation of the cell as given in the question shows that

Oxidation reaction at anode : $Ni (s) \rightarrow Ni^{2+} (aq) + 2e^-$
Reduction reaction at cathode : $Ag^+ (aq) + e^- \rightarrow Ag (s)] \times 2$

Cell reaction : $Ni (s) + 2Ag^+ (aq) \rightarrow Ni^{2+} (aq) + 2Ag (s)$

Since two electrons are involved in the cell reaction ($n = 2$), Nernst equation for this cell reaction can be written as :

$$E_{cell} = E^o_{cell} - \frac{0.0591}{2} \log \frac{\text{Product of the concentrations of the products}}{\text{Product of of concentrationsof the reactants}}$$

$$= E^o_{cell} - 0.0295 \log \frac{[Ni^{2+}]\ [Ag]^2}{[Ni]\ [Ag^+]^2}$$

$$= E^o_{cell} - 0.0295 \log \frac{[Ni^{2+}]}{[Ag^+]^2}$$

or

$$E_{cell} = E^o_{cell} - 0.0295 \log \frac{0.1}{(0.1)^2}$$

(*b*)

$$E^o_{cell} = (E^o_{red})_{cathode} - (E^o_{red})_{anode}$$

$$= E^o_{Ag^+/Ag} - E^o_{Ni^{2+}/Ni}$$

$$= 0.80 - (-0.25) = 1.05 \text{ V}$$

$$\therefore \quad E_{cell} = E^o_{cell} - 0.0295 \log \frac{0.1}{(0.1)^2}$$

$$= 1.05 - 0.0295 \times 1 = \mathbf{1.0205 \text{ V}}$$

Example 18.31. *A galvanic cell contains two hydrogen electrodes. The negative electrode is in contact with a solution of 10^{-6} M hydrogen ions. The e.m.f. of the cell is 0.118 volt at 25°C. Calculate the concentration of hydrogen ions at the positive electrode.* *(J.E.E. 1988)*

Solution. Let the concentration of H^+ ions in hydrogen electrode which acts as a positive electrode be c. When this hydrogen electrode is coupled with other hydrogen electrode which acts as a negative electrode and in which the concentration of H^+ ions is 10^{-6} M, the cell thus obtained is represented as :

$$\underset{\text{(Anode)}}{\underset{\text{Negative electrode}}{\text{Pt, } H_2\,(g) \mid H^+\,(aq,\, 10^{-6}\text{ M})}} \quad \| \quad \underset{\text{(Cathode)}}{\underset{\text{Positive electrode}}{H^+\,(aq,\, c)\,/\,H_2\,(g)\text{, Pt}}}$$

Obviously, $E^o_{cell} = (E^o_{red})_{cathode} - (E^o_{red})_{anode}$

$$= E^o_{2H^+/H_2} - E^o_{2H^+/H_2} = 0 - 0 = 0.0 \text{ V}$$

Oxidation reaction at anode : $H_2\,(g) \rightarrow 2H^+\,(aq,\, 10^{-6}) + 2e^-$

Reduction reaction at cathode : $2H^+\,(aq,\, c) + 2e^- \rightarrow H_2\,(g)$

Cell reaction : $H_2\,(g) + 2H^+\,(aq,\, c) \rightarrow 2H^+\,(aq,\, 10^{-6}) + H_2\,(g)$

Obviously $n = 2$

$$\therefore \quad E_{cell} = E^o_{cell} - \frac{0.0591}{2} \log \frac{[H^+]^2\,[H_2]}{[H_2]\,[H^+]^2}$$

$$= 0.0 - 0.0295 \log \frac{(10^{-6})^2 \times 1}{1 \times c^2}$$

$$\text{or} \quad 0.118 = -0.0295 \log \frac{10^{-12}}{c^2}$$

$$\frac{0.118}{0.295} = -\log \frac{10^{-12}}{c^2}$$

$$\text{or} \quad 4 = 12 + \log c^2 = \log_{10} + \log c^2 = 12 \times \log_{10} + \log c^2$$

$$\text{or} \quad -8 = \log c^2 \quad \text{or} \quad -8 = 2 \log c \quad \text{or} \quad -4 = \log c$$

$$\therefore \quad c = 10^{-4} \text{ M}$$

Thus the concentration of H^+ ions in the hydrogen electrode which constitutes the positive terminal of the cell is equal to **10^{-4} M.**

Example 18.32. *Select the correct answer. Tin metal will reduce Pb^{2+} ions, if the concentration of Pb^{2+} ions and Sn^{2+} ions is (a) 0.50 M and 0.01 M respectively (b) 0.01 M and 0.50 M respectively.*

Solution. The reduction of Pb^{2+} ions by Sn metal can be shown by the following equations :

(*i*) $Sn\,(s) + Pb^{2+}\,(0.50\text{ M}) \rightarrow Sn^{2+}\,(0.01\text{ M}) + Pb\,(s)$

(*ii*) $Sn\,(s) + Pb^{2+}\,(0.01\text{ M}) \rightarrow Sn^{2+}\,(0.50\text{ M}) + Pb\,(s)$

Obviously galvanic cell, in which the above reaction takes place, can be represented as :

$$\underset{\text{Anode}}{Sn\,(s)\,/\,Sn^{2+}\,(0.01\text{ M})} \quad \| \quad \underset{\text{Cathode}}{Pb^{2+}\,(0.50\text{ M})\,/\,Pb\,(s)} \quad \text{[For reaction (i)]}$$

$$\underset{\text{Anode}}{Sn\,(s)\,/\,Sn^{2+}\,(0.50\text{ M})} \quad \| \quad \underset{\text{Cathode}}{Pb^{2+}\,(0.01\text{ M})\,/\,Pb\,(s)} \quad \text{[For reaction (ii)]}$$

$$E^o_{cell} = (E^o_{red})_{cathode} - (E^o_{red})_{anode}$$

$$= E^o_{Pb^2/Pb} - E^o_{Sn^{2+}/Sn}$$

$$= -0.126 - (-0.136) = +0.010 \text{ V}$$

$$\therefore \quad \textit{For first cell :} \quad E_{cell} = E^o_{cell} - \frac{0.0592}{2} \log \frac{[Sn^{2+}]\,[Pb]}{[Pb^{2+}]\,[Sn]}$$

$$= 0.010 - 0.0295 \log \frac{0.01}{0.50} = \mathbf{+\,0.061\ V.}$$

Since E_{cell} is positive, Sn metal will reduce Pb^{2+} ions, having concentration equal to 0.50 M as shown in equation (*i*).

For second cell : $E_{cell} = E^o_{cell} - 0.0295 \log \frac{0.50}{0.01}$

$= 0.010 - 0.0295 \log 5 = -0.041$ V

Since the value of E_{cell} is negative, Sn will not reduce Pb^{2+} ions having concentration equal to 0.01 M, as shown in equation (*ii*). The above discussion shows that reaction (*i*) is feasible while reaction (*ii*) is not.

Example 18.33. *Consider the reaction, $2Ag^+ + Cd \rightarrow 2Ag + Cd^{2+}$. The standard electrode potentials for Ag^+ / Ag and Cd^{2+} / Cd couples are 0.80 V and 0.40 V respectively. (i) What is the standard potential for this reaction (ii) For the electrochemical cell in which the above reaction takes place which electrode is negative electrode (iii) Will the total e.m.f. of the given reaction be more negative or positive, if the concentration of Cd^{2+} ions is 0.1 M rather than 1.0 M.* *(Roorkee 82)*

Solution. (*i*) In the given reaction Cd is being oxidised to Cd^{2+} ions and Ag^+ ions are being reduced to Ag metal. Thus the galvanic cell in which the given reaction takes place can be represented as :

$$\underset{\text{Anode (negative electrode)}}{Cd\,(s) / Cd^{2+}\,(aq)} \quad || \quad \underset{\text{Cathode (positive electrode)}}{Ag^+\,(aq) / Ag\,(s)}$$

Thus the standard potential of the given reaction would be the standard e.m.f. of the above cell (E^o_{cell}) which is given by :

$$E^o_{cell} = (E^o_{red})_{cathode} - (E^o_{red})_{anode}$$

$$= E^o_{Ag^+/Ag} - E^o_{Cd^{2+}/Cd} = 0.80 - 0.40 = \mathbf{0.40\ V}$$

(*ii*) Cd (*s*) / Cd^{2+} (*aq*) electrode constitutes the negative electrode of the cell, since it is this electrode at which oxidation takes place.

(*iii*) The cell in which the concentration of Cd^{2+} ions is 0.1 M and not 1.0 M can be shown as :

$$Cd\,(s) / Cd^{2+}\,(aq,\ 0.1\ M)\ ||\ Ag^+\,(aq,\ 1.0\ M) / Ag\,(s)$$

For this cell, $E_{cell} = E^o_{cell} - \frac{0.05912}{2} \log \frac{[Cd^{2+}]}{[Ag^+]^2}$

$= 0.40 - 0.0295 \times \log \frac{0.1}{1.0}$

$= 0.40 - 0.0295 \log 10^{-1}$

$= 0.40 + 0.0295 \times 1 = \mathbf{+\ 0.4295\ V}$

Thus we see that the total e.m.f. of cell becomes more positive.

Example 18.34. *Write Nernst equation and calculate e.m.f. of the following cells at 298 K.*

(i) Al (s) / Al^{3+} (aq) || Ni^{2+} (aq) / Ni (s)

(ii) Fe (s) / Fe^{2+} (aq, 0.001 M) || H^+ (aq, 1 M) / H_2 (g, 1 atm), Pt

(iii) Mg (s) / Mg^{2+} (aq, 0.001 M) || Cu^{2+} (aq, 0.0001 M) / Cu (s).

Solution. (*i*) The given cell is composed of two electrodes which are Al (*s*) / Al^{3+} (*aq*) (anode) and Ni^{2+} (*aq*) / Ni (*s*) (cathode). Reactions taking place at the two electrodes are :

Oxidation reaction at anode : $Al\,(s) \rightarrow Al^{3+}\,(aq) + 3e^-] \times 2$

Reduction reaction at cathode : $Ni^{2+}\,(aq) + 2e^- \rightarrow Ni\,(s)\ ] \times 3$

Cell reaction : $2Al\,(s) + 3Ni^{2+}\,(aq) \rightarrow 2Al^{3+}\,(aq) + 2Ni\,(s)$

Here $n = 6$

Nernst equation for this cell reaction at 298 K is written as :

$$E_{cell} = E^o_{cell} - \frac{0.05912}{6} \log \frac{[Al^{3+}\,(aq)]^2\ [Ni\,(s)]^2}{[Al\,(s)]^2\ [Ni^{2+}\,(aq)]^3}$$

Now since Ni (*s*) and Al (*s*) both are solids, their concentrations are equal to unity [*i.e.*, [Ni (*s*)] = 1 and [Al (*s*)] =1] and hence the above equation becomes.

$$E_{cell} = E^o_{cell} - \frac{0.05912}{6} \log \frac{[Al^{3+}\,(aq)]^2}{[Ni^{2+}(aq)]^3}$$

(*ii*) Oxidation reaction anode : $Fe\ (s) \rightarrow Fe^{2+}\ (aq) + 2e^-$
Reduction reaction at cathode : $2H^+\ (aq) + 2e^- \rightarrow H_2\ (g)$

Cell reaction : $Fe\ (s) + 2H^+\ (aq) \rightarrow Fe^{2+}\ (aq) + H_2\ (g)$

Here $n = 2$

Nernst equation for this cell reaction at 298 K is :

$$E_{cell} = E^o_{cell} - \frac{0.0591}{2} \log \frac{[Fe^{2+}\ (aq)]\ [H_2\ (g)]}{[Fe\ (s)]\ [H^+\ (aq)]^2}$$

$$= (E^o_{2H^+/H_2} - E^o_{Fe^{2+}/Fe}) - \frac{0.0591}{2} \log \frac{[Fe^{2+}\ (aq)]\ 1}{1\ [H^+\ (aq)]^2}$$

$$= [0.0 - (-0.44)] - \frac{0.0591}{2} \log \frac{0.001}{1}$$

$$= 0.44 - \frac{0.0591}{2} \log 10^3 = \mathbf{0.528\ V}$$

(*iii*) Oxidation reaction at anode : $Mg\ (s) \rightarrow Mg^{2+}\ (aq) + 2e^-$
Reduction reaction at cathode : $Cu^{2+}\ (aq) + 2e^- \rightarrow Cu\ (s)$

Cell reaction : $Mg\ (s) + Cu^{2+}\ (aq) \rightarrow Mg^{2+}\ (aq) + Cu\ (s)$

Here $n = 2$

Nernst equation for this cell reaction at 298 K is given by :

$$E_{cell} = E^o_{cell} - \frac{0.05112}{2} \log \frac{[Mg^{2+}\ (aq)]\ [Cu\ (s)]}{[Mg\ (s)]\ [Cu^{2+}\ (aq)]}$$

$$= [E^o_{Cu^{2+}/Cu} - E^o_{Mg^{2+}/Mg}] - \frac{0.0591}{2} \log \frac{0.001 \times 1}{1 \times 0.0001}$$

$$= 0.34 - (-2.37) - \frac{0.0591}{2} \log 10$$

$$= 2.71 - 0.0295 = \mathbf{2.6805\ V}$$

Example 18.35. *A galvanic cell consists of metallic Zn-plate immersed in 0.1 M Zn $(NO_3)_2$ solution and metallic plate of lead in 0.02 M $Pb(NO_3)_2$ solution. Calculate e.m.f. of the cell. Write the chemical equations for the electrode reactions and represent the cell. Given that $E^o_{Zn^{2+}/Zn}$ = – 0.76 V and $E^o_{Pb^{2+}/Pb}$ = – 0.13 V.*

(A.I.S.E. 1985)

Solution. Obviously the two electrodes of which the cell is composed are $Zn\ (s)\ /\ Zn^{2+}\ (aq)$ and $Pb\ (s)\ /\ Pb^{2+}\ (aq)$. Since $E^o_{Pb^{2+}/Pb}$ (= – 0.13 V) is greater than $E^o_{Zn^{2+}/Zn}$ (= – 0.76 V) $Pb^{2+}\ (aq)/Pb\ (s)$ constitutes the cathode of the cell and $Zn\ (s)\ /\ Zn^{2+}\ (aq)$ forms the anode of the cell. Thus the cell can be represented as :

$$\underset{\text{Anode}}{Zn\ (s)\ /\ Zn^{2+}\ (aq,\ 0.1\ M)} \ ||\ \underset{\text{Cathode}}{Pb^{2+}\ (aq,\ 0.02\ M)\ /\ Pb\ (s)}$$

$$E^o_{Zn^{2+}/Zn} = -0.76\ V \qquad E^o_{Pb^{2+}/Pb} = -0.13\ V$$

Reactions taking place the two electrodes can be written as :

Oxidation reaction at anode : $Zn\ (s) \rightarrow Zn^{2+}\ (aq) + 2e^-$
Reduction reaction at cathode : $Pb^{2+}\ (aq) + 2e^- \rightarrow Pb\ (s)$

Cell reaction : $Zn\ (s) + Pb^{2+}\ (aq) \rightarrow Zn^{2+}\ (aq) + Pb\ (s)$

Here $n = 2$

Nernst equation for this cell reaction at 298 K is given by :

$$E_{cell} = E^o_{cell} - \frac{0.05912}{2} \log \frac{[Zn^{2+}\ (aq)]\ [Pb\ (s)]}{Zn\ (s)\ [Pb^{2+}\ (aq)]}$$

$$= (E^o_{Pb^{2+}/Pb} - E^o_{Zn^{2+}/Zn}) - \frac{0.0591}{2} \log \frac{0.1 \times 1}{1 \times 0.02}$$

$$= [-0.13 - (-0.76)] - \frac{0.0591}{2} \log \frac{0.1}{0.02}$$

$$= 0.63 - \frac{0.0591}{2} \log \frac{0.10}{0.02} = \mathbf{0.609\ V}$$

Example 18.36 : *The e.m.f. of the cell, Zn (s) / Zn^{2+} (aq 0.1 M) || Cd^{2+} (aq, M_1) / Cd (s) has been found to be equal to 0.3305 V at 298 K. Calculate the value of M_1. Given that $E^o_{Zn^{2+}/Zn}$ = – 0.76 V and $E^o_{Cd^{2+}/Cd}$ = – 0.40.*

Solution. It is obvious from the cell representation that Zn (*s*) / Zn^{2+} (*aq*) electrode acts as anode and Cd^{2+} (*aq*) / Cd (*s*) electrode behaves as cathode. The cell reaction is :

$$Zn\ (s) + Cd^{2+}\ (aq) \rightarrow Zn^{2+}\ (aq) + Cd\ (s)$$

Here $n = 2$

Thus from Nernst equation, E_{cell} at 298 K is given by :

$$E_{cell} = E^o_{cell} - \frac{0.05912}{2} \log \frac{[Zn^{2+}(aq)]\ [Cd\ (s)]}{[Zn\ (s)]\ [Cd^{2+}(aq)]}$$

$$= [E^o_{Cd^{2+}/Cd} - E^o_{Zn^{2+}/Zn}] - 0.0295 \log \frac{0.1 \times 1}{1 \times M_1}$$

or $$0.3305 = [-0.40 - (-0.76)] - 0.0295 \log \frac{0.1}{M_1}$$

or $$-0.0295 = -0.295 \log \frac{0.1}{M_1}$$

$\therefore$ $$M_1 = \mathbf{0.01\ M}$$

Example 18.37. *The standard reduction potential of Cu^{2+}/Cu and Ag^+/Ag electrodes are 0.337 and 0.799 volt respectively. Construct a galvanic cell using these electrodes so that its standard e.m.f. is positive. For what concentration of Ag^+ will the e.m.f. of the cell at 25°C be zero, if the concentration of Cu^{2+} is 0.01 M ?* **(I.I.T. 1990)**

Solution. Since $E^o_{Ag^{2+}/Ag}$ (= 0.799 V) is greater than $E^o_{Cd^{2+}/Cd}$ (= 0.337 V), Ag^+ (*aq*) / Ag (*s*) will form the cathode and Cu (*s*) / Cu (*aq*) will behave as anode of the galvanic cell. Thus the cell can be represented as given below. Here M_1 is the concentration of Ag^+ (*aq*) ions.

Cu (*s*) / Cu^{2+} (*aq*, 0.01 M) || Ag^+ (*aq*, M_1) / Ag (*s*)
Anode — Cathode

$E^o_{Cu^{2+}/Cu} = 0.337$ $\quad$ $E^o_{Ag^{2+}/Ag} = 0.799$ V

Cell reaction is : Cu (*s*) + $2Ag^+$ (*aq*) $\rightarrow$ Cu^{2+} (*aq*) + 2Ag (*s*)

Here $n = 2$

From Nernst equation at 298 K, we have

$$E_{cell} = E^o_{cell} - \frac{0.0591}{2} \log \frac{[Cu^{2+}(aq)]\ [Ag\ (s)]}{[Cu\ (s)]\ [Ag^+(aq)]^2}$$

$$= [E^o_{Ag^{2+}/Ag} - E^o_{Cu^{2+}/Cu}] - 0.0295 \log \frac{0.01 \times 1}{1 \times M_1}$$

$$0.0 = (0.799 - 0.337) - 0.0295 \log \frac{0.01}{M_1}$$

$$-0.462 = -0.0295 \log \frac{0.01}{M_1}$$

or $$\mathbf{M_1 = 1.48 \times 10^{-9}\ M}$$

Example 18.38. *A zinc rod is placed in 0.1 M solution of $ZnSO_4$ at 25°C. Assuming that the salt is dissociated to the extent of 95% at this dilution, calculate the potential of the electrode at this temperature. Given that $E^o_{Zn^{2+}/Zn} = -0.76$ volts.*

Solution. Concentration of $ZnSO_4$ solution = 0.1 M

Percentage of dissociation of $ZnSO_4$ solution = 95%

$\therefore$ Concentration of Zn^{2+} ions in $ZnSO_4$ solution $= \dfrac{0.1 \times 95}{100} = 0.095$ M

Thus the electrode can be represented as :

$Zn\,(s)\,/\,Zn^{2+}\,(aq,\ 0.095\ M)$

Reduction reaction taking place at this electrode is :

$Zn^{2+}\,(aq,\ 0.095\ M) + 2e^- \rightarrow Zn\,(s)$. (Here $n = 2$)

According to Nernst equation, the reduction potential of the above electrode $[(E_{red})_{elec}]$ is given by :

$$(E_{red})_{elec} = (E^o_{red})_{elec} - \frac{Rt}{nF}\ln\frac{\text{Concentration of the products}}{\text{Concentration of the reactants}}$$

or
$$(E_{red})_{elec} = (E^o_{red})_{elec} - \frac{2.303\ RT}{nF}\log\frac{\text{Concentration of the products}}{\text{Concentration of the reactants}}$$

or
$$E_{Zn^{2+}/Zn} = E^o_{Zn^{2+}/Zn} - \frac{2.303\ RT}{nF}\log\frac{[Zn]}{[Zn^{2+}(aq)]}$$

or
$$E_{Zn^{2+}/Zn} = -0.76 - \frac{2.303\ RT}{nF}\log\frac{1}{0.095}$$

$$= -0.76 + \frac{2.303\ RT}{nF}\log 0.095$$

$$= -0.76 + \frac{2.303 \times (8.31\ C.V.\ K^{-1}\ mol^{-1}) \times 298\ K}{2 \times (96500\ Cmol^{-1})}\log 0.095$$

or
$$E_{Zn^{2+}/Zn} = \mathbf{-0.79\ volt}$$

Example 18.39. *Calculate the reduction potential of the following electrode at 298 K.*

Pt, Cl_2(2.5 atm) / HCl (0 .01 M), $E^o_{Cl_2/2Cl^-} = +1.36$ V

Solution. The reduction reaction taking place at the given electrode can be written as :

$Cl_2\,(g) + 2e^- \rightarrow 2Cl^-\,(aq)$ (Here $n = 2$)

According to Nernst equation, the reduction potential of the given electrode ($E^o_{Cl_2/2Cl^-}$) is given by :

$$E_{Cl_2/2Cl^-} = E^o_{Cl_2/2Cl^-} - \frac{RT}{nF}\ln\frac{[\text{Concentration of } Cl^- \text{ ions}]^2}{\text{Pressure of } Cl_2 \text{ gas}}$$

$$= +1.36 - \frac{2.303\ RT}{2F}\log\frac{[0.01]^2}{2.5}$$

At 298 K, the above equation becomes

$$E^o_{Cl_2/2Cl^-} = +1.36 - \frac{2.303 \times (8.31\ C.V.\ K^{-1}\ mol^{-1}) \times 298\ K}{2 \times (96500\ C\ mol^{-1})}\log\frac{[0.01]^2}{2.5}$$

$$= +1.36 - 0.0295\log\frac{(0.01)^2}{2.5}$$

$$E^o_{Cl_2/2Cl^-} = \mathbf{1.49\ V}$$

Relation between free energy decrease (– ΔG) and cell potential (E_{cell})

It has been shown that if n is the number of electrons involved in a cell reaction, n faradays (*i.e.*, nF) of electricity will be generated in the reaction. If E_{cell} is the e.m.f. of the cell, then electrical energy supplied by the cell = $nF\,E_{cell}$. It has been pointed out by Gibbs that the electrical energy of a reversible cell (= $nF\,E_{cell}$) is equal to the decrease in the free energy ($-\Delta G$) of the reaction occurring in the cell, *i.e.*,

$$-\Delta G = nF\,E_{cell} \qquad ...(i)$$

Here negative sign indicates the decrease in free energy of the cell reaction.

Equation (*i*) gives us a relation between the free energy decrease ($-\Delta G$) and cell potential of a reversible cell (E_{cell}). If the reactants and products appearing in the cell reaction are in their standard states, then equation (*i*) becomes

$$-\Delta G^\circ = nF\,E^\circ_{cell} \qquad ...(ii)$$

ΔG° is called standard free energy change and E°_{cell} is the standard e.m.f. of the galvanic cell.

To find out a relation between the standard cell potential (E°_{cell}) and equilibrium constant (K_{eq}) of a cell reaction from Nernst equation

In the beginning of the operation of a galvanic cell, the reduction potential of the anode is less than that of cathode $[(E_{red})_{anode} < (E_{red})_{cathode}]$. As the reaction in the cell proceeds, the concentration of the metal ions round the anode increases due to oxidation of metal to metal ions while the concentration of these ions round the cathode decreases due to the reduction of metal ions to free metal. The net result is that the reduction potential of anode gradually increases while that of cathode decreases. This continues till a stage is reached when the reduction potentials of both the electrodes become equal and hence e.m.f. of the cell (E_{cell}) becomes equal to zero.

Let us consider the following redox reaction taking place in a galvanic cell. Let the number of electrons involved in this reaction be *n*.

$$aA + bB \rightleftharpoons cC + dD$$

Suppose the equilibrium constant of this reaction is K_{eq} whose value is given by :

$$K_{eq} = \frac{[C]^c\,[D]^d}{[A]^a\,[B]^b}$$

Nernst equation for the above reaction is given by :

$$E_{cell} = E^\circ_{cell} - \frac{2.303\,RT}{nF}\log\frac{[C]^c\,[D]^d}{[A]^a\,[B]^b} \qquad ...(i)$$

At equilibrium, since $E_{cell} = 0$ and $K_{eq} = \dfrac{[C]^c\,[D]^d}{[A]^a\,[B]^b}$, equation (*i*) becomes

$$0 = E^\circ_{cell} - \frac{2.303\,RT}{nF}\log K_{eq}$$

or $$E^\circ_{cell} = \frac{2.303\,RT}{nF}\log K_{eq} \qquad ...(ii)$$

At 25°C or 298 K, K_{eq} is given by :

$$E^\circ_{cell} = \frac{2.303 \times (8.31\text{ C.V. K}^{-1}\text{ mol}^{-1}) \times 298\text{ K}}{n \times 96500\text{ C mol}^{-1}}\log K_{eq}$$

or $$E^\circ_{cell} = \frac{0.05912}{n}\log K_{eq} \qquad ...(iii)$$

Equations (*ii*) or (*iii*) gives us a relation between the standard cell potential (E°_{cell}) and the equilibrium constant (K_{eq}) of the redox reaction occurring in the cell. If E°_{cell} is known, the value of K_{eq} can be calculated with the help of equation (*ii*) or (*iii*).

Relation between standard free energy decrease ($-\Delta G^\circ$) and equilibrium constant (K_{eq}) of a cell reaction

We have seen that the decrease in the standard free energy ($-\Delta G^\circ$) of a redox reaction taking place in a reversible cell and standard potential of the cell (E^o_{cell}) are related as :

$$-\Delta G^\circ = nF\,E^o_{cell} \quad ...(a)$$

We have also seen that at equilibrium, the equilibrium constant (K_{eq}) of the reaction and the standard potential of the cell (E^o_{cell}) are related as :

$$E^o_{cell} = \frac{2.303\,RT}{nF}\log K_{eq} \quad ...(b)$$

Put the value of E^o_{cell} from equation (*b*) in equation (*a*) to get :

$$-\Delta G^\circ = nF \times \frac{2.303\,RT}{nF}\log K_{eq}$$

or $$\Delta G^\circ = -2.303\,RT\log K_{eq} \quad ...(i)$$

The above equation gives us a relation between ΔG° and K_{eq}.

At 25°C or 298 K, ΔG° and K_{eq} are related as :

$$\Delta G^\circ = -2.303 \times (8.31\ JK^{-1}\ mol^{-1}) \times 298\ K \times \log K_{eq}$$

or $$\Delta G^\circ = -(5703.1031 \log K_{eq})\ J\ mol^{-1} \quad ...(ii)$$

If the value of ΔG° is known, the value of K_{eq} can be calculated from the above equation.

Relation between ΔG°, E^o_{cell} and K_{eq}

We have already shown that :

$$\Delta G^\circ = -nF\,E^o_{cell} \quad ...(a)$$

$$E^o_{cell} = \frac{2.303\,RT}{nF}\log K_{eq} \quad ...(b)$$

$$\Delta G^\circ = -2.303\,RT\log K_{eq} \quad ...(c)$$

When all the above relations are combined, we get :

$$\Delta G^\circ = -nF\,E^o_{cell} = -2.303\,RT\log K_{eq} \quad ...(i)$$

Equation (*i*) gives a relation between ΔG° and K_{eq}.

At 25°C or 298 K, equation (*i*) becomes

$$\Delta G^\circ = -nF\,E^o_{cell} = -2.303 \times (8.31\ C.V.\ K^{-1}\ mol^{-1}) \times 298\ K \times \log K_{eq}$$

$$= -2.303 \times (8.31\ JK^{-1}\ mol^{-1}) \times 298\ K \times \log K_{eq}$$

or $$\Delta G^\circ = -nF\,E^o_{cell} = -5703.1031 \log K_{eq}\ Jmol^{-1} \quad ...(ii)$$

($\because$ C.V. = J)

Significance of K_{eq} value

The value of K_{eq} gives the extent of the cell reaction. If the value of K_{eq} is very large, the equilibrium position of the reaction lies far to the right and, therefore, large amount of products is formed, *i.e.*, if the value of K_{eq} is very high, the reaction proceeds in the forward direction almost to completion.

Maximum work (W_{max}) that can be obtained from an electrochemical cell : Relation between W_{max}, ΔG° and E^o_{cell}

In an electrochemical cell, the system does work by transferring electrical charge through an external circuit. The magnitude of useful maximum work (W_{max}) that can be obtained from the reaction taking place in any electrochemical cell has been found to be equal to the change in free energy (ΔG) of the cell reaction, *i.e.*,

$$W_{max} = \Delta G \quad ...(i)$$

Now if F is the Faraday constant (the charge on one mole of electrons) and *n* is the number of moles of electrons transferred in the cell reaction, then

$$\text{Charge on } n \text{ moles of electrons} = nF \quad ...(ii)$$

Now if E_{cell} is the e.m.f. of the cell, then

$$W_{max} = -\text{(charge on } n \text{ moles of electrons)} \times E_{cell}$$

or $$W_{max} = -nF\,E_{cell} \quad ....(iii)$$

With the help of equations (*i*) and (*iii*), we can write .

$$W_{max} = \Delta G = -nF\,E_{cell} \quad ...(iv)$$

At standard conditions, equation (*iv*) becomes

$$W_{max} = \Delta G^\circ = -nF\,E^o_{cell}$$

Relation between W_{max}, ΔG°, E^o_{cell} and K_{eq}

We have already shown that ΔG°, E^o_{cell} and K_{eq} are related by equation (*i*) while W_{max}, ΔG° and E^o_{cell} are related by equation (*ii*).

$$\Delta G^\circ = -nF\,E^o_{cell} = -2.303\,RT \log K_{eq} \quad ...(i)$$

$$W_{max} = \Delta G^\circ = -nF\,E^o_{cell} \quad ...(ii)$$

With the help of equations (*i*) and (*ii*), we can write

$$W_{max} = \Delta G^\circ = -nF\,E^o_{cell} = -2.303\,RT \log K_{eq} \quad ...(iii)$$

At 25°C or 298 K, equation (*iii*) becomes

$$W_{max} = \Delta G^\circ = -nF\,E^o_{cell} = -2.303 \times (8.31\text{ C.V. K}^{-1}\text{ mol}^{-1}) \times (298\text{ K}) \times (\log K_{eq})$$

or $$W_{max} = \Delta G^\circ = -nF\,E^o_{cell} = -(5703.1031 \times \log K_{eq})\text{ J mol}^{-1} \quad ...(iv)$$

($\because$ C.V. = J)

Example 18.40. *Calculate the maximum work that can be obtained from the Daniell cell given below:*

$$Zn\,(s) / Zn^{2+}\,(aq) \,||\, Cu^{2+}\,(aq) / Cu\,(s)$$

Given that $E^o_{Zn^{2+}/Zn} = -0.76$ *V and* $E^o_{Cu^{2+}/Cu} = +0.34$ *V.*

Solution. Cell reaction is :

$$Zn\,(s) + Cu^{2+}\,(aq) \rightarrow Cu\,(s) + Zn^{2+}\,(aq)$$

Here $n = 2$

$$E^o_{cell} = (E^o_{red})_{cathode} - (E^o_{red})_{anode}$$

$$= E^o_{Cu^{2+}/Cu} - E^o_{Zn^{2+}/Zn}$$

$$= 0.34 - (-0.76) = 1.10\text{ V}$$

We know that :

$$W_{max} = \Delta G^\circ = -nF$$

$$= -(2\text{ mol}) \times (96500\text{ C mol}^{-1}) \times (1.10\text{ V})$$

$$= -212300\text{ C.V.}$$

$$= -212300\text{ J (C.V. = J)}$$

or $$W_{max} = \mathbf{-212.300\ kJ}$$

Thus maximum work that can be obtained by the operation of the given cell is equal to 212.3 kJ.

Example 18.41. *Calculate the standard e.m.f., standard free energy change and equilibrium constant for the following redox reaction taking place in a galvanic cell at 25°C.*

$$\frac{1}{2} Cu\,(s) + \frac{1}{2} Cl_2\,(g) \rightleftharpoons \frac{1}{2} Cu^{2+}\,(aq) +]Cl^-(aq)$$

Solution. The given reaction shows that Cu is being oxidised to Cu^{2+} and Cl_2 is being reduced to Cl^-. Hence Cu (*s*) / Cu^{2+} (*aq*) forms the anode of the cell (whose cell reaction is given in the question) and Pt, Cl_2 (*g*) / Cl^- (*aq*) acts as cathode. Thus the cell can be represented as :

$$\underset{\text{Anode}}{Cu\,(s) / Cu^{2+}\,(aq)} \,||\, \underset{\text{Cathode}}{Cl_2\,(g) / Cl^-\,(aq),\ Pt}$$

Calculation of standard e.m.f. of the cell (E^o_{cell})

$$E^o_{cell} = (E^o_{red})_{cathode} - (E^o_{red})_{anode}$$
$$= E^o_{Cl_2/2Cl^-} - E^o_{Cu^{2+}/Cu}$$
$$= 1.36 - 0.34 = \mathbf{1.02\ Volts}$$

Calculation of standard free energy change ($\Delta G°$) *of the reaction.*

Oxidation reaction at anode : $\frac{1}{2}Cu\,(s) \rightarrow \frac{1}{2}Cu^{2+}\,(aq) + e^-$

Reduction reaction at cathode : $\frac{1}{2}Cl_2\,(g) + e^- \rightarrow Cl^-\,(aq)$

Cell reaction : $\frac{1}{2}Cu\,(s) + \frac{1}{2}Cl_2\,(g) \rightarrow \frac{1}{2}\ Cu^{2+}\,(aq) + Cl^-\,(aq)\ (n = 1)$

Since $n = 1$ mol,

$$\Delta G° = -nFE^0_{cell}$$
$$= -(1\ mol) \times (96500\ C\ mol^{-1}) \times 1.02\ V$$
$$= -96500 \times 1.02\ C.V.\ mol^{-1}$$
$$= \mathbf{-96500 \times 1.02\ J\ mol^{-1}} \qquad (\because C.V. = J)$$

Calculation of equilibrium constant (K_{eq}) *of the reaction*

$$\Delta G° = -2.303\ RT \log K_{eq}$$
$$-96500 \times 1.02\ J\ mol^{-1} = -2.303 \times (8.31\ JK^{-1}\ mol^{-1}) \times (298\ K) \times \log$$

or
$$\log K_{eq} = \frac{-96500 \times 1.02\ J\ mol^{-1}}{-2.303 \times (8.31\ J\,K^{-1}\ mol^{-1}) \times 298\ K}$$
$$= \frac{96500 \times 1.02}{2.303 \times 8.31 \times 298}$$

$$\therefore \quad K_{eq} = \text{Antilog}\,\frac{96500 \times 1.02}{2.303 \times 8.31 \times 198}$$

or $K_{eq} = \mathbf{2 \times 10^{17}}$

Example 18.42. *Determine the equilibrium constant of the following reaction at 298 K.*

$$2Fe^{3+} + Sn^{2+} \rightarrow 2Fe^{2+} + Sn^{4+}$$

From the calculated value of equilibrium constant, predict whether Sn^{2+} ions can reduce Fe^{3+} ions to Fe^{2+} ions quantitively or not ($E^o_{Fe^{3+},\ Fe^{2+}/Pt} = +0.771$ V, $E^o_{Sn^{4+},\ Sn^{2+}/Pt} = +0.15$ V).

Solution. In the given reaction, Sn^{2+} ions are being oxidised to Sn^{4+} ions and hence Pt / Sn^{2+}, Sn^{4+} electrode will constitute the anode. Fe^{3+} ions are being reduced to Fe^{2+} ions and hence Fe^{3+}, Fe^{2+} / Pt electrode will constitute the cathode. Thus the cell in which the above reaction takes place can be represented as :

$$\underset{\text{Anode}}{Pt / Sn^{2+}, Sn^{4+}} \ \| \ \underset{\text{Cathode}}{Fe^{3+}, Fe^{2+} / Pt}$$

The oxidation and reduction reactions occurring at the two electrodes can be written as :

Oxidation reaction at anode : $Sn^{2+} \rightarrow Sn^{4+} + 2e^-$

Reduction reaction at cathode : $Fe^{3+} + e^- \rightarrow Fe^{2+}] \rightarrow 2$

Cell reaction : $Sn^{2+} + 2Fe^{3+} \rightarrow Sn^{4+} + 2Fe^{2+}\ (n = 2)$

The value of E^o_{cell} of the above cell is given by :

$$E^o_{cell} = (E^o_{red})_{cathode} - (E^o_{red})_{anode}$$
$$= E^o_{Fe^{3+},\ Fe^{2+}/Pt} - E^o_{Sn^{4+},\ Sn^{2+}/Pt}$$
$$= 0.771 - 0.150 = +0.621\ volt$$

We know that at 25°C or 298 K, E^o_{cell} and K_{eq} are related as :

$$E^o_{cell} = \frac{0.05912}{n} \log K_{eq}$$

$$0.621 = \frac{0.05912}{2} \log K_{eq}$$

or $$\log K_{eq} = \frac{2 \times 0621}{0.5912} = 21.00$$

$\therefore$ $$K_{eq} = \text{Antilog } (21.00) = \mathbf{1.0 \times 10^{21}}$$

Example 18.43. *Calculate the equilibrium constant for the reaction, $2H_2O \rightleftharpoons H_3O^+ + OH^-$ at 25°C, if the standard reduction potential at 25°C of the reaction, $2H_2O + 2e^- \rightarrow H_2 + 2OH^-$ is – 0.8277 V.* **(J.E.E. 1989)**

Solution. We know that at 25°C or 298 K, E^o_{cell} and K_{eq} are related as :

$$E^o_{cell} = \frac{0.5912}{n} \log K_{eq}$$

$$-0.8277 = \frac{0.05912}{2} \log K_{eq}$$

or $$\log K_{eq} = -\frac{2 \times 0.8277}{0.05912} = -28.00$$

$\therefore$ $$K_{eq} = \text{Antilog } (-28.00) = \mathbf{1.64 \times 10^{-28}}.$$

Example 18.44. *The zinc-silver oxide cell is used in hearing aids and electric watches. From the following data:*

(i) $Zn^{2+} + 2e^- \rightarrow Zn$ $E^o = -0.76$

(ii) $Ag_2O + H_2O + 2e^- \rightarrow 2Ag + 2OH^-$ $E^o = +0.344$ V

Answer the following :

(a) Construct the cell and what will be its e.m.f. ?

(b) What is its cell reaction and what will be the ΔG° value of this cell reaction ?

(c) Name the species which undergo oxidation and reduction when the cell is in operation.

Solution. (*a*) Both the given reactions are reduction reactions. The electrode corresponding to reaction (*i*) can be represented as Zn^{2+} / Zn having $E^o_{Zn^{2+}/Zn} = -0.76$ V and the electrode corresponding to reaction (*ii*) can be written as OH^- / Ag_2O/Ag which has its reduction electrode potential ($E^o_{OH^-/Ag_2O/Ag}$) equal to + 0.344 V.

We know that in a galvanic cell, the value of $(E^o_{red})_{cathode}$ is always greater than the value of $(E^o_{red})_{anode}$. Thus, since $E^o_{OH^-/Ag_2O/Ag}$ (= + 0.34 V) is greater than $E^o_{Zn^{2+}/Zn}$ (= – 0.76 V), OH^- (*aq*) / Ag_2O / Ag electrode will act as cathode and Zn^{2+}/Zn electrode will act as anode when these two electrodes are coupled together to form a galvanic cell. Thus the resulted cell can be represented as :

Zn / Zn^{2+} (*aq*)	‖	OH^- (*aq*) / Ag_2O / Ag
Negative terminal (anode)		Positive terminal (cathode)
$E^o_{Zn^{2+}/Zn} = -0.76$ V		$E^o_{OH^-/Ag_2O/Ag} = +0.344$ V

The value of standard e.m.f. of the above cell (E^o_{cell} value) is given by :

$$E^o_{cell} = (E^o_{red})_{cathode} - (E^o_{red})_{anode}$$

$$= +0.344 - (-0.76) = \mathbf{1.104\ V}$$

(*b*) Oxidation reaction at anode : $Zn \rightarrow Zn^{2+} + 2e^-$

Reduction reaction at cathode : $Ag_2O + H_2O + 2e^- \rightarrow 2Ag + 2OH^-$
(Ag = +1)   (Ag = 0)

Cell reaction : $Zn + Ag_2O + H_2O \rightarrow Zn^{2+} + 2Ag + 2OH^-$

This is the desired cell reaction. The value of ΔG° of the above cell reaction which involves two electrons ($n = 2$) is given by :

$$\Delta G^\circ = -nFE^\circ_{cell}$$
$$= -2 \times (96500\ \text{Cmol}^{-1}) \times (1.104\ \text{V})$$
$$= -213072\ \text{C.V. mol}^{-1}$$
$$= -213072\ \text{J mol}^{-1}\ (\text{J} = \text{C.V.})$$

or $\Delta G^\circ = \mathbf{-213.072\ kJ\ mol^{-1}}$

(*c*) From the reactions taking place at the two electrodes, it is evident that Zn is oxidised to Zn^{2+} and Ag_2O is reduced to Ag when the cell operates.

Example 18.45. *For the Daniell cell involving the cell reaction : Zn (s) + Cu^{2+} (aq) $\rightleftharpoons$ Zn^{2+} (aq) + Cu (s), the standard free energies of formation for Zn (s), Cu (s), Cu^{2+} (aq) and Zn^{2+} (aq) are 0, 0, 15.66 and – 35.14 Kcal mol^{-1} respectively. Calculate the standard e.m.f. of the cell.*

Solution. We know that $\Delta G^\circ = -nF\,E^\circ_{cell}$.

In order to know the value of E°_{cell}, we should know the values of ΔG°, n and F. Again we know that $n = 2$, F = 96500 C. The value of ΔG° is calculated as follows :

$$\Delta G^\circ = \Delta G^\circ\ (\text{products}) - \Delta G^\circ\ (\text{reactants})$$
$$= (-35.14 + 0) - (0 + 15.66)\ \text{Kcal} = -50.80\ \text{Kcal}$$
$$= -50.80 \times 4.184\ \text{kJ} = -212.55\ \text{kJ}$$
$$= -212550\ \text{volt-coulombs}$$

Now $\Delta G^\circ = -nFE^\circ_{cell}$

or -212550 volt–coulombs $= -2 \times 96500$ coulombs $\times E^\circ_{cell}$

$$\therefore \quad E^\circ_{cell} = \frac{-212550\ \text{volt coulombs}}{-2 \times 96500\ \text{coulombs}}$$
$$= \mathbf{1.101\ volts.}$$

To calculate the value of standard potential of a given half-cell reaction (E^0) from the values of standard potentials of other given half-cell reactions

When we are required to calculate E° value for a given half-cell reaction, we should remember that the free energy changes for each of the combining half-cell reactions can be added or subtracted as required. Free energy change (ΔG°) for a half-reaction is given by :

$$\Delta G^\circ = -nF\,E^\circ$$

where n = No. of electrons involved in the half-cell reaction, F = Faraday constant and E° = standard reduction electrode potential for the half-reaction. This method can be understood by considering the following examples.

Example 18.46. *Calculate the E° value for the half-cell reaction, $MnO_4^- + 4H^+ + 3e^- \rightarrow MnO_2 + 2H_2O$. Given that :*

$$MnO_4^- + 8H^+ + 5e^- \rightarrow Mn^{2+} + 4H_2O; \quad E^\circ_{red} = +1.51\ V$$
$$Mn^{2+} + 2H_2O \rightarrow MnO_2 + 4H^+ + 2e^-; \quad E^\circ_{ox} = -1.23\ V$$

Solution.

	Half-cell reactions	*n*	*E°*	$\Delta G^\circ = -nFE^\circ$
(*i*)	$MnO_4^- + 8H^+ + 5e^- \rightarrow Mn^{2+} + 2H_2O$	5	$E_{red} = +1.51$ V	$= -5F \times 1.51 = -7.55$ F
(*ii*)	$Mn^{2+} + 2H_2O \rightarrow MnO_2 + 4H^+ + 2e^-$	2	$E_{ox} = -1.23$ V	$= -2 \times F \times (-1.23) = 2.46$ F

On adding equations (*i*) and (*ii*), we get :

$$MnO_4^- + 4H^+ + 3e^- \rightarrow MnO_2 + 2H_2O\ (n = 3)$$

ΔG° for this reaction $= -7.55\ F + 2.46\ F = -5.09\ F$

Again we know that :

ΔG° for the above reaction $= -nFE^\circ = -3FE^\circ_{MnO_4^-/MnO_2} = -5.09$ F

$$\therefore \quad E^{o}_{MnO_4^-/MnO_2} = \frac{-5.09}{-3} = \mathbf{+1.69\ V}$$

Example 18.47. *Calculate E° value for the half-cell reaction,* $\frac{1}{2}Cl_2 + 3H_2O \rightarrow ClO_3^- + 6H^+ + 5e^-$. *Given that :*

(i) $Cl^- + 3H_2O \rightarrow ClO_3^- + 6H^+ + 6e^-, \quad E_{ox} = -1.45\ V$

(ii) $2Cl^- \rightarrow Cl_2 + 2e^-; \quad E_{ox} = -1.36\ V$

Solution. Dividing *(ii)* reaction by two, we get :

(iii) $Cl^- \rightarrow 1/2\ Cl_2 + e^-; \quad E_{ox} = -1.36\ V$

Half-cell reactions	*n°*	*E°*	*ΔG° = –nFE°*
(i) $Cl^- + 3H_2O \rightarrow ClO_3^- + 6H^+ + 6e^-$	6	$E^o_{ox} = -1.45\ V$	$= -6 \times F \times (-1.45) = +8.70\ F$
(iii) $Cl^- \rightarrow 1/2\ Cl_2 + e^-$	1	$E_{OX} = -1.36\ V$	$= -1 \times F \times (-1.36) = +1.36\ F$

On substracting *(iii)*, *(i)*, we get :

$$3H_2O \rightarrow ClO_3^- + 6H^+ + 5e^- - \tfrac{1}{2}Cl_2$$

or $$\tfrac{1}{2}Cl_2 + 3H_2O \rightarrow ClO_3^- + 6H^+ + 5e^- \qquad (n = 5)$$

$\Delta G°$ for this reaction = 8.70 F – 1.36 F = +7.24 F

or $$-5 \times F \times E^{o}_{\frac{1}{2}Cl_2/ClO_3^-} = +7.34\ F$$

$$\therefore \quad E^{o}_{\frac{1}{2}Cl_2/ClO_3^-} = \frac{+7.34}{-5} = \mathbf{-1.47\ V}$$

E.M.F. Diagrams

E.M.F. diagram is a short form in which standard reduction potentials of an element in different oxidation states are presented in a condensed manner. In an E.M.F. diagram, the species of the element having different oxidation states are arranged from left to right in the decreasing order of the oxidation number of the element and the values of standard reduction potential are inserted between the oxidised and reduced species. Thus the highest oxidation state is written on the left and the lowest at the right. E^o_A, E^o_B and E^o_N denote the values of standard reduction potentials in acidic, basic and neutral medium respectively.

Example 18.48. *Vanadium exists in 0, +2, +3, + 4 and +5 oxidation states. EMF diagram for this element in acidic medium in the above oxidation states is shown as :*

$$E^o_A\ (volts) \quad V(OH)_4^+ \xrightarrow{+1.00} VO^{2+} \xrightarrow{+0.359} V^{3+} \xrightarrow{-0.256} V^{2+} \xrightarrow{-1.186} V^{o}$$

(V = + 5) *(V = + 4)* *(V = + 3)* *(V = + 2)* *(V = 0)*

($V(OH)_4^+ \rightarrow V^o$: –0.254)

Solution. The reduction reactions corresponding for this diagram are given as :

$V(OH)_4^+ + 2H^+ + e^- \rightarrow VO^{2+} + 3H_2O; \quad E^o_A = +1.00\ V$

$VO^{2+} + 2H^+ + e^- \rightarrow V^{3+} + H_2O \quad ; \quad E^o_A = +0.359\ V$

$V^{3+} + e^- \rightarrow V^{2+} \quad ; \quad E^o_A = -0.256\ V$

$V^{2+} + 2e^- \rightarrow V \quad ; \quad E^o_A = -1.186\ V$

$[V(OH)_4]^+ + 4H^+ + 5e^- \rightarrow V + 4H_2O; \quad E^o_A = -0.254\ V$

Applications of E.M.F. diagrams

(i) The diagrams give us a general idea of the chemistry of the element.

(ii) They instantly reveal the comparative stability of different oxidation states of the element.

(iii) The standard electrode potentials given in the diagram can be used to calculate the value of standard reduction electrode potential of other reduction half-reaction. Since the EMF diagrams

refer to volts per electron, the value of E° for a new half-reaction can be obtained from the diagram through the sum of the intermediate couples, each multiplied by the number of the electrons (n) involved in each of the intermediate couples.

Example 18.49. *EMF diagram for Mn is given as :*

$$E^{\circ}_{A}\ (volts)\ MnO_4^- \xrightarrow{+0.56} MnO_4^{2-} \xrightarrow{+2.26} MnO_2 \xrightarrow{+0.95} Mn^{3+} \xrightarrow{+0.51} Mn^{2+} \xrightarrow{-1.81} Mn^0$$

($MnO_4^- \to Mn^{2+}$: +1.51; $MnO_4^- \to MnO_2$: +1.70; $MnO_2 \to Mn^{2+}$: +1.23)

Calculate E°_{A} value for $MnO_4^- - Mn^{3+}$ couple.

Solution. The half-reactions showing $MnO_4^- \to MnO_4^{2-}$, $MnO_4^{2-} \to MnO_2$ and $MnO_2 \to Mn^{3+}$ charges are :

$$MnO_4^- + e^- \to MnO_4^{2-}\ (n = 1) \qquad ; \qquad E^{\circ}_{A} = +\,0.56\ V$$

$$MnO_4^{2-} + 4H^+ + 2e^- \to + MnO_2 + 2H_2O\ \ (n = 2)\ ;\ E^{\circ}_{A} = +\,2.26\ V$$

$$MnO_2 + 4H^+ + e^- \to Mn^{3+} + 2H_2O\ ;\ E^{\circ}_{A} = +\,0.95\ V$$

adding the above equations, we get :

$$MnO_4^- + 8H^+ + 4e^- \to Mn^{3+} + 4H_2O$$

This is half-reaction for $MnO_4^- - Mn^{3+}$ couple.

$$\therefore\ E^{\circ}_{A}\ \text{value for}\ MnO_4^- - Mn^{3+}\ \text{couple} = \frac{0.56 \times 1 + 2.26 \times 2 + 0.95 \times 1}{1 + 2 + 1} = \mathbf{1.51\ V}$$

Example 18.50. *EMF diagram for iron is given as :*

$$E^{\circ}_{A}\ (volts)\ \ \underset{(Fe\ =\ +\ 6)}{FeO_4^{2-}} \xrightarrow{+2.20} \underset{(Fe\ =\ +\ 3)}{Fe^{3+}} \xrightarrow{+0.77} \underset{(Fe\ =\ +\ 2)}{Fe^{2+}} \xrightarrow{-0.445} \underset{(Fe\ =\ 0)}{Fe^{o}}$$

($FeO_4^{2-} \to Fe^{2+}$: +1.08)

Determine the value of $E^{o}_{FeO_4^{2-}/Fe^{2+}}$.

Solution. The half-reactions showing $FeO_4^{2-} \to Fe^{3+}$ and $Fe^{3+} \to Fe^{2+}$ changes are :

$$FeO_4^{2-} + 8H^+ + 3e^- \to Fe^{3+} + 4H_2O\ (n = 3);\ E^{\circ}_{A} = +\,2.20\ V$$

$$Fe^{3+} + e^- \to Fe^{2+}\ (n = 1);\ E^{\circ}_{A} = +\,0.77\ V$$

On adding the above two equations, we get :

$$FeO_4^{2-} + 8H^+ + 4e^- \to Fe^{2+} + 4H_2O\ (n = 4)$$

which represents $FeO_4^{2-} \to Fe^{2+}$ change.

$$\therefore\ E^{o}_{FeO_4^{2-}/Fe^{2+}} = \frac{2.20 \times 3 + 0.77 \times 1}{3 + 1} = \mathbf{+1.84\ V}$$

University Questions

1. Explain the differences between the following :
   (*i*) Electrode potential and standard electrode potential. *(Guwahati 91; Agra 92,)*
   (*ii*) Electrode potential and ionisation potential. *(E.C.C. Allahabad 94; Lucknow 95;)*
2. What are standard electrode potentials? How are they measured? Explain the process of determining the equilibrium constant of a redox reaction from standard electrode potentials. *(Guwahati 91)*

3. Wrtie short note on:

(*i*) Standard electrode potential and its applications. *(Allahabad 91; E.C.C. Allahabad 94; Lucknow 94)*

(*ii*) Electrochemical series. *(Avadh 93)*

(*iii*) (*a*) Nernst's equation (*b*) calomel electrode *(Meerut 94)*

(*iv*) Single electrode potential. *(Bahuguna 93)*

4. Show how the e.m.f. method is used in determining the pH of a solution and the solubility of a sparingly soluble electrolytes. *(Agra 92; Meerut 94)*

5. How will you determine the ionic product of water from E.M.F. measurements ? *(Agra 92)*

6. Select the possible reactions from the following :

(*i*) $Cu + Zn^{2+} \rightarrow Cu^{2+} + Zn$

(*ii*) $Co + Fe^{2+} \rightarrow Co^{2+} + Fe$

(*iii*) $Fe + Zn^{2+} \rightarrow Fe^{2+} + Zn$

(*iv*) $Zn + Cu^{2+} \rightarrow Zn^{2+} + Cu$

7. Explain the following :

(*i*) Cu-metal does not dissolve in dil. H_2SO_4 and H_2 gas is not evolved. *(I.T.T. 74)*

(*ii*) Cu and Ag do not liberate H_2 gas from the acids ?

(*iii*) H_2 gas is displaced from dil. acids by Zn and not by Cu. *(Madras 85)*

(*iv*) Zn and Fe can replace Cu from $CuSO_4$ solution while Pt and Au cannot do so. *(R.E.E. 76)*

(*v*) Zn rod and $CuSO_4$ solution will react but Cu rod and $ZnSO_4$ will not. *(Ruhelkhand 90)*

(*vi*) Fe displaces Cu from $CuSO_4$ solution but Ag does not. *(Ravishankar 93)*

(*vii*) Alkali metals are more reactive than coinage metals, although both have one electron in their outermost shell. (Given that : $E^o_{Li^+/Li} = -3.04$ V, $E^o_{Na^+/Na} = -2.71$ V, $E^o_{k^+/k} = +2.99$ V; $E^o_{Cu^+/Cu} = +0.52$ V, $E^o_{Ag^+/Ag} = +0.80$ V, $E^o_{Au^+/Au} = 1.70$ V). *(Allahabad 82)*

(viii)A Cu wire when placed in a solution of $AgNO_3$, Ag is precipitated. *(Guwahati 91)*

8. With the help of following reduction potential diagram applied to different manganese species in acidic medium, identify the species which would undergo disproportionation reactions, giving equations :

$$MnO_4^- \xrightarrow{+0.56\ V} MnO_4^{2-} \xrightarrow{+2.26\ V} MnO_2 \xrightarrow{+0.95\ V} Mn^{3+} \xrightarrow{+1.51\ V} Mn^{2+} \xrightarrow{+1.19\ V} Mn^o$$

(E.C.C. Allahabad 94)

9. In the following reduction potential diagram identity the species that will undergo disproportionation reaction and work out the reason for the same.

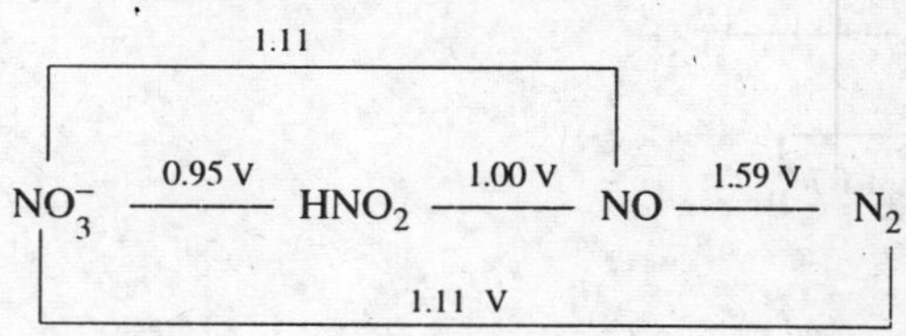

or

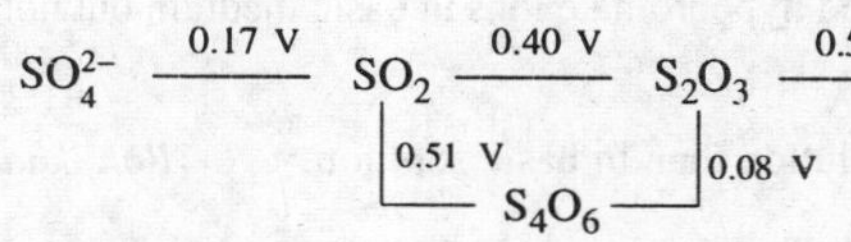

10. With the help of the reduction potential digram.

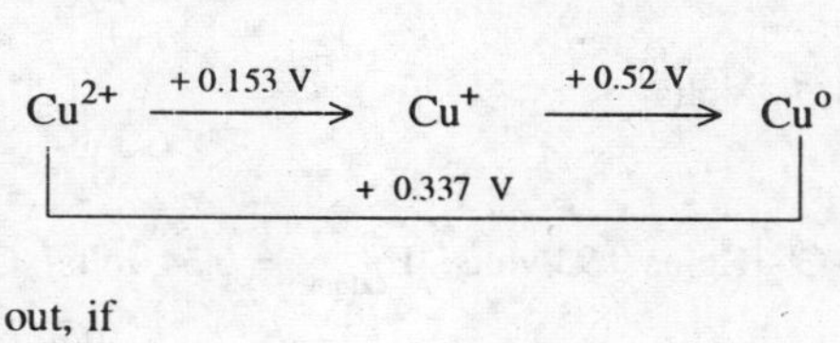

find out, if

(*a*) Cu^+ will readily change into Cu° .

(*b*) Cu^+ will undergo disproportionation action into Cu^{2+} and Cu°. *(Allahabad 1979)*

11. What informations do you get from the following E.M.F. diagram :

$$CuO^+ \xrightarrow{+0.18\ V} Cu^{2+} \xrightarrow{+0.15\ V} Cu^+ \xrightarrow{+0.50\ V} Cu$$

$$Cu^{2+} \xrightarrow{+0.34\ V} Cu$$

12. The reduction potential diagram for some sulphur species in basic solution is the following :

$$SO_4^{2-} \xrightarrow{-0.98\ V} SO_3^{2-} \xrightarrow{-0.58\ V} S_2O_3^{2-} \xrightarrow{-0.74\ V} S \xrightarrow{-0.51\ V} S^{2-}$$

Comment on the oxidising power of these species and the stability of $S_2O_3^{2-}$ ion. *(Allahabad 88)*

13. Calculate the electrode potential of a Zn^{2+} / Zn electrode in which the concentration of Zn^{2+} ions is 0.01 M. Given that $E^{2-}_{Zn/Zn^{2+}}$ = .076 V. *(Avadh 93)*

14. Differentiate between ionisation potential and electrode potential of a metal. *(VBS Purvanchal 2002)*

15. Give reason for the fact that E° of Cu is + 0.34 V while that of Zn is – 0.76 V. (Delhi 2002)

Answer. We know that a less reactive metal has higher value of E^O than a more reactive metal. Thus Cu has higher value of E^O (= + 0.34 V) than Zn (= – 0.76 V), since Cu is less reactive than Zn.

16. (*a*) What is electrode potential ? How does it differ from ionisation energy ?

(*b*) Predict whether the following reaction will occur spontaneously or not ?

$$Ni(s) + Cu^{2+}(aq) \longrightarrow Ni^{2+}(ag) + Cu\,(s)$$

Given that : $E^0_{Ni/Ni^{2+}} = +0.25\ V$ and $E^0_{Cu/Cu^{2+}} = -0.337\ V$.

(*c*) What is electrochemical series ? Describe its uses in explaining chemical reactions. *(VBS Purvanchal 2003)*

17. Consider the following EMF diagram for bromine and answer the following :

$$E_A^0 \quad BrO_3^- \xrightarrow{1.49} HOBr \xrightarrow{1.59} \tfrac{1}{2}Br_2\,(l) \xrightarrow{1.07} Br^-$$

$$BrO_3^- \xrightarrow{1.44} \tfrac{1}{2}Br_2\,(l); \qquad HOBr \xrightarrow{E} Br^-$$

$$E_A^0 \quad BrO_3^- \xrightarrow{0.54} BrO^- \xrightarrow{0.45} \tfrac{1}{2}Br_2\,(l) \xrightarrow{1.07} Br^-$$

$$BrO_3^- \xrightarrow{0.61} Br^-$$

(*a*) Calculate the value of E.

(*b*) State whether the following statements are true or false. Give reasons to your answer.

(*i*) Bromine disproportionates into bromide and hypobromate ions in basic medium but not in acidic medium.

(*ii*) BrO_3^- is better oxidizing agent in acid solution than in basic solution. *(Rohilkhand 2003)*

18. (*a*) Calculate the equilibrium constant for the reaction

$$Zn + Cu^{+2} \longrightarrow Cu + Zn^{+2} \text{ at } 25°\,C$$

$$E°_{Zn/Zn^{+2}} = 0.763\ \text{Volts},\ E°_{Cu/Cu^{+2}} = -0.337\ \text{Volts}$$

(*b*) The e.m.f. of following cell

$Pt, H_2\,|\,H^+\ a_{H^+} = 1\,|\text{Solution}|\ 0.1\ M\ KCl\ |Hg_2Cl_2|Hg$ is 0.93 volts. [$E_{calomel}$ = 0.34 volts]. Calculate the pH of the solution. *(Gujrat 2003)*

19

Concepts of Acids and Bases

Introduction

Various definitions of acids and bases have been proposed by different chemists. These definitions have tried to cover a large number of acid-base reactions, redox reactions and complex formation reactions.

Each definition is correct, when applied to the particular range for which this was meant. So it would be meaningless to try to select the "right" or "wrong" definition. However, following concepts of acids and bases are important.

Arrhenius Concept : Water System Concept (1884)

According to Arrhenius (Swedish chemist), an acid is defined as any species that gives H^+ ions (or better hydronium ions, H_3O^+) in water and a base is a substance which furnishes OH^- ions in water. Thus, according to this concept, the substances like HCl, HNO_3, H_2SO_4, CH_3COOH etc. are acids, since they give H^+ ions, when dissolved in water.

$$HCl \xrightleftharpoons{H_2O} H^+ + Cl^- \qquad ...(i)$$

H^+, ion produced in equation (*i*), gets hydrated by one molecule of water (H_2O) and gives hydronium ion, H_3O^+.

$$H^+ + H_2O \rightleftharpoons H_3O^+ \qquad ...(ii)$$

On adding the equations (*i*) and (*ii*), we get equation (*iii*) which represents the acidic behaviour of HCl in water.

$$HCl + H_2O \rightleftharpoons H_3O^+ + Cl^- \qquad ...(iii)$$

Equation (*iii*) can also be written as :

$$HCl + \text{water} \rightleftharpoons H^+ (aq) + Cl^- (aq)$$

Some chemists believe that H^+ produced in equation (*i*) gets associated with four molecules of water and gives $[H(H_2O)_4]^+$ or $[H_9O_4]^+$ ion.

$$H^+ + 4H_2O \longrightarrow [H(H_2O)_4]^+ \text{ or } [H_9O_4]^+$$

The substances like NaOH, KOH, $Ca(OH)_2$ etc., are bases, since they give OH^- ions in water.

$$NaOH \rightleftharpoons Na^+ + OH^-$$

$$NaOH + \text{water} \rightleftharpoons Na^+ (aq) + OH^- (aq)$$

Acid-base neutralisation reactions taking place in water

According to Arrhenius concept, acid-base neutralisation reaction taking place in water, is that in which a compound containing or making available H_3O^+ (or H^+) ions (acid) combines with a compound containing or making available OH^- ions (base) to form the salt and water, *e.g.*,

$$\underset{Acid}{HCl} + \underset{Base}{NaOH} \xrightarrow{H_2O} \underset{Salt}{NaCl} + \underset{Water}{H_2O}$$

Actually the above reaction takes place through the following steps :

$$HCl \xrightarrow{H_2O} H^+ + Cl^-$$
$$NaOH \xrightarrow{H_2O} Na^+ + OH^-$$

$$HCl + NaOH \xrightarrow{H_2O} Na^+Cl^- + H^+[OH]^-$$

or $$HCl + NaOH \xrightarrow{H_2O} NaCl + H_2O$$

The formation of salt (NaCl) and H_2O can also be shown as follows :

$$HCl \xrightarrow{H_2O} H^+ + Cl^-$$
$$H^+ + H_2O \xrightarrow{H_2O} H_3O^+$$
$$NaOH \xrightarrow{H_2O} Na^+ + OH^-$$

$$HCl + H_2O + NaOH \xrightarrow{H_2O} Na^+ Cl^- + [H_3O]^+ [OH]^-$$

or $$HCl + H_2O + NaOH \xrightarrow{H_2O} NaCl + 2H_2O$$

or $$HCl + NaOH \xrightarrow{H_2O} NaCl + H_2O$$

Due to the formation of a salt, acid-base neutralisation reaction is also called *salt formation reaction.* This shows that in neutralisation reaction, H_2O is produced by the combination of H^+ and OH^-. Thus we can say that neutralisation reaction, according to Arrhenius concept of acids and bases, is the combination of H^+ and OH^- ions, which are produced by the dissociation of the acid and base respectively in aqueous medium.

Applications (advantages) of Arrhenius concept

With the help of this concept, we can explain the following :

(*i*) Aqueous solution of non-metallic oxides (*e.g.*, CO_2, SO_2, SO_3, N_2O_3, N_2O_5, P_4O_6, P_4O_{10} etc.) is acidic, since it gives H^+ ions in water.

$$SO_3 + H_2O \rightleftharpoons H_2SO_4 \rightleftharpoons 2H^+ + SO_4^{2-}$$
$$N_2O_5 + H_2O \rightleftharpoons 2HNO_3 \rightleftharpoons 2H^+ + NO_3^-$$

(*ii*) Aqueous solution of metallic oxides (*e.g.*, CaO, Na_2O etc.,) and the compounds like NH_3, N_2H_4, NH_2OH etc. are basic, since these substances give OH^- ions in water.

$$CaO + H_2O \rightleftharpoons Ca(OH)_2 \rightleftharpoons Ca^{2+} + 2OH^-$$
$$NH_3 + H_2O \rightleftharpoons NH_4OH \rightleftharpoons NH_4^+ + OH^-$$
$$N_2H_4 + H_2O \rightleftharpoons [N_2H_5]^+[OH]^- \rightleftharpoons N_2H_5^+ + OH^-$$
$$NH_2OH + H_2O \rightleftharpoons [NH_3OH]^+[OH]^- \rightleftharpoons NH_3OH^+ + OH^-$$

(*iii*) This concept can explain the acid-base neutralisation reactions taking place in water.

(*iv*) The strength of an acid (HA) and a base (BOH) can be expressed quantitatively in terms of the ionisation (or dissociation) constant of the acid and base in aqueous solution.

$$\underset{\text{Acid}}{HA} \rightleftharpoons H^+ + A^-, \quad K_a = \frac{C_{H^+} \times C_{A^-}}{C_{HA}}$$

$$\underset{\text{Base}}{BOH} \rightleftharpoons B^+ + OH^-, \quad K_b = \frac{C_{B^+} \times C_{OH^-}}{C_{BOH}}$$

(*v*) The catalytic properties of acids in many reactions can be explained by saying that H^+ ions are available from the acid.

Limitations of Arrhenius concept

(*i*) According to this concept, the acid or base property of a substance is not supposed to be inherent in the substance itself, but is dependent on its aqueous solution, *e.g.*, according to this concept HCl is an acid, only when it is dissolved in water, but it is not considered as an acid when it is in the gaseous state.

(*ii*) According to this concept, acid-base neutralisation reactions take place only in water and hence cannot explain such reactions occurring in other solvents or in the gas phase, *e.g.*, the formation of NH_4Cl (*s*) by the combination of NH_3 (*g*) and HCl (*g*) cannot be explained by Arrhenius concept.

$$NH_3\,(g) + HCl\,(g) \rightleftharpoons NH_4Cl\,(s)$$

(*iii*) According to this concept, acids and bases undergo dissociation only in water (aqueous solvent). Thus, it cannot explain the dissociation of acids and bases in non-aqueous solvents like liq. NH_3, liq. SO_2, etc.

Bronsted-Lowry Concept : Protonic Concept (1923)

In 1923, J.N. Bronsted (Denmark) and J.M. Lowry (England) independently proposed a broader concept of acids and bases. According to it, an acid is a species (molecule or ion) that can lose a proton (H^+) and a base is a species (molecule or ion) that can accept a proton. In other words an acid is a proton donor and a base is a proton acceptor.

An acid qualifying Bronsted-Lowry concept is termed as a Bronsted-Lowry acid or simply as a Bronsted acid; similarly a base qualifying Bronsted-Lowry definition is called Bronsted base.

Illustrative examples (*i*) Since HCl can lose one proton (H^+) to give Cl^- ion, HCl, according to Bronsted - Lowry concept, is an acid.

$$\underset{\text{Bronsted acid}}{HCl} \xrightarrow{-H^+} Cl^- \qquad \ldots (i)$$

Similarly, since Cl^- ion can accept a proton (H^+) to form HCl, Cl^- ion acts as a base.

$$\underset{\text{Bronsted base}}{Cl^-} \xrightarrow{+H^+} HCl \qquad \ldots (ii)$$

On combining equations (*i*) and (*ii*), we get the following equation which shows that HCl is a Bronsted acid and Cl^- ion is a Bornsted base.

$$\underset{\text{Bronsted acid}}{HCl} \underset{+H^+}{\overset{-H^+}{\rightleftharpoons}} \underset{\text{Bronsted base}}{Cl^-}$$

(*ii*) The following reaction shows that NH_3 acts as a Bronsted base and NH_4^+ ion acts as a Bronsted acid.

$$\underset{\text{Bronsted base}}{NH_3} \underset{-H^+}{\overset{+H^+}{\rightleftharpoons}} \underset{\text{Bronsted acid}}{NH_4^+}$$

Examples of Bronsted acids and Bronsted bases

Examples of Bronsted acids and Bronsted bases are given below :

Bronsted acids *Bronsted bases*

(These can lose one or more protons) *(These can accept one or more protons)*

$$HX \underset{+H^+}{\overset{-H^+}{\rightleftharpoons}} X^-$$

$$NH_4^+ \underset{+H^+}{\overset{-H^+}{\rightleftharpoons}} NH_3$$

$$H_2S \underset{+H^+}{\overset{-H^+}{\rightleftharpoons}} HS^-$$

$$H_2S \underset{-2H^+}{\overset{-2H^+}{\rightleftharpoons}} S^{2-}$$

$$\begin{array}{c}COOH\\ |\\ COOH\end{array} \underset{+2H^+}{\overset{-2H^+}{\rightleftharpoons}} \begin{array}{c}COO^-\\ |\\ COO^-\end{array}$$

$$H_3O^+ \underset{+H^+}{\overset{-H^+}{\rightleftharpoons}} H_2O$$

$$H_2SO_4 \underset{+2H^+}{\overset{-2H^+}{\rightleftharpoons}} SO_4^{2-}$$

$$H_3SO_4^+ \underset{+H^+}{\overset{-H^+}{\rightleftharpoons}} H_2SO_4$$

$$H_3PO_4 \underset{+3H^+}{\overset{-3H^+}{\rightleftharpoons}} PO_4^{3-}$$

$$H_2CO_3 \underset{+H^+}{\overset{-H^+}{\rightleftharpoons}} HCO_3^-$$

$$H_2O \underset{+H^+}{\overset{-H^+}{\rightleftharpoons}} OH^-$$

$$CH_3COOH \underset{+H^+}{\overset{-H^+}{\rightleftharpoons}} CH_3COO^-$$

$$CH_3COOH_2^+ \underset{+H^+}{\overset{-H^+}{\rightleftharpoons}} CH_3COOH$$

$$NH_2CONH_2 \underset{+H^+}{\overset{-H^+}{\rightleftharpoons}} H_2N.\ CO.\ NH^-$$

$$NH_2CONH_3^+ \underset{+H^+}{\overset{-H^+}{\rightleftharpoons}} NH_2CONH_2$$

$$HNO_3 \underset{+H^+}{\overset{-H^+}{\rightleftharpoons}} NO_3^-$$

$$H_2NO_3^+ \underset{+H^+}{\overset{-H^+}{\rightleftharpoons}} HNO_3$$

Monoprotonic and polyprotonic Bronsted acids and Bronsted bases

Bronsted acid, which can lose one proton, is called *monoprotonic acid,* while the Bronsted acid which loses two or more protons is called *polyprotonic acid.* Examples are given below :

Monoprotonic acids	*Polyprotonic acids*
$HF \rightarrow H^- + F^-$	$H_2S \rightarrow 2H^+ + S^{2-}$
$H_3COOH \rightarrow CH_3COO^- + H^+$	$\begin{array}{c}COOH\\ \vert \\ COOH\end{array} \rightarrow 2H^+ + \begin{array}{c}COO^-\\ \vert \\ COO^-\end{array}$

Bronsted base that can accept one proton is called *monoprotonic base* while the Bronsted base that accepts two or more protons, is called *polyprotonic base.* Examples are:

Monoprotonic bases	*Polyprotonic bases*
$HS^- + H^+ \rightarrow H_2S$	$SO_4^{2-} + 2H^+ \rightarrow H_2SO_4$
$H_2O + H^+ \rightarrow H_3O^+$	$PO_4^{3-} + 3H^+ \rightarrow H_3PO_4$

All Arrhenius acids are also Bronsted acids, but all Arrhenius bases are not Bronsted bases

Let us consider the behaviour of HCl in the following two reactions :

(*i*) $HCl + H_2O \rightarrow H_3O^+ + Cl^-$ (*ii*) $HCl + NH_3 \rightarrow NH_4^+ + Cl^-$

Reaction (i) Since HCl gives H_3O^+ ion when dissolved in water, this acid acts as an Arrhenius acid in this reaction

$$HCl \rightarrow H^+ + Cl^-$$
$$H_2O + H^+ \rightarrow H_3O^+$$
$$\overline{HCl\ (\textit{Arrhenius acid}) + H_2O \rightarrow H_3O^+ + Cl^-}$$

Reaction (ii) In this reaction, since HCl loses a proton (H^+) which combines with NH_3 to form NH_4^+ ion, HCl acts as a Bronsted acid.

$$HCl \rightarrow H^+ + Cl^-$$
$$NH_3 + H^+ \rightarrow NH^4_+$$
$$\overline{HCl\ (\textit{Bronsted acid}) + NH_3 \rightarrow NH^4{}_3 + Cl^-}$$

H^+

The above discussion shows that in reaction (*i*) HCl acts as an Arrhenius acid; while in reaction (*ii*), the same substance (*i.e.,* HCl) behaves as a Bronsted acid. Thus, we can say that an Arrhenius acid also behaves as a Bronsted acid.

According to Arrhenius concept, a substance that gives OH^- ions in aqueous solution, is called Arrhenius base. Such a substance (*i.e.,* Arrhenius base) cannot accept a proton and hence, according to Bronsted-Lowry concept, cannot act as a Bronsted base. Thus, an Arrhenius base cannot act as a Bronsted base.

Conjugate acid-base pairs

In order to understand the meaning of conjugate acid-base pairs, let us consider the reaction between HCl (*aq*) and NH_3 (*aq*). When these substances react together, NH_4^+ and Cl^- ions are produced.

$$HCl + NH_3 \rightleftharpoons NH_4^+ + Cl^- \qquad ...(A)$$

Actually, the above reaction takes place through the following two steps :

(*i*) In this step HCl loses a proton and gives Cl^- ion. Thus HCl acts as an acid and Cl^- ion is its base, since this ion has been obtained by removing a proton from HCl.

$$\underset{\text{Acid}}{HCl} \rightleftharpoons \underset{\text{Base}}{Cl^-} + H^+ \qquad ...(i)$$

Here HCl (acid) - Cl^- (base) pair is called conjugate acid-base pair (conjugate = related).

(*ii*) The proton lost by HCl in step (*i*) is accepted by NH_3 molecule so that NH_4^+ ion is obtained. Thus here NH_3 acts as a base and NH_4^+ ion is its acid, since this ion has been obtained by adding a proton to NH_3.

$$\underset{\text{Base}}{NH_3} + H^+ \rightleftharpoons \underset{\text{Acid}}{NH_4^+} \qquad ...(ii)$$

Here NH_4^+ (acid) – NH_3 (base) pair is also called conjugate acid-base pair.

On adding the above two equations, we get :

$$\underset{\text{Acid}}{HCl} + \underset{\text{Base}}{NH_3} \rightleftharpoons \underset{\text{Acid}}{NH_4^+} + \underset{\text{Base}}{Cl^-}$$

This equation is the same as equation (A).

Reaction (A) shows that it consists of two conjugate acid-base pairs, which are HCl – Cl^- and NH_4^+ – NH_3. Thus, if in HCl – Cl^- pair, HCl is designated as $Acid_1$, then its conjugate base Cl^- may be designated as $Base_1$. Similarly, if in NH_4^+ – NH_3 pair, NH_4^+ is designated as $Acid_2$, then NH_3 may be represented as $Base_2$. Thus, equation (A) can be written as :

Conjugate acid-base pair ($Acid_1$ – $Base_1$)

Conjugate acid-base pair ($Base_2$ – $Acid_2$)

$$\underset{Acid_1}{HCl} + \underset{Base_2}{NH_3} \rightleftharpoons \underset{Acid_2}{NH_4^+} + \underset{Base_1}{Cl^-}$$

(HCl —H^+→ NH_3; Cl^- ←H^+— NH_4^+)

The above discussion shows that the general equation representing acid-base reaction can be written as :

$$Acid_1 + Base_2 \rightleftharpoons Acid_2 + Base_1$$

It also indicates that in order to get the conjugate base from a given acid, one proton has to be removed from that acid, *i.e.*,

$$\underset{\text{Acid}}{HCl} \xrightarrow{-H^+} \underset{\text{Conjugate base}}{Cl^- + H^+}$$

Similarly in order to get a conjugate acid from a given base one proton has to be added to that base, *i.e.*,

$$\underset{\text{Base}}{NH_3} \xrightarrow{+H^+} \underset{\text{Conjugate acid}}{NH_4^+}$$

Examples of conjugate acid-base pairs

Some examples of conjugate acid-base pairs are given in Table 19.1.

Acid-base neutralisation reaction according to Bronsted-Lowry concept

According to this concept, an acid-base neutralisation reaction is that in which a Bronsted acid reacts with Bronsted base and their conjugate base and conjugate acid respectively are produced. In the neutralisation reaction the acid (HA) gives up its proton (H^+) and produces its conjugate base (A^-). The proton given by HA is accepted by the base (B) which is thereby converted into its conjugate acid, BH^+. Thus, according to Bronsted-Lowry concept, the neutralisation reaction between the acid, HA and the base, B can be represented as :

$$\underset{\text{Acid}}{HA} + \underset{\text{Base}}{B} \xrightarrow{H^+} \underset{\text{Conjugate}}{BH^+} + \underset{\text{Conjugate}}{A^-}$$

Thus, we see that HA(acid) - A^- (base) and BH^+ (acid) - B (base) are two conjugate acid-base pairs in the above acid-base neutralisation reaction.

Table 19.1. *Some examples of conjugate acid-base pairs.*

$Acid_1$		$Base_2$		$Acid_2$ (Conjugate acid to $base_2$)		$Base_1$ (Conjugate base to $acid_1$)
HCO_3^-	+	H_2O	$\rightleftharpoons$	H_3O^+	+	CO_3^{2-}
HCN	+	H_2O	$\rightleftharpoons$	H_3O^+	+	CN^-
H_2S	+	H_2O	$\rightleftharpoons$	H_3O^+	+	HS^-
H_2O	+	CO_3^{2-}	$\rightleftharpoons$	HCO_3^-	+	OH^-
H_2O	+	H_2O	$\rightleftharpoons$	H_3O^+	+	OH^-
NH_4^+	+	H_2O	$\rightleftharpoons$	H_3O^+	+	NH_3
HCl	+	CH_3COOH	$\rightleftharpoons$	$CH_3COOH_2^+$	+	Cl^-
$HClO_4$	+	H_2SO_4	$\rightleftharpoons$	$H_3SO_4^+$	+	ClO_4^-
HCOOH	+	$H_2N.CO.NH_2$	$\rightleftharpoons$	$H_2NCONH_3^+$	+	$HCOO^-$

Amphiprotonic substances

Molecules or ions that can lose as well as accept proton(s) are called amphoprotonic substances, *i.e.*, the molecules or ions that can act as Bronsted acids (loss of proton) as well as Bronsted bases (gain of proton) are called amphoprotonic substances.

Illustration. As an example let us consider the behaviour of H_2O in the following acid-base reactions.

$$\underset{Acid_1}{\boxed{H_2O}} + \underset{Base_2}{NH_3} \xrightarrow{H^+} \underset{Acid_2}{NH_4^+} + \underset{Base_1}{OH^-} \quad ...(i)$$

$$HCl + \boxed{H_2O} \xrightarrow{H^+} H_3O^+ + Cl^- \quad ...(ii)$$

In reaction (*i*), H_2O molecule loses a proton ($H_2O \xrightarrow{-H^+} OH^-$) and hence acts as an acid while in reaction (*ii*) this molecule gains a proton from HCl ($H_2O \xrightarrow{+H^+} H_3O^+$) and hence behaves as a base. Thus since H_2O can lose as well as gain a proton, it behaves as an acid as well as a base. In other words, we can say that H_2O is an amphoteric substance.

Examples of amphoteric species (molecules and ions) are given in acid-base reactions, shown in Table 19.2. These substances have been shown in a rectangle.

Table 19.2 *Acid-base reactions in which the species shown in rectangle behave as amphoteric substances.*

	Acid$_1$		Base$_2$		Acid$_2$		Base$_1$
(*i*) (*a*)	$\boxed{CH_3COOH}$	+	H_2O	$\rightleftharpoons$	H_3O^+	+	CH_3COO^-
(*b*)	HF	+	$\boxed{CH_3COOH}$	$\rightleftharpoons$	$CH_3COOH_2^+$	+	F^-
(*ii*) (*a*)	$\boxed{H_2SO_4}$	+	$2H_2O$	$\rightleftharpoons$	$2H_3O^+$	+	SO_4^{2-}
(*b*)	HF	+	$\boxed{H_2SO_4}$	$\rightleftharpoons$	$H_3SO_4^+$	+	F^-
(*iii*) (*a*)	$\boxed{H_2N.CO.NH_2}$	+	NH_3	$\rightleftharpoons$	NH_4^+	+	$H_2N.CO.NH^-$
(*b*)	H_2SO_4	+	$\boxed{H_2N.CO.NH_2}$	$\rightleftharpoons$	$H_2N.CO.NH_3^+$	+	HSO_4^-
(*iv*) (*a*)	$\boxed{HNO_3}$	+	H_2O	$\rightleftharpoons$	H_3O^+	+	NO_3^-
(*b*)	HF	+	$\boxed{HNO_3}$	$\rightleftharpoons$	$H_2NO_3^+$	+	F^-
(*v*) (*a*)	HF	+	HNO_3	$\rightleftharpoons$	$H_2NO_3^-$	+	F^-
(*b*)	$HClO_4$	+	$\boxed{HF}$	$\rightleftharpoons$	H_2F^+	+	ClO_4^-
(*vi*) (*a*)	$\boxed{HSO_4^-}$	+	H_2O	$\rightleftharpoons$	H_3O^+	+	SO_4^{2-}
(*b*)	H_3O^+	+	$\boxed{HSO_4^-}$	$\rightleftharpoons$	H_2SO_4	+	H_2O
(*vii*) (*a*)	$\boxed{HS^-}$	+	NH_3	$\rightleftharpoons$	NH_4^+	+	S^{2-}
(*b*)	H_2O	+	$\boxed{HS^-}$	$\rightleftharpoons$	H_2S	+	OH^-

Applications of amphoprotic character of water

The tendency of water to show amphoprotic character (*i.e.*, in some acid-base reactions water behaves as an acid while in others it behaves as a base) has been utilised to explain the following:

(*i*) *Aqueons solution of NaHS is basic and gives the smell of H_2S.* When NaHS is dissolved into water, HS^- ion is obtained by the ionisation of NaHS.

$$NaHS \longrightarrow Na^+ + HS^-$$

Now since HS^- ion thus obtained has a stronger tendency than H_2O to accept a proton, this ion accepts the proton from H_2O molecule and is converted into H_2S and H_2O is changed into OH^- ion. Thus in this reaction, H_2O acts as an acid and HS^- ion behaves as a base

$$\underset{Acid}{H_2O} + \underset{Base}{HS^-} \rightleftharpoons \underset{Acid}{H_2S} + \underset{Base}{OH^-}$$

In other words we can say that NaHS behaves as a base in aqueous solution and due to the production of H_2S, aqueous solution of NaHS gives the smell of H_2S.

(*ii*) *Aqueous solution of Na_2CO_3 shows alkaline character.* When Na_2CO_3 is dissolved in H_2O, Na_2CO_3 and H_2O both get dissociated to give $CO_3{}^{2-}$ and H^+ ions respectively. These ions (*i.e.*, $CO_3{}^{2-}$ and H^+ ions) combine together to produce HCO_3 ions

$$Na_2CO_3 \rightleftharpoons 2Na^+ + CO_3{}^{2-}$$
$$H_2O \rightleftharpoons H^+ + OH^-$$
$$CO_3{}^{2-} + H^+ \rightleftharpoons HCO_3{}^-$$

$$Na_2CO_3 + H_2O \rightleftharpoons 2Na^+ + OH^- + HCO_3{}^-$$

or $$2Na^+ + CO_3{}^{2-} + H_2O \rightleftharpoons 2Na^+ + OH^- + HCO_3{}^-$$

or $$CO_3{}^{2-} + H_2O \rightleftharpoons OH^- + HCO_3{}^-$$

In the reation shown above, CO_3^{2-} ion acts as a base (proton acceptor) and H_2O behaves as an acid (proton looser). Due to the production of OH^- ions, the aqueous solution of Na_2CO_3 shows alkaline nature.

$$CO_3^{2-}\text{ (base)} + H_2O\text{ (acid)} \rightleftharpoons HCO_3^-\text{ (acid)} + OH^-\text{ (base)}$$

(*iii*) *Aqueous solution of the salts of Fe^{3+} and Al^{3+} shows acidic character.* Aqueous solution of the salts of Fe^{3+} and Al^{3+} from hexahydrated cations, $[M(H_2O)_6]^{3+}$ (M = Fe, Al) which act as acid in water, since they donate one proton to H_2O which is converted into H_3O^+. Due to the production of H_3O^+ ions, the aqueous solution of the salts of Fe^{3+} and Al^{3+} show acidic character.

$$\underset{\text{Acid}}{[M(H_2O)_6]^{3+}} + \underset{\text{Base}}{H_2O} \rightleftharpoons \underset{\text{Acid}}{H_3O^+} + \underset{\text{Base}}{[M(H_2O)_5(OH)]^{2+}}$$

Relative strength of Bronsted acids and Bronsted bases in aqueous medium

Relative strength of Bronsted acids and Bronsted bases can be determined by treating these substances with water. When an acid reacts with H_2O, water acts as a base in this reaction. In the reaction between a base and water, water acts as an acid. Here we shall consider the reaction of HCl (strong acid) and CH_3COOH (weak acid) with H_2O. The strenght of an acid depends on its ability to transfer its proton (H^+) to H_2O (which acts as a Bronsted base) to form its conjugate base. When a monoprotonic acid like HCl, CH_3COOH etc., reacts with H_2O, it transfers it proton to H_2O to form H_3O^+ ion and a conjugate base of the monoproton acid.

(*i*) *Reaction of HCl (strong acid) with H_2O (base).* Since HCl is a strong acid, it is nearly 100% ionised in water and hence this reaction proceeds almost completely towards right. This state has been shown by a longer half arrow, pointing towards right. Thus the reaction of HCl with H_2O can be depicted as :

Conjugate strong acid-weak base pair (HCl → Cl^-)

$$\underset{\text{Strong acid}}{HCl} + \underset{\text{Strong base}}{H_2O} \rightleftharpoons \underset{\text{Weak acid}}{H_3O^+} + \underset{\text{Weak base}}{Cl^-} \quad ...(i)$$

Conjugate weak acid-strong base pair (H_2O → H_3O^+)

In the above reaction the two acids *viz.*, HCl and H_3O^+ are competing for donating a proton to the base ($HCl \xrightarrow{-H^+} Cl^-$, $H_3O^+ \xrightarrow{-H^+} H_2O$). Since the equilibrium of this reaction lies towards right, HCl donates the proton more strongly than does H_3O^+ ion. Thus HCl is a stronger acid than H_3O^+ ion. In other wards, we can say that HCl is a strong acid and H_3O^+ ion is a weak acid.

In this reaction the two bases namely H_2O and Cl^- are competing for the gain of a proton from the acid ($H_2O \xrightarrow{+H^+} H_3O^+$, $Cl^- \xrightarrow{+H^+}$ HCl). Since the reaction proceeds towards right, H_2O molecule gains the proton more strongly than does Cl^- ion. Thus H_2O is stronger base than Cl^- ion, *i.e.*, H_2O is a strong base and Cl^- ion is a weak base.

(*ii*) *Reaction of CH_3COOH (weak acid) with H_2O (base).* Since CH_3COOH is a weak acid, it is less than 1% ionised in H_2O and hence this reaction proceeds mostly towards left. This state of affairs is represented by using a longer half arrow, pointing towards left. Thus, the reaction of CH_3COOH with H_2O can be shown as :

Conjugate weak acid-strong base pair (CH_3COOH → CH_3COO^-)

$$\underset{\text{Weak acid}}{CH_3\,COOH} + \underset{\text{Weak base}}{H_2O} \rightleftharpoons \underset{\text{Strong acid}}{H_3O^+} + \underset{\text{Strong base}}{CH_3\,COO^-} \quad ...(ii)$$

Conjugate strong acid-weak base pair (H_2O → H_3O^+)

In the above reaction, the two acids *viz.*, CH_3COOH and H_3O^+ are competing for donating a proton to the base ($CH_3COOH \xrightarrow{-H^+} CH_3COO^-$, $H_3O^+ \xrightarrow{-H^+} H_2O$). Since the equilibrium of the reaction lies towards the left, H_3O^+ ion donates the proton more strongly than does CH_3COOH molecule. Thus H_3O^+ ion is stronger acid than CH_3COOH, *i.e.*, H_3O^+ ion is a strong acid and CH_3COOH is a weak acid.

In the above reaction, the two bases, namely H_2O molecule and CH_3COO^- ion, are competing for the gain of a proton from the acid ($H_2O \xrightarrow{+H^+} H_3O^+$, $CH_3COO^- \xrightarrow{+H^+} CH_3COOH$). Since the reaction proceeds towards left, CH_3COO^- ion gains the proton more strongly than does H_2O. Thus CH_3COO^- ion is stronger base than H_2O, *i.e.*, CH_3COO^- ion is a strong base and H_2O is a weak base.

Comparison between the basic strength of the three bases namely H_2O, Cl^- and CH_3COO^-

We have seen above that in reaction (*i*), Cl^- ion is a weak base and H_2O is a strong base *i.e.*, $H_2O > Cl^-$. In reaction (*ii*), CH_3COO^- ion is a strong base and H_2O is a weak base *i.e.*, $CH_3COO^- > H_2O$. Thus the basic strength of these three bases is in the order $CH_3COO^- > H_2O > Cl^-$. That the basic strength of CH_3COO^- ion is greater than that of Cl^- ion can also be shown as follows. Since reaction (*i*) proceeds towards right and reaction (*ii*) proceeds towards left, the base namely Cl^- ion has lesser tendency than the base *viz.*, CH_3COO^- ion to accept the proton and hence is much weaker base than CH_3COO^- ion ($Cl^- < CH_3COO^-$).

Important conclusions

From the above study we can conclude that :

(*i*) In a given conjugate acid-base pair, if the acid is strong, its conjugate base will be weak. Similarly if the acid is weak, its conjugate base will be strong. This means, that only strong acid-weak base and weak acid-strong base pairs are allowable.

(*ii*) The acid-base reaction, in which a strong acid reacts with a strong base, proceeds completely towards right and hence weak acid and weak base are produced. On the other hand, the acid-base reaction in which a weak acid reacts with a weak base, proceeds towards left *i.e.*, if a weak acid reacts with a weak base, no forward reaction will take place. Thus :

$$\underset{(e.g.,\ HCl)}{\text{Strong acid}} + \underset{(e.g.,\ H_2O)}{\text{Strong base}} \rightleftharpoons \underset{(e.g.,\ H_3O^+)}{\text{Weak acid}} + \underset{(e.g.,\ Cl^-)}{\text{Weak base}}$$

$$\underset{(e.g.,\ CH_3COOH)}{\text{Weak acid}} + \underset{(e.g.,\ H_2O)}{\text{Weak base}} \rightleftharpoons \underset{(e.g.,\ H_3O^+)}{\text{Strong acid}} + \underset{(e.g.,\ CH_3COO^-)}{\text{Strong base}}$$

(*iii*) If an acid (HA) is stronger than H_3O^+, then the reaction of this acid with H_2O (which acts as a base) will proceed mostly towards right. On the other hand, if the acid (HA) is weaker than H_3O^+, then the reaction of this reaction with H_2O will proceed towards left.

$$\underset{\text{Strong acid}}{HA} + \underset{\text{Base}}{H_2O} \rightleftharpoons \underset{\text{Weak acid}}{H_3O^+} + OH^-$$

$$\underset{\text{Weak acid}}{HA} + \underset{\text{Base}}{H_2O} \rightleftharpoons \underset{\text{Strong acid}}{H_3O^+} + OH^-$$

(*iv*) On similar lines it may be shown that if a base (B) is stronger than OH^- ions, then the reaction of this base with H_2O (which acts as an acid) will proceed towards right. On the other hand if the base (B) is weaker than OH^- ions, then the equilibrium will lie towards left.

$$\underset{\text{Strong base}}{B} + \underset{\text{Acid}}{H_2O} \rightleftharpoons BH^+ + \underset{\text{Strong base}}{OH^-}$$

$$\underset{\text{Weak base}}{B} + \underset{\text{Acid}}{H_2O} \rightleftharpoons BH^+ + \underset{\text{Strong acid}}{OH^-}$$

Levelling and differentiating solvents

Let us consider the ionisation of HCl, HBr and HI in two different solvents like water and CH_3COOH.

(*i*) *Ionisation in H_2O.* When these acids are dissolved in water, they get ionised to the *same extent* and hence appear to be almost equally strong in this solvent. H^+ ion lost by these acids is accepted by H_2O (which acts as a base) to form H_3O^+ ion.

$$\underset{\text{acid}}{HX\ (X = Cl, Br, I)} + \underset{\text{base (solvent)}}{H_2O} \rightleftharpoons \underset{\text{acid}}{H_3O^+} + \underset{\text{base}}{X^-}$$

Thus, we find that water (solvent) brings all the three acids to the same level of their acidic strength and hence is called a levelling solvent for these acids. The phenomenon due to which the levelling solvent makes all the acids behave equally strong is called levelling effect. It is due to the levelling effect that the relative acid strength of the above acids cannot be measured in aqueous medium.

(*ii*) *Ionisation in CH_3COOH.* When the above said acids are dissolved in CH_3COOH, they are ionised to *different extents* and hence show different acidic strength in this solvent.

$$\underset{\text{acid}}{HX\ (X = Cl, Br, I)} + \underset{\text{base (solvent)}}{CH_3COOH} \rightleftharpoons \underset{\text{acid}}{CH_3COOH_2^+} + \underset{\text{base}}{X^-}$$

Since CH_3COOH (solvent) brings the acids to different level of their acid strength, this solvent is called a differentiating solvent for these acids.

The above two examples show that water acts as levelling solvent for HCl, HBr and HI, but CH_3COOH acts as a differentiating solvent for the same acids.

Binary hydride anions derived from binary hydrides are Bronsted bases

The hydride anions derived from the binary hydrides are Bronsted bases. Examples of such anions are CH_3^-, AH_2^-, (A = N, P, As, Sb, Bi), AH^- (A = O, S, Se, Te, Po), X^- (X = F, Cl, Br, I). These anions act as Bronsted bases because they have a tendency to accept a proton (H^+) to form their corresponding Bronsted acids (conjugate acid) as shown below :

Base (Hydride anion)		*Conjugate acid*	*Base (Hydride anion)*		*Conjugate acid*
CH_3^-	$\xrightarrow{+H^+}$	CH_4	AH^-	$\xrightarrow{+H^+}$	AH_2
			(A = O, S, Se, Te Po)		
AH_2^-	$\xrightarrow{+H^+}$	AH_3	X^-	$\xrightarrow{+H^+}$	HX
(A = N, P, As, Sb, Bi)			(X = F, Cl, Br, I)		

What is basicity or basic strength of a Bronsted base ?

According to Bronsted-Lowry concept of acids and bases, the basicity of a molecule or an ion that acts as a Bronsted base, is defined as its tendency to accept (add) a proton (H^+) to form its conjugate acid. If the given molecule or ion accepts the proton quite readily, it is said to have high basicity. On the other hand, if a molecule or an ion gains the proton less readily, it has low basicity.

For example, since F^- ion has greater power to accept a proton to form its conjugate acid, HF, while Cl^- ion has a lesser power to do so to form HCl (conjugate acid), F^- ion has higher basic strength (*i.e.,* proton accepting power) than Cl^- ion. ($F^- > Cl^-$). In short, we say that the basic strenght of a Bronsted base is defined *as its proton accepting power.*

Variation of the basicity of the hydride anions of the non-metals of the same period

Let us consider the anions namely CH_3^-, NH_2^-, OH^- and F^-. These anions have been derived from the hydrides CH_4, NH_3, H_2O and HF respectively. These anions act as Bronsted bases, since they can accept a proton to give their corresponding conjugate acids as shown below:

It has been found that the basicity of the given anions decreases as we move from left to right in the period, *i.e.* the basicity is in the order $CH_3^- > NH_2^- > OH^- > F^-$. This order can be explained as follows:

Base (Hydride ion)		*Conjugate acid*	*Base (Hydride ion)*		*Conjugate acid*
CH_3^-	$\xrightarrow{+H^+}$	CH_4	OH^-	$\xrightarrow{+H^+}$	OH_2
NH_2^-	$\xrightarrow{+H^+}$	NH_3	F^-	$\xrightarrow{+H^+}$	HF

(*i*) We know that in the formation of the given anions, H-atom overlaps with the negatively-charged central atom and hence some volume of the central atom is occupied by H-atoms and the remaining volume of the central atom is occupied by the electron due to which the central atom is carrying negative charge on it. Since the number of H-atoms overlapping the central atom is decreasing from CH_3^- to F^- ions, the volume occupied by H-atoms is also decreasing from CH_3^- to F^- ions and hence the volume occupied by the electron is increasing. In case of F^- ion, since no H-atom is present, the whole volume of this ion is occupied by electron. Thus :

Ions :	CH_3^-	NH_2^-	OH^-	F^-
No. of H-atoms :	3	2	1	zero
Volume occupied by H-atoms:	$1 \times \frac{3}{4} = \frac{3}{4}$	$1 \times \frac{2}{3} = \frac{2}{3}$	$1 \times \frac{1}{2} = \frac{1}{2}$	zero
Volume occupied by the electron:	$\left(1 - \frac{3}{4}\right) = \frac{1}{4}$	$\left(1 - \frac{2}{3}\right) = \frac{1}{3}$	$\left(1 - \frac{1}{2}\right) = \frac{1}{2}$	1 (whole volume)

The increase in the volume occupied by the electron produces an increase in the delocalisation of the electron and hence the electron density round the central atom decreases from CH_3^- to F^-. The decrease in the electron density decreases the tendency of the anion to accept the proton and hence the basicity of the anion decreases from CH_3^- to F^-. Thus we see that the basicity of the given anions decreases as $CH_3^- > NH_2^- > OH^- > F^-$. In general we can say that the basicity of the anions derived from the hydrides of the elements of the same period decreases as we move from left to right in the period.

(*ii*) We know that in a given conjugate acid-base pair, if the acid is strong, its conjugate base will be weak and *vice versa.* Thus since the acidic strength of CH_4, NH_3, H_2O and HF which are conjugate acids of CH_3^-, NH_2^-, OH^- and F^- ions respectively increases as $CH_4 < NH_3 < H_2O < HF$, the basic strength of the given bases should decrease in the same direction, *i.e.,* the basic strength of the hydride anions should be in the order $CH_3^- > NH_2^- > OH^- > F^-$.

Variation of the basicity of the hydride anions of the elements of groups VA, VIA and VIIA

The hydride anions of the elements of these groups are NH_2^-, PH_2^-, AsH_2^-, SbH_2^-, BiH_2^- (group VA), OH^-, SH^-, HSe^-, HTe^-, HPo^- (group VIA) and F^-, Cl^-, Br^-, I^- (group VIIA). These anions have a tendency to accept a proton and hence, according to Bronsted-Lowry concept of bases, act as Bronsted bases. After accepting the proton they give their corresponding conjugate acids.

It has been found that the basicity of the hydride anions of the elements of a given group decreases as we move down the group. This progressive decrease can be explained by using the same argument that has been used in explaining the successive decrease in the basicity of the hydride anions of the non-metals along the period (*e.g.*, $CH_3^- > NH_2^- > OH^- > F^-$). Thus the basicity of the hydride anions of the elements of groups VA, VIA and VIIA decrease in the following order :

Group VA : $NH_2^- > PH_2^- > AsH_2^- > SbH_2^- > BiH_2^-$

Group VIA : $OH^- > SH^- > SeH^- > TeH^- > PoH^-$

Group VIIA : $F^- > Cl^- > Br^- > I^-$

Summary. We have seen above that : (*i*) On proceeding from left to right in a given period, the basicity of the hydride anions of the non-metals decreases. (*ii*) On moving down a group, the basicity of the hydride anions of the elements decreases. On combining these two conclusions, we can say that the basicity of the hydride anions decreases, on proceeding from left to right in a period, as well as on moving down a group.

——— *Basicity decreases* ——→ (across); ↓ *Basicity decreases* (down)

CH_3^-	NH_2^-	OH^-	F^-
	PH_2^-	SH^-	Cl^-
	AsH_2^-	SeH^-	Br^-
	SbH_2^-	TeH^-	I^-
	BiH_2^-	PoH^-	

Hydrides of the elements of group VIA (H_2O, H_2S, H_2Se, H_2Te and H_2Po) and group VIIA (HF, HCl, HBr, HI) are Bronsted acids

The hydrides of the elements of group VIA are Bronsted acids, since they have a tendency to release a proton (H^+) to form their corresponding Bronsted base (conjugate base), e.g.,

$$H_2S \rightleftharpoons H^+ + HS^-; \qquad HS^- \rightleftharpoons H^+ + S^{2-}$$

The hydrides of the elements of group VIIA are essentially covalent in their gaseous state but in aqueous solution, they ionise to give hydrated proton, H^+ (*aq*) and hence act as Bronsted acids.

$$HX\ (g) + aq \rightarrow HX\ (aq) \xrightarrow{\text{Ionisation}} H^+\ (aq) + X^-\ (aq),$$

or
$$HX\ (g) + H_2O \longrightarrow H_3O^+ + X^-(aq)$$

What is acidity or acidic strength of a Bronsted acid ?

According to Bronsted-Lowry concept of acids and bases, the acidity or acidic strength of a molecule or an ion which acts as a Bronsted acid is defined as its tendency to lose (release) a proton (H^+) to form its conjugate base. If the given molecule or ion loses the proton quite readily, it is said to have higher acidity. On the other hand, if a molecule or an ion loses the proton less readily, it has low acidity. For example, since H_2O has lesser power to lose H^+ to form OH^- ions (conjugate base of H_2O) while H_2S has greater power to do so to give HS^- ions (conjugate base of H_2S), the former has lesser acidic strength than the latter ($H_2O < H_2S$). In short we say that the acidic strength of a Bronsted acid is defined as its *proton-releasing power.*

Variation of the acidity or acidic strength (proton releasing power) of the hydrides of the non-metals of the same period. Let us consider the hydrides of the non-metals of 2*nd* period of the periodic table namely CH_4, NH_3, H_2O and HF. These hydrides have a tendency to release a proton (H^+) and hence, according to Bronsted-Lowry concept of acids and bases, act as Bronsted acids. After losing the proton they give their conjugate bases as shown below:

Acid (Hydride molecule)		*Conjugate base*	*Acid (Hydride molecule)*		*Conjugate base*
CH_4	$\xrightarrow{-H^+}$	CH_3^-	H_2O	$\xrightarrow{-H^+}$	OH^-
NH_3	$\xrightarrow{-H^+}$	NH_2^-	HF	$\xrightarrow{-H^+}$	F^-

Experiments have shown that the acidity of CH_4, NH_3, H_2O and HF molecules increases as we move from left to right in the period *i.e.,* the acidity is in the order : $CH_4 < NH_3 < H_2O < HF$. This order can be explained in any of the following ways :

(*i*) *Explanation based on the electronegativity of the central atom of the molecule.* As we move from CH_4 to HF, the electrongativity of the central atom (A) increases and hence its tendency to pull the electron pair of A – H bond (A = C, N, O, F) towards itself also increases. This results in that the electronic charge density on H-atom of A –H bond is decreased. The decrease produced in the electron density on H-atom makes the release of H-atom as H^+ ion (proton) easy by the rupture of A – H bond. Thus we see that as the electronegativity of the central atom increases, the possibility of getting H^+ ions from the hydride molecule also increases and hence acidic strength of these molecules also increases from CH_4 to HF.

(*ii*) *Explanation based on Bronsted-Lowry concept of acids and bases.* Since CH_4, NH_3, H_2O, and HF molecules, after loosing a proton, give CH_3^-, NH_2^-, OH^- and F^- ions which are conjugate bases, CH_4 (acid)- CH_3^- (base), NH_3 (acid)- NH_2^- (base), H_2O (acid) - OH^- (base) and HF (acid) - F^- (base) are conjugate acid-base pairs. We have already stated that the basicity of CH_3^-, NH_2^-, OH^- and F^- ions (which are conjugate bases of CH_4, NH_3, H_2O and HF respectively) decreases as $CH_3^- > NH_2^- > OH^- > F^-$. Again we know that, according to Bronsted-Lowry concept, in a given conjugate acid-base pair, if the acid is strong, its conjugate base would be weak and *vice versa.* Thus since the basicity of CH_3^-, NH_2^-, OH^- and F^- ions is in the decreasing order ($CH_3^- > NH_2^- > OH^- > F^-$), the acidity of their conjugate acids (*viz.*, CH_4, NH_3, H_2O and HF respectively) would be in the increasing order ($CH_4 < NH_3 < H_2O < HF$).

Both the explanations show that the acidity of the hydrides of the non-metals of the same period (*e.g.*, CH_4, NH_3, H_2O, HF) increases as we pass from left to right in the given period.

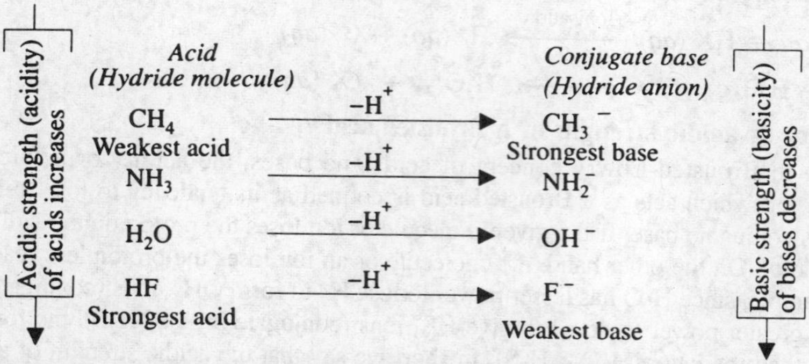

Relative order of the acidic strength of H_2O, H_2S, H_2Se and H_2Te

We have said above that the given hydrides have a tendency to lose a proton (H^+) and hence act as Bronsted-Lowry acids. It has been observed that their acidic strength (*i.e.,* proton releasing power) increases as we move down the group from H_2O to H_2Te. Thus the acidic strength is in the order $H_2O < H_2S < H_2Se < H_2Te$. This increasing order is also confirmed by the fact that the values of dissociation constant (K_a) for the given molecules also increase in the same order ($H_2O = 1.0 \times 10^{-14}$, $H_2S = 1.0 \times 10^{-7}$, $H_2Se = 1.7 \times 10^{-4}$, $H_2Te = 2.3 \times 10^{-3}$).

Explanation. The above increasing order of the acidic strength can be explained as follows :

(*a*) *Explanation based on the strength of A—H bond (A = O, S, Se, Te).* As we move down the group, the size of the central atom (A) of the hydride increases gradually and hence the length of

A—H bond also increases in the same direction. With the increase in A—H bond length, the strength of A—H bond decreases from O—A bond to O—Te bond. Consequently, the tendency of A—H bond to break to give H^+ ions increases as we move down the group *i.e.*, the acidic strength (*i.e.*, H^+ ion releasing power) increases as $H_2O < H_2S < H_2Se < H_2Te < H_2Po$.

(*b*) *Explanation based on the strength of conjugate acid-base pair.* Since the given hydride molecules, after losing a proton (H^+), give OH^-, SH^-, HSe^- and HTe^- ions respectively, which are their conjugate bases, H_2O (acid)- OH^- (base), H_2S (acid)- HS^- (base), H_2Se (acid)- HSe^- (base) and H_2Te (acid)- HTe^- (base) are conjugate acid-base pairs. We have already stated that since the basic strength (*i.e.*, proton accepting power) of OH^-, SH^-, SeH^- and TeH^- ions (which are conjugate bases of H_2O, H_2S, H_2Se and H_2Te molecules respectively) decreases as $OH^- < SH^- < SeH^- < TeH^-$, the acidic strength (*i.e.*, proton releasing power) of H_2O, H_2S, H_2Se and H_2Te which are conjugate acids of OH^-, SH^-, SeH^- and HTe^- ions respectively should increase as $H_2O < H_2S < H_2Se < H_2Te$.

Relative order of the acidity or acidic strength (*i.e.*, proton releasing ability) of HX molecules in aqueous solution

HX molecule in aquous solution is represented as HX (*aq*). We have already said that HX (*aq*) (X = F, Cl, Br, I) ionises and gives hydrated proton, H^+ (*aq*). It is due to the production of H^+ (*aq*) ions that HX molecules in aqueous solution behave as Bronsted acids. In the ionisation of HX (*aq*) molecule some energy is also released (evolved). This energy is called **heat of ionisation of HX (*aq*) molecule**. The magnitude of this heat is a measure of the acidic strength of HX (*aq*). Larger is the amount of this energy, greater is the degree of ionisation of HX (*aq*) and hence greater is the acidic strength of HX (*aq*). Since heat of ionisation is the heat released, it is represented by a negative (–) sign.

$$\underset{(X = F, Cl, Br, I)}{HX(aq)} \xrightarrow{\text{Ionisation}} \underset{\text{Hydrated proton}}{H^+(aq)} + X^-(aq)\ ; \quad \begin{aligned} &= \text{Heat released} \\ &= -\text{ Heat of ionisation of } HX(aq) \\ &= (\Delta H)_{HX(aq)} \end{aligned} \quad \text{.......................(A)}$$

$$\text{or} \qquad HX + H_2O \longrightarrow H_3O^+ + X^-(aq)$$

The values of heat of ionisation (in KJ/mole) of HF (*aq*), HCl (*aq*), HBr (*aq*) and HI (*aq*) are given below: (For the sake of convenience – sign has been omitted). These values show that the acidic strength of HX molecules in aqueous solution is in the order:

HX(aq) molecules :	HF (*aq*)	<	HI (*aq*)	<	HCl (*aq*)	< HBr (*aq*)
Heat of ionisation : (kJ/mole)	8.0		57.8		59.8	63.6
	Increasing →					
Acidic strength of HX (aq) molecules :	HF (*aq*)	<	HI (*aq*)	<	HCl (*aq*)	< HBr (*aq*)
	Increasing →					

Calculation of heat of ionisation for the ionisation reaction of HX (*aq*) molecules.

We have shown above that the ionisation of HX (*aq*) molecule is represented by equation (A). In this reaction heat equal to heat of ionisation of HX (*aq*) is released. Heat of ionisation of HX (*aq*) molecule is denoted by $(\Delta H)_{HX(aq)}$.

The value of $(\Delta H)_{HX(aq)}$ can be calculated by using the fact that ionisation reaction shown at (A) takes place through the following steps :

(*a*) *Dehydration of HX (aq) to HX (g).* HX (*aq*) molecule absorbs (+ sign) heat equal to its heat of dehydration ($+\Delta H_1$) and gets dehydrated to give HX (*g*).

$$HX\ (aq) \xrightarrow{\text{Dehydration}} HX\ (g)\ ;\ \Delta H = \text{Heat absorbed}$$
$$= +\ \text{Heat of dehydration of HX}\ (aq)$$
$$= +\Delta H_1$$

(*b*) *Dissociation of HX (g) molecule to H(g) and X (g) atoms.* In this step HX (*g*) molecule formed in step (*a*) absorbs (+ sign) heat equal to H–X bond dissociation energy of HX (*g*) molecule ($+\Delta H_2$) and gets dissociated into H(*g*) and X(*g*) atoms.

$$HX\ (aq) \xrightarrow{\text{Dissociation}} H\ (g) + X\ (g);\ \Delta H = \text{Heat absorbed}$$
$$= \text{Heat of dissociation of H – X bond}$$
$$= +\ \Delta H_2$$

(*c*) *Removal of an electron from H (g) atom to get $H^+(g)$ ion.* In this step H(*g*) atom formed in step (*b*) absorbs (+ sign) energy equal to its ionisation energy ($+\Delta H_3$) and is converted into H^+ (*g*) by losing an electron. The value of ionisation energy of H(*g*) atom (ΔH_3) has been found to be equal to 1312 kJ/mol. Thus $\Delta H_3 = +\ 1312$ kJ/mol.

$$H\ (g) \xrightarrow[\text{an electron}]{\text{Removal of}} H^+(g) + e^-;\ \Delta H = \text{Energy absorbed}$$
$$= \text{Ionisation energy of H}(g)$$
$$= +\ \Delta H_3 = +\ 1312\ \text{kJ/mol.}$$

(*d*) *Addition of electron to X(g) atom to get X–(g) ion.* X(*g*) atom formed in step (*b*) gains an electron obtained in step (*c*) and is converted into X^-(*g*) ion. In this step energy equal to electron affinity (ΔH_4) of X(*g*) atom is released (– sign).

$$X\ (g) + e^- \xrightarrow[\text{electron}]{\text{Addition of}} X^-(g)\ ;\ \Delta H = \text{Energy released}$$
$$= -\ \text{Electron affinity of X(g) atom}$$
$$= -\ \Delta H_4$$

Here it may be noted that although electron affinity is $= -\ \Delta H_4$ represented by + sign, but here it has been denoted by – sign.

(*e*) *Hydration of H^+ (g) ion to hydrated proton, $H^+(aq)$.* In this step, energy equal to hydration energy of H^+(*g*) ion (ΔH_5) is released (– sign). The value of ΔH_5 has been found to be equal to 1091 kJ/mol.

$$H^+\ (g) + aq \xrightarrow{\text{Hydration}} H^+(aq);\ \Delta H = \text{Heat released}$$
$$= -\ \text{Hydration energy of}\ H^+\ (g)\ \text{ion}$$
$$= -\ \Delta H_5 = -1091\ \text{kJ/mol}$$

(*f*) *Hydration of X^- (g) ion to X^- (aq) ion.* In this step, energy equal to hydration energy of X^- (*g*) ion (ΔH_6) is released (– sign)

$$X^-(g) + aq \xrightarrow{\text{Hydration}} X^-(aq);\ \Delta H = \text{Heat released}$$
$$= -\ \text{Hydration energy of}\ X^-\ (g)\ \text{ion}$$
$$= -\ \Delta H_6$$

If all the equations given at (*a*), (*b*), (*c*), (*d*), (*e*) and (*f*) are added, we get :

$$HX\ (aq) + 2\ aq \longrightarrow H^+\ (aq) + X^-\ (aq)$$

or $$HX\ (aq) \longrightarrow H^+\ (aq) + X^-\ (aq)$$

Since this equation is the same as equation (A), the sum of ΔH_1, ΔH_2, ΔH_3, ΔH_4, ΔH_5 and ΔH_6 should be equal to $(\Delta H)_{HX(aq)}$. Thus :

$$(\Delta H)_{HX(aq)} = \Delta H_1 + \Delta H_2 + \Delta H_3 + \Delta H_4 + \Delta H_5 + \Delta H_6$$
$$= \Delta H_1 + \Delta H_2 + 1312 - \Delta H_4 - 1091 - \Delta H_6$$

This equation can be used to calculate the value of $(\Delta H)_{HX(aq)}$ for the ionisation of HF(*aq*), HCl (*aq*), HBr(*aq*) and HI(aq) as shown in Table 19.3.

Table 19.3. Calculation of heat of ionisation, $(\Delta H)_{HX(aq)}$ for the ionisation reaction, $HX(aq) \longrightarrow H^+(aq) + X^-(aq)$

Ionisation reactions	Values of heat of ionisation, $(\Delta H)_{HX(aq)}$ (in kJ/mole)
$HF\ (aq) \longrightarrow H^+\ (aq) + F^-(aq)$	$(\Delta H)_{HF(aq)} = \Delta H_1 + \Delta H_2 + \Delta H_3 + \Delta H_4 + \Delta H_5 + \Delta H_6 = +48 + 566 + 131\ 2 - 328 - 1091 - 515 = -8.0$
$HCl\ (aq) \longrightarrow H^+\ (aq) + Cl^-(aq)$	$(\Delta H)_{HCl(aq)} = +18 + 431 + 1312 - 348.8 - 1091 - 381 = -59.8$
$HBr\ (aq) \longrightarrow H^+\ (aq) + Br^-(aq)$	$(\Delta H)_{HBr(aq)} = +21 + 366 + 1312 - 324.6 - 1091 - 347 = -63.6$
$HI\ (aq) \longrightarrow H^+\ (aq) + I^-(aq)$	$(\Delta H)_{HI(aq)} = +23 + 299 + 1312 - 295.4 - 1091 - 305 = -57.8$

It is evident from the values of $(\Delta H)_{HX(aq)}$ given in Table 19.3 that these values are in order : $HF(aq)$ (=8.0) < $HI(aq)$ (=57.8) < $HCl\ (aq)$ (=59.8) <$HBr\ (aq)$ (=63.6) (Negative sign has not been considered). All the steps given at (*A*), (*a*), (*b*), (*c*), (*d*), (*e*) and (*f*) can be shown in the form of a cycle known as Born-Haber cycle given in Fig.19.1.

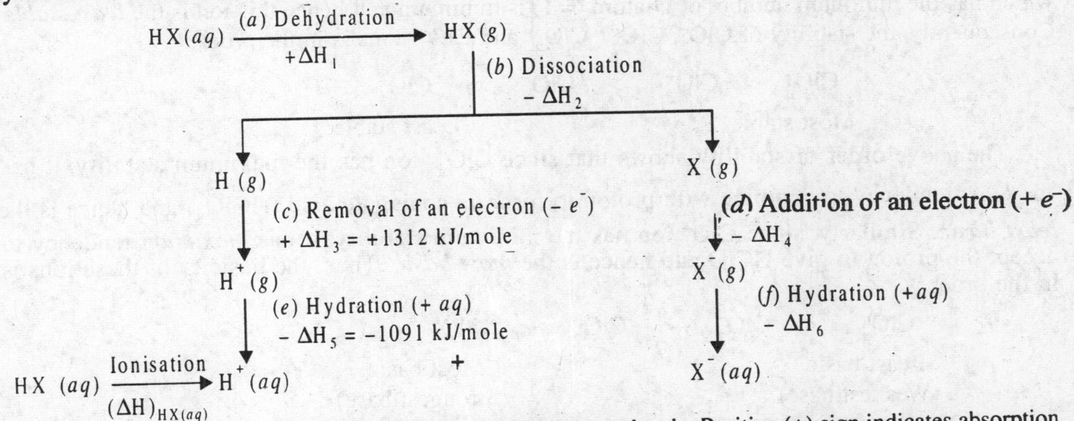

Fig. 19.1. Born-Haber cycle for the ionisation of $HX(aq)$ molecule. Positive (+) sign indicates absorption of energy while negative (–) sign indicates release of energy.

Why is the aqueous solution of HF weakest acid ?

We have said above that HF, in its aqueous solution, behaves as the weakest acid. Its weakest character is due to the fact that the dissociation energy of H—F bond in HF molecule is the highest (= 566 kJ/mol) and hence this molecule has least tendency to undergo ionisation to produce hydrated H^+ ions.

Relative order of acid strength of HX molecules in methanol solvent

We have seen that on the basis of H_I values the acid strength of HX molecules in aqueous solution should be in the order HF (*aq*) < HI (*aq*) < HCl (*aq*) < HBr (*aq*). However, the acid strength of these molecules in methanol solvent has been found in the order HF < HCl < HBr < HI. Thus we find that this order is not the same as that predicted by H_I values. In other words we say that H_I values are not able to explain the order of acid strength which has been found experimentally in methanol solution. This discrepancy has been explained on the basis of entropy changes (ΔS) accompanying the process of ionisation, the discussion of which is beyond the scope of this book.

Summary

We have seen above that (*i*) on proceeding from left to right in a given period, the acidity or acidic strength of the hydrides of non-metals increases (*ii*) on moving down a given group the acidity of the hydrides of the elements increases. On combining these two conclusions we can say that the acidity of the hydrides increases on proceeding from left to right in a given period as well as on moving down a given group. (See Fig. 19.2).

Relative order of the stability and basicity of the oxy-anions of a given element. As an example, let us find out the order of stability and basicity of the oxy-anions given by Cl-atom. The oxy-anions given by Cl-atom are ClO^-, ClO_2^-, ClO_3^- and ClO_4^-. Since O-atom is more electronegative than Cl-atom, the unit negative charge on each oxy-anion will be localised on O-atom rather than on Cl-atom. Since ClO^- ion has only one O-atom, the whole of unit negative charge will be localised on this O-atom. Since ClO_2^- ion has two O-atoms, the unit negative is now distributed on two O-atoms of ion. Similarly in case of ClO_3^- and ClO_4^- ions, the unit negative charge is distributed on three and four O-atoms, in these ions respectively. Thus, we see that since ClO_4^- ion has the maximum number of O-atoms (= 4), the distribution or spreading of the unit negative charge in this ion is maximum and hence this ion is the *most stable.* On the other hand, the distribution of unit negative charge in ClO^- ion, which has the minimum number of O-atom (= 1) is minimum and hence this ion is the *least stable.* Consequently, the stability of ClO^-, ClO_2^-, ClO_3^- and ClO_4^- ions is in the order :

Acidity increases (left to right) →; Acidity increases (top to bottom) ↓

CH_4	NH_3	H_2O	HF
	PH_3	H_2S	HCl
	AsH_3	H_2Se	HBr
	SbH_3	H_2Te	HI
	BiH_3	H_2Po	

Fig 19.2 : Variation of the acidity of the hydrides, on proceeding from left to right in a period and on proceeding from top to bottom, in a group.

$$ClO_4^- > ClO_3^- > ClO_2^- > ClO^-$$

Most stable — Least stable

The above order of stability shows that since ClO_4^- ion has the maximum stability, it has the *least* tendency to combine with proton to form its conjugate acid ($HClO_4$) and hence is the *least basic.* Similarly, since ClO^- ion has the minimum stability, it has *maximum* tendency to accept the proton to give HClO and hence is the *most basic.* Thus, the basicity of these ions is in the order :

$$ClO_4^- < ClO_3^- < ClO_2^- < ClO^-$$

Least basic (Weakest base) — Most basic (Strongest base)

or

$$ClO^- > ClO_2^- > ClO_3^- > ClO_4^-$$

Most basic (Strongest base) — Least basic (Weakest base)

Oxidation number of Cl-atom increases → (+1 → +3 → +5 → +7)

The above discussion can be concluded by saying that the basicity of oxy-anions, having the same central atom, decreases with the increase in the number of O-atoms present in the oxy-anion or with the increase in oxidation number of the central atom. Thus, the relative order of the basicity of some oxy-anions is as given below :

$$\overset{+1}{BrO^-} > \overset{+5}{BrO_3^-}$$

$$\overset{+1}{IO^-} > \overset{+4}{IO_3^-} > \overset{+7}{IO_4^-}$$

$$\overset{+3}{NO_2^-} > \overset{+5}{NO_3^-}$$

$$\overset{+4}{SO_4^{2-}} > \overset{+4}{SO_3^{2-}}$$

Relative order of the acidic strength of oxy-acids of the same element in different oxidation state

As an example, let us consider the oxy-acids of Cl-atom *viz.,* $\overset{+1}{HClO}$, $\overset{+3}{HClO_2}$, $\overset{+5}{HClO_3}$ and

$\overset{+7}{HClO_4}$. It has been shown earlier that the acidic strength of the given acids increases, as we move from HClO to $HClO_4$, ($HClO < HClO_2 < HClO_3 < HClO_4$). This order can be explained in any of the following ways:

(*i*) *Explanation based on oxidation number of central atom.* The structures of tnese acids can be written as :

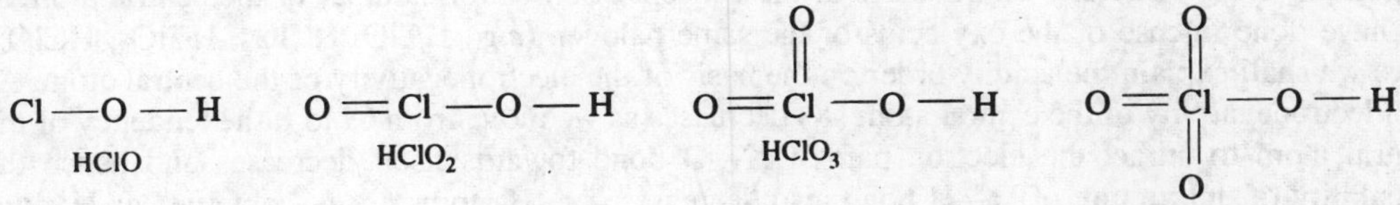

HClO $HClO_2$ $HClO_3$ $HClO_4$

These structures show that each acid has Cl—O—H bond. It may also be noted that the oxidation number of the central atom *viz.*, Cl-atom increases from +1 to +7 as we move from HClO to $HClO_4$ (HClO = +1, $HClO_2$ = +3, $HClO_3$ = +5, $HClO_4$ = +7). As the oxidation number of Cl-atom increases from +1 (in HClO) to +7 (in $HClO_4$), the tendency of Cl-atom to attract the electron pair of Cl—O bond towards itself increases and hence the rupture of O—H bond becomes easier, *i.e.*, with the increase in the oxidation number of Cl-atom, the possibiliy of getting H-atom as H^+ ion by the rupture of O—H bond increases. In other words we can say that the acidic character of the given acids increases from HClO to $HClO_4$ ($HClO < HClO_2 < HClO_3 < HClO_4$).

(*ii*) *Explanation based on Bronsted-Lowry concept of acids and bases.* According to this concept, ClO^-, ClO_2^-, ClO_3^- and ClO_4^- are the conjugate bases of HClO, $HClO_2$, $HClO_3$ and $HClO_4$ respectively. We know that the basicity of ClO^-, ClO_2^-, ClO_3^- and ClO_4^- ions decreases as we move from ClO^- to ClO_4^- ($ClO^- > ClO_2^- > ClO_3^- > ClO_4^-$). Again according to this concept, since the basicity of the anions decreases from ClO^- to ClO_4^- ($ClO^- > ClO_2^- > ClO_3^- > ClO_4^-$), the acidity of their conjugate acids (*viz.*, HClO, $HClO_2$, $HClO_3$ and $HClO_4$) will increase from HClO to $HClO_4$, *i.e.*, the acidity of HClO, $HClO_2$, $HClO_3$ and $HClO_4$ increases as $HClO < HClO_2 < HClO_3 < HClO_4$.

Summary

The above discussion shows that if an element forms more than one oxy-acids in which the element exhibts different oxidation states, then the acidic strength of different oxy-acids increases with the increase in the oxidation number of the element. Examples of some more such oxy-acids showing this trend of acidity are given below:

$\overset{+3}{HNO_2} < \overset{+5}{HNO_3}$ (*Group V A*)

$\overset{+4}{H_2SO_3} < \overset{+6}{H_2SO_4}$ (*Group VI A*)

$\overset{+1}{HBrO} < \overset{+3}{HBrO_2} < \overset{+5}{HBrO_3} < \overset{+7}{HBrO_4}$ } (*Group VIIA*)

$\overset{+1}{HIO} < \overset{+3}{HIO_2} < \overset{+5}{HIO_3} < \overset{+7}{HIO_4}$ }

$\overset{+6}{H_2MnO_4} < \overset{+7}{HMnO_4}$ (*Group VIII*)

Relative order of the acidic strength of oxy-acids of the elements of the same group and in the same oxidation state

As an example, let us consider halic acids of Cl, Br and I *viz.*, $HClO_3$, $HBrO_3$ and HIO_3. The acidic strength of these acids has been found to decrease as we move down the halogen group from Cl to I. ($HClO_3 > HBrO_3 > HIO_3$). This order can be explained as follows :

All these acids have the same structure *viz.*, $X(=O)_2—O—H$ (X = Cl, Br, I) and the oxidation number of the central atoms in each molecule is the same (= +5). We can, therefore, not explain decreasing order of acidity on the basis of the concept of oxidation number of the central atom as we have done in case of the oxy-acids of the same halogen (*e.g.*, $HClO$, $HClO_2$, $HClO_3$, $HClO_4$) Here we shall explain the acidity order on the basis of the electronegativity of the central atom. As the electronegativity of the central atom (X) decreases as we move from Cl to I, the tendency of the central atom to attract the electron pair of X—O bond towards itself decreases and hence the possibility of the rupture of O—H bond also decreases, *i.e.*, H-atoms are not obtained as H^+ ions quite easily. In other words we can say that the acid having the central atom with maximum electronegativity value will give H^+ ions quite easily while that having the central atom with minimum value of electronegativity will do so with great difficulty and hence acidic strength will decrease with the decrease in the electronegativity of the central atom ($HClO_3 > HBrO_3 > HIO_3$).

Summary

The above discussion shows that the acidic strength of oxy-acids of the elements of the same group and in the same oxidation, state decreases, as the electronegativity of the central atom decreases, on moving down the group. Examples of some more oxy-acids showing this trend in their acidic strength are given below :

$\overset{+5}{HNO_3} > \overset{+5}{HPO_3} > \overset{+5}{HAsO_3}$	*Group VA*
$\overset{+5}{H_3PO_4} > \overset{+5}{H_3AsO_4} > \overset{+5}{H_3SbO_4}$	
$\overset{+4}{H_2SO_3} > \overset{+4}{H_2SeO_4}$	*Group VIA*
$\overset{+6}{H_2SO_4} > \overset{+6}{H_2SeO_4} > \overset{+6}{H_2TeO_4}$	
$\overset{+1}{HClO} > \overset{+1}{HBrO} > \overset{+1}{HIO}$ (*hypohalous acids*)	*Group VIIA*
$\overset{+3}{HClO_2} > \overset{+3}{HBrO_2} > \overset{+3}{HIO_2}$ (*halous acids*)	
$\overset{+5}{HClO_4} > \overset{+5}{HBrO_4} > \overset{+5}{HIO_4}$ (*perhalic acids*)	

Acidic strength of H_3PO_2, H_3PO_3 and H_3PO_4 acids

We have said above that, in general, the acidic strength of oxy-acids of an element in different oxidation state increases with the increase in oxidation number of that element. But the acidic strength of the acids *viz.*, $\overset{+1}{H_3PO_2}$, $\overset{+3}{H_3PO_3}$ and $\overset{+5}{H_3PO_4}$ is not in the increasing order ($HPO_2 < HPO_3 < HPO_4$), although the oxidation number of P-atom increases from +1 to +5. The actual order has been found as $HPO_2 > H_3PO_3 > H_3PO_4$. This order has been explained on the basis of the number of OH groups attached directly with P-atom. These acids are monobasic, dibasic and tribasic and hence contain one, two and three OH groups respectively.

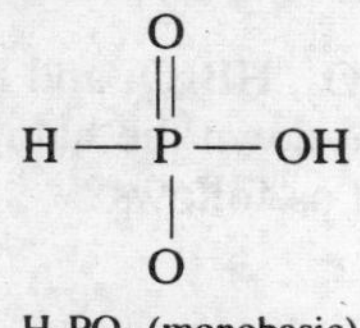
H_3PO_2 (monobasic)

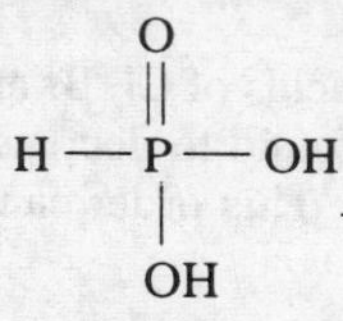
H_3PO_3 (dibasic)

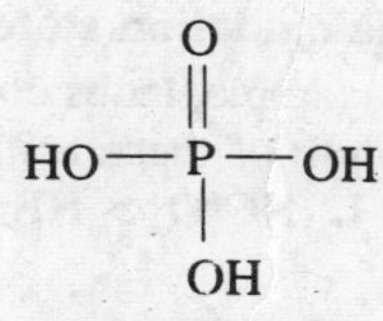
H_3PO_4 (tribasic)

The acidic strength decreases with the increase in the number of OH groups attached directly with P-atom. Thus the acidity order is $H_3PO_2 > H_3PO_3 > H_3PO_4$.

Advantages of Bronsted-Lowry concept

(*i*) This concept is superior to Arrhenius concept, since Arrhenius concept can explain the acidic/basic character of a substance only in aqueous medium (H_2O) while Bronsted-Lowry concept can explain the acidic/basic nature of a substance in aqueous (H_2O) as well as in other protonic solvents like liq. NH_3, liq. HF. For example, since $H_2N.CO.NH_2$ (urea) loses a proton in liq. NH_3, it behaves as a Bronsted acid in this solvent. On the other HNO_3 gains a proton in liq. HF, it behaves as a Bronsted base in this solvent.

$$\underset{Acid}{H_2N.CO.NH_2} + \underset{Base}{NH_3} \rightleftharpoons \underset{Acid}{NH_4^+} + \underset{Base}{H_2N.CO.NH^-}$$

$$HF + HNO_3 \rightleftharpoons H_2NO_3^+ + F^-$$

(*ii*) Acid-base reactions taking place in gaseous phase can also be explained by Bronsted-Lowry concept. For example in the reaction between HCl (*g*) and NH_3 (*g*), HCl (*g*) acts as an acid and NH_3 (*g*) behaves as a base.

$$\underset{Acid}{HCl\,(g)} + \underset{Base}{NH_3\,(g)} \rightleftharpoons \underset{Acid}{NH_4^+} + \underset{Base}{Cl^-} \text{ or } NH_4^+Cl^-$$

Limitations of Bronsted-Lowry concept

The most serious limitation of this concept is that it cannot explain the acid-base reactions taking place in non-protonic solvents, like liq. SO_2, liq. BF_3, BrF_3, $AlCl_3$, $POCl_3$ etc., in which no transfer of protons takes place.

$$\underset{Acid}{SO_2} + \underset{Base}{SO_2} \rightleftharpoons \underset{Acid}{SO^{2+}} + \underset{Base}{SO_4^{2-}}$$

$$BrF_3 + BrF_3 \rightleftharpoons BrF_4^+ + BrF_2^-$$

Solvent system (auto-ionisation) concept (1928)

This concept was introduced by Franklin (1905) and was extended by Cady-Esley (1928). The definition of acids and bases given by this concept can be applied for protonic (*e.g.*, H_2O, NH_3, N_2H_4, HF, H_2SO_4, CH_3COOH, HCN etc.) as well as for non-protonic (*e.g.*, SO_2, $COCl_2$, BrF_3, N_2O_4 etc.) solvents. According to this concept, the solvents undergo self ionisation (auto-ionisation) and give cations and anions which are called *solvent cations* and *solvent anions* respectively.

The definition of acids and bases given by this concept is based on the nature of the cations and anions which the acid or base produces, either by its self-ionisation or when it (*i.e.*, acid or base) is dissolved in the solvent. Thus, according to this concept, the substances which give solvent cations when dissolved in that solvent, are called *acids* while the substances which give solvent anions when dissolved in that solvent are called *bases*. It follows from this discussion that solvent cations can also be called *acid cations* and solvent anions can also be called *base anions*. The solvent system concept would be clear, if we consider the auto-ionisation of some solvents like water, liq. NH_3, liq. SO_2, liq. BrF_3, liq. HF, liq. N_2O_4, CH_3COOH, HCN and H_2SO_4 as discussed below :

Auto-ionisation of water

Water (H_2O) under-goes self-ionisation in the following three ways :

(*a*) $H_2O \rightleftharpoons \underset{\text{(solvent cations : acid ions)}}{H^+ \text{ (Hydrogen ions or protons)}} + \underset{\text{(solvent anions : base ions)}}{OH^- \text{ (Hydroxyl ions)}}$

(*b*)
$$H_2O \rightleftharpoons H^+ + OH^-$$
$$H^+ + H_2O \rightleftharpoons H_3O^+$$
$$\overline{2H_2O \rightleftharpoons \underset{\text{(solvent cations : acid ions)}}{H_3O^+ \text{ (Hydronium ions)}} + \underset{\text{(solvent anions : base ions)}}{OH^- \text{ (Hydroxyl ions)}}}$$

$$K_w = [H_3O^+][OH^-] = 1.0 \times 10^{-14} \text{ at } 25°C$$

(*c*) $3H_2O \rightleftharpoons 2H_3O^+$ (Hydronium ions) + O^{2-} (Oxide ions)

(solvent cations : acid ions) (solvent anions : base ions)

Different modes of ionisation given above suggest that, according to solvent system concept, the compounds that give H^+ or H_3O^+ ions in water act as acids in aqueous solution, while the compounds which furnish OH^- or O^{2-} ions in water, behave as bases in water. Thus, since HCl gives H^+ or H_3O^+ ions in water, it behaves as an acid in aqueous solution

$$HCl \xrightleftharpoons{\text{water}} H^+ + Cl^-$$

or $$HCl + H_2O \rightleftharpoons H_3O^+ + Cl^-$$

Similarly NaOH, which furnishes OH^- ions in its aqueous solution, acts as a base

$$NaOH \xrightleftharpoons{\text{water}} Na^+ + OH^-$$

Auto-ionisation of liq. NH_3

Liq. NH_3 ionises in the following three ways :

(*a*) $NH_3 \rightleftharpoons H^+ + NH_2^-$

$H^+ + NH_3 \rightleftharpoons NH_4^+$

$NH_3 + NH_3 \rightleftharpoons NH_4^+ + NH_2^-$

or $2NH_3 \rightleftharpoons NH_4^+$ (*Ammonium ions*) + NH_2^- (*Amide ions*)

Solvent (Solvent cations: acid ions) (Solvent anions : base ions)

$$K_b = [NH_4^+][NH_2^-] = 1.9 \times 10^{-33} \text{ at } -50°C.$$

(*b*) $3NH_3 \rightleftharpoons 2NH_4^+$ (*Ammonium cations*) + NH^{2-} (*Imide ions*)

Solvent (Solvent : cations : acid ions) (Solvent anions : base ions)

(*c*) $4NH_3 \rightleftharpoons 3NH_4^+$ (*Ammonium cations*) + N^{3-} (*Nitride ions*)

Solvent (Solvent cations : acid ions) (Solvent anions : base ions)

Different modes of auto-ionisation of liq. NH_3, given above, indicate that, according to solvent system concept, in liq. NH_3, any substance that gives NH_4^+ ions (which are solvent cations) will act as an acid in liq. NH_3, while that which produces NH_2^- or NH^{2-}or N^{3-} ions (which are solvent anions) will behave as a base in liq. NH_3. The compounds which give NH_4^+ ions in liq. NH_3 are called *ammono acids*, while those which give NH_2^- NH^{2-} or N^{3-} ions in this solvent are called *ammono bases*.

Examples of ammono acids

Ammonium salts [*e.g.*, NH_4Cl, NH_4Br, NH_4I, NH_4NO_3, $(NH_4)_2S$ etc.], organic amides (*e.g.*, urea, acetamide, sulphamide etc.), acetic acid, sulphamic acid etc., all act as ammono acids in liq. NH_3, since they produce NH_4^+ ions in this solvent, as shown below:

$$NH_4X \xrightleftharpoons{\text{liq. } NH_3} X^- + NH_4^+$$

(X = Cl, Br, I)

$$H_2N.CO.NH_2 + NH_3 \longrightarrow H_2N.CO.NH^- + NH_4^+$$

Urea

$$CH_3CO\ NH_2 + NH_3 \longrightarrow CH_3CONH^- + NH_4^+$$

Acetamide

$$H_2N.SO_2.NH_2 + NH_3 \longrightarrow H_2N.SO_2.NH^- + NH_4^+$$
Sulphamide

$$CH_3COOH + NH_3 \longrightarrow CH_3COO^- + NH_4^+$$
Acetic acid

$$H_2N.SO_2.OH + 2NH_3 \longrightarrow HN^-.SO_2.O^- + 2NH_4^+$$
Sulphamic acid
(Dibasic acid)

It may be noted that since $H_2N.SO_2.OH$ produces two ions, this acid behaves as a *dibasic acid* in liq. NH_3 .

The reaction between organic amides and liq. NH_3 is called *protolysis.*

The fact that ammonium salts behave as acids in liq. NH_3 is confirmed by an experimental observation that these salts are able to dissolve metallic oxides and hydroxides. For example :

$$2NH_4Cl + CaO \xrightarrow{\text{liq. } NH_3} Ca(NO_3)_2 + H_2O + 2NH_3$$

$$2NH_4Cl + Na_2O \xrightarrow{\text{liq. } NH_3} NaCl + H_2O + 2NH_3$$

$$NH_4Cl + NaOH \xrightarrow{\text{liq. } NH_3} NaCl + H_2O + NH_3$$

$$NH_4NO_3 + NaOH \xrightarrow{\text{liq. } NH_3} NaNO_3 + H_2O + NH_3$$

Examples of ammono bases

KNH_2, PbNH and BiN are the examples of ammono bases, since these compounds give NH_2^-, NH^{2-} and N^{3-} ions respectively in liq. NH_3.

$$KNH_2 \xrightarrow{\text{liq. } NH_3} K^+ + NH_2^- \quad (\textit{Amide ion})$$

$$PbNH \xrightarrow{\text{liq. } NH_3} Pb^{2+} + NH^{2-} \quad (\textit{Imide ion})$$

$$BiN \xrightarrow{\text{liq. } NH_3} Bi^{3+} + N^{3-} \quad (\textit{Nitride ion})$$

Reactions of ammono acids

(*i*) *Reaction between an ammono acid and a reactive metal.* When an ammono acid (*e.g.*, NH_4NO_3) reacts with a reactive metal like Na, Co etc., is liq. NH_3, H_2 gas is evolved.

$$2Na + 2NH_4NO_3 \xrightarrow{\text{liq. } NH_3} 2NaNO_3 + 2NH_3 + H_2$$
Ammono acid

$$\text{or} \quad 2Na + 2NH_4^+ \xrightarrow{\text{liq. } NH_3} 2Na^+ + 2NH_3 + H_2$$

$$Co + 2NH_4NO_3 \xrightarrow{\text{liq. } NH_3} Co(NO_3)_2 + 2NH_3 + H_2$$

$$\text{or} \quad Co + 2NH_4^+ \xrightarrow{\text{liq. } NH_3} Co^{2+} + 2NH_3 + H_2$$

The above reactions are analogous to the following reactions, which take place in aqueous medium.

$$2Na + H_2SO_4 \xrightarrow{H_2O} 2Na_2SO_4 + H_2$$
Acid

$$\text{or} \quad 2Na + 2H^+ \xrightarrow{H_2O} 2Na^+ + H_2$$

$$Co + H_2SO_4 \xrightarrow{H_2O} CoSO_4 + H_2$$
Acid

$$\text{or} \quad Co + 2H^+ \xrightarrow{H_2O} Co^{2+} + H_2$$

The evolution of H_2 gas by the reaction between an alkali metal (*e.g.,* Na) and an ammonium salt (ammono acid) in liq. NH_3, as shown above, indicates that ammonium salts decolourise blue solutions of alkali metals in liq. NH_3.

(*ii*) *Precipitation (Metathetical) reactions.* When ammono acids like NH_4X (X = Cl, Br, I), $(NH_4)_2S$ etc. are mixed with the liq. NH_3 solution of metallic salts like $Sr(NO_3)_2$, $Zn(NO_3)_2$, $AgNO_3$, $Cu(NO_3)_2$, $Ca(NO_3)_2$, KI etc., the halides or sulphides of the metal are precipitated. For example :

Ammono acid *Metallic salt*

$$2NH_4X + Sr(NO_3)_2 \xrightleftharpoons{\text{liq. } NH_3} SrX_2\downarrow + 2NH_4NO_3$$
(X = Cl, Br)

$$2NH_4I + Zn(NO_3)_2 \xrightleftharpoons{\text{liq. } NH_3} ZnI_2\downarrow + 2NH_4NO_3$$

$$(NH_4)_2S + 2AgNO_3 \xrightleftharpoons{\text{liq. } NH_3} Ag_2S\downarrow + 2NH_4NO_3$$

$$(NH_4)_2S + M(NO_3)_2 \xrightleftharpoons{\text{liq. } NH_3} MS\downarrow + 2NH_4NO_3$$
(M = Cu, Cd)

$$NH_4Cl + KI \xrightleftharpoons{\text{liq. } NH_3} KCl\downarrow + NH_4I$$

(*iii*) *Reaction between an ammono acid and a metallic oxide or hydroxide.* These substances get dissolved in ammono acid *e.g.,*

$$2NH_4Cl + Na_2O \xrightarrow{\text{liq. } NH_3} NaCl + H_2O + 2NH_3$$

$$2NH_4Cl + CaO \xrightarrow{\text{liq. } NH_3} Ca(NO_3)_2 + H_2O + 2NH_3$$

$$NH_4Cl + NaOH \xrightarrow{\text{liq. } NH_3} NaCl + H_2O + NH_3$$

$$NH_4NO_3 + NaOH \xrightarrow{\text{liq. } NH_3} NaNO_3 + H_2O + NH_3$$

Precipitation reactions shown by ammono bases

When an ammono base like KNH_2 is mixed with the liq. NH_3 solution of metallic salt like $AgNO_3$, PbI_2, BiI_3 etc., the amide, imide or nitride of the metal is precipitated. For example:

$$AgNO_3 + KNH_2 \xrightarrow{\text{liq. } NH_3} KNO_2 + AgNH_2\downarrow \text{ (Amide)}$$

$$PbI_2 + KNH_2 \xrightarrow{\text{liq. } NH_3} KI + HI + Pb(NH)\downarrow \text{ (Imide)}$$

$$BiI_3 + KNH_2 \xrightarrow{\text{liq. } NH_3} KI + 2HI + BiN\downarrow \text{ (Nitride)}$$

The above reactions are analogous to the following reactions, which take place in aqueous medium. In these reactions when the solution of a base like NaOH in aqueous medium is treated with a metallic salt, metallic oxide and hydroxides are precipitated. For example:

$$2AgNO_3 + 2NaOH \xrightarrow{H_2O} 2NaNO_3 + H_2O + Ag_2O\downarrow \text{ (Oxide)}$$

$$Pb(NO_3)_2 + 2NaOH \xrightarrow{H_2O} 2NaNO_3 + Pb(OH)_2\downarrow \text{ (Hydroxide)}$$

Reaction between an ammono acid and an ammono base in liq. NH_3 : Acid-base neutralisation reaction

When an ammono acid reacts with an ammono base in liq. NH_3, salt and the solvent (NH_3) are produced, *e.g.,*

$$\underset{\text{Ammono acid}}{NH_4Cl} + \underset{\text{Ammono base}}{KNH_2} \xrightarrow{\text{liq. } NH_3} \underset{\text{Salt}}{KCl} + \underset{\text{Solvent}}{2NH_3}$$

For details see "*Applications of solvent system concept*" (*Application No. 3*)

Auto-ionisation of liq. SO_2.

Self-ionisation of liq. SO_2 takes place as follows:

$$SO_2 \rightleftharpoons SO^{2+} + O^{2-}$$
$$SO_2 + O^{2-} \rightleftharpoons SO_3^{2-}$$
$$\underbrace{SO_2 + SO_2}_{\text{solvent}} \rightleftharpoons \underset{\text{(solvent cations : acid ions)}}{SO^{2+}\ (\textit{thionyl ions})} + \underset{\text{(solvent anions : base ions)}}{SO_3^{2-}\ (\textit{sulphite ions})}$$

When we compare the self-ionisation of liq. SO_2 with that of water and liq. NH_3, we find that SO^{2+} ion is analogous to H_3O^+ and NH_4^+ ions while SO_3^{2-} ion is analogous to OH^- and NH_2^- ion produced by the self-ionisation of H_2O and liq. NH_3 respectively. Thus all those compounds which contain or make available SO^{2+} ions (solvent cations-acid ions) will act as acids in liq. SO_2. Similarly all those compounds which contain or make available SO_3^{2-} ions, will behave as bases in liq. SO_2.

The above discussion shows that since $SOCl_2$, $SOBr_2$, $SO(SCN)_2$ etc., give SO^{2+} ions in liq. SO_2, they behave as acids in this solvent. On the other hand, the substances *viz*., Cs_2SO_3,$[N(CH_4)]_2SO_3$, K_2SO_3 which give SO_3^{2-} ions in this solvent act as bases.

Relation between the ions given by the auto-ionisation of H_2O, liq. NH_3 and liq. SO_2

The equations representing the auto-ionisation of H_2O, liq. NH_3 and liq. SO_2 indicate that NH_4^+ and SO^{2+} ions, which are obtained by the self-ionisation of liq. NH_3 and liq. SO_2 respectively, are analogous to H_3O^+ (or H^+) ions, which are obtained by the ionisation of H_2O. Similarly NH_2^- and NH^{2-} ions, obtained by the ionisation of liq. NH_3 and SO_3^{2-} ions (obtained from liq. SO_2), are analogous to OH^- ions obtained from H_2O. N^{3-} ions obtained form liq. NH_3 are equivalent to O^{2-} ions obtained from H_2O.

Auto-ionisation of liq. BrF_3

Liq. BrF_3 has high specific conductance and hence undergoes auto-ionisation to produce BrF_2^+ and BrF_4^- ions.

$$2BrF_3 \rightleftharpoons BrF_2^+ + BrF_4^-$$

Auto-ionisation of liq. BrF_3 and other items related with it have been discussed under *Interhalogen Compounds.*

Auto-ionisation of liq. HF

The high specific conductance of liq. HF suggests a relatively high degree of auto-ionisation of liq. HF shown below:

$$HF \rightleftharpoons H^+ + F^-$$
$$HF + H^+ \rightleftharpoons H_2F^+$$
$$HF + HF \rightleftharpoons \underset{\substack{\text{Solvent cations}\\\text{(acid ions)}}}{H_2F^+} + \underset{\substack{\text{Solvent anions}\\\text{(base ions)}}}{F^-}$$

or

$$3HF \rightleftharpoons \underset{\substack{\text{Solvent cations}\\\text{(acid ions)}}}{H_2F^+} + \underset{\substack{\text{Solvent anions}\\\text{(base ions)}}}{F^-.HF \text{ or } HF_2^-}$$

Thus any substance that can give H_2F^+ ions would behave as an acid in liq. HF and any substance that can furnish F^- or HF_2^- ions would act as a base in this solvent.

Examples. (*i*) CH_3COOH (which acts as a *weak acid* in water) acts as a base in liq. HF, because it gives F^- ions (solvent anions-base ions) when dissolved in liq. HF.

$$CH_3COOH\ (\text{Base}) + HF\ (\text{Solvent}) \longrightarrow CH_3COOH_2^+ + F^- \ [\textit{Solvent anions (base ions)}]$$

(*ii*) H_2SO_4 and HNO_3 both act as strong acids in aqueous medium, but show basic character in liq. HF, due to the production of F^- ions which are solvent anions (base ions).

$$HNO_3\ (\text{Base}) + HF\ (\text{Solvent}) \longrightarrow H_2NO_3^+ + F^- \ [\textit{Solvent anions (base ions)}]$$

$$HNO_3\ (\text{Base}) + HF\ (\text{Solvent}) \longrightarrow H_2NO_3^+ F^- \ [\textit{Solvent anions (base ions)}]$$

(*iii*) $HClO_4$ behaves as the *strongest acid* in water, but in liq. HF it acts as an *amphoteric substance*, since it produces both H_2F^+ (solvent cations-acid ions) and F^- (solvent anions-base ions) in this solvent

$$HClO_4 + HF\ (\text{Solvent}) \longrightarrow H_2ClO_4^+ + F^- \ [\textit{Solvent anions (base ions)}]$$

$$HClO_4 + HF\ (\text{Solvent}) \longrightarrow H_2F^+ + ClO_4^- \ [\textit{Solvent cations (acid ions)}]$$

Self-ionisation of some other non-aqueous solvents

The self-ionisation of liq. N_2O_4, CH_3COOH, HCN and H_2SO_4 is shown below :

(*a*) $2N_2O_4$ (*solvent*) $\rightleftharpoons$ $2NO^+$ (*nitrosyl ions : solvent cations-acid ions*) + $2NO_3^-$ (*nitrate ions : solvent anions-base ions*)

(*b*) $2CH_3COOH$ (*solvent*) $\rightleftharpoons$ CH_3COOH^+ (*solvent cations : acid ions*) + CH_3COO^- (*solvent anions : base ions*)

(*c*) 2HCN (*solvent*) $\rightleftharpoons$ H_2CN^+ (*solvent cations : acid ions*) + CN^- (*solvent anions : acid ions*)

(*d*) $2H_2SO_4$ (*solvent*) $\rightleftharpoons$ $H_3SO_4^+$ (*solvent cations : acid ions*) + HSO_4^- (*solvent anions : base ions*)

Applications of solvent system concept

With the help of solvent system concept of acids and bases we can explain the following :

1. To explain the acidic / basic nature of a given substance in a given solvent. With the help of solvent-system concept of acids and bases, we can predict whether a given substance will behave as an acid, as a base, as an amphoteric substance or as a neutral substance in a given solvent. The following examples illustrate this application.

(*i*) *Behaviour of* CH_3COOH *in* H_2O, *liq.* NH_3, *liq. HF and* H_2SO_4 *(a).* In water CH_3COOH ionises to a small extent to produce H_3O^+ ions (solvent cations-acid ions). Due to the feeble ionisation of CH_3COOH in water, the concentration of H_3O^+ ions obtained is very low and hence, according to solvent system concept, CH_3COOH acts as a weak acid in water.

$$\underset{\text{Weak acid}}{CH_3COOH} + \underset{\text{Solvent}}{H_2O} \rightleftharpoons \underset{\substack{\text{Solvent cations-acid ions} \\ \text{(low concentration)}}}{H_3O^+} + CH_3COO^-$$

The longer half arrow pointing towards left indicates that CH_3COOH undergoes partial ionisation in water or in other words the above equilibrium lies mostly towards the left.

(*b*) When dissolved in liq. NH_3, CH_3COOH is completely converted into NH_4^+ ions which are solvent cations or acid ions. CH_3COOH, therefore, behaves as a *strong acid* in liq. NH_3.

$$CH_3COOH\ (\textit{Strong acid}) + NH_3\ (\textit{Solvent}) \longrightarrow NH_4^+ + CH_3COO^-$$

[*Solvent cations (acid ions)*]

(*c*) When CH_3COOH is dissolved in liq. HF, it ionises to produce F^- ions (solvent anions-base ions) and hence behaves as a base in liq. HF

$$CH_3COOH\ (Base) + HF\ (Solvent) \longrightarrow CH_3COOH_2^+ + F^-$$

[*Solvent anions (base ions)*]

(*d*) When CH_3COOH is dissolved in H_2SO_4, the concentration of HSO_4^- ions (solvent anions base ions) is increased and hence CH_3COOH behaves as a base in H_2SO_4.

$$CH_3COOH\ (Base) + H_2SO_4\ (Solvent) \longrightarrow CH_3\ COOH_2^+ + HSO_4^-$$

[*Solvent anions (base ions)*]

It follows from the above discussion that according to solvent-system concept CH_3COOH acts as a weak acid in water; in liq. NH_3 it behaves as a strong acid but in liq. HF and H_2SO_4 both it shows basic character.

(*ii*) *Behaviour of H_2SO_4 in water and liq. HF.* (*a*) H_2SO_4 is completely ionised in water to produce H_3O^+ ions (solvent cations-acid ions) and hence, according to solvent-system concept this substance behaves as a strong acid in water.

$$H_2SO_4\ (Strong\ acid) + 2H_2O\ (Solvent) \longrightarrow 2H_3O^+ + SO_4^{2-}$$

[*Solvent cations (acid ions)*]

It may be noted from the above equation that H_2SO_4 behaves as a *dibasic acid in water* since it donates two protons to two water molecules.

(*b*) When H_2SO_4 is put into liq. HF, it is ionised to produce F^- ions (solvent anions-base ions) and hence, according to solvent-system concept, H_2SO_4 acts as a base in liq. HF

$$H_2SO_4\ (Base) + HF\ (Solvent) \longrightarrow H_3SO_4^+ + F^- \quad [Solvent\ anions\ (base\ ions)]$$

Whatever has been said at (*a*) and (*b*) can be summarised by saying that H_2SO_4 which behaves as a strong acid in aqueous solution acts as a base when dissolved in liq. HF.

(*iii*) *Behaviours of $HClO_4$ in water and liq. HF.* (*a*) $HClO_4$ ionises completely in H_2O and produces H_3O^+ ions (solvent cations-acid ions) in high concentration. $HClO_4$, therefore, acts as a *strong acid in aqueous solution.*

$$HClO_4\ (Strong\ acid) + H_2O\ (Solvent) \longrightarrow H_3O^+ + ClO_4^- \quad [Solvent\ cations\ (acid\ ions)]$$

(*b*) When $HClO_4$ is put into liq. HF, it is ionised to produce both H_2F^+ (which are solvent cations-acid ions) and F^- (which are solvent anions-base ions) and hence, according to solvent-system concept, $HClO_4$ behaves as an *amphoteric substance in liq. HF.*

$$HClO_4 + HF\ (Solvent) \longrightarrow H_2ClO_4^+ + F^- \quad [Solvent\ anions\ (base\ ions)]$$

$$HClO_4 + HF\ (Solvent) \longrightarrow H_2F^+ + ClO_4^- \quad [Solvent\ cations\ (acid\ ions)]$$

Thus, we see that $HClO_4$, in water, behaves as a strong acid but in liq. HF this substance shows amphoteric character.

(*iv*) *Behaviour of urea in water, liq. NH_3 and conc.H_2SO_4.* (*a*) Urea ($H_2N.CO.NH_2$) does not react with water to produce H_3O^+ or OH^- ions and hence behaves as a neutral substance in this solvent.

(*b*) When urea is dissolved in liq. NH_3, NH_4^+ ions (solvent cations-acid ions) are obtained in low concentration and hence urea behaves *as a weak acid in this solvent.*

$$H_2N.CO.NH_2\ (Weak\ acid) + NH_3\ (Solvent) \rightleftharpoons H_2N\ CONH^- + NH_4^+$$

[*Solvent cations-acid ions (Low concentration)*]

(*c*) In conc. H_2SO_4, urea produces HSO_4^- ions (solvent anions-base ions) in large concentration. Thus urea in this solvent behaves as a *strong base.*

$H_2N.CO.NH_2$ (*Strong base*) + H_2SO_4 (*Solvent*) $\longrightarrow$ $H_2N.CH.NH_3^+ + HSO_4^-$
[*Solvent anions-base ions (large concentration)*]

(*v*) *Behaviour of NHO_3 in water, liq. HF and H_2SO_4.* (*a*) HNO_3 is completely ionised in water to produce H_3O^+ ions (*solvent cations-acid ions*) in large concentration and, hence, according to solvent-system concept, HNO_3 behaves as a strong acid in water.

HNO_3 (*Strong acid*) + H_2O (*Solvent*) $\longrightarrow$ $H_3O^+ + NO_3^-$ [*Solvent cations (acid ions)*]

(*b*) When HNO_3 is put in liq. HF, it gets ionised to produce F^- ions (*solvent anions-base ions*) and hence HNO_3 behaves as a *base in liq.HF*.

HNO_3 (*Base*) + HF (*Solvent*) $\longrightarrow$ $H_2NO_3^+ + F^-$ [*Solvent anions (Base ions)*]

(*c*) When HNO_3 is dissolved in H_2SO_4, HSO_4^- ions which are solvent anions (base ions) are produced in high concentration. Consequently HNO_3 behaves as a *base in H_2SO_4*.

HNO_3 (*Base*) + H_2SO_4 (*Solvent*) $\longrightarrow$ $HSO_4^- + H_2NO_3^+$ [*Solvent anions (base ions)*]

(*vi*) *Behaviour of sulphamic acid ($H_2N.SO_2.OH$) in water and liq. NH_3.* Sulphamic acid donates only one proton (H^+) to H_2O in aqueous solution while it donates two protons to NH_3 in liq. NH_3. Thus $H_2N.SO_2.OH$ behaves as a monobasic acid in aqueous solution and as a dibasic acid in liq. NH_3.

$H_2N.SO_2.OH$ (*Monobasic acid*) + H_2O $\longrightarrow$ $H_2N.SO_2.O^- + H_3O^+$

$H_2N.SO_2.OH$ (*Dibasic acid*) + $2NH_3$ $\longrightarrow$ $\overline{H}N.SO_2.O^- + 2NH_4^+$

Table 19.4 : *Behaviour of different substances in different solvents.*

Substances	Solvents			
	H_2O	liq. NH_3	liq. HF	H_2SO_4
(*i*) CH_3COOH	Weak acid	Strong acid	Base	Base
(*ii*) H_2SO_4	Strong acid	—	Base	—
(*iii*) $HClO_4$	Strong acid	—	Amphoteric substance	—
(*iv*) Urea ($H_2N.CO.NH_2$)	Neutral	Weak acid	—	Strong base
(*v*) HNO_3	Strong acid	—	Base	Base
(*vi*) Sulphamic acid ($H_2N.SO_2.OH$)	Monobasic acid	Di basic acid	—	—

2. To explain the amphoteric behaviour of compounds in different solvents. A compound that reacts both with acids and bases is said to show amphoteric character. Let us explain the amphoteric character of substances in H_2O, liq. NH_3 etc.

(*i*) *To explain the amphoteric character of compounds in water.* In water, $Zn(OH)_2$ and $Al(OH)_3$ react both with acids and bases (alkalies) and hence are said to show amphoteric character in this medium.

$$Zn(OH)_2 + 2HCl\ (Acid) \xrightarrow{H_2O} ZnCl_2\ (Salt) + 2H_2O$$

or $$Zn(OH)_2 + 2H^+ \xrightarrow{H_2O} Zn^{2+} + 2H_2O$$

$$Zn(OH)_2 + 2NaOH\ (Base) \xrightarrow{H_2O} Na_2[Zn(OH)_4] \text{ or } Na_2[ZnO_2]\ (soluble\ complex) + 2H_2O$$

or $$Zn(OH)_2 + 2OH^- \xrightarrow{H_2O} [Zn(OH)_4]^{2-} \text{ or } [ZnO_2]^{2-} + 2H_2O$$

(ii) To explain amphoteric behaviour of compounds in liq. NH_3. In liq. NH_3, Zn (II) and Mg (II) salts [*e.g.*, $Zn(NO_3)_2$, $Zn(NH_2)_2$, $Mg(NH_2)_2$ etc.] react both with ammono acids and ammono bases, and hence are said to exhibit amphoteric character in this solvent. With ammono acids simple salts are obtained while with ammono bases soluble amido complex ions of the metals are produced.

(*a*) $Zn(NO_3)_2 + 2NH_4Cl$ (*Ammono acid*) $\xrightarrow{\text{liq. } NH_3}$ $ZnCl_2$ (Salt) + $2NH_4NO_3$

or $Zn(NO_3)_2 + 2NH_4^+ \xrightarrow{\text{liq. } NH_3} Zn^{2+} + 2NH_4NO_3$

$Zn(NO_3)_2 + 4KNH_2$ (*Ammono base*) $\xrightarrow{\text{liq. } NH_3}$ $K_2[Zn(NH_2)_4] + 2KNO_3$ [*Amido complex (soluble)*]

$Zn(NO_3)_2 + 4KNH_2$ (*Ammono base*) $\xrightarrow{\text{liq. } NH_3}$ $K_2[Zn(NH)_2] + 2NH_3 + 2KNO_3$ [*Imido complex (soluble)*]

or $Zn^{2+} + 4NH_2^- \xrightarrow{\text{liq. } NH_3} [Zn(NH_2)_4]^{2-}$

or $Zn^{2+} + 4NH_2^- \xrightarrow{\text{liq. } NH_3} [Zn(NH)_2]^{2-}$

(*b*) $M(NH_2)_2 + 2NH_4Cl$ (*Ammono acid*) $\xrightarrow{\text{liq. } NH_3}$ MCl_2 (Salt) + $4NH_3$
(M = Zn, Mg)

or $M(NH_2)_2 + 2NH_4^+ \xrightarrow{\text{liq. } NH_3} M^{2+} + 4NH_3$

$M(NH_2)_2 + 2NaNH_2$ (*Ammono base*) $\xrightarrow{\text{liq. } NH_3}$ $Na_2[M(NH_2)_4]$ [*Amido complex (soluble)*]

or $M(NH_2)_2 + 2NH_2^- \xrightarrow{\text{liq. } NH_3} [M(NH_2)_4]^{2-}$

$M(NH_2)_2 + 2NaNH_2 \xrightarrow{\text{liq. } NH_3}$ $Na_2[M(NH)_2]$ [*Imido complex (soluble)*] + $2NH_3$

or $M(NH_2)_2 + 2NH_2^- \xrightarrow{\text{liq. } NH_3} [M(NH)_2]^{2-} + 2NH_3$

(iii) To explain the amphoteric character of compounds in liq. SO_2. $AlCl_3$ and $GaCl_3$ both behave as amphoteric substances in liq. SO_2. This fact is evident from the reaction between $AlCl_3$ and tetramethyl ammonium sulphite $[N(CH_3)_4]_2SO_3$ which act as a base in liq. SO_2.

$2AlCl_3 + 3[N(CH_3)_4]_2SO_3$ (Base) $\xrightarrow{\text{liq. } SO_2}$ $6[N(CH_3)_4]Cl + Al_2(SO_3)_3$ (Gelatinous ppt.)

$Al_2(SO_3)_3 + 3[N(CH_3)_4]_2SO_3 \xrightarrow{\text{liq. } SO_2} 2[N(CH_3)_4]_3[Al(SO_3)_3]$
Soluble complex (solution)

On adding the above two equations, we get the final reaction represented by the equation:

$2AlCl_3 + 6[N(CH_3)_4]_2SO_3$ (Base (excess)) $\xrightarrow{\text{liq. } SO_2}$ $2[N(CH_3)_4]_3[Al(SO_3)_3]$ (Sulphite complex (soluble)) + $6[N(CH_3)_4]Cl$

From the soluble complex solution, $Al_2(SO_3)_3$ can be precipitated by adding $SOCl_2$ which acts as an acid in liq. SO_2.

$2[N(CH_3)_4]_3[Al(SO_3)_3]$ (Soluble complex (solution)) + $3SOCl_2$ (Acid) $\xrightarrow{\text{liq. } SO_2}$ $6[N(CH_3)_4]Cl + 6SO_2 + Al_2(SO_3)_3 \downarrow$ (ppt.)

In aqueous solution, an identical reaction takes place between $AlCl_3$ and NaOH which acts as a base in aqueous solution.

$$AlCl_3 + 3NaOH \text{ (Base)} \xrightarrow{H_2O} 3NaCl + Al(OH)_3 \text{ (\textit{Gelatinous ppt.})}$$

$$Al(OH)_3 + NaOH \xrightarrow{H_2O} Na[Al(OH)_4] \text{ [\textit{Soluble complex (solution)}]}$$

From the soluble complex solution, $Al(OH)_3$ can be precipitated by adding HCl, which acts as an acid in aqueous solution

$$\underset{\text{Soluble complex (solution)}}{Na[Al(OH)_4]} + \underset{\text{Acid}}{HCl} \longrightarrow NaCl + H_2O + \underset{\text{ppt.}}{Al(OH)_3 \downarrow}$$

Amphoteric character of $GaCl_3$ in liq. SO_2 can also be explained on similar lines.

3. To explain the acid-base neutralisation reactions in different solvents. Just as according to Arrhenius concept, and also according to solvent concept as well, an acid-base neutralisation reaction in a given solvent, is that in which an acid (an acid is a compound which contains or gives the solvent cation either by its direct ionisation or when dissolved in the given solvent) and a base (a base is a compound which contains or furnishes the solvent anion either by its direct ionisation or when dissolved in the given solvent) combine together in the presence of a given solvent and porduce the salt and the given solvent. Due to the formation of the salt, neutralisation reaction is also called *salt formation reaction.* In this reaction the solvent is produced by the combination of the solvent cation and solvent anion. Neutralisation reactions in H_2O, liq. NH_3 and liq. SO_2 are discussed below:

(i) Acid-base neutralisation reactions taking place in H_2O. In water, acid-base neutralisation reaction is that in which a compound containing or making available H_3O^+ or H^+ ions combines with a compound containing or making available OH^- ions to form the salt and the solvent (*i.e.*, H_2O), *e.g.*,

$$HCl \text{ (\textit{acid})} + KOH \text{ (\textit{base})} \xrightarrow{\text{Aqueous solution}} KCl \text{ (\textit{salt})} + H_2O \text{ (\textit{solvent})}$$

Actually the above reaction takes place through the following steps :

$$HCl \xrightarrow{H_2O} H^+ + Cl^- \quad ...(i)$$

$$(\text{or } HCl + H_2O \longrightarrow H_3O^+ + Cl^-)$$

$$KOH \xrightarrow{H_2O} K^+ + OH^- \quad ...(ii)$$

On adding (*i*) and (*ii*), we get :

$$HCl + KOH \xrightarrow{\text{Aqueous solution}} H^+ + Cl^- + K^+ + OH^-$$

$$\text{or} \quad HCl + KOH \xrightarrow{\text{Aqueous solution}} K^+ + Cl^- + H_2O$$

$$\text{or} \quad H^+ + Cl^- + K^+ + OH^- \longrightarrow K^+ + Cl^- + H_2O$$

$$\text{or} \quad H^+ \text{ (\textit{Solvent cation})} + OH^- \text{ (\textit{Solvent anion})} \longrightarrow H_2O \text{ (\textit{Solvent})} \quad ...(iii)$$

Equation (*iii*) indicates that neutralisation reaction in H_2O can be defined as the combination of solvent cation (H^+ or H_3O^+) and solvent anion (OH^-) to form the solvent (*i.e*,. H_2O).

(ii) Acid-base neutralisation reactions taking place in liq. NH_3. In liq. NH_3 acid-base neutralisation reaction is that in which a compound containing or making available NH_4^+ ion (*ammono acid*) combines with a compound containing or making available NH_2^-, NH^{2-} or N_3^{3-} ion (*ammono base*) to form the salt and the solvent (*i.e.*, NH_3), *e.g.*,

$$NH_4Cl \text{ (acid)} + KNH_2 \text{ (\textit{base})} \xrightarrow{\text{liq. } NH_3} KCl \text{ (\textit{salt})} + 2NH_3 \text{ (\textit{Solvent})}$$

Actually the above reaction takes place through the following steps :

$$NH_4Cl \xrightarrow{\text{liq. } NH_3} NH_4^+ + Cl^-$$

$$KNH_2 \xrightarrow{\text{liq. } NH_3} K^+ + NH_2^-$$

$$NH_4Cl + KNH_2 \xrightarrow{\text{liq. } NH_3} K^+Cl^- + NH_4^+ + NH_2^-$$

or $$NH_4Cl\ (\textit{acid}) + KNH_2\ (\textit{base}) \xrightarrow{\text{liq. } NH_3} KCl\ (\textit{salt}) + 2NH_3\ (\textit{solvent})$$

or $$NH_4^+ + Cl^- + K^+ + NH_2^- \xrightarrow{\text{liq. } NH_3} K^+ + Cl^- + 2NH_3$$

or $$NH_4^+\ (\textit{solvent cation}) + NH_2^-\ (\textit{solvent anion}) \xrightarrow{\text{liq. } NH_3} 2NH_3\ (\textit{solvent}) \quad ...(i)$$

Equation (*i*) indicates that neutralisation reaction in liq. NH_3 can be defined as the combination of solvent cation (NH_4^+) and solvent anion (NH_2^-) to form the un-ionised solvent (*i.e.*, liq. NH_3) and ammono salt.

Other examples of acid-base neutralisation reactions taking place in liq. NH_3 are :

Ammono acid		*Ammono base*		*Ammono salt*		*Solvent*
NH_4X ($X = Cl, NO_3$)	+	$NaNH_2$	$\xrightarrow{\text{liq. } NH_3}$	NaX	+	$2NH_3$
$2NH_4X$ ($X = Cl, I$)	+	$PbNH$	$\xrightarrow{\text{liq. } NH_3}$	PbX_2	+	$3NH_3$
$3NH_4X$ ($X = Cl, I$)	+	BiN	$\xrightarrow{\text{liq. } NH_3}$	BiX_3	+	$4NH_3$

(iii) Acid-base reactions taking place in liq. SO_2. In liq. SO_2, acid-base reaction is that in which a compound containing or making available SO_2^{2+} ion (acid) combines with a compound containing or making available SO_3^{2-} ion (base) to form the salt and the solvent (*i.e.*, SO_2), *e.g.*,

$$SOCl_2\ (\textit{acid}) + Cs_2SO_3\ (\textit{base}) \xrightarrow{\text{liq. } SO_2} 2CsCl\ (\textit{salt}) + 2SO_2\ (\textit{solvent})$$

Actually the above reaction takes place through the following steps :

$$SOCl_2 \xrightarrow{\text{liq. } SO_2} SO^{2+} + 2Cl^-$$

$$Cs_2SO_3 \xrightarrow{\text{liq. } SO_2} 2Cs^+ + SO_2^{2-}$$

$$SOCl_2 + Cs_2SO_3 \xrightarrow{\text{liq. } SO_2} 2Cs^+Cl^- + SO^{2+} + SO_3^{2-}$$

or $$SOCl_2\ (\textit{acid}) + Cs_2SO_3\ (\textit{base}) \xrightarrow{\text{liq. } SO_2} 2CsCl\ (\textit{salt}) + 2SO_2\ (\textit{solvent})$$

or $$SO^{2+} + 2Cl^- + 2Cs^+ + SO_3^{2-} \xrightarrow{\text{liq. } SO_2} 2Cs^+ + 2Cl^- + 2SO_2$$

or $$SO^{2+}\ (\textit{solvent cation}) + SO_3^{2-}\ (\textit{solvent anion}) \xrightarrow{\text{liq. } SO_2} 2SO_2\ (\textit{solvent}) \quad ...(i)$$

It may be seen from equation (*i*) that neutralisation reaction in liq. SO_2 is the combination of the solvent cation (SO^{2+}) and solvent anion (SO_3^{2-}) to produce the solvent (SO_2).

Examples of some other acid-base neutralisation reactions taking place in liq. SO_2 are given below:

Acid		Base		Salt		Solvent
$SO(SCN)_2$	+	K_2SO_3	$\xrightarrow{\text{liq. } SO_2}$	$2K(SCN)$	+	$2SO_2$
$SO\,Br_2$	+	$[N(CH_3)_4]2SO_3$	$\xrightarrow{\text{liq. } SO_2}$	$2[N(CH_3)_4]Br$	+	$2SO_2$

(iv) Acid-base reaction in other solvents. Acid-base reactions in solvents other than H2O, liq. NH3 and liq. SO2 are given below :

Acid		Base		Salt		Solvent
$NOCl$	+	$NaNO_3$	$\xrightarrow{\text{liq. } N_2O_4}$	$NaCl$	+	N_2O_4
HCl	+	CH_3COONa	$\xrightarrow{\text{liq. } CH_3COOH}$	$NaCl$	+	CH_3COOH
HCl	+	$NaCN$	$\xrightarrow{\text{liq. HCN}}$	$NaCl$	+	HCN

Advantage of solvent system concept

The definition of acids and bases given by solvent concept can be used for protonic (e.g., H_2O, NH_3, N_2H_4, HF, H_2SO_4, CH_3COOH, HCN etc.) as well as non-protonic (e.g., SO_2, $COCl_2$, BrF_3, N_2O_4 etc.) solvents i.e., this definition is applicable to aqueous (i.e., H_2O) as well as non-aqueous solvents (e.g., NH_3, N_2H_4, HF, H_2SO_4, CH_3COOH, HCN, $COCl_2$, BrF_3, N_2O_4 etc.).

Disadvantages of solvent system concept

(i) According to this concept, the definition of acids and bases is based on the nature of the solvent cation and solvent anion obtained by the auto-ionisation of the solvent.

(ii) According to this concept, acid-base reactions taking place in the absence of a solvent cannot be explained, i.e., acid-base reactions take place only in the presence of a solvent.

(iii) This concept cannot account for the acid-base reactions occurring in non-ionising solvents like C_6H_6, $CHCl_3$ etc.

Lewis concept : Electron pair acceptor-donor concept (1923)

In the early 1930, G.N. Lewis defined an acid as a molecule or an ion that can accept an electron pair from some other substance and a base as a molecule or an ion which can donate an electron pair (lone pair of electrons) to some other substance. In other words, a Lewis acid is an electron pair acceptor and a Lewis base is an electron pair donor.

Since Lewis acid is an electron pair acceptor, it is an electron-deficient species. Similarly, since Lewis base is an electron pair donor, it is an electron pair rich species.

Since a Lewis acid loves to accept the electron pair(s), it is also called an electrophilic (electro = electron, philic = lover) reagent or simply electrophile. Similarly, since a Lewis base loves to donate the electron pair(s) to a nucleus-in fact-to a positively-charged ion, it is a nucleus-lover or nucleus-seeker and hence is also called nucleophilic (nucleo = nucleus, philic = lover) reagent or simply nucleophile.

Lewis acid should contain one or more vacant orbitals into which electron pair(s) donated by the Lewis base can be accommodated.

An acid qualifying Lewis concept is called Lewis acid and a base qualifying Lewis concept is termed as Lewis base.

Neutralisation reaction according to Lewis concept

According to Lewis concept of acids and bases the neutralisation reaction is that in which a Lewis acid reacts with a Lewis base and forms a compound which is called an **adduct** or **complex compound.** This compound contains (Lewis base $\rightarrow$ Lewis acid) co-ordincate bond. For example :

Lewis acid		Lewis base		Adduct
BF_3	+	$:NH_3$	$\longrightarrow$	$[H_3N \rightarrow BF_3]$

(Lewis base $\rightarrow$ Lewis acid) co-ordinate bond results by the overlap of the filled orbital on the Lewis base with the vacant orbital on the Lewis acid.

Examples of Lewis acids

Following molecules/ions act as Lewis acids :

1. Molecules whose central atom has vacant p-orbitals or incomplete octet of electrons in its valence shell. Examples of such molecules are BeF_2, BX_3 (X = H, F, Cl, Me) etc. The valence shell

of Be-atom (central atom) in BeF2 molecule (F—Be—F) has only 4 electrons (incomplete octet of electrons). Similarly the valence-shell of B-atom is BX_3 molecule 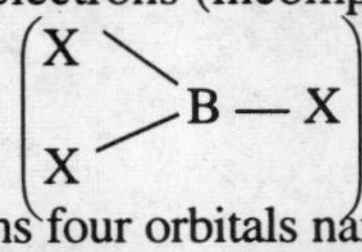 has only 6 electrons (incomplete octet). The central atom of these molecules contains four orbitals namely $2s$, $2p_x$, $2p_y$ and $2p_z$ but all these orbitals are not filled, *e.g.*, in BeF_2 molecule two orbitals *viz.*, $2p_y$ and $2p_z$ are vacant while in BX_3 molecule one orbital namely $2p_z$ is vacant. Thus these molecules are electron-deficient molecules. Due to the presence of vacant *p*-orbitals in the valence-shell of the central atom of these molecules, these molecules can accept one or more electron pairs from Lewis bases (Lewis bases are electron pair donor species) and hence act as Lewis acids. By accepting the electron pair(s) the central atom of the Lewis acid molecules completes its octet. Following Lewis acid—Lewis base reactions illustrate this point :

Lewis acid (Electron deficient molecules) *Lewis base (Electron pair donor species)* *Adduct*

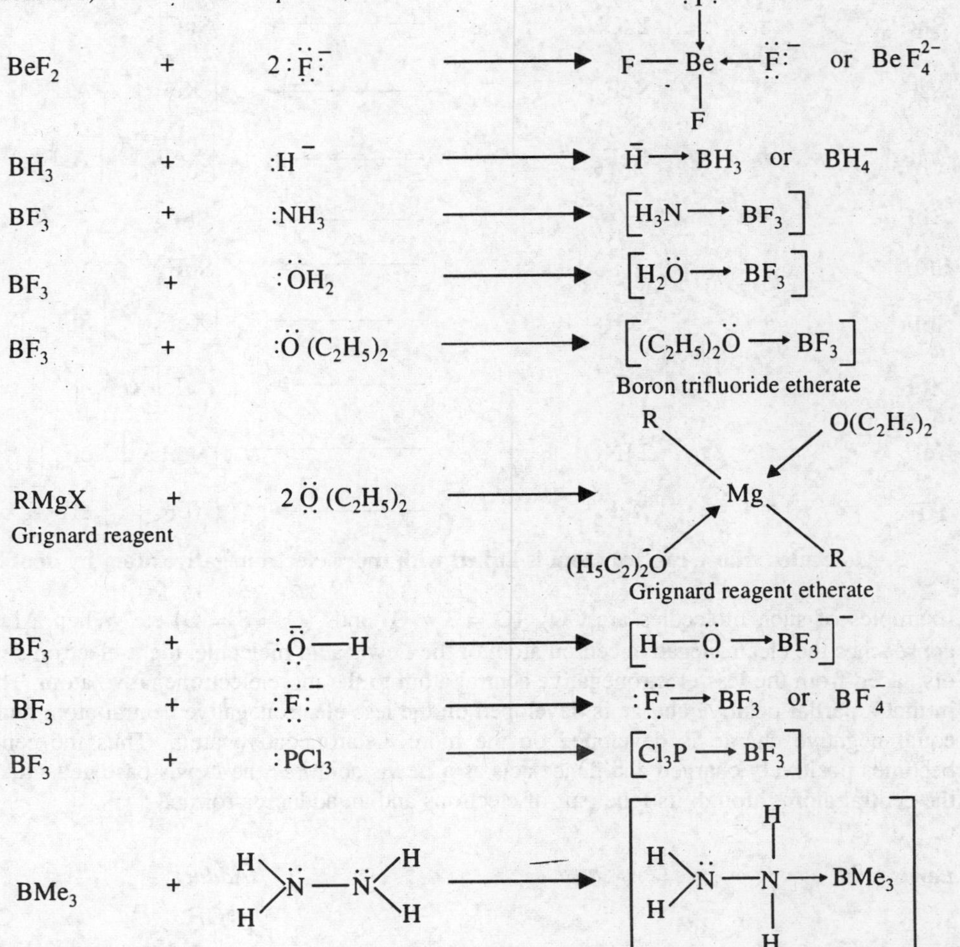

2. Molecules whose central atom has vacant d-orbitals in its valence-shell. Examples of such molecules are AlF_3, $AlCl_3$, $GaCl_3$, SiX_4, GeX_4, SnX_4, PX_3, PF_5, AsF_5, SbF_3, SbF_5, SF_4, $TeCl_4$, XeF6 etc. Due to the presence of vacant d-orbitals these molecules can accept the electron pair(s) donated by Lewis bases. By accepting the electron pair(s) the central atom gets more than eight electrons (expansion of octet). Following Lewis acid-Lewis base reactions illustrate this point:

Lewis acid		*Lewis base*		*Adduct*
AlF_3	+	$3:\ddot{F}:^-$	$\longrightarrow$	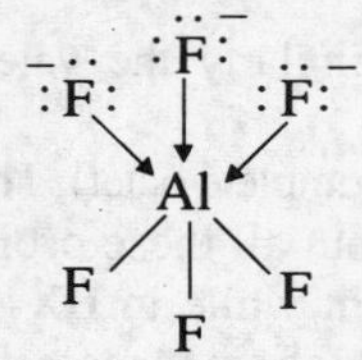 or AlF_6^{3-}
$AlCl_3$	+	$COCl_2$	$\longrightarrow$	$[COCl]^+[AlCl_4]^-$
$SnCl_3$	+	$:\ddot{Cl}:^-$	$\longrightarrow$	$GaCl_4^-$
SiF_4	+	$2:\ddot{F}:^-$	$\longrightarrow$	SiF_6^{2-}
$SnCl_4$	+	$2:\ddot{Cl}:^-$	$\longrightarrow$	$SnCl_6^{2-}$
AsF_5	+	XeF_2	$\longrightarrow$	$[XeF]^+[AsF_6]^-$
AsF_5	+	XeF_6	$\longrightarrow$	$[XeF_5]^+[AsF_6]^-$
SbF_3	+	$2:\ddot{F}:^-$	$\longrightarrow$	SbF_5^{2-}
SbF_5		$:\ddot{F}:^-$	$\longrightarrow$	SbF_6^-
SbF_5	+	XeF_4	$\longrightarrow$	$[XeF_3]^+[SbF_6]^-$
XeF_6	+	$2\ Cs^+F^-$	$\longrightarrow$	$Cs_2^+[XeF_8]^{2-}$
XeF_6	+	$2\ [NO]^+F^-$	$\longrightarrow$	$[NO]_2^+[XeF_8]^{2-}$
HF	+	XeF_6	$\longrightarrow$	$[XeF_5]^+[HF_2]^-$

3. Molecules whose central atom is linked with more electronegative atom by double bonds.

Examples of such molecules are CO_2 $(\overset{3.5}{O} = \overset{2.5}{S} = \overset{3.5}{O})$ and $(\overset{3.5}{O} = \overset{2.5}{S} = \overset{3.5}{O})$ etc. When a Lewis base approaches the electronegative central atom of the Lewis acid molecule, the π-electron density gets displaced from the less electronegative central atom to the more electronegative atom. This results in that a partial positive charge is developed on the less electronegative central atom and a partial equal negative charge is developed on the more electronegative atom. Thus the central atom becomes positively charged and hence acts as a Lewis acid *i.e.,* the Lewis base gets attached with the central atom through its lone pair of electrons and an adduct is formed :

Lewis acid *Lewis base* *Adduct*

or HCO_3^-

$$CO_2 + H^+[OH]^- \longrightarrow H^+[HCO_3]^- \text{ or } H_2CO_3$$

$$CO_2 + Li^+[OH]^- \longrightarrow Li^+[HCO_3]^- \text{ or } LiHCO_3$$

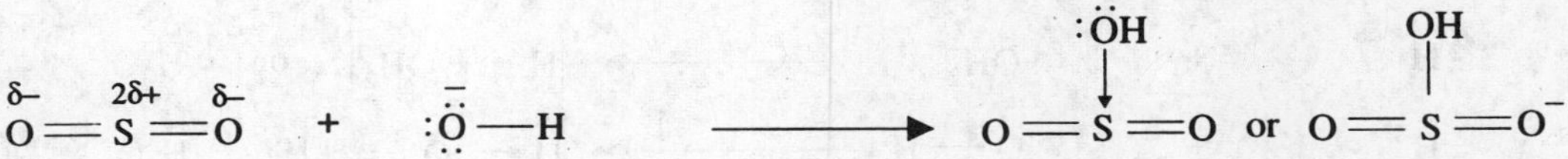

or HSO_3^-

$$SO_2 + Ca^{2+}O^{2-} \longrightarrow Ca^{2+}[SO_3]^{2-} \text{ or } CaSO_3$$

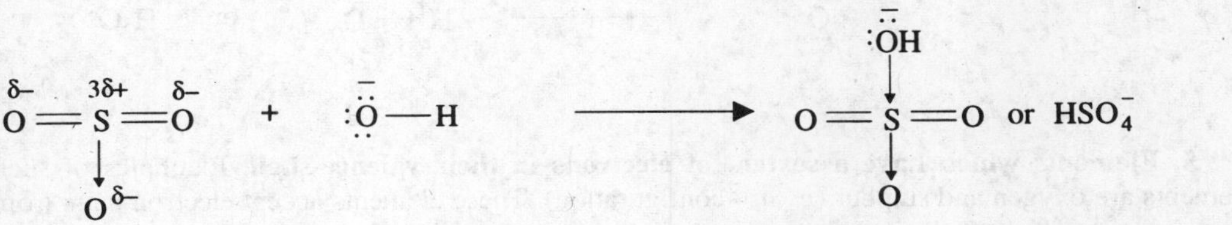

$$SO_3 + H^+[OH]^- \longrightarrow H^+[HSO_4]^- \text{ or } H_2SO_4$$

$$SO_3 + Na_2^+O^{3-} \longrightarrow Na_2^+[SO_4]^{2-} \text{ or } Na_2SO_4$$

$$SO_3 + Ca^{2+}O^{2-} \longrightarrow Ca^{2+}[SO_4]^{2-} \text{ or } CaSO_4$$

4. Simple cations, which have low lying empty orbital(s). When a simple metal cation (*e.g.*, Ag^+, Cu^{2+}, Cd^{2+} etc.) combines with a ligand, a complex compound is formed. In its formation, the ligand donates one or more electron pairs to the metal cation, which accommodates them in its vacant orbitals and (ligand → metal cation) co-ordinate bond is established. In this reaction, metal cation acts as a Lewis acid and the ligand behaves as a Lewis base. Examples of some Lewis acid-Lewis base reactions, in which metal cations act as Lewis acids are given below :

Lewis acid		*Lewis base*		*Complex ion (Adduct)*
Ag^+	+	$2:NH_3$	$\longrightarrow$	$H_3N \longrightarrow \overset{+}{Ag} \longleftarrow NH_3$ or $[Ag(NH_3)_2]^+$
Cu^{2+}	+	$4:NH_3$	$\longrightarrow$	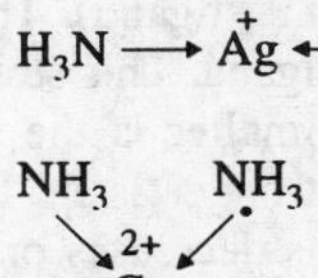 or $[Cu(NH_3)_4]^{2+}$
Cd^{2+}	+	$4:NH_3$	$\longrightarrow$	$[Cd(NH_3)_4]^{2+}$

Since H^+ ion has empty 1*s* orbital, it is able to accommodate an electron pair donated by the electron pair donors. Thus, in the following reactions H^+ ion acts as a Lewis acid.

Lewis acid		*Lewis base*		*Adduct*		
H^+	+	$:NH_3$	$\longrightarrow$	$[H^+ \leftarrow NH_3]$	or	NH_4^+
H^+	+	$:PH_3$	$\longrightarrow$	$[H^+ \leftarrow PH_3]$	or	PH_4^+
H^+	+	$:\ddot{O}H_2$	$\longrightarrow$	$[H^+ \leftarrow OH_2]$	or	H_3O^+
H^+	+	$:\ddot{\underset{..}{O}}^- — H$	$\longrightarrow$	$[H^+ \leftarrow \ddot{\underset{..}{O}} — H]$	or	H_2O
H^+	+	$:\ddot{\underset{..}{F}}:^-$	$\longrightarrow$	$[H^+ \leftarrow \ddot{\underset{..}{F}}:]$	or	HF
$2H^+$	+	$:\ddot{\underset{..}{O}}:^{2-}$	$\longrightarrow$	$H^+ \leftarrow \overset{2-}{\ddot{O}}: \rightarrow H^+$	or	H_2O

5. Elements which have a sextent of electrons in their valence-shell. Examples of such elements are oxygen and sulphur (ns^2np^4 configuration). These elements accept electron pairs from Lewis bases like SO_3^{2-} to complete their octet and hence act as Lewis acids. For example :

Lewis acid *Lewis base* *Complex ion (adduct)*

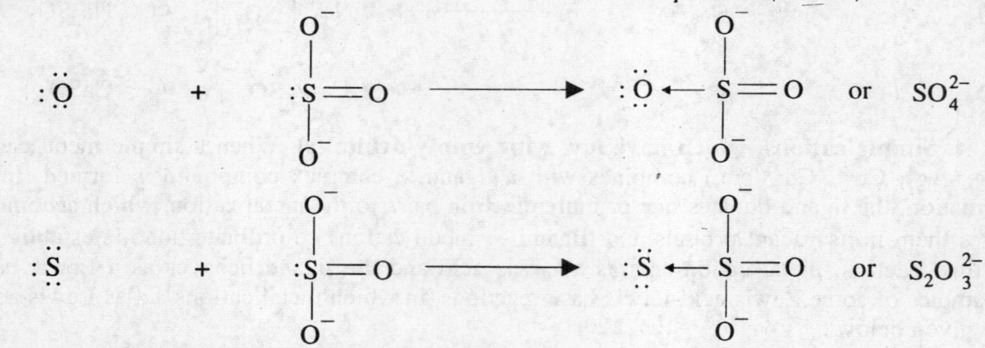

Variation of Lewis acid strength of simple cations

The Lewis acid strength of a cation is defined as its tendency to accept the electron pair(s) donated by the Lewis base (*i.e.*, ligand). Thus Lewis acid strength of a cation is its co-ordinating ability with a Lewis base or ligand. This strength depends on the following properties of the cation.

1. Size of the cation. Smaller is the size of a cation, greater is its tendency to accept the electron pair(s) and hence stronger is its Lewis acid strength or co-ordinating ability. Thus the Lewis acid strength of cations decreases on moving down the group and increases on proceeding from left to right in a period. For example :

$H^+ > Na^+ > K^+$ (cations of the same group)

$Li^+ < Be^{2+}$ (cations of the same period)

2. Oxidation state of the cation. The Lewis acid strength of the cations derived from the same metal atom increases with the increase in its oxidation number, since the size of the cation decreases with the increase in oxidation number. For example the Lewis acid strength of Fe^{2+} and Fe^{3+} ions is in the order: $Fe^{2+} < Fe^{3+}$.

Trihalides of boron are Lewis acids *i.e.*, trihalides of boron have Lewis acid character

We know that all trihalides of boron namely BF_3, BCl_3, BBr_3 and BI_3 are obtained by sp^2 hybridisation of B-atom (central atom) in its excited state ($B = 2s^1\ 2p_x^1\ 2p_y^1\ 2p_z^0$). Since B-atom in its excited state uses one $2s$ and only two $2p$ (say $2p_x$ and $2p_y$) orbitals in sp^2 hybridisation, $2p_z$ orbital remains vacant and unused on B-atom. Due to the presence of this vacant $2p_z$ orbital, the trihalide molecules can accept an electron pair donated by the donor molecules (*e.g.*, RHO, $(C_2H_5)_2O$, NH_3, H_2O etc.) or ions (*e.g.*, OH^-, F^-, Cl^- etc.) and can thus form adducts or addition compounds with these molecules or ions. Consequently in these adducts, the trihalide molecules act as Lewis acids and the molecules/ions having lone pairs of electrons act as Lewis bases. The above discussion shows that the trihalides of boron have a tendency to accept an electron pair donated by Lewis bases and hence are Lewis acids. Thus, on the basis of Lewis definition of acids, we can say that boron trihalides have Lewis acid character. The formation of some adducts is shown below :

Lewis acid		*Lewis base*		*Adduct*
BF_3	+	$:NH_3$	$\longrightarrow$	$H_3N \rightarrow BF_3$
BF_3	+	$:\ddot{O}H_2$	$\longrightarrow$	$H_2\ddot{O} \rightarrow BF_3$
BF_3	+	$:\ddot{O}(C_2H_5)_2$	$\longrightarrow$	$[(C_2H_5)_2\ddot{O} \rightarrow BF_3]$
BF_3	+	$:\ddot{O}-H^-$	$\longrightarrow$	$[H-\ddot{O} \rightarrow BF_3]^-$
BF_3	+	$:\ddot{F}:^-$	$\longrightarrow$	$:\ddot{F} \rightarrow BF_3$ or BF_4^-
BF_3	+	$:PCl_3$	$\longrightarrow$	$Cl_3P \rightarrow BF_3$

Relative order of Lewis acid character of BF_3, BCl_3, BBr_3 and BI_3 molecules

The measurement of the molar heats of formation of the solutions of BX_3 molecules in nitrobenzene and the dipole moments of their adducts like $H_3N \rightarrow BX_3$ have shown that the Lewis acid character of the given molecules is in the order $BF_3 < BCl_3 < BBr_3 < BI_3$. This order shows that BBr_3 will give more stable adducts than BCl_3, and BCl_3 will yield more stable adducts than BF_3.

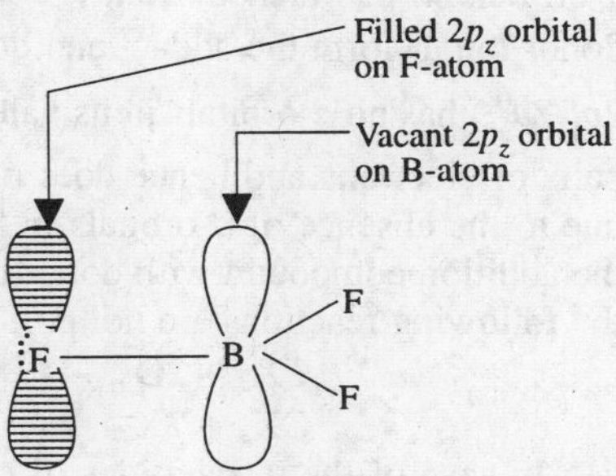

Fig. 19.3 : Structure of BF_3 molecule.

Explanation. The above order can be explained on the basis of boron-halogen π-back bonding. We know that each of the given molecules has a trigonal planar geometry which arises due to sp^2 hybridisation of B-atom in its excited state. In sp^2 hybridisation, one $2p$ orbital, say $2p_z$, of B-atom remains unhybridised and vacant. The halogen atom has valence-shell configuration as $ns^2\ np_x^2\ np_y^1\ np_z^2$. The singly filled np_y orbital of halogen atom overlaps with the singly-filled sp^2 hybrid orbital of B-atom and forms B-X σ-bond, while each of the remaining three orbitals of halogen atom contains one lone pair of electrons. Thus the structure of BF_3 molecule can be shown, as depicted in Fig. 19.3. Now the filed $2p_z$ orbital of F-atom makes a lateral overlap with the vacant $2p_z$ orbital of B-atom and gives rise to the formation of an additional F → B π–bond, called *p*π–*p*π back bond (See Fig. 19.4).

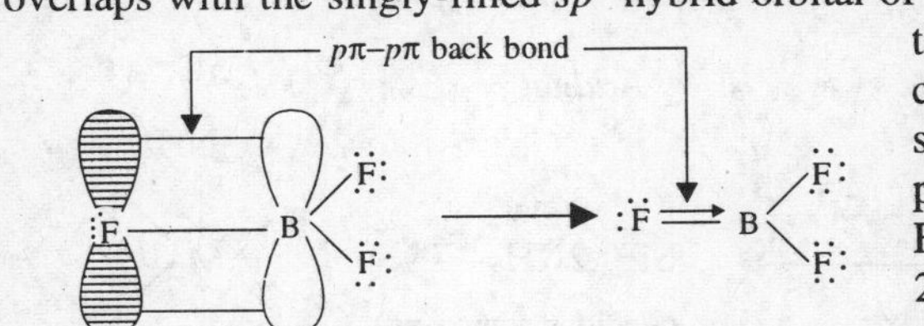

Fig. 19.4 : Formation of *p*π–*p*π back bond in BF_3 molecule by the lateral overlap between the filled $2p_z$ orbital of F-atom and vacant $2p_z$ orbital on B-atom.

The formation of *p*π–*p*π back bonding results in the following :

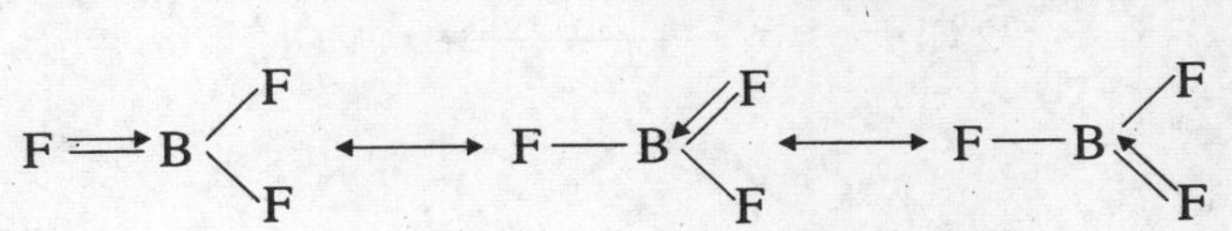

(*a*) Due to the formation of extra F $\rightarrow$ B π-bond, B—F bond acquires some double bond character. Thus B—F bond should be shown as $B \underset{\longleftarrow}{=\!=} F$. Since any one of the three F-atoms can take part in the formation of $B \underset{\longleftarrow}{=\!=} F$ bond, the structure of BF_3 can be supposed to be a resonance hybrid of the following three equivalent resonating structures.

On similar lines the resonating structures of BCl_3, BBr_3 and BI_3 molecules can also be written. It is because of the resonance that all the three B—X bond lengths in a given trihalide are equal.

(*b*) In case of BF_3 molecule F $\rightarrow$ Bπ-bond results by 2*p* (F) – 2*p* (B) overlap while in case of BCl_3, BBr_3 and BI_3 molecules this bond is produced by 3*p* (Cl) – 2*p* (B), 4*p* (Br) – 2*p* (B) and 5*p* (I) – 2*p* (B) overlaps respectively. Since in case of BF_3 molecule the overlap takes place between the orbitals which have the same energy and shape, their overlap occurs with great effectiveness. As we move from F to I, since the difference in energy between 3*p* (Cl) –2*p* (B), 4*p* (Br) – 2*p* (B) and 5*p* (I) – 2*p* (B) pairs of orbitals increases, the effectiveness with which these pairs of orbitals overlap decreases. Thus we see that the tendency of back bonding is maximum in BF_3 molecule. This tendency falls rapidly on passing from BCl_3 to BI_3. This means that the tendency of BF_3 molecule to accept electron pair given by Lewis base (*e.g.*, NH_3, PH_3, F^- etc.) is minimum and this tendency increases as we move from BF_3 to BI_3 *i.e.*, the Lewis acid character of BX_3 molecules is in the order :

$$BF_3 < BCl_3 < BBr_3 < BI_3$$

Tetrahalides of carbon donot behave as Lewis acids, while the tetrahalides of other elements of group IVA are Lewis acids

We have already said that if the central atom of a given molecule does not have vacant *d*-orbitals in its valence-shell, it cannot accept the lone pairs of electrons from a donor atom or a donor ion to form the adduct and does not show Lewis acid character. Now since C-atom ($2s^2\ 2p_x^1\ 2p_y^1\ 2p_z^0$) has no *d*-orbitals in its valence-shell, this atom in CX_4 molecule is not able to accept lone pairs of electrons and hence does not show Lewis acid character. In other words we can say that, due to the absence of *d*-orbitals in the valence-shell of C-atom, CX_4 molecule is not able to form the addition compound with donor molecules (D) or complex halide ion with halide ions (X^-), *i.e.*, the following reactions are not possible.

$$CX_4 + 2D \rightarrow CX_4.2D \text{ (addition compound)}$$
$$CX_4 + 2X^- \rightarrow [CX_6]^{2-} \text{ (complex hexahalide ion)}$$

In case of the tetrahalides of other elements, since the central atom (which may be Si, Ge, Sn or Pb) has vacant *d*-orbitals, these tetrahalides can accommodate the lone pairs of electrons donated by the donor molecule (*e.g.*, NH_3, H_2O etc.) or halide ions and can thus form the addition compound or complex halide ion. Thus the following Lewis acid-Lewis base reactions are possible.

Lewis acid		*Lewis base*		*Adduct*
MF_4 (M = Si, Ge)	+	2(:NH_3)	$\longrightarrow$	$SiF_4.2NH_3$ **(Addition compound)** or 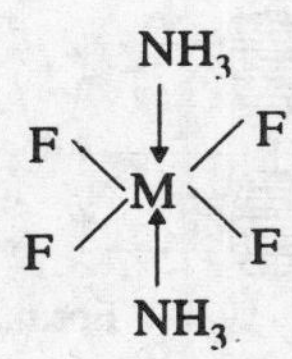
MCl_4 (M = Sn, Pb)	+	2 Cl^-	$\longrightarrow$	MCl_6^{2-} or 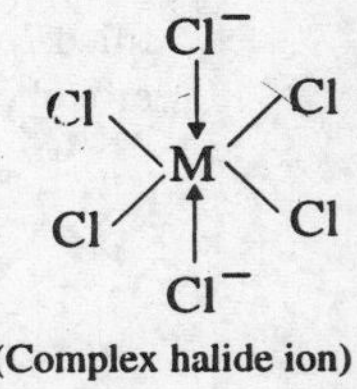

(Complex halide ion)

In the formation of addition compound or complex halide ion the central atom of the tetra-halide is sp^3d^2 hybridised.

Examples of Lewis bases :

Following molecules/ions act as Lewis bases

1. Molecules whose central atom has one or more unshared electron pairs (lone pairs of electrons). Examples of such molecules are the hydrides of the elements of groups VA (NH_3, PH_3, AsH_3 etc.), VIA (H_2O, H_2S etc.) and VIIA (HF, HCl etc.). Neon which has four lone pairs of electrons is also a Lewis base.

2. Almost all negative ions. Typical examples are F^-, Cl^-, I^-, OH^-, O^{2-}, SO_6^{2-} etc.

Lewis acid		*Lewis base*		*Adduct*
BF_3	+	$:\ddot{F}:^-$	$\longrightarrow$	$BF_3 \leftarrow :\ddot{F}:^-$ or BF_4^-
H^+	+	$:\ddot{F}:^-$	$\longrightarrow$	$H^+ \leftarrow \ddot{F}:^-$ or HF
$AlCl_3$	+	$:\ddot{Cl}:^-$	$\longrightarrow$	$Cl_3Al \leftarrow \ddot{Cl}:^-$ or $AlCl_4^-$
I_2	+	$:\ddot{I}:^-$	$\longrightarrow$	I_3^-
H^+	+	$:\ddot{O}^- —H$	$\longrightarrow$	$H^+ \leftarrow \ddot{O} — H$ or H_2O
$2H^+$	+	$:\ddot{O}:^{2-}$	$\longrightarrow$	$H^+ \leftarrow \ddot{O}:^{2-} — H$ (↓ H^+) or H_2O

3. Molecules containing C = C double bond. The donation of an electron pair from a π-bond to a Lewis acid like metal ion, BF_3, AlF_3 etc., results in the formation of π-complexes. In a number of complex compounds of ethylene with Ag, Pt and Rh, the metal atoms are bonded not to any single carbon atom but to the π-bond directly.

4. Halides. The halides that can make halide ion available behave as Lewis bases. Examples of such halides are XeF_2, XeF_4, XeF_6, CsF, NOF, $COCl_2$ etc. In the following reactions these halides act as Lewis bases :

Lewis acid		*Lewis base*		*Adduct*
AsF_5	+	XeF_2	$\longrightarrow$	$[XeF]^+ [AsF_6]^-$
SbF_5	+	XeF_4	$\longrightarrow$	$[XeF_3]^+ [SbF_6]^-$
AsF_6	+	XeF_6	$\longrightarrow$	$[XeF_5]^+ [AsF_6]^-$
HF	+	XeF_6	$\longrightarrow$	$[XeF_5]^+ [HF_2]^-$
XeF_6	+	2CsF	$\longrightarrow$	$Cs_2^+ [XeF_8]^{2-}$
$AlCl_3$	+	$COCl_2$	$\longrightarrow$	$[COCl]^+ [AlCl_4]^-$

Relative order of the basic strength of NH_3, H_2O, HF and Ne

We have already said that all these substances are Lewis bases. The basic strength (*i.e.*, electron-releasing power) of these substances depends on the number of lone pairs present in them. Greater is the number of lone pairs, lesser is the tendency of the substance to donate the electron pair and hence lesser is the basic strength of the substance. Thus since the number of lone pairs in these substances increases as we move from NH_3 to Ne (NH_3 = 1, H_2O = 2, HF = 3, Ne = 4), the basic strength decreases as $NH_3 > H_2O > HF > Ne$.

Relative order of the basic strength of NH_3, PH_3, AsH_3, SbH_3 and BiH_3

We know that the central atom in each of the hydrides has one lone pair of electrons. These molecules can donate this lone pair to the donor atom/molecule (Lewis acid) and hence act as Lewis base.

Lewis base		*Lewis acid*		
NH_3	+	H^+	$\longrightarrow$	$H_3N \rightarrow H^+$ or NH_4^+
NH_3	+	BF_3	$\longrightarrow$	$H_3N \rightarrow BF_3$

It has been found that the basic strength (*i.e.*, electron donating power) of these molecules decreases in the order $NH_3 > PH_3 > AsH_3 > SbH_3 > BiH_3$. Thus NH_3 is distinctly basic and forms salts with weak as well as strong acids, *e.g.*,

$$NH_3 + HCl \rightarrow NH_4Cl; \qquad 2NH_3 + H_2CO_3 \rightarrow (NH_4)_2CO_3$$

PH_3 is also a base but has less basic character than NH_3 while AsH_3, SbH_3 and BiH_3 have no basic character at all *i.e.*, AsH_3, SbH_3, and BiH_3 are neutral.

Explanation. The above order of basic character of AH_3 molecules has been explained on the basis of the relative size of the central atom, A. As we move down the group, the size of the central atom increases from N to Bi. Since N-atom is the smallest in size, the electron density of lone pair of electrons on it is concentrated over a small region. This electron pair can, therefore, be donated to the Lewis acid quite easily. As the size of the central atom increases down the group, the electron density of lone pair on the central atom gets diffused over a larger region and hence the ability of the central atom to donate the lone pair to the Lewis acid decreases. Thus we find that with the increase in the size of the central atom, the electron density of lone pair on the central atom decreases and hence the ability of the central atom to donate the lone pair to the Lewis acid also decreases, *i.e.*, the basic character of AH_3 molecules decreases as we move from NH_3 to BiH_3.

Comparison between acids and bases as defined by Arrhenius, Lowry-Bronsted and Lewis

The comparison between different definitions of acids and bases can be summarised as follows :

(*i*) (*a*) Arrhenius acid is a compound that gives H^+ or H_3O^+ ions in water and Arrhenius base is a compound that furnishes OH^- ions in water.

(*b*) Bronsted acid is a molecule or an ion that can donate one or more protons (H^+), *i.e.*, Bronsted acid is a proton-donor. Bronsted base is a molecule or an ion that can accept one or more protons, *i.e.*, Bronsted base is a *proton acceptor.*

(*c*) Lewis acid is a molecule or an ion that can accept one or more electron pairs, *i.e.*, a Lewis acid is an electron pair-acceptor. Lewis base is a molecule or an ion that can accept one or more electron pairs, *i.e.*, a Lewis base is an electron pair-donor.

(*ii*) Arrhenius acids and Lewis acids are the same substances, *i.e.*, the substances that act as Arrhenius acids also act as Lewis acids. This fact can be illustrated by considering the behaviour of HCl in the following two reactions :

(*a*) $HCl + H_2O \rightarrow H_3O^+ + Cl^-$ (*b*) $HCl + NH_3 \rightarrow NH_4^+ Cl^-$

Reaction. (*a*) Since HCl gives H_3O^+ ion when dissolved in water, this acid acts as an Arrhenius acid in this reaction.

$$HCl \rightarrow H^+ + Cl^-$$
$$H_2O + H^+ \rightarrow H_3O^+$$

$$HCl \text{ (Arrhenius acid)} + H_2O \rightarrow H_3O^+ + Cl^-$$

Reaction. (*b*) The formation of $NH_4^+Cl^-$ takes place through the following two steps :

$$HCl \rightarrow H^+ + Cl^-$$
$$H_3N: + H^+ \rightarrow H_3N \rightarrow H^+ \text{ or } NH_4^+$$

$$HCl \text{ (Bronsted acid)} + NH_3 \rightarrow NH_4^+ + Cl^- \text{ or } NH_4^+ Cl^-$$

In reaction (*b*) HCl undergoes ionisation and gives H^+ and Cl^- ions. H^+ ion thus obtained, has empty 1*s* orbital and hence accepts an electron pair from NH_3 and thus forms NH_4^+ ion. Now NH_4^+ and Cl^- ions combine together to form $NH_4^+ Cl^-$. Thus, we see that since H^+ ion obtained by the ionisation of HCl accepts an electron pair from NH_3 molecule, HCl, according to Lewis concept, acts as a Lewis acid.

The above discussion clearly shows that HCl, which acts as an Arrhenius acid, also behaves as a Lewis acid. In general, we can say that the substances that act as Arrhenius acids also act as Lewis acids, *i.e.,* Arrhenius acids and Lewis acids are the same substances.

(*iii*) Bronsted bases and Lewis bases are the same substances, *i.e.*, the substances that behave as Bronsted bases also behave as Lewis bases. This can be illustrated by considering the formation of NH_4^+ ion by the combination of H^+ ion and NH_3 molecule, according to Bronsted and Lewis concepts.

According to ***Bronsted concept***, the formation of NH_4^+ ion can be shown as :

$$H^+ + NH_3 \text{ (Bronsted base)} \longrightarrow NH_4^+$$

In this reaction, since NH_3 molecule accepts a proton (H^+) to form NH_4^+ ion, this molecule acts as a Bronsted base.

According to Lewis concept, the combination of H^+ and NH_3 takes place as follows :

$$H^+ + :NH_3 \text{ (Lewis base)} \longrightarrow [H^+ \leftarrow NH_3] \text{ or } NH_4^+$$

Here, since NH_3 molecule loses one electron pair, this molecule acts as a Lewis base.

The above discussion shows that NH_3 molecule which acts as a Bronsted base, also acts as a Lewis base. In general, we can say that the substances that act as Bronsted bases, also act as Lewis bases *i.e.*, Bronsted bases and Lewis bases are the same substances.

Pearson's classification of Lewis acids and Lewis bases into hard and soft acids and bases

R.G. Pearson (1963) has classified the Lewis acids and Lewis bases as hard and soft acids and bases. A third category whose characteristics are intermediate between those of hard and soft acids/bases are called borderline acids or borderline bases. The principal characteristics of hard and soft acids and bases along with their examples are given in Table 19.5.

Pearson's hard and soft acids and bases principle (HSAB principle)

On the basis of experimental data of various complexes obtained by the combination of Lewis acids and Lewis bases, pearson (1963) discovered a principle known as Hard and Soft Acids and Bases Principle (HSAB principle) (some chemists prefer the abbreviation SHAB instead of HSAB used by Pearson). This principle states that a hard Lewis acid prefers to combine with a hard Lewis base and similarly a soft Lewis acid prefers to combine with a soft Lewis base, since this type of combination gives a more stable product. Thus we can say that (hard acid + hard base) and (soft acid + soft base) combinations give more stable products than the (hard acid + soft base) and (soft acid + hard base) combinations. The combination of hard acid and hard base occurs mainly through ionic bonding as in $Mg(OH)_2$ (Mg^{2+} = hard acid, OH^- = hard base) and that of soft acid and soft base takes place mainly by covalent bonding as in HgI_2 (Hg^{2+} = soft acid, I^- = soft base).

Applications of HSAB principle

HSAB principle is extremely useful in explaining the following :

1. Stability of complex compounds, having the same ligands. This application can be understood by considering the following examples :

(*a*) AgI_2^- is stable while AgF_2^- does not exist. We know that Ag^+ is a soft acid, F^- ion is a hard base and I^- ion is a soft base. Thus, since AgI_2^- is obtained by the combination of a soft acid (Ag^+) and soft base (I^-) and AgF_2^- results by the interaction of a soft acid (Ag^+) and a hard base (F^-), AgI_2^-, ion is stable but AgF_2^- does not exist.

Table 19.5 : *Classification of Lewis acids and Lewis bases into hard, soft and borderline acids and bases.*

Lewis Acids (acceptors)

Hard acids [Ahrland and Chatt (1958) have arbitrarily called hard acids as class (*a*) metal ions or metal acceptors]	**Soft acids** [These have been called class (*b*) metal ions or metal acceptors]	**Borderline (intermediate) acids**
(*i*) They have acceptor metal atom of small size. (*ii*) They have acceptor with high positive charge (oxidation state). (*iii*) The valence-electrons of the acceptor atom of these acids cannot be polarised (or distorted or removed) easily (*i.e.*, they have low polarisability), since they are held strongly and it is for this reason that these Lewis acids have been called hard acids (or hard metal ions) by Pearson (1963).	(*i*) They have acceptor metal atom of large size. (*ii*) They have acceptor atom with low or zero positive charge. (*iii*) The valence-electrons of the acceptor atom of these acids can be polarised easily (*i.e.*, they have high polarisability), since they are held weakly and for this reason these Lewis acids have been called soft acids (or soft metal ions) by Pearson.	The characteristics of borderline acids are intermediate between those of hard acids and soft acids.
Examples: H^+, Li^+, Na^+, K^+, Be^{2+}, Mg^{2+}, Ca^{2+}, Sr^{2+}, Mn^{2+}, Ag^{2+}, Al^{3+}, Sc^{3+}, Ga^{3+}, In^{3+}, La^{3+}, N^{3+}, Cl^{3+}, Gd^{3+}, Lu^{3+}, Cr^{3+}. Co^{3+}, Fe^{3+}, As^{3+}, CH_3Sn^{3+}, Si^{4+}, Ti^{4+}, Zr^{4+}, Th^{4+}, Li^{4+}, Pu^{4+}, Ce^{3+}, Hf^{4+}, Wo^{4+}, Sn^{4+}, UO_2^{2+}, MoO^{3+}, $BeMe_2$, BF_2, $B(OR)_3$, $Al(CH_3)_3$, $AlCl_3$, AlH_3, RPo_2, SO_3, phenol, I^{7+}, I^{5+}, Cl^{7+}, Cr^{6+}, RCO^+, Fe^{6+}, Pt^{6+}, CO_2, NC^+, HX (hydrogen-bonding molecules)	**Examples:** Cu^+, Ag^+, Au^+, Tl^+, Hg^+, Cs^+, Pd^{2+}, Cd^{2+}, Pt^{2+}, Hg^{2+}, CH_3Hg^+, $Co(CN)_5^{2-}$, Pt^{4+}, Te^{4+}, Tl^{3+}, $Tl(CH_3)_3$, BH_3, $Ga(CH_3)_3$, $GaCl_3$, GaI_3, $InCl_3$, RS^+, RSe^+, RTe^+, I^+, Br^+, I_2, Br_2, ICN, trinitrobenzene, chloranil, quinones, tetracyanoethylene, O, Cl, Br, I, N, RO, RO_2, M (metal atoms), CH_2, carbenes.	**Examples:** Fe^{2+}, Co^{2+}, Ni^{2+}, Cu^{2+}, Zn^{2+}, Pb^{2+}, Sn^{2+}, Sb^{3+}, Bi^{3+}, Rh^{3+}, Ir^{3+}, $B(CH_3)_3$, SO_2, NO^+, Ru^{2+}, Os^{2+}, R_3C^+, C_6H^{5+}, GaH_3.

Lewis Bases (Donors or ligands)

Hard bases (*Hard ligands*)	**Soft bases** (*Soft ligands*)	**Borderline (intermediate) bases**
The donor atom of a hard base:	The donor atom of a soft base:	These bases have intermediate properties.
(*i*) has high electronegativity.	(*i*) has low electronegativity.	
(*ii*) holds its valence-electrons strongly and hence cannot be polarised (removed or deformed) easily, *i.e.*, the donor atom of hard base has low polarisability.	(*ii*) holds its valence-electrons weakly and hence can be polarised easily, *i.e.* the donor atom of a soft base has high polarisability.	
(*iii*) has filled orbitals	(*iii*) has partially filled orbitals.	
Examples: H_2O, OH^-, ROH, R_2O, RO^-, CH_3COO^-, PO_4^{3-}, SO_4^{2-}, RCO_2^-, CO_3^{2-}, ClO_4^-, NO_3^-, O^{2-}, $C_2O_4^{2-}$, (co-ordination through O-atom), NH_3, NR_3, NHR_2, NH_2R, N_2H_4, NCS^-(co-ordination through N-atom), F^-, Cl^-.	**Examples:** R_2S, RSH, RS^-, SCN^- (co-ordination through S-atom). S^{2-}, R_3P, R_3As, I^-, CN^-, H^-, R^-, $S_2O_3^{2-}$, $(RO)_3P$, RNC, CO, C_2H_4, C_6H_6, CH_3^-.	**Examples:** $C_6H_5NH_2$, C_5H_5N, N_3^-, Br^-, NO_2^-, SO_3^{2-}, N_2.

(*b*) CoF_6^{3-} (hard acid + hard base) is more stable than CoI_6^{3-} (hard acid + soft base).

(*c*) $(CH_3)_2$ N – PF_2 molecule acts as a bidentalte ligand, since it has two lone pairs of electrons one of which is on N-atom and the other is on P-atom. Both BH_3 and BF_3 molecules combine with this ligand and forms an adduct. With the help of HSAB principle we can predict the structure of this adduct. We know that BF_3 is a hard acid and $-\ddot{N}(CH_3)_2$ is a hard base. It is also known that BH_3 is a soft acid and $-\ddot{P}F_2$ is a *soft base*. On applying the principle that (hard acid + hard base) combinations and (soft acid + soft base) are preferred, the structure of the adduct should be that in which N-atom donates its lone pairs of electrons to B-atom of BF_3 molecule and P-atom donates its lone pair of electrons to B-atom of BH_3 molecule. Thus the structure of the adduct is :

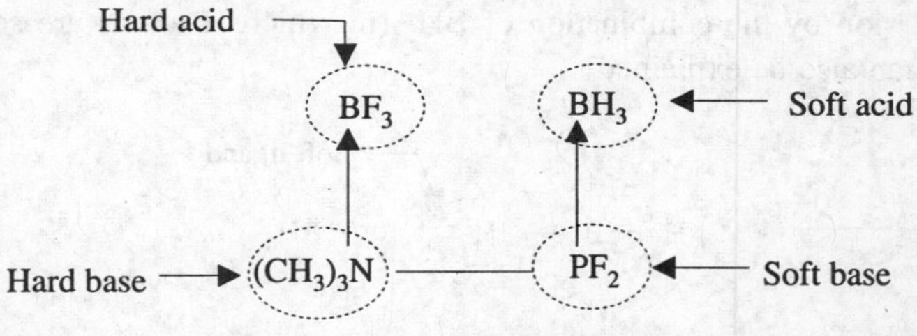

2. To predict the nature of bonding in complex ions given by ambidentate ligands (*a*) With the help of HSAB principle we can predict which atom of an ambidentate ligand will combine with metal ion to form the complex. SCN^- ion is an ambidentate ligand since it can co-ordinate to the metal ion either through its S-atom or through N-atom. It has been found that Co^{2+} and Pd^{2+} both combine with four SCN^- ligands to form the complex ion, $[M(SCN)_4]^{2-}$ (M = Co^{2+}, Pd^{2+}). With the help of HSAB principle it can be shown that in $[Co(SCN)_4]^{2-}$ ion, Co^{2+} is linked with the ligand through N-atom while in $[Pd(SCN)_4]^{2-}$ ion, Pd^{2+} is co-ordinated with the ligand through S-atom. Thus the complex ions given by Co^{2+} and Pd^{2+} ions should be represented as $[Co(NCS)_4]^{2-}$ and $[Pd(SCN)_4]^{2-}$ respectively. The reason for this is that since Co^{2+} ion is a hard acid, it prefers to co-ordinate with N-atom of the hard ligand, NCS^-. On the other, Pd^{2+} ion is a soft acid and hence combines with the S-atom of the soft ligand, SCN^-.

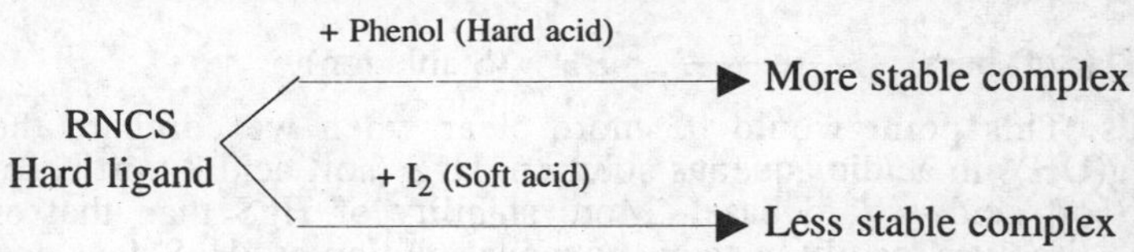

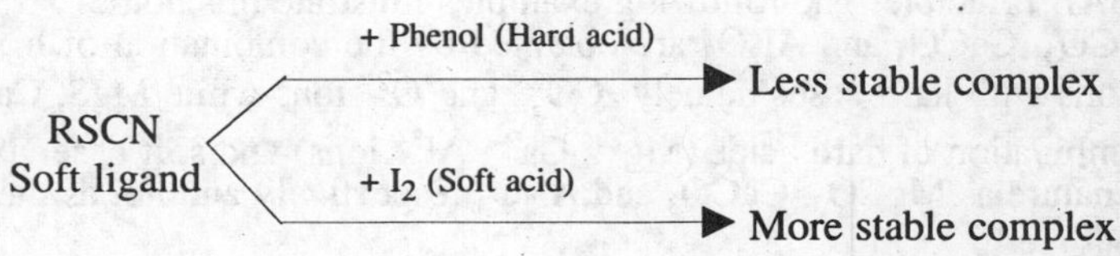

(*b*) We know that phenol is a hard acid and I_2 is a soft acid. It is also known that alkyl thiocyanate, RSCN (S-atom acting as a donor) is a soft ligand and alkyl iso-thiocyanate, RNCS (N-atom acting as a donor) is a hard ligand. Thus, if RSCN and RNCS are complexed with phenol and I_2, RNCS will form more stable complex with phenol due to hard acid (phenol) - hard ligand (RNCS) combination than with I_2. On the other hand RSCN will give more stable complex with I_2 due to soft acid (I_2) - soft ligand (RSCN) combination.

3. Stability of complex compounds having different ligands. Jorgensen has pointed out that in a complex compound having different ligands, if all the ligands are of the same nature, *i.e.*, if all the ligands are soft ligands or hard ligands, the complex compound will be stable. On the other hand, if the ligands are of different nature, the complex compound would be unstable. This point may be illustrated by the following examples:

(*a*) Since in $[Co(NH_3)_5F]^{2+}$ (I) both the ligands *viz.*, NH_3 molecule and F^- ion are hard ligands and in $[Co(NH_3)_5I]^{2+}$ (II) NH_3 is a hard ligand and I^- ion is a soft ligand, (I) is a stable complex ion while (II) is unstable.

(*b*) $[Co(CN)_5I]^{3-}$ (I) is more stable than $[Co(CN)_5F]^{3-}$ (II) because in (I) both the ligands are soft ligands while in (II) CN^- ions are soft ligands and F^- ion is a hard ligand.

4. Symbiosis. Soft ligands prefer to get attached with a centre which is already linked with soft ligands. Similarly hard ligands prefer to get attached with a centre which is already linked with hard ligands. This tendency of ligands is called **symbiosis** and can be explained by considering the formation of ($F_3B \leftarrow NH_3$) adduct and BH_4^- ion. Hard ligand like NH_3 co-ordinates with B-atom of BF_3 molecule to form ($F_3B \leftarrow NH_3$) adduct, since F^- ions which are already attached with B-atom in BF_3 molecule are also hard ligands. Thus:

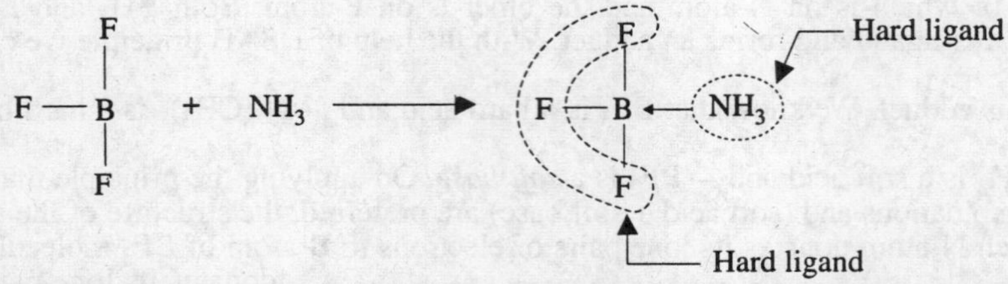

Similarly the formation of BH_4^- ion by the combination of BH_3 (in which H atoms are soft ligands) and H^- ions (soft ligands) can also be explained

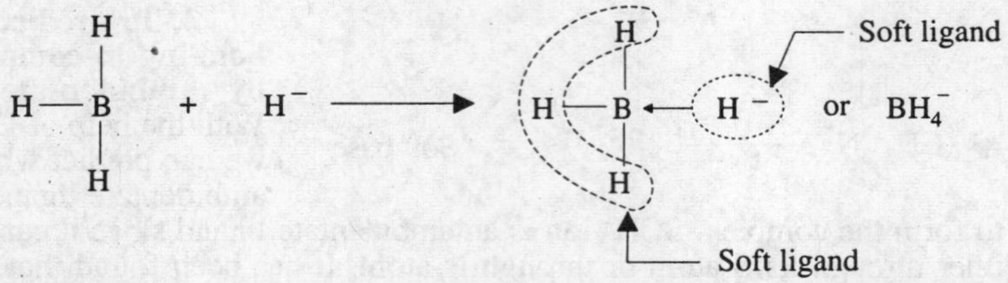

The formation of $F_3B \leftarrow NH_3$ adduct can also be explained on the basis of the fact that since BF_3 and NH_3 are hard acid and hard base respectively, they combine together to form a stable $F_3B \leftarrow NH_3$ adduct.

$$F_3B \text{ (Hard acid) } + NH_3 \text{ (Hard base) } \longrightarrow F_3B \leftarrow NH_3 \text{ (Stable adduct)}$$

Similarly since BH_3 is a soft acid and H^- ion is a soft base, their combination gives a stable BH_4^- ion.

$$BH_3 \text{ (Soft acid) } + H^- \text{ (Soft base) } \longrightarrow BH_4^- \text{ (Stable ion)}$$

5. Solubility of compounds. This point would be more clear when we compare the relative stability of HgS and $Hg(OH)_2$ in acidic aqueous solution. HgS (soft acid + soft base) is more stable than $Hg(OH)_2$ (soft acid + hard base). More stability of HgS than that of $Hg(OH)_2$ explains why $Hg(OH)_2$ dissolves readily in acidic aqueous solution but HgS does not.

6. Occurrence of metals in nature. The occurrence of some metals in nature as their ores can be explained with the help of HSAB principle. The following examples illustrate this point :

(*a*) We know that since $MgCO_3$, $CaCO_3$ and Al_2O_3 are obtained by the combination of hard acids *viz.*, Mg^{2+}, Ca^{2+} and Al^{3+} ions with hard bases namely CO_3^{2-} and O^{2-} ions while MgS, CaS and Al_2S_3 are obtained by the combination of hard acids (Mg^{2+}, Ca^{2+}, Al^{3+} ions) and soft base *viz.*, S^{2-} ion, Mg, Ca and Al occur in nature as $MgCO_3$, $CaCO_3$ and Al_2O_3 respectively and not as their sulphides (MgS, CaS and Al_2S_3).

(*b*) Since Cu_2S, Ag_2S and HgS are obtained by the combination of soft acids namely Cu^+, Ag^+ and Hg^{2+} ions and soft base *viz.*, S^{2-} ion while Cu_2CO_3, Ag_2CO_3 and $HgCO_3$ result by the interaction of soft acids (Cu^+, Ag^+, Hg^{2+}) and hard base *viz.*, CO_3^{2-}, Cu, Ag and Hg occur in nature as their sulphides (Cu_2S, Ag_2S and HgS) and not as their carbonates.

(*c*) Ni^{2+}, Cu^{2+} and Pb^{2+} ions which are borderline (intermediate) acids occur in nature both as carbonates and sulphides.

7. Jorginsen has also pointed that hard solvents tend to dissolve hard solutes and *vice versa*.

8. Course of reaction. The principle of (hard acid + hard base) and (soft acid + soft base) combination has also been used to predict the course of many reactions. For example :

LiI	+	CsF ⟶	LiF	+	CsI
(hard acid + soft base)		(soft acid + hard base)	(hard acid + hard base)		(soft acid + soft base)
HgF_2	+	BeJ_2 ⟶	BeF_2	+	HgI_2
(soft acid + hard base)		(hard acid + soft base)	(hard acid + hard base)		(soft acid + soft base)

Limitations of HSAB principle

Although (hard + hard) and (soft + soft) combination is a useful principle, yet many reactions cannot be explained with the help of this principle. For example in the reaction :

$$SO_3^{2-} + HF \longrightarrow HSO_3^- + F^-$$

or SO_3^{2-} + $H^+ F^-$ ⟶ $[H]^+ [SO_3]^{2-} + F^-$
soft base (hard acid + hard base) (hard acid + soft base)

which proceeds towards right, hard acid (H^+) combines with soft or borderline base (SO_3^{2-}) to form $[H^+] [SO_3^{2-}]$ or HSO_3^- ion which is a stable ion. (Hard acid + soft base) combination is against the HSAB principle.

Utility of Lewis concept

(*i*) This concept includes those reactions also in which no protons are involved.

(*ii*) Lewis concept is more general than the Bronsted-Lowry concept (*i.e.,* protonic concept) in that acid-base behaviour is not dependent on the presence of one particular element or on the presence or absence of a solvent.

(*iii*) It explains the long accepted basic properties of metallic oxides and acidic properties of non-metallic oxides.

(*iv*) This theory also includes many reactions, such as gasphase, high temperature and non-solvent reaction as neutralisation processes.

(*v*) The Lewis approach is, however, of great value in cases where the protonic concept is inapplicable, for example, in reaction between acidic and basic oxides in the fused state.

Limitations of Lewis concept

(*i*) Since the strength of Lewis acids and Lewis bases is found to depend on the type of reaction and not on the absolute power of accepting or donating a pair of electrons, it is not possible to arrange the Lewis acids and Lewis bases in any order of their acid or base strength. For example, since fluoride complex of Be^{2+} ion is more stable than the fluoride complex of Cu^{2+} ion, Be^{2+} ion is more stronger Lewis acid than Cu^{2+} ion. On the other hand, since ammine complex of Cu^{2+} ion is far more stable than that of Be^{2+} ion, Cu^{2+} ion is more stronger Lewis acid than Be^{2+} ion.

(*ii*) One of the starting points of Lewis concept was that Lewis acid-Lewis base reactions should be fast. But this is not so in actual practice for many reactions due to kinetic factors.

(*iii*) The reactions catalysed by Lewis acids are generally not catalysed by the protonic acids.

(*iv*) The conventional protonic acids *viz.,* H_2SO_4 and HCl are not covered under Lewis concept, as they do not establish a covalent bond by accepting a pair of electrons (which a Lewis acid ought to).

Usanovich Concept (1939)

M. Usanovich (1939), Russian chemist, proposed a very wide definition of acids and bases. The merits of this concept are: (*i*) It is the most general of all the different definitions. (*ii*) It includes all acids and bases, defined by Lewis. (*iii*) It states that the donation or acceptance of electrons need not take place as shared pairs. Thus, according to this concept, oxidation-reduction reactions are also classified as acid-base reactions.

According to this concept, an acid is any chemical species which (*i*) reacts with a base, (*ii*) accepts anions or electrons, (*iii*) furnishes cations. A base is any chemical species that (*i*) reacts with an acid, (*ii*) accepts cations, (*iii*) furnishes anions or electrons.

Some examples of acid-base reactions in terms of Usanovich concept are given below :

(*i*) SO_3 (Acid) + Na_2O (Base) ⟶ Na_2SO_4 (Salt)

This reaction takes place through the following stages :

Na_2O (Base) ⟶ $2Na^+ + O^{2-}$, Na_2O gives O^{2-} anion and hence acts as a base

SO_3 (Acid) + O^{2-} ⟶ SO_4^{2-}, SO_3 accepts O^{2-} anion and hence behaves as an acid

Na_2O (Base) + SO_3 (Acid) ⟶ $2Na^+ + SO_4^{2-}$ or Na_2SO_4 (Salt)

(*ii*) SO_2 (Acid) + Na_2O (Base) ⟶ Na_2SO_3 (Salt)

The formation of Na_2SO_3 (salt) by the combination of SO_2 (acid) and Na_2O (base) can also be explained as we have explained the formation of Na_2SO_4 by the combination of SO_3 and Na_2O.

(*iii*) Cl_2 (Acid) + 2Na (Base) ⟶ 2NaCl (Salt)

2Na (Base) $\xrightarrow{-2e^-}$ $2Na^+ + 2C^-$, since Na furnishes electrons, it acts as a base

2Cl $\xrightarrow{+2e^-}$ $2Cl^-$, Cl accepts electron and :s therefore an acid

2Na (Base) + 2Cl (Acid) ⟶ $2Na^+ + 2Cl^-$ or 2NaCl (Salt)

(*iv*) $Fe(CN)_2$ (Acid) + 4KCN (Base) ⟶ $K_4[Fe(CH)_6]$ (Salt)

4KCN (Base) ⟶ $4K^+ + 4CN^-$, KCN acts as a base, since it gives CN^- anions

$Fe(CN)_2$ (Acid) + $4CN^-$ ⟶ $[Fe(CN)_6]^{4-}$, $Fe(CN)_2$ behaves as an acid because it adds CN^- anions

4KCN (Base) + $Fe(CN)_2$ (Acid) ⟶ $4K^+[Fe(CN)_6]^{4-}$ or $K_4[Fe(CN)_6]$ (Salt)

Limitation of the concept

The only point that has been raised against this definition is that it is an extremely general concept, so general that almost all chemical reactions are considered as acid-base reactions in this concept.

University Questions

1. (*i*) What do you understand by soft and hard acids and bases ? Explain with suitable examples. Discuss the Pearson's principle and give a brief account of the important theories which favour this principle.
(*ii*) Explain why do Ca and Mg occur as oxides and carbonates but not as sulphides in nature. *(Lucknow 2002)*

2. Lewis strength of BBr_3 is greater than that of BCl_3 and BF_3. Explain. *(Lucknow 2002)*

3. Explain why all Bronsted bases are Lewis bases but all Bronsted acids are not Lewis acids. *(Lucknow 2002)*

4. (*a*) What are hard and soft acids and bases ? Explain with examples.
(*b*) On the basis of SHAB principle, explain why HgS is insoluble and $Hg(OH)_2$ is soluble in dil. HCl. *(Nagpur 2002)*

5. Discuss the Pearson's hard and soft acid and base principle for the classification of acids and bases. Explain theoretical behaviour of hardness and softness. Predict the relative strength of HF, HBr and HI in aqueous solution in the light of the above concept. *(VBS Purvanchal 2003)*

6. Explain Bronsted-Lowry concept of acids and bases with suitable examples. *(VBS Purvanchal 2003)*

7. Explain Lewis concept of acids and bases. *(VBS Purvanchal 2003)*

8. Discuss the concept of hard and soft acids and bases. *(Gujrat 2003)*

9. (*a*) What are Lewis acids and bases ? How are they classified ?
(*b*) Discuss the concept of the following, giving suitable examples : (*i*) Lux flood concept of acids and bases (*ii*) Hard and soft acids and bases. *(VBS Purvanchal 2003)*

10. (*a*) BCl_3 is stronger acid than BF_3. Explain.
(*b*) Which is a Lewis acid in the following ? (*i*) NaCl (*ii*) $BaCl_2$ (*iii*) $AlCl_3$. *(VBS Purvanchal 2003)*

11. (*a*) Enumerate the factors which govern the hardness and softness in acids and bases.
(*b*) Illustrate HSAB principle. Give reasons for the following on the basis of HSAB principle : (*i*) AgI_2^- is stable but AgF_2^- is not; (*ii*) CsF reacts with LiI but the reverse is not true.
(*c*) Identify Lewis acid or base in the following : Ag^+, BF_3, CO, PCl_3 *(Rohilkhand 2003)*

12. (*a*) Explain the terms conjugate acids and conjugate bases with the help of suitable examples.
(*b*) Discuss the effect of substituents on the strength of carboxylic acid.
(*c*) Which is the stronger acid of each pair. H_3O^+ or H_2O, NH_4^+ or NH_3, H_2S or HS^- and H_2O or OH? *(Delhi 2003)*

Answer.

(*b*) The presence of an electron withdrawing substituent like Cl in the carboxylic acid increases the acidity of carboxylic acid, since the electron withdrawing substituent withdraws electrons and stabilizes the anion by resonance. On the other hand, the presence of an electron releasing group like CH_3 in the acid decreases the acidity of the acid, since this group destabilise the anion.

13. (*a*) Apply SHAB principle in : (*i*) occurrence of metal minerals and (*ii*) predicting the stability of halogen acids.
(*b*) Classify the following as hard and soft acids : Ca^{2+} and Hg^{2+}. *(Nagpur 2003)*

14. (*a*) Discuss the Bronsted - Lowry concept of acids and bases. What are its applications and limitations?
(*b*) Explain the following :
(*i*) Arrhenius hypothesis (*ii*) Lux-Flood acid - base hypothesis.
(*c*) Explain the following :
(*i*) Hard acid and soft base (*ii*) Hard base and soft base
(*iii*) Lewis acid and Lewis base. *(Bhopal 2004)*

15. (*a*) Out of BCl_3 and BF_3 which is more acidic?
(*b*) BF_3 is a Lewis acid whereas CCl_4 is not a Lewis acid. Explain. *(GND Amritsar, 2004)*

20

General Principles of Extraction of Metals and Furnaces

Parts of Earth

Earth can be broadly divided into the following three parts :

1. Atmosphere. This part consists of a mixture of various gases. The main gases present in atmosphere are nitrogen (= 78.08%) and oxygen (= 21%).

2. Hydrosphere. It covers about 80% of the earth's surface and consists of rivers, lakes and oceans (sea).

3. Lithosphere. It is the solid part of the earth and consists of different types of rocks like igneous rocks, sedimentary rocks and metamorphic rocks.

Composition of lithosphere

According to Gold Schmidt, lithosphere consists of different layers of rocks. These rocks are formed by the cooling of the gaseous mixtures present in the atmosphere. All rocks are not formed in the same way. Depending on the way in which these rocks are formed, these rocks can be classified into the following three types :

1. Igneous rocks. We know that since the temperature, deep under the surface of the earth, is very high, the substances present deep down the earth's surface are extremely hot. These substances are called **magma.** The very high temperature prevailing deep down the earth, sometimes forces the magma to come out of the earth's surface in the form of volcanic erruptions. This results in that the magma spreads over the surface of the earth. Now when this magma cools gradually over the years, it gets solidified into rocks, which are called igneous rocks. The word igneous implies "*made by fire*". *Granite, pumice, feldspar, magnetite, odsiadian* etc., are the common examples of igneous rocks. The igneous rocks are generally the silicates and / or alumino silicates of Fe^{2+}, Mg^{2+}, Ca^{2+}, Na^{+} and K^{+}.

2. Sedimentary rocks. Wind, rain and flowing water slowly break the igneous rocks into small pieces, which are called **pebbles.** These pebbles are carried away by winds and rivers into the lakes and sea where they get deposited on the river-banks and at the bottom of the sea. The skeletons of sea animals also settle alongwith pebbles at the bottom of the sea. These are called sediments and the process of their formation is called **sedimentation.** With the passage of time different layers of sediments get deposited one upon each other and ultimately condense into a hard mass, which is called *sedimentary rock.* The common sedimentary rocks are *sand stone, lime stone, dolomite* etc.

3. Metamorphic rocks. These rocks are believed to be formed by the structural changes of igneous and sedimentary rocks, due to high temperatures and pressures. The word metaphoric implies "*change of form*". *Genesius, marble, slate* etc., are the common examples of metamorphic rocks.

Different layers of earth

According to Gold Schmidt, earth is composed of different layers or zones. There is a *metallic core* in the centre of the earth. This is called *central layer* or *siderophil*. This layer consists of the elements, which occur in their *native state*. This layer consists of the elements like Mn, Fe, Co, Ni, Cu, Ru, Rh, Pd, Ag, Re, Os, Ir, Pt, Au etc. The central layer is surrounded by a sulphide layer, which is called *chalcophil*. This layer consists of the elements, which have a tendency to form sulphides and is mainly composed of P, S, Zn, Ge, As, Se, Cd, In, Sb, Te, Hg, Tl, Pb, Bi. The outer layer, covering the sulphide layer, is called *silicate layer* or *lithophil*. This layer consists of chlorides, sulphates and carbonates of Li, Na, K, Rb, Cs, Mg, Ca, Sr, Ba etc. The silicate layer is surrounded by the solid crust, which we actually see. This solid crust is called *atmophile* and consists of the elements, which occur as gases (F, Cl, Br, I, B, C, Si, N, O, He, Ne, Ar, Kr, Xe).

The different layers, mentioned above, have been formed by the slow cooling of the magma coming out from deep inside the surface during the eruptions that are taking place from time to time.

Percentage distribution of different elements in the earth's crust

Silicon and oxygen together form 74.3% of the earth's crust (Si = 27.7%, O = 46.6%). Other elements present in the earth's crust are Al (8.3%), Fe (5.1%), Ca (3.6%), Na (2.8%), K (2.6%), Mg (2.1%), Ti (0.4%), Mn (0.1%), H (0.1%), P (0.1%), other elements (< 0.1%).

Chemical weathering of igneous rocks

We have already said that when magma cools gradually over the years, the magma gets solidified into rocks, which are called igneous rocks. We have also said that igneous rocks are silicates and / or alumino silicates of Fe^{2+}, Mg^{2+}, Ca^{2+}, Na^+ and K^+ ions. When rain water, containing CO_2 and other acids dissolved in it, comes in contact with these rocks, some of the ions like Fe^{2+}, Mg^{2+}, Ca^{2+}, Na^+ etc., are dissolved in the rain water and a residue, which contains TiO_2, Fe_2O_3 and SiO_2 is left behind. This process is called *chemical weathering* of rocks. Thus, weathering may be defined as the process of selective dissolution of some ions like Fe^{2+}, Mn^{2+}, Ca^{2+} etc., in rain water containing CO_2 and other acids and leaving behind insoluble residue, containing TiO_2, Fe_2O_3 and SiO_3.

As a matter of fact, the differential dissolution or extraction of various ions during weathering process depends on their charge / size ratio. For examples :

(*a*) The elements for which charge / size ratio for their cations is *less than four,* are easily extracted by water and become soluble. Examples of such ions are Na^+, Mg^{2+}, K^+ etc.

(*b*) The elements for which charge / size ratio for their cations is *greater than twelve* form oxoanions like sulphates and phosphates.

(*c*) The elements for which charge / size ratio lies between 4 and 12 dissolve under reducing media but precipitate in oxidising media.

The weathered particles are transporated to different parts of the earth's crust, by means of winds and flowing water and gradually get accumulated on the earth's surface, to form minerals.

Production of elements in sea water

Sea water contains a large number of elements. The most common of these elements are Na, K, Ca, Mg etc. These elements are present in sea water, in the form of their respective chlorides and sulphates. These salts are formed by the process which consists of the following steps :

(*i*) Igneous rocks consist of alumino silicates of Ca, Na and K. When these rocks come in contact with rain water, which contains CO_2, soluble bicarbonates of Na, K and Ca are formed. Being soluble, these bicarbonates pass into sea water, while a mixture of sand (SiO_2) and kaoline clay, called *detritus clay*, settles down.

$$\underset{\substack{\text{(Alumino silicates}\\ \text{of Ca, Na and K)}}}{\text{Igneous rock}} + \text{Rain water } (H_2O) \text{ containing } CO_2 \xrightarrow{H_2O\ +\ CO_2}$$

Sea water containing bicarbonates of K, Na and Ca (soluble) + Detritus clay containing sand (SiO_2) + kaoline clay

(*ii*) Kaoline clay, which is a alumino silicate reacts with sea water, containing K^+ and HCO_3^- ions, to form illite clay along with CO_2 and H_2O

$$\underset{\text{(Alumino silicate)}}{\text{Kaoline clay}} + \text{Sea water containing } K^+ \text{ and } HCO_3^- \text{ ions} \longrightarrow \text{Illite clay} + CO_2 + H_2O$$

(*iii*) Ca^{2+} and HCO_3^- ions combine to form $CaCO_3$ which forms sediments

$$Ca^{2+} + 2HCO_3^- \rightarrow CaCO_3 + H_2O + CO_2$$

(*iv*) As a result of the volcanic eruptions taking place in sea from time to time, HCl is formed. It reacts with HCO_3^- ions as follows :

$$HCl + HCO_3^- \rightarrow Cl^- + H_2O + CO_2$$

(*v*) The above reaction is followed by metaphorphism of clay to form alumino silicates of Ca, K and Na which are the constituents of igneous rocks. The clay reacts with interstial water at high temperature and pressure to form igneous rocks, HCl and alumino silicate.

$$\underset{\text{(Na, K, Ca)}}{\text{Clay}} + \text{Interstitial water} \xrightarrow[\text{pressure}]{\text{Heat}} \text{Igneous rock} + HCl + \text{Alumino silicate}$$

NaCl is produced by the combination of Na^+ and Cl^- ions.

$$Na^+ + Cl^- \rightarrow NaCl.$$

Although sea water contains a large number of elements, yet only a few elements, namely Cl, Br, Mg and Na are commercially recovered from sea water.

Metals, non-metals and metalloids

We know that all kinds of materials are made up of elements. These elements constitute the entire matter on this earth. The elements are classified as metals, non-metals and metalloids.

Metals are the elements which are generally lustrous, solids, malleable, ductile and good conductors of heat and electricity. Two-third of the known elements are metals. Examples of metals which are of common use in our daily life are Fe, Cu, Al, Pb, Au, Ag, Hg etc.

Non-metals are the elements which are generally non-lustrous, brittle and poor conductor of electricity (graphite is an exception). Some of the common non-metals are O, H, C, S, P etc.

Metalloids are the elements which have common characteristics of metals and non-metals. Examples are antimony, tin, arsenic, bismuth etc.

Occurrence of elements in nature

Elements occur in nature in the following two states :

1. Native or free (uncombined) state. The elements are said to occur in the native state, when they are found in their elementary form. Generally less active elements like Au, Ag, Cu, Pt (metals), O, N, noble gases (non-metals) occur in the native state.

2. Combined state. The elements are said to occur in the combined state when they are found in the form of their compounds. Generally the reative metals occur as oxides, sulphides, sulphates, silicates, carbonates, phosphates, chlorides, nitrates etc.

Minerals and ores

Minerals are the compounds of the metals which are found in the earth and in which the metals are present in the combined state. Minerals have chemical composition and structure. Ores are those minerals from which the metals can be extracted economically and conveniently. Thus all minerals cannot be called ores but all ores are called minerals.

Minerals and ores differ from each other in the following respects : (*i*) The minerals contain a low percentage of the metal while the ores contain a large percentage of the metal. (*ii*) The metal cannot be extracted from the mineral. On the other hand ores can be used for the extraction of metals. (*iii*) All minerals are not ores but all the ores are minerals.

Types of ores

The ores can be divided into the following groups :

1. Free or native ores. These ores contain the metals in the free (native) or metallic state. Cu, Ag, Au, Hg, Bi and platinum metals (Ru, Pd, Os, Ir and Pt) are the examples of the elements which occur in the free state. These elements are generally found in the association with rock or alluvial impurities like clay, sand etc. Sometimes lumps of almost pure metals are also found. These are called *nuggets*.

2. Sulphide ores. Ag, Cu, Zn, Hg, Pb, Fe, Cd, Co, Ni, As etc., occur as sulphides. Examples of sulphide ores are Zinc blende (ZnS), Copper glance (Cu_2S), galena (Pbs), cinnabar (HgS), copper-iron pyrites or chalco-pyrite ($Cu_2S.Fe_2S_3$ or $CuFeS_2$), iron-pyrites (FeS_2) etc.

3. Oxide ores. Cu, Al, Zn, Sn, Fe, Mn, Cr etc., occur as oxides. Examples are bauxite ($Al_2O_3.2H_2O$), tin stone or cassiterite (SnO_2), cuprite (Cu_2O), haematite (Fe_2O) pyrolusite (Mn_2O_3), zincite (ZnO), chrome ochre (Cr_2O_3), magnetite (Fe_3O_4) etc.

4. Carbonate ores. Na, Ca, Mg, Fe, Zn, Cu, Ba, Pb etc., occur as carbonates. Examples are magnesite ($MgCO_3$), lime stone ($CaCO_3$), calamine ($ZnCO_3$), dolomite ($CaCO_3.MgCO_3$), siderite ($FeCO_3$); malachit [$CuCO_3$. $Cu(OH)_2$], cerussite (Pb CO_3) etc.

5. Halide ores. Na, K, Ag, Ca, Mg, Al etc., occur as halide ores. Examples are rock salt or common salt (NaCl), sylvine (KCl), horn silver (AgCl), flourspar (CaF_2), carnallite ($KCl.MgCl_2.6H_2O$), cryolitge ($AlF_3 3NaF$ or Na_3AlF_6) etc.

6. Sulphate ores. Ba; Pb Ca, Mg etc., occur as sulphates. Examples are barytes or heavy spar ($BaSO_4$), anglesite ($PbSO_4$), gypsum ($CaSO_4.2H_2O$), kieserite ($MgSO_4.H_2O$), epsom salt ($MgSO_4$. $7H_2O$). etc.

7. Silicate ores. Li, Al, Be, Zn etc., occur as silicates. Examples are spodumene [LiAl $(SiO_3)_2$], potash felspar ($K_2O.Al_2O_3.6SiO_2$), beryl ($3BeO.Al_2O_3.6SiO_2$), willemite (Zn_2SiO_4) etc.

8. Phosphate ores. Li, Ca, Fe etc., occur as phosphates. Examples are triphylite [(L, Na)$_3$ PO_4. (Fe, Mn)$_3$ $(PO_4)_2$], phosphorite [$Ca_3(PO_4)_2$].

9. Nitrate ores. Na, K, etc., occur as nitrates. Examples are chile salt petre or caliche ($NaNO_3$), nitre (KNO_3) etc.

Metallurgy

Metallurgy is a process of extracting a metal in its free state from its ore. The process of metallurgy depends on the nature of the ore and the impurities present in the ore. Thus, different procedures are adopted in case of different metals.

Gangue or matrix

The ore is generally associated with non-metallic impurities of earth and rocks. Mica, lime stone, sand etc., are the important impurities. These impurities are called **gangue or martix.**

Different Steps of Metallurgy

Different steps that are involved in the metallurgy of a metal from a given ore are discussed below :

I. Crushing and grinding of the ore : Pulverisation of the ore

Most of the ores occur in nature as huge lumps. These lumps have to be broken into small

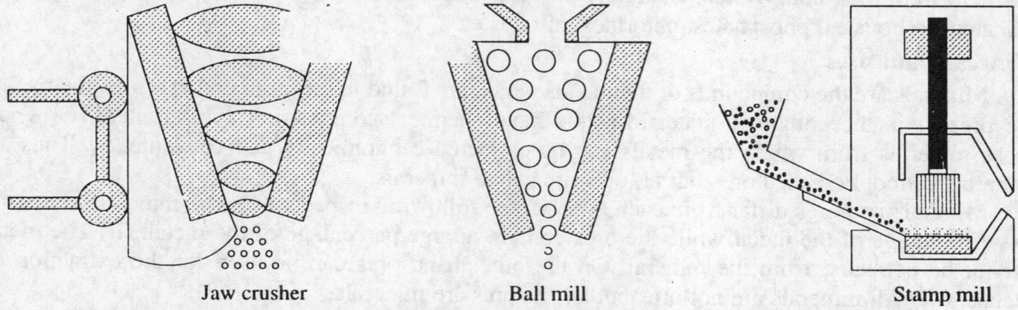

Fig. 20.1 : Crushing and grinding of the ore by jaw crusher, ball mill and stamp mill.

pieces, so that chemical changes, which take place at the later stages, become convenient. The huge lumps are broken into small pieces *in jaw crushers*. They are further converted into fine powder (pulverisation of the ore) in *ball mill* or *stamp mill* (See Fig. 20.1).

II. Concentration or dressing of the ore : Removal of impurities from the ore

Concentration is a process by which the impurities, like earthy matter, rocky matter, sand, lime stone etc., present in the ore are removed. Various methods used for the concentration of the ore are discussed below. The method used for the concentration of a given ore depends on the nature of the impurities to be removed and on the nature of the ore. Concentration of the ore is also called **benefication of the ore.**

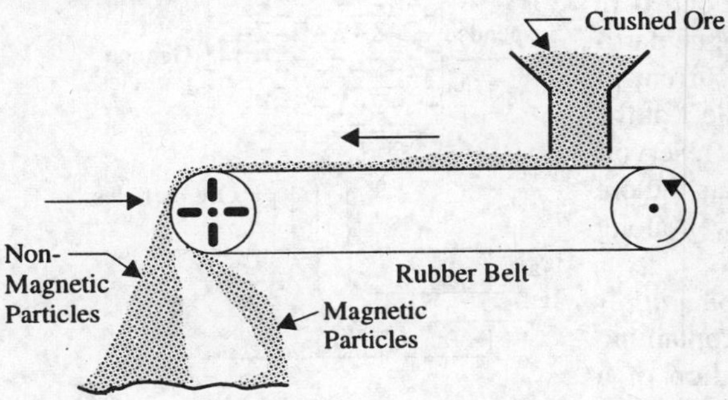

Fig. 20.2 : An electromagnetic separator, used to concentrate the ore.

1. Electromagnetic separation method : Removal of magnetic/non-magnetic impurities. If the impurity particles are magnetic and the ore particles are non-magnetic or *vice-versa,* then this method is used. In this method, an electromegnetic separator is used. The separator consists of a rubber belt which moves over two rollers, one of which encloses a magnet in it. The crushed ore is dropped through a hopper over the belt, when the magnetic particles are attracted by the magnet and hence fall just below the magnet, while the non-magnetic particles fall away from it (Fig. 20.2).

Examples. (*a*) Tin stone ore, SnO_2 (an ore of tin) is non-magnetic. This ore contains the impurity of wolframite, (Fe, Mn) WO_4 which is magnetic. This impurity of wolframite is removed from the ore by electromagnetic separation method.

(*b*) Chromite ore, $FeO.Cr_2O_3$ or Fe $(CrO_2)_2$ (an ore of Cr) is magnetic. This ore contains silicious impurities which ore non-magnetic. These non-magnetic impurities are removed from the ore by electromagnetic separation method.

2. Hydraulic washing method : Gravity separation method (Removal of lighter impurities). This method is used for the removal of those gangue particles which are lighter than the particles of the ore. This method is based on the fact that when the crushed and finely-powdered ore particles are treated with a strong and rapidly flowing current of water, the gangue particles, being lighter, are washed away by water and are thus removed while the ore particles, being heavier, settle down at the bottom.

The particles of oxide ores like tinstone (SnO_2), heamatite (Fe_2O_3), chromite ($FeO.Cr_2O_3$) etc., are heavier than many impurities present in them. Tin stone contains lighter impurities of silicious and earthy matter, heamatite contains lighter impurities of clay, sand etc., and chromite has also many highter impurities. All these lighter impurities are removed by gravity separation method.

The above discussion shows that gravity separation method makes use of the difference in the densities of the gangue (*i.e.,* impurity) particles and the ore particles.

Gravity separation method is of the following two types :

(*a*) *Wilfley table method (used for removing lighter impurities from tin stone*). In this method, a slopping table is used for the purpose. This table has a slanting floor on which long

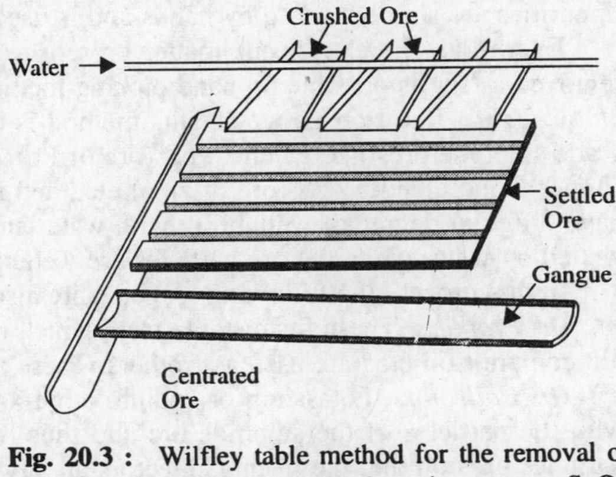

Fig. 20.3 : Wilfley table method for the removal of lighter impurities from tin stone ore, SnO_2.

wooden strips called *cleats* or *riffles* are fixed (Fig. 20.3). The crushed tin stone ore is fed on to the top of this table which is kept vibrating all the time. A running stream of water is made to flow across the table at right angles to the slope. The lighter silicious and earthy gangue particles rise up and are carried away with water while the heavier ore particles settle at the bottom. (Fig. 20.3).

(*b*) *Hydraulic classifier method (used for removing lighter impurities from tin stone)*. In this method, hydraulic classifier is used. It is a conical reservoir having a hopper at its top and a pipe in its bottom. Finely powdered tin stone ore is dropped in through the hopper and a powerful current of water is introduced in through the pipe fitted at the bottom. The gangue particles, being lighter, are carried away by the current of water and the ore particles, being heavier, collect at the apex of the cone. The conical shape of the reservoir helps in reducing the velocity of water and thus the ore particles are prevented from being carried away along with the current of water. (Fig. 20.4).

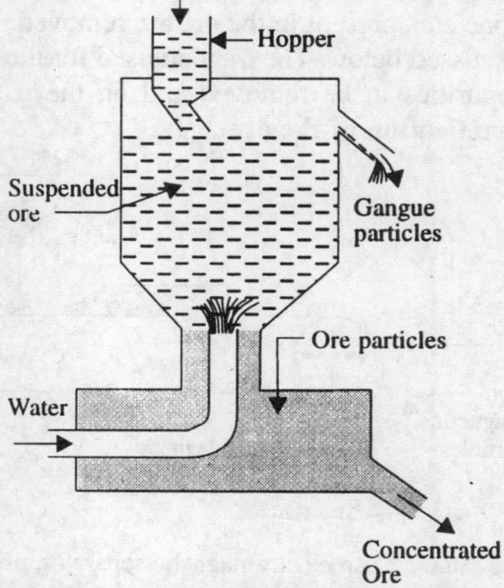

Fig. 20.4 : Hydraulic classifier method for removing lighter impurities from tin stone, SnO_2.

3. Leaching process. (Also called *hydrometallurgical* or *wet process*). In this process the ore containing impurities is leached with the aqueous solution of a suitable dissolving reagent so that the metal present in the ore is converted into its soluble simple salt or complex compound while the impurities remain undissolved in the reagent and are removed by filtration. The metal can be extracted from the soluble complex salt.

Example. Bauxite ore, $Al_2O_3.2H_2O$ (an ore of Al) contains the impurities of FeO, Fe_2O_3 and SiO_2. These impurities can be removed by Baeyer's process or by Hall's process both of which are leaching processes. The crushed bouxite ore is heated with a solution of NaOH (*Baeyer's process*) or Na_2CO_3 (*Holl's process– also called dry* or *fusion process*) in an autoclave for a few hours. Al present in the ore is converted into a *soluble* complex compound namely meta aluminate, $Na[AlO_2]$ while the impurities of FeO, Fe_2O_3 and SiO_2 remain undissolved in the reagent and hence are removed by filtration.

$$Al_2O_3.2H_2O + 2NaOH \longrightarrow 2Na[AlO_2] + 3H_2O$$

$$Al_2O_3.2H_2O + Na_2CO_3 \longrightarrow 2Na[AlO_2] + 2H_2O + CO_2\uparrow$$

The solution of $Na[AlO_2]$ obtained as above is used to extract Al metal by different methods.

4. Hand picking method. When impurities present in the ore are quite distinct from the ore and can easily be differentiated with nacked eyes, picking method is used. In this method the impurities are simply picked by hands and is used in the initial stage of the concentration.

Example. The extraneous matter present as impurity in heamatite ore, Fe_2O_3 (ore of Fe) is removed as for as possible by hand picking method.

5. Froth flotation process. This method is commonly used for the concentration of the low grade sulphide ores like galena, PbS (ore of Pb), copper-ion pyrites $Cu_2S.Fe_2S_3$ or $CuFeS_2$ (ore of copper), zinc blende, ZnS (ore of zinc) etc., and is based on the fact that gangue and ore particles have different degree of wettability with water and pine oil; the gangue particles are preferentially wetted by water while the ore particles are wetted by oil.

In this process the following reagents are also used :

(*a*) *Frothers*. These form stable froth which rises to the top of the flotation cell. Oils like pine oil, comphor oil etc., are used as frothers. These are added in small quantity.

(*b*) *Collectors*. Potassium or sodium ethyl xanthate is used as a collector. These get attached with the particles of the sulphide ore and thus make them water-repellant. Consequently the ore particles pass on into the froth. Collectors are always added in small quantity.

(*c*) *Activating and depressing agents.* When a mineral contains other minerals as impurities, the addition of these agents activates or depresses the flotation property of other minerals present as impurities and thus helps in separating the impurities. For example galena (PbS) usually contains the minerals namely ZnS and pyrites (FeS_2) as impurities. Flotation is carried out by using potassium ethyl xanthate (used as a collector) along with NaCN and Na_2CO_3 (used as depressing agent). The addition of NaCN and Na_2CO_3 depresses the flotation property of ZnS and FeS_2 grains, so mainly PbS passes into the froth when air is blown in. After PbS has been collected with the froth, the process is repeated by adding $CuSO_4$ (activator) which activates the flotation property of ZnS grains which are now removed with the froath. The acidification of the remaining material left in the flotation cell leads to the flotation of FeS_2.

Procedure. The powdered sulphide ore is mixed with water and the mixture thus obtained (called *slurry*) in taken in a flotation cell (Fig. 20.5). To this cell is added pine oil (frother), potassium or sodium ethyl xanthate (collector) and activating and / depressing agent, if necessary and then the mixture is agitated by blowing into it compressed air so that froth is formed and the ore particles which are preferentially wetted by pine oil rise upto the upper surface with the froth (foam) and the gengue particles (sand and earthy impurities) which are preferentially wetted by water remain in water below the froth at the bottom of the flotation cell. Thus the ore particles alongwith froth bubbles float on the top and can be separated.

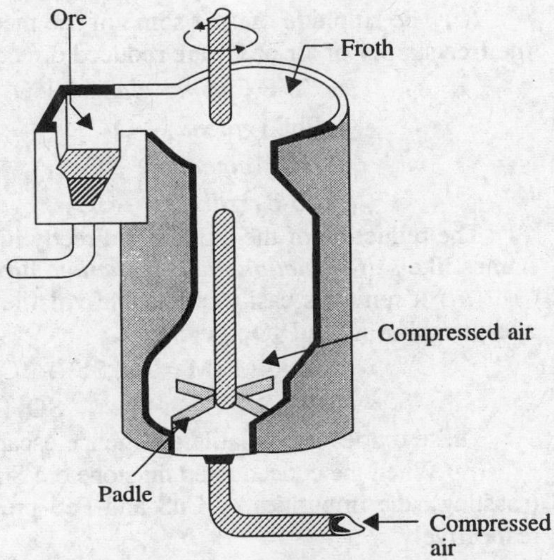

Fig. 20.5 : Flotation cell, used for the concentration of a sulphide ore by froth flotation process.

III. Calcination and roasting of the concentrated ore

Calcination. It is a process of heating the concentrated ore strongly in a limited supply of air or in the absence of air. The process of calcination brings about the following changes :

(*a*) The carbonate ore gets decomposed to form the oxide of the metal, *e.g.*,

$$FeCO_3 \xrightarrow{\Delta} FeO + CO_2 \; ; \quad PbCO_3 \xrightarrow{\Delta} PbO + CO_2$$

$$CaCO_3 \text{ (\textit{calcite} or \textit{lime stone})} \xrightarrow{\Delta} CaO + CO_2$$

$$ZnCO_3 \text{ (\textit{calamine})} \xrightarrow{\Delta} ZnO + CO_2$$

$$CuCO_3.Cu(OH)_2 \text{ (\textit{malachit})} \xrightarrow{\Delta} 2CuO + H_2O + CO_2$$

$$MgCO_3.CaCO_3 \text{ (\textit{dolomite})} \xrightarrow{\Delta} MgO + CaO + 2CO_2$$

(*b*) Water of crystallisation present in the hydrated oxide ore gets lost as moisture, *e.g.*,

$$2Fe_2O_3.3H_2O\ (g) \text{ (\textit{limonite})} \xrightarrow{\Delta} 2Fe_2O_3\ (s) + 3H_2O\ (g)\uparrow$$

$$Al_2O_3.\ 2H_2O \text{ (\textit{bauxite})} \xrightarrow{\Delta} Al_2O_3\ (s) + 2H_2O\ (g)\uparrow$$

(*c*) Organic matter, if present in the ore, gets expelled and the ore becomes porous. Volatile impurities are removed.

Roasting. It is a process of heating the concentrated ore (generally sulphide ore) strongly in the excess of air or O_2 below its melting point. The process of roasting does the following things :

(*i*) *Roasting at moderate temperature.* Some portion of the sulphide ores like galena (PbS), Zinc blende (ZnS) is first converted into metallic oxide and then the remaining portion is converted into metallic sulphate.

$$2PbS\ (\textit{some portion}) + 3O_2 \xrightarrow{\Delta} 2PbO + 2SO_2$$
$$PbS\ (\textit{remaining portion}) + 2O_2 \xrightarrow{\Delta} PbSO_4$$
$$2ZnS\ (\textit{some portion}) + 3O_2 \xrightarrow{\Delta} 2ZnO + 2SO_2$$
$$ZnS\ (\textit{remaining protion}) + 2O_2 \xrightarrow{\Delta} ZnSO_4$$

(*ii*) *Roasting at high temperature.* (*a*) Some sulphide ores, on being heated *strongly* in the presence of O_2, give metallic oxides.

$$2ZnS\ (\textit{zinc blende}) + 3O_2 \xrightarrow{\Delta} 2ZnO + 2SO_2$$
$$2Cu_2S\ (\textit{copper glance}) + 3O_2 \xrightarrow{\Delta} 2Cu_2O + SO_2$$

(*b*) The sulphide ores of some of the metals like Cu, Pb, Hg, Sb etc., when heated strongly in the free supply of air or O_2 are reduced directly to the metal rather than to the metallic oxides, *e.g.*,

$$Cu_2S\ (\textit{copper glance}) + O_2 \longrightarrow 2Cu + SO_2$$
$$PbS\ (\textit{galena}) + O_2 \longrightarrow Pb + SO_2$$
$$HgS\ (\textit{cinnabar}) + O_2 \longrightarrow Hg + SO_2$$
$$Sb_2S_3\ (\textit{stibnite}) + 3O_2 \longrightarrow 2Sb + 3SO_2$$

The reduction of the oxide ore directly into metal by heating it in air or O_2 is called by various names like *self-reduction, auto-reduction, air-reduction* etc.

(*iii*) It removes easily oxidisable volatile impurities like arsenic (as As_2O_3), sulphur (as SO_2) and phosphorus (as P_4O_{10}) etc.

$$4M\ (M = As, Sb) + 3O_2 \longrightarrow 2M_2O_3\uparrow$$
$$S + O_2 \longrightarrow SO_2\uparrow;\quad P_4 + 4O_2 \longrightarrow P_4O_{10}\uparrow$$

These oxides are volatile and hence escape as gases through the chimney.

(*iv*) When the concentrated tin stone ore SnO_2 (ore of Sn) is heated strongly in a free supply of air (roasting), the impurities of CuS and FeS present in the ore are converted into $CuSO_4$ and $FeSO_4$ respectively.

$$CuS + 2O_2 \xrightarrow{\Delta} CuSO_4\ ;\ FeS + 2O_2 \xrightarrow{\Delta} FeSO_4$$

Both calcination and roasting are generally carried out in a reverberatory furnace. In case of roasting, the air holes are kept open while they are partially or completely closed during calcination.

The differences between calcination and roasting are shown in the following Table.

Calcination	Roasting
(*i*) This process is used for hydrated oxide or carbonates ores.	(*i*) It is used for sulphide ores.
(*ii*) In calcination, the ore is strongly heated in the absence of air.	(*ii*) In this process the ore is heated in the presence of air.
(*iii*) Calcination converts the hydrated oxide or carbonate into metallic oxide.	(*iii*) In this process the sulphide ore is converted into metallic oxide or sulphate or into metal directly.

IV. Reduction of the compounds obtained after calcination or roasting : Pyrometallurgical process

In order to extract metals from the ores obtained after calcination or roasting, these are reduced by various reducing agents. Some of the methods used for reduction are given below. Here we shall consider the reduction of *metallic oxides, metallic sulphides, metallic halides* etc., to the free metal.

1. Reduction of metallic oxides. The reduction of metallic oxides to the free metal can be carried in the following ways :

(*a*) *Reduction by Al-powder: Gold Schmidt's aluminothermic process.* Metallic oxides can be reduced to free metals by Al-powder. This type of reduction is called *Gold Schmidt's aluminothermic process.* Equations representing this process are given below :

$$3MnO_2 + 4Al \rightarrow 3Mn + 2Al_2O_3$$
$$3Mn_3O_4 + 8Al \rightarrow 9Mn + 4Al_2O_3$$ *(Extraction of Mn)*

$$Cr_2O_3 + 2Al \rightarrow 2Cr + Al_2O_3$$ *(Extraction of Cr)*

$$Fe_2O_3 + 2Al \rightarrow 2Fe + Al_2O_3$$ *(Extraction of Fe and thermite welding process)*

$$B_2O_3 + 2Al \rightarrow 2B + Al_2O_3$$ *(Extraction of boron)*

$$3V_2O_5 + 10Al \rightarrow 6V + 5Al_2O_3$$ *(Extraction of vanadium)*

$$3U_3O_8 + 16Al \rightarrow 9U + 8Al_2O_3$$
$$3UO_2 + 4Al \rightarrow 3U + 2Al_2O_3$$ *(Extraction of uranium)*

The reaction between Al-powder and metallic oxides is exothermic and heat evolved produces a high temperature due to which the metals obtained by this process are in the molten state.

Reduction of Cr_2O_3 to Cr-metal by Al-powder. Cr_2O_3 is mixed with requisite amount of Al-powder (this mixture is called *thermite mixture*) and is placed in a large fire-clay crucible. An intimate mixture of Na_2O_2 or BaO_2 and Mg powder (called *ignition mixture* or *igniter*) is placed in a small depression made in the thermite mixture. The crucible is surrounded by sand which prevents the loss of heat by radiation. A piece of Mg ribbon is struck into the ignition mixture and the charge is covered by a layer of fluorspar (CaF_2) which acts as a heat insulator. Now Mg-ribbon is ignited so that ignition mixture catches fire and flame is produced, leading to a violent reaction between Mg and BaO_2 with the evolution of large amount of heat.

$$Mg + BaO_2 \rightarrow BaO + MgO + Heat$$

Heat produced in the above reaction makes Cr_2O_3 and Al-powder react together.

$$Cr_2O_3 + Al \rightarrow 2Cr\ (molten) + Al_2O_3 + 112\ Kcal.$$

Molten Cr-metal formed settles down at the bottom of the crucible (Fig. 20.6).

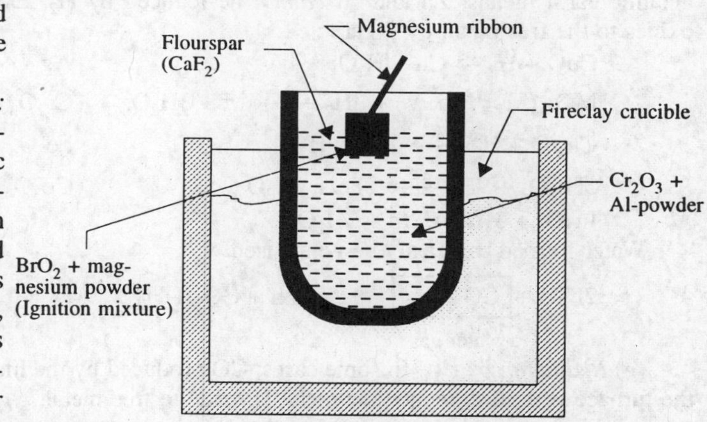

Fig. 20.6 : Reduction of Cr_2O_3 by Al-powder (Aluminothermic process).

Thermite welding process: An application of aluminothermic process. Aluminothermic process has been used for joining the broken pieces of iron (welding). This process is called *thermite welding process.* It was discovered by German chemist, Gold Schmidt in 1905. In this process thermite mixture consisting of Fe_2O_3 and Al-powder in 3:1 ratio is placed in a funnel-shaped crucible lined internally with magnesite and having a plug hole at its bottom. The thermite mixture is covered with a mixture of BaO_2 plus Mg-powder (ignition mixture) in which a piece of Mg ribbon is inserted. The ends of the iron pieces to be welded are throughly cleaned and surrounded by a fire-clay mould. When Mg ribbon is ignited, ignition mixture catches fire and Fe_2O_3 gets reduced to Fe by Al-powder.

$$2Al + Fe_2O_3 \rightarrow Al_2O_3 + 2Fe\ (molten) + 3230\ kJ$$

Due to the production of large amount of heat iron melts and hence is obtained in the molten state. Molten iron is allowed to fall through the hole into the gap between the broken iron pieces to be welded (Fig. 20.7).

(b) Reduction by Na-metal. The following example is important.

$$B_2O_3 + 6Na \rightarrow 2B + 3Na_2O$$ *(Extraction of boron)*

(c) Reduction by K-metal. Boron is extracted by reducing B_2O_3 by K.

$$B_2O_3 + 6K \rightarrow 2B + 3K_2O$$

(*d*) *Reduction by Mg-metal.* Examples are given below :

$$Rb_2O_3 + 3Mg \xrightarrow{\Delta} 2Rb + 3MgO$$
(*Extraction of Rb*)

$$B_2O_3 + 3Mg \rightarrow 2B + 3MgO$$
(*Extraction of boron*)

(*e*) *Reduction by Ca-metal.* Vanadium can be obtained by Beard and Gook's method, in which V_2O_5 is heated with excess of Ca metal, in presence of $CaCl_2$ and iodine, in a magnesite crucible at 900 – 950°C. In this reaction V_2O_5 gets reduced to vanadinum metal.

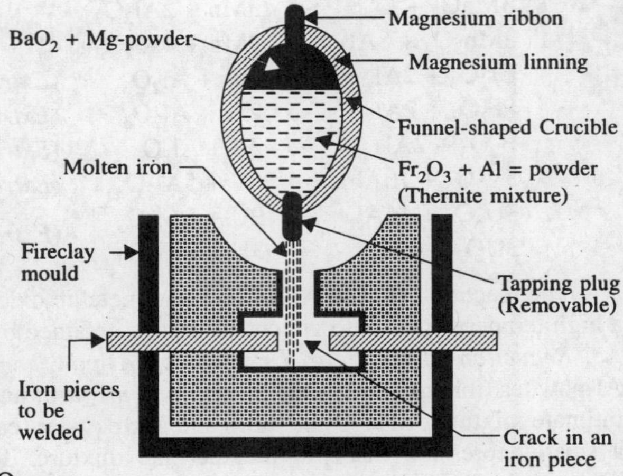

Fig. 20.7 : Thermite welding process.

$$V_2O_5 + 5Ca\ (excess) \xrightarrow{\Delta} 2V + 5CaO$$

When ThO_2 is heated at 950°C with Ca in a steel bomb in presence of $CaCl_2$ which acts as a flux, thorium is obtained.

$$ThO_2 + 2Ca \xrightarrow[\text{in presence of } CaCl_2 \text{ (flux)}]{950^\circ C} Th + 2CaO$$

(*f*) *Reduction by H_2 or water gas, (H_2 + CO).* The oxides of those metals which are less electropositive than hydrogen can be reduced by H_2 when heated. The oxides of alkali metals, alkaline earth metals, Zn and Al *cannot* be reduced by H_2. Examples showing the reduction of oxides to the free metal by H_2 are :

$$CuO + H_2 \rightarrow Cu + H_2O;$$
$$Ag_2O + H_2 \rightarrow 2Ag + H_2O$$
$$MO_3\ (M = Mo, V) + 3H_2 \xrightarrow{\Delta} M + 3H_2O;$$
$$Cr_2O_3 + 3H_2 \xrightarrow{\Delta} 2Cr + 3H_2O$$
$$Cu_2.O + H_2 \xrightarrow{\Delta} 2Cu + H_2O\ ;$$
$$Fe_3O_4 + 4H_2 \xrightarrow{\Delta} 3Fe + 4H_2O$$
$$3Fe_2O_3 + 9H_2 \xrightarrow{\Delta} 6Fe + 9H_2O\ ;$$
$$Co_3O_4 + 4H_2 \rightleftharpoons 3Co + 4H_2O$$
$$In_2O_3 + 3H_2 \rightarrow 2In + 3H_2O$$

Water formed can easily be volatalised.

$$2NiO + \underbrace{\boxed{CO + H_2}}_{\text{water gas}} \rightarrow 2Ni + CO_2 + H_2O$$

(*g*) *Reduction by CO.* In some cases, CO produced by the incomplete combustion of carbon in the furnace is used to reduce the metallic oxide to free metal, *e.g.*,

$$PbO + CO \rightarrow Pb + CO_2$$
$$FeO + CO \rightarrow Fe + CO_2$$
$$Fe_2O_3 + 3CO \rightarrow 2Fe + 3CO_2$$
$$Fe_3O_4 + 4CO \rightarrow 3Fe + 4CO_2$$
$$NiO + CO \rightarrow Ni + CO_2$$
$$CuO + CO \rightarrow Cu + CO_2$$

(*h*) *Reduction by coke or coal.* This process is also called *carbon reduction process.* The oxides of the metals like Zn, Pb, Sn, Fe etc., which do not form carbides with carbon are reduced to free metal by heating them with coke or coal. The metals obtained by this reduction process are either, in the form of *vapours* which are condensed (*e.g.*, Zn) or are in the *molten state* (*e.g.*, Sn, Fe). For reducing tin stone (SnO_2) anthracite is used. Examples showing carbon reduction process are given below :

$$ZnO + C \xrightarrow{\Delta} Zn + CO\ (\textit{Extraction of Zn})$$
$$NiO + C \xrightarrow{\Delta} Ni + CO\ (\textit{Extraction of Ni})$$

$PbO + C \longrightarrow Pb + CO$ (*Extraction of Pb*)

$Cr_2O_3 + 3C$ (*sugar charcoal*) $\xrightarrow{\Delta} 2Cr + 3CO$ (*Davill's process*) (*Extraction of Cr*)

$SnO_2 + 2C$ (*anthracite*) $\xrightarrow{\Delta} Sn + 2CO$ (*Extraction of Sn*)

$Fe_2O_3 + 3C \xrightarrow{\Delta} 2Fe + 3CO$ (*Extraction of Fe*)

$MnO_2 + 2C \xrightarrow{\Delta} Mn + 2CO$
$Mn_3O_4 + 4C \xrightarrow{\Delta} 3Mn + 4CO$ (*Extraction of Mn*)

$BeO + C \xrightarrow{\Delta} Be + CO$ (*Extraction of Be*)

Carbon reduction process is usually carried out in a *blast furnace.*

Some of the disadvantage of using carbon as reducing agent are given below :

(*i*) Some metallic oxides (*e.g.*, CaO) give metallic carbides instead of metals, *e.g.*,

$$CaO + 3C \xrightarrow{\Delta} CaC_2 + CO$$

(*ii*) In some cases reversible reaction takes place and hence during the cooling of the products, reformation of the oxide and carbon may take place, *e.g.*,

$$MgO + C \rightleftharpoons Mg + CO$$

Some of these disadvantages have been overcome by forming alloys (*e.g.*, ferromanganese, ferrotitanium etc.) instead of metals.

Smelting, flux and slag. Some of the oxide ores like heamatite (Fe_2O_3), tin stone (SnO_2), even after concentration, still contain some infusible earthy impurities which may be basic impurities (*i.e.*, metallic oxides like CaO, FeO, MgO etc.,) or acidic impurities (*i.e.*, non-metallic oxides like SiO_2, P_2O_5 etc.).

In order to remove the acidic or basic impurities the concentrated oxide ore is mixed with coke and other substance and the mixture thus obtained is heated. The other substance which is added is called **flux** and the process of heating the mixture (oxide ore + coke + flux) in this manner is called **smelting**. Reactions taking place in smelting are :

(*i*) Coke or CO reduces the oxide ore to free metal. Examples are given below :

(*a*) $PbO + C \rightarrow Pb$ (*molten lead*) $+ CO$

$PbO + CO \xrightarrow{\text{High temperature}} Pb$ (*molten lead*) $+ CO_2$

PbO used in the above reactions is obtained from galena ore (PbS) by roasting it.

(*b*) $2C + O_2 \rightarrow 2CO$

Fe_2O_3 (*haematite ore*) $+ 3CO \rightarrow 2Fe$ (*spongy iron*) $+ 3CO_2$

(*c*) SnO_2 (*tin stone*) $+ 2C \rightarrow Sn$ (*molten tin*) $+ 2CO\uparrow$

(*ii*) The flux combines with the infusible acidic or basic impurity present in the oxide ore and forms a fusible mass which is called **slag.** Thus the impurity is removed as a fusible slag. The formation of slag can be shown as :

Flux + Infusible acidic or basic impurity present in the oxide ore $\xrightarrow{\text{Heat}}$ Fusible slag

Thus a flux can be defined as *a substance which is added during the smelting of an oxide ore to remove infusible acidic or basic impurities as a fusible mass known as slag.*

Types of flux. Fluxes are of two types *viz., acidic flux* and *basic flux.*

(*i*) *Acidic flux.* It is an acidic oxide (oxide of a non-metal) like SiO_2, P_2O_5, B_2O_3 (from borax). It is used to remove the basic impurity like CaO, FeO, MgO etc. The acidic flux combines with the basic impurity and forms a slag,

$$SiO_2 \text{ (acidic flux)} + FeO \text{ (basic impurity)} \rightarrow FeSiO_3 \text{ (slag)}$$

(*ii*) *Basic flux.* It is a basic oxide (*i.e.*, oxide of a metal) like CaO (obtained from lime stone, $CaCO_3$), MgO (from magnesite, $MgCO_3$), haematite (Fe_2O_3) etc., It is used to remove the acidic impurity like SiO_2, P_2O_5 etc. The basic flux combines with the acidic impurity and forms a slag.

$$CaO \text{ (basic flux)} + SiO_2 \text{ (acidic impurity)} \rightarrow CaSiO_3 \text{ (slag)}$$

Thus, slag can be defined *as a fusible mass, which is obtained when a flux reacts with an infusible acidic or basic impurity present in the oxide ore.*

Properties of a slag. (*i*) Slag is a fusible mass. (*ii*) It has low melting point. (*iii*) It is lighter than and immiscible with the molten metal. It is due to these impurities that the slag floats as a separate layer on the molten metal and can thus be easily separated from the metal. The layer of the slag on the molten metal prevents the metal from being oxidised.

Example of slag formation. (*i*) Tin is generally extracted from tin stone (SnO_2). This ore contains SiO_2 as an acidic impurity. In order to remove this impurity and to reduce tin stone (SnO_2) to tin metal, the oxide ore is mixed with anthracite coal and a basic flux like lime (CaO). The mixture is heated so that SnO_2 gets reduced to molten tin metal and CaO combines with the impurity of SiO_2 and forms a slag of $CaSiO_3$. Thus SiO_2 is removed as a slag.

$$SnO_2 + 2C \text{ (\textit{anthracite coal})} \xrightarrow{\Delta} Sn \text{ (\textit{molten tin metal})} + 2CO\uparrow$$

$$CaO \text{ (\textit{basic flux})} + SiO_2 \text{ (\textit{acidic impurity})} \longrightarrow CaSiO_3 \text{ (\textit{slag})}$$

The slag of $CaSiO_3$, being lighter, floats on the molten tin metal as an upper layer which can easily be removed.

(*ii*) In the manufacture of steel from molten pig iron, if the pig iron contains phosphorus as impurity, the molten pig iron is taken in a Bessemer converter lined with lime, CaO (Basic Bessemer process) and a blast of air is passed into it. Phosphorus combines with O_2 to form P_2O_5 which forms a slag of $Ca_3(PO_4)_2$ with the lining of CaO (basic flux). The slag of $Ca_3(PO_4)_2$ is called *Thomas slag*.

$$4P + 5O_2 \rightarrow 2P_2O_5$$

$$P_2O_5 \text{ (\textit{acidic impurity})} + 3CaO \text{ (\textit{basic flux})} \rightarrow Ca_3(PO_4)_2 \text{ (\textit{slag})}$$

(*iii*) In the extraction of copper from copper-iron pyrites ore ($CuFeS_2$), the ore is roasted in a free supply of air so that a mixture of Cu_2S and FeS (called copper matte) is obtained.

$$2CuFeS_2 + O_2 \xrightarrow{\Delta} \underbrace{\boxed{Cu_2S + 2FeS}}_{\text{Copper matte}} + SO_2\uparrow$$

Then the copper matte is mixed with SiO_2 (acidic flux), put into a Bessemer's converter and a blast of hot air is blown into it. Her, oxygen converts FeS to FeO and reduces Cu_2S to molten copper metal.

$$2FeS + 3O_2 \rightarrow 2FeO + 2SO_2\uparrow$$

$$Cu_2S + O_2 \rightarrow 2Cu \text{ (\textit{molten metal})} + SO_2$$

Molten Cu-metal obtained contains FeO as basic impurity which reacts with SiO_2 (acidic flux) and forms a slag of $FeSiO_3$.

$$FeO \text{ (\textit{basic impurity})} + SiO_2 \text{ (\textit{acidic flux})} \rightarrow FeSiO_3 \text{ (\textit{slag})}$$

Thus we see that the basic impurity of FeO is removed as a slag.

2. Reduction of carbonate. The carbonate ore is first heated strongly in the absence of air (calcination) so that the ore loses moisture, CO_2 and other volatile impurities, and is ultimately converted into metallic oxide which, on being reduced by a suitable reducing agent, gives the metal. Examples are :

(*a*) $CuCO_3.Cu(OH)_2$ (*Malachit*) $\rightarrow 2CuO + CO_2\uparrow + H_2O\uparrow$
$CuO + CO \rightarrow Cu + CO_2$

(*b*) $ZnCO_3$ (*Calamine*) $\xrightarrow{\Delta} ZnO + CO_2$
$ZnO + C \rightarrow Zn + CO\uparrow$

3. Reduction of metallic sulphides. The reduction of metallic sulphides to free metals can be done either by air or O_2 or by iron.

(*a*) *Reduction by air or O_2–called auto-reduction or self-reduction or air-reduction.* The metallic sulphides like Cu_2S, PbS, HgS, Sb_2S_3 etc., when heated strongly in a free supply of air or O_2 (roasting), get reduced directly to the free metal. This is called *air-reduction, auto-reduction* or *self-reduction.* Examples have already been given in the discussion of "*Roasting*".

(*b*) *Reduction by iron.* Sulphides like Sb_2S_3 (stibnite ore), PbS (galena), when heated with iron, are reduced to free metals.

$$Sb_2S_3 + 3Fe \xrightarrow{\Delta} 2Sb + 2FeS \; ; \quad PbS + Fe \xrightarrow{\Delta} Pb + FeS$$

4. Reduction of metallic sulphates. The metallic sulphate is first reduced to metallic sulphide by heating it with carbon and then the sulphide thus obtained is roasted to get the metallic oxide. The oxide, on being reduced by a suitable metal, gives the metal, *e.g.*,

$$BaSO_4 \text{ (Barytes ore)} + 4C \xrightarrow[\text{Reduction}]{\Delta} BaS + 4CO$$

$$2BaS + 3O_2 \xrightarrow[\text{Reduction}]{\Delta} 2BaO + 2SO_2$$

$$3BaO + 2Al \xrightarrow[\text{Reduction}]{\Delta} 3Ba + Al_2O_3$$

5. Reduction of metallic halides. Metallic halides are usually reduced by heating with Al, Mg, Na, K, H_2 etc. The reduction is extremely violent in many cases. The advantage of using the above mentioned metals as reducing agent is that these metals have greater reducing power and the halides of these metals formed are soluble in water so that metals obtained can easily be purified simply by washing them with water. Examples, showing the extraction of metals by reducing their halides with Al, Mg etc., are given below.

(*a*) *Reduction by Mg*. Examples are given below :

$$TiCl_4 \text{ (vapours)} + 2Mg \text{ (molten)} \xrightarrow[\text{argon}]{800° C} Ti \text{ (spongy)} + 2MgCl_2$$ (*Extraction of titanium – Kroll process*)

$$2K[BF_4] + 3Mg \longrightarrow 2B + 2KF + 3MgF_2$$ (*Extraction of boron*)

$$VCl_4 + 2Mg \longrightarrow V + 2MgCl_2$$ (*Extraction of vanadium*)

$$2VCl_3 + 3Mg \longrightarrow 2V + 3MgCl_2$$ (*Extraction of vanadium*)

$$UF_4 + 2Mg \longrightarrow O + 2MgF_2$$ (*Extraction of uranium*)

(*b*) *Reduction by Na.* Examples are given below :

$$TiCl_4 + 4Na \xrightarrow[\text{argon}]{800° C} Ti + 4NaCl$$ (*Extraction of titanium–I.C.I. process*)

$$BeCl_2 + 2Na \xrightarrow{\Delta} Be + 2NaCl$$

$$K_2[BeF_4] + 2Na \xrightarrow{\Delta} Be + 2NaF + 2KF$$

$$K_2[ZrF_6] + 4Na \xrightarrow{\Delta} Zr + 2NaF + 2KF$$

$$VCl_3 + 3Na \xrightarrow{\Delta} V + 3NaCl$$ (*Extraction of vanadium*)

$$UCl_4 + 4Na \longrightarrow U + 4NaCl$$ (*Extraction of uranium*)

$$ThCl_4 + 4Na \longrightarrow Th + 4NaCl$$ (*Extraction of thorium*)

(*c*) *Reduction by K.* Examples are :

$$CaI_2 + 2K \xrightarrow{\Delta} Ca + 2KI$$

$$K_2[TiF_6] + 4K \xrightarrow{\Delta} Ti + 6KF$$

$$K[BF_4] + 3K \xrightarrow{\Delta} B + 4KF$$

(*d*) *Reduction by Al.* Examples are given below :

$$K_2[ZrF_6] + 2Al \xrightarrow{\Delta} Zr + 2K + 2AlF_3$$ (*Extraction of Zr*)

$$K[BF_4] + Al \xrightarrow{\Delta} B + KF + AlF_3$$ (*Extraction of boron*)

(*e*) *Reduction by H_2*. Examples are :

$$2VCl_4 + 4H_2 \xrightarrow{600° C} 2V + 8HCl$$ (*Extraction of vanadium : Jantesh method*)

$$2VCl_3 + 3H_2 \longrightarrow 2V + 6HCl$$ (*Extraction of vanadium*)

(*f*) *Rreduction by Ca.* Examples are given below :

$UF_4 + 2Ca \longrightarrow U + CaF_2$ (*slag*) ; $ThF_4 + 2Ca \longrightarrow Th + 2CaF_2$

6. Reduction of molten (fused) compounds by electricity : Electrolytic reduction (Also called electrometallurgy). The oxides, chlorides, hydroxides etc., of more reactive metals like Na, K, Mg, Ca, Al, lanthanides etc., are very stable and hence cannot be reduced to the free metals either by carbon or by aluminium. However the oxides of the above metals can be reduced by carbon at a very high temperature, but as soon as the metal is formed, that metal reacts with carbon and metallic carbide is formed. Therefore, in such cases the metals are obtained by reducing their suitable molten (fused) salts like chlorides, hydroxides etc., (containing some other substance) by electricity, called *electrolytic reduction* or *electrometallurgy.* The metal is liberated at the cathode. It should be noted that the metals cannot be obtained by the electrolytic reduction of the aqueous solution of their salts, since the metal obtained at cathode reacts with H_2O (present in the solution) and metal hydroxide and H_2 ore formed. In order to make the electrolysis of the molten salts successful, the following conditions most be satisfied.

(*a*) The liberated metal should not be soluble (miscible) in the melt so that it can be collected neatly from the electrolytic cell. Thus potassium cannot be obtained by the electrolysis of ($KCl + CaCl_2$) melt, since liquid potassium is soluble in $CaCl_2$.

(*b*) The products obtained as a result of electrolysis may react together and hence a suitable arrangement has to be made to keep them separate.

(*c*) Some other salt may have to be added to lower the m.pt. of the compound taken for electrolysis. The decrease in m.pt. makes the process of electrolysis take place at a lower temperature.

(*d*) The cell and electrodes should be of such materials which should not be affected by the electrolyte or the products obtained after electrolysis.

The following examples of the extraction of metals by the electrolysis of their appropriate fused salts may be studied.

(*a*) *Extraction of Al.* Pure Al_2O_3 is dissolved in cryolite solvent ($3NaF.AlF_3$ or Na_3AlF_6) and then the solution is melted in an electric furnace. The fused mixture of Al_2O_3 and Na_3AlF_6 is taken in an iron tank lined with carbon from inside. This carbon lining acts as cathode and carbon rods which are dipped into the molten mass of Al_2O_3 and Na_3AlF_6 act as anode. On passing electric current the following reactions take place :

Ionisation of Al_2O_3 : $Al_2O_3 \rightleftharpoons 2Al^{3+} + 3O^{2-}$

Formation of Al-metal at cathode : $Al^{3+} + 3e^- \rightarrow Al$ (*Reduction*)

Liberation of CO and CO_2 *gases at carbon anode* : $O^{2-} - 2e^- \rightarrow O$ (*Oxidation*)

$O + O \rightarrow O_2$

$4C$ (from carbon anode) $+ 3O_2 \rightarrow 2CO + 2CO_2$

(*b*) *Extraction of Na.* The fused mixture of NaCl and $CaCl_2$ is taken in Down's cell which consists of circular iron cathode and carbon anode. On passing the electric current the following reactions take place :

Ionisation of NaCl : $NaCl \rightleftharpoons Na^+ + Cl^-$

Collection of Na at cathode : $Na^+ + e^- \rightarrow Na$ (*Reduction*)

Collection of Cl_2 *at anode* : $Cl^- + e^- \rightarrow Cl$ (*Oxidation*), $Cl + Cl \rightarrow Cl_2$

Na can also be obtained by electrolysing molten NaOH in Castmer's cell.

(*c*) *Extraction of K.* It is obtained by the electrolysis of fused KCN.

(*d*) *Extraction of Ca.* It is obtained by electrolysing molten $CaCl_2$ containing some CaF_2.

(*c*) *Extraction of magnesium.* Mg metal is obtained by electrolysing magnesia (MgO) dissolved is a fused mixture of MgF_2, BaF_2 and NaF in a steel cell (cathode = cast iron, anode = carbon, temperature = 950°C). Molten Mg is collected at cathode.

Mg can also be obtained by electrolysing fused $MgCl_2$ containing some NaCl.

V. Purification or refining of impure metals

The metals obtained by any of the methods discussed above, contain a number of impurities and hence need to be purified *i.e.,* the impurities should be removed from the metals. The impurities include other metals (produced by the simultaneous reduction of their respective oxides, originally present in the ore), non-metals (*e.g.,* Si, P, S etc.), unreduced oxides and sulphides of the metals, slags, dissolved gases etc. Which method is to be used for the purification of a particular metal depends on the nature of the metal to be purified and the impurities to be removed and the purpose for which the metal is to be used. The methods used for refining the impure (crude) metals can be grouped into the following four catagories (A), (B), (C) and (D).

(A) **Physical methods.** These methods include the following processes :

1. Liquation process. This process is used for the purification of the metal, which itself is readily fusible, but the impurities present in it are not, *i.e.,* the impurities are infusible. In other words, we can say that the melting point of the metal to be purified should be lower than that of each of the impurities associated with the metal. This process is used for the purification of Sn and Zn, and for removing Pb from Zn-Ag alloy, which is obtained at the end of Parke's process and contains Pb as impurity.

Examples. *(a) Purification of impure tin metal.* The impure tin metal contains Cu, Fe, W etc., as impurities. This metal is placed on the slopping heart of a reverberatory furnace and gently heated. When the temperature of the furnace reaches the melting point of tin metal, this metal, on account of its lower melting point melts earlier than the impurities and hence flows down the inclined hearth and the solid infusible (non-fusible) impurities (called *dross*) are left behind on the hearth. The pure tin metal is collected in a cast iron vessel in the molten state. The metal obtained in this manner is called *pig tin.* The purification of this metal has been illustrated in Fig. 20.8.

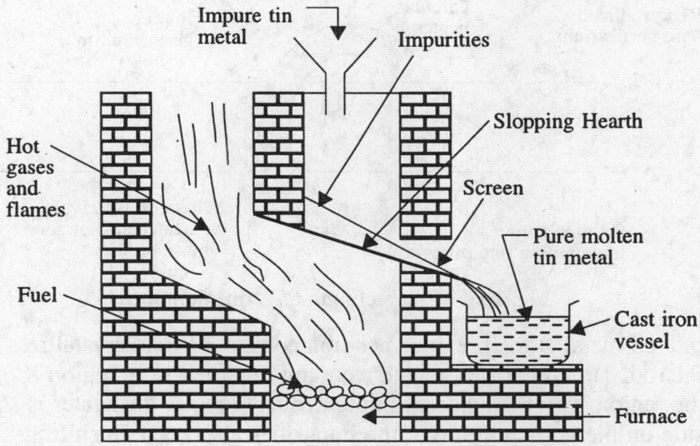

Fig. 20.8 : Purification of tin metal by liquation process.

(b) Purification of crude zinc. The crude zinc or the spelter is melted on the slopping hearth of a reverberatory furnace. Molten zinc flows down while the non-fusible impurities are left on the hearth.

(c) Removal of lead from Zn–Ag alloy obtained at the end of Parke's process. See "Combined discussion of softening processes, Parke's process, liquation process etc."

2. Fractional distillation process. This process is used to purify those metals which themselves are volatile and the impurities in them are non-volatile and *vice-versa.* Zn, Cd and Hg are purified by this process.

In this method the impure metal is heated in a vessel so that the metal is converted into its vapours while the impurities remain in the vessel in the solid state. The vapours of the metal are condensed in a separate vessel. The non-volatile impurities are left behind in the vessel.

Examples. *(a) Purification of impure zinc.* The purification of impure zinc (called *spelter* or *commercial zinc*) which contains As, Cd, Pb and Fe as impurities is done by distillation process. The removal of these impurities by the method of fractional distillation is based on the following facts : (*i*) As volatilises simply by heating (*ii*) Cd and Zn vaporise between 767 – 907°C, since the melting points of these metals are 767°C and 907°C respectively. (*iii*) Pb boils at 1600°C while Fe boils at 3000°C. In this process, the spelter is distilled between 897 – 997°C. First As volatialises away and soon after the mixture of Zn and Cd distills over, leaving behind Fe and Pb. The mixed vapours of Zn and Cd are led through a column maintained at 797°C, where Zn condenses and vapours of Cd pass over. Hundred percent pure Zn is obtained, after a number of such distillations.

(*b*) *To get Ag from Zn–Ag alloy.* Distillation process can also be employed to get Ag from Zn–Ag alloy. For this, see "*Combined discussion of softening processes, Parke's process, liquation process, distillation process and cupellation process*".

3. Zone refining method (Also called **fractional crystallisation method**). This process is used when metals are required in very high purity, for specific applications. For example pure Si and Ge are used in semiconductors and hence are purified by this method. Zone refining method is based on the principle that an impure molten metal on gradual cooling will deposit crystals of the pure metal, while the impurities will be left in the remaining part of the molten metal. Germanium metal, which is used in semiconductor devices, is refined (purified) by the zone refining method. The impure germanium metal to be refined is taken in the form of a rod. A circular heater H is fitted around this rod (Fig. 20.9) and this heater is slowly moved along the length of the rod. When the heater is at the extreme left end of the impure germanium rod, it melts a narrow zone (narrow region) of the germanium rod at that place. Now, when the heater moves on a little to the right side, then the molten metal at the previous position cools down and crystallizes to give pure metal at region X of the rod. The impurities, which were initially present in region X of germanium rod, now pass on to the region Y in the adjacent molten zone. Now, as the heater is shifted more and more to the right side on the germanium rod, the impurities also keep on shifting to the right side in the newer and newer molten zones. Ultimately, the impurities reach the extreme right end Z of the germanium rod. This end Z of the germanium rod containing all the impurities is then discarded. The remaining rod is now of highly pure germanium metal. In addition to germanium, silicon and gallium used as semiconductors are also refined by the zone refining method.

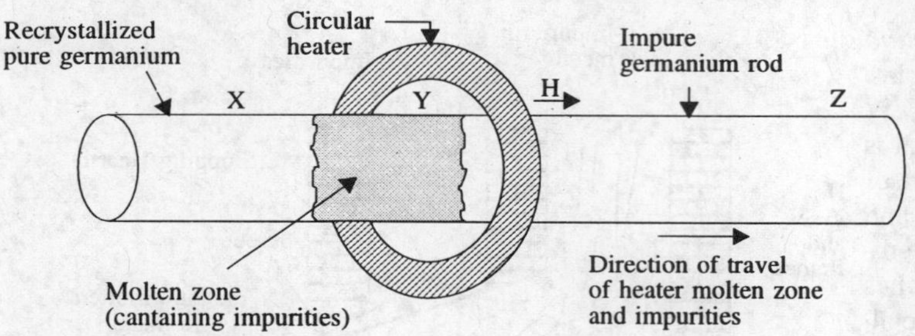

Fig. 20.9 : Zone refining of germanium metal.

4. Parke's process. For this, see "*Combined discussion of softening processes, Parke's process*".

(B) Chemical methods. These methods include the following methods :

1. Oxidative process. This process is used for removing the metallic (*e.g.*, Mn, Cu, Pb, Sn, Fe, Ag etc.) as well as non-metallic (*e.g.*, C, P, S, Si etc.,) elements present as impurities. This process is based on the fact that when O_2 or air is passed through the impure molten metal, the impurities are easily oxidised into their oxides. Then these oxides may be removed as follows :

(*a*) These oxides may form a scum on the surface of the metal. This scum can easily be removed by skimming.

(*b*) If the oxides are volatile, they escape from the mouth of the furnace.

(*c*) The oxides may form a slag with the lining on the inside surface of the furnace and may thus be removed. In the formation of the slag, the lining acts as a flux.

Sometimes O_2, needed for converting the impurities into their oxides, is supplied by the oxide ore of the impure metal. The oxide ore is used either as a lining of the furnace (see Manufacture of wrought iron from cast iron by puddling process) or is added as an extra substance from outside.

The oxidation of impurities can be done through the following processes :

(*i*) *Softening process. See "Combined discussion of softening processes, Parke's process etc."*

(*ii*) *Cupellation process.* Ag obtained by the distillation of Zn–Ag alloy contains traces of Pb as impurity. Similarly Au obtained from the auriferrous quartz and alluvial deposits also contains lead as impurity. Cupellation process has been used to remove lead from Ag and Au. For the details of this process see "*Combined discussion of softening process, Parke's process, liquation process, distillation process and cupellation process*".

(*iii*) *Bessemer's process.* (Also called *Bessemerisation process*). In this process the material containing the impurities is taken into a Bessemer's converter containing acidic or basic lining and blast of hot air is sent into it. The air converts the impurities into oxides. If these oxides are volatile, they get volatilised and thus escape from the mouth of the converter. The oxides may form a slag with the lining of the Bessemer's converter and are thus removed

Examples. (*a*) *Manufacture of steel from cast iron by Bessemer's process.* In the manufacture of steel from cast iron by Bessemer's process, the impurities of C, S, Si, Mn and P present in the cast iron react with the hot air and are converted into their oxides *viz.*, CO, SO_2, SiO_2, MnO and P_2O_5 respectively. SO_2 escapes from the mouth of the converter while CO burns with a blue flame at its mouth. SiO_2, MnO and P_2O_5 react with the lining of the Bessemer's converter to form the slags and thus are removed as slags. In the formation of slag the material used for the lining of the converter acts as a flux.

$$MnO + SiO_2 \text{ (acidic lining)} \rightarrow MnSiO_3 \text{ (slag)}$$
$$P_2O_5 + 3CaO \text{ (basic lining)} \rightarrow Ca_3(PO_4)_2 \text{ (slag)}$$
$$SiO_2 + CaO \text{ (basic lining)} \rightarrow CaSiO_3 \text{ (slag)}$$

(*b*) *Removal of sulphur and FeS from the molten mass obtained during the extraction of nickel from pentlandite ore. (Ni, Cu, Fe)*S by *Bessemerisation process.* Pentlandite ore, (Ni, Cu, Fe)S (an ore of Ni), after being roasted and smelted, gives a molten mass which contains NiS, CuS, S and FeS. Sulphur and FeS are removed form this molten mass by the *Bessemerisation process* which has been discussed under "*Extraction of nickel.*"

(*c*) *Removal of FeS from the copper matte obtained during the extraction of copper from copper-iron pyrites* ($CuFeS_2$) *by Bessemerisation process.* Copper-iron pyrites ($CuFeS_2$), after being roasted and smelted, gives a molten mass which is called *copper matte.* This copper matte contains a mixture of Cu_2S and FeS. FeS is removed from this mixture (copper matte) by Bessemerisation process.

(*iv*) *Puddling process.* This process is used for the manufacture of wrought iron from cast iron. We know that cast iron contains the impurities of C, S, Si, Mn, and P. When these impurities are removed from cast iron, we get wrought iron. In this process the impurities are oxidised to their oxides *not* by blast of air but by the haematite (Fe_2O_3) lining of the furnace. (See "*Manufacture of wrought iron from cast iron by puddling process*").

Combined discussion of softening processes; Parke's process, liquation process, distillation process and cupellation process. Lead obtained from galena ore (PbS) by air-reduction or carbon-reduction process, contains base metals (Cu, Bi, As, Sn, Ag, Sn, Zn etc.) as impurities. It is due to the presence of these impurities that lead becomes hard and brittle. The removal of these impurities is done by the process, which is called **softening process**, since the removal of impurities makes lead soft. This process consists of heating the impure lead on the hearth of a shallow flattened reversible furnace, in a blast of air. The impurities of the base metals (except Ag) are oxidised to their respective oxides, which form a scum on the surface of lead. This scum is removed by skimming. Lead obtained by this method still contains Ag as impurities and is called *argentiferrous lead* or *commercial lead.* The removal of the impurities of Ag from the commercial lead is called *desilverisation of lead* and is done by **Parke's process.** Thus, Parke's process is the desilverisation of lead. In Parke's process, the commercial lead, which contains Ag as impurity, is melted in iron pots and 1% of Zn is added to it. The molten mass is throroughly agitated. Since Ag is about 300 times more soluble in Zn than in Pb, most of the Ag present in the commercial lead as impurity mixes with Zn, to form Zn-Ag alloy. When the whole mass is cooled, two layers are obtained. The upper layer contains Zn-Ag alloy in the solid state, while the lower layer has lead in the molten state. This lead contains only 0.0004% of Ag and hence is almost pure. Lead obtained after removing most of Ag from it (desilverisation of lead) by Parke's process, is called *desilverised lead.* This lead contains the impurities of metals like Zn, Au, Sb etc. These metal impurities are removed from desilverised lead by **Bett's electrolytic process,** which will be discussed later on, in this chapter.

Zn-Ag alloy, formed in the upper layer, is skimmed off from the surface of the molten lead by perforated ladles. This alloy contains lead as impurity. This impurity of Pb is removed from the alloy by **liquation process**, in which Zn-Ag alloy is heated in a slopping furnace, when the impurity of Pb melts and hence drains away from the solid alloy. Thus purified Zn-Ag is obtained. Now Ag can be obtained from this purified Zn-Ag alloy by **distillation process**, in which the alloy is heated strongly in presence of little carbon in a fire-clay retort. Zn, being more volatile, distills off while Ag remains in the retort, carbon used in the process reduces the oxide of Zn, if formed. Ag obtained from Zn-Ag alloy is contaminated with a little of Pb as impurity. This impurity of Pb in removed by **cupellation process.** In this process, Ag containing the traces of Pb as impurity, is placed in a cupel (cupel is a boat-shaped dish made of bone ash which is porous in nature) in a reverberatory furnace and heated in the presence of air. By doing so, lead (impurity) is oxidised to PbO (litharge) which volatilises and pure Ag is left behind in the cupel. Last traces of PbO are absorbed by the porous mass of the cupel.

2. Poling process. This process is used for the purification of copper and tin.

Examples. (*a*) *Purification of impure copper.* Impure copper contains a number of impurities like S, Sb, As, Fe, CuO etc. In this process, impure copper is remelted in a reverberatory furnace lined with SiO_2 and a blast of O_2 is blown into the furnace. O_2 oxidises S, Sb and As to their respective oxides which, being, volatile, get volatalised and are thus removed. Fe is oxidised to FeO which forms a slag of $FeSiO_3$ with SiO_2 lining of the furnace. Molten copper left behind contains CuO as impurity. This molten copper is treated with powdered anthracite and then stirred with a pole of green wood. Green wood, at high temperature, liberates hydrocarbon gases, which are converted into methane (CH_4). Methane thus obtained reduces CuO to free Cu-metal, which is about 99.5% pure and is called *tough pitch copper.*

$$\text{Green wood} \rightarrow \text{Hydrocarbons} \rightarrow CH_4$$

$$4CuO + CH_4 \rightarrow 4Cu\ (\textit{pure metal}) + CO_2 + 2H_2O$$

(*b*) *Purification of impure tin.* Impure tin metal contains the impurities of Cu, Fe, W and SnO_2. The impurity of SnO_2 is due to the incomplete reduction of tin stone ore (SnO_2) during smelting. In order to remove these impurities, the impure molten tin metal is taken in a big pot and stirred with a pole of green wood. Green wood, at high temperature liberates hydrocarbon gases, which are converted into methane CH_4. Methane, thus produced, reduces SnO_2 to pure Sn metal while the impurities of Cu, Fe, W etc., come up to the surface, where they come in contact with air and are oxidised to their respective oxides. The oxides form a scum on the surface of pure tin metal. This scum is removed from the surface. Tin metal obtained by this method is 99% pure.

$$\text{Green wood} \rightarrow \text{Hydrocarbon} \rightarrow CH_4$$

$$2SnO_2 + CH_4 \rightarrow 2Sn + CO_2 + 2H_2O$$

3. Thermal decomposition method. In this method, the impure metal is converted into a suitable volatile compound which, on being heated, gets decomposed, to give pure metal. Some examples of thermal decomposition method are given below :

(*i*) *Mond's process (Also called carbonyl process)* : *Purification of nickel.* This is the most important process for getting 99.9% pure nickel metal from crude nickel. This process was discovered by German chemist, Ludvig in 1893. For the details of this process, see "*Extraction of nickel from pentlandite ore, (Ni, Cu, Fe)S*".

(*ii*) *Van-Arkelde Boer's process: Also called iodine refining process (Ti, Zr, etc.)*. In this process, the impure metal is heated with iodine, in a sealed vessel, so that the metal forms volatile metallic iodide and impurities do not react with iodine. The metallic iodide, so formed, is decomposed to get the pure metal and iodine. The decomposition of the metallic iodide to pure metal is done by passing it over a thin tungsten or tantalum wire (filament) placed in the vessel and electrically heated to a high temperature, when pure metal gets deposited on the surface of the hot wire and iodine vapour gets condensed. The pure metal can be removed from the surface of the tungsten wire. This method is very expensive and hence is used only when very pure metal is required for special uses. The only requirement to make this method successful is that the metal should not volatalise on the temperature at which the metallic iodide decomposes, *i.e.*, the decomposition temperature of the metallic iodide should be less than the melting point of the metal (*e.g.*, decomposition temperature of ZrI_4 = 1800°C, while the melting point of Zr-metal = 1857°C).

Van-Arkel process has been used to purify Ti, Zr, V and Th metals. Equation showing the purification of Ti-metal by Van-Arkel process can be written as :

$$\text{Ti}\ (s)\ (\textit{Impure metal}) + 2I_2\ (g) \xrightarrow{150\ -\ 250°\text{ C}} \text{TiI}_4\ (g) \xrightarrow[\text{Tungsten filament}]{1400°\text{ C}} \text{Ti}\ (s)\ (\textit{pure metal}) + 2I_2\ (g)$$

Similar equation for the purification of Th, Zr and V can also be written

$$\text{Th}\ (\textit{crude}) + 2I_2\ (g) \rightarrow \text{ThI}_4\ (g) \xrightarrow{1100\ -\ 1200°\text{ C}} \text{Th}\ (\textit{pure}) + 2I_2\ (g)$$

(C) Amalgamation process. This process is used for the extraction of silver and gold and is based on the fact that when Ag or Au reacts with Hg, Au-Hg / Ag – Hg amalgam is formed. When this amalgam is allowed to distill in an iron retort, Hg, being more volatile, distills off and Au / Ag metal is left behind in the retort. The following examples explain this process.

(*a*) *Extraction of Ag from silver glance, Ag_2S.* The ore is crushed and treated with a solution of $CuCl_2$, so that a slime of AgCl is obtained.

$$Ag_2S + CuCl_2 \longrightarrow 2AgS + CuS$$

Hg is then added to the product when Ag set free gets dissolved in the excess of Hg due to the formation of Ag–Hg amalgam.

$$2AgCl + 2Hg \longrightarrow Hg_2Cl_2 + 2Ag\downarrow$$

$$Ag + Hg \longrightarrow (Ag - Hg)\ \text{amalgam}$$

(Ag – Hg) amalgam is washed and allowed to distill in an iron retort. Hg, being more volatile, distills off and Ag metal is left behind in the retort.

(*b*) *Extraction of gold from auriferous quartz rocks.* The auriferous quartz rocks are broken into smaller pieces by means of rock crushers. The smaller pieces, thus obtained are finely powdered in big stamp mills. This is called *pulverisation* of the quartz. The pulverised ore is carried off by a current of H_2O through a screen, at the other end of the stamp mill and thus slurry (*i.e.*, mixture of finer particles of quartz and H_2O) is obtained. The course particles, if any, are retained back and ore again pulverised.

Now the slurry is allowed to flow over Cu-plates, which are coated with Hg and are placed in a slopping position. The particles of Au-metal present in the slurry form Au-Hg amalgam with Hg, which is retained on the Cu-plates. The amalgam is scrapped off from plates and distilled in iron retorts, when Hg, being more volatile than Au, distills off and Au-metal is left behind in the retort. Nearly 75-80% of total gold originally present in the quartz rock is thus recovered and the remaining 25-20% gold remains associated with the rock particles. These rock particles are called **tailings,** from which gold can be recovered by *Mac-Arthur Forest cyanide process* (See *Leaching process*).

(D) Electrolytic process. Sn, Pb, Cu, Ag, Ni, Zn, Cr etc., are purified by this process. This process gives metals of high purity. In this process, a thick plate of the impure metal to be purified is made the anode (*i.e.*, positive electrode) of the electrolytic cell and hence is connected to the positive terminal of the battery. A thin plate of pure metal is made cathode (*i.e.*, negative electrode) of the cell and hence is connected to the negative terminal of the battery. An aqueous solution of a suitable simple salt or complex salt of the metal having some corresponding acid, if necessary, is used as an electrolytic solution. When electric current of appropriate strength is passed through the electrolytic solution, metal from the impure plate (anode) is transferred to the pure plate (cathode), where it gets deposited. The soluble impurities go into the solution, while the insoluble impurities settle down at the bottom, below the anode. The matter settled below the anode is called *anode mud* or *anode sludge.* Sometimes the anode mud contains valueable metals which can be extracted from it, *e.g.*, the anode mud obtained in the electrolytic refining of lead contains Ag, Au, Sb, Cu etc.

Examples. (*a*) *Purification of blister copper.* Copper obtained from copper-iron pyrites, $CuFeS_2$, is called blister copper, which contains many impurities like Fe, Ni, Zn, Ag, Au etc., and hence is only 98% pure. For refining blister copper, a *thick* plate of *impure copper* is made anode and a *thin* plate of *pure copper* is made cathode. Cathode is coated with graphite and oil, which helps in scrapping the copper easily that gets deposited on the cathode. 15% $CuSO_4$ containing 5% H_2SO_4 is used as an electrolyte. On passing electric current of 0.1 – 0.3 volt at 40–60°C through the electrolyte, ionisation of $CuSO_4$ takes place ($CuSO_4 \rightleftharpoons Cu^{2+} + SO_4^{2-}$) and Cu^{2+} ions, thus obtained, move towards the cathode, where they accept electrons, to become neutral Cu atoms, which deposit as free metal on the cathode. Thus cathode (pure Cu plate) goes on becoming thicker and thicker in size, as the reaction goes on.

$$Cu^{2+} + 2e^- \xrightarrow{\text{Reduction}} Cu^\circ \text{ (\textit{gets deposited on cathode})}$$

This equation represents the cathode reaction (reduction of Cu^{2+} ions to Cu atoms).

SO_4^{2-} ions move towards the anode, where they lose electrons and are converted into sulphate redicals (SO_4). These radicals have a strong tendency to combine with Cu-atoms of Cu-anode (impure Cu-plate) to form $CuSO_4$ molecules, which go into the solution in which they are again ionised to give Cu^{2+} ions. These Cu^{2+} ions travel through the electrolytic solution and get deposited on cathode as free Cu-metal.

$$SO_4^{2-} \longrightarrow SO_4 + 2e^- \quad \text{... (\textit{i})}$$

$$SO_4 + Cu \text{ (\textit{from Cu-anode})} \longrightarrow CuSO_4 \text{ (\textit{go into solution})} \quad \text{...(\textit{ii})}$$

$$CuSO_4 \longrightarrow Cu^{2+} + SO_4^{2-} \quad \text{...(\textit{iii})}$$

On adding equations (*i*), (*ii*) and (*iii*), we get

$$Cu \xrightarrow{\text{Oxidation}} Cu^{2+} + 2e^-$$

or $\quad Cu \text{ (From impure Cu-plate : Cu anode)} \xrightarrow{\text{Oxidation}} Cu^{2+} \text{ (\textit{go into solution})} + 2e^-$

This equation represents the anode reaction (oxidation of Cu-atoms to Cu^{2+} ions). This equation indicates that Cu dissolves out of the impure Cu-plate (anode) and is deposited on pure Cu-plate (cathode). Thus, anode goes on growing thiner and thinner in size as the reaction proceeds. From the above discussion it follows that pure Cu is transferred from impure Cu-plate (anode) to pure Cu-plate (cathode) on which Cu-gets deposited.

The soluble impurities, like Fe, Ni, Zn, pass into solution as their sulphates, while the insoluble impurities (*e.g.*, Ag, Au), which are not affected by H_2SO_4 – $CuSO_4$ solution, settle down below the anode, as *anode mud* or *anode sludge* (See Fig. 20.10).

(*b*) *Purification of silver.* Silver obtained by any of the methods may contain the traces of Zn, Cu and Au as impurities. This is purified by electrolytic method. A block of impure Ag is made anode, while the cathode is a thin sheet of pure Ag. 6% solution of $AgNO_3$ containing 1% HNO_3 is used as an electrolytic solution. On passing electric current, pure Ag is deposited on the cathode (sheet of pure Ag). Zn and Cu, if present, pass into solution, while Au, if any, falls down as the anode mud. Reactions taking place are :

Ionisation : $AgNO_3 \; Ag^+ + NO_3^-$

At cathode : $Ag^+ + e^- \xrightarrow{\text{Reduction}} Ag^\circ$ (*get deposited on cathode*)

At anode : Ag (from impure Cu-block : Cu anode) $\rightarrow Ag^+$ (*go in solution*) + e^-

(*c*) *Purification of nickel.* Although impure nickel can be purified by Mond's process, nickel obtained by this method, is only 92.8% pure and hence needs further purification. This is done by electrolytic method. The crude nickel metal is melted and cast into anodes. Cathodes are sheets of pure nickel metal and nickel (II); ammonium sulphate solution is used as electrolyte. On passing electric current, pure nickel gets deposited on the cathode sheets. The anode mud obtained below the anode, contains Ag, Au and platinum metals, which are obtained as by-products.

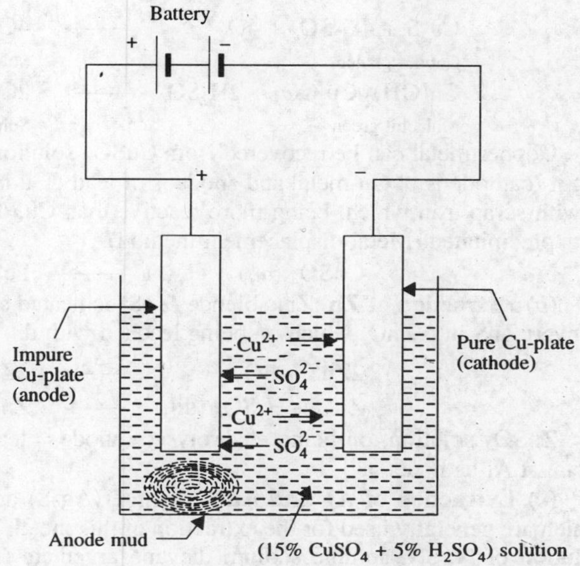

Fig 20.10 : Electrolytic refining of blister copper.

(*d*) *Purification of impure tin.* Tin extracted from tin stone (SnO_2) contains the impurities of Cu, Fe, W etc. These impurities can be removed by electrolytic process. In this process, blocks of impure tin metal are suspended in the electrolytic bath, containing hydrofluosilicic acid (H_2SiF_6), tin sulphate and H_2SO_4 as electrolyte.

Impure tin metal is made the anode, while the pure tin plate or wire acts as cathode. On passing the electric current through the electrolytic solution, tin dissolves from the anode and is deposited on the cathode. Tin metal thus obtained is 99.9% pure. Impurities settle down on the bottom of the electrolytic tank and collect as mud near the anode. This mud is called *anode mud.*

(*e*) *Removal of metal impurities (e.g., Zn, Au, Sb etc.) from the desilverised lead by Bett's electrolytic process.* We have seen during the study of Parke's process that lead, from which the impurity of Ag has been removed by Parke's process (lead from which the impurity of Ag has been removed is called *desilverised lead*) still contains metallic impurities like Zn, Au, Sb etc. These impurities are removed from the desilverised lead by Bett's electrolytic process. In this process desilverised lead is made the anode while the electrolyte is lead silicic fluoride, $PbSiF_6$ containing 8-12% of hydrofluosilicic acid (H_2SiF_6). HCl and H_2SO_4 cannot be used because they form sparingly soluble lead salts. In the presence of gelatin which acts as an addition agent, the deposit of pure lead on the cathode which are of pure lead is more smooth and uniform. The metallic impurities which are more electropositive than lead (*e.g.*, Zn) etc. remain in solution while less electropositive metallic impurities (*e.g.*, Au, Sb etc.) fall down as anode mud.

More about hydro-metallurgical process

This process is also called **leaching process** or **wet process.** We have already said that this process is used for the concentration of ores. In addition to this application, this process is also used for the extraction of metals like Zn, Cu, Ag, Au etc. In this process, the powdered ore is leached with some suitable dissolving reagent (*e.g.*, dil. H_2SO_4, moist air, NaCN solution etc.) whereby the metal present in the ore is converted into a soluble simple salt (see the extraction of Cu and Zn given below) or soluble complex compound (see the extraction of Ag and Au). Now the metal can be recovered from the salt / complex compound solution by electrolysing it (*electrolytic method*) or by treating it with some more reactive metal (*metal displacement method*). Extraction of Cu, Zn, Ag and Au with the help of leaching process has been discussed below :

(*a*) **Extraction of copper.** Leaching of cuprite (Cu_2O) or copper glance (Cu_2S) (both are the ores of copper) with dil. H_2SO_4 in presence of air, gives a solution of $CuSO_4$ and the impurities present in the ores remain undissolved in the acid. Leaching of malachite green, $Cu(OH)_2.CuCO_3$ (it is also an ore of copper) with dil. H_2SO_4 also gives a solution of $CuSO_4$.

$$\underset{\text{Cuprite}}{2Cu_2O} + 4H_2SO_4 + O_2 \longrightarrow \underset{\text{Solution}}{4CuSO_4} + 4H_2O$$

$$\underset{\text{Copper glance}}{Cu_2S + 4H_2SO_4 + 4O_2} \longrightarrow \underset{\text{Solution}}{4CuSO_4 + 4H_2O + 2SO_2\uparrow}$$

$$\underset{\text{Malachit green}}{Cu(OH)_2.Cu\ CO_3 + 2H_2SO_4} \longrightarrow \underset{\text{Solution}}{2CuSO_4 + 3H_2O + CO_2\uparrow}$$

Copper metal can be recovered from $CuSO_4$ solution (obtained as above) either by electrolysing it (cathode is of Cu-metal and anode is of lead ; Cu-metal is collected at cathode) or by treating it with scrap iron which, being more reactive than Cu, displaces Cu from $CuSO_4$ solution and Cu gets precipitated (Metal displacement method).

$$CuSO_4\ (aq) + Fe\ (s) \longrightarrow FeSO_4\ (aq) + Cu\ (s)\downarrow$$

(*b*) **Extraction of Zn**. Zinc blende (ZnS) is heated strongly in the presence of air (roasting) to convert ZnS into ZnO, which on being leached with dil. H_2SO_4, forms a solution of $ZnSO_4$.

$$2ZnS + 3O_2 \xrightarrow{\Delta} 2ZnO + 2SO_2$$

$$ZnO + H_2SO_4\ (dil.) \longrightarrow ZnSO_4 + H_2O$$

$ZnSO_4$ solution, on being electrolysed (anode = lead, cathode = aluminium sheet) gives Zn-metal at Al-cathode.

(*c*) **Extraction of Ag and Au.** Argentite (Ag_2S) and horn silver (AgCl) are the ores of Ag which are generally used for the extraction of this metal. When AgCl ore is treated with an aqueous solution of NaCN, soluble sodium dicyano argentate (I), $Na[Ag(CN)_2]$ (complex compound) is obtained.

$$AgCl + 2NaCN \longrightarrow Na\ [Ag\ (CN)_2] + NaCl$$

or $$Ag^+ + CN^- \longrightarrow [Ag\ (CN)_2]^-$$

The same complex compound is also obtained when Ag_2S is treated with aqueous solution of NaCN in presence of a current of hot air.

$$Ag_2S + 4NaCN + 2O_2\ \text{(From air)} \longrightarrow 2Na\ [Ag\ (CN)_2] + Na_2SO$$

or $$Ag_2S + 4CN^- + 2O_2 \longrightarrow 2[Ag\ (CN)_2]^- + SO_4^{2-}$$

The tailings obtained at the end of the amalgamation process (see the extraction of gold from auriferous quartz rocks by amalgamation process) also contain gold. When these tailings are leached with a dilute aqueous solution of NaCN (0.3%) in presence of air, gold gives soluble complex compound, $Na[Au(CN)_2]$.

$$4Au + 8NaCN + 2H_2O + O_2 \longrightarrow 4Na[Au(CN)_2] + 4NaOH$$

Ag and Au can be precipitated from their respective complex compounds with the help of Zn-scrap.

$$\underset{(M = Ag\ or\ Au)}{2Na\ [M(CN)_2] + Zn} \longrightarrow Na_2\ [Zn(CN)_4] + 2M\downarrow$$

The method mentioned above for the extraction of Ag and Au is also called by other names like *Mac-Arthur Forest cyanide process, cyanide process, complex formation followed by metal displacement method etc.*

Types of furnaces

Different types of furnaces are used in different metallurgical operations. Important furnaces which are generally used for the extraction of metals, are discussed below :

1. Reverberatory furnace

This furnace has been used for smelting (reducing) and roasting the ores. The reduction of the ore is done by the use of some reducing agent like anthracite coal (as in case of tin) while the oxidation (roasting) is done by means of air. This furnace has been used (*i*) for reducing (smelting) the roasted tin stone, SnO_2 (an ore of tin) to molten tin metal by the use of anthracite coal, (*ii*) for roasting the concentrated galena ore, PbS (an ore of Pb) to convert into PbO and $PbSO_4$ by means of air, (*iii*) for roasting the concentrated copper pyrites ore, $CuFeS_2$ (an ore of Cu) to convert it into Cu_2O and FeO by means of air.

Construction and functioning of the furnace. This furnace consists of three parts namely (*a*) *fire-place* where fuel is burnt (*b*) *bed* or *hearth* where charge is placed and heated (*c*) *chimney,* from which the waste gases escape. The fire place is built at one end of the furnace, at a slightly lower level than that of the hearth and is closed at the bottom by fire-bars, on which the fuel is placed and burnt. It is covered by a low roof, which is made slanting towards the other end and connects with chimney. The hot gases (waste gases) from the fire place are reflected by the concave ceiling over the hearth. The furnace is surrounded on all sides by walls of fire-bricks. The sketch of a reverberatory furnace, which is used for roasting the concentrated galena ore, PbS (an ore of Pb), has been shown in Fig. 20.11. The charge is placed on the hearth and the fuel is placed on the fire-place. Air enters from the fire-place, through the iron grates and waste gases (hot gases) escape out of chimney. There are doors on one side of the furnace for introducing the charge and tapping holes on the other side, to remove the charge.

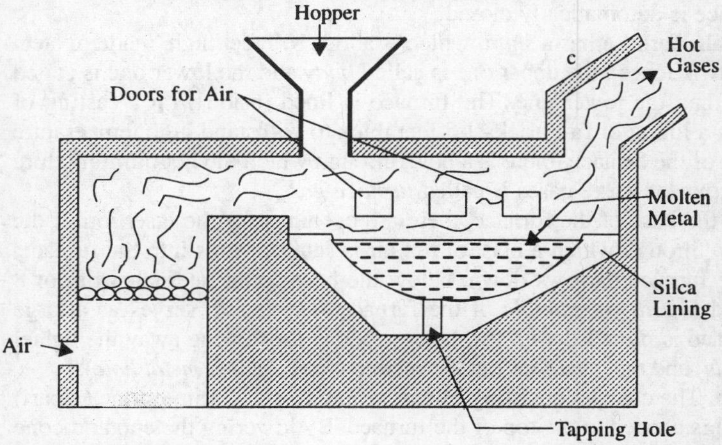

Fig. 20.11 : Reverberatory furnace, used for roasting the concentrated galena ore, PbS.

2. Blast furnace

This furnace is used for the manufacture of cast iron, for smelting roasted copper pyrites ore, $CuFeS_2$ in the extraction of copper metal and for smelting PbO in the extraction of lead metal.

Construction of the blast furnace. It is a tall, cylindrical furnace, the dimension and construction of which varies from place to place. It is from 80 to 120 feet in diameter. Following are its main parts :

(*a*) *A double cup and cone arrangement.* At the top, the furnace carries a hopper, which rests over a funnel-shape double cup and cone arrangement. This arrangement gets lowered by a lever arrangement, worked out by chains and wheels (not fully shown in

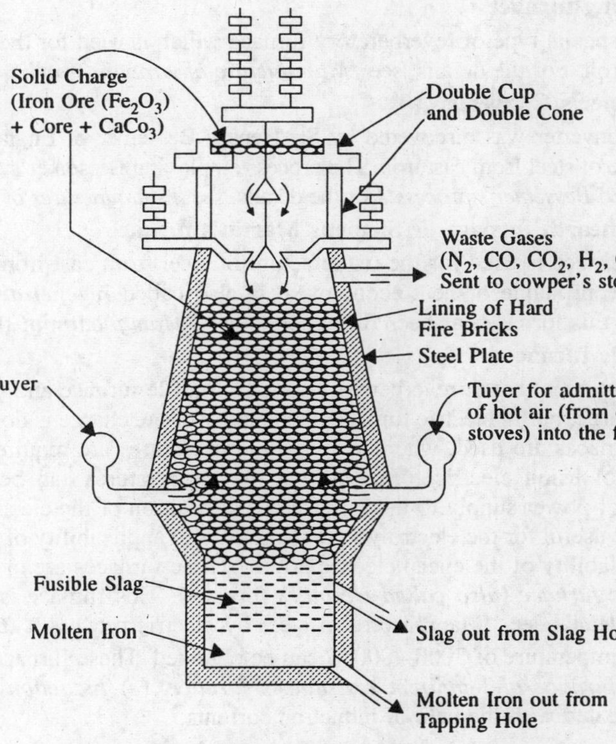

Fig. 20.12 : Blast furnace for the manufacture of cast iron.

Fig. 20.12). By lowering it, a fresh charge is introduced into the furnace. This arrangement prevents the exit of gases during the charging of the furnace. The charge is fed at the cup and when it is full, the cone is depressed, so that the charge automatically goes down into the furnace and subsequently the throat of the furnace is automatically closed.

(*b*) *Body, bosh and tuyers.* The furnace has a *shaft,* which is about 100 feet high, made of steel plates and is formed by joining two cones : the upper one is called *body* and the lower one is called *bosh*. The upper one is longer than the lower one. The furnace is lined inside with a casting of ordinary bricks and then there is a lining of fire bricks 1.5 feet thick to withstand high temperature of the furnace. At the upper part of the furnace, there is a hole, through which the hot burnt mixture of waste gases escape. This mixture of waste gases is called *furnace gas*.

Between 6 and 8 feet from the base of the furnace, there are openings for the insertion of the water-cooled pipes called *tuyers,* through which a blast of hot air is sent upwards into the furnace.

(*c*) *Hearth or crucible.* The furnace narrows down, below the bosh to an internal width of 8 feet at the base, which is called *hearth* or *crucible* of the furnace. The hearth serves as a large crucible to hold the molten iron and is provided with two holes – one is for flowing away the lighter slag (called *slag notch* or *slag hole*) and the other taps out the molten metal (called *metal hole*).

Functioning of the furnace. The charge consisting of roasted ore (8 parts), lime stone (1 part) and desulphurised coke (4 parts) is carried to the top of the furnace. By lowering the cup and cone arrangement, the charge is introduced into the furnace. At the same time, the furnace is lit and a blast of hot air is sent upwards, through the *tuyers.* The blast of hot air for this purpose is obtained by heating air in a series of *cowper stoves* or *preheaters* and is conveyed through a steel pipe, lined with fire-bricks, to the tuyers, from where it is discharged into the furnace. As the charge moves down the furnace, it meets high temperature and various gases are liberated, as a result of the chemical reaction between the charge and the air blast rising from the below. The waste gases leave the furnace through the out let near the top of the furnace. The molten metal, together with the slag collects on the hearth of the furnace, from where these are removed by their respective outlets.

3. Puddling furnace

It is a special type of reverberatory furnace which is used for the manufacture of wrought iron from cast iron. For the details, see "*Manufacture of wrought iron*"

4. Bessemer's Converter

This converter was discovered by Sir Henery Bessemer of England in 1856. It is used for the manufacture of steel from cast iron. The process employing Bessemer's converter in the manufacture of steel is called *Bessemer's process.* For the details, see "*Manufactures of Steel by Bessemer's Process*".

5. Open-hearth furnace or Siemens Martin's furnace

This furnace is used for the manufacture of steel from cast iron. Since this furnace works on regenerative principle of heat economy, it is also called *regenerative furnace.* Construction and working of this furnace has been discussed under "*Manufacture of steel by open-hearth process*".

6. Electric furnaces

We have seen that in reverberatory furnace, muffle furnace and blast furnace, a fuel is used to heat the charge, but in electric furnace, the heating of the charge is done with the help of electricity. Electric furnaces are used, where very high tamperatures are required. Further, it is an important advantage of using electric furnaces, that the temperatures can be easily controlled by simple regulation of power supply to the furnace or the position of the electrodes etc., These furnaces are particularly useful for the electrolytic reductions. The applicability of electric furnaces is dependent on the availability of the cheap electric power. These furnaces are of three types :

(*a*) *Arc furnace* (*also called Heroult's furnace*). This furnace is used for the manufacture of steel. For details see "Manufacture of steel by electric process". (*b*) *Resistance furnace.* In this furnace a temperature of 3000-4000°C can be obtained. These furnaces are of two types *viz., direct resistance furnace and indirect resistances furnace.* (*c*) *Induction furnace.* In this furnace the charge is heated with the help of induction currents.

Unversity Questions

1. Write short notes on :

(*i*) Refining of metals. *(Lucknow 93)*
(*ii*) (*a*) Ores and minerals (*b*) Calcination (*c*) Roasting (*d*) Smelting (*e*) Poling. *(Agra 92)*
(*iii*) Gangue *(Meerut 89)*
(*iv*) Aluminothermic process *(Meerut 89)*
(*v*) Use of electricity in the extraction of metals *(MLNR 85)*
(*vi*) Froath flotation process *(Meerut 89)*
(*vii*) Different types of furnaces, used in metallurgy (Draw labelled sketch of each) *(Agra 85; MLNR 85)*
(*viii*) Electrolytic refining of metals *(Meerut 90)*
(*ix*) Liquation process of refining metals *(Delhi 89)*
(*x*) Cupellation *(Delhi 89)*
(*xi*) Dressing of the ore *(Meerut 90)*
(*xii*) Roasting (give examples) *(Meerut 89)*
(*xiii*) Hydrometallurgy *(Meerut 90)*
(*xiv*) Electrometallurgy *(Meerut 90)*

2. What methods are available for the extraction of a metal from its oxide and sulphide ores. Illustrate your answer with examples. *(Meerut 91)*

3. What are the differences between :

(*i*) Minerals and ores *(Delhi 89)*
(*ii*) Calcination and roasting *(Delhi 82)*
(*iii*) Roasting, calcination and smelting *(Agra 82)*
(*iv*) Slag and flux *(I.I.T. 78)*

4. What is the role of acidic and basic flux in the extractive metallurgy ? *(Delhi Hons. 89)*

5. Explain the following terms with examples :

(*i*) Electrolytic refining of metals *(Meerut 89)*
(*ii*) Liquation process for refining of metals *(Delhi 89)*
(*iii*) Cupellation *(Delhi 89)*
(*iv*) Van Arkel de Boer process *(REE 85)*

6. Explain the chemical principles involved in the extraction of metals from their oxide, sulphide and halide ores. *(Lucknow 95)*

7. Wrıte a note on "Calcinatıon and Roasting". *(Lucknow 2002)*

8. (*a*) Name the various metallurgical operations used in the extraction of metal from the ore. Discuss the various methods used in refining and purification of metal.

(*b*) Write a note on Blast Furnace.

(*c*) Distinguish between roasting and calcination. *(Nagpur 2002)*

9. How is reverberatory furnace used both for reduction and oxidation processes ? *(Nagpur 2002)*

10. (*a*) Explain Froth flotation process along with a neat diagram.

(*b*) Describe any *two* of the following : (*i*) Calcination and Roasting; (*ii*) Zone Refining; (*iii*) Mond's process for Nickel. *(Delhi 2002)*

11. Write different methods for the purification of metals. *(Rohilkhand 2003)*

12. (*a*) Explain the terms calcination and roasting.

(*b*) What are the steps involved in metallurgy ? Describe "Froth Flotation Process" for concentration of ores.

(*c*) Write a note on Reverberatory Furnace.

(*d*) Draw a well labelled diagram of "Bessemer Convertor". *(Nagpur 2003)*

13. Explain the following terms with a suitable examples : *(Banglore, 2004)*

(*i*) Roasting (*ii*) Slag

14. List the different methods of refining of a crude metal. *(Banglore, 2004)*

15. Name an acidic flux. *(Banglore, 2004)*

16. Give an example of a metal refined by Van Arkel method. *(Banglore, 2004)*

17. What are the differences between roasting and calcination ? *(Banglore, 2004)*

21

Nomenclature of Inorganic Compounds

Introduction

When comparatively few inorganic compounds were known, there was no difficulty in naming them. But with the increase in the number of these compounds, the systematic nomenclature presented many difficulties. Some of which are given below:

(*i*) One difficulty is that many of the compounds had become so familiar that it is difficult to discard their familiar names. For example although ferrocyanides and ferricyanides should be named as hexacyanoferrate (II) and hexacynoferrate (III) respectively, yet industry still calls them *yellow prussiate* and *red prussiate* respectively.

(*ii*) The other difficulty in systematic nomenclature for inorganic compounds is that the structure of the compounds depends on the state of aggregation. For example:

(*a*) In the vapour state PCl_5 exists as a discrete molecular species and hence is correctly named as phosphorus pentachloride, but in the solid state, since it consists of $[PCl_4]^+$ $[PCl_6]^-$ ion pairs, it is named as tetrachlorophosphonium (V) hexachlorophosphate (V).

(*b*) In the vapour state PBr_5 is correctly called phosphorus pentabromide, but in the solid state, since it exists as $[PBr_4]^+$ Br^-, it is named as tetrabromophosphonium (V) bromide.

(*c*) Phosphorus pentaoxide has become so familiar name for P_4O_{10} molecule that no body likes to call it as tetraphosphorus decaoxide.

The problem of nomenclature of inorganic compounds was ultimately examined by the *International Union of Pure and Applied Chemistry* (IUPAC) and some rules were given by it. These rules are sometimes not completely acceptable to all chemists. Hence these should not be viewed as a rigid code. These are simply an attempt to classify the process of naming.

Terms used in the nomenclature of inorganic compounds

Before discussing the actual nomenclature of various kinds of ions and inorganic compounds, we should be fimiliar with various terms like numerical prefixes, Arabic numericals, square brackets etc., which are used in the nomenclature of inorganic compounds. These terms are discussed below :

1. Greek numerical prefixes. (*a*) Simple numerical prefixes like *mono, di, tri, tetra, penta, hexa* etc. are used to indicate the following :

(*i*) *Stoichiometric proportions.* Examples are CO (carbon monoxide), CO_2 (carbon dioxide), SO_2 (sulphur dioxide), SO_3 (sulphur trioxide) etc.

(*ii*) *Extent of substitution.* For example $SiCl_2H_2$ (dichlorosilane), $PO_2S_2^{2-}$ (dithiophosphate ion) etc.

(*iii*) *Number of identical ligands in co-ordinated ions.* Examples are $[Co(NH_3)_4Cl_2]^+$ is called tetrammine dichlorocobalt (III) ion.

(*iv*) *Number of identical atoms in condensed acids.* Examples are H_3PO_4 (monop hosphoric acid. This acid is also called simply phosphoric acid), $H_4P_2O_7$ (diphosphoric acid), $H_2S_3O_{10}$ (trisulphuric acid), $H_5P_3O_{10}$ (triphosphoric acid).

(*v*) *Number of atoms of the same element forming the skeleton of molecules or ions.* This point may be illustrated by the following examples.

(*a*) Si_2H_6 (disilane), B_4H_{10} (tetraborane), $S_4O_6^{2-}$ (tetrathionate ion) etc.

(*b*) If a compound contains organic radicals or groups, their numbers two, three, four, five etc., are indicated by the prefixes like *bis, tris, tetrakis, pentakis* etc., respectively. The name of the organic radical or group is written in parenthesis. The following examples illustrate this point.

$(ClCH_2CH_2)_2$ NH.... *bis* (2-chloroethyl) amine
$(Cr(C_2O_4)_3]^{3-}$........... *tris* (oxalate) chromate (III) ion
$[Pt\ (CH_2OH)_4]^+$ ..... *tetrakis* (hydroxy methyl) phosphonium ion.

2. Structural abbreviations. In order to indicate the structure of a given inorganic compound, some specific abbreviations are used. For example *cis* is used to indicate two groups occupying adjacent positions while *trans* is used to indicate two groups occupying the diagonal (*i.e.*, opposite) positions in a square plane. Some structural abbreviations and their meaning are given below :

pentaprismo ... 10 atoms are bound into a pentagonal prism.
antiprismo ... 8 atoms are bound into a rectangular antiprism
hexaprismo ... 12 atoms are bound into a hexagonal prism
triprismo ... 6 atoms are bound into a triangular prism
dodecahedro ... 8 atoms are bound into a dodecahedron with triangular faces.
hexahedro ... 8 atoms are bound into a hexahedron (*i.e.*, a cube)
icosahedro ... 12 atoms are bound into a triangular icosahedron
octahedro ... 6 atoms are bound into an octahedron
quadro ... 4 atoms are bound into a quadrangle (*e.g.*, a square)
tetrahedro ... 4 atoms are bound into a tetrahedron
triangulo ... 3 atoms are bound into a triangle.
sym and *asym* ... symmetric (symmetrical) and asymmetric or unsymmetrical structure respectively
catena ... a chain structure
closo ... a cage or closed structure
cyclo .. a ring or cyclic structure
fac ... 3 groups are occupying the corners of the same face of an octahedron
mer (*meridional*) ... 3 groups are on an octahedron in such a way that one is *cis* to the other two which are *trans* to each other
nido ... a nest like structure
η (*eta*) ... number of contiguous atoms of the group that are attached to a metal
μ (*mu*) ... the group so designated bridges two centres of coordination
σ (*sigma*) ... one atom of the group is attached to a metal
(+) *or d-form* ... dextro-rotatory form
(–) *or l-form* ... levo (or laevo) rotatory form
dl or (±) *form* ... recemic form
(A A) ... A symmetrical bidentate ligand having two A-atoms as donor atoms
(AB) ... A symmetrical bidentate ligand having A and B as donor atoms

3. Arabic numbers. Arabic numbers are used in inorganic compounds

(*a*) to indicate the atoms at which there is a substituent or addition in a chain. For example $\overset{1}{H_3Si}—\overset{2}{ClHSi}—\overset{3}{SiH_2}—\overset{4}{SiH_3}$ is called *2-chlorotetrasilane* and $\overset{1}{CH_3NH}—\overset{2}{NHCH_3}$ is called *1, 2-dimethylhydrazine.*

(*b*) in place of simple Greek numerals prefixes. For example :

B_6H_{10} ... Hexaborane (IV) (in place of hexaboron decahydride)

MgSO4.7H_2O ... Magnesium sulphate 7-hydrate (in place of magnesium sulphate heptahdrate)
AlK$(SO_4)_2$.12H_2O ... Aluminium potassium sulphate 12-hydrate (in place of aluminium potassium sulphate dodecahydrate)

(*c*) to indicate the number of positive and negative charges on a free ion [*e.g.*, Fe^{2+}, SO_4^{2-} etc.] or on a complex ion [*e.g.*, $[Fe(CN)_6]^{3-}$ $[Co(NH_3)_6]^{3-}$ etc.].

4. Square brackets. These are used:

(*a*) To enclose sets of atoms or ligards in the formula of a compound, a complex ion or a neutral complex compound, *e.g.*, $Al_2(SO_4)_3$, $[Cu(CN)_6]^{4-}$, $[Co(NH_3)_6]^{3-}$, $[Co(NH_3)_3 (NO_2)_3]^0$ etc.

(*b*) To enclose the names of organic radicals or groups acting as ligands in complex compounds. Before the square bracket the prefixes like *bis, tris, tetrakis* etc., are used to indicate their number. For example $[Co(NH_3)_2 (en)_2]^{3+}$ is called diammine *bis* (ethylene diammine) cobalt (III) ion.

5. Symbol of an element in italics. The symbol of an element printed in italics is used :

(*a*) To indicate the element in a compound at which there is a substitution. For example $(CH_3)ONH_2$ (*O*-methyl hydroxylamine), $HONH(CH_3)$ (*N*-methyl hydroxylamine), CH_3ONH_2-CH_3 (*O*-methyl hydroxylamine-borane) etc.

(*b*) To indicate the element of an ambidentate ligand co-ordinated to the central metal atom. For example $[Co(NH_3)_5(ONO)]^{2+}$ is called *O*-nitropentammine cobalt (III) ion while $[Co(NH_3)_5NO_2]^{2+}$ is called *N*-nitropentammine cobalt (III) ion. Similarly $[Hg(SCN)_4]^{2-}$ is called *S*-tetrathiocyanato mercurate (II) ion while $[Hg (NCS)_4]^{2-}$ is called *N*-tetrathiocyanato mercurate (II) ion.

Now let discuss the rules for naming various types of ions and compounds.

Representation of the valence state of the element

The termination *ous* and *ic* are used for naming the compounds of an element which show only two valence states. For example : $FeCl_2$ ferrous chloride, $FeCl_3$ ferric chloride.

Sometimes, as suggested by Stock (1919), a Roman numeral (I, II, III etc.) in parentheses is used to denote the valence state of the element. For example : CuCl copper (I) chloride, $CuCl_2$ copper (II) chloride, FeO iron (II) oxide, Fe_2O_3 iron (III) oxide, Fe_3O_4 (FeO.Fe_2O_3) iron (II, III) oxide, $PbCl_2$ lead (II) chloride, $PbCl_4$ lead (IV) chloride.

The *cipher* (0) is used to denote an oxidation state of zero. A negative oxidation state is indicated by the use of the negative sign with a Roman numeral.

Stock system can also be used for some ternary compounds. Examples are : $Cr(CH_3COO)_2$ chromium (II) acetate, Ni $(CN)_2$ nickel (II) cyanide.

Stock system cannot be used for naming many compounds. For example (*i*) the name nitrogen (IV) oxide for NO_2 does not distinguish it from N_2O_4. (*ii*) the name lead (II, IV) oxide for Pb_2O_3 (PbO.PbO_2) does not distinguish it from Pb_3O_4(2PbO.PbO_2). In such cases the use of Greek or Latin numeral prefixes [*mono, di, tri, tetra, penta, hexa, hepta, octa, nona* (*ennea*), *deca, undeca* (*hendeca*) and *dodeca*] preceding the name of the elements, without hyphens, are used. Examples are : N_2O_4 dinitrogen tetraoxide, Pb_3O_4 trilead tetraoxide, S_2Cl_2 disulphur dichloride.

The prefix *mono* is generally omitted. For example : $SiCl_4$ silicon tetrachloride, NO nitrogen oxide, NO_2 nitrogen dioxide, N_2O dinitrogen oxide.

Nomenclature of cations

(*a*) *Monoatomic cations* like Cu^+, Cu^{2+}, Cr^{3+}, Sn^{4+} etc., are named as their elements with their oxidation state shown by Roman numerals (such as I, II, III, IV etc.) in parentheses. Examples are Cu^+ copper (I) ion ; Cu^{2+} copper (II) ion ; Fe^{3+} iron (III) ion, Sn^{4+} tin (IV) ion.

(*b*) The *polyatomic cations* like IH_2^+, SH_3^+, PH_4^+, AsH_4^+ etc., are obtained by adding a proton (H^+) to the neutral molecules *viz*., HI, H_2S, PH_3 and AsH_3. The names of these cations end in *-onium.* Thus IH_2^+, SH_3^+, PH_4^+ and AsH_4^+ are called *iodonium sulphonium, phosphonium* and *arsonium* respectively.

(*c*) *Certain polyatomic cations* are called by their trivial names. For example NH_4^+, $H_3O_2^+$ and NH_2OH^+ are named as *ammonium, hydronium* and *hydroxylammonium* respectively.

(*d*) *Some polyatomic cations* are called by special names *e.g.,* NO^+ (nitrosyl ion), VO_2^{2+} (vanadyl ion), NO_2^+ (nitronium ion), UO_2^{2+} [uranyl (VI) ion], CrO_2^{2+} (chromyl ion) etc.

Nomenclature of anions

(*i*) *Monoatomic anions* are named by adding *ide* to the stem name of the element. Examples are : H^- (hydride ion), F^- (fluoride ion), Se^{2-} (selenide ion), N^{3-} (nitride ion), P^{3-} (phosphide ion), O^{2-} (oxide ion) etc.

(*b*) *Certain polyatomic anions* have their names ending in *ide.* Examples are OH^- (hydroxide ion), O_2^{2-} (peroxide ion), HO_2^- (hydrogen peroxide ion), O_2^- (superoxide ion), O_3^- (ozonide ion), NH_2^- (amide ion), NH^{2-} (imide ion), N_3^- (azide ion), CN^- (cyanide ion), HS^- (hydrogen sulphide ion), S_2^{2-} (disulphide ion), I_3^- (triodide ion), HF_2^- (hydrogen difluoride ion) etc.

(*c*) The names of *polyatomic anions containing oxygen* end in – *ite.* – ite denotes lower oxidation state. Examples are NO_2^{2-} (nitrite ion), $N_2O_2^{2-}$ (hyponitrite ion), AsO_3^{3-} (arsenite ion), SO_3^{2-} (sulphite ion), $S_2O_6^{2-}$ (disulphite ion), $S_2O_4^{2-}$ (dithionite ion), SeO_3^{2-} (selenite ion), ClO^- (hypochlorite ion), ClO_2^- (chlorite ion), IO^- (hypoiodite ion) etc. It may be noted that AsO_3^{3-} is called arsenite while PO_3^{3-} is called phosphonate and not phosphite.

Nomenclature of isopoly anions

Isopoly acids are obtained by the condensation of molecules of the same acid. The anions derived from isopoly acids are called isopoly anions. These anions are named by indicating the number of atoms of the characteristic element with Greek number prefixes. For example $S_2O_7^{2-}$ (disulphate ion), $P_3O_{10}^{5-}$ (triphosphate ion), $B_4O_7^{2-}$ (tetraborate ion), $Mo_7O_{24}^{6-}$ (heptamolybdate ion), $Cr_4O_{18}^{2-}$ (tetrachromate ion) etc.

Isopoly anions having *cyclic* (*ring*) and *straight chain structures* are distinguished by using the prefixes *cyclo* and *catena* respectively. The prefix catena may also be omitted. Examples are:

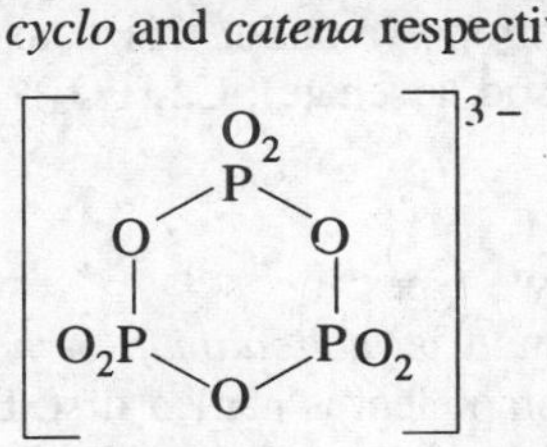

Cyclo triphosphate ion, $[(PO_3)_3]^{3-}$

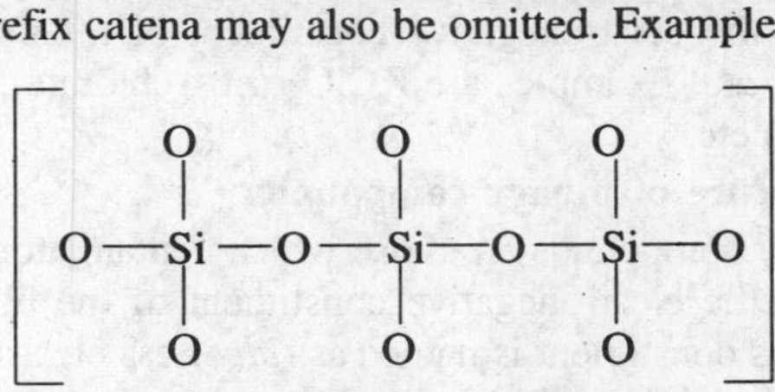

Catena trisilicate ion, $[Si_3O_{10}]^{6-}$

Nomenclature of heteropoly anions

(*a*) *Dinuclear anions* are named by considering the anion which occurs first in alphabetical order as the ligand on the atoms of second anion. For example $[O_3P - O - SO_3]^{3-}$ is called phosphato sulphate (– 3) and $[O_3As—O—PO_3]^{4-}$ is called arsenato phosphate (– 4). (– 3) and (– 4) indicate the oxidation state of the anion.

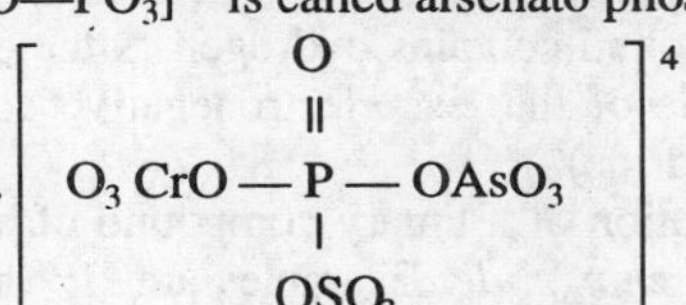

is called arsenato chromate sulphato phosphate (– 4).

(*b*) *Cyclic heteropoly anions* are named in a manner similar to that used for naming chain structures. The starting point and the direction of citation of various units are taken in alphabetical order. For example :

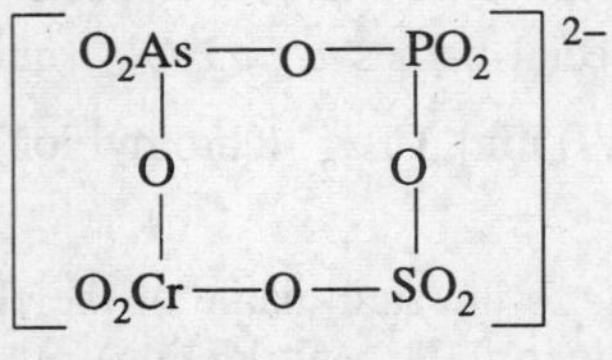

cyclo arsenato chromate
sulphate phosphate (–2)

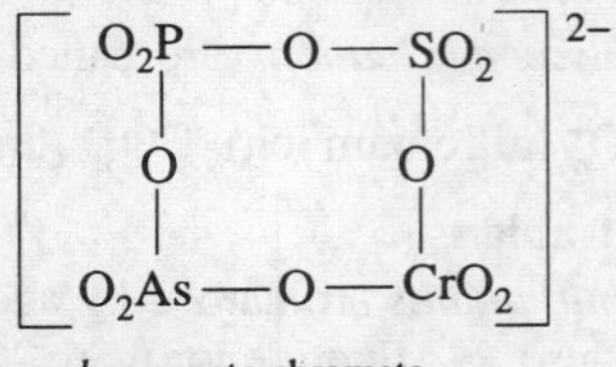

cyclo arsenato chromate
phosphato sulphato chromate (–2)

(*c*) *Condensed heteropoly anions* consisting of three-dimensional frameworks of tetrahedra of linked MoO_6, WO_2 etc. and surrounding a central atom are called by using the prefixes *molybdo, tungto* etc. Examples are $P_2Mo_{18}O_{62}^{-6}$ is called 18-molybdodiphosphate (–6) and $PW_{12}O_{40}^{-3}$ is called dodecatunsto phosphate (– 3).

Nomenclature of neutral and cationic radicals

The names of certain neutral and cationic radicals end in *–yl.* Examples are OH (hydroxyl), CO (carbonyl), NO (nitrosyl), NO_2 (nitryl), PO (phosphoryl), SO (sulphinyl or thionyl), SO_2 (sulphonyl or sulpheryl), S_2O_5 (disulphuryl or pyrosulphuryl), SeO (seleninyl), SeO_2 (selenonyl), ClO (chlorosyl), ClO_2 (chloryl), ClO_3 (perchloryl), VO (vanadyl), CrO_2 (chromyl), UO_2 (uranyl), NpO_2 (neptunyl), PS (thiophosphoryl), CSe (seleno carbonyl) etc.

Nomenclature of compounds in general

(*a*) When we are required to write the formula of an inorganic salt, the cation is written first and then the anion is written. The oxidation state of the cation is written in Roman numeral in parentheses. For example NO_2CO_3 (sodium carbonate), $CaCN_2$ (calcium cyanamide), $VOSO_4$ (vanadyl sulphate), $FeCl_3$ [iron (III) chloride], $NaNH_2$ (sodium amide), CrO_2Cl_2 (chromyl chloride).

(*b*) Salts containing acid hydrogen (*e.g.*, $NaHCO_3$, LiH_2PO_4 etc.) are named by writing "*hydrogen*" before the name of the anion. Number of H-atoms is indicated by *di, tri, tetra* etc. For example $NaHCO_3$ (sodium hydrogen carbonate), KHS (potassium hydrogen sulphide), LiH_2PO_4 (lithium dihydrogen phosphate), Na_2HPO_4 (disodium hydrogen phosphate) etc.

(*c*) If the electronegative constituent of a compound is hetropolyatomic, this constituent is named as *-ate.* Examples are $KClO_3$ (pot. chlorate), Na_3AsO_4 (sod. arsenate), $Ca_3(PO_4)_2$ (calcium phosphate) etc.

Nomenclature of binary compounds

Binary compounds are those which contain atoms of only two types:

(*a*) If the electronegative constituent of the binary compound is *monoatomic* or *homopolyatomic*, this constituent is named as *-ide.* Less electronegative component is named first. Examples are CaC_2 (calcium carbide), Mg_3N_2 (magnesium nitride), AlN (aluminium nitride), NaCl (sodium chloride) etc.

(*b*) When we are required to write the formula of a binary compound between non-metals, the constituent which is written earlier in the series : Rn, Xe, Kr, B, Si, C, Sb, As, P, N, H, Te, Se, Ta, S, At, I, Br, Cl, O, F should be written first. Examples are XeF_2, NH_3, H_2S, S_2Cl_2, Cl_2O SiC, OF_2 etc.

(*c*) If the binary compound contains hydrogen as the electronegative constituent, binary compound is named as hydride of the less electronegative constituent. For example CaH_2 (calcium hydride), NaH (sodium hydride).

(*d*) If the aqueous solution of a binary compound of hydrogen acts as an acid, it is named as *hydro.......ic acid* or *hydrogen........ide.* Examples are HF (hydrofluoric acid or hydrogen fluoride), HCl (hydrochloric acid or hydrogen chloride), H_2S (hydrogen sulphide), H_2Te (hydrogen telluride), N_3H (hydrazoic acid or hydrogen azide).

(*e*) There are many binary compounds of hydrogen for which the trivial name is used. For example H_2O (water), BH_3 (borane), SiH_4 (silane), GeH_4 (germane), NH_3 (ammonia), PH_3 (phosphine), AsH_3 (arsine), SbH_3 (stibine), N_2H_4 or $NH_2.NH_2$ (hydrazine or diammine) etc.

Nomenclature of oxy-acids

(*a*) *Hypo, ous, ic and per acids.* Different oxidation states of the central atom are distinguished by using the notations *hypo, ous, ic* and *per.* Hypo indicates the lowest oxidation state of the central metal atom while per indicates the highest oxidation state. *Ous* indicates an oxidation state higher than that of hypo acid while *ic* indicates an oxidation state lower than that of per acid. Thus the oxidation state increases as *hypo* → *ous* → *ic* → *per.*

Examples are (*i*) hyponitrous acid, $H_2N_2O_2$ (N = + 1); nitrous acid, HNO_2 (N = + 3); nitric acid, HNO_3 (N = + 5); pernitric acid, HNO_4 (N = + 7), (*ii*) hypophosphorus acid, H_3PO_2 (P = + 1); phosphorus acid, H_3PO_3 (P = + 3); phosphoric acid (also called orthophosphoric acid), H_3PO_4 (P = + 5); perphosphoric, HPO_5 (P = + 7). (*iii*) sulphurous acid, H_2SO_3 (S = + 4); sulphuric acid, H_2SO_4 (S = + 6), (*iv*) hypophosric acid, $H_4P_2O_6$ (P = + 4); phosphoric acid, H_3PO_4 (P = + 5), (*v*) dithionous acid, $H_2S_2O_4$ (S = + 3); dithionic acid, $H_2S_2O_6$ (S = + 5), (*vi*) hypochlorous acid, HClO; chlorous acid, $HClO_2$; chloric acid, $HClO_3$; perchloric acid, $HClO_4$.

(*b*) *Ortho, meta and pyroacids.* In some cases an element forms more than one oxyacids in which the oxidation state of the element is different. A group of such oxyacids can be distinguished by the use of the prefixes *ortho, meta* and *pyro.* The prefix *ortho* is used for the *most hydroxylated acid* in the set, known in the free state, in salt or in esters.

An acid which may be regarded as being formed by the removal of one H_2O molecule from *one* molecule of an ortho acid is called the **meta acid** while that formed by the removal of one H_2O molecule from *two* molecules of ortho acid is called the **pyro acid.**

For example:

$$\underset{\text{One molecule of orthophosphoric acid}}{H_3PO_4} \xrightarrow{-H_2O} \underset{\text{Metaphosphoric acid}}{HPO_3} \xrightarrow{\text{polymerisation}} \underset{\text{Polymerised molecule}}{(HPO_3)_n}$$

$$\underset{\text{One molecule of ortho boric acid}}{H_3BO_3} \xrightarrow{-H_2O} \underset{\text{Meta boric acid}}{HBO_2} \xrightarrow{\text{polymerisation}} \underset{\text{Polymerised molecule}}{(HBO_2)_n}$$

$$\underset{\text{One molecule of ortho silicic acid}}{H_4SiO_4} \xrightarrow{-H_2O} \underset{\text{Meta silicic acid}}{H_2SiO_3} \xrightarrow{\text{polymerisation}} \underset{\text{Polymerised molecule}}{(H_2SiO_3)_n}$$

$$2H_3PO_4 \text{ (Two molecules of ortho phosphoric acid)} \xrightarrow{-H_2O} H_4P_2O_7 \text{ (Pyrophosphoric acid)}$$

Only a few meta acids are known, since they are generally polymerised species.

Sometimes the prefix *di* is preferred to the prefix *pyro.* For example pyrophosphoric acid ($H_4P_2O_7$) is also called *disphosphoric acid.* Some other examples of diacids are given below :

$$2H_2SO_4 \text{ (Two molecules of sulphuric acid)} \xrightarrow{-H_2O} H_2S_2O_7 \text{ (Disulphuric acid)}$$

$$2H_2SO_3 \text{ (Two molecules of sulphorus acid)} \xrightarrow{-H_2O} H_2S_2O_5 \text{ (Disulphorus acid)}$$

$$2H_2CrO_4 \text{ (Two molecules of chromic acid)} \xrightarrow{-H_2O} H_2Cr_2O_7 \text{ (Dichromic acid)}$$

(*c*) *Peroxo (peroxy) acids.* Peroxo acids can be regarded as having been obtained by removing one or two H_2O molecules from the combination of one H_2O_2 molecule and one or two molecules of oxyacid. For example.

(*i*) 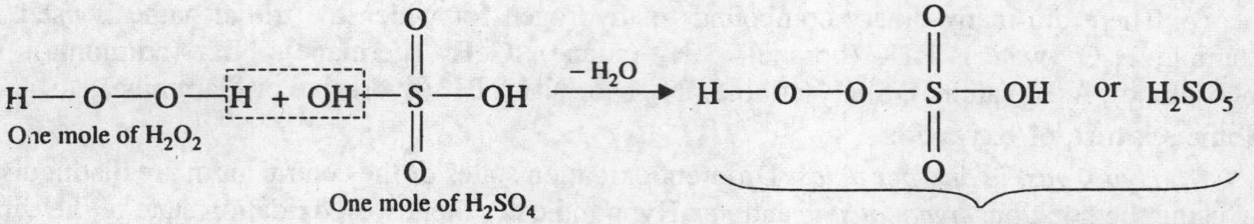

Peroxomonosulphuric acid (also called permonosulphuric acid or Caro's acid)

(*ii*) 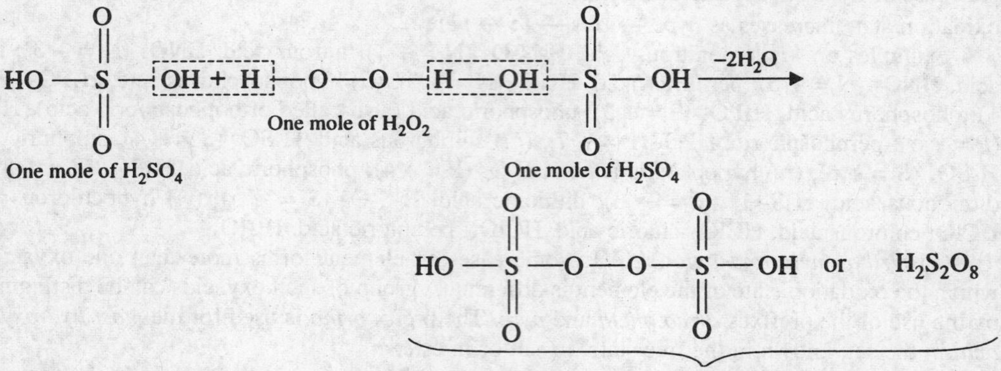

Peroxodisulphuric acid (also called perdisulphuric acid or Marshall's acid).

(*iii*) 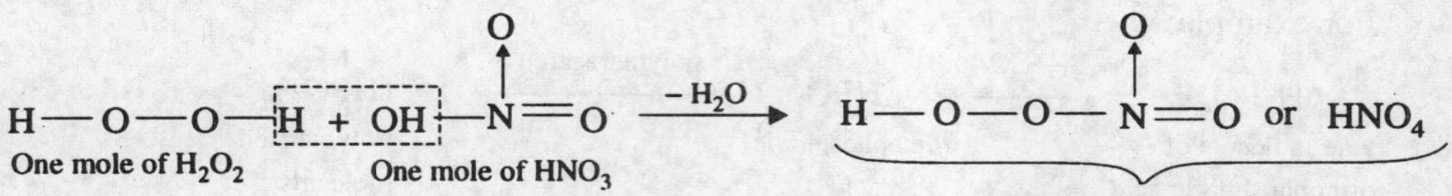

Peroxomononitric acid

(*iv*) 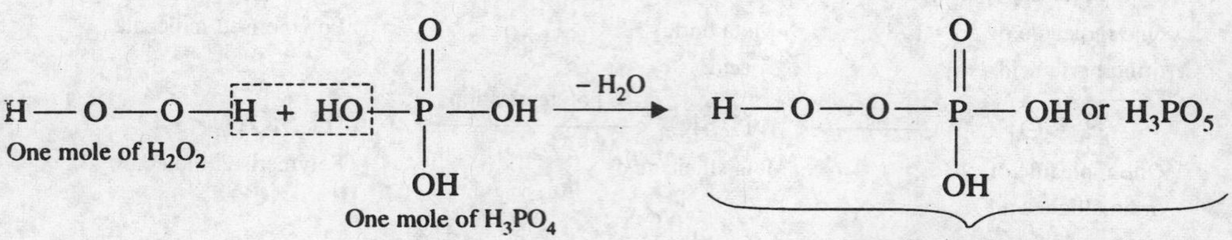

Peroxomonophosphoric acid

(*v*) 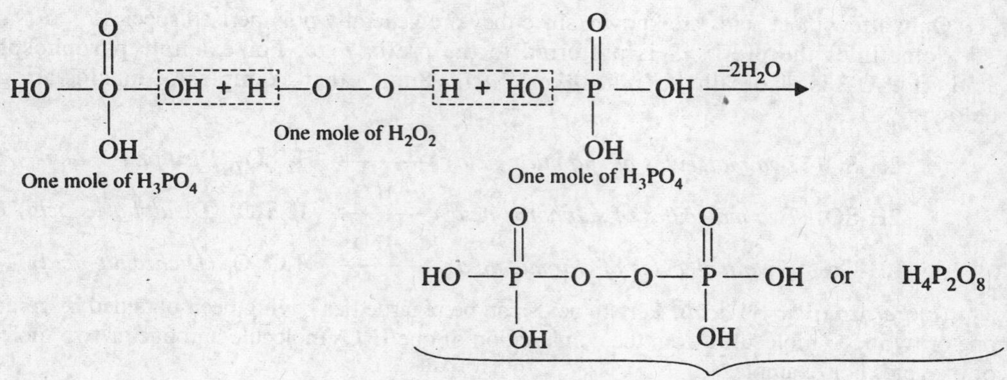

Peroxodiphosphoric acid

(*vi*) $\underset{\text{One mole of } H_2O_2}{H—O—O—H} + \underset{\text{One mole of } CH_3COOH}{OH—\overset{\overset{\large O}{\|}}{C}—CH_3} \xrightarrow{-H_2O} \underbrace{H—O—O—\overset{\overset{\large O}{\|}}{C}—CH_3 \text{ or } CH_3COO_2H}_{\text{Peroxomonacetic acid}}$

(*d*) *Thio acids.* Thio acids are obtained from oxyacids by replacing one or more oxygen atoms by sulphur atoms. For example : (*i*) H_2CO_3 (carbonic acid), H_2CS_3 (trithio carbonic acid) (*ii*) H_3PO_4 (phosphoric acid), H_3PO_3S (thiophosphoric acid), $H_3PO_2S_2$ (dithiophosphoric acid) (*iii*) H_2SO_4 (sulphuric acid), $H_2S_2O_3$ (thiosulphuric acid) (*iv*) H_2SO_3 (sulphurous acid), $H_2S_2O_2$ (thiosulphurous acid) (*v*) HOCN (cyanic acid), HSCN (thiocyanic acid).

Nomenclature of hydrates, ammoniates and peroxy hydrates

See chapter (*Basic Concepts of Co-ordination Chemistry: Volume II*)

Nomenclature of double salts containing more than one cation or more than one anion

See chapter (*Basic Concepts of Co-ordination Chemistry: Volume II*)

Nomenclature of hydrolysed or Basic salts

See chapter (*Basic Concepts of Co-ordination Chemistry: Volume II*)

Nomenclature of co-ordination compounds.

See chapter (*Basic Concepts of Co-ordination Chemistry: Volume II*)

University Questions

1. Write short notes on:
   (*i*) Nomenclature of inorganic compounds by IUPAC syste'
   (*ii*) Nomenclature of oxyacids by IUPAC system.
   (*iii*) The systematic nomenclature of inorganic salts containing two or more cations.
   (*iv*) Main rules of inorganic compounds nomenclature. (*Lucknow, 79, 80; Kerala, 82; Meerut, 85*)
2. Give the IUPAC names of the following compounds: $NaAlO_3$, $BeCl_2$, $H_4O_2O_7$, N_2O_3, P_2O_5, LiCl, $MgBr_2$, $H_4P_2O_7$, $H_2S_2O_8$, N_2O_5, $FeCl_2$, $FeCl_3$, N_2O_4, $CuSO_5.5H_2O$, H_2O_2, HNO_3, H_3PO_4, P_4O_6, $H_2S_2O_6$, $HClO_2$, NO_2, SO_2, $(HF)_n$, $CuCl_2$. (*Lucknow, 80, 81, 85 ; Allahabad, 86*)
3. Discuss the nomenclature of complex compounds by IUPAC system. (*Raj, 84 : Meerut, M.Sc. 82*)
4. Name the following species according to IUPAC nomenclature:
   IH_2^+, PH_4^+, SH_3^+, AsH_4^+, S_2^{2-}, $N_2H_3^-$, C_2^{2-}, SeO_3^{2-}, NO, NO_2, SeO_2, UO_2, ClO_3^-, $S_2O_5^-$, $Mo_7O_{24}^{6-}$, $Cr_4O_{18}^{2-}$, $[O_2P–O–SO_3]^{3-}$, $[O_3As–O–PO_3]^4$, $[P_2Mo_{18}O_{62}]^{6-}$ $[PW_{12}O_{40}]^{3-}$.
5. Name the following compounds according to IUPAC nomenclature:
   H_2SO_5, $H_2S_2O_8$, $H_2S_2O_6$, $H_5P_8O_{10}$, $H_4P_2O_6$, H_4SiO_4, $CaCl_2.8NH_3$, $3CdSO_4.8H_2O$, $AlCl_3.4C_2H_5OH$, $Na_2CO_3.10H_2O$, $FeSO_4.(NH_4)_2SO_4.6H_2O$, $NH_3.BF_3$, $8H_2S.46H_2O$, $K_2SO_4.Al_2(SO_4)_3.24H_2O$. B_2H_6, B_4H_6, B_4H_{10}, B_5H_2, B_6H_{10}, $B_{10}H_{14}$, $B_{20}H_{16}$.
6. Name at least 10 structural affixes. Give the specific structures which they signify.
7. Which of the following is the correct name for F_2O ?
   (*a*) Fluorine oxide (*b*) Oxygen difluoride.
8. Which of the following is the correct name for $[PW_{12}O_{14}]^{3-}$?
   (*a*) Dodecatungsto phosphate (3–) (*b*) Phosphododeca tungstate (3–)
9. Which of the following is the correct name for $H_5P_3O_{10}$?
   (*a*) Pyrophosphoric acid (*b*) Orthophosphoric acid (*c*) Triphosphoric acid.
10. Which of the following is the correct name for $H_2S_2O_4$?
    (*a*) Dithionic acid (*b*) Thiosulphuric acid (*c*) Persulphuric acid (*d*) Dithionous acid.
11. Which of the following is the correct name for $((CH_3)_3Si)_2N–NH_2$?
    (*a*) bis (trimethylsilyl) hydrazine (*b*) 1, 1-bis (trimethylsilyl) hydrazine (*c*) *N, N*-bis (trimethylsilyl) hydrazine.

[**Ans :** 7 (*b*), 8 (*a*), 9 (*c*), 10 (*d*), 11 (*b*)]

Part II : Chemistry of Main Group Elements and their Compounds (*s*- and *p*-Block Elements)

22

General Characteristics of Main Group Elements (*s*- and *p*- Block Elements)

Introduction

The elements of groups IA(1), IIA (2), IIIA (13), IVA (14), VA (15), VIA (16), VIIA (17) and zero (18) are called *main group elements.* Out of these the elements of groups IA, IIA, IIIA, IVA, VA, VIA and VIIA are called *normal or representative elements* and those of zero group are called *noble gases* or *inert gases.* Thus we see that the representative elements and noble gases are main group elements. On the basis of the orbital into which the last electron goes, main group elements have been grouped into the following categories : (*a*) *s-block elements*. In the atoms of these elements the last electron goes to *ns* orbital. Thus the elements of groups IA and IIA are *s*-block elements. (*b*) *p-block elements.* In the atoms of these elements the last electron enters *np* orbital. Thus the elements of groups IIIA, IVA, VA, VIA and VIIA are *p*-block elements. *s*- and *p*-block elements are collectively called main group elements.

General characteristics of main group elements

1. Electronic configuration. Electronic configuration of the atoms of main group elements, according to their position in different periods and in different groups of the periodic table, has already been discussed in chapter 4.

2. Atomic and ionic radii. The atomic and ionic radii of main group elements generally decrease in moving from left to right in a given period and increase in moving down the group. Some irregular trends in these have also been observed. An explanation of irregular trends of atomic and ionic radii has already been given in chapter 5.

3. Ionisation energy (IE). Generally the values of first IE increase on crossing a period from left to right and decrease on moving down a group. A detailed explanation of these trends has already been given in chapter 5.

From these values as given in chapter 5, it may be seen that the ionisation energies decrease regularly as we move down the groups (groups IA and IIA) and increase across a period of *s*-block elements, but some *p*-block elements show irregular trends in these values. For example in the elements of *2nd* and *3rd* periods, the ionisation energy values do not increase regularly as we move from left to right in the given period. This irregular variation of ionisation energy has already been explained in chapter 5.

4. Electron affinity (EA). Generally electron affinity values increase as we move from left to right in a given period (*i.e.,* from alkali metals to halogens) and decrease in going down a group. The reason for this variation has already been given in chapter 5. The elements of group IIA have either zero or negative values for their electron affinity. EA values of N and P are very low. Noble gaves have either zero or negative EA values. The reasons for these abnormal values have already been discussed at the appropriate pages of chapter 5.

5. Electronegativity. According to pauling, electronegativity of an atom of an element in a molecule is its relative tendency (or power or ability) to attract the shared electron pair towards itself. Electronegativity values of main group elements are given in chapter 5. These values increase on moving from left the right in a period and decrease in moving down a group of *s*-block elements, but these values do not follow this trend in the elements of groups IIIA and IVA which belong to *p*-block (*See chapter 5*).

6. Metallic (electropositive) and non-metallic (electronegative) character. The variation of these properties on moving from left to right and down a given group of main group elements has already been discussed in chapter 5.

7. Density of elements in the solid state (*Also see chapter 5*). Densities (g cm^{-3}) of main group elements in the solid state are given in Table 5.8 of chapter 5. It may be seen from the table that the densities do not show any regular trend as we move from left to right in a given period. These values generally increase as we move down a given group.

8. Atomic volumes of elements (*Also see chapter 5*). The atomic volumes (in cm^3) of main group elements are given in Table 5.9 of chapter 5. Atomic volumes increase in moving down the group. This increase is due to the increase in atomic radii. Atomic volumes show no regular trend in a period.

9. Melting and boiling points (*Also see chapter 5*). The values of melting and boiling points (in °C) of the main group elements are given in Table 5.10 of chapter 5. These values generally decrease on moving down the group. The melting and boiling points generally increase across a period and attain a maximum value in the middle and, thereafter, decrease until we reach noble gas element.

Explanation. The explanation of the variation of melting and boiling points along a group and a period has already been given in chapter 5.

10. Electrode potential of elements (*Also see chapter 5*). The values of standard reduction electrode potential (in volts) of some main group elements are given in Table 5.11 of chapter 5. From these values a number of important points may be noted. These points have been discussed in chapter 5.

11. Oxidising and reducing properties of the elements in aqueous solution. *See chapter 5.*

12. Oxidation states (*Also see chapter 5*). Oxidation states shown by main group elements are given in Table 5.12 of chapter 5. It may be noted that *s*-block elements show only one positive oxidation state which is equal to their group number. On the other hand most of the *p*-block elements show more than one positive as well as negative oxidation states.

13. Diagonal relationship or diagonal similarity. *See chapter 10.*

14. Reluctance of the heavier elements to form (*p* – *p*) π-bonds. The lighter non-metals of *p*-block elements (*e.g.*, C, N, O etc.) frequently give structures which contain one or more (*p* — *p*) π-bonds. On the other hand, the heavier *p*-block elements like Si, P, S etc., give structures which contain only single σ-bonds. For example :

(*i*) While carbon forms double bond between carbon and oxygen atoms in CO_2 molecule, Si – atom forms four single Si — O σ-bonds in SiO_2 molecule. CO_2 molecule has a *linear monomeric* structure (O = C = O) in which C—O π-bonds are obtained by the *p* — *p* overlap. On the other hand SiO_2 does not have the structure *viz.*, O = Si = O; rather it has an *infinite three dimensional polymeric* structure in which a large number of SiO_4 tetrahedrons are linked together. The linking of many SiO_4 tetrahedrons gives giant molecular (*i.e.*, polymeric) structure to SiO_2 molecule. In each SiO_4 tetrahedron Si-atom is *covalently* linked with four O-atoms. Thus we see Si-atom does not form (*p* — *p*) π-bonds with O-atoms in SiO_2 molecule as carbon atom does in CO_2 molecule.

Reason. Since the size and energy of *2p* orbitals of carbon and oxygen atoms are nearly the same, they overlap effectively and form C—O π-bond in CO_2 molecule. In case of SiO_2 molecule, since the 3*p* orbitals of Si-atom and *2p* orbitals of O-atoms have appreciable difference in their size and energy, they cannot overlap effectively to form Si—O π-bond. The tetracovalency of Si-atom is satisfied by forming four single Si—O σ bonds.

(*ii*) Nitrogen exists as a *diatomic linear molecule,* N_2 ($N \equiv N$) which contains two (p — p) π-bonds. On the other hand phosphorus exists as a discrete *tetrahedral tetra atomic molecule,* P_4 in which each P-atom is linked with each of the three remaining P-atoms by single P—P σ-bonds. P_4 molecule has a tetrahedral structure in which the four P-atoms lie at the corners of a regular tetrahedron.

(*iii*) Oxygen exists as a *diatomic molecule,* O_2 ($O = O$) which contains one (p —p) π-bond. On the other hand sulphur exists as S_8 which has a *puckered ring structure* in which each S-atom is linked with two S-atoms by single S — S σ-bonds.

Reason. The probable reason for the non-formation of (p—p) π-bond between two P-atoms in P_4 molecule and between two S-atoms in S_8 molecule can be given as follows :

We know that as we move from N to P or from O to S, the covalent radius of the elements increases (N = 0.75 Å, P = 1.06 Å; O = 0.73 Å, S = 1.02 Å) and hence the length of P—P σ-bond which is equal to double the covalent radius of P-atom becomes greater than that of N—N σ-bond. Similarly the length of S—S σ-bond is also greater than that of O—O σ-bond. It is due to greater P—P σ-bond or S—S σ-bond that the p-orbitals of the two heavier elements (*i.e.*, two P-atoms or two S-atoms) are at a greater distance from each other and hence are not able to overlap effectively to form (p—p) π-bond(s). This is illustrated in Fig. 22.1, the numerical values shown are the covalent radii of the atoms of the elements. Covalent radius is equal to half of the length of M—M σ-bond (M = N or P).

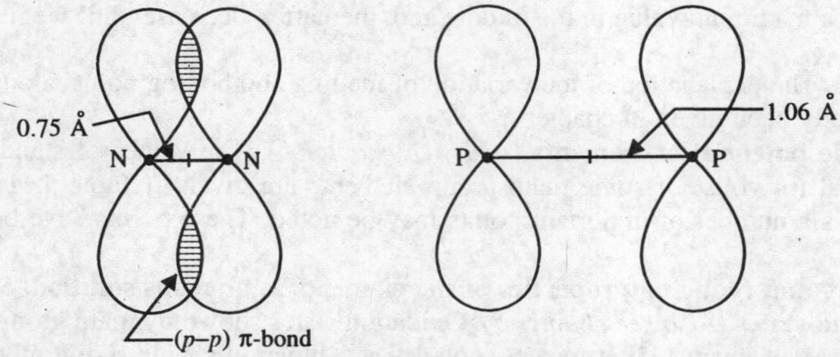

Fig. 22.1 : P—P σ-bond length is so large that p-orbitals of two P-atoms are not able to overlap effectively and no P—P π-bond is formed in P_4 molecule.

Conclusion. The above discussion shows that the lighter non-metals like C, N, O etc., have a tendency to form (p — p) π-bond(s) while the heavier elements (*e.g.*, Si, P, S etc.) show a reluctance to form (p — p) π-bond(s).

15. Important compounds of main group elements. Main group elements form a large variety of compounds. Out of these, oxides, hydrides, halides and oxy halides are the must important. Detailed discussion of these compounds has been given in their respective chapters.

University Questions

1. What are *s*- and *p*-block elements. Write down their electronic configurations and discuss their general characteristics. *(Avadh, 93)*

2. Write a note on "Representative Elements". *(Rohilkhand, 91)*

3. What are main group elements ? Discuss briefly their general characteristics.

4. Explain the following, giving appropriate reasons.

(*i*) The oxidising character of elements increases and reducing character decreases as we move from left to right in a period.

(*ii*) The oxidising character and reducing character of elements increases as we move down a group.

(*iii*) Metallic character of *p*-block elements decreases along a period but increases down a group.

(*iv*) Electron affinity generally decreases down a group.

(*v*) Ionisation energy increases along a period and decreases down a group.

(*vi*) Melting and boiling points decrease down a group.
(*vii*) The heavier *p*- block elements are reluctant to show maximum oxidation states.
(*viii*) The heavier *p*-block elements are reluctant to involve *p*-orbitals for pi bonding.
(*ix*) CO_2 is a monomer while SiO_2 is a three-dimensional giant molecule.
(*x*) The elements of *2nd* period show a covalency of 4 while the elements of *3rd* period and higher periods show covalency more than 4.
(*xi*) CCl_4 does not undergo hydrolysis while $SiCl_4$ and PCl_5 are easily hydrolysed.

5. Discuss the variation of the following properties as we move along a period and down a group of main group elements. (*i*) Atomic and ionic radii (*ii*) Ionisation energy (*iii*) Electron affinity (*iv*) Electronegativity (*v*) Metallic and non-metallic character (*vi*) Oxidising and reducing properties (*vii*) Redox potentials (*viii*) Oxidation states (*ix*) Acidic and basic character of oxides (*x*) Ionic and covalent character of halides.

23

Group Discussion of Main Group Elements (*s*- and *p*- Block Elements)

Introduction

Main group elements have been grouped into 8 vertical columns, called groups. These groups are IA (1), IIA (2), IIIA (13), IV (14), VA (15), VIA (16), VIIA (17) and zero (18). These elements are placed at the extreme left and extreme right of the periodic table as shown in Fig. 23.1. Here we shall make a study of the general characteristics and variations of these properties in moving down each of the above said groups.

← *s*-block elements →			← *p*-block elements →					
IA (1)	IIA (2)		IIIA (13)	IVA (14)	VA (15)	VIA (16)	VIIA (17)	Zero (18)
H_1								He_2
Li_3	Be_4		B_5	C_6	N_7	O_8	F_9	Ne_{10}
Na_{11}	Mg_{12}		Al_{13}	Si_{14}	P_{15}	S_{16}	Cl_{17}	Ar_{18}
K_{19}	Ca_{20}	⟷	Ga_{31}	Ge_{32}	As_{33}	Se_{34}	Br_{35}	Kr_{36}
Rb_{37}	Sr_{38}	*d*-Block Element	In_{49}	Sn_{50}	Sb_{51}	Te_{52}	I_{53}	Xe_{54}
Cs_{55}	Ba_{56}		Tl_{81}	Pb_{82}	Bi_{83}	Po_{84}	At_{85}	Rn_{86}
Fr_{87}	Ra_{88}							

Fig. 23.1 : Location of *s*-and *p*-block elements (main group elements) in the periodic table.

Group IA (1) : H, Li, Na, K, Rb, Cs, Fr (Alkali Metals)

Group IA (1) of the periodic table contains six elements (excluding hydrogen, H_1) which are *lithium* (Li_3), *sodium* (Na_{11}), *potassium* (K_{19}), *rubedium* (Rb_{37}), *caesium* (Cs_{55}) and *francium* (Fr_{87}). These elements are collectively called alkali metals, since they form strongly alkaline oxides and hydroxides. Fr is a radioactive elements.

Gradation in physical properties

Some of the important physical properties of these elements are given in Table 23.1 and have been discussed below:

Table 23.1 : *Some important physical properties of the elements of group IA (1)*

Properties	Lithium (Li_3)	Sodium (Na_{11})	Potassium (K_{19})	Rubedium (Rb_{37})	Caesium (Cs_{55})
Electronic configuration	$[He]_2\ 2s^1$ or 2, 1	$[Ne]_{10}\ 3s^1$ or, 2, 8, 1	$[Ar]_{18}\ 4s^1$ or, 2, 8, 8, 1	$[Kr]_{36}\ 5s^1$ or 2, 8, 18, 8, 1	$[Xe]_{54}\ 6s^1$ or 2, 8, 18, 18, 8, 1
Atomic mass	6.94	22.9	39.10	85.48	132.91
Single bond covalent radius (Å)	1.23	1.54	2.03	2.16	2.35
Radius of M^+ ion (Å)	0.68	0.95	1.33	1.48	1.69
Atomic volume (cm^3)	12.97	23.68	45.36	55.80	69.95
Density ($g\ cm^{-3}$)	0.53	0.97	0.86	1.53	1.90
Specific heat at 0°C ($cal\ g^{-1}$)	0.941	0.293	0.17	0.08	0.049
Melting point (°C)	181	97.8	64.0	39.0	28.5
Boiling point (°C)	1330	888	756.0	688.0	669.0
Heat of atomisation ($kJ\ mol^{-1}$)	159.0	108.1	90.2	82.1	78.2
Ionisation energy ($kJ\ mol^{-1}$)	520.3	495.8	418.9	403.0	375.7
Oxidation state	+1	+1	+1	+1	+1
Flame colouration	Crimson red	Golden yellow	Violet	Violet	Violet
Hydration energies of M^+ cations ($kJ\ mol^{-1}$)	531.4	422.6	338.9	318.9	280.3
Size of hydrated M^+ ions, *i.e.*, size of $[M\ (aq)]^+$ ions (Å)	3.40	2.76	2.32	2.28	2.28
Ionic conductance or mobility (electrical conductivity) of $[M\ (aq)]^+$ ions ($ohm^{-1}\ cm^2\ equiv^{-1}$)	33.5	43.5	64.6	67.5	68.0
Standard reduction potential for $M^+ (aq) + e^- \rightleftharpoons M\ (s)$ (volts)	–3.05	–2.71	–2.93	–2.93	–2.93
Electronegativity	1.0	0.9	0.8	0.8	0.7
Abundance in earth's crust (ppm)	6	28,300	25,900	310	7
Electron affinity ($kJ\ mol^{-1}$)	59.8	52.7	48.36	46.89	45.50
Electropositive character	Increases →				

1. Electronic configuration. See chapter 4.

2. Physical state. With the exception of Li, the alkali metals are *extremely soft* and are *readily fused.* They are highly *malleable* (*i.e.*, can be pressed out into sheets) and *ductile* (*i.e.*, can be drawn into wires). When freshly cut, they have a bright lustre which is quickly tarnished as soon as metal comes in contact with atmosphere.

3. Atomic volume, atomic and ionic radii. See chapter 5.

4. Density. See chapter 5.

5. Melting and boiling points. See chapter 5.

6. Heats of atomisation. Alkali metals have *low heats* of atomisation which are indicative of the fact that there are weak interatomic bonds in alkali metals.

7. Ionisation energy. See chapter 5.

8. Tendency of the elements to form M^+ and M^{2+} ions. See "*Ionisation energy*" discussed in chapter 5.

9. Tendency of the elements to form ionic / covalent compounds. Due to the low value of ionisation potential, alkali metals can readily transfer their ns^1 electron to the highly electronegative elements and can thus form ionic compounds. Li^+ ion, due to its smallest size and hence high electron density round its nucleus, can polarise the anions most strongly and hence gives covalent compounds (*e.g.*, LiCl, LiBr, LiI, $LiNO_3$ etc.). The covalent character of these compounds is also supported by the fact that these compounds are soluble in organic solvents, while the corresponding compounds of other alkali metals are insoluble. Due to the larger size of M^+ cations of other alkali metals, these cations give largely ionic compounds.

Alkali metals also form covalent diatomic molecules of A_2 type (A = Li, Na etc.) in which the two atoms are linked together by a covalent bond. Li also forms organo metallic compounds which are covalent in nature.

10. Electropositive (metallic) character. See chapter 5.

11. Oxidation states of elements in ionic compounds. See chapter 5.

12. Alkali metals impart characteristic colour to the flame. When alkali metals or their compounds are put in Bunsen flame, they give characteristic colours. The reason is that the atoms of alkali metals absorb energy from the flame and the outermost electron gets promoted to higher energy levels. When these electrons, which are present in higher energy levels, come back to lower energy levels, the energy which they had absorbed earlier is given out by them in the form of light in the visible region of the spectrum. The amount of energy released is in the order : Li < Na < K < Rb < Cs. As a result, the frequency of light emitted is also in the same order. Consequently the colours exhibited by the alkali metals are Li-*crimson red* ; Na-*golden yellow* ; K, Rb and Cs-*violet.*

13. Electrical conductivity of metals. The individual atoms in the crystal structure of the metals (*i.e.*, metal structure) are held together by the valence-electrons. In alkali metals, this binding is relatively weak, since the valence-electrons are loosely held. These loosely held electrons are, therefore, free to move throughout the metal structure with the result that the alkali metals have high conductivity, *i.e.*, these are good conductors of electricity and heat.

14. Hydration energies of M^+ cations : Size and electrical conductivity of hydrated alkali metal cations, $[M\ (aq)]^+$. See chapter 16 (*Structure of solids*).

15. Alkali metals do not occur free in nature. Alkali metals, being very reactive, react with O_2, moisture and CO_2 of the atmosphere to form NaOH and Na_2CO_3. It is for this reason that these metals do not occur free in nature.

16. Alkali metals cannot be obtained by the electrolysis of the aqueous solution of their salts. As an example let us consider the extraction of Na metal by the electrolysis of an aqueous solution of NaCl. The aqueous solution of NaCl contains Na^+, Cl^-, H^+ and OH^- ions (NaCl $\rightleftharpoons$ $Na^+ + Cl^-$, $H_2O \rightleftharpoons H^+ + OH^-$). On electrolysing this solution, the cations namely Na^+ and H^+ move towards cathode.

Now since H^+ ions lose their charge much more readily than Na^+ ions. H^+ ions are discharged in preference to Na^+ ions to form H_2 gas at cathode, while OH^- ions remain in solution. Thus, instead of getting Na at the cathode, we get H_2 at this electrode. Cl^- and OH^- ions move towards anode at which Cl^- ions are discharged in preference to OH^- ions to form Cl_2 gas while OH^- ions remain in solution. Thus Cl_2 gas is collected at anode. Na^+ and OH^- ions which are present in solution combine together to form NaOH. Thus we see that instead of getting Na metal, we get H_2 (at cathode), Cl_2 (at anode) and NaOH (in solution).

17. Reducing property of alkali metals in aqueous solution. See chapter 18. (*Electrode Potentials and Electrochemical series*).

18. Lattice energies of alkali metal salt. See "*Factors affecting the magnitude of lattice energy of ionic solids*" discussed in chapter 16. (*Structure of Solids*).

Gradation in chemical properties

Some of the important chemical properties of alkali metals and gradation in them are discussed below:

1. Reactivity. We have seen that the first ionisation energies and electronegativities of alkali metals are the lowest of all the elements and their heats of atomisation are relatively low. These properties make the alkali metals highly reactive especially towards highly electronegative elements like Cl_2 and O_2 with which they give *ionic compounds. The reactivity of these metals increases with the increase of their atomic number,* since their is a steady decrease in the ionisation energies, *e.g.,* Li reacts *slowly* with H_2 ; Na reacts *vigorously* ; K, Rb and Cs react with *increasing violence.*

2. Formation of oxides : Reaction with O_2 or air. See chapter 28.

3. Formation of hydroxides. *Preparation.* Hydroxides of alkali metals can be prepared by treating the normal oxides or the metals with H_2O.

$$Na_2O + H_2O \longrightarrow 2NaOH \; ; \quad 2Li + 2H_2O \longrightarrow 2LiOH + H_2$$

Properties. Alkali metal hydroxides are crystalline solids. These are ionic compounds and are highly soluble in water and alcohol. Since the alkali metals are highly electropositive, their hydroxides are the strongest bases known. *Since the electropositive character* of the metals increases from Li to *Cs, the basic character of their hydroxides increases from LiOH to CsOH* . All the hydroxides are stable towards heat and *their thermal stability increases from* LiOH *to* CsOH, *e.g.,* LiOH decomposes at red heat, forming Li_2O while the other hydroxides do not decompose even at red heat. They melt, without decomposing and are good conductors of electricity in the fused state. At 400°C, the hydroxides consist of dimers, $(MOH)_2$. With the increase in the size of the cation, the internuclear distance between the oxygen of OH^- ion and the metal cation also increases and hence the ionisation of the hydroxides increases from LiOH to CsOH.

4. Formation of hydrides : Action of H_2. See chapter 25.

5. Formation of halides (MX) : Action of halogens. All the alkali metals combine directly with halogens and form halides of MX type. For the preparation and properties of these halides (See chapter 29).

6. Action of heat on carbonates, bicarbonates and nitrates. The carbonates (M_2CO_3) and bicarbonates ($MHCO_3$) are highlty stable to heat. *With increase of electropositive character from Li to Cs, the stability of these salts increases.* Their nitrates decompose on strong heating to the corresponding nitrite and O_2 (Exception is $LiNO_3$).

$$2NaNO_3 \longrightarrow 2NaNO_2 + O_2\uparrow$$

7. Solubility of alkali metals in mercury. Alkali metals dissolve readily in mercury forming *amalgams.* The process of the formation of amalgams is highly exothermic.

8. Formation of nitrides. See chapter 27.

Anomalous properties of lithium : Dissimilarities with other alkali metals

From the general properties of alkali metals discussed above it may be seen that Li differs in many properties from the other alkali metals. The main reasons due to which it shows anomalous properties are : (*a*) The size of Li atom and Li^+ ion is the smallest of all the alkali metal ions. (*b*) The polarising power of Li^+ ion is the greatest of all the alkali metal ions. (*c*) Li has the highest ionisation energy and electronegativity as compared to other alkali metals.

Points of differences. Following are the main points of differences between Li and other alkali metals:

1. The greatest polarising power of Li^+ ion results in its tendency towards *solvation* and *covalent bond formation.* Thus the salts of Li are covalent and are, therefore, more soluble in non-polar solvents (*e.g.,* organic solvents). However, the salts of other alkali metals are insoluble in organic solvents.

2. Li is the only alkali metal which reacts with nitrogen and silicon to form ionic lithium nitride (Li_3N) and lithium silicide (Li_6Si_2). Li_6Si_2 is a dark violet and hygroscopic compound and is formed when Li and Si are heated together.

3. Li is much *harder* and *lighter* than the other alkali metals.

4. Its melting and boiling points are rather high and it can be melted in dry air without losing its brilliancy.

5. Due to its high ionisation energy, Li *does not react* with O_2 below 0°C ; reacts *slowly* with H_2O to liberate H_2 or with liquid Br_2.

6. Li is the *least reactive* metal of all the alkali metals.

7. Li^+ ion shows a tendency to form complexes with NH_3, H_2O (to form hydrates) and forms a large number of organolithium compounds.

8. When burnt in air, Li gives only the *monoxide,* Li_2O (O^{2-}) while other alkali metals from *peroxide* (O_2^{2-}) and *super oxides* (O_2^-).

9. When acetylene gas is passed over heated Li metal, it does not form lithium acetylide, while other alkali metals do form metal acetylides.

10. Li when heated in NH_3 forms imide, Li_2NH while other alkali metals form amides, MNH_2.

11. Li is the only alkali metal whose salts (*e.g.*, LiCl) may undergo hydrolysis.

12. Due to the *low electropositive character* of Li, its various salts (*e.g.*, Li_2CO_3, $LiNO_3$) are *less stable* and, therefore, *decompose* to give oxide, *e.g.*,

$$2\,LiOH \xrightarrow{\text{Read Heat}} Li_2O + H_2O \;; \quad 4\,LiNO_3 \longrightarrow 2\,Li_2O + 4\,NO_2 + O_2$$

$$Li_2CO_3 \longrightarrow Li_2O + CO_2$$

13. Some of the salts of Li like Li_2CO_3, Li_3PO_4 and LiF are *insoluble* in H_2O while the corresponding salts of other alkali metals are *soluble.* The solubility of the salts of other alkali metals is due to the weak electrostatic force of attraction.

14. Li_2O dissolves in H_2O quietly while other oxides do so more energetically.

15. LiOH is considerably *less soluble* in H_2O and a *much weaker base* than NaOH or KOH.

LiOH decomposes at red heat forming Li_2O, whereas other alkali metal hydroxides sublime unchanged as $(MOH)_2$.

16. LiCl is deliquescent and is soluble in alcohol as well as in pyridine. It forms the hydrate, $LiCl.2H_2O$. Chlorides of other alkali metals do not show this behaviour.

17. $LiCO_3$ decomposes on heating to from Li_2O and CO_2 while other carbonates do not decompose. The small size of Li^+ ion makes Li_2O lattice *more stable* than Li_2CO_3 lattice. However, the large size of other alkali metal ions make M_2O lattices *less stable* than M_2CO_3 lattice and hence they decompose.

18. $LiNO_3$, on heating, forms Li_2O, NO_2 and O_2 while other alkali metal nitrates form nitrites and O_2.

$$4LINO_3 \longrightarrow 2Li_2O + 4NO_2 + O_2 \;; \quad NaNO_3 \longrightarrow 2NaNO_2 + O_2$$

19. Lithium hydrosulphide, LiSH is thermally unstable while the hydrosulphides of other alkali metals are so stable that they can be fused without extensive deomposition.

20. Lithium per chlorate, $LiClO_4$ is remarkably soluble in alcohol, acetone and ethyl acetone, while the per chlorates of other alkali metals are almost insoluble in these solvents. The solubility of $LiClO_4$ is due to strong solvation of Li^+ ion.

21. Li_2SO_4 is the only alkali metal sulphate which does not form double salts such as alums.

Diagonal relationship of Li with Mg : Similarities between Li and Mg

It has been observed that Li resembles more in its many properties with the element Mg, lying at its right in 3rd period (*i.e.,* diagonally opposite element) than with other members of its own sub-group IA. This type of similarity between Li and Mg is called diagonal relationship.

The points which show that Li resembles Mg in many respects, *i.e.*, Li and Mg have diagonal relationship have already been discussed in chapter 10.

Group IIA (2) : Be, Mg, Ca, Sr, Ba, Ra (Alkaline Earth Metals)

Group IIA (2) of the periodic table contains six elements which are *beryllium* (Be_4), *magnesium* (Mg_{12}), *calcium* (Ca_{20}), *stronsium* (Sr_{38}), *barium* (Ba_{56}) and *radium* (Ra_{88}). These elements are also called alkaline earth metals, since their oxides occur in the earth's crust and their hydroxides are strong alkalies. Although Ra has similar properties as alkaline earth metals, but, being radioactive, it is studied separately alongwith other radioactive elements.

Table 23.2 : *Some important physical properties of the elements of group IIA (2).*

Properties	Beryllium (Be_4)	Magnesium (Mg_{12})	Calcium (Ca_{20})	Stronsium (Sr_{38})	Barium (Ba_{56})
Electronic configuration	$[He]_2\ 2s^2$ or 2, 2	$[Ne]_{10}\ 3s^2$ or 2, 8, 2	$[Ar]_{18}\ 4s^2$ or 2, 8, 8, 2	$[Kr]_{36}\ 5s^2$ or 2, 8, 18, 8, 2	$[Xe]_{54}\ 6s^2$ or 2, 8, 18, 18, 8, 2
Atomic mass	9.01	24.31	40.08	87.67	137.34
Density (g cm^{-3})	1.84	1.74	1.55	2.54	3.75
Melting point (°C)	1277	650	838	768	714
Boiling point (°C)	2770	1107	1440	1380	1640
Atomic volume (cm^3)	4.90	13.97	25.91	34.54	36.72
Single bond covalent radius (Å)	0.90	1.36	1.74	1.91	1.98
Radius of M^{2+} cation (Å)	0.31	0.65	0.99	1.13	1.35
Colour of M^{2+} ions	Colourless	Colourless	Colourless	Colourless	Colourless
Magnetic property of M^{2+} ion	Diamagnetic	Diamagnetic	Diamagnetic	Diamagnetic	Diamagnetic
1st ionisation energy (kJ mol^{-1})	899.5	737.7	589.8	549.5	502.9
2nd ionisation energy (kJ mol^{-1})	1757.1	1450.7	1146.0	1046.0	965.0
Hydration energy of M^{2+} ion (kJ mol^{-1})	2494.0	1953.9	1577.0	1443.0	1352.0
Standard reduction potential (volts) for $M^{2+}(aq) + 2e^- \rightleftharpoons M(s)$	–1.70	–2.38	–2.76	–2.89	–2.90
Electronegativity	1.5	1.2	1.0	1.0	0.9
Heat of atomisation (kJ mol^{-1})	78.2	35.2	42.5	39.1	42.2
Flame colouration	None	None	Brick red	Crimson	Grassy green
Oxidation state	+2	+2	+2	+2	+2
Abundance in earth's crust (p.p.m.)	6	20,000	36,300	300	250
Electron affinity (kJ mol^{-1})	< 0	< 0	< 0	< 0	< 0
Electropositive character	Increases →				

Gradation in physical properties

Some of the important physical properties of these elements are given in Table 23.2 and have been discussed below:

1. Electronic configuration. See chapter 4.

2. Physical state. The elements of this group have a greyish whiter lustre when freshly cut and are malleable and ductile.

3. Density. See chapter 5.

4. Melting and boiling points. See chapter 5.

5. Atomic volume, atomic and ionic radii. See chapter 5.

6. Colour and magnetic properties of M^{2+} ions. Since the outer shell electronic configurations of M^{2+} ions are similar to those of inert gases (s^2 or s^2p^6 configuration), these ions have *no unpaired electrons* and consequently are *diamaganetic and colourless.*

7. Ionisation energy. See chapter 5.

8. Formation of M^+, M^{2+} and M^{3+} ions. See chapter 5.

9. Metallic or electropositive character. See chapter 5.

10. Hydration energies of M^{2+}cations, electrical conductivity of hydrated cations and formation of hydrated salts. Due to the increase in the size of M^{2+} cations on moving down the group from Be^{2+} to Ba^{2+}, the hydration energies of M^{2+} cations decrease in the same direction, *i.e.*, the hydration energies of M^{2+} cations are in the decreasing order, as shown in bracket (in KJ/mole): Be^{2+} (= 2494) > Mg^{2+} (= 1953.9) > Ca^{2+} (= 1577) > Sr^{2+} (= 1443) > Ba^{2+} (= 1352). Because of the decrease in the magnitude of hydration energies from Be^{2+} to Ba^{2+}, the ionic mobility or ionic conductance of M^{2+} cations (electrical conductivity) increases in the same direction.

On comparing the values of hydration energies of M^{2+} ions with those of M^+ ions (alkali metal ions), we find that M^{2+} ions have high values. It is because of these higher values that the compounds of M^{2+} ions are extensively hydrated while those of M^+ ions are not, *i.e.*, M^{2+} ions give hydrated salts while M^+ ions form anhydrous salts (*e.g.*, $MgCl_2.6H_2O$, NaCl; $CaCl_2.6H_2O$, KCl).

11. Standard reduction potentials and reducing property of metals in aqueous solution. See chapter 18.

12. Nature of compounds given by M^{2+} cations. We know from Fajans' rules that smaller is the size of a cation, more strongly or effectively it polarises a given anion and hence greater is the tendency of the cation to form covalent compounds. Thus, since Be^{+2} ion is the smallest of all the M^{2+} cations, it polarises a given anion to the maximum extent and hence forms covalent compounds, particularly halides. Consequently the compounds given by beryllium are covalent while those formed by other alkaline earth metals are ionic.

13. Electronegativity. These elements have small values of electronegativity. These values decrease as we move down the group from Be to Ba, since the size of the atoms of the elements increase in the same direction.

14. Flame colouration. Only Ca, Sr, Ba and Ra metals and their salts give a characteristic colour to the Bunsen flame. The colours given by them are : Ca-*brick red* ; Sr-*crimson* ; Ba-*apple-green* ; Ra-*red.* On heating these metals or their salts in a Bunsen flame, the electrons absorb energy from the flame and are excited to the higher energy levels. When these excited electrons return to their normal or ground states, the energy already absorbed by them is now released in the form of visible light, which imparts a characteristic colour to the flame. Due to the small size of Be and Mg atoms, the electrons in these atoms are held to the nucleus quite strongly and hence are not excited to the higher energy levels by the energy absorbed by them from the flame. However, these elements absorb radiation in the more energetic ultra violet range

15. Conductivity. Since two valence electrons are free to move throughout the metal structure. the alkaline earth metals are good conductors of heat and electricity.

Gradation in chemical properties

Some chemical properties and gradation in them are given below:

1. Formation of oxides. See chapter 28.

2. Formation of hydroxides. *Preparation.* The hydroxides, M $(OH)_2$ are obtained:

(*a*) by the reaction of the metals with H_2O, *e.g.*,

$$Ca + 2H_2O \longrightarrow Ca(OH)_2 + O_2$$

Be does not react with H_2O even at elevated temperature ; Mg reacts only with steam, while other metals react with cold water as well.

(*b*) by the reaction of the oxides (excepting MgO) with H_2O, *e.g.*,

$$MO + H_2O \longrightarrow M(OH)_2$$

Properties. (*i*) The heats of formation of hydroxides increase from $Be(OH)_2$ to $Ba(OH)_2$.

(*ii*) $Be(OH)_2$ is *not at all basic* ; in fact it is *amphoteric,* since it reacts with acids to form salts and with alkalies to give beryllates.

$$Be(OH)_2 + 2HCl \longrightarrow BeCl_2 + 2H_2O$$

$$Be(OH)_2 + 2NaOH \longrightarrow Na_2BeO_2 + 2H_2O$$

The hydroxides of other metals are basic in character. Their basic character increases on moving down the group. Thus $Mg(OH)_2$ is *weakly basic* while $Ba(OH)_2$ is the *strongest base.* The increase in basic character of the hydroxides on moving down the group is due to the fact that with the increase in size of M^{2+} cation both the *polarity* of M—OH bond and the *internuclear distance* between oxygen of OH^- ion and the metal atom *increase.* As a result of this, there is greater ionisation of $M(OH)_2$ and hence basic character increases.

(*iii*) Due to *high polarising power of small* Be^{2+} *ion,* $Be(OH)_2$ is *covalent* while other hydroxides are *ionic.*

(*iv*) $Be(OH)_2$ and $Mg(OH)_2$ are almost insoluble in H_2O while the hydroxides of other metals are slightly soluble. Their solubility increases on moving down the group as shown by the increasing value of the solubility products of these hydroxides. $Be(OH)_2 = 1.6 \times 10^{-26}$; $Mg(OH)_2 = 8.9 \times 10^{-12}$; $Ca(OH)_2 = 1.3 \times 10^{-4}$; $Sr(OH)_2 = 3.2 \times 10^{-4}$; $Ba(OH)_2 = 5.4 \times 10^{-3}$.

3. Formation of halides. See chapter 29.

4. Formation of hydrides. See chapter 25.

5. Formation of carbonates. (*i*) All the carbonates decompose on heating at appropriate temperature, evolving CO_2.

$$MCO_3 \xrightarrow{\Delta} CO_2 + MO$$

The stability of the carbonates of these metals *increases on moving down the group.* This is illustrated by the values of the decomposition temperatures of these carbonates given as: $BeCO_3$ = 25°C, $MgCO_3$ = 540°C, $CaCO_3$ = 900°C, $SrCO_3$ = 1290°C, $BaCO_3$ = 1360°C. The marked instability of $BeCO_3$ is due to *strong polarising effect of small* Be^{2+} *cation* on the large polarisable CO_3^{2-} ion.

(*ii*) Carbonates are insoluble in water and, therefore, occur as solid rock minerals in nature. However, they dissolve in H_2O containing CO_2 due to the formation of bicarbonates.

$$CaCO_3 \text{ (insoluble)} + CO_2 + H_2 \longrightarrow Ca(HCO_3)_2 \text{ (soluble)}$$

6. Formation of nitrides. See chapter 27.

7. Formation of carbides When the elements from Mg to Ba or their oxides are heated with carbon, carbides (MC_2) are obtained, *e.g.*,

$$CaO + 3C \longrightarrow CaC_2 + CO$$

All these carbides are *ionic in nature* and have NaCl type structure with M^{2+} replacing Na^+ and $[C \equiv C]^{2-}$ replacing Cl^-. On treatment with H_2O, they liberate *acetylene,* $CH \equiv CH$.

$$CaC_2 + 2H_2O \longrightarrow Ca(OH)_2 + CH \equiv CH$$

When Be is heated with carbon, it gives the carbide, Be_2C and not BeC_2. This carbide with H_2O gives *methane,* CH_4.

$$Be_2C + 2H_2O \longrightarrow 2BeO + CH_4$$

8. Formation of complexes. The tendency of M^{2+} ions to form the complexes *decreases with the increases of the size of* M^{2+} *ions.* Thus Be^{2+} ion, on account of its small size, forms many stable complexes such as $[BeF_3]^-$, $[BeF_4]^{2-}$, while Ba^{2+} forms *very few complexes.*

9. Solubility of sulphates and chromates. Calcium sulphate is sufficiently soluble in H_2O while stronsium and barium sulphates are almost insoluble, but calcium and stroncium chromates are soluble in CH_3COOH while barium chromate is insoluble.

Anomalous behaviour of beryllium : Dissimilarities with other alkaline earth metals

Beryllium, the first element of the group differs from rest of alkaline earth metals due to its small atomic size (0.90 Å) and comparatively high electronegativity (1.5). The main points of difference are:

1. Hardness. Beryllium is the hardest of all the elements of its group.

2. Melting and boiling points. The melting and boiling points of beryllium are the highest.

3. Formation of covalent compounds. Beryllium has a tendency to form covalent compounds. Thus, when it reacts with another element, the electronegativity difference is not so large and the bond is, therefore, covalent.

4. Reaction with water. Beryllium does not react with water even at high temperatures. Other alkaline earth metals decompose water liberating H_2 gas.

$$Mg + H_2O \longrightarrow MgO + H_2$$

5. Reaction with hydrogen. Beryllium does not react with hydrogen directly to form its hydride. Its hydride, however, has been prepared indirectly. The rest of the alkaline earth metals combine with hydrogen to form hydrides. The hydrides of Be and Mg are covalent, whereas the hydrides of other metals are ionic.

6. Reaction with alkalies. Beryllium reacts with alkalies to form hydrogen and sod. beryllate, Na_2BeO_2.

$$Be + 2NaOH \longrightarrow Na_2BeO_2 + H_2 \uparrow$$

Other alkaline earth metals do not react with alkalies.

7. Behaviour of oxides and hydroxides. The oxides and hydroxides of beryllium are amphoteric, *i.e.*, dissolve in both acids and alkalies to form salts.

$$BeO + H_2SO_4 \longrightarrow BeSO_4 + H_2O$$

The oxides and hydroxides of other alkaline earth metals are basic.

8. Behaviour of carbides. Beryllium carbide is decomposed by water to form methane (CH_4).

$$Be_2C + 4H_2O \longrightarrow 2Be(OH)_2 + CH_4$$

The carbides of other alkaline earth metals are decomposed by water to form acetylene (C_2H_2). For example:

$$Ca_2C + 2H_2O \longrightarrow Ca(OH)_2 + C_2H_2$$

9. Behaviour of nitrides. Be_3N_2 is volatile while the nitrides of other alkaline earth metals are nonvolatile.

10. No. of molecules of water of crystallisation. The salts of Be^{2+} ion cannot have more than four molecules of water of crystallisation while other alkaline earth metals have more than four molecules of water of crystallisation. This is explained as follows. In case of Be^{2+} ion there are only four orbitals (namely one 2*s* and three 2*p* orbitals) available in its outermost shell. These four orbitals can accept lone pairs of electrons denoted by O-atom on each of the water molecules as shown below.

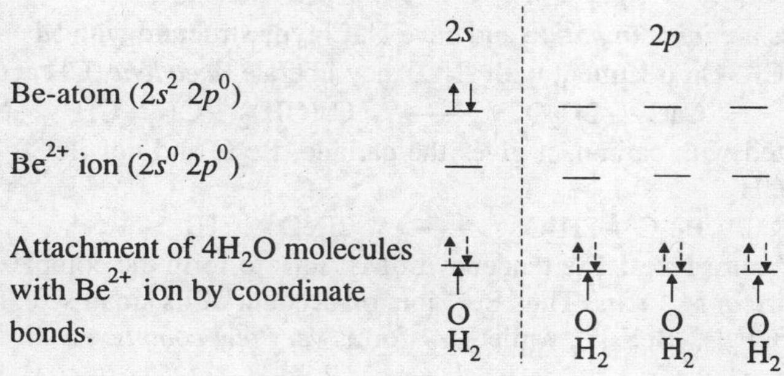

Dotted electrons indicate the electron pairs donated by $4H_2O$ molecules to the vacant orbitals on Be^{2+} ion.

On the other hand, other alkaline earth metals like Mg can extend their coordination number to six by using one 3*s*, three 3*p* and two 3*d* orbitals belonging to their outermost sheil.

11. Formation of complex compounds. Be^{2+} ion, on account of its small size, forms many stable complex compounds like $[BeF_3]^-$, $[BeF_4]^{2-}$ while M^{2+} ions derived from other alkaline earth metals form very few complex compounds.

Comparison of Be and Mg with other alkaline earth metals (Ca, Sr, and Ba)

Beryllium and magnesium, although placed in the same group, differ from rest of the alkaline earth metals in a number of properties. Their similarities and dissimilarities with each other are given below.

Similarities. (*i*) Both Be, Mg and Ca, Sr, Ba have two electrons in their outermost shells. (*ii*) Both have valency equal to two. (*iii*) Both react with halogens to form halides of the general formula, MX_2. (*iv*) Both combine directly with nitrogen to form nitrides. (*v*) Both react with dilute acids to form hydrogen.

Dissimilarities: These are given in the following table:

	Be, Mg		Ca, Sr, Ba
1.	They are hard metals, Be is the hardest.	1.	They are soft metals, Ba is the softest.
2.	They do not give any colour to the flame.	2.	They give characteristic colours to the flame.
3.	They do not react with hydrogen.	3.	They react with hydrogen to form hydrides.
4.	They are not affected by dry air.	4.	They are affected by dry air to form oxides.
5.	They do not react with water at ordinary temperature.	5.	They react with water to form hydrogen.
6.	Their hydroxides are insoluble in water and are amphoteric or weakly basic.	6.	Their hydroxides are soluble in water and are strongly alkaline.
7.	Their carbonates are not precipitated in the presence of $NH_4Cl + NH_4OH$.	7.	Their carbonates are precipitated in the presence of $NH_4Cl + NH_4OH$.
8.	Their sulphates are soluble in water.	8.	Their sulphates are insoluble in water.
9.	Their sulphides are soluble in water and are not hydrolysed.	9.	Their sulphides are soluble in water and are hydrolysed.

Diagonal relationship between Be and Al : Similarities between Be and Al

Since Be is placed diagonally opposite to Al in the periodic table, it shows diagonal relationship with Al, *i.e.*, both these elements show many similar properties. In fact the similarity between Be and Al is so striking that Be — Al pair is considered to be an ideal diagonal pair in the periodic table and hence the similarity between Be and Al is the best example of diagonal relationship.

The cause of diagonal relationship and the points showing the diagonal relationship between Be and Al have already been given in chapter 10.

Comparison of alkaline earth metals with alkali metals

(1) The metals calcium, barium and strontium have been isolated from their stable salts by similar electrolytic methods which are applied to the isolation of alkali metals. Beryllium and magnesium are also prepared by the electrolysis of their chlorides.

(2) These metals are, likewise, soft with silvery lustre and very light.

(3) They also *react with water* or *oxygen* with readiness and show much activity towards other substances.

(4) Magnesium and beryllium are comparatively *harder* and *more stable* metals, but in absorbing nitrogen, magnesium shows an analogy to lithium. They possess higher density and higher melting and boiling points.

(5) The *electropositive nature* continuously increases from beryllium to radium in the second group, as in the first group from lithium to caesium.

(6) The *oxides* of alkaline earth metals have less affinity towards water when compared to the oxides of alkali metals but they also dissolve in water to a fair extent (though much less when compared to Na_2O) with the evolution of a considerable amount of heat, and the solution obtained is alkaline. Their *hydroxides*, like alkali hydroxides, are good absorbents for carbon dioxide forming carbonates. But there is a difference between alkali *carbonates* and alkaline earth carbonates. With the exception of lithium carbonate, other alkali carbonates dissolve fairly in water and produce mild alkali solutions. But the carbonates of magnesium, calcium, strontium and barium are only sparingly soluble and come down as precipitates whenever they are formed in solution.

(7) Another contrast is in the behaviour of *bicarbonates*. Sodium and potassium bicarbonates are comparatively less soluble than their carbonates, but, on the contrary, magnesium and calcium bicarbonates are more soluble than the carbonates, and therefore, a fresh precipitate of magnesium or calcium carbonate can be redissolved by passing an excess of carbon dioxide.

(8) Alkali *sulphates* and *phosphates* are also soluble in water but calcium, strontium and barium sulphates and phosphates are much less soluble, and so is the case with their oxalates.

(9) In other salts as chlorides, acetates, nitrates, cyanides etc., the two series of salts are similar. But all these salts are not so important and not so extensively used as the alkali salts.

Magnesium acting as a bridge element between groups IIA and IIB elements

The subgroup A of the second group of periodic classification consists of the elements Be, Mg, Ca, Sr, Ba and Ra, whereas subgroup B consists of Zn, Cd and Hg. The branching of the two subgroups occurs at magnesium as shown below:

Ca — Sr — Ba — Ra ... Group IIA
Be — Mg <
Zn — Cd — Hg ... Group IIB

As we have said in the case of the first group, the metals of the subgroup B are the transitional ones of a long series, and the first metal in the second group of this transition is zinc and not magnesium, and necessary, therefore, magnesium does not belong to subgroup B. It is a member of the main group, but being at the position where the branches of two subgroups just begin, it shows similarity with calcium as well as with zinc. In fact *it resembles zinc more than it resembles calcium* (*See below*). It is also seen that the electropositive character of the elements increases from magnesium to radium whereas decreases gradually from magnesium to mercury. In other words, it gradually decreases from radium to mercury *via* magnesium. Magnesium thus links the two sub-groups and is, therefore, regarded as a **bridge element** between them.

A. Magnesium resembles calcium. Both the metals resemble in the following respects : (1) Both of them have similar electronic configuration, *i.e.*, they have two electrons in their ultimate shell and eight electrons in their penultimate shell. [Mg_{12} = 2, 8, 2 ; Ca_{20} = 2, 8, 8, 2]. The metals are, therefore, highly electropositive and bivalent. (2) Both are light silvery and white light metals which are maleable and ductile. (3) Both of them occur as carbonates and sulphates in nature. (4) Both form a number of deliquescent salts. (5) When these metals are heated in a current of nitrogen, both form nitrides (Mg_3N_2 and Ca_3N_2). (6) Both liberate H_2 when treated with very dil. HNO_3. (7) Sulphides of both the metals are soluble in water and are not precipitated by passing H_2S in their salt solutions. (8) Nitrates of both give oxide when heated. (9) The oxides of both the metals (MgO and CaO) are basic in nature and stable towards heat.

B. Magnesium resembles zinc. Both of these metals resemble in the following respects : (1) Both are maleable and ductile metals. (2) Both Mg and Zn *slowly* tarnish in air while Ca *quickly* tarnishes in air. (3) Both have the same electronic configuration, *i.e.,* they have two electrons in their outermost shell. [Mg_{12} = 2, 8, 2; Zn_{30} = 2, 8, 18, 2]. These metals are, therefore, highly electropositive and bivalent. (4) Both have *no action on cold water.* They, however, react with steam to give metal oxides and H_2. Calcium, however, *decomposes cold water.* (5) Salts of both the metals are *readily hydrolysed.* $MgCl_2 + H_2O \longrightarrow Mg(OH)Cl + HCl$;

$ZnCl_2 + H_2O \longrightarrow Zn(OH)Cl + HCl$. Calcium salts are *quite stable towards water.* (6) Their sulphates are isomorphous (*e.g.*, $MgSO_4.7H_2O$ and $ZnSO_4.7H_2O$). These sulphates are *readily soluble in water* and *decompose on strong heating* [$2MgSO_4 \longrightarrow 2MgO + 2SO_2 + O_2$; $2ZnSO_4 \longrightarrow 2ZnO + 2SO_2 + O_2$]. The sulphates also form double salts of the type $K_2SO_4.MgSO_4.6H_2O$ and $K_2SO_4.ZnSO_2.6H_2O$.

The sulphate of calcium, $CaSO_4.2H_2O$ is *insoluble in water* and gives the anhydrous sulphate, $CaSO_4$ on heating at 400°. Moreover, it does not form double salts as $MgSO_4$ and $ZnSO_4$ do. (7) Both form double phosphates, *e.g.*, $Mg(NH_4)PO_4$ and $Zn(NH_4)PO_4$ These are crystalline compounds, insoluble in water and give pyrophosphates on heating. [$2Mg(NH_4)PO_4 \longrightarrow Mg_2P_2O_7 + H_2O + 2NH_3$; $2Zn(NH_4)PO_4 \longrightarrow Zn_2P_2O_7 + H_2O + 2NH_3$]. Calcium does not form such double phosphates. (8) Their salts do not impart any characteristic colour to the flame while calcium salts give brick red colour to the flame. (9) They do not form hydrides with hydrogen while calcium unites directly with hydrogen to form a salt-like hydride, CaH_2.

Group IIIA (13) : B, Al, Ga, In, Tl

Group IIIA (13) consists of five elements namely *boron* (B_5) *aluminium* (Al_{13}), *gallium* (Ga_{31}), *indium* (In_{49}) and *thallium* (Tl_{81}). These elements are collectively called *boron group elements.*

Gradation in physical properties

Some important physical properties of these elements are given in Table 23.3 and have been discussed below:

1. Electronic configuration. See chapter 4.

2. Atomic volume, atomic and ionic radii and density. See chapter 5.

3. Melting and boiling points. See chapter 5.

4. Heats of sublimation. Heats of sublimation show a continuous *decrease.* The successive decrease in heat of sublimation indicates that the atoms of these elements are held less and less closely as we move from B to Tl.

5. Ionisation energies. See chapter 5.

6. Metallic and nonmetallic character. See chapter 5.

7. Oxidation states. See chapter 5.

8. Explanation of the relative stability of + 1 and + 3 oxidation states of the elements on the basis of inert pair effect. See "*Inert Pair Effect*" discussed in chapter 10.

9. Standard reduction electrode potential values and reducing property. See chapter 5.

10. Nature of M^{3+} compounds. The nature of the compounds of M^{3+} ions is decided by Fajans' rule, according to which *the smaller the cation, the greater is its tendency to form covalent compounds.* Thus with the *increase* of the size of M^{3+} ions from B^{3+} to Tl^3, the tendency of these ions to form covalent compounds *decreases.* Consequently the compounds of B^{3+} are predominantly covalent while Al^{3+}, Ga^{3+}, In^{3+} and Tl^{3+} give ionic compounds as well.

Gradation in chemical properties

Some chemical properties of these elements and gradation in them are given below:

1. Formation of oxides and hydroxides. All the elements of this group form sesquoxides of general formula, M_2O_3. These elements also form trihydroxides of $M(OH)_3$ type. It has been found that, on moving down the group, there is a gradual change from acidic to amphoteric and then to basic character of the oxides and hydroxides.

Explanation of the variation. The explanation for the variation from the acidic to amphoteric and then to basic character can be explained on the basis of polarising power of M^{3+} ions (See Fajans' rules discussed in chapter 10).

2. Formation of halides. See chapter 29.

3. Formation of hydrides. The elements of this group form various types of hydrides. See chapter 25.

Table 23.3 : *Some important physical properties of the elements of group IIIA (13).*

Property	Boron (B_5)	Aluminium (Al_{13})	Gallium (Ga_{31})	Indium (In_{49})	Thallium (Tl_{81})
Electronic configuration	$[He]_2\ 2s^2\ 2p^1$ or 2, 3	$[Ne]_{10}\ 3s^2\ 3p^1$ or 2, 8, 3	$[Ar]_{18}\ 4s^2\ 3d^{10}\ 4p^1$ or 2, 8, 18, 3	$[Kr]_{36}\ 5s^2\ 4d^{10}\ 5p^1$ or 2, 8, 18, 18, 3	$[Xe]_{54}\ 6s^2\ 4f^{14}\ 5d^{10}\ 6p^1$ or 2, 8, 18, 32, 18, 3
Atomic volume (cm^3)	4.62	10.0	11.8	15.1	17.2
Single bond covalent radius (Å)	0.85	1.43	1.35	1.67	1.70
Radius of M^{3+} ion (Å)	0.20	0.50	0.62	0.81	0.95
Radius of M^+ ion (Å)	—	—	1.13	1.32	1.40
Density ($g\ cm^{-3}$)	2.34	2.70	5.91	7.31	11.85
Heat of sublimation (kJ/mol)	5.648	324.3	273.2	241.4	179.1
Melting point (°C)	2030	659	28.8	156.4	303
Boiling point (°C)	2550	2450	2237	2000	1457
1st ionisation energy (kJ/mol)	800.6	577.6	578.8	558.3	589.3
Stable oxidation states	+3	+3	+3, +1	+3, +1	+3, +1
Standard reduction potential for $M^{3+}(aq) + 3e^- \rightleftharpoons M(s)$ reaction (volts)	– 0.87	– 1.68	– 0.52	– 0.34	– 0.34
Electron affinity ($kJ\ mol^{-1}$)	26.8	42.6	29.0	29.0	29.0
Atomic mass	10.81	26.98	69.72	114.82	204.37
Electronegativity	2.0	1.5	1.6	1.7	1.8
Abundance in earth's crust p.p.m.)	3	81,300	15	0.1	≈ 2
tallic character	Metalloid	←	Atomic same metallic character		→

malous properties of boron : Dissimilarities of boron with other elements of group IIIA

n many properties boron differs from the other elements of its own subgroup. The main points erence are. (1) B has a very small atomic radius. The hypothetical B^{3+} ion has very small gh charge and hence high charge density. This value is so high that B^{3+} ion does not exist. ompounds of boron are, therefore, covalent. (2) Boron has less than four valence-electrons ives it a great electron accepting power, *i.e.*, its compounds behave as strong Lewis acids forms a large number of complex compounds exhibiting its tendency to acquire a stable oron shows a maximum covalency of four while all other elements show a covalency of (4) It does not exhibit inert pair effect.

between boron and aluminium

larities. (*i*) *Electronic configuration.* Both have similar electronic configurations ectrons in their outermost shells. [B(Z = 5) = $1s^2\ 2s^2\ 2p^1$ (2, 3) ; Al(Z = 13) = $1s^2$ (2, 8, 3)].

ee Both the elements are trivalent. They do not show variable valency because their s are complete and stable. Both have generally no tendency to form B^{+3} and Al^{+3}

ions and, therefore, form tricovalent compounds. Aluminium, however, may also form electrovalent compounds with strong electron accepting atoms or groups.

(*iii*) *Oxidation state.* The common oxidation state of these elements is + 3. Boron, however, also exhibits an oxidation state of – 3 in the metal borides.

(*iv*) *Action of conc. H_2SO_4.* Both react with conc. H_2SO_4 to form SO_2.

$$2B + 3H_2SO_4 \longrightarrow 2H_3BO_3 + 3SO_2 \uparrow$$
$$2Al + 6H_2SO_4 \longrightarrow Al_2(SO_4)_3 + 3SO_2 \uparrow$$

(*v*) *Action of alkalies.* They react with alkalies to form H_2.

$$2B + 6NaOH \longrightarrow 2Na_3BO_3 + 3H_2 \uparrow$$
$$2Al + 2NaOH + 2H_2O \longrightarrow 2NaAlO_2 + 3H_2 \uparrow$$

(*vi*) *Formation of oxides.* Both form oxides when heated with oxygen at high temperature.

$$4B + 3O_2 \xrightarrow{700^\circ C} 2B_2O_3 \ ; \quad 4Al + 4O_2 \xrightarrow{800^\circ C} 2Al_2O_3$$

These oxides are amphoteric and dissolve in acids as well as in alkalies to form salts.

$$B_2O_3 + 6HCl \longrightarrow 2BCl_3 + 3H_2O$$
$$B_2O_3 + 6NaOH \longrightarrow 2Na_3BO_3 + 3H_2O$$
$$Al_2O_3 + 6HCl \longrightarrow 2AlCl_3 + 3H_2O$$
$$Al_2O_3 + 2NaOH \longrightarrow 2NaAlO_2 + H_2O$$

(*viii*) *Formation of nitrides.* Both, when heated with nitrogen or ammonia, form nitrides.

$$2B + N_2 \longrightarrow 2BN; \quad 2Al + N_2 \longrightarrow 2AlN$$
$$2B + 2NH_3 \longrightarrow 2BN + 3H_2 \ ; \quad 2Al + 2NH_3 \longrightarrow 2AlN + 3H_2$$

These nitrides are decomposed by steam to form ammonia.

$$BN + 3H_2O \longrightarrow H_3BO_3 + NH_3 \ ; \quad AlN + 3H_2O \longrightarrow Al(OH)_3 + NH_3$$

(*viii*) *Formation of sulphides.* Both from sulphides on heating with sulphur at high temperatures. These are hydrolysed by water.

$$B_2S_3 + 6H_2O \longrightarrow 2H_3BO_3 + 3H_2S \ ; \quad Al_2S_3 + 6H_2O \longrightarrow 2Al(OH)_3 + 3H_2S$$

(*ix*) *Formation of chlorides.* They form trichlorides by direct combination with Cl_2 or by passing Cl_2 over heated mixture of their oxides and charcoal.

$$2B + 3Cl_2 \longrightarrow 2BCl_3; \quad B_2O_3 + 3C + 3Cl_2 \longrightarrow 2BCl_3 + 3CO$$
$$2Al + 3Cl_2 \longrightarrow 2AlCl_3; \quad Al_2O_3 + 3C + 3Cl_2 \longrightarrow 2AlCl_3 + 3CO$$

These trichlorides are covalent and are hydrolysed by water.

$$BCl_3 + 3H_2O \longrightarrow H_3BO_3 + 3HCl \ ; \quad AlCl_3 + 3H_2O \longrightarrow Al\,(OH)_3 + 3HCl$$

(*x*) *Formation of alkyl compounds.* Both form similar types of alkyl compounds.

$$2BCl_3 + 3Zn\,(CH_3)_2 \longrightarrow 2B\,(CH_3)_3 + 3ZnCl_2 \ ;$$
$$2Al + 3Hg\,(CH_3)_2 \longrightarrow 2Al\,(CH_3)_3 + 3Hg$$

(B) Dissimilarities. Although B and Al both belong to the same group of the periodic table, yet they show many dissimilarities. The difference in their porperties is due to the big difference in their size and ionisation potential. The points showing dissimilarities between boron and aluminium are given in the table.

Diagonal relationship between B and Si : Similarities between B and Si

Boron resembles silicon in many of its properties with which it is diagonally related. Some common properties showing the similarities between B and Si have already been given in chapter 10.

	Boron	Aluminium
1.	There are two electrons in the penultimate shell (2, 3).	There are eight electrons in the penultimate shell (2, 8, 3).
2.	It is a non-metal and is a bad conductor of heat and electricity.	It is a metal and is a good conductor of heat and electricity.
3.	It has high melting point (2300°C).	It has low melting point (660°C).
4.	It exhibits allotropy and is found in two forms *viz.*, crystalline and amorphous.	It does not exhibit allotropy.
5.	Its crystalline form is very hard.	It is sufficiently soft.
6.	It forms covalent compounds only.	It forms both covalent and electrovalent compounds.
7.	It reacts with hydrogen to form a number of hydrides like B_2H_6, B_5H_9 etc.	It does not form any stable hydride.
8.	It does not decompose water or steam.	It decomposes steam to form hydrogen. $2Al + 6H_2O \rightarrow 2Al(OH)_3 + 3H_2$
9.	It does not react with dil. acids.	It reacts with dil. acids to form hydrogen. $2Al + 3H_2SO_4 \rightarrow Al_2(SO_4)_3 + 3H_2 \uparrow$
10.	It is oxidised by conc. HNO_3 to form boric acid. $B + 3HNO_3 \rightarrow H_3BO_3 + 2NO_2$	It becomes passive with conc. HNO_3.
11.	It dissolves in fused alkalies to form hydrogen. $2B + 6NaOH \rightarrow 2Na_3BO_3 + 3H_2$	It dissolves in hot alkalies to form hydrogen. $2Al + 2NaOH + 2H_2O \rightarrow 2NaAlO_2 + 3H_2$
12.	Borates are very stable.	Aluminates are less stable.
13.	$B(OH)_3$, *i.e.*, H_3BO_3 is weakly acidic.	$Al(OH)_3$ is amphoteric.
14.	BCl_3 is a fuming liquid.	$AlCl_3$ is a solid.
15.	It combines with metals to form borides. $3Mg + 2B \rightarrow Mg_3B_2$	It combines with metals to form alloys.
16.	Its maximum covalency is 4.	Its maximum covalency is 6.

Comparison between boron and silicon

(A) Similarities. For this see diagonal relationship between boron and silicon discussed in chapter 10.

(B) Dissimilarities. The following points show the dissimilarities between boron and silicon.

	Boron	Silicon
1.	Boron has three electrons in its valence-shell (2, 3).	Silicon has four electrons in its valance shell (2, 8, 4).
2.	Boron in the form of minerals constitutes only 0.001% of the earth's crust.	Silicon is the second most abundant element and constitutes 26% of the earth's crust.
3.	Boron is tricovalent in its compounds.	Silicon is tetracovalent in its compounds.
4.	Boron on heating with N_2 gives BN. $2B + N_2 \rightarrow 2BN$	Silicon does not combine with N_2.
5.	Boron is oxidised by hot conc. H_2SO_4 or conc. HNO_3 to form H_3BO_3. $2B + 3H_2SO_4 \rightarrow 2H_3BO_3 + 3SO_2$; $B + 3HNO_3 \rightarrow H_3BO_3 + 3NO_2$	Silicon is attached only by HF, forming hydrofluosilicic acid, H_2SiF_6. $Si + 6HF \rightarrow H_2SiF_6 + 2H_2$
6.	B_2O_3 is amphoteric in nature.	SiO_2 is acidic in nature.

Position of boron in the periodic table

We know that boron is present in group IIIA of the periodic table. The element lying below it is aluminium and the element which is diagonally related to boron is silicon which is present in group IVA (period number = 3) as shown in the margin.

III A	IV A
B ↘	C
Al	→ Si
Ga	Ge

Position of boron in the periodic table

Thus the position of boron in the periodic table is justified by discussing (1) *Similarities between B and Al,* and (2) *Similarities between B and Si.* These points have already been discussed.

Group IVA(14) : C, Si, Ge, Sn, Pb

Group IVA (14) of the periodic table contains five elements which are *carbon* (C_6), *silicon* (Si_{14}), *germanium* (Ge_{32}), *tin* (Sn_{50}) and *lead* (Pb_{82}). Carbon is an essential constituent of all organic matter while silicon is the main constituent of inorganic matter.

Some important physical properties of these elements are given in Table 23.4 and have been discussed below:

Table 23.4 : *Some important physical properties of the elements of group IVA (14).*

Property	Carbon (C_6)	Silicon (Si_{14})	Germanium (Ge_{32})	Tin (Sn_{50})	Lead (Pb_{82})
Electronic configuration	$[He]_2\ 2s^2\ 2p^2$ or 2, 4	$[Ne]_{10}\ 3s^2\ 3p^2$ or 2,8, 4	$[Ar]_{18}\ 4s^2\ 3d^{10}\ 4p^2$ or 2, 8, 18, 4	$[Kr]_{36}\ 5s^2\ 4d^{10}\ 5p^2$ or 2, 8, 18, 18, 4	$[Xe]_{54}\ 6s^2\ 4f^{14}\ 5d^{10}\ 6p^2$ or 2, 8, 18, 32, 18, 4
Atomic weight	12.000	28.086	72.590	118.890	207.190
Occurrence	Free state and combined state	Found in rocks as silicon or silicate	Rare element	Found as tinstone (SnO_2)	Found as galena (PbS)
Allotropic forms	Crystalline (*e.g.*, diamond and graphite) as well as amorphous forms	Crystalline as well as amorphous forms	Two crystalline forms	White tin, grey tin and rhombic tin	No allotropic forms
Atomic volume (cm^3)	3.4	11.4	13.6	16.3	18.27
Density	2.26	2.49	5.32	7.30	11.40
Atomic and ionic radii (Å)	0.77 0.15 (M^{4+}) 2.60 (M^{4-})	1.11 0.41 (M^{4+}) 2.71 (M^-)	1.22 0.53 (M^{4+}) 0.93 (M^{2+})	1.41 0.71 (M^{4+}) 1.12 (M^{2+})	1.47 0.84 (M^{4+}) 1.20 (M^{2+})
Melting point (°C)	3727	1410	937.5	231.8	327.4
Boiling point (°C)	4830	2680	2830.0	2270.0	1725.0
Ionisation energy (kJ/mol)	1086.4	786.5	762.2	708.6	715.5
Eelctronegativity	2.5	1.8	2.0	1.9	2.3
Metallic/nonmetallic character	Nonmetal	Metalloid	Metalloid	Metal	Metal
Stable oxidation states	+4	+4	+2, +4	+2, +4	+2, +4
M—M bond energy (kJ/mol)	348.0	222.0	167.0	155.0	—
Electron affinity (kJ mol^{-1})	122.3	133.6	116.0	121.0	35.2
Standard reduction electrode potential for $M^{2+} + 2e^- \rightarrow M$ reaction (V)	—	—	—	– 0.14	– 0.13

Gradation in physical properties

1. Electronic configuration : See chapter 4.

2. Occurrence. Carbon is the only element of this group which occurs in the *free state as diamond, graphite* and *coal.* In the combined state it occurs as *hydrocarbons, carbohydrates* and *carbon dioxide* in the atmosphere. Silicon is found as a major constituent in rocks, either in the form of *silica* or *silicates.* Germanium is a *rare element.* Tin is mainly found in the form of *tinstone,* SnO_2 while lead occurs as mineral *galena,* PbS.

3. Allotropy. Excepting Pb, all other elements of this group show *allotropy, i.e.,* these elements exist in different allotropic forms. For example : (*i*) Carbon exists in the following *two* allotropic forms. (*a*) Crystalline forms : diamond and graphite. (*b*) Amorphous forms which are microcrystalline forms, *e.g.,* coal, coke, charcoal etc. (*ii*) Silicon exists in two forms. (*a*) Crystalline form. (*b*) Amorphous form. It is the common variety of silicon. (*iii*) Germanium exists in two crystalline forms. (*iv*) Tin exists in three allotropic forms. (*a*) White tin (*b*) Grey tin (*c*) Rhombic tin.

4. Atomic volume, density and atomic and ionic radii. See chapter 5.

5. Melting and boiling points. See chapter 5.

6. Ionisation energies. See chapter 5.

7. Electronegativity. See chapter 5.

8. Metallic and nonmetallic character. See chapter 5.

9. Explanation of the relative stability of + 2 and + 4 oxidation states of the elements on the bases of inert pair effect. See "*Inert Pair Effect*" discussed in chapter 10.

10. Oxidation states. (*a*) *Positive oxidation states.* See chapter 5 and "*Inert Pair Effect*" (chapter 10).

(*b*) *Negative oxidation states.* See chapter 5.

11. Covalency of the elements. See "*Rule of Maximum Covalency*" discussed in chapter 10.

12. Tendency of the elements to form homo chains (M—M chains) : Catenation or self-linkage. All the elements of this group have a tendency to link with each other and thus form long chains of identical atoms (M—M chains). Such long chains of identical atoms are called *homo chains.* This type of linking of identical atoms with each other is called *catenation* or *self-linkage.*

The tendency of an element M towards catenation is related with the magnitude of M—M bond energy. Greater is the magnitude of M—M bond energy, greater is the instability associated with M—M bond and hence greater will be the tendency of M atom to show the property of catenation. Thus, since the values of M—M bond energies decrease in the order : C—C (= 348 KJ/mole) > Si—Si (= 222 KJ/mole) > Ge—Ge (= 167 KJ/mole) > Sn—Sn (= 155 KJ/mole), C—C bond is the most unstable and hence C-atom shows maximum tendency for catenation. Carbon atoms, by uniting with other C-atoms, can form long C—C chains of any length. These chains may be straight chains, branched chains and cyclic chains. C—C bonds are highly stable in the compounds containing C—C long chains. The property of catenation shown by other elements of this group decreases as the values of M—M bond energies decrease on moving down the group from carbon to tin. Thus it is due to the maximum property of catenation of carbon atom that it is easy to prepare compounds containing C—C chains upto twenty carbon atoms, while for Si and Ge it is not possible to extend Si—Si or Ge—Ge chains beyond six units. For Sn and Pb it is not possible to extend the Sn—Sn or Pb—Pb chain even beyond two units.

13. Nature of $SnCl_2$-$SnCl_4$ and $PbCl_2$-$PbCl_4$ halides. See "*Fajans' Rules*" discussed in chapter 10.

Gradation in chemical properties

Some chemical properties of these elements and gradation in them are given below:

1. Formation of oxides. See chapter 28.

2. Formation of hydrides. See chapter 25.

3. Formation of tetrahalides, MX_4. For the general methods of preparation and general properties (see chapter 29).

4. Formation of oxy acids. Carbon forms carbonic acid (H_2CO_3). Silicon gives orthosilicic acid (H_4SiO_4) and metasilicic acid (H_2SiO_3). H_4SiO_4 is obtained as a gelatinous precipitate when SiF_4 or $SiCl_4$ is passed into H_2O.

$$3SiF_4 + 4H_2O \longrightarrow H_2SiO_4 + 2H_2SiF_6$$

H_2SiO_3 is obtained as a gelatinous ppt. by adding HCl to a conc. solution of an alkali silicate.

$$Na_2SiO_3 + 2HCl \longrightarrow 2NaCl + H_2SiO_3$$

Germanium forms germanic acid, H_2GeO_3.

Tin gives α-stannic acid, H_2SnO_3 and β-stannic acid (also called meta stannic acid), $H_2Sn_5O_{11}.4H_2O$ or $(H_2SnO_3)_5$. H_2SnO_3 is obtained as a white gelatinous precipitate when $SnCl_4$ solution is treated with NH_4OH and the precipitate of $Sn(OH)_4$ is dried at 100°C.

$$SnCl_4 + 4NH_4OH \longrightarrow Sn(OH)_4 + 4NH_4Cl$$

$$Sn(OH)_4 \longrightarrow H_2SnO_3 + H_2O$$

H_2SnO_3 can also be obtained by the hydrolysis of Na_2SnO_3 with dil. HCl

$$Na_2SnO_3 + 2HCl \longrightarrow H_2SnO_3 + 2NaCl$$

β-stannic acid is obtained (*a*) when α-stannic acid is allowed to stand at ordinary temperature or when α-stannic acid is boiled with water (*b*) when fairly conc. HNO_3 is treated with Sn metal.

Both α- and β-stannic acids are soluble in alkali hydroxide and carbonate solutions. When β-stannic acid is boiled for a long time with conc. HCl, it gradually changes into α-stannic acid. Lead forms metaplumbic acid, H_2PbO_2. H_2CO_3, H_2SiO_3, H_2GeO_3 and H_2SnO_3 can be represented by the structure $O = M\langle^{OH}_{OH}$. Thus these acids can also be formulated as $MO(OH)_2$.

5. Action of caustic alkalies. Except C, all are attacked by caustic alkalies with the evolution of H_2, *e.g.*,

$$Si + 2NaOH + H_2O \longrightarrow Na_2SiO_3 + 2H_2\uparrow$$

$$Ge + 2NaOH + H_2O \longrightarrow K_2GeO_3 + 2H_2\uparrow$$

$$Sn + 2NaOH + H_2O \longrightarrow Na_2SnO_3 + 2H_2\uparrow$$

$$Pb + 2NaOH \longrightarrow Na_2PbO_2 + H_2\uparrow$$

Differences of carbon and silicon from other elements of group IVA

Carbon and silicon differ from the remaining members of group IVA in the following respects:

1. Electronic configuration. Both carbon and silicon have respectively 2 and 8 electrons (configuration of inert gas) in their penultimate shell while the other elements of this group possess 18 electrons in their penultimate shell.

2. Metallic and nonmetallic character. Both carbon and silicon are nonmetals while Ge is a metalloid and Sn and Pb are metals.

3. Valency. Both C and Si show tetravalency and do not give tetravalent or bivalent ions (M^{4+} or M^{2+} ions). On the other hand Ge, Sn and Pb show tetracovalency as well as form bivalent ions *viz.*, Ge^{2+}, Sn^{2+} and Pb^{2+}.

4. Oxidation states. Both C and Si show an oxidation state of +4 only while Ge, Sn and Pb show two types of oxidation states which are +2 and +4.

5. Catenation. Carbon and, to some extent, silicon have a tendency for catenation while other elements do not have this property to a great extent.

6. Nature of dioxide. CO_2 and SiO_2 are acidic while other dioxides (GeO_2, SnO_2 and PbO_2) are amphoteric.

7. Formation of acids. Carbon and silicon form the acids of the type, HOOC—COOH (oxalic acid), HOOSi—SiOOH (silicon oxalic acid) while other elements do not give the acids of this type.

Comparison between carbon and silicon

(A) Similarities. Both these elements belong to the same subgroup (group IVA) of the periodic table and have the same number of electrons (= 4) in their valence shell. It is for this reason that these elements have many similarities as discussed below:

(*i*) *Electronic configuration.* Both the elements have 4 electrons in their valence-shell [C_4 = 2, 4,; Si_{14} = 2, 8, 4].

(*ii*) *Occurrence.* Both are widely distributed in nature. Carbon is an essential constituent of vegetable and animal kingdom, *i.e.*, of the organic compounds. Silicon is an essential constituent of mineral kingdom, *i.e.*, of the inorganic compounds.

(*iii*) *Valency.* Both are tetravalent. These elements show a covalency of four and form compounds by sharing of electrons.

(*iv*) *Allotropy.* Both show allotropy, *i.e.*, they exist in more than one form. Thus, carbon occurs in two crystalline forms namely diamond and graphite and a number of amorphous forms such as coal, coke, charcoal etc. Silicon exists in two forms *viz.*, crystalline and amorphous.

(*v*) *Formation of oxides.* Both burn in oxygen to form acidic oxides.

$$CO_2 + 2NaOH \longrightarrow Na_2CO_3 + H_2O \quad ; \quad SiO_2 + 2NaOH \longrightarrow Na_2SiO_3 + H_2O$$

(*vi*) *Formation of hydrides.* Both form a number of covalent hydrides.

CH_4	C_2H_6	C_3H_6	C_4H_{10}
SiH_4	Si_2H_6	Si_3H_8	Si_4H_{10}

These are all colourless gases and burn in air to form oxides.

$$CH_4 + 2O_2 \longrightarrow CO_2 + 2H_2O \; ; \quad SiH_4 + 2O_2 \longrightarrow SiO_2 + 2H_2O$$

Both CH_4 and SiH_4 can be prepared by similar methods, *i.e.*, by the action of water or acids on carbides and silicides, respectively.

$$Be_2C + 4HCl \longrightarrow 2BCl_2 + CH_4 ; \quad Mg_2Si + 4HCl \longrightarrow 2MgCl_2 + SiH_4$$

(*vii*) *Formation of halides.* Both form similar types of covalent halides.

CF_4	CCl_4	CBr_4	$CHCl_3$
SiF_4	$SiCl_4$	$SiBr_4$	$SiHCl_3$

	Carbon	Silicon
1.	It has two electrons in the penultimate shell.	It has eight electrons in the penultimate shell.
2.	It is a black solid, melting point 3500°C.	It is a reddish brown solid, melting point 1410°.
3.	Some varieties of carbon, *i.e.*, graphite are good conductors of heat and electricity.	Bad conductor of heat and electricity.
4.	It has a tendency to form long chains of identical atoms.	It has a lesser tendency to form such chains.
5.	No action with alkalies.	It reacts with fused alkalies to form H_2. $Si + 2NaOH + H_2O \rightarrow Na_2SiO_3 + 2H_2$
6.	CO_2 is a gas. It has a linear structure. The carbon atom is attached to two oxygen atoms by double covalent bond. O = C = O	SiO_2 is a solid. It has a giant structure. The silicon atom is attached with oxygen atoms by single covalent bonds. 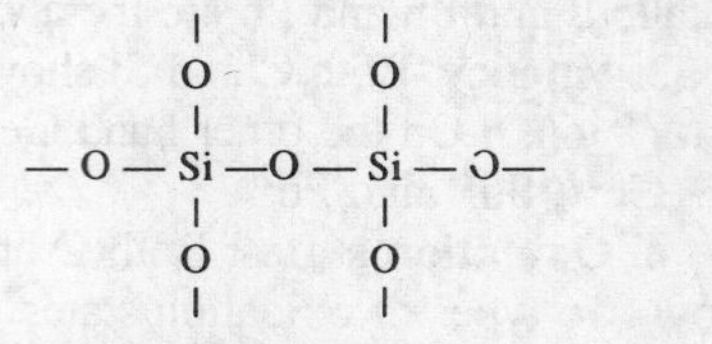
7.	The lower oxide (CO) is known.	The lower oxide SiO is unknown.
8.	It forms a large number of stable hydrides known as hydrocarbons.	It forms lesser number of hydrides which are unstable.
9.	It does not react with Cl_2 or Br_2 directly.	It reacts with Cl_2 or Br_2 at 500°C. $Si + 2Cl_2 \xrightarrow{500^\circ C} SiCl_4$
10.	CCl_4 is quite stable and is not hydrolysed by water.	$SiCl_4$ is less stable and hence is hydrolysed by water. $SiCl_4 + 3H_2O \rightarrow H_2SiO_3 + 4HCl$
11.	The oxy acid H_2CO_3 is unstable.	The oxy acid H_2SiO_3 is stable.
12.	It does not form complex compounds.	It forms a number of complex compounds because of availability of *d*-orbitals in the valency shell.

(*viii*) *Formation of acids,* Both form similar types of acids, *e.g.*, HOOC—COOH (*oxalic acid*) and HOOSi—SiOOH (*silicon oxalic acid*).

(B) Dissimilarities. In some properties these elements differ from each other. The dissimilarity is due to the difference in the number of electrons in the penultimate shell (carbon has 2 electrons while silicon has 8 electrons). The points showing dissimilarities between these elements are given in the table.

Comparison between silicon and tin

(A) Similarities. The following similarities are important:

(*i*) *Allotropy.* Both tin and silicon show allotropy.

(*ii*) *Oxides.* Oxides of both (SiO_2 and SnO_2) are acidic.

(*iii*) *Hydrides.* Both form hydrides which are gases. Tin gives a hydride SnH_4 by the action of HCl on an alloy of tin and magnesium. The hydride of silicon, SiH_4 is, however, more stable.

(*iv*) *Halides.* Both give similar halides. Their chlorides are colourless, volatile liquids which are easily hydrolysed. The fluorides of both, SiF_4 and SnF_4 are known to form similar silicofluorides, K_2SiF_6 and stannifluorides, K_2SnF_6.

(*v*) *Action of alkalies.* Both of them react with hot alkalies solution and liberate hydrogen.

$$Si + 2NaOH + H_2O \longrightarrow Na_2SiO_3 + 2H_2$$

$$Sn + 2NaOH + H_2O \longrightarrow Na_2SnO_3 + 2H_2$$

(*vi*) A similarity also exists between the silicates and metastannates.

(B) Dissimilarities. The following dissimilarities may be noted:

(*i*) *Metallic character.* Tin is metalloid whereas silicon is a non-metal.

(*ii*) *Melting point.* Tin metals at 232°C whereas the melting point of silicon is very high.

(*iii*) *Action of dilute acids.* Tin reacts with dilute acids to liberate hydrogen whereas silicon has no such action.

(*iv*) *Salts.* Nitrates, sulphates and other salts of tin are known but such compounds are rare in the case of silicon. Many varieties of silicon are found in nature (*olivine* Mg_2SiO_4) and can also be prepared but orthostannates are almost unknown (with the exception of cobalt stannate Co_2SnO_4). Most of the stannates are meta as Na_2SnO_3.

Similarities between tin and lead

The following points show the similarities between the two elements:

(*i*) Although both the elements have electropositive character, yet, being in the middle of the periodic table, they also show feeble electronegative character in some of its compounds like halides, metastannates, metaplumbates etc.

(*ii*) Both the elements form highly unstable hydrides.

(*iii*) They form quite stable halides. The formation of such halides indicates that these elements have feeble electronegative character.

(*iv*) Both the elements form similar complexes namely stannates and plumbates respectively.

(*v*) When caustic alkali is added to a Sn (+2) salt solution in the absence of air, Sn $(OH)_2$ is first precipitated which dissolves in the excess of alkali, forming stannite (SnO_2^{2-}). Similarly a plumbite (PbO_2^{2-}) is formed by the addition of an excess of an alkali to a Pb (+2) salt solution.

$$Sn^{2+} + 2NaOH \longrightarrow Sn(OH)_2\ (ppt.) + 2Na^+$$

$$Sn(OH)_2 + 2NaOH \longrightarrow Na_2SnO_2\ (Sod.\ stannite) + 2H_2O$$

$$Pb^{2+} + 2NaOH \longrightarrow Pb(OH)_2\ (ppt.) + 2Na^+$$

$$Pb(OH)_2 + 2NaOH \longrightarrow Na_2PbO_2\ (Sod.\ plumbite) + 2H_2O$$

(*vi*) Metastannates (*e.g.*, Na_2SnO_3) and metaplumbates (*e.g.*, Na_2PbO_3) are stable and well-defined compounds.

Abnormal behaviour of carbon

We have studied that the first member of all the three groups (groups IA, IIA and IIIA) shows many different properties from the remaining members of its own group. Carbon is no exception, since this element also differs in properties from the remaining members of group IVA.

The abnormal behaviour of carbon is due to its small size, its high electronegativity, absence of *d*-orbitals in its valence-shell (C = $2s^2\ 2p^2$) and its tendency to show maximum catenation.

The abnormal behaviour of carbon is supported by the following properties:

1. Carbon is quite hard. Diamond which is an allotropic form of carbon is the hardest substance known. The other members of this group are comparatively soft.
2. Its melting and boiling points are higher than the rest of the members
3. CCl_4 cannot be easily hydrolysed while tetrahalides of other elements of this group can be hydrolysed.
4. CF_4 cannot form the complex ion, CF_6^{2-} while SiF_4 can form the complex ion, SiF_6^{2-}.

Table 23.5 : *Some important physical properties of the elements of group VA (15).*

Properties	Nitrogen (N_7)	Phosphorus (P_{15})	Arsenic (As_{33})	Antimony (Sb_{51})	Bismuth (Bi_{83})
Electronic configuration	$[He]_2\ 2s^2\ 2p^3$ or 2, 5	$[Ne]_{10}\ 3s^2\ 3p^3$ or 2, 8, 5	$[Ar]_{18}\ 3d^{10}\ 4s^2\ 4p^3$ or 2, 8, 18, 5	$[Kr]_{36}\ 4d^{10}\ 5s^2\ 5p^3$ or 2, 8, 18, 18, 5	$[Xe]_{54}\ 4f^{14}\ 5d^{10}\ 6s^2\ 6p^3$ or 2, 8, 18, 32, 18, 5
Standard reduction electrode potential (V) for the reaction $M^{3+} + 3e^- \rightarrow M$	—	—	+ 0.25	+ 0.21	+ 0.23
Atomic mass	14.007	30.970	74.922	121.750	208.980
Density (g cm^3)	0.81 (liquid)	1.82	5.72	6.62	9.80
Electron affinity (kJ mol^{-1})	≤ 0.0	71.7	77.0	101.0	91.0
Atomic volume (cm^3)	17.3	17.0	13.1	18.4	21.3
van der Waals radius (Å)	1.5	1.85	2.00	2.20	—
Atomic radius (Å)	0.75	1.06	1.20	1.40	1.46
Radius of M^{3+} ion (Å)	—	—	—	—	1.20
Radius of M^{3-} ion (Å)	1.71	2.12	2.22	2.45	
Radius of M^{5+} ion (Å)	0.11	0.34	0.47	0.62	0.74
Ionisation energies (kJ mol^{-1})					
IE_1	1402.3	1011.8	944.0	831.6	703.3
IE_2	2856.1	1903.0	1798.0	1595.0	1610.0
IE_3	4578.1	2910.0	2736.0	2443.0	2466.0
Metallic/nonmetallic character	Nonmetal	Nonmetal	Metalloid	Metalloid	Metal
Oxidation states	−3, −2, −1, +1, +2, +3, +4, +5	−3, +3, +5	−3, +3, +5	−3, +3, +5	+3, +5
Electronegativity	3.0	2.1	2.0	1.8	1.7
Melting point (°C)	−210.0	44.1 (white phosphorus)	817 (at 36 atm.)	630.5	271.3
Boiling point (°C)	−195.8	280.0	613 (sublimes)	1440	20
Heat of fusion (kJ mol^{-1})	0.36	0.63	27.7	19.9	10.9
M—M bond energy of M_2 molecule (kJ mol^{-1})	163.1	200.8	146.8	—	—

5. CCl_4 cannot form the addition compounds like $CCl_4.2H_2O$, $CCl_4.\ 2NH_3$ while $SiCl_4$ can form the addition compounds like $SiCl_4.2H_2O$, $SiCl_4.2NH_3$.

6. Carbon shows a covalency of 4 only while other members also show higher covalencies in addition to that of 4.

7. CO_2 is a gas at room temperature while the dioxides of other member are crystalline solids.

8. Carbon has the maximum tendency to show catenation while other members have comparatively little tendency.

9. Carbon has great tendency to form multiple bonds with other C-atoms or other different atoms like N, O, S etc. (*e.g.*, C = C, C ≡ C, C = O, C = N, C ≡ N, C = S etc.). The remaining elements have little tendency to form multiple bonds.

Group VA (15) : N, P, As, Sb, Bi

Group VA (15) contains five elements *viz.*, *nitrogen* (N_7), *phosphorus* (P_{15}), *arsenic* (As_{33}), *antimony* (Sb_{51}) and *bismuth* (Bi_{83}). These elements are also called **nitrogen family elements.**

Gradation in physical properties

Some of the important physical properties of these elements are given in Table 23.5 and have been discussed below :

1. Occurrence. Nitrogen is found in free as well as in combined state. Air is the most abundant source of nitrogen. In the combined state, this element occurs as nitrates like Chile salt petre ($NaNO_3$), Indian salt petre (KNO_3) etc., and as ammonium compounds. Nitrogen is the essential constituent of all living things (animals and vegetables) in which it is found in the form of proteins.

Phosphorus, being an active element, is not found in the free state. It is found in the form of phosphates in rocks and in the soil. Important phosphate minerals are *phosphorite* [$Ca_3(PO_4)_2$], *fluorapatite* [$3Ca_3(PO_4)_2.CaF_2$] and *chlorapatite* [$3Ca_3(PO_4)_2.CaCl_2$]. Phosphorus also occurs as phosphoproteins in all living beings. It is an essential constituent of bones, teeth, blood and nervous tissues.

As, Sb and Bi are rarely found in the free state. These elements occurs as sulphides or oxides.

2. Reactivity. Nitrogen is chemically inert under ordinary atmospheric conditions. This is due to high dissociation energy of N = N bond in N_2 molecule. Other elements are quite reactive as they have single A—A bonds in their A_4 molecules (A = P, As, Sb), *i.e.*, bond dissociation energy of A — A bond is very low.

3. Allotropy. Excepting Bi, all the remaining elements show allotropy. Different allotropic forms of different elements are given below:

Nitrogen (*Two forms*): α-nitrogen with cubic crystalline structure and β-nitrogen having hexagonal crystalline structure. Transition temperature is –238.5°C. Both these forms are solid allotropic forms

Phosphorus: White, red, scarlet, metallic or α-black, β-black and violet.

Arsenic (*Three forms*): Grey, yellow and black.

Antimony (*Three forms*): Metallic, yellow or α-antimony and explosive.

4. Tendency of the elements to act as electron pair donor (Lewis base). All the elements in their compounds like AH_3, A_2H_4 etc., have a lone pair of electrons which can be donated by them easily to a Lewis acid. Thus these elements can act as electron pair donor (Lewis base). Nitrogen shows this property to the maximum extent. H_3N: (*Lewis base*) + H^+ (*Lewis acid*) → $H_3N \rightarrow H^+$ or NH_4^+.

5. Tendency of the elements to act as electron pair acceptor (Lewis acid). Nitrogen does not have *d*-orbitals in its valence-shell and hence cannot accept electron pair(s) donated by a Lewis base. Thus N-atom cannot act as an electron pair acceptor. On the other hand, since other elements have vacant *d*-orbitals in their valence-shell, they can accept electron pair(s) donated by a Lewis base and hence can act as electron pair acceptor. This means that nitrogen cannot form complexes by accepting electron pairs while other elements can form complexes.

6. Electronic configuration. The electronic configuration of the atoms of the elements of this group have already been discussed in chapter 4.

7. Physical state. Nitrogen (first element) is a gas while phosphorus (second element) is a waxy solid which can readily pass into vapour state. The remaining elements are hard solids and possess metallic lustre. Thus we see that there is a gradual change in physical state of these elements as we move down the group from N to Bi.

8. Formation of linear A_2 or tetrahedral A_4 type molecules: Atomicity. Nitrogen exists as a *diatomic linear molecule,* N_2 ($N \equiv N$) which contains one N— N σ-bond and two N—N π– bonds. Each of these bonds is formed by (*p* — *p*) overlap. On the other hand P, As and Sb all exist as *discrete tetrahedral tetratomic molecule,* A_4 (A = P, As and Sb) in which each A-atom is linked with each of the remaining three A-atoms by single A — A σ-bond. Thus A_4 molecules exist as solids. In A_4 molecule each A-atom lies at the corner of a regular tetrahedron. In A_4 molecule each A atom completes its octet.

Explanation. The non-existence of A_2 type molecules in case of P, As and Sb is explained on the basis of the fact that these elements are not capable of forming multiple bonds ($A \equiv A$ bonds) due to the repulsion between non-bonded electrons of the inner core. Also see chapter on "*General Characteristics of Main Group Elements.*"

9. Oxidation states. (Also see "*oxidation states*" discussed in chapter 5 and "*Inert Pair Effect*" discussed in chapter 10). Nitrogen shows a large number of oxidation states from +5 to –3 [+5(in N_2O_5), +4 (N_2O_4), +3(N_2O_3), +2 (NO), +1(N_2O), –1 (NH_2OH), –2 (N_2H_4), –3(NH_3)]. Other elements of this group do not show such a large variety of oxidation states. These elements show only a limited number of oxidation states from +5 to –3.

10. Covalency. See chapter 10.

11. Metallic and non-metallic character. See chapter 5.

12. Atomic volume and density. See chapter 5.

13. Atomic and ionic radii. See chapter 5.

14. Ionisation potential. See chapter 5.

15. Melting and boiling points. See chapter 5.

16. Electronegativity. See chapter 5.

17. Formation of M^{3+} and M^{3-} ions. Being smallest in size and most electronegative, N-atom can accept three electrons from highly electropositive metals and can thus form N^{3-} ions quite readily. The tendency of phosphorus to form P^{3-} ion is much less and the tendency of forming M^{3-} ions is absent in other elements. Sb and Bi form Sb^{3+} and Bi^{3+} ions but these ions do not appear to be stable, since these are rapidly hydrolysed to form antimonyl (SbO^+) and bismuthyl (BiO^+) ions.

$$M^{3+} + 3H_2O \longrightarrow MO^+ + 2H_3O^+$$

18. Tendency of elements to form covalent compounds. Covalent compounds of N are more common. The electronegativities of other elements are so low that they generally form covalent compounds even with highly electropositive elements. Bismuth compounds, although covalent, do not give ionic species in polar solvents like H_2O. This tendency decreases on moving in the group from Bi to P through Sb and As.

19. Property of catenation. Although the property of catenation is considered to be a property of carbon in particular, the elements of this group which lie close to carbon in the periodic table also show this property to some extent. Nitrogen which has the minimum value for its N—N bond energy (= 163.1 kJ mol^{-1}) shows this property to the maximum extent, although this property of nitrogen is much smaller then that of carbon (C—C = 353.9 kJ mol^{-1}). Thus N-atom can form many compounds which contain straight chains of two or more N-atoms. These compounds are called straight chain compounds. Examples of such compounds are given below : (R = organic alkyl or aryl group or H-atom).

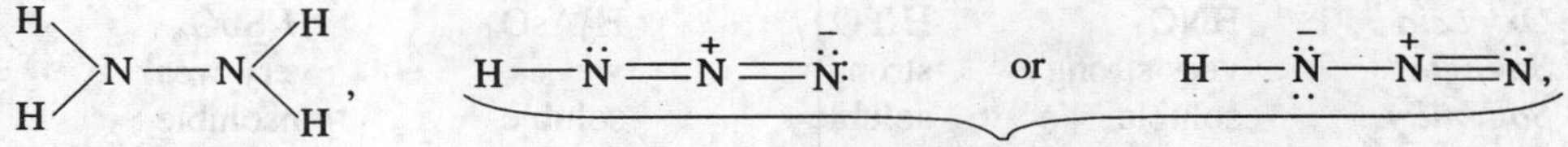

(N_2H_4 molecule) (HN_3 molecule)

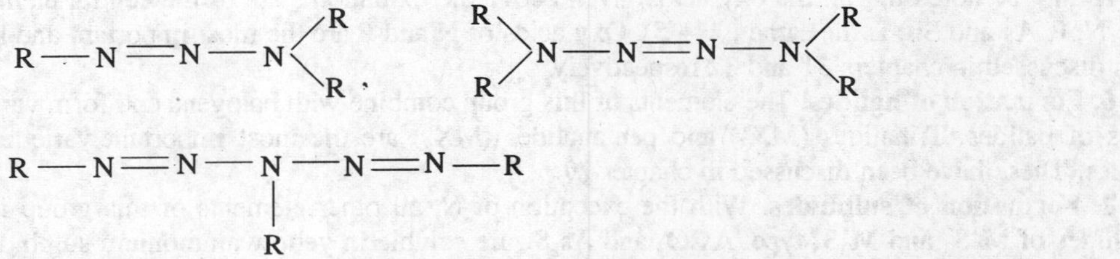

Some cyclic compounds containing rings with upto five consecutive N-atoms have been reported. These cyclic compounds are not very stable, since the bond energies of N—N and N = N bonds are very low. In case of other elements of this group the catenation is limited to two atoms only (*e.g.*, P_2H_4, As_2H_4 etc.).

20. Hydrogen bonding. On account of high electronegativity of N-atom, this element enters hydrogen bond formation in its hydrogen compounds like NH_3. On the other hand, because of their comparatively low electronegativity, other elements do not form H-bonds. It is because of H-bonding between many NH_3 molecules that NH_3 exists as an associated molecule, $(NH_3)_n$. It is because of the formation of the associated molecule of NH_3 that NH_3 has higher boiling point than PH_3 (NH_3 = –34.5°C, PH_3 = – 87.5°C).

Gradation in chemical properties of the elements and their compounds

These elements can form many compounds. Some of these compounds are discussed below:

1. Compounds formed by the combination of the elements with metals. N, P, As, Sb and Bi combine with metals to form *nitrides* (*e.g.*, Mg_3N_2), *phosphides* (*e.g.*, Ca_3P_2), *arsenides* (*e.g.*, Na_3As), *antimonides* (*e.g.*, Mg_3Sb_2) and *bismuthanides* (*i.e.*, Mg_3Bi_2) respectively.

2. Compounds formed by the combination of the elements with alkyl radicals. All the five elements combine with one or more alkyl radicals and form organic compounds. Examples of such compounds are $NH_2(CH_3)$, $N(CH_3)_3$, $P(CH_3)_3$, $Bi(CH_3)_3$ etc.

3. Formation of trihydrides of AH_3 type (A = N, P, As, Sb, Bi). All the elements of this group form trihydrides of AH_3 type. All these hydrides are volatile. These are *ammonia* (NH_3), *phosphine* (PH_3), *arsine* (AsH_3), *stibine* (SbH_3) and *bismuthine* (BiH_3). General methods of their preparation, their general properties etc., have been discussed in chapter 25.

4. Formation of oxides. See chapter 28.

5. Formation of oxy acids. All the elements of this group form oxyacids. Oxy acids formed by nitrogen are hyponitrous acid ($H_2N_2O_2$), nitrous acid (HNO_2), nitric acid (HNO_3) and peroxynitric acid (HNO_4). Phosphorus forms larger number of oxy-acids. Important ones are hypophosphorus acid (H_3PO_2), phosphorus acid (H_3PO_3), pyrophosphorus acid or diphosphorus acid ($H_4P_2O_5$), metaphosphorus acid (HPO_2), hypophosphoric acid ($H_4P_2O_6$), orthophosphoric acid or phosphoric acid (H_3PO_4), metaphosphoric acid (HPO_3), pyrophosphoric acid or diphosphoric acid ($H_4P_2O_7$).

Arsenic forms two oxy acids *viz.*, arsenous acid (H_3AsO_3) and arsenic acid (H_3AsO_4). Antimony gives meta-antimonic acid ($HSbO_3$), orthoantimonic acid (H_3SbO_4) and pyroantimonic acid ($H_4Sb_2O_7$). Bismuth gives metabismuthic acid ($HBiO_3$) and orthobismuthic acid (H_3BiO_4). Oxy-acids of Sb and Bi are unstable and hence exist as their salts.

The strength, stability and solubility of oxy acids of different elements in the same oxidation state decrease gradually as we move down the group, *e.g.*,

Oxy acid	:	HNO_3	H_3PO_4	H_3AsO_4	H_3SbO_4
Strength	:	very strong	strong	weak	very weak
Solubility	:	soluble	soluble	soluble	insoluble
Stability	:	——— *Decreases* ———→			

It may be noted that in the oxy acids given above the oxidation state of the central elements (*i.e.*, N, P, As and Sb) is the same (= + 5). Oxy acids of N and P are the most important and have been discussed in chapters 31 and 32 respectively.

6. Formation of halides. The elements of this group combine with halogens and form various types of halides. Trihalides (MX_3) and pentahalides (MX_5) are the most important varieties of halides. These have been discussed in chapter 29.

7. Formation of sulphides. With the exception of N, all other elements of this group form sulphides of M_2S_3 and M_2S_5 type. As_2O_3 and As_2S_5 are soluble in yellow ammonium sulphide or ammonium sulphide, $(NH_4)_2S$ and form ammonium thioarsenite, $(NH_4)_3AsS_3$ and ammonium thioarsenate, $(NH_4)_3AsS_4$ respectively.

$$As_2S_3 + 3(NH_4)_2S \longrightarrow 2(NH_4)_3\ AsS_3$$

$$As_2S_5 + 3(NH_4)_2S \longrightarrow 2(NH_4)_3\ AsS_4$$

Sb_2S_3 and Sb_2S_5 also form similar thio compounds with $(NH_4)_2S$. As_2S_3 is soluble in NaOH and forms a mixture of sodium arsenite (Na_3AsO_3) and sodium thioarsenite (Na_3AsS_3).

Differences between nitrogen and the remaining elements of group VA: Abnormal behaviour of nitrogen

Nitrogen which is the first element of group VA differs considerably from the remaining elements of this group. These differences are due to the *small size, high electronegativity* and *non-availability of d-orbitals* in the valence-shell of N-atom (N = $2s^2\ 2p^3$).

Main differences are as follows :

(*i*) Nitrogen occurs in the free as well as in the combined state while other elements are found only in the combined state.

(*ii*) Since N-atom does not have *d*-orbitals in its valence-shell, it cannot form complexes by accepting electron pairs while other elements, due to the presence of vacant *d*-orbitals in their valence shell, can accept electron pairs and can thus form complexes.

(*iii*) N is a gas at room temperature while all the remaining elements are solids.

(*iv*) Nitrogen molecule is a diatomic molecule (N_2) while the molecules of other elements are tetratomic molecules such as P_4, As_4 and Sb_4.

(*v*) Nitrogen exhibits a large number of oxidation states such as + 5, + 4, + 3, + 2, + 1, – 1, – 2 and – 3. Other elements show only a limited number of oxidation states from + 5 to – 3.

(*vi*) Since N-atom does not have *d*-orbitals in its valence-shell, its maximum covalency is 4 while other elements can show a maximum covalency of 5. This means that N-atom cannot show pentacovalency while other elements can exhibit pentacovalency.

(*vii*) On account of high value of electronegativity, N behaves as a typical nonmetal while As, and Sb are metalloids and Bi is a metal.

(*viii*) Due to small size and high electronegativity, N can from the nitride ion (N^{3-}). This tendency is less in phosphorus and absent in other elements.

(*ix*) The catenation property is more pronounced in nitrogen. Chain compounds having 8 nitrogen atoms are known, but in case of other elements catenation is limited to two atoms only.

(*x*) Hydride of nitrogen (NH_3) is stable and non-poisonous while the hydrides of other elements are less stable and poisonous. Due to high electronegativity of N-atom, many NH_3 molecules can polymerise together by H-bonding to give associated molecule, $(NH_3)_n$. On the other hand, due to the small electronegativity of other elements, the hydrides of these elements cannot be associated together through H-bonding and hence cannot form $(AH_3)_n$ molecules.

(*xi*) (*a*) All the elements form the oxides of A_2O_3, A_2O_4 and A_2O_5 type. In addition to these oxides, N also forms three more oxides *viz.*, N_2O, NO, NO_3 or N_2O_6. Bi also forms BiO.

(*b*) Oxides of N are monomeric while many oxides of other elements are dimeric, *e.g.*, $(P_2O_3)_2$ or P_4O_6, $(P_2O_4)_2$ or P_4O_8, $(P_2O_5)_2$ or P_4O_{10} etc.

(*xii*) Since N≡N bond energy in N_2 molecule is very high, N_2 is chemically inert under ordinary conditions. On the other hand, due to the low A — A bond energy in A_4 molecule (A = P, As, Sb), other molecules are quite reactive .

(*xiii*) N does not form sulphide while all other elements form sulphides of M_2S_3 and M_2S_5 type

(*xiv*) (*a*) Except NF_3, other trihalides of N (NCl_3, NBr_3 and NI_3) are unstable and explosive. The trihalides of other elements are stable.

(*b*) N can form only trihalides and not pentahalides while P, As, and Sb can form both the types of halides.

Table 23.6 : *Some physical properties of the elements of group VIA (16)*

Property	Oxygen (O_8)	Sulphur (S_{16})	Selenium (Se_{34})	Tellurium (Te_{52})	Polonium (Po_{84})
Electronic configuration	$[He]_2\ 2s^2\ 2p^4$ or 2,6	$[Ne]_{10}\ 3s^2\ 3p^4$ or 2, 8, 6	$[Ar]_{18}\ 4s^2\ 3d^{10}\ 4p^4$ or 2, 8, 18, 6	$[Kr]_{36}\ 5s^2\ 4d^{10}\ 5p^4$ or 2, 8, 18, 18, 6	$[Xe]_{54}\ 6s^2\ 5d^{10}\ 4f^{14}\ 6p^4$ or 2, 8, 18, 32, 18, 6
Density (g cm^{-3})	1.14 (in liquid state)	2.7	2.79	6.24	9.2
Atomic volume (cm^3)	14.0	15.5	16.5	20.5	22.7
Atomic and ionic radii (Å)	0.73 0.09 (+ 6) 1.42 (– 2)	1.02 0.29 (+ 6) 1.84 (– 2)	1.16 0.42 (+ 6) 1.98 (– 2)	1.36 0.56 (+ 6) 2.21 (– 2)	1.46 — —
van der Waals radii (Å)	1.40	1.80	1.90	2.10	—
Melting points (°C)	– 218.8	119.0	217.0	449.5	254.0
Boiling points (°C)	– 183.0	444.6	685.0	989.8	962.0
Ionisation energy (kJ mol^{-1})	1314.00	999.62	940.90	869.30	812.00
Atomic mass	15.999	32.064	78.96	127.60	(210)
Electronegativity	3.5	2.5	2.4	2.1	2.0
Heat of fusion (kJ mol^{-1})	0.22	1.42	5.3	17.7	11.0
Heat of atomisation (kJ mol^{-1})	250	280	207	197	145
Oxidation states	– 1,– 2, + 2	– 2, + 2, + 4, + 6	– 2, + 2, + 4, + 6	– 2, + 2, + 4, + 6	+ 2, + 4
Abundance in earth's crust (p.p.m.)	46,600	520	0.09	0.002	traces
M—M bond energy in M_2 molecules (kJ mol^{-1})	142	226	172	126	—
Electron affinity (kJ mol^{-1}) EA_1 EA_2	 141.1 844.0	 200.43 333.00	 194.97 420.00	 190.16 —	 — —
Metallic/nonmetallic character	Nonmetal	Nonmetal	Nonmetal	Nonmetal	Metal
E^0_{red} (volts)	+ 0.44 (O_2/OH^-)	+ 0.49 (S/S^{2-})	—	—	—

Group VIA (16) : O, S, Se, Te, Po

Group VIA (16) of the periodic table consists of *five* elements *viz.*, *oxygen* (O_8), *sulphur* (S_{16}), *selenium* (Se_{34}), *tellurium* (Te_{52}) *and polonium* (Po_{84}). Oxygen and sulphur are the *typical elements* of this group. The elements O, S, Se and Te are often collectively called *chalcogens, i.e.*, the ore forming elements.

Gradation in physical properties

Some of the physical properties of these elements are given in Table 23.6, and are discussed as follows:

1. Electronic configuration. See chapter 4.

2. Density, atomic volume, atomic radii, ionic radii and van der Waals radii. See chapter 5.

3. Melting and boiling points. See chapter 5.

4. Ionisation energies. See chapter 5.

5. Metallic and nonmetallic character. See chapter 5.

6. Molecular structure (atomicity) and physical state. Oxygen is a *diatomic molecule* (O=O) and is, therefore, a gas while other elements have complex structures and hence are solids. Sulphur and selenium exist as *octa-atomic molecules* (S_8 and Se_8 molecules) at room temperature and have the structures in which the atoms are linked by covalent bonds, leading to the formation of either puckered ring structure or chain polymers (See Fig. 23.2).

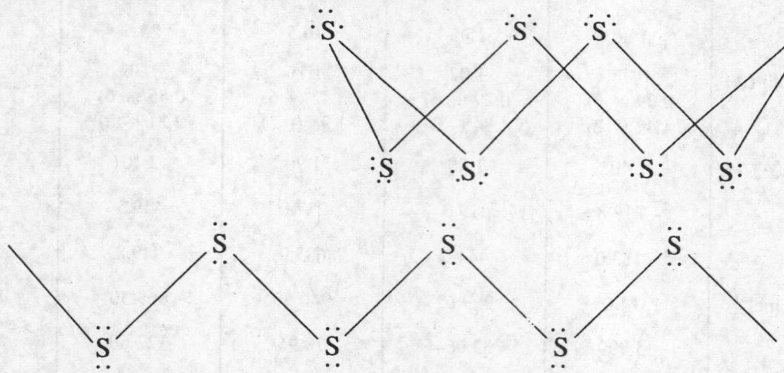

Fig. 23.2 : Puckered ring or polymeric chain of S_8 molecule.

7. Allotropy. All the elements of this group exist in different allotropic forms. Oxygen exists in two metallic forms which are oxygen (O_2) and zone (O_3). Sulphur exists in several allotrophic forms all of which are nonmetallic. These forms have been discussed in chapter 16. Se has two forms : *red* (*nonmetallic*) and *grey* (*metallic*). Te exists in two forms *nonmetallic and metallic* (more stable). Po also exists in two forms α and β (*both metallic*).

8. Catenation. Oxygen and to a greater extent sulphur shows the property of catenation *i.e.*, they have a tendency to form chains of identical atoms *e.g.*, chains of sulphur atoms are present in persulphides, polythionic acids and in S_8 molecule. Oxygen gives polyoxides similar to poly sulphides given by sulphur. Examples are : H_2O_2, H—O—O—H; H_2S_2, H—S—S—H; H_2S_3, H—S—S—S—H.

9. Oxidation states. See chapter 5.

10. Covalency of elements. See chapter 10.

Gradation in chemical properties

Some chemical properties of these elements and gradation in them are discussed below:

1. Formation of oxides. S, Se, Te and Po form various types of oxides. Their classification, methods of preparation and properties have been given in chapter 28 while the structure of the oxides has been discussed in chapter 15.

2. Formation of halides. All the elements of this group form various types of halides. The classification, methods of preparation and properties of these halides have been given in the chapter 29 while the structure of these halides has been discussed in chapter 14.

3. Formation of hydrides. All the elements of this group form hydrides which may be of H_2A (A = O, S, Se, Te, Po) and H_2A_2 (A = O, S) type. A comparative study of these hydrides has been given in chapter 25.

4. Formation of oxy acids. Sulphur, selenium and tellurium form a number of oxy acids.

(*a*) Oxy-acids formed by sulphur exist either in the free state or as their solution or as their salts. Sulphur also forms thio acids which have S—S linkage. Oxy acids formed by sulphur can be grouped into five categories. Classification, methods of preparation and properties of important oxy acids of sulphur have been discussed in chapter 33. (*b*) Selenium forms *selenious acid* (H_2SeO_3) and *selenic acid* (H_2SeO_4). (*c*) Tellurium forms *tellurous acid* (H_2TeO_3 or $TeO_2.H_2O$), *orthotelluric* or *telluric acid* (H_6TeO_3 or H_6TeO_6) and *metatelluric acid* (H_2TeO_4 or $TeO_3.H_2O$).

5. Compounds with carbon. All the elements of this group form compounds with carbon. Examples are CO_2, CS_2, CSe_2 and CTe_2. All these compounds are covalent in nature.

Differences between oxygen and the remaining elements of group VIA: Abnormal behaviour of oxygen

Oxygen which is the first element of group VIA differs considerably from the rest of the elements of this group. These differences are due to the *small size high electronegativity* and *non-availablity of d-orbitals in the valence-shell of O-atom* (O = $2s^2\ 2p^4$) : Main points of difference are given below:

1. Atomicity. Oxygen is a diatomic molecule (O = O) and is, therefore, a gas while other elements have complex structures and hence exist as solids. For example sulphur and selenium exist as S_8 and Se_8 at room temperature.

2. Formation of M^{2-} ions. Because of high electronegativity, O-atom forms O^{2-} ion readily while other elements have a little tendency to form M^{2-} ions (M = S, Se and Te).

3. Oxidation states. Because of high electronegativity, O-atom shows negative oxidation state equal to – 2. This element does not show any positive oxidation state except in OF_2 (F = –1, O = +2). Other elements show +2, +4 and +6 oxidation states in addition to – 2 oxidation state (Polonium does not show –2 oxidation state).

4. Maximum covalency. Maximum covalency of O-atom is 2, since it has only two unpaired and no *d*- orbitals in its valence-shell. Due to the presence of *d*-orbitals in the valence-shell of other elements, their *np* and *ns* paired electrons can be promoted to *nd* orbitals and thus more number of unpaired electrons may be obtained. So, other elements show covalencies of 2, 4 and 6 corresponding to 2, 4 and 6 unpaired electrons and sp^3, sp^3d and sp^3d^2 hybridisation of the atom.

5. Hydrogen bonding. On account of high electronegativity of O-atom, this element forms hydrogen bonds in its hydrogen compounds like H_2O. On the other hand, because of comparatively low electronegativity, other elements do not form H-bonds. It is because of H-bonding between many H_2O molecules that H_2O exists as an associated molecule, $(H_2O)_n$. It is because of the formation of the associated molecule of H_2O that H_2O is a liquid and has high boiling point (=100°C) while the hydrides of other elements (*i.e.*, H_2S, H_2Se, H_2Te) are gases and have low boiling points. (H_2S = – 60.75°C, H_2Se = – 41.5°C, H_2Te = – 1.8°C).

6. Tendency to form (*p*—*p*) π-bonds. Oxygen forms (*p*—*p*) π-bond with the elements having similar size (*e.g.*, C, N). Other elements show rebuctance to form multiple bonds *i.e.*, other elements do not have a tendency to form (*p*—*p*) π-bonds.

7. Magnetic property. Molecular oxygen in the gaseous, liquid and solid states is paramagnetic while the molecules of other elements are diamagnetic in nature.

Comparison between oxygen and sulphur

These are the typical elements of group VIA and show many similar and dissimilar properties.

(*a*) **Similarities.** (*i*) Both are non-metals. (*ii*) They have six electrons in their outermost shell (O_8 → 2, 6 ; S_{16} → 2, 8, 6). (*iii*) Oxygen shows analogy with sulphur in compounds like (*a*) C_2H_5OH and C_2H_5SH. (*b*) CO_2 and CS_2. (*c*) $(C_2H_5)_2O$ and $(C_2H_2)_2S$. (*d*) H_2O and $H_2.S$ (*iv*) O and S combine with H_2 to form the hydrides *viz.* H_2O and H_2S (*v*) They both form oxides of RO_2 type. *e.g.*, O_3 (O O_2) and SO_2. (*vi*) Both the elements are obtained by heating their higher oxides and sulphides.

$$2BaO_2 \rightarrow 2BaO + O_2\uparrow;\ FeS_2 \rightarrow FeS + S\uparrow$$

(*vii*) They combine with carbon on heating to give CO_2 and CS_2.

(*b*) **Dissimilarities.** (*i*) O_2 is colourless and odourless gas while sulphur is a pale yellow solid with a faint odour, (*ii*) Oxygen is more widely distributed in nature than sulphur, (*iii*) Maximum covalency of oxygen is 2 while that of sulphur is 6, (*iv*) Oxygen does not react with conc. HNO_3 while sulphur is oxidised by conc. HNO_3, (*v*) Oxygen, due to non-availability of *d*-orbitals, does not form co-ordination compounds while S gives a large number of coordination compounds.

Group VIIA (17) : F, Cl, Br, I, At (Halogens)

Group VIIA (17) of the periodic table consists of five elements *viz., fluorine* (F), *chlorine* (Cl), *bromine* (Br), *iodine* (I) and *astatine* (At). The elements F, Cl, Br and I are collectively called *Halogens.* The term *halogen* is derived from the Greek *halos* (= salt) and *genes* (= born) meaning salt producers because most of these elements exist in sea water, notably in the form of their sodium compounds. Astatine was discovered in 1940 and is an unstable element of radioactive origin. The other four elements are stable and resemble each other in physical and chemical properties.

Gradation in physical properties

Some of the physical properties of these elements are given in Table 23.7 and are discussed as follows:

1. Electronic configuration. See chapter 4.

2. Occurrence. These elements do not occur in the free state. But in the combined state they occur fairly abundantly in nature.

Table 23.7 : *Some physical properties of the elements of group VIIA (17)*

Property	Fluorine (F_9)	Chlorine (Cl_{17})	Bromine (Br_{35})	Iodine (I_{53})
Electronic configuration	$[He]_2\ 2s^2\ 2p^5$ or 2, 7	$[Ne]_{10}\ 3s^2\ 3p^5$ or 2, 8,7	$[Ar]_{18}\ 3d^{10}\ 4s^2\ 4p^5$ or 2, 8, 18, 7	$[Kr]_{36}\ 5d^{10}\ 6s^2\ 6p^5$ or 2, 8, 18, 18, 7
Heat of dissociation (dissociation energy of X_2 molecules) (KJ mol^{-1})	159	243	193	151
Hydration energy of X^- (*g*) ions (KJ mol^{-1})	515.0	381.0	347.0	305.0
Density (g cm^{-3})	1.11 (liquid)	1.56 (liquid)	3.12 (liquid)	4.94 (solid)
Physical state	Gas	Gas	Liquid	Solid
Atomic radius (Å)	0.72	0.99	1.14	1.33
Radii of ions (Å)	1.36 (–1)	1.81 (–1)	1.95 (–1)	2.16 (–1)
	—	0.26 (+7)	0.39 (+7)	0.50 (+7)
van der Walls radii (Å)	1.35	1.80	1.95	2.15
Atomic volume (cm^3)	17.1	18.7	23.5	25.7
Electronegativity	4.0	3.0	2.8	2.4
Atomic weight	18.998	35.453	79.909	126.904
Melting point (°C)	–218.6	–101.0	–7.2	113.5

Boiling point (°C)	– 188.1	– 34.6	59.5	185.2
Colour of vapour	Pale yellow	Greenish yellow	Orange-red	Violet
Colour of liquid	Clear yellow	Amber yellow	Reddish brown	Shining dark solids
Ionisation energy (kJ mol^{-1}) IE_1 IE_2 IE_3 IE_4	1681.0 3374.2 6050.4 7450.0	1251.1 2260 3390 4590	1139.9 2080 3090 4140	1008.4 1800 3010 3610
X—X bond length in X_2 molecule (Å)	1.43	1.99	2.28	2.66
Standard reduction electrode potential for $X_2\ (g) + 2e^- \rightleftharpoons 2X^-\ (aq)$ reaction (volts)	+ 2.85	+ 1.36	+ 1.06	+ 0.53
Oxidation states	–1	–1, +1, +3, + 4, +5, + 6, +7	–1, +1, +3, + 4, + 5, + 6	–1, +1, +3, + 4, + 5, + 6, + 7
Covalencies	1	1, 3, 5, 7	1, 3, 5, 7	1, 3, 5, 7
Nonmetallic character	Decreases →			
Abundance in earth's crust (%)	0.075	0.04	0.002	0.0003
Polarisability (CC/atom)	1.04×10^{-24}	3.66×10^{-24}	4.77×10^{-24}	7.10×10^{-24}
Reactivity of X_2 molecules	Decreases →			
Heat of fusion (+Δ H) of solid X_2 molecules $[X_2\ (s) \rightarrow X_2\ (l)]$ (kJ mol^{-1})	—	—	—	+ 16
Heat of vaporisation (+ΔH) of liquid X_2 molecules $[X_2\ (l) \rightarrow X_2\ (g)]$ (kJ mol^{-1})	—	—	+ 30	+ 44
Heat released (kJ mol^{-1}) in the reaction, X $(g) + e^- \rightarrow X^-\ (g)$ (Electron affinity)	328	348.8	324.6	– 295.4
Energy liberated (– Δ H) in the reduction reaction , $1/2\ X_2\ (s, l, g) + e^- + aq \rightarrow X^-\ (aq)$ (kJ mol^{-1})	– 763.5	– 608.3	– 575.1	– 524.9
Relative order of oxidising power of X_2 molecules in aqueous solution	Decreases →			

3. Diatomic nature. Heat capacity and combining volume data have shown that all the halogens exist as covalent diatomic molecules (X_2 molecules) in the gaseous, liquid and crystalline states. These discrete X_2 molecules are held together by weak van der Waals forces which explain the volatile nature of these elements.

4. Heat of dissociation (Dissociation energy) of X_2 molecules and reactivity of halogens. Heat of dissociation, ΔH_d of X_2 molecule is the amount of heat required to dissociate X_2 molecule in the gaseous state into X gaseous atoms or to break X—X bond in X_2 molecule. Thus ΔH_d can be defined by the equation, $X_2\ (g) + \Delta H_d \rightarrow 2X\ (g)$. Obviously ΔH_d is equal to X — X bond energy. Dissociation energies of X_2 molecules as given in Table 23.7 show that these values are very low.

Why is the dissociation energy of F_2 less than that of Cl_2? The lower value of heat of dissociation of F_2 molecule as compared to Cl_2 molecule is due to the reasons namely: (*i*) Due to very small F—F bond length in F_2 molecule, repulsion between the nonbonding electrons is very large as compared to other X—X bond lengths (X = Cl, Br, I). This makes the F-atoms in F_2

molecule repel each other and helps the dissociation of F_2 molecule into F atoms. (*ii*) X—X bond in Cl_2, Br_2 and I_2 molecules is stronger than F— F bond in F_2 molecule. This is due to the possibility of the existence of multiple bonding in X—X bond involving *d*-orbitals.

We have said above that X_2 molecules have low values of their dissociation energies. It is due to the low values that X_2 molecules can easily be dissociated into X-atoms and are, therefore, very reactive. The reactivity of halogens decreases as we move down the group from F_2 to I_2. Due to the repulsion between the nonbonding electrons in F_2 molecule, F — F bond dissociation energy is very low and hence F — F bond is very weak. Due to the weak nature of F — F bond, F_2 molecule is the most reactive among the halogens. Maximum reactivity of F_2 is evident from the following reactions:

(*a*) *Decomposition of water by halogens.* F_2 decomposes H_2O vigorously even in the dark, Cl_2 decomposes H_2O in sunlight, Br_2 also decomposes H_2O in sunlight but very slowly while I_2 does not decompose H_2O at all.

$$2F_2 + 2H_2O \rightarrow 4HF + O_2 \; ; \qquad 3F_2 + 3H_2O \rightarrow 6HF + O_3$$

$$Cl_2 + H_2O \rightarrow HClO + HCl \; ; \qquad I_2 + H_2O \rightarrow \text{No reaction}$$

(*b*) *Combination of halogens with H_2.* All the halogens combine with H_2 to form hydrogen halides (HX). The tendency of halogens to combine with H_2 decreases from F_2 to I_2. Thus F_2 combines with H_2 explosively and even in the dark, Cl_2 reacts with H_2 in sunlight or ultraviolet light, Br_2 reacts with H_2 only on heating above 500°C while I_2 reacts with H_2 only on heating and in presence of Pt as a catalyst. The reaction of I_2 with H_2 is reversible and endothermic.

5. Hydration energy of X^- (*g*) ions (*Also see chapter 16.*) Hydration energy of X^- ion is defined as the amount of energy liberated in converting X^- (*g*) ion into hydrated X^- (*g*) ion, X^- (*aq*) [X^- (*g*) + H_2O (*l*) or *aq* → X^- (*aq*) + Hydration energy]. Hydration energies of X^- ions as given in Table 23.7 show that these values decrease as the size of X^- ions increases from F^- to I^- ions.

6. Physical state and density. There is a change of state from gas to solid and hence density increases as we move down the group. (See chapter 5).

7. Atomic and ionic radii. See chapter 5.

8. Nonmetallic (basic) character. See chapter 5.

9. Electronegativity. See chapter 5.

10. Ionisation energies. See chapter 5.

11. Electron affinity. See chapter 5.

12. Melting and boiling points. See chapter 5.

13. Colour of vapour. All the halogens are coloured. Their colour is due to the fact that their molecules absorb light in the visible region by means of which the outer electrons are excited to higher energy levels. The amount of energy required for the excitation (*i.e., excitation energy*) decreases progressively form fluorine to iodine as the size of the atom increases. Thus there is a progressive deepening of colour of the halogens from fluorine to iodine. Fluorine absorbs violet light (higher excitation energy) and hence appears *yellow,* while iodine vapour absorbs yellow light (lower energy) and appears *deep violet.* Similarly we can account of *greenish yellow* colour of chlorine and *orange red* colour of bromine.

14. Odour. All the halogens have very pungent and unpleasant odours. If inhaled even in very small quantities, they cause inflammation of the mucous membranes of the throat and nose and serious irritation of respiratory ducts. Large quantities may have fatal effect.

15. Solubility. The halogens are slightly soluble in water and their solubility decreases from Cl to I. F cannot be dissolved in H_2O, since it decomposes H_2O liberating O_2 and some O_3. The halogens dissolve much better than H_2O in many organic solvents like carbon disulphide, chloroform, ether etc. Their highter solubility in organic solvents has been utilised to extract them from aqueous solutions. In organic solvents, iodine gives violet colour, bromine yields brown colour and chlorine produces yellow colour. It is assumed that in organic solvents halogens exist as free molecule as in the gas phase.

In nucleophilic (electron donating) polar solvents like alcohols, ketones, liquid SO_2, the halogens give coloured solutions whose colours are due to the formation of solvent → halogen type complexes. These complexes are charge-transfer compounds. In such compounds an electron located mainly in an orbital of the solvent molecule jumps into orbital located mainly on the halogen and absorbs energy from the light, causing the compound to look coloured. Such compounds have been isolated at low temperatures.

16. Oxidation states. See chapter 5.

17. Covalencies shown by halogens. See chapter 10.

18. Formation of clathrates. Water solutions of the halogens on cooling give the crystals of their clathrates of $X_2 . xH_2O$ type. Examples are : $Cl_2 . 7.3\ H_2O$, $Br_2 . 10H_2O$, $I_2 . xH_2O$ etc.

19. Oxidising property of halogen molecules (X_2) and reducing power of halide ions (X^-) in aqueous solution. See chapter on *"Electrode Potentials and Electrochemical Series"* (Chapter 18).

Gradation in chemical properties

Some chemical properties of these elements and gradation in them are discussed below:

1. Formation of oxy acids. Excepting fluorine, all the remaining halogens form oxy-acids. Classification, general methods of preparation and general properties of these acids have been discussed in the chapter on "*Oxy acids of Halogens*" (Chapter 34).

2. Formation of halogen-oxygen binary compounds. Halogens form a large number of halogen-oxygen binary compounds. Examples are OF_2, O_2F_2 (called fluorides of oxygen), Cl_2O, ClO_2, Cl_2O_6, Cl_2O_7, ClO_4 (called oxides of Cl), Br_2O, BrO_2, BrO_3 (called oxides of Br), I_2O_4, I_4O_9, I_2O_5 (called oxides of I). These compounds cannot be obtained by the direct combination of halogen and O_2. These are obtained by indirect methods. A detailed study of these compounds has been given in chapter 28.

3. Formation of hydrogen halides. (Also called *halogen acids* or *hydroacids or hydrohalic acids*). Each of the halogens combines directly with H_2 to form hydrogen halides *viz*., HF, HCl, HBr and HI. General methods of preparation and general properties of these acids have been discussed in chapter 29.

4. Formation of interhalogen compounds. Two different halogen atoms, due to the difference in their electronegativity values, combine with one another to form a number of covalent compounds which are called *interhalogen compounds.* These are of four types namely AB (*e.g.*, ClF, BrF etc.); AB_3 (*e.g.*, ClF_3, BrF_3, ICl_3); AB_5 (*e.g.*, BrF_5, IF_5) and AB_7 (*e.g.*, IF_7) where A and B are two different halogen atoms. A detailed study of interhalogen compounds is given in chapter 29.

5. Formation of polyhalides. Two or more similar or different halogen atoms combine together and form *polyhalide ions* which may be cations or anions. Examples of such ions are Cl_2^+, Br_2^+, I_2^+, I_3^-, ICl_2^+, IF_4^+, $ClBrI^-$ etc. These ions combine with monovalent metallic atoms and form the compounds which are called *polyhalides.* Examples of such compounds are $K[I_3]$, $Cs[ClF_4]$, $Na[IBr_2]$ etc. A detailed study of polyhalides has been given in chapter 29.

6. Action on alkalies. The reaction of Cl_2, Br_2 and I_2 with cold and dil. solution of an alkali gives halide (X^-) and hypohalide (XO^-) while the reaction of these halogens with hot and concentrated solution of an alkali gives halide (X^-) and halate (XO_3^-). Thus:

X_2 + 2NaOH (*dil. and cold solution*) ⟶ $NaX + NaXO + H_2O$

(X = Cl , Br. I)

$3X_2$ + 6NaOH (*conc. and hot solution*) ⟶ $5NaX + NaXO_3 + 3H_2O$

F_2 reacts with dil. alkalies (2% NaOH) and forms NaF and OF_2 in place of NaFO

$2F_2$ + 2NaOH (*2% solution*) ⟶ $2NaF + OF_2 + H_2O$

F_2 also reacts with conc. alkalies and gives NaF and O_2 in place of $NaFO_3$.

$2F_2$ + 4NaOH (*conc. solution*) ⟶ $4NaF + O_2 + 2H_2O$

The non-formation of $NaFO$ and $NaFO_3$ is due to the fact that the oxy acids of these salts (*i.e.*, HFO and HFO_3 respectively) do not exist.

7. Action on NH_3. All the four halogens react with NH_3 and form different products; *e.g.*,

(*a*) F_2 reacts with NH_3 to give N_2 and some NF_3

$$3F_2 + 2NH_3 \longrightarrow N_2 + 6HF; \quad 3F_2 + NH_3 \longrightarrow NF_3 + 3HF$$

(*b*) Cl_2 reacts with excess of NH_3 to form N_2 and NH_4Cl. In this reaction Cl_2 oxidises NH_3 to N_2

$$3Cl_2 + 8NH_3\ (excess) \longrightarrow N_2 + 6NH_4Cl$$

Excess of Cl_2 reacts with NH_3, forming NCl_3 (explosive material)

$$3Cl_2\ (excess) + NH_3 \longrightarrow NCl_3 + 3HCl$$

(*c*) Br_2 oxidises NH_3 to N_2

$$3Br_2 + 8NH_3 \longrightarrow N_2 + 6NH_4Br$$

(*d*) I_2 reacts with aqueous solution of NH_3 to form an explosive powder of $NI_3 . NH_3$

$$3I_2 + 2NH_3 \longrightarrow NI_3 . NH_3 + 3HI$$

8. Action on organic compounds. (*a*) F_2 reacts explosively with most organic compounds, forming *fluorocarbons* like CF_4, C_6F_{12} etc. (Also see chapter 38)

$$CH_4 + 4F_2 \longrightarrow CF_4 + 4HCl$$

$$C_6H_6 + 9F_2 \xrightarrow[265°\,C]{Cu\text{–}guage} C_6F_{12} + 6HF$$

(*b*) Cl_2, Br_2 and I_2 form substitution and addition compounds with saturated (*e.g.*, CH_4) and unsaturated (*e.g.*, C_2H_4) organic compounds.

$$\underset{(X = Cl,\ Br)}{CH_4 + X_2} \longrightarrow CH_3X + HX\ (substitution)$$

$$C_2H_6 + I_2 \longrightarrow C_2H_5I + HI\ \ (substitution)$$

$$C_2H_4 + X_2 \longrightarrow C_2H_4.X_2\ \ (addition)$$

9. Action on the aqueous solution of halides. (*a*) F_2 can displace Cl_2, Br_2 and I_2 from the aqueous solution of chlorides, bromides and iodides respectively.

$$\underset{(X = Cl,\ Br,\ I)}{F_2 + 2KX} \longrightarrow 2KF + X_2$$

or $$F_2 + 2X^- \longrightarrow 2F^- + X_2$$

In this reaction X^- ion is oxidised to X_2.

(*b*) Cl_2 can displace Br_2 and I_2 from the aqueous solution of bromides and iodides respectively but it cannot displace F_2 from a fluoride, *i.e.*, only reaction (A) takes place while reaction (B) does not occur.

$$\left.\begin{aligned} &\underset{(X = Br,\ I)}{Cl_2 + 2KX} \longrightarrow 2KCl + X_2 \\ \text{or}\quad &Cl_2 + 2X^- \longrightarrow 2Cl^- + X_2 \end{aligned}\right\}\ \ldots (A)$$

$$\left.\begin{aligned} &Cl_2 + 2KF \longrightarrow 2KCl + F_2 \\ \text{or}\quad &Cl_2 + 2F^- \longrightarrow 2Cl^- + F_2 \end{aligned}\right\}\ (Does\ not\ occur)\ \ldots(B)$$

(*c*) Br_2 can displace I_2 from the aqueous solution of an iodide but it cannot displace F_2 from KF, Cl_2 from KCl and I_2 from KI. Thus only reaction (A) occurs and reactions shown at (B) do not take place

$$\left.\begin{aligned} &Br_2 + 2KI \longrightarrow 2KBr + I_2 \\ \text{or}\quad &Br_2 + 2I \longrightarrow 2Br^- + I_2 \end{aligned}\right\}\ \ldots(A)$$

$$\left.\begin{aligned} &\underset{(X = F,\ Cl,\ Br)}{Br_2 + 2KX} \longrightarrow 2KBr + X_2 \\ \text{or}\quad &Br_2 + 2X^- \longrightarrow 2Br^- + X_2 \end{aligned}\right\}\ (Does\ not\ occur)\ \ldots(B)$$

(*d*) I_2 cannot displace any of the halogens from the aqueous solution of their respective halides, *i.e.*, none of the following reactions takes place.

$$I_2 + 2KX \longrightarrow 2KI + X_2 \text{ (Do not occur)}$$
$$(X = F, Cl, Br)$$

10. Decomposition of water by halogens. It has already been discussed.

11. Combination of halogens with H_2. It has already been discussed.

12. Action on elements. (*a*) F_2 combines with metals (*e.g.*, Na, Ca, Mg, Al, Ni, Pt etc.), nonmetals (*e.g.*, H_2, C, S, P, Cl_2, Br_2, I_2 etc.) and metalloids (*e.g.*, As, Sb) to form the fluorides of these elements. F_2 combines with xenon to form three different fluorides *viz.*, XeF_2 , XeF_4 and XeF_6, depending on the temperature, pressure and mixing ratio of F_2 and Xe. In case of Cu and Hg a protective layer of fluoride is formed which prevents further action.

(*b*) Na, Sb powder and copper leaves form their respective chlorides when thrown in a jar of Cl_2, but Al, Fe and other metals form their chlorides only when they are heated in the current of Cl_2.

$$2Sb + 3Cl_2 \longrightarrow 2SbCl_3 \; ; \qquad 2Al + 3Cl_2 \xrightarrow{\Delta} 2AlCl_3$$

When a metal forms two chlorides, it is usually higher chloride which is generally formed, *e.g.*, Fe, Cu and Hg give $FeCl_3$ (not $FeCl_2$), $CuCl_2$ (not Cu_2Cl_2) and $HgCl_2$ (not Hg_2Cl_2).

$$2Fe + 3Cl_2 \longrightarrow 2FeCl_3 \; ; \qquad M(M = Cu, Hg) + Cl_2 \longrightarrow MCl_2$$

Cl_2 combines with P and S to form PCl_5 and S_2Cl_2 respectively.

(*c*) Br_2 reacts with most metals and produces bromides. Al reacts with Br_2 vigorously with the emission of light and K reacts explosively. As and Sb powder catches fire in a jar full of Br_2 vapours.

$$2M\ (M = As, Al) + 3Br_2 \longrightarrow 2MBr_3 \; ; \qquad 2K + Br_2 \longrightarrow 2KBr$$

Pb, Ni and Mg are not attacked by Br_2 and hence may be used as containers for Br_2 liquid. In presence of moisture Fe and Zn are readily corroded. Br_2 also reacts with many nonmetals like S and P, forming their corresponding bromides. Br_2 has no action on C, O_2 and N_2.

(*d*) I_2 combines with most of the metals except Ag, Au and Pt. When Sb powder is dropped in a jar of I_2 vapours, it catches fire. P and K react explosively with I_2. On rubbing Hg and I_2 in a mortar, Hg_2I_2 (green) is formed. If Hg is taken in excess, HgI_2 (red) is formed. I_2 has no action on O_2, N_2 and C. Only P and halogens combine directly with I_2.

Differences between fluorine and the remaining elements of group VIIA: Abnormal behaviour of fluorine

As usual, fluorine which is the first element of this group, shows many differences from the remaining halogens. These differences are due to small size, high electronegativity and absence of *d*-orbitals in the valence-shell of F-atom. ($F = 2s^2\ 2p^5$).

Main points of difference are given below:

1. Formation of oxides. Since F shows only one oxidation state equal to –1, this element forms only one oxide namely F_2O (F = –1, O = +2). Other halogens show many oxidation states and hence form many oxides in +1 (Cl_2O, Br_2O), + 4 (ClO_2, BrO_2, I_2O_4), + 5 (I_2O_5), + 6 (Cl_2O_6, BrO_3) and +7 (Cl_2O_7, I_2O_7) oxidation states. Fluorine also forms F_2O_2 which is unstable.

2. Formation of interhalogen compounds. Due to high electronegativity of F-atom, this element gives the maximum number of interhalogen compounds. Other halogens give less number of interhalogen compounds.

3. Formation of oxy acids. Since F is more electronegative than oxygen, it does not form oxy acids. On the other hand, since the remaining halogens are less electronegative than oxygen, they give various types of oxy acids. Electronegativity values of F, Cl, Br, I and O are as F = 4.0, O = 3.5, Cl = 3.0, Br = 2.8 and I = 2.5.

4. Oxidation states. Since F is the most electronegative element, F shows only negative oxidation state equal to –1. This element does not show any positive oxidation state. In contrast, other halogens show + 1, +2, + 3, + 4, + 5, and + 6 oxidation states (positive oxidation states) in addition to – 1 oxidation state.

5. Maximum covalency. Maximum covalency of F is 1, since it contains only one electron and no *d*-orbitals in its valence-shell. Due to the presence of *d*-orbitals in the valence-shell of other elements, their *np* and *ns* paired electrons can be promoted to *nd* orbitals and thus more number of unpaired electrons may be obtained. So, other elements show covalencies of 3, 5, and 7 corresponding to 3, 5 and 7 unpaired electrons and sp^3d, sp^3d^2 and sp^3d^3 hybridisation of the atom.

6. F—F bond dissociation energy and reactivity of F_2 molecule. Due to the repulsion between nonbonding electrons in F_2 molecule, F—F bond dissociation energy is very low and hence F—F bond is very weak. Due to the weak nature of F—F bond, F_2 molecule is the most reactive among halogens.

7. Formation of polyhalide ions. Because of the absence of *d*-orbitals in the valence-shell of F-atom, this element cannot accept the lone pairs of electron(s) donated by the halide ions and hence cannot form polyhalide ions. On the other hand, other halogens, because of the presence of vacant *d*-orbitals in their valence-shell, accept the lone pairs of electrons and give rise to the formation of polyhalide ions like ClF_2^-, $BrCl_2^-$, ICl_2^-, Cl_3^-, Br_3^-, I_3^-. Thus F_3^- ion is not formed.

8. Formation of sulphur hexahalides. Since F-atom has a high value of electronegativity, this element readily transforms sulphur into + 6 oxidation state by forming stable sulphur hexafluoride, SF_6. Other halogens do not form any sulphur hexahalide *i.e.*, SF_6 exists but SCl_6, SBr_6 and SI_6 are not formed.

9. Action on alkalies. Cl_2, Br_2, and I_2 give the same types of products while F_2 gives different products.

Comparison between fluorine and oxygen

Fluorine resembles oxygen which is the first member of the neighbouring group *viz.*, VIA in some respects on account of the following factors: (*a*) *The atomic and ionic radii of both the elements are very similar* (*b*) *Both are highly electronegative.*

(A) Similarities. (*i*) Both of them exist in gaseous state at ordinary temperature (*ii*) Both react with metals (*e.g.*, Na) and nonmetals (*e.g*, H, C, Si, S etc.) to give fluorides and oxides which resemble each other in solubility, volatility, fusibility etc. *e.g.*, HF and H_2O resemble each other in the following respects : (*a*) Both possess associated molecules, $(H_2O)_x$ and $(HF)_x$ respectively due to hydrogen bonding. (*b*) The boiling points of both are abnormally high. (*c*) HF like H_2O is a weak acid and both are ionising solvents. (*d*) Heats of formation of HF and H_2O are close to each other. (*iii*) Certain elements like sulphur exhibit higher valencies in both fluorides and oxides, *e.g.*, the valency of sulphur in SF_6 and SO_3 is 6.

(B) Dissimilarities. However these elements differ from each other in the following respects : (*i*) F is always monovalent while oxygen is bivalent (*ii*) F is more reactive than oxygen.

Zero Group (18) : He, Ne, Ar, Kr, Xe, Rn (Noble Gases)

Zero group of the periodic table consists of six elements namely *helium,* He (Z = 2) ; *neon,* Ne (Z = 10) ; *argon,* Ar (Z = 18) ; *krypton,* Kr (Z = 36) ; *xenon,* Xe (Z = 54) and *radon,* Rn (Z = 86). These elements are also called by other names like *inactive gases, inert gases, rare gases* and *noble gases.*

Physical properties of noble gases

Some physical properties of noble gases are given in Table 23.8. These are discussed below:

1. Occurrence. Excepting radon, all other noble gases occur throughout the universe in atomic state. Since radon is a radioactive element and hence decays rapidly, it does not occur in the free state. Other rare gases occur in the following sources:

(*i*) *In the stars and interstellar space.* The masses of the major planets Jupiter and Saturn are estimated to be due to 23% and 27% of helium respectively. The atmosphere of Mars consists of mostly N_2 and argon.

(*ii*) *In the earth's atmosphere.* The nonradioactive rare gases are present in the atmosphere to the extent of 1% by volume. The volume of each rare gas in 100 litres of this is as : He = 0.524 c.c., Ne = 1.82 c.c., Ar = 9.34 c.c., Kr = 0.114 c.c., Xe = 0.0087 c.c.

Table 23.8 : *Some physical properties of noble gases*

Property	Helium (He_2)	Neon (Ne_{10})	Argon (Ar_{18})	Krypton (Kr_{36})	Xenon (Xe_{54})	Radon (Rn_{86})
Electronic configuration	2 ($1s^2$)	2, 8 ($2s^2\,p^6$)	2, 8, 8 ($3s^2\,p^6$)	2, 8, 18, 8 ($3d^{10}\,4s^2\,p^6$)	2, 8, 18, 18, 8 ($4d^{10}\,5s^2\,p^6$)	2, 8, 18, 32, 18, 8 ($4f^{14}\,5d^{10}\,6s^2\,p^6$)
Atomic weight	4.003	20.183	39.948	83.30	131.30	222
Density of liquid gas (g cm^3)	—	1.20	1.40	2.60	3.06	—
Melting points (°C)	– 272.1	– 246.6	–189.3	–157.2	–118.8	–71.0
Boiling points (°C)	– 268.9	– 246.0	–185.8	–153.0	–108.1	– 62.0
Ionisation energy (KJ mol^{-1})	2372.3	2080.7	1520.5	1350.7	1170.4	1037.0
Heat of vaporisation (KJ mol^{-1})	0.08	1.74	6.52	9.05	12.65	18.1
van der Waals radii (Å)	0.93	1.31	1.74	1.89	2.09	2.14
Electronegativity	zero	zero	zero	zero	zero	zero
Critical temperature (°C)	– 267.9	– 228.7	– 122.4	– 62.5	+ 16.6	+ 14.5
Critical pressure (atm)	2.26	26.9	50.0	54.3	58.2	62.4
$\gamma = C_p / C_v$	1.652	1.642	1.60	1.689	1.60	—
Atomicity	Monoatomic	Monoatomic	Monoatomic	Monoatomic	Monoatomic	Monoatomic
Absorption coefficient in water at 25°C	0.0097	0.0114	0.0053	0.1105	0.2420	0.5100
Electron affinity (KJ mol^{-1})	≤ 0	≤ 0	≤ 0	≤ 0	≤ 0	≤ 0
Relative abundance in earth's crust (ppm)	5.2	18.2	93.4	1.1	0.09	traces

(*iii*) *In natural gas.* Helium and traces of neon are found in natural gas which is found in the wells of certain oil fields in U.S.A. At present most of helium is isolated from natural gas as the concentration of helium in it varies from 2 to 7%.

(*iv*) *In minerals and spring waters.* Noble gases are also found in certain minerals. Except helium, the concentration of other noble gases in minerals is quite low. For example, the concentration of helium in thorianite is 10.5 c.c. gm. Certain spring waters give off gases which contain helium and argon. The percentage of these gases in spring water is so low that these cannot be isolated.

2. Electronic configuration. See chapter 4.

3. Melting and boiling points. See chapter 5.

4. Solidification of noble gases. Excepting Rn, all other noble gases solidify at 2 to 9°C below their normal boiling points. Helium is most difficult to be solidified. It solidifies only when it is cooled to about 1 K under a pressure of 25 atmospheres. Liquid helium is, therefore, used as an ideal cooling agent for research work at low temperatures.

5. Liquefaction. Noble gases can be liquefied with great difficulty. It is because of the existence of weak van der Waals forces in these gases.

6. Heat of vaporisation and critical temperature. Heats of vaporisation of these gases are very low. These values increase as we move down the group. Critical temperatures increase from He to Xe.

7. Solubility of noble gases in water. These gases are slightly soluble in water. Solubility generally increases with increasing atomic weight.

8. Ionisation energies. See chapter 5.

9. Electron affinity. See chapter 5.

10. Monoatomicity. Noble gases exist as monoatomic molecules, since the atoms are not capable of combining even amongst themselves.

11. Characteristic spectra. Noble gases give characteristic spectra by means of which they can be identified. Thus helium gives characteristic spectrum containing green and yellow lines while neon gives a characteristic spectrum with orange and red lines and so on.

Position of inert gases in the periodic table

The inert gases were not known at the time of Mendeleef when he gave the original classification of elements. He did not leave gaps for these gases although he left gaps for several other unknown elements. After their discovery a new group was introducted into the period table. In the modern periodic table, these may be placed in the zero group as well as in the VIII group because of the following reasons.

1. Reasons for placing in zero group. (*i*) *Chemical inertness.* These elements are chemically inert. The existence of such a group is expected from the fact that there must be a separating group between strongly electronegative halogens of VIIA group and strongly electropositive alkali metals of IA group. Therefore these elements should be placed in between VIIA group and IA group, *i.e.*, in the zero group.

(*ii*) *Valency.* Valency is defined as the combining capacity of the element. The electronic configurations of these elements given as : He $\rightarrow$ 2 (Is), Ne $\rightarrow$ 2, 8 ($2s^2p^6$), Ar $\rightarrow$ 2, 8, 8 ($3s^2p^6$), Kr $\rightarrow$ 2, 8, 18, 8 ($4s^2p^6$), Xe $\rightarrow$ 2, 8, 18, 18, 8 ($5s^2p^6$), Rn $\rightarrow$ 2, 8, 18, 32, 18, 8 ($6s^2p^6$) indicate that the atoms of these elements have saturated shell. They have, therefore, no tendency either to lose, gain or share electrons with the atoms of other elements. In other words their combining capacity or valency is zero. Therefore, these elements should be placed in the zero group.

2. Reasons for placing in VIII group. (*i*) *Electronic configuration.* Electronic configurations of inert gases indicate that, except helium which has $1s^2$ configuration, all the elements have eight electrons in the outermost shell. Therefore, on this basis they should be placed in the VIII group.

(*ii*) *Analogy with other groups.* In the modern periodic table, all the groups, except VIII, are divided into two sub-groups A and B each. The elements of A subgroups are normal elements, whereas those of B subgroups are transitional elements. If the VIII group is also considered parallel to others, it should also be divided into two subgroups *viz.*, VIIIA and VIIIB. The VIIIA group should contain inert elements and VIIIB group should contain transitional elements. In such a case, the periodic table will be more symmetrical. Thus, from the above discussion it is clear that the inert gases may be placed in both the groups, *i.e.*, zero and VIII. However, for the general reference they are still considered as the members of the zero group. The position of these gases is further justified because of the facts namely: (*a*) They are all colourless, odourless and tasteless gases, under ordinary conditions, (*b*) The specific heat ratio, *i.e.*, C_p / C_v is the same (1.66) for all gases, which indicates their monoatomic character, (*c*) They show a regular gradation in their physical and chemical properties, (*d*) They do not have any resemblance with the elements of their neighbouring groups, (*e*) They are all chemically inert.

Why are the rare gases chemically inert?

The chemical inertness of these elements is due to the following reasons:

(*i*) We have already seen that the atoms of rare gases have saturated shells. They all have eigh electrons in their outermost shells, except helium, which has two. They have, therefore, no ten dency either to lose, gain or share electrons with atoms of other elements. These elements, thus, d not enter into chemical combination.

(*ii*) The atoms of these elements have all the electrons paired. Thus, the electrons are not available for sharing with the electrons of other atoms. Hence, they cannot form covalent compounds.

(*iii*) The electronegativity and electron affinity of these elements are zero. Thus, these elements have no tendency to attract or accept electrons and hence, are unable to form compounds.

(*iv*) The ionization potentials of these gases are very high. Hence, the removal of the electron from the outermost shell is very difficult.

History of discovery of rare gases

The credit for the discovery of rare gases goes to Lord Rayleigh and William Ramsay.

1. Discovery of argon. Cavendish in 1785, repeatedly sparked a mixture of purified air and excess of oxygen and absorbed the resulting oxides of nitrogen and unused oxygen in caustic potash and potassium pentasulphide solutions, respectively. He found that a small volume of the gas which was about 1/120 part of the original air, was left. It was neither nitrogen nor oxygen. Cavendish, however, could not draw any conclusion regarding the nature of the residual gas.

The experimental observation of Cavendish was not given any importance for a long time. Rayleigh in 1892, observed that the density of nitrogen isolated from air was 0.5% more than the density of nitrogen obtained by chemical means, *i.e.*, by the decomposition of urea of ammonium nitrite. Ramsay suggested that this difference of density was due to the presence of some heavier gas of the atmosphere. Rayleigh and Ramsay repeated Cavendish's experiment in a modified apparatus and found that the residual gas was chemically inert. They also removed various constituents of air. The residual gas thus obtained was the same as obtained by Cavendish's experiment. This gas was completely unreactive and different from all the known gases. On sparking in a discharge tube, it gave red light and a spectral line never observed before. This indicated the presence of a new element. It was named argon derived from the Greek word argos, meaning lazy.

2. Discovery of helium. Lockeyer and Frankland in 1868, observed a new yellow line in the chromosphere of the sun, during total solar eclipse. This line did not coincide with the D_1 and D_2 lines of sodium and Lockeyer called it D_3 line. This indicated the presence of a new element. It was named helium derived from the Greek word helios, meaning sun.

Hillebrand in 1888, observed that uranium minerals such as clevite, on treatment with dil. H_2SO_4 gave out a gas which he thought to be nitrogen. Ramsay who was constantly in the search of some other sources of argon, suspected the same gas to be argon. He examined the spectrum of the gas which indicated the presence of helium, the element already discovered by Lockeyer and Frankland in the chromosphere of the sun.

3. Discovery of neon, krypton and xenon. After the discovery of argon and helium, Ramsay and Travers collected large amounts of argon by the fractional evaporation of liquid air. This argon was then subjected to systematic fractional evaporation under different reduced pressures. From the first fraction, they isolated a new element which gave a number of red lines in the spectrum. It was named neon (meaning new). From the final fraction they isolated two more inert gases *viz.*, krypton (meaning hidden) which gave green spectral lines and xenon (meaning strange) which gave a blue glow.

4. Discovery of radon. Dorn in 1900, discovered radon as one of the disintegration products of radium.

$$^{226}_{88}Ra \xrightarrow{-^4_2He} {}^{222}_{86}Rn$$

Isolation of rare gases from dry air (*Chemical Method*)

The method used for the isolation of rare gases from dry air is the chemical method. This method consists of the following two steps:

1. Isolation of inert gases mixture from the dry air. Different constituents of the air *viz.*, CO_2, water vapour, O_2 and N_2 are removed from the air by any of the following methods and thus inert gases mixture which is free from CO_2, water vapour, O_2 and N_2 is obtained.

(*i*) *Rayleigh and Ramsay first method.* In this method, the different constituents of air, *i.e.*, carbon dioxide, water vapour, oxygen and nitrogen are removed, one by one, as shown below:

(*a*) *Carbon dioxide.* By passing over soda lime, [$2NaOH + CO_2 \longrightarrow Na_2CO_3 + H_2O$]

(*b*) *Water vapour.* By passing over [P_2O_5. $P_2O_5 + 3H_2O \longrightarrow 2H_3PO_4$]

(*c*) *Oxygen.* By passing over red hot Cu. [$2Cu + O_2 \longrightarrow 2CuO$]

(*d*) *Nitrogen.* By passing over heated Mg. [$3Mg + N_2 \longrightarrow Mg_3N_2$]

The process of passing air successively over red hot copper and heated magnesium is repeated till it is completely free from nitrogen and oxygen. The residual gas is a mixture of inert gases.

(*ii*) *Rayleigh and Ramsay second method.* In this method 9 volumes of air and 11 volumes of oxygen are introduced into a glass globe of 50 litres capacity. This glass globe is fitted with two heavy platinum electrodes through which a transformer discharge on 600-800 volts is passed (Fig. 23.3). The nitrogen of the air combines with excess of oxygen to form oxides of nitrogen. Now

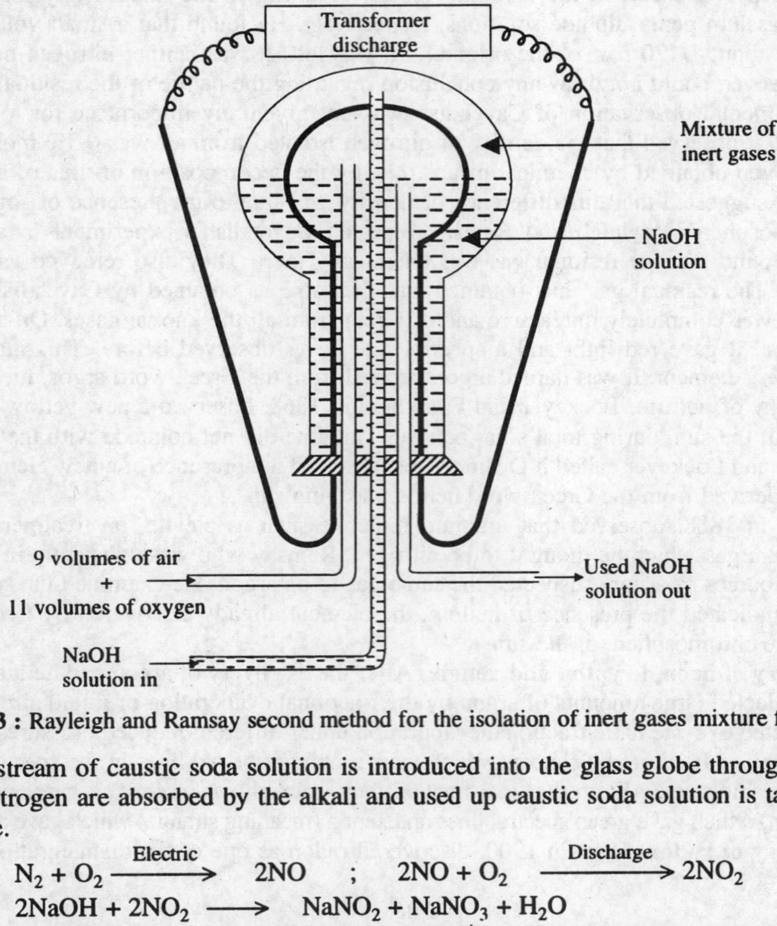

Fig. 23.3 : Rayleigh and Ramsay second method for the isolation of inert gases mixture from dry air.

continuous stream of caustic soda solution is introduced into the glass globe through a tube. The oxides of nitrogen are absorbed by the alkali and used up caustic soda solution is taken away by another tube.

$$N_2 + O_2 \xrightarrow{\text{Electric}} 2NO \quad ; \quad 2NO + O_2 \xrightarrow{\text{Discharge}} 2NO_2$$

$$2NaOH + 2NO_2 \longrightarrow NaNO_2 + NaNO_3 + H_2O$$

As a result of the above method, most of O_2 and N_2 is removed from the air in the glass globe. The mixture of residual gases contains noble gases with traces of O_2. The mixture of residual gases is passed through tubes containing alkaline pyrogallol in which O_2 gas gets absorbed and the mixture of rare gases which is free from O_2 gas comes out.

(*iii*) *Fisher Ringe's method.* In this method, the air free from moisture and carbon dioxide is passed over a heated (800°C) mixture containing 90% calcium carbide and 10% calcium chloride in an iron tube, when the following reactions take place:

$$CaC_2 + N_2 \xrightarrow{800^\circ C} CaCN_2 \text{ (\textit{calcium cyanamide})} + C \; ; \quad 2C + O_2 \longrightarrow 2CO$$

$$C + O_2 \longrightarrow CO_2 \; ; \quad 2CaC_2 + 3CO_2 \longrightarrow 2CaCO_3 + 5C$$

The mixture of gases containing CO and CO_2 so obtained is passed over heated copper oxide when CO is oxidized to CO_2 and CO_2 thus formed is absorbed by caustic potash solution.

$$CuO + CO \longrightarrow Cu + CO_2; \quad 2KOH + CO_2 \longrightarrow K_2CO_3 + H_2O$$

The individual members of the inert gases mixture are dried hy bubbling the mixute through concentrated H_2SO_4 and are finally passed through tubes packed with P_4O_{10}. The gaseous mixture thus obtained contains only noble gases.

2. Separation of the individual gases from the inert gases mixture. The separation of the individual gases from the inert gass mixture can be done by *Dewar's coconut charcoal method* which is based on the fact that coconut charcoal absorbs different inert gases at different temperatures and, therefore, the separation of these gases may be done. The mixture of inert gases is passed over coconut charcoal which is contained in a double walled bulb (called Dewar's flask) maintained at –100°C by liquid air (Fig. 23.4). He and Ne remain unadsorbed on the coconut charcoal and hence are pumped out of the bulb while Ar, Kr and Xe are adsorbed by the coconut charcoal. The mixture of He and Ne is then introduced into bulb containing coconut charcoal and maintained at –180°C when Ne gets adsorbed on the charcoal while He remains unadsorbed and hence is pumped out. In order to recover Ne the charcoal is heated.

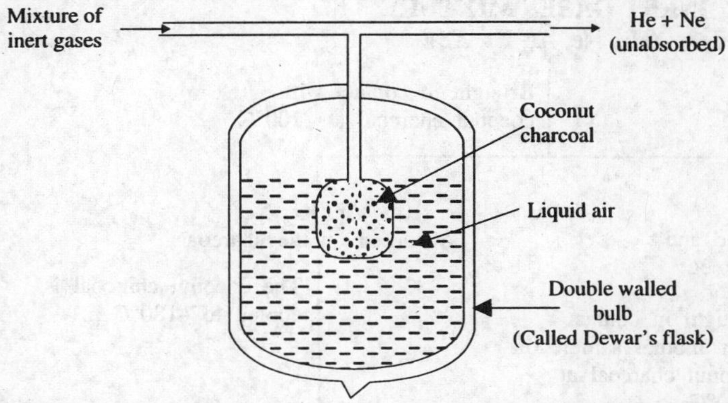

Fig. 23.4 : Dewar's coconut charcoal method for the separation of inert gases from the inert gases mixture.

The charcoal in which Ar, Kr and Xe had been adsorbed is cooled to liquid air temperature (–180°C). Ar gets liquefied and is thus separate while Kr and Xe remain adsorbed on the charcoal which is slightly warmed to a temperature equal to –90°C. At this temperature Kr remains unadsorbed and hence comes out of the charcoal. Xe remains adsorbed on the charcoal which, on being heated at 0°C, gives Xe. Various steps of this method are illustrated in flow sheet given on the next page.

Isolation of inert gases from liquid air (*Physical Method*)

Liquid air contains He, Ne, Ar, Kr, Xe, N_2 and O_2. Inert gases can be obtained from this liquid air by its fractional evaporation. The method used

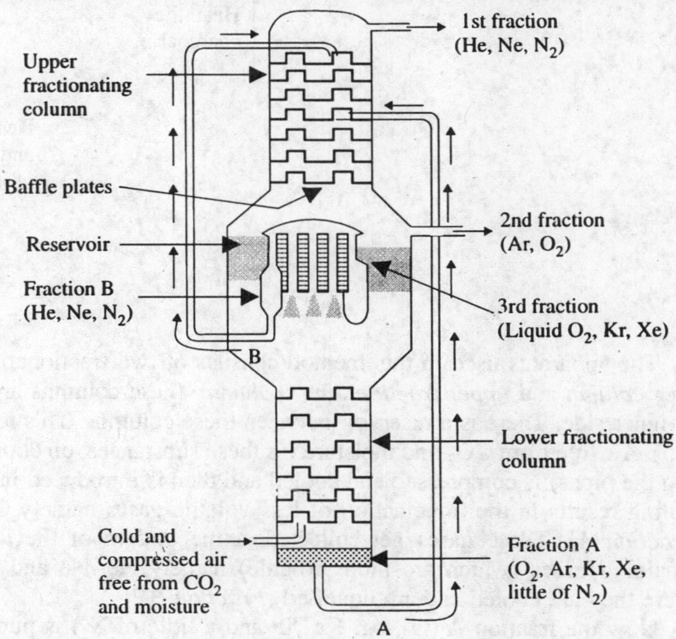

Fig. 23.5 : Isolation of inert gases from liquid air.

is based on the fact that the boiling points of the various constituents of liquid air widely differ from each other as shown : He = – 269°C, Ne = – 246°C, N_2 = –195.7°C, Ar = –186°C, O_2 = –183°C, Kr = –152°C, Xe = –107°C. Important conclusions drawn from these values can be summarised as : (*i*) He and Ne have very low boiling points and hence during the process of liquefying the air, He and Ne will remain as gases while other constituents will change into liquids. (*ii*) There is little difference between the boiling points of Ar and O_2 and hence these gases will be collected more or less as a single fraction. (*iii*) Kr and Xe have fairly high boiling points as compared to those of N_2, Ar and O_2. Therefore, Kr and Xe are always found in the less volatile portion of the liquid air and are always easily separated from other gases.

Flow sheet for the separation of inert gases by Dewar's coconut charcoal method

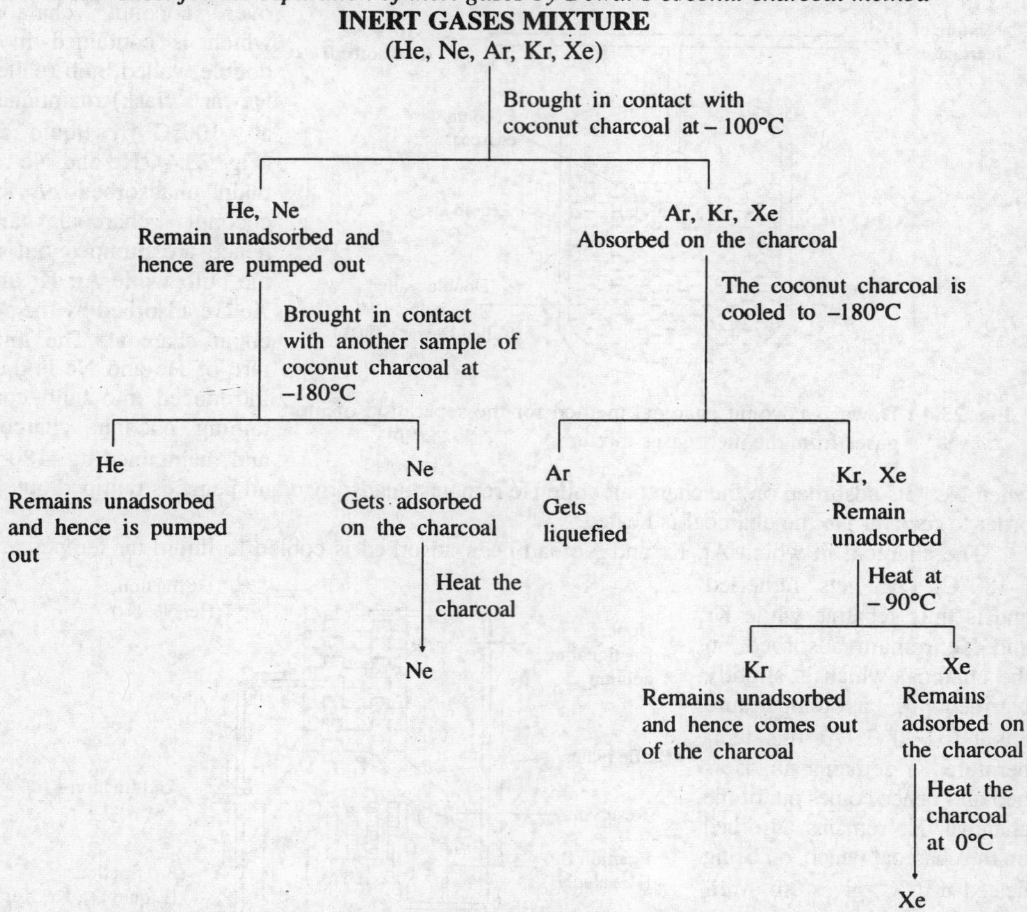

The apparatus used in this method consists of two fractionating columns *viz., lower fractionating column* and *upper fractionating column*. These columns are provided with baffle plates on the inner side. There is a reservoir between these columns. This reservoir contains liquid O_2. (Fig. 23.5). Air free from CO_2 and moisture (as these substances, on cooling, change into solid and hence clog the pipes) is compressed and cooled and then is introduced into the fractionating column. The cooling results in the liquefaction of less volatile gases namely O_2, Ar, Kr, Xe and a little of N_2 (*Fraction A*). These gases get collected at the bottom of the lower fractionating column. The remaining gases (which are more volatile) namely He, Ne and N_2 rise upwards in the column where they are cooled and are liquefied (*Fraction B*).

Now the fraction A (O_2, Ar, Kr, Xe and a little of N_2) is pumped up from the bottom and is dropped from the middle of the upper fractionating column. The fraction B (He, Ne and N_2) is also pumped and is dropped from the top of the upper fractionating column. By this operation the liquid air gets separated into three fractions *viz.,* (*a*) *1st fraction.* This fraction contains most volatile gases namely He, Ne and N_2. The mixture of these gases escapes from the top of the upper fractionating column. (*b*) *2nd fraction.* This fraction contains Ar and O_2. The mixture of these gases escapes from the top of central reservoir. (*c*) *3rd fraction.* This fraction contains liquid O_2, Kr and Xe. The mixture of these gases gets collected at the bottom of the reservoir.

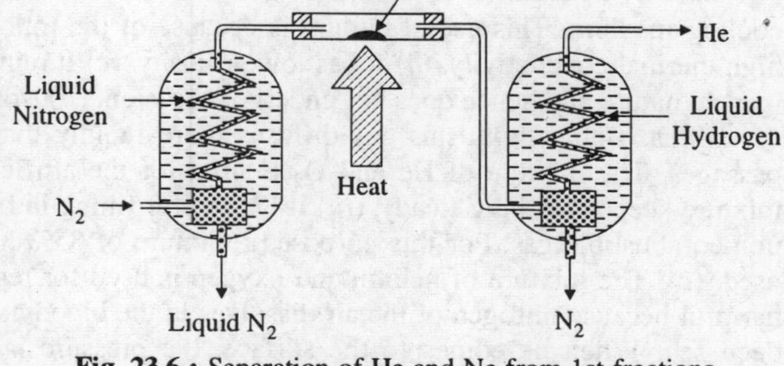

Fig. 23.6 : Separation of He and Ne from 1st fractions.

Now from these fractions the individual inert gases are obtained as follows:

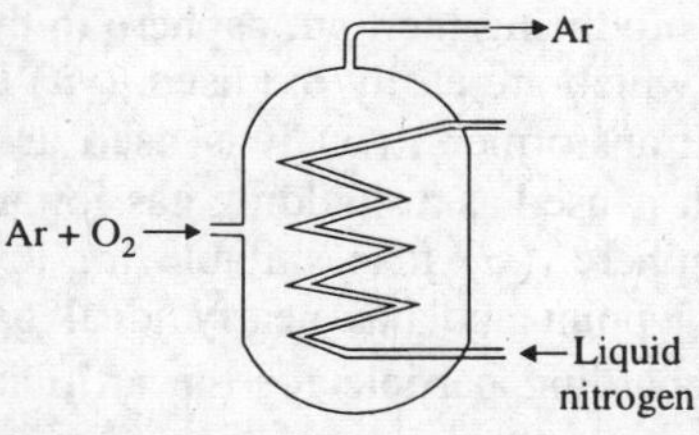

Fig. 23.7 : Separation of Ar from the 2nd.

(*i*) *Separation of helium and neon form the 1st fraction.* The first fraction containing He, Ne and N_2 is passed through a spiral tube cooled by liquid nitrogen. Here most of the nitrogen is condensed and a mixture of He, Ne and some N_2 passes out. The remaining N_2 is removed by passing the mixture over beated calcium carbide.

$$CaC_2 + N_2 \rightarrow CaCN_2 + C$$

The gaseous mixture containing He and Ne is then passed through a spiral tube cooled by liquid H_2 at –253°C when Ne solidifies (boiling point –246°C) and is collected at the bottom, while helium in the gaseous state escapes from the top (Fig. 23.6).

(*ii*) *Separation of argon from the 2nd fraction.* The second fraction containing argon and oxygen is passed through a column which is provided with a coil in which liquid N_2 is being circulated. Here most of the oxygen is condensed while more of argon escapes at the top (Fig. 23.7). The argon, so obtained, still contains some oxygen which is removed by passing over heated copper.

$$2Cu + O_2 \longrightarrow 2CuO$$

(*iii*) *Separation of krypton and xenon from the 3rd fraction.* From the third fraction containing liquid O_2, Kr and Xe, oxygen is removed by evaporation. The remaining liquid becomes richer in krypton and xenon. These two may be separated easily by fractional evaporation as there is a wide difference in their boiling points (Kr = –152°C, Xe = –107°C) (Fig. 23.8).

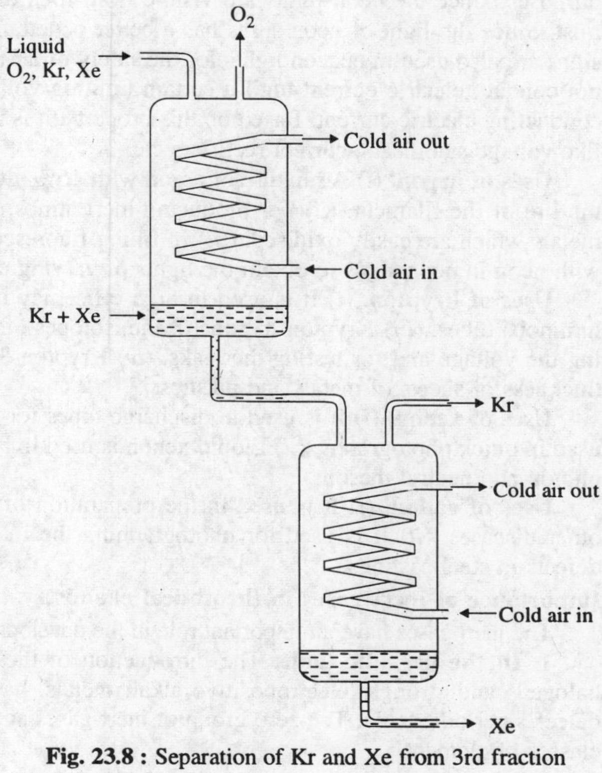

Fig. 23.8 : Separation of Kr and Xe from 3rd fraction

Uses of noble gases

Of all the noble gases, only He and Ar which are available easily and in plenty have got many uses. The uses are due to their chemical inertness and low boiling points. Important uses of noble gases are given below:

Uses of helium. (*i*) It is used in gas cooled atomic reactors as a heat transfer gas, *i.e.*, as a cooling medium. This use of helium is because of the following properties of this gas: (*a*) It has high thermal conductivity. (*b*) It has low viscosity. (*c*) It remains unaffected by irradiation. (*d*) It is inert in nature and hence does not undergo corrosion. (*ii*) Mixed with O_2, it is used in the treatment of asthma. Being light, this gas diffuses more rapidly than air through the partly choked lung passages. The mixture of He and O_2 is used for the artificial breathing of asthma patients. This mixture keeps breathing steady. (*iii*) It is used for filling in balloons and airships as it is a light and non-combustible gas. For this purpose, a mixture of 85% helium and 15% hydrogen is generally used. (*iv*) The mixture of helium and oxygen is used for respiration by sea divers. The air proves harmful because nitrogen of the air dissolves in the blood at high pressure when the diver is under deep sea. When he comes to the surface, the pressure is suddenly decreased. The bubbles of nitrogen gas now escape from the blood and are collected at joints which cause pains. Helium taken in the place of air is much less soluble in blood at high pressure. (*v*) Liquid helium is used for producing low temperatures required for research work. (*vi*) For producing inert atmosphere in the welding and metallurgy of certain metals (*e.g.*, Mg, Al, Ti etc.) which are easily oxidised. (*vii*) In the preservation of food. (*viii*) It is used for filling electrical transformers. (*ix*) It is used as a pressuring agent in rockets to expel liquid O_2 and liquid H_2. (*x*) It is used as a shielding gas for arc welding as it protects both the arc and the metal from atmosphere. (*xi*) It is suitable for low temperature gas thermometry because this gas has low boiling point and has nearly ideal gas behaviour. (*xii*) Helium nucleus (α-particles) is used as a bombarding particle for the artificial disintegration of atoms.

Uses of neon. (*i*) It is used in neon discharge lamps and signs which are used for advertising purposes, since the neon lights are visible from long distances and even visible through fog and mist. Since the light of neon signs has a better penetrating power through fog and mist, the neon signs are also used in beacon lights for the safety of air navigation. (*ii*) Mixture of He and Ne does not conduct electric current until a certain limiting voltage is applied to it and thereafter it starts conducting electric current. Based on this property, it is used in certain protective electrical devices like voltage stabilisers, current rectifiers etc.

Uses of argon. (*i*) A mixture of argon with 15% nitrogen is filled in electric bulbs to increase the life of the filament. (*ii*) For producing inert atmosphere in welding and metallurgy of certain metals which are easily oxidised. (*iii*) In filling fluorescent tubes and radio-valves. (*iv*) It is mixed with neon in neon signs to obtain the lights of varying colours.

Uses of krypton. (*i*) It is used in high efficiency miner's cap lamps. (*ii*) It is used in filling luminous tubes. (*iii*) Krypton-85, one of the isotopes of Kr, is used in electronic tubes for regulating the voltage and for testing the leaks. (*iv*) Krypton-85 is also used for the measurement of the thickness of sheets of metals and plastics.

Uses of xenon. (*i*) It is used in discharge tubes for producing high speed flash of bluish light used in quick photography. (*ii*) Liquid xenon is used in bubble chamber for the detection of gamma photon and neutral mesons.

Uses of radon. (*i*) It is used in the preparation of ointment for the treatment of canner and other diseases. (*ii*) It is used for photographing the interior of opaque meterials for locating the defects in steel castings.

Importance of inert gases in theoretical chemistry.

The inert gases have an important role in the development of theoretical chemistry as follows:

1. In the periodic table. The introduction of these gases between strongly electronegative halogens and strongly electropositive alkali metals, has removed one of the anomalies of Mendeleef's periodic table. The zero group of inert gases acts as a separating group between these two classes of element.

2. In the elucidation of distribution of electrons in atoms. The outermost shell of the atoms of inert gases has eight electrons, except helium which has two. This configuration accounts for the inertness of these elements. On this basis Bohr and Bury postulated that outermost shells of the atoms cannot have more than eight electrons.

3. In the development of electronic theory of valency. The modern concept of valency is based upon the structure of inert gases. According to the concepts of Kossel and Lewis, the atoms tend to acquire inert gas type of electronic configuration by transference of electrons from one atom to another (electrovalency), by sharing of electrons between two atoms (covalency) and by the combination of both the processes, *i.e.*, transference and sharing (coordinate).

4. In the discovery of isotopes. Neon was the first nonradioactive substance whose isotopes were first of all separated. On this basis, it was concluded that other elements may also consist of a mixture of two or more isotopes and thereafter, a number of isotopes of other elements have been identified and separated.

5. In radioactivity. Helium is one of the disintegration products of some radioactive elements. This fact has been useful in the development of the theory of the radioactive disintegration and group displacement law.

Conditions and types of compounds formed by inert gases

The atoms of inert gases have saturated shells and, therefore, they are chemically inert. Recent researches have shown that under certain specified conditions, they enter into chemical combinations and form some rare chemical compounds. The specified conditions and types of compounds formed by these gases are described below :

1. Compounds formed under excited conditions: Compounds of helium. These compounds known as halides are formed by sparking helium at low pressure in presence of mercury, tungsten etc., to form $HgHe_2$, $HgHe_{10}$, WHe_2. In addition, metal electrodes in the discharged tubes also form compounds with helium like $BiHe_2$, FeHe, Pt_3He, PdHe. In these compounds, it is supposed that helium is adsorbed on the surface of the metal and, therefore, these compounds are not considered as true chemical compounds.

2. Compounds formed through coordination: Compounds of argon. Booth and Wilson in 1935 reported that argon forms a number of unstable compounds with varying numbers of BF_3 molecules, *e.g.*, $Ar.BF_3$, $Ar.2BF_3$, $Ar.3BF_3$, $Ar.6BF_3$, $Ar.8BF_3$ and $Ar.16BF_3$. In these compounds, argon atom donates a pair of electrons to boron atom. In the case of higher compounds, the fluorine atoms also donate pair of electrons. The structure of $Ar.8BF_3$ is given in the margin.

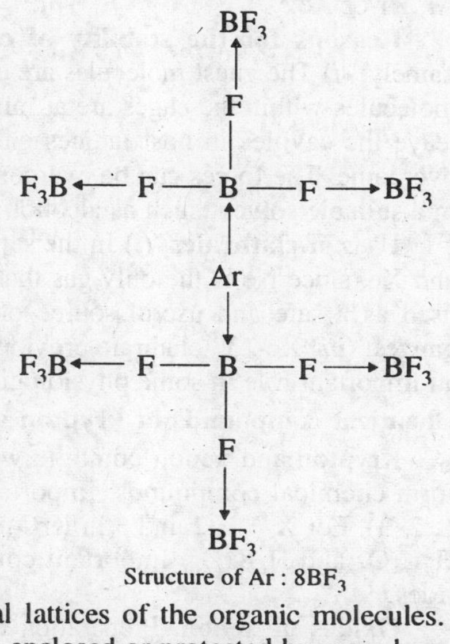

Structure of Ar : $8BF_3$

3. Compounds of inert gasses with water: Formation of hydrates. The inert gases Ar, Kr and Xe form compounds with water at low temperature and high pressure. These compounds are known as hydrates. The compounds like $Ar.6H_2O$, $Kr.6H_2O$ and $Xe.6H_2O$ have been obtained and studied.

4. Compounds formed by physical trapping : Clathrate compounds. The inert gases argon, krypton and xenon form solid compounds with certain organic molecules such as phenol, hydroquinone, $C_6H_4(OH)_2$ under pressure. In these compounds the inert gases are enclosed into the crystal lattices of the organic molecules. These compounds are known as clathrates. (Latin : clathrote = enclosed or protected by cross bars or grating). These are also referred to as cage compounds.

In clathrates the noble gases are trapped into the cavities of crystal lattices of certain organic and inorganic compounds.

The organic or inorganic substance having the cavities of crystal lattices is called the 'host' while the atom of noble gases entrapped in it is called the 'guest'.

In these compounds the guest components are held by van der Waals forces in crystalline cages formed by the host substances and consequently there is no chemical bonding involved in these compounds. It is not necessary that all the cavities in the host structure are filled. As a result the clathrates are normally nonstoichiometric compounds. The 'guest' can escape from the 'host' by heating the crystals or by dissolving them in a suitable solvent such as alcohol.

Conditions for the formation of clathrates. (*i*) The crystal lattice of the host component should contain cavities of appropriate size. (*ii*) Size of the guest component should be such that it should fit in the cavities of the host component without bringing any atoms closer together than which corresponds to the van der Waals radii of the atoms.

Types and preparation of clathrates. (*i*) *Gas hydrates.* Clathrates of Ar, Kr and Xe in which water molecule is the host component are called gas hydrates. These have been prepared by solidifying water in presence of Ar, Kr and Xe. The general formula of these is 5.75 H_2O.G where G = Ar, Kr or Xe. No gas hydrate of He and Ne has yet been prepared even under high pressure due to their small size and highly volatile nature. The stability of the gas hydrates increases as the atomic number of the 'guest' atom increases. In noble gas hydrates the water molecules are linked together by hydrogen bonds, leaving roughly spherical voids which are occupied by gas atoms.

(*ii*) *Quinol clathrates.* Clathrates of Ar, Kr and Xe have been obtained by dissolving any of these in aqueous saturated solution of quinol under a pressure of 10-40 atmospheres followed by slow cooling.

These compounds are crystalline and can be stored for several years. However, when heated or dissolved in water, the gas atoms escape and quinol remains behind as such. Quinol molecules which form the cavities are held together by H-bonds.

(*iii*) *Phenol and β-hydroquionone clathrates.* Clathrates given by phenol and β-hydroquinone with Ar, Kr and Xe have been prepared by dissolving any of these gases in phenol or β-hydroquinone followed by slow crystallisation. these clathrates are also quite stable. Clathrate of β-hydroquinone can be represented by the formula, 3. β-hydroquinone. G or 3.β-$C_6H_4(OH)_2$.G where G = Ar, Kr or Xe.

Reasons for the stability of clathrates. The stability of clathrates is due to the reasons namely: (*i*) The guest molecules are tightly held in the cavities of the host molecules. (*ii*) The guest molecules within the cages are at minimum potential energy. The guest molecules can be made to leave the cavities in host lattices only when the forces holding the molecular cages together are overcome. The forces can be overcome either by heating the crystals, or by dissolving the crystals in a suitable solvent such as alcohol.

Uses of clathrates. (*i*) In the separation of noble gases, *e.g.*, Ne can be separated from Ar, Kr and Xe, since Ne is the only gas that does not form a clathrate with quinol. (*ii*) Kr-85 clathrate is used as a safe and useful source of beta radiations which are used for measuring thickness of gauges. (*iii*) Xe-133 clathrate provides a compact source for gamma radiations. (*iv*) Clathrates play an important role in some physiological actions.

Chemical compounds of krypton and xenon

Krypton and xenon combine with strong electron-acceptors viz., oxygen and fluorine and form chemical compounds. Important compounds of Kr are KrF_2, KrF_4 and KrO_3. x H_2O (x = 1, 2, 3). For X = 1, 2 and 3 in KrO_3. x H_2O, we get the compounds having the formula H_2KrO_4, H_4KrO_5 and H_6KrO_6. Important compounds formed by Xe are oxides, fluorides and oxy fluorides.

Oxides of xenon. Preparation and properties of the oxides have been given in chapter on "Oxides" (chapter 28) while their structure has been discussed in chapter 15.

Fluorides of xenon. Stable fluorides formed by xenon are XeF_2, XeF_4 and XeF_6. Their preparation and properties have been given in chapter on "Halides" (chapter 29) while their structure/geometry has been discussed in chapter 14.

Oxy fluorides of xenon. Xenon forms *xenon oxy difluoride* ($XeOF_2$), *xenon oxy tetrafluoride* (XeO_2F_2), *xenon trioxydifluoride* (XeO_3F_2) and *xenon dioxytetrafluoride* (XeO_2F_4). Preparation and properties of these compounds are given in chapter on "Halides" (chapter 29) and their structure has been discussed in chapter 15.

University Questions

1. Explain why CCl_4 and $SiCl_4$ are different in their chemical behaviour ? *(VBS Phurvanchal 2002)*
2. Write a note on "Rhombic sulphur". *(VBS Purvanchal 2002)*
3. Justify the inclusion of oxygen, sulphur, selenium, tellurium and polonium in the same group of the periodic table. *(VBS Purvanchal 2002)*
4. Discuss the main points of similarity and dissimilarity between oxygen and sulphur. *(VBS Purvanchal 2002)*
5. (*a*) Give an account of the discovery of rare gases of the atmosphere. How can they be separated from each other ?
   (*b*) Comment on the statement that there is little likelihood of the formation of compounds of He and Ne. *(Lucknow 2002)*
6. Explain the necessary conditions for the formation of compounds of inert gases. Discuss the structure and geometry of XeF_6. (Nagpur 2002)
7. Why is iodine more basic than other halogens ? Explain. *(Nagpur 2002)*
8. (*a*) Explain why the solubilities of alkaline earth metal sulphates decrease down the group.
   (*b*) What is catenation ? Explain with reference to carbon group elements. *(Delhi 2002)*
9. Name the elements of carbon family and discuss the variation in their metallic properties on going down the group. *(VBS Purvanchal 2003)*
10. Explain why CO_2 is a gas while SiO_2 is a solid.
11. Writ a note on allotropy of sulphur. *(VBS Purvanchal 2003)*
12. Why does fluorine differ in its properties from other halogens ? *(VBS Purvanchal 2003)*
13. What happens when Cl_2 is passed through dil. solution of NaOH ? *(Delhi 2003)*
14. Explain giving reasons :
   (*i*) Fluorine exists only in –1 oxidation state whereas other halogens exhibit negative as well as positive oxidation states.
   (*ii*) Iodine shows metallic character. *(Nagpur 2003)*
15. (*i*) Oxygen exhibits – 2 oxidation state but other members of the group show oxidation states of + 4 and + 6 also. Explain.
   (*ii*) Compare the elements of nitrogen family with respect to the following : (*a*) Electronic configuration (*b*) Oxidation state (*c*) Electronegative character (*d*) Halides (*e*) Hydrides. *(Rohilkhand 2003)*
16. (*a*) Discuss the positive and negative oxidation state in nitrogen family.
   (*b*) Explain the oxidizing power of halogens.
   (*c*) Explain the abnormal behaviour of oxygen. *(Nagpur 2003)*
17. Explain why F exhibits only –1 oxidation state while other halogens also exhibit –1, +1, +3, +5, +7 oxidation states. *(Bhopal 2004)*
18. Discuss the variation in properties of group 15 elements hydrides with reference to :
   (*i*) Basicity (*ii*) Stability (*iii*) Reducing character (*iv*) Bond angle. *(Bhopal 2004)*
19. What you understand by the electropositive character of iodine? *(GND Amritsar, 2004)*
20. (*a*) Why is BF_3 is a weaker Lewis acid than BCl_3 and BBr_3?
   (*b*) What type of oxides are formed by alkali metals. Explain with examples.
   (*c*) Why do alkali metals give coloured solution in liq. NH_3? *(GND Amritsar, 2004)*
21. (*a*) Which out of Li and Na has greater value of hydration energy?
   (*b*) Nitrogen is a gas but phosphorus is a solid. Explain.
   (*c*) Nitrogen does not form pentahalides. Explain.
   (*d*) What led Ramsay to name elements of group 18 to call them as inert gases? *(GND Amritsar, 2004)*

24

Occurrence, Extraction / Isolation, Properties and Uses of Some Main Group Elements

Reactive Forms of Hydrogen

Following are the reactive forms of hydrogen:

1. Nascent hydrogen

If hydrogen gas is bubbled through acidified potassium permanganate, colour is not discharged. To the same if we add some zinc pieces, tiny bubbles of hydrogen are found to rise and potassium permanganate is decolorised. Hydrogen produced *in situ* in the second case is more active and is called **nascent hydrogen** (*newly born*).

$$Zn + H_2SO_4 \longrightarrow ZnSO_4 + \quad + 2H \text{ (Nascent hydrogen)}$$

$$2KMnO_4 + 3H_2SO_4 + 10H \longrightarrow K_2SO_4 + 2MnSO_4 + 8H_2O$$

Ferric chloride solution (*yellow*) is decolorised by adding zinc pieces and hydrochloric acid. This is also due to reduction of ferric chloride to ferrous chloride (*faint green solution*) by nascent hydrogen.

$$FeCl_3 + H \longrightarrow FeCl_2 + HCl$$

Nascent hydrogen produced from Zn and dil. H_2SO_4 can reduce $KClO_3$ to KCl and $AsCl_3$ to AsH_3.

$$KClO_3 + 6H \longrightarrow KCl + 3H_2O$$

$$AsCl_3 + 6H \longrightarrow AsH_3 + 3HCl$$

Preparation of nascent hydrogen. Nascent hydrogen can be prepared by the following methods:

(*i*) *By the action of Zn on* H_2SO_4

$$Zn + H_2SO_4 \longrightarrow ZnSO_4 + 2H$$

(*ii*) *By the action of water on sodium amalgam.*

$$Na + H_2O \longrightarrow NaOH + H$$

(*iii*) *A mixture of Sn and conc. HCl or Fe and water acidified with a little of HCl also gives nascent hydrogen which is used for reducing aromatic compounds.*

$$Sn + 2HCl \longrightarrow SnCl_2 + 2H$$

(*iv*) *Metallic sodium and absolute alcohol also give nascent hydrogen.*

$$Na + C_2H_5OH \longrightarrow C_2H_5ONa + H$$

(*v*) *By the action of NaOH on Zn or Al*

$$Zn + 2NaOH \longrightarrow Na_2ZnO_2 + 2H$$

Theories to explain the reactivity of nascent hydrogen. (*i*) *Atomic hydrogen theory.* In one of the explanations it was suggested that nascent hydrogen at the moment of its liberation exists as single atoms while in ordinary hydrogen these atoms have combined to form molecule, H_2. Its activity is due to its atomic state. Single atoms of nascent hydrogen are ready to react. On the other hand, some work is required to be done on molecules of ordinary hydrogen to split them into single atoms before they can react.

This theory fails to account for a number of facts such as the different degrees of reactivity of nascent hydrogen from different reagents. For example, nascent hydrogen from zinc and dilute sulphuric acid can reduce chlorates to chlorides but one from sodium amalgam and water cannot.

(*ii*) *Energy theory.* In another explanation, it is believed that difference in the degrees of activity of nascent hydrogen obtained from different sources is due to different amounts of energy associated with it. Some of the energy liberated as a result of chemical reaction appears as heat energy while the rest becomes associated with hydrogen.

(*iii*) *Internal pressure theory.* In another suggested explanation hydrogen at the moment of its production is believed to be in the form of very minute bubbles with great internal pressure on account of their small size. It has been shown experimentally that activity of ordinary hydrogen can be increased by applying great pressure.

(*iv*) *Theory of association of energy.* Recently, an attempt has been made to explain its activity on the basis of association of energy when electrons are transferred from metals of low electronegativity to H^+ ions in order to change its valency from + 1 to zero.

$$Zn \longrightarrow Zn^{2+} + 2e^- \; ; \quad Na \longrightarrow Na^+ + e^- \; ; \quad Al \longrightarrow Al^{3+} + 3e^-$$

The hydrogen in solution accepts these electrons : $H^+ + e^- \longrightarrow H$

As the energy liberated in the transfer of electrons is different for different reactions, the reducing capacity of various reactions producing nascent hydrogen would be different.

2. Adsorbed or occluded hydrogen

Some metals (*e.g.*, Pd, Pt, Au, Fe, Ni) can adsorb under certain conditions relatively large volumes of hydrogen. For example, Graham (1867) found that one volume of palladium adsorbs 935 volumes of hydrogen while cooling from red heat. The gas thus adsorbed by metals is given off when the metal is heated especially under reduced pressure. The phenomenon was termed **occlusion** or **adsorption** by Grahm. The gas so adsorbed by the metal is termed **occluded** or **adsorbed hydrogen.** *Occluded hydrogen is more active and a stronger reducing agent than ordinary hydrogen.* For example : (*i*) It reduces ferrous salts. (*ii*) It combines with chlorine and iodine even in the dark. (*iii*) It precipitates mercury from mercuric chloride.

Theories to explain the nature of occluded hydrogen. Different views have been put forward from time to time to explain the nature of occluded hydrogen. For example: (*i*) Formerly it was believed that occluded hydrogen forms a chemical compound with the metal. (*ii*) Holt, Edgar and Firth (1913) suggested that a part of occluded hydrogen is present as a condensed layer on the surface (adsorption) while the rest of it dissolves in the interior of the metal. (*iii*) According to Gillespie and Hall (1926) there was evidence for the existence of two immiscible solid solutions. They thought that one of these might be a compound. Pd_2H. (*iv*) At present it is believed that occluded hydrogen gives an *interstitial compound* in which hydrogen occupies the holes in the crystal lattice of the metal.

3. Active hydrogen

When ordinary hydrogen at room temperature is subjected to the action of silent electric discharge at an electrical pressure more than 30,000 volts, it changes into an active variety of hydrogen. This is called active hydrogen. Active hydrogen is **very reactive** as shown by the reactions *viz.*, (*i*) It combines directly with lead or sulphur to form hydrides. The combination occurs at the room temperature. (*ii*) It reduces cupric and ferric oxides to corresponding metals even in the cold.

Theories tr explain the nature of active hydrogen. Nature of active hydrogen is not clearly understood. Two views put forward are : (*i*) It consists of molecules H_3 (*cf.* formula of ozone, O_3 which is produced by passing silent electric discharge through ordinary oxygen). (*ii*) It is atomic hydrogen. Neither of these explanations has been universally accepted. Active hydrogen is at present regarded merely as an active form of ordinary hydrogen, H_2 (*cf.* active nitrogen).

4. Atomic hydrogen

In 1922, Langmuir showed that when molecular hydrogen (H_2) in contact with a tungsten or platinum wire was heated by an electric current at low pressure, it is dissociated into atomic hydrogen (2H). The dissociation of molecular hydrogen into atomic hydrogen is an endothermic reaction.

$$H_2 \longrightarrow H + H - 103.7 \text{ Kcals}$$

The atomic hydrogen so produced was found to be more reactive than ordinary, nascent and adsorbed hydrogen.

Preparation of atomic hydrogen. (*i*) *By high temperature dissociation method.* This method is based on the observation made by *Langmuir* who showed that wires of W, Pt and Pd heated to temperature of 1000-2000°C in molecular hydrogen, at pressure of 0.01 mm or less, produce atomic hydrogen, by absorbing a large amount of energy. This method consists in blowing a current of hydrogen across an arc (20 amp. at 300-800 volts) struck between tungsten electrodes in an atomosphere of hydrogen. The percentage of degree of dissociation (D) of molecular hydrogen into atomic hydrogen at different temperatures is as: at 1000°K, D = 0.00000371%; at 2000°K, D = 0.122%; at 3000°K, D = 9.03%; at 4000°K, D = 62.50% and at 5000°K, D = 94.69%.

(*ii*) *By discharge tube method.* Wood in 1922 showed that active hydrogen is formed when a glow discharge is passed through hydrogen at a pressure of 0.1 to 1.0 mm.

(*a*) *Bonhoeffer's method.* Bonhoeffer designed an apparatus for the production of atomic hydrogen by Wood's method and for studying its effect on various substances. The apparatus contains a S-shaped discharge tube, 2 metres long and about 2 centimetres in diameter, fitted with two cylindrical electrodes of sheet aluminium. It is connected with a U-tube on one side and to a trap bulb through an experimental tube on the other side. A very slow current of hydrogen is introduced into the U-tube which is surrounded by liquid air to remove water vapour from it. A pressure (0.1 to 1 mm) is 300 to 500 maintained in the discharge tube where the gas is subjected to an electric discharge of volts when atomic hydrogen is formed. The action of atomic hydrogen on various substances is studied by introducing them in the current of atomic hydrogen coming out of the discharge tube (See Fig. 24.1).

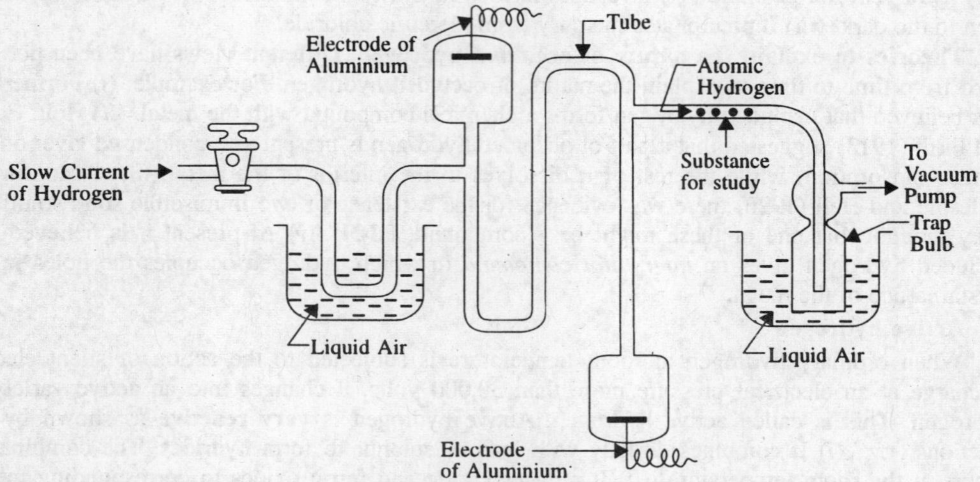

Fig. 24.1 : Bonhoeffer's apparatus for the production of atomic hydrogen.

(*b*) *Harteck's method.* It is a modification of Bonhoeffer's method. In this method a mixture of hydrogen and neon is fed into the discharge tube at pressures upto 20 mm. The presence of neon facilitates the dissociation of hydrogen into atoms.

(*iii*) *By photochemical method* (Cario and Frank-1922). Ordinary hydrogen mixed with mercury vapour is subjected to the resonance radiation of a mercury arc. The method gives larger yields of atomic hydrogen and at comparatively higher pressure (upto 0.5 atmosphere).

Properties of atomic hydrogen. (*i*) *Stability.* The life period of atomic hydrogen is only 1/3rd of a second but can be extended under special circumstances to 10 seconds.

(*ii*) *Recombination of atomic hydrogen to give molecular hydrogen.* Atomic hydrogen is extremely unstable (Life period = $\frac{1}{3}$ second) and quickly recombines to give molecular hydrogen. Metals like Pt, Pd etc., accelerate this recombination.

$$H + H \rightleftharpoons H_2 + 103.7 \text{ Kcals}$$

This exothermic reaction is used in *atomic hydrogen welding torch* which is employed for welding the cracked metals like Mn, Cr, W etc. In atomic hydrogen welding torch when molecular hydrogen is passed through an electric arc struck between two tungsten electrodes, the hydrogen molecules absorb heat and are split into atomic hydrogen.

$$H_2 + \text{Heat} \longrightarrow H + H$$

Just beyond the arc these atoms recombine to form H_2 molecules and evolve the heat which they previously absorbed.

$$H + H \longrightarrow H_2 + \text{Heat}$$

This heat plus the heat of burning hydrogen creates a high temperature (= 5000°C). This heat is used for welding the cracked metals. (See Fig. 24.2) and for melting some of the most refractory materials such as tungsten, tantalum and thorium oxide.

(*iii*) *Formation of hydrides.* It combines with metals (*e.g.*, Na, K, Li) and nonmetals (*e.g.*, S, P, As, Sb) to give hydrides.

$$Li + H \longrightarrow LiH$$
$$S + 2H \longrightarrow H_2S$$
$$P + 3H \longrightarrow PH_3$$
$$As + 3H \longrightarrow AsH_3$$
$$Sb + 3H \longrightarrow SbH_3$$

(*iv*) *Reducing action.* It is an extremely powerful reducing agent and reduces oxides, chlorides and sulphides of Cu, Ag, Hg etc., to metal.

$$HgO + 2H \longrightarrow Hg + H_2O$$

(*v*) *Combination with oxygen.* It combines with oxygen to give hydrogen peroxide.

$$2H + O_2 \longrightarrow H_2O_2$$

(*vi*) *Luminiscent effect.* A mixture of atomic hydrogen and vapour of alkali metals or mercury gives characteristic glows.

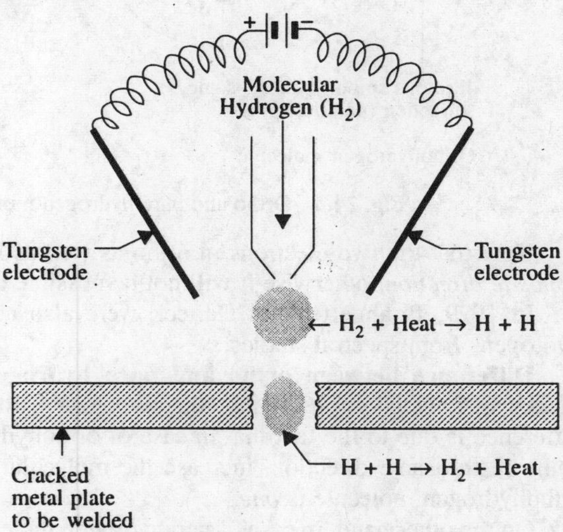

Fig. 24.2 : Atomic hydrogen welding torch.

(*vii*) *Combination with organic compounds.* It reduces unsaturated organic compounds to saturated compounds, *e.g.*, C_2H_4 is converted into C_2H_6 and oleic acid changes into stearic acid.

$$\underset{\text{Ethylene}}{C_2H_4} + 2H \longrightarrow \underset{\text{Ethane}}{C_2H_6}$$

$$\underset{\text{Oleic acid}}{C_{17}H_{33}COOH} + 2H \longrightarrow \underset{\text{Stearic acid}}{C_{17}H_{35}COOH}$$

Uses of atomic hydrogen. Atomic hydrogen is used for the production of atomic hydrogen welding torch which is employed for welding and cutting metals as described above.

5. Ortho and para hydrogen molecules

We know that a hydrogen molecule is composed of two hydrogen atoms. Each of these two H-atoms of H_2 molecule has one proton in its nucleus and one electron revolving round the nucleus. Thus H_2 molecule which is formed from two H-atoms has two protons and two electrons in it. The two electrons in these two H-atoms always spin in *opposite direction* in accordance with Pauli's principle.

$$H(\uparrow) + H(\downarrow) \rightarrow H\uparrow \quad \downarrow H \text{ or } H—H \text{ or } H_2$$

However, the two protons present in the nucleus of both the H-atoms may spin either in the *same direction* (*i.e., parallel direction*) or in *opposite direction* (*i.e., antiparallel direction*).

Thus, depending on the direction of the two protons in the nucleus, we have the following two types of hydrogen molecule:

(*i*) **Orthohydrogen molecule.** *Hydrogen molecule in which both the protons in the nuclei of both H-atoms spin in the same direction (i.e., parallel direction) is called orthohydrogen* (Fig. 24.3). In this from the resultant nuclear spin is *one.*

(*ii*) **Parahydrogen molecule.** *Hydrogen molecule in which both the protons in the nuclei of both H-atoms spin in opposite direction (i.e., antiparallel direction) is called parahydrogen* (Fig. 24.3). In this form the resultant nuclear spin is *zero.*

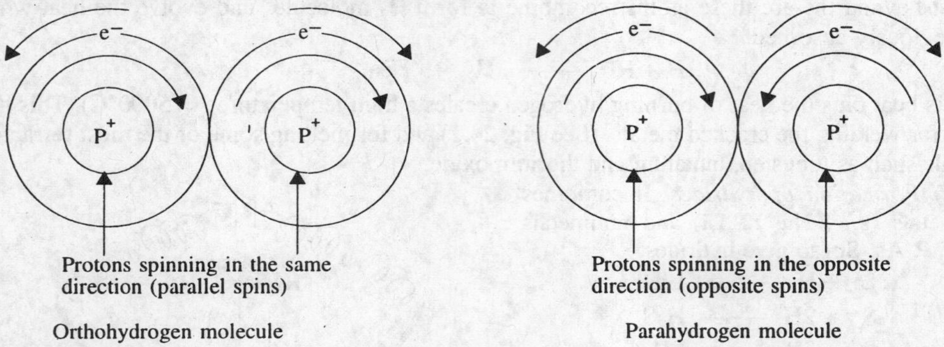

Fig. 24.3 : Ortho and para hydrogen molecules. Here P^+ indicates proton.

Note that the two electrons in ortho as well as in para hydrogen molecule *always spin in the opposite direction,* otherwise it will not be possible to get stable hydrogen molecule.

In 1929, Bonhoeffer and Harteck were also able to show the existence of ortho and para hydrogens from spectral studies.

Difference between ortho and para hydrogen molecules. (*i*) *Internal molecular energy.* Mecke discovered in 1934 that the two forms differ in their internal molecular energy. This difference is due to the fact that in case of orthohydrogen molecule, the spins of the two protons, being in the same direction, increase the molecular energy of ortho form. The resultant spin of orthohydrogen molecule is *one.*

On the other hand, in case of parahydrogen molecule, the spins of the two protons, being in the opposite direction, neutralise (*i.e.*, cancel) each other and thus make the resultant spin equal to *zero.* This neutralisation of spins decreases the molecular energy of parahydrogen molecule.

Thus we see that *para form has lower internal energy than the ortho form.*

(*ii*) The ortho form is more stable than the para form and hence the latter (*i.e.*, para form) has a tendency to change to the ortho form.

(*iii*) *Effect of temperature on the relative proportions of the two forms in ordinary hydrogen.* It has been shown that the relative proportion of ortho and para hydrogens (*i.e.*, ortho-para ratio) present in ordinary molecular hydrogen is dependent on temperature. *At absolute zero molecular hydrogen consists* only of para form (*i.e.*, ratio of para and ortho forms = 1 : 0) which has lower internal energy. *As the temperature increases, the proportion of ortho form increases and that of para form decreases. At the temperature of liquefaction of air, the ratio of para and ortho forms is 1 : 1,* while at room temperature and even at elevated temperatures this ratio is 1 : 3. Thus:

Temperature	% of para H_2	% of ortho H_2	Ratio (para : ortho)
Absolute zero	100%	Nil	1 : 0
Temperature of liquefaction of air	50%	50%	1 : 1
Room temperature and even at elevated temperatures.	25%	75%	1 : 3

It is quite evident that at room temperature the ordinary molecular hydrogen is a mixture of para and ortho hydrogen which are present in 1 : 3 ratio.

(*iv*) *Both the forms have different band spectrum.* This fact was used to demonstrate the existence of ortho and para hydrogens.

(*v*) *Physical properties.* Physical properties like melting point, boiling point, thermal conductivity and specific heat of both the forms are different from each other.

Some physical constants of ortho and para forms are given below. Ortho form is difficult to obtain in a pure state and as such its constants are left blank.

	Ortho form	*Para form*
Boiling point at 760 mm	—	20.26°C
Vapour pressure at 20 – 39° A :	751 mm (Extrapolated)	737 mm
Triple point	—	13.83°C

Due to the neutralisation of nuclear spins, the magnetic moment of parahydrogen is *zero* while that of ortho form, as extrapolated from the data for known ortho-para mixture, is approximately twice the momentum of a proton.

The appreciable difference in the thermal conductivities of the two forms has been used to estimate the percentage of ortho and para forms in a given sample of hydrogen.

Although the physical properties of the two forms are different, their chemical properties are similar, because both the forms have same covalent bonding.

Separation of the two forms from ordinary hydrogen : Preparation of pure parahydrogen. (*i*) We know that ordinary molecular hydrogen at room temperature consists of para and ortho hydrogens in 1 : 3 ratio. In order to separate para form from the ordinary hydrogen, the ordinary hydrogen is passed through a tube packed with activated charcoal and cooled in liquid air or liquid hydrogen for about four hours. The para form thus obtained is pumped off and is about 99.5% pure. It can be stored for a week in glass vessels at room temperature without appreciable change into ortho form. (*ii*) The two forms can be separated from each other by gas chromatography.

Conversion of parahydrogen into orthohydrogen. Orthohydrogen is more stable than the para form and the latter has a tendency to change to the ortho form. This process is accelerated under certain conditions and it is possible to convert the para form into ortho. (*i*) *By heating it to 800°C or above.* (*ii*) *By passing an electric discharge through it.* (*iii*) *By treatment with catalysts like platinum.* (*iv*) *By mixing it with paramagnetic molecules like oxygen or nitrogen peroxide.* (*v*) *By mixing with atomic hydrogen.* It was shown by Geib and Herteck that when atomic hydrogen collides with parahydrogen molecule, the parahydrogen molecule is converted into ortho molecule as depicted below in Fig. 24.4.

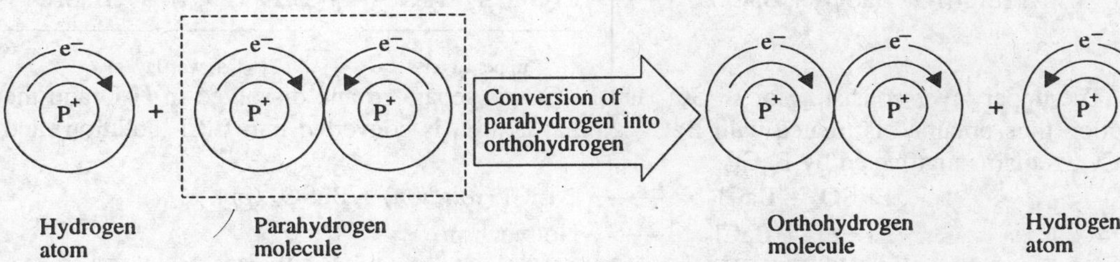

Fig. 24.4 : Conversion of parahydrogen into orthohydrogen.

Isotopes of Hydrogen

Hydrogen exists in three isotopes *viz.*, (*i*) *Hydrogen* or *protium* (^1_1H or H) (*ii*) *Heavy hydrogen* or *deuterium* (^2_1H or ^2_1D) (*iii*) *Tritium* (^3_1H or ^3_1T). Detailed study of these isotopes has already been given in chapter 7.

Lithium

Occurrence

Lithium does not occur in nature in the free state. In the combined state, it occurs as silicates and phosphates like (*i*) *Lepidolite* or *lithia mica,* (Li, Na, K)$_2$ Al$_2$ (SiO$_3$)$_3$ (F, OH)$_2$ (*ii*) *Triphyllite,* (Li, Na)$_3$ PO$_4$. (Fe, Mn)$_3$ (PO$_4$)$_2$ (*iii*) *Petalite,* LiAl (Si$_2$O$_5$)$_2$ (*iv*) *Spodumene,* LiAl (SiO$_3$)$_2$ (*v*) *Amblygonite,* LiAl (F, PO$_4$). In India lepidolite is found in Bihar, Kashmir, Karnataka and Rajasthan.

Extraction of lithium from spodumene and triphyllite ores

The extraction of lithium metal from these ores consists of the following steps:

1. Preparation of LiCl. (*i*) *From spodumene, LiAl (SiO$_3$)$_2$.* LiCl can be obtained from this ore by any of the following two methods:

(*a*) *Acid treatment method.* The finely-powdered spodumene ore is digested (boiled) with conc. H_2SO_4 and the digested mass is evaporated to dryness and extracted with H_2O when SiO_2 is obtained as residue and the filtrate contains Li_2SO_4, $Al_2(SO_4)_3$, $FeSO_4$ and $MgSO_4$. $FeSO_4$ and $MgSO_4$ are present as impurities in spodumene ore.

$$\underset{\text{spodumene}}{2LiAl\,(SiO_3)_2} + \underset{\text{(conc.)}}{4H_2SO_4} + \text{Impurities of } FeSO_4 \text{ and } MgSO_4$$
$$\longrightarrow \underbrace{Li_2SO_4 + Al_2(SO_4)_3 + FeSO_4 + MgSO_4}_{\text{solution}} + 4H_2O + 4\,SiO_2\,(\textit{residue})$$

The residue of SiO_2 is removed by filtration and the solution containing Li_2SO_4, $Al_2(SO_4)_3$, $FeSO_4$ and $MgSO_4$ is treated with Na_2CO_3 so that $Al_2(SO_4)_3$, $FeSO_4$ and $MgSO_4$ are converted into their basic carbonates (residue) and Li_2SO_4 remains in solution.

$$Li_2SO_4 + Al_2(SO_4)_3 + FeSO_4 + MgSO_4 + Na_2CO_3 \rightarrow Li_2SO_4\,(\textit{solution})$$
$$\boxed{\text{Basic carbonates of Al, Fe and Mg}}\ (\textit{residue}) + Li_2SO_4\,(\textit{solution})$$

The solution containing only Li_2SO_4 is concentrated and then treated with excess of Na_2CO_3 when the precipitate of Li_2CO_3 is obtained

$$Li_2SO_4 + Na_2CO_3 \longrightarrow Li_2CO_3\,(\textit{ppt.}) + Na_2SO_4$$

The precipitate of Li_2CO_3 is treated with HCl whereby LiCl is obtained.

$$Li_2CO_3 + 2HCl \longrightarrow 2LiCl + H_2O + CO_2$$

(*b*) *Fusion method.* The finely-powdered spodumene ore is fused with a mixture of $BaCO_3$ and $BaSO_4$ when a fused mass is obtained. The fused mass gets separated into two layers *viz.*, upper layer which consists of Li_2SO_4 and Li_2CO_3 and lower layer which consists of $Al_2\,(SiO_3)_3$ and $BaSiO_3$.

$$4LiAl\,(SiO_3)_2 + BaSO_4 + BaCO_3 \xrightarrow{\text{Fuse}} \underbrace{Li_2SO_4 + Li_2\,CO_3}_{\text{upper layer}} + \underbrace{2Al_2\,(SiO_3)_3 + 2BaSiO_3}_{\text{lower layer}}$$

The upper layer consisting of Li_2SO_4 and Li_2CO_3 is separated and dissolved in H_2O and the solution thus obtained is treated with $BaCl_2$ so that Li_2SO_4 is converted into LiCl (solution) and Li_2CO_3 remains unaffected by $BaCl_2$.

$$Li_2SO_4 + BaCl_2 \longrightarrow 2\,LiCl\,(\textit{solution}) + BaSO_4\,(\textit{ppt.})$$
$$Li_2CO_3 + BaCl_2 \longrightarrow \text{No reaction.}$$

(*ii*) *From triphyllite, (Li, Na)$_3$ PO$_4$. (Fe, Mn)$_3$ (PO$_4$)$_2$.* The finely-powdered ore is digested with conc. HCl and then boiled with some conc. HNO_3 to oxidise ferrous into ferric state. The solution is neutralised with NH_3 and then with solid CH_3COONa and CH_3COOH. The resulting solution is boiled with neutral $FeCl_3$ solution to get the precipitate of $FePO_4$ and basic ferric acetate and the solution contains the salts of Li, Na and Mn. This solution is treated with HCl to get the chlorides of Li, Na and Mn. Now when the solution which contains LiCl, NaCl and $MnCl_2$ is treated with BaS, MnS gets precipitated and LiCl, NaCl and $BaCl_2$ remain in solution.

$$MnCl_2 + BaS \longrightarrow MnS\ (ppt.) + BaCl_2\ (solution)$$

The solution containing LiCl, NaCl and $BaCl_2$ is treated with dil. H_2SO_4 so that Li_2SO_4 and Na_2SO_4 are obtained in solution and $BaSO_4$ gets precipitated.

$$2LiCl + H_2SO_4 \longrightarrow Li_2SO_4\ (solution) + 2HCl$$

$$2NaCl + H_2SO_4 \longrightarrow Na_2SO_4\ (solution) + 2HCl$$

$$BaCl_2 + H_2SO_4 \longrightarrow BaSO_4\ (ppt.) + 2HCl$$

The solution containing Li_2SO_4 and Na_2SO_4 is evaporated with oxalic acid ($H_2C_2O_4$) so that lithium oxalate ($Li_2C_2O_4$) and sodium oxalate ($Na_2C_2O_4$) are formed.

$$Li_2SO_4 + H_2C_2O_4 \longrightarrow Li_2C_2O_4 + H_2SO_4$$

$$Na_2SO_4 + H_2C_2O_4 \longrightarrow Na_2C_2O_4 + H_2SO_4$$

These oxalates, on being ignited, give Li_2CO_3, Na_2CO_3 and CO.

$$Li_2C_2O_4 \xrightarrow{\Delta} Li_2CO_3 + CO\uparrow$$

$$Na_2C_2O_4 \xrightarrow{\Delta} Na_2CO_3 + CO\uparrow$$

Now when the mixture of Li_2CO_3 and Na_2CO_3 is treated with HCl, a mixture of LiCl and NaCl is obtained. From this mixture LiCl is dissolved out in pyridine (since LiCl is a covalent compound) while NaCl remains insoluble. Now when LiCl-pyridine solution is distilled, LiCl is obtained.

2. To get lithium metal from LiCl by its electrolysis. Li-metal is obtained by the electrolysis of a fused mixture containing equal amounts of LiCl (obtained as above) and KCl. The addition of KCl increases the conductivity of the fused mixture and lowers the melting point (fusion temperature) of LiCl from 613° to 400°C. The mixture is electrolysed in an electrolytic cell having a graphite anode and a steel cathode. (Fig. 24.5). The cathode is surrounded by cast iron enclosure having iron guaze in its lower portion so that Li and Cl_2 fromed in the cell may not come in contact with each other. The cell is operated at a voltage of 8 – 9 volts and at 400 – 420°C. The reactions taking place in the cell are:

Ionisation of LiCl : $LiCl \quad Li^+ + Cl^-$

Collection of Li metal at cathode : $Li^+ + e^- \longrightarrow Li$

Liberation of Cl_2 gas at anode : $2Cl^- - 2e^- \longrightarrow Cl_2 + 2e^-$

Li metal is collected at cathode which is covered to protect the metal from air. Cl_2 gas is obtained at anode as a valuable by-product.

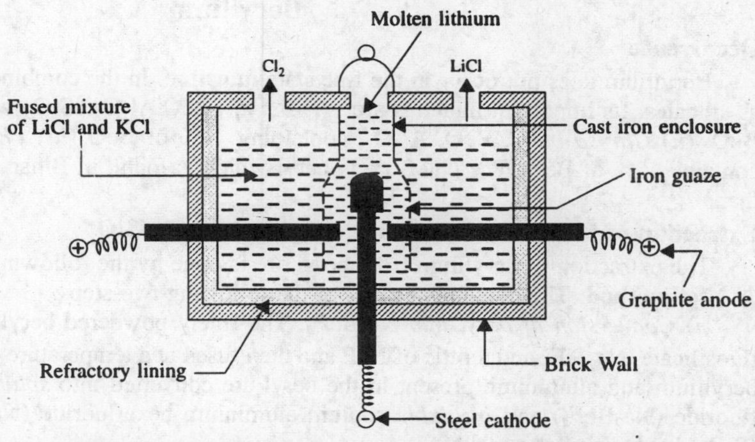

Fig. 24.5 : Preparation of lithium metal by the electrolysis of fused mixture of LiCl and KCl.

Properties of lithium

(*a*) **Physical.** Lithium is a silvery white metal with melting point = 186°C. It is harder than sodium or potassium but is extremely light. Its density which is 0.534 at 20°C is the lowest of all the metals. When volatalised, it imparts a red colour to the flame. It is a good conductor of heat and electricity. It forms an amalgam and alloys with a number of other metals. It has the highest specific heat of any element (0.9408).

(*b*) **Chemical.** (*i*) *Formation of compounds.* When heated above its fusion point, it combines with hydrogen, nitrogen, sulphur, halogens and carbon dioxide.

$$2Li + H_2 \longrightarrow 2LiH \quad (\textit{Lithium hydride})$$
$$6Li + N_2 \longrightarrow 2Li_3N \quad (\textit{Lithium nitride})$$
$$2Li + Cl_2 \longrightarrow 2LiCl \quad (\textit{Lithium chloride})$$

(*ii*) *Action of water.* It reacts with water forming the hydroxide and liberating hydrogen but unlike sodium or potassium it does not fuse on water as its melting point is higher.

$$2Li + 2H_2O \longrightarrow 2LiOH + H_2$$

(*iii*) *Action of air.* It is not affected in dry air but is oxidised when moisture is present. Heated in air above 180°C, it burns with a brilliant light forming lithium oxide and lithium nitride.

$$4Li + O_2 \longrightarrow 2Li_2O \quad (\textit{Lithium oxide})$$
$$6Li + N_2 \longrightarrow 2Li_3N \quad (\textit{Lithium nitride})$$

(*iv*) *Action of acids.* Lithium is a strongly electropositive metal and it reacts with dilute and conc. hydrochloric acid and dilute sulphuric acid liberating hydrogen. The reaction with conc. sulphuric acid is slow. Nitric acid attacks it so violently that the metal melts and catches fire.

(*v*) *Action of alkalies.* Alkalies have no action on lithium.

Uses of lithium and its compounds

(*i*) *In making alloys.* The most important use of lithium is in the manufacture of alloys. It is added to the alloys of magnesium to improve their tensile strength and resistance to corrosion. A lead-lithium alloy is used for making bearings and sheaths of electric cables.

(*ii*) *In the manufacture of glass and pottery.* Lithium compounds are employed to increase the fluidity of glass. Its carbonate is employed to increase the strength and resistance of glass. Its salts are used in ceramics to prevent surface cracking of pottery.

(*iii*) *As deoxidiser.* It is employed as a deoxidiser in the purification of nickel and copper.

(*iv*) *In medicine.* Its citrate and salicylate are employed in medicine for relieving gout since lithium urate is fairly soluble in water. Its compounds have also been used in the synthesis of vitamin A and some pharmaceuticals.

(*v*) *In lubricants.* Its hydroxide is used in the manufacture of high quality lubricating greases which can withstand extreme variations of temperature.

(*vi*) *In thermonuclear energy.* It is finding increasing use as an ingredient of high energy fuels for propulsion of intercontinental rockets.

Beryllium

Occurrence

Beryllium does not occur in the free state in nature. In the combined state this metal is found as silicates. Its important minerals are : (*i*) *Beryl,* $3BeO.Al_2O_3.6SiO_2$ which contains 11 – 15% of BeO. (*ii*) *Chryso beryl,* BeO, Al_2O_3, containing 19% of BeO (*iii*) *Phenacite,* $2BeO.SiO_2$ which contains 45% of BeO. The chief ore is beryl which is found in Bihar, Orissa, Madras and Rajasthan.

Extraction of beryllium from beryl ore, $3BeO.Al_2O_3.6SiO_2$

The extraction of beryllium from beryl ore is done by the following two methods:

1st method: This method consists of the following two steps:

(*i*) *Conversion of beryl into Na_2BeF_4.* The finely powdered beryl ore is mixed with sodium fluosilicate, Na_2SiF_6 and a little of NaF and then fused at a temperature of 800 - 900°C. As a result, beryllium and aluminium present in the beryl are converted into *soluble* sodium beryllium tetrafluoride (Na_2BeF_4) and *insoluble* sodium aluminium hexafluoride (Na_3AlF_6) respectively. Silica (SiO_2) is obtained as a residue.

SiF_4 formed in the above reaction reacts with the added NaF to form sodium fluosilicate, Na_2SiF_6 which can again be used to fuse beryl ore.

$$SiF_4 + 2NaF \longrightarrow Na_2SiF_6$$

The fused mass which consists of Na_2BeF_4, Na_3AlF_6 and SiO_2 is boiled with H_2O and filtered. The filtrate obtained contains Na_2BeF_4 while the residue contains Na_3AlF_6 and SiO_2.

(ii) Electrolysis of Na_2BeF_4. The filtrate containing Na_2BeF_4 is evaporated to dryness and the residue of Na_2BeF_4 so obtained is mixed with NaF and BaF_2. The fluorides of Na and Ba are added to increase the conductivity and to decrease the volatility of the fused mass at high temperatures. The mixture is fused at 1400°C and electrolysed in a graphite pot which itself acts as anode and a water-cooled iron pipe dipping in the melt acts as cathode. Beryllium metal is deposited on the cathode. The fluorides of Na and Ba are added at intervals to maintain the concentration of beryllium constant in the electrolyte.

2nd method: This method consists of the following steps:

(i) Conversion of beryl into BeO. The finely powdered beryl is fused with K_2CO_3 at about 1500°C in an electric furnace and then evaporated with conc. H_2SO_4. The resulting mass is extracted with boiled H_2O and fitered. The filtrate contains a mixture of $BeSO_4$, $Al_2(SO_4)_3$ and K_2SO_4 while the residue contains silica (SiO_2)

$$\underset{\text{Beryl ore}}{3BeO.Al_2O_3.6SiO_2} + 6K_2CO_3 \longrightarrow 3BeO + Al_2O_3 + 6K_2SiO_3 + 6CO_2\uparrow$$

$$3BeO + Al_2O_3 + 6K_2SiO_3 + 12H_2SO_4 \longrightarrow 3BeSO_4 + Al_2(SO_4)_3 + 6K_2SO_4 + 12H_2O \text{ (solution)} + 6SiO_2 \textit{ (residue)}$$

The fitrate containing a mixture of $BeSO_4$, $Al_2(SO_4)_3$ and K_2SO_4 is concentrated when all K_2SO_4 and some $Al_2(SO_4)_3$ combine together to give the crystals of potash alum, $K_2SO_4.Al_2(SO_4)_3.24H_2O$. $BeSO_4$ and the remaining $Al_2(SO_4)_3$ remain as such in solution.

$$K_2SO_4 \textit{ (all)} + Al_2(SO_4)_3 \textit{ (some quantity)} + 24H_2O \longrightarrow K_2SO_4.Al_2(SO_4)_3.24H_2O$$

Now the solution containing $BeSO_4$ and remaining $Al_2(SO_4)_3$ is treated with hot and conc. solution of $(NH_4)_2CO_3$ so that the precipitate of $Al(OH)_3$ is obtained and $BeSO_4$ is converted into a solution of basic beryllium carbonate, $Be(OH)_2.BeCO_3$.

$$Al_2(SO_4)_3 \textit{ (remaining quantity)} + 3(NH_4)_2CO_3 + 3H_2O \longrightarrow \underset{\text{(ppt.)}}{2Al(OH)_3} + 3(NH_4)_2SO_4 + 3CO_2\uparrow$$

$$2BeSO_4 + (NH_4)_2CO_3 + H_2O \longrightarrow \underbrace{Be(OH)_2.BeCO_3 + (NH_4)_2SO_4 + SO_3}_{\textit{solution}}$$

The solution containing $Be(OH)_2.BeCO_3$, on concentration, gives the crystals of $Be(OH)_2.BeCO_3$ which, on being ignited, gives BeO.

$$Be(OH)_2.BeCO_3 \xrightarrow{\text{Ignite}} 2BeO + H_2O + CO_2\uparrow$$

(ii) Electrolytic reduction of BeO to beryllium metal. BeO obtained as above is mixed with powdered coke and put into an alternating current arc furnace (Fig. 24.6). The furnace consists of a big carbon crucible provided with a tapping hole and is enclosed in the fire-bricks.

The crucible acts as one of the electrodes. The other carbon electrode dips in the charge and is held vertically. On striking the electric arc between the electrodes, BeO is reduced to beryllium which, in the molten state, is taken away through the tapping hole.

$$BeO + C \longrightarrow Be + CO\uparrow$$

Properties

(a) **Physical.** Beryllium is silvery white in colour, malleable, easily forged and takes a high polish. It is a light metal (S.G. = 1.842) and possesses a high melting point (1280°). Its specific heat is probably the highest of all the known metals and this accounts for the nonmetallic nature of the metal (*cf. Boron*).

(b) **Chemical.** *(i) Action of air.* The metal is more permanent in the air than magnesium, but is oxidised in the moist air. When heated to redness in oxygen, it forms a protective layer of the oxide round the metal. The metal powder, however, burns in air brilliantly forming the oxide, BeO.

(ii) Action of water. Water or steam is not decomposed even at red heat (compare with magnesium).

(iii) Combination with elements. It does not directly combine with sulphur but unites with halogens (difficulty with iodine, but with incandescence with chlorine and bromine). It unites directly with carbon and boron and gives alloys with metals.

(iv) Action of acids. It is easily acted upon by hydrochloric acid and dilute sulphuric acid, evolving hydrogen. It liberates sulphur dioxide with strong sulphuric acid.

Flow sheet showing the extraction of beryllium from beryl ore, $3BeO.\ Al_2O_3.\ 6SiO_2$ by 2nd method

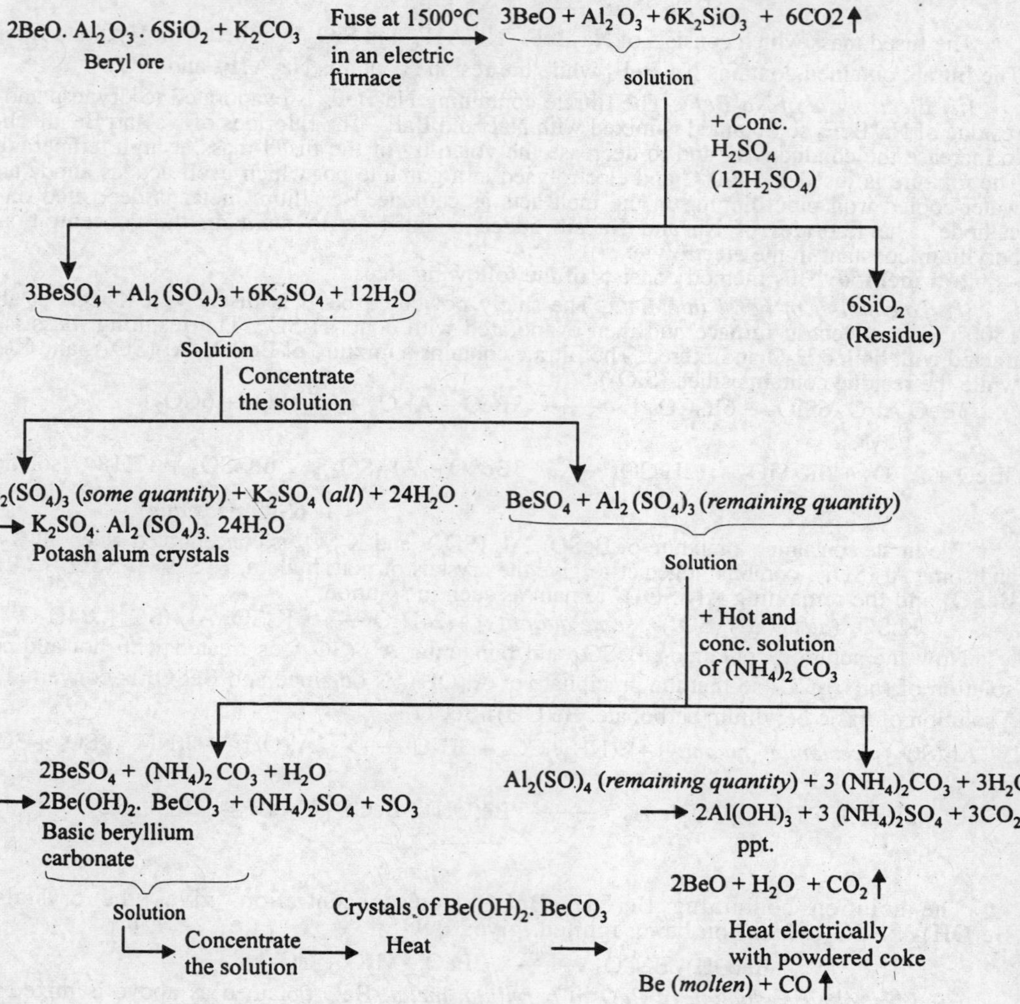

$$Be + 2H_2SO_4 \longrightarrow BeSO_4 + SO_2 + 2H_2O$$

Cold concentrated nitric acid makes it passive while hot acid produces nitrogen peroxide.

$$Be + 4HNO_3 \longrightarrow Be(NO_3)_2 + 2H_2O + 2NO_2$$

(*v*) *Action of alkalies.* It is attacked readily by alkalies with the formation of *beryllates* and liberation of hydrogen.

$$Be + 2NaOH \longrightarrow Na_2BeO_2 + H_2$$

Uses

(*i*) *Beryllium bronzes,* which contain 2.25 to 2.5% of beryllium are hard and can be tempered just like steel. These are being used for making springs and slip-rings for high speed motors. (*ii*) Because of its high transparency to X-rays, it is used for making "windows" of X-ray tubes. (*iii*) It is also used for making electrodes of neon signs. (*iv*) *It is used as a moderator in nuclear reactors.* This use depends on the fact that beryllium absorbs fewer neutrons than any other known structural meterial. (*v*) Beryllium also finds special applications in the fields of developing rockets, missiles and high speed aircraft. (*vi*) Recently beryllium has been used as an ingredient of a new textile fibre of rayon type.

Radium

Occurrence

Radium is found in all those minerals which contain uranium. Some important minerals of

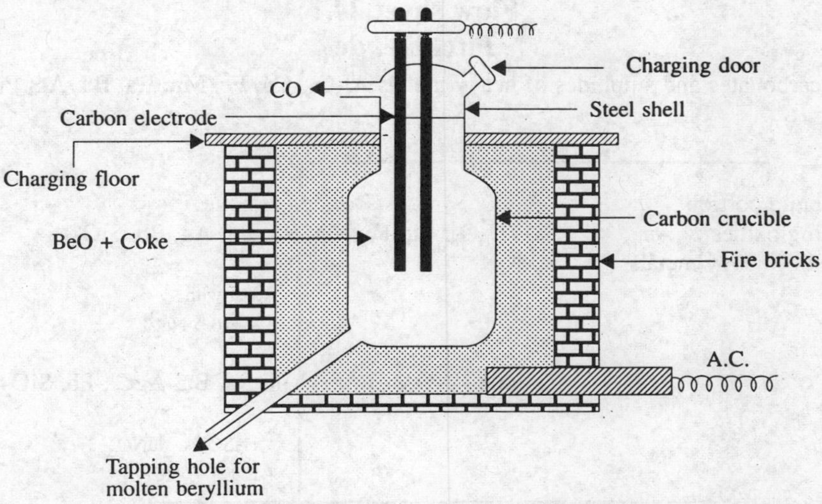

Fig. 24.6 : Preparation of beryllium metal by the electrolytic reduction of BeO.

radium are : (*i*) *Pitchblende* or *uranite* which is chiefly U_3O_8. It is found in Congo and Canada. (*ii*) *Carnotite* which is potassium uranyl vanadate, $K(UO_2)(VO_4).4H_2O$. This ore is found in Katanga mines of Congo, U.S.A. and Russia. (*iii*) *Autunite* or *uranium mica* which is calcium uranyl phosphate, $Ca(UO_2)_2(PO_4)_2.8H_2O$. (*iv*) *Calcolite* which is copper uranyl phosphate, $Cu(UO_2)_2(PO_4)_2.8H_2O$. Radium is also found in traces in certain mineral springs and sea water.

Extraction of radium

Radium is generally extracted from pitchblende or carnotite ores. The method consists of the following two steps:

1. Preparation of $RaCl_2$. (*a*) *From pitchblende*. (*i*) *By old method* (*Curie's method*). Pitchblende consists of traces of Ra, Ba, SiO_2, S, As, Sb, Ag, clay etc., as impurities. First of all, the ore is concentrated by gravity separation method to remove clay etc., and then is roasted in a reverberatory furnace in the excess of air to remove S, As and Sb as their volatile oxides. Ag is removed as insoluble AgCl by heating the roasted ore with NaCl at 800°C. Now the roasted ore becomes free from S, As, Sb, Ag and clay, and contains U_3O_8, RaO, BaO and SiO_2 in the from of a residue. This residue is leached with H_2SO_4 when BaO and RaO get precipitated as $BaSO_4$ and $RaSO_4$ respectively [$BaO + H_2SO_4 \rightarrow BaSO_4$ (*ppt.*) + H_2O; $RaO + H_2SO_4 \rightarrow RaSO_4$ (*ppt.*) + H_2O] while SiO_2 remains as such U_3O_3 is converted into uranyl sulphate, $UO_2(SO_4)$ which goes into solution [$U_3O_8 + 3H_2SO_4 + \frac{1}{2}O_2 \rightarrow 3UO_2(SO_4) + 3H_2O$]. This solution can be used for the extraction of uranium. Now the residue containing $BaSO_4$, $RaSO_4$ and SiO_2 is fused with Na_2CO_3 so that $BaSO_4$ and $RaSO_4$ are converted into $BaCO_3$ and $RaCO_3$ respectively while SiO_2 remains as such [$BaSO_4 + Na_2CO_3 \rightarrow BaCO_3 + Na_2SO_4$; $RaSO_4 + Na_2CO_3 \rightarrow RaCO_3 + Na_2SO_4$]. The mass containing $BaCO_3$, $RaCO_3$ and SiO_2 is treated with HCl so that $BaCO_3$ and $RaCO_3$ are converted into $BaCl_2$ and $RaCl_2$ (in solution) while SiO_2 remains unaffected and hence as residue [$RaCO_3 + 2HCl \rightarrow RaCl_2$ (*solution*) $+ H_2O + CO_2\uparrow$; $BaCO_3 + 2HCl \rightarrow BaCl_2$ (*solution*) $+ H_2O + CO_2\uparrow$]. Now when the solution containing $RaCl_2$ and $BaCl_2$ is concentrated and cooled, less soluble $RaCl_2$ crystallises out while more soluble $BaCl_2$ remains dissolved in solution (fractional crystallisation).

(*ii*) *By modern method.* Different steps of this method have been shown in flow sheet 24.1.

(*b*) *From carnotite.* In addition to uranium, carnotite also contains Ba, V, Fe, Al, Ra etc. This ore is heated with some HNO_3 at 100°C. The heating is done by the steam. Most of the ore dissolves in the acid and the solution thus obtained is neutralised by NaOH and then treated with $BaCl_2$ and H_2SO_4 whereby Ba and Ra are precipitated as $BaSO_4$ and $RaSO_4$ and U and V go into solution as UO_2SO_4 and $VOSO_4$. This solution can be used for the extraction of uranium and vanadium. The precipitate containing $BaSO_4$ and $RaSO_4$ is fused with Na_2CO_3 and then treated with HCl so that a solution containing $BaCl_2$ and $RaCl_2$ is obtained. From this solution, $RaCl_2$ is separated as crystals by fractional crystallisation.

2. To get radium metal by the electrolysis of $RaCl_2$ solution. When the solution of $RaCl_2$ is electrolysed in a cell (cathode: Hg, anode: Pt-Ir alloy), radium is collected at cathode with which it forms Ra-Hg amalgam. When this amalgam is heated in a retort at 400 - 700°C in an atmosphere of H_2, Hg gets volatilised and radium is left as a white metal in the retort. The electrolytic cell used for the purpose is shown in Fig. 24.7.

Flow sheet 24.1
Pitchblende

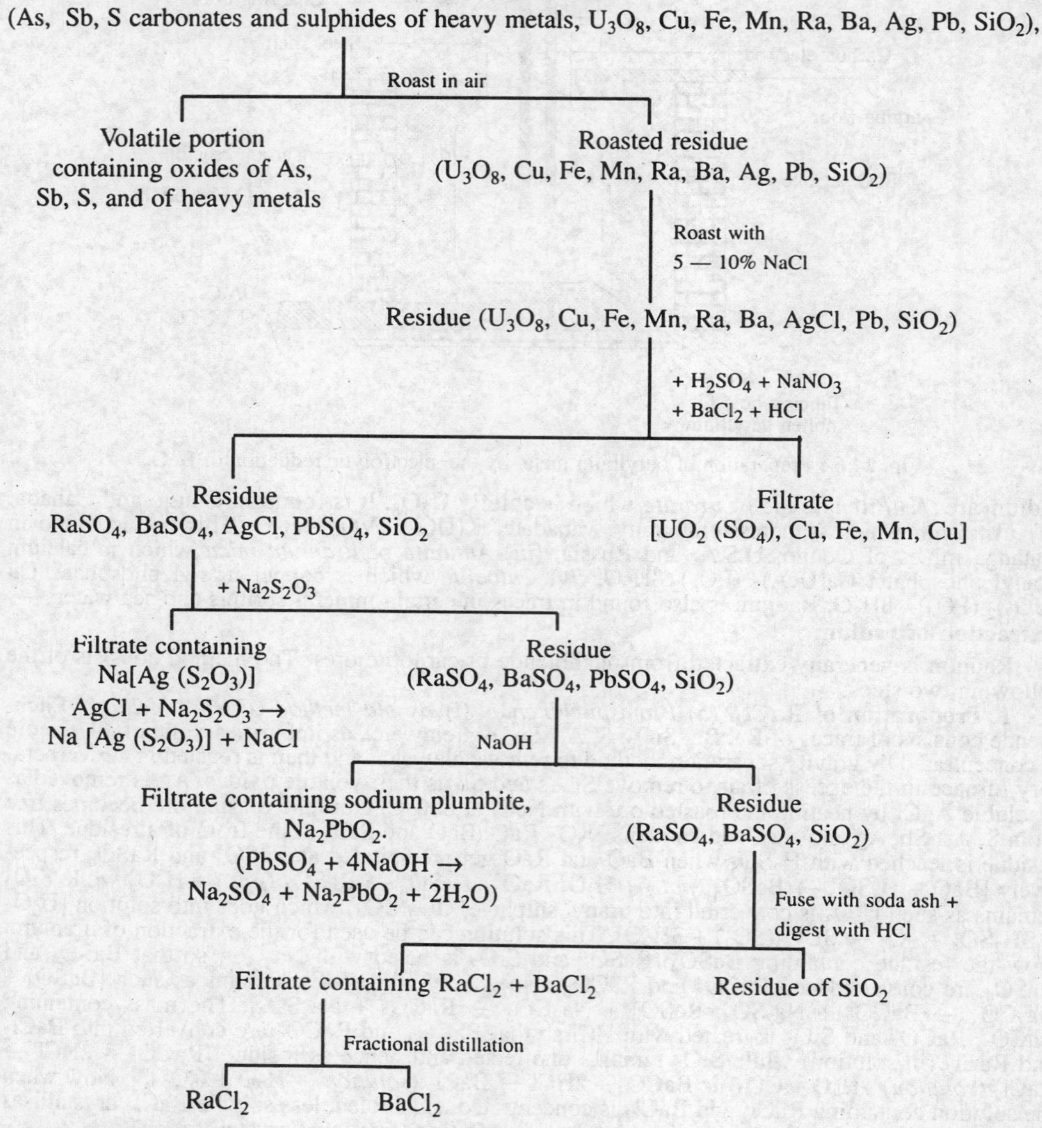

Properties of radium

Radium is a white metal which melts at 700°C, volatilises at 960°C and boils at 1140°C. When exposed to air, it blackens due to the formation of a nitride. It is luminescent and its blue luminescence becomes more pronounced when barium is present as impurity in it. It is acted upon by water, when its hydroxide is obtained. It is soluble in hydrochloric acid, forming the chloride. It gives out alpha, beta, and gamma rays and this property of radioactivity remains unaffected between temperatures 20000°C and –250°C, and under other physical conditions. the radiations affect the photographic plate and discharge the electroscope. Radium salts give a caramine colour to the flame.

Uses of radium salts

(*i*) In the manufacture of *luminous paints* used for painting clock and watch dials, push bottons etc. The paint consists of a mixture of sulphides of Zn, Cu, Mn etc., containing a trace of radium bromide (*ii*) In medicine in the treatment of cancer.

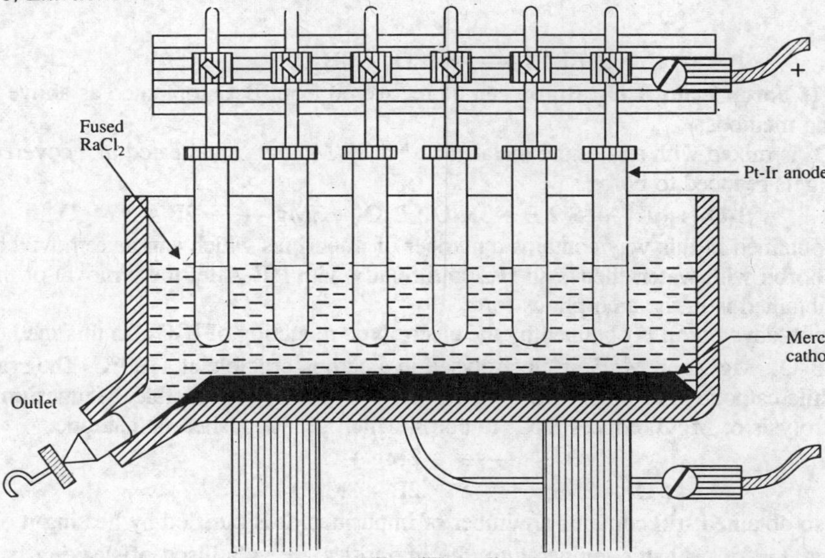

Fig. 24.7 : Electrolytic cell used for the extraction of radium metal by the electrolysis of $RaCl_2$ solutio

Boron

Occurrence

Boron does not occur in the free state in nature. In the combined state it occurs as: (*i*) *Borax* (also called *tincal* or *suhaga*), $Na_2B_4O_7.10H_2O$ which is found in the dried-up lakes of Tibet, Ceylon, California (U.S.A.) and India (Ladakh and Puja Valley of Kashmir). (*ii*) *Colemanite,* $Ca_2B_6O_{11}.5H_2O$ which is found in California and South Africa (*iii*) *Karnite* or *razorite,* $Na_2B_4O_7.4H_2O$. (*iv*) *Boracite,* $2Mg_3B_8O_{15}.MgCl_2$. (*v*) *Borocalcite.* $CaB_4O_7.4H_2O$. (*vi*) *Boric acid,* H_3BO_3. Boron is also present in some plants and sometimes in coal ash.

Extraction of boron from borax and colemanite ores

The method consists of the following steps:

1. Preparation of boric acid, H_3BO_3. (*a*) *From borax, $Na_2B_4O_7.10H_2O$.* Borax is finely powdered and then boiled with conc. HCl or conc. H_2SO_4 so that sparingly soluble H_3BO_3 separates out as crystals on cooling the solution.

$$Na_2B_4O_7.10H_2O + 2HCl \xrightarrow{\text{Boil}} 2NaCl + 4H_3BO_3 + 5H_2O$$

$$Na_2B_4O_7.10H_2O + H_2SO_4 \xrightarrow{\text{Boil}} Na_2SO_4 + 4H_3BO_3 + 5H_2O$$

(*b*) *From colemanite ore, $Ca_2B_6O_{11}.5H_2O$.* Boric acid can also be obtained from this ore by any of the following methods:

(*i*) The finely-powdered colemanite ore is dissolved in boiling water and then SO_2 gas is passed through this solution when highly soluble calcium bisulphite, $Ca(HSO_3)_2$ remains in solution while sparingly soluble H_3BO_3 separates out on cooling.

$$\underset{\text{Colemanite ore}}{Ca_2B_6O_{11}.5H_2O} + 4SO_2 + 6H_2O \longrightarrow \underset{\text{(in solution)}}{2Ca(HSO_3)_2} + \underset{\text{crystals}}{6H_3BO_3}$$

(*ii*) The finely powdered colemanite ore is boiled with Na_2CO_3 solution, so that $CaCO_3$ is obtained as a precipitate while borax ($Na_2B_4O_7$) and sodium metaborate ($NaBO_2$) go into solution.

$$Ca_2B_6O_{11}.5H_2O + 2Na_2CO_3 \xrightarrow{\text{Boil}}$$

$$2CaCO_3\ (ppt.) + Na_2B_4O_7 + 2NaBO_2 + 5H_2O \quad (solution)$$

On filtration, $CaCO_3$ is separated and on concentrating the solution, crystals of $Na_2B_4O_7$ separate out, but $NaBO_2$ still remains in solution. Now a current of CO_2 is passed in the solution when $NaBO_2$ present in the solution is converted into $Na_2B_4O_7$ which is separated by crystallisation.

$$4NaBO_2 + CO_2 \longrightarrow Na_2CO_3 + Na_2B_4O_7$$

The crystals of $Na_2B_4O_7$ obtained as above can be converted into H_3BO_3 by boiling it with conc. HCl or H_2SO_4 as shown at (*a*) above.

2. Conversion of H_3BO_3 into boric anhydride, B_2O_3. When H_3BO_3 is heated, B_2O_3 is obtained.

$$2H_3BO_3 \xrightarrow{\Delta} B_2O_3 + 3H_2O.$$

3. To get boron from B_2O_3. Boron can be recovered from B_2O_3 obtained as above by any of the following methods:

(*a*) B_2O_3 is mixed with any of the metals like Na, K, Mg etc., and heated in a covered crucible when B_2O_3 gets reduced to boron.

$$B_2O_3 + 6K \longrightarrow 2B + 3K_2O \; ; \; B_2O_3 + 3Mg \longrightarrow 2B + 3MgO$$

Boron obtained in this way contains a number of impurities which can be removed by boiling the impure boron with water, then with HCl and finally with HF. After the removal of the impurities boron obtained is in the amorphous form.

(*b*) Now-a-days boron is obtained by the electrolytic reduction of B_2O_3. In this method a fused mixture of B_2O_3, MgO and MgF_2 is electrolysed in a carbon crucible at 1100°C. The crucible acts as anode while cathode is of iron which remains in the centre of the crucible. Magnesium obtained by the electrolysis of MgO reduces B_2O_3 to boron which gets deposited on cathode.

$$2MgO \longrightarrow 2Mg + O_2$$

$$B_2O_3 + 3Mg \longrightarrow 2B + 3MgO$$

Boron so obtained still contains a number of impurities. It is purified by heating it electrically in vacuum at 1100°C. At this temperature, the impurities are volatilised off leaving behind pure boron.

Preparation of crystalline boron

(*i*) Crystalline boron can be obtained by the reduction of B_2O_3 or KBF_4 by Al

$$B_2O_3 + 2Al \longrightarrow Al_2O_3 + 2B$$

$$KBF_4 + Al \longrightarrow AlF_3 + KF + 2B$$

The crystals of boron are treated with NaOH solution to remove the impurity of Al.

(*ii*) Crystalline boron can also be prepared by dropping amorphous boron in molten Al at 1500°C when it appears to dissolve. On cooling, yellow coloured crystals deposit on the surface of Al. The mass is broken up and heated with HCl to dissolve away Al when very hard crystals are left behind. Boron obtained by this method is not pure since it contains some aluminium as impurity.

Properties of boron

(*a*) **Physical.** Boron exists in two allotropic forms *viz.*, *crystalline* or *adamantine* and *amorphous boron* Crystalline boron is black and possesses metallic lustre. It is extremely inert. It is scarcely attacked by conc. HNO_3 or H_2SO_4 and not at all by boiling conc. HCl or HF. It is, however, vigorously attacked by fused sodium peroxide and by fused oxidizing fusion mixture. Crystalline boron is very hard and is second to diamond in hardness. Its specific gravity is 3.3. Both needle-like and plate-like hexagonal crystals of crystalline boron are known. It is a bad conductor of electricity. Amorphous boron is brown powder of density 2.45. It is very hard to fuse. The melting point and boiling point are 2300°C and 2550°C respectively.

(*b*) **Chemical.** (*i*) *Action of air and O_2.* Amorphous boron is very reactive. When it is heated at 700°C in air or O_2, it burns with a reddish flame forming a mixture of oxide and nitride.

$$4B + 3O_2 \longrightarrow 2B_2O_3; \qquad 2B + N_2 \longrightarrow 2BN$$

(*ii*) *Action of acids.* Halogen acids have no action on boron but it is attacked by oxidising acids like HNO_3 and H_2SO_4,

$$B + 3HNO_3 \longrightarrow H_3BO_3 + 3NO_2; \qquad 2B + 3H_2SO_4 \longrightarrow 2H_2BO_3 + 3SO_2$$

(*iii*) *As reducing agent.* Boron is powerful reducing agent, since it can replace carbon from CO_2, silicon from SiO_2 and N_2 from NO.

$$4B + 3CO_2 \longrightarrow 2B_2O_3 + 3C; \qquad 4B + 3SiO_2 \longrightarrow 2B_2O_3 + 3Si$$

$$4B + 6NO \longrightarrow 2B_2O_3 + 3N_2$$

(*iv*) *Action of alkalies.* It is vigorously attacked by fused alkalies or by fusion with oxidising fusion mixture.

$$2B + 6NaOH \longrightarrow 2Na_3BO_3 + 3H_2\uparrow$$

(*v*) *Action of metals.* Excepting with Cu, Ag and Au, boron combines with other metals at high temperatures to form borides. These borides are extremely hard substances. Mg_3B_2 and Be_3B_2 are attacked by acids with the formation of boron hydrides.

(*vi*) *Action of nonmetals.* Boron also reacts with nonmetals. For example : (*a*) Boron combines with N_2 at a temperature greater than 900°C to form boron nitride, BN. (*b*) Boron forms boron sulphide B_2S_3 when heated with S. (*c*) On heating with carbon in an electric furnace an extremely hard substance namely boron carbide, B_4C is formed.

(*vii*) *Action of steam.* Boron reacts with steam at red heat liberating H_2.

$$2B + 3H_2O \longrightarrow B_2O_3 + 3H_2\uparrow$$

Uses of boron and its compounds

Boron is used: (*i*) as a deoxidiser in the casting of copper (*ii*) for making boron steels which are very hard and are used as control rods in atomic reactors (*iii*) the traces of boron compounds are essential for plant metabolism and for the fertility of soil (*iv*) as a catalytic agent (*v*) in the manufacture of thermometers and thermoregulators (*vi*) as a scavanger to remove gaseous impurities from molten copper and as a semiconducting material (*vii*) the hydrides of boron are used as fuels for rockets because of their very high energy/weight ratio (*viii*) metallic borides and boron carbides are used as abrasives.

Gallium

Occurrence

Gallium is one of the rarest elements and is widely distributed in the earth's crust in minute quantities. In zinc and germanium ores, gallium may occur as the sulphide. Thus zinc blende contains some amount of gallium. The residues of zinc furnaces obtained after the distillation of zinc also have 0.002 g of gallium per 10 kg of zinc.

Extraction of gallium from zinc blende

Gallium is generally extracted from zinc blende. The method consists of many steps which have been shown in Flow sheet 24.2.

Properties

(*a*) **Physical.** Gallium is a silvery-white, hard and brittle metal. It melts at 29.8°C and boils at 2237°C. The liquid may be cooled to almost 0°C without solidifying. The specific gravity is 5.9 (25°C). It dissolves in caustic alkalies as well as in HCl, Hg and H_2SO_4 but not in HNO_3. HNO_3 renders it *passive*. It is stable in dry air and does not decompose water, hot or cold in the absence of free O_2.

(*b*) **Chemical.** (*i*) Its chemistry is similar to that of Al. Important exception is the non-formation of Ga_4C_3. It combines with Cl_2, Br_2 and I_2 at cold temperature in the decreasing order of activity forming $GaCl_2$, $GaCl_3$, $GaBr_2$ and GaI_2.

(*ii*) The metal gets dissolved in cold and hot HCl, evolving H_2 gas

$$Ga + 2HCl \longrightarrow GaCl_2 + H_2\uparrow$$

(*iii*) Gallium dissolves in a solution of KOH, liberating H_2 and forming gallates (GaO_3^{3-}).

$$Ga + 2KOH \longrightarrow K_2GaO_2 + H_2$$

$$2Ga + 6KOH \longrightarrow 2K_3GaO_3 + 3H_2O$$

(*iv*) The salts of gallium are readily hydrolysed by water to form basic salts.

Uses.

(*i*) Since, at ordinary conditions of pressure, gallium is found as a liquid over a wide range of temperature [between 29.8°C (m.pt.) and 2237°C (b.pt.)], it is used in quartz thermostats for measuring high temperatures. Liquid gallium, like mercury, does not wet the glass or quartz. (*ii*) It is also used in arc lamps employed for spectroscopic analysis just as mercury is used in mercury lamps. (*iii*) Gallium is used in the production of semiconducting compounds. For many years, this technology was dominated by silicon and germanium. (*iv*) It is also used in low resistance contact electrodes.

Flow sheet 24.2

Zinc blende

↓ Aqua regia

Solution of zinc blende

↓ Treat the solution with strips of metallic zinc

- Precipitate containing Cu, Cd, Pb, Hg, Ag (Impurities)
- Filtrate

Filtrate ↓ Boil the filtrate with excess of zinc strips

White ppt. containing basic salts of Zn, Ga, Fe, Cr and a mixture of Al_2O_3 and SiO_2

↓ Dissolve the white ppt. in HCl and saturate with H_2S

- Precipitate if any. Reject it by filtration
- Filtrate

Filtrate ↓ Neutralise the filtrate with NH_4OH and then acidify with a mixture of CH_3COOH and CH_3COONH_4

Solution

↓ Saturate the solution with H_2S

- Filtrate. Reject it.
- Precipitate of ZnS and Ga_2S_3

Precipitate of ZnS and Ga_2S_3 ↓ Dissolve the ppt. in HCl

Solution containing $ZnCl_2$ and $GaCl_3$

↓ Add Na_2CO_3 solution

Precipitate of basic gallium carbonate $\xrightarrow{\text{Add } H_2SO_4}$ Soluble $Ga_2(SO_4)_3$

Soluble $Ga_2(SO_4)_3$ ↓ Boil with excess of water

White ppt. of basic gallium sulphate

↓ Dissolve the white ppt. in H_2SO_4 and excess of KOH

Solution containing potassium gallate $\xrightarrow[\text{the solution}]{\text{Electrolyse}}$ Gallium metal gets deposited on Pt-cathode

Thallium

Occurrence

This element does not occur in the free state in nature. Traces of this element are found in copper, iron and zinc blendes and pyrites and in mica, carnallite, allumite and other minerals. Thallium is also found in **flue dust** which contains Tl_2O along with CuO, PbO_2, As_2O_3, Bi_2O_3, Fe_2O_3 etc. Flue dust is obtained during the manufacture of H_2SO_4. Important minerals of thallium are : (*i*) *crookesite,* $(Cu, Tl, Ag)_2$ Se (*ii*) *hutchinsonite* $(Cu, Tl, Ag)_2S.PbS.2Ag_2S_2$ (*iii*) *lorandite,* $TlAsS_2$ (*iv*) *urabaite,* $TlAs_2$. SbS_3.

Extraction of thallium from flue dust

We have said above that flue dust contains Tl_2O along with CuO, PbO_2, As_2O_3, Bi_2O_3 and Fe_2O_3. It is dissolved in *aqua regia* so that we get a solution containing TlCl, $CuCl_2$, $PbCl_4$, $AsCl_3$, $BiCl_3$ and $FeCl_3$ [$Tl_2O + 2HCl$ (*from aqua regia*) $\rightarrow 2TlCl + H_2O$; $CuO + 2HCl \rightarrow CuCl_2 + H_2O$; $PbO_2 + 4HCl \rightarrow PbCl_4 + 2H_2O$; M_2O_3 (M = As, Bi, Fe) + $6HCl \rightarrow 2MCl_3 + 3H_2O$]. The solution containing the chlorides is evaporated to drive away the excess of HNO_3 and then treated with H_2S gas. By this operation, the chlorides namely $CuCl_2$, $PbCl_4$, $AsCl_3$ and $BiCl_3$ are converted into their corresponding sulphides while TlCl and $FeCl_3$ remain as such in solution. The solution containing TlCl and $FeCl_3$ is heated with NH_4OH to get $Fe(OH)_3$ as a precipitate and TlOH goes in solution. [$FeCl_3 + 3NH_4OH \xrightarrow{\Delta} Fe(OH)_3$ (*ppt.*) + NH_4Cl; $TlCl + NH_4OH \xrightarrow{\Delta} TlOH$ (*solution*) + NH_4Cl]. The solution containing TlOH is treated with KI solution to get TlI as a yellow precipitate. The precipitate of TlI is treated with conc. H_2SO_4 to get Tl_2SO_4. Now Tl_2SO_4 can be used in the following two different ways to get Tl metal. (*i*) When the solution of Tl_2SO_4 is electrolysed, Tl metal gets deposited on cathode. (*ii*) When Tl_2SO_4 solution is treated with Zn-metal, Tl metal gets precipitated [Tl_2SO_4 (*solution*) + Zn $\rightarrow$ $ZnSO_4$ (*solution*) + $2Tl\downarrow$].

Properties of thallium

(*a*) **Physical.** Thallium is a soft grey metal. Its melting and boiling points are 303° and 1457°C. It is so soft that it can be scratched even by finger nails. It is malleable but not ductile. It exists in two allotropic forms, namely α and β forms. It alloys with a number of metals like Cu, Ag, Au, Zn, Cd, etc.

(*b*) **Chemical.** Tl is more reactive than In and Ga due to its tendency to form unipositive ion, Tl^+. It oxidises in moist air and decomposes steam at red heat. It dissolves readily in dil. mineral acids except HCl giving Tl (I) compounds. It does not dissolve readily in HCl because TlCl is insoluble. It combines with halogens at ordinary temperature and with S, Se and Te on heating. It does not combine with B, C, N_2 and H_2 even on heating.

Similarities between thallium and other elements outside its group

(*a*) **Similarities with alkali metals.** Some of Tl (I) compounds are similar to the corresponding compounds of alkali metals. This similarity is due to the fact that Tl^+ ion has the same charge and approximately the same size as the ions derived from the heavier alkali metals. Some points showing the similarity are : (*i*) Tl (+1) cyanide, perchlorate, sulphate, phosphate and carbonate are stable salts and are isomorphous with alkali salts. For example Tl_2SO_4, $TlClO_4$ and Tl_2PtCl_6 are isomorphous with K_2SO_4, $KClO_4$ and K_2PtCl_6 respectively. (*ii*) TlOH is soluble in water giving strong alkaline solution which is very similar in behaviour to NaOH. (*iii*) The aqueous solution of Tl_2CO_3 is strongly alkaline like that of Na_2CO_3. (*iv*) TlF has distorted NaCl type structure while other Tl (+1) halides crystallise with CsCl crystal.

(*b*) **Similarities with lead.** The cause of similarity between Tl (I) and Pb (II) is that Tl^+ and Pb^{2+} ions are isoelectronic, *i.e.*, these ions have identical electronic configuration. Some points showing the similarity are : (*i*) Thallium and lead both are white metals with a bluish type. (*ii*) Both the metals are so soft that they can be cut with a knife and scratched with a finger's nail. (*iii*) Both the metals mark the paper and have low melting points. (*iv*) The halides, chromates and sulphides of both the elements are formed by the same method. These salts have the same appearance and are sparingly soluble in H_2O.TlCl, like $PbCl_2$, is soluble in hot water.

(*c*) **Similarities with silver.** (*i*) Chlorides, bromides and iodides of Tl (I) and Ag (I) have similar colour and are sparingly soluble in H_2O. (*ii*) Like AgCl, TlCl turns violet on exposure to light.

Uses of thallium

Thallium does not have any worthwhile application as a metal. However, its salts are very useful. Some uses of its salts are : (*i*) Thallium salts are toxic and hence are used as fungicide, germicide, and for killing rats and destroying vermin. (*ii*) Thallium salts have been used in ointments for the treatment of ring worms. (*iii*) Thallium compounds have unusually high refractive index and are, therefore, used for the preparation of optical glasses like Crookes glasses. (*iv*) Tl_2CO_3 has been recommended as a standard in acidimetry. (*v*) Thallium oxy sulphide salts have been used as depilatories. (*vii*) Alloys of Tl with Ag are lustrous and remain unattacked in air. An alloy containing Tl = 10%, Al = 10% and Ag = 80% is resistant to the action of H_2S.

Germanium

Occurrence

Germanium is a rare element and occurs in nature in very few minerals. Important minerals of this element are : (*i*) *Argyrodite,* $4Ag_2S \cdot GeS_2$ or $Ag_8[GeS_8]$. This mineral is mostly found in Germany. (*ii*) *Germanite,* 7CuS. FeS. GeS_2. This mineral is found in South Africa. This mineral also contains 1.8% gallium (*iii*) *Confielddite,* $4Ag_2S$ (Ge, Sn)S_2. Other sources from which germanium can be obtained include coke furnace, flue dust and gas liquor. Germanium is also sometimes found in the residue from zinc and lead smelters.

Extraction

Germanium is usually extracted from argyrodite and germanite ores. The method consists of the following steps:

1. Preparation of germanium dioxide, GeO_2. (*a*) *From argyrodite ore by Winkler's method.* The finely powered ore is fused with Na_2CO_3 and sulphur. The fused mass is extracted with H_2O and treated with dil. H_2SO_4 and filtered. The precipitate obtained contains sulphur, As_2S_3 and Sb_2S_3 and is rejected while the filtrate contains soluble germanium salts. This filtrate is treated with excess of HCl and is saturated with H_2S. Germanium is precipitated as germanium sulphide (GeS_2). This precipitate of GeS_2 is washed, roasted with HNO_3 and ignited when Ge_2S is converted into GeO_2.

(*b*) *From germanite ore.* The powered ore is dried in air and then heated to 350°C in an atmosphere of Cl_2 in a chlorination chamber so that $GeCl_4$, $AsCl_3$ and S_2Cl_2 are formed. Now these chlorides are subjected to fractional distillation whereby $GeCl_4$ is distilled over 83°C and is thus separated from $AsCl_3$ (boiling point = 130°C) and S_2Cl_2 (boiling point = 138°C). The clear distillate which contains only $GeCl_4$ is poured with vigorous stirring over crushed ice when $GeCl_4$ gets hydrolysed and is converted into light yellow suspension of hydrated GeO_2. Now the hydrated GeO_2 is saturated with Cl_2, distilled and hydrated with ice. The white precipitate is dried and ignited to get GeO_2.

2. To get germanium metal from GeO_2. When GeO_2 obtained as above is reduced, germanium metal is obtained. The reduction of GeO_2 to free Ge-metal can be done by any of the following methods:

(*a*) by heating GeO_2 to red heat in a current of H_2. (*Winkler's method*)

$$GeO_2 + 2H_2 \xrightarrow{540°C} Ge + 2H_2O$$

(*b*) by reducing GeO_2 with carbon under a reflux of NaCl.

$$GeO_2 + C \longrightarrow Ge + 2CO$$

(*c*) by heating GeO_2 with Al-metal.

$$3GeO_2 + 4Al \longrightarrow 3Ge + 2Al_2O_3$$

Properties

(*a*) **Physical.** Germanium is a greyish-white brittle metal with 937°C as its melting point and 2700°C as its boiling point.

(*b*) **Chemical.** (*i*) *Oxidation.* The metal is stable in air and undergoes oxidation only at red heat to form a thin film of GeO_2

(*ii*) *Decomposition of water.* Germanium does not decompose water below 200°C. However, at higher temperature water gets decomposed as shown below:

$$Ge + 2H_2O \xrightarrow{\text{High temperature}} GeO_2 + 2H_2$$

(*iii*) *Action of acids.* The metal is not attacked by dil. HCl or H_2SO_4 or by conc. H_2SO_4 at 90°C. However, the metal is attacked by hot conc. H_2SO_4 and dil. HNO_3

$$2Ge + 2HNO_3 \longrightarrow 2GeO_2 + 2NO + H_2$$

(*iv*) *Action of alkalies.* Molten alkalies easily dissolve the metal, evolving H_2.

$$Ge + 2KOH + H_2O \longrightarrow K_2GeO_3 + 2H_2\uparrow$$

(*v*) *Action of H_2 and Cl_2.* When heated above 700°C, Ge combines with H_2 to torm the volatile hydride, GeH_4. Finely powdered Ge metal ignites in Cl_2.

Uses of germanium

(*i*) Ge is now widely used in transistors for reactifying and amplifying alternating current. Most of the metal is used in semiconductor engineering. (*ii*) Ge has been used in germanium resistance thermometers which possess sensitivity and exceptional stability in the temperature range near absolute zero. (*iii*) It is used in mercury discharge lamps for supplying red rays. (*iv*) It is also used in television receivers, transistors and appliances for the transmission of infrared rays.

Tin

Occurrence

Tin does not occur free in nature. Its important ore is *cassiterite* (also called *tin stone*). SnO_2. This ore contains 0.5%-10% of the metal as SnO_2, the rest being the impurities of pyrities (*i.e.*, sulphides) of Fe, Cu and *wolframite* [which is an isomorphous mixture of the tungstates of Fe (+2) and Mn (+2), $FeWO_4 + MnWO_4$], As, sand, silicious matter, earth matter etc. The miners call tin stone as *tin* or *black tin* to distinguish it from *white tin* which is the name given to the metal. Its important deposits occur in Malaysia, Indonesia, Bolivia (South America), China, Burma, Thailand (Siam) and Nigeria (Africa). In India small quantities of tin stone occur in Hazrat Bagh (Bihar) and Orissa. Tin also occurs as *stannite* (*tin pyrites*), $CuS.FeS.SnS_2$.

Extraction of tin from tin stone ore, SnO_2

The extraction of tin from tin stone consists of the following steps:

1. Crushing and concentration of the ore. First of all, the ore is crushed and finally powdered in ball mills. The powdered ore contains the impurity of walframite, (Fe, Mn) WO_4 and silicons and earthy matter. Walframite is magnetic in nature and silicons and earthy matter are lighter than the ore of tin stone. These impurities are removed by the following methods:

(*a*) *Removal of magnetic impurity of walframite, (Fe, Mn) WO_4*. We know that tin stone itself is non-magnetic but the impurity of walframite, (Fe, Mn)WO_4 is magnetic. This impurity of walframite can be removed from tin stone by electromagnetic separation method. The details of this method have already been given in chapter 20.

(*b*) *Removal of lighter impurities of silicous and earthy matter.* We know that tin stone contains the impurities of silicous and earthy matter which are lighter than the ore of tin stone. These impurities are, therefore, removed by Wilfley table method and hydraulic classifier method. Both these methods are two forms of gravity separation method. These have already been discussed in chapter 20.

2. Roasting. In this process the concentrated ore is heated strongly in an inclined revolving furnace in a free supply of air, when the impurities of S, As, Sb etc., are converted to their oxides and volatilise away and the impurities of the sulphides of Cu and Fe present in the ore are oxidised to their respective sulphates.

$$S + O_2 \longrightarrow SO_2\uparrow$$

$$4M + 3O_2 \longrightarrow 2M_2O_3$$
(M = As, Sb)
$$MS + 2O_2 \longrightarrow MSO_4$$
(M = Cu, Fe)

Now the roasted mass which contains SnO_2 (tin stone ore), $CuSO_4$ and $FeSO_4$ is extracted with dil. H_2SO_4, when $CuSO_4$ and $FeSO_4$ dissolve in it and SnO_2 remains as residue.

3. Smelting: Reduction of SnO_2 to Sn. The black tin is mixed with anthracite coal and heated in a reverberatory furnace at 1200-1309°C. If silica (SiO_2) is present as impurity, a little lime (CaO) or fluorspar is also added as a flux. SnO_2 is reduced to molten tin metal which collects in the lower layer and is tapped down through the hole and cast into *blocks* or *ignots.*

$$SnO_2 + 2C \longrightarrow Sn\ (\textit{Molten tin}) + 2CO\uparrow$$

Molten metal obtained thus is called *block tin* and contains 70% of the metallic tin.

Some SnO is also formed during the reduction of SnO_2 to Sn.

$$SnO_2 + C \longrightarrow SnO + CO\uparrow$$

SnO thus formed and CaO (lime) react with silica (SiO_2) to form slag of $SnSiO_3$ and $CaSiO_3$.

$$SnO + SiO_2 \longrightarrow SnSiO_3\ (\textit{slag})$$
$$CaO\ (\textit{basic flux}) + SiO_2\ (\textit{acidic impurity}) \longrightarrow CaSiO_3\ (\textit{slag})$$

The slag, being lighter, floats on molten tin metal and thus forms the upper layer.

Tin metal can also be recovered from the slag, $SnSiO_3$ by melting it with lime stone ($CaCO_3$) in a reverberatory furnace.

$$CaCO_3 \longrightarrow CaO + CO_2$$
$$SnSiO_3 + CaO + C \longrightarrow Sn + CaSiO_3 + CO\uparrow$$

Sometimes in place of lime stone, some scrap iron may also be used.

$$SnSiO_3 + Fe \longrightarrow Sn + FeSiO_3$$

4. Refining of impure tin. Tin obtained as above contains the impurities of Cu, Fe, W etc., and SnO_2 left unreduced during smelting. These impurities are removed by the following methods: (*a*) *Liquation process.* (*b*) *Polling process.* (*c*) *Electrolytic process.* See chapter 20.

Properties of tin

(*a*) **Physical.** Tin is a lustrous white metal with bluish tinge, whiter than Ag and Zn. Its melting point is 232°C and boiling point is 2275°C. The metal is soft, can easily be cut with knife. It is highly malleable and ductile and can be beaten into very thin sheets known as tin foil. The ductility is maximum at 100°C. At 200°C it becomes brittle and falls into powder. If a sheet or bar of tin is bent, it produces a specific sound known as *'cry of tin'*. Tin shows enantiotropy among its three allotropic forms:

$$\underset{\text{sp. gr.} = 5.8}{\textit{Grey tin}} \underset{}{\overset{18°C}{\rightleftharpoons}} \underset{\text{sp. gr.} = 7.8}{\textit{White tin}} \overset{170°C}{\rightleftharpoons} \underset{\text{sp. gr.} = 7.56}{\textit{Rhombic tin}}$$

White tin is not only the heaviest but also the most stable variety. The conversion of white tin into grey tin takes place very slowly at 18° which is the transition temperature. However, if the temperature is as low as –50° and a small amount of grey tin is already present, the conversion becomes very rapid. This conversion is known as *tin pest* or *tin disease* or *tin plague.* White tin is malleable and not ductile. It becomes brittle when heated to 200°C.

(*b*) **Chemical.** When heated to whiteness (1500-1600°C) in the presence of O_2, tin burns with a bright flame, forming SnO_2. If the metal is heated in the atmosphere of Cl_2 or sulphur, it forms $SnCl_4$ and SnS_2.

$$Sn + 2Cl_2 \longrightarrow SnCl_2\ ; \qquad Sn + 2S \longrightarrow SnS_2$$

On being heated with conc. HCl, Sn forms $SnCl_2$ and H_2 is evolved. With conc. H_2SO_4, it gives $SnSO_4$ and SO_2 is evolved. Dil. HNO_3 attacks the metal forming $Sn(NO_3)_2$ and $(NH_4)_2NO_3$. Hot conc. HNO_3 produces copious fumes of NO_2 and metastannic acid, H_2SnO_3. Organic acids have no action on tin metal.

$$Sn + 2HCl \longrightarrow SnCl_2 + H_2$$
$$Sn + 2H_2SO_4 \longrightarrow SnSO_4 + SO_2 + 2H_2O$$
$$4Sn + 10HNO_3 \longrightarrow 4Sn(NO_3)_2 + NH_4NO_3 + 3H_2O$$
$$Sn + 4HNO_3 \longrightarrow H_2SnO_3 + 4NO_2 + H_2O$$

When the metal is heated with alkalies, H_2 gas is evolved and stannate is formed.

$$Sn + 2NaOH + H_2O \longrightarrow Na_2SnO_3\ (\textit{Sod. stannate}) + 2H_2\uparrow$$

Uses of tin. (*i*) Tin is used in the preparation of a number of alloys which are very useful. Some important alloys of tin and their uses are given in Table 24.1

Table 24.1 : *Important alloys of tin*

	Name	Percentage composition	Uses
1.	Solder	Sn = 67%, Pb = 33%	In soldering.
2.	Pewter	Sn = 75%, Pb = 25%	In making cups, mugs, and other utensils.
3.	Babbit metal	Sn = 90%, Sb = 7%, Cu = 3%	For making bearings for machinery
4.	Britannia metal	Sn = 90%, Sb = 8%, Cu = 2%	Being silvery white, it is also used for making table wares.
5.	White metal	Sn = 82%, Sb = 12%, Cu = 6%	For making crockery and table wares.
6.	Bell metal	Sn = 25%, Cu = 75%	For making bells, gongs, etc.
7.	Rose metal	Sn = 28%, Pb = 22%, Bi = 50%	For electric fuses.
8.	Type metal	Pb = 75%, Sn = 5%, Sb = 20%	Printing types.
9.	Wood's metal (fusible alloy)	Pb = 25%, Bi = 50%, Sn = 12.5%, Cd = 12.5%	Melting point being very low (71°C), it is used in fuse wires and safety plugs.

(*ii*) It is used in the preparation of collapsible tubes for tooth pastes and various ointments. (*iii*) *Tinning of household utensils.* Copper or brass utensils are protected from poisonous corrosion by atomsphere or organic acids present in the foodstuffs by tinning. Utensils are cleaned well by rubbing with sand or clay and heated. Some NH_4Cl is sprinkled on the hot surface which liberates HCl to remove any oxide film present on the surface.

$$NH_4Cl \longrightarrow NH_3 + HCl$$

A little tin alongwith some NH_4Cl powder is rubbed on the entire surface to provide a thin coating of the metal on the utensil. (*iv*) *Tin plating of iron sheets.* It is largely used in tin plating of iron sheets. In this process a protective coating of tin metal is made over iron sheets. The process of tin plating consists of dipping the clean sheets of iron or steel in a bath of molten tin and passing through hot pairs of rollers. A thin coherent film of tin metal is deposited on the iron sheets. Such plates are used in the manufacture of tin cans, kerosene oil containers and other similar articles. (*v*) Tin amalgam is used for making mirrors. (*vi*) Compounds of tin are used as mordants in dyeing and calico-printing.

Lead

Occurrence

The chief ore of lead is *galena,* PbS which is associated with zinc blende, iron pyrites and traces of silver and is found in Spain, North America, New South Wales, Mexico and Burma. Other minerals containing lead are *anglestite* ($PbSO_4$) and *cerrusite* ($PbCO_3$)

Occurrence in India

Galena occurs in Kurnool, Guntur and Cuddapah districts of Andhra Pradesh. Galena is found in Singhbhum, Ranchi, Hazaribagh districts of Bihar. It also occurs in Kulu and Kangra districts of Himachal Pradesh, Jammu and Kashmir (Baramula district). Lead ores are found on a fairly large scale in Madhya Pradesh, Mysore (Metri) and Tehri Garhwal in U.P.

Extraction of lead from galena ore, PbS

Lead is mainly extracted from *galena* ore, PbS. The extraction of lead from galena is carried out by the following two processes:

1. Air reduction process. This method is also called **auto-reduction** or **self-reduction process.** This method is used to extract lead from that galena ore which is *rich* in lead content. The concentrated galena ore (PbS) (concentrated by forth flotation process) is transferred to the hearth of a reverberatory furnace where it is heated at *moderate temperature* in the presence of *controlled quantity of air* (Roasting). Here some portion of galena is converted into lead oxide (PbO) and the remaining portion of galena into $PbSO_4$.

$$2PbS\ (\textit{some portion}) + 3O_2 \xrightarrow[\text{temp.}]{\text{Moderate}} 2PbO + 2SO_2$$

$$2PbS\ (\textit{remaining portion}) + 2O_2 \longrightarrow PbSO_4$$

Now the quantity of air is reduced and excess of galena is added to the charge and the temperature is raised. Under this condition PbO and $PbSO_4$ formed as above are reduced to molten lead (*Reduction*).

$$2PbO + PbS \longrightarrow 3Pb\ (\textit{molten}) + SO_2\ ; \quad PbSO_4 + PbS \longrightarrow 2Pb\ (\textit{molten}) + 2SO_2$$

Molten lead is drawn off from the bottom of the furnace.

If galena is heated strongly in the free supply of air or O_2, the galena ore is directly reduced to lead metal

$$PbS\ (\textit{galena}) + O_2 \xrightarrow[\text{strongly}]{\text{Heat}} Pb + SO_2$$

2. Carbon reduction process. This process is used to extract lead from that galena ore which is even *poor* in lead content. This process consists of the following steps:

(*i*) *Concentration of galena ore by froth flotation process.* See chapter 20.

(*ii*) *Roasting and smelting.* The concentrated galena ore is roasted in presence of air in a *sinterer* made of sheet metal and lined inside with fire bricks. The sinterer has a grating near the bottom and a hood at the top. Mounted on trunions, it can be brought in various positions for charging and discharging. The concentrated galena ore is placed on the coal fire at the grating and air is sucked in from the pipe by suction. Galena is oxidised to PbO

$$2PbS + 3O_2 \longrightarrow 2PbO + 2SO_2$$

PbO thus obtained usually contains silica (SiO_2) as impurity (*acidic impurity*). PbO is mixed with coke and lime, CaO (*basic flux*). Coke reduces PbO to molten Pb metal while CaO combines with SiO_2 to form the slag of $CaSiO_3$ which is removed away. Thus CaO acts as a flux.

$$2PbO + C \longrightarrow 2Pb + CO_2$$

$$CaO\ (\textit{basic flux}) + SiO_2\ (\textit{acidic impurity}) \longrightarrow CaSiO_3\ (\textit{slag})$$

Molten lead and slag ($CaSiO_3$) get collected in the lowest part of the furnace. The slag floats on the surface of the molten lead and is drawn out through slag hole while lead is removed through the tapping hole near the bottom of the furnace.

Here it should be noted that since CaO (flux) is more basic than PbO, the latter (*i.e.*, PbO) does not combine with SiO_2 (impurity) to form the slag of $PbSiO_3$, *i.e.*, the formation of $PbSiO_3$ is prevented

$$PbO + SiO_2 \longrightarrow PbSiO_3\ (\textit{Does not occur})$$

Purification of lead

Lead obtained by any of the above two methods contains the impurities of metals like Cu, Fe, Sn, Ag, Zn, Au, Sb etc. It is due to the presence of these impurities that impure lead is hard and brittle. When these impurities are removed, lead becomes soft. In order to remove the metallic impurities from the impure lead the following methods are used:

(*i*) *Removal of base metal impurities* (*Cu, Fe, Sn etc.*) *by softening process.* Base metals which are present as impurities in lead can be removed from the metal by softening process whose details are given in chapter 20.

(*ii*) *Removal of Ag from argentiferrous lead by Parke's process : Desilverisation of lead.* See chapter 20.

(*iii*) *Removal of metals (e.g., Zn, Au, Sb) from the desilverised lead by Bett's electrolytic process.* Desilverised lead obtained by Parke's process still contains the impurities of metals like Zn, Au, Sb etc. These metal impurities are removed by Bett's electrolytic process whose details are given in chapter 20.

Properties of lead

(*a*) **Physical.** Lead is a soft metal with a bluish grey colour. Its freshly cut surface shows a bright lustre. It leaves mark on paper. The metal is malleable but not very ductile. The presence of other metals makes it brittle. It has m.pt. = 32°, b.pt. = 1620° and sp. gr. = 11.34.

Lead is readily obtained in crystalline state by the process of precipitation. If a zinc rod be suspended in a weak solution of lead acetate, the metallic lead is thrown as a mass of arboraceous crystals, known as the *lead tree.*

$$(CH_3COO)_2Pb + Zn \longrightarrow Pb + (CH_3COO)_2Zn$$

(*b*) **Chemical.** (*i*) *Action of air.* Lead is unaffected in dry air but in the presence of moisture, it tarnishes due to the formation of a thin film of the hydroxide and finally carbonate. This layer protects the metal from further action of air.

Heated in air or oxygen, it is slowly oxidised to litharge, PbO, and at a high temperature (400-500°C) to red lead, Pb_3O_4.

$$2PbO + O_2 \longrightarrow 2PbO\ ; \qquad 6PbO + O_2 \longrightarrow 2Pb_3O_4$$

(*ii*) *Action of water.* Lead is not attacked by pure water in the absence of air but water containing dissolved air has a solvent action on it due to the formation of lead hydroxide. This phenomenon is called *plumbo solvency.*

$$2Pb + O_2 + 2H_2O \longrightarrow 2Pb(OH)_2$$

(*iii*) *Action of acids.* Dilute hydrochloric and sulphuric acids have practically no action on lead. Hot concentrated hydrochloric acid forms lead chloride with the evolution of hydrogen.

Hot concentrated sulphuric acid librerates sulphur dioxide but the reaction is retarded by the formation of an insoluble layer of lead sulphate.

$$Pb + 2H_2SO_4 \longrightarrow PbSO_4 + 2H_2O + SO_2$$

Dilute and concentrated nitric acid dissolves lead forming lead nitrate and the oxides of nitrogen.

$$3Pb + 8HNO_3\ (\text{dilute}) \longrightarrow 3Pb\ (NO_3)_2 + 2NO + 4H_2O$$

$$Pb + 4HNO_3\ (\text{concentrated}) \longrightarrow Pb\ (NO_3)_2 + 2NO_2 + 2H_2O$$

Organic acids dissolve lead. A considerable danger is thus involved in using cooking vessels tinned with tin containing lead. In the presence of air, it dissolves in acetic acid forming lead acetate.

$$Pb + 2CH_3COOH + \tfrac{1}{2}O_2 \longrightarrow (CH_3COO)_2Pb + H_2O$$

(*iv*) *Action of chlorine and sulphur.* Lead combines with chlorine and sulphur on heating.

$$Pb + S \longrightarrow PbS\ ; \qquad Pb + 2Cl_2 \longrightarrow PbCl_4$$

Uses

Lead is used for making cable covering, protective sheets for roofs and drains, water pipes and for lining the chambers in sulphuric acid manufacture. It is employed in making lead accumulator plates, lead shot, fuse wire and compounds like red lead, litharge, lead tetraethyl and white lead. It is an excellent protection from radiation in atomic work and its alloys find numerous applications. Some of the alloys of lead are: *type metal* (Pb-Sb-Sn in the ratio 60 : 30 : 10) used for making type for printing ; *cable sheet* (Pb = 29 ; Sb = 1) used for making cable covering and *solder* (Pb = 50 ; Sn = 50) used for soldering.

Selenium

Occurrence

Its minerals are mostly selenides and a few selenites. The important minerals are : *Berze-*

hamite, Cu_2Se; *Zorgite,* $Cu_2Se.PbSe$; *Clausthelite,* PbSe; *Chalcomerite,* $CuSeO_3.2H_2O$ and *Selenolite,* SeO_2. Selenium is also found in teethbones, lavas of vulcanoes, some vegetables and gold and copper ores.

Extraction of selenium from anode mud of copper refineries

Selenium is generally extracted from the anode mud of copper refineries. During the process of refining the crude copper by electrolytic method, a deposit is formed round the anode and settles down on the bottom. This deposit is called *anode mud.* This mud contains Se, Te, Cu, Ag, etc. Selenium may be obtained from this mud as follows: The mud is roasted under oxidising conditions when Cu and Se are converted into CuO and SeO_2 respectively. SeO_2 gets volatilised and deposited in the flues alongwith TeO_2.The deposit which contains SeO_2 and TeO_2 is dissolved in H_2O and the solution so obtained is filtered. SeO_2 goes into water as selenious acid, H_2SeO_3 while TeO_2, being sparingly soluble, remains undissolved. The solution of H_2SeO_3 is reduced with SO_2 when Se is precipitated as a grey powder.

$$SeO_2 + H_2O \longrightarrow H_2SeO_3 \text{ (Selenious acid)}$$

$$H_2SeO_3 + 2SO_2 + H_2O \longrightarrow Se\downarrow \text{ (Grey ppt.)} + 2H_2SO_4$$

Grey precipitate of selenium is washed with hot water and dried.

Properties

(*a*) **Physical.** Selenium, like sulphur, exists in allotropic forms. The well known allotropic forms are given below:

(*i*) *Amorphous selenium.* It is obtained by the reduction of selenious acid. It is a bright red powder first formed but when heated to 50° it gives darker clots of soft mass similar to vitreous selenium. If allowed to stand in contact with carbon disulphide, benzene or chloroform, it gives red crystals.

(*ii*) *Vitreous selenium.* It is prepared by heating amorphous form above 217°C, and cooling rapidly. It is glassy, brittle and black in lump, but the thin sheets have a ruby red colour. It is a dielectric like glass and can be thus electrified by friction. Sp. gr. is equal to 4.2.

(*iii*) *Soluble selenium.* It is also prepared by the reduction of selenious acid solution. In this form, the element is completely soluble in water, forming red fluorescent solution. This may be supposed to be colloidal selenium. It is precipitated from the solution on the addition of acids or salts.

(*iv*) *Red crystalline monoclinic selenium.* When amorphous selenium is crystallised from *hot* carbon disulphide, dark red translucent crystals are obtained, but if crystallised from *cold,* the crystals have orange red colour. Sp. gr. is equal to 4.4 – 4.5.

(*v*) *Metallic or grey selenium.* It is obtained by cooling molten selenium to 210°C and allowing it to stand at this temperature for sometime. It is also obtained by heating any other form to high temperature. This is stable form of selenium, melting at 217°C and has sp. gravity equal to 4.8.

(*b*) **Chemical.** (*i*) *Action of acids.* It is soluble in conc. H_2SO_4, giving a green solution. When boiled with HNO_3, it forms H_2SeO_3.

$$Se + 4HNO_3 \longrightarrow H_2SeO_3 + 4NO_2 + H_2O$$

It is not attacked by HCl. When HI is heated with Se in a sealed tube, H_2Se is obtained.

$$Se + 2HI \rightleftharpoons H_2Se + I_2$$

On cooling the reverse reaction takes place.

(*ii*) *Action of heat.* On being heated in air, it burns with a blue flame and gives fumes of SeO_2. Above 200°C, the metal becomes a good conductor.

(*iii*) *Action of nonmetals.* (*a*) With H_2 and N_2, H_2Se and NSe respectively are obtained.

(*b*) It combines with Cl_2 to form $SeCl_2$ (brown liquid) and $SeCl_4$ (white solid) and with F_2 in cold to give SeF_4 and SeF_6. When heated in sulphur, various sulphides are obtained.

(*iv*) *Action of metals.* (*a*) With alkali metals monomeric and polymeric selenides (*e.g.,* Na_2Se, Na_2Se_2, Na_2Se_3 etc.) are formed.

(*b*) With most of the remaining metals, only monoselenides are obtained. (*e.g.,* CdSe).

(*v*) *Action of compounds.* (*a*) When heated with $SOCl_2$ or SO_2Cl_2, $SeCl_4$ is obtained.

$$2SOCl_2 + Se \longrightarrow SeCl_4 + SO_2 + S$$

$$2SO_2Cl_2 + Se \longrightarrow SeCl_4 + 2SO_2$$

(*b*) With $AgNO_3$ solution, the reaction proceeds as:

$$3Se + 4AgNO_3 + 3H_2O \longrightarrow 2Ag_2Se + H_2SeO_3 + 4HNO_3$$

(*vi*) Selenium imparts various colours (*e.g.*, yellow, reddish, violet, brown etc.) to glass.

(*vii*) *Action of liq. SO_2.* Selenium dissolves in liq. SO_2 to form $SeSO_3$.

(*viii*) *Toxicity.* Selenium and its compounds are highly toxic. The salts of selenium on contact with skin causes dermatitis. Selenium when taken orally or inhalation of its vapours causes gastro-intestinal troubles and it has a disastrous effect on the lungs, liver, kidney and other organs.

Uses

Selenium is used (*i*) in glass industry as a deolouriser, in the production of various coloured glasses (like ruby glass) and in glazes and enamels (*ii*) in the rubber industry. Selenised rubber is used as a fire-proof material for cables (*iii*) as a semiconductor in photoelectric cells, optical warning instruments, indicators, rectifiers, television and sound pictures and picture telegraphy.

Tellurium

Occurrence

Generally it is found in combination with metals like Ag, Pb, Bi, Au etc., as tellurides. Some important telluride minerals are : *Salvanite* or *graphic tellurium,* (Ag, Au) Te_2 ; *Tetradymite,* Bi_2, Te_3 ; *Petzite,* $(Ag, Au)_2$ Te ; *Hessite,* Ag_2Te.

Extraction

Tellurium may be extracted either from the anode mud of copper refineries or from tetradynite ore, Bi_2Te_3.

(*i*) **From anode mud of copper refineries.** We have seen that in the extraction of selenium from the anode mud of copper refineries, TeO_2, being sparingly soluble, remains undissolved and hence is obtained as a precipitate by filtration. The residue of TeO_2 thus obtained is used to get tellurium metal by any of the following methods:

(*a*) The residue of TeO_2 is treated with a solution of NaOH to get sod. tellusite (Na_3TeO_3) in solution

$$TeO_2 + 2NaOH \longrightarrow Na_2TeO_3 + H_2O$$

The solution containing Na_2TeO_3 is electrolysed in a cell having stainless steel electrodes. Tellurium metal gets deposited on the cathode from where it can be removed, washed and dried.

(*b*) TeO_2 is dried and reduced to Te-metal by heating with charcoal in a furnace at 480°C

$$TeO_2 + 2C\ (charcoal) \longrightarrow Te + 4CO$$

(*ii*) **From tetradymite ore, Bi_2Te_3.** The Finely-powdered ore is made into a paste with Na_2CO_3 and then heated to white heat so that sod. telluride (Na_2Te) is obtained.

$$Bi_2Te_3 + 3Na_2CO_3 \longrightarrow 3Na_2Te + Bi_2O_3 + 3CO_2$$

The resulting mass which contains Na_2Te is washed with air-free water and then air is passed into the filtrate, when Te-metal gets precipitated.

$$2Na_2Te + 2H_2O + O_2 \longrightarrow 4NaOH + 2Te\downarrow$$

Properties

Like sulphur and selenium, tellurium also exists in various allotropic forms out of which the following are important:

(*i*) *Amorphous tellurium.* It is a grey powder having sp. gravity = 6.0 and m.pt. = 444.8°C. It is obtained by the reduction of tellurous acid (H_2TeO_3) or telluric acid (H_6TeO_6) with SO_2. This form, on heating, is converted into the crystalline or metallic form.

(*ii*) *Colloidal tellurium.* It is red coloured variety which melts at 45°C. It is prepared by the reduction of telluric acid or by electrolysing a solution of tellurium in HNO_3. The colour of the colloidal solution is dark brown.

(iii) Crystalline (also called metallic or hexagonal) tellurium. It is obtained by heating amorphous tellurium. It is silvery-white in colour with hexagonal crystals and has a metallic lustre. It is brittle and can be powdered easily. It is a poor conductor of electricity and heat. It is insoluble in H_2O and CS_2 but soluble in HNO_3. It melts at 453°C and boils at 1400°C.

It burns in air with a blue flame, forming white vapours of TeO_2.F_2 gives TeF_4, Cl_2 and Br_2 give di- and tetrahalides while I_2 does not give any iodide even on fusion.

It is oxidised by HNO_3 and H_2SO_4. With alkali metals it forms dark red poly tellurides and with many other metals monotellurides are formed. Powdered tellurium reacts with $AgNO_3$ and $AuCl_3$ in the following manner:

$$3Te + 4AgNO_3 + 3H_2O \xrightarrow{100°C} 2Ag_2Te + H_2TeO_3 + HNO_3$$

$$3Te + 4uCl_3 \longrightarrow 3TeCl_4 + 4Au$$

Uses

(i) It is used in making alloys which are very useful. For example, since Te-Pb alloy is resistant to chemical attack and has increased tensile strength, it is used in making water pipes which do not burst when H_2O freezes in them. Te-Al alloy is very ductile and hence is very important. Te-Ag alloy is used in dental fillings. *(ii)* Steel having small quantity of Te is used in making cutting tools. *(iii)* Te is also used in rubber industry. *(iv)* Colloidal tellurium is used as a fungicide. *(v)* It is used in glass and ceramic industry. *(vi)* It is also used in photography for imparting brown tone to silver prints. *(vii)* The use of Te has been suggested for the treatment of cancer and syphilis. *(viii)* Te can be used as an indicating electrode for potentiometric titration even in nonaqueous solvent.

Fluorine

History of its isolation

It was known to *Schwankhard* (1670) that glass is attacked when exposed to the fumes obtained by heating fluorspar with conc. H_2SO_4. In 1771 *Scheela* heated fluorspar with conc. H_2SO_4 in a tin retort and obtained an acid solution. By his experiment he concluded that fluorspar is the calcium salt of this acid solution which he named *fluoric acid. Gay Lussac* and *Thenard* in 1809 prepared this acid and showed that the acid resembles hydrochloric acid in many properties and on analogy they named it *hydrofluoric acid.* In 1813 *Davy* showed that hydrofluoric acid was a compound of H_2 and an unknown element which he named *fluorine.* Davy tried to prepare fluorine by the electrolysis of aqueous solution of hydrofluoric acid just as Cl_2 is obtained of the electrolysis of aqueous solution of HCl. To his great surprise he obtained H_2 and ozone mixed with O_2, instead of fluorine, at the anode. For being most resistant to oxidation, H_2O gets oxidised at the anode liberating O_3 mixed with O_2.

$$2H_2O \longrightarrow 4H^+ + O_2 + 4e^-$$

$$3H_2O \longrightarrow 6H^+ + O_3 + 6e^-$$

Gore then attempted to electrolyse the anhydrous hydrofluoric acid to isolate fluorine from it, but he found this acid to be nonconductor and hence failed to isolate fluorine from it. In 1860 he showed that the anhydrous acid dissolves in KHF_2 and gives a conducting solution. But another problem was that of great chemical reactivity of fluorine. It attacked the apparatus whether made of glass, carbon and platinum. When platinum vessel was used, it gave a chocolate powder of PtF_4 while with carbon it gave the gaseous CF_4. It was in 1886 that the long standing problem was finally solved by *Moisson.* He prepared fluorine by electrolysing a solution of KHF_2 (1 part) in anhydrous hydrofluoric acid (5 parts) in a U-tube made of Pt-Ir alloy using electrodes of the same alloy. He reduced the great chemical reactivity of fluorine by keeping the bath and collecting the gas at very low temperature.

What were the difficulties encountered in the discovery and isolation of fluorine?

For a long time a number of difficulties were experienced in the discovery and isolation of fluorine. The main difficulties (or causes) for the late discovery and isolation of fluorine are *(i) Extreme reactivity of fluorine.* Fluorine is so reactive that it reacts with the material of all the vessels, *e.g.*, F_2 attacks carbon and silicon present in the glass vessels with the formation of CF_4 and SiF_4 respectively. Similarly F_2 also attacks the electrodes and vessels made of Pt with the formation of chocolate coloured powder, PtF_4. Pb, Fe and Sn vessels were also tried and found to be unsuitable for the preparation of fluorine. Thus no metal could be employed for the preparation of F_2 as vessel or as electrode. (*ii*) *Greater stability and nonconducting nature of hydrofluoric acid.* Hydrofluoric acid is very stable, highly poisonous and corrosive in nature. It is so stable that all attemps to isolate F_2 from hydrofluoric acid by oxidising agents failed. Moreover, electrolysis of aqueous solution of hydrofluoric acid gives H_2 and ozone mixed O_2 while anhydrous hydrofluoric acid in nonconductor.

Occurrence

Fluorine is very reactive and hence does not occur free in nature but is widely found in the combined state as fluorides. Its chief minerals are : *fluorspar*, CaF_2 ; *cryolite*, Na_3AlF_6 and *fluorapatite*, $CaF_2.3Ca_3(PO_4)_2$. Small quantities of fluorine are present as fluoride in plant-ash, soil, sea-water, river water plants, bones and teeth of animals. One litre of sea-water contains about 2 mg of fluorine. Fluorine is also found in the *phosphate rocks. Fluorspar,* CaF_2 is the principal source of fluorine and its compounds. A six mile long belt of fluorspar has been recently discovered in Durgapur (Rajasthan) and it is considered to be sufficient to meet the need of the country.

Isolation

Different methods employed for the isolation of F_2 are given below:

1. Moissan's method. (*Historical method – 1886*). This is a historical method, since this method was used for the first time by Moissan to isolate fluorine. Essential parts of the electrolytic cell used by Moissan are : (*a*) *Electrolytic cell.* Electrolytic cell consists of a U-shaped tube made of Pt-Ir alloy cooled at – 23°C by placing it in a bath of CH_3Cl (b.pt. = – 23°C) which was constantly renewed. (*b*) *Electrodes.* These are made of Pt-Ir alloy and are insulated with fluorspar stoppers covered with shellac. (*c*) *Electrolyte.* Electrolyte consists of a solution of *potassium hydrogen fluoride*, KHF_2 (1 part) in *anhydrous hydrofluoric acid,* H_2F_2 (5 parts).

Process. Moissan took the solution consisting of KHF_2 and anhydrous H_2F_2 (in 1 : 5 ratio) in a U-shaped electrolytic cell at a temperature of –23°C by placing the cell in a bath of CH_3Cl and electrolysed this solution (Fig. 24.8). On electrolysis H_2 was liberated at the cathode and F_2 at the

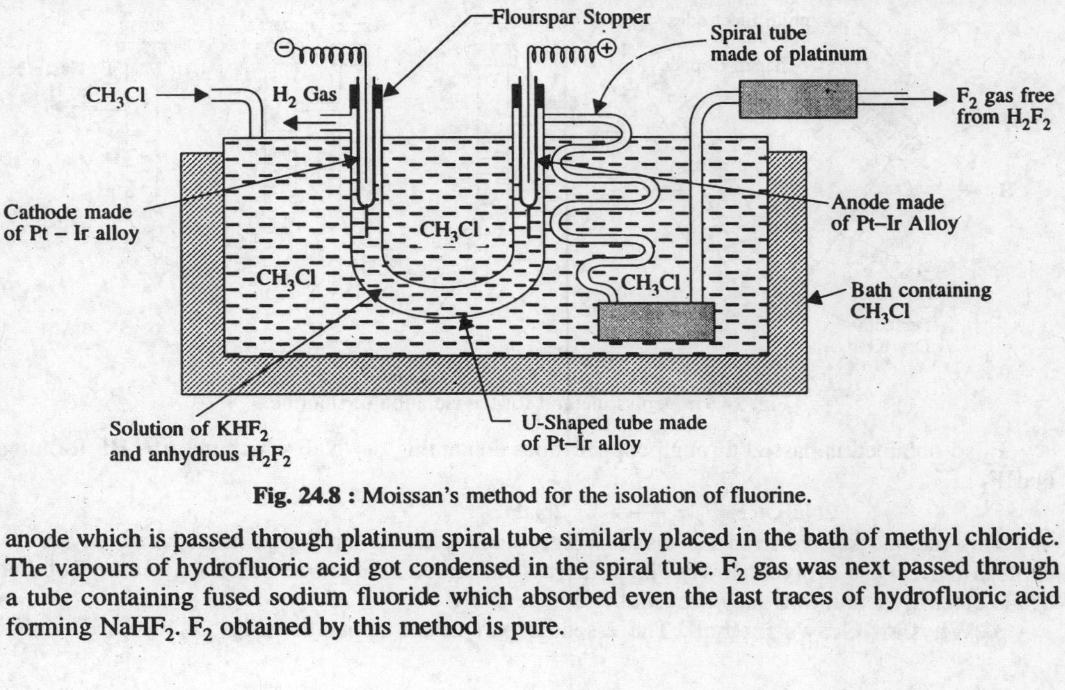

Fig. 24.8 : Moissan's method for the isolation of fluorine.

anode which is passed through platinum spiral tube similarly placed in the bath of methyl chloride. The vapours of hydrofluoric acid got condensed in the spiral tube. F_2 gas was next passed through a tube containing fused sodium fluoride which absorbed even the last traces of hydrofluoric acid forming $NaHF_2$. F_2 obtained by this method is pure.

$$2NaF + H_2F_2 \longrightarrow 2NaHF_2$$
$$2KHF_2 \longrightarrow 2KF + H_2F_2$$
$$H_2F_2 \rightleftharpoons 2H^+ + 2F^-$$

At anode : $F^- - e^- \longrightarrow F;\quad F + F \longrightarrow F_2$

At cathode : $H^+ + e^- \longrightarrow H;\quad H + H \longrightarrow H_2$

In spite of low temperature *viz.*, – 23°C employed in this method, there was severe corrosion and wastage of about 5 grams of platinum for the preparation of one gm of F_2. In 1899 he observed that a Cu-tube could be used in place of Pt-Ir tube, since Cu-tube gets coated with a projective layer of CuF_2 which prevents further action of F_2 on Cu, but he continued the use of Pt-Ir electrodes.

2. Deniss method. This method was devised by *Deniss, Veeder* and *Rochow* in 1931. Essential parts of the electrolytic cell used in this method are : (*a*) *Electrolytic cell.* Electrolytic cell consists of V-shaped tube made of Cu about 2" wide and fitted with copper caps into which graphite electrodes are fixed with bakelite cement. The V-shaped copper tube is thickly covered with an insulting layer of asbestos cement. This layer is wound by a resistance wire for heating the V-shaped tube electrically. The whole apparatus is then thickly lagged to prevent the loss of heat. (*b*) *Electrolyte.* Pure and dry KHF_2 (m.pt. = 227°C) is used as an electrolyte. KHF_2 should be perfectly dry, since the presence of traces of moisture results in the formation of O_2 or OF_2. KHF_2 is usually dried in a current of air at 140° – 150°C. (*c*) *Electrodes.* Graphite electrodes are fixed into the V-shaped copper tube by bakelite cement are used.

Process. When a current of 5 amperes and 12 volts is passed through perfectly dry electrolyte of KHF_2.H_2 is liberated at the cathode and F_2 at the anode and passes out through an exit near the anode (Feg. 24.9). F_2 so produced attacks the copper walls forming the layer of CuF_2 which protects the deeper layers of copper from further attack. The reactions taking place are :

$$KHF_2 \longrightarrow KF + HF;\quad H_2F_2 \longrightarrow 2H^+ + 2F^-$$

At cathode : $2H^+ + 2e^- \longrightarrow H_2$

At anode : $2F^- - 2e^- \longrightarrow F_2$

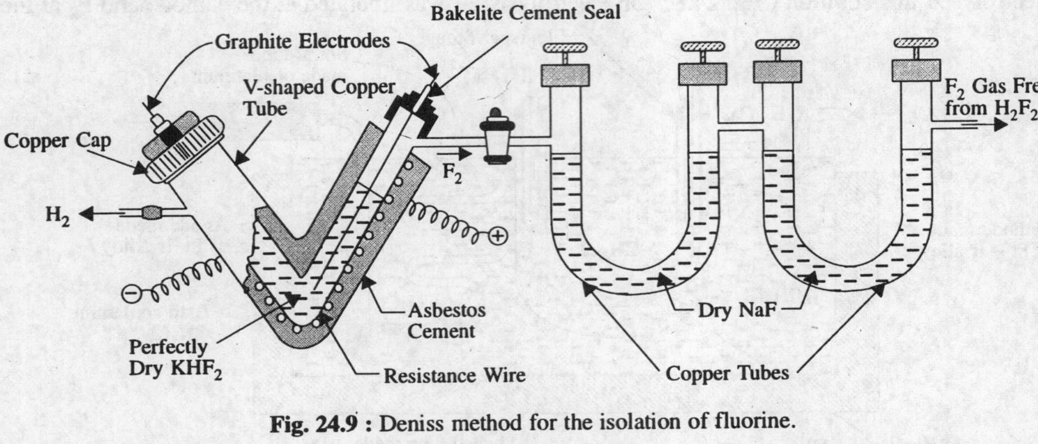

Fig. 24.9 : Deniss method for the isolation of fluorine.

F_2 so obtained is passed through copper tubes containing dry NaF which absorbs HF forming $NaHF_2$

$$2NaF + H_2F_2 \longrightarrow 2NaHF_2$$

This method suffers from the disadvantages namely (*i*) H_2 and F_2 get mixed and may cause explosion. (*ii*) F_2 is evolved slowly due to narrow exit around the anode which may cause frothing of the electrolyte and thus may prevent the escape of F_2.

3. Whytlaw-Gray's method. The essential part of the apparatus (Fig. 24.10) used in this method is a copper cell wound with resistance wire for electrical heating. It also acts as the cathode resistance and has an outlet for hydrogen near the top.

The anode is a rod of graphite enclosed in a copper cylinder. This cylinder has an outlet near the top for the escape of fluorine gas and is perforated near the bottom.

The electrolyte is fused potassium hydrogen fluoride (KHF_2). It undergoes electrolysis to give hydrogen and fluorine.

$$2KHF_2 \longrightarrow 2KF + F_2 + H_2$$

F_2 and H_2 gases escape out of the cell as shown without any possibility of mixing with each other as the anode and the cathode are separated by the cylindrical diaphragm.

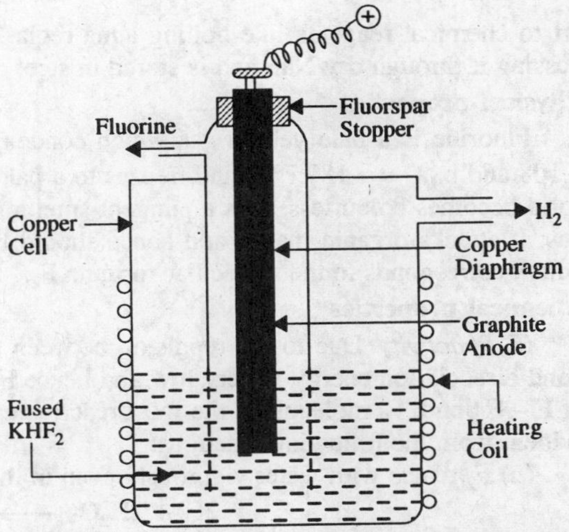

Fig. 24.10 : Whytlaw Gray's apparatus for the isolation of fluorine.

Precautions. (*i*) The electrolyte must be absolutely dry, because the presence of moisture hinders electrolysis and the fluorine evolved may react with it to produce oxygen and ozone. All parts of the apparatus which come into contact with fluorine must also be free from oil and grease. (*ii*) The gas must be freed from HF by using NaF. If HF is present, it will attack the glass vessels in which the fluorine is kept. (*iii*) The vessels in which fluorine is collected must also be absolutely dry. (*iv*) As F_2 attacks oraganic matter of all types, it should be very carefully handled.

Advantages of this method. This method is superior to the above methods in the following respects : (*i*) Copper cylinder (diaphragm) separates the anode from the cathode with the result that the liberated H_2 and F_2 are prevented from mixing with each other and thus the possibility of explosions is avoided. (*ii*) There is no chance of explosion due to the frothing of the electrolyte by the accumulation of gases as the exits for gases are quite wide to allow the gases to escape out quite promptly.

4. Modern method. The manufacture of F_2 by modern method consists of electrolysing fused KHF_2 mixed with 2 to 3 moles of H_2F_2 at 70 – 80°C in a rectangle steel vessel using a current of 1000-2000 amperes at 8.5 to 11 volts. The steel vessel acts as a cathode and the anode was earlier made of special graphite form of carbon. Now anodes of pertroleum coke impregnated with copper are preferred. A steel cylinder attached to the lid acts as a diaphragm (Fig. 24.11) which does not allow F_2 and H_2 to come in contact with each other and thus the two gases are led out through separate exits. Consequently the possibility of explosion is avoided. The valves used are of monel metal or nickel with packing of teflon, $(C_2F_4)_x$ which is in-

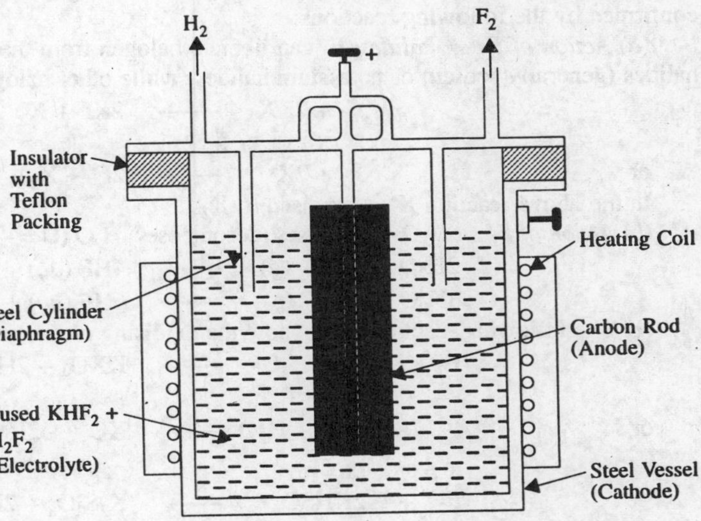

Fig. 24.11 : Manufacture of fluorine by modern method.

ert to chemical reagents like boiling aqua regia. F_2 obtained at the anode is freed from H_2F_2 by passing it through dry NaF and is stored in steel cylinders in compressed conditions.

Physical properties

Fluorine is a pale yellow gas which condenses to a clear yellow liquid having a density of 1.108 and b.pt. = – 187.7°C, and freezes to a pale yellow solid (m.pt. = – 223°C). At – 252°C the solid becomes colourless. It is a pungent smelling and poisonous gas but less poisonous than HF gas. It attacks organic matter and hence should be handled carefully. Its vapour density is 18.91 which corresponds to the molecular formula F_2.

Chemical properties

(*i*) *Reactivity.* Due to the repulsion between the nonbonding electrons in F_2 molecule, F – F bond dissociation energy is very low and hence F – F bond is very weak. Due to the weak nature of F – F bond, F_2 molecule is the most reactive among the halogens. Maximum reactivity of F_2 is evident from the following reactions:

(*a*) F_2 reacts with water vigorously even in the dark, giving O_2 and O_3

$$2F_2 + 2H_2O \longrightarrow 4HF + O_2$$
$$3F_2 + 3H_2O \longrightarrow 6HF + O_3$$

(*b*) F_2 combines with H_2 explosively even in the dark, forming HF

$$F_2 + H_2 \longrightarrow 2HF$$

(*ii*) *Combination with elements.* F_2 combines with metals (*e.g.*, Na, Ca, Mg, Al, Ni, Pt etc.), nonmetals (*e.g.*, H_2, C, S, P, Cl_2, Br_2, I_2 etc.) and metalloids (*e.g.*, As, Sb) to form the fluorides of these elements. F_2 also combines with xenon to form three different fluorides namely XeF_2, XeF_4 and XeF_6. Which fluoride will be formed depends on the temperature, pressure and mixing ratio of F_2 and Xe.

(*iii*) *Action on alkalies.* Unlike Cl_2, Br_2 and I_2, F_2 reacts with dil. alkalies (2% solution) and forms NaF and OF_2 (not NaFO). With conc. alkalies F_2 gives NaF and O_2 (not $NaFO_3$).

$$2F_2 + 2NaOH\ (2\%\ solution) \longrightarrow 2NaF + OF_2 + H_2O$$
$$2F_2 + 4NaOH\ (conc.\ solution) \longrightarrow 4NaF + O_2 + 2H_2O$$

(*iv*) *Oxidising property (Also see chapter 18).* Since F_2 has the largest positive value of standard reduction electrode potential for the reduction reaction,

$$F_2\ (g) + aq + e^- \longrightarrow 2F^-,$$

F_2 is the strongest oxidising agent among all the halogens. Maximum oxidising power of F_2 is confirmed by the following reactions:

(*a*) *Action of F_2 on halides.* F_2 can liberate halogen from the aqueous solution of all the three halides (generally sodium or potassium halides) while other halogens cannot do so.

$$F_2 + 2KX \longrightarrow 2KF + X_2$$
$$(X = Cl, Br, I)$$

or $$F_2 + 2X^- \longrightarrow 2F^- + X_2$$

In the above reaction X^- is oxidised to X_2.

(*b*) *Action of F_2 on H_2O.* F_2 oxidises (decomposes) H_2O (O = – 2) to O_2 (O = 0) and O_3 (O = 0)

$$2H_2O\ (l) + 2F_2\ (g) \longrightarrow 4HF\ (aq) + O_2\ (g)$$
$$3H_2O\ (l) + 3F_2\ (g) \longrightarrow 6HF\ (aq) + O_3\ (g)$$

(*c*) The following reactions also show the oxidising property of F_2

$$F_2 + KXO_3 + H_2O \longrightarrow KXO_4 + 2HF$$
$$(X = Cl, I)$$

or $$\underset{(F = 0)}{F_2} + \underset{(X = +5)}{XO_3^-} + H_2O \longrightarrow \underset{(X = +7)}{XO_4^-} + 2H^+ + \underset{(F = -1)}{2F^-}$$

$$F_2 + 2KHSO_4 \longrightarrow K_2S_2O_8 + 2HF$$

or $$\underset{(F = 0)}{F_2} + \underset{(S = +6)}{2HSO_4^-} \longrightarrow \underset{(S = +7)}{S_2O_8^{2-}} + 2H^+ + \underset{(F = -1)}{2F^-}$$

$$\underset{(F = 0)}{2F_2} + \underset{(Cr = +3)}{2Cr^{3+}} + 7H_2O \longrightarrow \underset{(Cr = +6)}{Cr_2O_7^{2-}} + 14H^+ + \underset{(F = -1)}{6F^-}$$

(*v*) *Action on* NH_3. F_2 reacts with NH_3 to give N_2 and some NF_3

$$3F_2 + 2NH_3 \longrightarrow N_2 + 6HF$$

$$3F_2 + NH_3 \longrightarrow NF_3 + 3HF$$

(*vi*) *Action on* H_2S. H_2S burns in an atmosphere of F_2, forming SF_6 and HF

$$H_2S + 4F_2 \longrightarrow SF_6 + 2HF$$

(*vii*) *Action on organic compounds*: *Formation of fluorocarbons.* F_2 reacts explosively with most organic compounds, forming fluorocarbons like CF_4, C_6F_{12} etc.

$$CH_4 + 4F_2 \longrightarrow CF_4 + 4HCl$$

$$C_6H_6 + 9F_2 \xrightarrow[265°C]{\text{Cu–guage}} C_6F_{12} + 6HF$$

(*viii*) *Action on glass*. If F_2 is free from HF, it has no action on glass. Below 100°C, the reaction between F_2 and glass is slow. Above 100°C, F_2 reacts with glass and forms SiF_4.

Uses of fluorine and its compounds

(*i*) F_2 is used as an insecticide. (*ii*) F_2 gives a number of compounds which are very useful. Some of the more important compounds and their uses are: (*a*) Freon-12 (CF_2Cl_2) is used in refrigeration and air conditioning in place of NH_3 and SO_2. It is also used as a solvent for D.D.T. and other insecticides. (*b*) Teflon, $(CF_2 = CF_2)_n$ is used as an insulating material in cables. (*c*) UF_6 is used in the separation of U^{235} from natural uranium which is used in atom and hydrogen bombs. (*c*) SF_6 has a remakable property of being a high voltage insulation material and hence is used as an insulating gas in nuclear physics and electrical and X-ray works. (*d*) *Freon*, CCl_2F_2 (*dichlorofluoromethane*) is used in refrigerators and cold storage plants. (*e*) H_2F_2 is used for etching glass, in brewing industry and for removing SiO_2 from iron castings. (*f*) *Cryolite* (Na_3AlF_6) and *fluorspar* (CaF_2) are used in the extraction of Al. (*g*) *Fluorspar* is used in the preparation of H_2F_2. (*h*) NaF and Na_3AlF_6 in powdered form are used as insecticides. (*i*) *Presence of small quantities of fluorides*, less than one part per million, in drinking water prevents dental decay. (*j*) *Antimony fluoride* is used in the manufacture of pottery and porcelain and *sodium antimony fluoride* which is a double fluoride of Na and Sb is used as a mordant in the textile industry. (*k*) *Sodium fluoro acetate* finds use as rat poison. (*l*) CuF_2 is used in the ceramic industry and as a flux in soldering, welding and blazing. (*m*) MgF_2 is used in optical industry. (*n*) KF is widely used in the preparation of fluorocarbons. (*o*) *Many inorganic fluorine compounds* are used in industry and technology. D.D.F.T. which is similar to D.D.T. is an extremely active fungicide and fumigant.

University Questions

1. How will you get radium from pitch blende ore ? Discuss the radioactive properties of radium. *(Lucknow, 95)*
2. How is thallium extracted from flue dust? Discuss its thallous and thallic compounds and compare them with Pb, Ag and K compounds. Give its properties and uses. *(Meerut, 95 ; Lucknow, 95)*
3. Give the important minerals of Li. How will you extract Li from its alumino-silicate ore? *(Meerut, 95)*
4. Mention the main sources of boron. How is it obtained from these minerals? Describe briefly the chemical characteristics of boron and its compounds. *(Meerut, 95)*
5. Discuss the position of beryllium in the P.T. How will you extract it from its minerals? *(Meerut, 96)*
6. Write a note on preparation of ultra pure germanium. *(Allahabad, 87)*
7. Give the chemical principles involved in the extraction of germanium. *(Allahabad, 88)*
8. Give two uses of germanium metal. *(G.N.D., 86)*
9. How is tin obtained from its ores? Give the industrial uses of tin. *(Meerut, 96)*

10. Outline the principle of the electrolytic process used for the preparation of F_2. *(Osmania, 93)*

11. Give occurrence and isolation of F_2. What are the reasons of its late discovery? Compare the properties of F_2 with O_2. *(Meerut, 97)*

12. Justify the position of boron in P.T. and explain in detail the properties in which boron differs from the other members of the group. *(Meerut, 97)*

13. Discuss the position of radium in P.T. How is it obtained from its ores? *(Meerut, 97)*

14. Give the names of important ores of lead. How is it obtained from sulphide ores? *(Meerut, 97)*

15. What happens when Sn reacts with HNO_3? *(Meerut, 97)*

16. How is lithium obtained from its phosphate mineral? *(Meerut, 97)*

17. Describe the methods of extraction and isolation of radium from pitch blende ore. *(Lucknow 2002)*

18. (*a*) Give the names and formulae of two ores of beryllium. Describe the method of extraction of beryllium from any one of them.

(*b*) How does lithium occur in nature ? Using flow sheet diagram give one method of its extraction. *(Rohilkhand 2003)*

19. (*a*) How does fluorine occur in nature ? What were the difficulties in the isolation of fluorine ?

(*b*) Describe the modern method for the isolation of fluorine. *(Rohilkhand 2003)*

20. Write a note on ortho and parahydrogen. *(Gujrat 2003)*

21. (*a*) Give an account of general method of purification of metals. Discuss one of them in brief.

(*b*) What is calcination ? How is it used in concentrating an ore ? Give example.

(*c*) Write down ores along with their composition of any two metals of your choice. *(Delhi 2003)*

22. Name the principle ore of lead. How is pure lead extracted from it ? *(Banglore, 2004)*

23. Explain the cause of chemical inertness of noble gases. *(Bhopal 2004)*

24. Explain the structure of XeF_6. *(Bhopal 2004)*

25. (*a*) Why inert gases react with F_2 and O_2 only?

(*b*) Discuss the factors which led to the late discovery of compounds of noble gases.

(*c*) Briefly discuss the clathrates of noble gases.

(*d*) How will you prepare XeF_2 and XeF_4?

(*e*) Discuss the structure of XeF_6, $XeOF_2$, $XeOF_4$ and XeO_3.

(*f*) Complete the reactions :

(*i*) $XeO_2F_2 + H_2O \longrightarrow \ldots\ldots$ (*ii*)

$XeF_6 + H_2O \longrightarrow \ldots\ldots$

(GND Amritsar, 2004)

25

Hydrides of Main Group Elements

Classification of Hydirdes

Hydrogen combines with a number of metals and nonmetals to form the compounds which are called *hydrides*. Based on their physical and chemical properties, the hydrides have been classified into the following four categories:

I. Ionic hydrides

These hydrides are given by the elements of group IA (alkali metals), Ca, Sr and Ba (alkaline earth metals). It may be noted that the hydrides of Be and Mg (*i.e.*, BeH_2 and MgH_2) are not ionic; rather these are covalent hydrides.

Preperation. (*i*) *By heating metal in a current of* H_2. The alkali and alkaline earth metals give hydrides when heated at different temperatures in a current of H_2. For example:

$$2M\ (M = Na, K\ etc.) + H_2 \xrightarrow{400°C} 2MH$$

$$M\ (M = Ca, Sr, Ba) + H_2 \xrightarrow{150-300°C} MH_2$$

(*ii*) *By heating carbonates in a current of* H_2. RbH and CsH are prepared by heating their carbonates in a current of H_2 at 620 – 650°C. In this reaction the presence of Mg is essential.

$$Rb_2CO_3 + 2H_2 \longrightarrow 2RbH + CO_2 + H_2O$$

(*iii*) *By heating metal nitride in a current of* H_2. By this method LiH can be prepared.

$$2Li_3N + 3H_2 \longrightarrow 6LiH + N_2$$

(*iv*) *By Alexander's method.* CaH_2 is prepared by heating CaO with metallic Mg in a current of H_2 at 250°C and 50 cm pressure of hydrogen. A yield of 99.2 % of CaH_2 is obtained in 2 hours.

$$CaO + Mg + H_2 \longrightarrow CaH_2 + MgO$$

Properties. (*i*) (*a*) Alkali metal hydrides are colourless crystalline solids having ionic lattices

(*b*) CaH_2, SrH_2 and BaH_2 which are ionic hydrides of alkaline earth metals are white powder with ionic lattices.

(*c*) As ionic compounds, they have high melting and boiling points and conduct electricity in their fused state liberating H_2 at anode.

(*d*) Their density is higher than that of the metal.

(*ii*) *Ionic/covalent character.* The ionic character of M—H bond in the hydrides of the elements of group IA and IIA increase on moving down the group. For example:

(*a*) The ionic character of the hydrides of alkali metals increases as LiH < NaH < KH < RbH < CsH.

(*b*) Among the hydrides of alkaline earth metals (MH_2 type hydrides) BeH_2 is covalent while MgH_2 is partly covalent and partly ionic. The remaining hydrides (CaH_2, SrH_2 and BaH_2) are largely ionic. Due to small size and high charge on Be^{2+} ion, this ion has high polarising power. Thus Be^{2+} ion polarises the electron cloud of H^- ion in BeH_2 molecule to the greatest extent. This makes BeH_2 molecule covalent. In case of MgH_2 molecule, the polarisation of the electron cloud of H^- ion by Mg^{2+} ion (larger ion) is to lesser extent and hence MgH_2 molecule is partly covalent and partly ionic. Ca^{2+}, Sr^{2+} and Ba^{2+} ions have weak polarising power and hence the hydrides of these cations have little or no covalent character, *i.e.*, these hydrides are largely ionic. Thus we see that the ionic character in MH_2 hydrides increases down the group.

(*iii*) *Order of thermal stability, reactivity and heats of formation.* The thermal stability of the hydrides of the elements of groups IA and IIA decreases down the group and hence the reactivity increases on moving downwards. The heats of formation (ΔH_f) also decrease in moving downwards. Here we have not considered BeH_2 and MgH_2 since these hydrides are covalent in nature.

	Group IA hydrides	*Group II A hydrides*
Thermal stability	LiH > NaH > KH > RbH > CsH	$CaH_2 > SrH_2 > BaH_2$
Reactivity	LiH < NaH < KH < RbH < CsH	$CaH_2 < SrH_2 < BaH_2$
Heats of formation $(-\Delta H_f)$ *(Kcal/mol)*	21.6 13.9 14.4 13.5 13.0	45.1 42.3 40.9

The heat of formation of KH is an exception. The greatest stability of LiH among alkali metal hydrides is due to its highest heat of formation. The decreasing order of the stability or the increasing order of the reactivity of the hydrides of alkali metals is evident from the fact that LiH is stable in dry air upto red heat. NaH inflames in O_2 at 500 K while KH does so even at room temperature. Similarly LiH does not react with Cl_2, NaH reacts only slowly whereas KH inflames.

$$KH + Cl_2 \longrightarrow KCl + HCl.$$

The order of thermal stability and reactivity of CaH_2, SrH_2 and BaH_2 as given above indicate that CaH_2 has the maximum stability and hence is the least reactive. Heat of formation of CaH_2 is maximum (= – 45.1 Kcal mol^{-1}).

(*iv*) *Electrolysis.* When ionic hydrides in the fused state are electrolysed, metal goes to cathode and H_2 is liberated at anode. Thus in these hydrides metal is present as M^+ (in MH type hydrides) or M^{2+} (in MH_2 type hydrides) and hydrogen is present as H^- ion.

(*v*) *Action of air.* They are oxidised by air, some burn spontaneously in air on heating on account of their decomposition into H_2 which is inflammable, *e.g.*,

$$CaH_2 \xrightarrow{\Delta} Ca + H_2$$

(*vi*) *Decomposition.* Excepting LiH, they decompose on fusion.

(*vii*) *Formation of complex hydrides.* (*a*) When ionic hydrides react in ether solution of covalent hydrides (*e.g.*, Br_2H_6, AlH_3, GaH_3) and covalent halides (*e.g.*, BF_3, $AlCl_3$), complex hydrides are obtained. For example:

$$B_2H_6 + \underset{(M = Li, Na)}{2MH} \xrightarrow{Ether} 2M[BH_4]$$

$$MH_3\ (M = Al, Ga) + LiH \longrightarrow Li\,[MH_4]$$

$$BF_3 + 4NaH \longrightarrow Na\,[BH_4] + 3NaF$$

$$\underset{(anhydrous)}{AlCl_3} + 4LiH \longrightarrow Li\,[AlH_4] + 3LiCl$$

(*b*) When NaH reacts with $B(OCH_3)_3$, $Na[BH_4]$ is obtained.

$$4NaH + B.(OCH_3)_3 \longrightarrow Na[BH_4] + 3NaOCH_3$$

(*viii*) *Reaction with N_2.* (*a*) LiH is oxidised by N_2 forming $LiNH_2$ (amide), Li_2NH (imide) and Li_3N (nitride).

$$LiH \xrightarrow{+N_2} LiNH_2 \xrightarrow{+N_2} Li_2NH \xrightarrow{+N_2} Li_3N$$

(*b*) CsH, which is the most reactive hydride among the alkali metal hydrides, gives $CsNH_2$ when heated with N_2.

$$Cs \xrightarrow{+N_2\,(\Delta)} CsNH_2$$

(*c*) CaH_2 gives Ca_3N_2 with N_2

$$3CaH_2 + N_2 \longrightarrow Ca_3N_2 + 3H_2$$

(ix) Decomposition by SO_2. KH and CaH_2 both are decomposed by SO_2, forming the metallic hyposulphite ($S_2O_4^{2-}$).

$$2KH + 2SO_2 \longrightarrow K_2S_2O_4 + H_2\uparrow$$
$$CaH_2 + 2SO_2 \longrightarrow CaS_2O_4 + H_2\uparrow$$

(x) Action of protonic solvents and reducing properties. These hydrides react with protonic solvents (*i.e.*, solvents giving H^+ ions) such as H_2O, NH_3 and C_2H_5OH to form H_2 and a base.

$$LiH + H_2O \longrightarrow LiOH \quad + H_2$$
$$LiH + NH_3 \longrightarrow LiNH_2 \quad H_2$$
$$LiH + C_2H_5OH \longrightarrow LiOC_2H_5 \quad H_2$$
$$CaH_2 + 2H_2O \longrightarrow Ca(OH)_2 \quad 2H_2$$

Due to the production of H_2 as shown above, these hydrides reduce many substances. For example:

(*a*) LiH, NaH and KH reduce CO_2 to metallic formate, HCOOM

$$MH + CO_2 \longrightarrow HCOOM$$
(M = Li, Na, K)

(*b*) NaH reduces CO to carbon, $SiCl_4$ to SiH_4 and Fe_3O_4 to Fe.

$$NaH + 2CO \longrightarrow HCOONa + C$$
$$4NaH + SiCl_4 \longrightarrow SiH_4 + 4NaCl$$
$$NaH + Fe_3O_4 \longrightarrow 3Fe + 4NaOH.$$

(*c*) CaH_2 reduces $PbSO_4$ to PbS and CO_2 to $(HCOO)_2Ca$

$$2CaH_2 + PbSO_4 \longrightarrow PbS + 2Ca(OH)_2$$
$$CaH_2 + 2CO_2 \longrightarrow (HCOO)_2Ca$$

(xi) Action of BCl_3. When the ionic hydrides like NaH, CaH_2 etc., are treated with BCl_3, B_2H_6 is obtained.

$$2BCl_3 + 6NaH \longrightarrow 6NaCl + B_2H_6$$
$$2BCl_3 + 3CaH_2 \longrightarrow 3CaCl_2 + B_2H_6$$

Structure. We have seen above that in these hydrides metal is present as M^+ or M^{2+} and hydrogen is present as H^-, *i.e.*, ionic hydrides are composed of M^+ or M^{2+} and H^- ions. Since these hydrides are ionic in nature, their bonds are not directed in space and hence the molecules of these hydrides have no geometry. Alkali metal hydrides possess face-centred cubic lattice while alkaline earth metal hydrides possess orthorhombic lattice. The radius of H^- ion in these hydrides decreases with the decrease of the electropositive character of the metal. The metal-hydrogen distance has been found accurately by X-ray diffraction technique. X-ray diffraction has shown that in the CaH_2 lattice, Ca^{2+} ions have slightly distorted hexagonally close-packed arrangement. The unit cell of CaH_2 is orthorhombic.

II. Covalent or molecular hydrides

These hydrides are given by most of the *p*-block elements. These have molecular lattices and hence are called molecular hydrides.

Preparation. (*i*) By the direct union of the free element with H_2 at elevated temperature.

$$N_2 + 3H_2 \longrightarrow 2NH_3; \qquad 2H_2 + O_2 \longrightarrow 2H_2O$$

(*ii*) By the hydrolysis of compounds like borides, silicides, and similar compounds. The hydrolysis may be affected either by H_2O only or by dil. acid solutions.

$$Mg_3B_2 + 6HCl \longrightarrow 3MgCl_2 + B_2H_6; \qquad CaC_2 + 2H_2O \longrightarrow Ca(OH)_2 + C_2H_2$$
$$Al_4C_3 + 6H_2O \longrightarrow 2Al_2O_3 + 3CH_4; \qquad Mg_2Si + 4HCl \longrightarrow 2MgCl_2 + SiH_4$$
$$Mg_3N_2 + 6H_2O \longrightarrow 3Mg(OH)_2 + NH_3; \qquad Ca_3P_2 + 6H_2O \longrightarrow 3Ca(OH)_2 + 2PH_3$$
$$Zn_3As_2 + 6HCl \longrightarrow 3ZnCl_2 + 2AsH_3; \qquad FeS + 2HCl \longrightarrow FeCl_2 + H_2S$$

(*iii*) By the reduction of appropriate anhydrous chlorides in ether solution by $Li[AlH_4]$.

$$4BCl_3 + 3Li[AlH_4] \longrightarrow 2B_2H_6 + 3LiCl + 3AlCl_3$$
$$MCl_4 + Li[AlH_4] \longrightarrow LiCl + AlCl_3 + MH_4$$
(M = Si, Ge, Sn)

(*iv*) By the decomposition of a carbide with alkalies.

$$Be_2C + 4KOH \longrightarrow 2Be(OK)_2 + CH_4$$

(*v*) Halogen hydrides, HX, may be obtained by heating double compounds such as HF. KF

$$HF.KF \xrightarrow{heat} HF + KF$$

(*vi*) *Paneth and Rabino-Witsch's method.* These persons prepared SnH_4 by electrolysing a sulphuric acid solution of $SnSO_4$ contaning glucose. The cell used had Pb-electrodes and the temperature kept was below 50 – 60°C.

Properties of covalent hydrides. (*i*) *Physical state.* In groups VA, VIA and VIIA, the hydride of the first element (*i.e.*, NH_3, H_2O and HF) is a liquid while each of the hydrides of the remaining elements of each of the above groups is a gas. For *explanation.* See "*H-bonding*".

(*ii*) *Boiling points.* (*a*) In groups VA, VIA and VIIA, NH_3, H_2O and HF have abnormally high boiling points in their respective groups. For explanation *see H-bonding.*

(*b*) Since the electronegativity difference between C and H is very small (C = 2.5, H = 2.1), there is no H-bonding in CH_4 molecule and hence CH_4 molecules cannot get associated to give clustered molecule, $(CH_4)_n$. Due to the non-formation of clustered molecule, CH_4 among the hydrides of group IVA has the lowest boiling point. As expected the boiling points of MH_4 hydrides (M = C, Si, Ge, Sn) are in the increasing order : CH_4 (= –161.3°C) < SiH_4 (= –111.9°C) < GeH_4 (= – 88.1°C) < SnH_4 (= – 53°C).

(*iii*) *Relative order of the stability of hydrides.* As we move down a group, the size of the central atom of the hydrides increases and hence M—H bond length also increases. With the increase in M—H bond length, the strength of M—H bond decreases. Wtih decrease of M—H bond length the stability of M—H bond also decreases.

Thus the stability of the hydrides of the elements of group IVA and VA decreases as CH_4 > SiH_4 > GeH_4 > SnH_4 > PbH_4 (group IVA); NH_3 > PH_3 > AsH_3 > SbH_3 > BiH_3 (group VA). These orders show that PbH_4 and BiH_3 are extremely unstable and can be prepared only in traces.

The thermal stability of the hydrides of the elements of *2nd* period increases from left to right with the increase in the electronegativity of the central atom in the hydride, *e.g.*, $CH_4 < NH_3 < H_2O$ < HF.

(*iv*) *Monomeric and polymeric nature of the hydrides namely NH_3, H_2O and HF.* In groups VA, VIA and VIIA, the hydride of the first element (*i.e.*, NH_3, H_2, and HF) exists as a polymerised molecule *i.e.*, $(NH_3)_n$, $(H_2O)_n$ and $(HF)_n$ respectively. On the other hand the remaining hydrides of each group exist as monomeric molecules. For explanation *see H-bonding.*

(*v*) *Relative order of basic strength of the hydrides.* The trihydrides of the elements of group VA (*i.e.*, NH_3, PH_3, AsH_3, SbH_3 and BiH_3) show basic property which decreases on moving down the group, *i.e.*, the basic strength decreases as NH_3 > PH_3 > AsH_3 > SbH_3 > BiH_3. For explanation see. "*General properties of hydrides of the elements of group VA*".

The basic strength of the covalent hydrides of the elements of *2nd* period decreases on moving from left to right. Thus the basic strength of NH_3, H_2O and HF molecules varies as $NH_3 > H_2O$ > HF.

(*vi*) *Volatility.* The volatility is decreased with those hydrides whose molecules are linked together by H-bonds (*e.g.*, NH_3, H_2O HF) or which are dimeric (*e.g.*, B_2H_6, Ga_2H_6) or polymeric [*e.g.*, $(AlH_3)_n$, $(InH_3)_n$].

(*vii*) *Property of acting as Lewis acids: Formation of addition compounds.* The hydrides of the elements of group IIIA [*e.g.*, B_2H_6, $(AlH_3)_n$ etc.] are electron-deficient molecules and hence have a tendency to accept an electron pair donated by a donor molecule like NR_3, PF_3, CO etc., and thus form addition compounds. In the formation of addition compounds the hydrides act as Lewis acids and donor molecules act as Lewis bases.

Lewis acid		*Lewis base*		*Addition compound*
B_2H_6	+	$2 : NR_3$	$\longrightarrow$	$2[H_3B \leftarrow NR_3]$
B_2H_6	+	$2 : PF_3$	$\longrightarrow$	$2[H_3B \leftarrow PF_3]$
B_2H_6	+	$2 : CO$	$\longrightarrow$	$2[H_3B \leftarrow CO]$
$(AlH_3)_n$	+	$n : NR_3$	$\longrightarrow$	$n[H_3Al \leftarrow NR_3]$

(*viii*) *Hydridic and protonic properties.* (*a*) The covalent hydrides of the elements having low electronegativity (*e.g.*, B_2H_6) react with H_2O to produce H_2 (*Hydrolysis*), *i.e.*, B_2H_6 shows *hydridic properties.*

$$B_2H_6 + 6H_2O \longrightarrow 2H_3BO_3 + 6H_2$$

(*b*) The covalent hydrides of the elements having high electronegativity (*e.g.*, HCl) react with a base (*e.g.*, NH_4OH) to form a salt, *i.e.*, HCl shows *protonic properties.*

$$HCl + NH_4OH \longrightarrow NH_4Cl + H_2O$$

(*c*) The hydrides of the elements having intermediate electronegativity (*e.g.*, hydrides of the elements of group IVA) can show hydridic as well as protonic properties. For example when GeH_4 reacts with HBr, H_2 is produced and hence GeH_4 is said to show hydridic properties. On the other hand when GeH_4 reacts with $NaNH_2$ (base), sodium germyl, $NaGeH_3$ (salt) is formed, showing that GeH_4 has protonic character.

$$GeH_4 + HBr \longrightarrow GeH_3\,Br + H_2$$

$$GeH_4 + NaNH_2 \longrightarrow Na\,[GeH_3] + NH_3$$

(*ix*) *Formation of complex hydrides.* The covalent hydrides (*e.g.*, B_2H_6, AlH_3, GaH_3) react with ionic hydrides like LiH, NaH etc., in etheral solution and give rise to the formation of complex hydrides. Thus:

$$B_2H_6 + 2MH\ (M = Li, Na) \xrightarrow{\text{Ether}} 2M\,[BH_4]$$

$$MH_3\ (M = Al, Ga) + LiH \xrightarrow{\text{Ether}} Li\,[MH_4]$$

Structure of covalent hydrides. In covalent hydrides the bonds formed are mostly covalent in character. These hydrides consist of discrete molecules. These molecules are held together either by weak van der Waals forces or by H-bonds (*e.g.*, NH_3, H_2O, HF). As a result of this type of bonding, the covalent hydrides are gases, liquids or (in a few cases) solids of low melting and boiling points. In polymeric covalent hydrides like $(BeH_2)_n$, $(Mg\,H_2)_n$, B_2H_6, $(AlH_3)_n$, $(InH_3)_2$ etc., the metal nonmetal atoms are linked together by H-bridges. See the structure of $(BeH_2)_n$, B_2H_6 etc.

III. Interstitial or metallic hydrides

These hydrides are also called *alloy-type hydrides* and are mostly given by transition metals and have, therefore, been discussed in the chapter on "*General Characteristic of Transition Elements*".

IV. Polymeric or polynuclear hydrides

Examples of such hydrides are $(BeH_2)_n$, $(MgH_2)_n$ and $(AlH_3)_n$. These hydrides are solids. These are covalent molecules linked together through bridging hydrogen bonds. Thus the structures of $(BeH_2)_n$ and $(AlH_3)_n$ can be shown as:

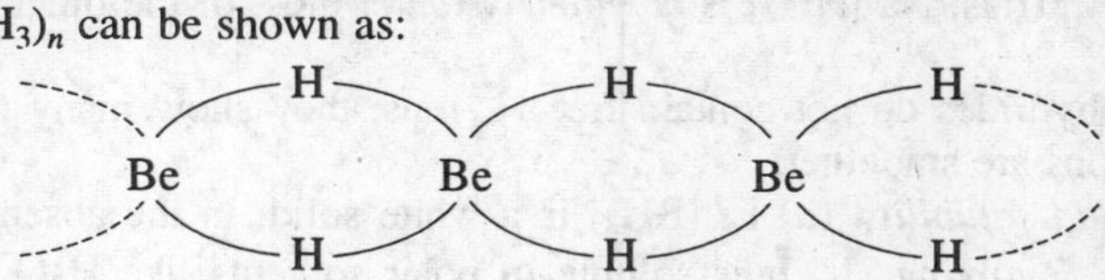

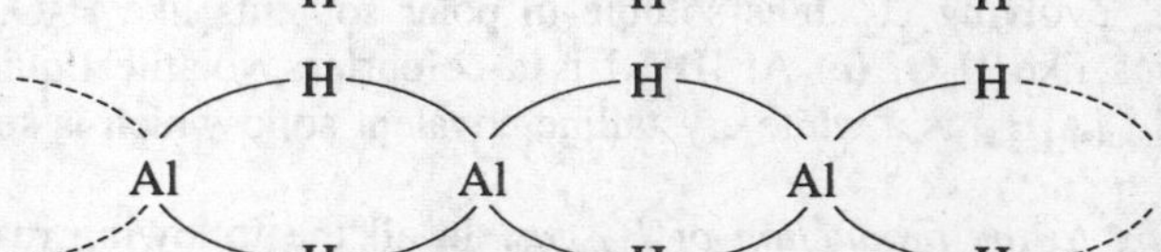

$(MgH_2)_n$ has the same structure as $(BeH_2)_n$ has.

V. Complex hydrides

Boron, aluminium and gallium form complex hydrides which contain $[XH_4]^-$ ion (X = B, Al, Ga). The complex hydrides containing BH_4^- ion are called *boro hydrides*, those having AlH_4^- ion are called *aluminium hydrides* while those having GaH_4^- ion are known as *gallium hydrides*. Examples of complex hydrides are $Li[BH_4]$, $Na[BH_4]$, $Be[BH_4]_2$, $Al[BH_4]_3$, $Li[AlH_4]$, $Na[AlH_4]$, $Li[GaH_4]$ etc.

Preparation. (*i*) These can be prepared by the action of covalent hydrides (*e.g.*, B_2H_6, AlH_3, GaH_3) or covalent halides (*e.g.*, BF_3, $AlCl_3$) on the ionic hydrides (*e.g.*, NaH, LiH). The reaction should be carried out in ether solution.

$$B_2H_6 + 2MH\ (M = Li, Na) \xrightarrow{\text{Ether}} 2M[BH_4]$$
$$MH_3\ (M = Al, Ga) + LiH \longrightarrow Li[MH_4]$$
$$BF_3 + 4NaH \longrightarrow Na[BH_4] + 3NaF$$
$$\underset{\text{(anhydrous)}}{AlCl_3} + 4LiH \longrightarrow \underset{\text{solution}}{Li[AlH_4]} + \underset{\text{ppt.}}{3LiCl\downarrow}$$
$$AlBr_3 + 4LiH \longrightarrow \underset{\text{solution}}{Li[AlH_4]} + \underset{\text{ppt.}}{3LiBr\downarrow}$$

In the preparation of $Li[AlH_4]$ by the action of LiH on $AlCl_3$ or $AlBr_3$, LiCl or LiBr is obtained as precipitate which is removed by filtration and the filtrate which contains $Li[AlH_4]$, on evaporation in the absence of air, gives $Li[AlH_4]$ in the solid state.

(*ii*) $Li[AlH_4]$ and $Na[AlH_4]$ both can be prepared by the direct combination of the constituent elements under high pressure in the presence of ether (*Industrial preparation*)

$$Li + Al + 2H_2 \xrightarrow[140°C,\ 350\ atm]{\text{Ether}} Li[AlH_4]$$

$$Na + Al + 2H_2 \xrightarrow[150°C,\ 136\ atm]{\text{Ether}} Na[AlH]_4$$

(*iii*) $Na[BH_4]$ can be obtained by treating NaH with methyl borate, $B(OCH_3)_3$ in tetrahydrofuran (THF) solvent.

$$4NaH + 4B(OCH_3)_3 \xrightarrow{\text{THF}} Na[BH_4] + 3Na(OCH_3)$$

(*iv*) $Al[BH_4]_3$ can be prepared by the following reactions :

$$Al(CH_3)_3 + 2B_2H_6 \longrightarrow B(CH_3)_3 + Al[BH_4]_3$$
$$AlCl_3 + 3Li[BH_4] \xrightarrow{\text{Warm}} 3LiCl + Al[BH_4]_3$$

(*v*) $Li[BH_4]$ can also be prepared by the reactions represented by the following equations:

(*a*) $LiCl + Na[BH_4] \longrightarrow Li[BH_4] + NaCl$

(*b*) $2B_2H_6 + 2LiC_2H_5 \xrightarrow{\text{Ether}} 2Li\ [BH_4] + B(C_2H_5)_3$

(*c*) $Al[BH_4]_3 + 3LiC_2H_5 \xrightarrow{\text{Benzene}} 3Li[BH_4] + Al(C_2H_5)_3$

Properties of $M^+[XH_4]^-$ hydrides. 1. The properties of complex hydrides depends on the nature of the cation, M^+. Thus, Na $[NH_4]$ is *completely ionic*, most of the other borohydrides are largely covalent.

2. Although these hydrides do not contain free H^- ions, they show many reactions of ionic hydrides, but the reactions are smoother.

3. *Physical state and solubility.* (*a*) Li $[BH_4]$ is a white solid. In the absence of air it melts (decomposes) at 275°C, evolving H_2. It is soluble in polar solvents like H_2O. (*b*) Na $[BH_4]$ is soluble in polar solvents like H_2O. (*c*) Al $[BH_4]$ is a colourless volatile liquid. It dissolves in nonpolar solvents. (*d*) Li $[AlH_4]$ is a white crystalline covalent solid which is soluble in nonpolar solvents like ether.

4. *Reactions leading to the production of H_2 gas.* In all the following reactions, H_2 gas is produced.

(*a*) $Li[AlH_4]$ reacts with excess of H_2O and C_2H_5OH, evolving H_2 gas.

$$Li[AlH_4] + 4H.OH \longrightarrow LiOH + Al(OH)_3 + 4H_2\uparrow$$

$$Li[AlH_4] + 4C_2H_5OH \longrightarrow Li(OC_2H_5) + Al(OC_2H_5)_3 + 4H_2\uparrow$$

(*b*) $Al[BH_4]_3$ reacts with H_2O and HCl. In both the reactions H_2 is produced.

$$Al[BH_4]_3 + 12H_2O \longrightarrow 3B(OH)_3 + Al(OH)_3 + 12H_2\uparrow$$

$$2Al[BH_4]_3 + 6HCl \xrightarrow{-60°C} 2B_2H_6 + 2AlCl_3 + 3H_2\uparrow$$

(*c*) $Li[BH_4]$ reacts with H_2O even in cold, with CH_3OH at 100°C and with HCl at – 80°C. In all these reactions H_2 is evolved. When a mixture of $Li[BH_4]$ and NH_4Cl is heated in vacuum at 230°C, H_2 is evolved.

$$Li[BH_4] + 2H_2O \xrightarrow{\text{cold}} LiBO_2 + 4H_2\uparrow$$

$$Li[BH_4] + 4CH_3OH \xrightarrow{100°C} Li(OCH_3) + B(OCH_3)_3 + 4H_2\uparrow$$

$$2Li[BH_4] + 2HCl \xrightarrow{-80°C} 2LiCl + B_2H_6 + 2H_2\uparrow$$

$$3Li[BH_4] + 3NH_4Cl \xrightarrow{230°C} B_3N_3H_6 + 3LiCl + 9H_2\uparrow$$

(*d*) $Na[BH_4]$ reacts with HCl and I_2 and produces H_2 in both the cases.

$$2Na[BH_4] + 2HCl \longrightarrow B_2H_6 + 2NaCl + 2H_2\uparrow$$

$$2Na[BH_4] + I_2 \xrightarrow{\text{Diglyme}} B_2H_6 + 2NaI + H_2\uparrow$$

In the absence of diglyme, BI_3 is produced.

$$Na[BH_4] + 4I_2 \longrightarrow BI_3 + 4HI + NaI$$

(*e*) NH_3, PH_3, primary amines ($R.NH_2$) and secondary amines (R_2NH) react with $Li[AlH_4]$ and H_2 is evolved.

$$Li[AlH_4] + 4NH_3 \longrightarrow Li[Al(NH_2)_4] + 4H_2\uparrow$$

$$Li[AlH_4] + 4PH_3 \longrightarrow Li[Al(PH_2)_4]$$

$$Li[AlH_4] + 4R.NH_2 \longrightarrow Li[Al(R.NH)_4] + 4H_2\uparrow$$

$$Li[AlH_4] + 4R_2.NH \longrightarrow Li[Al(R_2N)_4] + 4H_2\uparrow$$

However, with tertiary amines, LiH is produced.

$$Li[AlH_4] + (CH_3)_3N \longrightarrow (CH_3)_3\ [Al\ (NH_3)] + LiH$$

(*f*) When heated to 150°C, $Li[AlH_4]$ gets decompose to give LiH, Al and H_2 is evolved

$$2Li[AlH_4] \xrightarrow{150°C} 2LiH + 2Al + 3H_2$$

5. *Reducing properties.* $Li[AlH_4]$, $Li[BH_4]$ etc., are good reducing agents. Thus:

(*a*) The etheral solution of $Li[AlH_4]$ reduces aldehydes, ketones, carboxylic acids and acid chlorides to the corresponding alcohols.

$$CH_3CHO\ (\textit{Acetaldehyde}) \xrightarrow[\text{Li [AlH}_4]]{+2H} CH_3.CH_2.OH\ (\textit{Ethanol})$$

$$(CH_3)_2\ CO\ (\textit{Acetone}) \xrightarrow{+2H} (CH_3)_2CHOH\ (\textit{Isopropyl})$$

$$CH_3COOH \xrightarrow{+4H} CH_3CH_2OH + H_2O$$

$$CH_3COCl\ (\text{Acetyl chloride}) \xrightarrow{+4H} CH_3CH_2OH + HCl$$

$Li[AlH_4]$ also reduces amides, anilides, aliphatic nitrocompounds and nitriles to amines

$$CH_3CONH_2\ (\textit{Acetamide}) \xrightarrow{+4H} CH_3\ CH_2\ NH_2\ (\textit{Ethyl amine}) + H_2O$$

$$C_6H_5.NHCOCH_3\ (\textit{Acetanilide}) \xrightarrow{+4H} C_6H_5\ NH\ CH_2CH_3\ (\textit{N-Ethyl aniline}) + H_2O$$

$$\underset{\substack{|\\ NO_2}}{CH_3.CH}.CH_2.CH_3\ (\textit{2-nitro butane}) \xrightarrow{+6H} CH_3 — \underset{\substack{|\\ NH_2}}{CH}.CH_2.CH_3 + 2H_2O\ (\textit{2-amino butane})$$

$$R-C\equiv N\ (\textit{Nitrile}) \xrightarrow{+4H} R-CH_2-NH_2\ (\textit{primary amine})$$

(*iii*) Ether solution of $Li[AlH_4]$ reduces covalent chlorides [*e.g.*, BCl_3, $AlCl_3$, $SiCl_4$, $SnCl_4$, ACl_3 (A = N, P, As, Sb)] to their corresponding hydrides

$$2MCl_2\ (M = Be, Mg) + Li\ [AlH_4] \longrightarrow 2MH_2 + LiCl + AlCl_3$$
$$4BCl_3 + 3Li\ [AlH_4] \longrightarrow 2B_2H_6 + 3LiCl + 3AlCl_3$$
$$AlCl_3 + Li\ [AlH_4] \longrightarrow 4AlH_3 + 3LiCl$$
$$MCl_4\ (M = Si, Sn) + Li[AlH_4] \longrightarrow MH_4 + LiCl + AlCl_3$$
$$ACl_3\ (A = N, P, As, Sb) + 3Li\ [AlH_4] \longrightarrow AH_3 + 3LiCl + 3AlH_3$$

The above reactions have been used to prepare the hydrides of the elements of groups IIIA, IVA and VA.

(*b*) $Li[BH_4]$ reduces $FeCl_3$ to free Fe metal and nonmetallic halides (*e.g.*, $SiCl_4$) to their hydrides. B, B, B-trichloro borazine ($Cl_3B_3N_3H_3$) is reduced to borazine, $B_3N_3H_6$.

$$6Li[BH_4] + 2FeCl_3 \longrightarrow 6LiCl + 2Fe + 2B_2H_6 + 3H_2$$
$$4Li[BH_4] + SiCl_4 \longrightarrow 4LiCl + 2B_2H_6 + SiH_4$$
$$3Cl_3B_3N_3H_3 + 6Li[BH_4] \xrightarrow{\text{Ether}} 2B_3N_3H_6 + 6LiCl + 2B_2H_6$$

6. *Reactions leading to the formation of B_2H_6.* (*a*) When borohydrides like $Li[BH_4]$, $Na[BH_4]$ and $Al[BH_4]_3$ react with BF_3, HCl, trifluoro ether complex [$(C_2H_5)_2O.BF_3$] or I_2, B_2H_6 is obtained.

$$BF_3 + 3M\ [BH_4]\ (M = Li, Na) \longrightarrow 3MF + 2B_2H_6$$
$$2HCl + 2M[BH_4] \longrightarrow 2MCl + B_2H_6 + 2H_2\uparrow$$
$$(C_2H_5)_2O.BF_3 + 3M[BH_4] \longrightarrow 2MF + 2B_2H_6 + (C_2H_5)_2O$$
$$6HCl + 2Al[BH_4]_3 \xrightarrow{-60°C} 2AlCl_3 + 3B_2H_6 + 6H_2\uparrow$$
$$I_2 + 2Na[BH_4] \xrightarrow{\text{Diglyme}} B_2H_6 + 2NaI + H_2\uparrow$$

(*b*) When $Li[AlH_4]$ reacts with BCl_3, B_2H_6 is produced.

$$4BCl_3 + 3Li[AlH_4] \longrightarrow 3LiCl + 3AlCl_3 + 2B_2H_6$$

Structure of $[XH_4]^-$ ion. The structure of this ion can be written as

$$\begin{array}{c} H \\ | \\ H—X\leftarrow H^- \\ | \\ H \end{array}$$

This structure shows that $[XH_4]^-$ can be supposed to have been formed by the combination of XH_3 molecule and H^- ion

$$MH_3 + H^- \longrightarrow H_3X \leftarrow H^- \text{ or } XH_4^-$$

We know that XH_3 molecule results by sp^2 hybridisation of X-atom. In sp^2 hybridisation one *p*-orbital remains vacant and unhybridised. This vacant *p*-orbital accepts an electron pair residing in 1*s* orbital of H^- ion and $H_3X \leftarrow H^-$ coordinate bond is formed. Evidently in the formation of XH_4^- ion, XH_3 molecule acts as an acceptor (Lewis acid) and H^- ion acts as a donor (Lewis base). In XH_4^- ion X-atom is sp^3 hybridised and this ion has tetrahedral geometry. For details see the structure of $[BH_4]^-$ ion as an example discussed in chapter 14. (p. 419)

Uses. The complex hydrides particularly $Li[AlH_4]$ and $Na[AlH_4]$ are used as reducing agents. These reduce the aldehydes and ketones to alcohols and nitrates to amines.

Lithium aluminium hydride, Li $[AlH_4]$

It is a complex hydride and its IUPAC name is *lithium tetrahydro aluminate* (III).

Preparation. See "*Complex hydrides*".

Properties. (*i*) It is a white crystalline solid which is soluble in ether.

(*ii*) *Reducing properties.* See "*Complex hydrides*".

(*iii*) *Reactions leading to the production of H_2.* See "*Complex hydrides*".

(*iv*) *Action of B_2H_6.* See "*Diborane*".

(*v*) *Action of dimethyl beryllium, $(CH_3)_2Be$.* BeH_2 is formed.

$$Li[AlH_4] + (CH_3)_2Be \longrightarrow BeH_2 + Li[AlH_2(CH_3)_2]$$

(*vi*) *Action of acetylene, $HC \equiv CH$.* $Li[AlH(CH = CH_2)_3]$ is obtained

$$Li[AlH_4] + 3CH \equiv CH \longrightarrow Li[AlH(CH = CH_2)_3]$$

Uses. $Li[AlH_4]$ is used: (*i*) as a reducing agent in organic chemistry (see reducing properties discussed above), (*ii*) in the preparation of covalent hydrides like MH_2 (M = Be, Mg), B_2H_6, AlH_3, AH_3 (A = N, P, As, Sb), MH_4 (M = Si, Sn) etc., which are otherwise difficult to prepare. See "*Reducing Properties of Complex Hydrides*".

Hydrides of the elements of group IA (alkali metals)

All the alkali metals react with H_2 and form the hydrides of MH type. Li combines with H_2 at 700 – 800°C while other alkali metals do so at 350 – 430°C. Thus Li is not so reactive towards H_2 as other alkali metals are.

Since the electropositive character of alkali metals increases from Li to Cs, the ease with which these metals react with H_2 to form the hydrides decreases from Li to Cs, *i.e.*, Li has the maximum tendency to form hydride while Cs has the minimum tendency to do so.

Properties and structure. See "*Ionic Hydrides*".

Among the hydrides of alkali metals LiH is very important and is discussed below:

Lithium hydride, LiH

It is an ionic hydride.

Preparation. See "*General Preparation of Ionic Hydrides*".

Properties. (*i*) It is a hard white solid with melting point = 680°C.

(*ii*) *Thermal stability.* Among alkali metal hydrides, LiH has the greatest thermal stability which is due its highest heat of formation, ($-\Delta H_f = -21.6$ Kcal mol^{-1}). The greatest stability of LiH is evident from the facts that it remains stable in dry air upto red heat and also does not react with Cl_2.

(*iii*) *Electrolysis.* See "*Ionic Hydrides*".

(*iv*) *Decomposition.* LiH is not decomposed on fusion.

(*v*) *Reaction with covalent hydrides and covalent halides : Formation of complex hydrides.* See "*Ionic hydrides*".

(*vi*) It turns blue on exposure to light without any change in composition.

(*vii*) *Action of protonic solvents (e.g., H_2O, NH_3, C_2H_5OH) and reducing properties.* See "*Ionic hydrides*".

(*viii*) *Action of N_2.* See "*Ionic hydrides*".

Uses. It is used (*i*) as a strong reducing agent in synthetic organic chemistry (*ii*) for the preparation of complex hydrides like $Li[AlH_4]$, $Li[BH_4]$ etc. $Li[AlH_4]$ is a valuable reducing agent and $Li[BH_4]$ is used in jet propulsion (*iii*) as a source of H_2 which is used in air-sea rescue balloons

Structure. See "*Ionic hydrides*".

Hydrides of the elements of group IIA

Be and Mg give polymeric hydrides namely $(BeH_2)_n$ and $(MgH_2)_n$ while the remaining elements give monomeric hydrides of MH_2 type. Thus Ca, Sr and Ba give the hydrides having the formula CaH_2, SrH_2 and BaH_2 respectively.

Preparation. Excepting Be, all the remaining elements combine directly with H_2, on heating and give their hydrides. BeH_2 is produced by reducing $BeCl_2$ with $LiAlH_4$ or LiH.

$$2BeCl_2 + LiAlH_4 \longrightarrow 2BeH_2 + LiCl + AlCl_3$$

$$BeCl_2 + 2LiH \xrightarrow{\text{Ether}} BeH_2 + 2LiCl$$

MgH_2 can be obtained by the direct combination of Mg and H_2, if the combination is carried out at elevated temperature and pressure. It is more conveniently prepared by the action of $LiAlH_4$ on $MgCl_2$ in etheral solution.

$$2MgCl_2 + LiAlH_4 \longrightarrow 2MgH_2 + LiCl + AlCl_3$$

Properties. (*i*) BeH_2 is covalent, MgH_2 is partly covalent and partly ionic while the remaining hydrides (*i.e.*, CaH_2, SrH_2 and BaH_2) are largely ionic.

(*ii*) BeH_2 and MgH_2 both react with etheral solution of B_2H_6 and form borohydride, $[BH_4]^-$.

$$MH_2\ (M = Be, Mg) + B_2H_6 \longrightarrow M\,[BH_4]_2$$

(*iii*) BeH_2 is a nonvolatile white solid which is stable at room temperature but dissociates into its elements at 125°C.

(*iv*) MgH_2 is light grey powder and is thermally stable. Its decomposition temperature is equal to 280°C.

(*v*) All the hydrides react with H_2O and H_2 is liberated. Thus these hydrides are used as reducing agents.

$$MH_2 + 2H_2O \longrightarrow M\,(OH)_2 + 2H_2$$

Structure. The structures of polymeric hydrides namely $(BeH_2)_n$ and $(MgH_2)_n$ have already been discussed in this chapter. Since CaH_2, SrH_2 and BaH_2 are ionic solids, they contain M^{2+} and H^- ions.

Calcium hydride, CaH_2 *(Also called hydrolith).*

It is also an ionic hydride.

Preparation. CaH_2 can be prepared by heating Ca-metal in a current of H_2 or by heating CaO mixed with Mg in a current of H_2 at 250°C and 50 cm pressure of hydrogen.

$$Ca + H_2 \xrightarrow{\Delta} CaH_2; \qquad CaO + Mg + H_2 \xrightarrow{\Delta} CaH_2 + MgO$$

Properties. (*i*) It is a colourless crystalline compound of density 1.90 which is higher than that of Ca-metal.

(*ii*) *Thermal stability and reactivity.* See "*Ionic hydrides*".

(*iii*) *Electrolysis.* See "*Ionic hydrides*".

(*iv*) *Action of air on heating.* See "*Ionic hydrides*".

(*v*) *Action of N_2.* See "*Ionic hydrides*".

(*vi*) *Decomposition by SO_2.* See "*Ionic hydrides*".

(*vii*) *Action of H_2O and reducing properties.* See "*Ionic hydrides*".

Structure. See "*Ionic hydrides*".

Uses. It is used (*a*) as a source of H_2 for balloons and hence the name *hydrolith* for CaH_2 (*b*) as a reducing agent in metallurgical processes in organic chemistry.

Hydrides of the elements of group IIIA

Hydrides formed by the elements of group IIIA can be classified as:

(*i*) *Boranes.* Boron does not combine directly with hydrogen to form the hydrides. Nevertheless, this element gives a large number of stable covalent hydrides which are called *boranes*. These are of two types : (*a*) *Boranes of B_mH_{m+4} type.* Examples are B_2H_6, B_5H_9, B_6H_{10}, B_8H_{12}, $B_{10}H_{14}$ etc. (*b*) *Boranes of B_mH_{m+6} type.* Examples are B_4H_{10}, B_5H_{11}, B_6H_{12}, B_8H_{14}, B_9H_{15}, $B_{10}H_{16}$ etc. Boranes are electron-deficient compounds. B_2H_6 is the most important borane.

Nomenclature of boranes. Boranes are named by indicating the number of boron atoms by *Greek numerical prefixes* (*e.g.*, di, tetra, penta etc.) and the number of H-atoms are indicated by *Arabic numerals* (*e.g.*, 6, 9, 10 etc.) written at the end of the name of the borane. For example B_2H_6, diborane-6 or simply diborane ; B_4H_{10}, tetraborane-10 ; B_5H_9, pentaborane-9 ; B_5H_{11}, pentaborane-11 ; B_6H_{10}, hexaborane-10 ; B_6H_{12}, hexaborane-12 ; B_8H_{12}, octaborane-12 ; B_8H_{14}, octaborane-14 ; B_9H_{15}, nonaborane-15 ; $B_{10}H_{14}$, decaborane-14 ; $B_{10}H_{16}$, decaborane-16.

(*ii*) *Polymeric hydrides.* Al, Ga and In give high molecular weight polymeric hydrides *viz.*, $(AlH_3)_n$, $(GaH_3)_2$ or Ga_2H_6 (*digallane*) and $(InH_3)_n$ respectively. Tl does not form a hydride. $(AlH_3)_n$ is insoluble and decomposes on heating above 200°C into its elements. $(AlH_3)_n$ forms an addition compound with a variety of donor molecules like NR_3.

$$(AlH_3)_n + nNR_3 \longrightarrow n(AlH_3 . NR_3)$$

The structure of $(AlH_3)_n$ has already been explained in this chapter.

(*iii*) *Complex hydrides*. B, Al and Ga also form complex hydrides which contain $[AH_4]^-$ ion (A = B, Al, Ga). Examples are $Na[BH_4]$ (sodium borohydride), $Li[AlH_4]$ (lithium aluminium hydride). $Li[GaH_4]$ (lithium gallium hydride).

The hydrides of the elements of this group are electron-deficient and hence act as Lewis acids See "*General properties of covalent hydrides*". Also see "*Complex hydrides*".

Diborane, B_2H_6

Preparation. B_2H_6 can be prepared:

(*i*) by the action of ionic hydrides like NaH and CaH_2 on BCl_3 (See *ionic hydrides*).

(*ii*) by the action of borohydrides like $Li[BH_4]$, $Na[BH_4]$ and $Al[BH_4]$ on BF_3, HCl, I_2 or $(C_2H_5)_2O.BF_3$) (See *complex hydrides*).

(*iii*) by the action of $Li[AlH_4]$ on BCl_3. (See *complex hydrides*).

(*iv*) by the action of dil. HCl on magnesium boride, Mg_3B_2 (*Stock's method*). Stock treated Mg_3B_2 with a dilute aqueous solution of HCl and obtained a volatile mixture of various boranes. This mixture contains mainly B_4H_{10} which, on being heated, gets decomposed to give B_2H_6.

$$2Mg_3B_2 + 12HCl \longrightarrow 6MgCl_2 + B_4H_{10} + H_2$$

$$B_4H_{10} + H_2 \xrightarrow{\Delta} 2B_2H_6$$

(*v*) by passing electric discharge at 10 mm pressure through a mixture of BCl_3 or BBr_3 and excess of H_2.

$$2BX_3\ (X = Cl, Br) + 6H_2\ (excess) \xrightarrow[\text{discharge}]{\text{Electric}} B_2H_6 + 6HX$$

Properties. (*i*) it is a colourless gas, has sweet odour and is extremely toxic. It is extremely reactive gas. The melting and boiling points of B_2H_6 are – 165°C and – 92.5°C respectively.

(*ii*) *Action of heat.* (*a*) B_2H_6 is stable only at low temperatures in the absence of moisture.

(*b*) At higher temperatures, B_2H_6 gives higher boranes like B_5H_{11}, B_4H_{10}, $B_{10}H_{14}$ etc. In all the reactions H_2 is evolved.

$$5B_2H_6 \xrightarrow[\text{(U-tube)}]{115°C} B_5H_{11} + 4H_2$$

$$2B_2H_6 \xrightarrow{-78 - 100°C} B_4H_{10} + H_2$$

$$5B_2H_6 \xrightarrow[\text{(Seated tube)}]{100°C} B_{10}H_{14} + 8H_2$$

$$5B_2H_6 \xrightarrow{200 - 250°C} 2B_5H_9 + 6H_2$$

$$5B_2H_6 \xrightarrow[\text{pressure (tube)}]{250°C,\ 120\ mm} B_5H_9 + B_5H_{11} + 5H_2$$

(*c*) At red heat B_2H_6 gets decomposed into its constituent elements.

$$B_2H_6 \xrightarrow{\text{Red heat}} 2B + 3H_2$$

(*d*) When B_2H_6 mixed with H_2 is sent through a glass tube at 200–250°C, B_5H_{11} is obtained in large quantity.

(*iii*) *Action of H_2O and aqueous solution of an alkali (Hydrolysis).* B_2H_6 is hydrolysed by H_2O (H.OH) and aqueous solution of an alkali. H_2 is evolved in both the reactions.

$$B_2H_6 + 6H.OH \longrightarrow 2H_3BO_3 + 6H_2\uparrow$$

$$B_2H_6 + \underset{(M = Na,\ K)}{2MOH} + 2H_2O \longrightarrow \underset{\text{Metaborate}}{2MBO_2} + 6H_2\uparrow$$

(*iv*) *Action of O_2 or air.* Pure B_2H_6 undergoes no reaction when mixed with dry air or O_2 at room temperature, but it ignites (*i.e.*, burns) when it is impure. In this reaction B_2O_3 is obtained and energy equal to 2165 kJ is produced (*exothermic reaction*).

$$B_2H_6\ (impure) + 3O_2 \longrightarrow B_2O_3 + 3H_2O\ ; \quad \Delta H = -2165\ kJ\ mol^{-1}$$

(*v*) *Action of silent electric discharge.* Silent electric discharge in presence of an inert gas decomposes B_2H_6 into a mixture which contains B_4H_{10} = 40%, B_5H_9 = 20%, B_5H_{11} = 30% and a small quantity of B_9H_{15}. Different components can be separated by the fractional distillation of the mixture under reduced pressure.

(*vi*) *Formation of metallic borohydrides.* (*a*) When B_2H_6 reacts with ionic hydrides (*e.g.*, LiH, NaH) or LiC_2H_5 in ether solution, $Al(CH_3)_3$ etc., borohydride of the metal is obtained.

$$B_2H_6 + 2MH\ (M = Li, Na) \xrightarrow{\text{Ether}} 2M[BH_4]$$

$$2B_2H_6 + 3LiC_2H_5 \xrightarrow{\text{Ether}} 3Li[BH_4] + B(C_2H_5)_3$$

$$2B_2H_6 + Al(CH_3)_3 \longrightarrow Al[BH_4]_3 + B(CH_3)_3$$

(*b*) When B_2H_6 reacts with $Li[AlH_4]$, borohydrides of Li and Al are obtained.

$$Li[AlH_4] + 2B_2H_6 \longrightarrow Li[BH_4] + Al[BH_4]_3$$

(*c*) BeH_2 and MgH_2 both react with ether solution of B_2H_6 and form borohydride of Be and Mg respectively.

$$MH_2\ (M = Be, Mg) + B_2H_6 \xrightarrow{\text{Ether}} M[BH_4]_2$$

(*vii*) *Action of halogens.* B_2H_6 reacts with Cl_2 (at 25°C) and Br_2 (at 100°C) to form HCl and HBr respectively.

$$B_2H_6 + 6Cl_2 \xrightarrow{25^\circ C} 2BCl_3 + 6HCl\ (\textit{vigorous reaction})$$

$$B_2H_6 + Br_2 \xrightarrow{25^\circ C} B_2H_5Br\ (\textit{Bromo diborane}) + HBr\ (\textit{slow reaction})$$

(*viii*) *Action of hydrogen halides, HX (X = Cl, Br, I).* B_2H_6 reacts with HCl readily in the presence of a catalyst (*e.g.*, $AlCl_3$) to form chlorodiborane, B_2H_5 Cl. With HBr it reacts at 90°C in presence of $AlBr_3$ to give C_2H_5Br while with HI it reacts at 50°C in the absence of catalyst to form $B_2H_5I.H_2$ gas is evolved in all the reactions.

$$B_2H_6 + HX(X{=}Cl, Br, I) \xrightarrow{AlCl_3} B_2H_5X + H_2\uparrow$$

(*ix*) *Formation of addition compounds (adducts).* (*a*) When B_2H_6 is treated with Na or K amalgam (reducing agents) in cold, crystalline adduct, $B_2H_6.2M$ (M = Na or K) is formed.

$$\underset{(M = Na\ or\ K)}{2M\ (\textit{in the form of amalgam})} + B_2H_6 \longrightarrow B_2H_6.2M$$

(*b*) When B_2H_6 reacts with *excess* of NH_3 at *low temperature* (–120°C), addition compound, $B_2H_6.2NH_3$ (called *diammoniate of diborane) is obtained.*

$$B_2H_6 + 2NH_3\ (\textit{excess}) \xrightarrow{-120^\circ C} B_2H_6.2NH_3 \text{ or } [BH_2(NH_3)_2]^+ [BH_4]^-$$

$B_2H_6.2NH_3$ is a white nonvolatile solid and is soluble in water. Due to its solubility in water, it is supposed to be an ionic compound which can be represented as $[BH_2\ (NH_3)_2]^+$ $[BH_4]^-$.

(*c*) B_2H_6 reacts with a number of donor molecules like $(CH_3)_2O$ (*ether*), $(CH_3)_2$ S (*thioether*), CO etc., to give addition compounds. In these reactions. B_2H_6 acts as an acceptor (Lewis acid) and the other molecules behave as donor (Lewis base).

Lewis acid (Acceptor)		*Lewis base (Donor)*		*Adduct*
$\frac{1}{2}B_2H_6$	+	$(CH_3)_2O$	$\longrightarrow$	$(CH_3)_2O \rightarrow BH_3$ (*Ether borane*)
$\frac{1}{2}B_2H_6$	+	$(CH_3)_2S$	$\longrightarrow$	$(CH_3)_2S \rightarrow BH_3$ (*Thioether borane*)
$\frac{1}{2}B_2H_6$	+	CO	$\xrightarrow{200^\circ C,\ 120\ mm}$	$CO \rightarrow BH_3$ (*Carbonyl borane*)

(*d*) B_2H_6 reacts with PH_3 at –110°C and forms an unstable addition compound, $PH_3.BH_3$.

$$\frac{1}{2}B_2H_6 + PH_3 \xrightarrow{-110^\circ C} PH_3.BH_3\ (\textit{Unstable addition compound})$$

(*e*) When B_2H_6 reacts with $PH(CH_3)_2$, an addition compound having the composition $BH_3.PH(CH_3)_2$ is obtained.

$$\frac{1}{2}B_2H_6 + PH(CH_3)_2 \longrightarrow PH(CH_3)_2, BH_3$$

When this addition compound is heated to 150°C, its three molecules polymerise and form a cyclic compound, $(BH_3)_3P_3(CH_3)_6$ with the evolution of H_2 gas. In this compound P-atom is tricovalent, since it has one lone pair of electrons on it.

$$3[PH(CH_3)_2.BH_3] \xrightarrow[\text{Polymerisation}]{150°C} (CH_3)_6P_3(BH_3)_3 + \frac{3}{2}H_2$$

(*x*) *Methylation of* B_2H_6 *by trimethyl borane, $B(CH_3)_3$.* B_2H_6 reacts with $B(CH_3)_3$ at ordinary temperature and produces *four* methyl derivatives which are called *methyl diboranes.* In the formation of these derivatives only four terminal H-atoms of B_2H_6 are replaced by CH_3 groups while the two bridging H-atoms remain as such. Thus in this reaction both three-centre-two-electron bonds (B—H—B) bonds remain undisturbed.

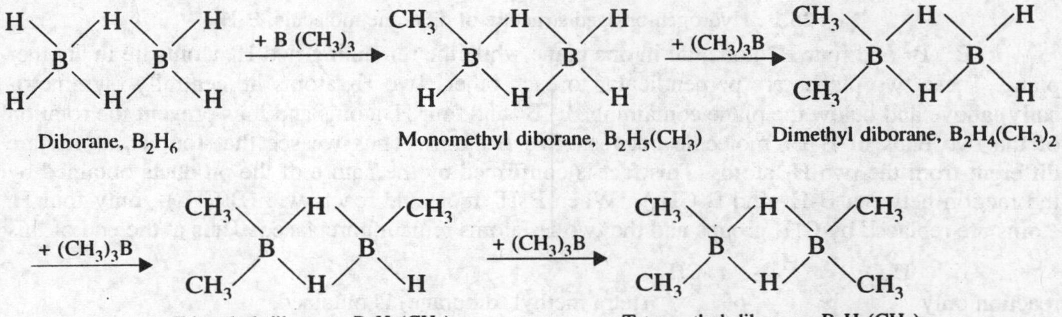

Diborane, B_2H_6 · Monomethyl diborane, $B_2H_5(CH_3)$ · Dimethyl diborane, $B_2H_4(CH_3)_2$ · Trimethyl diborane, $B_2H_3(CH_3)_3$ · Tetramethyl diborane, $B_2H_2(CH_3)_4$

The replacement of only four H-atoms by CH_3 groups shows that four H-atoms in B_2H_6 molecule are different from the remaining two H-atoms.

(*xi*) *Action of alkenes and alkynes (Hydroboration reaction).* B_2H_6 adds to alkenes and alkynes in ether at room temperature and gives organoboranes.

$$6RCH = CH_2 + B_2H_6 \longrightarrow 2B(CH_2CH_2R)_3$$

The above reaction is of great synthetic utility in organic chemistry.

(*xii*) *Action of NH_3.* (*a*) When B_2H_6 reacts with *excess* of NH_3 at –120°C, addition compound, $B_2H_6.2NH_3$ is formed. See "*Formation of addition compounds*".

(*b*) When B_2H_6 reacts with excess of NH_3 at high temperature, boron nitride (BN) is obtained and H_2 is evolved.

$$B_2H_6 + 2NH_3\ (excess) \xrightarrow{\text{High temperature}} 2BN + 6H_2$$

(*c*) When one mole of B_2H_6 reacts with two moles of NH_3 at high temperature, an addition compound, $B_2H_6.2NH_3$ is formed. This addition compound gets decomposed when heated in a closed tube at 200°C and forms *borazine* (also called *borazole*), $(BH)_3\ (NH)_3$ or $B_3N_3H_6$

$$B_2H_6 + 2NH_3 \xrightarrow{\text{High temperature}} B_2H_6.2NH_3 \xrightarrow[\text{vessel at 200°C}]{\text{Heat in a closed}} \frac{2}{3}B_3N_3H_6 + 4H_2$$

Uses. B_2H_6 is used: (*i*) as a rocket fuel; (*ii*) for the preparation of high energy fuels and propellents; (*iii*) for the preparation of higher boranes, organic boron compounds and metallic borohydrides like $Li[BH_4]$, $Na[BH_4]$, $Al[BH_4]_3$ etc.; (*iv*) as a reducing agent in organic chemistry; (*v*) as a catalyst in polymerisation reactions; (*vi*) for welding torches.

Structure and bonding in B_2H_6 molecule. Electron diffraction study of B_2H_6 molecule has shown that this molecule has hydrogen-bridged structure shown in Fig. 25.1.

From the structure shown in Fig. 25.1 the following points may be noted: (*i*) Two boron atoms have been represented as B^1 and B^2. Four H-atoms represented as H_t are called *terminal hydrogen atoms* while the remaining two H-atoms represented as H_b, are called *bridging hydrogen atoms.*

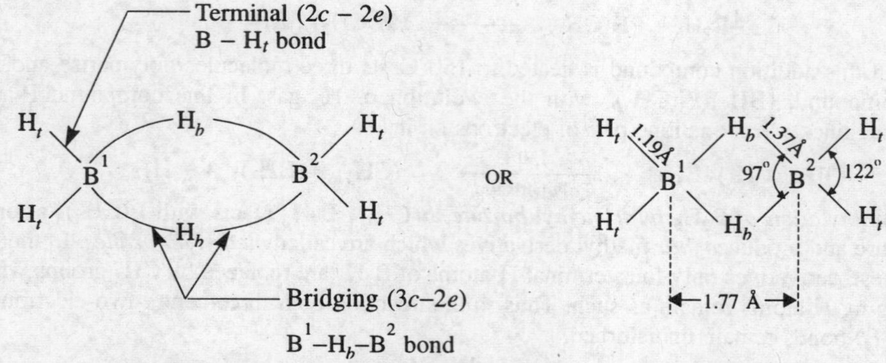

Fig. 25.1 : Hydrogen-bridged structure of diborane molecule, B_2H_6.

(*ii*) B^1, B^2 and four H_t atoms lie in one plane while the remaining two H_b atoms lie in another plane. These two planes are perpendicular to each other. Two H_b atoms lie centrally (symmetrically) above and below the plane containing B^1, B^2 and four H_t atoms and thus prevent the rotation of the two ends of B_2H_6 molecule between two B-atoms. Thus we see that four H_t atoms are different from the two H_b atoms. This fact is confirmed by the nature of the products obtained by the raction between B_2H_6 and $B(CH_3)_3$. When B_2H_6 molecule reacts with $B(CH_3)_3$, only four H_t atoms are replaced by CH_3 groups and the two H_b atoms remain unreplaced. Thus at the end of this reaction only

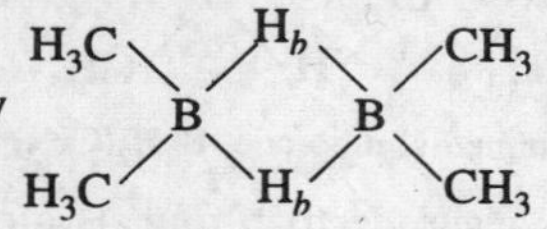

(tetra methyl diborane) is obtained.

(*iii*) $B^1 - H_t$ bond length = $B^2 - H_t$ bond length = 1.19 Å, $B^1 - H_b$ bond length = $B^2 - H_b$ bond length = 1.37 Å and $B^1 - B^2$ bond length = 1.77 Å.

(*iv*) $H_t B^1 H_t$ bond angle = $H_t B^2 H_t$ bond angle = 122° and $H_b B^1 H_b$ bond angle = $H_b B^2 H_b$ bond angle = 97°.

(*v*) B_2H_6 molecule has two types of bonds discussed below:

(*a*) *Four terminal two-centre-two-electron B – H_t bonds.* Each of these bonds is represented as (2*c* – 2*e*) B – H_t bond and is formed by the sharing of two electrons (one electron pair) between two atoms (*i.e.*, nuclei) *viz.*, B^1 or B^2 atom and H_t atom. It is a normal covalent bond. As is evident, this bond holds two atoms (or nuclei) namely B^1 or B^2 and H_t together.

(*b*) *Two bridging or open three-centre-two-electron $B^1 - H_b - B^2$ bonds.* Each of these bonds is repesented as (3*c* – 2*e*) $B^1 - H_b - B^2$ or (3*c* – 2*e*) $B^1...H_b...B^2$ and is formed by the sharing of two electrons (one electron pair) between three atoms (*i.e.*, nuclei) *viz.*, B^1, H_b and B^2 atoms. It is not a normal covalent bond. As is evident, this bond holds three atoms (nuclei) namely B^1, B_b and B^2 together. H_b atom acts as a bridge between B^1 and B^2 atoms. Since $B^1 - H_b - B^2$ bond or molecular orbital has banana shape, this bond is also called *banana bond.*

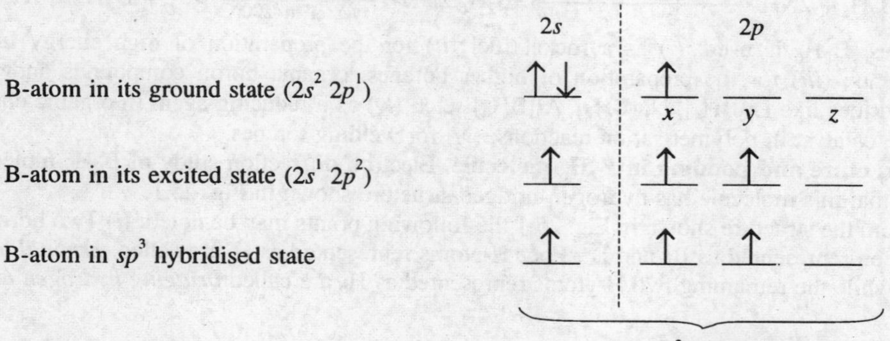

How are two bridging (3*c* – 2*e*) $B^1 – H_b – B^2$ bonds and four terminal (2*c* – 2*e*) B – H_t bonds formed. In B_2H_6 molecule each B-atom in its excited state is sp^3 hybridised as shown on page 832.

As is evident from the above scheme, three sp^3 hybrid orbitals contain one electron each while the fourth sp^3 hybrid orbital is empty. Now one singly-filled sp^3 hybrid orbital on B^1-atom, singly-filled 1*s* orbital on one H_b atom and empty sp^3 hybrid orbital on B^2-atom overlap together and give rise to the formation of one bridging (3*c*—2*e*) B^1 — H_b — B^2 bond or molecular orbital. Similarly the formation of the other (3*c*—2*e*) B^1—H_b—B^2 bond takes place. Thus in all two bridging (3*c* — 2*e*) B^1—H_b—B^2 bonds are obtained in B_2H_6 molecule.

It may be seen that after the formation of two (3*c*—2*e*) B^1—H_b—B^2 bonds has taken place, each B-atom has two singly-filled sp^3 hybrid orbitals which have not participated in the formation of the bonds. Each of these two singly-filled sp^3 hybrid orbitals on each B-atom overlaps with the singly-filled 1*s* orbital on H_t atom and gives rise to the formation of terminal (2*c*—2*e*) B—H_t σ-bond. Thus in all four terminal (2*c*—2*e*) B—H_t σ-bonds are obtained in B_2H_6 molecule. The formation of both the types of bonds has been shown in Fig. 25.2.

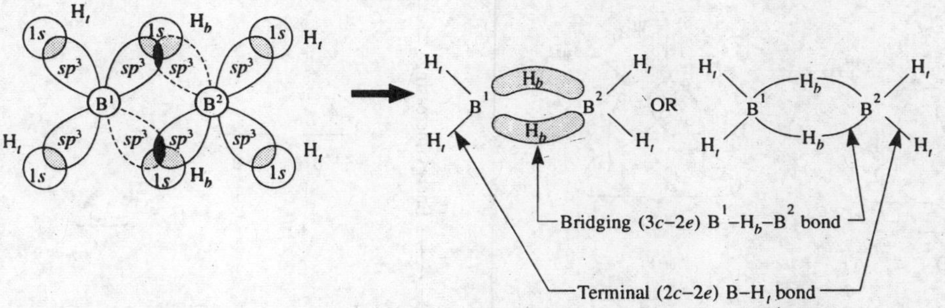

Fig. 25.2 : Formation of two bridging (3*c* - 2*e*) B^1 - H_b - B^2 and four terminal (2*c*—2*e*) B—H_t bonds lin B_2H_6 nolecule. Empty sp^3 hybrid orbitals have been shown by dotted lines.

Molecular orbital treatment of the formation of bridging (3*c*—2*e*) B^1—H_b—B^2 bonds in B_2H_6 molecule. In terms of molecular orbital theory (3*c*—2*e*) bond is called 3-centre molecular orbital. We have shown above that each B-atom in B_2H_6 molecule has four sp^3 hybrid orbitals. One of the four hybrid orbitals is empty while each of the remaining three is singly-filled. Let the wave function of the *singly-filled* sp^3 hybrid orbital on B^1 atom and that of the *empty* sp^3 hybrid orbital on B^2 atom be represented as ψ_{B_1} and ψ_{B_2} respectively. Let the wave function of 1*s* orbital (singly-filled) of bridging H-atom be denoted as ψ_H. When ψ_{B_1} (singly-filled), ψ_{B_2} (empty) and ψ_H (singly-filled) combine together, three molecular orbitals are obtained. These molecular orbitals are *bonding molecular orbital* (ψ^b), *antibonding molecular orbital* (ψ^*) and *nonbonding molecular orbital* (ψ_n). The combination of ψ_{B_1}, ψ_{B_2} and ψ_H leading to the formation of ψ^b, ψ^* and ψ_n takes place through the following steps:

(*i*) First of all ψ_{B_1} (singly-filled) and ψ_{B_2} (empty) make additive (+ +) and subtractive (+ –) combinations and form two 2-centre molecular orbitals whose wave functions are represented as ψ_{++} and ψ_{+-} respectively. Thus:

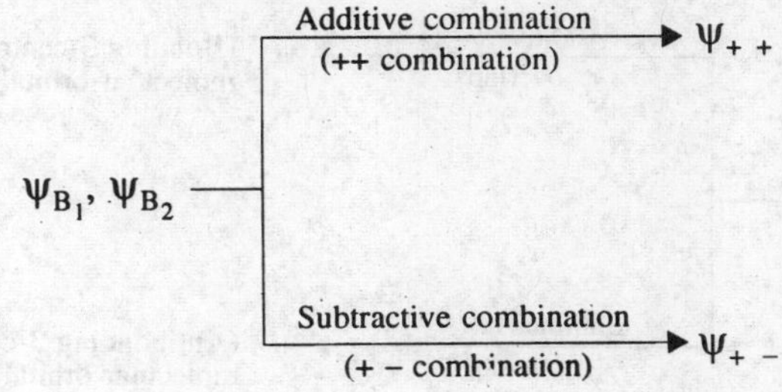

The formation of ψ_{++} and ψ_{+-} molecular orbitals has been shown in Fig. 25.3 at (*a*) and (*b*) respectively.

Regarding ψ_{++} and ψ_{+-} molecular orbitals the following points may be noted:

(*a*) Both the molecular orbitals are 2-centre molecular orbitals, since each of these are formed by the combination of two orbitals *viz*., ψ_{B_1} and ψ_{B_2}

(*b*) Since the additive combination of ψ_{B_1} and ψ_{B_2} results in a decrease of energy, ψ_{++} molecular orbital has lower energy than both the wave functions (*i.e.*, ψ_{B_1} and ψ_{B_2}) from which this molecular orbital has resulted. On the other hand, the subtractive combination of ψ_{B_1} and ψ_{B_2} results in an increase of energy and hence ψ_{+-} molecular has higher energy than ψ_{B_1} and ψ_{B_2} both. Thus the relative order of the energies of ψ_{B_1}, ψ_{B_2}, ψ_{++} and ψ_{+-} can be shown as depicted in Fig. 25.4.

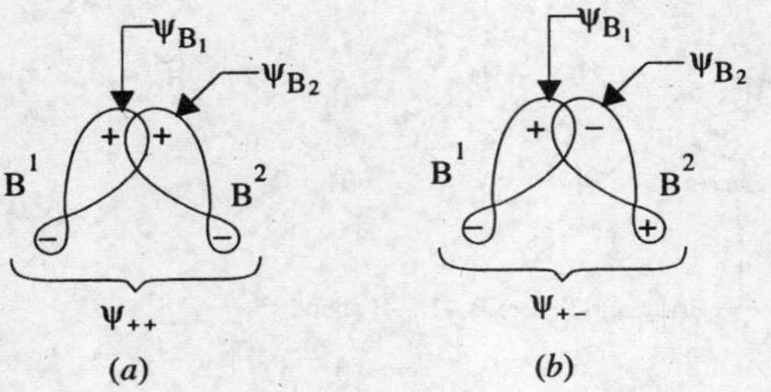

Fig. 25.3 : Formation of (*a*) ψ_{++} by the additive (+ +) combination of ψ_{B_1} and ψ_{B_2} (*b*) ψ_{+-} by the subtractive (+ –) combination of ψ_{B_1} and ψ_{B_2}.

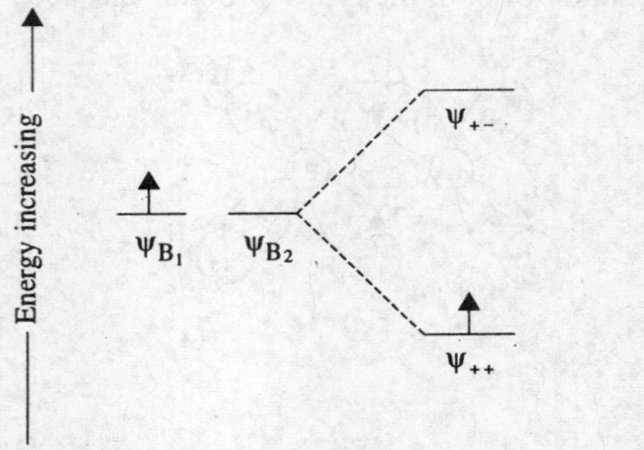

Fig. 25.4 : Relative order of the energies of ψ_{B_1}, ψ_{B_2}, ψ_{++} and ψ_{+-}.

The single electron residing initially in ψ_{B_1} atomic orbital now occupies ψ_{++} molecular orbital, since this molecular orbital has lower energy than ψ_{+-} molecular orbital.

(*c*) ψ_{++} and ψ_{+-} molecular orbitals in their normalised form can be represented by the equations:

$$\psi_{++} = \frac{1}{\sqrt{2}}(\psi_{B_1} + \psi_{B_2})\ ;\ \psi_{+-} = \frac{1}{\sqrt{2}}(\psi_{B_1} - \psi_{B_2}).$$

Obviously the sum of the contributions made by ψ_{B_1} and ψ_{B_2} in the formation of ψ_{++} is equal to $\left(\frac{1}{\sqrt{2}}\right)^2 + \left(\frac{1}{\sqrt{2}}\right)^2 = 1$ (unity). Similarly the sum of the contributions made by ψ_{B_1} and ψ_{B_2} in the formation of ψ_{+-} is also equal to unity $\left[\left(\frac{1}{\sqrt{2}}\right)^2 + \left(\frac{1}{\sqrt{2}}\right)^2 = 1\right]$.

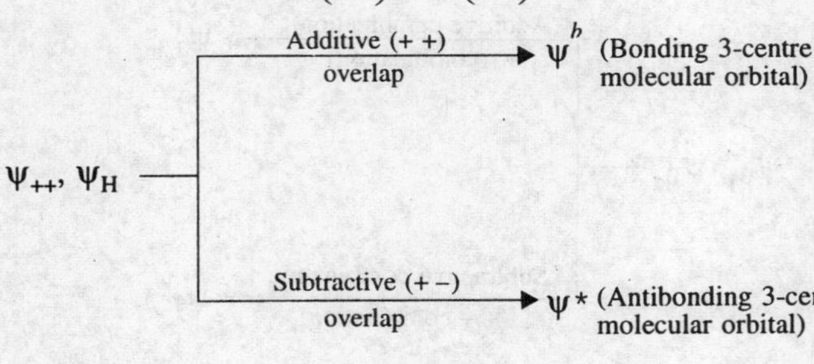

(*ii*) In the second step ψ_{++} molecular orbital makes an additive (+ +) overlap with ψ_H and gives *3-centre bonding molecular orbital* (ψ^b). Similarly when ψ_{++} molecular orbital makes a subtractive (+ –) overlap with ψ_H, *3-centre anti-bonding molecular orbital* (ψ^*) is obtained.

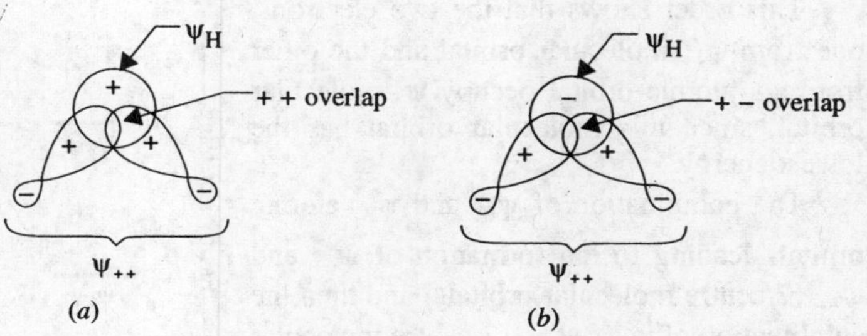

Fig. 25.5 : Formation of (*a*) ψ^b molecular orbital by the additive (+ +) overlap of ψ_{++} and ψ_H (*b*) ψ molecular orbital by the subtractive (+ –) overlap of ψ_{++} and ψ_H.

The formation of ψ^b and ψ^* molecular orbitals by the combination of ψ_{++} and ψ_H has been shown in Fig. 25.5.

The following points may be noted:

(*a*) ψ^b and ψ^* both are 3-centre molecular orbitals, since each of these is formed by the combination of one 2-centre molecular orbital *viz*., ψ_{++} and one atomic orbital namely ψ_H. Note that ψ_{++} is obtained by the combination of two atomic orbitals namely ψ_{B_1} and ψ_{B_2}.

(*b*) ψ^b and ψ^* molecular orbitals in their normalised form can be represented by the equations:

$$\psi^b = \frac{1}{2}(\psi_{B_1} + \psi_{B_2}) + \frac{1}{\sqrt{2}}\psi_H$$

$$\psi^* = \frac{1}{2}(\psi_{B_1} + \psi_{B_2}) - \frac{1}{\sqrt{2}}\psi_H$$

Obviously the sum of the contributions made by ψ_{B_1}, ψ_{B_2} and ψ_H in the formation of ψ^b and ψ^* molecular orbitals is equal to $\left(\frac{1}{2}\right)^2 + \left(\frac{1}{2}\right)^2 + \left(\frac{1}{\sqrt{2}}\right)^2 = 1$ (unity).

(*iii*) In the third step, ψ_{+-} molecular orbital makes an additive and subtractive overlaps with ψ_H atomic orbital as shown at (*a*) and (*b*) respectively of Fig. 25.6. It may be seen from Fig. 25.6 (*a*) that in the additive overlap, + + overlap is cancelled by + – overlap. Similarly it can be seen from Fig. 25.6 (*b*) that in subtractive combination + – overlap is cancelled by – – overlap. Thus ψ_{+-} molecular orbital cannot combine with ψ_H atomic orbital either by addition or by subtraction. In other wards we say that the combination of ψ_{+-} melecular orbital with ψ_H atomic orbital gives non - bonding molecular orbital (ψ^n).

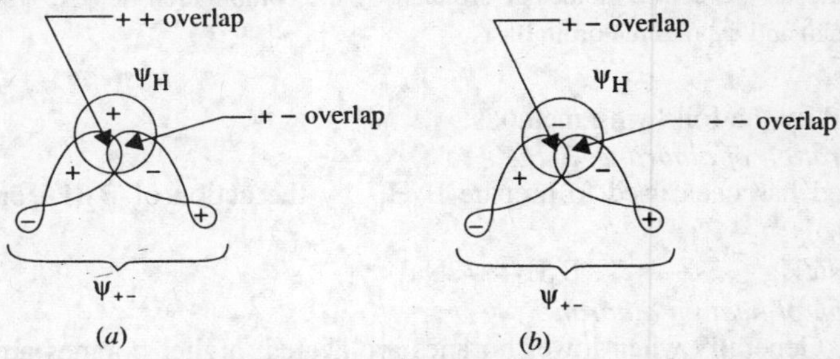

Fig. 25.6 : Formation of ψ^n molecular orbital by (*a*) the additive overlap of ψ_{+-} and ψ_H (*b*) the subtractive overlap of ψ_{+-} and ψ_H.

$$\psi_{+-}, \psi_H \xrightarrow[\text{Subtractive}(+-)\text{ combination}]{\text{Additive }(++)\text{ or}} \psi^n \text{ (Non-bonding 3-centre molecular orbital)}$$

The above discussion makes it evident that the combination of ψ_{++}, ψ_{+-} and ψ_H gives rise to the formation of three 3-centre molecular orbitals namely ψ^b, ψ^* and ψ^n.

It should be noted that ψ_{++} molecular orbital has one electron, ψ_{+-} molecular orbital is empty while ψ_H atomic orbital also contains one electron. The energy of ψ^b, ψ^n and ψ^* molecular orbitals varies in the order $\psi^b < \psi^n < \psi^*$ (See Fig. 25.7).

This order shows that the two electrons, one from ψ_{++} molecular orbital and the other from ψ_H atomic orbital occupy ψ^b molecular orbital, since this molecular orbital has the lowest energy.

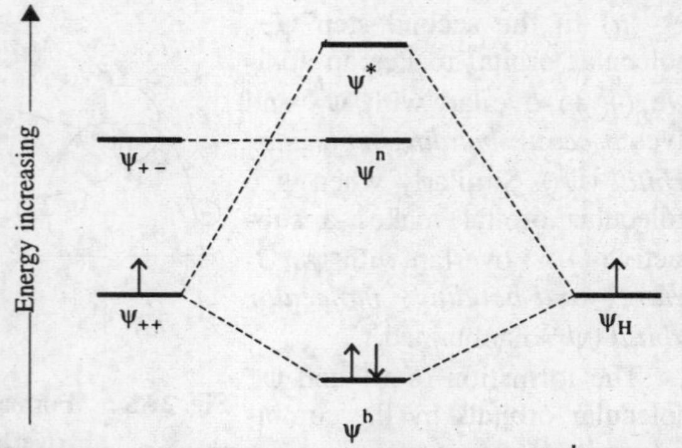

Fig 25.7: Relative order of the energy of Ψ^b, Ψ^n and Ψ^* molecular orbitals.

The combination of ψ_{B_1} and ψ_{B_1} atomic orbitals leading to the formation of ψ_{++} and ψ_{+-} (2-centre molecular orbitals) and then the combination of ψ_{++}, ψ_{+-} (2-centre molecular orbitals) and ψ_H (atomic orbital) leading to the generation of ψ^b, ψ^n and ψ^* (3-centre molecular orbitals) has been shown in Fig. 25.8. Thus Fig. 25.8 is a combination of Figs. 25.4 and 25.7. Fig. 25.8 shows that ψ^b, ψ^n and ψ^* molecular orbitals can be regarded as having been formed by the combination of ψ_{B_1}, ψ_{B_2} and ψ_H atomic orbitals.

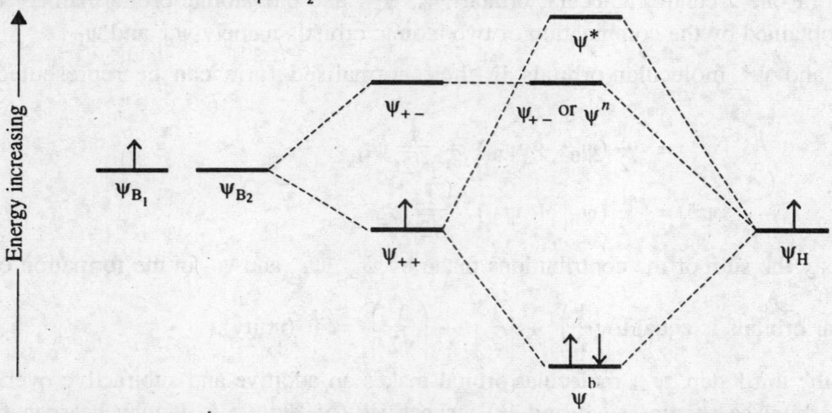

Fig 25.8 : Formation of ψ^b, ψ^n and ψ^* (3-centre molecular orbitals) by the combination of ψ_{++}, ψ_{+-} (2-centre molecular orbitals) and ψ_H (atomic orbital).

Higher boranes

Higher boranes can be prepared by the following methods:

(*i*) *Stock's method.* See "*Preparation of diborane, B_2H_6*".

(*ii*) *Wurt's method.* This method has been used to prepare B_4H_{10} by the action of B_2H_5I on sodium amalgam (Na / Hg).

$$2B_2H_5I + 2Na/Hg \longrightarrow B_4H_{10} + 2NaI$$

(*iii*) *From diborane.* See "*Action of heat on diborane*".

(*iv*) *By heating lower boranes.* Generally when lower boranes are heated, higher boranes are obtained. For example:

(*a*) $B_4H_{10} \xrightarrow{95°C} B_{10}H_{14} \xrightarrow{95°C} B_5H_9 \xrightarrow[\text{discharge}]{\text{Electric}} B_{10}H_{16}$

(*b*) $B_4H_{10} \xrightarrow{60°C} B_5H_{11} \xrightarrow{250°C} B_{10}H_{14}$

(*c*) $B_4H_{10} \xrightarrow{200°C} B_5H_9$

General properties. (*i*) *Physical state.* Boranes are volatile gases, liquids or solids.

(*ii*) *Electron-deficient compounds.* All the boranes are electron-deficient compounds, since total number of atomic orbitals (n_o) available on all boron and hydrogen atoms present in the borane is more than the total number of valence-electrons (n_e) present in the valence-shell of boron and H-atoms, *i.e.*, in these compounds there are not enough valence-shell electrons to form the expected number of 2-centre-2-electron B—H or B—B bonds. For example, in B_2H_6 molecule, since each B-atom has four atomic orbitals namely $2s$, $2p_x$, $2p_y$ and $2p_z$ and each H-atom has one $1s$ orbital, total number of orbitals (n_o) on two B-atoms and six H-atoms is equal to $4 \times 2 + 1 \times 6 = 14$. Total number of valence-shell electrons (n_e) in B_2H_6 molecule is equal to $3 \times 2 + 1 \times 6 = 12$ [B = $2s^2 2p^1$ (3), H = $1s^1$ (1)]. Since n_o is more than n_e, B_2H_6 molecule is an electron-deficient molecule. On the same lines of argument it can be shown that all the boranes are electron-dificient compounds.

(iii) Stability, melting and boiling points. Boranes of B_nH_{m+4} (*e.g.*, B_2H_6, B_5H_9, B_6H_{10} etc.) are quite stable and hence have high melting points while those of B_mH_{m+6} type (*e.g.*, B_4H_{10}, B_5H_{11}, B_6H_{12} etc.) are unstable and hence have low melting points. The melting points and boiling points (in °C) of some boranes of both types are given below :

B_mH_{m+4} type	melting points	boiling points	B_mH_{m+6} type	melting points	boiling points
B_2H_6	–165.5	– 92.5	B_4H_{10}	– 120	18
B_5H_9	– 46.6	48.0	B_5H_{11}	– 123	– 63
B_6H_{10}	– 65.0	—	B_6H_{12}	– 82.3	80 – 90
$B_{10}H_{14}$	– 99.7	213			

(iv) Action of heat. (a) Action of heat on diborane (B_2H_6) has already been discussed.

(b) Generally when lower boranes are heated, higher boranes are obtained. See "*General methods for the preparation of higher boranes*".

(v) Action of H_2O and aqueous solution of an alkali. (a) The action of H_2O and aqueous solution of NaOH or KOH on B_2H_6 has already been discussed.

(b) All the higher boranes are hydrolysed by H_2O and aqueous solution of KOH or NaOH. H_2 gas is evolved in all the cases.

$$B_4H_{10} + 12H_2O \longrightarrow 4H_3BO_3 + 11H_2\uparrow$$
$$B_4H_{10} + 4KOH + 4H_2O \longrightarrow 4KBO_2 + 11H_2\uparrow$$
$$B_6H_{12} + 18H_2O \longrightarrow 6H_3BO_3 + 15H_2\uparrow$$
$$B_6H_{12} + 6KOH + 6H_2O \longrightarrow 6KBO_2 + 15H_2\uparrow$$

(vi) Action O_2 or air. (a) Action of air on impure B_2H_6 has already been discussed.

(b) All the boranes are easily oxidised by O_2 to form coloured products. A mixture of the gaseous boranes with air is highly explosive in the presence of traces of moisture.

(vii) Formation of addition compounds (adducts). Since all the boranes, as mentioned above, are electron-deficient compounds, these compounds react with a number of donor molecules and give rise to the formation of addition compounds or adducts. In the formation of addition compounds donor molecules acts as *Lewis bases* while the boranes act as *Lewis acids or acceptors.*

(viii) Action of NH_3. All boranes react with NH_3 to give various products, depending on the experimental conditions.

Uses. Boranes are used as fuels for rockets because these compounds have high energy/weight ratio.

Types of bonds found in higher boranes

Higher boranes may contain some or all of the following types of bonds. Each of these bonds is formed by the sharing of two electrons between two atoms or three atoms as the case may be.

(i) Terminal (2c—2e) B — H bond. It is a normal covalent bond and is represented as B—H. It is formed by the overlap of one singly-filled sp^3 hybrid orbital of B-atom and singly-filled $1s$ orbital on H-atom.

(ii) (2c—2e) B — B bond. This bond links two B-atoms together and is a normal covalent bond. It is formed by the overlap of one singly-filled sp^3 hybrid orbital of one B-atom with the singly-filled sp^3 hybrid orbital of the other B-atom.

(iii) (3c—2e) bridging or bent B — H — B bond. This bond links two B-atoms and one H-atom

together and is formed by the sharing of two electrons. It is represented as 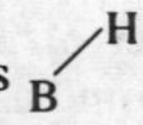 or 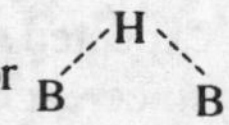

and is formed by the overlap between singly-filled sp^3 hybrid orbital of one B-atom, empty sp^3 hybrid orbital of the other B-atom and singly-filled 1*s* orbital of H-atom.

(*iv*) *(3c—2e) closed or triply bridged B—B—B bond.* This bond links three B-atoms and is represented as 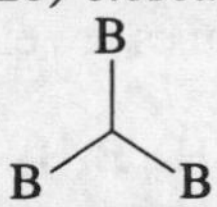. It is formed by the sharing of two electrons and results by the overlap of three sp^3 hybrid orbitals on three B-atoms. sp^3 hybrid orbital of one B-atom should be vacant while this hybrid orbital of both the remaining B-atoms should be singly-filled. The three sp^3 hybrid orbitals are situated at the three corners of an equilateral triangle. Since this bond contains two electrons, each boron atom contributes 2/3 electrons for the formation of this bond.

Structure and bonding in higher boranes

Now let us discuss the bonding in some higher boranes such as B_4H_{10}, B_5H_9, B_5H_{11}, B_6H_{10} and $B_{10}H_{14}$.

1. Tetraborane, B_4H_{10}. In this molecule four B-atoms may be regarded as a portion of a slightly distorted octahedron. The structure of this molecule has been shown in Fig. 25.9 which indicates that this molecule contains : (*i*) Four bridging (3*c* — 2*e*) *B — H — B bonds viz.*, B^1 — H — B^3, B^2 — H — B^3, B^1 — H — B^4 and B^2 — H — B^4 (*ii*) one direct (2*c* — 2*e*) B — B bond (B^1 - B^2 bond) (*iii*) six terminal (2*c* — 2*e*) *B — H bonds* namely B^3 — H, B^3 — H, B^2 — H, B^2 — H, B^4 — H, and B^4 — H bonds.

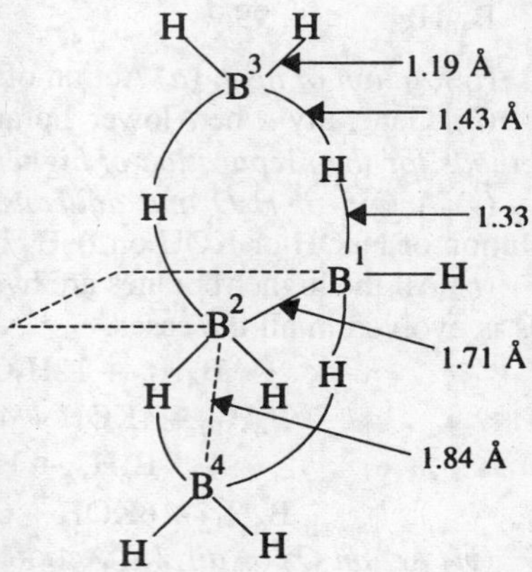

Fig. 25.9 : Structure of tetraborane, B_4H_{10}.

2. Pentaborane–9, B_5H_9. In this molecule five B-atoms occupy the five corners of a square pyramid. Each boron atom is linked with one H-atom by a terminal (2*c*—2*e*) B—H bond. Four B-atoms located in the base of the square pyramid (*i.e.*, basal B-atoms namely B^1, B^2, B^3 and B^4 atoms) are bonded to each other by four bridging (3*c*—2*e*) B—H—B bonds while the B-atom (namely B^5 atom) located at the apex (apical B-atom) of the pyramid is bonded to the basal two B-atoms (namely B^1 and B^4 atoms) by two (2*c*—2*e*) B—B bonds. This molecule also contains one closed (3*c*—2*e*) B—B—B bond. Thus the structure of this molecule can be shown as depicted in Fig. 25.10. This figure indicates that B_5H_9 molecule contains five terminal B—H bonds (*viz.*, B^1—H, B^4—H, B^2—H, B^3—H, and B^5—H bonds), four bridging B—H—B bonds (namely B^1—H—B^2, B^2—H—B^3, B^3—H—B^4, and B^4—H—B^1 bonds), two B—B bonds (B^1—B^5 and B^5—B^4 bonds) and one closed (3*c*—2*e*) B^5—B^3—B^2 bond. Thus this molecule contains all the types of bonds found in higher boranes.

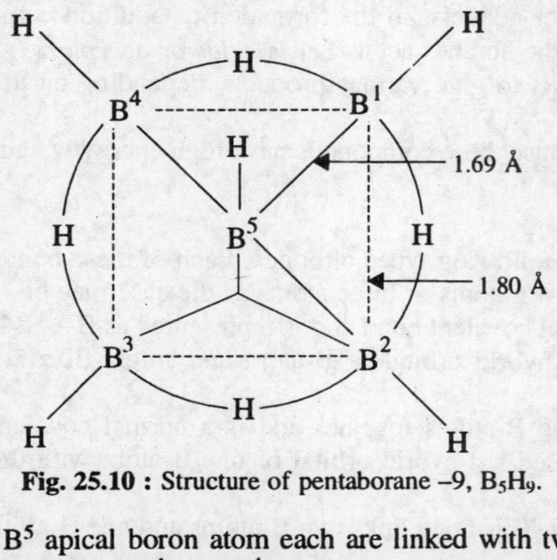

Fig. 25.10 : Structure of pentaborane –9, B_5H_9.

3. Pentaborane–11, B_5H_{11}. This molecule has an unsymmetrical square pyramidal structure. Like five B-atoms of B_5H_9 molecule, the five B-atoms of B_5H_{11} molecule also occupy the five corners of a square pyramid. B^2 and B^3 basal boron atoms and B^5 apical boron atom each are linked with two H-atoms by two terminal B—H bonds while the remaining B^1 and B^4 basal boron atoms each are linked with one H-atom by one terminal B—H

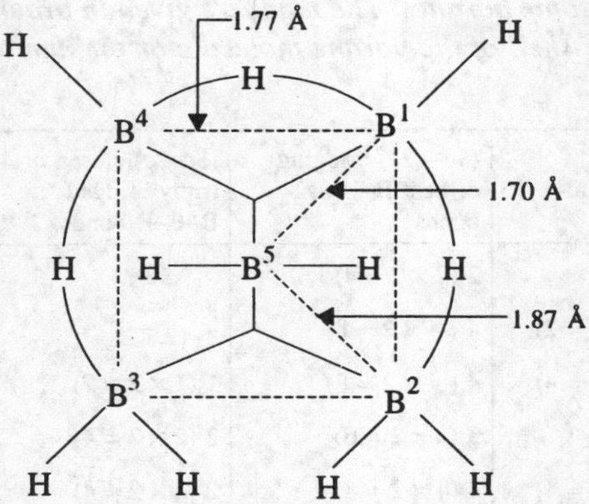

Fig. 25.11 : Structure of pentaborane-11, B_5H_{11}.

bond. This molecule also contains three bridging (3*c*—2*e*) B—H—B bonds and two closed (3*c*—2*e*) B—B—B bonds. Thus this molecule contains eight terminal B—H bonds (*viz.*, B^1 —H, B^2 —H, B^2 —H, B^3—H, B^3—H, B^4— H, B^5—H and B^5 —H bonds), three bridging (3*c*—2*e*) B—H—B bonds (namely B^1—H—B^4, B^4 —H—B^3 and B^2—H—B^1 bonds) and two closed (3*c*—2*e*) B—B—B bonds (B^1—B^4—B^5 and B^2—B^3—B^5 bonds). (Fig. 25.11)

4. Hexaborane–10, B_6H_{10}. In this molecule the six boron atoms lie at the corners of a pentagonal bipyramid. Each boron atom is linked with one H-atom by terminal B—H bond. Thus this molecule has six terminal B—H bonds namely B^1—H, B^2—H, B^3 —H, B^4—H, B^5—H and B^6 —H bonds. This molecule also contains four bridging (3*c* — 2*e*) B—H—B bonds (*viz.*, B^1—H—B^2, B^2—H—B^3, B^2 —H—B^4 and B^4—H—B^5 bonds). Two B—B bonds (B^1—B^5 and B^3—B^6 bonds) and two closed (3*c*—2*e*) B—B—B bonds (B^4—B^5—B^6 and B^1—B^2—B^6 bonds). All these bonds have been shown in Fig. 25.12.

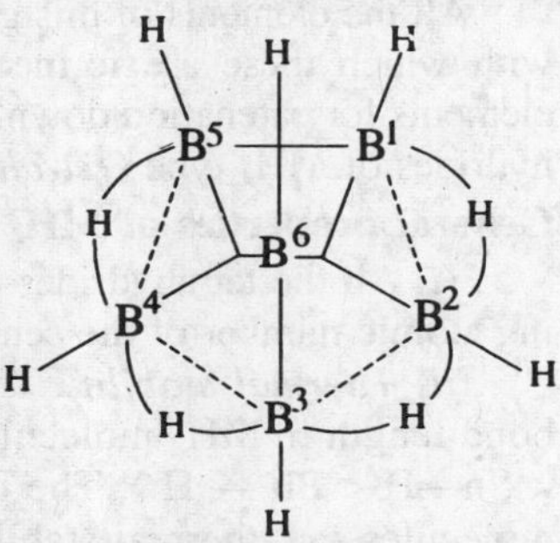

Fig. 25.12 : Structure of hexaborane–10, B_6H_{10}.

5. Decaborane–14, $B_{10}H_{14}$. The structure of this molecule has been determined by X-ray analysis. The boron atoms are arranged in two pentagonal pyramids with an edge common to both the pentagonal pyramids. This molecule has four bridging (3*c*—2*e*) B—H—B bonds namely B^5—H—B^6, B^6—H—B^7, B^8—H—B^9 and B^9—H—B^{10} in the open face. Each of B-atoms is linked with one H-atom by terminal B—H bonds. Thus the molecule has ten terminal B—H bonds *viz.*, B^1—H, B^2—H, B^3—H, B^4—H, B^5—H, B^6—H, B^7 —H, B^8—H, B^9—H and B^{10}—H bonds. In addition to these bonds, the molecule also has four B—B bonds (B^2—B^5, B^2—B^7, B^4—B^8, B^4—B^{10}) and four closed (3*c*—2*e*) B—B—B bonds (B^1—B^2—B^3, B^1—B^3—B^4, B^1—B^5—B^{10} and B^3—B^7—B^8 bonds). (See Fig. 25.13).

Summary. Number and types of different bonds in B_2H_6 and higher boranes like B_4H_{10}, B_5H_9, B_5H_{11}, B_6H_{10} and $B_{10}H_{14}$ have been shown in Table 25.1.

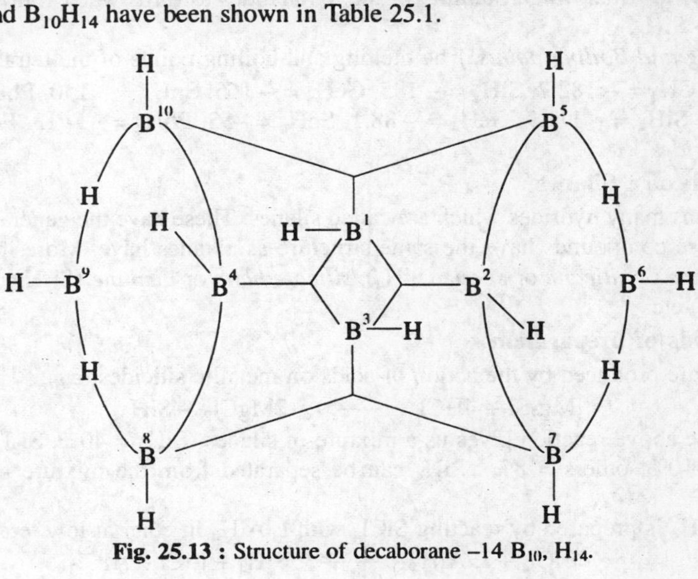

Fig. 25.13 : Structure of decaborane –14 B_{10}, H_{14}.

Table 25.1 : *Number and types of bonds found in some boranes. The numbers given in bracket indicate the number of electrons present in the valence-shell of the borane molecule or the number of electrons involved in the formation of bonds.*

Borane molecule	Terminal (2*c*–2*e*) B–H bonds	(2*c* – 2*e*) B–B bonds	(3*c* – 2*e*) bridging or bent B–H–B bonds	(3*c*–2*e*) closed or triply bridged B–B–B bonds
Diborane, B_2H_6 (12)	4 (4 × 2 = 8)	—	2 (2 × 2 = 4)	—
Tetraborane, B_4H_{10} (22)	6 (6 × 2 = 12)	1 (1 × 2 = 2)	4 (4 × 2 = 8)	—
Pentaborane-9 B_5H_9 (24)	5 (5 × 2 = 10)	2 (2 × 2 = 4)	4 (4 × 2 = 8)	1 (1 × 2 = 2)
Pentaborane-11, B_5H_{11} (26)	8 (8 × 2 =16)	—	3 (3 × 2 = 6)	2 (2 × 2 = 4)
Hexaborane-10, B_6H_{10} (28)	6 (6 × 2 = 12)	2 (2 × 2 = 4)	4 (4 × 2 = 8)	2 (2 × 2 = 4)
Decaborane-14 $B_{10}H_{14}$ (44)	10 (10 × 2 = 20)	4 (4 × 2 = 8)	4 (4 × 2 = 8)	4 (4 × 2 = 8)

Hydrides of the elements of group IVA

All the elements of this group form *covalent* hydrides but the number of hydrides and the ease with which these are formed decreases from C to Pb due to the diminishing tendency of the elements for catenation down the group. It may be seen that all the elements of group IVA form the hydrides of MH_4 type (*Tetrahydrides*).

General properties of MH_4 type hydrides

(*i*) All the tetrahydrides are gases. The volatility of these hydrides decreases with the increasing atomic number of the cental atom.

(*ii*) *Thermal stability.* As the size of the central atom, M increases from C to Pb, H—X bond length in MH_4 molecules also increases from C—H to Pb — H (C—H < Si—H < Ge—H < Sn—H < Pb — H). The increase in M—H bond length *decreases* thermal stability of MH_4 molecules *i.e.,* thermal stability of MH_4 molecules is in the order $CH_4 > SiH_4 > GeH_4 > SnH_4 > PbH_4$. This order of thermal stability is evidenced by the fact that the decomposition temperatures of these hydrides also decreases from CH_4 to PbH_4 (CH_4 = 800°C, SiH_4 = 450°C, GeH_4 = 285°C, SnH_4 = 150°C and PbH_4 = 0°C). PbH_4 is so unstable that it has been obtained only in traces.

(*iii*) *Stability towards water.* Stability of the tetrahalides towards water follow the order $CH_4 > GeH_4 > SnH_4 > SiH_4$.

(*iv*) *Melting and Boiling points.* The melting and boiling points of the tetrahydrides are given as : *M.pts(°C)*: CH_4 = – 182.7, SiH_4 = – 185, GeH_4 = – 165, SnH_4 = – 150, PbH_4 = –; *B.pts.(°C)*: CH_4 = – 161.3, SiH_4 = – 111.9, GeH_4 = – 88.1, SnH_4 = –53, PbH_4 = ~ – 13. For explanation see "*Covalent Hydrides*".

Hydrides of silicon : Silanes

Silicon forms many hydrides which are called silanes. These have the general formula Si_nH_{2n+2} (*n* upto 7). These compounds have the same structure as alkanes have. More important members are *silico-methane* or *silicane* or *silane* (SiH_4), *silico-ethane* or *disilane* (Si_2H_6), *silico-propane* or *trisilane* (Si_3H_8) etc.

General methods of preparation

(*i*) Silanes are produced by the action of acids on metallic silicides, *e.g.,*

$$Mg_2Si + 4HCl \longrightarrow 2MgCl_2 + SiH$$

Actually the above reaction gives us a mixture of silanes (SiH_4 = 40%, Si_2H_6 = 30%, Si_3H_8 = 12%, Si_4H_{10} = 10%, others = 8%). SiH_4 can be separated from the mixture by fractionating in vacuum.

(*ii*) Pure SiH_4 is prepared by reacting $SiCl_4$ with $LiAlH_4$ in ether at low temperature.

$$SiCl_4 + LiAlH_4 \longrightarrow SiH_4 + LiCl + AlCl_3$$

General properties. (*i*) All the silanes are colourless volatile covalent compounds.

(*ii*) The first two slanes are colourless gases while other members are liquids.

(*iii*) The melting and boiling points (°C) are given as:

	Melting point	*Boiling points*		*Melting point*	*Boiling points*
SiH_4	– 185	111.9	Si_4H_{10}	84.3	107.4
Si_2H_6	–132.5	–14.5	Si_5H_{12}	—	100
Si_3H_8	–117.4	52.9	Si_6H_{14}	—	100

(*iv*) *Thermal stability and reactivity.* The silanes are much less stable than the corresponding alkanes. All the silanes are extremely reactive and are spontaneously inflammable in air. The lower members decompose more rapidly than the higher ones. Thus SiH_4 decomposes spontaneously on coming in contact with air.

$$SiH_4 + 2O_2 \longrightarrow SiO_2 + 2H_2O$$

(*v*) *Hydrolysis*. Silanes are rapidly hydrolysed by water. In this reaction they decompose water when brought in contact with water, *e.g.*,

$$SiH_4 + 3H_2O \longrightarrow SiO_2.H_2O + 4H_2$$

or

$$SiH_4 + 3H_2O \longrightarrow H_2SiO_3 + 4H_2$$

$$Si_2H_6 + (2n + 4)\ H_2O \longrightarrow 2SiO_2.nH_2O + 7H_2$$

(*vi*) *Action of alkali solution.* Silanes are soluble in strong alkali solutions in presence of air and evolve H_2, *e.g.*,

$$2SiH_4 + 4NaOH + O_2 \longrightarrow 2Na_2SiO_3 + 6H_2$$

$$Si_2H_6 + 4KOH + O_2 \longrightarrow 2K_2SiO_3 + 5H_2$$

(*vii*) *Reducing properties.* Silanes are reducing agents and hence reduce Cl_2 and Br_2 to HCl and HBr, *e.g.*,

$$SiH_4 + 4Cl_2 \longrightarrow SiH_4 + 4HCl.$$

(*viii*) *Action of HCl and HBr.* The treatment of SiH_4 with HCl or HBr at 100°C in presence of Al_2Cl_3 as a catalyst results in the substitution of H-atom by Cl or Br atom.

$$SiH_4 + HCl \xrightarrow{Al_2Cl_3} SiH_3Cl + H_2.$$

(*ix*) *Action of heat.* When heated to 670 K, SiH_4 breaks into its elements. Similarly Si_2H_6 breaks into its elements at 470 K.

$$SiH_4 \xrightarrow{\Delta} Si + 2H_2; \qquad Si_2H_6 \xrightarrow{\Delta} 2Si + 3H_2$$

(*x*) *Action of salts.* (*a*) When solid AgCl reacts with SiH_4 in a heated flow reactor, Ag is precipitated.

$$SiH_4 + 2AgCl \xrightarrow{200°C} SiH_3Cl + HCl + 2Ag$$

(*b*) With $AgNO_3$ the reaction proceeds as

$$SiH_4 + 4AgNO_3 \longrightarrow Si + 4Ag + 4HNO_3$$

(*c*) SiH_4 gives a ppt. of Cu_2Si when passed in a solution of $CuSO_4$

$$2CuSO_4 + SiH_4 \longrightarrow Cu_2Si + 2H_2SO_4$$

(*xi*) *Action of $CHCl_3$ and CCl_4 on Si_2H_6.*

$$Si_2H_6 + 3CHCl_3 \longrightarrow 2SiCl_4 + HCl + 4H_2 + 3C$$

$$Si_2H_6 + 2CCl_4 \longrightarrow 2SiCl_4 + 3H_2 + 2C$$

Structure of silanes

These compounds have the structures which are similar to those of the corresponding saturated hydrocarbons. The structure of these compounds are given below.

Difference between alkanes and silanes

(*i*) In the hydrocarbon series, the number of C-atoms in a chain can apparently have any value but in case of silanes the maximum number of Si-atoms is seven which is found in hepta silance (Si_7H_{14}). This is because of the fact that relatively weak Si—Si bond makes unstable silanes linked with large number of Si-atoms. (*ii*) SiH_4 gets hydrolysed while CH_4 does not. This is because

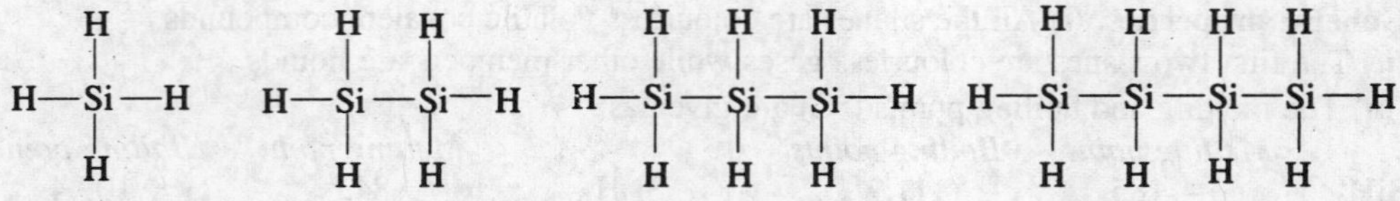

Monosilane (SiH_4) Disilane (Si_2H_6) Trisilane (Si_3H_8) Tetrasilane (Si_4H_{10})

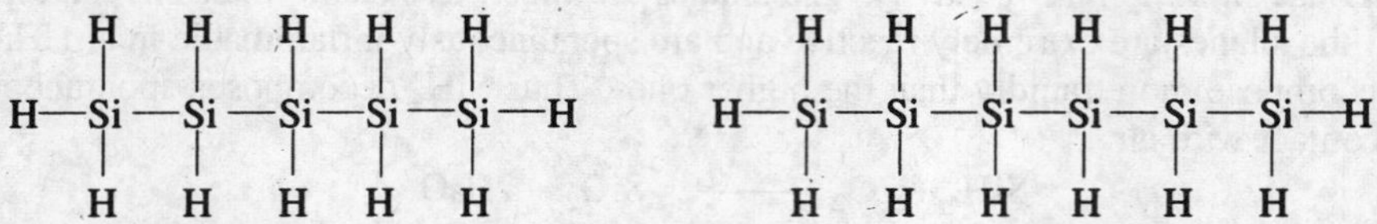

Pentasilane (Si_5H_{12}) Hexasilane (Si_6H_{14})

maximum covalency of carbon is four which is attained in CH_4. Thus CH_4 is resistant to the attack of electron-donating molecules. The maximum covalency of silicon atom is six. In SiH_4 molecule only four covalent bonds are present. It is, therefore, possible for SiH_4 to accept lone pair of electrons from the donor group such as water.

Comparison between silanes and germanes

(*i*) Germanes, like silanes, are colourless volatile covalent compounds. (*ii*) Just as with silanes, thermal stability of germanes decreases with increasing molecular weight. (*iii*) The germanes are less volatile and less reactive than the corresponding silanes. For example in comparison to SiH_4, GeH_4 does not ignite when brought in contact with air and also is far more stable towards hydrolysis. GeH_4, unlike SiH_4, is not attacked by 30% NaOH or kOH solution. (*iv*) Halogenation reactions are analogous to those of silanes. But the reaction rates are slower. (*v*) The boiling points of germanes are higher than those of the corresponding silanes.

Hydrides of the elements of group VA

Various types of hydrides given by the elements of group VA are given below:

Nitrogen	*Phosphorus*	*Arsenic*	*Antimony*	*Bismuth*
NH_3	PH_3	AsH_3	SbH_3	BiH_3
N_2H_4	P_2H_4	As_2H_4	—	—
HN_3	—	—	—	—
NH_2OH	—	—	—	—

Comparative study of NH_3, PH_3 ,AsH_3, SbH_3 and BiH_3

Preparation. (*i*) These can be prepared by the action of H_2O or dil. acids on binary metal compounds like Mg_3N_2, Ca_3P_2, Zn_3A_2 (A = As, Sb, Bi), Mg_3Sb_2 and Mg_3Bi_2.

$$Mg_3N_2 + 6H_2O \longrightarrow 3Mg(OH)_3 + 2NH_3$$
$$Mg_3N_2 + 6HCl \longrightarrow 3MgCl_2 + 2NH_3$$
$$Zn_3A_2 \text{ (A = As, Sb, Bi)} + 6HCl \longrightarrow 3ZnCl_2 + 2AH_3$$
$$Ca_3A_2 \text{ (A = N, P)} + 6H_2O \longrightarrow Ca(OH)_2 + 2AH_3$$

(*ii*) These can also be prepared by the reduction of trichlorides or oxides of these elements with nascent hydrogen or $LiAlH_4$.

$$A_2O_3 + 12H \longrightarrow 2AsH_3 + 3H_2O$$
$$ACl_3 \text{ (A = As, Sb)} + 6H \longrightarrow AH_3 + 3HCl$$
$$ACl_3 \text{ (A = N, P, As, Sb)} + 3LiAlH_4 \longrightarrow AH_3 + 3LiCl + 3AlH_3$$

Nascent hydrogen is obtained by the action of Zn on H_2SO_4.

(*iii*) NH_3 is prepared by the direct combination of N_2 and H_2. A mixture of N_2 and H_2 is now-a-days obtained by partial oxidation of hydrocarbons (derived from naptha, natural gas or coal gas) by oxygen-enriched air in the presence of a catalyst. The best catalyst is highly porous finely-divided iron containing small amounts of promoters usually molybdenum or oxides of K and Al.

$$\text{Hydrocarbons} + O_2 + \underbrace{O_2 + N_2}_{\text{air}} \xrightarrow{\text{catalyst}} x\,CO + y\,H_2 + N_2$$

$$N_2 + 3H_2 \rightleftharpoons 2NH_3, \qquad \Delta H = -92.0 \text{ kJ}$$

The process which consists of the formation of NH_3 by the direct combination of N_2 and H_2 is called Haber's process. The above reaction is reversible, exothermic and proceeds forward with a decrease in volume. Hence, according to Le Chatelier's principle, low temperature and high pressure are favourable conditions for better yield of NH_3. In actual practice the above reaction is allowed to take place at 450 – 550°C and under a pressure of 100 – 1000 atmospheres.

(*iv*) PH_3 is generally prepared by boiling white phosphorus with a conc. solution of NaOH in an inert atmosphere of CO_2, oil gas or H_2

$$P_4 + 3NaOH + 3H_2O \longrightarrow 3NaH_2PO_2 + PH_3.$$

(*v*) BiH_3 is extremely unstable, since it decomposes even at ordinary temperature into Bi and H_2. Therefore, BiH_3 is difficult to prepare.

General properties. General properties of trihydrides, AH_3 have been summarised in Table 25.2 and have been discussed below:

1. (*a*) **Physical state.** NH_3 is a liquid at ordinary temperature while other hydrides exist as gases (colourless) at room temperature.

Reason. See "*Hydrogen Bonding*" (chapter 10).

(*b*) **Smell.** The smell of these hydrides becomes more and more disagreeable as the atomic number increases from N to Bi.

Table 25.2 : *Some properties of the trihydrides of the elements of group VA.*

Property	NH_3	PH_3	AsH_3	SbH_3
Melting point (°C)	– 77.5	– 133.5	– 116.3	– 88.0
Boiling point (°C)	– 34.5	– 87.5	– 62.4	– 18.4
HAH bond angle (°)	107.5	93.5	91.8	91.3
	Decreasing →			
Shape	Trigonal pyramidal	Trigonal pyramidal	Trigonal pyramidal	Trigonal pyramidal
Heat of formation (kJ mol^{-1})	– 46.1	– 9.6	66.4	145.1
Physical state	Liquid	Gas	Gas	Gas
H—A bond length (Å)	1.02	1.42	1.52	1.70
	Increasing →			
Strength of H—A bond	Decreasing →			
Thermal stability	Decreasing →			
Dissociation temperature (°C)	1300	440	28	Room temperature
Reducing power	Increasing →			
Ionic character in A—H bond	Decreasing →			
Covalent character in AH_3 molecules	Increasing →			
Basic character of AH_3 molecules	Decreasing →			

(*c*) **Poisonous nature.** All the hydrides are poisonous. Their poisonous nature increases from NH_3 to BiH_3.

(*d*) **Solubility.** NH_3 is highly soluble in water but other hydrides are less soluble. The high solubility of NH_3 is due to the H-bonding between NH_3 and H_2O molecules as shown in the margin.

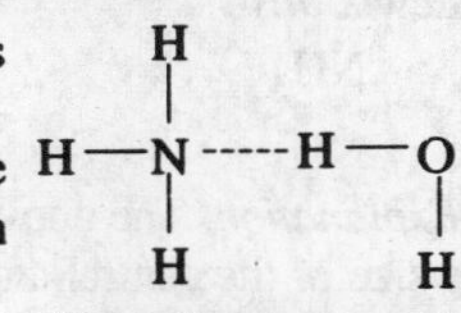

H-bonding between NH_3 and H_2O molecules.

2. H—A bond length, strength of H—A bond (H—A bond energy) and thermal stability of AH_3 molecules. As the size of the central atom A (A = N, P, As, Sb, Bi) increases from N to Bi, H—A bond length in AH_3 molecules increases from H—N to H—Sb (H—N = 1.02 Å, H—P = 1.42 Å, H—As = 1.52 Å, H—Sb = 1.71 Å). The increase in H—A bond length decreases the strength of H—A bond from H—N to H—Sb. Due to the successive decrease in the strength of H—A bond from H—N to H—Sb, the cleavage of A — H bond in AH_3 molecules becomes easier as we move down the group from NH_3 to SbH_3, *i.e.*, thermal stability of AH_3 molecules decreases from NH_3 to SbH_3. ($NH_3 > PH_3 > AsH_3 > SbH_3$). The decreasing order of thermal stability is evident from the fact that NH_3 dissociates at 1300°C, PH_3 at 440°C (in an inert atmosphere). AsH_3 at 280°C, SbH_3 dissociates at room temperature while BiH_3 is very unstable. Due to unstable nature of BiH_3, this substance has not been obtained in the pure form. It has been obtained only in traces.

3. Reducing power. All the molecules have a tendency to lose hydrogen and hence act as reducing agents. The reducing power of a given AH_3 molecule is expressed in terms of its ability to lose hydrogen. Easier is the cleavage of A—H bond, greater will be the tendency of AH_3 molecule to lose hydrogen and hence greater will be the reducing power of AH_3 molecule. Since, as we have seen above, the stability of AH_3 molecules decreases down the group, the ease with which A—H bond is broken to produce hydrogen increases and hence the reducing power of AH_3 molecules increases *i.e.*, the reducing power of AH_3 molecules is in the order $NH_3 < PH_3 < AsH_3 < SbH_3 < BiH_3$.

NH_3 is a mild reducing agent and reduces heated metallic oxides to their corresponding metal and is itself oxidised to N_2, *e.g.*,

$$2NH_3 + 3CuO\ (\textit{heated}) \longrightarrow 3Cu + 3H_2O + N_2$$

PH_3, AsH_3 and SbH_3 are powerful reducing agents and reduce $AgNO_3$ to metallic silver (*black ppt.*), *e.g.*,

$$AsH_3 + 6AgNO_3 + 3H_2O \longrightarrow 6Ag + H_3AsO_3 + 6HNO_3$$

4. Boiling points. Since as we move down the group from NH_3 to SbH_3, the magnitude of van der Waals forces of attraction increases and hence the boiling points of the hydrides should also increase in the same direction ($NH_3 < PH_3 < AsH_3 < SbH_3$). But actually what happens is that the boiling point decreases from NH_3 to PH_3 ($NH_3 > PH_3$) and then increases until we reach SbH_3 ($PH_3 < AsH_3 < SbH_3$). Why the boiling point of NH_3 is higher than that of PH_3 has been explained on the basis of inter-molecular H-bonding occurring between several NH_3 molecules. See "*H-bonding*" discussed in chapter 10.

5. Variation in the ionic/covalent character in NH_3, PH_3, AsH_3 and SbH_3 molecules. It has been found that the magnitude of ionic character of H—A bond in the given molecules decreases as H—N > H—P > H—As > H—Sb. Thus NH_3 is the most ionic (*i.e., least covalent*) and SbH_3 is the least ionic (most covalent).

Reason. The above order has been explained on the basis of electronegativity difference, ($x_A - x_H$). For this see "Polar and Non-polar covalent Bonds" (chapter 10).

6. Relative order of basic strength (electron-donating capacity). Since the central atom of each hydride has a lone pair of electrons, the hydride can donate this lone pair to the donor molecule (*e.g.*, BF_3) or ion (*e.g.*, H^+, Cu^{2+} etc.) and hence act as Lewis base. The basic strength has been found in the order $NH_3 > PH_3 > AsH_3 > SbH_3 > BiH_3$. This order is evident from the fact that NH_3 (Lewis base) readily co-ordinates with proton, H^+ (Lewis acid) to give stable ammonium ion, NH_4^+ while phosphonium ion, PH_4^+ is much less stable and the corresponding ions for the remaining elements are nonexistent.

Lewis base		*Lewis acid*		*Adduct*
NH_3	+	H^+	$\longrightarrow$	$H_3N \rightarrow H^+$ or NH_4^+
NH_3	+	BF_3	$\longrightarrow$	$H_3N \rightarrow BF_3$

Explanation. The above order of basic strength of AH_3 molecules is due to the successive increase in the size of the central atom, A as we move down the group from N to Bi. (*Also see* chapter 19).

The donor ability of AH_3 molecules is increased by the replacement of H-atoms by alkyl groups. thus trimethyl phosphine, $P(CH_3)_3$ is stronger base than phosphine, PH_3. The hydrides and alkyl-substituted hydrides, especially of N and P, show donor properties not only towards protonic acids but also form coordination compounds with a number of electron acceptors (Lewis acids) like BF_3 and transition metal ions.

Due to the maximum basic strength of NH_3, its aqueous solution forms stable complex ions with many transition metal cations like Ag^+, Cu^{2+}, Cd^{2+} etc., while PH_3, AsH_3, SbH_3 and BiH_3 donot form any complexes with the transition metal cations.

$$Ag^+ + 2NH_3 \rightleftharpoons [Ag(NH_3)_2]^+$$

$$M^{2+} (M^{2+} = Cu^{2+}, Cd^{2+}) + 4NH_3 \longrightarrow [M(NH_3)_4]^{2+}$$

NH_3 is highly basic, PH_3 is less basic than NH_3 while AsH_3, SbH_3 and BiH_3 are neutral.

7. Precipitation reactions. NH_3 precipitates heavy metals (*e.g.*, Fe) as their hydroxides from their aqueous salt solutions.

$$FeCl_3 + 3NH_3 + 3H_2O \longrightarrow Fe(OH)_3\ (ppt.) + 3NH_4Cl$$

or

$$Fe^{3+} + 3NH_3 + 3H_2O \longrightarrow Fe(OH)_3 \downarrow + 3NH_4^+$$

PH_3, AsH_3, SbH_3 and AsH_3 precipitate metals (*e.g.*, Ag, Cu) as their *phosphides* (P^{3-}), *arsenides* (As^{3-}), *antimonides* (Sb^{3-}) and *bismuthides* (Bi^{3-}) respectively when these gaseous hydrides are passed through the aqueous solutions of the salts of these metals.

$$2CuSO_4 + 2PH_3 \longrightarrow Cu_3P_2\downarrow + 3H_2SO_4$$

$$3AgNO_3 + AsH_3 \longrightarrow Ag_3As \downarrow + 3HNO_3$$

$$3AgNO_3 + SbH_3 \longrightarrow Ag_3Sb \downarrow + 3HNO_3$$

$$3CuSO_4 + 2BiH_3 \longrightarrow Cu_3Bi_2 \downarrow 3H_2SO_4$$

8. Hydrogen bonding and formation of $(AH_3)_x$ cluster. Since N-atom has much more electronegativity than H-atom and the electronegativity values of P, As and Sb are close to that of H-atom (N = 3.0, H = 2.1, P = 2.1, As = 2.0, Sb = 1.9), NH_3 molecule undergoes intermolecular hydrogen binding to form its cluster, $(NH_3)_x$ while the occurrence of H-bonding and hence the formation of $(AH_3)_x$ cluster (A = P, As, Sb) are not possible in case of PH_3, AsH_3 and SbH_3 molecules. Thus NH_3 molecule shows H-bonding while other molecules do not show H-bonding.

9. Shape of AH_3 molecules and variation in HAH bond angles. The shape of all the AH_3 molecules is *trigonal pyramidal* which results from sp^3 hybridisation of the central atom, A (A = N, P, As, Sb). As an example the structure of NH_3 molecule can be studied from the chapter on "*Hybridisation*".

HAH bond angles in the molecules are as : NH_3 = 107.5°, PH_3 = 93.5°, AsH_3 = 91.8° and SbH_3 = 91.3°. These values decrease as we move down the group from NH_3 to SbH_3. The cause of the decrease has been given under "*Bond Angles*".

Hydrazine or diammine, N_2H_4

Preparation. (*i*) *Rasching's method.* This method is used to prepare N_2H_4 in the laboratory as well as on commercial scale. In this method N_2H_4 is prepared by boiling a strong aqueous solution of NH_3 with sodium hypochlorite (NaOCl) in presence of little glue or gelatin. The reaction between NH_3 and NaOCl takes place through the following steps:

$$NH_3 + NaOCl \longrightarrow NaOH + NH_2Cl\ (\textit{Chloramine})$$

$$NH_3 + NaOH + NH_2Cl \longrightarrow N_2H_4 + NaCl + H_2O$$

$$2NH_3\ (N = -3) + NaOCl \longrightarrow N_2H_4\ (N = -2) + NaCl + H_2$$

Why is glue or gelatin used in the above reaction? N_2H_4 formed as above may react with NH_2Cl and oxidise it to N_2

$$2NH_2Cl + N_2H_4\ (N = -2) \longrightarrow 2NH_4Cl + N_2\ (N = 0)$$

The production of N_2 decreases the yield of N_2H_4. The addition of glue or gelatin minimises the oxidation of N_2H_4 by NH_2Cl to N_2. Thus the presence of glue or gelatin increases the yield of N_2H_4.

(*ii*) *Anhydrous* N_2H_4 is prepared by adding a solution of sodium methoxide (CH_3ONa) in CH_3OH to a solution of hydrazinium monochloride, $[N_2H_5]^+Cl^-$ dissolved in CH_3OH. The reaction between $[N_2H_5]$ Cl and CH_3ONa gives a solution which contains N_2H_4 and CH_3OH. when this solution is distilled under pressure, CH_3OH gets distilled first and then N_2H_4.

$$(N_2H_5)\boxed{Cl + Na}CH_3O \longrightarrow \boxed{N_2H_4 + CH_3OH}\ (\textit{Solution}) + NaCl\ (\textit{ppt})$$

(*iii*) N_2H_4 can also be prepared by reducing hydrazoic acid (HN_3) in presence of sodium amalgam or finely divided platinum or by reducing NO containing solution of K_2SO_3 by sodium amalgam and H_2O.

$$2HN_3\ (N = -1/3) + 10H \xrightarrow{Na/Hg} 3N_2H_4\ (N = -2)$$

$$K_2SO_3\ (S = +4) + 2NO\ (N = +2) + 6H \xrightarrow{Na.Hg/H_2O} K_2SO_4\ (S = +6) + N_2H_4\ (N = -2) + H_2O$$

Properties. Physical. N_2H_4 is a colourless fuming liquid whose melting point = 113.5°C. On cooling, it solidifies to a colourless solid having melting point 1.4°C. It is readily soluble in water and alcohol. It is strongly hygroscopic and poisonous. It dissolves S, As, Se and phosphorus in it.

Chemical. 1. *Action of heat : Decomposition.* On being heated alone, N_2H_4 decomposes to give NH_3 and N_2.

$$3N_2H_4 \xrightarrow{\Delta} N_2 + 4NH_3$$

2. *Oxidation: Reducing properties.* Oxidation reaction of N_2H_4 by various substances can be classified into the following groups:

(*a*) In these reactions, N_2H_4 (N = – 2) is oxidised to HN_3 (N = –1/3). For example :

$$N_2H_4 + HNO_2 \longrightarrow 2HN_3 + 2H_2O$$

$$3N_2H_4 + 5O_3 \longrightarrow 2HN_3 + 5H_2O + O_2$$

$$N_2H_4 + NCl_3 \xrightarrow{Benzene} HN_3 + 3HCl$$

$$3N_2H_4 + 5H_2O_2 \longrightarrow 2HN_3 + 10H_2O$$

These reactions are used to prepare HN_3.

(*b*) In the following reactions N_2H_4(N = – 2) is oxidised to N_2 (N = 0)

$$N_2H_4 + O_2\ (\textit{From air}) \xrightarrow{\Delta} N_2 + 2H_2O + \textit{Heat}\ (= 622\ kJ)$$

$$N_2H_4 + 2X_2\ (X = Cl, Br, I) \longrightarrow N_2 + 4HX$$

$$5N_2H_4 + 4KMnO_4 + 6H_2SO_4 \longrightarrow 5N_2 + 2K_2SO_4 + 4MnSO_4 + 16H_2O$$

or $$5N_2H_4 + 4MnO_4^- + 12H^+ \longrightarrow 5N_2 + 2Mn^{2+} + 16H_2O$$

This reaction can be used for the titration of hydrazine and its salts.

$$N_2H_4 + 4FCl_3 \longrightarrow N_2 + 4FCl_2 + 4HCl$$

or $$N_2H_4 + 4Fe^{3+} \longrightarrow N_2 + 4Fe^{2+} + 4H^+$$

$$3N_2H_4 + 2KIO_3 \longrightarrow 3N_2 + 2KI + 6H_2O$$

or $$3N_2H_4 + 2IO_3^- \longrightarrow 3N_2 + 2I^- + 6H_2O$$

$$N_2H_4 + 4CuO\ (\textit{Fehling's solution}) \longrightarrow N_2 + 2H_2O + 2Cu_2O\ (\textit{red ppt.})$$

$$2N_2H_4 + 4AuCl_3 \longrightarrow 3N_2 + 4Au + 12HCl$$

$$N_2H_4 + PtCl_4 \longrightarrow N_2 + Pt + 4HCl$$

$$N_2H_4 + 2HgCl_2 \longrightarrow N_2 + 2Hg + 4HCl.$$

$$N_2H_4 + 4AgNO_3 \longrightarrow N_2 + 4Ag + 4HNO_3$$

$$N_2H_4 + 4K_3[Fe(CN)_6] + 4KOH \longrightarrow N_2 + 4H_2O + 4K_4[Fe(CN)_6]$$

or $$N_2H_4 + 4[Fe(CN)_6]^{3-} + 4OH^- \longrightarrow N_2 + 4H^+ + 4OH^- + 4[Fe(CN)_6]^{4-}$$

or $$N_2H_4 + 4[Fe(CN)_6]^{3-} \longrightarrow N_2 + 4H^+ + 4[Fe(CN)_6]^{4-}$$

$$N_2H_4 + H_2[PtCl_6] \longrightarrow N_2 + 6HCl + Pt\ (\textit{colloidal platinum})$$

$$N_2H_4\ (l) + 2H_2O\ (l) \xrightarrow{Cu\ (II)\ Catalyst} N_2 + 4H_2O + Heat.$$

3. *Action of alkali metals: Formation of metal hydrazides.* N_2H_4 dissolves in alkali metals to give blue solution which slowly gives hydrazides, $M^+[N_2H_3]^-$ and H_2, *e.g.*,

$$N_2H_4 + Na \longrightarrow Na^+ [N_2H_3]^- \text{ (sodium hydrazide)} + 1/2H_2$$

4. *Reduction* . In presence of Zn and HCl, N_2H_4 is reduced to NH_3.

$$N_2H_4\ (N = -2) + 2H \xrightarrow{Zn/HCl} 2NH_3\ (N = -3)$$

5. *Diacidic basic character: Formation of hydrazinium salts with acids.* An inspection of the structure of N_2H_4 molecule reveals that both N-atoms have one lone pair of electron which resides in sp^3 hybrid orbital. Therefore, when N_2H_4 reacts with an acid like HCl, HN_3, H_2SO_4 etc., one or both lone pair of electrons on N-atoms are denoted to the H^+ ion obtained by the ionisation of the acid and $N_2H_5^+$ or $N_2H_6^{2+}$ ions (called *hydrazinium* or *hydrazonium ions*) are obtained. Thus N_2H_4 can give rise to the formation of two types of salts (called *hydrazinium* or *hydrazonium salts*) which contain $N_2H_5^+$ or $N_2H_6^{2+}$ ion. In other words we say that N_2H_4 molecule behaves as a *diacidic base.*

The formation of hydrazinium monochloride, $[N_2H_5]Cl$ and hydrazinium dichloride, $[N_2H_6]Cl_2$ can be explained as follows :

$$HCl \longrightarrow H^+ + Cl^-$$

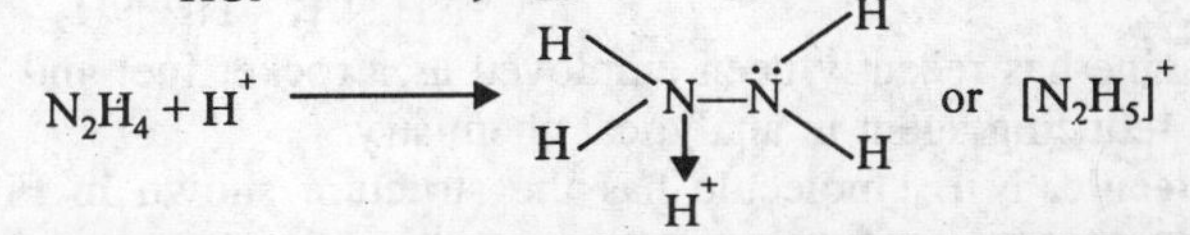

$$HCl + N_2H_4 \longrightarrow [N_2H_5]^+ + Cl^- \longrightarrow [N_2H_5]Cl^-$$

$$2HCl \longrightarrow 2H^+ + 2Cl^-$$

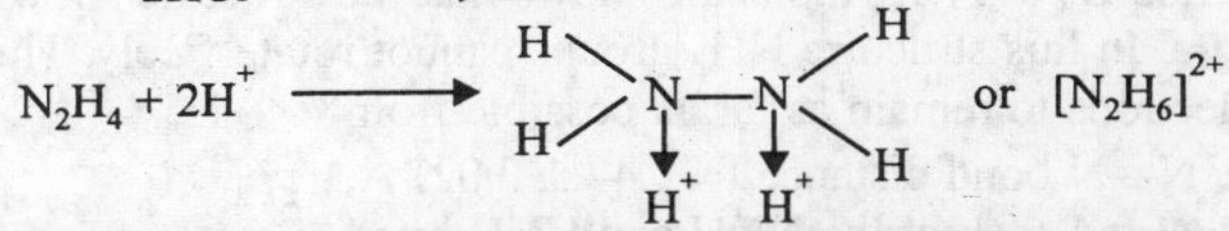

$$2HCl + N_2H_4 \longrightarrow [N_2H_6]^{2+} + 2Cl^- \longrightarrow [N_2H_6]Cl_2$$

On similar lines, the formation of hydrazinium azide, $[N_2H_5]N_3$ and hydrazinium sulphate, $[N_2H_6]SO_4$ can be shown as:

$$N_2H_4 + HN_3 \longrightarrow [N_2H_5]N_3$$
$$N_2H_4 + H_2SO_4 \longrightarrow [N_2H_6]SO_4$$

6. *Formation of complex compounds.* Although $\ddot{N}H_2 — \ddot{N}H_2$ has two lone pairs of electrons on two N-atoms, it cannot serve as a bidentate ligand, since it cannot get coordinated to the same metal atom at two sites, *i.e.*, the attachment of $NH_2 — NH_2$ molecule with the metal atom as shown in Fig. 25.14 (*a*) is *not possible.* This is because of the fact that a strained three-membered ring is formed at the metal atom. This ring is unstable.

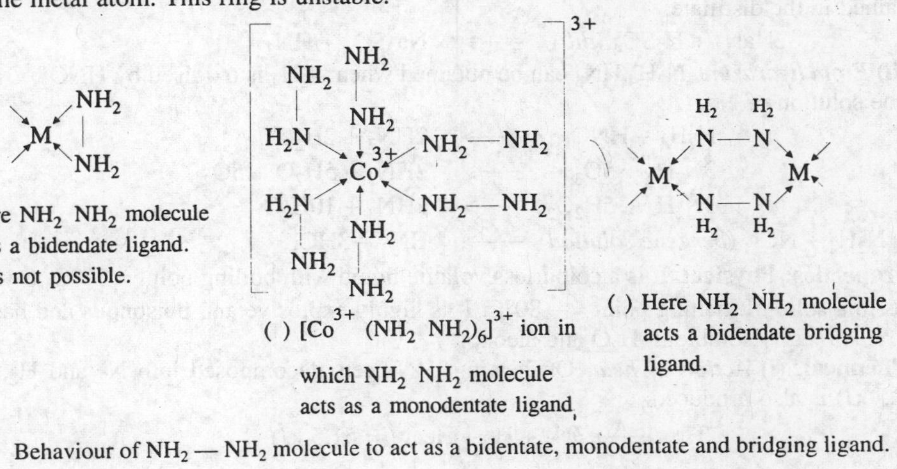

( ) Here NH_2 NH_2 molecule acts a bidendate ligand. It is not possible.

( ) $[Co^{3+}(NH_2\ NH_2)_6]^{3+}$ ion in which NH_2 NH_2 molecule acts as a monodentate ligand.

( ) Here NH_2 NH_2 molecule acts a bidendate bridging ligand.

Fig. 25.14 : Behaviour of $NH_2 — NH_2$ molecule to act as a bidentate, monodentate and bridging ligand.

$NH_2 — NH_2$ molecule, however, can act either as a monodentate ligand [Fig. 25.14 (*b*) or as a bidentate bridging ligand [Fig. 25.14 (*c*)].

7. *Condensation reactions.* N_2H_4 reacts with acetaldehyde (CH_3CHO) and acetone, $(CH_3)_2CO$ to form hydrazone

$$CH_3 . CH = \boxed{O + H_2} N. NH_2 \xrightarrow{-H_2O} CH_3 . CH = N. NH_2 \text{ (\textit{Acetaldehyde hydrazone})} + H_2O$$

$$(CH_3)_2 . C = \boxed{O + H_2} N. NH_2 \xrightarrow{-H_2O} (CH_3)_2 . C = N. NH_2 \text{ (\textit{Acetone hydrazone})} + H_2O$$

8. *Action of cynogen,* $N \equiv C—C \equiv N$. When aqueous solution of N_2H_4 reacts with cyanogen, oxaldimidic acid dihydrazide (I) is obtained.

Uses. (*i*) An organic derivative of hydrazine known as *phenyl hydrazine*, $C_6H_5NH.NH_2$ has been largely used in the preparation of hydrazones, osazones and nitrogen ring compounds in

$$N{\equiv}C—C{\equiv}N + 2\,NH_2 . NH_2\ (aq) \longrightarrow \underset{\text{NH.NH}_2}{HN{=}C} — \underset{\text{NH.NH}_2}{C{=}NH} \quad \text{(I)}$$

organic chemistry. (*ii*) Hydrazine has recently been employed as a rocket fuel and as an anti T.B. drug. (*iii*) It is also used as a reducing agent in analytical chemistry.

Structure of N_2H_4 molecule. N_2H_4 molecule has the structure shown in Fig. 25.15. This structure indicates that, since each N-atom is surrounded by three σ-bonding electron pairs (*σ-bps* = 3) and one lone pair of electrons (*lp* = 1), each N-atom is sp^3 hybridised. Raman spectra and high value of dipole moment (= 1.85 D) of N_2H_4 molecule shows that the structure shown in Fig. 25.15 is an *unsymmetrical structure.* In this structure NH_2 groups cannot rotate freely. This is due to the tendency of lone pairs of electrons to remain as far as possible from each other. In this molecule N—N bond distance (= 1.47 ± 0.02) Å, N—H bond distance = 1.04 Å and each of the NNH and HNH bond angles = 110°. The above structure can easily explain the formation of two types of salts *viz., monohydrazinium salts* (*e.g.,* N_2H_5Cl) and *dihydrazinium salts* (*e.g.,* $N_2H_6Cl_2$).

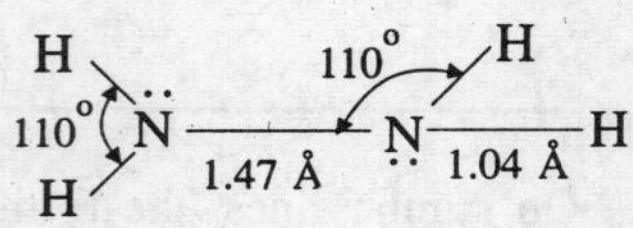

Fig. 25.15 : Unsymmetrical structure of N_2H_4 molecule.

Hydrazoic acid or hydrogen azide, HN_3

Preparation. (*i*) *From sodamide, $NaNH_2$.* First of all sodium azide (NaN_3) is obtained by treating $NaNH_2$ with N_2O or $NaNO_3$.

$$3NaNH_2 + N_2O \xrightarrow{200°C} NaN_3 + NaOH + NH_3$$

$$3NaNH_2 + NaNO_3 \xrightarrow{175°C} NaN_3 + 3NaOH + NH_3$$

NaN_3 is now distilled with dil. H_2SO_4 under reduced pressure so that aqueous solution of HN_3 is obtained in the distillate.

$$2NaN_3 + H_2SO_4\ (dil.) \longrightarrow Na_2SO_4 + HN_3$$

(*ii*) *From hydrazine.* $N_2H_4.HN_3$ can be obtained when N_2H_4 is oxidised by HNO_2, O_3, H_2O_2 or benzene solution of NCl_3.

$$N_2H_4 + HNO_2 \longrightarrow 2HN_3 + 2H_2O$$

$$3N_2H_4 + 5O_3 \longrightarrow 2HN_3 + 5H_2O + 5O_2$$

$$3N_2H_4 + 5H_2O_2 \longrightarrow 2HN_3 + 10H_2O$$

$$N_2H_4 + NCl_3\ (\textit{benzene solution}) \longrightarrow HN_3 + 3HCl$$

Properties: Physical. It is a colourless volatile liquid with boiling point = 37°C. On cooling it freezes to a solid of melting point = – 80°C. It is highly explosive and poisonous and has pungent odour. It is readily soluble in H_2O and alcohol.

Chemical. (*i*) *Action of heat.* On heating, HN_3 gets decomposed into N_2 and H_2, and heat (= 51.5 kJ) is also produced.

$$2HN_3 \xrightarrow{\Delta} 3N_2 + H_2 + \text{Heat } (= 51.5 \text{ kJ})$$

(*ii*) *Action of HCl and HI acids*. HCl and HI both decompose HN_3 into NH_3, N_2 and X_2.

$$HN_3 + 2HX(X = Cl, I) \longrightarrow NH_3 + N_2 + X_2$$

(*iii*) *Reduction.* HN_3 (N = –1/3) is reduced by sodium amalgam to NH_3 (N = –1/3) or hydrazine, N_2H_4 (N = –2).

$$HN_3 + 4H_2 \longrightarrow 3NH_3 \;; \qquad 2HN_3 + 5H_2 \longrightarrow 2N_2H_4$$

(*iv*) *Oxidation : Reducing properties.* HN_3 gets oxidised to N_2 by HNO_2, HOCl and acidified $KMnO_4$ solution and reduces them to N_2O, HCl and $MnSO_4$ respectively.

$$HN_3 + HNO_2 \longrightarrow N_2 + N_2O + H_2O$$

$$HN_3 + HOCl \longrightarrow 3N_2 + HCl + H_2O$$

$$2KMnO_4 + 3H_2SO_4 + 10HN_3 \longrightarrow K_2SO_4 + 2MnSO_4 + 8H_2O + 15N_2$$

or

$$2MnO_4^- + 6H^+ + 10HN_3 \longrightarrow 2Mn^{2+} + 8H_2O + 15N_2$$

(*v*) *Acidic nature and formation of azides.* HN_3 is a weak monobasic acid and dissociates to give azide ions (N_3^-) and H^+ ions.

$$HN_3\,(aq) \rightleftharpoons H^+\,(aq) + N_3^-\,(aq), \quad K_a = 1.8 \times 10^{-5}$$

Its acidic strength is equal to that of CH_3COOH ($K_a = 1.8 \times 10^{-5}$). The salts containing the azide ions (N_3^-) are called *azides*. The azides can be obtained by treating HN_3 with active metals (*e.g.*, Zn, Cu, Fe, Mg etc.), metallic salts (*e.g.*, AlH_3, $AgNO_3$ etc.), alkalies, NH_3, hydrazine (N_2H_4) etc.

$$3HN_3\,(aq) + M(M = Zn, Cu, Fe) \longrightarrow M(N_3)_2 + N_2 + NH_3$$

$$2HN_3\,(aq) + Mg \longrightarrow Mg(N_3)_2 + H_2$$

$$3HN_3\,(\textit{in ether}) + AlH_3 \longrightarrow Al(N_3)_3 + 3H_2$$

$$HN_3\,(\textit{vapour}) + AgNO_3\,(aq) \longrightarrow AgN_3 + HNO_3$$

$$HN_3 + NaOH \longrightarrow NaN_3 + H_2O$$

$$HN_3 + NH_3 \longrightarrow [NH_4]\,N_3 \;(\textit{Ammonium azide})$$

$$HN_3 + N_2H_4 \longrightarrow [N_2H_5]N_3 \;(\textit{Hydrazonium azide})$$

(*vi*) *Action of cyanogen, $(CN)_2$.* $(CN)_2$ reacts with HN_3 to form a cyclic compound, called 5-cyanotetrazole (I).

Uses. (*i*) The azides of alkali metals and of barium are used for the preparation of N_2 of high

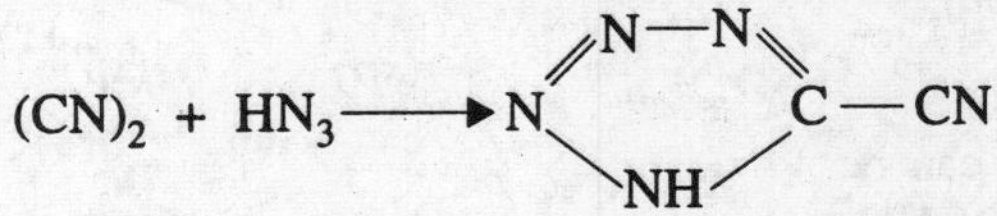

(I)

purity, *e.g.*, $2NaN_3\,(s) \xrightarrow{300°C} 2Na\,(l) + 3N_2\,(g)$. (*ii*) The azides of heavy metals (*e.g.*, Pb, Hg and Ba) are used as detonators in bombs, since these azides explode when struck or heated. (*iii*) Many azides are used as organic dyestuffs.

Structure of HN_3 molecule. Infrared studies of HN_3 molecule have shown that this molecule which is covalent in nature has an *unsymmetrical bent structure* in which all the three N-atoms lie in the same line but H-atom lies out of line. Theoretically, the true structure of HN_3 molecule should be a resonance hybrid of the following three structures *viz.*, (I), (II), and (III) but structure (III) cannot be considered to be the contributing structure, since two adjacent N-atoms namely N_a and N_b in this structure have the same charge (+ charge) which is against Pauling's adjacent charge rule. Consequently the true structure of HN_3 molecule is a resonance hybrid of structures (I) and (II) only.

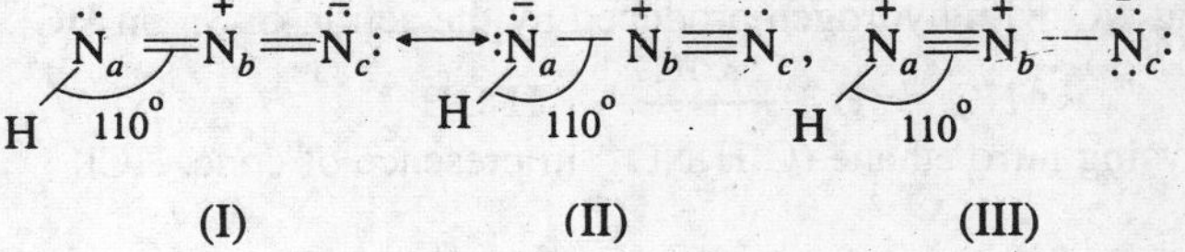

(I) (II) (III)

Bond distances are N_a—N_b = 1.24Å, N_b—N_c = 1.13 Å.

Structure of N_3^- ion. X-ray and electron diffraction studies of the metallic azides have indicated that the azide ion (N_3^-) has *linear symmetrical structure* in which all the three N-atoms lie in the same line. Both N—N bond distances are equal to 1.15 Å. Its true structure is supposed to be a resonance hybrid of the following three structures *viz.*, (IV), (V) and (VI).

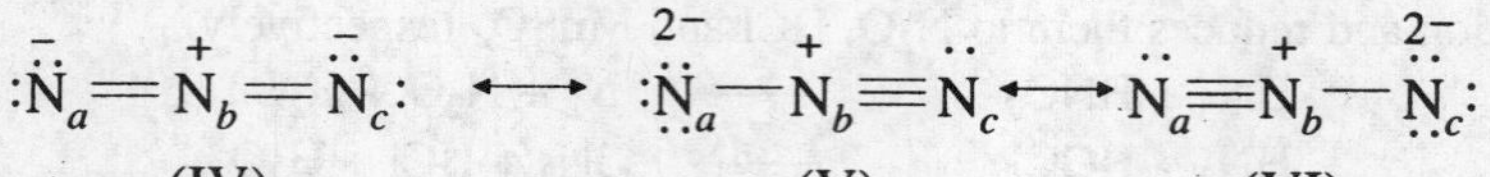

(IV) (V) (VI)

In structure (IV) the central N-atom (*i.e.*, N_b^+) in its excited state is *sp* hybridised while the terminal N-atoms remain unhybridised (See Fig. 25.16). The formation of different bonds in N_3^- ion has been shown in Fig. 25.17.

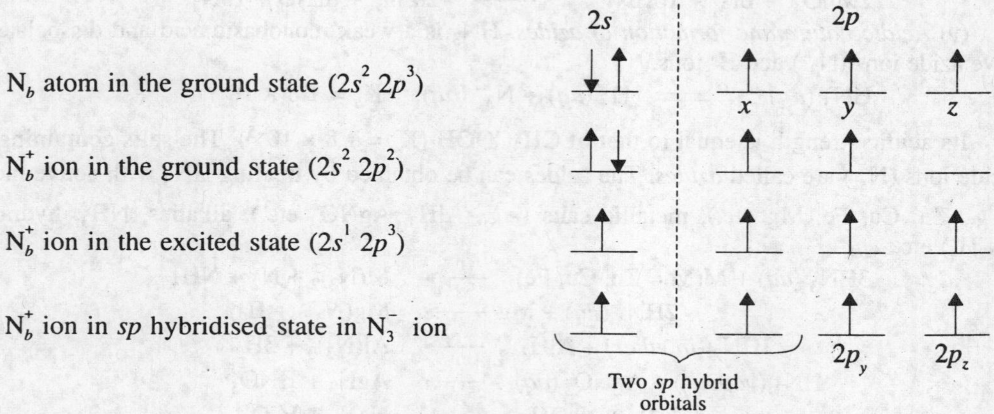

Fig. 25.16 : *sp* hybridisation of N_b^+ in structure (IV), $:\ddot{N}_a = N_b^+ = \ddot{N}:_c$

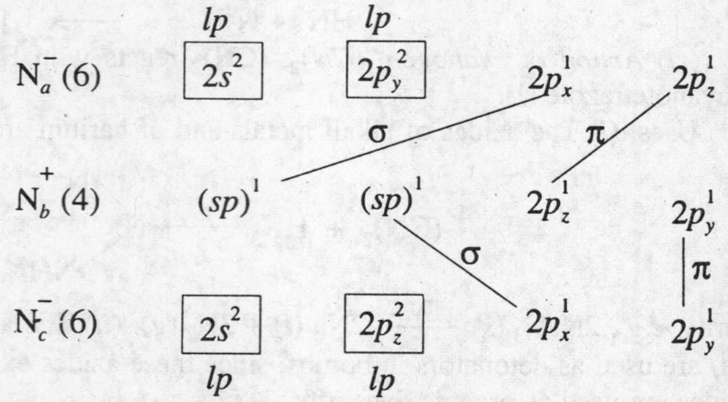

Fig. 25.17 : Formation of σ- and π-bonds in N_3^- ion (*sp* hybridisation).

Hydroxyl amine, NH_2OH

Preparation. NH_2OH can be prepared: (*i*) by reducing $NaNO_2$ with SO_2. A concentrated solution of $NaNO_2$ is mixed with a solution of Na_2CO_3 and SO_2 gas is passed through it till the solution becomes just acidic.

$$Na_2CO_3 + SO_2 + H_2O \longrightarrow NaHSO_3 + NaHCO_3$$

$$3NaHSO_3 + NaNO_2 \longrightarrow (SO_2.ONa)_2N.OH + Na_2SO_3 + H_2O$$

$(SO_2.ONa)_2N.OH$ formed as above is warmed with dil. H_2SO_4 at 90°C so that it undergoes hydrolysis to form NH_2OH.

$$(SO_2.ONa)_2N.OH + 2H_2O \xrightarrow[(90°C)]{\text{dil. } H_2SO_4} NH_2OH + 2NaHSO_4$$

(*ii*) by reducing NO with hydrogen produced by the action of Sn on HCl.

$$NO + 3H \xrightarrow{Sn/HCl} NH_2OH$$

(*iii*) by hydrolysing nitro ethane ($C_2H_5NO_2$) in presence of conc. HCl.

$$C_2H_5NO_2 + H_2O \xrightarrow[\Delta]{\text{conc. HCl}} CH_3COOH + NH_2OH$$

(*iv*) by the electrolytic reduction of HNO_3 in presence of 5% H_2SO_4 at amalgamated lead cathode (cathodic reduction at lead cathode).

$$O_2N.OH + 6H^+ + 6e^- \rightarrow NH_2OH + 2H_2O$$

(*v*) *Anhydrous* NH_2OH is obtained by heating hydroxylamine orthophosphate, $(NH_2OH)_3.H_3PO_4$ to 125°C under low pressure.

$$(NH_2OH)_3.H_3PO_4 \xrightarrow[\text{low pressure}]{125°C} 3NH_2OH + H_3PO_4$$

Properties: Physical. NH_2OH is a white crystalline solid with melting point = 33°C and boiling point = 58°C (at 22 mm pressure). It is deliquescent in air and can be crystallised from its etheral solution.

Chemical. 1. *Stability: Disproportionation of NH_2OH.* It is unstable and hence decomposes slowly above 15°C, giving NH_3 and N_2. Thus, on being heated, NH_2OH undergoes disproportionation.

$$3NH_2OH \xrightarrow{\Delta} NH_3 + N_2 + 3H_2O$$

2. *Action of HNO_2 and HNO_3.* With HNO_2, NH_2OH gives nitrous oxide (N_2O) and with HNO_3, nitric oxide (NO) is obtained.

$$NH_2OH + HNO_2 \longrightarrow N_2O + 2H_2O$$
$$NH_2OH + HNO_3 \longrightarrow 2NO + 2H_2O$$

3. *Monoacidic basic character. Formation of hydroxylammonium salts with acids.* An inspection of the structure of NH_2OH shows that N-atom has one lone pair of electrons which resides in *sp^3 hybrid orbital.* Therefore, when NH_2OH reacts with an acid like HCl, HNO_3, H_2SO_4 etc., the lone pair of electrons is donated to H^+ ion obtained by the ionisation of the acid and $[NH_3OH]^+$ ion (called *hydroxyl ammonium ion*) is obtained. Thus NH_2OH forms the salts with acids which are called *hydroxyl ammonium salts.* In other words, we say that NH_2OH behaves as a *monoacidic base.* The formation of hydroxyammonium salts with HCl, HNO_3 and H_2SO_4 has been shown below:

$$HCl + H_2N-OH \longrightarrow [NH_3OH]^+Cl^-$$
$$HNO_3 + H_2N-OH \longrightarrow [NH_3OH]^+ NO_3^-$$
$$H_2SO_4 + 2NH_2OH \longrightarrow [NH_3OH]_2^{2+} SO_4^{2-}$$

4. *Formation of complex compounds.* Due to the presence of one lone pair of electrons on N-atom, NH_2OH molecule can function as a neutral monodentate ligand and thus can form the complex compounds. $[Pt^{2+}(NH_3)(NH_2OH)(py)(NO_2)]^+$ is an important example of square planar complex compound in which NH_2OH acts as a monodentate ligand. $[Ca(NH_2OH)_6]Cl_3$ is an example of octahedral complex. In complex compounds NH_2OH links to the central metal atom through its N-atom.

5. *Condensation reactions.* NH_2OH reacts with aldehydes and ketones, giving oximes, *e.g.*,

$$\underset{\text{Acetaldehyde}}{CH_3.CH} = \boxed{O + H_2}\,N.OH \xrightarrow{-H_2O} \underset{\text{Acetaldoxime}}{CH_3.CH = N.OH} + H_2O$$

$$\underset{\text{Acetone}}{(CH_3)_2C} = \boxed{O + H_2}\,N.OH \xrightarrow{-H_2O} \underset{\text{Acetoxime}}{(CH_3)_2C = N.OH} + H_2O$$

6. *Action of cyanogen, $(CN)_2$.* Aqueous solution of NH_2OH reacts with $(CN)_2$ and forms oxaldihydroxamide (I).

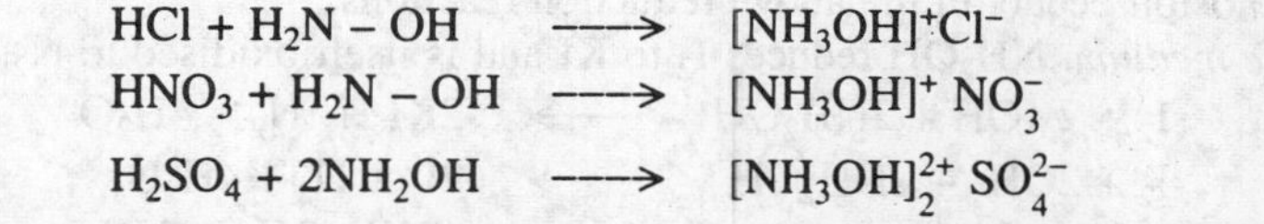

7. *Oxidising properties.* (*a*) *In alkaline medium.* NH_2OH acts as an oxidising agent according

to the equation $\overset{-1}{N}H_2OH \longrightarrow \overset{-3}{N}H_3 + O$ which indicates that NH_2OH is reduced to NH_3. Thus NH_2OH oxidises $Fe(OH)_2$ to $Fe(OH)_3$, nitrites (*e.g.*, $NaNO_2$) to nitrates, arsenites (*e.g.*, Na_3AsO_3) to arsenates and H_2O_2 to O_2.

$$2Fe(OH)_2 + NH_2OH + H_2O \longrightarrow 2Fe(OH)_3 + NH_3$$
$$NaNO_2 + NH_2OH \longrightarrow NaNO_3 + NH_3$$
$$Na_3AsO_3 + NH_2OH \longrightarrow Na_3AsO_4 + NH_3$$
$$H_2O_2 + NH_2OH \longrightarrow H_2O + O_2 + NH_3$$

(*b*) *In acidic medium*, NH_2OH oxidises $FeSO_4$ to $Fe_2(SO_4)_3$ and is itself reduced to NH_4^+ salt.

$$4FeSO_4 + 3H_2SO_4 + \underset{(N = -1)}{2NH_2OH} \longrightarrow 2Fe_2(SO_4)_3 + \underset{(N = -3)}{(NH_4)_2SO_4} + 2H_2O$$

or
$$4Fe^{2+} + 6H^+ + \underset{(N = -1)}{2NH_2OH} \longrightarrow 4Fe^{3+} + \underset{(N = -3)}{2[NH_4]^+} + 2H_2O$$

8. *Reducing properties.* (*a*) *In acidic medium.* NH_2OH acts as a reducing agent according to the equation $2\overset{-1}{N}H_2OH \rightarrow \overset{+1}{N_2}O + H_2O + 4H$ which indicates that NH_2OH is oxidised to N_2O. Thus NH_2OH reduces Fehling solution (CuO) to red ppt. of Cu_2O, halogens (X_2) to halogen acids (HX), $FeCl_3$ to $FeCl_2$ and metallic salts like $AuCl_3$, $PtCl_4$, $HgCl_2$, $AgNO_3$ to free metals.

$$4CuO + 2NH_2OH \longrightarrow 2Cu_2O\ (\textit{red ppt.}) + N_2O + 3H_2O$$
$$2X_2\ (X = Cl, Br) + 2NH_2OH \longrightarrow 4HX + N_2O + H_2O$$
$$4FeCl_3 + 2NH_2OH \longrightarrow 4FeCl_2 + N_2O + H_2O + 4HCl$$
$$4AuCl_3 + 6NH_2OH \longrightarrow 4Au + 3N_2O + 3H_2O + 12HCl$$
$$PtCl_4 + 2NH_2OH \longrightarrow Pt + N_2O + H_2O + 4HCl$$
$$2HgCl_2 + 2NH_2OH \longrightarrow 2Hg + N_2O + H_2O + 4HCl$$
$$4AgNO_3 + 2NH_2OH \longrightarrow 4Ag + N_2O + H_2O + 4HNO_3$$

Generally explosion occurs in the above reduction reactions.

(*b*) *In alkaline medium.* NH_2OH reduces I_2 to KI and is itself oxidised to N_2.

$$I_2 + 2KOH + 2NH_2OH \longrightarrow 2KI + N_2 + 4H_2O$$

or
$$I_2 + 2OH^- + 2NH_2OH \longrightarrow 2I^- + N_2 + 4H_2O$$

9. *Comparison between the base strength of NH_3 and NH_2OH.* In NH_3 molecule N-atom is attached with three H-atoms while in NH_2OH molecule N-atom is attached with two H-atoms and one O-atom of OH group. Since O-atom is more electronegative than H-atom, the base strength of NH_2OH is less than that of NH_3. This fact is also supported by K_B values of these substances (K_B for $NH_2OH \approx 6.0 \times 10^{-6}$, K_B for $NH_3 \approx 10^{-5}$).

Uses. (*i*) NH_2OH and its salts are used as reducing agents (*ii*) NH_2OH is used as a reagent in organic chemistry for the identification of aldehydes and ketones.

Structure. NH_2OH molecule can be regarded as a hydroxy derivative of NH_3. Thus NH_2OH may be obtained by replacing one H-atom in NH_3 molecule by OH group. NH_3 and NH_2OH both have one lone pair of electrons on their N-atom. N-atom which is the central atom is sp^3 hybridised in both the molecules.

N-atom in the ground state ($2s^2\ 2p^3$)
N-atom in sp^3 hybridised state in NH_2OH molecule.

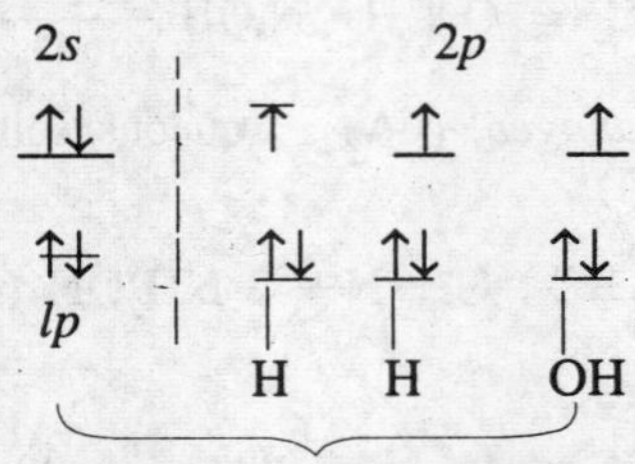

sp^3 hybridisation of N-atom in NH_2OH molecule.

One of the sp^3 hybrid orbitals contains a lone pair of electrons and the remaining three singly-filled sp^3 hybrid orbitals form N — H, N— H and N — OH bonds. Due to the presence of one lone pair of electrons on N-atom, NH_2OH molecule, like NH_3 molecule, has trigonal pyramidal shape (Fig. 25.18).

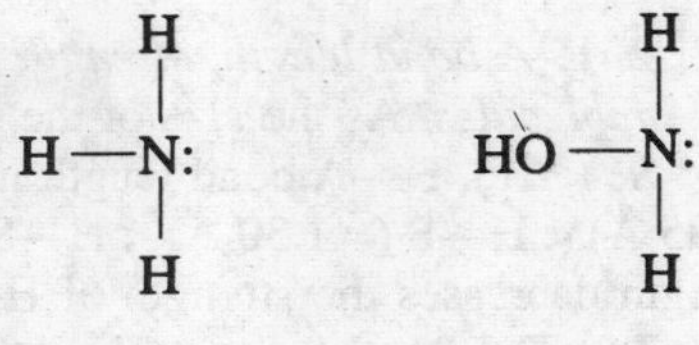

(*a*) (*b*)

Fig. 25.18 : Pyramidal shape of (*a*) NH_3 molecule and (*b*) NH_2OH molecule.

Hydrides of the elements of group VIA

The elements of group VIA give the following types of hydrides. (*a*) *Hydrides of H_2A type (A = O, S, Se, Te, Po).* All the members of this group form hydrides of this type. (*b*) *Hydrides of H_2A_2 type (A = O, S).* It is only oxygen and sulphur which form hydrides of this type.

Comparative study of H_2O, H_2S, H_2Se and H_2Te hydrides

Preparation. H_2O is prepared by burning H_2 in an atmosphere of O_2 while the hydrides of S, Se and Te are obtained by the action of acids on metal sulphides (S^{2-}), selenides (Se^{2-}) and tellurides (Te^{2-}) respectively.

$$FeS + H_2SO_4 \longrightarrow FeSO_4 + H_2S$$
$$K_2Se + H_2SO_4 \longrightarrow K_2SO_4 + H_2Se$$

H_2Po has been prepared only in traces by dissolving Mg foil plated with Po in 0.2 N HCl.

General properties. General properties of the hydrides of H_2A type have been summarised in Table 25.3 and are discussed below:

1. *Physical state.* H_2O exists as a liquid (colourless and odourless) at ordinary temperature while the hydrides of S, Se and Te exist as gases (colourless, poisonous and with unpleasant odour) at room temperature. These are inflammable gases which burn with a blue flame:

Explanation. See "*H-bonding*" discussed in chapter 10.

Table 25.3 : *Some properties of the hydrides of H_2A type (A=O, S, Se, Te).*

Property	H_2O	H_2S	H_2Se	H_2Te
Physical state	Liquid	Gas	Gas	Gas
H—A bond length (Å)	0.95	1.30	1.45	1.72
	Increasing ⟶			
Strength of H—A bond (*i.e.*, H—A bond dissociation energy)	Decreasing ⟶			
Thermal stability of H2A molecules	Decreasing ⟶			
Dissociation temperature (°C)	2000	400–600	150	Room temperature
	Decreasing ⟶			
Readucing power of H2A molecules	Increasing ⟶			
	H_2O is not a reducing agent			
Boiling points (°C) [H2O > H2S < H2Se < H2Te]	100 (Maximum)	– 60.4	– 41.5	– 1.8
Melting point (°C)	0.0	– 85.5	– 65.7	– 51.1
Volatility	($H_2O < H_2S > H_2Se > H_2Te$)			
Ionic character of A–H bond	Decreasing ⟶			
Covalent character of A–H bond	Increasing ⟶			
Tendency of A2A molecules to release H^+ ions	Increasing ⟶			
Acidic strength of H2A molecules	Increasing ⟶			
Dissociation constant as an acid at 25°C	1.0×10^{-14}	1.0×10^{-7}	1.7×10^{-4}	2.3×10^{-3}
	Increasing ⟶			
Shape of H2A molecules	V-shape	V-Shape	V-shape	V-shape
HAH bond angles	104.5°	92.2°	91.0°	89.5°
	Decreasing ⟶			

2. *H–A bond length, strength of H — A bond (H — A bond energy) and thermal stability of H_2A molecules.* As the size of the central atom, A (A = O, S, Se, Te) increases from O to Te (O < S < Se < Te), H—A bond length in H_2A molecules also increases from H—O to H—Te [H—O = 0.95 Å) < H—S (= 1.30 Å) < H—Se (= 1.45 Å) < H—Te (= 1.72 Å)]. The increase in H—A bond length decreases the strength of H—A bond from H—O to H—Te (H—O > H—S > H—Se > H—Te). Due to the successive decrease in the strength of H—A bond from H—O to H—Te, thermal stability of H_2A molecules also decreases from H_2O to H_2Te ($H_2O > H_2S > H_2Se > H_2Te$). This decreasing order of thermal stability is evident from the fact that H_2O dissociates at 2000°C, H_2S dissociates at 400—600°C, H_2Te undergoes dissociation at 150°C while H_2Te gets dissociated even at room temperature.

3. *Reducing power.* We have seen above that thermal stability of the hydrides decreases from H_2O to H_2Te ($H_2O > H_2S > H_2Te$). Due to the successive decrease in thermal stability, the ease with which H_2A molecules can lose H-atoms increases from H_2O to H_2Te and hence the reducing power (reducing power of a molecule is its tendency to lose H atoms) of the given molecules also increases in the same direction, *i.e.,* the reducing power is in the order $H_2O < H_2S < H_2Se < H_2Te$. The order shows that H_2S is a better reducing agent than H_2O. As a matter of fact H_2O never acts as a reducing agent.

4. *Boiling points.* Since as we move down the group from H_2O to H_2Te, the magnitude of van der Waals forces increases and hence the boiling points of the hydrides should increase as $H_2O < H_2S < H_2Se < H_2Te$. But, the boiling point decreases from H_2O to H_2S ($H_2O > H_2S$) and then increases until we reach H_2Te ($H_2S < H_2Se < H_2Te$). Why the boiling point of H_2O is higher than that of H_2S has been explained on the basis of intermolecular H-bonding occurring between several H_2O molecules. See "H-bonding" discussed in chapter 10.

5. *Volatility.* Since boiling point decreases from H_2O to H_2S and then increases from H_2S to H_2Te, the volatility of these hydrides increases from H_2O to H_2S and then decreases. Since H_2O has the highest boiling point, it has the minimum volatility. Abnormal low volatility of H_2O (*i.e.,* high boiling point) is due to its association as a result of hydrogen bonding both in solid and liquid states. As a consequence H_2O has the highest boiling point and melting point among the hydrides. Due to lower electronegativity of S-atom, H-bonding in H_2S is rare and hence the association of H_2S molecule does not occur, but H_2S gas remains as unassociated molecule. This makes H_2S to have more volatility than H_2O. In going down the group the volatility of H_2A hydrides decreases from $H_2S \rightarrow H_2Se \rightarrow H_2Te$ due to the increase in molecular weights of the hydrides.

6. *Nature of H — A bond: Ionic / covalent character of H — A bond in H_2A molecules.* Since the electronegativity values of O, S, Se and Te decreases as we move from O to Te, the value of electronegativity difference, ($x_A - x_H$) (A = O, S, Se, Te) will also decrease in the same direction. Since these values decrease, the amount of ionic character in A — H bonds in H_2A molecules also decreases from H_2O to H_2Te ($H_2O > H_2S > H_2Se > H_2Te$). Thus the covalent character of A — H bonds in H_2A molecules increases from H_2O to H_2Te ($H_2O < H_2S < H_2Se < H_2Te$). H_2O is the least covalent and H_2Te is the most covalent. The increase in covalent character from H_2O to H_2Te can also be explained on the basis of one of the Fajans' rules according to which as the size of A^{-2} ions in H_2A molecules increases from O^{2-} to Te^{2-}, the tendency of these ions to form covalent H_2A molecules increases

7. *Relative order of acidic strength.* Since these hydrides have a tendency to give a proton (H^+), they behave as Bronsted acids, *e.g.,*

$$H_2S \rightleftharpoons H^+ + HS^-; \qquad HS \rightleftharpoons H^+ + S^{2-}$$

The acidic stength (*i.e.,* proton releasing power) increases in the order $H_2O < H_2Se < H_2Se < H_2Te$. This order is confirmed by the increasing order of the values of dissociation constant of these hydrides as acid at 25°C. ($H_2O = 1.0 \times 10^{-14}$, $H_2S = 1.0 \times 10^{-7}$, $H_2Se = 1.7 \times 10^{-4}$, $H_2Te = 2.3 \times 10^{-3}$).

Explanation. The above order can be explained on the basis of the strength of A—H bond (A = O, S, Se, Te) and on the basis of the strength of conjugate acid-base pair. Both these explanations have been discussed in the chapter on "*Modern concept of Acids and Bases* "*(Chapter 19).*

8. Hydrogen bonding and formation of $(H_2A)_x$ *cluster.* Since O-atom has much more electronegativity than H-atom and the electronegativity values of S, Se and Te are close to that of H-atom (O = 3.5, H = 2.1, S = 2.5, Se = 2.5, Te = 2.1), H_2O molecule undergoes intermolecular H-bonding to form its cluster, $(H_2O)_x$ while the occurrence of H-bonding and hence the formation of $(AH_2)_x$ cluster (A—S, Se, Te) are not possible in case of H_2S, H_2Se and H_2Te molecules. Thus H_2O molecule shows H-bonding while other molecules do not show H-bonding.

9. *Shape of AH_3 molecules and variation in HAH bond angles.* All H_2A molecules have *V-shape* (*angular or bent shape*) which results from sp^3 hybridisation of the central atom. The central atom has two lone pairs of electrons and hence the shape of the molecule gets distorted from tetrahedral and becomes **V-shape.** As an example, the structure of H_2O molecule can be studied from chapter 14.

HAH bond angles are H_2O = 104.5°, H_2S = 92.2°, H_2Se = 91.0° and H_2Te = 89.5°. These values decrease as we move down the group. The cause of the decrease has been given under "*Bond Angles*" (Chapter 10).

H_2A_2 type hydrides

H_2O_2 and H_2S_2 are examples of the hydrides of this type. Both these molecules have similar nonpolar structures as shown below:

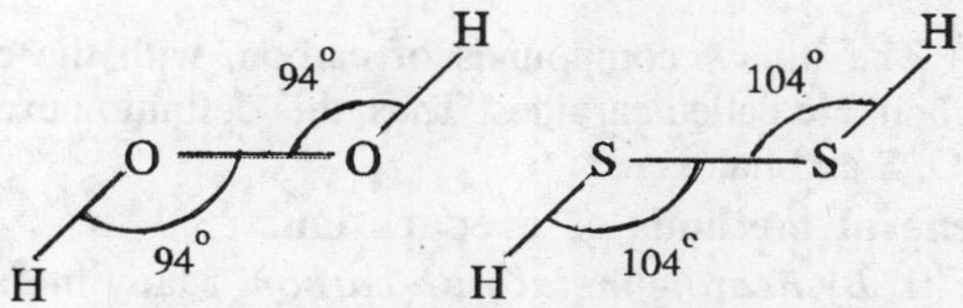

H_2O_2 molecule H_2S_2 molecule

These molecules differ from each other in the following properties. (*i*) H_2O_2 is fairly stable (except in the presence of certain impurities like finely-divided Cu, Pt etc.) while H_2S_2 is unstable. The unstability of H_2S_2 molecule is evident from the fact that this molecule decomposes readily in presence of OH ions to produce S and H_2S ($2H_2S_2 \xrightarrow{OH^-} H_2S + S$). (*ii*) H_2O_2 acts as a strong oxidising agent while H_2S_2 does not. (*iii*) On account of H-bonding occurring in H_2O_2 molecule, this molecule gets associated to give the associated molecule, $(H_2O_2)_x$ while H_2S_2 gives only discrete molecules.

University Questions

1. Give a comparative discussion of the properties of the trihalides of N, P, As, Sb and Bi. *(VSB Purvanchal 2002)*
2 Write methods of preparation, properties and uses of $LiAlH_4$. *(Lucknow 2002)*
3 Write a detailed note on "structure of diborane". *(Lucknow 2002)*
4 Discuss the structure and bonding in diborane (B_2H_6) molecule. *(Nagpur 2002)*
5. Write a note on "Bonding in hydrides". *(Delhi 2002)*
6. (*a*) Give two methods for the preparation of hydrazoic acid.
7. (*b*) Give the reaction of hydrazoic acid with (*i*) Hydrogen; (*ii*) Acidified $KMnO_4$; (*iii*) Hydrochloric
8. acid; (*iv*) Silver nitrate solution. *(Delhi 2002)*
.Discuss the variation in bond angle and reducing properties ot NH_3, PH_3 and ASH_3 molecules. *(VBS Purvanchal 2003)*
9. Explain the structure of diborane *(Nagpur 2003)*
Write note on ortho and para hydrogen *(VBS Purvanchal 2002, 2003)*
10. .What are hydrides and how are they classified ? *(VBS Purvanchal 2003)*
11. Give one method of preparation of the following compounds : (*i*) Diborane; (*ii*) Hydrazine;
12 (*iii*) Hydroxylamine. *(Rohilkhand 2003)*
13. .How is diborane prepared in the laboratory ? Discuss the structure of diborane. *(Rohilkhand 2003)*
Discuss three centered bound in diborane. *(Gujrat 2003)*
13 (*a*) Draw the structure of : (*i*) Hydrazine and (*ii*) Phosphine.
14. (*b*) What are boranes ? Discuss the structure and bonding in diborane. *(Nagphr 2003)*
15. Discuss the bridge structure of diborane. *(Bhopal 2004)*
16. Hydriedes of alkali metals are very reactive towards water. Why? *(GND Amritsar, 2004)*

26

Carbides of Main Group Elements

The binary compounds of carbon, with the elements which are more electropositive than carbon, are called carbides. Thus, this definition excludes the binary compounds of carbon with N, P, O, S and halogens.

General methods of preparation.

1. *By heating metal with carbon.* Many metals like Be, Ag, Mn, Li, Ta, V etc., combine with carbon, at high temperature, to form carbides. For example :

$Be + 2C \longrightarrow BeC_2$; $2Ag + 2C \longrightarrow Ag_2C_2$

$3Mn + C \longrightarrow Mn_3C$; $2Li + C \longrightarrow Li_2C_2$

$3Fe + C \longrightarrow Fe_3C$; $3Mn + C \longrightarrow Mn_3C$

2. *By heating oxides hydroxides, nitrides, phosphates, sulphates etc. with carbon.* For example :

$MgO + 3C \longrightarrow MgC_2 + CO\uparrow$; $CeO_2 + 4C \longrightarrow CeC_2 + 2CO\uparrow$

$2Al_2O_3 + 9C \longrightarrow Al_4C_3 + 6CO\uparrow$; $SiO_2 + 3C \longrightarrow SiC + 2CO\uparrow$

$U_3O_8 + 14C \longrightarrow 3UC_2 + 8CO\uparrow$; $2Mg(OH)_2 + 4C \longrightarrow 2MgC_2 + 2H_2O + O_2$

$Ca_3N_2 + 6C \longrightarrow 3CaC_2 + N_2$; $Ca_3(PO_4)_2 + 14C \longrightarrow 3CaC_2 + 2P + SCO$

$4AlPO_4 + 19C \longrightarrow Al_4C_3 + 4P + 16CO$; $BaSO_4 + 2C \longrightarrow BaC_2 + SO_2 + O_2$

3. *By passing acetylene (C_2H_2) or ethylene (C_2H_4) on heated element or its oxide.* For example :

$4Li + C_2H_2 \longrightarrow Li_2C_2 + 2LiH$; $2Si + C_2H_2 \longrightarrow 2SiC + H_2$

$2Ag_2O + C_2H_2 \longrightarrow Ag_2C_2 + H_2O$; $6Li + C_2H_4 \longrightarrow Li_2C_2 + 4LiH$

4. *By heating the metallic oxide or halide with CaC_2.* For example :

$MgO + CaC_2 \longrightarrow MgC_2 + CaO$; $CuCl_2 + CaC_2 \longrightarrow CuC_2 + CaCl_2$

$MgF_2 + CaC_2 \longrightarrow MgC_2 + CaF_2$

5. *By passing CO over metal, at high temperature.*

$2Li + 2CO \longrightarrow Li_2C_2 + O_2$; $6Al + 3CO \longrightarrow Al_4C_3 + Al_2O_3$

6. *By passing acetylene (C_2H_2) through ammoniacal solution of the metal salt.*

$$2CuCl + 2NH_4OH + C_2H_2 \longrightarrow Cu_2C_2 + 2NH_4Cl + 2H_2O$$

$$2AgNO_3 + 2NH_4OH + C_2H_2 \longrightarrow Ag_2C_2 + 2NH_4NO_3 + 2H_2O$$

7. *By heating metallic carbonate with carbon, in presence of an oxidisable metal.*

$$BaCO_3 + C + 3Mg \longrightarrow 3MgO + BaC_2$$

General properties

(*i*) Generally, carbides are transparent crystalline solids. In the solid state, they are non-conductors of electricity.

(*ii*) *Colour.* Carbides of alkali metals and of Ca, Sr and Ba only are colourless, while most of the remaining carbides are coloured.

(*iii*) *Softness and hardness.* Alkali metal carbides are soft while others are usually hard. For example Be_2C and UC_2 are so hard that they can scratch glass and quartz.

(*iv*) *Explosive nature.* Carbides of U, Cu, Au, Hg (*ic*) etc., are explosive substances, *e.g.*, Hg (*ic*) carbide explodes on rapid heating. Uranium carbide emits sparks when struck and takes fire even when powdered quickly.

(*v*) *Reducing property.* The carbides of alkali metals and of Ca, Sr and Ba are strong reducing agents, *e.g.*, MgO and $MgCl_2$ are reduced to the metals, on heating with CaC_2.

$$3MgO + CaC_2 \longrightarrow 3Mg + CaO + 2CO\uparrow$$

$$MgCl_2 + CaC_2 \longrightarrow Mg + CaCl_2 + 2C$$

(*vi*) *Hydrolysis.* Ionic carbides can easily be hydrolysed by water or dil. acids, with the formation of different types of hydrocarbons.

Classification of carbides

Carbides are classified into four types, depending on the nature of bonding between carbon atoms and other elements. These four types are : (1) *Ionic or salt like carbides.* (2) *Interstitial or metallic carbides.* (3) *Iron type or borderline carbides. (4) Covalent carbides.*

Ionic Carbides

These carbides are given by the metals of groups IA, IIA and IIIA (except boron), coinage metals, Zn,Cd and some lanthanides.

Preparation

These are obtained by heating the metal or its oxide with carbon, CO or a hydrocarbon like acetylene

$$2Be + C \longrightarrow Be_2C; \qquad CaO + 3C \longrightarrow CaC_2 + CO$$

$$4Al + 3C \longrightarrow Al_4C_3\ ; \qquad 2Al_2O_3 + 9C \longrightarrow Al_4C_3 + 6CO$$

$$2Li + 2CO \longrightarrow Li_3C_2 + O_2\ ; \qquad 6Al + 3CO \longrightarrow Al_4C_3 + Al_2O_3$$

$$4Li + C_2H_2 \longrightarrow Li_2C_2 + 2LiH\ ; \qquad 2Ag_2O + C_2H_2 \longrightarrow Ag_2C_2 + H_2O$$

Properties. (*i*) These carbides form transparent, colourless crystals with ionic lattices, containing metal cations in the interstices between carbon anions.

(*ii*) These are non-conductors of electricity.

(*iii*) *Hydrolysis.* These are easily hydrolysed by water or dil. acids, to give different hydrocarbons. Depending on the nature of the hydrocarbon formed during the hydrolysis of ionic carbides by water, these carbides can be classified into the following types :

(*a*) *Acetylides.* These are ionic carbides which give acetylene (H — C≡ C — H) on hydrolysis. These carbides are, therefore, regarded as derivatives of acetylene and are called *acetylides.* Examples of such carbides are BeC_2, MgC_2, CaC_2, $Al_2(C_2)_3$ etc. Calcium carbide ($Ca\dot{C}_2$) which is called *calcium acetylide*, is the most important carbide.

Hydrolysis of CaC_2 has been shown as follows :

$$CaC_2 + 2H_2O \longrightarrow Ca(OH)_2 + H—C\equiv C—H \text{ or } C_2H_2\ (\textit{Acetylene})$$

Some of the carbides, like Cu_2C_2, Ag_2C_2 etc., do not hydrolyse to give acetylene (C_2H_2) and are, therefore, not regarded as true acetylides.

Structure of acetylide ion, C_2^{2-}. Since this ion is regarded as having been derived from acetylene, H—C≡ C—H by removing two H-atoms, its structure is represented as $:\overline{C} \equiv \overline{C}:$ or $(:C\equiv C:)^{2-}$. In this structure, due to the presence of one negative charge, each C-atom has five electrons (instead of four) in its valence-shell ($C = 2s^2p^3$). Due to the presence of one unshared electron pair on each C-atom and one C—C σ-bond , both C-atoms are *sp*-hybridised. Lone pair of electrons resides in one *sp* hybrid orbital. C—C σ-bond results by *sp* — *sp* overlap, while each of the two C—C π-bonds results from *p* — *p* overlap.

(*b*) *Methanides.* These are the ionic carbides, which give methane (CH_4) on hydrolysis. These carbides are regarded as derivatives of methane and hence are said to contain C^{4-} ions. These carbides are called *methanides.*

$$Al_4C_3 + 12H_2O \longrightarrow 4Al(OH)_3 + 3CH_4\uparrow$$
$$Be_4C_2 \text{ or } Be_2C + 4H_2O \longrightarrow 2Be(OH)_2 + CH_4\uparrow$$

Mn_3C probably has a related character, since its hydrolysis also gives CH_4.

$$Mn_3C + 6H_2O \longrightarrow 3Mn(OH)_2 + CH_4 + H_2$$

(*c*) *Allylides.* These are the ionic carbides which give allylene ($H_2C = C = CH_2$) on hydrolysis. These carbides are regarded as derivatives of allylene and hence are said to contain C_3^{-4} ion.

$$Mg_4(C_3)_2 \text{ or } Mg_2C_3 + 4H_2O \longrightarrow 2Mg(OH)_2 + C_3H_4 \text{ or } H_2C = C = CH_2$$

Structure of allylide ion, C_3^{4-}. Since this ion is regarded as a derivative of allylene, $H_2C = C = CH_2$ by removing four H-atoms, its structure is represented as $:\ddot{C}^{2-} = C = \ddot{C}:^{2-}$ or $[:\ddot{C} = C = \ddot{C}:]^{4-}$.

(*d*) *Mixed carbides.* A few carbides, like those of Th and U of the formula MC_2, give, on hydrolysis, a mixture of several hydrocarbons, including acetylene, olefins and H_2. These carbides are related in structure to inter-metallic compounds.

Structure of ionic carbides

Ionic carbides have ionic lattices, in which the metallic cations are packed into the cavities between C-atoms of acetylide (C_2^{2-}), methide (C^{-4}) and allylide (C_3^{-4}) ions.

Interstitial or Metallic Carbides

These carbides are given mostly by transition metals and hence have been discussed in chapter on "*General Characteristics of Transition Elements*".

Borderline Carbides

Fe, Mn, Co, Ni and Cr have atomic radii less than 1.3 (Mn is an exception) and hence octahedral holes in the metallic lattices of these metals are too small to accommodate C-atoms in them, without producing a great deal of distortion in the metallic lattice (or metal structure). Here the metallic lattice is distorted and these carbides are intermediate in properties between the ionic and interstitial carbides, but their radius ratio, namely R_C / R_M, has been found to be equal to 0.60 – 0.61, which place these carbides closer to the interstitial carbides.

These carbides are relatively less stable and liberate a mixture of hydrocarbons and H_2, on hydrolysis with H_2O or acids.

In these carbides, C-atoms are located inside the trigonal prism, formed by metal atoms. The metal lattice is distorted and long chains of C-atoms are running through distorted metal structures. C — C distance is almost equal to 1.65 Å.

Covalent Carbides

The only covalent carbides are those of silicon and boron (SiC and B_4C).

Preparation. SiC and B_4C are prepared by reducing their oxides with carbon in an electric furnace.

$$SiO_2 + 3C \longrightarrow SiC + 2CO\uparrow \; ; \quad 2B_2O_3 + 7C \longrightarrow B_4C + 6CO\uparrow$$

Properties. These are not attacked by H_2O, dil. and conc. acids. These are extremely hard and decompose at high temperatures. Because of their hardness they are used for cutting and as abrasives.

Uses of carbides

(*i*) Many metal carbides have been used as reducing agents in metallurgical and electrothermal process. (*ii*) In general, carbides are very hard and used as efficient abrasives, drilling tools and cutting tools. (*iii*) Carbides can withstand high temperature and hence are used widely in furnace lining. (*iv*) The uses of calcium carbide (CaC_2), boron carbide (B_4C) and aluminium carbide (Al_4C_3) can be seen at their appropriate places. (*v*) Tungsten carbide, WC is used for the manufacture of high speed tools. (*vi*) ZrC is used (*a*) as a component of cemented hard metal cutting tools, (*b*) as an electrode material, for refractory crucibles and as incandescent filament, (*c*) as a coating (1-2 mm thick) to delay the release of fission products from uranium carbide fuel elements.

Some Important Carbides

1. Calcium carbide (also called calcium acetylide), CaC_2

It is an ionic carbide and is represented as $Ca^{2+} (C_2)^{2-}$.

Preparation. This compound is prepared by the action of carbon on Ca_2N_2 or $Ca_3(PO_4)_2$ at 2000°C.

$$Ca_2N_2 + 6C \longrightarrow 3CaC_2 + N_2 \uparrow$$

$$Ca_3(PO_4) + 14C \longrightarrow 3CaC_2 + 2P + 8CO\uparrow$$

Manufacture by Moissan's process. Calcium carbide is manufactured by heating a mixture of calcium oxide (lime) and powdered coke in an electric furnace at a temperature above 2000°C.

$$CaO + 3C \longrightarrow CaC_2 + CO$$

The furnace, which is shown in Fig. 26.1 is made of fire-clay bricks and has a thick layer of graphite at the bottom. This layer of graphite acts as one electrode. A block of graphite hangs down from the top of the furnace and this block acts as another electrode. Lime and coke are introduced into the furnace from the top. Calcium carbide is formed in the molten state and taken out from the tapping hole from time to time.

Production in India. In India CaC_2 is being manufactured in three units namely (*i*) *M/S Birla Jute Manufacturing Co. Ltd., Birlapur (West Bengal)* (*ii*) *M/S Industrial Chemical Ltd.,* Talaiyuthu (Tamil Nadu) and (*iii*) *M/S Calico Mills, Trombay.* Imports have been stopped from 1961 and two other schemes have been licensed for the production of CaC_2, while the existing units are increasing their production.

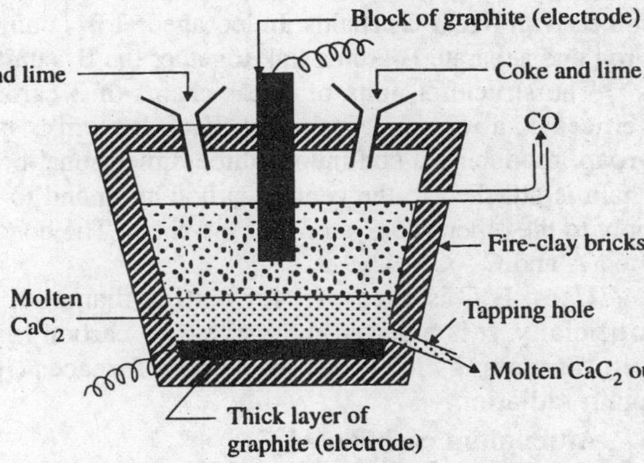

Fig. 26.1 : Manufacture of calcium carbide (CaC_2) by heating a mixture of calcium oxide (lime) and coke (*Moissan's process*).

Properties. (*i*) Pure calcium carbide forms colourless crystals. The commercial product is a hard stone-like mass, having a grey colour. Its melting point is equal to 2300°C.

(*ii*) *Action of water.* It is readily attacked by water to produce acetylene gas and slaked lime, $Ca(OH)_2$.

$$CaC_2 + 2H_2O \longrightarrow Ca(OH)_2 + C_2H_2 \text{ (acetylene)}$$

(*iii*) *Action of nitrogen.* When heated in the atmosphere of nitrogen, it gives calcium cyanamide, $CaCN_2$.

$$CaC_2 + N_2 \longrightarrow CaCN_2 + C$$

The mixture of calcium cyanamide and carbon is sold in the market under the name of *nitrolim* which is used as a nitrogenous fertiliser, since it liberates NH_3 when treated with H_2O.

$$CaCN_2 + 3H_2O \longrightarrow CaCO_2 + 2NH_3.$$

(*iv*) *Reducing property.* CaC_2 is an active reducing agent, since a powdered mixture of CaC_2, Fe_2O_3 and $FeCl_2$ burns violently on ignition and produces fused metallic iron. CaC_2 reduces MgO to Mg metal.

$$3MgO + CaC_2 \longrightarrow 3Mg + CaO + 2CO\uparrow$$

Uses. It is used (*i*) in the manufacture of calcium cyanamide which is employed as a nitrogenous fertiliser and source of NH_3, (*ii*) in the preparation of acetylene which is employed (*a*) in oxy-acetylene blow pipes for welding purposes, (*b*) for illuminating purposes, (*c*) for the manufacture of a large number of organic compounds like acetaldehyde, acetic acid etc., (*iii*) as a reducing agent.

Structure . See *"Structure of AB_2 type ionic crystals"*, discussed in chapter 16.

2. Boron Carbide, B_4C

It is a colvalent carbide. It is prepared : (*i*) by heating boron or boron trioxide (B_2O_3) with carbon, in an electric furnace, at about 2500°C.

$$4B + C \longrightarrow B_4C \qquad 2B_2O_3 + 7C \longrightarrow B_4C + 6CO$$

(*ii*) by dissolving carbon and boron in certain metals like Cu or Ag in an electric furnace. The mass obtained is then treated with HNO_3 to dissolve the metal;

(*iii*) by passing a mixture of H_2 and BCl_3 through the glass globe at 2500°. B_4C formed is deposited on the water cooled brass tubes. In the glass globe, there is carbon strip with brass tubes and water cooling arrangement.

Properties. It forms black lustrous crystals of density 2.52. Its melting point in 5350°C. It is completely resistant to attack to $KClO_3$ and HNO_3 but C is slowly attacked by O_2 and Cl_2 below 1000°C. It is a hard substance, even harder than SiC. It shows considerable electrical conductivity.

Structure. In his publication appearing in 1966. Lipp stated that the unit cell of B_4C contains three B_4C molecules and hence boron carbide should be assigned the formula $B_{12}C_3$. It has a complicated structure, in which icosahedral of 12 boron atoms (*i.e.*, icosahedral B_{12} units) are linked with linear C_3 chains. In icosahedral B_{12} unit, 12 B-atoms link together, to form an icosahedron and separate B-atoms link together the B_{12} units.

The structural units of linear chains of 3 carbon atoms and B_{12} groups are arranged at the vertices of a regular icosahedral. Each B-atom is bonded to five other B-atoms in the same B_{12} group, producing a continuous three dimensional boron network. A carbon at the end of a carbon chain is attached to the central carbon atom and to three B-atoms, the central C-atom is attached only to the other two C-atoms of the chain. The bond distances are B—B = 1.74 – 180 Å, B—C = 1.64 Å and C—C = 1.39 Å.

Uses. B_4C is used : (*i*) for cutting diamond, since it is the hardest abrasive ever made artificially. It is a better abrasive than carbon rundum (SiC) (*ii*) for drilling holes in rockets (*iii*) for making electrodes for electric furnaces (*iv*) for making lamp filaments (*v*) as a shield again radiation.

3. Aluminium carbide, Al_4C_3

It is also an ionic carbide. This carbide is also called *aluminium methanide.*

Preparation. It is prepared : (*i*) by heating a mixture of carbon and aluminium at a temperature above 1000°C.

$$4Al + 3C \longrightarrow Al_4C_3$$

The product is extracted with the cooled conc. HCl which dissolves unreacted Al metal. It is then dried over ether.

(*ii*) by heating alumina (Al_2O_3) or $AlPO_4$ with carbon in an electric furnace or by heating Al_2O_3 with calcium carbide, CaC_2.

$$2Al_2O_3 + 9C \longrightarrow Al_4C_3 + 6CO\uparrow$$
$$4AlPO_4 + 19C \longrightarrow Al_4C_3 + 4P + 16CO\uparrow$$
$$2Al_2O_3 + 3CaC_2 \longrightarrow Al_4C_3 + 3CaO + 3CO\uparrow$$

(*iii*) by the action of CO on Al metal at high temperature.

$$6Al + 3CO \longrightarrow Al_4C_3 + Al_2O_3$$

Properties. (*i*) Al_4C_3 forms yellow hexagonal crystals. It sublimes at 2200°C without decomposition.

(*ii*) It is hydrolysed by H_2O on warming, giving methane.

$$Al_4C_3 + 12H_2O \longrightarrow 4Al(OH)_3 + 3CH_4\uparrow$$

This reaction has been used for the preparation of CH_4.

(*iii*) With N_2 at 1800°C, it forms Al_5C_3N.

(*iv*) With O_2 it gives alumina, Al_2O_3

$$Al_4C_3 + 6O_2 \longrightarrow 2Al_2O_3 + 3CO_2\uparrow$$

Uses. Al_4C_3 is used for the preparation of methane.

Structure. The crystals of Al_4C_3 are regular hexagons. In the space lattice of the crystal of this compound, each Al atom is surrounded by three C-atoms and each C-atom is surrounded by four Al atoms. C—Al distance is 1.90 to 2.22 while the shortest C—C distance is 3.16. Therefore C-atoms are not linked directly to one another.

4. Silicon carbide or silicon silicide, SiC (also called carborundum)

It is also a covalent carbide.

Preparation. It is prepared by heating a mixture of coke and SiO_2 in an electric furnace at 2000°C or by passing acetylene on heated silicon.

$$SiO_2 + 3C \longrightarrow SiC + 2CO; \qquad 2Si + C_2H_2 \longrightarrow 2SiC + H_2$$

Manufacture. SiC is manufactured by Acheson's process. In this process a mixture of sand (54%), coke (34%), sawdust (10%) and salt (2%) is heated in an electric furnace made of fire bricks to 1550-2200°C. The bed of the furnace and the end walls of it are permanent while the side walls are built up with the charge and are pulled down after the completion of the process to take out the prouduct. It is provided at each end with carbon to take out the product. It is provided at each end with carbon electrodes consisting of sixty rods of carbon. A heavy current is passed for 36 hours whereby a high temperature is rapidly reached. At this high temperature, the following reaction occurs resulting in the formation of SiC.

$$SiO_2 + 3C \longrightarrow SiC + 2CO\uparrow$$

For about two hours in the beginning, the e.m.f. gradually decreases from 165 volts to 125 volts and the current increases from 1700 amperes to 6000 amperes, due to the gradual decrease in resistance. These conditions persist for the remaining period of time. At the end of the operation, the side walls are pulled down and the dark coloured mass of black crystals of SiC is crushed and washed successively with H_2SO_4 and NaOH solution, to remove the impurities. It is finally dried in kilns and graded into various portions, according to the size of the particles.

The salt acts as a flux, while saw dust increases the porosity of the charge, which enables a continuous escape of CO, that burns at the top of the charge.

Properties. (*a*) Physical. (*i*) Pure SiC is a colourless crystalline solid. The commercial sample is yellow, green or blue.

(*ii*) It is nearly as hard as diamond.

(*iii*) Of all the known binary compounds of Si, SiC has the highest thermal stability. It decomposes with a loss of silicon only at 2700°C.

(*iv*) It is insoluble in H_2O and alcohol.

(*b*) **Chemical.** (*i*) *Chemical stability.* Chemically, SiC is extremely inert and even at high temperatures, it is not attacked even by HF, HCl, O_2 or S, but it reacts with H_3PO_4. Even a mixture of fuming HNO_3 and HF has no action on SiC.

(*ii*) *Oxidation.* SiC is slightly oxidised by air at about 1000°C. Silica (SiO_2) formed during oxidation,forms a protective layer on the surface of SiC. This layer of SiO_2 stops further oxidation of SiC.

(*iii*) *Action of molten alkalies.* SiC reacts with molten NaOH, in presence of O_2, to form Na_2SiO_3.

$$SiC + 2NaOH + 2O_2 \xrightarrow{\text{Fuse}} Na_2SiO_3 + H_2O + CO_2$$

(*iv*) *Action of Cl_2.* $SiCl_4$ and CCl_4 are formed.

$$SiC + 2Cl_2 \xrightarrow{100° C} SiCl_4 + C \;; \quad SiC + 4Cl_2 \xrightarrow{1000° C} SiCl_4 + CCl_4$$

Uses. (*i*) Due to its extreme hardness, SiC is used as an abrasive for cutting and grinding glasses (*ii*) SiC has very low coefficient of expansion (6×10^{-6}) and very high thermal and chemical stability. It also has high thermal conductivity. Due to these properties, it is used as a refractory material for making crucibles, which are used for melting metals. (*iii*) It has a very high melting point and is, therefore, used in furnace lining. (*iv*) It is used as carbon rods in resistant heaters. (*v*) It is also used for making wheels, hones, wheatstones by mixing SiC with moistened china clay and felspar, moulding under pressure and firing the article in a kiln. (*vi*) It is also used as a deoxidant in metallurgy and as a resistor for electrical furnaces. (*vii*) SiC is used as high-temperature semiconductor in transistor diode rectifiers. (*viii*) It is mostly used in the fabrication of grinding wheels for sharpening metallic tools. (*ix*) SiC is also employed in atomic reactors, since it serves as an excellent container for fuel elements at high temperatures.

Structure. SiC exists in three forms, which are related to one another as diamond, zinc blende and wurtzite. These forms are different combinations of "layers" corresponding to zinc blende (*a*) and wurtzite (*b*) structures. These three forms are : (*i*) *Carborundum I* represented as *aaa bb* (*ii*) *Carborundum II* represented as *a a a b b b* (*iii*) *Carborundeum III* represented as *aa bb*. The structure of each form of SiC has infinite three-dimensional array of silicon and carbon atoms in which each C-atom is tetrahedrally surrounded by four Si-atoms and each Si-atom is surrounded tetrahedrally by four C-atoms. Thus, in a simplified way, the structure of SiC can be depicted as shown in Fig. 26.2.

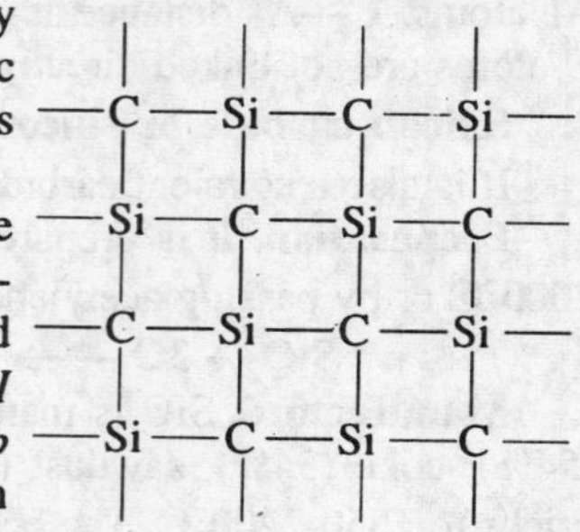

Fig. 26.2 : Structure of SiC.

University Questions

1. Write a general account of carbides, giving their classification, preparation, properties, structure and uses. *(Meerut M.Sc. 94)*
2. Give the preparation, properties, uses and structure of the following
   (*i*) Calcium carbide *(Meerut 95)*
   (*ii*) Boron carbide *(Madras 86)*
   (*iii*) Carborundum *(Rajasthan 87)*
3. The number and type of bonds between two C-atoms in CaC_2 are :
   (*a*) one σ and one π-bonds
   (*b*) one σ and two π-bonds
   (*c*) one σ and one a half π-bonds
   (*d*) one σ-bond
   Ans.: (*b*) *(I.I.T. 96)*
4. Discuss the classification of carbides. Give the action of water on different types of carbides. *(Rohilkhand 90)*
5. Write a short note on "Salt-like carbides". *(Lucknow 2002)*
6. How is silicon carbide prepared ? Give its important applications. *(Nagpur 2003)*
7. What is corundum ? *(Banglore, 2004)*
8. Saw dust is used in the manufacture of silicon carbide. *(Banglore, 2004)*

27

Nitrides of Main Group Elements

The binary compounds of nitrogen, with electropositive elements or with the elements which are less electronegative than nitrogen, are called nitrides.

This definition does not include the salts of hydrazoic acid, HN_3 (*e.g.*, NaN_3). Rather the salts of hydrazoic acid are called *azides*. Thus, NaN_3, AgN_3 etc., are the examples of azides, while Na_3N, Ag_3N etc., are the examples of nitrides.

General methods of preparation

1. *By direct combination of elements.* Combination is brought about, generally, by heating the element in a current of nitrogen to a suitable temperature.

In the case of alkaline earth metals, amalgam may be used in place of the pure metal.

$$6Li + N_2 \longrightarrow 2Li_3N; \qquad 3Mg + N_2 \longrightarrow Mg_2N_2$$

$$2Al + N_2 \longrightarrow 2AlN$$

In some cases, as for instance aluminium and chromium, the combination is effected by striking an electric arc between metal electrodes, in an atomsphere of nitrogen. In the case of the alkali metals, *active nitrogen* is used, in place of ordinary nitrogen.

2. *By heating the element, its oxide or chloride with NH_3.* The elements or their oxides are heated with NH_3 at 300-400°C to get their nitrides. The nitrides of Mg, Zn, Ag, B, Al, Fe, Co, Ni etc., can be obtained by this method.

$$3Mg + 2NH_3 \longrightarrow Mg_3N_2 + 3H_2; \qquad 6Cu + 2NH_3 \longrightarrow 2Cu_3N + 3H_2$$

$$3Fe + 2NH_3 \longrightarrow Fe_3N_2 + 3H_2; \qquad 3ZnO + 2NH_3 \longrightarrow Zn_3N_2 + 3H_2O$$

$$3Ag_2O + 2NH_3 \longrightarrow Ag_3N + 3H_2O; \qquad 3CuO + 2NH_3 \longrightarrow Cu_3N_2 + 3H_2O$$

$$B_2O_3 + 2NH_3 \longrightarrow 2BN + 3H_2O$$

The nitrides of the elements like Ti, Zr, Th, Cr, Mo, W, U, B etc., have been prepared by passing their chloride vapour with NH_3 through a tube, heated to 800°C.

$$BCl_3 + NH_3 \xrightarrow{800°\,C} BN + 3HCl$$

3. *By the action of nitrogen, on a mixture of metallic oxide and carbon.* The reaction is carried out at a high temperature, in an electric furnace. At the high temperature, carbon reduces the oxide to the metal, which reacts with nitrogen, to form the nitride. Nitrides of aluminium, zirconium, scandium, niobium and boron are prepared in this way.

$$Al_2O_3 + 3C + N_2 \longrightarrow 2AlN + 3CO\uparrow; \qquad B_2O_3 + 3C + N_2 \longrightarrow 2BN + 3CO\uparrow$$

$$ZrO + C + N_2 \longrightarrow ZrN_2 + 3CO\uparrow$$

4. *By heating imides and amides.* Imides and amides, on heating, decompose to give the nitrides. For example :

$$3Ba(NH_2)_2 \xrightarrow{\Delta} 4Ba_3N_2 + NH_3\uparrow; \qquad 3Ca(NH_2)_2 \xrightarrow{\Delta} Ca_3N_2 + 4NH_3\uparrow$$

$3Zn(NH_2)_2 \xrightarrow{\Delta} Zn_3N_2 + 4NH_3\uparrow$; $B_2(NH)_3 \xrightarrow{\Delta} 2BN + NH_3\uparrow$

5. *By heating the element with NO.* Under suitable conditions, certain elements combine with NO, to form nitrides.

$$5B + 3NO \longrightarrow BN + B_2O_3$$

6. *By the action of N_2 on hydrides.* The nitrides of certain metals can be prepared by heating their hydrides in a current of N_2.

$3MgH_2 + 2N_2 \longrightarrow Mg_3N_2 + 2NH_3\uparrow$; $3BaH_2 + 2N_2 \longrightarrow Ba_3N_2 + 2NH_3\uparrow$

7. *By the action of a halide on amide.* When halides and amides of certain metals are mixed together, nitride of the metals (whose halide has been taken) is formed.

$$BiCl_3 + 3KNH_2 \longrightarrow BiN + 3KCl + 2NH_3$$

8. *By the action of cyanides on nitrates.* Some nitrides have been obtained by the action of eyanides on nitrates.

$$3Cu(CN)_2 + 12NH_4NO_3 \longrightarrow Cu_3N_2 + 24H_2O + 6CO_2 + 14N_2$$

General Properties

1. *Action of water.* Nitrides react with water to produce ammonia gas. The rate of hydrolysis varies with the nature of the nitride and some of them (*e.g.*, alkaline earth metals, Mg and Zn nitrides) are decomposed even by the moisture present in air and the reaction is quite vigorous with water. Whereas Mn_4N_3 and W_2N_3 are only slowly hydrolysed by water. CrN is not decomposed even by steam at 220°C.

$Li_3N + 3H_2O \longrightarrow 3LiOH + NH_3$; $Mg_3N_2 + 6H_2O \longrightarrow 3Mg(OH)_2 + 2NH_3$

$AlN + 3H_2O \longrightarrow Al(OH)_3 + NH_3$; $BN + 3H_2O \longrightarrow B(OH)_3$ or $H_3BO_3 + NH_3$

2. *Action of hydrogen.* Nitrides are reduced by hydrogen with the formation of hydrides and evolution of ammonia.

$Mg_3N_2 + 6H_2 \longrightarrow 3MgH_2 + 2NH_3$; $CeN + 3H_2 \longrightarrow CeH_3 + NH_3$

3. *Action of acids.* Nitrides are decomposed by dilute acids, producing the salt of the metal atom present in the nitride.

$$2Cu_3N + 8HCl \longrightarrow 2NH_4Cl + 6CuCl$$

$$2CeN + 4H_2SO_4 \longrightarrow Ce_2(SO_4)_3 + (NH_4)_2\,SO_4$$

4. *Action of alkalies.* Caustic alkalies decompose nitrides liberating ammonia gas.

$AlN + 3KOH \longrightarrow K_3AlO_3 + NH_3\uparrow$; $BN + 3KOH \longrightarrow K_3BO_3 + NH_3\uparrow$

5. *Action of halogens.* Nitrides react with halogens at 700°C giving N_2 gas.

$$2AlN + 3Cl_2 \longrightarrow 2AlCl_3 + N_2\uparrow$$

6. *Action of CO and CO_2.* CO and CO_2 decompose nitrides, at about 200°C, to form metal oxides. For example :

$$Mg_3N_2 + 3CO \longrightarrow 3MgO + 3C + N_2$$

$$Mg_3N_2 + 3CO_2 \longrightarrow 3NgO + 3CO + N_2$$

7. *Sublimation.* There are many nitrides which, like oxides, sublime. For example, gallium nitride sublimes without decomposition above 800°C.

Classification of nitrides

The nitrides have been classified into three types depending on the nature of bonding between nitrogen and the other element. These three types are : (1) *Ionic or salt-like nitrides,* (2) *Covalent nitrides,* (3) *Interstitial or metallic or alloy nitrides.*

Ionic Nitrides

These are ionic in nature and contain the nitride ion, N^{3-}. These are given by the elements of groups IA (alkali metals), IIA (alkaline earth metals), Zn, Cd, Cu (*ous*), Ag.

Preparation. Ionic nitrides are prepared : (*i*) by direct union of the elements.

$6Li + N_2 \longrightarrow 2Li_3$; $3Mg + N_2 \longrightarrow Mg_3N$

(*ii*) by passing NH_3 over heated metal or its oxide.

$$3Mg + 2NH_3 \longrightarrow Mg_3N_2 + 3H_2\uparrow \; ; \; 3Ag_2O + 2NH_3 \longrightarrow 2Ag_3N + 3H_2O$$

(*iii*) by decomposing the amide at elevated temperature.

$$3Ba(NH_2)_2 \xrightarrow{\Delta} Ba_3N_2 + 4NH_3$$

Properties. These are crystalline, colourless compounds, having high melting points. These are hydrolysed by H_2O, to form the hydroxides and NH_3.

$$Li_3N + 3H_2O \longrightarrow 3LiOH + NH_3$$
$$Ca_3N_2 + 6H_2O \longrightarrow 3Ca(OH)_2 + 2NH_3$$

When heated in a current of CO_2 at 200°C, N_2 gas is evolved.

$$Mg_3N + 3CO_2 \longrightarrow 3MgO + 3CO + N_2$$

When heated in a current of H_2S, sulphides are obtained.

$$Mg_2N_2 + H_2S \longrightarrow 3MgS + (NH_4)_2S$$

Structure. Ionic nitrides contain metal ions in combination with nitride ion, N^{3-}. The structure of Li_3N is written as $Li^+[Li^+ - N^{3-} - Li^+]^-$. The nitride has a hexagonal structure.

The nitrides of M_3N_2 types, are *anti-isomorphous* with oxides of M_2O_3 type, metal atoms occupying oxygen positions and N-atoms occupying metal positions in the oxide structure.

Covalent Nitrides

These nitrides are formed by the sharing of electrons between nitrogen and the metal or nonmetal. These types of nitrides includes some volatile nitrides of non-metals, such as H, C, Si, P, Cl, F etc., and some nonvolatile nitrides of the elements of group IIIA like B, Al, Ga and In. Thus, the compounds like NH_3, cyanogen and oxides and halides of nitrogen are the examples of volatile covalent nitrides, while BN, AIN, Si_3N_4, P_3N_5 are non-volatile covalent nitrides. The non-volatile nature of the nitrides is attributed to their macromolecular structures. For example, BN has a layer-like structure, which is analogous to that of graphite and crystal structure of AIN is very similar to that of diamond. Non-volatile covalent nitrides undergo hydrolysis, to give NH_3.

$$BN + 3H_2O \longrightarrow B(OH)_3 \text{ or } H_3BO_3 + NH_3 \; ; \quad AIN + 3H_2O \longrightarrow Al(OH)_3 + NH_3$$

Intestitial Nitrides

These nitrides are formed by transition metals and hence have been discussed under "*General Characteristics of Transition Elements.*"

Uses of nitrides

(*i*) BN possesses same hardness as diamond and can withstand temperature of more than 300°C. Due to this property, it is used for coating crucible lining. (*ii*) The formation of AIN during the purification of bauxite by Serpeck's process, is utilised for the manufacture of NH_3. (*iii*) Mg_3N_2 and AlN are decomposed by water, to give NH_3, which is used for the preparation of nitrogenous fertilisers.

Some Important Nitrides

A. Nitrides of Group IA elements

Lithium nitride, Li_3N is obtained by the action of N_2 on Li metal. The reaction occurs slowly at ordinary temperature, but rapidly on heating.

$$6Li + N_2 \longrightarrow 2Li_3N$$

Li_3N is a red hygroscopic solid and gets hydrolysed by H_2O.

$$Li_3N + 3H_2O \longrightarrow 3LiON + NH_3$$

Li_3N has a hexagonal structure with one-third of Li^+ ions having two N-atoms at 1.94 Å and six at 2.11 Å.

Nitrides of sodium, potassium and rubidium. (Na_3N, K_3N and Rb_3N) are formed by striking an electric are between platinum cathode and the alkali metal (Na, K or Rb) as the anode under liquid nitrogen. These nitrides are unstable and decompose with explosion.

B. Nitrides of group IIA elements

All the elements burn in nitrogen to form nitrides, M_3N_2, *e.g.*,

$$3Ca + N_2 \longrightarrow Ca_3N_2$$

These react with H_2O to liberate NH_3 *e.g.*,

$$Ca_3N_2 + 6H_2O \longrightarrow 3Ca(OH)_2 + 2NH_3$$

Be_3N_2 is *volatile* while other nitrides are *not so*. Now let us discuss Be_3N_2, Mg_3N_2, Ca_3N_2 and Ra_3N_2.

1. Beryllium nitride, Be_3N_2.

Preparation. It is prepared by : (*i*) heating the metal above 900°C in the atmosphere of N_2 (*ii*) by heating a mixture of BeO and carbon in N_2, at 1900°C, (*iii*) by reducing Be_2C with N_2 at 1250°C or with NH_3 at 950-100°C.

$$3Be_2C + 2N_2 \longrightarrow 2Be_3N_2 + 3C$$

Properties. It forms white crystalline powder or colourless crystals. It fuses at 2200°C. It is stable in air but is hydrolysed by boiling water or alkalies, liberating NH_3. It is volatile.

$$Be_3N_2 + 3H_2O \longrightarrow 3B(OH)_2 + 2NH_3$$

Uses. It is used as a source of radioactive carbon (C^{14}).

2. Magnesium nitride, Mg_3N_2

Preparation. It is obtained when Mg is heated in N_2 or in NH_3 gas.

$$3Mg + 2NH_3 \longrightarrow Mg_3N_2 + 3H_2$$

Properties. It is a yellow powder and gets decomposed by water, evolving NH_3.

$$Mg_3N_2 + 6H_2O \longrightarrow 3Mg(OH)_2 + 2NH_3$$

3. Calcium nitride, Ca_3N_2

Preparation. It is prepared by heating Ca-metal in an atmosphere of N_2 at 440°C or by heating $Ca(NH_2)_2$.

$$3Ca + N_2 \longrightarrow Ca_3N_2; \qquad 3Ca(NH_2)_2 \longrightarrow Ca_3N_2 + 4NH_3$$

Properties. (*i*) It is a brownish-yellow solid of melting point = 1195°C.

(*ii*) It burns in O_2 giving CaO and N_2.

$$2Ca_3N_2 + 3O_2 \longrightarrow 6CaO + 2N_2$$

(*iii*) It is decomposed by H_2O forming $Ca(OH)_2$ and NH_3.

$$Ca_3N_2 + 6H_2O \longrightarrow 3Ca(OH)_2 + 2NH_3$$

(*iv*) When heated in H_2, it gives CaH_2

$$Ca_3N_2 + 6H_2 \longrightarrow 3CaH_2 + 2NH_3$$

4. Radium nitride, Ra_3N_2

It is prepared by the absorption of N_2 by radium. It is a black compounds.

C. Nitrides of group IIIA element

Boron and aluminium both form nitrides of MN type (M = B, Al). Both these compounds can be obtained by heating the element in the atmosphere of N_2.

$$2B + N_2 \longrightarrow 2BN\ ; \qquad 2Al + N_2 \xrightarrow{800°\ C} 2AlN$$

Both the nitrides are covalent compounds. BN remains unchanged by aqueous acids or alkalies but gets decomposed when fused with alkalies, giving borate (BO_3^{3-}).

$$BN + 3NaOH \xrightarrow{\text{Fuse}} Na_3BO_3 + NH_3$$

It is hydrolysed by steam evolving NH_3.

$$2BN + 3H_2O \longrightarrow B_2O_3 + 2NH_3$$

AlN, unlike BN, is readily hydrolysed by cold water or hot dil. alkalies.

$$AlN + 3H_2O \longrightarrow Al(OH)_3 + NH_3$$
$$AlN + NaOH + H_2O \longrightarrow NaAlO_2 + NH_3$$

Now let us discuss these nitrides in detail.

1. Boron nitride. BN

Preparation. It is formed (*i*) by heating boron to a white heat, in an atmosphere of nitrogen, NO or NH_3.

$$2B + N_2 \longrightarrow 2BN; \qquad 5B + 3NO \longrightarrow 3BN + B_2O_3$$
$$2B + 2NH_3 \longrightarrow 2BN + 3H_2$$

(*ii*) by heating B_2O_3 with $Hg(CN)_2$, KCN or NH_4Cl.

$$B_2O_3 + Hg(CN)_2 \longrightarrow 2BN + CO + CO_2 + Hg$$
$$B_2O_3 + 2KCN \longrightarrow 2BN + K_2O + 2CO$$
$$B_2O_3 + 2NH_4Cl \longrightarrow 2BN + 3H_2O + 2HCl$$

(*iii*) when perfectly anhydrous borax ($Na_2B_4O_7$) is heated with dry NH_4Cl to a red heat in a platinum crucible.

$$Na_2B_4O_7 + 2NH_4Cl \longrightarrow 2BN + 2NaCl + B_2O_3 + 4H_2O$$

The resulted mass is treated with dil. HCl which dissolves B_2O_3 and NaCl, leaving behind BN. It is repeatedly washed with water and dried.

(*iv*) *Pure.* BN is best prepared by heating boron amide, $B(NH_2)_3$ or boron imide, $B_2(NH)_3$ or by the action of NH_3 on BCl_3

$$2B(NH_2)_3 \xrightarrow{\Delta} 2BN + 4NH_3; \qquad B_2(NH)_3 \xrightarrow{\Delta} 2BN + NH_3$$
$$2BCl_3 + 8NH_3 \longrightarrow 2BN + 6NH_4Cl$$

(*v*) BN can also be prepared by passing N_2 gas through a mixture of B_2O_3 and carbon heated in an electric furnace.

$$B_2O_3 + 3C + N_2 \longrightarrow 2BN + 3CO\uparrow$$

(*vi*) When diborane (B_2H_6) reacts with *excess* of NH_3 at high temperature, BN is obtained.

$$B_2H_6 + 2NH_3\ (excess) \longrightarrow 2BN + 6H_2$$

Properties. (*i*) Boron nitride is a white powder of density 2.34. It melts under pressure at 3000°C. It is very stable and unreactive substance. It remains unaffected by mineral acids, solution of alkalies and Cl_2 at red heat.

(*ii*) *Decomposition.* (*a*) It gets decomposed when heated in steam, evolving NH_3.

$$BN + 3H_2O \longrightarrow B(OH)_3 \text{ or } H_3BO_3 + NH_3$$

(*b*) It is decomposed, but solwly, by HF, forming ammonium borofluoride, NH_4BF_4.

$$BN + 4HF \longrightarrow NH_4BF_4$$

(*c*) It is also decomposed when fused with KOH.

$$BN + 3KOH \longrightarrow K_3BO_3 + NH_3\uparrow$$

(*iii*) *Reaction with K_2CO_3.* When BN is fused with K_2CO_3, potassium cyanate (KCNO) and potassium metaborate (KBO_2) are obtained.

$$BN + K_2CO_3 \longrightarrow KCNO + KBO_2$$

Uses. BN possesses same hardness as diamond and can withstand temperature of more than 300°C. Due to this property it is used for coating crucible linings.

Structure. The lattice of boron nitride consists of different layers. Each layer has a hexagonal arrangement of B and N atoms. Different layers are arranged in such a manner that B-atoms in one layer are immediately above the N-atoms in the adjascent layer. The (B—N) distances are equal to 1.45 Å and the distance between the two layers is 8.33 Å. The (B—N) bonds in boron nitride are formed by the overlapping of sp^2 hybrid orbitals of boron and nitrogen atoms. The remaining electrons form π-bonds.

BN molecule is iso-electronic with two adjascent C-atoms in graphite (BN = 3 + 5 = 8, C_2 = 4 + 4 = 8) and hence the structure of $(BN)_n$ is similar to that of graphite, C_n (See Fig. 27.1). In graphite π-electrons are *completely-delocalised* over all the C-atoms of the lattice while in boron nitride the π-electrons are only *partially delocalised* over B-atoms, leading to unequal (B—N) bond lengths. Boron nitride, like graphite, is able to act as host, *i.e.*, it is able to occlude oxides, sulphides, oxy-halides and chlorides in the same way as graphite does. Boron nitride, when heated to 3000°C at 70,000 atm pressure gives a cubic form of boron nitride which has diamond like structure in which all the B and N-atoms attain tetrahedral (four) coordination. This cubic form is superior to diamond, in its mechanical strength and hardness, and suffers only superficial oxidation in air at high temperatures.

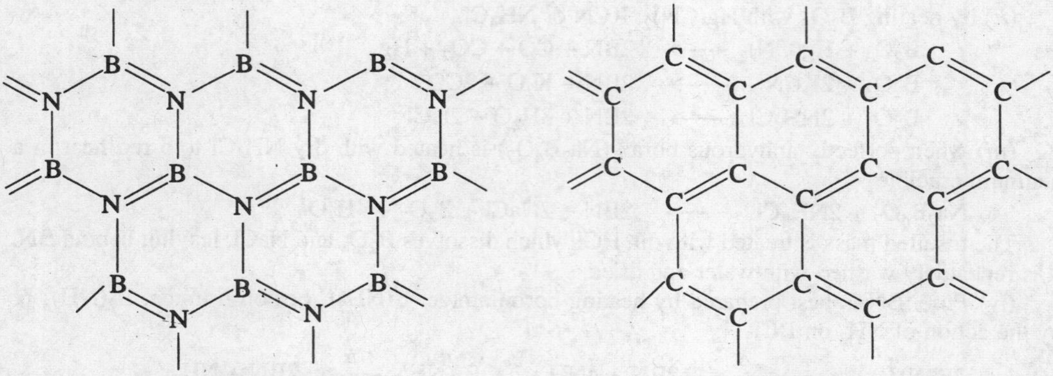

Boron nitride, $(BN)_n$ Graphite, C_n

Fig. 27.1 : Similarity between the structures of boron nitride and graphite.

2. Aluminium nitride, AlN

Preparation. AlN is prepared by heating strongly at 800°C Al-powder in an atmosphere of nitrogen or by heating, at 1800°C, mixed with coke in a current of N_2.

$$2Al + N_2 \xrightarrow{800°C} 2AlN; \qquad Al_2O_3 + 3C + N_2 \xrightarrow{1800°C} 2N + 3CO$$

Properties. It is a colourless crystalline solid which is decomposed by water or alkalies.

$$AlN + 3H_2O \longrightarrow Al(OH)_3 + NH_3$$

It is thermally stable upto 800°C. When heated with HCl at 900°C and with Cl_2 at 300°, it gives AlC_3.

$$AlN + 3HCl \xrightarrow{900°C} AlCl_3 + NH_3; \qquad AlN + 3Cl_2 \xrightarrow{300°C} 2AlCl_3 + N_2$$

Uses. Its formation during the purification of bauxite by Serpeck's process is utilised for the manufacture of NH_3.

Structure. Crystalline AlN has wurtzite structure.

D. Silicon nitride, Si_3N_4

Preparation. SiN_4 is prepared (*i*) by treating $SiCl_4$ with NH_3 and decomposing silicon amide, $Si(NH_2)_4$ by heat.

$$SiCl_4 + 4NH_3 \longrightarrow Si(NH_2)_4 + 4HCl$$

$$3Si(NH_2)_4 \xrightarrow{\Delta} Si_3N_4 + 8\,NH_3$$

(*ii*) by passing N_2 over silicon at 1400°C

$$3Si + 2N_2 \xrightarrow{1400°C} Si_3N_4$$

Properties. It is an extremely stable compound. With alkalies it gives NH_3 and silicate.

Structure. It has the phenacite structure.

E. Nitrides of sulphur

Sulphur forms many nitrides, which are polymeric in nature. These are covalent compounds and have industrial and technical importance. Examples are *tetrasulphur tetranitride* S_4N_4 or

$(SN)_4$; *disulphurdinitride,* S_2N_2. etc. These compounds have been discussed in the chapter on "*Inorganic Polymers*" (chapter 36).

University Questions

1. How are nitrides classified ? Give a brief account of interstitial nitrides.
2. Write the action of H_2O, H_2, acids, alkalies, halogens, CO, CO_2 on nitrides.
3. Describe the preparation, properties, uses and structure of boron nitride. *(Agra 84)*
4. How does BN resemble with and differ from graphite. *(Bharathiar 86)*
5. Explain why Li forms a nitride while other alkali metals donot. *(Delhi 2002)*

28

Oxides of Main Group Elements

Introduction

The binary compounds of oxygen, with other elements, are called oxides. The binary oxygen-fluorine compounds (*e.g.*, OF_2, O_2F_2) are not called oxides of fluorine, but are called fluorides of oxygen, since F is more electronegative than oxygen.

General methods of preparation of oxides

(*i*) By burning the element, like C, P, S, Na, K, Mg etc., in air or O_2. For example :

$$C + O_2 \longrightarrow CO_2 \, ; \qquad S + O_2 \longrightarrow SO_2$$
$$4Na + O_2 \longrightarrow 2Na_2O \, ; \qquad 4K + O_2 \longrightarrow 2K_2O$$
$$2Mg + O_2 \longrightarrow 2MgO$$

(*ii*) By strongly heating carbonates, hydroxides or nitrates. For example :

$$CaCO_3 \xrightarrow{\Delta} CaO + CO_2 \, ; \qquad MgCO_3 \xrightarrow{\Delta} MgO + CO_2$$
$$CuCO_3 \xrightarrow{\Delta} CuO + CO_2 \, ; \qquad 2Al(OH)_3 \xrightarrow{\Delta} Al_2O_3 + 3H_2O$$
$$Cu(OH)_2 \xrightarrow{\Delta} CuO + H_2O \, ; \qquad 2Pb(NO_3)_2 \xrightarrow{\Delta} 2PbO + 4NO_2 + O_2$$
$$2Ba(NO_3)_2 \xrightarrow{\Delta} 2BaO + 4NO_2 + O_2$$

Notes : (*a*) The carbonates and hydroxides of alkali metals do not decompose on heating (*b*) The nitrates of alkali metals give nitrites on heating.

(*iii*) By the oxidation of the non-metal or metalloid with conc. HNO_3. Here HNO_3 acts as an oxidising agent and oxidises the element to its corresponding oxide. HNO_3 is reduced to NO_2 in each case. Examples are given below :

$$6HNO_3 + S \longrightarrow 3H_2O + 6NO_2 + SO_3.$$
$$M\ (M = C, Si) + 4HNO_3 \longrightarrow MO_2 + 4NO_2 + 2H_2O$$
$$2P + 10HNO_3 \longrightarrow P_2O_5 + 18NO_2 + 5H_2O$$
$$I_2 + 10HNO_3 \longrightarrow I_2O_5 + 10NO_2 + 5H_2O$$
$$2As + 10HNO_3 \longrightarrow As_2O_5 + 10NO_2 + 5H_2O$$
$$Sn + 4HNO_3 \longrightarrow SnO_2 + 4NO_2 + 2H_2O$$

(*iv*) By passing steam on red hot element, *e.g.*,

$$3Fe + 4H_2O\ (steam) \longrightarrow Fe_3O_4 + 4H_2 \, ; \qquad C + H_2O\ (steam) \longrightarrow CO + H_2$$

(*v*) By the action of HNO_3 on metals, their oxides, hydroxides or carbonates *e.g.*,

$$Sn + 4HNO_3 \longrightarrow 4NO_2 + SnO_2 + 2H_2O$$
$$2BaCO_3 + 4HNO_3 \longrightarrow 4NO_2 + 2BaO + 2H_2O + 2CO_2 + O_2$$

Types of oxides

Oxides of various elements may be of any of the following types :

1. Acidic oxides. These oxides are generally formed by nonmetals (*e.g.*, B, C, Si, N, P, S, Cl, I etc.) and by metals in higher oxidation state [*e.g.*, Sn (+ 4), V (+ 5), Cr (+ 6), Mn (+ 7)]. Example of acidic oxides are B_2O_3, CO_2, N_2O_3, NO_2, N_2O_5, P_2O_3, P_2O_4, P_2O_5, SO_2, SO_3, Cl_2O_7 (non-metallic oxides), SnO_2, V_2O_5, CrO_3, Mn_2O_7 (metallic oxides). These oxides have the following characteristics :

(*i*) They dissolve in water to form the oxy-acids and hence are also called *acid anhydrides* of the corresponding oxy acids.

$$CO_2 + H_2O \longrightarrow H_2CO_3\ ; \qquad N_2O_3 + H_2O \longrightarrow 2HNO_2$$
$$N_2O_5 + H_2O \longrightarrow 2HNO_3\ ; \qquad P_2O_3 + 3H_2O \longrightarrow 2H_3PO_3$$
$$P_2O_5 + 3H_2O \longrightarrow 2H_3PO_4\ ; \qquad SO_2 + H_2O \longrightarrow H_2SO_3$$
$$SO_3 + H_2O \longrightarrow H_2SO_4\ ; \qquad Cl_2O_7 + H_2O \longrightarrow 2HClO_4$$
$$SnO_2 + H_2O \longrightarrow H_2SnO_3; \qquad CrO_3 + H_2O \longrightarrow H_2CrO_4$$
$$Mn_2O_7 + H_2O \longrightarrow 2HMnO_4$$

Some of the acidic oxides combine with water and form two types of acids. Such oxides are, therefore, called *mixed anhydrides*. Examples are NO_2 and P_2O_4.

$$2NO_2 + H_2O \longrightarrow HNO_2 + HNO_3\ ; \qquad P_2O_4 + 3H_2O \longrightarrow H_3PO_3 + H_3PO_4$$

(*ii*) They combine with bases to form salts and water, *e.g.*,

Acidic oxide		*Base*		*Salt*		*Water*
SiO_2	+	2NaOH	$\longrightarrow$	Na_2SiO_3	+	H_2O
B_2O_3	+	6NaOH	$\longrightarrow$	$2Na_3BO_3$	+	$3H_2O$
P_2O_5	+	6NaOH	$\longrightarrow$	$2Na_3PO_4$	+	$3H_2O$

(*iii*) They combine with basic oxides to form salt.

Acidic oxide		*Basic oxide*		*Salt*
SiO_2	+	CaO	$\longrightarrow$	$CaSiO_3$
SO_3	+	PbO	$\longrightarrow$	$PbSO_4$

(*iv*) These oxides are generally covalent in nature.

2. Basic oxides. These are generally formed by metals. Example are Na_2O, CaO, PbO, CuO, FeO, BaO etc. These oxides have the following characteristics :

(*i*) They dissolve in water to form alkalies, *e.g.*,

$$Na_2O + H_2O \longrightarrow 2NaOH\ ; \qquad CaO + H_2O \longrightarrow Ca(OH)_2$$

(*ii*) They combine with acids to form salts and water, *e.g.*,

$$CuO + H_2SO_4 \longrightarrow CuSO_4 + H_2O\ ; \qquad FeO + H_2SO_4 \longrightarrow FeSO_4 + H_2O$$
$$BaO + 2HCl \longrightarrow BaCl + H_2O$$

(*iii*) They combine with acidic oxides to form salts, *e.g.*,

Basic oxide		*Acidic oxide*		*Salt*
CaO	+	SiO_2	$\longrightarrow$	$CaSiO_3$
PbO	+	SO_3	$\longrightarrow$	$PbSO_4$

(*iv*) These oxides may be ionic or covalent in nature. Mostly they are ionic.

3. Neutral oxides. These oxides combine neither with bases (characteristic of acidic oxides) nor with acids (characteristic of basic oxides) to form salt and water. Examples of such oxides are CO, H_2O, N_2O, NO etc.

4. Amphoteric oxides. These oxides react with bases (acidic character) and acids (basic character) to form salt and water. Examples are ZnO, Al_2O_3, BeO, Sb_2O_3, Cr_2O_3, PbO, PbO_2 etc.

$$Al_2O_3 + 2NaOH \longrightarrow 2NaAlO_2 + H_2O \quad \textit{(Acidic behaviour)}$$
$$Al_2O_3 + 6HCl \longrightarrow 2AlCl_3 + 3H_2O \quad \textit{(Basic behaviour)}$$
$$PbO + 2NaOH \longrightarrow Na_2PbO_2 + H_2O$$
$$PbO + H_2SO_4 \longrightarrow PbSO_4 + H_2O$$
$$ZnO + 2NaOH \longrightarrow Na_2ZnO_2 + H_2O$$

$$ZnO + 2HCl \longrightarrow ZnCl_2 + H_2O$$
$$Cr_2O_3 + 2NaOH \longrightarrow Na_2Cr_2O_4 + H_2O$$
$$Cr_2O_3 + 3H_2SO_4 \longrightarrow Cr_2(SO_4)_3 + 3H_2O$$

5. Mixed or compound metallic oxides. Such oxides are the combination of two simple oxides. These two simple oxides may be of the same metal or of different metal in different oxidation state. Examples of some mixed oxides are : (*a*) Pb_3O_4 (*red lead*). This oxide is a combination of 2PbO and PbO_2. Thus Pb_3O_4 can be represented as $2Pb^{2+}O.Pb^{4+}O_2$. (*b*) Mn_3O_4 (*pyrolusite*) $\equiv 2Mn^{2+}O.Mn^{4+}O_2$. (*c*) Fe_3O_4 (*magnetite*) $\equiv FeO.Fe_2^{3+}O_3$ or $Fe^{2+}Fe_2^{3+}O_4$ (*d*) Mg Al_2O_4 $\equiv MgO.Al_2^{3+}O_3$. (*e*) Zn Fe_2O_4 (*Zinc ferrite*) $\equiv ZnO.Fe_2^{3+}O_3$. (*f*) $MgFe_2O_4$ (*Magnesium ferrite*) $\equiv$ $MgO.Fe_2^{3+}O_3$.

$MgAl_2O_4$ and $ZnFe_2O_4$ have normal spinel structure, while Fe_3O_4 or Fe Fe_2O_4 and $MgFe_2O_4$ have inverse spinel structure.

It may be noted that the mixed oxides, having normal spinel or inverse spinel structure, can be represented by the general formula $A^{2+}B_2^{3+}O_4$. In Fe_3O_4 or $Fe^{2+}Fe_2^{3+}O_4$, $A^{2+} = Fe^{2+}$ and $B^{3+} = Fe^{3+}$, *i.e.,* in Fe_3O_4 A and B are the same metal. For the details of the spinel and inverse spinel structure of the mixed oxides of AB_2O_4 type (See chapter 16).

6. Normal oxides (O^{2-}) . These oxides contain the oxide ion (O^{2-}). Examples of such oxides are Na_2O, MgO, PbO_2 (Pb = + 4), MnO_2 (Mn = + 4) etc.

7. Peroxides (O_2^{2-}). Peroxides contain peroxide ion, O_2^{2-}. This ion can be regarded as having been derived from H_2O_2 molecule by removing both H-atoms. All the alkali metals form peroxides which can be formulated as M_2O_2. Excepting Be all the remaining alkaline earth metals form peroxides of MO_2 type.

Preparation. Peroxides of alkali metals and alkaline earth metals can be prepared :

(*a*) by heating the metal in the presence of excess of O_2, *e.g.,*

$$2Na + O_2\ (excess) \xrightarrow{300°C} Na_2O_2$$

(*b*) by heating the monoxide of the metal alone or in presence of O_2 or air, *e.g.,*

$$2Na_2O \xrightarrow{7400°C} Na_2O_2 + 2Na\ (vapour)\ ; \qquad 2BaO + O_2 \xrightarrow{500°C} 2BaO_2$$

(*c*) by the action of O_2 or air on the metal dissolved in liq. NH_3. Thus method has been used for the preparation of K_2O_2, Rb_2O_2 and Cs_2O_2.

(*d*) by heating dioxides (MO_2). The peroxides of K, Rb and Cs can be obtained by heating carefully the dioxides of these metals.

(*d*) by the action of H_2O_2 on metallic salt solution in presence of an alkali. For example

$$Ca(NO_3)_2 + H_2O_2 + 2NaOH \longrightarrow CaO_2 + 2NaNO_3 + 2H_2O$$
$$BaCl_2 + H_2O_2 + 2KOH \longrightarrow BaO_2 + 2KCl + 2H_2O$$

Properties. (*i*) *Action of dil. acids.* When treated with dil. acids, peroxides give H_2O_2. For example :

$$Na_2O_2 + H_2SO_4\ (dil.) \longrightarrow Na_2SO_4 + H_2O_2$$
$$BaO_2 + H_2SO_4\ (dil.) \longrightarrow BaSO_4 + H_2O_2$$

(*ii*) *Oxidising property.* Peroxides give O_2 on heating and hence act as oxidising agents. For example : (*a*) Na_2O_2 oxidises $Cr(OH)_3$ (Cr = +3) to soluble Na_2CrO_4 (Cr = +6) and Na_2O_2 (O = –1) is reduced to NaOH (O = – 2).

$$3Na_2O_2 + 2Cr(OH)_3 \longrightarrow 2Na_2CrO_4 + 2NaOH + 2H_2O$$

or $$3O_2^{2-} + 2Cr(OH)_3 \longrightarrow 2CrO_4^{2-} + 2OH^- + 2H_2O$$

(*b*) Na_2O_2 oxidises $Mn(OH)_2$ (Mn = +2) to $NaMnO_4$ (Mn = +7).

$$Mn(OH)_2 + 2Na_2O_2 \longrightarrow Na_2MnO_4 + 2NaOH$$

or $$Mn(OH)_2 + 2O_2^{2-} \longrightarrow MnO_4^{2-} + 2OH^-$$

It may be noted that although PbO_2 is a strong oxidising agent, yet it is not a peroxide, since this oxide does not give H_2O_2 on treatment with dil. acids.

(*iii*) *Stability.* The stability of peroxides increases with the increase of the electropositive character of the metal. Generally, peroxides, in the dry state, are more stable than when they are in the solution state.

(*iv*) Many peroxides are highly hydrated due to hydrogen bonding. Examples of hydrated peroxides are $Na_2O_2.8H_2O$, $CaO_2 - 8H_2O$ etc., (octahydrates).

Structure of the peroxides of alkali metals and alkaline earth metals. Alkali metal peroxide molecule, M_2O_2 can be regarded as having been obtained by replacing the two H-atoms of one H_2O_2 molecule, by two alkali metal atoms (M), while the peroxide molecule of an alkaline earth metal (MO_2) can be supposed to have been formed by replacing both H-atoms by one M-atom. Thus, the structure of M_2O_2 molecule, can be written as M — O — O — M (in which M is in + oxidation state) and that of MO_2 molecule can be written as $M\langle$ O | O (in which M is in +2 oxidation state).

Structure of peroxide ion. (*i*) *Structure based on valence-bond theory.* O^{2-} ion has 6 + 6 + 2 = 14 electrons, in its valence–shall and hence its Lewis structure can be written as $:\ddot{O}-\ddot{O}:$

This structure shows that each O-atom has *σ-bps* = 1 and *lps* = 3, *i.e.*, each O^- is surrounded by 1+ 3 = 4 electron pairs and hence is sp^3 hybridised. Thus ,each O^- has three sp^3 hybrid orbitals, three of which contain lone pair of electrons and the fourth is singly-filled.

	2s	2p		
O-atom in its ground state ($2s^2\ 2p^4$)	↑↓	↑↓	↑	↑
O^- in its ground state ($2s^2\ 2p^5$)	↑↓	↑↓	↑↓	↑
O^- in sp^3 hybridised state in O_2^{2-} ion	↑↓	↑↓	↑↓	↑
	lp	*lp*	*lp*	

Now the singly-filled sp^3 hybrid orbitals on both O^- ions overlap together and give rise to the formation of a $\overline{O} - \overline{O}$ σ-bond in O_2^{2-} ion. Three sp^3 hybrid orbitals, having lone pairs of electrons, do not participate in bond formation. Thus, each O-atom in O_2^{2-} ion has three lone pairs of electrons (See Fig. 28.1). This structure has no unpaired electrons. Peroxides are, therefore, diamagnetic.

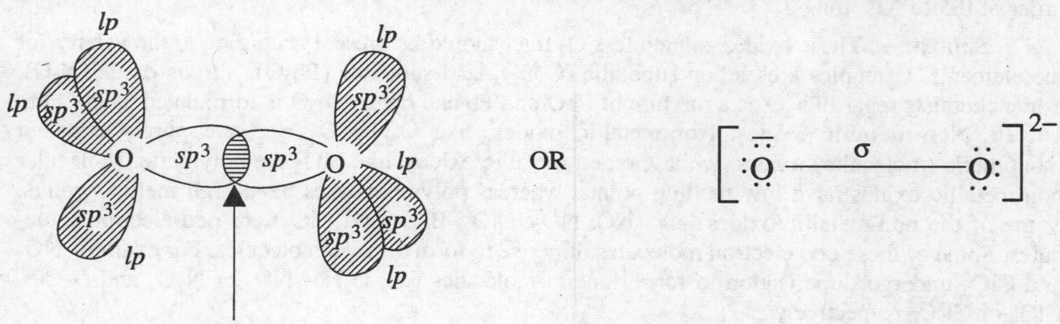

Fig. 28.1 : Structure of peroxide ion, O_2^{2-} on the basis of valence-bond theory. Each of the shaded sp^3 hybrid orbital contains a lone pair of electrons (*lp*).

(*b*) *Structure based on molecular orbital theory.* (See chapter 11). Molecular orbital theory shows that all the 14 electrons present in O_2^{2-} ion are in the paired state and hence this ion is diamagnetic. This theory also reveals the fact that the bond order of O_2^{2-} ion is equal to (8 – 6) / 2 = 1, implying that the two O-atoms in this ion, are linked together by a single σ-bond. Thus, the structure of O_2^{2-} ion is $:\ddot{\bar{O}}—\ddot{\bar{O}}:$

8. Superoxides (O_2^-). Superoxides contain superoxide ion, O_2^-. The known superoxides are KO_2, RbO_2 and CsO_2.

Preparation. These superoxides can be obtained by burning the metal in the excess of O_2 or air.

$$M + O_2\ (excess) \longrightarrow MO_2\ (M = K, Rb, Cs)$$

Properties. (*i*) These are yellow solids. (*ii*) The stability of these superoxides is in the order $KO_2 < RbO_2 < CsO_2$. (*iii*) These are strong oxidising agents. (*iv*) These oxides are hydrolysed by H_2O, forming H_2O_2 and O_2. For example :

$$2KO_2 + 2H_2O \longrightarrow 2KOH + H_2O_2 + O_2$$

or

$$2O_2^- + 2H_2O \longrightarrow 2OH^- + H_2O_2 + O_2$$

(*v*) Superoxides are paramagnetic in nature due to the presence of one unpaired electron in O_2^- ion.

Structure of superoxide ion, O_2^-. Superoxides are ionnic oxides, $M^+O_2^-$. Thus, these oxides contain superoxide ion, O_2^-. The negative charge on the ion is due to the electron transferred from the metal. O—O bond length, in this ion = 125 pm and bond dissociation energy = 394 KJ mol^{-1}. This ion has (6 + 6 + 1) = 13 electrons in its valence-shell.

(*a*) *Structure based on valence-bond theory.* According to valence-bond theory, in this ion, the two O-atoms are linked together by a two-electron bond (normal covalent bond) and a 3-electron bond. Thus, the structure of O_2^- ion can be written as $\left[:\ddot{O}\overset{\cdots}{—}\underset{\cdot}{\overset{\cdot\cdot}{O}}:\right]^-$. The presence of one un-paired electron in 3-electron bond explains the paramagnetic character of the superoxides.

(*b*) *Structure based on molecular orbital theory.* The structure of O_2^- ion has also been explained on the basis of molecular orbital theory (See chapter 11). According to this theory, bond order of this ion is equal to (8 – 5) / 2 = 1.5. The presence of one 3-electron bond gives a bond order of 0.5 to O_2^- ion.

9. Suboxides. These oxides contain less O_2 than should be present according to the valency of the elements. Examples are carbon suboxide (C_3O_2), lead suboxide (Pb_2O), nitrous oxide (N_2O). Some chemists regard Pb_2O as a mixture of PbO and Pb and hence Pb_2O is formulated as PbO.Pb.

10. Non-metallic oxides. Non-metallic oxides, like CO_2, SO_2 etc., are purely covalent monomeric (molecular) oxides, while the non-metallic oxides, like SiO_2 are polymeric. Molecular non-metallic oxides have low melting points, whereas polymeric ones have high melting points. Some of the non-metallic oxides [*e.g.*, NO, NO_2, ClO_2, BrO_2 and ClO_3) are odd-electron molecules. Some of these odd electron molecules dimerise to form dimeric molecules. For example NO_2 and ClO_3 undergo dimerisation to form dimeric molecules *viz.*, $O_2N – NO_2$ or N_2O_4 and $O_3Cl – ClO_3$ or Cl_2O_6 respectively.

11. Non-stoichiometric oxides. In these oxides, the oxide ions have close-packed structure and the holes present in the structure are occupied by the metal ions of suitable size. Oxides of transition metals like Fe, Ti, Mn etc., are nonstoichiometric oxides.

Periodic variation of acidic, basic and amphoteric character of oxides

(*a*) **In a period.** Generally, the oxides of the elements lying in the beginning of the period are strongly basic while those of the elements lying at the end of the period are strongly acidic. Thus, on moving across the period, the basic character first changes into amphoteric and finally into acidic character. This may be illustrated by considering the oxides of the elements of 3*rd* period.

Na_2O	MgO	Al_2O_3	SiO_2	P_2O_5	SO_3	Cl_2O_7
Strongly basic	Less basic	Amphoteric	Weakly acidic	Acidic	Strongly acidic	Even more acidic

(*b*) In a group. On moving down a group it is observed that the acidic character first changes to amphoteric and finally to basic character. This may be illustrated by considering the oxides of group IIIA elements.

B_2O_3	Al_2O_3, Ga_2O_3	In_2O_3	Tl_2O_3
Weakly acidic	Amphoteric	Basic	More basic

The weakly acidic character of B_2O_3 is confirmed by the fact that it reacts with NaOH solution to produce sodium metaborate, $NaBO_2$.

$$B_2O_3 + 2NaOH \longrightarrow 2NaBO_2 + H_2O$$

The amphoteric character of Al_2O_3 is confirmed by its ability to react with both acids and bases.

$$Al_2O_3 + 3H_2SO_4 \longrightarrow Al_2(SO_4) + 3H_2O$$
$$Al_2O_3 + 2NaOH \longrightarrow 2NaAlO_2 + H_2O$$

Explanation. The change from acidic to amphoteric and then to basic character can be explained on the basis of the polarising power of M^{3+} cations ($M^{3+} = B^{3+}, Al^{3+}, Ga^{3+}, In^{3+}$ and Tl^{3+}). For details see "*Applications of Fajans Rules : Consequences of Polarisation of Ions.*"

Stability of oxides

With the exception of halogens, N and noble gases, all other main group elements form at least one oxide, which has a negative value of its equivalent enthalpy of formation and hence is stable. The equivalent enthalpy of formation of an oxide, M_xO_y is obtained by dividing its molecular enthalpy of formation by 2*y*. These values, for various oxides, show that since F_2O has a positive value of equivalent enthalpy of formation, it is the *least stable oxide*. On the other hand, the oxides *viz.,* Li_2O, MgO, Al_2O_3, SiO_2, P_4O_{10}, SO_3 and Cl_2O_7 have negative value of their equivalent enthalpy of formation and hence are *stable oxides.* Greater is the negative value of equivalent enthalpy of formation of an oxide, greater is its stability. On the basis of the values of equivalent enthalpy of formation, the stability of the above oxides has been found in the order : $Li_2O > MgO > Al_2O_3 > SiO_2 > P_4O_{10} > SO_3 > Cl_2O_7$.

Structure of oxides

The structure of oxides depends on the electronegativity difference between the element and oxygen. Thus :

(*i*) If the electronegativity difference is *large,* the oxide generally has *ionic structure.* For example, the oxides of alkali metals (M_2O) are *ionic* and have *antifluorite structure, i.e.,* in the structure of alkali metal oxides, each metal ion is surrounded by four oxide ions and each O^{2-} ion is surrounded by eight metal ions. The oxides of alkaline earth metals (MO) are also *ionic* and have *rock salt-like structure.* Ionic oxides have high melting points and are good conductors in the fused state.

(*ii*) If the electronegativity difference is *small*, the oxides are generally *covalent.* For example, the oxides of higher non-metals (*e.g.,* NO, F_2O etc.) whose electronegativity is almost the same as that of oxygen, exist as discrete molecules and not as ions. These oxides have low melting points.

(*iii*) The oxides of *heavier non-metals* and *semimetals* (*e.g.,* SiO_2, GeO_2, SnO_2, oxides of Sb and Bi) have complex crystal structure in the solid state. For example :

(*a*) CO_2 molecule exists as a discrete molecule, *i.e.,* it has *monomeric structure* (O = C = O). On the other hand, SiO_2 has an *infinite three-dimensional polymeric structure,* in which each Si-atom is linked covalently with four O-atoms. GeO_2 and SnO_2 also have three dimensional structure containing Ge—O and Sn—O covalent bonds respectively.

(*b*) Oxides of N, P and As have simple molecular structures, whereas oxides of Sb and Bi have complex three-dimensional structures containing Sb—O and Bi—O covalent bonds.

(*c*) Oxides of sulphur have simple molecular structures. On the other hand SeO_2 and TeO_2 have infinitely long chain structures.

Oxides of hydrogen and deuterium

Hydrogen (^1_1H) forms two oxides, namely hydrogen oxide (called water), H_2O and hydrogen peroxide, H_2O_2. Deuterium (^2_1D) forms only one oxide, which is called tenterium oxide or heavy water and is formulated as D_2O. Here we shall discuss only H_2O_2 and D_2O in some detail.

Hydrogen peroxide, H_2O_2

Occurrence of H_2O_2

Hydrogen peroxide occurs in minute quantities in air, water, snow and juices of certain plants. It was first prepared by Thenard, a French chemist, in 1818 by the action of dilute acids on barium peroxide.

Laboratory preparation of H_2O_2

H_2O_2 can be prepared in the laboratory by the following methods :

(*i*) By adding Na_2O_2 with stirring to ice-cold dil. H_2SO_4 (20%)

$$Na_2O_2 + H_2SO_4 \longrightarrow Na_2SO_4 + H_2O_2$$

Na_2SO_4 is removed by cooling it when the crystals of $Na_2SO_4.10H_2O$ are obtained.

(*ii*) By passing CO_2 into cold water to which BaO_2 is slowly added (*Merk's method*)

$$BaO_2 + CO_2 + H_2O \longrightarrow BaCO_3\ (ppt.) + H_2O_2$$

(*iii*) By the action of cold dil. H_2SO_4 or H_3PO_4 on hydrated borium oxide, $BaO_2.8H_2O$.

$$BaO_2.8H_2O + H_2SO_4 \longrightarrow BaSO_4 \downarrow + H_2O_2 + 8H_2O$$

If cold dil. H_2SO_4 acts on anhydrous BaO_2, $BaSO_4$ formed gets deposited on the surface of BaO_2, and hence further reaction between BaO_2 and H_2SO_4 stops.

If H_3PO_4 is used in place of H_2SO_4, $Ba_3(PO_4)_2$ is obtained as precipitate. This precipitate is then decomposed by dil. H_2SO_4 to get H_3PO_4 which may again be used.

$$3(BaO_2.8H_2O) + 2H_3PO_4 \longrightarrow Ba_3(PO_4)_2 + 3H_2O_2 + 24H_2O$$

$$Ba_3(PO_4)_2 + 3H_2SO_4 \longrightarrow 3BaSO_4\ (ppt.) + 2H_3PO_4$$

(*iv*) By the hydrolysis of Caro's acid (H_2SO_5). When H_2SO_5 is distilled with dil. H_2SO_4, H_2O_2 is obtained.

$$H_2SO_5 + H_2O \xrightarrow{\text{dil. } H_2SO_4} H_2SO_4 + H_2O_2$$

Manufacture of H_2O_2

(*i*) By the electrolysis *or anodic oxidation of 50-60% solution of H_2SO_4*. When 50-60% solution of H_2SO_4 is electrolysed in a large test tube, in which a glass tube opened at both the ends is placed, H_2SO_4 gets ionised, to give bisulphate ions (HSO_4^-), which unite together at anode (Pt-wire) to form Marshall's acid, $H_2S_2O_8$. H_2 gas is given out at the cathode.

$$H_2SO_4 \rightleftharpoons 2H^+ + 2HSO_4^-\ (Ionisation)$$

$$2HSO_4^- \xrightarrow{\text{Oxidation}} H_2S_2O_8 + 2e^-$$
(*At anode*)

$$2H^+ + 2e^- \xrightarrow{\text{Reduction}} H_2(g)$$
(*At cathode*)

Electrolytic cell. The cell used for the preparation of $H_2S_2O_8$ consists of a fine Pt-wire, which acts as anode (+), while a Cu-coil, wound round the inner glass tube, acts as cathode (–). The large test tube is cooled by keeping it in a beaker containing ice and a current of 1–2 amperes is passed. (See Fig. 28.2).

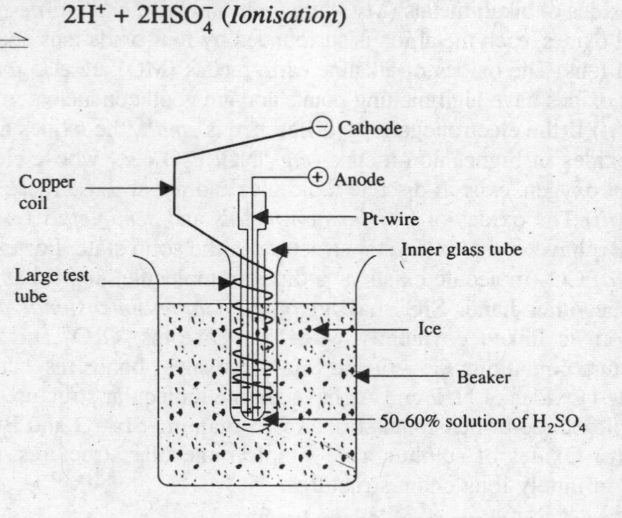

Fig. 28.2 : Preparation of H_2O_2 and $H_2S_2O_8$ by the anodic oxidation of 50-60% solution of H_2SO_4.

Now, when $H_2S_2O_8$ obtained as above at anode is

distilled (heated) with dil. H_2SO_4, a mixture of H_2SO_4 and H_2O_2 is obtained, due to the hydrolysis of $H_2S_2O_8$.

$$H_2S_2O_8 + 2H_2O \xrightarrow{\text{dil. } H_2SO_4} 2H_2SO_4 + H_2O_2 \ \textit{(mixture)}$$

When the mixture containing (H_2SO_4 + H_2O_2) is distilled, H_2O_2 gets separated from the mixture.

(*ii*) *By the electrolysis of NH_4HSO_4 in dil. H_2SO_4*. When a solution of NH_4HSO_4 in dil. H_2SO_4 is electrolysed at 0°C between Pt-electrodes, ammonium peroxodisulphate, $(NH_4)_2S_2O_8$ is obtained at anode.

$$2NH_4HSO_4 \rightleftharpoons 2H^+ + 2NH_4SO_4^- \ (\textit{Ionisation})$$

$$2NH_4SO_4^- \xrightarrow{\text{Oxidation}} (NH_4)_2S_2O_8 + 2e^- \ (\textit{At anode})$$

$$2H^+ + 2e^- \xrightarrow{\text{Reduction}} H_2(g) \ (\textit{At cathode})$$

The steam distillation of $(NH_4)_2S_2O_8$ (obtained as above at anode) in dil. H_2SO_4 at low pressure and high temperature gives $H_2S_2O_8$ which, on hydrolysis, gives H_2O_2.

$$(NH_4)_2S_2O_8 + H_2SO_4 \longrightarrow H_2S_2O_8 + (NH_4)_2SO_4$$

$$H_2S_2O_8 + 2H_2O \xrightarrow{\text{dil. } H_2SO_4} 2H_2SO_4 + H_2O_2$$

(*iii*) *By the oxidation of 2-ethyl anthraquinone (Modern method)*. This is the modern method and is used in America. Here 2-ethyl anthraquinone is reduced by H_2 to 2-ethyl anthraquinol in presence of Pd catalyst. 2-ethyl anthraquinol, on being oxidised by air, produces 2-ethyl anthraquinone along with H_2O_2. When H_2O_2 is distilled with H_2O, 20% solution of H_2O_2 is obtained.

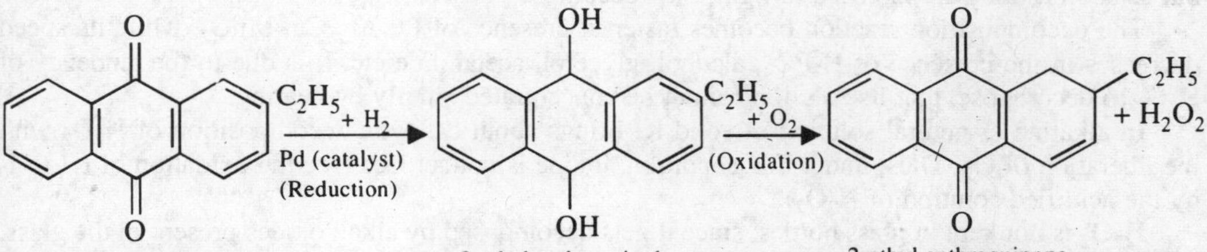

2-ethyl anthraquinone 2-ethyl anthraquinol 2-ethyl anthraquinone

(*iv*) *By the oxidation of 2-propanol, $(CH_3)_2$ CH.OH*. When $(CH_3)_2$ CH.OH is oxidised at high pressure, acetone, $(CH_3)_2CO$ and H_2O_2 are obtained.

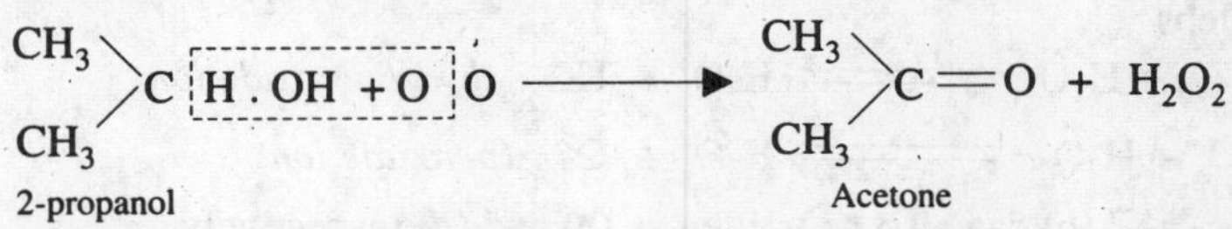

2-propanol Acetone

Preparation of pure H_2O_2 by direct synthesis

Exceptionally pure H_2O_2 can be obtained in small quantity by saturating (H_2 + O_2) mixture (in 19:1 ratio) with H_2O and subjecting it to high voltage and high frequency electric discharge.

Methods for concentrating H_2O_2 solution

Hydrogen peroxide, obtained by the above methods, is always in the form of a dilute solution. The concentration of hydrogen peroxide is a dangerous operation, as impurties, particularly organic material or metallic ions, may catalyse its explosive decomposition. The Germans had many serious explosions with the 80% solution produced by the auto-oxidation method. The dilute solution obtained in any of the above methods is concentrated as follows :

(*i*) *By careful evaporation of the solution.* The dilute solution, obtained as above, is carefully evaporated on a water bath, preferably under reduced pressure and using a fractionating column. Hydrogen peroxide, being appreciably less volatile than water, loses water by evaporation, till a 20% solution is obtained. Since it results in concentration of impurities in the product and many of these impurities catalyse decomposition, it is not advisable to concentrate it beyond about 30 per cent (*ii*) *By distillation under reduced pressure.* The above solution is heated in a distillation flask upto 45°C, the pressure being reduced to 15 mm. At first water comes over, as it is more volatile. At 70°C, hydrogen peroxide begins to distil over. The sample obtained is upto 90% pure.

(*iii*) *By crystallisation.* Further concentration of the peroxide is possible, if solid carbon dioxide and ether are both employed. The whole mass, at first freezes, but if, a little of the frozen solid be dropped in a concentrated solution of the peroxide, needle-shaped crystals of 100% pure hydrogen peroxide separate out.

Physical properties of H_2O_2

Anhydrous H_2O_2 is a colourless, syrupy liquid; in thick layers, it gives a bluish tinge. It produces blisters in contact with skin. It boils at 85°C, under a pressure of 68 mm and forms prismatic crystals at – 2°C. It melts at – 9.91°C. Its density, at 0°C, is 1.64. It is not possible to determine its boiling point due to its decomposition. It is miscible with water.

It is regarded as a feebly acidic substance. Anhydrous H_2O_2, however, does not turn blue litmus red.

Chemical properties of H_2O_2

(1) *Decomposition : Auto-oxidation-reduction of H_2O_2.* H_2O_2 is unstable, since it gets decomposed into H_2O and O_2, either on heating or on long standing.

$$\underset{(O=-1)}{2H_2O_2} \longrightarrow \underset{(O=-2)}{2H_2O} + \underset{(O=0)}{O_2}$$

In the decomposition of H_2O_2 shown above, H_2O_2 (O = –1) disproportionates into H_2O (O = –2) and O_2 (O = 0). The decomposition of H_2O_2 is, therefore, an example of auto-oxidation-reduction (disproportionation) reaction. About 3% solution of H_2O_2 keeps well in dark coloured bottle, but as soon as they are exposed to light, they decompse.

The decomposition reaction becomes faster in presence of Pt, Au, Cu, MnO_2, while its speed decreases in the presence of H_3PO_4, alcohol, glycerol, acetanilide etc. It is due to the tendency of H_2O_2 to decompose, that its solution cannot be concentrated simply by heating.

In alkaline or neutral solution of solid KI brings about catalytic decomposition of H_2O_2 with the liberation of O_2. Thus, under this condition, iodine is not set free (see the oxidation of KI to I_2 by the acidified solution of H_2O_2).

H_2O_2 is not kept in glass bottles, since it gets decomposed by alkali oxides present in the glass. H_2O_2 is, therefore, kept in bottles coated with paraffin wax. H_2O_2 gets decomposed into H_2O and O_2 when it is added to a solution of Na_2CO_3.

(2) *Acidic nature.* The aqueous solution of H_2O_2 acts as a weak acid. It ionises in water in the following two steps :

$$H_2O_2 + H_2O \rightleftharpoons H_3O^+ + HO_2^- \text{ (hydroperoxide ion)} \quad ...(a)$$

$$HO_2^- + H_2O \rightleftharpoons H_3O^+ + O_2^{2-} \text{ (peroxide ion)} \quad ...(b)$$

Equations (*a*) and (*b*) can also be written as (*c*) and (*d*) respectively.

$$H_2O_2 \rightleftharpoons H^+ + HO_2^- \quad ...(c)$$

$$HO_2^- \rightleftharpoons H^+ + O_2^{2-} \quad ...(d)$$

The above equations indicate that H_2O_2 acts as a dibasic acid and hence gives two types of salts *viz., hydroperoxide* (*e.g.,* $NaHO_2$) and *peroxide* (*e.g.,* Na_2O_2). The acidic character of the aqueous solution of H_2O_2 is confirmed by the fact that it reacts with bases like NaOH, $Ba(OH)_2$, Na_2CO_3 etc., to form the salt (peroxide).

Acid		*Base*		*Salt* (*peroxide*)		
H_2O_2	+	2NaOH	$\longrightarrow$	Na_2O_2	+	$2H_2O$
H_2O_2	+	$Ba(OH)_2$	$\longrightarrow$	BaO_2	+	$2H_2O$
H_2O_2	+	Na_2CO_3	$\longrightarrow$	Na_2O_2	+	$H_2O + CO_2$

(3) *Oxidising properties.* H_2O_2 acts as an oxidising agent (electron acceptor) in acidic as well as in alkaline medium. Oxidation is slow in acidic medium but rapid in alkaline solution.

(*a*) *In acidic medium.* In acidic medium, H_2O_2 (O = –1) is reduced to H_2O (O = –2)

$$\underset{(O=-1)}{H_2O_2} \xrightarrow{\text{Reduction}} \underset{(O=-2)}{H_2O} + O \text{ (Molecular equation)}$$

$$\underset{(O=-1)}{H_2O_2 + 2H^+ + 2e^-} \xrightarrow{\text{Reduction}} \underset{(O=-2)}{2H_2O} \text{ (Ionic equation)}$$

The value of E° for the ionic reaction is equal to + 1.77 volt. Thus, in acidic medium H_2O_2 oxidises :

(*i*) *black PbS(S = +2) to white $PbSO_4$ (S = +6)*

$$4H_2O_2 + PbS \longrightarrow PbSO_4 + 4H_2O \text{ (Molecular equation)}$$

$$4H_2O_2 + S^{2-} \longrightarrow SO_2^{2-} + 4H_2O \text{ (Ionic equation)}$$

The above reaction is employed in restoring the colour of old oil paintings, which turn black by the action of atmospheric H_2S on the white pigment, consisting of basic lead carbonate.

(*ii*) *arsenite, AsO_3^{3-} (As = +3) to arsenate, AsO_4^{3-} (As = +5)*

$$Na_3AsO_3 + H_2O_2 \longrightarrow Na_3AsO_4 + H_2O \text{ (Molecular equation)}$$

$$H_2O_2 + AsO_3^{3-} \longrightarrow H_2O + AsO_4^{3-} \text{ (Ionic equation)}$$

(*iii*) *sulphite, SO_3^{2-} (S = + 4) to sulphate, SO_4^{2-} (S = +6)*

$$Na_2SO_3 + H_2O_2 \longrightarrow Na_2SO_4 + H_2O \text{ (Molecular equation)}$$

$$H_2O_2 + SO_3^{3-} \longrightarrow H_2O + SO_4^{2-} \text{ (Ionic equation)}$$

(*iv*) *nitrite, NO_2^- (N = +3) to nitrate, NO_3^- (N = +5)*

$$KNO_2 + H_2O_2 \longrightarrow KNO_3 + H_2O \text{ (Molecular equation)}$$

$$H_2O_2 + NO_2^- \longrightarrow H_2O + NO_3^- \text{ (Ionic equation)}$$

(*v*) *H_2S (S = –2) to S (S = 0)*

$$H_2S + H_2O_2 \longrightarrow S + 2H_2O \text{ (Molecular equation)}$$

$$H_2O_2 + H_2S \longrightarrow S + 2H_2O \text{ (Ionic equation)}$$

(*vi*) *acidified solution of KI (I = –1) to I_2 (I = 0)*

$$2KI + H_2O_2 + H_2SO_4 \longrightarrow K_2SO_4 + I_2 + 2H_2O \text{ (Molecular equation)}$$

$$H_2O_2 + 2H^+ + 2I^- \longrightarrow 2H_2O + I_2 \text{ (Ionic equation)}$$

(*vii*) *ferrous salts to ferric salts ($Fe^{2+} \rightarrow Fe^{3+}$)*

$$2FeSO_4 + H_2SO_4 + H_2O_2 \longrightarrow Fe_2(SO_4) + 2H_2O \text{ (Molecular equation)}$$

$$H_2O_2 + 2Fe^{2+} + 2H^+ \longrightarrow 2H_2O + 2Fe^{3+} \text{ (Ionic equation)}$$

(*viii*) *Hg(Hg = 0) to HgO (Hg = +2) in presence of H_2SO_4*

$$Hg + H_2O_2 \xrightarrow{H_2SO_4} HgO + H_2O$$

(*ix*) *$S_2O_3^{2-}$ (S = +2) to $S_3O_6^{2-}$ (S = +10 / 3) and SO_4^{2-}(S = + 6)*

$$2Na_2S_2O_3 + 4H_2O_2 \longrightarrow Na_2S_3O_6 + Na_2SO_4 + 4H_2O$$

(*x*) *hydrazine, N_2H_4 (N = –2) to N_2 (N = 0) in presence of Cu (II) catalyst.*

$$3N_2H_4(l) + 2H_2O_2(l) \xrightarrow[\text{catalyst}]{\text{Cu (II)}} N_2\,(g) + 4H_2O + \text{Heat.}$$

(*xi*) *C_6H_6 to phenol (C_6H_5OH) in presence of $FeSO_4$*

$$C_6H_6 + H_2O_2 \xrightarrow{FeSO_4} C_6H_5OH + H_2O$$

(*xii*) *acidified solution of $[Fe(CN)_6]^{4-}$ (Fe = +2) to $[Fe(CN)_6]^{3-}$ (Fe = +3)*

$$2K_4\,[Fe\,(CN)_6] + H_2SO_4 + H_2O_2 \longrightarrow 2K_3\,[\,Fe(CM)_6] + K_2SO_4 + 2H_2O \text{ (Molecular equation)}$$

$$H_2O_2 + 2H^+ + 2[Fe(CN)_6]^{4-} \longrightarrow 2H_2O + 2[Fe(CN)_6]^{3-} \text{ (Ionic equation)}$$

(*b*) *In alkaline medium.* In alkaline medium, $H_2O_2(O = -1)$ gets reduced to OH^- ions $(O = -2)$.

$$\underset{(O=-1)}{H_2O_2} + 2e^- \longrightarrow \underset{(O=-2)}{2OH^-} \quad (\textit{Ionic equation})$$

The value of E° for the above reaction is equal to 0.87 volt. Sometimes H_2O_2 (H = +1) is reduced to $H_2(H = 0)$.

$$H_2O_2 \longrightarrow H_2 + 2O$$

Thus H_2O_2 oxidises formaldehyde (HCHO) to formic acid (HCOOH) in presence of pyrogallol

$$\begin{aligned} H_2O_2 &\longrightarrow H_2 + 2O \\ 2HCHO + 2O &\longrightarrow 2HCOOH \\ \hline 2HCHO + H_2O_2 &\longrightarrow H_2 + 2HCOOH \end{aligned}$$

This reaction in the dark is accompanied by the emission of light (chemiluminescence).

(4) *Bleaching properties.* We have seen above that H_2O_2 is a powerful oxidising agent. It is due to its oxidising property that H_2O_2 shows bleaching properties and hence can bleach straw, glue, textiles, fats and oils, silk, wool, hair, starch, leather, ivory etc., and is sold in the market under the name of *perhydrol.* The above substances cannot be bleached by chlorine.

(5) *Action of chlorosulphonic acid, $ClSO_2OH$.* (*a*) When *one* mole of $ClSO_2OH$ reacts with *one* mole of anhydrous H_2O_2, Caro's acid (H_2SO_5) is obtained.

$$H{-}O{-}O{-}\boxed{H + Cl}\,SO_2.OH \xrightarrow{-HCl} \underset{\text{Caro's acid } (H_2SO_5)}{H{-}O{-}O.SO_2.OH} + HCl$$

(*b*) When *two* moles of $ClSO_2OH$ react with *one* mole of anhydrous H_2O_2, Marshall's acid $(H_2S_2O_8)$ is obtained.

$$HO.SO_2.\boxed{Cl + H}{-}O{-}O{-}\boxed{H + Cl}.\,SO_2.\,OH \xrightarrow{-2HCl} 2HCl + \underset{\text{Marshall's acid } (H_2S_2O_8)}{HO.\,SO_2{-}O{-}O{-}SO_2.OH}$$

(6) *Action of diethyl sulphate, $(C_2H_5)_2SO_4$.* When H_2O_2 reacts with diethyl sulphate, diethyl peroxide, $(C_2H_5)_2O_2$ is formed.

$$H{-}O{-}O{-}H + (C_2H_5)_2SO_4 \longrightarrow \underset{\text{Diethyl peroxide}}{H_5C_2{-}O{-}O{-}C_2H_5} + H_2SO_4$$

(7) *Action of Caro's acid, H_2SO_5.* When H_2O_2 is warmed with H_2SO_5, the latter loses O_2 and is converted into H_2SO_4.

$$H_2SO_5 + H_2O_2 \xrightarrow{\Delta} H_2SO_4 + H_2O + O_2$$

(8) *Action of SO_3.* When anhydrous H_2O_2 reacts with SO_3, Caro's acid (H_2SO_5) is obtained.

$$H_2O_2 + SO_3 \longrightarrow H_2SO_5$$

(9) *Action of titanium dioxide (TiO_2) in conc. H_2SO_4 or titanium sulphate, $Ti(SO_4)_2$.* When H_2O_2 is added to the solution of TiO_2 in conc. H_2SO_4 or to the solution of $Ti(SO_4)_2$, blue coloured solution is produced. This colour is due to the formation of pertitanic acid, H_2TiO_4.

$$Ti(SO_4)_2 + H_2O_2 + 2H_2O \longrightarrow H_2TiO_4 + 2H_2SO_4$$

or $$Ti^{+4} + H_2O_2 + 2H_2O \longrightarrow H_2TiO_4 + 4H^+$$

H_2TiO_4, being unstable, decomposes to give bright yellow titanium trioxide, TiO_3.

$$H_2TiO_4 \longrightarrow TiO_3 + H_2O.$$

This reaction is used for the test of H_2O_2.

(10) *Action of acidic solution of K_2CrO_4 or $K_2Cr_2O_7$.* When H_2O_2 is added to a solution of K_2CrO_4 or $K_2Cr_2O_7$ containing dil. H_2SO_4 and a small amount of ether, an intense blue colour is produced which persists in ethereal layer. This blue colour is due to the formation of chromium diperoxide (also called chromic peroxide), $Cr(O_2)_2O$ or CrO_5.

$$K_2CrO_4 + H_2SO_4 + 2H_2O_2 \xrightarrow{\text{Ether}} K_2SO_4 + CrO_5 + 3H_2O$$

or $$CrO_4^{2-} + 2H^+ + 2H_2O_2 \xrightarrow{\text{Ether}} CrO_5 + 3H_2O$$

$$K_2Cr_2O_7 + H_2SO_4 + 4H_2O_2 \xrightarrow{\text{Ether}} K_2SO_4 + 2CrO_5 + 5H_2O$$

or $$Cr_2O_7^{2-} + 2H^+ + 4H_2O_2 \xrightarrow{\text{Ether}} 2CrO_5 + 5H_2O$$

This reaction is used for testing H_2O_2.

(11) *Reducing properties.* H_2O_2 also acts as reducing agent (electron donor) in acidic as well as in alkaline medium.

(*a*) *In acidic medium.* In acidic medium H_2O_2 (O = –1) is oxidised to O_2(O = 0).

$$H_2O_2 \xrightarrow{\text{Oxidation}} O_2 + 2H \text{ (Molecular equation)}$$

$$\underset{(O = -1)}{H_2O_2} \xrightarrow{\text{Oxidation}} \underset{(O = 0)}{O_2} + 2H^+ + 2e^- \text{ (Ionic equation)}$$

Thus is acidic medium, H_2O_2 reduces :

(*i*) *Ag_2O (Ag = +1) to Ag (Ag = 0)*

$$Ag_2O + H_2O_2 \longrightarrow 2Ag + H_2O + O_2$$

(*ii*) *PbO_2 (Pb = +4) to Pb(Pb = 0)*

$$PbO_2 + 2H_2O_2 \longrightarrow Pb + 2H_2O + 2O_2$$

(*iii*) *Au_2O_3 (Au = +3) to Au (Au = 0)*

$$Au_2O_3 + 3H_2O_2 \longrightarrow 2Au + 3H_2O + 3O_2$$

(*iv*) *Halogen (e.g., Cl_2, Br_2) to halogen acid (HCl or HBr) ($X_2^0 \longrightarrow X^-$)*

$$X_2 \text{ (X = Cl, Br)} + H_2O \longrightarrow 2HX + O_2 \text{ (Molecular equation)}$$

$$H_2O_2 + X_2 \longrightarrow O_2 + 2X^- + 2H^+ \text{ (Ionic equation)}$$

(*v*) *acidified solution of MnO_2 (Mn = +4) to Mn (II) salt and Pb_3O_4 (Pb = +8 / 3) to Pb (II salt).*

$$MnO_2 + H_2SO_4 + H_2O_2 \longrightarrow MnSO_4 + 2H_2O + O_2 \text{ (Molecular equation)}$$

$$H_2O_2 + MnO_2 + 2H^+ \longrightarrow Mn^{2+} + 2H_2O + O_2 \text{ (Ionic equation)}$$

$$Pb_3O_4 + 6HNO_3 + H_2O_2 \longrightarrow 3Pb(NO_3)_2 + 4H_2O + O_2 \text{ (Molecular equation)}$$

$$H_2O_2 + Pb_3O_4 + 6H^+ \longrightarrow 3Pb^{2+} + 4H_2O + O_2 \text{ (Ionic equation)}$$

(*vi*) *acidified solution of KnO_4 (Mn = +7) to Mn(II) salt.*

$$2KMnO_4 + 3H_2SO_4 + 5H_2O_2 \longrightarrow K_2SO_4 + 2MnSO_4 + 8H_2O + 5O_2 \text{ (Molecular equation)}$$

$$2MnO_4^- + 5H_2O_2 + 6H^+ \longrightarrow 2Mn^{2+} + 8H_2O + 5O_2 \text{ (Ionic equation)}$$

(*vii*) *hypochlorites, ClO^- (Cl = +1) and hypobromites, BrO^- (Br = +1) to their corresponding halides, Cl^- and Br^- (X = –1).*

$$\underset{(X = Cl, Br)}{NaXO} + H_2O_2 \longrightarrow HX + H_2O + O_2 \text{ (Molecular equation)}$$

$$H_2O_2 + XO^- \longrightarrow O_2 + X^- + H_2O \text{ (Ionic equation)}$$

(*viii*) *O_3 to O_2.* In this reaction H_2O_2 (O = –1) is oxidised to O_2 (O = 0) and O_3(O = 0) is reduced to H_2O (O = –2).

$$H_2O_2 + O_3 \longrightarrow H_2O + 2O_2$$

(*ix*) *hydroxylamine, NH_2OH (N = –1) to NH_3 (N = –3)*

$$H_2O_2 + NH_2OH \longrightarrow O_2 + NH_3 + H_2O$$

(*b*) *In alkaline medium.* In alkaline medium, H_2O_2 (O = –1) is oxidised to O_2 (O = 0)

$$H_2O_2 + O \longrightarrow H_2O + O_2 \text{ (Molecular equation)}$$

$$H_2O_2 + 2OH^- \longrightarrow 2H_2O + O_2 + 2e^- \text{ (Ionic equation)}$$

(*i*) In alkaline medium H_2O_2 reduces $K_3[Fe(CN)_6]$ to $K_4[Fe(CN)_6]$ and is itself oxidised to O_2.

$$H_2O_2 + 2OH^- + 2[Fe(CN)_6]^{3-} \longrightarrow 2H_2O + O_2 + 2[Fe(CN)_6]^{4-} \text{ (Ionic equation)}$$

(*ii*) In alkaline medium, H_2O_2 reduces $KMnO_4$ (Mn = +7) to MnO_2 (Mn = +4) (*ppt.*)

$$2KMnO_4 + 3H_2O_2 \longrightarrow 2MnO_2 + 3O_2 + 2H_2O + 2KOH$$

(12) *Formation of addition compounds.* Hydrogen peroxide forms addition compounds with some inorganic salts, just as water forms hydrates and also with a few organic compounds. For

example, $Na_2HPO_4.H_2O_2$; $(NH_4)_2SO_4.H_2O_2$ and $Co(NH_2).H_2O_2$ (trade name *hyperol* or *ortizon*). They give hydrogen peroxide on the addition of water.

Concentration or strength of H_2O_2 solution

The concentration or strength of H_2O_2 solution is expressed in the following ways :

1. In terms of volume of O_2 gas. The volume of O_2 gas obtained at N.T.P. by heating one volume of H_2O_2 solution is called the concentration of H_2O_2 solution. For example, 1 ml of a 10 volume solution of H_2O_2 will give, on heating, 10 ml of O_2 gas. Similarly 10 ml of a 10 volume solution of H_2O_2 will give $10 \times 10 = 100$ ml of O_2 gas at N.T.P. 10 ml of a 20 volume solution of H_2O_2 will liberate $10 \times 20 = 200$ ml of O_2 gas at N.T.P. Thus, the concentration of 10 volume and 20 volume solution of H_2O_2 is 10 ml and 20 ml respectively.

2. In terms of weight of pure H_2O_2 in a given solution of H_2O_2. The weight of pure H_2O_2 in a given sample of H_2O_2 solution may be expressed in precentage weight (*i.e.*, g / 100 ml) or in g/litre. For example a solution of H_2O_2 whose concentration is 5% has 5 g of pure H_2O_2 in 100 ml of the solution.

Now let us calculate the concentration of a given 10 volume solution of H_2O_2 in terms of percentage weight and in g/litre. We know that :

$$\underset{\substack{2[2+32]\text{g} \\ =68\text{ g}}}{2H_2O_2} \longrightarrow 2H_2O + \underset{\substack{32\text{ g at N.T.P. or} \\ 22400\text{ ml. at N.T.P.}}}{O_2}$$

It is clear form the above equation that 22400 ml of O_2 gas are obtained at N.T.P. by 68 g of 10 volume H_2O_2 solution. It is also known that 1 ml of a 10 volume solution of H_2O_2 gives $1 \times 10 = 10$ ml of O_2 gas at N.T.P. Thus :

$\because$ 22400 ml of O_2 at N.T.P. are obtained from 68 g of pure H_2O_2 present in 1 ml of 10 volume H_2O_2 solution.

$\therefore$ 10 ml of O_2 at N.T.P. are obtained from $\frac{68 \times 10}{22400}$ g = 0.03035 g of pure H_2O_2 present in 1 ml of 10 volume H_2O_2 solution.

$\therefore$ Wt. of pure H_2O_2 in 1 ml of 10 volume H_2O_2 solution = 0.03035 g and hence wt. of pure H_2O_2 in 100 ml of 100 volume H_2O_2 solution.

$$= 0.03035 \times 100 \text{ g}$$
$$= 3.035 \text{ g}$$

$\therefore$ Concentration of 10 volume H_2O_2 solution

$$= 3.035\%$$

or Concentration of 10 volume H_2O_2 solution

$$= 3.035 \times 10 \text{ g / litre}$$
$$= 30.35 \text{ g / litre.}$$

3. As normality. In order to calculate the normality we have to determine the equivalent weight of H_2O_2. This is done as follow :

$$\underset{\substack{2 \times 34 = 68 \text{ parts} \\ \text{by weight}}}{2H_2O_2} \longrightarrow 2H_2O + \underset{\substack{16 \times 2 = 32 \text{ parts} \\ \text{by weight}}}{O_2}$$

Thus :

32 parts by weight of O_2 are obtained from 68 parts by weight of H_2O_2.

$\therefore$ 8 parts by weight O_2 are obtained from $\frac{68 \times 8}{32} =$ 17 parts by weight of H_2O_2.

$\therefore$ Equivalent weight of H_2O_2 = 17

and hence, normality of 10 volume H_2O_2 solution

$$= \frac{\text{Strength of 10 volume } H_2O_2 \text{ solution in g / litre}}{\text{Equivalent weight of } H_2O_2} = \frac{30.35}{17} = 1.785.$$

Uses of hydrogen peroxide

(*i*) It finds use in medicine as an antiseptic for washing wounds, teeth and ears. (*ii*) An an oxidising agent in the laboratory. (*iii*) For bleaching hair, straw, silk, wool and ivory. (*iv*) For restoring the colour of blackened lead paintings. (*v*) As a propellent or fuel in submarines and rockets. Large volume of gaseous products, exothermic nature of its decomposition, and a rapid combustion of fuel in the available oxygen make it a good propellant.

Structure of hydrogen peroxide

Molecular formula. Thenard found that 17 parts by weight of hydrogen peroxide decompose to give 8 parts by weight of oxygen and 9 parts by weight of water. This corresponds to the empirical formula HO for hydrogen peroxide. The molecular weight of hydrogen peroxide, as determined by Victor Meyer method (its V.D. at 90° = 17) and depression of freezing point method is 34. Hence its molecular formula is H_2O_2.

Constitution. The two formulae suggested for hydrogen peroxide are :

(Kingzett's formula, 1884) (I) *(Bayer's formula : dihydroxyl formula, 1900)* (II)

Evidence in favour of structure I. This structure was suggested by Kingzett in 1884. This structure accounts for the oxidising properties of H_2O_2 in which oxygen atom attached by a co-ordinate bond is lost, *i.e.*, one of the oxygen atoms is unstable. Explained on the basis of modern electronic concepts, this structure corresponds to :

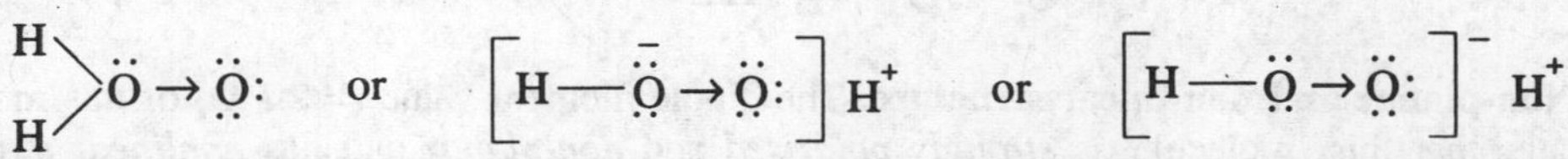

This explains why hydrogen peroxide in some respects behaves as a weak acid, dissociating into H^+ and $[O_2H]^-$ ions.

Evidence in favour of structure II. (*i*) H_2O_2 behaves as a weak dibasic acid and forms two series of salts, *e.g.*, Na—O—O—H and Na—O—O—Na. The formation of such salts indicates that H_2O_2 must contain two hydroxyl groups.

(*ii*) Hydrogen peroxide reacts with diethyl sulphate to give diethyl peroxide which on reduction with zinc and acetic acid gives ethyl alcohol. These reactions are explained only on the basis of formula II.

$$\underset{}{HO—OH} + \underset{\text{Diethyl sulphate}}{(C_2H_5)_2SO_4} \longrightarrow \underset{\text{Diethyl peroxide}}{C_2H_5O—OC_2H_5} + H_2SO_4$$

$$C_2H_5O—OC_2H_5 + 2H \xrightarrow{Zn/CH_3COOH} \underset{\text{Ethyl alcohol}}{2C_2H_5OH}$$

Taking I as the structure of hydrogen peroxide the product should be diethyl ether and not alcohol. As a matter of fact H_2O_2 gives diethyl ether.

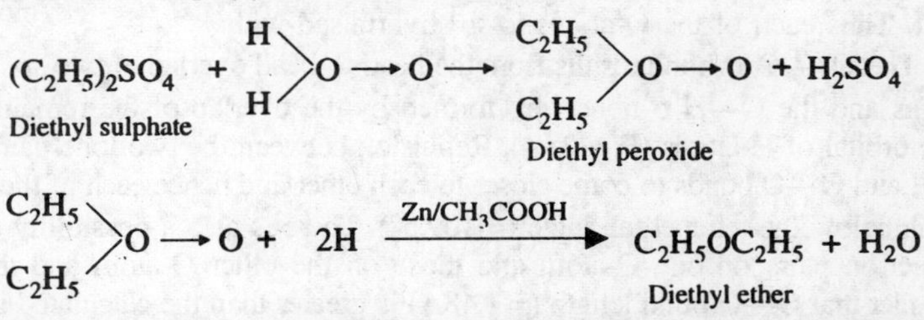

(*iii*) Perdisulphuric acid is formed when anyhydrous hydrogen peroxide reacts with chlorosulphonic acid. Per-disulphuric acid has a peroxide structure, as revealed by the X-ray analysis of the potassium persulphate crystals.

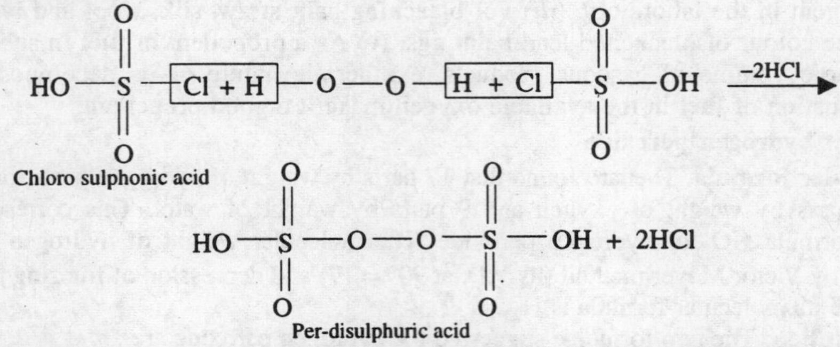

Therefore, hydrogen peroxide, of which perdisulphuric acid is a derivative, must also possess the corresponding dihydroxyl structure.

Equilibrium mixture view. The reported discovery of a new variety of hydrogen peroxide, produced by the action of atomic hydrogen on oxygen, at the liquid air temperature, suggests the existence of molecules of both kinds in hydrogen peroxide, because this form passes into the ordinary form at – 115°C, with some *decomposition into water and oxygen.*

It is suggested, therefore, that hydrogen peroxide is a tautomeric mixture of the two forms in equilibrium in aqueous solution as shown below :

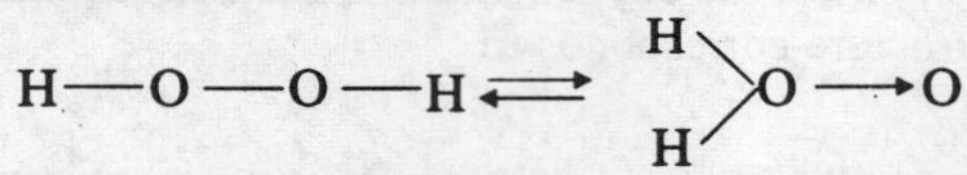

Non-planar and non-linear structure. The dipole moment value (= 2.1 D) of this molecule indicates that this 'molecule' is *strongly polarised* and *non-planar* and has *nonlinear* structure. X-ray study of the solid H_2O_2 molecule has also indicated that this molecule has a *non-planar* and *non-linear* structure in which the two H—O bonds are asymmetrically distributed, *i.e.,* all the four atoms are situated in different planes or are not lying in the same line. According to X-ray study H_2O_2 molecule can be best pictured as shown in Fig. 28.3 in which the two O-atoms can be considered to be lying on the spine of a book opened at an angle of 94°. The H-atoms are placed one on each cover. Each of the H— O—O bond angles is about 97°. The length of O—O bond is about 1.48Å while that of each of O—H bonds is 0.97Å. (See Fig. 28.3 shown in the margin).

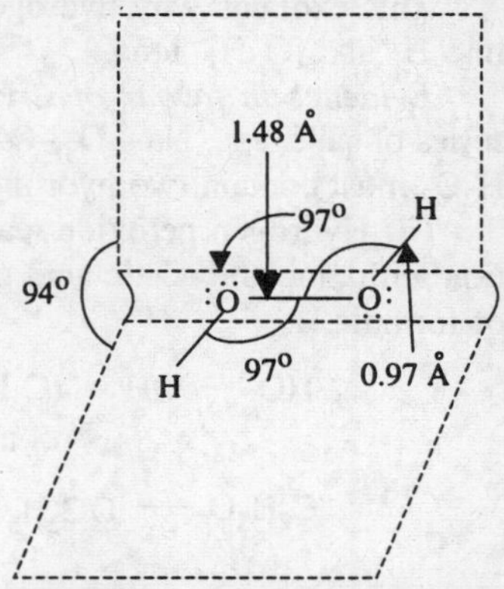

Fig. 28.3 : Nonlinear and nonplanar structure of H_2O_2 molecule.

***sp*^{3} hybridisation in H_2O_2 molecule.** Lewis structure of H_2O_2, $:\mathrm{H}—\ddot{\mathrm{O}}—\ddot{\mathrm{O}}—\mathrm{H}$, indicates that each of the two oxygen atoms is linked with H-atoms by two σ-bonds and has two lone pairs on it. Thus each of the O-atoms is sp^3 hybridised with two lone pairs. The O — O σ-bond results from the head-to-head overlap of two sp^3 hybrid orbitals on both the O-atoms and the O—H σ-bonds are formed by the overlap of the remaining sp^3 hybrid of O-atom with 1*s* orbital of H-atoms (Fig. 28.4). Repulsion between the two lone pairs on each O-atom forces the O—H and O—O bonds to come closer to each other and hence each of the H—O—O-bonds angles (= 97°) is below the tetrahedral angle (= 109.5°). There is also a possibility of some repulsion between the electron pairs on one O-atom and those on the other O-atom and this repulsion may account for the fact that O—O bond length (= 1.48Å) is greater than the calculated single bond O—O distance which is equal to the double of the radius of O-atom (= 1.46 = 2*r*(O) = 2 × 0.73 Å).

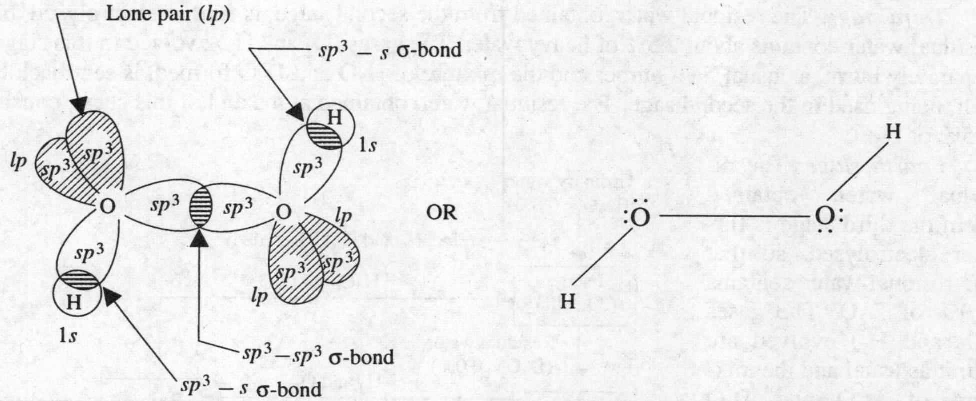

Fig. 28.4 : Structure of H_2O_2 molecule involving sp^3 hybridisation of both oxygen atoms. Each of the shaded sp^3 hybrids contains a lone pair of electrons.

Deuterium oxide or heavy water, D_2O

Occurrence: Washburn and Urey (1932) showed that 6000 parts of ordinary water contains one part of D_2O. D_2O is also found in minute quantities on the leaves of banyan trees, in rain water and in the last remains obtained by the melting of snow by sun on the hills, as in Himalayans.

Preparation. (*i*) *By the multistage electrolysis of water : Taylor, Eyring and Frost process (1933).* D_2O is obtained by the electrolysis of ordinary water, containing calculated quantity of NaOH. The electrolytic cell used in this process is shown in Fig. 28.5 and is, as designed by Brown, Degget and Urey. It consists of a steel cell (45 cm long and 10 cm in diameter). The cell itself acts as a cathode while a cylindrical sheet of nickel having a number of holes punched into it (*i.e.,* a perforated cylin drical sheet of nickel) serves as the anode.

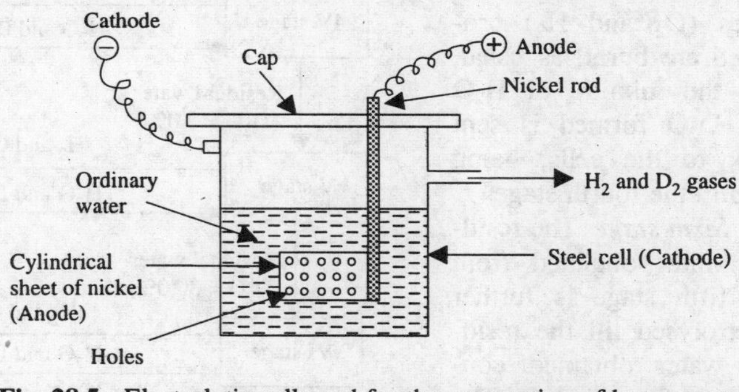

Fig. 28.5 : Electrolytic cell used for the preparation of heavy water.

In actual practice a large number of such cells is used so that the electrolysis of ordinary water may be done in several stages *i.e., the electrolysis of water is continuous and prolonged.* Thus by the continuous and prolonged electrolysis, the concentration of heavy water goes on increasing gradually in the residual water left behind at the end of each stage of electrolysis. Usually the process of electrolysis is complete in seven stages as discussed below :

First stage. Thirty electrolytic cells, joined together in series, are used in this stage. Each cell is charged with 77 litres of a 8% solution of NaOH in ordinary water, *i.e.,* 2310 litres of ordinary water are electrolysed in this stage. A current of 110 volts is used and the electrolysis is carried out for about 72 hours. The gases liberated during the electrolysis at the end of this stage contain protium (H_2) and O_2 which are discarded. At the end of 72 hours, the total volume of the charge reduces to about 1/6th of its original volume. The alkali present is neutralised by passing CO_2 in it.

Second stage. The residual water obtained from the first stage, is again electrolysed, using six cells. the gases *viz.,* O_2 and H_2 and evolved during electrolysis in this stage, are separately burnt in a burner and the mixture of H_2O and D_2O formed by the burning of gases, is sent back to the cells, being used in the first stage. The process of electrolysis in this stage is continued till the residual water contains about 0.5% of D_2O.

Third stage. The residual water, obtained from the second stage, is further electrolysed, till the residual water contains about 2.5% of heavy water. The gases (O_2 and H_2) evolved in this stage, are separately burnt, as usual, in a burner and the mixture of H_2O and D_2O formed, is sent back to the cell, being used in the second sage. The residual water, obtained at the end of this stage, consists of 2.5% of D_2O.

Fourth stage. The residual water, obtained from the third stage is further electrolysed, so that the residual water contains 8.0% of D_2O. The gases (O_2 and H_2) evolved are burnt as usual and the mixture of H_2O and D_2O formed is sent back to the cells, being used in the third stage.

Fifth stage. The residual water obtained from the fourth stage is further electrolysed, so that the residual water contains about 30.0% of D_2O. The gases (O_2 and H_2) produced are burnt, as usual, and the mixture of H_2O and D_2O formed is sent back to the cells, being used in the fourth stage.

Sixth stage. The residual water obtained from the fifth stage is further electrolysed till the residual water obtained contains 93.0% of D_2O. The gases (O_2 and H_2) produced are burnt as usual and the mixture of H_2O and D_2O obtained is sent back to the cells being used in fifth stage.

Seventh stage. The residual water obtained from the sixth stage is further electrolysed so that the residual water obtained at the end of this stage contains 99% of D_2O. The gases (O_2 and H_2) produced here are burnt as usual and the mixture of H_2O and D_2O obtained is

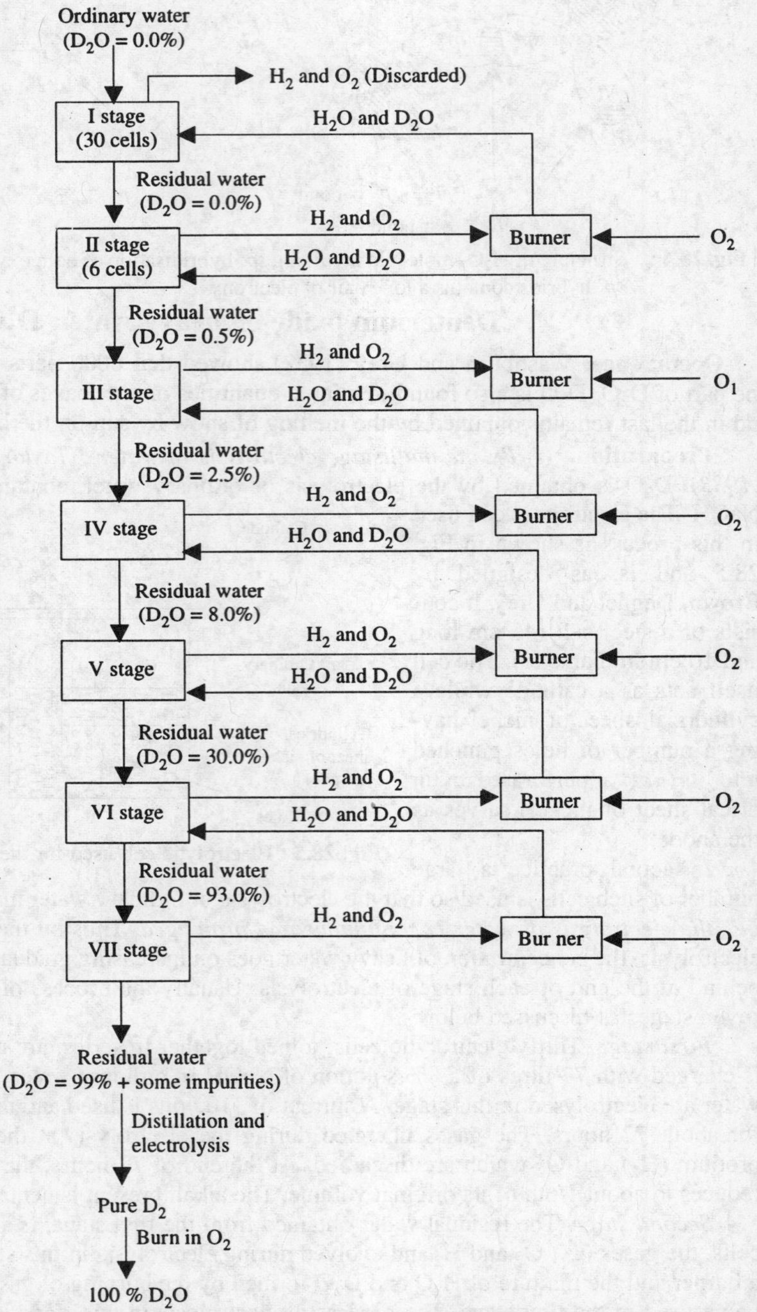

Fig. 28.6 : Flow sheet diagram showing the preparation of heavy water (D_2O) by multistage electrolysis of ordinary water.

sent back into the cells being used in the sixth stage. 99% pure D_2O obtained in this stage, also contains some impurities like alkali metals, ordinary water etc. Therefore, this heavy water is distilled and the distillate obtained is then electrolysed to get pure deuterium. The deuterium is then separately burnt in O_2 to get 100% pure D_2O.

A flow sheet diagram, showing the preparation of D_2O by the multistage electrolysis of ordinary water, is given in Fig. 28.6.

Taylor, Eyring and Frost started the electrolysis taking 2310 litres of ordinary water and obtained about 82 ml of D_2O.

(ii) By fractional distillation of ordinary water. Heavy water has been separated from ordinary water by the method, which takes the advantage of the difference in the boiling points of ordinary water (100°C) and heavy water (= 101.42°C). Since the difference between the boiling points of these two liquids is very small, the fractional distillation of ordinary water is done in a very long fractionating column (13 metre long) and the process is repeated several times. The lighter fraction (*i.e.*, H_2O), which has lower boiling point, is distilled first, while the heavier fraction (*i.e.*, D_2O), which has higher boiling point, is left behind. The heavier fraction is richer in D_2O.

(iii) By chemical exchange method (Spevack's method). When H_2S gas is passed through ordinary water at 100°C, D_2O present in ordinary water reacts with H_2S and a mixture of D_2S and H_2O is obtained by the exchange reaction.

$$D_2O\ (\textit{present in ordinary water}) + H_2S \rightleftharpoons D_2S + H_2O \quad ...(i)$$

Now when the mixture of D_2S and H_2O is passed into cold water, the opposite reaction takes place

$$D_2S + H_2O\ (\textit{cold}) \rightleftharpoons D_2O + H_2S \quad ...(ii)$$

Thus cold water becomes enriched with D_2O. On repeating the processes represented by (*i*) and (*ii*), cold water goes on becoming more and more enriched with D_2O. Now D_2O can be obtained from the fractional distillation of cold water which is enriched with D_2O.

(iv) From deuterium. Pure D_2O is obtained by the action of water vapours on deuterium at 200 - 400°C in presence of ZnO or Al_2O_3.

$$H_2O\ (\textit{vapours}) + D_2 \xrightarrow[200-400°C]{ZnO\ \text{or}\ Al_2O_3} D_2O + H_2$$

Manufacture of D_2O in India

In India, *the Fertilizer Corporation of India* set up a heavy water plant at Nangal (Punjab) in 1962. Since here the electric supply is cheap, the method being used is of continuous and prolonged electrolysis of ordinary water. D_2O manufactured here is being supplied to the Atomic Energy Commission. Plants at Rourkela, Trombay. Namrup, Neyveli and Naharkatiya have also been set up.

Physical properties

(i) Similarities with H_2O. Like ordinary water, heavy water is a colourless, odourless and tasteless mobile liquid.

(ii) Dissimilarities with H_2O. Since molecular weight of D_2O is higher than that of H_2O (D_2O = 20.03, H_2O = 18.0), there are slightly stronger van der Waals forces acting between D_2O molecules. Owing to this, D_2O has a higher density at 20°C (D_2O = 1.1059 g/cm^3, H_2O = 0.9982 g / cm^3), melting point or freezing point (D_2O = 3.82°C, H_2O = 0.00°C), boiling point (D_2O = 101.42°C, H_2O = 100.00°C) and specific heat (D_2O = 1.018, H_2O = 1.00). D_2O has higher dielectric constant than H_2O (D_2O = 82, H_2O = 80.5). Ionic compounds are less soluble in D_2O than in H_2O, *e.g.*, solubility of NaCl is 305 g / lit and 359g / lit in D_2O and H_2O respectively.

Chemical properties

(i) Electrolysis. A solution of D_2O, containing P_2O_5 or Na_2CO_3. (The addition of P_2O_5 or Na_2CO_3 makes D_2O a conductor), on electrolysis, decomposes into D_2 and O_2, which are evolved at cathode and anode respectively.

$$2D_2O \xrightarrow{\text{Electrolysis}} 2D_2 \text{ (At cathode)} + O_2 \text{ (At anode)}$$

(*ii*) *Action of metals.* D_2O reacts with alkali and alkaline earth metals to liberate D_2 and heavy alkalies (*i.e.*, metallic deuteroxides) are formed.

$$2Na + 2D_2O \longrightarrow 2NaOD + D_2; \qquad Ca + 2D_2O \longrightarrow Ca(OD)_2 + D_2$$

(*iii*) *Action of metallic oxides (i.e., basic oxides) and nonmetallic oxides (i.e., acid anhydrides).* With metallic oxides like Na_2O, CaO etc., heavy alkalies are formed.

$$Na_2O + D_2O \longrightarrow 2NaOD; \qquad CaO + D_2O \longrightarrow Ca(OD)_2$$

With nonmetallic oxides, deutero acids are obtained, *e.g.*,

$$Na_2O_5 + D_2O \longrightarrow 2DNO_3 \text{ (deutero nitric acid)}$$
$$SO_3 + D_2O \longrightarrow D_2SO_4 \text{ (deutero sulphuric acid)}$$
$$P_2O_5 + 3D_2O \longrightarrow 2D_3PO_4 \text{ (deutero phosphoric acid)}$$
$$CO_2 + D_2O \longrightarrow D_2CO_3 \text{ (deutero carbonic acid)}$$

(*iv*) *Action of metallic nitrides, phosphides, arsenides and carbides.* D_2O decomposes metallic nitrides, phosphides, arsenides and carbides to form the corresponding deutero compounds, *e.g.*,

$$2AIN + 6D_2O \longrightarrow 2Al(OD)_3 + 2ND_3 \text{ (Deuteroammonia)}$$
$$Mg_3N_2 + 6D_2O \longrightarrow 3Mg(OD)_2 + 2ND_3$$
$$Ca_3P_2 + 6D_2O \longrightarrow 3Ca(OD)_3 + 2PD_3 \text{ (Deutero phosphine)}$$
$$Na_3A + 6D_2O \longrightarrow 3NaOD + AsD_3 \text{ (Deutero arsine)}$$
$$Al_4C_3 + 12D_2O \longrightarrow 4Al(OD)_3 + 2CD_4 \text{ (Deutero methane)}$$
$$CaC_2 + 2D_2O \longrightarrow Ca(OD)_2 + C_2D_2 \text{ (Deutero acetylene)}$$

(*v*) *Formation of deuterates.* Like water, heavy water (D_2O) also gets associated with salts to form crystalline salts called *deuterates or deutero-hydrates.* (water gives hydrates) Examples of some deuterates are : $BeCl_2.4D_2O$; $CuSO_4.5D_2O$; $C_2Cl_2.6D_2O$; $NiCl_2.6D_2O$; $MgSO_4.7D_2O$. In these deuterates D_2O acts as heavy water of crystallisation. These deuterates are obtained when solution of salts in D_2O are crystallised. For example :

$$CuSO_4.5H_2O + 5D_2O \xrightarrow[-5H_2O]{\text{Crystallisation}} CuSO_4.5D_2O$$

(*vi*) *Action of compounds containing ionic or non-ionic hydrogen : Exchange reactions.* D_2O reacts with a number of compounds containing liable hydrogen atoms. In these reactions H atoms are partially or completely replaced by deuterium. Such reactions are called *exchange reactions.*

If the compounds contain ionic (*i.e.*, polar) hydrogen atoms, the exchange reactions occur more readily, *e.g.*,

$$NaOH + D_2O \rightleftharpoons NaOD + HOD$$
$$HCl\ (aq) + D_2O \rightleftharpoons DCl + HOD$$
$$NH_4Cl + 4D_2O \rightleftharpoons ND_4Cl + 4HOD$$

If the compounds contain non-ionic hydrogen atoms, the exchange reactions take place slowly and require the presence of a catalyst, *e.g.*, C_6H_6 when treated with D_2O using nickel dispersed on kieselguhr at a temperature of 200°C leads to the formation of pure *hexa deutero-benzene.*

$$C_6H_6 + 3D_2O \rightleftharpoons C_6D_6 + 3H_2O$$

Hexammine cobaltic chloride, $[Co(NH_3)_6]Cl_3$ shows a slow exchange reaction.

$$[CO(NH_3)_6]Cl_3 + D_2O \longrightarrow [CO(NH_3)_5(NH_2D)]Cl_3 + HOD$$

Exchange reactions have been used in determining the number of ionic and non-ionic hydrogen atoms present in a given compound, *e.g.*, treatment of hypophosphorus acid (H_3PO_2) with D_2O brings about the exchange of only one H atom of H_3PO_2 with deuterium.

$$H_3PO_2 + D_2O \longrightarrow H_2DPO_2 + HDO$$

This reaction shows the presence of only one ionic hydrogen atom in H_3PO_2, *i.e.*, H_3PO_2 is a *mono basic acid.*

(*vii*) *Physiological effects.* Heavy water of high concentration retards the growth of plants and animals. Lewis has shown that tobacco seeds do not grow in D_2O, though under the same conditions, they develop well in ordinary water. Pure D_2O kills small fish, tadpoles and mice, when fed on them. Mice and rats, on drinking D_2O, feel more and more thirsty. According to Taylor, it is a germicide and a bactericide. While water containing higher concentration of D_2O is poisonous, water containing small quantity of D_2O acts as a tonic and stimulates vegetable growth. Although D_2O is injurious to animal life, certain moulds have been found to flourish better in solutions richer in D_2O than in ordinary aqueous solutions.

(*viii*) *Deuterolysis*. Water brings about hydrolysis of certain inorgonic chlorides and some other salts. D_2O also brings about a similar reaction, which is called *deuterolysis*.

$AlCl_3 + 3D_2O \longrightarrow Al(OD)_3 + 3DCl$; $BaS + 2D_2O \longrightarrow Ba(OD)_2 + D_2S$

$BiCl_3 + D2O \longrightarrow BiOCl + 2DCl$; $SiCl_4 + 4D_2O \longrightarrow D_4SiO_4 + 4DCl$

$SOCl_3 + 2D_2O \longrightarrow D_4SO_3 + 2DCl$

Uses of D_2O. Important uses of D_2O are :

(*i*) *As a moderator of neutrons in nuclear reactors.* Neuterons are used for bringing about the fission of uranium atoms. For this purpose the neutrons of *slow speed* are required. The speed of neutrons is slowed down by passing them through D_2O. Such substances which are used for slowing down the speed of neutrons are called *moderators.* Obviously D_2O here acts as a mederator.

(*ii*) *For the preparation of* D_2. D_2 can be prepared either by the electrolysis of D_2O or by the action of sodium metal on D_2O.

$$2D_2O \xrightarrow{\text{Electrolysis}} 2D_2 + O$$

$$2D_2O + 2Na \longrightarrow 2NaOD + D_2$$

(*iii*) *As a tracer compound.* D_2O is commonly used as a tracer compound for studying reaction mechanism in organic chemistry. *e.g.*, D_2O has helped to elucidate the mechanism of electrophilic aromatic substitution reactions.

D_2O is also used as tracer in metabolic processes, *e.g.*, when mice fed with labelled butyric acid, within hours all the deuterium was eliminated from the body indicating that it is a ready source of energy.

(*iv*) Although D_2O injurious to animal life, yet certain moulds have been found to flourish better in solutions rich in D_2O than in ordinary aqueous solution.

(*v*) The ability of D_2O is to undergo exchange reactions has been used for studying the structure of some oxy-acids of phosphorus such as H_3PO_2 and H_3PO_3 by finding out the number of ionic hydrogen atoms in these compounds.

(*vi*) D_2O is also used for absorbing heat produced in nuclear reactors, *i.e.*, D_2O is also used as a coolant.

Oxides of Group IA Elements

When alkali metals are burnt in air, they form oxides. Depending on the nature of the alkali metal, different kinds of oxides are obtained.

A. Monoxides or normal oxides, M_2O.

Lithium forms monoxide, Li_2O.

$$2Li + O_2 \longrightarrow 2Li_2O$$

Normal oxide of sodium (Na_2O) is obtained by indirect methods, *e.g.*

$$Na_2O_2 + 2Na \longrightarrow 2Na_2O$$

$$2NaNO_2 + 6Na \longrightarrow 4Na_2O + N_2$$

Properties (*i*) *Solubility and basic nature:* Monoxides are highly soluble and strongly basic. They give strongly alkaline solution due to the formation of hydroxides, M^+OH^- which are very largely ionised.

$$M_2O + H_2O \longrightarrow 2MOH$$

or $$O^{2-} + H_2O \longrightarrow 2OH^-$$

(*ii*) *Stability.* Li_2O is the most stable oxide of all the monoxides of other alkali metals.

(*iii*) *Ionic / covalent character.* Li_2O is covalent in nature while other monoxides have almost entirely ionic character.

Explanation. See "*Applications of Fajans' Rules*" discussed in chapter 10.

Structure of monoxides. These oxides are ionic oxides and hence are represented as $M_2^+ O^{2-}$. Thus these oxides contain the oxide ion, O^{2-}. These oxides have *antifluorite structure* which is the reverse of fluorite structure (structure of CaF_2 crystal is called fluorite structure). For details see "*Structure of A_2B type ionic crystal*" discussed in chapter 16 (Structure of solids).

B. Superoxides, MO_2.

Excepting Li and Na, the remaining elements for superoxides. Thus the known superoxides are KO_2, RbO_2 and ClO_2. These are obtained when the metal is allowed to react with O_2.

$$M\ (M = K, Rb, Cs) + O_2 \longrightarrow MO_2$$

Superoxides contains superoxide ion, O^{2-} which has a *three-electron bond.* Thus this ion can be represented as $\left[:\overset{\cdot}{\underset{\cdot\cdot}{O}}\cdots\underset{\cdot\cdot}{O}:\right]$. The presence of one unpaired electron in three electron bond makes this ion *paramagnetic* and *coloured.*

C. Peroxides, M_2O_2

All the alkali metals form peroxides. These oxides are strong oxidising agents and react with H_2O to give H_2O_2 and O_2.

$$O_2^{2-}\ (\textit{peroxide}) + 2H_2O \longrightarrow OH^- + H_2O_2$$

$$2O_2^{2-} + 2H_2O \longrightarrow 4OH^- + O_2$$

Peroxides contain peroxide ion, O_2^{2-} or $[—O—O—]^{2-}$

1. Lithium peroxide, Li_2O_2. It is formed : (*a*) by heating Li_2O above 400°C.

$$2Li_2O \xrightarrow{7400^\circ C} Li_2O_2 + 2Li$$

(*b*) by treating a solution of LiOH with H_2O_2 and alcohol and then drying the white crystalline precipitate obtained in vacuum over P_2O_5.

$$2LiOH + 2H_2O_2 \xrightarrow{\text{Alcohol}} Li_2O_2.\ H_2O_2.3H_2O\ (\textit{white crystalline precipitate})$$

$$\xrightarrow[\text{over } P_2O_5]{\text{Dry in vacuum}} Li_2O_2\ (\textit{anhydrous})$$

Properties. (*i*) Li_2O_2 is a white solid and is not very stable.

(*ii*) When heated to 300°C in vacuum, Li_2O is obtained.

$$2Li_2O_2 \xrightarrow{300^\circ C} 2Li_2O + O_2$$

2. Sodium peroxide, Na_2O_2. This oxide is also called *oxone,* since it is used as a source of O_2.

Preparation. Na_2O_2 is prepared : (*a*) by heating Na_2O above 400°C.

$$2Na_2O \xrightarrow{>400^\circ C} Na_2O_2 + 2Na\ (\textit{vapour})$$

(*b*) When Na metal is burnt in the excess of O_2 at 300°C. Oxygen should be free from CO_2 and moisture (*Large scale preparation*)

$$2Na + O_2\ (\textit{excess}) \xrightarrow{300^\circ C} Na_2O$$

Properties. (*i*) It is a pale yellow solid which is stable in air.

(*ii*) *Action of moist air.* When exposed to moist air, Na_2O_2 turns white due to the formation of NaOH and Na_2CO_3.

$$Na_2O_2 + 2H_2O\ (\textit{From moist air}) \longrightarrow 2NaOH + H_2O_2$$

$$2NaOH + CO_2\ (\textit{From moist air}) \longrightarrow Na_2CO_3 + H_2O$$

(*iii*) *Action of heat.* When heated strongly, it evolves O_2.

$$2Na_2O_2 \xrightarrow{\Delta} Na_2O + O_2$$

(*iv*) *Action of water.* (*a*) Na_2O_2 dissolves in *ice cold water* (0°C) forming H_2O_2.

$$Na_2O_2 + 2H_2O \xrightarrow{0°\,C} 2NaOH + H_2O_2$$

Due to the formation of NaOH the solution is alkaline in nature. From this solution, a crystalline hydrate, $Na_2O_2.8H_2O$ can be isolated.

(*b*) On warming the solution (*i.e.*, at room temperature), O_2 is evolved.

$$2Na_2O_2 + 2H_2O \xrightarrow{\text{room temperature}} 4NaOH + O_2$$

(*v*) *Action of acids.* (*a*) With *ice cold dil. mineral acids,* H_2O_2 is produced.

$$Na_2O_2 + 2HCl\ (dil.) \longrightarrow 2NaCl + H_2O_2$$

$$Na_2O_2 + H_2SO_4\ (dil.) \longrightarrow Na_2SO_4 + H_2O_2$$

Second reaction is used for the preparation of H_2O_2.

(*b*) With *hot acids,* O_2 is evolved

$$Na_2O_2 + 4HCl\ (hot) \longrightarrow 4NaCl + 2H_2O + O_2$$

$$2Na_2O_2 + 2H_2SO_4\ (hot) \longrightarrow 2Na_2SO_4 + 2H_2O + O_2$$

(*vi*) *Action of Na metal.* When Na_2O_2 is heated with sodium metal, Na_2O is obtained.

$$Na_2O_2 + 2Na \xrightarrow{\Delta} 2Na_2O$$

(*vii*) *Action of CO.* Na_2O_2 absorbs CO, forming Na_2CO_3.

$$Na_2O_2 + CO \longrightarrow Na_2CO_3$$

(*viii*) *Action of CO_2.* O_2 gas is liberated.

$$2Na_2O_2 + 2CO_2 \longrightarrow 2Na_2CO_3 + O_2$$

It is due to this property that Na_2O_2 is used for purifying air in confined places like submarines, hospitals etc.

(*ix*) *Action of absolute alcohol (C_2H_5OH).* At 0°C, a white powder of sodium hydrogen peroxide ($NaHO_2$) is formed. This substance explodes on heating.

$$Na_2O_2 + C_2H_5\,OH \longrightarrow NaHO_2 + C_2H_5ONa$$

(*x*) *Oxidising properties.* Since Na_2O_2 gives O_2 on heating, it acts as on oxidising agent. Thus Na_2O_2.

(*a*) oxidises moist $Cr(OH)_3$ (Cr = +3) to soluble Na_2CrO_4(Cr = + 6). Na_2O_2 (O = –1) is reduced to NaOH (O = – 2)

$$[Na_2O_2 + H_2O \longrightarrow 2NaOH + O] \times 3$$

$$2Cr(OH)_3 + 4NaOH + 3O \longrightarrow 2Na_2CrO_4 + 5H_2O$$

$$3Na_2O_2 + 2Cr(OH)_3 \longrightarrow 2Na_2CrO_4 + 2NaOH + 2H_2O$$

or $$2Cr(OH)_3 + 3O_2^{2-} \longrightarrow 2CrO_4^{2-} + 2OH^- + 2H_2O$$

(*b*) oxidises $Mn(OH)_2$ (Mn = +2) to $NaMnO_4$ (Mn = +7)

$$Mn(OH)_2 + 2Na_2O_2 \longrightarrow Na_2MnO_4 + 2NaOH$$

or $$Mn(OH)_2 + 2O_2^{2-} \longrightarrow MnO_4^{2-} + 2OH^-$$

(*c*) in the fused state, Na_2O_2 oxidises *chromite* ore (*chrome iron stone*), $FeCr_2O_4$ or FeO. Cr_2O_3.

(Fe = +2, Cr = +3) to Fe_2O_3 (Fe = +3) and soluble Na_2CrO_4(Cr = + 6). Na_2O_2 (O = –1) is reduced to Na_2O (O = – 2)

$$2[FeO.\ Cr_2O_3] + 7Na_2O_2 \longrightarrow Fe_2O_3 + 4Na_2CrO_4 + 3Na_2O$$

(*d*) oxidises iron pyrites, FeS_2 (Fe = +2) to Fe_2O_3(Fe = +3) and Na_2SO_4.

$$2FeS_2 + 15Na_2O_2 \longrightarrow Fe_2O_3 + 4Na_2SO_4 + 11Na_2O$$

(*e*) oxidises charcoal (C = 0) at 300 – 400°C to Na_2CO_3 (C = + 4)

$$3Na_2O_2 + 2C \longrightarrow 2Na_2CO_3 + 2Na$$

(*f*) oxidises SO_2 (S = + 4) to sulphate, SO_4^{2-} (S = + 6)

$$Na_2O_2 + SO_2 \longrightarrow Na_2SO_4$$

(*xi*) *Action of slaked lime, $Ca(OH)_2$.* When a mixture of $Ca(OH)_2$ and Na_2O_2 is compressed, calcium peroxide (CaO_2) is obtained (*commercial preparation of CaO_2*).

$$Ca(OH)_2 + Na_2O_2 \longrightarrow CaO_2 + 2NaOH$$

Uses. It is used (*i*) as a source of O_2 under the name of *oxone* in the form of CuO and Na_2O_2; (*ii*) for the purification of air in confined places such as hospitals, submarines, diving bells, theatres, aeroplanes etc.; (*iii*) as an oxidising agent in laboratory; (*iv*) for bleaching horn, bone, ivory, sponges, wood, silk, wool, straw, hair, rayon etc. in industry. For this purpose a mixture of Na_2O_2 and dil. H_2SO_4 is used under the name of *soda bleach*; (*v*) in the manufacture of H_2O_2, sodium per borate, dyes and benzoyl peroxide; (*vi*) in the preparation of calcium peroxide (CaO_2. $8H_2O$) which is used for the activation of yeast, as an antiseptic and for bleaching; (*vii*) in the manufacture of washing and soap powders.

3. Rb_2O_2 and Cs_2O_2. These peroxides are obtained by heating the dioxides (MO_2) in a vacuum Rb_2O_2 is yellow-white and Cs_2O_2 is pale-yellow.

Oxides of group IIA elements

The elements of this group form two types of oxides *viz.*, monoxides (MO) (also called normal oxides) and peroxides (MO_2). Monoxides are formed by all the elements while peroxides are given by Mg, Ca, Sr and Ba only (Be does not form peroxide).

A. Monoxides, MO (M = Be, Mg, Ca, Sr, Ba)

These oxides are formed by all the elements. CaO, SrO and BaO were known much earlier than the metals themselves. These were called alkaline earths, since they were alkaline and existed in the earth.

Preparation. Monoxides are obtained by heating the metals in O_2 or by heating their carbonates strongly at high temperatures, *e.g.*,

$$2M\ (M = Be, Ca) + O_2 \xrightarrow{\Delta} 2MO\ ; \qquad MCO_3(M = Ca, Ba) \xrightarrow{\Delta} MO + CO_2$$

General properties. (*i*) *Action of carbon.* BeO, when heated with carbon, is converted into beryllium carbide with the formula, Be_2C while the monoxides of other metals under similar conditions form the carbides with the formula. MC_2. Thus :

$$2BeO + 3C \xrightarrow{\Delta} Be_2C + 2CO\ ; \quad MO\ (M = Mg, Ca, Sr, Ba) + 3C \xrightarrow{\Delta} MC_2 + CO$$

(*ii*) *Action of water.* The values of crystal lattice energies of the oxides are large. These values for BeO and MgO are so high that these oxides are almost insoluble in water. On the other hand, the oxides of the remaining elements slake (*i.e.*, react with H_2O) to form the hydroxides which are strongly basic in character.

$$CaO + H_2O \longrightarrow Ca(OH)_2\ (\textit{slaked lime})\ ; \quad BaO + H_2O \longrightarrow Ba(OH)_2$$

(*iii*) *Amphoteric / basic character.* The basic character of the oxides gradually increases as we move down the group. Thus BeO is amphoteric, MgO is weakly basic while SrO and BaO are strongly basic. Being amphoteric BeO dissolves in acids to form salts and in alkalies to form beryllates (BeO_2^{2-}).

$$BeO + 2HCl \longrightarrow BeCl_2 + H_2O\ ; \quad BeO + 2NaOH \longrightarrow Na_2BeO_2 + H_2O$$

(*iv*) These oxides are extremely stable white crystalline solids due to their high crystal lattice energy obtained by packing *doubly-charged ions* (M^{2+} and O^{2-} ions) in a sodium chloride type lattice

(*v*) Due to small size of Be^{2+} ion, BeO is *covalent* while the oxides of other metals are *ionic.* Although BeO is covalent, yet it has a higher melting point and is harder than other oxides, as it is *polymeric.* In polymerised molecule each Be-atom is tetrahedrally coordinated to four O-atoms.

Uses. MgO and CaO (lime) are used as basic lining in furnaces.

Structure of oxides. BeO has a wurtzite (ZnS) structure with 4 : 4 (four-fold) coordination while the monoxides of other elements have NaCl type structure with 6 : 6 (six-fold) coordination.

B. Peroxides, MO_2 (M = Mg, Ca, Sr, Ba)

Excepting Be, all the remaining metals form peroxides. The ease with which these elements form peroxides increases as we move down the group.

Preparation. Peroxides of heavier elements (Mg, Ca, Sr, Ba) can be obtained by heating the normal oxides with O_2 at high temperature, *e.g.*,

$$2BaO + O_2 \longrightarrow 2BaO_2$$

Properties. (*i*) These oxides are white ionic solids having peroxide ion, $[O—O]^{2-}$.

(*ii*) *Action of dil. acids.* These oxides react with dil. acids and produce H_2O_2.

$$MO_2 + H_2SO_4 \longrightarrow MSO_4 + H_2O_2$$

(*ii*) *Action of heat.* On heating in air, they lose O_2 and give monoxides (MO). Their thermal stability towards dissociation increases with increasing cation size on descending the group.

Now let us discuss some peroxides in detail.

1. Magnesium peroxides, MgO_2. A precipitate of the composition MgO. MgO_2 is obtained when a solution of $MgSO_4$ containing H_2O_2 is treated with NaOH. It is an unstable compound and gradually loses O_2. MgO_2 is used as an antiseptic in tooth pastes and as a bleaching agent since MgO_2 gradually loses O_2.

2. Calcium peroxide, $CaO_2.8H_2O$. It is obtained as a white crystalline precipitate :

(*a*) by adding H_2O_2 to lime water, $Ca(OH)_2$.

$$H_2O_2 + Ca(OH)_2 \longrightarrow CaO_2 + 2H_2O$$

(*b*) by compressing a mixture of $Ca(OH)_2$ (*slaked lime*) and Na_2O_2 (*commercial preparation*)

$$Ca(OH)_2 + Na_2O_2 \longrightarrow CuO_2 + 2NaOH.$$

Properties. (*i*) It is a white crystalline substance.

(*ii*) It is sparingly soluble in water and the resulting solution is alkaline due to hydrolysis.

(*iii*) It is similar to BaO_2 in properties but less stable.

(*iv*) When it is warmed alone or with 30% H_2O_2, a yellow powder of calcium tetroxide (CaO_4) is obtained.

(*v*) When treated with acids, H_2O_2 is evolved.

Uses. It is used as an antiseptic and for bleaching.

3. Strontium peroxide, SrO_2. Preparation. (*a*) When H_2O_2 is added to $Sr(OH)_2$ solution, a crystalline octahydrate, $SrO_2.8H_2O$ is precipitated as pearly scales. This precipitate on gentle heating gives SrO_2.

$$Sr(OH)_2 + H_2O_2 + 6H_2O \longrightarrow SrO_2.8H_2O \xrightarrow[-8H_2O]{\Delta} SrO_2 + 8H_2O$$

(*b*) SrO_2 can also be prepared by the combination of SrO with O_2 of the air under 150 atom pressure and 500°C.

$$2Sr + O_2 \xrightarrow[500°\ C]{150\ atm} 2SrO_2$$

Properties. (*i*) SrO_2 is a white powder which is sparingly soluble in water.

(*ii*) When heated, it gets decomposed into SrO.

$$2SrO_2 \xrightarrow{\Delta} 2SrO + O_2$$

4. Barium peroxide, BaO_2. It is prepared :

(*a*) by heating BaO in O_2 or air (free from CO_2 and moisture) at about 500°C.

$$2BaO + O_2 \xrightarrow{500°\ C} 2BaO_2$$

(*b*) by the action of H_2O_2 on $BaCl_2$ solution in presence of KOH.

$$BaCl_2 + H_2O_2 + 2KOH \longrightarrow BaO_2 + 2KCl + 2H_2O$$

Properties. (*i*) It is a white powder which is sparingly soluble in water and forms octahydrate, $BaO_2.8H_2O$. It is insoluble in alcohol and ether.

(*ii*) *Action of ice-cold dil. acids and CO_2.* When a paste of BaO_2 in ice-cold water is treated with dil. acid or CO_2, H_2O_2 is formed.

$$3BaO_2 + 2H_3PO_4 \longrightarrow Ba_3(PO_4)_2 + 3H_2O_2$$
$$BaO_2 + H_2SO_4\ (dil.) \longrightarrow BaSO_4 + H_2O_2$$
$$BaO_2 + CO_2 + H_2O \longrightarrow BaCO_3 + H_2O_2$$

The reaction between anhydrous BaO and dil. H_2SO_4 is slow and ceases after sometime, because $BaSO_4$ produced in the reaction makes a protective coating on BaO_2. In order to get H_2O_2, we, therefore, use hydrated barium peroxide ($BaO_2.8H_2O$) in place of BaO_2.

$$BaO_2.8H_2O + H_2SO_4 \longrightarrow BaSO_4 + H_2O_2 + 8H_2O$$

(*iii*) *Action of conc. acids and atmospheric CO_2.* BaO_2 absorbs CO_2 from the atmosphere and liberates O_2. Concentrated acids also give O_2.

$$2BaO_2 + 2CO_2\ (From\ atmosphere) \longrightarrow 2BaCO_3 + O_2$$
$$2BaO_2 + 2H_2SO_4\ (conc.) \longrightarrow 2BaSO_4 + 2H_2O + O_2$$

When CO_2 is passed through a cold pasty solution of BaO_2 in water, H_2O_2 is obtained.

$$BaO_2 + H_2O + CO_2 \longrightarrow BaCO_3 + H_2O_2$$

This equation represents Merck's process for the preparation of H_2O_2.

(*iv*) *Action of H_2O_2.* BaO_2 forms a double compound, $BaO_2.H_2O_2$, when treated with H_2O_2.

Uses. It is used (*i*) for the menufacture of H_2O_2 (*ii*) in bleaching silk, vegetable fibres and straw. (*iii*) in *Gold Schmidt's thermite process* for preparing ignitor pellets which consists of a mixture of BaO_2 and Mg powder (*iv*) as a disinfectant (*v*) as a decolourising agent for lead glass.

Oxides of group IIIA elements

All the elements of this group form the oxides of M_2O_3 type (M = B, Al, Ga, In and Tl) in which the element is in +3 oxidation state. These oxides are called *sesquioxides.* In addition to Tl_2O_3, thallium also forms thallium (I) oxide, Tl_2O in which Tl is in +1 oxidation state.

Acidic, amphoteric and basic character of $\mathbf{M_2O_3}$ type oxides

B_2O_3 is an acidic oxide although it also shows feebly basic character. Al_2O_3 and Ga_2O_3 are amphoteric oxides while In_2O_3 and Tl_2O_3 are basic in character. Thus we see that on moving down the group there is a gradual change from acidic to amphoteric and then to basic character of the oxides.

Explanation of the variation. See "Apprications of Fajans' Rules" discussed in chapter 10.

Oxides of thallium

Thallium forms two oxides *viz.*, thallium monoxide or thallous oxide (Tl_2O) and dithallium trioxide or thallic oxide or thallium sesquioxide (Tl_2O_3). In Tl_2O and Tl_2O_3, Th is in +1 and +3 oxidation state respectively.

1. Thallium oxide or thallous oxide, $\mathbf{Tl_2O}$. Preparation. It is obtained :

(*a*) as a black very hygroscopic solid by heating TlOH to 100°C out of contact of air.

$$2TlOH \xrightarrow{\Delta} Tl_2O + H_2O$$

(*b*) by heating Tl_2O_3

$$Tl_2O_3 \xrightarrow{\Delta} Tl_2O + O_2$$

(*c*) by the direct combination of Tl and O_2.

$$4Tl + O_2 \longrightarrow 2Tl_2O$$

Properties. (*i*) Tl_2O reacts with water to form colourless solution of TlOH.

$$Tl_2O + H_2O \longrightarrow 2TlOH$$

(*ii*) Tl_2O is a basic oxide and reacts with acids to form salts.

2. Dithallium trioxide or thallic oxide or thallium sesquioxide, $\mathbf{Tl_2O_3}$. It is obtained by heating Tl in the excess of O_2.

$$4Tl + 3O_2\ (excess) \longrightarrow 2Tl_2O_3$$

In this reaction some Tl_2O is also obtained.

Tl_2O_3 can also be obtained by heating $Tl(OH)_3$ or $TlNO_3$.

$$2Tl(OH)_3 \xrightarrow{\Delta} Tl_2O_3 + 3H_2O \ ; \quad 2TlNO_3 \xrightarrow{\Delta} Tl_2O_3 + N_2O_3$$

Properties. (*i*) Tl_2O_3 is brown or black in colour, insoluble in water but soluble in acids to give thallic salts.

(*ii*) On being heated Tl_2O_3 gets decomposed into Tl_2O and O_2 is evolved.

$$Tl_2O_3 \xrightarrow{\Delta} Tl_2O + O_2$$

(*iii*) When Tl_2O_3 is treated with dil. H_2SO_4 and the solution is allowed to crystallise, crystals of thallic sulphate, $Tl_2(SO_4)_3.7H_2O$ are obtained.

$$Tl_2O_3 + 3H_2SO_4 \longrightarrow Tl_2(SO_4)_3 + 3H_2O$$

Oxides of group IVA elements

The well known oxides formed by the elements of group IVA are (*i*) monoxides, MO (M = C, Si, Ge, Sn, Pb) (*ii*) dioxides, MO_2 (M = C, Si, Ge, Sn, Pb) (*iii*) suboxides, *e.g.*, C_3O_2 and Pb_2O. (*iv*) other oxides like Pb_2O_3 and Pb_3O_4.

A. Monoxides MO (M = C, Si, Ge, Sn, Pb)

These are given by all the elements. Except SiO and GeO, all the remaining monoxides are stable. CO is *neutral*, GeO is *basic* while SnO and PbO are *amphoteric*.

1. Germanium monoxide or germanous oxide, GeO. It is prepared : (*a*) by dehydrating $Ge(OH)_2$.

$$Ge(OH)_2 \xrightarrow[(-H_2O)]{\text{Dehydration}} GeO + H_2$$

(*b*) by heating GeO_2 with Ge at about 1000°C.

$$GeO_2 + Ge \xrightarrow{1000^\circ C} 2GeO$$

(*c*) by reducing GeO_2 with CO at 900°C.

$$GeO_2 + CO \longrightarrow GeO + CO_2$$

(*d*) by heating Ge metal in a current of CO_2 at 700-900°C.

$$Ge + CO_2 \longrightarrow GeO + CO$$

GeO is a black coloured powder. It is stable in air and is scarcely attacked by acid or alkali. On heating to 700°C, GeO undergoes disproportionation into Ge and GeO_2. GeO is a basic oxide.

$$\underset{(Ge = +2)}{2\,GeO} \xrightarrow{700^\circ C} \underset{(Ge = 0)}{Ge} + \underset{(Ge = +4)}{GeO_2}$$

2. Tin monoxide or stannous oxide, SnO. Preparation. It is obtained by dehydrating $Sn(OH)_2$ or by heating $Sn(COO)_2$.

$$Sn(OH)_2 \xrightarrow{-H_2O} SnO + H_2O$$

$$Sn(COO)_2 \xrightarrow{\Delta} SnO + CO + CO_2$$

Properties. (*i*) It is a dark grey or black powder.

(*ii*) *Burning.* SnO burns in air to form SnO_2.

$$2SnO + O_2 \longrightarrow 2SnO_2$$

(*iii*) *Amphoteric nature.* SnO is an amphoteric oxide, since it reacts with acids as well as alkalies, forming stannous salts and stannites (SnO_2^{2-}) respectively.

$$SnO + 2HCl \longrightarrow SnCl_2 + H_2O \ ; \quad SnO + 2NaOH \longrightarrow Na_2SnO_2 + H_2O$$

The stannites are known only in solution and absorb O_2 form air to form stannate (SnO_3^{2-}).

$$2Na_2SnO_2 + O_2 \longrightarrow 2Na_2SnO_3$$

(*iv*) *Action of conc. NaOH solution.* Stannate (SnO_3^{2-}) and spongy Sn are formed.

$$2SnO + 2NaOH\ (conc.) \longrightarrow Na_2SnO_3 + Sn + H_2O$$

3. Lead monoxide or plumbous oxide, PbO [Also called **litharge** or **massicot** (*murdasang*)].

Preparation. (*i*) This oxide exists in two forms *viz.*, massicot (yellow powder) and litharge (reddish-yellow crystalline form). Massicot is obtained when Pb is heated in air to about 300°C. At high temperature (877°C) it fuses to give litharge.

$$2Pb + O_2 \xrightarrow{300°\,C} 2PbO\ (\textit{Massicot})\ ; \quad 2Pb + O_2 \xrightarrow{877°\,C} 2PbO\ (\textit{Litharge})$$

(*ii*) When PbO_2 is heated at higher temperature, PbO is obtained.

$$2PbO_2 \xrightarrow{\Delta} 2PbO + O_2$$

Properties. (*i*) PbO varies in colour from yellow (massicot) to reddish-yellow (litharge). It fuses at 877°C and is volatile at red heat.

(*ii*) It is slightly soluble in water, probably forming $Pb(OH)_2$.

$$PbO + H_2O \longrightarrow Pb(OH)_2$$

(*iii*) *Amphoteric nature.* PbO is an amphoteric oxide, since it dissolves in acids as well as in alkalies, forming plumbous salts and plumbites (PbO_2^{2-}) respectively.

$$PbO + 2HNO_3 \longrightarrow Pb(NO_3)_2 + H_2O\ ; \quad PbO + 2NaOH \longrightarrow Na_2PbO_2 + H_2O$$

(*iv*) *Reduction.* PbO is reduced to Pb metal by heating with C, H_2 and CO.

$$PbO + C \longrightarrow Pb + CO\ ;\ PbO + H_2 \longrightarrow Pb + H_2O\ ;\ PbO + CO \longrightarrow Pb + CO_2$$

(*v*) *Fusion with $KClO_3$ and nitre.* PbO_2 is obtained.

$$3PbO + KClO_3 \xrightarrow[\text{of nitre }(KNO_3)]{\text{Fuse in presence}} 3PbO_2 + KCl$$

(*vi*) *Action of sodium hypochlorite, NaOCl.* When NaOCl is added to a cold solution of PbO in NaOH, yellow ppt. of lead sesquioxide (Pb_2O_3) is obtained.

$$2PbO + NaOCl \longrightarrow Pb_2O_3 + NaCl$$

(*vii*) *Action of heat.* When massicot (yellow lead monoxide) is heated in a current of air to about 340°C for several hours, red lead (Pb_3O_4) is obtained.

$$6PbO + O_2 \xrightarrow{340°\,C} 2Pb_3O_4$$

Uses. (*a*) Litharge is used in making flint glass, as a drier in paints and varnishes, in preparing some lead salts and in glazing pottery, (*b*) massicot when mixed with glycerine is used as a cement for glass and stone.

B. Dioxides, MO_2 (M = C, Si, Ge, Sn, Pb)

All the elements of group IVA form dioxides. The acidic character of these oxides decreases as we move down the group. Thus CO_2 and SiO_2 are acidic while GeO_2, SnO_2 and PbO_2 are amphoteric. All these oxides dissolve in alkalies, giving carbonate (CO_3^{2-}), silicate (SiO_3^{2-}), germanate (GeO_3^{2-}), stannate (SnO_3^{2-}) and metaplumbate (PbO_3^{2-}) respectively.

$$MO_2 + 2NaOH \longrightarrow Na_2MO_3 + H_2O$$

or

$$MO_2 + 2OH^- \longrightarrow MO_3^{2-} + H_2O$$

1. Solid carbon dioxide or dry ice. Gaseous CO_2 can be liquefied by cooling it under a pressure of 70 atmosphere at 20°C. When liquid CO_2 is allowed to escape out of a cylinder into a canvas bag tied to its nozzle, solid CO_2 is produced as a result of intense cooling.

Solid CO_2 is manufactured by evaporating liquid CO_2 at 20°C and 70 atmospheric pressure to a pressure of less than 5.2 atmosphere.

Properties. Solid CO_2 sublimes at –78.5°C and it is technically known as dry ice. Solid CO_2 is a soft white snow-like substance having its sp. gravity = 1.53 and melting point = –56.6°C at 2 atmospheres. It evaporates in air without melting, if the pressure is 5.2 atmospheres or more. Solid CO_2 like other solids will melt to a liquid. It dissolves in ether, accetone or alcohol and gives freezing mixtures which have exceedingly low temperatures. Ether-solid CO_2 mixture produces a temperature of –110°C and is used as an excellent freezing mixture.

Uses. Solid CO_2 is used (*i*) as a refrigerant under the commercial name *dricold* as it leaves no residue and produces much low temperature, (*ii*) as a freezing mixture since the solid CO_2-ether mixture produces exceedingly low temperature equal to –110°C, (*iii*) in hospitals for surgical operations of sores, (*iv*) as a substitute of ice in cold drinks and in the preparation of ice-creams.

2. Silicon dioxide or silica, SiO_2. *Occurrence.* Silica is widely distributed in nature. In the free state, it occurs in several crystalline and amorphous forms. Silica also occurs in vegetable and animal organisms. The stems and exterior coating of straw and bamboo, the quills of feathers, the claws of the animals and the finger nails contain considerable amounts of silica.

Different forms of silica. As already stated, silica exists in nature in several crystalline as well as amorphous forms. These forms have been discussed below :

(*i*) *Crystalline forms.* There are three main crystalline forms of silica which are :

(*a*) *Quartz.* Quartz is the most common form of silica. Its specific gravity is 2.648 and it melts in an electric furnace at 1740°C. There are many varieties of quartz which have different colours due to the presence of impurities in them. Some of these varieties are *amethyst* (purple), *jaspar* (red), *opal* (white), *caringorm* (yellow) etc. All these varieties are used as *gems.*

Quartz is more transparent to ultra violet rays than is glass. It is, therefore, used in all apparatuses where ultra violet hight is to be used.

Sand is the crushed form of quartz and is produced as a result of weathering or disintegration of rocks and further breaking down by violent movement of water. Sandstone consists of sand particles along with Fe_2O_3.

(*b*) *Tridynite.* It occurs more rarely than quartz in minute crystals. Its specific gravity is 2.26. It is in the hexagonal crystalline form.

(*c*) *Crystobalite.* It is a cubic crystalline form and its specific gravity = 2.32 and melting point = 1705°C.

It is formed by heating powdered silica glass at 1500°C.

Each of the above three crystalline forms has two allotropic modifications which have difinite transition temperatures as shown below in Fig. 28.7.

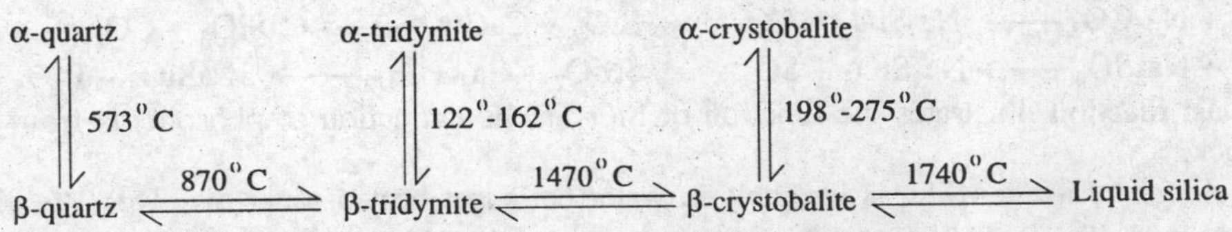

Fig. 28.7 : Allotropic modifications of three crystalline forms of silica.

It may be seen from Fig. 28.7 that quartz is stable below 870° above which it changes into tridynite. Tridynite is stable between 870° and 1470°C above which it changes into cristobalite. Cristobalite melts when heated to 1740°C.

(*ii*) *Amorphous forms.* Silica occurs in nature associated with 3-12% of water in the mineral *opal* which may be colourless and contains iron oxide, organic matter etc. *Flint, chalcedony, chent, agate and onyx* are the amorphous forms of silica.

Preparation of artificial silica (amorphous form). It can be formed :

(*i*) by the action of water on $SiCl_4$ or SiF_4

$SiCl_4 + 4H_2O \longrightarrow H_4SiO_4$ [*orthosilicic acid (gelatinous mass)*] + 4HCl

$SiF_4 + 4H_2O \longrightarrow H_4SiO_4 + 2H_2SiF_6$ (*Hydrofluosilicic acid*)

The gelatinous mass of H_4SiO_4 obtained as above is heated to 1000°C to get pure SiO_2.

$$H_4SiO_4 \xrightarrow{1000^\circ\,C} SiO_2 + 2H_2O$$

(*ii*) by burning silicon in air or better in O_2 at 400°C.

$$Si + O_2 \xrightarrow{400^\circ\,C} SiO_2$$

(*iii*) by adding HCl or solution of $(NH_4)_2SO_4$ to that of Na_2SiO_3 when gelatinous mass of silicic acid (H_2SiO_3) is obtained. This acid, on being heated to redness in a Pt dish, gives a white powder of SiO_2.

$$2HCl + Na_2SiO_3 \longrightarrow H_2SiO_3 + 2NaCl$$

$$(NH_4)_2SO_4 + Na_2SiO_3 \longrightarrow H_2SiO_3 + Na_2SO_4 + 2NH_3$$

$$H_2SiO_3 \xrightarrow{\text{Heat to redness}} SiO_2 + H_2O$$

Properties. (*i*) Silica is a *hard solid with high melting point.* (Note that CO_2 is a gas). All the varieties of silica soften below 1610°C and melt at about 1710°C in oxygen-hydrogen flame to a colourless glass known as *quartz glass* or *silica glass.* These varieties boil in electric furnace at 2230°C.

(*ii*) *Comparison between quartz glass and ordinary glass.* (*a*) Quartz glass has specific gravity = 2.22 and low coefficient of thermal expansion (0.0000005). It is due to this low value of thermal expansion that when red hot quartz is plunged in water, it does not shatter into small pieces.

(*b*) Quartz glass is transparent to ultraviolet rays while ordinary glass is opaque.

(*iii*) *Action of heat.* When quartz glass is heated to about 1100°C, it passes into crystalline form and becomes opaque.

(*iv*) *Reduction.* When heated with Mg in an electric furnace, SiO_2 is reduced to silicon

$$SiO_2 + 2Mg \longrightarrow Si + 2MgO$$

(*v*) *Action of carbon.* When heated with carbon in an electric furnace, SiC (carborandum) is obtained.

$$SiO_2 + 3C \longrightarrow SiC + 2CO$$

(*vi*) *Action of metallic oxides.* SiO_2 acts as a powerful acid anhydride at high temperatures and combines with bases and many metallic oxides (*e.g.*, Na_2O, CaO etc.) to form silicates.

$$SiO_2 + Na_2O \longrightarrow Na_2SiO_3\ ; \qquad SiO_2 + CaO \longrightarrow CuSiO_3$$

These reactions show that SiO_2 has acidic nature.

(*vii*) *Action of metallic salts.* When SiO_2 is heated with metallic salts like Na_2CO_3, $CaCO_3$, Na_2SO_4, $Ca_3(PO_4)_2$ etc., silicates (SiO_3^{2-}) of the metals are obtained.

$$SiO_2 + Na_2CO_3 \longrightarrow Na_2SiO_3 + CO_2\ ; \qquad SiO_2 + CaCO_3 \longrightarrow CaSiO_3 + CO_2$$

$$SiO_2 + Na_2SO_4 \longrightarrow Na_2SiO_3 + SO_3\ ; \qquad 3SiO_2 + Ca_3(PO_4)_2 \longrightarrow 3CaSiO_3 + P_2O_5$$

The last reaction illustrates the function of SiO_2 in the extraction of phosphorus from phosphorite rock.

(*viii*) *Action of water.* SiO_2 is insoluble in water but super heated water dissolves 1% silica at 300°C and gives *silicic acid,* H_2SiO_3 which is a very weak acid. This acid is more stable than H_2CO_3.

$$SiO_2 + 2H_2O \longrightarrow H_2SiO_3$$

(*ix*) *Action of acids.* SiO_2 is insoluble in all acids except in HF with which it forms SiF_4.

$$SiO_2 + 4HF \longrightarrow SiF_4 + 2H_2O$$

(*x*) *Action of conc. and hot. NaOH.* Soluble Na_2SiO_3 is formed.

$$SiO_2 + 2NaOH \longrightarrow Na_2SiO_3 + H_2O$$

The addition of HCl to the reaction mixture gives the ppt. of SiO_2.

$$Na_2SiO_3 + 2HCl \longrightarrow 2NaCl + H_2O + SiO_2$$

Uses. The important uses of various forms of silica are summarised as follows : (*i*) Transparent varieties of quartz are used in the manufacture of prisms in optical instruments and lenses. (*ii*) Opaque varieties of quartz are used in making vessels which are acid and alkali resistant and can stand sudden changes of temperatures and, therefore, superior to glass vessels. (*iii*) Coloured varieties of quartz are used as gems. (*iv*) Sand is used in making glass, glaze porcelain, mortar, silica bricks etc., and as building material when mixed intimately with cement or lime. Silica bricks made from sand are used for lining the furnaces employed in the manufacture of steel by acid process. (*v*) Finely ground sand is used in the extraction of phosphorus from phosphates. (*vi*) Sandstone is used in making mill stones etc.

Structure of CO_2 molecule and its comparison with that of SiO_2 molecule. See chapter 15.

3. Germanium dioxide or germanic oxide, GeO_2. Preparation. It is prepared :

(*a*) by the oxidation of $GeCl_2$ to $GeCl_4$.

$$2GeCl_2 + O_2 \longrightarrow GeO_2 + GeCl_4$$

(*b*) by hydrolysing GeF_4 with H_2O

$$3GeF_4 + 2H_2O \longrightarrow GeO_2 + 2H_2GeF_4 \text{ (Fluoro germanic acid)}$$

(*c*) by the disproportionation of GeO at 700°C.

$$2GeO \xrightarrow{700° C} Ge + GeO_2$$

(*d*) by the action of HNO_3 on GeS_2.

(*e*) by the action of an alkali on $GeCl_4$.

Properties. (*i*) It is a dense white powder soluble in hot water, forming an acidic solution.

(*ii*) It is easily soluble in an alkali and dissolves in HCl with some difficulty. GeO_2 is an amphoteric oxide. With alkali germanate (GeO_3^{2-}) is formed.

$$GeO_2 + 2NaOH \longrightarrow Na_2GeO_3 + H_2O$$

(*iii*) *Action of HF.* When GeO_2 is treated with HF, GeF_4 is obtained.

$$GeO_2 + 4HF \longrightarrow GeF_4 + 2H_2$$

(*iv*) *Reduction.* GeO_2 can be reduced to Ge metal with carbon or H_2 at red heat and by metallic aluminium.

$$GeO_2 + 2C \longrightarrow Ge + 2CO \; ; \qquad GeO_2 + 2H_2 \longrightarrow Ge + 2H_2O$$

$$3GeO_2 + 4Al \longrightarrow 3Ge + 2Al_2O_3$$

Uses. GeO_2 is used in making optical glasses.

4. Tin dioxide or stannic oxide, SnO_2 (Also called **cassiterite** or **tin stone**). **Preparation.** SnO_2 is obtained : (*a*) by heating tin in air [$Sn + O_2 \longrightarrow SnO_2$].

(*b*) by heating meta stannic acid, H_2SnO_3 or SnO_2. H_2O which is produced by the action of moderately HNO_3 (sp. gr. = 1.24) on tin metal.

$$Sn + 4HNO_3 \longrightarrow H_2SnO_3 + 4NO_2 + H_2O \; ; \quad H_2SnO_3 \xrightarrow{\Delta} SnO_2 + H_2O$$

Properties. (*i*) It is soft white powder which sublimes above 1800° without melting.

(*ii*) It is insoluble in water.

(*iii*) *Action of acids.* SnO_2 is not acted by acids except conc. H_2SO_4 which gives unstable stannic sulphate, $Sn(SO_4)_2$.

$$SnO_2 + 2H_2SO_4 \text{ (conc.)} \longrightarrow Sn(SO_4)_2 + 2H_2O$$

(*iv*) *Amphoteric nature.* SnO_2 is an ampholeric oxide with predominant acidic nature. When fused with alkalies, it gives stannate (SnO_3^{2-}).

$$SnO_2 + 2NaOH \longrightarrow Na_2SnO_3 + H_2O$$

(*v*) *Reduction.* SnO_2 is reduced to Sn metal by heating it with carbon or a current of H_2.

$$SnO_2 + C \longrightarrow Sn + CO_2 \; ; \qquad SnO_2 + 2H_2 \longrightarrow Sn + 2H_2O$$

(*vi*) *Action of Cl_2 or a mixture of (Cl_2 + S_2Cl_2).* In both cases, $SnCl_4$ is obtained.

$$SnO_2 + 2Cl_2 \longrightarrow SnCl_4 + O_2$$

$$2SnO_2 + 3Cl_2 + S_2Cl_2 \longrightarrow 2SnCl_4 + 2SO_2$$

Uses. SnO_2 is used (*i*) as a polishing powder under the name *putty powder* (*ii*) for making opaque milk-glass shades, white glaze for tiles etc.

5. Lead dioxide or plumbic oxide, PbO_2. Preparation. PbO_2 is prepared.

(*a*) by boiling red lead (Pb_3O_4) with dil. HNO_3.

$$Pb_3O_4 + 4HNO_3 \longrightarrow PbO_2 + 2Pb(NO_3)_2 + 2H_2O$$

(*b*) by heating Pb (II) salts with alkalies in presence of NaOCl.

$$(CH_3COO)_2Pb + NaOCl + 2NaOH \longrightarrow PbO_2 + 2CH_3COONa + NaCl + H_2O$$

(*c*) by fusing PbO with $KClO_3$ and KNO_3 (*nitre*).

$$3PbO + KClO_3 \xrightarrow{\text{Fuse with } KNO_3} 3PbO_2 + KCl$$

(*d*) by heating the solution of $(CH_3COO)_2Pb$ with excess of bleaching powder ($CaOCl_2$) on a water bath. First of all, $Ca(OH)_2$ which is present as impurity in bleaching powder reacts with $(CH_3COO)_2Pb$ forming $Ca(OH)_2$ which reacts with $CaOCl_2$ and gives PbO_2

$$(CH_3COO)_2Pb + Ca(OH)_2 \longrightarrow Pb(OH)_2 + (CH_3COO)_2Ca$$
$$Pb(OH)_2 + CaOCl_2 \longrightarrow PbO_2 + CaCl_2 + H_2O$$

$$(CH_3COO)_2Pb + Ca(OH)_2 + CaOCl_2 \longrightarrow (CH_3COO)_2Pb + PbO_2 + CaCl_2 + H_2O$$

(*c*) by treating red lead (Pb_3O_4) with cold conc. or hot dil. HNO_3. In this reaction a chocolate brown ppt. of PbO_2 and $Pb(NO_3)_2$ in solution are obtained.

$$Pb_3O_4 + 4HNO_3 \longrightarrow PbO_2\ (ppt.) + 2Pb\,(NO_3)_2\ (solution) + 2H_2O$$

Properties. (*i*) PbO_2 is a chocolate brown powder which is insoluble in water.

(*ii*) *Stability.* It is stable upto 300°C but at higher temperatures, it decomposes into PbO and O_2.

$$2PbO_2 \xrightarrow{\Delta} 2PbO + O_2$$

(*iii*) *Oxidising property.* PbO_2 is a powerful oxidising agent, since it gives up O_2 on heating (at higher temperatures) alone or with conc. H_2SO_4.

$$2PbO_2 \xrightarrow{\Delta} 2PbO + O_2$$
$$2PbO_2 + 2H_2SO_4\ (conc.) \longrightarrow 2PbSO_4 + 2H_2O + O_2$$

Thus :

(*a*) PbO_2 oxidises conc. HCl to Cl_2 and is itself reduced to $PbCl_2$.

$$\underset{(Pb\,=\,+4)}{PbO_2} + \underset{(Cl\,=\,-1)}{4HCl\ (conc.)} \longrightarrow \underset{(Pb\,=\,+2)}{PbCl_2} + 2H_2O + \underset{(Cl\,=\,0)}{Cl_2}$$

(*b*) When a manganeous salt (*e.g.*, $MnSO_4$) is boiled with PbO_2 in presence of HNO_3, the salt is oxidised to permanganic acid, $HMnO_4$ and PbO_2 is reduced to Pb (+2) salt.

$$\underset{(Mn\,=\,+2)}{2MnSO_4} + 6HNO_3 + \underset{(Pb\,=\,+4)}{5PbO_2} \longrightarrow \underset{(Mn\,=\,+7)}{2HMnO_4} + 2H_2O + \underbrace{2PbSO_4 + 3Pb(NO_3)_2}_{(Pb\,=\,+2)}$$

(*c*) When SO_2 is passed over PbO_2, $PbSO_4$ is formed and the whole mass becomes red hot.

$$\underset{(Pb\,=\,+4)}{PbO_2} + \underset{(S\,=\,+4)}{SO_2} \longrightarrow \underset{(Pb\,=\,+2,\ S\,=\,+6)}{PbSO_4}$$

(*d*) $Cr(OH)_3$ in presence of an alkali is oxidised to a chromate (CrO_2^{2-}) and PbO_2 is reduced to plumbite (PbO_2^{2-}).

$$2Cr(OH)_3 + 10KOH + 3PbO_2 \longrightarrow 2K_2CrO_4 + 8H_2O + 3K_2PbO_2$$

(*iv*) *Action of cold conc. HCl.* PbO_2 reacts with cold conc. HCl, forming $PbCl_4$.

$$PbO_2 + 4HCl\ (cold\ and\ conc.) \longrightarrow PbCl_4 + 2H_2O$$

(*v*) *Action of hot conc. H_2SO_4.* O_2 is evolved.

$$2PbO_2 + 2H_2SO_4 \longrightarrow 2PbSO_4 + 2H_2O + O_2$$

(*vi*) *Action of hot conc. solution of an alkali.* Metaplumbite (PbO_3^{2-}) is formed.

$$PbO_2 + 2KOH \longrightarrow K_2PbO_3 + H_2O$$

(*vii*) *Action of H_2O_2.* PbO_2 is reduced to lead monoxide, PbO (litharge) by H_2O_2.

$$PbO_2 + H_2O_2 \longrightarrow PbO + O_2 + H_2O$$

Structure. PbO_2 molecule can be represented by the structure O = Pb = O.PbO_2. When treated with dil. acids, does not give H_2O_2. This oxide is, therefore, not regarded as lead peroxide.

C. Other oxides

Lead, in addition to monoxide (PbO), dioxide (PbO_2) and suboxide (Pb_2O) also forms lead sesquioxide (Pb_2O_3) and triplumbric tetroxide (Pb_3O_4). Pb_3O_4 is also called red lead or minimum (*sandur*). Here we shall discuss Pb_2O_3 and Pb_3O_4.

1. Lead sesquioxide, Pb_2O_3. Preparation. Pb_2O_3 is obtained : (*a*) as a yellow precipitate by adding sodium hypochlorite (NaOCl) to a cold solution of litharge (PbO) in NaOH.

$$2PbO + NaOCl \longrightarrow Pb_2O_3 + NaCl$$

(*b*) when solutions of sodium plumbite (Na_2PbO_2) sodium metaplumbate (Na_2PbO_3) are mixed.

$$Na_2PbO_2 + Na_2PbO_3 \longrightarrow Pb_2O_3 + 2Na_2O$$

Properties. (*i*) It is a yellow powder.

(*ii*) *Action of dil. HNO_3.* It is decomposed by dil. HNO_3 into $Pb(NO_3)_2$ and PbO_2

$$Pb_2O_3 + 2HNO_3 \longrightarrow Pb(NO_3)_2 + PbO_2 + H_2O$$

Structure of Pb_2O_3 molecule. The molecule is composed of one molecule of lead monoxide (PbO) and one molecule of lead dioxide (PbO_2). Thus Pb_2O_3 is represented as $PbO.PbO_2$. It is better represented as *plumbous metaplumbate,* $Pb^{2+}[Pb^{4+}O_3]$ in which lead is in +2 and + 4 oxidation state.

2. Triplumbic tetroxide, Pb_3O_4 [Also called *red lead* or *minimum* (*sandur*)]. **Preparation.** Pb_3O_4 is obtained when massicot (yellow lead monoxide) is heated in a current of air to about 340°C for several hours.

$$6PbO + O_2 \xrightarrow{340° C} 2Pb_3O_4$$

The temperature must be kept near 340°C, since Pb_3O_4 decomposes into PbO and O_2 above 950°C.

Properties. (*i*) It is a scarlet powder.

(*ii*) It is insoluble in water but soluble in molten KNO_3.

(*iii*) *Action of heat.* On heating, Pb_3O_4 acquires a deeper and deeper tint as time goes on and appears to become violet and finally black. But on cooling, it regains its original scarlet colour. When heated above 450°C, it gets decomposed into PbO and O_2.

$$2Pb_3O_4 \xrightarrow{> 450° C} 6PbO + O_2$$

(*iv*) *Action of HNO_3* When Pb_3O_4 is treated with cold conc. or hot dil. HNO_3, a chocolate brown ppt. of PbO_2 and $Pb(NO_3)_2$ (*in solution*) are obtained.

$$Pb_3O_4 + 4HNO_3 \longrightarrow PbO_2\ (ppt.) + 2Pb(NO_3)_2\ (solution) + 2H_2O$$

(*v*) *Action of HCl.* When treated with conc. HCl, $PbCl_2$ is formed and O_2 is evolved.

$$Pb_3O_4 + 8HCl \longrightarrow 3PbCl_2 + 4H_2O + Cl_2$$

(*vi*) *Action of H_2SO_4.* With hot and conc. H_2SO_4, O_2 is evolved.

$$2Pb_3O_4 + 6H_2SO_4 \longrightarrow 6PbSO_4 + 6H_2O + O_2$$

(*vii*) *Reduction.* It is reduced to metallic lead on heating with carbon or CO.

$$Pb_3O_4 + 4C \xrightarrow{\Delta} 3Pb + 4CO\ ; \qquad Pb_3O_4 + 4CO \longrightarrow 3Pb + 4CO_2$$

Uses. Pb_3O_4 is used : (*i*) in the manufacture of flint glass (*ii*) as a red pigment (*iii*) as a drier in oil paints (*iv*) in the manufacture of matches (*v*) for making glazes for pottery (*vi*) in making anticorrosion paint for iron (*vii*) as an oxidising agent in laboratory.

Structure of Pb_3O_4 molecule. This molecule is composed of two molecules of lead monoxide (2PbO) and one molecule of lead dioxide (PbO_2). Thus Pb_3O_4 is represented as $2PbO.PbO_2$. It is better regarded as plumbous orthoplumbate, $Pb_2^{2+}\ [Pb^{4+}\ O_4]$ which contains Pb in +2 as well as in + 4 oxidation state.

Oxides of group VA elements

The well known oxides given by the elements of group VA can be mainly of M_2O_3, M_2O_4 and M_2O_5 type. Besides these three types of oxides, N also forms three more oxides *viz*., N_2O, NO and NO_3 or N_2O_6. Bismuth also forms BiO. It may be seen that nitrogen forms oxides like

N_2O, NO and N_2O_4 but other elements do not form such oxides. Oxides of nitrogen are monomeric, while many oxides of other elements are dimeric *e.g.*, $(P_2O_3)_2$ or P_4O_6, $(P_2O_4)_2$ or P_4O_8, $(P_2O_5)_2$ or P_4O_{10} etc. The trioxides and pentoxides of nitrogen are *monomeric* (N_2O_3 and N_2O_5) while such oxides of other elements are dimeric (*e.g.*, P_4O_6, As_4O_6, Sb_4O_6, P_4O_{10}). The tetroxide of phosphorus, P_2O_4 is also dimeric and hence is represented as $(P_2O_4)_2$ or P_4O_8.

Various oxides of nitrogen can be prepared by the reactions given below :

$$NH_4NO_3 \xrightarrow{220^\circ C} N_2O \text{ (\textit{Nitrous oxide or laughing gas})} + 2H_2O$$

$$2NaNO_2 + 2FeSO_4 + 3H_2SO_4 \longrightarrow Fe_2(SO_4)_3 + 2NaHSO_4 + 2H_2O + 2NO \text{ (\textit{Nitric oxide})}$$

$$2NO + N_2O_4 \xrightarrow{250\text{ K}} 2N_2O_3 \text{ (\textit{Dinitrogen trioxide})}$$

$$2Pb(NO_3)_2 \xrightarrow{673\text{ K}} 4NO_2 \text{ (\textit{Nitrogen dioxide})} + 2PbO + O_2$$

$$2NO_2 \underset{\text{Heat}}{\overset{\text{Cool}}{\rightleftharpoons}} N_2O_4 \text{ (\textit{Dinitrogen tetroxide})}$$

Important oxides of phosphorus are phosphorus (III) oxide (P_4O_6) and phosphorus (V) oxide (P_4O_{10}). These are prepared by the reactions given below :

$$P_4 + 3O_2 \text{ (\textit{limited quantity})} \xrightarrow{\Delta} P_4O_6\,; \quad P_4 + 5O_2 \text{ (\textit{excess})} \xrightarrow{\Delta} P_4O_{10}$$

N_2O and NO are neutral oxides.

A. Trioxides, M_2O_3 (Also called sesquioxide)

All the elements form trioxides.

Preparation. (*i*) N_2O_3 is obtained: (*a*) by distilling 50% HNO_3 with As_2O_3 or starch ($C_6H_{10}O_5$).

$$As_2O_3 + 4HNO_3 + 2H_2O \longrightarrow 2N_2O_3 + 2H_3AsO_4$$

$$2C_6H_{10}O_5 + 18HNO_3 \longrightarrow 9N_2O_3 + 6H_2C_2O_4 + 13H_2O$$

(*b*) by the action of 5N HNO_3 on metallic Cu.

$$2Cu + 6HNO_3 \longrightarrow 2Cu(NO_3)_2 + 3H_2O + N_2O_3$$

(*c*) by the action of NO on solid NO_2 cooled in liq. air.

$$NO + NO_2 \rightleftharpoons N_2O_3$$

(*ii*) P_2O_3 is prepared by the action of phosphorus vapour on N_2O at 600°C at low pressure.

$$P_4 + 6N_2O \longrightarrow 2P_2O_3 + 6N_2$$

(*iii*) As_2O_3 is manufactured by roasting arsenical pyrites (FeAsS) or cobaltite (CoAsS).

$$2FeAsS + 5O_2 \xrightarrow{\text{Roasting}} Fe_2O_3 + 2SO_2 + As_2O_3.$$

$$4CoAs\,S + 9O_2 \xrightarrow{\text{Roasting}} 4CoO + 4SO_2 + 2As_2O_3$$

The roasting is done in a revolving calcining furnace.

(*iv*) Sb_2O_3 is obtained. (*a*) by burning Sb or SbH_3 (*stibine*) in air or by roasting Sb_2S_3 (*stibinite*).

$$4Sb + 3O_2 \longrightarrow 2Sb_2O_3\ ; \qquad 2SbH_3 + 3O_2 \longrightarrow Sb_2O_3 + 3H_2O$$

$$2Sb_2S_3 + 9O_2 \longrightarrow 2Sb_2O_3 + 6SO_2$$

(*b*) by hydrolysing $SbCl_3$ with hot water in presence of Na_2CO_3

$$2SbCl_3 + 3H_2O \xrightleftharpoons{Na_2CO_3} Sb_2O_3 + 6HCl$$

(*v*) Bi_2O_3 is obtained by heating Bi, $Bi(OH)_3$, $(BiO)_2CO_3$ or $Bi(NO_3)_3$ to redness in air.

$$4Bi + 3O_2 \longrightarrow 2Bi_2O_3\ ; \qquad 2Bi(OH)_3 \longrightarrow Bi_2O_3 + 3H_2O$$

$$(BiO)_2CO_3 \longrightarrow Bi_2O_3 + CO_2; \qquad 4Bi(NO_3)_3 \longrightarrow 2Bi_2O_3 + 12NO_2 + 3O_2$$

Properties. (*i*) *Physical state.* Pure N_2O_3 exists only as a solid at low temperature. The solid is pale blue and melts at –102°C.

Pure P_2O_3 is a colourless solid having m.pt. = 23.8°C and b.pt. = 173.1C. When impure, it is waxy.

As_2O_3 exists in three forms which are : (*a*) *amorphous glass* which is a colourless non-crystalline form. It has glassy appearance and is, therefore, called *bitreous arsenic.* (*b*) *octahedral form* which is a common form as it is stable under ordinary conditions and is crystalline, (*c*) *monoclinic form* which is a crystalline form.

Sb_2O_3 is a white powder which turns yellow on heating and pale buff on cooling.

Bi_2O_3 is a pale yellow fusible powder. Impure form of Bi_2O_3 is greenish-grey in colour. It exists in three crystalline modifications.

(*ii*) *Thermal stability.* Thermal stability of trioxides increases down the group. The stability of Bi_2O_3 is by far the most pronounced.

In the liquid state, N_2O_3 decomposes into NO and NO_2 giving reddish-brown fumes.

$$N_2O_3 \rightleftharpoons NO + NO_2$$

The decomposition starts at –21°C and is almost complete at about 100°C.

The vapour density shows that the gas is an equimolecular mixture of NO and NO_2, so that N_2O_3 in the gaseous state is almost completely dissociated into NO and NO_2.

When P_2O_3 is heated in a sealed tube at 440°C, phosphorus tetroxide, P_2O_4 is formed.

$$4P_2O_3 \xrightarrow[\text{tube at 440° C}]{\text{Heat in a sealed}} 3P_2O_4 + 2P\ (\textit{red})$$

P_2O_4 appears as a crystalline sublimate while red phosphorus remains as a residue.

When heated to 100°C, Sb_2O_3 is converted into Sb_2O_4 which decomposes into Sb_2O_3 at 900°C.

$$2Sb_2O_3 + O_2 \underset{900°C}{\overset{400°C}{\rightleftharpoons}} 2Sb_2O_4$$

(*iii*) *Acidic character.* On moving down the group, it is observed that the acidic character of trioxides first changes to amphoteric and finally to basic. Thus N_2O_3 and P_2O_3 are exclusively acidic, since they give with water nitrous acid (HNO_2) and phosphorus acid (H_3PO_3) respectively

$$N_2O_3 + H_2O \xrightarrow{0°C} 2HNO_2\ ;\quad P_2O_3 + 3H_2O\ (\textit{cold}) \longrightarrow 2H_3PO_3$$

Thus N_2O_3 and P_2O_3 are anhydrides of HNO_2 and H_3PO_3 respectively.

With hot water, P_2O_3 gives H_3PO_4 and PH_3.

$$2P_2O_3 + 6H_2O\ (\textit{hot}) \longrightarrow 3H_3PO_4 + PH_3$$

With an alkali, N_2O_3 gives a nitrite

$$N_2O_3 + 2NaOH \longrightarrow 2NaNO_2 + H_2O$$

As_2O_3 and Sb_2O_3 are amphoteric because they form salts with acids as well as alkalies.

$$As_2O_3 + 6HCl \longrightarrow 2AsCl_3 + 3H_2O$$

$$As_2O_3 + 6NaOH \longrightarrow 2Na_3AsO_3\ (\textit{aresenite}) + 3H_2O$$

$$Sb_2O_3 + 6HCl \longrightarrow 2SbCl_3 + 3H_2O$$

$$Sb_2O_3 + 2NaOH \longrightarrow 2NaSbO_2\ (\textit{metantimonite}) + H_2O$$

$$Sb_2O_3 + 6NaOH \longrightarrow 2Na_3SbO_3 + 3H_2O$$

Bi_2O_3 is also amphoteric oxide but is predominantly basic. It dissolves in acids to form bismuth salts and in alkalies to produce *bismuthites* (BiO_2^-).

$$Bi_2O_3 + 6HCl \longrightarrow 2BiCl_3 + 3H_2O$$

$$Bi_2O_3 + 2NaOH \longrightarrow 2NaBiO_2\ (\textit{Bismuthite}) + H_2O$$

$$Bi_2O_3 + 6NaOH \longrightarrow 2Na_3BiO_3 + 3H_2O$$

The decrease in acidic character of trioxides down the group is consistant with the increasing metallic character of the elements in passing from nitrogen to bismuth.

(*iv*) *Oxidation.* P_2O_3 is oxidised to P_2O_5 in air. It inflames at 70°C in air and at 50°C in O_2.

$$P_2O_3 + O_2 \longrightarrow P_2O_5$$

As_2O_3 (As = +3) is easily oxidised to As_2O_5 (As = + 5) or arsenic acid, H_3AsO_4 (As = +5) by H_2O_2, Cl_2, HNO_3, $FeCl_3$ etc.

$$As_2O_3 + 2H_2O_2 \longrightarrow As_2O_5 + 2H_2O$$

$$As_2O_3 + 2Cl_2 + 5H_2O \longrightarrow 2H_3AsO_4 + 4HCl$$

$$As_2O_3 + 2HNO_3 + 2H_2O \longrightarrow 2H_3AsO_4 + N_2O_3$$
$$As_2O_3 + 4FeCl_3 + 5H_2O \longrightarrow 2H_3AsO_4 + 4HCl + 4FeCl_3$$

(*v*) *Reduction*. As_3O_3 is easily reduced to arsenic as is evident from the following reactions :

$$As_2O_3 + 3C \xrightarrow{\Delta} 2As + 3CO$$
$$As_2O_3 + 3SnCl_2 + 6HCl \longrightarrow 2As + 3SnCl_4 + 3H_2O$$
$$As_2O_3 + 6HCl + 6Cu \xrightarrow{\text{Boil}} 2As + 6CuCl + 3H_2O$$

Cu-foil becomes grey due to the deposition of As.

Like As_2O_3, Bi_2O_3 is also reduced to bismuth by carbon or H_2.

$$Bi_2O_3 + 3C \xrightarrow{\Delta} 2Bi + 3CO$$

Structure. The structure of N_2O_3 and P_2O_3 molecules have already been discussed in chapter 15.

Molecular determination of arsenic trioxide in nitrobenzene has shown that this oxide is dimeric and hence should be formulated as $(As_2O_3)_2$ or As_4O_6. The structure of this oxide in the solid as well as in the vapour state is analogous to that of $(P_2O_3)_2$ or P_4O_6.

Antimony sesquoxide is dimorphic, $(Sb_2O_3)_2$ or Sb_4O_6.

Bismuth trioxide (Bi_2O_3) crystallises in several forms. One of these forms is cubic and has an ionic structre.

B. Tetroxides, M_2O_4

The acidic nature of these oxides is less than that of M_2O_5 type oxides and more than that of M_2O_3 type oxides. Acidic nature of M_2O_4 type oxides also decreases from N_2O_4 to Bi_2O_4. The decrease in acidic nature of these oxides is due to the increase in metallic nature of the elements from N to Bi.N_2O_4, P_2O_4 and As_2O_4 are acidic and hence react with water to produce a mixture of oxy-acids.

$$N_2O_4 + H_2O \longrightarrow HNO_2 + HNO_3 \; ; \qquad P_2O_4 + 3H_2O \longrightarrow H_3PO_3 + H_3PO_4$$

These oxides give salts with alkalies.

Sb_2O_4 when fused with alkalies form soluble hypoantimonates, *e.g.*, $K_2Sb_2O_5$ or $K_2O.\ Sb_2O_4$.

$$Sb_2O_4 + 2KOH \longrightarrow K_2O.Sb_2O_4 + H_2O$$

Bi_2O_4 is not affected by alkalies but gives Bi (III) salts with acids *e.g.*,

$$Bi_2O_4 + 8HCl \longrightarrow 2BiCl_3 + 4H_2O + Cl_2$$

C. Pentoxides, M_2O_5

All the elements form pentoxides, M_2O_5.

Preparation. (*i*) N_2O_5 is formed : (*a*) by dehydrating conc. HNO_3 by P_2O_5.

$$2HNO_3\ (conc.) + P_2O_5 \longrightarrow N_2O_5 + 2HPO_3$$

(*b*) by the action of dry Cl_2 on gently heated $AgNO_3$.

$$2Cl_2\ (dry) + 4AgNO_3\ (heated) \longrightarrow 4AgCl + 2N_2O_5 + O_2$$

(*c*) by passing ozonised O_2 into cooled liquid N_2O_4.

$$O_3 + N_2O_4 \longrightarrow N_2O_5 + O_2$$

(*ii*) P_2O_5 is obtained by burning P_4 in an excess of dry air or O_2.

$$P_4 + 5O_2 \longrightarrow 2P_2O_5$$

(*iii*) As_2O_5 is obtained by the oxidation of As_2O_3 by HNO_3, Cl_2 etc. The resulting solution of H_3AsO_4 (orthoarsenic acid) is concentrated to crystallisation and then heated to 200°C.

$$As_2O_3 + 2HNO_3 + 2H_2O \longrightarrow 2H_3AsO_4 + N_2O_3$$
$$As_2O_3 + 2Cl_2 + 5H_2O \longrightarrow 2H_3AsO_4 + 4HCl$$
$$2H_3AsO_4 \xrightarrow{200^\circ C} As_2O_5 + 3H_2O$$

(*iv*) Sb_2O_5 is obtained : (*a*) by heating antimony with conc. HNO_3, evaporating the solution and then gently heating the residue of $HSbO_3$ (*meta antimonic acid*)

$$Sb + 5HNO_3 \longrightarrow HSbO_3 + 2H_2O + 5NO_2$$
$$2HSbO_3 \xrightarrow{\Delta} Sb_2O_5 + H_2O$$

(*b*) by hydrolysing $SbCl_5$

$$2SbCl_5 + 5H_2O \longrightarrow Sb_2O_5 + 10HCl$$

(*v*) Bi_2O_5 has not been obtained in the pure form, since the pentapositive state of Bi has strong oxidising properties. However, it can be obtained by the following reactions :

$$Bi_2O_3 + 2KOH \longrightarrow 2KBiO_2 + H_2O$$

$$KBiO_2 + H_2O + Cl_2 \xrightarrow{\text{Oxidation}} KBiO_3 + 2HCl$$

$$KBiO_3 + HNO_3 \longrightarrow HBiO_3 + KNO_3$$

$$2HBiO_3 \xrightarrow[H_2SO_4\ (-H_2O)]{\text{Dry over}} Bi_2O_5 + H_2O$$

Properties. (*i*) *Physical state.* N_2O_5 forms white highly hygroscopic crystals. P_2O_5 is a white solid which exists in three crystalline polymorphs, in an amorphous form and also in a glassy form.

As_2O_5 is a colourless deliquescent solid. Sb_2O_5 is a yellow powder. Bi_2O_5 is an unstable brown powder.

(*ii*) *Reduction.* N_2O_5 acts as a strong oxidising agent and attacks cork, rubber etc. Hence glass apparatus is used in its preparation and storage.

P_2O_5 is reduced to red phosphorus when heated with carbon.

$$P_2O_5 + 5C \longrightarrow 2P + 5CO$$

Bi_2O_5 is reduced to Bi (III) salts when warmed with conc. HCl or H_2SO_4.

$$Bi_2O_5 + 10HCl \longrightarrow 2BiCl_3 + 5H_2O + 2Cl_2$$

$$Bi_2O_5 + 3H_2SO_4 \longrightarrow Bi_2(SO_4)_3 + 3H_2O + O_2$$

(*iii*) *Acidic character.* All the pentoxides are acidic and their acidic character diminishes quite sharply as we move from N_2O_5 to Bi_2O_5. Thus N_2O_5, P_2O_5 and As_2O_5 readily dissolve in water to form their corresponding acids.

$$N_2O_5 + H_2O \longrightarrow 2HNO_3\ (\textit{Nitric acid})$$

$$P_2O_5 + 3H_2O\ (\textit{hot}) \longrightarrow 2H_3PO_4\ (\textit{Ortho phosphoric acid})$$

$$As_2O_5 + 3H_2O \longrightarrow 2H_3AsO_4\ (\textit{Arsenic acid})$$

Actually the formation of H_3PO_4 by the action of P_2O_5 on H_2O takes place through the following steps :

(*a*) $P_2O_5 + H_2O \longrightarrow 2HPO_3$ (*metaphosphoric acid*)

(*b*) $2HPO_3 + H_2O \longrightarrow H_4P_2O_7$ (*pyrophosphoric acid*)

(*c*) $H_4P_2O_7 + H_2O \longrightarrow 2H_3PO_4$ (*orthophosphoric acid*)

On adding (*a*), (*b*) and (*c*) we get :

$$P_2O_5 + 3H_2O \longrightarrow 2HPO_4$$

The formation of H_3PO_4 indicates that P_2O_5 is called *phosphoric anhydride.* Reactions (*a*), (*b*) and (*c*) take place more rapidly, if *hot* water is used.

Sb_2O_5 is acidic and forms salts called *antimonates* (SbO_4^{3-}).

$$S_2O_5 + 6KOH \longrightarrow 2K_3SbO_4 + 3H_2O$$

Bi_2O_5 is feebly acidic and forms salts when fused with alkalies called *bismuthates* (BiO_3^-)

$$Bi_2O_5 + 2KOH \longrightarrow 2K\ BiO_3 + H_2O\ ; \qquad Bi_2O_5 + 6KOH \longrightarrow 2K_3BiO_4 + 3H_2O$$

(*vi*) *Thermal stability and oxidising properties.* Pentoxides vary considerably in their thermal stability. N_2O_5 is dissociated into N_2O_4 and O_2 even at room temperature.

$$2N_2O_5 \longrightarrow 2N_2O_4 + O_2$$

P_2O_5 is thermally stable and sublimes on heating. When heated to red heat, As_2O_5 decomposes into As_2O_3 and O_2.

$$As_2O_5 \xrightarrow{\text{Red heat}} As_2O_3 + O_2$$

When heated above 440°C, Sb_2O_5 decomposes to give Sb_2O_3 and O_2.

$$Sb_2O_5 \longrightarrow Sb_2O_3 + O_2$$

Bi_2O_5 is extremely unstable and readily loses O_2 to form Bi_2O_3.

$$Bi_2O_5 \longrightarrow Bi_2O_3 + O_2$$

Since all the pentoxides, except P_2O_5, can readily lose O_2 and form lower oxides, they are strong oxidising agents. Among these oxides, N_2O_5 is the strongest oxidising agent.

Acidic nature of oxides of a given element.

The acidic nature of the oxides of a given element in different oxidation states increases as the percentage of oxygen or the oxidation state of the element increases. For example the acidic nature of N_2O_3, N_2O_4 and N_2O_5 is in the order $N_2O_5 > N_2O_4 > N_2O_3$. Similarly the acidic character of P_2O_3, P_2O_4 and P_2O_5 varies as $P_2O_5 > P_2O_4 > P_2O_3$. These orders are confirmed by the nature of oxy-acids formed in the following reactions :

$$N_2O_3 + H_2O \longrightarrow 2HNO_2 ; \qquad P_2O_3 + 3H_2O \longrightarrow 2H_3PO_3$$
$$N_2O_4 + H_2O \longrightarrow HNO_2 + HNO_3 ; \qquad P_2O_4 + 3H_2O \longrightarrow H_3PO_4 + H_3PO_3$$
$$N_2O_5 + H_2O \longrightarrow 2HNO_3 ; \qquad P_2O_5 + 3H_2O \longrightarrow 2H_3PO_4$$

HNO_3 is stronger than HNO_2; H_3PO_4 is stronger than H_3PO_3

Oxides of Nitrogen

Following are important oxides of nitrogen. Here we shall study their various methods of preparation, properties and uses. Their structures have been discussed in chapter 15.

1. Nitrous oxide, N_2O

Preparation. It may be prepared by reducing NO with moist iron filings or by heating NH_4NO_3.

$$Fe + H_2O + 2NO \longrightarrow N_2O + Fe(OH)_2$$
$$NH_4NO_3 \xrightarrow{220°C} N_2O + 2H_2O$$

It may also be prepared by heating a mixture of $NaNO_3$ and $(NH_4)_2SO_4$ or NH_4Cl

$$2NaNO_3 + (NH_4)_2SO_4 \longrightarrow N_2O + Na_2SO_4 + 4H_2O$$
$$NaNO_3 + NH_4Cl \longrightarrow N_2O + NaCl + 2H_2O$$

Properties. (*i*) Davy (1799) has called this gas *laughing gas* as it produces hysteric laughter. On inhalation but continued inhalation causes insensibility to pain and finally death.

(*ii*) It is a colourless gas with a faint sweetish odour and taste (melting point = – 90.8°C and boiling point = – 88.7°C).

(*iii*) It is appreciably soluble in water and more so in alcohol but is insoluble in hot water. The aqueous solution is neutral and does not behave as the anhydride of hyponitrous acid ($H_2N_2O_2$), although it is formed by its decomposition.

$$H_2N_2O_2 \longrightarrow N_2O + H_2O$$

N_2O is a neutral oxide.

(*iv*) *Decomposition.* When heated, N_2O gets decomposed into N_2 and O_2.

$$2N_2O \xrightarrow[\text{temp.}]{\text{Low}} 2N_2 + O_2 ; \qquad 2N_2O \xrightarrow{1300°\,C} 2NO + N_2$$

The gas is also decomposed by electric sparks

(*v*) *Reduction.* Burning pieces of P, C, S etc., burn brilliantly in N_2O gas and free N_2 is evolved *i.e.*, N_2O is reduced to N_2.

$$P + 5N_2O \longrightarrow P_2O_5 + 5N_2 ; \qquad C + 2N_2O \longrightarrow CO_2 + 2N_2$$
$$4P\,(vapour) + 6N_2O \xrightarrow{600°\,C} 2P_2O_3 + 6N_2 ; \qquad S + 2N_2O \longrightarrow SO_2 + 2N_2$$

When passed over hot Cu, N_2O is reduced to N_2

$$Cu + N_2O \longrightarrow CuO + N_2$$

Uses. N_2O is used (*i*) as a propellant gas in "*whipped*" cream bombs (*ii*) as a mild anaesthetic in dental and other minor surgical operations when mixed with 10% O_2.

2. Nitric oxide, NO.

Preparation. (*i*) In the laboratory, NO is obtained by the action of cold dil. HNO_3 on Cu.

$$3Cu + 8HNO_3 \longrightarrow 3Cu(NO_3)_2 + 4H_2O + 2NO$$

(*ii*) Pure NO is obtained by the action of $NaNO_2$ solution on an acidified solution of KI.

$$2NaNO_2 + 2KI + 4HCl \longrightarrow 2NaCl + 2KCl + I_2 + 2H_2O + 2NO$$

or by dropping a solution of $K_4[Fe(CN)_6]$ and $NaNO_2$ into dil. CH_3COOH.

$$2K_4[Fe(CN)_6] + KNO_2 + 2CH_3COOH \longrightarrow 2K_3[Fe(CN)_6] + 2CH_3COOK + H_2O + NO$$

Properties. (*i*) NO is a colourless gas which is slightly heavier than air.

(*ii*) It is only sparingly soluble in water and is neutral oxide like N_2O.

(*iii*) Liquid NO is blue in colour and boils at –151.7°C. At –163.6°C it freezes to a blue solid.

(*iv*) *Action of solid NO_2*. The action of solid NO_2 cooled in liquid air gives N_2O_3.

$$NO_2 + NO \rightleftharpoons N_2O_3$$

(*v*) *Reduction.* (*a*) Burning phosphorus, CS_2 vapour, H_2 and heated metals (*e.g.*,, Cu, Fe, Ni, Na etc.) reduce NO (N = +2) to N_2 (N = 0).

$$4P + 10NO \longrightarrow 2P_2O_5 + 5N_2 \; ; \quad 2CS_2 + 10NO \longrightarrow 2CO + 4SO_2 + 5N_2$$

$$2H_2 + 2NO \xrightarrow{\text{Explode}} 2H_2O + N_2 \; ; \; 2Cu + 2NO \xrightarrow{\Delta} 2CuO + N_2$$

(*b*) H_2SO_3 and moist Fe filings reduce NO (N = +2) to N_2O (N = +1).

$$H_2SO_3 + 2NO \longrightarrow N_2O + H_2SO_4 \; ; \quad Fe + H_2O + 2NO \longrightarrow N_2O + Fe(OH)_2$$

(*c*) When heated with H_2 in presence of Pt metal, NO (N = +2) is reduced to NH_3 (N = – 3).

$$5H_2 + 2NO \xrightarrow{Pt} 2NH_3 + 2H_2O$$

(*d*) Chromous ion (Cr^{2+}) reduces NO (N = +2) to NH_2OH (N = –1).

$$6H + 2NO \xrightarrow{Cr^{2+}\text{ ion}} 2NH_2OH.$$

(*e*) Alkaline solution of $SnCl_2$ reduces NO to potassium hyponitrite ($K_2N_2O_2$).

$$2NO + SnCl_2 + 6KOH \longrightarrow K_2N_2O_2 + K_2SnO_3 + 2KCl + 3H_2O$$

(*vi*) *Oxidation.* (*a*) Acidified solution of $KMnO_4$ and dil. solution of I_2 oxidise NO(N = +2) to HNO_3 (N = +5).

$$2KMnO_4 + 3H_2SO_4 \longrightarrow K_2SO_4 + 3MnSO_4 + 3H_2O + 5O$$

$$2NO + 3O + H_2O \longrightarrow 2HNO_3$$

This reaction is quantitative and is used for the estimation of NO.

$$2NO + 4H_2O + 3I_2 \longrightarrow 2HNO_3 + 6HI$$

(*b*) Conc. HNO_3 and O_2 oxidise NO (N = +2) to NO_2 (N = +4)

$$2HNO_3 + NO \rightleftharpoons 3NO_2 + H_2O \; ; \quad O_2 + 2NO \longrightarrow 2NO_2$$

(*vii*) *Absorption by $FeSO_4$.* $FeSO_4$ solution absorbs NO, forming brown nitrosoferrous sulphate, $(NO)FeSO_4$.

$$FeSO_4 + NO \rightleftharpoons (NO)FeSO_4$$

(*viii*) *Action of halogens.* Excepting I_2, F_2, Cl_2 and Br_2 react with NO and give corresponding nitrosyl halides, NOX.

$$2NO + X_2 \; (X = F, Cl, Br) \longrightarrow 2NOX$$

Use. NO is uses (*i*) in the manufacture of HNO_3 (*ii*) as a catalyst in lead chamber process for the manufacture of H_2SO_4 (*iii*) to distinguish O_2 from N_2O.

3. Dinitrogen trioxide, N_2O_3

Preparation. See *"Trioxides"*.

Properties. (*i*) *Physical state.* See *"Trioxides"*.

(*ii*) *Action of water.* See *"Trioxides"*.

(*iii*) *Absorption by H_2SO_4.* H_2SO_4 absorbs NO, forming nitrososulphuric acid, $(NO)HSO_4$

$$N_2O_3 + 2H_2SO_4 \longrightarrow 2(NO)HSO_4 + H_2O$$

4. Nitrogen dioxide (NO_2) and dinitrogen tetroxide (N_2O_4)

Preparation. NO_2 is prepared : (*a*) by mixing NO with O_2

$$2NO + O_2 \longrightarrow 2NO_2$$

(*b*) by heating dry powdered Pb $(NO_3)_2$ in a hard glass tube (*Laboratory method*)

$$2Pb(NO_3)_2 \xrightarrow{\Delta} 2PbO + O_2 + 4NO_2$$

(*c*) by the action of conc. HNO_3 on Cu or Bi.

$$Cu + 4HNO_3 \longrightarrow Cu(NO_3)_2 + 2H_2O + 2NO_2$$

(*d*) by heating nitroso sulphuric acid, $(NO)HSO_4$ with dry KNO_3 .

$$(NO)HSO_4 + KNO_3 \xrightarrow{\Delta} KHSO_4 + 2NO_2$$

Properties. (*i*) NO_2 is brown while N_2O_4 is colourless.

(*ii*) *Action of heat.* NO_2 and N_2O_4 exist in a temperature dependent equilibrium.

$$2NO_2 \text{ (Brown and paramagnetic)} \underset{140^\circ C}{\overset{\text{On cooling}}{\rightleftharpoons}} N_2O_4 \text{ (Colourless and diamagnetic)}$$

In the solid state, the oxide is entirely N_2O_4 which appears to be colourless. At –11.2°C the solid N_2O_4 melts and is converted into N_2O_4 liquid. At 21.5°C, N_2O_4 is present as reddish, brown vapour. At 140°C the oxide is present as NO_2. At 620°C, NO_2 dissociates into NO and O_2. Reverse changes take place on cooling.

$$\underset{\text{colourless}}{N_2O_4 \text{ (solid)}} \overset{-11.2^\circ C}{\rightleftharpoons} N_2O_4 \text{ (liquid)} \overset{-21.5^\circ C}{\rightleftharpoons} N_2O_4 \text{ (vapour)} \underset{\text{on cooling}}{\overset{-140^\circ C}{\rightleftharpoons}} 2NO_2 \overset{620^\circ C}{\rightleftharpoons} 2NO + O_2$$

The above scheme of changes shows that when NO is cooled, it gets dimerised into N_2O_4 (vapour) which, on being heated to 140°C, gets dissociated into NO_2.

(*iii*) *Action of metals and non-metals.* Heated sodium burns in NO_2 while potassium inflames spontaneously in the gas.

$$K + 2NO_2 \longrightarrow KNO_3 + NO$$

Strongly burning P and C burn in it because at high temperature, NO_2 dissociates into NO and O_2.

(*iv*) *Acidic nature.* N_2O_4 is an acidic oxide, since it dissolves in H_2O and gives a mixture of HNO_2 and HNO_3. Thus N_2O_4 is called a *mixed acid anhydride.*

$$3 \times [N_2O_4 + H_2O \longrightarrow HNO_2 + HNO_3] \quad ...(a)$$

HNO_2 produced as above decomposes into HNO_3 and NO.

$$3HNO_2 \longrightarrow HNO_3 + H_2O + 2NO \quad ...(b)$$

Add equations (*a*) and (*b*) to get :

$$3N_2O_4 + 2H_2O \longrightarrow 4HNO_3 + 2NO$$

N_2O_4 is absorbed by an alkali, forming a mixture of nitrite and nitrate.

$$N_2O_4 + 2KOH \longrightarrow KNO_2 + KNO_3 + H_2O$$

(*v*) *Action of conc. H_2SO_4.* Conc. H_2SO_4 absorbs N_2O_4 and forms nitroso sulphuric acid, $(NO)HSO_4$ and HNO_3

$$N_2O_4 + H_2SO_4 \rightleftharpoons (NO)HSO_4 + HNO_3$$

(*vi*) *Oxidising properties (Reduction).* (*a*) NO_2 (N = + 4) oxidises SO_2 to H_2SO_4 in presence of H_2O, H_2S to S, CO to CO_2 and liberates I_2 from KI solution and is itself reduced to NO (N = +2).

$$NO_2 + SO_2 + H_2O \longrightarrow H_2SO_4 + NO \ ; \qquad NO_2 + H_2S \longrightarrow S + NO + H_2O$$

$$NO_2 + CO \longrightarrow CO_2 + NO; \qquad NO_2 + 2KI + H_2O \longrightarrow I_2 + NO + 2KOH$$

(*b*) NO_2 (N = + 4) oxidises H_2(H = 0) to H_2O (H = +1) and is reduced to NH_3 (N = –3).

$$2NO_2 + 7H_2 \xrightarrow[\text{Pt. metal}]{\text{Heated}} 2NH_3 + 4H_2O$$

(*c*) NO_2 (N = + 4) oxidises Cu (Cu = 0) heated to bright redness to CuO (Cu = +1) and is itself reduced to N_2(N = 0).

$$4Cu + 2NO_2 \longrightarrow 4CuO + N_2$$

(*vii*) *Oxidation.* NO_2 (N = + 4) is oxidised to HNO_3 (N = +5) by means of acidified $KMnO_4$ solution.

$$10NO_2 + 2KMnO_4 + 3H_2SO_4 + 2H_2O \longrightarrow 10HNO_3 + 2MnSO_4 + K_2SO_4$$

(*viii*) *Action of ozonised O_2.* When ozonised O_2 (*i.e.*, O_3) is passed into cooled liquid N_2O_4, N_2O_5 is obtained.

$$N_2O_4 + O_3 \longrightarrow N_2O_5 + O_2$$

5. Dinitrogen pentoxide, N_2O_5

Preparation. See "*Pentoxides*".

Properties. (*i*) N_2O_5 forms white crystals which are highly hygroscopic and are suitable below 0°C but decompose slowly into N_2O_4 and O_2 and become yellow at ordinary temperature.

$$2N_2O_5 \longrightarrow 2N_2O_4 + O_2$$

(*ii*) *Acidic nature.* N_2O_5 dissolves in H_2O with a hissing noise giving HNO_3. Thus N_2O_5 is called *nitric anhydride.*

$$N_2O_5 + H_2O \longrightarrow 2HNO_3$$

With alkalies, N_2O_5 gives nitrates, *e.g.*, $N_2O_5 + 2NaOH \longrightarrow 2NaNO_3 + H_2O$.

(*iii*) *Action of P, K, C etc.* P and K burn in liquid N_2O_5, if slightly warmed. Charcoal is not attacked by liquid N_2O_5 even on boiling but burns brilliantly, if previously kindled.

(*iv*) *Oxidising action.* N_2O_5 is a powerful oxidising agent and attacks cork, rubber etc. Hence glass apparatus is used for its preparation and storage.

(*v*) *Action of anhydrous H_2O_2.* When N_2O_5 is dissolved in anhydrous H_2O_2, peroxynitric acid (HNO_4) is obtained.

$$N_2O_5 + H_2O_2 \longrightarrow HNO_4 + HNO_3$$

Structure of the oxides of nitrogen and ions derived from them

The structure and shape/geometry of the oxides and oxy-ions formed by nitrogen have already been discussed in chapter 15. From this discussion it may be seen that in all the species, N-atom (central atom) is *sp* or *sp*^{2} hybridised and all contain *sigma* as well as *pi* bonds. Each of the π-bonds is a $(p - p)$ π-bond.

Oxides of Phosphorus

Following are the important oxides of phosphorus. Here we shall study their preparation, properties, uses and structure.

1. Phosphorus trioxide or phosphorus sesquioxide, P_2O_3 or P_4O_6

The oxide is also called *anhydride of phosphorus acid* (H_3PO_3), since it may be regarded as having been formed by removing $3H_2O$ molecules from $2H_3PO_3$ molecules.

$$2H_3PO_4 \xrightarrow{-3H_2O} P_2O_3 + 3H_2O$$

Preparation. P_2O_3 is prepared : (*a*) by the combustion of yellow or white phosphorus in a limited quantity of air.

$$4P + 3O_2 \longrightarrow 2P_2O_3$$

(*b*) by the action of nitrous oxide (N_2O) on phosphorus vapours at low pressure at 600°C.

$$4P\ (vapours) + 6N_2O \xrightarrow[600°\ C]{\text{Low pressure}} 2P_2O_3 + 6N_2$$

P_2O_3 obtained by both these methods contains some phosphorus as impurity. This impurity can be removed by repeated crystallisation from carbon disulphide.

Properties. (*i*) Pure P_2O_3 is a colourless solid (m.p = 23.8°C and b.p. = 173.1°C). When impure it is snowwhite waxy solid. It smells like garlic and is very posinuous.

(*ii*) *Oxidation by air.* See "*Trioxides*".

(*iii*) *Acidic nature.* See "*Trioxides*".

(*iv*) *Action of C_2H_5OH.* P_2O_3 ignites in C_2H_5OH at ordinary temperature, but if it is mixed with C_2H_5OH at 273 K, diethyl phosphite, $(C_2H_5)_2HPO_3$ is formed.

$$2P_2O_3 + 8C_2H_5OH \xrightarrow{\text{Mix}} 4(C_2H_5)_2HPO_3 + 2H_2O$$

(*v*) *Action of heat.* See "*Trioxides*".

(*vi*) *Action of P_2O_5.* When P_2O_3 is heated with P_2O_5 in a sealed tube at 290°C, P_2O_4 is obtained.

$$P_2O_3 + P_2O_5 \xrightarrow{\Delta} 2P_2O_4$$

(*vii*) *Action of Cl_2.* P_2O_3 reacts vigorously with Cl_2 forming phosphoryl chloride ($POCl_3$) and metaphosphoryl chloride (PO_2Cl).

$$P_2O_3 + 2Cl_2 \longrightarrow POCl_3 + PO_2Cl$$

(*viii*) *Action of sulphur.* P_2O_3 combines violently with sulphur forming crystalline solid having the composition, $P_4O_6S_4$.

$$2P_2O_3 + 4S \longrightarrow P_4O_6S_4$$

Uses. P_2O_3 is used for preparing phosphorus acid (H_3PO_3) and phosphites which are used as good reducing agents.

2. Phosphorus tetroxide, P_2O_4 or $(P_2O_4)_2$ or P_4O_8

Preparation. It is formed by heating P_2O_3 in a sealed tube at 440°C or with P_2O_5 in a sealed tube at 290°C.

$$4P_2O_3 \xrightarrow[\text{tube at } 440^\circ C]{\text{Heat in a sealed}} 3P_2O_4 + 2P\ (\textit{red})$$

$$P_2O_3 + P_2O_5 \xrightarrow[\text{tube at } 290^\circ C]{\text{Heat in a sealed}} 2P_2O_4$$

Properties. (*i*) P_2O_4 is a colourless crystalline solid which sublimes in vacuum at 180°.

(*ii*) *Action of H_2O.* Mixture of H_3PO_3 and H_3PO_4 is obtained.

$$P_2O_4 + 3H_2O \longrightarrow H_3PO_3 + H_3PO_4$$

Thus like N_2O_4, P_2O_4 can be considered to be a *mixed anhydride of H_3PO_3 and H_3PO_4.*

3. Phosphorus pentoxide, P_2O_5 or P_4O_{10}

Preparation. It is obtained by burning phosphorus in an excess of dry air or O_2.

$$4P + 5O_2 \xrightarrow{\text{Burn}} 2P_2O_5$$

Properties. (*i*) *Physical state.* See "*Pentoxides*".

(*ii*) Pure P_2O_5 is odourless while impure form has a garlic odour due to the presence of P_2O_3.

(*iii*) *Action of water : Acidic character.* See "*Pentoxides*".

(*iv*) *Dehydrating property.* Because of the fact that P_2O_5 can combine with one molecule of water (H_2O) and form HPO_3, this oxide removes H_2O molecule from inorganic and organic compounds and renders them anhydrous. Thus, P_2O_5 can remove H_2O molecule from H_2SO_4, HNO_3, CH_3CONH_2 (acetamide), CH_3COOH, C_2H_5OH etc., and thus acts as a dehydrating agent. Examples are given below:

$$H_2SO_4 \xrightarrow{-H_2O} SO_3 + H_2O$$

$$P_2O_5 + H_2O \longrightarrow 2HPO_3$$

$$\overline{H_2SO_4 + P_2O_5 \longrightarrow SO_3 + 2HPO_3}$$

$$2HNO_3 + P_2O_5 \longrightarrow N_2O_5 + 2HPO_3$$

$$CH_3CONH_2 + P_2O_5 \longrightarrow CH_3CN + HPO_3$$

$$2CH_3COOH + P_2O_5 \longrightarrow (CH_3CO)O + 2HPO_3$$

$$C_2H_5OH + P_2O_5 \longrightarrow H_2C = CH_2 + 2HPO_3$$

$$2HClO_4 + P_2O_5 \longrightarrow Cl_2O_7 + 2HPO_3$$

The last reaction has been used for the preparation of Cl_2O_7.

It is due to the dehydrating property that P_2O_5 chars wood, paper, sugar etc. It should be noted that since P_2O_5 reacts with NH_4OH to form NH_4PO_3, it cannot be used for dehydrating NH_4OH.

(*v*) *Action of NH_4OH. Ammonium metaphosphate, NH_4PO_3* is formed.

$$P_2O_5 + 2NH_4OH \longrightarrow 2NH_4PO_3 + H_2O$$

(*vi*) *Reduction by carbon.* See "Pentaoxides".

(*vii*) *Action of basic oxides.* When P_2O_5 reacts with basic oxides like CaO, phosphate is obtained, *e.g.*,

$$3\,CaO + P_2O_5 \longrightarrow Ca_3(PO_4)_2\,; \qquad 3Na_2O + P_2O_5 \longrightarrow 2Na_3PO_4$$

(*viii*) *Action of malonic acid and its ester.* When a large excess of P_2O_5 reacts with malonic acid [$CH_2(COOH)_2$] or its ester like diethyl malonate at 300°C under 12 mm pressure, carbon suboxide (C_3O_2) is formed.

$$CH_2(COOH)_2 + 2P_2O_5 \longrightarrow C_3O_2 + 4HPO_3$$

$$CH_2(COOC_2H_5)_2 + 2P_2O_5 \longrightarrow C_3O_2 + 4HPO_3 + 2C_2H_4$$

(*xi*) *Action of H_2O_2.* When P_2O_5 is treated with 30% H_2O_2 at low temperature and then the reaction mixture is diluted with H_2O and cooled, a solution of peroxy-monophosphoric acid, H_3PO_5 is obtained.

$$P_2O_5 + 2H_2O_2 + H_2O \xrightarrow[\text{temperature}]{\text{Low}} 2H_3PO_5 \ (\textit{solution})$$

Uses. P_2O_5 is used (*i*) a drying and dehydrating agent (*ii*) for the preparation of orthophosphoric and (H_3PO_4) and peroxy-monophosphoric acid (H_3PO_5).

Structure of the oxides of phosphorus. The measurement of vapour density of the oxides of phosphorus has shown that phosphorus trioxide (P_2O_3) and phosphorus pentoxide (P_2O_5) exist as dimers and hence should be formulated as $(P_2O_3)_2$ or P_4O_6 and $(P_2O_5)_2$ or P_4O_{10} respectively. The vapour density of phosphorus tetroxide at 1800°C $(P_2O_4)_4$ indicates that this oxide exists as a tetramer, $(P_2O_4)_4$ or P_8O_{16}. Each of these oxides has tetrahedral shape. The four corners of the tetrahedron are occupied by four P-atoms. Six O-atoms in all the three oxides make a bridge between two P-atoms of P_4 tetrahedron. In all the three oxides P-atom is sp^3 hybridised. The structure of these oxides has been explained in detail in chapter 15.

Oxides of group VIA elements

Excepting Se, all the elements form monoxides. Dioxides are given by all the elements and with the exception of Po all the remaining elements give trioxides.

1. Monoxides (MO)

Excepting Se, all other elements give monoxides (SO, TeO, PoO). SO is produced by passing slow discharge through SO_2 while TeO and PoO are obtained by heating $TeSO_3$ and $PoSO_3$ respectively.

2. Dioxides (MO_2)

Preparation. These are prepared as follows :

$$Cu + 2H_2SO_4 \longrightarrow CuSO_4 + 2H_2O + SO_2\,; \qquad S + O_2 \longrightarrow SO_2$$

$$\underset{\text{Iron pyrites}}{4FeS_2} + 11O_2 \longrightarrow 2Fe_2O_3 + 8SO_2\,; \qquad \underset{\text{Zinc blende}}{2ZnS} + 3O_2 \longrightarrow 2ZnO + 2SO_2$$

$$Se + 4HNO_3 \longrightarrow SeO_2 + 2H_2O + 4NO_2\,; \qquad Te + O_2 \longrightarrow TeO_2$$

Properties. (*i*) *Physical state.* SO_2 is a gas (b.pt. = –10°C). SeO_2 is a white volatile solid which sublimes on heating (sublimation temperature = 315°C). TeO_2 and PoO_2 are nonvolatile crystalline solids and occur in two crystalline forms each.

(*ii*) *Stability.* SO_2 is more stable than SeO_2.

(*iii*) *Acidic/amphoteric character.* SO_2 and SeO_2 both are soluble in water and give acids.

$$H_2O + SO_2 \rightleftharpoons H_2SO_3 \ (\textit{Sulphurous acid})$$

$$H_2O + SeO_2 \longrightarrow H_2SeO_3 \ (\textit{Selenious acid})$$

H_2SO_3 exists only in solution and hence cannot be isolated. H_2SeO_3 can be isolated in the crystalline state. Due to the production of H_2SO_3 and H_2SeO_3, SO_2 and SeO_3 both show acidic character and are known as anhydrides of H_2SO_3 and H_2SeO_3 respectively. TeO_2 and PoO_2 are insoluble in water. TeO_2 dissolves in alkalies to form tellurites (TeO_3^{2-}) and in acids TeO_2 gives basic salts [*e.g.*, $2TeO_2.HNO_3$ or $Te_2O_3(OH)NO_3$]. This shows that TeO_2 has amphoteric character.

PoO_2 also has amphoteric character. Thus we see that the acidic character of the dioxides decreases as we move in the series $SO_2 \rightarrow SeO_2 \rightarrow TeO_2 \rightarrow PoO_2$.

Trioxides, MO_3

Preparation. SO_3 is prepared by heating conc. H_2SO_4 with a large excess of P_2O_5 or by distilling fuming sulphuric acid, $H_2S_2O_7$.

$$H_2SO_4\ (conc.) + P_2O_5 \longrightarrow SO_3 + 2HPO_3$$

$$H_2S_2O_7 \rightleftharpoons H_2SO_4 + SO_3$$

SeO_3 is obtained by subliming a mixture of selenic acid (H_2SeO_4) and P_2O_5 in vacuum.

TeO_3 is formed by heating telluric acid (H_6TeO_6) to about 360°C but below red heat.

$$H_6TeO_6 \xrightarrow{\Delta} TeO_3 + 3H_2O$$

Properties. (*i*) *Action of water.* SO_3 dissolves in H_2O with a loud hissing noise and evolution of heat forming H_2SO_4.

$$SO_3 + H_2O \rightleftharpoons H_2SO_4$$

SO_3 is, therefore, regarded as the anhydride of H_2SO_4.

SeO_3 dissolves in water to form selenic acid (H_2SeO_4). Hence it is known as an anhydride of H_2SeO_4.

$$H_2O + SeO_3 \longrightarrow H_2SeO_4$$

Thus SO_3 and SeO_3 both are acidic oxides. TeO_3 is insoluble in water.

(*ii*) *Thermal stability.* SO_3 is much more thermally stable than SeO_3 and TeO_3. Thus whereas SO_3 is only slightly decomposed into SO_2 and O_2 at 400°C, SeO_3 and TeO_3 are appreciably dissociated to MO_2 and O_2 around 200°C.

$$2MO_3 \xrightarrow{\Delta} 2MO_2 + O_2$$

Oxides of sulphur

Main oxides formed by sulphur are discussed below :

Sulphur sesquioxide, S_2O_3

It is obtained by dissolving sulphur in molten SO_3 at 15°C.

$$S + SO_3\ (molten) \longrightarrow S_2O_3$$

S_2O_3 is a bluish-green crystalline solid. It slowly decomposes giving S and SO_2.

$$2S_2O_3 \longrightarrow 2 + 3SO_2$$

With water it gives S and a mixture of H_2SO_3 and H_2SO_4.

The fact that S_2O_3 does not give hyposulphurous acid ($H_2S_2O_4$) with water does not prove that S_2O_3 is not an anhydride of $H_2S_2O_4$, since the latter (*i.e.*, $H_2S_2O_4$) decomposes, giving S_2O_3 and H_2O.

$$H_2S_2O_4 \longrightarrow S_2O_3 + H_2O$$

Sulphur dioxide, SO_2.

Preparation. SO_2 is prepared : (*i*) by burning sulphur or sulphide ores like iron phyrites (FeS_2), zinc blende (ZnS) etc. in the excess of air. (*Commercial method*).

$$S + O_2 \longrightarrow SO_2$$

$$4FeS_2 + 11O_2 \longrightarrow 2Fe_2O_3 + 8SO_2$$

$$2ZnS + 3O_2 \longrightarrow 2ZnO + 2SO_2$$

(*ii*) by reducing conc. H_2SO_4 by Cu turnings (*laboratory method*), Ag, Hg, S or charcoal. In these reactions H_2SO_4 is reduced to SO_2 and Cu, Ag and Hg are oxidised to their respective sulphates while S and charcoal are oxidised to SO_2 and CO_2 respectively. Thus H_2SO_4 acts as an oxidising agent.

(*a*) $Cu + 2H_2SO_4 \longrightarrow CuSO_4 + SO_2 + 2H_2O$

or $SO_4^{2-} + 4H^+ + Cu \longrightarrow SO_2 + 2H_2O + Cu^{2+}$

(*b*) $2Ag + 2H_2SO_4 \longrightarrow Ag_2SO_4 + SO_2 + 2H_2O$

or $2Ag + SO_4^{2-} + 4H^+ \longrightarrow 2Ag^+ + SO_2 + 2H_2O$

(*c*) $Hg + 2H_2SO_4 \longrightarrow HgSO_4 + 2H_2O + SO_2$

or $Hg + 4H^+ + SO_4^{2-} \longrightarrow Hg^{2+} + 2H_2O + SO_2$

(*d*) $2H_2SO_4 + S \longrightarrow 2H_2O + 3SO_2$

or $SO_4^{2-} + 4H^+ + S \longrightarrow 3SO_2 + 2H_2O$

(*e*) $C + 2H_2SO_4 \longrightarrow 2SO_2 + CO_2 + 2H_2O$

or $2SO_4^{2-} + 4H^+ + C \longrightarrow 2SO_2 + CO_2 + 2H_2O$

(*iii*) by heating gypsum ($CaSO_4.2H_2O$) or anhydrite ($CaSO_4$) with carbon at about 1000°C. (*Commercial method*)

$$2CaSO_4 + C \longrightarrow 2CaO + CO_2 + 2SO_2$$

This reaction is used at those places where gypsum or anhydrite is found in abundence.

(*iv*) by treating Na_2SO_3 or $NaHSO_3$ with dil. HCl or conc. H_2SO_4.

$Na_2SO_3 + 2HCl \longrightarrow 2NaCl + H_2O + SO_2$ or $SO_3^{2-} + 2H^+ \longrightarrow H_2O + SO_2$

$NaHSO_3 + HCl \longrightarrow NaCl + H_2O + SO_2$ or $HSO_3^- + H^+ \longrightarrow H_2O + SO_2$

$Na_2SO_3 + H_2SO_4 \longrightarrow Na_2SO_4 + H_2O + SO_2$ or $SO_3^{2-} + 2H^+ \longrightarrow H_2O + SO_2$

$NaHSO_3 + H_2SO_4 \longrightarrow NaHSO_3 + H_2O + SO_2$ or $HSO_3^- + H^+ \longrightarrow H_2O + SO_2$

Properties. (*i*) SO_2 is a colourless gas which has suffocating odour. It is heavier than air. It can be liquefied to a colourless liquid at –10°C and into a snow-like solid at – 76°C.

(*ii*) It neither burns nor helps in burning. However heated K, Mg, Sn and Fe burn in the gas and are oxidised.

$4K + 3SO_2 \longrightarrow K_2SO_3 + K_2S_2O_3$; $3M$ (M = Mg, Fe) $+ SO_2 \longrightarrow 2MO + MS$.

(*iii*) *Action of lime water, $Ca(OH)_2$.* When SO_2 is passed into lime water, the latter becomes milky due to the formation of the ppt. of $CaSO_3$.

$$Ca(OH)_2 + SO_2 \longrightarrow CaSO_3 \text{ (white ppt.)} + H_2O$$

(*iv*) *Decomposition.* (*a*) When heated to 1200°C, it gets decomposed to an appreciable extent, giving SO_3 and S.

$$3SO_2 \xrightarrow{1200^\circ C} 2SO_3 + S$$

(*b*) When acted upon by electric discharge at low temperature, it dissociates into sulphur monoxide (SO) and O_2.

$$2SO_2 \rightleftharpoons 2SO + O_2$$

(*v*) *Acidic nature.* SO_2 is an acidic oxide, since its aqueous solution shows acidic properties due to the formation of sulphurous acid, H_2SO_3.

$$SO_2 + H_2O \rightleftharpoons H_2SO_3$$

SO_2 is, therefore, known as *sulphurous anhydride.* Being acidic, SO_2 neutralises alkalies and decomposes carbonates forming sulphites (SO_3^{2-}) and bisulphites (HSO_3^-) in both the cases. It also reacts with metallic (basic) oxides to form sulphites.

$SO_2 + 2NaOH \longrightarrow Na_2SO_3 + H_2O$; $SO_2 + Na_2CO_3 \longrightarrow Na_2SO_3 + CO_2$

$2SO_2 + Na_2CO_3 + H_2O \longrightarrow 2NaHSO_3 + CO_2$; $SO_2 + CaO \longrightarrow CaSO_3$

(*vi*) *Formation of addition compounds.* (*a*) We know that maximum covalency of S-atom is six. In SO_2 molecule S-atom is only tetracovalent. In order to attain its maximum covalency of six, SO_2 directly combines with O_2, Cl_2 and lead dioxide (PbO_2) or barium peroxide (BaO_2), giving addition compounds.

$2SO_2 + O_2 \xrightleftharpoons{\text{Pt or } V_2O_5} 2SO_3$; $SO_2 + Cl_2 \xrightarrow[\text{or camphor}]{\text{Sunlight}} SO_2Cl_2$

$$SO_2 + PbO_2 \xrightarrow{\Delta} PbSO_4$$

(*b*) Liquid SO_2 also forms addition compounds with NaI, KCNS, $AlCl_3$ etc. Examples are NaI.$2SO_2$, NaI.$4SO_2$, 9KCNS.SO_2, $AlCl_3.SO_2$ etc.

(*vii*) *Reducing properties.* In presence of moisture SO_2 can liberate nascent hydrogen and hence acts as a reducing agent. When SO_2 reduces a certain substance, it gets oxidised to SO_3 or H_2SO_4.

$$SO_2 + H_2O \longrightarrow SO_3 + 2H$$
or $$SO_2 + H_2O \longrightarrow SO_3 + 2H^+ + 2e^-$$
$$SO_2 + 2H_2O \longrightarrow H_2SO_4 + 2H$$
or $$SO_2 + 2H_2O \longrightarrow SO_4^{2-} + 4H^+ + 2e^-$$

Thus :

(*a*) SO_2 reduces halogens (X_2) to their corresponding hydracids (HX).

$$SO_2 + 2H_2O + X_2 \longrightarrow H_2SO_4 + 2HX$$
or $$SO_2 + 2H_2O + X_2 \longrightarrow SO_4^{2-} + 2X^- + 4H^+$$

(*b*) SO_2 reduces acidified solution of $KMnO_4$ to $MnSO_4$.

$$2KMnO_4 + 5SO_2 + 2H_2O \longrightarrow K_2SO_4 + 2MnSO_4 + 2H_2SO_4$$
or $$2MnO_4^- + 5SO_2 + 2H_2O \longrightarrow 5SO_4^{2-} + 4H^+ + 2Mn^{2+}$$

(*c*) SO_2 reduces acidified $K_2Cr_2O_7$ to green $Cr_2(SO_4)_3$.

$$K_2Cr_2O_7 + 3SO_2 + H_2SO_4 \longrightarrow K_2SO_4 + Cr_2(SO_4)_3 + H_2O$$
or $$Cr_2O_7^{2-} + 3SO_2 + 2H^+ \longrightarrow 3SO_4^{2-} + H_2O + 2Cr^{3+}$$

(*d*) SO_2 reduces $Fe_2(SO_4)_3$ to $FeSO_4$

$$Fe_2(SO_4)_3 + SO_2 + 2H_2O \longrightarrow 2FeSO_4 + 2H_2SO_4$$
or $$2Fe^{3+} + SO_2 + 2H_2O \longrightarrow 2Fe^{2+} + SO_7^{2-} + 4H^+$$

(*e*) Iodic acid (HIO_3) or an acidified solution of an iodate (IO_3^-) is reduced to I_2 Chlorate (ClO_3^-) is not reduced to Cl_2 but to a chloride (Cl^-). SO_2 is oxidised to H_2SO_4 in both the cases.

$$5SO_2 + 2KIO_3 + 4H_2O \longrightarrow 4H_2SO_4 + K_2SO_4 + I_2$$
or $$5SO_2 + 2IO_3^- + 4H_2O \longrightarrow 5SO_4^{2-} + 8H^+ + I_2$$
$$3SO_2 + NaClO_3 + 3H_2O \longrightarrow 3H_2SO_4 + NaCl$$
or $$3SO_2 + ClO_3^- + 3H_2O \longrightarrow 3SO_4^{2-} + 6H^+ + Cl^-$$

(*f*) SO_2 reduces $Hg_2(NO_3)_2$ to metallic Hg.

$$Hg_2(NO_3)_2 + SO_2 + 2H_2O \longrightarrow 2Hg + 2HNO_3 + H_2SO_4$$
or $$Hg_2^{2+} + SO_2 + 2H_2O \longrightarrow 2Hg + 4H^+ + SO_4^{2-}$$

In the following examples SO_2 is oxidised to SO_3.

(*a*) H_2O_2 is reduced to H_2O.

$$SO_2 + H_2O_2 \longrightarrow SO_3 + H_2O$$

(*b*) HNO_3 is reduced to NO_2.

$$SO_2 + 2HNO_3 \longrightarrow SO_3 + H_2O + 2NO_2$$

Above reactions have been used for the manufacture of SO_3.

SO_2 can also be oxidised to SO_3 by the action of atmospheric O_2 in presence of V_2O_5 (catalyst) at 400°C and by the action of O_3 on SO_2.

$$2SO_2 + O_2 \text{ (from air)} \xrightarrow[400°\,C]{V_2O_5} 2SO_3\ ; \qquad 3SO_2 + O_3 \longrightarrow 3SO_3$$

(*viii*) *Oxidising properties.* SO_2 (S = + 4) oxidises particularly stronger reducing agents. When it oxidises a certain substance, it is reduced to sulphur (S = 0) in most cases. In some cases it is also reduced to sulphide (S^{2-}) or thiosulphate, $S_2O_3^{2-}$ (S = +2).

$$SO_2 + 4H^+ + 4e^- \longrightarrow S + 2H_2O$$
$$SO_2 + 4H^+ + 6e^- \longrightarrow S^{2-} + 2H_2O$$
$$2SO_2 + 2H^+ + 4e^- \longrightarrow S_2O_3^{2-} + H_2O$$

Thus :

(*a*) It oxidises H_2S to S, CO to CO_2 H_2 to H_2O (at 1000°C), $SnCl_2$ to $SnCl_4$ (in presence of HCl) and Hg_2Cl_2 to $HgCl_2$ (in presence of HCl). SO_2 is reduced to sulphur in each case.

$$2H_2S + SO_2 \longrightarrow 2H_2O + 3S\ (colloidal) \quad \text{or} \quad 2S^{2-} + SO_2 + 4H^+ \longrightarrow 2H_2O + 3S$$

$$2CO + SO_2 \longrightarrow 2CO_2 + S; \quad 2H_2 + SO_2 \xrightarrow{1000°\ C} 2H_2O + S$$

$$2SnCl_2 + SO_2 + 4HCl \longrightarrow 4SnCl_4 + 2H_2O + S$$

or $$2Sn^{2+} + SO_2 + 4H^+ \longrightarrow 4Sn^{4+} + 2H_2O + S$$

$$2Hg_2Cl_2 + SO_2 + 4HCl \longrightarrow 4HgCl_2 + 2H_2O + S$$

or $$2Hg_2^{2+} + SO_2 + 4H^+ \longrightarrow 4Hg^{2+} + 2H_2O + S$$

(*b*) It oxidises heated Mg and Fe to MgO and FeO respectively and is itself reduced to S^{2-}.

$$3M\ (M = Mg, Fe) + SO_2 \longrightarrow 2MO + MS$$

or $$3M + SO_2 + 4H^+ \longrightarrow 3M^{2+} + S^{2-} + 2H_2O$$

(*c*) It oxidises burning K to K_2SO_3 (K = +1) and is itself reduced to thiosulphate, $S_2O_3^{2-}$ (S = +2).

$$4K + 3SO_2 \longrightarrow K_2SO_3 + K_2S_2O_3 \quad \text{or} \quad 4K + 2SO_2 + 2H^+ \longrightarrow 4K^+ + S_2O_3^{2-} + H_2O$$

(*ix*) *Bleaching property.* We have said above that, in presence of moisture, SO_2 can liberate nascent hydrogen.

$$SO_2 + 2H_2O \longrightarrow H_2SO_4 + 2H$$

This nascent hydrogen reduces the coloured vegetable matter to colourless reduction matter. Thus SO_2 in presence of moisture, acts as a bleaching agent.

$$\text{Coloured vegetable matter} + H \xrightarrow{\text{Reduction}} \text{Colourless reduction matter (Bleached matter).}$$

On standing in air, the colourless reduction matter is reoxidised by atmospheric O_2 and hence becomes coloured. Thus the bleaching done by SO_2 is temperatory.

$$\text{Colourless reduction matter} \xrightarrow[\text{atmospheric } O_2]{\text{oxidation by}} \text{Coloured vegetable matter.}$$

In certain cases, SO_2 forms a colourless addition product with coloured vegetable matter. If the colourless addition product is treated with H_2SO_4, it gains it original colour and SO_2 is liberated.

$$\text{Coloured vegetable matter} \underset{\text{Decomposition by } H_2SO_4}{\overset{\text{Addition of } SO_2}{\rightleftharpoons}} \text{Colourless addition product}$$

(*x*) *Action of Zn-dust.* When SO_2 is passed through a suspension of Zn-dust in absolute alcohol, zinc hyposulphite (ZnS_2O_4) is obtained.

$$2SO_2 + Zn\ (in\ alcohol) \longrightarrow ZnS_2O_4$$

Uses. SO_2 is used : (*i*) in the manufacture of SO_3 and H_2SO_4. (see *Reducing properties of SO_2*) (*ii*) as a bleaching agent for delicate articles like wool, silk and straw (*iii*) in refiring of petroleum and sugar. (*iv*) as a disinfactant for killing desease germs, fungi and certain moulds. (*v*) in the preparation of $NaHSO_3$ and $Ca(HSO_3)_2$ which are used in paper industry (*vi*) as a refrigerant in the form of liquid SO_2 (*vii*) as an antichlor for removing Cl_2 from a fabric after bleaching.

Sulphur trioxide, SO_3

Preparation. SO_3 is prepared : (*i*) by oxidising SO_2 to SO_3. SO_2 can be oxidised to SO_3.

(*a*) by passing a dry mixture of SO_2 and O_2 over spongy platinum or platinised asbestos heated to 400°-450°C in a glass tube. The reaction is exothermic.

$$2SO_2 + O_2 \rightleftharpoons 2SO_3, \Delta H = -196.6 \text{ kJ}$$

SO_3 formed is collected in a receiver cooled in a freezing mixture.

(*b*) by Cl_2 or Br_2 in presence of moisture.

$$X_2 \text{ (X = Cl, Br)} + H_2O + SO_2 \longrightarrow SO_3 + 2HX$$

(*c*) by H_2O_2

$$H_2O_2 + SO_2 \longrightarrow SO_3 + H_2O$$

(*d*) by O_3

$$O_3 + 3SO_2 \longrightarrow 3SO_3$$

In this reaction, the whole O_2 is used up in the process of oxidation.

(*e*) by HNO_3

$$2HNO_3 + SO_2 \longrightarrow 2NO_2 + SO_3 + H_2O$$

(*f*) by NO_2

$$NO_2 + SO_2 \longrightarrow SO_3 + NO$$

(*ii*) by dehydrating conc. H_2SO_4 with excess of P_2O_5.

$$H_2SO_4 + P_2O_5 \longrightarrow SO_3 + 2HPO_3$$

(*iii*) by heating $Fe_2(SO_4)_3$ or dry $FeSO_4$.

$$Fe_2(SO_4)_3 \xrightarrow{\Delta} Fe_2O_3 + 3SO_3 \ ; \ 2FeSO_4 \xrightarrow{\Delta} Fe_2O_3 + SO_3 + SO_2$$

(*iv*) by distilling fuming sulphuric acid, $H_2S_2O_7$.

$$H_2S_2O_7 \rightleftharpoons H_2SO_4 + SO_3$$

(*v*) by heating $NaHSO_4$ at 300°C.

$$2NaHSO_4 \xrightarrow[-H_2O]{\Delta} Na_2S_2O_7 \xrightarrow{\Delta} Na_2SO_4 + SO_3$$

Properties. (*i*) *Allotropic forms.* SO_3 exists in three allotropic forms *viz.*, (*a*) *α-form.* It is chemically active form and forms long transparent ice-like crystals. Its m.pt. = 17°C. In the solid state, this form exists as a trimer, $(SO_3)_3$. (*b*) *β-form.* It is dimeric, $(SO_3)_2$. It forms needle-like silky white crystals. It melts at 32.5°C. Above 50°C, it changes to α-form. In the solid state this form has linear layer structure. (*c*) *γ-form.* It is like β-form and is obtained by completely drying β-form. It melts at 62.2°C under 2 atmospheric pressure. In the solid state, this form, like β-form, has linear layer structure.

(*ii*) *Acidic nature.* SO_3 is an acidic oxide, since it dissolves in H_2O, forming sulphuric acid (H_2SO_4).

$$SO_3 + H_2O \longrightarrow H_2SO_4$$

SO_3 is, therefore, called *sulphuric anhydride.*

Acidic nature of SO_3 is also supported by the fact that it forms sulphates with basic (metallic) oxides. and carbonates. $SO_3 + CaO \rightarrow CaSO_4$; $SO_3 + Na_2O \rightarrow Na_2SO_4$; $Na_2CO_3 + SO_3 \rightarrow Na_2SO_4 + CO_2$

With BaO so much heat is evolved that the mass becomes incandescent.

$$SO_3 + BaO \longrightarrow BaSO_4 + \text{Heat}$$

(*iii*) *Action of conc. HCl.* Chlorosulphonic acid [$(OH)SO_2Cl$] is obtained.

$$SO_3 + HCl \longrightarrow (OH)SO_2Cl$$

(*iv*) *Action of conc. H_2SO_4.* SO_3 dissolves in conc. H_2SO_4 and forms oleum, $H_2S_2O_7$ which is also called fuming sulphuric acid or pyrosulphuric acid.

$$SO_3 + H_2SO_4 \longrightarrow H_2S_2O_7$$

(*v*) *Action of heat.* SO_3 dissociates on heating at 1000°C into SO_2 and O_2.

$$2SO_3 \xrightarrow{1000° C} 2SO_2 + O_2$$

(*vi*) *Oxidising property.* SO_3 also acts as an oxidising agent. When it oxidises a certain substance, it gets reduced to SO_2. Examples are :

$$5SO_3 + 2P \longrightarrow 5SO_2 + P_2O_5; \qquad SO_3 + PCl_5 \longrightarrow SO_2 + Cl_2 + POCl_3$$

$$SO_3 + 2HBr \longrightarrow SO_2 + Br_2 + H_2; \qquad 2SO_3 + S \xrightarrow{1000^\circ C} 3SO_2$$

(*vii*) *Action of anhydrous H_2O_2*. Peroxy monosulphuric acid (H_2SO_5) is formed.

$$H_2O_2\ (anhydrous) + SO_3 \longrightarrow H_2SO_5$$

Uses. SO_3 is used : (i) in the manufacture of H_2SO_4 and oleum ($H_2S_2O_7$), (ii) as a drying agent for a gases.

Sulphur heptoxide, S_2O_7

It is formed by passing a silent electric discharge in a mixture of SO_2 and O_2.

$$4SO_2 + 3O_2 \longrightarrow 2S_2O_7$$

S_2O_7 is a liquid and liberates O_2 from H_2O.

$$2S_2O_7 + 4H_2O \longrightarrow 4H_2SO_4 + O_2$$

Structure of dioxide and trioxide molecules (*a*) *Structure of dioxide molecules, MO_2.* Structure of SO_2, SeO_2, TeO_2 and PoO_2 molecules have been discussed in chapter 15.

(*b*) *Structure of trioxide molecules MO_3.* MO_3 molecules in the gaseous state exists as monomers and have trigonal planar geometry. In the solid state, α-form of SO_3 exists as a trimer, $(SO_3)_3$ which has ring structure. On the other hand β-and γ-forms of SO_3 solid have linear layer structure.

In the solid state, SeO_3 exists as a tetramer, $(SeO_3)_4$ which has a cyclic structure. For the detailed structure of MO_3 molecules (See chapter 15).

Binary halogen-oxygen binary compounds

Halogens form a large number of halogen-oxygen binary compounds. The binary compounds of F_2 with O_2 are called *fluorides of oxygen* because of the greater electronegativity of fluorine as compared to that the oxygen while those of Cl, Br and I are called *oxides of halogens,* since oxygen is more electronegative than each of the three halogens.

Fluorides of oxygen have been discussed along with "*Halides of group VIA elements*". The oxides of Cl, Br and I are discussed here.

Oxides of chlorine

The known oxides given by chlorine are discussed below :

1. **Chlorine monoxide, Cl_2O**

It is formed by passing dry Cl_2 over freshly-precipitated yellow HgO (previously heated to 300 - 400°C) contained in a cooled tube.

$$2HgO + 2Cl_2 \longrightarrow HgCl_2.HgO + Cl_2O$$

Properties. (*i*) It is a brownish yellow gas having a penetrating odour. It condenses in a freezing mixture to an orange coloured liquid (b.pt. = 2°C and m.pt. = –120.6°C). Unlike OF_2, it is very unstable and explosive.

(*ii*) *Exploding property.* It explodes readily on heating or sparking giving Cl_2 and O_2.

$$2Cl_2O \longrightarrow 2Cl_2 + O_2$$

Liquid Cl_2O may explode, if the tube containing it is screatched with a file.

Its mixture with NH_3 explodes violently. The overall reaction may be represented as :

$$3Cl_2O + 10NH_3 \longrightarrow 2N_2 + 6NH_4Cl + 3H_2O$$

(*iii*) *Action of H_2O.* Unlike OF_2, Cl_2O dissolves in water, forming a golden yellow solution of hypochlorous acid (HClO).

$$Cl_2O + H_2O \rightleftharpoons 2HClO$$

Cl_2O is, therefore, called anhydride of HClO. On cooling the aqueous solution strongly, a crystalline hydrate, $Cl_2O.H_2O$ (m.pt. = –36°C) is obtained.

(*iv*) *Oxidising property.* Cl_2O is a powerful oxidising agent. For example it oxidises HCl (Cl = –1) to Cl_2 (Cl = 0).

$$Cl_2O + 2HCl \longrightarrow H_2O + 2Cl_2$$

2. Chlorine dioxide, ClO_2

It is obtained when a mixture of powdered $KClO_3$ and well cooled conc. H_2SO_4 is *cautiously* warmed or when $KClO_3$ is treated with $H_2C_2O_4$.

$$2KClO_3 + 3H_2SO_4 \longrightarrow 3KHSO_4 + HClO_4 + H_2O + 2ClO_2$$
$$2KClO_3 + 2H_2C_2O_4 \longrightarrow K_2C_2O_4 + 2CO_2 + 2H_2O + 2ClO_2$$

(*b*) *Pure.* ClO_2 is obtained when dry Cl_2 is passed over $AgClO_3$ heated to 90°C.

$$Cl_2 + 2AgClO_3 \longrightarrow 2AgCl + O_2 + 2Cl_2O$$

Properties. (*i*) ClO_2 is a dark yellow highly unstable pungent smelling gas and heavier than air.

(*ii*) The gaseous ClO_2 can be condensed to a dark red liquid, which freezes at –59°C to an orange-red crystalline solid.

(*iii*) *Decomposition.* The liquid and solid ClO_2 are slowly decomposed, when exposed to light.

$$2ClO_2 + H_2O \longrightarrow HClO_2 + HClO_3$$

(*iv*) *Action of alkalies.* With alkalies, ClO_2 gives a mixture of chlorite and chlorate.

$$2ClO_2 + 2KOH \longrightarrow KClO_2 + KClO_3 + H_2O$$

(*v*) *Oxidising properties.* ClO_2 is a powerful oxidising agent in acidic medium.

$$ClO_2 + 4H^+ + 5e^- \longrightarrow Cl^- + 2H_2O$$

(*vi*) P, S and many organic compounds spontaneously catch fire in the gas.

(*vii*) *Explosion.* The liquid is violently explosive. The gas explodes by an electric spark, on heating with a glass rod or hot wire and in contact with terpentine, alcohol or ether. On heating, it gives, Cl_2 and O_2.

$$2ClO_2 \longrightarrow Cl_2 + 2O_2$$

(*viii*) *Action of H_2O_2 or Na_2O_2.* When ClO_2 is acted upon by H_2O_2 or Na_2O_2, chlorous acid ($HClO_2$) or sodium chlorite ($NaClO_2$) is obtained.

$$2ClO_2 + H_2O_2 \longrightarrow 2HClO_2 + O_2; \qquad 2ClO_2 + Na_2O_2 \longrightarrow 2NaClO_2 + O_2$$

(*ix*) *Action of water.* ClO_2 dissolves in water and gives a mixture of chlorom acid ($HClO_2$) and chloric acid ($HClO_3$).

$$2ClO_2 + H_2O \longrightarrow HClO_2 + HClO_3$$

Thus ClO_2 can be considered to be a mixed anhydride of $HClO_2$ and $HClO_3$.

(*x*) *Action of O_3.* Cl_2O_6 is formed.

$$2ClO_2 + 2O_3 \xrightarrow{0°C} Cl_2O_6 + 2O_2$$

Uses. ClO_2 is used : (*i*) for bleaching the highest grade paper pulp, (*ii*) in water purification and for odour control, (*iii*) for improving the quality of low-grade fats and oils, (*iv*) in making the wool unshrinkable.

3. Chlorine hexoxide, Cl_2O_6

It is prepared by the action of ClO_2 (at 0°C) or Cl_2 on O_3.

$$2ClO_2 + 2O_3 \xrightarrow{0°C} Cl_2O_6 + 2O_2\ ; \qquad Cl_2 + 4O_3 \longrightarrow Cl_2O_6 + 3O_2$$

Properties. (*i*) It is a dark red oily liquid (density 1.65) which melts at 3.5°C.

(*ii*) It is less explosive than ClO_2. It reacts explosively with organic matter like stop cock grease and other reducing agents.

(*iii*) *Decomposition.* It is unstable and decomposes into ClO_2 and O_2 even at its melting point.

$$Cl_2O_6 \longrightarrow 2ClO_2 + O_2$$

(*iv*) *Action of H_2O and alkalies.* With cooled water vapour, a mixture of per chloric acid monohydrate; $HClO_4.H_2O$ (also called hydronium perchlorate, $[H_3O][ClO_4]$ and chloric acid ($HClO_3$) is obtained.

$$Cl_2O_6 + 2H_2O \longrightarrow HClO_4.H_2O + HClO_3$$

or

$$Cl_2O_6 + 2H_2O \longrightarrow [H_3O]\,[ClO_4] + HClO_3$$

The formation of ClO_4^- and ClO_3^- ions by the action of H_2O shows that Cl_2O_6 is a mixed anhydride of $HClO_4$ and $HClO_3$.

Cl_2O gives a mixture of chlorate (ClO_3^-) and perchlorate (ClO_4^-) ions with OH^- ions.

$$Cl_2O_6 + 2OH^- \longrightarrow ClO_3^- + ClO_4^- + H_2O$$

(*v*) *Action of HF.* With HF, Cl_2O_6 reacts as follows :

$$Cl_2O_6 + HF \longrightarrow HClO_4 + ClO_2F$$

(*vi*) Liquid Cl_2O_6 has no unpaired electrons and hence is diamagnetic. However, its vapours are paramagnetic due to the dissociation of Cl_2O_6 molecule into chlorine trioxide (ClO_3) which has one unpaired electron (odd-electron molecule) and hence is paramagnetic. Thus the following equilibrium exists between Cl_2O_6 and ClO_3.

$$Cl_2O_6 \text{ (Liquid : Diamagnetic)} \rightleftharpoons 2ClO_3 \text{ (Vapour: Paramagnetic)}$$

Structure of ClO_3 and Cl_2O_6 molecules. See "*Odd electron molecules/ion*" discussed in chapter 10.

4. Chlorine heptoxide, Cl_2O_7

It is prepared by the dehydration of anhydrous $HClO_4$ by P_2O_5 at –10°C followed by vacuum distillation with precautions against explosion.

$$2HClO_4 + P_2O_5 \longrightarrow Cl_2O_7 + 2HPO_3$$

Properties. (*i*) It is a colourless oily liquid (b.p = 82°C).

(*ii*) It is more stable than Cl_2O or ClO_2 and may be poured on P, S and wood or paper without explosion.

(*iii*) *Action of H_2O and alkalies.* It forms $HClO_4$

$$Cl_2O_7 + H_2O \longrightarrow 2HClO_4\ ; \qquad Cl_2O_7 + 2OH^- \longrightarrow 2ClO_4^- + H_2O$$

Thus Cl_2O_7 is an anhydride of $HClO_4$.

(*iv*) Cl_2O_7 is the only exothermic oxide of chlorine and is thermodynamically more stable than any of the other oxides of chlorine. (*v*) Cl_2O_7 is not so strong an oxidising agent as the other oxides of chlorine. (*vi*) Cl_2O_7 is less reactive than the lower oxides of chlorine.

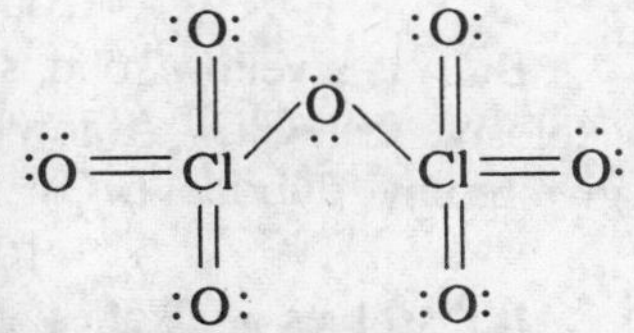

Fig. 28.8 : Electron dot structure of Cl_2O_7 molecule.

Structure. Electron dot structure of Cl_2O_7 molecule is as shown in Fig. 28.8. Electron diffraction analysis reveals that two tetrahedra are linked through an O-atom in its structure. ClOCl and OClO bond angles are equal to 115° and 97° respectively. (See Fig. 28.9). Thus Cl_2O_7 can be represented as $O_3Cl—O—ClO_3$. Cl_2O_7 molecule is polar having $\mu = 0.72$.

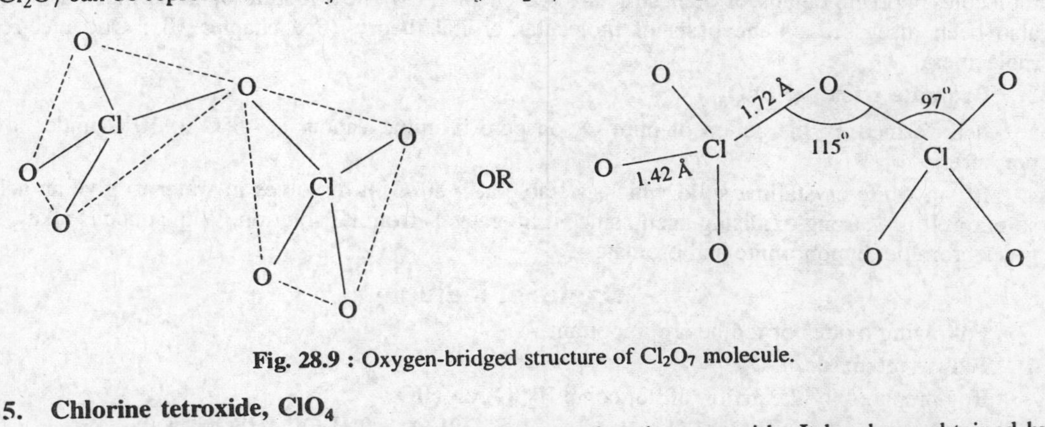

Fig. 28.9 : Oxygen-bridged structure of Cl_2O_7 molecule.

5. Chlorine tetroxide, ClO_4

This oxide should be regarded as a free radical rather than an oxide. It has been obtained by the action of I_2 on an etheral solution of $AgClO_4$.

$$2AgClO_4 + I_2 \longrightarrow 2Ag\downarrow + 2ClO_4$$

When the solution obtained after filtering AgI is shaken with water, it gets hydrolysed to produce $HClO_4$.

$$4ClO_4 + 2H_2O \longrightarrow 4HClO_4 + O_2$$

Oxides of bromine

Following are the oxides of bromine.

1. Bromine monoxide, Br_2O

It can be prepared by the action of Br_2 vapour on completely dry HgO at 50-60°C.

$$2Br_2\ (vapour) + 2HgO\ (dry) \xrightarrow{50-60^\circ} HgO.\ HgBr_2 + Br_2O$$

It is a drak brown liquid which freezes at –17.5°. It is the anhydride of HBrO.

$$Br_2O + H_2O \longrightarrow 2HBrO$$

BrO_2 reacts with an alkali and gives a mixture of bromide and bromate. Thus in this reaction BrO_2 undergoes disproportionation.

$$\underset{(Br = +4)}{6BrO_2} + 6NaOH \longrightarrow \underset{(Br = -1)}{NaBr} + \underset{(Br = +5)}{5NaBrO_3} + 3H_2O$$

BrO_2 oxidises I_2 to I_2O_5 and is itself reduced to Br_2.

$$5BrO_2 + I_2 \longrightarrow I_2O_5 + 5Br_2$$

Structure. The structure of Br_2O is similar to that of Cl_2O and OF_2. The structure of all three monoxides *viz*., OF_2, Cl_2O and Br_2O has been explained at a single place in chapter 14.

2. Bromine dioxide, BrO_2

It can be prepared by low temperature ozonalysis of Br_2.

$$Br_2 + 4O_3 \xrightarrow{-78^\circ} 2BrO_2 + 4O_2$$

BrO_2 is a yellow solid, stable below – 40°C. At 0°C it decomposes into its elements. It is less explosive than ClO_2. At low temperatures, it reacts with strong fluorinating agents such as BrF_5 to give bromyl fluoride, BrO_2F.

$$10BrO_2 + 2BrF_5 \longrightarrow Br_2 + 10BrO_2F$$

It dissolves in alkalies, gives bromides and bromates.

$$6BrO_2 + 6OH^- \longrightarrow Br^- + 5BrO_3^- + 3H_2O$$

Structure of ClO_2 and BrO_2 molecules. The structure of both these molecules is the same as that of NO_2 molecule. Like NO_2, both these molecules have angular geometry and are odd-electron molecules. For the details of their structure see chapter 15. The structure of these molecules has also been discussed on the basis of molecular orbital theory. (See chapter 10 : Odd electron molecules)]

3. Bromine trioxide, BrO_3

It is formed by the action of pure O_3 on pure bromine vapour at –5°C to 10°C under low pressure.

It is a white crystalline solid which is stable at – 80°C. It dissolves in water to give an acid solution. It is a strong oxidising agent, since it liberates I_2 from KI solution. With aqueous alkali, it gives bromide, hypobromite and bromate.

Oxides of iodine

Following oxides of iodine are important :

1. Iodine tetroxide, I_2O_4

It,is prepared by the action of hot conc. H_2SO_4 on HIO_3.

$$4HIO_3 + 2H_2SO_4\ (conc.) \longrightarrow 2I_2O_4 + 2[H_3O]^+\ [HSO_4]^- + O_2$$

or

$$4HIO_3 \xrightarrow{conc.\ H_2SO_4} 2I_2O_4 + O_2 + 2H_2O$$

Properties. (*i*) It is yellow solid.

(*ii*) *Action of H_2O.* It is sparingly soluble in cold water. With hot water it gives a solution containing HIO_3 and I_2 is liberated.

$$5I_2O_4 + 4H_2O\ (hot) \longrightarrow 8\ HIO_3 + I_2$$

(*iii*) *Action of alkalies.* It slowly dissolves in KOH solution to give a mixture of I_2 and iodate.

$$3I_2O_4 + 6KOH \longrightarrow KI + 5KIO_3 + 3H_2O$$

(*iv*) *Action of heat.* When heated to 130°C, it is decomposed to produce I_2O_5 and I_2.

$$5I_2O_4 \xrightarrow{\Delta} 4I_2O_5 + I_2$$

(*v*) *Action of HCl.* It gives Cl_2 and ICl.

$$I_2O_4 + 8HCl \longrightarrow 3Cl_2 + 2ICl + 4H_2O$$

Structure. It is an ionic compound and is represented as iodosyl iodate, $[IO]^+ [IO_3]^-$

2. Iodine iodate, I $(IO_3)_3$ or I_4O_9

It is prepared by the action of O_3 on dry I_2.

$$2I_2\ (dry) + 9O_3 \longrightarrow I_4O_9 + 9O_2$$

It is a pale yellow solid, deliquescent in nature. It reacts with H_2O, giving I_2 and HIO_3.

$$5I_4O_9 + 9H_2O \longrightarrow I_2 + 18HIO_3$$

When heated to 75°C, it is decomposed into I_2O_5, I_2 and O_2.

$$4I_4O_9 \xrightarrow{\Delta} 6I_2O_5 + 2I_2 + 3O_2$$

I_4O_9 is regarded as iodine iodate, $I^+ [IO_3]_3^-$.

3. Iodine pentoxide, I_2O_5

Preparation. It is prepared : (*a*) by heating I_4O_9 at 75°C.

$$4I_4O_9 \xrightarrow{75°\ C} 6I_2O_5 + 2I_2 + 3O_2$$

(*b*) by heating HIO_3 to about 240°C.

$$2HIO_3 \xrightarrow{240°\ C} I_2O_5 + H_2O$$

(*c*) by heating I_2 with conc. HNO_3.

$$I_2 + 10\ HNO_3 \longrightarrow I_2O_5 + 10\ NO_2 + 5H_2O$$

(*d*) by heating para periodic acid (H_5IO_6) to 138°C.

$$2H_5IO_6 \xrightarrow{138°\ C} I_2O_5 + 5H_2O + O_2$$

Properties. (*i*) It is a white solid stable upto 300°C. Beyond this temperature I_2O_5 melts and gets decomposed into I_2 and O_2.

$$2I_2O_5 \xrightarrow{> 300°\ C} 2I_2 + 5O_2$$

(*ii*) *Acidic nature.* I_2O_5 readily dissolves in water forming iodic acid (HIO_3).

$$I_2O_5 + H_2O \longrightarrow 2HIO_3$$

I_2O_5 is, therefore, an anhydride of HIO_3. I_2O_5 is, however, insoluble in organic solvents.

(*iii*) *Oxidising properties.* I_2O_5 is are oxidising agent. Thus :

(*a*) I_2O_5 oxidises CO to CO_2 and is itself reduced to I_2.

$$I_2O_5 + 5CO \longrightarrow 5CO_2 + I_2$$

It is due to this property of I_2O_5 that this oxide is used for the estimation of CO.

(*b*) I_2O_5 liberates I_2 from KI solution. This liberated iodine can be estimated by titrating it with standard solution of $Na_2S_2O_3$.

(*c*) I_2O_5 oxidises H_2S to S and is itself reduced to I_2.

$$I_2O_5 + 5H_2S \longrightarrow 5S + I_2 + 5H_2O$$

(*d*) I_2O_5 oxidises HCl to Cl_2 and is itself reduced to ICl_3.

$$10HCl + I_2O_5 \longrightarrow 2Cl_2 + 2ICl_3 + 5H_2O$$

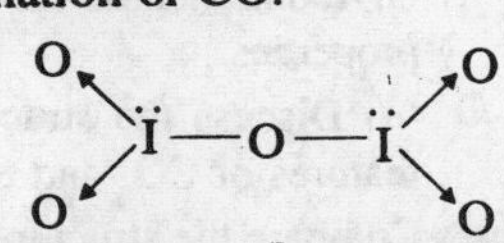

Fig. 28.10 : Oxygen-bridged structure of I_2O_5 molecule.

Uses. I_2O_5 is used (*i*) as an oxidising agent (*ii*) for the estimation of CO.

Structure. The infrared spectrum of I_2O_5 molecule indicates that it has a planar structure which has been shown in Fig. 28.10. Thus I_2O_5 has oxygen-bridged structure which can be represented as $O_2I—O—IO_2$. (See Fig. 28.10).

Oxides of xenon

Xenon forms two oxides namely xenon trioxide (XeO_3) and xenon tetraoxide (XeO_4). In these oxides Xe is present in + 6 and + 8 oxidation state respectively.

1. Xenon trioxide, XeO_3 (Xe = + 6)

XeO_3 is formed (*a*) when XeF_4 or XeF_6 is hydrolysed.

$$2XeF_4 + 3H_2O \longrightarrow Xe\,(g) + XeO_3\,(s) + 6HF\,(aq) + F_2\,(g)$$
$$XeF_6 + 3H_2O \longrightarrow XeO_3 + 6HF$$

(*b*) when sod. perxenate, $Na_4\,XeO_6.8H_2O$ is dissolved in an acid solution.

$$2Na_4\,XeO_6 + 8HCl \longrightarrow 8\,NaCl + 2XeO_3 + 4H_2O + O_2$$

Properties. (*i*) It is a white non-volatile in water.

(*ii*) XeO_3 is an endothermic compound and has a large positive value of its enthalpy of formation, Δ Hf (= + 401.7 kJ mol^{-1}).

(*iii*) When in aqueous solution, it is quite stable, but it is highly explosive when dry. It can be detonated by simple rubbing, pressing or slight heating and produces about the same effect as TNT does. Thus, its explosive power can be comparred to that of TNT.

(*iv*) *Oxidising action.* XeO_3 is a powerful oxidising agent. Thus it oxidises Pu^{3+} to Pu^{4+} in acidic solution and is itself reduced to xenon gas.

$$6Pu^{3+} + XeO_3 + 6H^+ \longrightarrow 6Pu^{4+} + Xe + H_2O$$

(*v*) *Nature of aqueous solution.* Aqueous solution of XeO_3 shows weakly acidic character, since it gives xenic acid (H_2XeO_4) in aqueous solution.

$$XeO_3 + H_2O \longrightarrow H_2XeO_4 \longrightarrow H^+ + HXeO_4^-$$

(*vi*) *Action of an alkali.* When XeO_3 is dissolved in an alkali like NaOH, it undergoes disproportionation and produces sod. per xenate, $Na_4X_2O_6$. $8H_2O$.

$$2XeO_3 + 4NaOH + 6H_2O \longrightarrow Xe + O_2 + Na_4XeO_6.8H_2O$$

or

$$2XeO_3 + 4OH^- \longrightarrow Xe + O_2 + XeO_6^{4-} + 2H_2O$$

2. Xenon tetraoxide, XeO_4 (Xe = + 8)

It can be prepared : (*a*) by the action of anhydrous H_2SO_4 on barium perxenate, Ba_2XeO_6.

$$Ba_2XeO_6 + 2H_2SO_4\,(anhydrous) \longrightarrow XeO_4 + 2BaSO_4 + 2H_2O$$

(*b*) by the action of conc. H_2SO_4 on Na_4XeO_6.

$$Na_4\,XeO_6 + 2H_2SO_4\,(conc.) \longrightarrow XeO_4 + 2Na_2\,SO_4 + 2H_2O$$

XeO_4 is an unstable gas at room temperature and is a yellow solid at low temperature. It decomposes to give xenon and O_2.

Structure of xenon oxides. The structure/geometry of XeO_3 and XeO_4 has already been explained in chapter 15.

University Questions

1. Describe how heavy water is prepared by Taylor, Erying and Frost method. *(Nagpur 2002, 2003)*
2. Give one method for the preparation of H_2O_2. Write one example each for its oxidising and reducing properties. *(VBS Purvanchal 2003)*
3. (*a*) Discuss the structures of N_2O_3 and N_2O_5 molecules. (*b*) Compare and contrast the significant features of CO_2 and SiO_2. *(Rohilkhand 2003)*
4. Compare the structure and bonding involved in N_2O and NO_2. *(Delhi 2003)*
5. Name the various oxides of sulphur.
6. Explain why CO_2 is a gas while SiO_2 is a solid. 3. Draw the structure of P_4O_{10}. *(Bhopal 2004, GND Amritsar 2004)*

29

Halides and Oxy-halides of Main Group Elements

The salts given by halogen acids *viz*., HF, HCl, HBr and HI are called *"halides"*.

Types of halides and oxy-halides formed by main group elements

I. Ionic or salt-like halides

These halides exist as ionic or semi-ionic crystals. Since alkali and alkaline earth metals have relatively low ionisation energy, the halides of these metals are ionic in nature. Some of the metals, in their lower oxidation states (*e.g.*, Fe^{2+}, Mn^{2+} etc.), also give essentially ionic halides. Due to high lattice energies, almost all the metallic fluorides are ionic substances.

Preparation. Ionic halides may be prepared by heating the metal, metal oxide, metal carbonate, with the corresponding halogen acid, *e.g.*,

$$Zn + 2HF \longrightarrow ZnF_2 + H_2$$

$$CaO + 2HCl \longrightarrow CaCl_2 + H_2O$$

$$Na_2CO_3 + 2HF \longrightarrow 2NaF + CO_2 + H_2O$$

Properties. (1) Ionic halides have high melting points and boiling points, because the forces holding together the individual MX units in the crystal lattice, are stronger. The fluoride has generally the highest melting point and boiling point, in a series of ionic halides of a given element. Ionic halides conduct electricity, in fused state. Ionic halides do not undergo extensive hydrolysis.

(2) *Covalent / ionic character and related properties.* The halides of alkali metals and alkaline earth metals are truely ionic in nature. The anhydrous halides of other elements are more covalent than ionic in the solid state, although they are soluble in polar solvents and given ionic solutions. The following points may be noted :

(*i*) If we consider the halides of the main group elements, belonging to the same period and having different oxidation state (*i.e.*, positive charge), it may be seen that the covalent character of these halides increases with the increase in the oxidation state of the cation. Due to the increase in covalent character, the melting points of the halides decrease in the same direction. For example, in the chlorides of Na^+, Mg^{2+} and Al^{3+} (3rd period), NaCl is ionic, while $AlCl_3$ is essentially covalent. Melting points of these chlorides are as: NaCl = 800° C, $MgCl_2$ = 712°C, $AlCl_3$ sublimes at 180°C.

Explanation. See Fajans' rules.

(*ii*) The covalent character, in the halides of a given element belonging to main group elements, increases with the increase in the size and decrease in the electronegativity of the halogen. For example :

(*a*) The covalent character of the halides of Li increases as LiF < LiCl < < LiBr < LiI. Due to the successive increase in the covalent character, the melting points of these halides show the expected decrease, as shown in bracket (LiF = 870° C, LiCl = 613°C, LiBr = 547°C, LiI = 446°C).

(*b*) Among the trihalides of Al, the covalent character increases, on moving from AlF_3 to AlI_3. Thus, AlF_3 is essentially ionic (least covalent), $AlCl_3$ dinner has intermediate character (but more covalent than ionic) while the dinners of $AlBr_3$ and AlI_3 show predominantly covalent character. This increasing order of covalent character is supported by the fact that the melting points of these halides decrease on moving from AlF_3 to AlI_3 (AlF_3 = 1270°C, $AlCl_3$ sablimes at 180°C, $AlBr_3$ = 975°C, AlI_3 = 383° C).

Explanation. See *"Fajans' rules"*.

(*iii*) If a given metal belonging to the main group elements forms two halides, the halide of the metal in higher oxidation state is largely covalent, while that of the same metal in lower oxidation state, is largely ionic. For example, while $PbCl_4$ (Pb = + 4) is largely covalent, $PbCl_2$ (Pb = +2) is largely ionic.

Explanation. See Fajans' rules.

(*iv*) The fluorides of metallic elements are essentially ionic, whereas the other halides may or may not be so.

(3) *Solubility of halides of different metals in water.* (*i*) Due to large values of the lattice energy of the fluorides of Li and alkaline earth metals (*i.e.*, LiF, CaF_2, BaF_2), these are insoluble in water. (*ii*) With the increase in the size of the halide ion, the solubility of the halides of a given alkali metal also increases, in the order MCl < MBr < MI. (*iii*) All the halides of Na are soluble. (*iv*) CaF_2 is insoluble, while $CaCl_2$, $CaBr_2$ and CaI_2 are soluble. (*v*) BaF_2 is insoluble, while $BaCl_2$, Ba Br_2 and Ba I_2 are soluble. (*vi*) $PbCl_2$ is sparingly soluble (*vii*) Cuprous chloride (Cu_2Cl_2) and mercurous chloride (Hg_2Cl_2) both are insoluble. (*viii*) AgF_2 is soluble, while AgCl, AgBr and AgI are insoluble.

(4) *Oxidation of metallic halides to halogens.* See *"Oxidising properties of halogens"* discussed in chapter 18.

II. Covalent or acidic halides

These halides are given by non-metals, as well as by metals with high charge / size ratio. Examples are F_4, CCl_4, $SiCl_4$, NF_3, PCl_3, PBr_3, IF_3, BrF_3, ICl_3, $COCl_2$, NOCl, $POCl_3$, etc.

Preparation. These are conveniently prepared by direct combination of the constituent elements, *e.g.*,

$$P_4 + 6Br_2 \longrightarrow 4PBr_3$$

Some of the chlorides are obtained by passing dry Cl_2 over a heated mixture of non-metallic oxide and carbon, *e.g.*,

$$SiO_2 + 2C + 2Cl_2 \longrightarrow SiCl_4 + 2CO$$

Properties. (*i*) Covalent halides generally exist as discrete molecules. These halides have low melting points and boiling points. The iodide generally has the highest melting point and boiling point, in a series of covalent halides of a given element. This is because of larger molecular weight and consequently high van der Waals attraction. In the halides of amphoteric elements, such as Al (III), Sn (IV) etc., the fluorides generally have the highest melting and boiling point, with a sharp drop to the chloride and then a slow increase to the iodide *via* bromide (See Fajans' rules).

(*ii*) These halides are volatile liquids, which are non-conductors of electricity. Some non-metallic halides, like BrF_3, ICl_3, $COCl_2$, NOCl, $POCl_3$, conduct electricity due to auto-ionisation process.

$$COCl_2 \rightleftharpoons COCl^+ + Cl^- ; \qquad 2BrF_3 \rightleftharpoons BrF_2^+ + BrF_4^-$$

(*iii*) *Hydrolysis.* (*a*) When treated with H_2O, the halides undergo hydrolysis, to form oxy-acid of the element and halogen acid (HX). The oxidation state of the element in the halide and in the oxy acid is the same. Thus :

$$\underset{(B = +3)}{BX_3} \text{ (X = Cl, Br, I)} + 3H_2O \longrightarrow \underset{(B = +3)}{HBO_3} + 3HX$$

$$\underset{(Si = +4)}{SiCl_4} + 4H_2O \longrightarrow \underset{(Si = +4)}{H_4SiO_4} + 4HCl$$

$$\underset{(P = +3)}{PCl_3} + 3H_2O \longrightarrow \underset{(P = +3)}{H_3PO_3} + 3HCl$$

(*b*) NCl_3, NBr_3 and NI_3 are completely hydrolysed, giving hypohalons acid, HOX (oxy-acid) and NH_3 (in place of halogen acid)

$$NX_3\ (X = Cl, Br, I) + 3H_2O \longrightarrow 3HOX + NH_3$$

(*c*) The hydrolysis of BF_3 and SiF_4 takes place in similar manner. Here, if BF_3 or SiF_4 gas is passed in excess into water, it dissolves in HF to form $H[BF_4]$ or $H_2[SiF_6]$ respectively.

$$BF_3 + 3H_2O \longrightarrow H_3BO_3 + 3HF$$
$$3 \times [BF_3 + HF] \longrightarrow H[BF_4]$$

$$4BF_3\ (excess) + 3H_2O \longrightarrow H_3BO_3 + 3H[BF_4]$$
$$3\ SiF_4 + 4H_2O \longrightarrow H_4\ SiO_4 + 4HF$$
$$2 \times [SiF_4 + 2HF \longrightarrow H_2\ [SiF_6]$$

$$3SiF_4\ (excess) + 4H_2O \longrightarrow H_4SiO_4 + H_2[SiF_6]$$

H_4SiF_6, being unstable, dissociates to give the ppt. of SiO_2. Thus, the above equation can be written as :

$$3SiF_4\ (excess) + 4H_2O \longrightarrow SiO_2.\ 2H_2O + H_2[Si\ F_6]$$

or $$3SiF_4\ (excess) + 2H_2O \longrightarrow SiO_2 + H_2[Si\ F_6]$$

(*d*) In some cases the oxy-acid, being unstable, gets dissociated to give the ppt. of the oxide of the element.

$$[\underset{(As = +3)}{AsCl_3} + 3H_2O \longrightarrow \underset{(As = +3)}{H_3AsO_3} + [3HCl] \times 2$$
$$2H_3AsO_3 \longrightarrow As_2O_3 + 3H_2O$$

$$2AsCl_3 + 6H_2O \longrightarrow As_2O_3 + 6HCl$$

(*e*) Sometimes the hydrolysis of halides gives halogen acid and oxy-halide of the element.

$$SbCl_3 + H_2O \rightleftharpoons SbOCl + 2HCl$$
$$BiCl_3 + H_2O \rightleftharpoons BiOCl + 2HCl$$

(*f*) Some of the halides of the elements of 2*nd* period (*e.g.*, CF_4, CCl_4, NF_3 etc.) are resistant to hydrolysis. This is because of the fact in a covalent halide, hydrolysis occurs as a result of coordination of a water molecule to the central atom of the halide molecule. Since the central atom, being a member of 2*nd* period, has no *d*-orbitals in its valence-shell, it cannot accept the electron pair donated by H_2O molecule and hence cannot expand its covalency beyond 4. On the other hand, the halides of the elements of higher periods can easily be hydrolysed, since the central element, due to the presence of vacant *d*-orbitals in its valence shell, can accept the electron pair from H_2O molecules and can thus increase its covalency beyond 4.

(*g*) Dimeric monohalides (*e.g.*, S_2F_2, S_2Cl_2 and Se_2Cl_2) and dihalides (*e.g.*, SCl_2, $TeCl_2$ and $TeBr_2$) are readily hydrolysed by H_2O. Their hydrolysis gives similar products.

$$\underset{(X = F, Cl)}{2S_2\ X_2} + 3H_2O \longrightarrow 3S + H_2SO_3 + 4HX$$
$$2Se_2Cl_2 + 3H_2O \longrightarrow 3Se_2 + H_2SeO_3 + 4HCl$$
$$2SCl_2 + 3H_2O \longrightarrow S + H_2SO_3 + 4HCl$$
$$\underset{(X = Cl, Br)}{2TeX_2} + 3H_2O \longrightarrow Te + H_2TeO_3 + 4HX$$

(*h*) $SeBr_4$ and TeX_4 (X = Cl, Br, I) are readily hydrolysed to give oxide of the element and halogen acid.

$$SeBr_4 + 2H_2O \longrightarrow SeO_2 + 4HBr\ ; \qquad TeX_4 + 2H_2O \longrightarrow TeO_2 + 4HX$$

(*i*) SF_6 and SeF_6 resist hydrolysis but TeF_6 is completely hydrolysed by H_2O, giving orthotelluric acid, $Te(OH)_6$ or H_6TeO_6 and HF.

$$TeF_6 + 6H_2O \longrightarrow H_6TeO_6 + 6HF$$

Also see "*Rule of maximum covalency*" discussed in chapter 10.

III. Halides of intermediate character

Certain halides like $FeCl_3$ are volatile and are easily hydrolysed (covalent properties) and are good conductor in the fused state (ionic properties). Such halides have intermediate character.

IV. Complex halides

The halides, whose central atom has vacant *p*-orbitals and/or vacant *d*-orbitals (*e.g.*, BeF_2, BF_3, SiF_4, $SnCl_4$ etc.) can accommodate lone pairs of electrons donated by the halide ions and can thus form complex halide ions. Examples of important complex halide ions are, BeF_3^-, BeF_4^{2-}, BF_4^-, AlF_4^-, AlF_6^{3-}, $GaCl_4^-$, $TlBr_4^-$, SiF_6^{2-}, $SnCl_4^{2-}$, PCl_6^-, SbF_5^{2-}, SbF_6^{3-}, $TeCl_5^-$, TeF_6^{2-} etc.

F^- ion forms stable complex halide ions with smaller cations like B^{3+}, Al^{3+}, Si^{4+}, because of strong electrostatic and high energies of the resulting bonds. Other halide ions cannot form such complex halide ions with smaller cations. For example, F^- ion can form the complex hexafluoride ion, AlF_6^{3-}, while other halide ions are not able to form AlX_6^{3-} ion (X = Cl, Br, I).

F^- ion does not give stable complex halide ions with larger cations. The stability of the complex halide ions formed by Cl^-, Br^- and I^- ions with larger cations increases in the order $Cl^- < Br^- < I^-$. These complex halide ions can easily be polarised.

Halides of hydrogen : Halogen acids

Introduction

Hydrogen combines directly with halogens (X) and gives binary compounds of HX type (X = F, Cl, Br, I). Since H-atom has less electronegativity than each of the halogens (H = 2.1, F = 4.0, Cl = 3.0, Br = 2.8, I = 2.5), H-atom in these compounds has positive oxidation state equal to + 1 and each of the halogen atoms has a negative oxidation state equal to –1. It is due to the positive oxidation state of H-atom and negative oxidation state of halogen atom that these compounds are called hydrogen halides (H^+X^-).

Since the aqueous solution of these compounds ionise to give hydrated proton, H^+ (*aq*) or H_3O^+, these compounds are also called "halogen acids".

$$HX + aq \longrightarrow H^+ (aq) + X^- (aq)$$

or

$$HX + H_2O \longrightarrow H_3O^+ + X^-$$

These compounds are also called by other names like hydroacids, hydrohalic acids etc.

We have the following four halogen acids (*i*) *Hydrofluoric acid* or *hydrogen fluoride*, HF or H_2F_2. (*ii*) *Hydrochloric acid* or *hydrogen chloride*, HCl (*iii*) *Hydrobromic acid* or *hydrogen bromide*, HBr (*iv*) *Hydroiodic acid* or *hydrogen iodide*, HI.

General methods for the preparation of halogen acids

Following are the important methods :

1. By the direct combination of elements : By synthesis. In this method, H_2 and X_2 (X = Cl, Br, I) combine together under different conditions and give different halogen acids, *e.g.*,

(*a*) HCl is produced by burning Cl_2 in the excess of H_2 or by the action of H_2 on Cl_2 in presence of sunlight.

$$Cl_2 + H_2\,(excess) \xrightarrow{\text{Burn}} 2HCl$$

$$Cl_2 + H_2 \xrightarrow{\text{Sunlight}} 2HCl + \text{Heat}$$

(*b*) HBr can be prepared by passing a mixture of H_2 and Br_2 over a platinum spiral heated to redness by an electric current.

$$H_2 + Br_2 \xrightarrow[\text{spiral}]{\text{Heat Pt.}} 2HBr$$

(*c*) HI can be prepared in small quantities by passing H_2 and I_2 vapours over red hot fine platinum.

$$H_2 + I_2 \xrightleftharpoons{\text{Heat pt. (450°C)}} 2HI$$

2. By heating an appropriate halide with conc. H_2SO_4 or conc. H_3PO_4. HF and HCl can be prepared by heating CaF_2 and NaCl respectively with conc. H_2SO_4.

$$CaF_2 + H_2SO_4 \xrightarrow{\Delta} CaSO_4 + 2HF; \quad 2NaCl + H_2SO_4 \xrightarrow{\Delta} Na_2SO_4 + 2HCl$$

When NaBr (or KBr) and NaI (or KI) are heated with conc. H_2SO_4, HBr and HI are produced respectively.

$$2MBr\ (M = Na, K) + H_2SO_4 \xrightarrow{\Delta} M_2SO_4 + 2HBr$$

$$2MI\ (M = Na, K) + H_2SO_4 \xrightarrow{\Delta} M_2SO_4 + 2HI$$

Now since HBr and HI, formed as above, react with conc. H_2SO_4 and are oxidised to Br_2 and I_2 respectively, these acids are consumed and hence are not obtained. In other words, we say that HBr and HI cannot be prepared by the action of their halides on conc. H_2SO_4. The oxidation of HBr and HI to Br_2 and I_2 respectively by H_2SO_4, can be shown by the following equations. In these reactions H_2SO_4 is reduced to SO_2 by HBr and SO_2, S and H_2S by HI.

$$2HBr + H_2SO_4 \longrightarrow SO_2 + 2H_2O + Br_2$$

or $$2Br^- + 4H^+ + SO_4^{2-} \longrightarrow SO_2 + 2H_2O + Br_2$$

$$2HI + H_2SO_4 \longrightarrow SO_2 + 2H_2O + I_2$$

or $$2I^- + 4H^+ + SO_4^{2-} \longrightarrow SO_2 + 2H_2O + I_2$$

$$6HI + H_2SO_4 \longrightarrow S + 4H_2O + 3I_2$$

or $$6I^- + 8H^+ + SO_4^{2-} \longrightarrow S + 4H_2O + 3I_2$$

$$8HI + H_2SO_4 \longrightarrow H_2S + 4H_2O + 4I_2$$

or $$8I^- + 10H^+ + SO_4^{2-} \longrightarrow H_2S + 4H_2O + 4I_2$$

Since HBr and HI are not able to reduce conc. H_3PO_4, these acids are obtained by heating NaBr and NaI respectively with conc. H_3PO_4.

$$3NaBr + H_3PO_4 \xrightarrow{\Delta} Na_3PO_4 + 3HBr; \quad 3NaI + H_3PO_4 \xrightarrow{\Delta} Na_3PO_4 + 3HI$$

3. By hydrolysing non-metallic halides with H_2O: Preparation of HCl, HBr and HI. HCl, HBr and HI may be obtained when chlorides, bromides or iodides of non-metals (*e.g.*, PCl_3, $SiCl_4$, PBr_3, PI_3 etc.) are hydrolysed by H_2O.

$$PCl_3 + 3H_2O \longrightarrow H_3PO_3 + 3HCl; \quad SiCl_4 + 3H_2O \longrightarrow H_2SiO_3 + 4HCl$$

$$PBr_3 + 3H_2O \longrightarrow H_3PO_3 + 3HBr; \quad PI_3 + 3H_2O \longrightarrow H_3PO_3 + 3HI$$

4. By oxidising H_2S and SO_2 by halogen: Preparation of HBr and HI. HBr and HI can be prepared by oxidising H_2S (to S) and SO_2 (to H_2SO_4) by Br_2 or I_2 as the case may be. When H_2S or SO_2 is bubbled through Br_2 water or iodine solution, dilute solution of HBr or HI is obtained.

$$Br_2 + H_2S \longrightarrow 2HBr + S \quad \text{or} \quad Br_2 + S^{2-} \longrightarrow 2Br^- + S$$

$$Br_2 + SO_2 + 2H_2O \longrightarrow 2HBr + H_2SO_4 \quad \text{or} \quad Br_2 + SO_2 + 2H_2O \longrightarrow 2Br^- + 4H^+ + SO_4^{2-}$$

$$I_2 + H_2S \longrightarrow 2HI + S \quad \text{or} \quad I_2 + S^{2-} \longrightarrow 2I^- + S$$

$$I_2 + SO_2 + 2H_2O \longrightarrow 2HI + H_2SO_4 \quad \text{or} \quad I_2 + SO_2 + 2H_2O \longrightarrow 2I^- + 4H^+ + SO_4^{2-}$$

5. By the action of water on a mixture of red phosphorus and Br_2 or I_2 : Preparation of HBr and HI. HBr and HI can be prepared in the laboratory by slowly pouring water from a dropping funnel, fitted to a flask containing a mixture of red phosphorus (1 part) and Br_2 or I_2 (20 parts).

$$8H_2O + 2P + 5Br_2 \rightarrow 10\ HBr + 2H_3PO_4; \quad 8H_2O + 2P + 5I_2 \rightarrow 10\ HI + 2H_3PO_4$$

Specific methods for the preparation of individual halogen acids

(*a*) **Preparation of H_2F_2.** (*i*) *Pure anhydrous* H_2F_2 is obtained by heating KHF_2 in a platinum retort (Laboratory preparation) or by passing H_2 over dry AgF.

$$2KHF_2 \xrightarrow{\Delta} K_2F_2 + H_2F_2; \quad 2AgF\ (dry) + H_2 \longrightarrow 2Ag + H_2F_2$$

The vapours of H_2F_2 are condensed in a platinum condenser and collected in a platinum receiver. In place of Pt, lead apparatus can also be used.

(*ii*) *The aqueous solution* of H_2F_2 can be obtained by heating CaF_2 with 90% conc. H_2SO_4 in a lead retord. The vapours are condensed in a strongly cooled lead receiver. Aqueous acid of about 40% strength can be obtained by passing the vapours through water in a lead receiver.

$$CaF_2 + H_2SO_4 \longrightarrow CaSO_4 + H_2F_2$$

The aqueous solution of H_2F_2 is kept in guatta-percha bottles, when dilute and in wax bottles, when concentrated. The anhydrous acid is stored in well-cooled Pt, Au or Ag vessels or steel cylinders.

(*b*) **Preparation of HCl.** (*i*) As already mentioned, HCl can be prepared in the laboratory by heating NaCl with conc. H_2SO_4.

$$2NaCl + H_2SO_4 \xrightarrow{\Delta} Na_2SO_4 + 2HCl$$

The gas of HCl obtained as above cannot be dried over P_2O_5 or quick lime (CaO), since HCl reacts with both of these substances.

$$2P_2O_5 + 3HCl \rightarrow POCl_3 + 3HPO_3 ; \qquad CaO + 2HCl \rightarrow CaCl_2 + H_2O$$

(*ii*) Pure HCl is obtained by the action of water on $SiCl_4$. (hydrolysis $SiCl_4$).

$$SiCl_4 + 2H_2O \longrightarrow SiO_2 + 4HCl$$

(*iii*) Considerable quantities of HCl are obtained as a by-product in the manufacture of Na_2CO_3 from NaCl by Le Blanc process.

(*c*) **Preparation of HBr.** HBr can be obtained by the action of Br_2 on well cooled C_6H_6 in presence of Al or iron powder (*catalyst*).

$$C_6H_6 + Br_2 \xrightarrow{Al} C_6H_5Br + HBr$$

General characteristics of halogen acids

Following are the important characteristics of halogen acids (Also see Table 29.1).

1. Mono-meric and di-meric nature of HX molecules. Hydrofluoric acid exists as a dimeric molecule, H_2F_2 even in the gaseous state. Due to its dimeric nature, H_2F_2 gives two types of salts which contain HF_2^- and F_2^{2-} ions.

$$2HF \rightleftharpoons HF_2^- + H^+; \qquad HF_2^- \rightleftharpoons F_2^{2-} + H^+$$

Salts (*e.g.*, Na_2F_2 or NaF), which contain F_2^{2-} or F^- ions, are called "normal salts", while those (*e.g.*, $NaHF_2$), which contain HF_2^- ions, are known as "acid salts". As an acid, H_2F_2 is weaker than H_3PO_4, but stronger than HNO_2.

Other HX molecules (X = Cl, Br, I) are monomeric and hence form only one type of salts, which contain X^- ion.

$$HX\ (X = Cl, Br, I) \rightleftharpoons H^+ + H^-$$

Explanation. Why hydro-fluoric acid exists as a dimeric molecule (H_2F_2) and other halides exist as monomeric molecules (HX) has already been explained under "*H-Bonding*", discussed in chapter 10.

2. Physical state. Anhydrons HF is a liquid at ordinary temperature and fumes strongly in air, while each of the remaining hydrogen halides, is a colourless gas, with pungent smell and acidic taste.

Reason. See "*Hydrogen Bonding*" (chapter 10).

3. H—X bond length, strength of H—X bond (H—X bond energy) and thermal stability of HX molecules. As the size of the halogen atom increases from F to I, H—X bond length in HX molecules also increases from H—F to H—I (H—F < H—Cl < H—Br < H—I). The increase in H—X bond length decreases the strength of H—X bond from H—F to H—I (H—F > H—Cl > H—Br > H—I). The decrease in the strength of H—X bond is evident from the fact that H—X bond dissociation energies decrease from H—F to H—I (H—F = 574.0 kJ/mol, H—Cl = 428.1 kJ/mol, HBr = 362.5 kJ/mol, H—I = 294.6 kJ/mol). Due to successive decrease in the strength of H—X bond from H—F to H—I, thermal stability of HX molecules also decreases from HF to HI (HF > HCl > HBr > HI). This order of thermal stability is evident from the fact that HF does not dissociate even at high temperature, HCl is decomposed into its elements at 1500° C (*i.e.*, by electricity).

HBr gets partially dissociated into its elements, when heated to 800°C, while HI, being the least stable, gets decomposed into its elements, even in sunlight or on heating to 180°C.

Maximum thermal stability of HF is also evident from the fact that this acid is not oxidised even by strong oxidising agents.

4. Boiling points. Since as we move down the group from HF to HI, the magnitude of van der Waals forces increases and hence the b.pts. of the hydrides should increase as HF < HCl < HBr < HI. But, as is evident from Table 29.1, the boiling point decreases from HF to HCl (HF > HCl) and then increases until we reach HI (HCl < HBr < HI). Why the boiling point of HF is higher than that of HCl has been explained on the basis of *H-bonding*, occurring in several HF molecules. (See "*H-bonding*" discussed in chapter 10).

5. Volatility. Since HF has the maximum boiling point and HCl has the minimum boiling point, HF is the least volatile and HCl is the most volatile. The least volatility of HF molecule is due to much stronger H—F bond and association of HF molecules to form dimer, H_2F_2 through H-bonding. The maximum volatility of HCl is due to its being a monomer acid and less strength of H—Cl bond in HCl molecule.

6. Solubility in water and formation of azeotropic mixtures. All the hydrogen halides are highly soluble in water and give a constant boiling mixture, which is called azotropic mixture. The azeotropic mixture of each acid contains a particular weight-percentage of the acid and boils at a fixed temperature. The dilute solution of the acid cannot be concentrated by boiling it beyond the weight-percentage of the acid. Dilute solution of HCl cannot be concentrated by boiling beyond 20.2%.

7. Dipole moment. Dipole moment values for HF, HCl, HBr and HI molecules, which are assumed to be completely ionic, are given in Table 29.1. The topic "*Dipole Moment*" given in chapter 10 can be consulted for the method for calculating the dipole moment values. Calculated values increase from HF to HI, while the experimental values decrease in the same direction.

8. Combustibility. All the halogen acids are neither combustible nor supporter of combustion, *i.e.*, they are incombustible and non-supporter of combustion.

9. Covalent / ionic character of H—X bond in gaseous HX molecules. H—X bond in gaseous HX molecules is largely covalent and has very small amount of ionic character. Thus, gaseous HX molecules are mostly covalent and have only partial ionic character. The covalent character of H—X bond is also confirmed by the fact that melting and boiling points of hydrogen halides are low and pure liquid halides are poor conductors of electricity.

The percentage of ionic character in HX molecules has been calculated with the help of Hanny and Smith equation and dipole moment values of HX molecules. Both these methods have been discussed in chapter 10. The values of percentage of ionic character, as calculated by these methods, are given in Table 29.1. These values show that the ionic character of H—X bond in gaseous HX molecules is in the order : H—F > H—Cl > H—Br > H—I. This order shows that HF is the most ionic and HI is the least so.

Explanation of the above order. The above order of the ionic character of H—X bond in gaseous HX molecules has been explained as follows :

(*a*) *Explanation based on the electronegativity difference value, (x_X — x_H).* See "*Polar and Non-polar Covalent Bonds*" in chapter 10.

(*b*) *Explanation based on dipole moment values of HX molecules.* See "*Polar and Non-polar Covalent Bonds*" in chapter 10.

(*c*) *Explanation based on Fajans' rules.* See "*Fajans' Rules*" (chapter 10).

10. Acidic nature of aqueous solution of HX molecules. HX molecules in the gaseous state are essentially covalent and hence are not able to ionise to yield proton (H^+) but in aqueous solution they give hydrated proton, H^+ (*aq*) and hence act as Bronsted acids.

$$HX(g) + aq \rightarrow HX(aq) \longrightarrow H^+(aq) + X(aq) \text{ or } HX(g) + H_2O \longrightarrow H_3O^+ + X^-(aq)$$

The acidic strength (*i.e.,* proton releasing power) of halogen acids in aqueous solution is in the order :

$$HF\ (aq) < HI\ (aq) < HCl\ (aq) < HBr\ (aq)$$

This above order has been explained in chapter 19 (*Modern Concepts of Acids and Bases*). The relative order of the acidic strength of HX molecules, in methanol solvent, has been found in the order :

$$HF < HCl < HBr < HI$$

Anhydrous hydrofluoric acid does not show acidic property and hence does not attack metals except potassium. However, its concentrated solution shows acidic character. It is a weak dibasic acid (even weaker than H_3PO_4 but stronger than HNO_2) and hence forms two types of salts which contain HF_2^- and F_2^{2-} ions. A concentrated solution of H_2F_2 reacts with metals (*e.g.*, Zn), their hydroxides, carbonates, oxides etc., to form metallic fluorides.

$$Zn + H_2F_2 \longrightarrow ZnF_2 + H_2\ ; \qquad 2NaOH + H_2F_2 \longrightarrow Na_2F_2 + 2H_2O$$

$$CaCO_3 + H_2F_2 \longrightarrow CaF_2 + H_2O + CO_2\ ; \qquad B_2O_3 + 3H_2F_2 \longrightarrow 2BF_3 + 3H_2O$$

The aqueous solution of HCl, HBr and HI show acidic nature and hence turn blue litmus red. Each of them reacts with metals, their hydroxides, carbonates, oxides etc., to form metallic halides.

$$Zn + 2HX\ (X = Cl, Br, I) \longrightarrow ZnX_2 + H_2\ ; \qquad NaOH + HX \longrightarrow NaX + H_2O$$

$$Na_2CO_3 + 2HX \longrightarrow 2NaX + CO_2 + H_2O$$

11. Action on silica (SiO_2) and glass. HCl, HBr and HI do not react with SiO_2 and glass. HF attacks on SiO_2 and gives white fumes of SiF_4.

$$SiO_2 + 4HF \longrightarrow SiF_4 + 2H_2O$$

The white fumes of SiF_4 form a glatinous mass of H_2SiO_3 or H_4SiO_4 and hydro-fluosilicic acid (H_2SiF_6) with a drop of water held in these fumes.

$$3SiF_4 + 3H_2O \longrightarrow H_2SiO_3 + 2H_2SiF_6$$

or $$3SiF_4 + 4H_2O \longrightarrow H_2SiO_3.H_2O \text{ or } H_4SiO_4 + 2H_2SiF_6$$

Now, we know that glass is composed of Na_2SiO_3 and $CaSiO_3$. HF reacts with these silicates and form sodium and calcium fluosilicate (Na_2SiF_6 and $CaSiF_6$) respectively.

$$Na_2SiO_3 + 6HF \rightarrow Na_2SiF_6 + 3H_2O\ ; \qquad CaSiO_3 + 6HF \rightarrow CaSiF_6 + 3H_2O$$

Since HF reacts with glass as shown above, it should not be kept in glass vessels; it should be stored in waxed glass vessels because wax prevents the action of HF on glass. The property of HF to act on glass has been utilised in the etching of glass and in making scales on the glass instruments. For etching glass, commercial H_2F_2 (40-60%) is used.

12. Action on salts. (*a*) *Action on salts of weaker acids.* HCl decomposes the salts of weaker acids like carbonates, bicarbonates, sulphides, sulphites, thiosulphates and nitrites.

$$2Na_2CO_3 + 2HCl \longrightarrow 2NaCl + H_2O + CO_2 \quad \text{or} \quad CO_3^{2-} + 2H^+ \longrightarrow H_2O + CO_2$$

$$NaHCO_3 + HCl \longrightarrow NaCl + H_2O + CO_2 \quad \text{or} \quad HCO_3^- + H^+ \longrightarrow H_2O + CO_2$$

$$Na_2S + 2HCl \longrightarrow 2NaCl + H_2S \quad \text{or} \quad S^{2-} + 2H^+ \longrightarrow H_2S$$

$$Na_2SO_3 + 2HCl \longrightarrow 2NaCl + H_2O + SO_2 \quad \text{or} \quad SO_3^{2-} + 2H^+ \longrightarrow H_2O + SO_2$$

$$Na_2S_2O_3 + 2HCl \longrightarrow 2NaCl + S + H_2O + SO_2 \quad \text{or} \quad S_2O_3^{2-} + 2H^+ \longrightarrow S + H_2O + SO_2$$

$$2NaNO_2 + 2HCl \longrightarrow 2NaCl + H_2O + NO + NO_2 \quad \text{or} \quad 2NO_2^- + 2H^+ \longrightarrow H_2O + NO + NO_2$$

With metallic oxides and hydroxides HCl forms salt and water.

(*b*) *Action on $AgNO_3$ solution.* With $AgNO_3$ solution, HF forms AgF which becomes soluble in water. On the other hand HCl, HBr and HI give the ppt. of AgCl (*white*), AgBr (*pale yellow*) and AgI (*yellow*) respectively.

$$AgNO_3 + HF \longrightarrow \underset{\text{soluble}}{AgF} + HNO_3\ ; \quad AgNO_3 + HX\ (X = Cl, Br, I) \longrightarrow \underset{\text{ppt.}}{AgX} + HNO_3$$

The ppt. of AgCl is insoluble in HNO_3 but soluble in NH_4OH. AgBr is soluble in HNO_3 but sparingly so in NH_4OH and AgI is soluble in both HNO_3 and NH_4OH.

(*c*) *Action of Pb (II) salts*. HF does not give any ppt. with the solution of Pb (II) salts like $Pb(NO_3)_2$, $(CH_3COO)_2$ Pb etc., HCl, HBr and HI give the ppt. of $PbCl_2$ (*white*) $PbBr_2$ (*white*) and PbI_2 (*yellow*) respectively.

$$Pb(NO_3)_2 + 2HX\ (X = Cl, Br, I) \longrightarrow PbX_2 + 2HNO_3$$

$$(CH_3COO)_2Pb + 2HX \longrightarrow PbX_2 + 2CH_3COOH$$

All the three precipitates (*i.e.*, $PbCl_2$ $PbBr_2$ and PbI_2) are soluble in hot water.

(*d*) *Action of mercurous and mercuric salts*. HF does not give any ppt. with these salts. HCl forms a white ppt. of Hg_2Cl_2 with $Hg_2\ (NO_3)_2$ solution.

$$2HCl + Hg_2\ (NO_3)_2 \longrightarrow Hg_2Cl_2 + 2HNO_3$$

Hg_2Cl_2 is soluble in aqua regia. HI forms scarlet ppt. of HgI_2 with $HgCl_2$ solution.

$$2HI + HgCl_2 \longrightarrow HgI_2 + 2HCl \quad \text{or} \quad 2I^- + HgCl_2 \longrightarrow HgI_2 + 2Cl^-$$

(*e*) *Action of $CuSO_4$ solution.* HF, HCl and HBr do not react with $CuSO_4$ solution, but HI gives CuI_2 which, being unstable, decomposes into CuI and I_2.

$$2HI + CuSO_4 \longrightarrow CuI_2 + H_2SO_4\ ; \qquad 2CuI_2 \longrightarrow 2CuI + I_2$$

(*f*) *Action of $BaCl_2$, $SrCl_2$ and $CaCl_2$*. With the solution of these salts, H_2F_2 forms white ppt. of BaF_2, SrF_2 and CaF_2 respectively. Other acids do not give any ppt., since MX_2 (M = Ba, Sr, Ca; X = Br, I) are soluble.

$$\underset{(M = Ba,\ Sr,\ Ca)}{MCl_2 + H_2F_2} \longrightarrow \underset{ppt.}{MF_2} + 2HCl\ ; \qquad \underset{(M = Ba,\ Sr,\ Ca)}{MCl_2 + 2HX} \longrightarrow \underset{soluble}{MX_2} + 2HCl$$

13. Action on metals. (*a*) *Anhydrous.* HF does not attack any metal under ordinary conditions except potassium while aqueous concentrated acid reacts with many metals to form their fluorides with the evolution of H_2. For example :

$$Zn + 2HF \longrightarrow ZnF_2 + H_2\ ; \qquad Mg + 2HF \longrightarrow MgF_2 + H_2$$

In these reactions the metal is oxidised and HF is reduced to H_2.

(*b*) *Gaseous.* HCl reacts with heated metals while aqueous HCl reacts with most metals in cold. The reaction between aqueous HCl and Pt, Au, Ag and Hg is not perceptible.

Aqueous HCl reacts with Ag in presence of air.

$$4Ag + 4HCl + O_2 \longrightarrow 4AgCl + 2H_2O$$

Cu dissolves in conc. HCl. Fe reacts with HCl and gives lower chloride *viz.*, $FeCl_2$ (not $FeCl_3$)

$$Fe + 2HCl \longrightarrow FeCl_2 + H_2$$

Dilute and concentrated HCl forms $ZnCl_2$ and evolves H_2 with zinc metal.

$$Zn + 2HCl \longrightarrow ZnCl_2 + H_2$$

(*c*) HBr dissolves Fe, Zn, Sn, Cu, Ag, Pb etc., with the liberation of H_2 and formation of bromides. HBr attacks Hg very slowly, forming H_2 and $HgBr_2$.

14. Reducing properties of HX molecules or X^- ions : Oxidation of HX molecules or X^- ions. When HX molecule or X^- ion reacts with an oxidising agent, it reduces the oxidising agent and is itself oxidised to X_2 molecule. Thus in this reaction HX molecule or X^- ion acts as a reducing agent.

$$2HX + O_2 \longrightarrow 2H_2O + 2X_2 \qquad \textit{(Molecular equation)}$$

$$\text{or} \qquad 2X^-\ (X = -1) \xrightarrow{\text{oxidation}} X_2 + 2e^-\ (X = 0)\ \textit{(Ionic equation)}$$

O_2 shown in the molecular equation is provided by the oxidising agent. Two electrons shown in the ionic equation are lost by X^- ion and thus X^- ion acts as a reducing agent. These electrons are accepted by the oxidising agent.

It has been found that the tendency of X^- ions to lose electrons to undergo oxidation to form X_2 molecule increases from F^- to I^- ion. Accordingly the reducing power of X^- ions or HX molecules increases from F^- to I^- ions or from HF to HI molecules.

$$F^- < Cl^- < Br^- < I^- \qquad \text{or} \qquad HF < HCl < HBr < HI.$$

As a matter of fact, HF or F^- does not show reducing properties at all.

Reason. The increasing order of the reducing power of HX molecules or X^- ions can be explained on the basis of electrode potential values of $\frac{1}{2}X_2/X^-$ electrode and on the basis of strength of H — X bond in HX molecule. For details, see chapter 18 (*Electrode potentials and electrochemical series*).

Reactions showing the reducing properties of HX molecules or X^- ions. We have seen above that the reducing power of HX molecules or X^- ions is in the order

$$HF < HCl < HBr < HI \quad \text{or} \quad F^- < Cl^- < Br^- < I^-$$

Thus, HF or F^- is the weakest reducing agent. It is so weak as a reducing agent, that it does not reduce even very strong oxidising agents. For example, HF molecule is not able to reduce H_2SO_4 to SO_2 as HBr and HI do, *i.e.*, the following reaction is not possible :

$$2HF + H_2SO_4 \longrightarrow SO_2 + 2H_2O + F_2$$

or

$$2F^- + SO_4^{2-} + 4H^+ \longrightarrow SO_2 + 2H_2O + F_2 \quad \text{(Not possible)}$$

HF or F^- is also not able to reduce any of the halogens (X_2) to its corresponding halide ion (X^-), *i.e.*, none of the following reactions is possible.

$$2HF + Cl_2 \longrightarrow 2HCl + F_2 \quad \text{or} \quad 2F^- + Cl_2 \longrightarrow 2Cl^- + F_2$$
$$2HF + Br_2 \longrightarrow 2HBr + F_2 \quad \text{or} \quad 2F^- + Br_2 \longrightarrow 2Br^- + F_2$$
$$2HF + I_2 \longrightarrow 2HI + F_2 \quad \text{or} \quad 2F^- + I_2 \longrightarrow 2I^- + F_2 \quad \text{(Not possible)}$$

As reducing agent, HCl molecule or Cl^- ion is weaker than HBr and HI both and hence reduces only strong oxidising agents and is itself oxidised to Cl_2. Thus HCl reduces :

(*i*) MnO_2 (acidified with conc. H_2SO_4) to $MnSO_4$.

$$MnO_2 + H_2SO_4 + 2HCl \longrightarrow MnSO_4 + 2H_2O + Cl_2$$

or

$$MnO_2 + 4H^+ + 2Cl^- \longrightarrow Mn^{2+} + 2H_2O + Cl_2$$

(*ii*) MnO_2 to $MnCl_2$.

$$MnO_2 + 4HCl\ (conc.) \longrightarrow MnCl_2 + 2H_2O + Cl_2$$

or

$$MnO_2 + 4H^+ + 2Cl^- \longrightarrow Mn^{2+} + 2H_2O + Cl_2$$

(*iii*) $KMnO_4$ to $MnCl_2$

$$2KMnO_4 + 16HCl\ (conc.) \longrightarrow 2KCl + 2MnCl_2 + 8H_2O + 5Cl_2$$

or

$$2MnO_4^- + 16H^+ + 10Cl^- \longrightarrow 2Mn^{2+} + 8H_2O + 5Cl_2$$

(*iv*) $K_2Cr_2O_7$ to $CrCl_3$

$$K_2Cr_2O_7 + 14HCl \longrightarrow 2KCl + 2CrCl_3 + 7H_2O + 3Cl_2$$

or

$$Cr_2O_7^{2-} + 14H^+ + 6Cl^- \longrightarrow 2Cr^{3+} + 7H_2O + 3Cl_2$$

(*v*). PbO_2 to $PbCl_2$

$$PbO_2 + 4HCl\ (hot\ and\ conc.) \longrightarrow PbCl_2 + 2H_2O + Cl_2$$

or

$$PbO_2 + 4H^+ 2Cl^- \longrightarrow Pb^{2+} + 2H_2O + Cl_2$$

(*vi*) Pb_3O_4 (*red lead*) to $PbCl_2$.

$$Pb_3O_4 + 8HCl \longrightarrow 3PbCl_2 + 4H_2O + Cl_2 \quad \text{or} \quad Pb_3O_4 + 8H^+ + 2Cl^- \longrightarrow 3Pb^{2+} + 4H_2O + Cl_2$$

(*vii*) O_2 to H_2O in presence of Cu-salts (catalyst).

$$O_2 + 4HCl \longrightarrow 2H_2O + 2Cl_2 \quad \text{or} \quad O_2 + 4Cl^- \longrightarrow 2O^{2-} + 2Cl_2$$

(*viii*) Bleaching power ($CaOCl_2$) to $CaCl_2$.

$$CaOCl_2 + 2HCl \longrightarrow CaCl_2 + H_2O + Cl_2$$

(*ix*) F_2 to HF but neither Br_2 to HBr nor I_2 to HI

$$2HCl + F_2 \longrightarrow 2HF + Cl_2 \quad \text{or} \quad 2Cl^- + F_2 \longrightarrow 2F^- + Cl_2 \quad \text{(Possible)}$$
$$2HCl + Br_2 \longrightarrow 2HBr + Cl_2 \quad \text{or} \quad 2Cl^- + Br_2 \longrightarrow 2Br^- + Cl_2$$
$$2HCl + I_2 \longrightarrow 2HI + Cl_2 \quad \text{or} \quad 2Cl^- + I_2 \longrightarrow 2I^- + Cl_2 \quad \text{(Not possible)}$$

HCl cannot reduce H_2SO_4 to SO_2, *i.e.*, HCl is not oxidised to Cl_2 by H_2SO_4.

$$2HCl + H_2SO_4 \longrightarrow SO_2 + 2H_2O + Cl_2$$

or

$$2Cl^- + 4H^+ + SO_4^{2-} \longrightarrow SO_2 + 2H_2O + Cl_2 \quad \text{(Not possible)}$$

Since HBr or Br^- is a weaker reducing agent than HI or I^-, it reduces some oxidising agents and is itself oxidised to Br_2. Thus, HBr reduces :

(*i*) H_2SO_4 to SO_2.

$2HBr + H_2SO_4 \longrightarrow SO_2 + 2H_2O + Br_2$ or $2Br^- + 4H^+ + SO_4^{2-} \longrightarrow SO_2 + 2H_2O + Br_2$

(*ii*) atmospheric O_2 to H_2O.

$4HBr + O_2 \longrightarrow 2H_2O + 2Br_2$ or $4Br^- + O_2 \longrightarrow 2O^{2-} + 2Br_2$

(*iii*) acidified solution of $KMnO_4$ to $MnSO_4$.

$2KMnO_4 + 3H_2SO_4 + 10HBr \longrightarrow K_2SO_4 + 2MnSO_4 + 8H_2O + 5Br_2$

or $2MnO_4^- + 16H^+ + 10Br^- \longrightarrow 2Mn^{2+} + 8H_2O + 5Br_2$

(*iv*) acidified solution of $K_2Cr_2O_7$ to $Cr_2(SO_4)_3$.

$K_2Cr_2O_7 + 4H_2SO_4 + 6HBr \longrightarrow K_2SO_4 + Cr_2(SO_4)_3 + 7H_2O + 3Br_2$

or $Cr_2O_7^{2-} + 14H^+ + 6Br^- \longrightarrow 2Cr^{3+} + 7H_2O + 3Br_2$

(*v*) H_2O_2 to H_2O.

$2HBr + H_2O_2 \longrightarrow 2H_2O + Br_2$ or $2Br^- + 2H^+ + H_2O_2 \longrightarrow 2H_2O + Br_2$

(*vi*) $FeCl_3$ to $FeCl_2$.

$FeCl_3 + 2HBr \longrightarrow 2FeCl_2 + 2HCl + Br_2$ or $2Br^- + 2Fe^{3+} \longrightarrow 2Fe^{2+} + Br_2$

(*vii*) F_2 to HF and Cl_2 to HCl but not I_2 to HI.

$2HBr + F_2 \longrightarrow 2HF + Br_2$ or $2Br^- + F_2 \longrightarrow 2F^- + Br_2$

$2HBr + Cl_2 \longrightarrow 2HCl + Br_2$ or $2Br^- + Cl_2 \longrightarrow 2Cl^- + Br_2$ } *(Not possible)*

$2HBr + I_2 \longrightarrow 2HI + Br_2$ or $2Br^- + I_2 \longrightarrow 2I^- + Br_2$ *(Not possible)*

(*viii*) MnO_2 (acidified with conc. H_2SO_4) to $MnSO_4$.

$MnO_2 + H_2SO_4 + 2HBr \longrightarrow MnSO_4 + 2H_2O + Br_2$

or $MnO_2 + 4H^+ + 2Br^- \longrightarrow Mn^{2+} + 2H_2O + Br_2$

Being the strongest reducing agent, HI or I^- reduces even very strong oxidising agents and is itself oxidised to I_2. Thus, HI or I^- reduces :

(*i*) H_2SO_4 to SO_2, S or H_2S (or S^{2-})

$2HI + H_2SO_4 \longrightarrow SO_2 + 2H_2O + I_2$ or $2I^- + SO_4^{2-} + 4H^+ \longrightarrow SO_2 + 2H_2O + I_2$

$6HI + H_2SO_4 \longrightarrow S + 4H_2O + 3I_2$ or $6I^- + SO_4^{2-} + 8H^+ \longrightarrow S + 4H_2O + 3I_2$

$8HI + H_2SO_4 \longrightarrow H_2S + 4H_2O + 4I_2$ or $8H^+ + 8I^- + SO_4^{2-} \longrightarrow S^{2-} + 4H_2O + 4I_2$

(*ii*) HNO_3 to NO or NO_2

$2HNO_3 + 6HI \longrightarrow 2NO + 4H_2O + 3I_2$ or $2NO_3^- + 8H^+ + 6I^- \longrightarrow 2NO + 4H_2O + 3I_2$

$2HNO_3 + 2HI \longrightarrow 2NO_2 + 2H_2O + I_2$ or $2NO_3^- + 2I + 4H^+ \longrightarrow 2NO_2 + 2H_2O + I_2$

(*iii*) HNO_2 to NO

$2HNO_2 + 2HI \longrightarrow 2NO + 2H_2O + I_2$ or $2NO_2^- + 4H^+ + 2I^- \longrightarrow 2NO + 2H_2O + I_2$

(*iv*) $FeCl_3$ to $FeCl_2$.

$2FeCl_3 + 2HI \longrightarrow 2FeCl_2 + 2HCl + I_2$ or $2Fe^{3+} + 2I^- \longrightarrow 2Fe^{2+} + I_2$

(*v*) $CuSO_4$ to CuI.

$2CuSO_4 + 4HI \longrightarrow 2CuI + 2H_2SO_4 + I_2$ or $2Cu^{2+} + 2I^- \longrightarrow 2Cu^+ + I_2$

(*vi*) atmospheric O_2 to H_2O.

$4HI + O_2 \longrightarrow 2H_2O + 2I_2$

It is due to the liberation of free I_2 that the solution of HI turns brown, on keeping in air.

(*vii*) MnO_2 (acidifed with conc. H_2SO_4) to $MnSO_4$.

$MnO_2 + H_2SO_4 + 2HI \longrightarrow MnSO_4 + 2H_2O + I_2$

or $MnO_2 + 4H^+ + 2I^- \longrightarrow Mn^{2+} + 2H_2O + I_2$

(*viii*) iodic acid (HIO_3) to I_2.

$$5HI + HIO_3 \longrightarrow 3H_2O + 3I_2$$

or $$5I^- + 6H^+ + IO_3^- \longrightarrow 3H_2O + 3I_2$$

(*ix*) arsenate (AsO_4^{3-}) to arsenite (AsO_3^{3-}).

$$2I^- + AsO_4^{3-} + 2H^+ \longrightarrow AsO_3^{3-} + H_2O + I_2$$

(*x*) acidified solution of $KMnO_4$ to $MnSO_4$.

$$10HI + 2KMnO_4 + 3H_2SO_4 \longrightarrow K_2SO_4 + 2MnSO_4 + 8H_2O + 5I_2$$

or $$16H^+ + 10I^- + 2MnO_4^- \longrightarrow 2Mn^{2+} + 8H_2O + 5I_2$$

(*xi*) acidified solution of $K_2Cr_2O_7$ to $Cr_2(SO_4)_3$.

$$6HI + K_2Cr_2O_7 + 4H_2SO_4 \longrightarrow K_2SO_4 + Cr_2(SO_4)_3 + 7H_2O + 3I_2$$

or $$14H^+ + 6I^- + Cr_2O_7^{2-} \longrightarrow 2Cr^{3+} + 7H_2O + 3I_2$$

(*xii*) all the three halogens *viz.*, F_2, Cl_2 and Br_2 to their corresponding halogen acids, *i.e.*, all the following reactions take place :

$$2HI + F_2 \longrightarrow 2HF + I_2 \quad \text{or} \quad 2I^- + F_2 \longrightarrow 2F^- + I_2$$
$$2HI + Cl_2 \longrightarrow 2HCl + I_2 \quad \text{or} \quad 2I^- + Cl_2 \longrightarrow 2Cl^- + I_2$$
$$2HI + Br_2 \longrightarrow 2HBr + I_2 \quad \text{or} \quad 2I^- + Br_2 \longrightarrow 2Br^- + I_2$$

When we consider the reduction of a halogen molecule (X_2) by a halogen acid (HX) or halide ion (X^-) as discussed above, we find that HF or F^-, being the weakest reducing agent, cannot reduce any of the halogens *viz.*, Cl_2, Br_2 and I_2, HCl or Cl^- can reduce only F_2 but neither Br_2 nor I_2; HBr or Br^- can reduce only F_2 and Cl_2 but not I_2; HI or I^-, being the strongest reducing agent, can reduce all the three halogen molecules (F_2, Cl_2 and Br_2) to their corresponding halide ions. This discussion can be summarised by saying that *a given halogen acid can reduce only that halogen molecule which lies above it in the halogen group.*

15. Action on halogens.(Also see *reducing properties of HX molecules or X^- ions*). A given HX molecule may or may not react with all X_2 molecules. When a given HX molecule reacts with X_2 molecule, it gets oxidised and X_2 molecule is reduced. Thus :

(*a*) HF does not react with any of the halogens.

(*b*) When HCl reacts with F_2, Cl_2 is liberated.

$$2HCl + F_2 \longrightarrow 2HF + Cl_2$$

Br_2 and I_2 do not react with HCl.

(*c*) On passing F_2 or Cl_2 through HBr solution, Br_2 is liberated.

$$2HBr + F_2 \longrightarrow 2HF + Br_2 \; ; \qquad 2HBr + Cl_2 \longrightarrow 2HCl + Br_2$$

I_2 does not react with HBr.

(*d*) When F_2, Cl_2 or Br_2 is passed in an aqueous solution of HI, I_2 is liberated.

$$2HI + F_2 \longrightarrow 2HF + I_2 \; ; \qquad 2HI + Cl_2 \longrightarrow 2HCl + I_2$$
$$2HI + Br_2 \longrightarrow 2HBr + I_2$$

16. Basicity of acids and nature of the salts given by them. We have seen above that hydro-fluoric acid is formulated as H_2F_2 (dimeric formula) while other acids are formulated as HCl, HBr and HI (monomeric formula). Thus H_2F_2 gives two types of salts *viz.*, acid salts [*e.g.*, $K^+[HF_2]^-$] and normal salts (*e.g.*, K_2F_2) while other acids give only one type of salts (*e.g.*, KX where X = Cl, Br, I). Thus hydrofluoric acid is dibasic and other acids are monobasic.

17. Action of NH_3. HF has no action on NH_3 while HCl, HBr and HI give white thick fumes of NH_4Cl, NH_4Br and NH_4I respectively.

$$HF + NH_3 \longrightarrow \text{No reaction} \; ; \qquad HX\ (X = Cl, Br, I) + NH_3 \longrightarrow NH_4X$$

18. Hydrogen bonding and formation of $(HF)_x$ cluster. Since F-atom has much more electronegativity than H-atom (as a matter of fact, F-atom has the maximum electronegativity among all the elements in the periodic table) and the electronegativity values of Cl, Br and I are close to that of H-atom (H = 2.1, F = 4.0, Cl = 3.0, Br = 2.8, I = 2.5), HF molecule undergoes intermolecular H-bonding to form its cluster, $(HF)_x$ while the occurrence of H-bonding and hence the formation of $(HX)_x$ cluster (X = Cl, Br, I) is not possible in case of HCl, HBr and HI molecules. Thus HF molecule shows H-bonding while other molecules do not show H-bonding.

19. Formation of aqua regia. When one part of conc. HNO_3 is mixed with three parts of conc. HCl, we get aqua regia. Aqua regia produces nascent chlorine (Cl) alongwith nitrosyl chloride (NOCl).

$$HNO_3 + 3HCl \longrightarrow NOCl + 2H_2O + 2Cl$$

According to another view, in place of NOCl, nitric oxide (NO) is generated.

$$HNO_3 + 3HCl \longrightarrow NO + 2H_2O + 3Cl$$

Due to the production of nascent chlorine, aqua regia is very reactive. For example, Au and Pt (noble metals) which are not acted upon by HNO_3 get dissolved in aqua regia due to the formation soluble complexes namely $H[AuCl_4]$ and $H_2[PtCl_6]$ respectively.

$$2Au + 3HNO_3 + 11HCl \longrightarrow H[AuCl_4] + 3NOCl + 6H_2O$$

This equation can also be written as :

$$Au + HNO_3 + 4HCl \longrightarrow H[AuCl_4] + NO + 2H_2O$$

$$Pt + 2HNO_3 + 8HCl \longrightarrow H_2[PtCl_6] + 2NOCl + 4H_2O$$

This equation can also be written as :

$$3Pt + 4HNO_3 + 18\ HCl \longrightarrow 4NO + 3H_2\ [PtCl_6] + 8H_2O$$

A mixture of HNO_3 and HBr acts like aqua regia.

20. Dissociation. H_2F_2 does not dissociate on heating while other halogen acids dissociate at definite temperatures.

21. Action of mixture of MnO_2 and H_2SO_4. When H_2F_2 is heated with a mixture of MnO_2 and H_2SO_4, no gas is evolved. On the other hand in case of HCl, HBr and HI acids, Cl_2 (greenish-yellow), Br_2 (radish brown vapour) and I_2 (violet vapour) gas are evolved.

$$2HF + MnO_2 + H_2SO_4 \longrightarrow \text{No gas is evolved}$$

$$2HX\ (X = Cl, Br, I) + MnO_2 + H_2SO_4 \longrightarrow X_2 + 2H_2O + MnSO_4$$

22. Solubility of metallic halides. The metallic fluorides differ from the other halides in solubility. For example : (*a*) AgF is soluble and other Ag halides are insoluble. (*b*) CaF_2 is insoluble whereas other calcium halides are soluble.

23. Nature of halides. Most of the fluorides are ionic and hence contain F^- ions. This behaviour of fluorides is due to the smaller size of F^- ion. Chlorides, bromides and iodides have frequently molecular lattices.

24. Formation of complexes by X^- ions. Being the smallest in size of all the halide ions, F^- ion has the greatest tendency to form complex ions like AlF_6^{3-}, FeF_6^{3-} etc., in which the metal atom exhibits a higher covalency. Other halide ions, being bigger in size, show this tendency to a much smaller extent.

25. Combination of HX molecules with X^- ions to form HX_2^- ion. Because of high electronegativity of F^- ion, this ion can combine with HF molecule and can form HF_2^- ion. On the other hand, due to smaller electronegativity of Cl^-, Br^- and I^- ions, these ions are not able to combine with HCl, HBr and HI molecules respectively and hence cannot form HCl_2^-, HBr_2^- and HI_2^- ions.

Reason. See "*Hydrogen Bond*".

26. Some extra properties of hydrogen halides. (*a*) **Hydrogen fluoride, HF.** (*i*) HF is highly poisonous, very corrosive to skin and is, therefore, dangerous to handle. Anhydrous HF is a bad conductor of electricity. It destroys organic matter but does not attack gutta-percha and crasin wax.

(*ii*) *Basic or amphoteric behaviour of substances in liq. HF.* See "*Auto-ionisation of liq. HF*" discussed in chapter 19 (*Modern Concepts of Acids and Bases*).

(*iii*) Anhydrous HF is a colourless, strongly firming liquid which boils at 19.4°C and freezes at – 83°C.

(*iv*) *Action on hydrocarbons : Formation of fluorocarbons.* See "*Fluorocarbons*" (chapter 39).

(*b*) **Hydrogen chloride, HCl.** (*i*) It fumes in moist air and is extremely soluble in water. (*ii*) It is heavier than air. (*iii*) It can easily be liquefied to colourless liquid and frozen to a white crystalline solid.

(*c*) **Hydrogen bromide, HBr.** (*i*) It fumes in moise air and is highly soluble in water. (*ii*) It is heavier than air and hence is collected by the upward displacement of air.

(*d*) **Hydrogen iodide, HI.** (*i*) It fumes in air and is highly soluble in water. (*ii*) It is heavier than air (*iii*) It can easily be liquefied and solidified.

Comparison of hydrofluoric acid with other halogen acids

Hydrofluoric acid differs from the remaining halogen acids. The points showing differences have already been discussed under "*General Characteristics of Halogen Acids*" in this chapter.

Similarities between HF and H_2O. (*i*) Both the compounds resemble each other in solubility, fusibility etc. (*ii*) Both form associated molecules, $(HF)_x$ and $(H_2O)_z$. The formation of such molecules is due to hydrogen bonding in HF and H_2O molecules. (*iii*) The boiling point of HF is exceptionally high in comparison to those of other hydrogen halides. (H_2F_2 = 19.4° C, HCl = – 85°C, HBr = – 67°C, HI = – 35°C). Similarly the boiling point of H_2O is also exceptionally high in comparison to those of H_2S and H_2Se (H_2O = 100°C, H_2S = – 61.8°C, H_2S = – 61.8° C, H_2Se = – 42°C). Just as the exceptionally high boiling point of HF is ascribed due to strong H-bonds between HF molecules, the high boiling point of H_2O is also attributed to the H-bonds between H_2O molecules. (*iv*) Like H_2O, H_2F_2 is a weak acid and both are ionising solvents. (*v*) Heats of formation of HF and H_2O are close to each other.

Etching of Glass by H_2F_2

In etching, glass surface is covered with a thin film of paraffin wax and the design to be produced, is scratched through the film, with a sharp needle. The surface is then exposed to the action of the gas or of an aqueous solution (40-60%) of the acid. The paraffin coating protects the unexposed glass. Burettes, thermometers and other glass wares are engraved in this manner. Etching with solution is clear, but it is opaque with the gas.

Uses of halogen acids

Halogen acids are very useful compounds. Some of their uses are given below :

(*a*) Hydrofluoic acid. *H_2F_2 is used:* (*i*) In the etching of glass and making scales on the glass instruments. (*ii*) As an antiseptic in dyeing and brewing industries, since its effect on yeast is very little as compared with its action on other organisms. (*iii*) In the manufacture of fluorine. (*iv*) In the manufacture of a class of fluorine-containing dyes that have proved very fast and remarkably resistant to light. (*v*) In the pickling of alloy steels. (*vi*) In the manufacture of fluorine-containing plastics like Teflon, freon-12 and Kel-F which are characterised by extreme chemical inertness, high thermal stability and very good dielectric properties. (*vii*) Because of its action on silica. H_2F_2 is extensively used for removing silica from artificial graphite and Swedish filter paper. (*viii*) By the action of H_2F_2 on sodium aluminate, synthetic (artificial) cryolite is prepared. H_2F_2 is also used for the preparation of AlF_3. Both these substances are used in the metallurgy of Al. (*ix*) H_2F_2 is also used in the production of uranium hexafluoride (UF_6) which, because of its thermal stability and sufficiently high vapour pressure at the room temperature, is employed for the separation of uranium isotopes by the gaseous diffusion process. UF_6 is prepared by the action of H_2F_2 on U_2O_3 to get UF_3 which is then treated with F_2 to get UF_6. (*x*) Fluorides of alkali metals and antimony are used as mordant in dyeing industry. (*xi*) H_2F_2 is also used as a refrigerant in many of the cooling process.

(*b*) Hydrochloric acid. (*i*) Hydrochloric acid is used in the manufacture of chlorides and chlorine. (*ii*) It is used in the textile and dyeing industries, as also in tanning. (*iii*) In the form of aqua regia it is used for dissolving metals like gold and platinum. (*iv*) It is used as an important reagent in the laboratory. (*v*) It is also used in medicine. (*vi*) It is also used for cleaning iron sheets during tin plating, galvanisation etc. (*vii*) It is also used for the extraction of glue from animal tissues and bones.

(*c*) Hydrobromic acid. HBr is used as an important laboratory reagent and in the preparation of derivatives of unsaturated organic compounds and some important bromides. AgBr and KBr are used in photography and in medicines.

(*d*) Hydroiodic acid. HI is used as a reducing agent in organic chemistry. KI is used in medicines and AgI is used in photography.

Table 29.1 : *Properties of hydrogen halides (Hydrides of halogens)*

Property	HF	HCl	HBr	HI
Monomeric or dimeric	Dimeric (H_2F_2)	Monomeric (HCl)	Monomeric (HBr)	Monomeric (HI)
Physical state at 15°C	Liquid	Gas	Gas	Gas
H—X bond length (Å)	0.92	1.27	1.41	1.61
	Increases →			
Strength of H—X bond or thermal stability of HX molecule	Maximum	High	Not so high	Very little
	Decreases →			
H—X bond dissociation energy or heat of dissociation (kJ mol^{-1})	+ 574.0	+ 428.1	+362.5	+ 294.6
	Decreases →			
Dissociation temperature (°C)	Does not dissociate	1500	800	180
	Decreases →			
Melting point (°C)	– 83	–111	– 86	–50.8
Boiling point (°C)	+ 19.4 (Maximum)	– 85 (Minimum)	– 67	– 35.0
Hydrogen bonding	Maximum	Little	Very little	Very little
Volatility	Minimum	Maximum	—	—
Solubility in water at 0°C (g/litre)	85.3	42.0	49.0	57.1
Boiling point of constant boiling mixture (°C)	120	110	126	127
Composition of constant boiling mixture (weight % of the acid)	36.0 HF	20.4 HCl	48.0 HBr	57.0 HI
Dipole moment values when HX molecules are assumed to be completely ionic [calculated values (C.m.)]	1.47×10^{-29}	2.03×10^{-29}	2.25×10^{-29}	2.57×10^{-29}
	Increase →			
Experimental values of dipole moment (C.m.)	0.640×10^{-29}	0.343×10^{-29}	0.263×10^{-29}	0.127×10^{-29}
	Decrease →			
% of ionic character in the gaseous HX molecules.				
(*a*) as calculated from their dipole moment values	43.5	16.8	11.6	4.9
	Decrease →			
(*b*) as calculated from Hanny-Smith's equation	40.14	19.5	15.04	8.87
	Decrease →			
Heat evolved (kJ mol^{-1}) in the reaction, HX (*g*) + aq → HX (*aq*) → H^+ (*aq*) + X^- (*aq*) or HX (*g*) + H_2O → H_3O^+ + X^- (*aq*) (called heat of ionisation)	– 9.0 weakest acid	– 60.8 very strong acid	– 64.6 strongest acid	–58.4 strong acid
Reducing property	Not a reducing agent	Mild reducing agent	Strong reducing agent	Strongest reducing agent
	Increases →			
Heat of formation (kJ mol^{-1}) [$\frac{1}{2}X_2$ (*s, l, g*) + $\frac{1}{2}H_2$ (*g*) → HX (*g*); Δ H = heat required (+) or released (–)]	– 271.1 (*g*)	– 92.3 (*g*)	– 36.4 (*g*)	+25.4 (*g*)
	Increases →			

(*Contd.*)

Heat of dehydration (kJ mol^{-1}) [HX (*aq*) + Heat → HX (*g*)]	+ 48.0	+18.0	+21.0	+23.0
Heat of neutralisation with NaOH (kJ mol^{-1})	+ 68.2	+ 57.3	+ 57.9	+ 57.6
Apparent degree of dissociation in 0.1 N solution at room temperature	0.08	0.93 Increases →	0.94	0.95
Solubility of silver salt (g/100 g)	172	0.00154	0.000084	0.000028
Solubility of calcium salt (g/100 g)	0.0016	42.7 Increases →	58.8	67.6

Halides of alkali metals (Group IA)

Preparaction. All the alkali metals combine directly with halogens and form the halides of MX type. As the electropositive character of alkali metals increases from Li to Cs, the ease with which these metals form halides increases from Li to Cs.

Properties. (*i*) These are crystalline solids with high melting points. and b.pts., volatile without decomposition at high temperatures; are good conductors of electricity in the fused state.

(*ii*) *Ionic/covalent character.* With the exception of LiBr and LiI, other halides have essentially ionic character. In LiBr and LiI, the large halide ions *viz.*, Br^- and I^- ions are strongly polarised by the small Li^+ ion (See Fajans rules) and hence these molecules have covalent character.

(*iii*) *Solubility.* With the exception of LiF, all other halides are soluble in water. LiF is insoluble because of its large lattice energy. High lattice energy of LiF is due to the fact that both the ions present in LiF (*i.e.*, Li^+ and F^- ions) are small in size. Due to the presence of covalent character in LiBr and LiI, these halides are soluble in organic solvents. The corresponding halides of Na and K are not soluble. LiCl is soluble in alcohol while the halides of other alkali metals are insoluble in alcohol.

(*iv*) *Formation of hydrates.* LiCl separates out from the aqueous solution as $LiCl.2H_2O$. Sodium and other alkali metal chlorides do not form hydrates at ordinary temperature. LiCl is dehiescent.

(*v*) *Hydrolysis.* LiCl undergoes hydrolysis in hot water to a small extent but NaCl and KCl do not undergo hydrolysis at all.

(*vi*) *Structure.* The structure of the halides is determined by radius ratio (r_+/r_-) and lattice energy of the halides. (See chapter 16).

(*vii*) *Formation of polyhalides.* KI, rubedium and caesium halides have a tendency to form simple and mixed polyhalides. Examples are KI_3, CsI_3, RbI_3 (*simple polyhalides*), CsI_2Cl, $CsBr_2Cl$, $RbIBr_2$, $RbBrCl_2$, $RbClI_4$ etc., (*mixed polyhalides*). The polyhaldes of Cs are thermally more stable than those of Rb.

Here we shall discuss LiCl, KCl, and KI in detail.

1. Lithium chloride, LiCl

Preparation. (*i*) LiCl can be prepared by burning Li-metal in Cl_2 gas or by the action of HCl on LiOH or Li_2CO_3 or by the action of $BaCl_2$ solution on Li_2SO_4 solution.

$$2Li + Cl_2\ (gas) \longrightarrow 2LiCl\ ; \qquad LiOH + HCl \longrightarrow LiCl + H_2O$$

$$Li_2CO_3 + 2HCl \longrightarrow 2LICl + H_2O + CO_2\uparrow$$

$$Li_2SO_4\ (solution) + BaCl_2\ (solution) \longrightarrow 2LiCl\ (solution) + BaSO_4\downarrow$$

(*ii*) *Manufacture.* LiCl is manufactured from spodumene and triphylite minerals as discussed under "Extraction of Lithium".

Properties. (*i*) It is a white crystalline solid and is the most deliquescent substance known. It is soluble in H_2O and organic solvents like alcohol, pyridine etc. Its solubility in H_2O and organic solvents both indicates that LiCl is both ionic and covalent. It is more volatile than any other alkali metal chloride.

(*ii*) *Action of heat.* When heated above its fusion temperature, Cl_2 gas is obtained.

$$2LiCl \xrightarrow[\text{fusion temp.}]{\text{Heat above}} 2Li + Cl_2$$

(*iii*) This compound crystallises as an anhydrous salt from its aqueous solution above 98°C, but at lower temperature, it crystallises as hydrated salts like $LiCl.H_2O$, $LiCl.2H_2O$ and $LiCl.3H_2O$.

(*iv*) *Action of NH_3.* LiCl absorbs NH_3 both in the solution state and dry state, and gives ammoniates like LiCl, nNH_3 where n = 1, 2, 3, 4.

(*v*) *Action of $(NH_4)_2CO_3$.* When treated with $(NH_4)_2CO_3$, Li_2CO_3 is obtained.

$$2LiCl + (NH_4)_2CO_3 \longrightarrow Li_2CO_3 + 2NH_4Cl$$

Uses. *LiCl is used* : (*i*) in the preparation of Li-metal by the electrolysis (*ii*) in air-conditioning plants in the form of its 35-40% aqueous solution for controlling the humidity of air (*iii*) as a constituent of welding fluxes (*iv*) in Ni-Fe accumulators.

Structure. LiCl is primarily an ionic compound and hence is represented as Li^+ Cl^-. The formation of Li^+Cl^- molecule can be represented by the following three steps :

(*a*) $Li\ (2, 1) \xrightarrow{-e^-} Li^+\ (2) + e^-$ (*b*) $Cl(2, 8, 7) + e^- \longrightarrow Cl^-\ (2, 8, 8)$

(*c*) $Li^+\ (2) + Cl^-\ (2, 8, 8) \longrightarrow Li^+\ Cl^-$

2. Potassium chloride, KCl

It occurs in Stassfurt deposits (America) as sylvine (KCl) and as carnallite ($KCl.MgCl_2.6H_2O$).

Preparation. (*i*) *From sylvine, KCl.* Sylvine is a mixture of KCl and NaCl. When the boiling hot saturated solution of the mixed salts in water is cooled, KCl separates out and NaCl is left behind in the solution.

(*ii*) *From carnallite, $KCl.MgCl_2.6H_2O$.* This ore is always found mixed with NaCl and $MgSO_4$. In order to remove NaCl and $MgSO_4$, the ore is ground and extracted with a hot 20% solution of $MgCl_2$. Carnallite dissolves while NaCl and $MgSO_4$ remain undissolved. These are filtered off and the solution which contains only carnallite is crystallised so that cubic crystals of KCl separate out leaving behind $MgCl_2$ in the solution.

Properties. It is a colourless crystalline solid which possesses cubic crystals. Its melting point and boiling point are 768°C and 1411°C respectively. It is extremely soluble in water. It closely resembles NaCl in most of its properties, except that its solubility increases rapidly with the temperature and is more readily fusible.

Uses. It is used : (*i*) in the manufacture of potassium and its compounds (*ii*) as a potassium fertiliser (K-type fertiliser), since it supplies potassium (as K_2O) to the soil. The quantity of K_2O found in KCl varies according to the purity of KCl. Usually 50-55% potassium content is found in the salt.

3. Potassium iodide, KI

Preparation. KI is prepared : (*a*) *by the action of HI on KOH or K_2CO_3.*

$$KOH + HI \longrightarrow KI + H_2O\ ; \qquad K_2CO_3 + 2HI \longrightarrow 2KI + CO_2 + H_2O$$

(*b*) *by heating I_2 with hot and conc. solution of KOH.*

$$3I_2 + 6KOH \longrightarrow \underbrace{\boxed{5KI + KIO_3}}_{\text{solution}} + 3H_2O$$

The solution containing KI and KIO_3 is evaporated to dryness so that a solid residue is obtained. When this residue is ignited with prowdered charcoal, KIO_3 present in the residue is reduced to KI.

$$KIO_3 + 3C \xrightarrow{\text{Reduction}} KI + 3CO$$

The mass is extracted with H_2O, filtered and the clear solution is evaporated to dryness to get the crystals of KI.

(*c*) *by the action of ferro-ferric iodide (FeI_2. $2FeI_3$) on K_2CO_3.* (*Large scale preparation*). The solution containing $FeI_2.2FeI_3$ and K_2CO_3 is boiled.

$$4K_2CO_3 + FeI_2.2FeI_3 + 4H_2O \longrightarrow 8KI + 4CO_2 + Fe(OH)_2.2Fe(OH)_3 \ (ppt.)$$

$FeI_2.2FeI_3$ required in the above reaction is prepared by adding I_2 to iron filings in water.

$$3Fe \xrightarrow{+3I_2} 3FeI_2 \xrightarrow{+I_2} FeI_2.2FeI_3.$$

Properties. (*i*) It is a white crystalline solid having cubic crystals with melting point = 6.80°. It is soluble in water and alcohol.

(*ii*) *Action of I_2.* Its concentrated aqueous solution dissolves I_2 to form potassium tridide, KI_3 which is a complex compound.

$$KI + I_2 \rightleftharpoons KI_3$$

KI_3 is an unstable compound and decomposes into KI and I_2 in all its chemical reactions.

(*iii*) *Action of salts. (a) Action of Pb (II) salts.* When KI is added to the aqueous solution of Pb (II) salts like $PbCl_2$ (in hot water, since $PbCl_2$ is soluble only in hot water), $Pb(NO_3)_2$, $(CH_3COO)_2Pb$ etc., yellow ppt. of PbI_2 is obtained which gets dissolved in the excess of KI due to the formation of potassium-tetra iodo plumbate (II), $K_2[PbI_4]$.

$$Pb^{2+} + 2KI \longrightarrow PbI_2 + 2K^+ \ ; \ PbI_2 + 2KI \ (excess) \longrightarrow K_2[PbI_4] \ [colourless \ (in \ solution)]$$

The above reaction has been utilised for the identification of Pb^{2+} ions present in an inorganic mixture.

(*b*) *Action of $AgNO_3$.* A yellow ppt. of AgI is obtained with $AgNO_3$.

$$AgNO_3 + KI \longrightarrow AgI\downarrow + KNO_3$$

The ppt. of AgI dissolves in the excess of KI, forming the complex compound, $K[AgI_2]$ (*red*).

$$AgI + 2KI \ (excess) \longrightarrow K[AgI_2] \ (red)$$

(*c*) *Action of $HgCl_2$.* When KI is treated with $HgCl_2$ solution, a scarlet (red) ppt. of HgI_2 is obtained. This ppt. dissolves in the excess of KI forming complex compound, $K_2[HgI_4]$ which is called potassium tetra-iodomercurate.

$$HgCl_2 + 2KI \longrightarrow HgI_2 + 2KCl \ ; \qquad HgI_2 + 2KI \longrightarrow K_2[HgI_4]$$

The alkaline solution of $K_2[HgI_4]$ is known as Nessler's reagent which is used for testing even the traces of NH_3 in solution.

(*iv*) *Oxidation of KI to I_2 : Reducing properties.* Many oxidising agents can oxidise KI to I_2, *i.e.*, many oxidising agents can liberate I_2 from KI. When KI reacts with an oxidising agent, it behaves as a reducing agent.

$$2I^- \longrightarrow I_2 + 2e^-$$

Examples are given below :

(*a*) Acidified solution of $KMnO_4$ oxidises KI to I_2 and is itself reduced to $MnSO_4$.

$$2KMnO_4 + 10KI + 8H_2SO_4 \longrightarrow 6K_2SO_4 + 2MnSO_4 + 8H_2O + 5I_2$$

or $$2MnO_4^- + 10I^- + 16H^+ \longrightarrow 2Mn^{2+} + 8H_2O + 5I_2$$

(*b*) $CuSO_4$ liberates I_2 from KI and is itself reduced to CuI.

$$2CuSO_4 + 4KI \longrightarrow 2K_2SO_4 + 2CuI + I_2$$

or $$2Cu^{2+} + 2I^- \longrightarrow 2Cu^+ + I_2$$

(*c*) Acidified solution of $K_2Cr_2O_7$ liberates I_2 from KI and is itself reduced to $Cr_2(SO_4)_3$.

$$6KI + K_2Cr_2O_7 + 7H_2SO_4 \longrightarrow 4K_2SO_4 + Cr_2(SO_4)_3 + 3I_2 + 7H_2O$$

or $$6I^- + Cr_2O_7^{2-} + 14H^+ \longrightarrow 2Cr^{3+} + 3I_2 + 7H_2O$$

All the above three reactions are used in iodometric titrations.

(*d*) Cl_2 liberates I_2 from KI.

$$2KI + Cl_2 \longrightarrow 2KCl + I_2 \quad \text{or} \quad 2I^- + Cl_2 \longrightarrow 2Cl^- + I_2$$

(*e*) H_2SO_4 liberates I_2 from KI and is itself reduced to SO_2, S or H_2S depending on experimental conditions.

$$2H_2SO_4 + 2KI \longrightarrow SO_2 + 2H_2O + K_2SO_4 + I_2 \text{ or } SO_4^{2-} + 4H^+ + 2I^- \longrightarrow SO_2 + 2H_2O + I_2$$

$$4H_2SO_4 + 6KI \longrightarrow S + 4H_2O + 3K_2SO_4 + 3I_2 \text{ or } SO_4^{2-} + 8H^+ + 6I^- \longrightarrow S + 4H_2O + 3I_2$$

$5H_2SO_4 + 8KI \longrightarrow H_2S + 4K_2SO_4 + 4H_2O + 4I_2$ or $SO_4^{2-} + 10H^+ + 8I^- \longrightarrow H_2S + 4H_2O + 4I_2$

(*f*) HNO_3 also liberates I_2 from KI solution and is itself reduced to NO.

$8HNO_3 + 6KI \longrightarrow 2NO + 4H_2O + 6KNO_3 + 3I_2$ or $2NO_3^- + 8H + 6I^- \longrightarrow 2NO + 4H_2O + 3I_2$

(*g*) Ozone liberates I_2 from KI solution and is itself reduced to O_2.

$$2KI + H_2O + O_3 \longrightarrow 2KOH + I_2 + O_2$$

(*h*) The following reactions also show reducing properties of KI, *i.e.*, in these reactions I_2 is liberated.

$$KIO_3 + 3H_2SO_4 + 5KI \longrightarrow 3K_2SO_4 + 3H_2O + 3I_2$$

or $$IO_3^- + 6H^+ + 5I^- \longrightarrow 3H_2O + 3I_2$$

$$H_2O_2 + 2H^+ + 2I^- \longrightarrow K_2SO_4 + 2H_2O + I_2$$

or $$H_2O_2 + 2H^+ + 2I^- \longrightarrow 2H_2O + I_2$$

$$2HNO_2 + H_2SO_4 + 2KI \longrightarrow K_2SO_4 + 2H_2O + 2NO + I_2$$

or $$2NO_2^- + 4H^+ + 2I^- \longrightarrow 2H_2O + 2NO + I_2$$

$CaOCl_2$ (*Bleacing power*) $+ 2CH_3COOH + 2KI \longrightarrow (CH_3COO)_2Ca + 2KCl + H_2O + I_2$

The liberated iodine in the above reactions can be titrated with $Na_2S_2O_3$ solution (intermediate solution) and thus the strength of the solution of an oxidising agent can be determined indirectly by using KI solution.

$$I_2 + Na_2S_2O_3 \longrightarrow Na_2S_4O_6 + 2NaI$$

In these titrations freshly-prepared starch solution is used as an internal indicator. At the end point, the blue colour disappears by the addition of $Na_2S_2O_3$ solution from the burette. This indirect method of determining the strength of the solution of an oxidising agent is called *iodometric litration.*

(*v*) *Oxidation of KI to KIO_3 and KIO_4.* Alkaline solution of KI is oxidised to KIO_3 (pot. iodate) and KIO_4 (pot. periodate) by O_3.

$$KI + 3O_3 \longrightarrow KIO_3 + 3O_2$$

$$KI + 4O_3 \longrightarrow KIO_4 + 4O_2$$

Uses. KI is used : (*i*) as a reagent in laboratory (*ii*) in photography (*iii*) in medicine (*iv*) for making iodine solutions for chemical analysis (*v*) for making Nesseler's reagent in the laboratory.

Halides of alkaline earth metals (Group IIA)

All the elements of group IIA can form the halides of MX_2 type. These halides can be obtained either by heating the metal with halogen at appropriate temperature or by reacting the metal carbonates with dil. halogen acid.

MX_2 molecules have linear shape. As an example, the geometry of BeF_2 molecule has already been discussed in chapter 14.

Be^{2+} ion, on account of its smaller size, also forms stable complex halide ions, like $[BeF_3]^-$ and $[BeF_4]^-$. These ions can be regarded as having been formed by the combination of one BeF_2 molecule and one or two F^- ions.

$$BeF_2 + F^- \longrightarrow BeF_3^- ; \qquad BeF_2 + 2F^- \longrightarrow BeF_4^{2-}$$

BeF_3^- has trigonal planar geometry which is due to sp^2 hybridisation of Be atom in its excited state (Be = $2s^1\ 2p^1$). BeF_4^{2-} ion has tetrahedral shape which is due to sp^3 hybridisation of Be-atom in its excited state.

Properties. (*i*) *Covalent / ionic character.* Due to the smallest size and high charge density on Be^{2+} ion, this ion has the maximum polarising power and hence polarises effectively the electron charge cloud of halide ion (X^-) in BeX_2 molecule. Consequently BeX_2 molecule becomes covalent. Mg^{2+}, Ca^{2+}, Sr^{2+} etc., ions have very weak polarising power so that little or no covalent character is imparted to the halides of these cations, *i.e.*, beryllium halides are covalent, while the halides of other alkaline earth metals are ionic compounds and are soluble in water. The ionic character increases with the increase in ionic size of the metal ion. The solubility of the halides decreases as the atomic number increases. The halides of Be do not conduct electricity, in fused state. Halides of other metals are good conductors of electricity in the fused as well as in the dissolved state.

(*ii*) *Ignition of hydrated chlorides.* The hydrated magnesium chloride, $MgCl_2.6H_2O$, on ignition, gives a residue of MgO.

$$MgCl_2.6H_2O \longrightarrow MgO + 2HCl + 5H_2O$$

No such reaction is shown by the hydrated chlorides of other alkaline earth metals.

(*iii*) Some of the halides of MX_2 type are hydrated while others are not. The chlorides of all metals are hydrated. For example $BeCl_2.4H_2O$, $MgCl_2.6H_2O$, $CaCl_2.6H_2O$, $SrCl_2$, $6H_2O$, $BaCl_2$, $2H_2O$, $RaCl_2.2H_2O$. Radium bromide is also hydrated ($RaBr_2.2H_2O$). The fluorides are non-hydrated, *e.g.*, BeF_2, MgF_2, CaF_2 etc. Most of the anhydrous halides of group IIA metals crystallise in the fluorite true structure.

(*iv*) *Hydrolysis.* Hydrolysis of $BeCl_2$, $MgCl_2$ and $CaCl_2$ takes place as follows :

$$BeCl_2 + H_2O \longrightarrow BeO + 2HCl \; ; \qquad BeCl_2 + 2H_2O \longrightarrow Be(OH)_2 + 2H_2O$$

$$MgCl_2 + H_2O \longrightarrow Mg(OH)Cl + HCl; \qquad CaCl_2 + H_2O \longrightarrow CaO + 2HCl$$

Halides of boron group elements (Group III A)

Various halides given by the elements of group IIIA can be grouped into the following catagories: (*a*) *Monohalides given by thallium.* Examples are TlF, TlCl, TlBr and TlI. (*b*) *Trihalides given by all elements (MX_3).* (*c*) *Dinuclear tetrahalides given by boron.* Examples are B_2F_4, B_2Cl_4, B_2Br_4 and B_2I_4. (*d*) *Dimeric trihalides given by Al, Ga,* etc. The trihalides of Al, Ga etc., also dimerise to give dimeric molecules namely $(MX_3)_2$ or M_2X_6 both in the vapour state and when the trihalides are dissolved in non-polar solvents like C_6H_6 etc. In the formation of dimeric molecules, the central atom (Al, Ga, In) makes use of its vacant sp^3 hybrid orbital (sp^3 hybridisation). Due to small size of B-atom, this atom is not able to coordinate to other halogen atoms and hence cannot form the dimeric molecules, $(BX_3)_2$ or B_2X_6. (*e*) *Complex halides.* All the elements of group III A form complex halides. Examples are BF_4^-, AlF_4^-, AlF_6^{3-}, $GaCl_4^-$, $GaCl_6^{3-}$, $InCl_6^{3-}$, $TlBr_4^-$, $TlCl_6^{3-}$ etc.

General characteristics of trihalides

(*i*) *Covalent and ionic character.* Boron trihalides are covalent in character due to the smaller size and high charge density of B^{3+} ion.

The fluorides of Al, Ga, In and Tl are ionic and have high melting points, while the other halides of these elements are covalent when anhydrous. For example, anhydrous $AlCl_3$ is covalent, but in water it gets hydrolysed to give Al^{3+} (*aq*) ions. This change from a covalent compound in an aqueous solution is mainly due to the high hydration energy of Al^{3+} ions. However, BCl_3, on hydrolysis, does not give B^{3+} ions, because the ionisation energy of B^{3+} ion is very large.

(*ii*) *Geometry (i.e., shape) of trihalides.* The normal trihalides are planar triangular in shape which results from sp^2 hybridisation of the central atom. The central atom in these trihalides has a vacant *p*-orbital which remains unhybridised. (See the structure of BCl_3 molecule explained in chapter 14.

(*iii*) *Formation of dimeric trihalides, $(MX_3)_2$ or M_2X_6.* Boron trihalides exist as monomeric molecules (*i.e.*, as BX_3 molecule) while the trihalides of other elements exist as dimeric molecules, namely $(MX_3)_2$ or M_2X_6, both in the vapour state and when the trihalides are dissolved in non-polar solvents like C_6H_6 etc. The existence of monomeric nature of the trihalides of boron is due to the smaller size of B-atom. Due to smaller size of B-atom, this atom is not able to coordinate to other halogen atoms and hence cannot form the dimeric molecules, $(BX_3)_2$ or B_2X_6.

MX_3 molecules have planar trigonal shape which results from sp^2 hybridisation of M-atom in its excited state. On the other hand, the addition compounds, like $H_3N \rightarrow BH_3$, $[BF_4]^-$, have tetrahedral shape, corresponding to sp^3 hybridisation of M-atom, since *np* orbital which remains unhybridised in the formation of MX_3 molecules (sp^2 hybridisation) participates in sp^3 hybridisation. The tendency of MX_3 molecules to act as Lewis acids goes on decreasing, as we pass from BX_3 to InX_3 molecule. Thus, the tendency of these molecules to act as Lewis acids is in the order : $BX_3 > AlX_3 > GaX_3 > InX_3$.

It may be seen, that BF_3 can combine only with one F^- ion to from $[BF_4]^-$ ion (in which the coordination number of B-atoms is 4), while the trihalides of other elements can combine with three halide ions to form $[AlF_6]^{3-}$, $[GaCl_3]^{3-}$ etc. (in which the coordination number of the metal atom is 6). Boron atom, which belongs to 2nd period of the periodic table, cannot have coordination number more than 4, due to the non-availability of *d*-orbital in its valence-shell, while other elements which belong to higher periods, can have co-ordination number upto 6, due to the presence of *d*-orbitals in their valence-shell (Also see "*Rule of Maximum Covalency*" discussed in chapter 10).

Lewis acid or acceptor molecule		*Lewis base or donor molecule or ion*		*Addition compound or adduct*
BF_3	+	$:NH_3$	$\longrightarrow$	$H_3N \longrightarrow BH_3$
BF_3	+	$:\ddot{F}:^-$	$\longrightarrow$	$F^- \longrightarrow BF_3$ or $[BF_4]^-$
AlF_3	+	$3:\ddot{F}:^-$	$\longrightarrow$	$[AlF_6]^-$
AlF_3	+	$3:\ddot{Cl}:^-$	$\longrightarrow$	$[GaCl_6]^-$
$InCl_3$	+	$3:\ddot{Cl}:^-$	$\longrightarrow$	$[InCl_6]^-$

Boron trihalides (BF_3, BCl_3, BBr_3, BI_3)

Preparation. BF_3 can be obtained : (*a*) by heating B_2O_3 with a fluoride (*e.g.*, CaF_2, NaF) or ammonium fluoborate (NH_4BF_4) and conc. H_2SO_4.

$$B_2O_3 + 3CaF_2 + CH_2SO_4 \longrightarrow 3CaSO_4 + 3H_2O + 2BF_3$$
$$B_2O_3 + 6NaF + 3H_2SO_4 \longrightarrow 3Na_2SO_4 + 3H_2O + 2BF_3$$
$$B_2O_3 + 6NH_4BF_4 + 6H_2SO_4 \longrightarrow 6NH_4HSO_4 + 3H_2O + 8BF_3$$

(*b*) by heating borax ($Na_2B_4O_7$) with CaF_2 and conc. H_2SO_4.

$$Na_2B_4O_7 + 6CaF_2 + 8H_2SO_4 \longrightarrow 2NaHSO_4 + 6CaSO_4 + 7H_2O + 4BF_3$$

(*c*) by the action of HF or KBF_4 on B_2O_3 or by heating potassium fluoborate (KBF_4).

$$2B_2O_3 + KBF_4 \longrightarrow K[B_4O_6F] + BF_3 \;; \quad B_2O_3 + 6HF \longrightarrow 3H_2O + 3BF_3$$
$$KBF_4 \xrightarrow{\Delta} KF + BF_3$$

(*d*) *Pure BF_3* is obtained by the thermal decomposition of diazonium tetrafluoroborate, $C_6H_5N_2BF_4$

$$C_6H_5N_2BF_4 \longrightarrow C_6H_5F + N_2 + BF_3$$

BCl_3 and BBr_3 both can be prepared by passing Cl_2 or Br_2 over a heated mixture of B_2O_3 and powderd charcoal.

$$B_2O_3 + 3C + 3X_2 \; (X = Cl, Br) \xrightarrow{\Delta} 2BX_3 + 3CO$$

BCl_3 and BBr_3 can also be prepared by heating B_2O_3 with PCl_5 or PBr_5.

$$B_2O_3 + 3PX_5 \; (X = Cl, Br) \xrightarrow{\Delta} 2BX_3 + 3POX_3$$

BI_3 is obtained by the action of $Na[BH_4]$ on I_2 or by the action of HI on BCl_3.

$$4I_2 + Na[BH_4] \longrightarrow BI_3 + 4HI + NaI \;; \quad BCl_3 + 3HI \longrightarrow BI_3 + 3HCl.$$

Since BF_3 is soluble in water, it is collected over mercury.

Properties of trihalides

(*i*) *Physical state.* The physical state of the halides becomes progressively more condensed from fluoride to iodide. Thus, BF_3 is a colourless pungent gas (m.pt. = –127.1°C, b.pt. = –99.9°C); BCl_3 is a colourless liquid (b.pt. = 12.5°C, freezing pt. = –107°C), BFr_3 is a colourless viscous liquid (b.pt. = 91.3° C, freezing point = – 46° C) and BI_3 is a white crystalline solid with m.pt. = 49.9° C.

(*ii*) All the trihalides of boron one volatile, highly reactive and monomeric.

(*iii*) They do not conduct electricity in the liquid state.

(*iv*) *Melting and boiling points.* The melting and boiling points of trihalides increases regularly from BF_3 to BI_3.

(*v*) *Formation of hydrates.* With limited amount of water, BF_3 gives two hydrates namely $BF_3.H_2O$ and $BF_3.\ 2H_2O$.

(*vi*) *Hydrolysis.* (*a*) BF_3 is hydrolysed in water to form orthoboric acid (H_3BO_3) and HF.

$$BF_3 + 3H_2O \longrightarrow H_3BO_3 + 3HF \quad \text{... }(i)$$

If BF_3 gas (BF_3 is a gas) is passed in excess in water, it dissolves in HF (formed in the above reaction) to form hydrofluoboric acid, $H[BF_4]$.

$$BF_3 + HF \longrightarrow H[BF_4] \quad \text{...}(ii)$$

Multiply equation (*ii*) by 3 and then add it to equation (*i*) to get :

$$4BF_3\ (excess) + 3H_2O \longrightarrow H_3BO_3 + 3H[BF_4]$$

The above equation shows that BF_3, on hydrolysis, gives H_3BO_3 and $H[BF_4]$.

(*b*) BCl_3, BBr_3 and BI_3 are completely hydrolysed by water, giving H_3BO_3 and the corresponding halogen acids (HX).

$$BX_3 + 3H_2O \longrightarrow H_3BO_3 + 3HX\ (X = Cl, Br, I)$$

(*vii*) *Lewis acid character (Electron accepting property).* All the trihalides of boron act as Lewis acids (electron-deficient molecules), *i.e.* they can accept the electron pair donated by donor molecules [*e.g.*, NH_3, H_2O, $O(C_2H_5)_2$ etc.] and donor ions (*e.g.*, F^-, OH^- etc.) to form addition compounds (adducts). In the formation of addition compounds the donor molecules/ions act as Lewis bases.

Explanation. See chapter 19 (*Modern Concept of Acids and Bases*).

(*viii*) *Monomeric structure of BX_3 molecules.* Boron trihalides exist as monomeric discrete molecules (*i.e.*, as BX_3 molecules). This is because of the smaller size of B-atom. Due to its smaller size B-atom can coordinate only to three halide ions and not to four halide ions and hence cannot form the dimeric molecules, $(BX_3)_2$ or B_2X_6.

(*ix*) *Covalent character.* Boron trihalides are covalent in character due to the small size and high charge density on B^{3+} ion. Some other points in favour of the covalent character of boron trihalides are : (*a*) As liquids they do not conduct electricity *i.e.*, they are non-electrolytes. (*b*) Their boiling points are very low as compared to the halides of the elements of group IA and IIA. (*c*) They exist as discrete molecular species (*i.e.*, monomeric molecules) in all the states. (*d*) The boiling points of boron halides increase as the atomic number of the halogen atoms increases. This behaviour is very similar to that shown by covalent halogen compounds of carbon.

(*x*) *Solvolytic reactions.* While BF_3 gives addition compounds with certain donor molecules having labile hydrogen, the other trihalides undergo solvolytic reactions with such donor molecules. Some solvolytic reactions shown by BCl_3 are given below :

$$BCl_3 + 3C_2H_5OH \longrightarrow B(C_2H_5O)_3 + 3HCl$$

$$BCl_3 + 6NH_3 \longrightarrow B(NH_2)_3 + 3NH_4Cl \longrightarrow BN + 2NH_3 + 3NH_4Cl$$

$$BCl_3 + 6N_2H_4 \longrightarrow B(N_2H_3)_3 + 2N_2H_5Cl \longrightarrow B(N_2H) + 2N_2H_4 + 2N_2H_5Cl$$

$$BCl_3 + 6C_6H_5NH_2\ (In\ benzene) \rightarrow B(C_6H_5NH)_3 + 3C_6H_5\ NH_2.HCl$$

(*xi*) *Action of heat in air.* When BI_3 is heated in air, it burns with a brilliant flame.

$$4BI_3 + 3O_2 \longrightarrow 2B_2O_3 + 6I_2$$

(*xii*) *Reduction with hydrogen.* When reduced with H_2, BCl_3 gives free boron.

$$2BCl_3 + 3H_2 \longrightarrow 2B + 6HCl$$

Uses of BF_3

It is due to the tendency of BF_3 molecule to accept a lone pair of electrons from a donor molecule or ion that BF_3 is used as a catalyst in a number of reactions, such as alkylation or acylation in organic chemistry.

Geometry of BX_3 molecules. (X = F, Cl, Br, I)

Each of the trihalide molecules has trigonal planar geometry with each of the three XBX angles, equal to 120°. Each of the three B—X bonds is of equal length. This geometry arises due to sp^2 hybridisation of B-atom (central atom) in its excited state. As an example, the geometry of BF_3 molecule has already been discussed in chapter 14.

Resonance in BX_3 molecules

See chapter 19.

Variation in the covalent / ionic character of the trihalides of Al

The covalent character of the trihalides of Al increases, on moving from AlF_3 to AlI_3. Thus, AlF_3 is essentially ionic (least covalent), $AlCl_3$ dimer has intermediate character (more covalent than ionic) while $AlBr_3$ and AlI_3 dimers are predominantly covalent. This increasing order of covalent character is confirmed by the decreasing order of their melting points (AlF_3 = 1270°C, $AlCl_3$ sublimes at 180°C, $AlBr_3$ = 975°C, AlI_3 = 383°C). For the explanation of this variation see *Fajans' rules.*

Hydrofluoboric acid, $H[BF_4]$

Preparation. It can be prepared by : hydrolysing BF_3 with water or by dissolving HBO_3 in conc. HF.

$$4BF_3 + 3H_2O \longrightarrow H_3BO_3 + 3H[BF_4]; \quad H_3BO_3 + 4HF \longrightarrow H[BF_4] + 3H_2O$$

Properties. It is a strong acid, stronger than HF. It is decomposed when heated with water.

Fluoborates or borofluorides (BF_4^-)

The salts of $H[BF_4]$ are called fluoborates.

Preparation. These salts are prepared.

(*a*) by dissolving the corresponding metallic oxides, hydroxides or carbonates in aqueous $H[BF_4]$.

(*b*) in solution by treating the alkali acid fluorides with H_3BO_3.

$$2NaHF_2 + H_3BO_3 \longrightarrow Na[BF_4] + NaOH + 2H_2O$$

$$2NH_4HF_2 + H_3BO_3 \longrightarrow NH_4[BF_4] + NH_4OH + 2H_2O$$

(*c*) by treating alkali metal borohydride like $Na[BH_4]$ with BF_3 in diglyme.

$$3Na[BH_4] + 4BF_3 \xrightarrow{\text{diglyme}} 3Ma[BF_4] + 2B_2H_6$$

Properties. (*i*) The fluorates are crystalline solids and are fairly soluble in water, except $K[BF_4]$.

(*ii*) They decompose on heating, *e.g.*,

$$K[BF_4] \xrightarrow{\Delta} KF + BF_3$$

(*iii*) When $K[BF_4]$ is heated with K, Na or Mg, boron (98.3% pure) is obtained as amorphous boron or black powder

$$K[BF_4] + 3K \longrightarrow B + 4KF; \quad 2K\,[BF_4] + 3Mg \longrightarrow 2B + 2KF + 3MgF_2$$

(*iv*) It may be noted that the covalency of boron in BF_4^- ion is 4 which is the maximum covalency of boron.

Uses of $H[BF_4]$ and its salts. (*i*) $H[BF_4]$ is used as a catalyst in organic chemistry and in the preparation of its salts. (*ii*) $H[BF_4]$ and its salts are used in electroplating. (*iii*) $K[BF_4]$ is used for the extraction of boron from its minerals.

Halides of carbon group elements (Group IVA)

The elements of group IVA give various types of halides like *dihalides, tetrahalides, catenated halides* and *complex halides.*

Dihalide

The dihalides of Ge, Sn and Pb are much less volatile than their corresponding tetrahalides. The stability of the dihalides, increases in going from Ge to Pb, due to inert pair effect. The dihalides of Ge_2 and Sn, but not of Pb, are reducing agents. (See inert pair effect).

Tetrahalides, MX_4

Preparation. The tetrahalides may be prepared, either by heating the element with the appropriate halogen, or by passing dry halogen over a heated mixture of the dioxide and carbon.

$$M + 2X_2 \longrightarrow MX_4$$

$$MO_2 \text{ (M = C, Si etc.)} + 2C + 2X_2 \longrightarrow MX_4 + 2CO$$

Properties. (*i*) *Covalent / ionic character.* SnF_4 definitely and PbF_4 probably have ionic character. Other tetrahalides are covalent.

(*ii*) *Tharmal stability.* Thermal stability of tetrahalides decreases from CX_4 to PbX_4 and for a given central atom decreases from fluoride to iodide. The fluorides of carbon (fluorocarbons) are therefore, the most stable and hence are inert.

(*iii*) *Formation of addition compound of $MX_4.2D$ (D = donor molecule) type and complex hexahalide ions of $[MX_6]^{2-}$ type.* The tetrahalides of carbon are not able to form any of these compouds while the tetrahalides of other elements of group IVA can form such compounds.

Explanation. The explanation of the difference between the behaviour of the tetrahalides of carbon and other elements has been given in chapter 10 (Rule of Maximum Covalency) and chapter 19 (Lewis Acids and Bases).

(*iv*) *Hydrolysis.* While the tetrahalides of cabon are not hydrolysed by water under ordinary condictions, the tetrahalides of other elements of group IVA are readily hydrolysed.

Explanation. The explanation of the above difference has been given in chapter 10 (Rule of Maximum Covalency).

The ease with which the the tetrahalides are hydrolysed by H_2O decreases from Si to Sn, as the metallic character of the central atom increases in this order. Thus, GeX_4 and SnX_4 tetrahalides are less readily hydrolysed than SiX_4 tetrahalides.

It may be mentioned here that empty orbitals are always available with any atom and they can be utilised, if sufficient energy is provided for the reaction to occur, *e.g.*, CCl_4 undergoes hydrolysis, when superheated steam is used.

$$CCl_4 + H_2O \text{ (superheated steam)} \longrightarrow COCl_2 \text{ (Phosgene)} + 2HCl$$

Hydrolysis of tetrahalides of Pb follows essentially the same pattern, but due to the instability of tetravalent compounds of Pb, some decomposition of $PbCl_4$ to $PbCl_2$ also takes place.

$$PCl_4 \longrightarrow PbCl_2 + Cl_2$$

$PbCl_4$ is hydrolysed by H_2O as follows :

$$PbCl_4 + 2H_2O \longrightarrow PbO_2 + 4HCl$$

(*v*) *Lewis acid character.* Tetrahalides of carbon do not behave as Lewis acid, while the tetrahalides of other elements are Lewis acids.

Explanation. The explanation of the difference between the tetrahalides of carbon and those of other elements is given in chapter 19 (Modern Concepts of Acids and Bases).

(*vi*) *Solvolysis : Solvolytic reactions.* With the exception of tetrafluorides, other tetrahalides undergo solvolysis with donor molecules containing labile hydrogen, such as N_2H_4, C_2H_5OH, CH_3COOH etc. Solvolysis increases with time, temperature and concentration of the donor molecules and in going from tetrachloride to tetraiodide.

$$SiCl_4 + 4C_2H_5OH \longrightarrow Si(OC_2H_5)_4 + 4HCl$$

$$GeCl_4 + 2CH_3COOH \longrightarrow Ge(CH_3COO)_2Cl_2 + 2HCl$$

$$SnCl_4 + 2N_2H_4 \longrightarrow Sn(N_2H_3)_2Cl_2 + 2N_2H_5Cl$$

(*vii*) *Boiling points.* CCl_4, CBr_4 and Cl_4 have boiling points which are unexpectedly higher than the corresponding silicon halides. This anomaly has been explained on the basis of partial multiple bonding of silicon and a consequent decrease in inter-molecular forces. *d*-orbital in Si-atom can participate in an overlap with a filled orbital of Cl, Br or I. But carbon cannot do this.

Hydrofluosilicic acid, H_2SiF_6

It is a complex halide of silicon, in which silicon atom shows hexacovalency.

Preparation. (*i*) It is made in solution by dissolving SiO_2 in aqueous HF.

$$SiO_2 + 6HF \longrightarrow H_2SiF_6 + 2H_2O$$

(*ii*) An aqueous solution of H_2SiF_6 is obtained by passing SiF_6 into H_2O (Hydrolysis of SiF_4)

$$3SiF_4 + 2H_2O \longrightarrow SiO_2\ (ppt.) + 2H_2SiF_6$$

Properties. (*i*) The concentrated solution of H_2SiF_6 is a colourless fuming liquid, which decomposes on evaporation into HF and SiF_4.

$$H_2SiF_6 \rightleftharpoons 3HF + SiF_4$$

(*ii*) It is a strong acid and is highly iomised in solution as is H_2SO_4.

$$H_2SiF_6 \rightleftharpoons 2H^+ + SiF_6^{2-}$$

(*iii*) *Formation of fluosilicates or silicofluorides.* With alkalies, it gives salts called fluosilicates or silicofluorides.

$$H_2SiF_6 + 2NaOH \longrightarrow Na_2SiF_6 + 2H_2O$$

Most of these fluosilicates are soluble in water but the fluosilicates of Na, K, Ba etc., dissolve with difficulty.

(*iv*) *Action of $KClO_3$.* When H_2SiF_6 reacts with $KClO_3$, chloric acid ($HClO_3$) is obtained.

$$2KClO_3 + H_2SIF_6 \longrightarrow K_2SiF_6\ (ppt.) + 3HClO_3$$

The reaction has been used for the preparation of $HClO_3$.

Uses : (*i*) The acid is used (*a*) as an antiseptic in medicines, (*b*) for testing barium salts, (*c*) for estimating potassium, (*d*) for giving wood a stone-like surface, (*e*) in the refining of lead metal by electrolytic method, (*f*) for the preparation of $HClO_3$. (*ii*) Na_2SiF_6 is used as an insecticide for chickens. (*iii*) $MgSiF_6$ is used for hardening and polishing the surface of concrete floors.

Halides of tin

Tin forms the following types of halides : (*a*) *Stannous halides or dihalides.* (SnX_2) in which Sn is in +2 oxidation state. All the four stannous halides are known (SnF_2, $SnCl_2$, $SnBr_2$, SnI_2). (*b*) *Stannic halides or tetrahalides* (SnX_4) in which Sn is in +4 oxidation state. All the four stamic halides are known (SnF_4, $SnCl_4$, $SnBr_4$, SnI_4). As expected from Fajans' rules, stannous halides are ionic and hence contain Sn^{2+} ions while stannic halides are mostly covalent. On the basis of inert pair effect, Sn (+ 2) halides are less stable than Sn (+ 4) halides ($Sn^{2+} < Sn^{4+}$). Being less stable, Sn (+ 2) halides are readily oxidised to Sn (+ 4) halides and are good reducing agents.

$$Sn^{2+}\ (\textit{Less stable : Reducing agent}) \longrightarrow Sn^{4+}\ (\textit{More stable}) + 2e^-$$

(*c*) *Complex halides.* Both Sn^{2+} and Sn^{4+} ions have a tendency to form complex halide anions like $SnCl_4^{2-}$ (containing Sn^{2+} ions) and $SnCl_6^{2-}$, SnF_6^{2-} which contain Sn^{4+} ions. $SnCl_6^{2-}$ ion is found in hydrogen stannic acid, $H_2[SnCl_6]$ and its salts like ammonium chlorostannate, $(NH_4)_2\ [SnCl_6]$. SnF_6^{2-} ion is found in $H_2[SnF_6]$ and its salts like $K_2[SnF_6]$.

1. Stannous fluoride, SnF_2

It is obtained by crystalline it from a solution of SnO in HF.

$$SnO + 2HF \longrightarrow SnF_2 + H_2O$$

SnF_4 forms white monoclinic prisms, which are very soluble in water.

2. Stannic fluoride, SnF_4

It can be prepared by gently heating $SnCl_4$ with excess of HF, until no more HCl is liberated

$$SnCl_4 + 4HF \longrightarrow SnF_4 + 4HCl$$

It is a white crystalline solid with boiling point = 705°C. It combines with H_2O with a lissing noise because large amount of heat energy is generated. It forms complex salts like $K_2[SnF_6]$. These salts are called fluostannates, which are analogous to fluosilicates, SiF_6^{2-}. SnF_4 absorbs NH_3, forming double compound, $SnF_4.NH_3$.

3. Hydrated stannous chloride, $SnCl_2.2H_2O$

It is obtained by dissolving Sn in hot conc. HCl.

$$Sn + 2HCl \longrightarrow SnCl_2 + H_2$$

The solution on vaporation gives monoclinic crystals of $SnCl_2.2H_2O$ which are called "*tin salt*".

Properties. (*i*) Hydrated stannous chloride forms transparent monoclinic crystals which melt at 40°C.

(*ii*) *Action of heat.* On heating $SnCl_2.2H_2O$ loses HCl, giving stannous oxy-chloride, Sn(OH)Cl.

$$SnCl_2.2H_2O \xrightarrow{\Delta} HCl + Sn(OH)Cl + \frac{1}{2} H_2$$

Hence anhydrous stannous chloride ($SnCl_2$) cannot be prepared simply by heating the hydrated salt, $SnCl_2.2H_2O$.

(*iii*) *Action of water.* (*a*) *With small quantity of H_2O.* Small quantity of water dissolves $SnCl_2$ without any noticeable change.

(*b*) *With excess of H_2O.* With excess of water, $SnCl_2$ undergoes hydrolysis and Sn(OH)Cl is precipitated.

$$SnCl_2 + H_2O\ (excess) \longrightarrow Sn\ (OH)\ Cl + HCl$$

(*iv*) *Action of alkalies.* When treated with an alkali, white ppt. of Sn $(OH)_2$ is obtained which dissolves in the excess of alkali due to the formation of sad. stannite, $Na[SnO_2]$ which is a complex compound.

$$SnCl_2 + 2NaOH \longrightarrow Sn\ (OH)_2 + 2NaCl\ ;\ Sn(OH)_2 + 2NaOH \longrightarrow Na_2[SnO_2] + 2H_2O$$

(*v*) *Action of H_2S.* With H_2S a dark brown ppt. of SnS is formed. This ppt. is soluble in yellow ammonium sulphide, $(NH_4)_2S_2$ due to the formation of ammonium thiostannate, $(NH_4)_2\ [SnS_3]$.

$$SnCl_2 + H_2S \longrightarrow SnS + 2HCl\ ; \qquad SnS + (NH_4)_2\ S_2 \longrightarrow (NH_4)_2\ [SnS_3]$$

(*vi*) *Reducing properties.* $SnCl_2$ is an active reducing agent and reduces a number of substances. In these reactions Sn^{2+} (stannous) ions lose two electrons and are oxidised to Sn^{4+} (stannic) ions.

$$Sn^{2+} \longrightarrow Sn^{4+} + 2e^-$$

The electrons lost by Sn^{2+} ions are accepted by oxidising agents. Thus :

(*a*) $SnCl_2$ reduces $HgCl_2$ to Hg_2Cl_2 (*white ppt.*) and finally to metallic Hg (*dark grey or black*)

$$2Hg^{2+} + Sn^{2+} \longrightarrow Hg_2^{2+} + Sn^{4+}; \qquad Hg_2^{2+} + Sn^{2+} \longrightarrow 2Hg \downarrow + Sn^{4+}$$

(*b*) It reduces ferric salts to ferrous salts and cupric salts to cuprous salts.

$$2Fe^{3+} + Sn^{2+} \longrightarrow 2Fe^{2+} + Sn^{4+}; \qquad 2Cu^{2+} + Sn^{2+} \longrightarrow 2Cu^+ + Sn^{4+}$$

(*c*) It reduces $AuCl_3$ to the ppt. of colloidal Au.

$$2\ AuCl_3 + 3SnCl_2 \longrightarrow 3SnCl_4 + 2Au \quad \text{or} \quad 2\ Au^{3+} + 3\ Sn^{2+} \longrightarrow 3Sn^{4+} + 2Au \qquad ...(i)$$

$SnCl_4$ obtained as above in equation (*i*) gets hydrolysed by H_2O to form $Sn(OH)_4$ or $SnO_2.2H_2O$ which absorbs the *ppt.* of colloidal gold and gives a solution having purple colour. This solution is called *purple of cassius.*

$$3 \times [SnCl_4 + 4H_2O \rightarrow Sn(OH)_4 \quad \text{or} \quad SnO_2.2H_2O + 4HCl] \qquad ...(ii)$$

On multiplying equation (*ii*) by 3 and adding it to equation (*i*), we get :

$$2AuCl_3 + 3SnCl_2 + 12H_2O \longrightarrow 2\ Au + 3Sn\ (OH)_4 \text{ or } 3(SnO_2.\ 2H_2O) + 12\ HCl$$

$$\text{or} \quad 2AuCl_3 + 3\ SnCl_2 + 6H_2O \longrightarrow \boxed{2\ Au + 3\ SnO_2} + 12HCl$$

Purple of cassius

Thus purple of cassius is the colloidal form of $Sn(OH)_4$ or SnO_2 containing absorbed colloidal gold.

(*d*) It decolourises I_2 and can, therefore, be estimated volumetrically against iodine solution.

$$Sn^{2+} + I_2 \longrightarrow Sn^{4+} + 2I^-$$

(*e*) Acidified solution of $KMNO_4$ (Mn = +7) is decolourised due to its reduction to Mn^{2+} ion.

$$2MnO_4^- + 5\ Sn^{2+} + 16H^+ \longrightarrow 2Mn^{2+} + 5Sn^{4+} + 8H_2O$$

(*f*) Golden yellow acidified solution of $K_2Cr_2O_7$ (Cr = + 6) changes green, due to its reduction to Cr^{3+} ion.

$$Cr_2O_7^{2-} + 3\ Sn^{2+} + 14\ H^+ \longrightarrow 2Cr^{3+} + 3Sn^{4+} + 8H_2O$$

(*g*) It reduces HNO_3 (N = +5) to NH_2OH (N = –1) in presence of HCl.

$$HNO_3 + 3\ SnCl_2 + 6HCl \longrightarrow NH_2OH + 3SnCl_4 + 2H_2O$$

or $$NO_3^- + 3\ Sn^{2+} + 7H^+ \longrightarrow NH_2OH + 3Sn^{4+} + 2H_2O$$

(*h*) Alkaline solution of $SnCl_2$ reduces NO to pot. hyponitrite ($K_2N_2O_2$).

$$\underset{(N = +2)}{2NO} + \underset{(Sn = +2)}{SnCl_2} + 6KOH \longrightarrow \underset{(N = +1)}{K_2N_2O_2} + \underset{(Sn = +4)}{K_2SnO_3} + 2KCl + 3H_2O$$

(*i*) It reduces nitro compounds in acidic medium to the corresponding amino compounds, *e.g.*, $C_6H_5NO_2$ is reduced to $C_6H_5\ NH_2$.

$$C_6H_5NO_2 + 6H^+ + 3Sn^{2+} \longrightarrow C_6H_5NH_2 + 2H_2O + 3Sn^{4+}$$

Anhydrous stannous chloride, $SnCl_2$

As already stated, anhydrous $SnCl_2$ cannot be prepared by heating the hydrated salt, $SnCl_2.2H_2O$. The anhydrous salts is preparred by heating tin in a stream of HCl or by heating a mixture of metallic tin (excess) with $HgCl_2$ when Hg volatilises and $SnCl_2$ remains behind in solution.

$$Sn + Hg\ Cl_2 \longrightarrow SnCl_2 + Hg$$

Properties of anhydrous $SnCl_2$. (*i*) It is a covalent compound and is soluble in organic solvents like alcohol and ether.

(*ii*) Formation of addition compounds. $SnCl_2$ combines with NH_3, forming various types of addition compounds depending on the temperature of the reaction. Example of some addition compounds are $SnCl_2.2NH_3$ (below 0°C), $SnCl_2.2NH_3$ and $SnCl_2.NH_3$ (at ordinary temperature), $SnCl_2\ NH_3$ (at 100° C), 3 $SnCl_2.2NH_3$ (above 100° C), etc. The last one is most stable.

Uses. It is used : (*i*) as a mordant in dying under the name "*tin salt*", (*ii*) as a reducing agent for organic nitro, ago and diazonium compounds, (*iii*) in the preparation of *purple of cessium*, (*iv*) in the preparation of lake colours.

4. Stannic chloride, $SnCl_4$

It is prepared : (*i*) by passing Cl_2 over molten tin kept in a recort [$Sn + 2Cl_2 \longrightarrow SnCl_4$]. (*ii*) by distilling tin with excess of $HgCl_2$. [$Sn + 2HgCl_2 \rightarrow SnCl_4 + 2Hg$]. (*iii*) by passing Cl_2 and S_2Cl_2 vapour over heated SnO_2. [$2SnO_2 + 3Cl_2 + S_2Cl_2 \rightarrow 2SnCl_4 + 2SO_2$].

Properties. (*i*) $SnCl_4$ is colourless liquid with boiling point = 114.1°C. It is a covalent compound as is indicated by its volatility, solubility in organic solvents like, benzene and very negligible electrical conductivity. It is missible with (S_2 in all proportions and can dissolve phosphorus, sulphur, iodine etc.). It catches fire with turpentine.

(*ii*) *Action of water. (a) With small quantity of H_2O.* $SnCl_4$ dissolves in a small quantity of H_2O and heat is generated. The clear solution thus obtained can be used to get the hydrates like $SnCl_4.3H_2O$, $SnCl_4.5H_2O$, $SnCl_4.6H_2O$ etc. Pentahydrate, $SnCl_4$, $5H_2O$ is called "*Oxy-muriated of tin*" or "*butter of tin*".

(*b*) *With excess of H_2O: Hydrolysis* with excess of H_2O, $SnCl_4$ undergoes hydrolysis, forming a basic chloride, $Sn(OH)Cl_3$ which reacts with H_2O forming a colloidal solution of stannic acid, $Sn(OH)_4$ or H_4SnO_4.

$$SnCl_4 + H_2O \rightleftharpoons Sn(OH)\ Cl_3 + HCl$$

$$Sn(OH)Cl_3 + 3H_2O \rightleftharpoons Sn\ (OH)_4 + 3HCl$$

$$SnCl_4 + 4H_2O \longrightarrow Sn\ (OH)_4 \text{ or } H_4SnO_4 + 4H_2O$$

The hydrolysis reaction, shown above, is slow and reversible.

Some chemists say that $SnCl_4$, on hydrolysis, gives the complex compound, $H_2[SnCl_3(OH)_3]$.

$$SnCl_4 + 3H_2O \longrightarrow H_2\ [SnCl_3\ (OH)_3] + HCl$$

(*iii*) *Action of HCl.* When $SnCl_4$ is mixed with HCl, solution containing stannic acid, $H_2[SnCl_6]$ is obtained.

$$SnCl_4 + 2HCl \longrightarrow H_2[SnCl_6]\ (\textit{solution})$$

On evaporating the solution carefully, colourless crystals of $H_2[SnCl_6]6H_2O$ are obtained.

(*iv*) *Action of NH_4Cl.* When $SnCl_4$ is mixed with a solution of NH_4Cl, ammonium chlorostannate, $(NH_4)_2$ $[SnCl_6]$ is obtained. This salt is called pink salt.

$$SnCl_4 + 2NH_4Cl \longrightarrow (NH_4)_2\,[SnCl_6]$$

(*v*) *Formation of double salts.* See *"Differences between the tetrahalides of carbon and tetrahalides of other elements of group IVA".*

Uses. (*i*) The pentahydrated, $SnCl_4.5H_2O$ is used as a mordant especially for silk and in "*weighting*" silk (*ii*) Ammonium chlorostannate, $(MH_4)_2$ $[SnCl_6]$ is used as a mordant, under the name "*pink salt*".

5. Ammonium chlorostannate, $(NH_4)_2$ $[SnCl_6]$

The salts derived from hydrogen stannic acid, $H_2[SnCl_6]$ are called chlorostannates. Ammonium chloro stannate, $(NH_4)_2$ $[SnCl_6]$ can be obtained by mixing $SnCl_4$ with a solution of NH_4Cl.

$$SnCl_4 + 2NH_4Cl \longrightarrow (NH_4)_2\,[SnCl_6]$$

This salt is called *pink salt.*

Halides of lead

Lead forms the following types of halides : (*a*) *Plumbous halides or dihalides* (PbX_2) in which Pb is in +2 oxidation state. All the four plombous halides are known [*e.g.*, PbF_2, $PbCl_2$, $PbBr_2$, PbI_2). (*b*) *Plumbic halides* or *tetrahalides* (PbX_4) in which Pb shows + 4 oxidation state. Only tetrafluoride (PbF_4) and tetrachloride (PCl_4) are known when plumbous or plumbic is not mentioned with the halide, plumbous is implied, *e.g.*, lead chloride means plumbous chloride, $PbCl_2$. As expected from Fajan's rules, plumbous halides are ionic and hence contain plumbous (Pb^{2+}) ions while plumbic halides are mostly covalent. On the basis of inert pair effect, Pb (+ 2) halides are more stable than Pb (+ 4) halides ($Pb^{2+} > Pb^{4+}$). It is due to the less stability of Pb (+ 4) halides that these are readily reduced to Pb (+2) halides and hence are good oxidising agents.

$$Pb^{4+}(\textit{Less stable : oxidising agent}) + 2e^- \xrightarrow{\text{reduction}} Pb^{2+}\ (\textit{More stable})$$

Plumbic halides are either insoluble or readily hydrolysed by water to give PbO_2. Dihalides of Pb, excepting PbI_2, are colourless. (*c*) *Complex halides.* Both Pb^{2+} and Pb^{4+} ions have a tendency to form complex halide anions like $PbCl_3^-$ (*which contains Pb^{2+}*) and PbX_6^{2-} (*which contains Pb^{4+}*) (X = F, Cl).

1. Lead difluoride, PbF_2

Preparation. It is formed : (*a*) as a white ppt. when a soluble fluoride is added to a lead salt solution. (*b*) by the action of HF on PbO or $Pb(CO_3)_2$.

Properties. It is a white powder with melting point = 818°C and boiling point = 1292°C. It is least stable of Pb (II) halides with the exception of PbI_2. It dissolves in HCl or HNO_3.

2. Lead tetrafluoride, PbF_4

It is prepared by : (*a*) passing F_2 over PbF_2 above 250°C. [$PbF_2 + F_2 \xrightarrow{250°\,C} PbF_4$]. (*b*) by dissolving red lead or freshly prepared PbO_2 in BrF_3.[$3PBO_2 + 4BrF_3 \longrightarrow 3PbF_4 + 2Br_2 + 3O_2$]. It forms colourless needles. It is stablished by the formation of complex fluoplumbate anion, $[PbF_6]^2$.

3. Lead dichloride, $PbCl_2$

Preparation. It is formed : (*a*) by heating Pb-metals in Cl_2. (*b*) by the action of boiling conc. HCl on Pb, its oxide or carbonate. [$Pb + 2HCl \longrightarrow PbCl_2 + H_2$]. (*c*) as a white ppt. by adding HCl or a soluble chloride to a lead salt solution.

Properties. It forms white crystals with silky lustre, melting point. = 498°C and boiling point = 956°C. It is sparingly soluble in cold water (0.91%) but more in boiling water (3.2%) from which it separates in needles on cooling. When heated in air, it suffers hydrolysis through the action of moisture present in air. It is reduced to metal when heated in H_2 or with carbon in presence of water vapour. When treated with conc. HCl, complex halides are formed.

$$HCl \rightleftharpoons H^+ + Cl^- ;\quad PbCl^2 + Cl^- \longrightarrow [PbCl_3]^-;\quad PbCl_3 + Cl^- \longrightarrow [PbCl_4]^-$$

With NH_4OH, white ppt. of lead oxy-chloride, Pb_2OCl_2 is formed.

$$2PbCl_2 + 2NH_4OH \longrightarrow Pb_2OCl_2 \downarrow + 2NH_4Cl + H_2O$$

4. Lead tetrachloride, $PbCl_4$.

Preparation. It is formed : (*a*) by dissolving PbO_2 in ice cold conc. HCl.[$PbO_2 + 4HCl \longrightarrow PbCl_4 + 2H_2O$]. (*b*) When ammonium hexachloroplumbate, $(NH_4)_2[PbCl_6]$ is added to well cooled concentrated H_2SO_4. [$(NH_4)_2[PbCl_6] + H_2SO_4 \longrightarrow PbCl_4 + (NH_4)_2SO_4 + 2HCl$].

Properties. It is a yellow oily liquid. It freezes at –15°C to a crystalline mass. It is a covalent compound and hence is soluble in organic solvents and is a non-conductor of electricity. It readily decomposes on warming (sometimes. explosively) with the evolution of Cl_2.

$$PbCl_4 \longrightarrow PbCl_2 + Cl_2$$

It fumes strongly in air since it is hydrolysed by water to form lead dioxide and HCl.

$$PbCl_4 + 2H_2O \longrightarrow PbO_2 + 4HCl$$

With NH_4Cl, $PbCl_4$ forms a complex halide namely ammonium chloroplumbate, $(NH_4)_2[PbCl_6]_n$.

$$2NH_4Cl + PbCl_4 \longrightarrow (NH_4)_2[PbCl_6]$$

This compound is fairly stable and resembles ammonium stannic chloride, $(NH_4)_2[SnCl_6]$.

5. Lead dibromide, $PbBr_2$

It is formed : (*a*) as a white ppt. by adding a soluble bromide to a lead salt solution. (*b*) by dissolving lead dioxide or carbonate in HBr. $PbBr_2$ forms white needles. It melts at 373°C to a red liquid which solidifies again on cooling to a horny mass. It boils at 916°C. It is approximately as soluble as $PbCl_2$ in cold water. It slowly blackens when exposed to light due to the deposition of the metal.

6. Lead di-iodide, PbI_2

It is formed : (*a*) by dissolving lead, its oxide or carbonate in HI. (*b*) by adding a soluble iodide to a lead salt solution. It forms golden-yellow crystals with melting point. = 412°C and melting point = 900°C. On heating it first, turns brick red and then brown red but regains its original colour on cooling. It dissolves in a large excess of KI forming $K[PI_3]$ which is decomposed on dilution with water depositing PbI_2 again.

Halides and oxy-halides of nitrogen group elements (Group VA)

The elements of group VA give various types of halides and oxy-halides. *Trihauaes* (MX_3) and *pentahalides* (MX_5) are important halides. Nitrogen cannot form pentahalides because nitrogen does not have *d*-orbitals in its valence-shell and therefore, cannot expand its octet. Phosphorus gives the pentahalides with all the four halogens. As gives only AsF_5 while Sb forms SbF_5 and $SbCl_5$. Bi does not give any penta halide. The formation of pentahalides in case of P, As and Sb is because of the presence of vacant *d*-orbitals in their valence-shell. Bi does not form pentahalide because, due to inert pair effect, + 3 oxidation state of Bi is more stable than its +5 oxidation state, *i.e.*, Bi prefers to give trihalides rather than pentahalides. Oxy-halides are given by nitrogen and phosphorus only. Examples are NOX (X = F, Cl, Br), NO_2X (X = F, Cl) and POX_3 (X = F, Cl, Br).

General properties of trihalides

(*i*) (*a*) **Ionic/covalent nature.** Trihalides are mostly covalent and only partly ionic. The ionic character increases as we move down the group. Thus $BiCl_3$ and BiF_3 are predominantly ionic.

(*b*) **Physical state.** Physical state of trichlorides is given below :

NCl_3	PCl_3	$AsCl_3$	$SbCl_3$	$BiCl_3$
Explosive liquid	Liquid	Liquid	Solid	Solid

(*ii*) **Hydrolysis.** The trichlorides (MCl_3) of all the elements are hydrolysed. The tendency towards hydrolysis gradually decreases with increasing metallic character of the element (M) *e.g.*, NCl_3 and PCl_3 are completely and irreversibly hydrolysed by water. Excepting NF_3, all the remaining trihalides of nitrogen are completely hydrolysed, giving NH_3 and hypohalous acid (HOX).

$$NX_3\ (X = Cl, Br, I) + 3H_2O \longrightarrow NH_3 + 3HOX$$

NF_3 reacts with water vapour, on sparking, forming N_2O_3 and HF.

$$2NF_3 + 3H_2O\ (vapour) \longrightarrow N_2O_3 + 6HF$$

$$2PCl_3 + 5H_2O \longrightarrow 6HCl + H_4P_2O_5 \text{ (Pyrophosphorus acid)}$$

$$2AsCl_3 + 3H_2O \longrightarrow 6HCl + As_2O_3$$

$SbCl_3$ and $BiCl_3$ are partly and reversibly hydrolysed to give HCl and oxy-chlorides of the metal.

$$SbCl_3 + H_2O \rightleftharpoons SbOCl + 2HCl$$

$$BiCl_3 + H_2O \rightleftharpoons BiOCl + 2HCl$$

The addition of excess of HCl suppresses the hydrolysis by shifting the equilibrium to the left and giving a clear solution.

The hydrolysis reaction shown for $SbCl_3$ takes place with little quantity of water. With larger amount of water, $Sb_4O_5Cl_2$ is formed.

$$4SbCl_3 + 5H_2O \rightleftharpoons Sb_4O_5Cl_2 + 10HCl$$

***(iii)* Electron-pair acceptor (Lewis acid) properties.** Due to the absence of *d*-orbitals in the valence-shell in N-atom, nitrogen trihalides are not able to give the complex anions like NX_5^{2-} while, by virtue of vacant *d*-orbitals, the trihalides of other elements can accept the lone pairs of electrons donated by halide ions and can thus form complex halide anions of MX_5^{2-} type. For ample :

$$SbF_3 + 2F^- \longrightarrow SbF_5^{2-}; \qquad SbCl_3 + 2Cl^- \longrightarrow SbCl_5^{2-}$$

The anions arise due to sp^3d^2 hybridisation. Due to the presence of one lone pair of electrons in one of the six sp^3d^2 hybrid orbitals, these ions have square pyramidal structure.

***(iv)* Electron-pair donating (Lewis base) property.** The central atom of the trihalides can donate the lone pair of electrons present in one of the four sp^3 hybrid orbitals to the Lewis acid (*e.g.*, BF_3) and can form the adduct. For example

$$BF_3\ (Lewis\ acid) + PCl_3\ (Lewis\ base) \longrightarrow Cl_3P \longrightarrow BF_3\ (Adduct)$$

***(v)* Stability.** Except NF_3, the remaining trihalides of nitrogen (NCl_3, NBr_3 and NI_3) are unstable and highly explosive. On the other hand, all the trihalides of the remaining elements are fairly stable. The unstability associated with the trihalides of nitrogen is because of high electronegativity of N-atom.

***(vi)* Geometry.** All the trihalides have pyramidal geometry which arises due to sp^3 hybridisation of the central atom and due to the presence of one lone pair of electrons in one of the four sp^3 hybrid orbitals.

General properties of pentahalides

***(i)* Hydrolysis.** Phosphorus pentahalides are rapidly hydrolysed with the *excess* of water to give H_3PO_4 and HX.

$$PX_5\ (X = F, Cl) + 4H_2O \longrightarrow H_3PO_4 + 5HX$$

With *calculated amount* of H_2O, POX_3 and HX are obtained, *e.g.*,

$$PX_5\ (X = Cl, Br) + H_2O \longrightarrow POX_3 + 2HX$$

Hydrolysis of pentahalides of Sb with the excess of hot water deposits hydrated pentoxide, $Sb_2O_5.2H_2O$ or $H_4Sb_2O_7$ (*antimonic acid*).

$$2SbCl_5 + 7H_2O \longrightarrow H_4Sb_2O_7 + 10\ HCl \quad \text{or} \quad 2SbCl_5 + 7H_2O \longrightarrow Sb_2O_5.2H_2O + 10HCl$$

***(ii)* Thermal dissociation.** The pentafluorides are fairly stable but the pentachlorides and bromides undergo thermal dissociation into trihalide and halogen. For example :

(*a*) The vapour of PCl_5 is partially dissociated into PCl_3 and Cl_2.

$$PCl_5 \rightleftharpoons PCl_3 + Cl_2$$

Above 300°C the dissociation is complete.

(*b*) The vapour of $SbCl_5$ at 150°C is slightly dissociated into $SbCl_3$ and Cl_2.

$$SbCl_5 \rightleftharpoons SbCl_3 + Cl_2$$

***(iii)* Electron pair accepting (Lewis acid) property.** The pentahalides like AsF_5, SbF_5 etc., act as electron pair acceptors from halide ions because of the presence of vacant *d*-orbital in the valence-shell of the central atom. Thus these molecules can form the complex halide ions of MX_6^- which have octahedral geometry due to sp^3d^2 hybridisation. For example :

Lewis acid		*Lewis base*		*Adduct*
AsF_5	+	XeF_2	$\longrightarrow$	$[XeF]^+ [AsF_6]^-$
AsF_5	+	XeF_6	$\longrightarrow$	$[XeF_5]^+ [AsF_6]^-$
SbF_5	+	$:\ddot{F}:^-$	$\longrightarrow$	SbF_6^-
SbF_5	+	XeF_4	$\longrightarrow$	$[XeF_3]^+ [SbF_6]^-$
PCl_5	+	$:\ddot{Cl}:^-$	$\longrightarrow$	PCl_6^-
PF_5	+	$:\ddot{F}:^-$	$\longrightarrow$	PF_6^-

***(iv)* Action of oxalic acid, $H_2C_2O_4$.** PCl_5 and PBr_5 both react with $H_2C_2O_4$ and the corresponding oxy-halides of phosphorus are obtained.

$$PCl_5 + H_2C_2O_4 \longrightarrow POCl_3 + CO + CO_2 + 2HCl$$
$$PBr_5 + H_2C_2O_4 \longrightarrow POBr_3 + CO + CO_2 + 2HBr$$

***(v)* Action of P_2O_5.** When a mixture of PCl_5 or PBr_5 and P_2O_5 is located in a sealed tube, oxy-halides are obtained.

$$3PCl_5 + P_2O_5 \longrightarrow 5POCl_3; \qquad 3PBr_5 + P_2O_5 \longrightarrow 5POBr_3$$

Structure. *(a)* In the vapour state, pentahalides have the trigonal bipyramidal structure corresponding to sp^3d hybridisation. In the solid state, PCl_5 consists of an ionic lattice which has tetrahedral $[PCl_4]^+$ and octahedral $[PCl_6]^-$ ions. PBr_5 and PI_5 exist as ionic lattices which can be represented as $[PBr_4]^+ Br^-$ and $[PI_4]^+ I^-$ respectively.

Nitrosyl halides, NOX (X = F, Cl, Br). Only NOF, NOCl and NOBr are known. NOI does not exist, since I^- ion is oxidised to I_2 by nitrosonium ion, NO^+

$$2I^- + 2NO^+ \longrightarrow 2NO + I_2$$

Nitrosyl halides are hightly coloured and volatile compounds. They are easily hydrolysed by water to give HX and HNO_2.

$$NOX + H_2O \longrightarrow HX + HNO_2$$

Although these compounds are covalent, they have polar character. Here, we shall discuss only NOCl.

Nitrosyl chloride, NOCl

Preparation. *(i)* It is formed by warming aqua regia (a mixture of 1 vol. of conc. NHO_3 and 3 vols. of conc. HCl) when an orange-yellow mixture of NOCl and Cl_2 is evolved.

$$HNO_3 + 3HI \longrightarrow 2H_2O + NOCl + Cl_2$$

The gaseous mixture is dried by $CaCl_2$ and passed into conc. H_2SO_4 which absorbs NOCl as nitroso sulphuric acid ($HSO_4.NO$) and Cl_2 gas passes out.

$$H_2SO_4 + NOCl \longrightarrow HSO_4.NO + HCl$$

Now, when $HSO_4.NO$ is heated with NaCl, pure NOCl is obtained.

$$HSO_4.NO + NaCl \xrightarrow{\Delta} NaHSO_4 + NOCl$$

(ii) It is also formed by the direct combination of NO and Cl_2 in bright sunlight or in the presence of animal charocoal at 40-50°C.

$$2NO + Cl_2 \xrightarrow{40–50°\text{ C}} 2NOCl$$

(iii) It is also prepared by the action of PCl_5 on KNO_2.

$$PCl_5 + KNO_2 \longrightarrow NOCl + POCl_3 + KCl$$

Properties. *(i)* It is an orange-yellow gas, with suffocating odour. It can be condensed to a ruby-red liquid (density = 1.4 and boiling point. = –55°C) in a freezing mixture, which freezes to a red solid (melting point. = – 61.5°C). The solid becomes lemon yellow at liquid air temperature.

(*ii*) *Decomposition.* (*a*) It is stable and decomposes into NO and Cl_2 above 160°C only.

$$2NOCl \rightleftharpoons 2NO + Cl_2$$

(*b*) It is readily decomposed by water and also by alkalies.

$$NOCl + H_2O \rightleftharpoons HNO_2 + HCl$$
$$NOCl + 2NaOH \longrightarrow NaNO_2 + NaCl + H_2O$$

HNO_2 and HCl then interact to form NO and Cl_2.

$$2HNO_2 + 2HCl \longrightarrow 2NO + Cl_2 + 2H_2O \; ; \; NOCl + 2NaOH \longrightarrow NaNO_2 + NaCl + H_2O$$

(*iii*) *Action of metals.* It has no action on Au or Pt but attacks Hg and many other metals forming metallic chlorides. For example :

$$2Hg + 2NOCl \longrightarrow Hg_2Cl_2 + 2NO$$

(*iv*) *Formation of addition compounds.* NOCl gives addition compounds (adducts) with many metal chlorides. For example :

$$NOCl + CuCl_2 \longrightarrow No^+[CuCl_3]^-$$
$$NOCl + SbCl_5 \longrightarrow NO^+SbCl_6^-$$
$$2NOCl + SnCl_4 \longrightarrow (NO^+)_2\,(SnCl_6^-)$$

(*v*) *Addition to unsaturated linkage.* NOCl adds to unsaturated linkage to form addition product. This reaction is called Tilden reaction and NOCl is called Tilden reagent.

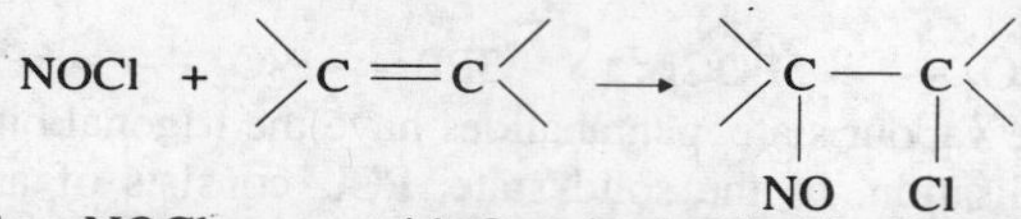

(*vi*) *Action of O_3.* When NOCl reacts with O_3, nitryl chloride (NO_2Cl) is obtained.

$$NOCl + O_3 \longrightarrow NO_2Cl + O_2$$

Uses. NOCl which is also known as Tilden reagent is used for knowing the presence and number of double bonds. This reagent also gives an idea about the nature of carbon atoms (*i.e.*, whether carbon atom is secondary or tertiary) having double bonds.

Nitryl halides, NO_2X (X = F, Cl)

These are the halides of HNO_3 or $NO_2(OH)$ in which OH group is replaced by halogens. Only NO_2F and NO_2Cl are known.

NO_2F is prepared by the action of NO on the excess of F_2 at liquid O_2 temperature or by the action of F_2 on heated $NaNO_2$.

$$4NO + F_2 \longrightarrow 2NO_2F + N_2 \; ; \qquad NaNO_2 + F_2 \longrightarrow NO_2F + NaF$$

NO_2F is a colourless gas having boiling point = –72.4°C and melting point = – 166°C. It is highly reactive.

NO_2Cl is obtained when O_3 reacts with NOCl.

$$NOCl + O_3 \longrightarrow NO_2Cl + O_2$$

NO_2Cl is a colourless gas of boiling point. = –15°C and melting point. = – 145°C. It is stable upto 100° C but decomposes at higher temperatures.

Phosphorus trichloride, PCl_3

Preparation. It is prepared : (*a*) by passing a current of dry Cl_2 over white or red phosphorus in a retort.

$$2P + 3Cl_2 \longrightarrow 2PCl_3$$

(*b*) by the action of phosphorus on $SOCl_2$

$$2P + 4SOCl_2 \longrightarrow 2PCl_3 + 2SO_2 + S_2Cl_2$$

Properties. (*i*) It is a colourless liquid with pungent odour having b.pt. = 75.0°C. It fumes in moist air.

(*ii*) *Hydrolysis.* See "*General Properties of Trihalides*".

(*iii*) *Reaction with hydroxy compounds.* PCl_3 reacts with organic compounds having OH group like C_2H_5OH, CH_3COOH etc. In these reactions OH group is replaced by Cl-atom. Thus :

$$P\boxed{Cl_3 + 3C_2H_5}OH \longrightarrow 3C_2H_5Cl + H_3PO_3$$

$$P\boxed{Cl_3 + 3CH_3CO}OH \longrightarrow 3CH_3COCl + H_3PO_3$$

(*iv*) *Decomposition.* PCl_3 is decomposed by conc. H_2SO_4 ($OH.SO_2.OH$) to give P_2O_5, SO_2 etc.

$$4HO.SO_2.OH + 2PCl_3 \longrightarrow 2Cl.SO_2.OH + 4HCl + P_2O_5 + 2SO_2 + H_2O$$

(*v*) *Reactions leading to the formation of PCl_5.* The action of excess of dry Cl_2, S_2Cl_2 and SO_2Cl_2 converts PCl_3 into PCl_5.

$$PCl_3 + Cl_2 \longrightarrow PCl_5\ ; \qquad PCl_3 + SCl_2 \longrightarrow PCl_5 + 2PSCl_3$$

$$PCl_3 + SO_2Cl_2 \longrightarrow PCl_5 + SO_2$$

(*vi*) *Combination with O_2, O_3 and SO_3.* PCl_3 combines with O_2, O_3 or SO_3, forming $POCl_3$ in each case. In these reactions PCl_3 (P = +3) is oxidised to $POCl_3$ (P = +5).

$$2PCl_3 + O_2 \longrightarrow 2POCl_3\ ; \qquad PCl_3 + O_3 \longrightarrow POCl_3 + O_2$$

$$PCl_3 + SO_3 \longrightarrow POCl_3 + SO_2$$

(*vii*) *Formation of adduct.* With BBr_3, adduct is formed.

$$PCl_3 + BBr_3 \longrightarrow [BBr_3\,(PCl_3)]$$

(*viii*) *Substitution reactions.* PCl_3 undergoes many substitution reactions some of which are shown below (R = alkyl group).

$$PCl_3 \xrightarrow{ROH} P(OR)_3\ ; \qquad PCl_3 \xrightarrow{NH_3} P(NH_2)_3$$

$$PCl_3 \xrightarrow{RSH} P(SR)_3\ ; \qquad PCl_3 \xrightarrow{R_2NH} P(NR_2)_3$$

$$PCl_3 \xrightarrow{AgCN} P(CN)_3\ ; \qquad PCl_3 \xrightarrow{AgCNO} P(CNO)_3$$

$$PCl_3 \xrightarrow{PF_3} PCl_2F + PClF_2.$$

(*ix*) *Action of oxalic acid crystals.* When PCl_3 is heated with the crystals of oxalic acid ($H_2C_2O_4$), phosphorus acid (H_3PO_3) is obtained.

$$PCl_3 + 3H_2C_2O_4 \xrightarrow{\Delta} H_3PO_3 + 3CO + 3CO_2 + 3HCl$$

Uses. PCl_3 is used as a chlorinating agent in organic chemistry.

Structure. PCl_3 molecule has trigonal pyramidal shape, which results due to sp^3 hybridisation of P-atom (central atom) and one lone pair of electrons present in one of the four sp^3 hybrid orbitals. The shape of PCl_3 molecule has already been explained.

Phosphorus pentachloride, PCl_5

Preparation. It is prepared by the action of excess of dry Cl_2 or S_2Cl_2 or SO_2Cl_2 on PCl_3 or by the action of SO_2Cl_2 on phosphorus.

$$PCl_3 + Cl_2 \longrightarrow PCl_5\ ; \qquad 2PCl_3 + S_2Cl_2 \longrightarrow PCl_5 + 2PSCl_3$$

$$PCl_3 + SO_2Cl_2 \longrightarrow PCl_5 + SO_2\ ; \qquad 2P + 5SO_2Cl_2 \longrightarrow 2PCl_5 + 5SO_2$$

Properties. (*i*) Pure PCl_5 is a colourless solid having sharp odour but generally it is a pale greenish yellow in colour.

(*ii*) *Thermal decomposition.* See "*General Properties of Pentahalides*".

(*iii*) *Hydrolysis.* See "*General Properties of Pentahalides*".

(*iv*) *Reaction with compounds containing OH groups.* Chloro derivatives are obtained by replacing OH group by Cl-atom.

$$CH_3COOH + PCl_5 \longrightarrow CH_3COCl + POCl_3 + HCl$$

$$C_2H_5OH + PCl_5 \longrightarrow C_2H_5Cl + POCl_3 + HCl$$

$$\underset{\text{Nitric acid}}{O_2N.OH} + PCl_5 \longrightarrow NO_2Cl + POCl_3 + HCl$$

$$HO.SO_2.OH + 2PCl_5 \longrightarrow Cl.SO_2.Cl + 2POCl_3 + 2HCl$$

(*v*) *Action of metals.* Metals like Cd, Au, Zn, Pt etc., give their chlorides, *e.g.*,

$$PCl_5 + Zn \longrightarrow PCl_3 + ZnCl_2$$

(*vi*) *Action of P_4O_{10} and P_4S_{10}.* With P_4O_{10}, $POCl_3$ is fomed while P_4S_{10} gives $PSCl_3$.

$$6\,PCl_5 + P_4O_{10} \longrightarrow 10\,POCl_3; \qquad 6PCl_5 + P_4S_{10} \longrightarrow 10PSCl_3$$

(*vii*) *Reactions leading to the formation of polymeric phosphonitrilic chlorides, $(NPCl_2)_n$.*

(*a*) When PCl_5 is allowed to undergo ammonolysis with NH_3, $(NPCl_2)_3$ and $(NPCl_2)_4$ are obtained.

$$3PCl_5 + 3NH_3 \longrightarrow (NPCl_2)_3 + 9HCl\ ;\ 4PCl_5 + 4NH_3 \longrightarrow (NPCl_2)_4 + 12HCl$$

(*b*) When PCl_5 is heated with solid NH_4Cl at 145 – 160°C, $(NPCl_2)_n$ is obtained :

$$nPCl_5 + nNH_4Cl\ (s) \xrightarrow{145-160°\,C} (NPCl_2)_n + 4nHCl$$

(*viii*) *Action of Cl^- ion acceptor covalent chloride molecules.* The molecules like BCl_3, $NbCl_4$ etc., accept Cl^- ion donated by PCl_5 and thus convert PCl_5 molecule into tetrahedral PCl_4^+ ion.

$$PCl_5 + BCl_3 \longrightarrow [PCl_4]^+\,[BCl_4]^-\ ; \qquad PCl_5 + NbCl_4 \longrightarrow [PCl_4]^+\,[NbCl_5]^-$$

(*ix*) *Action of KF.* $K^+\,[PF_6]^-$ is formed.

$$PCl_5 + 6KF \longrightarrow K^+\,[PF_6]^- + 6KCl$$

(*x*) *Reduction by H_2.* PCl_5 (P = +5) is reduced to PCl_3 (P = +3).

$$PCl_5 + H_2 \longrightarrow PCl_3 + 2HCl$$

(*xi*) *Action of H_3BO_3, H_3PO_3, HPO_3, SO_2 and Na_2SO_3.* In all these reactions $POCl_3$ is obtained.

$$3PCl_5 + 2H_3BO_3 \longrightarrow 3POCl_3 + B_2O_3 + 6HCl\ ;\ 3PCl_5 + H_3PO_3 \longrightarrow 3POCl_3 + PCl_3 + 3HCl$$

$$PCl_5 + SO_2 \longrightarrow SOCl_2\ (\text{b.pt.} = 78°\,C) + POCl_3\ (\text{b.pt.} = 107°C)$$

$$2PCl_5 + Na_2SO_3 \longrightarrow SOCl_2 + 2POCl_3 + 2NaCl\ ;\ 2PCl_5 + HPO_3 \longrightarrow 3POCl_3 + HCl$$

(*xii*) *Action of B_2O_3.* When B_2O_3 is heated with PCl_5 in a sealed tube at 150°C for a number of days, BCl_3 is formed.

$$B_2O_3 + PCl_5 \xrightarrow{\Delta} 2BCl_3 + 3POCl_3$$

Uses. PCl_5 is used as a chlorinating agent in Organic Chemistry.

Structure. See "*General Properties of Pentahalides.*"

Phosphorus oxy-trichloride or phosphoryl chloride, $POCl_3$

Preperation. It is prepared : (*a*) by the action of O_2, O_3 or SO_3 on PCl_3. In these reactions, PCl_3 (P = +3) is oxidised to $POCl_3$ (P = +5).

$$2PCl_3 + O_2 \longrightarrow 2POCl_3\ ; \qquad 2PCl_3 + O_3 \longrightarrow POCl_3 + O_2$$

$$PCl_3 + SO_3 \longrightarrow POCl_3 + SO_2$$

(*b*) by the action of H_2O (calculated amount), P_4O_{10}, $H_2C_2O_4$, H_3BO_3 and B_2O_3 on PCl_5.

$$PCl_5 + H_2O \longrightarrow POCl_3 + 2HCl\ ; \quad 3PCl_5 + P_4O_{10} \longrightarrow 10\,POCl_3$$

$$PCl_5 + H_2C_2O_4 \longrightarrow POCl_3 + CO + CO_2 + 2HCl$$

$$3PCl_5 + 2H_3BO_3 \longrightarrow 3POCl_3 + B_2O_3 + 6HCl$$

$$3PCl_5 + B_2O_3 \xrightarrow{\Delta} 3POCl_3 + 2BCl_3$$

Properties. (*i*) It is a colourless fuming liquid of specific gravity = 1.721 (at 0°C), boiling point. = 107.2° C and melting point. = 1.35°C.

(*ii*) It is slowly hydrlysed by H_2O, forming H_3PO_4.

$$POCl_3 + 3H_2O \longrightarrow H_3PO_4 + 3HCl$$

(*iii*) It also reacts with organic compounds having OH group. In these reactions OH group is replaced by Cl-atom. Thus :

$$3ROH + POCl_3 \longrightarrow 3RCl + H_3PO_4 \quad \text{or} \quad 3HO\boxed{R + Cl_3}OP \longrightarrow 3RCl + H_3PO_4$$

(*iv*) *Action of H_3PO_4 and $H_4P_2O_7$.* In both cases, metaphosphoric acid (HPO_3) is obtained.

$$POCl_3 + 2H_3PO_4 \longrightarrow 2HPO_3 + 3HCl\ ; \qquad POCl_3 + 2H_4P_2O_7 \longrightarrow 5HPO_3 + 3HCl.$$

Structure of $POCl_3$ molecule. The structure of $POCl_3$ molecule has been explained earlier. (chapter 15)

Halides and oxy-halides of oxygen group elements (Group VIA)

Oxygen forms a number of oxy-halogen binary compounds with halogens. Since F is more electronegative than oxygen (F = 4.0, O = 3.5), the oxygen-fluorine binary compounds are called fluorides of oxygen and not oxides of fluorine. Thus OF_2 is called oxygen difluoride and not fluorine oxide. On the other hand, since Cl, Br and I are less electronegative than oxygen (Cl = 3.0, Br = 2.8, I = 2.5), the oxygen-chlorine, oxygen-bromine and oxygen-iodine binary compounds are called oxides of chlorine, oxides of bromine and oxides of iodine respectively and not chlorides, bromides or iodides of oxygen. S, Se, Te and PO also form binary compounds with halogens and these compounds are called halides of S, Se, Te and Po respectively.

Fluorides of oxygen

Following are the important fluorides :

1. Oxygen difluoride, OF_2

Preparation. OF_2 is prepared by passing F_2 gas through NaOH solution.

$$2F_2 + 2NaOH \longrightarrow F_2O + 2NaF + H_2O$$

It may be collected over water.

Properties. (*i*) It is pale yellow gas which is highly poisonous (b.pt. = – 45° C, m.pt. = – 223.8°C). It is very slightly soluble in water.

(*ii*) *Action of H_2O.* OF_2 decomposes in presence of water, giving HF and O_2.

$$F_2O + H_2O \longrightarrow 2HF + O_2$$

(*iii*) *Action of heat.* On heating, it decomposes into its elements.

(*iv*) *Hydrolysis by base.* It is hydrolysed by a base.

$$OF_2 + 2OH^- \longrightarrow O_2 + 2F^- + H_2O$$

(*v*) *Oxidising property.* It is a powerful oxidising agent and hence liberates I_2 from KI solution.

$$OF_2 + 4KI + H_2O \longrightarrow 2KF + 2KOH + 2I_2$$

(*vi*) *Action of metals / non-metals.* Metals and non-metals are oxidised / or fluorinated by OF_2.

Uses. OF_2 is used as a rocket fuel.

Structure. See chapter 14.

2. Dioxygen difluoride, O_2F_2

Peparation. It has been isolated as a yellow orange solid (melting point = –163.5°C) by the action of an electric discharge on a mixture of F_2 and O_2 at low temperature and pressure.

Properties. Above –100°C, it decomposes into F_2 and O_2. It is a very good fluorinating and oxidising agent.

Uses. It is used for oxidising primary aliphatic amines to corresponding nitroso compounds.

Halides of S, Se, Te and Po

I. Dimeric monohalides, M_2X_2. (Examples : S_2F_2, S_2Cl_2, S_2Br_2, Se_2F_2, Se_2Cl_2, Se_2Br_2).

The known monohalides can be obtained by the following methods :

(*a*) S_2F_2 is obtained by the action of AgF on molten sulphur : $2AgF + 3S$ (*molten*) $\rightarrow Ag_2S + S_2F_2$.

(*b*) S_2Cl_2 can be obtained by passing Cl_2 over fused sulphur : $2S$ (*fused*) $+ Cl_2 \rightarrow S_2Cl_2$.

S_2Cl_2 can also be prepared by heating sulphur with thionyl chloride ($SOCl_2$) or sulphuryl chloride (SO_2Cl_2).

$$3S + 2SOCl_2 \longrightarrow 2S_2Cl_2 + SO_2 ; \qquad 2S + SO_2Cl_2 \longrightarrow S_2Cl_2 + SO_2$$

On commercial scale, S_2Cl_2 is obtained as a by-product in the manufacture of CCl_4 by the action of Cl_2 on CS_2 in presence of iodine or iron (catalyst).

$$CS_2 + 3Cl_2 \longrightarrow \boxed{CCl_4 + S_2Cl_2} \text{ (\textit{Liquid mixture})}$$

The liquid mixture thus produced is subjected to fractional distillation when CCl_4 distils over 348 K leaving behind S_2Cl_2 (boiling point = 411 K).

(*c*) S_2Br_2 is obtained by the combination of S and Br_2 : $2S + Br_2 \longrightarrow S_2Br_2$.

(*d*) Se_2Br_2 is obtained by reducing $SeBr_4$ with Se : $SeBr_4 + 3Se \longrightarrow 2Se_2Br_2$.

Properties. (*i*) With the exception of S_2F_2 which is a gas, all other monohalides are liquids.

(*ii*) *Hydrolysis.* All the monohalides are immediately decomposed by H_2O.

$$2S_2X_2 + 3H_2O \longrightarrow 3S + H_2SO_3 + 4HX \quad \text{or} \quad 2S_2X_2 + 2H_2O \longrightarrow 3S + SO_2 + 4HX$$

(X = F, Cl)

$$2Se_2Cl_2 + 3H_2O \longrightarrow 3Se + H_2SeO_3 + 4HCl$$

(*iii*) *Disproportionation reaction.* Monohalides also undergo disproportionation reaction to give free element and tetrahalide.

$$2Se_2Cl_2\ (Se = +1) \longrightarrow SeCl_4\ (Se = +4) + 3Se\ (Se = 0)$$

(*iv*) *Action of Cl_2.* When S_2Cl_2 reacts with Cl_2 at 20°C, SCl_2 is obtained : $S_2Cl_2 + Cl_2 \longrightarrow 2SCl_2$.

II. Dihalides, MX_2. (Examples : SCl_2, $SeCl_2$, $SeBr_2$, $TeCl_2$, $TeBr_2$, $PoCl_2$, $PoBr_2$).

Difluorides and diodides are not given by any of the elements. Amongst the dihalides the best known is SCl_2.

Preparation. SCl_2 is obtained as a red liquid by the action of Cl_2 on S_2Cl_2 at about 20°C. [$S_2Cl_2 + Cl_2 \rightarrow 2SCl_2$]

$TeCl_2$ and $TeBr_2$ both are formed by the direct combination of Te and halogen in suitable proportions. $TeCl_2$ can also be prepared by the reaction between $TeCl_4$ and Te, and $TeBr_2$ can be prepared by heating $TeBr_4$.

$$TeCl_4 + 2Te \rightarrow TeCl_2\ ;\ TeBr_4 \xrightarrow{\Delta} TeBr_2 + Br_2$$

$PoCl_2$ can be obtained by the action of HCl on polonium metal. $Po + 2HCl \longrightarrow PoCl_2 + H_2$

$PoCl_2$ can also be obtained by reducing $PoCl_4$ with SO_2.

Properties. (*i*) SCl_2, $TeCl_2$ and $TeBr_2$ are solids.

(*ii*) *Hydrolysis.* SCl_2, Te Cl_2 and $TeBr_2$ are readily hydrolysed by H_2O.

$$2SCl_2 + 2H_2O \longrightarrow S + SO_2 + 4HCl \quad \text{or} \quad 2SCl_2 + 3H_2O \longrightarrow S + H_2SO_3 + 4HCl$$

$$2\,TeX_2\ (X = Cl, Br) + 3H_2O \longrightarrow Te + H_2TeO_3 + 4HX$$

(*iii*) *Action of NaF.* When NaF reacts with SCl_2 in acetonitrile medium at 75°C, SF_4 is obtained :

$$3SCl_2 + 4NaF \xrightarrow{75°\,C} SF_4 + 4NaCl + S_2Cl_2$$

(*iv*) *Stability.* $SeCl_2$ and $SeBr_2$ are unstable and decompose in the vapour state. The stability of dihalides decreases from SX_2 to SeX_2 ($SCl_2 < SeCl_2 < TeCl_2 < PoCl_2$) and that of dihalides of the same element decreases from MCl_2 to MBr_2 (*e.g.*, $SeCl_2 > SeBr_2$; $TeCl_2 > TeBr_2$; $PoCl_2 > PoBr_2$).

III. Tetrahalides, MX_4. (Examples : SF_4, SCl_4, SeF_4, $SeCl_4$, $SeBr_4$, TeF_4, $TeCl_4$, $TeBr_4$, TeI_4, $PoCl_4$, $PoBr_4$ and PoI_4).

It may be seen that with the exception of SBr_4, SI_4 and SeI_4 all the tetrahalides are known.

Preparation. With the exception of tetrafluorides, all the tetrahalides can be prepared by the direct union of the elements.

$$M + 2Cl_2 \longrightarrow MCl_4\ (M = S, Se, Te)$$

SCl_4 may also be prepared by at direct chlorination of S_2Cl_2 at 251 K for a considerable time.

$$S_2Cl_2 + Cl_2 \longrightarrow 2SCl_4$$

SCl_4 can be prepared by the following reactions :

$$Se_2Cl_2 + 3Cl_2 \longrightarrow 2SeCl_4\ ; \qquad 6SeO_2 + 6PCl_5 \longrightarrow 6SeCl_4 + P_4O_{10} + 2PoCl_3$$

Tetrafluorides *cannot* be obtained by direct combination of the elements with F_2 because, on account of reactivity of F_2, the end product is always hexafluoride. The tetrafluorides can, however, be obtained by the following methods :

$$S + 4CoF_3 \longrightarrow SF_4 + 4CoF_2\ ; \qquad 3SCl_2 + 4NaF \xrightarrow{75°C} SF_4 + S_2Cl_2$$

$$SeCl_4 + 4AgF \longrightarrow SeF_4 + 4AgCl\ ; \qquad SF_4 + SeO_2 \xrightarrow{100°C} SeF_4 + SO_2$$
$$2SeF_4 + TeO_2 \longrightarrow TeF_4 + 2SeOF_2$$

Properties. (*i*) SF_4 is gaseous, SeF_4 is liquid, $SeCl_4$ is a yellow solid, $SeBr_4$ is a solid which sublimes to vapour state, TeF_4, $TeCl_4$, $TeBr_4$ and TeI_4 all are solids.

(*ii*) *Fluorinating agent.* SF_4 is used as a fluorinating agent. It converts C = O, COOH, P = O, P(O) (OH) into CF_2, CF_3, PF_2 and PF_3 respectively.

(*iii*) *Lewis acid character.* Tetrafluorides have Lewis acid character, since they readily accept electrons and are converted into hexafluorides : $SF_4 + F_2 \longrightarrow SF_6$.

SeF_4 and $TeCl_4$ combine with F^- and Cl^- ions respectively : $TeCl_4 \xrightarrow{+2\,Cl^-} [TeCl_6]^{2-}$.

(*iv*) *Formation of addition compounds.* SCl_4 forms addition compounds with metal chlorides. Examples are $SCl_4 . AlCl_3$, $SCl_4 . TlCl_3$, $SCl_4 . FeCl_3$. $SCl_4 . SbCl_5$. $2SCl_4 . SnCl_4$.

(*v*) *Hydrolysis.* $SeBr_4$ is readily hydrolysed by water, forming the oxide or oxy-acid and hydrogen bromide.

$$SeBr_4 + 2H_2O \longrightarrow SeO_2 + 4HBr \text{ or } SeBr_4 + 3H_2O \longrightarrow H_2SeO_3 + 4HBr$$

TeX_4 (X = Cl, Br, I) also hydrolysis in a similar manner

$$TeX_4 + 2H_2O \longrightarrow TeO_2 + 4HX$$

Tetrabromides of Po are fairly stable and hence do not undergo hydrolysis.

(*vi*) *Action of KI solution.* TeI_4 dissolves in the excess of KI solution, forming a deep red solution containing $K_2[TeI_6]$.

$$TeI_4 + 2KI \longrightarrow K_2[TeI_6]$$

(*vii*) $TeCl_4$ conducts electricity in the fused state probably because of the following type of ionisation.

$$2TeCl_4 \rightleftharpoons TeCl_3^+ + TeCl_5^-$$

(*viii*) *Thermal stability.* Tetrafluorides are more stable towards thermal decomposition than other tetrahalides. $TeCl_4$ is more stable than SCl_4 and $SeCl_4$.$SeBr_4$ is unstable and dissociates into Se_2Br_2 and Br_2.

$$2Se\,Br_4 \longrightarrow Se_2\,Br_2 + 3Br_2$$

IV. Dimeric sulphur pentafluoride, $(SF_5)_2$ or S_2F_{10}

Only sulphur pentafluoride is known and it is dimeric, S_2F_{10}. It is obtained in traces along with SF_6 when sulphur is directly fluorinated.

S_2F_{10} is a liquid. It is a quite stable compound but less so than SF_6. Like SF_6, it is an unreactive compound and hence is not acted upon by water or aqueous alkalies.

V. Hexafluorides, MF_6

S, Se and Te give only hexafluorides.

Preparation. F_2 directly combines with S, Se and Te to form the hexafluorides.

$$M + 3F_2 \longrightarrow MF_6 \text{ (M = S, Se, Te)}$$

Properties. (*i*) All the three hexafluorides are colourless gases.

(*ii*) *Reactivity and hydrolysis.* SF_6 is extremely stable and chemically inert. SeF_6 is slightly more reactive and TeF_6 is still more reactive ($SF_6 < SeF_6 < TeF_6$). Thus, we find that, while SF_6 and SeF_6 are inert towards water, TeF_6 is completely hydrolysed by H_2O, giving orthotelluric acid (H_6TeO_6) and HF.

$$TeF_6 + 6H_2O \longrightarrow H_6TeO_6 + 6HF$$

The explanation of the fact that SF_6 and SeF_6 are inert towards H_2O while TeF_6 can be hydrolysed has already been given in chapter 10.

(*iii*) *Formation of complex ions.* TeF_6 can add F^- ions to form complex fluorides like $[TeF_7]^-$ and $[TeF_8]^{2-}$.

(*iv*) *Covalent character.* These fluorides have low boiling points. (high volatility), indicating that these are highly covalent compounds.

(*v*) *Action of boiling sodium.* When SF_6 is passed through boiling sodium, it gets decomposed.

$$SF_6 + 8Na \longrightarrow Na_2S + 6NaF$$

(*vi*) *Action of* NH_3. Unlike SF_6, SeF_6 reacts with NH_3 at 200°C.

$$SeF_6 + 2NH_3 \longrightarrow N_2 + Se + 6HF$$

Structure of dimeric monohalide molecule, M_2X_2

Electron diffraction studies of O_2F_2, S_2F_2 and S_2Cl_2 have shown that these molecules can be represented by the covalent structures $F—\ddot{\underset{..}{O}}—\ddot{\underset{..}{O}}—F$, $F—\ddot{\underset{..}{S}}—\ddot{\underset{..}{S}}—F$ and $Cl—\ddot{\underset{..}{S}}—\ddot{\underset{..}{S}}—Cl$ respectively. These structures are the same as that of H_2O_2 molecule $\left(H—\ddot{\underset{..}{O}}—\ddot{\underset{..}{O}}—H\right)$. In these structures, one M—M bond and two M—X bonds, are lying in different planes, *i.e.*, these molecules, like H_2O_2, have *non-linear* and *non-planar* structures. The structures of O_2F_2, S_2F_2 and S_2Cl_2 are shown in Fig. 29.1.

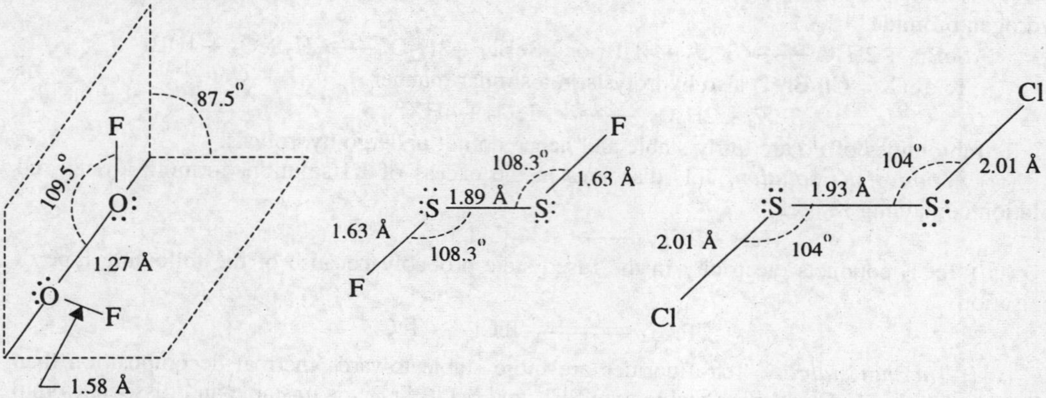

O_2F_2 molecule S_2F_2 molecule S_2Cl_2 molecule

Fig. 29.1 : Structures of dimeric monohalide molecules *viz.*, O_2F_2, S_2F_2 and S_2Cl_2.

The central M-atom (*i.e.*, O or S-atom) in each structure is sp^3 hybridised [O = $(sp^3)^2$ $(sp^3)^2$ $(sp^3)^1$ $(sp^3)^1$] and is surrounded by two lone pairs of electrons. Each of the two M—X bonds is obtained by the overlap between singly-filled *np* orbital of X-atom (X = ns^2 np_x^2 np_y^2 np_z^1) and singly-filled sp^3 hybrid orbital on M-atom while the overlap between the two singly-filled sp^3 hybrid orbitals on two M-atoms gives M—M bond.

It may be seen that in S_2F_2 and S_2Cl_2 molecules SSF and SSCl bond angles are less than the tetrahedral bond angle (= 109.5°). Their less values are due to the distortion caused in the molecules by the presence of two lone pairs of electrons round each S-atom.

It has been suggested that O_2F_2 molecule also has ionic structure which can be represented as $F—\overset{+}{O}=O\overline{F}$ or $\overline{F}\,O=\overset{+}{O}—F$.

It is proposed that S_2F_2 molecule also exists in an isomeric form which has the structure, $\begin{matrix} F \\ F \end{matrix}\!\!>\ddot{S}=\ddot{S}:$

Structure of dihalide molecule, MX_2

All the molecules of MX_2 type have angular (V-shaped) structure which is similar to that of H_2O molecule. The angular structure can be regarded as a distorted tetrahderal structure having two lone pairs of electrons. The angular structure of MX_2 molecules arises due to sp^3 hybridisation of the central atom (M-atom). The distortion in shape is due to the presence of two lone pairs of electrons. As an example, the structure of SCl_2 molecule has been explained in chapter 14. Other molecules, like $SeCl_2$, $SeBr_2$ etc., have similar structure.

Structure of tetrahalide molecule, MX_4

All the tetrahalides (MX_4) possess *see saw* geometry, which can be regarded as a distorted trigonal bipyramidal structure, having one lone pair (*lp*) of electrons in the basal (*i.e.,* equatorial) position of the trigonal bipyramid. See saw geometry of MX_4 molecules, arises due to the sp^3d hybridisation of the central atom (M-atom). The distortion in shape is due to the presence of one lone pair of electrons. Lone pair occupies the equatorial position of the trigonal bipyramid, since in this case the *lp* — *lp* repulsion is minimum. In MX_4 molecule the two axial M—X bonds are slightly bent towards the equatorial M—X bonds, showing the repulsion effect of lone pair. The axial M—X bonds are longer than the equatorial M—X bonds. As an example,the structure/geometry of SF_4 molecule has been explained in chapter 14.

Structure of dimeric sulphur pentafluoride, $(SF_5)_2$ or S_2F_{10}

S_2F_{10} molecule has two SF_5 units, linked together through S—S bond. Thus, this molecule has octahedral geometry. Each S-atom, in SF_5 unit, is sp^3d^2 hybridised. The structure of S_2F_{10} molecule is shown in Fig. 29.2, which shows that two octahedra are joined together.

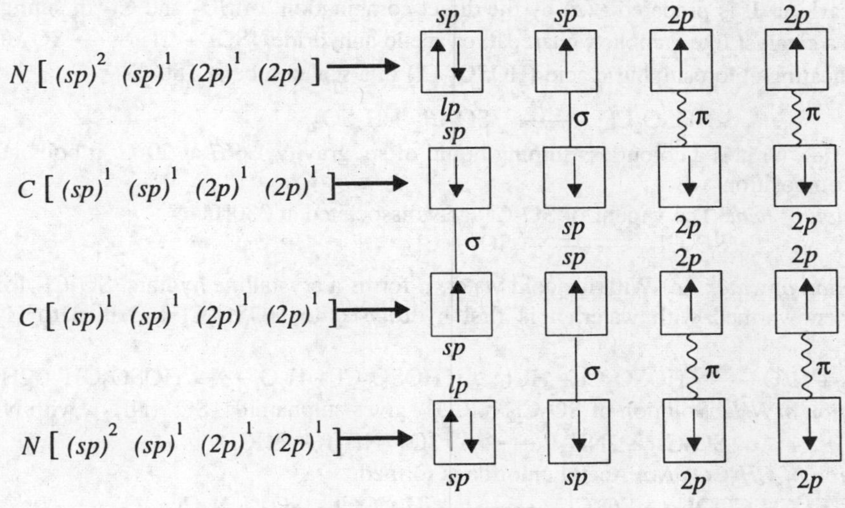

Fig. 29.2 : Structure of S_2F_{10} molecule.

Structure of hexafluoride molecule, MF_6

All the hexafluorides have regular octahedral geometry which results from sp^3d^2 hybridisation of the central atom, M (M = S, Se, Te). All M—F bonds are σ-bonds and all the six FMF bond angles and M—F bond lengths are equal. As an example the octahedral shape of SF_6 molecule has been explained in chapter 14.

Oxy-halides of sulphur

1. Thionyl chloride, $SOCl_2$

It is prepared :

(*a*) by the action of SO_2 or Na_2SO_3 on PCl_5.

$$SO_2 + PCl_5 \longrightarrow SOCl_2 \text{ (b.pt. = 7.8°C)} + POCl_3 \text{ (b.pt. = 107°C)}$$

$$Na_2SO_3 + 2PCl_5 \longrightarrow SOCl_2 + 2POCl_3 + 2NaCl$$

(*b*) by the action of excess of SO_2 on $COCl_2$: $SO_2 + COCl_2 \longrightarrow SOCl_2 + CO_2$.

(*c*) by adding sulphur to Cl_2O at –12° C: $S + Cl_2O \longrightarrow SOCl_2$.

(*d*) by the action of SO_3 on S_2Cl_2 at 75-80° C (*Commercial method*) : $SO_3 + S_2Cl_2 \longrightarrow SOCl_2 + SO_2 + S$.

Properties. (*i*) $SOCl_2$ is a colourless pungent smelling liquid, with boiling point. = 78°C and melting point. = –99.5° C.

(*ii*) It is decomposed by H_2O, forming H_2SO_3 and HCl : $SOCl_2 + 2H_2O \longrightarrow H_2SO_3 + 2HCl$.

(*iii*) It reacts with organic compounds having OH groups, *e.g.*, $ROH + SOCl_2 \longrightarrow RCl + SO_2 + HCl$.

(*iv*) It reacts with KBr, forming $SOBr_2$: $SOCl_2 + 2KBr \longrightarrow SOBr_2 + 2KCl$.

(*v*) It reacts with SbF_3 in presence of SbF_3 (Swart's reagent), forming SOF_2 : $3SOCl_2 + 2SbF_3 \longrightarrow 3SOF_2 + 2SbCl_3$.

Uses. It is used as a reagent for replacing OH groups by Cl-atom in organic compounds.

Structure. The structure and geometry of $SOCl_2$ molecule has been explained in chapter 15.

2. Sulphuryl chloride, SO_2Cl_2

Preparation. It is prepared : (*a*) by the direct combination of SO_2 and Cl_2 in sunlight or in presence of a catalyst like camphor, charcoal, or acetic anhydride. [$SO_2 + Cl_2 \longrightarrow SO_2Cl_2$]

(*b*) by heating chlorosulphuric acid ($HOSO_2Cl$) in a sealed tube at 180°C.

$$2HOSO_2Cl \xrightarrow{\Delta} SO_2Cl_2 + H_2SO_4$$

Properties. (*i*) It is a colourless fuming liquid of sp. gravity 1.667 at 20°C. It boils at 69.1°C without decomposition.

(*ii*) *Action of heat.* The vapour of SO_2Cl_2 gets dissociated at 330°C.

$$SO_2Cl_2 \rightleftharpoons SO_2 + Cl_2$$

(*iii*) *Action of water.* (*a*) With ice cold water, it forms a crystalline hydrate, $SO_2Cl_2.15H_2O$.

(*b*) When warmed with water, it is first hydrolysed to $HOSO_2Cl$ and then to H_2SO_4 or $(OH)_2SO_2$.

$SO_2Cl_2 + H_2O \longrightarrow HOSO_2Cl + HCl$; $HOSO_2Cl + H_2O \longrightarrow HOSO_2OH + 2HCl$

(*iv*) *Action of NH_3.* Solution of SO_2Cl_2 in C_6H_6 gives sulphamide, $SO_2(NH_2)_2$ with NH_3.

$$SO_2Cl_2 + 2NH_3 \longrightarrow SO_2(NH_2)_2 + 2HCl$$

(*v*) *Action of CH_3COONa.* Acetyl chloride is formed.

$$CH_3COONa + SO_2Cl_2 \longrightarrow CH_3COCl + SO_3 + NaCl$$

Uses. (*i*) Solution of SO_2Cl_2 in petrol is used in making the wood unshrinkable. (*ii*) SO_2Cl_2 is used as a chlorinating agent in orgnic chemistry.

Structure. See chapter 15.

Halides of halogens (Group VIIA)

Due to the difference in their electronegativity, two or more different halogens combine together and form the species which may be neutral molecules or ions. Examples of neutral molecules are ClF, ClF_3, IF_5, IF_7 etc., and the examples of ions are Cl_2^+, Cl_3^+, ICl_2^- (*cations*) I_3^-, I_5^-, ICl^- (*anions*) etc.

The neutral molecules are called **interhalogen compounds** and ions are called **interhalogen ions** or **polyhalide ions.** The compounds [*e.g.*, $K^+Cl_3^-$, $[NH_4]^+ I_5^-$, $[BrF_2]^+ [SbF_6]^-$ etc.] containing polyhalide ions are called **polyhalides.**

Interhalogen compounds

The known interhalogen compounds have been given in Table 29.2.

Table 29.2 : *Various types of inter-halogen compounds*

Element	Fluoride	Chloride	Bromide	Iodide
Fluorine	—	—	—	—
Chlorine	ClF, ClF_3, ClF_5	—	—	—
Bromine	BrF, BrF_3, BrF_5	$BrCl$	—	—
Iodine	IF, IF_3, IF_5, IF_7	ICl, ICl_3	IBr	—

From Table 29.2, the following points may be noted :

(*a*) An inter-halogen compound may be regarded as the halide of the more electropositive halogen, *i.e.*, a given halogen forms the interhalogen compound, only with that halogen, which has lower electropositive character. Thus, since the electropositive character of halogens is in the order $F < Cl < Br < I$, fluorine cannot form any interhalogen compound, while iodine has the maximum tendency to form inter-halogen compounds.

(*b*) Since F has the least electropositive character, all the halogens combine with it and give fluorides, *i.e.*, amongst interhalogen compounds, the fluorides are maximum in number.

(*c*) Interhaloges can be grouped into four catagories *viz.*, XY (*e.g.*, ClF, BrF, BrCl, ICl, IBr), XY_3 (*e.g.*, ClF_3, BrF_3, IF_3, ICl_3), XY_5 (*e.g.*, ClF_5, BrF_5, IF_5) and XY_7 (*e.g.*, IF_7). Here X halogen atom is more electropositive (*i.e.*, less electronegative) and larger in size than Y halogen atom.

(*d*) The oxidation state of atom X in XY, XY_3, XY_5 and XY_7 molecules is equal to +1, +3, +5 and +7 respectively.

(*e*) Y-atoms are present in odd number.

(*f*) As the ratio between the radii of X and Y atoms increases, the number of halogen atoms per interhalogen molecule increases. Obviously this ratio is maximum for IF_7 molecule.

Interhalogen compounds of XY type. (Diatomic interhalogens)

Preparation. (*i*) These are generally prepared by the direct combination of halogens *e.g.*.

$$\boxed{Cl_2 + F_2} \text{ (in equal amounts) } \xrightarrow[\text{Cu–vessel}]{250°\text{ C}} 2\,ClF$$

$$Br_2\,(g) + F_2\,(g) \longrightarrow 2BrF; \quad Br_2 + Cl_2 \longrightarrow 2BrCl$$

$$\boxed{I_2 + Cl_2\,(l)} \text{ (in equal amounts) } \longrightarrow 2ICl; \quad I_2 + F_2 \xrightarrow{-45°\text{ C}} 2IF$$

(*ii*) Some of them can be prepared by other methods, *e.g.*,

$$Cl_2 + ClF_3 \xrightarrow{300°\text{ C}} 3ClF\,; \quad Br_2 + BrF_3 \longrightarrow 3BrF$$

$$I_2 + AgF \longrightarrow IF + AgI; \quad I_2 + KClO_3 \xrightarrow{\Delta} ICl + KIO_2$$

Properties. (*i*) These are essentially covalent gases because of small difference in electronegativities between the two halogen atoms.

(*ii*) *Stability.* They differ in thermal stability. For example :

(*a*) ClF is extremely stable and dissociates only when heated strongly : $2ClF \rightarrow Cl_2 + F_2$.

(*b*) On heating IBr dissociates to a small extent : $2IBr \longrightarrow I_2 + Br_2$.

(*c*) BrCl decomposes reversibly at 10°C into Cl_2 and Br_2 : $2BrCl \rightleftharpoons Br_2 + Cl_2$.

(*d*) BrF and IF are unstable and undergo disproportionation rapidly.

$$3BrF \longrightarrow Br_2 + BrF_3\,; \quad 5IF \longrightarrow 2I_2 + IF_5$$

(*iii*) *Hydrolysis.* (*a*) BrCl gets hydrolysed to give oxy-acid and HCl : $BrCl + H_2O \longrightarrow HOBr + HCl$.

(*b*) ICl hydrolyses as : $5ICl + 3H_2O \longrightarrow HIO_3 + 5HCl + 2I_2$.

$$ICl + H_2O \longrightarrow HOI + HCl\,; \quad 3ICl + 3H_2O \longrightarrow HIO_3 + 3HCl + 3HI$$

(*iv*) *Action of metals and non-metals.* $Se + 4ClF \longrightarrow SeF_4 + 2Cl_2$.

$$ICl + 2Na \longrightarrow NaI + NaCl; \quad W + 6ClF \longrightarrow WF_6 + 3Cl_2$$

(*v*) *Action of alkali metal halides and olefines.* With alkali metal halides, polyhalides are formed while interhalogens add to olefins at double bond sites.

$$KI + ICl \longrightarrow K[ICl_2] \quad ; \qquad NaBr + IBr \longrightarrow Na[IBr_2]$$

$$CsI + BrCl \longrightarrow Cs\,[IBrCl]; \qquad CH_2 = CH_2 + BrCl \longrightarrow \underset{Br}{\underset{|}{CH_2}} - \underset{Cl}{\underset{|}{CH_2}}$$

$$—CH = CH— + ICl \longrightarrow —\underset{I}{\underset{|}{CH}} — \underset{Cl}{\underset{|}{CH}}—$$

(*vi*) *Stability*. The stability of XY type interhalogen compounds with respect to disproportionation is in the order ClF > ICl > IBr > BrCl > BrF.

(*vii*) *Lewis acid strength.* Lewis acid strength of some AB type interhalogens decreases in the order ICl >> BrCl > IBr > I_2 > Br > Cl_2. The order of Lewis acid strength of halogens has also been included for comparison.

(*viii*) *Reactivity.* The inter halogens of XY type are generally more reactive than each of the halogen molecules *viz.*, X_2 and Y_2. This is because bond energy of X—Y bond in XY interhalogen molecule is less than that of X—X bond in X_2 molecule and of Y—Y bond in Y_2 molecule.

1. Iodine monochloride, ICl. Preparation. It is formed : (*a*) by mixing I_2 and Cl_2 in equal amounts.

$$I_2 + Cl_2 \ (\textit{Equal amounts}) \longrightarrow 2ICl$$

(*b*) by heating I_2 with $KClO_3$.

$$KClO_3 + I_2 \longrightarrow KIO_3 + ICl$$

(*c*) by boiling I_2 in aqua regia or by heating ICl_3 at 68°C.

$$ICl_3 \xrightarrow{68^\circ C} ICl + Cl_2$$

Properties. (*i*) It is a dark red liquid which boils at 97.4°C.

(*ii*) *Solid modifications*. ICl exists in two solid modifications given below :

(*a*) *Solid form*. This form melts at 27.2°C and is obtained as red needle-like crystals on cooling the liquid rapidly. (*b*) *Metastable form.* This form melts at 14°C and is obtained as a black solid on cooling the liquid slowly at –10°C. This form changes to stable form on standing. In the gaseous state both the forms are identical and are composed of zig-zag chains.

(*iii*) *Hydrolysis*. ICl dissolves in water and gets hydrolysed as :

$$ICl + H_2O \rightleftharpoons IOH + HCl$$

$$3ICl + 3H_2O \rightleftharpoons HIO_3 + 3HCl + 3HI$$

$$5ICl + 3H_2O \rightleftharpoons HIO_3 + 5HCl + 2I_2$$

The hydrolysis can be prevented by adding HCl.

(*iv*) *Action of excess of KOH.* ICl is decomposed by the excess of KOH as :

$$3ICl + 6KOH \longrightarrow 3KCl + 2KI + KIO_3 + 3H_2O$$

(*v*) *Auto-iomisation of liquid ICl.* ICl in the liquid state undergoes auto-ionisation as :

$$2ICl \rightleftharpoons I^+ \ (\textit{Solvent cation}) + ICl_2^- \ (\textit{Solvent anion})$$

$AlCl_3$, MCl_4 (M = Si, Sn, Ti, V) and MCl_5 (M = Sb, Nb) give I^+ ions in liquid ICl and hence behave as acids in this solvent.

$$AlCl_3 + ICl \rightleftharpoons I^+ + AlCl_4^- \ ; \qquad MCl_4 + ICl \rightleftharpoons I^+ + MCl_5^-$$

$$MCl_5 + ICl \rightleftharpoons I^+ + MCl_6^-$$

Alkali metal chlorides, MCl (M = K, Rb, NH_4) give ICl_2^- ions in liquid ICl and, therefore, act as bases.

$$MCl + ICl \rightleftharpoons M^+ + ICl_2^-$$

Thus, neutralisation reactions in liquid ICl is that in which a compound containing I^+ ions reacts with that having ICl_2^- and a salt and ICl (*solvent*) are formed. For example :

$$I^+ [SbCl_6]^- \textit{(Acid)} + K^+[ICl_2]^- \textit{(Base)} \rightleftharpoons K^+ [SbCl_6]^- \textit{(Salt)} + 2ICl \textit{(Solvent)}$$

(*vi*) *Action of a bromide.* When a bromide like KBr, $[(CH_3)_4N]$ Br etc., reacts with ICl, $[ClBrI]^-$ ion is produced.

$$ICl + KBr \rightleftharpoons K^+ [ClBrI]^-$$

$$ICl + [(CH_3)_4N]Br \xrightarrow{CH_3COOH} [(CH_3)_4 N]^+ [ClBrI]^-$$

(*vii*) *Action of Cl_2.* Excess of Cl_2 converts ICl into ICl_3.

$$Cl_2 \textit{ (excess)} + ICl \xrightarrow{100° C} ICl_3.$$

(*viii*) *Electrolysis.* When molten ICl is electrolysed, a mixture of I_2 and Cl_2 is liberated at anode and only I_2 is liberated at cathode. This mode of electrolysis suggests that ICl ionises as :

$$2ICl \rightleftharpoons I^+ + ICl_2^-$$

$$2I^+ + 2e^- \longrightarrow I_2 \textit{ (at cathode)} ; \qquad 2ICl_2^- - 2e^- \longrightarrow I_2 + 2Cl_2 \textit{ (at anode)}$$

I^+ and ICl_2^- both have I^+. When ICl is dissolved in nitrobenzene or anhydrous CH_3COOH, the transport measurements support the above anode of ionisation of ICl.

Uses. (*i*) A solution of ICl in glacial acetic acid is used for determining the iodine values in oils and facts by Wiz's method (*ii*) Neutralisation reactions in liq. ICl have been used for the preparation of salts like $K[SbCl_6]$ (*iii*) The solution of ICl is commonly used as a catalyst in the oxidation of As (III) oxide by ceric sulphate. (*iv*) ICl has also been used for the preparation of polyhalides.

2. Chlorine monofluoride, ClF. Preparation. It is prepared by the action of Cl_2 on F_2 or ClF_3.

$$Cl_2 + F_2 \xrightarrow{250° C} 2ClF ; \qquad Cl_2 + ClF_3 \xrightarrow{300° C} 3ClF$$

Properties. (*i*) It is a colourless gas having melting point = – 156°C and boiling point = – 100.8°C.

(*ii*) It inflames cotton-wool and filter paper in the same way as F_2.

(*iii*) *Dissociation.* On heating it dissociates as : $3ClF \rightleftharpoons Cl_2 + F_2$

(*iv*) *Fluorination reactions.* It reacts with some metals more vigorously than F_2 itself. In these reactions the metals are converted into their fluorides.

$$Se + 4ClF \longrightarrow SeF_4 + 2Cl_2 \quad ; \qquad W + 6ClF \longrightarrow WF_6 + 3Cl_2$$

(*v*) *Chlorofluorinating ability.* ClF also acts as a chlorofluorinating agent.

$$CO + ClF \longrightarrow COFCl ; \qquad SF_4 + ClF \longrightarrow SF_5 Cl$$

$$SO_2 + ClF \longrightarrow SO_2F Cl$$

3. Iodine monobromide, IBr. It is obtained by the direct combination of the elements.

$$I_2 + Br_2 \longrightarrow 2IBr$$

Properties. (*i*) It is a hard, crystalline solid of grey-violet colour.

(*ii*) Its vapour dissociates to a small extent on heating : $2IBr \rightleftharpoons I_2 + Br_2$

(*iii*) IBr is an electrical conductor when in the molten state. It ionises as :

$$IBr \longrightarrow I^+ + Br^-$$

Thus, in IBr, iodine is present as I^+.

Interhalogens of XY_3 type (Tetra-atomic interhalogens)

(*i*) These are generally prepared by the direct combination of the elements under suitable conditions.

$$Cl_2 + 3F_2 \textit{ (excess)} \xrightarrow{200–300°C} 2ClF_3; \qquad Br_2 \textit{ (vapour)} + 3FS_2 \xrightarrow[\text{of } N_2]{\text{In presence}} 2BrF_3$$

$$3Cl_2 \textit{ (excess)} + I_2 \xrightarrow{100°C} 2ICl_3$$

(*ii*) Some of them may be prepared by other methods like

$$ClF_3 + Br_2 \longrightarrow BrF_3 + BrCl ; \qquad 3BrF \xrightarrow{\text{Disproportionation}} BrF_3 + Br_2$$

$$ICl + Cl_2 \longrightarrow ICl_3; \qquad I_2O_5 + 10 HCl \xrightarrow{\Delta} 2ICl_3 + 5H_2O.$$

Among these interhalogens, ClF_3 is the most reactive and reacts with many substances. Each of them has high electrical conductivity and hence undergoes self-ionisation.

$$2ClF_3 \rightleftharpoons ClF_2^+ + ClF_4^- ; \qquad 2BrF_3 \rightleftharpoons BrF_2^+ + BrF_4^-$$

$$2ICl_3 \rightleftharpoons ICl_2^+ + ICl_4^-$$

ClF_3 and BrF_3 are used as fluorinating agents.

1. Chlorine trifluoride, ClF_3. It is prepared by the action of Cl_2 on F_2 at 200-300°C in a Cu-vessel.

$$Cl_2 + 3F_2\ (excess) \xrightarrow{200-300° C} 2ClF_3$$

Properties. (*i*) It is a colourless gas which condenses to give a pale green liquid of boiling point = –12°C.

(*ii*) It is the most reactive of all the interhalogens of XY_3 type and reacts with many substances which otherwise are thought to be inert. It ignites materials like wood, abestos etc.

(*iii*) *Auto-ionisation of liquid ClF_3*. Liquid ClF_3 shows electrical conductance which suggests that liq. ClF_3 undergoes self ionisation as shown below :

$$2ClF_3 \rightleftharpoons ClF_2^+\ (Solvent\ cation) + CLF_4^-$$

It is obvious from the following reaction that MF_5 (M = As, Sb, V) acts as an acid in liquid BrF_3, since it produces ClF_2^+ ion (solvent cation) in this solvent.

$$MF_5 + ClF_3 \rightleftharpoons ClF_2^+\ (Solvent\ cation) + MF_6^-$$

(*iv*) *Fluorination reactions.* ClF_3 converts many metals (*e.g.*,uranium), metallic oxides (*e.g.*, NiO, Co_3O_4, MgO, CaO, Al_2O_3, MnO_2 etc.), metallic halides (*e.g.*, AgCl) into their fluorides.

$$U(s) + 3ClF_3 \xrightarrow{50-90° C} UF_6\ (l) + 3ClF\ (g); \qquad 6NiO + 4ClF_3 \longrightarrow 6Ni\ F_2 + 2Cl_2 + 3O_2$$

$$2Co_3O_4 + 6ClF_3 \longrightarrow 6CoF_3 + 3Cl_2 + 4O_2 ; \qquad 2AgCl + ClF_3 \longrightarrow 2AgF + Cl_2 + ClF$$

(*v*) *Hydrolysis.* It is hydrolysed by H_2O, forming ClOF: $ClF_3 + H_2O \longrightarrow ClOF + 2HF$.

(*vi*) *Action of NH_3 and N_2H_4.* HF is produced.

$$2NH_3 + 2ClF_3 \longrightarrow 6HF + N_2 + Cl_2 ; \qquad 2N_2H_4 + 4ClF_3 \longrightarrow 12HF + 3N_2 + 2Cl_2$$

(*vii*) *Action of F_2 and Br_2.* When ClF_3 reacts with F_2, it gets fluorinated with F_2, forming the higher inter halogen namely ClF_5 : $ClF_3 + F_2 \xrightarrow{hv} ClF_5$.

With Br_2, BrF_3 is obtained : $ClF_3 + Br_2 \xrightarrow{10° C} BrF_3 + BrCl$.

Uses. ClF_3 is used (*i*) as a fluorinating agent for the preparation of a number of metal fluorides, (*ii*) for the preparation of metallic polyhalides, (*iii*) as an incendiary (*iv*) in cutting oil well tubes.

2. Iodine trichloride, ICl_3. Preparation. (*i*) It is obtained by the action of excess of Cl_2 on I_2 or on ICI at 100°C.

$$3Cl_2\ (Excess) + I_2 \longrightarrow ICl_3 ; \qquad Cl_2\ (Excess) + ICl \longrightarrow ICl_3$$

(*ii*) It is also formed when dry HCl gas reacts with heated I_2O_5.

$$I_2O_5\ (heated) + 10\ HCl\ (dry\ gas) \longrightarrow 2ICl_3 + 5H_2O + 2Cl_2$$

(*iii*) It is also fomed along with ICI by the action of HI on Cl_2.

$$4HI + 4Cl_2 \longrightarrow 4HCl + I_2 + ICl + ICl_3$$

Properties. (*i*) It is a lemon-yellow crystalline solid which fumes readily. It is soluble in organic solvents, liq. NH_3 and liq. SO_2.

(*ii*) It forms crystalline additive compounds with metallic chlorides. These compounds dissociate on exposure to air, giving ICl_3 and MCl.

$$MCl + ICl_3 \rightleftharpoons MICl_4\ (Additive\ compound)$$

(*iii*) With fluorides of K, Rb, Cs, it gives polyhalides of $MICl_3F$ type.

(*iv*) ICl_3 is completely hydrolysed by H_2O : $2ICl_3 + 3H_2O \longrightarrow 5HCl + HIO_3 + ICl$.

(*v*) ICl_3 dissociates on heating at 68° giving ICl and Cl_2 : $ICl_3 \longrightarrow ICl + Cl_2$.

(*vi*) It reacts with CS_2 as : $4ICl_3 + 3CS_2 \longrightarrow CCl_4 + 2CSCl_2 + 2S_2Cl_2 + 2I_2$.

(*vii*) Like ICl, ICl_3 is also decomposed by KOH solution.

$$3ICl_3 + 12KOH \longrightarrow 9KCl + KI + 2KIO_3 + 6H_2O.$$

(*viii*) Molten ICl_3 has a high conductivity ($= 8.4 \times 10^{-3}$ ohm^{-1} cm^{-1}). The liberation of I_2 and Cl_2 at both electrodes indicates that ICl_3 is ionised into $[ICl_2]^+$ and $[ICl_4]^-$ ions.

$$2ICl_3 \rightleftharpoons [ICl_2]^+ + [ICl_4]^-$$

$$2[ICl_2]^+ + 2e^- \longrightarrow I_2 + Cl_2 \textit{ (At cathode)}; \; 2[ICl_4]^- - 2e^- \longrightarrow I_2 + 4Cl_2 \textit{ (At anode)}$$

ICl_2^+ and ICl_4^- both contain I^{3+} ions.

Uses. ICl_3 is used in medicines. It is also used in the preparation of polyhalides.

3. Bromine trifluoride, BrF_3. This compound was discovered by Lebean in 1905.

Preparation. It is obtained : (*i*) by mixing Br_2 vapour with F_2 in presence of N_2.

$$Br_2 \textit{ (vapour)} + 3F_2 \xrightarrow[\text{of } N_2]{\text{In presence}} 2BrF_3$$

(*ii*) by the action of ClF_3 on Br_2 at 10°C : $ClF_3 + Br_2 \xrightarrow{10°C} BrF_3 + BrCl$.

(*iii*) by the action of F_2 on HBr : $3F_2 + 2HBr \rightarrow 2BrF_3 + H_2$.

(*iv*) When BrF undergoes disproportionation.

$$3BrF \; (Br = +1) \longrightarrow BrF_3 \; (Br = +3) + Br_2 \; (Br = 0)$$

Properties. (*i*) BrF_3 is a straw coloured fuming liquid with boiling point = 125.8°C and melting point. = 8.8°C and is very reactive. It is a corrosive liquid and hence is used as a non-aqueous solvent to carry out many reactions. It is a non-protonic ionising solvent.

(*ii*) *Reaction with Br_2.* BrF_3 reacts with Br_2 to form BrF: $BrF_3 + Br_2 \rightarrow 3\,BrF$.

(*iii*) *Auto-ionisation of liquid BrF_3.* Liquid BrF_3 has high specific conductance ($= 0.8\ \Omega^{-1}\ m^{-1}$ at 298 K) and hence undergoes auto-ionisation which takes place in the following manner.

$$2BrF_3 \textit{ (Solvent)} \rightleftharpoons BrF_2^+ \textit{ (Solvent cations : acid ions)} + BrF_4^- \textit{ (Solvent anions : base ions)}$$

The above mode of auto-ionisation indicates that, according to solvent system concept of acids and bases, the substances containing or making available BrF_2^+ ions (solvent cations-acid ions) will act as acids in liq. BrF_3 and those which contain or give BrF_4^- ions in liq. BrF_3 behave as bases in this solvent. BrF_4^- ion is called *tetrafluorobromate.*

Examples of acids in liquid BrF_3. When the fluorides like AuF_3, MF_4 (M = Ge, Sn, Ti) and MF_5 (M = P, Sb, Bi, Nb, Ta) react with liquid BrF_3, BrF_2^+ ions are produced. Thus, these fluorides act as acids in liquid BrF_3.

$$AuF_3 + BrF_3 \rightleftharpoons BrF_2^+ + AuF_4^-$$

$$MF_4 \; (M = Ge, Sn, Ti)) + 2BrF_3 \rightleftharpoons 2BrF_2^+ + MF_6^{2-}$$

$$MF_5 + BrF_3 \; (M = P, Sb, Bi, Nb, Ta) \rightleftharpoons BrF_2^+ + MF_6^-$$

Examples of bases in liquid BrF_3. The metal fluorides like MF (M = Li, K, Ag), BaF_2, NOF etc., produce BrF_4^- ions (solvent anions) in liquid BrF_3 and hence act as bases in this solvent.

$$MF \; (M = Li, K, Ag) + BrF_3 \rightleftharpoons M^+ + BrF_4^-$$

$$CaF_2 + 2BrF_3 \rightleftharpoons Ca^{2+} + 2BrF_4^- ; \qquad NOF + BrF_3 \rightleftharpoons NO^+ + BrF_4^-$$

In the above reactions polyhalides like $[BrF_2]_2\,[MF_6]$, $[BrF_2]\,[MF_6]$, $M[BrF_4]$, $Ba[BrF_4]_2$, $[NO]\,[BrF_4]$ are produced.

(*iv*) *Neutralisation reactions involving acids and bases in liquid BrF_3.* The auto-ionisation of liq. BrF_3 as given above indicates that the compounds containing BrF_2^+ ions (*solvent cations-acid ions*) act as acids in liquid BrF_3 and those which have BrF_4^- ions (*solvent anions-base ions*) act as bases in this solvent. Thus, in liquid BrF_3, a neutralisation reaction is that in which a compound containing BrF_2^+ ions reacts with that having BrF_4^- ions and a salt and BrF_3 (*solvent*) are formed. Thus, the following reactions are acid-base reactions in liquid BrF_3:

Acid		*Base*		*Salt*		*Solvent*
$[BrF_2]_2^+ [SnF_6]^{2-}$	+	$2Ag^+ [BrF_4]^-$	$\longrightarrow$	$Ag_2 [SnF_6]$	+	$4BrF_3$
$[BrF_2]^+ [VF_6]^-$	+	$K^+ [BrF_4]^-$	$\longrightarrow$	$K[VF_6]$	+	$2BrF_3$
$[BrF_2]^+ [SbF_6]^-$	+	$Ag^+ [BrF_4]^-$	$\longrightarrow$	$Ag [SbF_6]$	+	$2BrF_3$
$[BrF_2]^+ [AuF_4]^-$	+	$Ag^+ [BrF_4]^-$	$\longrightarrow$	$Ag [AuF_4]$	+	$2BrF_3$
$[BrF_2]_2^+ [GeF_6]^{2-}$	+	$2NO^+ [BrF_4]^-$	$\longrightarrow$	$[NO]_2 [GeF_6]$	+	$4BrF_3$

The end point of acid base titrations in liquid BrF_3 may be determined either with the help of suitable indicators or may be followed conductometrically.

(v) Neutralisation reactions involving an acidic anhydride and a base or basic anhydride and an acid. Such reactions have also been reported to occur in liquid BrF_3. In these reactions salt and solvent (*i.e.,* BrF_3) are produced. For example :

$$VF_5 \ (\textit{Acidic anhydride}) + Ag^+ [BrF_4]^- \ (\textit{Base}) \xrightleftharpoons{\text{Liq. } BrF_3} Ag^+ [VF_6]^- \ (\textit{Salt}) + BrF_3 \ (\textit{Solvent})$$

$$LiF \ (\textit{Basic anhydride}) + [BrF_2]^+ [TaF_6]^- \ (\textit{Acid}) \xrightleftharpoons{\text{Liq. } BrF_3} Li^+ [TaF_6]^- \ (\textit{Salt}) + BrF_3 \ (\textit{Solvent})$$

(vi) Fluorination reactions. BrF_3 is a very reactive solvent and converts many metals, metallic and nonmetallic halides, oxides carbonates, iodates etc., into their fluorides. Thus BrF_3 is a useful *fluorinating agent.*

$$6M + 2BrF_3 \longrightarrow 6MF + Br_2 ; \qquad 3MCl + BrF_3 \longrightarrow 3MF + BrCl_3$$

$$3AO_2 \ (A = Si, Ge) + 4BrF_3 \longrightarrow 3AF_4 + 2Br_2 + 3O_2 ; 2B_2O_3 + 4BrF_3 \longrightarrow 4BF_3 + 2Br_2 + 3O_2$$

$$2WO_3 + 4BrF_3 \longrightarrow 2WF_6 + 2Br_2 + 3O_2 ; \qquad 6Sb_2O_3 + 20BrF_3 \longrightarrow 12\ SbF_5 + 10Br_2 + 9O_2$$

$$6CuO + 4\ BrF_3 \longrightarrow 6\ CuF_2 + 3Br_2 + 3O_2$$

BrF_3 reacts with SiO_2 and silica materials and sets fire to wood, paper etc.

(vii) Redox reactions. When alkali metals and other metals like Nb, Ta, Au, Ag etc., react liq. BrF_3, they are oxidised and BrF_3 (Br = +3) is reduced to Br_2 (Br = 0).

$$3K + 4BrF_3 \longrightarrow 3K^+ [BrF_4]^- + \frac{1}{2} Br_2$$

$$6Nb + 16\ BrF_3 \longrightarrow 6[BrF_2]^+ [NbF_6]^- + 5Br_2$$

(viii) Complex formation reactions. In presence of liq. BrF_3, fluorocomplexes of Au (III), Ge (IV), Sn (IV), Ti (IV), Mn (IV), Ru (IV), Os (IV), Ir (IV), Pd (IV), Pt (IV); As (V), V (V), Nb (V), Ta (V), Ru (V), Os (V), Ir (V) have been prepared. For example :

$$NO_2 + Au \xrightarrow{\text{liq. } BrF_3} [NO_2] [AuF_4] \ (\textit{Nitronium fluorocomplex})$$

$$NO_2 + As_2O_3 \xrightarrow{\text{liq. } BrF_3} [NO_2] [AsF_6] \ (\textit{Nitronium fluorocomplex})$$

$$NOCl + GeO_2 \xrightarrow{\text{liq. } BrF_3} [NO]_2 [GeF_6] \ (\textit{Nitrosonium fluorocomplex})$$

The following complex formation reaction has also been carried out in liq.BrF_3

$$SO_3 + K_2SO_4 \xrightarrow{\text{liq. } BrF_3} K[SO_3F] \ (\textit{Pot. fluorosulphonate})$$

(ix) Solvolysis reactions. In these reactions the solute reacts with liq. BrF_3 (*solvent*) and concentration of BrF_2^+ ions (*solvent cations*) and BrF_4^- ions (*solvent anions*) is increased. The increase in concentration is due to the auto-ionisation of liq. BrF_3. Examples are :

$$4BrF_3 \ (l) + K_2[TiF_6] \rightleftharpoons [BrF_2]_2 \ [TeF_6] + 2K[BrF_4]$$

$$2BrF_3 \ (l) + K[PF_6] \rightleftharpoons [BrF_2] \ [PF_6] + K[BrF_4]$$

(*x*) *Formation of adducts.* In BrF_3 solution, XeF_2 reacts with the fluoride like MF_5 (M = P, As, Sb etc.) and forms various adducts having the composition $XeF_2.MF_5$, XeF_2, $2MF_5$, $2XeF_2.MF_5$. These adducts are ionic compounds. The formation of $XeF_2.MF_5$ can be shown as :

$$XeF_2 \text{ (Lewis acid)} + MF_5 \text{ (Lewis base)} \longrightarrow XeF_2.MF_5 \text{ or } [XeF]^+ [MF_6]^- \text{ (adduct)}$$

(*xi*) *Hydrolysis.* BrF_3 is hydrolysed by water with the formation of bromine oxy-fluoride, $BrOF_3$.

$$BrF_3 + H_2O \longrightarrow BrOF_3 + H_2$$

(*xii*) *Action of SbF_3.* BrF_3 dissolves SbF_3 to form antimony hexafluoroborate, $Sb[BrF_6]$.

$$SbF_3 + BrF_3 \longrightarrow Sb\,[BrF_6]$$

Uses of liquid BrF_3. (*i*) The reaction of liq. BrF_3 with covalent fluorides / ionic halides gives polyhalides. Examples of the preparation of poly halides have already been given under "*Auto-ionisation of liq. BrF_3*" discussed on page 967 in this chapter. (*ii*) Acid-base (neutralisation) reactions carried out in liq. BrF_3 have been used to prepare a number of complex compounds which are difficult to prepare by other methods. Examples of such compounds have already been given on page 967 (same chapter). (*iii*) Due to the fact that liq. BrF_3 readily converts many metals, metallic and nonmetallic halides, oxides, carbonates, iodates etc., into their fluorides, this solvent is used as a fluorinating agent. Reactions showing the fluorinating ability of BrF_3 have already been given on page 968 (same chapter). (*iv*) In presence of liq. BrF_3, fluorocomplexes of many metals like Au (III), Ge (IV), As (V) etc., have been prepared.

Interhalogens of XY_5 type (Hexa-atomic interhalogens)

(*i*) These are prepared by the direct combination of the elements, *e.g.*,

$$Cl_2 + 5F_2 \xrightarrow[250 \text{ atm.}]{350°C} 2ClF_5 \; ; \qquad Br_2 + 5F_2 \text{ (excess)} \xrightarrow{>150° C} 2\,BrF_5$$

$$I_2 + 5F_2 \text{ (excess)} \longrightarrow 2IF_5$$

(*ii*) These can also be prepared by other methods, *e.g.*,

$$ClF_3 + F_2 \xrightarrow{h\nu} ClF_5 \; ; \qquad KBr + 3F_2 \xrightarrow{25° C} KF\,(s) + BrF_5$$

$$2I_2O_5 + 10F_2 \longrightarrow 4IF_5 + 5O_2 \; ; \qquad I_2 + 10\,AgF \longrightarrow 2IF_5 + 10Ag$$

1. Chlorine pentafluoride (ClF_5) and bromine pentafluoride (BrF_5). ClF_5 is prepared by the action of F_2 on Cl_2 or on ClF_3.

$$Cl_2 + 5F_2 \xrightarrow[250 \text{ atm.}]{350° C} 2ClF_5 \quad ; \qquad ClF_3 + F_2 \xrightarrow{h\nu} ClF_5$$

BrF_5 is obtained by the action of F_2 on Br_2 or on KBr.

$$5F_2 \text{ (excess)} + Br_2 \xrightarrow{>150° C} 2BrF_5 \; ; \qquad 3F_2 + KBr \xrightarrow{25° C} KF\,(s) + BrF_5$$

ClF_5 is a gas having melting point. = – 103°C and boiling point = – 13.1°C. It is a fluorinating agent. It is hydrolysed by water to liberate HF and $FClO_2$

$$ClF_5 + 2H_2O \longrightarrow FClO_2 + 4HF$$

ClF_5 ionises as : $2ClF_5 \rightleftharpoons ClF_4^+ + ClF_6^-$

With AsF_5 and SbF_5 it forms 1 : 1 adducts *viz.*, $ClF_5.AsF_5$ and $ClF_5.SbF_5$ which are ionic compounds and hence have the composition as $[ClF_4]^+ [AsF_6]^-$ and $[ClF_4]^+ [SbF_6]^-$ respectively.

BrF_5 reacts very violently. It is, therefore, generally diluted with N_2.BrF_5 is hydrolysed by water as : $BrF_5 + 3H_2O \longrightarrow 5HF + HBrO_2$.

BrF_5 is used in organic synthesis. It is also recommended as an oxidiser for propellants.

2. Iodine pentafluoride, IF_5. Preparation. It is formed : (*a*) by the direct combination of I_2 and F_2 (*excess*) : $I_2 + 5F_2 \text{ (excess)} \rightarrow 2IF_5$.

(*b*) by the action of F_2 on I_2O_5 : $10F_2 + 2I_2O_5 \longrightarrow 4IF_5 + 5O_2$.

(*c*) by heating I_2 with AgF : $I_2 + 10AgF \longrightarrow 2IF_5 + 10Ag$.

Properties. (*i*) It is a colourless liquid whose melting point. = 9.6°C.

(*ii*) *Self-ionisation.* It is a good conductor of electricity as it ionises as :

$$2IF_5 \rightleftharpoons IF_4^+ \text{ (solvent cation)} + IF_6^- \text{ (solvent anion)}$$

Thus, the substances that give IF_4^+ ions in liquid IF_5, act as acids in this solvent and those which produce IF_6^- ions, behave as bases in the solvent. The dielectric constant of IF_5 is extremely high, which makes it a good ionising solvent.

The following acid-base reaction which takes place in liquid IF_5, is important.

$[IF_4]^+ [SbF_6]^-$ (*Acid*) + $K^+ [IF_6]^-$ (*Base*) $\rightleftharpoons$ $K^+ [SbF_6]^-$ (*Salt*) + $2IF_5$ (*Solvent*)

(*iii*) *Action of I_2O_5.* IF_5 reacts with I_2O_5 to form iodine oxy-fluoride, IOF_3, which is a solid and decomposes on heating at 110°C.

$$3IF_5 + I_2O_5 \longrightarrow 5IOF_3 \quad ; \quad 2IOF_3 \xrightarrow{\Delta} IF_5 + IO_2F$$

(*iv*) *Hydrolysis.* Hydrolysis gives halogen acid and oxy-acid of the larger halogen atom.

$$IF_5 + 3H_2O \longrightarrow 5HF + HIO_3$$

(*v*) *Action of F_2.* When IF_5 is heated with F_2 and 250°-300°C, IF_7 is formed.

$$IF_5 + F_2 \xrightarrow{250° - 300° C} IF_7$$

Interhalogen compounds of XY_7 type (Octa-atomic interhalogens)

Only one compound of this type *viz.*, IF_7 (called iodine heptafluoride) has been formed and characterised. The reason of the formation of IF_7 molecule is that iodine atom has the largest size among halogens, while F atom has the smallest size.

IF_7 is prepared by the action of F_2 on IF_5, KI or PbI_2.

$$F_2 + IF_5 \xrightarrow{250° - 300°C} IF_7 \ ; \qquad F_2 + KI \xrightarrow{250°C} KF + IF_7$$

$$8F_2 + PbI_2 \longrightarrow PbF_2 + 2IF_7$$

Properties. (*i*) It is a gas at ordinary temperature. Like ClF_3 and BrF_3, it is also highly reactive.

(*ii*) *Action of SiO_2 and pyrex glass.* SiF_4 is formed.

$$2IF_7 + SiO_2 \xrightarrow{250° C} 2IOF_5 + SiF_4$$

(*iii*) *Hydrolysis.* Vapours of IF_7 are hydrolysed by water, giving HF and H_5IO_6.

$$IF_7 + H_2O \longrightarrow IOF_5 + 2HF$$
$$IOF_5 + 5H_2O \longrightarrow H_5IO_6 + 5HF$$
$$\overline{IF_7 + 6H_2O \longrightarrow H_5IO_6 + 7HF}$$

(*iv*) *Formation of adducts.* It forms adducts with SbF_5 and AsF_5. These adducts are ionic compounds having the structure $[IF_6]^+ [SbF_6]^-$ and $[IF_6]^+ [AsF_6]^-$ respectively.

Structure of interhalogen compounds

The structure and geometry of interhalogen molecules of all the four types *viz.*, XY, XY_3, XY_5 and XY_7 have already been discussed in chapter 14.

Polyhalide ions and polyhalides

What are polyhalide ions and polyhalides ?

The ions (cations or anions) composed of similar halogen atoms (*e.g.*, Cl_2^+, Br_2^+, I_2^+, Cl_3^+, Br_3^+, I_3^+, Br_5^+, I_5^+, I_3^-, I_4^-, I_5^-, I_7^-, I_8^- etc.) or dissimilar halogen atoms (*e.g.*, ICl_2^+, IF_4^+, ICl_2^-, $IBrCl^-$ etc.) are called **polyhalide ions.** These ions may be regarded as positively or negatively-charged interhalogen ions.

The formation of polyhalide ions was observed during the self-ionisation (auto-ionisation) of some interhalogen compounds, *e.g.*,

$$2ICl \rightleftharpoons I^+ + ICl_2^- \qquad 2BrF_3 \rightleftharpoons BrF_2^+ + BrF_4^-$$
$$2ClF_3 \rightleftharpoons ClF_2^+ + ClF_4^- \qquad 2ICl_3 \rightleftharpoons ICl_2^+ + ICl_4^-$$
$$2ClF_5 \rightleftharpoons ClF_4^+ + ClF_6^- \qquad 2IF_5 \rightleftharpoons IF_4^+ + IF_6^-$$
$$IF_7 \rightleftharpoons IF_6^+ + F^-$$

The ionic compounds containing either polyhalide cations or polyhalide anions are called **polyhalides.**

Examples of polyhalides, containing polyhalide anions, are $K^+ [Cl_3]^-$, $NH_4^+ [I_5]^-$, $Cs^+ [ClBrI]^-$, $M^+[ICl_2]^-$ (M = K, Rb, NH_4), $[(CH_3)_4 N]^+ [ClBrI]^-$, $[N(C_2H_5)_4]^+ [I_3]^-$, $[As(C_6H_5)_4]^+ [I_3]^-$, $Na^+ [IBr_2]^-$, $H^+[ICl_3]^-.4H_2O$, $H^+ [ICl_4]^-.4H_2O$ etc.

Example of polyhalides, containing polyhalide cations, are $[X_2]^+ [Sb_3F_{16}]^-$ (X = Cl, Br, I) $[Cl_3]^+ [AsF_6]^-$, $[I_5]^+ [AlCl_4]^-$, $[BrF_2]^+ [SbF_6]^-$, $[ClF_2]^+ [MF_6]^-$, $[ICl_2]^+ [MCl_4]^-$ (M = Sb, As), etc.

Iodine has maximum tendency to form polyhalide ions. It may be seen that Cl, Br and I atoms, can form the triatomic anions *viz.*, Cl_3^-, Br_3^- and I_3^- respectively, while F-atom is not able to form F_3^- ion. This is because of the fact that F-atom cannot show higher positive oxidation states (infact F-atom does not show any positive oxidation state) and has no *d*-orbitals in its valence-shell.

Preparation of polyhalides

These can be prepared by the following method :

(*i*) *By the action of halogens on metallic halides or other polyhalides.* Examples are given below:

(*a*) $Cl_2 + KCl \longrightarrow K[Cl_3]$; $Cl_2 + KI \longrightarrow K[ICl_2]$

$Br_2 + CsI \xrightarrow{C_2H_5OH} Cs [IBr_2]$; $Br_2 + KBr \longrightarrow K[Br_3]$

$I_2 + KI \longrightarrow K[I_3]$; $2I_2 + NH_4I \longrightarrow NH_4 [I_5]$

$$\left(\frac{n-1}{2}\right) I_2 + [(CH_3)_4N]I \xrightarrow{CH_3COOH} [(CH_3)_4N]^+ [I_n]^-$$

(*b*) $Cl_2 + K[ICl_2] \longrightarrow K[ICl_4]$; $Cl_2 + K[IBr_2] \longrightarrow K[ICl_2] + Br_2$

$I_2 + Cs[Br_3] \longrightarrow Cs[IBr_2] + IBr$

(*ii*) *By the action of interhalogens on metals, metallic halides or other polyhalides.* Examples are :

(*a*) $4BrF_3 + 3K \longrightarrow 3K [BrF_4] + \frac{1}{2} Br_2$; $16 BrF_3 + 6Nb \longrightarrow 6[BrF_2]^+ [NbF_6]^- + 5Br_2$

(*b*) $ICl + MCl$ (M = K, Rb, NH_4) $\longrightarrow M[ICl_2]$; $ICl + KI \longrightarrow K[ICl_2]$

$ICl + KBr \longrightarrow K[Cl BrI]$; $ICl + [(CH_3)_4 N]Br \xrightarrow{CH_3COOH} [(CH_3)_4 N]^+ [ClBrI]^-$

$IBr + NaBr \longrightarrow Na [IBr_2]$; $BrCl + CsI \longrightarrow Cs[IBrCl]$

$BrF_3 + MF$ (M = Li, K, Ag) $\rightleftharpoons M[BrF_4]$; $2BrF_3 + CaF_2 \rightleftharpoons Ba [BrF_4]$

$BrF_3 + NOF \rightleftharpoons [NO]^+ [BrF_4]^-$; $ClF_5 + AsF_5 \longrightarrow [ClF_4]^+ [AsF_6]^-$

$ClF_5 + SbF_5 \longrightarrow [ClF_4]^+ [SbF_6]^-$; $IF_7 + MF_5$ (M = As, Sb) $\longrightarrow [IF_6]^+ [MF_6]^-$

(*c*) $ICl + K [Cl Br I] \longrightarrow K[ICl_2] + IBr$

(*iii*) *By the action of I_2 and Cl_2 on the solution of metallic chlorides in HCl.* The following examples illustrate this method.

$I_2 + 3Cl_2 + 2MCl$ (M = Li, Na) $\xrightarrow{HCl} 2M^+ [ICl_4]^-$; $I_2 + Cl_2 + 2RbCl$ (in HCl) $\longrightarrow 2Rb^+[ICl_2]^-$

$I_2 + 3Cl_2 + MCl_2$ (M = Mg, Ca^{2+}) $\xrightarrow{HCl} M^{2+} [ICl_4]_2^-$

(*iv*) *By the action of an appropriate gaseous halogen on a metallic halide in the absence of a solvent, e.g.*, $2F_2 + CsCl \longrightarrow Cs[ClF_4]$

(*v*) I_2 dissolves in highly acidic oxidising media (*e.g.*, odeum) to give bright paramagnetic solutions which are believed to contain I_2^+ ion. Br_2^+ ion may be generated in a similar way. Cl_2^+ ion is known only in the gas phase.

The salts of X_2^+ ions (X = Cl, Br, I) are best prepared by the action of $S_2O_6F_2$ (used as an oxidising agent) on X_2 molecule and then SbF_5 is added.

$$X_2 + S_2O_6F_2 \xrightarrow{\text{Add SbF}_5} [X_2]^+ [Sb_3 F1_6]^-$$

$[I_2]^+ [Sb_2F_{11}]^-$ is obtained by the action of I_2 on SbF_5.

When a solution containing I_2^+ ions is cooled, brown I_4^+ ions are obtained.

(*vi*) X_3^+ (X = Cl, Br, I), Br_5^+ and I_5^+ ions are well established. Salts of these ions can be prepared by a number of methods. For example :

$$Cl_2 + ClF + AsF_5 \xrightarrow{-78°C} [Cl_3]^+ [AsF_6]^-$$

$$Au\ (s) + 4Br\ SO_3F \xrightarrow{\text{In BrSO}_3\text{F}} [Br_3]^+ [Au(SO_3F)_4]^- + \frac{1}{2} Br_2 \xrightarrow{+Br_2} [Br_5]^+ [Au\ (SO_3F)_4]$$

$$3I_2 + 3AsF_5 \xrightarrow{\text{In SO}_2} 2[I_3]^+ [AsF_6]^- + AsF_3$$

$$2I_2 + ICl + AlCl_3 \longrightarrow [I_5]^+ [AlCl_4]^-$$

Properties of polyhalides

(*i*) *Colour.* All the polyhalides are coloured compounds. The depth of the colour increases with the increase of the atomic number of the halogen atoms present. Thus the bromoiodides are red and chlorobromides are yellow.

(*ii*) *Dissociation in aqueous solution.* The polyhalides are usually very soluble in water, from which they can often be crystallised. Some of them get dissociated in water, to free halogen and halide ion (metallic monohalide). For example : $I_3^- \rightarrow I^- + I_2$

The stability of metallic trihalides of MX_3 type, having the same cation in the same oxidation state and different trihalide anions in aqueous solution, towards dissociation to free halogen and metallic monohalide (MX halide), is in the order : $MI_3 > MBr_3 > MCl_3$.

For the metallic trihalides, having the same trihalide anion and different cations in the same oxidation state, the stability increases with the increase of the size of the cation. For example : $NaI_3 < KI_3 < RbI_3 < CsI_3$. Note that the size of Na^+, Rb^+ and Cs^+ cations increases from Na^+ to Cs^+.

The polyhalides, containing ICl_4^- anion, dissociate in aqueous solution to metal chloride and ICl_3; the latter then hydrolysis to iodate, IO_3^-.

$$M^+ [ICl_4]^- \longrightarrow MCI + ICl_3] \times 5$$

$$5ICl_3 + 9H_2O \longrightarrow 3HIO_3 + 15\ HCl + I_2$$

$$5M^+ [ICl_4]^- + 9H_2O \longrightarrow 5MCl + 3HIO_3 + 15HCl + I_2$$

or

$$5ICl_4^- + 9H_2O \longrightarrow 20Cl^-\ 18H^+ + 3IO_3^- + I_2$$

(*iii*) *Formation of complexes.* Some of the polyhalide ions form complexes with organic donor molecules. Some examples of such complexes are : $LiI_3.4C_6H_5CN$, $MI_3.2C_6H_5CN$ (M = Na or K).

(*iv*) *Effect of heat : Thermal decomposition.* When the solid polyhalides are heated, they dissociate, more or less, readily. The ease of dissociation of a given polyhalide decreases with the increase of the size of the cation, with which the polyhalide anion is associated. This means that *the stability of the solid polyhalides towards dissociation is maximum, when the cation with which the polyhalide anion is associated, is large.*

On thermal dissociation, the polyhalides give the metal monohalides and halogen molecule or interhalogen molecule. For example :

$$CsI_3 \xrightarrow{\Delta} CsI + I_2\ ; \qquad RbICl_2 \xrightarrow{\Delta} RbCl + ICl$$

When a polyhalide, containing more than one type of halogen atoms, undergoes thermal dissociation, the metal monohalide formed is always that which has highest lattice energy. Thus, when $Cs^+ [ICl_2]^-$ undergoes thermal dissociation, the products formed are CsCl + ICl and not CsI + Cl_2, since CsCl has higher lattice energy than CsI.

For trihalide ions formed by the same metal, the order of thermal stability is : $I_3^- > IBr_2^- > ICl_2^- > I_2Br^- > Br_3^- > BrCl_2^- > Br_2Cl^-$.

(*v*) *Reaction with halogens.* Sometimes when a polyhalide is allowed to react with a halogen molecule, one or two atoms of halogen present in the polyhalide are substituted by one or two atoms of the reacting halogen, as the case may be.

$$KIBr_2 + Cl_2 \longrightarrow KICl_2 + Br_2 \;; \qquad CsBr_3 + I_2 \longrightarrow CsIBr_2 + IBr$$

Sometimes a higher polyhalide is formed. $KICl_2 + Cl_2 \longrightarrow KICl_4$

(*vi*) *Dissolution in liquid halogen or interhalogen.* Polyhalides may dissolve in liquid halogen or interhalogen to give solution from which the original polyhalide may be crystallised on evaporation. The polyhalides like $K^+[IBr_2]^-$, NH_4^+ $[IBr_2]^-$ and $Rb^+[IBr_2]^-$ form solution in bromine.

Structure of polyhalide ions

The structure/geometry of polyhalide ions like ICl_2^+, I_3^+, IF_4^+, I_5^+, IF_6^+ (cations), ICl_2^-, I_3^-, ICl_4^-, IF_6^- and I_8^{2-} (anions) has already been explained in chapter 14. The explanation is based on the concept of hybridisation. The structure of I_3^- ion has been explained with the help of molecular orbital theory.

Pseudohalides and pseudohalogens

There are certain uni-negative groups (*e.g.*, CN^-, OCN^-, N_3^- etc.), which are made up of two or more electronegative atoms and resemble halide ions in some respects. These uni-negative groups or ions are called *pseudohalide ions.* Examples of some important pseudohalide ions are : cyanide ion (CN^-), isocyanide ion (NC^-), fulminate ion (ONC^-), cyanate or oxy-cyanide ion (OCN^-), isocyanate ion (NCO^-), thiocyanate ion (SCN^-), isothiocyanate ion (NCS^-), tellurocyanate ion ($TeCN^-$), selenocyanate ion ($SeCN^-$), isoselenocyanate ion ($NCSe^-$), azide ion (N_3^-), azidothiocarbonate ion or azido carbondisulphide ion ($SCSN_3^-$). The salts (ionic compounds) containing pseudohalide ions (*e.g.*, AgCN, Pb $(CNS)_2$ etc.] are called *pseudohalides.*

As the dimers of halide ions are called *halogens*, the covalent dimers of the pseudohalide ions are called *pseudohalogens* or *halogenoids.* So far only a limited number of pseudohalogens, has been isolated and characterised. Examples of some pseudohalogen or halogenoids are cyanogen, $(CN)_2$; oxy-cynogen $(OCN)_2$; thiocynogen $(SeCN)_2$; azidocarbon disulphide, $(SCSN_3)_2$.

Similarities between halide ions and pseudohalide ions

Following are the important similarities between halide ions and pseudohalide ions :

1. Formation of ionic as well as covalent compounds. Like halide ions, pseudohalide ions also from ionic as well as covalent compounds. Ionic compounds are given by these ions with metals while covalent compounds are given by the combinations of these ions with nonmetals or metalloids. Examples are :

Ionic compounds : AgCl (AgCN), $PbCl_2$ $[Pb(CNS)_2]$, AgCl (HgCNS).

Covalent compounds : ICl(ICN), $SiCl_4[Si(NCS)_4]$ $COCl_2[CO(N_3)_2]$, SO_2Cl_2 $[SO_2(N_3)_2$.

The compounds given in parentheses are the analogous compounds given by pseudohalide ions.

2. Oxidation of hydracids. HX acids ($X = Cl^-, Br^-, I^-$) can be oxidised to free halogen (X_2). Similarly HY acids (Y = SCN CN etc.) can also be oxidised to free pseudohalogens (Y_2). For example :

$$4HCl + MnO_2 \longrightarrow MnCl_2 + 2H_2O + Cl_2 \;; 4HSCN + MnO_2 \longrightarrow Mn(SCN)_2 + 2H_2O + (SCN)_2$$

3. Thermal decomposition of Pb (IV) salts. On heating some of the Pb (IV) salts of halide ions (*e.g.*, $PbCl_4$) decompose to give Pb (II) salts and free halogen molecule. Similarly some Pb

(IV) salts of pseudohalide ions [*e.g.*, $Pb(SCN)_4$, $Pb(SeCN)_4$] decompose to give Pb (II) salts and free pseudohalogens. For example :

$$PbCl_4 \xrightarrow{\Delta} PbCl_2 + Cl_2 \ ; \quad Pb(SCN)_4 \xrightarrow{\Delta} Pb(SCN)_2 + (SCN)_2$$

4. Formation of complex ions. Like halide ions, pseudohalide ions also form complex ions with transition metal ions. Examples are given below :

Complex ions given by halide ions : $[FeF_6]^{3-}$, $[CoCl_4]^{2-}$, $[HgI_4]^{2-}$.

Complex ions given by pseudohalide ions $[Fe(CN)_6]^{3-}$, $[Co(SCN)_4]^{2-}$, $[Hg(CN)_4]^{2-}$

5. Formation of interpseudohalogen and interhalogen compounds. Halide ions can combine together to form interhalogen compounds like ClF, ICl, IBr etc. Similarly pseudohalide ions can also combine together to form interpseudohalogen compounds like $CN.N_3$, CN.SCN etc.

6. Formation of monobasic hydracids. Like halide ions, pseudohalide ions also combine with H_2 to form monobasic hydracids.

$$H_2 + 2Cl^- \longrightarrow 2HCl + 2e^- \ ; \quad H_2 + 2CN^- \longrightarrow 2HCN + 2e^-$$

7. Formation of insoluble salts. Like halide ions, pseudohalide ions give insoluble salts with Ag^+, Pb^{2+} and Hg^+ cations.

$Ag^+ + Cl^- \longrightarrow AgCl\downarrow$	$Pb^{2+} + 2Cl^- \longrightarrow PbCl_2\downarrow$	$Hg^+ + Cl^- \longrightarrow HgCl\downarrow$
$Ag^+ + CN^- \longrightarrow AgCN\downarrow$	$Pb^{2+} + 2CNS^- \longrightarrow Pb(CNS)_2\downarrow$	$Hg^+ + CNS^- \longrightarrow HgCNS\downarrow$

8. Acting as bridging ligands. Like halide ions, pseudohalide ions also have more than one pair of electrons and hence can coordinate with two metal ions simultaneously. Consequently like halide ions, pseudohalide ions also can act as bridging ligands. For example in $[R_2Au(CN)]_4$ which is a tetrameric compound, four CN^- ions act as bridging ligands. Similarly in $[R_2AuCl]_2$ which is a dimeric compound, two Cl^- ions act as bridging ligands.

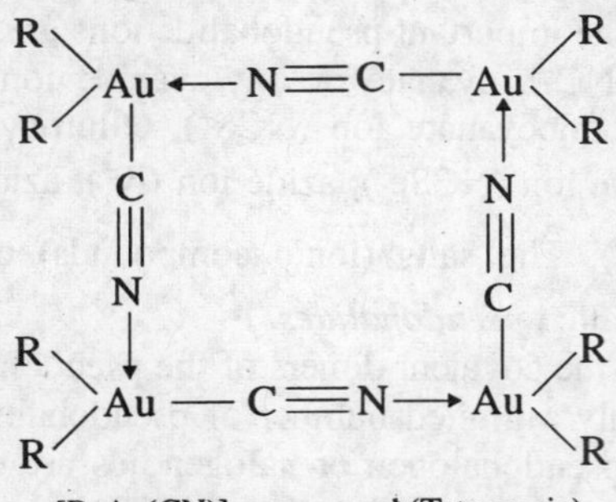

$[R_2Au(CN)]_4$ compound (Tetrameric)

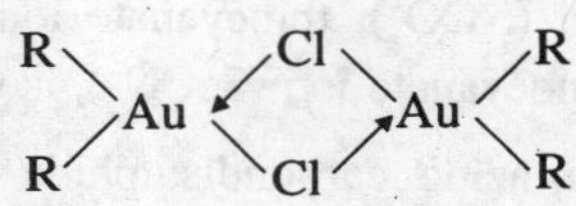

$[Ru_2AuCl]_2$ compound (Dimeric)

Note that in each of the above compounds, the geometry of Au atoms is square planar.

Dissimilarities between halide ions and pseudohalide ions

The differences between the two classes of ions are given as follows :

1. The hydracids, formed by the pseudohalide ions on combination with hydrogen, are relatively weaker than the analogous hydracids formed by the halide ions. For example, HCN is a very weak acid, while HCl is a very strong acid in aqueous medium. The weak nature of hydracids given by pseudohalide ions may be due to the poor electronegativity of the pseudohalogens.

2. Pseudohalide ions are stronger coordinating ligands than the halide ions and hence the complexes formed by pseudohalide ions are *low spin* (*i.e.*, inner-orbital involving d^2sp^3 hybridisation while those formed by halide ions are high spin (outer-orbital involving sp^3d^2 hybridisation). For example $[Fe(CN)_6]^{3-}$ is a low spin complex while $[FeF_6]^{3-}$ is a high spin complex.

That the pseudohalide ions are stronger ligands than the halide ions is due to the ability of the electron deficient pseudohalides to form σ-bond (from the pseudohalide to the metal) as well as π-bond (form the metal to the pseudohalide).

3. Pseudohalide ions, being made up of two hetero atoms can function as *ambidentate* ligands. For example class a metals form isothiocyanate complexes like $[Co\ (NCS)_4]^{2-}$ in which N-atom is

coordinated to Co^{2+} ion while class *B* metals form thiocyanate complexes like $[Pd(SCN)_4]^{2-}$ in which S atom is bonded to Pd^{2+} ion. Halide ions have no tendency to act as ambidentate ligands.

Similarities between halogens and pseudohalogens

Following are the points showing the similarities between the two classes of compounds :

1. Volatile and dimeric nature. Like halogens, pseudohalogens are also dimeric and fairly volatile (with the exception of polymeric thiocynogen) in the free state.

2. Isomorphous nature. Pseudohalogens are isomorphous to halogens when in the free or solid state. For example Cl_2 is isomorphous to $(CN)_2$ and similarly Br_2 is isomorphous with $(SCN)_2$.

3. Addition to ethylenic double bond. Like halogens pseudohalogens also add to ethylenic double bond linkage. For example :

$$H_2C=CH_2 + Cl_2 \rightarrow \underset{Cl}{H_2C}-\underset{Cl}{CH_2};\quad H_2C=CH_2 + (SCN)_2 \longrightarrow \underset{SCN}{H_2C}-\underset{SCN}{CH_2}$$

4. Reaction with alkalies. Like halogens, pseudohalogen also react with alkalies. For example :

(*i*) $(SCN)_2 + 2KOH$ (*cold and dilute*) $\longrightarrow KSCN + KOSCN + H_2O$

$Cl_2 + 2KOH$ (*cold and dilute*) $\longrightarrow KCl + KOCl + H_2O$

(*ii*) $3(SCN)_2 + 6KOH$ (*hot and concentrated*) $\longrightarrow 5KSCN + KSCNO_2 + 3H_2O$

$3Cl_2 + 6KOH$ (*hot and concentrated*) $\longrightarrow 5KCl + KClO_3 + 3H_2O$

5. Formation of monobasic hydracids. Like halogen pseudohalogens also combine with H_2 to form monobasic hydracids.

$$Cl_2 + H_2 \longrightarrow 2HCl\ ;\qquad (CN)_2 + H_2 \longrightarrow 2HCN$$

Dissimilarity between halogens and pseudohalogens

Pseudohalogens undergo polymerisation and form polymerised series. For example :

$$n(CN)_2 \xrightarrow{500°C} 2(CN)_n\ ;\quad n(SCN)_2 \xrightarrow{\text{Room temp.}} 2(SCN)_n$$

Halogens have no tendency to undergo polymerisation.

Some pseudohalogens

1. Cynogen, $(CN)_2$ or C_2N_2

This compound was discovered by Gay-Lussac in 1815.

Preparation. (*i*) It is formed by heating the cyanides of Hg (II), Ag (I) and Au (III). $Hg(CN)_2$ is heated to dull redness in a hard glass tube to get $(CN)_2$ gas which is collected over Hg.

$$Hg(CN)_2 \xrightarrow{\text{Dull redness}} Hg + (CN)_2$$

The yield of $(CN)_2$ obtained in this way is not quantitative, since $(CN)_2$ formed polymerises (at 500°) to insoluble paracynogen, $(CN)_x$.

$$x(CN)_2 \xrightarrow[500°C]{\text{Polymerisation}} 2(CN)_x$$

$(CN)_2$ is set free at a lower temperature and is obtained in better yield by heating a mixture of $Hg(CN)_2$ and $HgCl_2$.

$$Hg(CN)_2 + HgCl_2 \longrightarrow Hg_2Cl_2 + (CN)_2$$

(*ii*) $(CN)_2$ can also be prepared by the action of KCN on $CuSO_4$ solution. The reaction proceeds as :

$$CuSO_4\ (solution) + 2KCN \rightarrow K_2SO_4 + Cu(CN)_2$$

$$2Cu(CN)_2 \xrightarrow{\text{decomposition}} 2CuCN + (CN)_2$$

(*iii*) $(CN)_2$ can also be prepared by heating ammonium oxalate with P_2O_5.

$$\begin{matrix} COONH_4 \\ | \\ COONH_4 \end{matrix}\ (\textit{Ammonium oxalate}) \xrightarrow[2H_2O]{\text{Heat with } P_2O_5} \begin{matrix} CONH_2 \\ | \\ CONH_2 \end{matrix}\ (\textit{Oxamide}) \xrightarrow{-2H_2O} (CN)_2$$

Properties. (*i*) It is a colourless, extremely poisonous, flamable gas (boiling point = – 21.2°) and freezes to a solid (melting point = – 27.9°). It has a smell of bitter almonds.

(*ii*) *Burning in air.* It burns in air with a violet coloured flame to give CO_2 and N_2. During this reaction explosion takes place.

$$(CN)_2 + 2O_2 \longrightarrow 2CO_2 + N_2$$

(*iii*) *Polymerisation.* Although pure $(CN)_2$ is stable, the pure gas undergoes polymerisation on heating to 500° to an insoluble solid paracyanogen, $(CN)_n$ which, at 800°, regenerates $(CN)_2$ but decomposes above this temperature.

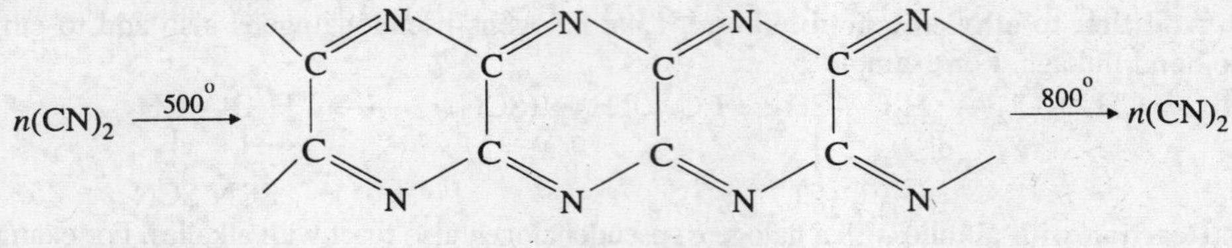

Paracynogen, $(CN)_n$

(*iv*) *Hydrolysis.* It is soluble in water and undergoes slow hydrolysis to give a number of products, the main products being ammonium formate, urea, oxamide and ammonium oxalate.

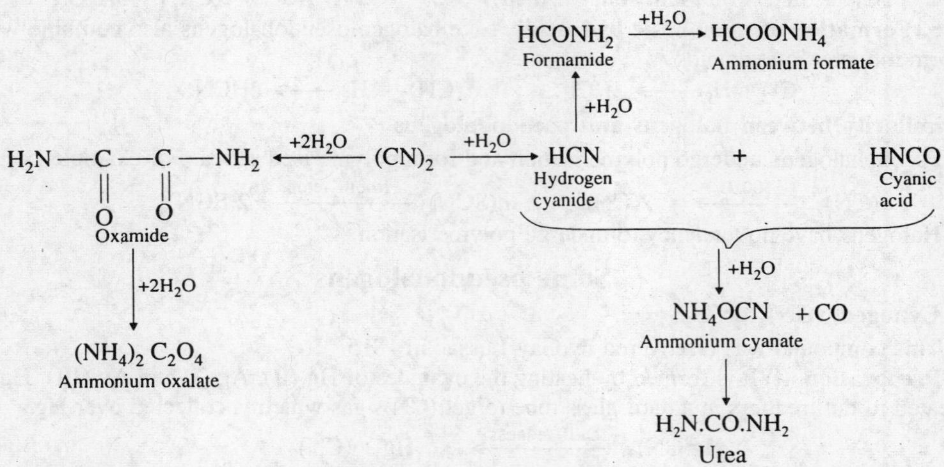

(*v*) *Action of alkalies.* It dissolves in alkalies (*e.g.*, KOH) to form the cyanide (KCN) and oxy-cyanide (KCNO).

$$(CN)_2 + 2KOH \longrightarrow KCN + KCNO + H_2O$$

(*vi*) *Action of alkali metals.* It combines directly with alkali metals to form their cyanides.

$$2K + (CN)_2 \longrightarrow 2KCN$$

(*vii*) *Addition reactions.* A number of reactions of $(CN)_2$ involve addition across one or both of the multiple bonds in its molecule. Thus, H_2S gives thiocynaformide and dithio-oxamide while aqueous NH_2OH gives oxaldihydroxamide.

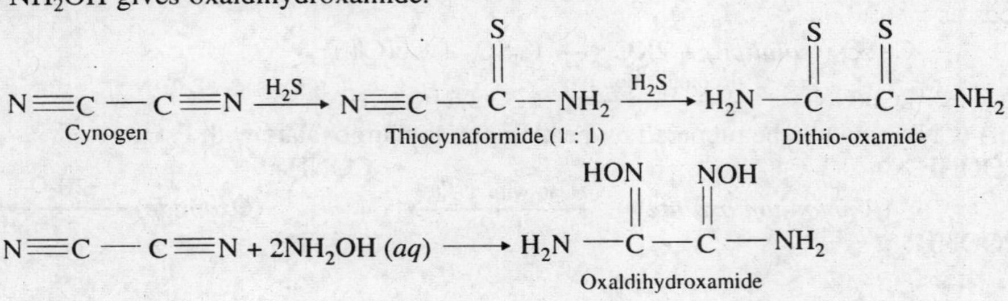

(viii) Reduction by hydrogen. When $(CN)_2$ is reduced by H_2, both the triple bonds are converted into single bonds by taking up hydrogen atoms and thus ethylenediamine, $H_2N—CH_2—CH_2—NH_2$ is formed.

$$N \equiv C — C \equiv N + 4H_2 \longrightarrow H_2N — CH_2 — CH_2 — NH_2$$

Under some conditions HCN is formed : $(CN)_2 + H_2 \longrightarrow 2HCN$

(ix) Flourination. When $(CN)_2$ is fluorinated by AgF_2 at ~ 110°, it gives $CF_2(N)_2(I)$, although other products may also be obtained.

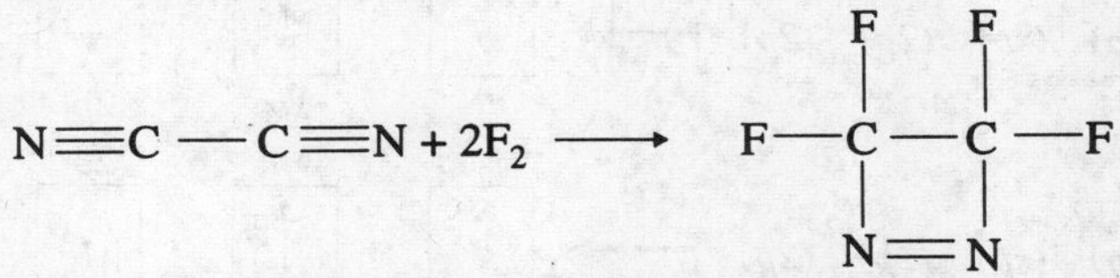

(x) Other reactions. Some other reactions of $(CN)_2$ are givn below :

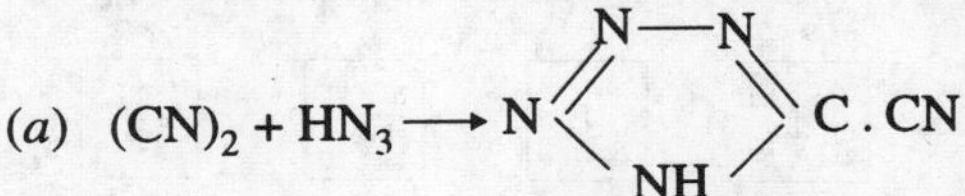

(5-cyanotetrazole)

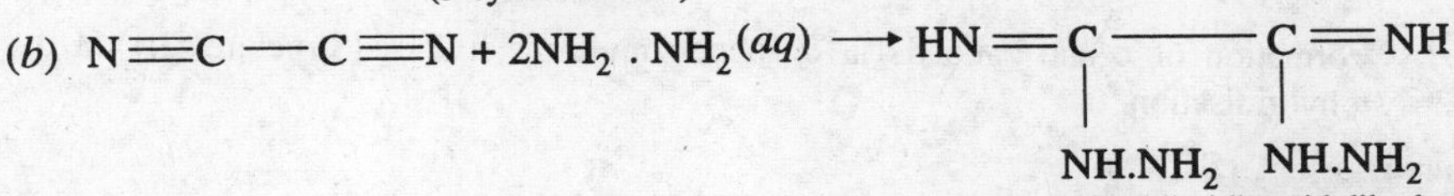

(Oxaldimidic acid dihydrazide)

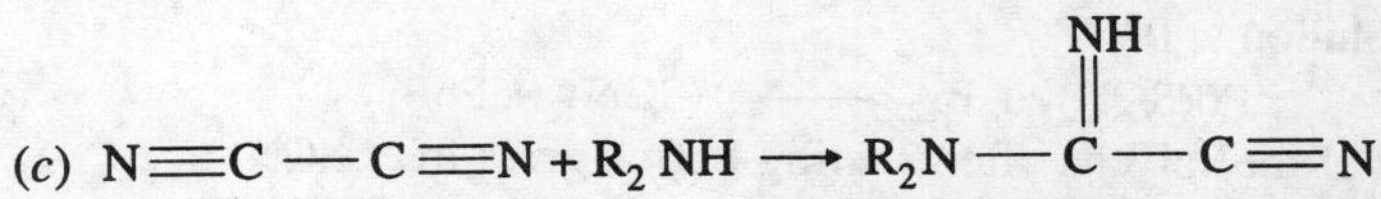

(N, N-dialkyl cyanotormamidine)

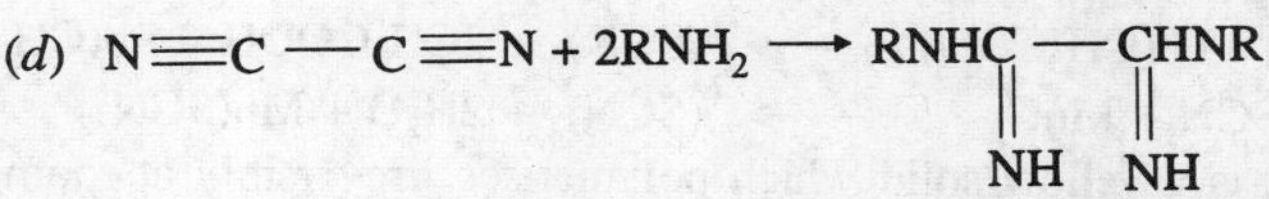

(Dialkyl oxamidines)

Uses. It is used in organic synthesis and as poisonous gas in wars.

Structure. The vapour density of the gas is 26 which shows that its formula is C_2N_2. The formation of ethylene diamine, $H_2N—CH_2—CH_2—NH_2$ on reduction and of oxamide, $H_2N—\underset{O}{\overset{C}{\|}}—\underset{O}{\overset{C}{\|}}—NH_2$ during hydrolysis shows that the order of atoms in cyanogen molecule should be N—C—C—N which is linear and symmetrical. Thus the structure of $(CN)_2$ is $\ddot{N} \equiv C—C \equiv \ddot{N}$. The N—C and C—C bond lengths which are equal to 1.16 Å and 1.37 Å respectively show that the true structure of the molecule is best represented as a resonance hybrid of the following three structures, structure (I) being the most contributing form.

$$\underset{(I)}{:N\equiv C — C\equiv N:} \longleftrightarrow \underset{(II)}{:\overset{+}{N}=C=C=\overset{-}{\ddot{N}}:} \longleftrightarrow \underset{(III)}{:\overset{-}{\ddot{N}}=C=C=\overset{+}{N}:}$$

In structure (I) both the carbon and both the nitrogen atoms are *sp* hybridised and each carbon and nitrogen pair form two π-bonds is between them. One of the two *sp*-hybrids on each N atoms contains a lone pair of electrons and other *sp*-hybrid orbitals has one electron. This singly-occupied *sp*-hybrid on each N atom overlaps with *sp*-hybrid on carbon atom to form *sp* — *sp* N — C σ-bond. *sp*-hybrid orbitals on each carbon atom overlap to form *sp* — *sp* C — C σ-bonds and four N — C *p* — *p* π-bonds, as shown in Fig. 29.3. Each N atom has a lone pair of electrons, residing in one of the two *sp*-hybrid orbitals.

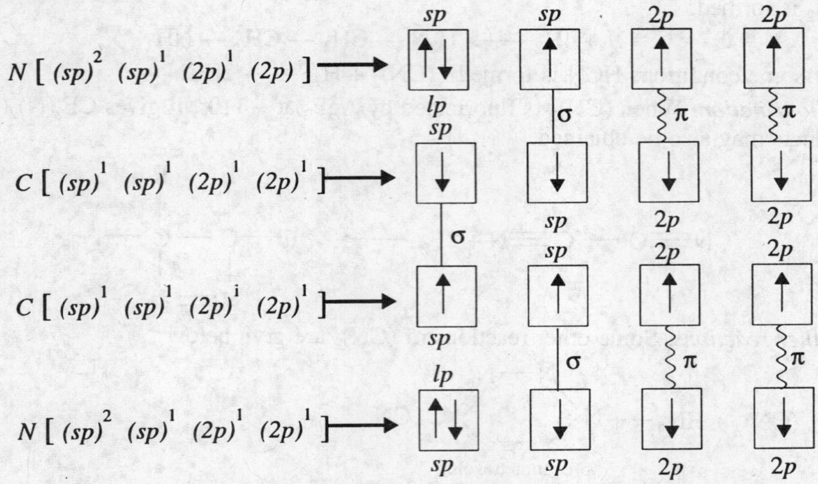

Fig. 29.3 : Formation of σ-and π-bonds in cyanogen molecule, $(CN)_2$ [Structure (I)] on the basis of *sp*-hybridisation.

2. Thiocyanogen, $(SCN)_2$

Preparation. It may be prepared : (*i*) by treating lead thiocyanate, $Pb(SCN)_2$ or AgSCN with Br_2 in the real solution at 0°C.

$$Pb(SCN)_2 + Br_2 \longrightarrow (SCN)_2 + PbBr_2 \downarrow$$

Evaporation of the solvent, after filtering $PbBr_2$, gives free $(SCN)_2$.

(*ii*) by electrolysing an alcoholic solution of NH_4SCN or KSCN.

(*iii*) by the oxidation of HSCN by $Pb(CH_3COO)_4$ or MnO_2.

$$2HSCN + Pb(CH_3COO)_4 \longrightarrow (SCN)_2 + 2CH_3COOH + Pb(CH_3COO)_2$$

$$4HSCN + MnO_2 \longrightarrow (SCN)_2 + 2H_2O + Mn(SCN)_2$$

Properties. (*i*) It is a yellow solid which polymerises irreversibly at room temperature to a red-brick amorphous solid which is parathiocyanogen, $(SCN)_n$ and insoluble in water.

(*ii*) It is most stable in CCl_4 or CH_3COOH solution in which it exists as NCSSCN.

(*iii*) It is rapidly hydrolysed by water to form thiocyanic acid (HSCN) and hydrocyanic acid (HCN).

(*iv*) Following reactions of $(SCN)_2$ indicate its oxidising property and similarities with halogens.

(*a*) A solution of $(SCN)_2$ in ether or in CS_2 liberates I_2 from an iodide.

$$(SCN)_2 + 2I^- \longrightarrow 2SCN^- + I_2$$

(*b*) It oxidises Cu (I) salts to Cu (II) salts.

$$(SCN)_2 + 2CuSCN \longrightarrow 2Cu^{2+} + 4SCN^-$$

(*c*) It combines with metals to form metallic salts.

$$(SCN)_2 + M(M = 0) \longrightarrow M(SCN)_2\ (M = +2)$$

(*d*) It oxidises thiosulphate ($S_2O_3^{2-}$) into tetrathionate ($S_4O_6^{2-}$).

$$(SCN)_2 + 2S_2O_3^{2-} \longrightarrow 2SCN^- + S_4O_6^{2-}$$

(*e*) Sulphur is precipitated from H_2S.

$$(SCN)_2 + H_2S \longrightarrow 2H^+ + 2SCN^- + S\downarrow$$

(*f*) Arsenite (AsO_3^{3-}) is oxidised to arsenate (AsO_4^{3-}).

$$(SCN)_2 + AsO_3^{3-} + H_2O \longrightarrow 2SCN^- + 2H^+ + AsO_4^{3-}.$$

(*v*) Some other reactions shown by $(SCN)_2$ can be represented by the following chemical equations :

(*a*) *Cold solution of* $(SCN)_2 + NO \longrightarrow$ NO.SCN (Nitrosyl thiocyanate)

(*b*) *1M cold solution of* $(SCN)_2 \xrightarrow{2HCl} (SCN)_2.\ 2HCl$ (*Dihydrochloride*) $\xrightarrow[(-2HCl)]{+H_2O}$ $(SCN)_2.H_2O$ (*Monohydrate*).

Uses. $(SCN)_2$ is used (*i*) as an oxidising agent (*ii*) for the determination of unsaturation in organic compounds.

Structure. $(SCN)_2$ molecule has tautomeric structures namely (*a*) and (*b*) shown below :

$$N\equiv C-S-S-C\equiv N$$

(*a*) Symmetrical structure

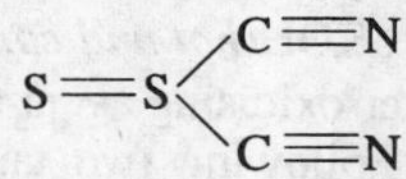

(*b*) Unsymmetrical structure

X-ray evidence points to a linear arrangement of S—C—N atoms and hence structure (*a*) is a symmetrical structure.

3. Selenocyanogen, $(SeCN)_2$

Preparation. This helegenoid may be prepared as follows :

(*i*) By the reaction of I_2 with silver selenocyanate, AgSeCN dissoved in ether.

$$2AgSeCN + I_2 \longrightarrow (SeCN)_2 + 2AgI\downarrow$$

(*ii*) When $Pb(CH_3COO)_4$ dissolved in $CHCl_3$ is treated with KSeCN in anhydrous acetone, $Pb(SeCN)_4$ is formed which is unstable and hence decomposes to give $Pb(SeCN)_2$ and $(SeCN)_2$.

$$Pb(CH_3COO)_4 + 4KSeCN \longrightarrow Pb(SeCN)_4 + 4CH_3COOK;\ Pb(SeCN)_4 \longrightarrow Pb(SeCN)_2 + (SeCN)_2$$

Properties. It is a yellow crystalline powder which turns red on standing. When kept dry and under pressure, it is quite stable. It is soluble in C_6H_6, $CHCl_3$ or CCl_4. On heating a solution of $(SeCN)_2$ in CS_2 under reflux, polyerisation occurs and crystalline compounds namely $Se_3(CN)_2$ and $Se(CN)_2$ are obtained. These compounds can be separated from each other on cooling. $(SeCN)_2$ reacts with H_2O forming hydrosentic acid, HSeCN.

$$2(SeCN)_2 + 3H_2O \longrightarrow H_2SeO_3 + 3HSeCN + HCN$$

It is slightly weaker in its oxidising properties than I_2.

Structure. Its molecular weight in C_6H_6 agrees with its dimeric formula, $(SeCN)_2$. Its structure is probably the same as that of $(SCN)_2$. Raman spectral studies have shown that $(SeCN)_2$ molecule has a linear structure which can be represented as $N\equiv C-\ddot{\underset{\cdot\cdot}{Se}}-\ddot{\underset{\cdot\cdot}{Se}}-C\equiv N$

4. Oxy-cyanogen, $(OCN)_2$

Preparation. (*i*) Lidov prepared it by the action of H_2O_2, CuO or sodium hypobromite on KOCN.

$$4KOCN + 2H_2O_2 \longrightarrow 2K_2CNO_2 + (OCN)_2 + 2H_2O$$

(*ii*) $(OCN)_2$ can also be prepared by the action of AgOCN on I_2 in CS_2 or CCl_4 solution.

$$2AgOCN + I_2 \longrightarrow 2AgI + (OCN)_2$$

(*iii*) $(OCN)_2$ can also be prepared by the action of cyanogen bromide (CNBr) on Ag_2O.

$$2CNBr + Ag_2O \longrightarrow 2AgBr + (OCN)_2$$

Properties. $(CNO)_2$ shows typical properties of pseudohalogens. It liberates I_2 from an iodide solution, reacts with metals, forms interpseudohalogen compounds.

5. Azido carbon disulphide, $(SCSN_3)_2$

Preparation. It may be prepared by oxidising potassium azidodithio carbonate, $KSCSN_3$ with $H_2O_2.I_2$, $FeCl_3$ etc.

$KSCSN_3$ required for the purpose is obtained by the interaction of KN_3 and CS_2 at 40°C.

$$KN_3 + CS_2 \longrightarrow KSCSN_3$$

$$2KSCSN_3 + H_2O_2 \longrightarrow (SCSN_3)_2 + 2KOH$$

$$2KSCSN_3 + I_2 \longrightarrow (SCSN_3)_2 + 2KI$$

$$2KSCSN_3 + 2FeCl_3 \longrightarrow (SCSN_3)_2 + 2FeCl_2 + 2KCl$$

Properties. It is a white crystalline solid and slightly soluble in water. It is unstable and decomposes violently automatically to give N_2, sulphur and $(SCN)_2$.

$$(SCSN_3)_2 \longrightarrow 2N_2 + 2S + (SCN)_2$$

It reacts slowly with dilute acid, and rapidly with concentrated acids to liberate sulphur. However, with HNO_3 no sulphur is precipitated.

In reaction with an alkali is resembles Cl_2 as shown below :

(*a*) $(SCSN_3)_2 + 2KOH$ (*cold and dilute*) $\longrightarrow KSCSN_3 + KOSCSN_3 + H_2O$

$Cl_2 + 2KOH$ (*cold and dilute*) $\longrightarrow KCl + KOCl + H_2O$

(*b*) $2(SCSN_3)_2 + 6KOH$ (*hot and concentrated*) $\longrightarrow 5KSCSN_3 + KSCSN_3O_3 + 3H_2O$

$3Cl_2 + 6KOH$(*hot and concentrated*) $\longrightarrow 5KCl + KClO_3 + 3H_2O$

It is a weaker oxidising agent than I_2.

Structure. Following two tautomeric forms have been suggested for the molecule. (I) is probably more likely.

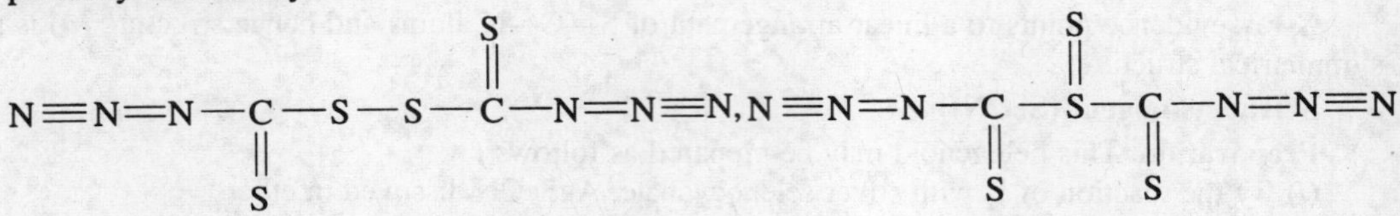

(I) (II)

Fluorides of xenon

The stable fluorides formed by xenon are xenon difluoride (XeF_2), xenor tetrafluoride (XeF_4) and xenon hexafluoride (XeF_6).XeF_2 and XeF_4 are more stable than XeF_6. In these fluorides Xe is in + 2, + 4, and + 6 oxidation state respectively.

Preparation of xenon fluorides

(*i*) These are generally prepared by mixing Xe and F_2 in different ratio in different conditions. For example :

$$\underbrace{Xe + F_2}_{2\,:\,1\text{ ratio}} \xrightarrow[\text{tube (400°C)}]{\text{Sealed Ni}} XeF_2 \ ; \quad \underbrace{Xe + F_2}_{2\,:\,1\text{ ratio}} \xrightarrow[\text{discharge}]{\text{Electric}} XeF_2$$

$$Xe + F_2 \xrightarrow[\text{(Hg vapours)}]{\text{Photochemical combination}} XeF_2 \ ; \quad \underbrace{Xe + 2F_2}_{1\,:\,5\text{ ratio}} \xrightarrow[\text{400°C, 5-6 atm}]{\text{Ni tube}} XeF_4$$

$$\underbrace{Xe + 2F_2}_{1\,:\,2\text{ ratio}} \xrightarrow[-\,80°C]{\text{Electric discharge}} XeF_4 \ ; \quad \underbrace{Xe + 3F_2}_{1\,:\,20\text{ ratio}} \xrightarrow[\text{250-300°C, 50-60 atm}]{\text{Ni vessel}} XeF_6$$

$$\underbrace{Xe + 3F_2}_{1\,:\,3\text{ ratio}} \xrightarrow[\text{low temp.}]{\text{Electric discharge}} XeF_6$$

(*ii*) The fluorides can also be prepared by the fluorination methods. For example :

(*a*) XeF_2 can be prepared by the fluorination of Xe by oxygen monofluoride, O_2F_2 at –178° C.

$$Xe + O_2F_2 \xrightarrow{-178°} XeF_2 + O_2$$

(*b*) XeF_6 can also be prepared by the fluorination of XeF_4 by O_2F_2 or by F_2 under pressure :

$$XeF_4 + O_2F_2 \longrightarrow XeF_6 + O_2 \ ; \qquad XeF_4 + F_2 \longrightarrow XeF_6$$

Properties of fluorides

(*i*) **Physical state.** All the three fluorides are colourless crystalline solids whose melting points. are as XeF_2 = 140°C, XeF_4 = 117.1°C, XeF_6 = 49.5°C. XeF_4 sublimes readily.

(*ii*) **Hydrolysis.** (*a*) XeF_2 is reduced by H_2O to Xe and H_2O is oxidised to O_2.

$$2XeF_2 + 2H_2O \longrightarrow 2Xe + 4FH + O_2$$

XeF_2 is rapidly hydrolysed by an aqueous solution of a base.

$$2XeF_2 + 4OH^- \longrightarrow 2Xe + 4F^- + 2H_2O + O_2$$

(*b*) XeF_4 undergoes disproportionation in water, giving XeO_3 which is a highly explosive solid.

$$2XeF_4 + 3H_2O \longrightarrow Xe\,(g) + XeO_3\,(s) + 6HF\,(aq) + F_2\,(g)$$

If the reaction is carried out at – 80° C, $XeOF_2$ is formed. In this reaction, slow and partial hydrolysis of XeF_4 takes place.

$$XeF_4 + H_2O \xrightarrow{-80^\circ} XeOF_2 + 2HF$$

(*c*) XeF_6 undergoes hydrolysis in water. The final product obtained is XeO_3 (an explosive solid).

$$XeF_6 + H_2O \longrightarrow XeOF_4 + 2HF\,; \qquad XeF_6 + 2H_2O \longrightarrow XeO_2F_2 + 4HF$$

$$XeF_6 + 3H_2O \longrightarrow XeO_3 + 6HF$$

In strongly basic solution, the reaction proceeds as :

$$XeF_6 + 7OH^- \longrightarrow 6F^- + 3H_2O + \underset{(Xe\,=\,+6)}{HXeO_4^-}\ (\textit{xenate ion}) \xrightarrow{+2OH^-} \frac{6}{4}H_2O + \frac{1}{4}Xe + \frac{3}{4}\underset{(Xe\,=\,+8)}{XeO_6^{-4}}\ (\textit{perxenate ion}).$$

In acidic solution, the main species is $XeOF_4$ (Xe = + 6) and H_6XeO_6 (Xe = + 6). H_6XeO_6 is called xenic acid.

(*iii*) **Fluorinating properties.** (*a*) XeF_2 is a mid fluorinating agent. Thus it reacts with C_6H_6 to give C_6H_5F.

(*b*) XeF_4 is a stronger fluorinating agent than XeF_2. Thus it converts Hg to HgF_2, Pt to PtF_4, NO to NOF, NO_2 to NO_2F, SF_4 to SF_6, BCl_3 to BF_3 etc.

$$2Hg + XeF_4 \longrightarrow Xe + 2HgF_2\,; \qquad Pt + XeF_4 \longrightarrow Xe + PtF_4$$

$$2SF_4 + XeF_4 \longrightarrow Xe + 2SF_6\,; \qquad 4BCl_3 + 3XeF_4 \longrightarrow 3X_3 + 4BF_3 + 6Cl_2$$

(*iv*) **Oxidising properties.** All the three fluorides can oxidise many substances and are themselves reduced to xenon. For example :

(*a*) $XeF_2 + H_2 \longrightarrow Xe + 2HF\,;$ $\qquad 2XeF_2 + 2H_2O \longrightarrow 2Xe + 4HF + O_2$

$XeF_2 + I_2 \xrightarrow{BF_3} Xe + 2IF\,;$ $\qquad XeF_2 + 2SO_3 \longrightarrow Xe + S_2O_6F_2$

$XeF_2 + 2HCl \longrightarrow Xe + 2HF + Cl_2\,;$ $\qquad XeF_2 + 2Ce^{3+} \longrightarrow 2Ce^{4+} + Xe + 2F^-$

(*b*) $XeF_4 + 2H_2 \longrightarrow Xe + 4HF;$ $\qquad 3XeF_4 + 4BCl_3 \longrightarrow 3X_3 + 4BF_3 + 6Cl_2$

$XeF_4 + 4I^- \longrightarrow Xe + 4F^- + 2I_2$

(*c*) $XeF_6 + 3H_2 \longrightarrow Xe + 6HF$

XeF_6 can also oxidise NH_3 (N = –3) to N_2 (N = 0) and HCl (Cl = –1) to Cl_2 (Cl = 0). In these reactions XeF_6 is reduced to Xe.

$$XeF_6 + 8NH_3 \longrightarrow Xe + 6NH_4F + N_2\,; \qquad XeF_6 + 6HCl \longrightarrow X_3 + 7HF + 3C_{12}$$

(*v*) **Formation of adducts.** All the three fluorides give adducts (addition compounds) with F^- ion acceptor fluorides like PF_5, AsF_5, SbF_5 etc.

(*a*) In BrF_3 solution, XeF_2 forms adducts with MF_5 molecules (M = P, As, Sb etc.).

$$XeF_2 + MF_5 \longrightarrow XeF_2.MF_5\,; \qquad XeF_2 + 2MF_5 \longrightarrow XeF_2.2MF_5$$

$$2XeF_2 + MF_5 \longrightarrow 2XeF_2.MF_5.$$

(*b*) XeF_4 dissolves in molten SbF_5 and gives the addition compound, $XeF_4.SbF_5$ which is an ionic compound and hence is represented as $[XeF_3]^+\,[SbF_6]^-$.

$$XeF_4 + SbF_5 \longrightarrow XeF_4.\,SbF_5 \longrightarrow [XeF_3]^+\,[SbF_6]^-$$

XeF_4 dissolves in molten TaF_5, giving a straw coloured compound, $Xe(TaF_6)_2$ which is discribed as an addition compound, $XeF_2.2TaF_5$.

$$XeF_4 + 2TaF_5 \longrightarrow Xe(TaF_6)_2 \quad \text{or} \quad XeF_2.2TaF_5 + F_2$$

(*c*) XeF_6 gives addition compound with AsF_5 and SbF_5 molecules.

$$XeF_6 + MF_5 \longrightarrow XeF_6.MF_5 \longrightarrow [XeF_5]^+ [MF_6]^-$$
(M = As, Sb)

In these reactions XeF_6 molecule acts as F^- ion donor while MF_5 molecule behaves as F^- ion acceptor.

XeF_6 also reacts with alkali metal fluorides (except LIF) and gives species containing XeF_7^- and XeF_8^{2-} ions. Thus in these reactions XeF_6 molecule acts as F^- ion acceptor and MF molecule (M = Na, K, Rb, (*s*) behaves as F^- ion donor.

$$XeF_6 + 2MF \longrightarrow M_2XeF_8 \text{ (Octafluoro-xenate)}$$
(M = Na, K)

$$XeF_6 + 2RbF \xrightarrow{200° C} 2Rb\,XeF_7 \xrightarrow{50° C} Rb_2\,XeF_8 + XeF_6$$
molten (Heptalfuoroxenate) (Octafluoroxenate)

(*vi*) **Action of HF.** XeF_2 and XeF_4 both dissolve in HF. On the other hand, XeF_6 dissolves in HF, giving a solution which is a good conductor of electricity. The electrical conductivity of the solution is because of the formation fo XeF_5^+ and HF_2^-.

$$XeF_6 + HF \longrightarrow XeF_5^+ + HF_2^-$$

(*vii*) **Heat of formation, Δ H*f*.** All the three fluorides are exothermic compounds and have (excepting XeF_2) large negative ΔHf values as shown $XeF_2 = -125.5$ kJ mol^{-1}, $XeF_4 = -410.0$ kJ mol^{-1}, $XeF_6 = -401.7$ kJ mol^{-1}.

(*viii*) **Some more properties of XeF_6.** (*a*) XeF_6 is the most volatile of all the fluorides. Its vapours have a greenish yellow colour.

(*b*) XeF_6 is extremely reactive. It cannot be stored in glass or quarts vessels, since it readily reacts with SiO_2 present in glass as one of its constituents. With SiO_2, XeF_6 gives dangerously explosive XeO_3 as a final product.

$$2XeF_6 + 3SiO_2 \longrightarrow 2XeO_3 + 3SiF_4$$

Structure and geometry of Xenon fluorides

The structure and shape (geometry) of XeF_2, XeF_4 and XeF_6 molecules have already been explained in chapter 14.

Oxy-fluorides of xenon

Xenon forms the following oxy-fluorides. $XeOF_2$ (xenon oxy-difluoride), $XeOF_4$ (xenon oxy-tetrafluoride), XeO_2F_2 (xenon dioxy-difluoride), $XeO_3\,F_2$ (xenon trioxy-difluoride) and XeO_2F_4 (xenon dioxy-tetrafluoride).

1. Xenon oxy-difluoride, $XeOF_2$ (Xe = +4)

It is formed by the partial and slow hydrolysis of XeF_4 at – 80°C.

$$XeF_4 + H_2O \xrightarrow{-80° C} XeOF_2 + 2HF$$

It is not a very stable compound.

2. Xenon oxy-tetrafluoride, $XeOF_4$ (Xe = +6)

It is obtained by the partial hydrolysis of XeF_6 with H_2O and by its action on SiO_2.

$$XeF_6 + H_2O \longrightarrow XeOF_4 + 2HF \; ; \qquad 2XeF_6 + SiO_2 \longrightarrow XeOF_4 + SiF_4$$

Properties. (*i*) It is a colourless volatile liquid which solidifies at about 245 K.

(*ii*) It reacts with SiO_2, forming $XeO_2\,F_2$.

$$2XeOF_4 + SiO_2 \longrightarrow XeO_2\,F_2 + SiF_4$$

It is due to this reaction that $XeOF_4$ is not kept in glass vessels, but in nickel vessels.

(*iii*) It is reduced to Xe by H_2

$$XeOF_4 + 3H_2 \longrightarrow Xe + H_2O + 4HF$$

(*iv*) $XeOF_4$ reacts with H_2O, giving XeO_3 as the final product.

$$XeOF_4\,(Xe = +6) + 2H_2O \longrightarrow XeO_3\,(Xe = +6) + 4HF$$

3. Xenon dioxy-difluoride, XeO_2F_2 (Xe = +6)

It is prepared by the action of SiO_2 on $XeOF_4$ or by hydrolysing it.

$$2XeOF_4 + SiO_2 \longrightarrow 2XeO_2F_2 + SiF_4\ ; \quad XeOF_4 + H_2O \longrightarrow XeO_2F_2 + 2HF$$

XeO_2F_2 is a colourless solid and melts at 29.8° C. It is easily hydrolysed by H_2O, giving XeO_3.

$$XeO_2F_2 + H_2O \longrightarrow XeO_3 + 2HF$$

Structure of xenon oxy-fluorides

The structure/geometry of $XeOF_2$, $XeOF_4$, XeO_2F_2, XeO_3F_2 and XeO_2F_4 molecules has already been explained in chapter 15.

University Questions

1. Explain the difference in the nature of bonding in LiF and LiI. *(I.I.T. 96)*
2. Which of the following statement is correct for $CsBr_3$?
   (*a*) It is a covalent compound.
   (*b*) It contains Cs^{3+} and Br^- ions
   (*c*) It contains Cs^+ and Br^- ions
   (*d*) It contains Cs^+, Br^- and lattice of Br_2 molecule. *(I.I.T., 96)*
3. Gradual addition of KI solution to $Bi(NO_3)_3$ solution initially produces a dark brown ppt., which dissolves in excess of KI, to give a clear yellow solution. Write chemical equations for the above reactions. *(IIT., 96)*
4. Complete the following chemical equation :
   $KI + Cl_2 \longrightarrow$ ...... *(I.I.T., 96)*
5. Discuss the structure of XeF_6. *(Usmania, 93)*
6. Give an account of the halides formed by S, Se, and Te. *(Madras, 86)*
7. **What are halides ? Discuss the properties of hydrides of halogen family in gaseous and aqueous solution with particular reference to acidic character.** ***(VBS Purvanchal 2002)***
8. **Write a note on Inter halogen compounds** ***(VBS Purvanchal 2002)***
9. **Explain why HCl in the gaseous state is covalent while in the aqueous solution it is ionic.** ***(VBS Purvanchal 2002)***
10. **Give an account of preparation and properties of xenon fluorides.** ***(VBS Purvanchal 2002)***
11. **(*a*) Write a note on pseudo halogens and pseudo halides.**
    (*b*) Explain why HF is a liquid while HCl is a gas. ***(Delhi 2002)***
12. **Explain why inter halogens are more reactive than halogens.** ***(VBS Purvanchal 2003)***
13. **Draw the structure of XeF_2, XeF_4 and XeF_6 and mention the type of hybridisation.** ***(VBS Purvanchal 2003)***
14. **Write a note on Fluorides of xenon.** **(Rohilkhand 2003)**
15. **Explain why HF is a weak acid as compared to HCl in water.** ***(Rohilkhand 2003)***
16. **(*i*) Write a short note on (*a*) Interhalogen compounds and polyhalides (*b*) Fluorides of xenon.**
    (*ii*) HF is a liquid whereas HCl, HBr and HI are gases. Explain. ***(Rohilkhand 2003)***
17. **(*a*) Give one method for the preparation and structure of XeF_2.**
    (*b*) What happens when XeF_6 reacts with HCl. ***(Delhi 2003)***
18. **(*a*) What are polyhalides ? How are they classified ? Discuss the structure and bonding in ICl_2^- ion.**
    (*b*) Discuss the structure and bonding in xenon difluoride and xenon hexafluoride.
    (*c*) Draw the structure of ICl_3 and BrF_3. ***(Nagpur 2003)***
19. **(*a*) Discuss the compound formation by inert gases under different conditions.**
    (*b*) Describe the hybridization involved and geometry of XeF_4.
    (*c*) Give general methods of preparation of interhalogens.
    (*d*) Explain back bonding in BF_3. ***(Nagpur 2003)***
20. Write a note on interhalogen compounds. *(Bhopal 2004)*
2. (*a*) Arrange the hydrogen halides in order of their increasing acidity, taking into account all the terms involved. (*b*) What are pseudohalogens? *(GND Amritsar, 2004)*

30

Oxy-acids of Boron (Boric acids) and their Salts

Boron forms several oxy-acids, which are called boric acids. All these acids may be regarded as having been obtained by adding one or more H_2O molecules to one or more B_2O_3 molecules, *e.g.*,

(*i*) $B_2O_3 \xrightarrow{+3H_2O} 2H_3BO_3$ or $2B(OH)_3$ (*Boric acid*, also called *orthoboric acid*)

(*ii*) $B_2O_3 \xrightarrow{+H_2O} 2HBO_2$ (*Metaboric acid*)

(*iii*) $2B_2O_3 \xrightarrow{+3H_2O} H_6B_4O_9$ (*Pyroboric acid*)

(*iv*) $2B_2O_3 \xrightarrow{+H_2O} H_2B_4O_7$ (*Tetraboric acid*)

(*v*) $3B_2O_3 \xrightarrow{+2H_2O} H_4B_6O_{11}$ (*Hexaboric acid*)

(*vi*) $4B_2O_3 \xrightarrow{+3H_2O} H_6B_8O_{15}$ (*Octaboric acid*)

Here we shall discuss orthoboric acid (H_3BO_3), metaboric acid (HBO_2) and pyroboric acid ($H_6B_4O_9$).

Boric acid, H_3BO_3 or $B(OH)_3$

Preparation. (*i*) *From borax, $Na_2B_4O_7$.* See "Extraction of Boron".

(*ii*) *From colemanite ore, $Ca_2B_6O_{11}$.* See "Extraction of Boron"

(*iii*) H_3BO_3 can also be prepared by the action of superheated water on boron nitride (BN) and boron sulphide (B_2S_3).

$$BN + 3H_2O \longrightarrow H_3BO_3 + NH_3 \quad ; \quad B_2S_3 + 6H_2O \longrightarrow 2H_3BO_3 + 3H_2S$$

(*iv*) H_3BO_3 is also obtained by the hydrolysis of BCl_3.

$$BCl_3 + 3H_2O \longrightarrow H_3BO_3 + 3HCl$$

Properties. 1. It forms soft pearly–white, needle like crystals, which have a greasy feel.

2. It is moderately soluble in H_2O. The dissolution is accompanied by the absorption of heat and hence its solubility increases with the increase in temperature.

3. *Acidic nature.* H_3BO_3 is a very weak acid and ionises as a monobasic acid and that, too, to a very limited extent. Although it is not a proton donor, but behaves as a Lewis acid, *i.e.*, accepts an electron pair from O – atom of OH^- ion (obtained from the ionisation of H_2O) and thus gets attached with it by B—OH^- coordinate bond, forming the tetrahedral metaborate ion, $[B(OH)_4]^-$.

$$H_2O \rightleftharpoons H^+ + OH^- \quad ...(i)$$

$$(OH)_3B + :\ddot{O}H^- \rightleftharpoons [(OH)_3B \leftarrow OH^-] \text{ or } [B(OH)_4]^- \quad ...(ii)$$

On adding (*i*) and (*ii*), we get

$$B(OH)_3 \text{ (or } H_3BO_3) \rightleftharpoons [B(OH)_4]^- + H^+ \text{ (pK = 9.00)}$$

The above equilibrium lies to the left, showing that H_3BO_3 is a weak acid. It is also evident from this equilibrium that H_3BO_3 behaves as a monobasic acid.

4. *Action of an alkali.* H_3BO_3 reacts with the aqueous solution of an alkali (*e.g.*, NaOH, KOH) and gives various types of polymeric metaborates like $NaBO_2.2H_2O$, $NaB_3O_5.2H_2O$, $Na_2B_4O_7.10H_2O$, $KB_5O_8.4H_2O$ etc. The formation of these metaborates is illustrated by the following equations.

$B(OH)_3$ + NaOH (*dil. solution*) $\longrightarrow$ $NaBO_2.2H_2O$ or $Na^+[B(OH)_4]^-$

$3B(OH)_3$ + NaOH (*conc. solution*) $\longrightarrow$ $Na_3B_5O_5.2H_2O + 3H_2O$

$4B(OH)_3$ + 2NaOH (*excess*) + $3H_2O$ $\longrightarrow$ $Na_2B_4O_7.10H_2O$

$5B(OH)_3$ + KOH $\longrightarrow$ $KB_5O_8.4H_2O + 4H_2O$

The titration between $B(OH)_3$ and a strong alkali solution (*e.g.*, NaOH] is not possible, because sodium metaborate, $Na^+[B(OH)_4]^-$ (or $NaBO_2.2H_2O$) formed by the reaction between $B(OH)_3$ and NaOH gets hydrolysed to regenerate $B(OH)_3$ and NaOH.

$$B(OH)_3 + NaOH \longrightarrow Na^+ [B(OH)_4]^- ; \qquad Na^+ [B(OH)_4]^- \longrightarrow B(OH)_3 + NaOH$$

However, if some quantity of polyhydroxy compound, like catechol, glycecol, mannitol, sugar etc., is added to the titratium solution, the polyhydroxy compound combines with H_3BO_3 or $B(OH)_3$ and forms chelated complex compound. This complex compound gives H^+ ions, which make H_3BO_3 to behave as a strong acid. The formation of complex compound by the action of two molecules of catechol / mannitol on one molecule of H_3BO_3 can be shown as :

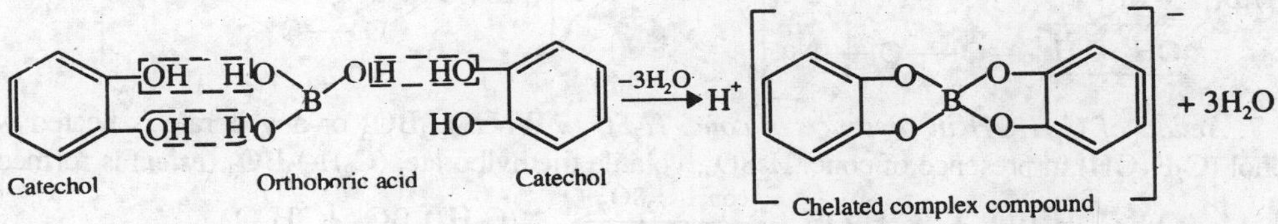

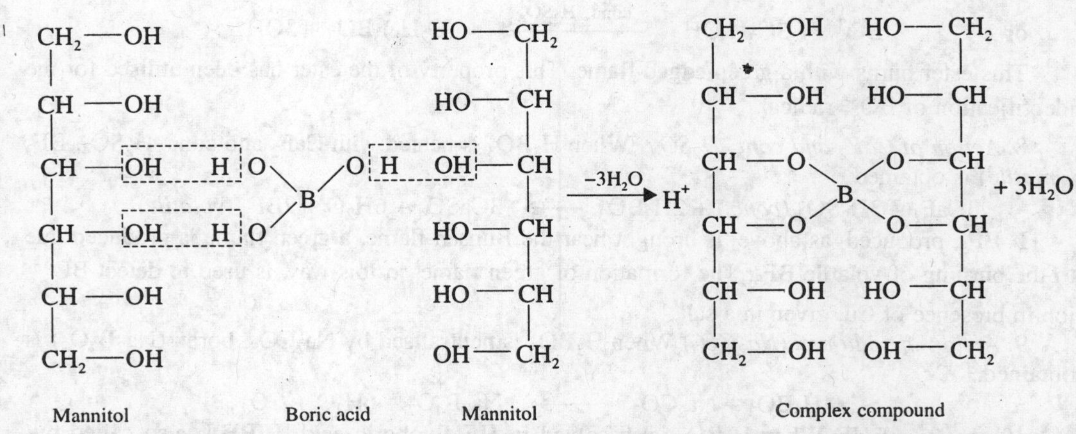

The titration between $B(OH)_3$ and strong alkali solution, in presence of polyhydroxy compound (phenolphthalein, is used as an indicator) makes the basis for the estimation of the acid and borax – phenolphthalein test of glycerol.

5. *Action of heat.* When heated to 100°C, H_3BO_3 loses H_2O to give metaboric acid (HBO_2) which, on being heated to 160°C, gives tetraboric acid ($H_2B_4O_7$). When $H_2B_4O_7$ is heated to redness, boron trioxide (B_2O_3) is obtained.

$$4H_3BO_3 \xrightarrow[-4H_2O]{100°C} 4HBO_2 \xrightarrow[-H_2O]{160°C} H_2B_4O_7 \xrightarrow[-H_2O]{\text{Red heat}} 2B_2O_3$$

6. *Action of sodium hydroperoxide, NaOOH.* When H_3BO_3 is treated with NaOOH, dihydrated sodium peroxyborate, $Na_2B_2O_6.2H_2O$ is formed.

$$2B(OH)_3 + 2NaOOH \longrightarrow Na_2B_2O_6.2H_2O + 2H_2O$$

Structure of $Na_2B_2O_6.2H_2O$. The crystal structure of this borate has shown that two H_2O molecules are a part of the anion of this borate *i.e.*, this borate is represented as $Na_2[B_2O_6.2H_2O]$. The anion, $[B_2O_6.2H_2O]^{2-}$ contains two peroxo groups (—O—O—) and four OH groups. Thus $[B_2O_6.2H_2O]^{2-}$ ion should be written as $[B_2(—O—O—)_2(OH)_4]^{2-}$.

In this ion, both B-atoms are negatively charged. The two peroxo groups make a bridge between two B-atoms and each B-atom is attached with two OH groups. Thus, the structure of $[B_2O_6.2H_2O]_2^-$ ion can be shown as :

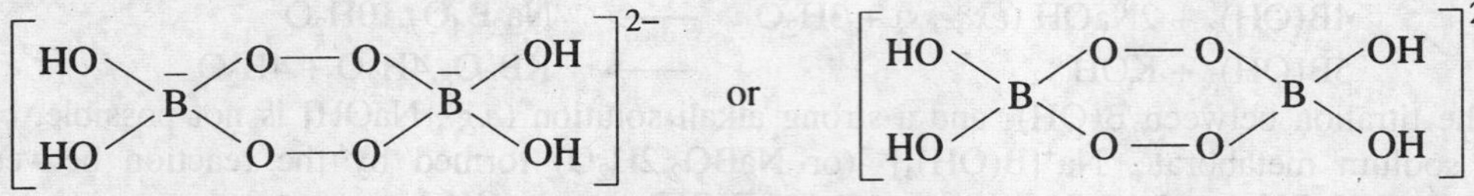

The above structure of $[B_2O_6.2H_2O]^{2-}$ confirms the fact that the formation of $Na_2^+[B_2O_6.2H_2O]^{2-}$ by the combination of two molecules of $B(OH)_3$ and two molecules of NaOOH can be shown as follows :

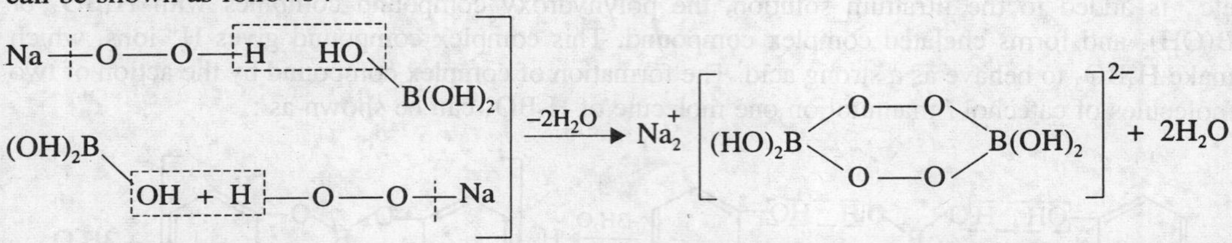

7. *Action of C_2H_5OH in presence of conc. H_2SO_4.* When H_3BO_3 or any borate is heated with alcohol (C_2H_5OH),in presence of conc. H_2SO_4, volatile triethylborate, $(C_2H_5)_3BO_3$ (*ester*) is formed.

$$H_3BO_3 + 3C_2H_5OH \xrightarrow{\text{conc. } H_2SO_4} (C_2H_5)_3BO_3 + 3H_2O$$

or

$$BO_3^{3-} + 3C_2H_5OH \xrightarrow{\text{conc. } H_2SO_4} (C_2H_5)_3BO_3 + 3OH^-$$

This ester burns with a green-edged flame. This property of the ester has been utilised for the identification of BO_3^{3-} radical.

8. *Action of CaF_2 and conc. H_2SO_4.* When H_3BO_3 is heated with CaF_2 and conc. H_2SO_4, BF_3 (*volatile*) is obtained.

$$3CaF_2 + 3H_2SO_4\ (\textit{conc.}) + 2H_3BO_3 \longrightarrow 3CaSO_4 + 6H_2O + 2BF_3\ (\textit{volatile})$$

If BF_3, produced as above, is brought near the Bunsen flame, a green flame is produced due to the burning of volatile BF_3. The formation of green flame, in this way, is used to detect BO_3^{3-} ion in presence of Cu, given in a salt.

9. *Action of soda-ash (Na_2CO_3)* When H_3BO_3 is neutralised by Na_2CO_3, borax ($Na_2B_4O_7$) is obtained.

$$4H_3BO_3 + Na_2CO_3 \longrightarrow Na_2B_4O_7 + 6H_2O + CO_2$$

10. *Action of HF.* When H_3BO_4 is dissolved in HF, fluoboric acid, $H[BF_4]_4$ also called hydrofluoboric acid) is obtained.

$$H_3BO_3 + 4HF \longrightarrow H^+[BF_4] + 3H_2O$$

11. *Action of alkali metal acid fluorides (MHF_2).* Alkali metal fluoborate, $M^+[BF_4]^-$ is obtained, *e.g.*, $2NaHF_2 + H_3BO_3 \longrightarrow Na^+[BF_4]^- + NaOH + 2H_2O$.

12. *Action of acetic anhydride, $(CH_3CO)_2O$.* When treated with $(CH_3CO)_2O$, boron acetate is obtained.

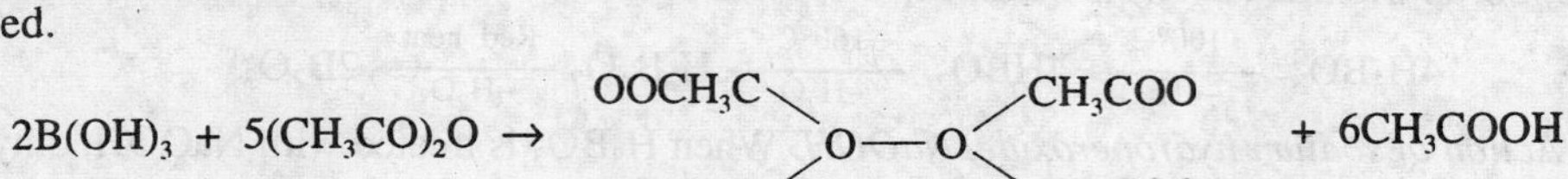

Boron acetate

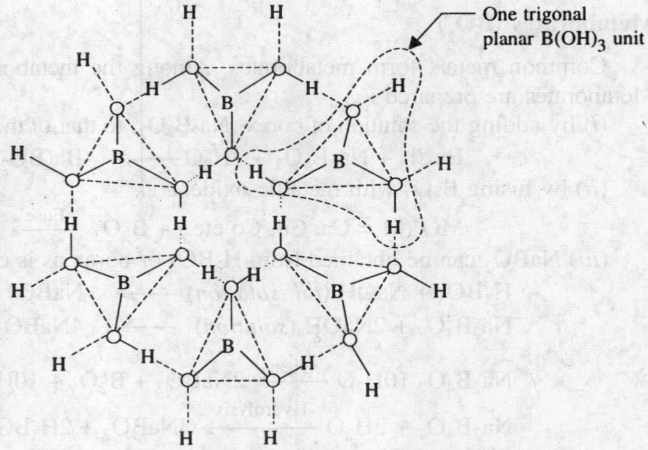

Fig. 30.1 : Layer structure of boric acid in the solid state.

Uses of boric acid. It is used : (*i*) as an anticeptic and as an eye-wash (eye-lotion) in medicines. (*ii*) in making enamels, borate glazes for pottery (borate glazes are more fusible than the silicate glazes), glasses, borax and pigments (*iii*) in the borax-phenolphthalein test for glycerol (*iv*) in tanning industry and in candle industry to stiffen wicks.

Structure of boric acid in the solid state. In the solid state, boric acid has layer structure, *i.e.,* boric acid consists of a number of layers, which are linked together by van der Waals forces and are at a distance of 318 pm from each other. Each layer is composed of various trigonal planar $B(OH)_3$ units. B-atom, in each $B(OH)_3$ unit, is linked with three OH groups, which are arranged at the corner of an equilateral triangle. The structure of one layer is shown in Fig. 30.1, from which it may be seen that H-atom acts as a bridge between the two O-atoms, belonging to two different $B(OH)_3$ units. In this figure, triangle closed in dotted lines contains B-atom at its centre and OH groups at its vertices, and thus represents triangular $B(OH)_3$ unit. The dotted lines between O-atom of one $B(OH)_3$ unit and H-atom of the other $B(OH)_3$ unit indicates hydrogen bonds. It is due to the layer structure of boric acid that this substance, in the solid state, consists of flakes, which are slippery to touch.

Metaboric acid, HBO_2

Preparation. It is obtained by heating H_3BO_3 at 100°C.

$$HBO_3 \xrightarrow{100°C} HBO_2 + H_2O$$

Properties. It is a solid. On being heated to 160°C, it gives tetraboric acid ($H_2B_4O_7$) which on being heated at red heat gives B_2O_3.

$$4HBO_2 \xrightarrow{160°C} H_2B_4O_7 + H_2O; \quad H_2B_4O_7 \xrightarrow{\text{Red heat}} 2B_2O_3 + H_2O$$

Thus: $4HBO_2 \xrightarrow{\text{Red heat}} 2B_2O_3 + 2H_2O$

When HBO_2 is dissolved in H_2O, it is changed into H_3BO_3.

$$HBO_2 + H_2O \longrightarrow H_3BO_3$$

Structure. This acid has the structure HO–B = O.

Pyroboric Acid, $H_6B_4O_9$

It is formed by heating H_3BO_3.

$$4H_3BO_3 \longrightarrow H_6H_4O_9 + 3H_2O$$

When dissolved in water, the acid gives H_3BO_3

$$4H_6B_4O_9 + 3H_2O \longrightarrow 4H_3BO_3$$

Salts of boric acids : Borates

The salts of boric acids are called borates. The borates of alkali metals are soluble in water and the resulting solutions have strongly basic reactions, due to hydrolysis. The borates of other metals are insoluble, but fairly soluble in acids. Among the orthoborates (BO_3^{3-}), $Mg_3(BO_3)_2$ and $(C_2H_5)_3BO_3$ are best known. Metaborates (BO_2^{2-}) are very common and most stable. Borax, $Na_2B_4O_7.10H_2O$, which is a salt of tetraboric acid ($H_2B_4O_7$) is an important borate.

Metaborates (BO_2^-)

Common metals form metaborates. Among the metaborates, $NaBO_2$ is the most important. Metaborates are prepared :

(*i*) by adding the solution of borax ($Na_2B_4O_7$) to that of the appropriate salt of the metal, *e.g.*,

$$BaCl_2 + Na_2B_4O_7 + 3H_2O \longrightarrow Ba(BO_2)_2\downarrow + 2H_3BO_3 + 2NaCl$$

(*ii*) by fusing B_2O_3 with metallic oxide, *e.g.*,

$$MO\ (M = Ca, Cu, Co\ etc.) + B_2O_3 \xrightarrow{Fuse} M(BO_2)_2$$

(*iii*) $NaBO_2$ can be obtained from H_3BO_3 or borax as is evident from the following equations:

$$H_3BO_3 + NaOH\ (\textit{dil. solution}) \longrightarrow NaBO_2 + 2H_2O$$

$$Na_2B_4O_7 + 2NaOH\ (\textit{solution}) \longrightarrow 4NaBO_2 + H_2O$$

$$Na_2B_4O_7.10H_2O \xrightarrow{Heat} 2NaBO_2 + B_2O_3 + 10H_2O$$

$$Na_2B_4O_7 + 3H_2O \xrightarrow{Hydrolysis} 3NaBO_2 + 2H_3BO_3$$

$$Na_2B_4O_7 + \underset{\text{Coloured salt}}{2MSO_4}\ (M = Cu, Co\ etc.) \xrightarrow{Heat} 2NaBO_2 + 10\ H_2O + SO_3 + M(BO_2)_2\ (\textit{Borax bead test})$$

$NaBO_2$ forms colourless needles. When a current of CO_2 is passed through a solution of $NaBO_2$, borax is obtained.

$$4NaBO_2 + CO_2 \longrightarrow Na_2CO_3 + Na_2B_4O_7.$$

The action of H_2O_2 on $NaBO_2$ gives sodium peroxy-borate ($NaBO_3.4H_2O$).

$$NaBO_2 + H_2O_2 + 3H_2O \longrightarrow NaBO_3.4H_2O$$

Sodium metaborate perhydrate, $NaBO_2.3H_2O.H_2O_2$

It is obtained by the electrolysis of a solution of borax and Na_2CO_3 at low temperature, using a platinum anode. It is a colourless crystalline solid. It is fairly soluble in water and its aqueous solution behaves as a mixture of borax and H_2O_2. Its aqueous solution shows alkaline behaviour towards litmus. When this substance, in the solid state or in its aqueous solution, is heated, O_2 gas and sodium metaborate ($NaBO_2$) are obtained.

$$NaBO_2.3H_2O.H_2O_2 \xrightarrow{\Delta} NaBO_2 + 4H_2O + \frac{1}{2}O_2$$

Due to evolution of O_2, this substance has strong oxidising and bleaching properties. It is, therefore, used in a number of bleaching and oxidising processes. It has also antiseptic properties and hence is used in mouth washes, toothpastes, soap powders, cosmetics etc.

Sodium pyroborate, $Na_6B_4O_9$

When borax ($Na_2B_4O_7$) is fused with Na_2CO_3, $Na_6B_4O_9$ is formed.

$$Na_2B_4O_7 + 2Na_2CO_3 \xrightarrow{Fuse} Na_6B_4O_9 + 2CO_2\uparrow$$

Sodium tetraborate or borax, $Na_2B_4O_7.10H_2O$

It is a sodium salt of tetraboric acid ($H_2B_4O_7$) and contains ten molecules of water as water of crystallisation (*decahydrate*).

Preparation. (*i*) *Form tincal mineral, $Na_2B_4O_7.10H_2O$.* Tincal is the crude form of borax, since it contains many impurities associated with it. The finely powdered tincal mineral is treated with hot water which dissolves borax and the impurities remain undissolved in water. The solution which contains borax only is now concentrated and cooled when the crystals of borax are obtained.

(*ii*) *From boric acid, H_3BO_3.* When H_3BO_3 is boiled with soda ash (Na_2CO_3) or excess of NaOH solution, $Na_2B_4O_7$ is obtained.

$$4H_3BO_3 + Na_2CO_3 \longrightarrow Na_2B_4O_7 + 6H_2O + CO_2\uparrow$$

$$4H_3BO_3 + 2NaOH\ (\textit{excess}) + 3H_2O \longrightarrow Na_2B_4O_7.10H_2O$$

(*iii*) *From colemanite mineral, $Ca_2B_6O_{11}.5H_2O$.* The mineral is finely powdered and then is boiled with calculated quantity of Na_2CO_3 solution. Borax goes into solution which, on concentration, gives the crystals of borax.

$$2Ca_2B_6O_{11} + 3Na_2CO_3 + H_2O \longrightarrow 3Na_2B_4O_7\ (solution) + 3CaCO_3\ (ppt.) + Ca(OH)_2$$

Properties. (*i*) Borax is a white crystalline solid and exists in three different forms *viz.*, (*a*) *ordinary* or *prismatic borax*, which is *decahydrate* or *monoclinic* and hence has the formula, $Na_2B_4O_7.10H_2O$. This is the common form of borax (*b*) *octahedral* or *Jeweller's borax*, which is pentahydrate ($Na_2B_4O_7.5H_2O$) and is obtained when the solution of $Na_2B_4O_7.10H_2O$ is crystallised above 60°C (*c*) *borax glass*, which is the anhydrous form ($Na_2B_4O_7$) and is obtained by heating $Na_2B_4O_7.10H_2O$ above its m.pt. It is a colourless glassy mass, absorbs moisture from air and is gradually converted into $Na_2B_4O_7.10H_2O$.

(*ii*) *Basic nature*. It is sparingly soluble in cold H_2O, but fairly soluble in hot water. In water it gets hydrolysed and gives a mixture of NaOH (*strong base*) and H_3BO_3 (*weak acid*). Due to the formation of strong base, namely NaOH, the aqueous solution of borax is alkaline in nature.

$$Na_2B_4O_7 + 7H_2O \rightleftharpoons 2NaOH\ (strong\ base) + 4H_3BO_3\ (weak\ acid)$$

(*iii*) *Reactions leading to the formation of metaborates.* (*a*) When the aqueous solution of borax is added to the salt of an appropriate metal, metaborate of the metal is obtained. For example :

$$BaCl_2 + Na_2B_4O_7 + 3H_2O \longrightarrow Ba(BO_2)_2\downarrow + 2NaCl + 2H_3BO_3$$

(*b*) Sodium metaborate, $NaBO_2$ can be obtained from borax, as is evident from the following equations :

$$Na_2B_4O_7 + 2NaOH\ (solution) \longrightarrow 4NaBO_2 + H_2O$$

$$Na_2B_4O_7.10H_2O \xrightarrow{\Delta} 2NaBO_2 + B_2O_3 + 10H_2O$$

$$Na_2B_4O_7 + 3H_2O \xrightarrow{\text{Hydrolysis}} 3NaBO_2 + 2H_3BO_3$$

$$Na_2B_4O_7.\ 10H_2O + \underset{\text{coloured salt}}{MSO_4\ (M = Cu, Co\ etc.)} \xrightarrow{\Delta} 2NaBO_2 + 10H_2O + SO_3 + M(BO_2)_2\ (Borax\ bead\ test)$$

(*iv*) *Action of Na_2CO_3.* When borax is fused with Na_2CO_3, sodium pyroborate ($Na_6B_4O_9$) is obtained.

$$Na_2B_4O_7 + 2Na_2CO_3 \longrightarrow Na_6B_4O_9 + 2CO_2\downarrow$$

(*v*) *Action of heat: Borax bead test.* On heating, borax first loses all the ten molecules of water of crystallisation and a white porous mass containing anhydrous borax ($Na_2B_4O_7$) is obtained. On further heating, the porous mass of $Na_2B_4O_7$ gives a glassy transparent mass of boric anhydride, B_2O_3.

$$Na_2B_4O_7.10H_2O \xrightarrow[(-10H_2O)]{\Delta} \underset{\text{White porous mass}}{Na_2B_4O_7} \xrightarrow{\Delta} 2NaBO_2 + \underset{\text{Glassy transparent mass}}{B_2O_3}$$

When coloured inorganic metallic salts (*e.g.*, $CoCO_3$, $CuSO_4$ etc.) are heated with borax ($Na_2B_4O_7.10H_2O$), the metallic oxide, obtained by the decomposition of metallic salts, combines with B_2O_3, obtained from $Na_2B_2O_7.10H_2O$ and forms coloured bead of metallic metaborates (BO_2^{2-}) which have specific colour, *e.g.*,

$$CuSO_4\ (metallic\ salt) \xrightarrow{\Delta} CuO + SO_3$$

$$Na_2B_4O_7.10H_2O \xrightarrow{\Delta} 2NaBO_2 + B_2O_3 + 10H_2O$$

$$CuO + B_2O_3 \longrightarrow \underset{\text{Coloured bead}}{Cu(BO_2)_2\ (bluish\ green)}$$

On adding the above equations, we get the final equation as :

$$CuSO_4 + Na_2B_4O_7.10H_2O \xrightarrow{\Delta} 2NaBO_2 + Cu\ (BO_2)_2\ (bluish\ green) + SO_3 + 10H_2O$$

On the basis of the colour of the metallic metaborates, we can identify the basic radical present in a given inorganic salt. The identification of basic radicals by this method is called *borax bead test.* The nature of the colour of the bead of metallic metaborate depends on whether the heating of the inorganic salt with borax has been done in oxidising flame or in reducing flame, as shown below :

Co'our of the bead of metallic metaborate in oxidising flame : Cu^{2+} – bluish green, Cr^{3+} – green, Co^{2+} – blue, Ni^{2+}– reddish brown, Mn^{2+}– violet, Fe^{2+}– yellow.

Colour of the bead of metallic metaborate in reducing flame : Cu^{2+} – red, Cr^{3+} – green, Co^{2+} – blue, Ni^{2+} – black, Mn^{2+} – colourless, Fe^{2+} – green.

In the reducing flame, the higher metaborates are reduced by carbon of the flame to the metaborates or metal and hence the colour of the bead is changed, *e.g.*,

$$2Cu(BO_2)_2\ (Cu = +2) + C \xrightarrow{\text{Reduction}} 2CuBO_2\ (Cu = +1)\ (\text{colourless}) + B_2O_3 + CO\uparrow$$

$$2CuBO_2\ (Cu = +1) + C \xrightarrow{\text{Reduction}} 2Cu\ (Cu = 0)\ (\text{Red bead}) + B_2O_3 + CO\uparrow$$

In cases where the metal metaborate is not reduced in reducing flame, the colour of the bead remains the same in both oxidising as well as reducing flames. For example Co (II) salts give blue bead in both types of flames. Similarly Cr (III) salts give green bead in both the types of flames.

$$CoO + B_2O_3 \longrightarrow Co(BO_2)_2 \quad (\textit{blue bead})$$

$$Cr_2O_3 + 3B_2O_3 \longrightarrow 2Cr(BO_2)_3 \quad (\textit{green bead})$$

Proceduce of borax bead test. The free end of a platinum wire is coiled into a small loop and heated in the Bunsen flame until it is red hot. It is then quickly dipped in borax and again heated when borax swells, loses all its water of crystallisation and ultimately gives a transparent, colourless glassy bead, which consists of $NaBO_2$ and B_2O_3. The bead is moistened with water and dipped in the powdered inorganic salt to be tested. Only a very small quantity of the inorganic salt should be allowed to adhere to the bead, otherwise the bead will become dark and opaque after heating. The bead is now heated first in the lower reducing flame and then in the lower oxidising flame and the colour of the bead in both types of flames is noted. From this colour, the nature of the basic radical present in the inorganic salt can be predicted.

(*vi*) *Action of acids.* $Na_2B_4O_7$ is acted upon by conc. HCl and conc. H_2SO_4 to form sparingly soluble boric acid, H_3BO_3.

$$Na_2B_4O_7 + 2HCl + 5H_2O \longrightarrow 2NaCl + 4H_3BO_3$$

$$Na_2B_4O_7 + H_2SO_4 + 5H_2O \longrightarrow Na_2SO_4 + 4H_3BO_3$$

(*vii*) *Acition of NH_4Cl.* When heated with NH_4Cl, boron nitride (BN) is obtained.

$$Na_2B_4O_7 + 2NH_4Cl \xrightarrow{\Delta} 2NaCl + 2NaCl + 2BN + B_2O_3 + 4H_2O$$

Uses. Borax is used : (*i*) in the manufacture of enamels, glazes (used in pottery), pyrex glass, soap, drying oils, washing powders, varnishes (*ii*) as a flux, since borax can dissolve may metal oxides to form borates of low melting points (*iii*) for the identification of some basic radicals (*e.g.*, Cu^{2+}, Co^{2+} etc.) by using borax bead test (*iv*) in water softening, since it precipitates Ca^{2+} ion (present in impure water) by forming insoluble calcium borate, CaB_4O_7. (*v*) As an antiseptic (*vi*) in leather and match industry (*vii*) for the preparation of metaborates like $NaBO_2$, $Ba(BO_2)_2$, $Cu(BO_2)_2$, $Co(BO_2)_2$ etc. (*viii*) for the preparation of H_3BO_3.

Structure of borates

The crystal structure of metallic borates shows that these salts consist of borate anions, which have cyclic (ring) or linear structure, which is obtained by the sharing of O-atoms between trigonal BO_3 or tetrahedral BO_4 units. The following examples illustrate this point.

1. Structure and geometry of borate ion, BO_3^{3-}. The structure of this ion has already been discussed in chapter 14. It may be seen that the central B-atom in its excited state ($2s^12p^2$) is sp^2 hybridised and BO_3^{3-} ion has symmetrical trigonal planar geometry with each of the three OBO bond angles equal to 120°.

2. Structure of sodium and potassium metaborates. Metaborates of Na and K, both exist as trimeric molecule, $M_3(BO_2)_3$ (M = Na, K). Thus, both these metaborates consist of M^+ (M^+ = Na^+, K^+) and $[(BO_2)_3]^{3-}$ or $B_3O_6^{3-}$ ions. $B_3O_6^{3-}$ ion is made up of three trigonal BO_3 units which are linked together through O-atoms, so that cyclic or ring structure is obtained (See Fig. 30.2).

3. Structure of calcium metaborate. Calcium metaborate consists of Ca^{2+} and $[(BO_2)_n]^{n-}$ ion.

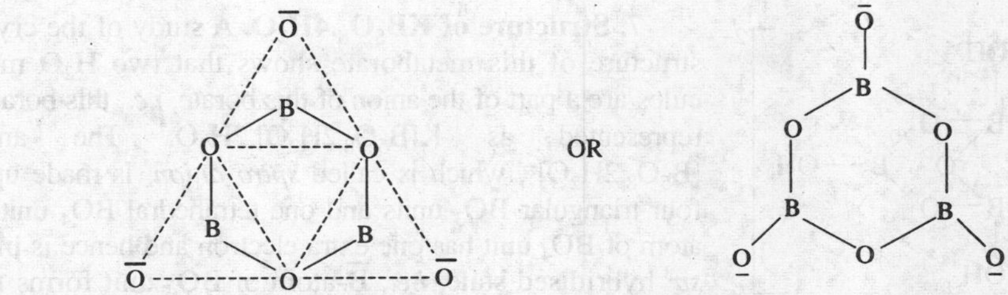

Fig 30.2 : Cyclic structure of $[(BO_2)_3]^{3-}$ or $B_3O_6^{3-}$ ion in the crystals of sodium and potassium metaborate.

$[(BO_2)_n]^{n-}$ ion is made up of triangular BO_3 units, which are joined together through O-atoms, in such a way, that the endless flat linear *zig-zag* chain is obtained. (See Fig. 30.3).

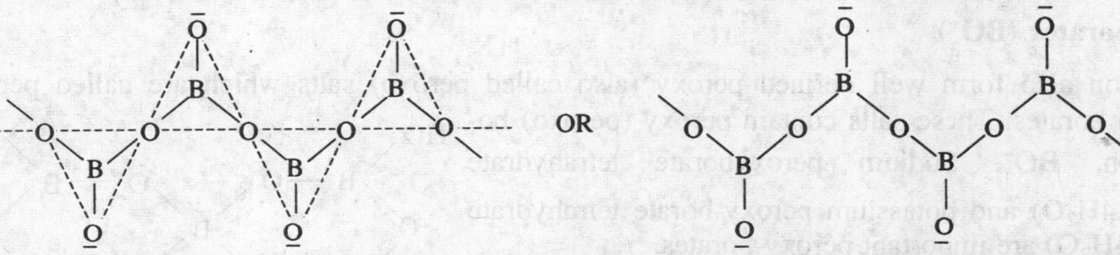

Fig 30.3: Linear structure of $[(BO_2)_n]^{n-}$ ion found in the crystals of calcium metaborate.

4. Structure of $NaBO_2.2H_2O$. The crystal structure of this metaborate shows that both H_2O molecules are a part of the anion in this borate, *i.e.*, this metaborate is represented as Na^+ $[BO_2.2H_2O]^-$. $[BO_2.2H_2O]^-$ or $[BO_4H_4]^-$ ion has tetrahedral structure, in which B-atom is negatively charged and with which four OH groups are attached by four σ-bonds. Thus $[BO_4H_4]^-$ ion should be represented as $[B(OH)_4]^-$ whose structure is

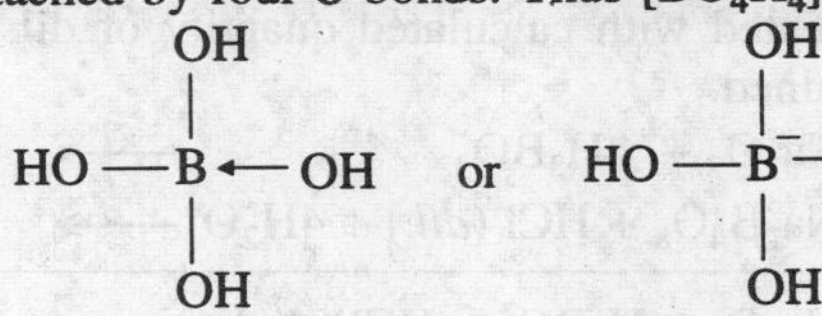

This ion has tetrahedral geometry.

5. Structure of $NaB_3O_5.2H_2O$. The crystal structure of this metaborate has reveals that both H_2O molecules participate in the constitution of the anion present in this metaborate, *i.e.*, this metaborate is represented as Na^+ $[B_3O_5.2H_2O]^-$. The anion *viz.*, $[B_3O_5.2H_2O]^-$ has two planar trigonal BO_3 units and one tetrahedral BO_4 unit. B-atom of BO_4 unit is negatively charged and hence is sp^3 hybridised. Each B-atom of both BO_3 units is linked with one OH group, while B-atom of BO_4 unit is linked with two OH groups. Thus, $[B_3O_5.2H_2O]^-$ ion has four OH groups and three bridging O-atoms and hence this ion should be represented as $[B_3O_3(OH)_4]^-$, which has the structure shown in Fig. 30.4.

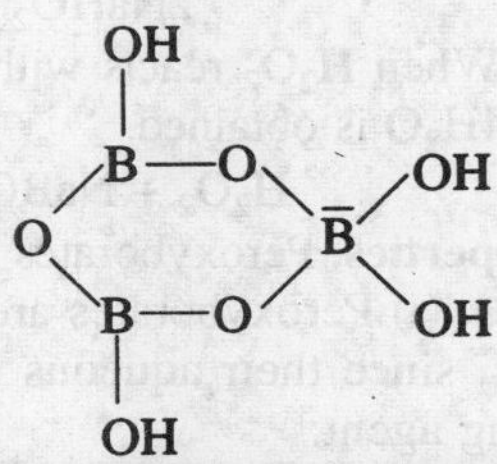

Fig. 30.4 : Structure of $[B_3O_5.2H_2O]^-$ ion found in Na^+ $[B_3O_5.2H_2O]$

6. Structure of borax, $Na_2B_4O_7.10H_2O$. The crystal structure study of this molecule has indicated that two water molecules participate in the constitution of the anion of this borate, *i.e.*, this borate is represented as Na_2 $[B_4O_7.2H_2O].8H_2O$. The anion, $[B_4O_7.2H_2O]^{2-}$ is made up of two triangular BO_3 units and two BO_4 units. Each boron atom of BO_4 unit has one extra electron and hence is sp^3 hybridised, *i.e.*, each boron atom of BO_4 unit forms four B — O σ-bonds. Each boron atom is linked with one OH group. O-atom is making a bridge between two B-atoms of two BO_4 units. Thus $[B_4O_7.2H_2O]^{2-}$ ion is represented as $[B_4O_5(OH)_4]^{2-}$ and consequently $Na_2[B_4O_7.2H_2O].8H_2O$ is represented as Na_2^+ $[B_4O_5(OH)_4]^{2-}.8H_2O$. The structure of $[B_4O_5(OH)_4]^{2-}$ ion is shown in Fig. 30.5.

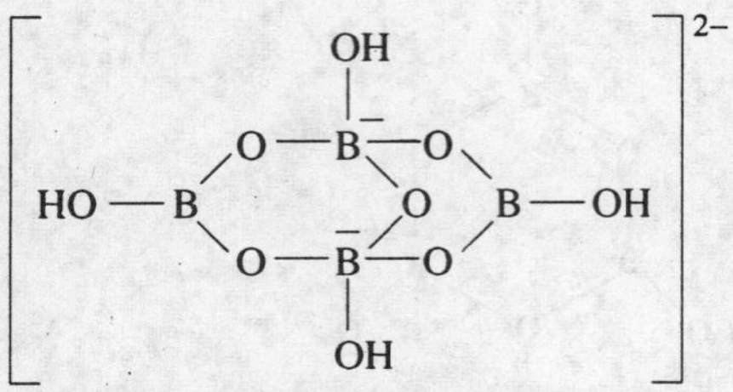

Fig. 30.5 : Structure of $(B_4O_7.2H_2O)^{2-}$ ion found in borax molecule, $Na_2B_4O_7.10H_2O$ or $Na_2[B_4O_7.2H_2O].8H_2O$.

7. Structure of $KB_5O_8.4H_2O$. A study of the crystal structure of this metaborate shows that two H_2O molecules are a part of the anion of this borate, *i.e.*, this borate is represented as $K[B_5O_8.2H_2O].2H_2O$. The anion, $[B_5O_8.2H_2O]^-$, which is called *spiro anion,* is made up of four triangular BO_3 units and one tetrahedral BO_4 unit. B-atom of BO_4 unit has one extra electron and hence is in the sp^3 hybridised state, *i.e.*, $\overline{B}$-atom of BO_4 unit forms four B — O σ-bonds. Each boron atom of BO_3 units is linked with one OH group while $\overline{B}$-atom of BO_4 unit is linked only with O-atoms. Thus $[B_5O_8.2H_2O]^-$ is represented as $[B_5O_6(OH)_4]^-$ which has the structure shown in Fig. 30.6.

Peroxyborates (BO_3^-)

Boron also form well defined peroxy (also called peroxo) salts which are called peroxy (peroxo) borates. These salts contain peroxy (peroxo) borate ion, BO_3^-. Sodium peroxyborate tetrahydrate ($NaBO_3.4H_2O$) and potassium peroxy-borate tetrahydrate ($KBO_3.4H_2O$) are important peroxy-borates.

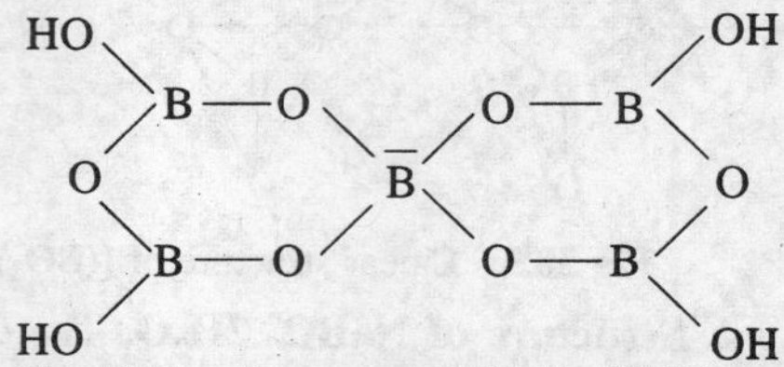

Fig. 30.6 : Structure of $[B_5O_8.2H_2O]^-$ or $[B_5O_6(OH)_4]^-$ ion found in $K^+[B_5O_8.2H_2O]^-.2H_2O$ or $K^+[B_5O_6(OH)_4]^-.2H_2O$.

Preparation. $NaBO_3$ is obtained by the following methods :

(*i*) When sodium peroxide (Na_2O_2) reacts with H_3BO_3 suspended in cold water, crystals of $Na_2B_4O_8$ are obtained on cooling the reaction mixture. When these crystals are treated with calculated quantity of dil. HCl, $NaBO_3$ is obtained.

$$Na_2O_2 + 4H_3BO_3 \longrightarrow Na_2B_4O_8 + 6H_2O$$
$$Na_2B_4O_8 + HCl\ (dil.) + 4H_2O \longrightarrow NaBO_3 + NaCl + 3H_3BO_3$$
$$\overline{Na_2O_2 + H_3BO_3 + HCl\ (dil.) \longrightarrow NaBO_3 + NaCl + 2H_2O}$$

(*ii*) When sodium hydroperoxide ($NaHO_2$) reacts with H_3BO_3, $NaBO_3$ is obtained.

$$NaHO_2 + H_3BO_3 \longrightarrow NaBO_3 + 2H_2O$$

or $$2NaHO_2 + 2H_3BO_3 \longrightarrow (NaBO_3.H_2O) + 2H_2O$$

(*c*) When H_2O_2 reacts with sodium metaborate ($NaBO_2$), sodium peroxy-borate tetrahydrate, $NaBO_3.4H_2O$ is obtained.

$$H_2O_2 + NaBO_2 + 3H_2O \longrightarrow NaBO_3.4H_2O$$

Properties. Peroxyborates are good oxidising agents.

Uses. (*i*) Peroxyborates are used as oxidising agents. (*ii*) Peroxyborates are used in washing powders, since their aqueous solution liberates H_2O_2 (*iii*) $NaBO_3$ is used as as cleansing and bleaching agent.

Structure of peroxyborates. Peroxyborates are extensively hydrated dimerised molecules, *e.g.*, sodium peroxyborate is a tetrahydrated dimerised molecule and hence is written as $(NaBO_3.4H_2O)_2$ or $Na_2B_2O_6.8H_2O$. The crystal structure determination of this dimerised molecule has shown that out of $8H_2O$ molecules, two H_2O molecules are a part of the anion present in this peroxyborate, *i.e.*, $Na_2B_2O_6.8H_2O$ should be formulated as $Na_2[B_2O_6.2H_2O].6H_2O$ or $Na_2[B_2O_8H_4].6H_2O$. The anion *viz.*, $[B_2O_8H_4]^{2-}$ consists of four OH^- groups and two peroxy groups, $(—O—O—)^{2-}$ or O_2^{2-}. Thus $[B_2O_8H_4]^{2-}$ ion should be represented as $[B_2(O_2)_2(OH)_4]^{4-}$. In this anion each

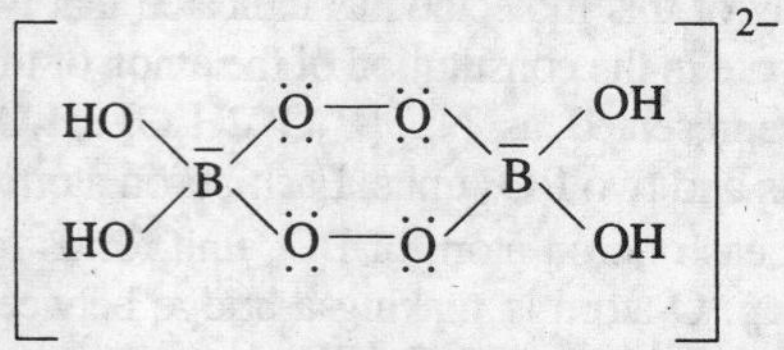

Fig. 30.7: Binuclear peroxy bridged structure of $[B_2O_8H_4]^{2-}$ ion in $(NaBO_3.4H_2O)_2$ *or* $Na_2[B_2O_8H_4].6H_2O$ molecule.

B-atom is negatively-charged and is attached with two OH^- groups and two O_2^{2-} groups. Thus two peroxy groups form two bridges between two B-atoms. Two OH groups act as terminal OH groups. Thus $[B_2O_8H_4]^{2-}$ or $[B_2(O_2)_2\,(OH)_4]^{2-}$ ion has a binuclear peroxy bridged structure shown in Fig. 30.7.

sp^3 hybridisation of B and O-atoms. Since each B-atom shows tetracovalency in the structure of $[B_2(O_2)_2\,(OH)_4]^{2-}$ ion it is negatively-charged and is sp^3 hybridised in its exited state .

B-atom in its ground state ($2s^2\,2p^1$)

B^- in its ground state ($2s^2\,2p^2$)

B^- in its excited state ($2s^1\,2p^3$)

B^- in sp^3 hybridised state

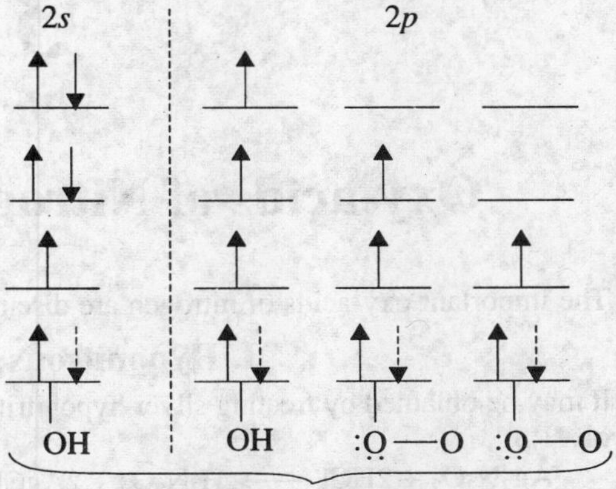

sp^3 hybridisation of $\bar{B}$ atom

Each O-atom of peroxide group is also sp^3 hybridised in its ground state. Two sp^3 hybrid orbitals contain two lone pairs of electrons each.

The above discussion shows that $[B_2(O_2)_2(OH)_4]^{2-}$ ion has two tetrahedral B-atoms. This discussion also indicates that the formation of sodium peroxyborate, $Na_2\,[B_2O_8H_4]$ or $Na_2[B_2(O_2)_2(OH)_4]$ by the combination of two $NaHO_2$ and two H_3BO_3 or $B(OH)_3$ molecules can be shown as :

O-atom in the ground state ($2s^2\,2p^4$)

O-atom in sp^3 hybridised state

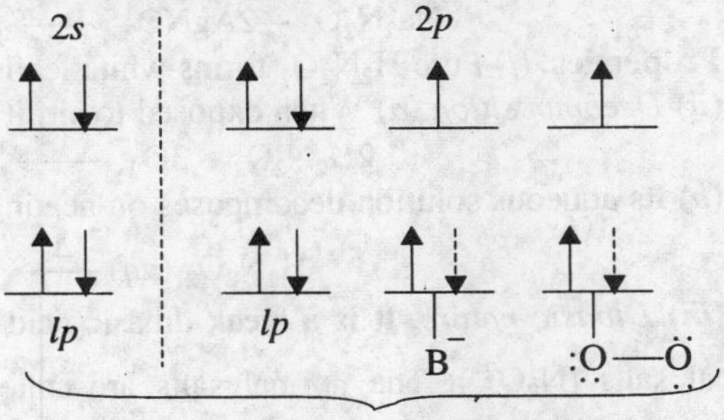

sp^3 hybridisation of O-atom

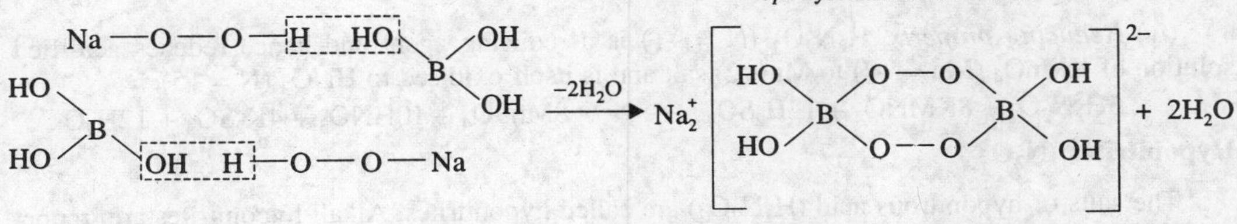

$$\text{or}\quad 2NaHO_2 + 2B(OH)_3 \xrightarrow{-2H_2O} Na_2\,[B(O_2)_2\,(OH)_4] + 2H_2O$$

$$\text{or}\quad 2NaHO_2 + 2B(OH)_3 \xrightarrow{-2H_2O} (NaBO_3.H_2O)_2 + 2H_2O$$

University Questions

1. Describe the preparation, properties and uses of borax. Starting from borax, how can you get : (*i*) BN (*ii*) B_2O_3 (*iii*) BCl_3. *(Meerut 95)*
2. Mention any two important chemical properties of boric acid. *(Osmania 93)*
3. Write a note on Borax bead test. *(Lucknow 2002)*

31

Oxy-acids of Nitrogen and their Salts

The important oxy-acids of nitrogen are discussed below:

I. Hyponitrous acid, $H_2N_2O_2$

It may be obtained by treating silver hyponitrite ($Ag_2N_2O_2$) with a solution of HCl in dry ether or with H_2S.

$$Ag_2N_2O_2 + 2HCl \longrightarrow H_2N_2O_2 + 2AgCl \; ; \qquad Ag_2N_2O_2 + H_2S \longrightarrow H_2N_2O_2 + Ag_2S$$

$Ag_2N_2O_2$ required in the above reactions is obtained by the following reactions:

$$2NaNO_2 + 4H \xrightarrow[\text{Na–Hg (Amalgam)}]{\text{Reduction by}} Na_2N_2O_2 + 2H_2O$$

When the above reaction which leads to the formation of $Na_2N_2O_2$ is complete, the resulting solution is neutralised with CH_3COOH and treated with $AgNO_3$ when yellow ppt. of $Ag_2N_2O_2$ is obtained.

$$Na_2N_2O_2 + 2AgNO_3 \longrightarrow Ag_2N_2O_2 + 2NaNO_3$$

Properties. (*i*) Pure $H_2N_2O_2$ forms white leaflets which explode when rubbed.

(*ii*) *Decomposition.* (*a*) When exposed to air, it decomposes showly to form HNO_2 and HNO_3.

$$2H_2N_2O_2 + 3O_2 \longrightarrow 2HNO_2 + 2HNO_3$$

(*b*) Its aqueous solution decomposes on heating evolving N_2O.

$$H_2N_2O_2\ (aq) \xrightarrow{\Delta} N_2O + H_2O$$

(*iii*) *Dibasic nature.* It is a weak dibasic acid and hence forms both acid salts ($HN_2O_2^-$) and normal salts ($N_2O_2^{2-}$). The normal salts are called hyponitrites which contain hyponitrite ion, $N_2O_2^{2-}$.

(*iv*) *Reducing property.* $H_2N_2O_2$ (N = +1) is a reducing agent and hence reduces acidified solution of $KMnO_4$ (Mn = +7) to Mn(+2) salt and is itself oxidised to HNO_3 (N = +5).

$$5H_2N_2O_2 + 8KMNO_4 + 12H_2SO_4 \longrightarrow 8MnSO_4 + 10HNO_3 + 4K_2SO_4 + 12H_2O$$

Hyponitrites ($N_2O_2^{2-}$)

The salts of hyponitrous acid ($H_2N_2O_2$) are called hyponitrites. Alkali hyponitrites are important and are prepared:

(*i*) by warming sodium hydroxylamine monosulphate, $OH.NH(SO_3Na)$ with NaOH

$$2OH.NH(SO_3Na) + 4NaOH \longrightarrow Na_2N_2O_2 + 2Na_2S_2O_3 + 4H_2O$$

(*ii*) by treating NH_2OH with amyl nitrite, $C_5H_{11}NO_2$ in ethyl alcohol containing sodium ethoxide, C_2H_5ONa.

$$NH_2OH + C_5H_{11}NO_2 + 2C_2H_5ONa \xrightarrow[\text{alcohol}]{\text{Ethyl}} Na_2N_2O_2 + C_5H_{11}OH + 2C_2H_5OH\ .$$

(*iii*) by the reduction of NO with an alkaline solution of $SnCl_2$.

$$\underset{(N = +2)}{2NO} + \underset{(Sn = +2)}{SnCl_2} + 6KOH \longrightarrow \underset{(N = +1)}{K_2N_2O_2} + \underset{(Sn = +4)}{K_2SnO_3} + 2KCl + 3H_2O$$

Properties. (*i*) Diethyl ester, $(C_2H_5)_2N_2O_2$ and sodium hyponitrite, $Na_2N_2O_2$ decompose on heating, giving N_2.

$$(C_2H_5)_2N_2O_2 \xrightarrow{\Delta} C_2H_5OH + CH_3CHO + N_2$$

$$3Na_2N_2O_2 \xrightarrow{\Delta} 2NaNO_2 + 2Na_2O + 2N_2$$

(*ii*) $(C_2H_5)_2N_2O_2$ on being reduced with sodium-amalgam and H_2O gives C_2H_5OH and N_2.

$$C_2H_5-\overset{H}{O}+N=N+\overset{H}{O}-C_2H_5 + 2H \xrightarrow{Na\text{-}Hg/H_2O} 2C_2H_5OH + N_2$$

(*iii*) Dipole moment of $(C_2H_5)_2N_2O_2$ has been found to be equal to zero.

Structure of $H_2N_2O_2$ molecule and $N_2O_2^{2-}$ ion. See chapter 15.

II. Nitrous acid, HNO_2

The acid does not exist in the free state and hence cannot be prepared in anhydrous state. However, the acid exists only in solution.

Preparation. Aqueous solution of HNO_2 can be obtained:

(*a*) by acidifying solutions of alkali nitrites with mineral acids.

$$2NaNO_2 + H_2SO_4 \longrightarrow Na_2SO_4 + 2HNO_2; \quad KNO_2 + HCl \longrightarrow KCl + HNO_2$$

(*b*) by dissolving N_2O_3 in water at 0°C.

$$N_2O_3 + H_2O \xrightarrow{0°C} 2HNO_2$$

(*c*) by adding calculated quantity of ice-cold H_2SO_4 to a well cooled solution of $Ba(NO_2)_2$ and filtering of the precipitate of $BaSO_4$.

$$H_2SO_4\ (\textit{ice-cold}) + Ba(NO_2)_2\ (\textit{cooled solution}) \longrightarrow BaSO_4\downarrow + 2HNO_2$$

Properties. (*i*) The aqueous solution of HNO_2 has pale blue colour due to the presence of N_2O_3 in it. The colour fades on standing for sometime.

(*ii*) *Acidic property.* It is a weak monobasic acid having its ionisation constant equal to 5.0×10^{-4} at 15°C. It reacts with alkalies to form nitrite.

$$HNO_2 + NaOH \longrightarrow NaNO_2 + H_2O$$

(*iii*) *Decomposition.* (*a*) Its dilute solution decomposes to give HNO_3 and NO.

$$3HNO_2 \longrightarrow 2NO + H_2O + HNO_3$$

(*b*) Its fairly concentrated solution decomposes into NO and NO_2.

$$2HNO_2 \rightleftharpoons H_2O + N_2O_3; \quad N_2O_3 \rightleftharpoons NO + NO_2$$

(*iv*) *Oxidising property.* HNO_2 oxidises many substances and in these reactions HNO_2 (N = +3) is reduced to NO (N = +2). Oxidising property of HNO_2 is due to the fact that this acid can readily decompose to give nascent oxygen.

$$2HNO_2 \longrightarrow 2NO + H_2O + O$$

The ionic equation showing the oxidising property (or reduction) of HNO_2 is:

$$NO_2^- + 2H^+ + e^- \xrightarrow{\text{Reduction}} NO + H_2O$$

Thus HNO_2

(*a*) liberates I_2 from the acidified solution of KI.

$$2KI + 4H_2SO_4 + 2HNO_2 \longrightarrow K_2SO_4 + 2NO + 2H_2O + I_2$$

or $$2I^- + 4H^+ + 2NO_2^- \longrightarrow 2NO + 2H_2O + I_2$$

(*b*) oxidises $SnCl_2$ (Sn = +2) to $SnCl_4$ (Sn = + 4).

$$SnCl_2 + 2HCl + 2HNO_2 \longrightarrow SnCl_4 + 2NO + 2H_2O$$

or $$Sn^{2+} + 4H^+ + 2NO_2^- \longrightarrow Sn^{4+} + 2NO + 2H_2O$$

(*c*) oxidises SO_2 (S = + 4) to H_2SO_4 (S = + 6).

$$SO_2 + 2HNO_2 \longrightarrow H_2SO_4 + 2NO \quad \text{or} \quad SO_2 + 2H^+ + 2NO_2^- \longrightarrow H_2SO_4 + 2NO$$

(*d*) oxidises H_2S (S = – 2) to sulphur (S = 0).

$$H_2S + 2HNO_2 \longrightarrow S + 2H_2O + 2NO \quad \text{or} \quad S^{2-} + 4H^+ + 2NO_2^- \longrightarrow S + 2H_2O + 2NO$$

(*e*) oxidises acidified $FeSO_4$ (Fe = +2) to $Fe_2(SO_4)_3$ (Fe = +3).

$$2FeSO_4 + H_2SO_4 + 2HNO_2 \longrightarrow Fe_2(SO_4)_3 + 2NO + 2H_2O$$

$$\text{or} \quad Fe^{2+} + 2H^+ + NO_2^- \longrightarrow Fe^{3+} + NO + H_2O$$

(*f*) oxidises Ti^{3+} to Ti^{4+} state.

$$Ti^{3+} + 2H^+ + NO_2^- \longrightarrow Ti^{4+} + NO + H_2O$$

(*g*) oxidises sod. arsenite, Na_3AsO_3 (As = +3) to sod. arsenate, Na_3AsO_4 (As = +5).

$$Na_3AsO_3 + 2HNO_2 \longrightarrow Na_3AsO_4 + 2NO + H_2O$$

$$\text{or} \quad AsO_3^{3-} + 2H^+ + 2NO_2^- \longrightarrow AsO_4^{3-} + 2NO + H_2O$$

(*v*) *Reducing property.* HNO_2 can also reduce many substances and in these reactions HNO_2 (N = +3) accepts nascent oxygen and is oxidised to HNO_3 (N = +5).

$$HNO_2 + O \longrightarrow HNO_3 \quad \text{or} \quad NO_2^- + H_2O \longrightarrow NO_3^- + 2H^+ + 2e^-$$

Thus HNO_2

(*a*) reduces Br_2 (Br = 0) to HBr (Br = –1).

$$Br_2 + H_2O + HNO_2 \longrightarrow 2HBr + HNO_3 \quad \text{or} \quad Br_2 + H_2O + NO_2^- \longrightarrow 2Br^- + NO_3^- + 2H^+$$

(*b*) reduces acidified $KMnO_4$ solution (Mn = +7) to $MnSO_4$ (Mn = +2).

$$2KMnO_4 + 3H_2SO_4 + 5HNO_2 \longrightarrow K_2SO_4 + 3H_2O + 2MnSO_4 + 5HNO_3$$

$$\text{or} \quad 2MnO_4^- + 6H^+ + 5NO_2^- \longrightarrow 3H_2O + 2Mn^{2+} + 5NO_3^-$$

(*c*) reduces acidified $K_2Cr_2O_7$ (Cr = + 6) to green $Cr_2(SO_4)_3$ (Cr = + 3).

$$K_2Cr_2O_7 + 4H_2SO_4 + 3HNO_2 \longrightarrow K_2SO_4 + Cr_2(SO_4)_3 + 3HNO_3 + 4H_2O$$

$$\text{or} \quad Cr_2O_7^{2-} + 8H^+ + 3NO_2^- \longrightarrow 2Cr^{3+} + 3NO_3^- + 4H_2O$$

(*d*) reduces H_2O_2 (O = –1) to H_2O (O = –2).

$$H_2O_2 + HNO_2 \longrightarrow HNO_3 + H_2O \quad \text{or} \quad O_2^{2-} + NO_2^- \longrightarrow NO_3^- + O^{2-}$$

(*vi*) *Action of NH_3, urea and other aliphatic amines.* In all these reactions N_2 is evolved.

$$NH_3 + HNO_2 \longrightarrow [NH_4NO_2]\ (\textit{Intermediate}) \longrightarrow N_2 + 2H_2O$$

$$NH_2CONH_2\ (\textit{Urea}) + 2HNO_2 \longrightarrow 2N_2 + CO_2 + 3H_2O$$

$$CH_3CONH_2 + HNO_2 \longrightarrow CH_3COOH + N_2 + H_2O$$

$$C_2H_5NH_2\ (\textit{Ethylamine}) + HO.NO \longrightarrow N_2 + C_2H_5OH + H_2O$$

In the last two reactions NH_2 group is replaced by OH group.

(*vii*) *Action of aniline and other aromatic amines.* At low temperatures, *diazo reaction* takes place *i.e.,* diazo salts are formed. For example:

$$\underset{\text{Aniline hydro chloride}}{C_6H_5NH_2.HCl} + HNO_2 \longrightarrow \underset{\text{Benzene diazonium chloride}}{C_6H_5 - N = N - Cl} + 2H_2O$$

(*viii*) *Action of primary alkyl alcohol.* When HNO_2 is treated with ROH, H-atom of HNO_2 molecule is replaced by alkyl group (R) and alkyl nitrite (R–O–N=O) is obtained.

$$R-\boxed{OH + H}-O-N=O \xrightarrow{-H_2O} R-O-N=O + H_2O$$

Uses. HNO_2 is used: (*i*) in organic chemistry for the preparation of diazo compounds which are the basic substances in the manufacture of azo dyes. (*ii*) as an oxidising as well as reducing agent in analytical chemistry (*iii*) for replacing – NH_2 group by – OH group in aliphatic amines.

Nitrites (NO_2^-)

We have already said that HNO_2 is a monobasic acid and hence gives only one type of salts which are called nitrites. Nitrites are quite stable compounds. $NaNO_2$ and KNO_2 are important. These may be prepared:

(*a*) by heating the nitrate.

$$2MNO_3\ (M = Na, K) \xrightarrow{\Delta} 2MNO_2 + O_2$$

(*b*) by the action of Pb on nitrate.

$$MNO_3 + Pb \longrightarrow MNO_2 + PbO$$

(*c*) by slowly adding sulphur to the melt containing $NaNO_3$ and NaOH.

$$3NaNO_3 + 2NaOH + S \longrightarrow 3NaNO_2 + Na_2SO_4 + H_2O$$

(*d*) by the action of N_2O_3 on NaOH.

$$N_2O_3 + 2NaOH \longrightarrow 2NaNO_2 + H_2O$$

Properties. (*i*) Nitrites of alkali-metals are yellow crystalline solids.

(*ii*) Excepting $AgNO_2$ which is sparingly soluble in water, all other nitrites, are soluble.

(*iii*) *Action of heat.* (*a*) All nitrites, except those of alkali metals, give NO and NO_2 on heating, *e.g.*,

$$Pb\,(NO_2)_2 \xrightarrow{\Delta} PbO + NO + NO_2$$

(*b*) NH_4NO_2 gives N_2 on heating

$$NH_4NO_2 \xrightarrow{\Delta} N_2 + 2H_2O$$

(*iv*) *Action of dil. H_2SO_4 or HCl.* Nitrites are decomposed by dil. H_2SO_4 or HCl with the evolution of reddish-brown fumes of NO_2 gas.

$$2NaNO_2 + 2H_2SO_4 \longrightarrow Na_2SO_4 + H_2O + \underset{\text{Colourless gas}}{NO} + \underset{\text{Reddish brown gas}}{NO_2}$$

or

$$NO_2^- + 2H^+ \longrightarrow H_2O + NO + NO_2$$

(*v*) *Action of solid NH_4Cl.* When aqueous solution of a nitrite is heated with solid NH_4Cl, effervescences of N_2 are produced.

$$NaNO_2 + NH_4Cl \xrightarrow{\Delta} NaCl + N_2 + 2H_2O \quad \text{or} \quad NO_2^- + NH_4^+ \xrightarrow{\Delta} N_2 + 2H_2O$$

(*vi*) *Action of $FeSO_4$ solution in presence of dil. H_2SO_4.* When $FeSO_4$ solution is added to a cold solution of a nitrite acidified by dil. H_2SO_4, brown black colour is produced. This colour is due to the formation of nitrosoferrous sulphate, $FeSO_4.NO$ or $[Fe(NO)]SO_4$.

$$2NaNO_2 + 2H_2SO_4 \longrightarrow Na_2SO_4 + H_2O + NO + NO_2 \text{ or } 2NO_2^- + 2H^+ \longrightarrow H_2O + NO + NO_2$$

$$FeSO_4 + NO \longrightarrow [Fe(NO)]\,SO_4\ (\textit{Brown black complex}) \text{ or } Fe^{2+} + NO \longrightarrow [Fe(NO)]^{2+}$$

Actually the complex obtained is $[Fe(NO)\,(H_2O)_5]^{2+}$

$$Fe^{2+} + NO + 5H_2O \longrightarrow [Fe(NO)\,(H_2O)_5]^{2+}$$

(*vii*) *Oxidising and reducing properties.* $NaNO_2$ acts both as an oxidising agent and as a reducing agent. These properties of $NaNO_2$ are the same as those of HNO_2. Thus $NaNO_2$ oxidises KI to I_2 and is itself reduced to NO in presence of dil. H_2SO_4.

$$2NaNO_2 + 2KI + 2H_2SO_4 \longrightarrow Na_2SO_4 + K_2SO_4 + 2NO + 2H_2O + I_2$$

or

$$2NO_2^- + 2I^- + 4H^+ \longrightarrow 2NO + 2H_2O + I_2$$

(*viii*) *Reduction of $NaNO_2$.* $NaNO_2$ (N = +3) can be reduced to sodium hydro nitrite, $Na_4N_2O_4$ (N = +2) by Na-metal in liq. NH_3.

$$2NaNO_2 + 2Na \xrightarrow{\text{Liq. } NH_3} Na_4N_2O_4$$

Tests of nitrite ion (NO_2^-)

(*i*) Warm some dil. H_2SO_4 in a test tube and then add to it a small amount of the given nitrite. Brown fumes of NO_2 are produced.

$$2NaNO_2 + H_2SO_4\ (dil.) \longrightarrow Na_2SO_4 + NO + NO_2 + H_2O$$

or $$2NO_2^- + 2H^+ \longrightarrow NO + NO_2 + H_2O$$

Now add $FeSO_4$ solution to the test tube, the solution becomes brown black due to formation of nitrosoferrous sulphate, $FeSO_4.NO$ (complex compound).

$$FeSO_4 + NO \longrightarrow [Fe(NO)]SO_4$$

Actually the above complex is $[Fe(NO)(H_2O)_5]SO_4$

$$FeSO_4 + NO + 5H_2O \longrightarrow [Fe(NO)(H_2O)_5]SO_4$$

or $$Fe^{2+} + NO + 5H_2O \longrightarrow [Fe(NO)(H_2O)_5]^{2+}$$

(*ii*) To the solution of a nitrite add a crystal of KI and some starch solution. Then add a few drops of dil. H_2SO_4. A deep blue colouraton confirms nitrite. Here $NaNO_2$ oxidises KI to I_2 and is itself reduced to NO in presence of dil. H_2SO_4.

$$NaNO_2 + 2KI + 2H_2SO_4 \longrightarrow Na_2SO_4 + K_2SO_4 + 2NO + 2H_2O + I_2$$

or $$NO_2^- + 2I^- + 4H^+ \longrightarrow 2NO + 2H_2O + I_2$$

I_2 which is liberated in the above reaction is absorbed by starch solution and a deep blue complex is formed.

$$I_2 + \text{Starch} \longrightarrow \text{Blue coloured complex}$$

(*iii*) To the aqueous solution of a nitrite add a pinch of thiourea (H_2NCSNH_2) and warm. Cool and then add a few drops of $FeCl_3$ solution. The production of blood red colour indicates the presence of NO_2^- ion. The production of blood red colour is due to the formation of $[Fe(CNS)]Cl_2$ (complex compound).

$$NaNO_2 + H_2NCSNH_2 \longrightarrow NaCNS + H_2O + N_2\ ;\ FeCl_3 + NaCNS \longrightarrow [Fe(CNS)]Cl_2 + NaCl$$

OR

$$NO_2^- + H_2NCSNH_2 \longrightarrow CNS^- + H_2O + N_2; \quad Fe^{3+} + CNS^- \longrightarrow [Fe(CNS)]^{2+}$$

Structure of HNO_2 molecule and NO_2^- ion. See chapter 15.

III. Nitric acid, HNO_3

Preparation. (*i*) In the laboratory, it is prepared by heating $NaNO_3$ or KNO_3 with conc. H_2SO_4. Reddish brown vapours of HNO_3 so produced are condensed in a water-cooled receiver.

$$NaNO_3 + H_2SO_4\ (conc.) \xrightarrow{\Delta} KHSO_4 + HNO_3$$

The acid thus produced is yellow in colour due to the presence of NO_2 formed by the decomposition of HNO_3.

$$4HNO_3 \longrightarrow 4NO_2 + 2H_2O + O_2$$

The acid is purified by redistillation and by passing a current of dry air or CO_2 through the warm acid.

(*ii*) *Manufacture.* On a large scale, the acid is generally prepared by *Ostwald's process* which consists of the following steps:

(*a*) NH_3 (N = – 3) obtained by Haber's process is oxidised to NO (N = + 2) by atmospheric O_2 (free from dust) in presence of platinum-guaze catalyst electrically heated to 750° – 900°C.

$$4NH_3 + 5O_2 \xrightarrow[750° - 900°C]{\text{Pt–guaze}} 4NO + 6H_2O;\ \Delta H = -90.0\ KJ$$

Since the reaction is exothermic, heat generated in the reaction maintains the temperature of the catalyst.

(*b*) NO produced as above is oxidised by atmospheric O_2 to NO_2 which is cooled to about 50°C and then absorbed in water (in the presence of air) to give HNO_3.

$$2NO + O_2 \longrightarrow 2NO_2; \quad 4NO_2 + O_2 + 2H_2O \longrightarrow 4HNO_3$$

The usual concentrated nitric acid contains 98% HNO_3 and its specific gravity = 1.5.

Properties. 1. Pure HNO_3 is colourless fuming liquid of sp. gravity = 1.56 (at 0°C). It boils at

86°C and freezes to a colourless solid having m.pt. = – 41.3°C. It is soluble in water in all proportions.

2. Corrosive nature. The acid is highly corrosive and produces painful blisters when it comes in contact with skin.

3. Colour. HNO_3 is colourless but often gradually turns yellow. Yellow colour is produced due to the fact that HNO_3, when exposed to light, undergoes decomposition to form NO_2 which gets dissolved in HNO_3 and gives it yellow colour.

$$4HNO_3 \xrightarrow{hv} 4NO_2 + O_2 + 2H_2O$$

The yellow colour of the acid can be removed by warming it to 60–80°C and bubbling dry air through it.

4. Decomposition. (*a*) At room temperature the pure acid undergoes dissociation to give N_2O_4 and O_2.

$$HNO_3 \xrightarrow[\text{temp.}]{\text{Room}} N_2O_5 + H_2O \;; \qquad N_2O_5 \rightleftharpoons N_2O_4 + \frac{1}{2}O_2$$

(*b*) On heating, HNO_3 decomposes to give O_2 and brown fumes of NO_2.

$$4HNO_3 \xrightarrow{\Delta} 4NO_2 + O_2 + 2H_2O$$

5. Acidic properties. HNO_3 is a very strong monobasic acid. It, therefore, reacts with basic oxides, carbonates, bicarbonates, sulphites and hydroxides, forming corresponding nitrates.

$$CaO + 2HNO_3 \longrightarrow Ca(NO_3)_2 + H_2O; \qquad Na_2CO_3 + 2HNO_3 \longrightarrow 2NaNO_3 + H_2O + CO_2$$

$$NaHCO_3 + HNO_3 \longrightarrow NaNO_3 + H_2O + CO_2; \quad NaOH + HNO_3 \longrightarrow NaNO_3 + H_2O$$

$$Na_2SO_3 + 2HNO_3 \longrightarrow 2NaNO_3 + SO_2 + H_2O; NaHSO_3 + HNO_3 \longrightarrow NaNO_3 + SO_2 + H_2O$$

Being a strong acid, HNO_3 is completely ionised in water as:

$$HNO_3 \longrightarrow H^+ + NO_3^-$$

6. Oxidising nature. HNO_3 is a very strong oxidising agent, since it can easily give nascent oxygen both in the conc. and dil. form. In these reactions HNO_3 is reduced to NO_2, O_2, NH_4NO_3 (or NH_4^+ ions), N_2O, NO, N_2 or NH_2OH depending on the nature of the substance to be oxidised, concentration of HNO_3 and the temperature employed.

Non-metals and metalloids (*e.g.*, C, Sn, P, As, Pb, S and I) are oxidised to their corresponding highest oxy-acid while the metals like Mg, Mn, Zn, Fe, Sn, Pb, Cu, Ag and Hg are oxidised to their corresponding nitrates.

The oxidation of various substances can be studied under the following heads:

A. Oxidation of nonmetals and metalloids. Examples of nonmetals and metalloids are C, Sn, P, As, Sb, S and I. These elements, when treated with HNO_3, are oxidsed to their corresponding highest oxy-acid and HNO_3 is reduced to NO_2. Thus, for example:

$$P + 5HNO_3 \longrightarrow H_3PO_4 + 5NO_2 + H_2O$$

Actually, the above reaction takes place through the following steps:

(*a*) HNO_3 loses nascent oxygen and is reduced to NO_2.

$$2HNO_3 \longrightarrow H_2O + 2NO_2 + O \quad \text{or} \quad 2H^+ + NO_3^- + e^- \longrightarrow H_2O + NO_2$$

(*b*) Nascent oxygen obtained as above oxidises the element to its highest oxide which reacts with H_2O and gives the oxy-acid, *e.g.*,

$$2P + 5O \longrightarrow P_2O_5$$

$$P_2O_5 + 3H_2O \longrightarrow 2H_3PO_4$$

$$\overline{2P + 5O + 3H_2O \longrightarrow 2H_3PO_4}$$

The following examples illustrate the above discription:

(*i*) *Carbon is oxidised to* H_2CO_3.

$$C + 4HNO_3 \longrightarrow H_2CO_3 + 4NO_2 + H_2O \text{ or } 2H^+ + 4NO_3^- + C \longrightarrow H_2O + 4NO_2 + CO_3^{2-}$$

(*ii*) *Tin is oxidised to metastannic acid,* H_2SnO_3 *by conc.* HNO_3.

$$Sn + 4HNO_3 \longrightarrow H_2SnO_3 + 4NO_2 + H_2O \text{ or } 2H^+ + 4NO_3^- + Sn \longrightarrow H_2O + 4NO_2 + SnO_3^{2-}$$

(iii) P, As and Sb are oxidised to H_3PO_4 (phosphoric acid), H_3AsO_4 (arsenic acid) and H_3SbO_4 (antimonic acid) respectively.

$$M\ (M = P, As, Sb) + 5HNO_3 \longrightarrow H_3MO_4 + 5NO_2 + H_2O$$

or

$$2H^+ + 5NO_3^- + M \longrightarrow H_2O + 5NO_2 + MO_4^{3-}$$

(iv) Sulphur is oxidised to H_2SO_4

$$S + 6HNO_3\ (\text{conc. and hot}) \longrightarrow H_2SO_4 + 6NO_2 + 2H_2O \text{ or } 4H^+ + 6NO_3^- + S \longrightarrow 2H_2O + 6NO_2 + SO_4^{2-}$$

(v) Iodine is oxidised to iodic acid, HIO_3.

$$I_2 + 10\ HNO_3 \longrightarrow 3HIO_3 + 10NO_2 + 4H_2O \text{ or } 8H^+ + 10NO_3^- + I_2 \longrightarrow 4H_2O + 10NO_2 + 2IO_3^-$$

This reaction has been used for the preparation of HIO_3.

B. Oxidation of metals which lie above hydrogen in the electrochemical series. Examples of such metals are Mg, Mn, Zn, Fe, Sn, Pb, etc.

(*i*) Mg and Mn liberate H_2, when treated with dil. HNO_3 (1–2%). In this reaction the metal is oxidised to M^{2+} and HNO_3 (H = +1) is reduced to H_2(H = 0).

$$M(M = Mg, Mn) + 2NHO_3 \longrightarrow M(NO_3)_2 + H_2 \quad \text{or} \quad M + 2H^+ \longrightarrow M^{2+} + H_2$$

(*ii*) Zn, Fe, Sn and Pb, when treated with HNO_3, are oxidised to their corresponding nitrates and HNO_3 is reduced to NH_4NO_3 (or NH_4^+ ion), N_2O, NO or NO_2 depending on the nature of the metal, concentration of the acid and temperature employed. Thus:

$$M\ (M{=}Zn, Fe, Sn, Pb) + 2HNO_3 \longrightarrow M(NO_3)_2 + NH_4NO_3, N_2O, NO \text{ or } NO_2 + H_2O$$

Actually the above reaction takes place through the following steps:

(*a*) The metal (M) on being treated with HNO_3 is converted into metallic nitrate and nascent hydrogen is produced.

$$M + 2HNO_3 \longrightarrow M(NO_3)_2 + 2H$$

(*b*) Nascent hydrogen produced as above reduces HNO_3 to NH_4NO_3 (or NH_4^+ ion), N_2O, NO or NO_2.

$$2HNO_3 + 8H \longrightarrow NH_4NO_3 + 3H_2O \quad \text{or} \quad NO_3^- + 10H^+ + 8e^- \longrightarrow NH_4^+ + 3H_2O$$

$$2HNO_3 + 8H \longrightarrow N_2O + 5H_2O \quad \text{or} \quad 2NO_3^- + 10H^+ + 8e^- \longrightarrow N_2O + 5H_2O$$

$$HNO_3 + H \longrightarrow NO_2 + H_2O \quad \text{or} \quad NO_3^- + 2H^+ + e^- \longrightarrow NO_2 + H_2O$$

$$HNO_3 + 3H \longrightarrow NO + 2H_2O \quad \text{or} \quad NO_3^- + 4H^+ + 3e^- \longrightarrow NO + 2H_2O$$

The following examples illustrate the above points:

Zinc. (*a*) Zn reacts with very dil. HNO_3 (6%) and is oxidised to $Zn(NO_3)_2$.HNO_3 is reduced to NH_4NO_3.

$$4Zn + 10\ HNO_3\ (\text{Very dilute}) \longrightarrow 4Zn(NO_3)_2 + NH_4NO_3 + 3H_2O$$

or

$$4Zn + NO_3^- + 10H^+ \longrightarrow 4Zn^{2+} + NH_4^+ + 3H_2O$$

(*b*) Zn reacts with cold and dil. HNO_3 (20%) and is oxidised to $Zn(NO_3)_2$. HNO_3 is reduced to N_2O.

$$4Zn + 10\ HNO_3\ (\text{Cold and dilute}) \longrightarrow 4Zn(NO_3)_2 + N_2O + 5H_2O$$

or

$$4Zn + 2NO_3^- + 10H^+ \longrightarrow 4Zn^{2+} + N_2O + 5H_2O$$

(*c*) Zn reacts with cold and conc. HNO_3 (70%) and is oxidised to $Zn(NO_3)_2$. HNO_3 is reduced to NO_2.

$$Zn + 4HNO_3\ (\text{Cold and conc.}) \longrightarrow Zn(NO_3)_2 + 2NO_2 + 2H_2O$$

or

$$Zn + 2NO_3^- + 4H^+ \longrightarrow Zn^{2+} + 2NO_2 + 2H_2O$$

(*d*) With cold and moderately conc. HNO_3 (50%), NO is evolved and $Zn(NO_3)_2$ is formed.

$$3Zn + 8HNO_3 \longrightarrow 3Zn(NO_3)_2 + 4H_2O + 2NO$$

or

$$3Zn + 2NO_3^- + 8H^+ \longrightarrow 3Zn^{2+} + 2NO + 4H_2O$$

Iron. (*a*) Fe reacts with very dil. HNO_3 and is oxidised to $Fe(NO_3)_2$. HNO_3 is reduced to NH_4NO_3.

$$4Fe + 10HNO_3\ (\textit{very dilute}) \longrightarrow 4Fe(NO_3)_2 + NH_4NO_3 + 3H_2O$$

or $$4Fe + NO_3^- + 10H^+ \longrightarrow 4Fe^{2+} + NH_4^+ + 3H_2O$$

(*b*) Fe is oxidised to $Fe(NO_3)_2$ by dil. HNO_3 and HNO_3 is reduced to N_2O.

$$4Fe + 10\ HNO_3\ (\textit{dilute}) \longrightarrow 4Fe(NO_3)_2 + N_2O + 5H_2O$$

or $$4Fe + 2NO_3^- + 10H^+ \longrightarrow 4Fe^{2+} + N_2O + 5H_2O$$

(*c*) Fe reacts with cold and conc. HNO_3 and is oxidised to $Fe(NO_3)_3$.HNO_3 is reduced to NO_2.

$$Fe + 6HNO_3\ (\textit{cold and conc.}) \longrightarrow Fe(NO_3)_3 + 3NO_2 + 3H_2O$$

or $$Fe + 3NO_3^- + 6H^- \longrightarrow Fe^{3+} + 3NO_2 + 3H_2O$$

(*d*) With hot and conc. HNO_3, Fe is oxidised to $Fe(NO_3)_2$ and HNO_3 is reduced to NO.

$$Fe + 4HNO_3\ (\textit{hot and conc.}) \longrightarrow Fe(NO_3)_3 + 2H_2O + NO$$

or $$Fe + NO_3^- + 4H^+ \longrightarrow Fe^{3+} + NO + 2H_2O$$

Note. Highly conc. HNO_3 renders iron passive (inert).

Lead. (*a*) With cold dil. HNO_3, $Pb(NO_3)_2$ is formed and NO is evolved.

$$3Pb + 8HNO_3\ (\textit{cold and dilute}) \longrightarrow 3Pb(NO_3)_2 + 2NO + 4H_2O$$

$$3Pb + 2NO_3^- + 8H^+ \longrightarrow 3Pb^{2+} + 2NO + 4H_2O$$

(*b*) With conc. HNO_3, $Pb(NO_3)_2$ is formed and NO_2 is evolved.

$$Pb + 4HNO_3\ (\textit{conc.}) \longrightarrow Pb(NO_3)_2 + 2NO_2 + 2H_2O$$

or $$Pb + 2NO_3^- + 4H^+ \longrightarrow Pb^{2+} + 2NO_2 + 2H_2O$$

Tin. (*a*) Dil. HNO_3 oxidises Sn to $Sn(NO_3)_2$ and is itself reduced to NH_4NO_3.

$$4Sn + 10HNO_3\ (\textit{dilute}) \longrightarrow 4Sn(NO_3)_2 + NH_4NO_3 + 3H_2O$$

or $$4Sn + NO_3^- + 10H^+ \longrightarrow 4Sn^{2+} + NH_4^+ + 3H_2O$$

(*b*) With conc. HNO_3 metastannic acid, H_2SnO_3 is formed and HNO_3 is reduced to NO_2.

$$Sn + 4HNO_3\ (\textit{conc.}) \longrightarrow H_2SnO_3 + 4NO_2 + H_2O$$

or $$Sn + 4NO_3^- + 2H^+ \longrightarrow SnO_3^{2-} + 4NO_2 + H_2O$$

C. Oxidation of metals which lie below hydrogen in the electrochemical series. Examples of such metals are Cu, Ag and Hg. These metals when treated with HNO_3, are oxidised to their corresponding nitrates and HNO_3 is reduced to N_2O, NO, NO_2 or N_2 depending on the nature of the metal, concentration of the acid and temperature employed. Thus:

$$M\ (M = Cu, Ag, Hg) + 2HNO_3 \longrightarrow M(NO_3)_2 + N_2O, NO, NO_2 \text{ or } N_2 + H_2O$$

Actually the above reaction takes place through the following steps:

(*a*) Since Cu, Ag and Hg lie below hydrogen in the electrochemical series, they are not able to liberate hydrogen from HNO_3 as Zn, Fe, Sn and Pb can do. In case of these metals HNO_3 acts as an oxidising agent, *i.e.*, HNO_3 gives nascent oxygen and is itself reduced to N_2O, NO, NO_2 or N_2.

$$2HNO_3 \longrightarrow N_2O + H_2O + 4O \quad \text{or} \quad 2NO_3^- + 10H^+ + 8e^- \longrightarrow N_2O + 5H_2O$$

$$2HNO_3 \longrightarrow 2NO + H_2O + 3O \quad \text{or} \quad NO_3^- + 4H^+ + 3e^- \longrightarrow NO + 2H_2O$$

$$2HNO_3 \longrightarrow 2NO_2 + H_2O + O \quad \text{or} \quad NO_3^- + 2H^+ + e^- \longrightarrow NO_2 + H_2O$$

$$2HNO_3 \longrightarrow N_2 + H_2O + 5O \quad \text{or} \quad NO_3^- + 6H^+ + 5e^- \longrightarrow N_2 + 3H_2O$$

(*b*) Nascent oxygen obtained as above, oxidises the metal to its oxide which on reacting with HNO_3, gives metallic nitrate.

$$M + O \longrightarrow MO\ (\textit{Metallic oxide})\ ; \quad MO + 2HNO_3 \longrightarrow M(NO_3)_2 + H_2O$$

The following examples illustrate the above discussion.

Copper. (*a*) Cold and dil. HNO_3 evolves N_2O.

$$10HNO_3\ (\textit{cold and dil.}) + 4Cu \longrightarrow N_2O + 5H_2O + 4Cu(NO_3)_2$$

or $$2NO_3^- + 10H^+ + 4Cu \longrightarrow N_2O + 5H_2O + 4Cu^{2+}$$

(*b*) Cold and moderately conc. HNO_3 evolves NO.

$$3Cu + 8HNO_3\ (\textit{cold and moderately conc.}) \longrightarrow 3Cu(NO_3)_2 + 2NO + 4H_2O$$

or $$3Cu + 2NO_3^- + 8H^+ \longrightarrow 3Cu^{2+} + 2NO + 4H_2O$$

(*c*) Cold and conc. HNO_3 evolves NO_2.

$$Cu + 4HNO_3\ (\textit{cold and conc.}) \longrightarrow Cu(NO_3)_2 + 2NO_2 + 2H_2O$$

or $$Cu + 2NO_3^- + 4H^+ \longrightarrow Cu^{2+} + 2NO_2 + 2H_2O$$

(*d*) Hot and conc. HNO_3 evolves N_2

$$12HNO_3 + 5Cu \rightarrow N_2 + 6H_2O + 5Cu(NO_3)_2 \text{ or } 2NO_3^- + 12H^+ + 5Cu \rightarrow N_2 + 6H_2O + 5Cu^{2+}$$

Silver. (*a*) With dil. HNO_3, $AgNO_3$ is obtained and NO is evolved.

$$3Ag + 4HNO_3\ (\textit{dilute}) \longrightarrow 3AgNO_3 + NO + 2H_2O$$

or $$3Ag + NO_3^- + 4H^+ \longrightarrow 3Ag^+ + NO + 2H_2O$$

(*b*) With conc. HNO_3, NO_2 is evolved.

$$Ag + 2HNO_3\ (\textit{concentrated}) \longrightarrow AgNO_3 + NO_2 + H_2O$$

or $$Ag + NO_3^- + 2H^+ \longrightarrow Ag^+ + NO_2 + H_2O$$

Mercury. (*a*) With dil. HNO_3, mercurous nitrate, $Hg_2(NO_3)_2$ is formed and NO is evolved.

$$6Hg + 8HNO_3\ (\textit{dilute}) \longrightarrow 3Hg_2(NO_3)_2 + 2NO + 4H_2O$$

or $$6Hg + 2NO_3^- + 8H^+ \longrightarrow 3Hg_2^{2+} + 2NO + 4H_2O$$

(*b*) With conc. HNO_3 mercuric nitrate, $Hg(NO_3)_2$ is obtained and NO_2 is evolved.

$$Hg + 4HNO_3\ (\textit{concentrated}) \longrightarrow Hg(NO_3)_2 + 2NO_2 + 2H_2O$$

or $$Hg + 2NO_3^- + 4H^+ \longrightarrow Hg^{2+} + 2NO_2 + 2H_2O$$

D. Oxidation of compounds. Both dil. and conc. HNO_3 can oxidise many compounds. In these reactions, dil. HNO_3 is reduced to NO while conc. HNO_3 is reduced to NO_2. When HNO_3 oxidises $SnCl_2$ to $SnCl_4$ in presence of HCl, it is reduced to NH_2OH. Thus:

$$2HNO_3\ (\textit{dil. acid}) \longrightarrow H_2O + 2NO + 3O \text{ or } NO_3^- + 4H^+ + 3e^- \longrightarrow NO + 2H_2O$$

$$2HNO_3\ (\textit{conc. acid}) \longrightarrow H_2O + 2NO_2 + O \text{ or } NO_3^- + 2H^+ + e^- \longrightarrow H_2O + NO_2$$

$$HNO_3 + 6H \longrightarrow 2H_2O + NH_2OH \text{ or } NO_3^- + 7H^+ + 6e^- \longrightarrow 2H_2O + NH_2OH$$

Examples are given below:

(*a*) I_2 is liberated from KI solution ($I^- \rightarrow I_2^0$).

$$8HNO_3\ (\textit{dil. acid}) + 6KI \longrightarrow 2NO + 4H_2O + 6KNO_3 + 3I_2$$

or $$2NO_3^- + 8H^+ + 6I^- \longrightarrow 2NO + 4H_2O + 3I_2$$

(*b*) CuS is oxidised to sulphur ($S^{2-} \rightarrow S$

$$8HNO_3\ (\textit{dil. acid}) + 3CuS \longrightarrow 2NO + 4H_2O + 3S + 3Cu(NO_3)_2$$

or $$2NO_3^- + 8H^+ + 3S^{2-} \longrightarrow 2NO + 4H_2O + 3S$$

(*c*) H_2S is oxidised to sulphur by both dil. and conc. HNO_3 ($S^{-2} \rightarrow S$).

$$2HNO_3\ (\textit{dil. acid}) + 3H_2S \longrightarrow 2NO + 4H_2O + 3S$$

or $$2NO_3^- + 8H^+ + 3S^{2-} \longrightarrow 2NO + 4H_2O + 3S$$

$$2HNO_3\ (\textit{conc. acid}) + H_2S \longrightarrow 2NO_2 + S + 2H_2O$$

or $$2NO_3^- + 4H^+ + S^{2-} \longrightarrow 2H_2O + 2NO_2 + S$$

(*d*) FeS is oxidised to $FeSO_4$ ($S^{2-} \rightarrow SO_4^{2-}$).

$$FeS + 8HNO_3\ (conc.\ acid) \longrightarrow FeSO_4 + 8NO_2 + 4H_2O$$

or $$8NO_3^- + 8H^+ + S^{2-} \longrightarrow 4H_2O + 8NO_2 + SO_4^{2-}$$

(*e*) SO_2 is oxidised to H_2SO_4 by both dilute and concentrated HNO_3 ($SO_2 \rightarrow SO_4^{2-}$).

$$SO_2 + 2HNO_3\ (conc.\ acid) \longrightarrow H_2SO_4 + 2NO_2 \text{ or } 2NO_3^- + SO_2 \longrightarrow 2NO_2 + SO_4^{2-}$$

$$2HNO_3\ (dil.acid) + 3SO_2 + 2H_2O \longrightarrow 2NO + 3H_2SO_4$$

or $$2NO_3^- + 2H_2O + 3SO_2 \longrightarrow 2NO + 3SO_4^{2-} + 4H^+$$

(*f*) Solution of $FeSO_4$ is oxidised to $Fe_2(SO_4)_3$ by both dil. and conc. HNO_3 in presence of H_2SO_4 ($Fe^{2+} \rightarrow Fe^{3+}$).

$$\underset{\text{Green solution}}{6FeSO_4} + 3H_2SO_4 + \underset{\text{dil.acid}}{2HNO_3} \longrightarrow \underset{\text{Yellow solution}}{3Fe_2(SO_4)_3} + 2NO + 4H_2O$$

or $$NO_3^- + 4H^+ + 3Fe^{2+} \longrightarrow NO + 2H_2O + 3Fe^{3+}$$

$$2FeSO_4 + \underset{\text{conc. acid}}{2HNO_3} + H_2SO_4 \longrightarrow Fe_2(SO_4)_3 + 2NO_2 + 2H_2O$$

or $$NO_3^- + 2H^+ + Fe^{2+} \longrightarrow H_2O + NO_2 + Fe^{3+}$$

(*g*) HBr and HI are oxidised to Br_2 and I_2 respectively ($Br^- \rightarrow Br_2$, $I^- \rightarrow I_2$).

$$2HNO_3\ (dil.\ acid) + 6HX \longrightarrow 2NO + 4H_2O + 3X_2$$

$$2NO_3^- + 8H^+ + 6X^- \longrightarrow 2NO + 4H_2O + 3X_2$$

$$2HNO_3\ (conc.\ acid) + 2HX \longrightarrow X_2 + 2NO_2 + 2H_2O$$

$$2NO_3^- + 4H^+ + 2X^- \longrightarrow 2H_2O + 2NO_2 + X_2 \qquad (X = Br, I)$$

(*h*) As_2O_3 is oxidised to H_3AsO_4 ($As^{3+} \rightarrow AsO_4^{3-}$).

$$2HNO_3 + As_2O_3 + H_2O \longrightarrow 4NO_2 + 2H_3AsO_4$$

or $$2NO_3^- + As^{3+} + 2H_2O \longrightarrow 2NO_2 + AsO_4^{3-} + 4H^+$$

(*i*) Cane sugar (sucrose), $C_{12}H_{22}O_{11}$ is oxidised to oxalic acid ($H_2C_2O_4$) by conc. HNO_3.

$$36HNO_3\ (conc.\ acid) + C_{12}H_{22}O_{11} \longrightarrow 23H_2O + 36NO_2 + 6H_2C_2O_4$$

(*j*) Toluene ($C_6H_5CH_3$) is oxidised to benzoic acid (C_6H_5COOH) by dil. HNO_3.

$$2HNO_3\ (\text{dil. acid}) + C_6H_5CH_3 \longrightarrow 2NO + 2H_2O + C_6H_5COOH$$

(*k*) $SnCl_2$ is oxidised to $SnCl_4$ in presence of HCl ($Sn^{2+} \rightarrow Sn^{4+}$).

$$3SnCl_2 + 6HCl + HNO_3 \longrightarrow 2H_2O + 3SnCl_4 + NH_2OH$$

or $$3Sn^{2+} + NO_3^- + 7H^+ \longrightarrow 3Sn^{4+} + 2H_2O + NH_2OH$$

7. Action of highly concentrated HNO_3 on Cr, Ni, Al, Fe etc. These metals become passive (inert) when treated with conc. HNO_3. For example, Fe displaces Cu from $CuSO_4$ solution but this property of Fe is lost when it is dipped in highly concentrated HNO_3.

Cause of passivity or inertness. It is assumed that when Fe reacts with conc. HNO_3, a thin film of the oxide of iron is formed and this film gets deposited on the surface of the metal and hence prevents the action of the acid on the metal.

8. Action on conc. HCl : Formation of aqua regia. When 1 part of conc. HNO_3 is mixed with 3 parts of conc. HCl, we get aqua regia. Aqua regia produces nascent chlorine (Cl) along with nitrosyl chloride (NOCl).

$$HNO_3 + 3HCl \longrightarrow NOCl + 2H_2O + 2Cl$$

According to another view, in place of NOCl, nitric oxide (NO) is generated.

$$HNO_3 + 3HCl \longrightarrow NO + 2H_2O + 3Cl$$

Due to production of nascent chlorine, aqua regia is very active. For example Au and Pt (noble metals) which are not acted upon by HNO_3 get dissolved in aqua regia due to the formation of soluble complexes namely $H[AuCl_4]$ and $H_2[PtCl_6]$ respectively.

$$2Au + 3HNO_3 + 11HCl \longrightarrow H[AuCl_4] + 2NOCl + 6H_2O$$

This equation can also be written as:

$$Au + HNO_3 + 4HCl \longrightarrow H[AuCl_4] + NO + 2H_2O$$

$$Pt + 2HNO_3 + 8HCl \longrightarrow H_2[PtCl_6] + 2NOCl + 4H_2O$$

This equation can be also written as :

$$3Pt + 4HNO_3 + 18HCl \longrightarrow 4NO + 3H_2[PtCl_6] + 8H_2O$$

9. Action of conc. HNO_3 on aromatic organic compounds in presence of conc. H_2SO_4. Conc. HNO_3 attacks many aromatic organic compounds (*e.g.*, C_6H_6, $C_6H_5\ CH_3$, C_6H_5OH etc.) in presence of conc. H_2SO_4 and substitutes one or more H-atoms in these organic compounds by nitro group, producing nitro compounds (nitration). Here H_2SO_4 absorbs water liberated in the reaction and also produces nitronium ion (NO_2^+) by reacting with HNO_3. This NO_2^+ ion makes the nitration of aromatic compounds.

(*a*)

$$HNO_3 + H_2SO_4 \longrightarrow HSO_4^- + NO_2^+ + H_2O$$

$$C_6H_6 + NO_2^+ \longrightarrow C_6H_5NO_2 + H^+$$

$$HNO_3 + H_2SO_4 + C_6H_6 \longrightarrow C_6H_5NO_2 + H^+ + HSO_4^- + H_2O$$

or

$$HNO_3 + C_6H_6 \xrightarrow{\text{conc. } H_2SO_4} \underset{\text{Nitrobenzene}}{C_6H_5NO_2} + H_2O$$

(*b*)

$$[HNO_3 + H_2SO_4 \longrightarrow HSO_4^- + NO_2^+ + H_2O] \times 3$$

$$C_6H_5CH_3 + 3NO_2^+ \longrightarrow C_6H_2\ (NO_2)_3\ CH_3 + 3H^+$$

$$3HNO_3 + 3H_2SO_4 + C_6H_5\ CH_3 \longrightarrow 3HSO_4^- + 3H_2O + C_6H_2(NO_2)_3CH_3 + 3H^+$$

or

$$3HNO_3 + C_6H_5\ CH_3 \xrightarrow{\text{conc. } H_2SO_4} \underset{\text{Trinitro toluene (TNT)}}{C_6H_2(NO_2)_3CH_3} + 3H_2O$$

(*c*)

$$[HNO_3 + H_2SO_4 \longrightarrow HSO_4^- + NO_2^+ + H_2O] \times 3$$

$$C_6H_5OH + 3NO_2^+ \longrightarrow C_6H_2\ (NO_2)_3OH + 3H^+$$

$$3HNO_3 + 3H_2SO_4 + C_6H_5OH \longrightarrow 3HSO_4^- + 3H_2O + C_6H_2\ (NO_2)_3OH + 3H^+$$

or

$$3HNO_3 + \underset{\text{Phenol}}{C_6H_5OH} \xrightarrow{\text{conc. } H_2SO_4} \underset{\text{Picric acid}}{C_6H_2(NO_2)_3OH} + 3H_2O$$

(*d*) Glycerol with conc. HNO_3 in presence of conc. H_2SO_4 below 25°C forms glycerol trinitrate (nitroglycerine).

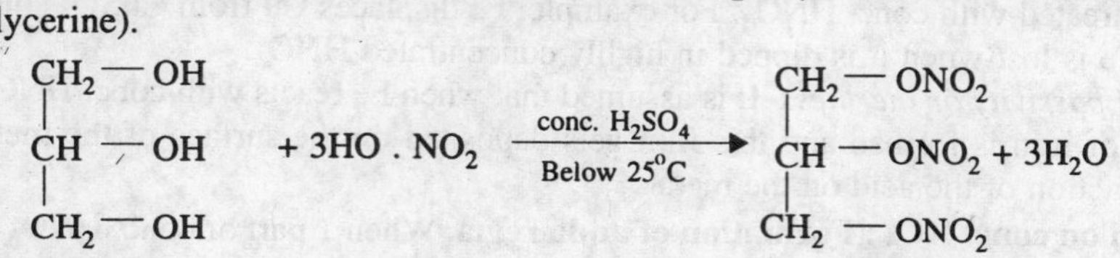

Glycerol Nitroglycerine

With protein, HNO_3 gives yellow coloured *xanthoprotein*. The acid produces yellow stain on skin and wood, since the skin and wood both contain protein.

Uses. HNO_3 is used (*i*) in the manufacture of (*a*) fertilisers like basic calcium nitrate [$Ca(NO_3)_2\ CaO$], ammonium nitrate etc. (*b*) explosives like TNT, gun cotton, nitroglycerine, picric acid etc. (*c*) H_2SO_4 by lead chamber process (*d*) dyes, perfumes, artifical silk, drugs etc. (*e*) nitrates of Ag, Na K etc. (*ii*) in electroplating (for refining the metal surface) (*iii*) in making batteries (*iv*) in the purification of Ag and Au by aqua regia (*v*) as an oxidising and nitrating agent in laboratory.

Nitrates (NO_3^-)

We have already said that HNO_3 is a monobasic acid and hence gives only one type of salts which are called nitrates. These salts contain nitrate ion, NO_3^-.

Preparation. $NaNO_3$ and KNO_3 occur in nature while other nitrates can be prepared by the action of HNO_3 on metal, metallic oxide, hydroxide, carbonate, bicarbonate or sulphite.

$$Ag + 2HNO_3 \longrightarrow AgNO_3 + NO_2 + H_2O$$
$$CaO + 2HNO_3 \longrightarrow Ca(NO_3)_2 + H_2O$$
$$KOH + HNO_3 \longrightarrow KNO_3 + H_2O$$
$$CaCO_3 + 2HNO_3 \longrightarrow Ca(NO_3)_2 + H_2O$$
$$(NH_4)_2CO_3 + 2HNO_3 \longrightarrow 2NH_4NO_3 + H_2O + CO_2$$
$$NaHCO_3 + HNO_3 \longrightarrow NaNO_3 + H_2O + CO_2$$
$$Na_2SO_3 + 2HNO_3 \longrightarrow 2NaNO_3 + SO_2 + H_2O$$

Properties. (*i*) *Solubility.* All nitrates are soluble in water.

(*ii*) *Action of heat.* (*a*) Alkali metal nitrates form nitrites and O_2 on heating.

$$2MNO_3 \text{ (M = Na, K)} \xrightarrow{\Delta} 2MNO_2 + O_2$$

(*b*) Nitrates of alkaline earth metals and heavy metals decompose to give a mixture of NO_2 and O_2 alongwith a solid residue of metallic oxide.

$$2M(NO_3)_2 \text{ (M = Ca, Pb, Cu)} \xrightarrow{\Delta} 2MO + 4NO_2 + O_2$$

(*c*) Nitrates of metals having low activity (*e.g.*, Ag, Hg etc.) decompose into the metal and a mixture of NO_2 and O_2.

$$2AgNO_3 \xrightarrow{\Delta} 2Ag + 2NO_2 + O_2\ ; \qquad Hg(NO_3)_2 \xrightarrow{\Delta} Hg + 2NO_2 + O_2$$

(*d*) NH_4NO_3 gives N_2O on heating.

$$NH_4NO_3 \xrightarrow{\Delta} N_2O + 2H_2O$$

(*iii*) *Action of conc. H_2SO_4 (Test of NO_3^- ion).* When heated with conc. H_2SO_4, nitrates undergo double decomposition reaction, forming free HNO_3. Some HNO_3 gets decomposed to give brown fumes of NO_2.

$$KNO_3 + H_2SO_4 \longrightarrow KHSO_4 + HNO_3$$
$$4HNO_3 \longrightarrow 4NO_2 + 2H_2O + O_2$$
or
$$4H^+ + 4NO_3^- \longrightarrow 4NO_2 + 2H_2O + O_2$$

The evolution of NO_2 increases on heating the reaction mixture with Cu turnings or paper pellets.

$$Cu + 4HNO_3 \longrightarrow Cu(NO_3)_2 + 2H_2O + 2NO_2$$
or
$$Cu + 4H^+ + 2NO_3^- \longrightarrow Cu^{2+} + 2H_2O + 2NO_2$$
$$C\ (\textit{from paper pellets}) + 4NHO_3 \longrightarrow 2H_2O + 4NO_2 + CO_2$$
or
$$C + 4H^+ + 4NO_3^- \longrightarrow 2H_2O + 4NO_2 + CO_2$$

(*iv*) *Oxidising nature.* Nitrates are good oxidising agents.

(*v*) *Action of freshly prepared $FeSO_4$ solution in presence of conc. H_2SO_4 (Ring test of NO_3^- ion).* When conc. H_2SO_4 is added carefully by the sides of the test tube containing a nitrate and freshly prepared solution of $FeSO_4$, a brown ring is formed at the junction of the two liquids. Brown ring is due to the formation of complex ion, $[Fe(NO)]^{2+}$ (called nitrosoferrous ion). The formation of $[Fe(NO)]^{2+}$ takes place through the following steps:

(*a*) First of all the nitrate reacts with conc. H_2SO_4 and HNO_3 is produced.

$$NaNO_3 + H_2SO_4 \longrightarrow NaHSO_4 + HNO_3 \quad \text{or} \quad NO_3^- + H^+ \longrightarrow HNO_3$$

(*b*) HNO_3 produced as above oxidises $FeSO_4$ to $Fe_2(SO_4)_3$ in presence of conc. H_2SO_4 and is itself reduced to NO.

$$6FeSO_4 + 2HNO_3 + 3H_2SO_4 \longrightarrow 3Fe_2(SO_4)_3 + 2NO + 4H_2O$$

or
$$3Fe^{2+} + NO_3^- + 4H^+ \longrightarrow 3Fe^{3+} + NO + 2H_2O$$

(*c*) NO evolved in the above reaction reacts with $FeSO_4$ solution and produces a brown ring at the junction of the two liquids. This ring is due to the formation of nitrosoferrous sulphate, $[Fe(NO)]SO_4$. (*Complex compound*).

$$FeSO_4 + NO \longrightarrow [Fe(NO)]\ SO_4 \quad \text{or} \quad Fe^{2+} + NO \longrightarrow [Fe(NO)]^{2+}$$

Actually the complex compound is $[Fe(NO)\ (H_2O)_5]^{2+}$.

$$FeSO_4 + NO + 5H_2O \longrightarrow [Fe(NO)\ (H_2O)_5]SO_4 \quad \text{or} \quad Fe^{2+} + NO + 5H_2O \longrightarrow [Fe(NO)\ (H_2O)_5]^{2+}$$

Structure of HNO_3 molecule of NO_3^- ion. See chapter 15.

Some important nitrates

Here we shall discuss the preparation, properties and uses of some important nitrates. These nitrates are used for various purposes.

1. Sodium nitrate, $NaNO_3$. This salt is commonly known as *Chile saltpetre* or Chile nitre as it occurs in the native state in large quantities in the rainless districts of Chile. Crude chile salt is called *caliche*. Caliche is usually impure as it contains sand, clay and other salts (*e.g.*, NaCl, $NaIO_3$, $NaClO_3$, KNO_3, $NaClO_4$ etc.) as impurities. Crude chile salt contains 20-50% of $NaNO_3$.

Preparation. (*i*) *From Caliche.* We have said above that caliche contains $NaNO_3$ and the impurities of other salts. This is crushed, luxated with hot water. From this solution $NaNO_3$ is separated by crystallisation.

(*ii*) $NaNO_3$ can also be obtained by the action of HNO_3 on NaOH or Na_2CO_3.

$$NaOH + HNO_3 \longrightarrow NaNO_3 + H_2O; \quad Na_2CO_3 + 2HNO_3 \longrightarrow 2NaNO_3 + H_2O + CO_2$$

Properties. (*i*) It is a colourless crystalline substance, deliquescent, m.pt. = 316°C, density = 2.26. It has a cold bitter taste. It is very soluble in water and the solubility increases with the rise of temperature. It gets decomposed into O_2 and $NaNO_2$.

$$NaNO_3 \xrightarrow{\Delta} NaNO_2 + \frac{1}{2}O_2$$

In presence of Cu or Pb this type of decompositon takes place at a low temperature.

$$NaNO_3 + Pb \longrightarrow NaNO_2 + PbO$$

Uses. It is used (*i*) as a nitrogenous fertiliser (*ii*) as a raw material in the manufacture of HNO_3, $NaNO_2$, KNO_3, Na_3AsO_4, glass and fire work. (*iii*) in dynamite (*iv*) for curing meat.

2. Potassium nitrate, KNO_3. This salt is also called by many other names like nitre, *Indian salt petre, Bengal saltpetre, Shora or Suverchi.* This is an excellent fertiliser, since it contains both nitrogen and potassium. It contains 50% K_2O. It occurs in large quantities in the soil of India (U.P., Bihar, Orissa, Bengal and Punjab), Iran, Peru, Chile etc.

Now it is prepared by mixing saturated hot sulutions of $NaNO_3$ and KCl. NaCl, being less soluble is separated from KNO_3 by fractional crystallisation.

$$NaNO_3 + KCl \longrightarrow NaCl\ (\textit{Less soluble}) + KNO_3$$

3. Ammonium nitrate, NH_4NO_3. Preparation. (*i*) It is prepared by neutralising HNO_3 with NH_3 or $(NH_4)_2CO_3$.

$$HNO_3 + NH_3 \longrightarrow NH_4NO_3 \quad ; \quad 2HNO_3 + (NH_4)_2CO_3 \longrightarrow 2NH_4\ NO_3 + H_2O + CO_2$$

(*ii*) On a large scale, it is obtained : (*a*) by passing NH_3 gas into 60% HNO_3. (*b*) by double decomposition of $NaNO_3$ and $(NH_4)_2SO_4$ in solution

$$NaNO_3 + (NH_4)_2SO_4 \longrightarrow 2NH_4NO_3 + Na_2SO_4\ (\textit{Least soluble})$$

$Na_2SO_4.10H_2O$, being least soluble at the temperature of hot solution, crystallises out first and NH_4NO_3 is obtained by evaporating the mother liquor.

Properties. (*i*) It is a colourless crystalline substance which deliquescences in moist air.

(*ii*) It dissolves readily in water with the absorption of heat. It is also soluble in alcohol.

(*iii*) *Action of heat.* (*a*) The anhydrous salt melts at 169.5° and on further heating at nearly 220°C, it gets decomposed into nitrous oxide (N_2O) and H_2O.

$$NH_4NO_3 \xrightarrow{220°C} N_2O + 2H_2O$$

(*b*) On being heated above 250°C, it deflagrates with a yellow flame.

$$2NH_4NO_3 \xrightarrow{>250°C} N_2 + 2NO + 4H_2O$$

(*c*) At higher temperature it detonates as:

$$2NH_3NO_3 \xrightarrow{\text{Higher temp.}} 2N_2 + O_2 + 4H_2O$$

Under ordinary conditions, NH_4NO_3 is not explosive but when stored in bulk, it may detonate with disastrous results.

(*iv*) *Action of trinitrotoluene, $C_6H_2(CH_3)(NO_2)_3$.* When mixed with 20% trinitrotoluene, amatol which is an explosive material, is obtained.

$$2C_6H_2(CH_3)(NO_2)_3 + 7NH_4NO_3 \longrightarrow 10N_2 + 14CO + 19H_2O$$

(*v*) *Action of Al-metal.* When NH_4NO_3 is mixed with a small quantity of Al-metal, ammonal which like amatol is an explosive substance is obtained.

$$3NH_4NO_3 + 2Al \longrightarrow 3N_2 + Al_2O_3 + 6H_2O$$

Uses. NH_4NO_3 is used. (*i*) in the preparation of N_2O (*ii*) for the preparation of calcium ammonqium nitrate, $Ca(NO_3)_2.NH_4NO_3$ and ammonium sulphate nitrate, $(NH_4)_2SO_4.2NH_4NO_3$ both of which are used as fertilisers (*iii*) in making explosives like amatol and ammonal (*iv*) in freezing mixtures.

4. Calcium ammonium nitrate (CAN) or nitrolime stone or Nangal fertiliser, $Ca(NO_3)_2$ NH_4NO_3. This compound is manufactured at Nangal in Punjab. The process used consists of the following steps:

(*a*) NH_3 is produced by Habber's process

$$N_2 + 3H_2 \xrightarrow[(Fe_2O_3\ +\ Al_2O_3\ +\ K_2O)]{450-500°C,\ 150-200\ \text{atm pressure}} 2NH_3$$

(*b*) Some portion of NH_3 obtained as above is converted into HNO_3 by Ostwald's process.

$$4NH_3 + 5O_2\ (\textit{from air}) \xrightarrow[750\ -\ 900°C]{Pt\ -\ \text{guage}} 4NO + 6H_2O$$

$$2NO + O_2\ (\textit{from air}) \longrightarrow 2NO_2$$

$$4NO_2 + O_2 + 2H_2O \longrightarrow 4NHO_3$$

(*c*) Now the remaining part of NH_3 is absorbed by the excess of HNO_3 to get a solution of NH_4NO_3.

$$NH_3 + HNO_3\ (\textit{excess}) \longrightarrow NH_4NO_3\ (\textit{solution})$$

The solution of NH_4NO_3 obtained as above also contains some unreacted HNO_3. This HNO_3 is neutralised by limestone $(CaCO_3)$ when $Ca(NO_3)_2$ is obtained.

$$2HNO_3 + CaCO_3 \longrightarrow Ca(NO_3)_2 + H_2O + CO_2\uparrow$$

(*d*) $Ca(NO_3)_2$ and NH_4NO_3 obtained as above at step (*c*) combine together to form the pellets of CAN.

$$Ca(NO_3)_2 + NH_4NO_3 \longrightarrow Ca(NO_3)_2.NH_4NO_3\ (CAN)$$

Limestone is added as a filler. Its function is to render NH_4NO_3 safe to handle, since NH_4NO_3 is an explosive substance. Since CAN is very hygroscopic, it has to be protected from the atmospheric moisture during its transportation and storage. For this the pellets of CAN are stirred with concentrated solution of finely powdered soap stone (sodium silicate). This treatment produces a thin layer of soap stone on the surface of pellets. The pellets are dried and packed in polythene lined bags.

Use of CAN as fertiliser. CAN is highly soluble in water and hence it can be assimilated by the plants very easily. It contains about 20% nitrogen. It is directly assimilated by plants and does not undergo any change in the soil.

5. Ammonium sulphate nitrate, $(NH_4)_2SO_4.2NH_4NO_3$ (ASN). It is a double salt composed of $(NH_4)_2SO_4$ and NH_4NO_3. It is prepared by a process which consists of the following steps:

(*a*) NH_3 is prepared by Habber's process.

$$N_2 + 3H_2 \xrightarrow[(Fe_2O_3\ +\ Al_2O_3\ +\ K_2O)]{450-500°C,\ 150-200\ \text{atm pressure}} 2NH_3$$

(*b*) Some portion of NH_3 obtained as above is convered into HNO_3 by Ostwald's process.

$$4NH_3 + 5O_2\ (\textit{from air}) \xrightarrow[750\ -\ 900°C]{Pt\ -\ \text{guage}} 4NO + 6H_2O$$

$$2NO + O_2\ (\textit{fromair}) \longrightarrow 2NO_2\ ; \qquad 4NO_2 + O_2 + 2H_2O \longrightarrow 4NHO_3$$

(*c*) Now the remaining portion of NH_3 is absorbed by a mixture of HNO_3 (*obtained as above*) and H_2SO_4 so that we get $(NH_4)_2SO_4$. $2NH_4NO_3$.

$$NH_3 + HNO_3 \longrightarrow NH_4NO_2; \qquad 2NH_3 + H_2SO_4 \longrightarrow (NH_4)_2SO_4$$

$$2NH_4NO_3 + (NH_4)_2SO_4 \longrightarrow (NH_4)_2SO_4\ .\ 2NH_4NO_3$$

Uses. ASN is used as nitrogenous fertiliser for sugar cane crops.

6. Basic calcium nitrate, $Ca(NO_3)_2.CaO$. It is also called *Norwegian saltpetre* or *nitrate of lime.* The process used to prepare it consists of the following steps:

(*a*) Finely powder limestone ($CaCO_3$) is treated with HNO_3 so that crystals of $Ca(NO_3)_2$ are obtained.

$$CaCO_3 + HNO_3 \longrightarrow Ca(NO_3)_2 + H_2O + CO_2\uparrow$$

(*b*) Crystals of $Ca(NO_3)_2$ obtained as above are mixed with lime (CaO) to get basic calcium nitrate, $Ca(NO_3)_2$. CaO.

$$Ca(NO_3)_2 + CaO \longrightarrow Ca(NO_3)_2.CaO$$

$Ca(NO_3)_2.CaO$ is highly deliquescent and hence is packed in water proof bags. It is very soluble in water. It is source of calcium and nitrogen both and is basic in nature. It is directly assimilated by plants and does not undergo any change in the soil. It is less deliquescent than ordinary calcium nitrate, $Ca(NO_3)_2.4H_2O$ and hence is used in preference to $Ca(NO_3)_2.4H_2O$ as a fertiliser.

IV. Peroxynitric acid $(OH)NO_3$ or HNO_4

Preparation. It is prepared by dissolving N_2O_5 in anhydrous H_2O_2.

Properties. (*i*) It is unstable and explosive. (*ii*) It smells like bleaching powder. (*iii*) It is a powerful oxidising agent and oxidises aniline to nitrobenzene and liberates Br_2 from KBr solution. (*iv*) HNO_4 ionises as $NHO_4 \rightleftharpoons H^+ + NO_4^-$. This mode of ionisation shows that HNO_4 is a monobasic acid and hence gives only one type of salts, called peroxy nitrates. These salts contain peroxy nitrate ion, NO_4^-..

Peroxy nitrates (NO_4^-)

As already said HNO_4 gives only one type of salts which are called peroxy nitrates and contain peroxy nitrate ion, NO_4^-. Potassium peroxy nitrate (KNO_4) is the most important salt. It is prepared by cooling a concentrated solution of KNO_3 in a freezing mixture and adding to it Na_2O_2 gradually and alcohol.

$$KNO_3 + Na_2O_2 + H_2O \longrightarrow 2NaOH + KNO_4$$

Peroxy nitrates are fairly stable but get easily decomposed by water.

Structure of HNO_4 molecule and NO_4^- ion. See chapter 15

University Questions

1. Which compound does not dissolve in hot, dilute HNO_3? (*a*) HgS (*b*) PbS (*c*) CuS (*d*) CdS *(I.I.T., 96)*
2. Draw the structures of the oxy-acids of nitrogen. *(Meerut, 85)*
3. Discuss the structure and constitutions of hyponitrous acid. *(Allahabad, 87)*
4. Write balanced equations. (*i*) When urea is heated with HNO_2. (*ii*) for the preparation of (*a*) NO from HNO_3 (*b*) HNO_3 from NH_3 (*c*) HNO_3 from N_2 and O_2.

5. Name the catalyst used in the oxidation of NH_3 by Ostwald's process for the manufacture of HNO_3. *(MLNR, 86)*

6. Explain the following, giving appropriate reasons:

(*i*) HNO_3 is stronger acid than HNO_2.

(*ii*) Conc. HNO_3 turns yellow in sunlight.

(*iii*) Al-metal reacts radily with dil. HNO_3 to give $Al(NO_3)_3$. This does not happen when Al reacts with conc. HNO_3.

(*iv*) Fe is not attacked by conc. HNO_3.

(*v*) Fe previously dipped in conc. HNO_3 does not precipitate Cu from $CuSO_4$ solution.

(*vii*) HNO_3 acts only as an oxidising agent but HNO_2 can act both as an oxidising and reducing agent. *(I.I.T., 80; M.D., 84.)*

7. Write a note on "Pernitric acid". *(Madras, 86)*

32

Oxy-acids of Phosphorus and their Salts

Oxy-acids formed by phosphorus can be grouped into the following three series:

(A) Phosphorus acid series. This series contains the following acids: (*i*) *Hypophosphorus acid,* H_3PO_2(P = +1) (*ii*) *Phosphorus acid,* H_3PO_3 (P = +3) (*iii*) *Pyrophosphorus acid,* $H_4P_2O_5$. This acid is also called *diphosphorus acid,* since it may be regarded as having been obtained by eliminating one H_2O molecule from two molecules of phosphorus acid, H_3PO_3 ($2H_3PO_3 \xrightarrow{-H_2O} H_4P_2O_5$) (*iv*) *Metaphosphorus acid,* HPO_2. This acid may be supposed to have been derived by removing one H_2O molecule from one molecule of phosphorus acid, H_3PO_3 ($H_3PO_3 \xrightarrow{-H_2O} HPO_2$).

(B) Phosphoric acid series. This series consists of the following acids: (*i*) *Hypophosphoric acid,* $H_4P_2O_6$ (P = + 4) (*ii*) *Phosphoric or orthophosphoric acid,* H_3PO_4 (P = +5) (*iii*) *Poly (or condensed) phosphoric acids,* $H_{n+2}P_nO_{3n+1}$. These acids may be assumed to have been derived by eliminating $(n-1)H_2O$ molecules from nH_3PO_4 molecules.

$$nH_3PO_4 \xrightarrow{-(n-1)\,H_2O} H_{n+2}P_n\,O_{3n+1}$$

Important examples of these acids are *diphosphoric acid* (also called *pyrophosphoric acid, $H_4P_2O_7$), triphosphoric acid* ($H_5P_3O_{10}$) and *tetraphosphoric acid* ($H_6P_4O_{13}$)

$$2H_3PO_4 \xrightarrow{-H_2O} H_4P_2O_7 \text{ (Diphosphoric acid)} + H_2O$$

$$3H_3PO_4 \xrightarrow{-2H_2O} H_5P_3O_{10} \text{ (Triphosphoric acid)} + 2H_2O$$

$$4H_3PO_4 \xrightarrow{-3H_2O} H_6P_4O_{13} \text{ (Tetraphosphoric acid)} + 3H_2O$$

(*iv*) *Metaphosphoric acid, HPO_3.* This acid may be supposed to have been obtained by removing one H_2O molecule from one molecule of phosphoric acid, H_3PO_4 ($H_3PO_4 \xrightarrow{-H_2O} HPO_3$).

(C) Peroxy phosphoric acids. These are : (*i*) *Peroxy-monophosphoric acid, H_3PO_5* (*ii*) *Peroxy-diphosphoric acid, $H_4P_2O_8$.*

Hypophosphorus acid, H_3PO_2 (P = +1, monobasic acid)

Preparation. Commercially it is obtained by the following methods:

(*i*) *From barium hypophosphite, $Ba(H_2PO_2)_2$.* When yellow phosphorus is boiled with baryta solution [$Ba(OH)_2$], $Ba(H_2PO_2)_2$ is precipitated which is filtered off.

$$\text{P (yellow)} + 3Ba(OH)_2 + 6H_2O \longrightarrow 3Ba(H_2PO_2)_2\downarrow + 2PH_3$$

$Ba(H_2PO_2)_2$ obtained as above is treated with calculated quantity of H_2SO_4. The insoluble $BaSO_4$ is filtered out and H_3PO_2 is obtained in the filtrate.

$$Ba(H_2PO_2)_2 + H_2SO_4 \longrightarrow BaSO_4\downarrow + 2H_3PO_2 \text{ (filtrate)}$$

When the filtrate containing H_3PO_2 is evaporated, syrupy mass is obtained. This mass can be crystallised by careful cooling below 0°C.

(ii) From $Ca(H_2PO_2)_2$. When $Ca(H_2PO_2)_2$ [obtained by heating white phosphorus with milk of lime, $Ca(OH)_2$] is treated with calculated quantity of $H_2C_2O_4$, CaC_2O_4 is precipitated and H_3PO_2 goes into solution.

$$Ca(H_2PO_2)_2 + H_2C_2O_4 \longrightarrow CaC_2O_4\downarrow + 2H_3PO_2\ (\textit{Solution})$$

The solution containing H_3PO_2 is concentrated to crystallisation in vacuum pans to get the crystals of H_3PO_2.

Properties. *(i)* It is a colourless crystalline solid having sp. gravity = 1.492 and m.pt. = 26.5°C. It is readily soluble in water.

(ii) Acidic nature. It is *monobasic* acid and hence ionises to give only one H^+ ion.

$$H_3PO_2 \rightleftharpoons H^+ + H_2PO_2^-,\ K = 1 \times 10^{-2}$$

$H_2PO_2^-$ ion is called *hypophosphite ion.* Thus H_3PO_2 forms only one type of salts, called *hypophosphites.* These salts contain hypophosphite ion, $H_2PO_2^-$.

(iii) Action of heat: Disproportionation. When heated to 140°C, phosphorus acid (H_3PO_3) and PH_3 are obtained. When heated to 160°C, phosphoric acid (H_3PO_4) and PH_3 are obtained. In both cases, H_3PO_2 undergoes disproportionation.

$$3H_3PO_2\ (P = +1) \xrightarrow{140°C} 2H_3PO_3\ (P = +3) + PH_3\ (P = -3)$$

$$2H_3PO_2\ (P = +1) \xrightarrow{160°C} H_3PO_4\ (P = +5) + PH_3\ (P = -3)$$

(iv) Action of nascent hydrogen (H). It is reduced to PH_3 by nascent hydrogen.

$$H_3PO_2\ (P = +1) + 4H \xrightarrow{Zn\ +\ dil.\ H_2SO_4} PH_3\ (P = -3) + 2H_2O$$

(v) Action on active metals. When active metals like Zn, Mg etc. are treated with H_3PO_2, their corresponding hypophosphites are obtained.

$$M(M = Zn, Mg) + 2H_3PO_2 \longrightarrow M(H_2PO_2)_2 + H_2$$

(vi) Reducing nature. The acid is a powerful reducing agent, since it can easily lose nascent hydrogen. When H_3PO_2 reduces a certain substance, it gets oxidised to H_3PO_4.

$$H_3PO_2\ (P = +1) + 2H_2O \longrightarrow H_3PO_4\ (P = +5) + 4H$$

or $$H_2PO_2^-\ (P = +1) + 2H_2O \longrightarrow H_2PO_4^-\ (P = +5) + 4H^+ + 4e^-$$

Thus :

(a) H_3PO_2 precipitates Ag and Au from their salt solutions.

$$H_3PO_2 + 2H_2O + 4AgNO_3 \longrightarrow H_3PO_4 + 4Ag + 4HNO_3$$

or $$H_2PO_2^- + 2H_2O + 4Ag^+ \longrightarrow H_2PO_4^- + 4H^+ + 4Ag^o$$

Similarly, $$3H_3PO_2 + 6H_2O + 4AuCl_3 \longrightarrow 3H_3PO_4 + 4Au + 12HCl$$

or $$3H_2PO_2^- + 6H_2O + 4Au^{3+} \longrightarrow 3H_2PO_4^- + 4Au + 12H^+$$

(b) $HgCl_2$ is first reduced to Hg_2Cl_2 (*white ppt.*) and finally to metallic Hg (*grey ppt.*)

$$4HgCl_2 + H_3PO_2 + H_2O \longrightarrow 2Hg_2Cl_2 + 4HCl + H_3PO_4$$

$$2Hg_2Cl_2 + H_3PO_2 + 2H_2O \longrightarrow 4Hg + 4HCl + H_3PO_4$$

(c) Cl_2 and I_2 are reduced to their corresponding hydracids namely HCl and HI respectively.

$$H_3PO_2 + 2H_2O + 2X_2\ (X = Cl, I) \longrightarrow H_3PO_4 + 4HX$$

or $$H_2PO_2^- + 2H_2O + 2X_2 \longrightarrow H_2PO_4^- + 4H^+ + 4X^-$$

(d) $CuSO_4$ is reduced to CuH.

$$4CuSO_4 + 3H_3PO_2 + 6H_2O \longrightarrow 4CuH + 3H_3PO_4 + 4H_2SO_4$$

or $$2Cu^{2+} + H_2PO_2^- + 2H_2O \longrightarrow Cu^+ + H_2PO_4^- + 4H^+$$

Hypophosphites ($H_2PO_2^-$)

Hypophosphites are the salts of hypophosphorus acid (H_3PO_2) and contain hypophosphite ion, $H_2PO_2^-$. Most important of these salts are Na (H_2PO_2) and $Ca_2(H_2PO_2)_2$.

Preparation. (*i*) Hypophosphites of active metals like Zn, Mg etc., are obtained by the action of H_3PO_2 on the metal.

$$M(M = Zn, Mg) + 2H_3PO_2 \longrightarrow M(H_2PO_2)_2 + H_2$$

(*ii*) The hypophosphites of Na, K, Ca, Ba etc., can be obtained by heating white phosphorus with the hydroxide of the metal.

$$4P\ (white) + 3MOH + 3H_2O \xrightarrow{\Delta} 3M(H_2PO_2) + PH_3$$

$$8P\ (white) + 3M(OH)_2\ (M = Ca, Ba) + 6H_2O \xrightarrow{\Delta} 3M(H_2PO_2)_2 + 2PH_3$$

The solution on evaporation gives the crystals of hypophosphite

(*iii*) Hypophosphites of many metals can be prepared by the action of calcium hypophosphite, $Ca(H_2PO_2)_2$ on the carbonate of the metal, *e.g.*,

$$Ca(H_2PO_2)_2 + M_2CO_3\ (M = Na, K) \longrightarrow 2M(H_2PO_2)\ (solution) + CaCO_3\downarrow$$

The crystals of hypophosphite can be obtained by evaporating the solution.

Properties. (*i*) $Na(HPO_2)$ is a crystalline solid and is soluble in water and alcohol.

(*ii*) *Action of heat: Disproportionation.* On heating, hypophosphites decompose to give pyrophosphates ($P_2O_7^{-4}$) and PH_3 and steam (H_2O) are liberated.

$$4Na(H_2PO_2)\ (P = 1) \xrightarrow{\Delta} Na_4P_2O_7\ (P = +5) + 2PH_3\ (P = -3) + H_2O\ (steam)$$

$$2Ca(H_2PO_2)_2 \xrightarrow{\Delta} Ca_2P_2O_7 + 2PH_3 + H_2O\ (steam)$$

(*iii*) *Reducing properties.* Hypophosphites, like hypophosphorus acid, are reducing agents. When a hypophosphite reduces a certain substance, it gets oxidised to dihydrogen phosphate ($H_2PO_4^-$).

$$H_2PO_2^-\ (P = +1) + 2H_2O \longrightarrow H_2PO_4^-\ (P = +5) + 4H^+ + 4e^-$$

Thus hypophosphites reduce Ag (+1) salts to Ag metal (Ag = 0) and $HgCl_2$ (Hg = +2) to Hg_2Cl_2 (Hg = +1) and finally to Hg metal (Hg = 0). Equations representing the reduction of $AgNO_3$ to Ag and $HgCl_2$ to Hg_2Cl_2 and finally to Hg have already been given under the study of hypophosphorus acid.

Uses. (*i*) $Na(H_2PO_2)$ is used : (*a*) in medicines as a tonic (*b*) for bleaching wood to manufacture paper (*ii*) $Ca(H_2PO_2)_2$ is used for preparing hypophosphites of other metals.

Structure of H_3PO_2 molecule and HPO_2^- ion. See chapter 15.

Phosphorus acid, H_3PO_3 (P = +3, dibasic acid)

Preparation. It is prepared:

(*a*) *by dissolving P_2O_3 in cold water*

$$P_2O_3 + 3H_2O\ (cold) \longrightarrow 2H_3PO_3$$

(*b*) *by hydrolysing PCl_3.* When Cl_2 is passed over molten white phosphorus under water, PCl_3 is formed which instantaneously gets hydrolysed to form H_3PO_3.

$$2P\ (molten\ white) + 3Cl_2 \longrightarrow 2PCl_3\ ;\ PCl_3 + 3H_2O \longrightarrow H_3PO_3 + 3HCl$$

(*c*) *by heating PCl_3 with $H_2C_2O_4$ crystals.*

$$PCl_3 + 3H_2C_2O_4 \xrightarrow{\Delta} H_3PO_3 + 3CO + 3CO_2 + 3HCl.$$

Properties. (*i*) It is a white crystalline solid with m.pt. = 73.6°C. It is deliquescent and is extremely soluble in water.

(*ii*) *Acidic nature.* Although H_3PO_3 molecule has three H-atoms, the aqueous solution of the acid gives only two H^+ ions on ionisation.

$$H_3PO_3 \rightleftharpoons H^+ + H_2PO_3^-$$ (*Dihydrogen phosphite ion*), $K_a^I = 1.6 \times 10^{-3}$

$$H_2PO_3^- \rightleftharpoons H^+ + HPO_3^{2-}$$ (*Hydrogen phosphite ion*), $K_a^{II} = 6.3 \times 10^{-7}$

On adding the above two equations we get: $H_3PO_3 \rightleftharpoons 2H^+ + HPO_3^{2-}$

Due to the production of two H^+ ions, H_3PO_3 behaves as a dibasic acid and hence can be represented as $H_2[HPO_3]$ or $(OH)_2POH$.

Thus this acid gives two types of salts which are called *dihydrogen phosphites* (which contain $H_2PO_3^-$ ion) and *hydrogen phosphites* (which contain HPO_3^{2-} ion).

(*iii*) *Action of heat.* (*a*) At 160°C, H_3PO_4 and PH_3 are obtained. In this reaction H_3PO_3 undergoes disproportionation.

$$4H_3PO_3\ (P = +3) \xrightarrow{160°C} 3H_3PO_4\ (P = +5) + PH_3\ (P = -3)$$

(*b*) On gentle heating pyrophosphorus acid ($H_4P_2O_5$) is obtained while on strong heating metaphosphorus acid (HPO_2) is obtained. Thus here dehydration of H_3PO_3 takes place.

$$2H_3PO_3 \xrightarrow[(-H_2O)]{\text{Gentle heating}} H_4P_2O_5 + H_2O\ ;\quad H_3PO_3 \xrightarrow[(-H_2O)]{\text{Strong heating}} HPO_2 + H_2O$$

(*iv*) *Action of nascent hydrogen: Reduction.* H_3PO_3 gets reduced to PH_3 by nascent hydrogen (H) produced by the action of Zn on dil. H_2SO_4.

$$H_3PO_3\ (P = +3) + 6H \xrightarrow{Zn/H_2SO_4} PH_3\ (P = -3) + 3H_2$$

(*v*) *Reducing nature: Oxidation.* Due to the presence of one P–H bond in H_3PO_3 molecule, this acid can lose nascent hydrogen and hence acts as a powerful reducing agent. In these reactions the acid reduces other substances and is itself oxidised to orthophosphoric acid, H_3PO_4

$$H_3PO_3\ (P = +3) + H_2O \longrightarrow H_3PO_4\ (P = +5) + 2H$$

or $$HPO_3^{2-} + 3OH^- \longrightarrow PO_4^{3-} + 2H_2O + 2e^-$$

Thus the acid reduces halogens (X_2) to halogen acids (HX), metallic salts (*e.g.*, $AuCl_3$, $AgNO_3$, $HgCl_2$, $CuSO_4$) to free metal, SO_2 or H_2SO_3 to S, and acidified $KMnO_4$ solution to colourless $MnSO_4$ solution.

$$H_3PO_3 + H_2O + 2X_2 \longrightarrow H_3PO_4 + 2HX$$

$$2AuCl_3 + 3H_3PO_3 + 3H_2O \longrightarrow 2Au\downarrow + 3H_3PO_4 + 6HCl$$

$$2AgNO_3 + H_3PO_3 + H_2O \longrightarrow 2Ag\downarrow + H_3PO_4 + 2HNO_3$$

$$CuSO_4 + H_3PO_3 + H_2O \longrightarrow Cu\downarrow + H_3PO_4 + H_2SO_4$$

$$HgCl_2 + H_3PO_3 + H_2O \xrightarrow{\text{Boil}} Hg\downarrow + H_3PO_4 + 2HCl$$

$$SO_2 + 2H_3PO_3 \longrightarrow S\downarrow + 2H_3PO_4$$

$$H_2SO_3 + 2H_3PO_3 \longrightarrow S\downarrow + 2H_3PO_4 + H_2O$$

$$2KMnO_4 + 3H_2SO_4 + 5H_3PO_3 \longrightarrow 2MnSO_4 + 5H_3PO_4 + 3H_2O + K_2SO_4$$

(*vi*) *Action of PCl_5.* PCl_3 and $POCl_3$ are obtained.

$$H_3PO_3 + 3PCl_5 \longrightarrow PCl_3 + 3POCl_3 + 3HCl$$

Structure of H_3PO_3 molecule and HPO_3^{2-} ion. See chapter 15.

Pyrophosphorus acid or diphosphorus acid, $H_4P_2O_5$ (P = + 3, dibasic acid)

Preparation. The acid is obtained by the hydrolysis of PCl_3 by a little of water and evaporating the solution in a dessicater over KOH.

$$2PCl_3 + 5H_2O \longrightarrow H_4P_2O_5 + 6HCl$$

Properties. (*i*) This acid is obtained as needle-shaped crystals having m.pt. = 38°C.

(*ii*) *Action of H_2O.* With H_2O, phosphorus acid (H_3PO_3) is obtained.

$$H_4P_2O_5 + H_2O \longrightarrow 2H_3PO_3$$

(*iii*) *Action of NaOH.* With NaOH, disodium dihydrogen pyrophosphite, $Na_2H_2P_2O_5$ is obtained.

$$H_4P_2O_5 + 2NaOH \longrightarrow Na_2H_2P_2O_5 + 2H_2O$$

(*iv*) *Action of heat.* On heating strongly, metaphosphorus acid (HPO_2) is obtained.

$$H_4P_2O_5 \xrightarrow[(-H_2O)]{\text{Strong heating}} 2HPO_2 + H_2O$$

(*v*) *Acidic nature.* Although $H_4P_2O_5$ molecule has four H-atoms, the aqueous solution of the acid gives only two H^+ ions on ionisation.

$$H_4P_2O_5 \rightleftharpoons H^+ + [H_3P_2O_5]^- \text{ (trihydrogen pyrophosphite ion)}$$
$$[H_3P_2O_5]^- \rightleftharpoons H^+ + [H_2P_2O_5]^{2-} \text{ (dihydrogen pyrophosphite ion)}$$
i.e., $$H_4P_2O_5 \rightleftharpoons 2H^+ + [H_2P_2O_5]^{2-}$$

Due to the production of two H^+ ions, $H_4P_2O_5$ behaves as a dibasic acid and hence can be represented as $H_2[H_2P_2O_5]$. Thus the acid gives two types of salts called *trihydrogen pyrophosphites* ($[H_3P_2O_5]^-$) and *dihydrogen pyrophosphites* ($[H_2P_2O_5]^{2-}$). Examples are $NaH_3P_2O_5$ and $Na_2H_2P_2O_5$.

Structure of $H_4P_2O_5$ molecule and $H_2P_2O_5^{2-}$ ion. See chapter 15.

Metaphosphorus acid, HPO_2

This acid is obtained by mixing together dried PH_3 and O_2 or by the dehydration of phosphorus acid, HPO_3.

$$PH_3\ (dry) + O_2 \longrightarrow HPO_2 + H_2 \quad ; \quad H_3PO_3 \xrightarrow[(-H_2O)]{\text{Dehydration}} HPO_2 + H_2O$$

Metaphosphorus acid a white crystalline solid.

Structure of HPO_2 molecule. See chapter 15.

Hypophosphoric acid, $H_4P_2O_6$ (P = + 4, tetrabasic acid)

Preparation. When phosphorus is treated with a limited quantity of air in presence of H_2O, a solution containing a mixture of phosphorus acid (H_3PO_3) and hypophosphoric acid ($H_4P_2O_6$) is obtained.

$$2P_4 + 10H_2O + 7O_2 \longrightarrow 4H_3PO_3 + 2H_4P_2O_6$$

When the solution containing H_3PO_3 and $H_4P_2O_6$ is treated with $(CH_3COO)_2Pb$, $H_4P_2O_6$ present in the solution precipitates lead hypophosphate, $Pb_2P_2O_6$.

$$H_4P_2O_6 + 2(CH_3COO)_2Pb \longrightarrow Pb_2P_2O_6 \downarrow + 4CH_3COOH.$$

Now when the precipitate of $Pb_2P_2O_6$ is suspended in H_2O and H_2S gas is passed into this suspension, $H_4P_2O_6$ is obtained in solution.

$$Pb_2P_2O_6 + 2H_2S \longrightarrow 2PbS\downarrow + H_4P_2O_6 \text{ (solution)}$$

The solution containing $H_4P_2O_6$, on being concentrated, gives the crystals of $H_4P_2O_6$.

Properties. (*i*) The acid is a colourless, crystalline and hygroscopic solid. The anhydrous acid, $H_4P_2O_6$ has its m.pt. = 70°C while the dihydrated acid, $H_4P_2O_6.2H_2O$ which forms rhombic crystals melts at 62°C.

(*ii*) *Acidic nature.* It is a tetrabasic acid and hence dissociates in four stages as shown below:

$$H_4P_2O_6 \rightleftharpoons H^+ + H_3P_2O_6^- \quad ; \quad K_1 = 6.4 \times 10^{-3}$$
$$H_3P_2O_6^- \rightleftharpoons H^+ + H_2P_2O_6^{2-} \quad ; \quad K_2 = 1.5 \times 10^{-3}$$
$$H_2P_2O_6^{2-} \rightleftharpoons H^+ + HP_2O_6^{3-} \quad ; \quad K_3 = 5.4 \times 10^{-8}$$
$$HP_2O_6^{3-} \rightleftharpoons H^+ + P_2O_6^{4-} \quad ; \quad K_4 = 9.4 \times 10^{-11}$$

$P_2O_6^{4-}$ ion is called *hypophosphate ion.* Although the acid is tetrabasic as shown above, yet it can form only three types of salts like $Na_2[H_2P_2O_6]$, $(NH_4)_2[H_2P_2O_6]$, $Na_3[HP_2O_6]$, $Pb_2[P_2O_6]$, $Na_4[P_2O_6]$, $(C_2H_5)_4[P_2O_6]$. The last compound is tetraethyl ester.

(*iii*) *Action of heat.* (*a*) On being heated above its m.pt., the acid gets decomposed into phosphorus acid (H_3PO_3) and metaphosphoric acid (HPO_3) (*Disproportionation*).

$$H_4P_2O_6\ (P = +4) \xrightarrow{\text{Above m.pt.}} H_3PO_3\ (P = +3) + HPO_3 (P = +5)$$

(*b*) On strong heating at about 180°C, the acid gives H_3PO_4, HPO_3 and PH_3.

$$4H_4P_2O_6 \xrightarrow{\text{Above 180°C}} 3H_3PO_4 + 4HPO_3 + PH_3$$

(*iv*) *Hydrolysis.* When $H_4P_2O_6$ is hydrolysed by a mineral acid, a mixture of H_3PO_3 and H_3PO_4 is obtained.

$$H_4P_2O_6 + H_2O \longrightarrow H_3PO_3 + H_3PO_4$$

(*v*) *Reducing nature.* Since $H_4P_2O_6$ molecule has no P–H bond in its structure, it is not able to reduce the metallic salts to free metal.

(*vi*) *Reduction by nascent hydrogen.* $H_4P_2O_6$ cannot be reduced by nascent hydrogen. Recall that H_3PO_2 and H_3PO_3 both can be reduced to PH_3 by nascent hydrogen

Structure of $H_4P_2O_6$ molecule and $P_2O_6^{4-}$ ion. See chapter 15.

Orthophosphoric acid, H_3PO_4 (P = + 5, tribasic acid)

Preparation. (*i*) *Laboratory preparation.* In laboratory, H_3PO_4 is obtained by the action of conc. HNO_3 on red phosphorus. In this reaction HNO_3 acts as an oxidising agent. It oxidises P to H_3PO_4 and is itself reduced to NO_2. A crystal of iodine is also added to the reaction mixture. This crystal acts as a catalyst.

$$5HNO_3 + P \longrightarrow H_2O + 5NO_2 + H_3PO_4 \quad \text{or} \quad 2H^+ + 5NO_3^- + P \longrightarrow H_2O + 5NO_2 + PO_4^{3-}$$

(*ii*) *Commercial preparation.* On large scale H_3PO_4 is obtained by the following methods:

(*a*) Bone ash which contains $Ca_3(PO_4)_2$ is dissolved in minimum amount of HNO_3 and $(CH_3COO)_2Pb$ is added to get the ppt. of $Pb_3(PO_4)_2$. This ppt. is then decomposed by H_2S to get H_3PO_4.

$$Pb_3(PO_4)_2 + 3H_2S \longrightarrow 3PbS\downarrow + 2H_3PO_4\ (\textit{in solution})$$

The solution containing H_3PO_4 is concentrated till it becomes syrupy. It is cooled over H_2SO_4 in vacuum dessicator when crystals of H_3PO_4 are formed.

(*b*) H_3PO_4 is obtained by heating phosphate rock or bone ash each of which contains $Ca_3(PO_4)_2$ with calculated quantity of conc. H_2SO_4.

$$\underset{\text{Phosphate rock or bone ash}}{Ca_3(PO_4)_2} + 3H_2SO_4 + 6H_2O \xrightarrow{\Delta} 2H_3PO_4 + 3\underset{\text{Gypsum}}{[CaSO_4.2H_2O]}$$

(*c*) H_3PO_4 can also be obtained by hydrolysing. P_2O_5 or PCl_5.

$$P_2O_5 + 3H_2O\ (\textit{hot}) \longrightarrow 2H_3PO_4\ ; \quad PCl_5 + 4H_2O \longrightarrow H_3PO_4 + 5HCl$$

P_2O_5 used in the above reaction can be obtained by the combustion of phosphorus vapour in air or by heating bone ash with SiO_2 (sand) in an electric furnace.

$$4P + 5O_2 \longrightarrow 2P_2O_5\ ; \quad 2Ca_3(PO_4)_2 + 6SiO_2 \xrightarrow{\Delta} 6CaSiO_3 + P_4O_{10}$$

(*d*) H_3PO_4 is obtained when pyrophosphoric acid, $H_4P_2O_7$ is boiled with water.

$$H_4P_2O_7 + H_2O \xrightarrow{\text{Boil}} 2H_3PO_4$$

(*d*) When metaphosphoric acid (HPO_3) is treated with cold or hot water, H_3PO_4 is obtained.

$$HPO_3 + H_2O \longrightarrow H_3PO_4$$

With hot water the reaction takes place rapidly.

Properties. (*i*) It is a transparent deliquescent crystalline colourless solid having its m.pt. = 42.3°C.

(*ii*) It absorbs water and forms colourless syrupy liquid which is a semihydrate, $2H_3PO_4.H_2O$. This melts at 29°C.

(*iii*) The acid is very soluble in water.

(*iv*) *Acidic nature: Tribasic nature.* The aqueous solution of the acid, on ionisation, gives up all its three H^+ ions and hence behaves as an acid. Due to the production of three H^+ ions, the acid behaves like a tribasic acid. Thus H_3PO_4 molecule can be represented as $H_3[PO_4]$ or $(OH)_3PO$. The removal of three H^+ ions takes place in three steps shown below:

$$H_3PO_4 \rightleftharpoons H^+ + H_2PO_4^- \text{ (Dilydrogen phosphate ion)}; \quad K_1 = 7.5 \times 10^{-3}$$

$$H_2PO_4^- \rightleftharpoons H^+ + HPO_4^{2-} \text{ (Hydrogen phosphate ion)}; \quad K_2 = 6.2 \times 10^{-8}$$

$$HPO_4^{2-} \rightleftharpoons H^+ + PO_4^{3-} \text{ (Phosphate ion)}; \quad K_3 = 1.1 \times 10^{-12}$$

The reason for the successive decrease in the value of K in the successive ionisation steps is that in the first ionisation step H^+ ion is removed from a neutral molecule (*viz.*, H_3PO_4) more easily than from the singly and doubly-negatively charged species in the second and third steps respectively. The successive decrease in the values of K shows that the first ionisation takes place to a considerable extent, the second ionisation takes place only slightly while the third ionisation takes place to an extremely small extent. The ionisation of H_3PO_4 in the three steps also indicates that H_3PO_4 forms three types of salts (called phosphates) namely (*a*) *primary phosphates* which contain dihydrogen phosphate ion, $H_2PO_4^-$, (*b*) *secondary phosphates* which contain hydrogen phosphate ion, HPO_4^{2-}, (*c*) *normal* or *tertiary phosphates* (also called simply *phosphates*) containing the phosphate ion, PO_4^{3-}.

(*v*) *Action of heat.* (*a*) On being heated to 250°C, H_3PO_4 is converted into pyrophosphoric acid ($H_4P_2O_7$).

$$2H_3PO_4 \xrightarrow[(-H_2O)]{250°C} H_4P_2O_7 + H_2O$$

(*b*) When H_3PO_4 is heated at 350°C, metaphosphoric acid (HPO_3) is obtained.

$$H_3PO_4 \xrightarrow[(-H_2O)]{350°C} HPO_3 + H_2O$$

(*c*) When heated strongly, H_3PO_4 gives P_2O_5

$$2H_3PO_4 \xrightarrow[(-3H_2O)]{\text{Strong heating}} P_2O_5 + 3H_2O$$

All the above three equations can be combined and written as:

$$P_2O_5 \xleftarrow[(-3H_2O)]{\text{Strong heating}} 2H_3PO_4 \xrightarrow[(-H_2O)]{250°C} H_4P_2O_7 \xrightarrow[(-H_2O)]{350°C} 2HPO_3$$

(*vi*) *Action of the solution of $AgNO_3$, $BaCl_2$, metallic bromide or metallic iodide.* When H_3PO_4 is treated with the solution of $AgNO_3$, $BaCl_2$, metallic bromide or metallic iodide, the phosphate of the metal is obtained.

$$3AgNO_3 + H_3PO_4 \longrightarrow Ag_3PO_4 \text{ (yellow ppt.)} + 3HNO_3$$

$$3BaCl_2 + 2H_3PO_4 \longrightarrow Ba_3(PO_4)_2 \text{ (white ppt.)} + 6HCl$$

$$3NaBr + H_3PO_4 \longrightarrow Na_3PO_4 + 3HBr$$

$$3NaI + H_3PO_4 \longrightarrow Na_3PO_4 + 3HI$$

The last two reactions have been used for the preparation of HBr and HI in the laboratory.

(*vii*) *Esterification.* When H_3PO_4 is treated with an alcohol, ester is formed.

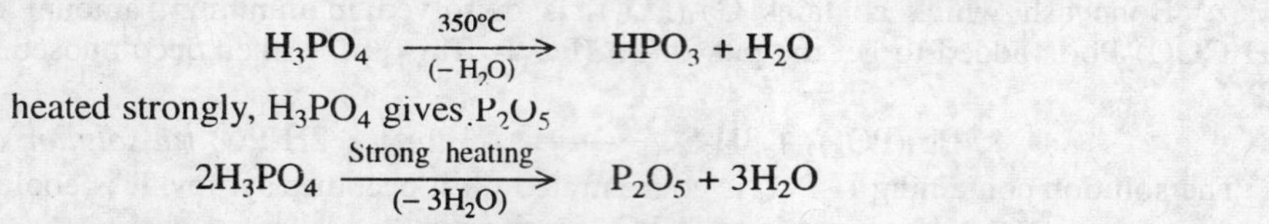

$$\text{or} \quad ROH + H_3PO_4 \xrightarrow{-H_2O} RH_2PO_4 \text{ (Ester)}$$

(*viii*) *Action of NH_3.* Depending on the quantity of NH_3 used, $(NH_4)H_2PO_4$ and $(NH_4)_2HPO_4$ are formed.

$$NH_3 + H_3PO_4 \longrightarrow (NH_4)H_2PO_4; \quad 2NH_3 \text{ (excess)} + H_3PO_4 \longrightarrow (NH_4)_2HPO_4$$

(*ix*) *Action of HPO_3.* When aqueous solution of H_3PO_4 and HPO_3 is heated at 100°C, pyrophosphoric acid ($H_4P_2O_7$) is obtained.

$$H_3PO_4 + HPO_3 \xrightarrow{100°C} H_4P_2O_7$$

(*x*) *Neutralisation by NaOH and Na_2CO_3.* When H_3PO_4 is just neutralised by NaOH or Na_2CO_3 (phenolphthalein is used as an indicator), disodium hydrogen phosphate (Na_2HPO_4) is obtained.

$$H_3PO_4 + 2NaOH \longrightarrow Na_2HPO_4 + 2H_2O$$
$$H_3PO_4 + Na_2CO_3 \longrightarrow Na_2HPO_4 + H_2O + CO_2$$

(*xi*) *Action of Na_2HPO_4.* When H_3PO_4 is added to a solution of $NaHPO_4$, NaH_2PO_4 is formed.

$$Na_2HPO_4 + H_3PO_4 \longrightarrow 2NaH_2PO_4$$

(*xii*) *Action of aqueous solution of H_3PO_4 on $Ca_3(PO_4)_2$, $Ca(HPO_4)$, $Ca(OH)_2$ and $CaCO_3$.* When the given salts are dissolved in the aqueous solution of H_3PO_4, calcium dihydrogen phosphate, $Ca(H_2PO_4)_2$ is obtained. This phosphate is also called *triple phosphate* and is used as a phosphatic fertiliser.

$$Ca_3(PO_4)_2 + 4H_3PO_4 \longrightarrow Ca(H_2PO_4)_2$$
$$Ca(HPO_4) + H_3PO_4 \longrightarrow Ca(H_2PO_4)_2$$
$$\underset{\text{slaked lime}}{Ca(OH)_2} + 2H_3PO_4 \longrightarrow Ca(H_2PO_4)_2 + H_2O + CO_2$$
$$\underset{\text{Limestone}}{CaCO_3} + 2H_3PO_4 \longrightarrow Ca(H_2PO_4) + H_2O + CO_2$$

(*xiii*) *Action of $POCl_3$.* H_3PO_4 reacts with $POCl_3$ in the following two different ways:

(*a*) When a mixture of H_3PO_4 crystals and $POCl_3$ are heated gently and evaporated in a vacuum dessicator and cooled at –10°C, pyrophosphoric acid ($H_4P_2O_7$) is obtained.

$$5H_3PO_4 + POCl_3 \longrightarrow 3H_4P_2O_7 + 3HCl$$

(*b*) When crystals of H_3PO_4 react with $POCl_3$ at ordinary temperature, metaphosphoric acid (HPO_3) is obtained.

$$2H_3PO_4 + POCl_3 \longrightarrow 3HPO_3 + 3HCl$$

(*xiv*) *Action of BaO_2.* When barium peroxide (BaO_2) is treated with H_3PO_4, H_2O_2 is obtained in solution.

$$3BaO_2 + 2H_3PO_4 \longrightarrow Ba_3(PO_4)_2 \downarrow + 3H_2O_2$$

The ppt. of $Ba_3(PO_4)_2$ is filtered off.

Uses. H_3PO_4 is used : (*i*) for the preparation of HBr and HI in the laboratory (*ii*) for the preparation of phosphates (*iii*) for the preparation of other oxy-acids of phosphorus [$2H_3PO_4 \xrightarrow{350°C} 2HPO_3 + 2H_2O$, $2HPO_3 + 5C \rightarrow 2P + 5CO + H_2O$]. (*v*) as a stabiliser for H_2O_2 (*vi*) in the preparation of nervous tonics (*vii*) for the manufacture of dyes, enamels, soaps, detergents and porcelain cement (*viii*) in dental filling (*ix*) for sugar refining.

Some important phosphates

Some of the phosphates and phosphates mixed with other salts have been found very useful. Here we shall discuss the preparation, properties and uses of such compounds.

1. Calcium dihydrogen phosphate, $Ca(H_2PO_4)_2.H_2O$

This compound is also called **triple superphosphate.**

Preparation. (*i*) It is obtained by dissolving $Ca_3(PO_4)_2$ or $Ca(HPO_4)$ in aqueous solution of H_3PO_4 and evaporating the resulting solution.

$$Ca_3(PO_4)_2 + 4H_3PO_4 \longrightarrow 3Ca(H_2PO_4)_2; \qquad Ca(HPO_4) + H_3PO_4 \longrightarrow Ca(H_2PO_4)_2$$

(*ii*) *On large scale,* pure $Ca(H_2PO_4)_2$ is made by dissolving slaked lime, $Ca(OH)_2$ or limestone, $Ca(CO_3)_2$ in H_3PO_4 and crystallising the resulting solution.

$$Ca(OH)_2 + 2H_3PO_4 \longrightarrow Ca(H_2PO_4)_2 + 2H_2O; \; CaCO_3 + 2H_3PO_4 \longrightarrow Ca(H_2PO_4)_2 + H_2O + CO_2$$

Properties. $Ca(H_2PO_4)_2$ is readily soluble in water while the other two phosphates of calcium *viz.,* $Ca(HPO_4)$ and $Ca_3(PO_4)_2$ are insoluble or sparingly soluble.

Uses. (*i*) It is used in the manufacture of superphosphate of lime, phosphorus and phosphoric acids. (*ii*) This compound makes a component of superphosphate of lime which is used as a phosphatic fertiliser. (*iii*) This compound contains 45-60% available phosphorus as P_2O_5 which is three times more concentrated than superphosphate of lime. Hence $Ca(H_2PO_4)_2$ is called triple superphosphate.(*iv*) This compound itself is also used as a phosphatic fertiliser. (*v*) Pure $Ca(H_2PO_4)_2$ is used in tooth paste and in American baking powder.

2. Ammonium dihydrogen phosphate, $NH_4(H_2PO_4)$

This phosphate is also called dihydrogen ammoniated phosphate or ammophos.

Preparation. (*i*) It is prepared by treating $Ca_3(PO_4)_2$ with a mixture containing $(NH_4)_2SO_4$ and H_2SO_4 (1:2 mixture).$(NH_4)(H_2PO_4)$ goes into solution.

$$Ca_3(PO_4)_2 + (NH_4)_2SO_4 + 2H_2SO_4 \longrightarrow 2(NH_4)(H_2PO_4)\ (solution) + 3CaSO_4\ (ppt.)$$

The solution containing $(NH_4)(H_2PO_4)$, on crystallisation, gives the crystals of $(NH_4)(H_2PO_4)$.

(*ii*) It is also obtained by the action of NH_3 on H_3PO_4.

$$NH_3 + H_3PO_4 \longrightarrow (NH_4)(H_2PO_4)$$

Properties. (*i*) It is non-hygroscopic and soluble in water.

(*ii*) On heating it gives NH_3 and H_3PO_4.

$$(NH_4)(H_2PO_4) \xrightarrow{\Delta} NH_3 + H_3PO_4$$

Uses. (*i*) It is used in fireproofing of fabrics (*ii*) It contains 12% nitrogen and 50-54% P_2O_5 and hence is used as a nitrogenous phosphorus fertiliser (NP fertiliser). It can supply both the essential elements (*viz.*, N and P) to the soil.

3. Diammonium monohydrogen phosphate, $(NH_4)_2(HPO_4)$

This is also called ordinary ammonium phosphate of commerce. It is obtained when H_3PO_4 containing excess of NH_3 is avaporated.

$$2NH_3\ (excess) + H_3PO_4 \longrightarrow (NH_4)_2(HPO_4)$$

It forms monoclinic crystals of density 1.62. These crystals effloresce in air evolving NH_3. It is very soluble in water. On heating is gives NH_3 and H_3PO_4.

$$(NH_4)_2(HPO_4) \xrightarrow{\Delta} 2NH_3 + H_3PO_4$$

It is used (*a*) as a reagent in laboratory (*b*) as a fireproofing of fabrics (*c*) in pharmacy (*d*) as a nitrogenous-phosphorus (NP) fertiliser, since it contains 21% nitrogen and 54% phosphorus as P_2O_5. It can be used as an NP-fertiliser in light as well as heavy soils.

4. Sodium ammonium hydrogen phosphate or microcosmic salt, $Na(NH_4)HPO_4.4H_2O$

This phosphate is also called microcosmic salt since it is present in the urine of the "*microcosm*" *i.e.*, man. It is prepared by dissolving molecular proportion of disodium hydrogen phosphate (Na_2HPO_4) and NH_4Cl in a little hot water, filtering off the precipitate of NaCl and crystallising the solution.

$$Na_2HPO_4 + NH_4Cl \longrightarrow Na(NH_4)HPO_4\ (solution) + NaCl\ (ppt.)$$

Properties and uses. (*i*) It is colourless and crystalline substance having needle-like crystals.

(*ii*) *Action of heat.* When heated strongly, it loses NH_3 and H_2O and forms a transparent glass beed of sodium metaphosphate, $NaPO_3$ in the molten state.

$$Na(NH_4)HPO_4 \xrightarrow{\Delta} NH_3 + H_2O + NaPO_3\ (in\ molten\ state)$$

When molten $NaPO_3$ is cooled rapidly, it gets polymerised and forms sodium hexametaphosphate, $(NaPO_3)_6$ or $Na_6P_6O_{18}$. (*calgon*).

$$3NaPO_3 \xrightarrow[\text{Polymerisation}]{\text{cool rapidly}} (NaPO_3)_6 \text{ or } Na_6P_6O_{18}$$

The bead of $NaPO_3$ obtained as above can combine with metallic oxides like CuO etc., and form coloured metallic sodium orthophosphate like $CuNaPO_4$.

$$CuO + NaPO_3\ (bead) \longrightarrow CuNaPO_4\ (blue)$$

The formation of coloured metallic sodium orthophosphate can be used to detect the coloured basic radicals like Cu^{2+}, Co^{2+} etc., present in a given inorganic mixture. Another use of $Na(NH_4)HPO_4$ is that it is also used for testing SiO_2 with which the salt gives a cloudy bead in which the particles of SiO_2 are floating. $Na(NH_4)HPO_4$ is also used for the preparation of sodium hexametaphosphate (calgon), $(NaPO_3)_6$ or $Na_6P_6O_{18}$ as shown above.

5. Trisodium phosphate dodecahydrate, $Na_3PO_4.12H_2O$

Preparation. It is prepared :

(*i*) by dissolving calculated amount of Na_2HPO_4 and NaOH in hot water and evaporating the resulted solution.

$$Na_2HPO_4 + NaOH \longrightarrow Na_3PO_4 + H_2O$$

On cooling the solution, the crystals of $Na_3PO_4.12H_2O$ separate out.

(*ii*) by fusing Na_2HPO_4 with Na_2CO_3. This method gives *anhydrous salt.*

$$2Na_2HPO_4 + Na_2CO_3 \longrightarrow 2Na_3PO_4 + H_2O + CO_2$$

(*c*) by fusing *wavellite* mineral ($AlPO_4$) with soda ash (Na_2CO_3).

$$AlPO_4 + 2Na_2CO_3 \longrightarrow NaAlO_2 + Na_3PO_4 + 2CO_2$$

Properties. (*i*) It is a white crystalline substance and is soluble in water.

(*ii*) The aqueous solution of the salt is alkaline due to the production of OH^- ions in hydrolysis.

$$PO_4^{3-} + H_2O \longrightarrow HPO_4^{2-} + OH^-$$

Uses. (*i*) It is used for softening water for boilers under the name of "*tripsa*". (*ii*) On account of its remarkable detergent properties, it is used either alone or as a constituent of soap powders, cleansers etc., for household purpose.

6. Ammonium dihydrogen phosphate sulphate, $NH_4(H_2PO_4).(NH_4)_2SO_4$ (Double salt)

This compound is also called *ammoniated sulphate* and is a double salt of $NH_4(H_2PO_4)$ and $(NH_4)_2SO_4$. It is obtained by neutralising a mixture of H_3PO_4 and H_2SO_4 by gaseous NH_3.

$$2NH_3 + H_2SO_4 \longrightarrow (NH_4)_2SO_4 \ ; \quad NH_3 + H_3PO_4 \longrightarrow (NH_4)(H_2PO_4)$$

$$(NH_4)_2SO_4 + (NH_4)(H_2PO_4) \longrightarrow (NH_4)(H_2PO_4).(NH_4)_2SO_4$$

The reaction is exothermic and the temperature rises to about 112°C. It contains 16% nitrogen and 20% phosphorus as P_2O_5 and hence is used as a nitrogenous-phosphorus fertiliser (NP fertiliser).

7. Calcium dihydrogen phosphate nitrate, $Ca(H_2PO_4)_2.2Ca(NO_3)_2$ (Double salt)

This compound is also called *calcium supersphosphate nitrate* or *nitrophosphate* and is a double salt containing one molecule of $Ca(H_2PO_4)$ and two molecules of $Ca(NO_3)_2$.

It is produced by the treatment of phosphate rock, $Ca_3(PO_4)_2$ with HNO_3.

$$Ca_3(PO_4)_2 + 4HNO_3 \longrightarrow \underbrace{Ca(H_2PO_4)_2 + 2Ca(NO_3)_2}_{\text{Cal. superphosphate nitrate}}$$

This compound is a mixed fertiliser (nitrogenous-phosphorus fertiliser), since it supplies the two essential elements *viz.*, nitrogen and phosphorus to the soil. $Ca(H_2PO_4)_2$ supplies phosphorus while $Ca(NO_3)_2$ supplies nitrogen to the plants and hence this fertiliser is superior to the ordinary superphosphate of lime, $[Ca(H_2PO_4)_2.H_2O + 2(CaSO_4.2H_2O)]$.

8. Calcium dihydrogen phosphate sulphate, $Ca(H_2PO_4)_2.H_2O + 2(CaSO_4.2H_2O)$ (Double salt)

This compound is also called *superphosphate of lime* or *calcium superphosphate* and is a mixture of $Ca(H_2PO_4).H_2O$ and two molecules of gypsum, $CaSO_4.2H_2O$.

Manufacture. It is manufactured by treating calcium phosphate, $Ca_3(PO_4)_2$ (obtained from phosphate rock or bone ash) with calculated quantity of commercial H_2SO_4 (*i.e.*, H_2SO_4 obtained from lead chamber process).

$$Ca_3(PO_4)_2 + 2H_2SO_4 + 5H_2O \longrightarrow [Ca(H_2PO_4)_2.H_2O + 2(CaSO_4.2H_2O)] \ (\textit{superphosphate of lime})$$

In the above reaction, heat is evolved and hence the reaction is exothermic in nature.

The plant used for the manufacture is shown in Fig. 32.1.

The phosphate rock is well powdered and is placed in a cast iron mixer. Now a calculated quantity of chamber H_2SO_4 is added to it. The whole mass is stirred by means of paddles for 2 to 5 minutes and then dumped mechanically into one of the brick-lined pits or dens D_1 or D_2 through the valves V_1 and V_2. Here the mixer is left for about 24 hours. The reaction which actually started in the mixer continues in the pot or den and the temperature gradually rises to 100-110°C because of the exothermic nature of the reaction.

Phosphate rock usually contains $CaCO_3$, CaF_2 and SiO_2 as impurities. The impurities of $CaCO_3$ and CaF_2 react with H_2SO_4 to produce CO_2 and HF respectively.

$$CaCO_3 + H_2SO_4 \longrightarrow CaSO_4 + H_2O + CO_2$$

$$CaF_2 + H_2SO_4 \longrightarrow CaSO_4 + 2HF \quad (poisonous)$$

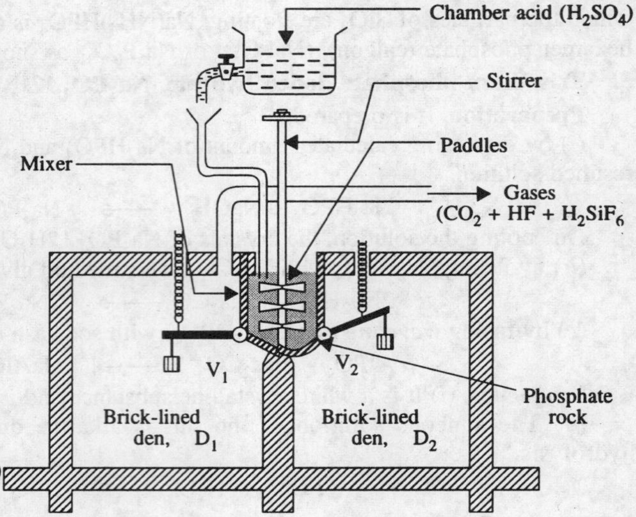

Fig. 32.1 : Plant used for the manufacture of superphosphate of lime, $Ca(H_2PO_4)_2\ H_2O + 2(CaSO_4.2H_2O)$.

HF produced in the above reaction reacts with SiO_2 (present as impurity in phosphate rock) to produce gaseous hydrofluosilicic acid, H_2SiF_6

$$6HF + SiO_2 \longrightarrow H_2SiF_6 + 2H_2O$$

The production of poisonous HF gas in the manufacture indicates that CaF_2 should not be present in large quantities in the phosphate rock. The presence of CaF_2 in large quantities also consumes huge quantities of extra H_2SO_4. The gases like CO_2, HF and H_2SiF_6 are led into a suitable absorption tower. As a result of the reaction, the mass becomes hard and is finally solidified to a porous dry solid due to the formation of gypsum. The mass is allowed to stand for a few days in the den until it becomes perfectly dry. The dried mass is removed from the den and is kept in the open for 4-6 weeks. During this period the excess of H_2SO_4 is completely consumed and the mass is completely dried. Finally the dry mass is crushed to a fine powder and is sold in the market under the name of *superphosphate of lime.*

Use of superphosphate of lime as a phosphatic fertiliser. It contains about 16-20% of P_2O_5, depending on the purity of the phosphate rock used in the manufacture of the fertiliser and its active ingredient is calcium dihydrogen phosphate which is soluble in water. Thus it provides phosphate in a water-soluble form which is readily assimilated by the plants for their growth. In this respect it has an advantage over the normal calcium phosphate which is insoluble in water and cannot be assimilated by the plants.

Superphosphate of lime has some disadvantages of its being used as a fertiliser. These disadvantages are : (*i*) It is relatively dilute fertiliser containing only 16-20% of P_2O_5 depending on the purity of the phosphate rock used in its manufacture. (*ii*) Being a dilute product, it requires relatively heavy freight and handling charges and a large number of expensive bags. (*iii*) It has a setting tendency and is converted into heavy lumps. (*iv*) The acidity in it causes the package to rot and contents to spill.

9. Phosphate slag, $Ca_3(PO_4)_2.CaSiO_3$ (Double salt)

Phosphate slag is probably a double salt of tricalcium phosphate, $Ca_3(PO_4)_2$ and calcium silicate, $CaSiO_3$ and hence is represented as $Ca_3(PO_4)_2$. $CaSiO_3$. Phosphate slag is also known by other names like *Thomas slag, Basic slag* and *Thomas meal.*

Phosphate slag is obtained as a by-product in the manufacture of steel from cast iron. When cost iron containing acidic impurities like phosphorus, silicon etc., is blasted (*i.e.*, burnt) with lime stone ($CaCO_3$) in presence of air, phosphorus and silicon are oxidised to P_2O_5 and SiO_2 respectively [$4P + 5O_2 \rightarrow 2P_2O_5$; $Si + O_2 \rightarrow SiO_2$] P_2O_5 and SiO_2 react with CaO obtained by the decomposition of $CaCO_3$ [$CaCO_3 \rightarrow CaO + CO_2\uparrow$] to form $Ca_3(PO_4)_2$ and $CaSiO_3$ [$3CaO + P_2O_5 \rightarrow Ca_3(PO_4)_2$; $CaO + SiO_2 \rightarrow CaSiO_3$] These salts mix together and form phosphate slag,

$Ca_3(PO_4)_2.CaSiO_3$ which, being lighter, rises on the top of the molten steel. The slag is separated from the top of the molten steel and cooled when it forms a solid which is ground to a fine powder and sold in the market under different names like *Thomas slag, Basic slag or Thomas meal.*

Manufacture in India. This fertiliser is obtained as a by-product in all the steel plants located at Jamshedpur, Burnpur, Rourkela, Durgapur, Bhillai and Bokaro.

Use of phosphate slag as a phosphatic fertiliser. (*i*) It contains 14-22% of P_2O_5 and hence is used as a phosphate fertiliser. It is crushed to a fine powder and is added to the soil. In the soil, it decomposes slowly in due course of time and liberates H_3PO_4 required for the growth of the plants. (*ii*) In the presence of salts and CO_2, the fertiliser dissolves in water. Thus it dissolves in soil water and is taken by the plants in the form of solution through their roots. (*iii*) This fertiliser also contains 40% of lime which is readily available for the neutralisation of acidity of the soil. (*iv*) This fertiliser is particularly useful for the production of hay.

Poly (or condensed) phosphoric acids, $H_{n+2}\, P_n\, O_{3n+1}$

Important examples of such acids are *diphosphoric acid,* $H_4P_2O_7$ (also called *pyrophosphoric acid* and *triphosphoric acid* ($H_5P_3O_{10}$).

1. Pyrophosphoric acid or diphosphoric acid, $H_4P_2O_7$ (P = +5, *tetrabasic acid*).

Preparation. This acid is obtained by heating H_3PO_4 alone or with $POCl_3$ or HPO_3.

$$2H_3PO_4 \xrightarrow{250°C} H_4P_2O_7 + H_2O$$

$$5H_3PO_4 + POCl_3 \xrightarrow{\Delta} 3H_4P_2O_7 + 3HCl$$

$$H_3PO_4 + HPO_3 \xrightarrow{100°C} H_4P_2O_7$$

(*b*) $H_4P_2O_7$ can also be obtained by the action of H_2S on lead pyrophosphate ($Pb_2P_2O_7$).

$$Pb_2P_2O_7 + 2H_2S \longrightarrow H_4P_2O_7 + 2PbS \text{ (black ppt.)}$$

$Pb_2P_2O_7$ used in the above reaction is obtained by the following reactions.

$$2Na_2HPO_4 \xrightarrow{>240°C} Na_4P_2O_7 + H_2O$$

$$Na_4P_2O_7 \text{ (solution)} + 2Pb(NO_3)_2 \longrightarrow Pb_2P_2O_7 \text{ (white ppt.)} + 4NaNO_3$$

Properties. (*i*) $H_4P_2O_7$ forms white grannular crystals which melt at 61°C. It is soluble in water.

(*ii*) *Acidic nature: Tetrabasic nature.* The aqueous solution of $H_4P_2O_7$, on ionisation, gives up all its four H^+ ions and hence behaves as a tetrabasic acid. Thus $H_4P_2O_7$ can be represented as $(OH)_4P_2O_3$. The removal of four H^+ ions takes place in four steps. The values of the four ionisation constants are $K_1 = 1.4 \times 10^{-1}$, $K_2 = 1.1 \times 10^{-2}$, $K_3 = 2.7 \times 10^{-7}$ and $K_4 = 2.4 \times 10^{-10}$. Although the acid is a tetrabasic acid, it gives only two types of salts namely *normal pyrophosphates* ($P_2O_7^{4-}$) and *dihydrogen pyrophosphates* ($H_2P_2O_7^{2-}$). $H_4P_2O_7$ is stronger than H_3PO_4.

(*iii*) *Action of water.* On boiling with water, $H_4P_2O_7$ is converted into H_3PO_4.

$$H_4P_2O_7 + H_2O \xrightarrow{\text{Boil}} 2H_3PO_4$$

(*iv*) *Action of heat.* When $H_4P_2O_7$ is heated alone at 350°C, metaphosphoric acid (HPO_3) is obtained.

$$H_4P_2O_7 \xrightarrow[-H_2O]{350°C} 2HPO_3 + H_2O$$

(*v*) Like phosphoric acid (H_3PO_4), $H_4P_2O_7$ also gives ammonium molybdate test.

(*vi*) *Action of H_2O_2.* When excess of $H_4P_2O_7$ is treated with H_2O_2, peroxy-diphosphoric acid ($H_4P_2O_8$) is obtained.

$$H_4P_2O_7 \text{ (excess)} + H_2O_2 \longrightarrow H_4P_2O_8 + H_2O$$

(*vii*) *Action of $POCl_3$ and PCl_5.* When $POCl_3$ and PCl_5 react with crystalline $H_4P_2O_7$, metaphosphoric acid (HPO_3) is formed.

$$2H_4P_2O_7 + POCl_3 \longrightarrow 5HPO_3 + 3HCl$$

$$H_4P_2O_7 + PCl_5 \longrightarrow 2HPO_3 + POCl_3 + 2HCl$$

(*viii*) *Action of $AgNO_3$ and $Co(NO_3)_2$ solutions.* White ppt. of $Ag_4P_2O_7$ and pink ppt. of $Co_2P_2O_7$ are obtained.

$$4AgNO_3 + H_4P_2O_7 \longrightarrow Ag_4P_2O_7 + 4HNO_3$$
$$2Co(NO_3) + H_4P_2O_7 \longrightarrow Co_2P_2O_7 + 4HNO_3$$

2. Triphosphoric acid, $H_5P_3O_{10}$

It is an unstable acid but its sodium salt namely *sodium triphosphate,* $Na_5P_3O_{11}$ is very stable.

Structure of polyphosphoric acids. The structure of diphosphoric acid ($H_4P_2O_7$), triphosphoric acid ($H_5P_3O_{10}$) and tetraphosphoric acid ($H_6P_4O_{13}$) have already been discussed in chapter 15.

Polyphosphates, $[P_nO_{3n+1}]^{-(n+2)}$

Among polyphosphates, diphosphates (pyrophosphates) like $Na_4P_2O_7.10H_2O$ and $Mg_2P_2O_7$ and triphosphates like $Na_5P_3O_{10}$ are very important.

1. Sodium diphosphate or sodium pyrophosphate, $Na_4P_2O_7.10H_2O$

It is prepared by heating disodium hydrogen phosphate (Na_2HPO_4) above 240°C when the salt fuses to a transparent glass which crystallises on cooling to white opaque mass which is $Na_4P_2O_7.10H_2O$.

$$2Na_2HPO_4 \xrightarrow{\text{Heat } (>240°C)} Na_4P_2O_7 + H_2O$$

It may also be obtained by heating sodium hypophosphite, NaH_2PO_2.

$$4NaH_2PO_2 \xrightarrow{\Delta} Na_4P_2O_7 + 2PH_3 + H_2O$$

$Na_2P_2O_7.10H_2O$ is a white crystalline substance which is soluble in water. Its aqueous solution is weakly alkaline due to hydrolysis.

When aqueous solution of $Na_4P_2O_7$ is treated with $Pb(NO_3)_2$ solution, white ppt. of lead pyrophosphate ($Pb_2P_2O_7$) is obtained.

$$Na_4P_2O_7 + 2Pb(NO_3)_2 \longrightarrow Pb_2P_2O_7 + 4NaNO_3$$

$Na_4P_2O_7$ is used in the preparation of $Pb_2P_2O_7$ which is employed for the preparation of pyrophosphoric acid, $H_4P_2O_7$.

2. Magnesium diphosphate, $Mg_2P_2O_7$ (Also called **magnesium pyrophosphate)**

When Mg $(NH_4)PO_4.6H_2O$ is heated to 105°C, it loses water and becomes anhydrous. On further heating to 800°C, it is converted into $Mg_2P_2O_7$ and NH_3 is evolved.

$$2[Mg(NH_4)PO_4.6H_2O] \xrightarrow[-12H_2O]{105°C} 2Mg(NH_4)PO_4 + 12H_2O \xrightarrow{800°C} Mg_2P_2O_7 + H_2O + 2NH_3$$

The use of this reaction is made in the estimation of Mg^{2+} and PO_4^{3-} radicals.

3. Sodium triphosphate, $Na_5P_3O_{10}$

It is a sodium salt of triphosphoric acid ($H_5P_3O_{10}$). This salt is prepared:

(*a*) by the hydrolysis of sodium trimetaphosphate, $(NaPO_3)_3$ or $Na_3P_3O_9$.

$$Na_3P_3O_9 + 2NaOH \longrightarrow Na_5P_3O_{10} + H_2O$$

(*b*) by fusing Na_2O with P_2O_5 followed by slow cooling.

$$5Na_2O + 3P_2O_5 \xrightarrow[\text{cool slowly}]{1000°C} 2Na_5P_3O_{10}$$

(*c*) by fusing appropriate amount of Na_2HPO_4 and NaH_2PO_4 followed by recrystallisation of the product.

$$2Na_2HPO_4 + NaH_2PO_4 \xrightarrow{450°C} Na_5P_3O_{10} + 2H_2O$$

$Na_5P_3O_{10}$ is used for softening water and often mixed with detergent and soaps in domestic washing powders.

Structure of polyphosphate ions. The structure of diphosphate ion (also called pyrophosphate ion) ($P_2O_7^{-4}$), triphosphate ion ($P_3O_{10}^{-5}$) and tetraphosphate ion ($P_4O_{13}^{-6}$) have already been discussed in chapter 15.

Metaphosphoric acid, HPO_3

This acid is also called *glacial phosphoric acid.*

Preparation. When $(NH_4)_3PO_4$ or H_3PO_4 or $H_4P_2O_7$ is heated to about 32°C, metaphosphoric acid is formed.

$$(NH_4)_3PO_4 \longrightarrow HPO_3 + 3NH_3 + H_2O \;; \quad H_3PO_6 \longrightarrow HPO_3 + H_2O$$

$$H_4P_2O_7 \longrightarrow 2HPO_3 + H_2O$$

Properties. (*a*) Physical. Metaphosphoric acid is a hard, sticky mass. The acid is, therefore, called *glacial phosphoric acid.* It dissolves in water with a cracking noise. It volatilises on heating. It melts at 38.4°C.

(*b*) **Chemical.** (*i*) *Action on silver nitrate.* Metaphosphoric acid gives a white precipitate with silver nitrate solution forming silver metaphosphate, $AgPO_3$.

$$AgNO_3 + HPO_3 \longrightarrow AgPO_3 + HNO_3$$

(*ii*) *Action on barium chloride.* Barium chloride gives a white precipitate in neutral or alkaline solutions.

$$BaCl_2 + 2HPO_3 \longrightarrow Ba(PO_3)_2 + 2HCl$$

The precipitate of barium phosphate is insoluble in acetic acid.

(*iii*) *Action on water.* Aqueous solution of metaphosphoric acid is gradually converted to orthophosphoric acid on prolonged standing. This conversion is accelerated in the presence of acids and on heating.

$$HPO_3 + H_2O \longrightarrow H_3PO_4$$

(*iv*) The alkali metaphosphate, $NaPO_3$, when heated with metal oxides, forms mixed phosphate. The mixed phosphates of heavy metals have characteristic colours.

$$NaPO_3 + CuO \longrightarrow NaCuPO_4$$

This reaction finds an application in the basic radical analysis as an alternative of the borax bead test. The mixed phosphates so formed are often referred to as microcosmic beads.

(*v*) Metaphosphoric acid or its salts are not present in the form of simple molecules but are often polymerised. In the vapour phase, metaphosphoric acid has been reported to have the formula $(HPO_3)_2$ but in solutions or in the melt, it is $(HPO_3)_n$ where $n = 3$ or 4.

(*vi*) *Acidic nature.* It is a monobasic acid and its salts are called *metaphosphates.*

(*vii*) Like other phosphoric acids, it also gives ammonium molybdate test.

(*viii*) Unlike other phosphoric acid, it at once coagulates albumin (white part of the egg).

Peroxyphosphoric acids

Phosphorus forms two peroxyphosphoric acids which are analogous to peroxysulphuric acids and are discussed below:

Peroxy monophosphoric acid, H_3PO_5

This acid is known in solution only. It is formed by the action of 30% H_2O_2 on P_2O_5 in acetonitrile solution at a low temperature.

$$P_2O_5 + 2H_2O_2 + H_2O \longrightarrow 2H_3PO_5$$

This acid is a powerful oxidising agent and oxidises manganous salts to permanganates.

Peroxy diphosphoric acid, $H_4P_2O_8$

This acid is prepared by the action of H_2O_2 on the excess of pyrophosphoric acid, $H_4P_2O_7$.

$$H_4P_2O_7 + H_2O_2 \longrightarrow H_4P_2O_8 + H_2O$$

Potassium salt of the acid is stable and is prepared by the electrolysis of potassium hydrogen phosphate in presence of KF. The solution of the acid oxidises aniline, nitrobenzene and nitrosobenzene. In acidic solution the acid is first hydrolysed to H_3PO_5 and finally to H_2O_2.

University Questions

1. Write structure and basicity of the following oxy acids : (*i*) Hypophosphorus acid (*ii*) Orthophosphoric acid (*iii*) Pyrophosphorus acid. *(Rohilkhand 2003)*
2. What are different types of oxyacids of phosphorus ? Discuss the structure of orthophosphric acid. *(Nagpur 2003)*
3. Explain the structure and geometry of orthophosphorous acid. *(Nagpur 2003)*

33

Oxy-acids of Sulphur and their Salts

A large number of oxy-acids of sulphur are known. These acids exist either in the free state or in the form of their solution or as their salts. Oxy-acids having S—S linkage are called **thio-acids.** These acids may be grouped into the following five series:

I. Sulphoxylic acid, H_2SO_2. It is an unstable acid.

II. Sulphurous acid series. (*i*) *Sulphurous acid,* H_2SO_3 (*ii*) *Thiosulphurous acid,* $H_2S_2O_2$ (*iii*) *Hyposulphurous acid,* $H_2S_2O_4$ (Also called *hydrosulphurous acid* or *dithionous acid*) (*iv*) *Pyrosulphurous acid* or *disulphorous acid,* $H_2S_2O_5$.

III. Sulphuric acid series. (*i*) *Sulphuric acid,* H_2SO_4 (*ii*) *Thiosulphuric acid,* $H_2S_2O_3$ (*iii*) *Pyro-sulphuric acid* or *disulphuric acid,* $H_2S_2O_7$ (also called *Oleum* or *fuming sulphuric acid*).

IV. Peroxy-sulphuric acid series. Two peroxy (or peroxo) sulphuric acids are known (*i*) *Peroxy-monosulphuric acid,* H_2SO_5 (also called *Caro's acid*) (*ii*) *Peroxy-disulphuric acid,* $H_2S_2O_8$ (also called *Marshall's acid*).

V. Thionic acids. These are of two types: (*i*) *Dithionic acid,* $H_2S_2O_6$ (*ii*) *Polythionic acids,* $H_2S_nO_6$ (n = 3 to 6). Different members of polythionic acids are : (*a*) *trithionic acid* ($H_2S_3O_6$) (*b*) *tetrathionic acid* ($H_2S_4O_6$) (*c*) *pentathionic acid* ($H_2S_5O_6$), and (*d*) *hexathionic acid* ($H_2S_6O_6$).

Here we shall discuss only a few acids and their salts.

Sulphurous acid, H_2SO_3 (S = + 4, Dibasic acid)

This acid is unstable and hence is unknown in the free state. It is known only in solution. In fact there is no direct evidence of the existence of the acid. Sulphurous acid is believed to be a solution of SO_2 in water. ($SO_2 + H_2O \equiv H_2SO_3$)

Preparation. (*a*) The solution of the acid is obtained by dissolving SO_2 in water.

$$SO_2 + H_2O \longrightarrow H_2SO_3$$

(*b*) H_2SO_3 can also be prepared by the action of water on thionyl chloride, $SOCl_2$.

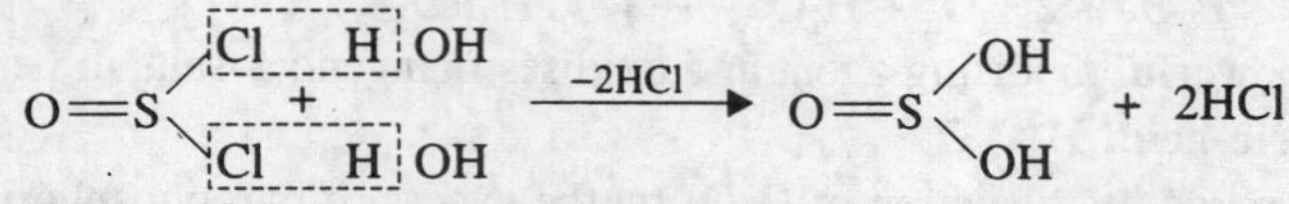

or $$SOCl_2 + 2H_2O \longrightarrow OS(OH)_2 \quad \text{or} \quad H_2SO_3 + 2HCl$$

Properties (*i*) The solution of the acid gives a smell of SO_2 which is evolved on heating the solution. Thus H_2SO_3 is present in equilibrium with the free SO_2 gas.

$$H_2SO_3 \rightleftharpoons H_2O + SO_2$$

(*ii*) When a saturated solution of SO_2 in H_2O is cooled to 3°C, crystals of the hexahydrate, $SO_2.6H_2O$ separate out.

(*iii*) *Acidic nature : Dibasic nature.* H_2SO_3 ionises in two stages and hence is a dibasic acid.

$$H_2SO_3 \rightleftharpoons H^+ + HSO_3^- \quad (\textit{Bisulphite ion}), \quad K_1 = 1.3 \times 10^{-2}$$

$$HSO_3^- \rightleftharpoons H^+ + SO_3^{2-}$$ (*Sulplite ion*), $K_2 = 6.24 \times 10^{-2}$

Thus H_2SO_3 forms two types of salts namely *bisulphites* which contain bisulphite ion, HSO_3^- (*e.g.*, $NaHSO_3$) and *sulphites* which have sulphite ion, SO_3^{2-} (*e.g.*, Na_2SO_3).

The ion obtained by removing one H_2O molecule from two bisulphite ions (HSO_3^-) is called *disulphite* or *pyrosulphite* or *metabisulphite ion, i.e.*,

$$2HSO_3^- \xrightarrow{-H_2O} S_2O_5^{2-} \text{ (disulphite ion)} + H_2O$$

It is incorrect to caii $S_2O_5^{2-}$ as pyrosulphite ion, since in the structure of this ion two S-atoms are not linked together through O-atom; rather they are connected directly with each other *i.e.*, the structure of $S_2O_5^{2-}$ ion is

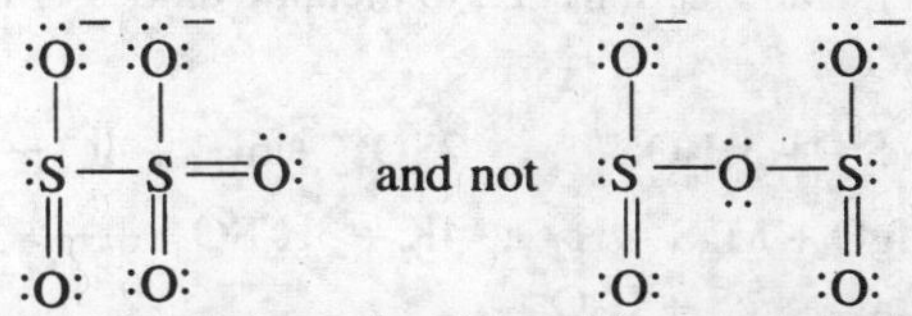

(*iv*) *Decomposition.* The acid solution when heated in a sealed tube at 150°C deposits sulphur.

$$3H_2SO_3 \xrightarrow{\Delta} 2H_2SO_4 + H_2O + S$$

(*v*) *Reducing properties.* H_2SO_3 (S = + 4) acts as a reducing agent. When it reduces a certain substance, it is oxidised to H_2SO_4(S = + 6) and gives nascent hydrogen.

$$H_2SO_3 + H_2O \longrightarrow H_2SO_4 + 2H \quad \text{or} \quad SO_3^{2-} + H_2O \longrightarrow SO_4^{2-} + 2H^+ + 2e^-$$

Thus:

(*a*) H_2SO_3 reduces halogens (X_2) to their corresponding hydracids (HX).

$$H_2SO_3 + H_2O + X_2 \longrightarrow H_2SO_4 + 2HX \quad \text{or} \quad SO_3^{2-} + H_2O + X_2 \longrightarrow SO_4^{2-} + 2H^+ + 2X^-$$

(*b*) Acidified solution of $KMnO_4$ is reduced to $MnSO_4$ solution.

$$2KMnO_4 + 5H_2SO_3 \longrightarrow K_2SO_4 + 2MnSO_4 + 2H_2SO_4 + 3H_2O$$

or $$2MnO_4^- + 5SO_3^{2-} + 6H^+ \longrightarrow 5SO_4^{2-} + 2Mn^{2+} + 3H_2O$$

(*c*) Acidified solution of $K_2Cr_2O_7$ is turned green due to the formation of $Cr_2(SO_4)_3$.

$$K_2Cr_2O_7 + H_2SO_4 + 3H_2SO_3 \longrightarrow K_2SO_4 + Cr_2(SO_4)_3 + 4H_2O$$

or $$Cr_2O_7^{2-} + 8H^+ + 3SO_3^{2-} \longrightarrow 3SO_4^{2-} + 2Cr^{3+} + 4H_2O$$

(*d*) $Fe_2(SO_4)_3$ is reduced to $FeSO_4$.

$$H_2SO_3 + H_2O + Fe_2(SO_4)_3 \longrightarrow 2H_2SO_4 + 2FeSO_4$$

or $$SO_3^{2-} + H_2O + 2Fe^{3+} \longrightarrow SO_4^{2-} + 2H^+ + 2Fe^{2+}$$

(*e*) It reduces acidified solution of iodate (*e.g.*, KIO_3) to I_2.

$$5H_2SO_3 + 2KIO_3 \longrightarrow 4H_2SO_4 + K_2SO_4 + H_2O + I_2$$

or $$5SO_3^{2-} + 2IO_3^- + 2H^+ \longrightarrow 5SO_4^{2-} + I_2 + H_2O$$

(*vi*) *Oxidising properties.* H_2SO_3 (S = + 4) also acts as an oxidising agent. When it oxidises a certain substance, it is reduced to sulphur (S = 0) in most cases. In some cases it is reduced to S^{2-} or $S_2O_3^{2-}$ (S = +2).

$$H_2SO_3 \longrightarrow S + H_2O + 2O \quad \text{or} \quad SO_3^{2-} + 6H^+ + 4e^- \longrightarrow S + 3H_2O$$

$$SO_3^{2-} + 6H^+ + 6e^- \longrightarrow S^{2-} + 3H_2O; \quad 2SO_3^{2-} + 6H^+ + 4e^- \longrightarrow S_2O_3^{2-} + 3H_2O$$

Thus :

(*a*) H_2SO_3 oxidises H_2S to S and is itself reduced to S.

$$H_2SO_3 + 2H_2S \longrightarrow 3S + 3H_2O \quad \text{or} \quad SO_3^{2-} + 6H^+ + 2S^{2-} \longrightarrow 3H_2O + 3S$$

(*b*) $SnCl_2$ and Hg_2Cl_2 are oxidised to $SnCl_4$ and $HgCl_4$ respectively. In both cases H_2SO_3 is reduced to sulphur.

$$H_2SO_3 + 2SnCl_2 + 4HCl \longrightarrow S + 3H_2O + 2SnCl_4$$

or $$SO_3^{2-} + 6H^+ + 2Sn^{2+} \longrightarrow S + 3H_2O + 2Sn^{4+}$$

$$H_2SO_3 + 2Hg_2Cl_2 + 4HCl \longrightarrow S + 3H_2O + 4HgCl_2$$

or $$SO_3^{2-} + 6H^+ + 2Hg_2^{2+} \longrightarrow S + 3H_2O + 4Hg^{2+}$$

(*c*) At 1000°C, H_2(H = 0) is oxidised to H_2O (H = +1) and H_2SO_3 is reduced to S.

$$H_2SO_3 + 2H_2 \longrightarrow S + 3H_2O \quad \text{or} \quad SO_3^{2-} + 6H^+ + 2H_2 \longrightarrow S + 3H_2O + 4H^+$$

(*d*) CO is oxidised to CO_2 and H_2SO_3 is reduced to sulphur.

$$H_2SO_3 + 2CO \longrightarrow S + H_2O + 2CO_2$$

(*e*) Metals are oxidised to metallic cations *i.e.*, to metallic oxides or metallic sulphides etc., and H_2SO_3 is reduced to S^{2-} or $S_2O_3^{2-}$.

$$SO_3^{2-} + 6H^+ + 6e^- \longrightarrow S^{2-} + 3H_2O; \qquad 2SO_3^{2-} + 6H^+ + 4e^- \longrightarrow S_2O_3^{2-} + 3H_2O$$

$$2Mg + H_2SO_3 \longrightarrow 2MgO + MgS + H_2O; \quad 4K + 3H_2SO_3 \longrightarrow K_2SO_3 + K_2S_2O_3 + 3H_2O$$

$$3Fe + H_2SO_3 \longrightarrow 2FeO + FeS + H_2O$$

(*f*) HI is oxidised to I_2. Here HI acts as a reducing agent.

$$H_2SO_3 + 4HI \longrightarrow S + 3H_2O + 2I_2 \quad \text{or} \quad SO_3^{2-} + 6H^+ + 4I^- \longrightarrow S + 3H_2O + 2I_2$$

This is why that in the titration between H_2SO_3 and I_2, the former must be run into I_2 solution and not *vice versa* otherwise HI produced will exert a reducing action on the excess of aqueous solution of SO_2 (*i.e.*, H_2SO_3).

(*viii*) *Bleaching properties.* We have said above that H_2SO_3 acts as a reducing agent, since it can liberate nascent hydrogen easily. This nascent hydrogen can render coloured matter to colourless matter. *i.e.*, H_2SO_3 acts as a bleaching agent.

$$H_2SO_3 + H_2O \longrightarrow H_2SO_4 + 2H$$

$$\text{Coloured matter} + H \longrightarrow \text{Colourless matter (Bleached matter)}$$

The bleaching done by H_2SO_3 is temporary, since the bleached matter (colourless matter) regains its colour due to oxidation when exposed to air.

$$\text{Colourless matter (Bleached matter)} + O \longrightarrow \text{Coloured matter}$$

(*viii*) H_2SO_3 molecule can readily take up oxygen and sulphur atom to form H_2SO_4 and $H_2S_2O_3$ (thiosulphuric acid).

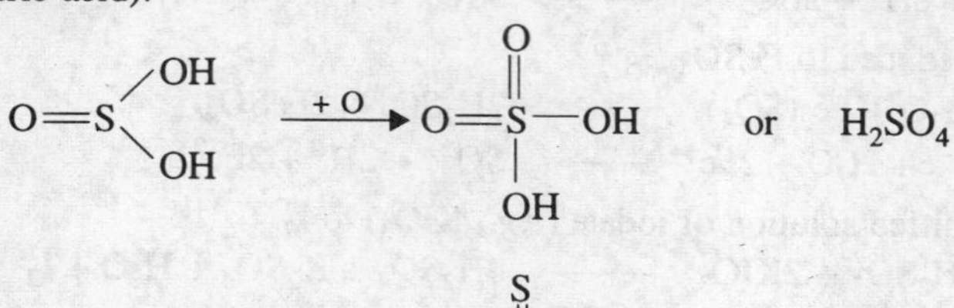

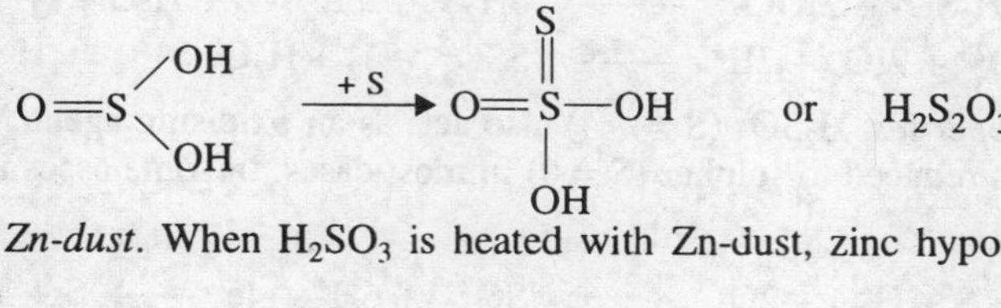

(*ix*) *Action of Zn-dust.* When H_2SO_3 is heated with Zn-dust, zinc hyposulphite (ZnS_2O_4) is obtained.

$$H_2SO_3 + Zn \xrightarrow{\Delta} ZnS_2O_4 + 2H_2O$$

Sulphites (SO_3^{2-}) and bisulphites (HSO_3^-)

Sulphites and bisulphites are the salts of H_2SO_3 and hence contain SO_3^{2-} and HSO_3^- ions respectively.

Preparation of sulphites. (*a*) Soluble sulphites are generally prepared by passing SO_2 into solutions of suspensions of metallic hydroxides or carbonates. For example:

$$4MOH + 2SO_2 \longrightarrow 2M_2SO_3 + 2H_2O \;; \quad M(OH)_2 + SO_2 \longrightarrow MSO_3 + H_2O$$

$$M_2CO_3 + SO_2 \longrightarrow M_2SO_3 + CO_2 \;; \quad MCO_3 + SO_2 \longrightarrow MSO_3 + CO_2$$

(*b*) Insoluble sulphites are formed by double decomposition between alkali sulphites and soluble salts of other metals.

Properties. (*i*) With the exception of alkali and ammonium sulphites, the normal sulphites are insoluble in water but dissolve in aqueous H_2SO_3. The aqueous solutions of alkali metal sulphites are stable and do not liberate SO_2. The solutions of alkali metal sulphites are alkaline towards litmus due to hydrolysis. Bisulphite solutions are acidic and smell of SO_2.

(*ii*) *Action of dil.* H_2SO_4 *and dil. HCl.* When a sulphite or a bisulphite is warmed with dil. H_2SO_4 or dil. HCl, suffocating colourless SO_2 gas is evolved. For example:

(*a*) $$Na_2SO_3 + H_2SO_4 \longrightarrow Na_2SO_4 + H_2O + SO_2$$

$$Na_2SO_3 + 2HCl \longrightarrow 2NaCl + H_2O + SO_2$$

or $$SO_3^{2-} + 2H^+ \longrightarrow H_2O + SO_2$$

(*b*) $$NaHSO_3 + HCl \longrightarrow NaCl + H_2O + SO_2$$

or $$HSO_3^- + H^+ \longrightarrow H_2O + SO_2$$

Both the above reactions have been used in the preparation of SO_2.

(*iii*) *Action of heat.* Sulphites, on being heated, give sulphide and sulphate, *e.g.*.

$$4Na_2SO_3 \xrightarrow{\Delta} Na_2S + 3Na_2SO_4$$

(*iv*) *Action of* $BaCl_2$ *and* $AgNO_3$ *solution.* An aqueous solution of a sulphite gives a white precipitate of $BaSO_3$ and Ag_2SO_3 respectively. Both these precipitates are soluble in HCl.

$$Na_2SO_3 + BaCl_2 \longrightarrow BaSO_3\downarrow + 2NaCl \quad \text{or} \quad SO_3^{2-} + Ba^{2+} \longrightarrow BaSO_3\downarrow$$

$$Na_2SO_3 + 2AgNO_3 \longrightarrow Ag_2SO_3\downarrow + 2NaNO_3 \quad \text{or} \quad SO_3^{2-} + 2Ag^+ \longrightarrow Ag_2SO_3\downarrow$$

(*v*) *Reducing properties.* The sulphites and bisulphites like H_2SO_3, show reducing properties. When they reduce any oxidising agent, they are oxidised to their corresponding sulphates by nascent oxygen given by the oxidising agent. For example:

$$BaSO_3 + O\ (\textit{Form oxidising agent}) \longrightarrow BaSO_4 \quad \text{or} \quad SO_3^{2-} + H_2O \longrightarrow SO_4^{2-} + 2H^+ + 2e^-$$

$$HSO_3^- + H_2O \longrightarrow SO_4^{2-} + 3H^+ + 2e^-$$

Thus sulphites reduce (*a*) Cl_2, Br_2 and I_2 water to their corresponding hydracids *viz.*, HCl, HBr and HI respectively (*b*) acidified $KMnO_4$ and $K_2Cr_2O_7$ solution to $MnSO_4$ and $Cr_2(SO_4)_3$ respectively. (*c*) acidified solution of iodic acid (HIO_3) or an iodate (IO_3^-) to I_2. Equations representing these reactions have already been given under the oxidising properties of H_2SO_3.

(*vi*) *Action of* H_2S. When H_2S reacts with a bisulphite (*e.g.*, $NaHSO_3$), sulphur is precipitated.

$$2NaHSO_3 + 2H_2S \longrightarrow Na_2SO_3 + 3S + 3H_2O$$

Hyposulphurous acid, $H_2S_2O_4$

This acid is also called *hydrosulphurous acid* or *dithionous acid.*

Preparation. The aqueous solution of zinc hyposulphite (ZnS_2O_4) is treated with lime [$Ca(OH)_2$] when $Zn(OH)_2$ gets precipitated and cal. hyposulphite (CaS_2O_4) remains in solution.

$$ZnS_2O_4\ (aq) + Ca(OH)_2\ (aq) \longrightarrow CaS_2O_4\ (\textit{solution}) + Zn(OH)_2\ (\textit{ppt.})$$

The solution containing CaS_2O_4 is treated with oxalic acid ($H_2C_2O_4$) when calcium oxalate (CaC_2O_4) is precipitated and $H_2S_2O_4$ is obtained in solution.

$$CaS_2O_4 + H_2C_2O_4 \longrightarrow H_2S_2O_4\ (\textit{solution}) + CaC_2O_4\ (\textit{ppt.})$$

ZnS_2O_4 solution needed in the above reaction can be obtained either by passing SO_2 gas through a suspension of Zn dust in absolute alcohol or by heating H_2SO_3 with Zn dust.

$$2SO_2 + Zn\ (\textit{in alcohol}) \longrightarrow ZnS_2O_4 \;; \quad H_2SO_3 + Zn \xrightarrow{\Delta} ZnS_2O_4 + 2H_2O$$

Properties. (*i*) *Decomposition.* The acid, even in solution, decomposes on heating to give first thiosulphurous acid ($H_2S_2O_3$) which decomposes to form H_2SO_3 and sulphur.

$$2H_2S_2O_4 \xrightarrow{\Delta} 2H_2S_2O_3 \; ; \quad H_2S_2O_3 \longrightarrow H_2SO_3 + S\downarrow$$

(*ii*) *Dibasic nature.* $H_2S_2O_4$ is a dibasic acid but gives only one type of salts, called *hyposulphites* or *hydrosulphites* which contain hyposulphite ion, $S_2O_4^{2-}$.

(*iii*) *Reducing properties.* The acid and its salts are strong reducing agent. They reduce the aqueous solution of the salts of Cu, Hg and Ag to free metals. They also reduce indigo and other vat dyes. When they reduce a certain substance, they are oxidised to sulphite (SO_3^{2-}) or bisulphite (HSO_3^-).

$$H_2S_2O_4 + 2H_2O \longrightarrow 2H_2SO_3 + 2H \text{ or } S_2O_4^{2-} + 2H_2O \longrightarrow 2SO_3^{2-} + 4H^+ + 2e^-$$

$$\text{or} \quad S_2O_4^{2-} + 2H_2O \longrightarrow 2HSO_3^- + 2H^+ + 2e^-$$

(*iv*) *Oxidation.* The aqueous solution of $H_2S_2O_4$ gets oxidised by atmospheric O_2 to form *pyro sulphurous* or *disulphurous acid* $H_2S_2O_5$.

$$2H_2S_2O_4 + O_2 \longrightarrow 2H_2S_2O_5$$

Uses. The acid and its salt *viz.*, $Na_2S_2O_4$ are used as reducing agents.

Sodium hyposulphite, $Na_2S_2O_4$

It is the most important salt of hyposulphurous acid, $H_2S_2O_4$.

Preparation. $Na_2S_2O_4$ is obtained by treating the aqueous solution of ZnS_2O_4 with the solution of soda ash (Na_2CO_3). $Na_2S_2O_4$ is obtained in solution.

$$ZnS_2O_4 + Na_2CO_3 \longrightarrow ZnCO_3\downarrow + Na_2S_2O_4 \; (solution)$$

The solution of ZnS_2O_4 needed in the above reaction is obtained either by passing SO_2 gas through a suspension of Zn dust in absolute alcohol or by heating H_2SO_3 with Zn dust.

$$2SO_2 + Zn \; (in\ alcohol) \longrightarrow ZnS_2O_4 \; ; \quad H_2SO_3 + Zn \xrightarrow{\Delta} ZnS_2O_4 + 2H_2O$$

Properties. (*i*) $Na_2S_2O_4$ is a colourless crystalline solid.

(*ii*) *Decomposition.* In solution, the salt decomposes to give $Na_2S_2O_3$ and $Na_2S_2O_5$ (called sod. pyrosulphite or sod. metabisulphite).

$$2Na_2S_2O_4 \longrightarrow Na_2S_2O_3 + Na_2S_2O_5$$

(*iii*) *Reducing properties.* It is a strong reducing agent. Its reducing action takes place according to the following equation :

$$Na_2S_2O_4 + 2H_2O \longrightarrow 2NaHSO_3 + 2H$$

Thus Cu, Hg and Ag are precipitated from the solutions of their salts, *e.g.*, when $Na_2S_2O_4$ reacts with Cu (+2) salt solution, a red colloidal solution of Cu-metal is obtained. This reaction has been used for the quantitative estimation of a hyposulphite. $Na_2S_2O_4$ also reduces indigo and other vat dyes.

(*iv*) *Oxidation.* The aqueous solution of $Na_2S_2O_4$ gets oxidised by the atmospheric O_2 to give sodium pyrosulphite (also called sodium metasulphite), $Na_2S_2O_5$.

$$2Na_2S_2O_4 + O_2 \longrightarrow 2Na_2S_2O_5$$

Uses. $Na_2S_2O_4$ is used (*i*) for reducing indigo and other vat dyes in industry (*ii*) for the quantitative estimation of a hyposulphite.

Sulphuric acid (Oil of vitriol), H_2SO_4 (S = + 6, Dibasic acid)

In ancient days, it was called oil of vitriol, since it was prepared by distilling green vitriol (ferrous sulphate), $FeSO_4.7H_2O$.

$$\underset{\text{Green vitriol}}{2[FeSO_4.7H_2O]} \longrightarrow Fe_2O_3 + SO_2 + 13H_2O + \underset{\text{Oil of vitriol}}{H_2SO_4}$$

Preparation. Sulphur is burnt in air so that it is oxidised to SO_2. SO_2 thus obtained is oxidised to SO_3 which, when dissolved in H_2O, gives H_2SO_4. The whole process can be summarised as:

$$S \xrightarrow{+O_2} SO_2 \xrightarrow{+\frac{1}{2}O_2} SO_3 \xrightarrow{+H_2O} H_2SO_4$$

SO_2 can be oxidised to SO_3 by various oxidising agents as discussed under "*preparation of SO_3*".

Manufacture of H_2SO_4

Sulphuric acid is manufactured these days by the following two processes:

1. Lead chamber process. In this process when the mixture containing SO_2, air and NO is treated with steam (H_2O), H_2SO_4 is formed. In this reaction NO acts as a catalyst.

$$2SO_2 + O_2 \text{ (from air)} + 2H_2O + \underset{\text{catalyst}}{[NO]} \longrightarrow 2H_2SO_4 + \underset{\text{catalyst}}{[NO]}$$

The following two mechanisms have been proposed to explain the formation of H_2SO_4.

(*i*) *Davy and Lunge mechanism.* According to this mechanism, the formation of H_2SO_4 takes place through the following steps:

(*a*) *Preparation of SO_2.* Sulphur or iron pyrites (FeS_2) is burnt in the excess of air to form SO_2.

$$S + O_2 \longrightarrow SO_2\ ; \quad 4FeS_2 + 11O_2 \longrightarrow 2Fe_2O_3 + 8\ SO_2$$

(*b*) *Oxidation of SO_2 to SO_3 by HNO_3.* SO_2 obtained as above is oxidised to SO_3 by HNO_3.HNO_3 is reduced to a mixture of NO and NO_2.

$$2HNO_3 + 2SO_2 \longrightarrow H_2O + NO + NO_2 + 2SO_3$$

HNO_3 required for the oxidation of SO_2 is obtained by the action of $NaNO_3$ on conc. H_2SO_4.

$$NaNO_3 + H_2SO_4 \text{ (conc.)} \longrightarrow KHSO_4 + HNO_3$$

(*c*) *Formation of nitroso sulphuric acid, NO.HSO_4 (nitrated acid).* NO and NO_2 formed in step (*b*) mix together to form N_2O_3 which reacts with SO_3 to to form NO.HSO_4.

$$NO + NO_2 \longrightarrow N_2O_3; \qquad 2SO_3 + N_2O_3 + H_2O \longrightarrow 2NO.HSO_4$$

(*d*) *Formation of H_2SO_4.* NO.HSO_4 reacts with H_2O and forms H_2SO_4. NO and NO_2 are again obtained.

$$2NO.HSO_4 + H_2O \longrightarrow 2H_2SO_4 + NO + NO_2$$

(*ii*) *Berzelius mechanism.* According to this mechanism, the formation of H_2SO_4 occurs through the following steps :

(*a*) *Preparation of SO_2.* SO_2 can be obtained by burning sulphur or iron pyrites (FeS_2) in the excess of air.

$$S + O_2 \longrightarrow SO_2; \qquad 4FeS_2 + 11O_2 \longrightarrow 2Fe_2O_3 + 8SO_2$$

(*b*) *Oxidation of SO_2 to SO_3.* SO_2 obtained as above is oxidised to SO_3 by NO_2 which is reduced to NO.

$$SO_2 + NO_2 \longrightarrow SO_3 + NO$$

NO_2 required for the oxidation of SO_2 to SO_3 is obtained by the oxidation of NO by atmospheric O_2.

$$2NO + O_2 \longrightarrow 2NO_2$$

(*c*) *Preparation of H_2SO_4.* SO_3 obtained in step (*b*) is allowed to react with steam (H_2O) to form H_2SO_4.

$$SO_3 + H_2O \text{ (steam)} \longrightarrow H_2SO_4$$

2. Contact process. This process involves the following steps:

(*a*) *Preparation of SO_2.* It can be obtained by burning sulphur or iron pyrites (FeS_2) in the excess of air.

$$S + O_2 \longrightarrow SO_2\ ; \quad 4FeS_2 + 11O_2 \longrightarrow 2Fe_2O_3 + 8SO_2$$

(*b*) *Oxidation of SO_2 to SO_3.* SO_2 is oxidised to SO_3 by the atmospheric O_2 in presence of a catalyst which may be Pt-asbestos, V_2O_5 or Fe_2O_3.

$$2SO_2 + O_2 \xrightleftharpoons{V_2O_5} 2SO_3; \qquad H = -196.6 \text{ KJ}$$

The gases (*i.e.*, SO_2 and O_2) must be purified before allowing them to combine to form SO_3. If they are not purified, the impurities will poison the catalyst used in the reaction. These days V_2O_5 is used as catalyst, as it is cheaper and not poisoned by impurities.

(*c*) *Dissolution of SO_3 in 98% H_2SO_4.* SO_3 is dissolved in 98% H_2SO_4 to get oleum, $H_2S_2O_7$ (*fuming sulphuric acid*)

$$SO_3 + H_2SO_4 \longrightarrow H_2S_2O_7$$

It may be noted that SO_3 cannot be dissolved in water because it produces a dense fog which does not easily condense.

(*d*) *Dilution of oleum by water.* Oleum is then diluted by water to get H_2SO_4 of any desired concentration.

$$H_2S_2O_7 + H_2O \longrightarrow 2H_2SO_4$$

The oxidation of SO_2 to SO_3 by atmospheric O_2 as shown at step (*b*) is a reversible and exothermic process. According to Le Chatelier's principle, the favourable conditions for the maximum yield of SO_2 are:

(*i*) *Low temperature.* Since the forward reaction is exothermic, low temperature favours the formation of SO_3 but a minimum temperature of 450°C, called *optimum temperature*, is needed to get the maximum yield of SO_3. Even at this temperature, the rate of reaction is slow and to accelerate the rate of reaction a suitable catalyst is required.

(*ii*) *High pressure.* Since the volume of the gaseous product (*i.e.*, SO_3) is less than that of the gaseous reactants (*i.e.*, SO_2 and O_2), high pressure will favour the oxidation of SO_2. But a very high pressure is likely to cause the corrosin of the vessel in which oxidation is carried out. The optimum pressure of 2-3 atmospheres is sufficient for oxidation.

(*iii*) *Excess of atmospheric O_2.* Maximum yield of SO_3 is obtained when SO_2 and atmospheric O_2 are taken in 2:3 ratio.

Fuming sulphuric acid is that which contains dissolved SO_3.

Physical properties

(*i*) Pure H_2SO_4 is a colourless viscous, heavy and syrupy oily liquid of density 1.84 (at 15°C) which does not fume. It forms a constant boiling mixture with water. This mixture contains 98% H_2SO_4 and its boiling point = 340°C which is usually given as the boiling point of H_2SO_4. 95% acid boils at 295°C. Pure H_2SO_4 is prepared by mixing 98% acid with calculated quantity of SO_3. When cooled with ice, it solidifies to colourless crystals which melt at 10.38°C. High boiling point and high viscosity of H_2SO_4 is due to the fact that H_2SO_4 molecules are associated together by H-bonding as shown below :

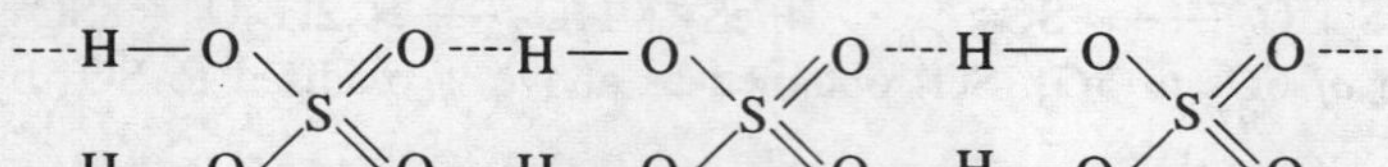

(*ii*) It is highly soluble in water. When water is added to the acid, a large amount of heat is produced and hence the temperature rises as high as 120°C. The heat produced is so large that the acid spurts out of the container. Therefore, if the acid is to be diluted, the acid should be added to water slowly with constant stirring and not water to the acid. The production of heat is due to the formation of hydrates like $H_2SO_4.H_2O$ (m.pt. = 8.5°C) and $H_2SO_4.2H_2O$ (m.pt. = 38°C). The formation of these hydrates is an exothermic process. In these hydrates both non-hydroxylic O-atoms of H_2SO_4 molecule are attached with H-atom of H_2O molecule by H-bonding.

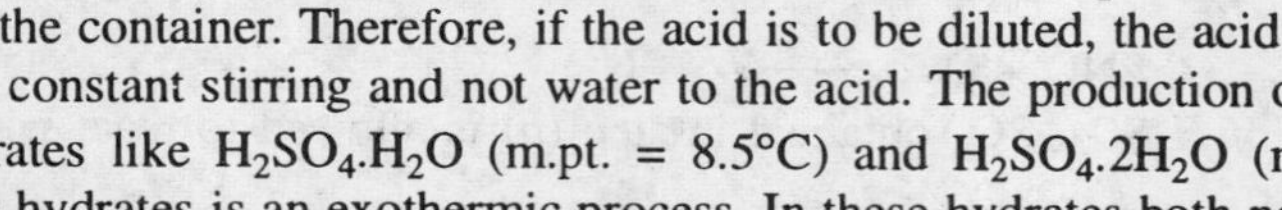

Monohydrate, $H_2SO_4.H_2O$

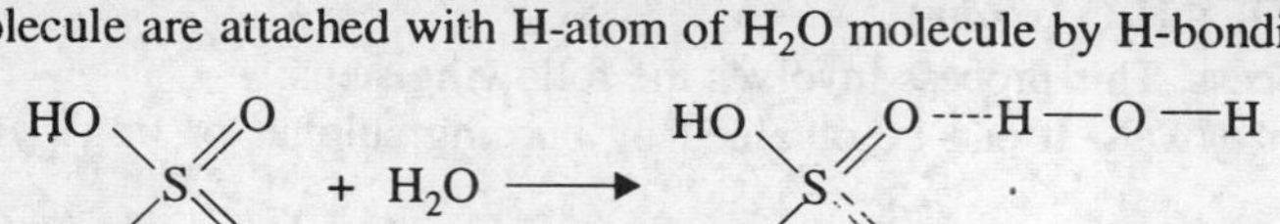

Dihydrate, $H_2SO_4.2H_2O$

(*iii*) It is a conductor of heat and electricity.

Chemical properties

1. Dissociation. (*a*) H_2SO_4, when heated undergoes dissociation into steam (H_2O) and SO_3.

$$H_2SO_3 \xrightleftharpoons{\Delta} H_2O\ (steam) + SO_3$$

The dissociation is almost complete at 444°C.

(*b*) when the acid vapour is passed through a strongly heated Pt or quartz tube, it is decomposed into SO_2, steam (H_2O) and O_2.

$$2H_2SO_4 \longrightarrow 2SO_2 + H_2O\ (steam) + O_2$$

2. Acidic nature. The aqueous solution of the acid behaves as a strong dibasic acid, since it ionises in two stages and gives two H^+ ions.

$$H_2SO_4 \rightleftharpoons H^+ + HSO_4^- \quad \textit{(bisulphate ion)}$$

$$HSO_4^- \rightleftharpoons H^+ + SO_4^{2-} \quad \textit{(sulphate ion)}$$

Thus H_2SO_4 forms two types of salts which are called *bisulphates* (*e.g.*, $NaHSO_4$) and *sulphates* (*e.g.*, Na_2SO_4). The acidic nature of H_2SO_4 is confirmed by the following reactions shown by H_2SO_4:

(*a*) H_2SO_4 is sour in taste and turns blue litmus red.

(*b*) It neutralises alkalies and forms bisulphates (HSO_4^-) and sulphates (SO_4^{2-}).

$$NaOH + H_2SO_4 \longrightarrow NaHSO_4 + H_2O; \quad 2NaOH + H_2SO_4 \longrightarrow Na_2SO_4 + 2H_2O$$

(*c*) It decomposes carbonates and bicarbonates into CO_2.

$$Na_2CO_3 + H_2SO_4 \longrightarrow Na_2SO_4 + H_2O + CO_2; \quad 2NaHCO_3 + H_2SO_4 \longrightarrow Na_2SO_4 + 2H_2O + 2CO_2$$

(*d*) Dilute H_2SO_4 usually reacts with all metals (except Pb) such as Al, Sn, Mn, Zn, Fe, Mg etc., which lie above hydrogen in the electrochemical series and H_2 gas is evolved.

$$M(M = Zn, Fe, Mg\ etc.) + H_2SO_4 \longrightarrow MSO_4 + H_2\uparrow$$

Metals like Cu, Ag, Hg, Pb etc., which lie below hydrogen in the series are not attacked by dil. H_2SO_4. However, these metals react with hot conc. H_2SO_4 (see *oxidising properties*).

3. Action on salts of more volatile acids. H_2SO_4 is a strong acid and hence decomposes the salts of more volatile acids (*e.g.*, chlorides, nitrates, sulphides, sulphites, carbonates etc.). The more volatile acid is liberated in each case. Examples are:

$$2NaCl + H_2SO_4 \longrightarrow Na_2SO_4 + 2HCl \quad \text{or} \quad 2Cl^- + H_2SO_4 \longrightarrow SO_4^{2-} + 2HCl$$

$$2KNO_3 + H_2SO_4 \longrightarrow K_2SO_4 + 2HNO_3 \quad \text{or} \quad 2NO_3^- + H_2SO_4 \longrightarrow SO_4^{2-} + 2HNO_3$$

$$FeS + H_2SO_4 \longrightarrow FeSO_4 + H_2S \quad \text{or} \quad S^{2-} + H_2SO_4 \longrightarrow SO_4^{2-} + H_2S$$

$$Ca_3(PO_4)_2 + 3H_2SO_4 \longrightarrow 3CaSO_4 + 2H_3PO_4$$

$$CaC_2O_4 + H_2SO_4 \longrightarrow CaSO_4 + H_2C_2O_4$$

$$2CH_3COONa + H_2SO_4 \longrightarrow Na_2SO_4 + 2CH_3COOH$$

$$CaF_2 + H_2SO_4 \longrightarrow CaSO_4 + 2HF$$

$$2NaNO_2 + H_2SO_4 \longrightarrow NaSO_4 + 2HNO_2$$

$$\downarrow$$

$$NO + NO_2 + H_2O$$

4. Oxidising properties. Hot concentrated H_2SO_4 (not dilute) acts as a powerful oxidising agent, since it can lose nascent oxygen quite readily. When H_2SO_4 oxidises any substance, it is reduced to SO_2.

$$\underset{\text{(Hot and concentrated)}}{H_2SO_4} \longrightarrow H_2O + SO_2 + O \quad \text{or} \quad SO_4^{2-} + 4H^+ + 2e^- \longrightarrow SO_2 + 2H_2O$$

Occasionally, H_2SO_4 is also reduced to sulphur, sulphide or H_2S.

$$SO_4^{2-} + 8H^+ + 6e^- \longrightarrow S + 4H_2O$$

$$SO_4^{2-} + 8H^+ + 8e^- \longrightarrow S^{2-} + 4H_2O$$

$$SO_4^{2-} + 10H^+ + 8e^- \longrightarrow H_2S + 4H_2O$$

(*i*) H_2SO_4 oxidises nonmetals like carbon and sulphur to their corresponding oxides *viz.*, CO_2 and SO_2. Phosphorus is oxidised to H_3PO_4, since P_2O_5 formed reacts with water to form H_3PO_4.

(*a*)
$$[H_2SO_4 \longrightarrow H_2O + SO_2 + O] \times 2$$
$$C + 2O \longrightarrow CO_2$$
$$2H_2SO_4 + C \longrightarrow 2H_2O + 2SO_2 + CO_2$$
or
$$2SO_4^{2-} + 4H^+ + C \longrightarrow 2SO_2 + 2H_2O + CO_2$$

(*b*)
$$2H_2SO_4 + S \longrightarrow 2H_2O + 3SO_2$$
or
$$2SO_4^{2-} + 4H^+ + S \longrightarrow 3SO_2 + 2H_2O$$

(*c*)
$$[H_2SO_4 \longrightarrow H_2O + SO_2 + O] \times 5$$
$$2P + 5O \longrightarrow P_2O_5$$
$$P_2O_5 + 3H_2O \longrightarrow 2H_3PO_4$$
$$2P + 5H_2SO_4 \longrightarrow 2H_3PO_4 + 5SO_2$$
or
$$5SO_4^{2-} + 4H^+ + 2P \longrightarrow 5SO_2 + 2PO_4^{3-} + 2H_2O$$

(*ii*) Metals like Cu, Ag, Hg, Pb etc., react with *hot conc.* H_2SO_4 and are oxidised to their oxides. These oxides then combine with acid and form their sulphates. Thus we find that the metals are oxidised to metallic cations and H_2SO_4 is reduced to SO_2. Examples are given below:

(*a*)
$$H_2SO_4 \longrightarrow H_2O + SO_2 + O$$
$$Cu + O \longrightarrow CuO$$
$$CuO + H_2SO_4 \longrightarrow CuSO_4 + H_2O$$
$$Cu + 2H_2SO_4 \longrightarrow CuSO_4 + SO_2 + 2H_2O$$
or
$$SO_4^{2-} + 4H^+ + Cu \longrightarrow SO_2 + 2H_2O + Cu^{2+}$$

(*b*)
$$2Ag + 2H_2SO_4 \longrightarrow Ag_2SO_4 + 2H_2O + SO_2$$
or
$$2Ag + SO_4^{2-} + 4H^+ \longrightarrow SO_2 + 2H_2O + 2Ag^+$$

(*c*)
$$Hg + 2H_2SO_4 \longrightarrow HgSO_4 + SO_2 + 2H_2O$$
or
$$Hg + 4H^+ + SO_4^{2-} \longrightarrow SO_2 + 2H_2O + Hg^{2+}$$

(*d*)
$$Pb + 2H_2SO_4 \longrightarrow PbSO_4 + 2H_2O + SO_2$$
or
$$Pb + 4H^+ + SO_4^{2-} \longrightarrow SO_2 + H_2O + Pb^{2+}$$

(*e*) Dil. H_2SO_4 reacts with commercial Zn, evolving H_2. Hot and conc. H_2SO_4 attacks Zn, giving off SO_2.

$$Zn + H_2SO_4 \longrightarrow ZnSO_4 + SO_2 + H_2O$$

Moderately conc. acid (20%) when heated with Zn metal gives H_2S and precipitates a little sulphur.

$$4Zn + 5H_2SO_4 \longrightarrow 4ZnSO_4 + H_2S + 4H_2O; \quad 3Zn + 4H_2SO_4 \longrightarrow 3ZnSO_4 + S + 4H_2O$$

(*iii*) H_2SO_4 cannot oxidise HF (or CaF_2) and HCl (or NaCl, KCl etc.) but it can oxidise HBr (or NaBr, KBr) and HI (or NaI, KI) to Br_2 and I_2 respectively. When H_2SO_4 oxidises HBr (NaBr or KBr), it is reduced to SO_2 but when the acid oxidises HI (or NaI or KI), it is reduced to SO_2, S or H_2S depending on the amount of HI used in the reaction.

$$H_2SO_4 \longrightarrow SO_2 + H_2O + O$$
$$2HBr + O \longrightarrow H_2O + Br_2$$
$$H_2SO_4 + 2HBr \longrightarrow SO_2 + 2H_2O + Br_2$$
or
$$SO_4^{2-} + 4H^+ + 2Br^- \longrightarrow SO_2 + 2H_2O + Br_2$$

Similarly, $H_2SO_4 + 2HI \longrightarrow SO_2 + 2H_2O + I_2$

or $SO_4^{2-} + 4H^+ + 2I^- \longrightarrow SO_2 + 2H_2O + I_2$

$H_2SO_4 + 6HI \longrightarrow S + 4H_2O + 3I_2$

or $SO_4^{2-} + 8H^+ + 6I^- \longrightarrow S + 4H_2O + 3I_2$

$H_2SO_4 + 8HI \longrightarrow H_2S + 4H_2O + 4I_2$

$SO_4^{2-} + 10H^+ + 8I^- \longrightarrow H_2S + 4H_2O + 4I_2$

Since HBr and HI are oxidised by H_2SO_4 to Br_2 and I_2 respectively, these acids cannot be prepared by the action of NaBr and NaI on H_2SO_4. Since HF and HCl are not oxidised by H_2SO_4, these acids can be obtained by heating CaF_2 and NaCl respectively with conc. H_2SO_4.

The equations representing the oxidation of KI and KBr to I_2 and Br_2 respectively can be written as follows:

$$H_2SO_4 \longrightarrow SO_2 + H_2O + O$$
$$2KI + H_2SO_4 \longrightarrow K_2SO_4 + 2HI$$
$$2HI + O \longrightarrow H_2O + I_2$$
$$\overline{2H_2SO_4 + 2KI \longrightarrow SO_2 + 2H_2O + K_2SO_4 + I_2}$$

or $SO_4^{2-} + 4H^+ + 2I^- \longrightarrow SO_2 + 2H_2O + I_2$

$$H_2SO_4 \longrightarrow S + H_2O + 3O$$
$$[2KI + H_2SO_4 \longrightarrow K_2SO_4 + 2HI] \times 3$$
$$[2HI + O \longrightarrow I_2 + H_2O] \times 3$$
$$\overline{4H_2SO_4 + 6KI \longrightarrow S + 4H_2O + 3K_2SO_4 + 3I_2}$$

or $SO_4^{2-} + 8H^+ + 6I^- \longrightarrow S + 4H_2O + 3I_2$

$$H_2SO_4 \longrightarrow H_2S + 4O$$
$$[2KI + H_2SO_4 \longrightarrow K_2SO_4 + 2HI] \times 4$$
$$[2HI + O \longrightarrow H_2O + I_2] \times 4$$
$$\overline{5H_2SO_4 + 8KI \longrightarrow H_2S + 4K_2SO_4 + 4H_2O + 4I_2}$$

or $SO_4^{2-} + 10H^+ + 8I^- \longrightarrow H_2S + 4H_2O + 4I_2$

Similar equations can be written for the oxidation of KBr to Br_2.

(*iv*) In presence of Hg (catalyst) naphthalene ($C_{10}H_8$) is oxidised to phthalic acid ($C_8H_6O_4$).

$$[H_2SO_4 \longrightarrow H_2O + SO_2 + O] \times 9$$
$$C_{10}H_8 + 9O \longrightarrow C_8H_6O_4 + 2CO_2 + H_2O$$
$$\overline{9H_2SO_4 + C_{10}H_8 \longrightarrow 10H_2O + 9SO_2 + C_8H_6O_4 + 2CO_2}$$

(*v*) H_2SO_4 oxidises H_2S to S

$$H_2SO_4 \longrightarrow H_2O + SO_2 + O$$
$$H_2S + O \longrightarrow H_2O + S$$
$$\overline{H_2SO_4 + H_2S \longrightarrow 2H_2O + SO_2 + S}$$

(*vi*) H_2SO_4 oxidises H_2 (H = 0) to H_2O (H = +1).

$$H_2SO_4 \longrightarrow H_2O + SO_2 + O$$
$$H_2 + O \longrightarrow H_2O$$
$$\overline{H_2SO_4 + H_2 \longrightarrow 2H_2O + SO_2}$$

5. Dehydrating properties. Conc. H_2SO_4 dissolves in water and forms sulphuric acid hydrates, $H_2SO_4.nH_2O$. Due to this property, conc. H_2SO_4 shows dehydrating properties. Thus it absorbs water from many organic compounds and dehydrates them. Absorption of water is highly exothermic. Water molecules obtained by the dehydration combine with H_2SO_4 and form $H_2SO_4.nH_2O$ hydrates. For example:

(*a*) Cane sugar ($C_{12}H_{22}O_{11}$), glucose ($C_6H_{12}O_6$) and starch ($C_6H_{10}O_5$) are dehydrated to carbon.

$$\underset{\text{Cane sugar}}{C_{12}H_{22}O_{11}} + H_2SO_4 \longrightarrow \underset{\text{Sugar charcoal}}{12C} + \underset{\text{Hydrate}}{[H_2SO_4.11H_2O]}$$

$$\underset{\text{Glucose}}{C_6H_{12}O_6} + H_2SO_4 \longrightarrow 6C + [H_2SO_4.6H_2O]$$

$$\underset{\text{Starch}}{(C_6H_{10}O_5)_n} + H_2SO_4 \longrightarrow 6nC + [5nH_2O\,.\,H_2SO_4]$$

(*b*) Oxalic acid, $(COOH)_2$ is dehydrated to CO and CO_2 and formic acid (HCOOH) gets dehydrated to CO only.

$$(COOH)_2 + H_2SO_4 \longrightarrow CO + CO_2 + [H_2SO_4.H_2O]$$

$$HCOOH + H_2SO_4 \longrightarrow CO + [H_2SO_4.H_2O]$$

(*c*) Ethyl alcohol (C_2H_5OH) is dehydrated to diethyl ether ($C_2H_5OC_2H_5$).

$$2C_2H_5OH + H_2SO_4 \longrightarrow C_2H_5OC_2H_5 + [H_2SO_4.H_2O]$$

(*d*) Dehydration of a mixture of chlorobenzene (C_6H_5Cl) and trichloroacetaldehyde (CCl_3CHO) (also called *chlorar*) gives D.D.T.

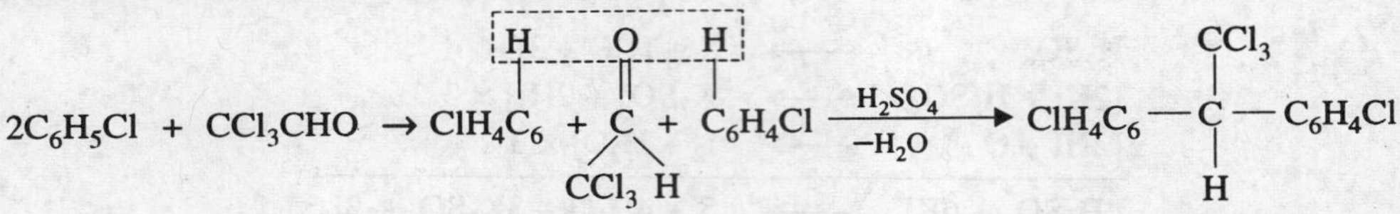

D.D.T.

It is due to the dehydrating property of conc. H_2SO_4 that cloth, wood, starch paper etc. [which are largely cellulose $(C_6H_{10}O_5)_x$ materials] get charred in this acid, *i.e.*, conc. H_2SO_4 removes water from these substances and thus chars them. The dehydrating property of conc. H_2SO_4 is used : (*i*) in drying the gases which do not react with the acid (*e.g.*, Cl_2, SO_2, HCl etc.) (*ii*) in many reactions like esterification etc. (*iii*) in the manufacture of dyes and explosives.

H_2SO_4 has very corrosive action on skin because it acts as dehydrating agent and absorption of water is accompanied by the release of heat.

6. Replacement of OH group(s). The molecule of H_2SO_4 contains two OH groups. One or both these groups can be replaced by other groups or atoms by treating H_2SO_4 with the appropriate compound. For example:

(*a*) When conc. H_2SO_4 is heated with C_6H_6, one OH group of H_2SO_4 is replaced by phenyl group (C_6H_5) and *benzene sulphonic acid,* $C_6H_5(OH)SO_2$ (also called *benzene sulphuric acid*) is obtained. Thus we see that benzene undergoes *sulphonation.*

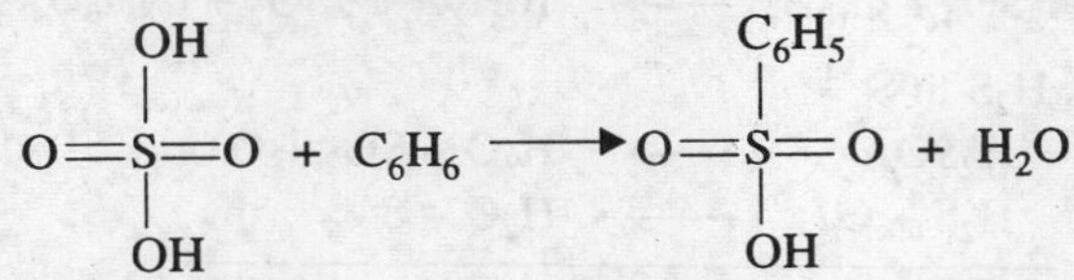

or $$O_2S(OH)_2 + C_6H_6 \longrightarrow O_2S(OH)\,(C_6H_5) + H_2O$$

(*b*) One or both OH groups of H_2SO_4 can be replaced by Cl-atom when H_2SO_4 is treated with PCl_5.

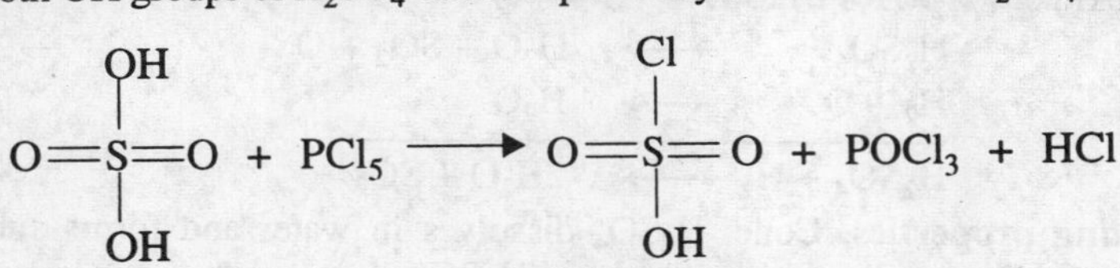

Chlorosulphonic acid

or $O_2S(OH)_2 + PCl_5 \longrightarrow Cl.SO_2.OH + POCl_3 + HCl$

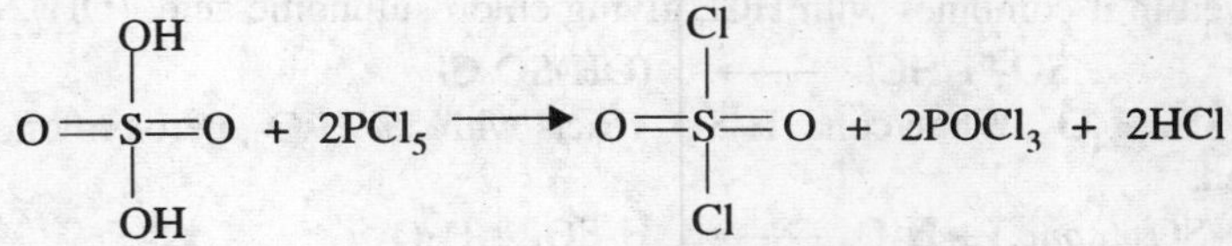

Sulphuryl chloride

or $O_2S(OH)_2 + 2PCl_5 \longrightarrow Cl.SO_2.Cl + 2POCl_3 + 2HCl$

(*c*) One Cl-atom can also be replaced by the action of $POCl_3$ on it. Thus the action of $POCl_3$ on H_2SO_4 produces chlorosulphonic acid.

$$2O_2S(OH)_2 + POCl_3 \longrightarrow 2Cl.SO_2.OH + HCl$$

7. Action on Ba (+2) and Pb (+2) salts. H_2SO_4 reacts with Ba (+2) and Pb (+2) salts and form the precipitate of $BaSO_4$ and $PbSO_4$ respectively. These sulphates are insoluble even in conc. HCl and conc. HNO_3.

$$BaCl_2 + H_2SO_4 \longrightarrow BaSO_4\downarrow + 2HCl \quad \text{or} \quad Ba^{2+} + SO_4^{2-} \longrightarrow BaSO_4\downarrow$$

$$Pb(NO_3)_2 + H_2SO_4 \longrightarrow PbSO_4\downarrow + 2HNO_3 \quad \text{or} \quad Pb^{2+} + SO_4^{2-} \longrightarrow PbSO_4\downarrow$$

8. Action of $KClO_3$. On heating $KClO_3$ with conc. H_2SO_4, ClO_2 is evolved with explosion.

$$3KClO_3 + 3H_2SO_4 \xrightarrow{\Delta} 3KHSO_4 + HClO_4 + 2ClO_2 + H_2O$$

9. Action of potassium ferrocyanide, $K_4[Fe(CN)_6]$. On heating $K_4[Fe(CN)_6]$ with conc. H_2SO_4, CO is evolved.

$$K_4[Fe(CN)_6] + 6H_2SO_4 + 6H_2O \xrightarrow{\Delta} 2K_2SO_4 + 6CO + FeSO_4 + 3(NH_4)_2SO_4$$

10. Action of P_2O_5. H_2SO_4 when treated with P_2O_5 loses water which combines with P_2O_5 and forms HPO_3.

$$H_2SO_4 \xrightarrow{-H_2O} H_2O + SO_3$$

$$H_2O + P_2O_5 \longrightarrow 2HPO_3$$

$$H_2SO_4 + P_2O_5 \longrightarrow SO_3 + 2HPO_3$$

11. Reactions leading to the formation of various types of fertilisers. H_2SO_4 reacts with many substances and forms various types of fertilisers. Examples are given below:

(*i*) When a mixture containing NH_3 and steam is passed through 60% H_2SO_4, crystals of $(NH_4)_2SO_4$ separate on cooling.

$$2NH_3 + H_2SO_4 \xrightarrow{\text{Cool}} (NH_4)_2SO_4\downarrow$$

(*ii*) $\underset{\text{Bone ash}}{Ca_3(PO_4)_2} + \underbrace{(NH_4)_2SO_4 + 2H_2SO_4}_{1:2 \text{ mixture}} \longrightarrow 3CaSO_4\downarrow + \underset{\text{Ammophos (solution)}}{2(NH_4)(H_2PO_4)}$

(*iii*) $3NH_3 + H_3PO_4 + H_2SO_4 \longrightarrow \underset{\text{Ammoniated phosphate sulphate}}{(NH_4)(H_2PO_4).(NH_4)_2SO_4}$

(*iv*) $\underset{\text{Bone ash}}{Ca_3(PO_4)_2} + 2H_2SO_4 + 5H_2O \longrightarrow \underset{\text{Superphosphate of lime}}{[Ca(H_2PO_4).H_2O + 2(CaSO_4.2H_2O)]}$

12. Action of $BaO_2.8H_2O$ When hydrated peroxide ($BaO_2.8H_2O$) is added to dil. H_2SO_4 at 0°C, H_2O_2 is obtained.

$$BaO_2.8H_2O + H_2SO_4 \longrightarrow BaSO_4\downarrow + H_2O_2 + 8H_2O$$

The precipitate of $BaSO_4$ is filtered off and H_2O_2 is obtained in solution.

13. Action on SO_3: Formation of oleum ($H_2S_2O_7$). H_2SO_4 dissolves SO_3, forming pyrosulphuric acid ($H_2S_2O_7$). This acid is also called *fuming sulphuric acid* or *oleum.*

$$H_2SO_4 + SO_3 \longrightarrow H_2S_2O_7$$

14. Electrolysis. When 50% H_2SO_4 is electrolysed, peroxydisulphuric acid ($H_2S_2O_8$) is obtained in solution at anode. For details see "*Preparation of Peroxydisulphuric Acid*".

15. Action of dry HCl gas on fuming H_2SO_4. When dry HCl gas is passed into fuming H_2SO_4, SO_3 dissolved in it combines with HCl, giving chlorosulphonic acid, $(OH).SO_2.Cl$.

$$SO_3 + HCl \longrightarrow (OH)SO_2Cl$$

16. Action of 5% H_2O_2. When conc. H_2SO_4 reacts with 5% H_2O_2, peroxymonosulphuric acid (H_2SO_5) is obtained.

$$H_2SO_4\,(conc.) + H_2O_2 \longrightarrow H_2SO_5 + H_2O$$

Uses and importance of H_2SO_4

So important is the acid and so varied are its uses that it is often called the *king of chemicals.* It is used: (*i*) in the manufacture of various fertilisers like $(NH_4)_2SO_4$, ammophos, ammoniated phosphate sulphate, superphosphate of lime, K_2SO_4 etc., (*ii*) in the manufacture of other acids such as HCl, H_3PO_4, HNO_3 etc., (*iii*) in the manufacture of explosives like nitroglycerine, trinitrotoluene (TNT), picric acid, gun cotton, dynamite etc., (*iv*) in the manufacture of many important chemicals like Na_2CO_3, alums, metal sulphates, ethers, phosphorus etc., (*v*) in the refining of petroleum, (*vi*) in many metallurgical processes for the purification of metals, (*vii*) in the manufacture of coaltar, dyes and a number of drugs and disinfectants, (*viii*) in the textile, paper and dyeing industries, (*iv*) in the leather industry for tanning, (*x*) for cleansing metals before electroplating, enameling, galvising etc., (pickling) (*xi*) in storage batteries, (*xii*) in the manufacture of paints and pigments, (*xiii*) as an important laboratory reagent, (*xiv*) for the synthesis of various organic compounds and sulphonation reactions, (*xv*) as an oxidising and dehydrating agent.

Sulphates (SO_4^{2-}) and bisulphates (HSO_4^-)

H_2SO_4 is a dibasic acid and hence forms two types of salts *viz.,* acid salts, called *bisulphates* (*e.g.*, $MHSO_4$) and normal sulphates, called *sulphates* (*e.g.*, M_2SO_4). Bisulphates are known only for alkali metals.

Preparation. These salts are prepared: (*i*) by the action on metals, their oxides, hydroxides, carbonates or other salts (*ii*) by the oxidation of sulphides or sulphites (*iii*) by decomposing the salts of more volatile acids by means of H_2SO_4.

Properties. (*i*) *Solubility.* Sulphates are mostly soluble in water and crystallise, forming hydrated sulphates. $CaSO_4$, $SrSO_4$ and $PbSO_4$ are *sparingly* soluble while $BaSO_4$ is *almost insoluble* in water.

(*ii*) *Action of heat.* Excepting $(NH_4)_2SO_4$, the sulphates of other alkali metals, of magnesium and of lead are stable towards heat, except at high temperatures. $CaSO_4$ is decomposed at high temperature while sulphates of Sr and Ba are stable. Sulphates of Zn, Fe, Cu etc., evolve SO_2 or SO_3 at high temperature.

$$MSO_4\ (M = Zn, Cu) \longrightarrow MO + SO_3$$
$$2FeSO_4 \longrightarrow Fe_2O_3 + SO_2 + SO_3$$
$$Fe_2(SO_4)_3 \longrightarrow Fe_2O_3 + 3SO_3$$

(*iii*) *Reduction.* Most of the sulphates can be reduced to sulphides when heated in a current of H_2 or with carbon (charcoal powder).

$$BaSO_4 + 4C \longrightarrow BaS + 4CO \quad \text{or} \quad SO_4^{2-} + 4C \longrightarrow S^{2-} + 4CO$$

(*iv*) *Reactions of SO_4^{2-} ions.* All soluble sulphates give white precipitates of $BaSO_4$ and $PbSO_4$ respectively with $BaCl_2$ and $Pb(NO_3)_2$. These precipitates are insoluble even in hot water, conc. HCl, conc. HNO_3 and aqua regia.

$$BaCl_2 + SO_4^{2-} \longrightarrow BaSO_4\downarrow + 2Cl^- \quad \text{or} \quad Ba^{2+} + SO_4^{2-} \longrightarrow BaSO_4\downarrow$$
$$Pb(NO_3)_2 + SO_4^{2-} \longrightarrow PbSO_4\downarrow + 2NO_3^- \quad \text{or} \quad Pb^{2+} + SO_4^{2-} \longrightarrow PbSO_4\downarrow$$

(*v*) *Action of HCl.* Many sulphates are converted into chlorides when heated in a current of HCl, *e.g.*,

$$CuSO_4 + 2HCl \longrightarrow CuCl_2 + H_2SO_4$$

Some important sulphates

Here we shall discuss $(NH_4)_2SO_4$, K_2SO_4, and double sulphates (also called alums) like $K_2SO_4.Al_2(SO_4)_3.24H_2O$ (called *potash alum*) and $K_2SO_4.Cr_2(SO_4)_3.24H_2O$ (*chrome alum*).

1. Ammonium sulphate, $(NH_4)_2SO_4$

It is the most important commercial salt of ammonium. This salt is used as a fertiliser. Large quantities of this salt are being manufactured in India at fertiliser factory at Sindri (Bihar). This salt is, therefore, called **Sindri fertiliser**. Tata Iron and Steel Company and Indian Iron and Steel Company are also manufacturing this salt.

Manufacture of $(NH_4)_2SO_4$ from ammonia liquor. Ammonia liquor is the by-product obtained during the dry distillation of coal. Besides organic compounds, this liquor also contains free NH_3 and many ammonium salts. These ammonium salts are of two types:

(*a*) *Volatile ammonium salts*. Examples of such salts are NH_4CN, NH_4HS etc. These salts are decomposed to give NH_3 simply by boiling, *e.g.*,

$$NH_4CN \xrightarrow{\text{Boil}} NH_3 + HCN \; ; \quad NH_4HS \xrightarrow{\text{Boil}} NH_3 + H_2S$$

(*b*) *Fixed ammonium salts*. Examples are $(NH_4)_2CO_3$, NH_4Cl etc. These are decomposed by lime water, $Ca(OH)_2$, *e.g.*,

$$(NH_4)_2CO_3 + Ca(OH)_2 \longrightarrow CaCO_3 + 2NH_3 + 2H_2O$$
$$2NH_4Cl + Ca(OH)_2 \longrightarrow CaCl_2 + 2NH_3 + 2H_2O$$

The liquor is heated in ammonia still first by steam to remove free NH_3 and to decompose volatile ammonium salts and then the residue which contains fixed salts is treated with lime water to decompose the fixed salts. The mixture of NH_3 and steam so produced is passed through a lead-lined tank containing 60% H_2SO_4 so that crystals of $(NH_4)_2SO_4$ separate on cooling.

$$2NH_3 + H_2SO_4 \longrightarrow (NH_4)_2SO_4$$

Manufacture of $(NH_4)_2SO_4$ at Sindri fertiliser factory. NH_3 is produced by Haber's process. This synthetic NH_3 is passed into a suspension of finely powdered $CaSO_4$ in water through which a stream of CO_2 is also passed.

$$2NH_3 + H_2O + CO_2 \longrightarrow (NH_4)_2CO_3 \xrightarrow{+\,CaSO_4} (NH_4)_2SO_4 + CaCO_3\downarrow$$

The insoluble $CaCO_3$ is filtered off and the filtrate is concentrated to crystallisation.

Properties. (*i*) $(NH_4)_2SO_4$ forms large colourless transparent rhombic crystals of density = 1.77. It is isomorphous with K_2SO_4 and is highly soluble in water.

(*ii*) *Action of heat.* On being heated, it gives NH_3 and NH_4HSO_4. Complete conversion of $(NH_4)_2SO_4$ into NH_4HSO_4 takes place at 300°C. On further heating, NH_4HSO_4 gives NH_3,N_2, SO_2 and steam.

$$(NH_4)_2SO_4 \xrightarrow{100\,-\,300^\circ C} NH_3 + NH_4HSO_4$$
$$3NH_4HSO_4 \xrightarrow{>\,300^\circ C} NH_3 + N_2 + 3SO_2 + 6H_2O\ (\textit{steam})$$

Uses. It is used: (*i*) as a fertiliser in agriculture (*ii*) as a reagent in laboratory (*iii*) in the preparation of other ammonium salts and alums.

Use of $(NH_4)_2SO_4$ as a fertiliser. This fertiliser contains 24-25% NH_3. This ammonia cannot be easily utilised by plants. When it is put into the soil, it is decomposed to give NH_3 by basic matter present in the soil. NH_3 thus obtained is converted into soluble nitrates by the nitrifying bacteria present in the soil. Nitrates thus obtained are taken up easily by plants from the soil.

$$(NH_4)_2SO_4 \xrightarrow[\text{in the soil}]{\text{Basic matter}} NH_3 \xrightarrow[\text{bacteria}]{\text{Nitrifying}} \text{soluble nitrates} \longrightarrow \text{Are taken up by plants.}$$

Methods of its application. The following precautions should be observed when $(NH_4)_2SO_4$ is being used as a fertilizer: (*i*) It should not be applied alongwith seeds as it is harmful to the germinating seeds. (*ii*) It should not be applied to wet clay soils, since in these soils the want of good aeration makes the nitrifying process slow and imperfect. (*iii*) It should not be used to soil soon after liming, since the fertilizer gets decomposed by lime. (*iv*) The fertilizer should not be used too frequently on soils which are poor in lime since loss of NH_3 makes the soil acidic. In such soils the fertilizer should be used in combination with ground limestone or chalk which neutralises the acidity of the soil.

$(NH_4)_2SO_4$ is less hygroscopic than urea and NH_4NO_3 and, therefore, has better storage properties. It can be applied prior to sowing or at the sowing time or during the growing period of the crop.

2. Potassium sulphate, K_2SO_4

It occurs in large quantities in the Stassfurt potash deposits as double salts like *Schonite* ($K_2SO_4.MgSO_4.6H_2O$) and *Kainite* ($K_2SO_4.MgSO_4.MgCl_2.6H_2O$).

Preparation. (*a*) When a saturated solution of kainite is allowed to cool, double sulphate *viz.*, $K_2SO_4.MgSO_4.6H_2O$ crystallises out.

$$\underset{\text{Kainite (Saturated solution)}}{K_2SO_4.MgSO_4.MgCl_2.6H_2O} \xrightarrow{\text{cool}} \underset{\text{Double sulphate (crystals)}}{K_2SO_4.MgSO_4.6H_2O} + \underset{\text{Soluble}}{MgCl_2}$$

Now when the double sulphate, $K_2SO_4.MgSO_4.6H_2O$ obtained as above is treated with a solution of KCl (sylvine) and concentrated, sparingly soluble K_2SO_4 crystallises out.

$$K_2SO_4.MgSO_4.6H_2O + 2KCl \longrightarrow \underset{\text{sparingly soluble}}{2K_2SO_4} + \underset{\text{soluble}}{MgCl_2.6H_2O}$$

(*b*) K_2SO_4 is also obtained by heating KCl with conc. H_2SO_4.

$$2KCl + H_2SO_4\ (conc.) \xrightarrow{\Delta} K_2SO_4 + 2HCl$$

Properties. (*i*) It forms colourless anhydrous rhombic crystals having melting point = 107.4°C and density = 2.67 (at 18°C).

(*ii*) *Formation of alums.* K_2SO_4 forms double sulphates (called alums) with the sulphates of trivalent metals. Examples are $K_2SO_4.Al_2(SO_4)_3.24H_2O$ (called potash alum), $K_2SO_4.Cr_2(SO_4)_3.24H_2O$ (called chrome alum), $K_2SO_4.Mn_2(SO_4)_3.24H_2O$ etc.

When calculated quantity of K_2SO_4 is added to the solution of $Al_2(SO_4)_3$, potash alum is obtained.

$$Al_2(SO_4)_3 + K_2SO_4 + 24H_2O \longrightarrow K_2SO_4.Al_2(SO_4)_3.24H_2O\ (\textit{Potash alum})$$

(*iii*) *Reduction to K_2S.* When heated with excess of carbon or in a current of H_2, K_2SO_4 gets reduced to K_2S.

$$K_2SO_4 + 4C \xrightarrow{\Delta} K_2S + 4CO; \qquad K_2SO_4 + 4H_2 \longrightarrow K_2S + 4H_2O$$

Uses. K_2SO_4 is used: (*i*) as a K-fertiliser, particularly for tabacco and wheat crops, and for the crops which are sensitive to Cl^- ions. It contains 50% K_2O. (*ii*) for the manufacture of potash alum and hard glass.

3. Double sulphates or alums

When hot solutions of equimolecular amounts of souble sulphates of monovalent and trivalent metallic cations are mixed, we get double sulphates which have the composition, $M_2^+SO_4.M_2^{3+}(SO_4)_3.24H_2O$ or $M^+ M^{3+} (SO_4)_2 .12H_2O$. Here $M^+ = Na^+, K^+, Rb^+, Cs^+, Ag^+, Tl^+$ (Li^+ is too small to be accomodated in the lattice) and $M^{3+} = Al^{3+}, Fe^{3+}, Cr^{3+}, Mn^{3+}, Co^{3+}, Ga^{3+}, V^{3+}, Ti^{3+}, Ir^{3+}, Rh^{3+}$ etc.

These double sulphates are called **alums.** Examples of some alums are:

Potash alum (commonly called alum), $K_2SO_4.Al_2(SO_4)_3.24H_2O$ or $KAl(SO_4)_2.12H_2O$

Ammonium alum, $(NH_4)_2SO_4.Al_2(SO_4)_3.24H_2O$ or $(NH_4)Al(SO_4)_2.12H_2O$

Ammonium iron alum, $(NH_4)_2SO_4.Fe_2(SO_4)_3.24H_2O$ or $(NH_4)Fe(SO_4)_2.12H_2O$

Chrome alum, $K_2SO_4.Cr_2(SO_4)_3.24H_2O$ or $KCr(SO_4)_2.12H_2O$

Magnetic alum, $K_2SO_4.Mn_2(SO_4)_3.24H_2O$ or $KMn(SO_4)_2.12H_2O$

Sodium alum, $Na_2SO_4.Al_2(SO_4)_3.24H_2O$ or $NaAl(SO_4)_2.12H_2O$

Preparation. Alums are generally obtained by mixing hot solutions of equimolar amounts of their constituent sulphates and by subjecting the resulted solution to crystallisation.

Properties. (*i*) Alums are crystalline compounds. (*ii*) In each alum, each metal ion is surrounded by six H_2O molecules. Thus the crystals of alums consist of $[M^+(H_2O)_6]^+$, $[M^{3+}(H_2O)_6]^{3+}$ and SO_4^{2-} ions. (*iii*) Alums are fairly soluble in hot water but less so in cold water. The aqueous solutions of alums are acidic and have stringent taste. The solution shows the properties of all the ions namely M^+, M^{3+} and SO_4^{2-} ions of the constituent salts. (*iv*) The alums are isomorphous to each other and form mixed crystals. (*v*) Each alum has different melting point. (*vi*) On heating strongly the alum swells up and a porous mass called **burnt alum** is left behind.

Uses. Alums act as coagulants and have germicide properties. Thus alums are used in the purification of water. They are also used in tanning of leather, as mordants in dyeing and as antiseptics.

Potash alum and chrome alum are important alums and hence are discussed below:

A. Potash alum, $K_2SO_4.Al_2(SO_4)_3.24H_2O$ or $KAl(SO_4)_2.12H_2O$. It is a double sulphate of K^+ and Al^{3+} ions. It is commonly called **alum.**

Preparation. (*i*) *From bouxite, $Al_2O_3.2H_2O$.* Bauxite is boiled with dil. H_2SO_4, to get $Al_2(SO_4)_3$ in solution. To this solution calculated quantity of K_2SO_4 is added and the resulted solution is concentrated and cooled. After sometime crystals of alum are obtained.

$$Al_2O_3.2H_2O + 3H_2SO_4 \xrightarrow{\text{Boil}} Al_2(SO_4)_3 + 5H_2O$$

$$Al_2(SO_4)_3 + K_2SO_4 + 24H_2O \longrightarrow \underset{\text{Alum}}{K_2SO_4.Al_2(SO_4)_3.24H_2O}$$

(*ii*) *From alum stone or alunite.* Alum stone is boiled with dil. H_2SO_4 so that $Al_2(SO_4)_3$ is obtained in solution. To this solution a calculated quantity of K_2SO_4 is added and the solution is concentrated and cooled. Crystals of alum are obtained.

$$\underset{\text{Alum stone or alunite}}{K_2SO_4.Al_2(SO_4)_3.4Al(OH)_3} + \underset{\text{Dilute}}{6H_2SO_4} \xrightarrow{\text{Boil}} K_2SO_4 + 3Al_2(SO_4)_3 + 12H_2O$$

$$Al_2(SO_4)_3 + K_2SO_4 + 24H_2O \longrightarrow K_2SO_4.Al_2(SO_4)_3.24H_2O$$

Properties. (*i*) It is a white crystalline compound.

(*ii*) It is soluble in water and its aqueous solution is acidic due to hydrolysis of $Al_2(SO_4)_3$. Its aqueous solution contains K^+, Al^{3+} and SO_4^{2-} ions and their usual tests can be performed.

(*iii*) *Action of heat.* On heating it swells up on account of elimination of all the water molecules.

$$K_2SO_4.Al_2(SO_4)_3.24H_2O \xrightarrow{200°C} K_2SO_4.\ Al_2(SO_4)_3 + 24H_2O$$

$$K_2SO_4.\ Al_2(SO_4)_3 \xrightarrow{\text{Red heat}} K_2SO_4 + Al_2O_3 + 3SO_3$$

(*iv*) *Action of NH_4OH.* $Al(OH)_3$ is precipitated.

$$K_2SO_4.Al_2(SO_4)_3.24H_2O + 6NH_4OH \longrightarrow K_2SO_4 + 3(NH_4)_2SO_4 + 24H_2O + 2Al(OH)_3\downarrow$$

Uses. It is used: (*i*) as a mordant is dyeing and calico printing, (*ii*) in sizing of cheap quality of paper, (*iii*) in the purification of water, (*iv*) as antiseptic and in stopping bleeding from cuts, (*v*) in leather tanning.

B. Potassium chromic sulphate or chrome alum, $K_2SO_4.Cr_2(SO_4)_3.24H_2O$ or $KCr(SO_4)_2.12H_2O$. It is a double sulphate of K^+ and Cr^{3+} cations.

Preparation. (*a*) In *laboratory*, it is prepared by reducing a solution of $K_2Cr_2O_7$ acidified with H_2SO_4 by SO_2.In this reaction SO_2 is oxidised to K_2SO_4.

$$K_2Cr_2O_7 + 3SO_2 + H_2SO_4 \longrightarrow K_2SO_4 + \underset{\text{green}}{Cr_2(SO_4)_3} + H_2O$$

or $$Cr_2O_7^{2-} + 2H^+ + 3SO_2 \longrightarrow 2Cr^{3+} + 3SO_4^{2-} + H_2O$$

On standing, beautiful dark purple octahedral crystals of chrome alum are separated.

(*b*) It is obtained as a by-product when anthracene is oxidised to anthraquinone by means of $K_2Cr_2O_7$ solution acidified with H_2SO_4.

Properties. It forms dark purple octahedral crystals which are isomorphous with other alums.

It is fairly soluble in water and forms a bluish-red solution. When this solution is heated to 90°C, it turns green due to the conversion of violet $Cr_2(SO_4)_3.18H_2O$ into the green form, $Cr_2(SO_4)_3.6H_2O$.

$$Cr_2(SO_4)_3.18H_2O \text{ (violet form)} \xrightarrow[-12H_2O]{90°C} Cr_2(SO_4)_3.6H_2O \text{ (green form)} + 12H_2O$$

Uses. It is used: (*i*) as a mordant in dyeing and calico printing, (*ii*) for tanning leather, (*iii*) in photography (during fixing) for hardening of the negatives.

Thiosulphuric acid, $H_2S_2O_3$

The free acid is quite unstable at ordinary temperature but has been isolated as an etherate at –78°C by the action of SO_3 on H_2S.

$$SO_3 + H_2S \longrightarrow H_2S_2O_3$$

It has also been obtained by the action of H_2S on chlorosulphonic acid, $O_2S(OH)Cl$ or by the action of very dil. HCl on sodium thiosulphate, $Na_2S_2O_3$.

$$O_2S\,(OH)\boxed{Cl\ +\ H}SH \xrightarrow{-HCl} O_2S(OH)-SH + HCl$$

or $$O_2S(OH)Cl + H_2S \xrightarrow{-HCl} H_2S_2O_3 + HCl$$

$$Na_2S_2O_3 + 2HCl \longrightarrow H_2S_2O_3 + 2NaCl$$

or $$S_2O_3^{2-} + 2H^+ \longrightarrow H_2S_2O_3$$

$H_2S_2O_3$, being unstable, decomposes to give sulphur and SO_2 (or H_2SO_3).

$$H_2S_2O_3 \longrightarrow S + SO_2 + H_2O \text{ or } H_2S_2O_3 \longrightarrow S + H_2SO_3$$

$H_2S_2O_3$ is a dibasic acid.

Thiosulphates ($S_2O_3^{2-}$)

Although $H_2S_2O_3$ is a dibasic acid, it gives only one series of salts of $M_2S_2O_3$ type. These salts are called thiosulphates which contain thiosulphate ion, $S_2O_3^{2-}$. These salts are fairly stable. Most thiosulphates except those of alkali metals are sparingly soluble in water but dissolve in alkali thiosulphate solutions forming complex anions, *e.g.*,

$$3Cu_2S_2O_3 + 2Na_2S_2O_3 \longrightarrow Na_4[Cu_6(S_2O_3)_5]$$

Thiosulphates are decomposed by acids, *e.g.*,

$$Na_2S_2O_3 + 2HCl \longrightarrow 2NaCl + H_2O + SO_2 + S\downarrow \quad \text{or} \quad S_2O_3^{2-} + 2H^+ \longrightarrow H_2O + SO_2 + S\downarrow$$

Among the thiosulphates, sodium thiosulphate (also called *hypo*), $Na_2S_2O_3$ is the most important and hence is described below:

Sodium thiosulphate, $Na_2S_2O_3.5H_2O$

Preparation. It is prepared:

(*i*) by boiling Na_2SO_3 solution with powdered sulphur till the solution is no more alkaline (*Laboratory preparation*).

$$Na_2SO_3 + S \longrightarrow Na_2S_2O_3$$

The unreacted sulphur is removed by filtration and the resulting solution, on concentration and cooling, gives monoclinic crystals of pentahydrate, $Na_2S_2O_3.5H_2O$.

(*ii*) by boiling sulphur with NaOH. In this reaction sulphur disproportionates to give S^{2-} (S = – 2) and $S_2O_3^{2-}$ (S = +2). Thus we see that in addition to $Na_2S_2O_3$, Na_2S is also formed.

$$6NaOH + 4S \longrightarrow Na_2S_2O_3 + 2Na_2S + 3H_2O \quad \text{or} \quad 6OH^- + 4S \longrightarrow S_2O_3^{2-} + 2S^{2-} + 3H_2O$$

If sulphur is taken in excess, in addition to Na_2S, sodium penta sulphide (Na_2S_5) is also formed. Na_2S and Na_2S_5 both are converted into $Na_2S_2O_3$ by atmospheric oxidation.

$$2Na_2S + H_2O + 2O_2 \longrightarrow Na_2S_2O_3 + 2NaOH$$

$$2Na_2S_5 + 3O_2 \longrightarrow 2Na_2S_2O_3 + 6S$$

(*iii*) by passing SO_2 gas into Na_2S solution or O_2 gas into NaHS solution.

$$3SO_2 + 2Na_2S \longrightarrow 2Na_2S_2O_3 + S$$

$$2O_2 + 2NaHS \longrightarrow Na_2S_2O_3 + H_2O$$

(*iv*) by decomposing sodium hyposulphite, $Na_2S_2O_4$.

$$2Na_2S_2O_4 \longrightarrow Na_2S_2O_3 + Na_2SO_3 + SO_2$$

(*v*) by adding calculated quantity of I_2 to a mixture of Na_2S and Na_2SO_3 solution (*Spring's method*).

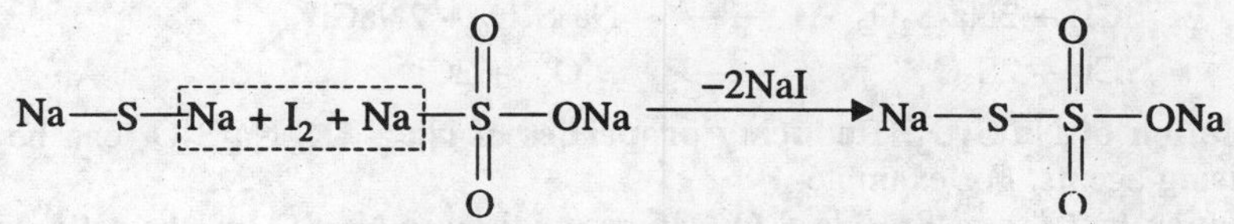

or $$Na_2S + I_2 + Na_2SO_3 \rightarrow Na_2S_2O_3 + 2NaI$$

or $$S^{2-} + I_2 + SO_3^{2-} \longrightarrow S_2O_3^{2-} + 2I^-$$

The resulting solution is concentrated and allowed to crystallise so that crystals of $Na_2S_2O_3.5H_2O$, being less soluble, separate out first.

(*vi*) by the action of sodium hydrosulphide (NaHS) on sodium bisulphite ($NaHSO_3$). This method gives pure $Na_2S_2O_3$.

$$2NaHS + 4NaHSO_3 \longrightarrow 3Na_2S_2O_3\ (pure) + 3H_2O$$

(*viii*) by passing SO_2 gas into waste liquor obtained in the manufacture of Na_2S (*Manufacture of* $Na_2S_2O_3$). Waste liquor contains Na_2S, Na_2CO_3 and Na_2SO_3.

$$\underbrace{2Na_2S + Na_2CO_3 + 4SO_2}_{\text{Waste liquor}} \longrightarrow 3Na_2S_2O_3 + CO_2$$

$$\underbrace{2Na_2S + Na_2SO_3 + 3SO_2}_{\text{Waste liquor}} \longrightarrow 3Na_2S_2O_3$$

The resulting solution is concentrated and allowed to crystallise so that the crystals of $Na_2S_2O_3.5H_2O$ which are less soluble separate out.

Properties. (1) This salt is a colourless crystalline solid, having melting point = 48°C. It is freely soluble in water. It contains five molecules of water as water of crystallisation.

(2) Action of heat. At 48°C the salt melts, at 215°C it loses all its five molecules of water and above 220°C it is converted into Na_2SO_4 and sodium pentasulphide (Na_2S_5).Na_2S_5 at high temperature loses sulphur and is converted into sodium tetrasulphide (Na_2S_4).

$$4[Na_2S_2O_3.5H_2O] \xrightarrow[(-\,20\,H_2O)]{250°C} 4Na_2S_2O_3 \xrightarrow{>220°C}$$

$$3Na_2SO_4 + \boxed{Na_2S_5} \xrightarrow{\text{High temerature}} Na_2S_4 + S$$

(3) Action of acids. (*a*) When $Na_2S_2O_3$ reacts with dilute acids (*e.g.*, H_2SO_4, HCl, HNO_3), SO_2 is evolved and colloidal sulphur (white or yellow turbidity) is obtained. Examples are:

$$H_2SO_4 + Na_2S_2O_3 \longrightarrow Na_2SO_4 + H_2O + SO_2 + S\downarrow$$

$$2HCl + Na_2S_2O_3 \longrightarrow 2NaCl + H_2O + SO_2 + S\downarrow$$

$$2HNO_3 + Na_2S_2O_3 \longrightarrow 2NaNO_3 + H_2O + SO_2 + S\downarrow$$

All the above reactions can be represented by the following ionic equation:

$$2H^+ + S_2O_3^{2-} \longrightarrow H_2O + SO_2 + S\downarrow$$

This reaction is used in the identification of $S_2O_3^{2-}$ radical.

(*b*) When a concentrated solution of $Na_2S_2O_3$ containing a little of sodium arsenite and cooled to –10°C is acted upon by conc. HCl, pentathionic acid ($H_2S_5O_6$), sodium pentathionate ($Na_2S_5O_6$) etc., are obtained.

$$5Na_2S_2O_3 + 8HCl\ (conc.) \longrightarrow Na_2S_5O_6 + H_2S_5O_6 + 8NaCl + 3H_2O$$

(4) Action of sulphur chlorides. When $Na_2S_2O_3$ reacts with sulphur monochloride (S_2Cl_2) and sulphur dichloride (SCl_2), tetrathionate ($S_4O_6^{2-}$) and pentathionate ($S_5O_6^{2-}$) are obtained respectively.

$$S_2Cl_2 + 2Na_2S_2O_3 \longrightarrow Na_2S_4O_6 + 2NaCl$$

or $$S_2Cl_2 + 2S_2O_3^{2-} \longrightarrow S_4O_6^{2-} + 2Cl^-$$

$$SCl_2 + 2Na_2S_2O_3 \longrightarrow Na_2S_5O_6 + 2NaCl$$

or $$SCl_2 + 2S_2O_3^{2-} \longrightarrow S_5O_6^{2-} + 2Cl^-$$

(5) Oxidation of $Na_2S_2O_3$: Reducing properties of $Na_2S_2O_3$. $Na_2S_2O_3$ can be oxidised by various oxidising agents. For example:

(*a*) *Oxidation by Cl_2 and Br_2.* $Na_2S_2O_3$ can be oxidised to Na_2SO_4 by the action of the excess of Cl_2 or Br_2 on it. These halogens are reduced to their hydracids *viz.*, HCl and HBr respectively, *e.g.*,

$$Na_2S_2O_3 + Cl_2 + H_2O \longrightarrow Na_2SO_4 + 2HCl + S$$

It is due to the above reaction that $Na_2S_2O_3$ is used as *anti-chlor, i.e.*, for removing the excess of Cl_2 sticking to the bleached articles in textile industry.

(*b*) *Oxidation by I_2.* I_2 oxidises $Na_2S_2O_3$ to sodium tetrathionate ($Na_2S_4O_6$) and is itself reduced to NaI.

$$2Na_2S_2O_3 + I_2 \longrightarrow Na_2S_4O_6 + 2NaI$$

or $$2S_2O_3^{2-} + I_2 \longrightarrow S_4O_6^{2-} + 2I^-$$

In the above reaction $Na_2S_2O_3$ acts as a reducing agent and I_2 behaves as an oxidising agent.

(*c*) *Oxidation by $FeCl_3$ solution.* When $Na_2S_2O_3$ is treated with a solution of $FeCl_3$, it gets oxidised to sodium tetrathionate ($Na_2S_4O_6$) and $FeCl_3$ is reduced to $FeCl_2$.

$$2Na_2S_2O_3 + 2FeCl_3 \longrightarrow Na_2S_4O_6 + 2FeCl_2 + 2NaCl$$

or $$2S_2O_3^{2-} + 2Fe^{3+} \longrightarrow S_4O_6^{2-} + 2Fe^{2+}$$

(*d*) *Oxidation by electricity.* When a neutral solution of $Na_2S_2O_3$ is allowed to undergo electrolytic oxidation, $Na_2S_4O_6$ is obtained.

$$2S_2O_3^{2-} \longrightarrow S_4O_6^{2-} + 2e^-$$

(*e*) *Oxidation by H_2O_2.* When H_2O_2 is added to a well-cooled saturated solution of $Na_2S_2O_3$ containing NaOH, the latter gets oxidised to sodium trithionate ($Na_2S_3O_6$) and the former (*i.e.*, H_2O_2) gets reduced to H_2O.

$$2Na_2S_2O_3 + 4H_2O_2 \xrightarrow{NaOH} Na_2S_3O_6 + 4H_2O + Na_2SO_4$$

or $$2S_2O_3^{2-} + 4H_2O_2 \xrightarrow{NaOH} S_3O_6^{2-} + 4H_2O + SO_4^{2-}$$

(*f*) *Oxidation by acidified solution of $KMnO_4$.* Acidified solution of $KMnO_4$ oxidises $Na_2S_2O_3$ to sodium dithionate ($Na_2S_2O_6$) and is itself reduced to Mn (+2) salt.

$$5Na_2S_2O_3 + 6KMnO_4 + 9H_2SO_4 \longrightarrow 5Na_2S_2O_6 + 3K_2SO_4 + 6MnSO_4 + 9H_2O$$

or $$S_2O_3^{2-} + 6MnO_4^- + 18H^+ \longrightarrow 5S_2O_6^{2-} + 6Mn^{2+} + 9H_2O$$

(*g*) *Oxidation by SO_2 gas.* When SO_2 gas is passed through a concentrated solution of $Na_2S_2O_3$, the latter is oxidised to sodium trithionate ($Na_2S_3O_6$) and the former gets reduced to sulphur.

$$2Na_2S_2O_3 + 3SO_2 \longrightarrow 2Na_2S_3O_6 + S\downarrow \quad \text{or} \quad S_2O_3^{2-} + 3SO_2 \longrightarrow 2S_3O_6^{2-} + S\downarrow$$

(*h*) *Oxidation by NaOH.* $Na_2S_2O_3$ is oxidised into Na_2SO_4 when it reacts with NaOH in presence of air.

$$Na_2S_2O_3 + 2NaOH + 2O_2 \longrightarrow 2Na_2SO_4 + H_2O$$

This reaction has been utilised in the extraction of Ag metal from silver glance ore (Ag_2S) by cyanide process (Mac Arthur Forest process).

(6) Action of salts. (*a*) *Action of $AgNO_3$ solution.* With dil. solution of $Na_2S_2O_3$, a white precipitate of silver thiosulphate ($Ag_2S_2O_3$) is formed which changes to yellow, orange, brown and finally to black pricipitate of Ag_2S.

$$2AgNO_3 + Na_2S_2O_3 \text{ (dil. solution)} \longrightarrow Ag_2S_2O_3 \text{ (white ppt.)} + 2NaNO_3$$

$$\text{or} \quad 2Ag^+ + S_2O_3^{2-} \longrightarrow Ag_2S_2O_3 \text{ (white ppt.)}$$

$$Ag_2S_2O_3 + H_2O \longrightarrow Ag_2S \text{ (black ppt.)} + H_2SO_4$$

The above reaction is used in the detection of $S_2O_3^{2-}$ radical given in an inorganic salt.

With conc. solution of $Na_2S_2O_3$, the white precipitate of $Ag_2S_2O_3$ formed as above dissolves in the excess of $Na_2S_2O_3$ solution by forming a colourless soluble complex compound namely sodium argento thiosulphate, $Na_3[Ag(S_2O_3)_2]$.

$$Ag_2S_2O_3 \text{ (white ppt.)} + 3Na_2S_2O_3 \text{ (excess)} \longrightarrow 2Na_3[Ag(S_2O_3)_2] \text{ (colourless)}$$

(*b*) *Action of $Pb(CH_3COO)_2$ solution.* A white precipitate of PbS_2O_3 is formed. This precipitate on boiling, turns black due to the formation of PbS.

$$Pb(CH_3COO)_2 + Na_2S_2O_3 \longrightarrow PbS_2O_3 \text{ (white ppt.)} + 2CH_3COONa$$

$$\text{or} \quad Pb^{2+} + S_2O_3^{2-} \longrightarrow PbS_2O_3 \text{ (white ppt.)}$$

$$PbS_2O_3 + H_2O \longrightarrow PbS \text{ (black ppt.)} + H_2SO_4$$

The above reaction is used for the detection of $S_2O_3^{2-}$ radical.

(*c*) *Action of silver halides.* Silver halides (AgCl, AgBr and AgI) get dissolved in the solution of $Na_2S_2O_3$ by forming colourless soluble complex compounds namely sodium argento thiosulphates whose composition depends on the number of $Na_2S_2O_3$ molecules used in the reaction.

$$AgCl + Na_2S_2O_3 \longrightarrow Na[Ag(S_2O_3)] + NaCl$$

$$AgCl + 1.5Na_2S_2O_3 \longrightarrow 0.5\ Na_4[Ag_2(S_2O_3)_3] + NaCl$$

$$AgCl + 2Na_2S_2O_3 \longrightarrow Na_3[Ag(S_2O_3)_2] + NaCl$$

$$AgBr + 2Na_2S_2O_3 \longrightarrow Na_3[Ag(S_2O_3)_2] + NaBr$$

The reaction between $Na_2S_2O_3$ and AgBr is utilised in the fixing of negatives and prints in photography, since $Na_2S_2O_3$ solution removes AgBr as a soluble colourless complex compound, $Na_3[Ag(S_2O_3)_2]$. Thus it is due to this reaction that $Na_2S_2O_3$ solution is used as a fixer in photography under the name of hypo. (see the uses of $Na_2S_2O_3$).

(*d*) *Action of cupric (Cu^{2+}) salts.* $Na_2S_2O_3$ solution reduces Cu^{2+} (cupric) salts to Cu^+ (cuprous) salts which dissolve in the excess of $Na_2S_2O_3$ solution by forming soluble sodium cuprous thiosulphate, $Na_4[Cu_6(S_2O_3)_5]$ (*complex compound*).

$$3 \times [2CuSO_4 + 3Na_2S_2O_3 \longrightarrow Cu_2S_2O_3 + 2Na_2SO_4 + Na_2S_4O_6$$

$$3Cu_2S_2O_3 + 2Na_2S_2O_3 \text{ (excess)} \longrightarrow Na_4[Cu_6(S_2O_3)_5]$$

$$6CuSO_4 + 11Na_2S_2O_3 \longrightarrow 6Na_2SO_4 + 3Na_2S_4O_6 + Na_4[Cu_6(S_2O_3)_5]$$

According to another view, cupric thiosulphate (CuS_2O_3) is first formed which gets dissolved in the excess of $Na_2S_2O_3$ solution by forming the soluble complex compound namely sodium cuprous thiosulphate, $Na[Cu(S_2O_3)]$.

$$[CuSO_4 + Na_2S_2O_3 \longrightarrow Na_2SO_4 + CuS_2O_3] \times 2$$

$$2Na_2S_2O_3 + 2CuS_2O_3 \longrightarrow 2Na[Cu(S_2O_3)] + Na_2S_4O_6$$

$$\underset{\text{Blue solution}}{2CuSO_4} + 4Na_2S_2O_3 \longrightarrow 2Na_2SO_4 + Na_2S_4O_6 + \underset{\text{Colourless}}{2Na[Cu(S_2O_3)]}$$

Thus the blue solution of $CuSO_4$ gets discharged due to the formation of $Na[Cu(S_2O_3)]$. $Na[Cu(S_2O_3)]$ is not very stable and hence, on boiling or on acidification, gives the precipitate of Cu_2S.

$$2Na[Cu(S_2O_3)] + H_2O \longrightarrow Cu_2S\downarrow + H_2SO_4 + Na_2S_2O_3$$

(*e*) *Action of auric (Au^{3+}) salts.* $Na_2S_2O_3$ solution reduces Au^{3+} salts to Au^+ (aurous) salts which dissolve in the excess of $Na_2S_2O_3$ solution by forming soluble sodium aurous thiosulphate, $Na_3[Au(S_2O_3)_2]$ (*complex compound*).

$$AuCl_3 + 2Na_2S_2O_3 \longrightarrow AuCl + Na_2S_4O_6 + 2NaCl$$
$$AuCl + 2Na_2S_2O_3\ (excess) \longrightarrow Na_3[Au(S_2O_3)_2] + NaCl$$
$$AuCl_3 + 4Na_2S_2O_3 \longrightarrow Na_2S_4O_6 + 3NaCl + Na_3[Au(S_2O_3)_2]$$

(*f*) *Action of $HgCl_2$ solution.* With excess of $HgCl_2$ solution, a white precipitate of an addition compound with the composition, $HgCl_2.2HgS$ is obtained.

$$3HgCl_2 + 2Na_2S_2O_3 + 2H_2O \longrightarrow HgCl_2.2HgS + 4NaCl + 2H_2SO_4$$

(*g*) *Action of bismuth salt solution.* With a solution of bismuth salt (*e.g.*, $BiCl_3$), a clear solution of sodium bismuth thiosulphate, $Na_3[Bi(S_2O_3)_3]$ is obtained.

$$3Na_2S_2O_3 + BiCl_3 \longrightarrow \underset{\text{Solution}}{Na_3[Bi(S_2O_3)_3]} + 3NaCl$$

The solution of $Na_3[Bi(S_2O_3)_3]$ quickly gets decomposed and gives black Bi_2S_3.

(*h*) *Action of $BaCl_2$ and $CaCl_2$ solution.* A moderately concentrated solution of $Na_2S_2O_3$ gives a white precipitate of BaS_2O_3 with $BaCl_2$ solution.

$$Ba^{2+} + S_2O_3^{2-} \longrightarrow BaS_2O_3\ (white\ ppt.)$$

With $CaCl_2$ no precipitate is obtained, since CaS_2O_3 formed becomes soluble in water.

Uses. $Na_2S_2O_3$ is used : (*i*) as an antichlor *i.e.*, for removing the excess of chlorine from the bleached articles in textile industry. (*ii*) for the determination of the strength of a given solution of iodine or for the determination of the strength of oxidising agents (*e.g.*, $K_2Cr_2O_7$, $KMnO_4$, $CuSO_4$), which on reacting with excess of KI solution (reducing agent), liberate iodine. The liberated iodine is titrated with standard solution of $Na_2S_2O_3$. (*iii*) in the extraction of Ag and Au (*iv*) for the preparation of $Na_2S_3O_6$ and $Na_2S_4O_6$ (*v*) *as a fixer in photography*. The negative plate or film obtained at the end of its development with reducing agent (*e.g.*, potassium ferrous oxalate, pyrogallol etc.) contains AgBr which has not been light-activated and reduced to Ag metal is treated with $Na_2S_2O_3$ solution so that AgBr is removed as a soluble complex compound, $Na_3[Ag(S_2O_3)_2]$. This is called fixing of the negative plate or film. (*vi*) in medicines.

Pyrosulphuric acid, $H_2S_2O_7$

This acid is also called *oleum, fuming sulphuric acid* or *disulphuric acid.*

Preparation. The acid is obtained:

(*a*) by dissolving SO_3 in conc. H_2SO_4.

$$H_2SO_4\ (conc.) + SO_3 \longrightarrow H_2S_2O_7$$

(*b*) by distilling $FeSO_4.H_2O$ crystals . The receiver should contain some conc. H_2SO_4.

$$6[FeSO_4.H_2O] \xrightarrow{\text{Distill}} Fe_2(SO_4)_3 + 2Fe_2O_3 + 3SO_2 + 6H_2O$$
$$Fe_2(SO_4)_3 \longrightarrow Fe_2O_3 + 3SO_3\ ; \qquad SO_3 + H_2SO_4\ (conc.) \longrightarrow H_2S_2O_7$$

Properties. (*i*) It is a viscous brownish liquid and is believed to be a solution of SO_3 in conc. H_2SO_4.

(*ii*) $H_2S_2O_7$ fumes in air and hence is also called *fuming sulphuric acid.*

(*iii*) *Action of heat.* On warming, $H_2S_2O_7$ changes into H_2SO_4, since the dissolved SO_3 escapes into the air.

$$H_2S_2O_7 \xrightarrow{\text{Warm}} H_2SO_4 + SO_3\uparrow$$

(*iv*) *Action of cooling.* $H_2S_2O_7$ solidifies on cooling below 0°C.

(*v*) *Dibasic nature.* Since the acid forms the salts of $M_2S_2O_7$ type, it is a dibasic acid. These salts are called *pyrosulphates* which contain pyrosulphate ion, $S_2O_7^{2-}$.

Uses. The acid is used: (*i*) in the manufacture of dyes and explosives (*ii*) in the refining of petroleum.

Peroxysulphuric acids

Sulphuric acid (H_2SO_4) gives two peroxy acids which are called peroxy sulphuric acids. These are *peroxymonosulphuric acid* (also called *Caro's acid*), H_2SO_5 and *peroxydisulphuric acid* (also called *Marshall's acid*), $H_2S_2O_8$.

Peroxy-monosulphuric acid or Caro's acid, H_2SO_5.

Preparation. (*i*) *Caro's method.* Caro in 1898 prepared the acid by mixing potassium peroxy-disulphate ($K_2S_2O_8$) with conc. H_2SO_4 and allowing the reaction mixture to stand for an hour and pouring on ice.

$$K_2S_2O_8 + H_2SO_4\ (conc.) \longrightarrow K_2SO_4 + H_2S_2O_8$$
$$H_2S_2O_8 + H_2O \longrightarrow H_2SO_5 + H_2SO_4$$
$$\overline{K_2S_2O_8 + H_2O \xrightarrow{\text{conc. } H_2SO_4} K_2SO_4 + H_2SO_5}$$

Unreacted H_2SO_4 is removed from the solution by shaking with excess of $Ba_3(PO_4)_2$ and filtering.

$$3H_2SO_4 + Ba_3(PO_4)_2 \longrightarrow 3BaSO_4\downarrow + 2H_3PO_4$$

(*ii*) *Ahrle's method.* Ahrle (1909) prepared nearly pure acid by the action of SO_3 or conc. H_2SO_4 on very concentrated H_2O_2. The reaction between conc. H_2SO_4 and H_2O_2 is reversible.

$$SO_3 + H_2O_2 \longrightarrow H_2SO_5$$
$$H_2SO_4 + H_2O_2 \rightleftharpoons H_2SO_5 + H_2O$$

(*iii*) *d'Ans and Friedrich method (1910).* Pure H_2SO_5 is obtained by the action of anhydrous H_2O_2 (1 mol) on chlorosulphuric acid (1 mol), $SO_2(OH)Cl$.

$$(HO)\ SO_2\boxed{Cl + H}\!-\!O\!-\!O\!-\!H \xrightarrow{-HCl} (HO)SO_2-O-O-H + HCl$$

or $$(HO)SO_2Cl + H_2O_2 \xrightarrow{-HCl} H_2SO_5 + HCl$$

Properties. (*i*) H_2SO_5 is a white, crystalline and hygroscopic solid with melting point = 45°C.

(*ii*) *Stability.* It is stable for some days but gradually loses ozonised O_2.

(*iii*) *Action of H_2O_2.* On warming with H_2O_2, H_2SO_5 loses O_2.

$$H_2SO_5 + H_2O_2 \xrightarrow{\Delta} H_2SO_4 + H_2O + O_2$$

(*iv*) *Action of dil. H_2SO_4 : Hydrolysis.* On distilling with dil. H_2SO_4, H_2O_2 is obtained.

$$H_2SO_5 + H_2O \xrightarrow{\text{Dil. } H_2SO_4} H_2SO_4 + H_2O_2$$

(*v*) *Oxidising properties.* H_2SO_5 is a powerful oxidising agent (more stronger than $H_2S_2O_8$). Thus it oxidises Fe^{2+} salts to Fe^{3+} salts, Mn^{2+} salts to MnO_2, Ag^+ salts to AgO, metals (*e.g.*, Zn, Cu) to their sulphates, SO_2 to SO_3 and aniline to nitrosobenzene and nitrobenzene (distinction from $H_2S_2O_8$) and liberates halogens from their respective halides. The liberation of I_2 from an iodide is instantaneous (distinction from $H_2S_2O_8$). When H_2SO_5 oxidises a certain substance, it gets reduced to sulphate.

$$H_2SO_5 \longrightarrow H_2SO_4 + O$$

Examples are given below:

$$H_2SO_5 \longrightarrow H_2SO_4 + O$$
$$2FeSO_4 + H_2SO_4 + O \longrightarrow Fe_2(SO_4)_3 + H_2O$$
$$\overline{2FeSO_4 + H_2SO_5 \longrightarrow Fe_2(SO_4)_3 + H_2O}$$

The following reactions can be written in a similar way:

$$Mn(OH)_2 + H_2SO_5 \longrightarrow MnO_2 + H_2SO_4 + H_2O$$
$$2AgNO_3 + H_2SO_5 + H_2O \longrightarrow 2AgO + H_2SO_4 + 2HNO_3$$
$$M(M = Zn, Cu) + H_2SO_5 \longrightarrow MSO_4 + H_2O$$
$$2KX\ (X = Cl, Br, I) + H_2SO_5 \longrightarrow K_2SO_4 + H_2O + X_2$$

(*vi*) *Monobasic nature.* H_2SO_5 does not give any salt of M_2SO_5 type. The non-formation of these salts shows that H_2SO_5 is not dibasic. It has been shown that when potassium chloro-sulphate, $O_2S(OK)(Cl)$ reacts with H_2O_2, potassium hydrogen peroxy-monosulphate ($KHSO_5$) is obtained.

$$O{=}\overset{\overset{O}{\|}}{\underset{\underset{OK}{|}}{S}}{-}\boxed{Cl + H}{-}O{-}O{-}H \xrightarrow{-HCl} O{=}\overset{\overset{O}{\|}}{\underset{\underset{OK}{|}}{S}}{-}O{-}O{-}H + HCl$$

or $\quad O_2S\,(OK)\,(Cl) + H_2O_2 \xrightarrow{-HCl} KHSO_5 + HCl$

The formation of $KHSO_5$ shows that H_2SO_5 is a monobasic acid and hence contains only one OH group attached directly with S-atom.

Uses. The acid is used as an oxidising agent and for the preparation of H_2O_2.

Peroxydisulphuric acid or Marshall's acid, $H_2S_2O_8$

Preparation. $H_2S_2O_8$ can be prepared:

(*i*) *By Berthelot's method.* Berthelot (1878) obtained $H_2S_2O_8$ by passing electric discharge through a mixture of SO_2 and O_2 and dissolving the product in H_2SO_4.

$$SO_2 + O_2 + H_2SO_4 \longrightarrow H_2S_2O_8$$

(*ii*) *By the electrolysis or anodic oxidation of 50-60% solution of H_2SO_4.* When 50-60% solution of H_2SO_4 is electrolysed in a large test tube in which a glass tube opened at both the ends is placed, H_2SO_4 gets ionised to give bisulphate ions (HSO_4^-) which unite together at anode (Pt wire) to form Marshall's acid, $H_2S_2O_8$.H_2 gas is given out at cathode.

$$H_2SO_4 \rightleftharpoons 2H^+ + 2HSO_4^- \quad (\textit{Ionisation})$$

$$2HSO_4^- \xrightarrow{\text{Oxidation}} H_2S_2O_8 + 2e^- \quad (\textit{At anode})$$

$$2H^+ + 2e^- \xrightarrow{\text{Reduction}} H_2\,(g) \quad (\textit{At cathode})$$

The anode is a platinum wire fused in a capillary tube and surrounded by a glass tube opened at both the ends. This tube serves as a diaphragm. A spiral of Cu wire outside the glass tube acts as a cathode.

(*iii*) *By the electrolysis of an aqueous solution of $(NH_4)_2SO_4$ containing H_2SO_4.* When aqueous solution of $(NH_4)_2SO_4$ and H_2SO_4 in molecular proportions is electrolysed at a low temperature in a cell provided with Pt anode and graphite cathode which are separated by a porous stoneware diaphragm, $(NH_4)_2S_2O_8$ is obtained at anode.

$$(NH_4)_2SO_4 + H_2SO_4 \longrightarrow 2NH_4HSO_4$$

$$NH_4HSO_4 \rightleftharpoons H^+ + [(NH_4)SO_4]^-$$

$$2[(NH_4)SO_4]^- \longrightarrow (NH_4)_2S_2O_8 + 2e^- \quad (\textit{At anode})$$

$$2H^+ + 2e^- \longrightarrow H_2\,(g) \quad (\textit{At cathode})$$

$(NH_4)_2S_2O_8$ formed in the anode compartment is quickly removed and is distilled with H_2SO_4 under reduced pressure so that $H_2S_2O_8$ is obtained.

$$(NH_4)_2S_2O_8 + H_2SO_4 \longrightarrow H_2S_2O_8 + (NH_4)_2SO_4$$

$(NH_4)_2S_2O_8$ required in the preparation of $H_2S_2O_8$ can also be prepared by electrolysing NH_4HSO_4 in dil. H_2SO_4 at 0°C between Pt-electrodes.

$$2NH_4HSO_4 \rightleftharpoons 2H^+ + 2NH_4SO_4^- \quad (\textit{Ionisation})$$

$$2NH_4SO_4^- \xrightarrow{\text{Oxidation}} (NH_4)_2S_2O_8 + 2e^- \quad (\textit{At anode})$$

$$2H^+ + 2e^- \xrightarrow{\text{Reduction}} H_2\,(g) \quad (\textit{At cathode})$$

(*iv*) *By d'Ans and Friedrich method (1910).* Pure $H_2S_2O_8$ is obtained by the action of anhydrous H_2O_2 (1 mol) on chlorosulphuric acid (2 mol), $SO_2(OH)Cl$.

$$(HO)\,SO_2\boxed{Cl + H}{-}O{-}O{-}\boxed{H + Cl}\,SO_2\,(OH) \xrightarrow{-2HCl} (HO)SO_2{-}O{-}O{-}SO_2(OH) + 2HCl$$

or $\quad (HO)SO_2Cl + H_2O_2 + (OH)SO_2Cl \xrightarrow{-2HCl} H_2S_2O_8 + 2HCl$

Properties. (*i*) $H_2S_2O_8$ is colourless, crystalline and hygroscopic solid with melting point = 65°C.

(*ii*) *Thermal stability.* It is fairly stable at ordinary temperature but on heating it loses O_2 and SO_3 gases.

$$2H_2S_2O_8 \xrightarrow{\Delta} 2H_2SO_4 + 2SO_3 + O_2$$

(*iii*) *Hydrolysis.* (*a*) On keeping the aquous solution of $H_2S_2O_8$, it gets hydrolysed to form H_2SO_4 and peroxymonosulphuric acid (H_2SO_5).

$$H_2S_2O_8 + H_2O \xrightarrow{\text{On standing}} H_2SO_4 + H_2SO_5$$

(*b*) When $H_2S_2O_8$ is distilled (heated) with dil. H_2SO_4, it gets hydrolysed to give 30% H_2O_2.

$$H_2S_2O_8 + 2H_2O \xrightarrow[\text{Heat}]{\text{Dil. } H_2SO_4} 2H_2SO_4 + H_2O_2$$

This reaction has been used to prepare H_2O_2 on a large scale.

(*iv*) *Oxidising properties.* The acid and its salts (which are called peroxydisulphates) both act as strong oxidising agents. These properties have been discussed under "*Peroxydisulphates*".

(*v*) $H_2S_2O_8$ bleaches indigo and oxidises aniline to aniline black but does not react with chromic acid or $KMnO_4$ solution.

Peroxydisulphates ($S_2O_8^{2-}$)

The salts of $H_2S_2O_8$ are called peroxydisulphates which contain peroxydisulphate ion, $S_2O_8^{2-}$. Two important peroxydisulphates are $K_2S_2O_8$ and $(NH_4)_2S_2O_8$. $K_2S_2O_8$ is obtained by the electrolysis of saturated solution of $KHSO_4$ while $(NH_4)_2S_2O_8$ is prepared by the electrolysis of a saturated solution of $(NH_4)_2SO_4$ in H_2SO_4.

$$2KHSO_4 \rightleftharpoons 2H^+ + 2KSO_4^-$$

$$2H^+ + 2e^- \longrightarrow H_2(g) \; (\textit{At cathode}); \quad 2KSO_4^- \longrightarrow K_2S_2O_8 + 2e^- \; (\textit{At anode})$$

$$(NH_4)_2SO_4 + H_2SO_4 \longrightarrow 2(NH_4)HSO_4$$

$$2NH_4HSO_4 \rightleftharpoons 2H^+ + 2NH_4SO_4^-$$

$$2H^+ + 2e^- \rightarrow H_2(g) \; (\textit{At cathode}); \quad 2NH_4SO_4^- \rightarrow (NH_4)_2S_2O_8 + 2e^- \; (\textit{At anode})$$

$K_2S_2O_8$ and $(NH_4)_2S_2O_8$ can also be obtained by the action of F_2 on $KHSO_4$ and $(NH_4)HSO_4$ respectively.

$$2KHSO_4 + F_2 \longrightarrow K_2S_2O_8 + 2HF; \quad 2(NH_4)HSO_4 + F_2 \longrightarrow (NH_4)_2S_2O_8 + 2HF$$

Peroxydisulphates of other metals can be obtained from $K_2S_2O_8$ and $(NH_4)_2S_2O_8$ by a process of double decomposition. For example BaS_2O_8 can be obtained by treating a solution of $(NH_4)_2S_2O_8$ with baryta, $Ba(OH)_2$.

$$(NH_4)_2S_2O_8 + Ba(OH)_2 \longrightarrow BaS_2O_8 + 4H_2O + 2NH_3$$

Properties. (*i*) $K_2S_2O_8$ is a colourless and crystalline solid. It is soluble in water.

(*ii*) *Action of heat.* (*a*) Peroxydisulphates decompose on heating into SO_3 and O_2.

$$2K_2S_2O_8 \xrightarrow{\Delta} 2K_2SO_4 + 2SO_3 + O_2$$

(*b*) When a solution a peroxydisulphate is heated, O_2 is evolved.

$$2K_2S_2O_8 + 2H_2O \xrightarrow{\Delta} 2K_2SO_4 + 2H_2SO_4 + O_2$$

(*iii*) (*a*) When peroxydisulphates are distilled with dil. H_2SO_4 under reduced pressure, H_2O_2 is obtained.

$$K_2S_2O_8 + 2H_2 \xrightarrow{\text{dil. } H_2SO_4} K_2SO_4 + H_2SO_4 + H_2O_2$$

Similarly, $(NH_4)_2S_2O_8 + 2H_2O \longrightarrow K_2SO_4 + H_2SO_4 + H_2O_2.$

(*b*) When a peroxy-disulphate is distilled with conc. H_2SO_4, peroxy-monosulphuric acid (H_2SO_5) is obtained (instead of H_2O_2).

$$K_2S_2O_8 + H_2O \xrightarrow{\text{conc. } H_2SO_4} K_2SO_4 + H_2SO_5$$

(*iv*) *Oxidising properties.* Like its parental acid *viz.*, $H_2S_2O_8$, $K_2S_2O_8$ also acts as an oxidising agent. When it oxidises a certain substances, it is reduced to SO_4^{2-} or HSO_4^-.

$$K_2S_2O_8 + H_2O \longrightarrow K_2SO_4 + H_2SO_4 + O$$

or $$S_2O_8 + 2e^- \longrightarrow 2SO_4^{2-}$$

$$K_2S_2O_8 + H_2O \longrightarrow 2KHSO_4 + O$$

or $$S_2O_8^{2-} + 2H^+ + 2e^- \longrightarrow 2HSO_4^-$$

Thus $K_2S_2O_8$ oxidises Fe^{2+} salts to Fe^{3+} salts. Mn^{2+} salts to MnO_2, Pb^{2+} salts to PbO_2, Ag^+ salts to AgO (*black ppt.*), metals to their sulphates, Cr to H_2CrO_4 and I_2 to iodic acid (HIO_3).

It also liberates halogens from their respective halides. The liberation of I_2 from an iodide is *slow* (Distinction from H_2SO_5).

$$2FeSO_4 + K_2S_2O_8 \longrightarrow Fe_2(SO_4)_3 + K_2SO_4$$

or $$2Fe^{2+} + S_2O_8^{2-} \longrightarrow 2Fe^{3+} + 2SO_4^{2-}$$

$$Mn(OH)_2 + K_2S_2O_8 \longrightarrow MnO_2 + 2KHSO_4$$

or $$Mn^{2+} + 2H_2O + S_2O_8^{2-} \longrightarrow MnO_2 + 4H^+ + 2SO_4^{2-}$$

$$Pb^{2+} + 2H_2O + S_2O_8^{2-} \longrightarrow PbO_2 + 4H^+ + SO_4^{2-}$$

$$2AgNO_3 + K_2S_2O_8 + 2H_2O \longrightarrow 2AgO + 2KHSO_4 + 2HNO_3$$

or $$2Ag^+ + S_2O_8^{2-} + 2H_2O \longrightarrow 2AgO + 2HSO_4^- + 2H^+$$

Here it may be noted that it is only $K_2S_2O_8$ and not $(NH_4)_2S_2O_8$ which gives a black precipitate of AgO with $AgNO_3$ solution.

$$M(M = Zn, Cu) + K_2S_2O_8 \longrightarrow MSO_4 + K_2SO_4$$

or $$M + S_2O_8^{2-} \longrightarrow M^{2+} + 2SO_4^{2-}$$

$$Cr + 3K_2S_2O_8 + 4H_2O \longrightarrow H_2CrO_4 + 3K_2SO_4 + 3H_2SO_4$$

or $$Cr + 3S_2O_8^{2-} + 4H_2O \longrightarrow CrO_4^{2-} + 6SO_4^{2-} + 8H^+$$

$$I_2 + 5K_2S_2O_8 + 6H_2O \longrightarrow 2HIO_3 + 5K_2SO_4 + 5H_2SO_4$$

or $$I_2 + 5S_2O_8^{2-} + 6H_2O \longrightarrow 2IO_3^- + 10SO_4^{2-} + 12H^+$$

$$K_2S_2O_8 + 2KI \longrightarrow 2K_2SO_4 + I_2 \quad \text{or} \quad S_2O_8^{2-} + 2I^- \longrightarrow 2SO_4^{2-} + I_2$$

Similar equations can be written for other halogens.

Uses. Peroxydisulphates are used : (*i*) as oxidising and bleaching agents (*ii*) for the preparation of H_2O_2 and in photography.

Table 33.1: *Comparison between H_2SO_5, $H_2S_2O_8$ and H_2O_2*

Reaction	H_2SO_5	$H_2S_2O_8$	H_2O_2
(*i*) With $KMnO_4$ solution acidified with dil. H_2SO_4.	No reaction, *i.e.*, acidified $KMnO_4$ solution does not get reduced.	NO reaction, *i.e.*, acidified $KMnO_4$ solution does not get reduced.	$KMnO_4$ solution gets decolourised and O_2 is evolved, *i.e.*, acidified $KMnO_4$ solution gets reduced.
(*ii*) With $K_2Cr_2O_7$ solution acidified by dil. H_2SO_4.	No reaction, *i.e.*, acidified $K_2Cr_2O_7$ solution does not get reduced.	No reaction, *i.e.* acidified $K_2Cr_2O_7$ solution does not get reduced.	Bright blue chromium diperoxide, CrO_5 is formed.
(*iii*) With aniline.	Nitrosobenzene and nitrobenzene are obtained.	Aniline black is formed.	No reaction.
(*iv*) With acidic solution of $Ti(SO_4)_2$.	No reaction.	No reaction.	Yellow or orange-red colouration, due to the formation of per titanic acid ($TiO_2.xH_2O$) is produced.

Dithionic acid, $H_2S_2O_6$

Preparation. Free acid is not known. It is known only in solution. The solution of the acid is obtained by treating barium dithionate (BaS_2O_6) with a calculated quantity of H_2SO_4.

$$BaS_2O_6\ (aq) + H_2SO_4\ (aq) \longrightarrow H_2S_2O_6\ (aq) + BaSO_4\downarrow$$

BaS_2O_6 needed in the above reaction is obtained by any of the following methods:

(*a*) A slow current of SO_2 is passed through a suspension of finely-powdered pyrollusite ore (MnO_2) in water kept in ice when MnS_2O_6 and $MnSO_3$ are formed. Now $Ba(OH)_2$ solution (baryta solution) is added to the reaction mixture so that BaS_2O_6 is formed in solution and $BaSO_3$ and $Mn(OH)_2$ are obtained as precipitate which can be removed by filtration.

$$2MnO_2 + 4SO_2 + H_2O \longrightarrow \underbrace{MnS_2O_6 + MnSO_3 + H_2SO_4}$$

$$\downarrow + 2Ba\ (OH)_2$$

$$\underset{\text{Solution}}{BaS_2O_6} + \underbrace{BaSO_3 + 2Mn(OH)_2}_{\text{ppt.}}$$

The solution containing BaS_2O_6, on concentration, gives the crystals of $BaS_2O_6.H_2O$.

(*b*) If we pass a slow current of SO_2 through $Fe(OH)_3$ (instead of MnO_2), the formation of BaS_2O_6 takes place through the following steps:

$$2Fe(OH)_3 + 3SO_2 \longrightarrow \underbrace{FeS_2O_6 + FeSO_3 + 3H_2O}$$

$$\downarrow + 2Ba(OH)_2$$

$$\underset{\text{Solution}}{BaS_2O_6} + \underbrace{BaSO_3 + 2Fe(OH)_2}_{\text{ppt.}}$$

Similar reactions take place when SO_2 gas is passed through $CO(OH)_3$ in place of MnO_2 or $Fe(OH)_3$.

Properties. The acid is available only in solution. On being concentrated, the acid decomposes to give SO_2 and H_2SO_4.

$$H_2S_2O_6 \longrightarrow SO_2 + H_2SO_4$$

The acid is stable towards oxidising agents. On hydrolysis, the acid gives H_2SO_4 and H_2SO_3.

$$H_2S_2O_6 + H_2O \longrightarrow H_2SO_4 + H_2SO_3$$

Although the acid is dibasic, it gives only one type of salts called dithionates (*e.g.*, $Na_2S_2O_6$) and the acid salts like $NaHS_2O_6$ are not known.

Uses. The acid and its salts (*i.e.*, dithionates) have been put to various uses.

Dithionates ($S_2O_6^{2-}$)

Although dithionic acid ($H_2S_2O_6$) is dibasic, it gives only one type of salts namely dithionates which contain dithionate ion, $S_2O_6^{2-}$. Acid salts like $NaHS_2O_6$ are unknown.

Preparation Dithionates can be obtained : (*a*) by decomposing BaS_2O_6 solution with a metal sulphate or metal carbonate, filtering off $BaSO_4$ or $BaCO_3$ and crystallising the solution.

(*b*) by the electrolytic oxidation of neutral or alkaline solution of sulphites.

$$2Na_2SO_3 + 2H_2O \longrightarrow Na_2S_2O_6 + 2NaOH + H_2$$

Properties. (*i*) All dithionates are soluble in water.

(*ii*) The dithionates of alkaline earth metals are very stable and hence their solutions may be boiled without decomposition.

(*iii*) Dithionates of heavy metals undergo hydrolysis in aqueous solution to give free $H_2S_2O_6$.

(*iv*) *Action of heat.* Dithionates decompose, on heating, to give sulphates and SO_2, *e.g.*,

$$K_2S_2O_6 \xrightarrow{\Delta} K_2SO_4 + SO_2 \quad \text{or} \quad S_2O_6^{2-} \xrightarrow{\Delta} SO_4^{2-} + SO_2$$

(*v*) *Action of sodium amalgam.* When aqueous solution of a dithionate is treated with Na-amalgam, it is reduced to sulphite, *e.g.*,

$$Na_2S_2O_6 + 2Na \longrightarrow 2Na_2SO_3$$

Barium dithionate, BaS_2O_6

Among dithionates BaS_2O_6 is the most important.

Preparation. Barium dithionate, BaS_2O_6 can be prepared by the methods which have already been discussed under "Preparation of dithionic acid".

Properties. (*i*) Solubility. It is soluble in water.

(*ii*) *Action of metal sulphate and metal carbonate.* When BaS_2O_6 solution is treated with a metal sulphate or a metal carbonate, dithionate of the other metal is obtained in solution and $BaSO_4$ and $BaCO_3$ are obtained as precipitate.

$$BaS_2O_6 + MSO_4 \longrightarrow MS_2O_6\ (solution) + BaSO_4\downarrow$$
$$BaS_2O_6 + MCO_3 \longrightarrow MS_2O_6\ (solution) + BaCO_3\downarrow$$

(*iii*) *Action of heat.* On being heated, it is converted into $BaSO_4$ and SO_2 gas is evolved.

$$BaS_2O_6 \xrightarrow{\Delta} BaSO_4 + SO_2$$

(*iv*) *Action of H_2SO_4.* When BaS_2O_6 solution is treated with a calculated quantity of H_2SO_4, dithionic acid ($H_2S_2O_6$) is obtained in solution.

$$BaS_2O_6(aq) + H_2SO_4\ (aq) \longrightarrow H_2S_2O_6\ (aq) + BaSO_4\downarrow$$

Uses. BaS_2O_6 is used for the preparation of $H_2S_2O_6$.

Polythionic acids, H_2SnO_6 (n = 3 to 6)

General methods for the preparation of polythionic acids and polythionates

Following methods are used:

(*i*) *By the hydrolysis of sulphur monochloride, S_2Cl_2.* When S_2Cl_2 is hydrolysed by H_2O, H_2SO_3, HCl and small quantity of polythionic acids like $H_2S_3O_6$, $H_2S_4O_6$ and $H_2S_5O_6$ are obtained.

$$S_2Cl_2 + 3H_2O \longrightarrow H_2SO_3 + H_2S + 2HCl$$

H_2SO_3 and H_2S formed above mix together and give $H_2S_3O_6$, $H_2S_4O_6$ and $H_2S_5O_6$.

$$6H_2SO_3 + 2H_2S \longrightarrow H_2S_3O_6 + H_2S_5O_6 + 6H_2O$$
$$3H_2SO_3 + H_2S \longrightarrow H_2S_4O_6 + 3H_2O$$

(*ii*) *By electrolytic oxidation.* Individual polythionates can be obtained by the oxidation of a thiosulphate (*e.g.*, $Na_2S_2O_3$) by means of H_2O_2, SO_2, I_2, $FeCl_3$ or electricity.

$$2Na_2S_2O_3\ (cooled\ solution) + 4H_2O_2 \xrightarrow{NaOH} Na_2S_3O_6 + 4H_2O + Na_2SO_4$$
or $$S_2O_3^{2-} + 4H_2O_2 \xrightarrow{NaOH} S_3O_3^{2-} + 4H_2O + SO_4^{2-}$$
$$2Na_2S_2O_3 + 3SO_2 \longrightarrow 2Na_2S_3O_6 + S\downarrow$$
or $$S_2O_3^{2-} + 3SO_2 \longrightarrow 2S_3O_6^{2-} + S\downarrow$$
$$2Na_2S_2O_3 + I_2 \longrightarrow Na_2S_4O_6 + 2NaI$$
or $$S_3O_6^{2-} + I_2 \longrightarrow S_4O_6^{2-} + 2I^-$$
$$2Na_2S_2O_3 + 2FeCl_3 \longrightarrow Na_2S_4O_6 + FeCl_2 + 2NaCl$$
or $$S_3O_3^{2-} + 2Fe^{3+} \longrightarrow S_4O_6^{2-} + 2Fe^{2+}$$
$$2S_2O_3^{2-} \xrightarrow[Oxidation]{Electrolytic} S_4O_3^{2-} + 2e^-$$

(*iii*) *From Wackenroder's solution : By passing H_2S gas through the solution of sulphurous acid.* When H_2S gas is passed through the solution of sulphurous acid at 0°C, a milky liquid is obtained which is called Wackenroder's solution. This solution contains colloidal sulphur, $H_2S_2O_3$ etc., alongwith $H_2S_4O_6$, $H_2S_5O_6$ and $H_2S_6O_6$. Polythionic acids in the form of their potassium salts can be obtained from this solution. The isolation of polythionic acids from this solution is based on the fact that potassium salts of polythionic acids have different solubility. The order of solubility is $K_2S_3O_6 > K_2S_6O_6 > K_2S_4O_6 = K_2S_5O_6$. Wackenroder's solution is first reacted with KNO_3 and concentrated so that sulphur separates out and is removed by filtration. The filtrate is neutralised by

CH_3COOK and allowed to evaporate whereby a mixture consisting of crystals of $K_2S_4O_6$ and $K_2S_5O_6$ is obtained. The mixture consisting of $K_2S_4O_6$ and $K_2S_5O_6$ crystals is suspended in a mixture of bromoform and xylene of specific gravity 2.2 in which $K_2S_5O_6$ floats and $K_2S_4O_6$ settles down to the bottom. The mother liquor obtained after the removal of $K_2S_4O_6$ which is recovered from this mother liquor by its spontaneous evaporation. The residual liquor is stirred with alcohol when the most soluble $K_2S_3O_6$ is precipitated.

(*iv*) Sodium tetrathionate and sodium pentathionate are obtained by the action of S_2Cl_2 and SCl_2 respectively on $Na_2S_2O_3$.

$$S_2Cl_2 + 2Na_2S_2O_3 \longrightarrow Na_2S_4O_6 + 2NaCl$$

or $$S_2Cl_2 + 2S_2O_3 \longrightarrow S_4O_6^{2-} + 2Cl^-$$

$$SCl_2 + 2Na_2S_2O_3 \longrightarrow Na_2S_5O_6 + 2NaCl$$

or $$SCl_2 + 2S_2O_3^{2-} \longrightarrow S_5O_6^{2-} + 2Cl^-$$

General properties of polythionates

(*i*) *Action of heat.* On heating, the solutions of all polythionates get decomposed to give sulphate, SO_2 and sulphur. For example:

$$K_2S_3O_6 \xrightarrow{\Delta} K_2SO_4 + SO_2 + S$$

or $$S_3O_6^{2-} \longrightarrow SO_4^{2-} + SO_2 + S$$

$$S_4O_6^{2-} \longrightarrow SO_4^{2-} + SO_2 + 2S$$

$$S_5O_6^{2-} \longrightarrow SO_4^{2-} + SO_2 + 3S$$

(*ii*) *Action of sodium or potassium amalgam.* (*a*) With sodium amalgam, $S_3O_6^{2-}$ gives thiosulphate ($S_2O_3^{2-}$) and sulphite (SO_3^{2-}) while $S_4O_6^{2-}$ gives only $S_2O_3^{2-}$.

$$S_3O_6^{2-} + 2Na \xrightarrow{Na/Hg} S_2O_3^{2-} + 2Na^+ + SO_3^{2-}$$

$$S_4O_6^{2-} + 2Na \xrightarrow{Na/Hg} S_2O_3^{2-} + 2Na^+$$

(*b*) with potassium amalgam, $S_5O_6^{2-}$ gives $S_2O_3^{2-}$ and S^{2-}.

$$S_5O_6^{2-} + 4K \longrightarrow 2S_2O_3^{2-} + S^{2-} + 4K^+$$

(*iii*) *Action of alkali metal sulphides (S^{2-}).* $S_3O_6^{2-}$ gives only $S_2O_6^{2-}$ while $S_4O_6^{2-}$, $S_5O_6^{2-}$ and $S_6O_6^{2-}$ give a mixture of $S_2O_3^{2-}$ and sulphur.

$$S_3O_6^{2-} + S^{2-} \longrightarrow 2S_2O_3^{2-}; \qquad S_4O_6^{2-} + S^{2-} \longrightarrow 2S_2O_3^{2-} + S$$

$$S_5O_6^{2-} + S^{2-} \longrightarrow 2S_2O_3^{2-} + 2S; \qquad S_6O_6^{2-} + S^{2-} \longrightarrow 2S_2O_3^{2-} + 3S$$

(*iv*) *Action of alkali metal sulphites (SO_3^{2-}).* Alkali metal sulphites have no action on trithionate ($S_3O_6^{2-}$) while $S_4O_6^{2-}$, $S_5O_6^{2-}$ and $S_6O_6^{2-}$ all give $S_2O_3^{2-}$ and lower polythionates. For example :

$$S_3O_6^{2-} + SO_3^{2-} \longrightarrow \text{No reaction}; \qquad S_4O_6^{2-} + SO_3^{2-} \longrightarrow S_2O_3^{2-} + S_3O_6^{2-}$$

$$S_5O_6^{2-} + SO_3^{2-} \longrightarrow S_2O_3^{2-} + S_4O_6^{2-}; \quad S_6O_6^{2-} + SO_3^{2-} \longrightarrow S_2O_3^{2-} + S_5O_6^{2-}$$

The above reactions show that higher polythionates are derived from trithionate.

(*v*) *Action of aqueous solution of KCN(CN^-).* On treatment with an aqueous solution of KCN, $S_3O_6^{2-}$ gives $S_2O_3^{2-}$, SO_4^{2-} and free HCN while $S_4O_6^{2-}$ and $S_5O_6^{2-}$ also give sulphocyanide (CNS^-) in addition to these products.

$$S_3O_6^{2-} + 2CN^- + H_2O \longrightarrow S_2O_3^{2-} + SO_4^{2-} + 2HCN$$

$$S_4O_6^{2-} + 3CN^- + H_2O \longrightarrow S_2O_3^{2-} + SO_4^{2-} + 2HCN + CNS^-$$

$$S_5O_6^{2-} + 4CN^- + H_2O \longrightarrow S_2O_3^{2-} + SO_4^{2-} + 2HCN + 2CNS^-$$

(*vi*) *Action of an alkali solution (OH^-).* $S_3O_6^{2-}$ and $S_4O_6^{2-}$, when treated with the aqueous solution of an alkali, decompose to give $S_2O_3^{2-}$ and SO_3^{2-} while $S_5O_6^{2-}$ gives only $S_2O_3^{2-}$.

$$2S_3O_6^{2-} + 6OH^- \longrightarrow S_2O_3^{2-} + 4SO_3^{2-} + 3H_2O$$

$$2S_4O_6^{2-} + 6OH^- \longrightarrow 3S_2O_3^{2-} + 2SO_3^{2-} + 3H_2O$$

$$2S_5O_6^{2-} + 6OH^- \longrightarrow 5S_2O_3^{2-} + 3H_2O$$

(*vii*) *Action of Cu^{2+} ions at 70°C.* It is only $S_3O_6^{2-}$ ion which reacts with Cu^{2+} ions at 70°C to give the ppt. of CuS.

$$S_3O_6^{2-} + 2H_2O + Cu^{2+} \longrightarrow CuS\downarrow + 2SO_4^{2-} + 4H^+$$

This reaction may be used for the estimation of trithionic acid in presence of other polythionic acids.

(*viii*) *Oxidation.* Aqueous solutions of $S_3O_6^{2-}$ and $S_4O_6^{2-}$ are oxidised to SO_4^{2-} and free H_2SO_4, when they are oxidised by air or an oxidising agent.

$$S_3O_6^{2-} + 2H_2O + 4O \longrightarrow SO_4^{2-} + 2H_2SO_4$$

$$S_4O_6^{2-} + 3H_2O + 7O \longrightarrow SO_4^{2-} + 2H_2SO_4$$

(*ix*) *Stability.* The alkali metal polythionates are quite stable in aqueous solution, those of alkaline earth metals are less stable while those of heavy metals are unstable. The stability of the solutions of polythionates increases by adding a little mineral acid while the same decreases when an alkali is added.

(*x*) *Dibasic nature.* All polythionic acids are dibasic in nature but they give only one type of salts like $Na_2S_3O_6$, $Na_2S_4O_6$ etc. The acids salt like $NaHS_3O_6$, $NaHS_4O_6$ etc., have not been prepared so far. The dibasic nature of these acids is because of the fact that each of the terminal S-atom is linked with one OH group.

(*xi*) *Solubility.* Excepting the polythionates of Ag and Hg, all other polythionates are soluble in H_2O. The order of the solubility of polythionates of potassium is $K_2S_3O_6 > K_2S_6O_6 > K_2S_4O_6 = K_2S_5O_6$. The difference in solubility of potassium polythionates has been utilised in the separation of a mixture of polythionates from Wackenroder's solution.

Trithionic acid, $H_2S_3O_6$

Preparation. Free acid in its solution form is obtained:

(*a*) by hydrolysing S_2Cl_2 by H_2O.

$$S_2Cl_2 + 3H_2O \longrightarrow H_2SO_3 + H_2S + 2HCl$$

H_2SO_3 and H_2S formed above mix together and give $H_2S_3O_6$ alongwith $H_2S_5O_6$.

$$6H_2SO_3 + 2H_2S \longrightarrow H_2S_3O_6 + H_2S_5O_6 + 6H_2O$$

(*b*) by treating potassium trithionate ($K_2S_3O_6$) with hydrofluosilicic acid, H_2SiF_6.

$$K_2S_3O_6 - H_2SiF_6 \longrightarrow H_2S_3O_6 + K_2SiF_6\downarrow$$

The precipitate of potassium fluosilicate (K_2SiF_6) is removed by filtration and $H_2S_3O_6$ is obtained in solution.

Properties. (*i*) On standing free acid decomposes into S, SO_2 and H_2SO_4.

$$H_2S_3O_6 \longrightarrow S + SO_2 + H_2SO_4$$

(*ii*) *Action of $AgNO_3$ solution.* With $AgNO_3$ solution, yellow precipitate of $Ag_2S_2O_3$ is obtained. This ppt. turns black on standing.

$$H_2S_2O_3 + 2AgNO_3 \longrightarrow Ag_2S_2O_3 \,(\textit{Yellow ppt.}) + 2HNO_3$$

(*iii*) *Action of $BaCl_2$ solution.* With $BaCl_2$ solution, no precipitate is obtained.

$$H_2S_2O_3 + BaCl_2 \longrightarrow \text{No precipitate}$$

Trithionates ($S_3O_6^{2-}$)

The salts of $H_2S_3O_6$ are called trithionates. Potassium trithionate ($K_2S_3O_6$) and sodium trithionate ($Na_2S_3O_6$) are only well-known trithionates. $K_2S_3O_6$ is obtained.

(*i*) by passing SO_2 gas into a saturated solution of $K_2S_2O_3$ until a yellow colour developed in solution no more fades out on standing.

$$2K_2S_2O_3 + 3SO_2 \longrightarrow 2K_2S_3O_6 + S$$

(*ii*) by warming a saturated solution of $KHSO_3$ with powdered sulphur.

$$6KHSO_3 + 2S \longrightarrow 2K_2S_3O_6 + K_2S_2O_3 + 3H_2O$$

(*iii*) by heating potassium silver thiosulphate, $KAgS_2O_3$ with H_2O.

$$2KAgS_2O_3 \xrightarrow{\text{Heat with } H_2O} K_2S_3O_6 + Ag_2S$$

$Na_2S_3O_6$ is obtained by oxidising the solution of $Na_2S_2O_3$ with H_2O_2 or SO_2.

$$2Na_2S_2O_3 + 4H_2O_2 \longrightarrow Na_2S_3O_6 + Na_2SO_4 + 4H_2O$$

$$2Na_2S_2O_3 + 3SO_2 \longrightarrow 2Na_2S_3O_6 + S\downarrow$$

Properties. All the trithionates, except those of Ag, Hg(ous) and Hg(ic) are soluble in H_2O.$K_2S_3O_6$ has the maximum solubility of all the remaining polythionates of potassium. For other properties see "General properties of polythionates".

Tetrathionic acid, $H_2S_4O_6$

Like $H_2S_3O_6$, this acid is also known only in solution. Free acid in its solution form is obtained:

(*a*) by hydrolysing S_2Cl_2 with H_2O.

$$S_2Cl_2 + 3H_2O \longrightarrow H_2SO_3 + H_2S + 2HCl$$

H_2SO_3 and H_2S obtained as above react together and give rise to the formation of $H_2S_4O_6$.

$$3H_2SO_3 + H_2S \longrightarrow H_2S_4O_6 + 3H_2O$$

(*b*) by treating a solution of barium tetrathionate (BaS_4O_6) or lead tetrathionate (PbS_4O_6) with calculated quantity of dil. H_2SO_4 and filtering off $BaSO_4$ or $PbSO_4$.

$$MS_4O_6 \text{ (M = Ba, Pb)} + H_2SO_4 \text{ (}dil.\text{)} \longrightarrow H_2S_4O_6 + MSO_4\downarrow$$

BaS_4O_6 and PbS_4O_6 used in the above reactions are obtained by the action of I_2 on a suspension of BaS_2O_3 or PbS_2O_3.

$$2MS_2O_3 \text{ (M = Ba, Pb)} + I_2 \longrightarrow MS_4O_6 + 2MI_2$$

Properties. The aqueous solution of $H_2S_4O_6$ is fairly stable and can be concentrated on a water bath and in vacuum over H_2SO_4 up to a point but decomposes as follows:

$$H_2S_4O_6 \longrightarrow H_2SO_4 + SO_2 + 2S$$

Tetrathionates ($S_4O_6^{2-}$)

The salts of tetrathionic acid ($H_2S_4O_6$) are called tetrathionates.

Preparation. (*i*) $Na_2S_4O_6$, BaS_4O_6 and PbS_4O_6 are obtained by the action of I_2 on the thiosulphates of Na, Ba and Pb.

$$2Na_2S_2O_3 + I_2 \longrightarrow Na_2S_4O_6 + 2NaI$$

$$2MS_2O_3 \text{ (M = Ba, Pb)} + I_2 \longrightarrow MS_4O_6 + 2MI_2$$

BaI_2 and unreacted iodine are removed by means of absolute alcohol in which both are soluble.

(*ii*) Cuprous tetrathionate ($Cu_2S_4O_6$) is obtained by the action of $CuSO_4$ on BaS_2O_3.

$$2CuSO_4 + 2BaS_2O_3 \longrightarrow Cu_2S_4O_6 + 2BaSO_4\downarrow$$

Properties. Tetrathionates are quite stable and soluble in water. Other properties–See "General properties of polythionates."

Sodium tetrathionate, $Na_2S_4O_6$

Among the tetrathionates, $Na_2S_4O_6$ is the most important.

Preparation. $Na_2S_4O_6$ is prepared by the action of $Na_2S_2O_3$ on I_2, $FeCl_3$ or H_2O_2.

$$Na_2S_2O_3 + I_2 \longrightarrow Na_2S_4O_6 + 2NaI$$

$$2Na_2S_2O_3 + 2FeCl_3 \longrightarrow Na_2S_4O_6 + 2FeCl_2 + 2NaCl$$

$$2Na_2S_2O_3 + H_2O_2 \longrightarrow Na_2S_4O_6 + 2NaOH$$

Properties. This salt is quite stable in aqueous solution. It is soluble in water, other properties. See "General properties of polythionates".

Uses. The formation of $Na_2S_4O_6$ by the action of $Na_2S_2O_3$ on I_2 has been utilised in the volumetric estimation of iodine (iodometric titration).

Pentathionic acid, $H_2S_5O_6$

Like other polythionic acids, this acid is also known only in solution.

Preparation. This acid is formed by the action of H_2S on H_2SO_3 in solution.

$$5H_2S + 10H_2SO_3 \longrightarrow 3H_2S_5O_6 + 12H_2O$$

Properties. The acid is known only in solution. Its aqueous solution can be concentrated on a water bath of density 1.3 and in vacuum desicator to density 1.6 but then decomposes into H_2SO_4, SO_2 and sulphur.

$$H_2S_5O_6 \longrightarrow H_2SO_4 + SO_2 + 3S$$

Pentathionates ($S_2O_5^{2-}$)

The salts of pentathionic acid are called pentathionates. $Na_2S_5O_6$ and $K_2S_5O_6.1.5H_2O$ are important pentathionates. $Na_2S_5O_6$ is obtained:

(*a*) by the action of conc. HCl on a concentrated solution of $Na_2S_2O_3$ containing a little sodium arsentie and cooled to –10°C.

$$5Na_2S_2O_3 + 8HCl\ (conc.) \longrightarrow Na_2S_5O_6 + H_2S_5O_6 + 8NaCl + 3H_2O$$

(*b*) by the action fo SCl_2 on $Na_2S_2O_3$.

$$SCl_2 + 2Na_2S_2O_3 \longrightarrow Na_2S_5O_6 + 2NaCl$$

or $$SCl_2 + 2S_2O_3^{2-} \longrightarrow S_5O_6^{2-} + 2Cl^-$$

The crystals of $K_2S_5O_6$. 1.5 H_2O are obtained by the action of CH_3COOK on $H_2S_5O_6$.

$$H_2S_5O_6 + 2CH_3COOK + 1.5H_2O \longrightarrow K_2S_5O_6.1.5\ H_2O + 2CH_3COOK$$

Pentathionate is finally crystallised from warm water acidified with H_2SO_4.

Properties of pentathionates. Pentathionates are not stable unless a little HCl or H_2SO_4 is added to the solution. The crystals of $K_2S_5O_6.1.5H_2O$ are rhombic.

Hexathionic acid ($H_2S_6O_6$) and hexathionates ($S_6O_6^{2-}$)

Free acid, like other polythionic acids, exists only in solution. Potassium hexathionate ($K_2S_6O_6$) is the important salt of this acid. This salt is obtained by adding a mixture of 1 mol of KNO_2 and 3 mols of $K_2S_2O_3$ to concentrated HCl at low temperature (–10 to – 15°C) in a flask. The oxides of nitrogen formed are removed by a current of air and KCl is removed by filtration. The filtrate, on being concentrated, gives $K_2S_6O_6$ in the form of white crystalline crust.

$K_2S_6O_6$ is quite stable in the solid state but readily decomposes in solution especially in the presence of alkali, giving lower polythionates.

Differences between $H_2S_2O_6$ and polythionic acids

$H_2S_2O_6$ is not regarded as a member of polythionic acids, since it differs from these acids in the following respects:

(*i*) Unlike polythionic acids, $H_2S_2O_6$ is oxidised only with difficulty, *e.g.*, boiling conc. HCl and potassium bromate slowly oxidises it to sulphate.

(*ii*) $H_2S_2O_6$ does not combine with sulphur as $H_2S_3O_6$ and $H_2S_4O_6$ do.

(*iii*) In $H_2S_2O_6$, two sulphonic acid groups ($O_2S.OH$) are attached together directly while in polythionic acids, these groups are attached together through one or more S-atoms which act as dinegative bridging atoms.

(*iv*) $H_2S_2O_6$ is prepared by the oxidation of H_2SO_3 while the polythionic acids are obtained by the reduction of H_2SO_3.

(*v*) $H_2S_2O_6$ is not formed in Wackenroder's solution while other polythionic acids like $H_2S_4O_6$, $H_2S_5O_6$ and $H_2S_6O_6$ can be obtained from this solution.

University Questions

1. Hydrolysis of one mole of $H_2S_2O_8$ produces. (*a*) two moles of H_2SO_4, (*b*) two moles of H_2SO_5, (*c*) one mole of H_2SO_4 and one mole of H_2SO_5 (*d*) one mole of H_2SO_4, one mole of H_2SO_5 and one mole of H_2O_2. *(I.I.T., 96)*
2. A dilute solution of Na_2SO_4 is electrolysed using platinum electrodes. The products obtained at anode and cathode are : (*a*) O_2 H_2, (*b*) $S_2O_8^{2-}$, Na, (*c*) O_2, Na, (*d*) $S_2O_8^{2-}$, H_2. *(I.I.T. 96)*
3. What will happen when chrome alum solution is treated with Na_2CO_3 solution?
Ans : Chrome alum, $K_2SO_4.Cr_2(SO_4)_3\ 24H_2O$ solution contains $[Cr(H_2O)_6]^{3+}$ ions. These ions react with CO_3^{2-} ions and $[Cr(H_2O)_3\ (OH)_3]^0$ gets precipitated and CO_2 gas is evolved.
$$2[Cr(H_2O)_6]^{3+} + 3CO_3^{2-} \rightleftharpoons 2[Cr(H_2O)_3\ (OH)_3] + 3CO_2 + 3H_2O$$
4. What will be the pH (acidic or alkaline) of chrome alum solution?
Ans : Chrome alum solution contains $[Cr(H_2O)_6]^{3+}$ ion. This ion reacts with H_2O and produces H_3O^+ ions due to which the solution of chrome alum is acidic in nature.
$$2[Cr(H_2O)_6]^{3+} + 3H_2O \rightleftharpoons 2[Cr(H_2O)_3\ (OH)_3]^0 + 3H_3O^+$$
5. What chromium ion is present in chrome alum?
Ans : Chrome alum contains hexaquo chromium (III) ion, $[Cr(H_2O)_6]^{3+}$.
6. Give the preparation and structure of peroxydisulphuric acid. *(Utkal, 91)*
7. Discuss the preparation, properties and structure of trithionic acid. *(Madurai, 93)*
8. Give equation to show the reaction of $H_2S_2O_8$ with water. *(Osmania, 93)*
9. Sodium thiosulphate is prepared by: (*i*) reducing Na_2SO_4 solution with H_2S, (*ii*) boiling Na_2SO_3 solution with sulphur in alkaline medium, (*iii*) neutralising $H_2S_2O_3$ solution with NaOH, (*iv*) boiling Na_2SO_3 solution with sulphur in acidic medium. *(I.I.T., 96)*
10. An aqueous solution of $FeSO_4$. $Al_2(SO_4)_3$ and chrome alum is heated with excess of Na_2O_2 and filtered. The materials obtained are : (*a*) a colourless filtrate and a green residue, (*b*) a yellow filtrate and a green residue, (*c*) a yellow filtrate and a brown residue, (*d*) a green filtrate and a brown residue. *(I.I.T., 96)*
11. Describe the preparation, properties, uses and structure of $Na_2S_2O_3.5H_2O$. *(Lucknow, 95; Meerut, 95)*
12. Explain why chrome alum and hexahydrate chromium (III) chloride have different colours.
Ans : Chrome alum has $[Cr^{3+}(H_2O)_6]^{3+}$ ions while $CrCl_3.6H_2O$ contains $[Cr^{3+}(H_2O)Cl]^+$ ions. Thus we see Cr^{3+} ion in both of these ions are surrounded by different ligands and have, therefore, different colours. Chrome alum is red violet while $CrCl_3.6H_2O$ is dark green.
13. Write a note on "Caro's and Marshall's acids". *(VBS Purvanchal 2002)*
14. (*a*) Give one method of preparation of Caro's acid and explain its structure.
(*b*) Draw the structure of Marshall's acid. *(Nagpur 2002)*
15. How is peroxodisulphuric acid prepared in the laboratory ? Draw a neat diagram. Give its reactions with (*i*) KI and (*ii*) H_2O. *(Delhi 2002)*
16. Describe the structures of H_2SO_5 and $H_2S_2O_8$. *(VBS Purvanchal 2003)*
17. Write structure and basicity of perdi sulphuric acid ($H_2S_2O_8$). *(Rohilkhand 2003)*
18. Name two peroxyacids of sulphur and give their structures. *(Rohilkhand 2003)*
19. Describe the preparation and uses of H_2SO_3 acid. Draw its geometrical shape. *(Delhi 2003)*
20. Name the various oxyacids of sulphur. Discuss their structure. *(GND Amritsar, 2004)*

34

Oxy-acids of Halogens and their Salts

Introduction

Since F-atom is more electronegative than O-atom, and does not have orbitals to receive the electron pair donated by back donation, it does not give any oxy-acids, while other halogen atoms form oxy-acids, which can be grouped into the following four types :

Halogen	Hypohalous acids, HXO (X = +1)	Halous acids, HXO_2 (X = +3)	Halic acids, HXO_3 (X = +51)	Perhalic acids, HXO_4 (X = +7)
Chlorine	HClO	$HClO_2$	$HClO_3$	$HClO_4$
Bromine	HBrO	—	$HBrO_3$	—
Iodinie	HIO	—	HIO_3	HIO_4, $HIO_4.H_2O$, $HIO_4.2H_2O$, $2HIO_4.H_2O$

The above table shows that the oxidation number of the central halogen atom, in hypohalous, halous, halic and perhalic acids of a given halogen, is +1, +3, +5 and +7 respectively. In other words, we can say, that the oxidation number of a given halogen atom, in all its four oxy acids, increases from +1 to +7, while the oxidation number of all the halogens in their oxy-acids of one type is the same, *e.g.*, oxidation number of Cl-atom, in all its four oxy-acids, *viz.*, HClO, $HClO_2$, $HClO_3$ and $HClO_4$ is +1, +3, +5, and +7 respectively *i.e.*, oxidation number of Cl-atom increases from +1 to +7. On the other hand, the oxidation number of Cl, Br and I in HClO, HBrO and HIO (all these acids are of one type, called *hypohalous acids*) is the same (= +1).

General properties of oxy-acids

Some of the general properties of oxy-acids are discussed below:

1. Thermal stability. Thermal stability depends on oxidation number and electronegativity of the central halogen atom. Greater is the oxidation number or electronegativity of the halogen atom, greater will be the thermal stability of the acid and *vice-versa*. Thus, the following two cases may be studied.

(*a*) In case of oxy-acids of a given halogen atom (*e.g.*, HClO, $HClO_2$, $HClO_3$, $HClO_4$), since the electronegativity of the central halogen atom *viz.*, Cl-atom remains the same, thermal stability of these acids depends on the oxidation number of the central atom. Thermal stability increases with the increase in oxidation number of the central atom, *i.e.*, thermal stability of the given acids is in the order : $HClO < HClO_2 < HClO_3 < HClO_4$

Explanation. As the oxidation number of the central atom increases, X—O bond in the acids becomes more and more covalent and hence thermal stability of the acids increases.

(*b*) In case of oxy-acids, having the same formula and different halogen atom (*e.g.*, HClO, HBrO, HIO), since the oxidation number of the halogen atom remains the same, thermal stability of the acids depends on the electronegativity of the halogen atom. Thermal stability decreases with the decrease in electronegativity of halogen atom, as we move down the group from Cl to I. Thus, thermal stability of the given acids is in the order : HClO > HBrO > HIO.

2. Oxidising power. (*a*) Oxidising power of oxy-acids of a given halogen atom (*e.g.*, HClO, $HClO_2$, $HClO_3$, $HClO_4$) decreases with the increase in the oxidation number of the central nalogen atom. Thus the oxidising power of the given acids is in the order : HClO > $HClO_2$ > $HClO_3$ > $HClO_4$.

Explanation. As the oxidation number of the central atom increases, X—O bond becomes more and more covalent and hence oxidising power of the acids decreases.

(*b*) In case of oxy-acids, having the same formula and different halogen atom, (*e.g.*, $HClO_3$, $HBrO_3$ and HIO_3), oxidising power decreases with the decrease in the electronegativity of the halogen atom, as we move down the group from Cl to I. Thus, oxidising power of the given acids is in the order : $HClO_3$ > $HBrO_3$ > HIO_3.

3. Stability and basicity of the oxy-anions of a given halogen atom. It has been observed that the oxy-anions, derived from the oxy-acids of the halogens, show basic character since they can accept a proton to form their conjugate acids. The stability and basicity of the oxy-anions of a given halogen atom (*e.g.*, ClO^-, ClO_2^-, ClO_3^- and ClO_4^-) have been found to be in the following order:

Stability : $ClO_4^- > ClO_3^- > ClO_2^- > ClO^-$ or $ClO^- < ClO_2^- < ClO_3^- < ClO_4^-$

Basicity : $ClO_4^- < ClO_3^- < ClO_2^- < ClO^-$ or $ClO^- > ClO_2^- > ClO_3^- > ClO_4^-$

For the explanation of these orders, see chapter on "*Modern Concept of Acids and Bases*" (chapter 19).

4. Acid strength (Acidity). (*a*) Acid strength of oxy-acids of a given halogen atom (*e.g.*, HClO, $HClO_2$, $HClO_3$ and $HClO_4$) increases with the increase of the oxidation state of the central halogen atom. Thus, the acid strength of the given acids increases in the order HClO < $HClO_2$ < $HClO_3$ < $HClO_4$. This increasing order can be explained on the basis of oxidation number of the central halogen atom or on the basis of Bronsted-Lowry concept of acids and bases. For this, see the chapter on "*Modern Concepts of Acids and Bases*" (chapter 19).

(*b*) The acid strength of the halogen acids of different halogen atoms, in the same oxidation state (*e.g.*, $HClO_3$, $HBrO_3$ and HIO_3) decreases, as we move down the halogen group. Thus, the acidic strength of the given acids is in the order $HClO_3$ > $HBrO_3$ > HIO_3. This order can be explained on the basis of electronegativity of the central halogen atom. For details see the chapter "*Modern Concepts of Acids and Bases*" (chapter 19).

Summary of the variation of different properties

The variation of different properties (general properties) of all the four types of oxy-acids has been summarised in Table 34.1.

Table 34.1: *Variation in the general properties of oxy-acids of halogens*

Halogen	Hypohalous acids (X = +1)	Halous acids (X = +3)	Halic acids (X = +5)	Perhalic acids (X = +7)
Cl	HClO	$HClO_2$	$HClO_3$	$HClO_4$
Br	HBrO		$HBrO_3$	
I	HIO		HIO_3	HIO_4

----- Oxidation number of the central atom increases (+1, +3, +5, +7) ---->

----- Thermal stability increases ---->

----- Covalent character of X—O bond increases ---->

——— Oxidising power decreses ———>

<---- Basicity decreases ----

----- Electronegativity of the central atom remains the same -----

↓ Electro-negativity of the central atom decreases ———

↓ Thermal stability decreases ———

↓ Oxidising power decreases ———

↓ Acidic strength decreases ———

↓ Oxidation number of the central atom remains the same ———

Oxy-acids of chlorine and their salts

Chlorine forms all the four oxy-acids, which are *hypochlorous acid* ($HCl^{+}O$), *chlorous acid* ($HCl^{+3}O_2$), *chloric acid* ($HCl^{+5}O_2$) and *perchloric acid* ($HCl^{+7}O_4$). It may be noted that Cl-atom is in +1, +3, +5 and +7 oxidation state respectively in these acids.

General properties

1. Thermal stability and oxidising power. Since the oxidation number of the central Cl-atom increases from +1 (in HClO) to +7 (in $HClO_4$), thermal stability of the acids increases while the oxidising power of the acids decreases on proceeding from HClO to $HClO_4$, *i.e.*, the thermal stability is in the order : $HCl^{+1}O < HCl^{+3} < HCl^{+5}O_3 < HCl^{+7}O_4$, while the oxidising power is in the order $HCl^{+1}O > HCl^{+3}O_2 > HCl^{+5}O_3 > HCl^{+7}O_4$.

Explanation. As the oxidation number of the central Cl-atom increases, Cl—O bond in the acids becomes more and more covalent and hence thermal stability increases and oxidising power decreases.

2. Acidity (acidic strength) and basicity. Acidic strength of the acids increases in the order $HClO < HClO_2 < HClO_3 < HClO_4$. Consequently the basicity of these acids decreases as $HClO > HClO_2 > HClO_3 > HClO_4$.

3. Monobasic nature. Since all the four acids contain one H-atom, all are monobasic acids. Their monobasic nature is confirmed by the fact that the acids contain only one OH group attached directly with the central Cl-atom.

Hypochlorous acid, HClO

This acid is known only in solution. HClO is prepared : (*a*) by passing Cl_2 into H_2O or into a suspension of $CaCO_3$ in water or into an aqueous solution of potassium hypochlorite (KClO) or bleaching powder ($CaOCl_2$).

$$Cl_2 + H_2O \longrightarrow HCl + HClO \quad \text{or} \quad Cl_2 + H_2O \longrightarrow Cl^- + ClO^- + 2H^+$$

In this reaction Cl_2 undergoes disproportionation into HCl and HClO.

$$CaCO_3 + H_2O + 2Cl_2 \longrightarrow CaCl_2 + CO_2 + 2HClO$$

$$KClO + Cl_2 + H_2O \longrightarrow KCl + 2HClO$$

$$CaOCl_2 + Cl_2 + H_2O \longrightarrow CaCl_2 + 2HClO$$

(*b*) by the action of atmospheric CO_2 on sodium hypochlorite, NaOCl.

$$NaOCl + CO_2 + H_2O \longrightarrow NaHCO_3 + HOCl.$$

(*c*) by distilling the aqueous solution of bleaching powder ($CaOCl_2$) with a calculated quantity of 5% HNO_3 or by passing CO_2 into aqueous solution of $CaOCl_2$ and then distilling.

$$2CaOCl_2 + 2HNO_3 \longrightarrow CaCl_2 + Ca(NO_3)_2 + 2HClO$$

$$CaOCl_2 + H_2O + CO_2 \longrightarrow CaCO_3 + 2HClO$$

(*d*) by shaking Cl_2 water with freshly precipitated HgO (General method for the preparation of hypohalous acids).

$$2Cl_2 + 2HgO + H_2O \longrightarrow HgCl_2.HgO\downarrow + 2HClO$$

The insoluble $HgCl_2.HgO$ is removed by filtration. The filtrate is distilled when dilute HClO passes over.

Properties. (*i*) The concentrated solution of HClO is yellow in colour, while the dilute solution is colourless. It is a weak acid, even weaker than H_2CO_3. Its dissociation constant is 3×10^{-8} at 20°C.

(*ii*) *Disproportionation of the acid.* (*a*) The dilute solution of the acid is farily stable in the dark, but when concentrated solution is exposed to light, it becomes unstable and hence undergoes disproportionation into Cl_2 (Cl = 0) and $HClO_3$ (Cl = + 5).

$$\underset{(Cl=+1)}{5HClO} \longrightarrow \underset{(Cl=0)}{2Cl_2} + \underset{(Cl=+5)}{HClO_3} + 2H_2O \ (\textit{Molecular equation})$$

$$\text{or} \quad 5ClO^- + 4H^+ \longrightarrow 2Cl_2 + ClO_3^- + 2H_2O \ (\textit{Ionic equation})$$

The decomposition is accelerated by Pt-black, MnO and CoO.

(*b*) When aqueous solution of HClO is heated, it undergoes disproportionation into HCl and $HClO_3$.

$$3HClO \xrightarrow{\Delta} 2HCl + HClO_3 \quad \text{or} \quad 3ClO^- \xrightarrow{\Delta} 2Cl^- + ClO_3^-$$

(*iii*) On distillation, HClO decomposes into H_2O and Cl_2O.

$$2HClO \longrightarrow Cl_2O + H_2O$$

(*iv*) *Action on metals.* HClO dissolves in Mg, with the evolution of H_2; in Fe and Al, with the evolution of H_2 and Cl_2, while with Co, Ni and Cu, Cl_2 and O_2 are evolved.

$$Mg + 2HClO \longrightarrow Mg(OCl)_2 + H_2$$

When HClO is shaken with Hg, a light brown ppt. of basic mercuric chloride, HgCl(OH), which is soluble in HCl, is obtained

$$2Hg + 2HClO \longrightarrow 2HgCl\,(OH)\ (\textit{Light brown ppt.})$$

(*v*) *Oxidising and bleaching properties.* The aqueous solution of HClO and its salts (*e.g.*, NaOCl) are oxidising and bleaching agents. This property is due to the fact that HClO or NaOCl decomposes to give nascant oxygen.

$$HClO \xrightarrow{\text{Reduction}} HCl + O$$

$$NaOCl \xrightarrow{\text{Reduction}} NaCl + O$$

$$ClO^- + 2H^+ + 2e^- \xrightarrow{\text{Reduction}} Cl^- + H_2O\ (\textit{Ion-electron equation})$$

Thus, we see that when OCl^- ion oxidises a certain substance, it gets reduced to Cl^- ion. See "*Oxidising properties of NaOCl*".

(*vi*) *Acidic nature : Monobasic nature.* HClO is a monobasic acid, since its aqueous solution gives only one H^+ ion on ionisation.

$$HClO(aq) \rightleftharpoons H^+\,(aq) + ClO^-\,(aq)$$

ClO^- ion is called **hypochlorite ion.**

Its monobasic nature shows that HClO molecule has one OH group attached directly with the central Cl-atom. Being an acid, HClO reacts with alkalies to form the salts which are called hypochlorites, *e.g.*,

$$NaOH + HClO \longrightarrow NaOCl + H_2O \quad \text{or} \quad OH^- + HClO \longrightarrow OCl^- + H_2O$$

(*vii*) *Action of $AgNO_3$.* HClO reacts with $AgNO_3$ and gives silver hypochlorite (AgClO). This compound is unstable and hence undergoes disproportionation into AgCl and $AgClO_3$.

$$HOCl + AgNO_3 \longrightarrow AgOCl + HNO_3; \quad 3AgOCl \longrightarrow AgCl + AgClO_3$$

Uses. HClO is used for bleaching paper pulp etc.

Hypochlorites (OCl^-)

We have already said the hypochlorous acid (HClO) is a monobasic acid and hence forms only one type of salts, which are called hypochlorites and contain hypochlorite ion (ClO^-). Sodium hypochlorite (NaOCl), calcium hypochlorite [$Ca(OCl)_2$] and calcium chlorohypochlorite, CaCl(OCl) or $CaOCl_2$ (called bleaching powder) are important hypochlorites. These are discussed below:

Sodium hypochlorite, NaOCl

Preparation. (*i*) NaOCl is obtained by passing Cl_2 into a cold and dilute solution of NaOH.

$$2NaOH\ (\textit{cold and dilute solution}) + Cl_2 \longrightarrow NaOCl + NaCl + H_2O$$

or

$$2OH^- + Cl_2 \longrightarrow OCl^- + Cl^- + H_2O$$

In this reaction Cl_2 molecule undergoes disproportionation into ClO_3^- and Cl^- ions. NaOCl formed in the above reaction goes into solution, but the salt cannot be obtained by concentrating the solution by evaporation, since the heating of NaOCl decomposes the salt into $NaClO_3$ and NaCl

$$3NaOCl \longrightarrow NaClO_3 + 2NaCl \quad \text{or} \quad 3OCl^- \longrightarrow ClO_3^- + 2Cl$$

In this reaction OCl^- ion undergoes disproportionation into ClO_3^- and Cl^- ions.

(*ii*) Many textile mills use NaOCl solution (called *bleaching solution*) for bleaching purpose. They prepare it by electrolysing a solution of brine (NaCl) in an undevided cell, which consists of two earthenware troughs (tanks), one placed inside the other. The inner trough is fitted with 25–30 vertical graphite electrodes. The outermost electrodes are connected with the positive and negative terminals of a 110 volt main. Each one of the outer electrodes acts as a cathode on one face and as anode on the other face, due to induction. Both the troughs are filled with brine solution. Cold water is circulated through a cooling pipe dipped in outer tank. This serves to keep the temperature of the electrolyte below 24°C. On passing electric current, NaOCl is obtained.

Reaction. (a) Ionisation of NaCl solution : $NaCl \rightleftharpoons Na^+ + Cl^-$

(*b*) *Formation of NaOH at cathode :* $2H_2O + 2e^- \xrightarrow{\text{Reduction}} H_2^0 + 2OH^-$

$$Na^+ + OH^- \longrightarrow NaOH$$

(*c*) *Formation of Cl_2 at anode :* $Cl^- \xrightarrow{\text{Oxidation}} Cl^0 + e^-,\ Cl^0 + Cl^0 \longrightarrow Cl_2$

(*d*) *Formation of NaOCl.* NaOH, obtained at cathode and Cl_2 liberated at anode, react together and give NaOCl.

$$2NaOH + Cl_2 \longrightarrow NaCl + NaOCl + H_2O$$

H_2 liberated at the cathode bubbles through the solution in the inner tank and thus over flows the outer tank.

Properties. (*i*) *Action of heat.* (*a*) When solution of NaOCl (Cl = +1) is heated, it undergoes disproportionation into $NaClO_3$ (Cl = +5) and NaCl (Cl = –1).

$$3NaClO \longrightarrow NaClO_3 + 2NaCl \quad \text{or} \quad 3ClO^- \longrightarrow ClO_3^- + 2Cl^-$$

(*b*) When NaOCl is heated with CaO in an alkaline solution, O_2 is liberated

$$2NaOCl \xrightarrow{\text{Heat in pressure of CaO}} 2NaCl + O_2$$

(*ii*) *Action of atmospheric CO_2 (or H_2CO_3).* Atmospheric CO_2 (or H_2CO_3) liberates HClO from NaOCl.

$$NaOCl + CO_2 + H_2O \longrightarrow NaHCO_3 + HOCl \quad \text{or} \quad NaOCl + H_2CO_3 \longrightarrow NaHCO_3 + HOCl$$

The above reaction has been used for the preparation of HClO.

(*iii*) *Oxidising and bleaching properties.* Since the aqueous solution of NaOCl produces nascent oxygen, NaOCl acts as an oxidising and bleaching agent. When NaOCl oxidises a certain substance, it is reduced to NaCl.

$$NaOCl\ (aq) \longrightarrow NaCl\ (aq) + O\ (aq)\ (\textit{Molecular equation})$$

$$OCl^- + 2H^+ + 2e^- \longrightarrow Cl^- + H_2O\ (\textit{Ion-electron equation})$$

Electrons which are accepted by OCl^- ion are furnished by reducing agent. The oxidising properties of NaOCl are evident from the following reactions :

(*a*) *NaOCl oxidises HCl (Cl = – 1) to Cl_2 (Cl = 0).*

$$2HCl + NaOCl \longrightarrow Cl_2 + NaCl + H_2O\ (\textit{Molecular equation})$$

or $$Cl^- + OCl^- + 2H^+ \longrightarrow Cl_2 + H_2O \quad (\textit{Ionic equation})$$

(*b*) *NaOCl oxidises AsO_3^{3-} (As = +3) to AsO_4^{3-} (As = +5)*

$$Na_3AsO_3 + NaOCl \longrightarrow Na_3AsO_4 + NaCl \quad (\textit{Molecular equation})$$

$$AsO_3^{3-} + ClO^- \longrightarrow AsO_4^{3-} + Cl^- \quad (\textit{Ionic equation})$$

(*c*) *NaOCl oxidises the alkaline solution of $Cr(OH)_3$ (Cr = +3) to sod. chromate, Na_2CrO_4*

$$2Cr(OH)_3 + 3NaOCl + 4NaOH \longrightarrow 2NaCrO_4 + 3NaCl + 5H_2O\ (\textit{Molecular equation})$$

or $$3OCl^- + 2Cr^{3+} + 5H_2O \longrightarrow 3Cl^- + 2CrO_4^{2-} + 10H^+ \quad (\textit{Ionic equation})$$

(*d*) *NaOCl oxidises $Mn(OH)_2$ (Mn = +2) to brown hydrated manganic oxide, $MnO_2.2H_2O$ or $Mn(OH)_4$ (Mn = +4).*

$$Mn(OH)_2 + NaOCl + H_2O \longrightarrow MnO_2.2H_2O + NaCl$$

or $$Mn(OH)_2 + NaOCl + H_2O \longrightarrow Mn(OH)_4 + NaCl$$

(e) NaOCl oxidises acidified solution of KI (I = –1) to I_2 (I = 0), i.e., I_2 is liberated from the acidified solution of KI.

$$2KI + NaOCl + 2HCl \longrightarrow I_2 + 2KCl + NaCl \quad \textit{(Molecular equation)}$$

$$OCl^- + 2H^+ + 2I^- \longrightarrow Cl^- + H_2O + I_2 \quad \textit{(Ionic equation)}$$

(f) NaOCl oxidises H_2O_2 (O = – 1) to O_2 *(O = 0)*

$$NaOCl + H_2O_2 \longrightarrow NaCl + H_2O + O_2 \quad \textit{(Molecular equation)}$$

or $$OCl^- + H_2O_2 \longrightarrow Cl^- + H_2O + O_2 \quad \textit{(Ionic equation)}$$

An important case. In all the reactions given above, OCl^- ion is reduced to Cl^-. But when OCl^- ion reacts with a Pb (+2) salt, it is reduced to Cl_2 (Cl = 0) and the salt is oxidised to PbO_2 (Pb = + 4).

$$(CH_3COO)_2Pb + 4NaOCl \longrightarrow 2PbO_2 + 2Cl_2 + 4CH_3COONa \quad \textit{(Molecular equation)}$$

or $$2OCl^- + Pb^{2+} \longrightarrow PbO_2 + Cl_2 \quad \textit{(Ionic equation)}$$

Uses. NaOCl is used : (*i*) as a disinfectant and deodorant in dairies, water supplies and sewage disposal (*ii*) as a bleaching agent for paper pulp, linen, cotton, rayon and jute.

Calcium hypochlorite, $Ca(OCl)_2$

It is also an important salt of hypochlorous acid and is prepared by treating lime slurry, $Ca(OH)_2$ with Cl_2.$Ca(OCl_2)$ formed is separated by adding common salt (*salting out*).

$$2Ca(OH)_2 + 2Cl_2 \longrightarrow Ca(OCl)_2 + CaCl_2 + 2H_2O$$

It is a stable substance, since it does not lose Cl_2 on standing. Its available chlorine content is very high and hence is used in place of bleaching powder under the commercial name *perchloron*. It is superior to blaching powder, since it is non-hygroscopic and gives a clear solution in water. When $Ca(OCl)_2$ reacts with calculated quantity of 5% HNO_3 or CO_2, HClO is obtained.

$$Ca(OCl)_2 + 2HNO_3 \longrightarrow Ca(NO_3)_2 + 2HClO; \quad Ca(OCl)_2 + CO_2 + H_2O \longrightarrow CaCO_3 + 2HClO$$

Bleaching powder, $CaCl(OCl).H_2O$ or $CaOCl_2.H_2O$

Manufacture. It is manufactured by the action of Cl_2 on dry slaked lime, $Ca(OH)_2$. Regarding the formation of bleaching powder, the following two views have been proposed.

(*i*) *Odling's view* (*1861*). Odling, in 1861, suggested that bleaching powder is a mixture of calcium hypochlorite, $Ca(ClO)_2$ and calcium chloride, $CaCl_2$. He, therefore, called it *calcium chlorohypochlorite*, CaCl(ClO) or *calcium oxy-chloride*, $CaOCl_2$. Thus, on this basis, the formation of bleaching powder can be shown as :

$$2Ca(OH)_2 + 2Cl_2 \longrightarrow \underbrace{Ca(ClO)_2 + CaCl_2}_{\text{Bleaching powder}} + 2H_2O$$

or $$2Ca(OH)_2 + 2Cl_2 \longrightarrow 2Ca(ClO)Cl + 2H_2O$$

or $$Ca(OH)_2 + Cl_2 \longrightarrow \underset{\text{Bleaching powder}}{Ca(ClO)Cl} + H_2O$$

(*ii*) *Bunn, Clark and Chifford's view (1935).* According to this view, bleaching powder is supposed to be a mixture of *calcium hypochlorite*, $Ca(ClO)_2$ and *basic calcium chloride*, $CaCl_2.Ca(OH)_2.H_2O$. Thus the formation of bleaching powder can be represented as :

$$2Cl_2 + 3Ca(OH)_2 \longrightarrow \underbrace{Ca(ClO)_2 + CaCl_2.Ca(OH)_2.H_2O}_{\text{Bleaching powder}} + H_2O$$

The manufacture of bleaching powder is carried out in any of the two plants namely *Hesenclever's plant* (*old process*) and *Backmann's plant* (*modern process*). Here we shall discuss the manufacture of bleaching powder by Backman's plant only. This plant consists of a chlorinating tower, which is made of cast iron and has a hopper at its top. There are two inlets, near the bottom of the tower, one for passing in Cl_2 gas and the other for passing in a current of hot air. There is also an exit near the top of the tower, through which the unreacted Cl_2 and air (waste gases) go out. There are a number of horizontal shelves inside the tower. Each shelf is provided with a rotating rake. These rakes are connected with a spindle which is rotated with the help of motor (Fig. 34.1).

The dry slaked lime is put into the hopper with the help of compressed air. From the hopper, slaked lime slowly drops into the chlorinating tower and falls on shelves. As the rakes rotate, the slaked lime moves downwards in the tower. Meanwhile Cl_2 gas and hot air are also being passed in, from near the bottom of the tower. As the slaked lime moves down, it meets the upcoming current of Cl_2 gas. $Ca(OH)_2$ reacts with Cl_2 gas to form $CaOCl_2$, which collects in the vessel placed at the bottom of the tower. Hot air is introducted to drive away the unreacted Cl_2 gas through the exit near the top of the tower.

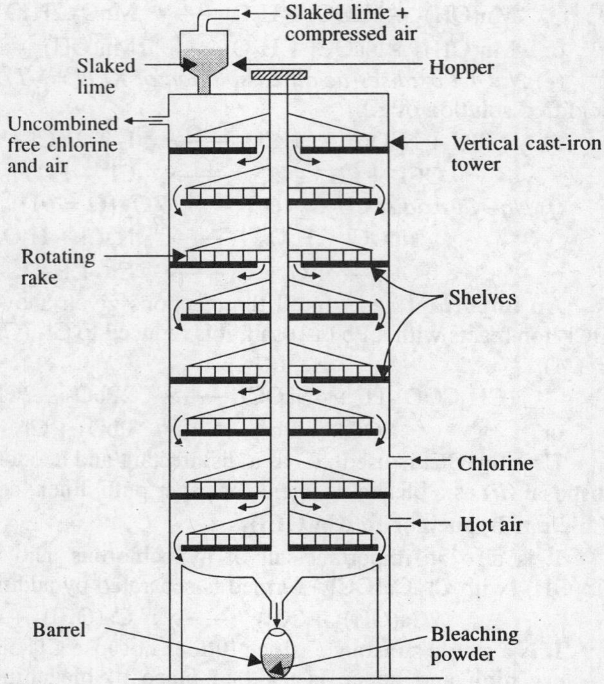

Fig. 34.1 : Backmann's plant used for the manufacture of bleaching powder, $CaOCl_2$.

Properties. **(1)** Bleaching powder is a pale yellow powder. It has strong smell of Cl_2. It is soluble in water but the solution is never clear due to the presence of impurities in it.

(2) Action of $CoCl_2$. In presence of $CoCl_2$, bleaching powder loses its oxygen.

$$2CaOCl_2 \xrightarrow{CoCl_2} 2CaCl_2 + O_2$$

(3) Action of acetone (CH_3COCH_3) and ethyl alcohol (C_2H_5OH). Both these organic compounds are converted into chloroform, $CHCl_3$.

$$4CaOCl_2 + 2CH_3COCH_3 \longrightarrow 2CHCl_3 + (CH_3COO)_2Ca + 2Ca(OH)_2$$

$$4CaOCl_2 + 2C_2H_5OH \longrightarrow 2CHCl_3 + (HCOO)_2Ca + Ca(OH)_2 + 2H_2O + CaCl_2$$

(4) Action of Cl_2. When Cl_2 is passed into an aqueous solution of $CaOCl_2$, HCl is obtained.

$$CaOCl_2 + Cl_2 + H_2O \longrightarrow CaCl_2 + 2HClO$$

(5) Action of 5% HNO_3. When an aqueous solution of $CaOCl_2$ is distilled with a calculated quantity of 5% HNO_3, HClO is obtained.

$$2CaOCl_2 + 2HNO_3 \longrightarrow CaCl_2 + Ca(NO_3)_2 + 2HClO$$

(6) Action of CO_2. (*a*) When dry CO_2 is passed into an aqueous solution of $CaOCl_2$, HClO is produced.

$$\underbrace{2CaOCl_2 + H_2O + CO_2}_{\text{Aqueous solution}} \longrightarrow CaCl_2 + CaCO_3 + 2HClO$$

(*b*) When moist CO_2 present in the atmosphere (*i.e.*, H_2CO_3) reacts with $CaOCl_2$, Cl_2 (called available chlorine) is liberated.

$$\underbrace{CaOCl_2.H_2O + (CO_2 + H_2O) \text{ or } H_2CO_3}_{\text{Moist } CO_2} \longrightarrow CaCO_3 + 2H_2O + Cl_2$$

(7) Oxidising and bleaching properties. In presence of *slight amount of dil. acids,* $CaOCl_2$ loses nascent oxygen.

$$2CaOCl_2 + H_2SO_4 \longrightarrow CaCl_2 + CaSO_4 + 2HCl + 2O$$

On account of the production of nascent oxygen, $CaOCl_2$ shows oxidising as well as bleaching properties.

(*a*) *Oxidising properties.* $CaOCl_2$ oxidises various substances in presence of dil. acids. When $CaOCl_2$ oxidises a certain substance, its hypochlorite ion (ClO^-) is reduced to Cl^- ion.

$$\underset{(Cl = +1)}{ClO^-} + 2e^- + 2H^+ \longrightarrow \underset{(Cl = -1)}{Cl^-} + H_2O$$

Thus Ca(ClO)Cl oxidises :

(*i*) KI solution acidified with CH_3COOH or HCl to I_2.

$$Ca(ClO)Cl + 2KI + 2CH_3COOH \longrightarrow 2KCl + I_2 + H_2O + Ca(CH_3COO)_2$$

$$Ca(ClO)Cl + 2KI + 2HCl \longrightarrow 2KCl + I_2 + H_2O + CaCl_2$$

(*ii*) Na_3AsO_3 to Na_3AsO_4

$$Na_3AsO_3 + Ca(ClO)Cl \longrightarrow Na_3AsO_4 + CaCl_2$$

or $$AsO_3^{3-} + ClO^- \longrightarrow AsO_4^{3-} + Cl^-$$

(*iii*) PbO to PbO_2 and MnO to MnO_2.

$$MO\ (M = Pb, Mn) + Ca(ClO)Cl \longrightarrow MO_2 + CaCl_2$$

or $$MO + ClO^- \longrightarrow MO_2 + Cl^-$$

(*iv*) H_2S to sulphur.

$$Ca(ClO)Cl + H_2S \longrightarrow CaCl_2 + H_2O + S \quad \text{or} \quad ClO^- + H_2S \longrightarrow Cl^- + H_2O + S$$

(*v*) Acidified solution of $FeSO_4$ to $Fe_2(SO_4)_3$.

$$Ca(ClO)Cl + H_2SO_4 + 2FeSO_4 \longrightarrow Fe_2(SO_4)_3 + CaCl_2 + H_2O$$

or $$ClO^- + 2H^+ + 2Fe^{2+} \longrightarrow 2Fe^{3+} + Cl^- + H_2O$$

(*vi*) KNO_2 to KNO_3.

$$Ca(ClO)Cl + KNO_2 \longrightarrow CaCl_2 + KNO_3 \quad \text{or} \quad ClO^- + NO_2^- \longrightarrow Cl^- + NO_3^-$$

(*vii*) NH_3 to N_2.

$$3Ca(ClO)Cl + 2NH_3 \longrightarrow 3CaCl_2 + 3H_2O + N_2 \quad \text{or} \quad 3ClO^- 2NH_3 \longrightarrow 3Cl^- + 3H_2O + N_2$$

(*b*) *Bleaching properties.* Nascent oxygen produced by the action of slight amount of dilute acids on bleaching powder renders the coloured matter colourless.

$$\text{Coloured matter} + O \longrightarrow \text{Colourless matter}$$

Thus we find that bleaching powder also shows bleaching properties.

(8) Action of excess of dilute acids : Available chlorine. When $CaOCl_2.H_2O$ is acted up by the *excess of dilute* acids like HCl, H_2SO_4, HNO_3, CH_3COOH etc., Cl_2 is liberated. Even H_2CO_3 (*i.e.,* moist CO_2 present in the atmosphere) also liberates Cl_2 from $CaOCl_2$.

$$\underset{(\text{Mol. wt.} = 145)}{CaOCl_2.H_2O} + 2HCl \longrightarrow CaCl_2 + 2H_2O + \underset{(\text{Mol. wt.} = 34.5 \times 2 = 71.0)}{Cl_2}$$

$$CaOCl_2.H_2O + H_2SO_4 \longrightarrow CaSO_4 + 2H_2O + Cl_2$$

$$CaOCl_2.H_2O + 2HNO_3 \longrightarrow Ca(NO_3)_2 + 2H_2O + Cl_2$$

$$CaOCl_2.H_2O + CH_3COOH \longrightarrow Ca(CH_3COO)_2 + 2H_2O + Cl_2$$

$$CaOCl_2.H_2O + \underbrace{H_2CO_3 \text{ or } (CO_2 + H_2O)}_{\text{Moist } CO_2} \longrightarrow CaCO_3 + 2H_2O + Cl_2$$

Amount of Cl_2 liberated by the action of excess of dilute acids or moist CO_2 present in the atmosphere on a sample of bleaching powder is called available chlorine of that sample. Generally available chlorine of a sample of bleaching powder is expressed as the amount of Cl_2 liberated by 100 g of the sample of bleaching powder.

The above equations show that since 145 g of bleaching powder gives 71 g of available chlorine, 100 g bleaching powder will give $\frac{145 \times 100}{71}$ = 49 g of available chlorine. Thus the value of available chlorine in a given sample of bleaching powder should be equal to 49%. Actually, the available chlorine in an old sample of bleaching powder has been found to vary between 35 to 40% and falls on keeping the sample for a long time. An old sample of bleaching powder may not have any available chlorine at all. The changes that are responsible for reducing the percentage of available chlorine content in an old sample of bleaching powder are given below :

(*i*) On long standing, $CaOCl_2.H_2O$ decomposes slowly to give a mixture of $Ca(ClO_3)_2$ and $CaCl_2$ or $CaCl_2$ and O_2.

$$6[CaOCl_2.H_2O] \longrightarrow Ca(ClO_3)_2 + 5CaCl_2 + 6H_2O$$
$$2[CaOCl_2.H_2O] \longrightarrow 2CaCl_2 + O_2 + 2H_2O$$

(*ii*) In presence of CO_2 or moist CO_2 present in the atmosphere (*i.e.*, H_2CO_3), $CaOCl_2$ decomposes to produce Cl_2.

$$CaOCl_2.H_2O + CO_2 \longrightarrow CaCO_3 + Cl_2 + 2H_2O$$
$$CaOCl_2.H_2O + \underbrace{(CO_2 + H_2O) \text{ or } H_2CO_3}_{\text{Moist } CO_2} \longrightarrow CaCO_3 + 2H_2O + Cl_2$$

The above reactions show that when bleaching powder is kept in an open bottle, available Cl_2 in bleaching powder is lost. So, bleaching powder should not be kept in an open bottle.

The power of a given sample of bleaching powder is evaluated on the basis of the percentage of available chlorine present in the sample.

Total chlorine. Bleaching powder contains free lime, water, $CaCl_2$ and $Ca(ClO_3)_2$, as impurities. These impurities do not liberate Cl_2 with weak acids like CH_3COOH. Total chlorine of a given sample of bleaching powder is the total amount of Cl_2 obtained by the action of an acid (*e.g.*, HCl) on the main constituent *viz.*, $CaOCl_2.H_2O$ of bleaching powder and the impurity like $Ca(ClO_3)_2$ present in the bleaching powder.

$$CaOCl_2.H_2O + 2HCl \longrightarrow CaCl_2 + 2H_2O + \boxed{Cl_2}$$
$$Ca(ClO_3)_2 + 12HCl \longrightarrow CaCl_2 + 6H_2O + \boxed{6Cl_2}$$

Total chlorine

Uses of bleaching powder. Bleaching powder is used : (*i*) as a disinfectant and germicide especially in the sterlisation of drinking water (*ii*) for the manufacture of chloroform ($CHCl_3$) by the interaction of bleaching powder and acetone or ethyl alcohol (*iii*) for making wool unshrinkable. (*iv*) as an oxidising agent in industry (*v*) mainly as bleaching agent for cotton, linen and wood pulp. However, delicate articles like straw, silk, ivory etc., are not bleached by bleaching powder.

Bleaching of cotton clothes. In textile industry, cotton clothes are mainly bleached with the help of $CaOCl_2$. Different steps of the procedure used for this purpose are as follows :

(*i*) Cloth is treated with a very dilute hot solution of NaOH, so that greasy matter is removed.

(*ii*) It is now treated with bleaching powder solution.

(*iii*) The cloth is next treated with a very dilute solution of HCl. Nascent oxygen and Cl_2 are produced by the interaction between $CaOCl_2$ and dil. HCl.

$$CaOCl_2 + HCl \longrightarrow CaCl_2 + HCl + O \quad \text{or} \quad CaOCl_2 \xrightarrow{\text{In presence of HCl}} CaCl_2 + O$$
$$CaOCl_2 + 2HCl \longrightarrow CaCl_2 + H_2O + Cl_2$$

Nascent oxygen, produced as above, bleaches the cloth. Cl_2, produced as above, sticks to the fibre and its removal is necessary, otherwise the cloth will be damaged.

(*iv*) In order to remove Cl_2, the cloth is treated with a solution of sodium thiosulphate, $Na_2S_2O_3$ (hypo) (use of $Na_2S_2O_3$ as anti-chlor).

$$Na_2S_2O_3 + Cl_2 + H_2O \longrightarrow Na_2SO_4 + 2HCl + S$$

(*v*) Finally the cloth is washed in running water, so that cloth is freed from all the sticking chemical substances.

Constitution of bleaching powder. In order to explain the constitution of bleaching powder, the following views have been put forward from time to time :

(*i*) *Dalton's view* (*1813*). Dalton regarded bleaching powder as a molecular compound of quick lime (CaO) and chlorine (Cl_2) and hence it is represented as $CaO.Cl_2$ (*chloride of lime*). But this view was rejected, since quick lime has no tendency to absorb Cl_2.

(*ii*) *Balard's view* (1835). Balard suggested that bleaching powder is a mixture of equi-molecular amounts of *calcium chloride* ($CaCl_2$) and *calcium hypochlorite* [$Ca(ClO)_2$]. Thus, according to this view, bleaching powder is represented as $CaCl_2 + Ca(ClO)_2$. This representation is based on the fact that the aqueous solution of bleaching powder gives the tests of both the compounds namely $CaCl_2$ and $Ca(ClO)_2$.

Objections. Balard's structure was rejected because it could not explain the following properties of bleaching powder :

(*a*) $CaCl_2$ is extremely deliquescent, while bleaching powder is not so.

(*b*) $CaCl_2$ is soluble in alcohol, but alcoholic extract of bleaching powder contains practically no $CaCl_2$.

(*c*) Moist CO_2 has no action on $CaCl_2$, while it liberates the whole of available Cl_2 from bleaching powder.

$$CaCl(ClO).H_2O + \underset{\text{Moist carbon dioxide}}{(CO_2 + H_2O)} \longrightarrow CaCO_3 + 2H_2O + \underset{\text{Available chlorine}}{Cl_2}$$

(*d*) Bleaching powder is not obtained by mixing $CaCl_2$ and $Ca(ClO)_2$ in equimolecular proportions.

(*iii*) *Odling's view* (1861). According to Odling, bleaching powder is a double salt, which is a compound of calcium chloride (C Cl_2) and calcium hypochlorite [$Ca(ClO)_2$], *i.e.,* bleaching powder is represented as $CaCl_2.Ca(ClO)_2$ or 2CaCl(ClO) or simply as CaCl(ClO). Thus, bleaching powder is *calcium chlorohypochl rite* or *calcium oxy-chloride.*

In other words, we can say that bleaching powder can be represented by the structure, $Ca^{2+}\langle^{Cl^-}_{ClO^-}$. This structure can well explain the formation of bleaching powder by the action of Cl_2 on slaked lime, $Ca(OH)_2$.

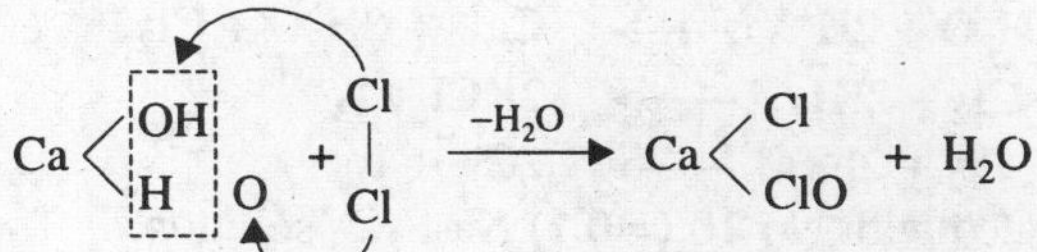

This structure can also explain the following properties of bleaching powder :

(*a*) On long standing, bleaching powder undergoes decomposition and gives a mixture of $CaCl_2$ and $Ca(ClO)_2$.$Ca(ClO)_2$ further gets decomposed into $Ca(ClO_3)_2$ and $CaCl_2$.

$$\left[2Ca\langle^{Cl}_{ClO} \longrightarrow CaCl_2 + Ca\,(ClO)_2\right] \times 3$$

$$3\,Ca\,(ClO)_2 \longrightarrow Ca(ClO_3)_2 + 2CaCl_2$$

$$6Ca\langle^{Cl}_{ClO} \longrightarrow 3CaCl_2 + Ca(ClO_3)_2 + 2CaCl_2$$

$$\text{or}\quad 6Ca\langle^{Cl}_{ClO} \longrightarrow Ca(ClO_3)_2 + 5CaCl_2$$

$$\text{or}\quad 6CaCl\,(ClO) \longrightarrow Ca(ClO_3)_2 + 5CaCl_2$$

The above reactions show that the final products obtained by the decomposition of bleaching powder are $Ca(ClO_3)_2$ and $CaCl_2$.

(*b*) In presence of water, bleaching powder gets decomposed into $CaCl_2$ and $Ca(ClO)_2$.

$$2Ca\langle^{Cl}_{ClO} \xrightarrow{H_2O} CaCl_2 + Ca(ClO)_2$$

(*iv*) *Bunn, Clark, and Clifford's view* (1935). These chemists proposed that when Cl_2 and slaked lime, $Ca(OH)_2$ combine together, a mixture of $Ca(ClO)_2$ and $CaCl_2$ is obtained.

$$2Ca(OH)_2 + 2Cl_2 \longrightarrow Ca(ClO)_2 + CaCl_2 + 2H_2O$$

$CaCl_2$ produced as above combines with $Ca(OH)_2$ and gives basic calcium chloride, $CaCl_2.Ca(OH)_2.H_2O$.

$$CaCl_2 + Ca(OH)_2 + H_2O \longrightarrow CaCl_2 . Ca(OH)_2 . H_2O$$

On combining the above two equations, we get :

$$3Ca(OH)_2 + 2Cl_2 \longrightarrow \underbrace{Ca(ClO)_2 + CaCl_2 . Ca(OH)_2 . H_2O}_{\text{Bleaching powder}} + H_2O$$

Thus, according to Bunn, Clark and Clifford, bleaching powder is a mixture of calcium hypochlorite, $Ca(ClO)_2$ and basic calcium chloride $CaCl_2.Ca(OH)_2$. This view is fully in accordance with the physical and chemical properties of bleaching powder and has been confirmed by physical rule study and by microscopic and X-ray analysis.

The following experimental facts also support this formula :

(*i*) Bleaching powder is neither soluble in alcohol nor is hygroscopic.

(*ii*) The solution of bleaching powder gives the test of Ca^{2+}, OCl^-, Cl^- and OH^- ions which are present in the compound.

(*iii*) The above formula is also confirmed by its method of formation as already shown.

To estimate the percentage of available chlorine, in bleaching powder, by titrating with N/10 sodium thiosulphate solution. The estimation of available chlorine, in a given sample of bleaching powder ($CaOCl_2.H_2O$), is based on the fact that when a fine suspension of bleaching powder in water is treated with an excess of KI solution and then acidified with glacial acetic acid, I_2 is liberated.

$$\underset{\text{(Mol.wt. = 145)}}{CaOCl_2.H_2O + 2CH_3COOH} \longrightarrow Ca(CH_3COO)_2 + 2H_2O + \underset{\text{(Mol.wt. = 2 × 34.5 = 71)}}{Cl_2 \textit{ (Available chlorine)}}$$

or $$CaOCl_2.H_2O + 2H^+ \longrightarrow Ca^{2+} + 2H_2O + Cl_2$$

$$Cl_2 + 2KI \longrightarrow 2KCl + I_2 \qquad ...(i)$$

or $$Cl_2 + 2I^- \longrightarrow 2Cl^- + I_2$$

The liberated I_2 is titrated with N / 10 (= 0.1) $Na_2S_2O_3$ solution.

$$I_2 + 2Na_2S_2O_3 \longrightarrow Na_2S_4O_6 + 2NaI \qquad ...(ii)$$

or $$I_2 + 2S_2O_3^{2-} \longrightarrow S_4O_6^{2-} + 2I^-$$

From equations (*i*) and (*ii*), it is clear that :

$$Cl_2 \equiv I_2 \equiv 2Na_2S_2O_3 \qquad ...(iii)$$

Procedure. Weigh out accurately about 2.5 g of bleaching powder sample and transfer it to a clean glazed porcelene mortar. Add some distilled water and rub the bleaching powder till a smooth paste is obtained. Transfer the paste to a 250 ml measuring flask through a funnel and add more water to it to make up the solution to 250 ml. Shake the solution thoroughly to ensure the homogeneity of the solution.

Rinse and fill the burette with 0.1 $Na_2S_2O_3$ solution (intermediate solution). Pipette out 25 ml of the solution of bleaching powder in a conical flask (titration flask) and dilute it with a small quantity of distilled water. Add 10 ml of a 10% solution of KI and 5 ml of glacial CH_3COOH. Titrate the liberated I_2 with 0.1 N-$Na_2S_2O_3$ solution. When the colour of the solution turns pale yellow, add 1 – 2 ml of starch solution. A blue colulur will be obtained. Continue the addition of

$Na_2S_2O_3$ solution till the blue colour is just discharged (end point). Repeat the titration till concordant values are obtained.

Observations and calculations. Suppose :

(*i*) Weight of bleaching powder dissolved to prepare 250 ml solution = W g

$\therefore$ Weight of bleaching powder dissolved to prepare

1000 ml solution = 4 × W g

or Strength of bleaching powder solution = 4 × W g/lit.

(*ii*) Volume of bleaching powder solution taken in each titration = 25 ml.

(*iii*) Volume of 0.1 N-$Na_2S_2O_3$ solution used to react with I_2 liberated by 25 ml solution of bleaching powder in each titration = V ml.

At the end point we have :

$$25 \times N_{I_2} = V \times 0.1 \qquad \text{or} \qquad N_{I_2} = \frac{0.1 \times V}{25}$$

or With the help of relation (*ii*) given above, the above relation can be written as :

$$N_{\text{available chlorine}} = \frac{0.1 \times V}{25}$$

$\therefore$ Strength (or amount) of available chlorine in 250 ml solution of bleaching powder

$$= \frac{0.1 \times V}{25} \times 34.5 \text{ g/lit}$$

Thus 4W g of bleaching powder contains available chlorine $= \dfrac{0.1 \times V \times 35.5}{25}$ g

$\therefore$ 100 g of bleaching powder contains available chlorine $= \dfrac{0.1 \times V \times 35.5}{25} \times \dfrac{100}{4W}$

$\therefore$ % of available chlorine $= \dfrac{0.1 \times V \times 35.5 \times 100}{25 \times 4W}$

$$= \frac{\text{(Normality of } Na_2S_2O_3 \text{ solution)} \times \text{(volume of } Na_2S_2O_3 \text{ solution needed for reacting with } I_2 \text{ liberated by 25 ml of bleaching powder solution)} \times 34.5 \times 100}{\text{(volume of bleaching powder taken in each titration)} \times \text{(weight of bleaching powder dissolved to prepare 1 litre solution)}}$$

Solved Examples

Example 34.1. *0.5 gm of bleaching powder was suspended in water and excess of KI added. On acidifying with dil. H_2SO_4, I_2 was liberated which required 50 ml of N/10-$Na_2S_2O_3.5H_2O$ solution for exact oxidation. Calculate the percentage of available chlorine in bleaching powder. Explain all the equations.* **(Meerut 1994)**

Solution. Percentage of available chlorine in bleaching powder

$$= \frac{0.1 \times 50 \times 35.5 \times 100}{25 \times 4 \times 0.5} = \mathbf{34.5\%}$$

Chlorous acid, $HClO_2$

Preparation. This acid is obtained in the aqueous solution by treating a suspension of barium chlorite, $Ba(ClO_2)_2$ with dil. H_2SO_4 and filtering off the precipitate of $BaSO_4$. The acid can also be obtained by the action of ClO_2 on H_2O_2.

$$Ba(ClO_2)_2 + H_2SO_4 \longrightarrow 2HClO_2 + BaSO_4\downarrow; \quad 2ClO_2 + H_2O \longrightarrow 2HClO_2 + O_2$$

Properties. (*i*) The freshly prepared solution of the acid is colourless but it soon decomposes to ClO_2 which colours the solution yellow.

(*ii*) The acid gives a violet colour with $FeSO_4$.

(*iii*) $HClO_2$ and its salts undergo disproportionation on heating in an alkaline solution.

$$\underset{(Cl = +3)}{3HClO_2} \longrightarrow \underset{(Cl = +5)}{2HClO_3} + \underset{(Cl = -1)}{HCl}$$

or $$3ClO_2^- \longrightarrow 2ClO_3^- + Cl^-$$

(*iv*) $HClO_2$ and its saits both show oxidising properties. When they oxidise a certain substance, they are reduced to chloride, *i.e.*,

$$HClO_2 \longrightarrow HCl + 2O \quad \text{or} \quad ClO_2^- + 4H^+ + 4e^- \longrightarrow Cl^- + 2H_2O$$

$HClO_2$ and its salts liberate I_2 from KI solution.

$$HClO_2 + 2H_2O + 4KI \longrightarrow 4KOH + HCl + I_2$$

$$4I^- + ClO_2^- + 4H^+ \longrightarrow Cl^- + 2H_2O + 2I_2$$

(*v*) In acid solution, $HClO_2$ rapidly decomposes according to the following equation

$$4HClO_2 \longrightarrow 2ClO_2 + HClO_3 + HCl + H_2O$$

Chlorites (ClO_2^-)

The salts formed by $HClO_2$ are called chlorites. These are formed alongwith chlorates (ClO_3^-) by the action of ClO_2 on alkali oxide or hydroxide, *e.g.*,

$$Na_2O_2 + 2ClO_2 \rightarrow 2NaClO_2 + O_2 \;;\quad 2KOH + 2Cl_2O \rightarrow KClO_2 + KClO_3 + H_2O$$

Pure chlorites are obtained by the action of alkali and H_2O_2 on a conc. solution of ClO_2.

Chlorites are much more stable in aqueous solution than their parental acid, *viz.*, $HClO_2$. These salts form yellow crystals and with the exception of $AgClO_2$ and $Pb(ClO_2)_2$, they are soluble in water. Alkaline solutions are much more stable, but decompose on boiling.

$$3HClO_2 \longrightarrow 2HClO_3 + HCl \quad \text{or} \quad 3ClO_2^- \xrightarrow{\text{Boil}} 2ClO_3^- + Cl^-$$

The acid solution decomposes as :

$$4HClO_2 \longrightarrow 2ClO_2 + HClO_3 + HCl + H_2O \quad \text{or} \quad 4ClO_2^- \longrightarrow 2ClO_2 + ClO_3^- + Cl^- + O^{2-}$$

Chlorites liberate I_2 from iodides, *e.g.*,

$$NaClO_2 + 2H_2O + 4KI \longrightarrow 2I_2 + 4KOH + NaCl \quad \text{or} \quad ClO_2^- + 4H^+ + 4I^- \longrightarrow 2I_2 + 2H_2O + Cl^-$$

The alkali chlorites have a caustic taste and bleach vegetable colours. They may be distinguished from hypochlorites by their bleaching action in presence of sodium arsenite.

Chloric acid, $HClO_3$

This acid is know only in solution. It is prepared :

(*i*) by the action of dil. H_2SO_4 on $Ba(ClO_3)_2$.

$$Ba(ClO_3)_2 + H_2SO_4 \longrightarrow BaSO_4\,(ppt.) + 2HClO_3$$

The precipitate of $BaSO_4$ is obtained by filtration. The unused H_2SO_4 is precipitated with baryta water. The filtrate is evaporated in a vacuum desicator over conc. H_2SO_4 until a 4% solution of $HClO_3$ is obtained.

Note. If the solution containing $HClO_3$ is evaporated further more, it gets decomposed into perchloric acid, $HClO_4$.

$$3HClO_3 \longrightarrow HClO_4 + Cl_2 + 2O_2 + H_2O$$

(*ii*) by the action of hydrofluosilicic acid (H_2SiF_6) on $KClO_3$.

$$2KClO_3 + H_2SiF_6 \longrightarrow K_2SiF_6\,(ppt.) + 2HClO_3$$

Properties. (*i*) Conc. solution of the acid is colourless and pungent smelling liquid.

(*ii*) It is fairly stable in dark. In light it decomposes and becomes yellow. On heating $HClO_3$ decomposes to give $HClO_4$.

$$3HClO_3 \xrightarrow{\Delta} HClO_4 + Cl_2 + 2O_2 + H_2O$$

(*iii*) When organic substances like cotton, wool, paper etc., come in contact with the acid, they catch fire.

(*iv*) The acid is a powerful oxidising and bleaching agent.

(*v*) When iodine is evaporated with 25% $HClO_3$, iodic acid (HIO_3) is obtained.

$$2HClO_2 + I_2 \longrightarrow 2HIO_3 + Cl_2$$

This reaction has been used for the preparation of HIO_3.

(*vi*) $HClO_3$ is a monobasic acid.

$$HClO_3 \rightleftharpoons H^+ + ClO_3^-$$

Chlorates (ClO_3^-)

We have, already said that chloric acid ($HClO_3$) is a monobasic acid and hence forms only one type of salts, which are called **chlorates.** These salts contain chlorate ion, ClO_3^-. Among the chlorates, potassium chlorate is the most important chlorate and is discussed below.

Potassium chlorate, $KClO_3$

(*i*) **Laboratory preparation.** In the laboratory, it is obtained by passing Cl_2 gas through a boiling concentrated solution of KOH.

$$3Cl_2 + 6KOH \longrightarrow KClO_3 + 5KCl + 3H_2O \quad \text{or} \quad 3Cl_2 + 6OH^- \longrightarrow ClO_3^- + 5Cl^- + 3H_2O$$

On concentrating and cooling the products mixture, the crystals of $KClO_3$ are obtained. It may be seen from the above equation that Cl_2 (Cl = 0) undergoes disproportionation reaction, since it is oxidised to $KClO_3$ (Cl = +5) and reduced to KCl (Cl = –1) simultaneously.

(*ii*) **Manufacture.** (*a*) *Old process.* Cl_2 is passed through hot milk of lime, $Ca(OH)_2$ kept agitated in a cast iron tank when $Ca(ClO_3)_2$ is produced.

$$6Cl_2 + 6Ca(OH)_2 \longrightarrow Ca(ClO_3)_2 + 5CaCl_2 + 6H_2O$$

or

$$3Cl_2 + 6OH^- \longrightarrow ClO_3^- + 5Cl^- + 3H_2O$$

The decanted liquor, which contains $Ca(ClO_3)_2$ is treated with KCl (*equivalent quantity*) when $KClO_3$, being least soluble, crystallises on cooling.

$$Ca(ClO_3)_2 \text{ (Mother liquir)} + 2KCl \longrightarrow 2KClO_3 + CaCl_2$$

(*b*) *Modern process : Electrolysis of an aqueous concentrated solution of NaCl (brine).* Modern process consists of electrolysing a hot conc. solution of NaCl (brine) in an undivided cell, in which steel tank acts as cathode and graphite rods act as anode. These rods are arranged very close to each other, so that NaOH (obtained at cathode) and Cl_2 (obtained at anode) mix together to give $NaClO_3$. The temperature is maintained at 40°C, by circulating H_2O through cooling pipes. Reactions taking place in the cell are :

(*a*) *Ionisation of NaCl solution :* As a result of ionisation of NaCl solution, Na^+ and Cl^- ions are obtained.

$$NaCl \rightleftharpoons Na^+ + Cl^-$$

(*b*) *Reactions taking place at cathode.* Water gets reduced to H_2 and OH^- ions are also produced. These OH^- ions combine with Na^+ ions, obtained by the ionisation of NaCl solution and produce NaOH.

$$\underset{(H = +1)}{2H_2O} + 2e^- \xrightarrow{\text{Reduction}} \underset{(H = 0)}{H_2} + 2OH^-$$

$$Na^+ + OH^- \longrightarrow NaOH.$$

(*c*) *Reactions taking place at anode.* Here Cl^- ions, obtained by the ionisation of NaCl, get oxidised to Cl-atoms. These Cl-atoms combine together and give Cl_2 molecules.

$$Cl^- \xrightarrow{\text{Reduction}} Cl^\circ + e^- ; \quad Cl^\circ + Cl^\circ \longrightarrow Cl_2$$

Now NaOH produced at cathode and Cl_2 obtained in anode, react together and give $NaClO_3$.

$$6NaOH + 3Cl_2 \longrightarrow NaClO_3 + 5NaCl + 3H_2O$$

The products mixture is treated with an equivalent amount of KCl, which reacts with $NaClO_3$ and $KClO_3$ is obtained along with NaCl.

$$NaClO_3 + KCl \longrightarrow KClO_3 + NaCl$$

On cooling the mixture, $KClO_3$ crystallises out.

Note. If, in place of NaCl solution, we electrolyse KCl solution, we will get $KClO_3$ directly, but since $KClO_3$ is sparingly soluble, it crystallises out during the course of electrolysis and creates trouble.

Properties. (*i*) $KClO_3$ is a white crystalline substance (m.pt. = 370°C). It is sparingly soluble in cold water and more so in hot water.

(*ii*) *Action of hydrofluosilicic acid, H_2SiF_6.* When H_2SiF_6 reacts with $KClO_3$.$HClO_3$ is obtained.

$$2KClO_3 + H_2SiF_6 \longrightarrow K_2SiF_6\ (ppt.) + 3HClO_3$$

This reaction has been used for the preparation of $HClO_3$

(*iii*) *Action of S, P or charcoal.* With sulphur, phosphorus or charcoal, $KClO_3$ gives an explosive mixture which is used in crackers.

(*iv*) *Action of heat.* (*a*) When $KClO_3$ is heated to 400°C, KCl and O_2 are produced.

$$3KClO_3 \xrightarrow{400°C} 2KCl + 3O_2$$

This reaction has been used for the laboratory preparation of O_2. (*b*) When $KClO_3$(Cl = +5) is heated at a high temperature, it undergoes disproportionation into KCl (Cl = –1) and $KClO_4$(Cl = +7).

$$\underset{(Cl\,=\,+5)}{4KClO_3} \xrightarrow{400°C} \underset{(Cl\,=\,-1)}{KCl} + \underset{(Cl\,=\,+7)}{3KClO_4}$$

or $$4ClO_3^- \longrightarrow Cl^- + 3ClO_4^-$$

(*v*) *Action of conc. H_2SO_4.* When $KClO_3$ is very gently heated with conc. H_2SO_4, loud explosions are produced and ClO_2 and $KClO_4$ are produced. Thus $KClO_3$ undergoes disproportionation reaction when heated with conc. H_2SO_4.

$$\underset{(Cl\,=\,+5)}{3KClO_3} + H_2SO_4 \longrightarrow \underset{(Cl\,=\,+4)}{2ClO_2} + \underset{(Cl\,=\,+7)}{KClO_4} + H_2O + K_2SO_4$$

or $$3ClO_3^- + 2H^+ \longrightarrow 2ClO_2 + H_2O + ClO_4^-$$

(*vi*) *Oxidising properties.* $KClO_3$ is a powerful oxidising agent, since it can lose nascent oxygen. Thus :

(*a*) $KClO_3$ oxidises hot H_2SO_3 to H_2SO_4 and is itself reduced to KCl.

$$KClO_3 + 3H_2SO_3 \longrightarrow KCl + 3H_2SO_4 \quad \text{or} \quad ClO_3^- + 3SO_3^{2-} \longrightarrow Cl^- + 3SO_4^{2-}$$

(*b*) When $KClO_3$ is heated with conc. HCl, it oxidises HCl to Cl_2 and is itself reduced to ClO_2. The mixture of ClO_2 and Cl_2 is yellow gaseous mixture and is called *euchlorine.*

$$2KClO_3 + 4HCl \longrightarrow 2KCl + 2ClO_2 + Cl_2 + 2H_2O$$

or $$4H^+ + 2Cl^- + 2ClO_3^- \longrightarrow 2ClO_2 + Cl_2 + 2H_2O$$

(*c*) It oxidises HCl to Cl_2 and is itself reduced to KCl.

$$KClO_3 + 6HCl \longrightarrow KCl + 3Cl_2 + 3H_2O$$

or $$6H^+ + ClO_3^- + 6Cl^- \longrightarrow 3Cl_2 + 3H_2O + Cl^-$$

(*d*) (*i*) A concentrated solution of $KClO_3$ oidises I_2 to KIO_3 in presence of little HNO_3 and is itself reduced to Cl_2.

$$2KClO_3 + I_2 \longrightarrow 2KIO_3 + Cl_2 \quad \text{or} \quad 2ClO_3^- + I_2 \longrightarrow Cl_2 + 2IO_3^-$$

This reaction has been used for the preparation of KIO_3.

(*ii*) Hot and concentrated solution of $KClO_3$ oxidises I_2 to HIO_3 and is itself reduced to KCl. Here the presence of HNO_3 is not essential.

$$5KClO_3 + 3I_2 + 3H_2O \longrightarrow 6HIO_3 + 5KCl$$

or $$3I_2 + 3H_2O + 5ClO_3^- \longrightarrow 6H^+ + 6IO_3^- + 5Cl^-$$

This reaction has been used for the preparation of HIO_3.

(*e*) In presence of H_2SO_4, $KClO_3$ oxidises oxalic acid ($H_2C_2O_4$) to CO_2 and is itself reduced to HCl.

$$KClO_3 + H_2SO_4 + 3H_2C_2O_4 \longrightarrow KHSO_4 + HCl + 3H_2O + 6CO_2$$

or $$6H^+ + ClO_3^- + 3C_2O_4^{2-} \longrightarrow Cl^- + 3H_2O + 6CO_2$$

(*f*) In presence of dil. H_2SO_4, a hot solution of $KClO_3$ oxidises $FeSO_4$ present in ferrous ammonium sulphate to $Fe_2(SO_4)_3$ and is itself reduced to KCl.

$$KClO_3 + 6FeSO_4 + 3H_2SO_4 \longrightarrow KCl + 3Fe_2(SO_4)_3 + 3H_2O$$

or $$ClO_3^- + 6H^+ + 6Fe^{2+} \longrightarrow 6Fe^{3+} + Cl^- + 3H_2O$$

Uses. $KClO_3$ is used : (*i*) in the manufacture of gun powder, matches, fire works and photographic flash powders and is preferred over $NaClO_3$ since the latter is hygroscopic (*ii*) as an antiseptic, especially in sore throat. A very dilute solution of the salt is used as a gargle for sore throat (*iii*) for the preparation of O_2 and potassium perchlorate ($KClO_4$) in the laboratory (*iv*) as an oxidising agent (*v*) in calico printing.

Perchloric acid, $HClO_4$

Preparation. (*i*) It is prepared by heating $HClO_3$.

$$3HClO_3 \xrightarrow{\Delta} HClO_4 + Cl_2 + 2O_2 + H_2O$$

(*ii*) *Anhydrous acid* is obtained by distilling a mixture of potassium perchlorate ($KClO_4$) with conc. H_2SO_4 under reduced pressure.

$$KClO_4 + H_2SO_4 \longrightarrow HClO_4 + KHSO_4$$

(*iii*) *An aqueous solution* of the acid is obtained : (*a*) by treating $Ba(ClO_4)_2$ with calculated quantity of dil. H_2SO_4 and then removing the insoluble $BaSO_4$ by filtration.

$$Ba(ClO_4)_2 + H_2SO_4 \longrightarrow 2HClO_4 + BaSO_4\downarrow$$

(*b*) by adding NH_4ClO_4 dissolved in conc. HCl to warm conc. HNO_3 and then evaporating.

$$NH_4ClO_4 + 8HCl + 3HNO_3 \longrightarrow HClO_4 + 2N_2O + 4Cl_2 + 7H_2O$$

Properties. (*i*) Anhydrous $HClO_4$ is a colourless mobile hygroscopic and oily liquid. It fumes strongly in moist air and dissolves in water with a hissing sound due to the liberation of much heat.

(*ii*) It forms hydrates with 1, 2, 2.5, 3 and 3.5 molecules of water of crystallisation.

(*iii*) It is unstable and decomposes with explosion on heating and sometimes merely on standing for a few days even in the dark. Aqueous solution of the acid is quite stable and does not decompose and hence can be kept indefinitely.

(*iv*) It is highly dangerous acid and produces severe wounds on the skin.

(*v*) It is a powerful oxidising agent and inflames paper and wood.

(*vi*) On dehydration with P_2O_5 at –10°C, it gives Cl_2O_7 which is the anhydride of perchloric acid.

$$2HClO_4 + P_2O_5 \longrightarrow Cl_2O_7 + 2HPO_3$$

This reaction has been used for the preparation of Cl_2O_7.

(*vii*) $HClO_4$ is the strongest acid of all the acids.

(*viii*) The metals like Zn, Fe etc., dissolve in the aqueous solution of the acid and form the soluble perchlorates.

$$Zn + 2HClO_4\ (aq) \longrightarrow \underset{\text{soluble}}{Zn(ClO_4)_2\ (aq)} + H_2\downarrow$$

(*ix*) The acid is not reduced by nascent hydrogen but gets reduced to chloride by strong reducing agents like $SnCl_2$, $CrCl_2$ etc.

(*x*) When a suspension of iodine is heated with $HClO_4$, paraperiodic acid (H_5IO_6) is obtained.

$$2HClO_4 + I_2 + 4H_2O \longrightarrow 2H_5IO_6 + Cl_2$$

Uses. The aqueous solution of the acid is used for the estimation of potassium grametrically.

Potassium perchlorate, $KClO_4$

It is the most important salt of perchloric acid ($HClO_4$).

Preparation. (*i*) When $KClO_3$ is heated at a high temperature, $KClO_4$ is obtained.

$$4KClO_3 \xrightarrow{\Delta} KCl + 3KClO_4$$

Since $KClO_4$ is sparingly soluble and NaCl is soluble in H_2O, $KClO_4$ can be separated from the product mixture by washing the mixture with H_2O.

(*ii*) When conc. solution of $NaClO_3$ undergoes electrolytic oxidation, $NaClO_4$ is obtained (Pt is used as anode).

$$NaClO_3 + O \longrightarrow NaClO_4 \; ; \qquad H_2O + ClO_3^- \longrightarrow ClO_4^- + 2H^+ + 2e^-$$

$NaClO_4$ obtained as above is treated with KCl so that $KClO_4$ is obtained.

$$NaClO_4 + KCl \longrightarrow KClO_4 + NaCl$$

On cooling the solution, $KClO_4$ being sparingly soluble, crystallises out.

(*iii*) When $KClO_3$ is very gently heated with conc. H_2SO_4, $KClO_4$ and ClO_2 are obtained. In this reaction disproportionation of $KClO_3$ (Cl = +5) into ClO_2 (Cl = + 4) and $KClO_4$(Cl = +7) takes place.

$$3KClO_3 + H_2SO_4 \xrightarrow{\Delta} 2ClO_2 + KClO_4 + H_2O + K_2SO_4 \quad \textit{(Molecular equation)}$$

or
$$3ClO_3^- + 2H^+ \longrightarrow 2ClO_2 + ClO_4^- + H_2O \quad \textit{(Ionic equation)}$$

Properties. (*i*) It is a colourless crystalline solid, sparingly soluble in water.

(*ii*) *Action of heat.* On being heated it gives O_2 gas.

$$KClO_4 \xrightarrow{\Delta} KCl + 2O_2$$

This reaction has been utilises for the formation of O_2 gas.

(*iii*) *Action of conc. H_2SO_4.* When $KClO_4$ is distilled with conc. H_2SO_4 under reduced pressure, *anhydrous* $HClO_4$ is obtained.

$$KClO_4 + H_2SO_4 \longrightarrow HClO_4 + KHSO_4$$

Uses. $KClO_4$ is used in fire works and explosives.

Oxy-acids of bromine and their salts

The known oxy-acids formed by bromine are *hypobromous acid* (HBrO) and *bromic acid* ($HBrO_3$). Bromous acid ($HBrO_2$) is unknown but its salts which are called *bromites* (BrO_2^-) are known. Perbromic acid ($HBrO_4$) and its salts *viz.*, perbromates are not formed.

Hypobromous acid, HBrO

Like HClO, HBrO is also known only in solution.

Preparation. (*i*) It is prepared by shaking Br_2 water with freshly precipitated HgO (general method) or Ag_2O.

$$2Br_2 + 2HgO + H_2O \longrightarrow HgBr_2.HgO \text{ (Insoluble)} + 2HBrO$$

$$2Br_2 + Ag_2O + H_2O \longrightarrow 2AgBr + 2HBrO$$

The insoluble $HgBr_2.HgO$ is filtered off and the solution is distilled in a vacuum at 40°C.

(*ii*) It is also obtained by distilling mixture of Br_2 and concentrated solution of $AgNO_3$.

$$Br_2 + AgNO_3 + H_2O \longrightarrow HOBr + HNO_3 + AgBr$$

Properties. (*i*) *Disproportionation reaction.* The aqueous solution of the acid is straw-yellow in colour and is always contaminated with the products which are obtained due to its decomposition, On decomposition, HBrO (Br = +1) gives Br_2 (Br = 0) and $HBrO_3$ (Br = +5). Thus during decomposition, HBrO undergoes *disproportionation* into Br_2 and $HBrO_3$.

$$5HBrO \longrightarrow 2Br_2 + HBrO_3 + 2H_2O \quad \text{or} \quad 4H^+ + 5BrO^- \longrightarrow 2Br_2 + BrO_3^- + 2H_2O$$

(*ii*) HBrO is a strong oxidising and bleaching agent. *e.g.*, HBrO oxidises H_2O_2 to O_2.

$$HOBr + H_2O_2 \longrightarrow H_2O + O_2 + HBr$$

Sodium hypobromite, NaOBr

It is an important salt of hypobromous acid, HOBr.

Preparation. It is prepared by passing I_2 into a cold and dilute solution of NaOH (*Disproportionation of Br_2*).

$$2NaOH + Br_2 \longrightarrow NaOBr + H_2O \quad \text{or} \quad 2OH^- + Br_2 \longrightarrow OBr^- + Br^- + H_2O$$

The solution should not be concentrated by evaporation, since NaOBr, on being heated, gets decomposed into $NaBrO_3$ and NaBr (*Disproportionation* of OBr^- ion).

$$3NaOBr \longrightarrow NaBrO_3 + 2NaBr \quad \text{or} \quad 3OBr^- \longrightarrow BrO_3^- + 2Br^-$$

NaOBr is a good oxidising agent and oxidises amines and amides to N_2. Thus it can be used to estimate urea, $(NH_2)_2CO$.

$$(NH_2)_2CO + 3NaOBr \longrightarrow 3NaBr + 2H_2O + CO_2 + N_2$$

Bromites (BrO_2^-)

We have already said that bromous acid, $HBrO_2$, is unknown, but its salts, *bromites* are known. These salts are formed in solution by the spontaneous decomposition of hypobromites (BrO^-).

$$2BrO^- \longrightarrow BrO_2^- + Br^-$$

Bromites are unstable and rapidly decompose as follows :

$$3BrO_2^- \longrightarrow 2BrO_3^- + Br^-$$

Bromic acid, $HBrO_3$

Like $HClO_3$, $HBrO_3$ is also known in solution only.

Preparation. An aqueous solution of $HBrO_3$ is obtained :

(*i*) by passing Cl_2 in bromine water.

$$Br_2 + 5Cl_3 + 6H_2O \longrightarrow 2HBrO_3 + 10HCl$$

(*ii*) by treating $Ba(BrO_3)_2$ with the calculated quantity of H_2SO_4.

$$Ba(BrO_3)_2 + H_2SO_4 \longrightarrow BaSO_4\ (ppt.) + 2HBrO_3$$

(*iii*) by shaking $AgBrO_3$ with bromine water.

$$5AgBrO_3 + 3Br_2 + 3H_2O \longrightarrow 5AgBr\downarrow + 6HBrO_3$$

Properties. (*i*) On evaporating the aqueous solution of the acid, it decomposes as follows :

$$4HBrO_3 \longrightarrow 3Br_2 + 5O_2 + 2H_2O$$

(*ii*) *Oxidising properties.* The aqueous solution of $HBrO_3$ decomposes on heating and gives Br_2 and nascent oxygen. It is due to the production of oxygen that $HBrO_3$ acts as a good oxidising agent. When this acid oxidises a certain substance, it gets reduced to Br_2. The oxidising action of the acid can thus be represented as :

$$2HBrO_3 \xrightarrow{\text{Reduction}} Br_2 + H_2O + 5O \quad \text{or} \quad 2BrO_3^- + 12H^+ + 10e^- \longrightarrow Br_2 + 6H_2O$$

Thus $HBrO_3$ oxidises :

(*a*) H_2S (S = –2) to sulphur (S = 0)

$$2HBrO_3 + 5H_2S \longrightarrow Br_2 + 6H_2O + 5S \quad (\textit{Molecular equation})$$

$$\text{or} \quad 2BrO_3^- + 12H^+ + 5S^{2-} \longrightarrow Br_2 + 6H_2O + 5S \quad (\textit{Ionic equation})$$

(*b*) SO_2 (S = + 4) to H_2SO_4 (S = +6).

$$2HBrO_3 + 5SO_2 + 4H_2O \longrightarrow Br_2 + 5H_2SO_4 \quad (\textit{Molecular equation})$$

$$\text{or} \quad 2BrO_3^- + 5SO_2 + 4H_2O \longrightarrow 8H^+ + Br_2 + 5SO_4^{2-} \quad (\textit{Ionic equation})$$

(*c*) HBr (Br = –1) to Br_2 (Br = 0).

$$HBrO_3 + 5HBr \longrightarrow 3Br_2 + 3H_2O \quad (\textit{Molecular equation})$$

$$\text{or} \quad BrO_3^- + 6H^+ + 5Br^- \longrightarrow 3Br_2 + 3H_2O \quad (\textit{Ionic equation})$$

Potassium bromate, $KBrO_3$

It is the most important salt of bromic acid.

Preparation. It is prepared.

(*i*) by adding a large quantity of Br_2 to a hot concentrated solution of an alkali

$$3Br_2 + 6KOH \longrightarrow 5KBr + KBrO_3 + 3H_2O \quad \text{or} \quad 3Br_2 + 6OH^- \longrightarrow 5Br^- + BrO_3^- + 3H_2O$$

In the above reaction Br_2 undergoes disproportionation into Br^- and BrO_3^- ions. $KBrO_3$ is much less soluble than KBr and hence can be separated by cooling.

(*ii*) by passing Cl_2 into an alkaline solution of KBr.

$$KBr + 6\ KOH + 3Cl_2 \longrightarrow KBrO_3 + 6KCl + 3H_2O$$

Properties. (*i*) It forms hexagonal crystals which are sparingly soluble in H_2O.

(*ii*) *Action of heat.* KIO_3 gets decomposed on heating and gives KBr and O_2, but never perbromate, since no perborates are yet known.

$$2KBrO_3 \longrightarrow 2KBr + 3O_2$$

(*iii*) *Action of KBr on $KBrO_3$ in presence of HCl.* When a mixture of $KBrO_3$ and a bromide (*e.g.*, KBr) is acidified by HCl, Br_2 is evolved.

$$KBrO_3 + 5KBr + 6HCl \longrightarrow 3Br_2 + 6KCl + 3H_2O$$

or $$BrO_3^- + 5Br^- + 6H^+ \longrightarrow 3Br_2 + 3H_2O$$

The above reaction makes the basis of preparing standard solution of bromine. Bromine is a volatile substance and hence its standard solution cannot be prepared merely by dissolving bromine in water. It is, therefore, prepared by dissolving $KBrO_3$ in H_2O and then treating the solution with excess of KBr in presence of HCl.

Oxy-acids of iodine and their salts

The known oxy-acids, formed by iodine, are hypoiodous acid (HIO), iodic acid (HIO_3) and periodic acids. Example of per iodic acids are metaperiodic and (HIO_4), mesoperiodic acid (HIO_4. H_2O or H_3IO_5), paraperiodic acid ($HIO_4.2H_2O$ or H_5IO_6) and dimesoperiodic acid ($2HIO_4.H_2O$ or $H_4I_2O_9$). Mesoperiodic acid (H_3IO_5) is unknown in the free state but its salts like Ag_3IO_5 are well known. In all these acids iodine is in +7 oxidation state. Among these acids, metaperiodic acid (HIO_4) and paraperiodic acid ($HIO_4.H_2O$ or H_5IO_6) are important acids.

Hypoiodous acid, HIO

Like HClO and HBrO, HIO is also known only in solution.

Preparation. It is prepared by shaking a solution or a suspension of finely powdered iodine with freshly precipitated HgO.

$$2I_2 + 2HgO + H_2O \longrightarrow HgI_2.\ HgO\ (\textit{Insoluble}) + 2HIO$$

Properties. It is a very weak acid. It is the weakest acid of all the hypohalous acid. This fact is evident from the values of dissociation constants of these acids $HOCl = 3 \times 10^{-8}$, $HOBr = 2 \times 10^{-9}$, $HOI = 4.5 \times 10^{-11}$.

The acid also acts as a very weak base, since it also gives OH^- ions.

$$HIO \rightleftharpoons I^+ + OH^-$$

Like HOCl and HOBr, HIO also undergoes disproportionation, farming I_2 and HIO_3.

$$5HIO \longrightarrow 2I_2 + HIO_3 + 2H_2O \quad \text{or} \quad 4H^+ + 5IO^- \longrightarrow 2I_2 + IO_3^- + 2H_2O$$

Sodium hypoiodıte, NaIO

It is an important salt of hypoiodous acid, HIO.

Preparation. NaIO is prepared by treating I_2 with a cold and dilute solution of NaOH, (Disproportionation of I_2 to IO^- and I^- ions).

$$2NaOH + I_2 \longrightarrow NaOI + NaI + H_2O \quad \text{or} \quad 2OH^- + I_2 \longrightarrow OI^- + I^- + H_2O$$

Properties. The freshly prepared solution of NaIO has oxidising and bleaching properties. Thus it oxidises AsO_3^{3-} to AsO_4^{3-}, ethyl alcohol to iodoform, H_2O_2 to O_2, $MnSO_4$ to $Mn(OH)_4$ and bleaches indigo solution.

Iodic acid, HIO_3. Of the halic acids, only this acid is known in the free state. It is the most stable oxy-acid of iodine.

Preparation. The acid is obtained :

(*i*) by oxidising iodine with boiling conc. HNO_3.

$$I_2 + 10\ HNO_3 \longrightarrow 3HIO_3 + 10NO_2 + 4H_2O \quad (\textit{Molecular equation})$$

or $$I_2 + 8H^+ + 10NO_3^- \longrightarrow 3IO_3^- + 10NO_2 + 4H_2O \quad (\textit{Ionic equation})$$

(*ii*) by passing Cl_2 into a suspension of iodine in water.

$$I_2 + 5Cl_2 + 6H_2O \longrightarrow 2HIO_3 + 10HCl$$

HCl is removed as AgCl by adding Ag_2O.

(*iii*) by dissolving I_2O_4 in hot water.

$$5I_2O_4 + 4H_2O \longrightarrow 8HIO_3 + I_2$$

(*iv*) by evaporating iodine with 25% $HClO_3$ or with a hot and concentrated solution of $KClO_3$.

$$2HClO_3 + I_2 \longrightarrow 2HIO_3 + Cl_2\ ;\quad 5KClO_3 + 3H_2O + 3I_2 \longrightarrow 6HIO_3 + 5KCl.$$

Preperties. (*i*) It is a colourless crystalline solid, very soluble in water but insoluble in alcohol. The aqueous solution first reddens and then bleaches litmus paper.

(*ii*) *Action of heat.* When heated to about 240°C, it decomposes to give I_2O_5.

$$2HIO_3 \xrightarrow{240°C} I_2O_5 + H_2O$$

The solid inflames when heated with powdered sulphur, phosphorus, charcoal or organic matter.

(*iii*) *Action of metals.* When Sn, Au, Pb, Pt, Fe, Al and Zn are heated with the acid, H_2 is evolved and iodate of the metal is obtained in solution, *e.g.*,

$$Zn + 2HIO_3 \longrightarrow Zn(IO_3)_2 + H_2$$

(*iv*) *Oxidising properties.* Like $HBrO_3$, it acts as an oxjdising agent. When it oxidises a certain substance, it is reduced to I_2. Thus HIO_3 oxidises H_2S to sulphur, SO_2 to H_2SO_4, HI to I_2 and H_2SO_3 to H_2SO_4.

$$2HIO_3 \xrightarrow{\text{Reduction}} I_2 + H_2O + 5O \quad \text{or} \quad 2IO_3^- + 12H^+ + 10e^- \longrightarrow I_2 + 6H_2O$$

(*a*) $5H_2S + 2HIO_3 \longrightarrow I_2 + 6H_2O + 5S$ or $5S^{2-} + 12H^+ + 2IO_3^- \longrightarrow I_2 + 6H_2O + 5S$

(*b*) $2HIO_3 + 5SO_2 + 4H_2O \longrightarrow I_2 + 5H_2SO_4$ or $2IO_3^- + 5SO_2 + 4H_2O \longrightarrow I_2 + 8H^+ + 5SO_4^{2-}$

(*c*) $2HIO_3 + 5HI \longrightarrow 3I_2 + 3H_2O$ or $IO_3^- + 6H^+ + 5I^- \longrightarrow 3I_2 + 3H_2O$

(*d*) When HIO_3 react with H_2SO_3, no reaction seams to take place but after a few seconds the solution turns brown due to the liberation of iodine. In this reaction H_2SO_3 is oxidised to H_2SO_4 and HIO_3 gets reduced to I_2.

$$6HIO_3 + 15H_2SO_3 \longrightarrow 15H_2SO_4 + 3I_2 + 3H_2O$$

or

$$2IO_3^- + 5SO_3^{2-} + 2H^+ \longrightarrow 5SO_4^{2-} + I_2 + H_2O$$

So long as H_2SO_3 is present, it removes iodine and no reaction appears to occur. After all H_2SO_3 has been used up, iodine separates out.

(*iv*) *Formation of salts.* HIO_3 is a monobasic acid and gives normal salts (*e.g.*, KIO_3) and acid salts. ($HClO_3$ and $HBrO_3$ do not give acid salts). Examples of acid salts given by HIO_3 are acid potassium iodate, KIO_3. HIO_3 or $KH(IO_3)_2$ (bi-iodate) and diacid potassium iodate, $KIO_3.2HIO_3$ or $KH_2(IO_3)_3$ (tri-iodate).

(*v*) When 50% solution of HIO_3 undergoes electrolytic oxidation (PbO_2 is used as an anode), para periodic acid, $HIO_4.2H_2O$ or H_5IO_6 is obtained. This property of HIO_3 has been utilised for the preparation of H_5IO_6.

(*vi*) *Action of conc. H_2SO_4.* When HIO_3 is heated with conc. H_2SO_4, I_2O_4 is obtained.

$$4HIO_3 + 2H_2SO_4 \longrightarrow 2I_2O_4 + 2[H_3O]^+ [HSO_4]^- + O_2$$

Potassium iodate, KIO_3

It is an important salt of iodic acid, HIO_3.

Preparation. It is prepared :

(*i*) by the action of I_2 on hot and conc. KOH solution.

$$3I_2 + 6KOH \longrightarrow 5KI + KIO_3 + 3H_2O \quad \text{or} \quad 3I_2 + 6OH^- \longrightarrow 5I^- + IO_3^- + 3H_2O$$

In the above reaction I_2 undergoes disproportionation into I^- and IO_3^- ions.

(*ii*) by heating a conc. solution of $KClO_3$ with I_2 in presence of a little HNO_3.

$$2KClO_3 + I_2 \longrightarrow 2KIO_3 + Cl_2 \quad \text{or} \quad 2ClO_3^- + I_2 \longrightarrow 2IO_3^- + Cl_2$$

Here ClO_3^- ion oxidises I_2 to IO_3^- ion and is itself reduced to Cl_2.

Properties. (*i*) It forms monoclinic (or cubic) crystals, which are sparingly soluble in cold water, but soluble in hot water.

(*ii*) *Action of KI on KIO_3 in presence of HCl.* When a mixture of KIO_3 and an iodide (*e.g.*, KI) is acidified by HCl, I_2 is liberated.

$$KIO_3 + 5KI + 6HCl \longrightarrow 3I_2 + 6KI + 3H_2O \quad \text{or} \quad IO_3^- + 5I^- + 6H^+ \longrightarrow 3I_2 + 3H_2O$$

The above reaction makes the basis for standardising sodium thiosulphate solution.

(*iii*) *Action of heat.* On heating it melts at 560°C and then decomposes at a somewhat higher temperature into KI and O_2.

$$2KIO_3 \longrightarrow KI + 3O_2$$

On thermal decomposition, KIO_3 also gives potassium metaperiodate (KIO_4) and potassium iodide (KI).

$$4KIO_3 \xrightarrow{\Delta} 3KIO_4 + KI$$

(*iv*) *Oxidising property.* Due to the production of O_2 when it is heated, KIO_3, like $KClO_3$ acts a good oxidising agent.

Metaperiodic acid, HIO_4

Preparation. It is obtained by heating para periodic acid, H_5IO_6.

$$H_5IO_6 \xrightarrow{100°C} HIO_4 + 2H_2O$$

Its salts which are called metaper iodates the prepared by the thermal decomposition of iodates, *e.g.*,

$$4KIO_3 \xrightarrow{\Delta} 3KIO_4 + KI$$

Properties. (*i*) It is a weak and dissociates as :

$$HIO_4 \rightleftharpoons H^+ + IO_4^-$$

(*ii*) When dissolved in water, it changes back to H_2IO_6.

$$HIO_4 + 2H_2O \longrightarrow H_5IO_6$$

(*iii*) HIO_4 and its salts are strong oxidising agents in acidic medium.

$$2IO_4^- + 16H^+ + 14e^- \longrightarrow I_2 + 8H_2O$$

Paraperiodic acid, $HIO_4.2H_2O$ or H_5IO_6

Preparation. It is obtained

(*i*) by the electrolytic oxidation of 50% iodic acid, HIO_3 on a PbO_2 anode.

(*ii*) by decomposing silver mesoperiodate, Ag_3IO_5 with Cl_2.

$$4Ag_3IO_5 + 6Cl_2 + 10H_2O \longrightarrow 4H_5IO_6 + 12AgCl + 3O_2$$

(*iii*) by heating a suspension of I_2 in $HClO_4$.

$$2HClO_4 + I_2 + 4H_2O \longrightarrow 2H_5IO_6 + Cl_2$$

(*iv*) by the action of calculated quantity of H_2SO_4 on barium paraperiodate, $Ba_5(IO_6)_2$.

$$Ba_5(IO_6)_2 + 5H_2SO_4 \longrightarrow 5BaSO_4 + 2H_5IO_6$$

(*v*) A stream of Cl_2 gas is passed into a boiling solution of I_2, in presence of excess of NaOH, when a white ppt. of sod. tribydrogen paraperiodate ($Na_2H_3IO_6$) is obtained.

$$7Cl_2 + I_2 + 18NaOH \longrightarrow 2Na_2H_3IO_6 \text{ (\textit{White ppt.})} + 14\,NaCl + 6H_2O$$

The white ppt. of $Na_2H_3IO_6$ is treated with $AgNO_3$ solution, so that a black ppt. of silver paraperiodate (Ag_5IO_6) is obtained.

$$Na_2H_3IO_6 + 5AgNO_3 \text{ (\textit{solution})} \longrightarrow Ag_5IO_6 \text{ (\textit{Black ppt.})} + 2NaNO_3 + 3HNO_3$$

Now Cl_2 gas is passed into a suspension of Ag_5IO_6, so that a solution containing H_5IO_6 is obtained. This solution, when evaporated, gives the colourless crystals of H_5IO_6.

$$4Ag_5IO_6 + 10Cl_2 + 10H_2O \longrightarrow 4H_5IO_6 \text{ (\textit{solution})} + 20AgCl + 5O_2$$

Properties. (*i*) It is a colourless crystalline solid and deliquescent in nature. It is readily soluble in water, but slightly so in alcohol and ether.

(*ii*) It is a fairly strong monobasic acid, but weak dibasic.

(*iii*) *Action of heat.* (*a*) When heated to 138°C, it gets decomposed into I_2O_5, H_2O and O_2.

$$2H_5IO_6 \xrightarrow{100°C} I_2O_5 + 5H_2O + O_2$$

(*b*) When heated in vacuum at 80°C, $H_4I_2O_9$ is obtained. On further heating at 100°C, HIO_4 is obtained. HIO_4, on being heated to 140°C, gives HIO_3 and O_2.

$$\underset{\text{Para-periodic acid}}{2H_5IO_6} \xrightarrow[-3H_2O]{80°C} \underset{\text{Dimeso-periodic acid}}{H_4I_2O_9} \xrightarrow[-H_2O]{100°C} \underset{\text{Metaper-iodic acid}}{2HIO_4} \xrightarrow{-40°C} \underset{\text{Iodic acid}}{2HIO_3} + O_2$$

(*iv*) The acid and its salts are powerful oxidising agents. Thus :

(*a*) They oxidise Mn^{2+} salts to permanganates.

(*b*) They liberte I_2 from KI in neutral or acidic medium.

$$H_5IO_6 + 7KI + 7HCl \longrightarrow 4I_2 + 7KCl + 6H_2O$$

or

$$24H^+ + 2\{IO_6\}^{5-} + 14I^- \longrightarrow 12H_2O + 8I_2$$

(*c*) The are reduced to I_2 by H_2O_2 and sulphites in acidic, neutral or alkaline solution.

Structure of paraperiodic acid molecules (H_5IO_6) and paraperiodate ion (IO_6^{5-})

X-ray study has shown that H_5IO_6 molecule, as well as IO_6^{5-} ion, have octahedral structure due to sp^3d^2 hybridisation of I-atom (central atom), which lies at the centre of the octahedron (See Fig. 34.2).

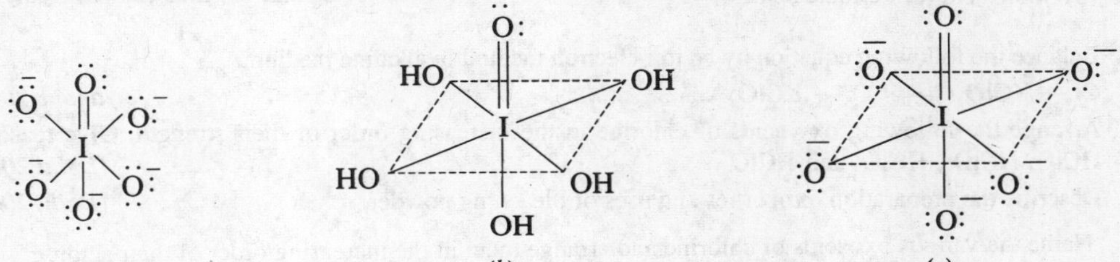

(*a*) (*b*) (*c*)

Fig. 34.2 : (*a*) Electron dot structure of IO_6^{5-} ion (*b*) Octahedral structure of H_5IO_6 molecule (*c*) Octahedral structure of IO_6^{5-} ion.

sp^3d^2 hybridisation of I-atom in IO_6^{5-} ion is, because of the fact that I-atom, in the electron dot structure of IO_6^{5-} ion, shown at (*a*) of Fig. 34.2, is surrounded by six σ-bonding electron pairs, used in the formation of five I — O^- σ-bonds and one I — O σ-bond. Seventh electron of I-atom (I-atom has seven electrons in its valence-shell), is used in forming I — O π-bond. Formation of different bonds in IO_6^{5-} ion has been shown in Fig. 34.3.

It may be seen from sp^3d^2 hybridisation scheme that I — O π-bond is a (*d*—*p*) π-bond.

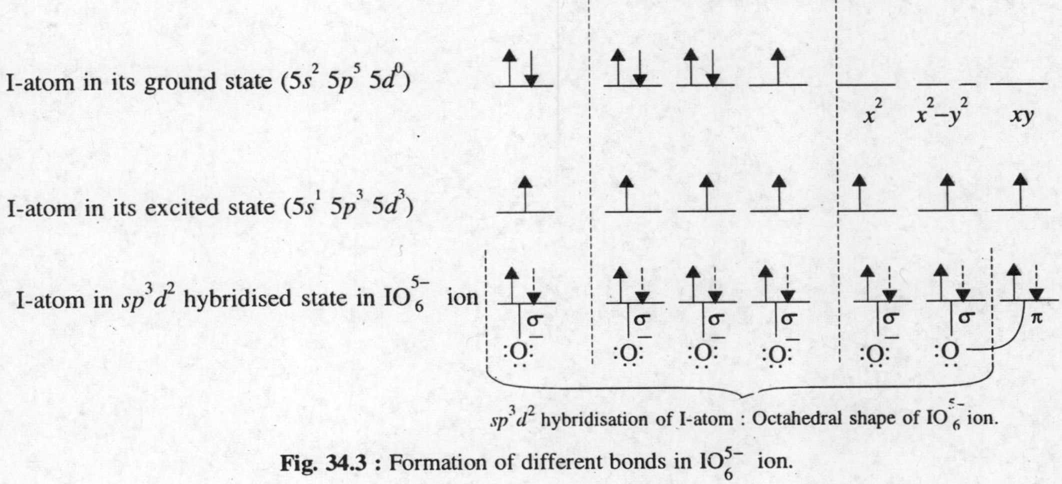

Fig. 34.3 : Formation of different bonds in IO_6^{5-} ion.

University Questions

1. Balance the following equation.

   $KClO_3 + I_2 \rightarrow ....$ *(I.I.T. 96)*

2. The oxy-acids of chlorine have been arranged in the order of decreasing acid strength. Identify to correct order.

   (*a*) I > II > III
   (*b*) II > I > III
   (*c*) III > II > I
   (*d*) I > III > II *(I.I.T. 96)*

3. Write a short note on oxy-acids of chlorine. *(Lucknow 95)*

4. What is available chlorine ? Describe the chemistry involved in the method for the determination of percentage of available chlorine in a given sample of bleaching powder. *(Meerut 96)*

5. Write a note on "Periodic acids". *(Rajasthan 84)*

6. Give the preparation, properties and uses of (*a*) Potassium chlorate (*b*) Perchloric acid (*c*) $KBrO_4$ (*d*) Iodic acid (*e*) Periodic acid. *(Calcutta 84, Banaras 85, Delhi 86)*

7. **Balance the following equation by an ion-electron method in alkaline medium.**

   $Cl_2 + KClO_3 \longrightarrow Cl^- + KClO_4$ .... ***(Nagpur 2002)***

8. **Arrange the following oxy acids of chlorine in the increasing order of their strength. Give reasons. $HClO$, $HClO_2$, $HClO_3$ and $HClO_4$.** ***(Delhi 2002)***

9. **Describe the preparation, properties and uses of bleaching powder.** ***(Delhi 2003)***

10. Name the various oxoacids of chlorine and arrange them in the increasing order of their acidity. *(GND Amritsar 2004)*

35

Fixation of Nitrogen and Fertilisers

Fixation of Nitrogen

What is fixation of nitrogen?

The conversion of atmospheric nitrogen into useful nitrogenous compounds, by natural or artificial methods, is called fixation of nitrogen. Nitrogen present in these nitrogenous compounds is called "fixed" or combined nitrogen.

Natural methods of the fixation of nitrogen

In nature, nitrogen is fixed by the following methods :

1. By lightning discharges. The nitrogen and oxygen, present in air, combine together to form nitric oxide, under the influence of lighting discharges. Nitric oxide is oxidised by excess of oxygen present in the atmosphere, to form nitrogen pentoxide, which further combines with water, to form nitric acid. This acid is washed down by rain into the soil, where it reacts with limestone and alkalies of the soil, to form nitrogen and is stored there as plant-food.

$$N_2 + O_2 \xrightarrow{\text{Lightning}} 2NO \quad (\textit{Nitric oxide})$$
$$4NO + 3O_2 \longrightarrow 2N_2O_5 \quad (\textit{Nitrogen pentoxide})$$
$$N_2O_5 + H_2O \longrightarrow 2HNO_3$$
$$2HNO_3 + CaCO_3 \longrightarrow Ca(NO_3)_2 + CO_2 + H_2O$$

About 250,000 tons of the acid are said to be formed in this way in 25 hours.

2. By symbiotic bacteria. The atmospheric nitrogen is being constantly transferred to the soil through the agency of certain bacteria called *symbiotic bacteria.* These grow in small nodules, in the roots of plants belonging to the family of *Leguminacea* (pea, gram etc.) which directly assimilate atmospheric nitrogen and convert it into products useful for plant growth.

Artificial methods used for the fixation of nitrogen

Following are the main artificial methods used for the **fixation of nitrogen.**

1. Fixation of nitrogen as HNO_3. Under the influence of high tension electric arc, where the temperature is high, nitrogen of the air combines with oxygen, to form nitric oxide. The equilibrium mixture is suddenly cooled when it combines with more of oxygen, to form nitrogen peroxide. This may be absorbed in water in presence of excess of air, to get nitric acid, which may be used for the manufacture of nitrogenous fertilisers.

$$N_2 + O_2 \rightleftharpoons 2NO \quad (\textit{Nitric oxide})$$
$$2NO + O_2 \longrightarrow 2NO_2 \quad (\textit{Nitrogen peroxide})$$
$$4NO_2 + 2H_2O + O_2 \ (\textit{From air}) \longrightarrow 4HNO_3$$

2. Fixation of nitrogen as NH_3 and ammonium salts. Under the influence of certain catalysts nitrogen combines with hydrogen to form ammonia. A mixture of nitrogen (manufactured by liquefaction of air) and hydrogen, in the ratio 1:3, is compressed to a pressure of 200-500 atmospheres and is passed over a catalyst (finely divided iron + molybdenum), heated to about 550°C. This is *Haber's process* for the manufacture of ammonia, which can be converted into ammonium salts by treatment with suitable acids.

$$N_2 + 3H_2 \underset{\text{High pressure}}{\overset{\text{Catalyst}}{\rightleftharpoons}} 2NH_3$$

3. Fixation of nitrogen as calcium cyanamide, $CaCN_2$. When nitrogen gas, obtained by the evaporation of liquid air, is passed over calcium carbide, heated to 800 – 1000°C, the two combine to form a mixture of calcium cyanamide and carbon, which is extensively used as a fertilizer under the name of *nitrolim.*

$$CaC_2 + N_2 \longrightarrow [CaCN_2 + C] \ (\textit{Nitrolim})$$

4. Fixation of nitrogen as nitrides. Nitrogen combines with magnesium and aluminium, at high temperature, to give nitrides, which are employed as a source of ammonia, since these nitrides are decomposed by H_2O and NH_3 is evolved.

$$3Mg + N_2 \longrightarrow Mg_3N_2 \quad (\textit{Magnesium nitride})$$
$$2Al + N_2 \longrightarrow 2AlN \quad (\textit{Aluminium nitride})$$
$$AlN + 3H_2O \longrightarrow Al(OH)_3 + NH_3$$

Fertilisers

Definition of plant nutrients, natural manures and artificial manures (fertilisers)

Just as we need various vitamins, minerals, proteins etc., for our healthy growth, in the same way plants also need certain elements for their growth. Examples of such elements are C, O, H, N, P, K, Ca, Mg, Fe, Cu, Co, Zn etc. These elements are called **plant nutrients.** Out of these elements, C, H, O and a part of N are taken up by the plants directly from the inexhaustible air and water.

The remaining elements, including N, are supplied to the plants by the soil itself. N, P and K are needed by the plants in large quantities and hence are called *primary* or *essential plant nutrients.* Other elements are required, in minute quantities, by plants and hence are termed as *secondary nutrients.* The soil is never deficient in secondary nutrients, but, after a few years of continuous cultivation, the soil becomes poorer and poorer in primary nutrients. If the soil is not compensated for what it has lost in time, it becomes less productive and ultimately it becomes infertile.

Thus, in order to maintain the fertility of the soil, it is necessary to add materials, containing N, P and K, to the soil. These materials are called **manures.** Manures are of two types *viz., natural manures* and *artificial* or *chemical manures*

1. Natural manures. These include animal dung, blood, bones, horns, hair locks, leather waste, fish scrap, decayed plants, plants burried in the soil etc. A natural manure is obtained by the decomposition of animal wastes and plant residues.

2. Artificial or chemical fertilisers. These are also called simply as **fertilisers.** These are chemical compounds obtained by artificial methods. A fertiliser is a chemical compound that is added to the soil to supply to it the essential nutrients (*i.e.*, N, P and K), so that the soil may become more fertile.

Function of essential plant nutrients

We have already said that N, P and K are the essential plant nutrients. Their functions are discussed below :

(*i*) **Nitrogen.** The addition of nitrogen to the soil helps in the rapid and healthy growth of the plants and thus improves the yields of the crops. It also increases the protein content in the crops and thus the food value of the crops is increased.

(*ii*) **Phosphorus.** Phosphorus promotes early growth, as well as, early maturity of the plants. Phosphorus also increases the resistance of the plants to frost and disease and helps in the formation of high-yielding seeds.

(*iii*) **Potassium.** Potassium provides a healthy root system to the plant. With the help of this system, the plants become capable of getting a regular supply of nutrients from the soil. The addition of potassium also helps in the formation of albuminoids and carbohydrates in various plants.

Essential qualities of a fertiliser

We have said above that N, P and K are the essential nutrients for a plant. But every compound of N, P or K cannot be used as a fertiliser. A compound containing N, P or K can act as a fertiliser only when it has the following qualities : (*i*) It should be soluble in water (*ii*) The nutrient elements (*viz.*, N, P and K) must be present in the compound, in such a form, that they can easily be assimilated by the plants. (*iii*) By rain or water, the compound must be converted into a form which the plant can take up easily. (*iv*) The compound should be stable, so that it may be available for a long time for the growing plant. (*v*) It should not be injurious to the plants. (*vi*) It should be able to correct the acidity of the soil. (*vii*) It should be cheap.

Classification of fertilisers

Depending on the nature of the essential element that a fertiliser can supply to the soil, the fertilisers have been classified into the following groups :

I. Nitrogenous fertilisers : N-type fertilisers

These fertilisers supply only nitrogen to the soil and hence are also called N-type fertilisers. These are single fertilisers, since they supply only one element (*viz.*, nitrogen) to the soil or plant. Some such fertilisers are discussed below :

1. Sodium nitrate, $NaNO_3$. (Also called *chile salt petre* or *chile nitre*). See "*Nitrates*" (chapter 31).

2. Calcium ammonium nitrate (CAN), $Ca(NO_3)_2.NH_4NO_3$. (Also called *nitrolime stone* or *Nangal fertiliser*). See "*Nitrates*".

3. Ammonium sulphate nitrate (ASN), $(NH_4)_2SO_4.2NH_4NO_3$. See "*Nitrates*".

4. Basic calcium nitrate, $Ca(NO_3)_2.CaO$ (Also called *Norwegian salt petre* or *nitrate of lime*). See "*Nitrates*" .

5. Ammonium sulphate, $(NH_4)_2SO_4$. (Also called *Sindri fertiliser*) See "*Sulphates*".

6. Calcium cyanamide or nitrolim, $CaCN_2$. The fertiliser is a derivative of cyanamic acid $H-N=C=N-H$ or H_2CN_2, in which two H-atoms have been replaced by one divalent Ca-atom.

Manufacture. It is manufactured by heating finely powdered CaC_2 with a little of $CaCl_2$ in the atmosphere of N_2 at 1000-1100°C, in an electric furnace. The presence of $CaCl_2$ accelerates the reaction. It is usually obtained as a dark grey mass. Its grey colour is due to the presence of carbon formed in the reaction. The grey mass, which is a mixture of $CaCN_2$ and carbon, is sold in the market under the name of *nitrolim*.

$$CaC_2 + N_2 \xrightarrow[(CaCl_2)]{1000-1100°C} [CaCN_2 + C] \text{ (\textit{Nitrolim})}$$

CaC_2, required in the above reaction, is obtained by heating a mixture of CaO (CaO is obtained by heating $CaCO_3$ in kilns) and coal in an electric furnace at 2500°C.

$$CaCO_3 \xrightarrow{\Delta} CaO + CO_2 \; ; \quad CaO + 3C \longrightarrow C_2 + CO$$

Since the formation of nitrolim is accompanied by the release of heat (exothermic reaction), the heating is to be carried out only at one point, so that the reaction further propagates by itself.

Properties. (*i*) Pure $CaCN_2$ forms colourless, rhombohedral crystals, which sublime at about 1300°C, without melting.

(*ii*) *Action of superheated steam.* When treated with superheated steam, under a pressure of 3-2 atmospheres, $CaCN_2$ is decomposed, to form NH_3.

$$CaCN_2 + 3H_2O \text{ (\textit{superheated steam})} \rightarrow CaCO_3 + 2NH_3$$

This reaction has been used to manufacture NH_3 on large scale by cyanamide process.

(*iii*) *Fusion with coke.* When pure $CaCN_2$ is fused with coke, calcium cyanide, $Ca(CN)_2$ is formed.

$$CaCN_2 + C \xrightarrow{\text{Fusion}} Ca(CN)_2$$

(*iv*) *Action of NaCl or Na_2CO_3.* When crude calcium cyanamide (*i.e.*, mixture of $CaCN_2$ and coke) is heated with NaCl or Na_2CO_3, NaCN is obtained.

$$(CaCN_2 + C) + 2NaCl \longrightarrow CaCl_2 + 2NaCN$$
$$(CaCN_2 + C) + Na_2CO_3 \longrightarrow CaCO_3 + 2NaCN$$

These reactions have been used to prepare NaCN.

(*v*) *Action of acids.* With acids like H_2SO_4, urea ($H_2N.CO.NH_2$) is obtained.

$$CaCN_2 + H_2SO_4 + H_2O \longrightarrow CaSO_4 + H_2N.CO.NH_2$$

(*vi*) *Action of atmospheric CO_2.* With atmospheric CO_2, $CaCN_2$ is converted into urea.

$$CaCN_2 + CO_2 + 2H_2O \longrightarrow CaCO_3 + H_2N.CO.NH_2$$

Uses. (*i*) Nitrolim is a very good nitrogenous fertiliser, since it provides both Ca and N_2 to the soil. It is added to the soil before sowing the surface of the soil and when the plants are actually growing. It is preferred to NH_4NO_3 and $(NH_4)_2SO_4$, as it is a slow-acting fertiliser. In the soil, it is converted into urea ($H_2N.CO.NH_2$) by the atmospheric CO_2 and moisture. Urea, thus formed, is decomposed into NH_3 which is converted into soluble nitrates by the nitrifying bacteria. These nitrates are assimilated by plants quite readily.

$$CaCN_2 + CO_2 + 2H_2O \longrightarrow CaCO_3 + H_2N.CO.NH_2$$
$$H_2N.CO.NH_2 + H_2O \longrightarrow CO_2 + 2NH_3$$
$$NH_3 \xrightarrow[\text{bacteria}]{\text{Nitrifying}} \textit{Soluble nitrates} \longrightarrow \text{Are assimilated by plants.}$$

Nitrolim is a slow-acting fertiliser, since the conversion of $CaCN_2$ into NH_3 and that of NH_3 into soluble nitrates by nitrifying bacteria takes place slowly.

(*ii*) It is used for the large scale preparation of NH_3 and NaCN.

(*iii*) It is also used in case-hardening of steel.

Structure of cyanamide ion, CN_2^{2-}. The structure of this ion can be represented as $:\bar{\ddot{N}}=C=\bar{\ddot{N}}:$ which shows that, due to the presence of one negative charge on each N-atom, each N-atom has six electrons (instead of five) in its valence-shell ($\bar{N} = 2s^2p^4$). C-atom ($2s^2p^2$) is *sp* hybridised, while N^- ion is sp^2 hybridised. The formation of different bonds has been shown in Fig. 35.1.

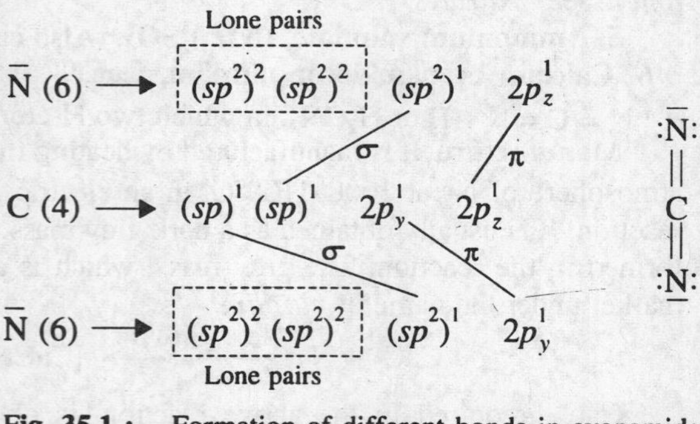

Fig. 35.1 : Formation of different bonds in cyanamide ion, CN_2^{2-}.

7. Urea, $H_2N.CO.NH_2$. It is the most important nitrogenous fertiliser.

Manufacture. The process used for the manufacture of urea consists of the following steps.

(*i*) Crude naptha obtained from oil refineries is subjected to partial combustion in specially designed burners to get a mixture of H_2 and CO.

$$\text{Crude naptha} \xrightarrow[\text{combustion}]{\text{Partial}} CO + H_2$$

(*ii*) The mixture of CO and H_2 obtained as above is made to react with steam (H_2O) in the presence of a Fe, Cr and CO (catalyst) when CO is oxidised to CO_2 and H_2 remains as such.

$$(CO + H_2) + H_2O\ (\textit{steam}) \xrightarrow{(Fe + Cr + Co)} CO_2 + 2H_2$$

CO_2 is separated from H_2 by washing with H_2O under pressure or by the treatment with $KHCO_3$ solution.

(*iii*) H_2 separated from the mixture, ($CO_2 + H_2$) is mixed with N_2. The mixture containing H_2 and N_2 is passed over heated (temperature = 450–500°C) Fe_2O_3 (catalyst) containing Cr_2O_3 (promoter) under 200 atmospheric pressure when NH_3 is obtained (Haber's process for the preparation of NH_3).

$$3H_2 + N_2 \longrightarrow 2NH_3$$

(*iv*) Now NH_3 obtained as above is allowed to react with gaseous CO_2 under 200 atmospheric pressure when unstable ammonium carbamate ($H_2NCOONH_4$) is obtained. $H_2NCOONH_4$, being unstable, decomposes to give urea.

$$2NH_3 + CO_2 \longrightarrow H_2NCOONH_4 \longrightarrow H_2NCONH_2\ (urea) + H_2O$$

Use of urea as a fertiliser. (*a*) *Advantages of its use.* (*i*) Urea has the highest nitrogen content, equal to 46.6%. This percentage of N_2 is much higher than that for $(NH_4)_2SO_4$ and CAN. Higher nitrogen content in urea results in economy in freight and packing. (*ii*) The cost of production of urea is relatively low since CO_2 required for its manufacture is obtained from crude naphtha. (*iii*) Urea is not subject to fire or explosion hazards, as is NH_4NO_3 and hence there is no risk in the storage of urea. (*iv*) It does not change the pH of the soil. (*v*) Urea can be used for all types of crops and soils, since after its assimilation by plants through the interaction of nitrifying bacteria, it leaves behind only CO_2 in the soil. This CO_2 does no harm to the soil.

$$H_2N.CO.NH_2 + H_2O \longrightarrow 2NH_3 + CO_2$$

On the other hand, $(NH_4)_2SO_4$ and CAN produce SO_4^{2-} and NO_3^- ions, which make the soil acidic.

(*b*) *Disadvantages of its use.* (*i*) Urea is very soluble in H_2O and hygroscopic and hence requires better quality package than $(NH_4)_2SO_4$. (*ii*) It is not so stable as other solid nitrogenous fertilisers are, since urea decomposes, even at lower temperatures, particularly in humid atmosphere, releasing NH_3 and CO_2. The formation of NH_3 and CO_2 results in serious loss. (*iii*) If urea contains impurities more than 2%, it cannot be used as a fertiliser, since the impurities are toxic to certain crops, particularly to citrus crops.

(*c*) *Methods of its application.* (*i*) Urea should be applied at the sowing time. It should not come in contact with the seeds. (*ii*) It can also be applied as a top dressing. (*iii*) Since urea is highly concentrated, it should be used in combination with earth or sand before its application. (*iv*) It should not be applied when the soil contains free water or is likely to remain wet for three or four days after urea application.

II. Phosphatic or phosphorus fertilisers : P-type fertilisers

These fertilisers supply only phosphorus to the soil and hence are also called P-type fertilisers. These are also single fertilisers, since they supply only one element (*viz.*, phosphorus) to the soil or plant. The minerals of phosphorus, like *phosphorite* $[Ca_3(PO_4)_2]$ and *apatite* $[3Ca_3(PO_4)_2.CaF_2]$, are sparingly soluble in water and hence are not able to supply phosphorus to the plants. These minerals are, therefore, converted into soluble materials, which act as good phosphatic fertilisers. Important phosphatic fertilisers are discussed below :

1. Calcium dihydrogen phosphate, $Ca(H_2PO_4)_2.H_2O$. (Also called *triple superphosphate*). See "*Phosphates*"

2. Calcium dihydrogen phosphate sulphate, $Ca(H_2PO_4)_2.H_2O.\ 2(CaSO_4.2H_2O)$ (Also called *superphosphate of lime* or *calcium superphosphate*) . See "*Phosphates*"

3. Phosphate slag, $Ca_3(PO_4)_2.CaSiO_3$. (Also called *Thomas slag, basic slag* or *Thomas meal*). See "*Phosphates*".

III. Potassium fertilisers : K-type fertilisers

These fertilisers supply potassium to the plants. This element is very useful for the growth of the plants like medowgrass, tobacco, cotton, coffee, potatoes and corn. All the these fertilisers occur in nature. These fertilisers, like N-type and P-type fertilisers, are also single fertilisers because they supply only one element *viz.*, potassium, to the soil or plants. The chief K-type fertilisers are discussed below :

1. Potassium chloride, KCl. See "*Halides of alkali metals*" in chapter 29.

2. Potassium sulphate, K_2SO_4. See "*Sulphates*" in chapter 33.

IV. Mixed fertilisers

Mixed fertilisers are those which can supply more than one essential element to the soil or plant. Depending on the nature of the essential element supplied by the fertiliser, mixed fertilisers can be classified into the following groups :

1. Nitrogen-phosphorus fertilisers : NP fertilisers. These fertilisers supply two essential elements *viz.*, nitrogen and phosphorus to the plants and hence are called nitrogen-phosphorus fertilises. Important examples of such fertilisers are :

(*a*) *Ammonium dihydrogen phosphate,* $(NH_4)(H_2PO_4)$. (Also called *dihydrogen ammoniated phosphate* or *ammophos*). See *"Phosphates"* (chapter 32).

(*b*) *Diammonium monohydrogen phosphate,* $(NH_4)_2(HPO_4)$. (Also called *ordinary ammonium phosphate of commerce*). See *"Phosphates"*.

(*c*) *Ammonium dihydrogen phosphate sulphate,* $NH_4(H_2PO_4).(NH_4)_2SO_4$. See *"Phosphates"*.

(*d*) *Calcium dihydrogen phosphate nitrate,* $Ca(H_2PO_4)_2.2Ca(NO_3)_2$. (Also called *calcium superphosphate nitrate or nitrophosphate*). See *"Phosphates"*.

2. Phosphorus-potassium fertilisers or PK-type fertilisers. PK-type fertiliser is a mixture of two compounds one containing phosphorus and the other containing potassium. For example, a mixture of triple superphosphate $[Ca(H_2PO_4)_2.H_2O]$ and potassium sulphate (K_2SO_4) is a PK-type fertiliser.

3. Potassium-nitrogen fertilisers : KN type fertilisers. Since KNO_3 contains potassium and nitrogen, it is a mixed fertiliser of KN-type. KNO_3 is also called by many other names like *nitre, Indian salt petre, Bengal salt petre, shora* or *suverchi.*

4. Nitrogen-phosphorus-potassium type or NPK type fertilisers : Complete fertilisers. These fertilisers supply all the three essential elements, namely N, P and K, to the soil. The percentage content of N, P and K, in these fertilisers, is usually indicated by numbers like 5-10-10, which means that NPK fertiliser contains 5% N_2, 10% P as P_2O_5 and 10% K as K_2O. These fertilisers are being produced by *Zuari Agro-chemical Ltd., Goa* and are obtained by mixing nitrogenous, phosphatic and potassium fertilisers in suitable proportion. For example, a mixture of $(NH_4)_2SO_4$ (N-type fertiliser), triple superphosphate, $Ca(H_2PO_4)_2.H_2O$ (P-type fertiliser) and K_2SO_4 (K-type fertiliser) gives a NPK-type fertiliser.

Manufacture of fertilisers in India

India is an agricultural country and requires fertilisers to increase food production. In our country, there are more than 60 factories, in private as well as in public sectors, which are manufacturing fertilisers. Some important fertiliser factories are located at Sindri (Bihar), Nangal (Punjab), Namrup (Assam), Gorakhpur (U.P.), Trombay (Maharashtra), Korba (M.P.), Barauni (Bihar), Talchar (Orissa), Durgapur (West Bengal), D.C.M. Chemical Works, Delhi and Kota (Rajasthan), Indian Explosive Ltd., Kanpur (U.P.), Zauri-Agro-Chemicals Ltd. (Goa), Phosphate Company Ltd. (Calcutta), Coromandal Fertiliser Ltd. (Visakhapatnam), Madras Fertilisers (Madras), The Fertiliser and Chemical Travancore, Ltd. (Alwaye), Rourkela (Orissa), Jamshedpur etc

Many new units have been setup to increase the production of fertilisers in India in recent years. Some of them are based on natural gas.

University Questions

1. Discuss the various methods for the fixation of nitrogen. *(Indore, 85)*
2. Give an account of various types of fertilisers. *(Madras, 86)*
3. Write a note on "Fertilisers and their manufacture in India". *(Bundelkhand, 88)*
4. Give the methods of industrial preparation, properties and uses of : (*i*) CAN (*ii*) Calcium cyanamide (*iii*) Urea (*iv*) Calcium superphosphate. *(Delhi, 89; Meerut, 89 S)*

36

Inorganic Polymers

Introduction

All the covalent macromolecules, which do not have carbon in their back bones, are considered to be inorganic polymers. Covalently-bonded crystals (*e.g.*, oxides and halides), condensed phosphates etc., are the examples of inorganic polymers. These polymers possess distinctive physico-chemical characteristics and unique physical, mechanical and electrical properties. These polymers are of extensive utility in everyday life, particularly in the area of engineering and technology.

The important point of similarity between inorganic and organic polymers is, that both can be prepared by the addition and condensation methods. The former method is used when polymers of higher molecular weight and greater mechanical strength are needed.

Clasification of inorganic polymers

There are different ways of classifying inorganic polymers.

1st classification. Depending on whether the polymer contains the atoms of only one element or of different elements in its backbone, the polymers are classified into the following two groups.

1. Homo-atomic polymers. These polymers contain the atoms of only one element in their back bones. Silicon, phosphorus, sulphur, germanium and tin form homo-atomic inorganic polymers. For example, sulphur has a tendency to form chain or rings in its elemental form (S_8) and in several compounds, like persulphides (H—S—S—H, H—S—S—S—H, H—S—S—S—S—H etc.), polythionic acids etc. Single bond strengths in homo-atomic inorganic polymers are 54 to 60 Kcal for sulphur, about 53 Kcal for silicon, 48-53 for phosphorus, about 45 for germanium and about 39 for tin.

2. Hetero-atomic polymers. These contain the atoms of different elements in their backbones.

2nd classification. Inorganic polymers can also be classified in another way, which is based on the type of reaction by which the polymers are formed. On this basis, inorganic polymers may be of the following types :

1. Condensation (polymerisation) polymers. Condensation polymers are those, which are formed by condensation (polymerisation) process. In this process, two or more simple molecules of the same substance polymerise together and form the condensation polymer. One or more H_2O, NH_3, H_2, HCl etc., molecules are also eliminated.

Examples. (*a*) Cross-linked silicone is obtained by the polymerisation of many $RSi(OH)_3$ molecules. Similarly, when many molecules of R_2 $Si(OH)_2$ undergo polymerisation, a straight chain (linear) or cyclic (ring) silicone is obtained. When two molecules of $R_3Si(OH)$ undergo polymerisation, a straight chain silicone (dimer) is obtained. For detail see "*Silicones*".

(*b*) When PCl_5 is partially hydrolysed by water, dichloro phosphoric acid, $PO(OH)Cl_2$ is obtained. When $PO(OH)Cl_2$ is heated, many molecules of this substance get polymerised and give rise to the formation of a condensation polymer. In this process HCl is eliminated.

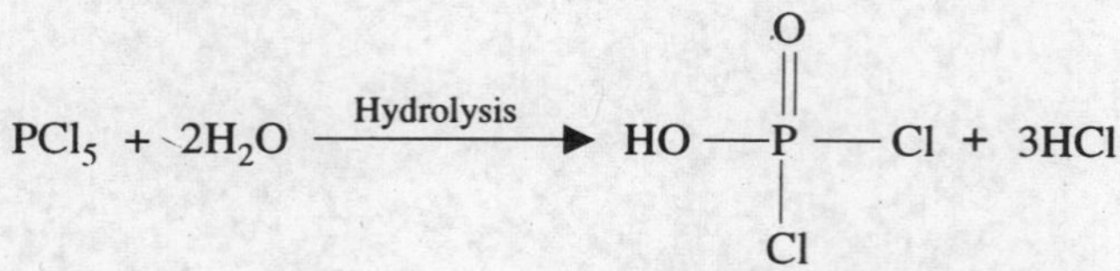

Dichlorophosphoric acid, $PO(OH)Cl_2$

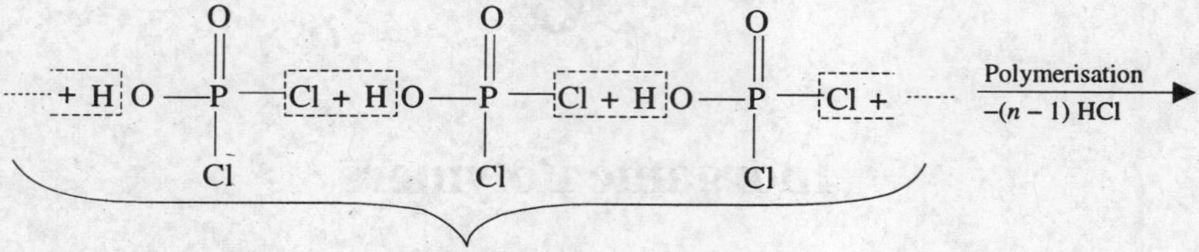

n molecules of $PO(OH)Cl_2$

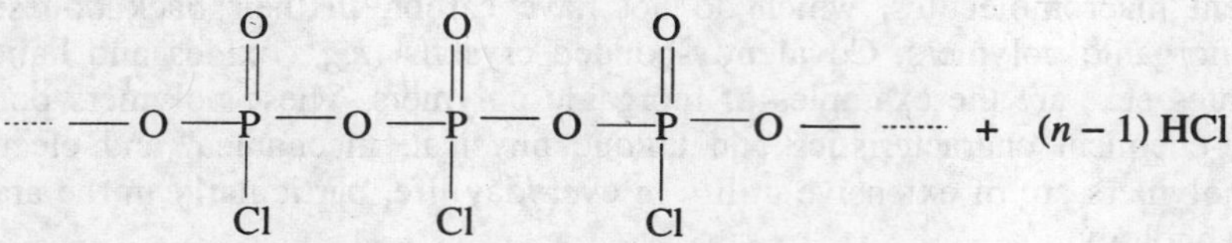

Condensation polymer
(Linear long chain polymer)

2. Addition polymers. These polymers are obtained, when many simple molecules (monomers) combine together.

Examples. (*a*) Many molecules of sulphur trioxide may be polymerised by the addition of a small amount of water. This gives addition polymer.

(*b*) When SO_2 reacts with propylene, $CH_3—CH = CH_2$, in presence of benzoyl peroxide, an addition polymer is obtained.

3. Coordination polymers. These are formed by the addition of saturated molecules to each other or by combining a ligand with a metal atom. These polymers contain chelated metal atoms or ions.

Types of coordination polymers. (*i*) In these polymers, the chelated metal atom (ion) is an integral part of the polymer framework.

(*ii*) In these polymers the metal atom (ion) remains bound to a polymeric ligand, which has donor groups in its framework.

Preparation of coordination polymers. (*i*) This method involves the combination of molecule/ion (ligand) with the metal atom/ion.

(*ii*) In this method, the metal ions are fixed into a polymeric ligand, which has donor atom in its framework.

(*iii*) Coordination polymers can be prepared by condensation polymerisation, which requires appropriate monomers.

(*iv*) Coordination polymers can also be obtained by using elimination-addition reactions.

3rd classification. According to this classification, the inorganic polymers can be classified into the following catagories :

1. *Polymers containing two bridging bonds per units, e.g.,* homo-atomic sulphur, selenium and teturium polymers.

2. *The alternating silicone-oxygen polymers.* Exmples are silicones and related compounds.

3. *The alternating phosphorus-nitrogen polymers.* Examples are phosphonitrilic halides, $(NPX_2)_n$.

4. *The alternating phosphorus-oxygen polymers.* Examples are metaphosphates, polyphosphates and cross-linked phosphates.

5. *The alternating sulphur-nitrogen polymers.* Examples are (*i*) polymeric nitrides of sulphur (*e.g.*, S_2N_2, S_4N_4, S_5N_2 etc.) (*ii*) thiazyl halides, [*e.g.*, $(NSF)_3$, $(NSF)_4$, $(NSCl)_3$] (*iii*) imides of sulphur [*e.g.*, $S_7(NH)$, $S_6(NH)_2$, $S_5(NH)_3$ and $S_4(NH)_4$].

4th classification. This classification is based on the element which forms inorganic polymers. Thus, we have :

1. *Polymers containing boron.* Examples are : (*a*) Borazine, $(BH)_3$ $(NH)_3$ or $B_3N_3H_6$. (*b*) Substituted borazines like (*i*) B-trimethyl borazine, $[B(CH_3)_3$ $(NH)_3]$ (*ii*) Boroxine, $(BH)_3O_3$ (*iii*) N-trimethyl borazine, $(BH)_3$ $[N(CH_3)]_3$ (*c*) Boron nitride, $(BN)_n$.

2. *Polymers containing silicon.* These are called silicones.

3. *Polymers containing phosphorus.* Examples are : (*a*) Metaphosphates, (*b*) Polyphosphates, (*c*) Cross-linked phosphates, (*d*) Phosphonitrilic halides, $[PNX_2]_n$.

4. *Polymeric compounds of sulphur.* Examples are nitrides of sulphur, thiazyl halides and imides of sulphur.

General properties of inorganic polymers

(*i*) Most of the inorganic polymers do not burn, but only soften or melt at high temperatures. Inorganic polymers, containing sulphur etc., are exceptions.

(*ii*) Inorganic polymers, having cross-linked structures with many covalent bonds, are generally stiffer and harder than the organic polymers.

(*iii*) Since most of the inorganic polymers are built up of highly polar repeat units, these polymers dissolve only in polar solvents. These polymers react with solvent molecules.

(*iv*) Inorganic polymers are generally less ductile than the organic polymers.

(*v*) Inorganic polymers have structures which are purely crystalline or purely amorphous while organic polymers have structures which are partly crystalline and parlty amorphous.

(*vi*) Inorganic polymers are generally stronger, harder and more brittle than the organic polymers.

Polymers containing boron

Borazine, substituted borazines and boron nitride are polymeric compounds of boron. Borazine is of special interest, because it is isoelectronic with benzene and some of its physical properties resemble those of benzene.

A. Borazine or borazole, $(BH)_3$ $(NH)_3$ or $B_3N_3H_6$

This compound is isoelectronic with benzene and hence has been called *Inorganic Benzene* by Weberg.

Preparation. Borazine can be prepared by the following methods :

(*i*) *By Stock and Pohland's method (1926).* Borazine was originally prepared by Stock and Pohland, in 1926, by the action of NH_3 on diborane (B_2H_6). The adduct, B_2H_6. $2NH_3$ is first formed, which then gets decomposed by heating in a closed tube at 200°C.

$$\underset{(\text{Ratio} = 1:2)}{3B_2H_6 + 6NH_3} \longrightarrow \underset{\text{Adduct}}{3[B_2H_6.2NH_3]} \xrightarrow{200^\circ C} 2B_3N_3H_6 + 12H_2$$

This method gives low yield of $B_3N_3H_6$ because of simultaneous formation of solid polymeric by-products.

(*ii*) *By heating BCl_3 with NH_4Cl.* When BCl_3 is heated with NH_4Cl in chlorobenzene (C_6H_5Cl) in the presence of Fe, Ni, or Co (used as catalyst) at about 140°C, B, B, B–trchloroborazine (one chloro group attached with each of the three B-atoms in $B_3N_3H_6$) is formed. This derivative, on being reduced by $LiBH_4$ in polyether, gives borazine, $B_3N_3H_6$.

$$3NH_4Cl + 3BCl_3 \xrightarrow[C_6H_5Cl \text{ and Fe})]{140^\circ C \text{ (In presence of}} \underset{\text{B, B, B-trichloroborazine}}{Cl_3B_3N_3H_3} + 9\,HCl$$

$$3Cl_3B_3N_3H_3 + 6LiBH_4 \xrightarrow{\text{Reduction}} 2B_3N_3H_6 + 6LiCl + 3B_2H_6$$

(iii) By heating a mixture of $LiBH_4$ and NH_4Cl (Laboratory method). $B_3N_3H_6$ can be prepared in the laboratory by heating a mixture of $LiBH_4$ and NH_4Cl in vacuum at 230°C.

$$3NH_4Cl + 3LiBH_4 \xrightarrow{230°C} B_3N_3H_6 + 3\,LiCl + 9H_2$$

This method gives 30% borazine.

Physical properties. *(i)* Barazine is a colourless mobile volatile liquid. It freezes at –58°C. It's b.pt. and m.pt. are 64.5°C and –58°C respectively.

(ii) Decomposition. Borazine decomposes slowly on storage even at – 80°C and small quantities of a white solid are deposited over a period of several days. At room temperature, decomposition occurs much more rapidly and H_2, B_2H_6 and other volatile products are formed.

(iii) Similarities with benzene. Since borazine is isoelectronic with benzene, some of its physical properties are similar to those of benzene. This is shown below :

		Mol wt.	*M.pt.* (C°)	*B.pt.* (°C)	ΔH_{vap} (kJ mole^{-1})	*Trouton's constant*
Borazine ($B_3N_3H_6$)	:	80.6	– 58	64.5	39.3	21.4
Benzene (C_6H_6)	:	78.0	+ 6	80.0	31.0	21.0

This similarity of physical properties has unfortunately been over emphasised and hence borazine has been called *Inorganic Benzene.* However, the fact is that the properties of borazine and benzene are quite different.

Chemical properties. Main chemical properties of borazine are given below. These properties are not shown by benzene.

(i) Addition reactions. (a) One molecule of $B_3N_3H_6$ adds three molecule of HCl or HBr in the cold, without a catalyst. These molecules get attached with all the three B-atoms of $B_3N_3H_6$ molecule, since B-atom is more negative than N-atom in B—N or B = N bond and hydrogen chloride derivative ($B_3N_3H_9Cl_3$) is obtained. This addition reaction is not shown by benzene.

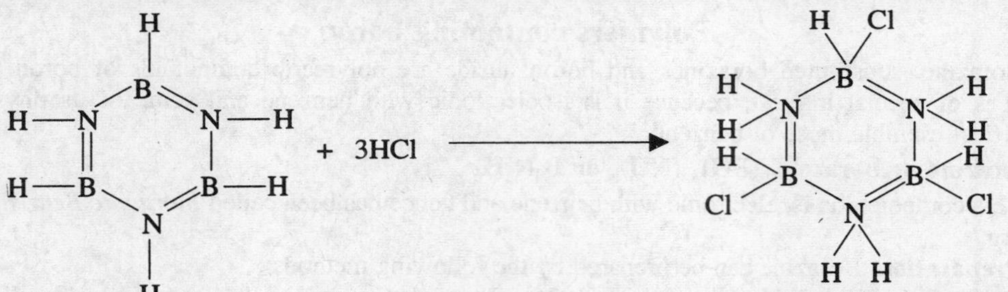

Hydrogen chloride derivative

or $\quad B_3N_3H_6 + 3HCl \longrightarrow B_3N_3H_9Cl_3$

When this derivative is heated to 50-100°C, it looses three H_2 molecules, to give B, B, B-trichloroborazine, $B_3N_3H_3Cl_3$.

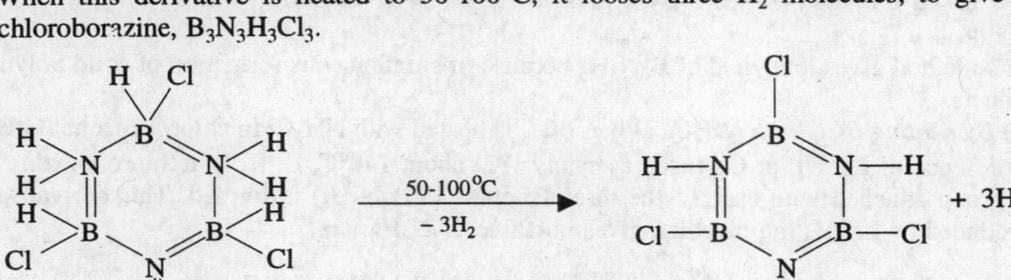

B, B, B-trichloroborazine

or $\quad B_3N_3H_9Cl_3 \xrightarrow[-\,3H_2]{50\text{-}100°C} B_3N_3H_3Cl_3 + 3H_2$

(*b*) One molecule of borazine also adds three molecules of Br_2 at 0°C and gives B-tribromo –N—tribromiborazine, which on being heated to 60° C, loses three molecules of HBr and forms B-tribromoborazine.

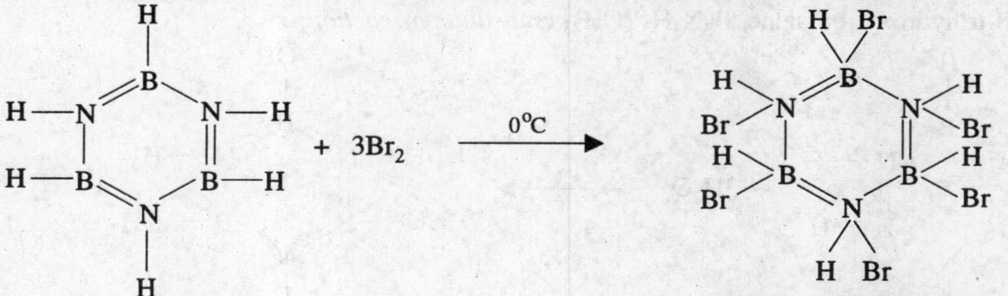

B-tribromo-N-tribromoborazine

$$\text{or} \quad \underset{\text{Borazine}}{B_3N_3H_6} + 3Br_2 \xrightarrow{0^{\circ}C} B_3N_3H_6Br_6$$

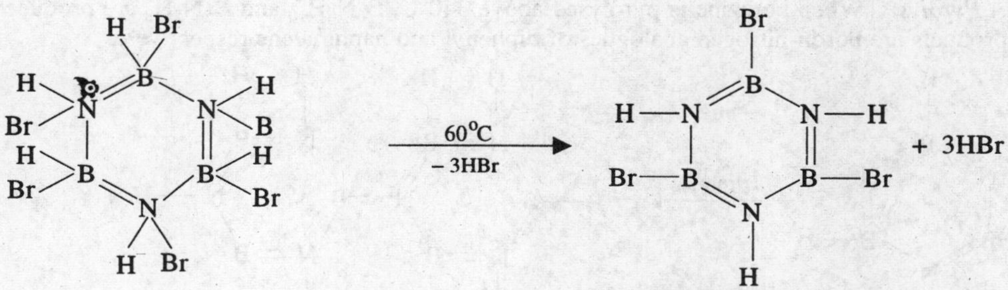

B-tribromoborazine

$$\text{or} \quad B_3N_3H_6Br_6 \xrightarrow[-\,3HBr]{60^{\circ}C} B_3N_3H_3Br_3 + 3HBr$$

Benzene shows substitution reaction and gives monobromobenzene, $C_6H_5\,Br$.

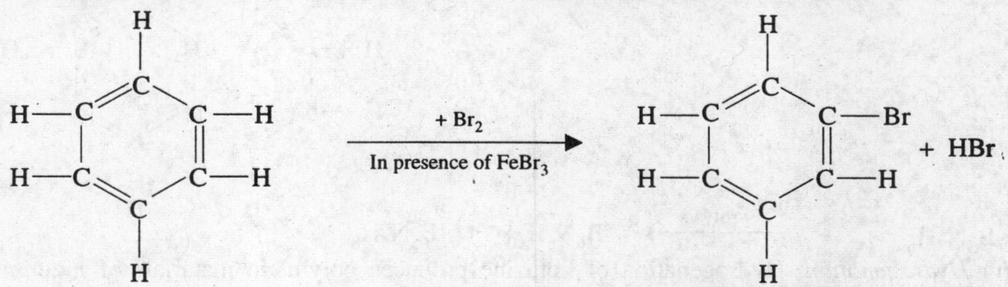

Bromobenzene

$$\text{or} \quad C_6H_6 \xrightarrow{+\,Br_2} C_6H_5\,Br + HBr$$

(*ii*) *Hydrolysis.* (*a*) Borazine gets slowly hydrolysed by water to produce boric acid [$B(OH)_3$ or H_3BO_3], NH_3 and H_2. Hydrolysis is favoured by the increase in temperature.

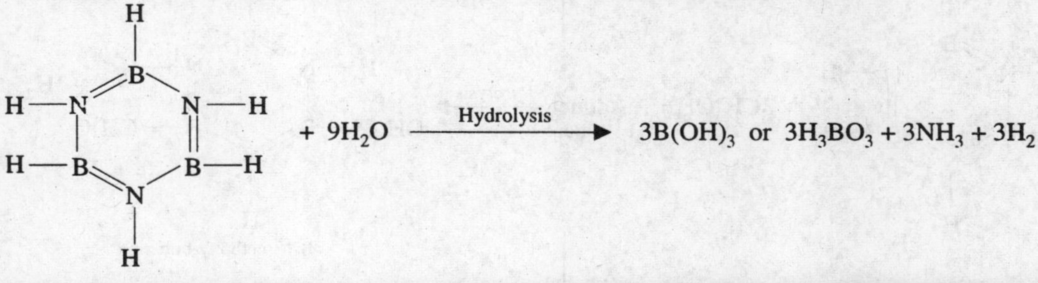

or $B_3N_3H_6 + 9H_2O \longrightarrow 3B(OH)_3 + 3NH_3 + 3H_2$

C_6H_6 does not show this type of reaction.

(*b*) It is reported that under proper conditions, borazine reacts with three molecules of water and gives B-trihydroxyl borazine, $B_3N_3H_3(OH)_3$ (*substitution reaction*).

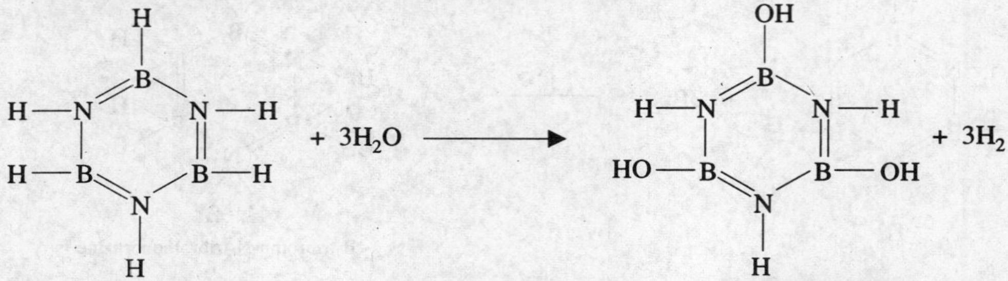

B-trihydroxyl borazine

or $B_3N_3H_6 + 3H_2O \longrightarrow B_3N_3H_3(OH)_3 + 3H_2$

(*iii*) *Pyrolysis.* When borazine is pyrolysed above 340°C, $B_6N_6H_{10}$ and $B_5N_5H_8$ are produced. These products are boron-nitrogen analogues of diphenyl and naphthalene respectively.

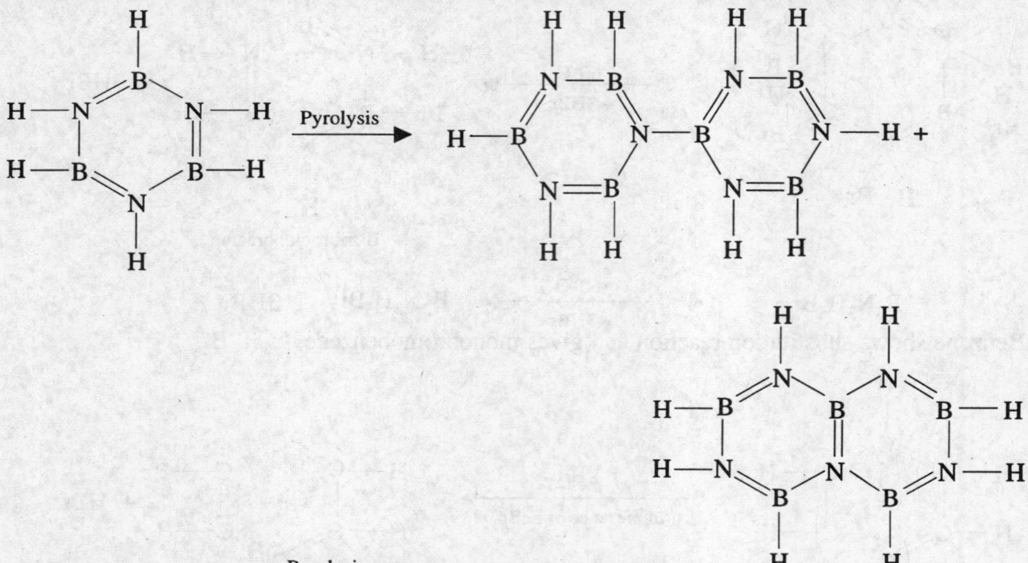

or $B_3N_3H_6 \xrightarrow{\text{Pyrolysis}} B_6N_6H_{10} + B_5N_5H_8$

(*iv*) *Hydrogenation.* Hydrogenation of borazine produces polymeric materials of indefinite composition.

(*v*) *Formation of adduct.* Borazine forms an adduct with CH_3OH. This adduct undergoes pyrolysis with the elimination of H_2 and gives B-trimethoxy-borazine.

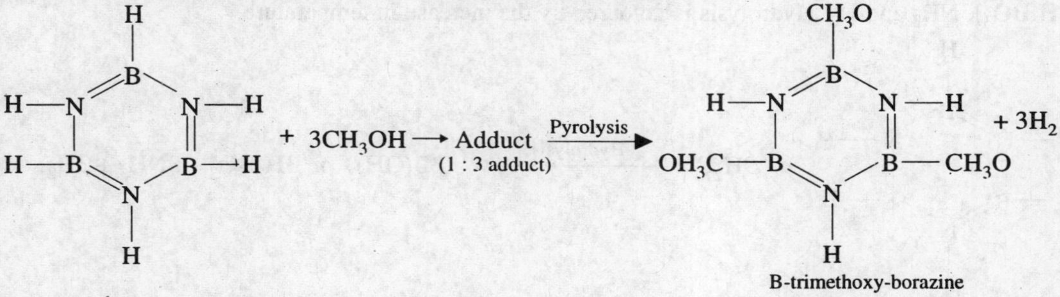

B-trimethoxy-borazine

(*vi*) *Reaction with aniline.* Borazine undergoes a strongly exothermic reaction with aniline, to produce tri-aminoborine,

H_2N and (C_6H_5) HN $>$ B—N—NH (C_6H_5)

Structure of borazine molecule. Various chemical reactions and electron diffraction study of borazine molecule have shown, that this molecule is isoelectronic with benzene and hence its structure is the same as that of benzene. Thus, like benzene, borazine has a *planar hexagonal structure*, containing six-membered ring, in which B and N atoms are arranged alternately (See Fig. 36.1). It is because of the similarity between the structures of borazine and benzene that borazine is called *inorganic benzene.*

In the structure of borazine, both B and N atoms are sp^2 hybridised. Each N-atom has one lone pair of electrons, while each B-atom has an empty *p*-orbital. (B—N) π-bond in borazine is a dative bond, which arises from the sidewise overlap, between the filled *p*-orbitals of N-atom and empty *p*-orbitals of B-atom.

Since borazine is isoelectronic with benzene, both the compounds have aromatic π-electron cloud (electrons in π-orbitals). Due to greater difference in the electronegativity values of B and N-atoms, the π-electron cloud in B_3N_3 ring of borazine molecule is *partially delocalised,* while in case of benzene ring, the π-electron cloud is *completely delocalised.* In fact, complete delocalisation of π-electron cloud in B_3N_3 ring in borazine molecule, cannot be expected, since N-π orbitals are of lower energy than the B-π orbitals. Molecular orbital calculations have indicated that π-electron drift from N to B is less than the σ-electron drift from B to N, due to greater electronegativity of N-atom. In C_6H_6 molecule, C = C bonds are *nonpolar,* while in case of $B_3N_3H_6$ molecule, due to the difference in electronegativities between B and N atoms, B—N bond is *polar.*

The ring structure of borazine molecule is the same as the layer lattice structure of boron nitride, $(BN)_n$. (See Fig. 36.1).

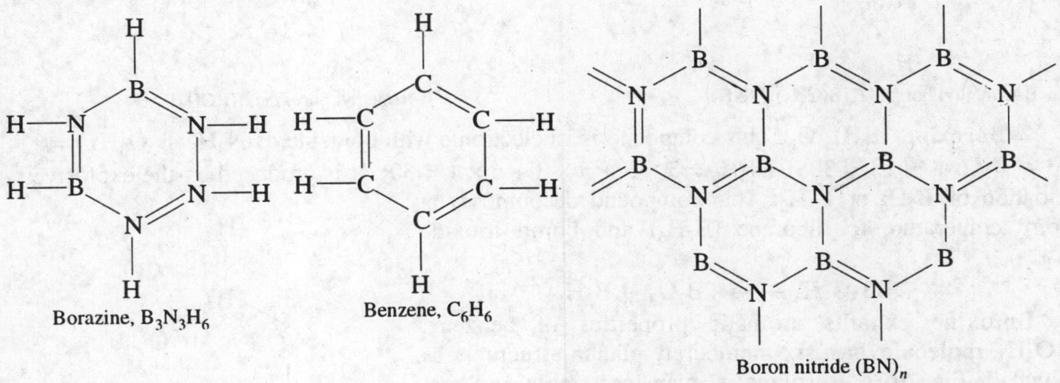

Fig. 36.1 : Comparison between the structures of borazine, benzene and boron nitride.

It is due to the partial delocalisation of the π-electron cloud that π-bonding in B_3N_3 ring is weakened. In addition, N-atom retains some of its basicity and boron atom retains some of its acidity. Polar species like HCl, therefore, attack the double bond between N and B. Thus, borazine, in contrast to C_6H_6, readily undergoes addition reactions

In these reactions, more electronegative atom (*e.g.,* Cl in HCl molecule) is generally attached with B-atom, which is less electronegative than N in B—N bond.

In borazine, B—N bond length is equal to 1.44 Å, which is between the calculated single B—N bond (= 1.54 Å) and double bond, B = N (= 1.36 Å) distances. The angles are equal to 120°. In benzene C—C bond length is equal to 1.42 Å.

B. Substituted borazines

Here, we shall discuss the following substituted borazines.

1. B-trimethyl borazine, $[B(CH_3)]_3\ (NH)_3$. It is prepared by heating $B(CH_3)_2$ with NH_3 at 320–340° at 20 atm. for 2 hours.

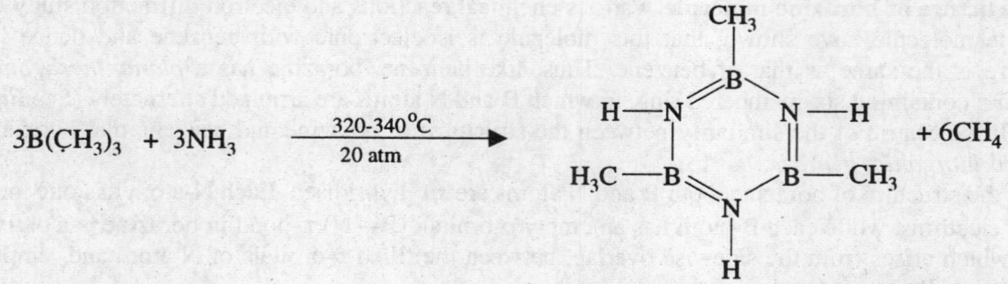

B-trimethyl borazine, $[B(CH_3)]_3\ (NH)_3$

At room temperature, this compound exists as colourless crystals. Its m.pt. and b.pt. are 31.8° and 127°C respectively. It is stable in vacuum upto 350°C. It is sensitive towards moisture, but insoluble in water. It is soluble in many organic solvents. At 100°C, water replaces the NH groups by O-atoms and gives B-trimethyl boroxine, $[B(CH_3)]_3\ O_3$.

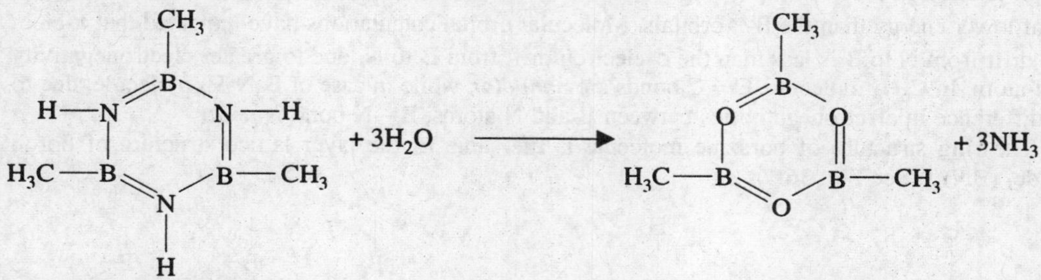

B-trimethyl borazine, $[B(CH_3)]_3\ (NH)_3$ B-trimethyl boroxine, $[B(CH_3)]_3\ O_3$

2. Boroxine, $(BH)_3\ O_3$. This compound is isoelectronic with borazine, $B_3N_3H_6$ ($B_3O_3H_3$) = 3 × 3 + 3 × 6 + 3 × 1 = 30, $B_3N_3H_6$ = 3 × 3 + 5 × 3 + 6 × 1 = 30). It is produced by the explosive oxidation of B_2H_6 or B_5H_9. This compound decomposes at room temperature to diborane (B_2H_6) and boron trioxide (B_2O_3).

$$2B_3O_3H_3 \longrightarrow 2B_2O_3 + B_2H_6$$

Boroxine exhibits aromatic properties of benzene. $B_3O_3H_3$ molecule has six-membered planar structure, as shown in Fig. 36.2. Boroxine is even less stable and presumably has less π-delocalisation than borazine. In this molecule, B—O bond distance is equal to 1.38 Å. The characteristic Roman frequency of the ring is at 807 cm^{-1}. B = O, double bond present in the structure is due to the donation of alone pair of electrons from O-atom to boron atom. This results in the development of a formal negative charge on B-atom and equal formal positive charge on O-atom. These charges have not been shown in the Fig. 36.2.

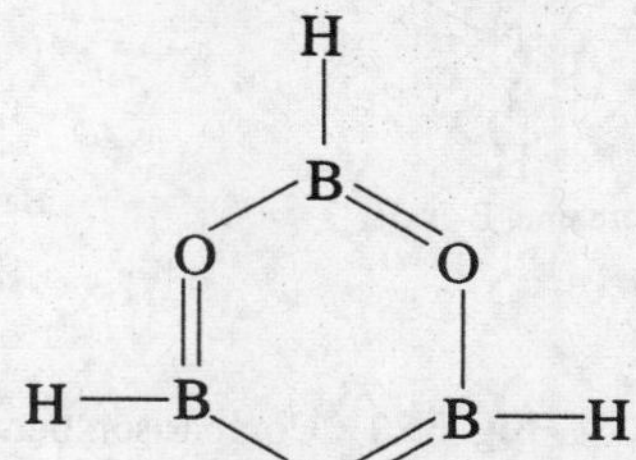

Fig. 36.2 : Six membered planar ring structure of boroxine molecule, $(BH)_3O_3$.

3. N-trimethyl borazine, $(BH)_3\ [N(CH_3)]_3$. It is obtained in 90% yield by heating a mixture of B_2H_6 and $NH_2.(CH_3)$ in the correct proportions at 180-200°C for 2 hours.

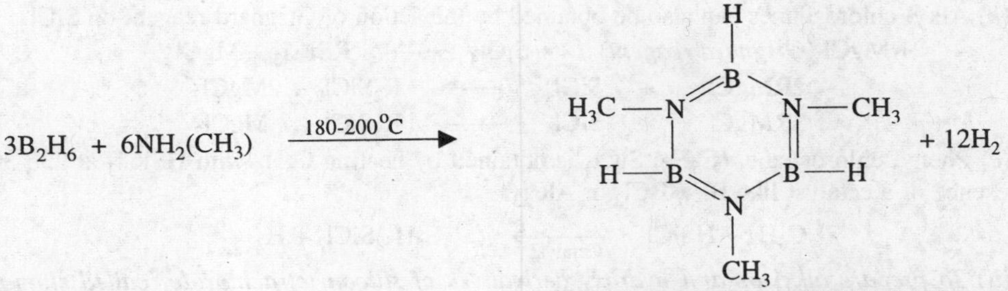

N-trimethyl borazine, $(BH)_3\ [N(CH_3)_3]$

This compound can also be prepared by reducing monomethyl ammonium chloride, $(CH_3)NH_3Cl$ with lithium borohydride, $LiBH_4$.

$$3(CH_3)NH_3Cl + 3LiBH_4 \longrightarrow (BH)_3\ [N(CH_3)]_3 + 3LiCl + 9H_2$$

It is a colourless mobile liquid having m.pt. = —9° C and b.pt. = 132°C. It is stable upto 500°C. At 60°C, H-atom attached with B-atom is replaced by OH group.

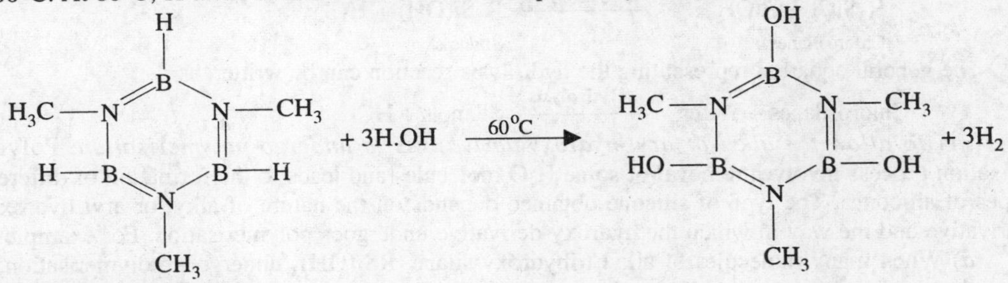

C. Boron nitride, BN

Methods of its preparation, properties, uses and structure have already been discussed in chapter 27.

Silicones : Polymers containing silicon

What are silicones?

Silicons are organo-silicon polymers containing ..... O—Si—O.... linkages. These may be *linear silicones cyclic silicones and cross-linked silicones.*

Preparation of silicones

These are prepared by the hydrolysis of alkyl or aryl derivatives of $SiCl_4$ like $RSiCl_3$, R_2SiCl_2 and R_3SiCl and polymerisation of alkyl or aryl hydroxy-derivatives obtained by hydrolysis. Thus, this method consists of the following steps :

(*i*) *To prepare alkyl or aryl derivatives of silicon tetrachloride.* Examples of such derivatives are $RSiCl_3$, R_2SiCl_2 and R_3SiCl where R is an alky (*e.g.*, CH_3, C_2H_5 etc.) or aryl (*e.g.*, C_6H_5) group. These derivatives are prepared by the following methods :

(*a*) Methyl chlorosilanes like $(CH_3)SiCl_3$, $(CH_3)_2SiCl_2$ and $(CH_3)_3SiCl$ are prepared by heating methyl chloride, $(CH_3)Cl$ with Si, catalysed by Cu, at 300°C. This reaction gives a mixture of methyl chlorosilanes.

$$2CH_3Cl + Si \xrightarrow[300°\ C]{Cu} \underbrace{(CH_3)\ SiCl_3 + (CH_3)_2\ SiCl_2 + (CH_3)_3\ SiCl}_{\text{Mixture of methyl chlorosilanes}} + \text{Other compounds}$$

The yield of $(CH_3)_2\ SiCl_2$ (b.pt. = 69.6°C) is over 50%. Careful fractionation is used to separate $(CH_3)_2SiCl_2$ from $(CH_3)\ SiCl_3$ (b.pt. = 66.9°C) and $(CH_3)_3\ SiCl$ (b.pt. = 87.7°C).

If a mixture of CH_3Cl and HCl is heated with Si, catalysed by Cu, at 300°C, then $(CH_3)\ SiCl_3$ is obtained as the main product.

$$CH_3Cl + 2HCl + Si \xrightarrow[300°\ C]{Cu} (CH_3)\ SiCl_3 + H_2$$

(*b*) Alkyl chlorosilanes can also be obtained by the action of Grignard reagent on $SiCl_4$.

$$RMgCl\ (\textit{Grignard reagent}) + SiCl_4 \longrightarrow RSiCl_3 + MgCl_2$$
$$2RMgCl + SiCl_4 \longrightarrow R_2SiCl_2 + 2MgCl_2$$
$$3RMgCl + SiCl_4 \longrightarrow R_3SiCl + 3\ MgCl_2$$

(*c*) Phenyl chlorosilane, $(C_6H_5)\ SiCl_3$ is obtained by heating C_6H_6 with $HSiCl_3$ at 230-300°C in presence of a catalyst like BF_3, BCl_3 or $AlCl_3$.

$$C_6H_6 + HSiCl_3 \xrightarrow[\text{Catalyst}]{\Delta} (C_6H_5)SiCl_3 + H_2$$

(*ii*) *To prepare alkyl or aryl hydroxy derivatives of silicon tetrachloride (called silanols or silandiols).* Examples of such silanols are R Si $(OH)_3$, $R_2Si(OH)_2$ and $R_3Si(OH)$. These silanols are obtained by the hydrolysis of $RSiCl_3$, R_2SiCl_2 and R_3SiCl respectively.

$$RSiCl_3 + 3H.OH \xrightarrow{\text{Hydrolysis}} RSi(OH)_3 + 3HCl$$
$$R_2SiCl_2 + 2H.OH \xrightarrow{\text{Hydrolysis}} R_2Si(OH)_2 + 2HCl$$
$$R_3SiCl + H.OH \xrightarrow{\text{Hydrolysis}} R_3Si(OH) + HCl$$

Chlorosilanes Silanols

The general equation representing the hydrolysis reaction can be written as :

$$\text{Chlorosilanes} + \text{Water} \xrightarrow{\text{Hydrolysis}} \text{Silanols} + HCl$$

(*iii*) *To allow the alkyl or aryl hydroxy derivatives to undergo polymerisation.* Polymerisation process involves removal of some H_2O molecules and leads to the formation of different types of silicones. The type of silicone obtained depends on the nature of alkyl or aryl hydroxy-derivative and the way in which the hydroxy-derivative undergoes polymerisation. For example :

(*a*) When many molecules of alkyl trihydroxy-silane, $RSi(OH)_3$ undergoes polymerisation, a cross-linked two dimensional silicone is obtained.

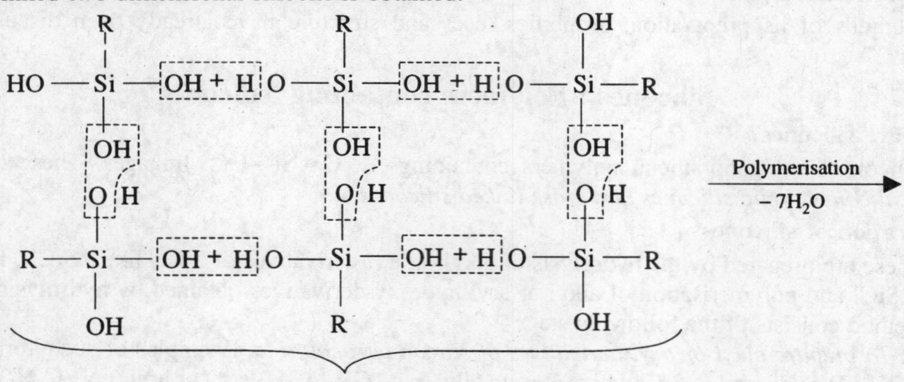

Six $RSi(OH)_3$ molecules

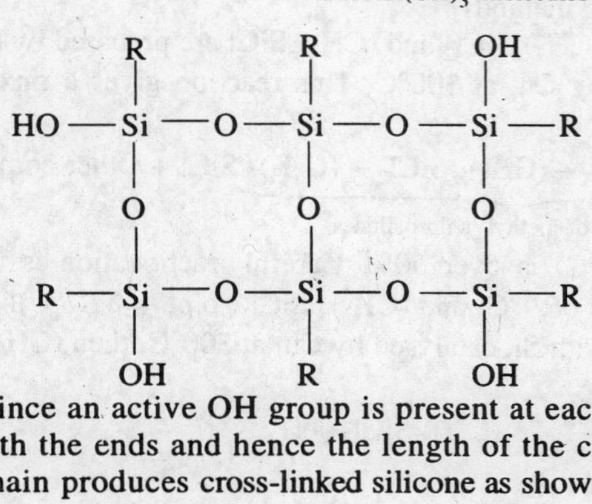

Since an active OH group is present at each end of the chain, polymerisation continues on both the ends and hence the length of the chain increases. The increase in the length of the chain produces cross-linked silicone as shown below :

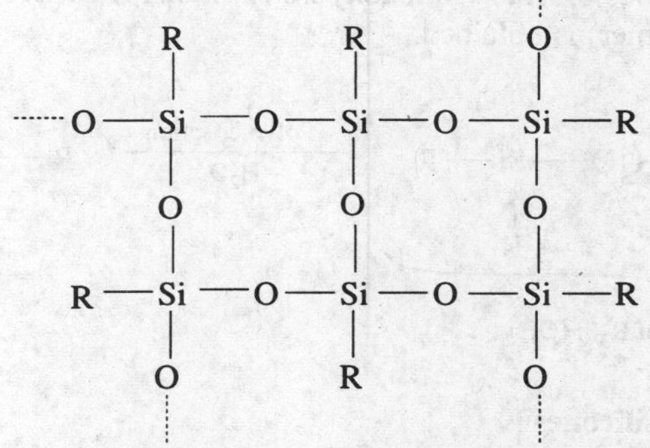

Two dimensional cross-linked silicone

(*b*) When many molecules of dialkyl dihydroxy-silane, $R_2Si(OH)_2$ undergo polymerisation, a straight chain (linear) or cyclic (ring) silicone is obtained.

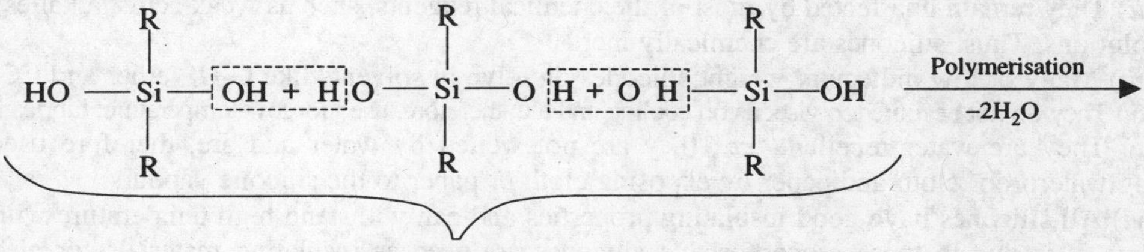

Three molecules of $R_2Si(OH)_2$

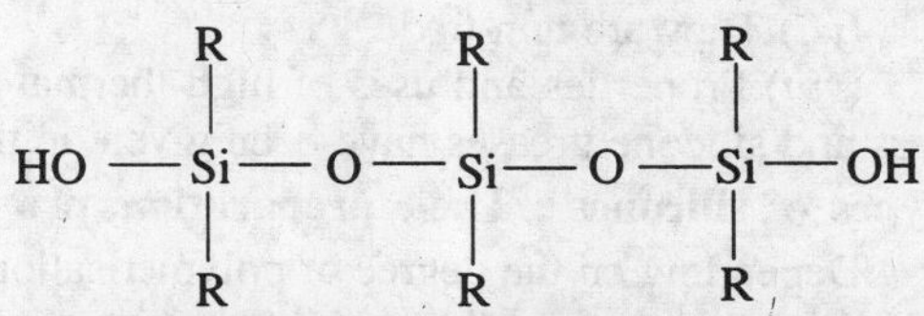

Since an active OH group is present at each end of the chain, polymerisation continues and hence the length of the chain increases and gives rise to the formation of long chain silicon, as shown below :

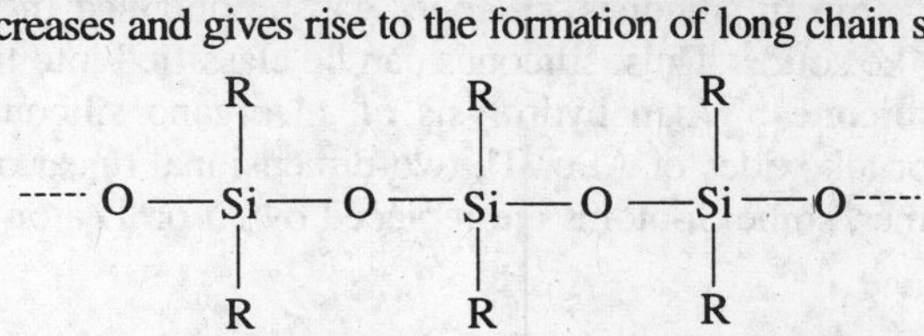

Linear or straight chain silicone (Thermoplastic polymer) (siloxane)

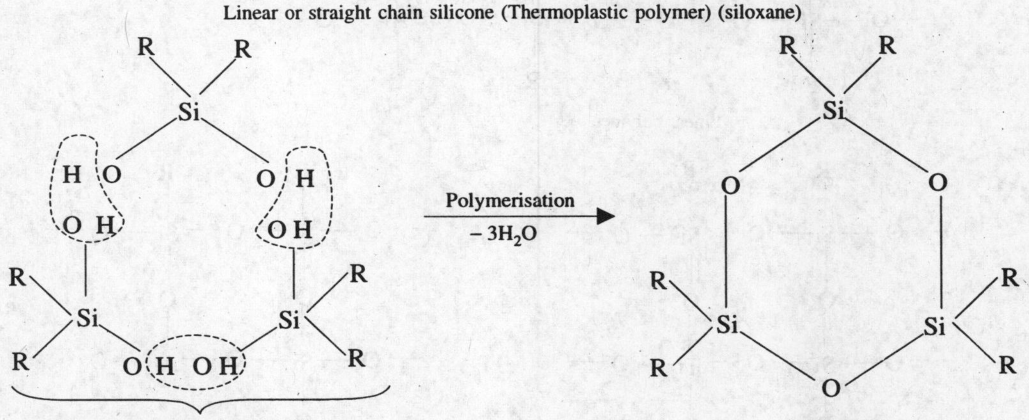

Three molecules of $R_2Si(OH)_2$ Cyclic trimer

(*c*) When two molecules of trialkyl monohydroxy-silane, $R_3Si(OH)$ undergo polymerisation, a straight chain silicone (dimer) is obtained.

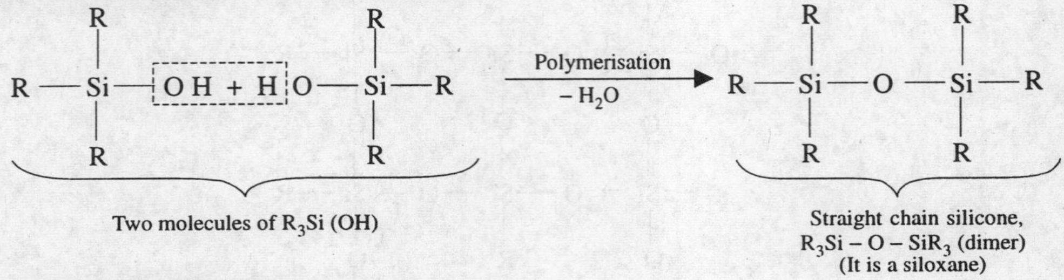

Properties and uses of silicones

The following properties and uses are common to all types of silicones :

(*i*) They have high thermal stability in the absence of air and withstand temperature upto 250–300°C.

(*ii*) They remain unaffected by most of the chemical reagents, such as weak acids, alkalies and salt solutions. Thus, silicones are chemically inert.

(*iii*) Many of low molecular weight silicones dissolve in solvents like C_6H_6, ether and CCl_4.

(*iv*) They do not become too viscous on cooling and are, therefore, used for low-temperature, lubrication.

(*v*) They are water repellent, *i.e.*, they are not wetted by water and are, therefore used in making waterproof cloth and paper by exposing cloth or paper to the silicone vapours.

(*vi*) All silicones have good insulating properties and can withstand high temperature without charring. It is due to these properties that silicones are used as insulating materials for electric motors and other electric appliances.

(*vii*) They are non-toxic.

(*viii*) Properties and uses of high-thermal silicones, silicone resins, silicone oils, silicone rubbers and silicone greases have been given on the subsequent pages.

Types of silicones : Their preparation, properties and uses

Depending on the degree of polymerisation undergone by the alkyl or aryl hydroxy-derivatives of $SiCl_4$, the length of the straight chain in cross-linked silicones and the nature of the alkyl or aryl groups attached with Si-atom in silicones, silicones can be obtained in the form of oils, viscous fluids, resins or rubber-like solids. Thus, silicones can be classified into the following types :

1. High thermal silicones. When hydrolysis of an organo silicon halide is carried out in presence of the halides or alkoxides of Al or Ti, two-dimensional linear or cyclic silicone polymer is obtained. In this polymer, some Si-atoms are replaced by Al or Ti atoms.

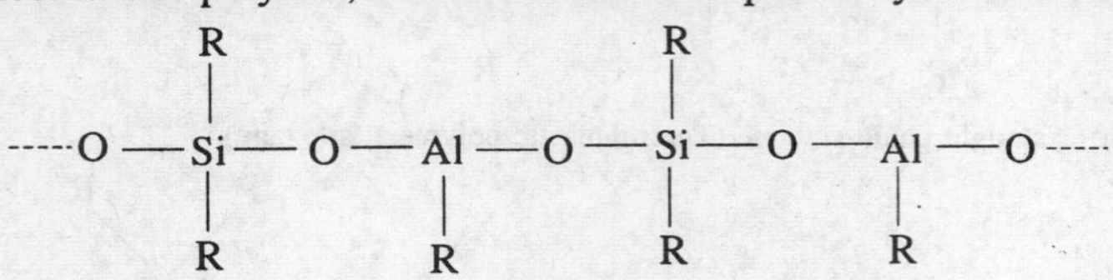

Linear polymer

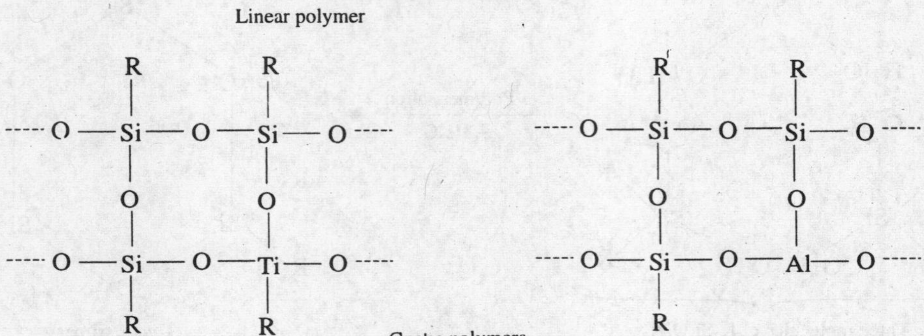

Cyclic polymers

High themal silicones have exceptionally high thermal stability and remain unchanged, even in contact with a white-hot electrically-heated wire. The presence of Al and Ti atoms in the structure of the polymers increases the thermal stability of the polymers.

In 1960, Brown *et al* reported the synthesis of a linear polymer. This polymer has exceptionally high thermal stability. Its structure can be represented as :

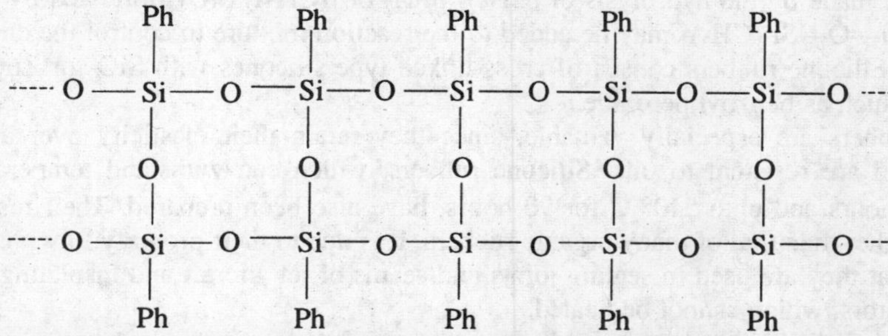

2. Silicone resins. Silicone resin is obtained by blending silicone with organic resin, such as acrylic ester. If the hydrolysis of $(CH_3)_2\ SiCl_2$ is carried out in presence of $(CH_3)SiCl_3$ and then polymerisation is allowed, a rigid silicone is obtained. This silicone is called *silicone resin.* These are of many types, like coating resins, laminating resins, release resins, water-repellent resins, molding resins and electrical resins. These are stable to heat, water-repellent and have good electrical conductivity, chemical inertness and weather resistance. To develop best properties in them, baking or heat treatment is given to them. These can withstand temperature as high as 250°C and coating made up of silicone resin plus organic resins and Al can withstand temperature upto 500°C.

Release resins are baked on metal surfaces, to prevent materials from sticking to these surfaces. Silicone resin solutions are mixed in paints, pigments and enamels, to make them resistent to the effects of high temperatures, sunlight, chemicals and damp. Silicone resin, mixed with the pigment like Al-flakes, can withstand 400°C. Silicone resins, having TiO_2 (TiO_2 is a pigment) resist dissociation, due to heat and weathering. These silicone resins are not affected by soap and fruit juices.

3. Silicone fluids or oils. A silicone fluid is a mixture of equal proportions of linear and cyclic polymers. These have low temperature-coefficient of viscosity, which makes them valuable hydraulic fluids. They have good insulating properties and high thermal stability. It is due to these properties that silicone fluids are used as dielectric fluids.

Emulsions of silicone fluids in water are made for convenience in applying small amounts to silicone textiles, paper or other surfaces. Silicone oils are highly stable and non-volatile, even on heating. They are, therefore, used for high temperature oil baths, high vacuum pumps and low temperature lubrication. Variation in the length of the chain gives silicone oils varying viscosities. The viscosity of silicone oils remains constant, even with the change of temperature and as such these silicones do not thicken in cold weather. Generally, low molecular weight silicones are oily liquids.

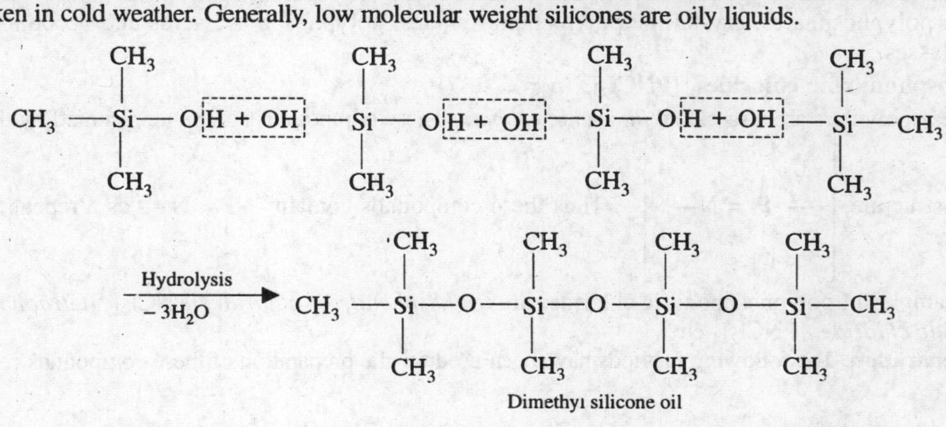

Dimethyl silicone oil

Preparation of dimethyl silicone oil. If a mixture containing $(CH_3)_3SiCl$ and $(CH_3)_2SiCl_2$, in proper ratio, is hydrolysed and the hydroxy-derivatives, thus obtained, are polymerised, then dimethyl silicone oil is obtained. This compound contains CH_3 group at each end of the chain. It is a linear silicone.

4. Silicone rubbers. These are long chain polymers, with some cross-linking between the chains. They are made by the hydrolysis of $(CH_3)_2\ SiCl_2$ or $[(CH_3)_2\ SiO]_4$ followed by polymerisation. $(CH_3)_3\ Si—O—Si\ (CH_3)_3$ may be added to the reaction mixture to control the chain length of the polymer. Silicone rubbers consist of cross-linked type silicones with SiO_2 or ZnO as filler and vulcanizer, such as benzoyl peroxide.

Silicone rubbers are especially valuable, since they retain their elasticity over a range of temperatures and are resistent to oils. Silicone rubbers, which can withstand temperature upto 400°C for 200 hours and upto 540° C for 90 hours, have also been prepared. Their resistance to heat is much higher than that of many organic rubbers. It is due to their property to withstand high temperatures, that they are used in sealing joints and seams of jet aircraft and insulating electrical parts like transistors, which cannot be heated.

Silicone rubber retains its shape and elasticity permanently, even after vulcanisation and has, therefore, been used in a number of ways. High molecular weight silicones, containing long chain or ring structure, are generally waxy or rubber-like solids.

5. Silicone greases. These are made by adding carbon black, special soaps or finely dispersed SiO_2 to silicone oils. Silicone grease, prepared from medium phenyl oil and lithium soap, is used for ball bearings operating at high speed and high temperature. Silicone greases are vaseline-like, These are used as lubricants in aeroplanes, since they do not freeze at as low temperature as – 40°C and do not melt at 200°C.

Polymers containing phosphorus

Depending on the type of bonding existing between their atoms, these polymers can be of the following two types :

1. Chain polymers. These polymers include (*i*) polyphosphazines, $[NPR_2]_n$ in which R = Cl, OCH_3 or OC_2H_5. Polyphosphazines may be of three types namely (*a*) polyphosphonitrilic chlorides, $[NPCl_2]_n$ (*b*) polydimethoxy-phosphazines, $[NP(OCH_3)_2]_n$ (*c*) poly diethoxy-phosphazines, $[NP(OC_2H_5)_2]_n$ (*ii*) polyphosphoryl chlorides, $P_n\ O_{2n-1}\ Cl_{n+2}$ (*iii*) vitreous polyphosphates (phosphate glasses) (*iv*) crystalline polymetaphosphates. Examples are Maddrell's and Kuroll's salts (*v*) platicised and flexible phosphates (*vi*) polyphosphoric acids.

2. Network polymers. These polymers include (*i*) polymeric phosphorus pentoxide (*ii*) polymeric boron orthophosphates, $[BPO_4]_n$ (*iii*) polymeric aluminium orthophosphate, $[AlPO_4]_n$ (*iv*) polymeric silver orthophosphate, $[Ag_3PO_4]_n$ (*v*) polymeric ferrous orthophosphate, $[Fe_3(PO_4)_2.8H_2O]_n$ (*vi*) sulphur-phosphorus polymers (*vii*) ultraphosphate glasses (*viii*) boron phosphate glasses. Here, we shall discuss only polyphosphonitrilic chlorides, vitreous polyphosphates, crystalline polymetaphosphates, polyphosphoric acids and borophosphate glasses.

Polyphosphonitrilic chlorides, $[PNCl_2]_n$ (n = 3 to 7)

These polymers are also called *inorganic rubbers.* These compounds can be represented by the general structure, $\left[\begin{array}{c} Cl \\ | \\ —P=N— \\ | \\ Cl \end{array}\right]_n$. Thus these compounds contain $\begin{array}{c} Cl \\ | \\ —P=N— \\ | \\ Cl \end{array}$ as a repeating unit. Examples of polyphosphonitric chlorides are *triphosphonitrilic chloride,* $[PNCl_2]_3$; *tetraphosphonitrilic chloride,* $[PNCl_2]_4$ etc.

Preparation. The following methods have been used for the preparation of these compounds :

(*i*) $(NPCl)_3$ and $(NPCl_2)_4$ can be prepared by the ammonolysis of PCl_5.

$$3PCl_5 + 3NH_3 \longrightarrow (NPCl_2)_3 + 9HCl\ ;\ 4PCl_5 + 4NH_3 \longrightarrow (NPCl_2)_4 + 12HCl$$

(*ii*) These compounds can be prepared by the reaction between PCl_5 and NH_4Cl in presence of $C_2H_4Cl_2$ or C_6H_5Cl or by heating PCl_5 with solid NH_4Cl at 145-160° C.

$$nPCl_5 + nNH_4Cl \xrightarrow[\text{or } C_6H_5Cl]{C_2H_4Cl_2} (NPCl_2)_n + 4nHCl$$

$$nPCl_5 + nNH_4Cl\ (s) \xrightarrow{145-160^\circ} (NPCl_2)_n + 4nHCl$$

Both the above reactions produce a mixture of various phosphonitrilic chlorides, but under controlled conditions, high yields of $(NPCl_2)_3$ and $(NPCl_2)_4$ can be obtained. These two compounds can be seperated from each other by using the fact that $(NPCl_2)_3$ sublimes in vacuum at 50°C as a white crystalline solid whereas $(NPCl_2)_4$ does not do so under these conditions.

The formation of $(NPCl_2)_n$ takes place through the following steps :

(*a*) $NH_4Cl + PCl_5 \longrightarrow NH_4\ PCl_6$ (*Ammonium hexachlorophosphate—an unstable compound*)

(*b*) $NH_4\ PCl_6 \xrightarrow{\text{Decomposition}} HN.\ PCl_3 + 3HCl$

(*c*) $nHN.\ PCl_3 \xrightarrow[(-\,nHCl)]{\text{Condensation}} (NPCl_2)_n + nHCl$

(*iii*) S_4N_4 reacts with $SOCl_2$ in PCl_3 to give $(NPCl_3)_2$.

$$S_4N_4 + 6SOCl_2 + 12PCl_3 \longrightarrow 4(NPCl_2)_3 + 10\ S + 3O_2 + 12Cl_2$$

Properties. (1) $(NPCl_2)_3$ melts at 114° C, boils at 256°C at 1 atm pressure and is readily soluble in benzene, ether and carbon tetrachloride. $(NPCl_2)_4$ melts at 123.5°C, boils at 328.5°C and has a lower solubility in benzene, ether and carbon tetrachloride than $(NPCl_2)_3$.

(2) $(NPCl_2)_3$ and $(NPCl_2)_4$ are irritating and some what toxic also.

(3) Cylic trimer, $[NPCl_2]_3$ or tetramer, $[NPCl_2]_4$, when heated in vacuum at 250°C gives a chain polymer.

$$n[NPCl_2]_3 \quad \text{or} \quad n[NPCl_2]_4 \xrightarrow{250^\circ C} Cl\!-\!\left[\begin{array}{c} Cl \\ | \\ P=N \\ | \\ Cl \end{array}\right]_n\!-\!H$$

chain polymer

Chain polymer obtained in this way is somewhat cross-linked and hence has low solubility in organic solvents.

(4) *Reactions involving replacement of Cl-atom of P—Cl bond.* Cl-atom in P–Cl bond in fairly reactive and can be replaced by many nucleophiles. The mechanism of this replacement is not well understood. However, it appears to involve S_{N^2} attack on P-atom. The following reactions illustrate the replacement of Cl-atom.

(*i*) $(NPCl_2)_3 + 6NaF \xrightarrow{\text{Acetonitrile}} (NPF_2)_3 + 6\ NaCl$

(*ii*) $[NPCl_2]_3 + 2NaOCH_3 \longrightarrow [NP\ (OCH_3)_2]_3 + 2NaCl$

Tri (dimethoxy) phosphazine

$[NPCl_2]_3 + 2NaC_2H_5 \longrightarrow [NP\ (OC_2H_5)_2]_3 + 2NaCl$

Tri (diethoxy) phosphazine

(*iii*) $[NPCl_2]_3 + 6C_6H_5MgBr \xrightarrow[\text{toluene}]{\text{Boiling}} [NP\ (C_6H_5)_2]_3 + 3MgCl_2 + 3MgBr_2$

Grignard reagent

(*iv*) $[NPCl_2]_3 + 3C_6H_4(OH)_2 \xrightarrow{Et_3\ N} [NP\ (C_6H_4O_2)]_3 + 6HCl$

1, 2-phenol

or 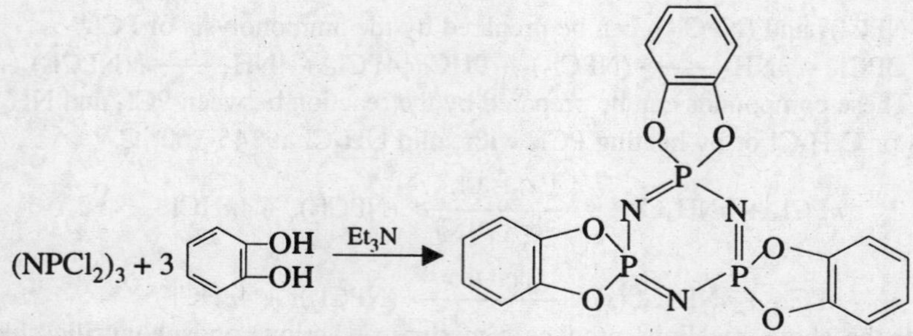

(*v*) *Hydrolysis.* (*a*) When $(NPCl_2)_3$ reacts with H_2O (aqueous ether), all Cl-atoms are replaced by OH groups.

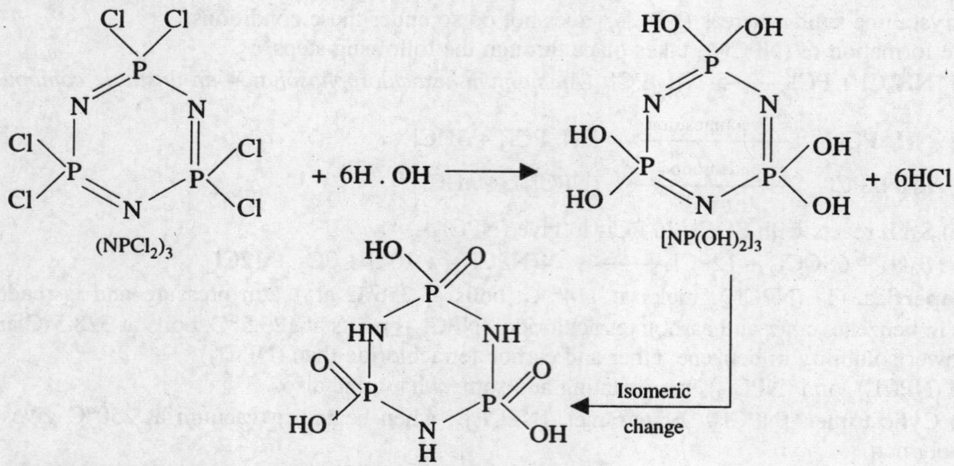

$(NPCl_2)_3$ in acid solution is hydrolysed to H_3PO_4 and NH_3.

(*b*) Hydrolysis of $(NPCl_2)_4$ in boiling water is rapid and a stable product is obtained.

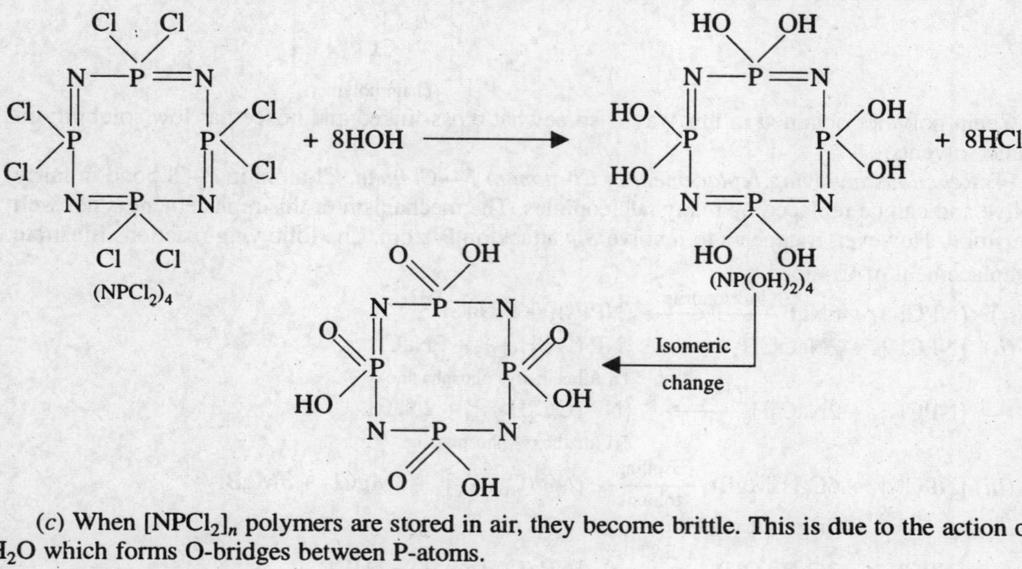

(*c*) When $[NPCl_2]_n$ polymers are stored in air, they become brittle. This is due to the action of H_2O which forms O-bridges between P-atoms.

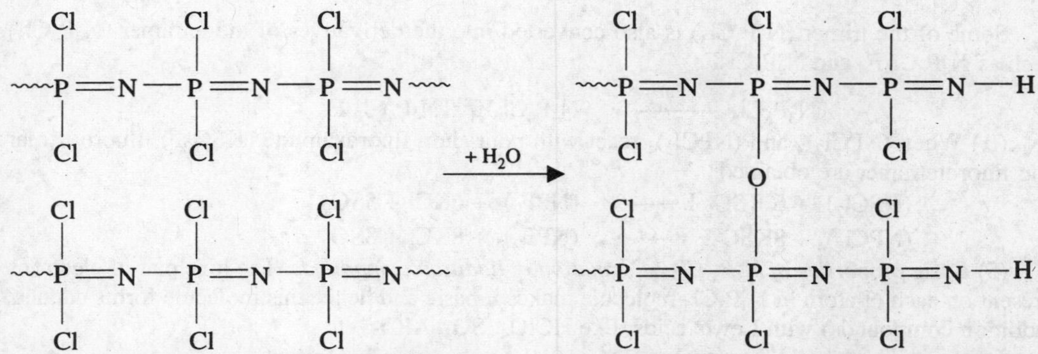

(*vi*) *Reaction with NH_3.* (*a*) When $[NPCl_2]_3$ reacts with *excess* of NH_3, all the six Cl-atoms are replaced by NH_2 groups and $N_3P_3(NH_2)_6$ is produced. This compound loses $3NH_3$ molecules, forming phospham, $N_3P_3\ (NH)_3$. $[NPCl_2]_4$ undergoes similar reaction with excess of NH_3.

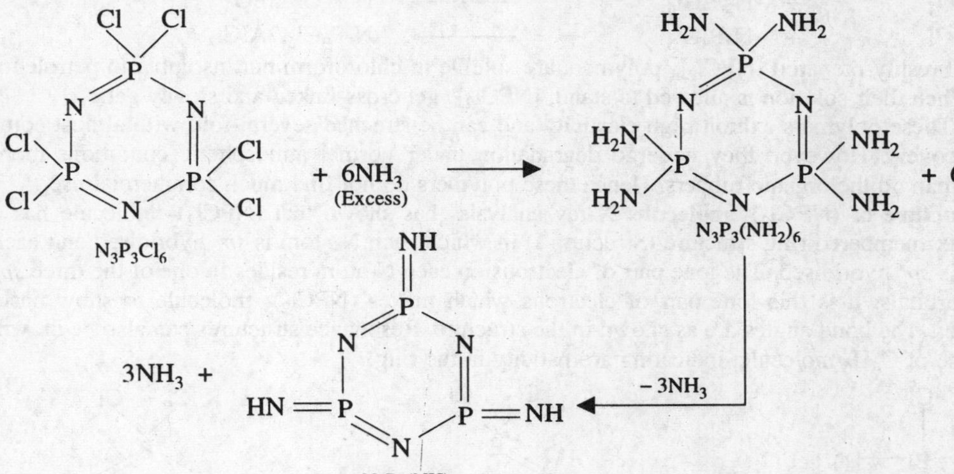

$N_3P_3Cl_6$ $N_3P_3(NH_2)_6$ $N_3P_3(NH)_3$

(*b*) When $[PNCl_2]_n$ reacts with NH_3 (*not excess*) and amines, cross-linked polymers are produced.

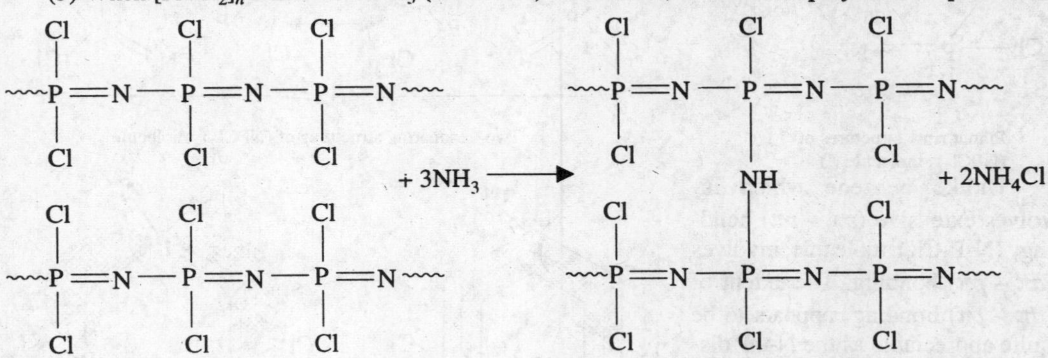

Cross-linked polymer

(*vii*) *Reaction with benzene, C_6H_6 (Friedel-Crafts reaction).* C_6H_6 reacts with $N_3P_3Cl_6$ in presence of $AlCl_3$ (Friedal-Crafts reaction) to form diphenyl derivative, $N_3P_3Cl_4\ (C_6H_5)_2$.

$$N_3P_3Cl_6 + 2\ C_6H_6 \xrightarrow{AlCl_3} N_3P_3\ Cl_4\ (C_6H_5)_2 + 2HCl$$

(*viii*) PbF_2 fluorinates $N_3P_3Cl_6$ (*trimer*), giving ultimately $N_3P_3F_6$.

$$N_3P_3Cl_6 + 3PbF_2 \longrightarrow N_3P_3F_6 + 3PbCl_2$$

Some of the trimer ($N_3P_3Cl_3$) is also converted into the derivatives of the tetramer ($N_4P_4Cl_8$) such as $N_4P_4Cl_4F_4$ and $N_4P_4Cl_2F_6$.

$$N_3P_3Cl_6 \xrightarrow{+PbF_2} N_4P_4Cl_4F_4 + N_4P_4Cl_2\,F_6$$

(*ix*) When $(NPCl_2)_3$ and $(NPCl_2)_4$ react with potassium fluorosulphite (KSO_2F), fluoro trimer and fluorotetramer are obtained.

$$(NPCl_2)_3 + 6\,KSO_2\,F \longrightarrow (NPF_2)_3 + 6KCl + 5SO_2$$
$$(NPCl_2)_4 + 8KSO_2F \longrightarrow (NPF_2)_4 + 8KCl + 8SO_2$$

(5) *Basic properties of $N_3P_3\,Cl_5$: Formation of addition compounds.* The lone pair of electrons present on each N-atom in $N_3P_3Cl_6$ molecule makes it basic and hence this molecule forms adducts (addition compounds) with Lewis acids, like $HClO_4$, SO_3, $AlCl_3$, etc.

Lewis acid		*Lewis base*		*Adduct*
$HClO_4$	+	$N_3P_3Cl_6$	$\longrightarrow$	$N_3P_3Cl_6.HClO_4$
$2HClO_4$	+	$N_3P_3Cl_6$	$\longrightarrow$	$N_3P_3Cl_6.2HClO_4$
$3\,SO_3$	+	$N_3P_3Cl_6$	$\longrightarrow$	$N_3P_3Cl_6.3SO_3$
$2AlCl_3$	+	$N_3P_3Cl_6$	$\longrightarrow$	$N_3P_3Cl_6.2AlCl_3$

(6) Freshly-prepared $[NPCl_2]_n$ polymers are soluble in chloroform but insoluble in petroleum ether. When their solution is allowed to stand, $[NPCl_2]_n$ get cross-linked and slowly gel.

(7) These polymers exhibit high elasticity and can be streched several-fold with almost complete recovery. However, they undergo degradation under normal atmospheric conditions more rapidly than do the organic rubbers. Hence these polymers do not find much commercial use.

Structure of $(NPCl_2)_3$ molecule. X-ray analysis has shown that $(NPCl_2)_3$ molecule has a planar six-membered ring structure (Structure I) in which each N-atom is sp^2 hybridised and each P-atom is sp^3 hybridised. The lone pair of electrons on each N-atom resides in one of the three sp^2 hybrid orbitals. It is this lone pair of electrons which makes $(NPCl_2)_3$ molecule to show basic properties. The bond angles are as shown in the structure. Resonance structures can also be drawn, as in case of C_6H_6 molecule, indicating aromaticity in the ring.

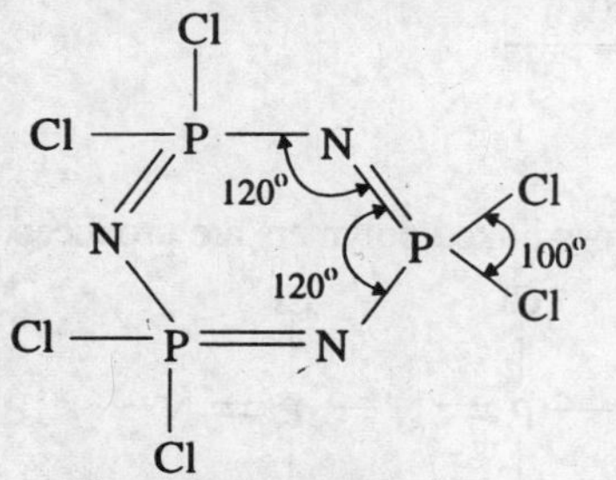

Planar ring structures of $(NPCl_2)_3$ molecule (I)

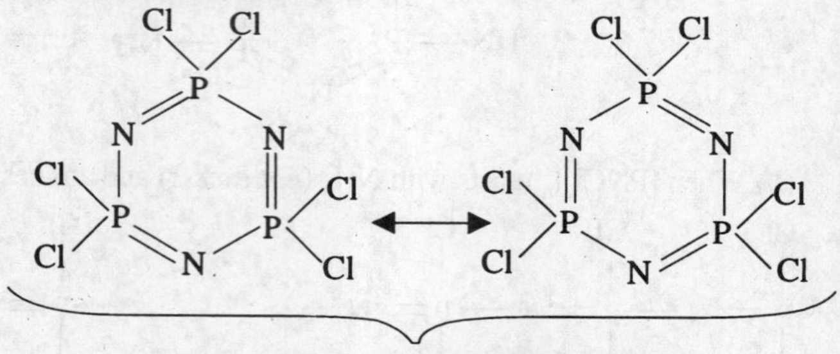

Two resonating structures of $(NPCl_2)_3$ molecule

Unlike benzene which involves extensive ($p\pi - p\pi$) bonding, ($N_3P_3Cl_6$) molecule involves ($d\pi - p\pi$) bonding. The extent of ($d\pi - p\pi$) bonding appears to be quite appreciable as the N—P distance (~ 1.6 Å) is considerably shorter than the N—P single bond distance (1.75 – 1.80 Å). Whether there is complete delocalisation of π-electron charge cloud on all the ring atoms as in C_6H_6 molecule or there are intensely-localised is-

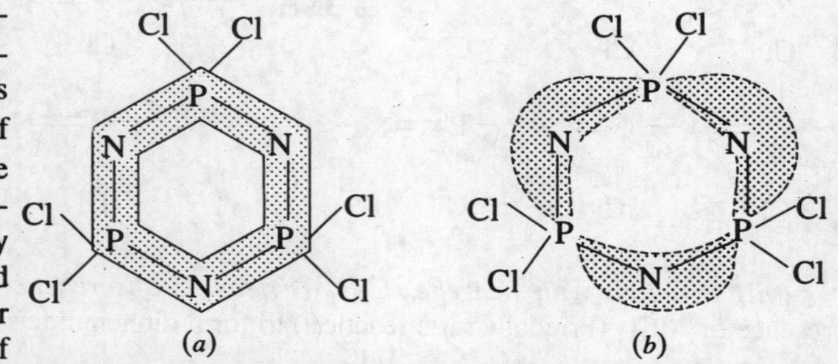

Fig. 36.3 : (*a*) Structure of $(NPCl_2)_3$ molecule having complete delocalisation of π-electron charge cloud on all the ring atoms (*b*) structure of $(NPCl_2)_3$ molecule having intensely-localised islands of π-electron charge cloud within the PNP segments.

lands of the electron cloud within the PNP segments (See Fig. 36.3) cannot be answered with certainity.

Structure of $(NPCl_2)_4$ molecule. This molecule has a tube-like puckered structure that has been shown in Fig. 36.4.

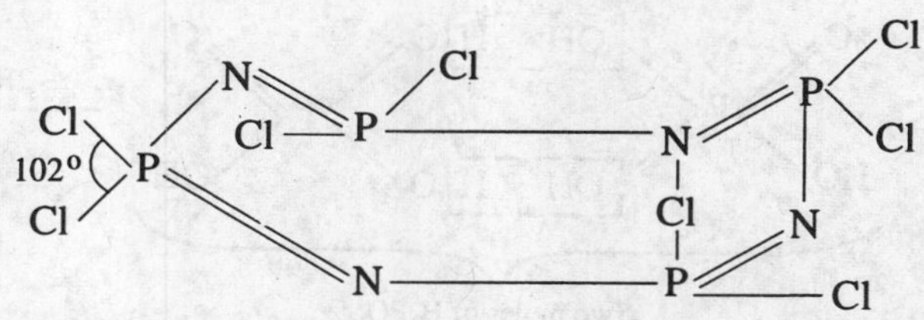

Fig. 36.4 : Tube-like packered structure of $(NPCl_2)_4$ molecule.

Linear phosphonitrilic halides or inorganic rubber. When PCl_5 and NH_4Cl are heated together in a refluxing S-tetrachloroethane, a product is obtained. Some portion of this product is insoluble in petroleum ether and has the composition, $[NPCl_2]_n$. PCl_5, where n = 10 to 15. This product is rubber-like and hence is called inorganic rubber. The structure of this compound can be represented as shown in the margin.

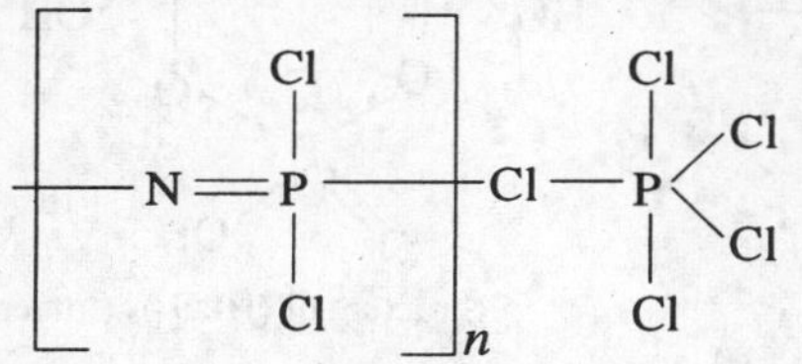

Structure of $[NPCl_2]_n$. PCl_5

This structure has been confirmed by the fact that ^{31}P *n.m.r.* spectrum of this compound has two peaks.

A similar product is formed when $(NPCl_2)_3$ is heated in a sealed tube at 350° with PCl_5. If inorganic rubber is kept in air for several hours, it becomes brittle due to hydrolytic condensation by atmospheric moisture and hence a chain-to-chain cross-linking through O-atoms takes place.

Uses of phosphonitrilic halides. These compounds find uses in flame-proofing of fabrics, as plasticisers and as catalysts in the manufacture of silicones.

Vitreous polyphosphates : Phosphate glasses

In these polymers, the negative charges present on the polyphosphate anions are balanced by the positive charges present on the cations like Na^+, K^+ etc., as shown below :

The polyphosphate anions present these polymers have long straight chains, but one P-atom out of 1000 P-atoms forms a cross-linked between the chains. These cross-links disappear within 12 hours when the polymer is dissolved in water or in a neutral aqueous salt solution.

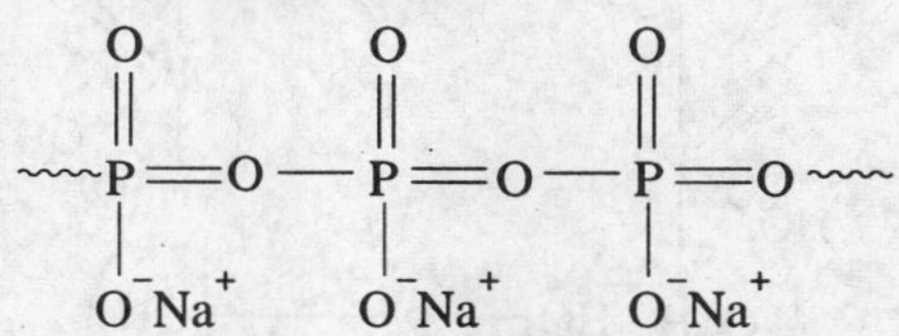

Vitreous sodium polyphosphate can be obtained by heating $Na_4P_2O_7$ or Na_3PO_4 with P_2O_5.

The short chain vitreous polyphosphate polymers are used for deflocculation of proteins. The long chain vitreous sodium phosphate polymers are used in food processing industry. Their use in this industry is based on the interaction of phosphates with proteins.

Crystalline polymetaphosphates

Polymetaphosphates are the salts of polymetaphosphoric acids. These acids are represented by the general formula $H_nP_nO_{3n}$ or $(HPO_3)_n$ and can be regarded as having been obtained by eliminating n moles of H_2O from n moles of H_3PO_4. Thus :

$$nH_3PO_4 \xrightarrow{-n\,H_2O} H_nP_n\,O_{3n} \quad \text{or} \quad (HPO_3)_n + nH_2O$$

The above discussion shows that $H_2P_2O_6$ or $(HPO_3)_2$, $H_3P_3O_9$ or $(HPO_3)_3$ and $H_4P_4O_{12}$ or $(HPO_3)_4$ can be obtained as :

$$2H_3PO_4 \xrightarrow{-2\,H_2O} H_2P_2O_6 \quad \text{or } (HPO_3)_2 \textit{ (dimetaphosphoric acid) (dimer)}$$

$$3H_3PO_4 \xrightarrow{-3\,H_2O} H_3P_3O_9 \quad \text{or } (HPO_3)_3 \textit{ (trimetaphosphoric acid) (trimer)}$$

$$4H_3PO_4 \xrightarrow{-4\,H_2O} H_4P_4O_{12} \quad \text{or } (HPO_3)_4 \textit{ (tetrametaphosphoric acid) (tetramer)}$$

Polymetaphosphoric acids and polymetaphosphates have cyclic structure. The structures of $(HPO_3)_2$ (dimer), $(HPO_3)_3$ (trimer) and $(HPO_3)_4$ (tetramer) are shown below in Fig. 36.5.

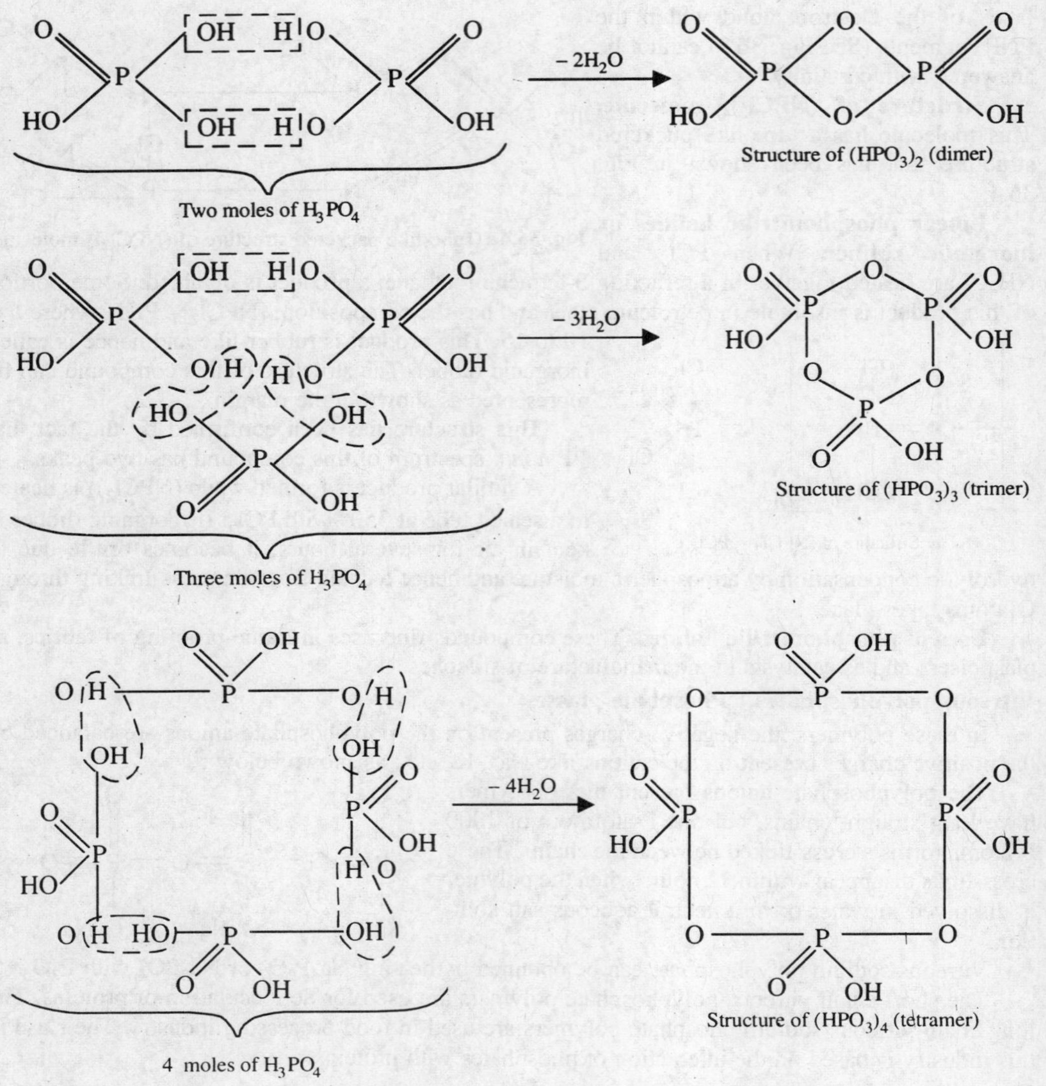

Fig. 36.5 : Structures of $(HPO_3)_2$, (HPO_3) and $(HPO_3)_4$ molecules.

The structure of trimetaphosphate ion, $[(PO_3)_3]^{3-}$ or $P_3O_9^{3-}$ has been shown in Fig. 36.6. The ring strucutre of $[(PO_3)_3]^{3-}$ ion has been established by the X-ray analysis and ^{31}P *n.m.r.* spectrum *of the trimetaphosphates of several cations. The* ^{31}P *n.m.r.* spectrum shows a single peak which is splitted into a triplet by two neighbouring phosphorus atoms in the ring, thereby showing that the three P-atoms are equivalent.

The structure of tetrametaphosphate ion, $[(PO_3)_4]^{4-}$ or $P_4O_{12}^{4-}$ can be obtained by removing 4H-atoms from 4OH groups present in $(HPO_3)_4$ molecule. Thus $[(PO_3)_4]^{4-}$ ion has the structure which has been shown in Fig. 36.6. It has been shown by X-ray diffraction study and ^{31}P *n.m.r.* spectrum of $Na_4P_4O_{12}$ that $P_4O_{12}^{4-}$ ion has a puckered eight-membered ring structure with O and P-atoms in alternate positions. Each P-atom is attached with more two O-atoms. $P_4O_{12}^{4-}$ ion can exist in both *boat* and *chair* forms.

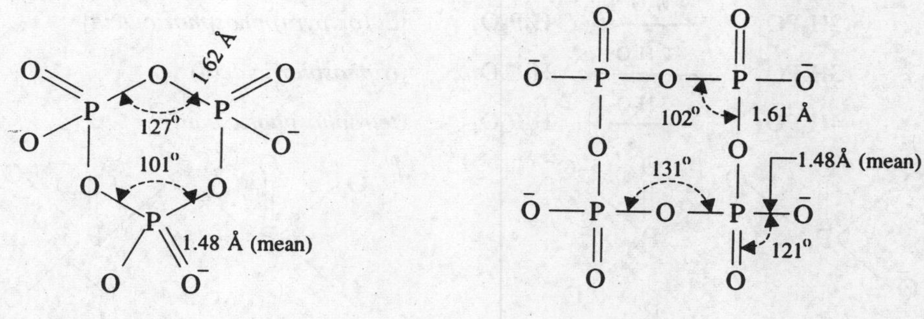

(*a*) (*b*)

Fig. 36.6 : (*a*) Structure of trimetaphosphate ion $[(PO_3)_3]^{3-}$ or $P_3O_9^{3-}$ (*b*) Structure of tetrametaphosphate ion, $[(PO_3)_4]^{4-}$ or $P_4O_{12}^{4-}$.

The above discussion shows that polymetaphosphate ions, $[(PO_3)_n]^{n-}$ have *cyclic (ring) structures* in which three or more PO_4 tetrahedra are linked together through two O-atoms to form a ring. The value of *n* varies from 3 to 8.

It is due to the cyclic structure of the metaphosphoric acids that these acids are also called *cyclic polymetaphosporic acids.*

Insoluble crystalline sodium or potassium polymetaphosphates have been prepared by the condensation of $Na_2H_2P_2O_7$ or $K_2H_2P_2O_7$. By using different condensation conditions, a number of polymetaphosphates having the same general formula, MPO_3 (M = Na or K) but different properties have been obtained. Important polymetaphosphates obtained in this way are Na (or K) PO_3–II, Na (or K) PO_3 – III (both these metaphosphates are called **Maddrell's salts)** and Na (or K) PO_3–IV (called **Kuroll's salt**). These metaphosphates are actually polymeric and hence are represented by the general formula, $[Na\ (or\ K)\ PO_3]_n$.

Sodium Maddrell's salts (*i.e.*, $NaPO_3$-II and $NaPO_3$-III) have negligible solubility in water and in saline solutions at room temperature. At higher temperatures and in the presence of salts of alkali metal ions (other than Na^+ ions) and NH_4^+ ions in solution, these salts dissolve at a higher rate. $NaPO_3$–II, dissolves more readily than $NaPO_3$-III. Under these conditions. Potassium Maddrell's salts (*i.e.*, KPO_3–II and KPO_3–III) also show similar behaviour.

Kuroll's salts, Na (or K) PO_3–IV are somewhat more soluble in pure water than the corresponding Maddrell's salt. When a Kuroll's salt is kept in ice-cold distilled water for a few days or in hot water for a few hours, its crystals swell and become gummy. The solubility of sodium Kroll's salt is increased by the presence of salts having cations other than Na^+ ions. Potassium Kuroll's salt also shows similar behaviour.

The process of dissolution of Kuroll's and Maddrell's salt, in solutions of simple salts containing cations other than Na^+ and K^+ ions, is an exchange process, taking place between cations of the salt and Na^+ or K^+ ions, present in Kuroll's and Maddrell's salts.

X-ray study of sodium Maddrell's salt, $NaPO_3$-II has shown that this polymer has long —P—O—P—chains which are composed of interconnected PO_4 groups. The pattern repeats itself after every three units and the chains do not spiral.

Sodium Maddrell's salt is used as a polishing agent in dental surgery while potassium Kuroll's salt is employed in the manufacture of sausages to prevent water loss.

Polyorthophosphoric acids (also called polyphosphoric acids)

Polyphosphoric acids can be regarded as having been obtained by eliminating (*n* – 1) moles of H_2O from *n* moles of H_3PO_4.

$$nH_3PO_4 \xrightarrow{-(n-1)\,H_2O} H_{n+2}\,P_n\,O_{3n+1} + (n-1)\,H_2O$$

Thus, these acids can be represented by the general formula, $H_{n+2}\,P_n\,O_{3n+1}$.

The above discussion shows that $H_4P_2O_7$, $H_5P_3O_{10}$, $H_6P_4O_{13}$ etc., can be obtained as :

$$2H_3PO_4 \xrightarrow{-H_2O} H_4P_2O_7 \quad \text{[di (or pyro) phosphoric acid]}$$

$$3H_3PO_4 \xrightarrow{-2H_2O} H_5P_3O_{10} \quad \text{(triphosphoric acid)}$$

$$4H_3PO_4 \xrightarrow{-3H_2O} H_6P_4O_{13} \quad \text{(tetraphosphoric acid)}$$

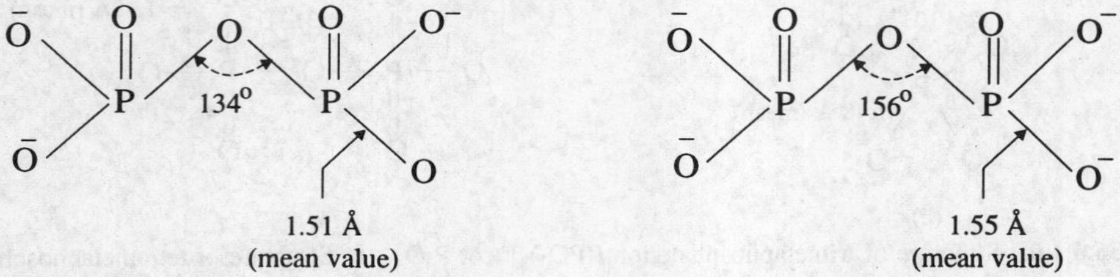

(*a*) (*b*)

Fig. 36.7 : Bond parameters of $P_2O_7^{4-}$ ion found in $Na_4P_2O_7 . 10H_2O$ [shown at (*a*)] and in α-$Mg_2P_2O_7$ [shown at (*b*)].

Polyphosphoric acids have linear chains and hence are also called *linear polyphosphoric acids.* In the structure of polyphosphate ion, $[P_n\ O_{3n+1}]^{-(n+2)}$ the linear chains are obtained by uniting PO_4 tetrahedrons together through one, two or more O-atoms. The structure of the ions are obtained by joining PO_4 tetrahedrons in such a way that one O-atom is common between two PO_4 tetrahedrons. The structure of pyrophosphate ion, $P_2O_7^{2-}$ found in $Na_4\ P_2O_7 . 10H_2O$ and in α-$Mg_2P_2O_7$ salts are shown in Fig. 36.7.

The structures of $H_5P_3O_{10}$ and $H_6P_4O_{13}$ molecules and the anions derived from them are given in Figs. 36.8 and 36.9 respectively.

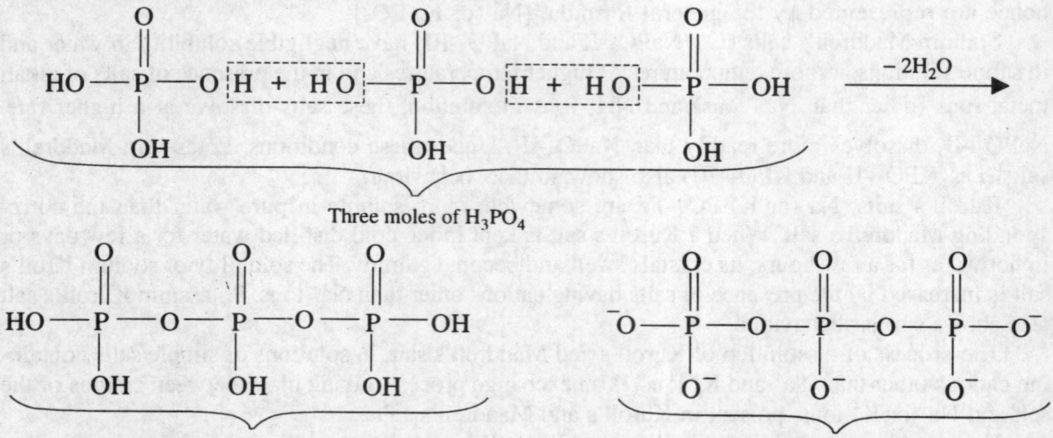

Structure of $H_5P_3O_{10}$ molecule Structure of $P_3O_{10}^{-5}$ ion

Fig. 36.8 : Structure of $H_5P_3O_{10}$ molecule and $P_3O_{10}^{5-}$ ion.

The linking of three PO_4 tetrahedrons through three O-atoms, to get the structure of $P_3O_{10}^{-5}$ ion has been shown in Fig. 36.10.

The structures of polyphosphate molecules and ions, as given above show that each P-atom in each structure is sp^3 hybridised.

Borophosphate glasses

These are formed by heating H_3PO_4, B_2O_3 and alkali metal carbonates or oxides at 700°C. Borophosphate glasses are of the following three types :

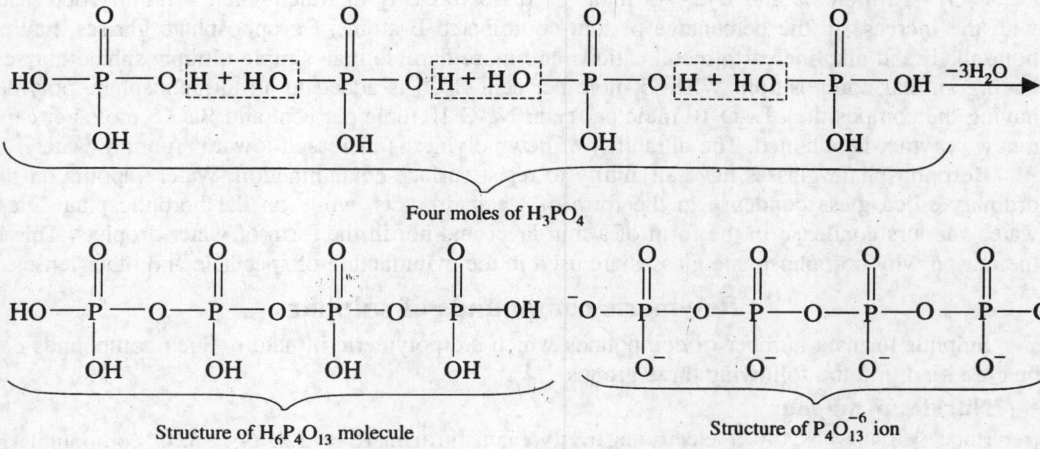

Fig. 36.9 : Structure of $H_6P_4O_{10}$ molecule and $P_4O_{13}^{-6}$ ion.

(*i*) Those which contain excess of alkali over P_2O_5. In this variety all the B-atoms are present as trigonal BO_3 groups.

(*ii*) Those which contain excess of P_2O_5 over alkali. These borophosphate glasses are called *acidic borophosphate glasses*. If there is less than 10 mole per cent of B_2O_3, all B-atoms are four coordinated.

(*iii*) Those which contain P_2O_5 and alkali in equivalent proportions. The number of 4-coordinated B-atoms decreases steadily with the increase in the content of B_2O_3 and becomes almost zero at about 47 mole per cent of B_2O_3.

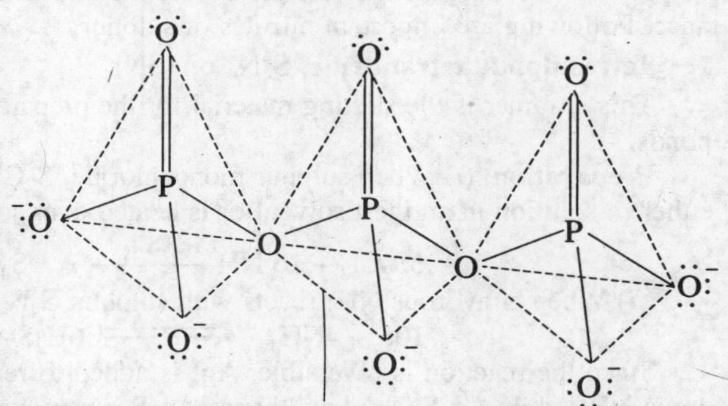

Fig. 36.10 : Linking of three PO_4 tetrahedrons through three O-atoms in the structure of $P_3O_{10}^{5-}$ ion.

Structures of borophosphate glass having three- and four-coordinated B-atoms are shown below in Fig. 36.11.

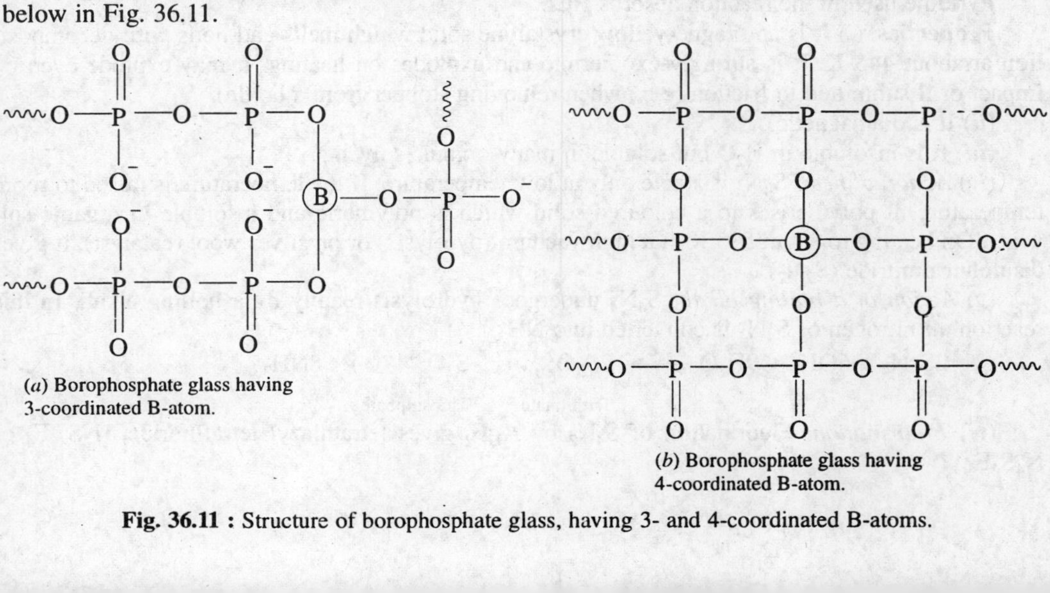

Fig. 36.11 : Structure of borophosphate glass, having 3- and 4-coordinated B-atoms.

P_2O_5 –80 mole % and B_2O_3 –5 mole %, dissolve easily in water. Their solubility decreases with the increase in the percentage of four-coordinated B-atoms. Borophosphate glasses, having both alkali and alkaline earth metal cations, are more durable than simple ultraphosphate glasses, having similar composition. When 5 mole per cent B_2O_3 is added to an ultraphosphate polymer having the composition, Li_2O 10 mole per cent, Na_2O 10 mole per cent and BaO 5 mole per cent, a new polymer is obtained. The durability of new polymer is increased towards running water.

Borophosphate glasses have an ability to resist surface contamination. Water vapours on the ordinary silica glass condense in the form of water droplets, while on the borophosphate glass water vapours condense in the form of a thin layer and not in the form of water droplets. This is the reason why borophosphate glasses are used in the manufacture of spectacle and other lenses.

Polymeric compounds of sulphur

Sulphur forms a number of compounds which are polymeric in nature. Such compounds can be classified into the following three groups :

I. Nitrides of sulphur

Since sulphur has lower electronegativity than nitrogen (S = 2.5, N = 3.0), compounds of sulphur with nitrogen are correctly called *nitrides* of *sulphur.* Sulphur forms many nitrides which are polymeric compounds. These are covalent compounds and have industrial and technical importance. Following are important nitrides of sulphur.

1. Tetrasulphur tetranitride, S_4N_4 or $(SN)_4$

This tetramer is the starting material for the preparation of many compounds containing S—N bonds.

Preparation. (*i*) When sulphur monochloride, S_2Cl_2 reacts with NH_3 (ammonolysis of S_2Cl_2) either in solution in an inert solvent or is heated over solid NH_4Cl, S_4N_4 is obtained.

$$6S_2Cl_2 + 16\,NH_3 \xrightarrow{CCl_4} S_4N_4 + S_8 + 12NH_4Cl$$

(*ii*) When anhydrous NH_3 reacts with sulphur, S_4N_4 is obtained.

$$10S + 4NH_3 \rightleftharpoons 6H_2S + S_4N_4$$

Since the reaction is reversible, AgI is added to remove H_2S. AgI is readily soluble in anhydrous NH_3 while Ag_2S is not and hence Ag_2S is removed from the reaction.

$$H_2S + 2AgI \longrightarrow Ag_2S + 2HI$$

(*iii*) When trithiazyl trichloride, $(NS)_3Cl_3$ is heated with tetrasulphur tetraimide, $S_4(NH)_4$ in presence of pyridine, $(NS)_3Cl_3$ undergoes reduction and tetrasulphur tetranitride (S_4N_4) is obtained.

$$4(NS)_3Cl_3 + 3S_4(NH)_4 \xrightarrow[\text{Pyridine}]{\Delta} 6S_4N_4 + 12\,HCl$$

Pyridine used in the reaction absorbs HCl.

Properties. (*i*) It is an oragne-yellow crystalline solid which melts and boils with decomposition at about 443 K. It is strongly exothermic and explodes on heating. It may explode even on impact or if submitted to friction (*e.g.*, when removing stopper from a bottle).

(*ii*) It is diamagnetic.

(*iii*) It is insoluble in H_2O but soluble in many organic solvents.

(*iv*) *Action of heat.* S_4N_4 is stable only at low temperature. If the temperature is raised to room temperature, it polymerises to a coloured solid which is polymeric and insoluble in organic solvents. On heating to about 575 K in a high vacuum (pyrolysis) or on silver wool (catalyst), it gives disulphur dinitride (S_2N_2).

(*v*) *Action of a boiling alkali.* S_4N_4 undergoes hydrolysis readily by a boiling alkali. In this reaction all nitrogen of S_4N_4 is converted into NH_3.

$$2S_4N_4 + 6OH^- + 9H_2O \longrightarrow \underset{\text{Trithionate}}{2S_3O_6^{2-}} + \underset{\text{Thiosulphate}}{S_2O_3^{2-}} + 8NH_3$$

(*vi*) *Fluorination.* Fluorination of S_4N_2 by AgF_2 gives tetrathiazyl tetrafluoride, $(NS)_4F_4$ or $N_4S_4F_4$.

$$S_4N_4 + AgF_2 \longrightarrow (NS)_4F_4 + 4AgF$$

(*vii*) *Reduction.* Reduction of S_4N_4 with $SnCl_2$ in ethanol gives tetrasulphur tetraimide, $S_4(NH)_4$ or $S_4N_4H_4$ which is a colourless compound.

$$S_4N_4 + 4H \xrightarrow{\text{Reduction by } SnCl_2 \text{ in } C_2H_5OH} S_4(NH)_4$$

(*viii*) *Oxidation.* Oxidation of S_4N_4 with Cl_2 gives trithiazyl trichloride, $(NS)_3Cl_3$.

$$3S_4N_4 + 6Cl_2 \longrightarrow 4(NS)_3Cl_3$$

(*ix*) *Bromination.* When S_4N_4 is brominated in a solvent, a bronze-coloured crystalline compound having the formula, NSBr is obtained. This compound is polymeric.

(*x*) *Reaction with S_2Cl_2.* When S_4N_4 is boiled with S_2Cl_2 in CCl_4 solution, thiotrithiazyl chloride, $[S(NS)_3]^+Cl^-$ or $[N_3S_4]^+Cl^-$ is obtained. As is evident, it is an ionic compound.

$$3S_4N_4 + 2S_2Cl_2 \xrightarrow[\text{Pyridine}]{\text{Boil}} 4[S(NS)_3]^+Cl^-$$

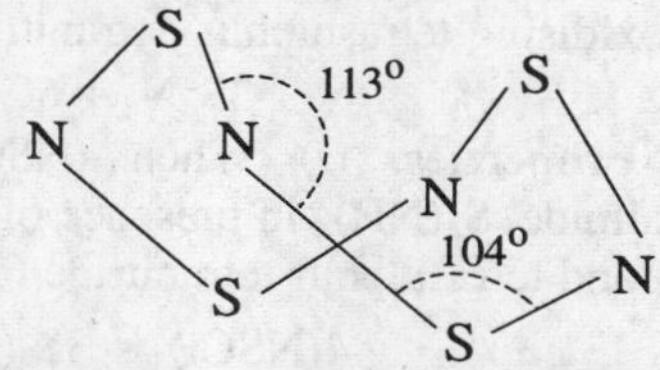

Fig. 36.12 : Eight-membered cradle ring (cyclic) structure of S_4N_4 molecule.

(*xi*) *Formation of adducts.* Due to the presence of lone pains of electrons on its N-atom, S_4N_4 acts as a Lewis base and hence forms adducts with several metal halides (Lewis acids). The adducts may be of S_4N_4. A (A = $SbCl_5$, $MoCl_5$, BF_3, WCl_4) or S_4N_4.2A (A = $SnCl_4$) type.

(*xii*) *Reaction with PCl_3.* S_4N_4 reacts with PCl_3 and forms an ionic compound whose formula is $[Cl_3P.N.PCl_2.N.PCl_3]^+[PCl_6]^-$.

Structure. Electron diffraction and X-ray measurements have shown that S_4N_4 molecule has an eight-membered cradle ring (cyclic) structure in which all the N—S bond lengths are equal to 1.62 Å. NSN bond angle = 104° and SNS bond angle = 113° (See Fig. 36.12).

2. Disulphur dinitride, S_2N_2. As already stated, when S_4N_4 is heated to about 575 K in a high vacuum or on silver wood, S_2N_2 is obtained. This compound forms highly volatile colourless crystals which explode when struck, rubbed or heated to about 303 K. They are soluble in organic solvents like benzene. On keeping, these crystals change to a dark blue polymer.

3. Pentasulphur dinitride, S_5N_2. When S_4N_4 is heated in solution, S_5N_2 is obtained. It is some what less stable than S_4N_4 and is a deep red oil which freezes to a grey solid at 284 K.

II. Thiazyl halides

Thiazyl halides can be represented by the general formula, $(NS)_nX_n$ or $(NSX)_n$ (X = F, Cl, Br, n = 3, 4 etc.). Examples of such compounds are trithiazyl trifluoride [$(NSF)_3$], tetrathiazyl tetrafluoride [$(NSF)_4$] and trithiazyl trichloride [$(NSCl)_3$].

1. Trithiazyl trifluoride, $(NS)_3F_3$ or $(NSF)_3$. It is prepared :

(*i*) by fluorinating trichloride, $(NSCl)_3$ with AgF_2 in CCl_4.

$$(NSCl)_3 \xrightarrow{AgF_2 \text{ in } CCl_4} (NSF)_3$$

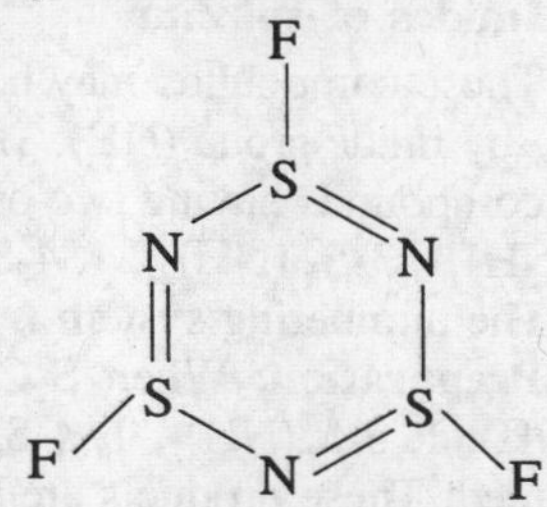

Fig. 36.13 : Six-membered ring structure of $(NSF)_3$ molecule.

(*ii*) by heating tetrathiazyl tetrafluoride, $(NSF)_4$ at 200-250°C which dissociates into NSF, followed by polymerisation.

$$(NSF)_4 \xrightarrow[200-250°\,C]{\text{Dissociation}} 4NSF; \quad 3NSF \xrightarrow{\text{Polymerisation}} (NSF)_3$$

The structure of $(NSF)_3$ molecule is not known with certainty. However, n.m.r. spectrum of this molecule indicates that all F-atoms are equivalent. $(NSF)_3$ molecule has a six-membered ring structure which is almost planar with all the F-atoms on one side of the ring in axial positions. All N—S distances are equal to 1.60 Å (See Fig. 36.13).

2. Tetrathaizyl tetrafluoride, $(NSF)_4$. It is prepared by fluorinating S_4N_4 with AgF_2.

$$S_4N_4 + 4AgF_2 \longrightarrow (NSF)_4 + 4AgF$$

It is a white sensitive solid. It dissociates at 200-250° C into NSF, which polymerises to trimer, $(NSF)_3$.

$$(NSF)_4 \xrightarrow{200\text{–}250°\,C} 4NSF\ ;\ 3NSF \longrightarrow (NSF)_3$$

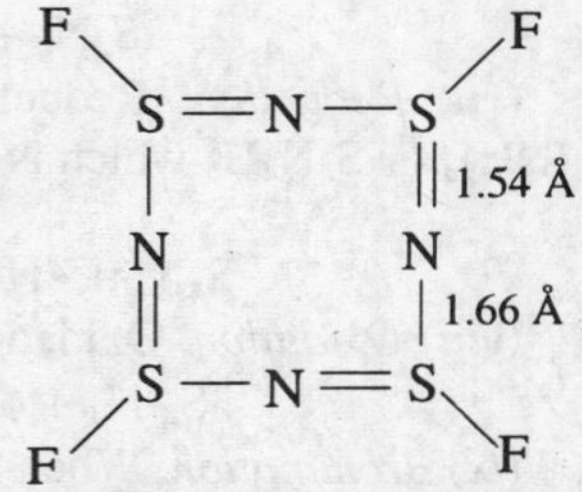

Fig. 36.14 : Eight-membered puckered ring structure of $(NSF)_4$ molecule.

The ^{19}F *n.m.r.* spectrum of tetramer, $(NSF)_4$ gives a single resonance peak which indicates that all the F-atoms are equivalent. $(NSF)_4$ molecule has a 8-membered puckered ring structure, with alternating S—N distances of 1.66 Å and 1.59 Å. (See Fig. 36.14).

3. Trithiazyl trichloride, $(NSCl)_3$. Preparation. It is prepared by oxidising tetrasulphur tetranitride (S_4N_4) by Cl_2.

$$3S_4N_4 + 6Cl_2 \longrightarrow 4\ (NSCl)_3$$

Properties. (*i*) When $(NSCl)_3$ is heated with tetrasulphur tetraimide, $S_4(NH)_4$ in presence of pyridine, $(NSCl)_3$ undergoes reduction and tetrasulphur tetranitride (S_4N_4) is obtained.

$$4(NSCl)_3 + 3S_4(NH)_4 \xrightarrow[\text{Pyridine}]{\Delta} 6S_4N_4 + 12HCl$$

Pyridine used in the above reaction absorbs HCl.

(*ii*) $(NSCl)_3$ is converted into $(NSF)_3$ by fluorinating it with AgF_2 in CCl_4.

$$(NSCl)_3 \xrightarrow{AgF_2 \text{ in } CCl_4} (NSF)_3$$

(*iii*) This compound forms an addition compound with three molecules of SO_3. The addition compound, on being heated to 425 K, givs sulphuranyl chloride.

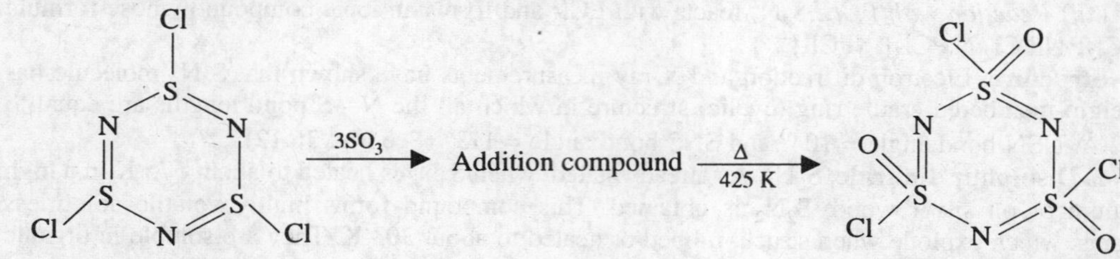

Trithiazyl trichloride

Sulphuranyl chloride

Structure. $(NSCl)_3$ molecule has the same structure as $(NSF)_3$ molecule has.

III. Imides of sulphur

These compounds may be regarded as having been obtained by replacing one or more S-atom in S_8 by imide group (NH). These compounds can be represented by the general formula, $S_n(NH)_{8-n}$. The compounds having two or more NH groups exist as isomers. Examples of such compounds are S_7 (NH) 1, 3-S_6 $(NH)_2$; 1, 4-S_6 $(NH)_2$; 1, 5-S_6 $(NH)_2$; 1, 3, 5-S_5 $(NH)_3$; 1, 3, 6-S_5 $(NH)_3$; $S_4(NH)_4$ etc. The numbering system is based on S_8 ring.

Preparation. When S_2Cl_2 reacts with NH_3 (ammonolysis of S_2Cl_2) in dimethyl formide, $S_7(NH)$, 1, 3-$S_6(NH)_2$, 1, 4-S_6 $(NH)_2$, 1, 5, S_6 $(NH)_2$, 1, 3, 5-S_5 $(NH)_3$, 1, 3, 6-S_5 $(NH)_3$ etc., are obtained. These products are separated from each other by chromatography and fractional crystallisation.

Tetrasulphur tetraimide, $S_4(NH)_4$ is prepared by reducing S_4N_4 with $SnCl_2$ in C_2H_5OH.

$$S_4N_4 + 4H \xrightarrow{\text{Reduction by } SnCl_2 \text{ in } C_2H_5OH} S_4(NH)_4$$

Properties of S_7 (NH). (*i*) The reaction of $S_7(NH)$ with C_6H_5Na gives $S_7(NNa)$.

$$S_7(NH) + C_6H_5Na \longrightarrow S_7(N\ Na) + C_6H_6$$

(*ii*) N—H bonds in S_7 (NH) can be acetylated and benzoylated.

(*iii*) The reaction of S_7 (NH) with the salts of Hg (II) and Hg (I) in non-aqueous solvents gives Hg $(NS_7)_2$ and Hg_2 $(NS_7)_2$ respectively.

Properties of $S_4(NH)_4$. (*i*) It is a colourless compound and reacts with C_6H_5Na to form $S_4(NNa)_4$.

$$S_4(NH)_4 + 4C_6H_5Na \longrightarrow S_4(NNa)_4 + 4C_6H_6$$

(*ii*) N—H bonds in $S_4(NH)_4$ molecule can be acetylated and benzoylated.

(*iii*) Air oxidation of $S_4(NH)_4$ at 100°C gives a tetrameric thionyl imide, $(OS)_4(NH)_4$ or $(OSNH)_4$.

Structure of imides of sulphur. None of the imides has N—N bonds. The structures of some imides are given in Fig. 36.15. $S_7(NH)$ does not have any isomer and hence, it is represented by the

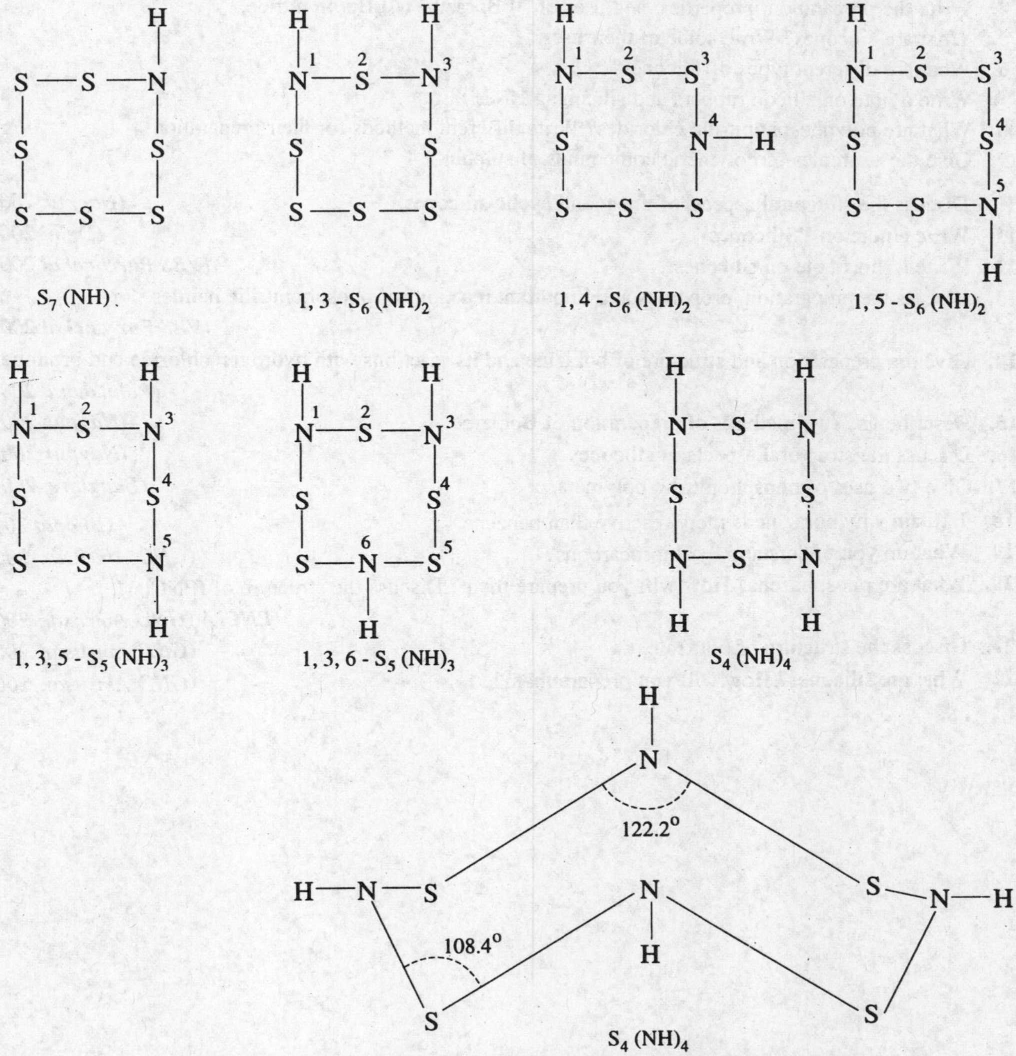

$S_7(NH)$ $1,3-S_6(NH)_2$ $1,4-S_6(NH)_2$ $1,5-S_6(NH)_2$

$1,3,5-S_5(NH)_3$ $1,3,6-S_5(NH)_3$ $S_4(NH)_4$

$S_4(NH)_4$

Fig. 36.15 : Structure of $S_7(NH)$, $S_6(NH)_2$, $S_5(NH)_3$ and $S_4(NH)_4$ molecules.

structure shown in the figure. It has one N—H bond. $S_6(NH)_2$ has three isomers namely 1, 3; 1, 4 and 1, 5 while $S_5(NH)_3$ exists in two isomeric forms which are 1, 3, 5 and 1, 3, 6. The structure of each isomer of $S_6(NH)_2$ and $S_5(NH)_3$ has been shown in the figure. The structure of $S_4(NH)_4$ is the same as that of S_8. NSN and SNS bond angles are 108.4° and 122.2°. Note that SSS bond angle in S_8 is 107.8°. All N—S bond lengths in $S_4(NH)_4$ are the same (= 1.67 Å). $S_4(NH)_4$ cannot be reduced to hydrazine. This shows no two N-atoms are directly linked to each other in $S_4(NH)_4$ molecule, *i.e.,* $S_4(NH)_4$ does not have any N—H bonds. $S_4(NH)_4$ molecule does not have any isomer and hence is represented by a single structure.

University Questions

1. What are inorganic polymers? Classify them?
2. What are coordination polymers? How will you prepare them?
3. Discuss general properties of inorganic polymers.
4. Write the preparation, properties, and uses of (*i*) Borazine (*ii*) Boron nitride.
5. What are silicones? Write some of their uses.
6. What are different types of silicones?
7. Write a note on silicon rubbers and silicone greases.
8. What are polyphosphonitrilic chlorides? Write different methods for their preparation.
9. Give the examples of polymeric compounds of sulphur.

10. **Discuss the structural aspects of linear and cyclic silicones.** ***(Nagpur 2002)***
11. **Write a note on "Silicones".** ***(Delhi 2002)***
12. **Write a short note on silicones.** ***(VBS Purvanchal 2003)***
13. **Discuss the preparation, properties and important reactions of phosphonitrilic halides.** ***(VBS Purvanchal 2003)***
14. **Give the preparation and structure of borazine and its reactions with hydrogen chloride and bromine.** ***(Rohilkhand 2003)***
15. **Describe any two methods of preparation of Borazine.** ***(Nagpur 2003)***
16. **Discuss the structural aspects of silicones.** ***(Nagpur 2003)***
17. **Give two uses of phosphonitrilic polymers.** ***(Banglore, 2004)***
18. Explain why borazine is more reactive than benzene. *(Bhopal 2004)*
19. What do you understand by fluorocarbons? *(GND Amritsar, 2004)*
20. What are phosphazens? How will you prepare them? Discuss the structure of $(PNCl_2)$? *$(PNCl_2)_3$(GND Amritsar, 2004)*
21. Discuss the structure of borazole. *(GND Amritsar, 2004)*
22. What are silicones? How will you prepare them? *(GND Amritsar, 2004)*

37

Important Compounds of Main Group Elements

Hydrides, carbides, nitrides, oxides, halides/oxy-halides and oxy-acids of main group elements have already been discussed in previous chapters. In this chapter, we shall discuss the methods of preparation, properties etc., of some other compounds, formed by main group elements.

Sodium amide or sodamide, $NaNH_2$

It is an important compound of sodium.

Preparation. It is formed when molten sodium is heated with NH_3 at 300°C.

$$2Na\ (molten) + 2NH_3 \xrightarrow{300°\ C} 2NaNH_2 + H_2$$

Properties. (*i*) It is a waxy white solid, with m.pt. = 210°C. It can be sublimed at high temperature in vaccum, but undergoes decomposition at about 500°C or even about 300°C in vacuum.

(*ii*) *Action of red hot charcoal.* When $NaNH_2$ is dropped on red hot charcoal (C), NaCN is obtained.

$$2NaNH_2 + 2C\ (red\ hot) \xrightarrow{300\text{–}800°\ C} 2NaCN + 2H_2$$

This reaction has been used to prepare NaOH by *Castner's process.*

(*iii*) *Action of water.* It gets hydrolysed by H_2O, giving NH_3.

$$NaNH_2 + H_2O \rightarrow NaOH + NH_3$$

(*iv*) *Action of nitrous oxide (N_2O) and $NaNO_3$. When $NaNH_2$* is heated in a current of H_2O at 200°C or is treated with $NaNO_3$ at 175°C, sodium azide (NaN_3) is obtained.

$$3NaNH_2 + N_2O \xrightarrow{200°\ C} NaN_3 + NH_3 + NaOH$$

$$3NaNH_2 + NaNO_3 \xrightarrow{175°\ C} NaN_3 + NH_3 + 3NaOH$$

These reactions have been used to prepare NaN_3.

(*v*) *Action of CO.* On heating with CO, NaCN is obtained.

$$2NaNH_2 + CO \xrightarrow{\Delta} NaCN + NaOH + NH_3$$

This reaction has been used to prepare NaCN by *Castner's Process.*

(*vi*) *Action of CO_2.* With CO_2, it gives cyanamide ($CN.NH_2$) and Na_2CO_3.

$$2NaNH_2 + 2CO_2 \longrightarrow Na_2CO_3 + H_2O + CN.NH_2$$

(*vii*) *Action of dry H_2.* It reacts with dry H_2 at 150-300°C to give NaH.

$$NaNH_2 + H_2\ (dry) \xrightarrow{150-300°\ C} NaH + NH_3$$

Uses. $NaNH_2$ is used (*i*) as a reducing as well as dehydrating agent in organic chemistry synthesis (*ii*) in the preparation of NaCN and NaN_3.

Structure of NH_2^- ion. $NaNH_2$ is an ionic molecule. NH_2^- ion has angular shape, which arises due to sp^3 hybridisation of N^- ion. For details see chapter 14.

Sodium cyanide, NaCN

Preparation. (*i*) *From sodamide, $NaNH_2$ (Castner's process).* First of all, $NaNH_2$ is obtained by the action of molten sodium on NH_3 at 300°C. Now, when $NaNH_2$ obtained as above, is dropped on red hot charcoal (C) at 300–800°C or is heated with CO, NaCN is obtained.

$$2Na\ (molten) + 2NH_3 \xrightarrow{300°\,C} 2NaNH_2 + H_2$$

$$2NaNH_2 + 2C\ (red\ hot) \xrightarrow{300\text{–}800°\,C} 2NaCN + 2H_2$$

$$2NaNH_2 + CO \xrightarrow{\Delta} NaCN + NaOH + NH_3$$

NaCN obtained by this method is impure. It is purified by dissolving NaCN into H_2O and then evaporating the solution.

(*ii*) *From crude calcium cyanamide.* Crude calcium cyanamide is a mixture of $CaCN_2$ and coke (C). When this crude mixture is fused with NaCl or Na_2CO_3, NaCN is produced

$$\underbrace{CaCN_2 + C}_{\text{Crude calcium cyanamide}} + 2NaCl \xrightarrow{\text{Fuse}} 2NaCN + CaCl_2$$

$$\underbrace{CaCN_2 + C}_{\text{Crude calcium cyanamide}} + Na_2CO_3 \xrightarrow{\text{Fuse}} 2NaCN + CaCO_3$$

(*iii*) *From soda ash.* Soda ash is a fused mixture of Na_2CO_3 and powdered charcoal. When NH_3 or N_2 is passed over this fused mixture, NaCN is obtained.

$$\underbrace{Na_2CO_3 + C}_{\text{soda ash}} + 2NH_3 \longrightarrow 2NaCN + 3H_2O$$

$$\underbrace{Na_2CO_3 + C}_{\text{soda ash}} + N_2 \xrightarrow[\text{divided Fe}]{\text{In pressure or}} 2NaCN + 3CO$$

(*iv*) *From coal gas.* Coal gas contains HCN and NH_3 as impurities. When coal gas is passed through an alkaline solution of $FeSO_4$, HCN forms $Na_4[Fe(CN)_6]$ and NH_3 remains unaffected.

$$6NaOH + 6HCN\ (present\ in\ coal\ gas) + FeSO_4 \longrightarrow 6H_2O + Na_2SO_4 + Na_4[Fe(CN)_6]$$

$Na_4[Fe(CN)_6]$ obtained as above is heated with metallic sodium in covered iron crucible, when NaCN and metallic iron are obtained.

$$Na_4[Fe(CN)_6] + 2Na \xrightarrow{\Delta} 6NaCN + Fe$$

Metallic iron mixed with NaCN is removed by filtration and NaCN thus obtained is cast into blocks.

Properties. (*i*) It is a colourless crystalline substance with a characteristic smell of HCN. It is soluble in alcohol and water and is highly poisonous.

(*ii*) *Alkaline nature.* The aqueous solution of NaCN is alkaline, since a strong base *viz.*, NaOH is obtained by the hydrolysis of NaCN. NaOH gives OH^- ions which make the solution alkaline HCN is a very weak acid.

$$NaCN + H_2O \rightleftharpoons \underset{\text{strong base}}{NaOH} + \underset{\text{weak acid}}{HCN} \rightleftharpoons Na^+ + OH^- + HCN$$

(*iii*) *Formation of soluble complex cyanides.* (*a*) when the salts of transition metals like Cu, Ag, Zn, Fe, Cd etc., react with NaCN, soluble metallic complex cyanides are obtained, *e.g.*,

$$ZnSO_4 + 4NaCN \longrightarrow Na_2[Zn(CN)_4] + Na_2SO_4 \quad \text{or} \quad Zn^{2+} + 4CN^- \longrightarrow [Zn(CN)_4]^{2-}$$

$$CdCl_2 + 4NaCN \longrightarrow Na_2[Cd(CN)_4] + 2NaCl \quad \text{or} \quad Cd^{2+} + 4CN^- \longrightarrow [Cd(CN)_4]^{2-}$$

When argentite (Ag_2S) is treated with aqueous solution of NaCN in presence of hot air, soluble complex cyanide *viz.*, sodium dicyno argentite (I), $Na[Ag(CN)_2]$ is formed.

$$Ag_2S + 4NaCN + 2O_2 \text{ (from air)} \dashrightarrow Na_2SO_4 + 2Na[Ag(CN)_2]$$

or $$Ag_2S + 4CN^- + 2O_2 \dashrightarrow 2[Ag(CN)_2]^- + SO_4^{2-}$$

When horn silver (AgCl) is acted upon by an aqueous solution of NaCN, soluble complex cyanide, $Na[Ag(CN)_2]$ is obtained.

$$AgCl \text{ (Horn silver)} + 2NaCN \longrightarrow Na[Ag(CN)_2] + NaCl \quad ...(i)$$

or $$Ag^+ + 2CN^- \longrightarrow [Ag(CN)_2]^- \quad ...(ii)$$

Reactions (*i*) and (*ii*) have been used in the extraction of Ag from argentite and horn silver respectively by *Mac-Arthur Forest cyanide process.*

(*b*) A dilute aqueous solution of NaCN (0.3%) dissolves Ag and Au metals in presence of air or O_2 and forms $Na[Ag(CN)_2]$ and $Na[Au(CN)_2]$ (soluble complexes) respectively.

$$4M \text{ (M = Ag, Au)} + 8\,NaCN + 2H_2O + O_2 \longrightarrow 4Na[M(CN)_2] + 4NaOH$$

This reaction has been used for getting Ag and Au from the tailings obtained at the end of amalgamation process which is used for the extraction of Au from auriferrous quartz rocks. For details see. "*Mac-Arthur Forest cyanide process*".

Uses. NaCN is used : (*i*) in the extraction of Ag and Au from *horn silver* (AgCl) and *argentite* (Ag_2S) *ores*. (*ii*) in getting Ag and Au from the tailings obtained at the end of amalgamation process used for the extraction of Au from auriferrous quartz rocks. (*iii*) in electroplating and galvanoplastics. (*iv*) as a laboratory reagent. (*v*) in photography.

Structure of NaCN molecule. NaCN is an ionic molecule and hence is represented as Na^+ CN^-. The structure of CN^- ion has been discussed in the chapters on "*Hybridisation*" and "*Molecular Orbital Theory*".

Potassium cyanide, KCN

Preparation. It is prepared : (*i*) By fusing potassium ferrocyanide, $K_4[Fe(CN)_6]$ with K_2CO_3.

$$K_4[Fe(CN)_6] + K_2CO_3 \longrightarrow 5KCN + KCNO + Fe + CO_2$$

KCNO produced in the reaction is reduced to KCN by adding a little carbon to the fused mass.

(*ii*) by fusing $K_4Fe(CN)_6$ with metallic K or Na.

$$K_4Fe(CN)_6 + 2K \longrightarrow 6KCN + Fe$$

$$K_4Fe(CN)_6 + 2Na \longrightarrow 4KCN + 2NaCN + Fe$$

(*iii*) *By Beilby's process.* In this process NH_3 gas is passed over a mixture of fused K_2CO_3 and carbon.

$$K_2CO_3 + C + 2NH_3 \longrightarrow 2KCN + 3H_2O$$

The fused KCN is then decanted and cystallised out by evaporation. *Very pure* KCN is obtained by recrystallising it from anhydrous liquid NH_3.

Properties. (*i*) It is colourless crystalline solid, highly hygroscopic, m.pt. = 634.5°. It is a *dead poison.*

(*ii*) It is very soluble in H_2O and alcohol. The aqueous solution is alkaline due to hydrolysis.

$$KCN + H_2O \longrightarrow K^+ + OH^- + HCN$$

(*iii*) *Formation of cyano complexes.* (*a*) Like NaCN, KCN also forms soluble cyano complexes with the salts of transition metals like Cu, Ag, Au, Zn, Cd, Fe, Co, Ni etc. For example :

$$2CuSO_4 + 10KCN \text{ (excess)} \longrightarrow 2K_2SO_4 + (CN)_2 + \underset{\text{Pot. cuprotetracyanide}}{2K_3[Cu(CN)_2]}$$

(*b*) When excess of KCN reacts with tetrammine cupric sulphate, $[Cu(NH_3)_4]SO_4$ and tetrammine cadmium sulphate, $[Cd(NH_3)_4]SO_4$, colourless tetracyano complexes are obtained in solution.

$$2[Cu(NH_3)_4]SO_4 + 10KCN + 8H_2O \longrightarrow 2K_3[Cu(CN)_4] + 2K_2SO_4 + 8NH_4OH + (CN)_2$$

$$[Cd(NH_3)_4]SO_4 + 4KCN + 4H_2O \longrightarrow K_2[Cd(CN)_4] + K_2SO_4 + 4NH_4OH$$

The formation of the tetracyano complexes of Cu^+ and Cd^{2+} has been used in separating Cu^{2+} from Cd^{2+} group IIA of qualitative analysis.

(*c*) When KCN solution is added, drop by drop, to the solution containing Ni^{2+} and Co^{2+} ions, colourless soluble cyano complexes are obtained in solution.

$$Ni^{2+} + 4KCN \longrightarrow K_2[Ni(CN)_4] \text{ [colourless (solution)]} + 2K^+$$
$$Co^{2+} + 6KCN \longrightarrow K_4[Co(CN)_6] \text{ (colourless solution)} + 2K^+$$

(*d*) A dilute solution (0.3%) of KCN dissolves Ag and Au metals in presence of air or O_2, forming respectively potassium argentocyanide, $K[Ag(CN)_2]$ and potassium aurocyanide, $K[Au(CN)_2]$.

$$4Ag + 8KCN + 2H_2O + O_2 \longrightarrow 4K[Ag(CN)_2] + 4KOH$$
$$4Au + 8KCN + 2H_2O + O_2 \longrightarrow 4K[Au(CN)_2] + 4KOH$$

Uses. It is used (*i*) in electroplating (*ii*) in the extraction of gold and silver (*iii*) as a reagent in the laboratory.

Basic beryllium acetate, $3Be(CH_3COO)_2.BeO$ or $Be_4O(CH_3COO)_6$

Preparation. This compound can be prepared by dissolving $Be(OH)_2$ or $BeCO_3$ in glacial acetic and. The resulting solution is diluted with water and then evaporated to dryness, so that a residue of $Be_4O(CH_3COO)_6$ is obtained. This residue is dissolved in chloroform. The chloroform solution, on evaporation, gives the crystals of $Be_4O(CH_3COO)_6$.

$$4Be(OH)_2 + 6CH_3COOH \longrightarrow Be_4O(CH_3COO)_6 + 7H_2O$$
$$4BeCO_3 + 6CH_3COOH \longrightarrow Be_4O(CH_3COO)_6 + 3H_2O + 4CO\uparrow$$

Properties. (*i*) $Be_4O\ (CH_3COO)_6$ is a crystalline solid with m.pt. = 283°C and b.pt. = 330°C.

(*ii*) It is insoluble in water but soluble in organic solvent like glacial acetic acid, chloroform and ether.

(*iii*) It is a covalent compound and is slowly hydrolysed by boiling water.

(*iv*) It reacts with mineral acids (*e.g.*, HCl, H_2SO_4 etc.) and forms the corresponding beryllium salts.

$$Be_4O(CH_3COO)_6 + 8HCl \longrightarrow 4BeCl_2 + 6CH_3COOH + H_2O$$
$$Be_4O(CH_3COO)_6 + 4H_2SO_4 \longrightarrow 4BeSO_4 + 6CH_3COOH + H_2O$$

Uses. $Be_4(CH_3COO)_6$ is used : (*i*) for the preparation of other beryllium salts (*ii*) for purifying beryllium metal by solvent extraction method.

Structure. X-ray diffraction of this salt has shown that $Be_4O(CH_3COO)_6$ molecule has tetrahedral shape whose centre is occupied by O-atom and the four corners are being occupied by four beryllium atoms. Six acetate groups, $\left[\begin{matrix} O \\ H_3C \end{matrix} \!\!>\! C = O\right]$ are arranged along the six edges of the tetrahedron (See Fig. 37.1).

The structure given in Fig. 37.1 clearly indicates that each Be-atom is covalently attached with O-atoms of three CH_3COO^- groups. The central O-atom is linked with four Be-atoms. Here it may be noted that although Be- and O-atoms are normally bicovalent, but in $Be_4O(CH_3COO)_6$ molecule, both these elements are *tetracovalent*.

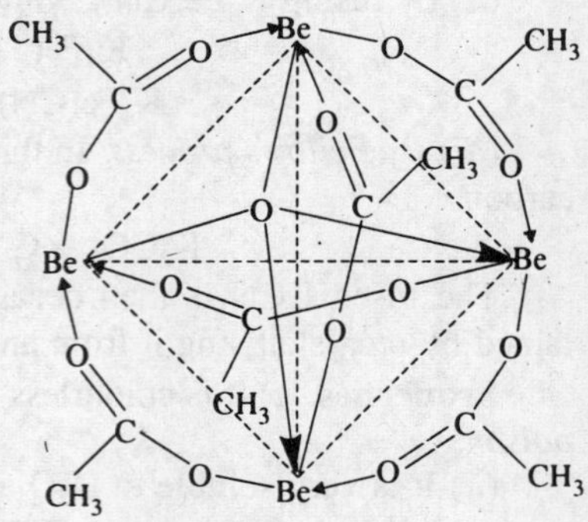

Fig. 37.1 : Tetrahedral shape of basic beryllium acetate molecule, $Be_4O\ (H_3COO)_6$.

Compounds of Radium

Radium is bivalent in its compounds. A number of compounds of radium, such as carbonate, chloride, bromide, sulphate, nitrate and azide, have been prepared and studied. They are all very similar to the corresponding compounds of barium. They are perfectly white when pure, but are pink if they contain any barium. As a rule, they are less soluble in water than the barium salt. They give a ceramine red colour to the non-luminous flame. All radium compounds are radioactive and are gradually decomposed by their own radiations. The chloride and the bromide continuously give out trace of chlorine and bromine respectively. As such, the radium salts are unstable. An aqueous solution of a radium salt gives off hydrogen and oxygen, due to the decomposition of water by the rays. These compounds produce phosphorescence and fluorescence in organic materials and gems.

1. Radium carbonate, $RaCO_3$. It is prepared either by fusing radium sulphate with sodium carbonate or by adding ammonium carbonate, to a radium salt solution. It is insoluble in water and forms salts with acids like hydrochloric etc.

2. Radium chloride, $RaCl_2.2H_2O$. It is prepared by dissolving radium carbonate in hydrochloric acid and crystallising it from the solution. It is isomorphous with $BaCl_2.2H_2O$. It is less soluble in water than barium chloride and can be precipitated from the solution by the addition of alcohol or strong hydrochloric acid. When heated to 200°C, it becomes anhydrous and on further heating, melts to a liquid, which solidifies to a glassy mass, shining with a bluish violet light.

3. Radium bromide, $RaBr_2.2H_2O$. It is prepared by dissolving radium carbonate in hydrobromic acid and crystallising the solution. It is a colourless, crystalline substance, isomorphous with barium bromide dihydrate. It is more soluble in water than the chloride. It is less volatile than barium bromide. It volatilizes at 820°C when exposed to air. It loses bromine and becomes alkaline after absorbing moisture and finally becomes radium carbonate after absorbing carbon dioxide. With barium bromide, it forms the double salt, $2BaBr_2RaBr_2.6H_2O$.

Note : *Radium fluoride and iodide have not been isolated.*

4. Radium sulphate, $RaSO_4$. It is obtained as a white precipitate by adding a soluble sulphate (or sulphuric acid) to a radium solution. It is a white substance, sparingly soluble in water (Its solubility is about 100 times less than that of barium sulphate). It is always carried down by barium sulphate, when the latter is precipitated from a solution containing barium and radium salts. This fact is utilized in the metallurgy of radium.

Basic lead carbonate (Also called "white lead"), $2PbCO_3.Pb(OH)_2$

It is manufactured by the following processes:

1. Dutch process. In this process, white lead is obtained by the action of the vapours of vinegar (CH_3COOH) on lead, in presence of O_2 from the air and CO_2 produced by the fermentation of tan bark.

$6Pb + 6CH_3COOH + 3O_2 + 2CO_2 \rightarrow Pb(OH)_2.2PbCO_3$ (*white lead*) $+ 3Pb(CH_3COO)_2 + 2H_2O$ .. (A)

Probably basic lead acetate $[Pb(OH)_2.Pb(CH_3COO)_2]$ is first formed by the action of CH_3COOH on Pb in presence of O_2. This basic lead acetate is then decomposed to white lead, $Pb(OH)_2.PbCO_3$ by the action of CO_2.

$3 \times [2Pb + 2CH_3COOH + O_2 \longrightarrow Pb(OH)_2.Pb(CH_3COO)_2$ (*basic lead acetate*)] ...(*i*)

$3[Pb(OH)_2.Pb(CH_3COO)_2] + 2CO_2 \longrightarrow Pb(OH)_2.2PbCO_3$ (*white lead*) $+ 3Pb(CH_3COO)_2 + 2H_2O$ ...(*ii*)

On multiplying equation (*i*) by 3 and then adding it to equation (*ii*), we get equation (A) which represents the formation of white lead. White lead, obtained by this method, is ground and washed with water to remove soluble $Pb(CH_3COO)_2$. It is then dried and the product is finely powdered. This process gives the best quality of white lead. The only drawback of this process is that it is a slow process.

2. Certer's process. In this process, superheated steam is passed over molten lead, so that molten lead is converted into very fine powder of lead. Now, this fine powder of lead, is placed in wooden cylinder (or drum), rotating slowly over horizontal axis and containing CH_3COOH in it. A mixture of CO_2 and air is introduced into the rotating drum, so that in two weeks' time lead is converted into white lead. In this process, no external heating is done, since the reaction is exothermic and the temperature remains nearly 40°C.

3. Electrolytic process. Solution of CH_3COONa is taken in the anodic chamber and solution of Na_2CO_3 is taken in the cathodic chamber of the electrolytic cell. Plates of lead and iron are dipped in CH_3COONa and Na_2CO_3 solutions respectively. Plate of lead acts as anode while that of iron acts as cathodes. On passing the electric current through the electrolyte, CH_3COO^- ions obtained the ionisation of CH_3COONa ($CH_3COONa \rightarrow CH_3COO^- + Na^+$) combine with Pb^{2+} ions, obtained from anode ($Pb \rightarrow Pb^{2+} + 2e^-$) to form $Pb(CH_3COO)_2$ in the electrolyte. $Pb(CH_3COO)_2$, thus obtained, reacts with Na_2CO_3 solution filled in the cathode chamber to form white lead.

$$3Pb(CH_3COO)_2 + 4Na_2CO_3 + 2H_2O \rightarrow 2PbCO_3.Pb(OH)_2 \text{ (white lead)} + 2NaHCO_3 + 6CH_3COONa$$

Very fine uniform grains of white lead are obtained by this process.

Properties. (*i*) It is a white amorphous powder and is insoluble in water. It is, however, soluble in linseed oil. Mixed with linseed oil, it forms a white paint. It is poisonous.

(*ii*) On heating to 43°C, it decomposes to give red lead (Pb_3O_4).

$$Pb(OH)_2.2PbCO_3 \xrightarrow{43°\,C} Pb_3O_4 + H_2O + CO_2$$

(*iii*) When placed in air, it gets blackened by H_2S gas (present in air) due to the formation of PbS.

(*iv*) When treated with CH_3COOH (45%), $Pb(CH_3COO)_2$ is obtained.

$$2PbCO_3.Pb(OH)_2 + 6CH_3COOH \longrightarrow 3Pb(CH_3COO)_2 + 4H_2O + 2CO_2$$

Uses. Mixed with linseed oil, it is used as a white paint, because of its great covering power. However, it has two drawbacks : it is poisonous and is blackened in air by certain gases like H_2S due to formation of PbS. It is also used in the preparation of pottery glazes.

Sublimed white lead or basic lead sulphate, $PbO.2PbSO_4$

Sublimed white lead is basic lead sulphate, containing some amount of ZnO. It is manufactured by burning galena (PbS) in the free supply of air. In this reaction, PbS (galena) is converted partly to $PbSO_4$ and partly PbO.

$$PbS + 2O_2 \longrightarrow PbSO_4\ ; \qquad 2PbS + 3O_2 \longrightarrow 2PbO + 2SO_2$$

Zn, which is present as impurity in galena, is oxidised to ZnO, which gets mixed with $PbO.2PbSO_4$. Thus sublimed white lead also contains some amount of ZnO. The approximate composition of sublimed white lead is $PbSO_4$ = 75%, PbO = 20% and ZnO = 5%.

Sublimed white lead is used as a *white pigment* and is superior to white lead [$(Pb(OH)_2).2PbCO_3$], since it is non-poisonous and does not get blackened by H_2S.

Super-sublimed white lead, $PbO.2PbSO_4$

Super-sublimed white lead is also basic lead sulphate ($PbO.2PbSO_4$) but does not contain any ZnO, as sublimed white lead has. It is prepared in a lead atomiser burner. In this method, molten lead is allowed to enter into the atomiser burner by means of compressed air and is allowed to burn with natural gas. The white flame, so produced, is directed into a chamber into which a regulated current of SO_2 is also fed in presence of O_2. The fumes of basic lead sulphate are collected in dust chambers.

$$2Pb\ (molten) + O_2 \longrightarrow 2PbO; \qquad 2Pb + 2SO_2 + O_2 \longrightarrow 2PbSO_4$$

$$PbO + 2PbSO_4 \longrightarrow PbO.2PbSO_4 \text{ (Super-sublimed white lead)}$$

Super-sublimed white lead is the brightest of all the white pigments and is whiter than any other lead pigment.

Potassium antimonyl tartrate, $K(SbO)[CH(OH)COO]_2$

This compound is the *potassium antimonyl salt* of *tartaric acid*,

$$\begin{array}{l} CH(OH)COOH \\ | \\ CH(OH)COOH \end{array}$$

or $[CH(OH)COOH]_2$ and hence is represented as

$$\begin{array}{l} CH(OH)COOK \\ | \\ CH(OH)COO(SbO) \end{array}$$

It is also called *tartar emetic*.

This compound is obtained by boiling antimony trioxide (Sb_2O_3) with water and pot. hydrogen tartrate (*cream of tartar*), until the solution becomes clear and then cooling the resulting solution.

$$2\begin{array}{l}\text{CH(OH)COOK}\\ |\\ \text{CH(OH)COOK}\end{array} + Sb_2O_3 \longrightarrow 2\begin{array}{l}\text{CH(OH)COOK}\\ |\\ \text{CH(OH)COO(SbO)}\end{array} + H_2O$$

Pot. hydrogen tartrate — Tartar emetic

Properties. It forms colourless octahedral crystals which are readily soluble in water but insoluble in alcohol. It has an unpleasant metallic taste. A small dose of this material (0.05 to 0.1g) produces vomiting; larger amounts are posinous.

Uses. It is used (*i*) as an emetic in medicines (*ii*) in the treatment of kala-azar and other deseases (*iii*) as a mordent in dying and calico-printing and in cheap enamels.

University Questions

1. Give the composition of White Lead. ***(Banglore, 2004)***

38

Miscellaneous Topics

Alkali metal complexes with crown ethers

Introduction

In 1967, a new type of ligands was discovered. These ligands were called *macrocyclic ethers* or *crown ethers* or simply *crowns* by C.J. Pederson. These ligands are polydentate ligands and are also called *polyether ligands.*

Nomenclature of polyether ligands

Trivial names have been suggested by Pedersen for these ligands. According to this nomenclature, the trivial name of a crown ether ligand consists of (*i*) number and name of the substituent groups on the polyether ring (*ii*) total number of atoms present in the ring (*iii*) class name *viz., crowns* (N) the number of ether oxygens present in the ring. On this basis, the trivial names and abbreviations used for some polyether ligands are given below. Thus, in dibenzo-18-crown-6, dibenzo are the substituent groups (two), 18 refers to the total number of atoms present in the ring, crown is the name for the class of ligands and 6 indicates the total number of ether oxygens present in the ring. These ligands are also called crown-4, crown-5, crown-6 etc. The structure of some polyether ligands is given below in Fig. 38.1.

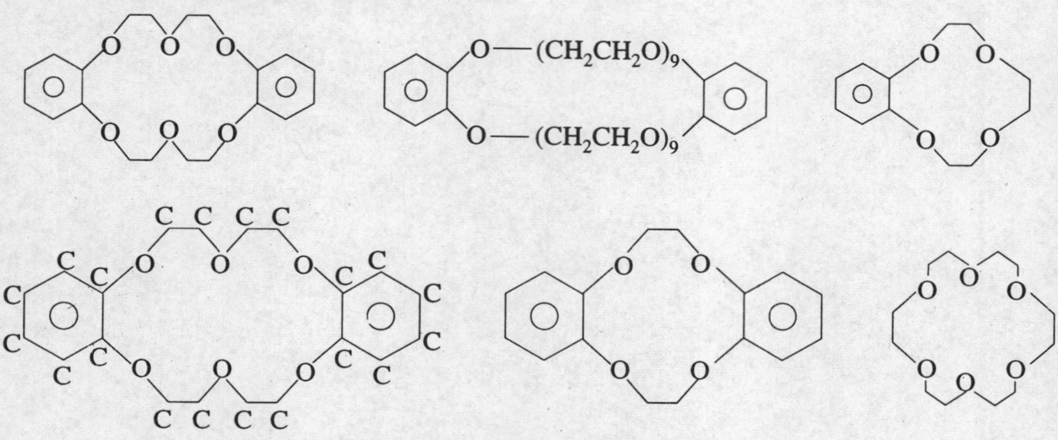

Fig. 38.1 : Structure of some polyether ligands.

Polyether complexes of alkali metals

Polyether ligands have an unusual property of forming stable complexes with alkali metal salts. Li^+ ion prefers to form complex compounds with crown-4, Na^+ ion with crown-5, while K^+ and Rb^+ ions form the polyether complexes with crown-6. Alkali metal salts are normally not soluble in organic solvents, but in the presence of these ligands, these salts become soluble, *e.g.*, $KMnO_4$ is soluble in C_6H_6, in the presence of dicyclohexyl-18-crown-6 (DC 18 C 6). Complexes of alkali metals with polyether ligands are called *crown complexes.*

Only that alkali metal ion, which exactly fits into the hole in the polyether ligand, forms stable complex. The cations which are either too small or too big to fit into the hole would not form stable complexes with polyether ligands. For example, Li^+ and Na^+ ions are too small in size and hence are not able to fit exactly into the hole of cryptate-222. Thus, these cations do not form stable complexes with cryptate-222. Rb^+ and Cs^+ cations are too big in size to fit into the hole of the ligand. Thus, these cations also do not form complexes with cryptate-222. K^+ ion is of the right size and hence it fits into the hole in the ligand. Thus, K^+ ion forms stable complex with cryptate-222. Thus we find that, among alkali metal cations, it is only K^+ which forms stable complex with cryptate-222.

Out of alkali metal cations, only Na^+ ion forms stable complex with cryptate-221. In the formation of complexes, polyether ligands wrap the metal cation completely and hence these complexes are also called *wrap around complexes.*

Synthesis of complexes

For the preparation of crown complexes of alkali metals, the desired alkali metal salt and crown ligand are dissolved in an appropriate warm, non-aqueous solvent (*e.g.*, ethanol). On cooling the solution, the crown complex compound gets precipitated. Examples of some crown complex compounds, given by benzo-15-crown-5 (B 15 C 5), dibenzo-18-crown-6 (DB 18 C 6) and dicyclohexyl-18-crown-6 (DC 18 C 6) ligands with alkali metal ions are (*i*) B 15 C5 $NaI.H_2O$, $(B\ 15\ C\ 5)_2.KI$ (*ii*) DB 18 C 6. RbNCS, DB 18 C 6. $NaNCS.H_2O$, DB 18 C 6.KNCS, DB 18C6.$NaBr.2H_2O$ (*iii*) DC 18 C 6. $NaBr.H_2O$.

Structure of complexes

In general, in the crystal lattice of 1 : 1 complex compounds, the metal ion is located at or near the centre of the oxygen atoms ring and the metal ion is linked with all O-atoms. The metal ion is also coordinated to the anion, in the direction perpendicular to the plane of the ring. The structure of DB 18 C 6, RbNCS complex compound is shown in Fig. 38.2.

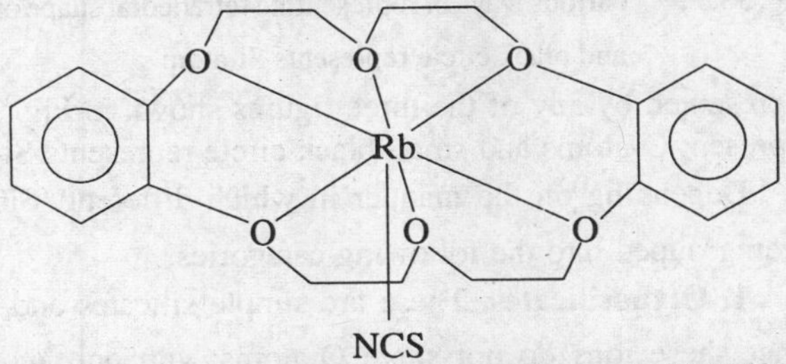

Fig. 38.2 : Structure of complex compound of RbNCS with dibenzo-18-crown-6 (DB 18 C 6. RbNCS complex compound).

Silicates

Introduction

Silicates are alumino-silicates of magnesium, calcium, iron etc. Quartz, mica, feld spar and zeolites are important examples of silicates. These are found in almost all rocks, clays and soils present in earth's crust. Glass and cement are two most useful silicates, which are the man-made silicates.

General properties of silicates

Silicates are of various types. Excepting Na_2SiO_3, which is soluble in water, all other silicates are insoluble in water. Si—O bond in silicates is very strong. This bond can be broken when silicates are treated with strong reagents like HF. Although Si—O bond is covalent, it has appreciable degree of ionic character. Ionic character is due to the fact that the electronegativity difference between oxygen and silicone atoms is equal to 3.5 – 1.8 = 1.7.

Structure and types of silicates

The structure of silicates has been found with the help of X-ray diffraction techniques. All silicates have tetrahedral SiO_4^{4-} ion, as a basic building unit, *i.e.*, all silicates are composed of many units. Tetrahedral shape of SiO_4^{4-} ion is due to sp^3 hybridisation of Si-atom, in its excited state. Valence-shell configuration of Si-atom is $3s^2\ 3p_x^1\ 3p_y^1\ 3p_z^0$, while this configuration, in the excited state is $3s^1\ 3p_x^1\ 3p_y^1\ 3p_z^1$. These four orbitals undergo sp^3 hybridisation and give rise to the formation of four singly-filled sp^3 hybrid orbitals. Each of the four Si—O-bonds in SiO_4^{4-} ion is [sp^3 (Si) — $2p$ (O^-)] σ-bond. Note that the valence-shell configuration of O^- has one singly-filled $2p$ orbital $\left[O = 2s^2\ 2p_x^1\ 2p_y^2\ 2p_z^1\ ;\ O^- = 2s^2\ 2p_x^1\ 2p_y^2\ 2p_z^2 \right]$. Tetrahedral shape of SiO_4^{4-} ion can be

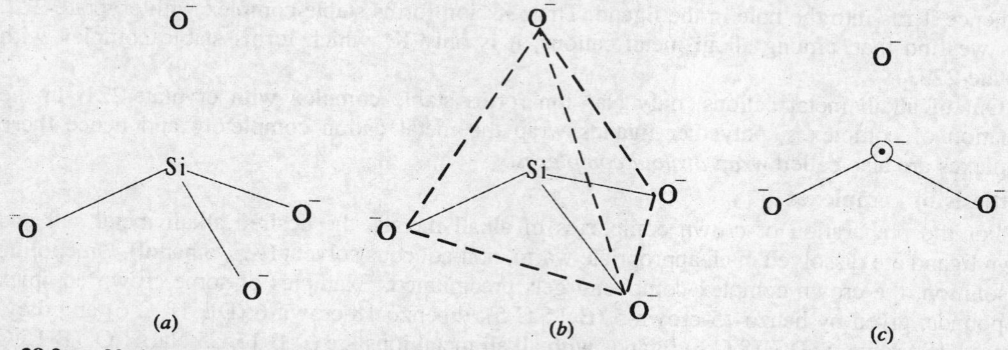

Fig. 38.3 : Various ways of representing tetrahedral shape of SiO_4^{4-} ion. In (*c*) white circles represent O-atoms and black circle represents Si-atom.

represented by any of the three figures shown in Fig. 38.3. In representation (*c*), the white circles represent O-atoms and small black circle represents Si-atom.

Depending on the manner in which different SiO_4^{4-} units are linked together, silicates have been grouped into the following catagories.

1. Orthosilicates. These are simple silicates and contain discrete orthosilicate (SiO_4^{4-}) ions, since these ions do not share O-atoms with one another. Examples of orthosilicates are *zircon* ($ZrSiO_4$), *wellimite* (Zn_2SiO_4), *phenacite* (Be_2SiO_4) and *garnets* [$M_3^{2+}\ M_2^{3+}(SiO_4)_3$] where M^{2+} = Ca^{2+}, Mg^{2+} or Fe^{2+} and M^{3+} = Al^{3+}, Cr^{3+} or Fe^{3+}.

2. Pyrosilicates or islands. The basic unit in these silicates in pyrosilicate ion, $Si_2O_7^{6-}$. This ion is formed when two SiO_4^{4-} tetrahedrons share one O-atom (bridging O-atom), *i.e.*, in this ion one O-atom acts as a bridge between two Si-atoms. The structure of this ion is shown in Fig. 38.4.

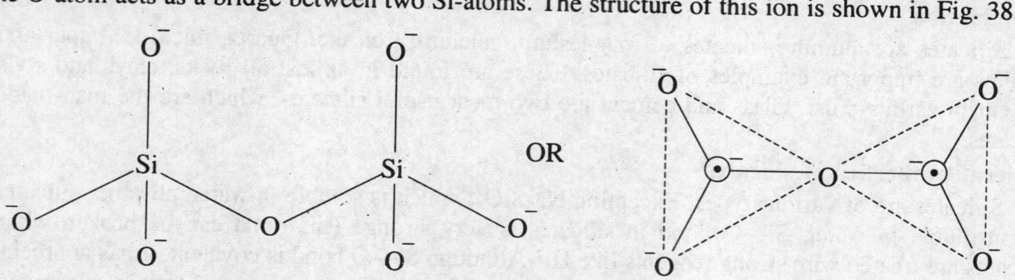

Fig. 38.4 : Structure of pyrosilicate ion, $Si_2O_7^{6-}$.

This structure shows that each Si-atom is surrounded by $3 + \frac{1}{2} = 3.5$ oxygen atoms. Thus the basic unit found in these silicates can be represented as $[SiO_{3.5}^{3-}]_2$ or $Si_2O_7^{6-}$ ion.

It may be noted from the structure of $Si_2O_7^{6-}$, ion that O-atom, which is shared by two SiO_4^{4-} units, does not carry any charge on it. O-atoms, forming no bridges between SiO_4^{4-} tetrahedrons, have negative charge. This negative charge is produced because each of these O-atoms picks up an electron from some metal. Examples of pyrosilicates are *thortveitite* ($Sc_2Si_2O_7$) and *hemimorphite* [$Zn_4(OH)\ Si_2O_7$].

3. Chain silicates. These silicates are of two types *viz., single chain silicates* and *double chain silicates.* Single chain silicates contain $[(SiO_3)_n]^{-2n}$ ion as the primary unit, while double chain silicates contain $[(Si_4O_{11})_n]^{-6n}$ as the primary unit. The structure of these ions is shown at (*a*) and (*b*) respectively of Fig. 38.5. $[(SiO_3)_n]^{-2n}$ ion is obtained when each tetrahedron shares two O-atoms (bridging O-atoms). $[Si_4O_{11})_n]^{-6n}$ ion is formed by further sharing of O-atom by half the Si-atom.

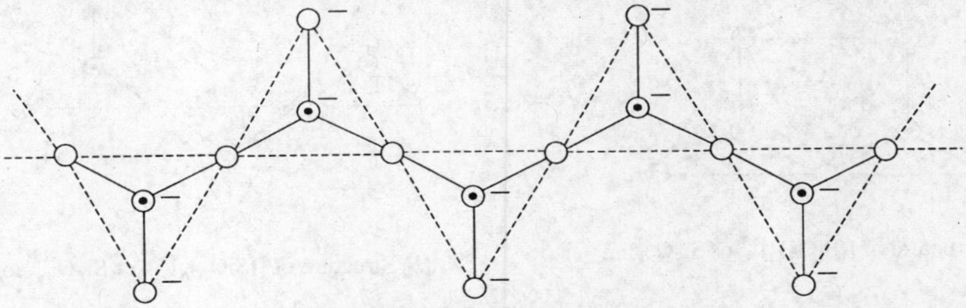

(*a*) **Structure of $[(SiO_3)_n]^{-2n}$ ion**

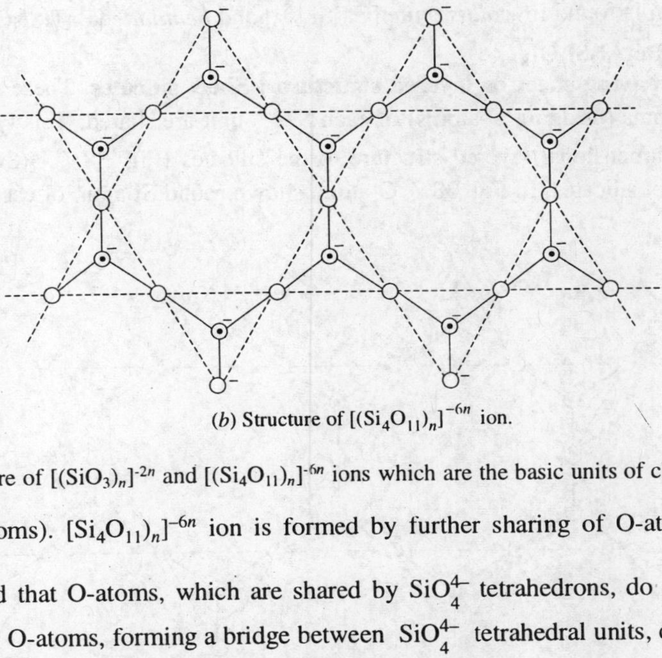

(*b*) **Structure of $[(Si_4O_{11})_n]^{-6n}$ ion.**

Fig. 38.5 : Structure of $[(SiO_3)_n]^{-2n}$ and $[(Si_4O_{11})_n]^{-6n}$ ions which are the basic units of chain silicates.

It may be noted that O-atoms, which are shared by SiO_4^{4-} tetrahedrons, do not contain any negative charge, *i.e.,* O-atoms, forming a bridge between SiO_4^{4-} tetrahedral units, do not carry any negative charge. O-atoms, forming no bridge, carry negative charge.

Common examples of single chain silicates are the minerals *pyroxenes,* like *enstatite* ($MgSiO_3$), *diopside* [$CaMg(SiO_3)_2$] and *spodumene* [$LiAl\ (SiO_3)_2$]. Common examples of double chain silicates are the minerals *amphiboles,* like *tremolite* [$Ca_2Mg_5(Si_4O_{11})_2\ (OH)_2$], *crocidolite* $\left[Na_2\ Fe_3^{2+}\ Fe_2^{3+}(Si_4O_{11})_2(OH)_2\right]$ etc.

4. Cyclic or ring silicates. These silicates have cyclic structure and contain $[(SiO_3)_n]^{-2n}$ ion as the primary unit. These silicates are formed when each SiO_4^{4-} tetrahedron shares two O-atoms (bridging O-atoms). The structures of cyclic ions *viz.*, $[(SiO_3)_3]^{-6}$ or $Si_3O_9^{-6}$ and $[(SiO_3)_6]^{-12}$ or $Si_6O_{18}^{-12}$ ions are shown at (*a*) and (*b*) respectively of Fig. 38.6.

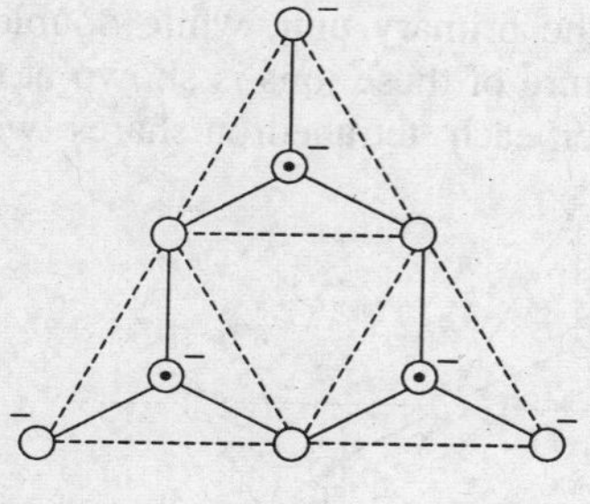

(*a*) Structure of $[(SiO_3)_3]^{-6}$ or $Si_3O_9^{-6}$ ion

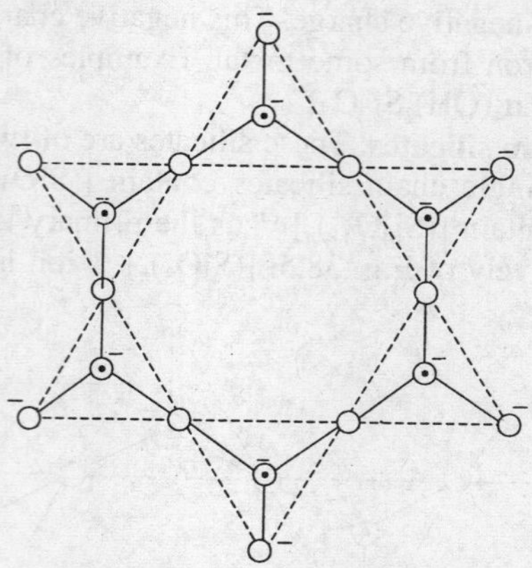

(*b*) Structure of $[(SiO_3)_6]^{-12}$ or $Si_6O_{18}^{-12}$ ion

Fig 38.6 : Structure of $Si_3O_9^{6-}$ and $Si_6O_{18}^{-12}$ ions.

$Si_3O_9^{6-}$ ion is found in *wollastonite* $[Ca_3Si_3O_9]$ and *bentonite* $[BaTiSi_3O_9]$, and $Si_6O_{18}^{-12}$ ion occurs in *beryl* $(Be_3Al_2Si_6O_{18})$.

5. Silicates having sheet or layered structure : Sheet silicates. These silicates are formed when three O-atoms (bridging O-atoms) of each SiO_4^{4-} unit are shared. This type of sharing gives an infinite two-dimensional layered structure to the silicates (Fig. 38.7). $[(Si_2O_5)_n]^{-2n}$ ion is the basic unit of sheet silicates. In Fig. 38.7, O-atom, shown round Si-atom of each SiO_4^{4-} tetrahedron

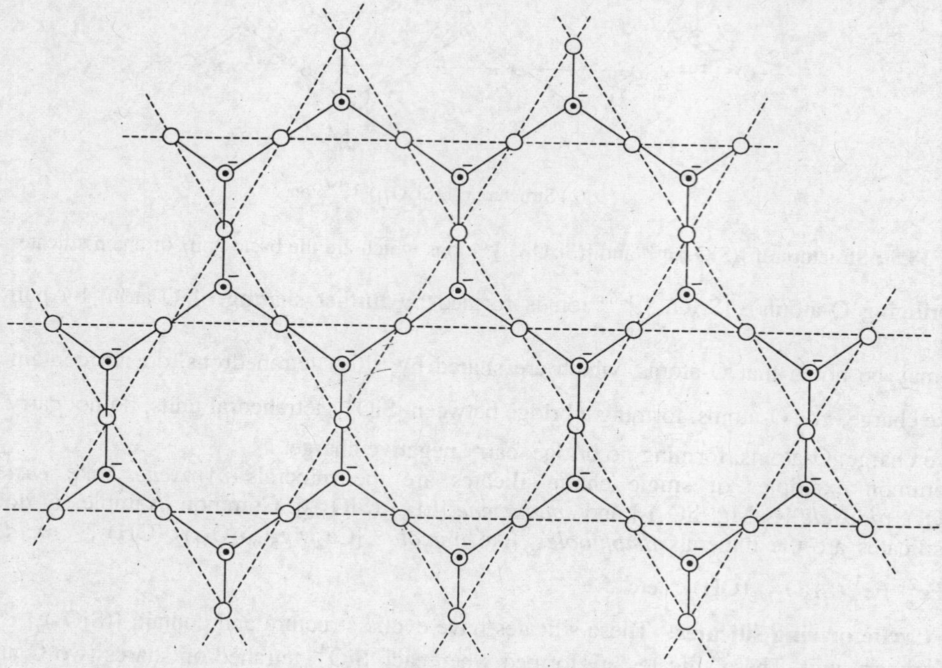

Fig. 38.7 : Structure of sheet silicates.

carries a negative charge, while the remaining three O-atoms, which act as bridging O-atoms, do not carry this charge. Negative charge on non-bridging O-atom is neutralised by the positive charge on the cations, which lie in between the sheets.

Examples of silicates having layered structure are given below. These are generally *aluminio silicates.*

(*i*) *Clays.* These are produced by weathering and decomposition of igneous rocks. They are generally composed of very fine particles. *Kaoline* (which is a china clay) is formulated as $Al_2O_3.2SiO_2.2H_2O$ or $Al_2(OH)_4Si_2O_5$ and has a sheet-structure, in which the sheets are held together by OH bridges.

When clay is mixed with water, it becomes plastic-soft and mouldable. Water of plasticity can be removed by heating at 100°C. Clay becomes rigid and brittle when water of plasticity has been eliminated. On heating beyond 100°C, structural water of crystallisation is removed and carbonaceous matter of the clay gets oxidised. At about 900°C, *multite* ($Al_6Si_2O_{13}$) begins to be formed. This new substance has *glassy appearance.* Clay is widely used in making chinaware, fire bricks and many other useful materials.

(*ii*) *Talc.* It is formulated as $Mg_3(OH)_2$ (Si_4O_{10}), which consists of electrically neutral sheets and has no cations in between them. It is soft and smoth. It is chemically inert. It is used as a dry lubricant, in ceramics, paper manufacture and cosmetics.

(*iii*) *Muscovite, $KAl_2(OH)_2$ (Si_3AlO_{10}).* Muscovite is an example of mica. It is obtained when one Si-atom in $Si_4O_{10}^{4-}$ ion is replaced by Al-atom. Thus we see, that in muscovite, the sheet composition is $(Si_3AlO_{10})^{5-}$. Two $(Si_3AlO_{10})^{5-}$ sheets, which have tetrahedral vertices inward, are linked by binding Al^{3+} cations, whose octahedral coordination is completed by OH^- ions. The double sheets, which have the composition, $[Si_3Al_3O_{10}(OH)_2]^-$, are stacked one upon another with sufficient number of K^+ cations. These cations are located in between two such double sheets and maintain electrical neutrality. The cleavage of mica is due to the weakening of these layers of K^+ ions. The binding Al^{3+} ions and the sandwiched K^+ ions can be replaced by Mg^{2+} and Na^+ ions respectively.

Micas are often tough, elastic and transparent. These are chemically inert, stable towards heat and have high dielectric constant. They are used in furnace windows and electrical applicances. The powdered mica is used as a filler for rubbers, plastics and in insulation boards.

6. Silicates having three dimensional structure. In these silicates, all the four O-atoms act as bridging atoms, *i.e.*, in these silicates, each SiO_4^{4-} tetrahedron shares all its four O-atoms with other SiO_4^{4-} tetrahedron. This type of sharing gives three-dimensional silicates, corresponding to the composition SiO_2 (*quartz, tridymite* or *crystobalite*). The replacement of Si^{4+} ions by a combination of Al^{3+} and other cations (to maintain electrical neutrality) gives minerals known as *feldspars, zeolites* and *ultramarines.*

Feldspars. These are also aluminio silicates. Almost 67% of the igneous rocks are composed of feldspars minerals. These have been grouped into two catagories namely (*i*) *Orthoclase feldspars.* Examples are *orthoclase* ($KAl\ Si_3O_8$), *celsian* ($BaAl_2Si_2O_8$) etc. (*ii*) *Plagioclase feldspars.* Examples are *albite* ($Na\ Al\ Si_3O_8$), *anorthite* ($CaAl_2Si_2O_8$) etc.

Orthoclase feldspars are more symmetrical than plagioclase feldspars, because the cations *viz.*, K^+ and Ba^{2+} present in them are of the right size to fit into the lattice. On the other hand in plagioclase feldspars the cations Na^+ and Ca^{2+} are smaller and hence produce distortion.

Zeolites. These are hydrated aluminosilicates and have three-dimensional structure. These may be either naturally occurring or artificially synthesised substances. These can be represented by the general formula, $M_x^{+n}\left[Al_x\,Si_y\,O_{2x+2y}\right]_n^{-x}.zH_2O$ or $M_x^{+n}\left[(AlO_2)_x\ (SiO_2)_y\right]_n^{-x}.zH_2O$ where n is the charge on metal cation M which may be Na^+, K^+, Ca^{2+} or Mg^{2+} and z is the number of moles of water of hydration. Examples of zeolites are (*i*) *Erionite* which may be formulated as $M^{2+}[(AlO_2)_2\ (SiO_2)_6].6H_2O$ ($M^{2+} = Ca^{2+}$ or Mg^{2+}) or $M_2^+\ [(AlO_2)_2\ (SiO_2)_6].6H_2O$ ($M^+ = Na^+$ or K^+)

(*ii*) *Gemelinite,* $Ca[(AlO_2)_2\ (SiO_2)_4].6H_2O$ or $Na_2[(AlO_2)_2\ (SiO_2)_4].6H_2O$ (*iii*) *Permutit,* $Na_2[(AlO_2)_2\ (SiO_2)_2].xH_2O$. It is an artifically prepared zeolite and can be obtained in the form of course sand by fusing together Na_2CO_3, Al_2O_3 and SiO_2. The product is washed with water to remove soluble impurities, leaving behind a porous mass of permutit.

Zeolites have SiO_4 and AlO_4 tetrahedrons, linked together in a three-dimensional open structure, in which four or six membered rings predominate. The open structure permits the formation of channels and cavities, having their diameter in the range 200-1100 pm. It is due to the presence of cavities in their open structure that zeolites can take up water and other small molecules (*e.g.*, NH_3, CO_2, C_2H_5OH etc.) in their cavities. Bigger or complex molecules, like methane and ethane, cannot be trapped in the cavities. Thus we find that zeolites can be used as molecular sieves. In zeolites, the cations move freely through the open structure and can be exchanged reversibly with other cations. For example, Na^+ ions in zeolites can be exchanged by Ca^{2+} or exchange their positions present in hard waters. Thus zeolites can also be used in softening water (water softeners). For example, permutit is used as a softener for removing Ca^{2+} / Mg^{2+} ions present in hard water. When hard water is passed through permutit, Na^+ ions present in permutit are exchanged by Ca^{2+}/ Mg^{2+} ions present in hard water.

$$Na_2[(AlO_2)_2(SiO_2)_2].xH_2O + M^{2+} \longrightarrow M^{2+}\ [(AlO_2)_2\ (SiO_2)_2.xH_2O + 2Na^+$$

Water coming out of permutit is soft because both Ca^{2+} and Mg^{2+} ions which are responsible for the hardness of water are retained by the permutit.

Regeneration of permutit. After about 12 hours, nearly the whole of permutit gets converted into calcium or magnesium zeolite $M^{2+}[(AlO_2)_2(SiO_2)_2].\ xH_2O$ ($M^{2+} = Ca^{2+}$ or Mg^{2+}), *i.e.*, permutit gets exhausted or spent. It may be regenerated [*i.e.,* it may be converted into $Na_2[(AlO_2)_2\ (SiO_2)_2].xH_2O]$ by passing a 10% solution of brine (NaCl) through it.

$$M^{2+}[(AlO_2)_2(SiO_2)_2].xH_2O + 2NaCl \longrightarrow Na_2[(AlO_2)_2(SiO_2)_2].xH_2O + MCl_2$$

$$\text{or}\quad M^{2+}\ [(AlO_2)_2\ (SiO_2)_2].x\ H_2O + 2Na^{2+} \longrightarrow \underset{\text{Permutit regenerated}}{Na_2[(AlO_2)_2(SiO_2)_2].xH_2O} + M^{2+}$$

The regenerated permutit can again be used for softening water.

Ultramarines. These are also aluminosilicates, but do not contain water of hydration. Some extra anions, like Cl^-, S_2^{2-} and SO_4^{2-}, may also be present in the cavities. Examples of ultramarines are *sodalite,* $Na_8(AlO_2)_6\ (SiO_2)_6Cl_2$; *ultramarine*, $Na_8(AlO_2)_6(SiO_2)_6S_2$; and *nosean,* $Na_8(AlO_2)_6\ (SiO_2)_6\ SO_4$. Many ultramarines are coloured and hence are used as pigments. When sodalite is fused with Na_2SO_4, nosean is obtained.

Glass

Composition of glass

Glass is an amorphous, transparent or translucent rigid super-cooled liquid of indefinite viscosity. Glass is not a single compound, but is a mixture of a number of metallic silicates. No definite formula can be assigned to glass and it has no definite melting point. Within certain limits, glass may be represented by the formula $x\ Na_2O$ (or $x\ K_2O$).yMO.6SiO_2, where M is a divalent metal like Ca, Ba, Pb, Zn etc., and x and y are the number of molecules. In other varieties of glass SiO_2 may be replaced by Al_2O_3, B_2O_3, P_2O_5 etc.

Raw materials required in the manufacture of glass

Raw materials required in the preparation of glass are (*i*) compounds of alkali metals like Na_2CO_3, Na_2SO_4, $NaNO_3$, K_2CO_3 and KNO_3. These compounds provide Na_2O or K_2O. (*ii*) compounds of alkaline earth metals such as $CaCO_3$, $BaCO_3$. These compounds provide CaO or BaO. CaO may be used as such. (*iii*) oxides of heavy metals like PbO, Pb_3O_4, ZnO. (*iv*) silica which is used in the form of quartz, ignited flint, white sand or ordinary sand. (*v*) $Ca_3(PO_4)_2$, oxides of As and Sb are used when opalescent glass is to be prepared. (*vi*) Raw materials used in the preparation of glass usually contain FeO and Fe_2O_3 as impurity. FeO gives a green colour to the glass whereas Fe_2O_3 imparts light yellow tint to the glass. In order to get colourless glass oxidising agents like MnO_2, KNO_3 etc., are also added along with raw materials. These oxidising agents are called *decolourisers.* (*vii*) In order to get coloured glass, certain metallic oxide is added to the molten mass. These oxides give specific colour to the glass. For example Fe_2O_3 is added to produce yellow colour in the glass, Cr_2O_3 produces green colour in the glass, MnO gives purple glass, CoO is used to get blue colour in the glass etc. (*viii*) Borax is used as a flux. Borax also lowers the coefficient of expansion of the glass and increases its durability. (*ix*) Alumina makes glass resistant to sudden changes of temperature. Alumina is often used in the form of feldspar. It is a naturally occurring mineral with a formula $Na_2O.Al_2O_3.6SiO_2$ (soda feldspar) or $K_2O.Al_2O_3.6SiO_2$ (potash feldspar). (*x*) Pieces of broken waste glass (scrap glass), called *cullets,* are added to the batch (Na_2CO_3 + $CaCO_3$ + SiO_2) during fusion. These pieces serve two purposes—fascilitate the fusion (flux) and secure utilisation of waste glass.

Manufacture of ordinary glass

We know that ordinary glass is soda lime glass and hence its approximate composition is $Na_2O.CaO.6SiO_2$. The manufacture of ordinary glass consists of the following steps :

1. Fusion of raw materials. Na_2CO_3, $CaCO_3$ and sand (SiO_2) are finely powdered and are mixed together in proper proportions. The resulting mixture, called **batch,** is mixed with broken pieces of scrap glass (called **cullets**), obtained from the previous run. Scrap glass fascilitates fusion and thus acts as a flux. The mixture of Na_2CO_3, $CaCO_3$, SiO_2 (batch) and scrap glass (cullet) is fused either in the pots of pot furnace or in the tank of tank furnace. The furnace is heated by producer gas, on the regenerative principle of heat economy. The following reaction takes place :

$$Na_2CO_3 + CaCO_3 + 6SiO_2 \longrightarrow \underset{\text{ordinary glass}}{Na_2O.CaO.6SiO_2} + 2CO_2 \uparrow$$

In the beginning, considerable frothing is produced on account of the formation of CO_2. On continued heating, the whole of CO_2 escapes out and a *clear, viscous, fused mass* is obtained. To get coloured glass, colouring material, which is usually a metallic oxide (*e.g.*, Fe_2O_3, CoO, Cr_2O_3, etc.) is added and the mass is heated again till it becomes homogeneous.

2. Shaping. The fused mass ,obtained as above, is allowed to cool down and then blown or cast into moulds, to get the article of the desired shape. If sheet glass is required, it is pressed between rollers.

3. Annealing. It is the process in which the newly shaped articles are cooled slowly. If the articles are cooled quickly, they become brittle on account of high internal strain. During annealing, the molecules of glass arrange themselves in such a way that there is no strain on them when the article is cooled. Thus, as soon as the article is ready, it is passed through a series of cooling chambers, where cooling is carried out, step by step.

4. Finishing. The articles so obtained require some finishing touch, such as cutting, grinding, polishing etc., after annealing. Only the finished products are sent to the market.

Varieties of glass

By varying the composition of bitch, several varieties of glass can be obtained. Some important varieties are discussed below :

1. Soda lime glass or soft glass or ordinary glass ($Na_2O.CaO.6SiO_2$) This is an ordinary variety of glass and is also called **window glass.** It is a mixture of sodium and calcium silicates. Its preparation has already been given. It is a cheap variety of glass. It is called, soft glass, because it softens at a relatively lower temperature and can easily be shaped into different shapes. It is very slowly attacked by chemical reagents and even by water. It is used in making glass tubes, window panes, bottles, dishes etc.

2. Potash lime glass or hard glass ($K_2O.CaO.6SiO_2$). It is also called **Boherian glass.** It is a mixture of potassium and calcium silicates. The batch used for its preparation consists of a mixture of K_2CO_3, limestone ($CaCO_3$) and sand (SiO_2). It fuses with difficulty and is quite resistant to the action of water and acids. It is used in making combustion tubes and chemical apparatus which can withstand high temperatures.

3. Potash-lead glass or flint glass ($K_2O.PbO.6SiO_2$). It is a mixture of potassium and lead silicates. The batch used for its preparation consists of K_2CO_3, red lead and sand. It is lustrous and transparent product and reflects light much better than other types of glass. It has high refractive index. It is soft and scratches easily. Because of its high refractive index, some varieties of flint glass are used in making artificial gems and decorative articles. It is also used in making prisms, lenses, optical instruments, electric bulbs etc.

4. Jena glass. It is mixture of barium and zinc boro-alumino silicates, *i.e.,* this glass contains BaO, ZnO Al_2O_3 and B_2O_3. In place of silica, Al_2O_3 is present in higher amount than in ordinary glass. This glass also contains alkali, which is present in low amount. It has low coefficient of expansion and is more resistant to heat, shock, acid and alkali.

5. Pyrex glass. It is mainly a mixture of boro and alumino silicates of Zn and Ba. Its approximate composition is SiO_2 = 80%, Na_2O = 4%, CaO = 0.5%, K_2O = 0.5%, B_2O_3 = 12%, Al_2O_3 = 3%. It can withstand sudden changes of temperature and is highly suited for laboratory glassware, such as flasks, beakers etc., which are to be heated to high temperatures.

6. Crooke's glass. In the preparation of this glass, a part of CaO is replaced by the oxides of Pb, Ba, Zn, Cd or Mg. Cerium oxide (CeO_2) is also added. This oxide cuts ultraviolet rays, which are harmful to the eyes. Thus, this glass is used in making special type of lenses.

7. Quartz glass or fused silica glass. This variety of glass is obtained by melting silica in the electric furnace and then made into tubes, basins, crucibles etc. Silica glass has the following characteristics, due to which silica glass articles have advantages over glass articles (*i*) silica glass has a very low coefficient of expansion and hence, a white hot article when put into water, does not break (*ii*)can withstand a temperature of over 1000°C (*iii*) it is less attacked by chemical reagents than glass. *Vitreosil* is an opalescent form of quartz glass and is cheaper than the transparent form.

8. Crown glass. A partial replacement of silica by phosphorus pentoxide gives crown glass.

9. Ground glass. It is prepared by grinding ordinary soda (soft) glass by emery and turpentine oil or by *sand blasting.*

10. Safety glass. It is used in making wind screens of automobiles, aeroplanes, railway trains, etc. It is prepared by placing a layer of *transparent plastic* (usually a sheet of *vinyl acetate resin*) between two layers of glass by means of a suitable adhesive. The three layers are joined together tightly by the action of heat and pressure.

This glass does not break easily under ordinary impact. This is the reason for the name "safety glass" or "unbreakable glass" given to it. However, it does break under heavy impact. But the particles of glass do not shatter because they are held by the plastic.

Enamel

When glass is rendered opaque by substances such as tin dioxide, antimony oxide, barium suphate etc., it is called enamel. We thus get white enamel. Like glass, enamel may be variously coloured by adding suitable colouring materials. Enamel is used for "*enameling*" iron articles, sign boards, name plates, gold and silver ornaments and various other materials.

Devitrification of glass

If glass articles are left unused for a long time or are cooled extremely slowly during annealing, they gradually lose their transparency and turn opaque, due to the reason that some of the metallic silicates, of which the glass is made, crystallize out. Glass is then said to have *devitrified.*

Transparent plastics or thermoplastic

Glass is unsuited for a number of purposes, since it is brittle and fragile. In recent years, it has been replaced for such uses by transparent plastics. A plastic consists of highly polymerised substances like glass, it softens on heating and assumes any given shape and, therefore, called thermosplastic. The interlinking of the molecules in the polymerised mass gives it hardness as well as complete insolubility.

Glass textiles

Recently, glass textiles have come into use on a large scale, because the filaments are now drawn much thinner than ever before, with a great increase in tensile strength. Fibres have been made 30 miles long without a break. The thinness of such filaments can be easily visualized by the fact that even two pounds of the finest filament would encircle the earth. Such filaments find a great market in insulating tapes, and for fluffy blanket, heat insulation of houses, ships and locomotives, and for dust removal from conditioned air. These fibres can reinforce plastics. These fibres are unsuitable for cloth making, since they slightly irritate the skin.

Eatching of glass

See "*Hydrofluoric Acid*".

Cement

Cement is an important building material, which is employed in the construction of buildings. It was introduced for the first time in England in 1824 by *Joseph Aspdin*. It is also called *Portland cement,* since it resembles with the famous building stone found near Portland in England. In India, *cement* came into use in 1914.

Composition of Portland cement

Different types of cement have different compositions of metal oxides. The average composition of different oxides found in Portland cement is *lime* (CaO) = 50–60%, *silica* (SiO_2) = 20–25%, *alumina* (Al_2O_3) = 5–10%, *magnesia* (MgO) = 2–3%, *ferric oxide* (Fe_2O_3) = 1–2%, *sulphur trioxide* (SO_3) = 1–2%, *sodium oxide* (Na_2O) = 1.0%, *potassium oxide* (K_2O) = 1.0%. The essential constituents of cement are lime (obtained from limestone), silica and alumina (present in clay). For the cement of good quality, the oxides should be kept in the following ratio :

$$\frac{\%\ SiO_2}{\%\ Al_2O_3} = 2.5 - 4.0, \qquad \frac{\%\ CaO}{\%\ SiO_2 + \%\ Al_2O_3 + \%\ Fe_2O_3} = 1.9 - 2.1$$

Cement containing no iron is white, but hard to burn. If less lime is present than given by the above ratio, the cement is low in strength and sets very soon. If more lime is there, the cement cracks. Excess of silica produces a slow-hardening cement, while excess of alumina gives a quick-setting product.

Raw materials used in the manufacture of cement

The important raw materials used in the manufacture of cement are :

1. Lime stone, $CaCO_3$. It provides lime, CaO. Chalk, marble and alkali waste can also be used, since these also supply CaO.

2. Clay. The composition of clay is $Al_2O_3.SiO_3.Fe_2O_3.2H_2O$. Clay supplies silica ($SiO_2$) and alumina ($Al_2O_3$) to the cement. Some of the clays do not contain Fe_2O_3 and the cement obtained in this case is white and hard to burn.

3. Gypsum, $CaSO_4.2H_2O$. The addition of gypsum decreases the setting time of cement.

Manufacture of cement

The manufacture of cement consists of the following steps :

1. Preparation of raw meal or slurry. The raw materials *viz.*, limestone and clay are mixed in proper ratio by any of the following methods :

(*a*) *Dry process.* This process is employed when the raw materials *viz.*, limestone and clay are hard. In this process the limestone is first broken into small pieces. It is then mixed with clay in the proper proportions and finally pulverised to such a fineness that 90-95% passes through a 100 mesh sieve. The mixture is made homogeneous to produce what is known as **raw meal.**

(*b*) *Wet process.* This process is used when the raw materials *viz.*, limestone and clay are soft, the climate is fairly damp and the fuel is cheap. In this process, the limestone is crushed to suitable size and the clay is washed with water in wash mills to remove foreign materials like flint. The powdered limestone is then mixed with the clay paste in the proper proportions (limestone 75%, clay 25%) and the mixture is finely ground and made homogeneous by means of a compressed air mixing arrangement. The resulting paste is known as **slurry.** The slurry contains about 40% water.

2. Burning (calcination) of raw meal or slurry in rotary kiln : Preparation of cement clinker. Raw meal or slurry prepared as above is introduced into the rotary kiln with the help of a screw conveyor. The rotary kiln consists of a long cylinder 6 to 8 feet in diameter and 100 to 250 feet in length. It is made of steel and is lined inside with fire bricks (Fig. 38.8). The kiln rotates on its axis at the rate of $\frac{1}{2}$ to 1 revolution per minute. As the kiln rotates, the charge slowly moves downwards, due to the rotary motion of the kiln. Here the charge is heated by burning coal dust, which enters at the lower end, with the help of a blower. As the charge moves forward, it meets higher temperatures. When it reaches the lowest part of the kiln, the temperature is about 1500°C, which is the maximum temperature in the kiln. The zone having this temperature is called *maximum temperature zone.* The charge takes 2–3 hours to cover the journey in the kiln.

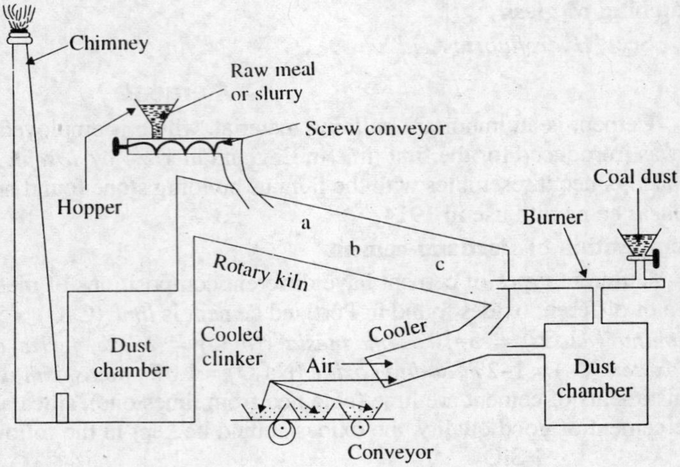

Fig. 38.8 : Rotary kiln used in the manufacture of cement. a = moderate temperature zone (800°C), b = average temperature zone (800-1000°C and c = maximum temperature zone (1000-15000°C).

Reactions taking place in the rotary kiln can be divided into the following three parts :

(*i*) *Reactions taking place in moderate temperature zone.* In this zone the temperature is upto 800°C. In this zone, free moisture is removed and clay ($Al_2O_3.2SiO_2Fe_2O_3.2H_2O$) is broken into Al_2O_3, SiO_2 and Fe_2O_3.

$$Al_2O_3.2SiO_2.Fe_2O_2.2H_2O \longrightarrow Al_2O_3 + 2SiO_2 + Fe_2O_3 + 2H_2O$$

(*ii*) *Reactions taking place in average temperature zone.* Here the temperature is 800–1000°C. In this zone, limestone ($CaCO_3$) decomposes into lime (CaO) and CO_2.

$$CaCO_3 \longrightarrow CaO + CO_2\uparrow$$

(*iii*) *Reactions taking place in maximum temperature zone.* In this zone, where the temperature is 1000–1500°C, the oxides *viz.*, CaO, $SiO_2.Al_2O_3$ and Fe_2O_3 combine together and form calcium silicates *viz.*, $2CaO.SiO_2$, $3CaO.SiO_2$; calcium aluminates *viz.*, $2CaO.Al_2O_3$; $3CaO.Al_2O_3$ and tetra-calcium alumino ferrite, $4CaO.Al_2O_3.Fe_2O_3$.

$$2CaO + SiO_2 \longrightarrow 2CaO.SiO_2 \quad (\textit{Dicalcium silicate})$$
$$3CaO + SiO_2 \longrightarrow 3CaO.SiO_2 \quad (\textit{Tricalcium silicate})$$
$$2CaO + Al_2O_3 \longrightarrow 2CaO.Al_2O_3 \quad (\textit{Dicalcium aluminate})$$
$$2CaO + Al_2O_3 \longrightarrow 3CaO.Al_2O_3 \quad (\textit{Tricalcium aluminate})$$
$$4CaO + Al_2O_3 + Fe_2O_3 \longrightarrow 4CaO.Al_2O_3.Fe_2O_3 \quad (\textit{Tetracalcium alumino ferrite})$$

The resulting product is known as **cement clinker** and as it comes out into the cooler part, it has the appearance of small greenish black or grey-coloured balls, varying in size from small nuts to peas.

Quite obviously the composition of cement clinker is : $2CaO.SiO_2$, $3CaO.SiO_2$, $2CaO.Al_2O_3$. $3CaO.Al_2O_3.4CaO.Al_2O_3.Fe_2O_3$.

3. Mixing the cement clinker with gypsum. The cement clinker is cooled in the coolers by cold air. The cooled clinker is mixed with 2–3% of its weight of gypsum ($CaSO_4.2H_2O$).3CaO. Al_2O_3, which is a fast setting constituent of the clinker, reacts with gypsum to form the crystals of calcium sulphoaluminate, $3CaO.Al_2O_2.3CaSO_4.2H_2O$.

$$3CaO.Al_2O + 3(CaSO_4.2H_2O) + 2H_2O \longrightarrow 3CaO.Al_2O_3.3CaSO_4.2H_2O + 6H_2O$$

Thus, we see that the addition of gypsum removes the fast setting $3CaO.Al_2O_3$ as $3CaO.Al_2O_3.3CaSO_4.2H_2O$ and hence the process of setting cement gets retarded and this results in better strength of the mass which sets.

Setting of cement

The use of cement in the construction of buildings is based on its property of setting to a hard mass, when its paste with water is allowed to stand for sometime. When the cement is mixed with water, it absorbs water and the mass becomes hard and very resistant to pressure. This is known as the *setting of cement.* The reactions involved in the setting of cement are :

(*i*) *Reactions taking place during first 24 hours.* A short time after the cement is mixed with H_2O, the following reactions take place :

(*a*) $3CaO.Al_2O_3$ absorbs H_2O (*hydration*) and forms a hydrated colloidal gel of tricalcium aluminate, $3CaO.Al_2O_3.6H_2O$.

$$3CaO.Al_2O_3 + 6H_2O \xrightarrow{\text{Hydration}} 3CaO.Al_2O_3.6H_2O$$

The gel of $3CaO.Al_2O_3.6H_2O$, so formed, starts crystallising slowly.

(*b*) $3CaO.Al_2O_3$, which is a fast-setting material, reacts with gypsum ($CaSO_4.2H_2O$), to form the crystals of calcium sulphoaluminate, $3CaO.Al_2O_3.3CaSO_4.2H_2O$.

$$3CaO.Al_2O_3 + 3(CaSO_4.2H_2O) + 2H_2O \longrightarrow 3CaO.Al_2O_3.3CaSO_4.2H_2O + 6H_2O$$

Thus we see that the addition of gypsum removes the fast setting $3CaO.Al_2O_3$ as $3CaO.Al_2O_3.3CaSO_4.2H_2O$ and hence the process of setting cement gets ratarded and this result in better strength of the mass which sets.

(*ii*) *Reactions taking place between 1 and 7 days.* $3CaO.SiO_2$ and $3CaO.Al_2O_2$ get hydrolysed (hydrolysis) according to the equations :

$$3CaO.SiO_2 + H_2O \xrightarrow{\text{Hydrolysis}} Ca(OH)_2 \downarrow + 2CaO.SiO_2 \ (\textit{colloidal gel})$$

$Ca(OH)_2$ formed starts changing into needle-shaped crystals which get studded in the colloidal gel, $2CaO.SiO_2$, formed as above, and thus impart strength to it.

$$3CaO.Al_2O_3 + 6H_2O \xrightarrow{\text{Hydrolysis}} 3Ca(OH)_2\downarrow + 2Al(OH)_3\downarrow$$

$Al(OH)_3$ so formed fills the interstices, resulting in hardening the mass.

(*iii*) *Reactions taking place between 7 and 28 days.* $2CaO.SiO_2$ begins to to hydrate (a very slow reaction) and forms the hydrated colloidal get of dicalcium silicate of the composition, $2CaO.SiO_2.xH_2O$.

$$2CaO.SiO_2 + xH_2O \xrightarrow{\text{Hydration}} 2CaO.SiO_2.xH_2O$$

Needles of $Ca(OH)_2$ formed above get studded in the hydrated colloidal gel of $2CaO.SiO_2.xH_2O$ and thus impart strength to it.

$4CaO.Al_2O_2.Fe_2O_3$ also gets hydrated to form a colloidal gel of the composition, $3CaO.Al_2O_3.6H_2O$.

$$4CaO.Al_2O_3.Fe_2O_3 + 6H_2O \xrightarrow{\text{Hydration}} 3CaO.Al_2O.6H_2O + CaO.Fe_2O_3$$

The gel formed as above starts losing water partly by evaporation and partly by forming hydrates with the hydrated constituents. Thus cement sets to a hard mass.

Manufacture of cement in India

In our country, at present, there are 56 factories in private as well as in public sectors, which are manufacturing cement. Following are the important factories running in the country.

1. Factories in public sector. The important factories of this type are : (*i*) *Cement Corporation of India.* This Corporation has set up the plants at Rajpur (M.P.), Kurkunta (Mysore) and Bokajan (Assam). (*ii*) *Assam Cement Corporation of India Ltd., Cherpaunji (Assam).* (*iii*) *U.P. State Cement Corporation Ltd., Dalla (U.P.)* (*iv*) *T.N. Industrial Development Corporations Ltd., Alangulam (Tamil Nadu).*

2. Factories in private sector. Important factories of this type are : (*i*) *Associated Cement Co. Ltd., Mumbai.* It is the biggest manufacturing unit in India, having 24 factories located in Madhya Pradesh (6), Gujarat (4), Bihar (3), Andhra Pradesh (3), Rajasthan (3), Mysore (2), Madras (1), Maharashtra (1) and Haryana (1). (*ii*) *Andhra Cement Co. Ltd., Vijayawada (Andhra Pradesh);* (*iii*) *Ashoka Cement Co. Ltd., Dalmianagar (Bihar);* (*iv*) *Bagalkot Udyog Co. Ltd., Bagalkot (Mysore);* (*v*) *Century Cement, Tilda (Madhya Pradesh);* (*vi*) *Dalmia Cement*

Ltd., Dalmiapuram (Tamil Nadu); (*vii*) *J.K. Synthetics Ltd., Nimbahera (Rajasthan);* (*viii*) *Jaipur Udyog Ltd., Sawai Modhopur and Beawar (Rajasthan)*; (*ix*) *Madras Cement Ltd., (Madras);* (*x*) *Mysore Cement Ltd., Bangalore;* (*xi*) *Orissa Cement Ltd., Sundergarh (Orissa);* (*xii*) *Punyam Cement and Minerals Industries Ltd., Cement Nagar (Andhra Pradesh);* (*xiii*) *Saurashtra Cement and Chemical Industries Ltd., Porbandar (Gujarat);* (*xiv*) *Shri Digvijay Cement Ltd., Digvijayagram, (Gujarat);* (*xv*) *Sone Valley Portland Cement Ltd., Japla (Bihar);* (*xvi*) *Travancore Cement Ltd., Kottayam (Kerala).*

Ceramics industry

Introduction

The art of making the articles from china clay or kaoline (*e.g.*, chinaware, earthware, stoneware etc.) comes under ceramics industry.

Raw materials required in ceramics industry

Following are the raw materials which are required in ceramics industry :

1. Clay. It is *hydrated aluminium silicate.* Some common clays are : (*i*) *Kaolin*, $Al_2O_3.2SiO_2.2H_2O$ (*ii*) *Bedellite*, $Al_2O_3.2SiO_2.H_2O$ (*iii*) *Halloysite*, $Al_2O_3.2SiO_2.3H_2O$. Clay is used for the manufacture of whiteware due to its following properties : (*i*) It is plastic and can be moulded when it is wet. (*ii*) It retains its shape when dried. (*iii*) When mould mass is heated to a high temperature, it sinters together forming a hard, coherent mass without losing its original colour.

2. Felspar or spar. It is double silicate of K or Ca with Al. Some common felspars are : (*i*) *Potash felspar*, $K_2O.Al_2O_3.6SiO_2$ (*ii*) *Soda felspar*, $Na_2O.Al_2O_3.6SiO_2$ (*iii*) *Lime felspar*, $CaO.Al_2O_3.6SiO_2$ Felspar is fusible and serves as a flux as well as a binder for the clay particles. It is not present in the naturally occurring clays and hence is added from outside.

3. Sand or flint. Sand (SiO_2) or flint is used as an opener for reducing the shrinkage of the clay.

4. Fluxes. Fluxes are easily fusible materials. Borax, boric acid, soda ash, pearl ash (K_2CO_3), fluourspar (CaF_2), cryolite (Na_3AlF_6), iron oxide, lead oxides etc., are used as fluxes,

5. Refractories. These are substances which can withstand high temperature (at least 600°) in the furnaces, without melting. Al_2O_3, FeO, Cr_2O_3, $MgCO_3$, TiO_2, SiC, ZrO_2 etc., are the important examples of refractories.

Manufacture of white ware

The manufacture of white ware consists of the following steps :

1. Preparation of slip. Clay, felspar, sand etc., are mixed with H_2O and stirred in a revolving *blunger*, to get a cream-like paste, which is known as *slip*. The slip, so prepared, is filtered through vibrating screen, to remove the coarse materials (*e.g.*, bits of wood, metals, coal etc.) present in it.

2. Filtering and aging. The slip obtained as above, is passed through the filter paper, so that almost all water is removed and a moist cake is obtained. The moist cake contains only 10–30% water. The moist cake is stored in damp cellars for a few weeks. This is called *aging of the slip*. This process increases the plasticity of the clay.

3. Shaping of the articles. The shaping of the articles is done by hand moulding, by potter's wheel or by die-pressing. When cast is required, Na_2CO_3 or K_2CO_3 or water glass is added to the clay.

4. Drying the articles. The articles are now dried in air or chambers, through which warm air circulates.

5. Firing. The dried articles are placed in clay boxes, called *saggers*, which are placed in the furnace, called *hovel oven*, so that the articles may not come in direct contact of the flames. During the process of firing the following changes occur : (*i*) At 150°- 600°C water is completely removed. (*ii*) At 800-900°C lime stone ($CaCO_3$), if any, gets decomposed ($CaCO_3 \rightarrow CaO + CO_2\uparrow$). (*iii*) At 350-900°C, oxidation of ferrous salts and organic matter present in the raw materials takes place. (*iv*) At 1200-1300°C, silicates of different metals may be formed by the combination of metallic oxides with silica (SiO_2).

The article, obtained after it has been fired, is porous like a biscuit and hence is called *biscuit.*

6. Glazing of the biscuit. In order to make the earthware impervious to liquids, glazes are applied on it. *Dipping* or *immersion, pouring, spraying, dusting, volatilization* etc., are the various methods of applying the glazes on the articles. The biscuit is again fired in a furnace at a temperature 700–990°C. During the firing, the article becomes smooth and shining in appearance. SiO_2, Al_2O_3, lead oxide, ceric oxide etc., are the examples of the glazes. The glazes contain at least one alkaline earth metal.

7. Colouring of the articles. The colouring agent is sometimes mixed in clay, felspar, sand etc., or in glaze components. The materials used as colouring agents are cobalt oxide (blue), copper oxide (green or red), chromium (green or red), iron oxide (yellow, orange or red), gold chloride (*pink*), MnO_2 (*violet*), TiO_2 (*faint yellow*) etc. The colours imparted by them are given in brackets.

Paints and pigments

Meaning of paints and pigments

Paints may be defined as liquid products, which hold in suspension solid colouring materials known as *pigments,* so blended that such mixtures can be applied evenly to a surface, for protective or decorative purposes.

Parts of a paint

There are essentially five parts of a paint which are discussed below :

1. Pigments. The colour of the paint depends upon the colour of the pigments used. These may be naturally occurring or manufactured. The pigments will be described in this book at their proper places.

2. Drying oils. The generally used drying oils are turpentine oil and tung oil (China wood oil). These act as the vehicles for the dispersion of the paint.

3. Driers. In order, that the oils may dry readily and properly, some additional substances, known as *driers* or *siccatives,* are added. They act as oxygen carriers to the oils and are added in quantities from 0.5% to 2% of the oil in the paint and are ususally metallic derivatives, particularly of cobalt, lead and manganese. MnO_2, litharge, lead acetate, manganese acetate and rosinates of these metals (obtained by the saponification of resin by NaOH and then the addition of lead, manganese or cobalt salts) are usually employed as driers. Cobalt linoleates and cobalt tungstates are also used.

4. Thinners. Thinning of the prepared paints is usually done by mixing solvents as petrol, benzene, naphtha and rosin spirit.

5. Fillers. These are inert soild substances, like china clay, talc, asbestos, $BaSO_4$ etc., which are added to the paints to increase their durability.

Requisites of a good paint

A good paint should have the requisites like : (*i*) It should be resistant to heat and light, *i.e.,* it should not peel off, on exposing to light or small changes of temperature. (*ii*) It should be resistant to the corrosive action of acids and alkalies. (*iii*) It should not be mutable by atmospheric changes, such as rains, winter, summer etc.

Classification of pigments

Pigments have been classified into the following two ways :

1st classification. This classification is based on the name of the element, which forms the pigment. This classification gives the following types of pigments :

1. Lead pigments. Examples are : (*a*) *Basic lead carbonate,* $Pb(OH)_2.2PbCO_3$ (also called *white lead*). It is used as a white pigment.

(*b*) *Basic lead sulphate* (also called ***sublimed white lead***), ***$PbO.2PbSO_4$***. **It is also a white** pigment and contains some ZnO.

(*c*) *Super-sublimed white lead,* $PbO.2PbSO_4$. It is also basic lead sulphate, but does not contain ZnO. It is whiter than any of the lead pigments.

(*d*) *Triplumbic tetroxide* (also called *red lead*), Pb_3O_4.

2. Chromium pigments. Several chromic compounds [*i.e.,* Cr (+3) compounds] are used as pigments in pottery glazes, building paints, printing colours etc. These pigments give fast and stable colours. Important compounds, which are used as pigments, are given below :

(*a*) *Lead chromate* (also called *chrome yellow*), $PbCrO_4$. It is used as a yellow pigment.

(*b*) *Basic lead chromate* (also called *chrome red*), $Pb(OH)_2.PbCrO_4$. It is a red pigment.

(*c*) *Hydrated chromic oxide,* $Cr_2O_3.2H_2O$*: Guigrets green.* It has a brilliant green colour and is in the powdered form. It is used as a green pigment.

(*d*) *Chrome green.* It is a green pigment and is obtained when chrome yellow ($PbCrO_4$) is mixed intimately with prussian blue.

(*e*) *Basic zinc chromate.* It is used as a yellow pigment.

3. Cobalt pigment. Examples are given below :

(*a*) *Thenard's blue,* $CoO.Al_2O_3$ *or* $Co(AlO_2)_2$. As is obvious, it is *cobalt meta-aluminate.* It is used as a blue pigment. This is obtained by heating CoO with Al_2O_3 or by heating cobalt phosphate or some other cobalt compound with freshly prepared $Al(OH)_3$.

(*b*) *Rinman's green, CoOZnO or* $CoZnO_2$. Obviously it is cobalt zincate. It is used as a green pigment. It is obtained when CoO combines with ZnO. It may also be prepared by heating a solution of $Co(NO_3)_2$ with ZnO.

$$2Co(NO_3)_2 \longrightarrow 2CoO + 4NO_2 + O_2 \;;\quad ZnO + CoO \longrightarrow ZnO.CaO \text{ or } CaZnO_2$$

(*c*) *Smelt,* $K_2O.CoO.3SiO_2$. It is used as a blue pigment and is obtained by fusing the mixture of CoO, SiO_2 and K_2CO_3.

4. Copper pigments. Examples are :

(*a*) *Basic cupric acetate* (also called *verdigris*), $Cu(OH)_2.(CH_3COO)_2Cu$. It is used as a green pigment.

(*b*) *Basic cupric carbonates.* Copper forms two basic cupric carbonates which are given below. Both these compounds are used as pigments. (*i*) *Malahit,* $Cu(OH)_2.CuCO_3$ *(green)* (*ii*) *Azurite,* $Cu(OH)_2.2CuCO_3$ *(blue).*

(*c*) *Cuprous oxide,* Cu_2O (also called *red oxide of copper).* It occurs in nature as *cuprite* or *red copper ore.* It is yellow or red in colour depending on the size of its particles. It is used to colour porcetain, in anti-rust paints and in making ruby glass.

5. Zinc pigments. These are used as white pigments. Examples are :

(*a*) *Lithopone,* $(BaSO_4 + ZnS)$. It is a white pigment and is a mixture of ZnS and $BaSO_4$. It is used in wall paints and cheap enamel paints.

(*b*) *Zinc oxide, ZnO* (also called *zinc white or chinese white*). It is used as a white pigment.

2nd classification. This classification is based on the colour of the pigment. According to this classification, the pigments are of the following types :

1. White pigments. Examples are :

(*a*) *Basic lead carbonate,* $Pb(OH)_2.\ 2PbCO_3$ (also called *white lead*)

(*b*) *Basic lead sulphate,* $PbO.2PbSO_4$ (also called *sublimed white lead*)

(*c*) *Lithopone,* $(BaSO_4 + ZnS)$

(*d*) *Zinc oxide, ZnO* (also called *zinc white* or *chinese white*)

(*e*) *Super-sublimed white lead,* $PbO.2PbSO_4$. It is the brightest of all the white pigments.

(*f*) *Titanium dioxide,* TiO_2. It is also called *titan white.* Mixed with $BaSO_4$ it is sold under the name of *titanox.*

2. Red pigments. Examples are :

(*a*) *Triplumbic tetroxide,* Pb_3O_4 (also called *red lead*)

(*b*) *Basic lead chromate,* $Pb(OH)_2.PbCrO_4$ (also called *chrome red*)

(*c*) *Ferric oxide,* Fe_2O_3. (also called *haematite*)

(*d*) *Cadmipone,* $(CdS + BaSO_4)$

(*e*) *Cuprous oxide* or *red oxide of copper,* Cu_2O.

3. Yellow pigments. Examples are :

(*a*) *Lead chromate,* $PbCrO_4$ (also called *chrome yellow*)

(*b*) *Basic zinc chromate,* $Zn(OH)_2.ZnCrO_4.H_2O$.

4. Green pigments. Examples are :

(*a*) *Guignet's green, $Cr_2O_3.2H_2O$*

(*b*) *Rinman's green, CoO.ZnO*

(*c*) *Basic cupric acetate, $Cu(OH)_2.(CH_3COO)_2Cu$* (also called *verdigris*)

(*d*) *Chromic oxide, Cr_2O_3* (also called *chrome green*)

(*e*) *Basic cupric carbonate,* $Cu(OH)_2.CuCO_3$ (called *malachit green*)

5. Blue pigments. Examples are :

(*a*) *Thenard's blue, $CoO.Al_2O_3$*

(*b*) *Smelt, $K_2o.CoO.3SiO_2$.*

(*c*) *Azurite, $Cu(OH)_2.2CuCO_3$.*

Fluorocarbons

What are fluorocarbons?

These are the compounds which can be regarded as having been obtained by replacing H-atoms of hydrocarbon or some other organic molecule, either completely or partially by F-atoms. Thus, C_8, F_{18}, C_6F_{12}, CCl F_3, CCl_2 F_2 (called freon 12), CCl_3F, $CHClF_2$, $CF_2 = CF_2$, $(CF_2 = CF_2)_n$ (called *teflon*) etc., are the examples of fluorocarbons.

Preparation

Fluorocarbons can be obtained by the following methods :

(*i*) *By the reactions in which F_2 is used.* (*a*) When C_6H_6 is heated with F_2 at 265°C in presence of inert gas such as N_2 or He and a catalyst of Cu-guaze, C_6F_{12} is obtained.

$$C_6H_6 + 9F_2 \xrightarrow[\text{Cu–guaze, 265° C}]{\text{In } N_2 \text{ or He}} C_6F_{12} + 6HF$$

(*b*) The action of F_2 on CH_4 gives CF_4.

$$CH_4 + 4F_2 \longrightarrow CF_4 + 4HCl$$

(*ii*) *By the reactions in which liq. HF is used.* (*a*) When organic compounds are electrolysed in liq. HF in steel cells with Ni-anodes and steel cathodes, fluorocarbons are obtained, *e.g.*,

$$(C_2H_5)_2O \longrightarrow (C_2F_5)_2O$$

$$(CH_3COOH) \longrightarrow CF_3COOF \xrightarrow{H_2O} CF_3COOH$$

$$C_8H_{18} \xrightarrow[\text{liq. H.F.}]{\text{Electron is in}} C_8F_{18} \; ; \quad (C_4H_9)_3N \longrightarrow (C_4F_9)_3N$$

This method has been used in laboratory and in industries.

(*b*) When anhydrous HF is heated with CCl_4 in presence of SbF_5 at 50–100°C temperature and 50–500 atmospheric pressure, **freon-12(CCl_2F_2)** is obtained.

$$CCl_4 + 2HF \xrightarrow[\text{50–100° C}]{SbF_5} CCl_2F_2 + 2HCl$$

50-500 at. pressure

In addition to freon-12, other freons like CCl_3F and $CClF_3$ are also formed in the above reaction.

(*c*) When HF reacts with $CHCl_3$ in presence of SbF_3, *teflon*, $(CF_2 = CF_2)_n$ is obtained.

$$CHCl_3 + 2HF \xrightarrow{SbF_3} CHF_2Cl + 2HCl$$

$$2CHF_2Cl \xrightarrow{850° C} CF_2 = CF_2 + 2HCl$$

Tetrafluoroethylene

$$n(CF_2 = CF_2) \xrightarrow{\text{Polymerisation}} (CF_2 = CF_2)_n \; (\textit{Teflone})$$

Teflon is a polymer of tetrafluoroethylene.

(*iii*) By *heating metal fluorides like LiF, NaF, CsF, KHF_2, AgF, AsF_3, HgF_2, ZnF_2, SbF_3 etc., with organic halides (RX).*

$$AgF + RCl \longrightarrow AgCl + RF \; (\textit{fluorocarbon})$$

$$AsF_3 + C_6H_5PCl_3 \xrightarrow{25° C} AsCl_3 + C_6H_5PF_3 \; (\textit{fluorocarbon})$$

The metal fluoride having bigger metal atom has been found to be a better fluorinating agent than the metal fluoride having smaller metal atom. Thus CsF is a better fluorinating agent than NaF.

(*iv*) *By heating CCl_4 with SbF_3, in presence of $SbCl_3$.* This reaction gives freon-12 (CCl_2F_2).

$$3CCl_4 + 2SbF_3 \xrightarrow[\text{Heat}]{SbCl_5} 2CCl_2F_2 + 2SbCl_3$$

(*v*) *By fluorinating organic halides, containging oxygen, with SF_4, e.g.,*

$$2\ R_2C=O + SF_4 \longrightarrow 2\ R_2CF_2 + SO_2$$

Properties. (*i*) Fluorocarbons are much more stable than the other halocarbons. Their higher stability is due to the following reasons :

(*a*) Since the difference between the sizes of H and F-atoms is smaller than that between the sizes of H and other halogen atoms, replacement of H-atoms by F-atoms does not produce appreciable strain on the resulting molecule and hence the fluorocarbons are much more stable than other halogen compounds.

(*b*) Since the energy of C—F bond (= 486 kJ mol^{-1}) is higher than the bond energies of C—H (= 415 kJ mol^{-1}) and C—Cl (= 332 kJ mol^{-1}) bonds, fluorocarbons are much more stable than the other halo compounds.

(*ii*) Fluorocarbons are not able to undergo hydrolysis, since C-atom does not have *d*-orbitals which could permit the formation of the 5-coordinated intermediate by the nucleophilic attack of H_2O molecule. (See the hydrolysis of $SiCl_4$).

(*iii*) Hydrocarbons are easily oxidised by atmospheric O_2 while fluorocarbons cannot be oxidised at all.

(*iv*) Fluorocarbons are not attacked by acids, alkalies and most chemical reagents. They are only by hot metals like molten sodium.

(*v*) Completely fluorinated compounds have very low boiling points.

(*vi*) Completely fluorinated compounds have very low coefficients of friction, since the intermolecular forces in these compounds are very low.

(*vii*) Certain perfluoro-organics are nontoxic and have the capacity to dissolve, transport and release O_2 by simple solubility.

(*viii*) Freon-12, (CCl_2F_2) is a colourless, non-toxic and non-inflammable gas having b.pt. = –29.6°C.

(*ix*) Teflon, $(CF_2 = CF_2)_n$ is a white plastic like material. It is insoluble in any solvent. It is resistant to the action of strong acids, alkalies and oxidising agents. Due to these properties, teflon is used in the construction of chemical plants. It has a very high electrical conductivity and is used as an insulating material in cables.

Uses. (*i*) Because of their very low coefficients of friction, fluorocarbons are used as subricating agents.

(*ii*) Chlorofluoro-carbons like CCl_3F and $CClF_3$ are used as refrigerants.

(*iii*) Freon-12. (CCl_2F_2) is used in refrigeration and in air-conditioning in place of NH_3 and SO_2. It is also used as a solvent for D.D.T. and other insecticides.

(*iv*) $CF_3CHBrCl$ and some similar compounds are used as anaesthetics

(*v*) $CHClF_2$ is used for the preparation of teflon, $(CF_2 = CF_2)_n$ as shown below :

$$2CHF_2Cl \xrightarrow{25°\,C} CF_2 = CF_2 + 2HCl$$

$$n(CF_2 = CF_2) \xrightarrow{\text{Polymerisation}} (CF_2 = CF_2)_n\ (\textit{Teflon})$$

(*vi*) Certain perfluoro-organics are used as synthetic blood for small mammals.

(*vii*) **Teflon** is used as an insulator in chemical plants such as in the manufacture of F_2.

Electroplating

What is electroplating?

Electroplating is a process in which a superior metal is deposited on an inferior metal (called base metal) by the electrolysis of the aqueous solution of a suitable electrolyte of the metal which has to be deposited on the base metal.

Purposes of electroplating

Electroplating is done for the following purposes :

(*i*) *For decorating the metals.* Au and Ag are deposited on base metals in order to increase the beauty of the latter.

(*ii*) *For protecting metals from corrosion.* Metals like Fe and Cu get easily corroded by the atmospheric air, moisture and CO_2 and hence are liable to be destroyed. Their life is considerably increased when they are coated electrolytically with Ni and Cr which protect them from corrosion and give them a good polish.

(*iii*) Sometimes suitable metal is deposited on the worn out parts of machinery to repair them.

Qualities of the metal coating

(*i*) The coating should be *adherent.* In order to get adherent coating the article to be plated should be free from oil, grease and corrosion products like rust, oxides, sulphides and other tarnishing films. Grease and oil on the surface are removed by treating the surface with hot alkali (*sponification*) or by using organic solvents. Tarnishing films are removed by treating the surface with conc. acids (*pickling*) or sand blasting.

(*ii*) The coating should be *coherent.* For coherent coatings, the deposit should be fine-grained and compact. For this, the current density and the temperature should have their optimum values for the given electrolyte. Form example, for $Na_3[Cu(CN)_4]$ bath, the optimum electron density is between 3 and 6 ampres per square foot and the optimum temperature is 45°C. When small quantities of organic substances like gelatin, glue etc., are added to the plating bath, cohesion is improved.

(*iii*) The coating should be of *uniform thickness* which is generally obtained by varying the shapes of the electrodes and distance between them.

(*iv*) Sometimes the coating is uncontinuous, since it has many pores. The number of pores can be decreased by increasing the thickness of the coating and by cleaning the surface of the base metal.

(*v*) Sometimes the coating is dull, though it is quite smoth and coherent. The dullness of the coating is due to (*a*) the use of excessive current as in Cu-plating from cyanide bath (*b*) low temperature as in case of Cr-plating (*c*) The change of bath, *e.g.,* in Zn-plating, the coating from the sulphate bath is more lustrous than from a cyanide bath. The opposite is true in Cu-plating.

In order to get brighter coating, suitable brighting agent should be added, *e.g.*, the addition of sodium formate and $CoSO_4$ brightens Ni-deposits. Similarly the addition of nickel salts brightens Cu-plating.

Apparatus

The apparatus used for electroplating is generally called *electrolytic* tank or *electrolytic bath.* It may be made of wood, iron, cement, stoneware etc., and may be lined inside with a suitable material like rubber, lead, etc. The plates are hung in the tank from rods which are placed horizontally at its top known as *bus bars.* Bus bars are insulated from the tank by means of hard rubber. A number of tanks arranged either in series or in parallels work together.

Requisites of a good electrolyte

The electrolyte used in the electrolytic bath depends on the nature of the metal to be deposited. For example in Cu-plating, when a course crystalline deposit is required, copper sulphate acidified with H_2SO_4 is used. But, for finer deposits, an alkaline solution of potassium tetracyano copper (I), $K_3[Cu(CN)_4]$ is used. The electrolyte used should be cheep, highly soluble, a good conductor, stable towards hydrolysis, oxidation and reduction, and should possess a good covering power. Generally a mixture of two or more substances is used as electrolyte

Procedure and theory

The article to the electroplated is made cathode (–) while the pure metal which is to be deposited on the article is made the anode (+). The following examples illustrate the procedure.

1. Nickel plating. In nickel plating of a steel article, an aqueous solution of $NiSO_4$ is used as an electrolyte, which is taken in the electrolytic tank. The steel article is made the cathode and the pure nickel metal plate is made the anode.

Reactions. When electric current of a suitable voltage is passed through the electrolytic solution, the following reactions take place :

(*a*) *Ionization of $NiSO_4$* : $NiSO_4 \rightleftharpoons Ni^{+2} + SO_4^{2-}$

(*b*) *Reactions taking place at cathode (steel article).* Ni^{+2} ions from the electrolyte move towards cathode (*i.e.*, towards the steel article), where they pick up electrons and are converted into neutral Ni atoms, which get deposited on the article (cathode).

$$Ni^{2+} + 2e^- \longrightarrow Ni \quad \textit{(gets deposited on the article)}$$

(*c*) *Reactions taking place at anode (pure nickel plate).* SO_4^{2-} ions lose their negative charges by losing electrons at the anode and become neutral SO_4 radicals, which combine with pure nickel metal of the anode, to form $NiSO_4$, which goes into the electrolytic solution.

$$SO_4^{2-} - 2e^- \longrightarrow SO_4$$

$$SO_4 + Ni \textit{ (From anode)} \longrightarrow NiSO_4 \textit{ (goes into electrolytic solution)}$$

Thus we see that Ni, from the electrolyte, gets deposited on the article (cathode) and an equivalent amount of Ni from the anode goes into solution of the electrolyte of $NiSO_4$ and hence the strength of $NiSO_4$ solution (electrolyte) remains unchanged. This process is continued as long as required.

2. Copper plating. When a course crystalline deposit of copper is required, $CuSO_4$ acidified with H_2SO_4 is used as an electrolyte, and if finer deposit of copper is needed, an alkaline solution of $K_3[Cu(CN)_4]$ is used as the electrolyte.

3. Silver plating. The article to be silver plated is made the cathode, while the anode consists of pure silver. $K[Ag(CN)_2]$ solution is used as the electrolyte. This complex compound dissociates in water as :

$$K[Ag(CN)_2] \longrightarrow K^+ + [Ag(CN_2]^-$$

Since $[Ag(CN)_2]^-$ ion formed as above is highly stable, it dissociates to a very small extent and hence gives Ag^+ ions of very low concentration. A coherent film of Ag gets deposited on the cathode (article).

$$[Ag(CN)_2]^- \rightleftharpoons Ag^+ + 2CN^-$$

$$Ag^+ + e^+ \longrightarrow Ag \textit{ (gets deposited on the article)}$$

This silver gets deposited more slowly and uniformly than that obtained from a solution of $AgNO_3$ when this solution is used as the electrolyte. The concentration of Ag^+ ions obtained from $[Ag(CN)_2]^-$ ion (which is very low) remains almost constant as the electrolysis proceeds.

4. Gold plating. In this plating, electrolyte used is the solution of $K[Au(CN)_2]$ or $K[Au(CN)_4]$. Like $[Ag(CN)_2]^-$ ions, $[Au(CN_2)]^-$ ions are also very stable and hence give Au^+ ions of very low concentration.

$$K[Au(CN)_2] \longrightarrow K^+ + [Au(CN)_2]^-$$

$$[Au(CN)_2]^- \rightleftharpoons Au^+ + 2CN^-$$

The concentration of Au^+ ions remains constant as the electrolysis proceeds.

Galvanising

Galvanising is a process in which iron articles are coated with a thin layer of zinc metal. On account of its resistance to rusting and corrosion, the layer of zinc prevents the iron articles from rusting. Zinc coating protects iron mechanically, as well as, chemically. The surface of iron does not come directly in touch with air and moisture, because of the layer of zinc. It prevents even that part of an article from rusting, from which the zinc has been slightly detached, due to wear and tear. This is due to the fact that zinc and iron will form a galvanic cell and zinc being more electropositive would react first and dissolve alone. Thus, galvanised iron remains free from rust, so long as all the zinc has not been detached.

Process of galvanising

The process of depositing a thin layer of zinc metal on iron articles is done by the following processes :

1. Galvanising. The iron articles, on which Zn coating is required, are first cleaned with hydrochloric acid. The process is known as "pickling". These are then thoroughly washed with water. Then these articles are dipped in molten zinc, covered with sal ammoniac. NH_4Cl, as flux and taken out immediately. The coating of zinc is made uniform by pressing through rollers.

2. Sherardising. In this process, iron articles, after cleaning with dilute hydrochloric acid, are heated for several hours, in closed drums and then zinc dust is sprayed on them, so that the upper surface of the article gets a coating of an alloy of zinc and iron. This coating is rust proof just like galvanisation.

3. Metallizing. It is not possible to impart a coating of zinc to large structures, like ships, tanks, bridges, by dipping them in molten zinc. Such objects are coated with zinc by spraying zinc over them.

Photography

What is photography?

Photography is a process or technique of getting an exact impression (photo) of an object, on an art paper, with the help of light.

Various steps of photography

The prcess of photography involves the following steps :

1. *To get sensitive photographic plate or thin celluloid film by coating it with an emulsion of AgBr.* In order to get a photographic plate, a glass plate or a thin strip (film) of celluloid is coated with a thin layer of the emulsion of AgBr, dispersed in gelatin. In order to get the emulsion of AgBr, a solution of $AgNO_3$ is mixed with the solution of NH_4Br, containing gelatin.

$$AgNO_3 + NH_4Br \xrightarrow{\text{Gelatin}} AgBr\ (\textit{Emulsion}) + NH_4NO_3$$

The mixture of AgBr and NH_4NO_3 is kept in a warm room, at the temperature of 45°C for some time. AgBr particles grow bigger in size and also become more sensitive to light. This process is called *ripening of the emulsion.* After sometime, some of the emulsion is cooled in ice, so that it solidifies to a jelly. It is then washed with water to remove NH_4NO_3. The solid jelly is melted and is spread over a glass plate or a thin celluloid film. All these operations are carried out in a dark and dustproof room.

2. *To get latent image of the object by exposing the sensitive photographic plate : To get exposed plate.* The sensitive photographic plate or film, obtained in step 1, is fixed in a camera and the image of the object to be photographed is focussed on it. Now the shutter of the camera lense is opened for a few seconds, so that the sensitive photographic plate or film is exposed to the light reflecting from the object.

Depending on the colour and intensity of the light falling on the plate, proportionate decomposition of AgBr to Ag particles takes place.

$$2AgBr \xrightarrow{\text{Exposure to light}} 2Ag + Br_2$$

Here some AgBr remains undecomposed and this undecomposed AgBr is present on the darker part of the plate. During the exposure, an inverted image of the object is formed on the plate. Since this image is not visible, it is called *latent image.*

3. *To get negative plate or negative picture, by developing the exposed plate : To get developed plate.* The latent image formed on the plate during the exposure of the sensitive plate is made visible by the process of developing. In this process, the exposed plate is immersed into the solution of a developer. The main function of the developer is to decompose (reduce) AgBr to black Ag that remained undecomposed during the process of exposure.

$$AgBr + \text{Developer} \longrightarrow Ag\ (\textit{Black}) + \text{Oxidised developer}$$

Developer is a weak reducing agent, such as potassium ferrous oxalate or an alkaline solution of quinol or pyrogellol. If quinol, $C_6H_4(OH)_2$ is used as a developer, the reduction of AgBr to Ag takes place as :

$$2AgBr + C_6H_4(OH)_2 \longrightarrow 2Ag + 2HBr + C_6H_4O_2$$

In the developed plate, the image of the object becomes visible, but the brighter parts of the picture become darker, due to greater decomposition of AgBr to Ag. Since the developed plate or developed picture has a negative relationship with the object, it is called *negative plate* or *negative picture.*

4. *To remove undecomposed AgBr by fixing the developed plate.* The developed plate still contains undecomposed AgBr. It is necessary to remove it, to get permanent image. In order to remove AgBr, the developed plate (*i.e.,* negative plate) is immersed in a solution of $Na_2S_2O_3$ (hypo), which is called *fixer* and the process is called *fixing*. $Na_2S_2O_3$ solution removes AgBr as a soluble complex compound, $Na_3[Ag(S_2O_3)_2]$.

$$2Na_2S_2O_3 + AgBr \longrightarrow Na_3[Ag(S_2O_3)_2] + NaBr \text{ or } 2S_2O_3^{2-} + AgBr \longrightarrow [Ag(S_2O_3)_2]^{3-} + Br^-$$

The process of fixing is also done in a dark room. After fixing, the plate is no longer affected by light, *i.e.,* it does not become black when brought to light.

5. *To print the fixed plate.* Printing out paper (P.O.P.) or bromide paper is generally used for printing the fixed paper. The fixed plate is placed on a printing out paper in a frame and exposed

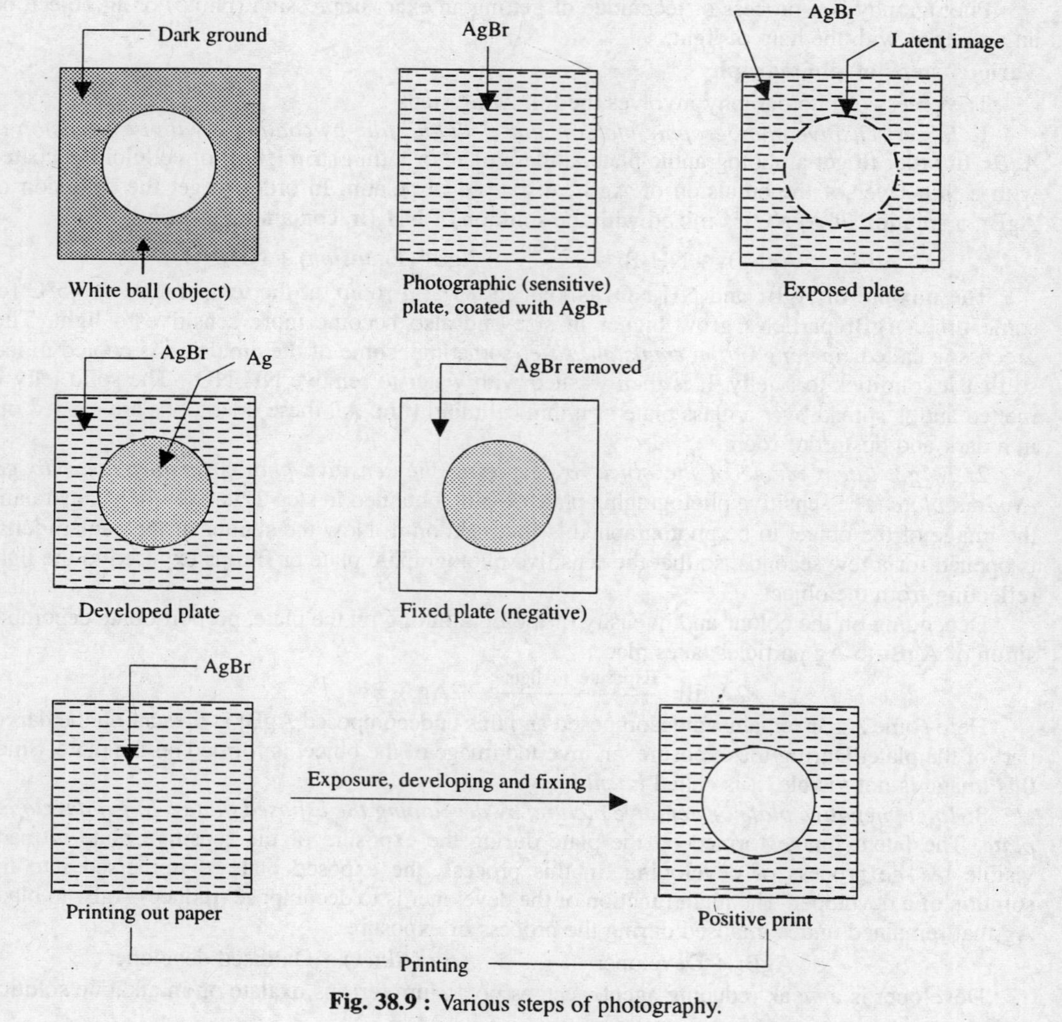

Fig. 38.9 : Various steps of photography.

to light for a few seconds. The printing paper is then subjected to developing and fixing as usual, whereby the shades of the negative are reversed on the printing paper. Thus a print, which is a negative of the negative plate is obtained. This negative has shades exactly similar to those of the object. This negative is called the positive or the actual photograph of the object.

6. *Toning*. The print is then dipped in $AuCl_3$ or $K_2[PtCl_6]$ solution when Ag partially goes in solution and beautiful purple tint or steel grey tint is imparted to the photograph due to gold and platinum particles respectively.

$$AuCl_3 + 3Ag \longrightarrow 3AgCl + Au\ (purple)\downarrow$$
$$K_2[PtCl_6] \longrightarrow PtCl_4 + 2KCl$$
$$PtCl_4 + 4Ag \longrightarrow 4AgCl + Pt\ (steel\ grey)\downarrow$$

All the five steps given above have been shown in Fig. 38.9.

University Questions

1. What are crown ethers? Illustrate your answer with two examples.
2. Give the structure of DB 18 C6, DB 14 C4, 18 C 6, DC 18C 6, DC 60 C20. B12 C4 ligands.
3. What are pyro and cyclic silicates? Give one example of each with its structure. *(Delhi, 86, 88)*
4. Give three examples of ultramarine. *(Madurai, 86)*
5. What are pyrex glass, flint glass, Crooke's glass and Jena glass? *(REE, 82; M.D. Rohtak, 85)*
6. Explain the following : (*i*) The inside surface of a glass bottle containing NaOH becomes dull (*ii*) Crooke's glass finds use in making spectacles glasses. *(Mysore, 80)*
7. Write note on : (*a*) Cement clinker (*b*) setting of cement. *(Delhi, 86; Madras, 85; Punjab, 87)*
8. Explain (*a*) changes taking place during the setting of cement (*b*) function of gypsum added to clinker before grinding *(Himachal, 85; Calicut, 84)*
9. Give the preparation, properties and uses of "ceramics". *(Meerut, 82, 88)*
10. Describe the composition of a paint. *(Madras, 1986)*
11. What do you understand by inorganic paints and pigments? Describe various inorganic pigments. *(Meerut, 87S)*
12. Write a short note on "freons" and "teflon". *(Meerut, 91)*
13. What are fluorocarbons? How are they prepared? What are their important properties? *(Madras, 85)*
14. Write a note on "Electroplating". *(Sagar, 85)*
15. Write a note on "Galvanising".
16. Write an essay on photography and explain the chemistry of process. *(M.D. Rohtak, 84; Bharthiar, 85)*
17. **Sketch the structure of $[(SiO_3)_3]^{6-}$ or $Si_3O_9^{6-}$ ion.** ***(VBS Purvanchal 2002)***
18. **Write a note on "Sheet silicates".** ***(Delhi 2002)***
19. **(*a*) Discuss the properties of glass.**
    (*b*) State ingredients you would use to make hard glass and flint glass. ***(Gujrat 2003)***
20. **Write a note on optical glass.** ***(Banglore, 2004)***
21. **Write a note on Fluorocarbos.** ***(Banglore, 2004)***
22. **What is the composition of hard glass ?** ***(Banglore, 2004)***
23. **What are the characteristics of a good paint ?** ***(Banglore, 2004)***
24. **Safety glass is used for wind screen i automobiles.** ***(Banglore, 2004)***
25. **Discuss the composition of cement. Highlight the theory of setting of cement.** *(GND Amritsar, 2004)*
26. **Write short notes on :**
    (*i*) Pyrosilicates and cyclic silicates (*ii*) Fluorocarbons. *(Bhopal 2004)*
27. **What are silicates? Name the different types of silicates known to you.** *(GND Amritsar, 2004)*

MULTIPLE CHOICE QUESTIONS

1

Quantum Mechanical Concepts of Atomic Structure : Rutherford's and Bohr's Atomic Models

MULTIPLE CHOICE QUESTIONS

1. Mass of one mole of electrons is :
(*a*) 0.20 mg (*b*) 0.02 mg
(*c*) 0.55 mg (*d*) 1.00 mg

2. The energy associated with Bohr's orbit in H-atom is given by $E_n = -\frac{13.6}{n^2}$ eV. The energy associated with the orbit having a radius $9r_1$ is equal to :
(*a*) – 13.6 eV (*b*) – 6.8 eV
(*c*) – 1.51 eV (*d*) zero

3. The wave number of the first line in the Balmer series of H-atom is 15200 cm^{-1}. The wave number of the first line in the Balmer series of Li^{2+} ion is
(*a*) 15200 cm^{-1} (*b*) 60800 cm^{-1}
(*c*) 76,000 cm^{-1} (*d*) 136,800 cm^{-1}
(I.I.T. Screening 1992)

4. The difference between 3*rd* and 2*nd* orbits in H-atom is :
(*a*) 2.116×10^{-8} cm
(*b*) 2.646×10^{-8} cm
(*c*) 0.529×10^{-8} cm
(*d*) 1.058×10^{-8} cm

5. What is degeneracy of the level of H-atom that has energy $= \frac{-R_H}{9}$?
(*a*) 0 (*b*) 3
(*c*) 7 (*d*) 9

6. If an electron in the ground state of H-atom is excited by monochromatic radiation of energy of 12.1 eV, then the number of lines emitted, according to Bohr's theory, is equal to :
(*a*) 1 (*b*) 2
(*c*) 3 (*d*) 4

7. The ratio of the difference between 2*nd* and 3*rd* Bohr's orbit energy to that between 3*rd* and 4*th* orbit energy is :
(*a*) 0.35 (*b*) 2.85
(*c*) 5.4 (*d*) 0.185

8. The number of elliptical orbits in N-shell of an atom is :
(*a*) 3 (*b*) 4
(*c*) 2 (*d*) 1

9. The IE of He^+ is 19.6×10^{-18} J $atom^{-1}$. The energy of the first stationary state of Li^{2+} is :
(*a*) 19.6×10^{-18} J $atom^{-1}$
(*b*) 4.41×10^{-18} J $atom^{-1}$
(*c*) 19.6×10^{-19} J $atom^{-1}$
(*d*) 4.41×10^{-17} J $atom^{-1}$
(Roorkee 1989, 1992)

10. The energy of an electron in the first Bohr orbit of H-atom is – 13.6 eV. The possible energy value (s) of the excited state (s) for the electron in Bohr orbits of H-atom is (are) :
(*a*) – 3.4 eV (*b*) – 4.2 eV
(*c*) – 6.8 eV (*d*) + 6.8 eV
(I.I.T. 1998)

11. What transition in H-spectrum would have the same wavelength as Balmer transition $n = 4$ to $n = 2$ in He^+ spectrum?

(*a*) $n = 4$ to $n = 1$ (*b*) $n = 3$ to $n = 2$
(*c*) $n = 3$ to $n = 1$ (*d*) $n = 2$ to $n = 1$

(I.I.T. 1993)

12. How many moles of electrons weigh one kilogram (Mass of electron = 9.108×10^{-31} Kg, Avogadro's number = 6.023×10^{23})

(*a*) 6.023×10^{-23} (*b*) $\frac{10^{31}}{9.108}$

(*c*) $\frac{6.023}{9.108} \times 10^{54}$ (*d*) $\frac{10^{8}}{9.108 \times 6.023}$

(I.I.T. 2002)

13. IE of H-atom is 13.6 eV. IE of He^+ is :

(*a*) 13.6 eV (*b*) 54.4 eV
(*c*) 122.4 eV (*d*) zero

(Roorkee 1986)

14. What is the wavelength of the radiation emitted producing a line in the Lyman series when an electron falls to 4*th* stationary state in H-atom ($R_H = 1.1 \times 10^7$ m^{-1}) ?

(*a*) 96.97 nm (*b*) 969.7 nm
(*c*) 9.697 cm (*d*) None

(Roorkee 1995)

15. Which of the following statement regarding spectral lines is incorrect ?

(*a*) The lines in the Balmer series correspond to the electronic transition from energy level having *n* higher than 1.
(*b*) Paschen series appear in the infrared region.
(*c*) The lines of Lyman series appear in the visible region.
(*d*) Transition from higher energy levels to 5*th* energy level produces *Pfund* series which falls in the infra red region.

(CEE, UP 1996)

16. If an electron falls from $n = 3$ to $n = 2$ level in H-atom, then energy emitted is :

(*a*) 10.2 eV (*b*) 12.09 eV
(*c*) 1.9 eV (*d*) 0.65 eV

(AFMC, Pune 1997)

17. If the ionisation energy of H-atom is W, then the energy of the electron in its first excited state is equal to :

(*a*) – W (*b*) – 2 W
(*c*) – W/2 (*d*) – W/4

18. Assuming that the energy of an electron is quantitised. Which of the following statement is not true ?

(*a*) The centrifugal force created by the electron's motion about the nucleus is counter balanced by the electrostatic attraction between the nucleus and the electron.
(*b*) The radius (R) of an electron's orbit about the nucleus is equal to $R = n^2 a_0/Z$ (Z = atomic number, a_0 = size of the orbit, *n* = principal quantum number).
(*c*) If the principal quantum number (*n*) of an electron in known, only the size of its orbit and energy of the electron can be calculated.
(*d*) As the principal quantum number increases, the difference in the energy between the levels increases.
(*e*) The energy (*E*) of an electron in a hydrogen-like ion is equal to $E = -K \times \frac{Z^2}{n^2}$ (*K* = a constant, *Z* = atomic number of the ion, *n* = principal quantum number).

19. Bohr's theory for hydrogen-like atoms consists of which of the following postulates ?

(*a*) The electron moves round the nucleus of charge +Ze in an elliptical orbit.
(*b*) The orbit has constant energy.
(*c*) Transitions between orbits generate spectral lines.
(*d*) The frequency of the spectral lines is given by $h/\Delta E$.

20. Bohr orbit radius of H-atom is approximately 0.53 A°. The radius for the first excited orbit is (in A°) :

(*a*) 0.13 (*b*) 1.06
(*c*) 4.77 (*d*) 2.12

(CBSE Med. 1998)

ANSWERS

1. (*c*)	**2.** (*c*)	**3.** (*d*)	**4.** (*b*)	**5.** (*d*)	**6.** (*c*)	**7.** (*b*)	**8.** (*a*)
9. (*d*)	**10.** (*a*)	**11.** (*d*)	**12.** (*d*)	**13.** (*b*)	**14.** (*c*)	**15.** (*c*)	**16.** (*c*)
17. (*d*)	**18.** (*c*)	**19.** (*b*),(*c*)	**20.** (*d*)				

HINTS AND EXPLANATIONS

1. Since mass of one electron = 9.1×10^{-31} kg = $9.1 \times 10^{-31} \times 10^3$ g = $9.1 \times 10^{-31} \times 10^3 \times 10^3$ mg,
mass of one mole of electrons = ($9.1 \times 10^{-31} \times 10^3 \times 10^3$ × Avogadro's number) mg

$$= 9.1 \times 10^{-31} \times 10^6 \times 6.022 \times 10^{23} \text{ mg}$$

$$= 9.1 \times 6.022 \times 10^{-2} = \frac{9.1 \times 6.022}{10^2} \text{ mg}$$

$$= \mathbf{0.548\ mg}$$

2. Radius of *n*th orbit (r_n) and that of 1st orbit (r_1) in H-atom are related as :

$$r_n = n^2 \times r_1$$

or $9r_1 = n^2 \times r_1$ or $n = 3$

Now, $E_n = \dfrac{-13.6}{n^2}\text{ eV} = \dfrac{-13.6}{(3)^2}\text{ eV} = \mathbf{-1.51\ eV.}$

3. We know that wave number ($\overline{v}$) of a particular line in Balmer series of H-atom is given by :

$$\overline{v}_H = R_H\left(\frac{1}{n_1^2} - \frac{1}{n_2^2}\right) = R_H\left(\frac{1}{2^2} - \frac{1}{n_2^2}\right) = 15,200 \text{ cm}^{-1}.$$

The wave number ($\overline{v}$) of a particular line in Balmer series of H-like species with atomic number Z is given by :

$$\overline{v} = \overline{v}_H \times Z^2$$

Since for Li^{2+} ion Z = 3, $\overline{v}_{Li^{2+}}$ is given by :

$$\overline{v}_{Li^{2+}} = \overline{v}_H \times (3)^2 = 15,200 \times 9 = \mathbf{13,6800\ cm^{-1}}$$

4. Radius of *n*th orbit in H-atom (r_n) is given by :

$$r_n = n^2 \times r_1 = n^2 \times 0.529 \text{ Å}. \qquad (\because r_1 = 0.529 \text{ Å})$$

$\therefore$ $r_3 = (3)^2 \times 0.529$ Å and $r_2 = (2)^2 \times 0.529$ Å

$\therefore$ $r_3 - r_2 = 0.529 \times [(3)^2 - (2)^2]$ Å $= 0.529 \times 5$ Å

$= 2.645$ Å $= \mathbf{2.645 \times 10^{-8}\ cm.}$

5. Energy of an electron in *n*th level of H-atom in given by :

$$E_n = \frac{-R_H}{n^2} = \frac{-R_H}{9} \text{ (Given)}$$

$\therefore$ $n = 3$

Now when $n = 3$, $l = 0, 1, 2$

For $l = 0$, $m = 0$ (3s orbital = 1 orbital)

For $l = 1$, $m = +1, 0, -1$ ($3p_x$, $3p_y$, $3p_z$ orbitals = 3 orbitals)

For $l = 2$, $m = +2, +1, 0, -1, -2$ ($3d_{xy}$, $3d_{yz}$, $3d_{xz}$, $3d_{z^2}$, $3d_{x^2-y^2}$ orbitals = 5 orbitals)

Thus $n = 3$ (3*rd*) level has 1 + 3 + 5 = 9 orbitals and hence degeneracy for 3*rd* level = 9.

6. Suppose the electron of H-atom in the ground state ($n = 1$) is excited to *n*th orbit ($n = n$). Then $E_n - E_1 = 12.1$

or $$\frac{-13.6}{n^2} - \left[\frac{-13.6}{(1)^2}\right] = 12.1$$

or $$12.1 = \frac{-13.6}{n^2} + 13.6$$

or $$12.1n^2 = -13.6 + 13.6n^2$$

$\therefore$ $$n^2 = \frac{-13.6}{-1.5} = +9.0$$

$\therefore$ $$n = 3$$

$\therefore$ $$\text{No. of lines} = \Sigma\Delta n = \Sigma(3-1) = \Sigma 2$$
$$= 2 + 1 = \mathbf{3.}$$

7. $E_2 = \frac{-13.6}{2^2} = -3.4 \text{ eV}$, $E_3 = \frac{-13.6}{3^2} = -1.51 \text{ eV}$ and $E_4 = \frac{-13.6}{4^2} = -0.85 \text{ eV}$.

$$\text{Ratio} = \frac{E_3 - E_2}{E_4 - E_3} = \frac{-1.51-(-3.4)}{-0.85-(-1.51)} = \frac{1.89}{0.66} = 2.85.$$

8. For N-shell $n = 4$ and hence for this shell the number of circular orbits = 1 and the number of elliptical orbits = 4 – 1 = 3.

9. I.E. of Li^{2+} = IE of H × Z^2 = $(IE)_H \times 3^2$

IE of He^+ = IE of H × Z^2 = $(IE)_H \times 2^2$

$\therefore$ $$\frac{\text{IE of } Li^{2+}}{\text{IE of } He^+} = \frac{(IE)_H \times 9}{(IE)_H \times 4}$$

$\therefore$ $$\text{IE of } Li^{2+} = \frac{9}{4} \times \text{IE of } He^+$$

$$= \frac{9}{4} \times 19.6 \times 10^{-18} \text{ J atom}^{-1}$$

$$= 4.41 \times 10^{-17} \text{ J atom}^{-1}.$$

10. The excited states for an electron in Bohr orbits have $n > 1$, *i.e.*, for excited states $n = 2, 3, 4$ ..... .

Thus $$E_2 = \frac{-13.6}{2^2} = \mathbf{-3.4\ eV}$$

11. Wave length for $n = 4$ to $n = 2$ transition in Balmer's series in He^+ spectrum (*Z* for He = 2) is given by :

$$\frac{1}{\lambda} = R_H\left[\frac{1}{2^2} - \frac{1}{4^2}\right] \times Z^2 = R_H \times \frac{3}{16} \times 4 = \frac{3R_H}{4}$$

$$\therefore \quad \lambda = \frac{4}{3R_H}$$

$$\frac{1}{\lambda_H} = R_H\left[\frac{1}{n_1^2} - \frac{1}{n_2^2}\right]$$

or $$\frac{1}{4/3R_H} = R_H\left[\frac{1}{n_1^2} - \frac{1}{n_2^2}\right]$$

or $$\frac{3R_H}{4} = R_H\left[\frac{1}{n_1^2} - \frac{1}{n_2^2}\right]$$

or $$\frac{3}{4} = \left[\frac{1}{n_1^2} - \frac{1}{n_2^2}\right]$$

Obviously $n_1 = 1$ and $n_2 = 2$.

12. $\because$ 1 electron weighs = 9.108×10^{-31} kg.

$\therefore$ 1 mole of electrons or 6.023×10^{23} electrons weigh = $9.108 \times 10^{-31} \times 6.023 \times 10^{-23}$ Kg

$= 9.108 \times 6.023 \times 10^{-8}$ kg

Thus $9.108 \times 6.023 \times 10^{-8}$ kg contain 1 mole of electrons.

$$1 \text{ kg contains} = \frac{1}{9.108 \times 6.023 \times 10^{-8}} \text{ moles of electrons}$$

$$= \frac{10^8}{9.108 \times 6.023} \text{ moles of electrons.}$$

13. We know that :

IE of H-like species with atomic number, Z = IE of H-atom × Z^2

$= 13.6 \times (2)^2 =$ **54.4 eV.**

14. For Lyman series $n_1 = 1$, $n_2 = 4$

$$\frac{1}{\lambda} = R_H\left[\frac{1}{n_1^2} - \frac{1}{n_2^2}\right] = 1.1 \times 10^7\left[\frac{1}{(1)^2} - \frac{1}{(4)^2}\right] m^{-1}$$

$$= 1.1 \times 10^7 \times \frac{15}{16} \text{ m}^{-1}$$

$$\therefore \quad \lambda = \frac{16}{1.1 \times 10^7 \times 15} \text{ m} = \frac{16 \times 10}{1.1 \times 10^7 \times 10 \times 15} \text{ m}$$

$$= \frac{160 \times 10^{-8}}{1.1 \times 15} \text{ m} = 9.69 \times 10^{-8} \text{ m}$$

$= 9.69$ nm.

15. (*a*) The spectral lines in Balmer series are produced when an electron falls from energy level with $n = 3, 4$ ... (n is higher than 1) to energy level with $n = 2$.

(*b*) Paschen series appears in IR region.

(*c*) The lines of Lyman series appear in UV region.

(*d*) The lines of *Pfund* series are produced when an electron falls from 6*th*, 7*th* ... energy level to 5*th* ($n = 5$) energy level. These lines fall in IR region.

16. $E_{emitted} = E_{3\to2} = E_3 - E_2$

$$= \frac{-13.6}{(3)^2} - \left[\frac{-13.6}{(2)^2}\right] \text{eV} = \left[\frac{-13.6}{9} + \frac{13.6}{4}\right] \text{eV}$$

$$= 13.6\left[\frac{1}{4} - \frac{1}{9}\right] \text{eV} = 13.6 \times \frac{5}{36} \text{eV} = 1.888 \text{ eV}$$

17. Ionisation energy of H-atom is equal to the difference, $(E_\infty - E_1)$. Thus :

IE of H–atom = W = $E_\infty - E_1$

$$= 0 - \left(-\frac{2\pi^2 me^4}{h^2} \times \frac{Z^2}{n^2}\right)$$

or $$W = +\frac{2\pi^2 me^4}{h^2} \times \frac{(1)^2}{(1)^2} \quad \text{... (i)}$$

Since first excited state is $n = 2$ state, energy of the electron in the first excited state of H-atom

$$= E_2 = \frac{-2\pi^2 me^4}{h^2} \times \frac{Z^2}{n^2}$$

$$= -\frac{2\pi^2 me^4}{h^2} \times \frac{(1)^2}{(2)^2}$$

= – W/4. [From equation (*i*)]

18. The statement given at (*c*) is wrong. The correct form of this statement is :

If the principal quantum number (n) of an electron is known, the size (*i.e.*, radius) of its orbit, energy of the electron and angular momentum of the electron all can be calculated.

19. Postulate (*a*) is wrong. The correct form of this postulate is "The electron moves round the nucleus of charge + Ze in a circular orbit".

Postulates (*b*) and (*b*) both are correct.

Postulate (*d*) is also wrong, since the frequency of the spectral lines is given by $\Delta E = hv$ or $v = \Delta E / h$.

20. For first excited state n = 2. Radians of first excited state ($n = 2$) of H-atom ($Z = 1$) is given by :

$$r_2 = \frac{h^2}{4\pi^2 me^2} \times \frac{n^2}{Z} = \frac{h^2}{4\pi^2 me^2} \times \frac{n^2}{1} = \frac{h^2}{4\pi^2 me^2} \times (2)^2$$

Now since Bohr radius for H-atom $= \frac{h^2}{4\pi^2 me^2} = 0.53$ A°,

$$r_2 = 0.53 \times (2)^2 \text{ A°} = 0.53 \times 4 \text{ A°} = 21.2 \text{ A°}.$$

2

Wave Mechanical Concept of Atomic Structure

MULTIPLE CHOICE QUESTIONS

1. The electrons, identified by quantum numbers n and l (*i*) $n = 4, l = 1$, (*ii*) $n = 4$, $l = 0$, (*iii*) $n = 3, l = 2$ and (*iv*) $n = 3, l = 1$ can be placed in order of increasing energy, from the lowest to highest, as :

(*a*) (*iv*) < (*ii*) < (*iii*) < (*i*)
(*b*) (*ii*) < (*iv*) < (*i*) < (*iii*)
(*c*) (*i*) < (*iii*) < (*ii*) < (*iv*)
(*d*) (*iii*) < (*i*) < (*iv*) < (*ii*)

2. The orbital diagram in which Aufbau principle is violated is :

	2*s*	2*p*		
(*a*)	↑↓	↑↓	↑	
(*b*)	↑	↑↓	↑	↑
(*c*)	↑↓	↑	↑	↑
(*d*)	↑↓	↑↓	↑↓	↑

3. Which of the following statements on radial probability distribution curves is wrong ?

(*a*) $n = 2, l = 0$, nodes = 1
(*b*) $n = 3, l = 0$, nodes = 2
(*c*) $n = 2, l = 1$, nodes = 0
(*d*) $n = 3, l = 2$, nodes = 1

4. Which one of the following statements related to radial probability distribution curves is false ?

(*a*) Most probable radius varies as $3s > 3p > 3d$.
(*b*) Most probable radius varies as $2s < 3s < 4s$.
(*c*) Most probable radius varies $2p < 3p < 4p$.
(*d*) Most probable radius varies as $1s < 1p < 1d$.

5. The electrons represented by the sets of quantum numbers (*i*) $n = 4, l = 0, m = 0, s = +\frac{1}{2}$ (*ii*) $n = 3, l = 1, m = 1, s = -\frac{1}{2}$ (*iii*) $n = 3, l = 2, m = 0, s = +\frac{1}{2}$ (*iv*) $n = 3, l = 0, m = 0, s = -\frac{1}{2}$ have the decreasing order of energy as :

(*a*) (*iii*) > (*i*) > (*ii*) > (*iv*)
(*b*) (*iv*) > (*i*) > (*iii*) > (*ii*)
(*c*) (*i*) > (*ii*) > (*iii*) > (*iv*)
(*d*) (*ii*) > (*iii*) > (*iv*) > (*i*)

6. Which of the following sets represents 4*d* electron ?

(*a*) 4, 3, 2, +1/2 (*b*) 4, 2, i, 0
(*c*) 4, 3, – 2, +1/2 (*d*) 4, 2, 1, – 1/2

7. Which of the following statements is not correct for an electron that has the quantum numbers $n = 4$ and $m = 2$?

(*a*) The electron may have $s = +1/2$.
(*b*) The electron may have $l = 2$.
(*c*) The electron may have $l = 3$.
(*d*) The electron may have $l = 0, 1, 2, 3, 4$.

8. Ground state electronic configuration of N-atom can be represented by :

	1*s*	2*s*	2*p*		
(*a*)	↑↓	↑↓	↑	↑	↑
(*b*)	↑↓	↑↓	↑	↓	↑

(*c*) [↑↓] [↑↓] [↑ | ↓ | ↓]

(*d*) [↑↓] [↑↓] [↓ | ↓ | ↓]

9. Which of the following sets of quantum numbers is(are) permitted ?

(*a*) $n = 3, l = 3, m = 0, s = +\frac{1}{2}$

(*b*) $n = 3, l = 2, m = +2, s = -\frac{1}{2}$

(*c*) $n = 3, l = 1, m = -2, s = -\frac{1}{2}$

(*d*) $n = 3, l = 0, m = 0, s = +\frac{1}{2}$

(*I.S.M. Dhanbad 1994*)

10. The set of quantum numbers not applicable for an electron in an atom is :

(*a*) $n = 1, l = 1, m = 1, s = +\frac{1}{2}$

(*b*) $n = 1, l = 0, m = 0, s = +\frac{1}{2}$

(*c*) $n = 1, l = 0, m = 0, s = -\frac{1}{2}$

(*d*) $n = 2, l = 0, m = 0, s = +\frac{1}{2}$

11. The orbital angular momentum of an electron in 2s orbital is :

(*a*) $+\frac{1}{2} \cdot \frac{h}{2\pi}$ (*b*) zero

(*c*) $\frac{h}{2\pi}$ (*d*) $\sqrt{2} \cdot \frac{h}{2\pi}$

12. The magnitude of spin angular momentum of an electron is equal to :

(*a*) $\sqrt{s(s+1)} \cdot \frac{h}{2\pi}$ (*b*) $s \cdot \frac{h}{2\pi}$

(*c*) $\pm\frac{1}{2} \cdot \frac{h}{2\pi}$ (*d*) $s(s+2) \cdot \frac{h}{2\pi}$

13. Total spin value for *n* unpaired electrons is equal to :

(*a*) $\sqrt{S(S+1)}$ (*b*) $\sqrt{s(s+1)}$

(*c*) $+\frac{1}{2}\sqrt{S(S+1)}$ (*d*) $-\frac{1}{2}\sqrt{s(s+1)}$

where $S = \frac{1}{2} \times n$ and $s = \pm\frac{1}{2}$.

14. The orbital diagrams in which, both Paulie's exclusion principle and Hund's rule are violated are :

(*a*) ↑↓ ¦ ↑↑ ↑ —

(*b*) ↑↓ ¦ ↑↓ ↑↓ —

(*c*) ↑↓ ¦ ↓ ↓ ↓

(*d*) ↑↓ ¦ ↑↓ ↑↓ ↑

15. State if the following statement is true or false. The electron density in *xy* plane in $3d_{x^2-y^2}$ orbital is zero.

16. Fill up the blanks :

(*a*) Wave function of electrons in atoms and molecules are called ..... .

(*b*) The $2p_x, 2p_y$ and $2p_z$ orbitals of atom have identical shapes but differ in their .... . (*I.I.T. 1993*)

17. The number of nodal planes in a p_x orbital is :

(*a*) 1 (*b*) 2

(*c*) three (*d*) zero (*I.I.T. 2000*)

18. The four quantum numbers of the valence shell electron of K atom are :

(*a*) 4, 1, 1, 1/2 (*b*) 4, 0, 0, 1/2

(*c*) 4, 1, 0, 1/2 (*d*) 4, 4, 0, 1/2

(*EAMCET 1998*)

19. Which of the following sets is not possible?

(*a*) $n = 3, l = 0, m = 0$

(*b*) $n = 3, l = 1, m = -1$

(*c*) $n = 2, l = 0, m = -1$

(*d*) $n = 2, l = 1, m = 0$ (*B.H.U. 2000*)

20. For how many orbitals the quantum numbers $n = 3$, $l = 2$ and $m = +2$ are possible ?

(*a*) 1 (*b*) 2

(*c*) 3 (*d*) 4

21. Nineteenth electron of Cr-atom has which of the following sets of quantum numbers.

	n	*l*	*m*	*s*
(*a*)	3	0	0	1/2
(*b*)	3	2	–2	1/2
(*c*)	4	0	0	1/2
(*d*)	4	1	– 1	1/2

(*DCE 2001*)

ANSWERS

1. (*a*) **2.** (*b*) **3.** (*d*) **4.** (*a*), (*b*), (*c*) **5.** (*a*) **6.** (*d*) **7.** (*d*)
8. (*a*), (*d*) **9.** (*b*),(*d*) **10.** (*a*) **11.** (*b*) **12.** (*a*) **13.** (*a*) **14.** (*a*),(*b*) **15.** False,
16. orbitals, orientation, **17.** (*a*) **18.** (*b*) **19.** (*c*) **20.** (*a*) **21.** (*c*)

HINTS AND EXPLANATIONS

1. For the given orbitals : (*i*) $(n + l) = 4 + 1 = 5$, (*ii*) $(n + l) = 4 + 0 = 4$, (*iii*) $(n + l) = 3 + 2 = 5$, (*iv*) $(n + l) = 3 + 1 = 4$. According to $(n + l)$ rules : (*a*) Energy of an orbital having higher value of $(n + l)$ has higher energy and (*b*) of the two or more orbitals having the same value of $(n + l)$, the orbital having higher value of n has higher energy. Thus on the basis of these rules the order of energy is :
(*i*), (*iii*) > (*ii*), (*iv*) or (*i*) > (*iii*) > (*ii*) > (*iv*) or (*iv*) < (*ii*) < (*iii*) < (*i*).

2. We know that the energy of 2*s* orbital is less than that of 2*p* orbital. 2*s* orbital is, therefore, filled up with two electrons before 2*p* orbitals. On the basis of this rule orbital diagram (*b*) is wrong. Now since 2*p* sub-shell has three orbitals *viz* $2p_x$, $2p_y$ and $2p_z$, all these three orbitals have equal energy (degenerate orbitals). On the basis of this rule orbital diagram (*a*) is not wrong. Thus (*b*) is the answer of this question.

3. We know that the number of spherical nodes for a given sub-shell is equal to $(n - l - 1)$. Thus the number of nodes for (*a*), (*b*) (*c*). and (*d*) are as : (*a*) = $(n - l - 1) = 2 - 0 - 1 = 1$, (*b*) = $3 - 0 - 1 = 2$, (*c*) $2 - 1 - 1 = 0$ and (*d*) = $3 - 2 - 1 = 0$. Thus we find that the number of nodes for (*d*) is 0 and not 1 as given in the question.

4. We know that 1*p* and 1*d* sub-shells, donot exist. Answer (*d*) is, therefore, a false answer. When we look at the radial probability distribution curves for 2*s*, 2*p*, 3*s*, 3*p*, 3*d*, 4*s* and 4*p* sub-shells, we find that answers (*a*), (*b*) and (*c*) are correct answers.

5. The orbitals represented by the given sets of quantum numbers are as (*i*) 4*s* ($n = 4, l = 0$), (*ii*) 3*p* ($n = 3, l = 1$), (*iii*) 3*d* ($n = 3, l = 2$) and (*iv*) 3*s* ($n = 3, l = 0$). $(n + l)$ values for these orbitals are (*i*) = $4 + 0 = 4$, (*ii*) = $3 + 1 = 4$, (*iii*) $3 + 2 = 5$ and (*iv*) = $3 + 0 = 3$. Thus on the basis of $(n + l)$ values the energy of the orbitals is in the order : (*iii*) > (*i*), (*ii*) > (*iv*) or (*iii*) > (*i*) (*ii*) > (*iv*)

6. We know that for an electron residing in 4d orbital, $n = 4$, $l = 2$, $m = +2, +1, 0, -1, -2$ and $s = +1/2$, or $-1/2$. Thus for this electron the correct set of quantum numbers is 4, 2, 1, – 1/2. Thus the correct answer is (*d*).

7. The shell having n = 4 is 4th shell which is divided into 4*s*, 4*p*, 4*d* and 4*f* sub-shells. The values of all the three quantum numbers (n, *l* and m) for these sub shells are given below. The value of s may be + 1/2 or –1/2.

4*s* → $n = 4, l = 0, m = 0$

4*p* → $n = 4, l = 1, m = +1, 0, -1$

4*d* → $n = 4, l = 2, m = +2, +1, 0, -1, -2$

4*f* → $n = 4, l = 3, m = +3, +2, +1, 0, -1, -2, -3$

Obviously an electron which has $n = 4$ and $m = 2$ is either a 4*d* electron ($n = 4, l = 2, m = +2, +1, 0, -1, -2$) or a 4*f* electron ($n = 4, l = 3, m = +3, +2, +1, 0, -1, -2, -3$). The electron residing in 4*d* orbital has $l = 2$ and that residing in 4*f* orbital has $l = 3$, *i.e.*, none of the electrons present in 4*d* or 4*f* orbitals has $l = 0, 1, 2, 3, 4$. Thus correct answer is (*d*).

8. Answer (*a*) and (*d*) are correct answers, since all the three electrons in 2*p* orbitals have their spins in the same direction (Hund's rule of maximum multiplicity). Answers (*b*) and (*c*) are wrong, since all the three electrons in 2*p* orbitals donot have their spins in the same direction.

9. (*a*) Since n cannot be equal to l, answer (*a*) is incorrect. Thus this set of quantum numbers is not permitted.

(*b*) $n = 3$ and $l = 2$ indicates 3*d* sub-shell. For this sub-shell $m = 0, \pm 1, \pm 2$. Thus (*b*) set is permitted.

(*c*) $n = 3$, $l = 1$ indicates 3*p* sub-shell. For this sub-shell $m = 0, \pm 1$. Thus set (*c*) is not permitted.

(*d*) $n = 3$, $l = 0$ indicates 3*s* sub-shell. For this sub-shell $m = 0$. Thus set (*d*) is permitted.

10. (*a*) In this set the values of n and l are equal. Hence this set is not applicable

(*b*) $n = 1, l = 0$ represents 1*s* sub-shell for which $m = 0$. Thus this set is applicable ($s = + 1/2$).

(*c*) $n = 1$, $l = 0$ represents 1*s* sub-shell. For this sub-shell $m = 0$. Thus this set is also applicable ($s = - 1/2$).

(*d*) $n = 2$, $l = 0$ indicates 2*s* sub-shell. For this sub-shell $m = 0$ and hence this set is also applicable.

11. The orbital angular momentum (L) of an electron in an orbital depends on the value of l of the orbital and is given by :

$$L = \sqrt{l\,(l+1)} \times \frac{h}{2\pi}.$$

Now since for 2*s* orbital, $l = 0$

$$L_{2s} = \sqrt{0 \times (0+1)} \times \frac{h}{2\pi} = 0$$

12. The spin angular momentum (*S*) of an electron in any orbital depends on the value s of that electron and is given by :

$$S = \sqrt{s(s+1)} \times \frac{h}{2\pi} = \sqrt{\frac{1}{2} \times \left(\frac{1}{2}+1\right)} \times \frac{h}{2\pi}$$

$$= \sqrt{\frac{1}{2} \times \frac{3}{2}} \times \frac{h}{2\pi} = \sqrt{\frac{3}{4}} \times \frac{h}{2\pi} = \frac{\sqrt{3}}{2} \times \frac{h}{2\pi}.$$

13. Total spin value for n unpaired electrons depends on the value of S and is equal to $\sqrt{S\,(S+1)}$ where $S = \frac{1}{2} \times n$.

14. (*a*) Three *p*-orbitals in this orbital diagram should be singly filled (Hund's rule of maximum multiplicity). Thus this diagram is incorrect. The correct diagram is :

↑↓ ¦ ↑ ↑ ↑

(*b*) This diagram is also incorrect. The correct diagram is :

↑↓ ¦ ↑↓ ↑ ↑

(*c*) This is correct diagram.

(*d*) This orbital diagram is also correct.

15. $3d_{x^2-y^2}$ orbital has four lobes two of which are along *x*-axis and the other two are along *y*-axis. All the four lobes lie in *xy* plane. Electron density is in the four lobes, *i.e.*, in *xy* plane. Thus electron density in *xy* plane in not zero and hence the given statement is false.

17. Number of nodal planes for an orbital is always equal to the value of l for that orbital, *i.e.* the number of nodal planes for an orbital = l. Now since for p_x orbital $l = 1$, number of nodal planes for this orbital = 1.

18. Valence-shell configuration of K ($Z = 19$) is $4s^1$. For $4s^1$ electron $n = 4$, $l = 0$, $m = 0$ and $s = +1/2$ or $-1/2$.

19. (*a*) When $n = 3$, $l = 0, 1, 2$

Now when $l = 0$, $m = \boxed{0}$. Thus this set is possible.

(*b*) When $n = 3$, $l = 0, 1, 2$

Now when $l = 1$, $m = +1, 0, \boxed{-1}$. Thus this set is possible.

(*c*) When $n = 3$, $l = 0, 1$

Now when $l = 0$, $m = \boxed{0}$. This set is not possible, since the given value of m is equal to -1.

(*d*) When $n = 2$, $l = 0, 1$

Now when $l = 1$, $m = +1, \boxed{0}, -1$. Thus this set is possible.

20. When $n = 3$, $l = 0, 1, 2$

Now when $l = 0$ (3*s*), $m = 0$ (one 3*s* orbital)

When $l = 1$ (3*p*), $m = +1, 0, -1$ (three 3*p* orbitals)

When $l = 2$ (3*d*), $m = +2, +1, 0, -1, -2$ (five 3*d* orbitals)

$m = +2$ represents one of the five 3*d* orbitals.

21. According to Aufbau principle, the orbitals are filled up with electrons in the order 1*s*, 2*s*, 2*p*, 3*s*, 3*p*, 3*p*, 4*s*, 3*d* ..... . Thus 19*th* electron of Cr atom will go to 4*s* orbital, since 18 electrons have already occupied 1*s*(2), 2*s*(2), 2*p*(6), 3*s*(2) and 3*p*(6) orbitals. For 4s orbital $n = 4$, $l = 0$, $m = 0$, $s = +\frac{1}{2}$ or $-\frac{1}{2}$.

3

Periodic Table

MULTIPLE CHOICE QUESTIONS

1. Metals belonging to the same group in the periodic table are :

(*a*) Mg, Na (*b*) Mg, Cu

(*c*) Mg, Ba (*d*) Mg, K

(MP, PMT 1998)

2. In the periodic table, the element with atomic number 38 belongs to :

(*a*) period IV and group II

(*b*) period IV and group IV

(*c*) period III and group IV

(*d*) period V and group II

3. The atomic numbers of four elements are given below. Which one is an alkaline earth metal ?

(*a*) 10 (*b*) 20

(*c*) 30 (*d*) 40

(JMC, AMC Aligarh 1992)

4. Which of the following is a defect in the Mendeleef's periodic table ?

(*a*) The position of alkali metals

(*b*) The position of transition metals

(*c*) The position of isotopes

(*d*) All of these.

5. The elements of a group of the periodic table have :

(*a*) the same number of protons

(*b*) the same valence shell

(*c*) the same valence electrons

(*d*) same electron affinity.

6. The tenth element in the periodic table resembles the element with atomic number

(*a*) 2 as well as 30

(*b*) 2 as well as 54

(*c*) 8 as well as 18

(*d*) 8 only.

7. Without looking at the periodic table, select from each of the following sets, the elements belonging to the same group :

(*a*) Z = 12, 38, 4, 88

(*b*) Z = 9, 16, 3, 35

(*c*) Z = 5, 11, 27, 19

(*d*) Z = 24, 47, 42, 55

8. 3*rd* period of the periodic table contains :

(*a*) 8 elements (*b*) 32 elements

(*c*) 3 elements (*d*) 18 elements.

9. Group number of elements in the periodic table indicates :

(*a*) valency with respect to hydrogen

(*b*) the atomicity

(*c*) the number of electrons in the outer most shell

(*d*) none of these.

10. Without looking at the periodic table, select the elements of group IIIA of the periodic table (Atomic numbers are given)

(*a*) 3, 11, 19, 37 (*b*) 5, 13, 21, 39

(*c*) 7, 15, 31, 49 (*d*) 5, 13, 31, 49

11. The elements which occupy positions in periodic table just after the members of zero group are :

(*a*) non-metals

(*b*) gases

(*c*) radioactive elements

(*d*) alkali metals

12. The statement that is not true for long form of periodic table is :

(*a*) It reflects the sequence of filling the electrons in the order of the sub-energy shells *s*, *p*, *d* and *f*.
(*b*) It helps to predict the stable valency states of the elements.
(*c*) It reflects trends in physical and chemical properties of the elements.
(*d*) It helps to predict the relative ionicity of the bond between any two elements.

13. Elements belonging to the same group of periodic table have generally the same :
(*a*) electronic configuration
(*b*) number of electrons in the outermost shell
(*c*) chemical properties
(*d*) physical properties.

14. The basis of periodic law presented by Mendeleef was :
(*a*) valency
(*b*) atomic mass
(*c*) atomic number
(*d*) atomic volume.

15. An element that belongs to third period and fifth group has the electronic configuration :
(*a*) $1s^2$, $2s^2\ 2p^6$, $3s^2\ 3p^3$
(*b*) $1s^2$, $2s^2\ 2p^3$
(*c*) $1s^2$, $2s^2\ 2p^6$, $3s^2\ 3p^4$
(*d*) None of these

16. The halogens are :
(*a*) normal elements
(*b*) noble elements
(*c*) transition elements
(*d*) inner transition elements.

17. Modern extended periodic table was given by :
(*a*) Bohr (*b*) Lother Meyer
(*c*) Mosley (*d*) None of these.

18. In the modern periodic table one of the following does not have appropriate position :
(*a*) inert gases
(*b*) inner-transition elements
(*c*) transition elements
(*d*) fluorine.

19. Which of the following is not a representative element ?
(*a*) Fe (*b*) K
(*c*) Ba (*d*) N

20. Elements of 2*nd* and 3*rd* periods of periodic table are called
(*a*) transition elements
(*b*) typical elements
(*c*) metalloids
(*d*) bridge elements.

ANSWERS

1. (*c*)	**2.** (*d*)	**3.** (*b*)	**4.** (*c*)	**5.** (*c*)	**6.** (*b*)	**7.** (*a*)	**8.** (*a*)
9. (*c*)	**10.** (*d*)	**11.** (*d*)	**12.** (*b*)	**13.** (*b*)	**14.** (*b*)	**15.** (*a*)	**16.** (*a*)
17. (*d*)	**18.** (*b*)	**19.** (*a*)	**20.** (*b*)				

HINTS AND EXPLANATIONS

1. Group numbers of different metals are as : (*a*) Mg – IIA, Na – IA (*b*) Mg – IIA, Cu – IB (*c*) Mg – IIA, Ba – IIA (*d*) Mg – IIA, K – IA. Obviously (*c*) is the correct answer.

2. Noble gases are He_2 (1*st* period), Ne_{10} (2*nd* period), Ar_{18} (3rd period), Kr_{36} (4*th* period), Xe_{54} (5*th* period) etc. Since the atomic number of the given element is 38, this element is present in 5*th* period and 2*nd* group in the periodic table.

3. We know that the atomic number of an alkaline earth metal present in *n*th period is equal to the atomic number of noble gas placed in (*n* – 1)*th* period plus 2. Thus the atomic number of different alkaline earth metals can be determined with the help of the following table which shows that is only 20 which is the atomic number of alkaline earth metal. This metal is Ca (Z = 20).

Period number	*Atomic number of noble gas*	*Atomic number of alkaline earth metal*
1	2(He)	
2	10 (Ne)	2 + 2 = 4 (Be)
3	18 (Ar)	10 + 2 = 12 (Mg)
4	36 (Kr)	18 + 2 = 20 (Ca)
5	54 (Xe)	36 + 2 = 38 (Sr)
6	86 (Rn)	54 + 2 = 56 (Ba)
7		86 + 2 = 88 (Ra)

4. Although isotopes of an element have different atomic weights, these have not been given different places in the Mendeleef's periodic table.

6. Tenth element in the periodic table is Ne_{10} which is a noble gas. This element resembles other noble gases which are He_2, Ar_{18}, Kr_{36}, Xe_{54} and Rn_{86}. Thus Ne_{10} resembles He ($Z = 2$) and Xe ($Z = 54$).

7. We know that the atomic numbers of noble gases and alkali metals are as :

Group IA (Alkali metals)	*Group Zero (Noble gases)*
H = 1	He = 2
Li = 3	Ne = 10
Na = 11	Ar = 18
K = 19	Kr = 36
Rb = 37	Xe = 54
Cs = 55	Rn = 86
Fr = 87	

On the basis of the above scheme the group number of different elements given in each set are :

(*a*) 12 (IIA), 38 (IIA), 4 (IIA), 88 (IIA) (*b*) 9 (VIIA), 16 (VIA), 3 (IA), 35(VIIA)

(*c*) 5 (IIIA), 11(IA), 27 (TE), 19 (IA) (*d*) 24(TE), 47(TE), 42(TE), 55(IA).

Obviously the elements given at (*a*) belong to the same group *viz* IIA.

8. We know that magic numbers are 2, 8, 8, 18, 18, 32. As a matter of fact these are the number of elements which are present in 1*st*, 2*nd*, 3*rd*, 4*th*, 5*th* and 6*th* periods respectively. Thus 3*rd* period contains 8 elements.

10. The atomic number of the elements of group IIIA should be equal to 2 + 2 + 1 = 5, 10 + 2 + 1 = 13, 18 + 2 + 10 + 1 = 31, 36 + 2 + 10 + 1 = 49 and 54 + 2 + 1 + 14 + 9 + 1 = 81. Here 2, 10, 18, 36 and 54 are the atomic numbers of noble gases *viz* He, Ne, Ar, Kr and Xe respectively.

14. Mendeleef proposed that the physical and chemical properties of the elements are a periodic function of their atomic weights (*i.e.*, atomic masses). He, therefore, arranged the elements in the increasing order of their atomic masses.

15. An element belonging to 3*rd* period and 5*th* group has three shells and five electrons in its 3*rd* shell (valence shell). Thus since configuration (*a*) has three shells and five electrons in its 3*rd* shell, configuration (*a*) is the correct configuration.

18. If we look at the modern periodic table, we find that inner-transition elements have not been placed in the main body of the periodic table. These elements have been put at a separate place at the bottom of the periodic table.

19. Representative elements are *s*-and *p*-block elements. Since K and Ba are *s*-block clements and N is a *p*-block element, these are representative elements. Fe is a *d*-block element (transition metal).

4

Electronic Configuration of Elements

MULTIPLE CHOICE QUESTIONS

1. Electronic configuration of Cr atom (Z = 24) is :

(a) $[Ar]_{18}$ $3d^5 4s^1$ (b) $[Ar]_{18}$ $3d^4 4s^2$
(c) $[Ar]_{18}$ $3d^6 4s^0$ (d) $[Ar]_{18}$ $3d^4 4s^1 4p^1$

2. General outer electronic configuration of transition metals is :

(a) $ns^2 nd^{1-10}$
(b) $ns^2 np^1$ $(n - 1)d^{1-10}$
(c) $ns^2 np^6$ $(n - 1)d^{1-10}$
(d) ns^{0-2} $(n - 1)d^{1-10}$

3. An element with atomic number 106 has been discovered recently. Its expected configuration is :

(a) [Rn] $5f^{14} 6d^4 7s^2$
(b) [Rn] $5f^{14} 6d^5 7s^1$
(c) [Rn] $5f^{14} 6d^6 7s^0$
(d) [Rn] $5f^{14} 6d^1 7s^2 7p^3$

4. The electronic configuration (EC) of an element which is just above the element with atomic number 43 in the same periodic group is :

(a) $1s^2$, $2s^2 2p^6$, $3s^2 3p^6 3d^5$, $4s^2$
(b) $1s^2$, $2s^2 2p^6$, $3s^2 3p^6 3d^{10}$, $4s^2 4p^5$
(c) $1s^2$, $2s^2 2p^6$, $3s^2 3p^6 3d^6$, $4s^1$
(d) $1s^2$, $2s^2 2p^6$, $3s^2 3p^6 3d^{10}$, $4s^1 4p^6$

(MLNR 1992)

5. Which of the following ions is colourless ?

(a) Cu^+ (b) Co^{2+}
(c) Ni^{2+} (d) Fe^{3+}

6. Which one of the following ionic species will impart colour to an aqueous solution?

(a) Ti^{4+} (b) Cu^+
(c) Zn^{2+} (d) Cr^{3+}

7. Which of the following has the maximum number of unpaired electrons ?

(a) Mg^{2+} (b) Ti^{3+}
(c) V^{3+} (d) Fe^{2+}

8. Which of the following electronic configurations belong s-block elements ?

(a) $1s^2$ $2s^2$ $2p^6$ $3s^2$ $3p^6$ $4s^1$
(b) $1s^2$ $2s^2$ $2p^6$ $3s^2$ $3p^6$ $3d^5$ $4s^1$
(c) $1s^2$ $2s^2$ $2p^6$ $3s^2$ $3p^6$ $4s^2$
(d) $1s^2$ $2s^2$ $2p^6$ $3s^2$ $3p^6$ $3d^{10}$ $4s^2$

9. The outer shell configuration and group number of the element with atomic number 107 is :

(a) $7s^2$, 7 (b) $6s^2$, 2
(c) $7s^2$ $4f^{14}$ $6d^5$, 7 (d) $7s^1$, 15.

10. Which of the following E.C. belongs to d-block element ?

(a) [Rn] $6s^2$ $7s^2$
(b) [Xe] $4f^1$ $5d^1$ $6s^2$
(c) [Xe] $4f^{14}$ $6d^1$ $6s^2$
(d) [Xe] $5d^1$ $6s^2$

11. Among KO_2, AlO_2^-, BaO_2, NO_2^+ unpaired electrons are present in :

(a) NO_2^+ and BaO_2
(b) KO_2 and AlO_2^-
(c) KO_2 only
(d) BaO_2 only

12. Which among the following is not isoelectronic ?

(*a*) NO^- (*b*) CN^-

(*c*) N_2 (*d*) O_2^{2+}

13. Which of the following are isoelectronic species ?

(*a*) CH_3^+, NH_2^-, NH_4^+

(*b*) NH_2^-, NH_4^+, NH_3

(*c*) CH_3^+, NH_2^-, NH_3

(*d*) NH_2^-, CH_3^+ (*CPMT 1999*)

14. The outer EC of Pd is $4d^{10}5s^0$. It belongs to :

(*a*) 4*th* period, group 11

(*b*) 6*th* period, group 9

(*c*) 5*th* period, group 10

(*d*) 6*th* period, group 10.

15. Among the following groupings which represents the collection of iso-electronic species ?

(*a*) NO^+, C_2^{2-}, O_2, CO

(*b*) N_2, C_2^{2-}, CO, NO

(*c*) $CO, N_2\ CN^-, C_2^{2-}$

(*d*) NO, CN^-, N_2, O_2^-

16. Which of the following species would be expected to be paramagnetic ?

(*a*) Copper crystal (*b*) Cu^+

(*c*) Cu^{2+} (*d*) H_2

17. Among the following, the lowest degree of paramagnetism per mole of the compound of 298 K will be shown by :

(*a*) $MnSO_4, 4H_2O$

(*b*) $CuSO_4 \cdot 5H_2O$

(*c*) $FeSO_4 \cdot 6H_2O$

(*d*) $Ni\ SO_4 \cdot 6H_2O$ (*I.I.T. 1988*)

18. Which of the following compounds is expected to be coloured ?

(*a*) Ag_2SO_4 (*b*) CuF_2

(*c*) MgF_2 (*d*) CuCl

(*I.I.T. 1997*)

19. The E.C. of four elements are given below. Which element does not belong to the same family ?

(*a*) [Xe] $4f^{14}\ 5d^{10}\ 6s^2$

(*b*) [Kr] $4d^{10}\ 5s^2$

(*c*) [Ne] $3s^2 3p^5$

(*d*) [Ar] $3d^{10}\ 4s^2$

20. Which of the following is not paramagnetic ?

(*a*) Cl^- (*b*) Be^-

(*c*) Ne^{2+} (*d*) As^+

(*CBSE 1998*)

ANSWERS

1. (*a*)	**2.** (*d*)	**3.** (*a*)	**4.** (*a*)	**5.** (*a*)	**6.** (*d*)	**7.** (*d*)	**8.** (*a*), (*c*)
9. (*c*)	**10.** (*d*)	**11.** (*c*)	**12.** (*a*)	**13.** (*b*)	**14.** (*c*)	**15.** (*c*)	**16.** (*c*)
17. (*b*)	**18.** (*b*)	**19.** (*c*)	**20.** (*a*).				

HINTS AND EXPLANATIONS

1. Since both the orbitals *viz.* 3*d* and 4*s* present in $[Ar]_{18}\ 3d^5\ 4s^1$ configuration are half-filled, $[Ar]_{18}\ 3d^5\ 4s^1$ configuration is more stable than each of the remaining configurations.

2. In transition metal atoms $(n-1)d$ orbitals are successively filled up with electrons. In some cases one or both electrons from *ns* orbital are shifted to $(n-1)d$ orbitals. Thus the general outer configuration of transition metal atoms becomes $(n-1)d^{1-10}\ ns^{0-2}$ or $ns^{0-2}\ (n-1)d^{1-10}$.

3. The element with $Z = 106$ will be a member of 6*d* series elements in 7*th* period and hence its expected configuration would be $[Rn]_{86}\ 7s^2\ 6d^1\ 5f^{14}\ 6d^3$ or $[Rn]_{86}\ 7s^2\ 6d^4\ 5f^{14}$.

4. The element with $Z = 43$ is placed in 5th period, since it comes after Kr_{36} which is present in 4*th* period. Thus EC of 43 element is $[Kr]_{36}\ 5s^2\ 4d^5$ and hence EC of the element which is just above 43 element in the same group would be $[Ar]_{18}$ $4s^2$ $3d^5$ or $1s^2$, $2s^2$ $2p^6$, $3s^2$ $3p^6$ + $4s^2$ + $3d^5$ or $1s^2$, $2s^2$ $2p^6$, $3s^2$ $3p^6$ $3d^5$, $4s^2$

5. Valence-shell configurations of the given ions are : (*a*) $Cu^+ = 3d^{10}$ ($n = 0$), (*b*) $Co^{2+} = 3d^7$ ($n = 3$), (*c*) $Ni^{2+} = 3d^8$ ($n = 2$) and (*d*) $Fe^{3+} = 3d^5$ ($n = 5$). It is only Cu^+ ion which has no unpaired electrons and hence is colourless in aqueous solution.

6. We know that only those ionic species which have one or more unpaired electrons ($n = 1$ or more) impart colour to the aqueous solution. The ionic species having all the electrons in the paired state (*i.e.*, having no unpaired electrons, $n = 0$) give no colour to the aqueous solution.

   Now (*a*) $Ti^{4+} = 3d^0\ 4s^0$ ($n = 0$), (*b*) $Cu^+ = 3d^{10}$ ($n = 0$) (*c*) $Zn^{2+} = 3d^{10}$ ($n = 0$), (*d*) $Cr^{3+} = 3d^4$ $4s^0$ ($n = 4$). Thus Cr^{3+} ion ($n = 4$) will give colour to the aqueous solution, *i.e.*, answer (*d*) is the correct answer.

7. In order to find out the number of unpaired electrons (n) in the given ions, we will have to write their valence-shell electronic configuration.

   (*a*) $Mg^{2+} = 3s^0$ ($n = 0$), (*b*) $Ti^{3+} = 3d^1 4s^0$ ($n = 1$)

   (*c*) $V^{3+} = 3d^2 4s^0$ ($n = 2$), (*d*) $Fe^{2+} = 3d^6\ 4s^0$ ($n = 4$)

   Thus the correct answer is (*d*).

8. On the basis of Aufbau principle the given configurations should be written as :

   (*a*) $1s^2\ 2s^2\ 2p^6\ 3s^2 3p^6\ 4s^1$

   (*b*) $1s^2\ 2s^2\ 2p^6\ 3s^2\ 3p^6\ 4s^2\ 3d^4$ or $1s^2\ 2s^2\ 2p^6\ 3s^2\ 3p^6\ 4s^1\ 3d^5$

   (*c*) $1s^2\ 2s^2\ 2p^6\ 3s^2\ 3p^6\ 4s^2$

   (*d*) $1s^2\ 2s^2\ 2p^6\ 3s^2\ 3p^6\ 4s^2\ 3d^{10}$

   These configurations show that is only (*a*) and (*c*) configurations, in which the extra electron enters 4*s* orbital. Thus (*a*) and (*c*) are *s*-block elements.

9. Since the element with atomic number 107 lies in 7*th* period and group 7 after Rn ($Z = 86$) in the periodic table, the valence shell E.C. of this element is $7s^2\ 4f^{14}\ 5d^5$.

10. (*a*) Since this element is placed after Rn ($Z = 86$) in the periodic table, its E.C. should be $[Rn]_{86}\ 7s^2\ 6d^1\ 5f^1$ or $[Rn]_{86}\ 7s^2\ 6d^2\ 5f^0$. Thus since the extra electron enters 5*f* orbitals, it is a *f*-block element.

    (*b*) The E.C. of this element should be $[Xe]_{54}\ 6s^2\ 5d^1\ 4f^1$. Thus it is also a *f*-block element.

    (*c*) The E.C. of this element should be $[Xe]_{54}\ 6s^2\ 5d^1\ 4f^{14}$. Thus it is also a *f*-block element.

    (*d*) The E.C. of this element is $[Xe]_{54}\ 6s^2\ 5d^1$. Since the extra electron goes to *d*-orbitals, it is a *d*-block element.

11. Since the orbitals present in the inner shells of each atom in the given species are completely-filled, we should find out the sum of the electrons present in the valence shell of each of the atoms of the given species. Thus :

    (*a*) $NO_2^+ = 2s^2\ 2p^3 + 2 \times 2s^2\ 2p^4 - 1 = 5 + 12 - 1 = 16\ (n = 0)$

    $BaO_2 = 6s^2 + 2 \times 2s^2 2p^4 = 2 + 12 = 14\ (n = 0)$

    (*b*) $KO_2 = 4s^1 + 2 \times 2s^2\ 2p^4 = 1 + 12 = 13\ (n = 1)$

    $AlO_2^- = 3s^2\ 3p^1 + 2 \times 2s^2\ 2p^4 + 1$

    $= 3 + 12 + 1 = 16\ (n = 0)$

    (*c*) $KO_2 = 13\ (n = 1)$

    (*d*) $BaO_2 = 14\ (n = 0)$

    Obviously, since KO_2 has unpaired electrons ($n = 1$), (*c*) is the answer.

12. Find out the sum of electrons present in the atoms in each species.

(*a*) $NO^- = 1s^2\,2s^2\,2p^3 + 1s^2\,2s^2\,2p^4 + 1 = 7 + 8 + 1 = 16$

(*b*) $CN^- = 1s^2\,2s^2\,2p^2 + 1s^2\,2s^2\,2p^3 + 1 = 6 + 7 + 1 = 14$

(*c*) $N_2 = 2 \times (1s^2\,2s^2\,2p^3) = 2 \times 7 = 14$

(*d*) $O_2^{2+} = 2 \times (1s^2\,2s^2\,2p^4) - 2 = 16 - 2 = 14$

Since NO^- has 16 electrons while other species have 14 electrons each, answer is (*a*).

13. Number of electrons in each species is as :

(*a*) $\overset{+}{C}H_3 = 6 + 3 + 1 = 8$, $NH_2^- = 7 + 2 + 1 = 10$, $NH_4^+ = 7 + 4 - 1 = 10$,

(*b*) $NH_2^- = 10$, $NH_4^+ = 10$, $NH_3 = 7 + 3 = 10$,

(*c*) $\overset{+}{C}H_3 = 8$, $NH_2^- = 10$, $NH_3 = 10$

(*d*) $NH_2^- = 10$, $\overset{+}{C}H_3 = 8$.

Thus only NH_2^-, NH_4^+ and NH_3 have the same number of electrons (= 10).

14. The expected outer E.C. of Pd is $5s^2\,4d^8$ while the observed E.C. is $5s^0\,4d^{10}$ which is obtained by shifting both $5s^2$ electrons to 4*d* orbitals to make these orbitals completely filled. Thus, since the extra electron enters 4*d* orbitals, Pd is a *d*-block element and hence is present in 5*th* period and group 10.

15. Find out the sum of the electrons present in both the atoms of the given species

(*a*) $NO^+ = 7 + 8 - 1 = 14$, $C_2^{2-} = 2 \times 6 + 2 = 14$, $O_2 = 2 \times 8 = 16$, $CO = 6 + 8 = 14$

(*b*) $N_2 = 2 \times 7 = 14$, $C_2^{2-} = 14$, $CO = 14$, $NO = 7 + 8 = 15$

(*c*) $CO = 14$, $N_2 = 14$, $CN^- = 6 + 7 + 1 = 14$, $C_2^{2-} = 14$

(*d*) $NO = 15$, $CN^- = 14$, $N_2 = 14$, $O_2^- = 2 \times 8 + 1 = 17$.

Obviously the species given in (*c*) have the same number of electrons and hence are isoelectronic.

16. Copper crystal has Cu^+ ion. Now write the E.C. of the given species and find out the number of unpaired electrons (*n*) in them.

(*a*) Cu^+ in copper crystal = $[Ar]_{18}\,3d^{10}\,4s^0$ (*n* = 0, diamagnetic), (*b*) $Cu^+ = [Ar]_{18}\,3d^{10}\,4s^0$ (*n* = 0, diamagnetic), (*c*) $Cu^{2+} = [Ar]_{18}\,3d^9 4s^0$ (*n* = 1, paramagnetic) and (*d*) $H_2 = 1s^1 + 1s^1 = 2$ (*n* = 0, diamagnetic).

17. The given compounds have Mn^{2+}, Cu^{2+}, Fe^{2+} and Ni^{2+} ions respectively. Write the E.C. of these ions and find out the number of unpaired electrons (*n*) in them.

(*a*) $Mn^{2+} = [Ar]_{18}\,3d^5\,4s^0$ (*n* = 5, paramagnetic)

(*b*) $Cu^{2+} = [Ar]_{18}\,3d^9\,4s^0$ (*n* = 1, paramagnetic)

(*c*) $Fe^{2+} = [Ar]_{18}\,3d^6\,4s^0$ (*n* = 4, paramagnetic)

(*d*) $Ni^{2+} = [Ar]_{18}\,3d^8\,4s^0$ (*n* = 2, paramagnetic)

Since Cu^{2+} ion has the lowest number of unpaired electrons (= 1), $CuSO_4 \cdot 5H_2O$ has the lowest degree of paramagnetism.

18. The given compounds have Ag^+, Cu^{2+}, Mg^{2+} and Cu^+ ions respectively. Write the E.C. of these ions and find out the number of unpaired electrons (*n*) in them. The ions having one or more unpaired electrons are coloured while those having $n = 0$ are colourless. $Ag^+ = [Kr]_{36}\ 3d^{10}\ 4s^0$ ($n = 0$, colourless), $Cu^{2+} = [Ar]_{18}\ 3d^9\ 4s^0$ ($n = 1$, coloured), $Mg^{2+} = [Ne]_{10}\ 4s^0$ ($n = 0$, colourless) and $Cu^+ = [Ar]_{18}\ 3d^{10}\ 4s^0$ ($n = 0$, colourless). Since Cu^{2+} ion has unpaired electrons ($n = 1$), CuF_2 is coloured.

19. (*a*) Since this element is placed after Xe ($Z = 54$) in the periodic table, its E.C. should be written as $[Xe]_{54}\ 6s^2\ 5d^1\ 4f^{14}$ + 9 electrons in 5*d* orbitals or $[Xe]_{54}\ 6s^2\ 4f^{14}\ 5d^{10}$. Since the extra electron enters 5*d* orbitals, this element belongs to d-block elements.

(*b*) Its E.C. is $[Kr]_{36}\ 5s^2\ 4d^{10}$ (*d-block element*)

(*c*) Its E.C. is $[Ne]_{10}\ 3s^2\ 3p^5$ (*p-block element*)

(*d*) Its E.C. is $[Ar]_{18}\ 4s^2\ 3d^{10}$ (*d- block element*)

Obviously (*c*) belongs to *p*-block elements but (*a*), (*b*) and (*d*) belong to *d*-block elements.

20. Electronic configurations of the given ions are :

(*a*) $Cl^- = [Ne]_{10}\ 3s^2 3p^6$ ($n = 0$, diamagnetic),

(*b*) $Be^- = [He]_2\ 2s^2\ 2p^1$ ($n = 1$, paramagnetic),

(*c*) $Ne^{2+} = [He]_2\ 2s^2 2p^4$ ($n = 2$, paramagnetic),

(*d*) $As^+ = [Ar]_{18}\ 4s^2\ 3d^{10}\ 4p^2$ ($n = 2$, paramagnetic).

5

Atomic Properties : Perodic Properties

MULTIPLE CHOICE QUESTIONS

Atomic and Ionic Radii

1. Size of species I, I^+ and I^- is in the order :
   (*a*) $I > I^+ > I^-$ (*b*) $I^- > I^+ > I$
   (*c*) $I^- > I > I^+$
   (*d*) All have the same size.
2. Which of the following has the smallest size ?
   (*a*) Al^{3+} (*b*) Al^{2+}
   (*c*) Al^+ (*d*) Al
3. The ionic radii of the following ions can be arranged as :
   (*a*) $K^+ < Ca^{2+} < S^{2-}$
   (*b*) $Be^{2+} < Mg^{2+} < Na^+$
   (*c*) $O^{2-} < F^- < N^{3-}$
   (*d*) $S^{2-} < O^{2-} < As^{3-}$
4. Which one of the following ions has the largest size ?
   (*a*) Rb^+ (*b*) Mg^{2+}
   (*c*) Li^+ (*d*) Ne^+
5. Atomic radii of F and Ne (in Angstrom units) respectively are :
   (*a*) 0.72, 1.60 (*b*) 1.60, 1.60
   (*c*) 0.72, 0.72 (*d*) None of these.
6. Covalent radius of Li is 123 pm. The crystal radius of Li will be :
   (*a*) > 123 pm (*b*) = 123 pm
   (*c*) < 123 pm (*d*) 123/2 pm
7. Which of the following statements is correct ?
   (*a*) X^- ion is larger in size than X atom.
   (*b*) X^+ ion is larger in size than X atom.
   (*c*) X^+ ion is larger in size than X^- ion.
   (*d*) X^+ and X^- ions are equal in size.
8. Which one of the following is correct order of increase in size ?
   (*a*) $Mg < Na^+ < F^- < Al$
   (*b*) $Na^+ < Al < Mg < F^-$
   (*c*) $Na^+ < F^- < Al < Mg$
   (*d*) $Na^+ < F^- < Mg < Al$
9. The correct order of ionic radii is :
   (*a*) $Ti^{4+} < Mn^{7+}$ (*b*) $^{35}Cl^- < {}^{37}Cl^-$
   (*c*) $K^+ > Cl^-$ (*d*) $P^{3+} > P^{5+}$
10. The correct order of radii is :
   (*a*) N < Be < B (*b*) $F^- < O^{-2} < N^{3-}$
   (*c*) Na < Li < K (*d*) $Fe^{3+} < Fe^{2+} < Fe^{4+}$

Ionisation Energy (IE)

11. Amongst the following configurations, the one having the highest ionisation energy is :
   (*a*) [Ne] $3s^2\ 3p^1$ (*b*) [Ne] $3s^2 3p^3$
   (*c*) [Ne] $3s^2\ 3p^2$ (*d*) [Ar] $3d^{10}\ 4s^2\ 4p^2$
12. The set representing the correct order of first ionisation potential is :
   (*a*) K > Na > Li (*b*) Be > Mg > Ca
   (*c*) B > C > N (*d*) Ge > Si > C
13. The decreasing order of 2nd ionisation energy of K, Ca and Ba is :
   (*a*) K > Ca > Ba (*b*) Ca > Ba > K
   (*c*) Ba > K > Ca (*d*) K > Ba > Ca
14. The incorrect statement among the following is :
   (*a*) IE_1 of Al is less than IE_1 of Mg.
   (*b*) IE_2 of Mg is greater than IE_2 of Na
   (*c*) IE_1 of Na is less than IE_1 of Mg.
   (*d*) IE_3 of Mg is greater than IE_3 of Al.
15. Which of the following configurations represents atom of the element having the highest second ionisation energy ?
   (*a*) $1s^2\ 2s^2\ 2p^4$ (*b*) $1s^2\ 2s^2\ 2p^6$

(c) $1s^2\ 2s^2\ 2p^6\ 3s^1$
(d) $1s^2\ 2s^2\ 2p^6\ 3s^2$

16. The correct sequence of elements in the decreasing order of IE is :
(a) Na > Mg > Al > Si
(b) Mg > Na > Al > Si
(c) Al > Mg > Na > Si
(d) Si > Mg > Al > Na

17. With reference to ionisation potential which one of the following sets is correct ?
(a) Li < K < Cs (b) B> Li > K
(c) Cs > Li > B (d) Cs < Li < K

18. Which of the following species has the highest ionisation potential ?
(a) Ne (b) Al^+
(c) Mg^+ (d) Li^+

19. In which of the following pairs, the ionisation energy of the first species is less than that of the second
(a) N, P (b) Be^{2+}, Be
(c) N, N^- (d) S, P

20. Ionisation energy decreases in moving down a group. This is due to :
(a) increase in nuclear charge.
(b) increase in atomic size and nuclear charge.
(c) increase in nuclear charge and decrease in shielding effect.
(d) increase in atomic size and also shielding effect.

Electron Affinity (EA)

21. The outermost electronic configuration of the element with highest value of electron affinity is :
(a) $ns^2\ np^3$ (b) ns^2np^5
(c) $ns^2\ np^4$ (d) $ns^2\ np^6$

22. The correct order of electron affinity of the elements of oxygen family in the periodic table is :
(a) O > S > Se (b) S > O > Se
(c) S > Se > O (d) Se > O > S

23. Which of the following elements has the highest electron affinity ?
(a) Ne (b) K
(c) Cl (d) F

24. The correct order of electron affinity among the following is :
(a) F > Cl > Br (b) Br > Cl > F
(c) Cl > F > Br (d) F > Br > Cl

25. Which of the following has least electron affinity ?
(a) O (b) N
(c) Ar (d) F

26. The correct order of electron affinity of B, C, N and O is :
(a) O>C>N>B (b) B>N>C>O
(c) O>C>B>N (d) O>B>C>N

27. The higher value of EA of Cl as compared to that of F is due to :
(a) Higher atomic radiaus of F.
(b) Smaller electronegativity of F.
(c) Weaker electron-electron repulsion in Cl.
(d) More vacant p-sub shell in Cl.

28. Electron affinity of an element depends on its :
(a) atomic size
(b) nuclear charge
(c) electronic configuration
(d) all these properties.

29. Which of the following processes involve absorption of energy ?
(a) $O \rightarrow O^-$ (b) $O^- \rightarrow O^{2-}$
(c) $O \rightarrow O^{2-}$ (d) $F \rightarrow F^-$

30. EA of K^+ ion would be numerically equal to :
(a) IE of K
(b) electronegativity of K
(c) IE of K^+
(d) none of the above.

Electronegativity

31. Two elements whose electronegativities are 1.2 and 3.0, the bond formed between them would be :
(a) ionic (b) covalent
(c) coordinate (d) metallic.

32. An element X has IE and EA respectively 275 and 1450 KJ mol^{-1}. The electronegativity of X, according to Pauling's scale, is :

(*a*) 2.4 (*b*) 2.8
(*c*) 3.2 (*d*) 4.0

33. If the electronegativity difference between two atoms A and B of a molecule is 2.0, then the percentage of covalent character in the molecule is :

(*a*) 54% (*b*) 46%
(*c*) 23% (*d*) 72%

34. Which of the following bonds will be the most polar ?

(*a*) N – Cl (*b*) O – F
(*c*) C – F (*d*) N – N

35. The electronegativity of F, Cl, Br and I are 4.0, 3.0, 2.8 and 2.5 respectively. The hydrogen halide with the highest degree of ionic character is

(*a*) HF (*b*) HCl
(*c*) HBr (*d*) HI

36. Which is true about electronegativity order of the following ?

(*a*) P > Si (*b*) C > N
(*c*) Br > Cl (*d*) Sr > Ca

37. The electronegativity of the following elements increases in the order :

(*a*) C, N, Si, P (*b*) N, Si, C, P
(*c*) Si, P, C, N (*d*) P, Si, N, C

38. The correct order of electronegativity of N, O, F and P is :

(*a*) F > N > P > O (*b*) F > O > P > N
(*c*) F > O > N > P (*d*) N > O > F > P

39. The attraction that an atom exerts on a pair of electrons that is being shared with another atom for forming covalent bond is referred to as its :

(*a*) Electron affinity
(*b*) Electronegativity
(*c*) Ionisation energy
(*d*) Valency

40. Electronegativity values of Fe, Fe^{2+} and Fe^{3+} are in the order :

(*a*) Fe > Fe^{2+} > Fe^{3+}
(*b*) Fe^{2+} > Fe > Fe^{3+}
(*c*) Fe^{3+} > Fe > Fe^{2+}
(*d*) Fe^{3+} > Fe^{2+} > Fe

ANSWERS

1. (*c*)	**2.** (*a*)	**3.** (*b*)	**4.** (*a*)	**5.** (*a*)	**6.** (*c*)	**7.** (*a*)	**8.** (*b*)
9. (*d*)	**10.** (*b*)	**11.** (*b*)	**12.** (*b*)	**13.** (*a*)	**14.** (*b*)	**15.** (*c*)	**16.** (*d*)
17. (*b*)	**18.** (*d*)	**19.** (*d*)	**20.** (*d*)	**21.** (*b*)	**22.** (*c*)	**23.** (*c*)	**24.** (*c*)
25. (*c*)	**26.** (*c*)	**27.** (*c*)	**28.** (*d*)	**29.** (*b*), (*c*)	**30.** (*a*)	**31.** (*a*)	**32.** (*c*)
33. (*a*)	**34.** (*c*)	**35.** (*a*)	**36.** (*a*)	**37.** (*c*)	**38.** (*c*)	**39.** (*b*)	**40.** (*d*)

HINTS AND EXPLANATIONS

1. Number of electrons in the given species is as : I (Z = 53) = 53, I^+ (Z = 53) = 53 – 1 = 52 and I^- (Z = 53) = 53 + 1 = 54. Smaller is the number of electrons in the species, greater is the nuclear attractive force acting on these electrons and hence smaller is the size of the species. Thus the size of the species decreases with the decrease in the number of electrons in the species.

2. Number of electrons (*e*) in each of the given species is as : (*a*) Al^{3+} (Z = 13), *e* = 13 – 3 = 10, (*b*) Al^{2+} (Z = 13), *e* = 13 – 2 = 11, (*c*) Al^+ (Z = 13), *e* = 13 – 1 = 12 and (*d*) Al (Z = 13), *e* = 13. Greater is the positive charge on the cations, smaller is the number of electrons in the cations and hence greater is the nuclear attractive force acting on the electrons and smaller is the size of the cation. Thus the size of cations increases with the decrease in positive charge on the cation [size of cation α 1/positive charge on cation]

3. (*a*) Number of electrons in the given ions is as : K^+ (Z = 19) = 19 – 1 = 18, Ca^{2+} (Z = 20) = 20 – 2 = 18 and S^{2-} (Z = 16) = 16 + 2 = 18. Since these ions are isoelectronic, their size will decrease with the increase of their nuclear charge (N.C.). Thus the size of K^+, Ca^{2+} and S^{2-} ions will decrease as S^{2-} (N.C. = + 16) > K^+ (N.C. = + 19) > Ca^{2+} (N.C. = + 20) or $Ca^{2+} < K^+ < S^{2-}$. Thus the increasing order given at (*a*) is wrong.

(*b*) Since Be^{2+} and Mg^{2+} ions belong to the same group (group 2), their radii are in the order $Be^{2+} < Mg^{2+}$. Na^+ and Mg^{2+} ions belong to the same period (3*rd* period), their radii are in the order $Na^+ > Mg^{2+}$. Thus the radii of Na^+, Mg^{2+} and Be^{2+} ions are in the order : $Na^+ > Mg^{2+} > Be^{2+}$ or $Be^{2+} < Mg^{2+} < Na^+$.

(*c*) Since O^{2-}, F^- and N^{3-} ions are isoelectronic (No. of electrons in each ion = 10), their size decreases with the increase in their nuclear charge (N.C.). Thus their radii are in the order : N^{3-} (N.C. = + 7) > O^{2-} (N.C. = + 8) > F^- (N.C. = + 9) or $F^- < O^{2-} < N^{3-}$. Thus the increasing order given at (*c*) is wrong.

(*d*) Valence shell E. Cs. of the given ions are as : $S^{2-} = 3s^2\,3p^6$, $O^{2-} = 2s^2\,2p^6$ and $As^{3-} = 4s^2\,4p^6$. These ions have noble gas configurations and hence their radii increase as $O^{2-} < S^{2-} < As^{3-}$ or $As^{3-} > S^{2-} > O^{2-}$. Thus the increasing order given at (*d*) is wrong.

4. Since Li^+, Na^+ and Rb^+ ions belong to the same group, their radii increase as $Li^+ < Na^+ < Rb^+$. Since Na^+ and Mg^{2+} ions belong to the same period, their radii are in the order : $Na^+ < Mg^{2+}$. Thus Rb^+ ion is the largest in size.

5. Since atomic radius of F atom is its covalent radius (r_{cov}) and that of Ne is its vander waals radius (r_{van}), r_{cov} is smaller than r_{van}.

6. Crystal radius of Li is the radius of Li^+ ion. Since a cation is smaller in size than its parental atom, Li^+ ion is smaller in size than Li atom. Thus Li^+ is smaller than 123 pm (Li^+ < 123 pm).

7. (*a*) Since an anion (X^-) is larger in size than its parental atom (X), statement (*a*) is correct.

(*b*) Since a cation (X^+) is smaller in size than its parental atom (X), statement (*b*) is wrong.

(*c*) and (*d*). Since X^+ ion has lesser number of electrons than X^- ion (both the ions have the same nuclear charge equal to + Z), this ion is smaller in size than X^-. Thus statements (*c*) and (*d*) are wrong.

8. Since Na^+ (Z = 11) and F^- (Z = 9) ions are iso-electronic (Number of electrons = 11 – 1 = 10, 9 + 1 = 10 respectively) and their nuclear charges (N.C.) are + 11 and + 9 respectively, their size is inversely proportional to their nuclear charge. Thus Na^+ ion is the smallest in size and F^- ion is the largest in size. Further since Mg and Al belong to the same period, Mg has larger size than Al (Mg > Al). Thus the overall order is $Na^+ < Al < Mg < F^-$.

9. (*a*) Since Ti^{4+} and Mn^{2+} ions both have the same number of electrons (Ti^{4+} = 22 – 4 = 18 and Mn^{7+} = 25 – 7 = 18), the ion having higher nuclear charge (N.C.) is smaller in size. Thus Mn^{7+} ion (N.C. = + 25) is smaller in size than Ti^{4+} ion (N.C. = + 22) ($Mn^{7+} < Ti^{4+}$).

(*b*) Both the ions are the same ions which have different mass number (A). A for $^{35}Cl^-$ = 35 and A for $^{37}Cl^-$ = 37. Mass number has no effect on the size of the ion. Thus the given (these ions are isotopes) have the same size.

(*c*) Since K^+ and Cl^- ions both have the same number of electrons (K^+ = 19 – 1 = 18 and Cl^- = 17 + 1 = 18), the ion having higher nuclear charge (N.C.) has smaller size. Thus K^+ ion (N.C. = + 19) is smaller in size than Cl^- (N.C. = + 17) ($K^+ < Cl^-$).

(*d*) If an element forms two cations having different positive charges, then the cation having higher positive charge is smaller in size than the cation having smaller positive charge. Thus P^{5+} (positive charges = + 5) should be smaller in size than P^{3+} (positive charges = + 3).

10. (*a*) Since N, Be and B are the elements of the same period, their radii decrease with the increase in their nuclear charge (N.C.). Thus the radii of these elements are in the order Be (N.C. = + 4) > B (N. C. = + 5) > N (N.C. = + 7) or N < B < Be. Thus order given at (*a*) is wrong.

(*b*) Since the given ions are iso-electronic ions (F^- = 9 + 1 = 10, O^{2-} = 8 + 2 = 10 and N^{3-} = 7 + 3 = 10), their radii decrease with the increase of their nuclear charge (N.C.). Thus N^{3-} (N.C. = + 7) > O^{2-} (N.C. = + 8) > F^- (N.C. = + 9) or F^- < O^{2-} < N^{3-}. Thus order given at (*b*) is correct.

(*c*) Since Na, Li and K are the elements of the same group, their radii should be in the order : Li < Na < K. Thus order given at (*c*) is wrong.

(*d*) If a metal forms many cations having different positive charges, then the radii of these cations decrease with the increase in the magnitude of positive charge on the cations. Thus the radii of Fe^{2+}, Fe^{3+} and Fe^{4+} are in the order : $Fe^{2+} > Fe^{3+} > Fe^{4+}$ or $Fe^{4+} < Fe^{3+} < Fe^{2+}$. Thus order given at (*d*) is wrong.

11. [Ne] $3s^2\ 3p^3$ configuration has half-filled 3*p* orbitals, this configuration is the most stable of all other configurations and hence the highest amount of energy will be needed to remove an electron from 3*p* orbitals.

12. Due to the increase in atomic size and screening effect of the inner electrons on the valence shell electron on moving down the group, IE values of the elements decrease down the group.

13. IE_2 value of an element (M) corresponds to the removal of an electron from its monovalent cation (M^+). E.C. of M^+ ions are $K^+ = [Ar]_{18}$, $Ca^+ = [Ar]_{18}\ 4s^1$ and $Ba^+ = [Xe]_{54}\ 6s^1$. Now since Ca^+ and Ba^+ ions belong to the same group, IE_2 of Ba is smaller than IE_2 of Ca (Ca > Ba). IE_2 of K is much higher than those of Ca and Ba, since in case of K, second electron is to be removed from $[Ar]_{18}$ configuration (stable configuration) while in case of Ca and Ba second electron is to be removed from 4s and 6s orbital respectively.

14. E.C. of Mg^+ and Na^+ ions are $Mg^+ = [Ne]_{10}\ 3s^1$ and $Na^+ = [Ne]_{10}$. IE_2 value of Mg corresponds to the removal of an electron from 3*s* orbital while IE_2 value of Na corresponds to the removal of an electron from neon configuration, $[Ne]_{10}$ which is a stable configuration.

15. IE_2 value of an element (M) is the energy needed to remove an electron from its M^+ cation. M^+ cations obtained from the given configurations are as : (*a*) $1s^2\ 2s^2\ 2p^3$, (*b*) $1s^2\ 2s^2\ 2p^5$, (*c*) $1s^2\ 2s^2\ 2p^6$ and (*d*) $1s^2\ 2s^2\ 2p^6\ 3s^1$. Since (*c*) has noble gas configuration, it has the highest value of IE_2.

16. Since Na, Mg, Al and Si are the elements of the same period (3rd period), their IE values should increase as Na < Mg < Al < Si. But since Mg has completely filled 3*s* orbital (Mg = $3s^2$), its IE is higher than that of Al and hence the correct order is as : Na < Al < Mg < Si or Si > Mg > Al > Na.

17. (*a*) Since Li, K and Cs are the elements of the same group, their IE values should decrease down the group as Li > K > Cs or Cs < K < Li. Thus (*a*) is incorrect order.

(*b*) Li > K (same group) and Li < B (same period). Hence B > Li > K. Thus (*b*) is the correct order.

(*c*) Li > Cs (same group) and Li < B (same period). Hence Cs < Li < B or B > Li > Cs. Thus (*c*) is incorrect order.

(*d*) Li > K > Cs (same group) or Cs < K < Li. Thus (*d*) is incorrect order.

18. Electronic configuration of the given species are : (*a*) Ne (Z = 10) = $[Ne]_{10} = 1s^2, 2s^2\ 2p^6$, (*b*) Al^+ (Z = 13) = $[Ne]_{10}\ 3s^2$, (*c*) Mg^+ (Z = 12) = $[Ne]_{10}\ 3s^1$ and (*d*) Li^+ (Z = 3) = $[He]_2 = 1s^2$. Since Ne and Li^+ both have noble gas configurations, these species have higher values of IE than the remaining two species. Now since $1s$ orbital present in Li^+ is nearest to the nucleus, highest amount of energy will be needed to remove an electron from this orbital.

19. (*a*) Since N and P belong to the same group, $(IE)_N > (IE)_P$.

(*b*) $Be^{2+} = 1s^2$, $Be = 1s^2, 2s^2$, $\therefore (IE)_{Be2+} > (IE)_{Be}$

(*c*) $N = 2s^2 2p^3$, $N^- = 2s^2\, 2p^4$, $\therefore (IE)_N > (IE)_{N^-}$

(*d*) $S = 3s^2 3p^4$, $P = 3s^2\, 3p^3$, $\therefore (IE)_P > (IE)_S$

20. Larger atoms can loose the valence-shell electron more easily and hence IE value decreases. Due to the increase in the magnitude of shielding effect caused by the inner electrons on the valence shell electron, the attraction between the valence-shell electron and nucleus decreases and hence IE values also decrease.

21. Due to the presence of half-filled and completely-filled *np* orbitals in configurations (*a*) and (*d*) respectively, these configurations are stable and hence donot accept the extra electron to be added to them. Thus these configurations have very low or approximately zero EA value. Configuration (*b*) will accept the extra electron readily, since the addition of an electron makes this configuration stable ($ns^2\ np^6$ configuration). Thus this configuration has the highest value of EA.

22. The given elements belong to the same group *viz* 6A which contains O, S, Se, Te and Po. Since EA values of the elements belonging to the same group generally decrease on moving down the group, EA values of O, S, Se, Te and Po should be in the order : O > S > Se > Te > Po. Due to the smallest size of O atom electron density on this atom is very high. Due to high electron density, O-atom shows the highest reluctance to accept the incoming electron and hence EA value of O-atom is the lowest amongst the elements of group 6A. Thus the actual order of EA values becomes as shown below :

$$\underbrace{O > S > Se > Te > Po}_{\text{Expected order}} \longrightarrow \underbrace{\underset{200.7}{S} > \underset{195.0}{Se} > \underset{190.0}{Te} > \underset{180.0}{Po} > \underset{140.9}{O}}_{\text{Actual order (Values are in KJ mol}^{-1}\text{)}}$$

Thus amongst oxygen family elements, O-atom has the lowest value of EA.

23. Halogens (ns^2np^5) have the highest EA values, since, by gaining one electron, they acquire the stable noble gas configuration ($ns^2\ np^6$). Thus F and Cl both have higher EA values than Na and K both. Now since F atom has smaller size than Cl atom and stronger electron-electron repulsion, this atom shows greater reluctance to accept the incoming electron and has lower value of EA than Cl-atom. Thus Cl atom has the highest value of EA.

24. Generally, EA values of elements decrease on moving down the groups. Thus these values for halogens should be in the order F > Cl > Br > I. Now since F atom is smaller in size than Cl atom, electron density on F atom is higher than that on Cl atom. Due to higher electron density, F atom shows greater reluctance to accept the incoming electron than Cl atom does and hence F atom has lower value of EA. Thus the actual order of EA values is as :

$$\underbrace{F > Cl > Br > I}_{\text{Expected order}} \longrightarrow \underbrace{\underset{348.5}{Cl} > \underset{332.6}{F} > \underset{324.7}{Br} > \underset{295.5}{I}}_{\text{Actual order (EA values are in KJ mol}^{-1}\text{)}}$$

Thus amongst halogens Cl has the highest EA.

25. Valence-shell electronic configurations of the given elements are $O = 2s^2\ 2p^4$, $N = 2s^2\ 2p^3$, $Ar = 3s^2 3p^6$ and $F = 2s^2\ 2p^5$. Since E.C. of Ar has an octet of electrons, it is a stable configuration and hence does not accept the incoming electron.

26. Since the given elements belong to the same period (2*nd* period), their EA values should increase as B < C < N < O. But since N-atom has stable half-filled 2*p* orbitals ($N = 2s^2\,2p^3$), this atom does not accept the additional electron and hence its EA is zero. Thus the actual order is N < B < C < O or O > C > B > N.

27. Since F-atom is smaller in size than Cl atom, 2*p* subshell in F atom ($F = 2s^2\,2p^5$) is relatively more compact than 3*p* sub shell in Cl atom ($Cl = 3s^2\,3p^5$). Due to compact nature of 2*p* sub-shell in F atom, the repulsion between the incoming electron and the electrons already present in 2*p* sub-shell in F atom is relatively stronger than that between the incoming electron and the electrons already present in 3*p* sub-shell of Cl atom. Due to stronger electron - electron repulsion in F-atom, this atom shows greater reluctance to accept the extra electron and hence has lower value of EA as compared to that of Cl atom.

28. With the increase in atomic size of the atom of an element, EA of the element decreases (Variation in a group). With the increase in the nuclear charge, EA increases (Variation in a period). Elements having half-filled or completely-filled orbitals have low or zero EA values. Thus EA value of an element depends on all the three properties *viz* atomic size, nuclear charge and electronic configuration.

29. (*a*) and (*d*). The conversion of O to O^- in process (*a*) is possible only when an electron is added to O-atom. In this process of adding an electron to a neutral atom energy is released. Similarly in process (*d*) energy is released.

$$O + e^- \rightarrow O^- + \text{Energy released } (= \text{EA of O atom})$$

$$F + e^- \rightarrow F^- + \text{Energy released } (= \text{EA of F atom})$$

(*b*) The conversion of O^- to O^{2-} is possible only when an electron is added to O^-. Since the electron is added against the electrostatic repulsion existing between the negative charge on electron and negative charge on O^- ion, energy is absorbed.

$$O^- + e^- + \text{Energy required } (= EA_2 \text{ of O-atom}) \rightarrow O^{2-}$$

(*c*) The conversion of O-atom to O^{2-} ion takes place through the following steps :

$$O + e^- \rightarrow O^-, \qquad \Delta H_1 = \text{Energy released } (-) = -EA_1 \text{ of O-atom}$$

$$O^- + e^- \rightarrow O^{2-}, \qquad \Delta H_2 = \text{Energy absorbed } (+) = +EA_2 \text{ of O-atom}$$

$$O + 2e^- \rightarrow O^{2-}, \qquad \Delta H = \Delta H_1 + \Delta H_2 = -EA_1 + EA_2$$

Now since $EA_2 > EA_1$, ΔH is a positive quantity, ..*e.*, in process (*c*) energy will be absorbed.

30. EA of K^+ ion is represented as :

$$K^+ + e^- \rightarrow K + \text{Energy released } (= \text{EA of } K^+).$$

IE of K atom is represented by the equation :

$$K + \text{Energy required (IE of K atom)} \rightarrow K^+ + e^-.$$

Since both the above equations are opposite to each other, EA of K^+ ion is the same as IE of K-atom

31. If the electronegativity difference of two elements is greater than 1.7, the bond between the elements is ionic. Here since the electronegativity difference is 3.0 – 1.2 = 1.8 which is greater than 1.7, the bond is ionic.

32. We know that if IE and EA of an element (M) are given in KJ mol^{-1}, then electronegativity of that element (x_M), on Pauling's scale, is given by :

$$(x_M)_{Pauling} = \frac{\text{IE of M+EA of M}}{193 \times 2.8}$$

$$= \frac{275+1450}{540.4} = 3.192 \approx 3.2$$

33. According to Hanny and Smith equation percentage of ionic character in A–B bond is given by :

Ionic character in A – B bond = $[16\ (x_A - x_B) + 3.5\ (x_A - x_B)^2]\%$

$= [16 \times 2 + 3.5 \times (2)^2]\%$

$= 46\%$

$\therefore$ Covalent character = (100 – 46)% = **54%**

34. Since F-atom which is the most electronegative atom is present in (*b*) and (*c*), the correct answer will be out of these two alternatives. Now since electronegativity of O-atom is greater than that of C-atom, $(x_F - x_C)$ is greater than $(x_F - x_O)$ and hence F – C bond will be the most polar.

35. Since F-atom has the maximum electronegativity, the electronegativity difference between halogen and hydrogen atom would be maximum when halogen atom is F, *i.e.*, $(x_F - x_H)$ would be the maximum for HF and hence HF would have maximum ionic character.

36. (*a*) Since P and Si are the members of the same period, their electronegativity order is Si < P or P > Si. Thus (*a*) is the correct answer.

(*b*) Since C and N are the members of the same period, their electronegativity order is C < N. Thus (*b*) is wrong answer.

(*c*) Since Br and Cl are the members of the same group, their electronegativity order is Br > Cl. Thus (*c*) is a wrong answer.

(*d*) Since Ca and Sr are the members of the same group, their electronegativity is in the order Ca > Sr or Sr < Ca. Thus answer (*d*) is a wrong answer.

37. We know that in a period electronegativity increases and in a group electronegativity decreases. On this basis we can write the following order of electronegativity values of the given elements (*i*) C < N (same period) (*ii*) Si < P (same period) (*iii*) C > Si (same group) (*iv*) N > P (same group).

On the basis of (*i*) and (*iii*) relations we can write Si < C < N. Similarly on the basis of (*ii*) and (*iv*) relations we can write Si < P < N. These relations show that Si has the lowest electronegativity and N has the highest electronegativity. Now since it is only (*c*) option in which Si has the lowest electronegativity and N has the highest electronegativity.

38. We know that in a period electronegativity increases and in a group electronegativity decreases. Thus on this basis we can write the relations : (*i*) N < O < F (same period) (*ii*) N > P or P < N (same group). On combining these relations we can write P < N < O < F or F > O > N > P.

39. The given statement is the definition of electronegativity of the atom of an element.

40. Since the size of the given species increases in the order : $Fe^{3+} < Fe^{2+} <$ Fe, electronegativity values of these species decrease as $Fe^{3+} > Fe^{2+} >$ Fe.

6

Radioactivity

MULTIPLE CHOICE QUESTIONS

1. An isotope of ^{231}Th can be converted into ^{227}Th by the emission of :

(a) One α-particle

(b) One β-particle

(c) two α and one β-particle

(d) one α and two β-particles

2. In $_{92}U^{238} \xrightarrow{-\alpha} A \xrightarrow{-\beta} {}_xB^y$, the value of x and y is :

(a) 90, 234 (b) 91, 234

(c) 92, 234 (d) 93, 234

3. β-particle is emitted in radioactivity by :

(a) conversion of proton to neutron

(b) from outermost orbit

(c) conversion of neutron to proton

(d) β-particle is not emitted.

4. Loss of a β-particle is equivalent to

(a) Increase of one proton only

(b) Decrease of one neutron only

(c) Both (a) and (b)

(d) None of these.

5. $_7N^{13}$ changes to $_6C^{13}$ by emission of :

(a) Electron (b) Neutron

(c) Positron (d) Proton.

6. A radioisotope has half-life period of 10 years. What percentage of the original amount of it would you expect to remain after 20 years ?

(a) 0 (b) 12.5

(c) 25 (d) 8

7. $_{91}Pa^{234}$ belongs to group 3 and emits a β-particle. The daughter nuclide thus obtained belongs to group :

(a) 2 (b) 4

(c) 1 (d) 3

8. In the radioactive change $^x_yP \longrightarrow {}^x_{y+1}Q \longrightarrow {}^{x-1}_{y-1}R \longrightarrow {}^{x-1}_{y-1}S$

The radiations emitted in the sequence are :

(a) α, β, γ (b) β, α, γ

(c) γ, α, β (d) β, γ, α

9. Thorium series emits :

(a) 6α and 4β particles

(b) 4α and 6β particles

(c) 3α and 2β particles

(d) 2β and 3α particles.

10. $_{84}Rn^{219}$ is a member of actinium series. Another member of the same series is :

(a) $_{92}U^{235}$ (b) $_{90}Th^{232}$

(c) $_{89}Ac^{225}$ (d) $_{15}P^{36}$

11. During the transformation of $_cX^a$ to $_dY^b$ by α and β-decay, the number of β-particles emitted are :

(a) $\frac{a-b}{4}$ (b) $d+\frac{a-b}{2}+c$

(c) $d+\frac{a-b}{2}-c$ (d) $2c-d+a-b$

12. $t_{0.5}$ of a radioactive isotope is 1.5 hours. How long will it take for its activity to be reduced to 1/16 of its original value ?

(a) 3 h (b) 360 min

(c) 4.5 h (d) 600 min

13. The activity of a sample of Ti ($Z = 22$) is decreased by 90% over a period of 10 years. The half-life period ($t_{0.5}$) of the sample is :

(*a*) 1 year (*b*) 2 years
(*c*) 3 years (*d*) 5 years.

14. Half-life of a radioactive substance is 15 seconds. After 45 seconds what percentage of activity will be left

(*a*) 15% (*b*) 18%
(*c*) 12.5% (*d*) 10%

15. Two radioactive nuclides A and B have half-life *t* and 2*t* respectively. If we start an experiment with one mole of each of them, what will be mole ratio after a time interval of 6*t* ?

(*a*) 1 : 2 (*b*) 1 : 8
(*c*) 1 : 6 (*d*) 1 : 1

16. The change $_{15}P^{30} \longrightarrow {}_{14}Si^{30}$ requires the emission of :

(*a*) α-particle (*b*) β-particle
(*c*) neutron (*d*) positron.

17. For a radioactive substance the values of $t_{0.5}$ and λ are identical in magnitude. The value of $t_{0.5}$ is :

(*a*) 1/0.693 (*b*) $(0.693)^2$
(*c*) 0.693 (*d*) $(0.693)^{1/2}$

18. If $_bX^a$ emits firstly a positron, then two α and two β and in the last one α is emitted and is converted into $_dY^c$. The correct relations are :

(*a*) $a = c + 12, d = b - 5$
(*b*) $a = c + 8, d = b - 1$
(*c*) $a = c + 6, d = b - 2$
(*d*) $A = c + 4, d = b - 2$

19. The half-life of a radioactive element is 30 min. One sixteenth of original quantity of element will be left after :

(*a*) 1 hr (*b*) 16 hrs
(*c*) 4 hrs (*d*) 2 hrs

20. After the emission of one α-particle followed by one β-particle from $_{92}X^{238}$, the number of neutrons in the daughter element will be :

(*a*) 142 (*b*) 146
(*c*) 144 (*d*) 143

21. Total number of electrons, protons and neutrons in the product formed by the loss of one α-particle from $_{92}U^{238}$ is :

(*a*) 326 (*b*) 333
(*c*) 324 (*d*) 332

ANSWERS

1. (*d*)	**2.** (*b*)	**3.** (*c*)	**4.** (*c*)	**5.** (*c*)	**6.** (*c*)	**7.** (*d*)	**8.** (*b*)
9. (*a*)	**10.** (*a*)	**11.** (*c*)	**12.** (*b*)	**13.** (*c*)	**14.** (*c*)	**15.** (*b*)	**16.** (*d*)
17. (*d*)	**18.** (*a*)	**19.** (*d*)	**20.** (*d*)	**21.** (*c*)			

HINTS AND EXPLANATIONS

1. The emission of one α ($_2He^4$) and two β ($_{-1}e^0$) particles from ^{231}Th produced ^{227}Th as shown below :

$$^{231}_{x}Th \xrightarrow{-\alpha \text{ or } _2He^4} {}^{231-4}_{x-2}X \text{ or } {}^{227}_{x-2}X \xrightarrow{-2\beta \text{ or } 2\ _{-1}e^0} {}^{227}_{x-2-(-1\times2)}X \text{ or } {}^{227}_{x}X$$

Since *x* for ^{231}Th and ^{227}Th are the same, ^{231}Th and ^{227}Th are isotopes with each other

2. $$_{92}U^{238} \xrightarrow{-\alpha \text{ or } _2He^4} {}_{92-2}A^{238-4} \text{ or } {}_{90}A^{234} \xrightarrow{-\beta \text{ or } _{-1}e^0} {}_{90-(-1)}B^{234} \text{ or } {}_{91}B^{234}$$

Thus $y = 234$ and $x = 91$.

3. If an isotope $_bX^a$ loses a β-particle, it is converted into $_{b+1}Y^a$.

$$_{b}X^{a} \xrightarrow{-\beta \text{ or } _{-1}e^{0}} {}_{b+1}Y^{a}$$

No. of protons in X = b No. of protons in Y = $b + 1$

No. of neutrons in X = $(a - b)$ No. of neutrons in Y = $a - (b + 1) = a - b - 1$

Obviously Y has one more proton and one less neutron than X. Thus one neutron of X is converted into proton.

4. If an isotope $_{b}X^{a}$ loses a β-particle, it is converted into $_{b+1}Y^{a}$.

$$_{b}X^{a} \xrightarrow{-\beta \text{ or } _{-1}e^{0}} {}_{b+1}Y^{a}$$

No. of protons in X = b No. of protons in Y = $b + 1$

No of neutron in X = $(a - b)$ No. of neutrons in Y = $a - (b + 1) = a - b - 1$

Obviously Y has one more proton (increase of one proton) and one less neutron (decrease of one neutron) than X.

5. Suppose $_{7}N^{13}$ changes to $_{6}C^{13}$ by emitting $_{b}X^{a}$ particle.

$$_{7}N^{13} \longrightarrow {}_{6}C^{13} + {}_{b}X^{a}$$

Obviously, $7 = 6 + b$, $\therefore\ b = 7 - 6 = 1$

$13 = 13 + a$, $\therefore\ a = 0$

Thus $_{b}X^{a}$ particle is $_{+1}e^{0}$ (positron).

6. We know that :

$$W_t \text{ or } W_n = W_o \times \left(\frac{1}{2}\right)^n$$

Here W_t or W_n = Amount of the radioisotope left undecayed after t years or after n half-lives.

W_o = Original amount of the radioisotope = 100 g (suppose)

$$n = \text{No. of half-lives in } t \text{ years } = \frac{t}{t_{1/2}} = \frac{20}{10} = 2.$$

$$\therefore\ W_t \text{ or } W_n = 100 \times \left(\frac{1}{2}\right)^2 = 100 \times \frac{1}{4} = \mathbf{25\%}$$

7. Obviously the daughter nuclide would be a nuclide having mass number = 234 and atomic number = 91 – (–1) = 92 as shown below :

$$_{91}Pa^{234} \xrightarrow{-\alpha \text{ or } _{-1}e^{0}} {}_{91-(-1)}X^{234} \quad \text{or} \quad {}_{92}X^{234}$$

Now since the elements with atomic number 90 to 103 belong to group 3 (actinides), the daughter nuclide whose atomic number is 92 will also belong to group 3.

8. Answer (*b*) is based on the facts that : (*i*) When a radioactive element emits an α-particle, the mass number decreases by 4 units and atomic number of the resulting nuclide decreases by 2 units. (*ii*) When a radioactive element emits a β-particle, the mass number of the resulting nuclide remains the same but the atomic number increases by one unit. (*iii*) The emission of a γ-particle does not change mass number nor atomic number.

9. Answer (*a*) is evident from thorium radioactive series given below $(\alpha = {}_2He^4$ and $\beta = {}_{-1}e^0)$

$$ {}_{90}Th^{232} \xrightarrow{-\alpha} {}_{88}Ra^{228} \xrightarrow{-\beta} {}_{89}Ac^{228} \xrightarrow{-\beta} {}_{90}Th^{228} $$

$$ \xrightarrow{-\alpha} {}_{88}Ra^{224} \xrightarrow{-\alpha} {}_{86}Rn^{220} \xrightarrow{-\alpha} $$

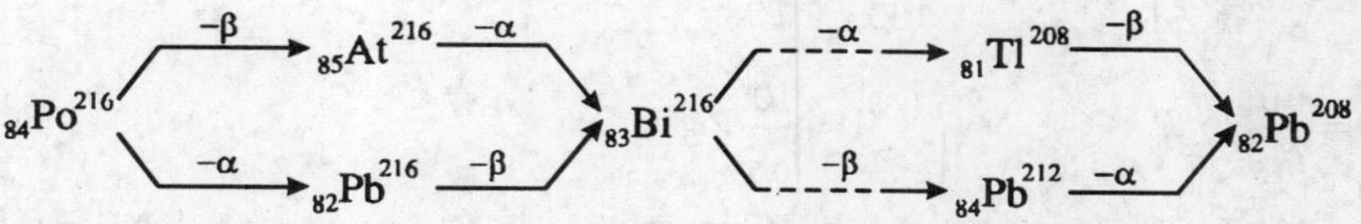

10. Actinium series is $4n + 3$ series which means that when the mass number of each element of this series is divided by 4, 3 is obtained as remainder (*r*). Thus for ${}_{92}U^{235}$, ${}_{90}Th^{232}$, ${}_{89}Ac^{225}$ and ${}_{15}P^{36}$ the values of r = 3, 0, 1 and 0 respectively. Thus it is only ${}_{92}U^{235}$ which is a member of actinium series.

11. Let the number of α and β particles emitted is x and y respectively. Then :

$$ {}_cX^a \longrightarrow {}_dY^b + x\ {}_2He^4 + y\ {}_{-1}e^0 $$

Thus the following relations can be written :

(*i*) $a = b + 4x + 0$ or $x = \dfrac{a-b}{4}$

(*ii*) $c = d + 2x - y$ or $c = d + 2 \times \dfrac{a-b}{4} - y$ or $y = d + \dfrac{a-b}{2} - c$

12. We know that :

$$ W_t \text{ or } W_n = W_o \times \left(\frac{1}{2}\right)^n $$

Here W_o = original amount of radioactive isotope (suppose 1 g)

$$ n = \text{No. of half-lives in } t \text{ years } = \frac{t}{t_{0.5}} = \frac{t}{1.5} $$

W_t or W_n = Amount of radioactive isotope left after t years (t_o be determined) or after n half-lives = 1/16

$$ \therefore \quad \frac{1}{16} = 1 \times \left(\frac{1}{2}\right)^{t/1.5} \quad \text{or} \quad \left(\frac{1}{2}\right)^4 = \left(\frac{1}{2}\right)^{t/1.5} $$

$$ \therefore \quad 4 = \frac{t}{1.5} \text{ or } t = 4 \times 1.5 \text{ hours} = 6 \text{ hours} = 6 \times 60 = 360 \text{ minutes} $$

13. We know that :

$$ a_t = a_o \times \left(\frac{1}{2}\right)^n $$

Here a_o = Activity of the sample in the beginning = 100 (suppose)

a_t = Activity of the sample left undecayed after 10 years = (100 – 90) = 10

n = No. of half-lives in 10 years $= \dfrac{10}{t_{0.5}}$

Hence $10 = 100 \times \left(\dfrac{1}{2}\right)^{\frac{10}{t_{0.5}}}$ or $\dfrac{10}{100} = \left(\dfrac{1}{2}\right)^{\frac{10}{t_{0.5}}}$ or $10 = (2)^{10/t_{0.5}}$

or $\log 10 = \dfrac{10}{t_{0.5}} \log 2$ or $1 = \dfrac{10}{t_{0.5}} \log 2$

or $t_{0.5} = 10 \times \log 2 = 10 \times 0.3010 = \mathbf{3.010\ years}$ $(\because \log 2 = 0.3010)$

14. No. of half-lives (n) in 45 seconds $= \dfrac{45}{15} = 3$

$\therefore \quad a_t = a_o \times \left(\dfrac{1}{2}\right)^n = a_o \times \left(\dfrac{1}{2}\right)^3 = a_o \times \dfrac{1}{8}$

$\therefore \quad \dfrac{a_t}{a_o} = \dfrac{1}{8} = 12.5\%$

15. *For nuclide A* : $(N_{6t})_A = (N_o)_A \times \left(\dfrac{1}{2}\right)^{6t/(t_{0.5})_A} = 1 \times \left(\dfrac{1}{2}\right)^{6t/t} = \left(\dfrac{1}{2}\right)^6$

For nuclide B : $(N_{6t})_B = (N_o)_B \times \left(\dfrac{1}{2}\right)^{6t/(t_{0.5})_B} = 1 \times \left(\dfrac{1}{2}\right)^{6t/2t} = \left(\dfrac{1}{2}\right)^3$

$$\therefore \quad \frac{(N_{6t})_A}{(N_{6t})_B} = \frac{\left(\frac{1}{2}\right)^6}{\left(\frac{1}{2}\right)^3} = \left(\frac{1}{2}\right)^3 = \frac{1}{2} \times \frac{1}{2} \times \frac{1}{2} = \frac{1}{8} = \mathbf{1:8}$$

16. If the particle which is emitted from ${}_{15}P^{30}$ to give ${}_{14}Si^{30}$ is supposed to be ${}_bX^a$, then :

${}_{15}P^{30} \longrightarrow {}_{14}Si^{30} + {}_bX^a$

Hence 30 = 30 + a, $\therefore$ $a = 0$ and 15 = 14 + b, $\therefore$ $b = 15 - 14 = 1$

Thus the particle emitted is positron $\left({}_{+1}e^0\right)$

17. We know that half-life period ($t_{0.5}$) and disintegration constant (λ) of a radioactive substance are related as :

$$t_{0.5} = \frac{0.693}{\lambda}$$

Now since $t_{0.5} = \lambda$, we can write :

$$t_{0.5} = \frac{0.693}{t_{0.5}} \text{ or } (t_{0.5})^2 = 0.693$$

$$\therefore \quad t_{0.5} = \sqrt{0.693}.$$

18. $_bX^a \longrightarrow {}_{+1}e^0 + 2\ {}_2He^4 + 2\ {}_{-1}e^0 + {}_2He^4 + {}_dY^c$

Positron α-particle β-particle α-particle

$\therefore\ a = 0 + 2 \times 4 + 2 \times 0 + 4 + c$ or $a = c + 12$

$b = +1 + 2 \times 2 + 2 \times (-1) + 2 + d$ or $b = d + 5$ or $d = b - 5$

19. $t_{0.5} = 30$ minutes, $a_0 = 1$ (suppose)

$$a_t = \frac{1}{16}, \quad t = ?$$

$$a_t = a_0 \times \left(\frac{1}{2}\right)^{t/t_{0.5}}$$

or $$\frac{1}{16} = 1 \times \left(\frac{1}{2}\right)^{t/30} \quad \text{or} \quad 16 = (2)^{t/30} \quad \text{or} \quad (2)^4 = (2)^{t/30}$$

$\therefore\ \frac{t}{30} = 4$ or $t = 30 \times 4 = 120$ minutes = **2 hours.**

20. $_{92}X^{238} \longrightarrow {}_2He^4 + {}_{-1}e^0 + {}_bY^a$

Obviously, $238 = 4 + 0 + a$, $\therefore\ a = 238 - 4 = 234$

$92 = 2 - 1 + b$, $\therefore\ b = 92 - 1 = 91$

Thus $_bY^a$ is $_{91}Y^{234}$ in which the number of neutrons = 234 – 91 = **143.**

21. The product formed is $_{90}Th^{234}$ as shown below :

$$_{92}U^{238} \xrightarrow{-\alpha \text{ or } _2He^4} {}_{90}Th^{234} + {}_2He^4$$

Obviously in $_{90}Th^{234}$ since the number of electrons = number of protons = 90 and number of protons + neutrons = 234, total number of electrons, protons end neutrons

= 90 + 234 = **324**.

7

Isotopes, Isotones, Isobars and Nuclear Isomers

MULTIPLE CHOICE QUESTIONS

1. Two elements X and Y are isotonic having atomic weight 54 and, 56 respectively. If the atomic number of X is 26, then the atomic number of Y is :

(*a*) 26 (*b*) 27
(*c*) 28 (*d*) 30

2. The activity of C–14 in a piece of ancient wood is only 12.5%. If half-life period of C–14 is 5760 years, the age of the piece of wood will be :

(*a*) 17.285×10^2 years
(*b*) 172.85×10^2 years
(*c*) 1.7285×10^2 years
(*d*) 1728.5×10^2 years

3. The radioactive isotope used to estimate the age of ancient geological formations is :

(*a*) O^{16} (*b*) Pb^{212}
(*c*) Fe^{59} (*d*) C^{14}

4. An isotone of $^{76}_{32}Ge$ is :

(*a*) $^{77}_{32}Ge$ (*b*) $^{77}_{33}As$
(*c*) $^{77}_{34}Se$ (*d*) $^{78}_{34}Se$

5. The analysis of a rock shows that relative number of Pb^{206} and U^{238} atoms = 0.25. If $t_{0.5}$ for $U^{238} \longrightarrow Pb^{206}$ is 4×10^9 years. The age of the rock is :

(*a*) $\frac{2.303}{0.693} \times (4 \times 10^9) \log 1.25$

(*b*) $\frac{2.303}{0.693} \times (4 \times 10^9) \log 0.25$

(*c*) $\frac{2.303}{0.693} \times (4 \times 10^9) \log 4$

(*d*) $\frac{2.303}{4 \times 10^9} \times 0.693 \times \log 4$

6. An isobar of $_{20}Ca^{40}$ is :

(*a*) $_{18}Ar^{40}$ (*b*) $_{20}Ca^{38}$
(*c*) $_{20}Ca^{40}$ (*d*) $_{18}Ar^{38}$

7. The two nuclides $_bA^a$ and $_{b-2}B^{a-4}$ are known as :

(*a*) isotones (*b*) nuclear isomers
(*c*) isosters (*d*) isodiaphers

8. Isobar of a nuclide is formed by :

(*a*) one α-emission
(*b*) one β-emission
(*c*) (α + β) emission
(*d*) (2α + 2β) emission.

9. An element X loses one α and two β particles in three successive stages. The resulting element will be :

(*a*) an isobar of X (*b*) an isotope of X
(*c*) X itself (*d*) an isotone of X

10. The composition of tritium is :

(*a*) 1 electron, 1 proton, 1 neutron
(*b*) 1 electron, 2 protons, 1 neutron
(*c*) 1 electron, 1 proton, 2 neutrons
(*d*) 1 electron, 1 proton, 3 neutrons

11. The triad of nuclei is isotonic in :

(*a*) $_6C^{14}$, $_7N^{15}$, $_9F^{17}$

(*b*) $_6C^{12}$, $_7N^{14}$, $_9F^{18}$

(*c*) $_6C^{14}$, $_7N^{14}$, $_9F^{17}$

(*d*) $_6C^{14}$, $_7N^{14}$, $_9F^{19}$

12. The amount of C–14 in a piece of wood is found to be 1/6 of the amount present in a fresh piece of wood. What is the age of the wood ? ($t_{0.5}$ of C–14 = 5770 years)
(*a*) 14921 years (*b*) 11540 years
(*c*) 5770 years (*d*) 17310 years

13. An archeological sample of wood has half C^{14} activity as compared to that found in fresh wood of the same plant. The half-life of C^{14} is 5770 years. The age of archiological sample of wood will be :
(*a*) 5772 years (*b*) 2885 years
(*c*) 11,540 years (*d*) 1945 years

14. Radium and uranium atoms in a sample of uranium mineral are in the ratio of 1 : 2.8×10^6. If half-life period of radium is 1620 years, the half-life period of uranium is :
(*a*) 45.36×10^9 years
(*b*) 4.536×10^{10} years
(*c*) 4.536×10^9 years
(*d*) 4.536×10^{10} years.

15. A wooden antifact and freshly cut tree show the activity of 7.8 and 15.6 disintegrations min^{-1} g^{-1} of carbon whose half-life is 5760 years. The age of the antifact is :
(*a*) 5762 years
(*b*) $5762 \times \frac{15.6}{7.8}$ years
(*c*) $5762 \times \frac{7.8}{15.6}$ years
(*d*) 5762 × (15.6 – 7.8) years.

16. The final product of U^{238} is Pb^{206}. A sample of pitchblende contains 0.0453 g of Pb^{206} for every gram of U^{238} present in it. Supposing that the mineral formed at the time of formation of earth didnot contain any Pb^{206}, the age of the earth is (Half-life period of U^{238} = 4.5×10^9 years) :
(*a*) 3.310×10^8 years
(*b*) 0.3310×10^8 years
(*c*) 33.10×10^8 years
(*d*) 33.10×10^8 years.

17. The ratio of the number of g atoms of Pb to the number of g atoms of U in a mineral is 0.33. Assuming that all the lead has come from uranium, the age of the mineral is ($K = 1.5 \times 10^{-10}$ $years^{-1}$) :
(*a*) 1.9×10^9 years
(*b*) 19×10^9 years
(*c*) 0.19×10^9 years
(*d*) 0.019×10^9 years.

18. The proper rays for radio carbon dating are :
(*a*) U V rays (*b*) I R rays
(*c*) Cosmic rays (*d*) X-rays.

19. The age of most of the ancient formations is estimated by :
(*a*) K–Ar method
(*b*) C–14 dating method.
(*c*) Ra–Si method
(*d*) U–Pb method

20. Which of the following element belongs to 4*n* series ?
(*a*) Pb-206 (*b*) Pb-207
(*c*) Pb-208 (*d*) Bi-209

ANSWERS

1. (*c*)	**2.** (*b*)	**3.** (*d*)	**4.** (*b*)	**5.** (*a*)	**6.** (*a*)	**7.** (*d*)	**8.** (*b*)
9. (*b*)	**10.** (*c*)	**11.** (*a*)	**12.** (*a*)	**13.** (*a*)	**14.** (*c*)	**15.** (*a*)	**16.** (*a*)
17. (*a*)	**18.** (*c*)	**19.** (*d*)	**20.** (*c*).				

HINTS AND EXPLANATIONS

1. Isotones are the elements in which the number of neutrons i.e., (A – Z) are equal. Thus if *x* is the atomic number of Y, then for ${}_{26}X^{54}$ and ${}_{x}Y^{56}$ we have :

$$54 - 26 = 56 - x, \quad \therefore \; x = 28$$

2. Age of the piece of wood (t) is given by :

$$t = \frac{2.303 \times t_{1/2} \text{ of C} - 14}{0.693} \times \log \frac{\text{Original activity of C} - 14 \text{ in the wood}}{\text{Activity of C} - 14 \text{ in the wood after time } t}$$

$$= \frac{2.303 \times 5760}{0.693} \times \log \frac{100}{12.5} = \frac{2.303 \times 5760}{0.693} \times \log 8$$

$$= 19141.818 \times 0.9030 = 17285.0 \text{ years} = 172.85 \times 10^2 \text{ years}$$

3. Radio-isotope used in radio carbon dating method (method used to estimate the age of ancient geological formation) is C^{14}.

4. Isotones are the elements which have the same number of neutrons, (A – Z). Thus $^{76}_{32}\text{Ge}$ (No. of neutrons = 76 – 32 = 44) and $^{77}_{33}\text{As}$ (No. of neutrons = 77 – 33 = 44) are isotones.

5. $\dfrac{\text{No. of atoms of Pb}^{206}}{\text{No. of atoms of U}^{238}} = 0.25 = \dfrac{0.25}{1}$

$\therefore$ $[N_o]_{U^{238}}$ = (No. of atoms of Pb^{206} + No. of atoms of U^{238}) = 0.25 + 1 = 1.25

$[N_t]_{U^{238}}$ = No. of atoms of U^{238} after time t = 1

t = Age of the mineral (to be determined), $t_{0.5} = 4 \times 10^9$ years.

$$\therefore \quad t = \frac{2.303 \times t_{0.5}}{0.693} \times \log \frac{[N_o]_{U^{238}}}{[N_t]_{U^{238}}} = \frac{2.303}{0.693} \times (4 \times 10^9) \times \log \frac{1.25}{1}$$

$$= \frac{2.303}{0.693} \times (4 \times 10^9) \times \log 1.25$$

6. Isobars are the elements which have same mass number (A). Thus $_{20}\text{Ca}^{40}$ and $_{18}\text{Ar}^{40}$ are isobars, since these have the same mass number, A (= 40).

7. The elements having the same value of (n – p) are called isodiaphers (n = No. of neutrons and p = No. of protons). Also the nuclide and its decay product obtained after α-particle ($_2\text{He}^4$) emission are known as isodiaphers to each other. Since $_{b-2}\text{B}^{a-4}$ is obtained after the emission of α-particle from $_b\text{A}^a$, these two nuclides are isodiaphers.

$$_b\text{A}^a \longrightarrow {}_2\text{He}^4 + {}_{b-2}\text{B}^{a-4}$$

8. Isobars are nuclides which have same mass number (A) but different atomic numbers (Z). During β-decay, no change in mass number takes place.

$$_b\text{X}^a \longrightarrow {}_{-1}\text{e}^0 \ (\beta\text{-particle}) + {}_{b+1}\text{Y}^a$$

9. When X loses and one α ($_2\text{He}^4$) and two β ($_{-1}\text{e}^0$) particles, the mass number of the daughter element is four units less than that of X and the atomic number is the same as that of X.

$$_b\text{X}^a \longrightarrow {}_2\text{He}^4 + 2{}_{-1}\text{e}^0 + {}_b\text{X}^{a-4}$$

Thus the daughter element is the isotope of X.

10. Tritrium is ${}_1H^3$ in which the number of electrons = number of protons = 1 and the number of neutrons = mass number – atomic number = 3 – 1 = 2.

11. Since ${}_6C^{14}$, ${}_7N^{15}$ and ${}_9F^{17}$ have the same number of neutrons equal to 8 (14 – 6 = 8, 15 – 7 = 8, 17 – 9 = 8), these nuclei are isotonic with each other.

12. Age of the wood (t) is given by :

$$t = \frac{2.303 \times t_{0.5} \text{ of } C-14}{0.693} \times \log \frac{\text{Weight of } C-14 \text{ in the fresh wood } (W_0)}{\text{Weight of } C-14 \text{ in wood after time } t\ (W_t)}$$

$$= \frac{2.303 \times 5770}{0.693} \times \log \frac{1}{1/16} = \frac{2.303 \times 5770}{0.693} \times \log 6 = 19175.051 \times 0.778 = \mathbf{14921 \text{ years.}}$$

13. The age (t) of archeological sample of wood is given by :

$$t = \frac{2.303 \times t_{0.5} \text{ of } C^{14}}{0.693} \times \log \frac{\text{Activity of } C^{14} \text{ in the fresh wood } (a_0)}{\text{Activity of } C^{14} \text{ in wood after time } t\ (a_t)}$$

$$= \frac{2.303 \times 5770}{0.693} \times \log \frac{1}{1/2} = \frac{2.303 \times 5770}{0.693} \times \log 2 = 19175.051 \times 0.301 = \mathbf{5772 \text{ years.}}$$

14. We know that :

$$\frac{N_{Ra}}{N_U} = \frac{(t_{0.5})_{Ra}}{(t_{0.5})_U} \quad \text{or} \quad \frac{1}{2.8 \times 10^6} = \frac{1620}{(t_{0.5})_U}$$

or $(t_{0.5})_U = 2.8 \times 10^6 \times 1620$ years $= 4536 \times 10^6$ years $= \mathbf{4.536 \times 10^9}$ **years.**

15. The age (t) of the antifact is given by :

$$t = \frac{2.303 \times t_{0.5} \text{ of carbon}}{0.693} \times \log \frac{\text{Activity of carbon in freshly cut tree}}{\text{Activity of carbon in wooden antifact}}$$

$$= \frac{2.303 \times 5760}{0.693} \times \log \frac{15.6}{7.8} = 19141.818 \times \log 2 = 19141.818 \times 0.3010 = \mathbf{5762.26 \text{ years.}}$$

16. Amount of U^{238} present in the rock in the beginning, $[N_o]_{U^{238}}$ = Weight of U^{238} in the rock

$$+ \frac{\text{A of } U^{238}}{\text{A of } Pb^{206}} \times \text{Weight of } Pb^{206} \text{ in the rock } = 1 + \frac{238}{206} \times 0.0453$$

$= 1 + 1.155 \times 0.0453 = 1.0523$ g

Amount of U^{238} present in the rock after time (t), $[N_t]_{U^{238}} = 1\,g$

$$\therefore\ t = \frac{2.303 \times (t_{0.5})_{U^{238}}}{0.693} \times \log \frac{[N_o]_{U^{238}}}{[N_t]_{U^{238}}} = \frac{2.303 \times 4.5 \times 10^9}{0.693} \times \log \frac{1.0523}{1}$$

$= 14.9545 \times 10^9 \times 0.0221 = 14.9545 \times 0.0221 \times 10 \times 10^8 = \mathbf{3.310 \times 10^8}$ **years.**

17. $$\frac{\text{No of g atoms of Pb}}{\text{No. of g atoms of U}} = \frac{0.33}{1}$$

$\therefore$ $[N_o]_U$ = (No. of g atoms of Pb + No. of g atoms of U) in the mineral

= 0.33 + 1 = 1.33 g atom

$[N_t]_U$ = No. of atoms of U after time t in the mineral = 1 g atom

t = Age of the mineral (to be determined)

$K = 1.5 \times 10^{-10}$ years^{-1}.

$$\text{Now,} \quad t = \frac{2.303}{K} \log \frac{[N_o]_U}{[N_t]_U} = \frac{2.303}{1.5 \times 10^{-10}} \log \frac{1.33}{1} = \frac{2.303 \times 10^{10}}{1.5} \times 0.1238$$

$$= \frac{2.303 \times 10 \times 0.1238}{1.5} \times 10^9 = \mathbf{1.9007 \times 10^9 \text{ years.}}$$

18. Cosmic ray neutrons (present in the upper atmosphere) convert $_7N^{14}$ to radioactive $_6C^{14}$.

$$_7N^{14} + {}_0n^1 \text{ (present in the cosmic rays)} \longrightarrow {}_6C^{14} \text{ (radioactive carbon)} + {}_1H^1$$

19. The age of ancient formation like earth, rock etc. is found by $_{92}U^{238}$ – $_{82}Pb^{206}$ method.

20. 4*n* series is that in which the mass members of all the members of the series are exactly divisible by 4. Now since it is only 208 which is completely divisible by 4, Pb-208 belongs to 4*n* series.

8

Nuclear Chemistry : Chemistry of the Nucleus

MULTIPLE CHOICE QUESTIONS

1. The mass of an atom of $_2He^4$ is 4.0026 amu. The mass of a neutron and a proton are 1.0087 amu and 1.0078 amu respectively. The nuclear binding energy per nucleon of $_2He^4$ is nearly :
(*a*) 7 MeV (*b*) 5 MeV
(*c*) 10 MeV (*d*) 14 MeV

2. The unstability of a nucleus is due to :
(*a*) higher electron-to-proton ratio
(*b*) high neutron-to-proton ratio
(*c*) low electron-to-proton ratio
(*d*) low neutron-to-electron ratio.

3. The isotopic mass of $_{92}U^{238}$ is 238.125 amu. Its packing fraction is :
(*a*) 5.25 (*b*) 0.125
(*c*) 12.5 (*d*) 1.25

4. Binding energy per nucleon of $_5B^{11}$ with mass defect 0.081 is :
(*a*) 6.85 MeV (*b*) 68.5 MeV
(*c*) 0.685 MeV (*d*) 685 MeV

5. Chlorine has two isotopes $_{17}Cl^{35}$ and $_{17}Cl^{37}$. The atomic weight of chlorine is 35.5. In what ratio Cl^{35} and Cl^{37} are present in chlorine ?
(*a*) 1 : 3 (*b*) 3 : 1
(*c*) 1 : 5 (*d*) 5 : 1

6. Which of the following statements is (are) true ?
(*a*) In α-emission there is a loss of two protons and gain of two neutrons.
(*b*) In β-emission one neutron is converted into one proton.
(*c*) In positron emission one proton is converted into one neutron.
(*d*) In electron capture one proton is converted into one neutron.

7. In terms of energy 1 amu is equal to :
(*a*) 100 J (*b*) 931.46 MeV
(*c*) 931.1 K cal (*d*) 10^7 ergs.

8. In terms of Kg, 1 amu is equal to :
(*a*) 1.66×10^{-27} Kg
(*b*) 16.6×10^{-27} Kg
(*c*) 166.0×10^{-27} Kg
(*d*) 0.166×10^{-27} Kg

9. If the mass defect of $_4X^9$ is 0.09 amu, then binding energy per nucleon is (1 amu = 931.5 MeV) :
(*a*) 9.315 MeV (*b*) 931.5 MeV
(*c*) 83.8 MeV (*d*) 8.38 MeV

10. Which of the following nuclei is unstable ?
(*a*) $_5B^{10}$ (*b*) $_4Be^{10}$
(*c*) $_7N^{14}$ (*d*) $_8O^{16}$

11. The number of electrons in the nucleus of Cl-12 is :
(*a*) 6 (*b*) 12
(*c*) 17 (*d*) 0

12. A radioisotope will emit :
(*a*) α-and β-particles simultaneously.
(*b*) β-and γ-rays simultaneously.
(*c*) γ-and α-rays simultaneously.
(*d*) α-or B-rays and then γ-rays.

13. Unstable substances exhibit higher radioactivity due to :

(*a*) low *p*/*n* ratio (*b*) high *p*/*n* ratio
(*c*) $p/n = 1$ (*d*) none of these.

14. A radio-active element X has atomic number of 100. It decays into an element Y which decays directly into an element Z. In both the processes charged particles are emitted. Which of the following statements could be true ?
(*a*) Y has atomic number 101.
(*b*) Z has atomic number 99.
(*c*) Y has atomic number 102.
(*d*) Z has atomic number 100.

15. If the actual isotopic mass of an isotope is M and the sum of the masses of protons, neutrons and electrons present in the isotope is N, then :
(*a*) M < N (*b*) M > N
(*c*) M = N
(*d*) No relation exists between M and N.

16. If the mass number of a nuclide is 27, the nuclear radius of the nuclide is (1 Fermi = 10^{-15} m) :
(*a*) 4.2 Fermi (*b*) 0.42 Fermi
(*c*) 0.042 Fermi (*d*) 42 Fermi

17. The mass defect of $_{18}Ar^{40}$ with atomic mass of 39.962384 amu is (mass of proton = 1.007825 amu and mass of neutron = 1.008665 amu)
(*a*) 0.369096 amu
(*b*) 3.69096 amu
(*c*) 0.0369096 amu
(*d*) 36.9096 amu

18. Decrease in atomic number is observed during :
(*a*) α-emission
(*b*) β-emission
(*c*) positron emission
(*d*) electron capture.

19. Loss of a β-particle is equivalent to :
(*a*) Increase of one proton only
(*b*) Decrease of one neutron only
(*c*) Both (*a*) and (*b*)
(*d*) None of these.

20. Loss of α-particle is equivalent to :
(*a*) Loss of two neutrons
(*b*) Loss of two protons
(*c*) Loss of two neutrons and two protons
(*d*) None of the above.

ANSWERS

1. (*a*) **2.** (*b*) **3.** (*a*) **4.** (*a*) **5.** (*b*) **6.** (*b*), (*c*), (*d*) **7.** (*b*)
8. (*a*) **9.** (*a*) **10.** (*b*) **11.** (*d*) **12.** (*d*) **13.** (*a*) **14.** (*a*), (*b*) **15.** (*a*)
16. (*a*) **17.** (*a*) **18.** (*a*), (*c*), (*d*) **19.** (*c*) **20.** (*c*)

HINTS AND EXPLANATIONS

1. Mass of $_2He^4$ = 4.0026 amu.

No. of protons in $_2He^4$ = 2 and number of neutrons in $_2He^4$ = 4 – 2 = 2

∴ Mass defect = (2 × mass of proton + 2 × mass of neutron) – mass of $_2He^4$
= (2 × 1.0078 + 2 × 1.0087) – 4.0026
= 2.0156 + 2.0174 – 4.0026
= 4.0330 – 4.0026 = 0.0304 amu

∴ Binding energy of $_2He^4$ = Mass defect × 931.5 MeV
= 0.0304 × 931.5 MeV
= 28.3176 MeV

$\therefore$ Binding energy per nucleon in ${}_2He^4$

$$= \frac{\text{Binding energy of } {}_2He^4}{A} \text{ MeV}$$

$$= \frac{28.317}{4} \text{ MeV} = \mathbf{7.079\ MeV}$$

2. When $n|p$ ratio of nuclei is high, the nuclei donot lie in the stability belt and hence are unstable. Being unstable such nuclei undergo spontaneous radioactive disintegration.

3. Packing fraction of ${}_{92}U^{238}$

$$= \frac{\text{Atomic mass of } {}_{92}U^{238} - A}{A} \times 10^4$$

$$= \frac{238.125 - 238}{238} \times 10^4$$

$$= \frac{1250}{238} = \mathbf{5.25}$$

4. Binding energy per nucleon of ${}_5B^{11} = \dfrac{\text{Mass defect} \times 931.5}{A}$ MeV

$$= \frac{0.081 \times 931.5}{11} \text{ MeV} = \mathbf{6.85\ MeV}$$

5. If an element has two isotopes in the ratio of R_1 and R_2 and their mass numbers are A_1 and A_2 respectively, then the atomic weight of the element is given by :

Atomic weight of the element $= \dfrac{R_1 A_1 + R_2 A_2}{R_1 + R_2}$

Thus atomic weight of Cl element $= \dfrac{R_1 \times 35 + R_2 \times 37}{R_1 + R_1}$

or $\quad 35.5 = \dfrac{R_1 \times 35 + R_2 \times 37}{R_1 + R_2}$

or $\quad 35.5\, R_1 + 35.5\, R_1 = R_1 \times 35 + R_2 \times 37$

or $\quad 35.5\, R_1 - 35 R_1 = 37 R_2 - 35.5 R_2$

or $\quad 0.5\, R_1 = 1.5\, R_2$

$\therefore \quad R_1 : R_2 = \dfrac{1.5}{0.5} = \mathbf{3:1}$

6. (*a*) *α-emission* $(-{}_2He^4)$:

$${}_bX^a \longrightarrow {}_{b-2}Y^{a-4} + {}_2He^4$$

$p = b \qquad\qquad p = b - 2$

$n = a - b \qquad\qquad n = a - 4 - b + 2$

$\qquad\qquad\qquad\quad = a - b - 2$

Thus we see that in α-emission two protons and two neutrons are lost.

(*b*) β-*emission* $(-\,_{-1}e^0)$:

$$_{b}X^{a} \longrightarrow \,_{b+1}Y^{a} + \,_{-1}e^{0}$$

$p = b,\ n = a - b \qquad p = b + 1$

$\qquad\qquad\qquad\qquad n = a - b - 1$

Since the number of protons increases by one and that of neutrons decreases by one, one neutron is converted into proton.

$$_{0}n^{1} \longrightarrow \,_{1}p^{1} + \,_{-1}e^{0} \text{ (β-particle)}$$

(*c*) *Positron emission* $(-\,_{+1}e^0)$

$$_{b}X^{a} \longrightarrow \,_{b-1}Y^{a} + \,_{+1}e^{0}$$

$p = b \qquad p = b - 1$

$n = a - b \qquad n = a - b + 1$

Since the number of protons decreases by one and that of neutrons increases by one, one proton is converted into a neutron.

$$_{1}p^{1} \longrightarrow \,_{0}n^{1} + \,_{+1}e^{0} \text{ (positron)}$$

(*d*) *Electron capture* $(+\,_{-1}e^0)$

$$_{b}X^{a} + \,_{-1}e^{0} \longrightarrow \,_{b-1}Y^{a} + \,_{0}v^{0} \text{ (Neutrino)}$$

$p = b \qquad p = b - 1$

$n = a - b \qquad n = a - b + 1$

Since the number of protons decreases by one and that of neutrons increases by one, one proton is converted into neutron

$$_{1}p^{1} + \,_{-1}e^{0} \text{ (electron)} \longrightarrow \,_{0}n^{1}.$$

7. Energy (E) released when 1 amu of mass is lost is given by Einsten equation, $E = mc^2$ in which E = energy in joules, m = mass (in kg) corresponding to 1 amu and c = velocity of light in ms^{-1}. Now we know that 1 amu = 0.16606×10^{-26} kg and $c = 2.9979 \times 10^{8}\ ms^{-1}$, energy (E) corresponding to 1 amu is given by :

$$E = (0.16606 \times 10^{-26}\ kg) \times (2.9979 \times 10^{8}\ ms^{-1})^2$$
$$= 0.16606 \times 2.9979 \times 2.9979 \times 10^{-10}\ kg\ m^2 s^{-2}$$
$$= 1.4924 \times 10^{-10}\ kg\ m^2\ s^{-2}$$
$$= 1.4924 \times 10^{-10}\ J \qquad (\because J = kg\ m^2 s^{-2})$$
$$= \frac{1.4924 \times 10^{-10}}{1.6022 \times 10^{-19}}\ eV \qquad (\because 1\ eV = 1.6022 \times 10^{-19}\ J)$$
$$= 0.93146 \times 10^{9}\ eV$$
$$= 931.46 \times 10^{6}\ eV$$
$$= \frac{931.46 \times 10^{6}}{10^{6}}\ MeV \qquad (\because 10^{6}\ eV = 1\ MeV)$$
$$= \mathbf{931.46\ MeV}$$

8. 1 amu $= \frac{1}{12}$ th of the mass of ${}_6C^{12}$ atom

$$= \frac{1}{12} \times \frac{12}{\text{Avogadro's number}}\, g = \frac{1}{6.022 \times 10^{23}}\, g$$

$$= \frac{1}{6.022} \times 10^{-23}\, g = \frac{1}{6.022} \times \frac{10^{-23}}{10^{3}}\ \text{Kg} = \frac{1}{6.022} \times 10^{-26}\ \text{Kg}$$

$$= 0.166 \times 10^{-26}\ \text{Kg} = 1.66 \times 10^{-27}\ \text{Kg}$$

9. Binding energy per nucleon of ${}_4X^9 = \frac{\text{Mass defect} \times 931.5}{\text{A}}$ MeV

$$= \frac{0.09 \times 931.5}{9}\ \text{MeV} = \mathbf{9.315\ MeV}$$

10. The nuclei having the same number of neutrons and protons are stable while the nuclei having different number of neutrons and protons are unstable. The number of neutrons (n) and protons (p) in the given nuclei are as :

(*a*) ${}_5B^{10} \longrightarrow p = 5, n = 10 - 5 = 5$, (*b*) ${}_4Be^{10} \longrightarrow p = 4, n = 10 - 4 = 6$,

(*c*) ${}_7N^{14} \longrightarrow p = 7, n = 14 - 7 = 7$, (*d*) ${}_8O^{16} \longrightarrow p = 8, n = 16 - 8 = 8$.

Thus ${}_5B^{10}$, ${}_7N^{14}$ and ${}_8O^{16}$ are stable nuclei and ${}_4Be^{10}$ is an unstable nucleus.

11. The nucleus of an atom contains no electrons. The nucleus contains only protons and neutrons. Electrons are present in various shells surrounding the nucleus.

12. A radioisotope first emits α or β particles and becomes unstable. Then it emits γ-rays

$${}_bX^a \xrightarrow{-{}_2He^4} {}_{b-2}X^{a-4} \xrightarrow{-{}_0\gamma^0} {}_{b-2}X^{a-4}$$

$${}_bX^a \xrightarrow{-{}_{-1}e^0} {}_{b+1}X^{a} \xrightarrow{-{}_0\gamma^0} {}_{b+1}X^{a}$$

13. The nuclei having high n/p ratio or low p/n ratio are unstable and hence are highly radioactive.

14. Charged particle may be an α-particle (${}_2He^4$) or a β-particle (${}_{-1}e^0$). Thus we have the following nuclear reactions :

$${}_{100}X \xrightarrow{-{}_2He^4} {}_{98}Y \xrightarrow{-{}_2He^4} {}_{96}Z$$

$${}_{100}X \xrightarrow{-{}_{-1}e^0} {}_{101}Y \xrightarrow{-{}_{-1}e^0} {}_{102}Z$$

$${}_{100}X \xrightarrow{-{}_2He^4} {}_{98}Y \xrightarrow{-{}_{-1}e^0} {}_{99}Z$$

$${}_{100}X \xrightarrow{-{}_{-1}e^0} {}_{101}Y \xrightarrow{-{}_2He^4} {}_{99}Z$$

Obviously (*a*) and (*b*) are the correct answers.

15. When the nucleons (*i.e.*, neutrons and protons) combine together to form the nucleus of an atom, mass equal to $(M'-M)$ which is called mass defect is lost.

16. Radius of the nucleus (nuclear radius), R of a nuclide is given by :

$$R = R_o \times A^{1/3}$$

where R_o = a constant = 1.4×10^{-15} m and A = Mass number of the nuclide = 27 (given)

$\therefore$ $R = 1.4 \times 10^{-15} \times (27)^{1/3}$ m $= 1.4 \times [(3)^3]^{1/3} \times 10^{-15}$ m $= 1.4 \times 3 \times 10^{-15}$ m

$$= \frac{1.4 \times 3 \times 10^{-15}}{10^{-15}} \text{ Fermi} \qquad (\because 1 \text{ Fermi} = 10^{-15} \text{ m})$$

$= 1.4 \times 3$ Fermi = **4.2 Fermi**

17. Number of protons in $_{18}Ar^{40} = 18$ and number of neutrons in $_{18}Ar^{40} = 40 - 18 = 22$

Mass defect = (Sum of the masses of 18 protons and 22 neutrons) – Atomic mass of $_{18}Ar^{40}$

$= (18 \times 1.007825 + 22 \times 1.008665) - 39.962384$

$= 40.33148 - 39.962384$

= **0.369096 amu**

18. (*a*) α-emission $(-\,_2He^4)$: $_bX^a \longrightarrow \,_{b-2}Y^{a-4} + \,_2He^4$

$Z = b \qquad Z = b - 2$

(*b*) β-emission $(-\,_{-1}e^0)$: $_bX^a \longrightarrow \,_{b+1}Y^a + \,_{-1}e^0$

$Z = b \qquad Z = b + 1$

(*c*) Positron emission $(-\,_{+1}e^0)$: $_bX^a \longrightarrow \,_{b-1}Y^a + \,_{+1}e^0$

$Z = b \qquad Z = b - 1$

(*d*) Electron capture $(+\,_{-1}e^0)$: $_bX^a + \,_{-1}e^0 \longrightarrow \,_{b-1}Y^a + \,_0\nu^0$

$Z = b \qquad Z = b - 1$

Only in (*b*), atomic number (*Z*) is increased while in other nuclear reactions *Z* is decreased.

19. $_bX^a \xrightarrow{-\beta \text{ or } _{-1}e^0} \,_{b+1}Y^a + \,_{-1}e^0$

$p = b \qquad p = b + 1$

$n = a - b \qquad n = a - b - 1$

Thus we see that the number of protons increases by one and that of neutrons decreases by one. Thus answer is (*c*). Since the number of neutrons decreases by one and that of protons increases by one, one neutron $(_1n^0)$ is converted into one proton $(_1n^1)$ during β-emission.

$$_1n^0 \longrightarrow \,_1p^1 + \,_{-1}e^0$$

20. $_bX^a \xrightarrow{-\alpha \text{ or } _2He^4} \,_{b-2}Y^{a-4} + \,_2He^4$

$p = b \qquad p = b - 2$

$n = a - b \qquad n = a - 4 - b + 2$

$= a - b - 2$

This nuclear reaction shows that the number of protons is decreased by two units $[b \rightarrow (b-2)]$. Similarly the number of neutrons is also decreased by two units $[(a-b) \rightarrow (a-b)-2]$

9

Nuclear Reactions

MULTIPLE CHOICE QUESTIONS

1. The nuclear reaction :

$$_{29}Cu^{63} + {}_2He^4 \longrightarrow {}_{17}Cl^{37} + 14\ {}_1H^1 + 16\ {}_0n^1$$

is referred to as :

(*a*) Spallation reaction
(*b*) Fusion reaction
(*c*) Fission reaction
(*d*) Chain reaction

2. The number of neutrons accompanying the formation of $_{54}Xe^{139}$ and $_{38}Sr^{94}$ from the absorption of slow neutron by $_{92}U^{235}$ followed by nuclear fission is :

(*a*) 0 (*b*) 2
(*c*) 1 (*d*) 3

3. The fissionable material commonly used in atomic bomb is (are) :

(*a*) $_{92}U^{235}$ (*b*) $_{92}U^{238}$
(*c*) $_{94}Pu^{239}$ (*d*) $_{90}Th^{232}$

4. The source of enormous energy of sun is :

(*a*) Fission of uranium
(*b*) Fusion of hydrogen nuclei to form helium nucleus.
(*c*) Fusion of deuterium and tritium
(*d*) Fission of tritium to form helium.

5. The radiant energy of sun is due to :

(*a*) Disintegration (*b*) Nuclear fission
(*c*) Nuclear fusion (*d*) Combustion.

6. Nuclear fusion is the source of energy in :

(*a*) Atomic bomb (*b*) Hydrogen bomb
(*c*) The sun (*d*) The moon.

7. Atom bomb is based on :

(*a*) Nuclear fusion
(*b*) Nuclear fission
(*c*) Induced radioactivity
(*d*) Disintegration

8. Nuclear reactors are based on :

(*a*) Nuclear fission
(*b*) Natural radioactivity
(*c*) Nuclear fusion
(*d*) Spontaneous chemical reaction

9. Which of the following is a thermonuclear reaction ?

(*a*) $_{92}U^{238} + {}_0n^1 \longrightarrow {}_{93}Np^{239} + {}_{-1}e^0$

(*b*) $4\ {}_1H^1 \longrightarrow {}_2He^4 + 2\ {}_{+1}e^0$

(*c*) $_{92}U^{238} + {}_6C^{12} \longrightarrow {}_{98}Cf^{246} + 4\ {}_0n^1$

(*d*) $_{13}Al^{27} + {}_2He^4 \longrightarrow {}_{15}P^{30} + {}_0n^1$

10. When $_{17}Cl^{35}$ undergoes (*n*, *p*) reaction, the radioactive isotope formed is :

(*a*) $_{15}P^{32}$ (*b*) $_{16}S^{35}$
(*c*) $_{16}S^{35}$ (*d*) $_{15}P^{34}$

11. Control rods used in the nuclear reactor are made of :

(*a*) Fe (*b*) Graphite
(*c*) Cd (*d*) B

12. The moderator used in nuclear reactor is :

(*a*) H_2O (*b*) D_2O
(*c*) Graphite (*d*) Boron

13. In a nuclear reactor heavy water is to :
(a) transfer the heat from the reactor
(b) provide high speed neutrons for fission reaction
(c) reduce the speed of fast moving neutrons
(d) increase the speed of neutrons.

14. Which of the following is a fission reaction ?
(a) ${}_1H^1 + {}_1H^1 + 2\ {}_0n^1 \longrightarrow {}_2He^4 + \text{Energy}$
(b) ${}_{92}U^{235} + {}_0n^1 \longrightarrow {}_{56}Ba^{141} + {}_{36}Kr^{92} + 3\ {}_0n^1$
(c) ${}_{92}U^{239} + {}_0n^1 \longrightarrow {}_{92}U^{239} + {}_0\gamma^0$
(d) ${}_{27}Co^{59} + {}_1H^2 \longrightarrow {}_{27}Co^{60} + {}_1H^1$

15. Which of the following is a fusion reaction ?
(a) ${}_{92}U^{235} + {}_0n^1 \longrightarrow {}_{56}Ba^{141} + {}_{36}Kr^{92} + 3\ {}_0n^1$
(b) ${}_{92}U^{239} + {}_0n^1 \longrightarrow {}_{92}U^{239} + {}_0\gamma^0$
(c) ${}_{26}Fe^{55} + {}_{-1}e^0 \longrightarrow {}_{25}Mn^{55}$
(d) ${}_1H^1 + {}_1H^1 + 2\ {}_0^1n \longrightarrow {}_2He^4 + \text{Energy}$

16. The nuclear reaction,
${}_1H^2 + {}_1H^2 \longrightarrow {}_2He^4$ is called :
(a) fission reaction
(b) fusion reaction
(c) chain reaction
(d) thermal reaction

17. Which of the following is correct ?
(a) Atom bomb and H-bomb both are examples of nuclear fission.
(b) Nuclear reactions are markedly affected by temperature.
(c) Electron and positron both have the same mass.
(d) The lesser is the half-life, the lesser is the number of atoms decaying in a given time.

18. The energy released when 1 g of ${}_{92}U^{235}$ undergoes fission in the fission reaction :
${}_{92}U^{235} + {}_0n^1 \longrightarrow$ Fission products + Newtrons + 3.20×10^{-11} J
is :
(a) 12.75×10^8 KJ
(b) 18.60×10^9 KJ
(c) 8.20×10^7 KJ
(d) 6.55×10^6 KJ.

19. The reaction,
${}_1D^2 + {}_1T^3 \longrightarrow {}_2He^4 + {}_0n^1$ is an example of :
(a) nuclear fission
(b) nuclear fussion
(c) artificial radioactivity
(d) radio disintegration

20. Nuclear reactions accompanied with emission of neutrons (s) :
(a) ${}_{13}Al^{27} + {}_2He^4 \longrightarrow {}_{15}P^{30}$
(b) ${}_6C^{12} + {}_1H^1 \longrightarrow {}_7N^{13}$
(c) ${}_{15}P^{30} \longrightarrow {}_{14}Si^{30} + {}_{+1}e^0$
(d) ${}_{96}Cm^{241} + {}_2He^4 \longrightarrow {}_{97}Bk^{244} + {}_{+1}e^0$

ANSWERS

1. (a) **2.** (d) **3.** (a), (c) **4.** (b) **5.** (c) **6.** (b), (c) **7.** (b) **8.** (a)
9. (b) **10.** (b) **11.** (c),(d) **12.** (b), (c) **13.** (a), (c) **14.** (b) **15.** (d) **16.** (b)
17. (c) **18.** (c) **19.** (b) **20.** (a), (d)

HINTS AND EXPLANATIONS

1. Since in the given nuclear reaction high energy α-particle, $_2He^4$ (Energy = 400 MeV) converts Cu^{63} nucleus into Cl^{37} nucleus and a large number of sub-atomic particles *viz* 14 protons ($14\ _1H^1$) and 16 neutrons ($16\ _0n^1$) are produced, it is a spallation reaction.
2. Let the number of neutrons released is x. Then we have :

$$_{92}U^{235} + {_0n^1} \longrightarrow {_{54}Xe^{139}} + {_{38}Sr^{94}} + x\ {_0n^1}$$

$$\therefore \quad 235 + 1 = 139 + 94 + x \times 1$$

$$\therefore \quad x = 236 - 139 - 94 = 236 - 233 = 3$$

3. The atomic bomb contains two pieces of U^{235} or Pu^{239} each of sub-critical mass. One piece is called wedge and the other is called target. The size of the fashionable material, U^{235} or Pu^{239} should not be less than the critical mass.
4. In the sun four hydrogen atoms (protons) get fused to give one $_2He^4$ nucleus with the emission of two positrons ($_{+1}e^0$). A large amount of energy is also produced.

$$4\ {_1H^1} \xrightarrow{\text{Fusion}} 2\ {_2He^4} + 2\ {_{+1}e^0} + 24.64 \text{ MeV}$$

5. The radiant energy of the sun is due to the fusion of four $_1H^1$ nuclei to form one $_2He^4$ nucleus with the emission two positrons ($_{+1}e^0$). In this reaction a large amount of energy is also produced.

$$4\ {_1H^1} \xrightarrow{\text{Fusion}} {_2He^4} + 2\ {_{+1}e^0} + 24.64 \text{ MeV}$$

6. Energy produced in hydrogen bomb is due to the nuclear fusion reaction between duterium ($_1H^2$) and tritium ($_1H^3$) nuclei. This fusion gives helium nucleus ($_2He^4$) and one neutron ($_0n^1$). In this reaction an uncontrolled amount of energy is also generated.

$$_1H^2 + {_1H^3} \xrightarrow{\text{Fusion}} {_2He^4} + {_0n^1} + 17.6 \text{ MeV}$$

$_1H^3$ used in the above reaction is generated by the following nucleus reaction.

$$_3Li^6 + {_0n^1} \longrightarrow {_1H^3} + {_2He^4}$$

Energy in the sun is produced due to the fusion of four $_1H^1$ nuclei to form one $_2He^4$ nucleus with the liberation of two positons ($_{+1}e^0$) and a large amount of energy is generated

$$4\ {_1H^1} \xrightarrow{\text{Fusion}} {_2He^4} + 2\ {_{+1}e^0} + 24.64 \text{ MeV}$$

7. The large amount of energy produced in atom bomb is due to the fission of large piece of $_{92}U^{235}$ by the neutrons obtained from Ra – Be source.

$$_{92}U^{235} + {_0n^1} \longrightarrow {_{56}Ba^{140}} + {_{36}Kr^{93}} + 3\ {_0n^1} + \text{ Huge amount of energy.}$$

8. In the nuclear reactor energy is produced by the fission of U^{235}. Since U^{235} is very costly, U^{235} enriched with U^{238} is used as the fuel element. Thus it is only U^{235} isotope alone that undergoes nuclear fission and hence acts as the real fuel element.

9. Since in nuclear reaction (*b*), lighter nuclei of $_1H^1$ fuse together to form heavier helium nucleous ($_2He^4$), it is a nuclear fusion reaction. Further, since nuclear reaction can take place only at extremely high temperatures only (> 10^6 K), this reaction is a thermo-nuclear reaction.

10. Suppose the radioactive isotope formed is $_bX^a$. Then :

$$_{17}Cl^{35} + {_0n^1} \longrightarrow {_1p^1} + {_bX^a}$$

$$\therefore \quad 35 + 1 = 1 + a, \ \therefore \ a = 35$$

$$17 + 0 = 1 + b, \ \therefore \ b = 16$$

Thus the radioactive isotope formed is $_{16}S^{35}$.

11. The function of control rods used in the nuclear reactor is to control the number of neutrons. This is achieved by using the rods of Cd or boron as control rods which absorb the neutrons and thus control their number. On absorbing neutrons, $_{48}Cd^{113}$ and $_5B^{10}$ are converted into non-radioactive isotopes *viz* $_{48}Cd^{114}$ and $_5B^{11}$ respectively.

$$_{48}Cd^{113} + {_0n^1} \longrightarrow {_{48}Cd^{114}} + {_0\gamma^0}$$

$$_5B^{10} + {_0n^1} \longrightarrow {_5B^{11}} + {_0\gamma^0}$$

Thus Cd or boron is used as control rods.

12. The function of moderator used in the nuclear reactor is to slow down the speed of the fast moving neutrons (without absorbing them) to such an extent that they can easily be captured by the fuel rods (U^{235}). Thus is achieved by surrounding the fuel rods by D_2O or graphite. Thus D_2O or graphite is used as moderator. The moderator not only slows down the speed of the neutrons but also transfers the heat from the reactor.

13. Heavy water is used as moderator. The function of a moderator is to reduce the speed of fast moving neutrons and transfers the heat from the reactor.

14. Since in the nuclear reaction (*b*), the heavier atom, $_{92}U^{235}$ gets split into lighter atoms *viz* $_{56}Ba^{141}$ and $_{36}Kr^{92}$ by the bombardment with neutrons, this reaction is a nuclear fission reaction.

15. Since in the nuclear reaction (*d*) two lighter nuclei *viz* $_1H^1$ get fused (combine) together to form a heavier nucleus ($_2He^4$), this reaction is a nuclear fusion reaction.

16. In the given reaction, since two higher nuclei *viz* $_1H^2$ get fused to form heavier nucleus *viz* $_2He^4$, the given reaction is a fusion reaction.

17. Electron and positron both have charges equal to – 1 and + 1 respectively. Each has mass equal to 9.1×10^{-31} Kg which is supposed to be equal to zero. Thus both are massless particles. These are, therefore, represented as $_{-1}e^0$ and $_{+1}e^0$ respectively.

18. No of atoms in 1 g of $U^{235} = \dfrac{\text{Avogadro's number}}{A}$

$$= \frac{6.02 \times 10^{23}}{235}$$

$\therefore$ Energy released when 1 g of U^{235} undergoes fission $= (3.20\times10^{-11})\times\frac{6.02\times10^{23}}{235}$ J

$$= \frac{3.20\times6.02\times10^{12}}{235} \text{ J} = \frac{3.20\times6.02\times10^{12}}{235\times10^{3}} \text{KJ} = \frac{3.20\times6.02}{235}\times10^{9} \text{ KJ}$$

$= 0.0819 \times 10^9$ KJ $= 8.19 \times 10^7$ KJ $=$ **8.20×10^7 KJ**

19. Since in the given reaction, two smaller nuclei viz $_1D^2$ and $_1T^3$ get fused together and give a bigger nucleus $(_2He^4)$, the given reaction is a example of nuclear fusion.

20. In reactions (*b*) and (*c*) the sum of the mass numbers of the reactants is the same as that of the products. Similarly the sum of the atomic numbers of the reactants and products is the same. Thus these reactions have been written correctly. In reaction (*a*), since the sum of mass numbers on LHS is one unit less and that of atomic number is the same, one $_0n^1$ should be added to RHS. Similar is the case of reactions (*d*).

$$_{13}Al^{27} + {}_2He^4 \longrightarrow {}_{15}P^{30} + {}_0n^1$$

$$_{96}Cm^{241} + {}_2He^4 \longrightarrow {}_{97}Bk^{244} + {}_{+1}e^0 + {}_0n^1$$

10

Different Types of Chemical Bonds

MULTIPLE CHOICE QUESTIONS

1. A diatomic molecule has a dipole moment equal to 1.2 D. If its bond distance is 1.0 A°, what fraction of an electronic charge exists in each atom ?

(*a*) 12% (*b*) 18%

(*c*) 25% (*d*) 29%

2. Which of the following molecules is polar ?

(*a*) SO_3 (*b*) SO_2

(*c*) SF_6 (*d*) All the polar.

3. NH_3 has measurable dipole moment while BF_3 has zero dipole moment, because :

(*a*) the N – H bond is polar but B – F bond is non-polar.

(*b*) BF_3 has planar symmetrical structure but NH_3 has a pyramidal structure.

(*c*) NH_3 is a base while BF_3 is a Lewis acid.

(*d*) NH_3 has a planar symmetrical structure but BF_3 has a pyramidal structure.

4. Which of the following has the highest dipole moment ?

(*a*) NH_3 (*b*) PH_3

(*c*) SbH_3 (*d*) As H_3

5. The molecules that will have dipole moment are :

(*a*) 2, 2-dimethyl propane

(*b*) trans-2-pentene

(*c*) *cis*-3-hexene

(*d*) 2, 2, 3, 3-tetra methyl butane

6. Which of the following factors produces covalent character in ionic bond in a molecule ?

(*a*) Small charge on ions

(*b*) Noble gas electronic configuration of the cation

(*c*) Large ionic radius

(*d*) High degree of polarisation

7. The correct order of the ionic character of PbF_2, $PbCl_2$, $PbBr_2$ and PbI_2 is :

(*a*) $PbF_2 > PbCl_2 > PbBr_2 > PbI_2$

(*b*) $PbF_2 > PbBr_2 > PbCl_2 > PbI_2$

(*c*) $PbF_2 < PbCl_2 < PbBr_2 < PbI_2$

(*d*) $PbF_2 < PbCl_2 > PbBr_2 > PbI_2$

8. Which of the following will be the least soluble in water ?

(*a*) NaCl (*b*) Na_2S

(*c*) $MgCl_2$ (*d*) MgS

9. Among the halides namely, NaF, NaCl, NaBr and NaI, NaF has the highest melting point, because NaF has :

(*a*) minimum ionic character

(*b*) maximum ionic character

(*c*) highest oxidising power

(*d*) lowest valency

10. Which of the following exists ?

(*a*) $K[HF_2]$ (*b*) $K[HCl_2]$

(*c*) $K[HBr_2]$ (*d*) $K[HI_2]$

11. The hydrogen bond is the strongest in :

(*a*) O – H ...... S (*b*) S – H ...... O

(*c*) F – H ...... F (*d*) F – H ...... O

12. NH_3 has much higher boiling point than PH_3 because

(*a*) NH_3 has larger molecular weight

(*b*) NH_3 undergoes umbrella inversion

(*c*) NH_3 forms hydrogen bond

(*d*) NH_3 contains ionic bonds whereas PH_3 contains covalent bonds.

13. Strength of hydrogen bonding in O, F, S, Cl, N is in the order :
(*a*) Cl < S < N < O < F
(*b*) Cl < F < S < N < O
(*c*) Cl < N < S < O < F
(*d*) Cl < O < S < N < F

14. What is the maximum number of H-bonds which a water molecule can form ?
(*a*) 1 (*b*) 2
(*c*) 3 (*d*) 4

15. The increasing order of C – C bond length in C_2H_6, C_2H_4 and C_2H_2 is :
(*a*) $C_2H_2 < C_2H_4 < C_2H_6$
(*b*) $C_2H_2 < C_2H_6 < C_2H_4$
(*c*) $C_2H_4 < C_2H_2 < C_2H_6$
(*d*) $C_2H_6 < C_2H_2 < C_2H_4$

16. What is not true about resonance ?
(*a*) The resonating structures is are hypothetical.
(*b*) The unpaired electrons in various resonating structures are the same.
(*c*) Hybrid structure is least stable.
(*d*) Hybrid structure is least energetic.

17. Resonance structures of a molecule should have :
(*a*) identical arrangement of atoms
(*b*) nearly the same energy content
(*c*) the same number of paired electrons
(*d*) identical bonding.

18. The correct order of increasing C – O bond length in CO, CO_3^{2-} and CO_2 is :
(*a*) $CO_3^{2-} < CO_2 < CO$
(*b*) $CO_2 < CO_3^{2-} < CO$
(*c*) $CO < CO_3^{2-} < CO_2$
(*d*) $CO < CO_2 < CO_3^{2-}$

19. The correct sequence of decrease in bond angles in NH_3, PH_3, AsH_3 and SbH_3 is
(*a*) $NH_3 > PH_3 > AsH_3 > SbH_3$
(*b*) $NH_3 > AsH_3 > PH_3 > SbH_3$
(*c*) $SbH_3 > AsH_3 > PH_3 > NH_3$
(*d*) $PH_3 > NH_3 > AsH_3 > SbH_3$

20. Which of the following let raiodides is the least stable and has doubtful existence ?
(*a*) CI_4 (*b*) GeI_4
(*c*) SnI_4 (*d*) PbI_4

21. Oxidising power of M^{4+} ions (M^{4+} = Ge^{4+}, Sn^{4+} and Pb^{4+}) is in the order :
(*a*) $Ge^{4+} < Sn^{4+} < Pb^{4+}$
(*b*) $Ge^{4+} < Pb^{4+} < Sn^{4+}$
(*c*) $Pb^{4+} < Ge^{4+} < Sn^{4+}$
(*d*) No order can be predicted.

22. Reducing power of M^{2+} ions (M^{2+} = Ge^{2+}, Sn^{2+} and Pb^{2+}) is in the order :
(*a*) $Ge^{2+} > Sn^{2+} > Pb^{2+}$
(*b*) $Sn^{2+} > Ge^{2+} > Pb^{2+}$
(*c*) $Pb^{2+} > Sn^{2+} > Ge^{2+}$
(*d*) $Sn^{2+} > Pb^{2+} > Ge^{2+}$

23. Which of the following statements is(are) false with respect to group 14 elements?
(*a*) The stability of + 4 oxidation state decreases as we move down the group.
(*b*) The + 4 oxidation state is more common in carbon and silicon.
(*c*) Pb^{4+} ion is more stable than Pb^{2+} ion.
(*d*) The stability of + 2 oxidations state increases on moving down the group.

24. Among the following ions $p\pi - p\pi$ overlap is present in :
(*a*) NO_3^- (*b*) CO_3^{2-}
(*c*) PO_4^{3-} (*d*) NO_2^-

25. The enolic form of acetone has :
(*a*) 9 sigma bonds, 1 pi bond and 2 lone pairs.
(*b*) 8 sigma bonds, 2 pi bonds and 2 lone pairs
(*c*) 10 sigma bonds, 1 pi bond and 1 lone pair
(*d*) 9 sigma bonds, 2 pi bonds and 1 lone pair.

26. NH_3 and BF_3 form adduct readily through :
(*a*) Ionic bond between BF_3 and NH_3
(*b*) Coordinate bond between B and N

(*c*) Covalent bond between B and N

(*d*) H-bonds between F atoms of BF_3 and H-atoms of NH_3.

27. One of the following statements is not the characteristic of a covalent compound.

(*a*) The compounds are generally poor conductors in the fused state or in solution.

(*b*) They possess low melting and boiling points.

(*c*) They usually show reactions which are instaneous.

(*d*) Mostly they exist in the solid state.

28. Which one of the following combinations gives the most ionic compound ?

(*a*) Na and H (*b*) Mg and O

(*c*) Br and F (*d*) Cs and F

29. The compound that contains both ionic and covalent bonds is :

(*a*) CH_4 (*b*) H_2

(*c*) KCN (*d*) KCl

30. A covalent bond is formed as a result of :

(*a*) transfer of electrons from one atom to the other.

(*b*) electrostatic attraction between two atoms.

(*c*) sharing of electrons contributed by both atoms.

(*d*) sharing of electrons contributed by only one atom.

31. Which of the following ions does not exist ?

(*a*) $[CCl_6]^{2-}$ (*b*) $[SiCl_6]^{2-}$

(*c*) $[GeCl_6]^{2-}$ (*d*) $[SnCl_5]^{2-}$

32. Which of the following has a coordinate bond ?

(*a*) NaOH (*b*) NH_4^+

(*c*) H_2O (*d*) NH_3

33. F_2 molecule is formed by the overlap of :

(*a*) *p–p* orbitals by side-to-side manner

(*b*) *p–p* orbitals by end-to-end manner

(*c*) *s–p* orbitals

(*d*) *s–s* orbitals

34. Which one of the following statements regarding a pi bond is not correct ?

(*a*) Pi bond results from the lateral overlap of atomic orbitals.

(*b*) Pi bond may be formed by (*p-p*) or (*p-d*) overlap.

(*c*) Pi bond is formed when a sigma bond already exists.

(*d*) Pi bond is formed from hybrid orbitals

35. The types of bonds present in $CuSO_4 . 5H_2O$ are

(*a*) Electrovalent and covalent only

(*b*) Electrovalent and coordinate covalent only

(*c*) Electrovalent, covalent, coordinate covalent and hydrogen bonds

(*d*) Covalent and coordinate covalent only.

36. Which type of bond is not present in HNO_2 molecule ?

(*a*) covalent (*b*) co-ordinate

(*c*) ionic

(*d*) ionic as well as coordinate

37. The nodal plane in the π-bond of ethene is located in :

(*a*) the molecular plane

(*b*) a plane parallel to the molecular plane

(*c*) a plane perpendicular to the molecular plane which bisects the carbon-carbon sigma bond at right angles.

(*d*) a plane perpendicular to the molecular plane which contains the carbon-carbon σ-bond.

38. The carbon atoms in calcium carbide are held by :

(*a*) ionic bonds (*b*) 2σ-bonds

(*c*) 2 covalent and one coordinate bonds

(*d*) 2π and 1 σ-bond

39. Number of sigma bonds in P_4O_{10} molecule is :

(*a*) 6 (*b*) 7

(*c*) 17 (*d*) 16

40. Hydrogen bonding is maximum in :

(*a*) Ethanol (*b*) Diethyl ether

(*c*) Ethyl chloride (*d*) Triethyl amine.

ANSWERS

1. (*c*)	**2.** (*b*)	**3.** (*a*), (*b*)	**4.** (*a*)	**5.** (*b*), (*c*)	**6.** (*d*)	**7.** (*a*)	**8.** (*d*)
9. (*b*)	**10.** (*a*)	**11.** (*c*)	**12.** (*c*)	**13.** (*a*)	**14.** (*d*)	**15.** (*a*)	**16.** (*c*)
17. (*a*), (*b*), (*c*)		**18.** (*d*)	**19.** (*a*)	**20.** (*d*)	**21.** (*a*)	**22.** (*a*)	**23.** (*c*)
24. (*c*)	**25.** (*a*)	**26.** (*b*)	**27.** (*c*)	**28.** (*d*)	**29.** (*c*)	**30.** (*c*)	**31.** (*a*)
32. (*b*)	**33.** (*b*)	**34.** (*d*)	**35.** (*c*)	**36.** (*d*)	**37.** (*a*)	**38.** (*d*)	**39.** (*d*)
40. (*a*)							

HINTS AND EXPLANATIONS

1. Bond distance, $d = 1.0$ A° $= 1.0 \times 10^{-8}$ cm (given)

Dipole moment, $\mu = 1.2$ D $= 1.2 \times 10^{-18}$ esu. cm(given)

Electronic charge $= 4.8 \times 10^{-18}$ esu. cm (given)

Fraction of electronic charge $= x$ (to be calculated)

Now we know that :

$$\mu = x \times d$$

1.2×10^{-18} esu. cm $= x \times 10^{-8}$ cm

$$\therefore \qquad x = \frac{1.2\times10^{-18}\text{ esu . cm}}{10^{-8}\text{ cm}} = 1.2\times10^{-10}\text{ esu}$$

Out of 4.8×10^{-10} esu. electronic charge, the electronic charge existing on each atom $= 1.2 \times 10^{-10}$ esu

$\therefore$ Out of 100 esu. electronic charge, the electronic charge existing on each atom

$$= \frac{1.2\times10^{-10}\text{ esu}}{4.8\times10^{-10}\text{ esu. cm}}\times100$$

$= 25\%$ (Answer)

2. Amongst the given molecules, it is only SO_2 which has irregular geometry (SO_2 has V-shaped geometry) while other molecules have regular geometry (SO_3–trigonal planar, SF_6–octahedral). Due to irregular geometry, SO_2 has $\mu > 0$. Now we know that a molecular having $\mu > 0$ is polar. Thus SO_2 is a polar molecule.

3. We know that a molecule having regular (symmetrical) geometry is non-polar and has $\mu = 0$. On the other hand, a molecule having irregular (unsymmetrical) geometry is polar and has $\mu > 1$. Now since BF_3 has $\mu = 0$, this molecule has symmetrical geometry (trigonal planar geometry) and B–F bands in it are non-polar. On the other hand, since NH_3 molecule has $\mu > 0$, it has unsymmetrical geometry (trigonal pyramidal geometry) and N – H bonds in it are polar.

4. NH_3 has the highest value of dipole moment. It can be explained as follows :

(*i*) Since N-atom (central atom) in NH_3 molecule has the highest electronegativity of the central atoms in PH_3, SbH_3 and AsH_3 respectively, N – H bond in NH_3 mulecule is the most polar. Now we know that with the increase in polarity, μ values increases. Thus NH_3 has the highest value of μ.

(*ii*) N-atom (central atom) in NH_3 has higher electronegativity than H-atom (N = 3.0, H = 2.1). In other molecules the central atom has equal or lower value of electronegativity than H-atom (P = 2.1, As = 2.0, Sb = 1.9, Bi = 1.9, H = 2.1). Since N-atom has higher electronegativity value, the bond moments of three N–H bonds in NH_3 are towards more electronegative N-atom $\left(\overset{\longmapsto}{\text{H}-\text{N}}\right)$, *i.e.*, the three N–H bond moments are in the same direction as that of *lp* on N-atom. Thus the resultant dipole moment of NH_3 is equal to the sum of three bond moments of three N–H bonds and that of *lp*. In case of other molecules, the resultant dipole moment is obtained by subtracting the moment of *lp* from the sum of bond moments of three M–H bonds (M = P, As, Sb).

5. (*a*) 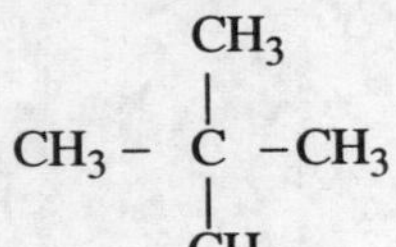 ... Symmetrical structure. Hence $\mu = 0$

(*b*) 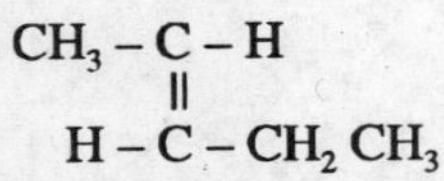 ... Unsymmetrical structure. Hence $\mu > 0$

(*c*) 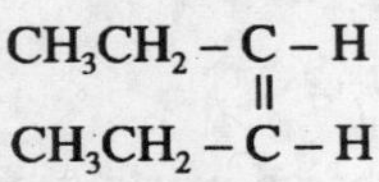 ... Unsymmetrical structure. Hence $\mu > 0$

(*d*) 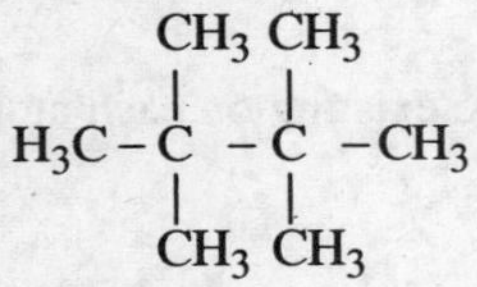 ... Symmetrical structure. Hence $\mu = 0$

Since (*a*) and (*d*) have symmetrical structures, these molecules have $\mu = 0$. On the other hand, since (*b*) and (*c*) have unsymmetrical structures, each of these molecules has $\mu > 0$.

6. The only condition necessary to produce covalent character in ionic bond in a molecule is that the degree of polarisation of anion should be high. To achieve high degree of polarisation (*i*) The size of cation should be small (*ii*) The size of anion should be large (*iii*) The charge on cation or anion should be large (*iv*) The cation should have 18-electron configuration and not 8-electron configuration (noble gas configuration).

7. Since the given dihalides contain the same cation viz Pb^{2+} but different anions (halide ions), the ionic character of these halides depends on the size of the halide ions. The ionic character decreases with the increase in the size of the halide ions. Thus since the size of halides ions increases as $F^- < Cl^- < Br^- < I^-$, the ionic character of Pb X_2 molecules decreases as $PbF_2 > PbCl_2 > PbBr_2 > Pb\,I_2$. Thus (*a*) is the correct answer.

8. We know that Na^+ ion is larger in size than Mg^{2+} ($Na^+ > Mg^{2+}$). Similarly, S^{2-} ion is also larger in size than Cl^- ion ($S^{2-} > Cl^{2-}$). Now MgS has smaller cation and larger anion, it would be the most covalent (Fajan's rules) and hence would be the least soluble in water.

9. Since the size of halide ions (X^-) increases as $F^- < Cl^- < Br^- < I^-$ the ionic character of MX molecules decreases as NaF > NaCl > NaBr > NaI. Being the most ionic, NaF has the highest melting point.

10. The formation of $K[HX_2]$ molecule takes place when KX combines with HX in aqueous solution.

$$KX + HX \longrightarrow K[HX_2]$$

Actually the formation of $K[HX_2]$ takes place through the following steps :

(*a*) $KX \longrightarrow K^+ + X^-$

(*b*) $X^- + H-X \longrightarrow X^-.....H-X$ or $[HX_2]^-$

On adding : $KX + HX \longrightarrow K^+ + [HX_2]^-$ or $K^+[HX_2]^-$

The formation of $[HX_2]^-$ ion as shown in step (*b*) takes place due to the attachment of X^- ion to H-atom of HX molecule through H-bonding. Due to smaller electronegativity of Cl^-, Br^- and I^- ions, these ions cannot combine with HCl, HBr and HI to form $[HCl_2]^-$, $[HBr_2]^-$ and $[HI_2]^-$ ions respectively. Due to non-formation of these ions, $K[HCl_2]$, $K[HBr_2]$ and $K[HI_2]$ donot exist. In other words we can say that KCl, KBr and KI cannot combine with HCl, HBr and HI and hence cannot form $K[HCl_2]$, $K[HBr_2]$ and $K[HI_2]$ respectively. This discussion also shows that $K[HF_2]$ contains K^+ and $[HF_2]^-$ ions and not K^+, H^- and F^- ions. Due to the non-formation of $K[HCl_2]$, $K[HBr_2]$ and $K[HI_2]$, these molecules donot contain K^+ and $[HX_2]^-$ ions. Nor these molecules contain K^+, H^+ and X^- ions.

11. The strength of a hydrogen bond depends on the magnitude of electronegativity of the atom attached with H-atom by H-bond. Greater is the electronegativity of the atom, greater is the strength of H-bond. Now since F has the maximum electronegativity, H .... F bond in F – H .... F has maximum strength, *i.e.*, H .... F bond in F – H .... .F is the strongest. Thus the correct answer is (*c*).

12. Due to H-bonding, several NH_3 molecules get associated together and form a cluster, $(NH_3)_x$. Due to lower electronegativity of P-atom compared to that of N-atom, PH_3 molecule is not able to form the cluster, $(PH_3)_x$. Due to the formation of $(NH_3)_x$ cluster, NH_3 has much higher boiling point than PH_3.

13. Strength of hydrogen bonding, H .... M increases with the increase in the electronegativity of M-atom. Out of the given atoms, Cl-does not form H-bond while the strength of H-bonding of other atoms is in the order S < N < O < F, since the electronegativity of these atoms also increases in the same order $\left(\begin{matrix} S < N < O < F \\ 2.5 \quad 3.0 \quad 3.5 \quad 4.0 \end{matrix}\right)$.

14. We know that H_2O molecule has the structure, 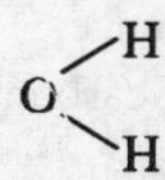 which shows that O-atom is linked with H atom by two O – H covalent bonds. In water, H_2O molecules are joined together due to inter-molecular H-bonding. In the joining process, O-atom of a given H_2O molecule which is already linked with two H-atoms by two O – H covalent bonds is also linked with two H-atoms belonging to other two H_2O molecules (In Fig. 10.1 these H_2O molecules have been represented as 1 and 2) by H-bonding. Each H-atom of the given H_2O molecule is also linked with O-atom belonging to other H_2O molecules by H-bonding. (These other H_2O molecules have been represented as 3 and 4 in Fig. 10.1). This description shows that each H_2O molecule is tetrahedrally surrounded by other four H_2O molecules. It is evident from Fig. 10.1 that in water each H_2O molecule can form maximum four H-bonds and O-atom of H_2O molecule can form two H-bonds.

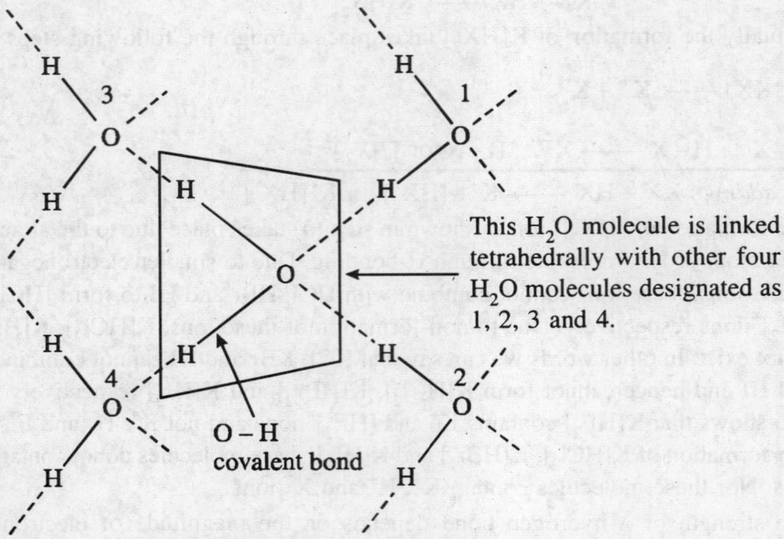

Fig. 10.1 Structure of water in liquid state.

15. The structure of C_2H_6, C_2H_4 and C_2H_2 molecules are given below :

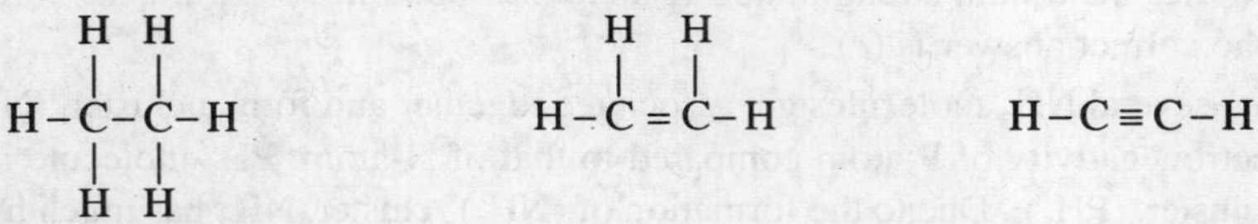

C_2H_6 [Multiplicity (m)] of carbon – carbon bond = 1 $\qquad$ C_2H_4 (m = 2) $\qquad$ C_2H_2 (m = 3)

Now since m of carbon – carbon bond in the given molecules decreases as C_2H_2 (m = 3) > C_2H_4 (m = 2) > C_2H_6 (m = 1), C – C bond length increases as $C_2H_2 < C_2H_4 < C_2H_6$.

16. Hybrid structure is the most stable.

17. (*a*), (*b*) and (*c*) are the conditions that have to be fulfilled by resonating structures of a molecule or ion showing resonance.

18. The structure and bond order of carbon – oxygen bond in the given molecules are given as :

$$\left(\text{B. O.} = \frac{\text{No. of } (\sigma + \pi) \text{ bonds}}{\text{No. of } \sigma \text{ bonds}}\right)$$

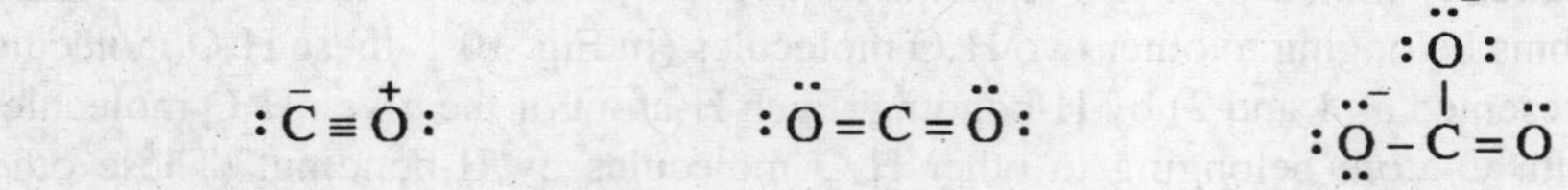

B.O. = 3/1 = 3 $\qquad$ B.O. = 4/2 = 2 $\qquad$ B.O. = 4/3 = 1.33

Since the bond order decreases as CO (= 3) > CO_2 (= 2) > CO_3^{2-} (= 1.3), C – O bond length increases as CO < CO_2 < CO_3^{2-}.

19. Since the given molecules have the same surrounding atoms linked with the central atom and the same geometry (trigonal pyramidal geometry) but different central atom, their HMH bond angles depend on the electronegativity of the central atoms (different atoms in each molecule). It has been found that HMH bond angle in these molecules decreases as the electronegativity of the central atom decreases from N to Bi (N = 3.0, P = 2.1, As = 2.0, Sb = 1.8 and Bi = 1.7).

20. The given tetraiodides contain C, Ge, Sn and Pb as C^{4+}, Ge^{4+}, Sn^{4+} and Pb^{4+} ions respectively. According to inert pair effect, the stability of these cations decreases as $C^{4+} > Ge^{4+} > Sn^{4+} > Pb^{4+}$. This order shows that since Pb^{4+} ion is the least stable, its tetraiodide, PbI_4 is also least stable and hence has doubtful existence.

21. According to inert pair effect, the stability of given ions is in the order $Ge^{4+} > Sn^{4+} > Pb^{4+}$. Now since Ge^{4+} ion is the most stable, this ion is reduced to Ge^{2+} ion with greatest difficulty and hence is the weakest oxidising agent *i.e.*, Ge^{4+} ion has the poorest oxidising power. On the other hand, since Pb^{4+} ion is the least stable, it is reduced to Pb^{2+} ion most easily and hence is the strongest oxidising agent. Thus the oxidising power of M^{4+} ions is in the order $Ge^{4+} < Sn^{4+} < Pb^{4+}$. Note that this order is opposite to the order of stability of M^{4+} ions.

22. According to inert pair effect, the stability of the given ions is in the order $Ge^{2+} < Sn^{2+} < Pb^{2+}$. Now since Ge^{2+} ion is the least stable, this ion is oxidised to Ge^{4+} ion most easily and hence this ion is the strongest reducing agent. On the other hand, since Pb^{2+} ion is the most stable, this ion is oxidised to Pb^{4+} ion with greatest difficulty and hence this ion is the weakest reducing agent. Thus the reducing power of M^{2+} ions is in the order $Ge^{2+} > Sn^{2+} > Pb^{2+}$. Note that this order is opposite to the order of stability of M^{2+} ions.

23. Since, due to inert pair effect, the stability of lower oxidation state (= + 2, M^{2+} ions) increases and that of higher oxidation state (= + 4, M^{4+} ions) decreases on moving down the group, Pb^{4+} is less stable than Pb^{2+} ion. Thus statement (*c*) is false.

24. The structures of the given species are given as :

(*a*) 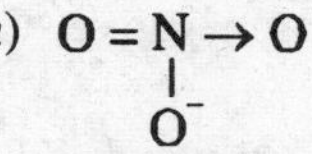

σ-*bps* = 3, *lps* = 0
Sum = 3 + 0 = 3 (sp^3).
(N–O) π bond is
[2p(N) – 2*p*(O)] or
(*p*π – *p*π) bond

(*b*) 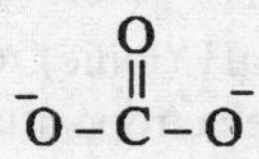

σ-*bps* = 3, *lps* = 0, Sum = 3 + 0 = 3
(sp^2) . (C – O) π bond is
[2*p*(C) – 2*p*(O)] or (*p*π – *p*π) bond

(*c*) 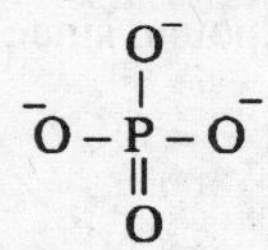

σ-*bps* = 4, *lps* = 0
Sum = 4 + 0 = 4 (sp^3).
(P–O) π bond is
[3*d*(P) – 2*p*(O)] or
(*d*π – *p*π) bond.

(*d*) 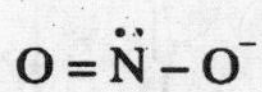

σ-*bps* = 2, *lps* = 1, Sum = 2 + 1 = 3
(sp^2) . (N – O) π bond is
[2*p*(N) – 2*p*(O)] or (*p*π – *p*π) bond

Obviously (*p*π – *d*π) overlap is present in PO_4^{3-} ion *i.e.* in (*c*).

25. Acetone $CH_3 - CO - CH_3$ exists in keto and enolic forms whose structures are given below :

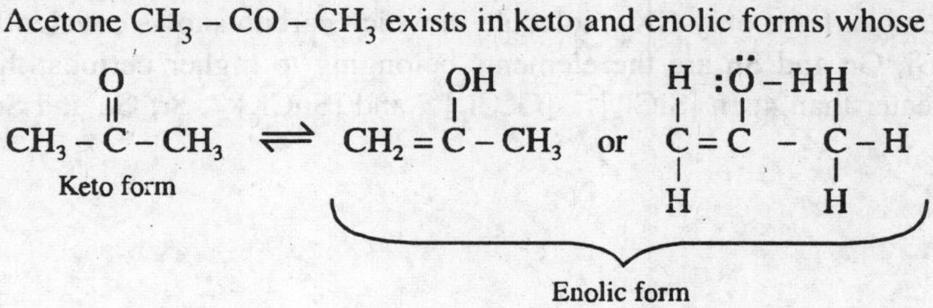

Keto form

Enolic form

The structure of enolic form shows that it has nine sigma bonds, [(C – C) σ-bonds = 2, (C – H) σ-bonds = 5, (C – O) σ-bond = 1, (O – H) σ bond = 1], one pi bond [(C – C) π-bond] and two lone pairs on O-atom.

26. In NH_3 molecule, N-atom has a lone pair of electrons and in BF_3 molecule B-atom in its excited state has one vacant 2*p* orbital ($B^* = 2s^1\ 2p^1\ 2p^1\ 2p^0$). Thus when NH_3 molecule combines with BF_3 molecule to form an adduct, lone pair of electrons on N-atom in NH_3 molecule is donated to the vacant 2*p* orbital on B-atom of BF_3 molecule and a N $\rightarrow$ B coordinate bond is formed.

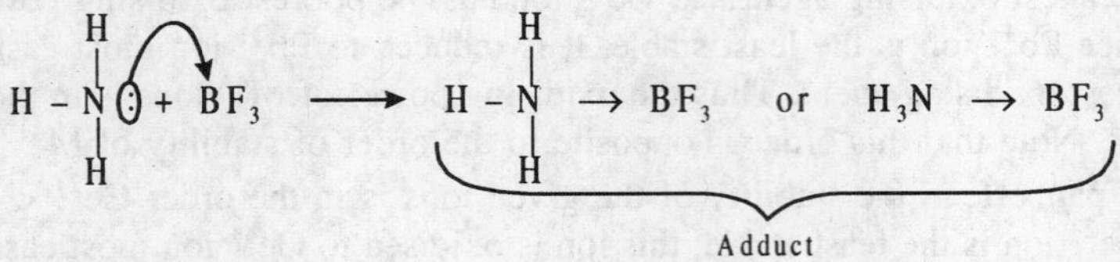

27. (*a*) Since these compounds donot contain free ions to conduct electricity, these compounds are poor conductors in the fused state or in solution.

(*b*) Since the molecules in covalent compounds are held together less tightly due to the presence of weak van der Walls forces, these have low melting and boiling points.

(*c*) Since these compounds are molecular and non-ionic, they show reactions which are molecular and not instaneous.

(*d*) Covalent compounds exist in all the three states *viz.* solid, liquid and gaseous. Mostly they exist in solid states.

28. Compounds formed are (*a*) NaH (ionic), (*b*) MgO (ionic), (*c*) BrF (covalent) and (*d*) CsF (ionic). Since Cs is the least electropositive and F is the most electronegative, CsF is the most ionic compound.

29. KCN is an ionic compound and hence can be represented as K^+CN^- in which K^+ and CN^- ions are linked together by ionic bond. In CN^- ion, C and N atoms are linked together by covalent bonds ($:\ddot{C}\equiv\bar{N}:$ or $:\bar{C}\equiv N:$). CH_4 and H_2 are covalent molecules, since they contain only covalent bonds $\left[\begin{array}{c} H \\ | \\ H-C-H, \ H-H \\ | \\ H \end{array}\right]$. KCl is an ionic compound since it contains only ionic bond between K^+ and Cl^- ($KCl \equiv K^+Cl^-$).

30. (*a*) Transfer of one or more electrons from one atom to the other gives ionic bond.

(*b*) Electrostatic attraction between two atoms is the cause of ionic bond.

(*c*) Sharing of two or more electrons by both the atoms gives one or more covalent bonds between the atoms.

(*d*) Sharing of electrons donated by only one atom gives coordinate bond.

31. Since carbon is an element of 2nd period, its maximum covalency is 4 (Rule of maximum covalency), *i.e.* carbon can form only CX_4 molecule in which carbon shows covalently of 4. On the other hand, Si, Ge and Sn are the elements belonging to higher periods, they can show a covalency greater than 4. In $[SiCl_6]^{2-}$, $[GeCl_6]^{2-}$ and $[SnCl_6]^{2-}$, Sr, Ge and Sn show covaleny of 6.

32. Structures of the given molecules are as : (*a*) $Na^+ OH^-$ (*b*) 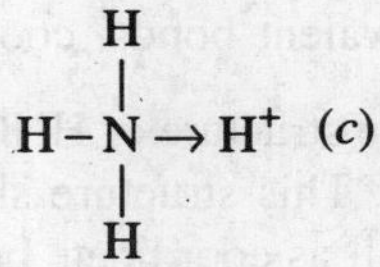 (*c*) 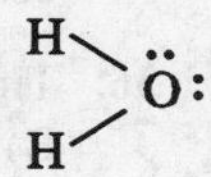

(*d*) 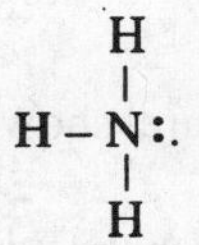. These structures show that it is only NH_4^+ ion which has a coordinate bond ($N \rightarrow H^+$ coordinate bond).

33. Valence shell electronic configuration of F atom is $2s^2\ 2p^5$ or $2s^2\ 2p_x^{\ 1}\ 2p_y^{\ 2}\ 2p_z^{\ 2}$ which shows that it is only $2p$ orbital which has unpaired electron (singly-filled). Now singly-filled $2p_x$ orbital of one F-atom overlaps with the singly-filled $2p_x$ orbital of another F-atom in end-to-end manner and thus forms [F(p) – F(p)] σ bond in F_2 molecule. Thus F – F bond in F_2 molecule results by *p–p* overlap in end-to-end manner.

34. (*a*) Pi bond is formed when two singly-filled *p*-orbitals overlap in side-to-side manner (lateral overlap) (*e.g.* O_2 molecule).

(*b*) Pi bond is formed by the overlap between two singly filled *p* orbitals [(*p* – *p*) overlap] or two singly-filled *p* and *d* orbitals [(*p* – *d*) overlap] (*e.g.* $SO_4^{\ 2-}$ ion).

(*c*) Pi bond is formed in those molecules or ions which already contain one or two σ-bonds.

(*d*) Overlap between hybrid orbitals gives σ-bond (*s*).

35. When $CuSO_4 \cdot 5H_2O$ (blue) is heated, four H_2O molecules are lost at 373 K to give copper sulphate monohydrate ($CuSO_4 \cdot H_2O$) while the fifth H_2O molecule is lost at 432 K to form anhydrous $CuSO_4$ (white)

$$CuSO_4 \cdot 5H_2O \text{ (Blue)} \xrightarrow[(-\,4H_2O)]{373\text{ K}} CuSO_4 \cdot H_2O \xrightarrow[(-\,H_2O)]{432\text{ K}} CuSO_4 \text{ (White)}$$

The action of heat on $CuSO_4 \cdot 5H_2O$ as given above shows that four H_2O molecules are different from the fifth H_2O molecule. In $CuSO_4 \cdot 5H_2O$ molecule, four H_2O molecules are coordinated to the central Cu^{2+} ion through coordinate bonds while the fifth H_2O molecule is attached to two O-atoms of $SO_4^{\ 2-}$ ion and two H-atoms of two H_2O molecules through H-bonding [See Fig. 10.2]. Thus $CuSO_4 \cdot 5H_2O$ should be formulated as $[Cu(H_2O)_4]SO_4 \cdot H_2O$ and should be named as ***tetraqua copper (II) sulphate monohydrate.*** Thus $CuSO_4 \cdot 5H_2O$ is a complex compound.

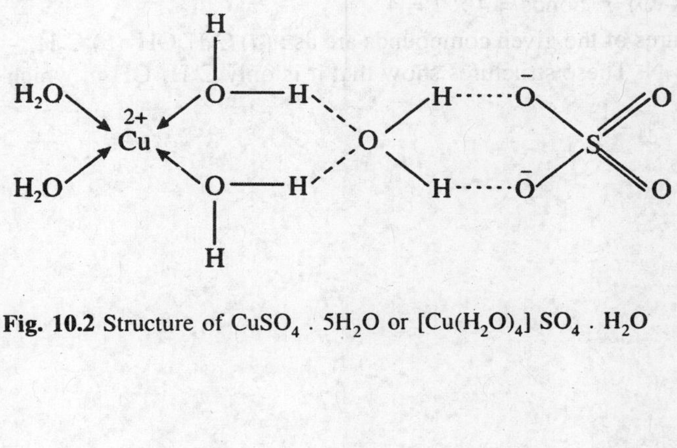

Fig. 10.2 Structure of $CuSO_4 \cdot 5H_2O$ or $[Cu(H_2O)_4]\ SO_4 \cdot H_2O$

The above structure of $CuSO_4 . 5H_2O$ shows that it contains ionic bonds (electrovalent), covalent bonds, coordinate bonds and H-bonds.

36. The structure of HNO_2 molecule is $H-O-\ddot{N}=O$ (structure A) or $H-O-\ddot{N}\rightarrow O$ (structure B). This structure shows that it contains only covalent bonds (structure A) or covalent as well as coordinate bonds (structure B). Thus HNO_2 molecule does not contain ionic as well as coordinate bonds.

37. The structure of ethene ($CH_2 = CH_2$) is $H_2C=CH_2$. This structure shows that the molecular plane does not have any π-electron density and hence the nodel plane in π-bond is located in the molecular plane. Note that nodal plane is that plane on which the electron density is zero.

38. On hydrolysis, since calcium carbide gives acetylene ($H - C \equiv C - H$), this compound has acetylide ion, C_2^{2-} whose structure is $:\bar{C} \equiv \bar{C}:$. Obviously $:\bar{C} \equiv \bar{C}:$ ion (acetylide) ion has one σ and two π bonds. Thus calcium carbide is represented as Ca^{2+} $(:C \equiv C:)^{2-}$ which contains one σ- and to π-bonds.

39. The structure of P_4O_{10} molecule is shown below in Fig. 10.3.

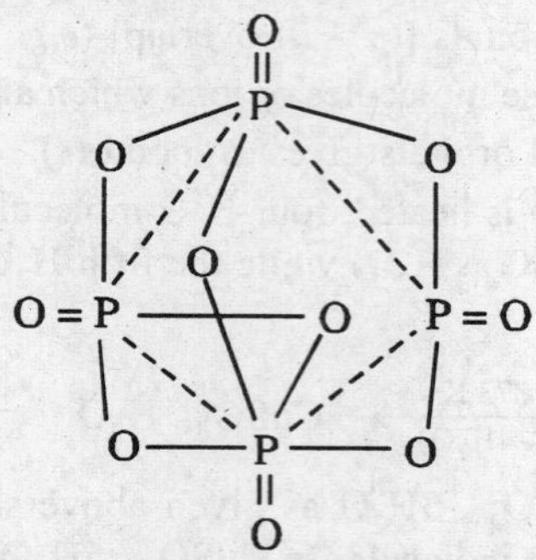

Fig. 10.3 Structure of P_4O_{10} molecule

This structure has four P-atoms. Since each P-atom is linked with three O-atoms by (P – O) single bonds and with one O-atom by (P = O) double bond.

No. of (P – O) single bonds or (P – O) σ bonds = $4 \times 4 = 16$

No. of (P = O) double bonds = $4 \times 1 = 4$

No. of (P – O) π bonds = $4 \times 1 = 4$

40. The structures of the given compounds are as : (*a*) C_2H_5OH (*b*) $C_2H_5 - O - C_2H_5$ (*c*) C_2H_5Cl (*d*) $(C_2H_5)_3N$. These structures show that it is only C_2H_5 OH in which hydrogen bonding is maximum.

11

Molecular Orbital Theory of Simple Diatomic Molecules and Ions

MULTIPLE CHOICE QUESTIONS

1. If x-axis is supposed to be the molecular axis of a diatomic molecule, then π molecule orbital is formed by the overlap of :

(*a*) s atomic orbitals

(*b*) s and p_x orbitals

(*c*) p_y and p_y orbitals

(*d*) p_x and p_z orbitals

2. Bond order of a diatomic molecule is given by:

(*a*) The difference between the number of electrons in bonding and antibonding molecular orbitals.

(*b*) Total number of electrons in bonding and antibonding molecular orbitals.

(*c*) Twice the difference between the number of electrons in bonding and antibonding molecular orbitals.

(*d*) Half the difference between the number of electrons in bonding and antibonding molecular orbitals.

3. In a homonuclear diatomic molecule which of the following sets of orbitals are degenerate ?

(*a*) σ_{2s} and σ_{1s} (*b*) π_{2p_x} and π_{2p_y}

(*c*) π_{2p_x} and σ_{2p_z} (*d*) σ_{2p_z} and $\overset{*}{\sigma}_{2p_z}$

4. Which of the following is non-existent according to molecular orbital theory ?

(*a*) H_2^- (*b*) O_2^-

(*c*) He_2 (*d*) O_2^+

5. According to MOT, the species O_2^+ possesses :

(*a*) bond order of 2.5

(*b*) three unpaired electrons

(*c*) diamagnetic character

(*d*) stability lower than O_2

6. N_2 and O_2 are converted to mono cations N_2^+ and O_2^+ respectively. Which is wrong ?

(*a*) In N_2^+, N – N bond weakens.

(*b*) In O_2^+, O – O bond order increases

(*c*) In O_2^+, paramagnetism decreases

(*d*) N_2^+ becomes diamagnetic

7. The number of antibonding electron pairs in O_2^{2-} ion, on the basis of MO theory, is :

(*a*) 4 (*b*) 3

(*c*) 2 (*d*) 5

8. N – O bond length in NO^+, NO and NO^- species is in the order:

(*a*) $NO^+ < NO < NO^-$

(*b*) $NO^+ > NO > NO^-$

(*c*) $NO^+ < NO^- < NO$

(*d*) $NO^+ > NO^- > NO$

9. Which of the following species is paramagnetic ?

(*a*) O_2^{-} (*b*) CN^-

(*c*) CO (*d*) NO^+

10. Which of the following is not paramagnetic ?

(*a*) S^{2-} (*b*) NO

(*c*) O_2^- (*d*) N_2^-

11. Which of the following have identical bond order ?

(*a*) CN^- (*b*) O_2^-

(*c*) NO^+ (*d*) CN^+

12. Among KO_2, AlO_2^-, BaO_2 and NO_2^+ unpaired electron is present in :

(*a*) NO_2^+ and BaO_2

(*b*) KO_2 and AlO_2^-

(*c*) KO_2 only

(*d*) BeO_2 only

13. O_2 molecule is paramagnetic because

(*a*) Bonding electrons are more than antibonding electrons

(*b*) It contains unpaired electrons

(*c*) Bonding electrons are less than antibonding electrons

(*d*) Bonding electrons are equal to antibonding electrons.

14. Which of the following molecular species has unpaired electrons(s) ?

(*a*) N_2 (*b*) F_2

(*c*) O_2^- (*d*) O_2^{2-}

15. Which among the following species donot exist (unstable) ?

(*a*) F_2^{2-} (*b*) Ar_2

(*c*) He_2^+ (*d*) H_2^+

16. The correct sequence of bond order is :

(*a*) $O_2^+ > O_2^- > O_2$

(*b*) $O_2 > O_2^- > O_2^+$

(*c*) $O_2^+ > O_2 > O_2^-$

(*d*) $O_2^- > O_2^+ > O_2$

17. Which of the following has unpaired electrons ?

(*a*) O_2^+ (*b*) O_2^-

(*c*) O_2 (*d*) O_2^{2-}

18. The common factures among the species CN^-, CO and NO^+ are :

(*a*) Bond order three and isoelectronic

(*b*) Bond order three and weak field legands

(*c*) Bond order three and π-acceptors

(*d*) Iso electronic and weak field ligands.

19. Which of the following species has the lowest first ionisation energy ?

(*a*) O (*b*) O_2

(*c*) O_2^+ (*d*) O_2^-

20. Combination of two AO's lead to the formation of :

(*a*) Two MO's (*b*) One MO

(*c*) Three MO's (*d*) Four MO's

ANSWERS

1. (*c*)	**2.** (*d*)	**3.** (*b*)	**4.** (*c*)	**5.** (*a*)	**6.** (*d*)	**7.** (*c*)	**8.** (*a*)
9. (*a*)	**10.** (*a*)	**11.** (*a*), (*c*)	**12.** (*c*)	**13.** (*b*)	**14.** (*c*)	**15.** (*a*), (*b*)	**16.** (*c*)
17. (*a*), (*b*), (*c*)		**18.** (*a*)	**19.** (*d*)	**20.** (*a*)			

HINTS AND EXPLANATIONS

1. (*a*) $(s - s)$ overlap always gives σ_s^b and σ_s^* molecular orbitals.

(*b*) $(s - p_x)$ overlap gives σ_{sp}^b and σ_{sp}^* molecular orbitals.

(*c*) $(p_y - p_y)$ overlap gives π_y^b and π_y^* molecular orbitals, since the two p_y orbitals have the same symmetry about the molecular (nuclear) axis.

(*d*) $(p_x - p_z)$ overlap will not take place, since + + overlap region is cancelled by + − overlap region as shown in Fig. Thus $(p_x - p_z)$ overlap does not give any molecular orbitals.

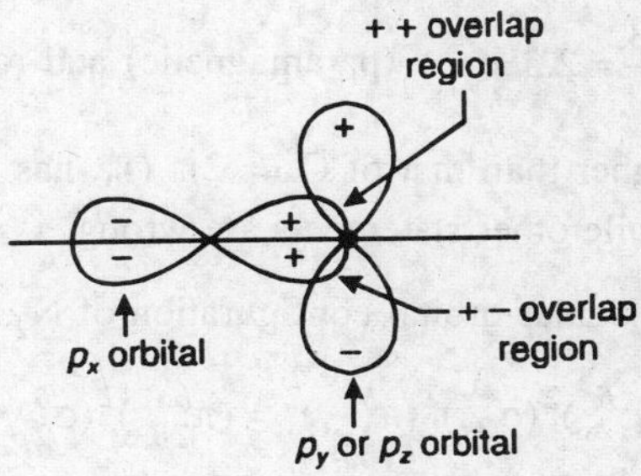

2. Bond order of a diatomic molecule is given by :

$$\text{Bond order} = \frac{N_b - N_a}{2}$$

Where N_b and N_a are the number of electrons present in bonding molecular orbitals and antibonding molecular orbitals respectively.

3. Energy of different molecular orbitals formed in a homonuclear diatomic molecule (*e.g.* Li_2, Be_2, B_2, C_2 and N_2) increases in the order :

Energy increases ⟶

$$\sigma_{1s}^b, \sigma_{1s}^*, \sigma_{2s}^b, \sigma_{2s}^*, \pi_{2p_y}^b = \pi_{2p_z}^b, \sigma_{2p_x}^b, \pi_{2p_y}^* = \pi_{2p_z}^*, \sigma_{2p_x}^*$$

This order shows that only π_{2p_x} and π_{2p_y} molecular orbitals are degenerate.

4. If x-axis is assumed to be the molecular axis, then MO configuration of the given species can be written as : (e = sum of the electrons in the valence shell of each atom)

(*a*) $H_2^-(e = 3) = (\sigma_{1s}^b)^2(\sigma_{1s}^*)^1$, $\quad\therefore\quad$ $\text{B.O.} = \frac{2-1}{2} = 1.5$

(*b*) $O_2^-(e = 13) = (\sigma_{2s}^b)^2(\sigma_{2s}^*)^2(\sigma_{2p_x}^b)^2 (\pi_{2p_y}^b)^2 = (\pi_{2p_z}^b)^2(\pi_{2p_y}^*)^2 = (\pi_{2pz}^*)^2$

$\therefore\quad \text{B.O.} = \frac{8-5}{2} = 1.5$

(*c*) $He_2(e = 4) = (\sigma_{1s}^b)(\sigma_{1s}^*)^2$, $\quad\therefore\quad$ $\text{B.O.} = \frac{2-2}{2} = 0$

(*d*) $O_2^+(e = 11) = (\sigma_{2s}^b)^2(\sigma_{2s}^*)^2(\sigma_{2p_x}^b)^2(\pi_{2p_y}^b)^2 = (\pi_{2p_z}^b)^2(\pi_{2p_z}^*)^1$ $\quad\therefore\quad$ $\text{B.O.} = \frac{8-3}{2} = 2.5$

Since it is only He_2 molecule whose bond order is zero, this molecule does not exist.

5. MO configuration of O_2^+ and O_2 are as :

$$O_2^+(e=11)=(\sigma_{2s}^b)^2(\sigma_{2s}^*)^2(\sigma_{2p_x}^b)^2(\pi_{2p_y}^b)^2=(\pi_{2p_z}^b)^2(\pi_{2p_y}^*)^1$$

$$O_2(e=12)=(\sigma_{2s}^b)^2(\sigma_{2s}^*)^2(\sigma_{2p_x}^b)^2(\pi_{2p_y}^b)^2=(\pi_{2p_z}^b)^2(\pi_{2p_y}^*)^1=(\pi_{2p_z}^*)^1$$

Thus for O_2^+, B.O. $=\dfrac{8-3}{2}=2.5, n=1$ (paramagnetic) and for O_2, B.O. $=\dfrac{8-4}{2}=2$. Since B.O. of O_2^+ (= 2.5) is higher than that of O_2(= 2), O_2^+ has higher stability than O_2. Thus statement (*a*) is correct while other statements are wrong.

6. Let us write down the molecular orbital configuration of N_2, N_2^+, O_2 and O_2^+ species.

$$N_2(e=10)=(\sigma_{2s}^b)^2(\sigma_{2s}^*)^2(\pi_{2p_y}^b)^2=(\pi_{2p_z}^b)^2(\sigma_{2p_x}^b)^2$$

$\therefore$ B.O. $=\dfrac{8-2}{2}=3$ and n = 0 (diamagnetic)

$$N_2^+(e=9)=(\sigma_{2s}^b)^2(\sigma_{2s}^*)^2(\pi_{2p_y}^b)^2=(\pi_{2p_z}^b)^2(\sigma_{2p_x}^b)^1$$

$\therefore$ B.O. $=\dfrac{7-2}{2}=2.5, n=1$ (paramagnetic)

$$O_2(e=12)=(\sigma_{2s}^b)^2(\sigma_{2s}^*)^2(\sigma_{2p_x}^b)^2(\pi_{2p_y}^b)^2=(\pi_{2p_z}^b)^2(\pi_{2p_y}^*)^1=(\pi_{2p_z}^*)^1$$

$\therefore$ B.O. $=\dfrac{8-4}{2}=2$ and n = 2 (paramagnetic)

$$O_2^+(e=11)=(\sigma_{2s}^b)^2(\sigma_{2s}^*)^2(\sigma_{2p_x}^b)^2(\pi_{2p_y}^b)^2=(\pi_{2p_z}^b)^2(\pi_{2p_y}^*)^1$$

$\therefore$ B.O. $=\dfrac{8-3}{2}=2.5$ and n = 1 (paramagnetic)

(*a*) Since N_2^+ has lower B.O. than $N_2(N_2^+=2.5, N_2=3)$, N – N bond in N_2^+ is weaker than N–N bond in N_2. Thus statement (*a*) is correct.

(*b*) Since B.O. of O_2^+ is higher than that of $O_2(O_2^+=2.5, O_2=2)$, statement (*b*) is correct.

(*c*) Since the number of unpaired electrons (n) in O_2^+ is lower than that in O_2 $(O_2^+=1, O_2=2)$, O_2^+ has less paramagnetic character than O_2. Thus statement (*c*) is correct.

(*d*) Since n = 1 in N_2^+, N_2^+ is paramagnetic and not diamagnetic. Thus statement (*d*) is wrong.

7. Molecular orbital configuration of O_2^{2-} is : (e = Total number of electrons in O_2^{2-})

$$O_2^{2-}(2\times8+2=18)=(\sigma_{1s}^b)^2,(\sigma_{1s}^*)^2,(\sigma_{2s}^b)^2,(\sigma_{2s}^*)^2,(\sigma_{2p_x}^b)^2,(\pi_{2p_y}^b)^2=(\pi_{2p_z}^b)^2,(\pi_{2p_y}^*)^2=(\pi_{2p_z}^*)^2$$

Obviously the number of electron pairs present in antibonding molecular orbitals is equal to

$$1 \text{ (in } \sigma_{1s}^{*}) + 1 \text{ (in } \sigma_{2s}^{*}) + 1 \text{ (in } \pi_{2p_y}^{*}) + 1 \text{ (in } \pi_{2p_z}^{*}) = 4$$

8. Since NO^+, NO and NO^- have 10, 11 and 12 electrons in their respective valence shells, their molecular orbitals configuration can be written as :

$$NO^{+}(e=10) = (\sigma_{2s}^{b})^2(\sigma_{2s}^{*})^2(\pi_{2p_y}^{b})^2 = (\pi_{2p_z}^{b})^2(\sigma_{2p_x}^{b})^2$$

$$\therefore \quad \text{B.O.} = \frac{8-2}{2} = 3$$

$$NO\ (e=11) = (\sigma_{2s}^{b})^2(\sigma_{2s}^{*})^2(\pi_{2p_y}^{b})^2 = (\pi_{2p_z}^{b})^2(\sigma_{2p_x}^{b})^2(\pi_{2p_y}^{*})^1$$

$$\therefore \quad \text{B.O.} = \frac{8-3}{2} = 2.5$$

$$NO^{-}(e=12) = (\sigma_{2s}^{b})^2(\sigma_{2s}^{*})^2(\pi_{2p_y}^{b})^2 = (\pi_{2p_z}^{b})^2(\sigma_{2p_x}^{b})^2(\pi_{2p_y}^{*})^1 = (\pi_{2p_z}^{*})^1$$

$$\therefore \quad \text{B.O.} = \frac{8-4}{2} = 2$$

Since bond orders of NO^+, NO and NO^- are in the order : NO^+ (= 3) > NO (= 2.5) > NO^- (= 2), N–N bond lengths in these species are in the reverse order *viz* NO^+ < NO < NO^-.

9. Method 1. Since O_2^- has 13 electrons in its valence shell, its MO configuration can be written as :

$$O_2^{-}(e=13) = (\sigma_{2s}^{b})^2(\sigma_{2s}^{*})^2(\sigma_{2p_x}^{b})^2(\pi_{2p_y}^{b})^2 = (\pi_{2p_z}^{b})^2(\pi_{2p_y}^{*})^2 = (\pi_{2p_z}^{*})^1$$

$\therefore$ n = 1 (paramagnetic)

Since CN^-, CO and NO^+ have the same number of electrons in their respective shells (= 10), they have the same MO configuration which is as :

$$(\sigma_{2s}^{b})^2(\sigma_{2s}^{*})^2(\pi_{2p_y}^{b})^2 = (\pi_{2p_z}^{b})^2(\sigma_{2p_x}^{b})^2$$

$\therefore$ n = 0 (diamagnetic)

Method 2. Number of electrons in the valence shell of the given species is as : $O_2^- = 2\times+1 = 13$ (odd number–paramagnetic), CN^- = 4 + 5 + 1 = 10 (Even number–diamagnetic), CO = 4 + 6 = 10 (Even number–diamagnetic), NO^+ = 5 + 6 – 1 = 10 (Even number – diamagnetic). Obviously it is only O_2^- ion which has odd number of electrons and hence is paramagnetic. Thus the correct answer is (*a*).

10. Method 1. Since S^{2-} is a monoatomic ion, its valence-shell electronic configuration has to be written to determine its magnetic property. Its valence-shell electronic configuration is $3s^2\ 3p^6$ which has n = 0. Thus S^{2-} ion is diamagnetic, *i.e.*, S^{2-} ion is not paramagnetic.

Since NO and N_2^- both have the same number of electrons (= 11) in their respective valence shell, their molecular orbital configurations are also the same which is :

$$(\sigma_{2s}^{b})^2(\sigma_{2s}^{*})^2(\pi_{2p_y}^{b})^2 = (\pi_{2p_z}^{b})^2(\sigma_{2p_x}^{b})^2(\pi_{2p_y}^{*})^1$$

which gives $n = 1$. Thus NO and N_2^- both are paramagnetic.

Now since O_2^- has 13 electrons in its valence shell, its molecular orbital configuration is :

$$(\sigma_{2s}^{b})^2(\sigma_{2s}^{*})^2(\sigma_{2p_x}^{b})^2(\pi_{2p_y}^{b})^2 = (\pi_{2p_z}^{b})^2(\pi_{2p_y}^{*})^2 = (\pi_{2p_z}^{*})^1$$

which gives $n = 1$. Thus O_2^- is also paramagnetic.

Method 2. Number of electrons in the valence shell of the given species is as : $S^{2-} = 6 \times 2 + 2 = 14$ (Even number–diamagnetic), NO = 5 + 6 = 11 (odd number–paramagnetic), $O_2^- = 6 \times 2 + 1 = 13$ (Odd number–paramagnetic), $N_2^- = 5 \times 2 + 1 = 11$ (Odd number–paramagnetic). Obviously S^{2-} is not paramagnetic while other species are paramagnetic. Thus correct answer is (*a*).

11. Number of electrons in the valence-shell of the given species is : (*a*) $CN^- = 4 + 5 + 1 = 10$, (*b*) $O_2^- = 6 \times 2 + 1 = 13$, (*c*) $NO^+ = 5 + 6 - 1 = 10$ and (*d*) $CN^+ = 4 + 5 - 1 = 8$. Now since CN^- and NO^+ have the same number of electrons (= 10) in their respective valence shell, they have the same molecular orbital configuration and hence same number of electrons in bonding molecular orbitals (N_b) and in antibonding molecular orbitals (N_a). Consequently these species (*i.e.*, CN^- and NO^+) have the same (identical) bond order.

12. MOT method. KO_2 has O_2^- ion which, according to molecular orbital theory, has one unpaired electron.

Alternative method. Number of electrons present in the valence-shell in the given species is as : $KO_2 = 1 + 6 \times 2 = 13$ (odd number–one unpaired electron), $AlO_2^- = 3 + 6 \times 2 + 1 = 16$ (even number–no unpaired electron), $BaO_2 = 2 + 6 \times 2 = 14$ (even number–no unpaired electron) and $NO_2^+ = 5 + 6 \times 2 - 1 = 16$ (even number–no unpaired electron). Obviously the correct answer is (*c*).

13. We know that the paramagnetic character of a molecular does not depend on the number of bonding and antibonding electrons. It depends on the number of unpaired electrons. Since O_2 ($e = 12$) has two unpaired electrons in $\pi_{2p_y}^{*}$ and $\pi_{2_z}^{*}$ molecular orbitals as shown below, it is paramagnetic.

$$O_2(e=12) : (\sigma_{2s}^{b})^2(\sigma_{2s}^{*})^2(\sigma_{2p_x}^{b})^2(\pi_{2p_y}^{b})^2 = (\pi_{2p_z}^{b})^2(\pi_{2p_y}^{*})^1 = (\pi_{2p_z}^{*})^1 (n = 2)$$

14. Method 1. Write the MO configurations of N_2 and O_2^- as follows :

$$N_2(e=10) : (\sigma_{2s}^{b})^2(\sigma_{2s}^{*})^2(\pi_{2p_y}^{b})^2 = (\pi_{2p_z}^{b})^2(\sigma_{2p_x}^{b})^2, \;\therefore n = 0$$

$$O_2^-(e=13) : (\sigma_{2s}^{b})^2(\sigma_{2s}^{*})^2(\sigma_{2p_x}^{b})^2(\pi_{2p_y}^{b}) = (\pi_{2p_z}^{b})^2(\pi_{2p_y}^{*})^2 = (\pi_{2p_z}^{*})^1, \quad \therefore n = 1$$

Now since F_2 and O_2^{2-} have the same number of electrons (= 14) in their respective valence-shell, their MO configuration is the same. This configuration is :

$$F_2 \text{ or } O_2^{2-}(e=14) : (\sigma_{2s}^{b})^2(\sigma_{2s}^{*})^2(\sigma_{2p_x}^{b})^2(\pi_{2p_y}^{b})^2 = (\pi_{2p_z}^{b})^2(\pi_{2p_y}^{*})^2 = (\pi_{2p_z}^{*})^2, \quad \therefore n = 0$$

Method 2. Number of electrons present in the valence shell of the given species is as :

(*a*) $N_2 = 5 \times 2 = 10$ (Even number–all electrons in the paired state)

(*b*) $F_2 = 7 \times 2 = 14$ (Even number–all electrons in the paired state)

(*c*) $O_2^- = 6 \times 2 + 1 = 13$ (Odd number–one unpaired electron)

(*d*) $O_2^{2-} = 6 \times 2 + 2 = 14$ (Even number–all electrons in the paired state).

15. The species having bond order (B.O.) equal to zero donot exist. Let us write the MO configuration of the given species :

(*a*) $F_2^{2-}(e = 16) = (\sigma_{2s}^b)^2(\sigma_{2s}^*)^2(\sigma_{2p_x}^b)^2(\pi_{2p_y}^b)^2 = (\pi_{2p_z}^b)^2(\pi_{2p_y}^*)^2 = (\pi_{2p_z}^*)^2(\sigma_{2p_x}^*)^2$

$\therefore$ B.O. $= \dfrac{8-8}{2} =$ zero

(*b*) $Ar_2(e = 16) = (\sigma_{3s}^b)^2(\sigma_{3s}^*)^2(\sigma_{3p_x}^b)^2(\pi_{3p_y}^b)^2 = (\pi_{3p_z}^b)^2(\pi_{3p_y}^*)^2 = (\pi_{3p_z}^*)^2(\sigma_{3p_x}^*)^2$

$\therefore$ B.O. $= \dfrac{8-8}{2} =$ zero

(*c*) $He_2^+(e = 3) = (\sigma_{1s}^b)^2(\sigma_{1s}^*)^1$, $\quad\therefore$ B.O. $= \dfrac{2-1}{2} = 0.5$

(*d*) $H_2^+(e = 1) = (\sigma_{1s}^b)^1$, $\quad\therefore$ B.O. $= \dfrac{1}{2} = 0.5$

Thus, since F_2^{2-} and Ar_2 have B.O. = zero, these species donot exist (unstable).

16. MO configurations of the given species are :

(*a*) $O_2^+(e = 11) = (\sigma_{2s}^b)^2(\sigma_{2s}^*)^2(\sigma_{2p_x}^b)^2(\pi_{2p_y}^b)^2 = (\pi_{2p_z}^b)^2(\pi_{2p_y}^*)^1$

$\therefore$ B.O. $= \dfrac{8-3}{2} = 2.5$

(*b*) $O_2^-(e = 13) = (\sigma_{2s}^b)^2\,(\sigma_{2s}^*)^2\,(\sigma_{2p_x}^b)^2\,(\pi_{2p_y}^b)^2 = (\pi_{2p_z}^b)^2\,(\pi_{2p_y}^*)^2 = (\pi_{2p_z}^*)^1$

$\therefore$ B.O. $= \dfrac{8-5}{2} = 1.5$

(*c*) $O_2(e = 12) = (\sigma_{2s}^b)^2\,(\sigma_{2s}^*)^2\,(\sigma_{2p_x}^b)^2\,(\pi_{2p_y}^b)^2 = (\pi_{2p_z}^b)^2\,(\pi_{2p_y}^*)^1 = (\pi_{2p_z}^*)^1$

$\therefore$ B.O. $= \dfrac{8-4}{2} = 2$

Thus correct sequence of bond order is $O_2^+(=2.5) > O_2(=2) > O_2^-(=1.5)$.

17. In order to find out the presence of unpaired electrons in the given species let us write the MO configuration of the given species.

(*a*) $O_2^+(e = 11) : (\sigma_{2s}^b)^2\,(\sigma_{2s}^*)^2\,(\sigma_{2p_x}^b)^2\,(\pi_{2p_y}^b)^2 = (\pi_{2p_z}^b)^2\,(\pi_{2p_y}^*)^1$, $\quad\therefore n = 1$

(*b*) $O_2^-(e=13):(\sigma_{2s}^b)^2\,(\sigma_{2s}^*)^2\,(\sigma_{2p_x}^b)^2\,(\pi_{2p_y}^b)^2=(\pi_{2p_z}^b)^2\,(\pi_{2p_y}^*)^2=(\pi_{2pz}^*)^1,\quad \therefore\ n=1$

(*c*) $O_2(e=12):(\sigma_{2s}^b)^2\,(\sigma_{2s}^*)^2\,(\sigma_{2p_x}^b)^2(\sigma_{2p_y}^b)^2\;=(\pi_{2p_z}^b)^2\,(\pi_{2p_y}^*)^1=(\pi_{2p_z}^*)^1,\quad \therefore\ n=2$

(*d*) $O_2^{2-}(e=14):(\sigma_{2s}^b)^2\,(\sigma_{2s}^*)^2\,(\sigma_{2p_x}^b)^2\,(\pi_{2p_y}^b)^2\;=(\pi_{2p_z}^b)^2\,(\pi_{2p_y}^*)^2=(\pi_{2p_z}^*)^2,\quad \therefore\ n=0$

18. Since all the given three species have the same number of electrons in their respective valence shell (= 10), these species are isoelectronic and have the same MO configuration which is as given below :

$$CN^-, CO, NO^+(e=10)=(\sigma_{2s}^b)^2\,(\sigma_{2s}^*)^2\,(\pi_{2p_y}^b)^2=(\pi_{2p_z}^b)^2(\sigma_{2p_x}^b)^2$$

$$\therefore\quad B.O.=\frac{8-2}{2}=3$$

19. Valence-shell electronic configuration of O-atom is $2s^2 2p^4$ and MO configurations of O_2, O_2^+ and O_2^- are given below :

$$O_2(e=12):(\sigma_{2s}^b)^2\,(\sigma_{2s}^*)^2\,(\sigma_{2p_x}^b)^2(\pi_{2p_y}^b)^2=(\pi_{2p_z}^b)^2\,(\pi_{2p_y}^*)^1=(\pi_{2p_z}^*)^1$$

$$\therefore\quad B.O.=\frac{8-4}{2}=2$$

$$O_2^+(e=11):(\sigma_{2s}^b)^2\,(\sigma_{2s}^*)^2\,(\sigma_{2p_x}^b)^2\,(\pi_{2p_y}^b)^2=(\pi_{2p_z}^b)^2\,(\pi_{2p_y}^*)^1$$

$$\therefore\quad B.O.=\frac{8-3}{2}=2.5$$

$$O_2^-(e=13):(\sigma_{2s}^b)^2\,(\sigma_{2s}^*)^2\,(\sigma_{2p_x}^b)^2\,(\pi_{2p_y}^b)^2=(\pi_{2p_z}^b)^2\,(\pi_{2p_y}^*)^2=(\pi_{2p_z}^*)^1$$

$$\therefore\quad B.O.=\frac{8-5}{2}=1.5$$

Since B.O. is in the order $O_2^+(B.O.=2.5)>O_2(=2)>O_2^-(=1.5)$, stability of O_2^+, O_2 and O_2^- is also in the same order. Thus O_2^- is the least stable and hence least amount of energy will be needed to remove an electron from $\pi_{2p_z}^*$ antibonding molecular orbital. Now since $\pi_{2p_z}^*$ antibonding molecular orbital in O_2^- ion has higher energy than 2*p* orbital of O-atom, this antibonding molecular orbital is less stable than 2*p* orbital. Consequently lesser amount of energy will be needed to remove an electron form $\pi_{2p_z}^*$ antibonding molecular orbital than 2*p* orbital. Thus O_2^- ion has the least ionisation energy.

20. According to MO theory, when two atomic orbitals combine together, two molecular orbitals are formed. One of these molecular orbitals is a σ or π bonding molecular orbital and the other is σ or π antibonding molecular orbital.

12

Valence-Shell Electron Pair Repulsion (VSEPR) Theory : Shape of Molecules and Ions

MULTIPLE CHOICE QUESTIONS

1. Which of the following is correct order of repulsive interactions ?

(*a*) $(lp - lp) > (lp - bp) > (bp - bp)$

(*b*) $(lp - bp) > (lp - lp) > (bp - bp)$

(*c*) $(bp - lp) > (lp - bp) > (lp - lp)$

(*d*) The relation depends on the nature of the molecule or ion.

2. The shape of AX_3 (*lp*) molecule is :

(*a*) Triangular

(*b*) T-shape

(*c*) Linear

(*d*) Trigonal pyramidal

3. Out of BF_3, CH_4 and SF_4 the molecules having regular geometry are :

(*a*) BF_3 and CH_4 (*b*) CH_4 and SF_4

(*c*) SF_4 and BF_3 (*d*) BF_3, CH_4 and SF_4

4. A molecule XY_2 contains two σ bonds, two π bonds and one *lp* in the valence shell of X. The arrangement of *lp* as well as bonding electron pairs is :

(*a*) Square pyramidal

(*b*) Linear

(*c*) Trigonal planar

(*d*) Unpredicted

5. Which of the following pairs contains iso-structural species ?

(*a*) CH_3^- and CH_3^+

(*b*) NH_4^+ and NH_3

(*c*) SO_4^{2-} and BF_4^-

(*d*) NH_2^- and BeF_2

6. BCl_3 molecule is planar while NCl_3 is pyramidal, because :

(*a*) BCl_3 does not have lone pair (*lp*) on B but NCl_3 has because

(*b*) B – Cl bond is more polar than N – Cl bond

(*c*) N atom is smaller than B

(*d*) N – Cl bond is more covalent than B – Cl bond.

7. Which of the following molecules is linear ?

(*a*) C_2H_2 (*b*) CH_4

(*c*) H_2O (*d*) NH_3

8. Which of the molecules/ions has triangular pyramidal shape ?

(*a*) BF_3 (*b*) NO_3^-

(*c*) H_3O^+ (*d*) CO_3^{2-}

9. The octahedral shape is associated with :

(*a*) PF_5 (*b*) SF_4

(*c*) TeF_6 (*d*) ClF_3

10. XeF_4 has a shape of :

(*a*) Sphere

(*b*) Trigonal bipyramidal

(*c*) Tetrahedral

(*d*) Square planar

11. Which pair among the following is isostructural ?

(*a*) XeF_2, IF_2^- (*b*) NH_3, BF_3

(*c*) CO_3^{2-}, SO_3^{2-} (*d*) PCl_5, ICl_5

12. Among the following species, identify the iso-structural pairs : NF_3, NO_3^-, BF_3, H_3O^+ and HN_3

(*a*) $[NF_3, NO_3^-]$ and $[BF_3, H_3^+O]$

(*b*) $[NF_3, HN_3]$ and $[NO_3^-, BF_3]$

(*c*) $[NF_3, H_3^+O]$ and $[NO_3^-, BF_3]$

(*d*) $[NF_3, H_3^+O]$ and $[HN_3, BF_3]$

13. Two lone pairs of electrons and two bond pairs of electrons are present in :

(*a*) NH_3 (*b*) BF_3
(*c*) H_2O (*d*) CO_2

14. In which of the following bond angle is the maximum ?

(*a*) NH_3 (*b*) NH_4^+
(*c*) PCl_5 (*d*) SCl_2

15. In NO_3^- ion, the number of bond pairs and lone pairs of electrons are respectively :

(*a*) 2, 2 (*b*) 3, 1
(*c*) 1, 3 (*d*) 4, 8

16. In OF_2 molecule, number of bond pairs and lone pairs of electrons are respectively :

(*a*) 2, 6 (*b*) 2, 8
(*c*) 2, 10 (*d*) 2, 9

17. Which of the following have linear shape ?

(*a*) N_2O (*b*) SO_2
(*c*) I_3^+ (*d*) I_3^-

18. Which pair among the following has same bond order and same shape.

(*a*) CO_2, N_3^- (*b*) CO_2, O_3
(*c*) O_3, NO_2^- (*d*) N_3^-, NO_2^-

19. The structure of $CH_2 = C = CH_2$ is :

(*a*) Linear (*b*) Planar
(*c*) Non-planar (*d*) Dissymmetric

20. Which of the following molecules has unequal bond lengths ?

(*a*) NF_3 (*b*) BF_3
(*c*) PCl_5 (*d*) SF_6

ANSWERS

1. (*a*)	**2.** (*d*)	**3.** (*a*)	**4.** (*c*)	**5.** (*c*)	**6.** (*a*)	**7.** (*a*)	**8.** (*c*)
9. (*c*)	**10.** (*d*)	**11.** (*a*)	**12.** (*c*)	**13.** (*c*)	**14.** (*b*)	**15.** (*d*)	**16.** (*b*)
17. (*a*), (*d*)		**18.** (*c*)	**19.** (*b*)	**20.** (*c*)			

HINTS AND EXPLANATIONS

1. If the central atom in a molecule or ion is surrounded by both one or more bonding electron pairs (*bp*) as well as by one or more lone pairs of electrons (*lp*), then (*lp* – *lp*) repulsion is greater than (*lp* – *bp*) and similarly (*lp* – *bp*) repulsion is greater than (*bp* – *bp*) repulsion, *i.e.,* (*lp* – *lp*) > (*lp* – *bp*) and (*lp* – *bp*) > (*bp* – *bp*). These relations are based on the fact that since a *lp* is under the influence of only one central atom while a *bp* is under the influence of two atoms, the orbital containing *lp* is more spread out in space than the orbital containing *bp*. On combining the above two relations we get the following repulsive interaction relation

$$(lp - lp) > (lp - bp) > (bp - bp)$$

2. In $AX_3(lp)$, since the central atom (A) is surrounded by three σ-*bps* and one *lp* and the sum of σ-*bps* and *lps* is equal to 3 + 1 = 4, this molecules, due to the presence of one *lp* on the central A atom, does not have the expected tetrahedral shape. Rather it has distorted tetrahedral (*i.e.,* trigonal pyramidal) shape.

3. The Lewis structures of the given molecules are :

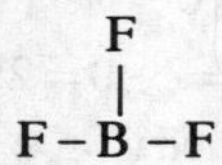 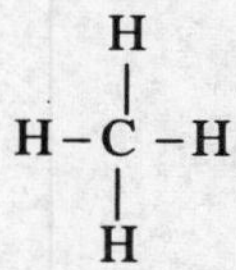 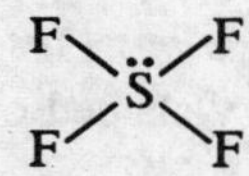

σ-*bps* = 3, *lp* = 0 σ-*bps* = 4, *lp* = 0 σ-*bps* = 4, *lp* = 1

Since BF_3 and CH_4 molecules donot contain any *lp*, these molecules have regular geometry. On the other hand, since SF_4 molecule has a *lp*, it has irregular geometry *viz* irregular tetrahedral or see-saw.

4. Since the sum of σ-*bps* and *lps* present in the valence-shell of the X (central atom) is 2 + 1 = 3, the arrangement of *lp* as well as *bps* round the central atom is trigonal planar. The presence of π bonds does not affect the arrangement of (*lps* + σ-*bps*).

5. The Lewis structures of each molecule/ion in the given pairs are :

(*a*) 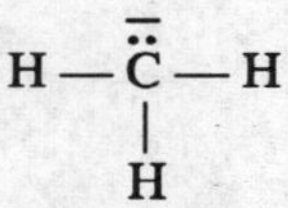 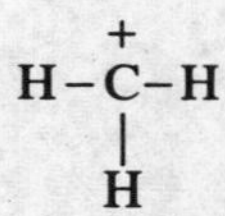

σ-*bps* = 3, *lps* = 1 σ-*bps* = 3, *lps* = 0

∴ σ-*bps* + *lps* = 3 + 1 = 4 ∴ σ-*bps* + *lps* = 3 + 0 = 3

∴ Trigonal pyramidal geometry ∴ Trigonal planar geometry

(*b*) 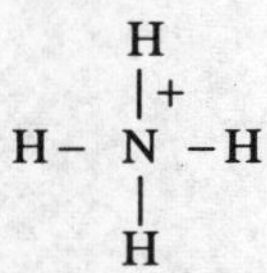 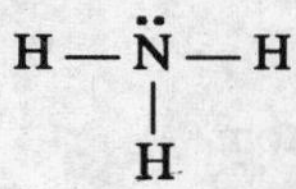

σ-*bps* = 4, *lps* = 0 σ-*bps* = 3, *lps* = 1

∴ σ-*bps* + *lps* = 4 + 0 = 4 ∴ σ-*bps* + *lps* = 3 + 1 = 4

∴ Tetrahedral geometry ∴ Trigonal pyramidal geometry

(*c*) 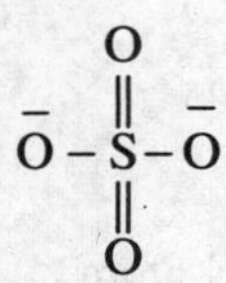 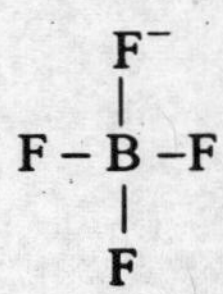

σ-*bps* = 4, *lps* = 0 σ-*bps* = 4, *lps* = 0

∴ σ-*bps* + *lps* = 4 + 0 = 4 ∴ σ-*bps* + *lps* = 4 + 0 = 4

∴ Tetrahedral geometry ∴ Tetrahedral geometry

(*d*) H — $\ddot{\text{N}}$ — H

σ-*bps* = 2, *lps* = 2

$\therefore$ σ-*bps* + *lps* = 2 + 2 = 4

$\therefore$ V-shaped or angular geometry

F – Be – F

σ-*bps* = 2, *lps* = 0

$\therefore$ σ-*bps* + *lps* = 2 + 0 = 2

$\therefore$ Linear geometry

Obviously SO_4^{2-} and BF_4^- ions given at (*c*) have the same geometry (tetrahedral geometry), *i.e.,* these ions are iso-structural ions.

6. The geometry of a molecule depends on the sum of σ-*bps* and *lps* residing on the central atom of the molecule. Now since B-atom (central atom) in BCl_3 molecule does not have any *lp* on it, BCl_3 molecule has the expected trigonal planar geometry. On the other hand, since N-atom in NCl_3 molecule has a *lp*, NCl_3 molecule has trigonal pyramidal geometry (not the expected tetrahedral geometry).

7. The structures of the given molecules are :

(*a*) H – C ≡ C – H

(σ-*bps* + *lps*) on each C-atom = 2 + 0 = 2

$\therefore$ Linear geometry

(*b*) H – C – H (with H above and H below C)

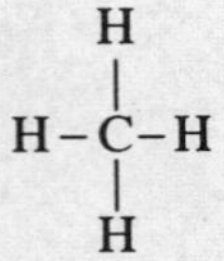

(σ-*bps* + *lps*) on C-atom = 4 + 0 = 4

$\therefore$ Tetrahedral geometry

(*c*) H — $\ddot{\underset{\cdot\cdot}{\text{O}}}$ — H

(σ-*bps* + *lps*) on O-atom = 2 + 2 = 4

$\therefore$ Angular geometry (due to the presence of two *lps* on O-atom)

(*d*) H — $\ddot{\text{N}}$ — H (with H below N)

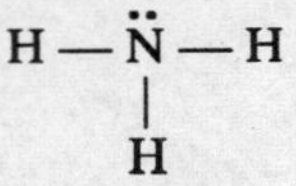

(σ-*bps* + *lps*) = 3 + 1 = 4

$\therefore$ Trigonal pyramidal geometry (due to the presence of one *lp* on N-atom)

8. The structures of the given molecules/ions are given below :

(*a*) F – B – F (with F above B)

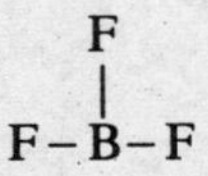

(σ-*bps* + *lps*) on B-atom = 3 + 0 = 3

$\therefore$ Trigonal planar geometry

(*b*) O = N → O (with O^- above N)

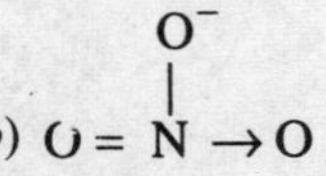

(σ-*bps* + *lps*) surrounding N-atom = 3 + 0 = 3

$\therefore$ Trigonal planar geometry

(*c*) 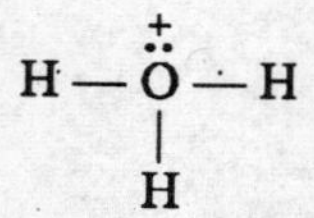

(σ-*bps* + *lps*) on O-atom = 3 + 1 = 4

∴ Triangular pyramidal geometry (due to the presence of one *lp* on O-atom)

(*d*) 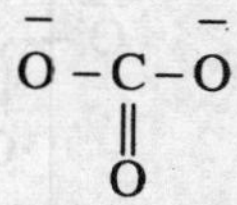

(σ-*bps* + *lps*) on C-atom = 3 + 0 = 3

∴ Trigonal planar geometry

9. The structures of the given species are given below :

(*a*) 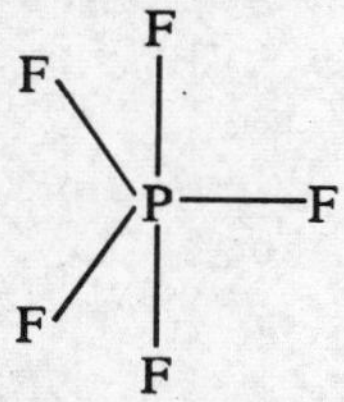

(σ-*bps* + *lps*) surrounding P-atom = 5 + 0 = 5

∴ Trigonal bipyramidal geometry.

(*b*) 

(σ-*bps* + *lps*) surrounding S-atom = 4 + 1 = 5

Sea-saw shape due to the presence of one *lp* on S-atom.

(*c*) 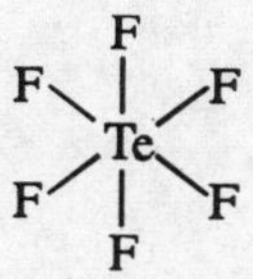

(σ-*bps* + *lps*) surrounding Te-atom = 6 + 0 = 6

∴ Octahedral geometry

(*d*) 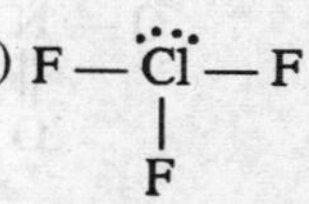

(σ-*bps* + *lps*) surrounding Cl-atom = 3 + 2 = 5

∴ T-shape of ClF_3 molecule due to the presence of two *lps* on Cl-atom.

10. Lewis structure of XeF_4 molecule is as shown in the margin. Since the sum of σ-*bps* and *lps* surrounding Xe-atom (central atom) is 4 + 2 = 6, XeF_4 molecule, due to the presence of two *lps* on Xe-atom, has square planar shape.

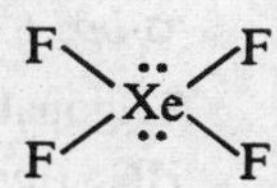

11. (*a*) 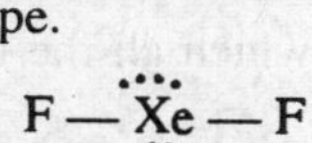

(σ-*bps* + *lps*) = 2 + 3 = 5

Due to the presence of three *lps* on Xe-atom, XeF_2 molecule has linear shape

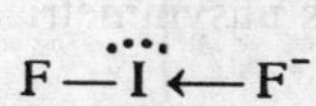

(σ-*bps* + *lps*) = 2 + 3 = 5

Linear shape

(*b*) 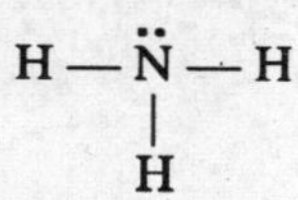

(σ-*bps* + *lps*) = 3 + 1 = 4

Due to the presence of one *lp* on N-atom, NH_3 molecule has trigonal pyramidal geometry

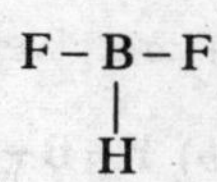

(σ-*bps* + *lps*) = 3 + 0 = 3

Trigonal planar geometry.

(*c*) 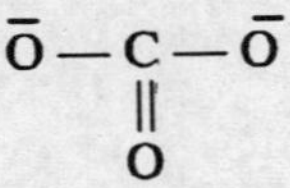

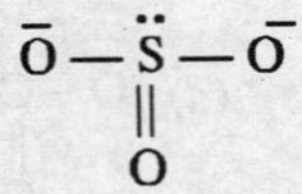

(σ-*bps* + *lps*) = 3 + 0 = 3

Trigonal planar geometry

(σ-*bps* + *lps*) = 3 + 1 = 4

Like NH_3, SO_3^{2-} ion also has trigonal pyramidal geometry.

(*d*) 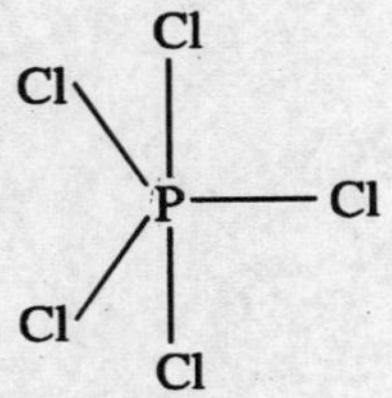

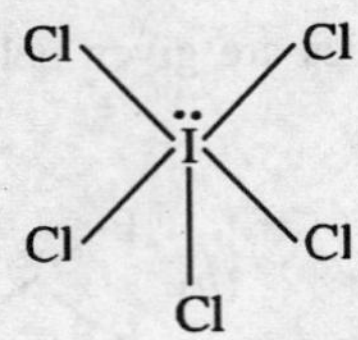

(σ – *bps* + *lps*) = 5 + 0 = 5

Trigonal bipyramidal geometry

(σ-*bps* + *lps*) = 5 + 1 = 6

Due to the presence of one *lp* on I-atom, ICl_5 molecule has square pyramidal shape.

Obviously, since XeF_2 and IF_2^- have the same geometry (linear), these are iso-structural.

12. The structures and geometries of the given species are given below :

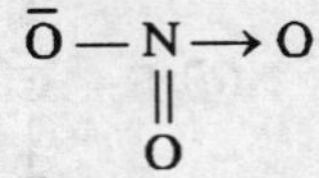

(σ-*bps* + *lps*) = 3 + 1 = 4

Trigonal pyramidal

(σ-*bps* + *lps*) = 3 + 0 = 3

Trigonal planar

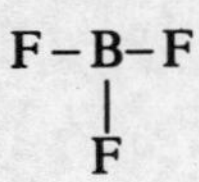

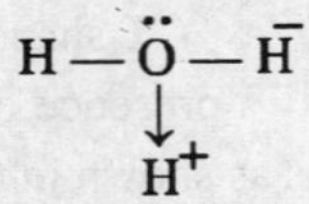

σ-*bps* + *lps* = 3 + 0 = 3

Trigonal planar

σ-*bps* + *lps* = 3 + 1 = 4

Trigonal pyramidal

HN_3 (hydrazoic acid) molecule has unsymmetrical bent structure in which all the three N-atoms lie in the same line but H-atom lies out of line 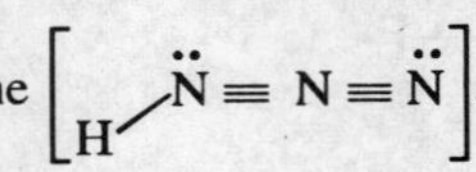. Azide ion (N_3^-) has linear symmetrical structure, since the central N-atom is surrounded by two σ-*bps* and no *lp* $[:\ddot{N}=N\equiv\ddot{N}]$.

13. Lewis structures of the given molecules are given below :

(*a*) 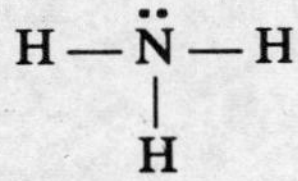

(*b*) 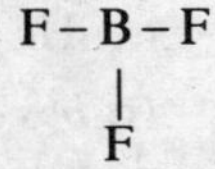

σ-*bps* = 3, *lp* = 1

σ-*bps* = 3, *lp* = 0

(*c*) H — $\ddot{\underset{\cdot\cdot}{O}}$ — H

σ-*bps* = 2, *lps* = 2

(*d*) O = C = O

σ-*bps* = 2, π-*bps* = 2, *lp* = 0

Thus it is only H_2O molecule whose central atom (O-atom) has two *bps* and two *lps*.

14. (*a*) 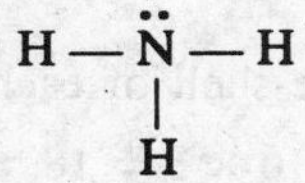

σ-*bps* + *lps* = 3 + 1 = 4

Due to the presence of one *lp*, NH_3 has trigonal pyramidal shape with HNH bond angle less than tetrahedral bond angle of 109.5°.

(*b*) 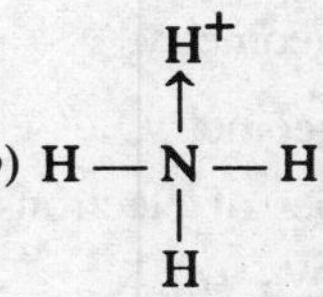

σ-*bps* + *lps* = 4 + 0 = 4

Tetrahedral shape with HNH bond angle equal to tetrahedral angle of 109.5°.

(*c*) 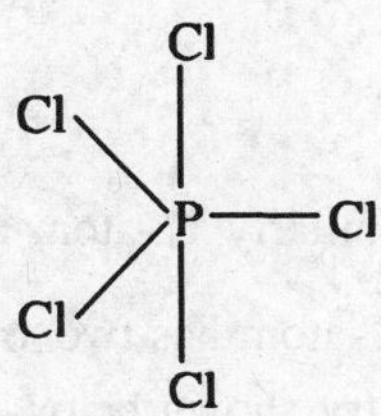

σ-*bps* + *lps* = 5 + 0 = 5

Trigonal bipyramidal shape with axial ClPCl bond angle = 90°

(*d*) Cl — $\ddot{\underset{\cdot\cdot}{S}}$ — Cl

σ-*bps* + *lps* = 2 + 2 = 4

Due to the presence of two *lps*, SCl_2 molecule has angular shape with ClSCl bod angle less than 109.5°.

Thus it is only NH_4^+ ion that has the maximum bond angle.

15. Lewis structure of NO_3^- ion is given in the margin. This structure shows that the number of *lps* on N-atom = zero, on the right hand side O-atom = 3, on the left hand side O-atom = 2 and on negatively-charged O-atom = 3. Thus total number of *lps* in NO_3^- ion = 0 + 3 + 2 + 3 = 8. Number of σ-*bps* and π-*bps* = 3 + 1 = 4.

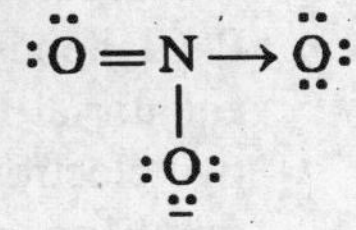

16. Lewis structure of OF_2 molecule as given in the margin shows that the number of *bps* = 2, number of *lps* on O-atom = 2 and number of *lps* on two F-atoms = 2 × 3 = 6. Thus total number of *lps* in OF_2 molecule = 2 + 6 = 8.

:$\ddot{\underset{\cdot\cdot}{F}}$ — $\ddot{\underset{\cdot\cdot}{O}}$ — $\ddot{\underset{\cdot\cdot}{F}}$:

17. Lewis structures of the given species are given below :

(*a*) :N ≡ N ⟶ $\ddot{\underset{\cdot\cdot}{O}}$:

(σ-*bps* + *lps*) surrounding the central N-atom = 2 + 0 = 2. Thus N_2O has linear shape

(*b*) O = $\ddot{S}$ = O

(σ-*bps* + *lps*) = 2 + 1 = 3. Due to the presence of one *lp* on S-atom, SO_2 has angular geometry.

(*c*) $I—\overset{+}{\ddot{\underset{\cdot\cdot}{I}}}—I$

(σ-*bps* + *lps*) = 2 + 2 = 4

Due to the presence of two *lps* on central I-atom, I_3^+ has angular geometry.

(*d*) $I—\ddot{\underset{\cdot\cdot}{I}}\leftarrow I$

(σ-*bps* + *lps*) = 2 + 3 = 5

Due to the presence of three *lps* shape of I_3^- ion is linear.

Thus N_2O and I_3^- have linear geometry.

18. Let us calculate the total number of electrons present in the valence-shell of each of the given species. CO_2 = 4 + 6 × 2 = 16, N_3^- = 5 × 3 + 1 = 16, O_3 = 6 × 3 = 18, NO_2^- = 5 + 6 × 2 + 1 = 18. We know that the species having the same number of electrons in their valence shell are called iso-electronic species. Such species have the same bond order and same shape. Thus the two species given at (*a*) (CO_2, N_3^-) have the same bond order and shape shape. Similarly the two species given at (*c*) (O_3 and NO_2^-) also have the same bond order and same shape.

19. The structue of $H_2C = C = CH_2$ can be written as :

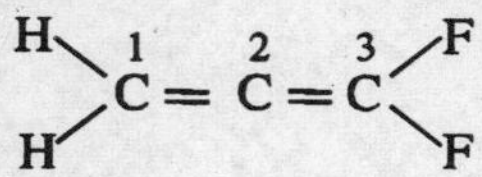

Since $\overset{1}{C}$-atom has three σ-*bps*, its structure is trigonal planar. Similarly $\overset{3}{C}$-atom also has three σ -*bps*. This atom, therefore, has trigonal geometry. Since $\overset{2}{C}$-atom has two σ -bps, it has linear geometry. Thus the given molecule has trigonal geometry shown below.

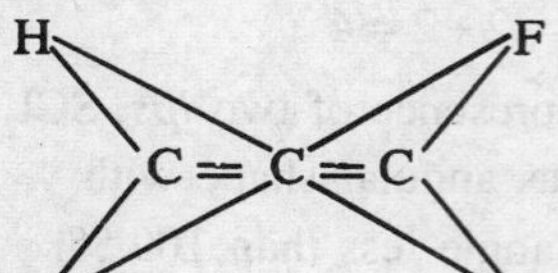

20. Since P-atom (central atom) in PCl_5 is surrounded by five σ -*bps* and no *lp*, PCl_5 molecule has trigonal bipyramidal geometry with two types of P–Cl covalent bonds. Two are axial P–Cl bonds and three are equatorial P–Cl bonds. Axial P–Cl bonds are larger than the equatorial P–Cl bonds. The reason is that the three equatorial P–Cl bonds are repelled by two electron pairs while the two axial P–Cl bonds are repelled by three electron pairs. Thus repulsion exerted on the axial bonds electron pairs is more and hence the axial bond lengths are larger.

13

Hybridisation and Its Different Types

MULTIPLE CHOICE QUESTIONS

1. Hybrid orbitals formed from *s* and *p* orbitals can contribute to :
   (*a*) σ-bond only
   (*b*) π-bond only
   (*c*) either σ and π bond
   (*d*) σ and π bonds both.
2. Increasing order (lowest first) of size of hybrid orbitals is :
   (*a*) sp, sp^2, sp^3 (*b*) sp^3, sp^2, sp
   (*c*) sp^2, sp^3, sp (*d*) sp^2, sp, sp^3
3. Directional characters of *sp*, sp^2 and sp^3 hybrid orbitals are related as :
   (*a*) $sp < sp^2 < sp^3$ (*b*) $sp^2 < sp < sp^3$
   (*c*) $sp^3 < sp < sp^2$ (*d*) $sp > sp^2 > sp^3$
4. Strength of the bond formed by the overlap between two *sp*, sp^2 or sp^3 hybrid orbitals is in the order :
   (*a*) $sp^3 - sp^3 > sp^2 - sp^2 > sp - sp$
   (*b*) $sp^2 - sp^2 > sp^3 - sp^3 > sp - sp$
   (*c*) $sp - sp > sp^2 - sp^2 > sp^3 - sp^3$
   (*d*) $sp^3 - sp^3 = sp^2 - sp^2 = sp - sp$
5. Hybridisation involves :
   (*a*) separation of atomic orbitals
   (*b*) overlapping of atomic orbitals
   (*c*) mixing of atomic orbitals
   (*d*) all of them
6. The equilateral shape has :
   (*a*) *sp* hybrid orbitals
   (*b*) sp^2 hybrid orbitals
   (*c*) sp^3 hybrid orbitals
   (*d*) sp^3d hybrid orbitals
7. sp^3d^2 hybrid orbitals are :
   (*a*) linear
   (*b*) pentagonal bipyramidal
   (*c*) hexagonal
   (*d*) octahedral
8. Which of the following types of hybridisation leads to two dimensional geometry of bonds around the atoms ?
   (*a*) *sp* (*b*) sp^2
   (*c*) sp^3 (*d*) None of these
9. Which of the following hybridisation results in non-planar orbitals ?
   (*a*) sp^3 (*b*) dsp^2
   (*c*) sp^2 (*d*) dsp^3
10. *d*-orbital involved in sp^3d hybridisation is :
    (*a*) $d_{x^2-y^2}$ (*b*) d_{xy}
    (*c*) d_{z^2} (*d*) d_{zx}
11. *d*-orbital involved in dsp^2 hybridisation is :
    (*a*) $d_{x^2-y^2}$ (*b*) d_{xy}
    (*c*) d_{z^2} (*d*) d_{zx}
12. Square planar complex is formed by the hybridisation of which atomic orbitals ?
    (*a*) s, p_x, p_y, p_z (*b*) $s, p_x, p_y, d_{x^2-y^2}$
    (*c*) s, p_x, p_y, d_{z^2} (*d*) s, p_x, p_z, d_{xy}
13. Which of the following statement is not correct ?
    (*a*) Hybridisation is the mixing of atomic orbitals and their simultaneous combination into molecular orbitals
    (*b*) sp^2 hybrid orbitals are formed from two *p*-orbitals and one *s*-orbital

(c) dsp^2 hybrid orbitals are all at 90° to one another

(d) d^2sp^3 hybrid orbitals are directed towards the corners of a regular octahedron.

14. Linear combination of two hybridised orbitals each having one electron and belonging to two different atoms leads to the formation of a :

(a) sigma bond (b) *pi* bond

(c) double bond (d) co-ordinate bond.

15. A molecular or ion is square planar with no lone pair of electrons. What type of hybridisation is associated with it ?

(a) sp^3d (b) sp^3d^2

(c) dsp^3 (d) dsp^2

16. AsF_5 molecule is trigonal bipyramidal. The orbitals used by As for hybridisation are :

(a) $d_{z^2}, s, p_x, p_y, p_z$

(b) $d_{x^2-y^2}, s, p_x, p_y, p_z$

(c) s, p_x, p_y, p_z, d_{xz}

(d) None

17. The bond angle around the atom which uses sp^2 hybrid orbitals is

(a) 120° (b) 180°

(c) 105 (d) 109° 28′.

18. Most efficient overlapping is :

(a) $sp^2 - sp^2$ (b) $s - s$

(c) $sp^2 - sp^3$ (d) $sp - sp$

19. On hybridisation of one *s* and one *p* orbital we get :

(a) two mutually perpendicular orbitals

(b) two orbitals at 180°

(c) four orbitals directed tetrahedrally

(d) three orbitals in a plane.

20. In which of the following the hybrid orbitals of the central atom have the same *s*-character ?

(a) CH_4 (b) $Ni(CO)_4$

(c) $[NiCl_4]^{2-}$ (d) $[Ni(CN)_4]^{2-}$

ANSWERS

1. (a)	**2.** (a)	**3.** (a)	**4.** (a)	**5.** (c)	**6.** (b)	**7.** (d)	**8.** (b)
9. (a), (d)		**10.** (c)	**11.** (a)	**12.** (b)	**13.** (c)	**14.** (a)	**15.** (d)
16. (a)	**17.** (a)	**18.** (c)	**19.** (b)	**20.** (a), (b), (c).			

HINTS AND EXPLANATIONS

1. Hybrid orbitals formed from *s* and *p* orbital are called *sp* hybrid orbitals. The two *sp* hybrid orbitals formed are directed in opposite direction along the nuclear axis and hence overlap along the nuclear axis (axial overlap) and consequently form σ-bond.

2. The size of a hybrid orbital depends on the % of *s*-character present in hybrid orbital. Greater is the % of *s*-character, smaller is the size of the hybrid orbital. Now since % of *s*-character in *sp*, sp^2 and sp^3 hybrid orbitals decreases as *sp* (= % of *s*-character $= \frac{100}{2} = 50\%$) $> sp^2 \left(= \frac{100}{3} = 33.3\%\right) > sp^3 \left(= \frac{100}{4} = 25\%\right)$, the size of these hybrid orbitals increases as $sp < sp^2 < sp^3$.

3. Greater is the % of *p*-character in *sp*, sp^2 and sp^3 hybrid orbitals, greater is the directional character of the hybrid orbitals. Thus since the % of *p*-character in *sp*, sp^2 and sp^3 hybrid orbitals increases as *sp* (% of *p*-character $= 100 - \frac{100}{2} = 50\%$) $< sp^2$

$\left(=100-\frac{100}{3}=66.6\%\right) < sp^3\left(100-\frac{100}{4}=75\%\right)$, ∴ directional character of these hybrid orbitals also increases in the same direction, *i.e.*, directional character of sp, sp^2 and sp^3 hybrid orbitals increases as $sp < sp^2 < sp^3$.

4. Since directional character of sp, sp^2 and sp^3 hybrid orbitals increases as $sp < sp^2 < sp^3$, the strength of bond formed by the overlap between two sp, two sp^2 or two sp^3 hybrid orbitals increases as $(sp - sp) < (sp^2 - sp^2) < (sp^3 - sp^3)$ or $(sp^3 - sp^3) > (sp^2 - sp^2) > (sp - sp)$.

5. Hybridisation is a process in which atomic orbitals on the same atom mix together so that their energies become equal due to the formation of hybrid orbitals.

6. The lobes of the three sp^2 hybrid orbitals are directed towards the corners of an equilateral triangle and are, therefore, inclined at an angle of 120° with each other.

7. sp^3d^2 hybrid orbitals are six in number. The six lobes of these hybrid orbitals are directed towards the six corners of a regular octahedron. Thus sp^3d^2 hybrid orbitals are octahedral.

8. sp^2 hybridisation gives trigonal planar geometry which is two dimensional geometry.

9. sp^3, dsp^2, sp^2 and dsp^3 hybridisations give respectively tetrahedral, square planar, trigonal planar and trigonal bipyramidal geometries. Out of these geometries, only tetrahedral (given by sp^3 hybridisation) and trigonal bipyramidal (given by dsp^3 hybridisation) geometries are non-planar geometries while square planar and trigonal planar are planar geometries.

10. sp^3d hybridisation gives five hybrid orbitals out of which three are planar triangular (equilateral) hybrid orbitals and two are axial hybrid orbitals. Thus the three planar equilateral hybrid orbitals may be assumed to have been formed by the mixing of one s and two p orbitals (sp^2 hybrid orbitals) and the remaining two axial hybrid orbitals may be assumed to have resulted by the mixing of one p and one d_{z^2} orbital (pd_{z^2} hybrid orbitals)

11. Since the four lobes of $d_{x^2-y^2}$ orbital lie along x and y axis, two p orbitals can combine with this orbital along x and y axes to form four dsp^2 hybrid orbitals.

12. Square planar complex is formed by dsp^2 hybridisation. This hybridisation requires the mixing of $(n-1)\,d_{x^2-y^2}$, ns, np_x and np_y orbitals.

13. dsp^2 hybridisation gives four hybrid orbitals which are lying in a square plane since their lobes are directed towards the four corners of a square. The angle between the two adjacent hybrid orbitals is 90° and that between the two opposite hybrid orbitals is 180°.

14. The linear overlap between two hybridised orbitals, each having one electron and belonging to two different atoms, form only σ-bonds and not π bonds.

15. If a molecule or an ion has no lone pair and has square planar geometry, this geometry results from dsp^2 hybridisation of the central atom or ion. For example $[Ni(CN)_4]^{2-}$ ion is obtained by dsp^2 hybridisation of Ni^{2+} ($Ni^{2+} = 3d^2\ 3d^2\ 3d^2\ 3d^2\ 3d^0$, $4s^\circ$, $4p^\circ\ 4p^\circ\ 4p^\circ$). $[Ni(CN)_4]^{2-}$ ion has square planar geometry due to dsp^2 hybridisation.

16. Since AsF_5 molecule has trigonal bipyramidal geometry, As atom (central atom) is sp^3d hybridised. Five hybrid orbitals obtained from sp^3d hybridisation contain three planar triangular (equilateral) hybrid orbitals and two axial hybrid orbitals. Thus the three planar equilateral hybrid orbitals may be assumed to have been formed by the combination of one s and two p orbitals (sp^2) and the remaining two axial hybrid orbitals may be assumed to have resulted from the mixing of one p and one d_{z^2} orbital (pd_{z^2}).

17. sp^2 hybrid orbitals are three in number. The lobes of these orbitals are directed towards the three corners of an equilateral triangle and hence the angle between two adjacent hybrid orbitals is of 120°.

18. The strength of a bond formed by the overlap of two atomic orbitals or two hybrid orbitals decreases with the decrease of the directional character of orbitals. Since directional character of s, p, sp, sp^2 and sp^3 orbitals decreases as $sp^3 > sp^2 > sp > p > s$, the strength of a bond formed by $sp^3 - sp^3$, $sp^2 - sp^2$, $sp - sp$, $p - p$, $s - s$ and $s - p$ overlaps decreases as $sp^3 - sp^3 > sp^2 - sp^2 > sp - sp > p - p > s - p > s - s$. Thus the bond formed by $sp^3 - sp^3$ overlap would be the strongest bond.

19. The hybridisation between one s and one p orbital gives two linear sp hybrid orbitals. The two lobes of these orbitals are inclined at an angle of 180°.

20. CH_4, $Ni(CO)_4$ and $[NiCl_4]^{2-}$ are tetrahedral species since their central atoms [C in CH_4, Ni in $Ni(CO)_4$ and Ni^{2+} in $(NiCl_4)^{2-}$] are sp^3 hybridised. On the other hand $[Ni(CN)_4]^{2-}$ ion is square planar, since Ni^{2+} ion is dsp^2 hybridised. Now since the type of hybridisation undergone by the central atoms in CH_4, $Ni(CO)_4$ and $[NiCl_4]^{2-}$ is the same, sp^3 hybrid orbitals formed in these species has the same s-character.

14

Structure and Shape of Molecules/Ions Containing only Sigma Bonds

MULTIPLE CHOICE QUESTIONS

1. The hybrid state of S in SO_3 is similar to that of carbon in :

(a) C_2H_2 (b) C_2H_4
(c) CH_4 (d) CO_2

2. In OF_2 the hybridisation of O-atom is :

(a) sp (b) sp^2
(c) sp^3 (d) None of these

3. Which of the following molecules contains an sp^2 hybridised carbon atom ?

(a) C_2H_5CHO (b) $CH_3C \equiv CH$
(c) CO_2 (d) $CH_3CH_2CH_3$

4. Carbon atom in CO_2 molecule is :

(a) sp hybridised (b) sp^3d hybridised
(c) sp^2 hybridised (d) sp^3 hybridised

5. The central atom does not assume sp^2 hybridisation in :

(a) PCl_3 (b) SO_3
(c) BF_3 (d) NO_3^-

6. The compound in which $\overset{*}{C}$-atom uses sp^3 hybrid orbitals for bond formation is :

(a) $H\overset{*}{C}OOH$ (b) $(NH_2)_2\overset{*}{C}O$
(c) $(CH_3)_2\overset{*}{C}OH$ (d) $CH_3\overset{*}{C}HO$

7. The type of hybrid orbitals used by Cl-atom in ClO_3^- ion is :

(a) sp^3 (b) sp^2
(c) sp (d) None of these

8. The hybridisation of atomic orbitals of nitrogen in NO_2^+, NO_3^- and NH_4^+ are respectively :

(a) sp, sp^3, sp^2 (b) sp, sp^2, sp^3
(c) sp^2, sp, sp^3 (d) sp^2, sp^3, sp^3.

9. The correct order of hybridisation of the central atom in NH_3, $[PtCl_4]^{2-}$, PCl_5 and BCl_3 is respectively :

(a) dsp^2, dsp^3, sp^2, sp^3
(b) sp^3, dsp^2, dsp^3, sp^2
(c) dsp^2, sp^2, sp^3, dsp^3
(d) dsp^2, sp^3, sp^2, dsp^3

10. Among the following compounds the one that is polar and has central atom with sp^2 hybridisation is :

(a) H_2CO_3 (b) SiF_4
(c) BF_3 (d) $HClO_2$

11. The maximum number of C-atoms arranged linearly in the molecule, $CH_3-C\equiv C-CH-CH_2$ is :

(a) 5 (b) 4
(c) 3 (d) 2

12. In the compound, $CH_2 = CH - CH_2 - CH_2 - C \equiv CH$, the $\overset{2}{C}-\overset{3}{C}$ bond is of the type :

(a) $sp - sp^2$ (b) $sp^3 - sp^3$
(c) $sp - sp^3$ (d) $sp^2 - sp^3$

13. Linear structure is assumed by :

(a) Isocyanate ion (b) Azide ion
(c) NO_2^+ (d) SO_2

14. Which of the following compounds has sp^2 hybridisation ?

(*a*) CO_2 (*b*) SO_2
(*c*) N_2O (*d*) CO

15. Specify the coordination geometry around and hybridisation of N and B atoms in a 1 : 1 complex of BF_3 and NH_3.

(*a*) N : tetrahedral, sp^3; B : tetrahedral, sp^3
(*b*) N : pyramidal, sp^3; B : pyramidal, sp^3
(*c*) N : pyramidal, sp^3; B : planar, sp^2
(*d*) N : pyramidal, sp^3; B : tetrahedral, sp^3

16. The molecule that has linear structure is :

(*a*) CO_2 (*b*) NO_2
(*c*) SO_2 (*d*) SiO_2

17. The C–H bond distance is the longest in :

(*a*) C_2H_2 (*b*) C_2H_4
(*c*) C_2H_6 (*d*) $C_2H_2Br_2$

18. The hybridisation of carbons involved in C–C single bond in $CH\equiv C-CH=CH_2$ is :

(*a*) $sp^3 - sp^2$ (*b*) $sp^3 - sp^3$
(*c*) $sp - sp^2$ (*d*) $sp - sp^3$

19. Which of the following has a bond formed by the overlap of $sp - sp^3$ hybrid orbitals ?

(*a*) $CH_3-C\equiv CH$
(*b*) $CH_3-CH=CH-CH_3$
(*c*) $CH_3=CH-CH=CH_2$
(*d*) $HC\equiv CH$

20. CO_2 is iso-structural with :

(*a*) $HgCl_2$ (*b*) $SnCl_2$
(*c*) C_2H_2 (*d*) NO_2

ANSWERS

1. (*b*) **2.** (*c*) **3.** (*a*) **4.** (*a*) **5.** (*a*) **6.** (*c*) **7.** (*a*) **8.** (*b*)
9. (*b*) **10.** (*a*) **11.** (*b*) **12.** (*d*) **13.** (*a*), (*b*), (*c*) **14.** (*b*) **15.** (*a*)
16. (*a*) **17.** (*c*) **18.** (*c*) **19.** (*a*) **20.** (*a*), (*c*)

HINTS AND EXPLANATIONS

1. Lewis structures of SO_3, C_2H_2, C_2H_4, CH_4 and CO_2 molecules and the type of hybridisation undergone by the central atom are shown below :

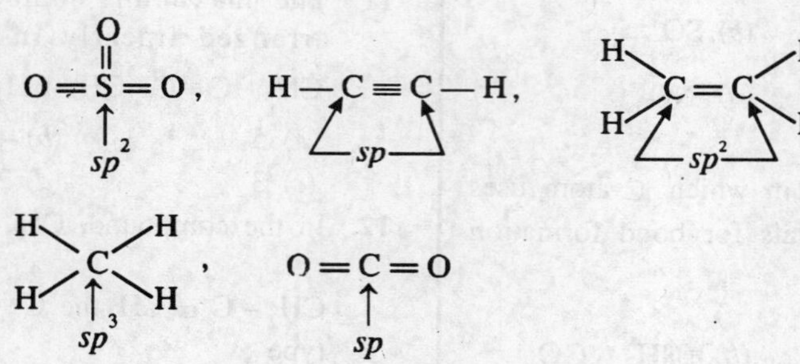

Obviously SO_2 and C_2H_4 have similar hybridisation.

2. Lewis structure of OF_2 molecule is $F - \ddot{\underset{\cdot\cdot}{O}} - F$ which shows that since O-atom (central atom) is surrounded by $(\sigma\text{-}bps + lp) = 2 + 2 = 4$, O-atom is sp^3 hybridised.

3. The structures of the given molecules and the type of hybridisation undergone by each C-atom are shown below :

(*a*) 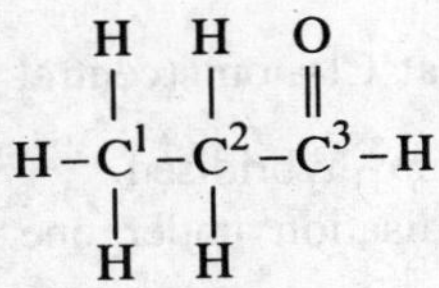

$C^1 = sp^3$, $C^2 = sp^3$, $C^3 = sp^2$

(*b*) 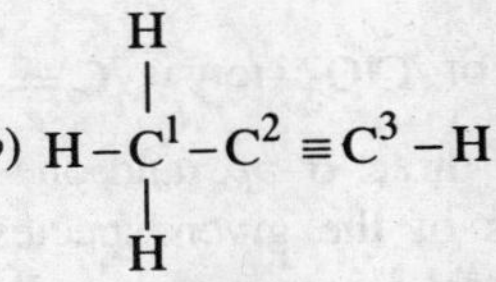

$C^1 = sp^3$, $C^2 = sp$, $C^3 = sp$

(*c*) O = C = O
↑
sp

(*d*) 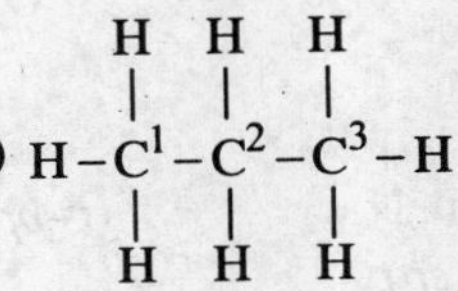

$C^1 = sp^3$, $C^2 = sp^3$, $C^3 = sp^3$

Obviously $\overset{3}{C}$-atom is C_2H_5CHO is sp^2 hybridised.

4. Lewis structure of CO_2 is O = C = O which shows that C-atom (central atom) is surrounded by (σ-*bps* + *lps*) = 2 + 0 = 2. C-atom is, therefore, *sp* hybridised in CO_2 molecule.

5. Lewis structures of the given molecules and the type of hybridisation undergone by their respective central atom have been shown below.

(*a*) 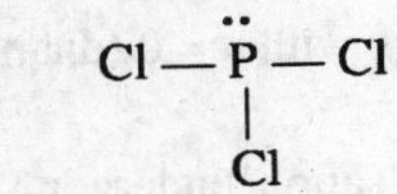

(σ-*bps* + *lps*) = 3 + 1 = 4
(sp^3)

(*b*) 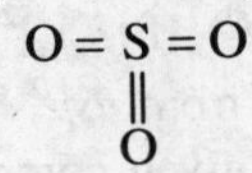

(σ-*bps* + *lps*) = 3 + 0 = 3
(sp^2)

(*c*) 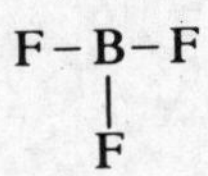

(σ-*bps* + *lps*) = 3 + 0 = 3
(sp^2)

(*d*) 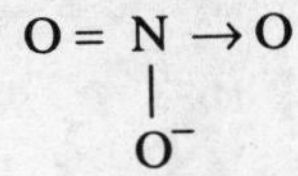

(σ-*bps* + *lps*) = 3 + 0 = 3
(sp^2)

6. Structures of the given molecules and the type of hybridisation undergone by $\overset{*}{C}$-atom are shown below.

(*a*) 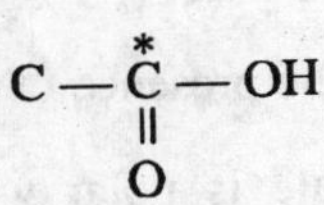

(σ-*bps* + *lps*) = 3 + 0 = 3
(sp^2)

(*b*) 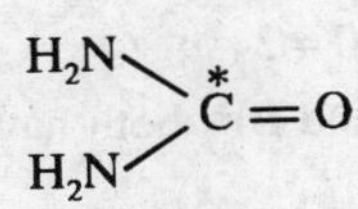

(σ-*bps* + *lps*) = 3 + 0 = 3
(sp^2)

(*c*) 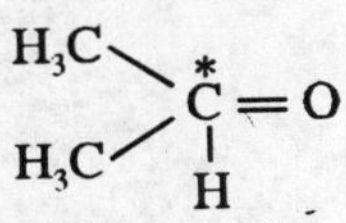

(σ-*bps* + *lps*) = 4 + 0 = 4
(sp^3)

(*d*) 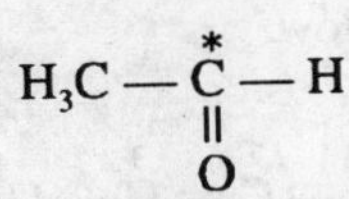

(σ-*bps* + *lps*) = 3 + 0 = 3
(sp^2)

7. The structure of ClO_3^- ion is 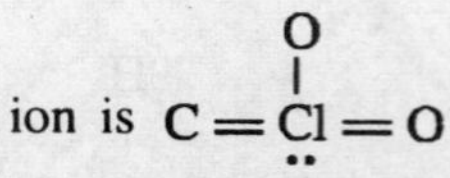 which shows that Cl-atom (central atom) is surrounded by three σ-*bps* and one *lp* and hence Cl-atom is sp^3 hybridised.

8. The structures of the given species and the type of hybridisation undergone by their respective central atoms are given below.

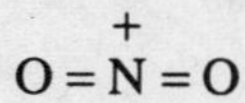

(σ-*bps* + *lps*) on $\overset{+}{N}$ = 2 + 0 = 2 (*sp*)

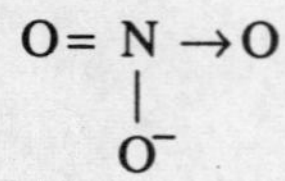

(σ-*bps* + *lps*) on N-atom = 3 + 0 = 3 (sp^2)

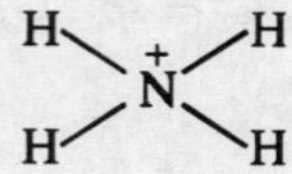

(σ-*bps* + *lps*) on $\overset{+}{N}$ = 4 + 0 = 4 (sp^3)

9. The structures of NH_3, PCl_5 and BCl_3 molecules are given below :

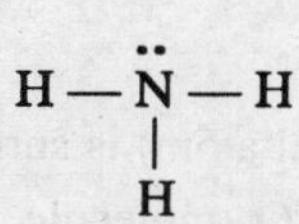

σ-*bps* + *lps* = 3 + 1 = 4 (sp^3)

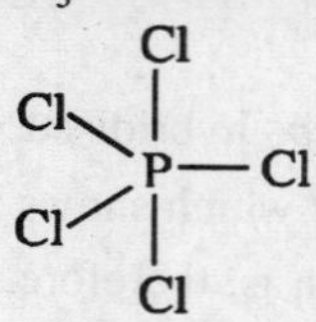

σ-*bps* + *lps* = 5 + 0 = 5 (sp^3d)

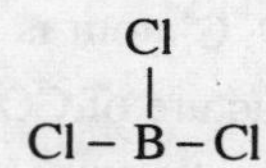

σ-*bps* + *lps* = 3 + 0 = 3(sp^2)

In $[Pt\,Cl_4]^{2-}$, Pt is present as Pt^{2+} whose valence shell electronic configuration is $5d^8\,6s^0\,6p^0$ (Pt = $5d^8\,6s^2$, $Pt^{2+} = 5d^8$). $[Pt\,Cl_4]^{2-}$ ion has square planar geometry with $n = 0$ (diamagnetic). This geometry results from dsp^2 hybridisation of Pt^{2+} ion.

10. The structures of the given compounds and the type of hybridisation undergone by their respective central atom are given below :

(*a*) 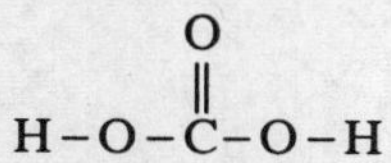

(σ-*bps* + *lps*) = 3 + 0 = 3 (sp^2)

(*b*) 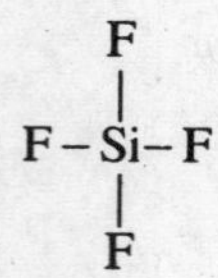

(σ-*bps* + *lps*) = 4 + 0 = 4 (sp^3)

(*c*) 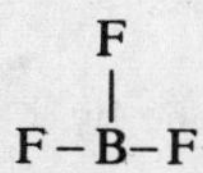

(σ-*bps* + *lps*) = 3 + 0 = 3 (sp^2)

(*d*) 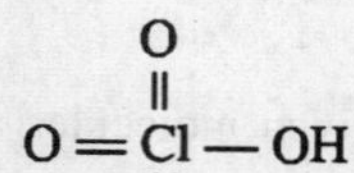

(σ-*bps* + *lps*) = 3 + 1 = 4 (sp^3)

Although H_2CO_3 and BF_3 both have sp^2 hybridisation, yet due to the presence of ionisable H-atom in H_2CO_3 ($H_2CO_3 \rightleftharpoons 2H^+ + CO_3^{2-}$), H_2CO_3 molecule is polar while BF_3 molecule is non polar.

11. The structure of the given molecule and the type of hybridisation undergone by each C-atom are shown below.

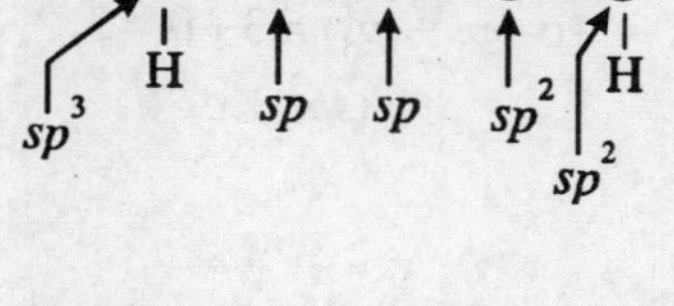

This structure shows that since $\overset{2}{C}$ atom is *sp* hybridised, $\overset{1}{C}, \overset{2}{C}$ and $\overset{3}{C}$ atoms are arranged linearly in this molecule. Similarly, since $\overset{3}{C}$ atom is also *sp* hybridised, $\overset{2}{C}, \overset{3}{C}$ and $\overset{4}{C}$ atoms are also arranged linearly in the molecule. Thus $\overset{1}{C}, \overset{2}{C}, \overset{3}{C}$ and $\overset{4}{C}$ atoms are arranged linearly. Consequently, the maximum number of C-atoms arranged linearly is 4.

12. When both double and triple bonds are present in a molecule and if there is choice, the numbering of C-atoms is done from that C-atom which is nearer to the double bond. Thus in the given molecule C-atoms are numbered as :

$$\overset{1}{C}H_2 = \overset{2}{C}H - \overset{3}{C}H_2 - \overset{4}{C}H_2 - \overset{5}{C} \equiv \overset{6}{C}H$$

The structure of the molecule can be written as :

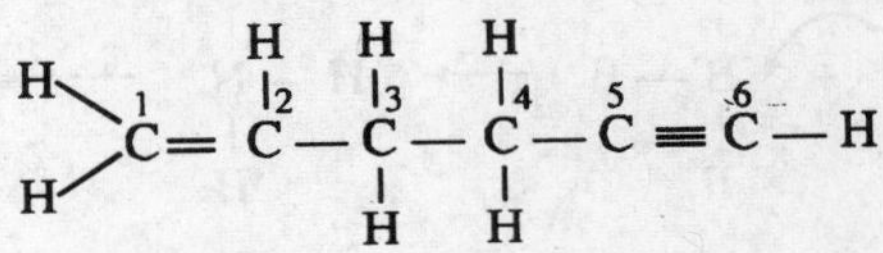

Obviously $\overset{2}{C}$ and $\overset{3}{C}$ atoms are sp^2 and sp^3 hybridised. Thus $\overset{2}{C}-\overset{3}{C}$ bond is obtained by $sp^2 - sp^3$ overlap.

13. The octet Lewis structure of the given species and the type of hybridisation involved in them are given below :

(*a*) $:\overset{-}{\underset{..}{N}} = C = \ddot{\underset{..}{O}}:$
Isocyanate ion

(σ-*bps* + *lps*) on C-atom = 2 + 0 = 2,

$\therefore$ C = *sp* (Linear geometry)

(*b*) $:\overset{-}{\underset{..}{N}} = \overset{+}{N} = \overset{-}{\underset{..}{N}}:$
Azide ion

(σ-*bps* + *lps*) on $\overset{+}{N}$ = 2 + 0 = 2,

$\therefore$ $\overset{+}{N}$ = *sp* (Linear geometry)

(*c*) $:\ddot{O} = \overset{+}{N} = \ddot{O}:$

(σ-*bps* + *lps*) on $\overset{+}{N}$ = 2 + 0 = 2,

$\therefore$ $\overset{+}{N}$ = *sp* (Linear geometry)

(*d*) $O = \ddot{S} = O$

(σ-*bps* + *lps*) on *S* = 2 + 1 = 3,

$\therefore$ S = sp^2 (Angular geometry)

Obviously, isocyanate ($\overset{-}{N}CO$), azide (N_3^-) and NO_2^+ ions have linear geometry.

14. (*a*) O = C = O

(σ-*bps* + *lps*) on C-atom
= 2 + 0 = 2, $\therefore$ C = *sp*

(*b*) $O = \ddot{S} = O$

(σ-*bps* + *lps*) on S-atom = 2 + 1 = 3

$\therefore$ S = sp^2

(*c*) $:\overset{-}{\ddot{N}} = \overset{+}{N} = \ddot{O}:$

(σ-*bps* + *lps*) on $\overset{+}{N}$

= 2 + 0 = 2, $\therefore$ $\overset{+}{N}$ = *sp*

(*d*) Lewis structure of CO molecule is $:\text{O}\equiv\text{O}:$ in which C and O both are *sp* hybridised. (C–O) σ bond is formed by (*sp* – *sp*) overlap.

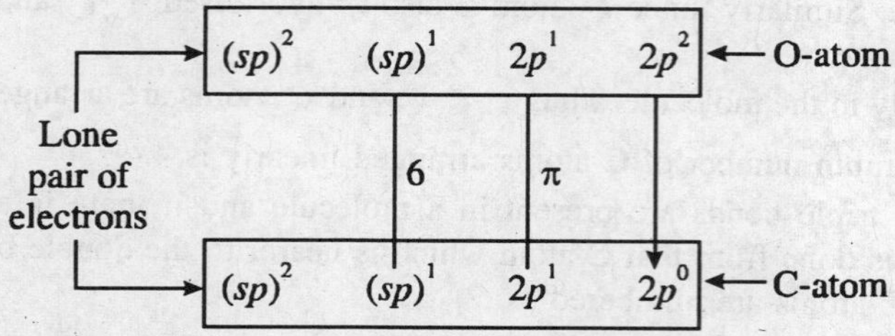

15. The formation of 1 : 1 complex obtained by the combination of BF_3 (Lewis acid) and NH_3 (Lewis base) can be represented as :

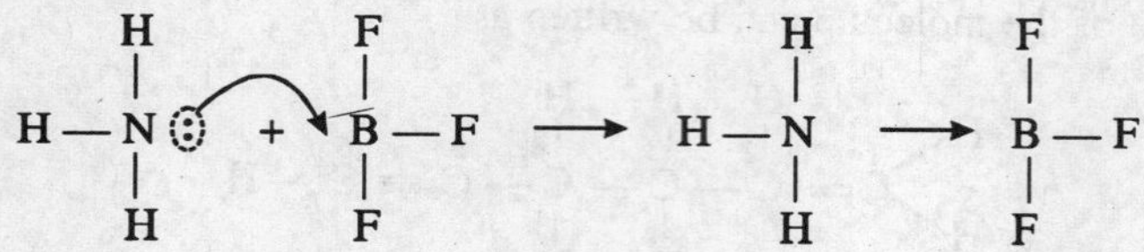

1 : 1 complex

The structure of 1 : 1 complex shows that N-atom is surrounded by four σ-*bps* and hence is sp^3 hybridised and has tetrahedral arrangement for the four surrounding atoms. Similarly B-atom is also sp^3 hybridised and the arrangement of four atoms (3F + N) round B-atom is tetrahedral.

16. Lewis structures of CO_2, NO_2 and SO_2 molecules and the type of hybridisation undergone by C, N and S atoms (central atoms) are given below.

(*a*) $:\ddot{\text{O}}=\text{C}=\ddot{\text{O}}:$

(σ-*bps* + *lps*) *C*-atom = 2 + 0 = 2

∴ C = *sp* (Linear shape)

(*b*) $:\overset{-}{\ddot{\text{O}}}-\overset{+}{\dot{\text{N}}}=\ddot{\text{O}}:$

Odd-electron molecule

(σ-*bps* + *lps*) around $\overset{+}{\text{N}} = 2 + \frac{1}{2} = 2.5$

∴ $\overset{+}{\text{N}} = sp^2$ (Angular geometry)

(*c*) $\text{O}=\ddot{\text{S}}=\text{O}$

(σ-*bps* + *lps*) around

S-atom = 2 + 1 = 3

∴ S = sp^2 (Angular geometry).

Unlike CO_2 (O = C = O), SiO_2 molecule does not have linear structure, O = Si =O; rather it has an infinite three-dimensional structure in which a large number of SiO_4 tetrahedrons are linked together through O-atoms. Thus SiO_2 does not have linear geometry.

17. The type of hybridisation undergone by C-atom of C–H bond in each of the given molecules has bean shown below :

(*a*) $\text{H}-\text{C}\equiv\text{C}-\text{H}$

Acetylene (C_2H_2)

σ-*bps* + *lps* = 2 + 0 = 2

(∴ *C* = *sp*)

(*b*)

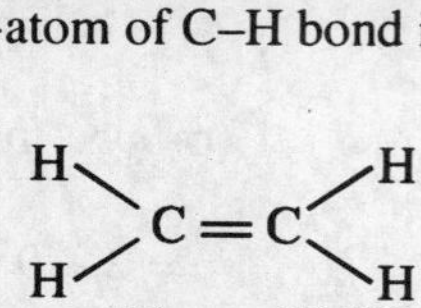

Ethene (C_2H_4)

σ-*bps* + *lps* = 3 + 0 = 3

(∴ $C = sp^2$)

(*c*) 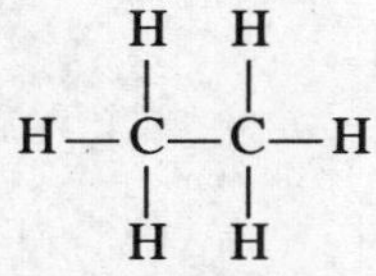

Ethane (C_2H_6)
σ-*bps* + *lps* = 4 + 0 = 4

($\therefore$ $C = sp^3$)

(*d*) 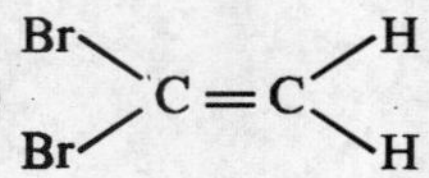

Dibromo ethene ($C_2H_2Br_2$)
σ-*bps* + *lps* = 3 + 0 = 3

($\therefore$ $C = sp^2$)

As the size of hybrid orbital of C-atom forming C–H bond decreases as $sp^3 > sp^2 > sp$, C–H bond distance in C_2H_6 molecule which has sp^3 hybridised C-atom is the largest.

18. The structure of the given molecule can be written as shown below.

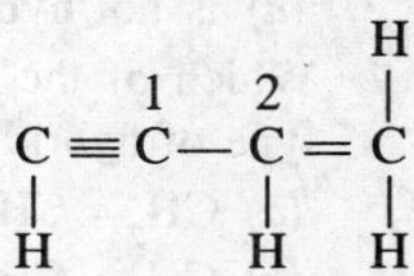

Obviously in the formation of C — C single bond, $\overset{1}{C}$ and $\overset{2}{C}$ atoms are involved. It is clear from the structure that $\overset{1}{C}$ is *sp* hybridised and $\overset{2}{C}$ is sp^2 hybridised. Thus $\overset{1}{C}$—$\overset{2}{C}$ single bond is formed by $sp - sp^2$ overlap.

19. The type of hybridisation undergone by each C-atom in each of the given molecules can be shown as :

(*a*) 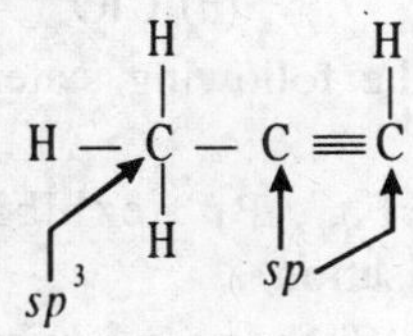

(*b*) 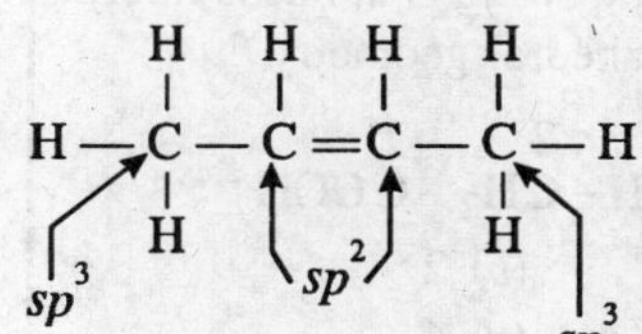

(*c*) 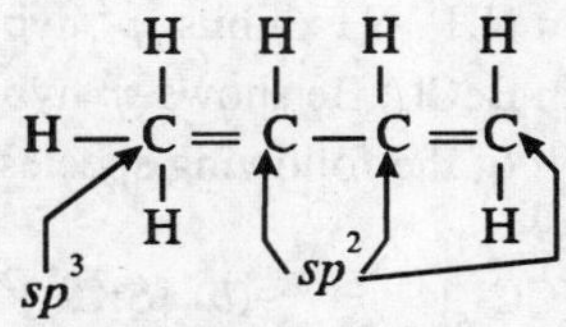

(*d*) 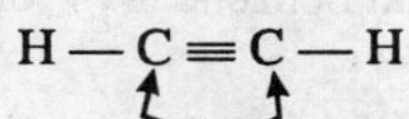

Obviously, it is only (*a*) in which a bond (C–C bond) is formed by $sp - sp^3$ overlap.

20. Lewis structure of CO_2 molecule is O = C = O which shows that since C-atom is *sp* hybridised, CO_2 molecule has linear shape.
The structures of $HgCl_2$, $SnCl_2$ C_2H_2 and NO_2 molecules are given below :

(*a*) **Cl — Hg — Cl.** Since (σ-*bps* + *lps*) around Hg atom (central atom) is 2 + 0 = 2, Hg is *sp* hybrised and hence $HgCl_2$ molecule has linear geometry.

(*b*) **Cl —** $\ddot{\text{Sn}}$ **— Cl.** Since (σ-*bps* + *lps*) around Sn atom is 2 + 1 = 3, Sn is sp^2 hybridised and hence $SnCl_2$ has angular geometry.

(*c*) **H – C ≡ C – H**. Here since each C-atom is *sp* hybridised, C_2H_2 molecule is linear.

(*d*) $:\overset{-}{\ddot{O}} = \overset{+}{\dot{N}} = \ddot{O}:$ Here since $\overset{+}{N}$-atom (central atom) is surrounded by (σ-*bps* + *lps*) $= 2 + \frac{1}{2} = 2.5$, $\overset{+}{N}$ is sp^2 hybridised and hence NO_2 molecule has angular geometry.

15

Structure and Shape of Molecules/Ions Containing Sigma as well as Pi Bonds

MULTIPLE CHOICE QUESTIONS

1. The type of hybrid orbitals used by Cl-atom in ClO_2^- is

(a) sp^3 (b) sp^2

(c) sp (d) None of these

2. The Cl–C–Cl angle in 1, 1, 2, 2 tetrachloro ethene and tetrachloro methane will be about :

(a) 120° and 109.5°

(b) 90° and 109.5°

(c) 109.5° and 90°

(d) 109.5° and 120°

3. Which of the following C-atoms is likely to possess tetrahedral geometry ?

$$\overset{4}{H_2C}=\overset{3}{C}H-\overset{2}{C}H_2-\overset{1}{C}OOH$$

(a) 1 (b) 2

(c) 3 (d) 4

4. Bond length of ethane (I), ethene (II), acetylene (III) and benzene (IV) follows the order :

(a) (I) > (II) > (III) > (IV)

(b) (I) > (II) > (IV) > (III)

(c) (I) > (IV) > (II) > (III)

(d) (III) > (IV) > (II) > (I)

5. Which of the following types of hybridisation is involved in the formation of cycloalkanes ?

(a) sp^3 (b) sp

(c) sp^2 (d) sp^3d^2

6. Each carbon atom in CaC_2 is :

(a) sp hybridised

(b) sp^2 hybridised

(c) sp^3 hybridised

(d) is not hybridised

7. Which of the following have the same type of hybrid carbon ?

(a) $CH_2=CH-CH=CH_2$

(b) $CH_3-CH_2-CH_2-CH_3$

(c) $CH_3-C\equiv C-CH_3$

(d) $HC\equiv C-C\equiv CH$

8. Which of the following molecules has linear geometry ?

(a) CO_2 (b) NO_2

(c) SO_2 (d) ClO_2

9. Which of the following statements are correct ?

(a) In BrF_3, Br exhibits dsp^3 hybridisation

(b) In SiF_4, Si shows sp^3 hybridisation

(c) In NH_3, N exhibits sp^2 hybridisation

(d) In $BeCl_2$, Be shows sp hybridisation

10. Which of the following species does not exist ?

(a) $[CCl_6]^{2-}$ (b) $[SiCl_6]^{2-}$

(c) $[GeCl_6]^{2-}$ (d) $[SnCl_6]^{2-}$

11. The linear species is :

(a) N_2O (b) SO_2

(c) I_3^+ (d) I_3^-

12. Electronegativity of carbon atom in CH_4, C_2H_4 and C_2H_2 molecules follows the order :

(a) $CH_4 > C_2H_4 < C_2H_2$

(b) $C_2H_4 < C_2H_2 < CH_4$

(c) $C_2H_2 < C_2H_4 < CH_4$

(d) $C_2H_2 < CH_4 < C_2H_4$

13. The acidic character of CH_4, C_2H_4 and C_2H_2 molecules follows the order :
(*a*) $CH_4 < C_2H_4 < C_2H_2$
(*b*) $C_2H_4 < C_2H_2 < CH_4$
(*c*) $C_2H_2 < C_2H_4 < CH_4$
(*d*) $C_2H_2 < CH_4 < C_2H_4$

14. In which of the following the central atom does not use sp^3 hybrid orbitals in its bonding ?
(*a*) BeF_3^- (*b*) OH_3^+
(*c*) NH_2^- (*d*) NF_3

15. Carbon atoms in allene molecule involve the hybrid orbitals.
(*a*) sp^2, sp, sp^2 (*b*) sp, sp^2, sp^2
(*c*) sp^2, sp, sp^3 (*d*) sp^2, sp^3, sp

16. In $XeOF_4$ the type of hybridisation is :
(*a*) sp^3d^3
(*b*) sp^3d^2
(*c*) sp^3d
(*d*) Compound is not possible.

17. The geometry and type of hybrid orbitals present about the central atom in BF_3 is :
(*a*) Linear, sp
(*b*) Trigonal planar, sp^2
(*c*) Tetrahedral, sp^3
(*d*) Pyramidal, sp^3

18. The hybridisation of carbon atoms in C – C single bond of $HC \equiv C - CH = CH_2$ is :
(*a*) $sp^3 - sp^3$ (*b*) $sp^2 - sp^3$
(*c*) $sp - sp^2$ (*d*) $sp^3 - sp$

19. In the compound, $CH_2 = CH - CH_2 - CH_2 - C \equiv CH$, $\overset{2}{C}-\overset{3}{C}$ bond is of the type :
(*a*) $sp - sp^2$ (*b*) $sp^3 - sp$
(*c*) $sp - sp^3$ (*d*) $sp^2 - sp^3$

20. The hybridisation of C-atom in diamond, graphite and acetylene is respectively :
(*a*) sp^3, sp^2, sp (*b*) sp, sp^3, sp^2
(*c*) sp^2, sp, sp^3 (*d*) sp, sp^2, sp^3

ANSWERS

1. (*a*)	**2.** (*a*)	**3.** (*b*)	**4.** (*c*)	**5.** (*a*)	**6.** (*a*)	**7.** (*a*), (*b*), (*d*)
8. (*a*)	**9.** (*b*), (*d*)	**10.** (*a*)	**11.** (*a*), (*d*)	**12.** (*a*)	**13.** (*a*)	**14.** (*a*)
15. (*a*)	**16.** (*b*)	**17.** (*b*)	**18.** (*c*)	**19.** (*d*)	**20.** (*a*)	

HINTS AND EXPLANATIONS

1. Lewis structure of ClO_2^- ion is $O = \underset{\cdot\cdot}{\overset{\cdot\cdot}{Cl}} - O^-$. Thus structure shows that since Cl-atom (central atom) is surrounded by $(\sigma\text{-}bps + lps) = 2 + 2 = 4$, this atom is sp^3 hybridised.

2. The structures of the given molecules are given below :

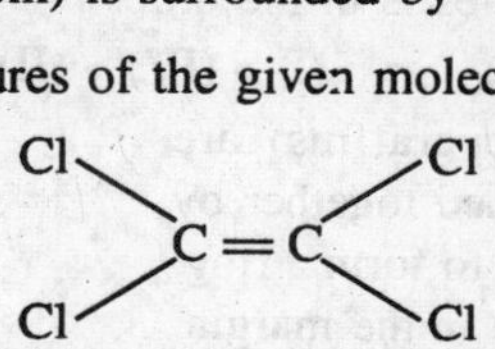

1, 1, 2, 2 - tetra chloro ethene.

Around each C-atom $(\sigma\text{-}bps + lps)$

$= 3 + 0 = 3$ ($\therefore$ $C = sp^2$: Angle = 120°)

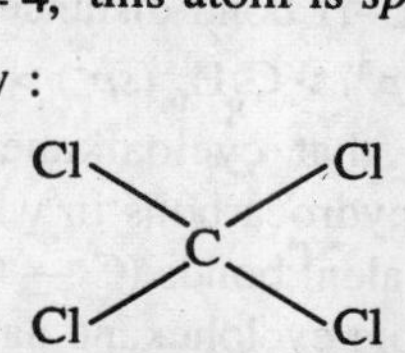

Tetrachloromethane. Around C-atom

$(\sigma\text{-}bps + lps) = 4 + 0 = 4$

($\therefore$ $C = sp^3$: Angle = 109.5°)

These structures show that since in these molecules C-atom is sp^2 and sp^3 hybridised respectively, Cl–C–Cl angle in these molecules is respectively 120° and 109.5°. Thus answer (*a*) is the correct answer.

3. The type of hybridisation undergone by the four C-atoms in the given molecule has been shown below. Obviously, since $\overset{2}{C}$-atom undergones sp^3 hybridisation, this atom has tetrahedral geometry.

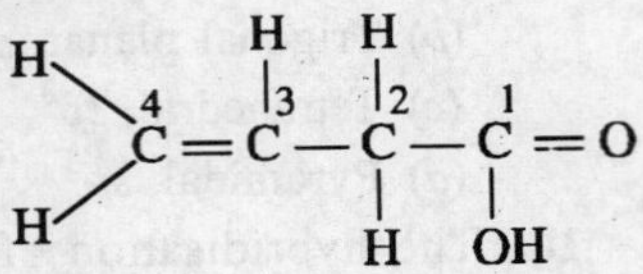

Obviously, $\overset{4}{C}: \sigma\text{-}bps + lps = 3 + 0 = 3\,(sp^2)$, $\overset{3}{C}: \sigma\text{-}bps + lps = 3 + 0 = 3\,(sp^2)$,

$\overset{2}{C}: \sigma\text{-}bps + lps = 4 + 0 = 4\,(sp^3)$ and $\overset{1}{C}: \sigma\text{-}bps + lps = 3 + 0 = 3\,(sp^2)$

4. The type of hybridisation undergone by C-atom of C — H bond in each of the given molecules has been shown below :

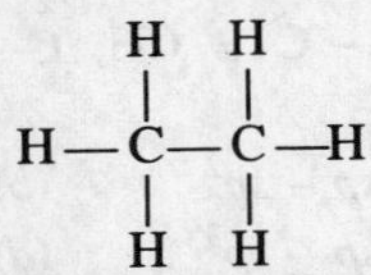

(I) Ethane (C_2H_6)

$(\sigma\text{-}bps + lps) = 4 + 0 = 4$

$(\therefore\ C = sp^3)$

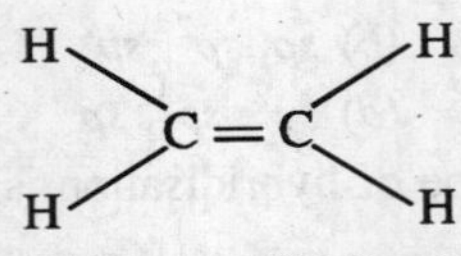

(II) Ethene ($H_2C = CH_2$)

$(\sigma\text{-}bps + lps) = 3 + 0 = 3$

$(\therefore\ C = sp^2)$

$H-C\equiv C-H$

(III) Acetylene (C_2H_2)

$(\sigma\text{-}bps + lps) = 2 + 0 = 2$ $(\therefore\ C = sp)$

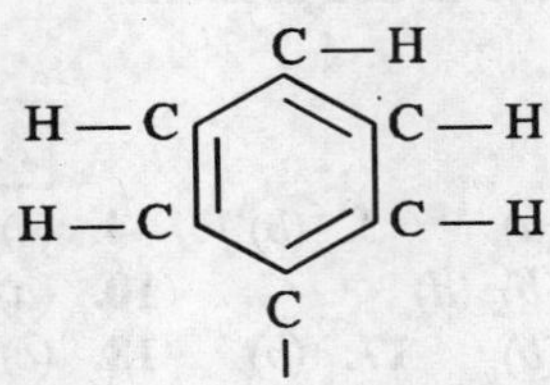

(IV) Benzene (C_6H_6)

$(\sigma\text{-}bps + lps) = 3 + 0 = 3$ $(\therefore\ C = sp^2)$

As the size of the hybrid orbital of C-atom forming C–H bond decreases in the order : $sp^3 > sp^2 > sp$, C–H bond distance in the given molecules is in the order :

$C_2H_6\,(sp^3) > C_6H_6\,(sp^2) > C_2H_4(sp^2) > C_2H_2(sp)$ or (I) > (IV) > (II) > (III)

5. We know that cycloalkanes (also called cycloparaffins) are saturated hydrocarbons in which C-atoms are joined together by single covalent bonds (C — C and C — H bonds) to form a ring. For example cyclohexane has the structure shown in the margin. Obviously each C-atom is sp^3 hybridised.

Cyclohexane

6. CaC_2 (Calcium carbide or calcium acetylide) is an ionic compound and hence is represented as $Ca^{2+}\,(C_2)^{2-}$. C_2^{2-} ion present in CaC_2 is called acetylide ion whose Lewis structure is $:\bar{C}\equiv\bar{C}:$. This structure shows that since each C-atom is surrounded by

(σ-*bps* + *lps*) = 1 + 1 = 2, C-atom is *sp* hybridised. Note C_2^{2-} ion has been derived by removing both H-atoms from acetylene molecule, H—C ≡ C—H

$$\underset{\text{Acetylene molecule}}{H—C \equiv C—H} \xrightarrow{-2H} 2H^+ + \underset{\text{Acetylide ion}}{:\bar{C} \equiv \bar{C}: \text{ or } C_2^{2-}}$$

7. The structures and the type of hybridisation of each C-atom are given below :

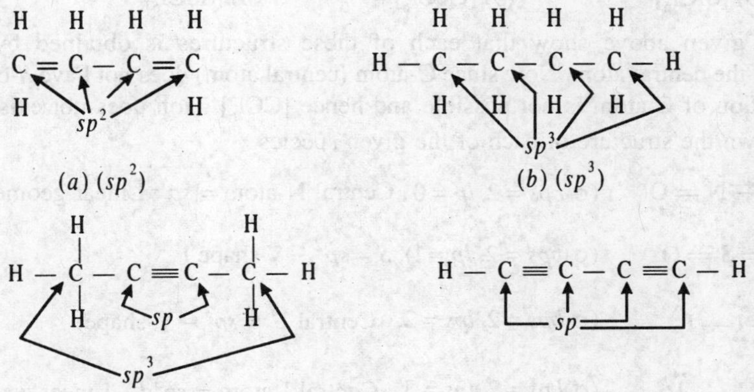

Obviously, C-atom in structures (*a*), (*b*) and (*d*) undergoes the same type of hybridisation.

8. The structures of the given molecules and their shape are given below.

(*a*) O = C = O

σ-*bps* + *lps* = 2 + 0 = 2

(C = sp-Linear shape)

(*b*) $:\ddot{O} \leftarrow \dot{N} = \ddot{O}$ or $:\ddot{O} — \overset{+}{N} = \ddot{O}:$

σ-*bps* + *lps* = $2 + \frac{1}{2} = 2\frac{1}{2}$

(N = sp^2 – Angular geometry)

(odd-electron molecule)

(*c*) $O = \ddot{S} = O$

σ-*bps* + *lps* = 2 + 1 = 3 (S = sp^2 – Angular geometry)

(*d*) $O = \ddot{Cl} = O$

σ-*bps* + *lps* = $2 + 1\frac{1}{2} = 3\frac{1}{2}$ (Cl = sp^3 – Angular geometry) (odd-electron molecule)

Thus it is only CO_2 which has linear shape.

9. Lewis structures and the type of hybridisation of the central atom in each molecule are given below :

(*a*) F—Br—F with F on Br and lone pairs on Br

σ-*bps* + *lps* = 3 + 2 = 5 (Br = sp^3d)

(*b*) SiF_4: F—Si—F with F above and below Si

σ-*bps* + *lps* = 4 + 0 = 4 (Si = sp^3)

(*c*) H—N—H with H on N and a lone pair on N

σ-*bps* + *lps* = 3 + 1 = 4(N = sp^3)

(*d*) Cl – Be – Cl

σ-*bps* + *lps* = 2 + 0 = 2 (Be = *sp*)

Obviously in (*b*) and (*d*) the central atom is sp^3 and *sp* hybridised respectively.

10. The structures of (*b*), (*c*) and (*d*) are given below :

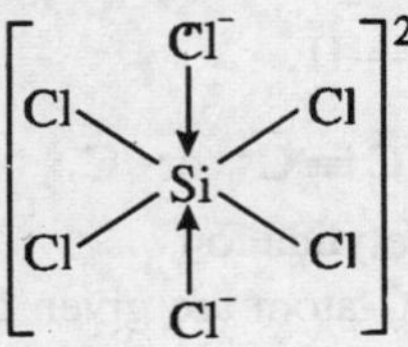

(*b*) $[SiCl_6]^{2-}$

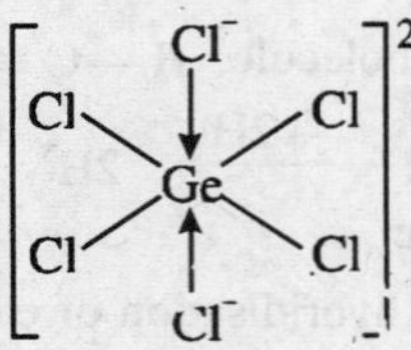

(*b*) $[GeCl_6]^{2-}$

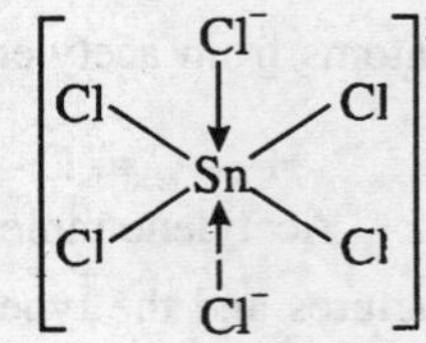

(*b*) $[SnCl_6]^{2-}$

The structures given above show that each of these structures is obtained by sp^3d^2 hybridisation of the central atom. Now since C-atom (central atom) does not have *d*-orbitals, sp^3d^2 hybridisation of C-atom is not possible and hence $[CCl_6]^{2-}$ ion does not exist.

11. Let us write down the structure of each of the given species.

(*a*) N_2O $\quad :\ddot{N}\leftrightharpoons N=\ddot{\underset{..}{O}}$ $\quad$ (σ-*bps* = 2, *lp* = 0, Central N-atom = *sp* : Linear geometry).

(*b*) SO_2 $\quad O=\ddot{S}=O$ $\quad$ (σ-*bps* = 2, *lp* = 1), $S = sp^2$ – V shape)

(*c*) I_3^+ $\quad I-\overset{+}{\underset{..}{\ddot{I}}}-I$ $\quad$ (σ-*bps* = 2, *lps* = 2, Central $I^+ = sp^3$ – V-shape)

(*d*) I_3^- $\quad I-I\leftarrow\overset{-}{\underset{..}{\ddot{I}}}:$ $\quad$ (σ-*bps* = 2, *lps* = 3 Central I-atom = sp^3d – Linear geometry)

These structures show that the species which have linear geometry are N_2O and I_3^-.

12. The structures of CH_4, C_2H_4 and C_2H_2 molecules given below show that C-atom in these molecules is sp^3, sp^2 and *sp* hybridised respectively. Now since the percentage of *s*-character in sp^3, sp^2 and *sp* hybrid orbitals increases as sp^3 (*s*-character = 25%) < sp^2 (*s*-character = 33%) < *sp* (*s*-character = 50%), the electronegativity of C-atom in CH_4, C_2H_4 and C_2H_2 molecules also increases in the same order, *i.e.*, electronegativity of C-atom increases as CH_4 (sp^3) < C_2H_4 (sp^2) < C_2H_2 (*sp*).

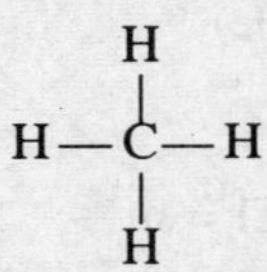

$CH_4(C = sp^3)$

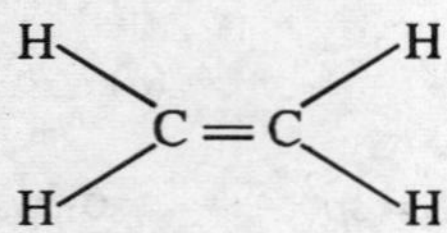

C_2H_4 (C = sp^2)

$H-C\equiv C-H$

C_2H_2 (C = *sp*)

13. We know that C-atom in CH_4, C_2H_4 and C_2H_2 molecules is sp^3, sp^2 and *sp* hybridised respectively. We also know that percentage of *s*-character in sp^3, sp^2 and *sp* hybrid orbitals increases as sp^3 (= 25%) < sp^2 (= 33%) < *sp* (= 50%). With the increase in *s*-character, the electronegativity of C-atom in the given molecules also increases as $CH_4 < C_2H_4 < C_2H_2$. With the increase in the electronegativity of C-atom, the electron pair-accepting power (acidic character) of C-atom in these molecules increases as $CH_4 < C_2H_4 < C_2H_2$.

14. Lewis structures of the given species and the type of hybridisation undergone by the central atom are given below.

(*a*) 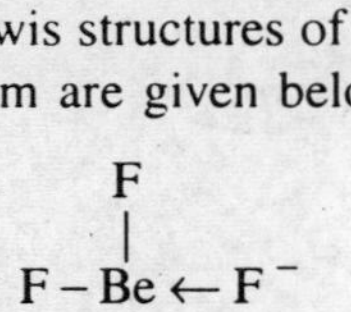

(σ-*bps* + *lps*) = 3 + 0 = 3

$\therefore$ Be = sp^2

(*b*) 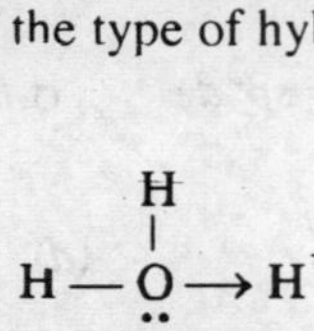

(σ-*bps* + *lps*) = 3 + 1 = 4

$\therefore$ O = sp^3

(*c*) 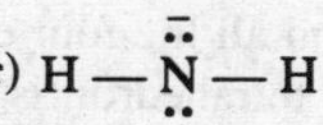

(σ-*bps* + *lps*) = 2 + 2 = 4

∴ N^- = sp^3

(*d*) 

(σ-*bps* + *lps*) = 3 + 1 = 4

∴ N = sp^3

Obviously Be atom in BeF_3^- does not use sp^3 hybridisation. The central atom in all other species uses sp^3 hybridisation.

15. Allene molecule is C_3H_4 whose structure is given in the margin. Carbon atoms have been designated as $\overset{1}{C}, \overset{2}{C}$ and $\overset{3}{C}$. Obviously $\overset{1}{C}, \overset{2}{C}$ and $\overset{3}{C}$ atoms are sp^2, *sp* and sp^2 hybridised.

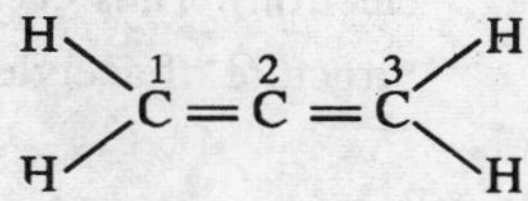

16. Lewis structure of $XeOF_4$ molecule given in the margin shows that since Xe-atom (central atom) is surrounded by (σ-*bps* + *lps*) = 5 + 1 = 6, Xe atom is sp^3d^2 hybridised. Thus correct answer is (*b*).

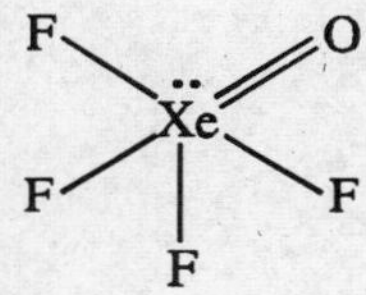

17. Lewis structure of BF_3 molecule is given in the margin. In this molecule, since B-atom (central atom) is surrounded only by three σ-*bps* and no *lp*, BF_3 molecule has trigonal planar geometry in accordance with VSEPR theory. Since (σ-*bps* + *lps*) round B-atom = 3 + 0 = 3, B-atom is sp^2 hybridised. Thus answer is (*b*).

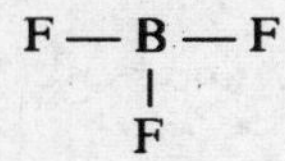

18. The type of hybridisation undergone by two C-atoms of C—C single bond in the given molecule has been shown below.

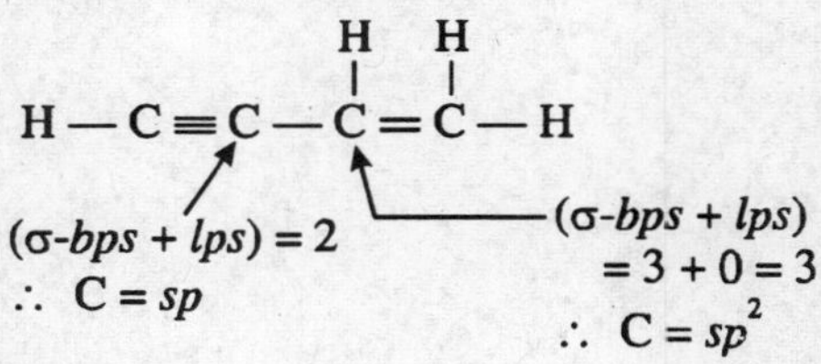

Thus answer is (*c*).

19. The numbering of various C-atoms in the given compound can be shown as :

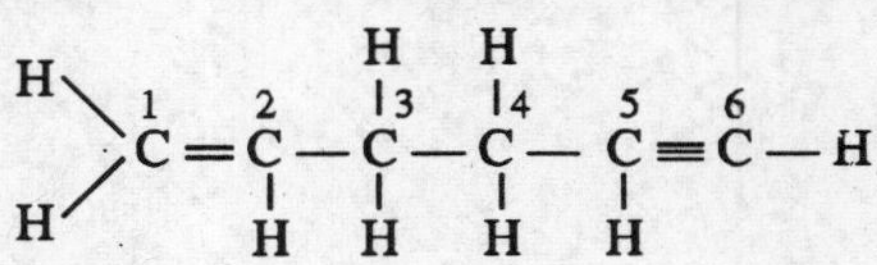

Since $\overset{2}{C}$-atom is surrounded by three σ-*bps* and no *lp*, this atom is sp^2 hybridised. Similarly, since $\overset{3}{C}$-atom is surrounded by four σ-*bps* and no *lp*, this atom is sp^3 hybridised. Thus $(\overset{2}{C}-\overset{3}{C})$ bond is obtained by $\left[sp^2\left(\overset{2}{C}\right) - sp^3\left(\overset{3}{C}\right)\right]$ overlap, *i.e.*, $\overset{2}{C}-\overset{3}{C}$ bond is of $sp^2 - sp^3$ type.

20. We know that the crystal of diamond consists of several tetrahedrons of C-atoms which are linked together. At the centre of each tetrahedron C-atom is situated and all the four corners of each tetrahedron contain four C-atoms. Thus each C-atom of each tetrahedron is linked with other four C-atoms by four (C — C) σ bonds. Thus each C-atom in diamond is sp^3 hybridised.

Graphite contains several parallel hexagonal layers (or sheets) of C-atoms. These layers are linked together by weak van der Waals forces. In each layer, each C-atom is linked with other three C-atoms by three (C — C) σ bonds and has one electron in the free state (free electron). Thus each C-atom in graphite is sp^2 hybridised.

Structure of acetylene is $H-C\equiv C-H$ which shows that each C-atom is *sp* hybridised.

16

Structure of Solids

MULTIPLE CHOICE QUESTIONS

1. In a face-centred cubic (*fcc*) arrangement, the number of atoms per unit cell is :

(*a*) 8 (*b*) 2
(*c*) 1 (*d*) 4

2. A compound formed by the elements X and Y crystallises in cubic structure in which X atoms are at the corners of the cube and Y atoms are at the centre of the face. The formula of the compound is :

(*a*) XY (*b*) XY_2
(*c*) X_2Y_2 (*d*) XY_3

3. A binary solid (A^+B^-) has zinc blende structure with B^- ions constituting the lattice and A^+ ions occupying 25% tetrahedral holes. The formula of the solid is :

(*a*) AB (*b*) A_2B
(*c*) AB_2 (*d*) AB_4

4. In a face-centred cubic arrangement of A and B atoms, A atoms are at the corners of the unit cell and B atoms are at the face centres. If one of the A atoms is missing from one corner in the unit cell, the simplest formula of the compound is :

(*a*) A_7B_3 (*b*) AB_3
(*c*) A_7B_{24} (*d*) $A_{7/8}B_3$

5. A substance A_xB_y crystallises in a face-centred cubic (FCC) lattice in which A atoms occupy each corner of the cube and B atoms occupy the centres of each face of the cube. Identify the correct composition of the substance A_xB_y

(*a*) AB_3 (*b*) A_4B_3
(*c*) A_3B
(*d*) Composition cannot be specified

6. In a solid AB having NaCl structure atoms occupy the corners of the cube unit cell. If all the face-centred atoms along one of the axis are removed, then the resultant stoichiometry of the solid is :

(*a*) AB_2 (*b*) A_2B
(*c*) A_4B_3 (*d*) A_3B_4

7. How many atoms are there in a cube-based unit cell having one atom on each corner and two atoms on each body diagonal of cube ?

(*a*) 8 (*b*) 6
(*c*) 4 (*d*) 9

8. A compound formed by element A and B crystallises in cubic structure in which A atoms are at the corners of the cube while B atoms are at the centre of the cube. The formula of the compound is :

(*a*) A_2B_3 (*b*) AB_2
(*c*) AB_3 (*d*) AB

9. An alloy of copper, silver and gold is found to contain Cu forming the simple cubic close-packed lattice. If Ag atoms occupy the face centres and Au atom is present at the body centre, then the formula of the alloy is :

(*a*) Cu_4Ag_4Au (*b*) Cu_4Ag_2Au
(*c*) Cu Ag Au (*d*) $CuAg_3Au$

10. An alloy of Cu, Ag and Au is found to have Cu constituting the CCP lattice. If Ag atoms occupy the edge centres and Au is present at body centre, the alloy has the formula :

(*a*) Cu_4Ag_2Au (*b*) Cu_4Ag_4Au
(*c*) Cu_4Ag_3Au (*d*) Cu Ag Au

11. A solid has bcc structure. If the distance of closest approach between the two atoms in 1.73 A°. The edge length of the cell is :

(*a*) 200 pm (*b*) $\frac{\sqrt{3}}{\sqrt{2}}$ pm
(c) 142.2 pm (*d*) $\sqrt{2}$ pm

12. In corrundum, oxide ions are arranged in hexagonal close packing and Al^{3+} ions occupy two-third of the octahedral voids. Then the formula of corundum is :

(*a*) Al_2O_3 (*b*) Al_2O
(c) AlO (*d*) AlO_3

13. In a solid, oxide ions are arranged in ccp. One-sixth of the tetrahedral voids are occupied by A cations and one-third of the octahedral voids are occupied by B cations. The formula of the solid is :

(*a*) ABO_3 (*b*) $A_2B_2O_3$
(c) ABO (*d*) ABO_2

14. A solid is made up of two elements P and Q. Atoms of Q are in *ccp* arrangement while atoms of P occupy all the tetrahedral sites. The formula of the compound is :

(*a*) P_2Q (*b*) PQ_2
(c) PQ (*d*) P_2Q_3

15. The ratio between the number of close-packed atoms and the number of tetrahedral holes in cubic close packing is :

(*a*) 1 : 1 (*b*) 1 : 2
(c) 1 : 3 (*d*) 2 : 1

16. The number of NaCl units in the unit cell of its crystal is :

(*a*) 2 (*b*) 4
(*c*) 6 (*d*) 8

17. In the crystal structure of sodium chloride, the arrangement of Cl^- ions is :

(*a*) *fcc* (*b*) *fcc* and *bcc*
(*c*) *bcc* (*d*) none of these.

18. In a crystalline solid, ions B are arranged in a cubic close packing. Cations A are equally distributed between octahedral and tetrahedral voids. If all the octahedral voids are occupied, the formula of the crystalline solid is :

(*a*) A_2B (*b*) AB_2
(*c*) A_3B_2 (*d*) A_2B_3

19. Which of the following statement (s) is (are) correct ?

(*a*) The coordination number of each type of ion in CsCl crystal is 8.
(*b*) A metal that crystallises in *bcc* structure has a coordination number of 12.
(*c*) A unit cell of an ionic crystal shares some of its ions with other unit cells.
(*d*) The length of unit cell in NaCl is 552 pm ($r_{Na^+} = 95$ pm and $r_{Cl^-} = 181$ pm)

20. The coordination number of a metal crystallising in a hexagonal close packing structure is :

(*a*) 12 (*b*) 4
(*c*) 8 (*d*) 6

ANSWERS

1. (*d*) **2.** (*d*) **3.** (*c*) **4.** (*c*) **5.** (*a*) **6.** (*c*) **7.** (*d*) **8.** (*d*)
9. (*d*) **10.** (*c*) **11.** (*a*) **12.** (*a*) **13.** (*a*) **14.** (*a*) **15.** (*b*) **16.** (*b*)
17. (*a*) **18.** (*a*) **19.** (*a*), (*c*), (*d*) **20.** (*a*)

HINTS AND EXPLANATIONS

1. A face-centred cubic unit cell consists of 8 atoms at the 8 corners of the cube and 6 atoms at the six centres of the six faces of the cube. Each of the 8 atoms placed at the 8 corners contributes $\frac{1}{8}$ *th* part of the atom to the whole unit cell and each of the 6 atoms placed at the 6 centres of the six faces contributes $\frac{1}{2}$ atom to the whole unit cell. Thus total number of atoms present in a face-centred cubic unit $= \frac{1}{8}$ × No. of atoms placed at 8 corners of the cube $+ \frac{1}{2} \times$ No. of atoms placed at the six centres of the six faces of the cube

$$= \frac{1}{8} \times 8 + \frac{1}{2} \times 6 = 1 + 3 = 4$$

2. In a cubic structure :

No. of X atoms from 8 corners of the unit cell $= \frac{1}{8} \times 8 = 1$

No. of Y atoms from 6 face centres of the unit cell $= \frac{1}{2} \times 6 = 3$

∴ Formula of the compound $= X_1Y_3 = XY_3$

3. We know that the unit cell of zinc blende of binary solid A^+B^- contains $4A^+$ ions and $14B^-$ ions. Four A^+ ions occupy 50% of the available tetrahedral sites (holes) created by the cubic-close packed arrangement of B^- ions. Thus since 50% tetrahedral holes are occupied by four A^+ ions, 25% tetrahedral holes will be occupied by $\frac{4}{2} = 2\ A^+$ ions. Thus the unit cell contains two A^+ ions.

We also know that since zinc blende structure has eight B^- ions at the eight corners of the cube and six B^- ions at the six centres of the six faces of the cube, the number of B^- ions present in a unit cell $= 8 \times \frac{1}{8} + 6 \times \frac{1}{2} = 1 + 3 = 4$. Thus the formula of the solid is $A_2B_4 = AB_2$.

4. Since one A atom is missing from one corner in the unit cell, the number of A atoms from 7 corners of the unit cell $= \frac{1}{8} \times 7 = \frac{7}{8}$.

No. of B atoms from 6 face centres of the unit cell $= \frac{1}{2} \times 6 = 3$

∴ Formula of the compound $A_{7/8}\ B_3 = A_7B_{24}$

5. In a face-centred structure :

No. of A atoms from 8 corners of the unit cell $= \frac{1}{8} \times 8 = 1$

No. of B atoms from 6 face centres of the unit cell $= \frac{1}{2} \times 6 = 3$

∴ Composition of the substance $A_xB_y = A_1B_3 = AB_3$

6. Since the solid AB has NaCl type structure, one A^+ cation is present at the centre of the cube unit cell and twelve A^+ cations are present at the twelve centres of 12 edges. Now since the single A^+ cation present at the centre of the cubic unit cell is not shared by any adjacent unit cell, the whole of A^+ cation contributes to the unit cell. Since each of the twelve A^+ cations located at the twelve centres of 12 edges is shared by 4 adjacent unit cells, each A^+ cation contributes only $\frac{1}{4}$ A^+ ion to the unit cell. Thus total number of A^+ cations contributed to the unit cell of AB solid $= 1 + 12 \times \frac{1}{4} = 1 + 3 = 4$

The unit cell of AB solid has eight B^- ions at the eight corners of the tube and six B^- ions at the six centres of the six faces of the cube. Now since each of the eight B^- ions plac the eight corners is being shared by eight surrounding unit cells, each B^- ion co only $\frac{1}{8}B^-$ ion to the unit cell. Similarly, since each of the six B^- ions placed at the centres is being shared by two surrounding unit cells, each B^- ion contributes $\frac{1}{2}B^-$ ion to the unit cell. Now we know that the number of face-centred B^- ions lying on axis is 2. After removing these two B^- ions, we are left with four face-centred B^- ions only. Thus total number of B^- ions contributed to the unit cell $= 8\times\frac{1}{8}+4\times\frac{1}{2}=1+2=3$. Thus the stoichiometry of AB solid = A_4B_3.

7. Each of the 8 atoms placed at the 8 corners of the cube contributes $\frac{1}{8}$th part of it to the whole unit cell and each of 8 atoms placed at the 4 body diagonals of the cube contributes the whole part of it to the whole unit cell. Thus total number of atoms present in the cube

$= \frac{1}{8}\times$ No. of atoms placed at 8 corners of the cube + 1 × No. of atoms placed at 4 body diagonals

$$= \frac{1}{8}\times 8 + 1\times 2\times 4 = 1+8 = 9$$

8. Since each of the eight A atoms placed at the eight corners of the cube is placed at the eight corners of the cube is shared by eight adjacent cubic unit cell, each A atom situated at the corner contributes only $\frac{1}{8}$th part of A atom to the cubic unit cell and hence the number of A atoms contributed by eight atoms to the unit cell $= \frac{1}{8}\times 8 = 1$. Thus the compound contains one A atom. Now since B atom is placed at the centre of the cube, this atom is not shared by any adjacent unit cells and hence belongs entirely to the unit cell. Thus the number of B atoms in the unit cell is equal to 1. Consequently the formula of the compound is AB.

9. In a simple cubic close packed structure, since Cu atoms are present as eight corners of the cube, each of the eight Cu atoms is shared by 8 neighboring unit cells and hence each Cu atom contributes $\frac{1}{8}$ th atom to the whole unit cell. Thus total number of Cu atoms present in the whole unit cell $= \frac{1}{8}\times 8 = 1$.

Since Ag atoms occupy the six face centres, each of the six Ag atoms is shared by two adjacent cubic unit cells and hence each Ag atom contributes $\frac{1}{2}$ atom to the whole unit cell. Thus total number of Ag atom present in the whole unit cell $= \frac{1}{2}\times 6 = 3$.

Since gold (Au) atom is present at the body centre of the cube, there is only one Au atom in the cube and this atom is not shared by any other adjacent cubic unit cells and hence this contributes the whole part of it to the whole unit cell. Thus total number of Au atoms present in the whole unit cell = 1.

Above discussion shows that the alloy contains one Cu atom, three Ag atoms and one Au atom and hence the formula of the alloy is $CuAg_3Au$.

10. In a lattice having *ccp* (cubic close-packed) structure. eight Cu atoms are placed at the eight corners of the cube and six Cu atoms are placed at six face centres of the cube. Thus total number of Cu atoms in the whole unit cell $= \frac{1}{8} \times 8 + \frac{1}{2} \times 6 = 1 + 3 = 4$

Since a cube has 12 edges, there are twelve Ag atoms at 12 edge centres. Total number of Ag atoms in the whole unit cell $= \frac{1}{4} \times 12 = 3$

Since a cube has only one body centre, there is only one Au atom in the whole unit cell.
Above discussion shows that the alloy contains four Cu atoms, three Ag atoms and one Au atom and hence the formula of the alloy is Cu_4Ag_3Cu.

11. We know that in a body-centred cubic (*bcc*) structure, the relation between the distance (*d*) between the two nearest atoms and the edge length (*a*) of the unit cell is given by :

$$d = \frac{\sqrt{3}}{2} \times a$$

or $\quad 1.73 = 0.866 \times a$

or $\quad a = \frac{1.73}{0.866}\ A° = 1.99 A° \approx 2 A° \approx 200\ \text{pm}$

12. If the number of O^{2-} ions arranged in hexagonal close packing = *a*, then the number of octahedral holes = *a*. Now since Al^{3+} ions occupy two-third of the octahedral holes, the

number of Al^{3+} ions $= \frac{2}{3} \times$ No. of octahedral holes $= \frac{2}{3} \times a = \frac{2a}{3}$

$\therefore$ Ratio $Al^{3+} : O^{2-} = \frac{2a}{3} : a = 2 : 3$

$\therefore$ Formula of corrundum = Al_2O_3.

13. Suppose the number of oxide (O^{2-}) ions arranged in cubic close packing = *a*. Then the number of octahedral holes = *a* and the number of tetrahedral holes = 2 × *a* = 2*a*. Thus :

No. of A cations $= \frac{1}{6} \times$ No. of tetrahedral holes

$$= \frac{1}{6} \times 2a = \frac{a}{3}$$

and No. of B cations $= \frac{1}{3} \times$ No. of octahedral holes

$$= \frac{1}{3} \times a = \frac{a}{3}$$

$\therefore$ Ratio $A:B:O^{2-} = \frac{a}{3}:\frac{a}{3}:a = 1:1:3$

$\therefore$ Formula of the solid = ABO_3.

14. Suppose the number of atoms of Q having *ccp* arrangement = *a*. Then the number of tetrahedral holes = 2 × *a* = 2*a*. Now since atoms of P occupy all the tetrahedral holes, then the number of atoms of P = 1 × 2*a* = 2*a*.

$\therefore$ Ratio P : Q = 2*a* : *a* = 2 : 1

$\therefore$ Formula of the compound = P_2Q.

15. We know that if the number of atoms having hexagonal (or cubic) close-packed arrangement is *a*, then the number of tetrahedral holes is 2*a*. Thus the ratio between the close-packed atoms and the number of tetrahedral holes is *a* : 2*a* = 1 : 2.

16. The unit cell of sodium chloride crystal contains 13 Na^+ ions and 14 Cl^- ions. Out of 13 Na^+ ions, one Na^+ ion (cation) is present at the centre (body) of the cube unit and twelve Na^+ ions are present at the twelve centres of 12 edges of the cube unit. Thus the total number of Na^+ ions present in the whole unit cell

$$= 1 + \frac{1}{4} \times 12 = 1 + 3 = 4.$$

Out of fourteen Cl^- ions, eight Cl^- ions are present at the eight corners and six Cl^- ions are placed at the six face centres. Thus the total number of Cl^- ions in the whole unit cell

$$= \frac{1}{8} \times 8 + \frac{1}{2} \times 6 = 1 + 3 = 4$$

It is clear from the above discussion that the unit cell of sodium chloride contains four Na^+ and four Cl^- ions ($4Na^+ 4Cl^-$ or $4Na^+ Cl^-$), *i.e.*, the unit cell of sodium chloride contains four NaCl units.

17. The unit cell of NaCl crystal has 14 Cl^- ions. Since eight Cl^- ions are present as the eight corners of the cube and six Cl^- ions are present at the six face centres of the cube, the fourteen Cl^- ions have face-central cubic (fcc) arrangement.

18. Suppose the number of B anions arranged in cubic close packing = *a*.

$\therefore$ No. of octahedral holes = *a*.

and, No. of tetrahedral holes = 2 × *a* = 2*a*.

Now we know that since 2 tetrahedral voids are occupied by one A cation, 2*a* tetrahedral holes will be occupied by $\frac{1}{2} \times 2a$ A cations = *a* A cations. Further since tetrahedral and octahedral holes are equally occupied by A cations, *a* octahedral holes will also be occupied by *a* A cations. Thus total number of A cations occupying tetrahedral and octahedral holes = *a* + *a* = 2*a*.

Hence ratio between the number of A cations and B anions = 2*a* : *a* = 2 : 1. Formula of the solid is, therefore, A_2B.

19. (*a*) The coordination number (C.N.) of each type of ions in CsCl crystal is 8, since each Cs^+ ion is surrounded by eight Cl^- ions and each Cl^- ion is surrounded by eight Cs^+ ions. Thus statement (*a*) is correct.

(*b*) The C.N. of each atom of a metal having *bcc* (body-central cubic) arrangement of atoms is 8 (not 12), since each atom is surrounded by 8 atoms. Thus statement (*b*) is wrong.

(*c*) Statement (*c*) is correct

(*d*) The length of unit cell of NaCl $= 2 \times (r_{Na^+} + r_{Cl^-}) = 2 \times (95 + 181)$ pm $= 2 \times 276$ pm $=$ 552 pm. Thus statement (*d*) is also correct.

20. In hexagonal close packing (*hcp*) structure, a sphere touches 6 other spheres in its own layer, 3 spheres in the layer above its own layer and 3 spheres in the layer below its own layer. Thus in hcp structure, a sphere touches 12 neighboring spheres and hence coordination number of this sphere is 12.

17

Redox Reactions

MULTIPLE CHOICE QUESTIONS

1. The equivalent weight of $MnSO_4$ is half of its molecular weight, when it is converted to :

(*a*) Mn_2O_3 (*b*) MnO_2

(*c*) MnO_4^- (*d*) MnO_4^{2-}

2. Equivalent mass of oxidising agent in the reaction,

$SO_2 + 2H_2S \longrightarrow 3S + 2H_2O$ is :

(*a*) 32 (*b*) 64

(*c*) 16 (*d*) 8

3. Oxidation state of A, B and C is + 2, + 5 and – 2 respectively. Possible formula of the compound is :

(*a*) $A_2(BC_2)_2$ (*b*) $A_2(BC_3)_2$

(*c*) $A_3(B_2C)_2$ (*d*) $A_3(BC_4)_2$

4. When $KMnO_4$ is titrated against ferrous ammonium sulphate, the equivalent weight of $KMnO_4$ is (M = Molecular weight) :

(*a*) $\frac{M}{10}$ (*b*) $\frac{M}{5}$

(*c*) $\frac{M}{2}$ (*d*) M

5. Amongst the following identify the species with an atom in + 6 oxidation state *(IIT Screening 2000)*

(*a*) MnO_4^- (*b*) $[Cr(CN)_6]^{3-}$

(*c*) NiF_6^{2-} (*d*) CrO_2Cl_2

6. O.N. of O-atom in O_2PtF_6 is :

(*a*) zero (*b*) +1/2

(*c*) + 1 (*d*) –1/2

7. If equal volumes of 1 M $KMnO_4$ and 1 M $K_2Cr_2O_7$ solutions are used to oxidise Fe^{2+} to Fe^{3+} in acidic medium, then Fe^{2+} will be oxidised :

(*a*) more by $K_2Cr_2O_7$

(*b*) more by $KMnO_4$

(*c*) to the equal extent in both the solutions

(*d*) the data is insufficient to predict the extent of oxidation

8. Number of electrons required for the reduction of one NO_3^- ion to hydrazine is :

(*a*) 8 (*b*) 7

(*c*) 5 (*d*) 3

9. Two Cl atoms in bleaching powder have O.N. respectively :

(*a*) + 2, – 2 (*b*) – 2, + 2

(*c*) – 1, + 1 (*d*) + 1, – 1

10. Consider the reaction,

$M^{n+} + MnO_4^- \longrightarrow MO_3^- + Mn^{2+}$

If one mole of MnO_4^- oxidises 1.67 moles of M^{n+} to MO_3^-, then the value of *n* in the reaction is :

(*a*) 2 (*b*) 3

(*c*) 4 (*d*) 5

11. O.N. of Fe atom in $Fe_{0.94}O$ is :

(*a*) + 2 (*b*) 94/200

(*c*) 200/94 (*d*) + 1

12. Sulphur has the highest oxidation state in :

(*a*) SO_2 (*b*) H_2SO_4

(*c*) $Na_2S_2O_3$ (*d*) $Na_2S_4O_6$

13. The O.N. of S-atom in $Na_2S_4O_6$ is
(*a*) + 2.5 for each S-atom
(*b*) + 2 for two S-atoms and + 3 for other two S-atoms
(*c*) + 2 for three S-atoms and + 3 for the other S-atom
(*d*) + 5 for two S-atoms and 0 for the other two S-atoms.

14. In the given reaction, $K_2Cr_2O_7 + XH_2SO_4 + YSO_2 \longrightarrow K_2SO_4 + Cr_2(SO_4)_3 + ZH_2O$
X, Y and Z are :
(*a*) 1, 3, 1 (*b*) 4, 1, 4
(*c*) 3, 2, 3 (*d*) 2, 1, 2

15. In the reaction,
$I_2 + 2S_2O_3^{2-} \longrightarrow 2I^- + S_4O_6^{2-}$,
the equivalent weight of I_2 will be equal to :
(*a*) its molecular weight (M.)
(*b*) $\frac{M}{2}$ (*c*) $\frac{M}{4}$
(*d*) 2 M

16. In acidic medium, equivalent weight of $K_2Cr_2O_7$ (Molecular weight = M) is :
(*a*) M/3 (*b*) M/4
(*c*) M/6 (*d*) M/2

17. In which of the following compounds Fe has the lowest oxidation state ?
(*a*) $K_4[Fe(CN)_6]$ (*b*) $Fe(CO)_5$
(*c*) Fe_2O (*d*) K_2FeO_4

18. A metal ion, M^{3+} loses 3 electrons, its O.N. will be :
(*a*) + 3 (*b*) + 6
(*c*) 0 (*d*) – 3

19. In which of the following pairs there is greatest difference in O.N. of the underlined elements ?
(*a*) $\underline{N}O_2$ and $\underline{N}_2O_4$ (*b*) $\underline{P}_2O_5$ and $\underline{P}_4O_{10}$
(*c*) $\underline{N}_2O$ and $\underline{N}O$ (*d*) $\underline{S}O_2$ and $\underline{S}O_3$

20. Oxidation number of Cr and Co in $Cr(CO)_6$ and $K[Co(CO)_4]$ is respectively :
(*a*) 0, – 1 (*b*) + 2, + 1
(*c*) – 2, + 3 (*d*) + 6, – 3

21. The O.N. of oxygen in OF_2 is :
(*a*) + 2 (*b*) – 2
(*c*) + 1 (*d*) – 1

22. Oxidation state of N-atom is correctly given for :
(*a*) $[Co(NH_3)_5Cl]\,Cl_2$ (0)
(*b*) NH_2OH (+ 1)
(*c*) $(N_2H_5)_2\,SO_4$ (+ 2)
(*d*) Mg_3N_2 (– 3)

23. In the reaction : $3Cl_2 + 6NaOH \longrightarrow NaClO_3 + 5NaCl + 3H_2O$
the element that loses as well as gains electrons is :
(*a*) Na (*b*) O
(*c*) Cl (*d*) None of these

24. Which of the following is not a redox reaction ?
(*a*) $CaCO_3 \longrightarrow CaO + CO_2$
(*b*) $O_2 + 2H_2 \longrightarrow 2H_2O$
(*c*) $Na + H_2O \longrightarrow NaOH + \frac{1}{2} H_2$
(*d*) $MnCl_3 \longrightarrow MnCl_2 + \frac{1}{2} Cl_2$

25. NH_3 or NH_4^+ ion is acting as a reducing agent in :
(*a*) $3\,CuO + NH_3 \longrightarrow Cu + 3H_2O + N_2$
(*b*) $NH_4Cl + NaNO_2 \longrightarrow N_2 + NaCl + 2H_2O$
(*c*) $NH_3 + HCl \longrightarrow NH_4Cl$

26. 2.5 moles of N_2H_4 lose 25 moles of electrons in being converted to a new compound X. Assuming that all the nitrogen appears in X, what is the oxidation state of N in X ?
(*a*) – 1 (*b*) – 2
(*c*) + 3 (*d*) + 4

27. A mole of N_2H_4 loses 10 moles of electrons to form a new compound Y. Assuming that all the N-atoms appear in the new compound, what is the oxidation state of N in Y.
(*a*) – 1 (*b*) – 3
(*c*) + 3 (*d*) + 5

28. Number of moles of $K_2Cr_2O_7$ reduced by one mole of Sn^{2+} ions is :
(*a*) 1/3 (*b*) 3
(*c*) 1/6 (*d*) 6

ANSWERS

1. (*b*)	**2.** (*c*)	**3.** (*d*)	**4.** (*b*)	**5.** (*d*)	**6.** (*b*)	**7.** (*a*)	**8.** (*b*)
9. (*c*)	**10.** (*a*)	**11.** (*c*)	**12.** (*b*)	**13.** (*d*)	**14.** (*a*)	**15.** (*b*)	**16.** (*c*)
17. (*b*)	**18.** (*b*)	**19.** (*d*)	**20.** (*a*)	**21.** (*a*)	**22.** (*d*)	**23.** (*c*)	**24.** (*a*)
25. (*a*), (*b*)		**26.** (*c*)	**27.** (*c*)	**28.** (*a*)			

HINTS AND EXPLANATIONS

1. Equivalent weight (E.W.) of $MnSO_4$ is given by :
E.W. of $MnSO_4$

$$= \frac{\text{Molecular weight (M.W.) of } MnSO_4}{\text{No. of electrons lost by one mole of } Mn^{2+} \text{ ion in the oxidation half-reaction}}$$

O.N. of Mn in $MnSO_4 = +2$.

(*a*) $\underset{(Mn\,=\,+2)}{2Mn^{2+}} + 3H_2O \xrightarrow{\text{Oxidation}} \underset{(Mn\,=\,+3)}{Mn_2O_3} + 6H^+ + 2e^-$

$$\therefore \quad \text{E.W. of } Mn^{2+} \text{ ion} = \frac{M.W/2}{2} = M.W.$$

(*b*) $\underset{(Mn\,=\,+2)}{Mn^{2+}} + 2H_2O \xrightarrow{\text{Oxidation}} \underset{(Mn\,=\,+4)}{MnO_2} + 4H^+ + 2e^-$

$$\therefore \text{E.W. of } Mn^{2+} \text{ ion} = \frac{M.W.}{2}$$

(*c*) $\underset{(Mn\,=\,+2)}{Mn^{2+}} + 4H_2O \xrightarrow{\text{Oxidation}} \underset{(Mn\,=\,+4)}{MnO_4^-} + 8H^+ + 5e^-$

$$\therefore \quad \text{E.W. of } Mn^{2+} \text{ ion} = \frac{M.W.}{5}$$

(*d*) $\underset{(Mn\,=\,+2)}{Mn^{2+}} + 4H_2O \xrightarrow{\text{Oxidation}} \underset{(Mn\,=\,+6)}{MnO_4^{2-}} + 8H^+ + 4e^-$

$$\therefore \text{E.W. of } Mn^{2+} \text{ ion} = \frac{M.W.}{4}$$

2. In the given reaction SO_2 (S = + 4) acts as an OA, since it is reduced to S (S = 0). H_2S (S = – 2) acts as a RA since it gets oxidised to S (S = 0). Thus the given reaction can be broken into the following half reactions :

$$\underset{(S\,=\,+4)}{SO_2} + 4H^+ + 4e^- \xrightarrow{\text{Reduction}} \underset{(S\,=\,0)}{S} + H_2O$$

$$\left[\underset{(S\,=\,-2)}{H_2S} \xrightarrow{\text{Oxidation}} \underset{(S\,=\,0)}{S} + 2H^+ + 2e^-\right] \times 2$$

on adding, we get :

$$\underset{\substack{(S = +4)\\(OA)}}{SO_2} + \underset{\substack{(S = -2)\\(RA)}}{2H_2S} \longrightarrow \underset{(S = 0)}{3S} + 2H_2O$$

∴ Equivalent mass of SO_2

$$= \frac{\text{Mol. wt of } SO_2}{\text{No. of electrons accepted by one } SO_2 \text{ molecule to get reduced to S}}$$

$$= \frac{32 + 16 \times 2}{4} = \frac{64}{4} = 16$$

3. (*a*) $A_2(BC_2)_2 = A_2B_2C_4$
Total charges $= 2 \times (+2) + 2 \times (+5) + 4 \times (-2)$
$= +4 + 10 - 8$
$= +6$ (not zero)

(*b*) $A_2(BC_3)_2 = A_2B_2C_6$
$= 2 \times (+2) + 2 \times (+5) + 6 \times (-2)$
$= +4 + 10 - 12$
$= -2$ (Not Zero)

(*c*) $A_3(B_2C)_2 = A_3B_4C_2$
$= 3 \times (+2) + 4 \times (+5) + 2 \times (-2)$
$= +6 + 20 - 4$
$= +22$ (Not zero)

(*d*) $A_3(BC_4)_2 = A_3B_2C_8$
$= 3 \times (+2) + 2 \times (+5)\ 8 \times (-2)$
$= +6 + 10 - 16$
$=$ Zero

4. In this titration, MnO_4^- ion oxidises Fe^{2+} to Fe^{3+} and is itself reduced to Mn^{2+}. In this reduction reaction five electrons are gained by one mole of MnO_4^-.

$$\underset{(Mn = +7)}{MnO_4^-} + 8H^+ + 5e^- \xrightarrow{\text{Reduction}} \underset{(Mn = +2)}{Mn^{2+}} + 4H_2O$$

∴ Equivalent weight of MnO_4^-

$$= \frac{\text{Molecular weight of } MnO_4^-}{\text{No. of electrons gained by one mole of } MnO_4^- \text{ in the reduction half-reaction}}$$

$$= \frac{M}{5}$$

5. (*a*) $\overset{x}{Mn}O_4^-$, $x + 4 \times (-2) = -1$ or $x = +7$

(*b*) $\left[\overset{x}{Cr}(CN)_6\right]^{3-}$, $x + 6 \times (-1) = -3$ or $x = +3$

(*c*) $\overset{x}{Ni}F_6^{2-}$, $x + 6 \times (-1) = -2$ or $x = +4$

(*d*) $\overset{x}{Cr}O_2Cl_2$, $x + 2 \times (-2) + 2 \times (-1) = 0$ or $x = +6$

6. O_2PtF_6 is an ionic compound and hence is represented as $O_2^+ [PtF_6]^-$ which shows that the O.N. of O-atom = + 1/2.

7. The oxidation of Fe^{2+} to Fe^{3+} in acidic medium can be shown by the following ionic equations :

$$MnO_4^- + 5Fe^{2+} + 8H^+ \longrightarrow Mn^{2+} + 5Fe^{3+} + 4H_2O$$

$$Cr_2O_7^{2-} + 6Fe^{2+} + 14H^+ \longrightarrow 2Cr^{3+} + 6Fe^{3+} + 7H_2O$$

These equations show that 1 mole of $K_2Cr_2O_7$ oxidises six moles of Fe^{2+} while 1 mole of $KMnO_4$ oxidises only 5 moles of Fe^{2+}. Thus $K_2Cr_2O_7$ oxidises more number of moles of Fe^{2+}.

8. The reduction of NO_3^- ion to hydrazine (N_2H_4) can be shown by the following reduction-half reaction.

$$2NO_3^- + 16H^+ + 14e^- \longrightarrow N_2H_4 + 6H_2O$$

This equation shows that :

For the reduction of two NO_3^- ions, the number of electrons required = 14

∴ For the reduction of one NO_3^- ion, the number of electrons required $= \frac{14}{2} = 7$

9. Bleaching powder ($CaOCl_2$) contains Ca^{2+}, Cl^- and OCl^- ions. Obviously in Cl^- and OCl^- ions, Cl has O.N. = –1 and + 1 respectively.

10. The given equation can be written as :

$$\underset{1.67\text{ moles}}{M^{n+}} + \underset{1\text{ mole}}{MnO_4^-} \longrightarrow MO_3^- + Mn^{2+}$$

This equation shows that 1.67 moles of M^{n+} are oxidised to MO_3^- (M = + 5) and 1 mole of MnO_4^- (Mn = + 7) is reduced to Mn^{2+} (Mn = + 2). For the reduction of 1 mole of MnO_4^- to Mn^{2+}, five electrons ($5e^-$) are required.

$$MnO_4^- + 8H^+ + 5e^- \xrightarrow{\text{Reduction}} Mn^{2+} + 4H_2O$$

Thus in the oxidation of 1.67 moles of M^{n+} to MO_3^-, $5e^-$ must be released. Thus 1.67 moles of M^{n+} release 5 electrons and hence 1 mole of M^{n+} releases $\frac{5}{1.67}$ electrons = 2.99 e^- = $3e^-$ (approximately). Thus the oxidation of 1 mole of M^{n+} to MO_3^- can be shown as :

$$\underset{1\text{ mole}}{M^{n+}} + 3H_2O \longrightarrow MO_3^- + 3e^- + 6H^+$$

This equation gives :

$$n = -1 - 3 + 6 = +2$$

11. $Fe_{0.94}O$

∴ $x \times 0.94 + 1 \times (-2) = 0$

∴ $x = \frac{2}{0.94} = \frac{200}{94}$

12. (*a*) S = + 4 (*b*) S = + 6 (*c*) S = – 2 (for terminal S-atom), + 6 (for central S-atom)
(*d*) S = + 5 (for each of the two terminal S-atoms), 0 (for each of the two bridging S-atoms).

(*c*) $Na_2S_2O_3$,

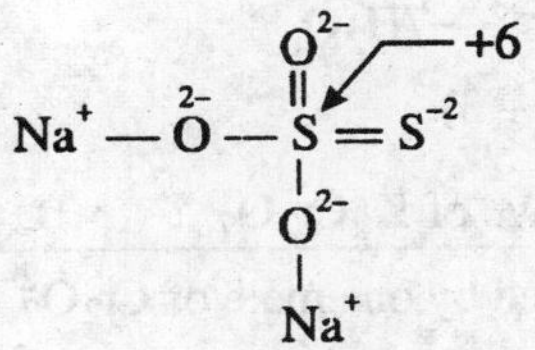

(*d*) $Na_2S_4O_6$,

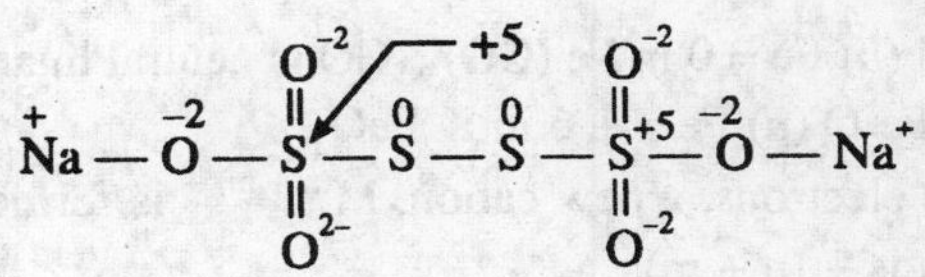

13.

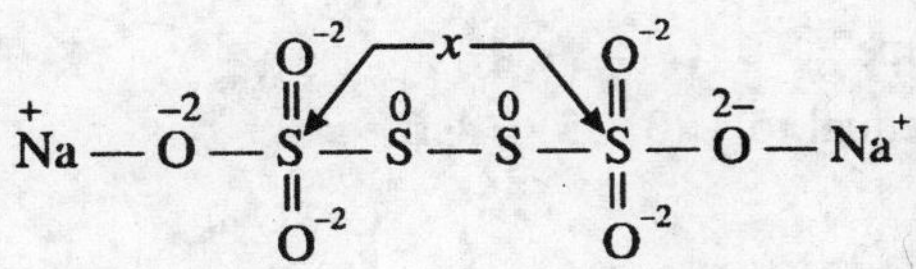

$+1 \times 2$ (For two Na-atoms) + $6 \times (-2)$ (For six O-atoms) + 2×0 (For two bridging S-atoms) + $2 \times x = 0$ (For two terminal S-atoms)

$\therefore \quad x = +5.$

Thus O.N. of each of the two terminal S-atoms = + 5

O.N. of each of the two bridging S-atoms = 0

14. In the given reaction, $Cr_2O_7^{2-}$ is reduced to Cr^{3+} and SO_2 is oxidised to SO_4^{2-}.

$$14\,H^+ + 6e^- + Cr_2O_4^{2-} \xrightarrow{\text{Reduction}} 2Cr^{3+} + 7H_2O$$

$$\left.SO_2 + 2H_2O \xrightarrow{\text{Oxidation}} SO_4^{2-} + 4H^+ + 2e^-\right] \times 3$$

$$2H^+ + Cr_2O_7^{2-} + 3SO_2 \longrightarrow 2Cr^{3+} + H_2O + 3SO_4^{2-}$$

or $\quad H_2SO_4 + K_2Cr_2O_7 + 3SO_2 \longrightarrow Cr_2(SO_4)_3 + H_2O + 3K_2SO_4$

$\therefore \quad$ X = 1, Y = 3 and Z = 1.

15. The given redox reaction can be broken into the following half-reactions.

$$2S_2O_3^{2-} \xrightarrow{\text{Oxidation}} S_4O_6^{2-} + 2e^-$$

$$I_2 + 2e^- \xrightarrow{\text{Raduction}} 2I^- + S_4O_6^{2-}$$

Thus :

Equivalent weight of iodine

$$= \frac{\text{Molecular weight of } I_2 \text{(M.W.)}}{\text{No. of electrons gained by } I_2 \text{ in the reduction half-reaction}}$$

$$= \frac{M}{2}$$

16. In acidic medium $Cr_2O_7^{2-}$ is reduced to Cr^{3+} and in this reduction reaction six electrons are gained by one mole of $Cr_2O_7^{2-}$ ions.

$$Cr_2O_7^{2-} + 14H^+ + 6e^- \longrightarrow 2Cr^{3+} + 7H_2O$$
$$(Cr = +6) \qquad\qquad (Cr = +3)$$

$$\therefore \text{Eq. wt. of } K_2Cr_2O_7 = \frac{\text{Mol. wt. of } K_2Cr_2O_7}{\text{No. of electrons gained by one mole of } Cr_2O_7^{2-} \text{ ions}} = \frac{M}{6}$$

Obviously answer is (*c*).

17. (*a*) Fe = + 2 in $K_4[Fe(CN)_6]$ (*b*) Fe = 0 in $Fe(CO)_5$ (CO is neutral ligand and hence O.N. of CO = zero). (*c*) Fe = + 1 in Fe_2O (*d*) Fe = + 6 in K_2FeO_4.

18. When a cation, M^{n+} loses *m* electrons, a new cation, $M^{(n+m)+}$ is formed. Thus O.N. of the new cation is + (n + m).

$$M^{n+} \longrightarrow M^{(n+m)} + me^-$$

In our case O.N. of the new cation will be equal to + (3 + 3) = + 6

$$M^{3+} \longrightarrow M^{6+} + 3e^-$$

19. (*a*) N = + 4 (in NO_2 and N_2O_4 both), ∴ Difference = zero.
(*b*) P = + 5 (in P_2O_5 and P_4O_{10} both), ∴ Difference = zero.
(*c*) N = + 1 (in N_2O), + 2 (in NO), ∴ Difference = 2 – 1 = 1.
(*d*) S = + 4 (in SO_2), + 6 (in SO_3), ∴ Difference = 6 – 4 = 2.

20. Since O.N. of CO molecule (neutral molecule) is zero, O.N. of Cr atom in $Cr(CO)_6$ is zero and that of Co atom in $K[Co(CO)_4]$ is – 1 as shown below :

$$x + 6 \times 0 = 0, \qquad \therefore x = 0$$
$$+1 + x + 4 \times 0 = 0, \qquad \therefore x = -1$$

21. In most of the compounds O.N. of O-atom in – 2 but in H_2O_2, N_2O_2, CaO_2 etc (peroxides) O.N. of O-atom is – 1. Now since F is more electronegative than O, O.N. of F atom is equal to – 1 and that of O-atom is equal to + 2 as shown below.

$$x + 2 \times (-1) = 0 \quad \text{or} \quad x = +2$$

22. Oxidation state of N-atom in the given compounds is as
(*a*) $[Co(NH_3)_5Cl]Cl = -3$ (*b*) $NH_2OH = -1$
(*c*) $(N_2H_5)_2SO_4 = -2$ (*d*) $Mg_3N_2 = -3$

23. The given reaction can be written as

Reduction (Gain of electrons)

$$3Cl_2 + 6OH^- \longrightarrow ClO_3^- + 3H_2O + 5Cl^-$$
$$(Cl = 0) \qquad (Cl = +5) \qquad (Cl = -1)$$

Oxidation
(Loss of electrons)

This reaction shows that Cl (Cl = 0) is oxidised to ClO_3^- (Cl = +5) (loss of electrons) and is also reduced to Cl^- ion (Cl = – 1) (gain of electrons)

24. In reaction (*a*) O.N. of Ca, C and O atoms remains the same on both sides.

25. The given reactions can be written as :

Reduction (Cu: +2 → 0)

(*a*) $3\,CuO + NH_3 \longrightarrow Cu + 3H_2O + N_2$

(Cu = +2) (OA); (N = – 3) (RA); (Cu = 0); (N = 0)

Oxidation (N: –3 → 0)

Thus NH_3 is acting as a RA.

Oxidation (N: –3 → 0)

(*b*) $NH_4Cl + NaNO_2 \longrightarrow N_2 + NaCl + 2H_2O$

(N = –3) (RA); (N = +3) (OA); (N = 0)

Reduction (N: +3 → 0)

Obviously, NH_4Cl or NH_4^+ ion is acting as a RA.

(*c*) $NH_3 + HCl \longrightarrow NH_4Cl$

(N = – 3) (Cl = –1) (N = – 3, Cl = – 1)

Since O.N. of N in NH_3 and in NH_4Cl is the same (= – 3), NH_3 acts neither as OA nor as RA.

26. Since 2.5 moles of N_2H_4 contains 2.5 × 2 = 5 atoms of N, total number of charges on five N atoms in 2.5 moles of N_2H_4 = – 2 × 5 = – 10

Now since 2.5 moles of N_2H_4 loses 25 moles of electrons ($25e^-$), total number of charges developed on five N-atoms in X = –10 + 25 = + 15. Thus total number of charges on one N-atom in $X = \frac{+15}{5} = +3$

27. We know that the loss of electrons from an atom or on ion increases the charge on the atom or ion. The total number of charges developed on the new species is equal to the sum of its original charges and the number of electrons lost by the atom or ion.

One mole of N_2H_4 contains two N-atoms. Total number of charges on two N-atoms in 1 mole of N_2H_4 = – 2 × 2 = – 4 (– 2 is the oxidation number of N-atom in N_2H_4 molecule).

Now since 1 mole of N_2H_4 loses 10 moles of electrons ($10e^-$), total number of charges developed on two N-atoms in Y = – 4 + 10 = + 6

$\therefore$ No. of charges on one N-atom in $Y = \frac{+6}{2} = +3$

Thus O.N. of N atom in Y = + 3.

28. In the reduc..ion of $Cr_2O_7^{2-}$ by Sn^{2+} involves the reduction of $Cr_2O_7^{2-}$ to Cr^{3+} and oxidation of Sn^{2+} to Sn^{4+}.

$$Cr_2O_7^{2-} + 14H^+ + 6e^- \xrightarrow{\text{Reduction}} 2Cr^{3+} + 7H_2O$$

$$\left[Sn^{2+} \xrightarrow{\text{Oxidation}} Sn^{4+} + 2e^-\right] \times 3$$

$$Cr_2O_7^{2-} + 14H^+ + 3Sn^{2+} \longrightarrow 2Cr^{3+} + 7H_2O + 3Sn^{4+}$$

This equation shows that :

$\therefore$ 3 moles of Sn^{2+} reduce 1 mole of $Cr_2O_7^{2-}$

$\therefore$ 1 mole of Sn^{2+} will reduce 1/3 moles of $Cr_2O_7^{2-}$

18

Electrode Potentials and Electrochemical Series

MULTIPLE CHOICE QUESTIONS

1. F_2 is the best oxidising agent because it has :

(*a*) highest electron affinity

(*b*) highest standard reduction potential

(*c*) highest standard oxidation potential

(*d*) lowest electron affinity

2. Given that :

$I_2 + 2e^- \longrightarrow 2I^-, \quad E^0 = 0.54$ V

$Br_2 + 2e^- \longrightarrow 2Br^-, \quad E^\circ = 1.09$ V

Predict which of the following is true

(*a*) I^- ions are able to reduce Br_2

(*b*) Br^- ions are able to reduce I_2

(*c*) I_2 is able to reduce Br_2

(*d*) Br_2 is able to reduce I^- ions.

3. A gas X at 1 atm is bubllet through a solution containing a mixture of 1 M Y^- and 1 M Z^- at 25° C. If the reduction potentials are in the order Z > Y > X, then

(*a*) Y will oxidise X and not Z

(*b*) Y will oxidise Z and not X

(*c*) Y will oxidise both X and Z

(*d*) Y will reduce both X and Z.

4. If the pressure of hydrogen gas is increased from 1 atm to 100 atm keeping the hydrogen ion concentration of 1 M, the voltage of hydrogen half-cell at 25°C will be : *(IAS 1996)*

(*a*) – 0.059 V (*b*) 0.59

(*c*) 0.295 V (*d*) 0.118 V

5. The reduction potential of hydrogen half-cell will be negative, if :

(*a*) P_{H_2} = 1 atm and $[H^+]$ = 1 M

(*b*) P_{H_2} = 1 atm and $[H^+]$ = 2 M

(*c*) P_{H_2} = 2 atm and $[H^+]$ = 1 M

(*d*) P_{H_2} = 2 atm and $[H^+]$ = 2 M

6. The oxidation potential of hydrogen electrode, $H_2(g)/H_3O^+$ (aq) will be greater than zero, if

(*a*) concentration of H_3O^+ ions is 1 M

(*b*) concentration of H_3O^+ ions is 2 M

(*c*) partial pressure of H_2 gas is 2 atm

(*d*) E_{ox} can never be + ive.

7. Given that $E^0_{Zn^{2+}/Zn} = -0.763$ V and $E^0_{Cd^{2+}/Cd} = -0.403$ V, e.m.f. of the cell,

$Zn|Zn^{+2}(c = 0.004\text{ M})||Cd^{2+}(c = 0.2\text{ M})|Cd$

Anode Cathode

is given by : *(IAS 1996)*

(*a*) $E_{cell} = -0.36 + \frac{0.059}{2} \log \frac{0.004}{0.2}$

(*b*) $E_{cell} = +0.36 - \frac{0.059}{2} \log \frac{0.004}{0.2}$

(*c*) $E_{cell} = -0.36 + \frac{0.059}{2} \log \frac{0.2}{0.004}$

(*d*) $E_{cell} = +0.36 + \frac{0.059}{n} \log \frac{0.2}{0.004}$

8. For the cell $Tl|Tl^+$ (0.001 M)| $|Cu^{2+}$ (0–1 M)| Cu, E_{cell} at 298 K is 0.83 V which can be increased by :

(*a*) increasing $[Cu^{2+}]$

(*b*) increasing $[Tl^+]$

(*c*) decreasing $[Cu^{2+}]$

(*d*) decreasing $[Tl^+]$

9. Which one of the following will increase the voltage of the cell reaction ?

$Sn(s) + 2Ag^+(aq) \longrightarrow Sn^{2+}(aq) + 2Ag(s)$

(*a*) increase in the size of silver rod

(*b*) increase in the concentration of Sn^{2+} ions

(*c*) increase in the concentration of Ag^+ ions

(*d*) none of the above.

10. e.m.f. of the following three galvanic cells :

I . Zn $|Zn^{2+}$ (1 M)$|$ $|Cu^{2+}$ (1 M)$|$ Cu

II. Zn $|Zn^{2+}$ (0.1 M)$|$ $|Cu^{2+}$ (1 M)$|$ Cu

III. Zn $|Zn^{2+}$ (1 M)$|$ $|Cu^{2+}$ (0.1 M)$|$ Cu

are represented by E_1, E_2 and E_3. Which of the following statements is true ?

(*a*) $E_1 > E_2 > E_3$ (*b*) $E_3 > E_2 > E_1$

(*c*) $E_3 > E_1 > E_2$ (*d*) $E_2 > E_1 > E_3$

11. For the reduction of MnO_4^- to Mn^{2+}, the correct form of the Nernst equation is :

(*a*) $E = E^0 + \dfrac{0.0591}{3}\log\dfrac{[Mn^{2+}]}{[MnO_4^-][H^+]^8}$

(*b*) $E = E^0 - \dfrac{0.0591}{5}\log\dfrac{[Mn^{2+}]}{[MnO_4^-][H^+]^8}$

(*c*) $E = E^0 - \dfrac{0.0591}{3}\log\dfrac{[Mn^{2+}]}{[MnO_4^-][H^+]^8}$

(*d*) $E = E^0 + \dfrac{0.0591}{5}\log\dfrac{[MnO_4^-][H^+]^8}{[Mn^{2+}]}$

12. Consider the two cells A and B given below :

Cell A : Fe(s) $|Fe^{2+}$ (1 M)$|$ $|H^+$ (1 M)$|$ H_2 (1 atm), Pt

Cell B : Fe (s) $|$ Fe^{2+} (0.001 M) $|$ $|H^+$ (1 M)$|$ H_2 (1 atm), Pt

The difference in e.m.f. of these cells is :

(*a*) 0.0885 V (*b*) 0.0295 V

(*c*) – 0.0885 V (*d*) – 0.0595 V

13. e.m.f of the following cell containing two hydrogen electrodes

Pt, $\frac{1}{2}H_2(g)\,|\,H^+\,(10^{-8}\,M)\,||$ $H^+\,(0.001\,M)\,|\,\frac{1}{2}H_2(g)$, Pt is :

(*a*) 0.295 V (*b*) 0.0295 V

(*c*) 0.0590 V (*d*) zero

14. For the electro chemical cell,

$M\,|\,M^+\,||\,X^-\,|\,X, E^0_{M^+/M} = 0.44$ V and $E^0_{X/X^-} = 0.33$ V. From these data one can deduce that :

(*a*) $M + X \longrightarrow M^+ + X^-$ is the spontaneous cell reaction

(*b*) $M^+ + X^- \longrightarrow M + X$ is the spontaneous cell reaction.

(*c*) $E^0_{cell} = 0.77$ V

(*d*) $E^0_{cell} = -0.77$ V

15. The standard reduction potential of PbO_2/Pb^{2+} is + 1.46 V and that for Fe^{3+}/Fe^{2+} is 0.77 V. The standard cell voltage for the redox reaction :

$PbO_2 + 4H^+ + 2Fe^{2+} \longrightarrow Pb^{2+} + 2H_2O + 2Fe^{3+}$ is

(*a*) – 2.33 V (*b*) + 0.69 V

(*c*) + 2.33 V (*d*) – 0.69 V

16. In a simple electro chemical cell, which is in standard state, half-cell reactions with their appropriate oxidation potentials are :

$Pb(s) - 2e^- \longrightarrow Pb^{2+}(aq)$; $E^0 = +0.13$ V

$Ag(s) - e^- \longrightarrow Ag^+(aq)$; $E^0 = -0.80$ V

which of the following reactions takes place ?

(*a*) $Pb^{2+}(aq) + 2Ag(s) \longrightarrow 2Ag^+(aq) + Pb(s)$

(*b*) $Pb^{2+}(aq) + Ag(s) \longrightarrow Ag^+(aq) + Pb(s)$

(*c*) $Ag^+(aq) + Pb(s) \longrightarrow Ag(s) + Pb^{2+}(aq)$

(*d*) $2Ag^+ (aq) + Pb(s) \longrightarrow 2Ag (s) + Pb^{2+}(aq)$

17. The spontaneous redox reaction is :

(*a*) $2Fe^{3+} + Fe \longrightarrow 3Fe^{2+}$

(*b*) $Hg_2^{2+} \longrightarrow Hg^{2+} + Hg$

(*c*) $H_2O_2 + Ni^{2+} \longrightarrow 2H^+ + O_2 + Ni$

(*d*) $3AgCl + NO + 2H_2O \longrightarrow 3Ag + 3Cl^- + NO_3^- + 4H^+$

Given that

$E^0_{Fe^{3+}/Fe^{2+}} = 0.77$ V, $E^0_{Fe^{2+}/Fe} = -0.44$ V,

$E^0_{Hg_2^{2+}/Hg} = 0.85$ V, $E^0_{Hg^{2+}/Hg_2^{2+}} = 0.92$ V,

$E^0_{Ni^{2+}/Ni} = -0.25$ V, $E^0_{O_2/H_2O_2} = 0.68$ V,

$E^0_{AgCl/Ag} = 0.22$ V and $E^0_{NO_3^-/NO} = 0.96$ V

18. e.m.f. of a cell having two electrodes A/A^+ ($E^0_{A/A^+} = +0.35$ V) and B^{2+}/B ($E^0_{B^{2+}/B} = -0.42$ V) is :

(*a*) + 0.07 V (*b*) – 0.07 V

(*c*) + 0.77 V (*d*) – 0.77 V

19. The hydrogen electrode is dipped in a solution of pH = 3.0 at 25°C. The potential of hydrogen electrode is :

(*a*) 0.177 V (*b*) – 0.177 V

(*c*) 0.087 V (*d*) 0.059 V

20. Nernst equation, $E = E^0 - \frac{RT}{nF} \ln Q$ indicates that the equilibrium constant (K_{eq}) will be equal to Q when :

(*a*) $E = E^0$ (*b*) $\frac{RT}{nF} = 1$

(*c*) E = zero (*d*) $E^0 = 1$

21. For the cell reaction.

$A(s) + B^{2+} (aq) \longrightarrow B(s) + A^{2+} (aq)$

equilibrium constant is 100. E^0_{Cell} at 25° C is :

(*a*) 0.0591 V (*b*) 0.0295 V

(*c*) 0.118 V (*d*) 0.590 V

22. For the cell reaction, $4Br^- + O_2 + 4H^+ \longrightarrow 2Br_2 + 2H_2O$, $E^0_{Cell} = 0.16$ V, the value of log K_{eq} is :

(*a*) 108 (*b*) 10.8

(*c*) 5.4 (*d*) 1080

23. E^0 for the cell Zn $|Zn^{2+}|$ $|Cu^{2+}/Cu$ is 1.10 V at 25°C. The equilibrium constant for the redox reaction :

$Zn + Cu^{2+} \longrightarrow Cu + Zn^{2+}$

is of the order :

(*a*) 10^{-28} (*b*) 10^{37}

(*c*) 10^{18} (*d*) 10^{17}

24. The value of E^0 for the redox reaction, $Cu^{2+} + Sn^{2+} \longrightarrow Cu + Sn^{4+}$ having equilibrium constant equal to 10^6 is

(*a*) 0.17773 V (*b*) 0.01773 V

(*c*) 0.2153 V (*d*) 0.02153 V

25. For the cell reaction,

$Zn(s) + Cu^{2+} (aq) \rightarrow Cu(s) + Zn^{2+} (aq)$

$E^0_{cell} = 1.10$ V, ΔG^0 is equal to :

(*a*) 2123 KJ (*b*) 106.05 KJ

(*c*) – 212.3 KJ (*d*) – 424.6 KJ

26. The standard electrode potential (E^0) for $OCl^- \mid Cl^-$ and $Cl^- \mid \frac{1}{2}Cl_2$ respectively are 0.94 V and – 1.36 V. The E° value for $OCl^- \left| \frac{1}{2} Cl_2 \right.$ will be :

(*a*) – 0.42 V (*b*) – 2.20 V

(*c*) 0.52 V (*d*) 1.04 V

27. $PbO_2 \longrightarrow PbO$, $\Delta G_{298} < 0$

$SnO_2 \longrightarrow SnO$, $\Delta G_{298} > 0$

Most probable oxidation state of Pb and Sn will be :

(*a*) Pb^{4+}, Sn^{4+} (*b*) Pb^{4+}, Sn^{2+}

(*c*) Pb^{2+}, Sn^{2+} (*d*) Pb^{2+}, Sn^{4+}

28. The standard reduction potential of Cu^{2+}/Cu and Cu^{2+}/Cu^+ are 0.337 V and 0.153 V respectively. The standard electrode potential of $Cu^+|Cu$ half-cell is

(*a*) 0.184 V (*b*) 0.827

(*c*) 0.521V (*d*) 0.490 V

29. $Cu^+(aq)$ is unstable in solution and undergoes simultaneous oxidation and reduction according to the reaction :

$$2Cu^+(aq) \rightleftharpoons Cu^{2+}(aq) + Cu(s)$$

Choose correct E^0 for the above reaction,

if $E^0_{Cu^{2+}/Cu} = 0.34$ V and

$$E^0_{Cu^{2+}/Cu^+} = 0.15 \text{ V}$$

(*a*) + 0.49 V (*b*) + 0.38 V

(*c*) – 0.19 V (*d*) – 0.38 V

30. Standard electrode potentials are : $Fe^{2+} \mid Fe$ = – 0.44 V and $Fe^{3+}| Fe^{2+}$ = + 0.77 V. Fe^{2+} and Fe^{3+} blocks are kept together, then

(*a*) Fe^{3+} increases

(*b*) Fe^{3+} decreases

(*c*) Fe^{2+}/Fe^{3+} remains unchanged

(*d*) Fe^{2+} decreases.

ANSWERS

1. (*b*)	**2.** (*a*)	**3.** (*a*)	**4.** (*a*)	**5.** (*c*)	**6.** (*c*)	**7.** (*b*)	**8.** (*a*), (*d*)
9. (*c*)	**10.** (*c*)	**11.** (*b*)	**12.** (*c*)	**13.** (*a*)	**14.** (*b*)	**15.** (*b*)	**16.** (*d*)
17. (*a*)	**18.** (*a*)	**19.** (*b*)	**20.** (*c*)	**21.** (*a*)	**22.** (*b*)	**23.** (*b*)	**24.** (*a*)
25. (*c*)	**26.** (*c*)	**27.** (*d*)	**28.** (*c*)	**29.** (*b*)	**30.** (*b*)		

HINTS AND EXPLANATIONS

1. Higher is the value of E^0_{red} of halogen (X_2), more readily it gets reduced to halide ion (X^-) and hence higher is its oxidising power.

2. We know that in a redox reaction : (*a*) Higher oxidises lower (*b*) Lower reduces higher. Thus on the basis of rule (*b*), since $E^0_{I_2/2I^-}$ is lower than $E^0_{Br_2/2Br^-}$, I^- ions are able to reduce Br_2 to Br^- ions.

3. We know that : (*a*) Higher oxidises lower, and (*b*) Lower reduces higher. On the basis of these rules, since Y has higher E^0 value than X, Y will oxidise X. Similarly, since Y has lower E^0 value than Z, Y will not oxidise Z. Since Y has lower E^0 value than Z, Y will reduce Z. Y has higher value of E^0 than X, Y will not reduce X.

4. $2H^+(1\text{ M}) + 2e^- \longrightarrow H_2\ (1g, 100\text{ atom})$

$$\therefore \quad E_{H^+/H_2} = E^0_{H^+/H_2} - \frac{0.0591}{n}\log\frac{p_{H_2}}{[H^+]^2}$$

$$= 0 - \frac{0.0591}{2}\log\frac{10^2}{(1)^2}$$

$$= \frac{-0.0591}{2} \times 2 \times \log 10$$

$$= -0.591 \text{ V}$$

5. Reduction reaction taking place in hydrogen half-cell is

$$2H^+(aq) + 2e^- \longrightarrow H_2\ (g)\ (n = 2)$$

Reduction potential of this half-cell is given by :

(*a*) $E_{H^+/H_2} = E^0_{H^+/H_2} - \dfrac{0.0531}{n}\log\dfrac{p_{H_2}}{[H^+]^2}$

$$= 0 - \frac{0.0591}{2} \log \frac{1}{(1)^2} = + 0.0591 \log 1 = \text{zero}$$

(*b*) $$E_{H^+/H_2} = -\frac{0.0591}{2} \log \frac{1}{(2)^2}$$

$$= -\frac{0.0591}{2}(\log 1 - 2\log 2)$$

$$= \frac{-0.0591}{2} \times (-2\log 2)$$

$$= +0.0591 \log 2$$

$$= \text{Positive value}$$

(*c*) $$E_{H^+/H_2} = \frac{-0.0591}{2} \log \frac{2}{(1)^2}$$

$$= -0.0295 [\log 2 - 2\log 1]$$

$$= -0.0295 \times \log 2$$

$$= \text{Negative value}$$

(*d*) $$E_{H^+/H_2} = \frac{-0.0591}{2} \log \frac{2}{(2)^2}$$

$$= -0.0295 \times (\log 2 - 2 \log 2)$$

$$= -0.0295 \times (-\log 2)$$

$$= +0.0295 \times \log 2$$

$$= \text{Positive value}$$

6. We know that E_{H^+/H_2} (reduction potential) is given by : $(2H^+ + 2e^- \rightarrow H_2)$

$$E_{H^+/H_2} = E^0_{H^+/H_2} - \frac{0.0591}{2} \log \frac{p_{H_2}}{[H^+]^2}$$

$$= 0 - 0.0295 \log \frac{p_{H_2}}{[H^+]^2}$$

or $E^0_{H_2/H^+}$ (Oxidation potential) $= +0.0295 \log \frac{p_{H_2}}{[H^+]^2}$

(*a*) $E_{H_2/H^+} = +0.0295 \times \log \frac{1}{(1)^2} = \text{zero}$

Thus answer (*a*) is wrong answer.

(*b*) $E_{H_2/H^+} = +0.0295 \times \log \frac{1}{(2)^2}$

$= + 0.0295 + \log (2)^{-2}$
$= - 2 \times 0.0295 \times \log 2$
= Negative value and hence less than zero

Thus answer (*b*) is also a wrong answer.

(*c*) $E_{H_2/H^+} = + 0.0295 \times \log \frac{2}{(1)^2} = + 0.0295 \times \log 2$

= Positive value and hence greater than zero.

Thus answer (*c*) is the correct answer.

(*d*) Calculations made at (*c*) show that answer (*d*) is wrong.

7. Cell reaction is :

$Zn + Cd^{2+}$ (0.2 M) $\longrightarrow$ Zn^{2+} (0.004 M) + Cd which gives $n = 2$.

Now according to Nernst equation,

$$E_{cell} = E^0_{cell} - \frac{0.059}{n} \log \frac{C_{anode}}{C_{cathode}}$$

$$= E^0_{cathode} - E^0_{anode} - \frac{0.059}{2} \log \frac{0.004}{0.2}$$

$$= - 0.403 - (- 0.763) - \frac{0.059}{2} \log \frac{0.004}{0.2}$$

$$= + 0.360 - \frac{0.059}{2} \log \frac{0.004}{0.2}$$

8. E_{cell} for the given cell is given by Nernst equation given below :

$$E_{cell} = E^0_{cell} - \frac{0.0591}{n} \log \frac{C_{anode}}{C_{cathode}}$$

or $$0.83 = E^0_{cell} - \frac{0.0591}{n} \log \frac{[Tl^+]}{[Cu^{2+}]}$$

This equation shows that the value of E_{cell} is inversely proportional to $[Tl^+]$ and directly proportional to $[Cu^{2+}]$. Thus the value of E_{cell} can be increased by decreasing $[Tl^+]$ [answer (*d*)] and by increasing $[Cu^{2+}]$ [answer (*a*)]

9. For the given redox reaction (cell reaction), E_{cell} is given by Nernst equation :

$$E_{cell} = E^0_{cell} - \frac{0.0591}{n} \log \frac{[Sn^{2+}(aq)]}{[Ag^+(aq)]^2}$$

This equation shows that the value of E_{cell} will increase with the increase in the magnitude of the concentration of Ag^+ (aq) ions. Thus answer (*c*) is the correct answer.

10. For galvanic cell I

$$E_{cell} = E_1 = \left(E^0_{cathode} - E^0_{anode}\right) - \frac{0.0591}{n} \log \frac{C_{anode}}{C_{cathode}}$$

$$= \left(E^0_{Cu^{2+}/Cu} - E^0_{Zn^{2+}/Zn}\right) - \frac{0.0591}{2}\log\frac{1}{1}$$

$$= \left(E^0_{Cu^{2+}/Cu} - E^0_{Zn^{2+}/Zn}\right) - 0$$

For galvanic cell II

$$E_{cell} = E_2 = \left(E^0_{Cu^{2+}/Cu} - E^0_{Zn^{2+}/Zn}\right) - \frac{0.0591}{2}\log\frac{0.1}{1}$$

$$= \left(E^0_{Cu^{2+}/Cu} - E^0_{Zn^{2+}/Zn}\right) - \frac{0.0591}{2}\times(-1)$$

$$= \left(E^0_{Cu^{2+}/Cu} - E^0_{Zn^{2+}/Zn}\right) + \frac{0.0591}{2}$$

For galvanic cell III

$$E_{cell} = E_3 = \left(E^0_{Cu^{2+}/Cu} - E^0_{Zn^{2+}/Zn}\right) - \frac{0.0591}{2}\log\frac{1}{0.1}$$

$$= \left(E^0_{Cu^{2+}/Cu} - E^0_{Zn^{2+}/Zn}\right) - \frac{0.0591}{2}\times\log 10$$

$$= \left(E^0_{Cu^{2+}/Cu} - E^0_{Zn^{2+}/Zn}\right) - \frac{0.0591}{2}$$

Obviously : $E_2 > E_1 > E_3$

11. The reduction of MnO_4^- to Mn^{2+} can be shown as :

$$MnO_4^- + 8H^+ + 5e^- \longrightarrow Mn^{2+} + 4H_2O\ (n = 5)$$

Obviously, $E_{redox\ reaction}$ is given by Nerst equation.

$$E_{redox\ reaction} = E^0_{redox\ reaction} - \frac{0.0591}{n}\log\frac{\text{Product of the concentration of products}}{\text{Product of the concentration of reactants}}$$

$$= E^0 - \frac{0.0591}{5}\log\frac{[Mn^{2+}][H_2O]^4}{[MnO_4^-][H^+]^8}$$

$$= E^0 - \frac{0.0591}{5}\log\frac{[Mn^{2+}]}{[MnO_4^-][H^+]^8}$$

12. *Cell A* : $Fe(s)\,|\,Fe^{2+}(1\ M)\,||\,H^+(1\ M)\,|\,H_2(atm), Pt$

Anode Cathode

Since the concentration of Fe^{2+} in anode and H^+ in cathode is unity.

$$(E_{cell})_A = E^0_{Cathode} - E^0_{Anode}$$

$$= E^0_{H^+/H_2} - E^0_{Fe^{2+}/Fe}$$

$$= 0 - E^0_{Fe^{2+}/Fe} = - E^0_{Fe^{2+}/Fe}$$

Cell B : $Fe(s)\,|\,Fe^{2+}\,(0.001\,M)\,||\,H^+\,(1\,M)\,||\,H_2\,(1\,atm)\,|,\,Pt$

Anode Cathode

Since the concentration of Fe^{2+} in anode is different from that of H^+ in cathode, $(E_{cell})_B$ is given by Nernst equation

$$(E_{cell})_B = E^0_{cell} - \frac{0.0591}{n} \log \frac{C_{Anode}}{C_{Cathode}}$$

$$= (E^0_{Cathode} - E^0_{Anode}) - \frac{0.0591}{2} \log \frac{0.001}{1}$$

$$= (E^0_{H^+/H_2} - E^0_{Fe^{2+}/Fe}) - \frac{0.0591}{2} \log 10^{-3}$$

$$= 0 - E^0_{Fe^{2+}/Fe} + \frac{3 \times 0.0591}{2} \times 1$$

$$= - E^0_{Fe^{2+}/Fe} + 0.0885$$

$$\therefore\ (E_{cell})_A - (E_{cell})_B = - E^0_{Fe^{2+}/Fe} - (-E^0_{Fe^{2+}/Fe} + 0.0885)$$

$$= - 0.0885\ V$$

13. Pt, $\frac{1}{2} H_2(g)\,|\,H^+\,(10^{-8} M)\,||\,H^+\,(0.001\,M)\,|\,\frac{1}{2} H_2\,(g)$, Pt

Anode Cathode

Since the concentration of H^+ in anode and H^+ in cathode is different, E_{cell} is given by Nernst equation.

$$E_{cell} = E^0_{cell} - \frac{0.0591}{n} \log \frac{C_{anode}}{C_{cathode}}$$

$$= (E^0_{cathode} - E^0_{anode}) - \frac{0.0591}{1} \log \frac{10^{-8}}{10^{-3}}$$

$$= (E^0_{H^+/H_2} - E^0_{H^+/H_2}) - 0.0591 \log 10^{-5}$$

$$= (0 - 0) - 0.0591 \times (-5)$$

$$= + 0.059 \times 5 = 0.2955\ V$$

14. Since $E^0_{M^+/M} > E^0_{X/X^-}$, M^+/M electrode should act as cathode and X^-/X electrode should act as anode

$X^-\,|\,X\,|$ $|\,M^+\,|\,M$

Anode Cathode

$$E^0_{anode} = E^0_{X/X^-} = 0.33\text{ V} \qquad E^0_{cathode} = E^0_{M^+/M} = 0.44\text{ V}$$

(*a*) $E^0_{reaction} = E^0_{OA} - E^0_{RA} = E^0_{X/X^-} - E^0_{M^+/M}$

$= 0.33 - 0.44 = -0.11$ V (Non spontaneous reaction)

(*b*) $E^0_{reaction} = E^0_{OA} - E^0_{RA} = E^0_{M^+/M} - E^0_{X/X^-}$

$= 0.44 - 0.33 = +0.11$ V (Spontaneous reaction)

(*c*), (*d*) $E^0_{cell} = E^0_{cell} = E^0_{cathode} - E^0_{anode} = 0.44 - 0.33 = +0.11\text{ V}$

Reduction ($\overset{4+}{Pb}O_2 \rightarrow Pb^{2+}$)

15. $\overset{4+}{PbO_2} + 4H^+ + 2Fe^{2+} \longrightarrow Pb^{2+} + 2H_2O + 2Fe^{3+}$

(OA) (RA)

Oxidation ($Fe^{2+} \rightarrow Fe^{3+}$)

$\therefore \quad E^0_{redox\ reaction} = E^0_{OA} - E^0_{RA}$

$= E^0_{PbO_2/Pb^{2+}} - E^0_{Fe^{3+}/Fe^{2+}}$

$= (1.46 - 0.77)\text{V} = 0.69\text{ V}$

16. Write the given half-reactions as reduction half-reactions

$Pb^{2+}(aq) + 2e^- \longrightarrow Pb(s), \qquad E^0_{Pb^{2+}/Pb} = -0.13\text{ V}$

$Ag^+(aq) + e^- \longrightarrow Ag(s), \qquad E^0_{Ag^+/Ag} = +0.80\text{ V}$

The value of $E^0_{reaction}$ for the given reaction is calculated as follows :

(*a*) $E^0_a = E^0_{OA} - E^0_{RA} = E^0_{Pb^{2+}/Pb} - E^0_{Ag^+/Ag}$

$= -0.13 - 0.80 = -1.13$ V (Non spontaneous reaction)

(*d*) $E^0_c = E^0_{OA} - E^0_{RA} = E^0_{Ag^+/Ag} - E^0_{Pb^{2+}/Pb}$

$= +0.80 - (-0.13) = +1.13$V (Spontaneous reaction)

(*b*), (*c*) In reactions (*b*) and (*c*), since positive charges on metallic ions are not balanced, these reactions will not take place.

Reduction ($Fe^{3+} \rightarrow Fe^{2+}$)

17. (*a*) $2Fe^{3+} + Fe \longrightarrow 2Fe^{2+} + Fe^{2+}$

(OA) (RA)

Oxidation ($Fe \rightarrow Fe^{2+}$)

$\therefore \quad E^0_{redox\ reaction} = E^0_{OA} - E^0_{RA} = E^0_{Fe^{3+}/Fe^{2+}} - E^0_{Fe^{2+}/Fe} = 0.77 - (-0.44) = +1.21\text{ V}$

Since $E^0_{\text{redox reaction}}$ is positive, reaction (*a*) is spontaneous.

Oxidation

$$(b)\quad \underset{\substack{(Hg = +1)\\(OA)}}{Hg_2^{2+}} + \underset{\substack{(Hg = +1)\\(RA)}}{Hg_2^{2+}} \longrightarrow \underset{(Hg = 0)}{2Hg} + \underset{(Hg = +2)}{2Hg^{2+}}$$

Reduction

$$\therefore\ E^0_{\text{redox reaction}} = E^0_{OA} - E^0_{RA} = E^0_{Hg_2^{2+}/Hg} - E^0_{Hg^{2+}/Hg_2^{2+}}$$

$$= 0.85 - 0.92 = -0.07\ V$$

Since $E^0_{\text{redox reaction}}$ is negative, reaction (*b*) is non spontaneous.

Reduction

$$(c)\quad \underset{\substack{(O = -1)\\(RA)}}{H_2O_2} + \underset{(OA)}{Ni^{2+}} \longrightarrow \underset{(O = 0)}{O_2} + 2H^+ + Ni$$

Oxidation

$$\therefore\quad E^0_{\text{redox reaction}} = E^0_{OA} - E^0_{RA}$$

$$= E^0_{Ni^{2+}/Ni} - E^0_{O_2/H_2O_2} = -0.25 - 0.68$$

$$= -0.93\ V$$

Since $E^0_{\text{redox reaction}}$ is negative, reaction (*c*) is non spontaneous.

Oxidation

$$(d)\quad \underset{\substack{(Ag = +1)\\(OA)}}{3AgCl} + \underset{\substack{(N = +2)\\(RA)}}{NO} + 2H_2O \longrightarrow \underset{(Ag = 0)}{3Ag} + 3Cl^- + \underset{(N = +5)}{NO_3^-} + 4H^+$$

Reduction

$$\therefore\quad E^0_{\text{redox reaction}} = E^0_{OA} - E^0_{RA} = E^0_{AgCl/Ag} - E^0_{NO_3^-/NO}$$

$$= 0.22 - 0.96 = -0.74\ V$$

Since $E^0_{\text{redox reaction}}$ is negative, reaction (*d*) is non spontaneous.

18. $\because\ E^0_{A/A^+} = 0.35\ V, E^0_{A^+/A} = -0.35$ and $E^0_{B^{2+}/B} = -0.42\ V.$

Now since $E^0_{A^+/A} (= -0.35\ V)$ is higher than $E^0_{B^{2+}/B} = (-0.42\ V)$, e.m.f of the cell $= E^0_{higher} - E^0_{lower}$

$= -0.35 - (-0.42)$

$= -0.35 + 0.42 = +0.07$ V

19. $E^0_{H^+/\frac{1}{2}H_2} = -0.0591\,pH = -0.0591 \times 3$

$= -0.177$ V

20. We know that when a redox reaction is at equilibrium, the electrode potential of the two electrodes constituting the redox reaction are equal and hence the potential of the cell is zero, *i.e.*, E = 0. Thus answer (*c*) is the correct answer.

21. To find out the value of number of electrons (*n*) involved in the given cell reaction break the reaction as :

$$A(s) \xrightarrow{\text{Oxidation}} A^{2+}(aq) + 2e^-$$

$$B^{2+}(aq) + 2e^- \xrightarrow{\text{Reduction}} B(s)$$

$$A(s) + B^{2+}(aq) \longrightarrow A^{2+}(aq) + B(s)$$

Obviously $n = 2$ and hence

$$E^0_{\text{cell reaction}} = \frac{0.0591}{n} \log K_{eq}$$

$$= \frac{0.0591}{2} \log 100$$

$$= \frac{0.0591}{2} \log 10^2$$

$$= 0.0591 \text{ V}$$

22. $$\left[2Br^- \xrightarrow{\text{Oxidation}} Br_2 + 2e^-\right] \times 2$$

$$O_2 + 4H^+ + 4e^- \xrightarrow{\text{Reduction}} 2H_2O$$

Cell reaction : $4Br^- + O_2 + 4H^+ \longrightarrow 2Br_2 + 2H_2O \ (n = 4)$

$$\therefore \quad E^0_{\text{cell reaction}} = \frac{0.0591}{n} \log K_{eq}$$

$$0.16 = \frac{0.0591}{4} \log K_{eq}$$

$$\therefore \quad \log K_{eq} = \frac{0.16 \times 4}{0.0591} = 10.8$$

23. $$E^0_{\text{redox reaction}} = \frac{0.0591}{n} \log K_{eq}$$

or $$1.1 = \frac{0.0591}{2} \log K_{eq}$$

$\therefore$ $$\log K_{eq} = \frac{2 \times 1.1}{0.0591} = 37.22$$

$\therefore$ $$K_{eq} = \text{Antilog } 37.2 = 10^{37}.$$

24. $$E^0_{\text{redox reaction}} = \frac{0.0591}{n} \log K_{eq}$$
$$= \frac{0.0591}{2} \log 10^6$$
$$= \frac{0.0591}{2} \times 6$$
$$= 0.1773 \text{ V}$$

25. Since $n = 2$ for the given cell reaction, $\Delta G°$ in given by :

$$\Delta G° = -nFE^0_{\text{cell reaction}}$$
$$= -2 \times 96500 \times 1.10 \text{ J}$$
$$= \frac{-2 \times 96500 \times 1.10}{1000} \text{ KJ} = 212.3 \text{ KJ}$$

26. Write the reduction half reactions corresponding to the given electrodes as :

(1) $$OCl^- + 2H^+ + 2e^- \xrightarrow[(n=2)]{\text{Reduction}} Cl^- + H_2O, \quad E^0_{OCl^-/Cl^-} = 0.94 \text{ V}$$
(Cl = + 1) (Cl = − 1)

(2) $$\frac{1}{2}Cl_2 + e^- \xrightarrow[(n=1)]{\text{Reduction}} Cl^-, \quad E^0_{Cl^-/\frac{1}{2}Cl_2} = -1.36 \text{ V} \quad \text{(Given)}$$
(Cl = 0) (Cl = − 1)

or $$E^0_{\frac{1}{2}Cl_2/Cl^-} = +1.36 \text{ V}$$

(3) $$OCl^- + 2H^+ + e^- \xrightarrow[(n=1)]{\text{Reduction}} \frac{1}{2}Cl_2 + H_2O, \quad E^0_{OCl^-/Cl_2} = ?$$
(Cl = + 1) (Cl = 0)

Since on subtracting equation (2) from (1) we get equation (3), we have :

$$\Delta G^0_3 = \Delta G^0_1 - \Delta G^0_2$$

or $$-nFE^0_{OCl^-/Cl_2} = -nFE^0_{OCl^-/Cl^-} - (-nF \times E^0_{\frac{1}{2}Cl_2/Cl^-})$$

or $$-1 \times F \times E^0_{OCl^-/Cl_2} = -2 \times F \times 0.94 + 1 \times F \times 1.36$$

$\therefore$ $$E^0_{OCl^-/Cl_2} = 2 \times 0.94 - 1.36 = 0.52 \text{ V}$$

27. $\overset{4+}{\text{Pb}}\,\text{O}_2 \longrightarrow \overset{2+}{\text{Pb}}\text{O}$, $\Delta G_{298} < 0$ (*i.e.* ΔG_{298} is negative)

$\overset{4+}{\text{Sn}}\,\text{O}_2 \longrightarrow \overset{2+}{\text{Sn}}\,\text{O}$, $\Delta G_{298} > 0$ (*i.e.* ΔG_{298} is positive)

Since for the process, $\overset{4+}{\text{Pb}}\,\text{O}_2 \longrightarrow \overset{2+}{\text{Pb}}\text{O}$, ΔG_{298} is negative, this process is spontaneous and hence Pb^{2+} is more stable than Pb^{4+}. Similarly, since for the process, $\overset{4+}{\text{Sn}}\,\text{O}_2 \longrightarrow \overset{2+}{\text{Sn}}\,\text{O}$, ΔG_{298} is positive, this process is not spontaneous and hence Sn^{4+} is more stable than Sn^{2+}.

28. (1) $Cu^{2+} + 2e^- \longrightarrow Cu\ (n = 2),\ E^0_{Cu^{2+}/Cu} = 0.337\ V$

(2) $Cu^{2+} + e^- \longrightarrow Cu^+\ (n = 1),\ E^0_{Cu^{2+}/Cu^+} = 0.153\ V$

(3) $Cu^+ + e^- \longrightarrow Cu\ (n = 1),\ E^0_{Cu^+/Cu} = ?$

Since (3) is obtained by subtracting (2) from (1), we have :

$$\Delta G_3^0 = \Delta G_1^0 - \Delta G_2^0$$

or $$-nF\,E^0_{Cu^+/Cu} = -nFE^0_{Cu^{2+}/Cu} - (-nFE^0_{Cu^{2+}/Cu^+})$$

or $$-1 \times F \times E^0_{Cu^+/Cu} = -2 \times F \times 0.337 - (-1 \times F \times 0.153)$$

or $$-E^0_{Cu^+/Cu} = -0.674F + 0.153$$

or $$E^0_{Cu^+/Cu} = -0.674 - 0.153 = +0.521$$

Oxidation (from the first Cu^+ to Cu^{2+})

29. $Cu^+\,(aq)$ (RA) $+$ $Cu^+\,(aq)$ (OA) $\rightleftharpoons$ $Cu^{2+}\,(aq) + Cu(s)$

Reduction (from the second Cu^+ to $Cu(s)$)

$$E^0_{\text{redox reaction}} = E^0_{OA} - E^0_{RA}$$

$$= E^0_{Cu^+/Cu} - E^0_{Cu^{2+}/Cu^+}$$

or $$E^o_{\text{redox reaction}} = E^0_{Cu^+/Cu} - 0.15 \qquad \text{... } (i)$$

In order to determine the value of $E^0_{\text{redox reaction}}$, the value of $E^0_{Cu^+/Cu}$ must be known. In order to find out the value of $E^0_{Cu^+/Cu}$ we proceed as follows :

(1) $Cu^{2+}(aq) + 2e^- \xrightarrow{\text{Reduction}} Cu(s)\ (n = 1),\ E^0_{Cu^{2+}/Cu} = 0.34\ V$

(2) $Cu^{2+}(aq) + e^- \xrightarrow{\text{Reduction}} Cu^+(aq)\ (n = 1),\ E^0_{Cu^{2+}/Cu^+} = 0.15\ V$

(3) $Cu^+(aq) + e^- \xrightarrow{\text{Reduction}} Cu(s)\ (n = 1),\ E^0_{Cu^+/Cu} = ?$

Since (3) is obtained by subtracting (2) from (1), we can write.

$$\Delta G_3^0 = \Delta G_1^0 - \Delta G_2$$

or $$-\ nFE^0_{Cu^+/Cu} = -nFE^0_{Cu^{2+}/Cu} - (-nFE^0_{Cu^{2+}/Cu^+})$$

or $$-1 \times F \times E^0_{Cu^+/Cu} = -2 \times F \times 0.34 - (-1 \times F \times 0.15)$$

or $$-E^0_{Cu^+/Cu} = -0.68 + 0.15$$

or $$E^0_{Cu^+/Cu} = 0.68 - 0.15 = 0.53\ V$$

Now put $E^0_{Cu^+/Cu} = 0.53$ V in equation (*i*) to get :

$$E^0_{\text{redox reaction}} = 0.53 - 0.15 = 0.38\ V$$

30. If Fe^{2+} | Fe electrode is made anode and Fe^{3+}| Fe^{2+} electrode is made cathode, then the cell is represented as :

A. Fe |Fe^{2+}| |Fe^{3+}| Fe^{2+}
Anode Cathode

On the other hand, if Fe^{2+} | Fe electrode is made cathode and Fe^{3+} | Fe^{2+} electrode is made anode, then the cell can be represented as :

B. Fe^{2+} |Fe^{3+}| |Fe^{2+}| Fe
Anode Cathode

Now, $$(E_{cell})_A = E^0_{Cathode} - E^0_{Anode}$$
$$= E^0_{Fe^{3+}/Fe^{2+}} - E^0_{Fe^{2+}/Fe}$$
$$= +\ 0.77 - (-\ 0.44)$$
$$= +\ 1.21\ V \text{ (Positive value)}$$

$$(E_{cell})_B = E^0_{Cathode} - E^0_{Anode}$$
$$= E^0_{Fe^{2+}/Fe} - E^0_{Fe^{3+}/Fe^{2+}}$$
$$= -\ 0.44 - 0.77$$
$$= -\ 1.21\ V \text{ (Negative value)}$$

Since $(E_{cell})_A$ is positive, cell A works and the cell reaction is the sum of following two half-reactions :

$$Fe \xrightarrow{\text{Oxidation}} Fe^{2+} + 2e^-$$

$$\left[Fe^{3+} + e^- \xrightarrow{\text{Reduction}} Fe^{2+}\right] \times 2$$

$$Fe + 2Fe^{3+} \longrightarrow 2Fe^{2+}$$

This equation shows that Fe^{3+} will decrease.

19

Concepts of Acids and Bases

MULTIPLE CHOICE QUESTIONS

1. The strongest base of the following species is
(a) NH^{2-} (b) OH^-
(c) O^{2-} (d) S^{2-}

2. Which of the following can act both as Bronsted acid and Bronsted base ?
(a) Na_2CO_3 (b) OH^-
(c) HCO_3^- (d) NH_3

3. Which of the following species can act both as Bronsted acid and as Bronsted base ?
(a) O^{2-} (b) HCl
(c) HSO_4^- (d) Na_2CO_3

4. The strongest conjugate base is :
(a) NO_3^- (b) Cl^-
(c) SO_4^{2-} (d) CH_3COO^-

5. Conjugate acid of NH_2^- is :
(a) NH_4^+ (b) NH_2OH
(c) NH_3 (d) N_2H_4

6. The following equilibrium is established when HCl is dissolved in CH_3COOH

$$HCl + CH_3COOH \rightleftharpoons Cl^- + CH_3COOH_2^+$$

The set that characterises the conjugate acid-base pair is :
(a) (HCl, CH_3COOH) and $(CH_3COOH_2^+, Cl^-)$
(b) $(HCl, CH_3COOH_2^+)$ and (CH_3COOH, Cl^-)
(c) $(CH_3COOH_2^+, HCl)$ and (Cl^-, CH_3COOH)
(d) (HCl, Cl^-) and $(CH_3COOH_2^+, CH_3COOH)$

7. Which of the following sets contains Lewis acids only ?
(a) BF_3, Ni^{2+}, $SnCl_4$, CO_2
(b) NH_3, C_5H_5N, H_2O, CO_2
(c) BF_3, NH_3, H_2O, CO_2
(d) $SnCl_4$, C_5H_5N, Ni^{2+}, CO_2

8. The acidic nature of CH_4, NH_3, H_2O and HF is in the order :
(a) $CH_4 < NH_3 < H_2O < HF$
(b) $CH_4 < H_2O < NH_3 < HF$
(c) $CH_4 > NH_3 > H_2O > HF$
(d) $CH_4 > H_2O > NH_3 > HF$

9. Conjugate base of hydrozoic acid is :
(a) HN_3^- (b) N_2^-
(c) N_3^- (d) N^{3-}

10. The decreasing order of strength of the bases $OH^-, NH_2^-, H-C\equiv\bar{C}$ and $CH_3-CH_2^-$ is :
(a) $CH_3-CH_2^- > NH_2^- > H-C\equiv\bar{C} > OH^-$
(b) $H-C\equiv\bar{C} > CH_3-CH_2^- > NH_2^- > OH^-$
(c) $OH^- > NH_2^- > H-C\equiv\bar{C} > CH_3-CH_2^-$
(d) $NH_2^- > H-C\equiv\bar{C} > OH^- > CH_3-CH_2^-$

11. NH_4^+ ion is :
(*a*) a conjagate acid
(*b*) a conjugate base
(*c*) neither an acid nor a base
(*d*) both an acid and a base

12. Which of the following is a Bronsted acid as well as Bronsted base ?
(*a*) Na_2CO_3 (*b*) H_2O
(*c*) NH_3 (*d*) BF_3

13. Of the given anions, the strongest Bronsted base is :
(*a*) ClO^- (*b*) ClO_2^-
(*c*) ClO_3^- (*d*) ClO_4^-

14. In the reaction, $I_2 + I^- \rightarrow I_3^-$ which species acts as Lewis base ?
(*a*) I_2 (*b*) I^-
(*c*) I_3^- (*d*) None of these

15. Which one of the following is the strongest acid ?
(*a*) $ClO_3(OH)$ (*b*) $ClO_2(OH)$
(*c*) $SO(OH)_2$ (*d*) $SO_2(OH)_2$

16. Which of the following is not a Lewis acid ?
(*a*) BF_3 (*b*) $AlCl_3$
(*c*) $FeCl_3$ (*d*) PH_3

17. Which of the following is not a Lewis base ?
(*a*) Ag^+ (*b*) H_2O
(*c*) CN^- (*d*) NH_3

18. Which of the following can be classified as a Bronsted base ?
(*a*) NO_3^- (*b*) H_3O^+
(*c*) CH_3COOH (*d*) NH_4^+

19. A compound having the formula, $NH_2 - CH_2 - COOH$ may behave :
(*a*) only as an acid
(*b*) only as a base
(*c*) both as an acid and a base
(*d*) neither as an acid nor as a base

20. Out of the equilibrium reactions :

(*i*) $NH_3 + CH_3COOH \rightleftharpoons CH_3COO^- + NH_4^+$

(*ii*) $H_2O + CH_3COOH \rightleftharpoons CH_3COO^- + H_3O^+$

(*iii*) $4NH_3 + Cu^{2+} \rightleftharpoons [Cu(NH_3)_4]^{2+}$

(*iv*) $HCl + CH_3COOH \rightleftharpoons CH_3COOH_2^+ + Cl^-$

the acid-base reactions, on the basis of Bronsted-Lowry concept, are :
(*a*) (*i*), (*ii*), (*iii*)
(*b*) (*i*), (*ii*), (*iv*)
(*c*) (*i*), (*ii*), (*iii*), (*iv*)
(*d*) None of the reactions

21. In HS^-, I^-, $R{-}NH_2$ and NH_3, the order of proton accepting tendency will be :
(*a*) $I^- > NH_3 > R - NH_2 > HS^-$
(*b*) $NH_3 > R - NH_2 > HS^- > I^-$
(*c*) $R - NH_2 > NH_3 > HS^- > I^-$
(*d*) $HS^- > R - NH_2 > NH_3 > I^-$

ANSWERS

1. (*a*)	**2.** (*c*)	**3.** (*c*)	**4.** (*d*)	**5.** (*c*)	**6.** (*d*)	**7.** (*a*)	**8.** (*a*)
9. (*c*)	**10.** (*a*)	**11.** (*a*)	**12.** (*b*)	**13.** (*a*)	**14.** (*b*)	**15.** (*a*)	**16.** (*d*)
17. (*a*)	**18.** (*a*)	**19.** (*c*)	**20.** (*b*)	**21.** (*c*)			

HINTS AND EXPLANATIONS

1. Since the electron density on N-atom in NH^{2-} ion is the highest, this ion can donate electron pair or can accept a proton most easily and hence is the strongest base.

2. It is only HCO_3^- ion which can accept (Bronsted base) as well as donate (Bronsted acid) a proton (H^+). When it accepts a proton, it forms its conjugate acid (H_2CO_3) and hence acts as

Bronsted base. On the other hand, when HCO_3^- ion donates a proton, it forms its conjugate base (CO_3^{2-}) and hence acts as Bronsted acid.

$$\underset{\text{Conjugate base}}{CO_3^{2-}} \xleftarrow[\text{Bronsted acid}]{\text{Donates a proton }(-H^+)} HCO_3^- \xrightarrow[\text{Bronsted base}]{\text{Accepts a proton }(+H^+)} \underset{\text{Conjugate acid}}{H_2CO_3}$$

3. It is only HSO_4^- ion which can accept (Bronsted base) as well as donate (Bronsted acid) a proton (H^+)

$$SO_4^{2-} \xleftarrow[\text{Bronsted acid}]{\text{Donates a proton }(-H^+)} HSO_4^- \xrightarrow[\text{Bronsted base}]{\text{Accepts a proton }(+H^+)} H_2SO_4$$

4. The conjugate acids of the given bases are HNO_3 ($NO_3^- + H^+ \rightarrow HNO_3$), HCl ($Cl^- + H^+ \rightarrow HCl$), H_2SO_4 ($SO_4^{2-} + 2H^+ \rightarrow H_2SO_4$) and CH_3COOH ($CH_3COO^- + H^+ \rightarrow CH_3COOH$) respectively. Now amongst the conjugate acids *viz* HNO_3, HCl, H_2SO_4 and CH_3COOH, since CH_3COOH is the weakest acid, its conjugate base namely CH_3COO^- is the strongest base.

5. Conjugate acid of a base is obtained by adding a proton (H^+) to the base.

$$\text{Conjugate base} + H^+ \longrightarrow \text{Conjugate acid}$$

Thus conjugate acid of NH_2^- is NH_3, since NH_3 is obtained by adding a proton (H^+ to NH_2^-.

$$\underset{\text{Conjugate base}}{NH_2^-} \xrightarrow{+H^+} \underset{\text{Conjugate acid}}{NH_3} + H^+$$

6. Since a conjugate acid-base pair differs only by a proton (H^+), HCl – Cl^- and $CH_3COOH_2^+$ – CH_3COOH are two conjugate acid-base pairs.

Conjugate acid-base pair (HCl → Cl^-)

$$\underset{\text{Acid}}{HCl} + \underset{\text{Base}}{CH_3COOH} \rightleftharpoons \underset{\text{Base}}{Cl^-} + \underset{\text{Acid}}{CH_3COOH_2^+}$$

Conjugate acid-base pair (CH_3COOH → $CH_3COOH_2^+$)

7. Since B-atom in BF_3 molecule has incomplete octet, this molecule can accept a pair of electrons and hence acts as a Lewis acid. Since Ni^{2+} ion has vacant *d*-orbitals, this ion can accept pairs of electrons and hence acts as a Lewis acid. Since Sn atom (central atom) in $SnCl_4$ molecule can expand its outer octet by taking up electrons in its empty *d*-orbitals, $SnCl_4$ is a Lewis acid. CO_2 molecule has two double bonds (multiple bonds) between C and O atoms (O = C = O) which have different electronegativities. Since C-atom has lower electronegativity than O-atom, this atom acquires a slight positive charge and hence can accept a pair of electrons. Thus CO_2 acts as a Lewis acid.

8. The given compounds are the hydrides of the elements of the same period (*2nd* period). Their acidic nature increases with the increase in electronegativity of their central atom. Thus since the electronegativities of C, N, O and F increase as C < N < O < F, the acidic nature of their hydrides also increases as $CH_4 < NH_3 < H_2O < HF$.

9. N_3^- ion is the conjugate base of hydrazoic acid (HN_3), since it is obtained by removing a proton (H^+) from HN_3

Conjugate acid — H^+ $\longrightarrow$ Conjugate base

or Conjugate acid $\longrightarrow$ Conjugate base + H^+

$$\underset{\text{Conjugate acid}}{HN_3} \longrightarrow \underset{\text{Conjugate base}}{N_3^-} + H^+$$

10. The conjugate acids corresponding to the given bases are H_2O, NH_3, $H-C\equiv C-H$ and CH_3-CH_3 respectively. Now since the acidic nature of these acids decreases as :

$$H_2O > H-C\equiv C-H > NH_3 > CH_3-CH_3$$

Strongest acid Weakest acid

the basic nature of the bases derived from these acids will be in the opposite order (Note that a strong acid has weak conjugate base) :

$$OH^- < H-C\equiv C-H < NH_2^- < CH_3-CH_2^-$$

Weakest base Strongest base

or

$$CH_3-CH_2^- > NH_2^- > H-C\equiv C-H > OH^-$$

Strongest base Weakest base

11. Since NH_4^+ ion can lose a proton to give NH_3, this ion is a conjugate acid of NH_3 which acts as a base

$$\underset{\text{Conjugate acid}}{NH_4^+} - H^+ \longrightarrow \underset{\text{Conjugate base}}{NH_3}$$

12. H_2O is a Bronsted acid, since it can lose a proton (H^+) as is evident from reaction (*i*). H_2O is also a Bronsted base, since it can also accept a proton as is evident from reaction (*ii*)

(*i*) $\underset{\text{Bronsted acid}}{H_2O} + CO_3^{2-} \rightleftharpoons OH^- + HCO_3^-$ (H^+ transferred from H_2O to CO_3^{2-})

(*ii*) $\underset{\text{Bronsted base}}{H_2O} + HCl \rightleftharpoons H_3O^+ + Cl^-$ (H^+ transferred from HCl to H_2O)

13. Explanation based on the distribution of unit negative charge on O-atoms. Since O-atom is more electronegative than Cl-atom, the unit negative charge on each of the oxy-anions is distributed (localised) on O-atoms rather than on Cl-atom. Since ClO^- ion has one O-atom, the whole of unit negative charge is distributed on this O-atom. Since ClO_2^- ion has two O-atoms, the unit negative charge is now distributed on two O-atoms of the ion. Similarly in

ClO_3^- and ClO_4^- ions, the unit negative charge is distributed on three and four O-atoms respectively. Thus we see that as we move from ClO^- to ClO_4^- ions, due to the increase in the number of O-atoms, the distribution of unit negative charge on these ions increases as : $ClO^- < ClO_2^- < ClO_3^- < ClO_4^-$. Thus the stability of these ions also increases as $ClO^- < ClO_2^- < ClO_3^- < ClO_4^-$. Now since ClO^- ion has the minimum stability, it has maximum tendency to accept the proton (H^+) to from its conjugate acid (HClO) and hence is the most basic.

Explanation based on acidic nature of conjugate acids of ClO^-, ClO_2^-, ClO_3^- and ClO_4^- ions. Conjugate acids corresponding to ClO^-, ClO_2^-, ClO_3^- and ClO_4^- ions are HClO, $HClO_2$, $HClO_3$ and $HClO_4$ respectively. Due to the increase of O.N. of the central Cl-atom in these acids, the acidic nature of the acids increases as HClO (O.N. of Cl = + 1) < $HClO_2$ (O.N. of Cl = + 3) < $HClO_3$. (O.N. of Cl = + 5) < $HClO_4$ (O.N. of Cl = + 7). Now we know that a weak acid has strong conjugate base and *vice versa*. Thus since HClO has the minimum acidic nature, its conjugate base *viz* ClO^- has the maximum basic character. Thus the basic character of the ions is in the order : $ClO^- > ClO_2^- > ClO_3^- > ClO_4^-$

14. I^- ion has 4 electron pairs ($I^- = 5s^2\ 5p^6$). When it reacts with I_2 molecule (I – I), this ion donates one of its electron pairs to I atom of I_2 to form $I - I \leftarrow I^-$ or I_3^- ion. Thus I^- ion acts as a Lewis base.

$$I - I + I^- \rightarrow I - I \leftarrow I^- \text{ or } I_3^-$$

15. The acidic nature of the given acids increases with the increase in the oxidation number of the central atom. Thus the acidic nature of the given acids increases as :
$\overset{+4}{H_2SO_3} < H\overset{+5}{Cl}O_3 < H_2\overset{+6}{Cl}O_4 < H\overset{+7}{Cl}O_4$. Consequently $HClO_4$ is the strongest acid.

16. Since BF_3, $AlCl_3$ and $FeCl_3$ are electron deficient compounds, these are Lewis acids. On the other hand, since PH_3 molecule has a lone pair of electrons on P-atom, this molecule is a Lewis base. Thus PH_3 is not a Lewis acid.

17. Since H_2O, CN^- and NH_3 have two (on O-atom), one (on C as well as O-atom, $:C \equiv N:$) and one (on N-atom) electron pairs, these species can donate these electron pairs to Lewis acids and hence act as Lewis bases. Since Ag^+ ion has vacant orbitals, this ion can accept electron pairs and hence acts as a Lewis acid, *i.e.,* Ag^+ ion is not a Lewis base.

18. Since NO_3^- ion can accept a proton (H^+) to give its conjugate acid (HNO_3), this ion is a Bronsted base. Since other species *viz* H_3O^+, CH_3COOH and NH_4^- are not able to accept a proton, these species are not Bronsted bases.

19. Since the given compound has a lone pair of electrons on N-atom and one carboxylic acid group (COOH), it can donate its lone pair of electrons to a Lewis acid and can give a proton

(H^+) to a Bronsted base. Thus this compound can act as a Lewis base and Bronsted acid. Consequently this compound can behave both as an acid (Bronsted acid) and as a base (Lewis base).

20. In reaction (*i*) since CH_3COOH gives a proton (H^+) to NH_3, it acts as a Bronsted acid and NH_3 acts as a Bronsted base. Similarly in reaction (*ii*) CH_3COOH acts as a Bronsted acid and H_2O acts aas a Bronsted base. In reaction (*iii*), since NH_3 molecule donates a lone pair of electrons to Cu^{2+} ion, this molecule acts as a Lewis base and Cu^{2+} ion behaves as a Lewis acid. In reaction (*iv*) since HCl gives a proton (H^+) to CH_3COOH, it acts as a Bronsted acid and CH_3COOH behaves as a Bronsted base. This discussion shows that reactions (*i*), (*ii*) and (*iv*) are acid-base reactions on the basis of Bronsted-Lowry concept. Thus the correct answer is (*b*).

21. Since R is electron releasing group, the electron pair on N-atom in $R - NH_2$ is the most easily available for protonation and hence $R - NH_2$ is the most basic.

20

General Principles of Extraction of Metals and Furnaces

MULTIPLE CHOICE QUESTIONS

1. In the manufacture of iron from haematite ore, lime stone is added to act as :

(*a*) flux
(*b*) slag
(*c*) a reducing agent
(*d*) an oxidising agent

2. Which of the following is not basic flux ?

(*a*) $CaCO_3$ (*b*) CaO
(*c*) MgO (*d*) SiO_2

3. Which of the following can be used as a flux for removing basic impurities during smelting ?

(*a*) H_2SO_4
(*b*) $CaCO_3$
(*c*) SiO_2
(*d*) Both CaO and SiO_2

4. The process of converting hydrated alumina into anhydrous alumina is called :

(*a*) Roasting (*b*) Smelting
(*c*) Dressing (*d*) Calcination

5. Froth flotation process is used for the metallurgy of :

(*a*) Chloride ores (*b*) Amalgams
(*c*) Oxide ores (*d*) Sulphide ores

6. The oil used in the flotation process for the purification of ores is :

(*a*) coconut oil (*b*) olive oil
(*c*) prime oil (*d*) mustard oil

7. The metals that cannot be obtained by the electrolysis of aqueous solution of their salts are :

(*a*) Ag (*b*) Mg
(*c*) Cu (*d*) Al

8. Froth flotation process may be used to increase the concentration of the mineral in :

(*a*) Chalcopyrites (*b*) Bauxite
(*c*) Haematite (*d*) Calamine

9. Which of the following processes is fused in the extractive metallurgy of Mg metal ?

(*a*) Fused salt electrolysis
(*b*) Self reduction
(*c*) Aqueous solution electrolysis
(*d*) Thermite reduction

10. Electrometallurgical process (electrolysis of fused salt) is employed to extract :

(*a*) Pb (*b*) Ag
(*c*) Na (*d*) None of these

11. The most electropositive metals are isolated from their ores by :

(*a*) high temperature reduction with carbon
(*b*) self reduction
(*c*) thermal decomposition
(*d*) electrolysis of fused ionic salts
(*e*) displacement method

12. Electrolytic reduction process is used for the extraction of :

(*a*) highly electronegative elements
(*b*) highly electropositive elements
(*c*) transition metals
(*d*) noble metals

13. The metals than can be extracted by carbon reduction process are :

(*a*) Pb, Fe (*b*) Na, K
(*c*) Pb, Na (*d*) Fe, K

14. Which of the following metals is obtained by leaching its ore with dilute cyanide solution ?
(*a*) Ag (*b*) Ti
(*c*) V (*d*) Zn

15. The metals that can be extracted by cyanide process are :
(*a*) Ag, Au (*b*) Cu, Fe
(*c*) Ag, Cu (*d*) Au Fe

16. The process of zone refining is used in the purification of :
(*a*) Si (*b*) Al
(*c*) Ag (*d*) Cu

17. Lead present as impurity in impure Ag is removed by :
(*a*) Polling (*b*) Cupellation
(*c*) Levigation (*d*) Distillation

18. In the manufacture of cast iron from haematite are (Fe_2O_3) the slag formed is :
(*a*) CO (*b*) $FeSiO_3$
(*c*) $MgSiO_3$ (*d*) $CaSiO_3$

19. Which of the following statements is correct regarding the slag obtained during the extraction of a metal like Cu or Fe ?
(*a*) The slag is lighter and has lower melting point than the metal
(*b*) The slag is ligher and has higher melting point than the metal
(*c*) The slag is heavier and has lower melting point than the metal
(*d*) The slag is heavier and has higher melting point than the metal

20. Heating pyrites to remove sulphur is called :
(*a*) Smelting (*b*) Calcination
(*c*) Liquation (*d*) Roasting

21. Nickel is purified by thermal dicomposition of its :
(*a*) Hydride (*b*) Chloride
(*c*) Carbonyl (*d*) Iodide

22. In the equation, $4\text{ M} + 8\text{ CN}^- + 2H_2O + O_2 \rightarrow 4[M(CN)_2]^- + 4OH^-$, the metal M is
(*a*) Copper (*b*) Iron
(*c*) Gold (*d*) Zinc

23. The common method of extraction of metals from oxide ores is :
(*a*) Reduction with carbon
(*b*) Reduction with hydrogen
(*c*) Reduction with aluminium
(*d*) Electrolytic method

24. Auto-reduction process is used for the extraction of :
(*a*) Hg (*b*) Fe
(*c*) Cu (*d*) Pb

25. Which method of purification is represented by the equations :

$$\text{Ti}(impure) + 2I_2 \xrightarrow{500\text{ K}} TiI_4$$

$$\xrightarrow{1675\text{ K}} \text{Ti }(pure) + 2I_2$$

(*a*) Cupellation (*b*) Poling
(*c*) Van Arkel (*d*) Zone refining

ANSWERS

1. (*a*)	**2.** (*d*)	**3.** (*c*)	**4.** (*d*)	**5.** (*d*)	**6.** (*c*)	**7.** (*b*), (*d*)	**8.** (*a*)
9. (*a*)	**10.** (*c*)	**11.** (*d*)	**12.** (*b*)	**13.** (*a*)	**14.** (*a*)	**15.** (*a*)	**16.** (*a*)
17. (*b*)	**18.** (*d*)	**19.** (*a*)	**20.** (*d*)	**21.** (*c*)	**22.** (*c*)	**23.** (*a*)	**24.** (*a*), (*c*), (*d*)
25. (*c*)							

HINTS AND EXPLANATIONS

1. Lime stone ($CaCO_3$) present in the charge (charge contains haematite ore, Fe_2O_3 and $CaCO_3$) decomposes at 1000° C to give CaO which acts as a basic flux, since it combines with silica (SiO_2) present as acidic impurity (gangue) to form a fusible slag of $CaSiO_3$

$$CaCO_3 \xrightarrow{1000°\text{ C}} CaO + CO_2$$

$$\underset{\text{Basic flux}}{CaO} + \underset{\text{Acidic impurity}}{SiO_2} \longrightarrow \underset{\text{Fusible slag}}{CaSiO_3}$$

2. Since CaO (obtained by the decomposition of $CaCO_3 \longrightarrow CaO + CO_2$), MgO and CaO all are basic oxides, these are basic fluxes. On the other hand, since SiO_2 is an acidic oxide, it is an acidic flux, *i.e.*, SiO_2 is not a basic flux.

3. Basic impurities are removed by using acidic flux. H_2SO_4 cannot be used as a flux. CaO obtained by the decomposition of $CaCO_3$ is a basic oxide and hence is used for removing acidic impurities. However, SiO_2 is an acidic oxide (or acidic flux) and hence is used for removing basic impurities.

$$\underset{\text{Acidic flux}}{SiO_2} + \text{Basic impurity} \longrightarrow \text{Slag}$$

4. We know that the process of converting hydrated oxides to anydrous oxides by heating is called calcination. For example :

$$2Fe_2O_3 . 3H_2O \text{ (Limonite)} \xrightarrow{\Delta} 2Fe_2O_3\,(s) + 3H_2O\,(g)$$

$$Al_2O_3 . 2H_2O \text{ (Hydrated alumina)} \xrightarrow{\Delta} Al_2O_3\,(s) + 2H_2O(g)$$

5. Froth flotation process is commonly used for the concentration of low grade sulphide ores like galena (PbS), copper-iron pyrites (Cu_2S. Fe_2S_3 or $CuFeS_2$), zinc blende (ZnS) etc.

6. Pine oil is used as a frother. It forms a stable froth which rises to the top of the flotation cell. It is added to the water solution of the sulphide ore in small quantity.

7. Cu and Ag can be obtained by electrolysing aqueous solution of $CuSO_4$ and $AgNO_3$ containing 10% HNO_3 respectively. Metals are obtained on cathode. On the other hand, Mg and Al can not be obtained by this process since these metals are very reactive and react with H_2O and evolve H_2 gas. Thus instead of metals, H_2 is discharged on cathode.

$$Mg + H_2O \rightarrow Mg\,(OH)_2 + H_2\uparrow, \qquad 2Al + 6H_2O \rightarrow 2Al(OH)_3 + 3H_2\uparrow$$

8. We know that froth flotation process is used to increase the concentration of sulphide ores only. Out of the given minerals the sulphide ore is only chalcopyrite (CuF. FeS or $CuFeS_2$). Other minerals are bauxite, Al_2O_3. $2H_2O$ (oxide ore) haematite, Fe_2O_3 (oxide ore) and calamine, $ZnCO_3$ (carbonate ore)

9. Since Mg is a highly electropositive metal, it is obtained by electrolysing the suitable fused salt of the metal. The salts of Mg cannot be reduced to free Mg metal by other metals.

10. Since Na is a highly electropositive metal and strong reducing agent, its salt cannot be easily reduced to the free metal by other metals. This metal can also not be obtained by displacing it from its salt aqueous solution by other metals. However this metal can be obtained by electrolysing its suitable fused salt.

11. (*a*) If oxides of highly electropositive metals are reduced by carbon at high temperature, the metal obtained combines with carbon and forms metallic carbide. Thus high temperature reduction with carbon process cannot be used to extract highly electropositive metals.

(*b*) Self-reduction process is used to extract less electropositive metals like Pb, Cu, Hg etc by reducing their sulphide ore by heating them in air. No external reducing agent is used.

(*c*) Thermal decomposition process is used for getting pure Ni by decomposing $Ni(CO)_4$ (volatile compound) at 180° C.

$$Ni(CO)_4 \xrightarrow{180^\circ C} Ni + 4CO$$

(*d*) Being highly electropositive in nature, these metals cannot obtained by displacing them from their salt aqueous solutions by other metals.

12. Since highly electropositive metals like Na, K, Mg, Ca, Al etc are strong reducing agents, their salts cannot be easily reduced to the free metals by other metals. These metals are, therefore, obtained by electrolytic reduction of their fused oxides (*e.g.*, MgO, Al_2O_3), hydroxides (*e.g.*, NaOH, KOH) and chlorides (*e.g.*, NaCl, KCl, $MgCl_2$, $CaCl_2$)

13. Since Na and K are highly electropositive metals, these metals are obtained by electrolytic reduction of their fused hydroxies (NaOH, KOH) or chlorides (NaCl, KCl). On the other hand, since Pb and Fe are less electropositive metals, these can be obtained by reducing their oxides with carbon (carbon reduction process)

$$PbO + C \rightarrow Pb + CO, \qquad PbO + CO \rightarrow Pb + CO_2$$

$$Fe_2O_3 + 3C \rightarrow 2Fe + 3CO, \qquad Fe_2O_3 + 3CO \rightarrow 2Fe + 3CO_2$$

14. Ag is obtained by leaching its ore *viz* argentite (Ag_2S) or horn silver (AgCl) with an aqueous solution of NaCN. When Ag_2S ore is treated (leached) with aqueous solution of NaCN in presence of a current of hot air, soluble sodium dicyano argentate (I), Na[Ag (CN_2)] (complex compound) is obtained

$$Ag_2S\ (\textit{Argentite ore}) + 4NaCN + 2O_2\ (\textit{From air}) \rightarrow 2Na\,[Ag\,(CN)_2] + Na_2SO_4$$

or $$Ag_2S + 4CN^- + 2O_2 \rightarrow 2[Ag\,(CN)_2]^- + SO_4^{2-}$$

When AgCl ore is treated with an aqueous solution of NaCN, the same complex is formed

$$AgCl + 2NaCN \rightarrow Na\,[Ag\,(CN)_2] + NaCl$$

or $$Ag^+ + 2CN^- \rightarrow [Ag\,(CN)_2]^-$$

Alternatively, Ag from its ore (Ag_2S or AgCl) reacts with the aqueous solution of NaCN in presence of hot air and forms the complex, Na[Ag $(CN)_2$]

$$4Ag + 8\,NaCN + 2H_2O + O_2 \rightarrow 4Na\,[Ag\,(CN)_2] + 4NaOH$$

or $$4Ag + 8\,CN^- + 2H_2O + O_2 \rightarrow 4[Ag(CN)_2]^- + 4OH^-$$

Ag can be precipitated from its complex compound, Na[Ag$(CN)_2$] by treating it with Zn dust. Since Zn is more electropositive than Ag, this metal displaces Ag from the complex compound.

$$2Na[Ag(CN)_2] + Zn \rightarrow Na_2\,[Zn(CN)_4] + 2Ag \downarrow$$

15. The process of extracting a metal by treating (leaching) its ore with an aqueous solution of NaCN is called by various names like *Mac-Arthur Forest cyanide process, Complex formation process* or simply by *cyanide process*. This process is used for the extraction of Ag and Au as shown below :

Extraction of Ag. See the answer of question number 14.

Extraction of Au. The tailings obtained at the end of the amalgamation process which is used for the extraction of Au from *hard auriferous quartz rocks* also contain some quantity of Au. When these tailings are treated with a dilute solution of NaCN (0.3%) in presence of air, gold gives soluble complex compound, Na[Au $(CN)_2$] (called sodium aurocyanide)

$$4Au + 8NaCN + 2H_2O + O_2 \rightarrow 4Na\,[Au\,(CN)_2] + 4NaOH$$

or $$4Au + 8CN^- + 2H_2O + O_2 \rightarrow 4\,[Au\,(CN)_2]^- + 4OH^-$$

Au can be precipitated from the complex compound by treating it with Zn dust

$$2Na\,[Au\,(CN)_2] + Zn \rightarrow Na_2\,[Zn\,(CN)_4] + 2Au \downarrow$$

16. The process of zone refining is used for the purification of semiconductors like silicon (Si), germanium (Ge) and gallium (Ga).

17. Since Pb which is present as impurity in Ag forms a volatile oxide (PbO) with a blast of air or O_2, this impurity is removed from the impure Ag by cupellation process. In this process the impure Ag is heated in a cupel in presence of a blast of air or O_2. The impurity of Pb is converted into volatile oxide *viz* hitharge (PbO) which escapes along with the blast of air or O_2 while pure Ag (shining mass) is left behind is the cupel.

18. The charge used in the manufacture of cast iron is a mixture of calcined haematite ore (Fe_2O_3), coke and lime stone ($CaCO_3$). Coke acts both as a fuel as well as a reducing agent while lime stone acts as a basic flux. Spongy iron is formed in the upper part of the furnace (blast furnace). This spongy iron contains SiO_2 as impurity. When impure spongy iron and $CaCO_3$ reach the middle part of the furnace, $CaCO_3$ gets decomposed to give lime (CaO). CaO thus formed acts as a basic flux and combines with the impurity of SiO_2 (acidic impurity) to form fusible slag of $CaSiO_3$

$$\underset{\text{Basic flux}}{CaO} + \underset{\text{Acidic impurity}}{SiO_2} \longrightarrow \underset{\text{Slag}}{CaSiO_3}$$

19. Since slag is lighter than the molten metal, it floats over the surface of the molten metal and hence can be skimmed off the surface from time to time.

20. The copper pyrites, CuS, FeS or $CuFeS_2$ (ore of Cu) contains the non-metallic impurities of S, P, As and Sb. When the concentrated pyrites ore is heated strongly in the presence of excess of air (Roasting), the following reactions take place :

(*a*) The impurities of S, P, As and Sb are removed as their volatile oxides *viz* SO_2, P_2O_5, As_2O_3 and Sb_2O_3

$$S + O_2 \longrightarrow SO_2, \quad 4P + 5O_2 \longrightarrow 2P_2O_5$$
$$4M\ (M = As, Sb) + 3O_2 \longrightarrow 2M_2O_3$$

(*b*) Copper pyrites is converted into a mixture of cuprous sulphide (Cu_2S) and iron sulphide (FeS) which, in turn, are partially, oxidised to cuprous oxide (Cu_2O) and ferrous oxide (FeO) respectively.

$$\underset{\text{Copper pyrites}}{2CuFeS_2} + O_2 \longrightarrow Cu_2S + 2FeS + 5O_2$$
$$2Cu_2S + 3O_2 \longrightarrow 2Cu_2O + 2SO_2$$
$$2FeS + 3O_2 \longrightarrow 2FeO + 2SO_2$$

21. Ni obtained from its ore *viz* pentlandite (Ni, Cu, Fe) S contains platinum metals (Ru, Rh, Pd, Os, Ir, Pt) and coinage metals (Cu, Ag, Au) etc. as impurities. In order to remove these impurities, crude nickel is treated with CO at 60-80°C. Ni from the crude matter forms volatile nickel tetracarbonyl, $Ni(CO)_4$ while Pt metals and coinage metals (impurities) are left as non-volatile residue. Vapours of Ni $(CO)_4$ get decomposed at 180° C to give pure Ni. [Thermal decomposition of $Ni(CO)_4$]

$$Ni(s)\ (\textit{Crude}) + 4\ CO\ (g) \xrightarrow{60\text{-}80°\ C} Ni(CO)_4\ (g)$$

$$Ni\ (CO)_4 \xrightarrow{180°\ C} Ni\ (\textit{Pure metal}) + 4CO$$

22. The given equation represents the formation of a complex compound of gold (Au) in the extraction of this metal by *cyanide process* (also called *Mac-Arthur Forest cyanide process or complex formation process*). In this process, Au metal from its ore reacts with an aqueous solution of NaCN in presence of hot air.

$$4Au + 8\ NaCN + 2H_2O + O_2 \rightarrow 4Na\,[Au(CN)_2] + 4NaOH\ (\textit{Molecular equation})$$

or $$4Au + 8CN^- + 2H_2O + O_2 \rightarrow 4[Au(CN)_2]^- + 4OH^-\ (\textit{Ionic equation})$$

23. The method used for the extraction of metals from their oxides consists of reducing their oxides to free metals. Reduction of oxides may be done with the help of various types of reducing agents such as C, CO, H_2, highly electropositive metals like Mg, Al etc. The common reduction method is the reduction by carbon. Examples are :

$$ZnO + C \rightarrow Zn + CO\uparrow, \qquad SnO_2 + C \rightarrow Sn + 2CO\uparrow$$

$$PbO + C \rightarrow Pb + CO\uparrow, \qquad Fe_2O_3 + 3C \rightarrow 2Fe + 3CO\uparrow$$

24. In some reactions the sulphides of less active metals like Pb, Cu, Mg etc. reduce the oxides of these meals to the free metals, when the sulphides are heated strongly in the excess of air (Roasting). In this reaction, first of all the sulphides are oxidised to the oxides or sulphates which are further reduced to the free metal by the excess of sulphides. Examples are :

(*a*) $$2PbS + 3O_2 \longrightarrow 2PbO + 2SO_2$$

$$2PbO + 2PbS \xrightarrow{\text{self-reduction}} 3Pb + SO_2$$

(*b*) $$2Cu_2S + 3O_2 \longrightarrow 2Cu_2O + 2SO_2$$

$$2Cu_2O + Cu_2S \xrightarrow{\text{self-reduction}} 6Cu + SO_2$$

(*c*) $$2HgS + 3O_2 \longrightarrow 2HgO + 2SO_2$$

$$2HgO + HgS \xrightarrow{\text{self-reduction}} 3Hg + SO_2$$

(*d*) $$PbS + 2O_2 \longrightarrow PbSO_4$$

$$PbSO_4 + PbS \xrightarrow{\text{self-reduction}} 2Pb + 2SO_2$$

The reduction of oxides (PbO, Cu_2O, HgO) or sulphate ($PbSO_4$) to the free metal by the excess of sulphides of these metals is called self-reduction or auto-reduction.

25. The method represented by the given equation is used for purifying impure Ti metal and is called Van Arkel or iodine refining process. In this process the impure Ti metal is heated with I_2 so that the metal is converted into volatile unstable Ti I_4 (*g*) and impurities are left behind. The unstable TiI_4 (*g*), on being heated, gives pure Ti metal.

21

Nomenclature of Inorganic Compounds

MULTIPLE CHOICE QUESTIONS

1. $H_4P_2O_7$ is called :
(*a*) Diphosphoric acid
(*b*) Triphosphoric acid
(*c*) Phosphoric acid
(*d*) Trisulphuric acid.

2. Which of the following is the correct name of F_2O ?
(*a*) Fluorine oxide
(*b*) Oxygen difluoride

3. Which of the following is hypochlorous acid ?
(*a*) $HClO$ (*b*) $HClO_2$
(*c*) $HClO_3$ (*d*) $HClO_4$

4. The acids obtained by removing one H_2O molecule from one and two molecules of H_3PO_4 are called :
(*a*) metaphosphoric acid and pyrophosphoric acid
(*b*) pyrophosphoric acid and metaphosphoric acid
(*c*) metaplosphoric acid and hypophosphoric acid
(*d*) hypophosphoric acid and metaphosphoric acid.

5. The acid obtained by the reaction between H_2O_2 and H_2SO_4 is called :
(*a*) Caros's acid
(*b*) Marshall's acid

6. Which of the following is the correct name for $H_5P_3O_{10}$?
(*a*) Pyrophosphoric acid
(*b*) Orthophosphoric acid
(*c*) Triphosphoric acid
(*d*) Metaphosphoric acid

7. Fill in the blanks :
(*a*) The cation obtained by adding a proton to PH_3 is called ....
(*b*) Pb_3O_4 is called ....
(*c*) NO^+ ion is called ..... .
(*d*) NO_2^+ ion is called ..... .
(*e*) $H_2S_2O_8$ is called ..... .

8. Which of the following is the correct name for $[P\ W_{12}\ O_{14}]^{3-}$?
(*a*) Dodecatungsto phosphate (3 –)
(*b*) Phosphododeca tungstate (3 –)

9. Which of the following is the correct name of $H_2S_2O_4$?
(*a*) Dithionic acid
(*b*) Thiosulphuric acid
(*c*) Persulphuric acid
(*d*) Dithionous acid.

10. Which of the following is the correct name of $[(CH_3)_3\ Si]_2\ N - NH_2$?
(*a*) *bis* (trimethylsilyl) hydrazine
(*b*) 1, 1 – *bis* (trimethylsilyl) hydrazine
(*c*) N, N – *bis* (trimethylsilyl) hydrazine

11. The ions obtained by removing two H-atoms from H_2O_2 molecule, K-atom from KO_2 molecule and H-atom from HN_3 molecule are called :
(*a*) Peroxide, superoxide and azide ions respectively
(*b*) Super oxide, peroxide and azide ions respectively
(*c*) Azide, peroxide and superoxide ions respectively
(*d*) None of these ions.

ANSWERS

1. (*a*) **2.** (*b*) **3.** (*a*) **4.** (*a*) **5.** (*a*) **6.** (*c*)

7. (*a*) phosphonium, (*b*) trilead tetroxide, (*c*) nitrosyl ion, (*d*) nitronium ion, (*e*) peroxodisulphuric acid,

8. (*a*) **9.** (*d*) **10.** (*b*) **11.** (*a*)

HINTS AND EXPLANATIONS

1. Since the given acid is a condensed acid and contains two P-atoms, it is called diphosphoric acid.

2. Since *F* is more electronegative than *O*, *F* has negative charge on it and *O* has positive charge ($F_2^- \, O^{2+}$). Due to negative charge on *F*-atom, the given compound is the fluoride of oxygen and not the oxide of fluorine. Thus it is called oxygen difluoride.

3. Oxidation state of Cl-atom in the given acids is +1, + 3, + 5 and + 7. We know that if an element forms more than one oxy acids having the element in different oxidation states, then the acid having the element in the lowest oxidation state is called hypo acid. Thus HClO (Cl = +1) is called hypochrous acid.

4. The acids obtained by removing one H_2O molecule from one and two molecules of H_3PO_4 (orthophosphoric acid) are called metaphosphoric acid (HPO_3) and pyrophosphoric acid ($H_4P_2O_7$) respectively.

$$H_3PO_4 \xrightarrow{-H_2O} HPO_3 \quad + H_2O$$

$$2H_3PO_4 \xrightarrow{-H_2O} H_4P_2O_7 \; + H_2O$$

5. The acid obtained by the reaction between H_2O_2 and H_2SO_4 gives peroxomonosuphuric acid, H_2SO_5 (also called Caro's acid)

$$H_2O_2 + H_2SO_4 \xrightarrow{-H_2O} H_2SO_5 \; + H_2O$$

or

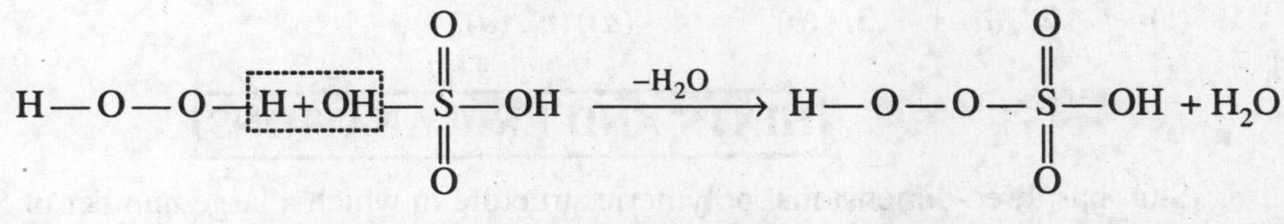

6. Since the given acid has been obtained by condensing three molecules of phosphoric acid (H_3PO_4), it is an isopoly acid and is called triphosphoric acid in which the number of atoms of phosphorus have been indeated by Greek number prefix.

11. H_2O_2 (*Hydrogen peroxide*) $\xrightarrow{-2H^+}$ O_2^{2-} (*Peroxide ion*)

KO_2 (*Potassium superoxide*) $\xrightarrow{-K^+}$ O_2^- (*Superoxide ion*)

HN_3 (*Hydrazoic acid*) $\xrightarrow{-H^+}$ N_3^- (*Azide ion*)

22

General Characteristics of Main Group Elements (*s*-and-*p*-Block Elements)

MULTIPLE CHOICE QUESTIONS

1. Main group elements are :
(*a*) *s*-block elements
(*b*) *p*-block elements
(*c*) *s*-as well as *p*-block element
(*d*) None of these

2. In the atoms of *p*-block elements the differentiating electron enters :
(*a*) *np* orbitals
(*b*) $(n-1)p$ orbitals
(*c*) *ns* orbitals
(*d*) *s* and *p* orbitals both

3. Which of the following has infinite three dimensional polymeric structure ?
(*a*) CO_2 (*b*) SiO_2
(*c*) CO (*d*) CO_3^{2-}

4. Which of the following statements is (are) correct ?
(*a*) The higher non-metals like C, N, O etc have a tendency to form $(p-p)\ \pi$ bond (*s*) while heavier elements like Si, P, S etc. show a reluctance to form $(p-p)\pi$ bond (*s*)
(*b*) The higher non-metals show a reluctance to form $(p-p)\pi$ bond (*s*) while heavier elements have a tendency to form $(p-p)\pi$ bond (*s*)
(*c*) CO_2 has linear structure
(*d*) S_8 has puckered ring structure

ANSWERS

1. (*c*) **2.** (*a*) **3.** (*b*) **4.** (*a*), (*c*), (*d*).

HINTS AND EXPLANATIONS

3. SiO_2 has three-dimensional polymeric structure in which a large number of SiO_4 tetrahedrons are linked together. The linking of many SiO_4 tetrahedrons gives giant molecular (polymeric) structure to SiO_2 molecule.

23

Group Discussion of Main Group elements (*s*-and-*p*-Block Elements)

MULTIPLE CHOICE QUESTIONS

1. The metal which is the best reducing agent in aqueous solution is :

(*a*) Na (*b*) K

(*c*) Rb (*d*) Li

2. A solution of Na metal in liq. NH_3 acts as a strong reducing agent due to the presence of :

(*a*) $NaNH_2$ (*b*) Na atom

(*c*) NaH

(*d*) Solvated electrons

3. The alkali metals dissolve in liq. NH_3 to give a blue solution which is conducting due to :

(*a*) ammoniated electrons

(*b*) loss of electrons

(*c*) ammoniated cations

(*d*) both (*a*) and (*c*)

4. The decrease in melting and boiling points of alkali metals with rise in atomic number is due to :

(*a*) the weakening of covalent bonding

(*b*) weak ionic bonding

(*c*) Van der Wall's forces

(*d*) the weakening of metallic bonding.

5. Which of the following alkali metals emits light of the largest wavelength in the flame test ?

(*a*) Na (*b*) Li

(*c*) K (*d*) Cs

6. Ionic mobility of Li^+ is less than that of Na^+ and K^+ ions both, because :

(*a*) Ionisation potential of Li is mall

(*b*) Charge density of Li^+ ion is high

(*c*) Hydration tendency of Li^+ ion is high

(*d*) Li^+ has two electrons

7. Pure anhydrous magnesium chloride ($MgCl_2$) can be prepared from hydrated salt ($MgCl_2 . 6H_2O$) by :

(*a*) heating the hydrate to red heat in the atmosphere of HCl gas

(*b*) melting the hydrate

(*c*) heating the hydrate with coke

(*d*) heating the hydrate with Mg ribbon.

8. Which of the following is least soluble in water ?

(*a*) $CaCO_3$ (*b*) $NaHCO_3$

(*c*) Na_2SO_4 (*d*) $KHCO_3$

9. $BeSO_4$ is soluble in water but $BaSO_4$ is insoluble, because :

(*a*) $BeSO_4$ is ionic and $BaSO_4$ is covalent.

(*b*) $BeSO_4$ is crystalline while $BaSO_4$ is amorphous.

(*c*) $BeSO_4$ has smaller lattice energy and high heat of hydration as compared to $BaSO_4$.

(*d*) None of these.

10. The sulphate of which metal has the highest solubility in water ?

(*a*) Ca (*b*) Ba

(*c*) Sr (*d*) Mg

11. The metal that is extracted from sea water is :

(*a*) Fe (*b*) Ca

(*c*) Mg (*d*) Au

12. Alkaline earth metals form M^{2+} ions instead of M^+ ions because :
(*a*) M^{2+} ions carry more charge
(*b*) their IE_2 values are not different from their IE_1 values.
(*c*) M^+ ions donot have stable configuration
(*d*) M^{2+} ions have more hydration energy.

13. Amongst the elements of group 3A (B, Al, Ga, In and Tl) the elements which have the highest and lowest melting points are respectively :
(*a*) B, Al (*b*) Ga, In
(*c*) Tl, B (*d*) B, Ga

14. The electropossible character of the elements of group 3A first increases from B to Al and then decreases from Al to Tl down the group because of :
(*a*) decrease in ionisation energy of the elements.
(*b*) increase in size of the elements
(*c*) ineffective shielding of nuclear charge by *d*-electrons in Ga, In and Tl.
(*d*) decrease in electronegativity of the elements

15. Boron does not form B^{3+} ion because
(*a*) Boron has small size and high ionisation energy
(*b*) Boron has high electronegativity
(*c*) Boron has high charge density
(*d*) None of the above

16. Which of the following elements does not show the group oxidation state in its compounds ?
(*a*) B (*b*) Al
(*c*) Ga (*d*) Tl

17. Amongst the elements of group 14 the one having the highest tendency to show stable + 2 oxidation state is :
(*a*) C (*b*) Pb
(*c*) Ge (*d*) Si

18. Amongst the elements of group 14 the oxidising power of he tetravalent species (M^{4+}) increases in the order :
(*a*) Ge < Pb < Sn
(*b*) Ge < Sn < Pb
(*c*) Pb > Ge > Sn
(*d*) None of the above.

19. The tendency towards catenation of group 14 elements down the group :
(*a*) increases
(*b*) decreases
(*c*) remains the same
(*d*) unpredictable

20. Density is the highest for :
(*a*) Pb (*b*) C
(*c*) Si (*d*) Ge

21. Which of the following statements is correct with respect to the property of elements of group 14 with the increase in atomic number ?
(*a*) Their metallic character decreases
(*b*) The stability of + 2 oxidation stable increases
(*c*) Their ionisation energies increase
(*d*) Their atomic size decreases

22. The most non-metallic element in group 15 is :
(*a*) N (*b*) Bi
(*c*) P (*d*) Sb

23. The element having a greater tendency to form multiple bond is :
(*a*) N (*b*) P
(*c*) Bi (*d*) Sb

24. Which of the following has $p\pi - p\pi$ bonding ?
(*a*) P_4 (*b*) As_4
(*c*) Sb_4 (*d*) N_2

25. Which element of the following has the lowest melting point ?
(*a*) N (*b*) As
(*c*) P (*d*) Sb

26. The low reactivity of nitrogen is due to :
(*a*) small atomic radius
(*b*) high electronegativity
(*c*) stable configuration
(*d*) high dissociation energy

27. Which of the elements has maximum tendency for catenation ?

(*a*) O (*b*) Se
(*c*) S (*d*) Te

28. SF_6 is known but SH_6 is not known, because :
(*a*) F is highly electronegative
(*b*) F is present in group 7A but H is present in group 1A.
(*c*) F has seven electrons in its valence shell but H has only one.
(*d*) H can show +1 as well as – 1 oxidation state while F shows only – 1 oxidation state.

29. Oxygen does not form OF_6 because :
(*a*) it has a small size
(*b*) there are no vacant *d*-orbitals available
(*c*) it has high ionisation energy
(*d*) it has large size

30. In its compounds, oxygen can show oxidation state (*s*) of :
(*a*) – 1 (*b*) – 2
(*c*) + 1 (*d*) + 2

31. Excepting Po, all other elements of group 6A are :
(*a*) non-metals (*b*) metalloids
(*c*) radioactive (*d*) polymorphic

32. Which of the following has the highest bond energy ?
(*a*) F_2 (*b*) Br_2
(*c*) I_2 (*d*) Cl_2

33. Which of the following halogens has the lowest bond dissociation energy ?
(*a*) F_2 (*b*) Cl_2
(*c*) Br_2 (*d*) I_2

34. Which of the following halogens has the highest bond dissociation energy ?
(*a*) F_2 (*b*) Cl_2
(*c*) Br_2 (*d*) I_2

35. Which of the following is reduced most readily ?
(*a*) I_2 (*b*) Cl_2
(*c*) F_2 (*d*) Br_2

36. Oxidising action of halogens increases in the order :
(*a*) Cl < Br < I < F
(*b*) Cl < I < Br< F
(*c*) I < F < Cl < Br
(*d*) I < Br < Cl < F

37. When I_2 is passed through KCl, KF and KBr solutions.
(*a*) Cl_2 and Br_2 are evolved
(*b*) Cl_2 is evolved
(*c*) Cl_2, Br_2 and F_2 are evolved
(*d*) None of the these

38. The ease of liquification of noble gases decreases in the order :
(*a*) He > Ne > Ar > Kr > Xe
(*b*) Xe > Kr > Ar > Ne > He
(*c*) Kr > Xe > He > Ar > Xe
(*d*) Ar > Kr > Xe > He > Ne

39. On moving down the group from He to Rn, the boiling points :
(*a*) increase
(*b*) decrease
(*c*) increase but irregularly
(*d*) decrease but irregularly

40. Xe shows oxidation state (s) of :
(*a*) + 2 (*b*) + 4
(*c*) + 6 (*d*) + 8

ANSWERS

1. (*d*) **2.** (*d*) **3.** (*d*) **4.** (*d*) **5.** (*b*) **6.** (*c*) **7.** (*a*) **8.** (*a*)
9. (*c*) **10.** (*d*) **11.** (*c*) **12.** (*d*) **13.** (*d*) **14.** (*c*) **15.** (*a*) **16.** (*d*)
17. (*b*) **18.** (*b*) **19.** (*b*) **20.** (*a*) **21.** (*b*) **22.** (*a*) **23.** (*a*) **24.** (*d*)
25. (*a*) **26.** (*d*) **27.** (*c*) **28.** (*a*) **29.** (*b*) **30.** (*a*), (*b*), (*d*) **31.** (*a*)
32. (*d*) **33.** (*d*) **34.** (*b*) **35.** (*c*) **36.** (*d*) **37.** (*d*) **38.** (*b*) **39.** (*a*)
40. (*a*), (*b*), (*c*), (*d*)

HINTS AND EXPLANATIONS

1. The metal which has the highest value of E^o_{ox} for the oxidation half-reaction, $M(s) \longrightarrow M^+(aq) + e^-$ is the best reducing agent. Now since Li has the highest value of E^o_{ox}, it is the best reducing agent (E^o_{ox} for Na = + 2.70 V, for K = +2.92 V, for Rb = +2.99 V and for Li = +3.04 V).
2. Due to the presence of free ammoniated electrons, $[e(NH_3)_y]^-$ (*i.e.,* solvated electrons) in alkali metal – NH_3 solution, this solution acts as a strong reducing agent.

$$\underset{\text{Alkali metal}}{M} + (x+y)NH_3 \longrightarrow [M(NH_3)_x]^+ + [e(NH_3)_y]^-$$

3. Alakali metal – NH_3 solution is highly conducting due to the formation of ammoniated electrons, $[e(NH_3)_y]^-$ and ammoniated alkali metal cations, $[M(NH_3)_x]^+$.

$$M \longrightarrow M^+ + e^- \quad \text{... }(i)$$

$$M^+ + xNH_3 \longrightarrow [M(NH_3)_x]^+ \quad \text{... }(ii)$$

$$e^- + yNH_3 \longrightarrow [e(NH_3)_y]^- \quad \text{... }(iii)$$

On adding (*i*), (*ii*) and (*iii*) we get :

$$M + (x+y)NH_3 \longrightarrow \underset{\text{Ammoniated cation}}{[M(NH_3)_x]^+} + \underset{\text{Ammoniated electron}}{[e(NH_3)_y]^-}$$

4. The melting and boiling points of alkali metals decrease down the group from Li to Cs because the bonding between the atoms of the metals (metallic bonding) in the solid state decreases down the group.
5. The colours emitted by the given metals are Na – Golden yellow, Li – Cimson red, K – Pale violet, Cs – Violet. Among these colours since red colour has the largest wave length, Li emits light of the largest wavelength. The magnitude of wavelength of the light emitted by a certain metal is directly proportional to its ionisation energy. Thus, since Li has the highest ionisation energy, the wavelength of the red light emitted by it would be the highest.
6. Since Li^+ ion is smaller than Ni^+ and K^+ ions both, this ion has the maximum hydration tendency and hence $[Li(aq)]^+$ ion is the largest. Due to largest size of $[Li(aq)]^+$ ion, ionic mobility of Li^+ ion is least, *i.e.,* it is less than Na^+ and K^+ ions both.
7. Anhydrous magnesium chloride ($MgCl_2$) is obtained by heating the hydrated magnesium chloride ($MgCl_2 . 6H_2O$) to red heat in the atmosphere of HCl gas.

$$MgCl_2 . 6H_2O \xrightarrow[\text{HCl }(g)]{\text{Heat to red heat}} MgCl_2 + 6H_2O\uparrow$$

If hydrated salt is heated in the absence of HCl gas, this salt undergoes hydrolysis with its own water of crystallization and MgO is obtained.

$$MgCl_2 . 6H_2O \longrightarrow Mg(OH)Cl + HCl + 5H_2O$$

$$Mg(OH)Cl \longrightarrow MgO + HCl.$$

The presence of HCl gas prevents the hydrolysis of $MgCl_2 . 6H_2O$ or the formation of MgO.

8. We know that bicarbonates of alkali metals are soluble in water. We also know that among the sulphates of alkali metals only Li_2SO_4 is insoluble while other sulphates (*e.g.*, Na_2SO_4, K_2SO_4 etc) are soluble. Thus $NaHCO_3$, $KHCO_3$ and Na_2SO_4 are soluble. Now since hydration energy of Ca^{2+} ion is low, solubility of $CaCO_3$ in water is also low. Thus $CaCO_3$ is the least soluble in water.

9. Solubility of a salt is high, if its hydration energy is high and lattice energy is low. Thus $BaSO_4$ is soluble because its hydration energy is higher and lattice energy is lower than that of $BaSO_4$.

10. All the given metals belong to the same group (group IIA). The variation in the solubility of their sulphates depends on the variation of hydration energy of M^{2+} ions. Due to the increase in the size of M^{2+} ions from Mg^{2+} to Ba^{2+} ($Mg^{2+} < Ca^{2+} < Sr^{2+} < Ba^{2+}$), hydration energies of these ions decrease as $Mg^{2+} > Ca^{2+} > Sr^{2+} > Ba^{2+}$. Due to the decrease in hydration energies, the solubility of sulphates also decreases ($MgSO_4 > CaSO_4 > SrSO_4 > BaSO_4$). Thus $MgSO_4$ has the highest solubility.

11. Sea water contains Mg in the form of $MgCl_2$ and $MgSO_4$. This water can be used to extract Mg metal as shown below.

Sea water containing $MgCl_2$ and $MgSO_4 + Ca(OH)_2 \longrightarrow \underset{\text{ppt}}{Mg(OH)_2}$

$$\underset{\text{Hydrated}}{MgCl_2 \cdot 6H_2O} \xrightarrow[\text{of HCl gas}]{\text{Heat in presence}} \underset{\text{Anhydrous}}{MgCl_2 + 6H_2O}$$

$$MgCl_2 + NaCl + CaCl_2 \xrightarrow{\text{Fuse}} \text{Fused mixture.}$$

$$\text{Fused mixture} \xrightarrow[\text{in presence of inert gas}]{\text{Electrolysis at 700°C}} \text{Mg metal is collected at cathode}$$

(Iron container – cathode, Graphite rod – anode)

12. Since M^{2+} ions are smaller in size than M^+ ions, the amount of energy released (hydration energy) when these ions are dissolved in water is greater than that which is released when M^+ ions are dissolved in water. M^{2+} ions derived from alkaline earth metals are more stable than M^+ ions due to their higher hydration energy.

13. B has the highest melting point because its crystal structure consists of icosahedral units with B-atoms at all the 12 corners and each B-atom is bonded to 5 equidistant neighbours. This structure of B suggests that this element has a giant molecular structure and hence its melting point is the highest.

Ga has the lowest melting point. Each Ga metal atom has one close neighbour at a distance of 2.43 A° and six more distant neighbours at distances between 2.70 and 2.79 A°. This unique structure of Ga suggests that Ga consists of almost discrete diatomic molecules and hence its melting point is excaptionally low.

14. The variation of electropositive character of the elements of group 3A can be shown as :

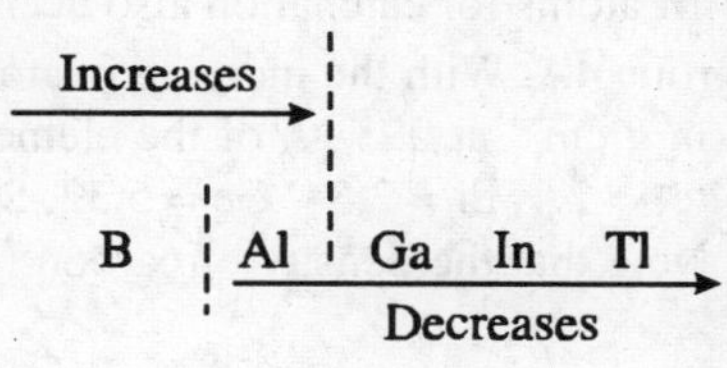

Since B-atom is the smallest in size and the sum of its three ionisation potentials (*i.e.*, IE_1 + IE_2 + IE_3) is the maximum, this atom has the least tendency to lose its valence electrons and hence is the least electropositive. Now since the size of Al atom is larger than that of B atom and the sum, (IE_1 + IE_2 + IE_3) for Al atom is much smaller than that of B atom [IE_1 + IE_2 + IE_3 for B = 800 + 2427 + 3659 = 6886 KJ mol^{-1}, for Al = 577 + 1816 + 2744 = 5137 KJ mol^{-1}], Al has higher tendency to lose electrons and hence has higher electropositive character than B.

The decrease in electropositive character from Al to Tl (Al > Ga > In > Tl) can be explained as follows :

(*a*) The decrease is due to the fact that the valence electrons in Ga, In and Tl are tightly held by the nucleus and hence are not easily removed. The tight linking of the valence electrons with the nucleus is because of the reason that the electrons present in *d* and *f* orbitals in Ga, In and Tl donot shield the valence electrons from the force of attraction of the nucleus effectively.

(*b*) Since the electrode potentials for the reaction, M^{3+} (*aq*) + $3e^-$ $\longrightarrow$ M(*s*) increases from Al to Tl (Al^{3+}/Al = – 1.66 V, Ga^{3+}/Ga = – 0.56 V, In^{3+}/In = – 0.34 V and Tl^{3+}/Tl = + 1.26 V), the electropositive character decreases accordingly.

15. Since B-atom has small size and the sum of its three ionisation energies (*i.e.*, IE_1 + IE_2 + IE_3) is the highest, this atom has the least tendency to lose all its three valence shell electrons to form B^{3+} ion.

16. Group valency (oxidation state) of the elements of group 13 is +3. Other oxidation state shown by these elements is + 1. For getting + 1 oxidation state (M^+ ion) np^1 electron is lost by M atom (M = ns^2 np^1, M^+ = ns^2 np^0) and + 3 oxidation state is shown by these elements when ns^2 electron pair present in M^+ ion is lost (M^{3+} = ns^0 np^0).

Now since $6s^2$ electron pair present in Tl^+ ion is the most stable (Maximum inert pair effect), Tl^+ ion does not lose its $6s^2$ electron pair and hence Tl^{3+} ion is not formed. Thus + 3 oxidation state which is group oxidation state (velency) is not shown by Tl.

17. Since the inert pair effect (*i.e.*, stability) of ns^2 electron pair present in C^{2+}, Si^{2+}, Ge^{2+} and Pb^{2+} ions increases down the group, the stability of M^{2+} ions also increases shown the group (C^{2+} < Si^{2+} < Ge^{2+} < Pb^{2+}). Thus, due to the maximum stability of Pb^{2+} ion, Pb has the highest tendency to show + 2 oxidation state.

18. We know that the stability of Ge^{4+}, Sn^{4+} and Pb^{4+} ions decreases as Ge^{4+} > Sn^{4+} > Pb^{4+}. Since Pb^{4+} ion is the least stable, this ion will reduce to Pb^{2+} ion most readily and hence has the highest oxidising power. Thus the oxidising power of M^{4+} ions decreases as Pb^{4+} > Sn^{4+} > Ge^{4+} or increases as Ge^{4+} < Sn^{4+} < Pb^{4+}.

19. The tendency of the elements for catenation depends on the size of the atoms of the elements. With the increase in the size of the atoms (M) the M – M bond lengths also increase. Due to the increase in M – M bond lengths, M – M bond energies decrease (C – C = 348 KJ mol^{-1}, Si – Si = 297 KJ mol^{-1}, Ge – Ge = 260 KJ mol^{-1} and Sn – Sn = 240 KJ mol^{-1}). With the decrease of M – M bond energies, the tendency of M atoms for catenation also decreases.

20. The given elements are the elements of group 4A. With the increase in atomic masses of the elements down the group, the densities (in g cm^{-1} at 293 K) of the elements also increases from carbon (graphite) to lead. [C(graphite) = 2.1, Si = 2.34, Ge = 5.32, Sn = 7.26 and Pb = 11.34]. Thus Pb has the highest density. Note that the density of carbon (diamond) is equal to 3.51 g cm^{-1} at 293 K.

21. (*a*) This statement is false, since the electropositive character of the elements of group 14 increases from B to Al and then decreases from Al to Tl.

(*b*) This statement is correct, since due to the increase in inert pair effect down the group, the stability of M^{2+} ions increases down the group, *i.e.*, the stability of + 2 oxidation state of the elements of group 14 increases down the group.

(*c*) It is a false statement, since, due to the increase in atomic size and screening effect of the inner electrons, IE values decrease from C to Sn and then increases from Sn to Pb. Thus the actual order of IE values (KJ mol^{-1}) is as :

$$C > Si > Ge > Sn < Pb$$

or

$$C > Si > Ge > Pb > Sn$$

$$1086 \quad 786 \quad 761 \quad 715 \quad 708 \text{ KJ mol}^{-1}.$$

(*d*) It is a false statement, because, due to the increase in the number of shells down the group, the atomic size of the elements increases.

22. As we move down the group 5A from N to Bi, the atomic size of the elements increases and hence the tendency of the elements to lose the electrons also increases. With the increase of this tendency, the electropositive character of the elements also increases. Thus N and P are non-metals, As and Sb are metalloids and Bi is a typical meal

N, P	As, Sb	Bi
Non-metals	Metalloids	Metal

——— Electropositive character increases ⟶

This order shows that N shows the least electropositive or most electronegative (or most non-metallic) character.

23. Amongst the elements of group 5A, nitrogen has the strongest tendency to form the multiple bonds with itself ($:N \equiv N:$) and with carbon and oxygen. On the other hand, the elements of this group have no tendency to form multiple bonds.

24. The structure of N_2 molecule is $:N \equiv N:$ which shows that this molecule has one σ and two π bonds. Each of the π bonds is a $(p\pi - p\pi)$ bond. On the other hand P_4, As_4 and Sb_4 all have tetrahedral structures in which the four M atoms (M = P, As or Sb) lie at the four corners of a regular tetrahedron. Each M atom is linked with other three M atoms by three M–M single bonds as shown in Fig. 23.1

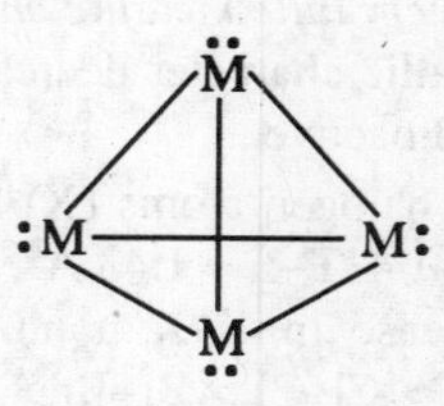

Fig. 23.1 Tetrahedral structure of M_4 molecule (M = P, As, Sb)

25. The given elements are the elements of group 5A. The melting points of these elements first increase from N to As and then decrease from As to Bi as shown below :

——— Increase ———→

N	P (white)	As (grey a-form)	Sb	Bi
63 K	317.1 K	1089 K	903.7 K	544.7 K

——— Decrease ———→

The increase in melting points from N to As is due to the increase in atomic size from N to As. Because of larger size of the atoms of Sb and Bi, the interatomic forces in these elements is weak and hence he melting points of these elements are lower.

Thus since N has the lowest atomic size, it has the lowest melting point.

26. We know that the structure of N_2 molecule is $N \equiv N$. Due to the presence of the triple bond between two N-atoms in N_2 molecule, the bond dissociation energy of N_2 molecule is very high (= 941.4 KJ mol^{-1}). Because of high value of bond dissociation energy, N_2 is inert and unreactive (low reactivity) in its elemental state. Thus low reactivity of nitrogen is due to its high bond dissociation energy.

27. O and S both have the tendency for catenation but Se and Te donot have this tendency. Since O — O bond dissociation energy is lesser than S — S bond dissociation energy, O has lesser tendency for catenation than S. Thus out of the given elements S has the maximum tendency for catenation.

28. Although in both the given compounds, S shows its highest oxidation state of + 6. We know that S can show its highest oxidation state only when it combines with highly electronegative element like F. Thus S can combine with F (highly electronegative element) to form SF_6 but not with H (less electronegative element) to form SH_6.

29. If it is supposed that OF_6 molecule is really formed, then its structure would be as shown in Fig. 23.2 In the structure of OF_6 molecule the central O-atom forms six (O — F) σ-bonds. For the formation of six (O — F) σ bonds, O-atom should have six unpaired electrons. As a matter of fact, O-atom has only two unpaired electrons and two paired electrons ($O = 2s^2\ 2p_x^2 2p_y^1 2p_z^1$). In order to get six unpaired electrons, the presence of two *d* orbitals in the valence shell of O-atom is essential. Due to the absence of *d*-orbitals in O-atom, the formation of OF_6 molecule is not possible.

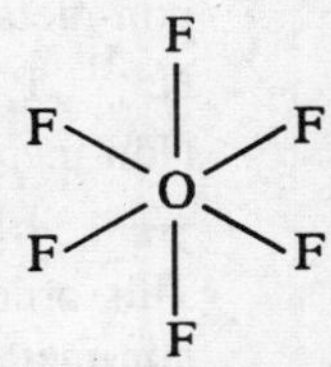

Fig. 23.2

30. In its compounds, O-atom shows –1, –2 and + 2 oxidation states. – 1 oxidation state is shown in H_2O_2 and metallic peroxides (*e.g.*, Na_2O_2), – 2 oxidation state is exhibited in its metallic oxides (*e.g.*, CaO) while + 2 oxidation state is shown by O-atom in its compounds in which this atom is linked with more electronegative atoms (more electronegative than O-atom) like F (*e.g.*, OF_2 molecule)

31. Since ionisation energies of O, S, Se and Te are very high, these elements donot lose the electrons easily and hence have *very little metallic character*, *i.e.*, these elements show non-metallic character. The non-metallic character decreases down the group from O to Te. Po has metallic character and is radioactive.

32. With the increase in the size of halogen atoms (X) from F to I, X – X bond length in X_2 molecules increases from F–F to I–I (F–F = 1.43 A°, Cl–Cl = 1.99 A°, Br–Br = 2.28 A° and I–I = 2.66 A°). With the increase in X–X bond lengths, the X–X bond dissociation energies should decrease as F–F > Cl–Cl > Br–Br > I–I. However, the experimental values of X–X bond dissociation energies (in KJ mol^{-1}) have been found to vary as : F–F (= 159 KJ mol^{-1}) < Cl–Cl (= 243) > Br–Br (= 193) > I–I (= 151). F_2 has lesser bond dissociation energy than Cl_2. Lesser value of bond energy of F_2 is due to greater inter electronic repulsions between the lone pairs of electrons on the two bonded F-atoms in F_2 molecule. Such repulsions are almost absent in other X_2 molecules due to larger X–X bond lengths in them. Experimental order of bond dissociation energy is Cl–Cl (= 243 KJ mol^{-1}) > Br – Br (= 193) > F – F (= 159) > I – I (= 151)

33. I_2 has the least bond dissociation energy due to the largest size of I-atom or largest I–I bond length in I_2 molecule.

34. Due to the increase in X – X bond length from F – F to I – I, the bond dissociation energy of X – X bond in X_2 molecules should decrease as F – F > Cl – Cl > Br – Br > I – I. Experiments have, however, shown that the actual order is Cl – Cl (= 243 KJ mol^{-1}) > Br – Br (= 193) > F – F (= 159) > I – I (151). Thus Cl_2 has the highest bond dissociation energy.

35. Since F_2 has the highest value for its reduction potential of all the halogens ($F_2/2F^-$ = + 2.87 V, $Cl_2/2Cl^-$ = + 1.36 V, $Br_2/2Br^-$ = + 1.08 V and $I_2/2I^-$ = + 0.54 V), it is reduced to F^- ion $\left(F_2 + 2e^- \xrightarrow{\text{Reduction}} 2F^-\right)$ most easily.

36. Since E^0 values of halogens decrease from F_2 to I_2, the oxidising power of halogens also decrease in the same sequence.

$$F_2 + 2e^- \longrightarrow 2F^- (E^0 = +2.87\text{ V})$$

Strongest OA — Weakest RA

$$Cl_2 + 2e^- \longrightarrow 2Cl^- (E^0 = +1.36\text{ V})$$

$$Br_2 + 2e^- \longrightarrow 2Br^- (E^0 = +1.08\text{ V})$$

$$I_2 + 2e^- \longrightarrow 2I^- (E^0 = +0.53\text{ V})$$

Weakest OA — Strongest RA

Oxidising power of X_2 molecules decreases ↓ ; Reducing power of X^- ions increases ↓

Thus oxidising power of halogens is in the order : $F_2 > Cl_2 > Br_2 > I_2$ or $I_2 < Br_2 < Cl_2 < F_2$

37. The probable reaction that may take place between I_2 and KCl/KF/KBr solutions can be represented by the following ionic equations.

$$\underset{\text{OA}}{I_2} + \underset{\text{RA}}{2Cl^-} \longrightarrow 2I^- + Cl_2$$

$$\underset{\text{OA}}{I_2} + \underset{\text{RA}}{2F^-} \longrightarrow 2I^- + F_2$$

$$\underset{\text{OA}}{I_2} + \underset{\text{RA}}{2Br^-} \longrightarrow 2I^- + Br_2$$

For the occurrence of these reactions the value of E° for I_2 (OA) should be greater than E° value for other halogens (RA). Now since E^0 value for I_2 is the lowest, I_2 cannot oxidise any X^- ion to X_2 molecule (E^0 values are : $F_2/2F^-$ = + 2.87 V, $Cl_2/2Cl^-$ = + 1.36 V, $Br_2/2Br^-$ = + 1.08 V and $I_2/2I^-$ = + 0.53 V)

38. With the increase in atomic size and atomic weights of noble gases down the group, the magnitude of intermolecular forces also increases. Due to the increase in intermolecular forces the ease of liquification of noble gases also increases down the group from He to Rn. (He < Ne < Ar < Kr < Xe < Rn)

39. Due to the increase in atomic size and atomic weights of noble gases down the group, the magnitude of van der Walls forces of attraction also increases. With the increase in the magnitude of van der Walls forces the boiling points (also melting points) of noble gases also increases down the group.

40. Xe shows oxidation states of + 2 (*e.g.*, XeF_2), + 4 (*e.g.*, XeF_4), + 6 (*e.g.*, XeF_6, XeO_3) and + 8 (*e.g.*, XeO_6^{4-})

24

Occurrence, Extraction/Isolation, Properties and Uses of Some Main Group Elements

MULTIPLE CHOICE QUESTIONS

1. Ortho and para hydrogens resemble in which of the following property ?
(*a*) Thermal conductivity
(*b*) Magnetic properties
(*c*) Chemical properties
(*d*) Heat capacity

2. Hydrogen behaves as an oxidising agent in its reaction with :
(*a*) Cl_2 (*b*) N_2
(*c*) Na (*d*) S

3. Hydrogen will not reduce :
(*a*) heated CuO (*b*) heated Fe_2O_3
(*c*) heated SnO_2 (*d*) heated Al_2O_3.

4. H_2 directly combines with :
(*a*) Au (*b*) Cu
(*c*) Ni (*d*) Ca

5. Which of the following is correct for hydrogen ?
(*a*) It can form compounds in + 1 as well as – 1 oxidation state.
(*b*) It is always collected at cathode.
(*c*) It has very high ionisation energy.
(*d*) It has same electronegativity as halogens

6. Which of the following does not react with aqueous NaOH ?
(*a*) B (*b*) Al
(*c*) Ga (*d*) Tl

7. Which of the following does not react with water ?
(*a*) B (*b*) Al
(*c*) Na (*d*) Tl

8. Lead dissolves most readily in :
(*a*) CH_3COOH (*b*) HNO_3
(*c*) H_2SO_4 (*d*) HCl

9. The ore of tin is :
(*a*) Galena (*b*) Cerussite
(*c*) Tin stone (*d*) Anglesite

10. Action of conc HNO_3 on metallic tin produces :
(*a*) Stanneous nitrate
(*b*) Staneous nitrate
(*c*) Stannic nitrate
(*d*) Hydrated stannic oxide

11. When tin is treated with conc. HNO_3,
(*a*) it is converted into staneous nitrate
(*b*) it is converted into stannic nitrate
(*c*) it is converted into meta (or β-) stannic acid
(*d*) it becomes passive.

12. Tin plague means :
(*a*) Conversion of stannous salt to stannic salt
(*b*) Conversion of white tin to grey tin
(*c*) Tin plating
(*d*) Emission of sound while bending a tin plate

13. Important ores of tin and lead are :
(*a*) Stannite and galena
(*b*) Cassiterite and stannite
(*c*) Cassiterite and galena
(*d*) Calamine and cryolite

14. Softening of lead means :
(*a*) Conversion of Pb into PbO

(*b*) Washing Pb with HNO_3 followed by dil. alkali solution
(*c*) Conversion of Pb into Pb_3O_4
(*d*) Removal of impurities of Cu, Ag, Fe from impure lead.

15. Lead pipes are not safe for transportation of drinking water, because :
(*a*) Lead forms lead oxide with water.
(*b*) Water containing dissolved air reacts slowly with lead forming $Pb(OH)_2$
(*c*) They are corroded by air
(*d*) They are slowly covered with the layer of $PbCO_3$.

16. Which of the following difficulties was/were encountered in Dennis method for the preparation of F_2 ?
(*a*) H_2 and F_2 get mixed together and produce explosion.
(*b*) F_2 is evolved slowly due to narrow exist around anode
(*c*) F_2 is obtained at anode
(*d*) None of these.

17. Which electrolyte is used in Dennis process for the preparation of fluorine ?
(*a*) KHF_2 solution in anhydrous HF
(*b*) Fused cryolite
(*c*) Pure dried KHF_2
(*d*) None of these

18. F_2 reacts with cold dilute NaOH solution to give :
(*a*) NaF and O_2
(*b*) NaF and OF_2
(*c*) NaF and H_2O_2
(*d*) NaF, H_2O_2 and F_2O

19. The manufacture of F_2 is done by :
(*a*) Heating anhydrous HF and MnO_2
(*b*) Electrolysing an aqueous solution of HF
(*c*) Electrolysing fused KHF_2 mixed with HF
(*d*) Heating a mixture of KF, MnO_2 and conc. H_2SO_4.

20. Which of the following halogens oxidises water vigorously to liberate O_2 exothermally ?
(*a*) F_2 (*b*) Cl_2
(*c*) Br_2 (*d*) I_2

ANSWERS

1. (*c*)	**2.** (*c*)	**3.** (*d*)	**4.** (*d*)	**5.** (*a*)	**6.** (*a*)	**7.** (*a*)	**8.** (*a*)
9. (*c*)	**10.** (*d*)	**11.** (*c*)	**12.** (*b*)	**13.** (*c*)	**14.** (*d*)	**15.** (*b*)	**16.** (*a*), (*b*)
17. (*c*)	**18.** (*b*)	**19.** (*c*)	**20.** (*a*)				

HINTS AND EXPLANATIONS

1. Since ortho and para hydrogens are allotropic forms of hydrogen, these resemble with each other in chemical properties.

2. When H_2 (O.S. = 0) reacts with Cl_2, N_2, Na and S, HCl (H = +1), NH_3 (H = + 1) NaH (H = – 1) and H_2S (H = +1) are formed respectively. It is only in NaH in which O.S. of H of zero is decreasesd to – 1. Thus in the reaction between H_2 and Na, H_2 acts as an oxidising agent.

3. H_2 can reduce the oxides only of those metals whose E° values are greater than zero (*i.e.*, positive). Since E° value for Al^{3+}/Al is less than zero (= – 1.66 V), H_2 will not reduce Al_2O_3.

4. H_2 combines only with active metals (highly electropositive) like Na, K, Ca etc. to form their corresponding hydrides under suitable conditions.

$$Ca + H_2 \longrightarrow CaH_2$$

5. (*a*) Like alkali metals, hydrogen shows + 1 oxidation state in its compounds (*e.g.*, H^+Cl^-, $Na^+ Cl^-$ etc). Like halogens H also shows – 1 oxidation state in its metallic hydrides (*e.g.*, $Na^+ H^-$, $Na^+ Cl^-$ etc.). Thus statement (*a*) is correct.

(*b*) When an aqueous solution of HCl is electrolysed, H_2 is collected at cathode. On the other hand when fused alkali metal hydride (*e.g.*, NaH) is electrolysed, H_2 is collected at anode. Thus statement (*b*) is wrong.

(*c*) IE of hydrogen is much higher than those of alkali metals but comparable to those of halogens. Thus statement (*c*) is wrong.

(*d*) Hydrogen has lesser value of its electronegativity than halogens. Thus statement (*d*) is also wrong.

6. Aqueous NaOH has no action on boron. However, boron dissolves in fused alkali due to the formation of sodium borate ($NaBO_3$)

$$2B + 6NaOH \text{ (Fused)} \rightarrow 2Na_3BO_3 + 3H_2 \uparrow$$

Al and Ga both get dissolved in aqueous as well as fused alkalies.

7. Boron does not react with water. However, it reacts with steam at red heat to liberate H_2 gas.

$$2B + 3H_2O \text{ (Steam)} \xrightarrow{\text{Red heat}} B_2O_3 + 3H_2 \uparrow$$

8. Pb dissolves most readily in CH_3COOH in presence of O_2 due to the formation of $(CH_3COO)_2Pb$.

$$Pb + 2CH_3COOH + \frac{1}{2}O_2 \longrightarrow (CH_3COO)_2Pb + H_2O$$

Dilute and concentrated HNO_3 also dissolves Pb forming $Pb(NO_3)_2$ and oxides of nitrogen *viz* NO and NO_2 respectively.

$$3Pb + 8HNO_3 \text{ (Dilute)} \longrightarrow 3Pb(NO_3)_2 + 2NO + 4H_2O$$

$$Pb + 4HNO_3 \text{ (Concentrated)} \longrightarrow Pb(NO_3)_2 + 2NO_2 + 2H_2O$$

Pb is slightly affected by dilute HCl or dilute H_2SO_4, since a layer of insoluble $PbCl_2$ or $PbSO_4$ formed on the surface of the metal prevents the further action of the acid on the metal.

$$Pb + 2H_2SO_4 \text{ (Dil.)} \longrightarrow PbSO_4 + 2H_2O + SO_2$$

$$Pb + 2HCl \text{ (Dil)} \longrightarrow PbCl_2 + H_2$$

9. It is only tin stone (SnO_2) which is the ore of tin. Other ores are of Pb as is evident from their formulae given as : galena (PbS), cerrusite ($PbCO_3$) and anglesite ($PbSO_4$).

10. $Sn + 4HNO_3 \text{ (conc)} \longrightarrow H_2O + 4NO_2 + \underset{\text{Hydrated stannic oxide}}{SnO_2 \cdot H_2O}$

11. $Sn + 4HNO_3 \text{ (conc)} \longrightarrow H_2O + 4NO_2 + \underset{\text{Metastannic acid}}{H_2SnO_3}$

12. Tin has three allotropic forms viz grey tin,, white tin and rhombic tin. White tin can be convected into grey tin. This conversion of white tin into grey tin is called *tin pest, tin desease* or *tin plague*. This conversion is accompanied by an increase in volume and the grey tin obtained is very brittle and hence crumbles down easily to a powder.

13. The ores of tin are stannite (also called tin pyrites) – CuS. FeS. SnS_2, cassiterite (also called tine stone) – SnO_2. Out of these cassiterite is the important ore of Sn. Galena (PbS) is the only ore of Pb. Calamine (Zn CO_3) is the ore of Zn and cryolite ($Na_3 AlF_6$) is the ore of Al.

14. Lead obtained from galena (PbS) either by air reduction process or by carbon reduction process contains base metals like Sn Bi, Cu, Ag, Fe Zn etc. as impurities. Due to the presence of base metals, impure Pb is hard and brittle. Since after, the removal of base metals, Pb becomes soft, the removal of base metals from impure Pb is called softening of lead and the process used for softening impure lead is called softening process.

15. Lead pipes should not be used for transporting drinking water, since water reacts with Pb slowly in presence of O_2 dissolved in it and forms soluble $Pb(OH)_2$ which is posinuous.

$$2Pb + 2H_2O + O_2 \longrightarrow 2Pb(OH)_2.$$

The formation of $Pb(OH)_2$ by the action of Pb on water in presence of dissolved O_2 is called plumbosolvency.

18. Unlike Cl_2, Br_2 and I_2, F_2 reacts with dil. alkalies (2% solution) to form NaF and OF_2 (not NaFO). However with cone. alkalies F_2 gives NaF and O_2 (not $NaFO_3$).

$$2F_2 + 2NaOH \text{ (2\% solution)} \longrightarrow 2NaF + OF_2 + H_2O$$

$$2F_2 + 4NaOH \text{ (Conc. solution)} \longrightarrow 4NaF + O_2 + 2H_2O$$

19. Modern method used for the manufacture of F_2 consists of electrolysing fused KHF_2 mixed with HF at 70–80°C.

Reactions taking place are :

$$KHF_2 \xrightarrow{\text{Electrolysis}} KF + HF$$

$$KF \rightleftharpoons K^+ + F^-$$

At cathode :

$$K^+ + e^- \longrightarrow K$$

$$K + HF \longrightarrow KF + H$$

$$H + H \longrightarrow H_2$$

At anode :

$$F^- \longrightarrow F + e^-$$

$$F + F \longrightarrow F_2$$

20. F_2 decomposes water very vigorously even at low temperature and in dark, forming a mixture of O_2 and O_3 (ozonised oxygen)

$$2F_2 + 2H_2O \longrightarrow 4HF + O_2$$

$$3F_2 + 3H_2O \longrightarrow 6HF + O_3$$

Cl_2 and Br_2 both decompose H_2O slowly in presence of sunlight to give HCl/HBr and O_2

$$2X_2 \ (X_2 = Cl_2, Br_2) + 2H_2O \longrightarrow 4HX \times O_2$$

I_2 has no action on water.

25

Hydrides of Main Group Elements

MULTIPLE CHOICE QUESTIONS

1. Ionic hydrides are usually :
   (*a*) good electrical conductors in the solid state
   (*b*) easily reduced
   (*c*) good reducing agents
   (*d*) liquid at room temperature
2. Alkali metal hydrides react with water to give :
   (*a*) acidic solution
   (*b*) basic solution
   (*c*) neutral solution
   (*d*) hydride ion
3. Electrolysis of fused NaH liberates H_2 at :
   (*a*) anode
   (*b*) cathode
   (*c*) cathode and anode both
   (*d*) none of these electrodes
4. Which of the following alkali metal hydrides is thermally most stable ?
   (*a*) LiH (*b*) NaH
   (*c*) KH (*d*) RbH
5. Which of the following hydrides has significant covalent character ?
   (*a*) BeH_2
   (*b*) MgH_2
   (*c*) BeH_2 and MgH_2 both
   (*d*) None of these
6. Which of the following statements is correct about B_2H_6 ?
   (*a*) There is a direct boron–boron bond
   (*b*) The boron atoms are linked together through H-bridges
   (*c*) The structure of B_2H_6 is similar to that of C_2H_6.
   (*d*) All the atoms of B_2H_6 are in the same plane
7. Which of the following statements is false about B_2H_6 ?
   (*a*) It contains two types of H atoms
   (*b*) It contains B–B covalent bond
   (*c*) Rotation of B–B axis is not possible
   (*d*) The two B-atoms linked with four H-atoms by B–H covalent bonds lie in the same plane.
8. $Li[AlH_4]$ is obtained by reacting excess of ... with an etheral solution of $AlCl_3$.
   (*a*) LiCl (*b*) LiH
   (*c*) Li (*d*) LiOH
9. Thermal stability of tetrahydrides, MH_4 (M = C, Si, Ge, Sn and Pb) is in the order :
   (*a*) $CH_4 > SiH_4 > GeH_4 > SnH_4 > PbH_4$
   (*b*) $CH_4 < SiH_4 < GeH_4 < SnH_4 < PbH_4$
   (*c*) $CH_4 > GeH_4 > SiH_4 > SnH_4 > PbH_4$
   (*d*) $CH_4 < GeH_4 < SiH_4 < SnH_4 < PbH_4$
10. Which hydride is the most volatile ?
    (*a*) CH_4 (*b*) SiH_4
    (*c*) GeH_4 (*d*) SnH_4
11. Bond angles in trihydrides of group 5A elements (MH_3) are in the order :
    (*a*) $NH_3 > PH_3 > AsH_3 > SbH_3$
    (*b*) $PH_3 > NH_3 > AsH_3 > SbH_3$
    (*c*) $NH_3 < PH_3 < AsH_3 < SbH_3$
    (*d*) $PH_3 < NH_3 < AsH_3 < SbH_3$
12. The boiling points of the following hydrides follow the order :
    (*a*) $SbH_3 > NH_3 > AsH_3 > PH_3$
    (*b*) $NH_3 > PH_3 > AsH_3 > SbH_3$

(*c*) $NH_3 > AsH_3 > PH_3 > SbH_3$
(*d*) $SbH_3 > AsH_3 > NH_3 > PH_3$

13. Thermal stability of trihydrides of group 5A elements is in the order :
(*a*) $NH_3 > PH_3 > AsH_3 > SbH_3 > BiH_3$
(*b*) $NH_3 < PH_3 < AsH_3 < SbH_3 < BiH_3$
(*c*) $NH_3 > AsH_3 > PH_3 > SbH_3 > BiH_3$
(*d*) $NH_3 < AsH_3 < PH_3 < SbH_3 < BiH_3$

14. The basic character of the hydrides of 15 group elements decreases in the order :
(*a*) $SbH_3 > PH_3 > AsH_3 > NH_3$
(*b*) $NH_3 > SbH_3 > PH_3 > AsH_3$
(*c*) $NH_3 > PH_3 > AsH_3 > SbH_3$
(*d*) $SbH_3 > AsH_3 > PH_3 > NH_3$

15. Which of the following hydrides has the largest bond angle ?
(*a*) H_2O (*b*) H_2S
(*c*) H_2Se (*d*) H_2Te

16. Bond angle is minimum for :
(*a*) H_2O (*b*) H_2S
(*c*) H_2Se (*d*) H_2Te

17. The boiling points of hydrides of group 16 (6A) elements are in the order :
(*a*) $H_2O > H_2Te > H_2S > H_2Se$
(*b*) $H_2O > H_2S > H_2Se > H_2Te$
(*c*) $H_2O > H_2Te > H_2Se > H_2S$
(*d*) $H_2O > H_2Se > H_2S > H_2O$

18. Which of the following hydrides is the least acidic ?
(*a*) H_2Te (*b*) H_2O
(*c*) H_2Se (*d*) H_2S

19. Which of the following hydrides is the least thermally stable ?
(*a*) H_2O (*b*) H_2S
(*c*) H_2Se (*d*) H_2Te

20. Stability of CH_4, NH_3, H_2O and HF is in the order :
(*a*) $CH_4 < NH_3 < H_2O < HF$
(*b*) $CH_4 < NH_3 < H_2O < HF$
(*c*) $CH_4 < NH_3 < HF < H_2O$
(*d*) No order exists.

ANSWERS

1. (*c*)	**2.** (*b*)	**3.** (*a*)	**4.** (*a*)	**5.** (*c*)	**6.** (*b*)	**7.** (*b*)	**8.** (*b*)
9. (*a*)	**10.** (*a*)	**11.** (*a*)	**12.** (*a*)	**13.** (*a*)	**14.** (*c*)	**15.** (*a*)	**16.** (*d*)
17. (*c*)	**18.** (*d*)	**19.** (*d*)	**20.** (*a*)				

HINTS AND EXPLANATIONS

1. Ionic hydrides react with protonic solvents (*i.e.*, solvents giving H^+ ions) such as H_2O, NH_3 and C_2H_5OH to given H_2 and a base

Li H (*Ionic hydride*) + H_2O $\longrightarrow$ Li OH + H_2

Due to the production of H_2 as shown above, ionic hydrides are usually good reducing agents, e.g.,

$$NaH + 2CO \longrightarrow HCOONa + C$$
$$NaH + Fe_3O_4 \longrightarrow 4NaOH + 3Fe$$

2. Alkali metal hydrides (MH) react with H_2O to give alkali metal hydroxide (MOH) and H_2. Since alkali metal hydroxides are strongly basic, the solution of alkali metal hydrides in water is basic.

$$MH + H_2O \longrightarrow MOH\ (\textit{Strongly basic}) + H_2$$

3. Since in NaH, hydrogen is present as H^- ion, the electrolysis of fused NaH gives H^- ion on anode.

$$NaH \rightleftharpoons Na^+ + H^-$$

At anode $H^- \rightarrow H + e^-$, $H + H \rightarrow H_2$

4. Since thermal stability of alkali metal hydrides (MH) decreases down the group (LiH > NaH > KH > RbH > GH), LiH is thermally most stable. Greatest stability of LiH is due to its highest heat of formation.
5. With the exception of Be and Mg, all other alkaline earth metals viz Ca, Sr and Ba give ionic hydrides. Be and Mg give covalent hydrides (BeH_2 and MgH_2).
6. There is no direct B – B bond in B_2H_6 molecule. The two B-atoms are linked together through two bridging H-atoms making two (3c – 2e) bonds. The structure of B_2H_6 cannot be similar to that of C_2H_6, since B-atom has only three electrons in its valence shell and hence cannot make four B–H bonds. Two B-atoms and four terminal H-atoms lie in one plane and the remaining two bridging H-atoms lie in a different plane. These two planes are perpendicular to each other.
7. Instead B_2H_6 molecule contains B ... H .... B banana - shaped bond which is called three-centre two-electron bond [(3c – 2e) bond].
8. $Li[AlH_4]$ is obtained by reacting an excess of ionic hydride, LiH with an ethereal solution of covalent halide, $AlCl_3$ (anhydrous)

$$4LiH + AlCl_3 \text{ (anhydrous)} \longrightarrow Li[AlH_4] \text{ (Solution)} + 3\,LiCl \downarrow$$

9. As the size of the central atom, M(M = C, Si, Ge, Sn and Pb) increases from C to Pb, M–H bond length in MH_4 molecules increases from C – H to Pb – H (C – H > Si – H > Ge – H > Sn – H > Pb – H). The increase in M – H bond length decreases the strength of M – H bond from C – H to Pb – H. Due to the successive decrease in the strength of M – H bond from C – H to Pb – H, the cleavage of M – H bond in MH_4 molecules becomes easier as we move down the group from CH_4 to PbH_4, *i.e.*, thermal stability of MH_4 molecules decreases from CH_4 to PbH_4 (CH_4 > SiH_4 > GeH_4 > SnH_4 > PbH_4). PbH_4 is so unstable that it has been obtained only in traces. In short we can say that the stability of MH_4 hydrides decreases with the increase of the size of M atoms down the group.
10. Out of the given hydride, CH_4 is the most volatile due to its lesser molecule mass.
11. The decrease in HMH bond angle from NH_3 to SbH_3 is due to the decrease in electronegativity of the central atom from N to Sb. Since the electronegativity of N atoms is the highest, two adjacent bonding electron pairs (*bps*) are nearer to N-atom (central atom) and hence (*bp* – *bp*) distance is minimum or (*bp* – *bp*) repulsion is maximum in NH_3 molecule. Due to maximum (*bp* – *bp*) repulsion, HMH bond angle in NH_3 molecule is also maximum. On the other hand, since the electronegativity of Sb atom is the lowest, the two *bps* are farthest away from Sb atom (central atom) and hence (*bp* – *bp*) distance is maximum or (*bp* – *bp*) repulsion is minimum in Sb H_3 molecule. Due to minimum (*bp* – *bp*) repulsion, HSbH bond angle in SbH_3 molecule is also minimum (See Fig. 25.1). Thus with the decrease in electronegativity of the central atom from N to Sb, (*bp* – *bp*) repulsion goes on decreasing and hence HMH bond angle also goes on decreasing.

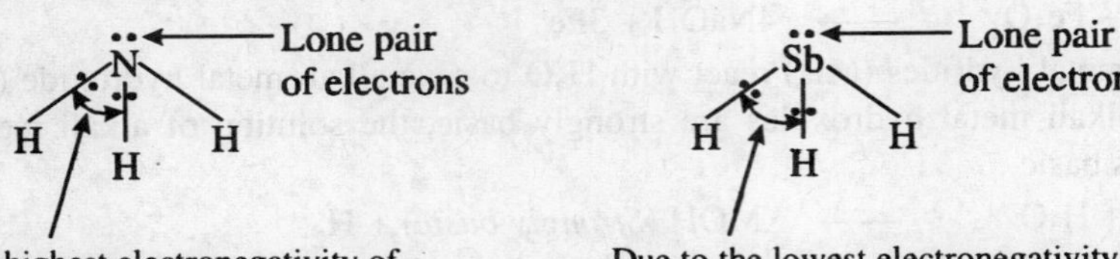

Due to the highest electronegativity of N-atom, (*bp* – *bp*) distance is minimum and hence (*bp* – *bp*) repulsion and HNH bond angle both are maximum.

Due to the lowest electronegativity of Sb atom, (*bp* – *bp*) distance is maximum and hence (*bp* – *bp*) repulsion and HSbH bond angle both are minimum.

Fig 25.1 Comparison of bond angles in NH_3 and SbH_3 molecules.

12. Since the molecular masses of the hydrides increases down the group from NH_3 to SbH_3 (NH_3 = 17, PH_3 = 34, AsH_3 = 78, SbH_3 = 125), the magnitude of van der wall's forces of attraction also increases in the same direction and hence the boiling points should also increase as : $NH_3 < PH_3 < AsH_3 < SbH_3$. However the actual order is $PH_3 < AsH_3 < NH_3 < SbH_3$ (PH_3 = 185.5 K, AsH_3 = 210.6 K, NH_3 = 238.5 K, SbH_3 = 254.6 K). Higher boiling point of NH_3 as compared to those of PH_3 and AsH_3 is because of the reason that, due to higher electronegativity of N-atom, NH_3 molecule undergoes strong inter-molecular H-bonding which increases the boiling point of NH_3.

13. The stability of MH_3 hydrides decreases down the group from NH_3 to BiH_3 with the increase of the size of M atoms (M = N, P, As, Sb and Bi) down the group.

14. Each of the given trihydrides has a lone pair of electrons in one of the four sp^3 hybrid orbitals on the central atom. It is due to the presence of lone pair of electrons that trihydrides show Lewis basic character. Basic character of the trihydrides decreases down the group as we move from NH_3 to BiH_3 ($NH_3 > AsH_3 > SbH_3 > BiH_3$). Thus NH_3 is distinctly basic and hence forms salts with strong and weak acids. PH_3 is a very weak base and forms salts only with anhydrous HCl. AsH_3, SbH_3 and BiH_3 are not at all basic. The decreases in basic character down the group can be explained on the basis of the size of the central atom. As the size of the central atom increases from N to Bi, the lone pair of electrons occupies larger and larger volume round the central atom. In other words, the electron density of the lone pair of electrons on the central atom decreases and consequently its tendency to donate the lone pair of electrons to a Lewis acid also decreases and hence basic strength of MH_3 molecules decreases as we move down the group from NH_3 to BiH_3.

15. The central atom present in the given hydrides in O, S, Se and Te respectively. The electronegativity values of these atoms decreases on moving down the group 6A from O to Te. Since O-atom has the highest electronegativity, the two adjascent bonding electron pairs (*bps*) involved in the formation of two O – H bonds in H_2O molecule come nearest to O-atom (central atom) and (*bp* – *bp*) distance is minimum or (*bp* – *bp*) repulsion is maximum. Due to maximum (*bp* – *bp*) repulsion, HOH bond angle in H_2O molecule is also maximum.

16. The central atom present in the given hydrides is O, S, Se and Te respectively. The electronegativity value of these atoms decreases on moving down the group 6 A from O to Te. Since Te atom has the lowest electronegativity, the two adjascent bonding electron pairs (*bps*) involved in the formation of two Te – H bonds in H_2Te molecule get shifted farthest away from Te atom (central atom) and (*bp* – *bp*) distance is maximum or (*bp* – *bp*) repulsion is minimum. Due to minimum (*bp* – *bp*) repulsion, H Te H bond angle in H_2Te molecule is also minimum.

17. Since the molecular masses of the the given hydrides increases down the group from H_2O to H_2Te (H_2O = 18, H_2S = 33, H_2Se = 81, H_2Te = 130), the magnitude of vander Waals forces of attraction also increases in the same direction and hence the boiling points should also increase as $H_2O < H_2S < H_2Se < H_2Te$. However the actual order is $H_2S < H_2Se < H_2Te < H_2O$ (H_2S = 231 K, H_2Se = 232 K, H_2Te = 269K, H_2O = 373K). The unexpected highest boiling point of H_2O is because of the reason that, due to the highest electronegativity of O-atom, H_2O molecule undergoes strong inter-molecular H-bonding which makes the boiling point of H_2O maximum. Since other elements (S, Se and Te) have much lower electronegativity, their hydrides donot undergo H-bonding.

18. Acidic nature (or acidic strength) of the given molecules increases down the group from H_2O to H_2Te ($H_2O < H_2S < H_2Se < H_2Te$). Since the size of the central atom A (A = O, S, Se, Te) increases from O to Te, A – H bond length also increases. With the increases of A – H bond length from O – H to Te – H, the ease with which A – H bond is broken to give H^+ ion also increases and hence acidic nature (or acidic strength) of AH_2 molecules also increases from H_2O to H_2Te ($H_2O < H_2S < H_2Se > H_2Te$). Since H_2O is neutral, H_2S is the least acidic and H_2Te is the most acidic.

19. Due to the increase in A – H bond length (A = O, S, Se, Te) in H_2A molecules from O – H to Te – H (O – H = 0.95 A°, S – H = 1.30 A°, Se – H = 1.45 A° and Te – H = 1.72 A°), A – H bond becomes weaker and weaker and hence its cleavage, on heating, becomes easier and easier. Thus thermal stability of A – H bonds or H_2A molecules decrease from H_2O to H_2Te [H_2O (*Most stable*) > H_2S > H_2Se > H_2Te (*Least stable*)]. Thus H_2O is the most stable and H_2Te is the least stable.

20. Since the given hydrides are the hydrides of the elements of the same period (*2nd* period), the stability of the hydrides increases with the increase of electronegativity of the central atom. Thus since the electronegativity of C, N, O and F increases as C < N < O < F, the stability of their hydrides also increases as $CH_4 < NH_3 < H_2O < HF$.

26

Carbides of Main Group Elements

MULTIPLE CHOICE QUESTIONS

1. Which of the following compounds will liberate methane upon reaction with water ?

(*a*) BeC_2 (*b*) Al_4C_3

(*c*) Be_2C (*d*) Both (*b*) and (*c*).

2. Which of the following compounds of boron is the hardest ?

(*a*) Boron oxide (*b*) Boron carbide

(*c*) Boron nitride (*d*) Boron hydroxide

3. Which of the following is methanide ion ?

(*a*) C^{2-} (*b*) C_2^{2-}

(*c*) C_3^{4-} (*d*) C^{4-}

4. Carborundum is the commercial name of :

(*a*) Al_2O_3 (*b*) SiC

(*c*) SiO_2 (*d*) HPO_3

5. Silicon carbide is used as :

(*a*) a solvent

(*b*) a dehydrating agent

(*c*) an abrasive

(*d*) a catalyst

6. Carborundum is :

(*a*) $Al_2(SO_4)_3$ (*b*) $Al_2O_3 . 2H_2O$

(*c*) $AlCl_3$ (*d*) SiC

7. Carborundum is obtained by :

(*a*) heating silica with coke

(*b*) heating silica with CO

(*c*) heating silica with CO_2

(*d*) passing acetylene on heated silicon

8. Magnesium carbide gives allylene on hydrolysis. The formula of this carbide is :

(*a*) MgC_2 (*b*) Mg_2C

(*c*) Mg_2C_3 (*d*) Mg_3C_2

9. Nitrolim is prepared by :

(*a*) Heating CaC_2 with N_2

(*b*) Heating CaC_2

(*c*) Hydrolysing CaC_2

(*d*) None of these methods.

10. Which of the following methods are used for the prepartion of Al_4C_3

(*a*) It can be prepared by heating Al with carbon

(*b*) It can be prepared by heating Al_2O_3 with carbon

(*c*) It can be prepared by passing CO on heated Al.

(*d*) It can be prepared by heating Al_2O_3 with CaC_2.

11. The number and type of bonds between two carbon atoms in CaC_2 are :

(*a*) one σ and π bonds

(*b*) one σ and two π bonds

(*c*) one σ and a half π bonds

(*d*) only one σ bond

12. Which of the following are covalent carbides ?

(*a*) CaC_2 (*b*) Be_2C

(*c*) B_4C (*d*) SiC

13. SiC reacts with molten NaOH in presence of O_2 to form :

(*a*) SiO_2 (*b*) Na_2SiO_3

(*c*) Si (*d*) None of these

ANSWERS

1. (*d*) **2.** (*b*) **3.** (*d*) **4.** (*b*), **5.** (*c*) **6.** (*d*) **7.** (*a*), (*d*) **8.** (*c*)
9. (*a*), **10.** (*a*), (*b*), (*c*), (*d*) **11.** (*b*) **12.** (*c*), (*d*) **13.** (*b*)

HINTS AND EXPLANATIONS

1. All the given compounds are ionic compounds. Hence their structures can be shown as : (*a*) $Be^{2+}(C_2)^{2-}$ (*b*) $(Al^{3+})_4(C^{4-})_3$ (*c*) $(Be^{2+})_2(C^{4-})$. These structures show that BeC_2 has acetylide ion ($C^- \equiv C^-$ or C_2^{2-}) and hence gives acetylene (C_2H_2) with H_2O.

$$BeC_2 + 2H_2O \longrightarrow Be(OH)_2 + H-C \equiv C-H \text{ or } C_2H_2.$$

Al_4C_3 and Be_2C both contain methanide ion (C^{4-}) and give methane with H_2O

$$Al_4C_3 + 12H_2O \longrightarrow 4Al(OH)_3 + 3CH_4\uparrow$$

$$Be_2C + 4H_2O \longrightarrow 2Be(OH)_2 + CH_4\uparrow$$

2. Boron carbide, B_4C (also called norbide) is the hardest substance.

3. Methanide ion is that which is obtained by removing all the four protons (H^+) from methane (CH_4) molecule.

$$CH_4(\textit{Methane}) \xrightarrow{-4H^+} C^{4-} \ (\textit{Methanide ion})$$

4. SiC is called carborundum while Al_2O_3 and SiO_2 are called corundum and silica respectively. HPO_3 is called metephosphoric acid, since it is obtained by removing one H_2O molecule from orthophosphoric acid, H_3PO_4.

$$H_3PO_4 \xrightarrow{-H_2O} HPO_3 + H_2O$$

5. Since silicon carbide is quite hard, it is used as an abrasive.

6. Carborundum is SiC. $Al_2O_3 \cdot 2H_2O$ (hydrated aluminium oxide) is called bauxite. $Al_2(SO_4)_3$ and $AlCl_3$ are sulfate and chloride of Al respectively.

7. Carborundum is silicon carbide (SiC) which may be obtained by heating silica (SiO_2) with coke (C) in an electric furnace at 2000°C or by passing acetylene (C_2H_2) on heated silicon (Si)

$$SiO_2 + 3C \longrightarrow SiC + 2CO, \qquad 2Si + C_2H_2 \longrightarrow 2SiC + H_2.$$

8. The structure of allylene is $H_2C = C = CH_2$ which, after losing four H-atoms, gives allylide ion, C_3^{4-} ($H_2C = C = CH_2 \xrightarrow{-4H^+} \overset{2-}{C} = C = \overset{2-}{C}$ or C_3^{4-}). Since magnesium carbide gives allylene on hydrolysis, it should contain allylide ion, C_3^{4-}. Thus the formula of magnesium carbide is Mg_2C_3.

$$Mg_2C_3 + 4H_2O \longrightarrow 2Mg(OH)_2 + H_2C = C = CH_2$$

9. Nitrolim is a mixture of calcium cyanamide ($CaCN_2$) and carbon. It is prepared by heating CaC_2 with N_2

$$CaC_2 + N_2 \xrightarrow{\Delta} \underbrace{CaCN_2 + C}_{\text{Nitrolim}}$$

Hydrolysies of CaC_2 gives acetylene (C_2H_2) and not nitrolim.

$$CaC_2 + 2H_2O \longrightarrow Ca(OH)_2 + C_2H_2$$

10. Al_4C_3 can be prepared by all the given methods.

(*a*) $4Al + 3C \rightarrow Al_4C_3$

(*b*) $2Al_2O_3 + 9C \rightarrow Al_4C_3 + 6CO\uparrow$

(*c*) $6Al + 3CO \rightarrow Al_4C_3 + Al_2O_3$

(*d*) $2Al_2O_3 + 3CaC_2 \rightarrow Al_4C_3 + 3CaO + 3CO\uparrow$

11. CaC_2 is an ionic carbide and is represented as $(Ca)^{2+}$ $(C_2)^{2-}$. Thus this carbide contains C_2^{2-} ion which is called acetylide ion, since it is obtained by removing both H-atoms from one molecule of acetylene, C_2H_2 $\left(H-C\equiv C-H \xrightarrow{-2H^+} C^-\equiv C^- \text{ or } C_2^{2-}\right)$. Thus the structure of acetylide ion (C_2^{2-}) is $C^-\equiv C^-$ which shows that this ion has one sigma and two pi bonds.

12. CaC_2 and Be_2C both are ionic carbides, since they give acetylene (C_2H_2) and methane (CH_4) on hydrolysis respectively.

$$CaC_2 + 2H_2O \longrightarrow Ca(OH)_2 + H-C\equiv C-H \text{ or } C_2H_2$$

$$Be_2C + 4H_2O \longrightarrow 2Be(OH)_2 + CH_4$$

B_4C and SiC both are not attacked by water.

13. Na_2SiO_3 is formed

$$SiC + 2NaOH + 2O_2 \xrightarrow{\text{Fuse}} Na_2SiO_3 + H_2O + CO_2$$

27

Nitrides of Main Group Elements

MULTIPLE CHOICE QUESTIONS

1. When NH_3 is passed over heated Mg, ... is formed.
   (a) Magnesium nitride
   (b) Magnesium hydride
   (c) Magnesium amide
   (d) Magnesium imide
2. BN gives NH_3 when it reacts with :
   (a) NaOH (b) H_2O
   (c) HF (d) KOH
3. Which of the sets contains only ionic nitrides ?
   (a) Li_3N, Mg_3N_2 (b) Mg_3N_2, BN
   (c) BN, Al N (d) AlN, Li_3 N
4. Which of the following nitrides is volatile ?
   (a) Be_3N_2 (b) Mg_3 N_2
   (c) Ca_3 N_2 (d) Ra_3 N_2
5. When BN reacts with HF,
   (a) NH_4BF_4 is formed
   (b) NH_3 gas is evolved
   (c) H_2 gas is evolved
   (d) F_2 gas is evolved
6. Cystalline BN has :
   (a) Wurtzite structure
   (b) Rutile structure
   (c) Fluorite structure
   (d) None of these

ANSWERS

1. (a) **2.** (a), (b), (d) **3.** (a) **4.** (a), **5.** (a) **6.** (a).

HINTS AND EXPLANATIONS

1. $3Mg + 2NH_3 \rightarrow Mg_3N$ (*Magnesium nitride*) $+ 3H_2 \uparrow$
2. (a) $BN + 3\,NaOH \longrightarrow Na_3BO_3 + NH_3$
   (b) $2BN + 3H_2O \longrightarrow B_2O_3 + 2NH_3$
   (c) $BN + 4HF \longrightarrow NH_4BF_4$
   (d) $BN + 3KOH \longrightarrow K_3BO_3 + NH_3$
3. The classification of the given nitrides into ionic and covalent is as Li_3N–ionic, Mg_3N_2 – ionic, BN – covalent and AlN – covalent. Both the nitrides given at (a) are ionic.
4. All the given nitrides are the nitrides of alkaline earth metals. It is only Be_3N_2 which is volatile.
5. Ammonium borofluoride (NH_4BF_4) is formed
   $$BN + 4HF \longrightarrow NH_4BF_4.$$

28

Oxides of Main Group Elements

MULTIPLE CHOICE QUESTIONS

1. The acidic character of trioxides of N, P, As, Sb and Bi is in the order :

(*a*) $N_2O_3 > P_2O_3 > As_2O_3 > Sb_2O_3 > Bi_2O_3$

(*b*) $N_2O_3 < P_2O_3 < As_2O_3 < Sb_2O_3 < Bi_2O_3$

(*c*) $N_2O_3 > As_2O_3 > P_2O_3 > Sb_2O_3 > Bi_2O_3$

(*d*) $N_2O_3 < As_2O_3 < P_2O_3 < Sb_2O_3 < Bi_2O_3$

2. The correct order of acidic strength is :

(*a*) $Cl_2O_7 > SO_2 > P_4O_{10}$

(*b*) $CO_2 > N_2O_5 > SO_3$

(*c*) $Na_2O > MgO > Al_2O_3$

(*d*) $K_2O > CaO > MgO$

3. The sequence of acidic character is :

(*a*) $SO_2 > CO_2 > CO > N_2O_5$

(*b*) $SO_2 > N_2O_5 > CO > CO_2$

(*c*) $N_2O_5 > SO_2 > CO > CO_2$

(*d*) $N_2O_5 > SO_2 > CO_2 > CO$

4. Which of the following oxides is basic in nature ?

(*a*) N_2O (*b*) P_4O_6

(*c*) As_2O_3 (*d*) Bi_2O_3

5. The basic character of Na_2O, MgO, CaO and Al_2O_3 follows the order :

(*a*) $Na_2O > CaO > MgO > Al_2O_3$

(*b*) $Na_2O < CaO < MgO < Al_2O_3$

(*c*) $Na_2O > MgO > CaO > Al_2O_3$

(*d*) $Na_2O < MgO < CaO < Al_2O_3$

6. Identify the correct order of acidic strength of CO_2, CuO, CaO and H_2O

(*a*) $CaO < CuO < H_2O < CO_2$

(*b*) $H_2O < CuO < CaO < CO_2$

(*c*) $CaO < H_2O < CuO < CO_2$

(*d*) $H_2O < CO_2 < CaO < CuO$

7. Which of the oxides is neutral ?

(*a*) CO (*b*) SnO_2

(*c*) ZnO (*d*) SiO_2

8. Which of the following oxides is called mixed anhydride ?

(*a*) N_2O (*b*) NO

(*c*) N_2O_3 (*d*) N_2O_5

9. Which of the following oxides has three - electron bond ?

(*a*) N_2O (*b*) NO

(*c*) N_2O_3 (*d*) N_2O_5

10. Which of the following oxides is (are) neutral ?

(*a*) N_2O (*b*) NO

(*c*) N_2O_3 (*d*) N_2O_5

11. The correct order for the acidic strength of oxides of nitrogen is :

(*a*) $N_2O < NO < N_2O_4 < N_2O_3 < N_2O_5$

(*b*) $N_2O < NO < N_2O_3 < N_2O_4 < N_2O_5$

(*c*) $N_2O < N_2O_3 < N_2O_4 < N_2O_5 < NO$

(*d*) $NO < N_2O < N_2O_3 < N_2O_4 < N_2O_5$

12. Which of the following pentaoxides is the most stable and does not show oxidising properties ?

(*a*) P_2O_5 (*b*) As_2O_5

(*c*) Sb_2O_5 (*d*) Bi_2O_5

13. Which of the following oxides is the most acidic ?

(*a*) Na_2O (*b*) NiO

(*c*) CaO (*d*) SnO

14. Which of the following oxides is different from the other three oxides ?

(*a*) MgO (*b*) ZnO

(*c*) SnO (*d*) PbO

Which of the following oxides is an amphoteric oxide ?

(*a*) MnO_2 (*b*) ZnO
(*c*) CaO (*d*) CO_2

16. HCl is added to the following oxides. Which oxide would give H_2O_2 ?

(*a*) BaO_2 (*b*) MnO_2
(*c*) PbO_2 (*d*) NO_2

17. Which of the following oxides is different from the other three oxides ?

(*a*) MgO (*b*) SnO
(*c*) ZnO (*d*) Cr_2O_3

18. The oxide that gives H_2O_2 on treatment with a dilute acid is :

(*a*) PbO_2 (*b*) MnO_2
(*c*) Na_2O_2 (*d*) TiO_2

19. Which of the following oxides is the most basic ?

(*a*) ZnO (*b*) MgO
(*c*) Al_2O_3 (*d*) N_2O_5

20. The oxide which is amphoteric is :

(*a*) H_2O (*b*) CaO
(*c*) Al_2O_3 (*d*) Cl_2O_7

21. Stannous oxide can be obtained by :

(*a*) heating tin strongly in air
(*b*) heating meta-stannic acid
(*c*) heating tin (II) oxalate
(*d*) None of these

22. A red crystalline solid A reacts with HNO_3 to form yellow powder B which is insoluble in water. Both A and B react with HCl to form $PbCl_2$. A and B respectively are :

(*a*) PbO, PbO_2 (*b*) Pb_3O_4, PbO_2
(*c*) PbO_2, Pb_3O_4 (*d*) PbO_2, PbO

23. Litharge is not commonly used in :

(*a*) The manufacture of special glasses
(*b*) Glazing pottery
(*c*) Preparing paints
(*d*) Lead storage batteries

24. Which of the following can be regarded as plumbous metaplumbate ?

(*a*) Pb_2O_3 (*b*) Pb_3O_4
(*c*) PbO (*d*) PbO_2

25. Lead (IV) oxide is obtained by :

(*a*) heating lead (II) oxide strongly in air.
(*b*) heating lead strongly in pure O_2
(*c*) oxidising Pb with conc. HNO_3
(*d*) None of these

26. Which of the following can be regarded as plumbous orthoplumbate ?

(*a*) Pb_2O_3 (*b*) Pb_3O_4
(*c*) PbO (*d*) PbO_2

27. Which of the following is used as red pigment ?

(*a*) Pb_3O_4 (*b*) PbO
(*c*) $PbCrO_4$ (*d*) $Pb(OH)_2$

28. When Pb_3O_4 is heated with HNO_3, we get a mixture of :

(*a*) $Pb(NO_3)_2$ and PbO_2
(*b*) PbO and NO_2
(*c*) PbO, PbO_2, NO_2 and O_2
(*d*) None of the above.

ANSWERS

1. (*a*)	**2.** (*a*)	**3.** (*d*)	**4.** (*d*)	**5.** (*a*)	**6.** (*a*)	**7.** (*a*)	**8.** (*d*)
9. (*b*)	**10.** (*a*),(*b*)	**11.** (*b*)	**12.** (*a*)	**13.** (*d*)	**14.** (*a*)	**15.** (*b*)	**16.** (*a*)
17. (*a*)	**18.** (*c*)	**19.** (*b*)	**20.** (*c*)	**21.** (*c*)	**22.** (*b*)	**23.** (*d*)	**24.** (*a*)
25. (*d*)	**26.** (*b*)	**27.** (*a*)	**28.** (*c*)				

HINTS AND EXPLANATIONS

1. Oxidation state of the central atom in the given oxides is the same (= + 3) and these oxides belong to the same group. Since metallic character of N, P, As, Sb and Bi increases down the group, the basic character of their oxides also increases down the group or their acidic character decreases. Thus N_2O_3 and P_2O_3 are strongly acidic, since they react with H_2O to give acids .

$N_2O_3 + H_2O \longrightarrow 2HNO_2$, $\quad P_2O_3 + 3H_2O \text{ (cold)} \rightarrow 2H_3PO_3$

As_2O_3 and Sb_2O_3 are amphoteric, since they react with acids as well as alkalies.

$As_2O_3 + 6HCl \longrightarrow 2AsCl_3 + 3H_2O$

$As_2O_3 + 6NaOH \longrightarrow 2Na_3AsO_3$ (*Arsenite*) $+ 3H_2O$

$Sb_2O_3 + 6HCl \longrightarrow 2SbCl_3 + 3H_2O$

$Sb_2O_3 + 6NaOH \longrightarrow 2Na_3SbO_3 + 3H_2O$

Bi_2O_3 is distinctly basic.

2. (*a*) All the given oxides in (*a*) are the oxides of P, S and Cl which are present in the same period (3rd period). Since the metallic character of P, S and Cl decreases from left to right (P $\rightarrow$ S $\rightarrow$ Cl), the basic character of their oxides also decreases as $P_2O_5 > SO_2 > Cl_2O_7$ or acidic character increases as $P_2O_5 < SO_2 < Cl_2O_7$. Thus the order given at (*a*) is correct.

(*b*) Since the oxides given at (*b*) are the oxides of non-metals, these are acidic oxides. The acids given by these oxides are H_2CO_3 ($CO_2 + H_2O \rightarrow H_2CO_3$), HNO_3 ($N_2O_5 + H_2O \rightarrow 2HNO_3$) and H_2SO_4 ($SO_3 + H_2O \rightarrow H_2SO_4$). Since the acidic strength of these acids is in the decreasing order ($HNO_3 > H_2SO_4 > H_2CO_3$), the acidic nature of N_2O_5, SO_3 and CO_2 is also in the decreasing order : $N_2O_5 > SO_3 > CO_2$. Thus the order give at (*b*) is wrong.

(*c*) Since Na_2O and MgO are the oxides of highly electropositive metals (Na = alkali metal, Mg = alkaline earth metal), these are highly basic oxides. Al_2O_3 is an amphoteric oxide. Now since Mg has lesser metallic (basic) character than Na, MgO is less basic than Na_2O or MgO is more acidic than Na_2O (MgO > Na_2O). Thus the acidic character of Na_2O, MgO and Al_2O_3 is in the order MgO > Na_2O > Al_2O_3. Consequently the order given at (*c*) is wrong.

(*d*) Since K_2O is the oxide of the most basic metal, K_2O is the most basic (least acidic). Since Ca has higher metallic (basic) character than Mg, CaO is more basic than MgO or CaO is less acidic than MgO. Thus the acidic character of K_2O, MgO and CaO is in the order MgO > CaO > K_2O. Consequently the order given at (*d*) is wrong.

3. CO is a neutral oxide, since it reacts neither with an acid nor with a base. N_2O_5, SO_2 and CO_2, on reacting with H_2O, give HNO_3, H_2SO_3 and H_2CO_3 respectively. Out of these acids HNO_3 is the strongest acid and H_2CO_3 is the weakest acid. Thus the acidic character of N_2O_5, SO_2 and CO_2 is in the order $N_2O_5 > SO_2 > CO_2$. Consequently, the acidic character of all the given four oxides is in the order : $N_2O_5 > SO_2 > CO_2 > CO$.

4. N_2O is a neutral oxide. The basic character of other oxides increases as $P_4O_6 < As_2O_3 < Bi_2O_3$, since these oxides are the oxides of the elements of the same group (group 5A). Thus Bi_2O_3 is the most basic oxide.

5. Since Na_2O, MgO and CaO are the oxides of highly electropositive alkali metals and alkaline earth metals, these oxides are basic in nature. Now since alkali metals are more electropositive (metallic or basic) than alkaline earth metals, Na_2O is more basic than MgO – CaO pair (Na_2O > MgO – CaO pair). Since Ca is more metallic (basic) than Mg. CaO is more basic than MgO (CaO > MgO). Thus the basic character of Na_2O, MgO and CaO is in the order : Na_2O > CaO > MgO. Since Al_2O_3 is amphoteric, the overall basic character of the four oxides follows the order Na_2O > CaO > MgO > Al_2O_3.

6. Since Ca and Cu are metals, CaO and CuO are basic oxides. Since CaO is an oxide of alkaline earth metal (highly electro positive metal) and CuO is an oxide of *d*-block element,

CaO is more basic than CuO or CaO is less acidic than CuO (CaO < CuO). H_2O is neutral and CO_2 is acidic oxide. Thus the acidic strength of the given oxides is in the order CaO < CuO < H_2O < CO_2

7. CO is a neutral oxide, since this oxide reacts neither with an acid nor with a base. SnO_2 and ZnO both are amphoteric oxides, since they react with an acid and base both to form salt and water.

$$SnO_2 + 2H_2SO_4\ (Conc) \xrightarrow{\Delta} Sn(SO_4)_2 + 2H_2O$$

$$SnO_2 + 2NaOH \xrightarrow{\text{Fuse}} Na_2SnO_3\ (Stannate) + H_2O$$

$$ZnO + 2HCl \longrightarrow ZnCl_2 + H_2O$$

$$ZnO + 2NaOH \longrightarrow Na_2ZnO_2\ (Zincate) + H_2O$$

SiO_2 is an acidic oxide, since it gives an acid with H_2O.

$$SiO_2 + H_2O \longrightarrow H_2SiO_3\ (Silicic\ acid)$$

8. (*a*) N_2O is a neutral oxide.

(*b*) NO is sparingly soluble in water. Like N_2O, NO is also a neutral oxide

(*c*) N_2O_3 dissolves in water to give HNO_2. Thus this oxide is called *nitrous anhydride.*

$$N_2O_3 + H_2O \xrightarrow{0°C} 2HNO_2$$

(*d*) NO_2 gives a mixture of two acids *viz* nitrous acid (HNO_2) and nitric acid (HNO_3) on dissolution with H_2O. Thus NO_2 is called a *mixed anhydride.*

$$2NO_2 + H_2O \longrightarrow HNO_2 + HNO_3$$

(*c*) N_2O_5 dissolves in water with a hissing sound to give only one acid *viz* HNO_3. Thus N_2O_5 is called *nitric anhydride.*

$$N_2O_5 + H_2O \longrightarrow 2HNO_3$$

9. The structure of the given oxides are given as :

(*a*) $:\ddot{N} = \overset{+}{N} = \ddot{O}:$ (N_2O)

(*b*) $:\dot{N} = \ddot{O}:$ (According to VBT)
$:N \equiv O:$ (According to MOT) } (NO)

(*c*) 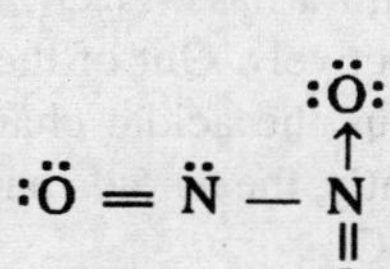 (N_2O_3)

(*d*) 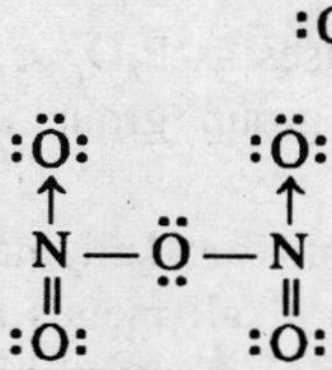 (N_2O_5)

Obviously it is only NO molecule which, according to MOT, contains one three – electron bond and two normal two – electron covalent bonds.

10. N_2O and NO both are neutral oxides since they donot react with water. N_2O_3 and N_2O_5 both are acidic oxides because both give acids on dissolution with water.

$$N_2O_3 + H_2O \longrightarrow 2HNO_2, \qquad N_2O_5 + H_2O \longrightarrow 2HNO_3$$

11. Since the given oxides are the oxides of the same element *viz* N, their acidic character increases with the increase of oxidation number of N-atom. Thus :

+1	+2	+3	+4	+5
N_2O	NO	N_2O	N_2O_4	N_2O_5
Neutral		Acidic	Strongly acidic	Very strongly acidic

—— Acidic character increases ⟶

12. P_2O_5 is thermally stable while all other pentoxides get dissociated into lower oxides and O_2

$$2N_2O_5 \xrightarrow{\text{Room temp}} 2N_2O_4 + O_2$$

$$As_2O_5 \xrightarrow{\text{Red heat}} As_2O_3 + O_2$$

$$Sb_2O_5 \xrightarrow{44°C} Sb_2O_3 + O_2$$

Bi_2O_5 is extremely unstable and readily forms Bi_2O_3.

$$Bi_2O_5 \longrightarrow Bi_2O_3 + O_2$$

P_2O_5, being thermally stable, does not give O_2 and hence does not show oxidising properties. Since other oxides give O_2, they are strong oxidising agents.

13. Since Na_2O and CaO are the oxides of highly electropositive (metallic or basic) metals *viz* Na and Ca, these oxides are highly basic in nature. NiO is a neutral oxide. SnO is an amphoteric oxide with predominant acidic character.

$$SnO + H_2SO_4 \longrightarrow SnSO_4 + H_2O \text{ (\textit{Basic character})}$$

$$SnO + 2NaOH \longrightarrow \underset{\text{Sod. stannite}}{Na_2SnO_2} + H_2O \text{ (\textit{Acidic character})}$$

Thus it can be said that SnO is the most acidic oxide.

14. Since MgO is an oxide of highly electropositive metal, it is a basic oxide. ZnO, SnO and PbO are amphoteric oxides, since each of them reacts with acids as well as bases

$$ZnO + 2HCl \longrightarrow ZnCl_2 + H_2O, \qquad ZnO + 2NaOH \longrightarrow \underset{\text{Zincate}}{Na_2ZnO_2} + H_2O$$

$$SnO + H_2SO_4 \longrightarrow SnSO_4 + H_2O, \qquad SnO + 2NaOH \longrightarrow \underset{\text{Stannite}}{Na_2SnO_2} + H_2O$$

$$PbO + H_2SO_4 \longrightarrow PbSO_4 + H_2O, \qquad PbO + 2NaOH \longrightarrow \underset{\text{Plumbite}}{Na_2PbO_2} + H_2O$$

15. (*a*) MnO_2 is insoluble and shows no acid-base properties. However, MnO_2 acts as a basic oxide towards cold and concentrated HCl, forming $MnCl_4$ (salt) and H_2O.

$$MnO_2 + 4HCl \text{ (\textit{Cold and concentrated})} \longrightarrow MnCl_4 + 2H_2O$$

(*b*) ZnO is an amphoteric oxide, since it reacts both with acids and bases to form salt and H_2O

$$ZnO + 2HCl \longrightarrow ZnCl_2 + H_2O, \qquad ZnO + 2NaOH \longrightarrow Na_2ZnO_2 + H_2O$$

(*c*) CaO is highly basic oxide. It gives $Ca(OH)_2$ with H_2O which is a strong hydroxide.

$$CaO + H_2O \longrightarrow Ca(OH)_2$$

(*d*) CO_2 is an acidic oxide, since it reacts with H_2O to give carbonic acid, H_2CO_3

$$CO_2 + H_2O \longrightarrow H_2CO_3.$$

16. (*a*) BaO_2 is a peroxide, since it gives salt and H_2O_2 with dilute acids like HCl, H_2SO_4 etc.

$BaO_2 + HCl\ (dil) \longrightarrow BaCl_2 + H_2O_2$, $BaO_2 + H_2SO_4\ (dil) \longrightarrow BaSO_4 + H_2O_2$

(*b*) MnO_2 is insoluble and shows no acid-base properties. However, MnO_2 acts as a basic oxide towards cold and conc. HCl.

$MnO_2 + 4HCl\ (\textit{Cold and concentrated}) \longrightarrow MnCl_4 + 2H_2O$

(*c*) PbO_2 gives Cl_2 with conc. HCl

$PbO_2 + 4HCl\ (\textit{conc}) \longrightarrow PbCl_2 + 2H_2O + Cl_2$

(*d*) NO_2 oxidises HCl to Cl_2

$$\underset{(N = +4)}{2NO_2} + \underset{(Cl = -1)}{4HCl} \longrightarrow \underset{(N = +3)}{2NOCl} + 2H_2O + \underset{(Cl = 0)}{Cl_2}$$

17. Since MgO is an oxide of highly electropositive metal, it is a basic oxide while other oxides are amphoteric. This fact is evident from the following reactions :

$$\left.\begin{aligned} MgO + H_2O &\longrightarrow Mg(OH)_2 \\ MgO + 2HCl &\longrightarrow MgCl_2 + H_2O \\ MgO + H_2SO_4 &\longrightarrow MgSO_4 + H_2O \end{aligned}\right\} \text{Basic nature of MgO}$$

$$\left.\begin{aligned} SnO + H_2SO_4 &\longrightarrow SnSO_4 + H_2O \\ SnO + 2NaOH &\longrightarrow \underset{\text{Sod. stanite}}{Na_2SnO_2} + H_2 \end{aligned}\right\} \text{Amphoteric nature of SnO}$$

$$\left.\begin{aligned} ZnO + 2HCl &\longrightarrow ZnCl_2 + H_2O \\ ZnO + 2NaOH &\longrightarrow Na_2ZnO_2 + H_2O \end{aligned}\right\} \text{Amphoteric nature of ZnO}$$

$$\left.\begin{aligned} Cr_2O_3 + 3H_2SO_4 &\longrightarrow Cr(SO_4)_3 + 3H_2O \\ Cr_2O_3 + 2NaOH &\longrightarrow \underset{\text{Sod. chromite}}{2NaCrO_2} + H_2O \end{aligned}\right\} \text{Amphoteric nature of } Cr_2O_3.$$

18. It is only Na_2O_2 which is a peroxide (sodium peroxide) and hence gives H_2O_2 on treatment with dil acids.

$Na_2O_2 + H_2SO_4\ (dil) \longrightarrow Na_2SO_4 + H_2O_2$

PbO_2 and MnO_2 are dioxides (covalent molecules) and give Cl_2 and O_2 respectively when treated with conc. HCl and conc. H_2SO_4.

$PbO_2 + 4HCl\ (\textit{conc}) \longrightarrow PbCl_2 + 2H_2O + Cl_2$

$MnO_2 + 2H_2SO_4\ (\textit{conc}) \longrightarrow 2MnSO_4 + 2H_2O + O_2$

TiO_2 is an acidic oxide because it gives H_2TiO_3 (acid) with H_2O.

$TiO_2 + H_2O \longrightarrow H_2TiO_3$

19. ZnO and Al_2O_3 both are amphoteric oxides, since both react with acids and bases to form salt and H_2O as shown below.

$ZnO + 2HCl \longrightarrow ZnCl_2 + H_2O$, $ZnO + 2NaOH \longrightarrow Na_2ZnO_2 + H_2O$

$Al_2O_3 + 3HCl \longrightarrow 2AlCl_3 + 3H_2O$, $Al_2O_3 + 2NaOH \longrightarrow 2NaAlO_2 + H_2O$

N_2O_5 is an acidic oxide, since it reacts with bases and H_2O.

$N_2O_5 + 2NaOH \longrightarrow 2NaNO_3 + H_2O$, $N_2O_5 + H_2O \longrightarrow 2HNO_3$

MgO is a basic oxide since it reacts with acids and H_2O.

$MgO + 2HCl \longrightarrow MgCl_2 + H_2O$, $MgO + H_2O \longrightarrow Mg(OH)_2$

20. (*a*) H_2O is a neutral oxide since it reacts neither with an acid nor with a base.

(*b*) Since CaO is an oxide of highly electropositive metal, it is a basic oxide. Being basic it reacts with an acids and H_2O

$$CaO + 2HCl \longrightarrow CaCl_2 + H_2O, \qquad CaO + H_2O \longrightarrow Ca(OH)_2$$

(*c*) Al_2O_3 is an amphoteric oxide, since it reacts with an acid and a base both to give salt and H_2O.

$$Al_2O_3 + 6HCl \longrightarrow 2AlCl_3 + 3H_2O$$
$$Al_2O_3 + 2NaOH \longrightarrow 2NaAlO_2 + H_2O$$

(*d*) Cl_2O_7 is an acidic oxide because its reacts with a base to form salt and H_2O and with H_2O to give an acid ($HClO_4$)

$$Cl_2O_7 + 2NaOH \longrightarrow 2NaClO_4 + H_2O$$
$$Cl_2O_7 + H_2O \longrightarrow 2HClO_4$$

21. Heating Sn strongly in air gives stannic oxide (SnO_2) and not stanneous oxide (SnO)

$$Sn + O_2 \xrightarrow{1800\text{ K}} SnO_2$$

(*b*) Heating tin (II) oxalate, $Sn(COO)_2$ gives stannous oxide (SnO)

$$Sn(COO)_2 \xrightarrow{\Delta} SnO + CO\uparrow + CO_2\uparrow$$

22. The reactions involved are :

$$\underset{\substack{\text{(A)}\\ \text{Red crystalline}\\ \text{solid}}}{Pb_3O_4} + 4HNO_3 \longrightarrow 2Pb(NO_3)_2 + 2H_2O + \underset{\substack{\text{(B)}\\ \text{Yellow}\\ \text{powder}}}{PbO_2}$$

$$\underset{\text{(A)}}{Pb_3O_4} + 8HCl \longrightarrow 3PbCl_2 + 4H_2O + Cl_2$$

$$\underset{\text{(B)}}{PbO_2} + 4HCl \longrightarrow PbCl_2 + 2H_2O + Cl_2$$

23. Litharge (PbO) is used in glazing pottery, in the manufacture of paints and special glasses (*e.g.* flint glass) but not in lead storage batteries.

It is lead dioxide (PbO_2) which is used in lead storage batteries.

24. Since metaplumbate ion is $[Pb^{4+}O_3]^{2-}$, plumbous metaplumbate is $Pb^{2+}[Pb^{4+}O_3]^{2-}$ or Pb_2O_3. This formula of Pb_2O_3 also suggests that in this oxide Pb is present in + 2 and + 4 oxidation state ($Pb_2O_3 \equiv \overset{2+}{Pb}O\,.\,\overset{4+}{Pb}O_2$)

25. (*a*) When lead (II) oxide, PbO is strongly heated in air, Pb_3O_4 is formed.

$$6PbO + O_2 \longrightarrow 2Pb_3O_4$$

(*b*) When lead is strongly heated in pure O_2, PbO (litharge) is formed but on further heating to 400 – 500° C, it gives red lead (Pb_3O_4)

$$2Pb + O_2 \longrightarrow 2PbO, \qquad 6PbO + O_2 \longrightarrow 2Pb_3O_4$$

(*c*) Oxidation of Pb by conc. HNO_3 gives $Pb(NO_3)_2$. In this reaction HNO_3 is reduced to NO_2NPPb + $4HNO_3$ (Conc) $\longrightarrow Pb(NO_3)_2 + 2NO_2 + 2H_2O$

(Pb = 0) (N = + 5) (Pb = + 2) (N = + 4)

26. Since orthoplumbate ion is $[Pb^{4+}O_4]^{4-}$, plumbous orthoplumbate is $Pb_2^{2+}[\overset{4+}{Pb}O_4]^{4-}$ or Pb_3O_4. This formula of Pb_3O_4 also suggests that in this oxide Pb is present in + 2 and + 4 oxidation state $(Pb_3O_4 \equiv 2\overset{2+}{Pb}O . \overset{4+}{Pb}O_2)$

27. Pb_3O_4 is called red lead, since it is used as a red pigment in the preparation of red protective paints for iron and steel structure.

28. We know that Pb_3O_4 is a mixed oxide which contains 2PbO and PbO_2 $(Pb_3O_4 \equiv 2PbO . PbO_2)$. It is only PbO which reacts with HNO_3 and is converted into $Pb(NO_3)_2$ while PbO_2 remains unaffected by HNO_3.

$$Pb_3O_4 \text{ or } 2PbO . PbO_2 + 4HNO_3 \longrightarrow 2Pb(NO_3)_2 + PbO_2 + 2H_2O$$

29

Halides and Oxy-halides of Main Group Elements

MULTIPLE CHOICE QUESTIONS

1. Which of the following halides is the most acidic ?

(*a*) PCl_3 (*b*) $SbCl_3$

(*c*) $BiCl_3$ (*d*) CCl_4

2. Which of the following statements is correct about NF_3 and NCl_3 ?

(*a*) NF_3 does not show basic character but NCl_3 acts as a base

(*b*) NCl_3 does not show basic character but NF_3 acts as a base.

(*c*) Neither NF_3 nor NCl_3 is a base

(*d*) NF_3 and NCl_3 both are acids.

3. Among the trihalides of nitrogen which one is the least basic ?

(*a*) NF_3 (*b*) NCl_3

(*c*) NBr_3 (*d*) NI_3

4. Basic strength of trihalides of nitrogen increases in the order :

(*a*) $NF_3 < NCl_3 < NBr_3 < NI_3$

(*b*) $NF_3 < NBr_3 < NCl_3 < NI_3$

(*c*) $NF_3 < NI_3 < NCl_3 < NBr_3$

(*d*) $NCl_3 < NBr_3 < NI_3 < NF_3$

5. The hydrolysis of PCl_3 produces :

(*a*) $H_3PO_3 + HClO$ (*b*) $H_3PO_3 + HCl$

(*c*) $H_3PO_4 + HCl$ (*d*) $PH_3 + HClO$

6. Which of the following trichlorides does not undergo hydrolysis ?

(*a*) $AsCl_3$ (*b*) $SbCl_3$

(*c*) PCl_3 (*d*) NF_3

7. PCl_5 is kept in well stoppered bottles because :

(*a*) it is highly volatile

(*b*) it reacts with O_2.

(*c*) it reacts with moisture

(*d*) it is explosive.

8. Solid PCl_5 exists as :

(*a*) PCl_5 molecules

(*b*) P_2Cl_6

(*c*) $[PCl_4]^+ [PCl_6]^-$

(*d*) None of these

9. $BiCl_3$, on hydrolysis, forms a white precipitate of :

(*a*) Bismuthio acid

(*b*) Bismuth oxychloride

(*c*) Bismuth pentechloride

(*d*) Bismuth hydroxide.

10. Which of the following pentahalides does not exist ?

(*a*) NCl_5 (*b*) PF_5

(*c*) AsF_5 (*d*) $SbCl_5$

11. Which of the following tetraiodides is the least stable and has doubtful existence ?

(*a*) CI_4 (*b*) GeI_4

(*c*) SnI_4 (*d*) PbI_4

12. Which of the following pentafluorides cannot be formed ?

(*a*) PF_5 (*b*) AsF_5

(*c*) SbF_5 (*d*) BiF_5

13. Which of the following is least basic ?

(*a*) NF_3 (*b*) NH_3

(*c*) NCl_3 (*d*) NI_3

14. The halide that is not hydrolysed is :

(*a*) $SiCl_4$ (*b*) SiF_4

(*c*) CCl_4 (*d*) $PbCl_4$.

15. Which of the following halides is the most stable ?

(*a*) CF_4 (*b*) CCl_4

(*c*) $C\,Br_4$ (*d*) CI_4

16. PbF_4 and $PbCl_4$ exist but $PbBr_4$ and PbI_4 donot exist, because of :

(*a*) Small size of Pb^{4+} ion and larger size of Br^- and I^- ions.

(*b*) High oxidising nature of Pb^{4+} ion

(*c*) High reducing nature of Br^- and I^- ions

(*d*) Both (*b*) and (*c*).

17. Which of the following complex hexahalide ions does not exist ?

(*a*) $[CCl_6]^{2-}$ (*b*) $[SiCl_6]^{2-}$

(*c*) $[GeF_6]^{2-}$ (*d*) $[SnCl_6]^{2-}$

18. The correct order of the ionic character of the following dihalides is :

(*a*) $PbF_2 > PbCl_2 > PbBr_2 > PbI_2$

(*b*) $PbF_2 > PbBr_2 > PbCl_2 > PbI_2$

(*c*) $PbF_2 < PbCl_2 < PbBr_2 < PbI_2$

(*d*) $PbF_2 < PbCl_2 > PbBr_2 > PbI_2$

19. The power of trihalides of boron to act as Lewis acids decreases in the order :

(*a*) $BF_3 > BCl_3 > BBr_3 > BI_3$

(*b*) $BI_3 > BBr_3 > BCl_3 > BF_3$

(*c*) $BCl_3 > BF_3 > PBr_3 > BI_3$

(*d*) $BBr_3 > BCl_3 > BF_3 > BI_3$

20. Lewis acid character of trichlorides of B, Al, Ga and In decreases in the order :

(*a*) $BCl_3 > AlCl_3 > GaCl_3 > InCl_3$

(*b*) $AlCl_3 > GaCl_3 > InCl_3 > BCl_3$

(*c*) $GaCl_3 > InCl_3 > AlCl_3 > BCl_3$

(*d*) $InCl_3 > GaCl_3 > AlCl_3 > BCl_3$

21. Which of the following statements is correct ?

(*a*) Both $SnCl_4$ and $SnBr_4$ are coloured

(*b*) All $SnCl_4$, $SnBr_4$ and SnI_4 are coloured

(*c*) Only $SnBr_4$ is coloured

(*d*) Only SnI_4 is coloured.

22. Tin (II) fluoride (anhydrous) can be obtained by :

(*a*) treating Sn with F_2

(*b*) treating Sn with HF

(*c*) dissolving SnO in HF

(*d*) None of these

23. $SnCl_2$ is used :

(*a*) as a mordant in dyeing

(*b*) as an oxidising agent

(*c*) as a reducing agent

(*d*) in the preparation of colloidal gold

24. Which of the following compounds of tin is a reducing agent ?

(*a*) $SnCl_2$ (*b*) $SnCl_4$

(*c*) SnO (*d*) None

25. When excess of $SnCl_2$ is added to a solution of $HgCl_2$, a white ppt. turning black is obtained. The black colour is due to the formation of :

(*a*) Hg_2Cl_2 (*b*) $SnCl_4$

(*c*) Sn (*d*) Hg

26. Tin (II) chloride (anhydrous) is obtained :

(*a*) by melting Sn in an atmosphere of dry Cl_2

(*b*) by treating Sn with dil. HCl and heating the product to dryness

(*c*) by treating Sn with conc. HCl and heating the product to dryness.

(*d*) by treating Sn with a current of dry HCl gas.

27. Tin (IV) chloride (anhydrous) can be obtained :

(*a*) by action of dry Cl_2 on molten tin

(*b*) by heating Sn with conc. HCl

(*c*) by treating tin with dil. HCl

(*d*) None of these

28. Sn (IV) chloride is :

(*a*) a covalent compound

(*b*) a volatile liquid at room temperature

(*c*) tetrahedral in shape

(*d*) an ionic compound

29. Butter of tin :

(*a*) $SnCl_4 \cdot 6H_2O$ (*b*) $SnCl_2 \cdot 6H_2O$

(*c*) $SnCl_4 \cdot 5H_2O$ (*d*) $SnCl_2 \cdot 5H_2O$

ANSWERS

1. (*c*)	**2.** (*a*)	**3.** (*a*)	**4.** (*a*)	**5.** (*b*)	**6.** (*d*)	**7.** (*c*)	**8.** (*c*)
9. (*b*)	**10.** (*a*)	**11.** (*d*)	**12.** (*d*)	**13.** (*a*)	**14.** (*c*)	**15.** (*a*)	**16.** (*b*), (*c*)
17. (*a*)	**18.** (*a*)	**19.** (*b*)	**20.** (*d*)	**21.** (*d*)	**22.** (*c*)	**23.** (*a*), (*c*), (*d*)	
24. (*a*)	**25.** (*d*)	**26.** (*d*)	**27.** (*a*)	**28.** (*a*), (*b*), (*c*)		**29.** (*c*)	

HINTS AND EXPLANATIONS

1. Since C-atom (central atom) in CCl_4 molecule has no vacant orbitals in its valence shell, this atom cannot accept electron pairs donoted by donor molecules or ions (Lewis bases) and hence does not show acidic properties. On the other hand, due to the presence of *d*-orbitals in the valence shell in P, Sb and Bi, PCl_3, $SbCl_3$ and $BiCl_3$ molecules can accept electron pair donated by donor molecules or ions (Lewis bases) and show acidic properties. Since Bi atom is the largest in size, $BiCl_3$ molecule will show maximum tendency to accept the electrons donated by a Lewis base. Thus $BiCl_3$ is the most acidic.

2. Since both the given molecules have a lone pair of electrons (*lp*) on N-atom (central atom), both the molecules should show basic character. In fact NF_3 does not show basic character while NCl_3 acts as a base.

Explanation. Since F-atom is much more electronegative than Cl-atom, the *lp* residing on N-atom in NF_3 molecule is attracted towards F-atom most strongly and hence electron density on N-atom in NF_3 molecule decreases to the maximum extent. Due to the decrease in electron density on N-atom, the electron pair on *N*-atom in NF_3 molecule is not available for being donated to Lewis acid (electron pair acceptor). Thus NF_3 molecule does not show basic character. Due to lesser electronegativity of Cl-atom, the *lp* of electrons on N-atom in NCl_3 molecule is easily available for being donated to Lewis acid. Thus NCl_3 molecule acts as a base.

3. As the electronegativity of halogens decreases from F to I (F > Cl > Br > I), the basic character of trihalides of nitrogen increases as $NF_3 < NCl_3 < NBr_3 < NI_3$. Thus NF_3 is the least basic.

Explanation. Each of the given molecules has a lone pair of electrons in sp^3 hybrid orbital on N-atom. The given trihalides are the trihalides of the same element *viz* N-atom and have different halogens. The basic character of these trihalides will, therefore, depend on the electronegativity of the halogen atom. Now since the electronegativity of F-atom is maximum, this halogen attracts the lone pair of electrons residing on N-atom towards itself more strongly than Cl, Br and I in NCl_3, NBr_3 and NI_3 molecules respectively. Thus lone pair of electrons present on N-atom in NF_3 molecule is least available to be donated to a Lewis acid. Thus NF_3 molecule is the poorest Lewis base, *i.e.*, NF_3 is the least basic.

4. As the electronegativity of halogen atom decreases from F to I (F > Cl > Br > I), the basic strength of trihalides of N-atom increases as $NF_3 < NCl_3 < NBr_3 < NI_3$.

Explanation. Since the central atom in the given trihalides and their geometry are the same (geometry is pyramidal), but the surrounding halogen atoms linked with the central N-atom are different, the basic character of trihalides depends on the electronegativity of the halogen atom. Since the electronegativity of F-atom is maximum, lone pair of electrons present on N-atom will be attracted towards F-atom most strongly and hence the electron density due to the *lp* of electrons will be decreased on N-atom. Due to the decreases in electron density on N-atom, NF_3 molecule will show minimum basic character. As we move from NF_3 to NI_3, electronegativity of halogen atom decreases and hence electron density on N-atom goes on increasing. Due to the increase in electron density on N-atom, the basic character of NX_3 molecules also increases from NF_3 to NI_3. (Greater is the electron density on the central N-atom, greater is the basic character of the trihalide of nitrogen).

5. PCl_3 reacts with water violently to produce phosphorus acid (H_3PO_3) and HCl

$$PCl_3 + 3H_2O \longrightarrow H_3PO_3 + 3HCl$$

6. NF_3 does not undergo hydrolysis due to the absence of vacant *d*-orbitals in the valence-shell of N-atom (central atom). All other trichlorides undergo hydrolysis as shown below :

$$PCl_3 + 3H_2O \longrightarrow H_3PO_3 + 3HCl$$
$$2AsCl_3 + 3H_2O \longrightarrow As_2O_3 + 6HCl$$
$$SbCl_3 + H_2O \longrightarrow [SbO]^+ Cl^- + 2HCl$$

7. PCl_5 reacts readily and violently with water to give posphoric acid (H_3PO_4) and HCl.

$$PCl_5 + 4H_2O \text{ (Excess)} \longrightarrow H_3PO_4 + 5HCl$$

8. X-ray study of the structure of PCl_5 in the solid state has shown that two PCl_5 molecules unite together and then get broken into $[PCl_4]^+$ and $[PCl_6]^-$ ions. Thus PCl_5, in the solid state, is an ionic compound which may be represented as $[PCl_4]^+ [PCl_6]^-$

$$PCl_5 + PCl_5 \longrightarrow 2PCl_5 \longrightarrow [PCl_4]^+ + [PCl_6]^- \longrightarrow [PCl_4]^+ [PCl_6]^-$$

Ionic compound. Ionic compound

9. When $BiCl_3$ is hydrolysed by water, white ppt. of bismuth oxy chloride (BiOCl) is obtained.

$$BiCl_3 + H_2O \longrightarrow \underset{\text{White ppt.}}{BiOCl} + 2HCl$$

10. NCl_5 does not exist, since N-atom cannot have a covalency of 5. This is because of the fact that N-atom cannot expand its octet due to the absence of *d*-orbitals in its valence shell.

11. In the given tetraiodides the central metallic atoms are in + 4 oxidation state and all the central atoms belong to group 4A. Now we know that, due to inert pair effect, the stability of + 4 oxidation state of C, Ge, Sn and Pb (group 4 A) decreases down the group. Thus + 4 oxidation state of Pb is the least stable. Consequently PbI_4 in which Pb shows + 4 oxidation state is the least stable and has, therefore, doubtful existence.

12. In the given pentafluorides the central atoms (*i.e.*, P, As, Sb and Bi) show + 5 oxidation state and belong to group 5A. We know that, due to inert pair effect, the stability of + 5 oxidation state of P, As, Sb and Bi (group 5A) decreases down the group. Thus + 5 oxidation state of Bi is unstable. Due to unstability of + 5 oxidation state, Bi cannot form BiF_5 molecule.

13. Trihalides of N (NF_3, NCl_3 and NI_3) are less basic than its hydride (NH_3), since electronegativity of each halogen atom is higher than that of H-atom (H = 2.1, F = 4.0, Cl = 3.0, I = 2.4). Out of different trihalides, NF_3 is the least basic because electronegativity of F-atom is higher than that of Cl and I both.

14. CCl_4, $SiCl_4$, SiF_4 and $PbCl_4$ molecules are formed by sp^3 hybridisation of the central atom. In sp^3 hybridisation *s*-and all the three *p*-orbitals are used up. In case of $SiCl_4$, SiF_4 and $PbCl_4$ molecules *d*-orbitals remain unused and vacant on the central atom but in case of CCl_4 molecule C-atom has no *d*-orbitals. Due to the presence of *d*-orbitals in the valence-shell of Si and Pb atoms, these atoms can accept the electron pair donated by H_2O molecule during hydrolysis to form $M \leftarrow O\genfrac{}{}{0pt}{}{H}{H}$ bond (M = Si, Pb). Thus $SiCl_4$, SiF_4 and $PbCl_4$ can undergo hydrolysis due to the presence of *d*-orbitals (vacant) in the valence shell of their central atom. On the other hand, since C-atom (central atom) in CCl_4 molecule has no *d*-orbitals, this atom cannot accept electron pair donated by H_2O molecule to form $C \leftarrow O \langle^{H}_{H}$ bond. Thus CCl_4 cannot undergo hydrolysis.

15. Since F atom is the smallest of all the halogens, C–F bond length is the smallest of all other C–X bond lengths (X = Cl, Br, I). Due to the smallest length of C–F bond, C–F bond is the strongest bond and hence CF_4 molecule is the most stable.

16. The statement that $PbBr_4$ and PbI_4 donot exist means that these halides are unstable compounds and get decomposed into ($PbBr_2 + Br_2$) and ($PbI_2 + I_2$) respectively. Similarly the statement that PbF_4 and $PbCl_4$ exist means that these halides are stable compounds and hence donot get decomposed to give ($PbF_2 + F_2$) and ($PbCl_2 + Cl_2$). We know that halide ions act as reducing agents and their reducing power increases as $F^- < Cl^- < Br^- < I^-$. Actually the conversion of $PbBr_4$ and PbI_4 into ($PbBr_2 + Br_2$) and ($PbI_2 + I_2$) takes place through the following steps :

$$PbBr_4 \longrightarrow Pb^{4+} + 4Br^-$$

$$\underset{\text{(OA)}}{Pb^{4+}} + \underset{\text{(Strong RA)}}{2Br^-} \longrightarrow Pb^{2+} + Br_2 \qquad \ldots (a)$$

$$PbBr_4 \longrightarrow Pb^{2+} + 2Br^- + Br_2$$

$$\text{or} \quad PbBr_4 \longrightarrow PbBr_2 + Br_2$$

$$PbI_4 \longrightarrow Pb^{4+} + 4I^-$$

$$\underset{\text{(OA)}}{Pb^{4+}} + \underset{\text{(Strongest RA)}}{2I^-} \longrightarrow Pb^{2+} + I_2 \qquad \ldots (b)$$

$$PbI_4 \longrightarrow Pb^{2+} + 2I^- + I_2$$

$$\text{or} \quad PbI_4 \longrightarrow PbI^2 + I_2$$

Since Br^- and I^- ions are strong reducing agents, they reduce Pb^{4+} ion to Pb^{2+} ion as shown at equations (*a*) and (*b*) respectively and are themselves oxidised to Br_2 and I_2. Hence $PbBr_2$ and PbI_2 are formed. On the other hand, since F^- and Cl^- ions are very weak reducing agents, they cannot reduce Pb^{4+} to Pb^{2+} and hence PbF_2 and $PbCl_2$ are not formed.

$$PbF_4 \longrightarrow Pb^{4+} + 4F^-$$

$$Pb^{4+} + \underset{\text{(Weakest RA)}}{2F^-} \longrightarrow \text{No reaction}$$

$$PbCl_4 \longrightarrow Pb^{4+} + 4Cl^-$$

$$Pb^{4+} + \underset{\text{(Weaker RA)}}{2Cl^-} \longrightarrow \text{No reaction}$$

17. The structure of $[SiCl_6]^{2-}$, $[GeF_6]^{2-}$ and $[SnCl_6]^{2-}$ ions is given below. The two electron pairs donated by two X^- ions to form two $X^- \longrightarrow M$ bonds are accommodated in two *d*-orbitals of the central M atom (sp^3d^2 hybridisation). Since C-atom (central atom) in $[CCl_6]^{2-}$ ion has no *d*-orbitals, the formation of $[CCl_6]^{2-}$ ion is not possible.

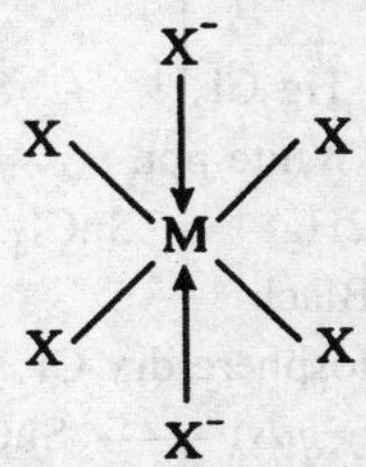

18. Since the central metallic ion in all the given dihalides is the same *viz* Pb^{2+}, the ionic character of dihalides depends on the size of halide ion (X^- ion). Since the size of X^- ion increases down the group from F^- to I^- ($F^- < Cl^- < Br^- < I^-$), the ionic character of PX_2 molecules decreases down the group from PbF_2 to PbI_2 ($PbF_2 > PbCl_2 > PbBr_2 > PbI_2$) in accordance with Fajan's rules.

19. Since all the given trihalides have the same central atom *viz* B and different halogen atoms, the variation of Lewis acid character of BX_3 molecules depends on the variation of the size of halogen atoms (F, Cl, Br, I). Lewis acid character increases with the increase in the size of halogen atom ($BF_3 < BCl_3 < BBr_3 < BI_3$)

Explanation. Since the size of halogen atoms increases from F to I, the electron density produced on B-atom in BX_3 molecules due to the transfer of electron pair from the filled *np* orbital of halogen atom, X (X = $ns^2\, np^1\, np^2\, np^2$) to the vacant 2*p* orbital on B-atom (B = $2s^2\, 2p^1\, 2p^1\, 2p^0\, 2p^0$, $B^* = 2s^1\, 2p^1\, 2p^1\, 2p^0$) decrease as we move from BF_3 to BI_3 molecule. With the decrease in electron density on B-atom on moving from BF_3 to BI_3, the tendency of B-atom to accept the electron pair donated by Lewis base (electron pair donor) increases and hence Lewis acid character of BX_3 molecules also increases from BF_3 to BI_3. Thus the Lewis acid character of BX_3 molecules increases as $BF_3 < BCl_3 < BBr_3 < BI_3$ or decreases in the order $BI_3 > BBr_3 > BCl_3 > BF_3$

20. Since all the given trichlorides have the same halide ion *viz* Cl^- ion, and different central atoms, the variation of Lewis acid character of trichlorides depends on the size of the central atoms *viz* B, Al, Ga and In. Since the size of central atoms increases down the group from B to In, the Lewis acid character of trichlorides of these elements also increases from BCl_3 to $InCl_3$ ($BCl_3 < AlCl_3 < GaCl_3 < InCl_3$) or decreases as $InCl_3 > GaCl_3 > AlCl_3 > BCl_3$.

22. $SnO + HF \longrightarrow SnF_2 + H_2O$

23. $SnCl_2$ is used as a mordant in dyeing and as a reducing agent. Its reducing property is used in preparing colloidal gold by reducing $AuCl_3$. In this reaction $AuCl_3$ is reduced to coiloidal particles of Au and $SnCl_2$ is oxidised to $SnCl_4$.

$$2AuCl_3 + 3SnCl_2 \longrightarrow 2Au + 3SnCl_4$$

24. $SnCl_2$ can act as a reducing agent, since it can easily be oxidised to $SnCl_4$.

$$Sn^{2+} \xrightarrow{\text{Oxidation}} Sn^{4+} + 2e^-$$

The solution of $SnCl_2$ in HCl reduces mercuric chloride ($HgCl_2$) to mercurous chloride (Hg_2Cl_2) and then to Hg. $SnCl_2$ is oxidised to $SnCl_4$.

$$2HgCl_2 + SnCl_2 \longrightarrow Hg_2Cl_2 + SnCl_4$$

$$Hg_2Cl_2 + SnCl_2 \longrightarrow 2Hg + SnCl_4$$

$SnCl_2$ also reduces $FeCl_3$ to $FeCl_2$ and is itself oxidised to $SnCl_4$.

$$2FeCl_3 + SnCl_2 \longrightarrow 2FeCl_2 + SnCl_4$$

25. When excess of $SnCl_2$ is added to a solution of $HgCl_2$, $HgCl_2$ is reduced by $SnCl_2$ to white ppt of Hg_2Cl_2 which is further reduced by $SnCl_2$ to black Hg. In this reaction $SnCl_2$ is oxidised to $SnCl_4$.

$$SnCl_2 + 2HgCl_2 \longrightarrow \underset{\text{White ppt.}}{Hg_2Cl_2\downarrow} + SnCl_4$$

$$SnCl_2 + Hg_2Cl_2 \longrightarrow \underset{\text{Black}}{2Hg} + SnCl_4.$$

26. (*a*) When Sn is melted in an atmosphere dry Cl_2 gas, $SnCl_4$ (instead of $SnCl_2$) is obtained.

Sn (*melted or molten*) + $2Cl_2$ (*dry gas*) $\longrightarrow$ $SnCl_4$ (*g*)

(*b*) and (*c*). The treatment of Sn with dil. HCl gives $SnCl_2$ while that of Sn with conc. HCl gives the crystals of tin dichloride dihydrate, $SnCl_2 \cdot 2H_2O$

$$Sn + 2HCl\ (dil) \longrightarrow SnCl_2 + H_2$$

$$Sn + 2HCl\ (conc) + 2H_2O \longrightarrow SnCl_2 \cdot 2H_2O + H_2$$

(*d*) When Sn is heated in a current of dry HCl gas, anhydrous $SnCl_2$ is obtained.

$$Sn(s) + 2HCl\ (g) \xrightarrow{\Delta} \underset{\text{Anhydrous}}{SnCl_2\ (s)} + H_2\ (g)$$

27. (*a*) The action of dry Cl_2 gas on molten (melted) Sn give $SnCl_4$

$$Sn\ (\text{molten}) + 2Cl_2\ (\text{dry gas}) \longrightarrow SnCl_4\ (g)$$

(*b*) The action of conc. HCl on Sn gives the crystals of $SnCl_2 \cdot 2H_2O$

$$Sn + 2HCl\ (\text{conc}) \xrightarrow{\Delta} 2H_2O \longrightarrow SnCl_2 \cdot 2H_2O + H_2$$

(*c*) The action of dil. HCl on Sn gives $SnCl_2$.

$$Sn + 2HCl\ (dil) \longrightarrow SnCl_2 + H_2$$

28. (*a*) Since Sn^{4+} ion has higher positive charge and is smaller in size, it polarises Cl^- ion in $SnCl_4$ molecule very strongly and hence $SnCl_4$ becomes covalent.

(*b*) $SnCl_4$ is a colourless volatile liquid which fumes in moist air due to hydrolysis

$$SnCl_4 + 2H_2O \longrightarrow SnO_2 + 4HCl$$

(*c*) Lewis structure of $SnCl_4$ is 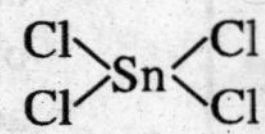 which shows that since Sn atom (central atom) is surrounded by four (Sn – Cl) σ bonding electron pairs and no lone pair of electrons, $SnCl_4$ molecule has tetrahedral shape.

29. Butter of tin is stannic chloride pentahydrate, $SnCl_4 \cdot 5H_2O$ (also called oxymuriate of tin). It is obtained when small amount of water is added to $SnCl_4$.

30

Oxy-Acid of Boron (Boric acids) and their Salts

MULTIPLE CHOICE QUESTIONS

1. Boric acid can be prepared :

(*a*) by hydrolysing BCl_3
(*b*) by hydrolysing B_2H_6
(*c*) by the action of H_2O on B_2O_3.
(*d*) by heating a mixture of H_2, B and O_2 in an inert atmosphere.

2. The action of NaOH on H_3BO_3 gives :

(*a*) Sodium borate (*b*) Boric anhydride
(*c*) Borax (*d*) None of these

3. The final product obtained by heating boric acid (H_3BO_3) is

(*a*) Metaboric acid
(*b*) Tetraboric acid
(*c*) Boric anhydride
(*d*) None of these

4. Which of the following statements is false about boric acid ?

(*a*) It acts as a tribasic acid
(*b*) It has a planar structure
(*c*) It acts as monobasic acid
(*d*) It is fairly soluble in hot water.

5. Which of the following statements is wrong about H_3BO_3 ?

(*a*) It is a strong tribasic acid
(*b*) It is prepared by acidifying an aqueous solution of borax
(*c*) It has a layer structure in which planar BO_3 units are joined together by H-bonds
(*d*) It does not act as a proton donor as it acts as a Lewis acid by accepting OH^- ions.

6. In the solid state, boric acid is polymeric due to :

(*a*) its acidic nature
(*b*) the presence of H-bonds
(*c*) its monobasic nature
(*d*) its geometry

7. Boric acid dissolves in oleum due to the formation of :

(*a*) $[B(HSO_4)_4]^-$ ion (*b*) B_2O_3
(*c*) H_2SO_4 (*d*) SO_3

8. Boric acid can be titrated against NaOH using methyl orange as indicator only in presence of catechol, because the addition of catechol :

(*a*) prevents the hydrolysis of $NaBO_2$ by forming a chelated complex compound with it.
(*b*) prevents the formation of NaOH
(*c*) increases the rate of formation of $NaBO_2$
(*d*) decreases the rate of formation of $NaBO_2$.

9. Which of the following compounds are formed when BCl_3 is hydrolysed by water ?

(*a*) H_3BO_3 + HCl (*b*) B_2H_6 + HCl
(*c*) B_2O_3 + HCl (*d*) None of these

10. Boric acid is prepared from borax by the action of :

(*a*) HCl (*b*) NaOH
(*c*) CO_2 (*d*) Na_2CO_3

11. $B(OH)_3$ is a

(*a*) base (*b*) monobasic acid
(*c*) dibasic acid (*d*) tribasic acid.

12. Borax is :

(*a*) decahydrated sodium salt of tetraboric acid
(*b*) a salt of boric acid

(*c*) $Ca_2B_6O_{11}$
(*d*) Sodium metaborate

13. Borax bead test is responded by :
(*a*) divalent metals
(*b*) heavy metals
(*c*) light metals
(*d*) metals which form coloured metaborates

14. Which of the following does not respond to borax bead test ?
(*a*) Nickel salts (*b*) Copper salts
(*c*) Cobalt salts (*d*) Aluminium salts

15. Borax is represented as $Na_2^+[B_4O_5(OH)_4]^{2-}.8H_2O$. The correct statements about $[B_4O_5(OH)_4]^{2-}$ ion are :
(*a*) Two B-atoms are sp^3 hybridised and two B-atoms are sp^2 hybridised.
(*b*) Two B-atoms have extra electron.
(*c*) There are five bridging O-atoms between two B-atoms
(*d*) Each B-atoms is sp^3 hybridised.

16. Borax is prepared by treating colemanite ore with :
(*a*) $NaNO_3$ (*b*) NaCl
(*c*) $NaHCO_3$ (*d*) Na_2CO_3

17. The colour of ferric metaborate is :
(*a*) yellow (*b*) blue
(*c*) green (*d*) grey

18. Which of the following is borax ?
(*a*) $Na_2B_4O_7.10H_2O$
(*b*) $Na_2B_4O_7.4H_2O$
(*c*) $Ca_2B_6O_{11}.5H_2O$
(*d*) CaB_4O_7

19. Which of the following statements is (are) wrong for borax ?
(*a*) Aqueous solution of borax is alkaline.
(*b*) Borax cannot be hydrolysed.
(*c*) Action of conc. H_2SO_4 on hot solution of borax gives metaboric acid (HBO_2)
(*d*) Borax has $[B_4O_5(OH)_3]^{2-}$ ion in its structure.

20. On the addition of a mineral acid to the aqueous solution of borax, the compound formed is :
(*a*) Boron hydride
(*b*) Orthoboric acid
(*c*) Metaboric acid
(*d*) Pyroboric acid

21. Ethyl borate, $(C_2H_5)_3BO_3$ burns with :
(*a*) yellow flame (*b*) green flame
(*c*) blue flame (*d*) crimson flame

22. Pick up the wrong statement.
(*a*) Borax is used in the manufacture of optical glasses
(*b*) Borax is used as a flux
(*c*) Borax is used as a water softener
(*d*) Borax is not used for glazing paper.

ANSWERS

1. (*a*), (*b*), (*c*)	**2.** (*a*)	**3.** (*c*)	**4.** (*a*)	**5.** (*a*)	**6.** (*b*)	**7.** (*a*)
8. (*a*)	**9.** (*a*)	**10.** (*a*)	**11.** (*b*)	**12.** (*a*)	**13.** (*d*)	**14.** (*d*)
15. (*a*), (*b*), (*c*)	**16.** (*d*)	**17.** (*a*)	**18.** (*a*)	**19.** (*b*), (*c*), (*d*)	**20.** (*b*)	**21.** (*b*)
22. (*d*)						

HINTS AND EXPLANATES

1. Boric acid (H_3BO_3) can be prepared by hydrolysing BCl_3 or B_2H_6 or by the action of H_2O on B_2O_3.

$$BCl_3 + 3H_2O \longrightarrow H_3BO_3 + 3HCl$$
$$B_2H_6 + 6H_2O \longrightarrow 2H_3BO_3 + 6H_2$$
$$B_2O_3 + 3H_2O \longrightarrow 2H_3BO_3$$

2. $H_3BO_3 + 3NaOH \longrightarrow Na_3BO_3$ (*Sodium borate*) $+ 3H_2O$

3. The final product obtained by heating boric acid (H_3BO_3) is B_2O_3 as shown below :

$$\underset{\text{Boric acid}}{4H_3BO_3} \xrightarrow[-4H_2O]{100°C} \underset{\text{Metaboric acid}}{4HBO_2} \xrightarrow[-H_2O]{160°C} \underset{\text{Tetraboric acid}}{H_2B_4O_7} \xrightarrow{\text{Red heat}} \underset{\text{Boric anhydride}}{2B_2O_3}$$

4. (*a*) H_3BO_3 or $B(OH)_3$ acts as a monobasic acid, since its aqueous solution, on ionisation, gives only one H^+ ion as shown below :

$$H_2O \longrightarrow H^+ + :\ddot{\underset{..}{O}}—H$$

$$(OH)_3B + :\ddot{\underset{..}{O}}—H \longrightarrow \underset{\text{Unstable ion}}{[(OH)_3B \leftarrow O—H]^- \text{ or } [B(OH)_4]^-}$$

$$[B(OH)_4]^- \xrightarrow{\text{Dissociation}} \underset{\text{Metaborate ion}}{BO_2^- + 2H_2O}$$

On adding : $(OH)_3B \rightarrow BO_2^- + H_2O + H^+$

(*b*) Lewis structure of $B(OH)_3$ is HO — B(HO)(OH) which shows that since B-atom (central atom) is linked with three OH^- ions by three σ-bonds, $B(OH)_3$ has trigonal planar geometry.

(*c*) $B(OH)_3$ acts as a monobasic acid as shown above at (*a*).

(*d*) $B(OH)_3$ is fairly soluble in hot water.

5. H_3BO_3 is a monobasic acid, since it gives only one H^+ ion on ionisation (See answer 4)

6. In the solid state, boric acid contains triangular planar $B(OH)_3$ units which are linked together through H-bonds. H-bond is formed between O-atom of one $B(OH)_3$ unit and H-atom of the other $B(OH)_3$ unit.

7. Boric acid (H_3BO_3) dissolves in oleum ($H_2S_2O_7$) due to the formation of *tetra kis* (*hydrogen sulphate*) *boric acid anion*, $[B(HSO_4)_4]^-$.

$$2H_3BO_3 + 3H_2S_2O_7 + 4H_2SO_4 \longrightarrow 2[B(HSO_4)_4]^- + 2H_3SO_4^+ + 3H_2O$$

8. In the titration of boric acid $[B(OH)_3]$ against NaOH, sodium metaborate ($NaBO_2$) is formed

$$B(OH)_3 + NaOH \longrightarrow NaBO_2 + 2H_2O \quad ...(a)$$

If catechol is not added to the titration solution, $NaBO_2$ formed as above gets hydrolysed to give back $B(OH)_3$ and NaOH.

$$NaBO_2 + 2H_2O \xrightarrow{\text{Hydrolysis}} B(OH)_3 + NaOH$$

The addition of catechol to the titration solution prevents the hydrolysis of $NaBO_2$ by forming a chalated complex compound of Na with $NaBO_2$ formed in reaction (*a*).

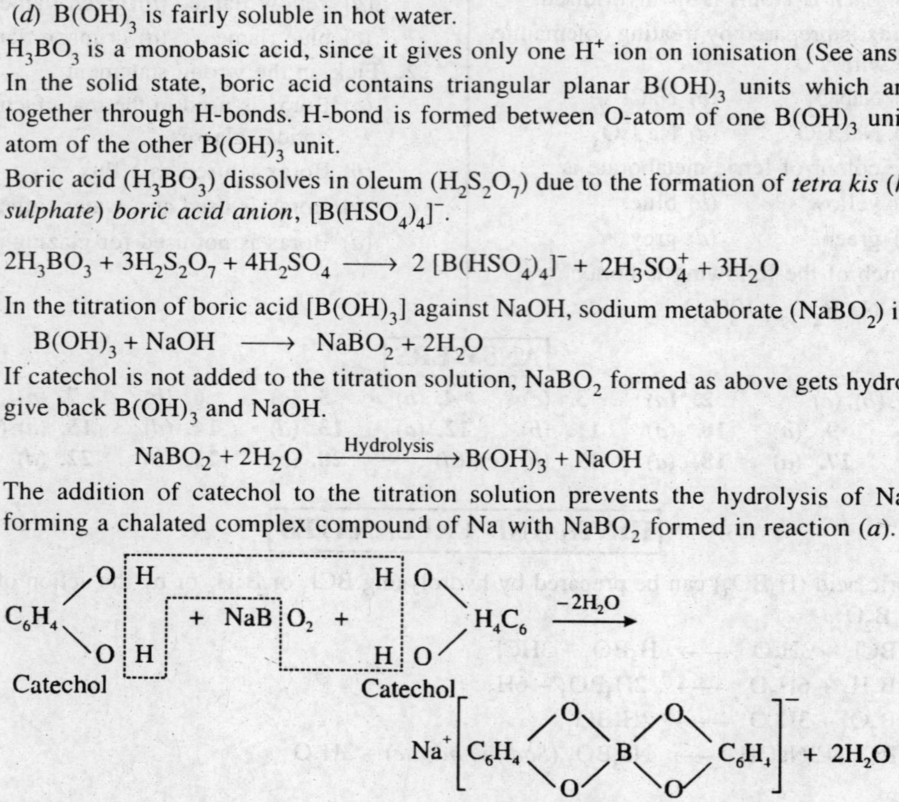

Catechol Catechol

Chelated complex compound

9. One of the methods for preparing boric acid (H_3BO_3) consists of hydrolysing BCl_3 with water.

$$BCl_3 + 3H_2O \longrightarrow H_3BO_3 + 3HCl$$

10. Boric acid (H_3BO_3) is obtained from borax ($Na_2B_4O_7$) by the action of HCl on it.

$$Na_2B_4O_7 + 2HCl + 5H_2O \longrightarrow 4H_3BO_3 + 2NaCl$$

11. Boric acid, $B(OH)_3$ is a monobasic acid, since it gives only one H^+ ion in water as shown below :

$$H_2O \longrightarrow H^+ + :\ddot{\underset{..}{O}}-H$$

$$(OH)_3B + :\ddot{\underset{..}{O}}-H \longrightarrow [(OH)_3B \leftarrow O-H]^- \text{ or } [B(OH)_4]^-$$

Unstable ion

$$[B(OH)_4]^- \xrightarrow{\text{Dissociation}} BO_2^- + 2H_2O$$

Metaborate ion

on adding : $(OH)_3B \longrightarrow BO_2^- + H_2O + H^+$

12. Tetraboric acid is $H_2B_4O_7$. Borax is sodium salt of this acid containing ten H_2O molecules as water of crystallisation, *i.e.*, borax is sodium tetraborate decahydrate, $Na_2B_4O_7 \cdot 10H_2O$. Salt of boric acid is called borate (*i.g.*, Na_3BO_3) . $Ca_2B_6O_{11}$ is colemanite ore and sodium metaborate is $NaBO_2$.

13. Example : $\underset{\text{Coloured salt}}{CuSO_4} \xrightarrow{\Delta} CuO + SO_3$

$$\underset{\text{Borax}}{Na_2B_4O_7} \xrightarrow{\Delta} B_2O_3 + 2NaBO_2$$

$$CuO + B_2O_3 \longrightarrow \underset{\text{Cupric borate (blue)}}{CuO \cdot B_2O_3 \text{ or } Cu(BO_2)_2}$$

14. Aluminium salts donot respond to borax bead test, since aluminium metaborate, $Al(BO_2)_3$ formed by these salts with B_2O_3 is not coloured.

15. The structure of $[B_4O_5(OH)_4]^{2-}$ ion is shown below. This structrue shows that :

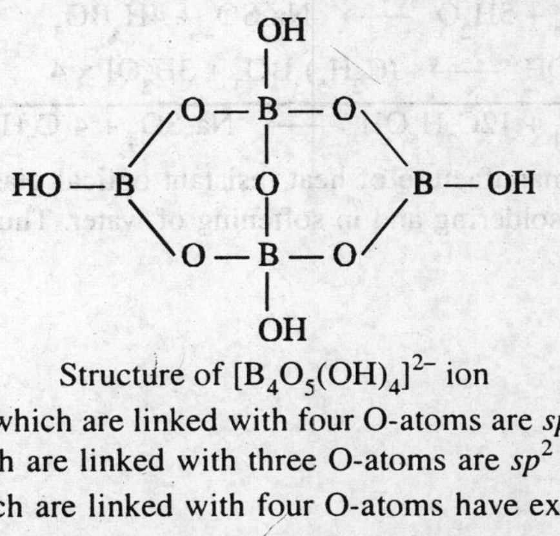

Structure of $[B_4O_5(OH)_4]^{2-}$ ion

(*a*) The two B-atoms which are linked with four O-atoms are sp^3 hybridised while the other two B-atoms which are linked with three O-atoms are sp^2 hybridised.

(*b*) Two B-atoms which are linked with four O-atoms have extra electron.

(*c*) The anion has nine O-atoms. Five O-atoms are making a bridge between two B-atoms while the remaining four O-atoms are attached with H-atoms of OH^- ions.

16. The preparation of borax ($Na_2B_4O_7$) by treating calemanite ore ($Ca_2B_6O_{11}$) with Na_2CO_3 solution can be shown as :

$$\underset{\text{Colemarite ore}}{Ca_2B_6O_{11}} + 2Na_2CO_3\,(aq) \xrightarrow{\text{Boil}} 2Ca\,CO_3\,(ppt) + \underbrace{Na_2B_4O_7 + 2NaBO_2}_{\text{Solution}}$$

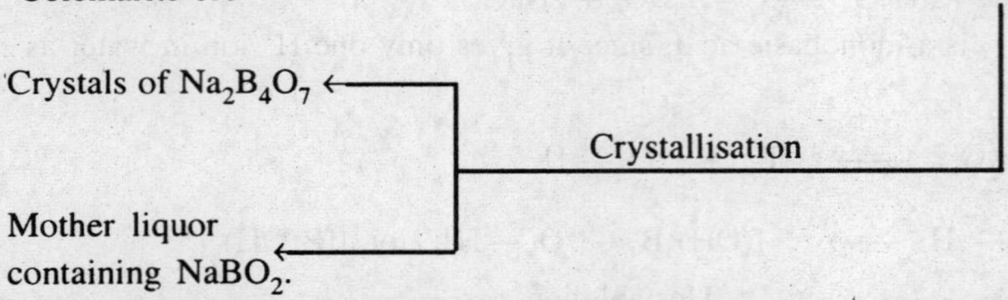

18. (*a*) Tetraboric acid is $H_2B_4O_7$. Boron is sodium salt of this acid containing ten H_2O molecules as water of crystallisation, *i.e.,* borax is sodium tetraborate decahydrate, $Na_2B_4O_7 . 10H_2O$

(*b*) $Na_2B_4O_7 . 4H_2O$ is sodium tetraborate tetrahydrate

(*c*) $Ca_2B_6O_{11} . 5H_2O$ is colemanite ore

(*d*) CaB_4O_7 is calcium tetraborate

19. (*a*) Since the aqueous solution of borax ($Na_2B_4O_7$) gives NaOH (strong alkali) and weak acid (H_3BO_3), this solution is alkaline in nature.

$$Na_2B_4O_7 + 7H_2O \longrightarrow \underset{\text{(strong alkali)}}{2NaOH} + \underset{\text{(weak acid)}}{4H_3BO_3}$$

(*b*) Borax can be hydrolysed by H_2O as shown above at (*a*).

(*c*) Action of conc. H_2SO_4 on hot solution of borax gives boric acid (H_3BO_3) and not metaboric acid (HBO_2)

$$Na_2B_4O_7 + H_2SO_4 + 5H_2O \longrightarrow Na_2SO_4 + 4H_3BO_3$$

(*d*) Borax can be represented by the structure, $Na_2^+[B_4O_7(OH)_4]^{2-}$

20. $Na_2B_4O_7$ (*Borax*) + $H_2SO_4 + 5H_2O \longrightarrow Na_2SO_4 + 4H_3BO_3$ (*Orthoboric acid*)

$Na_2B_4O_7$ (*Borax*) + $2HCl + 5H_2O \longrightarrow 2NaCl + 4H_3BO_3$ (*Orthoboric acid*)

21. When borax ($Na_2B_4O_7$) is heated with ethyl alcohol (C_2H_5OH) and conc. H_2SO_4, vapours of triethyl borate, $(C_2H_5)_3BO_3$ are produced. When ignited, these vapours burn with green *edged flame.*

$$Na_2B_4O_7 + H_2SO_4 + 5H_2O \longrightarrow Na_2SO_4 + 4H_3BO_3$$
$$[H_3BO_3 + 3C_2H_5OH \longrightarrow (C_2H_5)_3BO_3 + 3H_2O] \times 4$$
$$\overline{Na_2B_4O_7 + H_2SO_4 + 12C_2H_5OH \longrightarrow Na_2SO_4 + 4(C_2H_5)_3\,BO_3 + 7H_2O}$$

22. Borax is used in the manufacture of heat resistant optical glasses and glazing papers. It is also used as a flux is soldering and in softening of water. Thus statement (*d*) is wrong.

31

Oxyacids of Nitrogen and Their Salts

MULTIPLE CHOICE QUESTIONS

1. The salts of HNO_2 :
(*a*) act as reducing agent only
(*b*) act as oxidising agent only
(*c*) act both as reductants and oxidants.
(*d*) cannot exhibit redox properties.

2. Which of the following acids possesses oxidising, reducing as well as complex forming properties?
(*a*) H_2SO_4 (*b*) HNO_3
(*c*) HNO_2 (*d*) HCl

3. HNO_2 acts both as a reductant and as an oxidant while HNO_3 acts only as an oxidant. It is due to their :
(*a*) solubility ability
(*b*) maximum oxidation number
(*c*) minimum oxidation number
(*d*) minimum number of valence electrons

4. HNO_2 acts as an/a
(*a*) acid (*b*) oxidising agent
(*c*) reducing agent (*d*) all the three

5. Which gas is obtained when urea reacts with HNO_2 ?
(*a*) N_2 (*b*) NO
(*c*) N_2O (*d*) NO_2

6. N_2 is liberated by the thermal decomposition of :
(*a*) NH_4NO_2 (*b*) NaN_3
(*c*) $(NH_4)_2Cr_2O_7$ (*d*) All the three

7. N_2 is obtained by the thermal decomposition of :
(*a*) $Pb(NO_3)_2$ (*b*) $AgNO_3$
(*c*) NH_4NO_3 (*d*) NH_4NO_2

8. The gas obtained by heating NH_4NO_2 is :
(*a*) N_2O (*b*) N_2
(*c*) N_2O_3 (*d*) N_2O_4

9. In the laboratory HNO_3 is prepared by heating :
(*a*) $NaNO_3$ (*b*) KNO_3
(*c*) $KMnO_4$ (*d*) Both (a) and (b).

10. The starting material used in Birkland and Eyde process for the manufacture of HNO_3 is :
(*a*) NO_2 (*b*) air
(*c*) NH_3 (*d*) chile salt petre

11. NH_3 , on catalytic oxidation, gives an oxide from which NHO_3 is obtained. This oxide is :
(*a*) NO (*b*) NO_2
(*c*) N_2O_3 (*d*) N_2O_5

12. Which of the following is used to oxidise NH_3 in the manufacture of HNO_3 by Ostwald's process ?
(*a*) Pt (*b*) atmospheric O_2
(*c*) H_2O (*d*) NO_2

13. When conc. HNO_3 is heated, it decomposes to give :
(*a*) $N_2 + O_2$ (*b*) NO
(*c*) O_2 (*d*) $NO_2 + O_2$

14. The oxidising property of HNO_3 is due to :
(*a*) its concentration
(*b*) positive valency of N-atom in HNO_3
(*c*) its dilution
(*d*) unstability of +5 oxidation state of N-atom in HNO_3.

15. HNO_3 has the following properties :
(*a*) Oxidising and bleaching
(*b*) Acidic and oxidising
(*c*) Basic and reducing
(*d*) Reducing and bleaching.

16. When HNO_3 reacts with metals, NO_2 is usually evolved if the acid is :
(*a*) Dilute
(*b*) Very dilute
(*c*) Moderately strong
(*d*) Concentrated

17. Mg, on reaction with very dilute HNO_3, gives :
(*a*) NO (*b*) N_2O
(*c*) N_2 (*d*) NO_2

18. The metal that reacts with HNO_3 to give five different products depending on the concentration of the acid is :
(*a*) Cu (*b*) Fe
(*c*) Zn (*d*) Al

19. Which of the following metals behaves as a non-metal towards concentrated HNO_3 ?
(*a*) Cu (*b*) Fe
(*c*) Sn (*d*) Pb

20. Action of concentrated HNO_3 on metallic tin produces :
(*a*) Stanneous nitrite
(*b*) Staneous nitrate
(*c*) Stannic nitrate
(*d*) Hydrated stannic oxide.

21. Concentrated HNO_3 is stored in containers made of :
(*a*) Cu (*b*) Zn
(*c*) Al (*d*) Sn

22. When Zn reacts with very dilute HNO_3, it produces :
(*a*) NO (*b*) NH_4NO_3
(*c*) $NO_2 + N_2$ (*d*) O_2

23. Conc. HNO_3 reacts with I_2 to give :
(*a*) HI (*b*) HOI
(*c*) HIO_2 (*d*) HIO_3

24. Concentrated HNO_3 oxidises phosphorus to :
(*a*) H_3PO_4 (*b*) P_2O_3
(*c*) H_3PO_3 (*d*) $H_4P_2O_7$

25. Skin becomes yellow in concentrated HNO_3 because :
(*a*) The proteins are converted into xantho proteins.
(*b*) HNO_3 acts as a dehydrating agent.
(*c*) Nitrocellulose is formed
(*d*) HNO_3 acts as an oxidising agent

26. The oxide of nitrogen obtained by the dehydration of nitric acid by P_2O_5 is :
(*a*) N_2O_5 (*b*) N_2O_3
(*c*) N_2O (*d*) NO

27. Concentrated HNO_3 oxidises cane sugar to :
(*a*) CO_2 and H_2O
(*b*) CO and H_2O
(*c*) CO, CO_2 and H_2O
(*d*) oxalic acid and H_2O

28. The mixture of conc. HCl and HNO_3 taken in 3:1 ratio contains :
(*a*) ClO_2 (*b*) NOCl
(*c*) NCl_3 (*d*) N_2O_4

29. Gold and platinum dissolve in aqua regia due to the formation of :
(*a*) $AuCl_3$, $Pt\ Cl_4$
(*b*) AuCl, $PtCl_2$
(*c*) $HAuCl_4$, H_2PtCl_6
(*d*) $Au(NO_3)_3$, $Pt(NO_3)_4$

30. HNO_3 is used in the purification of :
(*a*) H_2 (*b*) Au
(*c*) Ag (*d*) both (*b*) and (*c*).

31. Nitrates of all alkati metals are :
(*a*) soluble
(*b*) insoluble
(*c*) partially soluble
(*d*) none of the above.

32. On strong heating lead nitrate gives :
(*a*) PbO, NO and O_2
(*b*) PbO, NO_2 and O_2
(*c*) PbO, NO and NO_2
(*d*) PbO_2, PbO and NO_2

33. Which of the following nitrates gives a mixture of two gases ?

(*a*) Ba $(NO_3)_2$ (*b*) Na NO_3
(*c*) KNO_3 (*d*) $RbNO_3$

34. NH_4NO_3 decomposes on heating to give :
(*a*) $NH_3 + HNO_3$ (*b*) $N_2 + H_2 + O_3$
(*c*) $NO + NO_2 + H_2$ (*d*) $N_2O + H_2O$

35. Nitrogen (I) oxide is produced by :
(*a*) thermal decomposition of NH_4NO_3
(*b*) thermal decomposition of NH_4NO_2
(*c*) disproportionation of N_2O_4
(*d*) Interaction between hydroxylamine and HNO_2

36. Nitrous oxide is prepared in the laboratory by :
(*a*) combination of N_2 and O_2
(*b*) reduction of NO
(*c*) reduction of HNO_3
(*d*) thermal decomposition of NH_4NO_3

37. Sodium nitrate decomposes above 800° C to give :
(*a*) N_2 (*b*) O_2
(*c*) NO_2 (*d*) Na_2O

38. When $AgNO_3$ is heated strongly, the products formed are :
(*a*) $NO + NO_2$ (*b*) $NO_2 + O_2$
(*c*) $NO_2 + N_2O$ (*d*) $NO + O_2$

39. In NO_3^- ion the number of bond pairs and lone pairs of electrons on N-atom are :
(*a*) 2, 2 (*b*) 3, 1
(*c*) 1, 3 (*d*) 4, 0

40. NO_2 gas cannot be obtained by heating :
(*a*) KNO_3 (*b*) Pb $(NO_3)_2$
(*c*) Cu $(NO_3)_2$ (*d*) Ag NO_3

41. Which of the following decompose easily on heating to give O_2 as the only gaseons product ?
(*a*) Pb $(NO_3)_2$ (*b*) $KClO_3$
(*c*) HgO (*d*) MnO_2

42. Pernitric acid is chemically :
(*a*) HNO_2 (*b*) HNO_3
(*c*) HNO_4 (*d*) $H_2N_2O_2$

ANSWERS

1. (*c*)	**2.** (*c*)	**3.** (*c*)	**4.** (*d*)	**5.** (*a*)	**6.** (*d*)	**7.** (*d*)	**8.** (*b*)
9. (*d*)	**10.** (*b*)	**11.** (*a*)	**12.** (*b*)	**13.** (*d*)	**14.** (*d*)	**15.** (*b*)	**16.** (*d*)
17. (*c*)	**18.** (*c*)	**19.** (*c*)	**20.** (*d*)	**21.** (*c*)	**22.** (*b*)	**23.** (*c*)	**24.** (*a*)
25. (*a*)	**26.** (*a*)	**27.** (*d*)	**28.** (*b*)	**29.** (*c*)	**30.** (*d*)	**31.** (*a*)	**32.** (*b*)
33. (*a*)	**34.** (*d*)	**35.** (*a, d*)	**36.** (*d*)	**37.** (*a,b,d*)	**38.** (*b*)	**39.** (*d*)	**40.** (*a*)
41. (*b,c,d*)	**42.** (*c*)						

HINTS AND EXPLANATIONS

1. In the salts of HNO_2 the oxidation state (O.S.) of N-atom is +3. When these salts act as oxidising agents, they are reduced to NO in which O.S. of N-atom is +2.

$$\underset{(N=+3)}{NO_2^-} + 2H^+ + 2e^- \xrightarrow{\text{Reduction}} \underset{(N=+2)}{NO} + H_2O$$

On the other hand when these salts act as reducing agents, these are oxidised to nitrates in which O.S. of N-atom is +5.

$$\underset{(N=+3)}{NO_2^-} + H_2O \xrightarrow{\text{Oxidation}} \underset{(N=+5)}{NO_3^-} + 2H^+ + 2e^-$$

Since +3 and +5 oxidation states are possible for N-atom, the salts of HNO_2 can act both as oxidising as well as reducing agents.

2. Oxidising property. HNO_2 oxidises KI to I_2, $SnCl_2$ to $SnCl_4$, SO_2 to H_2SO_4 etc and is itself reduced to NO

$$\underset{(I=-1)}{2I^-} + 4H^+ + \underset{(N=+3)}{2NO_2^-} \longrightarrow \underset{(N=+2)}{2NO} + 2H_2O + \underset{(I=0)}{I_2}$$

$$Sn^{2+} + 4H^+ + 2NO_2^- \longrightarrow 2NO + 2H_2O + Sn^{4+}$$

$$SO_2 + 2H^+ + 2NO_3^- \longrightarrow 2NO + H_2SO_4$$

Reducing property. HNO_2 reduces Br_2 to HBr, acidified $KMnO_4$ to $MnSO_4$ etc. and is itself oxidised to HNO_3.

$$Br_2 + NO_2^- + H_2O \longrightarrow 2Br^- + NO_3^- + 2H^+$$

$$2MnO_4^- + 5NO_2^- + 6H^+ \longrightarrow 2Mn^{2+} + 5NO_3^- + 3H_2O$$

Complex forming property. When $FeSO_4$ solution is added to *a* cold solution of HNO_2, brown black pentaqua nitrosonium iron (I) sulphate, $[Fe^+(NO^+)(H_2O)_5]SO_4$ is obtained which is a complex compound.

$$3HNO_2 \longrightarrow 2NO + H_2O + HNO_3$$

$$FeSO_4 + NO + 5H_2O \longrightarrow [Fe(NO)(H_2O)_5]SO_4$$

or $$Fe^{2+} + NO + 5H_2O \longrightarrow \left[Fe^+(NO^+)(H_2O)_5\right]^{2+}$$

3. In HNO_3, N-atom is in +5 oxidation state (O.S.) which is the highest O.S. exhibited by N-atom in its compounds. If HNO_3 acts as an oxidising agent (O.A.), this O.S. of +5 of N-atom will be decreased. The decrease in O.S. is possible, since N-atom also shows oxidation states lower than +5. If HNO_3 acts as a reducing agent, O.S. of +5 of N-atom will be increased. The increase in O.S. of N-atom is not possible, since N-atom is not able to show O.S. higher than +5. Thus HNO_3 can act only as an oxidising agent.

In HNO_2, N-atom is in +3 O.S. If HNO_2 acts as an oxidising agent, O.S. of +3 of N-atom will be decreased. The decrease in O.S. is possible, since N-atom also shows O.S. lower than +3. If HNO_2 acts as a reducing agent, O.S. of +3 of N-atom will be increased. The increase in O.S. is also possible, since N-atom also exhibits O.S. higher than +3. Thus HNO_2 can act both as an oxidising agent as well as a reducing agent.

4. HNO_2 is a weak monobasic acid. Hence it reacts with alkalies to form salt (nitrite) and H_2O, *e.g.*

$$HNO_2 + NaOH \longrightarrow NaNO_2 + H_2O$$

HNO_2 acts as an oxidising agent, since it is reduced to NO.

$$\underset{(N=+3)}{NO_2^-} + 2H^+ + 2e^- \xrightarrow{\text{Reduction}} \underset{(N=+2)}{NO} + H_2O$$

HNO_2 also acts as a reducing agent, since it is oxidised to HNO_3

$$HNO_2 + O \longrightarrow HNO_3$$

or $$NO_2^- + H_2O \xrightarrow{\text{Oxidation}} NO_3^- + 2H^+ + 2e^-$$

(N = +3) (N = +5)

5. HNO_2 decomposes urea ($NH_2CO\ NH_2$) and other aliphatic primary amines to form N_2.

$$NH_2CONH_2 + 2HNO_2 \longrightarrow CO_2 + 3H_2O + 2N_2$$

6.
$$NH_4NO_2 \xrightarrow{\Delta} 2H_2O + N_2$$

$$2NaN_3 \xrightarrow{\Delta} 2Na + N_2$$

$$(NH_4)_2Cr_2O_7 \xrightarrow{\Delta} Cr_2O_3 + 4H_2O + N_2$$

7. (*a*) $2Pb(NO_3)_2 \xrightarrow{\Delta} 2PbO + 4NO_2 + O_2$

(*b*) $2AgNO_3 \xrightarrow{\Delta} 2Ag + 2NO_2 + O_2$

(*c*) $NH_4NO_3 \xrightarrow{\Delta} N_2O + 2H_2O$

(*d*) $NH_4NO_2 \xrightarrow{\Delta} N_2 + 2H_2O$

8.
$$NH_4NO_2 \xrightarrow{\Delta} N_2 + 2H_2O$$

9. In the laboratory, HNO_3 is prepared by heating $NaNO_3$ or KNO_3 with conc. H_2SO_4.

$$MNO_3(M = Na, K) + H_2SO_4(conc) \longrightarrow MHSO_4 + HNO_3$$

10. The starting material required for the manufacture of HNO_3 by Birkland and Eyde process is air. Air is passed through an electric arc when N_2 and O_2 present in air combine to form NO

$$\underbrace{N_2 + O_2}_{\text{(From air)}} \rightleftharpoons 2NO$$

11. HNO_3 can be prepared by catalytic oxidation of NH_3 by a process which consists of the following steps :

(*i*)
$$4NH_3 + 5O_2 \xrightarrow[\substack{\text{oxidation of } NH_3 \\ (Pt,\ 1100\ K)}]{\text{Catalytic}} 4NO + 6H_2O, \quad \Delta H = -89.8\ KJ$$

(*ii*)
$$2NO + O_2 \xrightarrow[\text{of NO}]{\text{Oxidation}} 2NO_2$$

(*iii*)
$$4NO_2 + 2H_2O + O_2 \longrightarrow 4HNO_3$$

12. In the manufacture of HNO_3 by Ostwald's process NH_3 obtained by Haber's process is oxidised to NO by atomospharic O_2 (free from dust) in presence of Pt-gauze catalyst electrically heated to 750° – 900° C.

$$\underset{(N = -3)}{4NH_3} + 5O_2(\text{From air}) \xrightarrow[750° - 900°C]{\text{Pt-gauze}} \underset{(N = +2)}{4NO} + 6H_2O$$

13. When HNO_3 is heated, it gets decomposed to give NO_2, O_2 and H_2O

$$4HNO_3(aq) \xrightarrow{\Delta} 4NO_2(g) + O_2(g) + 2H_2O(l)$$

14. O.S. of N-atom in HNO_3 is +5. When HNO_3 acts as an oxidising agent, it gets reduced to various products in which unstable O.S. of N-atom of +5 in HNO_3 decreases to + 4 (NO_2),

+2 (NO), –1 (N_2O), 0 (H_2), –1 (NH_2OH) and –3 (NH_3 or NH_4^+ ion).

15. HNO_3 is a strong monobasic acid, since in aqueous solution it gives H_3O^+ ions

$$HNO_3(aq) + H_2O(l) \longrightarrow H_3O^+(aq) + NO_3^-(aq)$$

HNO_3 is a very strong oxidising agent, since it gives nescent oxygen both in the concentrated and in dilute form. This nescent oxygen oxidises many non-metals, compounds and metals

$$2HNO_3(conc) \longrightarrow 2H_2O + 2NO_2 + O$$

$$2HNO_3(dil.) \longrightarrow H_2O + 2NO + 3O$$

16. Many metals like Zn, Mg, Pb, Bi etc., on reacting with conc. HNO_3, produce their nitrates and NO_2 is evolved. For example:

$$M + 4HNO_3\ (\text{Conc.}) \longrightarrow M(NO_3)_2 + 2H_2O + 2NO_2$$

(M = Zn, Mg, Pb)

$$Bi + 6HNO_3\ (\text{Conc.}) \longrightarrow Bi\ (NO_3)_3 + 3H_2O + 3NO_2$$

17. Magnesium reacts with very dilute HNO_3 and forms H_2 gas

$$Mg + 2HNO_3 \longrightarrow Mg\ (NO_3)_2 + H_2$$

18. (*a*) Cu reacts with HNO_3 and gives N_2O, NO and NO_2 as the reduction products (three products)

(*i*) $Cu + 2HNO_3$ (cold and dilute) $\longrightarrow Cu\ (NO_3)_2 + 2H\ ] \times 4$

$2HNO_3 + 8H \longrightarrow N_2O + 5H_2O$

$4\ Cu + 10\ HNO_3$ (cold and dilute) $\longrightarrow 4\ Cu\ (NO_3)_2 + 5\ H_2O + N_2O$

or $4\ Cu + 2NO_3^- + 10\ H^+ \longrightarrow 4\ Cu^{2+} + 5\ H_2O + N_2O$

(*ii*) $Cu + 2\ HNO_3$ (Dilute) $\longrightarrow Cu(NO_3)_2 + 2H\ ] \times 3$

$HNO_3 + 3H \longrightarrow NO + 2\ H_2O\ ] \times 2$

$3\ Cu + 8\ HNO_3 \longrightarrow 3\ Cu\ (NO_3)_2 + 4\ H_2O + 2\ NO$

or $3\ Cu + 2NO_3^- + 8H^+ \longrightarrow 3\ Cu^{2+} + 4\ H_2O + 2\ NO$

(*iii*) $Cu + 2HNO_3$ (concentrated) $\longrightarrow Cu\ (NO_3) + 2H$

$HNO_3 + H \longrightarrow NO_2 + H_2O] \times 2$

$Cu + 4\ HNO_3$ (concentrated) $\longrightarrow Cu\ (NO_3)_2 + 2H_2O + 2NO_2$

or $Cu + 2NO_3^- + 4H^+ \longrightarrow Cu^{2+} + 2H_2O + 2NO_2$

(*b*) Fe reacts with cold and dilute HNO_3 and gives NH_4NO_3 but with conc. HNO_3, iron becomes passive due to the formation of a thin layer of Fe_3O_4 on the surface of Fe. This layer prevents further action of conc. HNO_3 on Fe.

$Fe + 2\ HNO_3$ (cold and dil.) $\longrightarrow Fe\ (NO_3)_2 + 2H] \times 4$

$HNO_3 + 8H \longrightarrow NH_3 + 3H_2O$

$NH_3 + HNO_3 \longrightarrow NH_4NO_3$

$4\ Fe + 10\ HNO_3$ (cold and dil.) $\longrightarrow 4\ Fe\ (NO_3)_2 + 3H_2O + NH_4NO_3$

or $4\ Fe + NO_3^- + 10\ H^+ \longrightarrow 4Fe^{2+} + 3\ H_2O + NH_4^+$

(*c*) Zn reacts with HNO_3 and gives five different reduction products namely NH_4NO_3 (or NH_4^+ ion), NH_3, N_2O, NO and NO_2 depending on the concentration of the acid and the temperature of the reaction. This is evident from the following reactions :

(*i*) Very dilute HNO_3 (N = +5) is reduced to NH_4NO_3 (N = +1) or NH_4^+ ion. (N = –3)

$$\begin{array}{rcl} Zn + 2\,HNO_3\text{ (very dil.)} & \longrightarrow & Zn(NO_3)_2 + 2H] \times 4 \\ HNO_3 + 8H & \longrightarrow & NH_3 + 3H_2O \\ NH_3 + HNO_3 & \longrightarrow & NH_4NO_3 \\ \hline 4\,Zn + 10\,HNO_3\text{ (very dil.)} & \longrightarrow & 4\,Zn(NO_3)_2 + 3\,H_2O + NH_4NO_3 \end{array}$$

or $4\,Zn + NO_3^- + 10\,H^+ \longrightarrow 4\,Zn^{2+} + 3H_2O + NH_4^+$

(*ii*) Dilute HNO_3 (N = +5) is reduced to NH_3 (N = –3) which reacts with HNO_3 to form NH_4NO_3 (N = +1). NH_4NO_3 gets decomposed to N_2O (N = +1)

$$\begin{array}{rcl} Zn + \text{(Dilute)} & \longrightarrow & Zn(NO_3)_2 + 2H] \times 4 \\ HNO_3 + 8H & \longrightarrow & 3H_2O + NH_3 \\ NH_3 + HNO_3 & \longrightarrow & NH_4NO_3 \\ NH_4NO_3 & \longrightarrow & N_2O + 2H_2O \\ \hline 4\,Zn + 10\,HNO_3\text{ (Dilute)} & \longrightarrow & 4\,Zn(NO_3)_2 + 5H_2O + N_2O \end{array}$$

or $4\,Zn + 2NO_3^- + 10\,H^+ \longrightarrow 4\,Zn^{2+} + 5H_2O + N_2O$

(*iii*) Cold and 50% concentrated HNO_3 (N = +5) is reduced to NO (N = +2)

$$\begin{array}{rcl} 2\,HNO_3\text{ (cold and 50\% concentrated)} & \longrightarrow & H_2O + 2NO + 3O \\ Zn + O + 2\,HNO_3 & \longrightarrow & Zn(NO_3)_2 + H_2O] \times 3 \\ \hline 3\,Zn + 8\,HNO_3\text{ (cold and 50\% concentrated)} & \longrightarrow & 3\,Zn(NO_3)_2 + 4H_2O + 2NO \end{array}$$

or $3\,Zn + 2NO_3^- + 8H^+ \longrightarrow 3\,Zn^{2+} + 4H_2O + 2NO$

(*iv*) Hot and conc. HNO_3 is reduced to NO_2

$$\begin{array}{rcl} Zn + 2\,HNO_3\text{ (Hot and conc.)} & \longrightarrow & Zn(NO_3)_2 + 2H \\ HNO_3 + H & \longrightarrow & H_2O + NO_2] \times 2 \\ \hline Zn + 4\,HNO_3\text{ (Hot and conc.)} & \longrightarrow & Zn(NO_3)_2 + 2H_2O + 2NO_2 \end{array}$$

or $Zn + 2NO_3^- + 4H^+ \longrightarrow Zn^{2+} + 2\,H_2O + NO_2$

(*d*) Al metal is rendered passive by the action of conc. HNO_3 on it. Passive nature of Al is due to the formation of a thin film of Al_2O_3 on its surface. This film prevents further action of HNO_3 on Al.

19. (*a*) Cu is oxidised to $Cu(NO_3)_2$ by means of concentrated HNO_3. HNO_3 is reduced to NO_2.

$$Cu + 4HNO_3\text{ (Conc.)} \longrightarrow Cu(NO_3)_2 + 2H_2O + 2NO_2$$

or $Cu + 2NO_3^- + 4H^+ \longrightarrow Cu^{2+} + 2\,H_2O + 2\,NO_2$

(*b*) When dipped in conc. HNO_3, Fe loses its normal activity and becomes passive. Its passivity is due to the formation of a thin protective layer of ferrosoferric oxide ($FeO.Fe_2O_3$ or Fe_3O_4) on the surface of iron.

(*c*) Sn reacts with conc. HNO_3 to form metastanic acid (H_2SnO_3) and NO_2.

$$Sn + 4HNO_3\text{ (Conc.)} \longrightarrow H_2SnO_3 + 4NO_2 + H_2O$$

or $Sn + 4NO_3^- + 2H^+ \longrightarrow SnO_3^{2-} + H_2O + 4NO_2$

In the above reaction, Sn behaves as a non-metal, since many non-metals like C, S, I, P, As and Sb also give their corresponding oxy acids (H_2CO_3, H_2SO_4, HIO_3, H_3PO_4, H_3AsO_4 and H_3SbO_4 respectively) and NO_2.

(*d*) Pb is also oxidised to $Pb(NO_3)$ and HNO_3 is reduced to NO_2.

$$Pb + 4HNO_3 \text{ (conc.)} \longrightarrow Pb(NO_3)_2 + 2H_2O + 2NO_2$$

or $$Pb + 2NO_3^- + 4H^+ \longrightarrow Pb^{2+} + 2H_2O + 2NO_2$$

20. Sn reacts with conc. HNO_3 but this metal is not oxidised to its nitrate. It is oxidised to hydrated stannic oxide ($SnO_2 . H_2O$) or metastanic acid, H_2SnO_3 (Sn = +4) and HNO_3 is reduced to NO_2.

$$2\ HNO_3 \longrightarrow H_2O + 2\ NO_2 + O] \times 2$$
$$Sn + H_2O + 2O \longrightarrow SnO_2 . H_2O \text{ or } H_2SnO_3$$
$$\overline{Sn + 4\ HNO_3 \text{ (conc.)} \longrightarrow H_2O + 4NO_2 + SnO_2 . H_2O \text{ or } H_2SnO_3}$$

OR

$$\underset{(N = +5)}{NO_3^-} + 2H^+ + e^- \xrightarrow{\text{Reduction}} \underset{(N = +4)}{NO_2} + H_2O] \times 4$$
$$\underset{(Sn = 0)}{Sn} + 3H_2O \xrightarrow{\text{Oxidation}} \underset{(Sn = +4)}{SnO_3^{2-}} + 6H^+ + 4e^-$$
$$\overline{\underset{(Sn = 0)}{Sn} + \underset{(N = +5)}{4\ NO_3^-} + 2H^+ \longrightarrow \underset{(Sn = +4)}{Sn\ O_3^{2-}} + H_2O + \underset{(N = +4)}{4\ NO_2}}$$

21. Since Al is rendered passive when put into conc. HNO_3, the acid can be stored in containers made of Al. Passive nature of Al is due to the formation of a thin layer of Al_2O_3 on its surface.

22. $$Zn + 2HNO_3 \text{ (Very dilute)} \longrightarrow Zn\ (NO_3)_2 + 2H] \times 4$$
$$HNO_3 + 8H \longrightarrow NH_3 + 3H_2O$$
$$NH_3 + HNO_3 \longrightarrow NH_4NO_3$$
$$\overline{4Zn + 10\ HNO_3 \text{ (Very dilute)} \longrightarrow 4\ Zn\ (NO_3)_2 + 3H_2O + NH_4NO_3}$$

or $$\underset{(Zn = 0)}{4\ Zn} + \underset{(N = +5)}{NO_3^-} \longrightarrow \underset{(Zn = +2)}{4\ Zn^{2+}} + 3\ H_2O + \underset{(N = -3)}{NH_4^+}$$

23. Conc. HNO_3 oxidises I_2 to I_2O_5 which gives HIO_3 with H_2O

$$[2HNO_3 \longrightarrow H_2O + 2NO_2 + O] \times 5$$
$$I_2 + 5\ O \longrightarrow I_2O_5$$
$$I_2O_5 + H_2O \longrightarrow 2\ HIO_3$$
$$\overline{I_2 + 10\ HNO_3 \longrightarrow 2\ HIO_3 + 10\ NO_2 + 4\ H_2O} \quad \textit{(Molecular equation)}$$

OR

$$\underset{(N = +5)}{[NO_3^-} + 2H^+ + e^- \xrightarrow{\text{Reduction}} H_2O + \underset{(N = +4)}{NO_2]} \times 10$$
$$\underset{(I = 0)}{I_2} + 6H_2O \xrightarrow{\text{Oxidation}} \underset{(I = +5)}{2IO_3^-} + 12H^+ + 10\ e^-$$
$$\overline{\underset{(N = +5)}{10\ NO_3^-} + 8\ H^+ + \underset{(I = 0)}{I_2} \longrightarrow 4H_2O + \underset{(N = +4)}{10\ NO_2} + \underset{(I = +5)}{2IO_3^-}} \quad \textit{(Ionic equation)}$$

24. Conc. HNO_3 oxidises phosphorus to P_2O_5 which reacts with water to give H_3PO_4. In this reaction HNO_3 is reduced to NO_2.

$$
\begin{array}{lll}
[2HNO_3 & \longrightarrow & H_2O + 2NO_2 + O\,] \times 5 \\
2P + 5O & \longrightarrow & P_2O_5 \\
P_2O_5 + 3H_2O & \longrightarrow & 2H_3PO_4 \\
\hline
P + 5\,HNO_3 & \longrightarrow & H_3PO_4 + 5\,NO_2 + H_2O \text{ (Molecular equation)}
\end{array}
$$

$$
\begin{array}{lll}
[2H^+ + NO_3^- + e^- & \longrightarrow & H_2O + NO_2] \times 5 \\
P + 4\,H_2O & \longrightarrow & PO_4^{3-} + 8\,H^+ + 5\,e^- \\
\hline
2H^+ + P + 5\,NO_3^- & \longrightarrow & H_2O + PO_4^{3-} + 5NO_2 \text{ (Ionic equation)}
\end{array}
$$

25. Concentrated HNO_3 attacks proteins of the skin, forming a yellow nitro compound with skin which is called xantho protein.

26. Since P_2O_5 is a powerful dehydrating agent, it extracts water from HNO_3 when treated with it.

$$2\,HNO_3 \xrightarrow[\text{of } P_2O_5]{\text{In presence}} H_2O + N_2O_5$$

Now since P_2O_5 has strong affinity for H_2O, it combines with H_2O lost by HNO_3 and forms metaphosphoric acid (HPO_3)

$$H_2O + P_2O_5 \longrightarrow 2\,HPO_3$$

On adding the above equations we get :

$$2\,HNO_3 + P_2O_5 \longrightarrow N_2O_5 + 2HPO_3$$

27.

$$
\begin{array}{lll}
[2HNO_3 & \longrightarrow & H_2O + 2\,NO_2 + O] \times 18 \\
C_{12}H_{22}O_{11} + 18\,O & \longrightarrow & 6\begin{array}{l}COOH\\ |\\ COOH\end{array} + 5\,H_2O \\
\text{Cane. sugar} & & \text{Oxalic acid} \\
(C = 0) & & (C = +3) \\
\hline
36HNO_3 + C_{12}H_{22}O_{11} & \longrightarrow & 23H_2O + 36\,NO_2 + 6\begin{array}{l}COOH\\ |\\ COOH\end{array}
\end{array}
$$

28. The mixture of conc. HCl and HNO_3 taken in 3:1 ratio contains NOCl (nitrosyl chloride) as is evident from the following chemical equation :

$$\underbrace{3HCl + HNO_3}_{(3:1\,\text{ratio})} \longrightarrow NOCl + 2H_2O + 2Cl$$

29.

$$\underbrace{3HCl + HNO_3}_{\text{Aqua regia}} \longrightarrow NOCl + 2H_2O + 2Cl$$

$$
\begin{array}{lll}
Au + 3Cl & \longrightarrow & AnCl_3 \\
AuCl_3 + HCl \text{ (excess)} & \longrightarrow & HAu\,Cl_4 \\
 & & \text{(Aurochloric acid)}
\end{array}
\quad\Bigg|\quad
\begin{array}{lll}
Pt + 4Cl & \longrightarrow & PtCl_4 \\
PtCl_4 + 2HCl \text{ (excess)} & \longrightarrow & H_2PtCl_6 \\
 & & \text{(Chloroplatinic acid)}
\end{array}
$$

30. HNO_3 is used in the purification of Ag and Au by aqna regia.

32. $2Pb(NO_3)_2 \xrightarrow{\Delta} 2PbO + 4NO_2 + O_2$

33. Since $NaNO_3$, KNO_3 and $RbNO_3$ are the nitrates of Na, K and Rb (alkali metals), they give MNO_2 and O_2 on heating.

$$2MNO_3 \text{ (M = Na, K, Rb)} \xrightarrow{\Delta} 2MNO_2 + O_2$$

$BaNO_3$ gives a mixture of two gases *viz* NO_2 and O_2.

$$2Ba(NO_3)_2 \xrightarrow{\Delta} 2BaO + 4NO_2 + O_2$$

34. $$NH_4NO_3 \xrightarrow{\Delta} N_2O + 2H_2O$$

35. (*a*) $$NH_4NO_3 \xrightarrow{\Delta} \underset{\text{Nitrogen (I) oxide}}{N_2O} + 2H_2O$$

(*b*) Decomposition of NH_4NO_2 gives N_2

$$NH_4NO_2 \longrightarrow N_2 + 2H_2O$$

(*c*) N_2O_4 (N = +4) undergoes disproportionation when dissolved in water, leading to the formation of HNO_2 (N = +3) and HNO_3 (N = +5) and not N_2O.

$$\underset{(N = +4)}{N_2O_4} + H_2O \longrightarrow \underset{(N = +3)}{HNO_2} + \underset{(N = +5)}{HNO_3}$$

(*d*) $$\underset{\text{Hydroxyl amine}}{NH_2OH} + HNO_2 \longrightarrow N_2O + 2H_2O$$

36. (*a*) When mixture of containing N_2 and O_2 is heated to 2273-3273 K in an electric furnace, NO is obtained. This reaction forms the basis for the manufacture of HNO_3 by Birkland and Eyde process.

$$N_2 + O_2 \rightleftharpoons 2NO$$

(*b*) Reduction of NO by burning P, H_2 and heated metals etc. gives N_2

$$10\,NO + 4P \longrightarrow 2P_2O_5 + 5N_2$$
$$2NO + 2H_2 \longrightarrow 2H_2O + N_2$$
$$2NO + 2Cu \xrightarrow{\Delta} 2CuO + N_2$$

(*c*) Reduction of HNO_3 gives different products like NO_2, O_2, NH_4NO_3 (or NH_4^+ ions), N_2O, NO, N_2 or NH_2OH depending on the nature of the substance to be oxidised, concentration of HNO_3 and the temperature employed.

(*d*) Thermal decomposition of NH_4NO_3 gives N_2O

$$NH_4NO_3 \xrightarrow{\Delta} N_2O + 2H_2O$$

37. $$2NaNO_3 \xrightarrow{\Delta} 2NaNO_2 + O_2$$

$$\downarrow \Delta$$

$$\underset{(d)}{Na_2O} + \underset{(a)}{N_2} + \underset{(b)}{\frac{3}{2}O_2}$$

38. $2AgNO_3 \xrightarrow{\Delta} 2Ag + 2NO_2 + O_2$

39. The structure of NO_3^- ion is given in the margin. This structure shows that since all the five electrons present in the valence shell of N-atom are used in forming bonds with O-atoms, N-atom has no lone pairs of electrons on it. Since the number of covalent bonds by which N-atom is linked with the surrounding O-atoms is four, N-atom has four bond pairs of electrons.

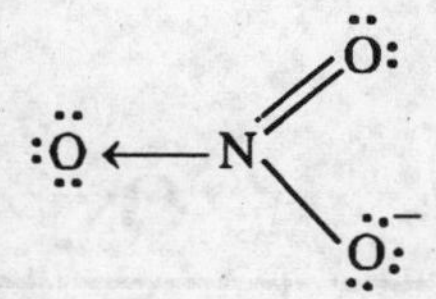

40. $2KNO_3 \xrightarrow{\Delta} 2KNO_2 + O_2$

$Pb(NO_3)_2 \xrightarrow{\Delta} 2PbO + 4NO_2 + O_2$

$2Cu(NO_3)_2 \xrightarrow{\Delta} 2CuO + 4NO_2 + O_2$

$2AgNO_3 \xrightarrow{\Delta} 2Ag + 2NO_2 + O_2$

Obviously since KNO_3 does not give NO_2, answer (*a*) is the correct answer.

41. $KClO_3$, HgO and MnO_2 give only O_2 as the gaseous product while $Pb(NO_3)_2$ gives two gaseous products *viz* O_2 and NO_2.

$2KClO_3 \xrightarrow{\Delta} 2KCl + 3O_2$

$2HgO \xrightarrow{\Delta} 2Hg + O_2$

$3MnO_2 \xrightarrow{>530°c} Mn_3O_4 + O_2$

$2Pb(NO_3)_2 \xrightarrow{\Delta} PbO + 4NO_2 + O_2$

42. Oxidation state of N-atom in the given acids is as +1 (in $H_2N_2O_2$), +3 (in HNO_2), +5 (in HNO_3) and +7 (in HNO_4). Now since oxidation state of N-atom increases as hypo-ous $\longrightarrow$ ous $\longrightarrow$ ic $\longrightarrow$ per-ic, these acids are called hyponitrous acid ($H_2N_2O_2$ with N = +1), nitrous acid (HNO_2 with N = +3), nitric acid (HNO_3 with N = +5) and pernitric acid (HNO_4 with N = +7). Thus HNO_4 is pernitric acid.

32

Oxy acids of Phosphorus and Their Salts

MULTIPLE CHOICE QUESTIONS

1. The oxide which is anhydride of orthophosphoric acid is :
(*a*) P_4O_{10} (*b*) P_2O_5
(*c*) P_3O_6 (*d*) P_2O_3

2. Basicity of orthophosphric acid is :
(*a*) 2 (*b*) 3
(*c*) 4 (*d*) 5

3. Which of the following is a tetrabasic acid ?
(*a*) Orthophosphorus acid
(*b*) Orthophosphoric acid
(*c*) Metaphosphoric acid
(*d*) Pyro (di) phosphoric acid

4. P – O – P bond is present in :
(*a*) $H_4P_2O_6$
(*b*) $H_4P_2O_5$
(*c*) Both (*a*) and (*b*)
(*d*) None of these

5. Which is wrongly matched ?
(*a*) H_3PO_4 : Orthophosphoric acid
(*b*) H_3PO_3 : Phosphorus acid
(*c*) $H_5P_3O_{10}$: Pyrophosphoric acid
(*d*) $H_6P_4O_{13}$: Tetrapoly phosphoric acid

6. Pure orthophosphoric acid exhibits syrupy nature because it :
(*a*) is a triprotic acid
(*b*) is hygroscopic
(*c*) contains PO_4 groups connected together by H-bonds
(*d*) is a strong acid

7. If phosphoric acid is allowed to react with sufficient quantity of NaOH, the product obtained is :
(*a*) Na_3PO_4 (*b*) NaH_2PO_4
(*c*) $Na_2H\,PO_4$ (*d*) $NaH\,PO_3$

8. Phosphorus has oxidation state of +3 in :
(*a*) Phosphorus acid
(*b*) Hypophosphorus acid
(*c*) Orthophosphoric acid
(*d*) Pyrophosphoric acid

9. The acid obtained by the action of cold water on P_4O_6 is :
(*a*) H_3PO_2 (*b*) H_3PO_3
(*c*) H_3PO_4 (*d*) $H_4P_2O_7$

10. Which of the following acids has maximum number of P – H bonds ?
(*a*) H_3PO_2 (*b*) H_3PO_4
(*c*) H_3PO_3 (*d*) $H_4P_2O_7$

11. The number of hydroxyl groups in pyrophosphoric acid is :
(*a*) 3 (*b*) 4
(*c*) 5 (*d*) 7

12. Which of the following oxy acids has maximum number of oxo groups ?
(*a*) H_2SO_4 (*b*) H_2SO_3
(*c*) H_3PO_3 (*d*) H_3PO_4

13. Which of the following salts does not exist ?
(*a*) NaH_2PO_2 (*b*) $NaH_3P_2O_7$
(*c*) NaH_2PO_3 (*d*) NaH_2PO_4

14. The acid that forms two series of salts is :
(*a*) H_3PO_4 (*b*) H_3PO_3
(*c*) H_3BO_3 (*d*) H_3PO_2

15. Which of the following is not an acid salt ?
(*a*) NaH_2PO_2 (*b*) NaH_2PO_3
(*c*) NaH_2PO_4 (*d*) $Na_2H\,PO_4$

16. One of the acids listed below is formed from P_2O_3 and the rest are formed from P_2O_5. The acids formed from P_2O_5 are :
(*a*) HPO_3 (*b*) $H_4P_2O_7$
(*c*) H_3PO_4 (*d*) H_3PO_3

17. When orthophosphoric acid is heated to 600° C, the product formed is :
(*a*) Phosphine
(*b*) Phosphorus pentoxide
(*c*) Phosphorus acid
(*d*) Metaphosphoric acid

18. The formula of micro cosmic salt is :
(*a*) $KBiO_3$
(*b*) $NaPO_3$
(*c*) $KHPO_4$
(*d*) $NaNH_4HPO_4.4H_2O$

19. There is very little difference in acid strength in the series H_3PO_2, H_3PO_3 and H_3PO_4, because :
(*a*) P-atom in these acids exists in different oxidation states.
(*b*) Number of unprotonated O-atoms remains the same.
(*c*) Phosphorus is not a highly electronegative element.
(*d*) Phosphorus oxides are less basic

20. The structural formula of hypophophorus acid is :

(*a*) 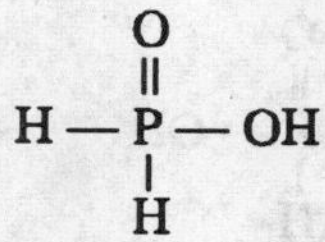

(*b*) 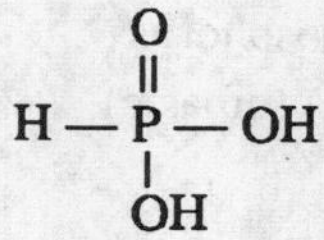

(*c*) 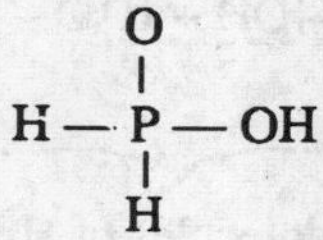

(*d*) 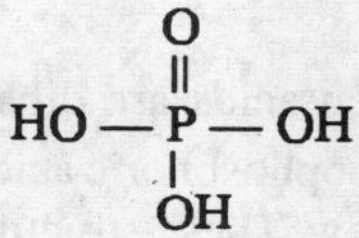

21. In PO_4^{3-} ion, the formal charge on each O-atom and P – O bond order respectively are :
(*a*) – 0.75, 1.25 (*b*) – 3, 1.25
(*c*) – 0.75, 2.0 (*d*) – 0.75, 0.6

22. The equivalent weight of H_3PO_4 in the reaction :

$$NaOH + H_3PO_4 \longrightarrow NaH_2PO_4 + H_2O$$ is :

(*a*) 25 (*b*) 49
(*c*) 59 (*d*) 98

ANSWERS

1. (*a*)	**2.** (*b*)	**3.** (*d*)	**4.** (*b*)	**5.** (*c*)	**6.** (*c*)	**7.** (*a*)	**8.** (*a*)
9. (*b*)	**10.** (*a*)	**11.** (*b*)	**12.** (*a*)	**13.** (*b*)	**14.** (*b*)	**15.** (*a*)	**16.** (*a,b,c*)
17. (*d*)	**18.** (*d*)	**19.** (*b*)	**20.** (*a*)	**21.** (*a*)	**22.** (*d*)		

HINTS AND EXPLANATIONS

1. The oxide that gives orthophosphoric acid (H_3PO_4) with H_2O is called anthydride of orthophosphoric acid. It is only P_4O_{10} that gives H_3PO_4 with H_2O

$$P_4O_{10} + 6H_2O \text{ (hot)} \longrightarrow 4H_3PO_4$$

P_2O_3 gives phosphorus acid (H_3PO_3) with H_2O.

$$P_2O_3 + 3H_2O \text{ (cold)} \longrightarrow 2H_3PO_3$$

2. The structure of orthophosphoric acid (H_3PO_4) is given on the next page. Since P-atom (central atom) is attached with three OH groups directly, H_3PO_4 is a tribasic acid. *i.e.* the basicity of H_3PO_4 is 3.

3. (*a*), (*b*), (*c*) and (*d*) have 2, 3, 1 and 4 OH groups attached directly with central P-atom as shown on the next page.

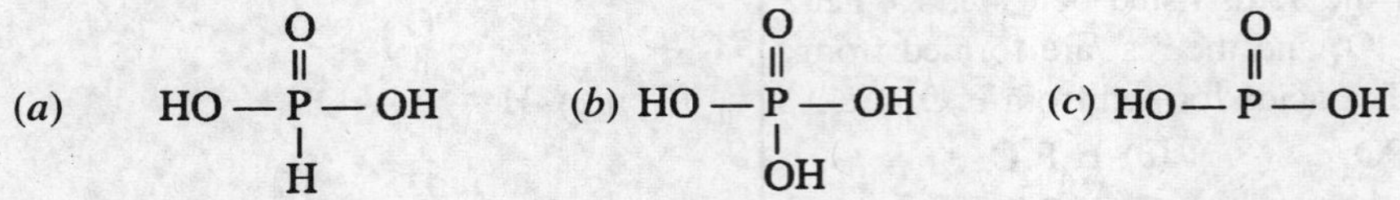

Orthoposphorus acid (H_3PO_3) (dibasic) — Orthophosphoric acid (H_3PO_4) (tribasic) — Metaphosphoric acid (HPO_3) (monobasic)

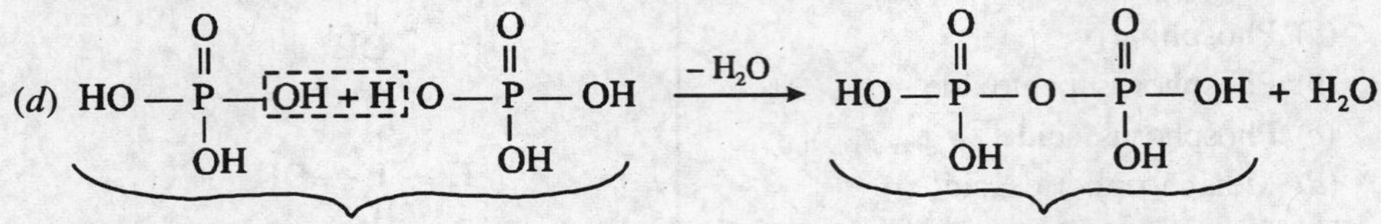

Two molecules of H_3PO_4 — Pyrophosphoric acid ($H_4P_2O_7$) (tetrabasic)

Thus the given acids are dibasic, tribasic, monobasic and tetrabasic respectively.

4. $H_4P_2O_6$ is hypophosphoric acid and $H_4P_2O_5$ is diphosphorus acid. The structures of these acids are given below. These structures show that it is only $H_4P_2O_5$ which contains P–O–P bond.

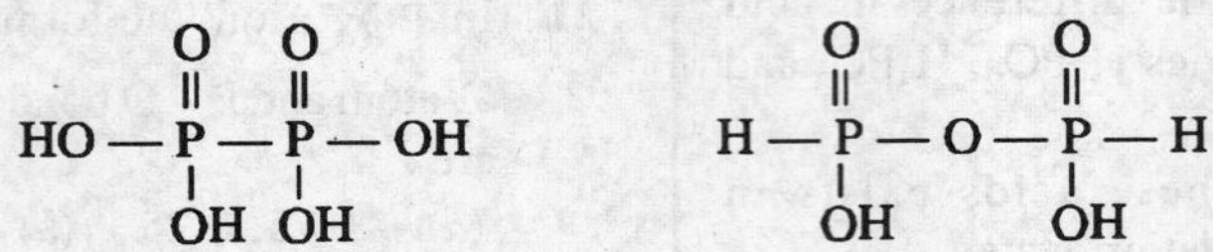

$H_4P_2O_6$ $H_4P_2O_5$

5. The formula of pyro (di) phosphoric acid is obtained by removing one H_2O molecule from two molecules of H_3PO_4.

$$2H_3PO_4 \xrightarrow{-H_2O} H_4P_2O_7 + H_2O$$

Thus the correct formula of this acid is $H_4P_2O_7$ and not $H_5P_3O_{10}$. In fact, $H_5P_3O_{10}$ is tripolyphosphoric acid as shown below :

$$3H_3PO_4 \xrightarrow{-2H_2O} H_5P_3O_{10} + 2H_2O$$

6. The structure of concentrated solution of H_3PO_4 contains many H_3PO_4 tetrahedral molecules connected together by H-bonds. The existance of H-bonds between different tetrahedral H_3PO_4 molecules increases the viscosity of the concentrated solution of the acid. The increase in viscocity makes the acid syrupy.

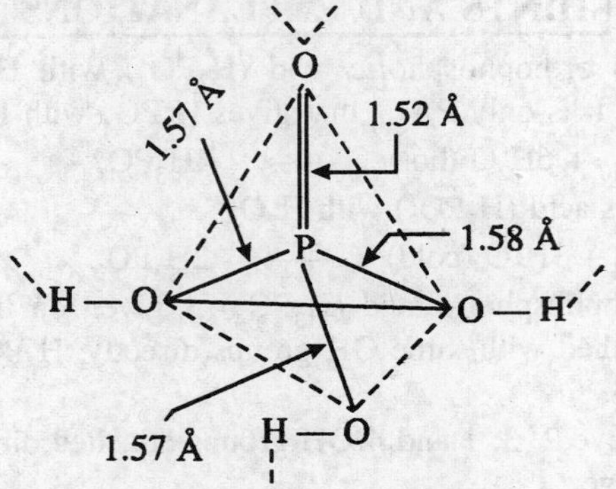

Fig. 31.1. Structure of H_3PO_4 in solid and concentrated solution form.

7. Phosphoric acid (H_3PO_4) is an acid (tribasic acid). Hence it reacts with NaOH to form Na_2HPO_4. With excess of NaOH, Na_3PO_4 is obtained.

$$H_3PO_4 + 2NaOH \longrightarrow Na_2HPO_4 + 2H_2O$$
$$Na_2HPO_4 + NaOH\ (Excess) \longrightarrow Na_3PO_4 + H_2O$$
$$\overline{H_3PO_4 + 3NaOH\ (Excess) \longrightarrow Na_3PO_4 + 3H_2O}$$

8. The formula of the given acids are (*a*) H_3PO_3 (phosphorus acid) (*b*) H_3PO_2 (hypophosphorus acid) (*c*) H_3PO_4 (orthophosphoric acid) and (*d*) $H_4P_2O_7$ (pyrophosphoric acid). Obviously oxidation state of P-atom in these acids is +3, +1, +5 and +5 respectively.

9. The acid obtained by the action of cold water on P_4O_6 is phosphorus acid (H_3PO_3)

$$P_4O_6 + 6H_2O\ (cold) \longrightarrow 4H_3PO_3$$

10. The structures of the given acids are given below :

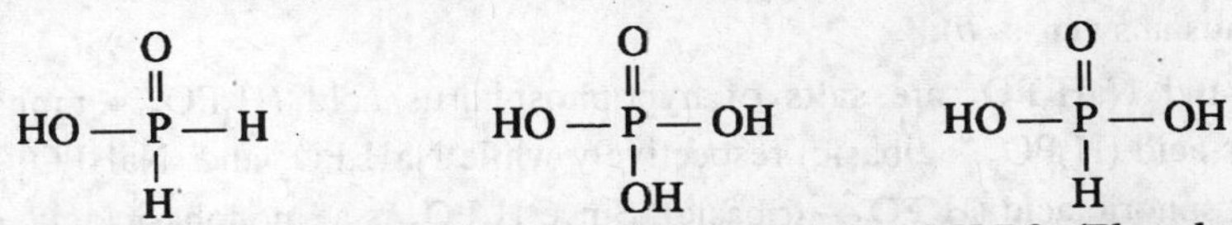

(*a*) H_3PO_2 (Hypophosphorus acid) (monobasic) (*b*) H_3PO_4 (Phosphoric acid) (tribasic) (*c*) H_3PO_3 (Phosphorus acid) (Dibasic)

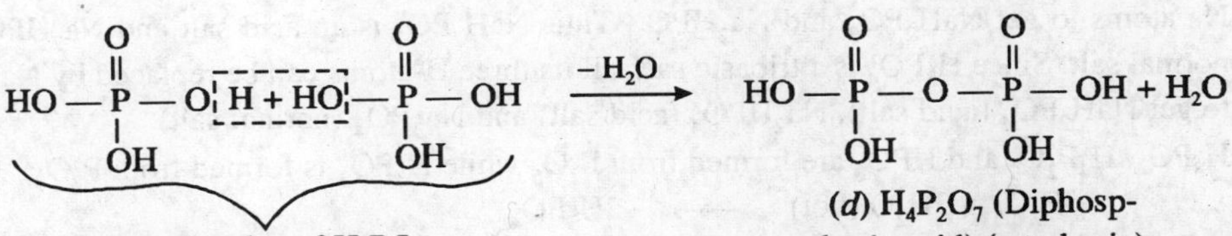

Two molecules of H_3PO_4

(*d*) $H_4P_2O_7$ (Diphosphoric acid) (tetrabasic)

It is obvious from the structures that H_3PO_2 has two P–H bonds while H_3PO_3 has only one P–H bond.

11. Pyrophosphoric acid is $H_4P_2O_7$. Since it is a tetrabasic acid as is evident from its structure given in below, it has four hydroxyl (OH) groups.

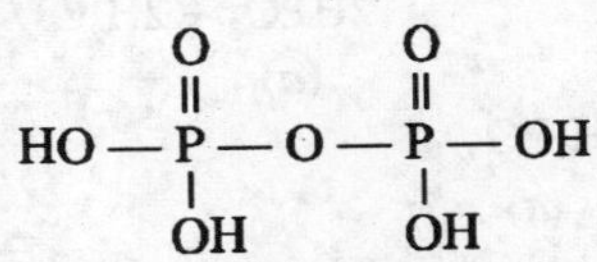

Structure of $H_4P_2O_7$ molecule

12. The structures of the given molecules are given below :

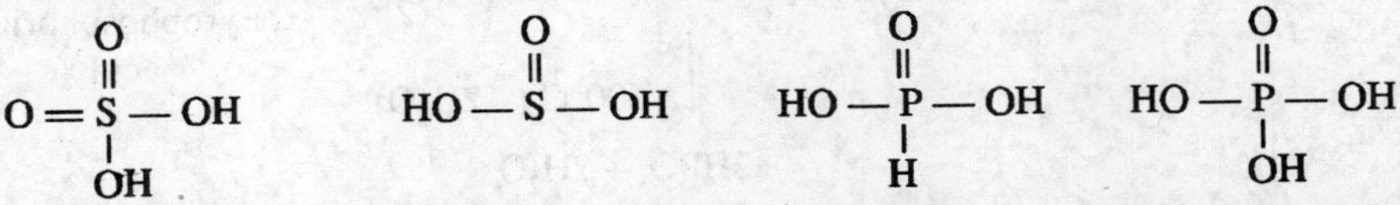

(*a*) H_2SO_4 (Dibasic) (*b*) H_2SO_3 (Dibasic) (*c*) H_3PO_3 (Dibasic) (*d*) H_3PO_4 (Tribasic)

These structures show that the number of unprotonated O-atoms attached directly with the central P or S atom in these molecules is 2, 1, 1 and 1 respectively.

13. The structures of the given salts show that the acids corresponding to these salts are H_3PO_2 (monobasic), $H_4P_2O_7$ (tetrabasic), H_3PO_3 (dibasic) and H_3PO_4 (tribasic) respectively. Thus, on the basis of the bascity of these acids, we can say NaH_2PO_2, NaH_2PO_3 and NaH_2PO_4 can be formed by H_3PO_2, H_3PO_3 and H_3PO_4 acids. Hence these salts exist.

Now since $H_4P_2O_7$ is a tetrabasic acid, it is expected to form four types of salts like $Na_3HP_2O_7$, $Na_2H_2P_2O_7$, $NaH_3P_2O_7$ and $Na_4P_2O_7$. Infact this acid gives only two types of salts like $Na_2H_2P_2O_7$ and $Na_4P_2O_7$. Thus $NaH_3P_2O_7$ cannot be formed by $H_2P_2O_7$.

14. (*a*) H_3PO_4 is phosphoric acid which is tribasic and hence forms three series of salts

(*b*) H_3PO_3 is phosphorus acid which is dibasic. Thus this acid gives two series of salts

(*c*) H_3BO_3 is boric acid which is a weak monobasic acid and hence gives only one type of salts (*e.g.* Na_3BO_3)

(*d*) H_3PO_2 is hypophosphorus acid which is monobasic. Thus this acid gives only one type of salts. Thus answer is (*b*).

15. NaH_2PO_2 and NaH_2PO_3 are salts of hypophosphorus acid (H_3PO_2 – monobasic) and phosphorus acid (H_3PO_3 – dibasic) respectively while NaH_2PO_4 and $NaHPO_4$ both are the salts of phosphoric acid (H_3PO_4 – tribasic). Since H_3PO_2 is a monobasic acid, only one H-atom can be repleced by Na atom to get NaH_2PO_2. Thus NaH_2PO_2 is a normal salt and not an acid salt of H_3PO_2. Since H_3PO_3 is a dibasic acid, only two H-atoms can be replaced by Na atoms to get NaH_2PO_3 and Na_2HPO_3. Thus NaH_2PO_3 is an acid salt and Na_2HPO_3 is a normal salt. Since H_3PO_4 is a tribasic salt, all its three H-atoms can be replaced by Na atoms to get NaH_2PO_4 (acid salt), Na_2HPO_4 (acid salt) and Na_3PO_4 (normal salt).

16. H_3PO_4, $H_4P_2O_7$ and HPO_3 are formed from P_2O_5 while H_3PO_3 is formed from P_2O_3.

$$P_2O_5 + 3H_2O \text{ (hot)} \longrightarrow \underset{(c)}{2H_3PO_4}$$

$$2H_3PO_4 \xrightarrow[(-H_2O)]{250°C} \underset{(b)}{H_4P_2O_7 + H_2O}$$

$$2H_3PO_4 \xrightarrow[(-2H_2O)]{600°C} \underset{(a)}{2HPO_3 + 2H_2O}$$

$$P_2O_3 + 3H_2O \text{ (cold)} \longrightarrow \underset{(d)}{2H_3PO_3}$$

17. The action of heat on H_3PO_4 on different temperatures can be shown as :

$$P_2O_5 + 3H_2O \xleftarrow[(-3H_2O)]{\text{Strong heating}} 2H_3PO_4 \xrightarrow[(-H_2O)]{250°C} \underset{\text{Pyrophosphoric acid}}{H_4P_2O_7} + H_2O$$

$$2H_3PO_4 \xrightarrow{600°C\ (-2H_2O)} \underset{\text{Metaphosphoric acid.}}{2HPO_3 + 2H_2O}$$

18. Microcosmic salt is a tetrahydrated salt of orthophosphoric acid (H_3PO_4) which is called sodium ammonium hydrogen phosphate tetrahydrate and hence is formulated as $Na(NH_4)\,HPO_4.\,4H_2O$.

19. Structures of the given acids are given below :

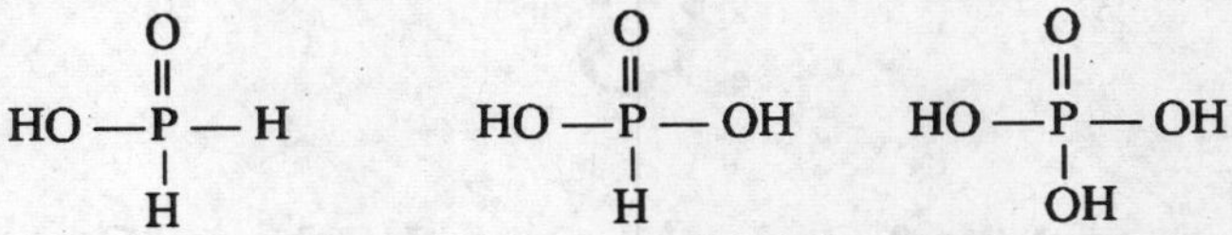

H_3PO_2 (Monobasic) H_2PO_3 (Dibasic) H_3PO_4 (Tribasic)

The acid strength of the given acids depends on the number of unprotonated O-atoms present in them. Since the number of unprotonated O-atoms in all the three acids is the same (equal to 1), the acidic strength of the acids is also almost the same.

20. The formulae corresponding to the given structures are H_3PO_2, H_3PO_3, H_3PO_2 and H_3PO_4 respectively. Since hypophosphorus acid is H_3PO_2 and is monobasic, its structure is (*a*). Structure (*c*) is wrong, since O-atom is attached with P-atom by P – O single bond and not by P = O double bond.

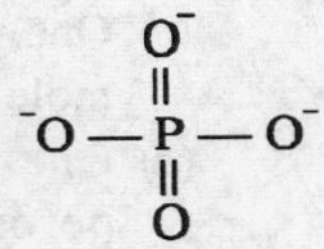

21. The structure of PO_4^{3-} ion is given in the margin. This structure shows that total charges on PO_4^{3-} ion = – 3 and the total number of O-atoms = 4. Thus :

$$\text{Formal charge on each O-atom} = \frac{\text{Total charges}}{\text{Total number of O-atoms}} = \frac{-3}{4} = -0.75$$

$$\text{Bond order of each P–O bond} = \frac{\text{Total number of P–O bonds}}{\text{Total number of O-atoms}} = \frac{5}{4} = 1.25$$

22. In the reaction : $NaOH + H_3PO_4 \longrightarrow NaH_2PO_4 + H_2O$, since one mole of H_3PO_4 reacts with one mole of NaOH, the equivalent weight of H_3PO_4 is its molecular weight which is equal to $1 \times 3 + 31 + 4 \times 16 = 98$

33

Oxy acids of Sulphur and Their Salts

MULTIPLE CHOICE QUESTIONS

1. Hydrolysis of one mole of peroxodisulphuric and produces:
(*a*) Two moles of sulphuric acid
(*b*) Two moles of peroxomonosulphuric acid
(*c*) One mole of sulphuric acid and one mole of peroxomonosulphuric acid.
(*d*) One mole each of sulphuric acid, peroxomonosulphuric acid and hydrogen peroxide.

2. There is S – S bond in :
(*a*) $S_2O_3^{2-}$ (*b*) $S_2O_4^{2-}$
(*c*) $S_2O_5^{2-}$ (*d*) $S_2O_7^{2-}$

3. $Fe_2(SO_4)_3$, on heating, gives:
(*a*) SO_2 and SO_3
(*b*) SO_2 only
(*c*) SO_3 only
(*d*) None of these oxides.

4. The reaction: $H_2S + H_2SO_4\,(conc) \longrightarrow A + B + 2H_2O$ is a redox reaction in which H_2SO_4 and H_2S act as oxidising and reading agents respectively. Identify *A* and *B*
(*a*) SO_2, S (*b*) SO_2, SO_3
(*c*) H_2, H_2SO_3 (*d*) SO_4^{2-}, S^{2-}

5. By the action of conc H_2SO_4, phosphorus changes to:
(a) Phosphorus acid
(*b*) Orthophosphoric acid
(*c*) Metaphosphoric acid
(*d*) Pyrophosphoric acid

6. S – S bond is present in :
(*a*) $S_2O_4^{2-}$ (*b*) $S_2O_7^{2-}$
(c) $S_2O_8^{2-}$ (*d*) S_3O_9

7. The number of S – S bonds in sulphur troxide trimer (S_3O_9) is :
(*a*) 3 (*b*) 2
(*c*) 1 (*d*) zero

8. The oxo acid (s) of sulphur containing – S – S – bond is (are):
(*a*) $H_2S_2O_3$ (*b*) $H_2S_2O_4$
(*c*) $H_2S_2O_6$ (*d*) $H_2S_2O_7$

9. Acid, having peroxy linkage are:
(*a*) H_2SO_5 (*b*) $H_2S_2O_6$
(*c*) $H_2S_2O_7$ (*d*) $H_2S_2O_8$

10. Oleum is chemically known as:
(*a*) H_2SO_3 (*b*) H_2SO_5
(*c*) $H_2S_2O_7$ (*d*) $H_2S_2O_8$

11. Caro's acid is:
(*a*) H_2SO_3 (*b*) $H_2S_2O_5$
(*c*) H_2SO_5 (*d*) $H_2S_2O_8$

12. What is the number of sigma (σ) and pi (π) bunds present in sulphuric acid molecule?
(a) $6\sigma, 2\pi$ (b) $6\sigma, 0\pi$
(c) $2\sigma, 4\pi$ (d) $2\sigma, 2\pi$

13. Which of the following ions does not contain S – S linkage?
(a) $S_2O_3^{2-}$ (*b*) $S_2O_4^{2-}$
(c) $S_2O_6^{2-}$ (*d*) $S_2O_8^{2-}$

14. S – S bond is not present in:

(*a*) $S_2O_3^{2-}$ (*b*) $S_2O_4^{2-}$

(*c*) $S_2O_5^{2-}$ (*d*) $S_2O_7^{2-}$

15. Which of the following species is basic and reducing?

(*a*) SO_3^{2-} (*b*) SO_4^{2-}

(*c*) $S_2O_4^{2-}$ (*d*) HSO_4^-

16. Hypo is used in photography because of its :

(*a*) Oxidising behaviour

(*b*) Reducing behaviour

(*c*) Complex forming behaviour

(*d*) Reaction with light

17. H_2SO_4 has very corrosive action on skin because

(*a*) It acts as dehydrating agent

(*b*) It reacts with proteins

(*c*) It acts as an oxidising agent.

(*d*) It acts as dehydrating agent and absorption of water is highly exothermic.

18. Which one of the following charrs when warmed with conc. H_2SO_4 ?

(*a*) Carbohydrates

(*b*) Proteins

(c) Fats

(*d*) Hydrocarbons

19. Which liberates SO_2 with dil. H_2SO_4 ?

(*a*) Na_2SO_4 (*b*) $NaHSO_4$

(*c*) Na_2SO_3 (*d*) Na_2S

20. Which of the following acids has the maximum number of oxo group (S = O or P = O groups) ?

(*a*) H_2SO_3 (*b*) H_2SO_4

(*c*) H_3PO_3 (*d*) H_3PO_4

21. The acid which contains a peroxo linkage is:

(*a*) Sulphurous acid

(*b*) Pyrosulphuric acid

(*c*) Dithionic acid

(*d*) Caro's acid

22. Acidified solution of $Na_2S_2O_3$ is unstable because in $Na_2S_2O_3$:

(*a*) The two S-atoms are in unstable oxidation state of + 2.

(*b*) The two S-atoms are in different oxidation states of + 6 and –2.

(*c*) The S – S bond in unstable

(*d*) Thio compounds contain S-atom in oxidation state of zero.

23. Oxidation number of S-atom in $H_2S_2O_8$ molecule is:

(a) + 2 (b) + 4

(c) + 6 (d) + 7

24. The oxidation states of the most electronegative element in the products of the reaction of BaO_2 with dilute H_2SO_4 are:

(*a*) 0 and – 1 (*b*) – 1 and – 2

(*c*) – 2 and 0 (*c*) – 2 and +1

25. The oxidation state (O.S.) of S-atom is – 1 in:

(*a*) FeS (*b*) FeS_2

(*c*)
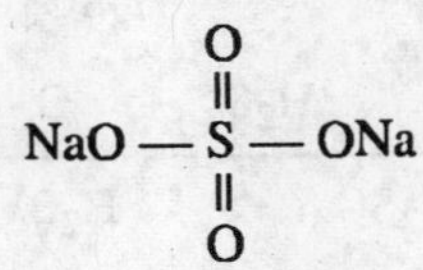

(*d*)
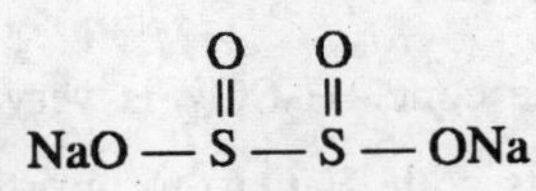

ANSWERS

1. (*c*) **2.** (*a, b*) **3.** (*c*) **4.** (*a*) **5.** (*b*) **6.** (*a*) **7.** (*d*) **8.** (*a,b,c*)
9. (*a,d*) **10.** (*c*) **11.** (*c*) **12.** (*a*) **13.** (*d*) **14.** (*c, d*) **15.** (*a*) **16.** (*c*)
17. (*c*) **18.** (*a*) **19.** (*c*) **20.** (*b*) **21.** (*d*) **22.** (*b*) **23.** (*c*) **24.** (*b*)
25. (*b*)

HINTS AND EXPLANATIONS

1. 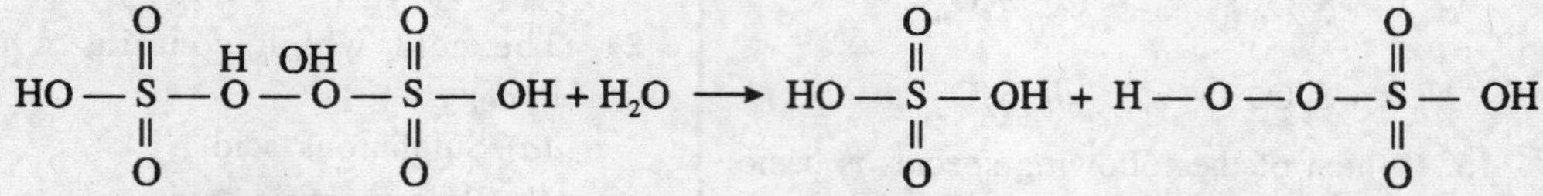

Peroxodisulphuric acid ($H_2S_2O_8$) + H_2O → Sulphuric acid (H_2SO_4) + Peroxomonosulphuric acid (H_2SO_5)

2. The struture of the given ions are given below. These structures show that $S_2O_3^{2-}$ and $S_2O_4^{2-}$ both contain S — S bond.

(a) 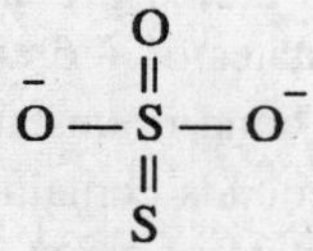

($S_2O_3^{2-}$ ion)

(b) 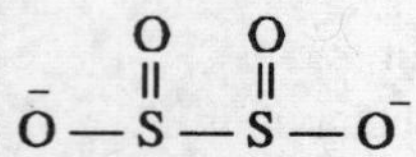

($S_2O_4^{2-}$ ion)

(c) 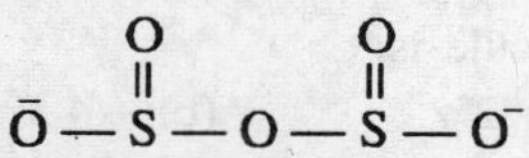

($S_2O_5^{2-}$ ion)

(d) 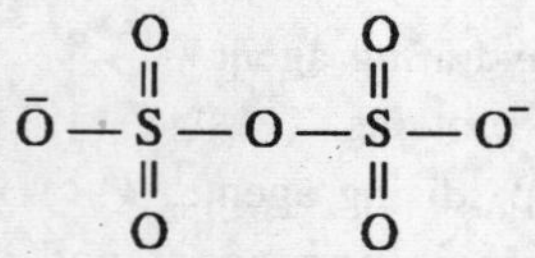

($S_2O_7^{2-}$ ion)

3. On heating, $Fe_2(SO_4)_3$ gives Fe_2O_3 and SO_3.

$$Fe_2(SO_4)_3 \xrightarrow{\Delta} Fe_2O_3 + 3SO_3$$

4. Since $H_2SO_4\,(S=+6)$ acts as an oxidising agent, it oxdises $H_2S(S=-2)$ to $S(S=O)$ and is itself reduced to $SO_2\,(S=+4)$.

$$H_2SO_4 \longrightarrow SO_2 + H_2O + O$$
$$H_2S + O \longrightarrow S + H_2O$$
$$\overline{H_2SO_4 + H_2S \longrightarrow \underset{(A)}{SO_2} + \underset{(B)}{S} + 2H_2O}$$

5. Since conc. H_2SO_4 is very good oxidising agent, it oxidises phosphorus to P_2O_5 which reacts with H_2O to form orthophosphoric acid (H_3PO_4)

$$[H_2SO_4(conc.) \longrightarrow H_2O + SO_2 + O] \times 5$$
$$2P + 5O \longrightarrow P_2O_5$$
$$[P_2O_5 + 3H_2O \longrightarrow 2H_3PO_4] \times 2$$
$$\overline{\underset{(P=0)}{2P} + \underset{(S=+6)}{5H_2SO_4} \longrightarrow \underset{(P=+5)}{2H_3PO_4} + \underset{(S=+4)}{5SO_2} + 2H_2O}$$

or $$2P + 5SO_4^{2-} + 4H^+ \longrightarrow 2PO_4^{3-} + 2H_2O + 5SO_2$$

Thus we see that P is oxidised to H_3PO_4 and H_2SO_4 is reduced to SO_2.

6. The structures of the given species are given below :

(*a*) 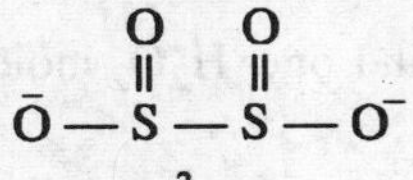

$S_2O_4^{2-}$ ion

(*b*) 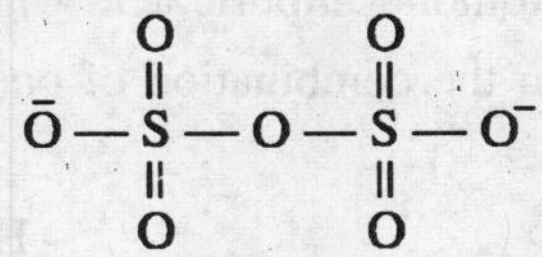

$S_2O_7^{2-}$ ion

(*c*) 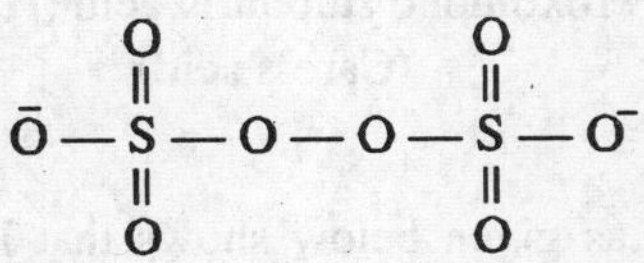

$S_2O_8^{2-}$ ion

(*d*) 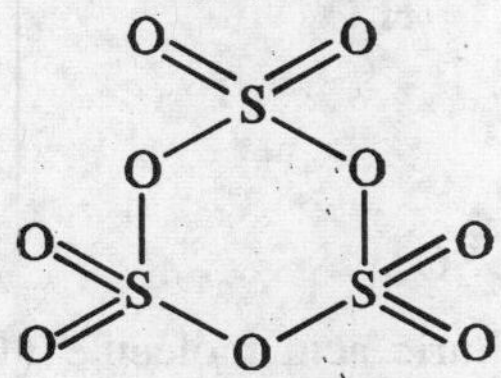

S_3O_9 molecule

These structures show that it is only $S_2O_4^{2-}$ ion which contains S – S bond.

7. The structure of sulphur trioxide trimer, $(SiO_3)_3$ or Si_3O_9 has already been given in the answer of question 6. This structure shows that Si_3O_9 molecule has no S – S bond.

8. The structures of the given acids are given below :

(*a*) 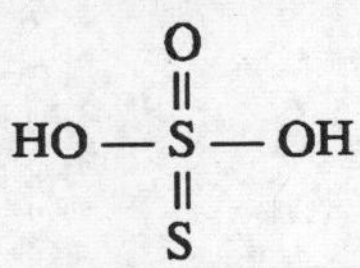

Thiosulphuric acid ($H_2S_2O_3$)

(*b*) 

Hyposulphurous acid ($H_2S_2O_4$)

(*c*) 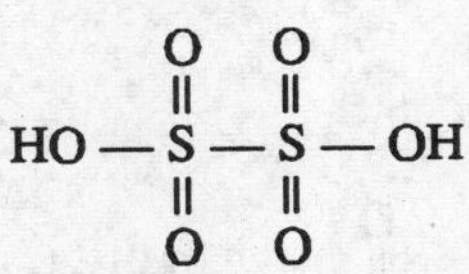

Dithionic acid ($H_2S_2O_6$)

(*d*) 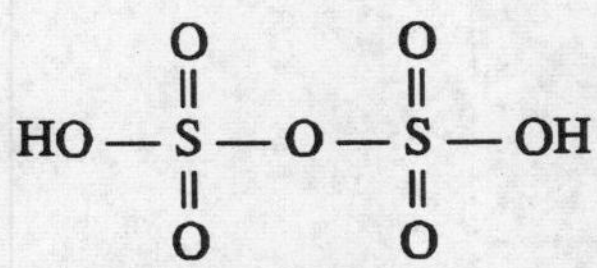

Di (pyro) sulphuric acid ($H_2S_2O_7$)

These strutures show that only $H_2S_2O_3$, $H_2S_2O_4$ and $H_2S_2O_6$ contain – S – S – O bond. $H_2S_2O_7$ has – S – O – S – bond.

9. The structures of the given acids are given below :

(*a*) 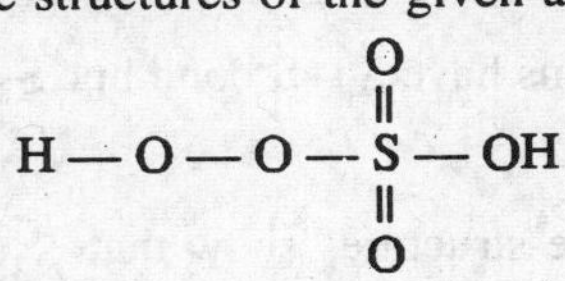

H_2SO_5

(*b*) 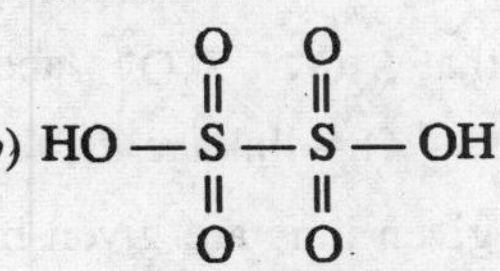

$H_2S_2O_6$

(*c*) 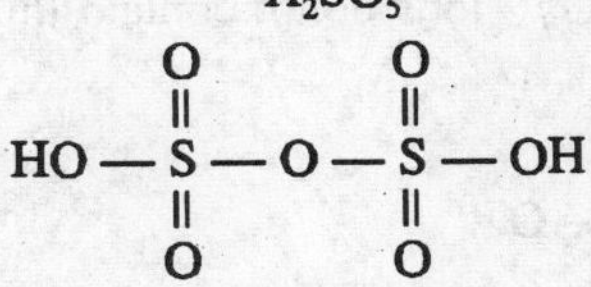

$H_2S_2O_7$

(*d*) 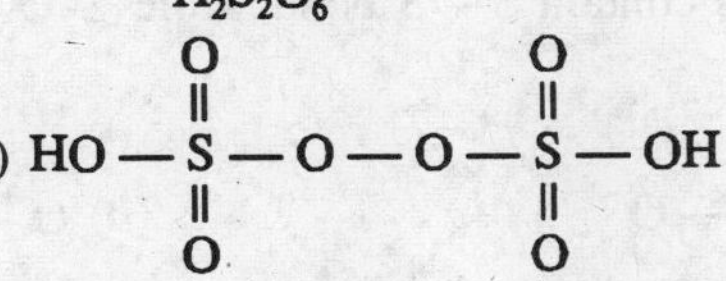

$H_2S_2O_8$

It is obvoous from these structures that only H_2SO_5 and $H_2S_2O_8$ contain peroxy linkage, – O – O – .

10. Since oleum is disulphuric acid (also called fuming sulphuric acid), its formula is $H_2S_2O_7$

$$2H_2SO_4 \xrightarrow{-H_2O} H_2S_2O_7 + H_2O$$

11. Caro's acid is peroxomonosullpuric acid whose structure can be obtained by removing one H_2O molecule from the combination of one H_2SO_4 molecule and one H_2O_2 molecule.

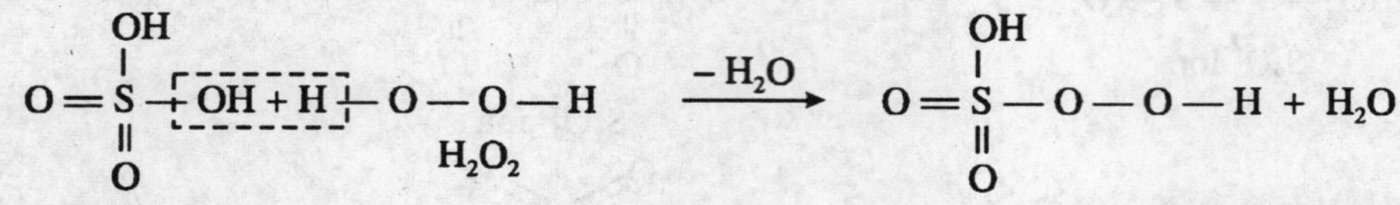

H_2SO_4

Peroxomono sulphuric acid (H_2SO_5)
(Caro's acid)

Thus Caro's acid is H_2SO_5.

12. The structure of sulphuric acid molecule (H_2SO_4) as given below shows that it has six sigma (σ) and two pi (π) bonds.

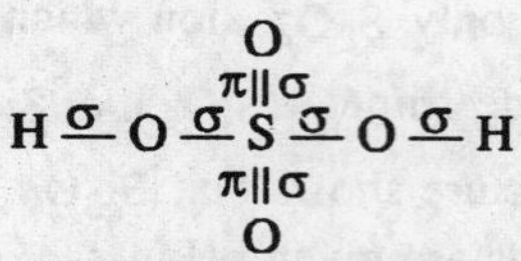

Structure of H_2SO_4 molecule

13. The structure of the given ions are given below:

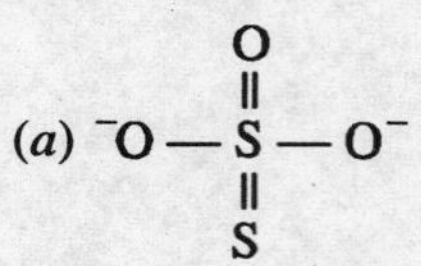

($S_2O_3^{2-}$)

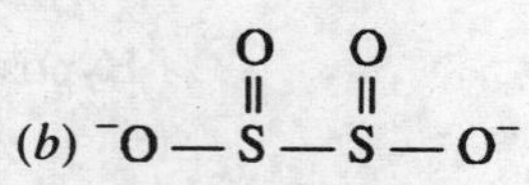

($S_2O_4^{2-}$)

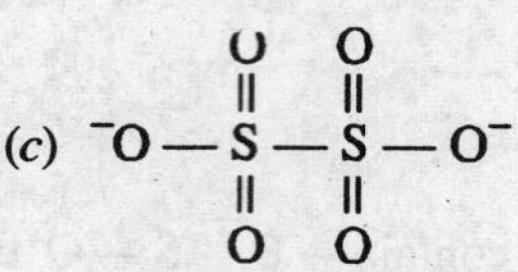

($S_2O_6^{2-}$)

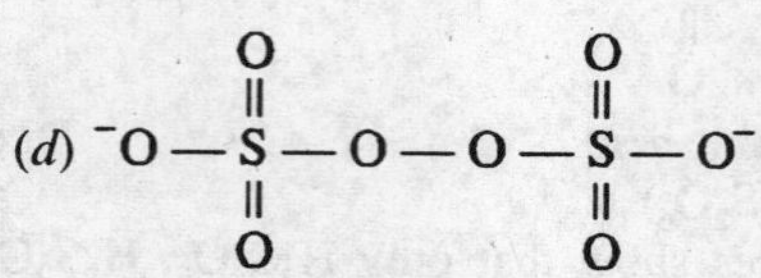

($S_2O_8^{2-}$)

These structures show that $S_2O_3^{2-}$, $S_2O_4^{2-}$ and $S_2O_6^{2-}$ ions have S – S bond but $S_2O_8^{2-}$ ion has peroxo linkage (– O – O – linkage)

14. The structures of the given ions are given below. These structures show that $S_2O_3^{2-}$ and $S_2O_4^{2-}$ ions contain S – S bond while $S_2O_5^{2-}$ and $S_2O_7^{2-}$ ions donot contain this bond.

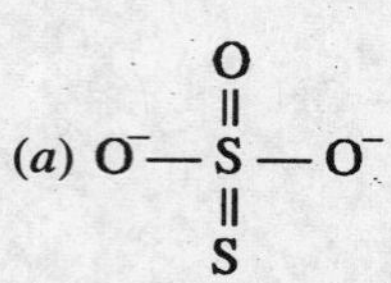

($S_2O_3^{2-}$ ion)

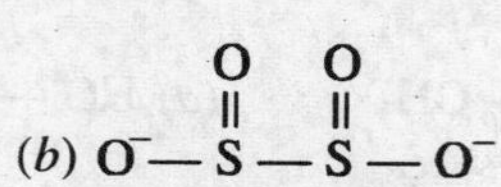

($S_2O_4^{2-}$ ion)

(c) 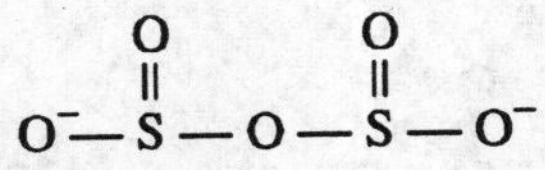

($S_2O_5^{2-}$ ion)

(d) 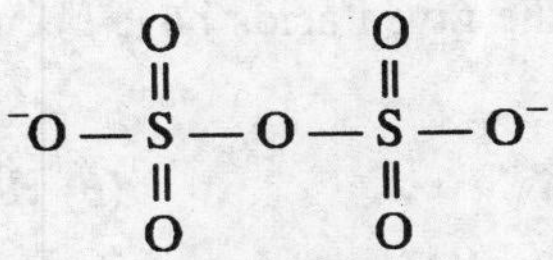

($S_2O_7^{2-}$ ion)

15. All the given species are basic since they can accept a proton (H^+), but it is only SO_3^{2-} ion that can act as a reducing agent because it can easily be oxidised to SO_4^{2-} ion.

$$SO_3^{2-} + H_2O \longrightarrow SO_4^{2-} + 2H^+ + 2e^-$$

16. Hypo $(Na_2S_2O_3)$ is used as photographer's fixer in photography because it removes the excess of unreacted (undecomposed) AgBr deposited on the surface of negative plate or film in the form of soluble colourless argento thiosulphate (complex compound), $Na_3[Ag(S_2O_3)_2]$

$$2Na_2S_2O_3 + AgBr \longrightarrow \underset{\text{Soluble complex}}{Na_3[Ag(S_2O_3)_2]} + NaBr$$

This hypo is used in photography for the fixation of picture.

17. H_2SO_4 has very corrosive action on skin because it dehydrates the skin and the absorption of water from the skin is accompanied by the release of heat (exothermic).

18. Conc. H_2SO_4 absorbs water from many coarbohydrates to from sulphuric acid hydrate, H_2SO_4, nH_2O.. Thus the acid dehydrates carbohydrates. It is due to the dehydrating property of H_2SO_4 that carbohyradtes get charred in this acid.

19. Dil. H_2SO_4 reacts with Na_2SO_3 to give H_2SO_3. Being unstable, H_2SO_3 gets decomposed to give SO_2.

$$H_2SO_4\ (\text{Dil.}) + Na_2SO_3 \longrightarrow Na_2SO_4 + H_2SO_3$$

$$H_2SO_3 \longrightarrow H_2O + SO_2$$

20. The structures of the given acids are given below:

(a) 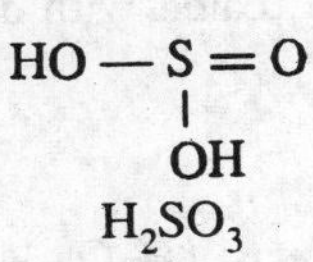

H_2SO_3

(b) 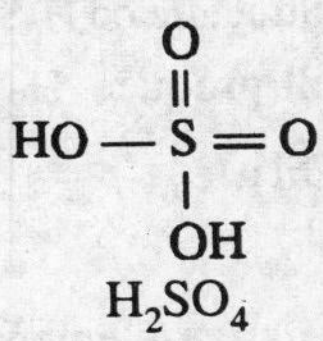

H_2SO_4

(c) 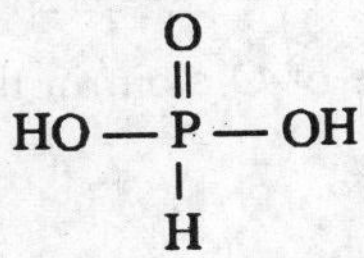

H_3PO_3

(d) 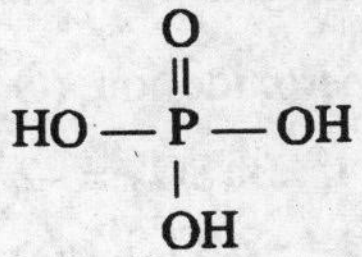

H_3PO_4

It is evident from the structures that H_2SO_4 has maximum number of oxo groups (M = O group) equal to 2. Other acids have one M = O group.

21. The structures of the given acids have been shown below:

(*a*) 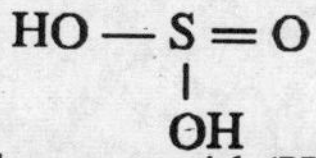

Sulphurous acid (H_2SO_3)

(*b*) 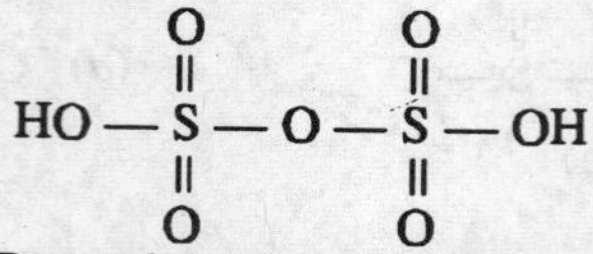

Pyrosulphuric acid ($H_2S_2O_7$)

(*c*) 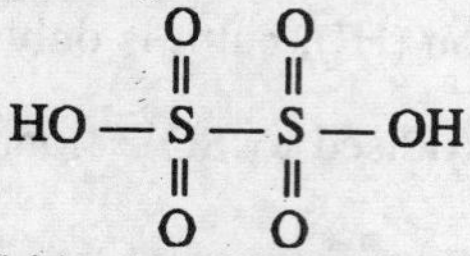

Dithionic acid ($H_2S_2O_6$)

(*d*) 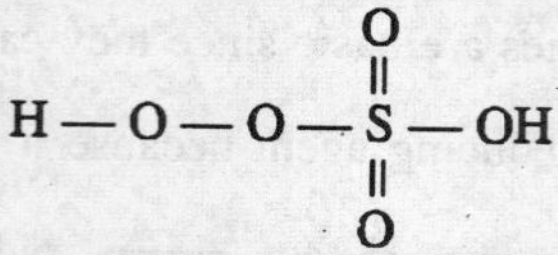

Caro's acid or peroxomono suplphuric acid (H_2SO_5)

Obviously it is only Caro's acid which contains peroxo linkage (– O – O – linkage).

22. In presence of dil. acids (e.g. HCl), $Na_2S_2O_3$ (S = + 6 and –2) gets decomposed to give SO_2 (S= + 4) and S(S = 0). This is because of the fact that the two S-atoms present in $Na_2 S_2O_3$ are in different oxidation states of + 6 (central S-atom) and – 2 (terminal S-atom) as shown in the structure of $Na_2S_2O_3$ molecule given in the margin.

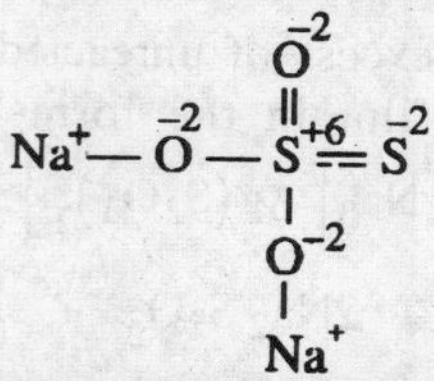

Fig. Oxidation states of two S-atom in $Na_2S_2O_3$ molecule

23. The structure of $H_2S_2O_8$ molecule is given below. Each of the four O-atoms present in four S = O bonds and each of the two O-atoms present in two O – H bonds has oxidation number equal to – 2. Since – O – O – is peroxide linkage, each O-atom in this linkage has oxidation number equal to – 1. Each H-atom has oxidation number equal to +1. Thus oxidation number of S-atom (= x) is given by:

$$6\times(-2)+2\times(-1)+2\times(+1)+2x=0$$

$$\therefore \quad x=+6$$

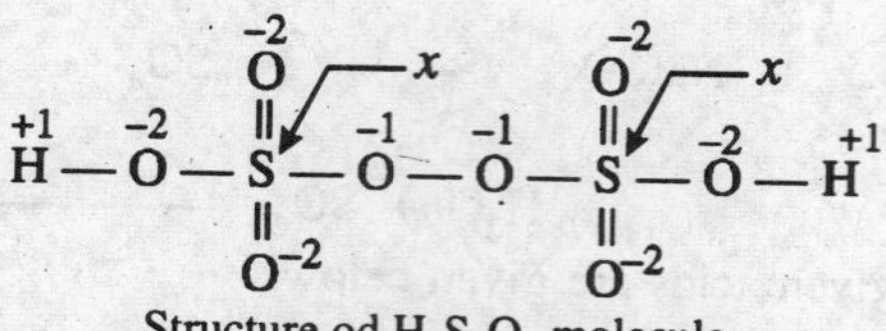

Structure od $H_2S_2O_8$ molecule.

24. Since BaO_2 is peroxide, it will produce H_2O_2 and $BaSO_4$ when treated with dil H_2SO_4.

$$BaO_2+H_2SO_4\,(\text{Dilute}) \longrightarrow \underbrace{H_2O_2 + BaSO_4}_{\text{Products}}$$

Obviously the most electronegative element present in the products $(H_2O_2 + BaSO_4)$ is O-atom. Since H_2O_2 has peroxide ion (O_2^{2-}), oxidation state of O-atom in this ion = –1. Oxidation state of O-atom in $BaSO_4$ = –2.

25. (*a*) In FeS, since O.S. of Fe = +2, O.S. of S-atom = – 2

(*b*) In FeS_2, if O.S. of Fe = +2, then O.S. of S-atom= – 1. On the other hand, if O.S. of Fe = +3, then O.S. of S-atom = – 3/2.

(*c*) The given salt is Na_2SO_4. Obviously O.S of S-atom = + 6

(*d*) The given salt is sodium hyposulphite ($Na_2S_2O_4$). Here O.S. of S-atom = +3.

34

Oxy acids of Halogens and Their Salts

MULTIPLE CHOICE QUESTIONS

1. Which of the following halogens does not form oxy acids?
   (a) F (b) Cl
   (c) Br (d) I
2. The correct order of oxidising power of oxy acids of chlorine is :
   (a) $HClO > HClO_2 > HClO_3 > HClO_4$
   (b) $HClO > HClO_3 > HClO_2 > HClO_4$
   (c) $HClO > HClO_4 > HClO_2 > HClO_3$
   (d) None of these.
3. The correct order of thermal stability of oxy acids of Cl, Br and I is :
   (a) $HClO > HBrO > HIO$
   (b) $HClO > HIO > HBrO$
   (c) $HBrO > HClO > HIO$
   (d) None of these.
4. The correct order of thermal stability of oxy acids of chlorine is :
   (a) $HClO < HClO_2 < HClO_3 < HClO_4$
   (b) $HClO > HClO_2 > HClO_3 > HClO_4$
   (c) $HClO < HClO_3 < HClO_2 < HClO_4$
   (d) $HClO > HClO_3 > HClO_2 > HClO_4$
5. Which of the following does not exist ?
   (a) $HClO$ (b) $HClO_2$
   (c) $HClO_3$ (d) $HClO_5$
6. In the oxy acids of Cl-atom, (Cl – O) π bond is :
   (a) $(p\pi - p\pi)$ π bond
   (b) $(d\pi - d\pi)$ π bond
   (c) $(d\pi - p\pi)$ π bond
   (d) None of these.
7. The acidic strength of HClO, HBrO and HIO decreases as :
   (a) $HBrO > HClO > HIO$
   (b) $HClO > HBrO > HIO$
   (c) $HIO > HBrO > HClO$
   (d) $HClO > HIO > HBrO$
8. The correct order of acid strength of the following oxy acids is :
   (a) $HClO_4 > HBrO_4 > HIO_4$
   (b) $HClO_4 < HBrO_4 < HIO_4$
   (c) $HClO_4 > HIO_4 > HBrO_4$
   (d) None of the above
9. The set with correct order of acidity is :
   (a) $HClO < HClO_2 < HClO_3 < HClO_4$
   (b) $HClO_4 < HClO_3 < HClO_2 < HClO$
   (c) $HClO < HClO_4 < HClO_3 < HClO_2$
   (d) $HClO_4 < HClO_2 < HClO_3 < HClO$
10. Which of the following is the strongest acid ?
    (a) $ClO_3(OH)$ (b) $ClO_2(OH)$
    (c) $SO(OH)_2$ (d) $SO_2(OH)_2$
11. Which of the following represents the correct order of increasing pK_a values of the given acids ?
    (a) $HClO_4 < HNO_3 < H_2CO_3 < B(OH)_3$
    (b) $HNO_3 < HClO_4 < B(OH)_3 < H_2CO_3$
    (c) $B(OH)_3 < H_2CO_3 < HClO_4 < HNO_3$
    (d) $HClO_4 < HNO_3 < B(OH)_3 < H_2CO_3$
    (*IAS Preliminary 1993*)
12. The correct order of stability of oxy anions of chlorine is :
    (a) $ClO^- < ClO_2^- < ClO_3^- < ClO_4^-$
    (b) $ClO^- < ClO_3^- < ClO_2^- < ClO_4^-$
    (c) $ClO^- < ClO_4^- < ClO_3^- < ClO_2^-$
    (d) None of these
13. The aqueous solution of which of the following has maximum pH value ?

(a) $NaClO$ (b) $NaClO_2$
(c) $NaClO_3$ (d) $NaClO_4$

14. The action of dil. and cold solution of NaOH on Cl_2 gas gives :
(a) $NaClO$ (b) $NaClO_2$
(c) $NaClO_3$ (d) None of these

15. The reaction :

$$3ClO^- (aq) \longrightarrow ClO_3^- (aq) + 2Cl^- (aq)$$

is an example of :
(a) oxidation reaction
(b) reduction reaction
(c) disproportionation reaction
(d) decomposition reaction

16. Ammonia, on heating with hypochlorite ion, can form :
(a) NO (b) NH_4Cl
(c) N_2H_4 (d) HNO_2

17. The anhydride of perchloric acid is :
(a) ClO_2 (b) Cl_2O
(c) Cl_2O_3 (d) Cl_2O_7

18. ClO_2 reacts with an alkali to give :
(a) Sodium chlorate
(b) Sodium chlorite
(c) Sodium chlorate and sodium chlorite
(d) None of the above.

19. When $KClO_3$ is heated at a high temperature, we get :
(a) $KCl + KClO_4$ (b) $KClO_2 + O_2$
(c) $KCl + O_2$ (d) $KCl + O_2 + O_3$

20. On boiling aqueous solution of $KClO_3$ with I_2, the product obtained is :
(a) KIO_3 (b) $KClO_4$
(c) KIO_4 (d) KCl

21. I_2 is liberated from sodium iodate by reacting with :
(a) Dil. H_2SO_4 (b) $KMnO_4$
(c) $NaHSO_3$ (d) HCl

22. Which of the following statements is correct for the reaction :

$$3NaClO \xrightarrow{\text{Heat}} NaClO_3 + 2NaCl$$

(a) It is a disproportionation reaction
(b) Oxidation state of Cl decreases as well as increases
(c) This reaction is used for the manufacture of halates
(d) All of the above.

23. When CO_2 is passed through a suspension of $CaOCl_2$, we get :
(a) $Ca(ClO)_2$ (b) $Ca(ClO_2)_2$
(c) $CaCl_2$ (d) Cl_2

24. Theoretical value of available chlorine in bleaching powder is :
(a) 56% (b) 50%
(c) 35% (d) 30%

25. Bleaching powder is an example if :
(a) A double salt (b) A complex salt
(c) An acidic salt (d) A mixed salt

26. Identify the false statement about bleaching powder.
(a) Amount of Cl_2 liberated when it is treated with excess of dil. acid is known as available chlorine.
(b) An old sample of bleaching powder contains 35-40% available chlorine.
(c) Bleaching powder is priced on the basis of its crystal structure.
(d) When stored for longer periods, bleaching powder changes to calcium chlorate, $Ca(ClO_3)_2$ and calcium chloride, $CaCl_2$.

27. Slaked lime reacts with Cl_2 to form :
(a) $Ca(OCl)_2$ (b) Ca(OCl) Cl
(c) $Ca(ClO_3)_2$ (d) $CaCl_2$

28. Bleaching powder reacts with dil. HCl to give :
(a) Cl_2 (b) $HClO_2$
(c) CaO (d) O_2

29. In presence of $CoCl_2$, bleaching powder decomposes to form
(a) $CaCO_3 + O_3$ (b) $ClO_2 + CaO$
(c) $Cl_2O + CaO$ (d) $CaCl_2 + O_2$

30. Bleaching properties of bleaching power are due to its :
(a) oxidising properties
(b) reducing properties
(c) basic prooperties
(d) disinfecting properties.

ANSWERS

1. (*a*)	**2.** (*a*)	**3.** (*c*)	**4.** (*a*)	**5.** (*d*)	**6.** (*c*)	**7.** (*b*)	**8.** (*a*)
9. (*a*)	**10.** (*a*)	**11.** (*a*)	**12.** (*a*)	**13.** (*a*)	**14.** (*a*)	**15.** (*c*)	**16.** (*b,c*)
17. (*d*)	**18.** (*c*)	**19.** (*c*)	**20.** (*a*)	**21.** (*c*)	**22.** (*d*)	**23.** (*d*)	**24.** (*a*)
25. (*d*)	**26.** (*c*)	**27.** (*b*)	**28.** (*a*)	**29.** (*d*)	**30.** (*a*)		

HINTS AND EXPLANATIONS

1. Since F-atom does not have *d*-orbitals, this atom cannot form $(d\pi - p\pi)$ π bond in its oxy acids.

2. Oxidising power of the oxy acids of halogens depends on the oxidation state (O.S.) or electronegativity of the central halogen atom. Since the electronegativity of Cl-atom (central halogen atom) in all the four acids remains the same, the oxidising power of the give acids decreases with the increase in O.S. of the central Cl-atom. Thus since O.S. of Cl-atom increases as +1 (in HClO) < +3 (in $HClO_2$) < +5 (in $HClO_3$) < +7 (in $HClO_4$), the oxidising power of the acids decreases as HClO > $HClO_2$ > $HClO_3$ > $HClO_4$. Thus $HClO_4$ is the weakest oxidising agent or least oxidising and HClO is the strongest oxidising agent or most oxidising.

3. Thermal stability of oxy acids of halogens depends on the O.S. or electronegativity of the cental halogen atom. Since in the given acids oxidation state of the central halogen atom remains the same (= +1), thermal stability of the given acids decreases with the decrease in electronegativity of halogen atom. Thus :

Oxy acids :	HClO		HBrO		HIO
Electronegativity of halogen atom :	Cl	>	Br	>	I
	——— Decreases ———→				
Thermal stability of oxy acids :	HClO	>	HBrO	>	HIO
	——— Decreases ———→				

4. Thermal stability of oxy acids of halogens depends on the O.S. or electronegativity of the central halogen atom. Since in the given acids electronegativity of Cl-atom remains the same, thermal stability of the given oxy acids increases with the increase of O.S. of Cl-atom from +1 (in HClO) to +7 (in $HClO_4$). Thus :

Oxy acids :	HClO	$HClO_2$	$HClO_3$	$HClO_4$
O.S. of Cl-atom :	+1	+3	+5	+7
	——— Increases ———→			

Thermal stability of oxy acids : HClO < $HClO_2$ < $HClO_3$ < $HClO_4$

——— Thermal stability increases ———→

Explanation. As oxidation state of Cl-atom increases, Cl — O bond in the acids becomes more and more covalent and hence thermal stability increases.

5. In HClO, $HClO_2$, $HClO_3$ and $HClO_5$ molecules, oxidation state (O.S.) of Cl-atom is +1, +3, +5 and +9 respectively. Since maximum oxidation state that Cl-atom can show is +7, $HClO_5$ molecule does not exist. Thus Cl-atom does not form $HClO_5$.

6. If we look at the structures of oxy acids of Cl atom, we find that $HClO_2$, $HClO_3$ and $HClO_4$ have one, two and three Cl = O bonds respectively. (Cl – O) σ bond is obtained by the overlap of sp^3 hybrid orbital of Cl-atom and 2p orbital of O-atom while (Cl–O) π bond results by the overlap of 3d orbital of Cl-atom and 2p orbital of O-atom. Thus (Cl–O) π bond is a $(d\pi - p\pi)$ π bond.

7. Acid strength of oxy acids of halogens depends on the oxidation state (O.S.) or electronegativity of the central halogen atom. Since the O.S. of halogen atom in the given acids remains the same equal to +1, the acid strength of the acids decreases with the decrease of electronegativity of halogen atom from Cl to I (Cl > Br > I). Thus the acid strength of the given acids decreases as HClO > HBrO > HIO.

8. Acid strength of oxy acids of halogens depends on the oxidation state (O.S.) or electronegativity of the central halogen atom. Since the O.S. of halogen atom in the given acids remains the same equal to +7, the acid strength of the acids decreases with the decrease in electronegativity of halogen atom from Cl to I (Cl > Br > I). Thus the acid strength of the given acids decreases as $HClO_4 > HBrO_4 > HIO_4$.

9. Acicity of oxy acids of halogens depends on the oxidation state (O.S.) or electronegativity of the central halogen atom. Since the electronegativity of the central Cl-atom in all the given acids remains the same, the acidity of the acids increases with the increase in O.S. of Cl atom in the given acids. Thus, since O.S. of Cl-atom increases as +1 (in HClO) < +3 (in $HClO_2$) < +5 (in $HClO_3$) < +7 (in $HClO_4$), the acidity of the acids also increases as HClO < $HClO_2 < HClO_3 < HClO_4$.

10. Since Cl is more electronegative than S, $HClO_4 - HClO_3$ pair is more acidic than $H_2SO_3 - H_2SO_4$ pair. Out of $HClO_4$ (O.S. of Cl = +7) and $HClO_3$ (O.S. of Cl = +5), $HClO_4$ is more acidic than $HClO_3$ as is evident from oxidation state of Cl-atom in these acids. Thus $HClO_4$ is the strongest acid.

11. Since O.S. of Cl-atom in $HClO_4$ is higher than that of N-atom in HNO_3 (Cl = +7 in $HClO_4$ and N = +5 in HNO_3), $HClO_4$ is stronger acid than HNO_3 ($HClO_4 > HNO_3$). Both H_2CO_3 and $B(OH)_3$ are weak acids. Now we know that CO_3^{2-} ion obtained from H_2CO_3 is stabilised due to resonnance and such type of resonance is not possible in $[B(OH)_4]^-$ ion obtained from $B(OH)_3$, H_2CO_3 is stronger acid than $B(OH)_3$

$$H_2CO_3\ (aq) \rightleftharpoons 2H^+\ (aq) + CO_3^{2-}\ (aq)$$

$$B(OH)_3\ (aq) + H_2O\ (l) \rightleftharpoons H^+\ (aq) + [B(OH)_4]^-\ (aq)$$

Thus the overall acid strength of the given acids is in the order :

$$\overset{3+}{B}(OH)_3 < H_2\overset{4+}{C}O_3 < H\overset{5+}{N}O_3 < H\overset{7+}{Cl}O_4$$

and hence pK_a values of these acids will be in the reverse order :

$$B(OH)_3 > H_2CO_3 > HNO_3 > HClO_4$$

12. Since the number of O-atoms present in the given oxyanions increases from ClO^- to ClO_4^- (ClO^- = 1, ClO_2^- =2, ClO_3^- =3 and ClO_4^- =4), the dispersal of negative charge on the oxy anions through ($p\pi-d\pi$) back bonding also increases from ClO^- to ClO_4^- ion and hence the stability of oxy anions also increases from ClO^- to ClO_4^-. Thus the stability of oxyanions increases as $ClO^- < ClO_2^- < ClO_3^- < ClO_4^-$

13. The oxy anions derived from the given salts are ClO^-, ClO_2^-, ClO_3^- and ClO_4^- respectively. All these oxy anions show basic character, since these anions can accept a proton (H^+) to form their corresponding conjugate acids namely HClO, $HClO_2$, $HClO_3$ and $HClO_4$ respectively. The basicity of these ions decreases with the increase of O.S. of Cl-atom from +1 (in ClO^-) to +7 (in ClO_4^-). Thus the basicity is in the order :

$$\overset{+1}{[ClO]^-} > \overset{+3}{[ClO_2]^-} > \overset{+5}{[ClO_3]^-} > \overset{+7}{[ClO_4]^-}$$

Thus since NaClO contains the most basic ClO^- ion, the aqueous sulution of this salt is the most basic and hence has maximum pH value.

14. Dil. and cold solution of NaOH gives sodium hypochlorite (NaClO) with Cl_2.

$$2NaOH \text{ (Dil. and cold solution)} + Cl_2 \longrightarrow NaCl + NaClO + H_2O$$

Sodium chlorate ($NaClO_3$) is obtained when conc. and hot solution of NaOH reacts with Cl_2.

$$6NaOH \text{ (Conc. and hot solution)} + 3Cl_2 \longrightarrow 5NaCl + NaClO_3 + 3H_2O$$

15. In this reaction, since O.S. of Cl-atom equal to +1 in ClO^- ion simultaneously increases from +1 to +5 in ClO_3^- ion and decreases from +1 to –1 in Cl^- ion, this reaction is a disproportionation reaction.

$$\underset{(Cl = +1)}{3ClO^- (aq)} \longrightarrow \underset{(Cl = +5)}{ClO_3^- (aq)} + \underset{(Cl = -1)}{2Cl^- (aq)}$$

16. Hypo chlorite ion (ClO^-) is an oxidising agent. When NH_3 is heated with ClO^- ion, NH_3 is oxidised to hydrazine ($NH_2 . NH_2$ or N_2H_4) and ClO^- ion is reduced to Cl^- ion (NH_4Cl)

$$\underset{(N = -3)}{3NH_3} + \underset{(Cl = +1)}{OCl^-} \longrightarrow \underset{(N = -2)}{N_2H_4} + OH^- + \underset{(Cl = -1)}{NH_4Cl}$$

Actually the above reaction takes place through the following steps :

(*i*) $NH_3 + OCl^- \longrightarrow OH^- + NH_2Cl$ (*chloramine*)

(*ii*) $NH_2Cl + 2NH_3 \longrightarrow N_2H_4 + NH_4Cl$

$$3NH_3 + OCl^- \longrightarrow \underset{(c)}{N_2H_4} + OH^- + \underset{(b)}{NH_4Cl.}$$

17. The anhydride of perchloric acid ($HClO_4$) is obtained by eliminating H_2O molecule from $HClO_4$.

$$2HClO_4 \xrightarrow{-H_2O} Cl_2O_7 + H_2O$$

Thus Cl_2O_7 is the anhydride of $HClO_4$. ClO_2 is the mixed anhydride of chlorous acid ($HClO_2$) and chloric acid ($HClO_3$), since this oxide gives the mixture of these acids with H_2O.

$$2ClO_2 + H_2O \longrightarrow HClO_2 + HClO_3$$

18. The action of ClO_2 on an alkati is used for the preparation chlorite along with chlorate.

$$ClO_2 + 2NaOH \longrightarrow \underset{\text{Sod. chlorite}}{NaClO_2} + \underset{\text{Sod. chlorate}}{NaClO_3} + H_2O$$

19. When $KClO_3$ (C. = +5) is heated at a high temperature, it undergoes disproportionation into KCl (Cl = –1) and $KClO_4$ (Cl = +7)

$$\underset{(Cl = +5)}{4KClO_3} \xrightarrow[\text{temperature}]{\text{Heat at high}} \underset{(Cl = -1)}{KCl} + \underset{(Cl = +7)}{3KClO_4}$$

or $$4ClO_3^- \longrightarrow Cl^- + ClO_4^-$$

However when heated to 400°C, KCl and O_2 are produced.

$$3KClO_3 \xrightarrow{400°\ C} 2KCl + 3O_2.$$

20. $KClO_3$ is a powerful oxidising agent. It oxidises I_2 to KIO_3 in presence of little HNO_3 and is itself reduced to Cl_2.

$$\underset{(Cl = +5)}{2KClO_3} + \underset{(I = 0)}{I_2} \longrightarrow \underset{(I = +5)}{2KIO_3} + \underset{(Cl = 0)}{Cl_2}$$

21. $$\underset{(I = +5)}{2NaIO_3} + \underset{(S = +4)}{5NaHSO_3} \longrightarrow \underset{(S = +6)}{3NaHSO_4} + H_2O + \underset{(S = +6)}{NaSO_4} + \underset{(I = 0)}{I_2}$$

In this reaction, $NaIO_3$ ($I = +5$) oxidises $NaHSO_3$ ($S = +4$) to $NaHSO_4$ ($S = +6$) and Na_2SO_4 ($S = +6$) and is itself reduced to I_2 ($I = 0$).

22. In the given reaction since O.S. of Cl-atom of +1 in NaClO is increased to +5 in $NaClO_3$ and is decreased to –1 in NaCl, this is a disproportionation reaction.

$$[\underset{(Cl = +1)}{NaClO} \longrightarrow \underset{(Cl = -1)}{NaCl} + O] \times 2$$

$$\underset{(Cl = +1)}{NaClO} + 2O \longrightarrow \underset{(Cl = +5)}{NaClO_3}$$

$$\underset{(Cl = +1)}{3NaClO} \xrightarrow{\Delta} \underset{(Cl = -1)}{2NaCl} + \underset{(Cl = +5)}{NaClO_3}$$

This reaction shows that chlorate (halate) can be obtained by heating hypochlorite (hypohalite).

23. $$CaOCl_2 + CO_2 \longrightarrow CaCO_3 + Cl_2$$

24. Theoretical value of available Cl_2 in bleaching powder is that weight of Cl_2 which is obtained when 100g of bleaching powder ($CaOCl_2$) reacts with dil. acids (*e.g.* H_2CO_3, H_2SO_4 etc.)

$$CaOCl_2 + H_2CO_3 \longrightarrow CaCO_3 + H_2O + Cl_2$$
$$CaOCl_2 + H_2SO_4 \longrightarrow CaSO_4 + H_2O + Cl_2$$

It is evident from the above equations that since $40 + 16 + 35.5 \times 2 = 127$g of $CaOCl_2$ gives $35.5 \times 2 = 71$g of available Cl_2, 100g of $CaOCl_2$ will give $\frac{71 \times 100}{127} \approx 56$g. Thus the theoretical value of available Cl_2 in bleaching powder is 56%.

25. According to Odling (1861), bleaching powder is a mixed salt of calcium chloride ($CaCl_2$) and calcium hypochlorite [$Ca(ClO)_2$].

$$CaCl_2 + Ca(ClO)_2 \longrightarrow Ca_2Cl_2(ClO)_2 \text{ or } 2CaCl(ClO) \text{ or } 2CaOCl_2$$

Thus bleaching powder is also called *calcium chloro hypochlorite*, CaCl (ClO) or *calcium oxy chloride*, $CaOCl_2$

26. (*a*) $CaOCl_2 + CH_3COOH \longrightarrow Ca(CH_3COO)_2 + H_2O + Cl_2$ (*Available chlorine*)

$CaOCl_2 + H_2CO_3 \longrightarrow CaCO_3 + H_2O + Cl_2$ (*Available chlorine*)

(*c*) Bleaching powder is priced on the basis of % of available Cl_2 obtained by the action of dil. acids on bleaching powder.

(*d*) $6CaOCl_2 \xrightarrow{\text{Decomposition}} Ca(ClO_3)_2 + 5CaCl_2$

27. Slaked lime, $Ca(OH)_2$ reacts with Cl_2 to give bleaching powder, CaCl(OCl) or $CaOCl_2$.

$$Ca(OH)_2 + Cl_2 \longrightarrow CaCl(ClO) \text{ or } CaOCl_2$$

28. When bleaching powder, $CaOCl_2$ is treated with dil. HCl, Cl_2 (called available chlorine) is obtained.

$$CaOCl_2 + 2HCl \longrightarrow CaCl_2 + H_2O + Cl_2$$

29. In presence of $CoCl_2$, bleaching powder ($CaOCl_2$) decomposes and gives $CaCl_2$ and O_2

$$2CaOCl_2 \xrightarrow[\text{in presence of } CoCl_2]{\text{Decomposition}} 2CaCl_2 + O_2$$

30. In presence of slight amount of dil. acids. $CaOCl_2$ loses nescent oxygen.

$$2CaOCl_2 + H_2SO_4 \longrightarrow CaCl_2 + CaSO_4 + 2HCl + 2O$$

Nescent oxygen thus produced renders the coloured matter colourless.

$$\text{Coloured matter} + O \longrightarrow \text{Colourless matter}$$

35

Fixation of Nitrogen and Fertilisers

MULTIPLE CHOICE QUESTIONS

1. Superphosphate of lime contains:

(*a*) $Ca_3(PO_4)_2$

(*b*) $CaHPO_4$

(*c*) $Ca_3(PO_4)_2 + H_3PO_4$

(*d*) $Ca(H_2PO_4)_2$

2. Which of the following is superphosphate of lime?

(*a*) $Ca(H_2PO_4)_2$

(*b*) $Ca(HPO_4)$

(*c*) $Ca_3(PO_4)_3$

(*d*) $Ca(H_2PO_4)_2 . H_2O + 2(CaSO_4 . 2H_2O)$

3. Fixation of nitrogen means:

(*a*) Reaction of N_2 with O_2

(*b*) Conversion of free atmospheric nitrogen into nitrogen compounds.

(*c*) Decomposition of nitrogenous compounds to free N_2.

(*d*) The action of denitrofying becteria on nitrogen compounds.

4. An important method for the fixation of nitrogen is:

(*a*) Haber's process

(*b*) Solvay's process

(*c*) Deacon's process

(*d*) Frasch process

5. Becteria convert molecular nitrogen into which of the following?

(*a*) HNO_3 (*b*) Amino acids

(*c*) NO_2 (*d*) NH_3

6. Superphosphate of lime is obtained from the reaction of:

(*a*) $CaCO_3$ with H_3PO_4

(*b*) $Ca_3(PO_4)_2$ with HCl

(*c*) $Ca_3(PO_4)_2$ with H_2SO_4

(*d*) Bones with gypsum

7. Superphosphate of lime is used in:

(*a*) Cement industry

(*b*) Glass industry

(*c*) Agriculture

(*d*) Metallurgy

8. Nitrolim is prepared by heating:

(*a*) CaO with N_2 (*b*) CaO with O_2

(*c*) CaC_2 with N_2 (*d*) CaC_2 with O_2

9. Superphosphate of lime is a mixture of $Ca(H_2PO_4)_2 . H_2O$ and

(*a*) $CaSO_4$

(*b*) $CaSO_4 . \frac{1}{2} H_2O$

(*c*) $CaSO_4 . H_2O$

(*d*) $CaSO_4 . 2H_2O$

10. The lightning bolts in the atmosphere cause the formation of:

(*a*) NO (*b*) NH_3

(c) NH_4OH (*d*) NH_2OH

11. In the soil, $CaCN_2$ which is used as a fertiliser, is converted into --- by atmospheric CO_2.

(*a*) Urea (*b*) NO_2

(*c*) NO (*d*) CO

12. Which of the following fertilisers is NPK fertiliser?
(*a*) Nitrogeneous (*b*) Phosphatic
(*c*) Mixed (*d*) All of these

13. Which of the following is a NP fertiliser?
(*a*) $(NH_4)_2(HPO_4)$
(*b*) $(NH_4)_2SO_4$
(*c*) $Ca(H_2PO_4)_2.H_2O$
(*d*) K_2SO_4

14. Which of the following nitrogen fertilisers has the highest nitrogen percentage?
(*a*) $CaCN_2$ (*b*) Urea
(*c*) NH_4NO_3 (*d*) $(NH_4)_2SO_4$

15. Which of the following nitrogen fertilisers has the lowest nitrogen percentage?
(*a*) Calcium ammonium nitrate (CAN), $Ca(NO_3)_2 . NH_4NO_3$
(*b*) Urea, $H_2N.CO.NH_2$
(*c*) Ammonium nitrate, NH_4NO_3
(*d*) Ammonium sulphate, $(NH_4)_2SO_4$

16. Which of the following nitrogenous fertilisers has greater nitrogen percentage than other fertilisers.
(*a*) $(NH_4)_2SO_4$
(*b*) $(NH_3)_3PO_4$
(*c*) $Ca(NO_3)_2.NH_4NO_3$
(*d*) Urea

17. Nitrophosphate is:
(*a*) $Ca(H_2PO_4)_2 + 2Ca(NO_3)_2$
(*b*) $Ca(H_2PO_4)_2.H_2O+(CaSO_4.2H_2O)$
(*c*) $Ca(H_2PO_4)_2.H_2O$
(*d*) $Ca_3(PO_4)_2$

18. Repeated use of which of the following fertilisers increases the acidity of the soil?
(*a*) $(NH_4)_2SO_4$
(*b*) $Ca(NO_3)_2.NH_4NO_3$
(*c*) Urea
(*d*) Superphosphate of lime

19. When lightning flash is produced through air, the compound (s) formed is (are):
(*a*) Nitrous oxide (*b*) Nitric oxide
(*c*) Nitric acid (*d*) Nitrogen dioxide

20. Calcium cyanamide on treatment with steam under pressure gives NH_3 and
(*a*) $CaCO_3$ (*b*) $Ca(OH)_2$
(*c*) CaO (*d*) $CaHCO_3$

21. When equal weights of urea and $(NH_4)_2SO_4$ are mixed, the mixture contains:
(*a*) Less nitrogen than $(NH_4)_2SO_4$
(*b*) As much nitrogen as $(NH_4)_2SO_4$
(*c*) Twice the amount of nitrogen present in $(NH_4)_2SO_4$
(*d*) More than twice the amount of nitrogen present in $(NH_4)_2SO_4$

22. The percentage of nitrogen in urea is about:
(*a*) 70 (*b*) 63
(*c*) 47 (d) 28

23. Complete manure is that which supplies...to the soil.
(*a*) S, K and N (*b*) N, K and P
(*c*) S and N (*d*) S, N and P

24. Which of the following is not used as a phosphatic fertiliser?
(*a*) Superphosphate of lime
(*b*) Guano
(*c*) Phosphatic slag
(*d*) Phosphorite mineral

25. The fertility of the soil can be increased by using
(*a*) Mixed fertilisers
(*b*) Nitrogenous fertilisers

(*c*) Phosphatic fertilisers
(*d*) None of these

26. Which of the following elements are used for fertilisers?
(*a*) N, P and K (*b*) Na, P and K
(*c*) Mg, P and K (*d*) Mg, Na and P

27. Which of the following is not a nitrogenous fertiliser?
(*a*) Superphosphate of lime
(*b*) CAN
(*c*) Urea
(*d*) Calcium cynamide

28. Triple superphosphate is prepared by the action of -----on phosphorite mineral:
(*a*) HCl (*b*) H_2SO_4
(*c*) H_3PO_4 (*d*) H_3PO_3

ANSWERS

1. (*d*)	**2.** (*d*)	**3.** (*b*)	**4.** (*a*)	**5.** (*d*)	**6.** (*c*)	**7.** (*c*)	**8.** (*c*)
9. (*d*)	**10.** (*a*)	**11.** (*a*)	**12.** (*c*)	**13.** (*a*)	**14.** (*b*)	**15.** (*d*)	**16.** (*d*)
17. (*a*)	**18.** (*a, b*)	**19.** (*b,c,d*)	**20.** (*a*)	**21.** (*d*)	**22.** (*c*)	**23.** (*b*)	**24.** (*d*)
25. (*a*)	**26.** (*a*)	**27.** (*a*)	**28.** (*c*)				

HINTS AND EXPLANATIONS

1. Superphosphate of lime is a fertiliser and is a mixture of one molecule of calcium dihydrogen monohydrate $\left[Ca(H_2PO_4)_2 \cdot H_2O\right]$ and two molecules of calcium sulphate dihydrate $\left[2(CaSO_4 . 2H_2O)\right]$ (gypsum). It is therefore, represented as $Ca(H_2PO_4)_2 . H_2O + 2(CaSO_4 . 2H_2O)$. Obviouslyj it contains $Ca(H_2PO_4)_2$.

2. See explanation of question number 1

3. Fixation of nitrogen is the process in which the atmospheric nitrogen is converted into its useful compounds either by natural methods or by artificial methods.

4. A large amount of atmospheric nitrogen is fixed in the form of NH_3 (i.e. is converted into NH_3) by Haber's process.

5. A large amount of atmospharic N_2 is converted into HNO_3 by lightning.

$$\underbrace{N_2 + O_2}_{\text{(From atmosphere)}} \xrightarrow[\text{lighting}]{\text{During}} 2NO$$

$$2NO + O_2 \longrightarrow 2NO_2$$

$$4NO_2 + O_2 + 2H_2O \longrightarrow 4HNO_3$$

HNO_3 thus obtained is carried by rain water to the soil where it combines with $CaCO_3$ and $MgCO_3$ present in the soil to form their respective nitrates. These nitrates are converted into NH_3 by de-nitrifying bacteria.

$$\text{Nitrate} \rightarrow \text{Nitr i te} \rightarrow NO_2 \rightarrow N_2 \rightarrow NH_3$$

NH_3 thus produced is used by plants for producing proteins.

6. Superphosphate of lime, $Ca(H_2PO4)_2 . H_2O + 2(CaSO_4 . 2H_2O)$ is obtained by treating calcium phosphate, $Ca_3(PO_4)_2$ (obtained from phosphate rock or bone ash) with calculated quantity of conc. H_2SO_4 (*i.e.* H_2SO_4 obtained from Lead chamber process).

$$\mathrm{Ca_3(PO_4)_2 + 2H_2SO_4 + 5H_2O \longrightarrow \underbrace{Ca(H_2PO_4)_2.H_2O + 2(CaSO_4.2H_2O)}_{\text{Superphosphate of lime}}}$$

7. Superphosphate of lime is used as a fertiliser in agriculture. Since $Ca(H_2PO_4)_2$ which is a constituent of the fertiliser is soluble in water, it $\left[i.e.\,Ca(H_2PO_4)_2\right]$ is provided to the soil in which it is assimilated readily.

8. Nitrolim is a mixture of calcium cyanamide $(CaCN_2)$ and carbon and is prepared by heating CaC_2 in an atmosphere of N_2.

$$\mathrm{CaC_2 + N_2 \xrightarrow{\Delta} \underbrace{CaCN_2 + C}_{\text{Nitrolim}}}$$

9. Superphosphate of lime is a mixture of $Ca(H_2PO_4)_2.H_2O$ and gypsum, $CaSO_4.2H_2O$. Hence it is representated as $\left[Ca(H_2PO_4)_2.H_2O + 2(CaSO_4.2H_2O)\right]$.

10. During the lightning bolts, atmospheric N_2 and O_2 combine together and form NO.

$$\underbrace{\mathrm{N_2 + O_2}}_{\text{(From atomosphere)}} \xrightarrow[\text{ting bolts}]{\text{During ligh-}} \mathrm{2NO}$$

11. $CaCN_2 + CO_2 + H_2O \longrightarrow CaCO_3 + H_2N{\cdot}CO{\cdot}NH_2$ (Urea)

12. Mixed fertiliser is that which can supply N, P and K to the soil. Thus mixed fertiliser is called NPK fertiliser.

13. It is only $(NH_4)_2(HPO_4)$ which contains N and P both. Thus it is a NP type fertiliser.

17. Nitrophosphate is a mixture of one molecule of $Ca(H_2PO_4)_2$ and two molecules of $Ca(NO_3)_2$. Thus it is $Ca(H_2PO_4)_2 + 2Ca(NO_3)_2$. It is also called superphosphate nitrate.

18. The repeated use of $(NH_4)_2SO_4$ and $Ca(NO_3)_2.NH_4NO_3$ produce SO_4^{2-} and NO_3^- ions respectively in the soil. These ions make the soil acidic.

19. When lightning flash is produced through air, N_2 and O_2 present in air combine to form nitric oxide (NO) which then combines with more O_2 to form NO_2 (Nitrogen dioxide).

$$\underbrace{\mathrm{N_2 + O_2}}_{\text{(From air)}} \xrightarrow{\Delta} \underset{(b)}{\mathrm{2NO}}, \quad \mathrm{2NO + O_2 \longrightarrow} \underset{(d)}{\mathrm{2NO_2}}$$

NO_2 thus produced combines with more O_2 in presence of H_2O to form nitric acid (HNO_3)

$$\mathrm{4NO_2 + O_2 + 2H_2O \longrightarrow} \underset{(c)}{\mathrm{4HNO_3}}$$

20. $\underset{\text{Cal. cyanamide}}{CaCN_2} + 3H_2O$ (Superheated steam) $\longrightarrow$ $CaCO_3 + 2NH_3$

22. Molecular mass of urea $(NH_2.CO.NH_2) = 14\times2+4\times1+12+16 = 60$. Thus since 60g of urea contains 28g of N_2, 100g of urea contains $\frac{28\times100}{60} = 46.66 \approx 47\%$

24. Phosphorite, $Ca_3(PO_4)_2$ is a mineral of phosphorus. Since it is sparingly soluble in water, it does not serve as a source of phosphorus for plants. It is, therefore, not used as a fertiliser of phosphorus. It is converted in soluble materials like superphosphate of lime etc which are used as phosphatic fertilisers. Guano is animal waste (natural manure).

25. Nitrogenous fertilisers supply only nitrogen to the soil or plants and phosphatic fertilisers supply only phosphorus to the plants while mixed fertilisers supply two or more essential elements (e.g. K,N,P) to the soil. Therefore mixed fertilisers should be used in suitable proportion to increase the fertility of the soil.

27. Since superphosphate of lime $[Ca(H_2PO_4)_2.H_2O + 2(CaSO_4.2H_2O)]$ supplies phosphorus to the soil, it is a phosphatic fertiliser. On the other hand, CAN (calcium amonium nitrate, $Ca(NO_3)_2.NH_4NO_3$), urea $(NH_2.CO.NH_2)$ and calcium cynamide $(CaCN_2)$ all supply nitrogen to the soil and hence are nitrogenous fertilisers.

28. Triple superphosphate is $Ca(H_2PO_4)_2$. It is prepared by the action of aqueous solution of H_3PO_4 on phosphorite mineral (Phosphate rock), $Ca_2(PO_4)_3$.

$$Ca_3(PO_4)_2 + 4H_3PO_4 \longrightarrow 3Ca(H_2PO_4)_2$$

36

Inorganic Polymers

MULTIPLE CHOICE QUESTIONS

1. Hydrolysis of borazine gives :

(*a*) B_2O_3 (*b*) $B(OH)_3$

(*c*) B_2H_6 (*d*) $H_3N \longrightarrow BH_3$

2. Which of the following bonds is present in silicones ?

(*a*) Si–Si–Si–Si (*b*) Si–C–Si–O–Si

(*c*) Si–C–Si–C–Si (*d*) –Si–O–Si–O–Si–

3. An alkyl halide obtained from CH_4 by monosubstitution reacts with a group 14 element Y at 570 K with Cu as a catalyst and produces a dialkyl dichloro compound Z. The compound Z on hydrolysis gives another compound which is water repellant and quite inert. The dioxide of Y is acidic in nature. The species Y and Z are :

(*a*) C, $C_2H_4Cl_2$

(*b*) Sn, $(CH_3)_2SnCl_2$

(*c*) $Si(C_2H_5)_2$, $SiCl_2$

(*d*) Si, $(CH_3)_2SiCl_2$

4. Hydrolysis of trialkyl chloro silane gives :

(*a*) R_3 Si–O–SiR_3 dimer

(*b*) Cyclic (ring) silicone

(*c*) Cross-linked silicone

(*d*) None of these.

5. The strength and inetness of silicones are related to :

(*a*) Stability of silica-like skeleton of – Si – O – Si – O – Si –

(*b*) Very high bond energy of Si – O bond

(*c*) High strength of Si–C bond

(*d*) All of the above.

6. Hydrolysis of dialkyl dichloro silane gives :

(*a*) Linear silicone

(*b*) SiO_2

(*c*) Cross-linked silicone

(*d*) None of these.

7. The number of P–O–P bonds in cyclic metaphosphoric acid is :

(*a*) Zero (*b*) 2

(*c*) 3 (*d*) 4.

8. Which of the following is a cyclic oxo acid ?

(*a*) $H_5P_3O_{10}$ (*b*) $H_6P_4O_{13}$

(*c*) $H_5P_5O_{15}$ (*d*) $H_7P_5O_{16}$

9. Sodium hexametaphosphate is used :

(*a*) As fertiliser

(*b*) For softening water

(*c*) In fruit ripening

(*d*) None of the above.

10. Which of the following is obtained when P_4O_{10} is is heated with Na_2O at 1000° C ?

(*a*) Sodium tripolyphosphate ($Na_5P_3O_{10}$)

(*b*) Sodium tripolyphosphate hexahydrate ($Na_5P_3O_{10}.6H_2O$)

(*c*) Sodium dihydrogen phosphate (NaH_2PO_4)

(*d*) Sodium hydrogen phosphate (Na_2HPO_4).

11. Sodium hexametaphosphate is known as :

(*a*) Calgon (*b*) Permutit

(*c*) Natalite (*d*) Nitrolim

12. Using chlorobenzene (C_6H_5Cl) as solvent, the reagents needed for the synthesis of borazine are :

(*a*) NH_4Cl, BCl_3

(*b*) NH_4Cl, BCl_3, $NaBH_4$

(*c*) NH_4Cl, $NaBH_4$

(*d*) NH_3, BCl_3

ANSWERS

1. (*b*) **2.** (*d*) **3.** (*d*) **4.** (*a*) **5.** (*d*) **6.** (*a*) **7.** (*c*) **8.** (*c*)
9. (*b*) **10.** (*a*) **11.** (*a*) **12.** (*b*)

HINTS AND EXPLANATIONS

1. Hydrolysis of borazine ($B_3N_3H_6$) gives $B(OH)_3$.

$$B_3N_3H_6 + 9H_2O \xrightarrow{\text{Hydrolysis}} 3B(OH)_3 + 3NH_3 + 3H_2$$

2. Silicones are organo-silicon compounds which contain repeated R_2SiO units (R = alkyl or phenyl group) which are linked together through O-atoms. Thus silicones have the following bonds.

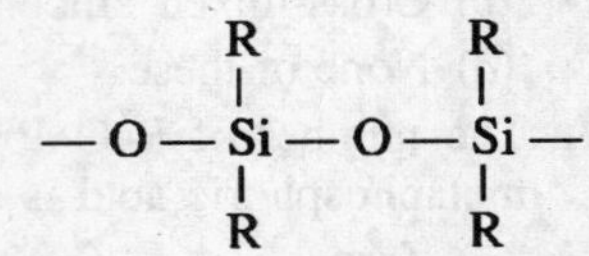

3. Since the alkyl halide has been obtained from CH_4 by monosubstitution, the alkyl halide may be CH_3F or CH_3Cl or CH_3 Br or CH_3I. Now since this alkyl halide reacts with group 14 element to give dialkyl dichloro compound, this alkyl halide must be methyl chloride (CH_3Cl) and dialkyl dichloro compound must be $(CH_3)_2SiCl_2$ (Z). Also group 14 element may be C or Si, since both these elements give acidic oxides namely CO_2 and SiO_2 respectively. Since hydrolysis of Z [*i.e.* $(CH_3)_2$ $SiCl_2$] gives a water repellent and chemically inert compound, this water repellent compound must be polymeric compound of Si. Carbon cannot give such a polymeric compound. Thus Y must be Si. The reactions involved can be written as :

$$2CH_4 + X_2 \ (X = F, Cl, Br, I) \longrightarrow 2CH_3X + H_2$$

$$2CH_3Cl + \underset{(Y)}{Si} \xrightarrow[\text{Cu as catalyst}]{570K} \underset{\substack{\text{Dimethyl dichloro silane} \\ (Z)}}{(CH_3)_2SiCl_2}$$

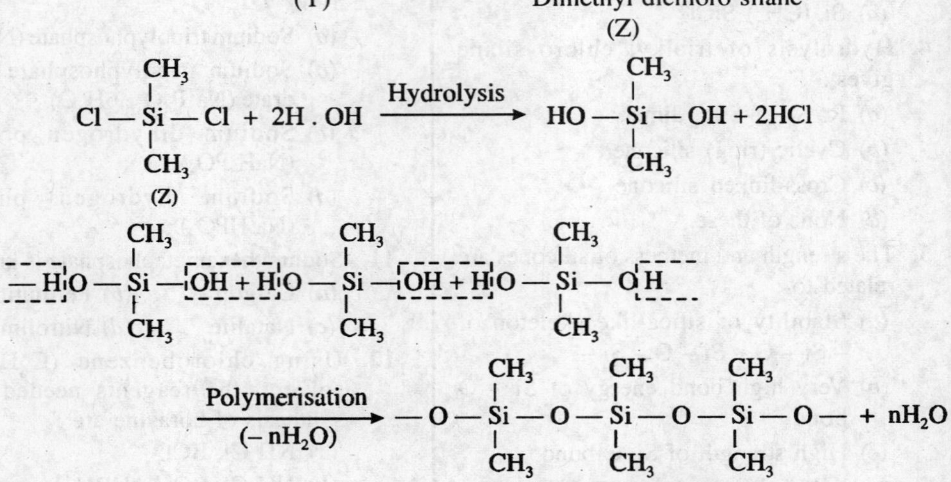

Polymeric compound
(Linear silicone)

4. Trialkyl chloro silane is R_3SiCl. The hydrolysis of this compound gives trialkyl silanol, R_3Si (OH) which gets dimerised to give the dimer, $R_3Si - O - SiR_3$.

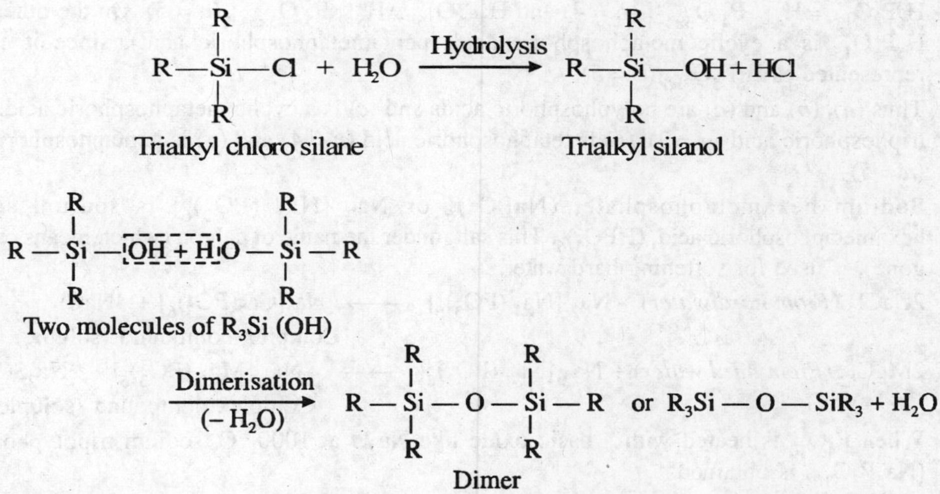

6. Dialkyl dichloro silane is R_2SiCl_2. The hydrolysis of this compound gives dialkyl silandiol, $R_2Si(OH)_2$ which gets polymerised to give the linear silicone.

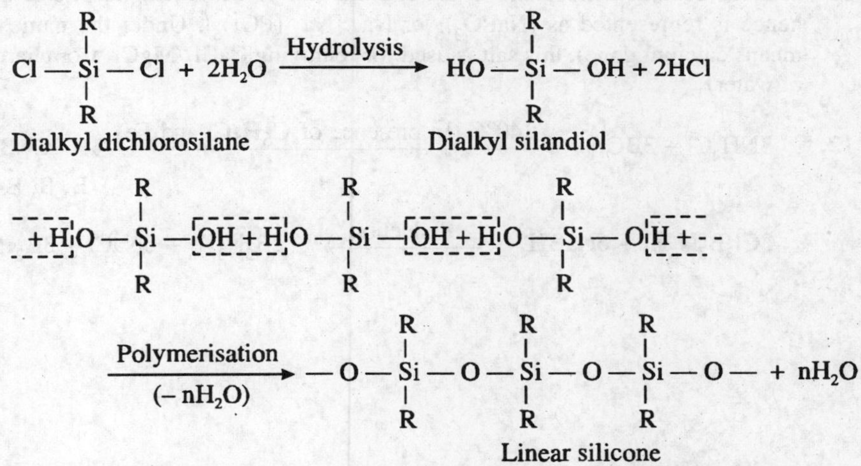

7. The cyclic metaphosphoric acid containing the smallest ring is trimetaphosphoric acid, $(HPO_3)_3$ whose structure is given in the margin. It is obvious form the structure that the acid has three P – O – P bonds.

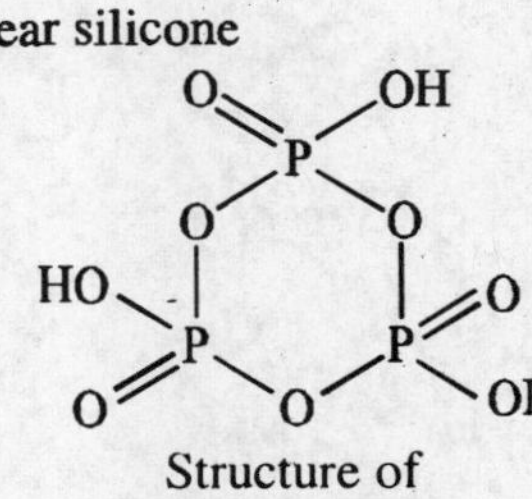

Structure of trimetaphsophoric acid, $(HPO_3)_3$

8. All the given acids are oxo acids of phosphorus. We know that cyclic phosphoric acids (also called cyclic metaphosphoric acids) are represented by the general formula $H_nP_nO_{3n}$ or $(HPO_3)_n$ while polyphosphoric acids (also called linear chain polyphosphoric acids) are represented by the general formula $H_{n+2} P_n O_{3n+1}$. Obviously $H_5 P_3 O_{10}$, $H_6P_4O_{13}$ and $H_7P_5O_{16}$ are polyphosphoric acids as shown $H_5P_3O_{10} \equiv H_{3+2} P_3 O_{3\times3+1}$ (n = 3),

$H_6P_4O_{13} \equiv H_{4+2}\,P_4\,O_{3\times4+1}$ (n = 4) and $H_7P_5O_{16} \equiv H_{5+2}\,P_5\,O_{3\times5+1}$ (*n* = 5). On the other hand $H_5P_5O_{15}$ is a cyclic metaphosphoric acid (pentametaphosphoric acid), since it can be represented as $H_5P_5O_{3\times5}$ (*n* = 5).

Thus (*a*), (*b*) and (*d*) are polyphosphoric acids and (*c*) is a cyclic metaphosphoric acid. (*a*) is triphosphoric acid (*n* = 3), (*b*) is tetaphosphoric acid (*n* = 4) and (*d*) is pentaphosphoric acid (*n* = 5)

9. Sodium hexametaphosphate, $(NaPO_3)_6$ or $Na_2\,[Na_4\,(PO_3)_6]$ is sodium salt of hexametaphosphoric acid, $(HPO_3)_6$. This salt, under the name of calgon (which means calcium gone), is used for softening hard water.

$2CaCl_2$ (*From hard water*) + $Na_2\,[Na_4\,(PO_3)_6] \longrightarrow Na_2[Ca_2(PO_3)_6] + 4NaCl$
Complex compound (soluble)

$2MgCl_2$ (*From hard water*) + $Na_2\,[Na_4\,(PO_3)_6] \longrightarrow Na_2\,[Mg_2\,(PO_3)_6] + 2Na_2SO_4$
Complex compound (soluble)

10. When P_4O_{10} is heated with a basic oxide like Na_2O at 1000° C, sodium tripolyphosphate ($Na_5P_3O_{10}$) is obtained.

$$3P_4O_{10} + 10Na_2O \xrightarrow{1000°C} 4Na_5P_3O_{10}$$

11. Sodium hexametaphosphate is the sodium salt of hexametaphosphoric acid, $(HPO_3)_6$ and hence is represented as $(NaPO_3)_6$ or $Na_2\,[Na_4\,(PO_3)_6]$. Under the name of calgon (which means calcium gone), this salt is used for removing $CaCl_2/MgCl_2$ from hard water (softening of water)

12. $$3NH_4Cl + 3BCl_3 \xrightarrow{140°C\text{ (In presence of }C_6H_5Cl\text{ and Fe)}} Cl_3B_3N_3H_3 + 9HCl$$
B, B, B-trichloroborazine

$$3Cl_3B_3N_3H_3 + 6NaBH_4 \xrightarrow{\text{Reduction}} 2B_3N_3H_6 + 6NaCl + 3B_2H_6$$

37

Important Compounds of Main Group Elements

MULTIPLE CHOICE QUESTIONS

1. Action of N_2O on sodamide gives :
(*a*) Ammonia only
(*b*) Sodium azide only
(*c*) Ammonia and sodium azide both
(*d*) None of these

2. $NaNH_3$ is obtained by :
(*a*) Treating molten Na with $(N_2 + H_2)$ mixture
(*b*) Treating molten Na with NH_3
(*c*) The action of NH_3 on NaH
(*d*) None of the above methods

3. When crude calcium cynamide is fused with NaCl, we get :
(*a*) Na_2CO_3 (*b*) NaCN
(*c*) CO (*d*) None of these

4. Which of the following complex ions is formed when argentite is treated with aqueous solution of NaCN in presence of hot air ?
(*a*) $[Ag(CN)_2]^-$ (*b*) $[Ag(CN)_4]^{3-}$
(*c*) $[Ag(CN)_3]^{2-}$ (*d*) None of these

5. Aqueous solution of NaCN is :
(*a*) weakly acidic (*b*) strongly acidic
(*c*) basic (*d*) neutral

6. On passing NH_3 gas over a mixture of K_2CO_3 and carbon, we get :
(*a*) KCN (*b*) $(NH_4)_2CO_3$
(*c*) CO_2 (*d*) $K(NH_4)CO_3$

7. Basic beryllium acetate has tetrahedral shape. The species placed at the centre, each of the four corners and each of the six edges are respectively :
(*a*) O, Be, CH_3COO^-
(*b*) Be, O, CH_3COO^-
(*c*) CH_3COO^-, Be, O
(*d*) CH_3COO^-, O, Be

8. Which of the following compounds is dissolved in glacial acitic acid to get basic beryllium acetate ?
(*a*) $Be(OH)_2$ (*b*) $BeCO_3$
(*c*) BeO (*d*) $BeHCO_3$

9. Which of the following is not a compound of lead ?
(*a*) Black lead (*b*) White lead
(*c*) Red lead (*d*) None

10. In Dutch process used for the preparation of white lead the vepours of vinegar are passed on lead in presence of O_2 and
(*a*) CO_2 (*b*) $Pb(OH)_2$
(*c*) PbO (*d*) PbO_2

11. In air, white lead gets blackend due to the formation of :
(*a*) Pb (*b*) PbS
(*c*) H_2S (*d*) C

12. Action of heat on white lead gives :
(*a*) $Pb(OH)_2$ (*b*) $PbCO_3$
(*c*) PbO (*d*) Pb_3O_4

13. Which of the following is sublimed white lead ?
(*a*) $PbO \cdot 2PbSO_4$
(*b*) $Pb(OH)_2 \cdot 2PbCO_3$
(*c*) Pb_3O_4
(*d*) PbO

14. Which of the following is (are) correct statement (s) ?
(*a*) White lead, sublimed white lead and super-sublimed white lead all are used as white pigments.

(*b*) Super-sublimed white lead is the brightest of all white pigments.

(*c*) The formulac of white lead and sublimed white lead are PbO $2PbSO_4$ and $Pb(OH)_2 \cdot 2PbCO_3$ respectively.

(*d*) Sublimed white lead contains ZnO as impurity but super-sublimed white lead does not contain this impurity.

15. White of the following oxides dissolves in potassium hydrogen tartrate (cream of tartar) to form tartar emetic ?

(*a*) Sb_2O_3

(*b*) Sb_2O_4

(*c*) Sb_2O_5

(*d*) None of these oxides

16. Which of the following is called sugar of lead ?

(*a*) $Pb(NO_3)_2$

(*b*) $PbCl_2$

(*c*) $PbCO_3 \cdot Pb(OH)_2$

(*d*) $Pb(CH_3COO)_2$.

17. Which compound of lead was used as anti-knocking agent ?

(*a*) Lead tetra acetate

(*b*) Basic lead sulphate

(*c*) Tetraethyl lead

(*d*) Sublimed white lead

ANSWERS

1. (*c*) **2.** (*b*) **3.** (*b*) **4.** (*a*) **5.** (*c*) **6.** (*a*) **7.** (*a*) **8.** (*a, b*)
9. (*a*) **10.** (*a*) **11.** (*b*) **12.** (*d*) **13.** (*a*) **14.** (*a,b,d*) **15.** (*a*) **16.** (*d*)
17. (*c*)

HINTS AND EXPLANATIONS

1. $3NaNH_2$ (*sodamide*) + N_2O $\longrightarrow$ NaN_3 (*sodium azide*) + NH_3 + NaOH.

2. 2Na (*molten*) + $2NH_3$ $\xrightarrow{300° C}$ $2NaNH_2 + H_2$

3. Crude calcium cynamide is a mixture of $CaCN_2$ and C. When this crude mixture is fused with NaCl, NaCN is obtained.

$$CaCN_2 + C + 2NaCl \xrightarrow{\text{Fuse}} 2NaCN + CaCl_2$$

4. Ag_2S (*Argentite*) + 4NaCN + $2O_2$ (*From air*) $\longrightarrow$ $Na_2SO_4 + 2Na[Ag(CN)_2]$

or $$Ag_2S + 4CN^- + 2O_2 \longrightarrow 2[Ag(CN)_2]^- + SO_4^{2-}$$

5. Since the aqueous solution of NaCN gives a strong base and weak acid *viz* NaOH and HCN respectively, it shows basic (alkaline) nature.

$$NaCN + H_2O \rightleftharpoons NaOH\ (\textit{strong base}) + HCN\ (\textit{weak acid})$$

$$NaOH \rightleftharpoons Na^+ + OH^-$$

6. $K_2CO_3 + C + 2NH_3 \longrightarrow 2KCN + 3H_2O$

This reaction represents Beilby's process for the preparation of KCN.

7. X-ray diffraction study of basic beryllium acetate, $Be_4O(CH_3CO_2)_6$ has shown that this molecule has tetrahedral geometry whose centre is occupied by O-atom and the four corners of the tetrahedron are ocunpied by four Be-atoms. Six acetate groups, 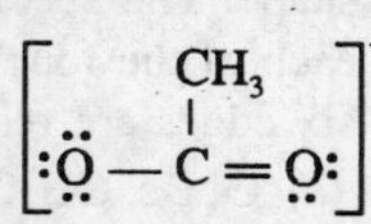 are arranged along six edges of the tetrahedron. One O-atom of CH_3COO^- ion is linked with

Be-atom placed on one corner and the other O-atom of this group is linked with Be-atom placed on the other corner.

8. Basic beryllium acetate, $Be_4O\ (CH_3COO)_6$ is obtained by dissolving Be $(OH)_2$ or $BeCO_3$ in glacial acetic acid, CH_3COOH.

$$4Be\ (OH)_2 + 6CH_3COOH \longrightarrow Be_4O\ (OH_3COO)_6 + 7H_2O$$

$$4BeCO_3 + 6CH_3COOH \longrightarrow Be_4O\ (CH_3COO)_6 + 3H_2O + 4CO \uparrow$$

9. Black lead is not a compound. Read lead is trilead tetroxide (Pb_3O_4) and white lead is basic lead carbonate, Pb $(OH)_2 \cdot 2PbCO_3$.

10. $$\underset{\text{Vinegar}}{6Pb + 6CH_3COOH} + 3O_2 + 2CO_2 \longrightarrow \underset{\text{White lead}}{Pb\ (OH)_2 \cdot 2PbCO_3} + 3Pb\ (CH_3COO)_2 + 2H_2O$$

Actually the above reaction takes place through the following steps :

(*i*) $$3 \times [2Pb + 2CH_3COOH + O_2 \longrightarrow \underset{\text{Basic lead acetate}}{Pb\ (OH)_2 \cdot Pb\ (CH_3COO)_2}]$$

(*ii*) $$3[Pb(OH)_2 \cdot Pb\ (CH_3COO)_2] + 2CO_2 \longrightarrow \underset{\text{White lead}}{Pb\ (OH)_2.2PbCO_3} + 3Pb\ (CH_3COO)_2 + 2H_2O$$

On adding : $$6Pb + 6CH_3COOH + 3O_2 + 2CO_2 \longrightarrow Pb\ (OH)_2 \cdot 2PbCO_3 + 3Pb\ (CH_3COO)_2 + 2H_2O$$

11. When placed in air, white lead gets blackened by H_2S gas present in air dur to the formation of PbS.

$$\underset{\text{White lead}}{2PbCO_3 \cdot Pb\ (OH)_2} + 3H_2S \longrightarrow \underset{\text{Black}}{3PbS} + 4H_2O + 2CO_2$$

12. $$\underset{\text{White lead}}{Pb\ (OH)_2 \cdot 2PbCO_3} \xrightarrow{43°\ C} Pb_3O_4 + H_2O + CO_2$$

13. $PbO \cdot 2PbSO_4$ is sublimed white lead, Pb $(OH)_2 \cdot 2PbCO_3$ is white lead, Pb_3O_4 is called red lead and PbO is called litharge.

14. Since white lead and sublimed white lead are basic lead carbonate and basic lead sulphate, their formulae are $2PbCO_3 \cdot Pb\ (OH)_2$ and $PbO \cdot 2PbSO_4$ respectively. Thus statement (*c*) is wrong.

15. $$2\ \begin{array}{l} CH(OH)\ COOK \\ | \\ CH(OH)\ COOK \end{array} + Sb_2O_3 \longrightarrow \begin{array}{l} CH(OH)COOK \\ | \\ CH(OH)COO(SbO) \end{array} + H_2O$$

Potassium hydrogen tartrate (Cream of tartar) — Tartar emetic

16. Lead acetate, Pb $(CH_3COO)_2$ is called sugar of lead, since it is sweet (but poisnous) in taste. It is obtained by dissolving litharge (PbO) or $PbCO_3$ in hot dil. CH_3COOH and crystallising.

$$PbO + 2CH_3COOH \longrightarrow (CH_3COO)_2\ Pb + H_2O$$

$$PbCO_3 + 2CH_3COOH \longrightarrow (CH_3COO)_2\ Pb + H_2O + CO_2$$

17. We know that CO is produced in the atmosphere due to the incomplete combustion of petrol in the internal engine of automobiles. Thus the automobiles exhaust contains CO which makes the air polluted. In order to minimise air pollution caused by CO, tetraethyl lead, $(C_2H_5)_4$ Pb is mixed with petrol. This compound provides more time to the petrol for its combustion and hence makes the petrol to burn at slow speed or more completely. Due to the slow burning of petrol, the quantity of CO released by the automobiles is decreased and hence the pollution of air caused by CO is also minimised.

38

Miscellaneous Topics

MULTIPLE CHOICE QUESTIONS

Alkali metal complexes with crown ethers

1. The crown ether ligand whose structure is given below is called :

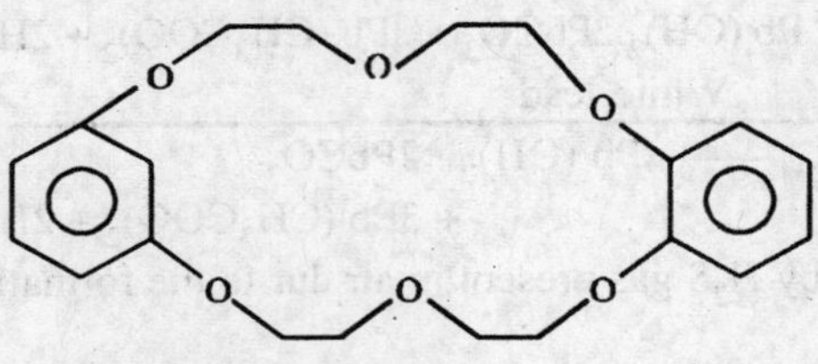

(a) Dibenzo-18-crown-6
(b) Dibenzo-16-crown-6
(c) Dibenzo-18-crown-5
(d) Benzo-18-crown-6

Silicates

2. Crown ethers are :
(a) Monodentate ligands
(b) Bidentate ligands
(c) Tridentate ligands
(d) Polydentate ligands

3. Which of the following is a cyclic silicate ?
(a) Zircon (b) Phenacite
(c) Wallastonite (d) Beryl

4. In silicates, silion atom is :
(a) sp^3 hybridised (b) sp^2 hybridised
(c) sp hybridised (d) dsp^2 hybridised

5. The basic structural unit present in all types of silicates is :
(a) SiO_2 unit (b) SiO_3^{2-} ion
(c) SiO_4^{4-} ion (d) SiO_6^{6-} ion

6. The geometry of SiO_4^{4-} ion is :
(a) Tetrahedral (b) Square planar
(c) Octahedral (d) Planar triangular

7. The silicates that contain discrete anions are :
(a) Sheet silicates
(b) Ortho silicates
(c) Three - dimensional silicates
(d) None of these

8. Which of the following is a chain silicate ?
(a) Beryl (b) Feld spar
(c) Spodumene (d) Zircon

9. The silicates which contain extended silicate ions are :
(a) Pyro silicates
(b) Three dimensional silicates
(c) Chain silicates
(d) Cyclic silicates

10. Quarts is a crystalline variety of :
(a) Silicon (b) Silica
(c) SiC (d) Na_2SiO_3

11. A silicate used in talcum powder :
(a) consists of planar sheets which can slip over one another
(b) is known as talc
(c) is a pure magnesium silicate which is represented as $3MgO \cdot 4SiO_2 \cdot H_2O$
(d) all of these

12. Which of the following does not represent silica ?
(a) Cristobalite (b) Quartz
(c) Tridymite (d) Zeolite

13. Kaolin and talc belong to :
(a) Orthosilicates
(b) Pyrosilicates
(c) Two-dimensional sheet silicates
(d) Three-dimensional sheet silicates.

14. Three dimensional silicate unit $(SiO_2)^0$ is present in :
(*a*) silica (*b*) asbestos
(*c*) beryl (*d*) mica

15. Pyrosilicate ion is :
(*a*) SiO_2^{2-} (*b*) SiO_4^{2-}
(*c*) SiO_3^{2-} (*d*) $Si_2O_7^{2-}$

16. The fibrous mineral which can withstand red hot flames without any damage is :
(*a*) Talc (*b*) Glass wool
(*c*) Asbestos (*d*) Soap stone

17. The structural representation given below is of :

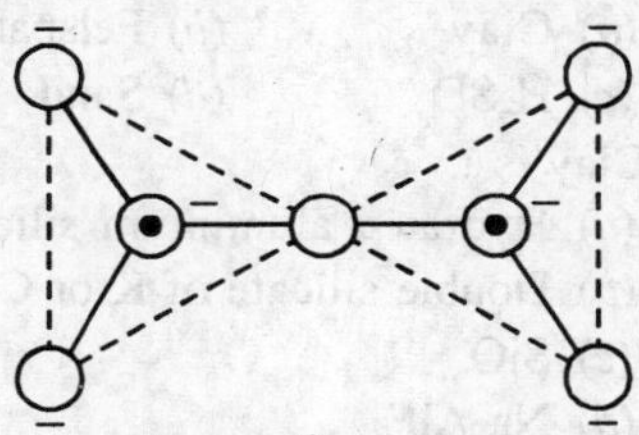

○ = Oxygen
● = Silcon

(*a*) Chain silicate ion
(*b*) Ortho silicate ion
(*c*) Pyrosilicate ion
(*d*) Sheet silicate ion

18. Pyrosilicates are the silicates in which two tetrahedral units are linked at :
(*a*) one point (*b*) two points
(*c*) three points (*d*) no point

Glass

19. Which glass has the highest percentage of lead ?
(*a*) Soda glass (*b*) Flint glass
(*c*) Jena glass (*d*) Pyrex glass

20. The colour producing oxide used to impart green colour to glass is :
(*a*) Cr_2O_3 (*b*) MnO_2
(*c*) CdS (*d*) Cu_2O

21. Ordinary glass is :
(*a*) Sodium silicate
(*b*) Calcium silicate
(*c*) Calcium and sodium silicate
(*d*) Copper silicate

22. The glass with the smallest coeffecient of expansion is :
(*a*) Soft glass (*b*) Soda lime glass
(*c*) Safety glass (*d*) Pyrex glass.

23. Which of the following is not a constituent of soda glass ?
(*a*) $CaSiO_3$ (*b*) Na_2SiO_3
(*c*) SiO_2 (*d*) H_2SiO_3

24. Ultraviolet rays can be checked by :
(*a*) Flint glass (*b*) Crooke's glass
(*c*) Soda glass (*d*) Pyrex glass.

25. Which of the following is used for the manufacture of optical glasses ?
(*a*) Sodium glass (*b*) Flint glass
(*c*) Ground glass (*d*) Quartz.

26. Glass is a :
(*a*) Liquid
(*b*) Solid
(*c*) Super-cooled liquid
(*d*) Transparent organic polymer.

27. The principal constituent of pyrex glass is :
(*a*) Zn (*b*) B
(*c*) Pb (*d*) Cl

28. Glass reacts with :
(*a*) Oleum (*b*) HF
(*c*) HNO_3 (*d*) $K_2Cr_2O_7$.

29. Ordinary glass contains :
(*a*) Silica, lime, NaCl
(*b*) Silica, lime, Na_2CO_3
(*c*) Silica, NaCl
(*d*) Lime, Na_2CO_3

30. Glass reacts with HF to produce :
(*a*) Na_3AlF_6 (*b*) H_2SiF_6
(*c*) H_2SiO_3 (*d*) SiF_4

31. Red colour can be imparted to glass by :
(*a*) Gold (*b*) Selenium
(*c*) Cuprous oxide (*d*) Iron (III) oxide.

Cement

32. Portland cement does not contain :
(*a*) $Ca_3Al_2O_6$ (*b*) Ca_3SiO_5
(*c*) Ca_2SiO_4 (*d*) $Ca_3(PO_4)_2$

33. A major constituent of Portland cement (except lime) is :
(*a*) Silica (*b*) Alumina
(*c*) Iron oxide (*d*) Magnesia

34. Cement contains mainly :

(*a*) Ca, Al, O (*b*) Ca, Si, O

(*c*) Ca, S, Al, O (*d*) Ca, Al, Si, O.

35. The raw materials used for the manufacture of cement are :

(*a*) $CaCO_3$ and SiO_2

(*b*) $CaCO_3$ and clay

(*c*) CaO and SiO_2

(*d*) $Ca(OH)_2$ and clay

36. Portland cement is manufactured by using :

(*a*) Lime stone, clay and sand

(*b*) Lime stone, gypsum and sand

(*c*) Lime stone, gypsum and alumina

(*d*) Lime stone, clay and gypsum.

37. Gypsum is added to clinker during cement manufacture to :

(*a*) decrease the rate of setting of cement

(*b*) bind the particles of calcium silicate

(*c*) fascilitate the formation of colloidal gel

(*d*) all the above.

38. Lime stone is used in the manufacture of :

(*a*) cement (*b*) fire bricks

(*c*) pigments (*d*) medicines.

39. The process of setting of cement under water is essentially :

(*a*) an oxidation process

(*b*) a reduction process

(*c*) a double decomposition process

(*d*) a hydration process.

40. Setting of cement is :

(*a*) an exothermic reaction

(*b*) an endothermic reaction

(*c*) neither exothermic nor endothermic

(*d*) None

41. Concrete is a mixture of :

(*a*) Cement, sand, gravel and water

(*b*) Cement, lime and water

(*c*) Cement, sand and water

(*d*) None.

Mortar

42. Mortar is a water solution containing slaked lime and sand in the ratio of :

(*a*) 3 : 1 (*b*) 1 : 3

(*c*) 2 : 3 (*d*) 1 : 4

43. The function of sand in mortar is :

(*a*) to decrease the hardness

(*b*) to prevent excessive shrinkage because of which cracks may result

(*c*) to increase hardness

(*d*) to make the mass compact

Ceramics industry

44. Which of the following is (are) used as raw materials in ceramics industry ?

(*a*) Clay (*b*) Felspar

(*c*) H_2SO_4 (*d*) Sand

45. Clay is :

(*a*) Hydrated aluminium silicate

(*b*) Double silicate of K or Ca with Al

(*c*) SiO_2

(*d*) Na_3AlF_6

46. The material used as colouring agent in ceramics industry is (are) :

(*a*) Gold chloride

(*b*) $Al_2O_3 . 2SiO_2 . 2H_2O$

(*c*) CaF_2

(*d*) MnO_2

Paints and pigments

47. Which of the following is used as red pigment ?

(*a*) Pb_3O_4 (*b*) PbO

(*c*) $PbCrO_4$ (*d*) $Pb(OH)_2$

48. The formula of chrome yellow is :

(*a*) K_2CrO_4 (*b*) $PbCrO_4$

(*c*) Na_2CrO_4 (*d*) $K_2Cr_2O_7$

49. Which of the following is not used as a pigment in paints ?

(*a*) White lead (*b*) Lead chromate

(*c*) Pb_3O_4 (*d*) PbO_2

Fluoro carbons

50. Freon-12 is used as a :

(*a*) Refrigerant (*b*) Insecticide

(*c*) Fungicide (*d*) Herbicide

51. Teflon is :

(*a*) CCl_2F_2

(*b*) $CF_2 = CF_2$

(*c*) CF_3COOH

(*d*) a polymer of the trafluoro ethylene.

Electoplating

52. Which of the following statement is (are) correct for electroplating ?

(*a*) Electroplating is a process in which a superior metal is deposited on an inferior metal (called base metal) by the electrolysis of the aqueous solution of a suitable electrolyte of the superior metal.

(*b*) The coating of the superior metal should be coherent.

(*c*) The coating should be of uniform thickness.

(*d*) The superior metal (pure metal) should be made anode (+) and the base metal should be made cathode (–).

53. Which of the following is used as an electrolyte in electroplating a base metal with gold ?

(*a*) $K[Au(CN)_2]$ (*b*) $AuCl_3$

(*c*) Au metal (*d*) None of the these

Galvanisation

54. Galvanisation of iron sheets is done by :

(*a*) Cu plating (*b*) Zn plating

(*c*) Ag plating (*d*) Tin plating

55. Galvanisation means :

(*a*) Deposition of Zn on Fe

(*b*) Deposition of Al on Fe

(*c*) Deposition of Sn on Fe

(*d*) Deposition of Cu on Fe

Photography

56. AgBr is used in photography, because AgBr :

(*a*) is most sensitive to light and hence undergoes photo reduction to metallic Ag instantaneously on exposure to light.

(*b*) is not reduced to Ag

(*c*) has high molecular weight

(*d*) is not affected by light.

57. As a fixing agent in photography, sodium thiosulphate is used for :

(*a*) Dissolving out unreacted AgBr.

(*b*) Converting silver halides to metallic Ag.

(*c*) Reducing solubility of AgBr.

(*d*) Preventing overdeveloping and fogging.

58. Hypo is used in photography for :

(*a*) Developing picture

(*b*) Picture printing

(*c*) Colouring picture

(*d*) The fixation of picture.

59. Sodium thiosulphate is used in photography to :

(*a*) Reduce AgBr to Ag

(*b*) Convert metallic Ag to silver salt.

(*c*) Remove and decompose AgBr as a soluble silver thiosulphate complex.

(*d*) Remove reduced Ag.

60. Which of the following chemical reactions represents developing of the film ?

(*a*) $NH_4Br + AgNO_3 \longrightarrow AgBr + NH_4NO_3$

(*b*) $2AgBr \xrightarrow{\text{Light}} 2Ag + Br_2$

(*c*) $\underset{\text{Quinol}}{C_6H_4(OH)_2} + 2AgBr \longrightarrow 2Ag + \underset{\text{Quinone}}{C_6H_4O_2} + 2HBr$

(*d*) $2Na_2S_2O_3 + AgBr \longrightarrow Na_3[Ag(S_2O_3)_2] + 2NaBr$

(e) $AuCl_3 + 3Ag \longrightarrow 3AgCl + Au$

61. Silver salt used in photography is ?

(*a*) AgCl (*b*) $AgNO_3$

(*c*) AgF (*d*) AgBr

62. Sodium thiosulphate is used in photography because of its :

(*a*) Oxidising behaviour

(*b*) Reducing behaviour

(*c*) Complexing behaviour

(*d*) Photo chemical behaviour

ANSWERS

1. (*a*)	**2.** (*d*)	**3.** (*c,d*)	**4.** (*a*)	**5.** (*c*)	**6.** (*c*)	**7.** (*b*)	**8.** (*c*)
9. (*c*)	**10.** (*b*)	**11.** (*d*)	**12.** (*d*)	**13.** (*c*)	**14.** (*a*)	**15.** (*d*)	**16.** (*c*)
17. (*c*)	**18.** (*a*)	**19.** (*b*)	**20.** (*a*)	**21.** (*c*)	**22.** (*d*)	**23.** (*d*)	**24.** (*b*)
25. (*b*)	**26.** (*c*)	**27.** (*b*)	**28.** (*b*)	**29.** (*b*)	**30.** (*b*)	**31.** (*a,b,c*)	**32.** (*d*)
33. (*a*)	**34.** (*d*)	**35.** (*b*)	**36.** (*d*)	**37.** (*a*)	**38.** (*a*)	**39.** (*d*)	**40.** (*a*)
41. (*a*)	**42.** (*a*)	**42.** (*b*)	**43.** (*b*)	**44.** (*a,b,d*)	**45.** (*a*)	**46.** (*a,d*)	**47.** (*a*)
48. (*b*)	**49.** (*d*)	**50.** (*a*)	**51.** (*d*)	**52.** (*a,b,c,d*)		**53.** (*a*)	**54.** (*b*)
55. (*a*)	**56.** (*a*)	**57.** (*a*)	**58.** (*d*)	**59.** (*c*)	**60.** (*c*)	**61.** (*d*)	**62.** (*c*)

HINTS AND EXPLANATIONS

1. Crown is the name for class of ligands. Since in the given ligand, there are two benzene groups (substituent groups), 18 atoms present in the ring and 6 ether oxygens, this ligand is called dibenzo-18-crown-6.

2. These are polydentate ligands, since in 1 : 1 complexes formed by these ligands, the central metallic ion located at or near the centre of oxygen atoms ring is attached with all O-atoms present in these ligands.

3. Since zircon ($ZrSiO_4$) and phenacite (Be_2SiO_4) both contain only one discrete silicate ion (SiO_4^{2-}), these are ortho silicates. Wallastonite (Ca_3SiO_9) and beryl ($Be_3Al_2Si_6O_{18}$) contain cyclic anions *viz.* $(SiO_3)_3^{6-}$ or $Si_3O_9^{6-}$ and $(SiO_3)_6^{12-}$ or $Si_6O_{18}^{12-}$ ions respectively. Thus both these silicates are cyclic silicates. These silicates contain $[SiO_3)_n]^{2n-}$ type ions in which n = 3 and 6 respectively.

4. Since in all types of silicates, Si-atom (central atom) is attached with four O-atoms by (Si – O) σ-bonds and has no lone pairs of electrons on it, this atom is sp^3 hybridised.

5. All types of silicates contain ortho silicate ion (SiO_4^{4-}) as the basic unit. Ortho silicates contain only one discrete SiO_4^{4-} ion while other silicates contain two or more SiO_4^{4-} ions which are linked together through bridging O-atoms (corners of different SiO_4^{4-} units).

6. In SiO_4^{4-} ion, since Si-atom (central atom) is linked with four O-atoms by four (Si – O) σ-bonds and has no lone pair of electrons on it, this ion has tetrahedral geometry.

7. It can be seen from the structure of ortho silicates that these silicates contain discrete ortho silicate ions, SiO_4^{4-}.

8. Spodumene, LiAl $(SiO_3)_2$ contains $[(SiO_3)_2]^{4-}$ ion which is a $[(SiO_3)_2]^{2n-}$ ion having $n = 2$. $[(SiO_3)_2]^{2-}$ ion is obtained when two O-atoms of each SiO_4^{4-} ion are shared with other SiO_4^{4-} ions in such a way that single chain is formed.

9. $Si_2O_6^{4-}$ and $Si_4O_{11}^{6-}$ ions are called extended silicate ions. Single chain silicates contain $Si_2O_6^{4-}$ ions while double chain silicates have $Si_4O_{11}^{6-}$ ions.

10. Quartz is a crystalline form of silicon dioxide (silica), SiO_2. This form is the purest form. Other forms of SiO_2 are tridymite and crystobalite.

11. The silicate used in talcum powder is magnesium silicate. Since it is a two-dimensional sheet silicate, it contains $(Si_2O_5)_n^{2n-}$ ion. Thus it is represented as 2Mg (Si_2O_5). Mg $(OH)_2$ or 3MgO . $4SiO_2$. H_2O.

12. Cristobalite (*a*), quartz (b) and tridymite (c) are three forms of silica. Zeolites are hydrated alumino silicates. Examples of zealites are :
$Na_2O. Al_2O_3. 2SiO_2 . xH_2O$ (permutit); $Na_2O . Al_2O_3 . 4SiO_2. 6H_2O$; $CaO . Al_2O_3 . 4SiO_2 . 6H_2O$; $M_2O . Al_2O_3. 6SiO_2 . 6H_2O$ (M = Na, K) and $MO . Al_2O_3 . 6SiO_2 . 6H_2O$ (M = Ca, Mg)

13. Kaolin is represented as $Al_2 (Si_2O_5) (OH)_4$ or $Al_2O_3 . 2SiO_2 . 2H_2O$ and talc is represented as $Mg_3 (Si_2O_5)_2 (OH)_2$ or $2Mg (Si_2O_5). Mg (OH)_2$ or $3MgO . 4SiO_2 . H_2O$. Since both these silicates contain $Si_2O_5^{2-}$ ions, these are two-dimensional sheet silicates.

15. $Si_2O_7^{2-}$ ion is pyro silicate ion. This ion is obtained when one O-atom (corner) is shared between two SiO_4^{4-} tetrahedral ions, *i.e.* this ion contains two tetrahedral SiO_4^{4-} ions linked together through O-atom.

17. In the given representation, since two SiO_4^{2-} units are connected together through one corner (*i.e.* through one O-atom), it is pyrosilicate ion, Si_2O_7.

18. Pyrosilicates are those silicates which contain pyrosilicate ion, $Si_2O_7^{6-}$. This ion is obtained when two tetrahedral SiO_4^{4-} units are linked together through one O-atom (one corner or one point) as shown in the figure given question 17.

19. (*a*) Soda glass has no lead, since its composition is :

$$Na_2O . CaO . 6SiO_2 \text{ or } Na_2SiO_3 . CaSiO_3 . 4SiO_2.$$

(*b*) Flint glass has lead as is evident from its composition which is $K_2O. PbO. 6SiO_2$.

(*c*) Jena glass contains BaO, ZnO, Al_2O_3 and B_2O_3

(*d*) The approximate composition of pyrex glass is SiO_2 = 80%, Na_2O = 4%, CaO = 0.5%, K_2O = 0.5%, B_2O_3 = 12% and Al_2O_3 = 3%.

20. In order to get coloured glasses, metallic oxides are added to the molten mass. For example Cr_2O_3 imparts green colour to the glass, MnO_2 gives purple colour, CdS gives lemon yellow colour and Cu_2O gives red colour.

21. Ordinary glass is a mixture of sodium and calcium silicates containing SiO_2. Thus it is represented as $Na_2SiO_3. CaSiO_3. 4SiO_2$ or $Na_2O. CaO. 6SiO_2$.

22. (*a*) Soft glass is a cheap variety of glass. It is called soft because it softens at a relatively lower temperature.

(*b*) Soda lime is another name of soft glass.

(*c*) Safety glass does not break under heavy impact.

(*d*) Quartz glass has low coeffecient of expansion. It can withstand high temperature.

23. Soda glass is a mixture of Na_2SiO_3, $CaSiO_3$ and SiO_2. Thus silicic acid (H_2SiO_3) is not a constituent of soda glass.

24. In the preparation of crooke's glass some CeO_2 is added. This oxide cuts ultraviolet rays which are harmful to the eyes.

26. Glass is an amorphous transparent or translucent rigid super-cooled liquid of indefinite viscosity :

27. Pyrex glass is a mixture of silicates of Na, K, Ca, B and Al, *i.e.* pyrex glass contains SiO_2 (= 80%), Na_2O (= 4%), K_2O (= 0.5%), CaO (= 0.5%), B_2O_3 (= 12%) and Al_2O_3 (= 3%). Thus pyrex glass has B_2O_3 in the maximum percentage.

28. The composition of common glass is $Na_2SiO_3. CaSiO_3. 4SiO_2$. Na_2SiO_3, $CaSiO_3$ and SiO_2 present in it react with HF to produce H_2SiF_6 as shown below :

$$Na_2SiO_3 + 8HF \longrightarrow 2NaF + H_2SiF_6 + 3H_2O$$

$$CaSiO_3 + 8HF \longrightarrow CaF_2 + H_2SiF_6 + 3H_2O$$
$$3SiO_2 + 12HF \longrightarrow H_2SiO_3 + 2H_2SiF_6 + 3H_2O$$

29. The composition of common glass is Na_2O. CaO (lime). $6SiO_2$ (silica). Na_2O is obtained from Na_2CO_3. Thus ordinary glass contains Na_2CO_3, CaO and SiO_2.

30. See the answer of question number 28.

31. Red colour can be imparted to glass by gold, selemium or cuprous oxide.

32. We know that cement is a grey heavy powder which contains dicalcium silicate (2CaO. SiO_2 or Ca_2SiO_4), tricalcium silicate (3CaO. SiO_2 or Ca_3SiO_5), tricalcium aluminate (3CaO, Al_2O_3 or $Ca_3\ Al_2O_6$), tricalcium aluminate ferrite (3CaO. Al_2O_3, Fe_2O_3 or $Ca_3\ Al_2Fe_2O_9$) and tetracalcium aluminate ferrate (4CaO. Al_2O_3, Fe_2O_3 or $Ca_4Al_2Fe_2O_{10}$) but not 3CaO. P_2O_5 or $Ca_3P_2O_8$ or $Ca_3(PO_4)_2$.

33. Major constituents of cement are lime (CaO = 50 - 60%), silica (SiO_2 = 20-25%), alumina (Al_2O_3 = 7.5%), magnesia (MgO = 2.5%) and iron oxide (Fe_2O_3 = 2.0%). Obviously excepting lime, since silica has the highest percentage (= 20 – 25%), it is the major constituent of cement.

34. We know that the major constituents of cement are lime (CaO = 50-60%), silica (SiO_2 = 20-25%) and alumina (Al_2O_3 = 7.5%). Obviously these substances contain Ca, O, Si and Al.

35. We know that the main constituents of cement are lime (CaO), silica (SiO_2) and alumina (Al_2O_3). CaO is provided by $CaCO_3$ and SiO_2 and Al_2O_3 both are provided by clay. Thus the raw materials used for the manufacture of cement are $CaCO_3$ and clay.

36. Raw materials needed for the manufacture of cement are (*i*) lime stone which provides lime (CaO), (*ii*) clay which provides alumina (Al_2O_3) and silica (SiO_2) (*iii*) gypsum which decreases the rate of setting of cement.

37. The addition of gypsum ($CaSO_4$. $2H_2O$) to clinker during cement manufacture removes the fast setting tricalcium aluminate (3CaO. Al_2O_3) as tricalcium sulpho aluminate crystals [3CaO. Al_2O_3 . $CaSO_4$. (*x*+2) H_2O] and thus the rate of setting of cement slows down.

$$3CaO.Al_2O_3 + CaSO_4.2H_2O + xH_2O \longrightarrow 3CaO.Al_2O_3 . CaSO_4. (x + 2)\ H_2O.$$

38. Raw materials used in the manufacture of cement are lime stone ($CaCO_3$), clay and gypsum ($CaSO_4.2H_2O$). Lime stone provides lime (CaO) and clay provides alumina (Al_2O_3) and silica (SiO_2).

39. When water is added to cement, cement starts setting. During the setting of cement, hydration of dicalcium silicate (2CaO. SiO_2), tricalcium aluminate (3CaO.Al_2O_3) and tetracalcium aluminate ferrite (4CaO. Al_2O_3. Fe_2O_3) to hydrated colloidal gel takes place.

$$2CaO.SiO_2 + xH_2O \xrightarrow{\text{Hydration}} \underset{\text{Hydrated colloidal gel}}{2CaO.SiO_2 . xH_2O}$$

$$3CaO.Al_2O_3 + 6H_2O \xrightarrow{\text{Hydration}} 3CaO.\ Al_2O_3.6H_2O$$

$$4CaO.Al_2O_3 . Fe_2O_3 + 6H_2O \xrightarrow{\text{Hydration}} 3CaO.Al_2O_3.6H_2O + CaO.Fe_2O_3.$$

40. Setting of cement is an exothermic process. Hence, cement structures have to be cooled during setting by sprinklinging water.

41. Concrete is a mixture of cement, sand, gravel and water. This mixture (*i.e.* concrete) sets to an extremely hard structure and is used in the construction of floors and roofs.

42. Mortar is a mixture of shaked lime and sand in 1 : 3 ratio. Mortar is used in building construction.

43. We know that mortar is a mixture containing slaked lime and sand. The presence of sand prevents the excessive shrinking of mortar which is used in building construction.

45. (*a*) Clay is hydrated aluminium silicate *e.g.* Al_2O_3. $2SiO_2.2H_2O$ (Kaolin), Al_2O_3. $2SiO_2.H_2O$ (bedellite) etc.

(*b*) Double silicate of K or Ca with Al is called felspar (*e.g.* $K_2O.Al_2O_3$. $6SiO_2$; $Na_2O.Al_2O_3$. $6SiO_2$; CaO. $Al_2O_3.6SiO_2$ etc.)

(*c*) SiO_2 is silica (sand)

(*d*) Na_3AlF_3 is cryolite and is used as a flux in ceramics industry.

46. Gold chloride and MnO_2 give pink and violet colour respectively to the articles. $Al_2O_3.2SiO_2$. $2H_2O$ is kaolin (clay) which is used as a raw material in ceramics industry. CaF_2 is flourspar and is used as a flux in ceramics industry.

47. Pb_3O_4 is a brilliant red powder insoluble in water and hence is called red lead. Due to its red colour it is used in the preparation of profective paints foı iron and steel structure.

48. Since $PbCrO_4$ is a yellow solid, it is used as a yellow pigment and hence is called chrome yellow.

49. (*a*) White lead is basic lead carbonate, $Pb(OH)_2$. $2PbCO_3$. It is used as a white pigment in paints.

(*b*) Lead chromate ($PbCrO_4$) is chrome yellow and is used as a yellow pigment in paints.

(*c*) Pb_3O_4 is used as a red pigment.

(*d*) PbO_2 is not used as a pigment in paints.

50. Freon –12 is dichlorodifluoro carbon, CCl_2F_2. It is used as a refrigerant and in air conditioning in place of NH_3 and SO_2.

51. (*a*) CCl_2F_2 is dichlorodifluoro carbon and is called freon –12.

(*b*) $CF_2 = CF_2$ is tetrafluoro ethylene

(*c*) CF_3 COOH is trifluoroacetic acid.

(*d*) Polymer of tetrafluoro ethylene is $(CF_2 = CF_2)_n$ which is called teflon.

$$n\,(CF_2 = CF_2) \xrightarrow{\text{Polymerisation}} \underset{\text{Teflon.}}{(CF_2 = CF_2)_n}$$

53. Since $[Au(CN)_6]^-$ ions are very stable, these ions give Au^+ ions of very low concentration.

$$K\,[Au\,(CN)_2] \longrightarrow K^+ + [Au\,(CN)_2]^-$$

$$[Au\,(CN)_2]^- \rightleftharpoons Au^+ + 2CN^-$$

54. Galvanisation of a metal is a process in which the metal surface is coated with a thin film of Zn metal.

55. See the answer of question number 54.

56. Out of all silver halides, AgBr is the most sensitive to light and hence undergoes photo-reduction to metallic Ag instataneously on exposure to light

$$2AgBr \xrightarrow{\text{Light(hv)}} 2Ag + Br_2$$

57. Sodium thiosulphate ($Na_2S_2O_3$) dissolves the undecomposed or unreacted AgBr as a soluble complex, $Na_3\,[Ag\,(S_2O_3)_2]$ and thus fixes the image.

$$AgBr\,(aq) + 2Na_2S_2O_3\,(aq) \longrightarrow \underset{\text{Soluble complex.}}{Na_3\,[Ag\,(S_2O_3)_2]\,(aq)} + NaBr\,(aq)$$

58. Hypo is used in photography as a fixer, *i.e.* for fixing the picture.

59. Sodium thiosulphate ($Na_2S_2O_3$) is used in photography to remove and decompose AgBr as a soluble complex compound, *viz* sodium dithiosulphato argentate (I), $Na_3[Ag(S_2O_3)_2]$

$$AgBr + 2Na_2S_2O_3 \longrightarrow Na_3[Ag(S_2O_3)_2] + NaBr$$

60. (*a*) Reaction given at (*a*) represents preparation of the sensitive plate or film.

(*b*) Reaction (*b*) represents the exposure of the plate or film to light.

(*c*) Reaction (*c*) represents developing of the film or plate.

(*d*) Reaction (*d*) represents the fixing of the negative (film).

(*e*) This reaction represents toning process.

61. Out of the given salts of Ag, Ag Br is the most sensitive to light and hence undergoes photo-reduction to metallic Ag on exposure to light.

$$2AgBr \xrightarrow{\text{Light}(h\nu)} 2Ag + Br_2$$

62. Because of complexing behaviour of sodium thiosulphate ($Na_2S_2O_3$), this salt dissolves the unreacted or undecomposed AgBr as soluble complex compound, $Na_3[Ag(S_2O_3)_2]$ and thus fixes the image.

$$AgBr + 2Na_2S_2O_3 \longrightarrow Na_3[Ag(S_2O_3)_2] + NaBr$$

SUBJECT INDEX

A

B

D

I

K

L

M

Q

R

T

U

V

W

X

Z

OTHER IMPORTANT BOOKS ON CHEMISTRY

ADVANCED PHYSICAL CHEMISTRY

D.N. Bajpai

For B.Sc. Part III students of chemistry, this second revised and enlarged edition of the book covers the newly revised syllabi of almost all the Indian universities. The subject matter is treated in simple and easily comprehensible language.

CONTENTS: Chemical Thermodynamics • Chemical Statistics or Statistical Thermodynamics • Phase Rule • Chemical Kinetics • Macromolecules • Electromotive Force • Electrochemistry • Dipole Moment • Molecular Structure • Interface Chemistry • Liquid State • Solid State: Chemical Crystallography • Environmental Chemistry • Wave Mechanics • Photo Chemistry • General Bibliography • Appendices • Logarithm Tables.

04 170 **ISBN:81-219-0408-0** **pp.1056**

ORGANIC CHEMISTRY— A SHORT COURSE

Harold Hart, David J. Hart & Leslie E. Craine

This book is designed to illustrate the practical applications of organic chemistry to everyday life and biological processes. There are about 925 problems and examples in this book, which appear at appropriate places within each chapter to help students develop problem solving skills.

CONTENTS: Introduction • Bonding and Isomerism • Alkanes and Cycloalkanes; Conformational and Geometric Isomerism • Alkenes and Alkynes • Aromatic Compounds • Ste reoisomerism • Organic Halogen Compounds; Substitution and Elimination Reactions • Alcohols, Phenols, and Thiols • Ethers and Epoxides • Aldehydes and Ketones • Carboxylic Acids and Their Derivatives • Amines and Related Nitrogen Compounds • Spectroscopy and Structure Determination • Heterocyclic Compounds • Synthetic Polymers • Lipids and Detergents • Carbohydrates • Amino Acids, Peptides and Proteins • Nucleotides and Nucleic Acids • Index.

04 223 **ISBN-81-219-1462-0** **pp. 594**

2000 SOLVED PROBLEMS IN ORGANIC CHEMISTRY

B. S. Bahl & Arun Bahl

The text, in the format of questions and answers, is a study guide for B.Sc. students. Problems answered in this guide have been carefully chosen from the final examination papers of more than 50 universities and include questions on every important principle, equation, theorem and concept. A clear, concise and correct answer is given to every problem. There are adequate illustrations to highlight important points and an extensive glossary is provided for quick cross-reference.

CONTENTS: Basic Principles: Principles of Bonding, Principles of Reactions, Isomerism • **Aliphatic Portion:** Alkanes, Petroleum, Alkenes, Dienes, Alkynes, Cycloalkanes, Alkyl Halides, Grignard Reagents and Lithium Reagents, Alcohols, Polyhydric Alcohols, Ethers and Epoxides, Mercaptans and Thioethers, Aldehydes and Ketones, Carboxylic Acids, Dicarboxylic Acids, Substituted Acids, Carboxylic Acid Derivatives, Urea and Related Compounds, Fats, Oils, and Lipids, Amines, Cyanides, Nitroalkanes and Diazomethane, Acetoacetic Ester and

Malonic Ester • **Aromatic Portion:** Aromatic Compounds, Benzene and Aromaticity, Directive Effect of Substituents, Alkylbenzenes, Aryl Halides, Benzenesulphonic Acid and Related Compounds, Nitrobenzene and Related Compounds, Aromatic Amines, Diazonium Salts, Phenols, Aromatic Aldehydes and Ketones, Aromatic Acids • **Special Topics:** Naphthalene, Anthracene, Hete-rocyclic Compounds, Carbohydrates, Amino Acids, Peptides and Proteins, Alkaloids, Terpenes, Dyes, Polymers • **Determination of Structures:** Analysis of Organic Compounds, Typical Numerical Problems, Reagents used in Organic Synthesis, Important Terms, Examination Guidelines, Index.

04 224 Ist Edn. Rep. 1997 ISBN:81-219-1466-3 pp. 836

A TEXTBOOK OF PHARMACEUTICAL CHEMISTRY

Jayashree Ghosh

22 022 2nd Edn. 1998 ISBN:81-219-1508-2 pp. 336

MODERN ORGANIC CHEMISTRY

S.P. Shukla & G.L. Trivedi

The book introduces the students of B.Sc. & M. Sc. to the topics in Organic Chemistry with U.G.C. Curriculum. The emphasis has been to help the reader obtain an understanding of the basic principles and concepts of Modern Organic Chemistry, with methodicity, simplicity and clarity of presentation being the watch word.

CONTENTS: Organic Reactions, Elucidations of their mechanisms, reagents and reaction intermediates • Substitution, elimination and addition reactions • Stereochemistry • Carbohydrates • Nitrogen Compounds • Amines • Aminoacids, peptides and proteins • Aliphatic diazocompounds, ketenes, carbenes and non-benzenoid compounds • Heterocyclic compounds • Synthetic Dyes • Index.

04 205 Ist Edn. 1997 ISBN:81-219-1497-3 pp. 800

आधुनिक भौतिक रसायन

फरीद खान

विशेष तौर पर भारतीय विश्वविद्यालयों की बी.एस.सी. स्नातक कक्षाओं के लिये लिखी गयी यह पुस्तक, अनेक आंकिक उदाहरणों से परिपूर्ण है जिससे विद्यार्थीगण विषय वस्तु को अधिक सरलता से आत्मसात कर सकें। इसके अलावा प्रत्येक अध्याय के अन्त में प्रश्नावली तथा प्रश्नों के उत्तर दिये गये हैं जो विद्यार्थियों के विषय परख के लिये उपयोगी सिद्ध होंगे।

विषय-सूची: प्रस्तावना • गणितीय रसायन • परमाणु संरचना • रेडियोऐक्टिवता • नाभिकीय रसायन • संयोजकता का इलेक्ट्रॉनिक सिद्धान्त • रासायनिक बन्धन: आर्बिटल संकल्पना • गैसीय अवस्था • रासायनिक साम्य • विलयन • तनु विलयन का सिद्धान्त • पदार्थ की कलिलीय अवस्था तथा वृहत् अणु • उत्प्रेरण • अधिशोषण • प्रावस्था नियम • वितरण नियम • रासायनिक बलगतिकी • प्रकाश रसायन • ऊष्मागतिकी का प्रथम नियम व ऊष्मा रसायन • ऊष्मागतिकी का द्वितीय नियम • भौतिक गुणधर्म एँव रासायनिक संगटन • ठोस एवं क्रिस्टल विज्ञान • आणविक सममिति • चुम्बकीय गुण • प्रकाश एव स्पेक्ट्रम विज्ञान • आयनिक साम्यावस्थाएं • अम्ल और क्षार - लवण जल अपघटन • विद्युत अपघटनी अभिगमन तथा चालकता • विद्युत वाहक बल • परिशिष्ट • पर्यावरण: प्रदूषण तथा नियँत्रण • ग्रीक-अक्षर • अनुक्रमाणिका।

16 211 Ist Edn. 1997 ISBN:81-219-1442-6 pp. 816

CHAND'S DICTIONARY OF CHEMISTRY

P.N. Arora & M.C. Arora (Editors)

04 206 Ist Edn. Rep. 1998 ISBN:81-219-0976-7 pp. 256

ESSENTIALS OF PHYSICAL CHEMISTRY

B.S. Bahl, G.D. Tuli & Arun Bahl

The book has completed 50 years of its successful and useful existence. The text has been re-written and enlarged so as to make it modern in approach. The latest trends have been integrated with the fundamental concepts and the contents matter made up-to-date.

CONTENTS: Structure of the Atom — Classical Mechanics • Structure of Atom–Wave Mechanical Approach • Isotopes, Isobars and Isotones • Nuclear Chemistry • Chemical Bonding–Lewis Theory • Chemical Bonding–Orbital Concept • First Law of Thermodynamics • Basic Concepts • Thermochemistry • Second Law of Thermodynamics • Gaseous State and the Kinetic Molecular Theory • Liquids • The Solid State • Physical Properties and Chemical Constitution • Solutions • Theory of Dilute Solutions • Osmosis and Osmotic Pressure • Chemical Equilibrium • Distribution Law • The Phase Rule • Chemical Kinetics • Catalysis • The Colloids • Absorption • Electrolysis and Electrical Conductance • Theory of Electrolytic Dissociation • Ionic Equilibria– Solubility Product • Acids and Bases • Salt Hydrolysis • Electromotive Force • Photochemistry • SI Units • Index.

04 028	**24th Edn. Rep. 1998**	**ISBN:81-219-0546-X**	**pp. 880**
04 204	**Hardbound**	**ISBN:81-219-1718-2**	**pp. 816**
04 030	**Hindi Edn.**	**ISBN:81-219-0022-0**	**pp. 790**

SELECTED TOPICS IN INORGANIC CHEMISTRY

W.U. Malik, G.D. Tuli & R.D. Madan

Most of the topics deal with the theoretical aspects of Inorganic Chemistry. The book elucidates the fundamentals and the approach to discussion is modern throughout the text which provides a clear understanding of the subject to the students.

CONTENTS: Structure of the Atom (Quantum, Mechanical and Wave Mechanical Approach) • Extended or Long Form of Periodic Table • Periodic Properties • Chemical Bonding • Nature of Covalent Bond and Shapes of Molecules • Modern Concepts of Acids and Bases • Non-aqueous Solvents • Transition Elements (d-Block Elements) • Lanthanides or Rare Earths (4f Block Elements) • Actinides (5f-block elements) • Basic Concepts of Co-ordination Chemistry • Nature of Metal-Ligand Bonding in Complexes • Isomerism among Inorganic Complexes • Stability of Complexes in Aqueous Solution • Ligand Substitution Reactions in Octahedral Complexes • Ligand Substitution Reactions in Square-planar Complexes • Electron-transfer (or oxidation-reduction) Reactions in Coordination Compounds • Electron-deficient Compounds: Boranes • Metallic Carbonyls and Metallic Nitrosyls • Basic Properties of Iodine, Interhalogen Compounds, Polyhalides and Pseudohalogens • Pollution and its Effects on the Living World • Subject Index.

04 039	**6th Edn. Rep. 1997**	**ISBN:81-219-0600-8**	**pp. 696**
04 207	**Hardbound**	**ISBN:81-219-0938-4**	**pp. 696**

ELEMENTARY ORGANIC SPECTROSCOPY (Principles and Chemical Applications)

Y.R. Sharma

Methods based on relationships between structural features and physical properties are not only making an ever increasing contribution but have also altered the scope and pace of organic chemistry due to the new technique of spectroscopy. The book as such deals with the application of these methods on selected representative molecules which provide insight into the subject for honours and postgraduate students of organic chemistry.

CONTENTS: Introduction • Ultra-violet and Visible Spectroscopy • Infrared Spectroscopy • Nuclear Magnetic Resonance Spectroscopy • Mass Spectrometry • Numerical Problems on UV, IR and NMR • Appendices.

04 073 2nd Edn. Rep. 1997 ISBN-81-219-0919-8 pp. 312

ADVANCED INORGANIC CHEMISTRY (VOL. I)

Satya Prakash, G.D. Tuli, S.K. Basu & R.D. Madan

This thoroughly revised and updated, seventeenth edition of the book is now divided into two volumes. Volume I covers the syllabus of all Indian Universities for the B. Sc. 1st and 2nd year-courses. It follows a new approach specially in the presentation of recent advances. As a matter of fact, the whole of the old text has been rewritten and re-arranged by Prof. R.D. Madan.

CONTENTS: Part I: Basic Principles of Chemistry • Quantum Mechanical Concepts of Atomic Structure: Rutherford's and Bohr's Atomic Models • Wave Mechanical Concept of Atomic Structure • Periodic Table • Electronic Configuration of Elements • Atomic Properties: Period Properties • Radioactivity • Isotopes, Isotones, Isobars and Nuclear Isomers • Nuclear Chemistry: Chemistry of the Nucleus • Nuclear Reactions • Different Types of Chemical Bonds • Molecular Orbital Theory of Simple Diatomic Molecules and Ions • Valence-shell Electron Pair Repulsion (VSERP) Theory: Shape of Molecules and Ions • Hybridisation and its Different Types • Structure and Shape of Molecules/Ions containing only Sigma Bonds • Structure and Shape of Molecules/Ions containing Sigma as well as Pi Bonds • Structure of Solids • Redox Reactions • Electrode Potentials and Electrochemical Series • Concepts of Acids and Bases • General Principles of Extraction of Metals and Furnaces • Nomenclature of Inorganic Compounds • **Part II: Chemistry of Main Group Elements and Their Compounds (*s*- And *p*-Block Elements)** • General Characteristic of Main Group Elements (*s*- and *p*- Block Elements) • Group Discussion of Main Group Elements (*s*- and *p*- Block Elements) • Occurrence, Extraction/ Isolation, Properties and Uses of Some Main Group Elements • Hydrides of Main Group Elements • Carbides of Main Group Elements • Nitrides of Main Group Elements • Oxides of Main Group Elements • Halides and Oxy-halides of Main Group Elements • Oxy-acids of Boron (Boric Acid) and their Salts • Oxy-acids of Nitrogen and their Salts • Oxy-acids of Sulphur and their Salts • Oxy-acids of Halogens and their Salts • Fixation of Nitrogen and Fertilisers • Inorganic Polymers • Important Compounds of Main Group Elements • Miscellaneous Topics • Index.

04 231 17 Rev. Edn. 1998 ISBN:81-219-0263-0 pp. 1156

ADVANCED INORGANIC CHEMISTRY (VOL. II)

Satya Prakash, G.D. Tuli, S.K. Basu & R.D. Madan

CONTENTS : Part I: Coordination Chemistry • **Part II:** Transition Metals (d-Block Elements) and Their Compounds • **Part III:** Inner Transition Elements and Their Compounds • **Part IV:** Miscellaneous Topics.

04 232 17th Edn. 1998 ISBN:81-219-1787-5 pp. 504

MODERN INORGANIC CHEMISTRY

R.D. Madan & Satya Prakash

This book explains the chemistry of elements on the basis of the latest theories of Chemistry. First few chapters of the book are devoted to the study of fundamentals of chemistry and others describe the extraction properties of elements along with the preparation, properties, uses and structure of their important compounds. Every chapter is followed by a large number of descriptive and objective type questions.